The Mathematica GuideBook

for Symbolics

Michael Trott

The Mathematica GuideBook
for Symbolics

With 848 Illustrations

INCLUDES DVD

Springer

Michael Trott
Wolfram Research
Champaign, Illinois

Mathematica is a registered trademark of Wolfram Research, Inc.

Library of Congress Control Number: 2005928496

ISBN-10: 0-387-95020-6   Printed on acid-free paper.
ISBN-13: 978-0387-95020-4
e-ISBN: 0-387-28815-5

Printed in the United States of America.   (HAM)

9 8 7 6 5 4 3 2 1

springeronline.com

Preface

Bei mathematischen Operationen kann sogar eine gänzliche Entlastung des Kopfes eintreten, indem man einmal ausgeführte Zähloperationen mit Zeichen symbolisiert und, statt die Hirnfunktion auf Wiederholung schon ausgeführter Operationen zu verschwenden, sie für wichtigere Fälle aufspart.

When doing mathematics, instead of burdening the brain with the repetitive job of redoing numerical operations which have already been done before, it's possible to save that brainpower for more important situations by using symbols, instead, to represent those numerical calculations.

— Ernst Mach (1883) [45]

Computer Mathematics and Mathematica

Computers were initially developed to expedite numerical calculations. A newer, and in the long run, very fruitful field is the manipulation of symbolic expressions. When these symbolic expressions represent mathematical entities, this field is generally called computer algebra [8]. Computer algebra begins with relatively elementary operations, such as addition and multiplication of symbolic expressions, and includes such things as factorization of integers and polynomials, exact linear algebra, solution of systems of equations, and logical operations. It also includes analysis operations, such as definite and indefinite integration, the solution of linear and nonlinear ordinary and partial differential equations, series expansions, and residue calculations. Today, with computer algebra systems, it is possible to calculate in minutes or hours the results that would (and did) take years to accomplish by paper and pencil. One classic example is the calculation of the orbit of the moon, which took the French astronomer Delaunay 20 years [12], [13], [14], [15], [11], [26], [27], [53], [16], [17], [25]. (The *Mathematica GuideBooks* cover the two other historic examples of calculations that, at the end of the 19th century, took researchers many years of hand calculations [1], [4], [38] and literally thousands of pages of paper.)

Along with the ability to do symbolic calculations, four other ingredients of modern general-purpose computer algebra systems prove to be of critical importance for solving scientific problems:

- a powerful high-level programming language to formulate complicated problems
- programmable two- and three-dimensional graphics
- robust, adaptive numerical methods, including arbitrary precision and interval arithmetic
- the ability to numerically evaluate and symbolically deal with the classical orthogonal polynomials and special functions of mathematical physics.

The most widely used, complete, and advanced general-purpose computer algebra system is *Mathematica*. *Mathematica* provides a variety of capabilities such as graphics, numerics, symbolics, standardized interfaces to other programs, a complete electronic document-creation environment (including a full-fledged mathematical typesetting system), and a variety of import and export capabilities. Most of these ingredients are necessary to coherently and exhaustively solve problems and model processes occurring in the natural sciences [41], [58], [21], [39] and other fields using constructive mathematics, as well as to properly represent the results. Consequently, *Mathematica*'s main areas of application are presently in the natural sciences, engineering, pure and applied mathematics, economics, finance, computer graphics, and computer science.

Mathematica is an ideal environment for doing general scientific and engineering calculations, for investigating and solving many different mathematically expressable problems, for visualizing them, and for writing notes, reports, and papers about them. Thus, *Mathematica* is an integrated computing environment, meaning it is what is also called a "problem-solving environment" [40], [23], [6], [48], [43], [50], [52].

Scope and Goals

The Mathematica GuideBooks are four independent books whose main focus is to show how to solve scientific problems with *Mathematica*. Each book addresses one of the four ingredients to solve nontrivial and real-life mathematically formulated problems: programming, graphics, numerics, and symbolics. The Programming and the Graphics volume were published in autumn 2004.

The four *Mathematica GuideBooks* discuss programming, two-dimensional, and three-dimensional graphics, numerics, and symbolics (including special functions). While the four books build on each other, each one is self-contained. Each book discusses the definition, use, and unique features of the corresponding *Mathematica* functions, gives small and large application examples with detailed references, and includes an extensive set of relevant exercises and solutions.

The *GuideBooks* have three primary goals:

- to give the reader a solid working knowledge of *Mathematica*
- to give the reader a detailed knowledge of key aspects of *Mathematica* needed to create the "best", fastest, shortest, and most elegant solutions to problems from the natural sciences
- to convince the reader that working with *Mathematica* can be a quite fruitful, enlightening, and joyful way of cooperation between a computer and a human.

Realizing these goals is achieved by understanding the unifying design and philosophy behind the *Mathematica* system through discussing and solving numerous example-type problems. While a variety of mathematics and physics problems are discussed, the *GuideBooks* are not mathematics or physics books (from the point of view of content and rigor; no proofs are typically involved), but rather the author builds on *Mathematica*'s mathematical and scientific knowledge to explore, solve, and visualize a variety of applied problems.

The focus on solving problems implies a focus on the computational engine of *Mathematica,* the kernel—rather than on the user interface of *Mathematica,* the front end. (Nevertheless, for a nicer presentation inside the electronic version, various front end features are used, but are not discussed in depth.)

The *Mathematica GuideBooks* go far beyond the scope of a pure introduction into *Mathematica*. The books also present instructive implementations, explanations, and examples that are, for the most part, original. The books also discuss some "classical" *Mathematica* implementations, explanations, and examples, partially available only in the original literature referenced or from newsgroups threads.

In addition to introducing *Mathematica*, the *GuideBooks* serve as a guide for generating fairly complicated graphics and for solving more advanced problems using graphical, numerical, and symbolical techniques in cooperative ways. The emphasis is on the *Mathematica* part of the solution, but the author employs examples that are not uninteresting from a content point of view. After studying the *GuideBooks*, the reader will be able to solve new and old scientific, engineering, and recreational mathematics problems faster and more completely with the help of *Mathematica*—at least, this is the author's goal. The author also hopes that the reader will enjoy using *Mathematica* for visualization of the results as much as the author does, as well as just studying *Mathematica* as a language on its own.

In the same way that computer algebra systems are not "proof machines" [46], [9], [37], [10], [54], [55], [56] such as might be used to establish the four-color theorem ([2], [22]), the Kepler [28], [19], [29], [30], [31], [32], [33], [34], [35], [36] or the Robbins ([44], [20]) conjectures, proving theorems is not the central theme of the *GuideBooks*. However, powerful and general proof machines [9], [42], [49], [24], [3], founded on *Mathematica'* s general programming paradigms and its mathematical capabilities, have been built (one such system is *Theorema* [7]). And, in the *GuideBooks*, we occasionally prove one theorem or another theorem.

In general, the author's aim is to present a realistic portrait of *Mathematica*: its use, its usefulness, and its strengths, including some current weak points and sometimes unexpected, but often nevertheless quite "thought through", behavior. *Mathematica* is not a universal tool to solve arbitrary problems which can be formulated mathematically—only a fraction of all mathematical problems can even be formulated in such a way to be efficiently expressed today in a way understandable to a computer. Rather, it is often necessary to do a certain amount of programming and occasionally give *Mathematica* some "help" instead of simply calling a single function like `Solve` to solve a system of equations. Because this will almost always be the case for "real-life" problems, we do not restrict ourselves only to "textbook" examples, where all goes smoothly without unexpected problems and obstacles. The reader will see that by employing *Mathematica's* programming, numeric, symbolic, and graphic power, *Mathematica* can offer more effective, complete, straightforward, reusable, and less likely erroneous solution methods for calculations than paper and pencil, or numerical programming languages.

Although the *Guidebooks* are large books, it is nevertheless impossible to discuss all of the 2,000+ built-in *Mathematica* commands. So, some simple as well as some more complicated commands have been omitted. For a full overview about *Mathematica*'s capabilities, it is necessary to study *The Mathematica Book* [60] in detail. The commands discussed in the *Guidebooks* are those that an scientist or research engineer needs for solving *typical* problems, if such a thing exists [18]. These subjects include a quite detailed discussion of the structure of *Mathematica* expressions, *Mathematica* input and output (important for the human–*Mathematica* interaction), graphics, numerical calculations, and calculations from classical analysis. Also, emphasis is given to the powerful algebraic manipulation functions. Interestingly, they frequently allow one to solve analysis problems in an algorithmic way [5]. These functions are typically not so well known because they are not taught in classical engineering or physics-mathematics courses, but with the advance of computers doing symbolic mathematics, their importance increases [47].

A thorough knowledge of:

- structural operations on polynomials, rational functions, and trigonometric functions
- algebraic operations on polynomial equations and inequalities
- process of compilation, its advantages and limits
- main operations of calculus—univariate and multivariate differentiation and integration
- solution of ordinary and partial differential equations is needed to put the heart of *Mathematica*—its symbolic capabilities—efficiently and successfully to work in the solution of model and real-life problems. *The Mathematica GuideBooks to Symbolics* discusses these subjects.

The current version of the *Mathematica GuideBooks* is tailored for *Mathematica* Version 5.1.

Content Overview

The Mathematica GuideBook for Symbolics has three chapters. Each chapter is subdivided into sections (which occasionally have subsections), exercises, solutions to the exercises, and references.

This fourth and last volume of the *GuideBooks* deals with *Mathematica*'s symbolic mathematical capabilities—the real heart of *Mathematica* and the ingredient of the *Mathematica* software system that makes it so unique and powerful. In addition, this volume discusses and employs the classical orthogonal polynomials and special functions of mathematical physics. To demonstrate the symbolic mathematics power, a variety of problems from mathematics and physics are discussed.

Chapter 1 starts with a discussion of the algebraic functions needed to carry out analysis problems effectively. Contrary to classical science/engineering mathematics education, using a computer algebra system makes it often a good idea to rephrase a problem—including when it is from analysis—in a polynomial way to allow for powerful algorithmic treatments. Gröbner bases play a central role in accomplishing this task. This volume discusses in detail the main functions to deal with structural operations on polynomials, polynomial equations and inequalities, and expressions containing quantified variables. Rational functions and expressions containing trigonometric functions are dealt with next.

Then the central problems of classical analysis—differentiation, integration, summation, series expansion, and limits—are discussed in detail. The symbolic solving of ordinary and partial differential equations is demonstrated in many examples.

As always, a variety of examples show how to employ the discussed functions in various mathematics or physics problems. The Symbolics volume emphasizes their main uses and discusses the specialities of these operations inside a computer algebra system, as compared to a "manual" calculation. Then, generalized functions and Fourier and Laplace transforms are discussed. The main part of the chapter culminates with three examples of larger symbolic calculations, two of them being classic problems. This chapter has more than 150 exercises and solutions treating a variety of symbolic computation examples from the sciences.

Chapters 2 and 3 discuss classical orthogonal polynomials and the special functions of mathematical physics. Because this volume is not a treatise on special functions, it is restricted to selected function groups and presents only their basic properties, associated differential equations, normalizations, series expansions, verification of various special cases, etc. The availability of nearly all of the special functions of mathematical physics for all possible arbitrary complex parameters opens new possibilities for the user, e.g., the use of closed formulas for the Green's functions of commonly occurring partial differential equations or for "experimental mathematics". These chapters focus on the use of the special functions in a number of physics-related applications in the text as well as in the exercises. The larger examples deal with are the quartic oscillator in the harmonic oscillator basis and the implementation of Felix Klein's method to solve quintic polynomials in Gauss hypergeometric functions $_2F_1$.

The Symbolics volume employs the built-in symbolic mathematics in a variety of examples. However, the underlying algorithms themselves are not discussed. Many of them are mathematically advanced and outside of the scope of the *GuideBooks*.

Throughout the Symbolics volume, the programming and graphics experience acquired in the first two volumes is used to visualize various mathematics and physics topics.

The Books and the Accompanying DVDs

Each of the *GuideBooks* comes with a multiplatform DVD. Each DVD contains the fourteen main notebooks, the hyperlinked table of contents and index, a navigation palette, and some utility notebooks and files. All notebooks are tailored for *Mathematica* 5.1. Each of the main notebooks corresponds to a chapter from the printed books. The notebooks have the look and feel of a printed book, containing structured units, typeset formulas, *Mathematica* code, and complete solutions to all exercises. The DVDs contain the fully evaluated notebooks corresponding to the chapters of the corresponding printed book (meaning these notebooks have text, inputs, outputs and graphics). The DVDs also include the unevaluated versions of the notebooks of the other three *GuideBooks* (meaning they contain all text and *Mathematica* code, but no outputs and graphics).

Although the *Mathematica GuideBooks* are printed, *Mathematica* is "a system for doing mathematics by computer" [59]. This was the lovely tagline of earlier versions of *Mathematica,* but because of its growing breadth (like data import, export and handling, operating system-independent file system operations, electronic publishing capabilities, web connectivity), nowadays *Mathematica* is called a "system for technical computing". The original tagline (that is more than ever valid today!) emphasized two points: doing mathematics and doing it on a computer. The approach and content of the *GuideBooks* are fully in the spirit of the original tagline: They are centered around *doing* mathematics. The second point of the tagline expresses that an electronic version of the *GuideBooks* is the more natural medium for *Mathematica*-related material. Long outputs returned by *Mathematica,* sequences of animations, thousands of web-retrievable references, a 10,000-entry hyperlinked index (that points more precisely than a printed index does) are space-consuming, and therefore not well suited for the printed book. As an interactive program, *Mathematica* is best learned, used, challenged, and enjoyed while sitting in front of a powerful computer (or by having a remote kernel connection to a powerful computer).

In addition to simply showing the printed book's text, the notebooks allow the reader to:

- experiment with, reuse, adapt, and extend functions and code
- investigate parameter dependencies
- annotate text, code, and formulas
- view graphics in color
- run animations.

The Accompanying Web Site

Why does a printed book need a home page? There are (in addition to being just trendy) two reasons for a printed book to have its fingerprints on the web. The first is for (*Mathematica*) users who have not seen the book so far. Having an outline and content sample on the web is easily accomplished, and shows the look and feel of the notebooks (including some animations). This is something that a printed book actually cannot do. The second reason is for readers of the book: *Mathematica* is a large modern software system. As such, it ages quickly in the sense that in the timescale of $10^{1.\,smallInteger}$ months, a new version will likely be available. The overwhelmingly large majority of *Mathematica* functions and programs will run unchanged in a new version. But occasionally, changes and adaptations might be needed. To accommodate this, the web site of this book—http://www.MathematicaGuideBooks.org—contains a list of changes relevant to the *GuideBooks*. In addition, like any larger software project, unavoidably, the *GuideBooks* will contain suboptimal implementations, mistakes, omissions, imperfections, and errors. As they come to his attention, the author will list them at the book's web site. Updates to references, corrections [51], hundreds of pages of additional exercises and solutions, improved code segments, and other relevant information will be on the web site as well. Also, information about OS-dependent and *Mathematica* version-related changes of the given *Mathematica* code will be available there.

Evolution of the Mathematica GuideBooks

A few words about the history and the original purpose of the *GuideBooks*: They started from lecture notes of an *Introductory Course in Mathematica 2* and an advanced course on the *Efficient Use of the Mathematica Programming System*, given in 1991/1992 at the Technical University of Ilmenau, Germany. Since then, after each release of a new version of *Mathematica*, the material has been updated to incorporate additional functionality. This electronic/printed publication contains text, unique graphics, editable formulas, runable, and modifiable programs, all made possible by the electronic publishing capabilities of *Mathematica*. However, because the structure, functions and examples of the original lecture notes have been kept, an abbreviated form of the *GuideBooks* is still suitable for courses.

Since 1992 the manuscript has grown in size from 1,600 pages to more than three times its original length, finally "weighing in" at nearly 5,000 printed book pages with more than:

- 18 gigabytes of accompanying *Mathematica* notebooks
- 22,000 *Mathematica* inputs with more than 13,000 code comments
- 11,000 references
- 4,000 graphics
- 1,000 fully solved exercises
- 150 animations.

This first edition of this book is the result of more than eleven years of writing and daily work with *Mathematica*. In these years, *Mathematica* gained hundreds of functions with increased functionality and power. A modern year-2005 computer equipped with *Mathematica* represents a computational power available only a few years ago to a select number of people [57] and allows one to carry out recreational or new computations and visualizations—unlimited in nature, scope, and complexity— quickly and easily. Over the years, the author has learned a lot of *Mathematica* and its current and potential applications, and has had a lot of fun, enlightening moments and satisfaction applying *Mathematica* to a variety of research and recreational areas, especially graphics. The author hopes the reader will have a similar experience.

Disclaimer

In addition to the usual disclaimer that neither the author nor the publisher guarantees the correctness of any formula, fitness, or reliability of any of the code pieces given in this book, another remark should be made. No guarantee is given that running the *Mathematica* code shown in the *GuideBooks* will give identical results to the printed ones. On the contrary, taking into account that *Mathematica* is a large and complicated software system which evolves with each released version, running the code with another version of *Mathematica* (or sometimes even on another operating system) will very likely result in different outputs for some inputs. And, as a consequence, if different outputs are generated early in a longer calculation, some functions might hang or return useless results.

The interpretations of *Mathematica* commands, their descriptions, and uses belong solely to the author. They are not claimed, supported, validated, or enforced by Wolfram Research. The reader will find that the author's view on *Mathematica* deviates sometimes considerably from those found in other books. The author's view is more on the formal than on the pragmatic side. The author does not hold the opinion that any *Mathematica* input has to have an immediate semantic meaning. *Mathematica* is an extremely rich system, especially from the language point of view. It is instructive, interesting, and fun to study the behavior of built-in *Mathematica* functions when called with a variety of arguments (like unevaluated, hold, including undercover zeros, etc.). It is the author's strong belief that doing this and being able to explain the observed behavior will be, in the long term, very fruitful for the reader because it develops the ability to recognize the uniformity of the principles underlying *Mathematica* and to make constructive, imaginative, and effective use of this uniformity. Also, some exercises ask the reader to investigate certain "unusual" inputs.

From time to time, the author makes use of undocumented features and/or functions from the `Developer`` and `Experimental`` contexts (in later versions of *Mathematica* these functions could exist in the `System`` context or could have different names). However, some such functions might no longer be supported or even exist in later versions of *Mathematica*.

Acknowledgements

Over the decade, the *GuideBooks* were in development, many people have seen parts of them and suggested useful changes, additions, and edits. I would like to thank Horst Finsterbusch, Gottfried Teichmann, Klaus Voss, Udo Krause, Jerry Keiper, David Withoff, and Yu He for their critical examination of early versions of the manuscript and their useful suggestions, and Sabine Trott for the first proofreading of the German manuscript. I also want to thank the participants of the original lectures for many useful discussions. My thanks go to the reviewers of this book: John Novak, Alec Schramm, Paul Abbott, Jim Feagin, Richard Palmer, Ward Hanson, Stan Wagon, and Markus van Almsick, for their suggestions and ideas for improvement. I thank Richard Crandall, Allan Hayes, Andrzej Kozlowski, Hartmut Wolf, Stephan Leibbrandt, George Kambouroglou, Domenico Minunni, Eric Weisstein, Andy Shiekh, Arthur G. Hubbard, Jay Warrendorff, Allan Cortzen, Ed Pegg, and Udo Krause for comments on the prepublication version of the *GuideBooks*. I thank Bobby R. Treat, Arthur G. Hubbard, Murray Eisenberg, Marvin Schaefer, Marek Duszynski, Daniel Lichtblau, Devendra Kapadia, Adam Strzebonski, Anton Antonov, and Brett Champion for useful comments on the *Mathematica* Version 5.1 tailored version of the *GuideBooks*.

My thanks are due to Gerhard Gobsch of the Institute for Physics of the Technical University in Ilmenau for the opportunity to develop and give these original lectures at the Institute, and to Stephen Wolfram who encouraged and supported me on this project.

Concerning the process of making the *Mathematica GuideBooks* from a set of lecture notes, I thank Glenn Scholebo for transforming notebooks to T_EX files, and Joe Kaiping for T_EX work related to the printed book. I thank John Novak and Jan Progen for putting all the material into good English style and grammar, John Bonadies for the chapter-opener graphics of the book, and Jean Buck for library work. I especially thank John Novak for the creation of *Mathematica* 3 notebooks from the T_EX files, and Andre Kuzniarek for his work on the stylesheet to give the notebooks a pleasing appearance. My thanks go to Andy Hunt who created a specialized stylesheet for the actual book printout and printed and formatted the 4×1000+ pages of the *Mathematica GuideBooks*. I thank Andy Hunt for making a first version of the homepage of the *GuideBooks* and Amy Young for creating the current version of the homepage of the *GuideBooks*. I thank Sophie Young for a final check of the English. My largest thanks go to Amy Young, who encouraged me to update the whole book over the years and who had a close look at all of my English writing and often improved it considerably. Despite reviews by many individuals any remaining mistakes or omissions, in the *Mathematica* code, in the mathematics, in the description of the *Mathematica* functions, in the English, or in the references, etc. are, of course, solely mine.

Let me take the opportunity to thank members of the Research and Development team of Wolfram Research whom I have met throughout the years, especially Victor Adamchik, Anton Antonov, Alexei Bocharov, Arnoud Buzing, Brett Champion, Matthew Cook, Todd Gayley, Darren Glosemeyer, Roger Germundsson, Unal Goktas, Yifan Hu, Devendra Kapadia, Zbigniew Leyk, David Librik, Daniel Lichtblau, Jerry Keiper, Robert Knapp, Roman Mäder, Oleg Marichev, John Novak, Peter Overmann, Oleksandr Pavlyk, Ulises Cervantes–Pimentel, Mark Sofroniou, Adam Strzebonski, Oyvind Tafjord, Robby Villegas, Tom Wickham–Jones, David Withoff, and Stephen Wolfram for numerous discussions about design principles, various small details, underlying algorithms, efficient implementation of various procedures, and tricks concerning *Mathematica*. The appearance of the notebooks profited from discussions with John Fultz, Paul Hinton, John Novak, Lou D'Andria, Theodore Gray, Andre Kuzniarek, Jason Harris, Andy Hunt, Christopher Carlson, Robert Raguet–Schofield, George Beck, Kai Xin, Chris Hill, and Neil Soiffer about front end, button, and typesetting issues.

It was an interesting and unique experience to work over the last 12 years with five editors: Allan Wylde, Paul Wellin, Maria Taylor, Wayne Yuhasz, and Ann Kostant, with whom the *GuideBooks* were finally published. Many book-related discussions that ultimately improved the *GuideBooks*, have been carried out with Jan Benes from TELOS and associates, Steven Pisano, Jenny Wolkowicki, Henry Krell, Fred Bartlett, Vaishali Damle, Ken Quinn, Jerry Lyons, and Rüdiger Gebauer from Springer New York.

The author hopes the *Mathematica GuideBooks* help the reader to discover, investigate, urbanize, and enjoy the computational paradise offered by *Mathematica*.

Wolfram Research, Inc.   Michael Trott
April 2005

References

1   A. Amthor. *Z. Math. Phys.* 25, 153 (1880).

2   K. Appel, W. Haken. *J. Math.* 21, 429 (1977).

3   A. Bauer, E. Clarke, X. Zhao. *J. Automat. Reasoning* 21, 295 (1998).

4   A. H. Bell. *Am. Math. Monthly* 2, 140 (1895).

5   M. Berz. *Adv. Imaging Electron Phys.* 108, 1 (2000).

6   R. F. Boisvert. *arXiv:cs.MS/*0004004 (2000).

7   B. Buchberger. *Theorema Project* (1997).   ftp://ftp.risc.uni-linz.ac.at/pub/techreports/1997/97-34/ed-media.nb

8   B. Buchberger. *SIGSAM Bull.* 36, 3 (2002).

9   S.-C. Chou, X.-S. Gao, J.-Z. Zhang. *Machine Proofs in Geometry*, World Scientific, Singapore, 1994.

10   A. M. Cohen. *Nieuw Archief Wiskunde* 14, 45 (1996).

11   A. Cook. *The Motion of the Moon*, Adam-Hilger, Bristol, 1988.

12   C. Delaunay. *Théorie du Mouvement de la Lune*, Gauthier-Villars, Paris, 1860.

13   C. Delaunay. *Mem. de l' Acad. des Sc. Paris* 28 (1860).

14   C. Delaunay. *Mem. de l' Acad. des Sc. Paris* 29 (1867).

15   A. Deprit, J. Henrard, A. Rom. *Astron. J.* 75, 747 (1970).

16   A. Deprit. *Science* 168, 1569 (1970).

17   A. Deprit, J. Henrard, A. Rom. *Astron. J.* 76, 273 (1971).

18   P. J. Dolan, Jr., D. S. Melichian. *Am. J. Phys.* 66, 11 (1998).

19   S. P. Ferguson, T. C. Hales. *arXiv:math.MG/* 9811072 (1998).

20   B. Fitelson. *Mathematica Educ. Res.* 7, n1, 17 (1998).

21   A. C. Fowler. *Mathematical Models in the Applied Sciences,* Cambridge University Press, Cambridge, 1997.

22   H. Fritsch, G. Fritsch. *The Four-Color Theorem*, Springer-Verlag, New York, 1998.

23   E. Gallopoulos, E. Houstis, J. R. Rice (eds.). *Future Research Directions in Problem Solving Environments for Computational Science: Report of a Workshop on Research Directions in Integrating Numerical Analysis, Symbolic Computing, Computational Geometry, and Artificial Intelligence for Computational Science*, 1991. http://www.cs.purdue.edu/research/cse/publications/tr/92/92-032.ps.gz

24   V. Gerdt, S. A. Gogilidze in V. G. Ganzha, E. W. Mayr, E. V. Vorozhtsov (eds.). *Computer Algebra in Scientific Computing*, Springer-Verlag, Berlin, 1999.

25   M. C. Gutzwiller, D. S. Schmidt. *Astronomical Papers: The Motion of the Moon as Computed by the Method of Hill, Brown, and Eckert*, U.S. Government Printing Office, Washington, 1986.

26   M. C. Gutzwiller. *Rev. Mod. Phys.* 70, 589 (1998).

27   Y. Hagihara. *Celestial Mechanics* vII/1, MIT Press, Cambridge, 1972.

28   T. C. Hales. *arXiv:math.MG/* 9811071 (1998).

29   T. C. Hales. *arXiv:math.MG/* 9811073 (1998).

30   T. C. Hales. *arXiv:math.MG/* 9811074 (1998).

31   T. C. Hales. *arXiv:math.MG/* 9811075 (1998).

32   T. C. Hales. *arXiv:math.MG/* 9811076 (1998).

33   T. C. Hales. *arXiv:math.MG/* 9811077 (1998).

34   T. C. Hales. *arXiv:math.MG/* 9811078 (1998).

35   T. C. Hales. *arXiv:math.MG/*0205208 (2002).

36   T. C. Hales in L. Tatsien (ed.). *Proceedings of the International Congress of Mathematicians* v. 3, Higher Education Press, Beijing, 2002.

37   J. Harrison. *Theorem Proving with the Real Numbers*, Springer-Verlag, London, 1998.

38   J. Hermes. *Nachrichten Königl. Gesell. Wiss. Göttingen* 170 (1894).

39   E. N. Houstis, J. R. Rice, E. Gallopoulos, R. Bramley (eds.). *Enabling Technologies for Computational Science*, Kluwer, Boston, 2000.

40   E. N. Houstis, J. R. Rice. *Math. Comput. Simul.* 54, 243 (2000).

41   M. S. Klamkin (eds.). *Mathematical Modelling*, SIAM, Philadelphia, 1996.

42   H. Koch, A. Schenkel, P. Wittwer. *SIAM Rev.* 38, 565 (1996).

43   Y. N. Lakshman, B. Char, J. Johnson in O. Gloor (ed.). *ISSAC 1998*, ACM Press, New York, 1998.

44   W. McCune. *Robbins Algebras Are Boolean*, 1997.   http://www.mcs.anl.gov/home/mccune/ar/robbins/

45   E. Mach (R. Wahsner, H.-H. von Borszeskowski eds.). *Die Mechanik in ihrer Entwicklung*, Akademie-Verlag, Berlin, 1988.

46   D. A. MacKenzie. *Mechanizing Proof: Computing, Risk, and Trust*, MIT Press, Cambridge, 2001.

47   B. M. McCoy. *arXiv:cond-mat*/0012193 (2000).

48   K. J. M. Moriarty, G. Murdeshwar, S. Sanielevici. *Comput. Phys. Commun.* 77, 325 (1993).

49   I. Nemes, M. Petkovšek, H. S. Wilf, D. Zeilberger. *Am. Math. Monthly* 104, 505 (1997).

50   W. H. Press, S. A. Teukolsky. *Comput. Phys.* 11, 417 (1997).

51   D. Rawlings. *Am. Math. Monthly* 108, 713 (2001).

52   *Problem Solving Environments Home Page.*   http://www.cs.purdue.edu/research/cse/pses

53   D. S. Schmidt in H. S. Dumas, K. R. Meyer, D. S. Schmidt (eds.). *Hamiltonian Dynamical Systems*, Springer-Verlag, New York, 1995.

54   S. Seiden. *SIGACT News* 32, 111 (2001).

55   S. Seiden. *Theor. Comput. Sc.* 282, 381 (2002).

56   C. Simpson. *arXiv:math.HO/*0311260 (2003).

57   A. M. Stoneham. *Phil. Trans. R. Soc. Lond.* A 360, 1107 (2002).

58   M. Tegmark. *Ann. Phys.* 270, 1 (1999).

59   S. Wolfram. *Mathematica: A System for Doing Mathematics by Computer*, Addison-Wesley, Redwood City, 1992.

60   S. Wolfram. *The Mathematica Book*, Wolfram Media, Champaign, 2003.

Contents

CHAPTER *2*

Classical Orthogonal Polynomials

CHAPTER *3*

Classical Special Functions

Introduction and Orientation to *The Mathematica GuideBooks*

0.1 Overview

0.1.1 Content Summaries

The *Mathematica GuideBooks* are published as four independent books: *The Mathematica GuideBook to Programming, The Mathematica GuideBook to Graphics, The Mathematica GuideBook to Numerics*, and *The Mathematica GuideBook to Symbolics*.

■ The Programming volume deals with the structure of *Mathematica* expressions and with *Mathematica* as a programming language. This volume includes the discussion of the hierarchical construction of all *Mathematica* objects out of symbolic expressions (all of the form *head* [*argument*]), the ultimate building blocks of expressions (numbers, symbols, and strings), the definition of functions, the application of rules, the recognition of patterns and their efficient application, the order of evaluation, program flows and program structure, the manipulation of lists (the universal container for *Mathematica* expressions of all kinds), as well as a number of topics specific to the *Mathematica* programming language. Various programming styles, especially *Mathematica*'s powerful functional programming constructs, are covered in detail.

■ The Graphics volume deals with *Mathematica*'s two-dimensional (2D) and three-dimensional (3D) graphics. The chapters of this volume give a detailed treatment on how to create images from graphics primitives, such as points, lines, and polygons. This volume also covers graphically displaying functions given either analytically or in discrete form. A number of images from the *Mathematica* Graphics Gallery are also reconstructed. Also discussed is the generation of pleasing scientific visualizations of functions, formulas, and algorithms. A variety of such examples are given.

■ The Numerics volume deals with *Mathematica*'s numerical mathematics capabilities—the indispensable sledgehammer tools for dealing with virtually any "real life" problem. The arithmetic types (fast machine, exact integer and rational, verified high-precision, and interval arithmetic) are carefully analyzed. Fundamental numerical operations, such as compilation of programs, numerical Fourier transforms, minimization, numerical solution of equations, and ordinary/partial differential equations are analyzed in detail and are applied to a large number of examples in the main text and in the solutions to the exercises.

■ The Symbolics volume deals with *Mathematica*'s symbolic mathematical capabilities—the real heart of *Mathematica* and the ingredient of the *Mathematica* software system that makes it so unique and powerful. Structural and mathematical operations on systems of polynomials are fundamental to many symbolic calculations and are covered in detail. The solution of equations and differential equations, as well as the classical calculus operations, are exhaustively treated. In addition, this volume discusses and employs the classical orthogonal polynomials and special functions of mathematical physics. To demonstrate the symbolic mathematics power, a variety of problems from mathematics and physics are discussed.

The four *GuideBooks* contain about 25,000 *Mathematica* inputs, representing more than 75,000 lines of commented *Mathematica* code. (For the reader already familiar with *Mathematica*, here is a more precise measure: The LeafCount of all inputs would be about 900,000 when collected in a list.) The *GuideBooks* also have more than 4,000 graphics, 150 animations, 11,000 references, and 1,000 exercises. More than 10,000 hyperlinked index entries and hundreds of hyperlinks from the overview sections connect all parts in a convenient way. The evaluated notebooks of all four volumes have a cumulative file size of about 20 GB. Although these numbers may sound large, the *Mathematica GuideBooks* actually cover only a portion of *Mathematica*'s functionality and features and give only a glimpse into the possibilities *Mathematica* offers to generate graphics, solve problems, model systems, and discover new identities, relations, and algorithms. The *Mathematica* code is explained in detail throughout all chapters. More than 13,000 comments are scattered throughout all inputs and code fragments.

0.1.2 Relation of the Four Volumes

The four volumes of the *GuideBooks* are basically independent, in the sense that readers familiar with *Mathematica* programming can read any of the other three volumes. But a solid working knowledge of the main topics discussed in *The Mathematica GuideBook to Programming*—symbolic expressions, pure functions, rules and replacements, and list manipulations—is required for the Graphics, Numerics, and Symbolics volumes. Compared to these three volumes, the Programming volume might appear to be a bit "dry". But similar to learning a foreign language, before being rewarded with the beauty of novels or a poem, one has to sweat and study. The whole suite of graphical capabilities and all of the mathematical knowledge in *Mathematica* are accessed and applied through lists, patterns, rules, and pure functions, the material discussed in the Programming volume.

Naturally, graphics are the center of attention of the *The Mathematica GuideBook to Graphics*. While in the Programming volume some plotting and graphics for visualization are used, graphics are not crucial for the Programming volume. The reader can safely skip the corresponding inputs to follow the main programming threads. The Numerics and Symbolics volumes, on the other hand, make heavy use of the graphics knowledge acquired in the Graphics volume. Hence, the prerequisites for the Numerics and Symbolics volumes are a good knowledge of *Mathematica*'s programming language and of its graphics system.

The Programming volume contains only a few percent of all graphics, the Graphics volume contains about two-thirds, and the Numerics and Symbolics volume, about one-third of the overall 4,000+ graphics. The Programming and Graphics volumes use some mathematical commands, but they restrict the use to a relatively small number (especially Expand, Factor, Integrate, Solve). And the use of the function N for numerical ization is unavoidable for virtually any "real life" application of *Mathematica*. The last functions allow us to treat some mathematically not uninteresting examples in the Programming and Graphics volumes. In addition to putting these functions to work for nontrivial problems, a detailed discussion of the mathematics functions of *Mathematica* takes place exclusively in the Numerics and Symbolics volumes.

The Programming and Graphics volumes contain a moderate amount of mathematics in the examples and exercises, and focus on programming and graphics issues. The Numerics and Symbolics volumes contain a substantially larger amount of mathematics.

Although printed as four books, the fourteen individual chapters (six in the Programming volume, three in the Graphics volume, two in the Numerics volume, and three in the Symbolics volume) of the *Mathematica Guide-Books* form one organic whole, and the author recommends a strictly sequential reading, starting from Chapter 1 of the Programming volume and ending with Chapter 3 of the Symbolics volume for gaining the maximum benefit. The electronic component of each book contains the text and inputs from all the four *GuideBooks*, together with a comprehensive hyperlinked index. The four volumes refer frequently to one another.

0.1.3 Chapter Structure

A rough outline of the content of a chapter is the following:

■ The main body discusses the *Mathematica* functions belonging to the chapter subject, as well their options and attributes. Generically, the author has attempted to introduce the functions in a "natural order". But surely, one cannot be axiomatic with respect to the order. (Such an order of the functions is not unique, and the author intentionally has "spread out" the introduction of various *Mathematica* functions across the four volumes.) With the introduction of a function, some small examples of how to use the functions and comparisons of this function with related ones are given. These examples typically (with the exception of some visualizations in the Programming volume) incorporate functions already discussed. The last section of a chapter often gives a larger example that makes heavy use of the functions discussed in the chapter.

■ A programmatically constructed overview of each chapter functions follows. The functions listed in this section are hyperlinked to their attributes and options, as well as to the corresponding reference guide entries of *The Mathematica Book*.

■ A set of exercises and potential solutions follow. Because learning *Mathematica* through examples is very efficient, the proposed solutions are quite detailed and form up to 50% of the material of a chapter.

■ References end the chapter.

Note that the first few chapters of the Programming volume deviate slightly from this structure. Chapter 1 of the Programming volume gives a general overview of the kind of problems dealt with in the four *GuideBooks*. The second, third, and fourth chapters of the Programming volume introduce the basics of programming in *Mathematica*. Starting with Chapters 5 of the Programming volume and throughout the Graphics, Numerics, and Symbolics volumes, the above-described structure applies.

In the 14 chapters of the *GuideBooks* the author has chosen a "we" style for the discussions of how to proceed in constructing programs and carrying out calculations to include the reader intimately.

0.1.4 Code Presentation Style

The typical style of a unit of the main part of a chapter is: Define a new function, discuss its arguments, options, and attributes, and then give examples of its usage. The examples are virtually always *Mathematica* inputs and outputs. The majority of inputs is in `InputForm` are the notebooks. On occasion `StandardForm` is also used. Although `StandardForm` mimics classical mathematics notation and makes short inputs more readable, for "program-like" inputs, `InputForm` is typically more readable and easier and more natural to align. For the outputs, `StandardForm` is used by default and occasionally the author has resorted to `InputForm` or `FullForm` to expose digits of numbers and to `TraditionalForm` for some formulas. Outputs are mostly not programs, but nearly always "results" (often mathematical expressions, formulas, identities, or lists of numbers rather than program constructs). The world of *Mathematica* users is divided into three groups, and each of them has a nearly religious opinion on how to format *Mathematica* code [1], [2]. The author follows the `InputForm` cult(ure) and hopes that the *Mathematica* users who do everything in either StandardForm or Traditional Form will bear with him. If the reader really wants to see all code in either StandardForm or Traditional Form, this can easily be done with the **Convert To** item from the **Cell** menu. (Note that the relation between InputForm and StandardForm is not symmetric. The InputForm cells of this book have been line-broken and aligned by hand. Transforming them into StandardForm or TraditionalForm cells works well because one typically does not line-break manually and align *Mathematica* code in these cell types. But converting StandardForm or TraditionalForm cells into InputForm cells results in much less pleasing results.)

In the inputs, special typeset symbols for *Mathematica* functions are typically avoided because they are not monospaced. But the author does occasionally compromise and use Greek, script, Gothic, and doublestruck characters.

In a book about a programming language, two other issues come always up: indentation and placement of the code.

- The code of the *GuideBooks* is largely consistently formatted and indented. There are no strict guidelines or even rules on how to format and indent *Mathematica* code. The author hopes the reader will find the book's formatting style readable. It is a compromise between readability (mental parsabililty) and space conservation, so that the printed version of the *Mathematica GuideBook* matches closely the electronic version.

- Because of the large number of examples, a rather imposing amount of *Mathematica* code is presented. Should this code be present only on the disk, or also in the printed book? If it is in the printed book, should it be at the position where the code is used or at the end of the book in an appendix? Many authors of *Mathematica* articles and books have strong opinions on this subject. Because the main emphasis of the *Mathematica GuideBooks* is on *solving* problems *with Mathematica* and not on the actual problems, the *GuideBooks* give all of the code at the point where it is needed in the printed book, rather than "hiding" it in packages and appendices. In addition to being more straightforward to read and conveniently allowing us to refer to elements of the code pieces, this placement makes the correspondence between the printed book and the notebooks close to 1:1, and so working back and forth between the printed book and the notebooks is as straightforward as possible.

0.2 Requirements

0.2.1 Hardware and Software

Throughout the *GuideBooks*, it is assumed that the reader has access to a computer running a current version of *Mathematica* (version 5.0/5.1 or newer). For readers without access to a licensed copy of *Mathematica*, it is possible to view all of the material on the disk using a trial version of *Mathematica*. (A trial version is downloadable from http://www.wolfram.com/products/mathematica/trial.cgi.)

The files of the *GuideBooks* are relatively large, altogether more than 20 GB. This is also the amount of hard disk space needed to store uncompressed versions of the notebooks. To view the notebooks comfortably, the reader's computer needs 128 MB RAM; to evaluate the evaluation units of the notebooks 1 GB RAM or more is recommended.

In the *GuideBooks*, a large number of animations are generated. Although they need more memory than single pictures, they are easy to create, to animate, and to store on typical year-2005 hardware, and they provide a lot of joy.

0.2.2 Reader Prerequisites

Although prior *Mathematica* knowledge is not needed to read *The Mathematica GuideBook to Programming*, it is assumed that the reader is familiar with basic actions in the *Mathematica* front end, including entering Greek characters using the keyboard, copying and pasting cells, and so on. Freely available tutorials on these (and other) subjects can be found at http://library.wolfram.com.

For a complete understanding of most of the *GuideBooks* examples, it is desirable to have a background in mathematics, science, or engineering at about the bachelor's level or above. Familiarity with mechanics and electrodynamics is assumed. Some examples and exercises are more specialized, for instance, from quantum mechanics, finite element analysis, statistical mechanics, solid state physics, number theory, and other areas. But the *GuideBooks* avoid very advanced (but tempting) topics such as renormalization groups [6], parquet approximations [27], and modular moonshines [14]. (Although *Mathematica* can deal with such topics, they do not fit the character of the *Mathematica GuideBooks* but rather the one of a *Mathematica Topographical Atlas* [a monumental work to be carried out by the *Mathematica*–Bourbakians of the 21st century]).

Each scientific application discussed has a set of references. The references should easily give the reader both an overview of the subject and pointers to further references.

0.3 What the GuideBooks Are and What They Are Not

0.3.1 Doing Computer Mathematics

As discussed in the Preface, the main goal of the *GuideBooks* is to demonstrate, showcase, teach, and exemplify scientific problem solving with *Mathematica*. An important step in achieving this goal is the discussion of *Mathematica* functions that allow readers to become fluent in programming when creating complicated graphics or solving scientific problems. This again means that the reader must become familiar with the most important programming, graphics, numerics, and symbolics functions, their arguments, options, attributes, and a few of their time and space complexities. And the reader must know which functions to use in each situation.

The *GuideBooks* treat only aspects of *Mathematica* that are ultimately related to "doing mathematics". This means that the *GuideBooks* focus on the functionalities of the kernel rather than on those of the front end. The knowledge required to use the front end to work with the notebooks can easily be gained by reading the corresponding chapters of the online documentation of *Mathematica*. Some of the subjects that are treated either lightly or not at all in the *GuideBooks* include the basic use of *Mathematica* (starting the program, features, and special properties of the notebook front end [16]), typesetting, the preparation of packages, external file operations, the communication of *Mathematica* with other programs via *MathLink*, special formatting and string manipulations, computer- and operating system-specific operations, audio generation, and commands available in various packages. "Packages" includes both, those distributed with *Mathematica* as well as those available from the *Mathematica* Information Center (http://library.wolfram.com/infocenter) and commercial sources, such as MathTensor for doing general relativity calculations (http://smc.vnet.net/MathTensor.html) or FeynCalc for doing high-energy physics calculations (http://www.feyncalc.org). This means, in particular, that probability and statistical calculations are barely touched on because most of the relevant commands are contained in the packages. The *GuideBooks* make little or no mention of the machine-dependent possibilities offered by the various *Mathematica* implementations. For this information, see the *Mathematica* documentation.

Mathematical and physical remarks introduce certain subjects and formulas to make the associated *Mathematica* implementations easier to understand. These remarks are not meant to provide a deep understanding of the (sometimes complicated) physical model or underlying mathematics; some of these remarks intentionally oversimplify matters.

The reader should examine all *Mathematica* inputs and outputs carefully. Sometimes, the inputs and outputs illustrate little-known or seldom-used aspects of *Mathematica* commands. Moreover, for the efficient use of *Mathematica*, it is very important to understand the possibilities and limits of the built-in commands. Many commands in *Mathematica* allow different numbers of arguments. When a given command is called with fewer than the maximum number of arguments, an internal (or user-defined) default value is used for the missing arguments. For most of the commands, the maximum number of arguments and default values are discussed.

When solving problems, the *GuideBooks* generically use a "straightforward" approach. This means they are not using particularly clever tricks to solve problems, but rather direct, possibly computationally more expensive, approaches. (From time to time, the *GuideBooks* even make use of a "brute force" approach.) The motivation is that when solving new "real life" problems a reader encounters in daily work, the "right mathematical trick" is seldom at hand. Nevertheless, the reader can more often than not rely on *Mathematica* being powerful enough to often succeed in using a straightforward approach. But attention is paid to *Mathematica*-specific issues to find time- and memory-efficient implementations—something that should be taken into account for any larger program.

As already mentioned, all larger pieces of code in this book have comments explaining the individual steps carried out in the calculations. Many smaller pieces of code have comments when needed to expedite the understanding of how they work. This enables the reader to easily change and adapt the code pieces. Sometimes, when the translation from traditional mathematics into *Mathematica* is trivial, or when the author wants to emphasize certain aspects of the code, we let the code "speak for itself". While paying attention to efficiency, the *GuideBooks* only occasionally go into the computational complexity ([8], [40], and [7]) of the given implementations. The implementation of very large, complicated suites of algorithms is not the purpose of the *GuideBooks*. The *Mathematica* packages included with *Mathematica* and the ones at *MathSource* (http://library.wolfram.com/database/MathSource) offer a rich variety of self-study material on building large programs. Most general guidelines for writing code for scientific calculations (like descriptive variable names and modularity of code; see, e.g., [19] for a review) apply also to *Mathematica* programs.

The programs given in a chapter typically make use of *Mathematica* functions discussed in earlier chapters. Using commands from later chapters would sometimes allow for more efficient techniques. Also, these programs emphasize the use of commands from the current chapter. So, for example, instead of list operation, from a complexity point of view, hashing techniques or tailored data structures might be preferable. All subsections and sections are "self-contained" (meaning that no other code than the one presented is needed to evaluate the subsections and sections). The price for this "self-containedness" is that from time to time some code has to be repeated (such as manipulating polygons or forming random permutations of lists) instead of delegating such programming constructs to a package. Because this repetition could be construed as boring, the author typically uses a slightly different implementation to achieve the same goal.

0.3.2 Programming Paradigms

In the *GuideBooks,* the author wants to show the reader that *Mathematica* supports various programming paradigms and also show that, depending on the problem under consideration and the goal (e.g., solution of a problem, test of an algorithm, development of a program), each style has its advantages and disadvantages. (For a general discussion concerning programming styles, see [3], [41], [23], [32], [15], and [9].) *Mathematica* supports a functional programming style. Thus, in addition to classical procedural programs (which are often less efficient and less elegant), programs using the functional style are also presented. In the first volume of the *Mathematica GuideBooks*, the programming style is usually dictated by the types of commands that have been discussed up to that point. A certain portion of the programs involve recursive, rule-based programming. The choice of programming style is, of course, partially (ultimately) a matter of personal preference. The *GuideBooks'* main aim is to explain the operation, limits, and efficient application of the various *Mathematica* commands. For certain commands, this dictates a certain style of programming. However, the various programming styles, with their advantages and disadvantages, are not the main concern of the *GuideBooks*. In working with *Mathematica*, the reader is likely to use different programming styles depending if one wants a quick one-time calculation or a routine that will be used repeatedly. So, for a given implementation, the program structure may not always be the most elegant, fastest, or "prettiest".

The *GuideBooks* are not a substitute for the study of *The Mathematica Book* [45] http://documents. wolfram.com/mathematica). It is impossible to acquire a deeper (full) understanding of *Mathematica* without a *thorough* study of this book (reading it twice from the first to the last page is highly recommended). It *defines* the language and the spirit of *Mathematica*. The reader will probably from time to time need to refer to parts of it, because not all commands are discussed in the *GuideBooks*. However, the story of what can be done with *Mathematica* does not end with the examples shown in *The Mathematica Book*. The *Mathematica GuideBooks* go beyond *The Mathematica Book*. They present larger programs for solving various problems and creating complicated graphics. In addition, the *GuideBooks* discuss a number of commands that are not or are only fleetingly mentioned in the manual (e.g., some specialized methods of mathematical functions and functions from the `Developer`` and `Experimental`` contexts), but which the author deems important. In the notebooks, the author gives special emphasis to discussions, remarks, and applications relating to several commands that are typical for *Mathematica* but not for most other programming languages, e.g., `Map`, `MapAt`, `MapIndexed`, `Distribute`, `Apply`, `Replace`, `ReplaceAll`, `Inner`, `Outer`, `Fold`, `Nest`, `Nest`. `List`, `FixedPoint`, `FixedPointList`, and `Function`. These commands allow to write exceptionally elegant, fast, and powerful programs. All of these commands are discussed in *The Mathematica Book* and others that deal with programming in *Mathematica* (e.g., [33], [34], and [42]). However, the author's experience suggests that a deeper understanding of these commands and their optimal applications comes only after working with *Mathematica* in the solution of more complicated problems.

Both the printed book and the electronic component contain material that is meant to teach in detail how to use *Mathematica* to solve problems, rather than to present the underlying details of the various scientific examples. It cannot be overemphasized that to master the use of *Mathematica,* its programming paradigms and individual functions, the reader must experiment; this is especially important, insightful, easily verifiable, and satisfying with graphics, which involve manipulating expressions, making small changes, and finding different approaches. Because the results can easily be visually checked, generating and modifying graphics is an ideal method to learn programming in *Mathematica*.

0.4 Exercises and Solutions

0.4.1 Exercises

Each chapter includes a set of exercises and a detailed solution proposal for each exercise. When possible, all of the purely *Mathematica*-programming related exercises (these are most of the exercises of the Programming volume) should be solved by every reader. The exercises coming from mathematics, physics, and engineering should be solved according to the reader's interest. The most important *Mathematica* functions needed to solve a given problem are generally those of the associated chapter.

For a rough orientation about the content of an exercise, the subject is included in its title. The relative degree of difficulty is indicated by level superscript of the exercise number (^{L1} indicates easy, ^{L2} indicates medium, and ^{L3} indicates difficult). The author's aim was to present understandable interesting examples that illustrate the *Mathematica* material discussed in the corresponding chapter. Some exercises were inspired by recent research problems; the references given allow the interested reader to dig deeper into the subject.

The exercises are intentionally not hyperlinked to the corresponding solution. The independent solving of the exercises is an important part of learning *Mathematica*.

0.4.2 Solutions

The *GuideBooks* contain solutions to each of the more than 1,000 exercises. Many of the techniques used in the solutions are not just one-line calls to built-in functions. It might well be that with further enhancements, a future version of *Mathematica* might be able to solve the problem more directly. (But due to different forms of some results returned by *Mathematica*, some problems might also become more challenging.) The author encourages the reader to try to find shorter, more clever, faster (in terms of runtime as well complexity), more general, and more elegant solutions. *Doing* various calculations is the most effective way to learn *Mathematica*. A proper *Mathematica* implementation of a function that solves a given problem often contains many different elements. The function(s) should have sensibly named and sensibly behaving options; for various (machine numeric, high-precision numeric, symbolic) inputs different steps might be required; shielding against inappropriate input might be needed; different parameter values might require different solution strategies and algorithms, helpful error and warning messages should be available. The returned data structure should be intuitive and easy to reuse; to achieve a good computational complexity, nontrivial data structures might be needed, etc. Most of the solutions do not deal with all of these issues, but only with selected ones and thereby leave plenty of room for more detailed treatments; as far as limit, boundary, and degenerate cases are concerned, they represent an outline of how to tackle the problem. Although the solutions do their job in general, they often allow considerable refinement and extension by the reader.

The reader should consider the given solution to a given exercise as a proposal; quite different approaches are often possible and sometimes even more efficient. The routines presented in the solutions are not the most general possible, because to make them foolproof for every possible input (sensible and nonsensical, evaluated and unevaluated, numerical and symbolical), the books would have had to go considerably beyond the mathematical and physical framework of the *GuideBooks*. In addition, few warnings are implemented for improper or improperly used arguments. The graphics provided in the solutions are mostly subject to a long list of refinements. Although the solutions do work, they are often sketchy and can be considerably refined and extended by the reader. This also means that the provided solutions to the exercises programs are not always very suitable for solving larger classes of problems. To increase their applicability would require considerably more code. Thus, it is not guaranteed that given routines will work correctly on related problems. To guarantee this generality and scalability, one would have to protect the variables better, implement formulas for more general or specialized cases, write functions to accept different numbers of variables, add type-checking and error-checking functions, and include corresponding error messages and warnings.

To simplify working through the solutions, the various steps of the solution are commented and are not always packed in a `Module` or `Block`. In general, only functions that are used later are packed. For longer calculations, such as those in some of the exercises, this was not feasible and intended. The arguments of the functions are not always checked for their appropriateness as is desirable for robust code. But, this makes it easier for the user to test and modify the code.

0.5 The Books Versus the Electronic Components

0.5.1 Working with the Notebooks

Each volume of the *GuideBooks* comes with a multiplatform DVD, containing fourteen main notebooks tailored for *Mathematica* 4 and compatible with *Mathematica* 5. Each notebook corresponds to a chapter from the printed books. (To avoid large file sizes of the notebooks, all animations are located in the Animations directory and not directly in the chapter notebooks.) The chapters (and so the corresponding notebooks) contain a detailed description and explanation of the *Mathematica* commands needed and used in applications of *Mathematica* to the sciences. Discussions on *Mathematica* functions are supplemented by a variety of mathematics, physics, and graphics examples. The notebooks also contain complete solutions to all exercises. Forming an electronic book, the notebooks also contain all text, as well as fully typeset formulas, and reader-editable and reader-changeable input. (Readers can copy, paste, and use the inputs in their notebooks.) In addition to the chapter notebooks, the DVD also includes a navigation palette and fully hyperlinked table of contents and index notebooks. The *Mathematica* notebooks corresponding to the printed book are fully evaluated. The evaluated chapter notebooks also come with hyperlinked overviews; these overviews are not in the printed book.

When reading the printed books, it might seem that some parts are longer than needed. The reader should keep in mind that the primary tool for working with the *Mathematica* kernel are *Mathematica* notebooks and that on a computer screen and there "length does not matter much". The *GuideBooks* are basically a printout of the notebooks, which makes going back and forth between the printed books and the notebooks very easy. The *GuideBooks* give large examples to encourage the reader to investigate various *Mathematica* functions and to become familiar with *Mathematica* as a system for doing mathematics, as well as a programming language. Investigating *Mathematica* in the accompanying notebooks is the best way to learn its details.

To start viewing the notebooks, open the table of contents notebook TableOfContents.nb. *Mathematica* notebooks can contain hyperlinks, and all entries of the table of contents are hyperlinked. Navigating through one of the chapters is convenient when done using the navigator palette GuideBooksNavigator.nb.

When opening a notebook, the front end minimizes the amount of memory needed to display the notebook by loading it incrementally. Depending on the reader's hardware, this might result in a slow scrolling speed. Clicking the "Load notebook cache" button of the GuideBooksNavigator palette speeds this up by loading the complete notebook into the front end.

For the vast majority of sections, subsections, and solutions of the exercises, the reader can just select such a structural unit and evaluate it (at once) on a year-2005 computer ($\geq$512 MB RAM) typically in a matter of minutes. Some sections and solutions containing many graphics may need hours of computation time. Also, more than 50 pieces of code run hours, even days. The inputs that are very memory intensive or produce large outputs and graphics are in inactive cells which can be activated by clicking the adjacent button. Because of potentially overlapping variable names between various sections and subsections, the author advises the reader not to evaluate an entire chapter at once.

Each smallest self-contained structural unit (a subsection, a section without subsections, or an exercise) should be evaluated within one *Mathematica* session starting with a freshly started kernel. At the end of each unit is an input cell. After evaluating all input cells of a unit in consecutive order, the input of this cell generates a short summary about the entire *Mathematica* session. It lists the number of evaluated inputs, the kernel CPU time, the wall clock time, and the maximal memory used to evaluate the inputs (excluding the resources needed to evaluate the Program cells). These numbers serve as a guide for the reader about the to-be-expected running times and memory needs. These numbers can deviate from run to run. The wall clock time can be substantially larger than the CPU time due to other processes running on the same computer and due to time needed to render graphics. The data shown in the evaluated notebooks came from a 2.5 GHz Linux computer. The CPU times are generically proportional to the computer clock speed, but can deviate within a small factor from operating system to operating system. In rare, randomly occurring cases slower computers can achieve smaller CPU and wall clock times than faster computers, due to internal time-constrained simplification processes in various symbolic mathematics functions (such as Integrate, Sum, DSolve, ...).

The Overview Section of the chapters is set up for a front end and kernel running on the same computer and having access to the same file system. When using a remote kernel, the directory specification for the package Overview.m must be changed accordingly.

References can be conveniently extracted from the main text by selecting the cell(s) that refer to them (or parts of a cell) and then clicking the "Extract References" button. A new notebook with the extracted references will then appear.

The notebooks contain color graphics. (To rerender the pictures with a greater color depth or at a larger size, choose **Rerender Graphics** from the **Cell** menu.) With some of the colors used, black-and-white printouts occasionally give low-contrast results. For better black-and-white printouts of these graphics, the author recommends setting the ColorOutput option of the relevant graphics function to GrayLevel. The notebooks with animations (in the printed book, animations are typically printed as an array of about 10 to 20 individual graphics) typically contain between 60 and 120 frames. Rerunning the corresponding code with a large number of frames will allow the reader to generate smoother and longer-running animations.

Because many cell styles used in the notebooks are unique to the *GuideBooks*, when copying expressions and cells from the *GuideBooks* notebooks to other notebooks, one should first attach the style sheet notebook GuideBooksStylesheet.nb to the destination notebook, or define the needed styles in the style sheet of the destination notebook.

0.5.2 Reproducibility of the Results

The 14 chapter notebooks contained in the electronic version of the *GuideBooks* were run mostly with *Mathematica* 5.1 on a 2 GHz Intel Linux computer with 2 GB RAM. They need more than 100 hours of evaluation time. (This does not include the evaluation of the currently unevaluatable parts of code after the **Make Input** buttons.) For most subsections and sections, 512 MB RAM are recommended for a fast and smooth evaluation "at once" (meaning the reader can select the section or subsection, and evaluate all inputs without running out of memory or clearing variables) and the rendering of the generated graphic in the front end. Some subsections and sections need more memory when run. To reduce these memory requirements, the author recommends restarting the *Mathematica* kernel inside these subsections and sections, evaluating the necessary definitions, and then continuing. This will allow the reader to evaluate all inputs.

In general, regardless of the computer, with the same version of *Mathematica*, the reader should get the same results as shown in the notebooks. (The author has tested the code on Sun and Intel-based Linux computers, but this does not mean that some code might not run as displayed (because of different configurations, stack size settings, etc., but the disclaimer from the Preface applies everywhere). If an input does not work on a particular machine, please inform the author. Some deviations from the results given may appear because of the following:
■ Inputs involving the function `Random[...]` in some form. (Often `SeedRandom` to allow for some kind of reproducibility and randomness at the same time is employed.)
■ *Mathematica* commands operating on the file system of the computer, or make use of the type of computer (such inputs need to be edited using the appropriate directory specifications).
■ Calculations showing some of the differences of floating-point numbers and the machine-dependent representation of these on various computers.
■ Pictures using various fonts and sizes because of their availability (or lack thereof) and shape on different computers.
■ Calculations involving `Timing` because of different clock speeds, architectures, operating systems, and libraries.
■ Formats of results depending on the actual window width and default font size. (Often, the corresponding inputs will contain `Short`.)

Using anything other than *Mathematica* Version 5.1 might also result in different outputs. Examples of results that change form, but are all mathematically correct and equivalent, are the parameter variables used in underdetermined systems of linear equations, the form of the results of an integral, and the internal form of functions like `InterpolatingFunction` and `CompiledFunction`. Some inputs might no longer evaluate the same way because functions from a package were used and these functions are potentially built-in functions in a later *Mathematica* version. *Mathematica* is a very large and complicated program that is constantly updated and improved. Some of these changes might be design changes, superseded functionality, or potentially regressions, and as a result, some of the inputs might not work at all or give unexpected results in future versions of *Mathematica*.

0.5.3 Earlier Versions of the Notebooks

The first printing of the Programming volume and the Graphics volumes of the *Mathematica GuideBooks* were published in October 2004. The electronic components of these two books contained the corresponding evaluated chapter notebooks as well as unevaluated versions of preversions of the notebooks belonging to the Numerics and Symbolics volumes. Similarly, the electronic components of the Numerics and Symbolics volume contain the corresponding evaluated chapter notebooks and unevaluated copies of the notebooks of the Programming and Graphics volumes. This allows the reader to follow cross-references and look up relevant concepts discussed in the other volumes. The author has tried to keep the notebooks of the *GuideBooks* as up-to-date as possible. (Meaning with respect to the efficient and appropriate use of the latest version of *Mathematica*, with respect to maintaining a list of references that contains new publications, and examples, and with respect to incorporating corrections to known problems, errors, and mistakes). As a result, the notebooks of all four volumes that come with later printings of the Programming and Graphics volumes, as well with the Numerics and Symbolics volumes will be different and supersede the earlier notebooks originally distributed with the Programming and Graphics volumes. The notebooks that come with the Numerics and Symbolics volumes are genuine *Mathematica* Version 5.1 notebooks. Because most advances in *Mathematica* Version 5 and 5.1 compared with *Mathematica* Version 4 occurred in functions carrying out numerical and symbolical calculations, the notebooks associated with Numerics and Symbolics volumes contain a substantial amount of changes and additions compared with their originally distributed version.

0.6 Style and Design Elements

0.6.1 Text and Code Formatting

The *GuideBooks* are divided into chapters. Each chapter consists of several sections, which frequently are further subdivided into subsections. General remarks about a chapter or a section are presented in the sections and subsections numbered 0. (These remarks usually discuss the structure of the following section and give teasers about the usefulness of the functions to be discussed.) Also, sometimes these sections serve to refresh the discussion of some functions already introduced earlier.

Following the style of *The Mathematica Book* [45], the *GuideBooks* use the following fonts: For the main text, Times; for *Mathematica* inputs and built-in *Mathematica* commands, Courier plain (like `Plot`); and for user-supplied arguments, Times italic (like *userArgument₁*). Built-in *Mathematica* functions are introduced in the following style:

```
MathematicaFunctionToBeIntroduced[typeIndicatingUserSuppliedArgument(s)]
```
is a description of the built-in command `MathematicaFunctionToBeIntroduced` upon its first appearance. A definition of the command, along with its parameters is given. Here, *typeIndicatingUserSupplied-Argument(s)* is one (or more) user-supplied expression(s) and may be written in an abbreviated form or in a different way for emphasis.

The actual *Mathematica* inputs and outputs appear in the following manner (as mentioned above, virtually all inputs are given in `InputForm`).

```
(* A comment. It will be/is ignored as Mathematica input:
   Return only one of the solutions *)
Last[Solve[{x^2 - y == 1, x - y^2 == 1}, {x, y}]]
```

When referring in text to variables of *Mathematica* inputs and outputs, the following convention is used: Fixed, nonpattern variables (including local variables) are printed in Courier plain (the equations solved above contained the variables x and y). User supplied arguments to built-in or defined functions with pattern variables are printed in Times italic. The next input defines a function generating a pair of polynomial equations in x and y.

```
equationPair[x_, y_] := {x^2 - y == 1, x - y^2 == 1}
```

x and y are pattern variables (usimng the same letters, but a different font from the actual code fragments x_ and y_) that can stand for any argument. Here we call the function equationPair with the two arguments u + v and w - z.

```
equationPair[u + v, w - z]
```

Occasionally, explanation about a mathematics or physics topic is given before the corresponding *Mathematica* implementation is discussed. These sections are marked as follows:

Mathematical Remark: Special Topic in Mathematics or Physics

A *short* summary or review of mathematical or physical ideas necessary for the following example(s).

From time to time, *Mathematica* is used to analyze expressions, algorithms, etc. In some cases, results in the form of English sentences are produced programmatically. To differentiate such automatically generated text from the main text, in most instances such text is prefaced by "○" (structurally the corresponding cells are of type "PrintText" versus "Text" for author-written cells).

Code pieces that either run for quite long, or need a lot of memory, or are tangent to the current discussion are displayed in the following manner.

```
mathematicaCodeWhichEitherRunsVeryLongOrThatIsVeryMemoryIntensive
OrThatProducesAVeryLargeGraphicOrThatIsASideTrackToTheSubjectUnder
Discussion
(* with some comments on how the code works *)
```

To run a code piece like this, click the **Make Input** button above it. This will generate the corresponding input cell that can be evaluated if the reader's computer has the necessary resources.

The reader is encouraged to add new inputs and annotations to the electronic notebooks. There are two styles for reader-added material: "ReaderInput" (a *Mathematica* input style and simultaneously the default style for a new cell) and "ReaderAnnotation" (a text-style cell type). They are primarily intended to be used in the Reading environment. These two styles are indented more than the default input and text cells, have a green left bar and a dingbat. To access the "ReaderInput" and "ReaderAnnotation" styles, press the system-dependent modifier key (such as Control or Command) and 9 and 7, respectively.

0.6.2 References

Because the *GuideBooks* are concerned with the solution of mathematical and physical problems using *Mathematica* and are not mathematics or physics monographs, the author did not attempt to give complete references for each of the applications discussed [38], [20]. The references cited in the text pertain mainly to the applications under discussion. Most of the citations are from the more recent literature; references to older publications can be found in the cited ones. Frequently URLs for downloading relevant or interesting information are given. (The URL addresses worked at the time of printing and, hopefully, will be still active when the reader tries them.) References for *Mathematica*, for algorithms used in computer algebra, and for applications of computer algebra are collected in the Appendix A.

The references are listed at the end of each chapter in alphabetical order. In the notebooks, the references are hyperlinked to all their occurrences in the main text. Multiple references for a subject are not cited in numerical order, but rather in the order of their importance, relevance, and suggested reading order for the implementation given.

In a few cases (e.g., pure functions in Chapter 3, some matrix operations in Chapter 6), references to the mathematical background for some built-in commands are given—mainly for commands in which the mathematics required extends beyond the familiarity commonly exhibited by non-mathematicians. The *GuideBooks* do not discuss the algorithms underlying such complicated functions, but sometimes use *Mathematica* to "monitor" the algorithms.

References of the form *abbreviationOfAScientificField/yearMonthPreprintNumber* (such as quant-ph/0012147) refer to the arXiv preprint server [43], [22], [30] at http://arXiv.org. When a paper appeared as a preprint and (later) in a journal, typically only the more accessible preprint reference is given. For the convenience of the reader, at the end of these references, there is a **Get Preprint** button. Click the button to display a palette notebook with hyperlinks to the corresponding preprint at the main preprint server and its mirror sites. (Some of the older journal articles can be downloaded free of charge from some of the digital mathematics library servers, such as http://gdz.sub.uni-goettingen.de, http://www.emis.de, http://www.numdam.org, and http://dieper.aib.uni-linz.ac.at.)

As much as available, recent journal articles are hyperlinked through their digital object identifiers (http://www.doi.org).

0.6.3 Variable Scoping, Input Numbering, and Warning Messages

Some of the *Mathematica* inputs intentionally cause error messages, infinite loops, and so on, to illustrate the operation of a *Mathematica* command. These messages also arise in the user's practical use of *Mathematica*. So, instead of presenting polished and perfected code, the author prefers to illustrate the potential problems and limitations associated with the use of *Mathematica* applied to "real life" problems. The one exception are the spelling warning messages General::spell and General::spell1 that would appear relatively frequently because "similar" names are used eventually. For easier and less defocused reading, these messages are turned off in the initialization cells. (When working with the notebooks, this means that the pop-up window asking the user "Do you want to automatically evaluate all the initialization cells in the notebook?" should be evaluated should always be answered with a "yes".) For the vast majority of graphics presented, the picture is the focus, not the returned *Mathematica* expression representing the picture. That is why the Graphics and Graphics3D output is suppressed in most situations.

To improve the code's readability, no attempt has been made to protect all variables that are used in the various examples. This protection could be done with Clear, Remove, Block, Module, With, and others. Not protecting the variables allows the reader to modify, in a somewhat easier manner, the values and definitions of variables, and to see the effects of these changes. On the other hand, there may be some interference between variable names and values used in the notebooks and those that might be introduced when experimenting with the code. When readers examine some of the code on a computer, reevaluate sections, and sometimes perform subsidiary calculations, they may introduce variables that might interfere with ones from the *GuideBooks*. To partially avoid this problem, and for the reader's convenience, sometimes Clear[*sequenceOfVariables*] and Remove[*sequenceOfVariables*] are sprinkled throughout the notebooks. This makes experimenting with these functions easier.

The numbering of the *Mathematica* inputs and outputs typically does not contain all consecutive integers. Some pieces of *Mathematica* code consist of multiple inputs per cell; so, therefore, the line numbering is incremented by more than just 1. As mentioned, *Mathematica* should be restarted at every section, or subsection or solution of an exercise, to make sure that no variables with values get reused. The author also explicitly asks the reader to restart *Mathematica* at some special positions inside sections. This removes previously introduced variables, eliminates all existing contexts, and returns *Mathematica* to the typical initial configuration to ensure reproduction of the results and to avoid using too much memory inside one session.

0.6.4 Graphics

In *Mathematica* 5.1, displayed graphics are side effects, not outputs. The actual output of an input producing a graphic is a single cell with the text ‑Graphics‑ or ‑Graphics3D‑ or ‑GraphicsArray‑ and so on. To save paper, these output cells have been deleted in the printed version of the *GuideBooks*.

Most graphics use an appropriate number of plot points and polygons to show the relevant features and details. Changing the number of plot points and polygons to a higher value to obtain higher resolution graphics can be done by changing the corresponding inputs.

The graphics of the printed book and the graphics in the notebooks are largely identical. Some printed book graphics use a different color scheme and different point sizes and line and edge thicknesses to enhance contrast and visibility. In addition, the font size has been reduced for the printed book in tick and axes labels.

The graphics shown in the notebooks are PostScript graphics. This means they can be resized and rerendered without loss of quality. To reduce file sizes, the reader can convert them to bitmap graphics using the Cell⟶ Convert To⟶Bitmap menu. The resulting bitmap graphics can no longer be resized or rerendered in the original resolution.

To reduce file sizes of the main content notebooks, the animations of the *GuideBooks* are not part of the chapter notebooks. They are contained in a separate directory.

0.6.5 Notations and Symbols

The symbols used in typeset mathematical formulas are not uniform and unique throughout the *GuideBooks*. Various mathematical and physical quantities (such as normals, rotation matrices, and field strengths) are used repeatedly in this book. Frequently the same notation is used for them, but depending on the context, also different ones are used, e.g. sometimes bold is used for a vector (such as **r**) and sometimes an arrow (such as $\vec{r}$). Matrices appear in bold or as doublestruck letters. Depending on the context and emphasis placed, different notations are used in display equations and in the *Mathematica* input form. For instance, for a time-dependent scalar quantity of one variable $\psi(t; x)$, we might use one of many patterns, such as $\psi[t][x]$ (for emphasizing a parametric t-dependence) or $\psi[t, x]$ (to treat t and x on an equal footing) or $\psi[t, \{x\}]$ (to emphasize the one-dimensionality of the space variable x).

Mathematical formulas use standard notation. To avoid confusion with *Mathematica* notations, the use of square brackets is minimized throughout. Following the conventions of mathematics notation, square brackets are used for three cases: a) Functionals, such as $\mathcal{F}_t[f(t)](\omega)$ for the Fourier transform of a function $f(t)$. b) Power series coefficients, $[x^k](f(x))$ denotes the coefficient of x^k of the power series expansion of $f(x)$ around $x = 0$. c) Closed intervals, like $[a, b]$ (open intervals are denoted by (a, b)). Grouping is exclusively done using parentheses. Upper-case double-struck letters denote domains of numbers, $\mathbb{Z}$ for integers, $\mathbb{N}$ for nonnegative integers, $\mathbb{Q}$ for rational numbers, $\mathbb{R}$ for reals, and $\mathbb{C}$ for complex numbers. Points in $\mathbb{R}^n$ (or $\mathbb{C}^n$) with explicitly given coordinates are indicated using curly braces $\{c_1, ..., c_n\}$. The symbols $\wedge$ and $\vee$ for And and Or are used in logical formulas.

For variable names in formula- and identity-like *Mathematica* code, the symbol (or small variations of it) traditionally used in mathematics or physics is used. In program-like *Mathematica* code, the author uses very descriptive, sometimes abbreviated, but sometimes also slightly longish, variable names, such as `buildBril` `louinZone` and `FibonacciChainMap`.

0.6.6 Units

In the examples involving concepts drawn from physics, the author tried to enhance the readability of the code (and execution speed) by not choosing systems of units involving numerical or unit-dependent quantities. (For more on the choice and treatment of units, see [39], [4], [5], [10], [13], [11], [12], [36], [35], [31], [37], [44], [21], [25], [18], [26], [24].) Although *Mathematica* can carry units along with the symbols representing the physical quantities in a calculation, this requires more programming and frequently diverts from the essence of the problem. Choosing a system of units that allows the equations to be written without (unneeded in computations) units often gives considerable insight into the importance of the various parts of the equations because the magnitudes of the explicitly appearing coefficients are more easily compared.

0.6.7 Cover Graphics

The cover graphics of the *GuideBooks* stem from the *Mathematica GuideBooks* themselves. The construction ideas and their implementation are discussed in detail in the corresponding *GuideBook*.

■ The cover graphic of the Programming volume shows 42 tori, 12 of which are in the dodecahedron's face planes and 30 which are in the planes perpendicular to the dodecahedron's edges. Subsections 1.2.4 of Chapter 1 discusses the implementation.

■ The cover graphic of the Graphics volume first subdivides the faces of a dodecahedron into small triangles and then rotates randomly selected triangles around the dodecahedron's edges. The proposed solution of Exercise 1b of Chapter 2 discusses the implementation.

■ The cover graphic of the Numerics volume visualizes the electric field lines of a symmetric arrangement of positive and negative charges. Subsection 1.11.1 discusses the implementation.

■ The cover graphic of the Symbolics volume visualizes the derivative of the Weierstrass $\wp'$ function over the Riemann sphere. The "threefold blossoms" arise from the poles at the centers of the periodic array of period parallelograms. Exercise 3j of Chapter 2 discusses the implementation.

■ The four spine graphics show the inverse elliptic nome function q^{-1}, a function defined in the unit disk with a boundary of analyticity mapped to a triangle, a square, a pentagon, and a hexagon. Exercise 16 of Chapter 2 of the Graphics volume discusses the implementation.

0.7 Production History

The original set of notebooks was developed in the 1991–1992 academic year on an Apple Macintosh IIfx with 20 MB RAM using *Mathematica* Version 2.1. Over the years, the notebooks were updated to *Mathematica* Version 2.2, then to Version 3, and finally for Version 4 for the first printed edition of the Programming and Graphics volumes of the *Mathematica GuideBooks* (published autumn 2004). For the Numerics and Symbolics volumes, the *GuideBooks* notebooks were updated to *Mathematica* Version 5 in the second half of 2004. Historically, the first step in creating the book was the translation of a set of Macintosh notebooks used for lecturing and written in German into English by Larry Shumaker. This was done primarily by a translation program and afterward by manually polishing the English version. Then the notebooks were transformed into T_EX files using the program nb2tex on a NeXT computer. The resulting files were manually edited, equations prepared in the original German notebooks were formatted with T_EX, and macros were added corresponding to the design of the book. (The translation to T_EX was necessary because *Mathematica* Version 2.2 did not allow for book-quality printouts.) They were updated and refined for nearly three years, and then *Mathematica* 3 notebooks were generated from the T_EX files using a preliminary version of the program tex2nb. Historically and technically, this was an important step because it transformed all of the material of the *GuideBooks* into *Mathematica* expressions and allowed for automated changes and updates in the various editing stages. (Using the *Mathematica* kernel allowed one to process and modify the notebook files of these books in a uniform and time-efficient manner.) Then, the notebooks were expanded in size and scope and updated to *Mathematica* 4. In the second half of the year 2003, and first half of the year 2004, the *Mathematica* programs of the notebooks were revised to be compatible with *Mathematica* 5. In October 2004, the Programming and the Graphics volumes were published. In the last quarter of 2004, all four volumes of the *GuideBooks* were updated to be tailored for *Mathematica* 5.1 A special set of styles was created to generate the actual PostScript as printouts from the notebooks. All inputs were evaluated with this style sheet, and the generated PostScript was directly used for the book production. Using a little *Mathematica* program, the index was generated from the notebooks (which are *Mathematica* expressions), containing all index entries as cell tags.

0.8 Four General Suggestions

A reader new to *Mathematica* should take into account these four suggestions.

■ There is usually more than one way to solve a given problem using *Mathematica*. If one approach does not work or returns the wrong answer or gives an error message, make every effort to understand what is happening. Even if the reader has succeeded with an alternative approach, it is important to try to understand why other attempts failed.

■ Mathematical formulas, algorithms, and so on, should be implemented as directly as possible, even if the resulting construction is somewhat "unusual" compared to that in other programming languages. In particular, the reader should not simply translate C, Pascal, Fortran, or other programs line-by-line into *Mathematica*, although this is indeed possible. Instead, the reader should instead reformulate the problem in a clear mathematical way. For example, Do, While, and For loops are frequently unnecessary, convergence (for instance, of sums) can be checked by *Mathematica*, and If tests can often be replaced by a corresponding pattern. The reader should start with an exact mathematical description of the problem [28], [29]. For example, it does not suffice to know which transformation formulas have to be used on certain functions; one also needs to know how to apply them. "The power of mathematics is in its precision. The precision of mathematics must be used precisely." [17]

■ If the exercises, examples, and calculation of the *GuideBooks* or the listing of calculation proposals from Exercise 1 of Chapter 1 of the Programming volume are not challenging enough or do not cover the reader's interests, consider the following idea, which provides a source for all kinds of interesting and difficult problems: The reader should select a built-in command and try to reconstruct it using other built-in commands and make it behave as close to the original as possible in its operation, speed, and domain of applicability, or even to surpass it. (Replicating the following functions is a serious challenge: N, Factor, FactorInteger, Integrate, NIntegrate, Solve, DSolve, NDSolve, Series, Sum, Limit, Root, Prime, or PrimeQ.)

■ If the reader tries to solve a smaller or larger problem in *Mathematica* and does not succeed, keep this problem on a "to do" list and periodically review this list and try again. Whenever the reader has a clear strategy to solve a problem, this strategy can be implemented in *Mathematica*. The implementation of the algorithm might require some programming skills, and by reading through this book, the reader will become able to code more sophisticated procedures and more efficient implementations. After the reader has acquired a certain amount of *Mathematica* programming familiarity, implementing virtually all "procedures" which the reader can (algorithmically) carry out with paper and pencil will become straightforward.

References

1   P. Abbott. *The Mathematica Journal* 4, 415 (2000).

2   P. Abbott. *The Mathematica Journal* 9, 31 (2003).

3   H. Abelson, G. Sussman. *Structure and Interpretation of Computer Programs*, MIT Press, Cambridge, MA, 1985.

4   G. I. Barenblatt. *Similarity, Self-Similarity, and Intermediate Asymptotics*, Consultants Bureau, New York, 1979.

5   F. A. Bender. *An Introduction to Mathematical Modeling*, Wiley, New York, 1978.

6   G. Benfatto, G. Gallavotti. *Renormalization Group*, Princeton University Press, Princeton, 1995.

7   L. Blum, F. Cucker, M. Shub, S. Smale. *Complexity and Real Computation*, Springer, New York, 1998.

8   P. Bürgisser, M. Clausen, M. A. Shokrollahi. *Algebraic Complexity Theory*, Springer, Berlin, 1997.

9   L. Cardelli, P. Wegner. *Comput. Surveys* 17, 471 (1985).

10 J. F. Carinena, M. Santander in P. W. Hawkes (ed.). *Advances in Electronics and Electron Physics* 72, Academic Press, New York, 1988.

11 E. A. Desloge. *Am. J. Phys.* 52, 312 (1984).

12 C. L. Dym, E. S. Ivey. *Principles of Mathematical Modelling*, Academic Press, New York, 1980.

13 A. C. Fowler. *Mathematical Models in the Applied Sciences,* Cambridge University Press, Cambridge, 1997.

14 T. Gannon. *arXiv:math.QA*/9906167 (1999).

15 R. J. Gaylord, S. N. Kamin, P. R. Wellin. *An Introduction to Programming with Mathematica*, TELOS/Springer-Verlag, Santa Clara, 1993.

16 J. Glynn, T. Gray. *The Beginner's Guide to Mathematica Version 3*, Cambridge University Press, Cambridge, 1997.

17 D. Greenspan in R. E. Mickens (ed.). *Mathematics and Science*, World Scientific, Singapore, 1990.

18 G. W. Hart. *Multidimensional Analysis*, Springer-Verlag, New York, 1995.

19 A. K. Hartman, H. Rieger. *arXiv:cond-mat*/0111531 (2001).

20 M. Hazewinkel. *arXiv:cs.IR*/0410055 (2004).

21 E. Isaacson, M. Isaacson. *Dimensional Methods in Engineering and Physics*, Edward Arnold, London, 1975.

22 A. Jackson. *Notices Am. Math. Soc.* 49, 23 (2002).

23 R. D. Jenks, B. M. Trager in J. von zur Gathen, M. Giesbrecht (eds.). *Symbolic and Algebraic Computation*, ACM Press, New York, 1994.

24 C. G. Jesudason. *arXiv:physics*/0403033 (2004).

25 C. Kauffmann in A. van der Burgh (ed.). *Topics in Engineering Mathematics*, Kluwer, Dordrecht, 1993.

26 R. Khanin in B. Mourrain (ed.). *ISSAC 2001*, ACM, Baltimore, 2001.

27 P. Kleinert, H. Schlegel. *Physica* A 218, 507 (1995).

28 D. E. Knuth. *Am. Math. Monthly* 81, 323 (1974).

29 D. E. Knuth. *Am. Math. Monthly* 92, 170 (1985).

30 G. Kuperberg. *arXiv:math.HO*/0210144 (2002).

31 J. D. Logan. *Applied Mathematics*, Wiley, New York, 1987.

32 K. C. Louden. *Programming Languages: Principles and Practice*, PWS-Kent, Boston, 1993.

33 R. Maeder. *Programming in Mathematica*, Addison-Wesley, Reading, 1997.

34 R. Maeder. *The Mathematica Programmer*, Academic Press, New York, 1993.

35   B. S. Massey. *Measures in Science and Engineering*, Wiley, New York, 1986.

36   G. Messina, S. Santangelo, A. Paoletti, A. Tucciarone. *Nuov. Cim.* D 17, 523 (1995).

37   J. Molenaar in A. van der Burgh, J. Simonis (eds.). *Topics in Engineering Mathematics*, Kluwer, Dordrecht, 1992.

38   E. Pascal. *Repertorium der höheren Mathematik* Theil 1/1 [page V, paragraph 3], Teubner, Leipzig, 1900.

39   S. H. Romer. *Am. J. Phys.* 67, 13 (1999).

40   R. Sedgewick, P. Flajolet. *Analysis of Algorithms*, Addison-Wesley, Reading, 1996.

41   R. Sethi. *Programming Languages: Concepts and Constructions*, Addison-Wesley, New York, 1989.

42   D. B. Wagner. *Power Programming with Mathematica: The Kernel*, McGraw-Hill, New York, 1996.

43   S. Warner. *arXiv:cs.DL*/0101027 (2001).

44   H. Whitney. *Am. Math. Monthly* 75, 115, 227 (1968).

45   S. Wolfram. *The Mathematica Book*, Wolfram Media, Champaign, 2003.

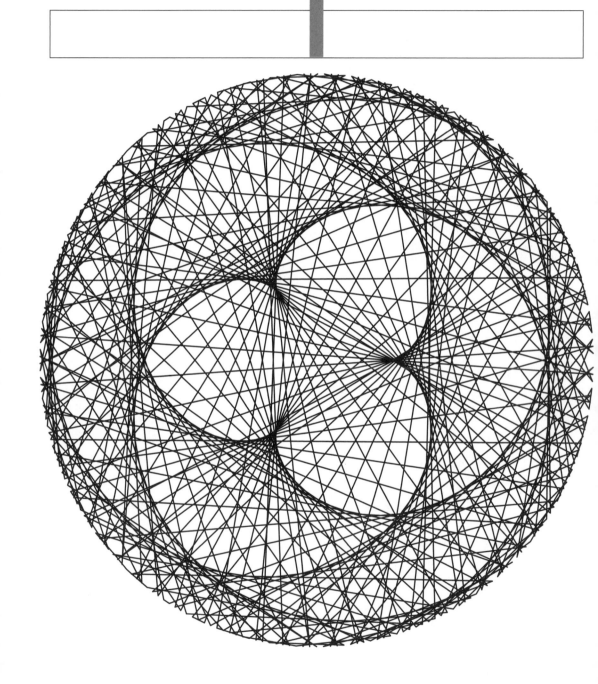

Symbolic Computations

1.0 Remarks

The functions discussed in this chapter form the most unique part of *Mathematica*'s mathematical capabilities. This explains the relatively large size of this chapter as well as the relatively large number of exercises. *Mathematica* can only be considered a complete system for doing various parts of mathematics and its applications on a computer if we include all of its graphical, pattern recognition, programming, numerical, and symbolic capabilities. On the other hand, capabilities such as factorization of polynomials, differentiation, integration, solution of equations and differential equations, computation of sums, products, and limit values can hardly be duplicated by other conventional programming systems that can handle graphics and numerical computations. So the commands for the operations `Factor`, `D`, `Integrate`, `FourierTransform`, `Solve`, `Reduce`, `Root`, `CylindricalDecomposition`, `Eliminate`, `Resultant`, `GroebnerBasis`, `DSolve`, `Limit`, and `Series` are in some sense the most important functions in *Mathematica* and the content of this chapter. As already discussed in the beginning (in Chapter 2 of the Programming volume [1735]), every object in *Mathematica* is an expression. This uniformity of the underlying data structures allows symbolic expressions to be integrated uniformly into the whole *Mathematica* system and to work naturally together with the graphical, pattern recognition, programming, and numerical functions of *Mathematica*.

A general remark about using the symbolic capabilities of *Mathematica* is in order. The functions discussed in this chapter are designed primarily for use with exact arguments and should, if possible, only be called with symbolic arguments, and not with numeric or mixed symbolic/numeric arguments. Although in most such instances *Mathematica* returns a sensible answer for numeric or mixed symbolic/numeric inputs, often in the case of numeric arguments the computation becomes considerably more complicated (for instance creating many-digit rational numbers by rationalizing approximate numbers is internally done) or even ill-defined.

1.1 Introduction

Before starting to discuss polynomials, the most important symbolic data structure, we will briefly discuss the meaning of a variable and assumptions and "types". *Mathematica* is mostly an untyped language. In nearly every function, a variable is interpreted and handled as a generic complex number of finite magnitude [1267]. Indefinite integrals $\int f(z)\,dz$ return a result that is correct almost everywhere in the complex z-plane (maybe with the exception of some zero-dimensional set with respect to the Lebesgue measure)—so from time to time the result of `Integrate` contains terms of the form $\sqrt[3]{z^3}$ and related forms that are sensible for complex numbers only, solutions to polynomial equations almost always contain complex numbers, and so on.

Some exceptions to the rule are that every variable is generically assumed to be complex are some logical functions. `LogicalExpand[a && Not[a]]` returns `False` and no error message saying that `Not[aSymbol-RepresentingAComplexNumber]` is not defined and comparisons, such as $x < 1$, do not automatically issue a warning message that a complex number is compared to 1, etc.

In symbolic summation, certain assumptions about the integer-valuedness of lower and upper limits are made, such as Sum[k, {k, 1, n}], which returns $1/2\,n\,(1+n)$ instead of If[Im[n] == 0 && n >= 0, 1/2 Floor[n] (1 + Floor[n]), Sum[k, {k, 1, n}]] which would correspond to the result returned by Sum for an explicit (real) *n*.

Variables that occur in inequalities will be considered as real-valued by many functions. For instance, for most functions a statement like $z^2 < -1$ will not include parts of the imaginary axis of the *z*-plane.

Many matrix operations, such as Cross[a, b] and Det[a] stay unevaluated for symbols a and b. Obviously, here a and b are not assumed to be complex numbers. There are some more exceptions, and we will encounter them in the following discussions.

Generically, the assumption that every variable is a complex one of finite size is very sensible. The complex numbers are an algebraically closed field and enable the inversion of polynomials and more complicated functions. Without using complex numbers, it would be, for instance, impossible to express the three real roots of $5\,x^3 - 9\,x^2 + x + 1 = 0$ in radicals without using $\sqrt{-1}$ explicitly. (See below for a more detailed discussion of this case.)

But in some instances one wants to make certain assumptions about the type of a variable, for example, when one wants to express that the parameter γ in $\int_{-\infty}^{\infty} e^{i\gamma x^2}\,dx$ is real so that the integral exists. A few *Mathematica* functions, notably Simplify, Integrate, Refine, and Assuming have currently the notion of a variable "type". We will discuss assumptions in Integrate in detail in Subsection 1.6.2. The function Simplify we discussed already in Section 3.5 of the Programming volume [1735], but not in its full generality. Because we will make use of it more frequently later, and because internally Simplify uses functions from all sections of this chapter, we will discuss all of its options now.

Simplify[*expression*, *assumptions*, *options*]

    tries to simplify *expression* under the assumptions *assumptions*.

We start with the last arguments of Simplify, its options.

In[1]:= **Options[Simplify]**

Out[1]= {Assumptions :→ $Assumptions, ComplexityFunction → Automatic,
      TimeConstraint → 300, TransformationFunctions → Automatic, Trig → True}

ComplexityFunction

    is an option of Simplify that determines what it means for an expression to be more "simple".

    Default:

        Automatic

    Admissible:

        userDefinedFunction

```
TimeConstraint
```
   is an option of `Simplify` that determines the maximal time (in seconds) a particular transformation rule is
   applied before it is aborted.
   Default:
```
      300
```

   Admissible:
   anyPositiveNumberOrInfinity

```
Trig
```
   is an option of `Simplify` that determines whether trigonometric identities should be used in the
   simplification of expressions.
   Default:
```
      True
```

   Admissible:
```
      False
```

```
TransformationFunctions
```
   is an option of `Simplify` that determines the transformation functions used in the simplification process.
   Default:
```
      Automatic
```

   Admissible:
   listOfUserGivenTransformationFunctions or

   {`Automatic`, *sequenceOfUserGivenTransformationFunctions*} (use built-in and user-added rules)

Let us give some examples. The meaning of `Automatic` in the `ComplexityFunction` option setting is basically to minimize the `LeafCount`. Some exceptions are made for numbers. Expressions of the form $\sqrt[integer_2]{integer_1{}^{integer_2}}$ autosimplify to *integer*$_1$. Such an operation might take awhile, but they will be carried out mostly automatically.

In[2]:= **(12345^45678)^(1/45678)**

Out[2]= 12345

The next expression, although being mathematically an integer, does not automatically evaluate to an integer.

In[3]:= **Exp[Log[12] + 3 (Sqrt[2] + 1)^2 Log[6] - 6 Sqrt[2] Log[6]]**

Out[3]= $12\ e^{-6\sqrt{2}\ \text{Log}[6]+3\left(1+\sqrt{2}\right)^2\ \text{Log}[6]}$

`Simplify` will simplify it to an explicit integer.

In[4]:= **Simplify[%]**

Out[4]= 120932352

Here are 13 expressions of this type.

In[5]:= ***g*[k_Integer] := Exp[Log[12] + k (Sqrt[2] + 1)^2 Log[6] -**
                                 2 k Sqrt[2] Log[6]]

```
gTab = Table[g[k], {k, 13}]
```

Out[7]= $\{12 \, e^{-2\sqrt{2}\,\text{Log}[6]+(1+\sqrt{2})^2\,\text{Log}[6]}, \; 12 \, e^{-4\sqrt{2}\,\text{Log}[6]+2\,(1+\sqrt{2})^2\,\text{Log}[6]}, \; 12 \, e^{-6\sqrt{2}\,\text{Log}[6]+3\,(1+\sqrt{2})^2\,\text{Log}[6]},$

$12 \, e^{-8\sqrt{2}\,\text{Log}[6]+4\,(1+\sqrt{2})^2\,\text{Log}[6]}, \; 12 \, e^{-10\sqrt{2}\,\text{Log}[6]+5\,(1+\sqrt{2})^2\,\text{Log}[6]}, \; 12 \, e^{-12\sqrt{2}\,\text{Log}[6]+6\,(1+\sqrt{2})^2\,\text{Log}[6]},$

$12 \, e^{-14\sqrt{2}\,\text{Log}[6]+7\,(1+\sqrt{2})^2\,\text{Log}[6]}, \; 12 \, e^{-16\sqrt{2}\,\text{Log}[6]+8\,(1+\sqrt{2})^2\,\text{Log}[6]}, \; 12 \, e^{-18\sqrt{2}\,\text{Log}[6]+9\,(1+\sqrt{2})^2\,\text{Log}[6]},$

$12 \, e^{-20\sqrt{2}\,\text{Log}[6]+10\,(1+\sqrt{2})^2\,\text{Log}[6]}, \; 12 \, e^{-22\sqrt{2}\,\text{Log}[6]+11\,(1+\sqrt{2})^2\,\text{Log}[6]},$

$12 \, e^{-24\sqrt{2}\,\text{Log}[6]+12\,(1+\sqrt{2})^2\,\text{Log}[6]}, \; 12 \, e^{-26\sqrt{2}\,\text{Log}[6]+13\,(1+\sqrt{2})^2\,\text{Log}[6]}\}$

Simplify's default measure for simplicity will not convert all of them into integers. Large integers are considered more complicated than certain symbolic expressions.

In[8]:= `Simplify[gTab]`

Out[8]= $\{2592, 559872, 120932352, 26121388032, 5642219814912, 1218719480020992,$
$263243407684534272, 56860576059859402752, 12281884428929630994432,$
$2652887036648800294797312, 573023599916140863676219392,$
$123773097581886426554063388672, 12 \, e^{-26\sqrt{2}\,\text{Log}[6]+13\,(1+\sqrt{2})^2\,\text{Log}[6]}\}$

Using LeafCount as the ComplexityFunction yields now 13 integers.

In[9]:= `Simplify[gTab, ComplexityFunction -> LeafCount]`

Out[9]= $\{2592, 559872, 120932352, 26121388032, 5642219814912, 1218719480020992,$
$263243407684534272, 56860576059859402752, 12281884428929630994432,$
$2652887036648800294797312, 573023599916140863676219392,$
$123773097581886426554063388672, 26734989077687468135677691953152\}$

Here is an expression.

In[10]:= `expr = (x - 8)^4 - (y + 4)^8`

Out[10]= $(-8 + x)^4 - (4 + y)^8$

Simplify will not convert it to the simple (appearing) original expression expr.

In[11]:= `Simplify[Expand[expr]]`

Out[11]= $-61440 - 2048\,x + 384\,x^2 - 32\,x^3 + x^4 - 131072\,y -$
$114688\,y^2 - 57344\,y^3 - 17920\,y^4 - 3584\,y^5 - 448\,y^6 - 32\,y^7 - y^8$

Simplify is not always able to find the expression that is "simplest" (for a discussion what "simplest" could mean, see [305]). There is no (and there cannot be a) guarantee that Simplify will find the simplest possible expression. Even for the case of (multivariate) polynomials, it will not always find the simplest form.

In[12]:= `{LeafCount[expr], LeafCount[Simplify[Expand[expr]]]}`

Out[12]= $\{13, 56\}$

There are no restrictions on the setting of the ComplexityFunction option. The following example uses a function that favors large expressions. As a side effect, the function blowItUp monitors all expressions generated in the simplification process. We collect all expressions generated in the list bag.

In[13]:= `blowItUp[expr_, n_] :=`
`    If[(counter = counter + 1) < n,`
`       AppendTo[bag, expr]; 1/LeafCount[expr], Abort[]]`

Now, we get a fairly large result.

```
In[14]:= counter = 0; bag = {};
        Simplify[Sin[x] - Cos[x] + (x - 1)^3/(x + 1)^3,
            ComplexityFunction -> (blowItUp[#, 10000]&)]
```

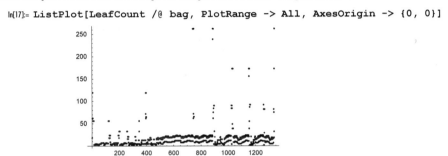

```
In[16]:= LeafCount[%]
Out[16]= 262
```

Here is a plot of the size of all expressions produced in the "simplification".

```
In[17]:= ListPlot[LeafCount /@ bag, PlotRange -> All, AxesOrigin -> {0, 0}]
```

Simplify always tries to minimize the value returned by the setting of the ComplexityFunction. This means that the functions Tan and Cot will not be present in the following result.

```
In[18]:= Simplify[Tan[x] - Cot[x], ComplexityFunction ->
            (Count[#, _Tan | _Cot, Infinity]&)]
Out[18]= -Cos[2 x] Csc[x] Sec[x]
```

Using all trigonometric functions as a measure, we do not obtain the expression rewritten in exponentials. The reason is that the function ExpToTrig (to be discussed below) is not automatically used.

```
In[19]:= Simplify[Tan[x] - Cot[x], ComplexityFunction ->
            (Count[#, _Tan | _Cot | _Sin | _Cos | _Csc | _Sec, Infinity]&)]
Out[19]= -Cot[x] + Tan[x]
```

When we explicitly include TrigToExp in the list of transformations to be applied, we get a result that does not contain any trig functions.

```
In[20]:= Simplify[Tan[x] - Cot[x], ComplexityFunction ->
            (Count[#, _Tan | _Cot | _Sin | _Cos | _Csc | _Sec, Infinity]&),
            TransformationFunctions -> {TrigToExp}]
```
$$Out[20]= \frac{i\,(e^{-ix} - e^{ix})}{e^{-ix} + e^{ix}} + \frac{i\,(e^{-ix} + e^{ix})}{e^{-ix} - e^{ix}}$$

If we allow additional built-in transformations, we obtain a simplified version of the last result.

```
In[21]:= Simplify[Tan[x] - Cot[x], ComplexityFunction ->
            (Count[#, _Tan | _Cot | _Sin | _Cos | _Csc | _Sec, Infinity]&),
            TransformationFunctions -> {Automatic, TrigToExp}]
```
$$Out[21]= \frac{2\,i\,(1 + e^{4ix})}{1 - e^{4ix}}$$

For all α, $\theta_j \neq \theta_k$, the following identity holds [292], [169]:

$$\frac{\sin(n\,\alpha)}{\sin(\alpha)} = \sum_{j=1}^{n} \prod_{\substack{k=1 \\ k \neq j}}^{n} \frac{\sin(\alpha + \theta_j - \theta_k)}{\sin(\theta_j - \theta_k)}.$$

```
In[22]:= sinIdentity[n_, {φ_, θ_, α_}] := Sin[n α]/Sin[α] -
            Sum[Product[If[k == j, 1, Sin[θ[j] - θ[k] + α]/Sin[θ[j] - θ[k]]],
                {k, n}], {j, n}]
```

Here is an example for *n* = 3.

```
In[23]:= sinIdentity[3, {φ, θ, α}]
```

```
Out[23]= Csc[α] Sin[3 α] -
            Csc[θ[1] - θ[2]] Csc[θ[1] - θ[3]] Sin[α + θ[1] - θ[2]] Sin[α + θ[1] - θ[3]] +
            Csc[θ[1] - θ[2]] Csc[θ[2] - θ[3]] Sin[α - θ[1] + θ[2]] Sin[α + θ[2] - θ[3]] -
            Csc[θ[1] - θ[3]] Csc[θ[2] - θ[3]] Sin[α - θ[1] + θ[3]] Sin[α - θ[2] + θ[3]]
```

`Simplify` can verify the identity.

```
In[24]:= Simplify[%]
```

```
Out[24]= 0
```

The running time of `Simplify` depends (often dramatically) on the size of the expression to be simplified.

```
In[25]:= Table[Timing[Simplify[sinIdentity[n, {φ, θ, α}]]], {n, 2, 6}]
```

```
Out[25]= {{0.01 Second, 0}, {0. Second, 0},
            {0.13 Second, 0}, {0.53 Second, 0}, {4.5 Second, 0}}
```

When the option setting `Trig` is set to `False`, the identities can no longer be verified.

```
In[26]:= Table[Timing[Simplify[sinIdentity[n, {φ, θ, α}],
                            Trig -> False]], {n, 2, 3}]
```

```
Out[26]= {{0.01 Second,
            Csc[α] Sin[2 α] + Csc[θ[1] - θ[2]] (-Sin[α + θ[1] - θ[2]] + Sin[α - θ[1] + θ[2]])},
            {0.04 Second, Csc[α] Sin[3 α] +
                Csc[θ[1] - θ[2]] (-Csc[θ[1] - θ[3]] Sin[α + θ[1] - θ[2]] Sin[α + θ[1] - θ[3]] +
                    Csc[θ[2] - θ[3]] Sin[α - θ[1] + θ[2]] Sin[α + θ[2] - θ[3]]) -
                Csc[θ[1] - θ[3]] Csc[θ[2] - θ[3]] Sin[α - θ[1] + θ[3]] Sin[α - θ[2] + θ[3]]}}
```

But we can instead use our own transformation functions for the simplification of trigonometric expressions—the functions `TrigExpand` and `TrigFactor` from Chapter 3 of the Programming volume [1735].

```
In[27]:= Table[Timing[
            Simplify[sinIdentity[n, {φ, θ, α}], Trig -> False,
                            TransformationFunctions -> {TrigExpand}]],
                {n, 2, 6}]
```

```
Out[27]= {{0. Second, 0}, {0.03 Second, 0},
            {0.12 Second, 0}, {0.51 Second, 0}, {4.38 Second, 0}}
```

The addition of other built-in transformation rules does not change the situation much.

```
In[28]:= Table[Timing[
            Simplify[sinIdentity[n, {φ, θ, α}], Trig -> False,
                            TransformationFunctions -> {Automatic,
                                                            TrigExpand}]],
                {n, 2, 6}]
```

Out[28]= {{0.01 Second, 0}, {0.03 Second, 0},
        {0.11 Second, 0}, {0.5 Second, 0}, {4.39 Second, 0}}

Here is an example that effectively uses the TransformationFunctions option to carry out a replacement rule. The momentum space volume element $d^n p$ is converted to a velocity space volume element $d^n v$ using $d^n p = \phi(v) \, d^n v$ [1140]. We calculate $\phi(v) = m^n \, \gamma^{n+2}$ and express the result in a compact way using $\gamma = 1 / (1 - \mathbf{v}.\mathbf{v} / c^2)^{1/2}$.

In[29]:= **Module[{v, p},**
        Table[(* velocity vector *)
                v = Table[v[k], {k, d}];
                (* momentum vector *)
                p = Table[m Sqrt[1/(1 - v.v/c^2)] v[k], {k, d}];
                (* simplify change-of-variables determinant *)
                {d, Simplify[Det[Outer[D, p, v]], c^2 > v.v && c > 0 && γ > 0,
                        TransformationFunctions ->
                    {(* use built-in rules *) **Automatic,**
                     (* express result using γ *) **(# /. v.v -> -c^2/γ^2 + c^2)&}]},**
                {d, 6}]]
Out[29]= {{1, m γ³}, {2, m² γ⁴}, {3, m³ γ⁵}, {4, m⁴ γ⁶}, {5, m⁵ γ⁷}, {6, m⁶ γ⁸}}

Simplify applies many transformation rules to an expression, but cannot guarantee finding the "simplest possible" expression. In the following input, we simplify (with respect to the LeafCount measure of complexity) the expanded form of $\sin(x)^n \sin(n\,x)$. While, on average, Simplify is be able to reduce the leaf count by a factor of 2, generically it does not reach the leaf count of 9 of $\sin(x)^n \sin(n\,x)$. Because zero testing is frequently considerably easier, in the following example Simplify always succeeds in showing that the difference between the expanded and the nonexpanded version is zero.

In[30]:= **Module[{s, t, x, y, o = 12},**
        Table[(* a larger trigonometric expression *)
                t = Sin[x]^n Sin[n x]; s = TrigExpand[t];
                (* measure sizes *)**{LeafCount[s],**
                LeafCount[Simplify[s, ComplexityFunction -> LeafCount]],
                LeafCount[t], (* difference is 0 *) **Simplify[s - t]}, {n, o}]]**
Out[30]= {{4, 4, 4, 0}, {21, 8, 9, 0}, {88, 13, 9, 0}, {103, 13, 9, 0},
        {204, 19, 9, 0}, {229, 17, 9, 0}, {368, 25, 9, 0}, {401, 21, 9, 0},
        {580, 31, 9, 0}, {621, 25, 9, 0}, {840, 37, 9, 0}, {889, 29, 9, 0}}

Now, let us discuss the possible assumptions for Simplify. The assumptions can be given in the form of equalities, inequalities, and type specification. The expression $\sqrt{x^2}$ does not simplify to x because it is not true in the whole complex x-plane.

In[31]:= **Simplify[Sqrt[x^2]]**

Out[31]= $\sqrt{x^2}$

Using an additional assumption, we can simplify this expression.

In[32]:= **Simplify[Sqrt[x^2], x > 3]**

Out[32]= x

Now, what are the "type" specifications available in *Mathematica*? The possible domains are listed below.

```
Integers
```
   represents the domain of integer numbers.
```
Rationals
```
   represents the domain of rational numbers.
```
Reals
```
   represents the domain of real numbers.
```
Complexes
```
   represents the domain of complex numbers.
```
Algebraics
```
   represents the domain of algebraic numbers (this means solutions of univariate polynomials).
```
Primes
```
   represents the domain of prime numbers.
```
Booleans
```
   represents the domain of booleans (True and False).

To use types, we need one more function: Element.

Element[*expression, domain*]

   asserts that the expression *expression* takes values from the domain *domain*.

Element[{*expression₁ , ..., expressionᵢ*} , *domain*]

   or

Element[{*expression₁ | ··· | expressionᵢ*} , *domain*]

   asserts that all of the expressions *expressionᵢ* take values from the domain *domain*.

For simple, decidable cases, Element[*expr, type*] will evaluate to True or False.

In[33]:= **Element[12, Integers]**

Out[33]= True

In[34]:= **Element[Pi, Reals]**

Out[34]= True

In[35]:= **Element[Pi + I E, Complexes]**

Out[35]= True

In[36]:= **Element[GoldenRatio, Algebraics]**

Out[36]= True

In[37]:= **Element[Pi, Rationals]**

Out[37]= False

In[38]:= **Element[Pi, Algebraics]**

Out[38]= False

Here are two more complicated examples [624].

In[39]:= **Element[Cos[Pi/111111] + Sin[Pi/22222], Algebraics]**

Out[39]= True

In[40]:= `Element[Cos[2], Algebraics]`

Out[40]= `False`

Still more complicated cases remain unevaluated.

In[41]:= `Element[Sin[Pi/24]^Log[2], Algebraics]`

Out[41]= $\mathrm{Sin}\left[\frac{\pi}{24}\right]^{\mathrm{Log}[2]} \in \mathrm{Algebraics}$

In[42]:= `Element[EulerGamma, Algebraics]`

Out[42]= $\mathrm{EulerGamma} \in \mathrm{Algebraics}$

`Element` deals exclusively with mathematical domains. It does not deal with explicitly given "sets". (In the next example, the function `MemberQ` would be more appropriate to test if the number 1 appears in the "set" {1, 2}.)

In[43]:= `Element[1, {1, 2}]`

Out[43]= $1 \in \{1, 2\}$

Using assumptions, we can now simplify expressions by giving additional type information. Here are some examples.

In[44]:= `Simplify[Sqrt[z^2], (* redundant *) Element[z, Complexes] &&`
              `Re[z] > 3]`

Out[44]= `z`

In[45]:= `Simplify[Sin[p Pi] + Cos[2 p Pi], Element[p, Primes]]`

Out[45]= `1`

In[46]:= `Simplify[a && (a || b) && b, Element[{a, b}, Booleans]]`

Out[46]= `a && b`

The first argument of `Element` can either be a single expression or a list of expressions, or it can have the head `Alternatives`.

In[47]:= `Simplify[Not[a] && Not[b || a], Element[a | b, Booleans]]`

Out[47]= `! a && ! b`

Type information can be supplemented with other assumptions.

In[48]:= `0^exp`

Out[48]= 0^{exp}

In[49]:= `Simplify[%, Element[exp, Reals] && exp > 2]`

Out[49]= `0`

Here are some more complicated examples using `Simplify`.

In[50]:= `Simplify[(4 (-1 + x^2))/((-1 + x) Sqrt[(1 + x)^2]*`
              `(2 + 3 x) + Sqrt[(-1 + x)^2] (-2 + x + 3 x^2)), x > 2]`

Out[50]= $\dfrac{2}{3\,x}$

In[51]:= `Simplify[(4 (-1 + x^2))/((-1 + x) Sqrt[(1 + x)^2]*`
              `(2 + 3 x) + Sqrt[(-1 + x)^2] (-2 + x + 3 x^2)), 0 < x < 1]`

Out[51]= `1`

In[52]:= `Simplify[(Sin[(3 n + 1) Pi] - Cos[(2 n + 1) Pi])*`
              `Sqrt[x^2] > 0, Element[n, Integers] && x > Pi]`

Out[52]= `True`

In[53]:= `Simplify[(-1)^(2 p) + Cos[2 Pi p], Element[p, Primes] && p > 3]`

Out[53]= 2

In[54]:= `Simplify[Log[Exp[x]] - Sqrt[x^2] - (x^4)^(1/4), x < -Pi]`

Out[54]= 3 x

When inequalities appear in the assumptions, `Simplify` assumes all variables that appear algebraically in these inequalities are real-valued. So the assumptions in the following input are contradictory and as a result, `True` is returned.

In[55]:= `Simplify[ω^4 > 1, Element[ω, Complexes] && ω^4 < -1]`

> Simplify::cas :
> Warning: Contradictory assumption(s) $\omega \in Complexes$ && $\omega^4 < -1$ encountered. More…

Out[55]= True

The second argument in `Simplify` can be used to give additional information about variables and symbols. This additional information often results in the possibility that more specialized transformation rules can be applied. But these more specialized transformation rules are applied only after more generically valid ones were already applied. So `Simplify` cannot (straightforwardly) be used to detect exceptions and special cases. So the next expression incorrectly simplifies to 0 (for generic x the numerator has been simplified to zero, resulting in 0 for the whole expression).

In[56]:= `Simplify[((x + 1)^2 - x^2 - 2x - 1)/x, x == 0]`

Out[56]= 0

Here is an even simpler example that gives incorrectly 0.

In[57]:= `Simplify[x Csc[x], x == 0]`

Out[57]= 0

And the following input returns `False`, although it is semantically meaningless for numbers.

In[58]:= `Simplify[cNumber && Not[cNumber], Element[cNumber, Complexes]]`

Out[58]= False

Here is another example of this kind. We use the `sinIdentity[3, {φ, θ, α}]` from above and give the conditions for degeneracy as assumptions. Nevertheless, `Simplify` returns the generically correct result 0.

In[59]:= `Simplify[sinIdentity[3, {φ, θ, α}],`
`            θ[1] == θ[2] && θ[1] == θ[3] && θ[2] == θ[3]]`

Out[59]= 0

Here is a slightly more complicated example. In the next inputs, we show that the given $\mathcal{U}(x)$ is a solution of $\mathcal{U}'(x) + \mathcal{U}(x)^2 = 2(\mathcal{V}(x) - \varepsilon)$ if the function $\psi(x)$ fulfills the Schrödinger equation $-\psi''(x)/2 + \mathcal{V}(x)\psi(x) = \varepsilon \psi(x)$ [1683].

In[60]:= `𝒰[x_] := D[Log[ψ[x]], x] + ψ[x]^-2/(λ + Integrate[ψ[x]^-2, x])`

`𝒰'[x] + 𝒰[x]^2 == 2 (𝒱[x] - ε) // Simplify[#,`
`    (* assumption for ψ[x] *) -ψ''[x]/2 + 𝒱[x] ψ[x] == ε ψ[x]]&`

Out[61]= True

`Simplify` will be a "strong" enough function for the nearly all applications treated in this chapter. In some situations (especially for nested radicals [927], [1112], expressions involving `Root`-objects, and special functions), `Simplify` will not achieve the "desired" form. In such cases, one can use the function `FullSim` plify. (We will discuss `FullSimplify` in greater detail in Chapter 3 when applying it to a variety of expressions containing special functions.) Here are two such cases.

In[62]:= `{Simplify[#], FullSimplify[#]}& @ ((2 + Sqrt[2])/(1 + Sqrt[2]))`

Out[62]= $\left\{ \dfrac{2 + \sqrt{2}}{1 + \sqrt{2}}, \ \sqrt{2} \right\}$

In[63]:= `{Simplify[#], FullSimplify[#]}& @ (Sqrt[5 + 2 Sqrt[6]])`

Out[63]= $\left\{ \sqrt{5 + 2\sqrt{6}}, \ \sqrt{2} + \sqrt{3} \right\}$

As mentioned, in general *Mathematica* does not make any assumptions about the type of variables occurring in expressions. This means that expressions such as $\sqrt{z^2}$ will by default stay in the given form to ensure correctness for any complex value of z. While `Simplify` will simplify such expressions (if possible) under further assumptions on the variables, `Simplify` is a relatively expensive operation and cannot always be afforded from a timing point of view.

In[64]:= `Do[Developer`ClearCache[];`
          `Simplify[Sum[Sqrt[z^k] + (z^k)^(1/k) + Log[z^k], {k, 12}],`
                   `z > Pi], {100}] // Timing`
Out[64]= `{4.75 Second, Null}`

A typically faster method to obtain an automatically simplified result is the use of the function `Refine`.

`Refine[`*expression*`,` *assumptions*`]`

   evaluates the expression *expression* under the assumptions *assumptions*.

The next input repeats the last evaluation and uses the function `Refine`.

In[65]:= `Do[Developer`ClearCache[];`
          `Refine[Sum[Sqrt[z^k] + (z^k)^(1/k) + Log[z^k], {k, 12}],`
                 `z > Pi], {100}] // Timing`
Out[65]= `{1.31 Second, Null}`

The last intended sense of the word "evaluate" needs some qualification. It means that the result is "equivalent" to the one that would obtained for numerical values satisfying the assumptions. Here are some examples.

In[66]:= `{#, Refine[#, z > 0]}&[Sqrt[z^2]]`
Out[66]= $\left\{ \sqrt{z^2}, \ z \right\}$

In[67]:= `{#, Refine[z < 0, z > 0]}&[Sqrt[z^2]]`
Out[67]= $\left\{ \sqrt{z^2}, \ \text{False} \right\}$

In[68]:= `{#, Refine[#, z == Pi/2]}&[Tan[z]]`
Out[68]= `{Tan[z], ComplexInfinity}`

In[69]:= `{#, Refine[#, z == E]}&[Round[z]]`
Out[69]= `{Round[z], 3}`

In[70]:= `{#, Refine[#, z < -1]}&[Log[z]]`
Out[70]= `{Log[z], i π + Log[-z]}`

But only "direct" inferences are carried out in `Reduce`. So the following call to `Refine` does not evaluate to `True`.

In[71]:= `{#, Refine[#, z > 20]}&[Log[z] > 1]`
Out[71]= `{Log[z] > 1, Log[z] > 1}`

If one has a larger program, one does not want to repeat the (frequently identical) assumptions in each call to functions like `Refine`, `Simplify`, or `Integrate`. The function `Assuming` allows stating the assumptions only once.

`Assuming[`*assumptions*`,` *expression*`]`

    evaluates the expression *expression* under the assumptions *assumptions*.

When `Refine` is called within `Assuming`, we do not have to supply the second argument. The inherited assumptions from the outer `Assuming` will be used.

In[72]:= **Assuming[z > 1, Refine[z > 0]]**

Out[72]= True

When the function `Assuming` appears in nested form, the assumptions are joined. Here is an example.

In[73]:= **Assuming[z > 1, Assuming[x < 0, Refine[z > 0 && x < 1]]]**

Out[73]= True

All currently active assumptions are stored in `$Assumptions`.

`$Assumptions`

    gives the currently active assumptions.

By default, `$Assumptions` has the value `True`. This means, nothing nontrivial can be derived.

In[74]:= **$Assumptions**

Out[74]= True

Here are the assumptions printed (using a `Print`-statement) that are active within the inner nested `Assuming`.

In[75]:= **Assuming[z > 1, Assuming[x < 0, Print[$Assumptions];**
                           Simplify[z > 0 && x < 1]]]

      $x < 0$ && $z > 1$

Out[75]= True

From contradictory assumptions (indicated when recognized as such), follow false statements.

In[76]:= **Assuming[z > 1, Assuming[z < -1, Simplify[-1/2 < z < 1/2]]]**

      $Assumptions::cas :
        Warning: Contradictory assumption(s) $z < -1$ && $z > 1$ encountered. More…

Out[76]= True

This was a short introduction into a directed use of `Simplify` and `Refine` using options and assumption specifications.

1.2 Operations on Polynomials

1.2.0 Remarks

Polynomials and polynomial systems play an extraordinary role in computational symbolic mathematics. In this section, we deal with three aspects of such systems: 1) Structural operations that express polynomials in various canonical forms, 2) manipulations of systems of polynomial equations and 3) manipulations of systems of polynomial inequations (meaning inequalities with Less and Greater, as well as Unequal as their heads). Explicit solutions of polynomial equations (which for most univariate polynomials of degree five or higher) cannot be given in radicals; their solutions will be discussed in Section 1.5. Here we largely focus on operations on polynomials that use their coefficients only.

1.2.1 Structural Manipulations on Polynomials

The two most important commands for manipulating polynomials, Expand and Factor, were already introduced in Chapter 3 of the Programming volume [1735]. Note that Factor also works for polynomials in several variables.

In[1]:= **Expand[(1 - x)^3 (3 + y - 2x)^2 (z^2 + 8y)]**

Out[1]= $72\,y - 312\,x\,y + 536\,x^2\,y - 456\,x^3\,y + 192\,x^4\,y - 32\,x^5\,y + 48\,y^2 -$
$176\,x\,y^2 + 240\,x^2\,y^2 - 144\,x^3\,y^2 + 32\,x^4\,y^2 + 8\,y^3 - 24\,x\,y^3 + 24\,x^2\,y^3 - 8\,x^3\,y^3 +$
$9\,z^2 - 39\,x\,z^2 + 67\,x^2\,z^2 - 57\,x^3\,z^2 + 24\,x^4\,z^2 - 4\,x^5\,z^2 + 6\,y\,z^2 - 22\,x\,y\,z^2 +$
$30\,x^2\,y\,z^2 - 18\,x^3\,y\,z^2 + 4\,x^4\,y\,z^2 + y^2\,z^2 - 3\,x\,y^2\,z^2 + 3\,x^2\,y^2\,z^2 - x^3\,y^2\,z^2$

In[2]:= **Factor[%]**

Out[2]= $-\,(-1+x)^3\,(3 - 2\,x + y)^2\,(8\,y + z^2)$

The following condition is often ignored.

> Factor works "properly" only for polynomials whose coefficients are exact (rational) numbers.

Thus, for instance, the following example does not work.

In[3]:= **Expand[(1.0 - x)^3 (3.0 + y - 2.0 x)^2 (z^2 + 8.0 y)] // Factor**

Out[3]= $-8.\,(-9.\,y + 39.\,x\,y - 67.\,x^2\,y + 57.\,x^3\,y - 24.\,x^4\,y + 4.\,x^5\,y - 6.\,y^2 + 22.\,x\,y^2 - 30.\,x^2\,y^2 +$
$18.\,x^3\,y^2 - 4.\,x^4\,y^2 - 1.\,y^3 + 3.\,x\,y^3 - 3.\,x^2\,y^3 + 1.\,x^3\,y^3 - 1.125\,z^2 + 4.875\,x\,z^2 -$
$8.375\,x^2\,z^2 + 7.125\,x^3\,z^2 - 3.\,x^4\,z^2 + 0.5\,x^5\,z^2 - 0.75\,y\,z^2 + 2.75\,x\,y\,z^2 - 3.75\,x^2\,y\,z^2 +$
$2.25\,x^3\,y\,z^2 - 0.5\,x^4\,y\,z^2 - 0.125\,y^2\,z^2 + 0.375\,x\,y^2\,z^2 - 0.375\,x^2\,y^2\,z^2 + 0.125\,x^3\,y^2\,z^2)$

The following simpler example works, but we highly discourage the use of inexact numbers inside Factor.

In[4]:= **x^2 - 5 x + 6. // Factor**

Out[4]= $1.\,(-3.+x)\,(-2.+x)$

Results such as the following are much better produced using NRoots or NSolve to achieve a factorization explicitly via solving for the roots.

In[5]:= **x^3 - x^2 - 5 x + 5.23 // Factor**

In[5]:= 1. (-2.19252 + x) (-1.05931 + x) (2.25183 + x)

In[6]:= (* better *)
        Times @@ (x - (x /. NSolve[x^3 - x^2 - 5 x + 5.23 == 0, x]))
Out[7]= (-2.19252 + x) (-1.05931 + x) (2.25183 + x)

Using the command `Rationalize` introduced in Chapter 1 of the Numerics volume [1737], we can convert approximate numbers to nearby rational numbers. (But be aware that for inputs with many-digit high-precision numbers, the functions myFactor might run a long time.)

In[8]:= myFactor[x_, opts___] :=
        N[Factor[MapAll[Rationalize[#, 0]&, x], opts],
           (* output precision = input precision *) Precision[x]]

        myFactor[%%%%]
Out[9]= 7.03682×10^{-24} $(-4.74876 \times 10^7 + 4.48287 \times 10^7 \, x)$
        $(-1.03683 \times 10^8 + 4.72897 \times 10^7 \, x)$ $(1.50951 \times 10^8 + 6.70348 \times 10^7 \, x)$

For nonexact integer exponents, Expand and Factor fail.

In[10]:= Expand[(1.0 - x)^3. (3.0 + y - 2.0 x)^2. (z^2 + 8.0 y)^2.]
Out[10]= $(1. - x)^{3.}$ $(3. - 2. \, x + y)^{2.}$ $(8. \, y + z^2)^{2.}$

But note that the application of N to a polynomial does not give numericalized exponents.

In[11]:= N[Expand[(1.0 - x)^3 (3.0 + y - 2.0 x)^2 (z^2 + 8.0 y)^2]]
Out[11]= $576. \, y^2 - 2496. \, x \, y^2 + 4288. \, x^2 \, y^2 - 3648. \, x^3 \, y^2 + 1536. \, x^4 \, y^2 - 256. \, x^5 \, y^2 +$
         $384. \, y^3 - 1408. \, x \, y^3 + 1920. \, x^2 \, y^3 - 1152. \, x^3 \, y^3 + 256. \, x^4 \, y^3 + 64. \, y^4 - 192. \, x \, y^4 +$
         $192. \, x^2 \, y^4 - 64. \, x^3 \, y^4 + 144. \, y \, z^2 - 624. \, x \, y \, z^2 + 1072. \, x^2 \, y \, z^2 - 912. \, x^3 \, y \, z^2 +$
         $384. \, x^4 \, y \, z^2 - 64. \, x^5 \, y \, z^2 + 96. \, y^2 \, z^2 - 352. \, x \, y^2 \, z^2 + 480. \, x^2 \, y^2 \, z^2 - 288. \, x^3 \, y^2 \, z^2 +$
         $64. \, x^4 \, y^2 \, z^2 + 16. \, y^3 \, z^2 - 48. \, x \, y^3 \, z^2 + 48. \, x^2 \, y^3 \, z^2 - 16. \, x^3 \, y^3 \, z^2 + 9. \, z^4 -$
         $39. \, x \, z^4 + 67. \, x^2 \, z^4 - 57. \, x^3 \, z^4 + 24. \, x^4 \, z^4 - 4. \, x^5 \, z^4 + 6. \, y \, z^4 - 22. \, x \, y \, z^4 +$
         $30. \, x^2 \, y \, z^4 - 18. \, x^3 \, y \, z^4 + 4. \, x^4 \, y \, z^4 + 1. \, y^2 \, z^4 - 3. \, x \, y^2 \, z^4 + 3. \, x^2 \, y^2 \, z^4 - 1. \, x^3 \, y^2 \, z^4$

Mathematica factorizes over the integers ("not over the rationals", and not over the algebraic numbers as long as they do not appear explicitly). This is not a big restriction for rational numbers. The following polynomial over the exact rationals is factored in such a way that the resulting polynomials have integer coefficients and is written out with a common denominator.

In[12]:= Expand[(1/4 - x)^3 (3/2 + y - 2x)^2 (z^2 + 8/5y)^2] // Factor
Out[12]= $-\dfrac{(-1 + 4 \, x)^3 \, (3 - 4 \, x + 2 \, y)^2 \, (8 \, y + 5 \, z^2)^2}{6400}$

An interesting theoretical question is the following: Given a polynomial $p = \sum_{k=0}^{d} c_k \, x^k$ of degree d with integer coefficients c_k in the range $-f \le c_k \le f$, what is the average number of factors of p [1413], [152], [1578], [525]? Here is a simulation for small d and f.

In[13]:= factorNumber[maxDegree_, maxCoefficient_] :=
         Module[{x}, (* count factors *)
                 If[Head[#] === Plus, 1, Length[#]]&[
                    (* factored random polynomial *) Factor[
                    Sum[Random[Integer, {-1, 1} maxCoefficient] x^i,
                        {i, 0, maxDegree}]]]]

In[14]:= Module[{n = 400, dMax = 12, fMax = 20, data},
                 (* use n random polynomials *)
                 data = Table[Plus @@ Table[factorNumber[d, f], {n}],
                              {d, 2, dMax}, {f, 1, fMax}]/n;

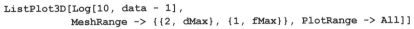

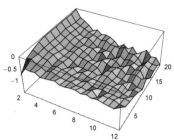

Polynomials that cannot be factored into multiple *x*-dependent factors are called irreducible [1439].

```
In[15]:= irreducibleQ[poly_, x_] :=
    With[{factors = Select[FactorList[poly], MemberQ[#, x, Infinity]&]},
        If[(* at least to x-containing factors exits *)
           Length[factors] > 1 || (* powers *) factors[[1, 2]] > 1,
           False, True]] /; PolynomialQ[poly, x] && Exponent[poly, x] > 0
```

"Most" univariate polynomials are irreducible. The following graphic shows the reducible polynomials of a quadratic, cubic and quartic polynomial over the plane of two coefficients. Reducible polynomials occur along certain lines.

```
In[16]:= Show[GraphicsArray[#]]& @
    Block[{o = 250, α, β},
       ListDensityPlot[Table[If[TrueQ[Not[irreducibleQ[#, x]]], 0, 1],
                       {α, -o, o}, {β, -o, o}],
                   Mesh -> False, MeshRange -> {{-o, o}, {-o, o}},
                   DisplayFunction -> Identity]& /@
       (* three polynomials with two parameters each *)
       {-2 + α x + β x^2, -4 + 3 x + α x^2 + β x^3,
        -4 - 3 x^2 + α x^3 + β x^4}]
```

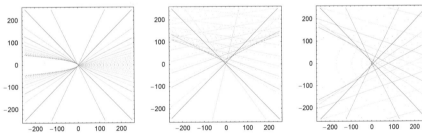

Given the digits d_k of an integer n in base b, we can naturally form the polynomial $p_b(n; x) = \sum_{k=0}^{\lceil \log_b n \rceil} d_k x^k$. Interestingly, when n is a prime number, the polynomial p is irreducible [1424], [937].

```
In[17]:= digitPolynomial[k_, b_, x_] :=
    Plus @@ MapIndexed[#1 x^(#2[[1]] - 1)&, Reverse[IntegerDigits[k, b]]]

In[18]:= (* checking a "random" prime in 100 bases *)
    Table[irreducibleQ[#, x]& @ digitPolynomial[Prime[123456789], b, x],
        {b, 2, 1001}] // Union

Out[19]= {True}
```

The following graphic shows the reducible $p_b(n; x)$ for 10000 values of n and 200 bases. The gray vertical lines indicate the integers n that are prime.

```
In[20]:= With[{nM = 10^6, δn = 9999, bM = 2, δb = 201},
    Show[Graphics[
        {{GrayLevel[0.8], (* gray stripes at position of primes *)
          Table[If[PrimeQ[n],
                Rectangle[{n - 1/2, bM - 1/2}, {n + 1/2, bM + δb + 1/2}],
    {}],
                {n, nM, nM + δn}]},
          {GrayLevel[0], (* black squares for reducible polynomials *)
           Table[If[Not[irreducibleQ[#, x]& @ digitPolynomial[n, b, x]],
                Rectangle[{n - 1/2, b - 1/2}, {n + 1/2, b + 1/2}], {}],
                {b, bM, bM + δb}, {n, nM, nM + δn}]}}],
        PlotRange -> All, AspectRatio -> 0.4, Frame -> True]]
```

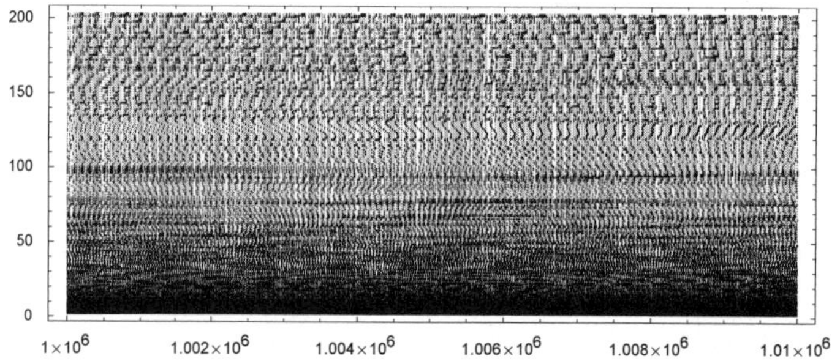

Because no algebraic numbers are computed in the factorization of a univariate polynomial with integer (rational) coefficients, we have the following behavior.

```
In[21]:= {Factor[x^2 - a^2], Factor[x^2 - 2^2], Factor[x^2 - Sqrt[2]^2]}
Out[21]= {-(a - x) (a + x), (-2 + x) (2 + x), -2 + x^2}
```

To factor polynomials over algebraic number fields, the option `Extension` can be used.

`Extension`

determines the generators of the field extension where the factorization is to be carried out.

Default:
    `None`

Admissible:
    `Automatic` or *explicitListOfAlgebraicNumbers*

With the default setting, `None` polynomials will not be factored over algebraic numbers.

```
In[22]:= Factor[x^2 - 11]
Out[22]= -11 + x^2
```

With the setting `Automatic`, polynomials will be factored over the field generated by the algebraic numbers explicitly present in the coefficients of the polynomial. This means for the following example that the polynomial will be factored over $\mathbb{Q}(\sqrt{2})$.

```
In[23]:= Factor[x^2 + 2 Sqrt[2] x + 2, Extension -> Automatic]
```

Out[23]= $\left(\sqrt{2} + x\right)^2$

By giving a list of algebraic numbers, one can explicitly specify the extension field. Here, a quartic polynomial is factored. Adjoining $\sqrt[4]{11}$ allows factoring $x^4 - 11$ into two linear factors with real roots and one quadratic factor with two complex roots.

In[24]:= `Factor[x^4 - 11, Extension -> (11)^(1/4)]`

Out[24]= $-\left(11^{1/4} - x\right)\left(11^{1/4} + x\right)\left(\sqrt{11} + x^2\right)$

Adding $\sqrt{-1}$ to the extensions allows for a complete factorization of $x^4 - 11$ into linear factors.

In[25]:= `Factor[x^4 - 11, Extension -> {(11)^(1/4), I}]`

Out[25]= $-\left(11^{1/4} - x\right)\left(11^{1/4} - i\,x\right)\left(11^{1/4} + i\,x\right)\left(11^{1/4} + x\right)$

Finding an extension such that a polynomial will factor is largely equivalent to solving *polynomial* = 0. Here is an unsuccessful trial to factor $3\,x^3 + 7\,x^2 - 9$.

In[26]:= `Factor[x^4 - 3x^3 + 7 x^2 - 9]`

Out[26]= $-9 + 7\,x^2 - 3\,x^3 + x^4$

Here, we use an extension such that the polynomial factors into one linear and one quadratic factor.

In[27]:= `Factor[3 x^3 + 7 x^2 - 9,`
`          Extension -> {(1501/2 - (27 Sqrt[2445])/2)^(1/3)}]`

Out[27]= $-\dfrac{1}{2490394032}\Big(\Big(-67228 + 4802\ 2^{2/3}\left(1501 - 27\sqrt{2445}\right)^{1/3} +$
$\qquad 1501\ 2^{1/3}\left(1501 - 27\sqrt{2445}\right)^{2/3} + 27\ 2^{1/3}\sqrt{2445}\left(1501 - 27\sqrt{2445}\right)^{2/3} - 86436\,x\Big)$
$\qquad \Big(11907\ 2^{2/3}\left(1501 - 27\sqrt{2445}\right)^{1/3} + 147\ 2^{2/3}\sqrt{2445}\left(1501 - 27\sqrt{2445}\right)^{1/3} +$
$\qquad 1701\ 2^{1/3}\left(1501 - 27\sqrt{2445}\right)^{2/3} + 21\ 2^{1/3}\sqrt{2445}\left(1501 - 27\sqrt{2445}\right)^{2/3} +$
$\qquad \Big(134456 + 4802\ 2^{2/3}\left(1501 - 27\sqrt{2445}\right)^{1/3} + 1501\ 2^{1/3}\left(1501 - 27\sqrt{2445}\right)^{2/3} +$
$\qquad 27\ 2^{1/3}\sqrt{2445}\left(1501 - 27\sqrt{2445}\right)^{2/3}\Big)\,x + 86436\,x^2\Big)\Big)$

Be aware that factoring of polynomials is a rather complex process (see the general references given in the appendix), which takes some time. Here, the timings for the expansion of $(C + 1)^i$ are compared with the timings for the factorization of the expanded object. (Be aware of the different degrees for `Expand` and `Factor`.)

In[28]:=
```
Show[GraphicsArray[
     ListPlot[(* reasonable units for a 2-GHz computer *)
             {#[[1]], 1000 #[[2, 1, 1]]}& /@ #[[1]],
             Frame -> True, PlotLabel -> #[[2]],
             FrameLabel -> {"degree", "milliSeconds"},
             DisplayFunction -> Identity]& /@
     (* Expand and Factor data *)
     {{(* clear caches for reliable timings *)
      Table[{i, Developer`ClearCache[];
             Timing[Expand[(C + 1)^i];]}, {i, 0, 6000, 50}], "Expand"},
      {Table[{i, Developer`ClearCache[];
             Timing[Factor[#]]&[Expand[(C + 1)^i]]}, {i, 300}], "Factor"}}]]
```

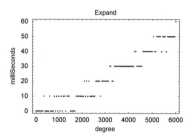

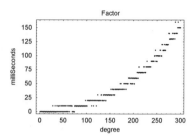

Let us take a graphical look at the result of expanding a power of a sum. Let the sum total be zero in the form $0 = \sum_{j=0}^{n-1} \exp(2\pi i j / n)$. Then the powers of this sum are also 0, and, by interpreting the partial sums of the expanded power as points in the complex plane, we get a closed path.

```
In[29]:= expandPicture[{n_, pow_}, opts___] :=
    Show[Graphics[{Thickness[0.002], Line[{Re[#], Im[#]}& /@
        (* form the partial sums *)
        FoldList[Plus, 0, N[(List @@
            (* first make list, and then replace to avoid reordering *)
            (* now comes the expansion *)
            Expand[Sum[C[i], {i, 0, n - 1}]^pow]) /. C[i_] -> Exp[i I 2Pi/n]]]]}],
        opts, AspectRatio -> Automatic, Frame -> True,
        PlotRange -> All, FrameTicks -> None];

In[30]:= Map[Show[GraphicsArray[
        expandPicture[#, DisplayFunction -> Identity]& /@ #]]&,
        (* the parameters for the pictures *)
        {Table[{3, i}, {i, 3, 21, 3}], Table[{4, i}, {i, 2, 20, 4}],
        Table[{5, i}, {i, 3, 15, 3}], Table[{6, i}, {i, 3, 12, 2}],
        Table[{8, i}, {i, 2, 8, 2}]}, {1}]
```

Here are three more complicated versions of such a graphic. We color the line segments from red to blue.

```
In[31]:= Show[GraphicsArray[
        expandPicture[#, DisplayFunction -> Identity] /.
            Line[l_] :> With[{n = Length[l]},
                MapIndexed[{Hue[0.78 #2[[1]]/n], Line[#1]}&,
                Partition[l, 2, 1]]]& /@ {{10, 10}, {16, 8}, {36, 4}}]]
```

Before discussing another algorithmically nontrivial operation for manipulating polynomials—namely, Decompose—we consider a method to rearrange an expression, if possible, into canonical polynomial form. PolynomialQ[*polynomial*, *var*] (we know this command from Chapter 5 of the Programming volume [1735]) tests whether *polynomial* is a polynomial in *var*. Be aware that PolynomialQ is a purely structural operation. While the expression $(\cos^2(1) + \sin^2(1) - 1) x^x + 2 x^2 - 1$ is mathematically a polynomial, structurally it is not. As a function ending with Q, PolynomialQ has to return True of False and cannot stay unevaluated. But it is always possible to construct terms of the form *hiddenZero nonPolynomialPart* so that it is algorithmically undecidable *hiddenZero* is zero (Richardson theorem [1482], [1484], [1485], [442]). From this, it follows that to guarantee not to get wrong answers from PolynomialQ is not doomed to give wrong results sometimes, it must be a purely structural function.

```
In[32]:= PolynomialQ[(Sin[1]^2 + Cos[1]^2 - 1) x^x + 2 x^2 - 1, x]
Out[32]= False
```

PolynomialQ[*polynomial*] tests if *polynomial* can be considered as a polynomial in at least one variable. Using Collect, we can now write an expression as an explicit polynomial in given variables.

Collect[*expression*, {*var₁*, *var₂*, ... , *varₙ*}, *function*]

writes *expression* recursively as a polynomial in the variables var_i ($i = 1, ..., n$) and applies the optional function *function* to the resulting coefficients. If *function* is omitted, the last argument is assumed to be Identity.

Here we again use our previous polynomial.

In[33]:= **polyInxInyInz = (1 - x)^3 (3 + y - 2x)^2 (z^2 + 8y)^2;**

Here is this expression as a polynomial in *x*.

In[34]:= **Collect[polyInxInyInz, x]**

Out[34]= $-4 x^5 (8 y + z^2)^2 + x^4 (24 + 4 y) (8 y + z^2)^2 + x (-39 - 22 y - 3 y^2) (8 y + z^2)^2 +$
$x^3 (-57 - 18 y - y^2) (8 y + z^2)^2 + (9 + 6 y + y^2) (8 y + z^2)^2 + x^2 (67 + 30 y + 3 y^2) (8 y + z^2)^2$

The result of Collect depends on the form of its input. Collect will not expand or factor the resulting coefficients by default.

In[35]:= **Collect[Expand[polyInxInyInz], x]**

Out[35]= $576 y^2 + 384 y^3 + 64 y^4 + 144 y z^2 + 96 y^2 z^2 + 16 y^3 z^2 + 9 z^4 + 6 y z^4 + y^2 z^4 +$
$x^5 (-256 y^2 - 64 y z^2 - 4 z^4) + x^4 (1536 y^2 + 256 y^3 + 384 y z^2 + 64 y^2 z^2 + 24 z^4 + 4 y z^4) +$
$x (-2496 y^2 - 1408 y^3 - 192 y^4 - 624 y z^2 - 352 y^2 z^2 - 48 y^3 z^2 - 39 z^4 - 22 y z^4 - 3 y^2 z^4) +$
$x^3 (-3648 y^2 - 1152 y^3 - 64 y^4 - 912 y z^2 - 288 y^2 z^2 - 16 y^3 z^2 - 57 z^4 - 18 y z^4 - y^2 z^4) +$
$x^2 (4288 y^2 + 1920 y^3 + 192 y^4 + 1072 y z^2 + 480 y^2 z^2 + 48 y^3 z^2 + 67 z^4 + 30 y z^4 + 3 y^2 z^4)$

Using the optional third argument of Collect, we can bring the coefficients to a canonical form.

In[36]:= **Collect[Expand[polyInxInyInz], x, Factor]**

Out[36]= $-4 x^5 (8 y + z^2)^2 + (3 + y)^2 (8 y + z^2)^2 +$
$4 x^4 (6 + y) (8 y + z^2)^2 - x (3 + y) (13 + 3 y) (8 y + z^2)^2 -$
$x^3 (57 + 18 y + y^2) (8 y + z^2)^2 + x^2 (67 + 30 y + 3 y^2) (8 y + z^2)^2$

Note that the individual terms are not strictly ordered; in particular, not all terms ($\propto x^0$) appear one after the other. This is a consequence of the Flat and Orderless attributes of Plus and the canonical order.

Here is the same expression as a polynomial in y.

In[37]:= **Collect[polyInxInyInz, y]**

Out[37]= $64 (1 - x)^3 y^4 + (1 - x)^3 y^3 (384 - 256 x + 16 z^2) +$
$(1 - x)^3 y^2 (576 - 768 x + 256 x^2 + 96 z^2 - 64 x z^2 + z^4) +$
$(1 - x)^3 y (144 z^2 - 192 x z^2 + 64 x^2 z^2 + 6 z^4 - 4 x z^4) + (1 - x)^3 (9 z^4 - 12 x z^4 + 4 x^2 z^4)$

Here it is again as a polynomial in z.

In[38]:= **Collect[polyInxInyInz, z]**

Out[38]= $64 (1 - x)^3 y^2 (3 - 2 x + y)^2 + 16 (1 - x)^3 y (3 - 2 x + y)^2 z^2 + (1 - x)^3 (3 - 2 x + y)^2 z^4$

Using as the second argument in Collect the list {x, y} results in a polynomial in x, whose coefficients are polynomials in y, whose coefficients are polynomials in z.

In[39]:= **Collect[polyInxInyInz, {x, y}]**

Out[39]= $64 y^4 + 9 z^4 + y^3 (384 + 16 z^2) +$
$x^5 (-256 y^2 - 64 y z^2 - 4 z^4) + y^2 (576 + 96 z^2 + z^4) + y (144 z^2 + 6 z^4) +$
$x (-192 y^4 - 39 z^4 + y^3 (-1408 - 48 z^2) + y (-624 z^2 - 22 z^4) + y^2 (-2496 - 352 z^2 - 3 z^4)) +$

$$x^3 (-64\ y^4 - 57\ z^4 + y^3 (-1152 - 16\ z^2) + y (-912\ z^2 - 18\ z^4) + y^2 (-3648 - 288\ z^2 - z^4)) +$$
$$x^4 (256\ y^3 + 24\ z^4 + y^2 (1536 + 64\ z^2) + y (384\ z^2 + 4\ z^4)) +$$
$$x^2 (192\ y^4 + 67\ z^4 + y^3 (1920 + 48\ z^2) + y^2 (4288 + 480\ z^2 + 3\ z^4) + y (1072\ z^2 + 30\ z^4))$$

Here we apply the function C to each of the coefficients in z.

In[40]:= `Collect[polyInxInyInz, {x, y}, C]`

Out[40]= $y^4\ C[64] + x^5 (y^2\ C[-256] + y\ C[-64\ z^2] + C[-4\ z^4]) +$
  $C[9\ z^4] + y^3\ C[384 + 16\ z^2] + x (y^4\ C[-192] + C[-39\ z^4] +$
    $y^3\ C[-1408 - 48\ z^2] + y\ C[-624\ z^2 - 22\ z^4] + y^2\ C[-2496 - 352\ z^2 - 3\ z^4]) +$
  $x^3 (y^4\ C[-64] + C[-57\ z^4] + y^3\ C[-1152 - 16\ z^2] + y\ C[-912\ z^2 - 18\ z^4] +$
    $y^2\ C[-3648 - 288\ z^2 - z^4]) + y^2\ C[576 + 96\ z^2 + z^4] +$
  $x^4 (y^3\ C[256] + C[24\ z^4] + y^2\ C[1536 + 64\ z^2] + y\ C[384\ z^2 + 4\ z^4]) +$
  $y\ C[144\ z^2 + 6\ z^4] + x^2 (y^4\ C[192] + C[67\ z^4] +$
    $y^3\ C[1920 + 48\ z^2] + y^2\ C[4288 + 480\ z^2 + 3\ z^4] + y\ C[1072\ z^2 + 30\ z^4])$

The second argument in `Collect` need not be an atomic expression, and thus the following expression will be written as a polynomial over `co[x]`.

In[41]:= `Collect[Expand[ (co[x] + 4 si[z] + 5 co[x]^3)^4], co[x]]`

Out[41]= $150\ co[x]^8 + 500\ co[x]^{10} + 625\ co[x]^{12} + 240\ co[x]^5\ si[z] + 1200\ co[x]^7\ si[z] +$
  $2000\ co[x]^9\ si[z] + 96\ co[x]^2\ si[z]^2 + 256\ co[x]\ si[z]^3 + 256\ si[z]^4 +$
  $co[x]^4 (1 + 960\ si[z]^2) + co[x]^6 (20 + 2400\ si[z]^2) + co[x]^3 (16\ si[z] + 1280\ si[z]^3)$

`Collect` only reorders. It does not carry out any "mathematical meaning- or content-dependent manipulations". It only looks at the syntactical structure of expressions. (This means that in the following example, `Cos[x]^2` is not rewritten as $1 - Sin[x]^2$.)

In[42]:= `Collect[Sin[x]^2 + (Cos[x]^2 + Sin[x]^2)^3 + 3 Sin[x]^3 + 7,`
         `Sin[x]]`

Out[42]= $7 + Cos[x]^6 + (1 + 3\ Cos[x]^4)\ Sin[x]^2 + 3\ Sin[x]^3 + 3\ Cos[x]^2\ Sin[x]^4 + Sin[x]^6$

Given an expression in several variables, we can use `Variables` to identify in which variables the expression is a polynomial.

`Variables[`*expression*`]`
   produces a list of the variables in which *expression* is a polynomial.

For the above `polyInxInyInz`, we get the expected result.

In[43]:= `Variables[polyInxInyInz]`

Out[43]= $\{x, y, z\}$

The following expression is a polynomial in x and y, but not in z.

In[44]:= `Variables[x^2 + 7 y + 5 z + Log[z]]`

Out[44]= $\{x, y, z, Log[z]\}$

The "variables" do not necessarily have to be atomic objects with head `Symbol`.

In[45]:= `Variables[Cos[x]^3 + Sin[y]^2 + Cos[x] Sin[y]]`

Out[45]= $\{Cos[x], Sin[y]\}$

The following expression is only a polynomial (of the trivial form $poly(x) = x$) "in itself".

In[46]:= `Variables[Cos[x^2 + 7 y + 5 z + Sqrt[z]]]`

Out[46]= $\{Cos[x^2 + 7 y + \sqrt{z} + 5 z]\}$

Variables does not pay attention to internal algebraic structures.

In[47]:= **Variables[f[Sqrt[z]] + f[z^(1/3)]^2 + Log[z^2]]**

Out[47]= $\{f[z^{1/3}], f[\sqrt{z}], Log[z^2]\}$

Variables uses a very general definition for a polynomial; it also considers Laurent and Puiseux polynomials. This means the terms can have the form $var^{negativeIntegerOrRational}$.

In[48]:= **Variables /@ {Sqrt[z] + 1, z^(1/3) + 1, z^-4 + 1,**
                      z^(1/3) + z^(-2/3) + 3, z^Pi + 1}

Out[48]= $\{\{z\}, \{z\}, \{z\}, \{z\}, \{z^{\pi}\}\}$

Be aware that Variables's interpretation of a polynomial does not agree with the one of PolynomialQ.

In[49]:= **PolynomialQ[#, z]& /@ {Sqrt[z] + 1, z^(1/3) + 1, z^-4 + 1, z^Pi + 1}**

Out[49]= $\{False, False, False, False\}$

The degree of a polynomial (i.e., the largest exponent) can be obtained using Exponent.

Exponent [*polynomial, var*]

    finds the degree of the polynomial *polynomial* in the variable *var*.

Here is a polynomial of twelfth degree in *x*.

In[50]:= **Exponent[(3 + 4 x + 5 x^4)^3, x]**

Out[50]= 12

Be aware that Exponent is in most situations a purely structural operation. For expanded polynomials, Exponent does not determine the mathematical degree of a polynomial, but rather finds the exponent of the highest power of *var* present in an expression (meaning the leading coefficient can be mathematically zero, but is not the explicit integer 0).

In[51]:= **zero0 = (Pi + 1)^2 - (Pi^2 + 2Pi + 1);**

    (* two linear polynomials in x; both disguised as quadratics *)
    {Exponent[zero0 x^2 + x + 1, x],
    Exponent[(Sin[α]^2 + Cos[α]^2 - 1) x^2 + x + 1, x]}

Out[53]= $\{2, 2\}$

If the polynomial is not expanded, Exponent will partially carry out nonstructural operations.

In[54]:= **(* three linear polynomials in x; all disguised as quadratics *)**
    {Exponent[(x + 1)^2 - x^2 - x - 1, x],
    Exponent[(zero0 x + 1)^2 - zero0 x^2 - x - 1, x],
    Exponent[(zero0 x + 1)^2 - (zero0 x)^2 - x - 1, x]}

Out[55]= $\{1, 2, 1\}$

For large polynomials with many terms, usually not all powers of the various variables occur (in fact, usually only a very small fraction of all possibilities is present). To find the powers that are involved, we use Coeffi⸫ cient.

Coefficient[*polynomial, var, degree*]

> finds the coefficient of var^{degree} in the polynomial *polynomial*. If no value is given for *degree*, the coefficient of $var^1 = var$ is found.

For typical cases, the result is obvious.

In[56]:= **Coefficient[12 x^3 + 23 x^2 + 45 x + 56, x]**

Out[56]= 45

In other cases, the output from Collect may not be so obvious. In the following, Coefficient uses a very liberal interpretation of a polynomial power.

In[57]:= **Coefficient[C x^Pi, x, Pi]**

Out[57]= C

Using a list in the second argument of Coefficient results in a list of the corresponding coefficients.

In[58]:= **Coefficient[polyInxInyInz, {x, y, z}]**

Out[58]= $\{(-39 - 22\ y - 3\ y^2)\ (8\ y + z^2)^2,\ (1 - x)^3\ (144\ z^2 - 192\ x\ z^2 + 64\ x^2\ z^2 + 6\ z^4 - 4\ x\ z^4),\ 0\}$

For polynomials in several variables (or for getting all coefficients in a univariate polynomial at once), we have the corresponding command CoefficientList.

CoefficientList[*polynomial*, {*var*$_1$, *var*$_2$, ... , *var*$_n$}]

> gives a list of *n*-dimensional tensor containing the coefficients of the powers products $var_1^{k_1} \cdots var_n^{k_n}$ in the polynomial *polynomial*.

Thus, CoefficientList finds a list of all coefficients appearing in the polynomial. Here is a list of the coefficients for a polynomial in one variable.

In[59]:= **CoefficientList[Sum[i x^i, {i, 0, 12}], {x}]**

Out[59]= {0, 1, 2, 3, 4, 5, 6, 7, 8, 9, 10, 11, 12}

For several variables, we get a list whose dimension is the number of variables. For the following polynomial in *x* and *y*, we get a 4×4 matrix.

In[60]:= **polyxy = a + b x + c y + d x^2 + e x y + f y^2 +**
          g x^3 + h x^2 y + i x y^2 + j y^3

Out[60]= $a + x\ b + y\ c + x^2\ d + x\ y\ e + y^2\ f + x^3\ g + x^2\ y\ h + x\ y^2\ i + y^3\ j$

In[61]:= **CoefficientList[polyxy, {x, y}]**

Out[61]= {{a, c, f, j}, {b, e, i, 0}, {d, h, 0, 0}, {g, 0, 0, 0}}

We can implement the above definition to get the following program. (In this code, we first look for all structures with powers of the form x^i y^j, and then separate out the x and y.)

In[62]:= **Table[**
        Which[i == 0 && j == 0, (* the constant term *)
                Cases[polyxy, _?(FreeQ[#, x] && FreeQ[#, y]&)],
                True, (* the coefficient of x^i y^j *)
                Cases[polyxy, _?(FreeQ[#, x] && FreeQ[#, y]&) x^i y^j |
                        x^i y^j]/(x^i y^j)],
              (* the exponents *)
              {i, 0, Exponent[polyxy, x]}, {j, 0, Exponent[polyxy, y]}] //.

$$\{\{\} \to 0, \{coeff\_\} \to coeff\} //.$$
$$(* \text{ all zero lists become empty lists } *) \ \{(0)..\} \to \{\}$$

Out[62]= {{a, c, f, j}, {b, e, i, 0}, {d, h, 0, 0}, {g, 0, 0, 0}}

Be aware that the result of CoefficientList[*poly, varList*] is always a rectangular array. For sparse polynomials of high degree, one will typically encounter many 0's in the result. Here is a typical example.

In[63]:= **CoefficientList[x^8 + y^6 - 7x - y, {x, y}]**

Out[63]= {{0, -1, 0, 0, 0, 0, 1}, {-7, 0, 0, 0, 0, 0, 0}, {0, 0, 0, 0, 0, 0, 0},
{0, 0, 0, 0, 0, 0, 0}, {0, 0, 0, 0, 0, 0, 0}, {0, 0, 0, 0, 0, 0, 0},
{0, 0, 0, 0, 0, 0, 0}, {0, 0, 0, 0, 0, 0, 0}, {1, 0, 0, 0, 0, 0, 0}}

Given the result of CoefficientList, it is straightforward to reconstruct the original polynomial. We just simply map the corresponding power product of the variables to the position of the coefficient and sum all terms. Here, this is done for the polynomial polyInxInyInz used above.

In[64]:= *cl* = **CoefficientList[polyInxInyInz, {x, y, z}]**

Out[64]= {{{0, 0, 0, 0, 9}, {0, 0, 144, 0, 6}, {576, 0, 96, 0, 1}, {384, 0, 16, 0, 0},
{64, 0, 0, 0, 0}}, {{0, 0, 0, 0, -39}, {0, 0, -624, 0, -22},
{-2496, 0, -352, 0, -3}, {-1408, 0, -48, 0, 0}, {-192, 0, 0, 0, 0}},
{{0, 0, 0, 0, 67}, {0, 0, 1072, 0, 30}, {4288, 0, 480, 0, 3}, {1920, 0, 48, 0, 0},
{192, 0, 0, 0, 0}}, {{0, 0, 0, 0, -57}, {0, 0, -912, 0, -18},
{-3648, 0, -288, 0, -1}, {-1152, 0, -16, 0, 0}, {-64, 0, 0, 0, 0}},
{{0, 0, 0, 0, 24}, {0, 0, 384, 0, 4}, {1536, 0, 64, 0, 0},
{256, 0, 0, 0, 0}, {0, 0, 0, 0, 0}}, {{0, 0, 0, 0, -4},
{0, 0, -64, 0, 0}, {-256, 0, 0, 0, 0}, {0, 0, 0, 0, 0}, {0, 0, 0, 0, 0}}}

Implementing a function fromCoefficientList that converts a nested list of coefficients to a polynomial is straightforward.

In[65]:= **fromCoefficientList[cl_, vars_] := Plus @@**
Flatten[MapIndexed[#1 Times @@ (vars^(#2 - 1))&, cl, {3}]]

In[66]:= **Expand[fromCoefficientList[*cl*, {x, y, z}] - polyInxInyInz]**

Out[66]= 0

For sparse multivariate high-degree polynomials, CoefficientList is an inefficient way to access the coefficients. To get the coefficients of such polynomials efficiently, we recommend the internal function Internal`DistributedTermsList. Internal`DistributedTermsList[*multivariatePolynomial, listOfVariables, options*] returns a list of the form {{{*listOfExponentsInVariables*$_1$, *coefficient*$_1$}, ..., {*listOfExponentsInVariables*$_n$, *coefficient*$_n$}}, *listOfVariables*} where n is the number of terms in the polynomial *multivariatePolynomial* collected with respect to the variables *listOfVariables*. Here is a simple example.

In[67]:= **Internal`DistributedTermsList[x^12 + 2 y^22 - 3 $\alpha\beta\gamma$ x^10 y^20, {x, y}]**

Out[67]= {{{{12, 0}, 1}, {{10, 20}, -3 $\alpha\beta\gamma$}, {{0, 22}, 2}}, {x, y}}

To recover the original polynomial, we can use the function Internal`FromDistributedTermsList.

In[68]:= **Internal`FromDistributedTermsList[%]**

Out[68]= $x^{12} + 2 y^{22} - 3 x^{10} y^{20} \alpha\beta\gamma$

Be aware that, in distinction to CoefficientList, the function Internal`DistributedTermsList returns also the original list of variables. The function Internal`DistributedTermsList has many Gröbner basis-like options (we will discuss some of these options in the next subsection).

In[69]:= **Options[Internal`DistributedTermsList]**

Out[69]= {CoefficientDomain → RationalFunctions,
         Modulus → 0, MonomialOrder → Lexicographic,
         ParameterVariables → {}, Sort → False, Tolerance → 0}

Their behavior is similar to the ones of other polynomial functions with the same options. The MonomialOr der order option, for instance, determines the order of the {*listOfExponentsInVariables$_k$, coefficient$_k$*}-sublists.

In[70]:= Internal`DistributedTermsList[
             x^12 + 2 y^22 - 3 αβγ x^10 y^20, {x, y},
             MonomialOrder -> DegreeReverseLexicographic]

Out[70]= {{{{10, 20}, -3 αβγ}, {{0, 22}, 2}, {{12, 0}, 1}}, {x, y}}

In addition to the command Factor discussed earlier, another important function can sometimes "simplify" a polynomial. Whereas Collect, Exponent, Coefficient and CoefficientList perform mainly structural operations, the following command carries out nontrivial calculations.

Decompose[*polynomial, var*]

> gives a list of the polynomials {*polynomial*$_1$(*var*), *polynomial*$_2$(*var*), ..., *polynomial*$_n$(*var*)}, such that *polynomial*$_1$(*polynomial*$_2$($\cdots$(*polynomial*$_n$(*var*) $\cdots$))) = *polynomial*.

To illustrate, we first create a "decomposable" polynomial.

In[71]:= Expand[Expand[(2x + 3)^2] /. {x -> (3x^2 + 5)^3}]

Out[71]= $64009 + 227700 x^2 + 339120 x^4 + 270324 x^6 + 121500 x^8 + 29160 x^{10} + 2916 x^{12}$

Now, we take it apart using Decompose.

In[72]:= dcp = Decompose[%, x]

Out[72]= $\{64009 + 9108 x + 324 x^2, 25 x + 15 x^2 + 3 x^3, x^2\}$

Here is a check.

In[73]:= (Composition @@ (* nest the polys *)
             (Function[x, #]& (* make Function from polys *)/@
                 Drop[dcp, -1]))[Last[dcp]] // Expand

Out[73]= $64009 + 227700 x^2 + 339120 x^4 + 270324 x^6 + 121500 x^8 + 29160 x^{10} + 2916 x^{12}$

The last example shows that the result of Decompose is not unique.

1.2.2 Polynomials in Equations

In this subsection, we will discuss various algebraic operations for sets of polynomials equated to zero. The most important functions of this kind are Resultant and GroebnerBasis. We will not elaborate on the intimately related subject of solving systems of polynomials (we will deal with this subject together with solving nonpolynomial equations when discussing the function Solve in the next section).

We start with a look at two functions for the direct manipulation of pairs of polynomials. They are important functions for algorithmic work on polynomials.

PolynomialQuotient[*polynomial$_1$*, *polynomial$_2$*, *var*]

finds the largest polynomial that can be divided out from $\frac{polynomial_1}{polynomial_2}$ with respect to the variable *var*.

PolynomialRemainder[*polynomial$_1$*, *polynomial$_2$*, *var*]

finds the remainder after dividing *polynomial$_1$* by *polynomial$_2$* with respect to the variable *var*.

We give just one simple example PolynomialQuotient. For the following example, the answer is obvious.

In[1]:= **Factor[PolynomialQuotient[**
Expand[(3 + 5 x^3 + 8 x^7)^6 (4 x^3 + 8 x^6)], (4 x^3 + 8 x^6), x]]

Out[1]= $(3 + 5\,x^3 + 8\,x^7)^6$

Some additional commands are useful for manipulating pairs of polynomials. These include PolynomialGCD, PolynomialLCM. Because these are of lesser importance in scientific and technical applications (they are more important in number theory and pure algebra), we do not use them here and in the following.

The next, more advanced command to be dealt with in this subsection is the resultant. Given two polynomials $p_1(z)$ and $p_2(z)$ (which might via their coefficients a_i and b_i also depend on further, not explicitly shown variables)

$$p_1(z) = \sum_{i=0}^{n} a_i z^i = \prod_{i=0}^{n} (z - \alpha_i) \quad p_2(z) = \sum_{i=0}^{m} b_i z^i = \prod_{i=0}^{m} (z - \beta_i)$$

the resultant $r(a_1, \ldots a_n; b_1, \ldots, b_m)$ of $p_1(z)$ and $p_2(z)$ is given by

$$r = a_n^m \prod_{i=0}^{n} p_2(\alpha_i) = (-1)^{mn} b_m^n \prod_{i=0}^{m} p_1(\beta_i) = a_n^m b_m^n \prod_{i=0}^{n} \prod_{j=0}^{n} (\alpha_i - \beta_j).$$

The resultant of two polynomials is an integral polynomial of their coefficients.

For nice reviews concerning properties and representations of the resultant, see, for instance, [1439], [708], [347], [1679], [1177], [441], [1823], [172], [286], [382], and [1824] (for subresultants, see [433]).

The *Mathematica* form for the resultant is Resultant.

Resultant[*polynomial$_1$*, *polynomial$_2$*, *var*]

calculates the resultant of the two polynomials *polynomial$_1$* and *polynomial$_2$* with respect to the variable *var*.

One important application of the resultant is the elimination of one variable from a system of two polynomial equations. As an example, let the two polynomials $p_1(z) = 0$ and $p_2(z) = 0$ both depend (in a polynomial manner) on the variable a.

In[2]:= **poly1[z_, a_] = Product[z - i a, {i, 1, 7, 2}] + 2**

Out[2]= $2 + (-7\,a + z)\,(-5\,a + z)\,(-3\,a + z)\,(-a + z)$

In[3]:= **poly2[z_, a_] = Product[z - i a, {i, 0, 6, 2}] + 3**

Out[3]= $3 + z\,(-6\,a + z)\,(-4\,a + z)\,(-2\,a + z)$

Then we can eliminate the variable z to get only one polynomial equation (which is, of course, much easier to solve) in the variable a.

In[4]:= **Resultant[poly1[z, a], poly2[z, a], z]**

Out[4]= $1 + 4260 \, a^4 - 26730 \, a^8 - 13500 \, a^{12} + 212625 \, a^{16}$

We, of course, could have also eliminated the variable a.

In[5]:= **Resultant[poly1[z, a], poly2[z, a], a]**

Out[5]= $93767625 - 168674724 \, z^4 - 23424810 \, z^8 - 298980 \, z^{12} + 2025 \, z^{16}$

This is the numerical solution of the above polynomial in a.

In[6]:= **aZeros = Chop[Last /@ (List @@ NRoots[%% == 0, a])]**

Out[6]= {-0.697333, -0.682883, -0.558864 - 0.558864 i, -0.558864 + 0.558864 i,
-0.0874929 - 0.0874929 i, -0.0874929 + 0.0874929 i, -0.697333 i, 0.697333 i,
-0.682883 i, 0.682883 i, 0.0874929 - 0.0874929 i, 0.0874929 + 0.0874929 i,
0.558864 - 0.558864 i, 0.558864 + 0.558864 i, 0.682883, 0.697333}

Because of the above formulas, these solutions are also solutions of $p_1(z) = 0$ and $p_2(z) = 0$ (or the highest coefficients a_n or b_n vanish for these roots). Plugging each root for a into the two polynomials and finding the root with respect to z that is the same for both polynomials will give all possible solutions. This is a first possibility to solve pairs of polynomial equations.

In[7]:= **Sort @ Table[{a -> aZeros[[i]], z -> Flatten[**
Outer[If[(* which are the same? *)
Chop[#1 - #2] === 0, Chop[#1], {}]&,
(* the zeros of poly1 with respect to z *)
Last /@ List @@ NRoots[poly1[z, aZeros[[i]]] == 0, z],
(* the zeros of poly2 with respect to z *)
Last /@ List @@ NRoots[poly2[z, aZeros[[i]]] == 0, z]]][[1]]},
{i, Length[aZeros]}]

Out[7]= {{a → -0.697333, z → -0.848507}, {a → -0.682883, z → -3.78763},
{a → -0.558864 - 0.558864 i, z → -1.88035 - 1.88035 i},
{a → -0.558864 + 0.558864 i, z → -1.88035 + 1.88035 i},
{a → -0.0874929 - 0.0874929 i, z → -1.21371 - 1.21371 i},
{a → -0.0874929 + 0.0874929 i, z → -1.21371 + 1.21371 i},
{a → -0.682883 i, z → -3.78763 i}, {a → 0.682883 i, z → 3.78763 i},
{a → -0.697333 i, z → -0.848507 i}, {a → 0.697333 i, z → 0.848507 i},
{a → 0.0874929 - 0.0874929 i, z → 1.21371 - 1.21371 i},
{a → 0.0874929 + 0.0874929 i, z → 1.21371 + 1.21371 i},
{a → 0.558864 - 0.558864 i, z → 1.88035 - 1.88035 i},
{a → 0.558864 + 0.558864 i, z → 1.88035 + 1.88035 i},
{a → 0.682883, z → 3.78763}, {a → 0.697333, z → 0.848507}}

A check in the original polynomials shows this.

In[8]:= **Chop[{poly1[z, a], poly2[z, a]} /. %]**

Out[8]= {{0, 0}, {0, 0}, {0, 0}, {0, 0}, {0, 0}, {0, 0}, {0, 0}, {0, 0},
{0, 0}, {0, 0}, {0, 0}, {0, 0}, {0, 0}, {0, 0}, {0, 0}, {0, 0}}

A comparison with the corresponding result of NSolve also confirms this.

In[9]:= **Sort @ Chop[NSolve[{poly1[z, a] == 0, poly2[z, a] == 0}, {a, z}]]**

Out[9]= {{a → -0.697333, z → -0.848507}, {a → -0.682883, z → -3.78763},
{a → -0.558864 - 0.558864 i, z → -1.88035 - 1.88035 i},
{a → -0.558864 + 0.558864 i, z → -1.88035 + 1.88035 i},
{a → -0.0874929 - 0.0874929 i, z → -1.21371 - 1.21371 i},
{a → -0.0874929 + 0.0874929 i, z → -1.21371 + 1.21371 i},

$\{a \to -0.682883 \, i, \ z \to -3.78763 \, i\}, \ \{a \to 0.682883 \, i, \ z \to 3.78763 \, i\},$
$\{a \to -0.697333 \, i, \ z \to -0.848507 \, i\}, \ \{a \to 0.697333 \, i, \ z \to 0.848507 \, i\},$
$\{a \to 0.0874929 - 0.0874929 \, i, \ z \to 1.21371 - 1.21371 \, i\},$
$\{a \to 0.0874929 + 0.0874929 \, i, \ z \to 1.21371 + 1.21371 \, i\},$
$\{a \to 0.558864 - 0.558864 \, i, \ z \to 1.88035 - 1.88035 \, i\},$
$\{a \to 0.558864 + 0.558864 \, i, \ z \to 1.88035 + 1.88035 \, i\},$
$\{a \to 0.682883, \ z \to 3.78763\}, \ \{a \to 0.697333, \ z \to 0.848507\}\}$

To calculate the resultant of two polynomials it is fortunately not necessary to calculate all roots of each polynomial and then form the product of all possible differences. Just carrying polynomial manipulations with the coefficients of both polynomials is enough and such an approach is much faster and guarantees that the resultant is an explicit polynomial in the coefficients of the two original polynomials. For the two polynomials at hand, an approach through the roots does not even succeed to calculate the resultant with respect to a as a polynomial in the coefficients.

```
In[10]:= Timing[Short[Simplify[Times @@ Flatten[
            Outer[Subtract, # /. Solve[poly1[z, a] == 0, {#}],
                             # /. Solve[poly2[z, a] == 0, {#}]]]], 2]]& /@ {z, a}
```

$\text{Out[10]= } \{\{0.22 \, \text{Second}, \ 1 + 4260 \, a^4 - 26730 \, a^8 - 13500 \, a^{12} + 212625 \, a^{16}\}, \ \{27.59 \, \text{Second},$

$$\frac{(286 \, z + \ll 7 \gg) \ll 10 \gg \left(\ll 5 \gg + 12 \sqrt{2} \, \sqrt{1724 \, z^2 - \frac{\ll 1 \gg}{\ll 1 \gg \ll 1 \gg} - 35 \ll 1 \gg + \frac{31616 \sqrt{2} \, z^3}{\sqrt{\ll 1 \gg}}} \right)}{40990810607230154227643842560000000000000} \}\}$$

An interesting representation of the resultant of two polynomials, due to Sylvester [32], is given by the determinant of the following square matrix.

```
In[11]:= MatrixALaSylvester[poly1_, poly2_, var_] :=
         Function[{coeffs1, coeffs2},
         With[{λ1 = Length[coeffs1], λ2 = Length[coeffs2]},
         (* form the matrix *)
         Join @@ (Table[Join[Table[0, {j}], #1,
                            Table[0, {#2 - 2 - j}]], {j, 0, #2 - 2}]& @@@
                    {{coeffs1, λ1}, {coeffs2, λ2}})]][
             (* the coefficients of the two polynomials *)
             Sequence @@ (Reverse[CoefficientList[#, var]]& /@ {poly1, poly2})]
```

Here, this representation is checked for the resultant of two degree 3 polynomials.

```
In[12]:= MatrixALaSylvester[a0 + a1 x + a2 x^2 + a3 x^3,
                            b0 + b1 x + b2 x^2 + b3 x^3, x] // MatrixForm
```

Out[12]//MatrixForm=

$$\begin{pmatrix} a3 & a2 & a1 & a0 & 0 & 0 \\ 0 & a3 & a2 & a1 & a0 & 0 \\ 0 & 0 & a3 & a2 & a1 & a0 \\ b3 & b2 & b1 & b0 & 0 & 0 \\ 0 & b3 & b2 & b1 & b0 & 0 \\ 0 & 0 & b3 & b2 & b1 & b0 \end{pmatrix}$$

```
In[13]:= Det[%]
```

Out[13]= $a3^3 \, b0^3 - a2 \, a3^2 \, b0^2 \, b1 + a1 \, a3^2 \, b0 \, b1^2 - a0 \, a3^2 \, b1^3 + a2^2 \, a3 \, b0^2 \, b2 - 2 \, a1 \, a3^2 \, b0^2 \, b2 -$
$a1 \, a2 \, a3 \, b0 \, b1 \, b2 + 3 \, a0 \, a3^2 \, b0 \, b1 \, b2 + a0 \, a2 \, a3 \, b1^2 \, b2 + a1^2 \, a3 \, b0 \, b2^2 -$
$2 \, a0 \, a2 \, a3 \, b0 \, b2^2 - a0 \, a1 \, a3 \, b1 \, b2^2 + a0^2 \, a3 \, b2^3 - a2^3 \, b0^2 \, b3 + 3 \, a1 \, a2 \, a3 \, b0^2 \, b3 -$
$3 \, a0 \, a3^2 \, b0^2 \, b3 + a1 \, a2^2 \, b0 \, b1 \, b3 - 2 \, a1^2 \, a3 \, b0 \, b1 \, b3 - a0 \, a2 \, a3 \, b0 \, b1 \, b3 - a0 \, a2^2 \, b1^2 \, b3 +$
$2 \, a0 \, a1 \, a3 \, b1^2 \, b3 - a1^2 \, a2 \, b0 \, b2 \, b3 + 2 \, a0 \, a2^2 \, b0 \, b2 \, b3 + a0 \, a1 \, a3 \, b0 \, b2 \, b3 +$
$a0 \, a1 \, a2 \, b1 \, b2 \, b3 - 3 \, a0^2 \, a3 \, b1 \, b2 \, b3 - a0^2 \, a2 \, b2^2 \, b3 + a1^3 \, b0 \, b3^2 - 3 \, a0 \, a1 \, a2 \, b0 \, b3^2 +$
$3 \, a0^2 \, a3 \, b0 \, b3^2 - a0 \, a1^2 \, b1 \, b3^2 + 2 \, a0^2 \, a2 \, b1 \, b3^2 + a0^2 \, a1 \, b2 \, b3^2 - a0^3 \, b3^3$

The last determinant agrees with the result from `Resultant`.

`In[14]:= Resultant[a0 + a1 x + a2 x^2 + a3 x^3,`
`                    b0 + b1 x + b2 x^2 + b3 x^3, x] === %`

`Out[14]= True`

(For similar determinant-based representations of eliminating n variables from $n+1$ polynomial equations, see [431]).

The function `Resultant` has the `Method` option. The following settings are possible: `Automatic`, `SylvesterMatrix`, `BezoutMatrix`, `Subresultants`, and `Modular`. The optimal choice depends dramatically on the concrete polynomial pair at hand and typically requires some experimentation. For higher-order univariate polynomials over the integers, the option setting `Modular` is frequently the fastest.

As an application of the use of the `Resultant` function, let us derive a nonlinear third-order differential equation for the function `EllipticNomeQ`.

`In[15]:= ?EllipticNomeQ`

> EllipticNomeQ[m] gives the nome q corresponding
> to the parameter m in an elliptic function. More...

This function $q(m)$ is an important function within the theory of elliptic functions (see Chapter 3) and theta functions. It can be expressed through complete elliptic integrals as $q(m) = \exp(-\pi\, K(1-m)/K(m))$, where $K(m)$ is the complete elliptic integral of the first kind (see Chapter 3).

We observe that the derivatives of the elliptic integrals of the first and second kind can be expressed rationally through $K(m)$ and $E(m)$. The elliptic integrals $K(m)$ and $E(m)$ are in *Mathematica* defined as `EllipticK[m]` and `EllipticE[m]`.

`In[16]:= {D[EllipticK[m], m], D[EllipticE[m], m]}`

$$Out[16]= \left\{ \frac{\text{EllipticE}[m] - (1-m)\,\text{EllipticK}[m]}{2\,(1-m)\,m},\ \frac{\text{EllipticE}[m] - \text{EllipticK}[m]}{2\,m} \right\}$$

We can differentiate $q(m)$ repeatedly and eliminate all occurrences of $K(m)$, $K(1-m)$, $E(m)$, and $E(1-m)$ to obtain a polynomial in the first four derivatives of $q(m)$ and m. Taking into account the so-called Legendre relation (see Chapter 1 of the Programming volume [1735])

$$E(m)\,K(1-m) + E(1-m)\,K(m) - K(1-m)\,K(m) = \pi/2$$

allows us to derive a differential equation of order three [163]. We will derive this differential equation now.

`In[17]:= legendreRelation =`
`         EllipticK[1 - m] EllipticE[m] + EllipticE[1 - m] EllipticK[m] -`
`         EllipticK[m] EllipticK[1 - m] - Pi/2;`

This is the set of equations obtained by differentiating $q(m)$ three times. For shorter, more readable output, we use one-letter symbols for the three functions `EllipticNomeQ`, `EllipticK`, and `EllipticE`.

`In[18]:= eqs = Table[Factor[Numerator[Together[`
`                D[q[m] - EllipticNomeQ[m], {m, i}]]]], {i, 3}] /.`
`                {EllipticNomeQ -> q, EllipticK -> 𝒦, EllipticE -> ℰ}`

$Out[18]= \{\pi\,q[m]\,ℰ[m]\,𝒦[1-m] + \pi\,q[m]\,ℰ[1-m]\,𝒦[m] - \pi\,q[m]\,𝒦[1-m]\,𝒦[m] -$
$2\,m\,𝒦[m]^2\,q'[m] + 2\,m^2\,𝒦[m]^2\,q'[m],\ -\pi^2\,q[m]\,ℰ[m]^2\,𝒦[1-m]^2 -$
$2\,\pi^2\,q[m]\,ℰ[1-m]\,ℰ[m]\,𝒦[1-m]\,𝒦[m] + 2\,\pi\,q[m]\,ℰ[m]^2\,𝒦[1-m]\,𝒦[m] +$
$2\,\pi^2\,q[m]\,ℰ[m]\,𝒦[1-m]^2\,𝒦[m] - \pi^2\,q[m]\,ℰ[1-m]^2\,𝒦[m]^2 + 2\,\pi\,q[m]\,ℰ[1-m]\,ℰ[m]\,𝒦[m]^2 +$

$$2 \pi^2 \, \mathsf{q}[m] \, \mathcal{E}[1-m] \, \mathcal{K}[1-m] \, \mathcal{K}[m]^2 - 2 \pi \, \mathsf{q}[m] \, \mathcal{E}[m] \, \mathcal{K}[1-m] \, \mathcal{K}[m]^2 -$$
$$2 \, m \, \pi \, \mathsf{q}[m] \, \mathcal{E}[m] \, \mathcal{K}[1-m] \, \mathcal{K}[m]^2 - \pi^2 \, \mathsf{q}[m] \, \mathcal{K}[1-m]^2 \, \mathcal{K}[m]^2 - 2 \, m \, \pi \, \mathsf{q}[m] \, \mathcal{E}[1-m] \, \mathcal{K}[m]^3 +$$
$$2 \, m \, \pi \, \mathsf{q}[m] \, \mathcal{K}[1-m] \, \mathcal{K}[m]^3 + 4 \, m^2 \, \mathcal{K}[m]^4 \, \mathsf{q}''[m] - 8 \, m^3 \, \mathcal{K}[m]^4 \, \mathsf{q}''[m] + 4 \, m^4 \, \mathcal{K}[m]^4 \, \mathsf{q}''[m],$$
$$\pi^3 \, \mathsf{q}[m] \, \mathcal{E}[m]^3 \, \mathcal{K}[1-m]^3 + 3 \, \pi^3 \, \mathsf{q}[m] \, \mathcal{E}[1-m] \, \mathcal{E}[m]^2 \, \mathcal{K}[1-m]^2 \, \mathcal{K}[m] -$$
$$6 \, \pi^2 \, \mathsf{q}[m] \, \mathcal{E}[m]^3 \, \mathcal{K}[1-m]^2 \, \mathcal{K}[m] - 3 \, \pi^3 \, \mathsf{q}[m] \, \mathcal{E}[m]^2 \, \mathcal{K}[1-m]^3 \, \mathcal{K}[m] +$$
$$3 \, \pi^3 \, \mathsf{q}[m] \, \mathcal{E}[1-m]^2 \, \mathcal{E}[m] \, \mathcal{K}[1-m] \, \mathcal{K}[m]^2 - 12 \, \pi^2 \, \mathsf{q}[m] \, \mathcal{E}[1-m] \, \mathcal{E}[m]^2 \, \mathcal{K}[1-m] \, \mathcal{K}[m]^2 +$$
$$6 \, \pi \, \mathsf{q}[m] \, \mathcal{E}[m]^3 \, \mathcal{K}[1-m] \, \mathcal{K}[m]^2 - 6 \, \pi^3 \, \mathsf{q}[m] \, \mathcal{E}[1-m] \, \mathcal{E}[m] \, \mathcal{K}[1-m]^2 \, \mathcal{K}[m]^2 +$$
$$12 \, \pi^2 \, \mathsf{q}[m] \, \mathcal{E}[m]^2 \, \mathcal{K}[1-m]^2 \, \mathcal{K}[m]^2 + 6 \, m \, \pi^2 \, \mathsf{q}[m] \, \mathcal{E}[m]^2 \, \mathcal{K}[1-m]^2 \, \mathcal{K}[m]^2 +$$
$$3 \, \pi^3 \, \mathsf{q}[m] \, \mathcal{E}[m] \, \mathcal{K}[1-m]^3 \, \mathcal{K}[m]^2 + \pi^3 \, \mathsf{q}[m] \, \mathcal{E}[1-m]^3 \, \mathcal{K}[m]^3 - 6 \, \pi^2 \, \mathsf{q}[m] \, \mathcal{E}[1-m]^2 \, \mathcal{E}[m] \, \mathcal{K}[m]^3 +$$
$$6 \, \pi \, \mathsf{q}[m] \, \mathcal{E}[1-m] \, \mathcal{E}[m]^2 \, \mathcal{K}[m]^3 - 3 \, \pi^3 \, \mathsf{q}[m] \, \mathcal{E}[1-m]^2 \, \mathcal{K}[1-m] \, \mathcal{K}[m]^3 +$$
$$12 \, \pi^2 \, \mathsf{q}[m] \, \mathcal{E}[1-m] \, \mathcal{E}[m] \, \mathcal{K}[1-m] \, \mathcal{K}[m]^3 + 12 \, m \, \pi^2 \, \mathsf{q}[m] \, \mathcal{E}[1-m] \, \mathcal{E}[m] \, \mathcal{K}[1-m] \, \mathcal{K}[m]^3 -$$
$$6 \, \pi \, \mathsf{q}[m] \, \mathcal{E}[m]^2 \, \mathcal{K}[1-m] \, \mathcal{K}[m]^3 - 12 \, m \, \pi \, \mathsf{q}[m] \, \mathcal{E}[m]^2 \, \mathcal{K}[1-m] \, \mathcal{K}[m]^3 +$$
$$3 \, \pi^3 \, \mathsf{q}[m] \, \mathcal{E}[1-m] \, \mathcal{K}[1-m]^2 \, \mathcal{K}[m]^3 - 6 \, \pi^2 \, \mathsf{q}[m] \, \mathcal{E}[m] \, \mathcal{K}[1-m]^2 \, \mathcal{K}[m]^3 -$$
$$12 \, m \, \pi^2 \, \mathsf{q}[m] \, \mathcal{E}[m] \, \mathcal{K}[1-m]^2 \, \mathcal{K}[m]^3 - \pi^3 \, \mathsf{q}[m] \, \mathcal{K}[1-m]^3 \, \mathcal{K}[m]^3 + 6 \, m \, \pi^2 \, \mathsf{q}[m] \, \mathcal{E}[1-m]^2 \, \mathcal{K}[m]^4 -$$
$$12 \, m \, \pi \, \mathsf{q}[m] \, \mathcal{E}[1-m] \, \mathcal{E}[m] \, \mathcal{K}[m]^4 - 12 \, m \, \pi^2 \, \mathsf{q}[m] \, \mathcal{E}[1-m] \, \mathcal{K}[1-m] \, \mathcal{K}[m]^4 +$$
$$2 \, \pi \, \mathsf{q}[m] \, \mathcal{E}[m] \, \mathcal{K}[1-m] \, \mathcal{K}[m]^4 + 10 \, m \, \pi \, \mathsf{q}[m] \, \mathcal{E}[m] \, \mathcal{K}[1-m] \, \mathcal{K}[m]^4 +$$
$$8 \, m^2 \, \pi \, \mathsf{q}[m] \, \mathcal{E}[m] \, \mathcal{K}[1-m] \, \mathcal{K}[m]^4 + 6 \, m \, \pi^2 \, \mathsf{q}[m] \, \mathcal{K}[1-m]^2 \, \mathcal{K}[m]^4 +$$
$$2 \, \pi \, \mathsf{q}[m] \, \mathcal{E}[1-m] \, \mathcal{K}[m]^5 - 2 \, m \, \pi \, \mathsf{q}[m] \, \mathcal{E}[1-m] \, \mathcal{K}[m]^5 + 8 \, m^2 \, \pi \, \mathsf{q}[m] \, \mathcal{E}[1-m] \, \mathcal{K}[m]^5 -$$
$$2 \, \pi \, \mathsf{q}[m] \, \mathcal{K}[1-m] \, \mathcal{K}[m]^5 + 2 \, m \, \pi \, \mathsf{q}[m] \, \mathcal{K}[1-m] \, \mathcal{K}[m]^5 - 8 \, m^2 \, \pi \, \mathsf{q}[m] \, \mathcal{K}[1-m] \, \mathcal{K}[m]^5 -$$
$$8 \, m^3 \, \mathcal{K}[m]^6 \, \mathsf{q}^{(3)}[m] + 24 \, m^4 \, \mathcal{K}[m]^6 \, \mathsf{q}^{(3)}[m] - 24 \, m^5 \, \mathcal{K}[m]^6 \, \mathsf{q}^{(3)}[m] + 8 \, m^6 \, \mathcal{K}[m]^6 \, \mathsf{q}^{(3)}[m] \}$$

To eliminate the four functions $K(m)$, $K(1-m)$, $E(m)$, and $E(1-m)$, we start with the Legendre relation. It is linear in each of the four functions. So, we solve, say, for $E(m)$ and substitute the result in eqs.

```
In[20]:= sol1 = Solve[legendreRelation == 0, EllipticE[m]] /.
                {EllipticNomeQ -> q, EllipticK -> 𝒦, EllipticE -> ℰ}
```

$$Out[20]= \left\{ \left\{ \mathcal{E}[m] \to \frac{\pi - 2 \, \mathcal{E}[1-m] \, \mathcal{K}[m] + 2 \, \mathcal{K}[1-m] \, \mathcal{K}[m]}{2 \, \mathcal{K}[1-m]} \right\} \right\}$$

```
In[22]:= eqs1 = Factor[Numerator[Together[eqs /. sol1[[1]]]]]
```

$$Out[22]= \{ \pi^2 \, \mathsf{q}[m] - 4 \, m \, \mathcal{K}[m]^2 \, \mathsf{q}'[m] + 4 \, m^2 \, \mathcal{K}[m]^2 \, \mathsf{q}'[m],$$
$$-\pi^4 \, \mathsf{q}[m] \, \mathcal{K}[1-m] + 2 \, \pi^3 \, \mathsf{q}[m] \, \mathcal{K}[m] - 4 \, \pi^2 \, \mathsf{q}[m] \, \mathcal{E}[1-m] \, \mathcal{K}[m]^2 +$$
$$4 \, \pi^2 \, \mathsf{q}[m] \, \mathcal{K}[1-m] \, \mathcal{K}[m]^2 - 4 \, m \, \pi^2 \, \mathsf{q}[m] \, \mathcal{K}[1-m] \, \mathcal{K}[m]^2 + 16 \, m^2 \, \mathcal{K}[1-m] \, \mathcal{K}[m]^4 \, \mathsf{q}''[m] -$$
$$32 \, m^3 \, \mathcal{K}[1-m] \, \mathcal{K}[m]^4 \, \mathsf{q}''[m] + 16 \, m^4 \, \mathcal{K}[1-m] \, \mathcal{K}[m]^4 \, \mathsf{q}''[m],$$
$$\pi^6 \, \mathsf{q}[m] \, \mathcal{K}[1-m]^2 - 6 \, \pi^5 \, \mathsf{q}[m] \, \mathcal{K}[1-m] \, \mathcal{K}[m] + 6 \, \pi^4 \, \mathsf{q}[m] \, \mathcal{K}[m]^2 +$$
$$12 \, \pi^4 \, \mathsf{q}[m] \, \mathcal{E}[1-m] \, \mathcal{K}[1-m] \, \mathcal{K}[m]^2 - 12 \, \pi^4 \, \mathsf{q}[m] \, \mathcal{K}[1-m]^2 \, \mathcal{K}[m]^2 +$$
$$12 \, m \, \pi^4 \, \mathsf{q}[m] \, \mathcal{K}[1-m]^2 \, \mathcal{K}[m]^2 - 24 \, \pi^3 \, \mathsf{q}[m] \, \mathcal{E}[1-m] \, \mathcal{K}[m]^3 +$$
$$24 \, \pi^3 \, \mathsf{q}[m] \, \mathcal{K}[1-m] \, \mathcal{K}[m]^3 - 24 \, m \, \pi^3 \, \mathsf{q}[m] \, \mathcal{K}[1-m] \, \mathcal{K}[m]^3 + 24 \, \pi^2 \, \mathsf{q}[m] \, \mathcal{E}[1-m]^2 \, \mathcal{K}[m]^4 -$$
$$48 \, \pi^2 \, \mathsf{q}[m] \, \mathcal{E}[1-m] \, \mathcal{K}[1-m] \, \mathcal{K}[m]^4 + 48 \, m \, \pi^2 \, \mathsf{q}[m] \, \mathcal{E}[1-m] \, \mathcal{K}[1-m] \, \mathcal{K}[m]^4 +$$
$$32 \, \pi^2 \, \mathsf{q}[m] \, \mathcal{K}[1-m]^2 \, \mathcal{K}[m]^4 - 56 \, m \, \pi^2 \, \mathsf{q}[m] \, \mathcal{K}[1-m]^2 \, \mathcal{K}[m]^4 + 32 \, m^2 \, \pi^2 \, \mathsf{q}[m] \, \mathcal{K}[1-m]^2 \, \mathcal{K}[m]^4 -$$
$$64 \, m^3 \, \mathcal{K}[1-m]^2 \, \mathcal{K}[m]^6 \, \mathsf{q}^{(3)}[m] + 192 \, m^4 \, \mathcal{K}[1-m]^2 \, \mathcal{K}[m]^6 \, \mathsf{q}^{(3)}[m] -$$
$$192 \, m^5 \, \mathcal{K}[1-m]^2 \, \mathcal{K}[m]^6 \, \mathsf{q}^{(3)}[m] + 64 \, m^6 \, \mathcal{K}[1-m]^2 \, \mathcal{K}[m]^6 \, \mathsf{q}^{(3)}[m] \}$$

The second of the resulting equations is linear in $E(1-m)$ and $K(1-m)$.

```
In[23]:= Exponent[#, {𝒦[m], ℰ[1 - m], 𝒦[1 - m]}]& /@ eqs1
Out[23]= {{2, 0, 0}, {4, 1, 1}, {6, 2, 2}}
```

Solving the second equation for $E(1-m)$ and substituting the result in the first and third equations gives the following two polynomials.

```
In[24]:= sol2 = Solve[eqs1[[2]] == 0, ℰ[1 - m]] /.
                {EllipticNomeQ -> q, EllipticK -> 𝒦, EllipticE -> ℰ}
```

Out[24]= $\{\{\mathcal{E}[1-m] \to \dfrac{1}{4\,\pi^2\,q[m]\,\mathcal{K}[m]^2}\,(-\pi^4\,q[m]\,\mathcal{K}[1-m] + 2\,\pi^3\,q[m]\,\mathcal{K}[m] +$

$4\,\pi^2\,q[m]\,\mathcal{K}[1-m]\,\mathcal{K}[m]^2 - 4\,m\,\pi^2\,q[m]\,\mathcal{K}[1-m]\,\mathcal{K}[m]^2 + 16\,m^2\,\mathcal{K}[1-m]\,\mathcal{K}[m]^4\,q''[m] -$

$32\,m^3\,\mathcal{K}[1-m]\,\mathcal{K}[m]^4\,q''[m] + 16\,m^4\,\mathcal{K}[1-m]\,\mathcal{K}[m]^4\,q''[m])\}\}$

In[26]:= **eqs2 = Factor[Numerator[Together[Delete[eqs1, 2] /. sol2[[1]]]]]**

Out[26]= $\{\pi^2\,q[m] - 4\,m\,\mathcal{K}[m]^2\,q'[m] + 4\,m^2\,\mathcal{K}[m]^2\,q'[m],$

$-\mathcal{K}[1-m]^2\,(\pi^8\,q[m]^2 - 16\,\pi^4\,q[m]^2\,\mathcal{K}[m]^4 + 16\,m\,\pi^4\,q[m]^2\,\mathcal{K}[m]^4 - 16\,m^2\,\pi^4\,q[m]^2\,\mathcal{K}[m]^4 -$

$768\,m^4\,\mathcal{K}[m]^8\,q''[m]^2 + 3072\,m^5\,\mathcal{K}[m]^8\,q''[m]^2 - 4608\,m^6\,\mathcal{K}[m]^8\,q''[m]^2 +$

$3072\,m^7\,\mathcal{K}[m]^8\,q''[m]^2 - 768\,m^8\,\mathcal{K}[m]^8\,q''[m]^2 + 128\,m^3\,\pi^2\,q[m]\,\mathcal{K}[m]^6\,q^{(3)}[m] - 384\,m^4\,\pi^2$

$q[m]\,\mathcal{K}[m]^6\,q^{(3)}[m] + 384\,m^5\,\pi^2\,q[m]\,\mathcal{K}[m]^6\,q^{(3)}[m] - 128\,m^6\,\pi^2\,q[m]\,\mathcal{K}[m]^6\,q^{(3)}[m])\}$

Now, the first of the resulting equations just contains $K(m)$ and the second equation contains a nontrivial factor that has only $K(m)$.

In[27]:= **eqs3 = {eqs2[[1]], eqs2[[2, -1]]}**

Out[27]= $\{\pi^2\,q[m] - 4\,m\,\mathcal{K}[m]^2\,q'[m] + 4\,m^2\,\mathcal{K}[m]^2\,q'[m],$

$\pi^8\,q[m]^2 - 16\,\pi^4\,q[m]^2\,\mathcal{K}[m]^4 + 16\,m\,\pi^4\,q[m]^2\,\mathcal{K}[m]^4 - 16\,m^2\,\pi^4\,q[m]^2\,\mathcal{K}[m]^4 -$

$768\,m^4\,\mathcal{K}[m]^8\,q''[m]^2 + 3072\,m^5\,\mathcal{K}[m]^8\,q''[m]^2 - 4608\,m^6\,\mathcal{K}[m]^8\,q''[m]^2 +$

$3072\,m^7\,\mathcal{K}[m]^8\,q''[m]^2 - 768\,m^8\,\mathcal{K}[m]^8\,q''[m]^2 + 128\,m^3\,\pi^2\,q[m]\,\mathcal{K}[m]^6\,q^{(3)}[m] -$

$384\,m^4\,\pi^2\,q[m]\,\mathcal{K}[m]^6\,q^{(3)}[m] + 384\,m^5\,\pi^2\,q[m]\,\mathcal{K}[m]^6\,q^{(3)}[m] - 128\,m^6\,\pi^2\,q[m]\,\mathcal{K}[m]^6\,q^{(3)}[m]\}$

Because both polynomials contain only even powers of $K(m)$ and we want to eliminate $K(m)$ anyway, we can reduce the degree with respect to $K(m)$ by introducing a new variable $\mathcal{K}2[m]$.

In[28]:= **eqs4 = eqs3 /. $\mathcal{K}$[m]^n_ -> $\mathcal{K}$2[m]^(n/2)**

Out[28]= $\{\pi^2\,q[m] - 4\,m\,\mathcal{K}2[m]\,q'[m] + 4\,m^2\,\mathcal{K}2[m]\,q'[m],$

$\pi^8\,q[m]^2 - 16\,\pi^4\,q[m]^2\,\mathcal{K}2[m]^2 + 16\,m\,\pi^4\,q[m]^2\,\mathcal{K}2[m]^2 -$

$16\,m^2\,\pi^4\,q[m]^2\,\mathcal{K}2[m]^2 - 768\,m^4\,\mathcal{K}2[m]^4\,q''[m]^2 + 3072\,m^5\,\mathcal{K}2[m]^4\,q''[m]^2 -$

$4608\,m^6\,\mathcal{K}2[m]^4\,q''[m]^2 + 3072\,m^7\,\mathcal{K}2[m]^4\,q''[m]^2 - 768\,m^8\,\mathcal{K}2[m]^4\,q''[m]^2 +$

$128\,m^3\,\pi^2\,q[m]\,\mathcal{K}2[m]^3\,q^{(3)}[m] - 384\,m^4\,\pi^2\,q[m]\,\mathcal{K}2[m]^3\,q^{(3)}[m] +$

$384\,m^5\,\pi^2\,q[m]\,\mathcal{K}2[m]^3\,q^{(3)}[m] - 128\,m^6\,\pi^2\,q[m]\,\mathcal{K}2[m]^3\,q^{(3)}[m]\}$

To eliminate $\mathcal{K}2[m]$, we finally use the function `Resultant`.

In[29]:= **res = Factor[Resultant[##, $\mathcal{K}$2[m]]& @@ eqs4]**

Out[29]= $256\,(-1+m)^2\,m^2\,\pi^8\,q[m]^2$

$(-q[m]^2\,q'[m]^2 + m\,q[m]^2\,q'[m]^2 - m^2\,q[m]^2\,q'[m]^2 + m^2\,q'[m]^4 - 2\,m^3\,q'[m]^4 +$

$m^4\,q'[m]^4 - 3\,m^2\,q[m]^2\,q''[m]^2 + 6\,m^3\,q[m]^2\,q''[m]^2 - 3\,m^4\,q[m]^2\,q''[m]^2 +$

$2\,m^2\,q[m]^2\,q'[m]\,q^{(3)}[m] - 4\,m^3\,q[m]^2\,q'[m]\,q^{(3)}[m] + 2\,m^4\,q[m]^2\,q'[m]\,q^{(3)}[m])$

The last term of `res` is the nontrivial one that is the differential equation we were looking for. (When carrying out repeated substitutions and resultant formations it happens frequently that the result will contain the "wanted" term, as well as other ones.)

In[30]:= **(Simplify[res[[-1]]]) // TraditionalForm**

Out[30]//TraditionalForm=

$(m-1)^2\,m^2\,q'(m)^4 - q(m)^2\,(3\,(m-1)^2\,q''(m)^2\,m^2 - 2\,(m-1)^2\,q'(m)\,q^{(3)}(m)\,m^2 + (m^2-m+1)\,q'(m)^2)$

In[31]:= **ellipticNomeQODE = %;**

Using the function `Together` (to be discussed in the next section), we succeed also symbolically in proving that the function `EllipticNomeQ[m]` is indeed a solution of the derived differential equation.

In[32]:= **Together[ellipticNomeQODE /. q -> EllipticNomeQ]**

Out[32]= 0

Now, we come to one of the most central functions of this chapter. `GroebnerBasis` has major applications to `Solve`, `NSolve`, etc.

GroebnerBasis[{*polynomial*$_1$, *polynomial*$_2$, ..., *polynomial*$_n$},
  {*var*$_1$, *var*$_2$, ... , *var*$_2$},
  {*elimVar*$_1$, *elimVar*$_2$, ... , *elimVar*$_2$}, *options*]
finds a Gröbner basis for the ideal generated by the polynomials *polynomial*$_i$ in the variables *var*$_i$ and
eliminates the variables *elimVar*$_i$.

Gröbner bases [1473], [198] play an important role in various applications, such as robotics [1661], [1528], [793], [1070], [69], [1068], [968], [414], and [298]; automatic geometric theorem proving [694], [1859], [1808], [359], [1860], [1807], [982], [979], [337], [1731], [693], [360], [1637], [877], [363], [1835], and [1090]; generating implicit representations of parametric curves and surfaces [1593], [1340], [960], [1067] and [1592]; in signal processing [1235]; and many more fields such as coding theory, Petri nets [328], computational group theory [515] and integer programming [400].

We do not give a precise definition of a Gröbner basis here; see the excellent treatments in [261], [414], [15], [1075], [669], [259], [415], [29], [258], [862], [1274], [135], [981], [765], [1138], and [564], and for a short introduction, see [782], [1315], [1121], and [827]. However, roughly speaking, a Gröbner basis for a set of polynomials is a set of "simpler" polynomials with the same common zeros as the original polynomial.

Because of its importance we will give a longer than usual list of examples in the following.

Here is a simple example for the following starting polynomials.

In[33]:= **eqs = {x^2 + 3 x + 4 y, 2 y^2 + 4 x + 5 y + 9}**

Out[33]= $\{3 x + x^2 + 4 y, 9 + 4 x + 5 y + 2 y^2\}$

These are the numerical solutions of `eqs`.

In[34]:= **NSolve[{eqs[[1]] == 0, eqs[[2]] == 0}, {x, y}]**

Out[34]= $\{\{x \to 1.2388 - 1.78102\,i, y \to -0.519754 + 2.43893\,i\}$,
  $\{x \to 1.2388 + 1.78102\,i, y \to -0.519754 - 2.43893\,i\}$,
  $\{x \to -2.6048, y \to 0.257352\}, \{x \to -5.8728, y \to -4.21784\}\}$

The polynomials computed by `GroebnerBasis` are not always shorter, but they have the same zeros and, in sequence, each of them depends on just one more variable than its predecessor in the list of polynomials (pseudotriangular form). (This form of a Gröbner basis, called a lexicographic Gröbner basis, is the default form generated by *Mathematica*. Other bases can be calculated too, see the option `MonomialOrder` below.)

In[35]:= **GroebnerBasis[eqs, {x, y}]**

Out[35]= $\{-27 + 94 y + 37 y^2 + 20 y^3 + 4 y^4, 9 + 4 x + 5 y + 2 y^2\}$

In[36]:= **NSolve[{%[[1]] == 0, %[[2]] == 0}, {x, y}]**

Out[36]= $\{\{x \to 1.2388 - 1.78102\,i, y \to -0.519754 + 2.43893\,i\}$,
  $\{x \to 1.2388 + 1.78102\,i, y \to -0.519754 - 2.43893\,i\}$,
  $\{x \to -2.6048, y \to 0.257352\}, \{x \to -5.8728, y \to -4.21784\}\}$

We give another example, this time with a plot. The Gröbner basis for the three polynomials $x^2 + y^2 - 1$, $x^2 + z^2 - 1$, and $y^2 + z^2 - 1$ is the following.

In[37]:= **GroebnerBasis[{x^2 + y^2 - 1, x^2 + z^2 - 1, y^2 + z^2 - 1}, {x, y, z}]**

Out[37]= $\{1 - 2 z^2, 1 - 2 y^2, 1 - 2 x^2\}$

The three polynomials produced by `GroebnerBasis` have the same zeros and are clearly simpler (in this example, also shorter). In this case, the set of zeros of the three polynomials can easily be plotted: They are the intersections of three cylinders surrounding the coordinate axes at $\{\pm2^{-1/2}, \pm2^{-1/2}, \pm2^{-1/2}\}$. In the following graphic, we display the three cylinders and mark the intersection points with little cubes.

```
In[38]:= (* cylinders parallel to the coordinate system axis *)
        {cylinderX, cylinderY, cylinderZ} =
        Cases[ParametricPlot3D[#, {φ, 0, 2Pi}, {t, -2, 2},
                        DisplayFunction -> Identity], _Polygon, Infinity]& /@
            {{t, Cos[φ], Sin[φ]}, {Cos[φ], t, Sin[φ]}, {Cos[φ], Sin[φ], t}};
```

```
In[40]:= (* display all three cylinders *)
        Show[Graphics3D[{ {EdgeForm[], Thickness[0.002],
            MapIndexed[{SurfaceColor[RGBColor @@
                    MapAt[1&, {0, 0, 0}, #2[[1]]]], #1}&,
                    {cylinderX, cylinderY, cylinderZ}]} /.
        Polygon[l_] :> (* making holes in cylinder polygons *)
        Module[{mp = Plus @@ l/Length[l]},
                ℓ = (mp + 0.8(# - mp))& /@ l;
                {MapThread[Polygon[Join[#1, Reverse[#2]]]&,
                Partition[Append[#, First[#]], 2, 1]& /@ {l, ℓ}]}],
            (* the intersection points as cubes *)
            {SurfaceColor[GrayLevel[0]],
            Cuboid[# - 0.1 {1, 1, 1}, # + 0.1 {1, 1, 1}]& /@
            (* intersection points of the three cylinders *)
                    ({x, y, z} /. NSolve[(0 == #)& /@
            {x^2 + y^2 - 1, x^2 + z^2 - 1, y^2 + z^2 - 1},
                            {x, y, z}])}}],
            DisplayFunction -> $DisplayFunction, PlotRange -> All]
```

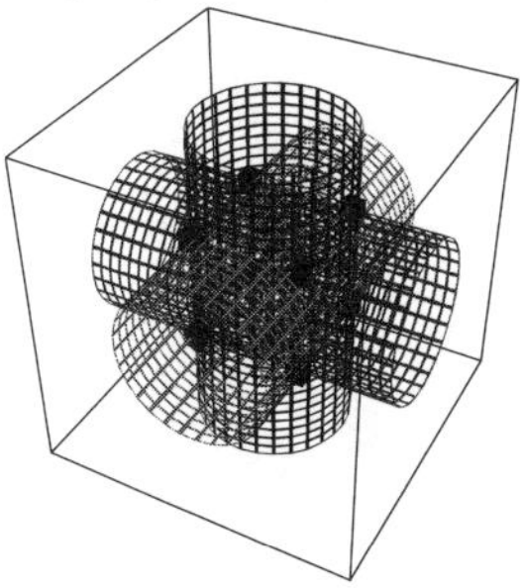

The Gröbner basis of an inconsistent polynomial system (this means the solution set is empty) is {1}.

```
In[42]:= GroebnerBasis[{x, x + 1, x - 1}, {x}]
```
```
Out[42]= {1}
```

`GroebnerBasis` has the following options.

```
In[43]:= Options[GroebnerBasis]
```
```
Out[43]= {CoefficientDomain → Automatic, Method → Buchberger,
        Modulus → 0, MonomialOrder → Lexicographic, ParameterVariables → {},
        SelectionStrategy → SugarCube, Sort → False, Tolerance → 0}
```

(For symmetric configurations of more cylinders, see [64].)

We will discuss some of these options; for more details, see [1141] and the Advanced Documentation in the online *Mathematica* documentation directory.

CoefficientDomain

    determines the domain of the coefficients of the polynomials.

    Default:

        Rationals

    Admissible:

        RationalFunctions or InexactNumbers or Integers or Polynomials[*vars*]

The coefficient domain influences the form of a Gröbner basis.

In[44]:= **GroebnerBasis[{x^2 - a y, y x - x^3 + a^3}, {x, y},**
                    CoefficientDomain -> RationalFunctions] // Expand

Out[44]= $\{a^5 - y^3 + 2\,a\,y^3 - a^2\,y^3,\ a^2\,x + y^2 - a\,y^2\}$

In[45]:= **GroebnerBasis[{x^2 - a y, y x - x^3 + a^3}, {x, y},**
                    CoefficientDomain -> Rationals] // Expand

Out[45]= $\{-a^6 + a\,y^3 - 2\,a^2\,y^3 + a^3\,y^3,\ a^3\,x + a\,y^2 - a^2\,y^2,\ a^3 + a^4 + a^5 + x\,y + a\,y^3 - a^2\,y^3,\ x^2 - a\,y\}$

In the last example, the variable *a* could be considered as a parameter. The option ParameterVariables allows to explicitly specify such parameters.

ParameterVariables

    declares certain variables as parameters with respect to the current monomial order.

    Default:

        {}

    Admissible:

        listOfVariables

Here we repeat the last example and specify *a* as a parameter.

In[46]:= **GroebnerBasis[{x^2 - a y, y x - x^3 + a^3}, {x, y},**
                    CoefficientDomain -> Rationals,
                    ParameterVariables -> {a}] // Expand

Out[46]= $\{-a^6 + a\,y^3 - 2\,a^2\,y^3 + a^3\,y^3,\ a^3\,x + a\,y^2 - a^2\,y^2,\ a^3 + a^4 + a^5 + x\,y + a\,y^3 - a^2\,y^3,\ x^2 - a\,y\}$

Again, both Gröbner bases have the same zeros with respect to *x* and *y*.

In[47]:= **Solve[# == 0& /@ %, {x, y}]**

Out[47]= $\left\{\left\{x \to \dfrac{-\dfrac{a^{10/3}}{(1-2\,a+a^2)^{2/3}} + \dfrac{a^{13/3}}{(1-2\,a+a^2)^{2/3}}}{a^2},\ y \to \dfrac{a^{5/3}}{(1-2\,a+a^2)^{1/3}}\right\},\right.$

$\left\{x \to \dfrac{\dfrac{(-1)^{1/3}\,a^{10/3}}{(1-2\,a+a^2)^{2/3}} - \dfrac{(-1)^{1/3}\,a^{13/3}}{(1-2\,a+a^2)^{2/3}}}{a^2},\ y \to \dfrac{(-1)^{2/3}\,a^{5/3}}{(1-2\,a+a^2)^{1/3}}\right\},$

$\left.\left\{x \to \dfrac{-\dfrac{(-1)^{2/3}\,a^{10/3}}{(1-2\,a+a^2)^{2/3}} + \dfrac{(-1)^{2/3}\,a^{13/3}}{(1-2\,a+a^2)^{2/3}}}{a^2},\ y \to -\dfrac{(-1)^{1/3}\,a^{5/3}}{(1-2\,a+a^2)^{1/3}}\right\}\right\}$

In[48]:= **Solve[# == 0& /@ %%%, {x, y}]**

Out[48]= $\left\{\left\{x \to \dfrac{-\dfrac{a^{10/3}}{(1-2\,a+a^2)^{2/3}}+\dfrac{a^{13/3}}{(1-2\,a+a^2)^{2/3}}}{a^2}\,,\ y \to \dfrac{a^{5/3}}{(1-2\,a+a^2)^{1/3}}\right\},\right.$

$\left\{x \to \dfrac{\dfrac{(-1)^{1/3}\,a^{10/3}}{(1-2\,a+a^2)^{2/3}}-\dfrac{(-1)^{1/3}\,a^{13/3}}{(1-2\,a+a^2)^{2/3}}}{a^2}\,,\ y \to \dfrac{(-1)^{2/3}\,a^{5/3}}{(1-2\,a+a^2)^{1/3}}\right\},$

$\left.\left\{x \to \dfrac{-\dfrac{(-1)^{2/3}\,a^{10/3}}{(1-2\,a+a^2)^{2/3}}+\dfrac{(-1)^{2/3}\,a^{13/3}}{(1-2\,a+a^2)^{2/3}}}{a^2}\,,\ y \to -\dfrac{(-1)^{1/3}\,a^{5/3}}{(1-2\,a+a^2)^{1/3}}\right\}\right\}$

Although in most cases a lexicographic basis will be the most useful one, it is at the same time typically the most difficult to compute. Here is a system of three equations in three variables.

In[49]:= **eqs = {2 x^4 y + y^3 x^3 - x z^2 + 1,**
x^2 + y^2 z^3 - 1,
x^2 y - 7 y^3 z^2 + y^2 z^3};

Calculating a lexicographic Gröbner basis takes just a few seconds.

In[50]:= **(gb = GroebnerBasis[eqs, {x, y, z}]); // Timing**

Out[50]= {28.21 Second, Null}

The list of polynomials gb contains three quite complicated polynomials. The first is a univariate polynomial of degree 44, and the largest integer coefficient that occurs has 251 digits.

In[51]:= **Exponent[#, {x, y, z}] & /@ gb**

Out[51]= {{0, 0, 44}, {0, 1, 43}, {1, 0, 43}}

In[52]:= **{Length /@ gb, {#, N[#]}& @ Max[Abs[Cases[gb, _Integer, Infinity]]]}**

Out[52]= {{42, 45, 45},
{660315050284902405127753569085965903934655262562978197853379515017909418018
283580174111147289043943242094943161981673659227156484042259064933530936400
123817867019162342716064243405446870093975459500383070825510773488184983110
22761249117137174194545028, 6.60315×10^{251}}}

Because of the large amount of calculations containing polynomials with large integer coefficients required in intermediate steps, calculating an exact Gröbner basis is an expensive process. Calculating a Gröbner basis with approximative numbers instead of with exact numbers is often much faster.

In[53]:= **(gbN = GroebnerBasis[N[eqs, 300], {x, y, z},**
CoefficientDomain -> InexactNumbers]); // Timing

Out[53]= {0.21 Second, Null}

We can avoid the explicit numericalization and let *Mathematica* do the work. The InexactNumbers option setting can have an argument that specifies the precision to be used.

In[54]:= **gbN == GroebnerBasis[eqs, {x, y, z},**
CoefficientDomain -> InexactNumbers[300]]

Out[54]= True

gbN has the same shape as gb.

In[55]:= **Exponent[#, {x, y, z}] & /@ gbN**

Out[55]= {{0, 0, 44}, {0, 1, 43}, {1, 0, 43}}

Actually, gbN is a very good numerical approximation of gb.

```
In[56]:= gb[[1]]/Coefficient[gb[[1]], z, 44] -
      gbN[[1]]/Coefficient[gbN[[1]], z, 44] //
             CoefficientList[#, z]& // Abs // Min
```

Out[56]= $0. \times 10^{-154}$

In calculating a Gröbner basis with the option setting `CoefficientDomain -> InexactNumbers`, it is important that the input equations have approximative numbers of high enough precision. Using only 200 digits instead of 300 digits results in a failure to calculate a Gröbner basis. (This is not unexpected; the largest numbers in the exact Gröbner basis had about 250 digits.)

```
In[57]:= (gbN = GroebnerBasis[N[eqs, 200], {x, y, z},
                       CoefficientDomain -> InexactNumbers]); // Timing
          GroebnerBasis::fltgb : Excessive loss of precision during computation. More…
```

Out[57]= {0.17 Second, Null}

When the coefficient domain of a Gröbner basis calculation is `Rationals`, often very large integers appear in the result and even larger integers are generated in intermediate stages of the calculation. Frequently the arithmetic of these large integers dominates the whole calculation time. Sometimes, for instance, when one is interested only in the shape of a Gröbner basis or when one attempts to show that a Gröbner basis is empty, it is sufficient to work over a smaller field than the integers. The option `Modulus` deals with this situation.

```
Modulus
```

    determines the modulus to be used in the calculations.

Default:

    0

Admissible:

    primeNumber

We again calculate the lexicographic basis for the set of polynomials `eqs`. But this time we treat all coefficients modulo 541. Calculating the Gröbner basis is now about 25 times faster and yields a Gröbner basis of the same shape as above.

```
In[58]:= (gb = GroebnerBasis[eqs, {x, y, z}, Modulus -> Prime[100]]); // Timing
```

Out[58]= {0.07 Second, Null}

```
In[59]:= Exponent[#, {x, y, z}] & /@ gb
```

Out[59]= {{0, 0, 44}, {0, 1, 43}, {1, 0, 43}}

So far, we have calculated only lexicographic Gröbner bases. The option influencing the kind of Gröbner basis calculated is `MonomialOrder`.

MonomialOrder

> determines the term order for the polynomials.

> Default:
> > Lexicographic

> Admissible:
> > Lexicographic or DegreeReverseLexicographic or EliminationOrder or *explicit-WeightMatrix*

We do not go into the details of term orderings here (see the above-mentioned references). A rough guide is the following: For solving polynomial systems (this means bringing them into [pseudo]triangular form), ones use the setting MonomialOrder -> Lexicographic. In the following example, the resulting basis contains one polynomial in z. It is easy to solve this univariate polynomial and substitute the solutions in the second polynomial to obtain a new univariate polynomial in y, which can be easily solved for, ...

In[60]:= `GroebnerBasis[{x + y + z, x - 2 y + z^3, x^2 - 2y^3 + z}, {x, y, z}]`

Out[60]= $\{27 z + 12 z^2 + 2 z^3 + 12 z^4 - 6 z^5 + 3 z^6 + 6 z^7 - 2 z^9, 3 y + z - z^3, 3 x + 2 z + z^3\}$

Reverting the order of the variables gives now one univariate polynomial in x.

In[61]:= `GroebnerBasis[{x + y + z, x - 2 y + z^3, x^2 - 2 y^3 + z}, {z, y, x}]`

Out[61]= $\{-27 x + 18 x^2 - 342 x^3 + 306 x^4 - 186 x^5 + 229 x^6 - 18 x^7 + 12 x^8 + 8 x^9,$
$-3469149 x + 826587 x^2 - 297774 x^3 + 1547550 x^4 + 1141206 x^5 + 33316 x^6 +$
$169480 x^7 + 48784 x^8 + 1194237 y, 4663386 x - 826587 x^2 + 297774 x^3 -$
$1547550 x^4 - 1141206 x^5 - 33316 x^6 - 169480 x^7 - 48784 x^8 + 1194237 z\}$

Calculating a Gröbner basis is typically a very time-consuming process for larger polynomial systems. In most cases, the calculation using the term order MonomialOrder -> DegreeReverseLexicographic is the fastest. Here is an example.

In[62]:=
```
{{tLex, gbLex}, {tDRLex, gbDegRevLex}} =
   Timing[(GroebnerBasis[{x^4 + y + z, x - 2 y^3 + 5 z^4,
                          x^2 - 7 y^3 + z^6}, {z, y, x},
                         MonomialOrder -> #])]& /@
                         {Lexicographic, DegreeReverseLexicographic};
```

```
(* the timings *)
{tLex, tDRLex}
```
Out[65]= $\{1.54 \text{ Second}, 0. \text{ Second}\}$

Both Gröbner bases consist of four polynomials.

In[66]:= `Map[Length, {gbLex, gbDegRevLex}]`

Out[66]= $\{4, 4\}$

But the one calculated with the option setting DegreeReverseLexicographic has many fewer terms and a much lower degree.

In[67]:= `Map[Length, {gbLex, gbDegRevLex}, {2}]`

Out[67]= $\{\{54, 70, 70, 3\}, \{3, 3, 4, 6\}\}$

In[68]:= `(Exponent[#, {x, y, z}]& /@ #)& /@ {gbLex, gbDegRevLex}`

Out[68]= $\{\{\{70, 0, 0\}, \{69, 1, 0\}, \{69, 3, 0\}, \{4, 1, 1\}\},$
$\{\{4, 1, 1\}, \{1, 3, 4\}, \{2, 3, 2\}, \{2, 6, 2\}\}\}$

The monomial ordering `DegreeReverseLexicographic` often results in the smallest runtime for `Groebn-erBasis`. The `DegreeReverseLexicographic` is not directly useful for equation solving (in some exercises we will use it as starting system for a Gröbner walk). But it is very useful for detecting an inconsistent system of polynomial equations. The next input shows that there are no 1 to 12-dimensional representations of the canonical commutation relations $\hat{p}\,\hat{q} - \hat{q}\,\hat{p} = $ id [807], [1686], [684], [725]. Although the last of the following Gröbner basis calculations contains 288 variables, the calculation is carried out quickly. (In this concrete example, a lexicographic ordering would need approximately the same time.)

```
In[69]:= Module[{p, q},
          Table[With[{p = Array[p, {d, d}], q = Array[q, {d, d}]},
              GroebnerBasis[Flatten[(* CCR *) p.q - q.p - IdentityMatrix[d]],
                  MonomialOrder -> DegreeReverseLexicographic]],
              {d, 1, 12}]] // Timing
Out[69]= {4.68 Second, {{1}, {1}, {1}, {1}, {1}, {1}, {1}, {1}, {1}, {1}, {1}, {1}}}
```

Using an additional variable $\mathfrak{t}$, we can deal with the situation $\hat{p}\,\hat{q} - \hat{q}\,\hat{p} = c$ id. The equation $1 - c\,\mathfrak{t}$ does not allow c to take the value 0.

```
In[70]:= Module[{p, q, c, t},
          Table[With[{p = Array[p, {d, d}], q = Array[q, {d, d}]},
              GroebnerBasis[Flatten[{(* CCR *) p.q - q.p - c IdentityMatrix[d],
                  (* c != 0 *) 1 - c t}],
                  MonomialOrder -> DegreeReverseLexicographic]],
              {d, 1, 12}]] // Timing
Out[70]= {5.2 Second, {{1}, {1}, {1}, {1}, {1}, {1}, {1}, {1}, {1}, {1}, {1}, {1}}}
```

As the last result suggests, there are no finite-dimensional representations of the canonical commutation relations [1568], [238], [1795], [527], [1409], [535], [1468], [465], [1685], [1447]. For another application of the term order `DegreeReverseLexicographic`, see below.

Finally, for eliminating variables, the term order `MonomialOrder -> EliminationOrder` is often the most appropriate one. Here, the parametrically defined surface $\{x, y, z\} = \{s\,t^2,\ s^2 - t^2,\ s^3 - t\}$ transformed into an implicit polynomial of the form $p(x, y, z) = 0$.

```
In[71]:= GroebnerBasis[{x - s t^2 + s, y - s^2 + t^2, z - s^3 + t},
              {z, y, x}, {s, t}, MonomialOrder -> EliminationOrder]
Out[71]= {4 x^4 + x^6 - 4 x^2 y + 7 x^4 y - 17 x^2 y^2 + 9 y^3 - 22 x^2 y^3 + 18 y^4 - 7 x^2 y^4 +
          15 y^5 + 6 y^6 + y^7 - 8 x^3 z - 6 x^5 z + 8 x y z - 16 x^3 y z + 14 x y^2 z + 6 x^3 y^2 z +
          18 x y^3 z + 2 x^3 y^3 z + 6 x y^4 z + 15 x^4 z^2 - 4 y z^2 + 6 x^2 y z^2 + 3 y^2 z^2 - 18 x^2 y^2 z^2 +
          12 y^3 z^2 - 6 x^2 y^3 z^2 + 13 y^4 z^2 + 6 y^5 z^2 + y^6 z^2 + 8 x z^3 - 20 x^3 z^3 + 8 x y z^3 +
          18 x y^2 z^3 + 6 x y^3 z^3 - 4 z^4 + 15 x^2 z^4 - 5 y z^4 - 6 y^2 z^4 - 2 y^3 z^4 - 6 x z^5 + z^6}
```

Here is a more complicated implicization problem. This is the parametrization of a surface [72], [73], [1723].

```
In[72]:= p[θ_, τ_] := 1/(1 + τ^2 - τ (Sin[2θ] - Cos[2θ])) *
          {(1 + 2τ) Cos[θ]/Sqrt[3], (1 - 2τ) Sin[θ]/Sqrt[3], 1};
```

The parameters belong to the domains $\theta \in [0, 2\pi)$, $\tau \in (-\infty, \infty)$.

```
In[73]:= Show[GraphicsArray[{#, (* show fixed z-range layer *)
          Append[#, PlotRange -> {All, All, {0, 1}}]}]]& @
          Graphics3D[{EdgeForm[], SurfaceColor[Hue[0.25], Hue[0.12], 2.8],
          Cases[ParametricPlot3D[Evaluate[
          p[θ, τ] /. (* change parametrization *) τ -> Tan[t]],
          {θ, 0, 2Pi}, {t, -Pi/2, Pi/2}, PlotPoints -> 80,
          DisplayFunction -> Identity], _Polygon, Infinity] /.
```

```
(* make diamonds *) Polygon[l_] :> Polygon[Plus @@@ Partition[
    Append[l, First[l]], 2, 1]/2]}, Boxed -> False]
```

To generate an implicit form of this surface, we transform the rational expression to a polynomial one, express $\cos(2\theta)$ and $\sin(2\theta)$ in $\cos(\theta)$ and $\sin(\theta)$ and supplement with the equation $\sin(\theta)^2 + \cos(\theta)^2 - 1$ [796].

```
In[74]:= GroebnerBasis[Append[TrigExpand[
            Numerator[Together[{x, y, z} - p[θ, τ]]]]], Cos[θ]^2 + Sin[θ]^2 - 1],
            {x, y, z}, {τ, Cos[θ], Sin[θ]},
            MonomialOrder -> EliminationOrder]
```

Out[74]= $\{-162\,x^8 - 648\,x^6\,y^2 - 972\,x^4\,y^4 - 648\,x^2\,y^6 - 162\,y^8 + 648\,x^6\,z + 1944\,x^4\,y^2\,z +$
$1944\,x^2\,y^4\,z + 648\,y^6\,z - 1008\,x^4\,z^2 - 378\,x^6\,z^2 - 216\,x^5\,y\,z^2 - 2016\,x^2\,y^2\,z^2 -$
$702\,x^4\,y^2\,z^2 - 1008\,y^4\,z^2 - 702\,x^2\,y^4\,z^2 + 216\,x\,y^5\,z^2 - 378\,y^6\,z^2 + 768\,x^2\,z^3 +$
$1224\,x^4\,z^3 + 288\,x^3\,y\,z^3 + 768\,y^2\,z^3 + 2448\,x^2\,y^2\,z^3 - 288\,x\,y^3\,z^3 + 1224\,y^4\,z^3 - 256\,z^4 -$
$1536\,x^2\,z^4 - 351\,x^4\,z^4 - 216\,x^3\,y\,z^4 - 1536\,y^2\,z^4 - 846\,x^2\,y^2\,z^4 + 216\,x\,y^3\,z^4 - 351\,y^4\,z^4 +$
$768\,z^5 + 1008\,x^2\,z^5 + 1008\,y^2\,z^5 - 864\,z^6 - 216\,x^2\,z^6 - 216\,y^2\,z^6 + 432\,z^7 - 81\,z^8\}$

Substituting the original parametrization shows the correctness of the last polynomial.

```
In[75]:= (% /. Thread[{x, y, z} -> p[θ, τ]]) // Simplify
```

Out[75]= $\{0\}$

Often one wants to solve a system of polynomials under the conditions that other polynomials do not vanish. Adding equations of the form $1 - Z\,q(x_1, \ldots, x_n)$ to the first set of polynomials ensures that $q(x_1, \ldots, x_n)$ will not vanish. Then one can use GroebnerBasis[*equations*, {…}, {Z}, MonomialOrder ->Elimination: Order] to eliminate the variable Z. The solutions of the resulting polynomials will then make $q(x_1, \ldots, x_n)$ nonvanishing generically (on lower dimensional solution sets, $q(x_1, \ldots, x_n)$ might vanish).

Next, we will use GroebnerBasis for a little geometry problem: Under which conditions are four points of an ellipse lying on a circle [155]? We use a rational parametrization for the ellipse. After writing all expressions over a common denominator, the rational functions will give polynomials and we can apply GroebnerBasis. $\{X,\ Y\}$ is the midpoint of the circle, and r is its radius. So, we have the following set of equations. ($s\,[i]$ stands for $\tan(t_i/2)$. The use of this parametrization is often of advantage compared with the use of $\cos(t_i)$ and $\sin(t_i)$ together with the equation $\cos^2(t_i) + \sin^2(t_i) - 1$ because it reduces the number of variables compared with other possibilities to introduce variables and supplementary equations.)

```
In[76]:= Table[Function[{x, y}, (x - X)^2 + (y - Y)^2 - r^2] @@
            ({a Cos[t], b Sin[t]} /. {Cos[t] -> (1 - s[i]^2)/(1 + s[i]^2),
                                      Sin[t] -> 2 s[i]/(1 + s[i]^2)}), {i, 4}]
```

Out[76]= $\{-r^2 + \left(-Y + \dfrac{2\,b\,s[1]}{1+s[1]^2}\right)^2 + \left(-X + \dfrac{a\,(1-s[1]^2)}{1+s[1]^2}\right)^2$,

$\quad -r^2 + \left(-Y + \dfrac{2\,b\,s[2]}{1+s[2]^2}\right)^2 + \left(-X + \dfrac{a\,(1-s[2]^2)}{1+s[2]^2}\right)^2$,

$\quad -r^2 + \left(-Y + \dfrac{2\,b\,s[3]}{1+s[3]^2}\right)^2 + \left(-X + \dfrac{a\,(1-s[3]^2)}{1+s[3]^2}\right)^2$,

$\quad -r^2 + \left(-Y + \dfrac{2\,b\,s[4]}{1+s[4]^2}\right)^2 + \left(-X + \dfrac{a\,(1-s[4]^2)}{1+s[4]^2}\right)^2\}$

To get a condition on the s[i], we form polynomials (by clearing denominators) and eliminate the midpoint and radius variables of the circle.

In[77]:= **GroebnerBasis[Numerator[Together[%]],**
{s[1], s[2], s[3], s[4], a, b}, {X, Y, r},
MonomialOrder -> EliminationOrder] // Factor

Out[77]= $\{(a-b)\,(a+b)\,(s[1]-s[2])\,(s[1]-s[3])\,(s[2]-s[3])$
$\quad (s[1]-s[4])\,(s[2]-s[4])\,(s[3]-s[4])\,(-s[1]-s[2]-s[3]+$
$\quad s[1]\,s[2]\,s[3]-s[4]+s[1]\,s[2]\,s[4]+s[1]\,s[3]\,s[4]+s[2]\,s[3]\,s[4])\}$

The last result shows the degenerate case that four points lie on a circle in case the ellipse degenerates to a circle or two of the four points coincide (the six factors s[i] − s[j]). The last factor, symmetric in the s[i], is the nontrivial condition we were looking for.

In[78]:= **Table[{(a Cos[t[i]] - X)^2 + (b Sin[t[i]] - Y)^2 - r^2,**
Cos[t[i]]^2 + Sin[t[i]]^2 - 1}, {i, 4}]

Out[78]= $\{\{-r^2 + (-X+a\,Cos[t[1]])^2 + (-Y+b\,Sin[t[1]])^2,\ -1+Cos[t[1]]^2+Sin[t[1]]^2\},$
$\quad \{-r^2 + (-X+a\,Cos[t[2]])^2 + (-Y+b\,Sin[t[2]])^2,\ -1+Cos[t[2]]^2+Sin[t[2]]^2\},$
$\quad \{-r^2 + (-X+a\,Cos[t[3]])^2 + (-Y+b\,Sin[t[3]])^2,\ -1+Cos[t[3]]^2+Sin[t[3]]^2\},$
$\quad \{-r^2 + (-X+a\,Cos[t[4]])^2 + (-Y+b\,Sin[t[4]])^2,\ -1+Cos[t[4]]^2+Sin[t[4]]^2\}\}$

Here is another application of the GroebnerBasis command. Let us calculate all square roots of the following matrix:

$$\mathbf{A} = \begin{pmatrix} 1 & 2 & 3 \\ 4 & 5 & 6 \\ 7 & 8 & 9 \end{pmatrix}.$$

This is the matrix **A** to be "square rooted".

In[79]:= **A = {{1, 2, 3}, {4, 5, 6}, {7, 8, 9}};**

Let **B** be a square root of the matrix **A**.

In[80]:= **B = Array[b, {3, 3}]**

Out[80]= $\{\{b[1, 1], b[1, 2], b[1, 3]\},$
$\quad \{b[2, 1], b[2, 2], b[2, 3]\}, \{b[3, 1], b[3, 2], b[3, 3]\}\}$

We set up all nine equations to be fulfilled by the elements of the matrix **B.B** and use GroebnerBasis to bring the equations to pseudotriangular form.

In[81]:= **gb = GroebnerBasis[Flatten[A - B.B], Flatten[B]]**

Out[81]= $\{10106041 - 7248120\,b[3, 3]^2 + 1986264\,b[3, 3]^4 - 246240\,b[3, 3]^6 + 11664\,b[3, 3]^8,$
$\quad 442720256\,b[3, 2] + 1487368267\,b[3, 3] - 1199777646\,b[3, 3]^3 +$
$\quad 246755268\,b[3, 3]^5 - 16420968\,b[3, 3]^7,\ 221360128\,b[3, 1] - 3812088313\,b[3, 3] +$
$\quad 1888967994\,b[3, 3]^3 - 338480748\,b[3, 3]^5 + 21253752\,b[3, 3]^7,$

$2434961408\,b[2, 3] + 15061017961\,b[3, 3] - 9776824218\,b[3, 3]^3 + 1910737836\,b[3, 3]^5 -$
$124612344\,b[3, 3]^7, 9739845632\,b[2, 2] - 168910878323\,b[3, 3] +$
$90441416958\,b[3, 3]^3 - 17000265636\,b[3, 3]^5 + 1090644264\,b[3, 3]^7,$
$4869922816\,b[2, 1] + 43883020907\,b[3, 3] - 25909742766\,b[3, 3]^3 +$
$4928643396\,b[3, 3]^5 - 317818728\,b[3, 3]^7, 1217480704\,b[1, 3] -$
$14085993229\,b[3, 3] + 7211276802\,b[3, 3]^3 - 1308060252\,b[3, 3]^5 + 82598616\,b[3, 3]^7,$
$1217480704\,b[1, 2] + 14411001473\,b[3, 3] - 8066459274\,b[3, 3]^3 +$
$1508952780\,b[3, 3]^5 - 96603192\,b[3, 3]^7, 1217480704\,b[1, 1] + 12543504281\,b[3, 3] -$
$6356094330\,b[3, 3]^3 + 1107167724\,b[3, 3]^5 - 68594040\,b[3, 3]^7\}$

Solving the equations yields the following four solutions.

```
In[82]:= sols = Solve[# == 0& /@ gb, Flatten[B]] // Union;
```

```
In[83]:= Length[sols]
```
Out[83]= 4

The solutions are relatively complicated. Here, we display the first one.

```
In[84]:= sols[[1]] // InputForm
```
Out[84]//InputForm=
```
{b[1, 1] -> (43*Sqrt[95/18 - ((8*I)/9)*Sqrt[2]] - (40*I)*Sqrt[2*(95/18 -
  ((8*I)/9)*Sqrt[2])])/187,
 b[1, 2] -> (2*(23*Sqrt[95/18 - ((8*I)/9)*Sqrt[2]] - (4*I)*Sqrt[2*(95/18 -
  ((8*I)/9)*Sqrt[2])]))/187,
 b[1, 3] -> (49*Sqrt[95/18 - ((8*I)/9)*Sqrt[2]] + (24*I)*Sqrt[2*(95/18 -
  ((8*I)/9)*Sqrt[2])])/187,
 b[2, 1] -> ((2*I)*Sqrt[95 - (16*I)*Sqrt[2]] + 41*Sqrt[2*(95 -
  (16*I)*Sqrt[2])])/561,
 b[2, 2] -> ((7*I)*Sqrt[95 - (16*I)*Sqrt[2]] + 50*Sqrt[2*(95 -
  (16*I)*Sqrt[2])])/561,
 b[2, 3] -> ((12*I)*Sqrt[95 - (16*I)*Sqrt[2]] + 59*Sqrt[2*(95 -
  (16*I)*Sqrt[2])])/561,
 b[3, 1] -> (11*Sqrt[95/18 - ((8*I)/9)*Sqrt[2]] + (4*I)*Sqrt[2*(95/18 -
  ((8*I)/9)*Sqrt[2])])/17,
 b[3, 2] -> ((2*I)*Sqrt[95 - (16*I)*Sqrt[2]] + 7*Sqrt[2*(95 -
  (16*I)*Sqrt[2])])/51,
 b[3, 3] -> Sqrt[95/18 - ((8*I)/9)*Sqrt[2]]}
```

Here is a quick check of the solutions.

```
In[85]:= A - B.B /. sols // Simplify
```
Out[85]= {{{0, 0, 0}, {0, 0, 0}, {0, 0, 0}}, {{0, 0, 0}, {0, 0, 0}, {0, 0, 0}},
      {{0, 0, 0}, {0, 0, 0}, {0, 0, 0}}, {{0, 0, 0}, {0, 0, 0}, {0, 0, 0}}}

(While a matrix typically has many square roots, only one is often the "right" one [939]; for nice closed forms, see [1436], [361].)

As a more complicated example of the use of the function GroebnerBasis, we will deal with an application from quantum mechanics, and we will calculate some exact radial wave functions $R_l(r)$ for the three-dimensional (3D) potential $V(r) = c^2\,r^{-6} + a^2\,r^2 + b\,r^{-4}$ in the radial Schrödinger equation

$$-R_l''(r) + \left(V(r) + \frac{l(l+1)}{r^2}\right)R_l(r) = \varepsilon_l\,R_l(r),$$

where l is a nonnegative integer [1912], [1796], [1912]. In [498], [500], [1089], [994], and [1221] the following ansatz was proposed for the radial wave function:

$$R_l(r) = r^\kappa \left(\beta r^2 + \alpha + \frac{\gamma}{r^2}\right) \exp\left(-\frac{1}{2}\left(a r^2 + \frac{c}{r^2}\right)\right).$$

α, β, γ, and κ are functions of a, b, c, and l. (For a more general ansatz, see [346], [1583], [503], and [1400].)

For brevity, we restrict ourselves in the following investigations to the search of nontrivial $R_l(r)$, meaning $\alpha \neq 0$, $\beta \neq 0$, and $\gamma \neq 0$.

Here is the potential $V(r)$ and the ansatz for the wave function $R_l(r)$.

In[86]:= `Clear[a, b, c, r, V, R, α, β, γ, κ, ε, "eqs*", "gb*"];`

    `V[r_] = a^2 r^2 + b r^-4 + c^2 r^-6;`

    `R[l_][r_] = r^κ (α + β r^2 + γ r^-2) Exp[-(a r^2 + c r^-2)/2];`

Substituting the ansatz in the radial Schrödinger equation, yields the following expression.

In[89]:= `SEquation = Factor[D[R[l][r], r, r] + (ε - V[r] - l(1 + 1)/r^2) R[l][r]]`

Out[89]= $-e^{-\frac{c}{2r^2} - \frac{a r^2}{2}} r^{-6+\kappa}$

$(b r^2 \alpha + 3 c r^2 \alpha + 2 a c r^4 \alpha + 1 r^4 \alpha + l^2 r^4 \alpha + a r^6 \alpha + b r^4 \beta - c r^4 \beta - 2 r^6 \beta + 2 a c r^6 \beta +$
$1 r^6 \beta + l^2 r^6 \beta + 5 a r^8 \beta + b \gamma + 7 c \gamma - 6 r^2 \gamma + 2 a c r^2 \gamma + 1 r^2 \gamma + l^2 r^2 \gamma -$
$3 a r^4 \gamma - r^6 \alpha \varepsilon - r^8 \beta \varepsilon - r^4 \gamma \varepsilon - 2 c r^2 \alpha \kappa + r^4 \alpha \kappa + 2 a r^6 \alpha \kappa - 2 c r^4 \beta \kappa -$
$3 r^6 \beta \kappa + 2 a r^8 \beta \kappa - 2 c \gamma \kappa + 5 r^2 \gamma \kappa + 2 a r^4 \gamma \kappa - r^4 \alpha \kappa^2 - r^6 \beta \kappa^2 - r^2 \gamma \kappa^2)$

For the last expression to be zero and the radial wave functions $R_l(r)$ to be physically sensible, the sum in the last expression must vanish identically. This means the individual coefficients of r in the last sum must vanish identically. We equate all powers of r to zero and obtain the following system of nonlinear equations for α, β, γ, and κ. The variables ε, a, b, and c are considered to be parameters, as is l.

In[90]:= `eqs = Factor[CoefficientList[SEquation[[-1]], r]]`

Out[90]= $\{\gamma (b + 7 c - 2 c \kappa), 0, b \alpha + 3 c \alpha - 6 \gamma + 2 a c \gamma + 1 \gamma + l^2 \gamma - 2 c \alpha \kappa + 5 \gamma \kappa - \gamma \kappa^2,$
$0, 2 a c \alpha + 1 \alpha + l^2 \alpha + b \beta - c \beta - 3 a \gamma - \gamma \varepsilon + \alpha \kappa - 2 c \beta \kappa + 2 a \gamma \kappa - \alpha \kappa^2, 0,$
$a \alpha - 2 \beta + 2 a c \beta + 1 \beta + l^2 \beta - \alpha \varepsilon + 2 a \alpha \kappa - 3 \beta \kappa - \beta \kappa^2, 0, \beta (5 a - \varepsilon + 2 a \kappa)\}$

Because we are only interested in generic solutions (none of the α, β, γ vanishing), we eliminate the obvious degenerate solution $\gamma = 0$, $\beta = 0$.

In[91]:= `eqs1 = MapAt[#/β&, MapAt[#/γ&, eqs, {1}], {-1}]`

Out[91]= $\{b + 7 c - 2 c \kappa, 0, b \alpha + 3 c \alpha - 6 \gamma + 2 a c \gamma + 1 \gamma + l^2 \gamma - 2 c \alpha \kappa + 5 \gamma \kappa - \gamma \kappa^2,$
$0, 2 a c \alpha + 1 \alpha + l^2 \alpha + b \beta - c \beta - 3 a \gamma - \gamma \varepsilon + \alpha \kappa - 2 c \beta \kappa + 2 a \gamma \kappa - \alpha \kappa^2, 0,$
$a \alpha - 2 \beta + 2 a c \beta + 1 \beta + l^2 \beta - \alpha \varepsilon + 2 a \alpha \kappa - 3 \beta \kappa - \beta \kappa^2, 0, 5 a - \varepsilon + 2 a \kappa\}$

We start by calculating a full Gröbner basis.

In[92]:= `gb1 = Factor[GroebnerBasis[eqs1, {α, β, γ, κ, ε},`
                                       `ParameterVariables -> {a, b, c, l}]];`

The resulting equations contain the variables α, β, γ, κ, ε, a, b, c, l in the following degrees.

In[93]:= `Exponent[#, {α, β, γ, κ, ε, a, b, c, l}]& /@ gb1 // Short[#, 4]&`

Out[93]//Short=
$\{\{0, 0, 0, 0, 1, 1, 1, 1, 0\}, \{0, 0, 0, 1, 0, 0, 1, 1, 0\},$
$\{0, 0, 0, 1, 1, 1, 0, 0, 0\}, \{0, 0, 1, 0, 0, 3, 6, 9, 6\},$

```
{0, 0, 1, 0, 1, 4, 5, 8, 6}, {0, 0, 1, 0, 2, 5, 4, 7, 6}, {0, 0, 1, 0, 3, 6, 4, 6, 6},
{0, 0, 1, 0, 4, 7, 3, 5, 6}, <<31>>, {0, 1, 1, 3, 3, 5, 1, 2, 4},
{0, 1, 1, 4, 2, 4, 1, 2, 4}, {1, 0, 1, 2, 0, 1, 0, 1, 2}, {1, 0, 1, 3, 0, 1, 1, 0, 2},
{1, 1, 0, 2, 0, 1, 0, 1, 2}, {1, 1, 0, 3, 1, 1, 1, 1, 2}, {1, 1, 1, 2, 1, 2, 0, 1, 2}}
```

As it turns out, gb1 still describes many degenerate solutions with $\gamma = 0$.

In[94]:= **Cases[gb1, γ _Plus, {1}] // Length**

Out[94]= 2

We can force generic solutions by adding an equation of the form $\alpha \beta \gamma z - 1 = 0$, where z is an auxiliary variable to the equations eqs1 (for $\alpha \beta \gamma z = 1$ to hold, the variables α, β, and γ cannot simultaneously vanish). We calculate the corresponding Gröbner basis. Because the auxiliary variable z is not needed later, we eliminate it.

In[95]:= **gb2 = Factor[GroebnerBasis[Append[eqs1, α β γ z - 1],**
                     {α, β, γ, κ, ε, a, b, c, l}, {z}]];

Now, the equations have a more interesting form. The first equation is an equation in the parameters only. The second equation is a linear equation in the energy ε.

In[96]:= **Exponent[#, {α, β, γ, κ, ε, a, b, c, l}]& /@ gb2 // Short[#, 4]&**

Out[96]//Short=

```
{{0, 0, 0, 0, 0, 3, 6, 9, 6}, {0, 0, 0, 0, 1, 1, 1, 1, 0},
 {0, 0, 0, 0, 1, 4, 5, 8, 6}, {0, 0, 0, 0, 2, 5, 4, 7, 6},
 {0, 0, 0, 0, 3, 6, 4, 6, 6}, {0, 0, 0, 0, 4, 7, 3, 5, 6}, {0, 0, 0, 0, 5, 8, 2, 4, 6},
 {0, 0, 0, 0, 6, 9, 2, 3, 6}, <<31>>, {0, 1, 1, 3, 3, 5, 1, 2, 4},
 {0, 1, 1, 4, 2, 4, 1, 2, 4}, {1, 0, 1, 2, 0, 1, 0, 1, 2}, {1, 0, 1, 3, 0, 1, 1, 0, 2},
 {1, 1, 0, 2, 0, 1, 0, 1, 2}, {1, 1, 0, 3, 1, 1, 1, 1, 2}, {1, 1, 1, 2, 1, 2, 0, 1, 2}}
```

The next output is the condition on the parameters a, b, and c (and l) that is needed for nondegenerate solutions. It is a nontrivial condition not easily obtained by using only paper and pencil.

In[97]:= **gb2[[1]]**

Out[97]= $-b^6 - 36\,b^5\,c - 505\,b^4\,c^2 - 3480\,b^3\,c^3 + 24\,a\,b^4\,c^3 - 12139\,b^2\,c^4 + 576\,a\,b^3\,c^4 - 19524\,b\,c^5 + 6160\,a\,b^2\,c^5 - 10395\,c^6 + 32448\,a\,b\,c^6 - 192\,a^2\,b^2\,c^6 + 83160\,a\,c^7 - 2304\,a^2\,b\,c^7 - 16960\,a^2\,c^8 + 512\,a^3\,c^9 + 12\,b^4\,c^2\,l + 288\,b^3\,c^3\,l + 2568\,b^2\,c^4\,l + 10080\,b\,c^5\,l - 192\,a\,b^2\,c^5\,l + 15468\,c^6\,l - 2304\,a\,b\,c^6\,l - 12864\,a\,c^7\,l + 768\,a^2\,c^8\,l + 12\,b^4\,c^2\,l^2 + 288\,b^3\,c^3\,l^2 + 2520\,b^2\,c^4\,l^2 + 9504\,b\,c^5\,l^2 - 192\,a\,b^2\,c^5\,l^2 + 13276\,c^6\,l^2 - 2304\,a\,b\,c^6\,l^2 - 12480\,a\,c^7\,l^2 + 768\,a^2\,c^8\,l^2 - 96\,b^2\,c^4\,l^3 - 1152\,b\,c^5\,l^3 - 4320\,c^6\,l^3 + 768\,a\,c^7\,l^3 - 48\,b^2\,c^4\,l^4 - 576\,b\,c^5\,l^4 - 2000\,c^6\,l^4 + 384\,a\,c^7\,l^4 + 192\,c^6\,l^5 + 64\,c^6\,l^6$

The generic solution of the system of equations under consideration is the following form.

In[98]:= **eqSol = Solve[# == 0& /@ gb2, {ε, κ, α, β, γ}]**

Solve::svars : Equations may not give solutions for all "solve" variables. More…

Out[98]= $\left\{\left\{\alpha \to \dfrac{1}{8192\,a\,c^8} \right.\right.$
$((-b^2 - 20\,b\,c - 99\,c^2 + 8\,a\,c^3 + 4\,c^2\,l + 4\,c^2\,l^2)\,(b^4 + 16\,b^3\,c + 86\,b^2\,c^2 + 176\,b\,c^3 - 16\,a\,b^2\,c^3 + 105\,c^4 - 128\,a\,b\,c^4 - 816\,a\,c^5 + 64\,a^2\,c^6 - 8\,b^2\,c^2\,l - 64\,b\,c^3\,l - 152\,c^4\,l + 64\,a\,c^5\,l - 8\,b^2\,c^2\,l^2 - 64\,b\,c^3\,l^2 - 136\,c^4\,l^2 + 64\,a\,c^5\,l^2 + 32\,c^4\,l^3 + 16\,c^4\,l^4)\,\gamma),$

$\beta \to \dfrac{1}{512\,c^6}\,((b^4 + 16\,b^3\,c + 86\,b^2\,c^2 + 176\,b\,c^3 - 16\,a\,b^2\,c^3 + 105\,c^4 - 128\,a\,b\,c^4 - 816\,a\,c^5 + 64\,a^2\,c^6 - 8\,b^2\,c^2\,l - 64\,b\,c^3\,l - 152\,c^4\,l + 64\,a\,c^5\,l - 8\,b^2\,c^2\,l^2 - 64\,b\,c^3\,l^2 - 136\,c^4\,l^2 + 64\,a\,c^5\,l^2 + 32\,c^4\,l^3 + 16\,c^4\,l^4)\,\gamma),\ \kappa \to \dfrac{b+7\,c}{2\,c},\ \varepsilon \to \dfrac{a\,(b+12\,c)}{c}\}\}$

In addition to the last result, the conditions on a, b, and c as expressed in `gb[[1]]` must be kept in mind. It is interesting to note that γ appears linearly on the right-hand side in the solutions for α and β. Let us establish a direct relation between the energy ε and the parameters a, b, and c (and l) by eliminating the variables α, β, γ, and κ from gb2.

In[99]:= `gbε = Factor[GroebnerBasis[gb2, {ε}, {α, β, γ, κ}]];`

In[100]:= `Exponent[gbε, ε]`

Out[100]= `{0, 1, 1, 2, 3, 4, 5, 6}`

The last Gröbner basis has eight elements. If we eliminate one additional parameter, say, a, we obtain one equation in ε, b, and c.

In[101]:= `Factor[GroebnerBasis[gbε, {ε}, {a}]] // Short[#, 4]&`

Out[101]//Short=

$\{-b^9 - 72\,b^8\,c - 2233\,b^7\,c^2 - 38940\,b^6\,c^3 - 417787\,b^5\,c^4 - 2832528\,b^4\,c^5 - 11970747\,b^3\,c^6 -$
$29784780\,b^2\,c^7 - 38228112\,b\,c^8 - 17962560\,c^9 + 12\,b^7\,c^2\,1 + 720\,b^6\,c^3\,1 + \ll77\gg +$
$9216\,b\,c^9\,1^4\,\varepsilon + 55296\,c^{10}\,1^4\,\varepsilon - 192\,b^3\,c^8\,\varepsilon^2 - 4608\,b^2\,c^9\,\varepsilon^2 - 44608\,b\,c^{10}\,\varepsilon^2 -$
$203520\,c^{11}\,\varepsilon^2 + 768\,b\,c^{10}\,1\,\varepsilon^2 + 9216\,c^{11}\,1\,\varepsilon^2 + 768\,b\,c^{10}\,1^2\,\varepsilon^2 + 9216\,c^{11}\,1^2\,\varepsilon^2 + 512\,c^{12}\,\varepsilon^3\}$

In a similar way, we can eliminate the parameter b or c.

In[102]:= `Factor[GroebnerBasis[gbε, {ε}, {b}]] // Short[#, 4]&`

Out[102]//Short=

$\{10395\,a^6 - 83160\,a^7\,c + 16960\,a^8\,c^2 - 512\,a^9\,c^3 - 15468\,a^6\,1 + 12864\,a^7\,c\,1 - 768\,a^8\,c^2\,1 -$
$13276\,a^6\,1^2 + 12480\,a^7\,c\,1^2 - 768\,a^8\,c^2\,1^2 + 4320\,a^6\,1^3 - 768\,a^7\,c\,1^3 + \ll31\gg +$
$192\,a^5\,c\,1^2\,\varepsilon^2 + 96\,a^4\,1^3\,\varepsilon^2 + 48\,a^4\,1^4\,\varepsilon^2 - 3480\,a^3\,\varepsilon^3 + 576\,a^4\,c\,\varepsilon^3 + 288\,a^3\,1\,\varepsilon^3 +$
$288\,a^3\,1^2\,\varepsilon^3 + 505\,a^2\,\varepsilon^4 - 24\,a^3\,c\,\varepsilon^4 - 12\,a^2\,1\,\varepsilon^4 - 12\,a^2\,1^2\,\varepsilon^4 - 36\,a\,\varepsilon^5 + \varepsilon^6\}$

In[103]:= `Factor[GroebnerBasis[gbε, {ε}, {c}]] // Short[#, 4]&`

Out[103]//Short=

$\{-17962560\,a^9 - 11975040\,a^{10}\,b - 203520\,a^{11}\,b^2 - 512\,a^{12}\,b^3 +$
$26728704\,a^9\,1 + 1852416\,a^{10}\,b\,1 + 9216\,a^{11}\,b^2\,1 + 22940928\,a^9\,1^2 +$
$1797120\,a^{10}\,b\,1^2 + \ll89\gg + 96\,a^4\,1^3\,\varepsilon^5 + 48\,a^4\,1^4\,\varepsilon^5 - 38940\,a^3\,\varepsilon^6 - 24\,a^4\,b\,\varepsilon^6 +$
$720\,a^3\,1\,\varepsilon^6 + 720\,a^3\,1^2\,\varepsilon^6 + 2233\,a^2\,\varepsilon^7 - 12\,a^2\,1\,\varepsilon^7 - 12\,a^2\,1^2\,\varepsilon^7 - 72\,a\,\varepsilon^8 + \varepsilon^9\}$

We could now proceed in a similar way for κ.

In[104]:= `gbκ = Factor[GroebnerBasis[gb2, {κ}, {α, β, γ, ε}]];`

In[105]:= `Exponent[gbκ, κ]`

Out[105]= `{0, 1, 1, 2, 3, 4, 5, 6}`

In[106]:= `Factor[GroebnerBasis[gbκ, {κ}, {a}]]`

Out[106]= `{-b - 7 c + 2 c κ}`

In[107]:= `Factor[GroebnerBasis[gbκ, {κ}, {b}]]`

Out[107]= $\{-280\,a\,c + 160\,a^2\,c^2 - 8\,a^3\,c^3 - 12\,1 + 96\,a\,c\,1 - 12\,a^2\,c^2\,1 - 4\,1^2 +$
$90\,a\,c\,1^2 - 12\,a^2\,c^2\,1^2 + 15\,1^3 - 12\,a\,c\,1^3 + 5\,1^4 - 6\,a\,c\,1^4 - 3\,1^5 - 1^6 - 12\,\kappa +$
$64\,a\,c\,\kappa - 12\,a^2\,c^2\,\kappa - 12\,a\,c\,1\,\kappa - 3\,1^2\,\kappa - 12\,a\,c\,1^2\,\kappa - 6\,1^3\,\kappa - 3\,1^4\,\kappa + 4\,\kappa^2 -$
$70\,a\,c\,\kappa^2 + 12\,a^2\,c^2\,\kappa^2 - 3\,1\,\kappa^2 + 12\,a\,c\,1\,\kappa^2 + 12\,a\,c\,1^2\,\kappa^2 + 6\,1^3\,\kappa^2 + 3\,1^4\,\kappa^2 +$
$15\,\kappa^3 + 12\,a\,c\,\kappa^3 + 6\,1\,\kappa^3 + 6\,1^2\,\kappa^3 - 5\,\kappa^4 - 6\,a\,c\,\kappa^4 - 3\,1\,\kappa^4 - 3\,1^2\,\kappa^4 - 3\,\kappa^5 + \kappa^6\}$

In[108]:= `Factor[GroebnerBasis[gbκ, {κ}, {c}]]`

Out[108]= $\{-13720\,a\,b - 1120\,a^2\,b^2 - 8\,a^3\,b^3 + 4116\,1 + 4704\,a\,b\,1 + 84\,a^2\,b^2\,1 + 1372\,1^2 +$
$4410\,a\,b\,1^2 + 84\,a^2\,b^2\,1^2 - 5145\,1^3 - 588\,a\,b\,1^3 - 1715\,1^4 - 294\,a\,b\,1^4 + 1029\,1^5 +$
$343\,1^6 + 4116\,\kappa + 10976\,a\,b\,\kappa + 404\,a^2\,b^2\,\kappa - 3528\,1\,\kappa - 3276\,a\,b\,1\,\kappa - 24\,a^2\,b^2\,1\,\kappa -$

$147\ 1^2\ \kappa - 3108\ a\ b\ 1^2\ \kappa - 24\ a^2\ b^2\ 1^2\ \kappa + 6468\ 1^3\ \kappa + 336\ a\ b\ 1^3\ \kappa + 2499\ 1^4\ \kappa +$
$168\ a\ b\ 1^4\ \kappa - 882\ 1^5\ \kappa - 294\ 1^6\ \kappa - 4900\ \kappa^2 - 6342\ a\ b\ \kappa^2 - 108\ a^2\ b^2\ \kappa^2 + 2037\ 1\ \kappa^2 +$
$1308\ a\ b\ 1\ \kappa^2 - 546\ 1^2\ \kappa^2 + 1284\ a\ b\ 1^2\ \kappa^2 - 5082\ 1^3\ \kappa^2 - 48\ a\ b\ 1^3\ \kappa^2 - 2331\ 1^4\ \kappa^2 -$
$24\ a\ b\ 1^4\ \kappa^2 + 252\ 1^5\ \kappa^2 + 84\ 1^6\ \kappa^2 - 2961\ \kappa^3 + 2804\ a\ b\ \kappa^3 + 24\ a^2\ b^2\ \kappa^3 - 3036\ 1\ \kappa^3 -$
$384\ a\ b\ 1\ \kappa^3 - 1838\ 1^2\ \kappa^3 - 384\ a\ b\ 1^2\ \kappa^3 + 2388\ 1^3\ \kappa^3 + 1174\ 1^4\ \kappa^3 - 24\ 1^5\ \kappa^3 - 8\ 1^6\ \kappa^3 +$
$5693\ \kappa^4 - 910\ a\ b\ \kappa^4 + 3045\ 1\ \kappa^4 + 48\ a\ b\ 1\ \kappa^4 + 2769\ 1^2\ \kappa^4 + 48\ a\ b\ 1^2\ \kappa^4 - 552\ 1^3\ \kappa^4 -$
$276\ 1^4\ \kappa^4 - 1669\ \kappa^5 + 216\ a\ b\ \kappa^5 - 1410\ 1\ \kappa^5 - 1386\ 1^2\ \kappa^5 + 48\ 1^3\ \kappa^5 + 24\ 1^4\ \kappa^5 -$
$685\ \kappa^6 - 24\ a\ b\ \kappa^6 + 300\ 1\ \kappa^6 + 300\ 1^2\ \kappa^6 + 506\ \kappa^7 - 24\ 1\ \kappa^7 - 24\ 1^2\ \kappa^7 - 108\ \kappa^8 + 8\ \kappa^9\}$

To make sure that the obtained solutions are physically sensible, we must require the above polynomial gb2[[1]] to have solutions for positive *a*, *b*, and *c*. Such solutions indeed exist, as shown by the following graphic for *l* = 1. Additionally, the corresponding r^κ should not be too singular. This is the case for the above solutions. The right graphic shows this.

```
In[109]:= Needs["Graphics`ContourPlot3D`"]
```

```
In[110]:= cp3D = ContourPlot3D[Evaluate[gb2[[1]] /. 1 -> 1],
                {a, 0, 5}, {b, 0, 5}, {c, 0, 5},
                DisplayFunction -> Identity,
                PlotPoints -> 22, MaxRecursion -> 0, Axes -> True];
```

```
In[111]:= Show[GraphicsArray[
        {cp3D, ListPlot[Sort[Apply[(#2 + #3)/(2 #3)&,
            DeleteCases[Level[Cases[cp3D, _Polygon, Infinity], {-2}],
                {_, _, _?(# == 0.&)}], {1}]],
                DisplayFunction -> Identity]}]]
```

Here is an explicit quick check of the solutions for *a* = *b* = 1, *l* = 2. We calculate the possible values of *c* from gb2[[1]] and substitute all solutions in the original equation. We see that within the precision used, the result is 0.

```
In[112]:= With[{r = Sequence[a -> 1, b -> 1, 1 -> 2]},
        With[{cs = Select[{ToRules[NRoots[(gb2[[1]] /. {r}) == 0, c, 60]]},
            Im[c /. #] == 0&]}, SEquation[[-1]] /. (Flatten[Join[eqSol[[1]] /.
                {r} /. #, {#, r}]]& /@ cs)]]
```

$\text{Out[112]= } \{0.\times10^{-58}\ \gamma + 0.\times10^{-56}\ r^2\ \gamma + 0.\times10^{-55}\ r^4\ \gamma + 0.\times10^{-55}\ r^6\ \gamma + 0.\times10^{-55}\ r^8\ \gamma,$
$0.\times10^{-58}\ \gamma + 0.\times10^{-55}\ r^2\ \gamma + 0.\times10^{-54}\ r^4\ \gamma + 0.\times10^{-54}\ r^6\ \gamma + 0.\times10^{-54}\ r^8\ \gamma,$
$0.\times10^{-58}\ \gamma + 0.\times10^{-54}\ r^2\ \gamma + 0.\times10^{-53}\ r^4\ \gamma + 0.\times10^{-53}\ r^6\ \gamma + 0.\times10^{-53}\ r^8\ \gamma,$
$0.\times10^{-58}\ \gamma + 0.\times10^{-53}\ r^2\ \gamma + 0.\times10^{-52}\ r^4\ \gamma + 0.\times10^{-52}\ r^6\ \gamma + 0.\times10^{-52}\ r^8\ \gamma,$
$0.\times10^{-59}\ \gamma + 0.\times10^{-57}\ r^2\ \gamma + 0.\times10^{-56}\ r^4\ \gamma + 0.\times10^{-56}\ r^6\ \gamma + 0.\times10^{-57}\ r^8\ \gamma,$
$0.\times10^{-58}\ \gamma + 0.\times10^{-57}\ r^2\ \gamma + 0.\times10^{-57}\ r^4\ \gamma + 0.\times10^{-57}\ r^6\ \gamma + 0.\times10^{-58}\ r^8\ \gamma,$
$0.\times10^{-57}\ \gamma + 0.\times10^{-57}\ r^2\ \gamma + 0.\times10^{-57}\ r^4\ \gamma + 0.\times10^{-57}\ r^6\ \gamma + 0.\times10^{-59}\ r^8\ \gamma\}$

For the corresponding two-dimensional problem, see [499], [501], and [502]; for the one-dimensional problem, see [111], [450], and [1400]; and for the *n*-dimensional problem, see [1310], [504], and [746]; for the use of Gröbner bases to find exact states in larger dimensions, see [713].

Let us use the option setting `MonomialOrder -> EliminationOrder` for another small, but interesting example: What is the largest area a hexagon of unit diameter can have [762], [1032]? Interestingly, the largest area is not the one of a regular hexagon. It can be shown that the hexagon we are looking for must have mirror symmetry and be of the following form. The black line indicates the unit diameters. Without loss of generality, we can use the following parametrization of the hexagon.

```
In[113]:= Clear["x*", "y*"];
```

```
        p1 = {0, 0}; p2 = { x2, y2}; p3 = { x3, y3};
        p4 = {0, 1}; p5 = {-x3, y3}; p6 = {-x2, y2};
```

```
In[116]:= Block[{x2 = 0.5, y2 = 0.402, x3 = 0.343, y3 = 0.939},
            Show[Graphics[{{Hue[0], Thickness[0.01],
                        (* outline *) Line[{p1, p2, p3, p4, p5, p6, p1}]},
                    {GrayLevel[0], Thickness[0.002],
                        (* diagonals *) Line /@ {{p1, p3}, {p1, p5},
                            {p2, p5}, {p3, p6}, {p1, p4}, {p2, p5}}}}],
                PlotRange -> All, Frame -> True, AspectRatio -> Automatic]]
```

The above hexagon has four degrees of freedom, the coordinates of the points p2 and p3. It follows from elementary geometry that the area of the hexagon is given by $x_3 - x_3 y_2 + x_2 y_3$. We take the unit diameter conditions into account using Lagrange multipliers λ_1, λ_2, and λ_3.

```
In[117]:= area = x3 - x3 y2 + x2 y3;
        L = area + λ1 (#.#&[p1 - p3] - 1) + λ2 (#.#&[p2 - p5] - 1) +
                λ3 (#.#&[p2 - p6] - 1);
```

To get the polynomial of the largest area `area`, we eliminate the coordinates and the Lagrange multipliers.

```
In[119]:= GroebnerBasis[
            {area - area, D[L, x2], D[L, y2], D[L, x3], D[L, y3],
                        D[L, λ1], D[L, λ2], D[L, λ3]},
            {area}, {x2, y2, x3, y3, λ1, λ2, λ3},
            MonomialOrder -> EliminationOrder] // Factor
```

$$
\begin{aligned}
Out[119]= \{ &(11993 + 78488\,area + 144464\,area^2 - 1232\,area^3 - 221360\,area^4 - \\
    &146496\,area^5 + 21056\,area^6 + 30848\,area^7 - 3008\,area^8 - 8192\,area^9 + 4096\,area^{10}) \\
    &(11993 - 78488\,area + 144464\,area^2 + 1232\,area^3 - 221360\,area^4 + 146496\,area^5 + \\
    &21056\,area^6 - 30848\,area^7 - 3008\,area^8 + 8192\,area^9 + 4096\,area^{10}) \}
\end{aligned}
$$

The first root larger than the area of a regular hexagon is the one we are looking for. It is nearly 4% larger than the area of a regular hexagon.

```
In[120]:= Select[NSolve[%[[1]] == 0, area], Im[(area /. #)] == 0 &&
            (* larger than regular hexagon *)
            3 Sqrt[3]/8 < (area /. #) &][[1]]
Out[120]= {area → 0.674981}
```

For the largest octagon, see [100], [101].

The calculation of a lexicographic Gröbner basis is sometimes a challenging undertaking and a direct call to `GroebnerBasis[eqs, vars, MonomialOrder -> Lexicographic]` might not finish in a reasonable amount of time. As mentioned, the calculation of a degree reverse lexicographic basis is often easier. But a degree reverse lexicographic basis is not directly usable for solving equations. Sometimes the function `Internal`GroebnerWalk` from the `Internal`` context can be very helpful here. For a parameter-free Gröbner basis this function will convert one monomial order to another. For most cases, we will be interested in a transformation from a degree-reverse lexicographic basis to a lexicographic basis. The default option settings of this function reflect these orders. (We only show the relevant options here.)

```
In[121]:= Cases[Options[Internal`GroebnerWalk],
            HoldPattern[_?(StringMatchQ[ToString[#], "*MonomialOrder*"] &) -> _]]
Out[121]= {GroebnerWalk`InitialMonomialOrder→ DegreeReverseLexicographic,
            MonomialOrder → Lexicographic}
```

As an example of the use of `Internal`GroebnerWalk`, let us consider the following system of equations.

```
In[122]:= (* individual equation *)
            tTerm[{{a__}, {b__}, {c__}}] :=
            16 + λ[a]^4 + (λ[b]^2 - λ[c]^2)^2 - 2 λ[a]^2 (λ[b]^2 + λ[c]^2)

            (* the equations *)
            eqs = tTerm /@
                    {{{1, 2}, {1, 3}, {2, 3}}, {{1, 2}, {1, 4}, {2, 4}},
                     {{1, 2}, {1, 5}, {2, 5}}, {{1, 3}, {1, 4}, {3, 4}},
                     {{1, 3}, {1, 5}, {3, 5}}, {{1, 4}, {1, 5}, {4, 5}},
                     {{2, 3}, {2, 4}, {3, 4}}, {{2, 3}, {2, 5}, {3, 5}},
                     {{2, 4}, {2, 5}, {4, 5}}, {{3, 4}, {3, 5}, {4, 5}}};

            Short[eqs, 3]
Out[128]//Short=
```
$$\{16 + \lambda[1, 2]^4 + (\lambda[1, 3]^2 - \lambda[2, 3]^2)^2 - 2 \lambda[1, 2]^2 (\lambda[1, 3]^2 + \lambda[2, 3]^2),$$
$$16 + \lambda[1, 2]^4 + (\lambda[1, 4]^2 - \lambda[2, 4]^2)^2 - 2 \lambda[1, 2]^2 (\lambda[1, 4]^2 + \lambda[2, 4]^2),$$
$$\ll 7 \gg, 16 + \lambda[3, 4]^4 + (\lambda[3, 5]^2 - \lambda[4, 5]^2)^2 - 2 \lambda[3, 4]^2 (\lambda[3, 5]^2 + \lambda[4, 5]^2)\}$$

These ten equations of total degree four each, look quite symmetric (they exhibit some permutation symmetries) and are not easy to solve. (The $\lambda[i, k]$ are the edge lengths of a 4D pentatope with unit 2D areas of the 2D faces.) The system has ten variables. This suggests a zero-dimensional solution.

```
In[129]:= (λs = Cases[eqs, _λ, Infinity] // Union) // Length
Out[129]= 10
```

While `GroebnerBasis[eqs, λs, MonomialOrder -> Lexicographic]` will not finish within several hours, the calculation of a degree-reverse lexicographic basis and its conversion to a lexicographic basis can be done in a couple of seconds.

```
In[130]:= (gb = GroebnerBasis[eqs, λs,
                    MonomialOrder -> DegreeReverseLexicographic];) // Timing
Out[130]= {13.83 Second, Null}
```

```
In[131]:= (gbLex = Internal`GroebnerWalk[gb, λs];) // Timing
Out[131]= {15.32 Second, Null}
```

In the lexicographic basis, we have one equation in only one variable.

```
In[132]:= Select[gbLex, Length[Union[Cases[#, _λ, Infinity]]] == 1&]
Out[132]= {12288 + 1280 λ[4, 5]⁴ - 752 λ[4, 5]⁸ + 15 λ[4, 5]¹²}
```

Solving this triquartic equation gives 12 possible values for $\lambda[4, 5]$, and, because of the symmetry of the system, for all λ's.

```
In[133]:= roots = λ[4, 5] /. Solve[gbLex[[1]] == 0, λ[4, 5]]
```

$$Out[133]= \left\{-2\left(-\frac{1}{5}\right)^{1/4}, 2\left(-\frac{1}{5}\right)^{1/4}, -\frac{2}{3^{1/4}}, -\frac{2\,i}{3^{1/4}}, \frac{2\,i}{3^{1/4}}, \frac{2}{3^{1/4}},\right.$$

$$\left. -2\,3^{1/4}, -2\,i\,3^{1/4}, 2\,i\,3^{1/4}, 2\,3^{1/4}, -\frac{2\,(-1)^{3/4}}{5^{1/4}}, \frac{2\,(-1)^{3/4}}{5^{1/4}}\right\}$$

To solve the whole system eqs we "just" try all 12^{10} possible combinations for the $\lambda_{i,j}$ and the roots and extract all correct tuples. Doing this in a recursive way eliminates most of these potential 10^{12} solutions. The function nextEquation extracts from a set of equations *eqs* and for a given set of λ's the equation with the fewest new λ's.

```
In[134]:= nextEquation[λs_, eqs_] :=
        eqs[[Position[#, Min[#], {1}, 1][[1, 1]]]]&[
            Length[(* all remaining λs *)
                Complement[#, λs]]& /@ (* all occurring λs *)
                (Union[Cases[#, _λ, Infinity]]& /@ eqs)]
```

We start with the first equation and find 112 possible solutions from the 1728 combinations tried. The expression $S[k]$ is a list of the solutions, after taking into account k equations.

```
In[135]:= (* the equation under consideration *)
        eqsActive = Expand[eqs[[1]]];
        (* the remaining equations *)
        remainingEqs = DeleteCases[Expand[eqs], eqsActive];
        (* the λs already used *)
        usedλs = {};
        (* the current λs to be solved for *)
        λsActive = Union[Cases[eqsActive, _λ, Infinity]];
        usedλs = Flatten[{usedλs, λsActive}];
        (* the 12^3 possible values for the first three λs *)
        start = Apply[Rule, Transpose[{λsActive, #}], {1}]& /@
                    Flatten[Outer[List, Sequence @@
                    Table[roots, {Length[λsActive]}]], Length[λsActive] - 1];
        (* the λ-values fulfilling the first equation *)
        (S[1] = Select[start, (eqsActive /. #) == 0&]) // Length
Out[147]= 112
```

Dealing now with one equation after another gives us after 10 steps $2^{16} = 65536$ solutions.

```
In[148]:= Do[eqsActive = nextEquation[λsActive, remainingEqs];
        remainingEqs = DeleteCases[remainingEqs, eqsActive];
        λsActive = Complement[Union[Cases[eqsActive, _λ, Infinity]], usedλs];
        usedλs = Flatten[{usedλs, λsActive}];
        (* rules for new λ-values *)
        newRules = Function[l, If[λsActive === {}, {{}},
            Apply[Rule, Transpose[{λsActive, #}], {1}]& /@
```

```
                Flatten[Outer[List, Sequence @@
                  Table[roots, {1}]], 1 - 1]]][Length[λsActive]];
       (* join already established values with potential new ones *)
       allRules = Flatten[Function[nR, Function[oR,
                          Join[nR, oR]] /@ S[k - 1]] /@ newRules, 1];
       (* the λ-values fulfilling the first k equations *)
       S[k] = Select[allRules, (eqsActive /. #) == 0&],
       {k, 2, 10}]
```

In[149]:= **Length[S[10]]**

Out[149]= 65536

26 solutions have all edge lengths positive.

In[150]:= **(posSol = Select[S[10], (And @@ (Positive /@ λs /. #))&]) // Length**

Out[150]= 26

A function that makes heavy use of Groebnerbasis is PolynomialReduce. To "simplify" a set of polynomials under the condition that another set of polynomial equations holds, one uses the function Polyno‐mialReduce.

PolynomialReduce[*polynomial*, {*reducingPolynomial*$_1$, *reducingPolynomial*$_2$, ...},
    varList, *options*]

reduces the polynomial *polynomial* with respect to the polynomials *reducingPolynomial*$_i$. The result is a list of the form {{*factor*$_1$, *factor*$_2$, ..., *factor*$_n$}, *reducedPolynomial*}, such that
polynomial = *reducedPolynomial* + $\sum_{k=1}^{n}$ *factor*$_k$ *reducingPolynomial*$_k$.

We implement the theorem of Vieta for an equation of fourth degree of the form $a_4 + a_3 x + a_2 x^2 + a_1 x^3 + x^4$. Given an expression containing the zeros x_1, x_2, x_3, x_4 of the polynomial, we want to replace the zeros by the coefficients a_1, a_2, a_3, a_4. (For multivariate versions of Vieta relations, see [1017] and [1636].)

In[151]:= **vietaEquations =**   **{x1 + x2 + x3 + x4 == -a1,**
     x1 x2 + x2 x3 + x1 x3 + x1 x4 + x2 x4 + x3 x4 == +a2,
       x1 x2 x3 + x1 x2 x4 + x1 x3 x4 + x2 x3 x4 == -a3,
               x1 x2 x3 x4 == +a4};

In applications, we often want to get rid of the xi-variables. We calculate a Gröbner basis using the term ordering DegreeReverseLexicographic (which, as mentioned, is often the fastest term ordering).

In[152]:= **gb = GroebnerBasis[Apply[Subtract, vietaEquations, {1}],**
                {x1, x2, x3, x4, a1, a2, a3, a4, x},
         MonomialOrder -> DegreeReverseLexicographic];

Using this Gröbner basis gb, we can reduce various polynomials quickly (this means to eliminate the xi).

In[153]:= **pR = PolynomialReduce[(x - x1)(x - x2)(x - x3)(x - x4), gb,**
                 {x1, x2, x3, x4},
            MonomialOrder -> DegreeReverseLexicographic]

Out[153]= $\{\{-x^3 + x^2 x2 + x^2 x3 - x\,x2\,x3 + x^2 x4 - x\,x2\,x4 - x\,x3\,x4 + x2\,x3\,x4,$
     $x^2 - x\,x3 - x\,x4 + x3\,x4, -x + x4, -1\}, a4 + a3\,x + a2\,x^2 + a1\,x^3 + x^4\}$

In[154]:= **PolynomialReduce[x1^8 + x2^8 + x3^8 + x4^8, gb, {x1, x2, x3, x4},**
              MonomialOrder -> DegreeReverseLexicographic][[-1]]

Out[154]= $a1^8 - 8\,a1^6 a2 + 20\,a1^4 a2^2 - 16\,a1^2 a2^3 + 2\,a2^4 + 8\,a1^5 a3 - 32\,a1^3 a2\,a3 + 24\,a1\,a2^2 a3 +$
     $12\,a1^2 a3^2 - 8\,a2\,a3^2 - 8\,a1^4 a4 + 24\,a1^2 a2\,a4 - 8\,a2^2 a4 - 16\,a1\,a3\,a4 + 4\,a4^2$

Be aware that the second element of `PolynomialReduce` must be a Gröbner basis. In the following input the second argument of `PolynomialReduce` is not a Gröbner basis and as a result the output still contains the variables x_i.

```
In[155]:= PolynomialReduce[(x - x1)(x - x2)(x - x3)(x - x4),
                Apply[Subtract, vietaEquations, {1}], {x1, x2, x3, x4}]
```
```
Out[155]= {{-x³ + x² x2 + x² x3 - x x2 x3 + x² x4 - x x2 x4 - x x3 x4 + x2 x3 x4, 0, 0, 0},
    a1 x³ + x⁴ - a1 x² x2 - x² x2² - a1 x² x3 + (a1 x - x²) x2 x3 + x x2² x3 - x² x3² + x x2 x3² -
    a1 x² x4 + (a1 x - x²) x2 x4 + x x2² x4 + (a1 x - x²) x3 x4 + (-a1 + 2 x) x2 x3 x4 -
    x2² x3 x4 + x x3² x4 - x2 x3² x4 - x² x4² + x x2 x4² + x x3 x4² - x2 x3 x4²}
```

Here is a small example concerning the use of `PolynomialReduce` in the univariate case. In [234], the following nice identity was given:

$$\alpha^{630} - 1 = \frac{(\alpha^{315} - 1)(\alpha^{210} - 1)(\alpha^{126} - 1)^2(\alpha^{90} - 1)(\alpha^3 - 1)^3(\alpha^2 - 1)^5(\alpha - 1)^3}{(\alpha^{35} - 1)(\alpha^{15} - 1)^2(\alpha^{14} - 1)^2(\alpha^5 - 1)^6\alpha^{68}}.$$

Here α is a solution of the equation $\alpha^{10} + \alpha^9 - \alpha^7 - \alpha^6 - \alpha^5 - \alpha^4 - \alpha^3 + \alpha + 1 = 0$. Numerically, the identity seems to hold.

```
In[156]:= poly1 = (α^630 - 1)(α^35 - 1)(α^15 - 1)^2(α^14 - 1)^2(α^5 - 1)^6 α^68 -
            (α^315 - 1)(α^210 - 1)(α^126 - 1)^2(α^90 - 1)(α^3 - 1)^3*
            (α^2 - 1)^5(α - 1)^3;
```
```
In[157]:= poly2 = α^10 + α^9 - α^7 - α^6 - α^5 - α^4 - α^3 + α + 1;
```
```
In[158]:= Solve[poly2 == 0 /. α^10 :> α10, α10]
```
```
Out[158]= {{α10 → -1 - α + α³ + α⁴ + α⁵ + α⁶ + α⁷ - α⁹}}
```
```
In[159]:= poly1 /. NSolve[poly2 == 0, α, 500]
```
```
Out[159]= {0. × 10⁻⁴⁹³ + 0. × 10⁻⁴⁹³ i, 0. × 10⁻⁴⁹³ + 0. × 10⁻⁴⁹³ i, 0. × 10⁻⁴⁹⁴ + 0. × 10⁻⁴⁹⁴ i,
    0. × 10⁻⁴⁹⁴ + 0. × 10⁻⁴⁹⁴ i, 0. × 10⁻⁴⁹² + 0. × 10⁻⁴⁹² i, 0. × 10⁻⁴⁹² + 0. × 10⁻⁴⁹² i,
    0. × 10⁻⁴⁹⁵ + 0. × 10⁻⁴⁹⁵ i, 0. × 10⁻⁴⁹⁵ + 0. × 10⁻⁴⁹⁵ i, 0. × 10⁻⁵⁰⁵, 0. × 10⁻⁴⁴¹}
```

An obvious way to prove the identity would be to repeatedly substitute $-1 - \alpha + \alpha^3 + \alpha^4 + \alpha^5 + \alpha^6 + \alpha^7 - \alpha^9$ for α^{10}. This is relatively time-consuming because many steps are needed. `PolynomialReduce` provides a proof of the above identity in less than a second.

```
In[160]:= PolynomialReduce[poly1, poly2][[2]] // Timing
```
```
Out[160]= {0.05 Second, 0}
```

Gröbner bases are among the most important tools to do symbolic mathematics. We will end here our short tour of *Mathematica*'s Gröbner basis command `GroebnerBasis` (we will repeatedly come back to it in the solutions of the exercises). Many more things could be said [1141], [1142]. We encourage the reader to experiment more with the Gröbner basis command. After a while, the reader will find it an indispensable tool for doing mathematics on the computer.

1.2.3 Polynomials in Inequalities

In this subsection, we will discuss various algebraic operations for systems of polynomial inequalities. The most important functions of this kind are `CylindricalDecomposition` and those functions related to quantifier elimination. While these functions can also deal with equations and inequations, their typical use is for inequalities and we discuss them in a separate subsection.

An advanced, quite useful function for dealing with polynomials is `CylindricalDecomposition`. Given a set of polynomials $p_i(x_1, x_2, ..., x_n), i = 1, ..., m$, the real vector space $\mathbb{R}^n$ can be naturally subdivided in cells, such that inside each cell the polynomials $p_i(x_1, x_2, ..., x_n)$ do not change signs (meaning they are either positive, negative, or zero). This forms a decomposition of the $\mathbb{R}^n$. Each cell can then be conveniently described in the form

$$x_1^{(l)} \lesssim x_1 \lesssim x_1^{(u)}$$

$$x_2^{(l)}(x_1) \lesssim x_2 \lesssim x_2^{(u)}(x_1)$$

...

$$x_m^{(l)}(x_1, x_2, ..., x_n) \lesssim x_m \lesssim x_m^{(u)}(x_1, x_2, ..., x_n)$$

(where all of the $\lesssim$ can be replaced by $<$, $\leq$ or $=$). Such a description is called a cylindrical decomposition [1275], [130], [960], [313], [935] (for more general decompositions, see [1253]).

The $x_i^{(l)}(x_1, x_2, ..., x_{i-1})$ are algebraic functions of the variables $x_1, x_2, ..., x_{i-1}$. This means they are either radicals or `Root`-objects. `Root[`*polynomialInFormOfAPureFunction*`, `*rootNumber*`]` is an inverse of the function *polynomialInFormOfAPureFunction*(z) (it is a unique one from the n possible inverses; where n is the degree of the univariate polynomial). `Root`-objects are generalizations of radicals to polynomials that cannot be uniquely inverted. We will discuss them in more detail in Section 1.5 as solutions of (parametrized) univariate polynomials, but we will encounter them repeatedly in some of the outputs in this subsection. The only operation we will carry out on them in this subsection is numericalization.

In most practical situations, one does not need all cells with all possible 3^n sign combinations, but rather one is interested in a cylindrical decomposition for a certain combination of signs only. In *Mathematica*, one can obtain a cylindrical decomposition by using the function `CylindricalDecomposition`.

> `CylindricalDecomposition[`*equalitiesAndInequalties*`, `*vars*`]`
>
> calculates a cylindrical algebraic decomposition of *equalitiesAndInequalties*. The result is returned in the form of a logical formula, and the inequalities with head `Inequality`.

The functions $x_i^{(l)}(...)$ and $x_i^{(u)}(...)$ are algebraic functions (meaning they contain potentially parametrized, radicals or `Root`-objects) of their arguments. Although `Root`-objects are typically not continuous functions of their arguments for all real values of the arguments, they are continuous functions within each region. (Inside each region returned by `CylindricalDecomposition` the sign of the polynomials is constant. But neighboring regions can have the same sign characteristic. For some purposes, it might be advantageous to join such regions.)

Here is a very simple example: The inside of a circle. A cylindrical description of this area is as follows. The following output has the head `And`.

In[1]:= `CylindricalDecomposition[x^2 + y^2 < 1, {x, y}]`

Out[1]= $-1 < x < 1$ && $-\sqrt{1 - x^2} < y < \sqrt{1 - x^2}$

The outside of a circle has a slightly more complicated description. Now the head of the output is `Or` and the second argument of this `Or` has the head `And` (and the second argument of this `And` has again the head `Or`).

In[2]:= `CylindricalDecomposition[x^2 + y^2 > 1, {x, y}]`

Out[2]= $x < -1$ || $\left(-1 \leq x \leq 1$ && $\left(y < -\sqrt{1 - x^2}$ || $y > \sqrt{1 - x^2}\right)\right)$ || $x > 1$

Here is the circle itself. The left and right "endpoints" appear separately.

In[3]:= `CylindricalDecomposition[x^2 + y^2 == 1, {x, y}]`

Out[3]= $-1 \leq x \leq 1 \;\&\&\; \left(y == -\sqrt{1 - x^2} \;||\; y == \sqrt{1 - x^2}\right)$

Reversing the variable ordering gives the top and bottom "endpoints" separately.

In[4]:= `CylindricalDecomposition[x^2 + y^2 == 1, {y, x}]`

Out[4]= $-1 \leq y \leq 1 \;\&\&\; \left(x == -\sqrt{1 - y^2} \;||\; x == \sqrt{1 - y^2}\right)$

Here is the outside of the circle, including the circle itself.

In[5]:= `CylindricalDecomposition[x^2 + y^2 >= 1, {x, y}]`

Out[5]= $x < -1 \;||\; \left(-1 \leq x \leq 1 \;\&\&\; \left(y \leq -\sqrt{1 - x^2} \;||\; y \geq \sqrt{1 - x^2}\right)\right) \;||\; x > 1$

The next result describes the lower half of two intersecting circles. A moment of reflection explains the result.

In[6]:= `CylindricalDecomposition[`
　　　　　　`(x - 1/2)^2 + y^2 < 1 && (x + 1/2)^2 + y^2 < 1 && y < 0, {x, y}]`

Out[6]= $\left(-\dfrac{1}{2} < x \leq 0 \;\&\&\; -\dfrac{1}{2}\sqrt{3 + 4x - 4x^2} < y < 0\right) \;||\; \left(0 < x < \dfrac{1}{2} \;\&\&\; -\dfrac{1}{2}\sqrt{3 - 4x - 4x^2} < y < 0\right)$

Here is a visualization of the last result and the circles involved.

In[7]:= `ContourPlot[Evaluate[If[%, 0, 1]], {x, -3/2, 3/2}, {y, -1, 1/2},`
　　　　　　`PlotPoints -> 240, AspectRatio -> Automatic,`
　　　　　　`Epilog -> {Hue[0], Line[{{-3/2, 0}, {3/2, 0}}],`
　　　　　　　　　　`Circle[{-1/2, 0}, 1], Circle[{ 1/2, 0}, 1]}]`

Now, let us look at a slightly more complicated example. We start with a torus in parametric form.

In[8]:= `torus[φ1_, φ2_, r1_, r2_] := {r1 Cos[φ1] + r2 Cos[φ1] Cos[φ2],`
　　　　　　　　　　　　　　　　`r1 Sin[φ1] + r2 Sin[φ1] Cos[φ2],`
　　　　　　　　　　　　　　　　`r2           Sin[φ2]};`

We squeeze the torus by making the larger radius a function of the angle $\varphi 1$. Using the `GroebnerBasis` function, we calculate an implicit representation of this squeezed torus.

In[9]:= `squeezedTorus[{x_, y_, z_}, r_, a_] =`
　　　　`GroebnerBasis[`
　　　　　　`Join[{x, y, z} - torus[φ1, φ2, r, a + Cos[φ1]],`
　　　　　　`{Cos[φ1]^2 + Sin[φ1]^2 - 1, Cos[φ2]^2 + Sin[φ2]^2 - 1}],`
　　　　　　`{x, y, z}, {Cos[φ1], Sin[φ1], Cos[φ2], Sin[φ2]},`
　　　　　　`MonomialOrder -> EliminationOrder][[1]]`

Out[9]= $x^4 - 2\,a^2\,x^4 + a^4\,x^4 - 2\,r^2\,x^4 - 2\,a^2\,r^2\,x^4 + r^4\,x^4 - 8\,a\,r\,x^5 - 2\,x^6 - 2\,a^2\,x^6 - 2\,r^2\,x^6 +$
$x^8 - 2\,a^2\,x^2\,y^2 + 2\,a^4\,x^2\,y^2 - 2\,r^2\,x^2\,y^2 - 4\,a^2\,r^2\,x^2\,y^2 + 2\,r^4\,x^2\,y^2 - 16\,a\,r\,x^3\,y^2 -$
$4\,x^4\,y^2 - 6\,a^2\,x^4\,y^2 - 6\,r^2\,x^4\,y^2 + 4\,x^6\,y^2 + a^4\,y^4 - 2\,a^2\,r^2\,y^4 + r^4\,y^4 - 8\,a\,r\,x\,y^4 -$

$$2\,x^2\,y^4 - 6\,a^2\,x^2\,y^4 - 6\,r^2\,x^2\,y^4 + 6\,x^4\,y^4 - 2\,a^2\,y^6 - 2\,r^2\,y^6 + 4\,x^2\,y^6 + y^8 - 2\,x^4\,z^2 -$$
$$2\,a^2\,x^4\,z^2 + 2\,r^2\,x^4\,z^2 + 2\,x^6\,z^2 - 2\,x^2\,y^2\,z^2 - 4\,a^2\,x^2\,y^2\,z^2 + 4\,r^2\,x^2\,y^2\,z^2 +$$
$$6\,x^4\,y^2\,z^2 - 2\,a^2\,y^4\,z^2 + 2\,r^2\,y^4\,z^2 + 6\,x^2\,y^4\,z^2 + 2\,y^6\,z^2 + x^4\,z^4 + 2\,x^2\,y^2\,z^4 + y^4\,z^4$$

Here the cross section of this torus in the x,y-plane is shown.

In[10]:= `ContourPlot[Evaluate[squeezedTorus[{x, y, 0}, 3, 7/10]],`
`{x, -5, 5}, {y, -5, 5},`
`PlotPoints -> 300, Contours -> {0}]`

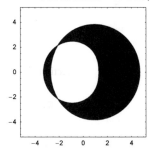

A cylindrical algebraic decomposition of this cross section can easily be calculated. The result looks quite complicated.

In[11]:= `(squeezedTorus2DCAD = CylindricalDecomposition[`
`squeezedTorus[{x, y, 0}, 3, 7/10] < 0, {x, y}]) // Short[#, 8]&`

Out[11]//Short=

$$\left| -\frac{33}{10} < x < -\frac{27}{10} \;\&\&\; -\sqrt{\frac{23}{200}\,\sqrt{529 - 400\,x} + \frac{1}{200}\,(529 - 200\,x - 200\,x^2)} \; < \right.$$

$$y < \sqrt{\frac{23}{200}\,\sqrt{529 - 400\,x} + \frac{1}{200}\,(529 - 200\,x - 200\,x^2)} \; \left. \right| \; | \,|$$

$$\left(-\frac{27}{10} \le x < -\frac{21}{10} \;\&\&\; \left(-\sqrt{\frac{23}{200}\,\sqrt{529 - 400\,x} + \frac{1}{200}\,(529 - 200\,x - 200\,x^2)} \; < \right. \right.$$

$$y < -\sqrt{\frac{37}{200}\,\sqrt{1369 + 400\,x} + \frac{1}{200}\,(1369 + 200\,x - 200\,x^2)} \; \left. \right| \,|$$

$$\sqrt{\frac{37}{200}\,\sqrt{1369 + 400\,x} + \frac{1}{200}\,(1369 + 200\,x - 200\,x^2)} \; < y <$$

$$\sqrt{\frac{23}{200}\,\sqrt{529 - 400\,x} + \frac{1}{200}\,(529 - 200\,x - 200\,x^2)} \left. \right) \left. \right) \; | \,|$$

$$\left(-\frac{21}{10} < x \le \frac{13}{10} \;\&\&\; \left(-\sqrt{\frac{37}{200}\,\sqrt{1369 + 400\,x} + \frac{1}{200}\,(1369 + 200\,x - 200\,x^2)} \; < \right. \right.$$

$$y < -\sqrt{\frac{23}{200}\,\sqrt{529 - 400\,x} + \frac{1}{200}\,(529 - 200\,x - 200\,x^2)} \; | \,|$$

$$\sqrt{\frac{23}{200}\,\sqrt{529 - 400\,x} + \frac{1}{200}\,(529 - 200\,x - 200\,x^2)} \; < y <$$

$$\sqrt{\frac{37}{200}\sqrt{1369+400\,x}+\frac{1}{200}\,(1369+200\,x-200\,x^2)}\Bigg)\Bigg)\,||$$

$$\left(\frac{13}{10}<x<\frac{47}{10}\,\&\&\,-\sqrt{\frac{37}{200}\sqrt{1369+400\,x}+\frac{1}{200}\,(1369+200\,x-200\,x^2)}\,<\right.$$

$$\left.y<\sqrt{\frac{37}{200}\sqrt{1369+400\,x}+\frac{1}{200}\,(1369+200\,x-200\,x^2)}\,\right)$$

Let us use the cylindrical algebraic decomposition to reproduce the contour plot of the cross section. We are interested only in the 2D regions; so, we only keep the finite *x*-intervals (meaning we omit single points). Then, we replace the Less by LessEqual. Last, we generate all cells by expanding the Ors in the second argument of the Ands.

```
In[12]:= squeezedTorus2DInequalities =
           Flatten[Cases[squeezedTorus2DCAD /. LessEqual -> Less,
             Inequality[_, Less, x, Less, _] && _] /.
               i1_Inequality && (i2_Inequality || i3_Inequality) :>
                                     {i1 && i2, i1 && i3}];
```

We generate polygons by traversing the boundaries of the cells of squeezedTorus2DInequalities.

```
In[13]:= makePolygon2D[Inequality[xl_, Less, x, Less, xu_] &&
                        Inequality[yl_, Less, y, Less, yu_], pp_] :=
         With[{ε = 1.`22*^-6},
         Polygon[Join[Table[{x, yl}, {x, xl + ε, xu - ε, (xu - xl - 2ε)/pp}],
         Block[{x = xu},
             Table[{x, y}, {y, yl + ε, yu - ε, (yu - yl - 2ε)/pp}]],
             Table[{x, yu}, {x, xu - ε, xl + ε, (xl - xu + 2ε)/pp}],
         Block[{x = xl},
             Table[{x, y}, {y, yu - ε, yl + ε, (yl - yu + 2ε)/pp}]]]]]
```

By coloring each of the resulting polygons in a different color, we finally arrive at the following picture.

```
In[14]:= Show[Graphics[{Hue[Random[]], makePolygon2D[#, 22]}& /@
                         squeezedTorus2DInequalities],
          AspectRatio -> Automatic, Frame -> True]
```

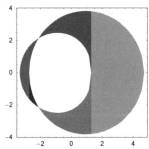

Next, we will use the function CylindricalDecomposition to calculate some exact iterations of the Arnold cat map [197] (see Section 2.4 of the Numerics volume [1737]). Given semialgebraic sets *parts* from within the unit square, the function catMapStep applies one step of the cat map and returns the resulting semialgebraic set. We realize the mod 1 operation by splitting the affine deformed semialgebraic set with respect to the six unit squares that are mapped back to $(0, 1)\times(0, 1)$.

```
In[15]:= catMapStep[parts_] :=
    Module[{r = 0, (* local counter *) r},
       (* stretch the given set made from the parts *)
       stretchedParts = parts /. {x -> 2 x - y, y -> -x + y};
       (* the six unit squares covered that cover any stretched parts *)
       squares = Outer[And, {0 < x < 1, 1 < x < 2},
                            {0 < y < 1, 1 < y < 2, 2 < y < 3}] // Flatten;
       (* the shifts corresponding to the six unit squares *)
       shifts = Outer[List, {0, 1}, {0, 1, 2}] // Flatten[#, 1]&;
       (* reduce all stretched parts with respect to the six squares *)
       Flatten[Table[
       Table[gcad = CylindricalDecomposition[
                       stretchedParts[[i, 2]] && squares[[j]], {x, y}] /.
                       (* keep 2D components *) LessEqual -> Less;
             part = Cases[{gcad}, _And, Infinity];
             If[part =!= {}, r = r++;
                C[r, Or @@ (modOne[# /. (* use Less instead Inequality *)
                       Inequality[a_, Less, b_, Less, c_] :> Less[a, b, c],
                             shifts[[j]]]& /@ part)], {}], {j, 6}],
             {i, Length[stretchedParts]}] /. (* keep all parts separated *)
                C[r_, or_Or] :> (C[r, #]& /@ (List @@ or))]]
In[16]:= (* shift semialgebraic set to unit square (0,1)×(0,1) *)
    modOne[(xMin_ < x < xMax_) && (yMin_ < y < yMax_), {ξ_, η_}] :=
       And @@ {(xMin - ξ) < x < (xMax - ξ),
               (yMin - η /. x -> x + ξ) < y < (yMax - η /. x -> x + ξ)}
```

The function `catMapGraphic` visualizes the semialgebraic sets resulting from applying the Arnold cat map. We color the parts according to the shift needed to carry out the mod 1 operation at each step.

```
In[18]:= catMapGraphic[parts_, pp_] :=
    Module[{μ = Max[Max[First /@ parts], 1]},
       (* avoid messages from numericalization at the edges *)
       Internal`DeactivateMessages[(* make colored polygons *)
          Graphics[{regionToBorderedPolygon[#2, Hue[0.8 #1/μ], pp]}& @@@
                     parts, PlotRange -> All, Frame -> True,
                     FrameTicks -> None, AspectRatio -> Automatic]]]
In[19]:= regionToBorderedPolygon[(xMin_ < x < xMax_) &&
                                 (yMin_ < y < yMax_), col_, pp_] :=
    Module[{iter = {x, xMin, xMax, (xMax - xMin)/pp}, lower, upper},
           {lower, upper} = {Table @@ {{x, yMin}, iter},
                             Table @@ {{x, yMax}, iter}} // N[#, 30]&;
           {(* colored polygon *)
            {col, Polygon[Join[lower, Reverse[upper]]]},
            (* thin black edge *)
            {GrayLevel[0], Thickness[0.002],
             Line[Join[lower, Reverse[upper], Take[lower, 1]]]}}]
```

We visualize the first few steps of repeatedly applying the cat map to the unit square $(0, 1) \times (0, 1)$ and a disk of radius $1/2$ in this unit square.

```
In[20]:= Function[initSAS,
           Show[GraphicsArray[catMapGraphic[#, 60]& /@
                             NestList[catMapStep, initSAS, 3]]]] /@
              {(* the unit square (0, 1)×(0, 1) *)
               {C[0, 0 < x < 1 && 0 < y < 1]},
```

```
(* radius 1/2 disk in the unit square (0, 1)×(0, 1) *)
{C[0, 0 < x < 1 && 1/2 - Sqrt[x - x^2] < y < 1/2 + Sqrt[x - x^2]]}}
```

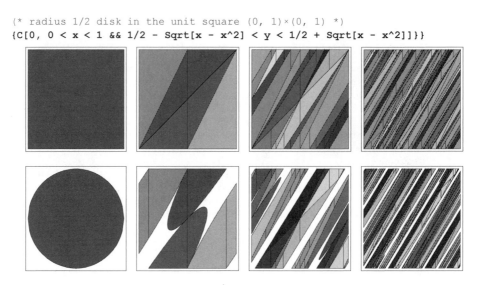

For the unit square itself, we apply the cat map five times. This results in 2636 polygons.

```
In[21]:= catSquare = Nest[catMapStep, {C[0, 0 < x < 1 && 0 < y < 1]}, 5];
        Length[catSquare]
Out[22]= 2636
```

As expected, the area of these 2636 polygons adds up to 1.

```
In[23]:= area[(xMin_ < x < xMax_) && (yMin_ < y < yMax_)] :=
        Module[{p1 = {xMin, yMin /. x -> xMin}, p2 = {xMax, yMin /. x -> xMax},
                p3 = {xMax, yMax /. x -> xMax}, p4 = {xMin, yMax /. x -> xMin}},
               area[{p1, p2, p3}] + area[{p3, p4, p1}]]

        area[{{x1_, y1_}, {x2_, y2_}, {x3_, y3_}}] :=
            Abs[x2 y1 - x3 y1 - x1 y2 + x3 y2 + x1 y3 - x2 y3]/2
In[26]:= allCatSquareAreas = area[#[[2]]]& /@ catSquare;
        Plus @@ allCatSquareAreas
Out[27]= 1
```

The following graphic shows the resulting polygons randomly colored (left). Along the main diagonal of the unit square, many small polygons exist. The right graphic shows the distribution of the (logarithm of the) sizes of the 2636 polygons.

```
In[28]:= Show[GraphicsArray[
        {catMapGraphic[catSquare, 2] /.
                        {p_Polygon :> {Hue[Random[]], p}, l_Line :> {}},
         ListPlot[N @ Log @ Sort[allCatSquareAreas], PlotRange -> All,
                  DisplayFunction -> Identity, Frame -> True, Axes -> False]}]]
```

 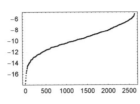

For 3D cat maps, see [338].

Let us now treat an equality. Here is a Lissajous curve.

```
In[29]:= ParametricPlot[{Cos[3θ], Sin[5θ]}, {θ, 0, 2Pi},
                        AspectRatio -> Automatic, Frame -> True, Axes -> False]
```

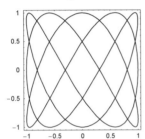

We calculate an implicit representation of this Lissajous curve using again the GroebnerBasis function.

```
In[30]:= lissajouCurve = GroebnerBasis[
    Join[TrigExpand[{x, y}  - {Cos[6θ], Sin[10θ]}],
        {Cos[θ]^2 + Sin[θ]^2 - 1}], {x, y}, {Cos[θ], Sin[θ]},
        MonomialOrder -> EliminationOrder][[1]]
```

$$Out[30]= \ -1 + 25\,x^2 - 200\,x^4 + 560\,x^6 - 640\,x^8 + 256\,x^{10} + 9\,y^2 - 24\,y^4 + 16\,y^6$$

Here is a sketch of the cylindrical algebraic decomposition of the Lissajous curve.

```
In[31]:= lissajouCAD = CylindricalDecomposition[lissajouCurve == 0, {x, y}];
```

```
In[32]:= Module[{i = 0}, InputForm[lissajouCAD /. _Root :> r[i++]]] // Short[#, 6]&
Out[32]//Short=
```

```
        (x == -1 && (y == -Sqrt[3]/2 || y == 0 || y == Sqrt[3]/2)) || (Inequality[-1,
          Less, x, Less, -Sqrt[(5 + Sqrt[5])/2]/2] && (y == -Sqrt[r[0]] || y == -Sqrt[
          r[1]] || y == -Sqrt[r[2]] || y == Sqrt[r[3]] || y == Sqrt[r[4]] || y == Sqrt[
          r[5]])) || (x == -Sqrt[(5 + Sqrt[5])/2]/2 && (y == -Sqrt[r[6]] || y == -Sqrt[
          r[7]] || y == Sqrt[r[8]] || y == Sqrt[r[9]])) || <<16>> || (Inequality[<<
          5>>] && <<1>>) || (x == 1 && (y == -Sqrt[3]/2 || y == 0 || y == Sqrt[3]/2))
```

Next, we separate the single points from the curve segments.

```
In[33]:= lissajouInequalities =
    Flatten[Cases[lissajouCAD /. LessEqual -> Less,
            Inequality[_, Less, x, Less, _] && _] /.
                i1_Inequality && or_Or :> ((i1 && #)& /@ (List @@ or))];
```

```
In[34]:= lissajouPoints = Point[{x, y}] /.
            (ToRules /@ Flatten[Cases[lissajouCAD, x == _ && _] /.
              x == x0_ && or_Or :>
              ((x == x0 && (# /. x -> x0))& /@ (List @@ or))]);
```

The cylindrical algebraic decomposition of the Lissajous curve consists of 60 line segments and 38 points.

```
In[35]:= Length /@ {lissajouInequalities, lissajouPoints}
Out[35]= {60, 42}
```

The next graphic shows the line segments as well as the single points. The single points are the leftmost and rightmost points (where the curve has a vertical tangent), the topmost and bottommost points (where the curve has a horizontal tangent), and the crossing points of the curve.

```
In[36]:= makeLine2D[Inequality[xl_, Less, x, Less, xu_] && y == yx_, pp_] :=
            With[{ε = N[10^-8, 22]},
              Line[Table[{x, yx}, {x, xl + ε, xu - ε, (xu - xl - 2ε)/pp}]]]

In[37]:= Show[Graphics[{
            {Thickness[0.002],
             {Hue[Random[]], makeLine2D[#, 22]}& /@ lissajouInequalities},
             {PointSize[0.02], GrayLevel[0], N @ lissajouPoints}}],
            AspectRatio -> Automatic, Frame -> True]
```

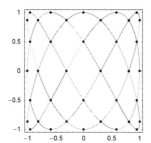

Mathematica can (given enough time and memory) calculate the cylindrical algebraic decomposition in any number of variables and (in)equalities (and as long as the arithmetic of the coefficients does not exceed the largest possible integers). Here is the cylindrical algebraic decomposition of a sphere.

```
In[38]:= CylindricalDecomposition[x^2 + y^2 + z^2 == 1, {x, y, z}]
```

$$Out[38]= \ (x == -1 \ \&\& \ y == 0 \ \&\& \ z == 0) \ ||$$
$$\left(-1 < x < 1 \ \&\& \ -\sqrt{1-x^2} \le y \le \sqrt{1-x^2} \ \&\& \ \left(z == -\sqrt{1-x^2-y^2} \ || \ z == \sqrt{1-x^2-y^2}\right)\right) \ ||$$
$$(x == 1 \ \&\& \ y == 0 \ \&\& \ z == 0)$$

As a more challenging example, let us deal with our squeezed torus from above. Its cylindrical algebraic decomposition is quite complicated.

```
In[39]:= squeezedTorus3DCAD = CylindricalDecomposition[
                                squeezedTorus[{x, y, z}, 3, 7/10] < 0, {x, y, z}];

          squeezedTorus3DCAD // Short[#, 8]&
```

Out[41]//Short=

$$\Biggl(-\frac{33}{10} < x < -\frac{27}{10} \;\&\&\; -\sqrt{\frac{23}{200}\sqrt{529 - 400\,x} + \frac{1}{200}\,(529 - 200\,x - 200\,x^2)} <$$

$$y < \sqrt{\frac{23}{200}\sqrt{529 - 400\,x} + \frac{1}{200}\,(529 - 200\,x - 200\,x^2)} \;\&\&\;$$

$$-\sqrt{\frac{-751\,x^2 - 100\,x^4 - 851\,y^2 - 200\,x^2\,y^2 - 100\,y^4}{100\,(x^2 + y^2)}} +$$

$$\frac{1}{5}\sqrt{\frac{49\,x^2 + 420\,x^3 + 900\,x^4 + 420\,x\,y^2 + 1800\,x^2\,y^2 + 900\,y^4}{x^2 + y^2}} <$$

$$z < \sqrt{\frac{-751\,x^2 - 100\,x^4 - 851\,y^2 - 200\,x^2\,y^2 - 100\,y^4}{100\,(x^2 + y^2)}} +$$

$$\frac{1}{5}\sqrt{\frac{49\,x^2 + 420\,x^3 + 900\,x^4 + 420\,x\,y^2 + 1800\,x^2\,y^2 + 900\,y^4}{x^2 + y^2}}\Biggr) \;\|\; \ll 1 \gg \; \ll 2 \gg$$

$$\ll 1 \gg \;\|\; \Biggl(\frac{13}{10} < x < \frac{47}{10} \;\&\&\; -\sqrt{\frac{37}{200}\sqrt{1369 + 400\,x} + \frac{1}{200}\,(1369 + 200\,x - 200\,x^2)} <$$

$$y < \sqrt{\frac{37}{200}\sqrt{1369 + 400\,x} + \frac{1}{200}\,(\ll 1 \gg)} \;\&\&\;$$

$$-\sqrt{\frac{-751\,x^2 - 100\,x^4 - 851\,y^2 - 200\,x^2\,y^2 - 100\,y^4}{100\,(x^2 + y^2)}} +$$

$$\frac{1}{5}\sqrt{\frac{49\,x^2 + 420\,x^3 + 900\,x^4 + 420\,x\,y^2 + 1800\,x^2\,y^2 + 900\,y^4}{x^2 + y^2}} <$$

$$z < \sqrt{\frac{-751\,x^2 - 100\,x^4 - 851\,y^2 - 200\,x^2\,y^2 - 100\,y^4}{100\,(x^2 + y^2)}} +$$

$$\frac{1}{5}\sqrt{\frac{49\,x^2 + 420\,x^3 + 900\,x^4 + 420\,x\,y^2 + 1800\,x^2\,y^2 + 900\,y^4}{x^2 + y^2}}\Biggr)$$

To better see the structure, we again use short-hand notations for the roots.

```
In[42]:= Module[{i = 0}, InputForm[squeezedTorus3DCAD /. _Root :> r[i++]]] //
                                                        Short[#, 8]&
```

Out[42]//Short=

```
(Inequality[-33/10, Less, x, Less, -27/10] && Inequality[-Sqrt[(23*Sqrt[
   529 - 400*x])/200 + (529 - 200*x - 200*x^2)/200], Less, y, Less,
   Sqrt[(23*Sqrt[529 - 400*x])/200 + (529 - 200*x - 200*x^2)/200]] &&
   Inequality[-Sqrt[(-751*x^2 - 100*x^4 - 851*y^2 - 200*x^2*y^2 - 100*y^
   4)/(100*(x^2 + y^2)) + Sqrt[(49*x^2 + 420*x^3 + 900*x^4 + 420*x*y^2 +
   1800*x^2*y^2 + 900*y^4)/(x^2 + y^2)]/5], Less, z, Less, Sqrt[(-751*x^
   2 - 100*x^4 - 851*y^2 - 200*x^2*y^2 - 100*y^4)/(100*(x^2 + y^2)) +
   Sqrt[(49*x^2 + 420*x^3 + 900*x^4 + 420*x*y^2 + 1800*x^2*y^2 + 900*y^
   4)/(x^2 + y^2)]/5]]) || <<2>> || (Inequality[13/10, <<4>>] && <<2>>)
```

Let us use `squeezedTorus3DCAD` to make a picture of the parts of $\mathbb{R}^3$ for which `squeezedTorus[{5, y,` `z}, 3, 7/10]` < 0. Similar to the 2D case, we select the inequalities in `squeezedTorus3DCAD` and expand remaining `Or`s.

```
In[43]:= squeezedTorus3DInequalities =
    Flatten[Cases[squeezedTorus3DCAD /. LessEqual -> Less,
                Inequality[_, Less, x, Less, _] && _] /.
        i1_Inequality && (a2_And || a3_And) :> {i1 && a2, i1 && a3}];
```

The function `makePolygons3D` generates polygons of the top and bottom boundaries for a given cell.

```
In[44]:= makePolygons3D[Inequality[xl_, Less, x, Less, xu_] &&
                Inequality[yl_, Less, y, Less, yu_] &&
                Inequality[zl_, Less, z, Less, zu_], pp_] :=
    Block[{x, y},
    Module[{ε = 10.^-12},
    (* avoid endpoints for numericalization *) δx = (xu - xl - 2ε)/pp;
    (* calculate points *)
    pointsl = Table[x = xl + ε + ix δx; δy = (yu - yl - 2ε)/pp;
            Table[y = yl + ε + iy δy; {x, y, Re[zl]},
                {iy, 0, pp}], {ix, 0, pp}];
    pointsu = Table[x = xl + ε + ix δx; δy = (yu - yl - 2ε)/pp;
            Table[y = yl + ε + iy δy; {x, y, Re[zu]},
                {iy, 0, pp}], {ix, 0, pp}];
    (* form polygons *)
    {Table[Polygon[{pointsl[[i, j]], pointsl[[i + 1, j]],
                    pointsl[[i + 1, j + 1]], pointsl[[i, j + 1]]}],
            {i, pp}, {j, pp}],
      Table[Polygon[{pointsu[[i, j]], pointsu[[i + 1, j]],
                    pointsu[[i + 1, j + 1]], pointsu[[i, j + 1]]}],
            {i, pp}, {j, pp}]}]]
```

Here, the pieces of the boundary of the 3D pieces of the cylindrical algebraic decomposition are randomly colored.

```
In[45]:= Show[Graphics3D[{EdgeForm[], {SurfaceColor[Hue[Random[]]],
            makePolygons3D[#, 24]}& /@ squeezedTorus3DInequalities}]]
```

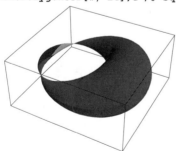

In many applications of cylindrical algebraic decompositions (for instance, for changing variables or the variable order in multidimensional integrals [1529]), one does not need the cells that are of lower dimensionality. This means only strong inequalities using `Less` and `Greater` are allowed. Fortunately, such a decomposition is also faster to calculate. The *Mathematica* command dealing especially with this case is `GenericCylindri\` `calAlgebraicDecomposition`. (This function comes from the `Experimental\`` context.)

GenericCylindricalAlgebraicDecomposition[*inequalties, vars*]

   calculates a generic cylindrical algebraic decomposition of *inequalties*. The result is returned in the form of a
   list. The first element is a logical formula describing the cells, and the second element describes the
   lower-dimensional parts.

Here is the difference between CylindricalDecomposition and GenericCylindricalAlgebraic⸱
Decomposition for a simple example—a circle and an ellipse.

In[46]:= **CylindricalDecomposition[**
$$(x^2 + y^2 - 1) (x^2/4 + y^2 - 1) <= 0, \{x, y, z\}]$$

Out[46]= $(x == -2 \&\& y == 0) \mid\mid \left(-2 < x < -1 \&\& -\frac{1}{2}\sqrt{4-x^2} \le y \le \frac{\sqrt{4-x^2}}{2}\right) \mid\mid$

$\left(-1 \le x < 0 \&\& \left(-\frac{1}{2}\sqrt{4-x^2} \le y \le -\sqrt{1-x^2} \mid\mid \sqrt{1-x^2} \le y \le \frac{\sqrt{4-x^2}}{2}\right)\right) \mid\mid$

$(x == 0 \&\& (y == -1 \mid\mid y == 1)) \mid\mid$

$\left(0 < x \le 1 \&\& \left(-\frac{1}{2}\sqrt{4-x^2} \le y \le -\sqrt{1-x^2} \mid\mid \sqrt{1-x^2} \le y \le \frac{\sqrt{4-x^2}}{2}\right)\right) \mid\mid$

$\left(1 < x < 2 \&\& -\frac{1}{2}\sqrt{4-x^2} \le y \le \frac{\sqrt{4-x^2}}{2}\right) \mid\mid (x == 2 \&\& y == 0)$

In[47]:= **Experimental`GenericCylindricalAlgebraicDecomposition[**
$$(x^2 + y^2 - 1) (x^2/4 + y^2 - 1) <= 0, \{x, y, z\}]$$

Out[47]= $\left\{\left(-2 < x < -1 \&\& -\frac{1}{2}\sqrt{4-x^2} < y < \frac{\sqrt{4-x^2}}{2}\right) \mid\mid\right.$

$\left(-1 < x < 1 \&\& \left(-\frac{1}{2}\sqrt{4-x^2} < y < -\sqrt{1-x^2} \mid\mid \sqrt{1-x^2} < y < \frac{\sqrt{4-x^2}}{2}\right)\right) \mid\mid$

$\left(1 < x < 2 \&\& -\frac{1}{2}\sqrt{4-x^2} < y < \frac{\sqrt{4-x^2}}{2}\right),$

$\left. -1 + x == 0 \mid\mid x == 0 \mid\mid 1 + x == 0 \mid\mid 4 - 5x^2 + x^4 - 8y^2 + 5x^2 y^2 + 4y^4 == 0\right\}$

The following output describes the regions in the *x,y*-plane where the product $p(x, y) = \prod_{i,j=-2}^{2}(x-i)(y-j)$ is
negative. Observe that the use of LessEqual does not cause the corresponding lines to be returned.

In[48]:= $p =$ **Product[x - i, {i, -2, 2}] Product[y - j, {j, -2, 2}];**
   Experimental`GenericCylindricalAlgebraicDecomposition[p <= 0,
$$\{x, y\}][[1]] // LogicalExpand$$

Out[49]= $(x > 2 \&\& y < -2) \mid\mid (y > 2 \&\& x < -2) \mid\mid (x > 2 \&\& -1 < y \&\& y < 0) \mid\mid (x > 2 \&\& 1 < y \&\& y < 2) \mid\mid$
$(y > 2 \&\& -1 < x \&\& x < 0) \mid\mid (y > 2 \&\& 1 < x \&\& x < 2) \mid\mid (-2 < x \&\& x < -1 \&\& y < -2) \mid\mid$
$(-2 < y \&\& x < -2 \&\& y < -1) \mid\mid (0 < x \&\& x < 1 \&\& y < -2) \mid\mid (0 < y \&\& x < -2 \&\& y < 1) \mid\mid$
$(-2 < x \&\& -1 < y \&\& x < -1 \&\& y < 0) \mid\mid (-2 < x \&\& 1 < y \&\& x < -1 \&\& y < 2) \mid\mid$
$(-2 < y \&\& -1 < x \&\& x < 0 \&\& y < -1) \mid\mid (-2 < y \&\& 1 < x \&\& x < 2 \&\& y < -1) \mid\mid$
$(-1 < x \&\& 0 < y \&\& x < 0 \&\& y < 1) \mid\mid (-1 < y \&\& 0 < x \&\& x < 1 \&\& y < 0) \mid\mid$
$(0 < x \&\& 1 < y \&\& x < 1 \&\& y < 2) \mid\mid (0 < y \&\& 1 < x \&\& x < 2 \&\& y < 1)$

We will make use of GenericCylindricalAlgebraicDecomposition for a nontrivial problem in
Subsection 1.9.1.

For details about the cylindrical algebraic decomposition method and its use, see [313], [260], [1081], [59],
[1206], [1225], [90], [91], [242], [89], [1275], [1123], and [941]. It should be mentioned that the algorithm
used to find a cylindrical algebraic decomposition is in the worst case double-exponential in the number of

variables. So whenever possible, the user should reduce the problem under consideration to one with fewer variables.

A subject intimately related to cylindrical algebraic decomposition is quantifier elimination. Given an expression e of the form

$$e = \exists_{qVars1_1} \forall_{qVars2_1} \exists_{qVars1_2} \forall_{qVars2_2} \cdots$$
$$\exists_{qVars1_n} \forall_{qVars2_n} L(p_1(qVars1, qVars2, vars), \ldots, p_n(qVars1, qVars2, vars))$$

where L is a logical expression (which can contain expressions with the head And, Or, Implies, Not) built from polynomial equations and inequations p_i (which can contain expressions with the head Equal, Unequal, Greater, Less, GreaterEqual, LessEqual, Inequality), it is in principle possible to eliminate the quantified variables and rewrite e in the form of a logical expression of polynomial equations and inequations p_i in the unquantified variables *vars* only [1703], [313]. The variables $qVars1 = \{qVars1_1, \ldots, qVars1_n\}$, $qVars2 = \{qVars2_1, \ldots, qVars2_n\}$ are quantified (and can be in principle eliminated) and the variables *vars* are free. All variables involved, $qVars1$, $qVars2$, and *vars*, will be assumed to be real.

The *Mathematica* form of the quantifiers $\exists$ and $\forall$ is as follows.

Exists[*listOfVariables*, *propertiesOfVariables*, *expression*]

>   represents the mathematical statement $\exists_{listOfVariables}$ *expression*. The variables fulfill the conditions *propertiesOf-Variables*.

ForAll[*listOfVariables*, *propertiesOfVariables*, *expression*]

>   represents the mathematical statement $\forall_{listOfVariables}$ *expression*. The variables fulfill the conditions *propertiesOf-Variables*.

Here are two very simple examples of expressions formed with the quantifiers Exists and ForAll. The first one qE1 states the existence of a real number x, $x > 5$, such that $x^3 = 200$.

In[50]:= **qE1 = Exists[x, Element[x, Reals] && x > 5, x^3 == 200]**

Out[50]= $\exists_{x, x \in Reals \&\& x>5} \; x^3 == 200$

The second statement qE2 represents the conjecture that for all positive integers n can be expressed as the sum of two squares.

In[51]:= **qE2 = ForAll[n, Element[n, Integers] && n > 0,**
                    Exists[{v1, v2}, Element[{v1, v2}, Integers],
                            n == v1^2 + v2^2]]

Out[51]= $\forall_{n, n \in Integers \&\& n>0} \; (\exists_{\{v1, v2\}, (v1|v2) \in Integers} \; n == v1^2 + v2^2)$

In[52]:= **Simplify[qE2]**

Out[52]= $\forall_{n, n \in Integers \&\& n>0} \; (\exists_{\{v1, v2\}, (v1|v2) \in Integers} \; n == v1^2 + v2^2)$

Both of the last inputs were not evaluated to True or False by *Mathematica*. To "evaluate" expressions containing quantifiers—this means to eliminate the quantifiers—we need one more function.

Resolve[*expressionContainingQuantifiedRealVariables*, *domain*]

>   eliminates the with Exists and ForAll quantified variables from the expression *expressionContainingQuantifiedRealVariables* assuming all occurring variables are from the domain *domain*.

The main function discussed in the previous subsection, `GroebnerBasis`, deals with polynomial equations, and the indeterminates are generically considered to be complex variables. The main functions of this subsection, `CylindricalDecomposition` and `Resolve`, deal generically with inequations of real variables.

`Resolve` will always carry out its calculation for polynomial equations and inequations in real variables (this means it will never, like `Integrate` does in case it cannot find a closed form of the integral, return unevaluated when properly called) and also complex variables, but the actual running time for some examples might be large. Contrary to many other *Mathematica* functions, `Resolve` does not make any genericity assumptions about variables. `Integrate[x^n, x]` returns the result $\frac{x^{n+1}}{n+1}$, which for $n = -1$ is ill-defined. x and n in this example are both assumed to be generic complex variables. `Resolve` will produce results where all "exceptions" are explicitly listed. This property of `Resolve` allows its use for proving a variety of theorems from arithmetic and geometry. Be aware that the running time of quantifier elimination problems depends dramatically on the number of variables to be dealt with (in the worse case, the dependence is double exponential in the number of variables).

Applying `Resolve` to the above expression qE1 yields `True`.

In[53]:= **Resolve[qE1]**

Out[53]= True

Applying `Resolve` to the above expression qE2 does not yield `False`. For quantified real variables, algorithms exist to carry out the elimination, but for integer-quantified variables, it might be algorithmically impossible to do this [1218].

In[54]:= **Resolve[qE2]**

Out[54]= $\forall_{n,\,n\in\text{Integers}\,\&\&\,n>0}\;(\exists_{\{v1,v2\},\,(v1|v2)\in\text{Integers}}\;n == v1^2 + v2^2)$

Now, let us consider some examples of the use of quantifier elimination. We will give a larger than usual amount of examples of these important tools for solving a variety of problems coming from different fields of applications. The first group of examples will be a set of theorems from elementary arithmetic and geometry [223], [1277], [267]. We start with the following theorem about the relation between the arithmetic mean and the geometric mean: $(x + y)/2 \geq \sqrt{x\,y}$ [268]. The statement does not hold for generic real x and y.

In[55]:= **ForAll[{x, y}, (x + y)/2 >= Sqrt[x y]] // Resolve**

Out[55]= False

But it does, however, hold for positive x and y.

In[56]:= **ForAll[{x, y}, x > 0 && y > 0, (x + y)/2 >= Sqrt[x y]] // Resolve**

Out[56]= True

Here is a similar statement about three variables: $(x + y + z)/3 \geq \sqrt[3]{x\,y\,z}$.

In[57]:= **Resolve[ForAll[{x, y, z}, x > 0 && y > 0 && z > 0,**
            (x + y + z)/3 >= (x y z)^(1/3)]]

Out[57]= True

This verifies the Cauchy inequality: $(a^2 + b^2)(c^2 + d^2) \geq (a\,c + b\,d)^2$.

In[58]:= **Resolve[ForAll[{a, b, c, d}, Element[{a, b, c, d}, Reals],**
            (a^2 + b^2)(c^2 + d^2) >= (a c + b d)^2]]

Out[58]= True

And the following input verifies the three-variable version of the Cauchy inequality:

$$(a_1^2 + a_2^2 + a_3^2)(b_1^2 + b_2^2 + b_3^2) \ge (a_1 b_1 + a_2 b_2 + a_3 b_3)^2.$$

```
In[59]:= Resolve[ForAll[{a1, a2, a3, b1, b2, b3},
                Element[{a1, a2, a3, b1, b2, b3}, Reals],
                (a1^2 + a2^2 + a3^2)(b1^2 + b2^2 + b3^2) >=
                (a1 b1 + a2 b2 + a3 b3)^2]]

Out[59]= True
```

An inequality from the year 2000 International Mathematics Olympiad (http://olympiads.win.tue.nl/imo/ imo2000/imo2000-problems.pdf) can be proven within a fraction of a second.

```
In[60]:= Timing @ Resolve[
        ForAll[{a, b, c}, Element[{a, b, c}, Reals] &&
                        a > 0 && b > 0 && c > 0 && a b c == 1,
            (a - 1 + 1/b) (b - 1 + 1/c) (c - 1 + 1/a) <= 1]]

Out[60]= {0.03 Second, True}
```

The following inequality from [943] can also be proved in less than a second.

```
In[61]:= Timing @ Resolve[
            Exists[{a, b, c},
                    Element[{a, b, c}, Reals] &&
                    0 < a < 1 && 0 < b < 1 && 0 < c < 1 &&
                    a b + a c + b c == 1,
                    a/(1 - a^2) + b/(1 - b^2) + c/(1 - c^2) == 3 Sqrt[3]/2]]

Out[61]= {0.52 Second, True}
```

Many inequalities concerning triangles exist. Using `Resolve`, it is straightforward to prove a large number of them. Here are some examples. In the following examples a, b, and c are the length of the sides of a triangle, s denotes its circumference, A its area, r is the radius of the inscribed circle, and R is the radius of the circumscribed circle.

```
In[62]:= s = a + b + c;
```

The conditions on the variables a, b, and c to be interpretable as the edge length of a triangle are as follows. (Note that we do not have to explicitly add `Element[{a, b, c}, Reals]`. Whenever a variable appears in an inequality inside `Exists` or `ForAll`, it is automatically assumed to be a real number.)

```
In[63]:= triangleConditions = a > 0 && b > 0 && c > 0 &&
                        a + b > c && a + c > b && b + c > a;
```

Here is an example from [1364].

```
In[64]:= Resolve[ForAll[{a, b, c}, triangleConditions,
                8 a b c <= (a + b) (b + c) (c + a)]]

Out[64]= True
```

And here is an example from [383], [488].

```
In[65]:= s = a + b + c;
```

```
In[66]:= Timing[Resolve @
        ForAll[{a, b, c}, triangleConditions,
                a b c < a^2 (s - a) + b^2 (s - b) + c^2 (s - c)]]

Out[66]= {0.04 Second, True}
```

Another solution is to quantify *s* itself.

```
In[67]:= Clear[s];

       Resolve @
       ForAll[{a, b, c}, triangleConditions, Exists[s, Element[s, Reals] &&
                   s == a + b + c,
             a b c < a^2(s - a) + b^2 (s - b) + c^2 (s - c)]] // Timing
Out[68]= {0.03 Second, True}
```

We take the above remark into account and evaluate s, A, and R in the following calculation outside of the ForAll instead of supplying them as equalities.

```
In[69]:= s = (a + b + c)/2;
       A = (* Heron's formula *) Sqrt[s (s - a) (s - b) (s - c)];
       R = a b c/(4 A);
       Timing[Resolve[ForAll[{a, b, c}, triangleConditions,
                   4 R^2 >= ((b^2 + c^2)/Sqrt[2 b^2 + 2 c^2 - a^2])^2]]]
Out[72]= {0.03 Second, True}
```

Here is a last, recently published inequality concerning triangles [175].

```
In[73]:= r = 2 A/(a + b + c);

       Resolve[ForAll[{a, b, c}, triangleConditions, 2 s^2 >= 27 R r]]
Out[74]= True
```

Using an algebraic representation for the maximum functions, we can also prove the inequality $\max(a, b, c) < (a + b + c)/2$.

```
In[75]:= max[x1_, x2_] := 1/2 (x1 + x2 + Sqrt[(x1 - x2)^2])
```

```
In[76]:= max[max[a, b], c] < 1/2 (a + b + c)
```

$$Out[76]= \frac{1}{2}\left(\frac{1}{2}\left(a + \sqrt{(a-b)^2} + b\right) + \sqrt{\left(\frac{1}{2}\left(a + \sqrt{(a-b)^2} + b\right) - c\right)^2} + c\right) < \frac{1}{2}(a + b + c)$$

```
In[77]:= Resolve[ForAll[{a, b, c}, triangleConditions, %]]
Out[77]= True
```

In the following example, the main statements are equations. Heron's formula is derived. We also get nondegener-
· acy conditions on the edge length such that Heron's formula holds.

```
In[78]:= Exists[{p2x, p3x, p3y},
                 (* coordinate and edge length constraints *)
                 Element[{p2x, p3x, p3y, area, 112, 123, 113}, Reals] &&
                       p2x >= 0 && p3x >= 0 && p3y >= 0 &&
                       123 >= 0 && 113 >= 0,
                 (* area and edge length expressions *)
                 area == p2x p3y/2 && 112 == p2x &&
                 123^2 == (p2x - p3x)^2 + p3y^2 &&
                 113^2 == p3x^2 + p3y^2] // Resolve
```

$Out[78]= (123 \geq 0 \,\&\&\, 113 = 123 \,\&\&\, 112 = 0 \,\&\&\, area = 0) \,||$
$(123 = 0 \,\&\&\, 113 > 0 \,\&\&\, 112 = 113 \,\&\&\, area = 0) \,||$
$\left(123 > 0 \,\&\&\, \left((113 = 0 \,\&\&\, 112 = 123 \,\&\&\, area = 0) \,||\, \left(0 < 113 < 123 \,\&\&\, \right.\right.\right.$
$\left(\left(112 = \sqrt{-113^2 + 123^2} \,\&\&\, area = \frac{112\,113}{2}\right) \,||\, \left(\sqrt{-113^2 + 123^2} < 112 < 113 + 123 \,\&\&\, \right.\right.$
$\left. area = \frac{1}{4}\sqrt{-112^4 + 2\,112^2\,113^2 - 113^4 + 2\,112^2\,123^2 + 2\,113^2\,123^2 - 123^4}\,\right) \,||$

$$\left.\left.\left.(112 == 113 + 123 \,\&\&\, \text{area} == 0)\right)\right) \,||\, \left(113 == 123 \,\&\&\, \left(\left(0 < 112 < 113 + 123 \,\&\&\, \right.\right.\right.\right.$$

$$\left.\text{area} == \frac{1}{4}\sqrt{-112^4 + 2\,112^2\,113^2 - 113^4 + 2\,112^2\,123^2 + 2\,113^2\,123^2 - 123^4}\right) \,||$$

$$\left.\left.\left.(112 == 113 + 123 \,\&\&\, \text{area} == 0)\right)\right) \,||\,\right.$$

$$\left(113 > 123 \,\&\&\, \left((112 == 113 - 123 \,\&\&\, \text{area} == 0) \,||\, \left(113 - 123 < 112 < 113 + 123 \,\&\&\,\right.\right.\right.$$

$$\left.\text{area} == \frac{1}{4}\sqrt{-112^4 + 2\,112^2\,113^2 - 113^4 + 2\,112^2\,123^2 + 2\,113^2\,123^2 - 123^4}\right) \,||$$

$$\left.\left.\left.\left.(112 == 113 + 123 \,\&\&\, \text{area} == 0)\right)\right)\right)\right)$$

The expression from which to eliminate quantifiers does not have to contain inequations at all. `Resolve` will also eliminate quantifiers from logical expressions formed purely by equations. Here is an example.

```
In[79]:= Exists[c, Element[c, Reals],
           ForAll[{a, b}, Element[{a, b, c, d}, Reals],
           Implies[(a == d && b == c) ||
                   (a == c && b == 1), a^6 == b c d]]] // Resolve
```

Out[79]= $d == 0 \,||\, d == 1$

Taking into account the above remark to express all formulas to be treated by `Resolve`, we can also prove many triangle-theorems containing angles. Using the cosine theorem, we express the values of the three cosines as rational functions in the side length.

```
In[80]:= cosα = -(a^2 - b^2 - c^2)/(2 b c);
         cosβ = -(b^2 - a^2 - c^2)/(2 a c);
         cosγ = -(c^2 - a^2 - b^2)/(2 a b);
```

Here are three theorems about angles in triangles proved:

$$0 < \cos(\alpha) + \cos(\beta) + \cos(\gamma) \le \frac{3}{2}$$

$$0 < \cos(\alpha) + \sqrt{2}\,(\cos(\beta) + \cos(\gamma)) \le 2$$

$$\cos(\alpha)\cos(\beta)\cos(\gamma) \le \frac{1}{8}.$$

```
In[83]:= ForAll[{a, b, c}, triangleConditions,
             0 < cosα + cosβ + cosγ <= 3/2] // Resolve
```

Out[83]= True

```
In[84]:= ForAll[{a, b, c}, triangleConditions,
             cosα + Sqrt[2](cosβ + cosγ) <= 2] // Resolve
```

Out[84]= True

```
In[85]:= ForAll[{a, b, c}, triangleConditions,
             cosα cosβ cosγ <= 1/8] // Resolve
```

Out[85]= True

But `Resolve` does not only allow us to prove theorems; it can also be used to derive new ones. Here, we derive lower and upper bounds for $\cos(\alpha) + 2\cos(\beta) + 3\cos(\gamma)$, where as always, α, β, and γ are the angles in a triangle. The new theorem is (the 4 on the right-hand side is the smallest possible value):

$$0 \le \cos(\alpha) + 2\cos(\beta) + 3\cos(\gamma) \le 4.$$

```
In[86]:= ForAll[{a, b, c}, triangleConditions && Element[{λ}, Reals],
             cosα + 2 cosβ + 3 cosγ <= λ] // Resolve
```

Out[86]= $\lambda \ge 4$

In[87]:= **ForAll[{a, b, c}, triangleConditions && Element[{λ}, Reals],**
            λ <= cosα + 2 cosβ + 3 cosγ] // Resolve

Out[87]= $\lambda \le 0$

Besides proving polynomial inequalities, quantifier elimination has a couple of other important applications. One of them is making certain statements about the location of zeros of polynomials. We start with a simple one: We prove that every cubic polynomial has at least one real zero.

In[88]:= **ForAll[{a, b}, Element[{a, b, c}, Reals],**
            Exists[x, Element[x, Reals],
                a + b x + c x^2 + x^3 == 0]] // Resolve

Out[88]= $c \in$ Reals

But not every quartic has real roots. (We normalize the leading coefficient to 1.)

In[89]:= **ForAll[{a, b, c}, Element[{a, b, c, d}, Reals],**
            Exists[x, Element[x, Reals],
                a + b x + c x^2 + d x^3 + x^4 == 0]] // Resolve

Out[89]= False

The coefficient of the cubic term was 1 in the last input. If we let the coefficient of the cubic term be the variable d, we get, as the condition for the statement to be correct, the nondegeneracy condition of the cubic polynomial.

In[90]:= **ForAll[{a, b}, Element[{a, b, c, d}, Reals],**
            Exists[x, Element[x, Reals],
                a + b x + c x^2 + d x^3 == 0]] // Resolve

Out[90]= $c \in$ Reals && (d < 0 || d > 0)

Here are the conditions for the coefficients a and b of a quadratic $a + b x + x^2$, such that this quadratic has at least one real root.

In[91]:= **Exists[{x}, Element[{x, a, b}, Reals],**
                a + b x + x^2 == 0] // Resolve

Out[91]= $b \in$ Reals && $a \le \dfrac{b^2}{4}$

To have two different roots, we get a strict inequality.

In[92]:= **Exists[{x1, x2}, Element[{x1, x2, a, b}, Reals],**
                (* different roots *) x1 != x2 &&
                a + b x1 + x1^2 == 0 && a + b x2 + x2^2 == 0] // Resolve

Out[92]= $b \in$ Reals && $a < \dfrac{b^2}{4}$

Using the last result, we can show that for any nondegenerate quadratic $x^2 + b x + a$ with real a and b and real roots, the following holds: $a + b + 1 \le 9/4 \max(a, b, 1)$ [1298].

In[93]:= **Resolve[ForAll[{a, b}, Element[{a, b, λ}, Reals] &&**
                a <= b^2/4, a + b + 1 <= λ Max[a, b, 1]]]

Out[93]= $\lambda \ge \dfrac{9}{4}$

Here are the more complicated conditions for the coefficients a, b, and c of the cubic $a + b x + c x^2 + x^3$ to have three different roots calculated. For calculating the probabilities for having k roots, see [1614].)

In[94]:= **Exists[{x1, x2, x3}, Element[{a, b, c, x1, x2, x3}, Reals],**
            (* roots are different *)
            x1 != x2 != x3 &&
            (* x1, x2, x3 _are_ roots *)

```
       a + b x1 + c x1^2 + x1^3 == 0 &&
       a + b x2 + c x2^2 + x2^3 == 0 &&
       a + b x3 + c x3^2 + x3^3 == 0] // Resolve
```

Out[94]= $c \in \text{Reals} \,\&\&\, b < \dfrac{c^2}{3} \,\&\&\, \dfrac{1}{27}(9\,b\,c - 2\,c^3) - \dfrac{2}{27}\sqrt{-27\,b^3 + 27\,b^2\,c^2 - 9\,b\,c^4 + c^6} <$

$a < \dfrac{1}{27}(9\,b\,c - 2\,c^3) + \dfrac{2}{27}\sqrt{-27\,b^3 + 27\,b^2\,c^2 - 9\,b\,c^4 + c^6}$

Here is a still more complicated example: For which a, b does the sextic $a^4 + a^2\,b\,x - a\,x^6$ have at least one real root?

```
In[95]:= Exists[{x}, Element[{x, a, b}, Reals],
            a^4 + a^2 b x - a x^6 == 0] // Resolve
```

Out[95]= $b \in \text{Reals} \,\&\&\, (a == 0 \;||$

$(a < 0 \,\&\&\, (b \le \text{Root}[46656\,a^9 + 3125\,\#1^6 \,\&,\, 1] \;||\; b \ge \text{Root}[46656\,a^9 + 3125\,\#1^6 \,\&,\, 2])) \;||$

$a \ge 0)$

Here is a numerical experiment. We randomly select 5000 pairs for a and b and make a point in the a,b-plane whenever the polynomial has at least one real root. The two lines are the curves described by the last result.

```
In[96]:= Show[Graphics[{{PointSize[0.01], Hue[0],
         Block[{a, b},
         Table[(* random values for a and b *)
                a = Random[Real, {-3, 3}]; b = Random[Real, {-5, 5}];
                (* check if at least one real root is present *)
                If[Cases[x /.
                    {ToRules[NRoots[a^4 + a^2 b x - a x^6 == 0, x]]}, _Real] =!= {},
                    Point[{a, b}], {}], {5000}]]},
         (* the lines *)
         {Thickness[0.002], GrayLevel[0], Table[(* two roots *)
                Line[Table[{a, Root[46656 a^9 + 3125 #^6&, j]},
                       {a, -((5 5^(2/9))/6^(2/3)), 0, 0.01}]], {j, 2}]}}],
         Frame -> True, PlotRange -> All]
```

Under which conditions on the real coefficients a_k can a cubic polynomial be expressed as the sum of three simple cubes $(x - \alpha_k)^3$ with real α_k?

$$x^3 + a_2\,x^2 + a_1\,x + a_0 = (x - \alpha_1)^3 + (x - \alpha_2)^3 + (x - \alpha_3)^3 = 0$$

```
In[97]:= Resolve[Exists[{α[1], α[2], α[3]},
             Element[{α[1], α[2], α[3], a[2], a[1], a[0]}, Reals],
                    (* from comparing coefficients of x0, x1, x2 *)
                    a[0] + α[1]^3/3 + α[2]^3/3 + α[3]^3/3 == 0 &&
                    a[1] - α[1]^2 - α[2]^2 - α[3]^2 == 0 &&
                    a[2] + α[1] + α[2] + α[3] == 0]]
```

Out[97]= $a[2] \in \text{Reals} \ \&\& \ \left(\left(a[1] == \frac{a[2]^2}{3} \ \&\& \ a[0] == \right.\right.$

$\left. \frac{1}{27} (9 a[1] a[2] - 2 a[2]^3) - \frac{\sqrt{27 a[1]^3 - 27 a[1]^2 a[2]^2 + 9 a[1] a[2]^4 - a[2]^6}}{27 \sqrt{2}} \right) \ || $

$\left(a[1] > \frac{a[2]^2}{3} \ \&\& \ \frac{1}{27} (9 a[1] a[2] - 2 a[2]^3) - \right.$

$\left. \frac{\sqrt{27 a[1]^3 - 27 a[1]^2 a[2]^2 + 9 a[1] a[2]^4 - a[2]^6}}{27 \sqrt{2}} \le a[0] \le \right.$

$\left.\left. \frac{1}{27} (9 a[1] a[2] - 2 a[2]^3) + \frac{\sqrt{27 a[1]^3 - 27 a[1]^2 a[2]^2 + 9 a[1] a[2]^4 - a[2]^6}}{27 \sqrt{2}} \right) \right)$

(For more concerning the connection between the number of roots and the coefficients, see [1874].)

A further very important application of quantifier elimination is the following: Given a polynomial, say, $a + b z + c z^2 + z^3$, under which conditions on the coefficients have all roots negative imaginary parts (or positive imaginary parts or are all inside or outside the unit circle or …) [692], [123]. This question has important application in control theory [1305], for the construction of stable finite difference schemes [1150], [871], [870], and more [960]. Because quantifier elimination techniques work only for real numbers, we write the condition for the complex roots z of $a + b z + c z^2 + z^3 = 0$ as two equations for the real and the imaginary parts.

In[98]:= **cubic = Expand[a + b z + c z^2 + z^3 /. z -> x + I y]**

Out[98]= $a + b x + c x^2 + x^3 + i b y + 2 i c x y + 3 i x^2 y - c y^2 - 3 x y^2 - i y^3$

In[99]:= **{im = Plus @@ (Cases[cubic, _Complex _]/I),**
       re = Expand[cubic - I im]}

Out[99]= $\{b y + 2 c x y + 3 x^2 y - y^3, \ a + b x + c x^2 + x^3 - c y^2 - 3 x y^2\}$

In[100]:= **ForAll[{x, y}, Element[{ x, y, a, b, c}, Reals],**
            Implies[re == 0 && im == 0, x < 0]] // Resolve

Out[100]= $c > 0 \ \&\& \ b > 0 \ \&\& \ 0 < a < b c$

Quantifier elimination can be used to solve many satisfiability questions. For example, given the two polynomials

$$x^2 + 2 y x - 3 = 0$$
$$y^2 - f x y - 5 = 0,$$

does an f exist, such that $-24/10 < x < -23/10$?

In[101]:= **Exists[{x, y}, Element[{x, y, f}, Reals],**
            x^2 + 2 x y - 3 == 0 && y^2 -f x y - 5 == 0 &&
            -24/10 < x < -23/10] // Resolve

Out[101]= $\frac{7471}{2208} < f < \frac{1005559}{242282}$

In[102]:= **N[%]**

Out[102]= $3.38361 < f < 4.15037$

Here is a graphical check for the last result.

In[103]:= **Show[Graphics[{**
        {Hue[0], (* bounding rectangle for x and f *)
        Polygon[{{1005559/242282, -23/10}, {1005559/242282, -24/10},
            {7471/2208, -24/10}, {7471/2208, -23/10}}]},
        {GrayLevel[0], PointSize[0.01],

```
Table[Point[{f, #}]& /@ (x /.
(* solve system numerically for given f *)
NSolve[{x^2 + 2 x y - 3 == 0, y^2 - f x y - 5 == 0},
      {x, y}]), {f, 3.3, 4.2, 0.05}]}],
   PlotRange -> {-2.2, -2.5}, Frame -> True]
```

The next input proves the Sendov–Iliev conjecture [1439] for quadratic polynomials. (The conjecture states that for univariate polynomials that have all of their roots in the unit circle, the maximal distance of a root of the differentiated polynomial from a root of the original polynomial is at most 1).

```
In[104]:= ForAll[{x1, x2, y2}, -1 < x1 < 1 &&  x2^2 + y2^2 < 1,
          Exists[{x, y, X, Y},
                 (* z == x + I y is a roots of (z - z1)(z - z2) == 0 *)
                 x1 x2 - x1 x - x2 x + x^2 + y2 y - y^2 == 0 &&
                 x1 y2 - y2 x - x1 y - x2 y + 2 x y == 0 &&
                 (* Z == X + I Y is a roots of 2 Z - (Z1 + Z2) == 0 *)
                 -x1 - x2 + 2 X == 0 && -y2 + 2 Y == 0,
                 (* bound distance of roots from derivative roots  *)
                 (x - X)^2 + Y^2 < ρ]] // Resolve[#, Reals]&
```

Out[104]= $\rho \geq 1$

We have a quick look at the options of `Resolve`.

In[105]:= `Options[Resolve]`

Out[105]= {Backsubstitution → False, Cubics → False, Quartics → False, WorkingPrecision → ∞}

The first three options we will discuss in Section 1.5 when we will deal with the solution of polynomial systems. The last option `WorkingPrecision` we already discussed in Chapter 1 of the Numerics volume [1737] for the numerical functions. The use of a finite working precision in `Resolve` avoids the potential coefficient swell that we already mentioned in connection with Gröbner basis calculations.

The next inputs calculate the parameter values a and b, such that the Hénon map $[x, y] \to \{a - x^2 + b\,y, x\}$ has period-three orbits [581]. This time, we use a finite working precision.

```
In[106]:= (* the Hénon map, parametrized by a and b *)
       HénonMap[{x_, y_}, {a_, b_}] := {a - x^2 + b y, x}
```

```
In[108]:= Resolve[Exists[{x0, y0}, Element[{x0, y0, a, b}, Reals],
                (* period three orbit *)
                And @@ Thread[{x0, y0} == Nest[HénonMap[#, {a, b}]&, {x0, y0}, 3]]],
                WorkingPrecision -> 100] // Timing
```

Out[108]= $\left\{88.03\,\text{Second},\ \left(b < \text{Root}[1 - 6\,\#1^3 + \#1^6\ \&,\ 1] \ \&\&\ a \geq \frac{1}{4}\,(-1 + 2\,b - b^2)\right)\ ||\right.$

$\left(b == \text{Root}[1 - 6\,\#1^3 + \#1^6\ \&,\ 1] \ \&\&\ a \geq \frac{1 - b^2 - 4\,b^3 - b^4 + b^6}{4\,b^2}\right)\ ||$

$$\left(\text{Root}[1 - 6\,\#1^3 + \#1^6\,\&,\,1] < b < \text{Root}[1 - 6\,\#1^3 + \#1^6\,\&,\,2]\,\&\&\,a \geq \frac{1}{4}\,(-1 + 2\,b - b^2)\right)\,||$$

$$\left(b == \text{Root}[1 - 6\,\#1^3 + \#1^6\,\&,\,2]\,\&\&\,a \geq \frac{1 - b^2 - 4\,b^3 - b^4 + b^6}{4\,b^2}\right)\,||$$

$$\left(b > \text{Root}[1 - 6\,\#1^3 + \#1^6\,\&,\,2]\,\&\&\,a \geq \frac{1}{4}\,(-1 + 2\,b - b^2)\right)\}$$

Here is the parameter domain that allows for period three orbits shown in black over the *a,b*-plane.

```
In[109]:= ContourPlot[Evaluate[If[%[[2]], 0, 1]], {a, -12, 12}, {b, -12, 12},
              PlotPoints -> 240]
```

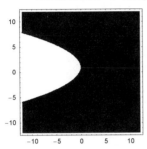

Let us now give two examples from quantum mechanics. The first example derives all nontrivial Bell and Clauser–Horne inequalities for an entangled two-particle system [1417], [1416], [102], [319], [913], [278], [19], [1565], [1833], [1834], [869], [771], [385], [253], [1163]. (For a *Mathematica* package to derive such inequalities, see [633].)

```
In[110]:= (* possible measurement outcomes *)
        elementaryEvents = {{𝓔[A1], 𝓔[A2]}, {𝓔[B1], 𝓔[B2]}};
        doubleEvents = Flatten[Outer[Join, Sequence @@ elementaryEvents]];
        allEvents = Flatten[{elementaryEvents, doubleEvents}]
Out[113]= {𝓔[A1], 𝓔[A2], 𝓔[B1], 𝓔[B2], 𝓔[A1, B1], 𝓔[A1, B2], 𝓔[A2, B1], 𝓔[A2, B2]}
```

```
In[114]:= (* vertices of the correlation polytope *)
        allPoints = Level[Table[Times @@@ allEvents,
                        Evaluate[Sequence @@ (({#[[1]], 0, 1}& /@
                        Cases[allEvents, 𝓔[_]]))]], {-2}];
```

```
In[116]:= (* express events as linear combinations of extremal points *)
        eventEquations = And @@ Thread[(𝓟 @@@ allEvents) ==
                Sum[λ[k] allPoints[[k]], {k, Length[allPoints]}]]
Out[117]= 𝓟[A1] == λ[9] + λ[10] + λ[11] + λ[12] + λ[13] + λ[14] + λ[15] + λ[16] &&
        𝓟[A2] == λ[5] + λ[6] + λ[7] + λ[8] + λ[13] + λ[14] + λ[15] + λ[16] &&
        𝓟[B1] == λ[3] + λ[4] + λ[7] + λ[8] + λ[11] + λ[12] + λ[15] + λ[16] &&
        𝓟[B2] == λ[2] + λ[4] + λ[6] + λ[8] + λ[10] + λ[12] + λ[14] + λ[16] &&
        𝓟[A1, B1] == λ[11] + λ[12] + λ[15] + λ[16] && 𝓟[A1, B2] == λ[10] + λ[12] + λ[14] + λ[16] &&
        𝓟[A2, B1] == λ[7] + λ[8] + λ[15] + λ[16] && 𝓟[A2, B2] == λ[6] + λ[8] + λ[14] + λ[16]
```

```
In[118]:= λs = Table[λ[k], {k, Length[allPoints]}];
        𝓟s = 𝓟 @@@ allEvents;
        (* do the work *)
        res = (Exists @@ {λs, Element[λs, Reals] && Element[𝓟s, Reals] &&
                (Sequence @@  (# > 0& /@ λs)) &&
                0 < (Plus @@ λs) < 1, eventEquations}) // Resolve;
```

```
In[122]:= (* select nontrivial inequalities *)
        Cases[(Select[Flatten[res /. {Or -> List, And -> List}],
```

```
                    Count[#, P[_, _], Infinity] == 4&] /. Less -> LessEqual) // Union,
              _? (MemberQ[Subtract @@ #, _Integer]&)]
Out[123]= {-1 + P[A2] + P[B2] + P[A1, B1] - P[A1, B2] - P[A2, B1] - P[A2, B2] ≤ 0,
          -1 + P[A2] + P[B1] - P[A1, B1] + P[A1, B2] - P[A2, B1] - P[A2, B2] ≤ 0,
          -1 + P[A1] + P[B2] - P[A1, B1] - P[A1, B2] + P[A2, B1] - P[A2, B2] ≤ 0,
          -1 + P[A1] + P[B1] - P[A1, B1] - P[A1, B2] - P[A2, B1] + P[A2, B2] ≤ 0}
```

Here is a more complicated example from quantum physics. The following proves that the four Bell-like inequalities in the second argument of the ForAll imply the existence of joint probability distributions for the three-particle Greenberger–Horne–Zeilinger state [449], [1248], [139], [102], [1154], [1898], [1687], [318].

```
In[124]:= ForAll[{p[A], p[B], p[C], p[D], p[ABC]},
          (* hidden variable probabilities for the state
           |ψ> = (|1;↑>|2;↑>|3;↑> - |1;↓>|2;↓>|3;↓>)/Sqrt[2])
           and the observables
           A == σ[1; x]⊗σ[2; y]⊗σ[3; y],
           B == σ[1; y]⊗σ[2; x]⊗σ[3; y],
           C == σ[1; y]⊗σ[2; y]⊗σ[3; x],
           D == A⊗B⊗C *)
          0 <= p[A] <= 1 && 0 <= p[B] <= 1 &&
          0 <= p[C] <= 1 && 0 <= p[D] <= 1 &&
          (* expectation values expressed through probabilities
           E[α] == p[α] - p[!α] == -1 + 2 p[α];
           Bell-like inequalities *)
          -2 <=  2 p[A] + 2 p[B] + 2 p[C] - 2 p[D] - 2 <= 2 &&
          -2 <= -2 p[A] + 2 p[B] + 2 p[C] - 2 p[D] <= 2 &&
          -2 <=  2 p[A] - 2 p[B] + 2 p[C] - 2 p[D] <= 2 &&
          -2 <=  2 p[A] + 2 p[B] - 2 p[C] - 2 p[D] <= 2,
          Exists[(* joint probability distributions *)
          {P[A, B, C], P[!A, B, C], P[A, !B, C], P[A, B, !C],
           P[!A, !B, C], P[!A, B, !C], P[A, !B, !C]},
          (* conditions on joint probability distributions *)
          0 <= P[A, B, C]   <= 1 && 0 <= P[!A, B, C]  <= 1 &&
          0 <= P[A, !B, C]  <= 1 && 0 <= P[A, B, !C]  <= 1 &&
          0 <= P[!A, !B, C] <= 1 && 0 <= P[!A, B, !C] <= 1 &&
          0 <= P[A, !B, !C] <= 1,
          (* locality assumptions *)
          p[A] == P[A, B, C] + P[A, B, !C] + P[A, !B, C] + P[A, !B, !C] &&
          p[B] == P[A, B, C] + P[A, B, !C] + P[!A, B, C] + P[!A, B, !C] &&
          p[C] == P[A, B, C] + P[A, !B, C] + P[!A, B, C] + P[!A, !B, C]]] //
                                                              Resolve
Out[124]= True
```

Quantifier elimination is also used in computer graphics. We start with a simple example. Given a rationally parametrized circle of radius r, what is a description of all points in maximal distance ε from the circle [1435]?

```
In[125]:= distanceSquared = #.#&[{x, y} - rρ {2t, t^2 - 1}/(t^2 + 1)]
```

$$Out[125]= \left(x - \frac{2\,t\,r\rho}{1 + t^2}\right)^2 + \left(y - \frac{(-1 + t^2)\,r\rho}{1 + t^2}\right)^2$$

```
In[126]:= Exists[t, Element[{t, x, y, rρ, εSquared}, Reals] && rρ > 0,
               distanceSquared < εSquared] // Resolve
```

$$Out[126]= (x \mid y) \in \text{Reals} \,\&\&\, r\rho > 0 \,\&\&\, \varepsilon\text{Squared} > x^2 + y^2 + r\rho^2 - 2\sqrt{x^2\,r\rho^2 + y^2\,r\rho^2}$$

Here is a visualization of the resulting geometric figure—a ring.

```
In[127]:= With[{εSquared = 0.1, r = 1},
             ContourPlot[r^2 + x^2 + y^2 - 2 Sqrt[r^2 x^2 + r^2 y^2],
```

```
                        {x, -2, 2}, {y, -2, 2},
                        PlotPoints -> 30, Contours -> {Sqrt[εSquared]}]]
```

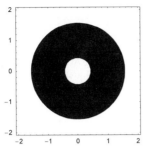

Next, we determine the conditions so that a disk and the inside of a parabola intersect.

```
In[128]:= Exists[{x, y}, Element[{x, y, ξ, η}, Reals],
               1 - x - 2 y^2 > 0 &&
               (x - ξ)^2 + (y - η)^2 < 1/64] // Resolve
```

Out[128]= $(\eta < 0 \,\&\&\, \xi < \text{Root}[45927 - 143424\,\eta^2 - 73728\,\eta^4 + 262144\,\eta^6 -$
$202176\,\#1 + 543744\,\eta^2\,\#1 - 196608\,\eta^4\,\#1 + 332352\,\#1^2 - 663552\,\eta^2\,\#1^2 +$
$262144\,\eta^4\,\#1^2 - 241664\,\#1^3 + 262144\,\eta^2\,\#1^3 + 65536\,\#1^4\,\&,\,2])\;||$
$\left(\eta == 0 \,\&\&\, \xi < \dfrac{9}{8}\right)\;||\;(\eta > 0 \,\&\&\, \xi < \text{Root}[45927 - 143424\,\eta^2 - 73728\,\eta^4 + 262144\,\eta^6 -$
$202176\,\#1 + 543744\,\eta^2\,\#1 - 196608\,\eta^4\,\#1 + 332352\,\#1^2 - 663552\,\eta^2\,\#1^2 +$
$262144\,\eta^4\,\#1^2 - 241664\,\#1^3 + 262144\,\eta^2\,\#1^3 + 65536\,\#1^4\,\&,\,2])$

Here are two examples of this situation.

```
In[129]:= Show[GraphicsArray[
     ContourPlot[1 - x - 2 y^2, {x, -2, 2}, {y, -2, 2},
                 Contours -> {0}, ContourLines -> False,
                 DisplayFunction -> Identity,
                 PlotPoints -> 50, Epilog -> {Hue[0], (* the disk *)
                   Disk[{%[[1, 2, 2]] /. η -> N[#], #}, 1/Sqrt[64]]}]& /@
                 {-1, 1/2}]]
```

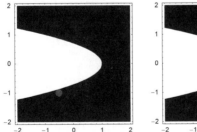

Here is another example. This time we start with a genuine 2D figure—a square. By taking again all points within a given distance from any square point, we smooth out the corners of the square.

```
In[130]:= Exists[{x, y}, Element[{x, y, ξ, η}, Reals],
                 (* the square *) -1 <= x <= 1 && -1 <= y <= 1 &&
                 (x - ξ)^2 + (y - η)^2 < 1/4] // Resolve
```

Out[130]= $\left(-\dfrac{3}{2} < \eta \le -1 \ \&\& \ -1 - \dfrac{1}{2}\sqrt{-3 - 8\,\eta - 4\,\eta^2} < \xi < 1 + \dfrac{1}{2}\sqrt{-3 - 8\,\eta - 4\,\eta^2}\right)\ ||$

$\left(-1 < \eta \le 1 \ \&\& \ -\dfrac{3}{2} < \xi < \dfrac{3}{2}\right)\ ||$

$\left(1 < \eta < \dfrac{3}{2} \ \&\& \ -1 - \dfrac{1}{2}\sqrt{-3 + 8\,\eta - 4\,\eta^2} < \xi < 1 + \dfrac{1}{2}\sqrt{-3 + 8\,\eta - 4\,\eta^2}\right)$

The following picture shows the resulting square with the smoothed corners as well circles of radius $1/2$ along the boundary points of the original square.

```
In[131]:= With[{pp = 12, qq = 8},
       Show[Graphics[
         {{Hue[0], (* the smoothed square *)
           Polygon[Join[Table[{η, -1 - Sqrt[-3 - 8 η - 4 η^2]/2},
                          {η, -3/2, -1, 1/2/pp}],
                     Reverse[Table[{η, 1 + Sqrt[-3 - 8 η - 4 η^2]/2},
                          {η, -3/2, -1, 1/2/pp}]]]],
           Polygon[Join[Table[{η, -3/2}, {η, -1, 1, 1/pp}],
                     Reverse[Table[{η, 3/2}, {η, -1, 1, 1/pp}]]]],
           Polygon[Join[Table[{η, -1 - Sqrt[-3 + 8 η - 4 η^2]/2},
                          {η, 1, 3/2, 1/2/pp}],
                     Reverse[Table[{η, 1 + Sqrt[-3 + 8 η - 4 η^2]/2},
                          {η, 1, 3/2, 1/2/pp}]]]]} // N,
         {GrayLevel[1/2], Thickness[0.002],
           (* circles of radius 1/2 along the
              boundary of the original square *)
           Circle[#, 1/2]& /@ Join[
           Table[{ x, -1}, {x, -1, 1, 2/qq}], Table[{ x,  1}, {x, -1, 1, 2/qq}],
           Table[{-1,  y}, {y, -1, 1, 2/qq}], Table[{ 1,  y}, {y, -1, 1, 2/qq}]]},
         {GrayLevel[0], (* the original square *) Rectangle[{-1, -1}, {1, 1}]}}],
           Frame -> True, PlotRange -> All, AspectRatio -> Automatic]]
```

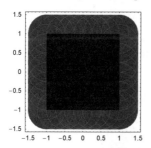

Another application of quantifier elimination are Minkowski sums and Minkowski products [612], [613], [614], [615], [616], [618]. Let us find all points $\{\xi, \eta\}$, such that $p(\xi + i\eta)$ is a point in the unit circle for a given polynomial p. toConditions expresses the real and imaginary parts of $p(X + iY)$ as a function of X and Y.

```
In[132]:= toConditions[p_, z_, {{x_, y_}, {X_, Y_}}] :=
       And @@ Thread[{x, y} == ComplexExpand[{Re[#], Im[#]}&[
                          p /. z -> X + I Y]]]
```

We will use the polynomial $p(z) = z^8 + z^7 + z^6 + z^5 + z^4 + z^3 + z^2 + z + 1$.

```
In[133]:= p =.; p[z_] := Sum[z^k, {k, 0, 8}];
```

The following is an implicit description of the points $\xi + i\eta$ such that $|p(\xi + i\eta)| \le 1$.

In[134]:= `Exists[{x, y}, Element[{x, y, ξ, η}, Reals],`
`        x^2 + y^2 <= 1 &&`
`        toConditions[p[z], z, {{x, y}, {ξ, η}}]] // Resolve`

Out[134]= $(\eta \mid \xi) \in$ Reals &&

$-\eta^2 + \eta^4 - \eta^6 + \eta^8 - \eta^{10} + \eta^{12} - \eta^{14} + \eta^{16} + 2\,\xi - 4\,\eta^2\,\xi + 6\,\eta^4\,\xi - 8\,\eta^6\,\xi - 8\,\eta^8\,\xi + 6\,\eta^{10}\,\xi -$
$4\,\eta^{12}\,\xi + 2\,\eta^{14}\,\xi + 3\,\xi^2 - 10\,\eta^2\,\xi^2 + 21\,\eta^4\,\xi^2 - 36\,\eta^6\,\xi^2 + 19\,\eta^8\,\xi^2 - 6\,\eta^{10}\,\xi^2 - 3\,\eta^{12}\,\xi^2 +$
$8\,\eta^{14}\,\xi^2 + 4\,\xi^3 - 20\,\eta^2\,\xi^3 + 56\,\eta^4\,\xi^3 + 48\,\eta^6\,\xi^3 - 2\,\eta^8\,\xi^3 - 16\,\eta^{10}\,\xi^3 + 14\,\eta^{12}\,\xi^3 + 5\,\xi^4 -$
$35\,\eta^2\,\xi^4 + 126\,\eta^4\,\xi^4 + 6\,\eta^6\,\xi^4 - 29\,\eta^8\,\xi^4 + 3\,\eta^{10}\,\xi^4 + 28\,\eta^{12}\,\xi^4 + 6\,\xi^5 - 56\,\eta^2\,\xi^5 - 36\,\eta^6\,\xi^5 -$
$20\,\eta^8\,\xi^5 + 42\,\eta^{10}\,\xi^5 + 7\,\xi^6 - 84\,\eta^2\,\xi^6 - 42\,\eta^4\,\xi^6 - 36\,\eta^6\,\xi^6 + 25\,\eta^8\,\xi^6 + 56\,\eta^{10}\,\xi^6 + 8\,\xi^7 -$
$48\,\eta^2\,\xi^7 - 36\,\eta^4\,\xi^7 + 70\,\eta^8\,\xi^7 + 9\,\xi^8 - 21\,\eta^2\,\xi^8 - 9\,\eta^4\,\xi^8 + 45\,\eta^6\,\xi^8 + 70\,\eta^8\,\xi^8 + 8\,\xi^9 -$
$2\,\eta^2\,\xi^9 + 20\,\eta^4\,\xi^9 + 70\,\eta^6\,\xi^9 + 7\,\xi^{10} + 10\,\eta^2\,\xi^{10} + 39\,\eta^4\,\xi^{10} + 56\,\eta^6\,\xi^{10} + 6\,\xi^{11} + 16\,\eta^2\,\xi^{11} +$
$42\,\eta^4\,\xi^{11} + 5\,\xi^{12} + 17\,\eta^2\,\xi^{12} + 28\,\eta^4\,\xi^{12} + 4\,\xi^{13} + 14\,\eta^2\,\xi^{13} + 3\,\xi^{14} + 8\,\eta^2\,\xi^{14} + 2\,\xi^{15} + \xi^{16} \le 0$

Here is this region visualized. The right picture displays the boundaries of the left region(s) under the mapping $z \to p(z)$ yields the unit circle.

In[135]:= `Show[GraphicsArray[`
`        Block[{$DisplayFunction = Identity, cp},`
`        {cp = ContourPlot[(* extract bivariate polynomial *)`
`                        Evaluate[Cases[%, _Plus, Infinity, 1][[1]]],`
`                        {ξ, -3/2, 3/2}, {η, -3/2, 3/2},`
`                        Contours -> {0}, PlotPoints -> 200],`
`            Show[Graphics[cp] /. (pl:(Polygon | Line))[1_] :>`
`                        (* map lines and polygons *)`
`                        pl[{Re[#], Im[#]}& /@ (p[{1, I}.#]& /@ 1)],`
`                PlotRange -> {{-3/2, 3/2}, {-3/2, 3/2}}]}]]]`

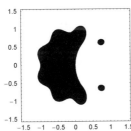
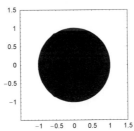

Still another area of application of quantifier elimination is the detection of collisions between (moving) bodies [1829]. The following input calculates the possible centers along the vertical axis of an ellipse moving in a circle.

In[136]:= `Clear[x, y, mx, r];`
`        ForAll[{x, y}, Element[{x, y, r, mx}, Reals] && r > 0,`
`                Implies[x^2/2^2 + y^2 - 1 < 0,`
`                        (x - mx)^2 + (y - 0)^2 < r^2]] // Resolve`

Out[137]= $(mx \le 0$ && $(r \le 0 \mid\mid r \ge 2 - mx)) \mid\mid (mx > 0$ && $(r \le 0 \mid\mid r \ge 2 + mx))$

Here is a more complicated example. A unit square is moving in the x,y-plane along the path parametrized by $\{t, t^2\}$. At the same time, a disk is moving along a straight path parametrized by $\{t, 2t + y_0\}$. Is there any time t where the square and the disk will collide? Here are the paths of the centers of the square and the disk.

In[138]:= `squarePath[t_] = {t, t^2};`
`        diskPath[t_, y0_] = {t, 2 t + y0};`

For $y_0 = -3$, the square and the disk will never collide, meaning there is no time t that the square and the circle overlap.

```
In[140]:= inSquare¿[{x_, y_}, t_] := -1 < x - squarePath[t][[1]] < 1 &&
                                      -1 < y - squarePath[t][[2]] < 1
```

```
In[141]:= collision[y0_] := Resolve[
              Exists[{t, x, y}, Element[{t, x, y}, Reals],
                  inSquare¿[{x, y}, t] && (#.#&[{x, y} - diskPath[t, y0]]) < 1]]
```

```
In[142]:= collision[-3]
```

```
Out[142]= False
```

For $y_0 = 3$, the square and the disk will collide.

```
In[143]:= collision[3]
```

```
Out[143]= True
```

A picture confirms these answer (time goes vertically). The $y_0 = 3$ is shown in the right graphic.

```
In[144]:= motionPicture[y0_, opts___] :=
              Graphics3D[{
                  (* the moving square *)
                  {SurfaceColor[Hue[0], Hue[0], 2.3],
                   Table[Polygon[Table[Append[squarePath[t], t] +
                           {Cos[φ], Sin[φ], 0}, {φ, Pi/4, 7 Pi/4, Pi/2}]],
                                 {t, -3, 3, 6/30}]},
                  (* the moving disk *)
                  {SurfaceColor[Hue[0.12], Hue[0.12], 2.3],
                   Table[Polygon[Table[Append[diskPath[t, y0], t] +
                           {Cos[φ], Sin[φ], 0}, {φ, Pi/4, 9 Pi/4, Pi/12}]],
                                 {t, -3, 3, 6/30}]}} // N, opts]
```

```
In[145]:= Show[GraphicsArray[{motionPicture[-3, ViewPoint -> {4, 2, 1}],
                              motionPicture[+3, ViewPoint -> {0, 2, 1}]}]]
```

By not eliminating the variable *t*, we obtain the time intervals of collision.

```
In[146]:= Exists[{x, y}, Element[{t, x, y}, Reals], inSquare¿[{x, y}, t] &&
              (#.#&[{x, y} - diskPath[t, 3]]) < 1] // Resolve
```

$$Out[146]= 1 - \sqrt{6} < t < 1 - \sqrt{2} \mid\mid 1 + \sqrt{2} < t < 1 + \sqrt{6}$$

Using a symbolic y_0, we obtain conditions on y_0 such that a collision occurs.

```
In[147]:= Exists[{x, y}, Element[{t, x, y, y0}, Reals], inSquare¿[{x, y}, t] &&
              (#.#&[{x, y} - diskPath[t, y0]]) < 1] // Resolve
```

$$Out[147]= t \in Reals \&\& -2 - 2 t + t^2 < y0 < 2 - 2 t + t^2$$

For y0 > -3, a collision will never happen.

```
In[148]:= Exists[{t, x, y}, Element[{t, x, y, y0}, Reals], inSquare¿[{x, y}, t] &&
              (#.#&[{x, y} - diskPath[t, y0]]) < 1] // Resolve
```

```
Out[148]= y0 > -3
```

For problems in which all quantified variables appear linearly, the execution speed of `Resolve` will be much larger and it is possible to deal with examples involving more variables. The next example [818] deals with the conditions on the sum of the rows, the columns, the diagonal, and the antidiagonal of a matrix **M**. We assume all matrix elements $m_{i,j}$ to have possible values from the interval $\{-1, 1\}$.

```
In[149]:= dim = 3;
          M = Table[m[i, j], {i, dim}, {j, dim}];
```

This is the set of conditions.

```
In[151]:= eqs = And @@ Join[
            (* row and column sums *)
            MapIndexed[((Plus @@ #1) == rowSum[#2[[1]]])&, M, {1}],
            MapIndexed[((Plus @@ #1) == columnSum[#2[[1]]])&, Transpose[M], {1}],
            (* diagonals *)
            {Tr[M] == diagonalSum, Tr[Reverse /@ M] == antiDiagonalSum},
            (* matrix element conditions *)
            Map[-1 < # < 1&, Flatten[M]]]
```

```
Out[151]= m[1, 1] + m[1, 2] + m[1, 3] == rowSum[1] && m[2, 1] + m[2, 2] + m[2, 3] == rowSum[2] &&
          m[3, 1] + m[3, 2] + m[3, 3] == rowSum[3] && m[1, 1] + m[2, 1] + m[3, 1] == columnSum[1] &&
          m[1, 2] + m[2, 2] + m[3, 2] == columnSum[2] &&
          m[1, 3] + m[2, 3] + m[3, 3] == columnSum[3] &&
          m[1, 1] + m[2, 2] + m[3, 3] == diagonalSum &&
          m[1, 3] + m[2, 2] + m[3, 1] == antiDiagonalSum && -1 < m[1, 1] < 1 &&
          -1 < m[1, 2] < 1 && -1 < m[1, 3] < 1 && -1 < m[2, 1] < 1 && -1 < m[2, 2] < 1 &&
          -1 < m[2, 3] < 1 && -1 < m[3, 1] < 1 && -1 < m[3, 2] < 1 && -1 < m[3, 3] < 1
```

```
In[152]:= sumVariables = Join[Table[rowSum[k], {k, dim}],
                              Table[columnSum[k], {k, dim}],
                              {diagonalSum, antiDiagonalSum}];
```

```
In[153]:= (res = Exists[Evaluate[Flatten[M]],
                        Element[Join[Flatten[M], sumVariables], Reals],
                        eqs] // Resolve);
```

The result is quite large; it consists of a logical And with 36 elements.

```
In[154]:= {Head[res], Length[DeleteCases[res, _Element]]}
```

```
Out[154]= {And, 36}
```

Here is the first of the resulting conditions on the various sums.

```
In[155]:= resDCs = DeleteCases[res, _Element][[1]]
```

```
Out[155]= 1 + columnSum[1] ≤ 0 && -3 - columnSum[1] < 0 &&
          3 + antiDiagonalSum + diagonalSum + 2 columnSum[1] - columnSum[3] - 2 rowSum[2] ≤ 0 &&
          -9 - antiDiagonalSum - diagonalSum - 2 columnSum[1] + columnSum[3] + 2 rowSum[2] < 0 &&
          2 antiDiagonalSum - diagonalSum + columnSum[1] + columnSum[3] - rowSum[2] ≤ 0 &&
          -6 - 2 antiDiagonalSum + diagonalSum - columnSum[1] - columnSum[3] + rowSum[2] < 0 &&
          -6 + antiDiagonalSum - 2 diagonalSum - columnSum[1] - columnSum[3] + rowSum[2] < 0 &&
          -antiDiagonalSum + 2 diagonalSum + columnSum[1] + columnSum[3] - rowSum[2] ≤ 0 &&
          -3 + antiDiagonalSum + diagonalSum - columnSum[1] - columnSum[3] + rowSum[2] < 0 &&
          -3 - antiDiagonalSum - diagonalSum + columnSum[1] + columnSum[3] - rowSum[2] < 0 &&
          columnSum[1] + columnSum[2] + columnSum[3] - rowSum[1] - rowSum[2] - rowSum[3] == 0 &&
          -3 + antiDiagonalSum - 2 diagonalSum - columnSum[1] - columnSum[3] +
            rowSum[2] + 3 rowSum[3] < 0 && -3 - antiDiagonalSum + 2 diagonalSum +
            columnSum[1] + columnSum[3] - rowSum[2] - 3 rowSum[3] ≤ 0 &&
```

$$-3 + 2\,\text{antiDiagonalSum} - \text{diagonalSum} - 2\,\text{columnSum}[1] - 3\,\text{columnSum}[2] -$$
$$2\,\text{columnSum}[3] + 2\,\text{rowSum}[2] + 3\,\text{rowSum}[3] < 0\ \&\&$$
$$-3 - 2\,\text{antiDiagonalSum} + \text{diagonalSum} + 2\,\text{columnSum}[1] +$$
$$3\,\text{columnSum}[2] + 2\,\text{columnSum}[3] - 2\,\text{rowSum}[2] - 3\,\text{rowSum}[3] \leq 0$$

Many more problems can be solved using quantifier elimination; see [494], [250], [493], [495], [806], [1677], [1678], [1305], [337], [55], [804], [805], [148], [1564], and [150] for many nice examples. For details about quantifier elimination in *Mathematica*, see [1674]. We will end with a last example showing a straightforward (although esoteric) way to calculate the limit of the rational function $(2\,x^3 - 4\,x + 6)/(7\,x^3 - 5\,x + 9)$ as x tends to $+\infty$ via the following definition of the limit λ [68]:

$$\forall_{\varepsilon,\varepsilon>0}\left(\exists_{X,X>0}\left(\forall_{x,x>X\wedge\lambda\in\mathbb{R}}\ \lambda-\varepsilon < \frac{2\,x^3 - 4\,x + 6}{7\,x^3 - 5\,x + 9} < \lambda+\varepsilon\right)\right).$$

In[156]:=
```
ForAll[ε, ε > 0,
    Exists[X, X > 0,
        ForAll[x, x > X && Element[limit, Reals],
            limit - ε < (2x^3 - 4x + 6)/(7x^3 - 5x + 9) < limit + ε]]] //
                                                              Resolve
```

Out[156]= $limit == \dfrac{2}{7}$

1.3 Operations on Rational Functions

The commands `Numerator` and `Denominator` introduced at the beginning (Subsection 2.4.1 of the Programming volume [1735]) work for rational numbers and for rational functions, that is, for fractions of polynomials. Here is an example.

In[1]:= `ratio = (3 + 6 x + 6 x^2)/(5 y + 6 y^3)`

Out[1]= $\dfrac{3 + 6\,x + 6\,x^2}{5\,y + 6\,y^3}$

In[2]:= `Numerator[ratio]`

Out[2]= $3 + 6\,x + 6\,x^2$

In[3]:= `Denominator[ratio]`

Out[3]= $5\,y + 6\,y^3$

The parts of a product that belong to the numerator and the parts that belong to the denominator are determined by the sign of the associated exponents after transformations of the form `1/k^(-1)` $\to$ `k^1`. Here is an expression that is a product of ten factors. After evaluation, eight have a positive exponent and four have a negative exponent. Some of the negative exponent terms are formatted in the denominator. (Be aware that `Exp[expr]` is rewritten is `Power[E, expr]` and the explicit formatting depends on *expr*.)

In[4]:= `expr = a b^2 c^-2 d^(4/3) e^-(5/6) 1/f^(-12/13) g^h i^-j 1/k^-1 Exp[-E^2]`

Out[4]= $\dfrac{a\,b^2\,d^{4/3}\,e^{-e^2}\,f^{12/13}\,g^h\,i^{-j}\,k^1}{c^2\,e^{5/6}}$

In[5]:= `{Numerator[expr], Denominator[expr]}`

Out[5]= $\left\{a\,b^2\,d^{4/3}\,f^{12/13}\,g^h\,k^1,\ c^2\,e^{5/6}\,e^{e^2}\,i^j\right\}$

Here is a nested fraction.

In[6]:= **nestedFraction = (a/(b + 1) + 2)/(c/(d + 3) + 4)**

Out[6]= $\dfrac{2 + \frac{a}{1+b}}{4 + \frac{c}{3+d}}$

For nested fractions, the functions Numerator and Denominator take into account only the "outermost" structure.

In[7]:= **{Numerator[#], Denominator[#]}&[nestedFraction]**

Out[7]= $\left\{2 + \dfrac{a}{1 + b},\ 4 + \dfrac{c}{3 + d}\right\}$

For fractions, the command Expand, which multiplies out polynomials, is divided into four parts to facilitate working on the numerators and denominators.

Expand[*rationalFunction*]

 multiplies out only the numerator of the *rationalFunction*, and divides all resulting terms by the (unchanged) denominator.

ExpandNumerator[*rationalFunction*]

 multiplies out only the numerator of the *rationalFunction*, and divides the result as a single expression by the (unchanged) denominator.

ExpandDenominator[*rationalFunction*]

 multiplies out only the denominator of the *rationalFunction*.

ExpandAll[*rationalFunction*]

 multiplies out the numerator and denominator of *rationalFunction*, and divides all resulting terms.

We now look at the effect of these four commands on the sum of two ratios of polynomials.

In[8]:= **ratio = (1 + 7 y^3)/(2 + 8 x^3)^2 + (1 + 6 x)^3/(1 - 4 y)^2**

Out[8]= $\dfrac{(1 + 6 x)^3}{(1 - 4 y)^2} + \dfrac{1 + 7 y^3}{(2 + 8 x^3)^2}$

Except for the lexicographic reordering of the two partial sums, *Mathematica* did not do anything nontrivial to this input automatically.

Now, all numerators are multiplied out, and all resulting parts are individually divided.

In[9]:= **Expand[ratio]**

Out[9]= $\dfrac{1}{(2 + 8 x^3)^2} + \dfrac{1}{(1 - 4 y)^2} + \dfrac{18 x}{(1 - 4 y)^2} + \dfrac{108 x^2}{(1 - 4 y)^2} + \dfrac{216 x^3}{(1 - 4 y)^2} + \dfrac{7 y^3}{(2 + 8 x^3)^2}$

ExpandNumerator also multiplies out, but does not divide the terms individually.

In[10]:= **ExpandNumerator[ratio]**

Out[10]= $\dfrac{1 + 18 x + 108 x^2 + 216 x^3}{(1 - 4 y)^2} + \dfrac{1 + 7 y^3}{(2 + 8 x^3)^2}$

With ExpandDenominator, the numerator remains unchanged, and only the denominator is multiplied out.

In[11]:= **ExpandDenominator[ratio]**

Out[11]= $\dfrac{(1 + 6 x)^3}{1 - 8 y + 16 y^2} + \dfrac{1 + 7 y^3}{4 + 32 x^3 + 64 x^6}$

Finally, we multiply everything out. ExpandAll typically produces the largest expressions. Now, the numerator is multiplied out and each of its terms is written over the expanded form of the denominator.

In[12]:= **ExpandAll[ratio]**

Out[12]= $\dfrac{1}{4 + 32\,x^3 + 64\,x^6} + \dfrac{7\,y^3}{4 + 32\,x^3 + 64\,x^6} +$

$\dfrac{1}{1 - 8\,y + 16\,y^2} + \dfrac{18\,x}{1 - 8\,y + 16\,y^2} + \dfrac{108\,x^2}{1 - 8\,y + 16\,y^2} + \dfrac{216\,x^3}{1 - 8\,y + 16\,y^2}$

In the following example, the same happens. Be aware that no common factors are cancelled.

In[13]:= `ExpandAll[(1 - x^4)/(1 + x^2)^2]`

Out[13]= $\dfrac{1}{1 + 2\,x^2 + x^4} - \dfrac{x^4}{1 + 2\,x^2 + x^4}$

In nested fractions, `ExpandAll` works again only the "outermost" structure.

In[14]:= `ExpandAll[nestedFraction]`

Out[14]= $\dfrac{2}{4 + \frac{c}{3+d}} + \dfrac{a}{(1 + b)\,\left(4 + \frac{c}{3+d}\right)}$

Mapping `ExpandAll` to all parts allows us to expand all parts.

In[15]:= `(* show all steps *)`
`FixedPointList[MapAll[ExpandAll, #]&, nestedFraction]`

Out[16]= $\left\{ \dfrac{2 + \frac{a}{1+b}}{4 + \frac{c}{3+d}},\ \dfrac{2}{4 + \frac{c}{3+d}} + \dfrac{a}{(1 + b)\,\left(4 + \frac{c}{3+d}\right)}, \right.$

$\left. \dfrac{2}{4 + \frac{c}{3+d}} + \dfrac{a}{4 + 4\,b + \frac{c}{3+d} + \frac{b\,c}{3+d}},\ \dfrac{2}{4 + \frac{c}{3+d}} + \dfrac{a}{4 + 4\,b + \frac{c}{3+d} + \frac{b\,c}{3+d}} \right\}$

Be aware that there is a difference between `Expand //@ expr` and `ExpandAll[expr]`. `ExpandAll` does not automatically recursively expanded in the inner levels of expressions with `Hold`-like attributes.

In[17]:= `{Expand //@ #, ExpandAll[#]}&[(α + β)^2 + Hold[(α + β)^2]]`

Out[17]= $\{\alpha^2 + 2\,\alpha\,\beta + \beta^2 + \text{Hold}[\text{Expand}[\text{Expand}[\text{Expand}[\alpha] + \text{Expand}[\beta]]^{\text{Expand}[2]}]],$
$\alpha^2 + 2\,\alpha\,\beta + \beta^2 + \text{Hold}[(\alpha + \beta)^2]\}$

Here is a small application of `ExpandAll`. Forming a matrix out of the first d rows of the Pascal triangle by right justifying them and filling in zeros into the empty spaces yields a matrix whose eigenvalues are modulo sign powers of the golden ratio. The next input is a one-liner that for a given positive d, returns the eigenvalues in the form $\pm\phi^k$ [1396]. The function `Eigenvalues` returns the eigenvalues as rational functions in $5^{1/2}$. We first approximate k numerically and then confirm the value of k using exact numerics.

In[18]:= `pascalMatrixEigenvalues[d_Integer?Positive] :=`
`    Function[ev, Sign[ev] GoldenRatio^If[`
`                  (* faster than cancelling *) ExpandAll[`
`                    Abs[ev]/((Sqrt[5] + 1)/2)^#] == 1, #]&[`
`                      Round[N[Log[GoldenRatio, Abs[ev]],`
`                             $MachinePrecision + 1]]]] /@`
`          (* the exact eigenvalues of the Pascal matrix *)`
`          Eigenvalues[Table[Binomial[n, k], {n, 0, d - 1}, {k, d - 1, 0, -1}]]`

Here is an example.

In[19]:= `pascalMatrixEigenvalues[16] // TraditionalForm`

Out[19]//TraditionalForm=

$\left\{\phi^{15}, -\phi^{13}, \phi^{11}, -\phi^9, \phi^7, -\phi^5, \phi^3, -\phi, \dfrac{1}{\phi}, -\dfrac{1}{\phi^3}, \dfrac{1}{\phi^5}, -\dfrac{1}{\phi^7}, \dfrac{1}{\phi^9}, -\dfrac{1}{\phi^{11}}, \dfrac{1}{\phi^{13}}, -\dfrac{1}{\phi^{15}}\right\}$

Because `ExpandAll` can greatly expand (the size of) an expression, sometimes we want to multiply out only certain parts of a rational expression. This can be accomplished with `ExpandAll` using its second argument.

> ExpandAll [*rationalFunction, relevantPattern*]
>
> multiplies out all those parts of the numerator and denominator of the *rationalFunction* that contain the pattern *relevantPattern*.

In the following example, only those terms (in either the numerator or the denominator) that contain z are multiplied out.

In[20]:= **ExpandAll[(1 + x^3)^3/(2 + z)^4 + (3 + z)^2/(r + 5)^2, z]**

Out[20]= $\dfrac{9}{(5+r)^2} + \dfrac{6z}{(5+r)^2} + \dfrac{z^2}{(5+r)^2} + \dfrac{(1+x^3)^3}{16+32z+24z^2+8z^3+z^4}$

There is still another important command for manipulating rational functions.

> Apart [*rationalFunction*]
>
> finds a decomposition of *rationalFunction* into partial fractions.
>
> Apart [*rationalFunction, var*]
>
> finds a decomposition of *rationalFunction* into partial fractions with respect to the variable *var*.

Apart leads to the simplest possible denominators, exactly those that are needed for the textbook integration of rational expressions. Here are three examples. The denominators are irreducible polynomials that are potentially raised to a power.

In[21]:= **Apart[(1 + 5 x + 6 x^2)/(1 - x^8)]**

Out[21]= $-\dfrac{3}{2(-1+x)} + \dfrac{1}{4(1+x)} + \dfrac{5(-1+x)}{4(1+x^2)} + \dfrac{1+5x+6x^2}{2(1+x^4)}$

In[22]:= **Apart[1/((x - 1)^12 (x - 100))]**

Out[22]= $\dfrac{1}{88638487171612928065801(-100+x)} - \dfrac{1}{99(-1+x)^{12}} - \dfrac{1}{9801(-1+x)^{11}} -$

$\dfrac{1}{970299(-1+x)^{10}} - \dfrac{1}{96059601(-1+x)^9} - \dfrac{1}{9509900499(-1+x)^8} -$

$\dfrac{1}{941480149401(-1+x)^7} - \dfrac{1}{93206534790699(-1+x)^6} - \dfrac{1}{9227446944279201(-1+x)^5} -$

$\dfrac{1}{913517247483640899(-1+x)^4} - \dfrac{1}{90438207500880449001(-1+x)^3} -$

$\dfrac{1}{8953382542587164451099(-1+x)^2} - \dfrac{1}{88638487171612928065801(-1+x)}$

In[23]:= **Apart[1/(1 - x) 1/(1 - x^2) 1/(1 - x^3)]**

Out[23]= $-\dfrac{1}{6(-1+x)^3} + \dfrac{1}{4(-1+x)^2} - \dfrac{17}{72(-1+x)} + \dfrac{1}{8(1+x)} + \dfrac{2+x}{9(1+x+x^2)}$

In the following example, the roots of the denominator polynomials are symbolic coefficients.

In[24]:= **Apart[(x - a1) (x - a2) (x - a3)/((x - b1) (x - b2) (x - b3) (x - b4)), x]**

Out[24]= $\dfrac{-a1\,a2\,a3 + a1\,a2\,b1 + a1\,a3\,b1 + a2\,a3\,b1 - a1\,b1^2 - a2\,b1^2 - a3\,b1^2 + b1^3}{(b1-b2)(b1-b3)(b1-b4)(-b1+x)} +$

$\dfrac{a1\,a2\,a3 - a1\,a2\,b2 - a1\,a3\,b2 - a2\,a3\,b2 + a1\,b2^2 + a2\,b2^2 + a3\,b2^2 - b2^3}{(b1-b2)(b2-b3)(b2-b4)(-b2+x)} -$

$\dfrac{a1\,a2\,a3 - a1\,a2\,b3 - a1\,a3\,b3 - a2\,a3\,b3 + a1\,b3^2 + a2\,b3^2 + a3\,b3^2 - b3^3}{(b1-b3)(b2-b3)(b3-b4)(-b3+x)} +$

$\dfrac{a1\,a2\,a3 - a1\,a2\,b4 - a1\,a3\,b4 - a2\,a3\,b4 + a1\,b4^2 + a2\,b4^2 + a3\,b4^2 - b4^3}{(b1-b4)(b2-b4)(b3-b4)(-b4+x)}$

So far, we have only expanded rational functions. We now look at making them (in most cases) smaller. For this the relevant commands are `Together`, `Cancel`, and `Factor`.

`Together [rationalFunction]`

> writes *rationalFunction* over a common denominator.

`Cancel [rationalFunction]`

> removes common factors from the numerator and denominator of *rationalFunction*.

`Factor [rationalFunction]`

> factors the numerator and denominator of *rationalFunction*, so that the final result is a single fraction.

Here are some examples with the rational function `ratio` defined earlier.

In[25]:= **ratio**

Out[25]= $\dfrac{(1 + 6\,x)^3}{(1 - 4\,y)^2} + \dfrac{1 + 7\,y^3}{(2 + 8\,x^3)^2}$

Over one denominator, we get the following result.

In[26]:= **Together[ratio]**

Out[26]= $\left(5 + 72\,x + 432\,x^2 + 896\,x^3 + 576\,x^4 + 3456\,x^5 + 6976\,x^6 + 1152\,x^7 + 6912\,x^8 + 13824\,x^9 - 8\,y + 16\,y^2 + 7\,y^3 - 56\,y^4 + 112\,y^5\right) \big/ \left(4\,(1 + 4\,x^3)^2\,(-1 + 4\,y)^2\right)$

Along with replacement rules, `Together` is a *very* important command for effective simplification of expressions. We now look at an example of a Cauchy determinant [1074]. Consider an $n \times n$ matrix with elements $1/(a_i + b_j)$.

$$\det\left(\frac{1}{a_i + b_j}\right) = \frac{\displaystyle\prod_{\substack{i,j=1 \\ i<j}}^{n} (a_i - a_j)(b_i - b_j)}{\displaystyle\prod_{i,j=1}^{n} (a_i + b_j)}$$

In[27]:= **CauchyDetIdentity[dim_] :=**
    Det[Outer[1/(#1 + #2)&, Array[a, dim], Array[b, dim]]] -
    Product[(a[i] - a[j])(b[i] - b[j]), {j, dim}, {i, j + 1, dim}]/
    Product[(a[i] + b[j]), {j, dim}, {i, dim}]

In[28]:= **Timing[Together[CauchyDetIdentity[4]]]**

Out[28]= {14.08 Second, 0}

Here is another example of the power of `Together`. In [909], the following identity for derivatives was given:

$$\sum_{k=0}^{n} \frac{1}{k+\alpha} \binom{n}{k} \frac{\partial^k f(x)^{-k-\alpha}}{\partial x^k} \frac{\partial^{n-k} f(x)^{k+\alpha}}{\partial x^{n-k}} = 0, \ n \in \mathbb{N}^+, \ \alpha \in \mathbb{C}.$$

We will check this identity explicitly for $n \le 12$. We measure the time `Together` needs as well as the leaf count of the expression.

In[29]:= **differentiationIdentity[n_] :=**
        Sum[1/(α + k) Binomial[n, k] D[f[x]^(-α - k), {x, k}]*
                        D[f[x]^(+α + k), {x, n - k}], {k, 0, n}]

```
In[30]:= Module[{sum},
         Table[sum = differentiationIdentity[n];
               {n, LeafCount[sum], Timing[Together[sum]]}, {n, 12}]]
Out[30]= {{1, 29, {0. Second, 0}}, {2, 114, {0. Second, 0}}, {3, 263, {0.02 Second, 0}},
         {4, 514, {0.02 Second, 0}}, {5, 907, {0.04 Second, 0}}, {6, 1556, {0.07 Second, 0}},
         {7, 2525, {0.14 Second, 0}}, {8, 4014, {0.24 Second, 0}},
         {9, 6177, {0.38 Second, 0}}, {10, 9346, {0.57 Second, 0}},
         {11, 13807, {1. Second, 0}}, {12, 20156, {1.68 Second, 0}}}
```

In the following simple identity, the left outer summation is carried out over all permutations S_n of the integers $1, ..., n$ [1167].

$$\sum_{\sigma \in S_n} \prod_{j=1}^{n} \frac{1}{\sum_{k=1}^{j} x_{\sigma(k)}} = \frac{1}{\prod_{j=1}^{n} x_j}$$

Checking this identity explicitly for small n is quite difficult (the identity can be easily checked by induction).

```
In[31]:= Table[Timing @ Together[Plus @@ (* form products of inverse sums *)
               ((Times @@ (1/(Plus @@@ NestList[Most, #, Length[#] - 1])))& /@
                         (* the permutations *) Permutations[Array[x, n]])],
               {n, 6}]
Out[31]= {{0. Second, 1/x[1]}, {0. Second, 1/(x[1] x[2])}, {0.01 Second, 1/(x[1] x[2] x[3])},
         {0.04 Second, 1/(x[1] x[2] x[3] x[4])}, {0.67 Second, 1/(x[1] x[2] x[3] x[4] x[5])},
         {11.81 Second, 1/(x[1] x[2] x[3] x[4] x[5] x[6])}}
```

Here is another stress test for `Together`: Consider the recursive definition (two term Gale–Robinson sequence [794], [685], [651], [43]) $y_k = (\alpha\, y_{k+p-n}\, y_{k-p} + \beta\, y_{k+q-n}\, y_{k-q})/y_{k-n}$ with $p \geq 1$, $q \geq p$, $n \geq 2q$. While the definition suggests that y_k is a rational function of the $n - 1$ initial values $y_j = Y_j$, $j = 1, ..., n - 1$, the y_k turn out to be Laurent polynomials in the Y_j. The next input implements a function to set up definitions for y_k for given integers p, q, and n. To write the y_k as a Laurent polynomial, we use `Together` to cancel common factors.

```
In[32]:= makeTwoTermGaleRobinsonDefinitions[{p_, q_, n_}, {η_, α_, β_, Y_}] :=
         ((* recursive definition *)
          η[k_] := η[k] = Expand[Together[(α η[k + p - n] η[k - p] +
                                  β η[k + q - n] η[k - q])/η[k - n]]];
          (* initial conditions *)
          η[k_?((0 <= # <= n - 1)&)] := η[k] = Y[k]) /;
          (* conditions *)
          (And @@ Positive[{p, q, n}]) && (And @@ (IntegerQ /@ {p, q, n})) &&
          q > p && n >= 2q
```

Here are the first seven y_k for $p = 1$, $q = 2$, and $n = 5$.

```
In[33]:= makeTwoTermGaleRobinsonDefinitions[{1, 2, 5}, {y, α, β, Y}]

In[34]:= Table[y[k], {k, 7}] // Together
```

Out[34]= $\left\{ Y[1], Y[2], Y[3], Y[4], \dfrac{\beta\, Y[2]\, Y[3] + \alpha\, Y[1]\, Y[4]}{Y[0]}, \right.$

$\dfrac{\alpha\, \beta\, Y[2]^2\, Y[3] + \alpha^2\, Y[1]\, Y[2]\, Y[4] + \beta\, Y[0]\, Y[3]\, Y[4]}{Y[0]\, Y[1]},$

$\dfrac{1}{Y[0]\, Y[1]\, Y[2]}\, (\alpha^2\, \beta\, Y[2]^2\, Y[3]^2 + \alpha^3\, Y[1]\, Y[2]\, Y[3]\, Y[4] +$

$\left. \beta^2\, Y[1]\, Y[2]\, Y[3]\, Y[4] + \alpha\, \beta\, Y[0]\, Y[3]^2\, Y[4] + \alpha\, \beta\, Y[1]^2\, Y[4]^2) \right\}$

Because higher y_k grow dramatically in size, we define a function `LaurentPolynomialQ` that checks if its argument is a Laurent polynomial (meaning no sums appear in denominators) instead of displaying them.

In[35]:= `LaurentPolynomialQ[expr_] := FreeQ[Head /@ Denominator[`
         `If[Head[#] === Plus, List @@ #, {#}]&[Expand[expr]]], Plus]`

The next input quickly checks whether the first 18 of the y_k are Laurent polynomials (y_{18} has more than 1000 terms).

In[36]:= `Table[{Length[y[k]], LaurentPolynomialQ[y[k]]}, {k, 18}]`

Out[36]= `{{1, True}, {1, True}, {1, True}, {1, True}, {2, True}, {3, True}, {5, True},`
         `{9, True}, {18, True}, {29, True}, {52, True}, {84, True}, {141, True},`
         `{218, True}, {341, True}, {496, True}, {748, True}, {1057, True}}`

Here is a hidden zero (the expression `zero[s]` is 0 for almost all complex s) [310], [1036], [149]. This time the expression is not a clean rational function, but it contains rational powers.

In[37]:= `zero[s_] = With[{x = s^(1/(s - 1)), y = s^(s/(s - 1))}, x^y - y^x]`

Out[37]= $\left(s^{\frac{1}{-1+s}}\right)^{s^{\frac{s}{-1+s}}} - \left(s^{\frac{s}{-1+s}}\right)^{s^{\frac{1}{-1+s}}}$

For certain rational values of s, the expression `zero[s]` does not autoevaluate to 0.

In[38]:= `zero[2/5]`

Out[38]= $-\left(\dfrac{5}{2}\right)^{\frac{5}{3}\left(\frac{5}{2}\right)^{2/3}} + \left(\dfrac{5}{2}\right)^{\left(\frac{5}{2}\right)^{2/3} + \frac{1}{3}\, 2^{1/3}\, 5^{2/3}}$

Applying `Together` yields 0.

In[39]:= `Together[%]`

Out[39]= `0`

`Together` is a very useful function for simplification. It is often faster and sometimes even more powerful than `Simplify`.

Now let us deal with the function `Cancel`. Common factors in numerators and denominators of rational functions are not removed, except when they appear already explicitly as factors.

In[40]:= `Expand[(x^2 - 1) (x^2 + 1)]/((x^2 - 1) (x^2 + 1))`

Out[40]= $\dfrac{-1 + x^4}{(-1 + x^2)\, (1 + x^2)}$

However, using `Cancel`, we can force the cancellation.

In[41]:= `Cancel[%]`

Out[41]= `1`

Common factors are not canceled without the explicit use of `Cancel` if they do not appear explicitly factored already. The two expressions above actually represent different *Mathematica* functions. The original is not even defined at $x = \pm 1$.

In[42]:= `%% /. x -> 1`

> Power::infy : Infinite expression $\frac{1}{0}$ encountered. More…

> ∞::indet : Indeterminate expression 0 ComplexInfinity encountered. More…

Out[42]= Indeterminate

Because this is a removable singularity, we are usually not too concerned about it. However, there is a theoretical difference between the functions $(x^4 - 1)/((x^2 - 1)(x^2 + 1))$ and 1; the first is from a computational point of view not defined at $x = 1$. But within *Mathematica*'s philosophy to carry out all operations that are correct for generic values of indeterminates, it is correct to cancel common factors. But because of efficiency reasons, this is not done automatically.

In[43]:= `Factor @ Cancel[(x^24 - y^24)/(x + y)]`

Out[43]= $(x - y)(x^2 + y^2)(x^2 - x y + y^2)(x^2 + x y + y^2)(x^4 + y^4)(x^4 - x^2 y^2 + y^4)(x^8 - x^4 y^4 + y^8)$

Finally, we give an example of the use of `Factor` with a rational function. Here, we get a single fraction with a factorized denominator and a factorized numerator. Terms that can be canceled from the numerator and denominator are removed.

In[44]:= `Factor[(1 - x^4)^2/(1 + x^8)^3 - (1 + x^8)^3/(1 - x^4)^2]`

Out[44]= $-\dfrac{x^4 (2 + 2 x^4 + 3 x^{12} + x^{20})(2 - 2 x^4 + 4 x^8 + 3 x^{16} + x^{24})}{(-1 + x)^2 (1 + x)^2 (1 + x^2)^2 (1 + x^8)^3}$

In the following example, the factor $1 + x^2$, common to numerator and denominator, is cancelled.

In[45]:= `Cancel[(1 - x^4)/(1 + x^2)^2]`

Out[45]= $\dfrac{1 - x^2}{1 + x^2}$

`Cancel` will not cancel common algebraic factors without being told.

In[46]:= `Expand[(x - Sqrt[2])(x - Sqrt[6])(x - Sqrt[12])]/(x - Sqrt[2]) // Cancel`

Out[46]= $\dfrac{12 - 6\sqrt{2}\, x - 2\sqrt{3}\, x - 2\sqrt{6}\, x + \sqrt{2}\, x^2 + 2\sqrt{3}\, x^2 + \sqrt{6}\, x^2 - x^3}{\sqrt{2} - x}$

Similar to `Factor`, `Cancel` has the option `Extension`.

In[47]:= `Options[Cancel]`

Out[47]= {Extension → None, Modulus → 0, Trig → False}

Using the `Extension` option (the possible settings are the same as in `Factor`), we can cancel the common factor.

In[48]:= `Expand[(x - Sqrt[2])(x - Sqrt[6])(x - Sqrt[12])]/(x - Sqrt[2]) //`
         `    Cancel[#, Extension -> Automatic]&`

Out[48]= $6\sqrt{2} - 2\sqrt{3}\, x - \sqrt{6}\, x + x^2$

Let us give a simple graphics-generating application of `Together`. We will visualize the interior of the well-known modular domain $\mathcal{FD} = \{z \mid -1/2 < \text{Re}(z) < 1/2 \wedge |z| > 1\}$ under repeated applications of the maps $z \to z \pm 1$ and $z \to -1/z$ (for the underlying mathematics, see, for instance, [608], [736], and [252]; for a detailed *Mathematica* implementation, see [1227]). Starting with the identity map $z \to z$, we will recursively generate all maps resulting from applying the translation and inversion operations. The first occurrences of the maps will be collected in lists *maps*[*k*]. `Together` is used to canonicalize the resulting rational functions of the form $(a + b z)/(c + d z)$.

In[49]:= `o = 12;`
         `maps[0] = {z};`

```
doneQ[z] = True;
Do[maps[k] = Complement[#, (* select new maps *)Select[#, doneQ]]&[
              Union[(* canonicalize *) Together[(* form next maps *)
              Join[(# + 1)& /@ #, (# - 1)& /@ #, (-1/#)& /@ #]]]&[
                                                   maps[k - 1]]];
     (doneQ[#] = True)& /@ maps[k], {k, o}]
```

The following list shows the number of newly occurring maps after k steps.

```
In[53]:= Table[{k, Length[maps[k]]}, {k, 0, o}]
Out[53]= {{0, 1}, {1, 3}, {2, 6}, {3, 10}, {4, 18}, {5, 28}, {6, 44},
          {7, 70}, {8, 112}, {9, 180}, {10, 290}, {11, 468}, {12, 756}}
```

Here are a few of the maps from the last calculated step.

```
In[54]:= {Take[maps[o], 5], Take[maps[o], -5]}
```

$$\text{Out[54]= } \left\{\left\{\frac{1}{-11-z}, \frac{1}{11-z}, -12+z, \frac{9-z}{-10+z}, \frac{-11+z}{-10+z}\right\},\right.$$
$$\left.\left\{\frac{-13+30\,z}{-3+7\,z}, \frac{-7+30\,z}{-3+13\,z}, \frac{-9+31\,z}{-2+7\,z}, \frac{-7+31\,z}{-2+9\,z}, \frac{-10+33\,z}{-3+10\,z}\right\}\right\}$$

We approximate the region $\mathcal{FD}$ as a polygon through the list $\mathcal{FD}$PolygonPoints (instead of letting the region extend vertically to ∞ we use a large upper vertex).

```
In[55]:= FDPolygonPoints =
         With[{pp = 20, ε = 10^-12}, Polygon[Join @@
         {(* left vertical boundary *)
         Reverse @ Table[1/2 + I Tan[ArcTan[(1 - t) Sqrt[3]/2] + t Pi/2],
                        {t, 0, 1 - ε, (1 - ε)/pp}],
         (* circular part of the boundary *)
                     Table[Exp[I Pi/3 (1 + t)], {t, 0, 1, 1/pp}],
         (* right vertical boundary *)
                     Table[-1/2 + I Tan[ArcTan[(1 - t) Sqrt[3]/2] + t Pi/2],
                        {t, 0, 1 - ε, (1 - ε)/pp}]} // N]];
```

Using now the maps from *maps*[k] to map the fundamental domain and coloring successive levels black and white, we obtain the following graphic. (We restrict the visible plot range to the neighborhood of the origin.) The graphic nicely shows how the images of the fundamental domain tessellate the upper half-plane.

```
In[56]:= mapFDPolygonPoints[f_] :=
         With[{f = Function @@ {z, f}}, Map[{Re[#], Im[#]}&,
             Map[f, FDPolygonPoints, {2}], {2}]]
```

```
In[57]:= FDG = Show[Graphics[{
         Table[{GrayLevel[Mod[k, 2]], mapFDPolygonPoints[#]& /@ maps[k]},
              {k, 0, o}]}], PlotRange -> {{-1, 1}, {0, 1}}, Frame -> True,
                   AspectRatio -> Automatic, FrameTicks -> False]
```

Applying a logarithmic transformation shows better the images of $\mathcal{FD}$ near the real axis (this time we use a random coloring). We also show a version resulting from mapping from the strip $-1/2 < \mathrm{Re}(z) < 1/2$ to the unit disk and from the x-unit interval to the circle.

```
In[58]:= (* remove far outside points *)
    FDGR[ξ_] := FDGR[ξ] = FDG /. Polygon[l_] :> Polygon[l /.
     {x_, y_} :> {Which[x < -ξ, -ξ, -ξ <= x <= ξ, x, x > ξ, ξ], y}];

In[60]:= {Show[#1], Show[GraphicsArray[{#2, #3}]]}& @@
    Block[{$DisplayFunction = Identity},
      {(* vertically stretched version *)
       Show[FDGR[1] /. Polygon[l_] :> {Hue[Random[]],
        (* stretch vertically *) Polygon[{#1, Log[#2]}& @@@ l]},
          PlotRange -> {All, {1, -5}}, AspectRatio -> 1/2],
        (* strip to unit disk *) Off[General::unfl];
       Show[FDGR[1/2] /. Polygon[l_] :> {(* map to unit circle *)
        Polygon[Exp[-2Pi #2]{Cos[2Pi #1], Sin[2Pi #1]}& @@@ l]},
          PlotRange -> All, AspectRatio -> 1],
        (* unit interval to unit disk *) On[General::unfl];
       Show[FDGR[1] /. Polygon[l_] :> Polygon[l /. (* map radially *)
        {x_, y_} :> {x, Which[y < -1, -1, -1 <= y <= 1, y, y > 1, 1]}] /.
        Polygon[l_] :> {Polygon[-Log[#2] {Cos[Pi #1], Sin[Pi #1]}& @@@ l]},
        PlotRange -> 4 {{-1, 1}, {-1, 1}}]}]
```

1.4 Operations on Trigonometric Expressions

Trigonometric functions are treated by `TrigFactor` and `TrigExpand` as discussed in Chapter 3 of the Programming volume [1735]. Thus, for example, `TrigFactor[`*trigExpression*`]` writes powers of the trigonometric functions in *trigExpression* as sums of trigonometric functions with sum arguments, and `Trig⌃ Expand[`*trigExpression*`]` writes trigonometric functions in *trigExpression* with sum arguments as sums of powers of trigonometric functions with simple variables.

Here is an example of multiplying out an expression.

In[1]:= **TrigExpand[Sin[2 x + 3 y]]**

Out[1]= $2 \, Cos[x] \, Cos[y]^3 \, Sin[x] + 3 \, Cos[x]^2 \, Cos[y]^2 \, Sin[y] - 3 \, Cos[y]^2 \, Sin[x]^2 \, Sin[y] -$
$6 \, Cos[x] \, Cos[y] \, Sin[x] \, Sin[y]^2 - Cos[x]^2 \, Sin[y]^3 + Sin[x]^2 \, Sin[y]^3$

We can factor to get back to our original expression.

In[2]:= **TrigFactor[%]**

Out[2]= $Sin[2 x + 3 y]$

The arguments of the trigonometric functions must have heads `Plus` or `Times` or `Power` (with an integer second argument) to allow for an expansion.

In[3]:= **TrigExpand[Sin[(x + y)^2]]**

Out[3]= $Sin[(x + y)^2]$

Trigonometric expressions will not be automatically simplified. This example remains unchanged.

In[4]:= **Sin[x]^2 + Cos[x]^2**

Out[4]= $Cos[x]^2 + Sin[x]^2$

We can force the simplification with `Simplify`.

In[5]:= **Simplify[%]**

Out[5]= 1

In this case, `TrigFactor` and `TrigExpand` would have led to the same result.

In[6]:= **{TrigExpand[#], TrigFactor[#]}& @ (Sin[x]^2 + Cos[x]^2)**

Out[6]= $\{1, 1\}$

In Chapter 2 of the Programming volume [1735], we have already seen that *trigFunction*[*inverseTrigFunction*[*var*]] will be automatically simplified.

In[7]:= **Cos[ArcSin[z]]**

Out[7]= $\sqrt{1 - z^2}$

However, frequently we have to deal with complicated expressions with a similar structure, such as the $\cos(5 \arcsin(x))$ example.

In[8]:= **Cos[5 ArcSin[z]]**

Out[8]= $Cos[5 \, ArcSin[z]]$

In[9]:= **TrigExpand[%]**

Out[9]= $5 \, z^4 \, \sqrt{1 - z^2} - 10 \, z^2 \, (1 - z^2)^{3/2} + (1 - z^2)^{5/2}$

`TrigExpand` can be used to generate the well-known addition formulas for trigonometric functions.

```
In[10]:= { (# == TrigExpand[#]) & [Sin[x + y]],
          (# == TrigExpand[#]) & [Tan[x + y]] }
```

$$\text{Out[10]= } \left\{ Sin[x + y] == Cos[y]\, Sin[x] + Cos[x]\, Sin[y], \right.$$

$$\left. Tan[x + y] == \frac{Cos[y]\, Sin[x]}{Cos[x]\, Cos[y] - Sin[x]\, Sin[y]} + \frac{Cos[x]\, Sin[y]}{Cos[x]\, Cos[y] - Sin[x]\, Sin[y]} \right\}$$

It would be nice to have the addition theorem for $\tan(x + y)$ containing only the functions $\tan(x)$, $\tan(y)$, and $\tan(x + y)$ and not sin and cos. Using the above-discussed function `GroebnerBasis`, it is straightforward to implement a little routine that generates an addition formula of this form.

```
In[11]:= makeAdditionTheorem[f_] :=
  Module[{trigFunction, vars, repRules, definitions,
          addFormula, toKeep, toEliminate, gb, revRepRules},
    (* the trigonometric function under consideration *)
    trigFunction = Head[f];
    (* the variables *)
    vars = Cases[f, _Symbol, {-1}];
    (* introduce dummies to avoid autoevaluation of 1/function *)
    repRules = {Sin -> sin, Cos -> cos, Tan -> tan,
                Cot -> cot, Sec -> sec, Csc -> csc};
    (* give dummies the same properties as the
       corresponding trigonometric functions;
       write identities in polynomial form *)
    definitions = Flatten[Numerator[Together[
      {sin[#]^2 + cos[#]^2 - 1,
       tan[#] - sin[#]/cos[#], cot[#] - cos[#]/sin[#],
       sec[#] - 1/cos[#], csc[#] - 1/sin[#]}]]& /@ vars];
    (* the addition formula *)
    addFormula = Numerator[Together[
      (f /. repRules) - (TrigExpand[f] /. repRules)]];
    (* the dummy trigonometric functions to keep *)
    toKeep = Flatten[{trigFunction /@ vars, f} /. repRules];
    (* the dummy trigonometric functions to eliminate *)
    toEliminate =
    Complement[Flatten[{sin[#], cos[#], tan[#],
                        cot[#], sec[#], csc[#]}& /@ vars], toKeep];
    (* do the elimination *)
    gb = GroebnerBasis[Join[{addFormula}, definitions],
                       toKeep, toEliminate,
                       MonomialOrder -> EliminationOrder];
    (* simplify result *)
    res = Simplify[First[gb]];
    (* replace dummies by original trigonometric functions *)
    revRepRules = Reverse /@ repRules;
    (* return held result to avoid autoevaluation *)
    HoldForm @@ {res /. revRepRules}]
```

Here are the addition formulas for the six trigonometric functions that contain only the function itself.

```
In[12]:= makeAdditionTheorem[#[x + y]]& /@
                {Sin, Cos, Tan, Cot, Sec, Csc}
```

$$\text{Out[12]= } \left\{ Sin[x]^4 + (Sin[y]^2 - Sin[x + y]^2)^2 + \right.$$

$$Sin[x]^2 \,(-2\, Sin[x + y]^2 + Sin[y]^2 \,(-2 + 4\, Sin[x + y]^2)),$$

$$1 - Cos[x]^2 - Cos[y]^2 + 2\, Cos[x]\, Cos[y]\, Cos[x + y] - Cos[x + y]^2,$$

$$\mathrm{Tan}[x] + \mathrm{Tan}[y] - \mathrm{Tan}[x + y] + \mathrm{Tan}[x]\,\mathrm{Tan}[y]\,\mathrm{Tan}[x + y],$$
$$-1 + \mathrm{Cot}[x]\,(\mathrm{Cot}[y] - \mathrm{Cot}[x + y]) - \mathrm{Cot}[y]\,\mathrm{Cot}[x + y],\ 2\,\mathrm{Sec}[x]\,\mathrm{Sec}[y]\,\mathrm{Sec}[x + y] -$$
$$\mathrm{Sec}[y]^2\,\mathrm{Sec}[x + y]^2 + \mathrm{Sec}[x]^2\,(-\mathrm{Sec}[x + y]^2 + \mathrm{Sec}[y]^2\,(-1 + \mathrm{Sec}[x + y]^2)),$$
$$\mathrm{Csc}[y]^4\,\mathrm{Csc}[x + y]^4 + \mathrm{Csc}[x]^4\,(\mathrm{Csc}[y]^2 - \mathrm{Csc}[x + y]^2)^2 -$$
$$2\,\mathrm{Csc}[x]^2\,\mathrm{Csc}[y]^2\,\mathrm{Csc}[x + y]^2\,(-2 + \mathrm{Csc}[y]^2 + \mathrm{Csc}[x + y]^2)\}$$

A third way to write trigonometric expressions is the use of arguments that are sums.

TrigReduce[*trigExpression*]

   rewrites *trigExpression* in the form of multiple angle arguments.

Both `TrigExpand` and `TrigFactor` do nothing with the following expression.

In[13]:= `{TrigExpand[Sin[x]^3 Cos[y]^4], TrigFactor[Sin[x]^3 Cos[y]^4]}`

Out[13]= $\{\dfrac{9\,\mathrm{Sin}[x]}{32} - \dfrac{9}{32}\,\mathrm{Cos}[x]^2\,\mathrm{Sin}[x] + \dfrac{3}{8}\,\mathrm{Cos}[y]^2\,\mathrm{Sin}[x] - \dfrac{3}{8}\,\mathrm{Cos}[x]^2\,\mathrm{Cos}[y]^2\,\mathrm{Sin}[x] +$

$\dfrac{3}{32}\,\mathrm{Cos}[y]^4\,\mathrm{Sin}[x] - \dfrac{3}{32}\,\mathrm{Cos}[x]^2\,\mathrm{Cos}[y]^4\,\mathrm{Sin}[x] + \dfrac{3\,\mathrm{Sin}[x]^3}{32} + \dfrac{1}{8}\,\mathrm{Cos}[y]^2\,\mathrm{Sin}[x]^3 +$

$\dfrac{1}{32}\,\mathrm{Cos}[y]^4\,\mathrm{Sin}[x]^3 - \dfrac{3}{8}\,\mathrm{Sin}[x]\,\mathrm{Sin}[y]^2 + \dfrac{3}{8}\,\mathrm{Cos}[x]^2\,\mathrm{Sin}[x]\,\mathrm{Sin}[y]^2 -$

$\dfrac{9}{16}\,\mathrm{Cos}[y]^2\,\mathrm{Sin}[x]\,\mathrm{Sin}[y]^2 + \dfrac{9}{16}\,\mathrm{Cos}[x]^2\,\mathrm{Cos}[y]^2\,\mathrm{Sin}[x]\,\mathrm{Sin}[y]^2 -$

$\dfrac{1}{8}\,\mathrm{Sin}[x]^3\,\mathrm{Sin}[y]^2 - \dfrac{3}{16}\,\mathrm{Cos}[y]^2\,\mathrm{Sin}[x]^3\,\mathrm{Sin}[y]^2 + \dfrac{3}{32}\,\mathrm{Sin}[x]\,\mathrm{Sin}[y]^4 -$

$\dfrac{3}{32}\,\mathrm{Cos}[x]^2\,\mathrm{Sin}[x]\,\mathrm{Sin}[y]^4 + \dfrac{1}{32}\,\mathrm{Sin}[x]^3\,\mathrm{Sin}[y]^4,\ \mathrm{Cos}[y]^4\,\mathrm{Sin}[x]^3\}$

`TrigReduce` rewrites the expression in a sum of trigonometric functions linear in x and y.

In[14]:= `TrigReduce[Sin[x]^3 Cos[y]^4]`

Out[14]= $\dfrac{1}{64}\,(18\,\mathrm{Sin}[x] - 6\,\mathrm{Sin}[3\,x] + 3\,\mathrm{Sin}[x - 4\,y] - \mathrm{Sin}[3\,x - 4\,y] + 12\,\mathrm{Sin}[x - 2\,y] -$

$4\,\mathrm{Sin}[3\,x - 2\,y] + 12\,\mathrm{Sin}[x + 2\,y] - 4\,\mathrm{Sin}[3\,x + 2\,y] + 3\,\mathrm{Sin}[x + 4\,y] - \mathrm{Sin}[3\,x + 4\,y])$

Often, it is very useful to convert from a trigonometric representation to an exponential one and vice versa. The functions `ExpToTrig` and `TrigToExp` carry out this transformation.

TrigToExp[*trigExpression*]

   rewrites all trigonometric functions in *trigExpression* as exponential ones.

ExpToTrig[*trigExpression*]

   rewrites all exponential functions in *trigExpression* as trigonometric ones.

Here are three examples. Inverse trigonometric and hyperbolic functions are rewritten as logarithms.

In[15]:= `TrigToExp[{Sin[z], Cosh[z], ArcTan[z], ArcCsch[z]}]`

Out[15]= $\{\dfrac{1}{2}\,i\,e^{-iz} - \dfrac{1}{2}\,i\,e^{iz},\ \dfrac{e^{-z}}{2} + \dfrac{e^{z}}{2},\ \dfrac{1}{2}\,i\,\mathrm{Log}[1 - i\,z] - \dfrac{1}{2}\,i\,\mathrm{Log}[1 + i\,z],\ \mathrm{Log}\Big[\sqrt{1 + \dfrac{1}{z^2}} + \dfrac{1}{z}\Big]\}$

`ExpToTrig` transforms exponentials back, but it leaves logarithms alone.

In[16]:= `ExpToTrig[%]`

Out[16]= $\left\{ \text{Sin}[z], \text{Cosh}[z], \frac{1}{2} \text{ i Log}[1 - \text{i } z] - \frac{1}{2} \text{ i Log}[1 + \text{i } z], \text{Log}\left[\sqrt{1 + \frac{1}{z^2}} + \frac{1}{z} \right] \right\}$

Single exponentials are converted to trigonometric or hyperbolic functions.

In[17]:= **ExpToTrig[{Exp[x], Exp[I x]}]**

Out[17]= $\{\text{Cosh}[x] + \text{Sinh}[x], \text{Cos}[x] + \text{i Sin}[x]\}$

TrigToExp can conveniently be used to prove the representations of π as sums of arctans. Here is a typical example.

In[18]:= **8 ArcTan[1/5] - 2 ArcTan[1/239]**

Out[18]= $-2 \text{ ArcTan}\left[\frac{1}{239}\right] + 8 \text{ ArcTan}\left[\frac{1}{5}\right]$

In[19]:= **Expand[TrigToExp[8 ArcTan[1/5] - 2 ArcTan[1/239]]]**

Out[19]= $-\text{i Log}\left[1 - \frac{\text{i}}{239}\right] + \text{i Log}\left[1 + \frac{\text{i}}{239}\right] + 4 \text{ i Log}\left[1 - \frac{\text{i}}{5}\right] - 4 \text{ i Log}\left[1 + \frac{\text{i}}{5}\right]$

After exponentiating this formula, we get just *i*.

In[20]:= **Together[Exp[I %]]**

Out[20]= i

Taking the logarithm and making sure by numerical evaluations that the two expressions corresponded to the same branch of the logarithm, this shows that $8 \tan^{-1}(\frac{1}{5}) - 2 \tan^{-1}(\frac{1}{239}) = \frac{\pi}{2}$.

TrigToExp also works for hyperbolic and inverse hyperbolic functions. Here, we use it to simplify $\sinh(2 \tanh^{-1}(\xi)) \sinh(\log(1/\xi))$.

In[21]:= **Sinh[2 ArcTanh[ξ]] Sinh[Log[1/ξ]] // TrigToExp**

Out[21]= $\frac{1}{4}\left(-1 + \frac{1}{\xi^2}\right) \xi \left(-\frac{1 - \xi}{1 + \xi} + \frac{1 + \xi}{1 - \xi}\right)$

In[22]:= **Together[%]**

Out[22]= 1

In dealing with trigonometric and hyperbolic functions, we frequently write the imaginary unit *i* explicitly, so that all variables involved are expected to be purely real-valued. However, *Mathematica* assumes automatically that all variables appearing in an expression can assume arbitrary complex values. This situation can be remedied with using ComplexExpand.

ComplexExpand[*expression*, {*var*$_1$, *var*$_2$, ..., *var*$_n$}]

 simplifies *expression* under the assumption that the variables *var*$_i$ can be complex values and all variables not mentioned can be only real values. If no variables are explicitly listed, it is assumed that all variables in the expression are real-valued.

Here is a simple example. The real and imaginary parts are given separately.

In[23]:= **ComplexExpand[Sin[x + I a] + Cos[(y + 5I s)^2] +**
       Exp[u + I o] + Re[t + o I]]

Out[23]= $t + e^u \text{ Cos}[o] + \text{Cos}[25 s^2 - y^2] \text{ Cosh}[10 s y] + \text{Cosh}[a] \text{ Sin}[x] +$
   $\text{i } (e^u \text{ Sin}[o] + \text{Cos}[x] \text{ Sinh}[a] + \text{Sin}[25 s^2 - y^2] \text{ Sinh}[10 s y])$

Here is another example that contains some different functions.

In[24]:= **ComplexExpand[Log[x + I y] + 1/Tan[x + I y] + (x + I y)^(x + I y)]**

Out[24]= $e^{-y \, Arg[x+i \, y]} \, (x^2 + y^2)^{x/2} \, Cos\left[x \, Arg[x + i \, y] + \frac{1}{2} \, y \, Log[x^2 + y^2]\right] +$

$\frac{1}{2} \, Log[x^2 + y^2] - \frac{Sin[2 \, x]}{Cos[2 \, x] - Cosh[2 \, y]} + i \left(Arg[x + i \, y] + \right.$

$\left. e^{-y \, Arg[x+i \, y]} \, (x^2 + y^2)^{x/2} \, Sin\left[x \, Arg[x + i \, y] + \frac{1}{2} \, y \, Log[x^2 + y^2]\right] + \frac{Sinh[2 \, y]}{Cos[2 \, x] - Cosh[2 \, y]} \right)$

The command `ComplexExpand` has one option.

In[25]:= `Options[ComplexExpand]`

Out[25]= {TargetFunctions → {Re, Im, Abs, Arg, Conjugate, Sign}}

`TargetFunctions`

> is an option of `ComplexExpand`. It determines which functions that return real numbers will be used to express the result.
> Default:
>> {Re, Im, Abs, Arg, Conjugate, Sign}
>
> Admissible:
>> arbitrary functions from the set {Re, Im, Abs, Arg, Conjugate, Sign}

Here, the expression $e^{i \, x} \ln(y)$ is decomposed in three different ways in terms of real and imaginary parts.

In[26]:= `ComplexExpand[Exp[I x] Log[y], TargetFunctions -> {Re, Im}]`

Out[26]= $\frac{1}{2} \, Cos[x] \, Log[y^2] - ArcTan[y, 0] \, Sin[x] + i \left(ArcTan[y, 0] \, Cos[x] + \frac{1}{2} \, Log[y^2] \, Sin[x]\right)$

In[27]:= `ComplexExpand[Exp[I x] Log[y], TargetFunctions -> {Abs, Arg}]`

Out[27]= $\frac{1}{2} \, Cos[x] \, Log[y^2] - Arg[y] \, Sin[x] + i \left(Arg[y] \, Cos[x] + \frac{1}{2} \, Log[y^2] \, Sin[x]\right)$

In[28]:= `ComplexExpand[Exp[I x] Log[y], TargetFunctions -> {Sign}]`

Out[28]= $\frac{1}{2} \, Cos[x] \, Log[y^2] + i \, Log[Sign[y]] \, Sin[x] +$

$i \left(-i \, Cos[x] \, Log[Sign[y]] + \frac{1}{2} \, Log[y^2] \, Sin[x]\right)$

Next, we try all possible 63 combinations of target functions. We obtain four different results from `ComplexExpand`.

In[29]:= `allCombinations = Join @@ Table[Flatten[Table[`
        `Table[{Re, Im, Abs, Arg, Conjugate, Sign}[[x[k]]], {k, n}],`
            `Evaluate[Sequence @@`
            `Table[{x[k], If[k === 1, 1, x[k - 1] + 1], 6}, {k, n}]]], n - 1],`
                `{n, 6}];`

In[30]:= `Short[allCombinations, 4]`

Out[30]//Short=
        {{Re}, {Im}, {Abs}, {Arg}, {Conjugate}, {Sign}, {Re, Im},
        {Re, Abs}, {Re, Arg}, ≪45≫, {Im, Arg, Conjugate, Sign},
        {Abs, Arg, Conjugate, Sign}, {Re, Im, Abs, Arg, Conjugate},
        {Re, Im, Abs, Arg, Sign}, {Re, Im, Abs, Conjugate, Sign},
        {Re, Im, Arg, Conjugate, Sign}, {Re, Abs, Arg, Conjugate, Sign},
        {Im, Abs, Arg, Conjugate, Sign}, {Re, Im, Abs, Arg, Conjugate, Sign}}

Here are these four results. The first argument of each sublist contains the target functions that produce the second argument.

```
In[31]:= {First /@ #, #[[1, 2]]}& /@
        Split[Sort[{#, ComplexExpand[Exp[I x] Log[y],
                            TargetFunctions -> #]}& /@ allCombinations,
                OrderedQ[{#1[[2]], #2[[2]]}]&], #1[[2]] == #2[[2]]&]
```

Out[31]= $\Big\{\{\{\{$Re$\}$, $\{$Im$\}$, $\{$Re, Im$\}$, $\{$Re, Abs$\}$, $\{$Re, Conjugate$\}$,

$\{$Re, Sign$\}$, $\{$Im, Abs$\}$, $\{$Im, Conjugate$\}$, $\{$Im, Sign$\}$, $\{$Re, Im, Abs$\}$,

$\{$Re, Im, Conjugate$\}$, $\{$Re, Im, Sign$\}$, $\{$Re, Abs, Conjugate$\}$,

$\{$Re, Abs, Sign$\}$, $\{$Re, Conjugate, Sign$\}$, $\{$Im, Abs, Conjugate$\}$,

$\{$Im, Abs, Sign$\}$, $\{$Im, Conjugate, Sign$\}$, $\{$Re, Im, Abs, Conjugate$\}$,

$\{$Re, Im, Abs, Sign$\}$, $\{$Re, Im, Conjugate, Sign$\}$, $\{$Re, Abs, Conjugate, Sign$\}$,

$\{$Im, Abs, Conjugate, Sign$\}$, $\{$Re, Im, Abs, Conjugate, Sign$\}\}$, $\frac{1}{2}$ Cos[x] Log[y^2] $-$

ArcTan[y, 0] Sin[x] $+ i \Big($ArcTan[y, 0] Cos[x] $+ \frac{1}{2}$ Log[y^2] Sin[x]$\Big)\Big\}$,

$\Big\{\{\{$Arg$\}$, $\{$Re, Arg$\}$, $\{$Im, Arg$\}$, $\{$Abs, Arg$\}$, $\{$Arg, Conjugate$\}$, $\{$Arg, Sign$\}$,

$\{$Re, Im, Arg$\}$, $\{$Re, Abs, Arg$\}$, $\{$Re, Arg, Conjugate$\}$, $\{$Re, Arg, Sign$\}$,

$\{$Im, Abs, Arg$\}$, $\{$Im, Arg, Conjugate$\}$, $\{$Im, Arg, Sign$\}$, $\{$Abs, Arg, Conjugate$\}$,

$\{$Abs, Arg, Sign$\}$, $\{$Arg, Conjugate, Sign$\}$, $\{$Re, Im, Abs, Arg$\}$,

$\{$Re, Im, Arg, Conjugate$\}$, $\{$Re, Im, Arg, Sign$\}$, $\{$Re, Abs, Arg, Conjugate$\}$,

$\{$Re, Abs, Arg, Sign$\}$, $\{$Re, Arg, Conjugate, Sign$\}$, $\{$Im, Abs, Arg, Conjugate$\}$,

$\{$Im, Abs, Arg, Sign$\}$, $\{$Im, Arg, Conjugate, Sign$\}$, $\{$Abs, Arg, Conjugate, Sign$\}$,

$\{$Re, Im, Abs, Arg, Conjugate$\}$, $\{$Re, Im, Abs, Arg, Sign$\}$,

$\{$Re, Im, Arg, Conjugate, Sign$\}$, $\{$Re, Abs, Arg, Conjugate, Sign$\}$,

$\{$Im, Abs, Arg, Conjugate, Sign$\}$, $\{$Re, Im, Abs, Arg, Conjugate, Sign$\}\}$,

$\frac{1}{2}$ Cos[x] Log[y^2] $-$ Arg[y] Sin[x] $+ i \Big($Arg[y] Cos[x] $+ \frac{1}{2}$ Log[y^2] Sin[x]$\Big)\Big\}$,

$\Big\{\{\{$Abs$\}$, $\{$Conjugate$\}$, $\{$Abs, Conjugate$\}\}$, $\frac{1}{2}$ Cos[x] Log[y^2] $+ i$ Log[y] Sin[x] $-$

$\frac{1}{2} i$ Log[y^2] Sin[x] $+ i \Big(-i$ Cos[x] Log[y] $+ \frac{1}{2} i$ Cos[x] Log[y^2] $+ \frac{1}{2}$ Log[y^2] Sin[x]$\Big)\Big\}$,

$\Big\{\{\{$Sign$\}$, $\{$Abs, Sign$\}$, $\{$Conjugate, Sign$\}$, $\{$Abs, Conjugate, Sign$\}\}$,

$\frac{1}{2}$ Cos[x] Log[y^2] $+ i$ Log[Sign[y]] Sin[x] $+$

$i \Big(-i$ Cos[x] Log[Sign[y]] $+ \frac{1}{2}$ Log[y^2] Sin[x]$\Big)\Big\}\Big\}$

When simplifying expressions involving a variable that takes only real values, we frequently want to reduce to x.

```
In[33]:= (x^2) ^ (1/2)
```

Out[33]= $\sqrt{x^2}$

This can be achieved using PowerExpand (or using corresponding replacement rules).

PowerExpand[*expression*, $\{var_1, var_2, \ldots, var_n\}$]

  simplifies roots and exponents under the assumption that $(a^b)^c = a^{bc}$ holds.

Note that the assumption $(a^b)^c = a^{bc}$ does not always hold (see the discussion in Subsection 2.2.6 of the Programming volume [1735]). So one must be careful when using PowerExpand.

Now, we get the "desired" result in the above root example.

```
In[34]:= PowerExpand[ (x^2) ^ (1/2) ]
```

Out[34]= x

Here is a demonstration of the effect of `PowerExpand` for general symbolic arguments.

In[35]:= **{ (x^a)^b, PowerExpand[(x^a)^b] }**

Out[35]= $\{ (x^a)^b, x^{ab} \}$

A second "simplification" carried out by `PowerExpand` is to rewrite a logarithm of a product as the sum of logarithms of the single factors. Be aware that for arbitrary complex numbers, this is not an identity.

In[36]:= **{Log[a b c], PowerExpand[Log[a b c]]}**

Out[36]= $\{ Log[a b c], Log[a] + Log[b] + Log[c] \}$

1.5 Solution of Equations

The most important built-in *Mathematica* functions for solving systems of linear and nonlinear equations are `Solve` and `Reduce`. (Also important for solving are the functions `Roots` and `FindInstance`; they will be discussed below. In Chapter 1 of the Numerics volume [1737], we discussed already the functions `NRoots`, `NSolve`, and `FindRoot` for the numerical solution of equations. In addition, the function `GroebnerBasis` is also useful for solving equations; see the remarks at the end of this section.) We start with the function `Solve`.

`Solve` [*equations, listOfVariablesToBeSolvedFor,*
    listOfVariablesToBeEliminated, options]

    solves (when it succeeds) the equations *equations* for the variables *listOfVariablesToBeSolvedFor* by simultaneous elimination of the variables *listOfVariablesToBeEliminated* using the options *options*.

`Solve` has already been briefly discussed in Chapter 6 of the Programming volume [1735] for linear systems. We now discuss the meaning of a "generic" solution. Let us choose a simple example: $ax + b = 0$. This is the solution calculated by *Mathematica* for this simple linear equation.

In[1]:= **Solve[a x == b, x]**

Out[1]= $\left\{ \left\{ x \to \dfrac{b}{a} \right\} \right\}$

However, this solution does not make sense for the special case $a = 0$, and solutions that would make sense in this setting are not produced by `Solve`!

Here, b/a is the "generic" solution of the equation $ax = b$. For almost all (in the sense of the corresponding Lebesgue measure for the complex planes of the variables involved), the solution $x = b/a$ is the correct solution. `Solve` will always return generic solutions. On lower-dimensional subsets (which have the corresponding Lebesgue measure 0), solutions might either be missing or after substitution of explicit numerical value for variables, indeterminate quantities may arise.

There is no guarantee that `Solve` will solve a given set of equations. However, for purely algebraic equations, there is a good chance that it will give a solution. For a polynomial system, it will virtually always find a solution, although it may take quite some time and memory.

Let us start with univariate polynomials. In general (provided no factors can be explicitly removed), we can find the zeros in closed form using radicals only for univariate polynomials up to (and including) the fourth degree (see, e.g., [1762]). *Mathematica* implements closed analytic solutions for polynomials up to the fourth degree; they can be expressed explicitly in terms of roots. Nevertheless, finding the four zeros of a general polynomial of fourth degree takes up around 50 kB.

```
In[2]:= {ByteCount[#], LeafCount[#], Depth[#]}&[
          Solve[a x^4 + b x^3 + c x^2 + d x + e == 0, x]]
Out[2]= {52104, 2885, 20}
```

Be aware of the following fact when solving cubic polynomials. In the case when a cubic polynomial has three real roots and does not factor, the answer returned by *Mathematica* will contain, when numericalized, a term of the form *verySmallNumber*×I. We will come back to this fact in a moment.

```
In[3]:= Solve[x^3 + x^2 - 12 x - 6 == 0, x] // N
Out[3]= {{x → 3.25417 + 0. i}, {x → -3.76437 + 0. i}, {x → -0.4898 + 0. i}}
```

Using high-precision arithmetic, we get *zeroWithFiniteAccuracy* I instead of *verySmallNumber*×I.

```
In[4]:= Solve[x^3 + x^2 - 12 x - 6 == 0, x] // N[#, 22]&
Out[4]= {{x → 3.254169471854462042793 + 0. × 10^-22 i},
         {x → -3.764369393648852639544 + 0. × 10^-22 i},
         {x → -0.4898000782056094032490 + 0. × 10^-23 i}}
```

NRoots gives in this case purely real solutions.

```
In[5]:= NRoots[x^3 + x^2 - 12 x - 6 == 0, x]
Out[5]= x == -3.76437 || x == -0.4898 || x == 3.25417
```

This fact also means that the *Mathematica* paradigm to numericalize real and imaginary parts independently is not always possible to fulfill. The three roots of the following cubic *are* real.

```
In[6]:= NRoots[x^3 - 12x + 5 == 0, x]
Out[6]= x == -3.65617 || x == 0.422973 || x == 3.23319
```

Here is the symbolic expression as returned by Solve for one of them.

```
In[7]:= Solve[x^3 - 12x + 5 == 0, x]
```

$$Out[7]= \left\{\left\{x \to \frac{4}{\left(\frac{1}{2}\left(-5 + i\sqrt{231}\right)\right)^{1/3}} + \left(\frac{1}{2}\left(-5 + i\sqrt{231}\right)\right)^{1/3}\right\},\right.$$

$$\left\{x \to -\frac{2\left(1 + i\sqrt{3}\right)}{\left(\frac{1}{2}\left(-5 + i\sqrt{231}\right)\right)^{1/3}} - \frac{1}{2}\left(1 - i\sqrt{3}\right)\left(\frac{1}{2}\left(-5 + i\sqrt{231}\right)\right)^{1/3}\right\},$$

$$\left.\left\{x \to -\frac{2\left(1 - i\sqrt{3}\right)}{\left(\frac{1}{2}\left(-5 + i\sqrt{231}\right)\right)^{1/3}} - \frac{1}{2}\left(1 + i\sqrt{3}\right)\left(\frac{1}{2}\left(-5 + i\sqrt{231}\right)\right)^{1/3}\right\}\right\}$$

Adding a purely real approximate number must result in a term *approximativeZero*×I.

```
In[8]:= root1 = 4/((-5 + I Sqrt[231])/2)^(1/3) + ((-5 + I Sqrt[231])/2)^(1/3);

        N[4, 100] + root1
Out[9]= 7.233193693489433088715431478758837920055527533867672575127066581330881255882ˋ
        9491441507436308794555 9 + 0. × 10^-116 i
```

Similarly, addition of an approximate real or imaginary zero will result in approximate imaginary and real parts.

```
In[10]:= (* add approximate real and complex zeros *)
         {root1 + 0., root1 + 0``100, root1 + 0``100 I}
Out[11]= {3.23319 - 4.44089×10^-16 i,
         3.233193693489433088715431478758837920055527533867672575127066581330881255888ˋ
         2949144150743630879455589 + 0. × 10^-117 i,
         3.233193693489433088715431478758837920055527533867672575127066581330881255888ˋ
         29491441507436308794555889470326832529375 + 0. × 10^-101 i}
```

As Abel, Ruffini, and Galois have shown [1718], [1422], it is generically impossible to express the roots of polynomials of degree five and higher in radicals (for the degree five and six exceptions, see [533] and [800]; for "closed form solvability" in general for equations, integrals, and differential equations, see [1018]). In *Mathematica*, irreducible higher-order polynomials will be solved for in Root-objects [1290]. We encountered Root-objects already in the last section. Now, let us deal with them in more detail.

Root [*irreduciblePolynomialOverTheIntergersAsAPureFunction*, *rootNumber*]

    represents the *rootNumber*th root of the equation *irreduciblePolynomialOverTheIntergersAsAPure-Function* [*x*] =0 with respect to *x*.

Here is an example.

In[12]:= `Solve[x^7 - 12x + 5 == 0, x]`

Out[12]= $\{\{x \to \text{Root}[5 - 12\,\#1 + \#1^7\,\&,\,1]\},$
$\{x \to \text{Root}[5 - 12\,\#1 + \#1^7\,\&,\,2]\},\ \{x \to \text{Root}[5 - 12\,\#1 + \#1^7\,\&,\,3]\},$
$\{x \to \text{Root}[5 - 12\,\#1 + \#1^7\,\&,\,4]\},\ \{x \to \text{Root}[5 - 12\,\#1 + \#1^7\,\&,\,5]\},$
$\{x \to \text{Root}[5 - 12\,\#1 + \#1^7\,\&,\,6]\},\ \{x \to \text{Root}[5 - 12\,\#1 + \#1^7\,\&,\,7]\}\}$

Root-objects are used to represent complex numbers. As such, they can be calculated to any required precision.

In[13]:= `N[%]`

Out[13]= $\{\{x \to -1.5735\},\ \{x \to 0.416849\},\ \{x \to 1.42858\},\ \{x \to -0.819692 - 1.31782\,i\},$
$\{x \to -0.819692 + 1.31782\,i\},\ \{x \to 0.683731 - 1.32212\,i\},\ \{x \to 0.683731 + 1.32212\,i\}\}$

In[14]:= `N[%%, 40]`

Out[14]= $\{\{x \to -1.573502020789201281884796361915924174292\},$
$\{x \to 0.4168489174904793059074502439012140522132\},$
$\{x \to 1.428575646587322164545991047112621100671\},$
$\{x \to -0.81969197825297226196468599028464333790\text{5}-$
$1.31781793599762931527222448155097155655\text{9}i\},$
$\{x \to -0.81969197825297226196468599028464333790\text{5}+$
$1.31781793599762931527222448155097155655\text{9}i\},$
$\{x \to 0.68373070660867216768036352573568784860\text{8}-$
$1.32211516505422041014930010984977950621\text{6}i\},$
$\{x \to 0.68373070660867216768036352573568784860\text{8}+$
$1.32211516505422041014930010984977950621\text{6}i\}\}$

Applying Re, Im, and Abs to a Root-object results in the automatic generation of a new Root-object. The following result shows that the first three roots were real.

In[15]:= `Re[x /. %%%]`

Out[15]= $\{\text{Root}[5 - 12\,\#1 + \#1^7\,\&,\,1],\ \text{Root}[5 - 12\,\#1 + \#1^7\,\&,\,2],\ \text{Root}[5 - 12\,\#1 + \#1^7\,\&,\,3],$
$\text{Root}[125 + 5400\,\#1 + 77760\,\#1^2 + 373248\,\#1^3 - 924800\,\#1^7 + 460800\,\#1^8 -$
$3907584\,\#1^9 - 4669440\,\#1^{14} + 9830400\,\#1^{15} + 2097152\,\#1^{21}\,\&,\,1],$
$\text{Root}[125 + 5400\,\#1 + 77760\,\#1^2 + 373248\,\#1^3 - 924800\,\#1^7 + 460800\,\#1^8 -$
$3907584\,\#1^9 - 4669440\,\#1^{14} + 9830400\,\#1^{15} + 2097152\,\#1^{21}\,\&,\,1],$
$\text{Root}[125 + 5400\,\#1 + 77760\,\#1^2 + 373248\,\#1^3 - 924800\,\#1^7 + 460800\,\#1^8 -$
$3907584\,\#1^9 - 4669440\,\#1^{14} + 9830400\,\#1^{15} + 2097152\,\#1^{21}\,\&,\,4],$
$\text{Root}[125 + 5400\,\#1 + 77760\,\#1^2 + 373248\,\#1^3 - 924800\,\#1^7 + 460800\,\#1^8 -$
$3907584\,\#1^9 - 4669440\,\#1^{14} + 9830400\,\#1^{15} + 2097152\,\#1^{21}\,\&,\,4]\}$

In[16]:= `Im[` (* for brevity, take first four roots *) `Take[x /. %%%%, 4]]`

Out[16]= {0, 0, 0,

Root[1658900974673 + 1143548280000 #1^2 − 16668733132800 #1^4 + 35143263977472 #1^6 +
17619114240000 #1^8 − 160168535654400 #1^{10} + 274207799771136 #1^{12} +
61441013760000 #1^{14} − 288898154496000 #1^{16} + 925736110129152 #1^{18} +
471300518707200 #1^{22} + 1038226739429376 #1^{24} − 80564191232000 #1^{28} −
414533063540736 #1^{30} + 11544872091648 #1^{36} + 4398046511104 #1^{42} &, 2]}

The absolute value of a complex-valued Root-object of degree d is typically the root of a polynomial of degree
$d(d-1)$.

In[17]:= **Abs[**(* for brevity, use last root only *) **Last[x /. %%%%]]**

Out[17]= Root[−15625 + 7500 #1^8 + 43200 #1^{10} + 248832 #1^{12} + 1875 #1^{14} + 7200 #1^{16} +
20736 #1^{18} − 600 #1^{22} − 3456 #1^{24} − 75 #1^{28} − 288 #1^{30} + 12 #1^{36} + #1^{42} &, 5]

The following function rootDegreeCount first forms all roots of irreducible polynomials of degree d and
integer coefficients from the interval $[-K, K]$ and then determines the degrees of the irreducible polynomial
obeyed by $f(root)$. It returns a list of the degrees and the number of $f(root)$ with this degree.

```
In[18]:= rootDegreeCount[f_, degree:d_?(# > 2&), maxCoefficient:K_Integer] :=
          Module[{c, poly, r, absr},
            {First[#], Length[#]}& /@ Split[Sort[DeleteCases[#, Null]& @ Flatten[
              Table[p = Sum[c[k] C^k, {k, 0, d}]; d = Exponent[p, C];
                If[d > 2,(* form Root-object *)
                    r = Root[Function[C, Evaluate[p]], Random[Integer, {1, d}]];
                    (* is r of degree d *)
                    If[Head[r] === Root && (Exponent[r[[1]][C], C] == d),
                      (* form f[r] and determine its degree *)
                      absr = RootReduce[f[r]];
                      Which[Head[absr] === Root, Exponent[absr[[1]][C], C],
                            IntegerQ[absr] || Head[absr] === Rational, 1, True, 2]]],
                (* iterator to form all polynomials *)
                Evaluate[Sequence @@ Table[{c[k], -K, K}, {k, 0, d}]]]]]]]]
```

The next input analyzes the real part, the imaginary part, the absolute value, the square, and the inverse of all
sextic polynomials with coefficients 0, or ±1.

In[19]:= **{#, rootDegreeCount[#, 6, 1]}& /@ {Re, Im, Abs, #^2&, 1/#&}**

Out[19]= {{Re, {{1, 7}, {3, 23}, {6, 115}, {9, 16}, {12, 30}, {15, 393}}},
{Im, {{1, 108}, {6, 27}, {12, 17}, {18, 16}, {24, 24}, {30, 392}}},
{Abs, {{1, 25}, {6, 105}, {12, 15}, {18, 16}, {24, 24}, {30, 399}}},
{#1^2 &, {{3, 24}, {6, 560}}}, {$\frac{1}{\#1}$ &, {{6, 584}}}}

Root-objects can be numericalized roots to any precision. The polynomial $x^n + c(\sum_{k=0}^{n-1} x^k) = 0$ has for integer c
one root that is near to an integer [524]. The following Root-object encodes this special root. We need the
Evaluate to evaluate the held argument of the pure function.

In[20]:= *nearlyIntegerRoot*[n_?(# > 1&), c_?(# > 3&)] :=
          Root[Function[Evaluate[#^n + c Sum[#^k, {k, 0, n - 1}]]], 1]

For $n = c = 60$, the resulting polynomial agrees with the nearest integer to more than 100 decimal digits.

In[21]:= **Round[-Log[10, (Abs[# - Round[#]])&[N[*nearlyIntegerRoot*[60, 60], 120]]]]**

Out[21]= 104

The numericalization of Root-objects can be done fast. The following shows that a root of an irreducible
quintic can be found to several hundred thousand digits in a few seconds.

```
In[22]:= Table[{100 2^k, Timing[N[Root[(#^5 + 4 # - 7)&, 1],
                            100 2^k]][[1, 1]]}, {k, 4, 12}]
```
Out[22]= {{1600, 0.}, {3200, 0.01}, {6400, 0.01}, {12800, 0.04}, {25600, 0.12},
    {51200, 0.34}, {102400, 0.91}, {204800, 2.1}, {409600, 4.78}}

```
In[23]:= (* a quick check of the last root approximation *)
      With[{x = N[Root[(#^5 + 4 # - 7)&, 1], 100 2^12]}, x^5 + 4x - 7]
```
Out[24]= $0. \times 10^{-409599}$

In Chapter 6 of the Programming volume [1735] we mentioned that the function Eigenvalues and Eigen⸱
system often return Root-objects for larger symbolic matrices. Let us give one example for such a situation.
Here is a symbolic 7×7 matrix with two indeterminates x and δ [1516].

```
In[25]:= (H[x_, δ_] = With[{d = 3},
        Table[If[i == j, Which[i < 0, x + i/d, i == 0, x,
                               i > 0, -x + i/d], δ],
              {i, -d, d}, {j, -d, d}]]) // TableForm[#, TableSpacing -> 1]&
```
Out[25]//TableForm=

$-1 + x$	δ	δ	δ	δ	δ	δ
δ	$-\frac{2}{3} + x$	δ	δ	δ	δ	δ
δ	δ	$-\frac{1}{3} + x$	δ	δ	δ	δ
δ	δ	δ	x	δ	δ	δ
δ	δ	δ	δ	$\frac{1}{3} - x$	δ	δ
δ	δ	δ	δ	δ	$\frac{2}{3} - x$	δ
δ	δ	δ	δ	δ	δ	$1 - x$

The eigenvalues of $H[x, \delta]$ are the roots of a septic polynomial. Here we encounter a Root-object with a
symbolic parameter.

```
In[26]:= evs[x_, δ_] = Eigenvalues[H[x, δ]];
      First[evs[x, δ]]
```
Out[27]= Root[$4x - 44x^2 + 193x^3 - 432x^4 + 522x^5 - 324x^6 + 81x^7 + 49x\delta^2 - 288x^2\delta^2 +$
    $648x^3\delta^2 - 648x^4\delta^2 + 243x^5\delta^2 - 98\delta^3 + 576x\delta^3 - 1296x^2\delta^3 + 1296x^3\delta^3 -$
    $486x^4\delta^3 - 378x\delta^4 + 972x^2\delta^4 - 729x^3\delta^4 + 504\delta^5 - 1296x\delta^5 + 972x^2\delta^5 +$
    $405x\delta^6 - 486\delta^7 - 4\#1 + 44x\#1 - 193x^2\#1 + 432x^3\#1 - 522x^4\#1 + 324x^5\#1 -$
    $81x^6\#1 - 147\delta^2\#1 + 864x\delta^2\#1 - 1944x^2\delta^2\#1 + 1944x^3\delta^2\#1 - 729x^4\delta^2\#1 -$
    $1008x\delta^3\#1 + 2592x^2\delta^3\#1 - 1944x^3\delta^3\#1 + 1890\delta^4\#1 - 4860x\delta^4\#1 +$
    $3645x^2\delta^4\#1 + 1944x\delta^5\#1 - 2835\delta^6\#1 - 49x\#1^2 + 288x^2\#1^2 - 648x^3\#1^2 +$
    $648x^4\#1^2 - 243x^5\#1^2 - 756x\delta^2\#1^2 + 1944x^2\delta^2\#1^2 - 1458x^3\delta^2\#1^2 + 2520\delta^3\#1^2 -$
    $6480x\delta^3\#1^2 + 4860x^2\delta^3\#1^2 + 3645x\delta^4\#1^2 - 6804\delta^5\#1^2 + 49\#1^3 - 288x\#1^3 +$
    $648x^2\#1^3 - 648x^3\#1^3 + 243x^4\#1^3 + 1260\delta^2\#1^3 - 3240x\delta^2\#1^3 + 2430x^2\delta^2\#1^3 +$
    $3240x\delta^3\#1^3 - 8505\delta^4\#1^3 + 126x\#1^4 - 324x^2\#1^4 + 243x^3\#1^4 + 1215x\delta^2\#1^4 -$
    $5670\delta^3\#1^4 - 126\#1^5 + 324x\#1^5 - 243x^2\#1^5 - 1701\delta^2\#1^5 - 81x\#1^6 + 81\#1^7$ &, 1]

```
In[28]:= Short[Expand[81 (Det[H[x, δ] - # IdentityMatrix[7]])], 6]
```
Out[28]//Short=
    $-4x + 44x^2 - 193x^3 + 432x^4 - 522x^5 + 324x^6 - 81x^7 - 49x\delta^2 + 288x^2\delta^2 - 648x^3\delta^2 +$
    $648x^4\delta^2 - 243x^5\delta^2 + 98\delta^3 - 576x\delta^3 + 1296x^2\delta^3 - 1296x^3\delta^3 + 486x^4\delta^3 +$
    $378x\delta^4 - 972x^2\delta^4 + 729x^3\delta^4 - 504\delta^5 + \ll53\gg + 6804\delta^5\#1^2 - 49\#1^3 + 288x\#1^3 -$
    $648x^2\#1^3 + 648x^3\#1^3 - 243x^4\#1^3 - 1260\delta^2\#1^3 + 3240x\delta^2\#1^3 - 2430x^2\delta^2\#1^3 -$
    $3240x\delta^3\#1^3 + 8505\delta^4\#1^3 - 126x\#1^4 + 324x^2\#1^4 - 243x^3\#1^4 - 1215x\delta^2\#1^4 +$
    $5670\delta^3\#1^4 + 126\#1^5 - 324x\#1^5 + 243x^2\#1^5 + 1701\delta^2\#1^5 + 81x\#1^6 - 81\#1^7$

While the matrix is diagonal for $\delta = 0$, the corresponding Root-objects do not autosimplify in the presence of the symbolic parameter x.

In[29]:= **Eigenvalues[$\mathcal{H}$[x, 0]]**

Out[29]= $\left\{ \dfrac{1}{3} - x, \ \dfrac{2}{3} - x, \ 1 - x, \ -1 + x, \ -\dfrac{2}{3} + x, \ -\dfrac{1}{3} + x, \ x \right\}$

In[30]:= **FreeQ[evs[x, 0], _Root, Infinity]**

Out[30]= False

Here are some visualizations of the eigenvalues of $\mathcal{H}[x, \delta]$ as a function of x for various δ. For small δ, one nicely sees the "avoided crossing "-behavior of the eigenvalues [826], [999], [1752], [1602], [1391], [1516], [1517], [1518], [470], [1184].

In[31]:= **Show[GraphicsArray[**
 Plot[Evaluate[evs[x, #]], {x, 0, 0.9},
 PlotRange -> All, Frame -> True,
 DisplayFunction -> Identity]& /@ {0, 0.01, 0.1, 1}]]

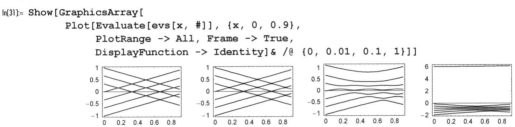

One side remark is in order at this point. When using the functions Eigenvalues, Eigenvectors, or Eigensystem with matrices containing symbolic entries, *Mathematica* has to repeatedly test if expressions are zero. These tests might fail if hidden algebraic dependencies exist between various symbolic parameters. So, it is frequently helpful to preprocess the initial matrix by introducing symbols for linear independent radicals, by introducing polynomial variables for exponentials, and by rewriting trigonometric expressions in exponential form. In addition, a carefully constructed function for the ZeroTest option might be useful.

It happens in calculations that one ends up with an expression containing more than one Root-object and/or radical. A very useful command for manipulating such expressions is RootReduce.

RootReduce [*expressionContainingRootObjects*]

    represents sums and products of roots and radicals in *expressionContainingRootObjects* as a canonical Root-object.

Here the sum of two Root-objects is reduced to one.

In[32]:= **sumOfTwoRoots = Root[5 - 12 #1 + #1^7 &, 1] + Root[5 - 12 #1 + #1^7 &, 2];**
 RootReduce[sumOfTwoRoots]

Out[33]= Root[$125 + 2700 \,\#1 + 19440 \,\#1^2 + 46656 \,\#1^3 -$
 $7225 \,\#1^7 + 1800 \,\#1^8 - 7632 \,\#1^9 - 285 \,\#1^{14} + 300 \,\#1^{15} + \#1^{21}$ &, 2]

The numerical value of the last result is equal to the numerical value of the sum of the two individual roots.

In[34]:= **{N[sumOfTwoRoots], N[%]}**

Out[34]= {-1.15665, -1.15665}

RootReduce can often be used to prove tricky (trigonometric and radical) identities such as the following.

In[35]:= **Tan[Pi/24] - (Sqrt[6] + Sqrt[2] - Sqrt[3] - 2)**

Out[35]= $2 - \sqrt{2} + \sqrt{3} - \sqrt{6} + \text{Tan}\left[\dfrac{\pi}{24}\right]$

First, we rewrite the `Tan` in exponential form.

In[36]:= **TrigToExp[%]**

Out[36]= $2 - \sqrt{2} + \sqrt{3} - \sqrt{6} + \dfrac{i\, e^{-\frac{i\pi}{24}}}{e^{-\frac{i\pi}{24}} + e^{\frac{i\pi}{24}}} - \dfrac{i\, e^{\frac{i\pi}{24}}}{e^{-\frac{i\pi}{24}} + e^{\frac{i\pi}{24}}}$

Now, we rewrite the exponentials `Exp[I rationalNumber]` in algebraic form $(-1)^{rationalNumber}$.

In[37]:= **Together[%]**

Out[37]= $\dfrac{(2+i) + (2-i)\,(-1)^{1/12} - \sqrt{2} - (-1)^{1/12}\sqrt{2} + \sqrt{3} + (-1)^{1/12}\sqrt{3} - \sqrt{6} - (-1)^{1/12}\sqrt{6}}{1 + (-1)^{1/12}}$

This expression is now in a form suitable for `RootReduce`.

In[38]:= **RootReduce[%]**

Out[38]= 0

`RootReduce` allows us to transform (nested) radicals into a canonical form.

In[39]:= **Nest[Sqrt[# + 1]&, 0, 6]**

Out[39]= $\sqrt{1 + \sqrt{1 + \sqrt{1 + \sqrt{1 + \sqrt{2}}}}}$

In[40]:= **RootReduce[%]**

Out[40]= Root$[-2 + 64\,\#1^8 - 128\,\#1^{10} - 160\,\#1^{12} + 736\,\#1^{14} - 700\,\#1^{16} - 384\,\#1^{18} +$
$1552\,\#1^{20} - 1744\,\#1^{22} + 1116\,\#1^{24} - 448\,\#1^{26} + 112\,\#1^{28} - 16\,\#1^{30} + \#1^{32}\ \&,\ 2]$

In the next input, we use `RootReduce` to show that some of the polynomial roots, of the irreducible (noncyclotomic) polynomial $z^6 - 2z^4 - 6z^3 - 2z^2 + 1$, have the interesting property that the product of two of them is again a root too [521].

In[41]:= **Module[{** (* the roots of the sextic *)
                roots = z /. Solve[z^6 - 2z^4 - 6z^3 - 2z^2 + 1 == 0, z]},
        Cases[Table[If[# === {}, {}, {i, j, #[[1, 1]]}]]&[
                (* is a products identical to a root? *)
                Position[roots, RootReduce[roots[[i]] roots[[j]]], {1}]],
                    {i, 6}, {j, i, 6}], {_, _, _}, {2}]]
Out[41]= {{1, 3, 5}, {1, 4, 6}, {2, 5, 3}, {2, 6, 4}, {3, 4, 2}, {5, 6, 1}}

For some `Root`-objects, it is possible to express them in radicals. The function `ToRadicals` is relevant here.

> `ToRadicals[rootObjects]`
>   tries to represent *rootObjects* as a (nested) radical expression.

Here are the above `Root`-object of degree 32 rewritten as a nested square root.

In[42]:= **ToRadicals[%]**

Out[42]= {{1, 3, 5}, {1, 4, 6}, {2, 5, 3}, {2, 6, 4}, {3, 4, 2}, {5, 6, 1}}

Sometimes it might be possible to express a `Root`-object in radicals, but *Mathematica* might not find it. Transforming a `Root`-object into radicals usually requires complicated Galois-group calculations [812], [811].

`Root`-objects can also contain parameters. In this case, such `Root`-objects represent algebraic functions, complex-valued, piecewise continuous functions of the parameters [1642], [1078], [1290], [1452], [421]. As such functions, `Root`-objects work nicely together with all other *Mathematica* functions. This means, for instance, one can differentiate them.

In[43]:= `root = Root[α - 12 #1 + #1^7&, 1];`
        `D[root, α]`

Out[44]= $-\dfrac{1}{-12 + 7\,\text{Root}[\alpha - 12\,\#1 + \#1^7\,\&,\,1]^6}$

One can expand them into a series (see Subsection 1.6.4 below) as long as the expansion point is not a branch point.

In[45]:= `Series[root, {α, 2, 2}] // N`

Out[45]= $-1.53924 - 0.0123312\,(\alpha - 2.) + 0.000340217\,(\alpha - 2.)^2 + O[\alpha - 2.]^3$

One can plot them.

In[46]:= `Plot3D[Evaluate[Re[root] /. α -> αr + I αi],`
        `{αr, -12, 12}, {αi, -12, 12}, PlotPoints -> 30]`

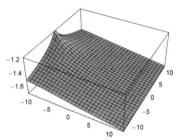

`RootReduce` does not handle `Root`-objects with parameters. `Root`-objects with symbolic parameters will have a complicated branch cut structure with respect to these parameters. This means that `Root`-objects as a function of a parameter are typically not smooth functions everywhere in the complex plane of the parameter(s). We show a contour plot of a `Root`-object as a function of a complex parameter. The branch cuts are easily identifiable as clusters of contour lines.

In[47]:= `ContourPlot[Im[Root[(αr + I αi) + 4 #1 +3 #1^7&, 1]],`
        `{αr, -5, 5}, {αi, -5, 5},`
        `PlotPoints -> 120, ColorFunction -> Hue, Contours -> 30]`

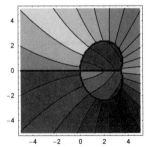

The following graphic visualizes the numbering of the `Root`-objects. We create an irreducible polynomial (by adding a small number) to the simple product $\prod_j (x - x_j)$. One sees that first come the real roots and then come the pairs of complex conjugate ones. We display the integer k at the position of the root with number k.

```
In[48]:= poly = 1/100 + Times @@
           ((x - #)& /@ Flatten[Table[x + I y, {x, -3, 3}, {y, -3, 3}]]);
```

```
In[49]:= sol = x /. Solve[poly == 0, x];
```

```
In[50]:= Take[sol, 3]
```

Out[50]= {Root[1 − 1918860503040000000000 #1 + 1498632389468928000000 #1^5 +
       435564030267789337600 #1^9 − 15037610025339776000 #1^{13} − 250937075173008000 #1^{17} −
       46934133632512000 #1^{21} − 439279531841500 #1^{25} + 3818649949800 #1^{29} −
       2212766500 #1^{33} + 515438000 #1^{37} + 7129500 #1^{41} + 49000 #1^{45} + 100 #1^{49} &, 1],
    Root[1 − 1918860503040000000000 #1 + 1498632389468928000000 #1^5 +
       435564030267789337600 #1^9 − 15037610025339776000 #1^{13} − 250937075173008000 #1^{17} −
       46934133632512000 #1^{21} − 439279531841500 #1^{25} + 3818649949800 #1^{29} −
       2212766500 #1^{33} + 515438000 #1^{37} + 7129500 #1^{41} + 49000 #1^{45} + 100 #1^{49} &, 2],
    Root[1 − 1918860503040000000000 #1 + 1498632389468928000000 #1^5 +
       435564030267789337600 #1^9 − 15037610025339776000 #1^{13} − 250937075173008000 #1^{17} −
       46934133632512000 #1^{21} − 439279531841500 #1^{25} + 3818649949800 #1^{29} −
       2212766500 #1^{33} + 515438000 #1^{37} + 7129500 #1^{41} + 49000 #1^{45} + 100 #1^{49} &, 3]}

```
In[51]:= Show[Graphics[MapIndexed[Text[#2[[1]], {Re[#1], Im[#1]}]& , N[sol]]],
           Frame -> True, PlotRange -> {{-4, 4}, {-4, 4}}];
```

Roots of polynomials have many interesting properties; for a bibliography, see [1233]; for more details about Root-objects in *Mathematica*, see [1671], [1672], and [1673].

Here is a small Galois-related application of Root-objects [93], [110], [297], [630], [1515], [1294], [635], [1718], and [1393] . We start with the roots of an irreducible cubic equation.

```
In[52]:= cubic = x^3 - x^2 - 2 // Factor
```

Out[52]= $-2 - x^2 + x^3$

```
In[53]:= roots = Last /@ List @@ Roots[cubic == 0, x, Cubics -> False]
```

Out[53]= {Root[−2 − #1^2 + #1^3 &, 1], Root[−2 − #1^2 + #1^3 &, 2], Root[−2 − #1^2 + #1^3 &, 3]}

Using the package NumberTheory`PrimitiveElement`, the three roots can be expressed as a polynomial of a root of a sextic.

```
In[54]:= Needs["NumberTheory`PrimitiveElement`"]
```

```
In[55]:= prim = PrimitiveElement[z, roots]
```

Out[55]= $\Big\{$Root[152 − 132 #1 + 193 #1^2 − 144 #1^3 + 58 #1^4 − 12 #1^5 + #1^6 &, 2],

$$\Big\{\frac{15}{14} + \frac{5z}{28} - \frac{67z^2}{112} + \frac{3z^3}{14} - \frac{3z^4}{112},$$
$$\frac{6}{7} - \frac{19z}{14} + \frac{67z^2}{56} - \frac{3z^3}{7} + \frac{3z^4}{56}, -\frac{13}{14} + \frac{33z}{28} - \frac{67z^2}{112} + \frac{3z^3}{14} - \frac{3z^4}{112}\Big\}\Big\}$$

We generate all six possible roots of the first element of the list `prim`.

```
In[56]:= pRoots = (Last /@ List @@ Roots[prim[[1, 1]][x] == 0, x,
                  Quartics -> False, Cubics -> False]) // RootReduce
```

$$Out[56]= \{Root[152 - 132 \#1 + 193 \#1^2 - 144 \#1^3 + 58 \#1^4 - 12 \#1^5 + \#1^6 \&, 3],$$
$$Root[152 - 132 \#1 + 193 \#1^2 - 144 \#1^3 + 58 \#1^4 - 12 \#1^5 + \#1^6 \&, 4],$$
$$Root[152 - 132 \#1 + 193 \#1^2 - 144 \#1^3 + 58 \#1^4 - 12 \#1^5 + \#1^6 \&, 2],$$
$$Root[152 - 132 \#1 + 193 \#1^2 - 144 \#1^3 + 58 \#1^4 - 12 \#1^5 + \#1^6 \&, 5],$$
$$Root[152 - 132 \#1 + 193 \#1^2 - 144 \#1^3 + 58 \#1^4 - 12 \#1^5 + \#1^6 \&, 1],$$
$$Root[152 - 132 \#1 + 193 \#1^2 - 144 \#1^3 + 58 \#1^4 - 12 \#1^5 + \#1^6 \&, 6]\}$$

Substituting the roots `pRoots` into the three elements of the second part of `prim` again yields all roots of the original cubic, this time in all possible orders [932].

```
In[57]:= RootReduce[prim[[2]] /. z -> #]& /@ pRoots
```

$$Out[57]= \{\{Root[-2 - \#1^2 + \#1^3 \&, 3], Root[-2 - \#1^2 + \#1^3 \&, 1], Root[-2 - \#1^2 + \#1^3 \&, 2]\},$$
$$\{Root[-2 - \#1^2 + \#1^3 \&, 2], Root[-2 - \#1^2 + \#1^3 \&, 1], Root[-2 - \#1^2 + \#1^3 \&, 3]\},$$
$$\{Root[-2 - \#1^2 + \#1^3 \&, 1], Root[-2 - \#1^2 + \#1^3 \&, 2], Root[-2 - \#1^2 + \#1^3 \&, 3]\},$$
$$\{Root[-2 - \#1^2 + \#1^3 \&, 3], Root[-2 - \#1^2 + \#1^3 \&, 2], Root[-2 - \#1^2 + \#1^3 \&, 1]\},$$
$$\{Root[-2 - \#1^2 + \#1^3 \&, 1], Root[-2 - \#1^2 + \#1^3 \&, 3], Root[-2 - \#1^2 + \#1^3 \&, 2]\},$$
$$\{Root[-2 - \#1^2 + \#1^3 \&, 2], Root[-2 - \#1^2 + \#1^3 \&, 3], Root[-2 - \#1^2 + \#1^3 \&, 1]\}\}$$

```
In[58]:= Sort[Map[#[[2]]&, %, {2}]] == Sort[Permutations[{1, 2, 3}]]
```

Out[58]= True

Similar to `NSolve` and `NRoots`, the function `Roots` calculates the roots of a univariate polynomial.

Roots[*polynomial*(*z*) == 0, *z*]

    finds the roots of the univariate polynomial equation *polynomial*(*z*)=0.

In distinction to `Solve`, the function `Roots` returns its answer in the form of `Or[solutions]`.

```
In[59]:= Roots[x^7 - 12x + 5 == 0, x]
```

$$Out[59]= x == Root[5 - 12 \#1 + \#1^7 \&, 1] \;||$$
$$x == Root[5 - 12 \#1 + \#1^7 \&, 2] \;||\; x == Root[5 - 12 \#1 + \#1^7 \&, 3] \;||$$
$$x == Root[5 - 12 \#1 + \#1^7 \&, 4] \;||\; x == Root[5 - 12 \#1 + \#1^7 \&, 5] \;||$$
$$x == Root[5 - 12 \#1 + \#1^7 \&, 6] \;||\; x == Root[5 - 12 \#1 + \#1^7 \&, 7]$$

`Roots` has the following options.

```
In[60]:= Options[Roots]
```

Out[60]= {Cubics → True, Eliminate → False, EquatedTo → Null,
         Modulus → 0, Multiplicity → 1, Quartics → True, Using → True}

`Cubics` is the option of most importance from the last list (`Quartic` is similar).

```
Cubics
```
> determines if the solutions to a cubic equation should be expressed in radicals (Cardano formula) or as
> Root-objects.
> **Default:**
> ```
> True
> ```
>
> **Admissible:**
> ```
> False
> ```

This solves a cubic equation in radicals.

In[61]:= `radicalCubeRoots = Roots[x^3 - 3 x + 5 == 0, x, Cubics -> True]`

Out[61]= $x == -\left(\dfrac{2}{5 - \sqrt{21}}\right)^{1/3} - \left(\dfrac{1}{2}(5 - \sqrt{21})\right)^{1/3}$ ||

$x == \dfrac{1}{2}(1 + i\sqrt{3})\left(\dfrac{1}{2}(5 - \sqrt{21})\right)^{1/3} + \dfrac{1 - i\sqrt{3}}{2^{2/3}(5 - \sqrt{21})^{1/3}}$ ||

$x == \dfrac{1}{2}(1 - i\sqrt{3})\left(\dfrac{1}{2}(5 - \sqrt{21})\right)^{1/3} + \dfrac{1 + i\sqrt{3}}{2^{2/3}(5 - \sqrt{21})^{1/3}}$

Here, the same cubic is solved in Root-objects.

In[62]:= `rootCubeRoots = Roots[x^3 - 3 x + 5 == 0, x, Cubics -> False]`

Out[62]= $x == \text{Root}[5 - 3\,\#1 + \#1^3\ \&,\ 1]$ || $x == \text{Root}[5 - 3\,\#1 + \#1^3\ \&,\ 2]$ || $x == \text{Root}[5 - 3\,\#1 + \#1^3\ \&,\ 3]$

Using RootReduce, we can establish equality between the roots.

In[63]:= `RootReduce[radicalCubeRoots[[1, 2]] - rootCubeRoots[[1, 2]]]`

Out[63]= 0

Let us take another cubic.

In[64]:= `radicalCubeRoots = Roots[x^3 - 19 x^2 + 76 x - 83 == 0, x,`
                       `Cubics -> True]`

Out[64]= $x == \dfrac{19}{3} + \dfrac{133}{3\left(\frac{1}{2}(2963 + 3\,i\,\sqrt{70131})\right)^{1/3}} + \dfrac{1}{3}\left(\dfrac{1}{2}(2963 + 3\,i\,\sqrt{70131})\right)^{1/3}$ ||

$x == \dfrac{19}{3} - \dfrac{1}{6}(1 + i\sqrt{3})\left(\dfrac{1}{2}(2963 + 3\,i\,\sqrt{70131})\right)^{1/3} - \dfrac{133(1 - i\sqrt{3})}{3\,2^{2/3}(2963 + 3\,i\,\sqrt{70131})^{1/3}}$ ||

$x == \dfrac{19}{3} - \dfrac{1}{6}(1 - i\sqrt{3})\left(\dfrac{1}{2}(2963 + 3\,i\,\sqrt{70131})\right)^{1/3} - \dfrac{133(1 + i\sqrt{3})}{3\,2^{2/3}(2963 + 3\,i\,\sqrt{70131})^{1/3}}$

In[65]:= `rootCubeRoots = Roots[x^3 - 19 x^2 + 76 x - 83 == 0, x, Cubics -> False]`

Out[65]= $x == \text{Root}[-83 + 76\,\#1 - 19\,\#1^2 + \#1^3\ \&,\ 1]$ ||
       $x == \text{Root}[-83 + 76\,\#1 - 19\,\#1^2 + \#1^3\ \&,\ 2]$ || $x == \text{Root}[-83 + 76\,\#1 - 19\,\#1^2 + \#1^3\ \&,\ 3]$

Although all three roots of this cubic are real, we see the symbol I in radicalCubeRoots, as already mentioned above. The numericalization of the Root-objects avoids these spurious imaginary parts.

In[66]:= `N[radicalCubeRoots]`

Out[66]= $x == 13.9924 + 0.\ i$ || $x == 3.08448 + 2.22045 \times 10^{-15}\ i$ || $x == 1.92311 - 2.22045 \times 10^{-15}\ i$

In[67]:= `N[rootCubeRoots]`

Out[67]= $x == 1.92311$ || $x == 3.08448$ || $x == 13.9924$

Because it is impossible to use only real expressions to represent the roots in the casus irreducibilis (see [867], [1035], [375], and [895]), the above occurrences of I are unavoidable from a mathematical point of view.

Roots deals with univariate polynomials over the integers.

```
In[68]:= Roots[Sqrt[2] x^5 - 2 x + 1 == 0, x]
```

$$Out[68]= \; x == \text{Root}[-1 + 4 \#1 - 4 \#1^2 + 2 \#1^{10} \&, 1] \; || $$
$$x == \text{Root}[-1 + 4 \#1 - 4 \#1^2 + 2 \#1^{10} \&, 7] \; || \; x == \text{Root}[-1 + 4 \#1 - 4 \#1^2 + 2 \#1^{10} \&, 8] \; ||$$
$$x == \text{Root}[-1 + 4 \#1 - 4 \#1^2 + 2 \#1^{10} \&, 3] \; || \; x == \text{Root}[-1 + 4 \#1 - 4 \#1^2 + 2 \#1^{10} \&, 4]$$

Solve will, of course, also solve this equation.

```
In[69]:= Solve[Sqrt[2] x^5 - 2 x + 1 == 0, x]
```

$$Out[69]= \; \{\{x \to \text{Root}[-1 + 4 \#1 - 4 \#1^2 + 2 \#1^{10} \&, 1]\},$$
$$\{x \to \text{Root}[-1 + 4 \#1 - 4 \#1^2 + 2 \#1^{10} \&, 3]\}, \{x \to \text{Root}[-1 + 4 \#1 - 4 \#1^2 + 2 \#1^{10} \&, 4]\},$$
$$\{x \to \text{Root}[-1 + 4 \#1 - 4 \#1^2 + 2 \#1^{10} \&, 7]\}, \{x \to \text{Root}[-1 + 4 \#1 - 4 \#1^2 + 2 \#1^{10} \&, 8]\}\}$$

After univariate polynomials, the second important set of equations that always will be solved by Solve are systems of univariate polynomials. Now, it is no longer easily predicted how many solutions we will have (see the remarks about this subject in the discussion of NSolve in Chapter 1 of the Numerics volume [1737].) Here are two examples of polynomial and algebraic equations and systems of equations that Solve can handle. Here are two examples. Often, the results returned by Solve for systems of polynomials are quite large. So, we will only display one of the solutions.

```
In[70]:= sol = Solve[{x + y + z == 1,
                x^2 + y^2 + z^2 == 4,
                x^3 + y^3 + z^3 == 9}, {x, y, z}];
```

```
In[71]:= Length[sol]
```

$$Out[71]= \; 6$$

```
In[72]:= sol[[1]]
```

$$Out[72]= \; \Big\{x \to \frac{1}{3} - \frac{1}{18} (5076 - 1134 \sqrt{14})^{1/3} + \frac{(94 - 21 \sqrt{14})^{1/3}}{6 \cdot 2^{2/3}} - $$
$$\frac{(94 + 21 \sqrt{14})^{1/3}}{6 \cdot 2^{2/3}} - \frac{1}{12} \sqrt{(88 - 3 (2 (94 - 21 \sqrt{14}))^{2/3} -}$$
$$3 (2 (94 + 21 \sqrt{14}))^{2/3} + 2 \cdot 2^{2/3} ((94 - 21 \sqrt{14}) (94 + 21 \sqrt{14}))^{1/3}),$$
$$y \to \frac{1}{72} \Big(24 - 6 (2 (94 - 21 \sqrt{14}))^{1/3} - 6 (2 (94 + 21 \sqrt{14}))^{1/3} +$$
$$6 \sqrt{(88 - 3 (2 (94 - 21 \sqrt{14}))^{2/3} - 3 (2 (94 + 21 \sqrt{14}))^{2/3} +}$$
$$2 \cdot 2^{2/3} ((94 - 21 \sqrt{14}) (94 + 21 \sqrt{14}))^{1/3})\Big),$$
$$z \to \frac{1}{3} + \frac{1}{18} (5076 - 1134 \sqrt{14})^{1/3} + \frac{(94 + 21 \sqrt{14})^{1/3}}{3 \cdot 2^{2/3}}\Big\}$$

```
In[73]:= sol = Solve[{x^3 + y^5 + x y == 3,
                x^3 + y^5 + x^2 y^2 == 5}, {x, y}];
```

```
In[74]:= Length[sol]
```

$$Out[74]= \; 16$$

```
In[75]:= sol[[1]]
```

$$Out[75]= \; \Big\{x \to \frac{1}{7892} \Big(648 + 264 \,\text{Root}[-1 - 4 \#1^3 + \#1^8 \&, 1] + 30141 \,\text{Root}[-1 - 4 \#1^3 + \#1^8 \&, 1]^2 +$$
$$2295 \,\text{Root}[-1 - 4 \#1^3 + \#1^8 \&, 1]^3 + 935 \,\text{Root}[-1 - 4 \#1^3 + \#1^8 \&, 1]^4 -$$
$$5465 \,\text{Root}[-1 - 4 \#1^3 + \#1^8 \&, 1]^5 - 1089 \,\text{Root}[-1 - 4 \#1^3 + \#1^8 \&, 1]^6 -$$

$$7678\ \text{Root}[-1 - 4\,\#1^3 + \#1^8\ \&,\ 1]^7 + 324\ \text{Root}[-1 - 4\,\#1^3 + \#1^8\ \&,\ 1]^8 +$$

$$132\ \text{Root}[-1 - 4\,\#1^3 + \#1^8\ \&,\ 1]^9 + 4219\ \text{Root}[-1 - 4\,\#1^3 + \#1^8\ \&,\ 1]^{10} +$$

$$297\ \text{Root}[-1 - 4\,\#1^3 + \#1^8\ \&,\ 1]^{11} + 121\ \text{Root}[-1 - 4\,\#1^3 + \#1^8\ \&,\ 1]^{12} -$$

$$243\ \text{Root}[-1 - 4\,\#1^3 + \#1^8\ \&,\ 1]^{13} - 99\ \text{Root}[-1 - 4\,\#1^3 + \#1^8\ \&,\ 1]^{14} -$$

$$698\ \text{Root}[-1 - 4\,\#1^3 + \#1^8\ \&,\ 1]^{15}),\ y \to \text{Root}[-1 - 4\,\#1^3 + \#1^8\ \&,\ 1]\}$$

Here is an example of a system of polynomial equations that can be solved by `Solve` but produces a fairly large output (of size about 4 MB). (For a "hand" solution, see [1665], [1351], [1493], and [1630]; this is the set of equations describing the circles of Apollonius shown in Chapter 1 of the Programming volume [1735].) The geometric interpretation of this system of equations as circles inside a triangle is obvious.

```
In[76]:= (* three circles touching a fourth one *)
        sol = Solve[{(x - x1)^2 + (y - y1)^2 == (r1 + r)^2,
                    (x - x2)^2 + (y - y2)^2 == (r2 + r)^2,
                    (x - x3)^2 + (y - y3)^2 == (r3 + r)^2}, {r, x, y}];

        LeafCount[sol]
Out[79]= 300381

In[80]:= sol // Short[#, 8] &
Out[80]//Short=
```

$$\left\{\left\{r \to \frac{1}{2\,r2 - 2\,r3}\left(-r2^2 + r3^2 + \ll8\gg + \frac{\ll1\gg}{\ll1\gg} - \right.\right.\right.$$

$$\frac{2\,x2\,(-r1^2\,r2 + \ll33\gg + \frac{r2\,y3\,(\ll1\gg)}{\ll1\gg + \ll93\gg + \ll1\gg})}{2\,r2\,x1 - 2\,r3\,x1 - 2\,r1\,x2 + \ll1\gg + 2\,r1\,x3 - 2\,r2\,x3} + \left(2\,x3\right.$$

$$\left(-r1^2\,r2 + r1\,r2^2 + r1^2\,r3 - r2^2\,r3 - r1\ \ll1\gg + \ll22\gg + \frac{\ll1\gg}{\ll1\gg} - \frac{r3\,y2\,(\ll1\gg)}{\ll1\gg} - \right.$$

$$\left.\left.\frac{r1\,y3\,(\ll1\gg)}{-4\,r2^2\,x1^2 + \ll93\gg + 4\,x2^2\,y3^2} + \frac{r2\,y3\,(\ll1\gg)}{-4\,r2^2\,x1^2 + \ll93\gg + 4\,x2^2\,y3^2}\right)\right)\bigg/$$

$$\left.\left(2\,r2\,x1 - 2\,r3\,x1 - 2\,r1\,x2 + 2\,r3\,x2 + 2\,r1\,x3 - 2\,r2\,x3\right)\right),$$

$$x \to \frac{-r1^2\,r2 + r1\,r2^2 + \ll32\gg + \frac{r2\,y3\,(\ll1\gg)}{-4\,r2^2\,x1^2 + \ll93\gg + 4\,x2^2\,y3^2}}{2\,r2\,x1 - 2\,r3\,x1 - 2\,r1\,x2 + 2\,r3\,x2 + 2\,r1\,x3 - 2\,r2\,x3},$$

$$y \to$$

$$\ll1\gg / (2\,(-4\,r2^2\,x1^2 + 8\,r2\,r3\,x1^2 - 4\,r3^2\,x1^2 + \ll84\gg +$$

$$8\,r1\,r2\,y3^2 - 4\,r2^2\,y3^2 + 4\,x1^2\,y3^2 - 8\,x1\,x2\,y3^2 + 4\,x2^2\,y3^2))\},$$

$$\left\{r \to \frac{-r2^2 + r3^2 + \ll11\gg + \frac{2\,x3\,(\ll1\gg)}{\ll1\gg}}{2\,r2 - 2\,r3},\ x \to \frac{\ll1\gg}{\ll1\gg},\right.$$

$$y \to$$

$$\left.\frac{\ll288\gg + \ll1\gg}{2\,(\ll1\gg)}\right\}\right\}$$

Note the form of the solutions produced by `Solve` for solutions with respect to more than one variable.

```
In[81]:= Solve[{x^2 == 4, y^2 == 9}, {x, y}]
Out[81]= {{x → -2, y → -3}, {x → -2, y → 3}, {x → 2, y → -3}, {x → 2, y → 3}}
```

> `Solve` produces a list of lists. The inner lists are the possible solutions and contain the values of the various variables to be solved for in the form of rules.

This is similar to the results of `NDSolve` that have already been discussed.

Solve does not always produce large expressions as in the example above. Here are two examples of polynomial equations whose inverses also have a polynomial structure.

In[82]:= **Solve[{X == x - 2(x z + y^2) y - (x z + y^2)^2 z,**
              Y == y + (x z + y^2) z,
              Z == z}, {x, y, z}]

Out[82]= $\{\{x \to X + 2 Y^3 + 2 X Y Z - Y^4 Z - 2 X Y^2 Z^2 - X^2 Z^3, z \to Z, y \to Y - Y^2 Z - X Z^2\}\}$

In[83]:= **Solve[{X == x - (z x + y w) w, Y == y + (z x + y w) z,**
              Z == z, W == w}, {x, y, z, w}]

Out[83]= $\{\{x \to X + W^2 Y + W X Z, y \to Y - W Y Z - X Z^2, w \to W, z \to Z\}\}$

This behavior is the exception, not the rule. For more on this situation, see [591], [1621], and [744].

Let us deal with a slightly more complicated example. Given the polynomial

$$(x - x_1)(x - x_2) \cdots (x - x_n) = \sum_{i=0}^{n} c_i x^i,$$

we have the following Vieta-relations: $c_n = 1$, $c_{n-1} = -x_1 - x_2 - \cdots - x_n$, ..., $c_0 = x_1 x_2 \cdots x_n$. Setting now $c_{n-1} = c_{n-2} = \cdots = c_1 = 0$ and $c_0 = 1$, the resulting polynomial is $x^n - 1 = 0$. The solutions of $x^n - 1 = 0$ are clearly also solutions for $c_{n-1} = c_{n-2} = \cdots = c_1 = 0$, $c_0 = 1$ [1862]. But given the system $0 = -x_1 - x_2 - \cdots - x_n$, ..., $1 = x_1 x_2 \cdots x_n$, we will solve it for the x_i. The function cyclicEquations produces the equation $c_{n-1} = c_{n-2} = \cdots = c_1 = 0$, $c_0 = 1$ for a given n.

In[84]:= **cyclicEquations[n_] :=**
         MapIndexed[If[#2 === {1}, #1 == 1, #1 == 0]&,
                        Drop[CoefficientList[Product[x - x[i], {i, n}], x], -1]]

Here are the resulting equations for $n = 3$.

In[85]:= **cyclicEquations[3]**

Out[85]= $\{-x[1] x[2] x[3] == 1, x[1] x[2] + x[1] x[3] + x[2] x[3] == 0, -x[1] - x[2] - x[3] == 0\}$

For $n = 2$, we have two solutions.

In[86]:= **Solve[cyclicEquations[2], Array[x, 2]] // Union**

Out[86]= $\{\{x[1] \to -i, x[2] \to i\}, \{x[1] \to i, x[2] \to -i\}\}$

In[87]:= **cyclicEquations[2] /. %**

Out[87]= {{True, True}, {True, True}}

For $n = 3$, we have six solutions. We turn off the N::meprec message because it would be issued in the solution verification process.

In[88]:= **Off[N::meprec]**
         csol3 = Solve[cyclicEquations[3], Array[x, 3]] // Union

Out[89]= $\{\{x[1] \to -1, x[2] \to 1 - (-1)^{1/3}, x[3] \to (-1)^{1/3}\},$
         $\{x[1] \to 1 - (-1)^{1/3}, x[2] \to -1, x[3] \to (-1)^{1/3}\},$
         $\{x[1] \to 1 - (-1)^{1/3}, x[2] \to (-1)^{1/3}, x[3] \to -1\},$
         $\{x[1] \to 1 + (-1)^{2/3}, x[2] \to -1, x[3] \to -(-1)^{2/3}\},$
         $\{x[1] \to 1 + (-1)^{2/3}, x[2] \to -(-1)^{2/3}, x[3] \to -1\},$
         $\{x[1] \to -(-1)^{1/3} + (-1)^{2/3}, x[2] \to (-1)^{1/3}, x[3] \to -(-1)^{2/3}\}\}$

Using RootReduce, we can rewrite the solutions in a slightly nicer form.

In[90]:= **RootReduce[csol3]**

Out[90]= $\{\{x[1] \to -1, x[2] \to \frac{1}{2}(1 - i\sqrt{3}), x[3] \to \frac{1}{2}(1 + i\sqrt{3})\},$

$\{x[1] \to \frac{1}{2}(1 - i\sqrt{3}), x[2] \to -1, x[3] \to \frac{1}{2}(1 + i\sqrt{3})\},$

$\{x[1] \to \frac{1}{2}(1 - i\sqrt{3}), x[2] \to \frac{1}{2}(1 + i\sqrt{3}), x[3] \to -1\},$

$\{x[1] \to \frac{1}{2}(1 + i\sqrt{3}), x[2] \to -1, x[3] \to \frac{1}{2}(1 - i\sqrt{3})\},$

$\{x[1] \to \frac{1}{2}(1 + i\sqrt{3}), x[2] \to \frac{1}{2}(1 - i\sqrt{3}), x[3] \to -1\},$

$\{x[1] \to -1, x[2] \to \frac{1}{2}(1 + i\sqrt{3}), x[3] \to \frac{1}{2}(1 - i\sqrt{3})\}\}$

As expected, the six solutions are just permutations of one fundamental solution.

In[91]:= `Sort[{x[1], x[2], x[3]} /. %] ===`
`    Sort[Permutations[{x[1], x[2], x[3]} /. %[[1]]]]`
Out[91]= `True`

Here is a quick numerical check for the solution.

In[92]:= `N[Apply[Subtract, cyclicEquations[3], {1}] /. csol3, 22] //`
`                                                    Abs // Max`
Out[92]= $0. \times 10^{-72}$

For $n = 4$, we have $4 != 24$ solutions.

In[93]:= `csol4 = Solve[cyclicEquations[4], Array[x, 4]] // Union;`

In[94]:= `Length[csol4]`
Out[94]= 24

Now, the solutions look a bit complicated.

In[95]:= `csol4[[1, 1]] // Simplify`

Out[95]= $x[1] \to \frac{1}{12(-5+3\sqrt{3})^{2/3}}\left((-5 - 5\,i)(-2)^{5/6} + (3 + 3\,i)(-2)^{5/6}\sqrt{3} - \right.$

$2(-1)^{5/12}2^{2/3}(-5+3\sqrt{3})^{1/3} - 4(-1)^{3/4}(-5+3\sqrt{3})^{2/3} + 2(-1)^{5/6}$

$\left. \sqrt{6\left((-1)^{1/6}2^{1/3}(-5+3\sqrt{3})^{2/3} + 2(-1)^{5/6}(-5+3\sqrt{3})^{4/3} + i\,2^{2/3}(-26+15\sqrt{3})\right)} \right)$

In[96]:= `(* restore message original state *)`
`        On[N::meprec]`

Instead of explicitly solving the equations, it is often useful to look at the structure of the solution. Inspecting a lexicographic Gröbner basis often gives valuable information. We see a quartic equation in x[4], a cubic in x[3], a quadratic in x[2], and a linear one in x[1].

In[98]:= `GroebnerBasis[Apply[Subtract, cyclicEquations[4], {1}], Array[x, 4]]`
Out[98]= $\{-1 - x[4]^4, x[3]^3 + x[3]^2 x[4] + x[3] x[4]^2 + x[4]^3,$
$-x[2]^2 - x[2] x[3] - x[3]^2 - x[2] x[4] - x[3] x[4] - x[4]^2, x[1] + x[2] + x[3] + x[4]\}$

Here is the structure of the Gröbner basis for $n = 8$. It is easy to see that solving this triangular system would result in $8 != 40320$ solutions, all permutations of a single solution [1861].

In[99]:= `Exponent[#, Array[x, 8]]& /@`
`        GroebnerBasis[Apply[Subtract, cyclicEquations[8], {1}], Array[x, 8]]`

Out[99]= {{0, 0, 0, 0, 0, 0, 0, 8}, {0, 0, 0, 0, 0, 0, 7, 7},
    {0, 0, 0, 0, 0, 6, 6, 6}, {0, 0, 0, 0, 5, 5, 5, 5}, {0, 0, 0, 4, 4, 4, 4, 4},
    {0, 0, 3, 3, 3, 3, 3, 3}, {0, 2, 2, 2, 2, 2, 2, 2}, {1, 1, 1, 1, 1, 1, 1, 1}}

The third class of equations Solve deals with systems of algebraic equations. By introducing auxiliary variables, it is always possible to rewrite such equations as systems of polynomials (the case we just discussed). Here is an example. The following pair of equations can be solved through a complicated sum of Root-objects. We have two solutions.

In[100]:= **sol = Solve[{x^2 + x^(1/2) == a^(1/3) x,**
                   a x + Sqrt[a x] == 2}, {a, x}];

In[101]:= **Length[sol]**

Out[101]= 2

In[102]:= **sol[[1]]**

Out[102]= $\left\{ a \rightarrow \frac{1}{981444059} \right.$

$\left(66190529663 - 173516674881 \, \text{Root}[-1 + \#1 - 6\,\#1^2 + 3\,\#1^3 - 2\,\#1^5 - 3\,\#1^6 + \#1^9 \,\&, \, 6] + \right.$

$509834475585 \, \text{Root}[-1 + \#1 - 6\,\#1^2 + 3\,\#1^3 - 2\,\#1^5 - 3\,\#1^6 + \#1^9 \,\&, \, 6]^2 -$

$824461295803 \, \text{Root}[-1 + \#1 - 6\,\#1^2 + 3\,\#1^3 - 2\,\#1^5 - 3\,\#1^6 + \#1^9 \,\&, \, 6]^3 +$

$350875125684 \, \text{Root}[-1 + \#1 - 6\,\#1^2 + 3\,\#1^3 - 2\,\#1^5 - 3\,\#1^6 + \#1^9 \,\&, \, 6]^4 +$

$103114768206 \, \text{Root}[-1 + \#1 - 6\,\#1^2 + 3\,\#1^3 - 2\,\#1^5 - 3\,\#1^6 + \#1^9 \,\&, \, 6]^5 -$

$201961328805 \, \text{Root}[-1 + \#1 - 6\,\#1^2 + 3\,\#1^3 - 2\,\#1^5 - 3\,\#1^6 + \#1^9 \,\&, \, 6]^6 -$

$253625653071 \, \text{Root}[-1 + \#1 - 6\,\#1^2 + 3\,\#1^3 - 2\,\#1^5 - 3\,\#1^6 + \#1^9 \,\&, \, 6]^7 +$

$81090341085 \, \text{Root}[-1 + \#1 - 6\,\#1^2 + 3\,\#1^3 - 2\,\#1^5 - 3\,\#1^6 + \#1^9 \,\&, \, 6]^8 -$

$141463345653 \, \text{Root}[-1 + \#1 - 6\,\#1^2 + 3\,\#1^3 - 2\,\#1^5 - 3\,\#1^6 + \#1^9 \,\&, \, 6]^9 -$

$2543840721 \, \text{Root}[-1 + \#1 - 6\,\#1^2 + 3\,\#1^3 - 2\,\#1^5 - 3\,\#1^6 + \#1^9 \,\&, \, 6]^{10} -$

$64898102529 \, \text{Root}[-1 + \#1 - 6\,\#1^2 + 3\,\#1^3 - 2\,\#1^5 - 3\,\#1^6 + \#1^9 \,\&, \, 6]^{11} +$

$1298518114 \, \text{Root}[-1 + \#1 - 6\,\#1^2 + 3\,\#1^3 - 2\,\#1^5 - 3\,\#1^6 + \#1^9 \,\&, \, 6]^{12} +$

$43860231228 \, \text{Root}[-1 + \#1 - 6\,\#1^2 + 3\,\#1^3 - 2\,\#1^5 - 3\,\#1^6 + \#1^9 \,\&, \, 6]^{13} +$

$25794421614 \, \text{Root}[-1 + \#1 - 6\,\#1^2 + 3\,\#1^3 - 2\,\#1^5 - 3\,\#1^6 + \#1^9 \,\&, \, 6]^{14} +$

$34464612 \, \text{Root}[-1 + \#1 - 6\,\#1^2 + 3\,\#1^3 - 2\,\#1^5 - 3\,\#1^6 + \#1^9 \,\&, \, 6]^{15} -$

$149063679 \, \text{Root}[-1 + \#1 - 6\,\#1^2 + 3\,\#1^3 - 2\,\#1^5 - 3\,\#1^6 + \#1^9 \,\&, \, 6]^{16} -$

$\left. 4295132733 \, \text{Root}[-1 + \#1 - 6\,\#1^2 + 3\,\#1^3 - 2\,\#1^5 - 3\,\#1^6 + \#1^9 \,\&, \, 6]^{17} \right),$

$\left. x \rightarrow \text{Root}[-1 + \#1 - 6\,\#1^2 + 3\,\#1^3 - 2\,\#1^5 - 3\,\#1^6 + \#1^9 \,\&, \, 6] \right\}$

A quick numerical check of the solutions is carried out by the following input.

In[103]:= **({x^2 + x^(1/2) - a^(1/3) x, a x + Sqrt[a x] - 2} /. sol) //**
                               SetPrecision[#, 100]&

Out[103]= {{$0. \times 10^{-99} + 0. \times 10^{-99}$ i, $0. \times 10^{-98} + 0. \times 10^{-98}$ i},
    {$0. \times 10^{-99} + 0. \times 10^{-99}$ i, $0. \times 10^{-98} + 0. \times 10^{-98}$ i}}

So far, we have discussed Solve for inputs where it will always (in principle) be successful. The algorithms built into Solve allow for an exhaustive treatment of polynomial inputs or algebraic input with no parameters. But the actual running times may sometimes be quite large. If the reader has a complicated system to be solved and Solve seems not able to find a solution in a reasonable amount of time, it is often useful to "massage" the system. There is no general rule of what kind of "massaging" might be useful (if there were one, *Mathematica* would do it). Typically, useful strategies are the elimination of linear variables "by hand" and appropriate

"Together-ing", the elimination of high multiplicities, precalculating a Gröbner basis using GroebnerBasis, and others. The reader will find at the end of this section some more detailed remarks.

Now, let us discuss the next set of equations: Transcendental equations. Here is one of the simplest examples.

In[104]:= **Solve[Sin[x] == a, x]**

>   Solve::ifun : Inverse functions are being used by Solve, so some solutions
>   may not be found; use Reduce for complete solution information. More…

Out[104]= {{x → ArcSin[a]}}

With trigonometric functions, this message is a typical "problem" with the use of Solve.

> For functions whose inverses are multivalued (with the exception of polynomials), Solve typically will not compute all solutions!

The last example issued a message. This message is connected to one of the options of Solve. Here are the options of Solve.

In[105]:= **Options[Solve]**

Out[105]= {InverseFunctions → Automatic, MakeRules → False, Method → 3, Mode → Generic,
        Sort → True, VerifySolutions → Automatic, WorkingPrecision → ∞}

The first option of Solve is InverseFunctions. It is this message and its setting True that caused the above message.

> InverseFunctions
>
>   is an option for Solve, and it determines whether inverse functions are to be used in finding the solution.
>
> **Default:**
>
>   True (uses inverse functions and gives a warning that all solutions may not have been found)
>
> **Admissible:**
>
>   False

With this option set to False, the following simple equation is no longer solved.

In[106]:= **Solve[Cos[x] == a, x, InverseFunctions -> False]**

>   Solve::ifun2 :
>   Cannot obtain solution with InverseFunctions->False option setting. More…

Out[106]= Solve[Cos[x] == a, x, InverseFunctions → False]

Here are three more examples. The first one is rather trivial; in the second example, inverse functions are needed three times (the variable to solve for is the difference in the three inputs); and in the third example, f is not found as an argument of a nonhead symbol.

In[107]:= **Solve[f[f[f[x]]] == a, {#}]& /@ {a, x, f}**

>   InverseFunction::ifun : Inverse functions are
>   being used. Values may be lost for multivalued inverses. More…
>   InverseFunction::ifun : Inverse functions are
>   being used. Values may be lost for multivalued inverses. More…
>   InverseFunction::ifun : Inverse functions are
>   being used. Values may be lost for multivalued inverses. More…
>   General::stop : Further output of
>   InverseFunction::ifun will be suppressed during this calculation. More…

In[107]:= `{{{a → f[f[f[f[x]]]]}}, {{x → f^(-1) [f^(-1) [f^(-1) [a]]]}}, {{}}}`

Mathematica can solve complicated equations and systems of equations involving trigonometric functions. Here, a slightly complicated equation in trigonometric functions is solved.

In[108]:= `sol = Solve[Sin[2 x] + Cos[3 x] + Tan[x/5] == 0, x];`

> Solve::ifun : Inverse functions are being used by Solve, so some solutions
> may not be found; use Reduce for complete solution information. More…

The individual solutions are expressed in Root-objects.

In[109]:= `sol[[1]]`

Out[109]= $\{x \rightarrow$

$-5 \, \text{ArcCos}\big[-\sqrt{\text{Root}[-1 - 19 \, \#1 + 465 \, \#1^2 - 15164 \, \#1^3 + 456192 \, \#1^4 - 7366272 \, \#1^5 + 69922304}$

$\#1^6 - 427648000 \, \#1^7 + 1782784000 \, \#1^8 - 5236326400 \, \#1^9 + 11024793600 \, \#1^{10} -$

$16712990720 \, \#1^{11} + 18087936000 \, \#1^{12} - 13631488000 \, \#1^{13} +$

$6794772480 \, \#1^{14} - 2013265920 \, \#1^{15} + 268435456 \, \#1^{16} \, \&, 1]\big]\}\}$

Here is a quick check that the solutions are correct.

In[110]:= `Sin[2 x] + Cos[3 x] + Tan[x/5] /. N[sol, 30] // Abs // Max`

Out[110]= $0. \times 10^{-28}$

In addition to trigonometric and inverse trigonometric equations, a large set of equations containing logarithms and exponentials that can be solved by the so-called ProductLog function (we will discuss this function in Chapter 3). ProductLog[z] = $W(z)$ is (one of) the solution(s) of the equation $z = W(z) \exp(W(z))$. Here are two examples of such equations.

In[111]:= `Solve[Log[x] == x^γ, x]`

> InverseFunction::ifun : Inverse functions are
> being used. Values may be lost for multivalued inverses. More…

> Solve::ifun : Inverse functions are being used by Solve, so some solutions
> may not be found; use Reduce for complete solution information. More…

Out[111]= $\left\{\left\{x \rightarrow \left(-\dfrac{\text{ProductLog}[-\gamma]}{\gamma}\right)^{\frac{1}{\gamma}}\right\}\right\}$

In[112]:= `Solve[Exp[x]^n Exp[α Exp[x]^m] == β, x]`

> InverseFunction::ifun : Inverse functions are
> being used. Values may be lost for multivalued inverses. More…

> Solve::ifun : Inverse functions are being used by Solve, so some solutions
> may not be found; use Reduce for complete solution information. More…

Out[112]= $\left\{\left\{x \rightarrow \dfrac{\text{Log}[\beta]}{n} - \dfrac{\text{ProductLog}\left[\frac{m \, \alpha \, \beta^{m/n}}{n}\right]}{m}\right\}\right\}$

Although *Mathematica* can (modulo a potentially long runtime) solve every system of polynomial equations, systems of transcendental equations are much more difficult to treat, and "helping" *Mathematica* is often required. Here is a system consisting of one polynomial and one transcendental equation. *Mathematica* cannot currently solve this system.

In[113]:= `Solve[{2^x + 4^y == 9, x == y}, {x, y}]`

> Solve::incnst : Inconsistent or redundant transcendental equation.
> After reduction, the bad equation is $\dfrac{\text{Log}[2^x]}{\text{Log}[2]} - \dfrac{\text{Log}[2^y]}{\text{Log}[2]} == 0$. More…

> Solve::incnst : Inconsistent or redundant transcendental equation.
> After reduction, the bad equation is $\dfrac{\text{Log}[2^x]}{\text{Log}[2]} - \dfrac{\text{Log}[2^y]}{\text{Log}[2]} == 0$. More…

```
Solve::ifun : Inverse functions are being used by Solve, so some solutions
    may not be found; use Reduce for complete solution information. More...

Solve::svars : Equations may not give solutions for all "solve" variables. More...
```

Out[113]= $\left\{\left\{x \to \dfrac{\text{Log}[2^y]}{\text{Log}[2]}\right\}\right\}$

Using the second equation in the "obvious" way results in one equation that *Mathematica* can now solve.

In[114]:= **Solve[2^x + 4^x == 9, x]**

```
Solve::ifun : Inverse functions are being used by Solve, so some solutions
    may not be found; use Reduce for complete solution information. More...
```

Out[114]= $\left\{\left\{x \to \dfrac{\text{Log}\left[\frac{1}{2}\left(-1+\sqrt{37}\right)\right]}{\text{Log}[2]}\right\}, \left\{x \to \dfrac{i\,\pi + \text{Log}\left[\frac{1}{2}\left(1+\sqrt{37}\right)\right]}{\text{Log}[2]}\right\}\right\}$

Another option of `Solve` is `WorkingPrecision`.

`WorkingPrecision`

> is an option for `Solve`, and it determines the number of digits to be used in carrying out the calculation.

Default:

> `Infinity` (symbolic computation)

Admissible:

> *arbitraryPositiveInteger*

With `WorkingPrecision` less than `Infinity`, we get a numerical result. In such cases, we would normally use NSolve.

In[115]:= **Solve[{x + y == 2, x + y^3 == 5}, {x, y}, WorkingPrecision -> 22]**

Out[115]= $\{\{x \to 0.32830011834283903 0252,\ y \to 1.67169988165716096975\},$
$\{x \to 2.83584994082858048487 - 1.04686931884998162528\,i,$
$\quad y \to -0.83584994082858048487 + 1.04686931884998162528\,i\},$
$\{x \to 2.83584994082858048487 + 1.04686931884998162528\,i,$
$\quad y \to -0.83584994082858048487 - 1.04686931884998162528\,i\}\}$

This gives the same result.

In[116]:= **NSolve[{x + y == 2, x + y^3 == 5}, {x, y}, WorkingPrecision -> 22]**

Out[116]= $\{\{x \to 2.83584994082858048 4874 - 1.046869318 84998 1625281\,i,$
$\quad y \to -0.83584994082858048 4874 + 1.0468693188499816 25281\,i\},$
$\{x \to 2.83584994082858048 4874 + 1.0468693188499816 25281\,i,$
$\quad y \to -0.83584994082858048 4874 - 1.0468693188499816 25281\,i\},$
$\{x \to 0.32830011834283903 0252,\ y \to 1.67169988165716096 9748\}\}$

In[117]:= **Sort[%] == Sort[%%]**

Out[117]= True

The next option that we want to discuss is `VerifySolutions`.

VerifySolutions

is an option for Solve, and instructs Solve to determine whether the solution found is really a solution of the given equation(s).

Default:

Automatic

Admissible:

True or False

The following situation can happen for nonpolynomial equations: Solve computes one or more "solutions" that not only fail to be solutions for some particular parameter value (i.e., they are not generic solutions), but also fail for a set of parameters of finite Lebesgue measure. (This is because certain nonequivalent transformations such as exponentiation are used which generate spurious solutions in the computation.) Here is an example of an algebraic equation containing the parameter a.

In[118]:= **Timing[(sol = Solve[x^(1/3) + x^(1/2) == a, x])];**

In[119]:= **Length[sol]**

Out[119]= 3

Here is one of the three solutions. It contains radicals of polynomials in a.

In[120]:= **sol[[1]]**

Out[120]= $\{x \rightarrow$

$\frac{1}{3} (1 - 6 a + 3 a^2) - (2^{1/3} (-1 + 12 a - 42 a^2 + 42 a^3)) / (3 (2 - 36 a + 234 a^2 - 666 a^3 + 783 a^4 -$

$216 a^5 + 3 \sqrt{3} \sqrt{-36 a^7 + 435 a^8 - 1552 a^9 + 1728 a^{10}})^{1/3}) +$

$\frac{1}{3\, 2^{1/3}} ((2 - 36 a + 234 a^2 - 666 a^3 + 783 a^4 - 216 a^5 +$

$3 \sqrt{3} \sqrt{-36 a^7 + 435 a^8 - 1552 a^9 + 1728 a^{10}})^{1/3})\}$

We look at the three solutions as a function of a by individually inserting them into the original equation. We see that the solutions depend explicitly on the value of a. Here, the behavior of the backsubstituted solution along the real axis is shown.

In[121]:= (* the three single solutions *)
        {pp[1, a_], pp[2, a_], pp[3, a_]} = (x /. sol);

        Module[{x},
        Show[GraphicsArray[Table[
        (* pictures of the three solutions used in the equation *)
        Plot[x = pp[i, a] // N;
            Chop[Abs[x^(1/3) + x^(1/2) - a]], {a, -1/2, 1/2},
            DisplayFunction -> Identity, PlotPoints -> 50,
            Axes -> False, PlotRange -> All, Frame -> True,
            PlotStyle -> {Thickness[0.002], Hue[0]}], {i, 3}]]]]

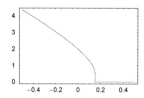 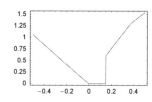 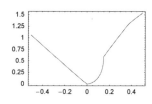

One clearly sees that none of the solutions is correct for negative a. A view of the backsubstituted solution in the complex a-plane shows a complicated behavior of the solutions.

```
In[124]:= Module[{x},
        Show[GraphicsArray[Table[
        (* pictures of the three solutions used in the equation *)
        Plot3D[x = pp[i, ar + I ai] // N;
              Chop[Abs[x^(1/3) + x^(1/2) - (ar + I ai)]],
              {ar, -1/2, 1/2}, {ai, -1/2, 1/2},
              DisplayFunction -> Identity, PlotPoints -> 90,
              Mesh -> False, PlotRange -> All], {i, 3}]]]]
```

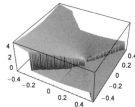 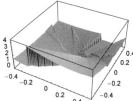 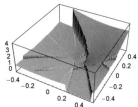

It is in such situations as shown in the last graphic, in which we have a complicated "parameter dependence", in which the option VerifySolutions becomes useful. With VerifySolutions -> True, a check of the correctness (as far as possible) automatically takes place. Here, the solutions are not verified.

```
In[125]:= With[{a = -3},
         Solve[x^(1/3) + x^(1/2) == a, x, VerifySolutions -> False]]
```

$$Out[125]= \left\{\left\{x \to \frac{46}{3} + \frac{1}{3}\left(\frac{136109}{2} - \frac{6561\sqrt{85}}{2}\right)^{1/3} + \frac{1}{3}\left(\frac{1}{2}(136109 + 6561\sqrt{85})\right)^{1/3}\right\},\right.$$

$$\left\{x \to \frac{46}{3} - \frac{1}{6}(1 + i\sqrt{3})\left(\frac{136109}{2} - \frac{6561\sqrt{85}}{2}\right)^{1/3} - \right.$$

$$\left.\frac{1}{6}(1 - i\sqrt{3})\left(\frac{1}{2}(136109 + 6561\sqrt{85})\right)^{1/3}\right\},$$

$$\left\{x \to \frac{46}{3} - \frac{1}{6}(1 - i\sqrt{3})\left(\frac{136109}{2} - \frac{6561\sqrt{85}}{2}\right)^{1/3} - \right.$$

$$\left.\frac{1}{6}(1 + i\sqrt{3})\left(\frac{1}{2}(136109 + 6561\sqrt{85})\right)^{1/3}\right\}\right\}$$

In the process of verifying the above three solutions, all three become rejected.

```
In[126]:= With[{a = -3},
         Solve[x^(1/3) + x^(1/2) == a, x, VerifySolutions -> True]]
Out[126]= {}
```

With the default setting VerifySolutions -> Automatic, no wrong solutions are returned for our example.

```
In[127]:= With[{a = -3},
            Solve[x^(1/3) + x^(1/2) == a, x, VerifySolutions -> Automatic]]
Out[127]= {}
```

Even with `VerifySolutions -> True`, one should be a bit skeptical of the results for equations containing radical expressions that have parameters, because it is very difficult to test with respect to all possible values of a.

```
In[128]:= Solve[x^(1/3) + x^(1/2) == a, x, VerifySolutions -> True] === sol
Out[128]= True
```

For the verification process, `Solve` will use numerical techniques. This means it can be fooled. So, for the following input of a cubic equation only one root is returned.

```
In[129]:= verySmallNumber =
            (Sqrt[2 - Sqrt[2 + Sqrt[2]]]/2 - Sin[Pi/16] + 10^-200);
            Solve[verySmallNumber x^3 + x + 1 == 0, x, VerifySolutions -> True]
Out[130]= {{x -> -1}}
```

> The solutions produced by `Solve` for equations containing multivalued functions should always be checked by substitution back into the original equation and then substitution of some random numerical values for the variables to check the solution!

At the end of our discussion of `Solve`, let us shortly discuss two limiting cases: Sometimes a system of equations has no solution at all.

```
In[131]:= Solve[x + 1 == x + 2, {x}]
Out[131]= {}
```

```
In[132]:= Solve[{x^2 + y^3 == 2, x^2 + y^3 == 3}, {x, y}]
Out[132]= {}
```

Note the difference between the case in which a system of equations has no solutions ({} is returned) and the case in which the equations are fulfilled for all values of the solved variables.

```
In[133]:= Solve[x + 1 == x + 1, {x}]
Out[133]= {{}}
```

> If a system of equations has no solution, `Solve` produces {}; if a system of equations is identically fulfilled, {{}} is returned.

While `Solve` tries to solve all solutions, it typically does not simplify them as much as it is possible. The process of simplification is potentially very time-consuming and therefore not carried out by default. Here is an example of a set of two transcendental equations for which `Solve` produces a really large solution. We consider the superposition of two damped oscillations, each of the form $f(x) = \exp(ax + b)\cos(cx + d)$. The superposition a $f(x + \alpha) + \mathrm{b}\, f(x + \beta)$ can be expressed in the form $\mathrm{c}\, f(x + \gamma)$ [1580]. We will determine c and γ as functions of a, b, c, d, a, α, b, β, and γ. Because $f(x)$ obeys a second-order linear differential equation, we match the function values and the first derivative at $x = 0$.

```
In[134]:= f[x_] := Exp[a x + b] Cos[c x + d];

            fSuperposition[x_] = a f[x + α] + b f[x + β] - c f[x + γ];

In[136]:= cγSol = Solve[{fSuperposition[0] == 0,
                         fSuperposition'[0] == 0}, {c, γ}];
```

```
Solve::eqf :  (≪15≫ + ≪14≫ == 0) == 0 is not a well-formed equation. More…
```

```
Solve::ifun :  Inverse functions are being used by Solve, so some solutions
    may not be found; use Reduce for complete solution information. More…
```

The solution consists of 16 branches.

In[137]:= `{Length[cγSol], ByteCount[cγSol]/10^6. "MB"}`

Out[137]= `{16, 384.033 MB}`

Applying `Simplify` results in a much smaller solution. (We carry out the simplification for the first solution only, for the other 15 solutions, one obtains similar results.)

In[138]:= `(cγSolS = Simplify[cγSol[[1]]]) // ByteCount`

Out[138]= `21360`

And applying the function `FullSimplify` gives a still smaller solution, which is small enough to display here.

In[139]:= `cγSolS // FullSimplify`

Out[139]= $\left\{ c \rightarrow \sqrt{e^{2\,a\,\alpha}\,a^2 + e^{2\,a\,\beta}\,b^2 + 2\,e^{a\,(\alpha+\beta)}\,a\,b\,\cos[c\,(\alpha-\beta)]} \right.$

$$\sqrt{\left(e^{a\,(\alpha+\beta)} \left(\cosh\left[\frac{1}{2}\,a\,(\alpha-\beta)\right] (a\,\sin[c\,\alpha] + b\,\sin[c\,\beta]) + (a\,\sin[c\,\alpha] - b\,\sin[c\,\beta])\,\sinh\left[\frac{1}{2}\,a\,(\alpha-\beta)\right] \right)^2 \right)}$$

$$\left. \left(\sqrt{\cosh\left[\frac{a\,\text{ArcSec}\left[-\frac{\sqrt{e^{2\,a\,\alpha}\,a^2 + e^{2\,a\,\beta}\,b^2 + 2\,e^{a\,(\alpha+\beta)}\,a\,b\,\cos[c\,(\alpha-\beta)]}\ (e^{a\,\alpha}\,a\,\sin[c\,\alpha] + e^{a\,\beta}\,b\,\sin[c\,\beta])}{(e^{a\,\alpha}\,a\,\cos[c\,\alpha] + e^{a\,\beta}\,b\,\cos[c\,\beta])\,\sqrt{(e^{a\,\alpha}\,a\,\sin[c\,\alpha] + e^{a\,\beta}\,b\,\sin[c\,\beta])^2}} \right]}{c} \right]}\ + \right.$$

$$\left. \sinh\left[\frac{a\,\text{ArcSec}\left[-\frac{\sqrt{e^{2\,a\,\alpha}\,a^2 + e^{2\,a\,\beta}\,b^2 + 2\,e^{a\,(\alpha+\beta)}\,a\,b\,\cos[c\,(\alpha-\beta)]}\ (e^{a\,\alpha}\,a\,\sin[c\,\alpha] + e^{a\,\beta}\,b\,\sin[c\,\beta])}{(e^{a\,\alpha}\,a\,\cos[c\,\alpha] + e^{a\,\beta}\,b\,\cos[c\,\beta])\,\sqrt{(e^{a\,\alpha}\,a\,\sin[c\,\alpha] + e^{a\,\beta}\,b\,\sin[c\,\beta])^2}} \right]}{c} \right] \right) \right/$$

$$(e^{a\,\alpha}\,a\,\sin[c\,\alpha] + e^{a\,\beta}\,b\,\sin[c\,\beta]),\ \gamma \rightarrow$$

$$-\frac{\text{ArcSec}\left[-\frac{\sqrt{e^{2\,a\,\alpha}\,a^2 + e^{2\,a\,\beta}\,b^2 + 2\,e^{a\,(\alpha+\beta)}\,a\,b\,\cos[c\,(\alpha-\beta)]}\ (e^{a\,\alpha}\,a\,\sin[c\,\alpha] + e^{a\,\beta}\,b\,\sin[c\,\beta])}{(e^{a\,\alpha}\,a\,\cos[c\,\alpha] + e^{a\,\beta}\,b\,\cos[c\,\beta])\,\sqrt{(e^{a\,\alpha}\,a\,\sin[c\,\alpha] + e^{a\,\beta}\,b\,\sin[c\,\beta])^2}} \right]}{c} \right\}$$

Here is an example for concrete parameter values. We show the actual superposition in black and the 16 solutions in gray. One of the solutions is the actual superposition.

In[140]:=
```
Block[{a = -1/3., b = 1., c = 2., d = 1.,
        a = -3., α = 1/2., b = -1/2., β = -2/3.},
    Plot[Evaluate[Flatten[{c f[x + γ] /. cγSol,
                        a f[x + α] + b f[x + β]}] // N],
        {x, 0, 6Pi}, AspectRatio -> 1/3,
        PlotRange -> All, Frame -> True, Axes -> False,
        (* solutions in gray; superposition in black *)
        PlotStyle -> Append[
            Table[{Thickness[0.008], GrayLevel[0.8]}, {Length[cγSol]}],
                {Thickness[0.002], GrayLevel[0]}]]];
```

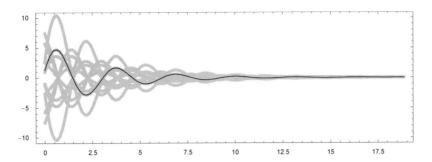

We now return to the problem of "generic" solutions mentioned in the beginning. If we need to find all solutions, and all exceptions to returned generic solutions—that is, including ones associated with special choices of the parameters—we can use Reduce.

Reduce [*equations*, *listOfVariablesToBeSolvedFor*, *domain*, *options*]

   solves (if possible) the equations *equations* for the variables *listOfVariablesToBeSolvedFor* over the domain *domain* using the options *options*. All solutions (including all special cases and degenerate cases) will be produced in the form of a logical Or construction.

Reduce is a very powerful function that deals with many different types of equations, inequations, and inequalities. Here we give some examples of its basic use. For many more details, see the Advanced Documentation.

Here again is the first equation from this section, but now all solutions will be simultaneously computed.

In[141]:= **Reduce [α x == β, x]**

Out[141]= $(\beta == 0 \text{ \&\& } \alpha == 0) \; || \; \left(\alpha \neq 0 \text{ \&\& } x == \dfrac{\beta}{\alpha}\right)$

Polynomial equations are also solved by Reduce.

In[142]:= **Reduce [{a x^2 + y == b, c x + d y == f}, {x, y}]**

Out[142]= $\left(a\,d \neq 0 \text{ \&\& } \left(x == -\dfrac{-c - \sqrt{c^2 + 4\,a\,b\,d^2 - 4\,a\,d\,f}}{2\,a\,d} \; || \; x == -\dfrac{-c + \sqrt{c^2 + 4\,a\,b\,d^2 - 4\,a\,d\,f}}{2\,a\,d}\right) \text{ \&\& }\right.$

$\left. y == b - a\,x^2\right) \; || \; \left(d == 0 \text{ \&\& } c \neq 0 \text{ \&\& } x == \dfrac{f}{c} \text{ \&\& } y == b - a\,x^2\right) \; ||$

$(f == 0 \text{ \&\& } d == 0 \text{ \&\& } c == 0 \text{ \&\& } y == b - a\,x^2) \; || \; \left(c == 0 \text{ \&\& } d \neq 0 \text{ \&\& } b == \dfrac{f}{d} \text{ \&\& } a == 0 \text{ \&\& } y == b\right) \; ||$

$\left(d \neq 0 \text{ \&\& } a == 0 \text{ \&\& } c \neq 0 \text{ \&\& } x == \dfrac{-b\,d + f}{c} \text{ \&\& } y == b\right)$

A convenient way to convert the results returned by Reduce into rules is the function ToRules.

ToRules [*logicalExpression*]

   converts the logical expression *logicalExpression* into a sequence of lists of rules.

The next input converts the last result into a set of rules. Be aware that the head of the next output is Sequence and that non-Equal results have been dropped.

In[143]:= **ToRules [%]**

Out[143]= $\text{Sequence}\Big[\Big\{x \to -\dfrac{-c - \sqrt{c^2 + 4\,a\,b\,d^2 - 4\,a\,d\,f}}{2\,a\,d}, \; y \to b - a\,x^2\Big\},$

$\Big\{x \to -\dfrac{-c + \sqrt{c^2 + 4\,a\,b\,d^2 - 4\,a\,d\,f}}{2\,a\,d}, \; y \to b - a\,x^2\Big\},$

$\Big\{d \to 0, \; x \to \dfrac{f}{c}, \; y \to b - a\,x^2\Big\}, \; \{f \to 0, \; d \to 0, \; c \to 0, \; y \to b - a\,x^2\},$

$\Big\{c \to 0, \; b \to \dfrac{f}{d}, \; a \to 0, \; y \to b\Big\}, \; \Big\{a \to 0, \; x \to \dfrac{-b\,d + f}{c}, \; y \to b\Big\}\Big]$

For the following equation with a rational function on the right-hand side, the situations of a vanishing denominator is treated especially. Here are four cases.

In[144]:= `Reduce[(x - a)/((x^2 - 1)(x - a b)) == 0, x]`

Out[144]= $x == a \;\&\&\; a - a^3 - a\,b + a^3\,b \neq 0$

In contrast, `Solve` finds only the one and two generic solutions for the last two examples.

In[145]:= `Solve[{a x^2 + y == b, c x + d y == f}, {x, y}] // Length`

Out[145]= 2

In[146]:= `Solve[(x - a)/((x^2 - 1)(x - a b)) == 0, x] // Length`

Out[146]= 1

Here is a similar example. We consider the matrix equation $\mathbf{A}.\mathbf{X} = \mathbf{X}.\mathbf{B}$ for the simplest case of 2×2 matrices. This matrix equation has nontrivial solutions only in the case if the matrices $\mathbf{A}$ and $\mathbf{B}$ have at least one common eigenvalue. This means that for generic matrices $\mathbf{A}$ and $\mathbf{B}$, `Solve` will only find the trivial solution because no generic solution exists.

In[147]:= `With[{n = 2},`
        `A = Table[Subscript[a, i, j], {i, n}, {j, n}];`
        `B = Table[Subscript[b, i, j], {i, n}, {j, n}];`
        `X = Table[Subscript[x, i, j], {i, n}, {j, n}]];`

In[148]:= `Solve[A.X == X.B, Flatten[X]]`

Out[148]= $\{\{x_{1,1} \to 0, \; x_{1,2} \to 0, \; x_{2,1} \to 0, \; x_{2,2} \to 0\}\}$

Under appropriate restrictions on the elements of the matrices $\mathbf{A}$ and $\mathbf{B}$, `Reduce` finds nontrivial solutions.

In[149]:= `(redSol = Reduce[A.X == X.B, Flatten[X]]) // Simplify;`
        `Length[redSol]`

Out[150]= 70

Here is one of these solutions.

In[151]:= `redSol[[-1]]`

Out[151]= $\Big(a_{2,2} == \dfrac{1}{2}\Big(b_{1,1} + b_{2,2} - \sqrt{b_{1,1}^2 + 4\,b_{1,2}\,b_{2,1} - 2\,b_{1,1}\,b_{2,2} + b_{2,2}^2}\Big) \;||$

$\quad a_{2,2} == \dfrac{1}{2}\Big(b_{1,1} + b_{2,2} + \sqrt{b_{1,1}^2 + 4\,b_{1,2}\,b_{2,1} - 2\,b_{1,1}\,b_{2,2} + b_{2,2}^2}\Big)\Big) \;\&\&$

$b_{1,1}^2 + 4\,b_{1,2}\,b_{2,1} - 2\,b_{1,1}\,b_{2,2} + b_{2,2}^2 \neq 0 \;\&\&$

$a_{1,1} == (2\,a_{1,2}\,a_{2,1}\,a_{2,2} - a_{1,2}\,a_{2,1}\,b_{1,1} - a_{2,2}\,b_{1,1}^2 + b_{1,1}^3 - 4\,a_{2,2}\,b_{1,2}\,b_{2,1} + 4\,b_{1,1}\,b_{1,2}\,b_{2,1} -$

$\quad a_{1,2}\,a_{2,1}\,b_{2,2} + 2\,a_{2,2}\,b_{1,1}\,b_{2,2} - b_{1,1}^2\,b_{2,2} + 4\,b_{1,2}\,b_{2,1}\,b_{2,2} - a_{2,2}\,b_{2,2}^2 - b_{1,1}\,b_{2,2}^2 + b_{2,2}^3)\;/$

$\quad (b_{1,1}^2 + 4\,b_{1,2}\,b_{2,1} - 2\,b_{1,1}\,b_{2,2} + b_{2,2}^2) \;\&\& -a_{2,2} + b_{1,1} \neq 0 \;\&\& x_{1,2} == \dfrac{b_{1,2}\,x_{1,1}}{-a_{2,2} + b_{1,1}} \;\&\&$

$-2\,a_{2,2} + b_{1,1} + b_{2,2} \neq 0 \;\&\& x_{2,1} == \dfrac{a_{2,1}\,x_{1,1}}{-2\,a_{2,2} + b_{1,1} + b_{2,2}} \;\&\& x_{1,1} \neq 0 \;\&\&$

$x_{2,2} == \dfrac{x_{1,2}\,x_{2,1}}{x_{1,1}} \;\&\& a_{1,2}\,a_{2,1} \neq 0$

As another small application of `Reduce` let us prove that there are no nonsingular real-valued 3×3 matrices $\mathcal{A}$ that fulfill the identity [938], [1867], [341]

$$\mathcal{A} \otimes_{\mathcal{H}} \left(\mathcal{A}^T\right)^{-1} = \frac{1}{3}\,\tilde{\mathbf{1}}_3.$$

Here $\otimes_{\mathcal{H}}$ denotes Hadamard multiplication (the default elementwise multiplication of *Mathematica*), $\left(\mathcal{A}^T\right)^{-1}$ is the inverse of the transpose of $\mathcal{A}$, and $\tilde{\mathbf{1}}_3$ is the 3×3 matrix with all elements 1. Here are the nine scalar equations representing the above matrix equation. We denote the elements of $\mathcal{A}$ by a_{ij}.

```
In[152]:= 𝒜 = Array[a, {3, 3}];
          oneMatrix[n_] := Table[1, {n}, {n}];
          (* the equations *)
          eqs = Factor[Numerator[Together[
               𝒜 Transpose[Inverse[𝒜]] - oneMatrix[3]/3]]] // Flatten;
          (* use subscripts *)
          eqs /. a[ij__] :> Subscript[a, ij]
```

Out[157]= $\{-a_{1,3}\,a_{2,2}\,a_{3,1} + a_{1,2}\,a_{2,3}\,a_{3,1} + a_{1,3}\,a_{2,1}\,a_{3,2} + 2\,a_{1,1}\,a_{2,3}\,a_{3,2} - a_{1,2}\,a_{2,1}\,a_{3,3} - 2\,a_{1,1}\,a_{2,2}\,a_{3,3},$

$a_{1,3}\,a_{2,2}\,a_{3,1} + 2\,a_{1,2}\,a_{2,3}\,a_{3,1} - a_{1,3}\,a_{2,1}\,a_{3,2} + a_{1,1}\,a_{2,3}\,a_{3,2} - 2\,a_{1,2}\,a_{2,1}\,a_{3,3} - a_{1,1}\,a_{2,2}\,a_{3,3},$

$2\,a_{1,3}\,a_{2,2}\,a_{3,1} + a_{1,2}\,a_{2,3}\,a_{3,1} - 2\,a_{1,3}\,a_{2,1}\,a_{3,2} - a_{1,1}\,a_{2,3}\,a_{3,2} - a_{1,2}\,a_{2,1}\,a_{3,3} + a_{1,1}\,a_{2,2}\,a_{3,3},$

$a_{1,3}\,a_{2,2}\,a_{3,1} - a_{1,2}\,a_{2,3}\,a_{3,1} + 2\,a_{1,3}\,a_{2,1}\,a_{3,2} + a_{1,1}\,a_{2,3}\,a_{3,2} - 2\,a_{1,2}\,a_{2,1}\,a_{3,3} - a_{1,1}\,a_{2,2}\,a_{3,3},$

$2\,a_{1,3}\,a_{2,2}\,a_{3,1} + a_{1,2}\,a_{2,3}\,a_{3,1} + a_{1,3}\,a_{2,1}\,a_{3,2} - a_{1,1}\,a_{2,3}\,a_{3,2} - a_{1,2}\,a_{2,1}\,a_{3,3} - 2\,a_{1,1}\,a_{2,2}\,a_{3,3},$

$a_{1,3}\,a_{2,2}\,a_{3,1} + 2\,a_{1,2}\,a_{2,3}\,a_{3,1} - a_{1,3}\,a_{2,1}\,a_{3,2} - 2\,a_{1,1}\,a_{2,3}\,a_{3,2} + a_{1,2}\,a_{2,1}\,a_{3,3} - a_{1,1}\,a_{2,2}\,a_{3,3},$

$2\,a_{1,3}\,a_{2,2}\,a_{3,1} - 2\,a_{1,2}\,a_{2,3}\,a_{3,1} + a_{1,3}\,a_{2,1}\,a_{3,2} - a_{1,1}\,a_{2,3}\,a_{3,2} - a_{1,2}\,a_{2,1}\,a_{3,3} + a_{1,1}\,a_{2,2}\,a_{3,3},$

$a_{1,3}\,a_{2,2}\,a_{3,1} - a_{1,2}\,a_{2,3}\,a_{3,1} + 2\,a_{1,3}\,a_{2,1}\,a_{3,2} - 2\,a_{1,1}\,a_{2,3}\,a_{3,2} + a_{1,2}\,a_{2,1}\,a_{3,3} - a_{1,1}\,a_{2,2}\,a_{3,3},$

$-a_{1,3}\,a_{2,2}\,a_{3,1} + a_{1,2}\,a_{2,3}\,a_{3,1} + a_{1,3}\,a_{2,1}\,a_{3,2} - a_{1,1}\,a_{2,3}\,a_{3,2} + 2\,a_{1,2}\,a_{2,1}\,a_{3,3} - 2\,a_{1,1}\,a_{2,2}\,a_{3,3}\}$

Adding the nonsingularity condition to the last nine equations and solving for the elements a_{ij} gives the following two solutions.

```
In[158]:= as = Flatten[𝒜];
          sol = (Reduce[Append[# == 0& /@ eqs,
                        (* 𝒜 is nonsingular *) Det[𝒜] != 0], as] //
                 (ExpandNumerator //@ #&)) //. a[ij__] :> Subscript[a, ij]
```

Out[159]= $a_{1,1} \neq 0$ && $\left(a_{2,2} == \dfrac{-a_{1,2}\,a_{2,1} - i\,\sqrt{3}\,a_{1,2}\,a_{2,1}}{2\,a_{1,1}} \;||\; a_{2,2} == \dfrac{-a_{1,2}\,a_{2,1} + i\,\sqrt{3}\,a_{1,2}\,a_{2,1}}{2\,a_{1,1}}\right)$ && $a_{1,2} \neq 0$ &&

$a_{2,3} == \dfrac{-a_{1,2}\,a_{1,3}\,a_{2,1} - a_{1,1}\,a_{1,3}\,a_{2,2}}{a_{1,1}\,a_{1,2}}$ && $a_{2,1} \neq 0$ && $a_{3,2} == \dfrac{-a_{1,2}\,a_{2,1}\,a_{3,1} - a_{1,1}\,a_{2,2}\,a_{3,1}}{a_{1,1}\,a_{2,1}}$ &&

$a_{3,3} == \dfrac{-a_{1,2}\,a_{1,3}\,a_{3,1} - a_{1,1}\,a_{1,3}\,a_{3,2}}{a_{1,1}\,a_{1,2}}$ && $a_{1,2}\,a_{1,3}\,a_{2,1}\,a_{3,1} + 2\,a_{1,1}\,a_{1,3}\,a_{2,2}\,a_{3,1} \neq 0$

Reduce calculates all complex solutions, but here we are interested in the real ones only. The a_{ij} were assumed real. A quick inspection of the last solution shows that to guarantee the real-valuedness of the left-hand sides of all solutions would require exactly these a_{ij} to vanish which according to the solution, cannot vanish.

The next equation contains radicals. The result `False` shows that no complex solution exists for this equation.

```
In[160]:= With[{a = -3}, Reduce[x^(1/3) + x^(1/2) == a, x, Complexes]]
```

Out[160]= False

For other values of the parameter a, `Reduce` shows that a solution exists.

```
In[161]:= With[{a = 8}, Reduce[x^(1/3) + x^(1/2) == a, x, Complexes]]
```

Out[161]= $x == \text{Root}[-262144 + 13312\,\#1 - 145\,\#1^2 + \#1^3 \,\&,\, 1]$

Reduce is a very general function used to solve a wide range of equations and inequations and restrict the solutions to be real or integer. In the following, we give a few examples.

The next pair of equations is solved for real numbers.

In[162]:= `Reduce[{Sqrt[a] + x == a x, a + Sqrt[x] == a x}, {a, x}, Reals]`

Out[162]= $(a == 0 \mid\mid a == \text{Root}[-1 + \#1 - 2\,\#1^2 + \#1^3\,\&,\,1])\,\&\&\,x == -\dfrac{\sqrt{a}}{1-a}$

Be aware that when solving over domains other than the complex numbers (Complexes), all intermediate results must be part of the specified domain too. So for the following equation no real-valued solutions for x are found because the intermediate expressions $x^{1/3}$ and $x^{2/3}$ would not be real for the "solution" $x = -1$.

In[163]:= `{(* no solution returned *) Reduce[x^(1/3) - x^(2/3) == 1, x, Reals],`
          `(* -1 is a "solution" *)    Simplify[x^(1/3) - x^(2/3) /. x -> -1]}`
Out[163]= `{False, 1}`

Here are the integer solutions of a diophantine equations.

In[164]:= `Reduce[{n^2 + 2 m - 3 == m^2 + 3 n - 2}, {m, n}, Integers]`
Out[164]= $(m == -1\,\&\&\,n == -1) \mid\mid (m == -1\,\&\&\,n == 4) \mid\mid (m == 1\,\&\&\,n == 0) \mid\mid$
          $(m == 1\,\&\&\,n == 3) \mid\mid (m == 3\,\&\&\,n == -1) \mid\mid (m == 3\,\&\&\,n == 4)$

The next transcendental equation has an infinite set of solutions. To represent the infinite set, the construction C[1]∈Integers is used.

In[165]:= `Reduce[1/Tan[x] == 0, x]`
Out[165]= $\text{C}[1] \in \text{Integers}\,\&\&\,x == \dfrac{\pi}{2} + \pi\,\text{C}[1]$

Here is a slightly more complicated example. We used this equation above for Solve.

In[166]:= `Reduce[Sin[2 x] + Cos[3 x] + Tan[x/5] == 0, x] //`
                          `LogicalExpand // #[[1]]&`
Out[166]= $\text{C}[1] \in \text{Integers}\,\&\&$
          $5\,(2\,\text{ArcTan}[\text{Root}[1 + 22\,\#1 - 436\,\#1^2 - 1030\,\#1^3 + 27840\,\#1^4 + 11254\,\#1^5 - 621180\,\#1^6 -$
          $20294\,\#1^7 + 6446700\,\#1^8 - 61970\,\#1^9 - 35897940\,\#1^{10} + 127906\,\#1^{11} +$
          $116538240\,\#1^{12} + 182510\,\#1^{13} - 231915900\,\#1^{14} - 205630\,\#1^{15} + 290845350\,\#1^{16} -$
          $205630\,\#1^{17} - 231915900\,\#1^{18} + 182510\,\#1^{19} + 116538240\,\#1^{20} + 127906\,\#1^{21} -$
          $35897940\,\#1^{22} - 61970\,\#1^{23} + 6446700\,\#1^{24} - 20294\,\#1^{25} - 621180\,\#1^{26} + 11254$
          $\#1^{27} + 27840\,\#1^{28} - 1030\,\#1^{29} - 436\,\#1^{30} + 22\,\#1^{31} + \#1^{32}\,\&,\,1]] + 2\,\pi\,\text{C}[1]) == x$

The equations can also contain logical conjunctions of equations and quantifiers. The next input establishes the existence of a 2D disk.

In[167]:= `Reduce[Exists[{x, y}, {x, y} ∈ Reals,  x^2 + y^2 <= 1]]`
Out[167]= `True`

In the next example, the inequalities are resolved and explicit intervals are given for x that fulfill the given inequalities.

In[168]:= `Reduce[-1 < Sum[k x^k, {k, -3, 3}] < 1, x]`
Out[168]= $\text{Root}[-3 - 2\,\#1 - \#1^2 + \#1^3 + \#1^4 + 2\,\#1^5 + 3\,\#1^6\,\&,\,1] <$
          $x < \text{Root}[-3 - 2\,\#1 - \#1^2 - \#1^3 + \#1^4 + 2\,\#1^5 + 3\,\#1^6\,\&,\,1] \mid\mid$
          $\text{Root}[-3 - 2\,\#1 - \#1^2 + \#1^3 + \#1^4 + 2\,\#1^5 + 3\,\#1^6\,\&,\,2] < x <$
          $\text{Root}[-3 - 2\,\#1 - \#1^2 - \#1^3 + \#1^4 + 2\,\#1^5 + 3\,\#1^6\,\&,\,2]$

And in the next input, we solve an inequality with a nonsmooth function.

In[169]:= **Reduce[Boole[Tan[x] > 1/2] == 1, x]**

Out[169]= $\frac{1}{2} + \frac{x}{\pi} \notin$ Integers && C[1] $\in$ Integers && ArcTan$\left[\frac{1}{2}\right] + \pi$ C[1] < x < $\frac{1}{2}$ (π + 2 π C[1])

Reduce has some options that are important for many uses.

In[170]:= **Options[Reduce]**

Out[170]= {Backsubstitution → False, Cubics → False, GeneratedParameters → C, Modulus → 0, Quartics → False, WorkingPrecision → ∞}

The Cubics and Quartics options are similar the identically named options of Roots. The GeneratedPa·rameters option setting was responsible for the C[1] that we obtained in the above example. In the next example, we generate uniquely named ck.

In[171]:= **counter = 0;**
Reduce[Sin[x] == 0 && Cos[y] == 1, {x, y},
    GeneratedParameters ->
        (ToExpression["c" <> ToString[counter++]]&)]

Out[172]= ((c0 | c1) $\in$ Integers && x == 2 π c0 && y == 2 π c1) ||
    ((c2 | c3) $\in$ Integers && x == π + 2 π c2 && y == 2 π c3)

The option Backsubstitution is especially relevant for polynomial systems. Here is a system of four equations in four variables.

In[173]:= **polys =**
    {32 - 216 v^2 - 216 v^3 - 216 x^2 + 648 v x^2 - 150 z +
    216 v^2 z + 216 x^2 z + 231 z^2 - 113 z^3 == 0,
    -432 u v - 648 u v^2 - 432 x + 1296 v x + 648 u x^2 +
    432 u v z + 432 x z == 0,
    -216 - 216 u^2 + 648 v - 648 u^2 v + 1296 u x + 216 z + 216 u^2 z == 0,
    648 u - 216 u^3 == 0};

With the default option setting of the Backsubstitution option, we get a solution where the values for z are explicit numbers, and the values for v are expressed implicitly through z.

In[174]:= **redSol1 = Reduce[polys, {x, z, u, v}]**

Out[174]= $\left(\left(z == 0 \mid\mid z == \frac{2}{3} \mid\mid z == 1\right) \&\& u == 0 \&\& v == \frac{1-z}{3}\right)$ ||

$\left(\left(z == 0 \mid\mid z == \frac{2}{3} \mid\mid z == 1\right) \&\& (u == -\sqrt{3} \mid\mid u == \sqrt{3}) \&\& v == \frac{1}{3}(-2 + 3ux + 2z)\right)$

With the option setting True of the Backsubstitution option, we get a solution where the values are as explicit as possible (the solutions depend parametrically on x).

In[175]:= **redSol2 = Reduce[polys, {x, z, u, v}, Backsubstitution -> True]**

Out[175]= $\left(z == 0 \&\& u == 0 \&\& v == \frac{1}{3}\right) \mid\mid \left(z == 0 \&\& u == -\sqrt{3} \&\& v == \frac{1}{3}(-2 - 3\sqrt{3}x)\right) \mid\mid$

$\left(z == 0 \&\& u == \sqrt{3} \&\& v == \frac{1}{3}(-2 + 3\sqrt{3}x)\right) \mid\mid \left(z == \frac{2}{3} \&\& u == 0 \&\& v == \frac{1}{9}\right) \mid\mid$

$\left(z == \frac{2}{3} \&\& u == -\sqrt{3} \&\& v == \frac{1}{9}(-2 - 9\sqrt{3}x)\right) \mid\mid \left(z == \frac{2}{3} \&\& u == \sqrt{3} \&\& v == \frac{1}{9}(-2 + 9\sqrt{3}x)\right) \mid\mid$

$(z == 1 \&\& u == 0 \&\& v == 0) \mid\mid (z == 1 \&\& u == -\sqrt{3} \&\& v == -\sqrt{3}x) \mid\mid (z == 1 \&\& u == \sqrt{3} \&\& v == \sqrt{3}x)$

For substitution of the solutions into equations, the second solution is more suited. (But frequently, the solutions obtained with Backsubstitution -> True are considerably larger.)

In[176]:= **polys /. {ToRules[redSol1]} // Simplify // Short[#, 4]&**

Out[176]//Short=

$$\{\{(-27\,x^2 + (-3 + z)^2)\,z == 0,\; x\,z == 0,\; z == 0,\; \text{True}\},$$
$$\{(243\,x^2 - (5 - 3\,z)^2)\,(-2 + 3\,z) == 0,\; 3\,x\,z == 2\,x,\; 3\,z == 2,\; \text{True}\},$$
$$\ll 6 \gg,\; \{216\,x^2\,(-2 + 3\,u\,x + 2\,z) == 8\,(-2 + 3\,u\,x + 2\,z)^3,$$
$$72\,(9\,\sqrt{3}\,x^2 + 6\,x\,(-2 + 3\,u\,x + 2\,z) - \sqrt{3}\,(-2 + 3\,u\,x + 2\,z)^2) == 0,$$
$$2 + 3\,\sqrt{3}\,x - 3\,u\,x == 2\,z,\; \text{True}\}\}$$

In[177]:= `polys /. {ToRules[redSol2]} // Simplify // Flatten // Union`

Out[177]= `{True}`

Sometimes one does not need a complete set of solutions for a set of equations or inequalities, but a single solution is already enough.

FindInstance [*equations, listOfVariables*]

  finds one possible solutions for the equations (and inequations and inequalities) equations.

Here is a very simple example. We are looking are a number that is larger than 0.

In[178]:= `FindInstance[x > 0, x]`

Out[178]= `{{x → 51}}`

The next input uses the optional third argument of `FindInstance` to specify that we are looking for a complex solution.

In[179]:= `FindInstance[I x > 0, x, Complexes]`

Out[179]= `{{x → -33 i}}`

And in the next input, we look for solutions of all possible types. The first argument of `FindInstance` contains an `Or`-expression to allow for solutions of all types.

In[180]:= `{#, FindInstance[x > 0 || I x > 0 || x == True, x, #]}& /@`
      `(* seven domains *) {Complexes, Reals, Algebraics,`
                  `Rationals, Integers, Primes, Booleans}`

Out[180]= `{{Complexes, {{x → 51}}}, {Reals, {{x → 51}}},`
      `{Algebraics, {{x → 51}}}, {Rationals, {{x → 51}}},`
      `{Integers, {{x → 51}}}, {Primes, {{x → 233}}}, {Booleans, {{x → True}}}}`

Often, we do not require solutions for all variables, but only for a prescribed subset of variables. This can be accomplished with `Eliminate`.

Eliminate [*equations, listOfVariablesToBeEliminated*]

  eliminates (when possible) the variables *listOfVariablesToBeEliminated* from the equations *equations*. The result of `Eliminate` is a logical expression of equations containing `Or` and `And`.

`Solve` offers a similar possibility for the cases in which we want to solve for some variables while at the same time eliminating others. This requires a third argument for `Solve`. Here is an example involving elimination of a variable from the following equations.

In[181]:= `equations = {a x^2 + y == 1, a x + y^2 == 1, a x y == 1}`

Out[181]= $\{a\,x^2 + y == 1,\; a\,x + y^2 == 1,\; a\,x\,y == 1\}$

The variable a is to be eliminated. The result of this operation is a logical `And` coupling the equations.

In[182]:= `Eliminate[equations, a]`

Out[182]= $-3 x + 2 x^2 + x^3 == -1$ && $y == 2 - 2 x - x^2$

In turn, it can be converted into a list of equations.

In[183]:= **List @@ %**

Out[183]= $\{-3 x + 2 x^2 + x^3 == -1, \ y == 2 - 2 x - x^2\}$

Now, we solve these equations for x and y after eliminating a.

In[184]:= **Solve[equations, {x, y}, a]**

Out[184]= $\left\{\left\{y \to \frac{1}{18} \left(52 \left(\frac{2}{97 - 3\sqrt{69}}\right)^{1/3} - \right.\right.\right.$

$\left. 338 \left(\frac{2}{97 - 3\sqrt{69}}\right)^{2/3} + 2 \cdot 2^{2/3} \left(97 - 3\sqrt{69}\right)^{1/3} - 2^{1/3} \left(97 - 3\sqrt{69}\right)^{2/3}\right),$

$\left. x \to \frac{1}{3} \left(-2 - 13 \left(\frac{2}{97 - 3\sqrt{69}}\right)^{1/3} - \left(\frac{1}{2} \left(97 - 3\sqrt{69}\right)\right)^{1/3}\right)\right\},$

$\left\{y \to -\frac{13}{9} \left(\frac{2}{97 - 3\sqrt{69}}\right)^{1/3} + \frac{13 i \left(\frac{2}{97 - 3\sqrt{69}}\right)^{1/3}}{3\sqrt{3}} - \frac{1}{3} \left(\frac{1}{2} \left(97 - 3\sqrt{69}\right)\right)^{1/3} - \right.$

$\frac{i \left(\frac{1}{2} \left(97 - 3\sqrt{69}\right)\right)^{1/3}}{3\sqrt{3}} + \frac{1}{18} \left(\frac{1}{2} \left(97 - 3\sqrt{69}\right)\right)^{2/3} + \frac{169}{9 \cdot 2^{1/3} \left(97 - 3\sqrt{69}\right)^{2/3}} +$

$\frac{169 i}{3 \cdot 2^{1/3} \sqrt{3} \left(97 - 3\sqrt{69}\right)^{2/3}} + \frac{1}{9} \cdot 2^{2/3} \left(97 - 3\sqrt{69}\right)^{1/3} - \frac{i \left(97 - 3\sqrt{69}\right)^{2/3}}{6 \cdot 2^{2/3} \sqrt{3}},$

$\left. x \to -\frac{2}{3} + \frac{1}{6} \left(1 + i\sqrt{3}\right) \left(\frac{1}{2} \left(97 - 3\sqrt{69}\right)\right)^{1/3} + \frac{13 \left(1 - i\sqrt{3}\right)}{3 \cdot 2^{2/3} \left(97 - 3\sqrt{69}\right)^{1/3}}\right\},$

$\left\{y \to -\frac{13}{9} \left(\frac{2}{97 - 3\sqrt{69}}\right)^{1/3} - \frac{13 i \left(\frac{2}{97 - 3\sqrt{69}}\right)^{1/3}}{3\sqrt{3}} - \frac{1}{3} \left(\frac{1}{2} \left(97 - 3\sqrt{69}\right)\right)^{1/3} + \right.$

$\frac{i \left(\frac{1}{2} \left(97 - 3\sqrt{69}\right)\right)^{1/3}}{3\sqrt{3}} + \frac{1}{18} \left(\frac{1}{2} \left(97 - 3\sqrt{69}\right)\right)^{2/3} + \frac{169}{9 \cdot 2^{1/3} \left(97 - 3\sqrt{69}\right)^{2/3}} -$

$\frac{169 i}{3 \cdot 2^{1/3} \sqrt{3} \left(97 - 3\sqrt{69}\right)^{2/3}} + \frac{1}{9} \cdot 2^{2/3} \left(97 - 3\sqrt{69}\right)^{1/3} + \frac{i \left(97 - 3\sqrt{69}\right)^{2/3}}{6 \cdot 2^{2/3} \sqrt{3}},$

$\left. x \to -\frac{2}{3} + \frac{1}{6} \left(1 - i\sqrt{3}\right) \left(\frac{1}{2} \left(97 - 3\sqrt{69}\right)\right)^{1/3} + \frac{13 \left(1 + i\sqrt{3}\right)}{3 \cdot 2^{2/3} \left(97 - 3\sqrt{69}\right)^{1/3}}\right\}\right\}$

Thus, Solve solves the equations produced by Eliminate for the desired variables.

In[185]:= **Solve[%%, {x, y}] == %**

Out[185]= True

For comparison, we give the result of Solve without a third argument (we would have three independent equations in two variables).

In[186]:= **Solve[equations, {x, y}]**

Out[186]= {}

The system of equations involves three equations in the two variables x and y, and in this case, there is typically no generic solution.

Let us use Eliminate to derive a fourth-order algebraic differential equation with the remarkable property that any continuous function can be approximated arbitrarily closely by a solution of this differential equation [1524], [219], [577], [241], [531] (for the construction of classes of universal differential equations, see [579]; for partial differential equations, see [264], and for a universal functional equation, see [578]).

In[187]:= `f[t_] = Integrate[Exp[-1/(1 - t^2)], t];`

In[188]:= `g[t_] = Exp[1/(1 - t^2)];`

In[189]:= ```
eqs = {y'[x] == Together[C α f'[t] g[t]],
       y''[x] == Together[C α^2 f''[t] g[t]],
       y'''[x] == Together[C α^3 f'''[t] g[t]],
       y''''[x] == Together[C α^4 f''''[t] g[t]]}
```

Out[189]= $\left\{ y'[x] == C\,\alpha,\ y''[x] == -\dfrac{2\,C\,t\,\alpha^2}{(-1+t^2)^2}, \right.$

$\left. y^{(3)}[x] == \dfrac{2\,C\,(-1+3\,t^4)\,\alpha^3}{(-1+t^2)^4},\ y^{(4)}[x] == -\dfrac{4\,C\,(3\,t-10\,t^3+3\,t^5+6\,t^7)\,\alpha^4}{(-1+t^2)^6} \right\}$

In[190]:= `uODE = Expand[Subtract @@ Eliminate[eqs, {C, α, t}]]`

Out[190]= $12\,y''[x]^7 - 29\,y'[x]^2\,y''[x]^3\,y^{(3)}[x]^2 - 12\,y'[x]^3\,y''[x]\,y^{(3)}[x]^3 + 24\,y'[x]^2\,y''[x]^4\,y^{(4)}[x] +$
$6\,y'[x]^3\,y''[x]^2\,y^{(3)}[x]\,y^{(4)}[x] - 4\,y'[x]^4\,y^{(3)}[x]^2\,y^{(4)}[x] + 3\,y'[x]^4\,y''[x]\,y^{(4)}[x]^2$

The next graphic shows a few of the possible solution curves for random complex initial conditions.

In[191]:= `SeedRandom[1];`

```
Module[{o = 24, r := Random[Complex, {-1 - I, 1 + I}], ndsols1, tRange},
  (* show all curves at once *)
  Show[DeleteCases[Internal`DeactivateMessages[
  Table[(* try to solve numerically for a few seconds *)
    TimeConstrained[ndsols =
    (* solve universal differential equation for random initial conditions *)
    NDSolve[{uODE == 0, y[0] == r, y'[0] == r, y''[0] == r, y'''[0] == r},
            {y}, {x, -10, 10}, PrecisionGoal -> 4, AccuracyGoal -> 4,
            Method -> StiffnessSwitching];
    (* plot solution over the solved x-range *)
    ParametricPlot[Evaluate[{Re[y[x]], Im[y[x]]} /. #],
                   Evaluate[{x, #[[1, 2, 1, 1, 1]], #[[1, 2, 1, 1, 2]]}],
                   DisplayFunction -> Identity]& /@ ndsols, 10], {o}]],
                   $Aborted] /. (* color each solution curve pair *)
            l_Line :> {Hue[Random[]], 1}, Frame -> True, Axes -> False,
    DisplayFunction -> $DisplayFunction]]
```

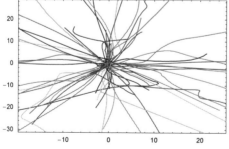

As another little application to `Eliminate`, let us calculate (non)linear differential equations with integer coefficients having $\cos(x) + \sin(2x)$ as their solutions. We construct these differential equations by eliminating all trigonometric functions from $\cos(x) + \sin(2x)$ and their derivatives.
```

In[193]:= `Flatten[Table[`
    `Factor[#] == 0& /@ Apply[Subtract,`
    `If[Head[#] === And, List @@ #, {#}]&[`
    `Eliminate[`(* eliminate trig functions *)
      `Join[Table[`(* table of derivatives *)
        `D[w[z], {z, i}] == TrigExpand[`
          (* rewrite all Sin, Cos of multiple arguments *)
          `D[Cos[x] + Sin[2 x], {x, i}]], {i, 0, j}],`
          (* algebraic relation between Sin[x] and Cos[x] *)
          `{Sin[x]^2 + Cos[x]^2 == 1}], {Sin[x], Cos[x]}]], {1}],`
          `{j, 1, 4}]] // TableForm`

Out[193]//TableForm=

$$27 - 207\,w[z]^2 + 64\,w[z]^4 - 27\,w'[z]^2 + 32\,w[z]^2\,w'[z]^2 + 4\,w'[z]^3 + 4\,w'[z]^4 == 0$$

$$-18711 - 15876\,w'[z] + 5859\,w'[z]^2 + 3024\,w'[z]^3 + 2259\,w''[z]^2 + 744\,w'[z]\,w''[z]^2 - 192\,w'[z]^2\,\textrm{v}$$

$$-9 + 84\,w[z]^2 - 12\,w'[z] - 3\,w'[z]^2 + 44\,w[z]\,w''[z] + 5\,w''[z]^2 == 0$$

$$-2079\,w[z] + 1008\,w[z]\,w'[z] - 99\,w''[z] + 246\,w'[z]\,w''[z] - 96\,w'[z]^2\,w''[z] + 64\,w[z]\,w''[z]^2 -$$
$$3267 + 1188\,w'[z] - 2367\,w'[z]^2 - 32\,w'[z]^3 + 256\,w'[z]^4 - 495\,w''[z]^2 - 24\,w'[z]\,w''[z]^2 + 128\,w'[$$

$$738 + 128\,w[z]^2 - 249\,w'[z] - 56\,w''[z]^2 - 57\,w^{(3)}[z] - 12\,w^{(3)}[z]^2 == 0$$

$$-33\,w[z] + 16\,w[z]\,w'[z] - 9\,w''[z] + 12\,w'[z]\,w''[z] + 2\,w''[z]\,w^{(3)}[z] == 0$$

$$-414 + 255\,w'[z] + 32\,w'[z]^2 + 24\,w''[z]^2 + 63\,w^{(3)}[z] + 4\,w^{(3)}[z]^2 == 0$$

$$-774 + 255\,w'[z] + 64\,w[z]\,w''[z] + 64\,w''[z]^2 + 63\,w^{(3)}[z] + 12\,w^{(3)}[z]^2 == 0$$

$$-401598 + 195075\,w'[z] + 24528\,w''[z]^2 + 4096\,w'[z]\,w''[z]^2 + 24003\,w^{(3)}[z] + 2560\,w''[z]^2\,w^{(3)}[\textrm{z}$$

$$63\,w[z] + 15\,w''[z] - 16\,w'[z]\,w''[z] + 8\,w[z]\,w^{(3)}[z] - 2\,w''[z]\,w^{(3)}[z] == 0$$

$$810 - 513\,w'[z] - 48\,w''[z]^2 - 129\,w^{(3)}[z] + 32\,w'[z]\,w^{(3)}[z] - 4\,w^{(3)}[z]^2 == 0$$

$$250047 - 34047\,w''[z]^2 + 1024\,w''[z]^4 - 8127\,w^{(3)}[z]^2 + 512\,w''[z]^2\,w^{(3)}[z]^2 - 16\,w^{(3)}[z]^3 + 64\,w^{(3)}$$

$$4\,w[z] + 5\,w''[z] + w^{(4)}[z] == 0$$

$$-3114 + 1023\,w'[z] + 255\,w^{(3)}[z] + 48\,w^{(3)}[z]^2 - 32\,w''[z]\,w^{(4)}[z] + 4\,w^{(4)}[z]^2 == 0$$

$$-99187011 + 12816144\,w^{(3)}[z] + 1568259\,w^{(3)}[z]^2 - 196416\,w^{(3)}[z]^3 + 589059\,w^{(4)}[z]^2 - 49056\,w$$

$$-189 + 1364\,w''[z]^2 + 48\,w^{(3)}[z] - 3\,w^{(3)}[z]^2 + 684\,w''[z]\,w^{(4)}[z] + 85\,w^{(4)}[z]^2 == 0$$

$$524799\,w''[z] + 65472\,w''[z]\,w^{(3)}[z] + 32319\,w^{(4)}[z] + 16344\,w^{(3)}[z]\,w^{(4)}[z] + 1536\,w^{(3)}[z]^2\,w^{(4)}$$

$$16579647 - 73872\,w^{(3)}[z] - 529407\,w^{(3)}[z]^2 + 128\,w^{(3)}[z]^3 + 4096\,w^{(3)}[z]^4 - 130815\,w^{(4)}[z]^2 + 9$$

Note, the last differential equation of order four is linear. Here, the results are checked.

In[194]:= `Simplify[% /. w -> Function[x, Cos[x] + Sin[2 x]]] // Union`

Out[194]= `{True}`

(For generalizations, see [1097].)

In this section, in Section 1.4 of the Numerics volume [1737], and in the above subsections on polynomial equations and inequalities, we have discussed many functions related to solving equations: `Solve`, `NSolve`, `Roots`, `NRoots`, `FindRoot`, `Eliminate`, `Reduce`, `Resultant`, `GroebnerBasis`, and `Cylindrical`· `Decomposition`. Together with the functions `SolveAlways`, `Experimental`Groebnerwalk`, and `Algebra`InequalitySolve`, this gives the user a broad set of tools to handle, rewrite, and solve equations and inequations. Which of the functions should be used under which circumstances? There is no one-sentence answer to this question. It depends on:

- Do we deal with equations, inequations, or inequalities?

- Do we deal with a univariate or a multivariate problem?

- Do we look for a numeric or symbolic solution?

- Are we interested in explicit or implicit solutions?

- Do we expect that the solutions are isolated or form finite-dimensional manifolds?

- Does the problem have parameters?

- Do (in the input problem) polynomial, rational or transcendental functions appear?

- Do we want all solutions or only some?

- Do we look for solutions correct only for special parameter values?

- How large and/or complicated is the system?

We cannot give exhaustive answers to all of the above questions, but in the following, we will give some rules, guidelines, recommendations, and tricks. The following discussion is restricted to methods using functions currently built into *Mathematica*. Further methods, such as multiresultants, can solve polynomial system efficiently.

If we want a numeric solution of a univariate polynomial equation of "reasonable" degree with rational coefficients, we can confidently use NRoots. If the coefficients contain exact numeric transcendental quantities (like E or Sqrt[5]), we can still use NRoots, but depending on the coefficients, we might want to numericalize them explicitly with a sufficiently high precision. The function NRoots will return all solutions for a univariate polynomial. Returning all solutions for large-degree univariate polynomial might take a considerable amount of time. If the polynomial is simple, a preprocessing call to Factor or Decompose may be useful. And if we are only interested in some solutions (often the smallest or largest root), then a call to FindRoot (with properly set parameters, especially WorkingPrecision and AccuracyGoal) is sometimes preferable. For some higher-order univariate polynomials with rational coefficients sometimes the numericalization of the exactly calculated is preferable.

If we want a symbolic solution of a univariate polynomial equation of "reasonable" degree with rational coefficients, we can use Roots. This will give all roots either as radicals or as Root-objects (depending on the degree of the polynomial and the setting of the Cubics and Quartics option). If the coefficients contain numeric transcendental quantities, we can still use Roots, but the result is more symbolic (meaning that not all operations that can, in principle, be carried out on Root-objects will work properly). (For most problems, a "reasonable" degree for numeric and symbolic univariate polynomials means less than or equal to a few hundred.)

After a univariate polynomial, the next complicated type would be a univariate rational function. NRoots will not deal with them, but NSolve will for numerical purposes. Similarly, for symbolic solutions, Solve is now the function of choice. Concerning the degree, and if all solutions are needed, the remarks from the last paragraph stay valid.

Inequalities and equalities with domain and/or arithmetic constraints on parameters and/or solutions are best handled by the function Reduce. Also, logical combinations of equations and inequations are handled by Reduce.

The most important type of equations (from the relevance in practical applications as well, from what can be dealt algorithmically) is a set of multivariate polynomial equations. If they are free of parameters, the set has only isolated solutions, and we want all of them, NSolve is the function of choice. If the set has parameters, they either have to be instantiated or they may be eliminated by using a three-argument form of NSolve. (Now that we have discussed Gröbner bases, let us add one sentence about what NSolve is doing with such a set of equations. For the most important case of *n* equations in *n* variables, it calculates a degree reverse-lexicographic basis using numericalized coefficients and then converts this basis to a lexicographic one using eigensystems of matrices [104]. A degree-reverse lexicographic basis is often much easier to calculate than a lexicographic one. But in the calculation and in the ongoing basis conversion, many digits can be lost due to cancellations; we saw examples in Chapter 1 of the Numerics volume [1737]. That is why it is important to use a sufficiently high working precision for such calculations.)

Now concerning the case where we want a symbolic solution: The function Solve is an appropriate choice in many cases. Solve will try to calculate a lexicographic Gröbner basis and then solve the resulting system of polynomials. Although this approach works theoretically always, the actual running time and memory needed depend crucially on the order of variables and certain internal choices. Solve has an elaborate set of heuristics to "do the right thing" and often will give a solution in a "reasonable" time. Similar statements hold for the function Reduce which will also care about degenerate cases.

If Solve does not return a solution after a "reasonable" amount of time (say after one day), the user can try to solve the system "manually". A first try would be a direct call to GroebnerBasis. (The order of the variables to be solved for might matter.) Often it is very helpful to eliminate certain parts of the solution the user is not interested in. Eliminating some variables (using GroebnerBasis with EliminationOrder, or Result‌ant on all pairs of equations) and then factoring the result sometimes achieves this. (But be aware that multiple uses of Resultant can give spurious solutions.) One could also try to calculate a degree-reverse lexicographic basis and then convert it to a lexicographic basis using the (undocumented and still experimental in Version 5.1) function Internal`Groebnerwalk [384], [54]. For systems that are linear in some of the variables, it often is a good idea to solve for these variables, and backsubstitute the solutions in the remaining equations. (Of course, care has to be taken with denominators that could vanish. Sometimes, a generalization of this approach, namely solving for the highest monomial x^d and replacing in other equations x^e ($e > d$) by $x^{\lfloor e/d \rfloor} x^{e - \lfloor e/d \rfloor}$ yields simpler equations.) Because the running time of GroebnerBasis is dramatically dependent on the number of variables, this method is sometimes of great advantage. As noted, NSolve will calculate an approximative Gröbner basis. Such a Gröbner basis can be also calculated directly using the CoefficientField -> InexactNumbers[*precision*] option setting. Because under this setting the coefficients cannot become arbitrary large numbers, GroebnerBasis runs often faster. And although numeric, the resulting basis can either be used to draw conclusions about the exact number of symbolic solutions of the system or sometimes even might be made symbolical (using Rationalize). In the solutions to the exercises of this chapter, we will demonstrate the mentioned possibilities on various example problems.

Often the just-mentioned alternatives to a direct call to Solve are also useful as processing steps to obtain numeric solutions. When we only want some solutions and we "roughly" know their values, FindRoot (again with properly set options) will come in handy. By starting with sufficiently many random starting values, we will sometimes also succeed to find all solutions for a polynomial system numerically.

Sometimes even a linear system with coefficients that are symbolic parameters may be a challenge to solve. For up to typically eight equations Solve often works fine, but for a larger system, it may need a long time and/or a lot of memory. Depending on the special characteristic of the system, solving one equation after the other and substituting the solutions may be an alternative. Especially if it is possible to simplify (using, e.g., Factor or Together) the coefficients after each step considerably (manual Gaussian elimination).

For a multivariate polynomial system that has finite-dimensional solutions, NSolve will resort to Solve. Either this works directly or we might have to resort to the above-mentioned manual alternatives and then numericalize the solutions.

Multivariate systems of rational functions formally reduce to a polynomial system after writing them over common denominators. Depending on the problem, care might have to be taken to ensure that denominators do not vanish and we do not get indeterminate expressions of the form $0/0$. In a direct call to `Solve`, this will be done automatically. When using the above-described manual approach, one either tests if the solutions are appropriate at the end or introduces in the beginning auxiliary equations of the form $1 - \xi_1 \, denominator_1$, $1 - \xi_2 \, denominator_2$, or $1 - \xi \, (denominator_1^2 + denominator_2^2 + \cdots)$ and eliminates the auxiliary variables ξ_k. These equations ensure that the denominators never vanish.

The most complicated equations are genuinely transcendental ones. For a few such equations, `Solve` will generate a solution. It will use heuristics to first find a variable substitution ($radicalOrRootOfOrder^n = radicalOrRootOfOrder$, like $y = \exp(i\,x)$ for trigonometric or $y = \exp(x)$ for hyperbolic equations) to transform the equations to a polynomial form. (Sometimes, operations on the whole equation, like exponentiating the equation when logarithmic or inverse hyperbolic functions exist, may be fruitful.) Then the described techniques will be used to solve these polynomials. Using the inverse functions, we get explicit solutions. Because the variable substitution is not an algorithmic process and not all possible inverse functions of transcendental functions are known to `Solve`, some manual work can lead to a success when `Solve` takes too long or fails.

Now, let us assume we have a polynomial system (or we have brought the problem into such a form). Solving a system is the most common operation, either with or without parameters. If we are interested in all special and degenerate solutions, `Reduce` should be used. Most remarks from above hold. If `Reduce` fails, we could try a manual approach. But great care has to be taken not to "lose" a branch. In the opposite case, where we are only interested in generic solutions, the function `SolveAlways` should be used.

Above, we used repeatedly the word "parameter", without formally defining it. Often it is clear from the problem at hand what are the solve variables and what are parameters. For `Solve`, parameters are all symbols that are not solve variables and that appear in the equations (the first argument of `Solve`). This means that potential coefficients such as `Pi`, `Sqrt[2]`, `Sin[Pi/20]` are parameters for `Solve`. Inside `GroebnerBasis`, parameters are either determined automatically or can be explicitly specified in the option `ParameterVari`⌇ `ables`. For radicals such as `Sqrt[2]`, the identity `Sqrt[2]^2 == 2` will automatically be used, but for transcendental constants such as `Sin[Pi/20]`, it is often advantageous to add the identity `Sin[Pi/20]^2 + Cos[Pi/20]^2 == 1` explicitly. Inside `GroebnerBasis` parameters could, of course, also be listed in the variable list; typically, they would be listed last. Finally, when parameters are present, the option setting `CoefficientDomain -> RationalFunctions` is sometimes useful.

Another frequently desired operation is the elimination of variables. Either a call to `Eliminate` or a call to `GroebnerBasis` with the option setting `MonomialOrder -> EliminationOrder` is advised here. Often this will succeed directly, but again, manual manipulations might dramatically accelerate an elimination process.

Solving systems containing inequations (such as `Solve[{x^2 ==1, x !=1}, x]`) reduces basically to the above cases. Either one finds first all solutions and then eliminates the unwanted ones or one introduces in the beginning auxiliary equations of the form $1 - \xi \, shouldNotVanish$.

Finally, sets of polynomial inequalities are either solved directly by using `Reduce` or by manually massaging the equations and then carrying out a call to the underlying `CylindricalDecomposition`.

> To summarize: If the problem under consideration is not very complicated, a direct call to `Solve`, `NSolve`, or `Reduce` will do the job. These functions make use of strong algorithmic techniques and elaborate heuristics. If a call to these functions does not succeed (from a timing or memory point of view), some manipulations of the equations and the use of numerical and symbolic techniques (and sometimes also graphical techniques to find appropriate starting values, for instance) will often make progress. (The most elaborate heuristics usually cannot compete with some "physical insight" into a

> problem of how many solutions and solutions of which kind one expects). In the exercises of this chapter, many of these solution strategies described will be put to work on concrete examples.

1.6 Classical Analysis

1.6.1 Differentiation

The important command D for differentiation has already been briefly discussed in Chapter 3 of the Programming volume [1735]. D has a more general syntax than was described there. D has one option.

> D[*toDifferentiate*, {*var₁*, *num₁*}, {*var₂*, *num₂*}, …, {*var_n*, *num_n*},
> NonConstants -> {*nonVar₁*, *nonVar₂*, …, *nonVar_n*}]
>
> differentiates *toDifferentiate num₁* times with respect to *var₁*, *num₂* times with respect to *var₂*, etc. Here, it is assumed that the variables *nonVar_i* depend implicitly on the variables for which the differentiation takes place.

Mathematica can of course compute the derivatives for all elementary functions and compositions of these functions, as well as for many special functions. This includes, for example, the Legendre functions (see Chapter 3).

In[1]:= **D[LegendreP[n, z], {z, 2}]**

$$\text{Out[1]= } -\frac{2 z \, (-n \, \text{LegendreP}[-1+n, z] + n z \, \text{LegendreP}[n, z])}{(-1+z^2)^2} + \frac{1}{-1+z^2}$$
$$\left(-\frac{n \, (-(-1+n)) \, \text{LegendreP}[-2+n, z] + (-1+n) z \, \text{LegendreP}[-1+n, z])}{-1+z^2} + \right.$$
$$\left. n \, \text{LegendreP}[n, z] + \frac{n z \, (-n \, \text{LegendreP}[-1+n, z] + n z \, \text{LegendreP}[n, z])}{-1+z^2} \right)$$

Here is the computation of the second derivatives of a function with respect to two different variables.

In[2]:= **D[Sin[x^3] y^2, {x, 2}, {y, 2}]**

Out[2]= $2 \, (6 \, x \, \text{Cos}[x^3] - 9 \, x^4 \, \text{Sin}[x^3])$

While D does not have the Listable attribute, it is automatically threaded over lists.

In[3]:= **D[{x, x^2, x^3}, x]**

Out[3]= $\{1, 2 \, x, 3 \, x^2\}$

Variables that do not explicitly depend on the variables of differentiation are treated as constants.

In[4]:= **D[x^2 z^z, {x, 2}]**

Out[4]= $2 \, z^z$

If z in this example also depends on x, we can take this into account in two different ways. First, we can explicitly state that x is an argument by writing z[x].

In[5]:= **ex1 = D[(x^2 z^z /. z -> z[x]), {x, 2}]**

$$\text{Out[5]= } 2 \, z[x]^{z[x]} + 4 \, x \, z[x]^{z[x]} \, (z'[x] + \text{Log}[z[x]] \, z'[x]) +$$
$$x^2 \left(z[x]^{z[x]} \, (z'[x] + \text{Log}[z[x]] \, z'[x])^2 + z[x]^{z[x]} \left(\frac{z'[x]^2}{z[x]} + z''[x] + \text{Log}[z[x]] \, z''[x] \right) \right)$$

Alternatively, we can tell D that z depends on x. (This can be useful for a proper treatment of total versus partial derivatives [369].)

In[6]:= **ex2 = D[x^2 z^z, {x, 2}, NonConstants -> z]**

Out[6]= $2 z^z + 4 x z^z$ (D[z, x, NonConstants → {z}] + D[z, x, NonConstants → {z}] Log[z]) +

$x^2 \left(z^z \text{ (D[z, x, NonConstants → {z}] + D[z, x, NonConstants → {z}] Log[z])}^2 + \right.$

$z^z \left(\text{D[z, x, NonConstants → {z}]} + \dfrac{\text{D[z, x, NonConstants → {z}]}^2}{z} + \right.$

$\left. \left. \text{D[z, x, NonConstants → {z}] Log[z]} \right) \right)$

In this case, the output form is different, but we can easily convert it back to the earlier one.

In[7]:= **ex3 = ex2 /. {z -> z[x],**
            D[z, x   , NonConstants -> {z}] -> z'[x],
            D[z, {x, 2}, NonConstants -> {z}] -> z''[x]}

Out[7]= $2 z[x]^{z[x]} + 4 x z[x]^{z[x]}$ (z'[x] + Log[z[x]] z'[x]) +

$x^2 \left(z[x]^{z[x]} \text{ (z'[x] + Log[z[x]] z'[x])}^2 + z[x]^{z[x]} \left(z'[x] + Log[z[x]] z'[x] + \dfrac{z'[x]^2}{z[x]} \right) \right)$

In[8]:= **Expand[ex3 - ex1]**

Out[8]= $x^2 z[x]^{z[x]} z'[x] + x^2 \text{ Log[z[x]] } z[x]^{z[x]} z'[x] -$
        $x^2 z[x]^{z[x]} z''[x] - x^2 \text{ Log[z[x]] } z[x]^{z[x]} z''[x]$

For a function of several variables, the derivative is not represented by ' in output form, but instead by using numbers in parentheses to specify how many times to differentiate with respect to the corresponding variable.

In[9]:= **D[func[t, t, t], {t, 2}, {t, 3}, {t, 3}]**

Out[9]= $func^{(2,3,3)} [t, t, t]$

Mathematica is able to explicitly differentiate nearly all special functions with respect to their "argument", but only a few special functions with respect to their "parameters". Here is the derivative of LegendreP with respect to its first argument.

In[10]:= **D[LegendreP[n, z], {n, 1}]**

Out[10]= $LegendreP^{(1,0)} [n, z]$

Numerically, these quantities can still be calculated.

In[11]:= **N[% /. {z -> 2.04, n -> 0.567}]**

Out[11]= 1.07925

Here is a high-precision evaluation of the last derivative.

In[12]:= **N[%% /. {z -> 204/100, n -> 567/1000}, 22]**

Out[12]= 1.079254609237523525024

Here, we have to make a remark about the numerical differentiation encountered in the last examples. Whereas the last derivative was evaluated "just fine" by *Mathematica*, the following "simple" derivative does "not work".

In[13]:= **Abs'[1.]**

Out[13]= Abs'[1.]

The reason Abs' [*inexactNumber*] does not evaluate is the fact that the derivative of Abs does not exist. The derivative is (by definition) the limit $\lim_{\delta \to 0} (f(z+\delta) - f(z))/\delta$ for *any complex* δ. But for Abs, the result depends on the direction of δ approaching 0. Let $z = 1$; then we have the following result. (Here, we use the soon-to-be-discussed function `Limit`.)

```
In[14]:= absDeriv[zr_, zi_, φ_] =
          With[{z = zr + I zi, δz = δ Exp[I φ]},
          Limit[ComplexExpand[(Abs[z + δz] - Abs[z])/δz,
                          TargetFunctions -> {Re, Im}], δ -> 0]]
```

$$Out[14]= \frac{(Cos[φ] - i\ Sin[φ])\ (zr\ Cos[φ] + zi\ Sin[φ])}{\sqrt{zi^2 + zr^2}}$$

Here is the direction dependence for $z = 1$ as a function of $\arg(z-1)$. The two curves show the real and the imaginary parts of `absDeriv[1, 0, φ]`.

```
In[15]:= Plot[{Re[absDeriv[1, 0, φ]], Im[absDeriv[1, 0, φ]]}, {φ, 0, 2Pi},
          PlotRange -> All, Frame -> True,
          Axes -> False, PlotStyle -> {{Thickness[0.005]},
                          {Thickness[0.005], Dashing[{0.01, 0.01}]}}]
```

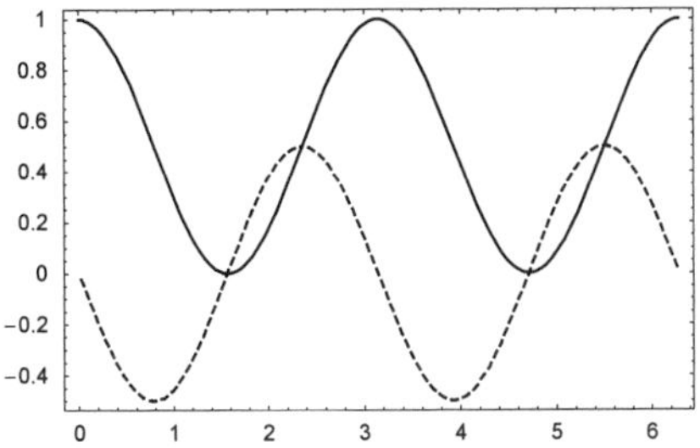

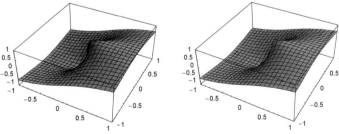

The numerical derivative is always taken for purely real δz. To use the numerical differentiation, we "hide" `Abs` in the following definition of `abs`.

```
In[16]:= (* make abs a numerical function for numerical arguments *)
          SetAttributes[abs, NumericFunction];
          abs[x_?InexactNumberQ] := Abs[x]
```

For purely real δz and complex z, we have the following behavior of the "derivative". The right picture shows the result of the numerical differentiation.

```
In[19]:= Show[GraphicsArray[
          Plot3D[#, {x, -1, 1}, {y, -1, 1}, DisplayFunction -> Identity,
                    PlotPoints -> 25]& /@ {absDeriv[x, y, 0], abs'[x + I y]}]]
```

While the last two pictures look "sensible", checking the difference between `absDeriv[x, y, 0]` and `abs'` more carefully, we see near the origin the effects of the numerical differentiation.

```
In[20]:= Plot3D[Abs[absDeriv[x, y, 0] - abs'[x + I y]],
            {x, -1/4, 1/4}, {y, -1/4, 1/4},
            PlotRange -> All, PlotPoints -> 50]
```

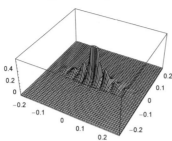

The following picture shows the points used in the numerical differentiation process near the high-precision number 1.

```
In[21]:= abs[x_?InexactNumberQ] := ((* collect values *) Sow[x]; Abs[x]);
         Show[Graphics[Line[{{#, 0}, {#, Abs[#]}}]& /@
               (* evaluate derivative and return sampled x-values *)
                                          Reap[abs'[1``100]][[2, 1]]],
             PlotRange -> All, Frame -> True]
```

From the last result, we conclude that we should not trust the numerical derivatives of discontinuous functions or quickly oscillating functions. Here is the numerical derivative of a step-like function (be aware that we display $f(x)$ and $f'(x)/60$).

```
In[23]:= SetAttributes[meander, NumericFunction];
         meander[x_?NumberQ] := Sin[x]/Sqrt[Sin[x]^2]

         Plot[{meander[x], meander'[SetPrecision[x, 100]]/60}, {x, 2, 4},
             PlotStyle -> {Hue[0], GrayLevel[0]}, PlotRange -> All,
             Compiled -> False]
```

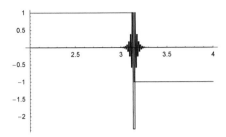

One word of caution is in order here: If very reliable high-precision numerical values of derivatives are needed, it is safer to use the function ND from the package NumericalMath`NLimit` instead of numericalizing symbolic expressions containing unevaluated derivatives using N. N[*unevaluatedDerivative*] has to choose a scale for sample points. Like other numerical functions, no symbolic analysis of *unevaluatedDerivative* is carried out, and as a result, the chosen scale may result in mathematically wrong values for the derivative (this is especially the case for values of derivatives near singularities; but in principle higher-order numerical differentiation is difficult [1331]).

We should make another remark concerning a potential pitfall when differentiating. *Mathematica* tacitly assumes that the differentiation with respect to different variables can be interchanged (a condition which is fulfilled for most functions used in practical calculations). This means that for the following well-known example strange⹁ Func, we do not get the "expected" derivatives for all values x and y if we specialize x- or y-values in intermediate steps.

```
In[26]:= strangeFunc[0, 0] = 0;
         strangeFunc[x_, y_] = x y (x^2 - y^2)/(x^2 + y^2);
```

Here is the function, its first derivative with respect to x, and its mixed second derivative are shown over the x,y-plane.

```
In[28]:= Show[GraphicsArray[
         Plot3D[Evaluate[D[strangeFunc[x, y], ##]], {x, -1, 1}, {y, -1, 1},
             PlotPoints -> 121, Mesh -> False, PlotRange -> All,
             DisplayFunction -> Identity]& @@@
             (* function, one first derivative, and mixed second derivative *)
                                              {{}, {x}, {x, y}}]]
```

First, we differentiate with respect to x and evaluate at $\{0, y\}$. And independently we first differentiate with respect to y and evaluate at $\{x, 0\}$.

```
In[29]:= {D[D[#, {x, 1}] /. {x -> 0}, y],
          D[D[#, {y, 1}] /. {y -> 0}, x]}& @ strangeFunc[x, y]
Out[29]= {-1, 1}
```

This means we have $\partial/\partial y(\partial/\partial x \, s(x, y)|_{x=0}) = -1$ and $\partial/\partial x(\partial/\partial y \, s(x, y)|_{y=0}) = +1$.

For comparison, generically, the results after differentiation are identical (this means for generic values of *x* and *y*, the order of differentiation does not matter). D uses the same kind of philosophy with respect to generic parameter values as already discussed for Solve.

In[30]:= `(D[#, {x, 1}, {y, 1}] == D[#, {y, 1}, {x, 1}]) & @ strangeFunc[x, y]`

Out[30]= True

Although the two calculated double derivatives were different at the origin, strangeFunc is differentiable at $\{0, 0\}$. An easy way to find out if a function $f(x_1, x_2, ..., x_n)$ is differentiable at a point $\{\tilde{x}_1, \tilde{x}_2, ..., \tilde{x}_n\}$ is to change from an *n*-dimensional Cartesian coordinate system to an *n*-dimensional polar coordinate system with the origin $\{\tilde{x}_1, \tilde{x}_2, ..., \tilde{x}_n\}$. If and only if $f(r, \vartheta_1, ..., \vartheta_n)$ in the limit $r \to 0$ is independent from all ϑ_i, the function f is differentiable at $\{\tilde{x}_1, \tilde{x}_2, ..., \tilde{x}_n\}$ [1780], [1232]. For the function strangeFunc, the limit $r \to 0$ exists and is 0.

In[31]:= `strangeFunc[r Cos[φ], r Sin[φ]] // Simplify`

Out[31]= $\frac{1}{4} r^2 \sin[4\varphi]$

But the second derivative under consideration does not have a φ-independent value in the limit $r \to 0$.

In[32]:= `D[strangeFunc[x, y], {y, 1}, {x, 1}] /.`
`              {x -> r Cos[φ], y -> r Sin[φ]} // Simplify`

Out[32]= $\frac{1}{2} (3 \cos[2\varphi] - \cos[6\varphi])$

For a collection of counter examples involving the interchangeability of differentiation, integration, and taking of limits, see [706], [1239], [1462], [1048], [1015], [820], [1048], [1179], and [1847]; for some vector analysis counter examples, see [1095].

Drawing attention to nongeneric points, it is even possible to encounter difficulties with functions of just one variable. Consider the following discontinuous function.

In[33]:= `Which[x <= 0, 0, x > 0, 1]`

Out[33]= Which$[x \le 0, 0, x > 0, 1]$

For some applications, the following result is not desired, although correct almost everywhere with respect to the 1D Lebesgue measure on the real axis.

In[34]:= `D[%, x]`

Out[34]= Which$[x \le 0, 0, x > 0, 0]$

Differentiating a univariate piecewise function (head Piecewise) gives the value Indeterminate at the position of the discontinuity.

In[35]:= `D[Piecewise[{{1, x > 1}}], x]`

Out[35]= $\begin{cases} 0 & x < 1 \,||\, x > 1 \\ \text{Indeterminate} & \text{True} \end{cases}$

Differentiating a multivariate piecewise function (head Piecewise) gives the result valid almost everywhere (including lower-dimensional curves where the result is not pointwise correct).

In[36]:= `D[Piecewise[{{1, x + y > 1}}], x]`

Out[36]= 0

The *Mathematica* kernel does not recognize the derivative of such discontinuous functions as proportional to Dirac delta function by default (but see Section 1.8). Other case-sensitive functions are differentiated in a similar way.

In[37]:= `D[If[true, false[unknownVariable], dontKnow[unknownVariable]],`
        `unknownVariable]`

Out[37]= `If[true, false'[unknownVariable], dontKnow'[unknownVariable]]`

This means that `D` does generically not act in a distributional sense, not for the above-mentioned case of piecewise-defined functions and not for other "closed-form" functions. The following two expressions are so-called differential algebraic constants.

In[38]:= `D[{(Sqrt[x^2] - x)/(2x), (Log[x^2] - 2 Log[x])}, x] // Together`

Out[38]= `{0, 0}`

Such constants are piecewise constants, but they may have jumps in the complex plane and/or the real axis.

In[39]:= `Show[GraphicsArray[`
        `Block[{$DisplayFunction = Identity},`
        `(* 2D and 3D plots of two piecewise constant functions *)`
        `{Plot[(Sqrt[x^2] - x)/(2x), {x, -1, 1}, Frame -> True, Axes -> False],`
        `Plot3D[Im[(Log[(x + I y)^2] - 2 Log[x + I y])],`
        `{x, -1, 1}, {y, -1, 1}, PlotPoints -> 30]}]]]`

As a result, for such functions the identity $\int_{x_0}^{x} \partial f(\xi)/\partial \xi \, d\xi = f(x) - f(x_0)$ does not hold for all generic x_0 and x. Using a generalized function `UnitStep` (to be discussed below) and its derivatives (Dirac δ function `Dirac` `Delta` and its derivative), the last identity hold almost everywhere.

In[40]:= `intDiffIdentity[f_, {x0_, x_}] :=`
                    `Integrate[D[f[ξ], ξ], {ξ, x0, x}] == f[x] - f[x0]`

In[41]:= `{(* generic, symbolic, not explicitly defined function *)`
        `intDiffIdentity[F, {x0, x}],`
        `(* distribution--can be differentiated freely *)`
        `intDiffIdentity[UnitStep, {-1, 1}],`
        `(* piecewise functions ignore jump contributions *)`
        `intDiffIdentity[Piecewise[{{1, # >= 0}}]&, {-1, 1}]}`

Out[41]= `{True, True, False}`

With respect to carrying out differentiations, the set system option should be mentioned.

In[42]:= `Cases[Developer`SystemOptions[],`
            `HoldPattern["DifferentiationOptions" -> _]]`

Out[42]= `{DifferentiationOptions → {AlwaysThreadGradients → False,`
            `DifferentiateHeads → True, DirectHighDerivatives → True,`
            `ExcludedFunctions → {Hold, HoldComplete, Less, LessEqual, Greater,`
            `GreaterEqual, Inequality, Unequal, Nand, Nor, Xor, Not, Element,`
            `Exists, ForAll, Implies, Positive, Negative, NonPositive, NonNegative},`
            `ExitOnFailure → False, SymbolicAutomaticDifferentiation → False}}`

The meaning of this system option setting is that multiple derivatives should be evaluated as efficiently (with respect to time) as possible. Here, we calculate the sixth derivative of the expression $(x + 1)^4 \exp(x^4)$.

```
In[43]:= fh = (x + 1)^4 Exp[x^4];
         D[fh, {x, 6}]
```

$$Out[44]= \; 360 \left(12\, e^{x^4}\, x^2 + 16\, e^{x^4}\, x^6\right) + 480\,(1+x)\left(24\, e^{x^4}\, x + 144\, e^{x^4}\, x^5 + 64\, e^{x^4}\, x^9\right) +$$
$$180\,(1+x)^2 \left(24\, e^{x^4} + 816\, e^{x^4}\, x^4 + 1152\, e^{x^4}\, x^8 + 256\, e^{x^4}\, x^{12}\right) +$$
$$24\,(1+x)^3 \left(3360\, e^{x^4}\, x^3 + 12480\, e^{x^4}\, x^7 + 7680\, e^{x^4}\, x^{11} + 1024\, e^{x^4}\, x^{15}\right) +$$
$$(1+x)^4 \left(10080\, e^{x^4}\, x^2 + 100800\, e^{x^4}\, x^6 + 134400\, e^{x^4}\, x^{10} + 46080\, e^{x^4}\, x^{14} + 4096\, e^{x^4}\, x^{18}\right)$$

Repeatedly differentiating $(x + 1)^4 \exp(x^4)$ gives a different-looking (although of course mathematically equivalent) expression.

```
In[45]:= Nest[D[#, x]&, fh, 6]
```

$$Out[45]= \; 4320\, e^{x^4}\, x^2 + 5760\, e^{x^4}\, x^6 + 11520\, e^{x^4}\, x\,(1+x) + 69120\, e^{x^4}\, x^5\,(1+x) + 30720\, e^{x^4}\, x^9\,(1+x) +$$
$$4320\, e^{x^4}\,(1+x)^2 + 146880\, e^{x^4}\, x^4\,(1+x)^2 + 207360\, e^{x^4}\, x^8\,(1+x)^2 + 46080\, e^{x^4}\, x^{12}\,(1+x)^2 +$$
$$80640\, e^{x^4}\, x^3\,(1+x)^3 + 299520\, e^{x^4}\, x^7\,(1+x)^3 + 184320\, e^{x^4}\, x^{11}\,(1+x)^3 +$$
$$24576\, e^{x^4}\, x^{15}\,(1+x)^3 + 10080\, e^{x^4}\, x^2\,(1+x)^4 + 100800\, e^{x^4}\, x^6\,(1+x)^4 +$$
$$134400\, e^{x^4}\, x^{10}\,(1+x)^4 + 46080\, e^{x^4}\, x^{14}\,(1+x)^4 + 4096\, e^{x^4}\, x^{18}\,(1+x)^4$$

To carry out multiple derivatives by repeated differentiation, we can use the system option setting `"Direct⋅ HighDerivatives"` -> `False`.

```
In[46]:= Developer`SetSystemOptions["DifferentiationOptions" ->
                              ("DirectHighDerivatives" -> False)];
         D[fh, {x, 6}]
```

$$Out[47]= \; 4320\, e^{x^4}\, x^2 + 5760\, e^{x^4}\, x^6 + 11520\, e^{x^4}\, x\,(1+x) + 69120\, e^{x^4}\, x^5\,(1+x) + 30720\, e^{x^4}\, x^9\,(1+x) +$$
$$4320\, e^{x^4}\,(1+x)^2 + 146880\, e^{x^4}\, x^4\,(1+x)^2 + 207360\, e^{x^4}\, x^8\,(1+x)^2 + 46080\, e^{x^4}\, x^{12}\,(1+x)^2 +$$
$$80640\, e^{x^4}\, x^3\,(1+x)^3 + 299520\, e^{x^4}\, x^7\,(1+x)^3 + 184320\, e^{x^4}\, x^{11}\,(1+x)^3 +$$
$$24576\, e^{x^4}\, x^{15}\,(1+x)^3 + 10080\, e^{x^4}\, x^2\,(1+x)^4 + 100800\, e^{x^4}\, x^6\,(1+x)^4 +$$
$$134400\, e^{x^4}\, x^{10}\,(1+x)^4 + 46080\, e^{x^4}\, x^{14}\,(1+x)^4 + 4096\, e^{x^4}\, x^{18}\,(1+x)^4$$

```
In[48]:= Expand[%%% - %]
```
```
Out[48]= 0
```

There is no general rule regarding what is preferable, the setting `"DirectHighDerivatives"` -> `False` or `"DirectHighDerivatives"` -> `True`. In many cases, `"DirectHighDerivatives"` -> `True` will be much faster, but will produce larger results. Here is a "typical" example.

```
In[49]:= With[{f = x^3 Log[x^5 + 1] Exp[-x^2], n = 50},
         {Developer`SetSystemOptions["DifferentiationOptions" ->
                              ("DirectHighDerivatives" -> True)];
          Timing[ByteCount[D[f, {x, n}]]],
          Developer`SetSystemOptions["DifferentiationOptions" ->
                              ("DirectHighDerivatives" -> False)];
          Timing[ByteCount[D[f, {x, n}]]]}]
```
```
Out[49]= {{0.06 Second, 934776}, {1.6 Second, 506008}}
```

For the following simple polynomial, it is fastest to not use special algorithms.

```
In[50]:= With[{f = (x + E)^400 - Expand[(x + E)^400], n = 500},
         {Developer`SetSystemOptions["DifferentiationOptions" ->
                              ("DirectHighDerivatives" -> True)];
```

```
          Timing[ByteCount[D[f, {x, n}]]],
          Developer`SetSystemOptions["DifferentiationOptions" ->
                                        ("DirectHighDerivatives" -> False)];
          Timing[ByteCount[D[f, {x, n}]]]}]
```
Out[50]= {{13.47 Second, 16}, {2.09 Second, 16}}

We restore the default setting.

In[51]:= **Developer`SetSystemOptions["DifferentiationOptions" ->**
                                        ("DirectHighDerivatives" -> True)];

Sometimes one does not want certain functions to be differentiated (for instance, when considering differentiation of a holomorphic functions and, say, Abs[z] or Conjugate[z] occurs). In the following input, the chain rule is used.

In[52]:= **D[f[g[x]], x]**

Out[52]= f'[g[x]] g'[x]

We add the function g to the list of functions that should not be differentiated (the list contains currently by default various comparison functions, but not Equal, which is a special case).

In[53]:= **Developer`SetSystemOptions[**
              "DifferentiationOptions" -> {"ExcludedFunctions" ->
              Append["ExcludedFunctions" /. ("DifferentiationOptions" /.
                  Developer`SystemOptions[]), g]}]

Out[53]= DifferentiationOptions → {AlwaysThreadGradients → False,
          DifferentiateHeads → True, DirectHighDerivatives → True,
          ExcludedFunctions → {Hold, HoldComplete, Less, LessEqual, Greater,
            GreaterEqual, Inequality, Unequal, Nand, Nor, Xor, Not, Element, Exists,
            ForAll, Implies, Positive, Negative, NonPositive, NonNegative, g},
          ExitOnFailure → False, SymbolicAutomaticDifferentiation → False}

Now the chain rule is still used, but no Derivative[][g] expression is formed from an expression of the form D[..., ...]. An expression of the form D[g[x], x] is returned instead.

In[54]:= **D[f[g[x]], x]**

Out[54]= $\partial_x g[x]$ f'[g[x]]

To avoid any partial differentiation in such cases, we can set the "ExitOnFailure" suboption to be True.

In[55]:= **Developer`SetSystemOptions[**
                  "DifferentiationOptions" -> {"ExitOnFailure" -> True}];

In[56]:= **D[f[g[x]], x]**

Out[56]= ∂_x f[g[x]]

In[57]:= **(* restore old settings *)**
          Developer`SetSystemOptions[
                  "DifferentiationOptions" -> {"ExitOnFailure" -> False}];

The next input makes use of fairly high derivatives. We visualize the (normalized) coefficients $c_k^{(n)}$ appearing in

$$\frac{\partial^n}{\partial z^n} \left(\frac{1}{\ln(z)} \right) = \frac{1}{z^n \ln^{n+1}(z)} \sum_{k=0}^{n-1} c_k^{(n)} \ln^k(z).$$

Here, we use $n = 1, 2, \ldots, 200$.

In[59]:= `Show[Graphics3D[`
   `Table[(* the derivative *) deriv = D[1/Log[z], {z, n}];`
        `(* the coefficients *) cl = CoefficientList[Expand[`
                `(-1)^n z^n Log[z]^(n + 1) deriv], Log[z]];`
        `(* colored line *)`
        `{Hue[n/ 250], Line[MapIndexed[{#2[[1]] - 1, n, #1}&,`
           `(* normalize coefficients *) cl/Max[Abs[cl]]]]]}, {n, 200}]],`
     `BoxRatios -> {1, 3/2, 0.5}, PlotRange -> All,`
     `Axes -> True, ViewPoint -> {0, 3, 1}]`

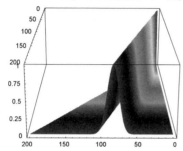

In addition to explicitly given functions, *Mathematica* is also able to differentiate certain abstract function(al)s of functions, for example, indefinite integrals or inverse functions [75], [942], [1724], [1622].

In[60]:= `Table[D[InverseFunction[Y][z], {z, i}], {i, 3}]`

Out[60]= $\left\{ \dfrac{1}{Y'[Y^{(-1)}[z]]},\ -\dfrac{Y''[Y^{(-1)}[z]]}{Y'[Y^{(-1)}[z]]^3},\ \dfrac{3\,Y''[Y^{(-1)}[z]]^2}{Y'[Y^{(-1)}[z]]^5} - \dfrac{Y^{(3)}[Y^{(-1)}[z]]}{Y'[Y^{(-1)}[z]]^4} \right\}$

Sometimes it is convenient to form derivatives of multivariate functions with respect to all independent variables at once (especially in carrying out vector analysis operations). This can be done with the following syntax.

> `D[`*toDifferentiate*`, {`*vector*`, n}]`
>
> differentiates *toDifferentiate* n times with respect to the vector variable *vector*. If the n is omitted, it is assumed to be 1.

Here are the first and second derivatives of the scalar function f, that depends on x, y, and z.

In[61]:= `D[f[x, y, z], {{x, y, z}, 1}]`

Out[61]= $\{ f^{(1,0,0)}[x, y, z],\ f^{(0,1,0)}[x, y, z],\ f^{(0,0,1)}[x, y, z] \}$

In[62]:= `D[f[x, y, z], {{x, y, z}, 2}]`

Out[62]= $\{\{ f^{(2,0,0)}[x, y, z],\ f^{(1,1,0)}[x, y, z],\ f^{(1,0,1)}[x, y, z]\},$
    $\{ f^{(1,1,0)}[x, y, z],\ f^{(0,2,0)}[x, y, z],\ f^{(0,1,1)}[x, y, z]\},$
    $\{ f^{(1,0,1)}[x, y, z],\ f^{(0,1,1)}[x, y, z],\ f^{(0,0,2)}[x, y, z]\}\}$

The last result is identical to carrying out two differentiation steps independently.

In[63]:= `Nest[D[#, {{x, y, z}, 1}]&, f[x, y, z], 2] === %`

Out[63]= `True`

For a function f that depends explicitly on a vector, the differentiation orders are also in lists.

In[64]:= `D[f[{x, y, z}], {{x, y, z}, 2}]`

Out[64]= $\{\{f^{(\{2,0,0\})}[\{x, y, z\}], f^{(\{1,1,0\})}[\{x, y, z\}], f^{(\{1,0,1\})}[\{x, y, z\}]\},$
$\{f^{(\{1,1,0\})}[\{x, y, z\}], f^{(\{0,2,0\})}[\{x, y, z\}], f^{(\{0,1,1\})}[\{x, y, z\}]\},$
$\{f^{(\{1,0,1\})}[\{x, y, z\}], f^{(\{0,1,1\})}[\{x, y, z\}], f^{(\{0,0,2\})}[\{x, y, z\}]\}\}$

The next function has three independent occurrences of the arguments x and y.

In[65]:= `D[f[{{x, y}}, {x, y}, x, y], {{x, y}, 1}]`

Out[65]= $\{f^{(\{\{0,0\}\},\{0,0\},1,0)}[\{\{x, y\}\}, \{x, y\}, x, y] + f^{(\{\{0,0\}\},\{1,0\},0,0)}[\{\{x, y\}\}, \{x, y\}, x, y] +$
$f^{(\{\{1,0\}\},\{0,0\},0,0)}[\{\{x, y\}\}, \{x, y\}, x, y], f^{(\{\{0,0\}\},\{0,0\},0,1)}[\{\{x, y\}\}, \{x, y\}, x, y] +$
$f^{(\{\{0,0\}\},\{0,1\},0,0)}[\{\{x, y\}\}, \{x, y\}, x, y] + f^{(\{\{0,1\}\},\{0,0\},0,0)}[\{\{x, y\}\}, \{x, y\}, x, y]\}$

While differentiation with respect to a vector argument works, differentiation with respect to a matrix does not.

In[66]:= `D[f[{{x1, x2}, {y1, y2}}], {{{x1, x2}, {y1, y2}}, 1}]`

General::ivar : {x1, x2} is not a valid variable. More...

Out[66]= $\partial_{\{\{\{x1,x2\},\{y1,y2\}\},1\}} f[\{\{x1, x2\}, \{y1, y2\}\}]$

As in the scalar case, for list arguments the differentiation is threaded over the arguments.

In[67]:= `D[{fx[x, y], fy[x, y]}, {{x, y}, 1}]`

Out[67]= $\{\{fx^{(1,0)}[x, y], fx^{(0,1)}[x, y]\}, \{fy^{(1,0)}[x, y], fy^{(0,1)}[x, y]\}\}$

We now turn to a topic that has not yet been discussed, the `FullForm` of derivatives. This came up in Chapter 1 of the Numerics volume [1737] in connection with the numerical solution of differential equations.

In[68]:= `FullForm[x'[t]]`

Out[68]//FullForm=

    `Derivative[1][\[DoubleStruckX]][\[DoubleStruckT]]`

In[69]:= `FullForm[x''[t]]`

Out[69]//FullForm=

    `Derivative[2][\[DoubleStruckX]][\[DoubleStruckT]]`

In[70]:= `FullForm[D[h[x, y], {x, 1}, {y, 2}]]`

Out[70]//FullForm=

    `Derivative[1, 2][\[DoubleStruckH]][\[DoubleStruckX], \[DoubleStruckY]]`

In this form, we have `Derivative`.

`Derivative[`i_1`, `i_2`, ..., `i_n`]`

    represents an operator that differentiates i_1 times with respect to the first variable, i_2 times with respect to the second variable, etc.

To use it, we write `Derivative[`i_1`, `i_2`, ..., `i_n`]`[*function*]`[`*argument*$_1$`, `*argument*$_2$`, ..., `*argument*$_n$`]`, where *function* must be a (pure) function. If possible, the differentiation will be carried out explicitly. The result of applying `Derivative` to a function is again a function; applying it to a (built-in) `Symbol` with function properties also yields a function (head `Function`).

In[71]:= `Derivative[1][Sin]`

Out[71]= `Cos[#1] &`

In[72]:= `Head[%]`

Out[72]= `Function`

From a mathematical point of view, this would have given the same thing.

In[73]:= `Derivative[1][Sin[#]&]`

Out[73]= `Cos[#1] &`

The following pure functions will also be differentiated.

In[74]:= `Derivative[1][Function[x, Sin[x]]]`

Out[74]= `Function[x, Cos[x]]`

In[75]:= `Derivative[1][Function[symbol, symbol^2]]`

Out[75]= `Function[symbol, 2 symbol]`

The following expression remains unchanged because *Mathematica* does not know any differentiation rules for `sin`.

In[76]:= `Derivative[1][sin[#]&]`

Out[76]= `sin'[#1] &`

The derivative function can then again be provided with variables.

In[77]:= `Derivative[1][#^2&][argument]`

Out[77]= `2 argument`

The next input generates an error message because the built-in `Sin` command has only one argument and in the process of calculating the derivative `Sin[#1, #2]` is evaluated..

In[78]:= `Derivative[1, 1][Sin]`

  Sin::argx : Sin called with 2 arguments; 1 argument is expected. More…

Out[78]= $Sin^{(1,1)}$

Because `Function` has the attribute `HoldAll`, their body will sometimes not be fully auto-evaluated after applying `Derivative`.

In[79]:= `{Nest[Derivative[1], Function[x, x Sin[x]], 5],`
     `Derivative[5][Function[x, x Sin[x]]]}`

Out[79]= `{Function[x, x Cos[x] + Sin[x] + Sin[x] + Sin[x] + Sin[x] + Sin[x]],`
     `Function[x, x Cos[x] + Sin[x] + 4 Sin[x]]}`

Here are some derivatives of simple pure functions.

In[80]:= `{Derivative[1, 1][Power], Derivative[1, 1][Times],`
     `Derivative[1, 2][Times]}`

Out[80]= $\{ \#1^{-1+\#2} + Log[\#1] \, \#1^{-1+\#2} \, \#2 \, \&, \, 1 \, \&, \, 0 \, \& \}$

The next input shows a higher-order derivative of the three-argument `Power`. (For brevity, we extract the body of the pure function and simplify the result).

In[81]:= `Derivative[1, 2, 3][Power][[1]] // Together`

Out[81]= $\dfrac{1}{\#1 \, \#2^2} \Big(6 \, Log[\#2] \, \#1^{\#2^{\#3}} \, \#2^{\#3} - 3 \, Log[\#2]^2 \, \#1^{\#2^{\#3}} \, \#2^{\#3} + 42 \, Log[\#1] \, Log[\#2] \, \#1^{\#2^{\#3}} \, \#2^{2 \, \#3} -$

$21 \, Log[\#1] \, Log[\#2]^2 \, \#1^{\#2^{\#3}} \, \#2^{2 \, \#3} + 36 \, Log[\#1]^2 \, Log[\#2] \, \#1^{\#2^{\#3}} \, \#2^{3 \, \#3} -$

$18 \, Log[\#1]^2 \, Log[\#2]^2 \, \#1^{\#2^{\#3}} \, \#2^{3 \, \#3} + 6 \, Log[\#1]^3 \, Log[\#2] \, \#1^{\#2^{\#3}} \, \#2^{4 \, \#3} -$

$3 \, Log[\#1]^3 \, Log[\#2]^2 \, \#1^{\#2^{\#3}} \, \#2^{4 \, \#3} + 6 \, Log[\#2]^2 \, \#1^{\#2^{\#3}} \, \#2^{\#3} \, \#3 - Log[\#2]^3 \, \#1^{\#2^{\#3}} \, \#2^{\#3} \, \#3 +$

$90 \, Log[\#1] \, Log[\#2]^2 \, \#1^{\#2^{\#3}} \, \#2^{2 \, \#3} \, \#3 - 15 \, Log[\#1] \, Log[\#2]^3 \, \#1^{\#2^{\#3}} \, \#2^{2 \, \#3} \, \#3 +$

$150 \, Log[\#1]^2 \, Log[\#2]^2 \, \#1^{\#2^{\#3}} \, \#2^{3 \, \#3} \, \#3 - 25 \, Log[\#1]^2 \, Log[\#2]^3 \, \#1^{\#2^{\#3}} \, \#2^{3 \, \#3} \, \#3 +$

$60 \, Log[\#1]^3 \, Log[\#2]^2 \, \#1^{\#2^{\#3}} \, \#2^{4 \, \#3} \, \#3 - 10 \, Log[\#1]^3 \, Log[\#2]^3 \, \#1^{\#2^{\#3}} \, \#2^{4 \, \#3} \, \#3 +$

$$6 \text{Log}[\#1]^4 \text{Log}[\#2]^2 \#1^{\#2^{\#3}} \#2^{5\,\#3} \#3 - \text{Log}[\#1]^4 \text{Log}[\#2]^3 \#1^{\#2^{\#3}} \#2^{5\,\#3} \#3 +$$
$$\text{Log}[\#2]^3 \#1^{\#2^{\#3}} \#2^{\#3} \#3^2 + 31 \text{Log}[\#1] \text{Log}[\#2]^3 \#1^{\#2^{\#3}} \#2^{2\,\#3} \#3^2 +$$
$$90 \text{Log}[\#1]^2 \text{Log}[\#2]^3 \#1^{\#2^{\#3}} \#2^{3\,\#3} \#3^2 + 65 \text{Log}[\#1]^3 \text{Log}[\#2]^3 \#1^{\#2^{\#3}} \#2^{4\,\#3} \#3^2 +$$
$$15 \text{Log}[\#1]^4 \text{Log}[\#2]^3 \#1^{\#2^{\#3}} \#2^{5\,\#3} \#3^2 + \text{Log}[\#1]^5 \text{Log}[\#2]^3 \#1^{\#2^{\#3}} \#2^{6\,\#3} \#3^2 \big)$$

Nesting the two argument `Power` gives the same result.

In[82]:= `% == (Derivative[1, 2, 3][Power[#1, Power[#2, #3]]&)[[1]] // Together)`

Out[82]= `True`

We have the following result for `Plus` and `Times` viewed as functions that can take multiple arguments.

In[83]:= `Derivative[1, 1] /@ {Times, Plus}`

Out[83]= `{1 &, 0 &}`

Because `Derivative` evaluates the function with symbolic arguments, the result for "program-like" functions is sometimes not so useful.

In[84]:= `fProgram[x_] := Module[{ξ = x}, Nest[#^2&, ξ, v]]`

In[85]:= `{fProgram[3], fProgram'}`

```
Nest::intnm :
  Non-negative machine-size integer expected at position 3 in Nest[#1² &, 3, v]. More…

Nest::intnm : Non-negative machine-size
  integer expected at position 3 in Nest[#1² &, #1, v]. More…
```

Out[85]= `{Nest[#1² &, 3, v], Nest⁽⁰,¹,⁰⁾[#1² &, #1, v] + 0 & Nest⁽¹,⁰,⁰⁾[#1² &, #1, v] &}`

Note the following difference between the assumptions on the dependence of the expressions considered for `D` and for `Derivative`. `D` acts on functions that were applied to arguments whereas `Derivative` acts on functions itself (like pure functions or symbols that have function definitions).

In[86]:= `{Derivative[1][fuga], Derivative[1][fuga][x], D[fuga, x]}`

Out[86]= `{fuga', fuga'[x], 0}`

This means the implicitly understood grouping is `((Derivative[1])[x])[t]`. The compound head `Derivative[1]` is applied to the (function) x and gives `Derivative[1][x]`. This expression is applied to the argument t.

`Derivative` can of course also calculate higher derivatives. Note that the body of the pure function was not fully evaluated (because the `HoldAll` attribute of `Function`).

In[87]:= `{#, Evaluate //@ #}&[Derivative[12][#1 Sin[#]&]]`

Out[87]= `{-11 Cos[#1] - Cos[#1] - Sin[#1] (-#1) &, -12 Cos[#1] + Sin[#1] #1 &}`

Note that `Derivative` is one of the few commands that does not carry the attribute `Protected`. (The `ReadProtected` attribute tries to hide some typesetting definitions for the function `Derivative`.)

In[88]:= `Attributes[Derivative]`

Out[88]= `{NHoldAll, ReadProtected}`

The reason for this is that derivatives can be user defined. The following attempts to make definitions for a second derivative of a function `bcr` both fail.

In[89]:= `{bcr''[x_] ^:= bcrss[x], bcr /: bcr''[x_] := bcrss[x]}`

```
UpSetDelayed::nosym :
  bcr"[x_] does not contain a symbol to which to attach a rule. More…
```

```
         TagSetDelayed::tagpos :
            Tag bcr in bcr''[x_] is too deep for an assigned rule to be found. More...
Out[89]= {$Failed, $Failed}
```

We recognize the reason for the above failure by looking at the `FullForm` of the expression. Thus, `bcr` is too deeply nested in `Derivative[2][bcr][Pattern[x, Blank[]]]` to associate a rule with it. (Remember, for `TagSet` to work, it was necessary for the symbol associated with the definition to appear in level 1.)

```
In[90]:= bcr''[x_] // FullForm
Out[90]//FullForm=
         Derivative[2][bcr][Pattern[x, Blank[]]]
```

Therefore, the rule has to be associated with `Derivative`, which does not carry the attribute `Protected`.

```
In[91]:= bcr''[x_] := bcrss[x]
```

Here is the information now associated with `Derivative`.

```
In[92]:= ClearAttributes[Derivative, ReadProtected]
```

```
In[93]:= SubValues[Derivative] // (* skip typeset definitions *) Last
Out[93]= HoldPattern[bcr''[x_]] :→ bcrss[x]
```

Another possibility to define a derivative is as a pure function. If we do not have an argument applied, the function is not too deep for `UpSetDelayed`.

```
In[94]:= (* remove last made definition *) bcr''[x_] =.

         (* make definition as a pure function *) bcr'' ^:= bcrss
```

Now the definition for the derivative is stored with the symbol `bcr`.

```
In[96]:= Definition[bcr]
Out[96]= bcr'' ^:= bcrss
```

The fact that derivatives are internally represented as `Derivative[n][f][arg]` allows us, in a convenient way, to replace the function f with an explicitly given function (head `Function`). The following function `odeForPolynomial` calculates the differential equation that annihilates an n-nomial (this is a sum of n different terms of the form x^m) [1525].

```
In[97]:= odeForPolynomial[y_, x_, numberOfTerms_] :=
            Det @ Table[Nest[D[#, x]&, Nest[(x D[#, x])&, y[x], j], i],
                       {i, 0, numberOfTerms}, {j, 0, numberOfTerms}]
```

Here is its explicit form for generic binomials, $y(x) = a x^p + b x^q$.

```
In[98]:= odeForPolynomial[y, x, 2]
```
$$Out[98]= 2\, y[x]\, y'[x]\, y''[x] - 2\, x\, y'[x]^2\, y''[x] - 2\, x\, y[x]\, y''[x]^2 + 3\, x^2\, y'[x]\, y''[x]^2 -$$
$$x^3\, y''[x]^3 + 4\, x\, y[x]\, y'[x]\, y^{(3)}[x] - 4\, x^2\, y'[x]^2\, y^{(3)}[x] + 2\, x^3\, y'[x]\, y''[x]\, y^{(3)}[x] -$$
$$x^3\, y[x]\, y^{(3)}[x]^2 + x^2\, y[x]\, y'[x]\, y^{(4)}[x] - x^3\, y'[x]^2\, y^{(4)}[x] + x^3\, y[x]\, y''[x]\, y^{(4)}[x]$$

Substituting this functions in the above expression gives zero for all p and q.

```
In[99]:= % /. y -> Function[x, a x^p + b x^q] // Expand
Out[99]= 0
```

Until now, we encountered derivatives of the form `Derivative[i_1, i_2, ..., i_n]`. When the arguments of a function have explicit list heads, the list structure is mirrored in the `Derivative` arguments. Some of the built-in special functions (to be discussed in Chapter 3), such as `HypergeometricPFQ` and `MeijerG`, have such arguments. Here is a simple example. The function f has a second argument that is a nested list.

```
In[100]:= Clear[f, x, a];
         D[f[x, {x, {x}}, a], x]
```
$$\text{Out[101]=}\ f^{(0,\{0,\{1\}\},0)}\ [x,\ \{x,\ \{x\}\},\ a]\ +\ f^{(0,\{1,\{0\}\},0)}\ [x,\ \{x,\ \{x\}\},\ a]\ +\ f^{(1,\{0,\{0\}\},0)}\ [x,\ \{x,\ \{x\}\},\ a]$$

Next, we use the function derivative to implement a generalized Taylor series expansion [884], [824], [1106], [754], [1646], [189], [1243]. In addition to the usual derivatives at the expansion point z_0, it also contains derivative contributions at z. (The two parameters m and n are nonnegative integers and R is the remainder term.)

$$f(z) = f(z_0) + \sum_{k=1}^{o} \frac{(m+n-k)!}{(m+n)!}\left(\binom{m}{k} f^{(k)}(z_0) - (-1)^k \binom{n}{k} f^{(k)}(z)\right)(z - z_0)^k + R_{o+1}[f](z, z_0)$$

```
In[102]:= GeneralizedTaylorSeries[f_, {m_, n_}, {z0_, z_}, o_] :=
         f[z0] + Sum[If[m + n - k < 0, 0, (m + n - k)!/(m + n)! *
            (Binomial[m, k] Derivative[k][f][z0] -
             (-1)^k Binomial[n, k] Derivative[k][f][z]) (z - z0)^k], {k, o}]
```

We will use $f(z) = \exp(z)$, $z_0 = 0$, and $o = 6$ in the following. For $n = 0$, we recover the ordinary Taylor series.

```
In[103]:= GeneralizedTaylorSeries[Exp, {6, 0}, {0, z}, 6]
```
$$\text{Out[103]=}\ 1 + z + \frac{z^2}{2} + \frac{z^3}{6} + \frac{z^4}{24} + \frac{z^5}{120} + \frac{z^6}{720}$$

For $m = n$, we obtain a series that contains the function $\exp(z)$ itself.

```
In[104]:= GeneralizedTaylorSeries[Exp, {6, 6}, {0, z}, 6]
```
$$\text{Out[104]=}\ 1 + \frac{1}{12}\ (6 + 6\ e^z)\ z + \frac{1}{132}\ (15 - 15\ e^z)\ z^2 +$$
$$\frac{(20 + 20\ e^z)\ z^3}{1320} + \frac{(15 - 15\ e^z)\ z^4}{11880} + \frac{(6 + 6\ e^z)\ z^5}{95040} + \frac{(1 - e^z)\ z^6}{665280}$$

Equating the last expression to $\exp(z)$ yields the diagonal Padé approximation of order 6.

```
In[105]:= Module[{exp}, Together //@
            Solve[(% == Exp[z] /. (* rename *) Exp[z] -> exp), exp] /.
            exp -> Exp[z]]
```
$$\text{Out[105]=}\ \left\{\left\{e^z \rightarrow \frac{665280 + 332640\ z + 75600\ z^2 + 10080\ z^3 + 840\ z^4 + 42\ z^5 + z^6}{665280 - 332640\ z + 75600\ z^2 - 10080\ z^3 + 840\ z^4 - 42\ z^5 + z^6}\right\}\right\}$$

```
In[106]:= (* load Padé package *)
         Needs["Calculus`Pade`"]
         Pade[Exp[x], {x, 0, 6, 6}] // Together
```
$$\text{Out[108]=}\ \frac{665280 + 332640\ x + 75600\ x^2 + 10080\ x^3 + 840\ x^4 + 42\ x^5 + x^6}{665280 - 332640\ x + 75600\ x^2 - 10080\ x^3 + 840\ x^4 - 42\ x^5 + x^6}$$

We calculate some more of the generalized expansion for $0 \le m, n \le 9$.

```
In[109]:= Do[exp[{m, n}, z_] = GeneralizedTaylorSeries[Exp, {m, n}, {0, z}, 6],
            {m, 0, 9}, {n, 0, 9}]
```

The left graphic shows the logarithm of the absolute error at $z = 1$ over the m, n-plane. The right graphic shows the logarithm of the absolute error of the 100 approximations as a function of real z. We color the curves according to their m,n-values. The two graphics show the error is smallest for $m \approx n$.

```
In[110]:= (* high-precision value for the absolute error *)
         δexpHP[{m_, n_}, z_?NumericQ] :=
            With[{(* make input exact *)ζ = SetPrecision[z, Infinity]},
```

```
            N[Abs[exp[{m, n}, ζ] - Exp[ζ]],
               (* guaranteed digits *) $MachinePrecision + 1]]
```

In[112]:= `Show[GraphicsArray[`
```
        (* avoid messages from numericalization for small quantities *)
        Internal`DeactivateMessages[
        Block[{$DisplayFunction = Identity},
          {(* at z = 1 over the n,m-plane *)
          ListPlot3D[Log[Table[δexpHP[{m, n}, 2],
                               {m, 0, 9}, {n, 0, 9}]]],
          (* as a function of real z *)
           Show[Table[Plot[Log[δexpHP[{m, n}, z]],
             {z, -2, 2}, PlotRange -> All, DisplayFunction -> Identity,
             PlotStyle -> RGBColor[m/9, 0, n/9]], {m, 0, 9}, {n, 0, 9}],
           Frame -> True, PlotRange -> All]}]]]]
```

Next, we discuss total differentials of a function of several variables [1124], [1179]. This is accomplished in *Mathematica* as follows.

$Dt[expr, var_1, var_2, \ldots, var_n, Constants \to \{varCon_1, varCon_2, \ldots, varCon_n\}]$

computes the total differential of *expr*, with respect to the variables $var_1, \ldots, var_n$. All derivatives with respect to the variables $varCon_i$ are assumed to vanish identically.

Here is the total differential of a function $z[x, y]$, where y is considered as a function of x.

In[113]:= `Clear[z, x, y];`
      `Dt[z[x, y], x]`
Out[114]= $Dt[y, x] z^{(0,1)}[x, y] + z^{(1,0)}[x, y]$

This behavior of `Dt` makes it very useful in changing variables in 1D integrals. If we have an integrand $f[s]$ and change the integration variable from s to $Sin[t]$, the new integrand can be obtained in the following way.

In[115]:= `(f[s] /. {s -> Sin[s]}) Dt[Sin[s], s]`
Out[115]= $Cos[s] f[Sin[s]]$

Here is an example of a total differential of order 2.

In[116]:= `Dt[z[x, y], x, z]`
Out[116]= $Dt[y, x, z] z^{(0,1)}[x, y] + Dt[y, z] z^{(1,1)}[x, y] +$
         $Dt[y, x] (Dt[y, z] z^{(0,2)}[x, y] + Dt[x, z] z^{(1,1)}[x, y]) + Dt[x, z] z^{(2,0)}[x, y]$

Of course, higher orders can be calculated [1107].

In[117]:= `Expand[Nest[Dt, f[x, y], 4]] // Short[#, 4]&`

Out[117]//Short=

Dt[Dt[Dt[Dt[y]]]] f$^{(0,1)}$[x, y] + 3 Dt[Dt[y]]^2 f$^{(0,2)}$[x, y] +
4 Dt[y] Dt[Dt[Dt[y]]] f$^{(0,2)}$[x, y] + 6 Dt[y]^2 Dt[Dt[y]] f$^{(0,3)}$[x, y] +
Dt[y]^4 f$^{(0,4)}$[x, y] + ≪10≫ + 6 Dt[x]^2 Dt[Dt[y]] f$^{(2,1)}$[x, y] +
6 Dt[x]^2 Dt[y]^2 f$^{(2,2)}$[x, y] + 6 Dt[x]^2 Dt[Dt[x]] f$^{(3,0)}$[x, y] +
4 Dt[x]^3 Dt[y] f$^{(3,1)}$[x, y] + Dt[x]^4 f$^{(4,0)}$[x, y]

Note the following difference in the assumption of the implicit dependence in D[...] compared with Dt[...]:

D[y, x]	gives	0
D[y, x, NonConstants -> y]	gives	D[y, x, NonConstants -> {y}]
Dt[y, x]	gives	Dt[y, x]
Dt[y, x, Constants -> y]	gives	0
Derivative[j][y]	gives	$y^{(j)}$

Although differentiation is generally an elementary process, the ability to do it on a computer is often a big advantage. We can illustrate this with a few examples from differential geometry.

Mathematical Remark: Metric Tensors, Christoffel Symbols, Geodesic Equations, and Riemann Curvature Tensors

Often, it is useful or necessary to work with coordinates other than the commonly used Cartesian coordinate system. For example, we may have to work with non-Euclidean geometry (e.g., in general relativity theory). For such calculations, the following three quantities play a fundamental role: the metric tensor, the Christoffel symbols, and the Riemann curvature tensor.

We now briefly discuss the basic relationships between these quantities. If we change from a coordinate system S with coordinates x to a new system S' with coordinates x' on a manifold, the local connection between the two is given by $A_{n'}^n = \partial x^n(x)/\partial x^{n'}$. Under such a coordinate transformation, the metric tensor $g_{mn} = \mathbf{e}_m(P)\,\mathbf{e}_n(P)$ where $\mathbf{e}_n(P)$ is the basis vector in the direction x^n at the point P, becomes $g_{n'm'} = g_{nm}\,A_{n'}^n\,A_{m'}^m$, where it is to be summed over repeated indices (Einstein's summation convention).

In addition to this covariant metric tensor (which is covariant because it has subscripts), its contravariant form g^{mn}, defined by $g_{mn}\,g^{np} = \delta_m^p$ (with δ_m^p the Kronecker symbol), also plays an important role. The Christoffel symbols Γ_{ab}^m can be computed from the metric tensor (which is in general x-dependent). They represent the changes in the basis vectors for infinitesimal movements:

$$\frac{\partial \mathbf{e}_i(P)}{\partial x^j} = \Gamma_{ij}^k\,\mathbf{e}_k(P).$$

The Christoffel symbols themselves are given by

$$\Gamma_{ab}^m = \frac{1}{2}\,g^{mn}\,(g_{an,b} + g_{bn,a} - g_{ab,n})$$

where f_x is the derivative of f with respect to x. Whether a space (or a geometry) is intrinsically curved, or whether only curvilinear coordinates are involved, can be decided on the basis of the Riemann curvature tensor R_{qkl}^i. This in turn can be found from the Christoffel symbols as follows:

$$R_{qkm}^i = \Gamma_{qk,m}^i - \Gamma_{qm,k}^i - \Gamma_{pk}^i\,\Gamma_{qm}^p + \Gamma_{pm}^i\,\Gamma_{qk}^p.$$

Geodesic equations are differential equations that describe the shortest path between two points (the path taken by a physical particle). They have the form:

$$\frac{d^2 x^m(\lambda)}{d\lambda^2} + \Gamma^m_{ab} \frac{dx^a(\lambda)}{d\lambda} \frac{dx^b(\lambda)}{d\lambda} = 0.$$

Here, λ is the parameter (e.g., time). (For more on differential geometry, see, for example, [1119], [1643], [1404], [358], [1224], [9], [1786], [528], [1148], [1880], [764], [687], and [1296].)

As a simple computational example, we now look at all of the above quantities for the case of 3D spherical coordinates. We begin with the metric tensor in Cartesian coordinates (we assume these are the initial coordinates). The variable u stands for upper and denotes an upper index; similarly, l denotes a lower index. (There is a more elegant way to do this than what we will do in the following program.) As is common in differential geometry, we will explicitly work with components and thus prefer to use g[n, m] rather than g[[n, m]]. (For a very elegant *Mathematica* implementation of upper and lower indices for tensor calculations, see [817].)

```
In[118]:= MetricTensorCart[1, 1] = IdentityMatrix[3];
          MetricTensorCart[1, 1, n_, m_] := MetricTensorCart[1, 1][[n, m]];
```

Here is the change of coordinates, where we express the old Cartesian coordinates in terms of new spherical coordinates.

```
In[120]:= oldVariable = {(* x = *)   r Cos[φ] Sin[θ],
                         (* y = *)   r Sin[φ] Sin[θ],
                         (* z = *)   r Cos[θ]};

          newVariable = {r, θ, φ};
```

The passage from the coordinates $x^{n'} = x^{n'}(x^n)$ (x^n are the three Cartesian coordinates, $x^{n'}$ are the three spherical coordinates) is described by the coordinate transformation matrix (the Jacobian) $A^n_{n'} = \partial x^n(x)/\partial x^{n'}$.

```
In[122]:= (* the whole matrix *)
          A[u, 1] = Outer[D, oldVariable, newVariable];
          (* the elements *)
          A[u, 1, n_, ns_] := A[u, 1][[n, ns]]
          A[u, 1] // TableForm
Out[126]//TableForm=
          Cos[φ] Sin[θ]        r Cos[θ] Cos[φ]        -r Sin[θ] Sin[φ]
          Sin[θ] Sin[φ]        r Cos[θ] Sin[φ]        r Cos[φ] Sin[θ]
          Cos[θ]               -r Sin[θ]              0
```

The metric tensor $g_{n'm'}$ in the new spherical coordinates becomes $g_{n'm'} = g_{nm} A^n_{n'} A^m_{m'}$. The replacement rules carry out specific simplification; they are significantly faster than Simplify.

```
In[127]:= MetricTensorSphere[1, 1] =
          Table[Sum[MetricTensorCart[1, 1, n, m] *
                    A[u, 1, n, ns] A[u, 1, m, ms],
                    {n, 3}, {m, 3}] //. {a_. Cos[φ_]^2 + a_. Sin[φ_]^2 -> a},
                {ns, 3}, {ms, 3}];

          MetricTensorSphere[1, 1, ns_, ms_] := MetricTensorSphere[1, 1][[ns, ms]]
```

Here is MetricTensorSphere evaluated.

```
In[129]:= MetricTensorSphere[1, 1] // TableForm
```

Out[129]//TableForm=

$$\begin{pmatrix} 1 & 0 & 0 \\ 0 & r^2 & 0 \\ 0 & 0 & r^2 \, \text{Sin}[\vartheta]^2 \end{pmatrix}$$

We still need the metric tensor with superscripts. The connection is given by $g_{mn}\, g^{np} = \delta_m^p$.

```
In[130]:= MetricTensorSphere[u, u] = Inverse[MetricTensorSphere[l, l]];
          MetricTensorSphere[u, u, ns_, ms_] := MetricTensorSphere[u, u][[ns, ms]]
```

```
In[132]:= MetricTensorSphere[u, u] // TableForm
```

Out[132]//TableForm=

$$\begin{pmatrix} 1 & 0 & 0 \\ 0 & \frac{1}{r^2} & 0 \\ 0 & 0 & \frac{\text{Csc}[\vartheta]^2}{r^2} \end{pmatrix}$$

We now compute the Christoffel symbols. A bit of work is involved in finding the derivatives.

```
In[133]:= Christoffel[u, l, l] =
          Table[1/2 Sum[  MetricTensorSphere[u, u, m, n]*
                  (D[MetricTensorSphere[l, l, a, n], newVariable[[b]]] +
                   D[MetricTensorSphere[l, l, b, n], newVariable[[a]]] -
                   D[MetricTensorSphere[l, l, a, b], newVariable[[n]]]),
                   {n, 3}], {m, 3}, {a, 3}, {b, 3}];
          (* the elements *)
          Christoffel[u, l, l, m_, a_, b_] := Christoffel[u, l, l][[m, a, b]]
```

We now display them explicitly.

```
In[136]:= Christoffel[u, l, l] //
                    TableForm[#, TableSpacing -> {4, 1, 1},
                              TableDirections -> {Row, Row, Column}]&
```

Out[136]//TableForm=

$$\begin{array}{ccc}
\begin{matrix} 0 & 0 & 0 \\ 0 & -r & 0 \\ 0 & 0 & -r\,\text{Sin}[\vartheta]^2 \end{matrix} &
\begin{matrix} 0 & \frac{1}{r} & 0 \\ \frac{1}{r} & 0 & 0 \\ 0 & 0 & -\text{Cos}[\vartheta]\,\text{Sin}[\vartheta] \end{matrix} &
\begin{matrix} 0 & 0 & \frac{1}{r} \\ 0 & 0 & \text{Cot}[\vartheta] \\ \frac{1}{r} & \text{Cot}[\vartheta] & 0 \end{matrix}
\end{array}$$

Next, we calculate the curvature tensor. For Euclidean space with a curvilinear coordinate system, all of these components are equal to 0.

```
In[137]:= CurvatureTensor[u, l, l, l] =
          Table[D[Christoffel[u, l, l, i, q, k], newVariable[[m]]] -
                 D[Christoffel[u, l, l, i, q, m], newVariable[[k]]]  -
                 Sum[Christoffel[u, l, l, i, p, k] *
                     Christoffel[u, l, l, p, q, m] -
                     Christoffel[u, l, l, i, p, m] *
                     Christoffel[u, l, l, p, q, k], {p, 3}],
                 {i, 3}, {q, 3}, {k, 3}, {m, 3}] // Simplify // Flatten // Union
```

Out[137]= {0}

To conclude, we also look at the geodesic equations. They have the following symbolic form:

$$\frac{dx^m(\lambda)}{d\lambda^2} + \Gamma_{ab}^m \frac{dx^a(\lambda)}{d\lambda}\, \frac{dx^b(\lambda)}{d\lambda} = 0$$

Because we need the explicit dependence of the variables r, φ, and ϑ on a parameter λ, we define an auxiliary function $var[n]$ that implements this.

```
In[138]:= var[n_] := var[n] = newVariable[[n]][λ]
```

Here is the result of applying var to 1, 2, and 3.

```
In[139]:= {var[1], var[2], var[3]}
Out[139]= {r[λ], ϑ[λ], φ[λ]}
```

```
In[140]:= GeodesicEquations =
    Table[D[var[m], {λ, 2}] +
        Sum[(Christoffel[u, 1, 1, m, a, b] /.
            (* replace variables with variables with arguments *)
                {r -> r[λ], φ -> φ[λ], ϑ -> ϑ[λ]})*
            D[var[a], λ]* D[var[b], λ],
        {a, 3}, {b, 3}] == 0, {m, 3}]
```

$$Out[140]= \left\{-r[\lambda]\,\vartheta'[\lambda]^2 - r[\lambda]\,\text{Sin}[\vartheta[\lambda]]^2\,\varphi'[\lambda]^2 + r''[\lambda] == 0,\right.$$

$$\frac{2\,r'[\lambda]\,\vartheta'[\lambda]}{r[\lambda]} - \text{Cos}[\vartheta[\lambda]]\,\text{Sin}[\vartheta[\lambda]]\,\varphi'[\lambda]^2 + \vartheta''[\lambda] == 0,$$

$$\left.\frac{2\,r'[\lambda]\,\varphi'[\lambda]}{r[\lambda]} + 2\,\text{Cot}[\vartheta[\lambda]]\,\vartheta'[\lambda]\,\varphi'[\lambda] + \varphi''[\lambda] == 0\right\}$$

In Euclidean space, geodesics (shortest paths) are straight lines. We quickly check this by substituting the equations for a parametrized line.

```
In[141]:= Block[{x, y, z, r, φ, ϑ},
            (* parametrized line *)
            x[t] = ax t + bx; y[t] = ay t + by; z[t] = az t + bz;
            (* express spherical coordinates through Cartesian coordinates *)
            r[t_] = Sqrt[x[t]^2 + y[t]^2 + z[t]^2];
            φ[t_] = ArcTan[x[t], y[t]];
            ϑ[t_] = ArcCos[z[t]/r[t]];
            (* simplify geodesic equations for the line *)
            GeodesicEquations // Simplify]
Out[141]= {True, True, True}
```

As a visually more interesting example than the just-calculated straight line, we now calculate some geodesics on a surface in $\mathbb{R}^3$. The implicit description of the surface is the following.

```
In[142]:= holedCube = (# @ x + # @ y + # @ z[x, y])&[(-8 #^4 + 8 #^2 - 1)&]
Out[142]= -3 + 8 x² - 8 x⁴ + 8 y² - 8 y⁴ + 8 z[x, y]² - 8 z[x, y]⁴
```

(We used this quartic describing a "holed cube" in Chapter 3 of the Graphics volume [1736].) Because in the above formulas, we need to express new coordinates through old ones, we view x and y as the old ones and $\{x, y, z(x, y)\}$ as the new ones (meaning, we consider x and y as the variables that parametrize the surface). To avoid ugly radicals and potential branch cut problems, we avoid solving explicitly for $z(x, y)$, but, rather express all quantities again in x, y, and $z(x, y)$. Differentiating the last equation repeatedly allows obtaining rational functions for the derivatives $z^{(i,j)}(x, y)$.

```
In[143]:= zDerivativeRules = Factor[Together[#]]& //@
        Solve[# == 0& /@ #[holedCube], #[z[x, y]]]&[
            (* all derivatives up to total order 2 *)
            Flatten[Table[If[0 < i + j < 3, D[#, {x, i}, {y, j}],
                        {}], {i, 0, 2}, {j, 0, 2}]]&][[1]];
    zDerivativeRules /. z[x, y] -> z // Simplify
```

Out[144]= $\{z^{(0,2)}[x, y] \to \dfrac{y^6 (4 - 24 z^2) + z^2 (1 - 2 z^2)^2 + 4 y^4 (-1 + 6 z^2) + y^2 (1 - 12 z^2 + 24 z^4 - 24 z^6)}{z^3 (-1 + 2 z^2)^3}$,

$z^{(1,1)}[x, y] \to - \dfrac{x (-1 + 2 x^2) y (-1 + 2 y^2) (-1 + 6 z^2)}{z^3 (-1 + 2 z^2)^3}$,

$z^{(2,0)}[x, y] \to \dfrac{x^6 (4 - 24 z^2) + z^2 (1 - 2 z^2)^2 + 4 x^4 (-1 + 6 z^2) + x^2 (1 - 12 z^2 + 24 z^4 - 24 z^6)}{z^3 (-1 + 2 z^2)^3}$,

$z^{(0,1)}[x, y] \to - \dfrac{y - 2 y^3}{z - 2 z^3}$, $z^{(1,0)}[x, y] \to - \dfrac{x - 2 x^3}{z - 2 z^3}\}$

Using these derivatives allows expressing g_{mn} and g^{mn} as rational functions in x, y, and $z(x, y)$ too.

```
In[145]:= xyz = {x, y, z[x, y]};
{xyzx, xyzy} = {{1, 0, D[z[x, y], x]},
                {0, 1, D[z[x, y], y]}} /. zDerivativeRules;

gll = {{xyzx.xyzx, xyzx.xyzy}, {xyzx.xyzy, xyzy.xyzy}};
guu = Factor //@ Together //@ Inverse[gll];
```

Next, we calculate the Christoffel symbols Γ_{abc} and Γ^c_{ab}. Again, they are rational functions in x, y, and z.

```
In[150]:= xy[1] = x; xy[2] = y;

Γlll[a_, b_, c_] := Γlll[a, b, c] =
  (* rewritten form of derivatives of gll *)
  Together[(D[xyz, xy[a], xy[b]].D[xyz, xy[c]]) /.
               zDerivativeRules] /. z[x, y] -> z;

Γllu[a_, b_, c_] := Γllu[a, b, c] = (* raise third index with g *)
Together[Sum[Γlll[a, b, d] guu[[c, d]], {d, 2}]] /. z[x, y] -> z
```

The two equations for the geodesics are of the form $x''(\tau) = F(x(\tau), y(\tau), z(\tau), x'(\tau), y'(\tau), z'(\tau))$ and similar for $y''(\tau)$. To make sure the geodesics stay on the original, implicitly defined surface, and to obtain three equations for the three coordinates, we supplement the two geodesic equations with the differentiated form of the implicit equation (meaning $(\text{grad } holedCube(x(\tau), y(\tau), z(\tau))).\{x'(\tau), y'(\tau), z'(\tau)\} = 0$ [332].

```
In[154]:= geodesicEquations = (# == 0)& /@ Append[
  (* second-order odes for x[τ] and y[τ] *)
  Table[D[xy[c][τ], τ, τ] +
  Sum[Γllu[a, b, c] D[xy[a][τ], τ] D[xy[b][τ], τ], {a, 2}, {b, 2}],
    {c, 2}] /. Derivative[n_][xy_][τ] :> Derivative[n][xy],
  (* differentiated form of the implicit equation of the surface *)
  D[holedCube /. {x -> x[τ], y -> y[τ], z[x, y] -> z[τ]}, τ] /.
       xyz_[τ] -> xyz] /. {x -> x[τ], y -> y[τ], z -> z[τ]} /.
       Derivative[n_][xy_[τ]] :> Derivative[n][xy][τ];

In[155]:= ((geodesicEquations // Simplify) /. {ξ_[τ] -> ξ}) //
                  Simplify // TraditionalForm
```

Out[155]//TraditionalForm=

$\{(2 x^2 y (2 y^2 - 1)(6 z^2 - 1) x' y' (1 - 2 x^2)^2 +$

$\quad x (2 x^2 - 1)(4 (6 z^2 - 1) x^6 + (4 - 24 z^2) x^4 + (24 z^6 - 24 z^4 + 12 z^2 - 1) x^2 - z^2 (1 - 2 z^2)^2)(x')^2 +$

$\quad x (2 x^2 - 1)(4 (6 z^2 - 1) y^6 + (4 - 24 z^2) y^4 + (24 z^6 - 24 z^4 + 12 z^2 - 1) y^2 - z^2 (1 - 2 z^2)^2)(y')^2 +$

$\quad z^2 (1 - 2 z^2)^2 (4 x^6 - 4 x^4 + x^2 + 4 y^6 - 4 y^4 + y^2 + z^2 (1 - 2 z^2)^2) x'')/$

$\quad ((2 z^3 - z)^2 (4 x^6 - 4 x^4 + x^2 + 4 y^6 + 4 z^6 - 4 y^4 - 4 z^4 + y^2 + z^2)) = 0,$

$(2 x (2 x^2 - 1) y^2 (6 z^2 - 1) x' y' (1 - 2 y^2)^2 +$

$$y(2y^2-1)\big(4(6z^2-1)x^6+(4-24z^2)x^4+(24z^6-24z^4+12z^2-1)x^2-z^2(1-2z^2)^2\big)(x')^2+$$
$$y(2y^2-1)\big(4(6z^2-1)y^6+(4-24z^2)y^4+(24z^6-24z^4+12z^2-1)y^2-z^2(1-2z^2)^2\big)(y')^2+$$
$$z^2(1-2z^2)^2\big(4x^6-4x^4+x^2+4y^6-4y^4+y^2+z^2(1-2z^2)^2\big)y''\big)\Big/$$
$$\big((2z^3-z)^2(4x^6-4x^4+x^2+4y^6+4z^6-4y^4-4z^4+y^2+z^2)\big)=0,$$
$$x(2x^2-1)x'+y(2y^2-1)y'+z(2z^2-1)z'=0\big\}$$

For a nicer visualization, we calculate a graphic of the surface.

```
In[156]:= Needs["Graphics`ContourPlot3D`"]
```

```
In[157]:= holedCubeGraphic3D = Graphics3D[{EdgeForm[],
          (* map in other positions *)
          Fold[Function[{p, r}, {p, Map[# r&, p, {-2}]}],
              (* 3D contour plot of holedCube in the first octant *)
              Cases[ContourPlot3D[Evaluate[holedCube /. z[x, y] -> z],
                  {x, 0, 1.1}, {y, 0, 1.1}, {z, 0, 1.1},
                  PlotPoints -> {{24, 2}, {20, 2}, {12, 2}},
                  MaxRecursion -> 1, DisplayFunction -> Identity], _Polygon,
                  {0, Infinity}] /. Polygon[l_] :> (* make diamonds *)
              Polygon[Plus @@@ Partition[Append[l, First[l]], 2, 1]/2],
                      {{-1, 1, 1}, {1, -1, 1}, {1, 1, -1}}]}];
```

We visualize the geodesics as lines on the surface. To avoid visually unpleasant intersections between the discretized surface and the discretized geodesics, we define a function liftUp that lifts the geodesics slightly in direction of the local surface normal.

```
In[158]:= normal[{x_, y_, z_}] = (* gradient gives the normal *)
          D[holedCube /. z[x, y] -> z, #]& /@ {x, y, z} // Expand;
```

```
In[159]:= liftUp[{x_, y_, z_}, ε_] = (* move in direction of normal *)
          {x, y, z} - ε #/Sqrt[#.#]&[normal[{x, y, z}]];
```

In the following graphic, we calculate 64 geodesics. We choose the starting points along the upper front "beam". The function rStart parametrizes the starting values.

```
In[160]:= rStart[φ_] := rStart[φ] = Module[{r}, r /. FindRoot[Evaluate[
          holedCube == 0 /. {x -> 0, y -> 0.7 + r Cos[φ],
                             z[x, y] -> 0.7 + r Sin[φ]}, {r, 0, 1/3}]]]
```

Here are the resulting geodesics. On the nearby smoothed corners of the cube, we see the to-be-expected caustics.

```
In[161]:= Module[{o = 128, T = 6, p, τ1, τ2},
          Show[{(* the surface *) holedCubeGraphic3D,
          Table[
          (* solve differential equations for geodesics *)
          (* avoid messages from caustics that run in problems *)
          Internal`DeactivateMessages[
          nsol = NDSolve[Join[geodesicEquations,
                      (* starting values *)
                      {x[0] == 0, y[0] == 0.7 + rStart[φ] Cos[φ],
                                  z[0] == 0.7 + rStart[φ] Sin[φ],
                        x'[0] == 1, y'[0] == 0}],
                  {x, y, z}, {τ, -T, T}, MaxSteps -> 2 10^4,
                  PrecisionGoal -> 6, AccuracyGoal -> 6,
                  (* use appropriate method *)
                  Method -> {"Projection", Method -> "StiffnessSwitching",
```

```
        (* stay on surface *) "Invariants" ->
        {holedCube /. {x -> x[τ], y -> y[τ], z[x, y] -> z[τ]}}}]];
(* parametrized geodesics *)
p[τ_] :=  (Append[liftUp[{x[τ], y[τ], z[τ]} /. nsol[[1]], 0.015],
                {Thickness[0.003], Hue[φ/(2Pi)]}]);
(* for larger T *) {τ1, τ2} = nsol[[1, 1, 2, 1, 1]];
(* show surface and geodesics *)
ParametricPlot3D[p[τ], {τ, τ1, τ2}, Compiled -> False,
                PlotPoints -> Round[200 (τ2 - τ1)],
                DisplayFunction -> Identity],
               {φ, 0, 2Pi (1 - 1/o), 2Pi/o}]},
       DisplayFunction -> $DisplayFunction, Boxed -> False,
       Axes -> False, ViewPoint -> {2.2, 2.4, 1.6}]]
```

We end here and leave it to the reader to calculate euthygrammes [885]. For large-scale calculations of this kind arising in general relativity (see [364]), we recommend the advanced (commercially available) *Mathematica* package *MathTensor* by L. Parker and S. Christensen [1379] (http://smc.vnet.net/MathSolutions.html); or the package *Cartan* by H. Soreng (http://store.wolfram.com/view/cartan). For the algorithmic simplification of tensor expressions, see [120] and [1430].

Next, we give an application of differentiation involving graphics: the evolute of a curve, the evolute of the evolute of a curve, the evolute of the evolute of the evolute of a curve, etc. [271].

Mathematical Remark: Evolutes

The evolute of a curve is the set of all centers of curvature associated with the curve. For a planar curve given in the parametric form $(x(t), y(t))$, the parametric representation of its evolute is:

$$\left(\frac{x(t) - y'(t)\,(x'(t)^2 + y'(t)^2)}{x'(t)\,y''(t) - y'(t)\,x''(t)}, \; \frac{y(t) + x'(t)\,(x'(t)^2 + y'(t)^2)}{x'(t)\,y''(t) - y'(t)\,x''(t)} \right).$$

For more on evolutes and related topics, see any textbook on differential geometry, for example, [1119], [1404], [492], [764], and [609]. For curves that are their own evolutes, see [1864].

Here we implement the definition directly. We apply `Together` to get one fraction. The optional function `simp` simplifies the resulting expressions in a user-specified way.

```
In[162]:= Evolute[{x_, y_}, t_, simp_:Identity] :=
        simp[{x - #2(#1^2 + #2^2)/(#1 #4 - #2 #3),
              y + #1(#1^2 + #2^2)/(#1 #4 - #2 #3)}&[
              (* compute all derivatives only once *)
```

```
                    D[x, t], D[y, t], D[x, {t, 2}], D[y, {t, 2}]] //
             (* avoid blow-up in size of iterated form *)  Together]
```

For a circle, the set of all centers of curvature is precisely the center of the circle.

In[163]:= **Evolute[{Cos[θ], Sin[θ]}, θ]**

Out[163]= {0, 0}

For an ellipse, we get a nontrivial parametrized curve.

In[164]:= **Evolute[{a Cos[θ], b Sin[θ]}, θ]**

Out[164]= $\left\{ \dfrac{a^2 \, Cos[\theta]^3 - b^2 \, Cos[\theta]^3}{a \, (Cos[\theta]^2 + Sin[\theta]^2)}, \dfrac{-a^2 \, Sin[\theta]^3 + b^2 \, Sin[\theta]^3}{b \, (Cos[\theta]^2 + Sin[\theta]^2)} \right\}$

This is a simplified form.

In[165]:= **Evolute[{a Cos[θ], b Sin[θ]}, θ,**
                (# /. {a_. Sin[θ]^2 + a_. Cos[θ]^2 -> a}) &]

Out[165]= $\left\{ \dfrac{a^2 \, Cos[\theta]^3 - b^2 \, Cos[\theta]^3}{a}, \dfrac{-a^2 \, Sin[\theta]^3 + b^2 \, Sin[\theta]^3}{b} \right\}$

We now iterate the formation of evolutes starting with an ellipse, and we graph the resulting evolutes. We use ten ellipses with different half-axes ratios. The right graphic shows a magnified view of the center region of the left graphic.

In[166]:= **Show[GraphicsArray[{Show[#],**
                        Show[#, PlotRange -> {{-3, 3}, {-3, 3}}]}]&[
         Table[ParametricPlot[Evaluate[(* nest forming the evolute *)
         NestList[Evolute[#, θ, (# /. {a_. Sin[θ]^2 + a_.Cos[θ]^2 -> a}) &]&,
         {α Cos[θ], 2 Sin[θ]}, 4]], {θ, 0, 2Pi},
                        PlotStyle -> {Hue[0.78 (α - 3/2)]},
                        Axes -> False, PlotRange -> All, AspectRatio -> 1,
                        DisplayFunction -> Identity, PlotPoints -> 140],
         (* values of α *) **{α, 3/2, 5/2, 1/11}]]]]**
```

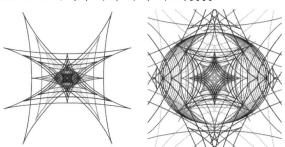

A rich field for the generation of nice curves is given by starting the above process, for instance, with Lissajous figures. We could now go on to the analogous situation for surfaces. Here, for an ellipsoid, we construct a picture with the two surfaces formed by going the amount of the principal radii of the curvature in the direction of the normal to the surface [1157].

In[167]:= **With[{**(* ellipsoid half axes *)**a = 1, b = 3/4, c = 5/4,**
                (* avoid 0/0 in calculations *) **ε = 10^-12},**
         **Module[{φ, θ, x, y, z, e, f, g, l, m, n, λ, v, μ, k, h,**
                **cross, normal1, normal, ellipsoid, makeAll},**
         (* parametrization of the ellipsoid *)
         **{x, y, z} = {a Cos[φ] Sin[θ], b Sin[φ] Sin[θ], c Cos[θ]};**
         (* E, F, G from differential geometry of surfaces *)
```

```
{e, g} = (D[x, #]^2 + D[y, #]^2 + D[z, #]^2)& /@ {φ, θ};
f = D[x, φ] D[x, θ] + D[y, φ] D[y, θ] + D[z, φ] D[z, θ];
(* L, M, N from differential geometry of surfaces *)
{l, n, m} =  Det[{{D[x, ##], D[y, ##], D[z, ##]},
                  D[#, φ]& /@ {x, y, z}, D[#, θ]& /@ {x, y, z}}]& @@@
                               {{φ, φ}, {θ, θ}, {φ, θ}};
{λ, ν, μ} = {l, m, n}/Sqrt[e g - f^2];
(* Gaussian curvature and mean curvature *)
k = (λ ν - μ^2)/(e g - f^2);
h = (g λ - 2 f μ + e ν)/(2 (e g - f^2));
(* normal on the ellipsoid *)
normal = #/Sqrt[#.#]&[Cross[D[{x, y, z}, φ], D[{x, y, z}, θ]]];
(* construct all pieces from the piece of one octant *)
makeAll[polys_] := Function[v, Map[v #&, polys, {-2}]] /@
  {{ 1,  1,  1}, { 1,  1, -1}, { 1, -1,  1}, {-1,  1,  1},
   {-1, -1,  1}, { 1, -1, -1}, {-1,  1, -1}, {-1, -1, -1}};
(* cut a hole in a polygon *)
makeHole[Polygon[l_], factor_] :=
Module[{mp = Plus @@ l/Length[l], L, nOld, nNew},
  L = (mp + factor(# - mp))& /@ l;
  {nOld, nNew} = Partition[Append[#, First[#]]&[#], 2, 1]& /@ {l, L};
  {MapThread[Polygon[Join[#1, Reverse[#2]]]&, {nOld, nNew}]}];
(* a sketch of the ellipsoid *)
ellipsoid = {Thickness[0.002],
(ParametricPlot3D[Evaluate[{x, y, z}],
   {φ, 0, 2Pi}, {θ, 0, Pi}, DisplayFunction -> Identity][[1]]) //.
                                  Polygon[l_] :> Line[l]};
(* surfaces of the centers of the principal curvatures *)
Show[GraphicsArray[
Graphics3D[{ellipsoid, {EdgeForm[],
makeAll @ ParametricPlot3D[#, {φ, ε, Pi/2 - ε}, {θ, ε, Pi/2 - ε},
               DisplayFunction -> Identity][[1]]}}, Boxed -> False,
               PlotRange -> All]& /@
(({x, y, z} + normal 1/(h + # Sqrt[h^2 - k]))& /@ {+1, -1}) ] /.
            p_Polygon :> makeHole[p, 0.7], GraphicsSpacing -> 0]]]
```

We give one more example illustrating the usefulness of symbolic differentiation.

Mathematical Remark: Phase Integral Approximation

Here, we are dealing with a method for the approximate solution of the ordinary differential equation (and associated eigenvalue problem):

$$y''(z) + R^2(z)\, y(z) = 0, \quad R^2(z) \gg 1.$$

If we assume $y(z)$ has the form

$$y(z) = q(z)^{-1/2} \exp\!\left(i \int^z q(z')\, dz' \right), \quad q(z) = Q(z)\, g(z)$$

where $Q(z)$ is " arbitrary", we get the following differential equation for g:

$$1 + \varepsilon(Q(\xi)) - g(\xi)^2 + g(\xi)^{1/2}\, \frac{d^2\, g(\xi)^{-1/2}}{d\xi^2} = 0$$

$$\xi = \xi(z) = \int^z Q(z')\, dz',$$

$$\varepsilon(Q) = \frac{(R(z) - Q^2(z))}{Q^2(z)} + Q(z)^{3/2}\, \frac{d^2\, Q(z)^{-1/2}}{dz^2}.$$

Introducing the parameter λ (which we will later set to 1) in the differential equation,

$$1 + \lambda^2\, \varepsilon(Q(\xi)) - g(\xi)^2 + \lambda^2\, g(\xi)^{1/2}\, \frac{d^2\, g(\xi)^{-1/2}}{d\xi^2} = 0$$

and expanding $g(z)$ in an infinite series in λ (with $Y_{2n+1} = 0$) $g(z) = \sum_{n=0}^{\infty} Y_{2n}(z)\, \lambda^{2n}$, we are led to the following recurrence formula for Y as a function of $\varepsilon(\xi)$ and its derivatives $\varepsilon'(\xi)$, $\varepsilon''(\xi)$, … :

$$Y_0 = 1$$

$$Y_{2n} = \frac{1}{2} \sum_{\substack{\alpha+\beta=n \\ 0 \le \alpha,\beta, \le n-1}} Y_{2\alpha}\, Y_{2\beta} - \frac{1}{2} \sum_{\substack{\alpha+\beta+\gamma+\delta=n \\ 0 \le \alpha,\beta,\gamma,\delta \le n-1}} Y_{2\alpha}\, Y_{2\beta}\, Y_{2\gamma}\, Y_{2\delta}$$

$$+ \frac{1}{2} \sum_{\substack{\alpha+\beta=n-1 \\ 0 \le \alpha,\beta \le n-1}} \left\{ \varepsilon\, Y_{2\alpha}\, Y_{2\beta} + \frac{3}{4}\, Y_{2\alpha}'\, Y_{2\beta}' - \frac{1}{4}\, (Y_{2\alpha}\, Y_{2\beta}'' + Y_{2\alpha}''\, Y_{2\beta}) \right\}$$

where $Y_\alpha' = Y_\alpha'(\xi) = dY_\alpha(\xi)/d\xi$ and $\xi = \xi(z)$.

$Y_2(\xi)$, $Y_4(\xi)$, and $Y_6(\xi)$ were found earlier by painful hand computations, but from $Y_8(z)$ on, computer algebra becomes necessary [296]. For more details on such asymptotic expansions, see [671], [487], [672], [428], [1006], [1623], [142], [1306], [1883], [1104], and [674]. For the corresponding supersymmetric problem, see [16], [673], [58], and [544].

Here, we want to find the first few nonvanishing $Y_{2\alpha}$. We now give an unrefined implementation of the above recurrence. (It is unrefined considering the restriction $\alpha + \beta + \gamma + \delta = n$, and the fourfold sum can be replaced by a threefold sum.)

```
In[168]:= Y[_Integer?OddQ] = 0;

        Y[0] = 1;

        Y[zn_Integer?EvenQ] := Y[zn] =
        Module[{n = zn/2},
        (* the If is the obvious implementation,
```

```
        but it requires summing over all variables *)
1/2 Sum[If[a + b == n, 1, 0] Y[2 a] Y[2 b],
        {a, 0, n - 1}, {b, 0, n - 1}] -
1/2 Sum[If[a + b + c + d == n, 1, 0] Y[2 a] Y[2 b] Y[2 c] Y[2 d],
        {a, 0, n - 1}, {b, 0, n - 1}, {c, 0, n - 1}, {d, 0, n - 1}] +
1/2 Sum[If[a + b == n - 1, 1, 0] *
        (ε[ξ] Y[2 a] Y[2 b] + 3/4 D[Y[2 a], ξ] D[Y[2 b], ξ] -
        1/4 (Y[2 a] D[Y[2 b], {ξ, 2}] + D[Y[2 a], {ξ, 2}] Y[2 b])),
        {a, 0, n - 1}, {b, 0, n - 1}] //
    (* keep the results as short as possible *) Expand // Factor]
```

We now look at the first few $Y_i(z)$.

In[171]:= **Y[2]**

Out[171]= $\dfrac{\varepsilon[\xi]}{2}$

In[172]:= **Y[4]**

Out[172]= $\dfrac{1}{8}\,(-\varepsilon[\xi]^2 - \varepsilon''[\xi])$

In[173]:= **Y[6]**

Out[173]= $\dfrac{1}{32}\,(2\,\varepsilon[\xi]^3 + 5\,\varepsilon'[\xi]^2 + 6\,\varepsilon[\xi]\,\varepsilon''[\xi] + \varepsilon^{(4)}[\xi])$

In[174]:= **Y[8]**

Out[174]= $\dfrac{1}{128}\,(-5\,\varepsilon[\xi]^4 - 50\,\varepsilon[\xi]\,\varepsilon'[\xi]^2 - 30\,\varepsilon[\xi]^2\,\varepsilon''[\xi] -$
$19\,\varepsilon''[\xi]^2 - 28\,\varepsilon'[\xi]\,\varepsilon^{(3)}[\xi] - 10\,\varepsilon[\xi]\,\varepsilon^{(4)}[\xi] - \varepsilon^{(6)}[\xi])$

The Y[i] for higher orders can also be found in a few seconds.

In[175]:= **Timing[Y[16]]**

Out[175]= $\{0.86\,\text{Second},$

$\dfrac{1}{32768}\,(-429\,\varepsilon[\xi]^8 - 60060\,\varepsilon[\xi]^5\,\varepsilon'[\xi]^2 - 316030\,\varepsilon[\xi]^2\,\varepsilon'[\xi]^4 - 12012\,\varepsilon[\xi]^6\,\varepsilon''[\xi] -$
$758472\,\varepsilon[\xi]^3\,\varepsilon'[\xi]^2\,\varepsilon''[\xi] - 496950\,\varepsilon'[\xi]^4\,\varepsilon''[\xi] - 114114\,\varepsilon[\xi]^4\,\varepsilon''[\xi]^2 -$
$1794156\,\varepsilon[\xi]\,\varepsilon'[\xi]^2\,\varepsilon''[\xi]^2 - 360932\,\varepsilon[\xi]^2\,\varepsilon''[\xi]^3 - 174317\,\varepsilon''[\xi]^4 -$
$168168\,\varepsilon[\xi]^4\,\varepsilon'[\xi]\,\varepsilon^{(3)}[\xi] - 877760\,\varepsilon[\xi]\,\varepsilon'[\xi]^3\,\varepsilon^{(3)}[\xi] -$
$1591304\,\varepsilon[\xi]^2\,\varepsilon'[\xi]\,\varepsilon''[\xi]\,\varepsilon^{(3)}[\xi] - 1533408\,\varepsilon'[\xi]\,\varepsilon''[\xi]^2\,\varepsilon^{(3)}[\xi] -$
$118404\,\varepsilon[\xi]^3\,\varepsilon^{(3)}[\xi]^2 - 562474\,\varepsilon'[\xi]^2\,\varepsilon^{(3)}[\xi]^2 - 684372\,\varepsilon[\xi]\,\varepsilon''[\xi]\,\varepsilon^{(3)}[\xi]^2 -$
$12012\,\varepsilon[\xi]^5\,\varepsilon^{(4)}[\xi] - 466180\,\varepsilon[\xi]^2\,\varepsilon'[\xi]^2\,\varepsilon^{(4)}[\xi] - 188760\,\varepsilon[\xi]^3\,\varepsilon''[\xi]\,\varepsilon^{(4)}[\xi] -$
$893724\,\varepsilon'[\xi]^2\,\varepsilon''[\xi]\,\varepsilon^{(4)}[\xi] - 543972\,\varepsilon[\xi]\,\varepsilon''[\xi]^2\,\varepsilon^{(4)}[\xi] -$
$800176\,\varepsilon[\xi]\,\varepsilon'[\xi]\,\varepsilon^{(3)}[\xi]\,\varepsilon^{(4)}[\xi] - 206138\,\varepsilon^{(3)}[\xi]^2\,\varepsilon^{(4)}[\xi] - 71786\,\varepsilon[\xi]^2\,\varepsilon^{(4)}[\xi]^2 -$
$163722\,\varepsilon''[\xi]\,\varepsilon^{(4)}[\xi]^2 - 92664\,\varepsilon[\xi]^3\,\varepsilon'[\xi]\,\varepsilon^{(5)}[\xi] - 144780\,\varepsilon'[\xi]^3\,\varepsilon^{(5)}[\xi] -$
$529776\,\varepsilon[\xi]\,\varepsilon'[\xi]\,\varepsilon''[\xi]\,\varepsilon^{(5)}[\xi] - 119548\,\varepsilon[\xi]^2\,\varepsilon^{(3)}[\xi]\,\varepsilon^{(5)}[\xi] -$
$272108\,\varepsilon''[\xi]\,\varepsilon^{(3)}[\xi]\,\varepsilon^{(5)}[\xi] - 159268\,\varepsilon'[\xi]\,\varepsilon^{(4)}[\xi]\,\varepsilon^{(5)}[\xi] - 23998\,\varepsilon[\xi]\,\varepsilon^{(5)}[\xi]^2 -$
$6006\,\varepsilon[\xi]^4\,\varepsilon^{(6)}[\xi] - 111020\,\varepsilon[\xi]\,\varepsilon'[\xi]^2\,\varepsilon^{(6)}[\xi] - 68068\,\varepsilon[\xi]^2\,\varepsilon''[\xi]\,\varepsilon^{(6)}[\xi] -$
$76986\,\varepsilon''[\xi]^2\,\varepsilon^{(6)}[\xi] - 113456\,\varepsilon'[\xi]\,\varepsilon^{(3)}[\xi]\,\varepsilon^{(6)}[\xi] - 41132\,\varepsilon[\xi]\,\varepsilon^{(4)}[\xi]\,\varepsilon^{(6)}[\xi] -$
$3431\,\varepsilon^{(6)}[\xi]^2 - 25168\,\varepsilon[\xi]^2\,\varepsilon'[\xi]\,\varepsilon^{(7)}[\xi] - 56328\,\varepsilon'[\xi]\,\varepsilon''[\xi]\,\varepsilon^{(7)}[\xi] -$
$25688\,\varepsilon[\xi]\,\varepsilon^{(3)}[\xi]\,\varepsilon^{(7)}[\xi] - 6004\,\varepsilon^{(5)}[\xi]\,\varepsilon^{(7)}[\xi] - 1716\,\varepsilon[\xi]^3\,\varepsilon^{(8)}[\xi] -$
$9210\,\varepsilon'[\xi]^2\,\varepsilon^{(8)}[\xi] - 11388\,\varepsilon[\xi]\,\varepsilon''[\xi]\,\varepsilon^{(8)}[\xi] - 4002\,\varepsilon^{(4)}[\xi]\,\varepsilon^{(8)}[\xi] -$
$3380\,\varepsilon[\xi]\,\varepsilon'[\xi]\,\varepsilon^{(9)}[\xi] - 2000\,\varepsilon^{(3)}[\xi]\,\varepsilon^{(9)}[\xi] - 286\,\varepsilon[\xi]^2\,\varepsilon^{(10)}[\xi] -$
$726\,\varepsilon''[\xi]\,\varepsilon^{(10)}[\xi] - 180\,\varepsilon'[\xi]\,\varepsilon^{(11)}[\xi] - 26\,\varepsilon[\xi]\,\varepsilon^{(12)}[\xi] - \varepsilon^{(14)}[\xi])\}$

1.6.2 Integration

Symbolic integration of functions is one of the most important capabilities of *Mathematica*. In contrast to many other operations (which can also be carried out by hand by the user, albeit more slowly and probably with more errors), and in addition to standard methods such as integration by parts, substitution, etc., *Mathematica* makes essential use of special algorithms for the determination of indefinite and definite integrals (see [244], the references cited in the appendix, Chapter 21 of [1917], and the very readable introductions of [1330], [990], [643], and [1205]).

Mathematica can find a great many integrals, including many not listed in tables. This holds primarily for integrands that are not special functions; but even for special functions, *Mathematica* is often able to find a closed-form answer. Nevertheless, once in a while, the user will have to refer to a book such as [1443] for complicated integrals. For most integrals, *Mathematica* works with algorithms rather than looking in tables. For indefinite integrals, these algorithms are based on the celebrated work by Risch and extensions by Trager and Bronstein [244]. Definite integrals are computed by using contour integration, by the Marichev–Adamchik reduction to Meijer-G functions [1560], [1561], [284], [1650], and [574], or by integration via differentiation (Cauchy contour integral formula).

We have already introduced the `Integrate` command for symbolic integration. In view of its extraordinary importance, we repeat it here.

`Integrate[`*integrand, var*`]`
    finds (if possible) the indefinite integral $\int^{var}$ *integrand dvar*.

`Integrate[`*integrand,* `{`*var, lowerLimit, upperLimit*`}]`
    finds (if possible) the definite integral $\int_{lowerLimit}^{upperLimit}$ *integrand dvar*.

Let us start with a remark similar to the one made in section dealing with the solution of equations using `Solve`. All variables in integrals are assumed to take generic complex values. So the result of the simple integral $\int^x x^n \, dx$ will be $1/(n+1) x^{n+1}$.

```
In[1]:= Integrate[x^n, x]
```

$$\text{Out[1]} = \frac{x^{1+n}}{1+n}$$

This is the correct answer for all complex x and for nearly all complex n (the exception, which is of Lebesgue measure zero with respect to dx, being $n = -1$). This assumption about all unspecified variables being generic can cause indeterminate expressions when substituting numerical values into the result of an integration containing parameters.

The integrand can either be given explicitly or unspecified. Here is such an example of the latter case.

```
In[2]:= Integrate[f'[x], x]
```
```
Out[2]= f[x]
```

```
In[3]:= Integrate[f'[x] f''[x], x]
```

$$\text{Out[3]} = \frac{1}{2} \, f'[x]^2$$

```
In[4]:= Integrate[f'[x] g[x] + f[x] g'[x], x]
```
```
Out[4]= f[x] g[x]
```

```
In[5]:= Integrate[Sin[f[x]] f'[x], x]
```

Out[5]= -Cos[f[x]]

Here is a slightly more complicated integral, the Bohlin constant of motion for a damped harmonic oscillator [720].

In[6]:= **Exp[Integrate[(λ1 - λ2) x'[t] (x''[t] + (λ1 + λ2) x'[t] + λ1 λ2 x[t])/**
                    ((x'[t] + λ1 x[t]) (x'[t] + λ2 x[t])), t]] // Together

Out[6]= $(\lambda 1\, x[t] + x'[t])^{\lambda 1}\, (\lambda 2\, x[t] + x'[t])^{-\lambda 2}$

The following product cannot be symbolically integrated without the result containing unevaluated integrals (which would cause recursion).

In[7]:= **Integrate[f'[x] g[x], x]**

Out[7]= $\int g[x]\, f'[x]\, dx$

Be aware that in distinction to NIntegrate, the function Integrate has no HoldAll attribute. This means that the scoping behavior in nested integrals is different. Whereas NIntegrate can treat its body as a black box that delivers values at given points (when the corresponding system option is set to avoid the evaluation of the body), the algorithms used in Integrate require unavoidably the evaluation of the integrand.

When Integrate carries out an indefinite integral, it does not return any explicit constants of integration. (Implicitly the result given amounts to selecting a concrete constant of integration.) So mathematically identical integrands can result in different indefinite integrals. The following polynomial $(x + 1)^4 + 1$ shows such a situation.

In[8]:= **(* integrals of original and expanded integrand and difference *)**
        {Integrate[#, x], Integrate[Expand[#], x],
         Expand[Integrate[#, x] - Integrate[Expand[#], x]]}&[(x + 1)^4 + 1]

Out[9]= $\left\{ x + \dfrac{1}{5}\, (1 + x)^5,\ 2\, x + 2\, x^2 + 2\, x^3 + x^4 + \dfrac{x^5}{5},\ \dfrac{1}{5} \right\}$

Mathematica's ability to integrate implicitly defined functions can be seen nicely in the following example. Suppose

$$G_0(x) = \frac{1}{2}$$

$$G_j(x) = \int^x (G'''_{j-1}(x) + 4\,u(x)\,G'_{j-1}(x) + 2\,u'(x)\,G_{j-1}(x))\,dx, \quad j = 1, 2, \dots.$$

These equations are of great practical importance for the construction of the Korteweg–de Vries equation hierarchy. Because the mathematical description of Lax pairs is slightly more complicated, we do not go into details here; see, however, [37], [719], [1204], [1442], [628], [1575], [1576], [325], [1782], [255], [1788], [717]. Note that $u(x)$ is not explicitly defined. We now implement the above definition of the $G_j(x)$.

In[10]:= $\mathcal{G}$**[0] = 1/2;**

        $\mathcal{G}$**[j_] := $\mathcal{G}$[j] = Integrate[D[$\mathcal{G}$[j - 1], {x, 3}] + 4 u[x] D[$\mathcal{G}$[j - 1], x] +**
                          2 u'[x] $\mathcal{G}$[j - 1], x] // Together // Numerator

We look at the first few $G_j(x)$; they are "completely" integrated.

In[12]:= **{$\mathcal{G}$[1], $\mathcal{G}$[2], $\mathcal{G}$[3]}**

Out[12]= $\{u[x],\ 3\, u[x]^2 + u''[x],\ 10\, u[x]^3 + 5\, u'[x]^2 + 10\, u[x]\, u''[x] + u^{(4)}[x]\}$

Here are the corresponding equations from the Korteweg–de Vries equation hierarchy. We denote the temporal derivative by u_t.

In[13]:= `KdVShortForm[j_] := Subscript[u, t] == Simplify[D[𝒢[j], x]] /.`
        `{u[x] -> u, Derivative[i_][u][x] -> Derivative[i][u]}`

In[14]:= `Table[KdVShortForm[k], {k, 3}]`

Out[14]= $\{u_t == u', \; u_t == 6\,u\,u' + u^{(3)}, \; u_t == 30\,u^2\,u' + 20\,u'\,u'' + 10\,u\,u^{(3)} + u^{(5)}\}$

Higher order equations become quickly quite complicated.

In[15]:= `KdVShortForm[6]`

Out[15]= $u_t == 2772\,u^5\,u' + 14784\,(u')^3\,u'' + 2310\,u^4\,u^{(3)} + 9702\,(u'')^2\,u^{(3)} +$
        $2838\,(u')^2\,u^{(5)} + 924\,u^{(4)}\,u^{(5)} + 924\,u^3\,(20\,u'\,u'' + u^{(5)}) + 660\,u^{(3)}\,u^{(6)} + 330\,u''\,u^{(7)} +$
        $198\,u^2\,(70\,(u')^3 + 70\,u''\,u^{(3)} + 42\,u'\,u^{(4)} + u^{(7)}) + 22\,u'\,(327\,(u^{(3)})^2 + 522\,u''\,u^{(4)} + 5\,u^{(8)}) +$
        $22\,u\,(966\,(u')^2\,u^{(3)} + 252\,u^{(3)}\,u^{(4)} + 168\,u''\,u^{(5)} + 6\,u'\,(217\,(u'')^2 + 12\,u^{(6)}) + u^{(9)}) + u^{(11)}$

Here are similar expressions for the Camassa–Holm equation [1129]. Again, the integrals can all be completely integrated.

In[16]:= `(* initial condition for the recursion downwards *)`
        `𝛿ℋCamassaHolm[-1][m_, x_] := 1/(2 Sqrt[m[x]])`

        `(* express -nth term as the integral over earlier`
        `   terms differentiated *)`
        `𝛿ℋCamassaHolm[n_][m_, x_] := 𝛿ℋCamassaHolm[n][m, x] =`
        `Expand[1/(2 Sqrt[m[x]]) Integrate[(D[𝛿ℋCamassaHolm[n + 1][m, x], x] -`
        `        D[𝛿ℋCamassaHolm[n + 1][m, x], {x, 3}])/Sqrt[m[x]], x]]`

In[20]:= `Table[𝛿ℋCamassaHolm[n][m, x], {n, -2, -4, -1}]`

Out[20]= $\Big\{ \dfrac{1}{8\,m[x]^{3/2}} - \dfrac{5\,m'[x]^2}{32\,m[x]^{7/2}} + \dfrac{m''[x]}{8\,m[x]^{5/2}},\; \dfrac{3}{64\,m[x]^{5/2}} - \dfrac{35\,m'[x]^2}{128\,m[x]^{9/2}} + \dfrac{1155\,m'[x]^4}{1024\,m[x]^{13/2}} +$

$\dfrac{5\,m''[x]}{32\,m[x]^{7/2}} - \dfrac{231\,m'[x]^2\,m''[x]}{128\,m[x]^{11/2}} + \dfrac{21\,m''[x]^2}{64\,m[x]^{9/2}} + \dfrac{7\,m'[x]\,m^{(3)}[x]}{16\,m[x]^{9/2}} - \dfrac{m^{(4)}[x]}{16\,m[x]^{7/2}},$

$\dfrac{5}{256\,m[x]^{7/2}} - \dfrac{315\,m'[x]^2}{1024\,m[x]^{11/2}} + \dfrac{15015\,m'[x]^4}{4096\,m[x]^{15/2}} - \dfrac{425425\,m'[x]^6}{16384\,m[x]^{19/2}} +$

$\dfrac{35\,m''[x]}{256\,m[x]^{9/2}} - \dfrac{2541\,m'[x]^2\,m''[x]}{512\,m[x]^{13/2}} + \dfrac{255255\,m'[x]^4\,m''[x]}{4096\,m[x]^{17/2}} +$

$\dfrac{189\,m''[x]^2}{256\,m[x]^{11/2}} - \dfrac{35607\,m'[x]^2\,m''[x]^2}{1024\,m[x]^{15/2}} + \dfrac{671\,m''[x]^3}{256\,m[x]^{13/2}} + \dfrac{63\,m'[x]\,m^{(3)}[x]}{64\,m[x]^{11/2}} -$

$\dfrac{2145\,m'[x]^3\,m^{(3)}[x]}{128\,m[x]^{15/2}} + \dfrac{1419\,m'[x]\,m''[x]\,m^{(3)}[x]}{128\,m[x]^{13/2}} - \dfrac{69\,m^{(3)}[x]^2}{128\,m[x]^{11/2}} - \dfrac{7\,m^{(4)}[x]}{64\,m[x]^{9/2}} +$

$\dfrac{825\,m'[x]^2\,m^{(4)}[x]}{256\,m[x]^{13/2}} - \dfrac{57\,m''[x]\,m^{(4)}[x]}{64\,m[x]^{11/2}} - \dfrac{27\,m'[x]\,m^{(5)}[x]}{64\,m[x]^{11/2}} + \dfrac{m^{(6)}[x]}{32\,m[x]^{9/2}} \Big\}$

Iterated integrals (which cannot be done) are left alone; they are not integrated by parts.

In[21]:= `Integrate[Integrate[f[x], x], x]`

Out[21]= $\displaystyle\int \left(\int f[x]\, dx \right) dx$

In dealing with nested integrals, the variables may be renamed to avoid interference of variables with one another. (The application of Set instead of SetDelayed in the following definition would lead to an error message concerning $RecursionLimit, as well as, an unexpected result; see also Exercise 3.)

In[22]:= `g[x_, 0] = x;`

        `g[x_, n_] := Integrate[g[x, n - 1], x]`

```
        {g[a, 0], g[b, 1], g[c, 2], g[z, 25]}
```
$$\text{Out[24]= } \left\{a, \frac{b^2}{2}, \frac{c^3}{6}, \frac{z^{26}}{4032914611266056355840000000}\right\}$$

Here is a small selection of indefinite integrals that *Mathematica* can compute.

In[25]:= `Integrate[1/x^2 1/(x^2 + 4), x]`

$$\text{Out[25]= } -\frac{1}{4x} - \frac{1}{8} \text{ArcTan}\left[\frac{x}{2}\right]$$

In[26]:= `Integrate[(1 + Sin[x])/(1 - Cos[x]), x]`

$$\text{Out[26]= } -\text{Cot}\left[\frac{x}{2}\right] + 2 \text{Log}\left[\text{Sin}\left[\frac{x}{2}\right]\right]$$

In[27]:= `Integrate[Sqrt[1/x^2 1/(x^2 + 4)], x]`

$$\text{Out[27]= } \frac{1}{2} x \sqrt{4 + x^2} \sqrt{\frac{1}{4x^2 + x^4}} \left(\text{Log}[x] - \text{Log}\left[2 + \sqrt{4 + x^2}\right]\right)$$

In[28]:= `Integrate[Exp[-x^2] x^3, x]`

$$\text{Out[28]= } \frac{1}{2} e^{-x^2} (-1 - x^2)$$

In[29]:= `Integrate[Tanh[x^2 + 3] x, x]`

$$\text{Out[29]= } \frac{1}{2} \text{Log}[\text{Cosh}[x^2] + e^6 \text{Cosh}[x^2] - \text{Sinh}[x^2] + e^6 \text{Sinh}[x^2]]$$

In[30]:= `Integrate[Cos[x + 3]/x^3 x, x]`

$$\text{Out[30]= } -\frac{\text{Cos}[3] \text{Cos}[x]}{x} - \text{CosIntegral}[x] \text{Sin}[3] + \frac{\text{Sin}[3] \text{Sin}[x]}{x} - \text{Cos}[3] \text{SinIntegral}[x]$$

In[31]:= `Integrate[Log[Log[x + 1]], x]`

$$\text{Out[31]= } (1 + x) \text{Log}[\text{Log}[1 + x]] - \text{LogIntegral}[1 + x]$$

In[32]:= `Integrate[ArcSin[x]^2, x]`

$$\text{Out[32]= } -2x + 2\sqrt{1 - x^2} \text{ArcSin}[x] + x \text{ArcSin}[x]^2$$

In[33]:= `Integrate[ArcSin[x]/x, x]`

$$\text{Out[33]= } \text{ArcSin}[x] \text{Log}[1 - e^{2 i \text{ArcSin}[x]}] - \frac{1}{2} i (\text{ArcSin}[x]^2 + \text{PolyLog}[2, e^{2 i \text{ArcSin}[x]}])$$

In[34]:= `Integrate[ArcCosh[Sec[ArcSinh[x]]], x]`

$$\text{Out[34]= } x \text{ArcCosh}[\text{Sec}[\text{ArcSinh}[x]]] - \left(\frac{1}{2} + \frac{i}{2}\right) e^{(-1+i) \text{ArcSinh}[x]} \text{Cot}\left[\frac{\text{ArcSinh}[x]}{2}\right]$$

$$\left(-i e^{2 \text{ArcSinh}[x]} \text{Hypergeometric2F1}\left[\frac{1}{2} - \frac{i}{2}, 1, \frac{3}{2} - \frac{i}{2}, -e^{2 i \text{ArcSinh}[x]}\right] + \right.$$

$$\left. \text{Hypergeometric2F1}\left[\frac{1}{2} + \frac{i}{2}, 1, \frac{3}{2} + \frac{i}{2}, -e^{2 i \text{ArcSinh}[x]}\right]\right) \sqrt{\text{Tan}\left[\frac{\text{ArcSinh}[x]}{2}\right]^2}$$

The following example is a complicated composition of two special functions (see Chapter 3 for Erf and ProductLog)—*Mathematica* is able to integrate it.

In[35]:= `Integrate[Erf[ProductLog[c1, x]], x]`

$$\text{Out[35]= } -\frac{1}{2} e^{1/4} \text{Erf}\left[-\frac{1}{2} + \text{ProductLog}[c1, x]\right] +$$

$$x \text{Erf}[\text{ProductLog}[c1, x]] + \frac{e^{-\text{ProductLog}[c1, x]^2} x}{\sqrt{\pi} \text{ProductLog}[c1, x]}$$

Using Simplify, it is straightforward to prove that the result was correct.

In[36]:= `D[%, x] // Simplify`

Out[36]= `Erf[ProductLog[c1, x]]`

There are, however, many integrals that in principle exist, but that *Mathematica* cannot find. Some of them are impossible to express in terms of finite sums of functions built into *Mathematica*, and some of them require algorithms that are just not implemented in Version 5.1.

In[37]:= `Integrate[Exp[1/(Log[x] + 1)], x]`

Out[37]= $\displaystyle\int e^{\frac{1}{1+\text{Log}[x]}} \, dx$

In[38]:= `Integrate[Cos[Exp[Sin[x + 1]]], x]`

Out[38]= $\displaystyle\int \text{Cos}[e^{\text{Sin}[1+x]}] \, dx$

Sometimes indefinite integrals can be expressed in quite different forms. The next integral contains a hypergeometric function.

In[39]:= `Integrate[1/(x^3 (a + b x)^(1/3)), x]`

Out[39]= $\displaystyle\frac{-3\,a^2 + a\,b\,x + 4\,b^2\,x^2 - 4\,b^2\,(1 + \frac{a}{b\,x})^{1/3}\,x^2\,\text{Hypergeometric2F1}[\frac{1}{3}, \frac{1}{3}, \frac{4}{3}, -\frac{a}{b\,x}]}{6\,a^2\,x^2\,(a + b\,x)^{1/3}}$

Sometimes *Mathematica* will produce answers in terms of special functions, also in cases in which it is possible to express the integral without a special function [931], [1768], [1769], [1190], [767], [1376], [526]. (Although the result becomes typically larger in this case.) Currently, no `Method` option is available for the function `Integrate` that could influence the mathematical form of the result.

In[40]:= `D[(* closed form elementary result *)`
`      (-2 b^2 x^2 Sqrt[3] Log[a^(1/3) (a + b x)^(2/3) +`
`      a^(2/3) (a + b x)^(1/3) + a] +`
`      4 b^2 x^2 Sqrt[3] Log[a^(2/3) (a + b x)^(1/3) - a] +`
`      12 b^2 x^2 ArcTan[(2 Sqrt[3] a^(2/3) (a + b x)^(1/3) +`
`      a Sqrt[3])/(3a)] + (12b x - 9a) Sqrt[3] a^(1/3) (a + b x)^(2/3))/`
`      (18 a^2 x^2 Sqrt[3] a^(1/3)), x] - 1/(x^3 (a + b x)^(1/3)) // Simplify`

Out[40]= 0

Mathematica can find the indefinite integral of every rational function over the rationals (this means every function of the form *polynomial1/polynomial2* in one indeterminate where both polynomials have integer or rational coefficients). Here is an example.

In[41]:= `Integrate[(x^3 - 4 x + 8)/(x^7 - 3 x^5 + 3 x^2 - x + 9), x]`

Out[41]= $\text{RootSum}\!\left[9 - \#1 + 3\,\#1^2 - 3\,\#1^5 + \#1^7 \,\&,\; \dfrac{8\,\text{Log}[x - \#1] - 4\,\text{Log}[x - \#1]\,\#1 + \text{Log}[x - \#1]\,\#1^3}{-1 + 6\,\#1 - 15\,\#1^4 + 7\,\#1^6} \,\&\right]$

In[42]:= `D[%, x]`

Out[42]= $\displaystyle\frac{8 - 4\,x + x^3}{9 - x + 3\,x^2 - 3\,x^5 + x^7}$

The above integral contained a `RootSum`. While the function `RootSum` is closely related to the function `Root`, it is of relevance mostly for integration and not for solving equations.

`RootSum[`*irreducibleUnivariatePolynomialAsAPureFunction, function*`]`

    represents the sum $\sum_i$ *function*(x_i). Here, the x_i are the roots of *irreducibleUnivariatePolynomialAsAPure-Function*$(x)=0$ and the sum extends over all roots x_i.

For rational functions as the second argument, `RootSum` can calculate closed-form expressions.

In[43]:= `RootSum[Function[x, x^7 - 4 x + 8], (#^2 + 1)/(# - 7)&]`

Out[43]= $-\dfrac{824323}{823523}$

We now turn to definite integrals. Again, a great many integrals can be computed, frequently in terms of special functions (see Chapter 3), especially (generalized) hypergeometric functions. We now look at a few examples.

In[44]:= **Integrate[Sin[x]/x, {x, -Infinity, Infinity}]**

Out[44]= π

In[45]:= **Integrate[Boole[x^2 - 1 < 0], {x, -Infinity, Infinity}]**

Out[45]= 2

In[46]:= **Integrate[Sin[x^2], {x, 0, 1}]**

Out[46]= $\sqrt{\dfrac{\pi}{2}}$ FresnelS$\left[\sqrt{\dfrac{2}{\pi}}\,\right]$

In[47]:= **Integrate[Log[Sin[x]]^2, {x, 0, Pi}]**

Out[47]= $\dfrac{\pi^3}{12} + \pi \, \mathrm{Log}[2]^2$

In[48]:= **Integrate[Exp[-x] Log[x]^6, {x, 0, Infinity}]**

Out[48]= $\mathrm{EulerGamma}^6 + \dfrac{5\,\mathrm{EulerGamma}^4\,\pi^2}{2} + \dfrac{9\,\mathrm{EulerGamma}^2\,\pi^4}{4} + \dfrac{61\,\pi^6}{168} +$
$40\,\mathrm{EulerGamma}^3\,\mathrm{Zeta}[3] + 40\,\mathrm{Zeta}[3]^2 + 4\,\mathrm{EulerGamma}\,(5\,\pi^2\,\mathrm{Zeta}[3] + 36\,\mathrm{Zeta}[5])$

In[49]:= **Integrate[x^n (1 - x^p)^o, {x, 0, 1}]**

Out[49]= $\mathrm{If}\left[\mathrm{Re}[n] > -1\ \&\&\ \mathrm{Re}[o] > -1\ \&\&\ \mathrm{Re}[p] > 0,\ \dfrac{\mathrm{Gamma}[1 + o]\ \mathrm{Gamma}\left[\frac{1+n}{p}\right]}{p\,\mathrm{Gamma}\left[\frac{1+n+o\,p}{p}\right]},\ \mathrm{Integrate}\left[\right.\right.$
$\left.\left. x^n\,(1 - x^p)^o,\ \{x, 0, 1\},\ \mathrm{Assumptions} \to !\ (\mathrm{Re}[n] > -1\ \&\&\ \mathrm{Re}[o] > -1\ \&\&\ \mathrm{Re}[p] > 0)\right]\right]$

In[50]:= **Integrate[(Exp[-x] - Exp[-z x])/x, {x, 0, Infinity}]**

Out[50]= $\mathrm{If}\left[\mathrm{Re}[z] > 0,\ \mathrm{Log}[z],\ \mathrm{Integrate}\left[\dfrac{e^{-x} - e^{-x\,z}}{x},\ \{x, 0, \infty\},\ \mathrm{Assumptions} \to \mathrm{Re}[z] \le 0\right]\right]$

In[51]:= **Integrate[(Sin[x] - x Cos[x])^6/x^4, {x, 0, Pi}]**

Out[51]= $\dfrac{2171\,\pi + 120\,\pi^3 + 72\,(5\,\mathrm{SinIntegral}[2\,\pi] - 4\,\mathrm{SinIntegral}[4\,\pi] + \mathrm{SinIntegral}[6\,\pi])}{1152}$

In[52]:= **Integrate[(Cos[n x] - Cos[n α])/(Cos[x] - Cos[α]), {x, 0, Pi}]**

Out[52]= $\mathrm{If}\left[\mathrm{Re}[\mathrm{Cos}[\alpha]] \le -1\ ||\ \mathrm{Im}[\mathrm{Cos}[\alpha]] \ne 0,\ -\dfrac{\pi\,\mathrm{Cos}[n\,\alpha]}{\sqrt{-1 - \mathrm{Cos}[\alpha]}\ \sqrt{1 - \mathrm{Cos}[\alpha]}}\ -\right.$
$\dfrac{\mathrm{HypergeometricPFQ}[\{\frac{1}{2}, 1, 1\}, \{1 - n, 1 + n\},\ \mathrm{Sec}[\frac{\alpha}{2}]^2]\ \mathrm{Sin}[n\,\pi]}{n + n\,\mathrm{Cos}[\alpha]},$
$\left.\mathrm{Integrate}\left[\dfrac{\mathrm{Cos}[n\,x]}{\mathrm{Cos}[x] - \mathrm{Cos}[\alpha]} - \dfrac{\mathrm{Cos}[n\,\alpha]}{\mathrm{Cos}[x] - \mathrm{Cos}[\alpha]},\ \{x, 0, \pi\},\ \mathrm{Assumptions} \to \mathrm{Cos}[\alpha] > -1\right]\right]$

In[53]:= **Integrate[r^-3 Sin[r ρ1] Sin[r ρ2]^4, {r, 0, Infinity}]**

Out[53]= $\mathrm{If}\left[\rho1 \in \mathrm{Reals}\ \&\&\ \rho2 \in \mathrm{Reals},\right.$
$\dfrac{1}{64}\,\pi\,(-6\,\rho1^2\,\mathrm{Sign}[\rho1] - (\rho1 - 4\,\rho2)^2\,\mathrm{Sign}[\rho1 - 4\,\rho2] + 4\,(\rho1 - 2\,\rho2)^2\,\mathrm{Sign}[\rho1 - 2\,\rho2] +$
$4\,(\rho1 + 2\,\rho2)^2\,\mathrm{Sign}[\rho1 + 2\,\rho2] - (\rho1 + 4\,\rho2)^2\,\mathrm{Sign}[\rho1 + 4\,\rho2]),$
$\left.\mathrm{Integrate}\left[\dfrac{\mathrm{Sin}[r\,\rho1]\,\mathrm{Sin}[r\,\rho2]^4}{r^3},\ \{r, 0, \infty\},\ \mathrm{Assumptions} \to \rho1 \notin \mathrm{Reals}\ ||\ \rho2 \notin \mathrm{Reals}\right]\right]$

In[54]:= **Integrate[Log[x]/(x^2 + a^2), {x, 0, 1}]**

Out[54]= If$\left[\text{Im}[a] \geq 1 \mid\mid 1 + \text{Im}[a] \leq 0 \mid\mid \text{Im}[a] == 0 \mid\mid \text{Re}[a] \neq 0,\right.$

$\quad -\dfrac{i \ (\text{PolyLog}[2, -\frac{i}{a}] - \text{PolyLog}[2, \frac{i}{a}])}{2\,a}$, Integrate$\left[\dfrac{\text{Log}[x]}{a^2 + x^2},\right.$

$\quad \{x, 0, 1\}$, Assumptions $\rightarrow (-1 < \text{Im}[a] < 0 \mid\mid 0 < \text{Im}[a] < 1)\ \&\&\ \text{Re}[a] == 0\big]\big]$

In[55]:= **Integrate[BesselJ[0, Sqrt[x]] BesselK[0, Sqrt[x]] Log[Sqrt[x]],**
$\qquad$ **{x, 0, Infinity}]**

Out[55]= $-$EulerGamma

In[56]:= **Integrate[x^2 Log[x]/(x^2 + 1)^(1/2), {x, 0, 1/(2 Sqrt[2])}]**

Out[56]= $\dfrac{1}{576}$

$\quad \left(\sqrt{2}\ \text{HypergeometricPFQ}\left[\{\frac{3}{2}, \frac{3}{2}, \frac{3}{2}\}, \{\frac{5}{2}, \frac{5}{2}\}, -\frac{1}{8}\right] + 18\ (1 + \text{Log}[8])\ (-3 + \text{Log}[16])\right)$

In[57]:= **Integrate[x^(-7/2) Log[x]^5, {x, 1, Infinity}]**

Out[57]= $\dfrac{1536}{3125}$

In[58]:= **Integrate[Sin[x^3], {x, 0, 1}]**

Out[58]= $\dfrac{1}{6}\ i\ \left(\text{ExpIntegralE}\left[\frac{2}{3}, -i\right] - \text{ExpIntegralE}\left[\frac{2}{3}, i\right] - i\ \text{Gamma}\left[\frac{1}{3}\right]\right)$

In[59]:= **Integrate[Gamma[l + m]/(Gamma[l] Gamma[m])***
$\qquad$ **x^(l - 1) (1 - x)^(m - 1) (a x + b (1 - x))^(-1 - m), {x, 0, 1}]**

Out[59]= $\dfrac{1}{\text{Gamma}[l]\ \text{Gamma}[m]}\ \left(\text{Gamma}[l + m]\ \text{If}\left[\text{Re}[b] > 0\ \&\&\ \text{Re}[l] > 0\ \&\&\ \text{Re}[m] > 0\ \&\&\right.\right.$

$\quad \left(\text{Re}\left[\frac{b}{a - b}\right] \leq -1 \mid\mid \text{Re}\left[\frac{b}{a - b}\right] \geq 0 \mid\mid \text{Im}\left[\frac{b}{a - b}\right] \neq 0\right), \dfrac{(\frac{a}{b})^{-1}\ b^{-1-m}\ \text{Gamma}[l]\ \text{Gamma}[m]}{\text{Gamma}[1 + m]},$

$\quad \text{Integrate}\left[(1 - x)^{-1+m}\ x^{-1+l}\ (b + a\,x - b\,x)^{-1-m}, \{x, 0, 1\}, \text{Assumptions} \rightarrow\right.$

$\quad \left(\text{Im}\left[\frac{b}{a - b}\right] == 0\ \&\&\ -1 < \text{Re}\left[\frac{b}{a - b}\right] < 0\right) \mid\mid \text{Re}[b] \leq 0 \mid\mid \text{Re}[l] \leq 0 \mid\mid \text{Re}[m] \leq 0\big]\big]\right)$

In[60]:= **Integrate[x^s Exp[-x]/Sqrt[x + z^2], {x, 0, Infinity}]**

Out[60]= If$\left[\text{Re}[s] > -1\ \&\&\ \text{Re}[z^2] > 0,\ \text{Gamma}\left[\frac{1}{2} + s\right]\ \text{Hypergeometric1F1}\left[\frac{1}{2}, \frac{1}{2} - s, z^2\right] +\right.$

$\quad \dfrac{(\frac{1}{z^2})^{-s}\ \sqrt{z^2}\ \text{Gamma}\left[-\frac{1}{2} - s\right]\ \text{Gamma}[1 + s]\ \text{Hypergeometric1F1}[1 + s, \frac{3}{2} + s, z^2]}{\sqrt{\pi}},$

$\quad \text{Integrate}\left[\dfrac{e^{-x}\ x^s}{\sqrt{x + z^2}}, \{x, 0, \infty\},\right.$

$\quad \text{Assumptions} \rightarrow \text{Re}[s] \leq -1 \mid\mid (\text{Re}[s] > -1\ \&\&\ \text{Re}[z^2] \leq 0)\big]\big]$

In[61]:= **Integrate[Csch[t]^2 - 1/t^2, {t, -Infinity, Infinity}]**

Out[61]= -2

The next integral extends over an infinite number of implicitly-described intervals.

In[62]:= **Integrate[Boole[Sin[x]^2 > 1/2] Exp[-x] Sin[x], {x, 0, Infinity}]**

Out[62]= $\dfrac{e^{3\pi/4}}{\sqrt{2}\ (1 + e^{\pi})}$

Here are two integrals of rational functions.

In[63]:= **Integrate[1/(x^4 + 3 x^2 + 1)^8, {x, 0, Infinity}]**

Out[63]= $\dfrac{21377637\ \pi}{160000000\ \sqrt{5}}$

```
In[64]:= largeResult = Integrate[1/(x^6 + 3 x^2 + 1)^2,
                                {x, -Infinity, Infinity}];

       Short[largeResult, 12]
```
Out[66]//Short=

$$\left(17 \text{ i Log}[\text{Root}[1 + 3 \#1 + \#1^3 \&, 3]] \right.$$

$$\sqrt{-\text{Root}[1 + 3 \#1 + \#1^3 \&, 1]} \text{ Root}[1 + 3 \#1 + \#1^3 \&, 2]^{5/2} + \ll 53 \gg +$$

$$2 \text{ i } \pi \left(17 - 6 \text{ Root}[1 + 3 \#1 + \#1^3 \&, 1] + 4 \text{ Root}[1 + 3 \#1 + \#1^3 \&, 1]^2 \right)$$

$$\left(\text{Root}[1 + 3 \#1 + \#1^3 \&, 2]^{5/2} \sqrt{\text{Root}[1 + 3 \#1 + \#1^3 \&, 3]} + \sqrt{\text{Root}[1 + 3 \#1 + \#1^3 \&, 2]} \right.$$

$$\text{Root}[1 + 3 \#1 + \#1^3 \&, 3]^{5/2} + \sqrt{\text{Root}[1 + 3 \#1 + \#1^3 \&, 2]} \text{ Root}[1 + 3 \#1 + \#1^3 \&, 3] +$$

$$\left. (\text{Root}[1 + 3 \#1 + \#1^3 \&, 2] \text{ Root}[1 + 3 \#1 + \#1^3 \&, 3])^{5/2} \right) \right) \Big/$$

$$\left(900 \sqrt{\text{Root}[1 + 3 \#1 + \#1^3 \&, 1]} \text{ Root}[1 + 3 \#1 + \#1^3 \&, 2] \text{ Root}[1 + 3 \#1 + \#1^3 \&, 3] \right)$$

The results returned by `Integrate` are typically not simplified. (It is always easily possible to apply a simplifying function to the result, but it would be impossible for a user to disable any built-in simplification if it would happen automatically.) Applying `RootReduce` to the last expression gives a much shorter answer.

```
In[67]:= Collect[RootReduce[largeResult], _Log, RootReduce]
```
Out[67]= π Root[−11449 + 17890956 #1^2 − 7103376000 #1^4 + 59049000000 #1^6 &, 2] +
        Log[Root[1 + 3 #1^2 + #1^6 &, 1]]
          Root[11449 + 17890956 #1^2 + 7103376000 #1^4 + 59049000000 #1^6 &, 1] +
        Log[Root[1 + 3 #1^2 + #1^6 &, 2]]
          Root[11449 + 17890956 #1^2 + 7103376000 #1^4 + 59049000000 #1^6 &, 2] +
        Log[Root[1 + 3 #1 + #1^3 &, 2]]
          Root[11449 + 71563824 #1^2 + 113654016000 #1^4 + 3779136000000 #1^6 &, 5] +
        Log[Root[1 + 3 #1 + #1^3 &, 3]]
          Root[11449 + 71563824 #1^2 + 113654016000 #1^4 + 3779136000000 #1^6 &, 6]

```
In[68]:= {LeafCount[%], LeafCount[largeResult]}
```
Out[68]= {191, 2958}

`Integrate` can, of course, also handle integrable singularities.

```
In[69]:= Integrate[1/Sqrt[x], {x, 0, 1}]
```
Out[69]= 2

The following two integrals contain Gamma functions. Here p is the probability distribution of the third power of a normal distributed variable $p(x) = \pi^{-1/2} \int_{-\infty}^{\infty} \delta(x - \xi^3) \exp(-x^2) d\xi$ [153]. Interestingly, the distribution q has the same moments as p. Because we cannot specify that n is an integer through `Assumptions` (we could, but `Integrate` would ignore this assumption), we use `Simplify` on the result.

```
In[70]:= p[x_] := 1/(3 Sqrt[Pi]) Abs[x]^-(2/3) Exp[-Abs[x]^(2/3)]

       q[x_, r_] := p[x] * (* the following factor is positive for r < 1/2 *)
                        (1 + r (Cos[Sqrt[3] Abs[x]^(2/3)] -
                            Sqrt[3] Sin[Sqrt[3] Abs[x]^(2/3)]))
```

```
In[72]:= (* moments of the two distributions *)
       {momentp[n_], momentq[n_, r_]} =
            Integrate[x^n {p[x], q[x, r]}, {x, -Infinity, Infinity},
                Assumptions -> Re[n] > 0]
```

Out[73]= $\left\{ \dfrac{(1 + (-1)^n)\ \text{Gamma}[\frac{1}{2} + \frac{3n}{2}]}{2\sqrt{\pi}},\ \dfrac{2^{-\frac{3}{2} - \frac{3n}{2}}\ (1 + (-1)^n)\ \text{Gamma}[\frac{1}{2} + \frac{3n}{2}]\ (2^{\frac{1}{2} + \frac{3n}{2}} - 2\ r\ \text{Sin}[\frac{n\pi}{2}])}{\sqrt{\pi}} \right\}$

Because of symmetry (and visible through the factors $(1 + (-1)^n)$), the odd moments vanish for p and q and the even moments agree.

In[74]:= `Simplify[momentp[n] == momentq[n, r], Element[n/2, Integers]]`

Out[74]= True

As some of the above examples show, sometimes *Mathematica* will produce `If` statements as results, where the first argument represents a set of conditions on parameters appearing in the integral such that the second argument of `If` is the integrated form. This form of the result allows giving sufficient conditions for the convergence of the integral depending on parameters appearing in the integrand (and potentially in the integration limits). The last argument contains the unevaluated form of the integral (which is possible because `If` has the `HoldRest` attribute) with the negated conditions. Here is an example.

In[75]:= `Integrate[Sin[a x] Cos[b x]/x, {x, 0, Infinity}]`

Out[75]= $\text{If}\Big[a - b \in \text{Reals}\ \&\&\ a + b \in \text{Reals},\ \dfrac{1}{4}\ \pi\ (\text{Sign}[a - b] + \text{Sign}[a + b]),$

$\qquad \text{Integrate}\Big[\dfrac{\text{Cos}[b\ x]\ \text{Sin}[a\ x]}{x},\ \{x, 0, \infty\},\ \text{Assumptions} \to a - b \notin \text{Reals}\ ||\ a + b \notin \text{Reals}\Big]\Big]$

For concrete values of the parameters, the last result could either use the calculated integral, or evaluate the last argument of the `If`-object. (In the case of the last result, the integral becomes nonconvergent for imaginary a.)

In[76]:= `Block[{a, b}, {{a, b} = {2, 1}; %, {a, b} = {1, I}; %}]`

$\qquad$ Integrate::idiv : Integral of $\dfrac{\text{Cosh}[x]\ \text{Sin}[x]}{x}$ does not converge on $\{0, \infty\}$. More…

Out[76]= $\left\{\dfrac{\pi}{2},\ \text{Integrate}\Big[\dfrac{\text{Cosh}[x]\ \text{Sin}[x]}{x},\ \{x, 0, \infty\},\ \text{Assumptions} \to \text{True}\Big]\right\}$

There are many integrands for which the indefinite integral cannot be expressed in closed form (using *Mathematica*'s built-in special functions), but certain definite integrals can be done. Here is an example.

In[77]:= `With[{f = Sqrt[1 - x^2] Exp[-x]},`
$\qquad$ `{Integrate[f, x], Integrate[f, {x, 0, 1}]}]`

Out[77]= $\left\{\displaystyle\int e^{-x}\ \sqrt{1 - x^2}\ dx,\ \dfrac{1}{2}\ \pi\ (\text{BesselI}[1, 1] - \text{StruveL}[1, 1])\right\}$

In the last example, it was crucial that the limits of the definite integral were 0 and 1. There are also integrals for which the indefinite integral cannot be done while the definite integral can be done for arbitrary limits. Here is such an integral.

In[78]:= `Integrate[Re[x], x]`

Out[78]= $\displaystyle\int \text{Re}[x]\ dx$

In[79]:= `Integrate[Re[x], {x, 0, ζ}]`

Out[79]= $\dfrac{1}{2}\ \zeta\ \text{Re}[\zeta]$

The indefinite integral could not be done because the integrand was not an analytic function. But still, along a straight line in the complex x-plane between 0 and ζ, the definite integral has a well-defined value.

The following integral is a nontrivial one; it arises in the calculation of the ground-state energy of the homogeneous electron gas [1126].

In[80]:= `Integrate[(1/x (1/x Log[(1 + x)/(1 - x)] - 2))^2, {x, 0, 1}]`

Out[80]= $\dfrac{2}{9}\ (-6 + \pi^2)$

Although *Mathematica* has powerful algorithms, some insight of the user in a given problem is often irreplaceable and allows a much faster, more compact solution. Sometimes it is the only way to obtain a solution at all. Here is such an example. *Mathematica* will not likely finish the evaluation of the following definite integral in a reasonable time.

```
In[81]:= int = Apply[Integrate,
         HoldForm @@ {{Function[ξ, 1/(5 - 27/(2^(2/3) ξ) (1 - I Sqrt[3]) -
                                   ξ/(2 2^(1/3)) (1 + I Sqrt[3]))^(1/8)][
            (277 + τ + Sqrt[-2003 + 554 τ + τ^2])^(1/3)],
            {τ, -277 - 162 Sqrt[3], 3}}}, {1}]
```

$$
\text{Out[81]= } \int_{-277-162\sqrt{3}}^{-3} \frac{1}{\left(5 - \dfrac{27\,(1-i\sqrt{3}\,)}{2^{2/3}\,(277+\tau+\sqrt{-2003+554\,\tau+\tau^2}\,)^{1/3}} - \dfrac{(1+i\sqrt{3}\,)\,(277+\tau+\sqrt{-2003+554\,\tau+\tau^2}\,)^{1/3}}{2\,2^{1/3}}\right)^{1/8}}\,d\tau
$$

```
In[82]:= TimeConstrained[ReleaseHold[int], 100]
```

Out[82]= $Aborted

Numerically, of course, it is easy to get an approximation of this integral.

```
In[83]:= ReleaseHold[int /. Integrate[l__] :> NIntegrate[l, PrecisionGoal -> 12]]
```

Out[83]= $469.197 - 1.33933 \times 10^{-13}\,i$

If one "knows" that the integrand has the form $(-p^{(-1)}(y))^{-1/2}$ where $p(x)$ is a triquartic polynomial $p(\xi) = (\xi^4)^3 - 15\,(\xi^4)^2 - 6\,(\xi^4)^1 + 3$ it is straightforward to carry out a change of variables $y \to p(y)$ manually, using [190], [1030], [1377], [1649] $\int g(p^{(-1)}(y))\,dy \longrightarrow \int g(y)\,p'(y)\,dy$.

```
In[84]:= p[x_] = (x^4)^3 - 15 (x^4)^2 - 6 (x^4)^1 + 3;
```

If one knows, in addition, that the limits of the integral are the values of the polynomial where its derivative vanishes, the calculation of the integral becomes trivial.

```
In[85]:= {x, p[x]} /. Solve[D[p[x], x] == 0, x] // Simplify // Union
```

$$
\text{Out[85]= } \left\{ \{0, 3\},\ \left\{-(5-3\sqrt{3}\,)^{1/4},\ -277+162\sqrt{3}\,\right\},\ \left\{-i\,(5-3\sqrt{3}\,)^{1/4},\ -277+162\sqrt{3}\,\right\}, \right.
$$
$$
\left\{i\,(5-3\sqrt{3}\,)^{1/4},\ -277+162\sqrt{3}\,\right\},\ \left\{(5-3\sqrt{3}\,)^{1/4},\ -277+162\sqrt{3}\,\right\},
$$
$$
\left\{-(5+3\sqrt{3}\,)^{1/4},\ -277-162\sqrt{3}\,\right\},\ \left\{-i\,(5+3\sqrt{3}\,)^{1/4},\ -277-162\sqrt{3}\,\right\},
$$
$$
\left.\left\{i\,(5+3\sqrt{3}\,)^{1/4},\ -277-162\sqrt{3}\,\right\},\ \left\{(5+3\sqrt{3}\,)^{1/4},\ -277-162\sqrt{3}\,\right\}\right\}
$$

Here is the exact symbolic result and its numerical value.

```
In[86]:= {Simplify[#], N[#]}&[
         ((# /. y -> 0) - (# /. y -> (5 + 3 Sqrt[3])^(1/4)))&[
             Simplify[Integrate[1/Sqrt[y] p'[y], y]]]]
```

$$
\text{Out[86]= } \left\{ \frac{128}{161}\,(5+3\sqrt{3}\,)^{7/8}\,(41+21\sqrt{3}\,),\ 469.197 \right\}
$$

(If we did not know the polynomial p, we could calculate it the following way.)

```
In[87]:= GroebnerBasis[Numerator[Together[
         {(5 - (27 (1 - I Sqrt[3]))/(2^(2/3) X) -
             ((1 + I Sqrt[3]) X)/(2 2^(1/3))) - y^4,
         (277 + τ + y) - X^3, -2003 + 554 τ + τ^2 - y^2}]], {τ, y}, {X, y}]
```

Out[87]= $\{3 - 6\,y^4 - 15\,y^8 + y^{12} - \tau\}$

While *Mathematica* tries hard to carry out an integral, it does not try all transformations of the integrand (for a nice collection of interesting integrals, see [212]). So sometimes rewriting and/or simplifying the integrand helps to carry out an integral. Here is a toy example demonstrating this. The integrand contains a term *hiddenZero* x^x. `Integrate` does not detect that *hiddenZero* is zero.

In[88]:= `Integrate[1 + x + x^x (Sqrt[2 - Sqrt[2 + Sqrt[2]]]/2 - Sin[Pi/16]), x]`

Out[88]= $\displaystyle\int \left(1 + x + x^x \left(\frac{1}{2}\sqrt{2 - \sqrt{2 + \sqrt{2}}} - \text{Sin}\left[\frac{\pi}{16}\right]\right)\right) dx$

But by applying various simplifications "manually" the *hiddenZero* can be reduced to zero.

In[89]:= `RootReduce //@ Simplify //@ (Together //@ TrigToExp //@ %)`

Out[89]= $\frac{1}{2}(2x + x^2)$

Using a sufficiently complicated transformation of the integration variable inside a known integral allows one easily to find examples of doable, but not done, integrals. Here is an example based on the Cauchy transformation $x \to \xi + 1/\xi$ [730].

In[90]:= `Integrate[Cos[x] Exp[-x^2] /. x -> x - 1/x, {x, -Infinity, Infinity}]`

Out[90]= $\displaystyle\int_{-\infty}^{\infty} e^{-\left(-\frac{1}{x}+x\right)^2} \text{Cos}\left[\frac{1}{x} - x\right] dx$

(For other possibilities to derive difficult integrals with known result, see [1582] and [906].)

Here is a transcendental integral. It can be carried out using Jensen's formula [933]. This is a quite nontrivial simplification.

In[91]:= `Module[{R = Pi},`
         `{Integrate[Log[Abs[Cos[R Exp[I φ]]]], {φ, 0, 2Pi}],`
          `{NIntegrate[Log[Abs[Cos[R Exp[I φ]]]], {φ, 0, 2Pi}],`
          `(* Jensen's formula; two simple zeros, no poles *)`
          `2Pi Log[Abs[Cos[0]] R^(2 - 0)/Abs[(-Pi/2) (Pi/2)]]} // N}]`

Out[91]= $\left\{\displaystyle\int_{0}^{2\pi} \text{Log}[\text{Abs}[\text{Cos}[e^{i\varphi}\pi]]] d\varphi, \{8.71034 + 0.\,i, 8.71034\}\right\}$

If one has a priori knowledge about the parameters, one can use them in `Integrate` via the `Assumptions` option, one of the three options of `Integrate`.

In[92]:= `Options[Integrate]`

Out[92]= `{Assumptions :→ $Assumptions,`
          `GenerateConditions → Automatic, PrincipalValue → False}`

`Assumptions`

   is an option of `Integrate` that specifies certain assumptions to be imposed on parameters appearing in the integral.

   Default:

   `$Assumptions` (the currently active assumptions)

   Admissible:

   listOfAssumptions or *logicalConjunctionOfAssumptions*

Without specifying the properties of a in the following example, a condition on a is generated.

In[93]:= `Integrate[Exp[a x^2], {x, -Infinity, Infinity}]`

Out[93]= $\text{If}\left[\text{Re}[a] < 0, \dfrac{\sqrt{\pi}}{\sqrt{-a}}, \text{Integrate}\left[e^{ax^2}, \{x, -\infty, \infty\}, \text{Assumptions} \to \text{Re}[a] \geq 0\right]\right]$

By specifying the properties of a, we do not get an `If` statement.

```
In[94]:= Integrate[Exp[a x^2], {x, -Infinity, Infinity},
              Assumptions -> {Re[a] < 0}]
```

$$Out[94]= \frac{\sqrt{\pi}}{\sqrt{-a}}$$

For real positive a, the integral is divergent.

```
In[95]:= Integrate[Exp[a x^2], {x, -Infinity, Infinity},
              Assumptions -> {Re[a] > 0}]
```

$$Out[95]= \text{Integrate}\left[e^{a\,x^2}, \{x, -\infty, \infty\}, \text{Assumptions} \to \{Re[a] > 0\}\right]$$

Here is a more complicated example—a nontrivial Lévy distribution [697].

```
In[96]:= Integrate[Exp[-q^(4/3)] Cos[q x], {q, 0, Infinity},
              Assumptions -> Im[x] == 0] // TraditionalForm
```

Out[96]//TraditionalForm=

$$\frac{9\,3^{3/4}\,\Gamma(-\frac{1}{4})\Gamma(\frac{13}{12})\Gamma(\frac{17}{12})\,_2F_2\left(\frac{13}{12}, \frac{17}{12}; \frac{5}{4}, \frac{3}{2}; \frac{27\,x^4}{256}\right)x^2 + 96\,\sqrt[4]{3}\,\Gamma(\frac{7}{12})\Gamma(\frac{11}{12})\Gamma(\frac{5}{4})\,_2F_2\left(\frac{7}{12}, \frac{11}{12}; \frac{1}{2}, \frac{3}{4}; \frac{27\,x^4}{256}\right)}{64\,\pi}$$

In the next example `Integrate`, inherits the currently active assumptions on x through `$Assumptions` that were created by the outer `Assuming`.

```
In[97]:= Assuming[Im[x] == 0, Print[$Assumptions]; TraditionalForm @
              Integrate[Exp[-q^(4/3)] Cos[q x], {q, 0, Infinity}]]
```

    Im[x] == 0

Out[97]//TraditionalForm=

$$\frac{9\,3^{3/4}\,\Gamma(-\frac{1}{4})\Gamma(\frac{13}{12})\Gamma(\frac{17}{12})\,_2F_2\left(\frac{13}{12}, \frac{17}{12}; \frac{5}{4}, \frac{3}{2}; \frac{27\,x^4}{256}\right)x^2 + 96\,\sqrt[4]{3}\,\Gamma(\frac{7}{12})\Gamma(\frac{11}{12})\Gamma(\frac{5}{4})\,_2F_2\left(\frac{7}{12}, \frac{11}{12}; \frac{1}{2}, \frac{3}{4}; \frac{27\,x^4}{256}\right)}{64\,\pi}$$

`Assumptions` can be given in the form of a `List` or as logical operators such as And and Or.

```
In[98]:= Integrate[Exp[a x^2 + b x^4], {x, -Infinity, Infinity},
              Assumptions -> {Re[a] < 0, Re[b] < 0}]
```

$$Out[98]= \frac{\sqrt{-a}\,e^{-\frac{a^2}{8b}}\,\text{BesselK}\left[\frac{1}{4}, -\frac{a^2}{8b}\right]}{2\sqrt{-b}}$$

```
In[99]:= Integrate[Exp[a x^2 + b x^4], {x, -Infinity, Infinity},
              Assumptions -> Re[a] < 0 && Re[b] < 0]
```

$$Out[99]= \frac{\sqrt{-a}\,e^{-\frac{a^2}{8b}}\,\text{BesselK}\left[\frac{1}{4}, -\frac{a^2}{8b}\right]}{2\sqrt{-b}}$$

Here is another example where giving assumptions allows giving a result (a Frullani-type integral [21], [912], [86] where for nonreal α and β the integral does not exist).

```
In[100]:= Integrate[(Sin[α x] - Sin[β x])^2/x^2, {x, 0, Infinity},
               Assumptions -> (Im[α] == 0 && Im[β] == 0)]
```

$$Out[100]= \frac{1}{2}\,\pi\,(\alpha\,\text{Sign}[\alpha] + (\alpha - \beta)\,\text{Sign}[\alpha - \beta] + \beta\,\text{Sign}[\beta] - (\alpha + \beta)\,\text{Sign}[\alpha + \beta])$$

Sometimes it can happen that the conditions returned by `Integrate` are not complete. The following integral converges for $-5 < \text{Re}(\alpha) < -1$. The lower bound on α is not found in the following input.

```
In[101]:= Integrate[x^α (Sin[x]/x - Cos[x])^2, {x, 0, Infinity}]
```

Out[101]= If$\left[\text{Re}\left[\alpha\right] < -1, \; -2^{-2-\alpha} \; (4 \text{ Gamma}[-1 + \alpha] + \text{Gamma}[1 + \alpha] + 4 \text{ Gamma}[\alpha, 0]) \; \text{Sin}\left[\frac{\pi \alpha}{2}\right], \right.$

$\left. \text{Integrate}[x^{-2+\alpha} \; (-x \text{ Cos}[x] + \text{Sin}[x])^2, \; \{x, 0, \infty\}, \; \text{Assumptions} \to \text{Re}[\alpha] \geq -1]\right]$

Assumptions in `Integrate` can currently have a form containing Re, Im, and equalities and inequalities. Specifying a parameter as integer is currently not very helpful (no algorithm will use such information literally, only the real-valuedness of the value can be inferred). Assumptions will be used in various stages in the integration algorithm to decide convergence and cases. They will also be used in the sense of `Refine` to allow certain evaluations which are correct under appropriate restrictions. Here is an example.

In[102]:= `Integrate[(1 + Cos[x]) Log[a + Cos[x]], {x, 0, Pi}, Assumptions -> a > 1]`

Out[102]= $-\pi \left(-a + \sqrt{-1 + a^2} + \text{Log}\left[\dfrac{2}{a + \sqrt{-1 + a^2}}\right]\right)$

The following half-sided Fourier transform of $(\exp(i x^3 /3) - 1) x^{-3/2}$ yields for positive ξ a complicated result involving generalized hypergeometric functions.

In[103]:= `Collect[#, Union[Cases[#, _HypergeometricPFQ, Infinity]], Simplify]& @`
`    Integrate[Exp[-I ξ x] (Exp[-I x^3/3] - 1) x^-(3/2),`
`        {x, 0, Infinity}, Assumptions -> ξ > 0] // TraditionalForm`

Out[103]//TraditionalForm=

$$\frac{(-1)^{11/12} \, \pi \, {}_1F_2\left(-\frac{1}{6}; \frac{1}{3}, \frac{2}{3}; \frac{\xi^3}{9}\right)}{3^{2/3} \, \Gamma\left(\frac{7}{6}\right)} + \frac{2\,(-1)^{7/12} \, 3^{2/3} \, \pi \, \xi \, {}_1F_2\left(\frac{1}{6}; \frac{2}{3}, \frac{4}{3}; \frac{\xi^3}{9}\right)}{\Gamma\left(-\frac{1}{6}\right)} + (1 + i) \sqrt{2\pi} \, \sqrt{\xi}$$

For analytic functions, the option `Assumptions` influences only definite integrals. For piecewise-defined functions, the option `Assumptions` allows also to evaluate some indefinite integrals along the real axis. Here is an example.

In[104]:= `{(* for generic complex x this integral stays undone *)`
`        Integrate[Floor[x - x^2 FractionalPart[x]], x],`
`        (* along the real axis for a finite x-range it can be done *)`
`        Integrate[Floor[x - x^2 FractionalPart[x]], x,`
`            Assumptions -> Element[x, Reals] && 0 < x < 2]}`

Out[104]= $\left\{\int \text{Floor}[x - x^2 \text{ FractionalPart}[x]] \, dx, \right.$

$$\left. \begin{cases} -2x & (x < 1 \;\&\&\; -x + x^3 > 1) \;||\; (x \geq 1 \;\&\&\; -x - x^2 + x^3 > 1) \\ -x & (x < 1 \;\&\&\; 0 < -x + x^3 \leq 1) \;||\; (x \geq 1 \;\&\&\; 0 < -x - x^2 + x^3 \leq 1) \\ x & x \geq 1 \;\&\&\; -x - x^2 + x^3 \leq -1 \end{cases} \right\}$$

The second option of `Integrate` is `GenerateConditions`.

`GenerateConditions`

   determines if conditions on the parameters should be generated that guarantee the convergence of the integral.

Default:

   Automatic

Admissible:

   True or False

In general, for a definite integral that contains parameters, `Integrate` will try to generate conditions on the convergence of the integral. When doing so, `Integrate` will return an `If`-statement with the first argument being the conditions, the second the result, and the third again an `Integrate`-object with negated conditions (the last argument of `If` is held, so this integral will not evaluate). Here is an example of an integrand containing three parameters. The integral returned is valid under conditions that are sufficient, but not necessary (exhaustive).

In[105]:= **Integrate[Exp[-α1 x]/x + Exp[-α2 x^2] + Exp[-α3 x^3], {x, 1, Infinity}]**

Out[105]= $\mathrm{If}\Big[\mathrm{Re}[\alpha1] > 0 \;\&\&\; \mathrm{Re}[\alpha2] > 0 \;\&\&\; \mathrm{Re}[\alpha3] > 0,$

$-\mathrm{CoshIntegral}[\alpha1] + \dfrac{\sqrt{\pi}\ \mathrm{Erfc}[\sqrt{\alpha2}\,]}{2\sqrt{\alpha2}} + \dfrac{1}{3}\ \mathrm{ExpIntegralE}\Big[\dfrac{2}{3},\ \alpha3\Big] + \mathrm{SinhIntegral}[\alpha1],$

$\mathrm{Integrate}\Big[e^{-x^2\,\alpha2} + e^{-x^3\,\alpha3} + \dfrac{e^{-x\,\alpha1}}{x},\ \{x,\,1,\,\infty\},$

$\mathrm{Assumptions} \to \mathrm{Re}[\alpha1] \le 0 \;||\; \mathrm{Re}[\alpha2] \le 0 \;||\; \mathrm{Re}[\alpha3] \le 0\Big]\Big]$

Specifying conditions that are also making the integral convergent but do not fall into the conditions covered by the generated conditions forces the third argument of the last If-statement to be evaluated and another result is returned.

In[106]:= **Block[{α1 = I}, %]**

Out[106]= $\mathrm{If}\Big[\mathrm{Re}[\alpha2] > 0 \;\&\&\; \mathrm{Re}[\alpha3] > 0,$

$-\mathrm{i}\,\pi + \dfrac{\sqrt{\pi}\ \mathrm{Erfc}[\sqrt{\alpha2}\,]}{2\sqrt{\alpha2}} + \dfrac{1}{3}\ \mathrm{ExpIntegralE}\Big[\dfrac{2}{3},\ \alpha3\Big] - \mathrm{ExpIntegralEi}[-\mathrm{i}],$

$\mathrm{Integrate}\Big[e^{-x^2\,\alpha2} + e^{-x^3\,\alpha3} + \dfrac{e^{-\mathrm{i}\,x}}{x},\ \{x,\,1,\,\infty\},\ \mathrm{Assumptions} \to \mathrm{Re}[\alpha3] \le 0 \;||\; \mathrm{Re}[\alpha2] \le 0\Big]\Big]$

Because in general there are many possible combinations of necessary conditions that guarantee the convergence of a given integral, it is too expensive to generate them all. So, if possible, Integrate returns just one set of sufficient convergence conditions and the corresponding value of the integral.

Sometimes one knows that a certain integral exists for certain parameters, but specifying the parameter ranges might be difficult. In this case, one uses the option setting GenerateConditions -> False. For the last example, we do not get an If statement.

In[107]:= **Integrate[Exp[a x^2 + b x^4], {x, -Infinity, Infinity},**
        GenerateConditions -> False]

Out[107]= $\dfrac{\sqrt{-a}\ e^{-\frac{a^2}{8b}}\ \mathrm{BesselK}\Big[\frac{1}{4},\ -\frac{a^2}{8b}\Big]}{2\sqrt{-b}}$

Be aware that the option setting GenerateConditions -> False results in two different behavior changes. One is the fact that the result will not be an If statement and no conditions are given. The second effect is that many divergent integrals will return finite values. Here is an example of an obviously divergent integral.

In[108]:= **Integrate[x^4 Sin[x], {x, 0, Infinity}]**

Out[108]= $\displaystyle\int_0^\infty x^4\,\mathrm{Sin}[x]\,dx$

Using GenerateConditions -> False gives a finite answer.

In[109]:= **Integrate[x^4 Sin[x], {x, 0, Infinity}, GenerateConditions -> False]**

Out[109]= 24

Giving finite values for divergent integrals is not a mathematical flaw in *Mathematica*, but for some kind of calculations a very useful property. The result 24 for the last integral can easily be understood when regularizing the integral with the convergence-implying factor $e^{-\varepsilon x}$.

In[110]:= **Integrate[x^4 Sin[x] Exp[-ε x], {x, 0, Infinity}, Assumptions -> ε > 0]**

Out[110]= $\dfrac{24\,(1 + 5\,\varepsilon^2\,(-2 + \varepsilon^2))}{(1 + \varepsilon^2)^5}$

In[111]:= **% /. ε -> 0**

Out[111]= 24

Here is another integral. Obviously, for large x, the integral $\int_0^x \xi^4 / (1 + e^{-\xi})\, d\xi$ diverges like $x^5/5$. So subtracting the diverging part yields a finite answer.

In[112]:= **Integrate[x^4/(1 + Exp[-x]) - x^4, {x, 0, Infinity}]**

Out[112]= $-\dfrac{45\, \text{Zeta}[5]}{2}$

Although the last divergent integral was "formally positive", the result was negative [1269], [1270], [1201]. The last result can be understood in a Hadamard(–Adhemar [273]) sense [978], [178], [179], [180], [1343], [576], [1337], [1757], [1758], [304], [1145] (for numerical applications, see [797], [1099], [882], [1701], [522]), which basically means to subtract all divergencies of the form $x^{positiveInteger}$ and logarithmic singularities (for generalizations, see [946]). Subtracting the divergent term, we get exactly the result from above.

Another way to understand the result $-45\,\zeta(5)/2$ is the so-called zeta function regularization technique [571], [572] (which is in most cases equivalent to the Hadamard regularization [1086]).

In[113]:= **Integrate[x^s/(1 + Exp[-x]), {x, 0, Infinity}, GenerateConditions -> False]**

Out[113]= $-(-2)^{-s}\,(-1 + 2^s)\,\text{Gamma}[1 + s]\,\text{Zeta}[1 + s]$

In[114]:= **% /. s -> 4**

Out[114]= $-\dfrac{45\, \text{Zeta}[5]}{2}$

Mathematica is often able to detect poles (and some branch points and branch cuts) along the path of integration.

In[115]:= **Integrate[1/x, {x, -1, 1}]**

Out[115]= $\displaystyle\int_{-1}^{1} \dfrac{1}{x}\, dx$

Often one wants such first-order pole singularities interpreted as Cauchy principal value integrals. The option `PrincipalValue` allows this (sometimes the limit $\lim_{\varepsilon \to 0} \left(\int_a^b f(x)\, dx + \int_b^c f(x)\, dx \right)$ of an integrand with any kind of singularity at $x = b$ is also called principal value integral).

`PrincipalValue`

    determines if a first-order pole in a definite integral is to be interpreted in the Cauchy principal value sense.

    Default:

        False

    Admissible:

        True

Now, we get the result 0.

In[116]:= **Integrate[1/x, {x, -1, 1}, PrincipalValue -> True]**

Out[116]= 0

Here is a slightly more complicated example.

In[117]:= **Integrate[1/x Sin[x], {x, -1/2, 1}, PrincipalValue -> True]**

Out[117]= $\text{SinIntegral}\left[\dfrac{1}{2}\right] + \text{SinIntegral}[1]$

For the following integral with a third-order pole, the classical Cauchy principal value is not defined.

In[118]:= **Integrate[1/x^3, {x, -1, 1}, PrincipalValue -> True]**

Out[118]= $\text{Integrate}\left[\dfrac{1}{x^3}, \{x, -1, 1\}, \text{PrincipalValue} \to \text{True}\right]$

Let us demonstrate the Kramers–Kronig transformation for the example function $\varepsilon(\omega) = 1/(\omega + i\beta)$. The real and imaginary parts of $\varepsilon(\omega)$ are then given by the following expressions.

In[119]:= **εr[ω_] = ω/(ω^2 + β^2);**
      εi[ω_] = -β/(ω^2 + β^2);

Here, the Kramers–Kronig transformation [1060], [915], [1402], [1434], [701], [1510], [619], [940] is carried out.

In[121]:= **(Integrate[#, {α, 0, Infinity}, Assumptions -> {ω > 0, β > 0},**
              PrincipalValue -> True] // Cancel) & /@
              {-2ω/Pi εi[α]/(ω^2 - α^2), 2/Pi εr[α] α/(ω^2 - α^2)}

Out[121]= $\{\dfrac{\omega}{\beta^2 + \omega^2}, -\dfrac{\beta}{\beta^2 + \omega^2}\}$

The next principal value integral is expressed in special functions.

In[122]:= **Integrate[Sin[x]/(x - P) , {x, 0, Infinity},**
              PrincipalValue -> True, Assumptions -> P > 0]

Out[122]= $-\text{CosIntegral}[P]\ \text{Sin}[P] + \dfrac{1}{2}\ \text{Cos}[P]\ (\pi + 2\ \text{SinIntegral}[P])$

Although *Mathematica* can integrate a huge amount of functions, there are many definite integrals that *Mathematica 5.1* cannot compute or that are impossible to express terms by using the built-in functions.

In[123]:= **Integrate[Sin[x^3]/(x + 5), {x, 0, 1}]**

Out[123]= $\displaystyle\int_0^1 \dfrac{\text{Sin}[x^3]}{5 + x}\ dx$

In[124]:= **Integrate[Sin[Exp[t] + Log[t]]/(t^6 + 7t^4 + 5t + 7), {t, 0, Infinity}]**

Out[124]= $\displaystyle\int_0^\infty \dfrac{\text{Sin}[e^t + \text{Log}[t]]}{7 + 5\,t + 7\,t^4 + t^6}\ dt$

We end now our discussions of univariate integration. To analytically compute multidimensional integrals, we use the standard iterator notation in `Integrate`.

`Integrate[integrand, {var₁, lowerLimit₁, upperLimit₁}, ..., {varₙ, lowerLimitₙ, upperLimitₙ}]`

finds (if possible) the multiple definite integral $\displaystyle\int_{lowerLimit_n}^{upperLimit_n} \cdots \int_{lowerLimit_1}^{upperLimit_1} integrand\ dvar_1 \cdots dvar_n$.

The following integrals will be integrated in the prescribed order. In contrast to numerical methods, no special methods are currently used in `Integrate`. Here are two examples for 2D integrals.

In[125]:= **Integrate[(x + y + z)^3 Exp[-x - y - z],**
              {x, 0, Infinity}, {y, 0, Infinity}]

Out[125]= $e^{-z}\ (24 + z\ (18 + z\ (6 + z)))$

In[126]:= **Integrate[1/(a y + b (x - y) + c (1 - x))^3, {x, 0, 1}, {y, 0, x},**
              GenerateConditions -> False]

Out[126]= $\dfrac{1}{2\ a\ b\ c}$

As for differentiation of nonintegrable functions (see the example with `strangeFunc` discussed above), for "nonnice" functions, the result of integration may depend on the order in which the integrals are calculated [706].

In[127]:= `{Integrate[(x^2 - y^2)/(x^2 + y^2)^2, {x, 0, 1}, {y, 0, 1}],`
`Integrate[(x^2 - y^2)/(x^2 + y^2)^2, {y, 0, 1}, {x, 0, 1}]}`

Out[127]= $\left\{\dfrac{\pi}{4}, -\dfrac{\pi}{4}\right\}$

The next input defines the function `dUnitSphereVolume` that calculates the volume of a *d*D unit sphere by integrating the volume element following from the implicit equation $x_1^2 + x_2^2 + \cdots + x_d^2 = 1$. We use the symmetry of the sphere to carry out the integral over the positive coordinate values only, and multiply the result with 2^d.

In[128]:= `dUnitSphereVolume[d_Integer?Positive]  :=`
`Module[{x}, If[d == 1, 2, (* number of d-orthant *) 2^d *`
`Integrate[Sqrt[1 - Sum[x[k]^2, {k, d - 1}]],`
`Sequence @@ (* limits of integration along the coordinate axes *)`
`Table[{x[j], 0, Sqrt[1 - Sum[x[k]^2, {k, j + 1, d - 1}]]},`
`{j, d - 1, 1, -1}]]]]`

Here are the resulting volume values and the known exact form for the volume.

In[129]:= `Table[{dUnitSphereVolume[d], Pi^(d/2)/(d/2)!}, {d, 5}]`

Out[129]= $\left\{\{2, 2\}, \{\pi, \pi\}, \left\{\dfrac{4\pi}{3}, \dfrac{4\pi}{3}\right\}, \left\{\dfrac{\pi^2}{2}, \dfrac{\pi^2}{2}\right\}, \left\{\dfrac{8\pi^2}{15}, \dfrac{8\pi^2}{15}\right\}\right\}$

Here is a more challenging 6D integral for the volume of a certain polytope [625]. Calculating the integral will take a few minutes.

In[130]:= `Integrate[Boole[x12 + x13 + 2 (x21  + x31) < 1 &&`
`x21 + x23 + 2 (x12  + x32) < 1 &&`
`x31 + x32 + 2 (x13  + x23) < 1],`
`{x12, 0, Infinity}, {x13, 0, Infinity},`
`{x21, 0, Infinity}, {x23, 0, Infinity},`
`{x31, 0, Infinity}, {x32, 0, Infinity}] // Timing`

Out[130]= $\left\{1264.51 \text{ Second}, \dfrac{1}{2880}\right\}$

Two remarks about multiple integrals are in order here:

`Integrate[`*integrand*`, {`*var*$_1$`, `*lowerLimit*$_1$`, `*upperLimit*$_1$`}, ..., {`*var*$_n$`, `*lowerLimit*$_n$`, `*upperLimit*$_n$`}]`   calculates a multiple integral, not a genuine multidimensional one. This means that it will first carry out the integration with respect to *var*$_1$, and then with respect to *var*$_2$, ... No change of variables will be carried out. This means that some "simple" multidimensional integrals will not be integrated in closed form. Here is an example (which is easily done in spherical coordinates).

In[131]:= `Integrate[Exp[-(x^2 + y^2 + z^2)]/(x^2 + y^2 + z^2),`
`{x, -Infinity, Infinity}, {y, -Infinity, Infinity},`
`{z, -Infinity, Infinity}]`

Out[131]= $\pi \displaystyle\int_{-\infty}^{\infty} \int_{-\infty}^{\infty} \dfrac{\text{Erfc}\left[\sqrt{x^2 + y^2}\right]}{\sqrt{x^2 + y^2}} \, dy \, dx$

As a little example showing *Mathematica*'s multiple integration capabilities, let us test a few special cases of Robbin's integral identity. The identity is [1494], [1894]:

$$
\int dx_1 \int dx_2 \cdots \int dx_n \atop {0 \le x_1 \le x_2 \le \dots \le x_{n-1} \le x_n}
\left\| \begin{pmatrix} x_1^{a_1-1} & x_1^{a_2-1} & \cdots & x_1^{a_n-1} \\ x_2^{a_1-1} & x_2^{a_2-1} & \cdots & x_2^{a_n-1} \\ \vdots & & \ddots & \vdots \\ x_n^{a_1-1} & \cdots & \cdots & x_n^{a_n-1} \end{pmatrix} \right\| =
$$

$$
\left(\prod_{i=1}^{n} \prod_{j=i+1}^{n} (a_j - a_i) \right) \left(\prod_{i=1}^{n} a_i \right)^{-1} \left(\prod_{i=1}^{n} \prod_{j=1}^{i-1} (a_j + a_i) \right)^{-1}.
$$

The *Mathematica* version of this theorem is as follows.

```
In[132]:= RobbinsIntegralIdentityTest[n_, a_Function] :=
    {Integrate[Det[Table[x[i]^(a[j] - 1), {i, n}, {j, n}]],
        Sequence @@ (Table[{x[i], x[i - 1], 1}, {i, n}] /. x[0] -> 0)],
    Product[a[j] - a[i], {i, n}, {j, i + 1, n}]/
    (Product[a[i], {i, n}] Product[a[j] + a[i], {i, n}, {j, i - 1}])}
```

The simplest possible choice for a_j is $a_j = j$.

```
In[133]:= RobbinsIntegralIdentityTest[3, #&]
```

$$
\text{Out[133]= } \left\{ \frac{1}{180}, \frac{1}{180} \right\}
$$

Here is a more difficult choice for the a_j. The resulting integrand is no longer a polynomial.

```
In[134]:= RobbinsIntegralIdentityTest[2, 1/#&]
```

$$
\text{Out[134]= } \left\{ -\frac{2}{3}, -\frac{2}{3} \right\}
$$

And here is a much larger example—120 five-dimensional examples have to be carried out.

```
In[135]:= RobbinsIntegralIdentityTest[5, # + 1&]
```

$$
\text{Out[135]= } \left\{ \frac{1}{2095632000}, \frac{1}{2095632000} \right\}
$$

The second remark concerns singularities. The option `GenerateConditions` is typically set to `False` in multiple integrals. As discussed above, this option setting results in certain regularizations of divergencies. As a result, one might get unexpected results.

Here is a 2D example of Hadamard-regularized integral. We have $\int_0^1 \int_0^1 |x - y|^\alpha \, dy \, dx = 2/((1 + \alpha)(2 + \alpha))$ for all α. *Mathematica* finds a condition for the existence of the integral in the classical sense.

```
In[136]:= Integrate[Sqrt[(x - y)^2]^α, {x, 0, 1}, {y, 0, 1}]
```

$$
\text{Out[136]= } \text{If}\left[\text{Re}[\alpha] > -1, \frac{2}{2 + 3\alpha + \alpha^2}, \right.
$$
$$
\left. \text{Integrate}\left[\text{Integrate}\left[((x - y)^2)^{\alpha/2}, \{x, 0, 1\}, \text{Assumptions} \to \text{Re}[\alpha] \le -1 \,\&\&\, 0 < y < 1\right], \right. \right.
$$
$$
\left. \left. \{y, 0, 1\}, \text{Assumptions} \to \text{Re}[\alpha] \le -1 \right]\right]
$$

One can understand this result by excluding a thin strip around the singular curve $x = y$. The resulting algebraic singular terms in ε are dropped within the Hadamard regularization. (Observe that for $-2 < \alpha < -1$ the regularized integral is negative despite the integrand being positive definite.)

```
In[137]:= Collect[#, ε, Together]& @ Simplify[#, ε > 0]& @ (Plus @@
    ((* carefully carrying out the pair of two nested integrals *)
    Fold[Integrate[##, GenerateConditions -> False]&,
        Sqrt[(x - y)^2]^α, {{y, x + ε, 1}, {x, 0, 1 - ε}}]& /@
        {{{y, x + ε, 1}, {x, 0, 1 - ε}}, {{y, x + e, 1}, {x, 0, 1 - e}}}))
```

Out[137]= $\dfrac{2}{2 + 3\,\alpha + \alpha^2} - \dfrac{2\,\varepsilon^{1+\alpha}}{1+\alpha} + \dfrac{2\,\varepsilon^{2+\alpha}}{2+\alpha}$

In computing definite integrals, whenever possible we should exploit *Mathematica*'s knowledge of definite integrals (although in general this is much slower than indefinite integration), rather than first finding indefinite integrals and inserting the limit values. In fact, in many cases, the latter approach can result in an incorrect value. For example, $\int_0^{4\pi} 1/(5 + \cos(x))\,dx$ has a strictly positive integrand, and hence, the integral must also be positive. But calculating the indefinite integral and substitution of the limits results in zero.

In[138]:= $cos[x\_]$ = **Integrate[1/(5 + Cos[x]), x]**

Out[138]= $\dfrac{\mathrm{ArcTan}\!\left[\sqrt{\tfrac{2}{3}}\ \mathrm{Tan}[\tfrac{x}{2}]\right]}{\sqrt{6}}$

In[139]:= $cos[4Pi] - cos[0]$

Out[139]= 0

Using *Mathematica*'s definite integration capabilities, we get the correct result.

In[140]:= **Integrate[1/(5 + Cos[x]), {x, 0, 4Pi}]**

Out[140]= $\sqrt{\dfrac{2}{3}}\ \pi$

A numerical check confirms the last result.

In[141]:= **{NIntegrate[1/(5 + Cos[x]), {x, 0, 4Pi}], N[%]}**

Out[141]= {2.5651, 2.5651}

The reason for the discrepancy in the above results is not because of any problem with the fundamental theorem of calculus [1311], of course; it is caused by the multivalued nature of the indefinite integral arctan. (Using nonstandard conventions for the branch cuts of arctan could make the result correct [585]. For the arithmetic nature of such integrals, see [1767].)

Let us have a look at the indefinite integral. The integral is correct, as can be shown by differentiation.

In[142]:= **Simplify[D[$cos[x]$, x]]**

Out[142]= $\dfrac{1}{5 + \mathrm{Cos}[x]}$

In[143]:= **Plot[$cos[x]$, {x, 0, 4Pi}];**

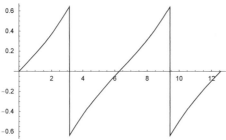

The indefinite integral is not a continuous function of x. The discontinuities are from the branch cuts that start and end from the simple poles of the integrand [1831] at $(2\,n + 1)\,\pi \pm i\,\mathrm{arccos}(5)$.

In[144]:= **Show[GraphicsArray[**
        Block[{$DisplayFunction = Identity, ε = 10^-10},
          {(* contour plot *)

```
ContourPlot[Evaluate[Re[cos[x + I y]]], {x, 0, 4Pi}, {y, -4, 4},
          PlotPoints -> 60, ContourShading -> False];
(* 3D plot made piecewise *)
Show[Table[Plot3D[Re[cos[x + I y]], {x, k Pi + ε, (k + 1) Pi - ε},
                {y, -4, 4}, PlotPoints -> {8, 60}], {k, 0, 3}],
    AxesLabel -> {"x", "y", None}]}]]]
```

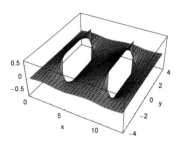

By adding the piecewise constant function, we can make the antiderivative a continuous function.

```
In[145]:= cosC[x_]  :=
        Which[x <  1  Pi // N, cos[x],
              x == 1 Pi // N, Pi/2/Sqrt[6],
              x <  3  Pi // N, cos[x] + Pi/Sqrt[6],
              x == 3 Pi // N, 3Pi/2/Sqrt[6],
              x <  5  Pi // N, cos[x] + 2 Pi/Sqrt[6]]
```

```
In[146]:= Plot[cosC[x], {x, 0, 4Pi}]
```

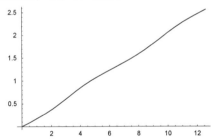

Now, the integral is given as the difference of the function value of the upper limit and the lower limit.

```
In[147]:= cosC[4Pi]  - cosC[0]
```

$$\text{Out[147]= } \sqrt{\frac{2}{3}}\,\pi$$

(For more details concerning such pitfalls, see [925], [923], and [926].)

Even for an everywhere smooth function, the indefinite integral returned by *Mathematica* might be discontinuous. The following plots show the real and imaginary parts of $(e^x - 1)/x$ and $\int (e^x - 1)/x\, dx$ along the real axis. The imaginary part (blue curves) of the indefinite integral is discontinuous at $x = 0$. (This integrand has the special property to give an integral that has a single line where its value is different from its left-side and right-side limits.)

```
In[148]:= Module[{f, F, x},
              f[x_] = (Exp[x] - 1)/x;  F[x_] = Integrate[f[x], x];
          Show[GraphicsArray[
```

```
(* show function and indefinite integral along real axis *)
Plot[{Re[#1[x]], Im[#1[x]]}, {x, -1, 1}, PlotLabel -> #2,
     PlotStyle -> {RGBColor[1, 0, 0], RGBColor[0, 0, 1]},
     DisplayFunction -> Identity, Frame -> True,
     PlotRange -> {{-1, 1}, {-3.5, 3.5}}]& @@@
{{f, "f[x]"}, {F, "∫f[x] dx"}}]]]
```

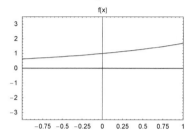

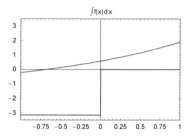

But the indefinite integral was nevertheless correct.

In[149]:= **D[Integrate[(Exp[x] - 1)/x, x], x] - (Exp[x] - 1)/x // Simplify**

Out[149]= 0

As an application of *Mathematica* integration capabilities, let us briefly discuss a class of parametrically describable minimal surfaces.

Mathematical Remark: Minimal Surfaces

Minimal surfaces are surfaces $z = f(x, y)$ that satisfy the differential equation

$$(1 + f_y^2) f_{xx} + 2 f_x f_y f_{xy} + (1 + f_x^2) f_{yy} = 0$$

or for surfaces given in parametric form $\{x(u, v), y(u, v), z(u, v)\}$, they satisfy $H = 0$ where (for brevity we suppress the u,v-dependence of all quantities)

$$H = \frac{(LG - 2\,FM + EN)}{2\,(EG - F^2)}$$

$$E = \left(\frac{\partial x}{\partial u}\right)^2 + \left(\frac{\partial y}{\partial u}\right)^2 + \left(\frac{\partial z}{\partial u}\right)^2$$

$$G = \left(\frac{\partial x}{\partial v}\right)^2 + \left(\frac{\partial y}{\partial v}\right)^2 + \left(\frac{\partial z}{\partial v}\right)^2$$

$$F = \left(\frac{\partial x}{\partial u}\right)\left(\frac{\partial x}{\partial v}\right) + \left(\frac{\partial y}{\partial u}\right)\left(\frac{\partial y}{\partial v}\right) + \left(\frac{\partial z}{\partial u}\right)\left(\frac{\partial z}{\partial v}\right)$$

$$\{L, N, M\} = \left\{\frac{l}{\sqrt{EG - F^2}}, \frac{n}{\sqrt{EG - F^2}}, \frac{m}{\sqrt{EG - F^2}}\right\}$$

$$\{l, n, m\} = \left\{ \begin{vmatrix} \dfrac{\partial^2 x}{\partial u^2} & \dfrac{\partial^2 y}{\partial u^2} & \dfrac{\partial^2 z}{\partial u^2} \\ \dfrac{\partial x}{\partial u} & \dfrac{\partial y}{\partial u} & \dfrac{\partial z}{\partial u} \\ \dfrac{\partial x}{\partial v} & \dfrac{\partial y}{\partial v} & \dfrac{\partial z}{\partial v} \end{vmatrix}, \begin{vmatrix} \dfrac{\partial^2 x}{\partial v^2} & \dfrac{\partial^2 y}{\partial v^2} & \dfrac{\partial^2 z}{\partial v^2} \\ \dfrac{\partial x}{\partial u} & \dfrac{\partial y}{\partial u} & \dfrac{\partial z}{\partial u} \\ \dfrac{\partial x}{\partial v} & \dfrac{\partial y}{\partial v} & \dfrac{\partial z}{\partial v} \end{vmatrix}, \begin{vmatrix} \dfrac{\partial^2 x}{\partial u \partial v} & \dfrac{\partial^2 y}{\partial u \partial v} & \dfrac{\partial^2 z}{\partial u \partial v} \\ \dfrac{\partial x}{\partial u} & \dfrac{\partial y}{\partial u} & \dfrac{\partial z}{\partial u} \\ \dfrac{\partial x}{\partial v} & \dfrac{\partial y}{\partial v} & \dfrac{\partial z}{\partial v} \end{vmatrix} \right\}$$

For given boundary curves, these are surfaces with minimal surface area. Here, H is the mean curvature.

It is remarkable that an infinite variety of such surfaces can be constructed relatively easily. (The general problem of more practical interest for computing the minimal surfaces corresponding to given boundary curves is called the Plateau–Douglas problem [650] and is far more difficult to solve.) Let $f(\xi)$ and $g(\xi)$ be two holomorphic functions with no common zeros in a connected domain of the complex $u + iv$-plane. Then, a parametric formula for a minimal surface is given by:

$$x(u, v) = x_0 + \text{Re} \int_{\xi_0}^{\xi(u,v)} (1 - g(\xi')^2)\, f(\xi')\, d\xi'$$

$$y(u, v) = y_0 + \text{Re} \int_{\xi_0}^{\xi(u,v)} i(1 + g(\xi')^2)\, f(\xi')\, d\xi'$$

$$z(u, v) = z_0 + \text{Re} \int_{\xi_0}^{\xi(u,v)} 2\, f(\xi')\, g(\xi')\, d\xi'.$$

Here, $\xi(u, v)$ is an arbitrary complex function of u and v (e.g., $u + i v$, $u e^{iv}$), whereas u and v are the two parameters in the parametric formula of the surface, and Re denotes the real part. For a proof of these formulas, see [1359] and [860].

We now implement the last formulas.

```
In[150]:= WeierstrassMinimalSurface[f_, g_, var_, pictVar_] :=
         (* pictVar for various coordinate systems;
            for instance, x + I y or r Exp[I φ] *)
         Re[Integrate[{(1 - g^2) f, I (1 + g^2) f, 2 f g}, var] /.
                                          var -> pictVar]
```

Here are two examples: the Enneper surface with $f(\xi) = 1$ and $g(\xi) = \xi$ and a Henneberg surface with $f(\xi) = -i/2\,(1 - \xi^{-4})$ and $g(\xi) = \xi$.

```
In[151]:= Show[GraphicsArray[
          Block[{$DisplayFunction = Identity,
                 opts = Sequence[Boxed -> False, Axes -> False, PlotRange -> All]},
            {(* Enneper surface *)
            ParametricPlot3D[Evaluate[
              WeierstrassMinimalSurface[1, ξ, ξ, r Exp[I φ]]],
                        {r, 0, 3}, {φ, 0, 2Pi},
                        PlotPoints -> {116, 80}, Evaluate[opts]],
            (* Henneberg surface *)
            ParametricPlot3D[Evaluate[
              WeierstrassMinimalSurface[-I/2 (1 - ξ^-4), ξ, ξ, r Exp[I φ]]],
                        {r, 0.72, 1}, {φ, 0, 2Pi},
                        PlotPoints -> {16, 40}, Evaluate[opts]]}]]]
```

Here is a spiraling minimal surface related to the behavior of a soap film near a boundary wire [236].

In[152]:= `Block[{γ = 0.02 Pi, wms},`
`    wms = WeierstrassMinimalSurface[`
`            I Exp[-w + I Pi w/(2 Cot[γ/2])], Exp[w], w, r Exp[I φ]];`
`    ParametricPlot3D[`
`            Evaluate[Append[wms, SurfaceColor[Hue[φ/(2 Pi)]]]],`
`                    {r, 0, 6}, {φ, 0, 2Pi},`
`                    PlotPoints -> {40, 160}, Boxed -> False,`
`                    Axes -> False, PlotRange -> All,`
`                    BoxRatios -> {1, 1, 2}]]`

We could plot many other such (generally unnamed) surfaces, for example, $f(\xi) = \xi^{1/4} + \xi^{1/3}$ and $g(\xi) = \xi$.

In[153]:= `wms = WeierstrassMinimalSurface[ξ^(1/4) + ξ^(1/3), ξ, ξ, r Exp[I φ]]`

Out[153]= $\left\{ -\frac{1}{260} \text{Re}\left[(e^{i\varphi} r)^{5/4} \left(-208 + 80 \, e^{2 i \varphi} r^2 - 195 \, (e^{i\varphi} r)^{1/12} + 78 \, (e^{i\varphi} r)^{25/12} \right) \right], \right.$

$-\frac{1}{260} \text{Im}\left[(e^{i\varphi} r)^{5/4} \left(208 + 80 \, e^{2 i \varphi} r^2 + 195 \, (e^{i\varphi} r)^{1/12} + 78 \, (e^{i\varphi} r)^{25/12} \right) \right],$

$\left. 2 \text{Re}\left[\frac{4}{9} (e^{i\varphi} r)^{9/4} + \frac{3}{7} (e^{i\varphi} r)^{7/3} \right] \right\}$

An initial attempt to plot this function does not produce a satisfactory result. (We use lines rather than polygons in the following graphic because the polygons touch each other often, and rendering the corresponding graphic takes a long time.)

In[154]:= `Show[Graphics3D[`
`    ParametricPlot3D[Evaluate[%], {r, 0.7, 0.75}, {φ, 0.001, 12Pi - 0.001},`
`                    PlotPoints -> {2, 600}, PlotRange -> All,`
`                    Axes -> False, DisplayFunction -> Identity][[1]] //.`
`            Polygon[l_] :> Line[Append[l, First[l]]]]]`

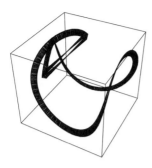

To get the "correct" function values for multivalued functions, we have to modify the results of the indefinite integration; in this case, we take the appropriate *n*th root. (If we would calculate the integrals through numerically solving a differential equation, we do not encounter such branch cut problems.) For ease of understanding, we view only a small strip.

```
In[155]:= ParametricPlot3D[Evaluate[(* analytically continue and add color *)
                  Append[wms /. {(r Exp[I φ])^n_ -> r^n Exp[I n φ]},
                  SurfaceColor[Hue[φ/(12 Pi)], Hue[φ/(12 Pi)], 2]]],
                  {r, 0.7, 0.75}, {φ, 0.01, 12Pi - 0.01},
                  PlotPoints -> {2, 600}, PlotRange -> All, Axes -> False]
```

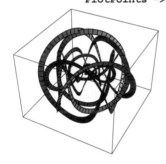

Many further examples of minimal surfaces exist and are easy to (re)produce in *Mathematica*. By changing the integrands in the three integrals in the Weierstrass representation from *integrand* to exp($i\vartheta$) *integrand*, we can (in dependence on ϑ) look at how a minimal surface evolves to its adjoint surface.

For additional examples of minimal surfaces, see [1586], [1164], [1344], [865], [481], [1228], [649], [1178], [1707], [863], [1848], [484], [864], [986], [642], [780], [874], [1698], [1423], [237], [1412], [1708], [985], [1513], [279], [1312], and [1709].

Remark: It is not necessary to use integrals when constructing minimal surfaces. If the above $g(\xi) \to \xi$ and $f(\xi) \to f'''(\xi)$, we can write

```
In[156]:= WeierstrassMinimalSurface[f'''[ξ], ξ, ξ, x] // TraditionalForm
```
Out[156]//TraditionalForm=

$$\{\mathrm{Re}(-f''(x)\,x^2 + 2\,f'(x)\,x - 2\,f(x) + f''(x)),$$
$$-2\,\mathrm{Im}(f(x)) + 2\,\mathrm{Im}(x\,f'(x)) - \mathrm{Im}(f''(x)) - \mathrm{Im}(x^2\,f''(x)), 2\,\mathrm{Re}(x\,f''(x) - f'(x))\}$$

As another application, let us deal with an infinite square grid electrical network. The resistance between the node {0, 0} and the node {m, n} in an infinite network is given by the following expression [645], [1896], [1897], [98], [1357] (each resistor has the value 1 Ohm).

In[157]:= `R[m_, n_] := R[m, n] = Integrate[(1 - ((t - I)/(t + I))^(m + n)*`
`((t - 1)/(t + 1))^Abs[m - n])/t,`
`{t, 0, Infinity}]/(2 Pi)`

It is a well-known result that the resistance between two neighboring nodes is 1/2.

In[158]:= `R[0, 1]`

Out[158]= $\dfrac{1}{2}$

Here are some more values of the resistance. The following second triple of values is equal because of symmetry.

In[159]:= `{{R[1, 1], R[1, 2], R[2, 4]}, {R[-1, -1], R[-1, 2], R[2, -4]}}`

Out[159]= $\left\{\left\{\dfrac{2}{\pi},\ \dfrac{8-\pi}{2\,\pi},\ \dfrac{-\frac{472}{15}+12\,\pi}{2\,\pi}\right\},\ \left\{\dfrac{2}{\pi},\ \dfrac{8-\pi}{2\,\pi},\ \dfrac{-\frac{472}{15}+12\,\pi}{2\,\pi}\right\}\right\}$

To avoid unnecessary calculations, we implement the symmetry of the underlying network and calculate the integral only once for every node distance.

In[160]:= `r /: DownValues[r] = (* order of the definitions matters *)`
`{HoldPattern[r[m_, n_]] :> (r[m, n] = R[m, n]) /;`
`NonNegative[m] && NonNegative[n] && m <= n,`
`HoldPattern[r[m_, n_]] :> r[n, m] /; NonNegative[m] && NonNegative[n],`
`HoldPattern[r[m_, n_]] :> r[-n, -m] /; Negative[m] && Negative[n],`
`HoldPattern[r[m_, n_]] :> r[n, -m] /; Negative[m],`
`HoldPattern[r[m_, n_]] :> r[-n, m] /; Negative[n]};`

Here is the resistance in the neighborhood of the origin.

In[161]:= `With[{n = 5},`
`ListPlot3D[Table[r[i, j], {i, -n, n}, {j, -n, n}],`
`MeshRange -> {{-n, n}, {-n, n}}, PlotRange -> All]];`

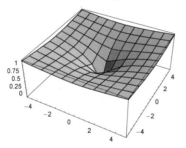

The function R makes heavy use of definite integration. For larger values of *n* and *m*, it becomes somewhat slow.

In[162]:= `{R[10, 10] // Timing, R[8, 12] // Timing}`

Out[162]= $\left\{\left\{0.43\,\text{Second},\ \dfrac{62075752}{14549535\,\pi}\right\},\ \left\{45.89\,\text{Second},\ \dfrac{-\frac{486215980256}{14549535}+10640\,\pi}{2\,\pi}\right\}\right\}$

Indefinite integration is often much faster than definite integration. As a result, it is sometimes advantageous to first calculate the indefinite integral and then substitute the integration limits. (Sometimes this might require the "manual" calculation of limits). For this procedure to be correct one must of course know that, inside the integration interval, the indefinite integral is a continuous function without any singularities. For the integrands under consideration this is actually the case, and we use the function Limit (to be discussed in the next subsection) to obtain the values at the integration end points. We also have to take care about contributions from branch cuts of the integral to make sure we use a continuous antiderivative.

```
In[163]:= RFast[m_, n_] :=
        Module[{upperLimitContribution, lowerLimitContribution,
                branchCutCorrection},
                (* the indefinite integral *)
                indefInt = Integrate[(1 - ((t - I)/(t + I))^(m + n)*
                                      ((t - 1)/(t + 1))^Abs[m - n])/t, t];
                (* contributions from the integration limits *)
                upperLimitContribution = Limit[indefInt, t -> Infinity];
                lowerLimitContribution = Limit[indefInt, t -> 0];
                (* contribution from making a continuous antiderivative *)
                branchCutCorrection = If[MemberQ[int, ArcTan[(1 + t)/(-1 + t)],
                                         Infinity], 2Pi, 0];
                (* simplify result *) Together @ ComplexExpand @ Re
                (upperLimitContribution + branchCutCorrection -
                                  lowerLimitContribution)/(2Pi)]
```

```
In[164]:= {RFast[10, 10] // Timing, RFast[8, 12] // Timing}
```

$$Out[164]= \left\{\left\{0.1 \text{ Second}, \frac{62075752 \text{ Re}}{14549535 \pi}\right\}, \left\{4.93 \text{ Second}, \frac{(-\frac{486215980256}{14549535} + 10640 \pi) \text{ Re}}{2\pi}\right\}\right\}$$

For the *n*-dimensional case of such resistor networks, see [419], [420], [1368], [930], [94]; for the continuous analog, see [1002], [1013]; for finite lattices, see [1857].

Mathematica can differentiate expressions arising from computations in which it is not able to explicitly integrate (meaning these expressions contain unevaluated integrals).

```
In[165]:= D[Integrate[f[x], y], y]
```

```
Out[165]= f[x]
```

This also works for integrals in which the variable of differentiation enters in a complicated way in the limits of integration (differentiation of parametric integrals).

```
In[166]:= Clear[f, x, y];
        D[Integrate[f[x], {x, 0, y}], y]
```

```
Out[167]= f[y]
```

```
In[168]:= D[Integrate[f[x], {x, -x, x}], x]
```

```
Out[168]= f[-x] + f[x]
```

```
In[169]:= Derivative[1, 0][Integrate[f[x], {x, #1, #2}]&][a, b]
```

```
Out[169]= -f[a]
```

```
In[170]:= Derivative[0, 1][Integrate[f[x], {x, #1, #2}]&][a, b]
```

```
Out[170]= f[b]
```

We now look at a somewhat more complicated expression: the d'Alembert solution of the one-dimensional wave equation.

Mathematical Remark: d'Alembert Solution of the One-Dimensional Wave Equation

Suppose we are given the following differential equation (wave equation)

$$\frac{\partial^2 u(x, t)}{\partial t^2} - a^2 \frac{\partial^2 u(x, t)}{\partial x^2} = f(x, t)$$

in $\mathbb{R}^1 \times \mathbb{R}^{1+}$. Here, $u(x, t)$ is the amplitude of the wave as a function of position x and time t, and a is the inverse phase velocity. The d'Alembert solution for prescribed $f(x, t)$ is:

$$u(x, t) = \frac{1}{2a} \int_0^t \int_{x-a(t-\tau)}^{x+a(t-\tau)} f(\xi, \tau)\, d\xi\, d\tau + \frac{1}{2a} \int_{x-at}^{x+at} u_1(\xi)\, dx + \frac{1}{2}\left(u_0(x+at) + u_0(x-at)\right).$$

Here, $u_0(x)$ is the initial position, and $u_1(x)$ is the initial velocity function; that is, $u(x, t = 0) = u_0(x)$ and $\partial u(x, t)/\partial t|_{t=0} = u_1(x)$.

For references, see any textbook on partial differential equations, for example, [1627] and [1047]. For some direct extensions, see [413] , [848], and [1853].

We now check this solution. The initial conditions are fulfilled.

```
In[171]:= u[x_, t_] =
          1/(2 a) Integrate[Integrate[f[ξ, τ], {ξ, x - a (t - τ), x + a (t - τ)}],
                 {τ, 0, t}] +
          1/(2 a) Integrate[u1[ξ], {ξ, x - a t, x + a t}] +
          1/2 (u0[x + a t] + u0[x - a t]);
```

```
In[172]:= {u[x, 0], D[u[x, t], t] /. t -> 0}
```

```
Out[172]= {u0[x], u1[x]}
```

A direct check of the differential equation fails.

```
In[173]:= D[u[x, t], {t, 2}] - a^2 D[u[x, t], {x, 2}] // Simplify
```

$$Out[173]= f[x, t] - \frac{1}{2}\, a \int_0^t (f^{(1,0)}[x + a\,(t - \tau), \tau] - f^{(1,0)}[-a\,t + x + a\,\tau, \tau])\, d\tau +$$

$$\frac{\int_0^t a^2\, (f^{(1,0)}[x + a\,(t - \tau), \tau] - f^{(1,0)}[-a\,t + x + a\,\tau, \tau])\, d\tau}{2\,a}$$

A quick glimpse at the result shows that the reason for the noncancellation is the presence of the constant a^2 under the second integral. So, we use a simplification function that pulls out constants.

```
In[174]:= (D[u[x, t], {t, 2}] - a^2 D[u[x, t], {x, 2}] //
          Simplify[#, (* pull out constants *) TransformationFunctions ->
             {Automatic, (# //. HoldPattern[Integrate[c_?(FreeQ[#, τ]&) r_, i_]] :>
                            c Integrate[r, i])&}]&)
```

```
Out[174]= f[x, t]
```

Here is a solution for the Schrödinger equation for a particle of mass $m(t)$ in a time-dependent linear potential $V(t)$ [627].

```
In[175]:= ψ[{x_, t_}, {B_, m_, V_}] =
          AiryAi[B (x + Integrate[1/m[τ] Integrate[V[σ], {σ, 0, τ}], {τ, 0, t}] -
             B^3/4 Integrate[1/m[τ], {τ, 0, t}]^2)]*
          Exp[I (B^3/2 Integrate[1/m[τ], {τ, 0, t}]*
             (x + Integrate[1/m[τ] Integrate[V[σ], {σ, 0, τ}], {τ, 0, t}] -
             B^3/6 Integrate[1/m[τ], {τ, 0, t}]^2) -
             1/2 Integrate[1/m[τ] Integrate[V[σ], {σ, 0, τ}], {τ, 0, t}]^2, {τ, 0, t}] -
             x Integrate[V[σ], {σ, 0, t}])]
```

$$Out[175]= e^{\mathrm{i}\left(\frac{1}{2} B^3 \left(\int_0^t \frac{1}{m[\tau]}\, d\tau\right)\left(x - \frac{1}{6} B^3 \left(\int_0^t \frac{1}{m[\tau]}\, d\tau\right)^2 + \int_0^t \frac{\int_0^\tau V[\sigma]\, d\sigma}{m[\tau]}\, d\tau\right) - \frac{1}{2}\int_0^t \frac{\left(\int_0^\tau V[\sigma]\, d\sigma\right)^2}{m[\tau]}\, d\tau - x \int_0^t V[\sigma]\, d\sigma\right)}$$

$$\text{AiryAi}\left[B \left(x - \frac{1}{4} B^3 \left(\int_0^t \frac{1}{m[\tau]}\, d\tau\right)^2 + \int_0^t \frac{\int_0^\tau V[\sigma]\, d\sigma}{m[\tau]}\, d\tau\right)\right]$$

The solution contains again unevaluated integrals and the Airy function Ai(z). We can verify that it is indeed a solution for any $m(t)$ and $\mathcal{V}(t)$.

```
In[176]:= With[{ψ = ψ[{x, t}, {B, m, V}]},
            I D[ψ, t] == -1/(2 m[t]) D[ψ, x, x] + V[t] x ψ]   // Simplify
Out[176]= True
```

As a related example, let us develop a series solution of the differential equation $z'(t) = f(z(t), t)$ for small t. We rewrite the differential equation as an integral equation $z(t) = z(0) + \int_0^t f(z(\tau), \tau)\, d\tau$ and calculate the series expansion of the right-hand side.

```
In[177]:= Together /@ Normal[Series[z[0] + Integrate[f[z[τ], τ], {τ, 0, t}],
                            {t, 0, 4}, Analytic -> True]] //.
           (* replace derivatives of z using the differential equations *)
              {Derivative[n_][z][0] :> (D[f[z[t], t], {t, n - 1}] /. t -> 0)}
```

$$\text{Out[177]= } t\, f[z[0], 0] + z[0] + \frac{1}{2} t^2\, (f^{(0,1)}[z[0], 0] + f[z[0], 0]\, f^{(1,0)}[z[0], 0]) +$$
$$\frac{1}{6} t^3\, (f^{(0,2)}[z[0], 0] + f^{(1,0)}[z[0], 0]\, (f^{(0,1)}[z[0], 0] + f[z[0], 0]\, f^{(1,0)}[z[0], 0]) +$$
$$2\, f[z[0], 0]\, f^{(1,1)}[z[0], 0] + f[z[0], 0]^2\, f^{(2,0)}[z[0], 0]) +$$
$$\frac{1}{24} t^4\, (f^{(0,3)}[z[0], 0] + 3\, (f^{(0,1)}[z[0], 0] + f[z[0], 0]\, f^{(1,0)}[z[0], 0])\, f^{(1,1)}[z[0], 0] +$$
$$3\, f[z[0], 0]\, f^{(1,2)}[z[0], 0] +$$
$$3\, f[z[0], 0]\, (f^{(0,1)}[z[0], 0] + f[z[0], 0]\, f^{(1,0)}[z[0], 0])\, f^{(2,0)}[z[0], 0] +$$
$$f^{(1,0)}[z[0], 0]\, (f^{(0,2)}[z[0], 0] + f^{(1,0)}[z[0], 0]$$
$$(f^{(0,1)}[z[0], 0] + f[z[0], 0]\, f^{(1,0)}[z[0], 0]) + f[z[0], 0]\, f^{(1,1)}[z[0], 0] +$$
$$f[z[0], 0]\, (f^{(1,1)}[z[0], 0] + f[z[0], 0]\, f^{(2,0)}[z[0], 0])) +$$
$$3\, f[z[0], 0]^2\, f^{(2,1)}[z[0], 0] + f[z[0], 0]^3\, f^{(3,0)}[z[0], 0])$$

Using $f(z, t) = z$, we get the series expansion of $\exp(t)$.

```
In[178]:= % /. f -> (#1&)
```

$$\text{Out[178]= } z[0] + t\, z[0] + \frac{1}{2} t^2\, z[0] + \frac{1}{6} t^3\, z[0] + \frac{1}{24} t^4\, z[0]$$

And using $f(z, t) = 2\,t\,z$, we get the series expansion of $\exp(t^2)$.

```
In[179]:= %% /. f -> (2 #1 #2&)
```

$$\text{Out[179]= } z[0] + t^2\, z[0] + \frac{1}{2} t^4\, z[0]$$

At this point, we mention that *Mathematica* can integrate a large class of functions whose antiderivatives can be expressed as elliptic integrals. Typically, such integrands contain roots of polynomials of third or fourth degree. Here are three examples.

```
In[180]:= Integrate[Sqrt[(b^2 - x^2)/(x^2 + a^2)], x]
```

$$\text{Out[180]= } \frac{\sqrt{\frac{b^2 - x^2}{a^2 + x^2}}\, \sqrt{1 + \frac{x^2}{a^2}}\, \text{EllipticE}\left[\text{ArcSin}\left[\sqrt{-\frac{1}{a^2}}\, x\right], -\frac{a^2}{b^2}\right]}{\sqrt{-\frac{1}{a^2}}\, \sqrt{1 - \frac{x^2}{b^2}}}$$

```
In[181]:= Integrate[Sqrt[(b^2 - x^2)/(x^2 + a^2)^3], x]
```

Out[181]= $\sqrt{\dfrac{b^2 - x^2}{(a^2 + x^2)^3}} \ (a^2 + x^2)$

$\left(\dfrac{x}{a^2} - \dfrac{1}{\sqrt{-\frac{1}{b^2}} \ (b^2 - x^2)} \left(i \ \sqrt{1 + \dfrac{x^2}{a^2}} \ \sqrt{1 - \dfrac{x^2}{b^2}} \left(\text{EllipticE}\left[i \ \text{ArcSinh}\left[\sqrt{-\dfrac{1}{b^2}} \ x \right], \ -\dfrac{b^2}{a^2} \right] - \right. \right. \right.$

$\left. \left. \left. \text{EllipticF}\left[i \ \text{ArcSinh}\left[\sqrt{-\dfrac{1}{b^2}} \ x \right], \ -\dfrac{b^2}{a^2} \right] \right) \right) \right)$

In[182]:= `Integrate[1/Sqrt[1 - x^3], x]`

Out[182]= $\dfrac{2 \ i \ \sqrt{(-1)^{5/6}} \ (-1 + x) \ \sqrt{1 + x + x^2} \ \text{EllipticF}\left[\text{ArcSin}\left[\frac{\sqrt{-(-1)^{5/6} - i \ x}}{3^{1/4}} \right], \ (-1)^{1/3} \right]}{3^{1/4} \ \sqrt{1 - x^3}}$

Note that sometimes *Mathematica* produces an incorrect result for a definite integral. Such cases usually involve integrands with symbolic parameters and branches. One possibility for checking the correctness of integrals is to compare the result of `Integrate` with that of `NIntegrate`. Here is an example: $\int_{1/10-i}^{1/10+i} \ln(z^2 - 1) \, dz$.

The integrand has a branch cut between -1 and 1. Here, the results of `Integrate` and `NIntegrate` do agree.

In[183]:= `Integrate[Log[z^2 - 1], {z, 1/10 - I, 1/10 + I}]`

Out[183]= $-\dfrac{1}{5} \ i \left(\text{ArcTan}\left[\dfrac{20}{199} \right] - 5 \left(-4 + \pi + \text{ArcTan}\left[\dfrac{400}{39999} \right] + \text{Log}\left[\dfrac{40001}{10000} \right] \right) \right)$

Here is an example where the two results do not agree. For generic endpoints of a definite integral, *Mathematica* must carry out the definite integral by first calculating the indefinite integral. Then it must find out if the straight line connecting the integration end points crosses any branch cuts of the antiderivative. In general, this means solving a transcendental equation and finding all relevant solutions. This is a very complicated step, and missing a crossed branch cut causes a different result from the one returned by `NIntegrate`.

In[184]:= `{N[Integrate[#, {z, -1 - I, -1 + I}]],`
`       NIntegrate[#, {z, -1 - I, -1 + I}]}&[`
`                   (1 + z^z (1 + Log[z]))/(z + z^z)] // Chop`

Out[184]= $\{4.80293 \ i, -1.48025 \ i\}$

1.6.3 Limits

Another important task of classical analysis that can be accomplished with *Mathematica* is the computation of limits. The appropriate command is `Limit`.

`Limit[`*function*`, `*var*` -> `*specificValue*`, `*options*`]`

finds the limit of *function* if *var* $\to$ *specificValue* taking into account the option settings *options*.

Here are four simple examples to start.

In[1]:= `Limit[Sin[x]/x, x -> 0]`

Out[1]= 1

In[2]:= `Limit[Exp[-x] x^2, x -> Infinity]`

Out[2]= 0

In[3]:= `Limit[((x + h)^(1/3) - x^(1/3))/h, h -> 0]`

Out[3]= $\dfrac{1}{3 \ x^{2/3}}$

In[4]:= `Limit[ (Tan[x]/x)^(1/x^2), x -> 0]`

Out[4]= $e^{1/3}$

Here are three slightly more complicated limits, two of the form ∞^0 [921] and one of the form 1^∞.

In[5]:= `Limit[ (1/x)^Tan[x], x -> 0]`

Out[5]= 1

In[6]:= `Limit[ (2 - 2 x)^Tan[Pi x], x -> 1/2]`

Out[6]= $e^{2/\pi}$

In[7]:= `Limit[ (((n - 1)^2 n^n)/(n^n - n))^((n - n^(2 - n))/(n - 1)^2),`
`        n -> Infinity]`

Out[7]= 1

A more complicated limit than contains a binomial coefficient.

In[8]:= `Limit[Binomial[n, k] (a/n)^k (1 - a/n)^(n - k), n -> Infinity]`

Out[8]= $\dfrac{a^k \, e^{-a}}{\text{Gamma}[1 + k]}$

The next limit is ∞.

In[9]:= `Limit[x^x - x^(Log[x]), x -> Infinity]`

Out[9]= ∞

The next limit shows how the logarithm $\ln(x)$ arises as the limit of a power function x^a. (For continuity, it follows from this that x^a and $\ln(x)$ should have the same branch cut structure.)

In[10]:= `Limit[Integrate[ξ^a, {ξ, 0, x}, Assumptions -> x > 0 && Re[a] > -1] -`
`        1/(1 + a), a -> -1]`

Out[10]= $\text{Log}[x]$

For functions whose limit values depend on the direction from which we approach *specificValue*, we can use the option `Direction`.

`Direction`

is an option for `Limit`, and it determines a direction for computing the limit.

Default:
1 (from the left)

Admissible:
−1 (from the right)

or

complexNumber (in direction *complexNumber*)

Here is an example of finding the limit of $\exp(1/x)$ as $x \to 0$. Using the `Direction` option in `Limit`, we can determine both limits.

In[11]:= `Limit[Exp[1/x], x -> 0, Direction -> #]& /@ {1, -1}`

Out[11]= $\{0, \infty\}$

Here, we look at limits of the analogous function on the imaginary axis.

In[12]:= `Limit[Exp[1/(I x)], x -> 0, Direction -> #]& /@ {-I, I}`

Out[12]= {0, ∞}

Here is another limit where the direction matters.

In[13]:= `Limit[(-1 + ε I)^λ, ε -> 0, Direction -> #]& /@ {+1, -1} //`
                                                    `ExpToTrig`

Out[13]= {Cos[π λ] - i Sin[π λ], Cos[π λ] + i Sin[π λ]}

`Limit` has two further options.

In[14]:= `Options[Limit]`

Out[14]= {Analytic → False, Assumptions :→ $Assumptions, Direction → Automatic}

For generic complex α, the following limit stays unevaluated.

In[15]:= `Limit[Exp[α/x], x -> 0, Direction -> #]& /@ {1, -1}`

Out[15]= $\left\{\text{Limit}\left[e^{\frac{\alpha}{x}}, x \to 0, \text{Direction} \to 1\right], \text{Limit}\left[e^{\frac{\alpha}{x}}, x \to 0, \text{Direction} \to -1\right]\right\}$

Under the assumption that the real part of α is positive, the last limit can be found by `Limit`.

In[16]:= `Assuming[Re[α] > 0,`
                      `Limit[Exp[α/x], x -> 0, Direction -> #]& /@ {1, -1}]`

Out[16]= {0, ∞}

Often, it is not possible to get an explicit value because the function does not approach a limit ($\sin(1/x)$ as x approaches 0 is a typical example). In cases in which the solution is bounded in an interval, the result will often be an `Interval`-object (we encountered these cases already in Chapter 1 of the Numerics volume [1737]). An example of such a situation is $\sin(1/x)$ as $x \to 0$.

In[17]:= `Limit[Sin[1/x], x -> 0]`

Out[17]= Interval[{-1, 1}]

The third option of `Limit` is `Analytic`.

`Analytic`

    is an option for `Limit`, and it describes whether the function whose limit value is to be found is analytic (in the sense of function theory).

    Default:

        False

    Admissible:

        True

Here is what happens with different option settings of `Analytic`. We use a function that is not explicitly defined.

In[18]:= `{Limit[f[x], x -> 1], Limit[f[x], x -> 1, Analytic -> True]}`

Out[18]= {Limit[f[x], x → 1], f[1]}

Almost all functions $f(z)$ of a complex variable z are nondifferentiable or even not continuous. As a result, the following limit stays unevaluated.

In[19]:= `Limit[(f[z + ε] - f[z])/ε, ε -> 0]`

Out[19]= $\text{Limit}\left[\frac{-f[z] + f[z + ε]}{ε}, ε \to 0\right]$

Assuming that $f(z)$ is an analytic function yields, as the result, the derivative $f'(z)$.

In[20]:= `Limit[ (f[z + ε] - f[z])/ε, ε -> 0, Analytic -> True]`

Out[20]= `f'[z]`

Here is a slightly more complicated limit.

In[21]:= `Limit[ (f[z + ε + ε^2] + f[z - ε - ε^3/4] - 2 f[z + ε^2/3])/ε^2,`
`        ε -> 0, Analytic -> True]`

Out[21]= $\dfrac{f'[z]}{3} + f''[z]$

Also in the following limit (that gives the Schwarz derivative $w^{(3)}/w' - 3/2\,w''^2/w'^2$) the option setting `Analytic -> True` is needed.

In[22]:= `Limit[6 D[Log[ (w[z] - w[ξ])/(z - ξ)], z, ξ], z -> ξ,`
`           Analytic -> True] // Expand`

Out[22]= $-\dfrac{3\,w''[ξ]^2}{2\,w'[ξ]^2} + \dfrac{w^{(3)}[ξ]}{w'[ξ]}$

The following input reduces also to the Schwarzian derivative [1370], [1371], [1342]. Because `w[ξ]` appears multiplicative in this expression, this time the option setting `Analytic -> True` is not needed.

In[23]:= `Limit[Derivative[3][Function[z, (z - w[z]/w'[z])/2]][ξ], w[ξ] -> 0] // Expand`

Out[23]= $-\dfrac{3\,w''[ξ]^2}{2\,w'[ξ]^2} + \dfrac{w^{(3)}[ξ]}{w'[ξ]}$

The next input represents a discrete approximation to the *n*th derivative (*n* a nonnegative integer) of a function *f* at $ξ$ [1842].

In[24]:= `derivativeApproximation[f_, n_, ξ_, ε_] :=`
`           Sum[ (-1)^k Binomial[n, k] f[ξ + (n - 2k)/2 ε], {k, 0, n}]/ε^n`

In the limit $ε \to \infty$, we get the explicit derivative for explicit nonnegative integer *n*.

In[25]:= `Table[Limit[derivativeApproximation[f, n, x0, ε],`
`             Analytic -> True], {n, 0, 6}]`

Out[25]= `{f[x0], f'[x0], f''[x0], f^(3)[x0], f^(4)[x0], f^(5)[x0], f^(6)[x0]}`

A generalized limit for expressing Euler's constant *e* [423].

In[26]:= `expr = ((n + 3)^(n + 3) - (n + 2)^(n + 2))/`
`               ((n + 2)^(n + 2) - (n + 1)^(n + 1)) -`
`               ((n + 2)^(n + 2) - (n + 1)^(n + 1))/`
`               ((n + 1)^(n + 1) - (n + 0)^(n + 0));`

`        Limit[expr, n -> Infinity]`

Out[28]= `e`

Subtracting the value of the limit allows finding the next terms as a correction term for large, but finite *n*.

In[29]:= `(* coefficient of 1/n term vanishes *)`
`        Limit[ (expr - E) n, n -> Infinity]`

Out[30]= `0`

In[31]:= `(* coefficient of 1/n^2 term is finite *)`
`        Limit[ (expr - E) n^2, n -> Infinity]`

Out[32]= $-1 + \dfrac{e}{24}$

In[33]:= `(* coefficient of 1/n^3 term is finite *)`
`        Limit[ (expr - E  - (* last term *) (E/24 - 1)/n^2) n^3, n -> Infinity]`

Out[34]= $2 - \dfrac{2}{e} - \dfrac{e}{6}$

`Limit` assumes that its variable approaches the limit point in a continuous manner. This means limits such as the following will stay unevaluated.

In[35]:= `Limit[Nest[Sqrt[5 + #]&, 5, n], n -> Infinity]`

> Nest::intnm : Non-negative machine-size
>   integer expected at position 3 in $Nest[\sqrt{5 + \#1} \ \&, \ 5, \ n]$. More...

Out[35]= $Limit\left[Nest\left[\sqrt{5 + \#1} \ \&, \ 5, \ n\right], \ n \to \infty\right]$

In[36]:= `Limit[Nest[1 + 1/#&, 1, n], n -> Infinity]`

> Nest::intnm : Non-negative machine-size
>   integer expected at position 3 in $Nest\left[1 + \frac{1}{\#1} \ \&, \ 1, \ n\right]$. More...

Out[36]= $Limit\left[Nest\left[1 + \frac{1}{\#1} \ \&, \ 1, \ n\right], \ n \to \infty\right]$

In[37]:= `Limit[Prime[n]/Exp[n], n -> Infinity]`

Out[37]= $Limit[e^{-n} \ Prime[n], \ n \to \infty]$

To compute limits when several variables are simultaneously tending toward given values, we have to apply `Limit` repeatedly. However, constructions of the form `Limit[f(a, b), a->a₀, b->b₀]` are not allowed. Here is a function, with two different limit values, that depends on the order in which `Limit` is applied.

In[38]:= `(* use different variable ordering *)`
   `{Limit[Limit[(x^2 - y^2)/(x^2 + y^2), x -> 0], y -> 0],`
   `Limit[Limit[(x^2 - y^2)/(x^2 + y^2), y -> 0], x -> 0]}`

Out[39]= `{-1, 1}`

The following input does not work (but be aware of the different message).

In[40]:= `Limit[(x^2 - y^2)/(x^2 + y^2), x -> 0, y -> 0]`

> Limit::optx : Unknown option y in $Limit\left[\frac{x^2 - y^2}{x^2 + y^2}, \ x \to 0, \ y \to 0\right]$. More...

Out[40]= $Limit\left[\frac{x^2 - y^2}{x^2 + y^2}, \ x \to 0, \ y \to 0\right]$

To conclude this section, we now present a tiny application of `Limit` concerning the computation of a 2D rotation matrix from infinitesimals [922]: An infinitesimal rotation by an angle φ_ϵ around the z-axis can be described (which is easily seen from the geometry) by $x' = x + \varphi_\epsilon \, y$, $y' = -\varphi_\epsilon \, x' + y$.

Here, x and y are the coordinates of a point before the rotation, and x' and y' are the coordinates after the rotation.

In matrix form, this is

$$\begin{pmatrix} x' \\ y' \end{pmatrix} = \begin{pmatrix} 1 & \varphi_\epsilon \\ -\varphi_\epsilon & 1 \end{pmatrix}\begin{pmatrix} x \\ y \end{pmatrix}.$$

Here, φ_ϵ is the infinitesimal angle of rotation. A finite rotation by an angle φ can be obtained by n-fold repetition of this small rotation, where $n \, \varphi_\epsilon = \varphi$. Here is the limit as $n \to \infty$.

In[41]:= `MatrixPower[{{1, φ/n}, {-φ/n, 1}}, n]`

Out[41]= $\left\{\left\{\frac{1}{2}\left(\frac{n - i \, \varphi}{n}\right)^n + \frac{1}{2}\left(\frac{n + i \, \varphi}{n}\right)^n, \ \frac{1}{2} i \left(\frac{n - i \, \varphi}{n}\right)^n - \frac{1}{2} i \left(\frac{n + i \, \varphi}{n}\right)^n\right\},\right.$
$\left.\left\{-\frac{1}{2} i \left(\frac{n - i \, \varphi}{n}\right)^n + \frac{1}{2} i \left(\frac{n + i \, \varphi}{n}\right)^n, \ \frac{1}{2}\left(\frac{n - i \, \varphi}{n}\right)^n + \frac{1}{2}\left(\frac{n + i \, \varphi}{n}\right)^n\right\}\right\}$

This is what we get after some reorganization.

In[42]:= **ComplexExpand[Map[Limit[#, n -> Infinity]&, %, {2}]] // Simplify**

Out[42]= $\{\{Cos[\varphi], Sin[\varphi]\}, \{-Sin[\varphi], Cos[\varphi]\}\}$

This is the well-known 2D rotation matrix for a given angle φ of rotation.

1.6.4 Series Expansions

Mathematica can compute

- Taylor series　　$(a_0 + a_1 x + a_2 x^2 + \cdots)$
- Laurent series　$(\cdots + a_{-2} x^{-2} + a_{-1} x^{-1} + a_0 x + a_1 x + a_2 x^2 + \cdots)$
- Puiseux series　expansions around branch points where functions behave like radicals or logarithms

(See [1445] and [1044] for more on the not-so-widely-known Puiseux series expansions.)

For all standard functions and compositions of functions, and for a multitude of special functions, the functions to generate a series expansion is `Series`.

Series [*function*, {*var*, *expansionPoint*, *order*}]

　　expands the function *function* around the point *var* = *expansionPoint* in a series up to order *order*.

We now look at some simple examples of series expansions.

In[1]:= **Series[Sin[x], {x, 0, 3}]**

Out[1]= $x - \dfrac{x^3}{6} + O[x]^4$

Here, we saw the remainder term $O[x]$ to represent a term of order n. Let us look at its `FullForm`.

In[2]:= **FullForm[O[x]^3]**

Out[2]//FullForm=

　　　　SeriesData[x, 0, List[], 3, 3, 1]

Written out, the above series has this form.

In[3]:= **Series[Sin[x], {x, 0, 3}] // FullForm**

Out[3]//FullForm=

　　　　SeriesData[x, 0, List[1, 0, Rational[-1, 6]], 1, 4, 1]

The object `SeriesData` that appears in this `FullForm` output has the following meaning.

SeriesData [*var*, *expansionPoint*, {$a_0, a_1, \ldots, a_n$}, *minExp*, *maxExp*, *denominator*]

　　represents a series expansion in the variables *var* around the point *expansionPoint*. The series has the form
　　$a_0 \, var^{minExp/denominator} + a_1 \, var^{(minExp+1)/denominator} + \cdots + a_n \, var^{(maxExp-1)/denominator}$.

The $O[x]$-form results in easier readability and inputting of a series. Using $O[x]^n$ often is a convenient way to indicate a series.

In[4]:= **FullForm[O[z, z0]]**

Out[4]//FullForm=

　　　　SeriesData[z, z0, List[], 1, 1, 1]

O[z, *ord*] evaluated to a SeriesData-object. Parsed, but unevaluated O is just a symbol.

In[5]:= **FullForm[Hold[z + O[z, z0]]]**

Out[5]//FullForm=

    Hold[Plus[z, O[z, z0]]]

In general, the expression O[z, *z0*] evaluates to SeriesData[z, z0, {}, 1, 1, 1]. If O[z, *z0*] is present in another expression, this SeriesData causes (if possible) the whole expression to be converted to a Series: Data. Here is an example.

In[6]:= **Cos[z] + O[z, Pi/2]^4**

Out[6]= $-\left(z - \dfrac{\pi}{2}\right) + \dfrac{1}{6}\left(z - \dfrac{\pi}{2}\right)^3 + O\left[z - \dfrac{\pi}{2}\right]^4$

We now look at a sample of computable series expansions.

■ **Taylor Series [963]**

Here is a rational function.

In[7]:= **Series[(a x^6 + b x^4 + c x^2 + 4)/**
    (g x^23 + 78 h x^12 + 456), {x, 0, 4}]

Out[7]= $\dfrac{1}{114} + \dfrac{c\,x^2}{456} + \dfrac{b\,x^4}{456} + O[x]^5$

We would have obtained the same result by just adding O[x]^5 to the rational expression (which is sometimes preferable because of readability of the code).

In[8]:= **(a x^6 + b x^4 + c x^2 + 4)/(g x^23 + 78 h x^12 + 456) + O[x]^5**

Out[8]= $\dfrac{1}{114} + \dfrac{c\,x^2}{456} + \dfrac{b\,x^4}{456} + O[x]^5$

Here is a Taylor series of a nested trigonometric function.

In[9]:= **Series[Tan[Sin[x]], {x, 0, 5}]**

Out[9]= $x + \dfrac{x^3}{6} - \dfrac{x^5}{40} + O[x]^6$

The next function is a trigonometric function with an algebraic argument.

In[10]:= **Series[Sin[Sqrt[z]]/Sqrt[z], {z, 0, 5}]**

Out[10]= $1 - \dfrac{z}{6} + \dfrac{z^2}{120} - \dfrac{z^3}{5040} + \dfrac{z^4}{362880} - \dfrac{z^5}{39916800} + O[z]^6$

Here is a rational function of a trigonometric function expanded in a series. To rewrite the result more elegantly, we use TrigFactor.

In[11]:= **(* rewrite the coefficients in a short form *)**
    TrigFactor /@ Collect[Normal[Series[
        (a Cos[x] - a^2)/(1 - 2a Cos[x] + a^2), {a, 0, 12}]], a]

Out[12]= a Cos[x] + a³ Cos[x] (−1 + 2 Cos[2 x]) +
    a⁵ Cos[x] (1 − 2 Cos[2 x] + 2 Cos[4 x]) + a⁹ Cos[x] (−1 + 2 Cos[2 x]) (−1 + 2 Cos[6 x]) +
    a⁷ Cos[x] (−1 + 2 Cos[2 x] − 2 Cos[4 x] + 2 Cos[6 x]) +
    a¹¹ Cos[x] (−1 + 2 Cos[2 x] − 2 Cos[4 x] + 2 Cos[6 x] − 2 Cos[8 x] + 2 Cos[10 x]) +
    a² (Cos[x] − Sin[x]) (Cos[x] + Sin[x]) + a⁴ (Cos[2 x] − Sin[2 x]) (Cos[2 x] + Sin[2 x]) −
    a⁶ (Cos[x] − Sin[x]) (Cos[x] + Sin[x]) (−1 + 2 Sin[2 x]) (1 + 2 Sin[2 x]) +
    a¹⁰ (Cos[x] − Sin[x]) (Cos[x] + Sin[x]) (−1 + 2 Cos[4 x] − 2 Sin[2 x])
    (−1 + 2 Cos[4 x] + 2 Sin[2 x]) + a⁸ (Cos[4 x] − Sin[4 x]) (Cos[4 x] + Sin[4 x]) −
    a¹² (Cos[2 x] − Sin[2 x]) (Cos[2 x] + Sin[2 x]) (−1 + 2 Sin[4 x]) (1 + 2 Sin[4 x])

Here is the series expansion of the sin iteration discussed graphically in Chapter 1 of the Graphics volume [1736].

```
In[13]:= Series[Nest[Sin, x, #], {x, 0, 5}]& /@ {1, 10, 100}
```

$$Out[13]= \left\{x - \frac{x^3}{6} + \frac{x^5}{120} + O[x]^6, \ x - \frac{5\,x^3}{3} + \frac{23\,x^5}{6} + O[x]^6, \ x - \frac{50\,x^3}{3} + \frac{1240\,x^5}{3} + O[x]^6\right\}$$

For a detailed analytical treatment of this example, see [453].

If the coefficients of a series are integers or rational numbers, it is frequently possible to calculate hundreds of terms in seconds. Here is an example.

```
In[14]:= Series[Exp[Tan[Sin[z]]], {z, 0, 100}] // Timing // Short[#, 6]&
```

Out[14]//Short=

$$\left\{3.96\,\text{Second},\ 1 + z + \frac{z^2}{2} + \ll145\gg +\right.$$

(11506912858774915055781664793422818662710748249729608435979487881910954354) ⋮
4282757730243546981255810241323201358626103391192121532266213z^{100}) /
9332621544394415268169923885626670049071596826438162146859296389521759999932 ⋮
2991560894146397615651828625369792082722375825118521091686400000000000000000 ⋮
000000000 + O[z]^{101}}

Here is a slightly more complicated example: The following continued fraction due to Ramanujan has the fascinating property that as $n \to \infty$, it approaches three different limit points [1459], [62], [224], [225].

$$1 + \cfrac{1}{-(1+q) + \cfrac{1}{+(1+q^2) + \ddots}}$$

$$+ \cfrac{1}{(-1)^{k-1}\left(1+q^{k-1}\right)+ \cfrac{1}{(-1)^k\left(1+q^k+\alpha\right)}}$$

Forming the explicit fraction for $n = 20, \ldots, 25$ and looking at the series terms for small q clearly shows how the three limit points are approached cyclically.

```
In[15]:= Table[{n, Series[FromContinuedFraction[
            Table[(-1)^k (1 + q^k + Which[k == 0, -1, k == n, α, True, 0]),
              {k, 1, n}]], {q, 0, 1}]}, {n, 20, 25}] // (Cancel //@ #)&
```

$$Out[15]= \left\{\left\{20,\ -\frac{\alpha}{1+\alpha} + \left(\frac{1}{-1-\alpha} - \frac{\alpha}{1+\alpha}\right)q + O[q]^2\right\},\ \left\{21,\ \frac{1}{\alpha} - q + O[q]^2\right\},\right.$$
$$\left\{22,\ (-1-\alpha) - q + O[q]^2\right\},\ \left\{23,\ -\frac{\alpha}{1+\alpha} + \left(\frac{1}{-1-\alpha} - \frac{\alpha}{1+\alpha}\right)q + O[q]^2\right\},$$
$$\left\{24,\ \frac{1}{\alpha} - q + O[q]^2\right\},\ \left\{25,\ (-1-\alpha) - q + O[q]^2\right\}\right\}$$

In the next input, we use `Series` to write the kth positive power of $\cos(x)$, $-\pi/2 \le x \le \pi/2$ as a decorated Gaussian.

```
In[16]:= expK[k_, x_] = With[{o = 20},
                    Exp[Expand[Normal[Log[Series[Cos[x], {x, 0, o}]^k]]]]]
```

$$Out[16]=\ e^{-\frac{k\,x^2}{2} - \frac{k\,x^4}{12} - \frac{k\,x^6}{45} - \frac{17\,k\,x^8}{2520} - \frac{31\,k\,x^{10}}{14175} - \frac{691\,k\,x^{12}}{935550} - \frac{10922\,k\,x^{14}}{42567525} - \frac{929569\,k\,x^{16}}{10216206000} - \frac{3202291\,k\,x^{18}}{97692469875} - \frac{221930581\,k\,x^{20}}{18561569276250}}$$

The first eleven nonvanishing terms of $\cos(x)$ are faithfully reproduced by `expK[1, x]`.

```
In[17]:= Cos[x] - expK[1, x] + O[x]^24
```

Out[17]= $-\dfrac{9444233042\, x^{22}}{2143861251406875} + O[x]^{24}$

The next graphics show the first powers of cos(x) (which asymptotically approach a Gaussian), the difference to the asymptotic Gaussian, and the logarithm of the absolute value of the difference to expK[k, x].

```
δcosExpK[k_, x_?NumberQ] :=
        With[{ξ = SetPrecision[x, Infinity], κ = Round[k]},
            N[Log[10, Abs[Cos[ξ]^ κ - expK[κ, ξ]]], 12]]
```

```
With[{o = 24},
Show[GraphicsArray[
  Block[{$DisplayFunction = Identity},
  (Plot[Evaluate[Table[#1, {k, o}]], {x, -Pi/2, Pi/2},
        PlotStyle  -> Table[{Thickness[0.002], Hue[0.8 k/o]}, {k, o}],
        PlotRange -> #2,  Frame -> True, Axes -> False,
        DisplayFunction -> Identity] //  Internal`DeactivateMessages) & @@@
      (* power and difference of powers to Gaussian and decorated  Gaussian *)
        {{Cos[x]^k, All}, {Cos[x]^k - Exp[-k/2 x^2], All},
        { δcosExpK[k, x], {0, -50}}}]]]]
```

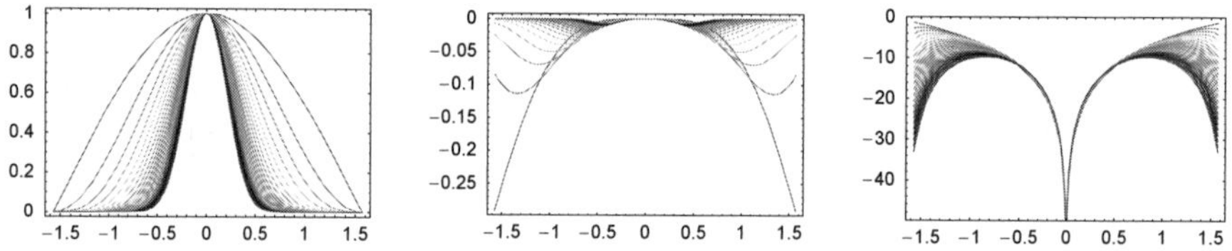

■ **Laurent Series**

Now, we have terms with negative powers of *x*. (Within *Mathematica*, it is a series with positive powers of $1/x$.)

In[21]:= `Series[1/(x^2 + a^2), {x, Infinity, 3}]`

Out[21]= $\left(\dfrac{1}{x}\right)^2 + O\left[\dfrac{1}{x}\right]^4$

In[22]:= `Series[Sin[x]^-1, {x, 0, 4}]`

Out[22]= $\dfrac{1}{x} + \dfrac{x}{6} + \dfrac{7\,x^3}{360} + O[x]^5$

Note the O[x] terms in the following two examples.

In[23]:= `Series[x^-6, {x, 0, 4}]`

Out[23]= $\dfrac{1}{x^6} + O[x]^5$

In[24]:= `Series[(1/Sin[x])^4, {x, 0, 4}]`

Out[24]= $\dfrac{1}{x^4} + \dfrac{2}{3\,x^2} + \dfrac{11}{45} + \dfrac{62\,x^2}{945} + \dfrac{41\,x^4}{2835} + O[x]^5$

The next series has no nonvanishing terms up to order x^4. And the result returned by Series indicates that the first nonvanishing coefficient might appear earliest at order x^{10}.

In[25]:= `Series[(x^2 + 3)/(x^12 - 17), {x, Infinity, 4}]`

Out[25]= $O\left[\dfrac{1}{x}\right]^{10}$

To get a nontrivial term for the last series, we must calculate more terms.

In[26]:= `Series[(x^2 + 3)/(x^12 - 17), {x, Infinity, 12}]`

Out[26]= $\left(\dfrac{1}{x}\right)^{10} + 3\left(\dfrac{1}{x}\right)^{12} + O\left[\dfrac{1}{x}\right]^{13}$

In case we have a series with many negative power terms and are only interested in the leading terms, we can use a negative value for *order*.

In[27]:= **Series[(1/Sin[x])^1000, {x, 0, -995}]**

Out[27]= $\dfrac{1}{x^{1000}} + \dfrac{500}{3\,x^{998}} + \dfrac{125050}{9\,x^{996}} + \dfrac{1}{O[x]^{994}}$

The trigonometric functions csc(z) and cot(z) have Laurent expansions around $z = 0$. The next input shows that the function Series is effectively behaving like a listable function (because its second argument is a list, Series cannot carry the Listable attribute).

In[28]:= **Series[{Csc[z], Cot[z]}, {z, 0, 3}]**

Out[28]= $\left\{ \dfrac{1}{z} + \dfrac{z}{6} + \dfrac{7\,z^3}{360} + O[z]^4, \dfrac{1}{z} - \dfrac{z}{3} - \dfrac{z^3}{45} + O[z]^4 \right\}$

Here is a series of a special function (to be discussed in Chapter 3). We use an approximate expansion point to force the numericalization of the resulting coefficients.

In[29]:= **Series[Gamma[z], {z, 1/2., 8}]**

Out[29]= $1.77245 - 3.48023\,(z - 0.5) + 7.79009\,(z - 0.5)^2 -$
$15.7948\,(z - 0.5)^3 + 31.8788\,(z - 0.5)^4 - 63.9127\,(z - 0.5)^5 +$
$127.943\,(z - 0.5)^6 - 255.961\,(z - 0.5)^7 + 511.974\,(z - 0.5)^8 + O[z - 0.5]^9$

Here are two series expansions for expressions that tend to e.

In[30]:= **(* expand one time at zero and one time at infinity *)**
    {Series[(1 + 1/n)^n, {n, Infinity, 2}],
     Series[(1 + n)^(1/n), {n, 0, 2}]}

Out[31]= $\left\{ e - \dfrac{e}{2\,n} + \dfrac{11}{24}\,e\left(\dfrac{1}{n}\right)^2 + O\left[\dfrac{1}{n}\right]^3, e - \dfrac{e\,n}{2} + \dfrac{11\,e\,n^2}{24} + O[n]^3 \right\}$

■ **Puiseux Series**

The expression $\sqrt{z}$ is an independent term in a Puiseux series. The $O[x]^{13/2}$ term arises from the order 6 of the series requested and the fact that the nonvanishing terms have fractional exponents with denominator 2.

In[32]:= **Series[Sqrt[x], {x, 0, 6}]**

Out[32]= $\sqrt{x} + O[x]^{13/2}$

The next series can be expressed in powers of $x^{1/2}$. The last argument of the SeriesData-object is 2, meaning that the increments in the powers of the expansion variable are $1/2$.

In[33]:= **Series[1 x^(1/2) + 3 x^(3/2) + 5 x^(5/2), {x, 0, 6}]**

Out[33]= $\sqrt{x} + 3\,x^{3/2} + 5\,x^{5/2} + O[x]^{13/2}$

In[34]:= **InputForm[%]**

Out[34]//InputForm=
        SeriesData[x, 0, {1, 0, 3, 0, 5}, 1, 13, 2]

Similarly, the *O*-term in the following has the value $7 \times (1/7) + 1/7 = 50/7$.

In[35]:= **Series[x^(1/7), {x, 0, 7}]**

Out[35]= $x^{1/7} + O[x]^{50/7}$

For large denominators, the third argument of the underlying SeriesData-object can become a long list.

In[36]:= **Series[x^(1/2000) + x^2, {x, 0, 2}][[3]] // Length**

Out[36]= 4000

The next two series expansions contain logarithms.

In[37]:= **Series[x^x, {x, 0, 4}]**

Out[37]= $1 + \text{Log}[x] \, x + \frac{1}{2} \text{Log}[x]^2 x^2 + \frac{1}{6} \text{Log}[x]^3 x^3 + \frac{1}{24} \text{Log}[x]^4 x^4 + O[x]^5$

In[38]:= **Series[x^(x^2), {x, 0, 3}]**

Out[38]= $1 + \text{Log}[x] \, x^2 + O[x]^4$

The last example contained a term of the form $\ln(x)\, x^2$. Logarithmic factors appear in the third argument of the underlying SeriesData-object.

In[39]:= **FullForm[%]**

Out[39]//FullForm=

     SeriesData[x, 0, List[1, 0, Log[x]], 0, 4, 1]

The function arcsin(z) has three branch points: two square-root–like branch points at ± 1 and a logarithmic branch point at ∞. Looking at the series expansion of ArcSin, these two different types of branch points are clearly visible.

In[40]:= **Series[ArcSin[z], {z, Infinity, 3}]**

Out[40]= $\left(\frac{\pi}{2} - \frac{1}{2} \, i \, \text{Log}[4] + i \, \text{Log}\left[\frac{1}{z}\right] \right) + \frac{1}{4} \, i \, \left(\frac{1}{z}\right)^2 + O\left[\frac{1}{z}\right]^4$

In[41]:= **Series[ArcSin[z], {z, -1, 3}]**

Out[41]= $-\frac{\pi}{2} + \sqrt{2} \, \sqrt{z+1} + \frac{(z+1)^{3/2}}{6\sqrt{2}} + \frac{3\,(z+1)^{5/2}}{80\sqrt{2}} + \frac{5\,(z+1)^{7/2}}{448\sqrt{2}} + O[z+1]^4$

In[42]:= **Series[ArcSin[z], {z, +1, 3}]**

Out[42]= $\frac{\pi}{2} + (-1)^{\text{Floor}\left[-\frac{\text{Arg}[-1+z]}{2\pi}\right]}$

$$\left(-i\sqrt{2}\,\sqrt{z-1} + \frac{i\,(z-1)^{3/2}}{6\sqrt{2}} + -\frac{3\,i\,(z-1)^{5/2}}{80\sqrt{2}} + \frac{5\,i\,(z-1)^{7/2}}{448\sqrt{2}} + O[z-1]^4 \right)$$

The last expansion at the branch point $z = 1$ shows the slightly unusual prefactor $(-1)^{\lfloor -\arg(z-1)/(2\pi)\rfloor}$. We will encounter such-type factors frequently when expanding analytic functions on branch points and branch cuts. Such factors ensure that the resulting series expansions are correct in any direction from the expansion point. The discontinuous function $\lfloor -\arg(z-1)/(2\pi)\rfloor$ reflects the fact that the original function arcsin(z) has a line of discontinuity (a branch cut) emerging from the point $z = +1$. The next input shows that in the last example, the factor is needed to get the sign of the imaginary part just above the branch cut corrected.

In[43]:= (* function, naive series, and corrected series *)
     {ArcSin[z], Pi/2 - I Sqrt[2] Sqrt[z - 1],
     Pi/2 - (-1)^Floor[-(Arg[z - 1]/(2 Pi))] I Sqrt[2] Sqrt[z - 1]} /.
         z -> 1 + 10^-3 + (* above branch cut *) 10^-10 I // N

Out[44]= {1.5708 + 0.0447176 i, 1.5708 - 0.0447214 i, 1.5708 + 0.0447214 i}

Here is a more complicated series of a complete elliptic integral. The piecewise constant term Floor[-Arg[-1+z]/(2 Pi)] appears outside of the SeriesData-object.

In[45]:= **Series[EllipticK[z], {z, 1, 1}]** // (* avoid PolyGamma functions *)
                    FunctionExpand // (Together //@ #) &

Out[45]= $\frac{1}{2} \left(\text{Floor}\left[-\frac{\text{Arg}[-1+z]}{2\pi}\right] \left(2\,i\,\pi + -\frac{1}{2}\,i\,\pi\,(z-1) + O[z-1]^2 \right) + \text{Floor}\left[-\frac{\text{Arg}[-1+z]}{2\pi}\right] \right.$

$(-4\,i\,\pi + i\,\pi\,(z-1) + O[z-1]^2) + \left(-i\,(\pi + 2\,i\,\text{Log}[4] - i\,\text{Log}[-1+z]) + \right.$

$\left. \left. \frac{1}{4}\,i\,(-2\,i + \pi + 2\,i\,\text{Log}[4] - i\,\text{Log}[-1+z])\,(z-1) + O[z-1]^2 \right) \right)$

Carrying out an expansion on a branch cut results in direction-dependent terms in a pure Taylor series.

In[46]:= **Series[ArcSin[z], {z, -2, 1}]**

Out[46]= $\pi \, \text{Floor}\left[\dfrac{\text{Arg}[2+z]}{2\,\pi}\right] + (-1)^{\text{Floor}\left[\frac{\text{Arg}[2+z]}{2\pi}\right]} \left(-\text{ArcSin}[2] + -\dfrac{i\,(z+2)}{\sqrt{3}} + O[z+2]^2\right)$

Here are the series expansions of a Bessel function. The result is a product of a Taylor series and the term z^ν. Without any information on ν, it is not possible to determine of z^ν will tend to zero or to infinity as z tends to zero.

In[47]:= **Series[BesselJ[ν, z], {z, 0, 4}]**

Out[47]= $z^\nu \left(\dfrac{2^{-\nu}}{\text{Gamma}[1+\nu]} - \dfrac{2^{-2-\nu}\,z^2}{(1+\nu)\,\text{Gamma}[1+\nu]} + \dfrac{2^{-5-\nu}\,z^4}{(1+\nu)\,(2+\nu)\,\text{Gamma}[1+\nu]} + O[z]^5\right)$

There are also some functions in which series expansions are (currently) not computable. These mostly involve essential singularities (in the function-theoretic sense) at the point of expansion, or they involve derivatives that currently cannot be expressed in closed form (for an extension of *Mathematica*'s capabilities for calculating series, see [1040], [1041], and [1043]).

In[48]:= **Series[{Exp[x], Sin[x], Zeta[x]}, {x, Infinity, 2}]**

    Series::esss : Essential singularity encountered in $e^{\frac{1}{x}+O\left[\frac{1}{x}\right]^3}$. More…

    Series::esss : Essential singularity encountered in $e^{\frac{1}{x}+O[x]^3}$. More…

    Series::esss : Essential singularity encountered in $\text{Sin}\left[\frac{1}{\frac{1}{x}}+O\left[\frac{1}{x}\right]^3\right]$. More…

    General::stop :
      Further output of Series::esss will be suppressed during this calculation. More…

Out[48]= $\{e^x,\ \text{Sin}[x],\ \text{Zeta}[x]\}$

Using the above-described function `SeriesData` such series cannot work. Where should, say, exponential terms be kept? In the following example, the result does not have the head `SeriesData`, but rather the head `Times`.

In[49]:= **Series[Gamma[z], {z, Infinity, 1}]**

Out[49]= $e^{-z}\,z^z \left(\sqrt{2\,\pi}\,\sqrt{\dfrac{1}{z}} + O\left[\dfrac{1}{z}\right]^{3/2}\right)$

In[50]:= **FullForm[%]**

Out[50]//FullForm=
    Times[Power[E, Times[-1, z]], Power[z, z], SeriesData[z,
      DirectedInfinity[1], List[Power[Times[2, Pi], Rational[1, 2]]], 1, 3, 2]]

Already simple arithmetic operations with the above expression will issue messages because a new series expansion will be attempted at an essential singularity.

In[51]:= **z %**

    Series::esss : Essential singularity encountered in $e^{-\frac{1}{z}+O\left[\frac{1}{z}\right]^3}$. More…

    Series::esss : Essential singularity encountered in $e^{-\frac{1}{z}+O[z]^3}$. More…

    Series::esss : Essential singularity encountered in $e^{-\frac{1}{z}+O\left[\frac{1}{z}\right]^3}$. More…

    General::stop :
      Further output of Series::esss will be suppressed during this calculation. More…

Out[51]= $e^{-z} z^z \left(\dfrac{\sqrt{2\pi}}{\sqrt{\frac{1}{z}}} + \sqrt{O\left[\frac{1}{z}\right]} \right)$

In the next example the expansion is done in a symbolic way; no explicit functions for `Derivative[i, 0][BesselJ]` are given (which is not so easy to do).

In[52]:= `Series[BesselJ[v, z], {v, 0, 3}]`

Out[52]= $\text{BesselJ}[0, z] + \text{BesselJ}^{(1,0)}[0, z]\, v +$

$\dfrac{1}{2}\, \text{BesselJ}^{(2,0)}[0, z]\, v^2 + \dfrac{1}{6}\, \text{BesselJ}^{(3,0)}[0, z]\, v^3 + O[v]^4$

A multitude of series expansions can also be found in [813].

For functions that are not given analytically, we can only differentiate the functions symbolically. Here is another such case, this time we use a nonbuilt-in function.

In[53]:= `Series[f[u], {u, v, 3}]`

Out[53]= $f[v] + f'[v]\, (u-v) + \dfrac{1}{2}\, f''[v]\, (u-v)^2 + \dfrac{1}{6}\, f^{(3)}[v]\, (u-v)^3 + O[u-v]^4$

Here is a small application of forming a series expansion of not explicitly given functions. The combined Numerov–Mickens nonstandard finite difference scheme discretizes the differential equation $\psi''(x) = \mathcal{V}(x)\,\psi(x)$ through the set of coupled difference equations $-u_{n-1} + 2 d_n u_n - u_{n+1} = 0$ where $u_n = 1 - h^2\,\mathcal{V}(x_n)/12$, $d_n = \cosh\!\left(h\,\sqrt{\mathcal{V}(x_n)}\right)$ and $x_n = x_0 + n\,h$ [340], [1252].

Expanding the difference equation around a fixed x_n in a power series in h and assuming $\psi''(x) = \mathcal{V}(x)\,\psi(x)$ shows that the first nonvanishing term is of order $O(h^6)$.

In[54]:= `Module[{x, u, d, discreteEq, ser},`
      `(* definitions *)`
      `x[n_] = x + n h;`
      `u[n_] = (1 - h^2 V[x[n]]/12) ψ[x[n]];`
      `d[n_] = Cosh[h Sqrt[V[x[n]]]];`
      `(* difference equation *)`
      `discreteEq[n_] = -u[n - 1] + 2 d[n] u[n] - u[n + 1];`
      `(* series expansion around x *)`
      `ser = Series[discreteEq[1], {h, 0, 6}];`
      `(* use differential equation ψ''[x] = V[x] ψ[x] *)`
      `ser //. Derivative[n_ /; n >= 2][ψ][x_] :> D[V[x] ψ[x],`
                                `{x, n - 2}] // Simplify]`

Out[54]= $\dfrac{1}{240}\, (2\, \psi'[x]\, (3\,\mathcal{V}[x]\,\mathcal{V}'[x] + 2\,\mathcal{V}^{(3)}[x]) + \psi[x]\, (4\,\mathcal{V}'[x]^2 + 7\,\mathcal{V}[x]\,\mathcal{V}''[x] + \mathcal{V}^{(4)}[x]))\, h^6 + O[h]^7$

`Series` can work with exact and approximate coefficients. `Series` applies zero testing to the series coefficients and drops terms that are considered to be zero. Here are the first terms of a function with the neat property $f^{(-1)}(z) = -f(z)^{-1}$.

In[55]:= `ffInv[z_] = 1 + Table[z^k, {k, 20}].` `(* series coefficient *)`
    `{ 0.11201933586799460636,`   `0.04013232971298552835,`   `0.05017599118780908587,`
      `0.02331170198490686166,`   `0.04169759366964792265,`   `0.01434818305566120933,`
      `0.04341898484131755962,`   `0.00427268422270174173,`   `0.05316437195644763211,`
    `-0.01227684573369744105,`   `0.07511431972356796605,` `-0.04373397070311347075,`
      `0.12019355271066564345,` `-0.10687118497186786129,`   `0.21222687421724158176,`
    `-0.23683787138794410136,`   `0.40230256474819762393,` `-0.50866679958475735404,`
      `0.80042942843989834362,`   `1.73706479076791285289};`

Calculating $f(-1/f(z))$ yields $z + O(z)^{20}$.

In[56]:= `ffInv[-1/ffInv[z]] + O[z]^21`

Out[56]= $z + 25.186282715164607 z^{20} + O[z]^{21}$

Calculating the same expression to polynomial and rational algebra we see that the coefficients of z^j, $j = 2, ..., 19$ are high-precision zeros that have been converted to exact zeros by `Series`.

In[57]:= `Take[Expand[ffInv[Normal[-1/ffInv[z] + O[z]^21]]], 21]`

Out[57]= $0. \times 10^{-19} + 1.00000000000000000 z + 0. \times 10^{-18} z^2 + 0. \times 10^{-18} z^3 + 0. \times 10^{-18} z^4 +$
$0. \times 10^{-18} z^5 + 0. \times 10^{-17} z^6 + 0. \times 10^{-17} z^7 + 0. \times 10^{-17} z^8 + 0. \times 10^{-17} z^9 + 0. \times 10^{-17} z^{10} +$
$0. \times 10^{-17} z^{11} + 0. \times 10^{-17} z^{12} + 0. \times 10^{-17} z^{13} + 0. \times 10^{-17} z^{14} + 0. \times 10^{-17} z^{15} +$
$0. \times 10^{-16} z^{16} + 0. \times 10^{-16} z^{17} + 0. \times 10^{-16} z^{18} + 0. \times 10^{-16} z^{19} + 25.186282715164607 z^{20}$

In[58]:= `(* for comparison: make coefficients exact *)`
`ffInvInf[z_] = SetPrecision[ffInv[z], Infinity];`
`SetPrecision[Normal[ffInvInf[-1/ffInvInf[z]] + O[z]^21], 2]`

Out[60]= $6.1 \times 10^{-20} + 1.0 z + 6.3 \times 10^{-20} z^2 + 3.7 \times 10^{-21} z^3 + 7.1 \times 10^{-21} z^4 - 1.1 \times 10^{-20} z^5 -$
$4.5 \times 10^{-20} z^6 - 2.4 \times 10^{-21} z^7 + 5.5 \times 10^{-20} z^8 - 1.1 \times 10^{-19} z^9 + 2.6 \times 10^{-19} z^{10} -$
$1.7 \times 10^{-19} z^{11} + 1.3 \times 10^{-19} z^{12} - 1.3 \times 10^{-19} z^{13} + 1.7 \times 10^{-19} z^{14} - 2.0 \times 10^{-19} z^{15} +$
$2.4 \times 10^{-19} z^{16} - 4.1 \times 10^{-19} z^{17} + 6.8 \times 10^{-19} z^{18} - 9.5 \times 10^{-19} z^{19} + 25. z^{20}$

`Series` can often be successfully used to calculate leading terms of very large, unexpanded polynomials. In the following input, we display the number of possibilities $c_n^{(F)}$ to represent an integer n as a sum of Fibonacci numbers $n = \sum_{j=0}^{\infty} \varepsilon_j F_j$ where ε_j is 0 or 1 [308], [1828], [558]. Using the obvious generating function for this problem, we obtain the following self-similar plot for $c_n^{(F)}$. (As we see from the following example, the function `CoefficientList` works not only for polynomials, but also for `SeriesData`.)

In[61]:= `ListPlot[Rest[CoefficientList[Series[Product[1 + z^Fibonacci[k],`
`{k, 21}] - 1, {z, 0, 10946}], z]]]`

Using symbolic exponents of another variable in addition to the explicit exponents allows obtaining the explicit form of the sums. Here are the 820 possibilities for representing the first 100 integers as sums of Fibonacci numbers calculated and some of them shown.

In[62]:= `fibonacci01Partitions100 =`
`    MapIndexed[((* make equations *)`
`                Equal @@ Flatten[{#2[[1]], #1}]) &,`
`        (* separate powers of ζ *)`
`        List @@@ Expand[Rest[CoefficientList[`
`            Series[Product[1 + z^Fibonacci[k] ζ^Subscript[F, k],`
`                {k, 12}] - 1, {z, 0, 100}], z]]] /.`
`                (* powers to exponents *) ζ^e_. :> e];`

In[63]:= `Short[fibonacci01Partitions100, 16] // TraditionalForm`

Out[63]//TraditionalForm=

$$\{1 = F_1 = F_2, 2 = F_1 + F_2 = F_3, 3 = F_1 + F_3 = F_2 + F_3 = F_4, 4 = F_1 + F_2 + F_3 = F_1 + F_4 = F_2 + F_4,$$
$$5 = F_1 + F_2 + F_4 = F_3 + F_4 = F_5, 6 = F_1 + F_3 + F_4 = F_2 + F_3 + F_4 = F_1 + F_5 = F_2 + F_5,$$
$$\ll 89 \gg, 96 = F_1 + F_2 + F_3 + F_4 + F_5 + F_6 + F_8 + F_{10} = F_1 + F_2 + F_3 + F_4 + F_7 + F_8 + F_{10} =$$
$$F_1 + F_2 + F_5 + F_7 + F_8 + F_{10} = F_3 + F_5 + F_7 + F_8 + F_{10} = F_1 + F_2 + F_3 + F_4 + F_9 + F_{10} =$$
$$F_1 + F_2 + F_5 + F_9 + F_{10} = F_3 + F_5 + F_9 + F_{10} = F_1 + F_2 + F_3 + F_4 + F_{11} = F_1 + F_2 + F_5 + F_{11} =$$
$$F_3 + F_5 + F_{11}, 97 = F_1 + F_3 + F_5 + F_7 + F_8 + F_{10} = F_2 + F_3 + F_5 + F_7 + F_8 + F_{10} = F_4 + F_5 + F_7 + F_8 + F_{10} =$$
$$F_6 + F_7 + F_8 + F_{10} = F_1 + F_3 + F_5 + F_9 + F_{10} = F_2 + F_3 + F_5 + F_9 + F_{10} = F_4 + F_5 + F_9 + F_{10} =$$
$$F_6 + F_9 + F_{10} = F_1 + F_3 + F_5 + F_{11} = F_2 + F_3 + F_5 + F_{11} = F_4 + F_5 + F_{11} = F_6 + F_{11},$$
$$98 = F_1 + F_2 + F_3 + F_5 + F_7 + F_8 + F_{10} = F_1 + F_4 + F_5 + F_7 + F_8 + F_{10} = F_2 + F_4 + F_5 + F_7 + F_8 + F_{10} =$$
$$F_1 + F_6 + F_7 + F_8 + F_{10} = F_2 + F_6 + F_7 + F_8 + F_{10} = F_1 + F_2 + F_3 + F_5 + F_9 + F_{10} =$$
$$F_1 + F_4 + F_5 + F_9 + F_{10} = F_2 + F_4 + F_5 + F_9 + F_{10} = F_1 + F_6 + F_9 + F_{10} = F_2 + F_6 + F_9 + F_{10} =$$
$$F_1 + F_2 + F_3 + F_5 + F_{11} = F_1 + F_4 + F_5 + F_{11} = F_2 + F_4 + F_5 + F_{11} = F_1 + F_6 + F_{11} = F_2 + F_6 + F_{11},$$
$$99 = F_1 + F_2 + F_4 + F_5 + F_7 + F_8 + F_{10} = F_3 + F_4 + F_5 + F_7 + F_8 + F_{10} =$$
$$F_1 + F_2 + F_6 + F_7 + F_8 + F_{10} = F_3 + F_6 + F_7 + F_8 + F_{10} = F_1 + F_2 + F_4 + F_5 + F_9 + F_{10} =$$
$$F_3 + F_4 + F_5 + F_9 + F_{10} = F_1 + F_2 + F_6 + F_9 + F_{10} = F_3 + F_6 + F_9 + F_{10} =$$
$$F_1 + F_2 + F_4 + F_5 + F_{11} = F_3 + F_4 + F_5 + F_{11} = F_1 + F_2 + F_6 + F_{11} = F_3 + F_6 + F_{11},$$
$$100 = F_1 + F_3 + F_4 + F_5 + F_7 + F_8 + F_{10} = F_2 + F_3 + F_4 + F_5 + F_7 + F_8 + F_{10} =$$
$$F_1 + F_3 + F_6 + F_7 + F_8 + F_{10} = F_2 + F_3 + F_6 + F_7 + F_8 + F_{10} = F_4 + F_6 + F_7 + F_8 + F_{10} =$$
$$F_1 + F_3 + F_4 + F_5 + F_9 + F_{10} = F_2 + F_3 + F_4 + F_5 + F_9 + F_{10} = F_1 + F_3 + F_6 + F_9 + F_{10} =$$
$$F_2 + F_3 + F_6 + F_9 + F_{10} = F_4 + F_6 + F_9 + F_{10} = F_1 + F_3 + F_4 + F_5 + F_{11} =$$
$$F_2 + F_3 + F_4 + F_5 + F_{11} = F_1 + F_3 + F_6 + F_{11} = F_2 + F_3 + F_6 + F_{11} = F_4 + F_6 + F_{11}\}$$

Next, we use `Series` to determine the limit of the reordered sequence $(-1)^k k^{-1}$. We reorder the series in such a way, that after p positive terms q negative ones follow [1441], [632], [1567], [1131] (for instance, for $p = 3$, $q = 2$ this means $1^{-1} + 3^{-1} + 5^{-1} - 2^{-1} - 4^{-1} + 7^{-1} + 9^{-1} + 11^{-1} - \cdots$). We sum the first o terms and determine the behavior around $o = \infty$. The resulting sum contains Polygamma functions.

```
In[64]:= (* p/(p + q) o terms of the first o are positive (for large o) *)
         Sum[1/j, {j, 1, p/(p + q) o, 2}] -
         (* q/(p + q) o terms of the first o are negative (for large o) *)
         Sum[1/j, {j, 2, q/(p + q) o, 2}] /. Floor[x_] :> x
```

$$\text{Out[65]}= -\frac{\text{EulerGamma}}{2} + \frac{1}{2} (\text{EulerGamma} + \text{Log}[4]) +$$
$$\frac{1}{2} \text{PolyGamma}\left[0, \frac{3}{2} + \frac{1}{2}\left(-1 + \frac{o\,p}{p + q}\right)\right] - \frac{1}{2} \text{PolyGamma}\left[0, 2 + \frac{1}{2}\left(-2 + \frac{o\,q}{p + q}\right)\right]$$

```
In[66]:= Series[%, (* take many terms of the sum *) {o, Infinity, 0}] //
                                              Simplify[#, p > 0 && q > 0]&
```

$$\text{Out[66]}= \frac{1}{2} \text{Log}\left[\frac{4\,p}{q}\right] + O\left[\frac{1}{o}\right]^1$$

If a function of several variables is to be expanded around a point in multidimensional space, we can use the standard iterator notation. The computation proceeds from right to left.

`Series` allows expanding with respect to two or more variables. The result is again a `SeriesData` object with further `SeriesData` in its third argument.

```
In[67]:= Series[Exp[x y], {x, 0, 2}, {y, 0, 2}] // InputForm
```

Out[67]//InputForm=

```
         SeriesData[x, 0, {1, SeriesData[y, 0, {1}, 1, 3, 1], SeriesData[y, 0, {1/2},
         2, 3, 1]}, 0, 3, 1]
```

```
In[68]:= {% === Series[Series[Exp[x y], {y, 0, 2}], {x, 0, 2}],
         % === Series[Series[Exp[x y], {x, 0, 2}], {y, 0, 2}]}
```

Out[68]= {False, True}

Note that the explicit form of multivariate series depends on the order in which the expansion is done. One has to specify orders for each variables; *Mathematica* does not allow for a total order with respect to all variables.

In[69]:= **sxy = Series[Sin[x y], {x, 0, 3}, {y, Pi/2, 3}]**

Out[69]= $\left(\frac{\pi}{2} + \left(y - \frac{\pi}{2}\right) + O\left[y - \frac{\pi}{2}\right]^4\right) x +$

$\left(-\frac{\pi^3}{48} - \frac{1}{8}\pi^2\left(y - \frac{\pi}{2}\right) - \frac{1}{4}\pi\left(y - \frac{\pi}{2}\right)^2 - \frac{1}{6}\left(y - \frac{\pi}{2}\right)^3 + O\left[y - \frac{\pi}{2}\right]^4\right) x^3 + O[x]^4$

In[70]:= **syx = Series[Sin[x y], {y, Pi/2, 3}, {x, 0, 3}]**

Out[70]= $\left(\frac{\pi x}{2} - \frac{\pi^3 x^3}{48} + O[x]^4\right) + \left(x - \frac{\pi^2 x^3}{8} + O[x]^4\right)\left(y - \frac{\pi}{2}\right) +$

$\left(-\frac{\pi x^3}{4} + O[x]^4\right)\left(y - \frac{\pi}{2}\right)^2 + \left(-\frac{x^3}{6} + O[x]^4\right)\left(y - \frac{\pi}{2}\right)^3 + O\left[y - \frac{\pi}{2}\right]^4$

But mathematically the last two results are identical. The function normal converts a SeriesData-object to a polynomial by explicitly forming the approximating polynomial.

In[71]:= **normal[expr_] := expr //.**
HoldPattern[SeriesData[x_, x0_, coeffs_, e1_, e2_, den_]] :>
coeffs.Take[Table[(x - x0)^(e/den), {e, e1, e2, den}],
Length[coeffs]]

In[72]:= **normal[sxy] - normal[syx] // Together**

Out[72]= 0

Sometimes one wants a multivariate series expansion up to a given total degree. This can be conveniently done by using the univariate built-in Series command by homogenizing the expression. The following function dSeries shows this.

In[73]:= **dSeries[f_, {vars_List, points_List, d_}] :=**
Module[{τ}, (Normal[Series[f /. Thread[vars -> τ vars + points],
{τ, 0, d}]] /. τ -> 1) + (* for correct form of the result *)
O[Times @@ (vars - points)]^d]

Here is an example that shows dSeries in action. Because the argument of O has head Times, we cannot use the result directly for further manipulations. The built-in code for manipulating SeriesData-objects has no rules to deal with such terms.

In[74]:= **dSeries[Sin[x] Cos[y] + Tan[z]^2, {{x, y, z}, {X, Pi/2, 0}, 4}]**

Out[74]= $\left(z^2 + \frac{2 z^4}{3} - x y \cos[X] + \frac{1}{6} x^3 y \cos[X] + \frac{1}{6} x y^3 \cos[X] -\right.$

$\left. y \sin[X] + \frac{1}{2} x^2 y \sin[X] + \frac{1}{6} y^3 \sin[X]\right) + O\left[(x - X)\left(-\frac{\pi}{2} + y\right) z\right]^4$

(For recovering the original function from the higher-order series terms, see [846].)

One word of caution is in order when working with multivariate SeriesData objects. Because constants with respect to the expansion variable do not form SeriesData objects, some orders might become wrong. So it is recommended to ensure that each of the coefficients of the outer SeriesData object is itself a SeriesData object. Here is an example that demonstrates the difference. The first series does not contain a SeriesData-object in the variable *x*.

In[75]:= **{Series[1 + y, {x, 0, 4}, {y, 0, 4}] + x^2 y^2,**
Series[1 + y + x^2 y^2, {x, 0, 4}, {y, 0, 4}]}

Out[75]= $\{1 + y + x^2 y^2 + O[y]^5, (1 + y + O[y]^5) + (y^2 + O[y]^5) x^2 + O[x]^5\}$

In[76]:= `Subtract @@ normal[%^3]`

Out[76]= $- (3 x^2 + 3 x^4) y^4 + (2 x^2 + x^4 + x^2 (1 + 2 x^2)) y^4$

`Series` also works in conjunction with other commands. The functions in the following example cannot be integrated explicitly; so `Series` operates inside the integral.

In[77]:= `Series[Integrate[g[y] f[x], {x, -1, 1}], {y, 0, 2}]`

Out[77]= $\left(\int_{-1}^{1} f[x] \, dx \right) g[0] + \left(\int_{-1}^{1} f[x] \, dx \right) g'[0] \, y + \frac{1}{2} \left(\int_{-1}^{1} f[x] \, dx \right) g''[0] \, y^2 + O[y]^3$

To transform a `SeriesData` expression into an "ordinary" *Mathematica* expression, we can use `Normal`. (The function `normal` from above is a crude version of the built-in function `Normal`.)

`Normal[`*series*`]`

transforms the `SeriesData` expression *series* into a sum or single term, respectively.

Now, the `O[n]` disappears.

In[78]:= `Series[Sin[x y], {x, 0, 3}, {y, Pi/2, 2}] // Normal`

Out[78]= $\frac{\pi x}{2} - \frac{\pi^3 x^3}{48} + \left(x - \frac{\pi^2 x^3}{8} \right) \left(-\frac{\pi}{2} + y \right) - \frac{1}{4} \pi x^3 \left(-\frac{\pi}{2} + y \right)^2$

The two series above agree up to their total order.

In[79]:= `Series[Sin[x y], {x, 0, 3}, {y, Pi/2, 2}] // Normal // Simplify`

Out[79]= $-\frac{1}{48} \pi^3 x^3 + x y + \frac{1}{8} \pi^2 x^3 y - \frac{1}{4} \pi x^3 y^2$

In[80]:= `Series[Sin[x y], {y, Pi/2, 2}, {x, 0, 3}] // Normal // Simplify`

Out[80]= $-\frac{1}{48} \pi^3 x^3 + x y + \frac{1}{8} \pi^2 x^3 y - \frac{1}{4} \pi x^3 y^2$

For `O[x^v]`, the result is trivially 0.

In[81]:= `Normal /@ {O[x^3], O[x^v], O[x]^3, O[x]^v,`
        `(* negative order *) SeriesData[x, 0, {}, -90, -90, 1]}`

Out[81]= $\{0, 0, 0, 0, 0\}$

One reason that the operation `Normal` is frequently needed can be seen from the following example. We define a function generated by a series expansion.

In[82]:= `serFunction1[x_] = Series[x, {x, 0, 3}]`

Out[82]= $x + O[x]^4$

We are not able to substitute numerical values for the expansion variable in series expansion, because for `SeriesData`-objects, numbers cannot be substituted for the variable.

In[83]:= `serFunction1[1]`

              SeriesData::ssdn :
                Attempt to evaluate a series at the number 1; returning Indeterminate. More…

Out[83]= Indeterminate

We can, of course, proceed as follows.

In[84]:= `serFunction2[x_] = Series[x, {x, 0, 3}] // Normal;`

       `serFunction2[1]`

Out[85]= 1

In the next input, we first form a series, then convert it to a normal expression and substitute another series for all occurrences of n. (This gives an expansion due to Ramanujan of the harmonic numbers H_n in terms of $m = n(n + 1)/2$ [1783].)

```
In[86]:= Simplify[Normal[Series[HarmonicNumber[n], {n, Infinity, 16}]] /.
            n -> Series[(Sqrt[1 + 8 m] - 1)/2, {m, Infinity, 6}]]
```

$$Out[86]= \left(EulerGamma + \frac{Log[2]}{2} - \frac{1}{2} Log\left[\frac{1}{m}\right] \right) + \frac{1}{12\,m} - \frac{1}{120}\left(\frac{1}{m}\right)^2 + \frac{1}{630}\left(\frac{1}{m}\right)^3 - \frac{\left(\frac{1}{m}\right)^4}{1680} + \frac{\left(\frac{1}{m}\right)^5}{2310} - \frac{191\left(\frac{1}{m}\right)^6}{360360} + O\left[\frac{1}{m}\right]^{13/2}$$

We now look at the zeros in the complex plane of the series expansion of e^z about $z = 0$ as a function of the order [438], [1803]. Because e^z has no finite zeros in the plane, the zeros in the following plot appear as smaller and smaller points moving outward for increasing orders. (Other than the variables for the polynomial, we do not need any further variables.) We implement a function `seriesZerosGraphics` that generates a graphic of the zeros of the series expansion of f to order o.

```
In[87]:= seriesZerosGraphics[f_, z_, {o_, n_}, opts___] :=
        Show[Graphics[Reverse[(* make colored points from zeros *)
        MapIndexed[{PointSize[0.001 #2[[1]]], Hue[0.8 (#2[[1]] - 1)/n], #1}&,
        Map[Point[{Re[#], Im[#]}]&,
        (* solve the polynomials *) (z /. NSolve[# == 0, z])& /@
        (* make all polynomials by series truncation *)
        NestList[Drop[#, -1]&, (* calculate series only once *)
                Normal[Series[f, {z, 0, o}]], n], {-1}]]]],
        opts, AspectRatio -> Automatic, Frame -> True,
        PlotRange -> All, FrameLabel -> {"Re[z]", "Im[z]"}]
```

```
In[88]:= seriesZerosGraphics[Exp[z], z, {35, 34}]
```

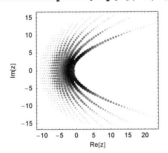

The zeros of the scaled partial sums $z \to order\,z$ cluster asymptotically at the curve $|z\exp(1 - z)| = 1$.

```
In[89]:= Module[{o = 60, z},
        Show[{(* region Abs[z Exp[1 - z]] < 1 *)
            ContourPlot[Abs[(x + I y) Exp[1 - x - I y]] - 1,
                    {x, -1, 1}, {y, -3/4, 3/4}, Contours -> {0},
                    PlotPoints -> 200, DisplayFunction -> Identity],
            (* zeros of the rescaled partial sums *)
            Graphics[Table[{Hue[0.8 n/o], Point[{Re[#], Im[#]}]& /@
                        N[z /. Solve[(Normal[Series[Exp[z], {z, 0, n}]] /.
                        (* rescale *) z -> n z) == 0, z]]}, {n, o}]]},
            AspectRatio -> Automatic, DisplayFunction -> $DisplayFunction]]
```

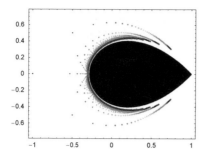

Let us use `seriesZerosGraphics` to visualize the zeros of some more polynomials that arise from a series. For sin(z), the zeros of the polynomials have to accumulate near the zeros of sin(z) at $k\pi$, $k \in \mathbb{Z}$ [1765], [1775].

In[90]:= `seriesZerosGraphics[Sin[z], z, {40, 19}]`

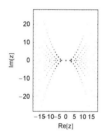

The Taylor series expansion of $(1 + z)^{1/2}$ has the finite convergence radius 1. Now, the zeros accumulate at the boundary of convergence.

In[91]:= `seriesZerosGraphics[Sqrt[1 + z], z, {35, 34}]`

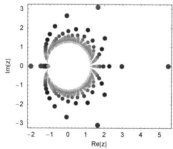

The zeros of the partial sums of an asymptotic series (the series of the Gamma function at infinity) are shown.

In[92]:=
```
Show[Graphics[Reverse[
    MapIndexed[{PointSize[0.0006 #2[[1]]], Hue[(#2[[1]] - 1)/60], #1}&,
    Map[Point[{Re[#], Im[#]}]&, (z /. NSolve[# == 0, z])& /@
    Rest[Table[Numerator @ Together @ Expand[Cancel[Normal[
    (* the series; divide out Puiseux part *)
    Series[Gamma[z], {z, Infinity, k}]]/
        (z^z Exp[-z] Sqrt[1/z] Sqrt[Pi/2])]], {k, 50}]], {-1}]]]],
    AspectRatio -> Automatic, Frame -> True,
    PlotRange -> All, FrameLabel -> {"Re[z]", "Im[z]"}]
```

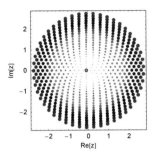

Here are the zeros of the series of a truncated polynomial [1609].

In[93]:= `seriesZerosGraphics[(1 - z)^-50, z, {40, 19}]`

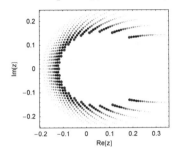

Here are some more pictures of this kind. The next example shows the zeros of the partial sums of the divergent series $\int_0^\infty e^{-t}(t\,z+1)^{-1}\,dt \propto \sum_{n=0}^\infty (-1)^n\,n!\,z^n$.

In[94]:= `Integrate[Normal[Series[Exp[-t]/(1 + x t), {x, 0, 9}]],`
                            `{t, 0, Infinity}]`

Out[94]= $1 - x + 2\,x^2 - 6\,x^3 + 24\,x^4 - 120\,x^5 + 720\,x^6 - 5040\,x^7 + 40320\,x^8 - 362880\,x^9$

In[95]:= `seriesZerosGraphics[Sum[(-1)^n n! z^n, {n, 0, 40}], z, {30, 22}]`

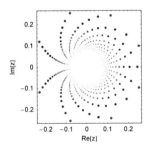

The last picture of this series shows the zeros of the partial sums of the series of the first seven derivatives of $(1 - z)^{-1}$.

In[96]:= `Show[GraphicsArray[#]]& /@ Partition[`
        `Table[seriesZerosGraphics[D[1/(1 - z), {z, n}], z, {30, 23},`
                                `DisplayFunction -> Identity],`
              `{n, 0, 7}], 4]`

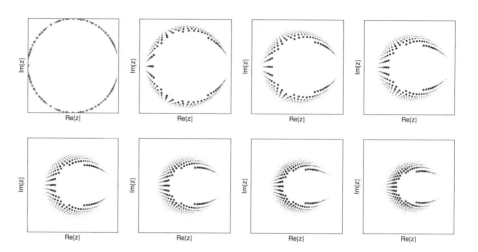

For the behavior of partial sums of other divergent series, see [486]. For mathematical details on the behavior of partial sum approximations of power series, see [1334], [557], [1773], [1774], and [218].

Let us use the `Series` command to build a more complicated series, a q-Taylor series [914], [587], [1094], [677], [962], [1682], [1453]. For any function $f(x)$ analytic at x_0, the series

$$f_n(x) = \sum_{k=0}^{n} \frac{(x, x_0; q)_n}{(k!)_q} \frac{d_q f(x)}{d_q x}$$

converges to $f(x)$ as $n \to \infty$ for all $|q| \neq 1$. Here $(k!)_q = \prod_{k=1}^{n} (q^k - 1)/(q - 1)$, $(x, a; q)_n = \prod_{k=0}^{n-1} (x - a\,q^k)$ and d_q is the q-differential operator

$$\frac{d_q f(x)}{d_q x} = \frac{f(x) - f(q\,x)}{(1 - q)\,x}.$$

It is straightforward to implement the last definitions.

```
In[97]:= (* q-factorial function *)
        qFactorial[q_][n_] := Product[(q^k - 1)/(q - 1), {k, 1, n}];
        qFactorial[q][0] = 1;

        (* q-differential operator; defined recursively *)
        qD[q_][f_, x_] := (f - (f /. x -> q x))/((1 - q) x);
        qD[q_][f_, {x_, 0 }] := f
        qD[q_][f_, {x_, 1 }] := qD[q][f, x]
        qD[q_][f_, {x_, n_}] := qD[q][f, {x, n}] =
                    Together[qD[q][qD[q][f, {x, n - 1}], x]]

        (* q product *)
        qProduct[q_][n_, x_, a_] := Product[x - a q^k, {k, 0, n - 1}]

        (* q-Taylor series *)
        qTaylorSeries[q_][f_, {x_, x0_, n_}] :=
        Sum[Together[qProduct[q][k, x, x0]/qFactorial[q][k] *
```

Here are the first two terms of the q-Taylor series expansion of an abstract function $f(x)$ around x_0.

In[110]:= **qTaylorSeries[q][f[x], {x, x0, 2}]**

Out[110]= $f[x0] - \dfrac{(x - x0)\ (f[x0] - f[q\,x0])}{(-1 + q)\ x0} -$

$\dfrac{(x - x0)\ (-x + q\,x0)\ (q\ f[x0] - f[q\,x0] - q\ f[q\,x0] + f[q^2\,x0])}{(-1 + q)^2\ (q + q^2)\ x0^2}$

In[111]:= **Series[%, {q, 1, 1}] // Simplify**

Out[111]= $\left(f[x0] + \dfrac{1}{2}\ (x - x0)\ (2\ f'[x0] + (x - x0)\ f''[x0])\right) +$

$\dfrac{1}{2}\ (x - x0)^2\ x0\ f^{(3)}[x0]\ (q - 1) + O[q - 1]^2$

For $x_0 = 0$, the q-Taylor series agrees with the ordinary Taylor series.

In[112]:= **qTaylorSeries[q][Cos[x], {x, 0, 6}]**

Out[112]= $1 - \dfrac{x^2}{2} + \dfrac{x^4}{24} - \dfrac{x^6}{720}$

But for general q, the q-Taylor series is typically far more complicated.

In[113]:= **qTaylorSeries[q][Cos[x], {x, 2Pi, 3}]**

Out[113]= $1 - \dfrac{(2\,\pi - x)\ (-1 + \text{Cos}[2\,\pi\,q])}{2\,\pi\ (-1 + q)} -$

$\dfrac{(2\,\pi - x)\ (2\,\pi\,q - x)\ (-q + \text{Cos}[2\,\pi\,q] + q\,\text{Cos}[2\,\pi\,q] - \text{Cos}[2\,\pi\,q^2])}{4\,\pi^2\ (-1 + q)^2\ (q + q^2)} -$

$((2\,\pi - x)\ (2\,\pi\,q - x)\ (2\,\pi\,q^2 - x)$

$(-q^3 + q\,\text{Cos}[2\,\pi\,q] + q^2\,\text{Cos}[2\,\pi\,q] + q^3\,\text{Cos}[2\,\pi\,q] - \text{Cos}[2\,\pi\,q^2] - q\,\text{Cos}[2\,\pi\,q^2] -$

$q^2\,\text{Cos}[2\,\pi\,q^2] + \text{Cos}[2\,\pi\,q^3]))\,/\,\left(8\,\pi^3\ (-q + q^2)^3\ (1 + 2\,q + 2\,q^2 + q^3)\right)$

Similarly to the above investigations, we could visualize the behavior of the zeros of $f_n(x)$. This time, we fix n and study the dependence on q. Here this is done for a simple polynomial $x^{50} - 1$ with $x_0 = 1$, $n = 25$ and q varying between -1 and 1.

In[114]:= **qft[x_] = qTaylorSeries[q][x^50 - 1, {x, 1, 25}];**

In[115]:= **Show[Graphics[{PointSize[0.003],**
**   Table[{Hue[0.8 (q0 - 1)/2], Point[{Re[#], Im[#]}]}& /@**
**   Cases[NRoots[qft[x] == 0 /. q -> N[q0], x], _?NumberQ, {-1}]},**
**   {q0, -1, 1, 2/501}]}],**
**   Frame -> True, PlotRange -> 3/2 {{-1, 1}, {-1, 1}}]**

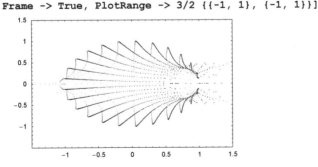

Here are three examples for the q-Taylor series of random polynomials.

In[116]:= **Show[GraphicsArray[**
**   (SeedRandom[#];** (* q-Taylor series *)

```
qft[x_] = qTaylorSeries[q][(* random polynomial *)
            Sum[Random[Integer, 10 {-1, 1}] x^k,
            {k, 10}], {x, 1, 9}];
(* make graphics *)
Graphics[{PointSize[0.003],
Table[{Hue[0.8 (q0 - 1)/2], Point[{Re[#], Im[#]}]& /@
        Cases[NRoots[qft[x] == 0 /. q -> N[Exp[I q0]], x],
        _?NumberQ, {-1}]},
    {q0, -1, 1, 2/1001}]}, Frame -> True, PlotRange -> All,
    PlotRange -> 3/2 {{-1, 1}, {-1, 1}}, FrameTicks -> False])& /@
    (* the seeds *)
    {51879769166741319900, 67570387056305739148}]]
```

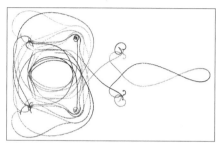 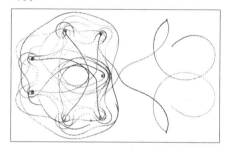

Many operations that apply to "ordinary functions" can also be applied to series. These include, for example, D, Integrate, Plus, Times, Power, etc. The results are SeriesData-objects (this is similar to interval arithmetic, in which the result has the head Interval). We present a few examples. Here is the initial object.

In[117]:= `ser = Series[Tan[Sin[x]] - Sin[Tan[x]], {x, 0, 10}]`

Out[117]= $\dfrac{x^7}{30} + \dfrac{29\,x^9}{756} + O[x]^{11}$

Here is its derivative.

In[118]:= `D[ser, x]`

Out[118]= $\dfrac{7\,x^6}{30} + \dfrac{29\,x^8}{84} + O[x]^{10}$

This is its indefinite integral with respect to x.

In[119]:= `Integrate[ser, x]`

Out[119]= $\dfrac{x^8}{240} + \dfrac{29\,x^{10}}{7560} + O[x]^{12}$

Here is an algebraic expression formed from it.

In[120]:= `ser + 5 ser^2 + Sqrt[ser]`

Out[120]= $\dfrac{x^{7/2}}{\sqrt{30}} + \dfrac{29}{252}\sqrt{\dfrac{5}{6}}\,x^{11/2} + \dfrac{x^7}{30} + O[x]^{15/2}$

In the next input ser is raised to its own power.

In[121]:= `ser^ser`

Out[121]= $1 + \left(-\dfrac{\text{Log}[30]}{30} + \dfrac{7\,\text{Log}[x]}{30}\right)x^7 + \left(\dfrac{29}{756} - \dfrac{29\,\text{Log}[30]}{756} + \dfrac{29\,\text{Log}[x]}{108}\right)x^9 + O[x]^{11}$

If SeriesData-objects are used as arguments of functions, a new SeriesData-object that has the corresponding order is automatically formed.

In[122]:= **Exp[ser]**

Out[122]= $1 + \dfrac{x^7}{30} + \dfrac{29\,x^9}{756} + O[x]^{11}$

In[123]:= **AiryAi[ser]**

Out[123]= $\dfrac{1}{3^{2/3}\,\text{Gamma}\,[\frac{2}{3}]} - \dfrac{x^7}{30\,(3^{1/3}\,\text{Gamma}\,[\frac{1}{3}])} - \dfrac{29\,x^9}{756\,(3^{1/3}\,\text{Gamma}\,[\frac{1}{3}])} + O[x]^{11}$

However, for a function that is not explicitly defined, nothing happens.

In[124]:= **f[ser]**

Out[124]= $f\left[\dfrac{x^7}{30} + \dfrac{29\,x^9}{756} + O[x]^{11}\right]$

If two series expansions are such that all of their terms coincide up to a certain order, their difference may still not be 0. They could be series expansions of two distinct functions that just happen to agree up to that order.

In[125]:= **s1 = Series[Sin[x], {x, 0, 3}]**

Out[125]= $x - \dfrac{x^3}{6} + O[x]^4$

In[126]:= **s2 = s1; s1 - s2**

Out[126]= $O[x]^4$

One must be careful with the following subtractions of two `SeriesData`-objects. The following result is also not 0. (Because evaluation of the summands happens before the actual summation this is exactly the same input as the last one.)

In[127]:= **s1 - s1**

Out[127]= $O[x]^4$

And the result of the following is not 1.

In[128]:= **s1/s1**

Out[128]= $1 + O[x]^3$

This happens because *Mathematica* computes the two arguments before subtracting, and at the point when the subtraction takes place, it no longer knows anything about the previous equality of the two arguments.

As a little application of `Series`, we will check the first few instances of the following neat identity [1841]. Let the (formal) series in x be $f = 1 + a_1 x + a_2 x^2 + \cdots$. Then the determinant of the matrix C with entries $c_{ij} = [x^i]\,(f(x)^j)$ is given by $\det(C) = a_1^{n(n+1)/2}$.

In[129]:= **With[{n = 7},**
             Module[{f = 1 + Sum[a[i] x^i, {i, n}] + O[x]^(n + 1)},
                 Table[Expand[Det[Table[If[i == 0, If[j == 0, 1, 0],
                         SeriesCoefficient[f^i, j]], {i, 0, k}, {j, 0, k}]]],
                 {k, 0, n}]]]
Out[129]= $\{1,\ a[1],\ a[1]^3,\ a[1]^6,\ a[1]^{10},\ a[1]^{15},\ a[1]^{21},\ a[1]^{28}\}$

A similar identity holds for $f = x + a_2 x^2 + a_3 x^3 + \cdots$ [1000]. Then, the determinant of the matrix C with entries $c_{ij} = [x^{i+1}]\,(f^{(j)})$, where $f^{(i)} = f(f^{(i-1)})$ and $f^{(0)} = x$ is given by:

$$\det(C) = \left(\prod_{k=0}^{n} k!\right) a_2^{n(n+1)/2}.$$

Here is a check of the last theorem for $n = 6$.

```
In[130]:= With[{n = 6},
          Module[{f},
                 f[x_] = x + Sum[a[i] x^i, {i, 2, n }];
                 Table[Expand[Det[Table[SeriesCoefficient[
                                        Nest[f, x + O[x]^(n + 1), i], j + 1],
                                        {i, 0, k}, {j, 0, k}]]], {k, 0, 5}]]]
```
Out[130]= $\{1,\ a[2],\ 2\,a[2]^3,\ 12\,a[2]^6,\ 288\,a[2]^{10},\ 34560\,a[2]^{15}\}$

```
In[131]:= Table[Product[k!, {k, 0, n}] a[2]^(n (n + 1)/2), {n, 0, 5}]
```
Out[131]= $\{1,\ a[2],\ 2\,a[2]^3,\ 12\,a[2]^6,\ 288\,a[2]^{10},\ 34560\,a[2]^{15}\}$

Sometimes one does not need the whole series, but rather a single term. SeriesCoefficient extracts a specific term from a series.

> SeriesCoefficient[*series*, *order*]
>
>   gives the term of order var^{order} of the series (head SeriesData) *series*.

```
In[132]:= SeriesCoefficient[Series[Exp[x], {x, 0, 10}], 10]
```
Out[132]= $\dfrac{1}{3628800}$

When the series does not have enough terms, Indeterminate is returned.

```
In[133]:= SeriesCoefficient[Series[Exp[x], {x, 0, 4}], 10]
```
Out[133]= Indeterminate

Here is a little application of SeriesCoefficient. How many ways are there to have change for $1? Using the generating function approach [1113], [866], it is straightforward to write down the following answer [1732], [759].

```
In[134]:= NumberOfDifferentChanges[coins_List, money_] :=
          Module[{z}, SeriesCoefficient[Series[1/(Times @@ (1 - z^coins)),
                                        {z, 0, money}], money]]
```

Using standard U.S. coins, we have 293 possibilities for $1.

```
In[135]:= NumberOfDifferentChanges[{1, 5, 10, 25, 50, 100}, 100]
```
Out[135]= 293

And we have 3954 possibilities for $1 if we would add a two-cent coin.

```
In[136]:= NumberOfDifferentChanges[{1, 2, 5, 10, 25, 50, 100}, 100]
```
Out[136]= 3954

Finding the number of ways to express $10 requires the calculation of the series up to order 1000. Nevertheless, this can be carried out quickly in a fraction of a minute only.

```
In[137]:= NumberOfDifferentChanges[{1, 5, 10, 25, 50, 100}, 1000] // Timing
```
Out[137]= {0.36 Second, 2103596}

The complexity of NumberOfDifferentChanges is $O(moneyInCent^2)$. This means that calculating that there are 139946140451 possibilities for $100 takes about 100 times as long as the last example.

```
In[138]:= NumberOfDifferentChanges[{1, 5, 10, 25, 50, 100}, 10000] // Timing
```
Out[138]= {45.52 Second, 139946140451}

The function `NumberOfDifferentChanges` can be extended to deal with the case of representing *money* through using exactly *number* coins *coins*.

```
In[139]:= NumberOfDifferentChanges[coins_List, money_, number_] :=
        Module[{z, ζ},
           SeriesCoefficient[Series[
             SeriesCoefficient[Series[1/(Times @@ (1 - ζ z^coins)),
                        {z, 0, money}], money], {ζ, 0, number}], number]]
```

Here is the number of possibilities to express \$1 using 1, 2, 5, 10, 25, 50 cent coins.

```
In[140]:= {#, NumberOfDifferentChanges[{1, 5, 10, 25, 50, 100}, 100, #]}& /@
               {1, 2, 5, 10, 20, 50, 100}
Out[140]= {{1, 1}, {2, 1}, {5, 1}, {10, 6}, {20, 8}, {50, 2}, {100, 1}}
```

For \$2, we see already a smooth dependence on the number of coins emerge.

```
In[141]:= ListPlot[
        Table[{number,
              NumberOfDifferentChanges[{1, 5, 10, 25, 50, 100}, 200, number]},
                {number, 200}], PlotRange -> All, Frame -> True,
                           Axes -> False, PlotStyle -> {PointSize[0.01]}]
```

For the optimal coin-set values, see [1607].

Here is another counting-related application of `SeriesCoefficient`. The following small program counts the number of different series-parallel networks with 1, 2, ..., n edges http://www.research.att.com/projects/-OEIS?Anum=A000084, [1156], [741], [48] (for random hierarchical series-parallel networks, see [809].)

```
In[142]:= SeriesParallelNetworksList[n_] :=
        Module[{S, τ}, Rest[#[[3]]]& @
                Fold[(#1/(1 - S^#2)^SeriesCoefficient[##])&,
                     1/(1 - (S = τ + O[τ]^(n + 1))), Range[2, n]]]

In[143]:= SeriesParallelNetworksList[12]
Out[143]= {1, 2, 4, 10, 24, 66, 180, 522, 1532, 4624, 14136, 43930}
```

Asymptotically the number of networks is proportional $n^{-3/2} \times 0.28083 \ldots^n$ [634].

```
In[144]:= ListPlot[MapIndexed[0.41276 #2[[1]]^(-3/2) 0.28083^(-#2[[1]])/#1&,
                          SeriesParallelNetworksList[200]]]
```

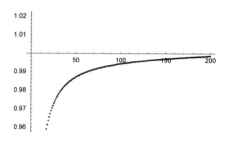

Here is a stress test for the calculation of Laurent series of rational functions. The (proved) Chan conjecture states that the iterated constant terms [1697] of a certain rational function equals a product of Catalan numbers [327], [1895], [1869]. Here is the precise statement.

$$[x_1^0][x_2^0] \cdots [x_n^0] \left(\frac{1}{\prod\limits_{i=1}^{n} (1 - x_i)} \frac{1}{\prod\limits_{j=1}^{n} \prod\limits_{i=1}^{j-1} (x_j - x_i)} \right) = \prod_{k=1}^{n-1} \frac{1}{k+1} \binom{2k}{k}$$

It is straightforward to implement a function `constantTerm` that returns the constant term of a Laurent series expansion.

```
In[145]:= (* the constant term with respect to x of a function *)
         constantTerm[expr_, x_] := SeriesCoefficient[Series[expr, {x, 0, 0}], 0]
```

So, we can code the left-hand side of Chan's conjecture through the function `ChanConjectureLHS`.

```
In[147]:= iteratedConstantTerms[f_, vars_] := (* iterated constant terms *)
              Fold[Together[constantTerm[##]]&, f, vars]
```

```
In[148]:= ChanConjectureLHS[n_] :=
              iteratedConstantTerms[1/Product[1 - x[i], {i, n}]*
                               1/Product[x[j] - x[i], {j, n}, {i, j - 1}],
                               Array[x, n]]
```

And in about a minute, we can verify the Chan conjecture up to order 8.

```
In[149]:= Table[ChanConjectureLHS[n], {n, 8}] // Timing
Out[149]= {44.49 Second, {1, 1, 2, 10, 140, 5880, 776160, 332972640}}
```

```
In[150]:= Table[Product[Binomial[2k, k]/(k + 1), {k, n - 1}], {n, 8}]
Out[150]= {1, 1, 2, 10, 140, 5880, 776160, 332972640}
```

Using series, we can perform another important operation: inversion. Thus, from a series $r(x)$, we obtain a new series $s(x)$ such that $r(s(x)) = \mathrm{id}(x)$, or in *Mathematica* semantics, `InverseSeries`.

> `InverseSeries[series, var]`
>   finds the inverse series to the series *series* with respect to the variables *var*.

We look at an example.

```
In[151]:= ser = Series[Exp[-Exp[x]], {x, 0, 4}]
```

$$Out[151]= \frac{1}{e} - \frac{x}{e} + \frac{x^3}{6\,e} + \frac{x^4}{24\,e} + O[x]^5$$

In[152]:= `invSer = InverseSeries[ser]`

Out[152]= $-e\left(x-\frac{1}{e}\right)-\frac{1}{6}e^3\left(x-\frac{1}{e}\right)^3+\frac{1}{24}e^4\left(x-\frac{1}{e}\right)^4+O\left[x-\frac{1}{e}\right]^5$

In[153]:= `ser /. x -> invSer`

Out[153]= $\frac{1}{e}+\left(x-\frac{1}{e}\right)+O\left[x-\frac{1}{e}\right]^5$

In[154]:= `Normal[%]`

Out[154]= x

The converse also holds.

In[155]:= `invSer /. x -> ser`

Out[155]= $x+O[x]^5$

We now invert the series $\sum_{i=0}^{n} a_i x^i$ with symbolic a_i. The coefficients of the resulting series grow quickly in size.

In[156]:= `Together /@ Normal[InverseSeries[`
`         (* the series of the form a0 + a1 x + a2 x^2 ... *)`
`                 Array[Subscript[a, #] ξ^#&, 5, 0, Plus] + O[ξ]^6]]`

Out[156]= $\dfrac{\xi-a_0}{a_1}-\dfrac{(\xi-a_0)^2\,a_2}{a_1^3}-\dfrac{(\xi-a_0)^3\,(-2\,a_2^2+a_1\,a_3)}{a_1^5}-$

$\dfrac{(\xi-a_0)^4\,(5\,a_2^3-5\,a_1\,a_2\,a_3+a_1^2\,a_4)}{a_1^7}+\dfrac{(\xi-a_0)^5\,(14\,a_2^4-21\,a_1\,a_2^2\,a_3+3\,a_1^2\,a_3^2+6\,a_1^2\,a_2\,a_4)}{a_1^9}$

Here is a series with the property that the coefficients of both, the direct and the inverse series are integers [1915].

In[157]:= `iSeries[z_, o_] :=`
`    z Exp[3 Sum[(3k)!/k!^3 Sum[1/j, {j, k + 1, 3k}] z^k, {k, o}]/`
`            Sum[(3k)!/k!^3 z^k, {k, 0, o}] + O[z]^o]`

In[158]:= `{iSeries[z, 5], InverseSeries[iSeries[z, 5]]}`

Out[158]= $\{z+15\,z^2+279\,z^3+5729\,z^4+124554\,z^5+O[z]^6,$

$z-15\,z^2+171\,z^3-1679\,z^4+15054\,z^5+O[z]^6\}$

Calculating the first 100 terms of the inverse series takes just a few seconds on a 2 GHz computer.

In[159]:= `(* check that all coefficients are integer and return maximal size *)`
`        {Union[Head /@ #], N[Max[Abs[#]]]}&[`
`                InverseSeries[iSeries[z, 100]][[3]]] // Timing`

Out[160]= $\{3.07\,\text{Second},\ \{\{\text{Integer}\},\ 3.094\times10^{81}\}\}$

The next input calculates the series expansion of the implicitly defined inverse function of $F(x)=\int_{\xi}^{x}f(t)\,dt$ (assuming $f(\xi)\neq0$) [497].

In[161]:= `InverseSeries[Series[Integrate[f[t], {t, ξ, x}], {x, ξ, 5}]]`

Out[161]= $\xi+\dfrac{x}{f[\xi]}-\dfrac{f'[\xi]\,x^2}{2\,f[\xi]^3}+\dfrac{(3\,f'[\xi]^2-f[\xi]\,f''[\xi])\,x^3}{6\,f[\xi]^5}+$

$\dfrac{(-15\,f'[\xi]^3+10\,f[\xi]\,f'[\xi]\,f''[\xi]-f[\xi]^2\,f^{(3)}[\xi])\,x^4}{24\,f[\xi]^7}+$

$\dfrac{1}{120\,f[\xi]^9}\,((105\,f'[\xi]^4-105\,f[\xi]\,f'[\xi]^2\,f''[\xi]+$

$10\,f[\xi]^2\,f''[\xi]^2+15\,f[\xi]^2\,f'[\xi]\,f^{(3)}[\xi]-f[\xi]^3\,f^{(4)}[\xi])\,x^5)+O[x]^6$

Using `InverseSeries`, we can easily compute the higher terms needed for finding zeros numerically using the Chebyshev method [335], [1370], [836], [1307], [1395], and [1408]. The idea behind this method is quite simple. Make a series expansion of an equation $f(x) = 0$ around an approximation $x \approx a$ of a root, invert this series, then use the fact that at the root, we have $f(x) = 0$.

```
In[162]:= C = Together /@ (Normal[InverseSeries[(* symbolic series; f not specified *)
              Series[f[x], {x, a, 5}]]] /. x -> 0)
```

$$\text{Out[162]= } a - \frac{f[a]}{f'[a]} - \frac{f[a]^2\, f''[a]}{2\, f'[a]^3} + \frac{f[a]^3\, (-3\, f''[a]^2 + f'[a]\, f^{(3)}[a])}{6\, f'[a]^5} - $$

$$\frac{f[a]^4\, (15\, f''[a]^3 - 10\, f'[a]\, f''[a]\, f^{(3)}[a] + f'[a]^2\, f^{(4)}[a])}{24\, f'[a]^7} + $$

$$\frac{1}{120\, f'[a]^9}\left(f[a]^5\left(-105\, f''[a]^4 + 105\, f'[a]\, f''[a]^2\, f^{(3)}[a] - \right.\right.$$

$$\left.\left. 10\, f'[a]^2\, f^{(3)}[a]^2 - 15\, f'[a]^2\, f''[a]\, f^{(4)}[a] + f'[a]^3\, f^{(5)}[a]\right)\right)$$

Here, a is the approximate value of the zero. We clearly see the Newton method in the first two terms. Here is a simple example that shows how the higher-order terms improve the solution. We use the function $f(x) = \cos(x^2) - x$ and $a = 1$.

```
In[163]:= Block[{λ = List @@ C, f = Cos[#^2] - #&, a = 1.},
           f /@ Rest[FoldList[Plus, 0, List @@ λ]]]
Out[163]= {-0.459698, -0.0553014, -0.0129139, -0.0025797, -0.000495504, -0.0000578252}
```

In addition to the wanted roots, the higher-order methods have other zeros. Here is the example $f(x) = x^2 - x$ used.

```
In[164]:= Solve[a == Take[C, 4] /. f -> ((#^2 - x)&), a]
```

$$\text{Out[164]= } \left\{\{a \to -\sqrt{x}\,\}, \{a \to \sqrt{x}\,\}, \left\{a \to -\frac{\sqrt{2\,x - i\,\sqrt{7}\,x}}{\sqrt{11}}\right\},\right.$$

$$\left\{a \to \frac{\sqrt{2\,x - i\,\sqrt{7}\,x}}{\sqrt{11}}\right\}, \left\{a \to -\frac{\sqrt{2\,x + i\,\sqrt{7}\,x}}{\sqrt{11}}\right\}, \left.\left\{a \to \frac{\sqrt{2\,x + i\,\sqrt{7}\,x}}{\sqrt{11}}\right\}\right\}$$

Let us remark here that such higher-order root finding procedures can be calculated using Schröder's formula [829], [39], [514], [937], [1326] without using explicit series inversion.

```
In[165]:= SchröderIterationFunction[f_, z_, n_] :=
           z + Sum[1/k! Nest[(1/f'[z] D[#, z])&, 1/f'[z], k - 1] (-f[z])^k,
              {k, n - 1}]

In[166]:= SchröderIterationFunction[f, a, 6] - C // Together
Out[166]= 0
```

Let us also remark that a truncated version of the continued fraction form of the first four terms of C yields Halley's root finding method [928].

Let us use `Series` to solve equations in an example from iteration theory. Let $f^{(k)}(x)$ stand for f applied k times to x. If $f(0) = 0$ for small x, under certain conditions the function $f^{(k)}(x)$ can be approximated through the following function $F[k, x, \{a, b, c\}]$ [778]. Here a, b, c are uniquely defined parameters.

```
In[167]:= F[k_, x_, {a_, b_, c_}] :=
              x/(1 - a k x) + (x/(1 - a k x))^2 b Log[1 - a k x] +
              (x/(1 - a k x))^3 b^2 (Log[1 - a k x]^2 -
              Log[1 - a k x] + c k x)
```

Let $f(x)$ be the relatively complicated function $(1 - x)\tan(x)\exp(x^2)/(\sin(x) + 1)$.

```
In[168]:= f[x_] := (1 - x) Tan[x] Exp[x^2]/(1 + Sin[x]);

           f[k_, x_] := Nest[f, x, k]
```

Fixing k allows to uniquely determine the parameters a, b, and c.

```
In[170]:= Module[{o = 6, ser1, ser2},
            Union @ (* one unique solution *)
              Table[ser1 = Series[f[k, x], {x, 0, o}];
                ser2 = Series[F[k, x, {a, b, c}], {x, 0, o}];
                Solve[(# == 0) & /@ Take[
                  DeleteCases[CoefficientList[ser1 - ser2, x], 0], 3],
                    {a, b, c}][[1]], {k, 2, 6}]]
```

$$Out[170]= \left\{\left\{c \to \frac{7}{2},\ b \to -\frac{1}{3},\ a \to -2\right\}\right\}$$

This yields the following approximation for $f^{(k)}(x)$.

```
In[171]:= Ff[k_, x_] = F[k, x, {a, b, c}] /. %[[1]]
```

$$Out[171]= \frac{x}{1 + 2\,k\,x} - \frac{x^2\,\text{Log}[1 + 2\,k\,x]}{3\,(1 + 2\,k\,x)^2} + \frac{x^3\,\left(\frac{7\,k\,x}{2} - \text{Log}[1 + 2\,k\,x] + \text{Log}[1 + 2\,k\,x]^2\right)}{9\,(1 + 2\,k\,x)^3}$$

The next graphic shows the relative difference between $f^{(1000)}(x)$ and the calculated approximation along the real and imaginary axes. The two functions agree quite well for small x.

```
In[172]:= With[{k = 1000, ε = 10.^-6},
          Show[GraphicsArray[
          Block[{$DisplayFunction = Identity},
              {(* along real axis *)
              ListPlot[Table[Log[10, Abs[Ff[k, x]/f[k, x] - 1]],
                          {x, ε, 1 - ε, (2 - ε)/200}],
                    Frame -> True, Axes -> False, PlotRange -> All],
                (* along imaginary axis *)
                ListPlot[Table[Log[10, Abs[Ff[k, I x]/f[k, I x] - 1]],
                          {x, -1 + ε, 1 - ε, (2 - ε)/400}],
                    Frame -> True, Axes -> False]}]]]]
```

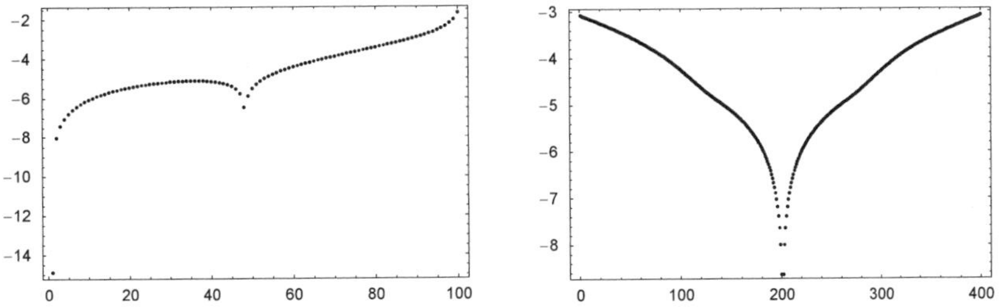

The error is a decreasing function of k.

```
In[173]:= With[{x = N[1/3, 100], k = 10000},
          ListPlot[Log[10, Abs[Rest[NestList[f, x, k]]/
                  Table[Ff[j, x], {j, k}] - 1]],
                PlotRange -> All, Frame -> True, Axes -> False]]
```

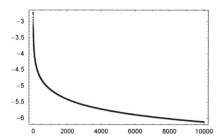

Next, we discuss a peculiarity of `Solve`. When applied with series (with the head `SeriesData`) as arguments in the `Equal`-objects of the first argument of `Solve`: `Solve[series₁ == series₂, {var₁, var₂, ..., varₙ}]` solves the equations resulting from comparing the coefficients in *series₁* and *series₂* for the variables *varᵢ*. This allows the convenient solution of differential equations by assuming a power series expansion for the solution. We now look at an example involving the (nonlinear!) differential equation of a planar mathematical pendulum $y''(t) = \sin(y(t))$.

In[174]:= **ySeries = Sum[a[i] t^i, {i, 0, 6}] + O[t]^7**

Out[174]= $a[0] + a[1] \ t + a[2] \ t^2 + a[3] \ t^3 + a[4] \ t^4 + a[5] \ t^5 + a[6] \ t^6 + O[t]^7$

In[175]:= **ySeriesDiff = D[ySeries, {t, 2}]**

Out[175]= $2 \ a[2] + 6 \ a[3] \ t + 12 \ a[4] \ t^2 + 20 \ a[5] \ t^3 + 30 \ a[6] \ t^4 + O[t]^5$

Now, we express the constants `a[2]` and `a[3]` in terms of `a[0]` and `a[1]` (which are to be determined by the initial conditions) using `Solve`.

In[176]:= **Solve[ySeriesDiff == Sin[ySeries], {a[2], a[3], a[4], a[5], a[6]}]**

Out[176]= $\left\{ \left\{ a[5] \rightarrow \frac{1}{120} \ (-a[1] - a[1]^3 \ \text{Cos}[a[0]] + 2 \ a[1] \ \text{Cos}[2 \ a[0]]), \right. \right.$

$a[6] \rightarrow \frac{1}{720} \ (-1 + a[1]^4 - 11 \ a[1]^2 \ \text{Cos}[a[0]] + 2 \ \text{Cos}[2 \ a[0]]) \ \text{Sin}[a[0]],$

$a[4] \rightarrow \frac{1}{48} \ (-2 \ a[1]^2 \ \text{Sin}[a[0]] + \text{Sin}[2 \ a[0]]),$

$\left. \left. a[3] \rightarrow \frac{1}{6} \ a[1] \ \text{Cos}[a[0]], \ a[2] \rightarrow \frac{1}{2} \ \text{Sin}[a[0]] \right\} \right\}$

This leads us to the following solution for $y(t)$.

In[177]:= **ySeries /. %**

Out[177]= $\left\{ a[0] + a[1] \ t + \frac{1}{2} \ \text{Sin}[a[0]] \ t^2 + \right.$

$\frac{1}{6} \ a[1] \ \text{Cos}[a[0]] \ t^3 + \frac{1}{48} \ (-2 \ a[1]^2 \ \text{Sin}[a[0]] + \text{Sin}[2 \ a[0]]) \ t^4 +$

$\frac{1}{120} \ (-a[1] - a[1]^3 \ \text{Cos}[a[0]] + 2 \ a[1] \ \text{Cos}[2 \ a[0]]) \ t^5 +$

$\left. \frac{1}{720} \ (-1 + a[1]^4 - 11 \ a[1]^2 \ \text{Cos}[a[0]] + 2 \ \text{Cos}[2 \ a[0]]) \ \text{Sin}[a[0]] \ t^6 + O[t]^7 \right\}$

In[178]:= **(* quick check of the solution *)**
D[%%, {t, 2}] == Sin[%%] // Simplify

Out[179]= True

Although `Series` alone can deal with Laurent and Puiseux series as well as Taylor series, this may not be the case when it is used in conjunction with `Solve`. In the following example, we expand `y[x]` and `y''[x]` in a series about $x = 0$, and compare the coefficients of the x^j. *Mathematica* does not currently handle this kind of series in `Solve`. Here is the case of a pure Taylor series.

In[180]:= `Solve[Series[y''[x] +  y[x] + 2 + x^2, {x, 0, 4}] == 0,`
            `Table[Derivative[k][y][0], {k, 2, 6}]]`

Out[180]= $\{\{y''[0] \to -2 - y[0], y^{(3)}[0] \to -y'[0], y^{(4)}[0] \to y[0], y^{(5)}[0] \to y'[0], y^{(6)}[0] \to -y[0]\}\}$

This leads to the following first terms of the solution.

In[181]:= `Series[y[x], {x, 0, 6}] /. %`

Out[181]= $\{y[0] + y'[0] x + \frac{1}{2} (-2 - y[0]) x^2 -$
          $\frac{1}{6} y'[0] x^3 + \frac{1}{24} y[0] x^4 + \frac{1}{120} y'[0] x^5 - \frac{1}{720} y[0] x^6 + O[x]^7\}$

In the following two examples, we would have to use a Laurent and Puiseux series, respectively, for y[x].

In[182]:= `Solve[Series[y''[x] +  y[x] + 2 + x^(-1), {x, 0, 2}] == 0,`
            `{y''''[0], y'''[0], y''[0]}]& /@ {x^(-1), x^(7/4)}`

Out[182]= `{{}, {}}`

Next, we express the coefficients c_{ijl} of the series

$$\sum_{i,j,k=0}^{n} c_{ijk} s_a^i s_b^j s_c^k = \ln\left(\sum_{i,j,k=0}^{n} m_{ijk} s_a^i s_b^j s_c^k\right)$$

in terms of the m_{ijl}. (This is a so-called cumulant expansion; see [1240], [1308], [1763], [696], [1219], [1029], and [1056].) Let $n = 2$.

In[183]:= `n = 2;`

            `(momentSum = Sum[m[i, j, l] sa^i sb^j sc^l,`
                        `{i, 0, n}, {j, 0, n}, {l, 0, n}] /. m[0, 0, 0] -> 1);`

In[185]:= `logMomentSum = Series[Log[momentSum], {sa, 0, n}, {sb, 0, n}, {sc, 0, n}];`

In[186]:= `(cumulantSum = Sum[c[i, j, l] sa^i sb^j sc^l,`
            `{i, 0, n}, {j, 0, n}, {l, 0, 2}] +` (* make Series *)
            `O[sa]^(n + 1) + O[sb]^(n + 1) + O[sc]^(n + 1)) // Short[#, 3]&`

Out[186]//Short=
        $((c[0, 0, 0] + c[0, 0, 1] sc + c[0, 0, 2] sc^2 + O[sc]^3) +$
          $(c[0, 1, 0] + sc\, c[0, 1, 1] + sc^2\, c[0, 1, 2]) sb +$
          $(c[0, 2, 0] + sc\, c[0, 2, 1] + sc^2\, c[0, 2, 2]) sb^2 + O[sb]^3) +$
          $(c[1, 0, 0] + sc\, c[1, 0, 1] + sc^2\, c[1, 0, 2] + \ll 3 \gg + sb^2\, c[1, 2, 0] +$
          $sb^2\, sc\, c[1, 2, 1] + sb^2\, sc^2\, c[1, 2, 2]) sa + \ll 1 \gg + O[sa]^3$

Calculating the cumulants can be easily done by a single call to `Solve`. Because of space limitations, we do not look at all k_{ijl}, but only at k_{222}, and this only in short form.

In[187]:= `Solve[logMomentSum == cumulantSum, {c[2, 2, 2]}] // Simplify // Short[#, 4]&`

Out[187]//Short=
        $\{\{c[2, 2, 2] \to -\frac{3}{2} m[0, 1, 1]^2 m[1, 0, 0]^2 -$
          $3 m[0, 1, 0] m[0, 1, 2] m[1, 0, 0]^2 + \ll 46 \gg + m[2, 2, 2]\}\}$

If we need the equations for comparison of the coefficients of equal powers themselves, we can use the command `LogicalExpand`, which we have already discussed in Chapter 5 of the Programming volume [1735] in connection with logical coupling.

LogicalExpand[*series*₁ == *series*₂]

gives the equations arising in the comparison of corresponding powers of the two series *series*₁ and *series*₂ in the form And[*coefficientEquations*].

Here are the resulting equations for the differential equation $y''(x) = y'(x) + 3\,y(x) + 4\,x + 5$ using a series expansion of the form $y(x) = \sum_{j=0}^{8} c_j\, x^j$.

In[188]:= `LogicalExpand[ (D[#, {x, 2}]  == D[#, x] + 3 # + 4 x + 5)&[`
`                        Sum[c[j] x^j, {j, 0, 8}] + O[x]^9]]`

Out[188]= $5 + 3\,c[0] + c[1] - 2\,c[2] = 0 \,\&\&\, 4 + 3\,c[1] + 2\,c[2] - 6\,c[3] = 0 \,\&\&$
$3\,c[2] + 3\,c[3] - 12\,c[4] = 0 \,\&\&\, 3\,c[3] + 4\,c[4] - 20\,c[5] = 0 \,\&\&$
$3\,c[4] + 5\,c[5] - 30\,c[6] = 0 \,\&\&\, 3\,c[5] + 6\,c[6] - 42\,c[7] = 0 \,\&\&\, 3\,c[6] + 7\,c[7] - 56\,c[8] = 0$

We could operate further on this system of equations, for example, with `Solve`. (We could plug in the last series directly into `Solve`, because, as just discussed, `Solve` accepts in addition to list of equations or a logical combination of equations as `SeriesData`-objects.)

In[189]:= `Solve[%, Array[c, 7, 2]] // Short[#, 6]&`
Out[189]//Short=

$$\left\{\left\{c[2] \rightarrow \frac{1}{2}\,(5 + 3\,c[0] + c[1]),\right.\right.$$

$$c[3] \rightarrow \frac{1}{6}\,(9 + 3\,c[0] + 4\,c[1]),\ c[4] \rightarrow \frac{1}{24}\,(24 + 12\,c[0] + 7\,c[1]),$$

$$c[5] \rightarrow \frac{1}{120}\,(51 + 21\,c[0] + 19\,c[1]),\ c[6] \rightarrow \frac{1}{720}\,(123 + 57\,c[0] + 40\,c[1]),$$

$$\left.\left.c[7] \rightarrow \frac{276 + 120\,c[0] + 97\,c[1]}{5040},\ c[8] \rightarrow \frac{645 + 291\,c[0] + 217\,c[1]}{40320}\right\}\right\}$$

Here is a quick check of the solution.

In[190]:= `(D[#, {x, 2}]  - (D[#, x] + 3 # + 4 x + 5))&[`
`          (Sum[c[j] x^j, {j, 0, 8}] /. %[[1]]]) + O[x]^9]`
Out[190]= $O[x]^7$

There is no built-in way of using series in `InverseFunction`. For example, if we try to find the series expansion of the inverse function for $y = x^5 + x$ about $y = \infty$, the following implementation fails. (`Inverse⁚Function[Function[x, x^5 + x]]` is not explicitly computed.)

In[191]:= `Series[InverseFunction[Function[x, x^5 + x]][y], {y, Infinity, 3}]`
Out[191]= `InverseFunction[Function[x, $x^5 + x$]][y]`

However, with `InverseSeries`, we can easily find the (more correctly—one of the) desired series representation.

In[192]:= `InverseSeries[Series[x^5 + x, {x, Infinity, 50}]]`

Out[192]= $\dfrac{1}{\left(\frac{1}{x}\right)^{1/5}} - \dfrac{1}{5}\left(\dfrac{1}{x}\right)^{3/5} - \dfrac{1}{25}\left(\dfrac{1}{x}\right)^{7/5} - \dfrac{1}{125}\left(\dfrac{1}{x}\right)^{11/5} + \dfrac{21\left(\frac{1}{x}\right)^{19/5}}{15625} +$

$\dfrac{78\left(\frac{1}{x}\right)^{23/5}}{78125} + \dfrac{187\left(\frac{1}{x}\right)^{27/5}}{390625} + \dfrac{286\left(\frac{1}{x}\right)^{31/5}}{1953125} - \dfrac{9367\left(\frac{1}{x}\right)^{39/5}}{244140625} -$

$\dfrac{39767\left(\frac{1}{x}\right)^{43/5}}{1220703125} - \dfrac{105672\left(\frac{1}{x}\right)^{47/5}}{6103515625} - \dfrac{175398\left(\frac{1}{x}\right)^{51/5}}{30517578125} + O\left[\dfrac{1}{x}\right]^{11}$

Here is a check for the correctness of the last result.

In[193]:= `#^5 + #&[%] // Normal`

Out[193]= x

While `Solve[Series[...] == 0, ...]` works, it is frequently more efficient to not generate too many terms of a series that have symbolic expressions because they frequently grow quickly in size. Instead solving for the highest term recursively is in many cases more efficient. Here is an example. The function $\psi(z) = z + \sum_{k=0}^{\infty} \beta_m z^{-m}$ defined implicitly through $p_n(\psi(z)) = z^{2^n} + O(z)$ for large z, maps the exterior of unit disk to the exterior of the Mandelbrot set [599]. Here $p_n(z) = p_{n-1}(z)^2 + z$, $p_0(z) = z$. The function `makeMandenbrotFunction[n]` calculates the approximation $\psi_n(z) = z + \sum_{k=0}^{2^n-2} \beta_m z^{-m}$. We use approximate series coefficients to avoid building fractions with large integers for the higher-order β_k.

```
In[194]:= makeMandenbrotFunction[n_Integer?Positive] :=
  Module[{},
      (* classical Mandelbrot recursion *)
      p[0][z_] := z;
      p[v_][z_] := p[v - 1][z]^2 + z;
      (* recursion start *)
      g[z_] = 1. z; L = Table[0, {2^n}]; L[[1]] = 1.;
      (* loop over terms of the series until order 2^n-2 *)
      Do[ψ = Series[g[z] + β[j]/z^j, {z, Infinity, j}];
          eqs = Chop @ Normal[p[n][ψ] - z^(2^n)] == 0;
          sol = Solve[eqs, β[j]];
          g[z_] = g[z] + β[j]/z^j /. sol[[1]]; L[[2 + j]] = sol[[1, 1, 2]],
          {j, 0, 2^n - 2}];
      (* return series as pure function and coefficient list *)
      {Function[Evaluate[g[#]]], L}]
```

Here is a shortened version of the resulting series.

```
In[195]:= Short[(mBF = makeMandenbrotFunction[8])[[1]][z], 6] // Timing
```

$$
\begin{aligned}
\text{Out[195]}= \Big\{ & 30.98\,\text{Second}, -0.5 - \frac{0.00230159}{z^{254}} - \frac{0.000258146}{z^{253}} - \frac{0.000107904}{z^{252}} - \frac{0.0012089}{z^{251}} + \\
& \frac{0.00022949}{z^{250}} - \frac{0.000188484}{z^{249}} - \frac{0.00031564}{z^{248}} - \frac{0.000225405}{z^{247}} - \frac{3.81051\times10^{-6}}{z^{246}} - \\
& \frac{0.000441422}{z^{245}} - \frac{0.00102796}{z^{244}} + \ll 360 \gg + \frac{0.03125}{z^{10}} - \frac{0.0140114}{z^{9}} + \frac{0.}{z^{8}} + \\
& \frac{0.0301208}{z^{7}} - \frac{0.0625}{z^{6}} - \frac{0.0458984}{z^{5}} + \frac{0.}{z^{4}} + \frac{0.117188}{z^{3}} - \frac{0.25}{z^{2}} + \frac{0.125}{z} + 1.\,z \Big\}
\end{aligned}
$$

Graphing the map of the unit circle under the ψ_k for the $k \le 2^8 - 2$ shows how the classical Mandelbrot shape arises.

```
In[196]:= Show[Graphics[
    (* color each ψ_k differently *)
    MapIndexed[{Hue[0.78 #2[[1]]]/Length[mBF[[2]]]], Line[#]}&,
    (* form curves efficiently *)
    Map[{Re[#], Im[#]}&, Transpose[Rest[FoldList[Plus, 0, #]]& /@
        Table[Table[Exp[I φ]^j, {j, 1, -Length[mBF[[2]]] + 2, -1}] mBF[[2]],
            (* use enough plotpoints *) {φ, 0., 2.Pi, 2Pi/1200}]], {2}]]]]
```

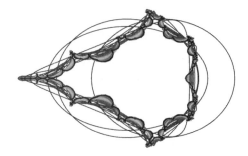

As an application of `Series`, we will deal with the discrete approximation of arbitrary linear functionals on equally spaced data points. This very elegant and useful application is from [381].

Mathematical Remark: Linear Functionals

A functional is a mapping $L : G \to \mathbb{R}$ (or $G \to \mathbb{C}$) of a linear space G (numbers, function space, etc.) into the real or complex numbers. A functional is linear, provided that $L[\lambda_1 g_1 + \lambda_2 g_2] = \lambda_1 L[g_1] + \lambda_2 L[g_2]$ for $\lambda_i \in \mathbb{R}$ or $\lambda_i \in \mathbb{C}$ and $g_i \in G$.

Examples of linear functionals are integrals, differential quotients at given points, etc. To work with functionals numerically, it is necessary to discretize them; that is, $L(g(x))$ should be approximated by an expression of the form $L[g(x)] \approx \sum_{i=0}^{n} w_i\, g(x_i)$, where $g(x)$ is a function of a real variable x. Here, $x_i = x_0 + i\,h$ are data points and w_i are weights. For n equally spaced data points with spacing h, the requirement that the approximation of the functional should reproduce exactly the moments $m_i = L[x^i]$ leads to the following result. If we expand $W(\xi) = \xi^{-x_0/h} \sum_{j=0}^{n-1} m_j / (j!\, h^j) \ln^j \xi$ in a Taylor series about $\xi = 1$ up to order $n - 1$ with respect to the variables ξ, the factors appearing in front of the ξ^i are the desired weights (the proof proceeds by substitution and explicit computation) $w_i = [\xi^i]\,(W(\xi))$. (For the remainder estimation, see [326].)

We now give a straightforward implementation.

```
In[197]:= weights[functional_, order_Integer, {x0_, h_}] :=
            Module[{X, var}, CoefficientList[Series[X^(-x0/h) *
                        Sum[functional[var^j, var] Log[X]^j/j!/h^j,
                            {j, 0, order - 1}], {X, 1, order - 1}], X]]
```

We assume that the functional being approximated is given in the form *functional* [*function*, *variable*]. We now look at a few examples. First, we consider the approximation of a first derivative of a function at $x = 0$ using a finite number of equally spaced points (with distance 1) lying symmetrically around $x_0 = 0$.

```
In[198]:= (* or, in a more functional style: Derivative[1][Function[#2, #1]][0]& *)
            der1[f_, var_] := D[f, var] /. var -> 0;

            Do[Print["i = ", i, "  ", weights[der1, i, {-i/2 + 1/2, 1}]], {i, 3, 11, 2}]
```

$$i = 3 \quad \left\{ -\frac{1}{2},\ 0,\ \frac{1}{2} \right\}$$

$$i = 5 \quad \left\{ \frac{1}{12},\ -\frac{2}{3},\ 0,\ \frac{2}{3},\ -\frac{1}{12} \right\}$$

$$i = 7 \quad \left\{ -\frac{1}{60},\ \frac{3}{20},\ -\frac{3}{4},\ 0,\ \frac{3}{4},\ -\frac{3}{20},\ \frac{1}{60} \right\}$$

$$i = 9 \quad \left\{ \frac{1}{280},\ -\frac{4}{105},\ \frac{1}{5},\ -\frac{4}{5},\ 0,\ \frac{4}{5},\ -\frac{1}{5},\ \frac{4}{105},\ -\frac{1}{280} \right\}$$

$$i = 11 \quad \{-\frac{1}{1260}, \frac{5}{504}, -\frac{5}{84}, \frac{5}{21}, -\frac{5}{6}, 0, \frac{5}{6}, -\frac{5}{21}, \frac{5}{84}, -\frac{5}{504}, \frac{1}{1260}\}$$

Here is the analog for second derivatives.

```
In[201]:= (* or, in a more functional style: Derivative[2][Function[#2, #1]][0]& *)
          der2[f_, var_] := D[f, {var, 2}] /. var -> 0;

          Do[Print["i = ", i, "  ", weights[der2, i, {-i/2 + 1/2, 1}]], {i, 3, 11, 2}]
```

$$i = 3 \quad \{1, -2, 1\}$$

$$i = 5 \quad \{-\frac{1}{12}, \frac{4}{3}, -\frac{5}{2}, \frac{4}{3}, -\frac{1}{12}\}$$

$$i = 7 \quad \{\frac{1}{90}, -\frac{3}{20}, \frac{3}{2}, -\frac{49}{18}, \frac{3}{2}, -\frac{3}{20}, \frac{1}{90}\}$$

$$i = 9 \quad \{-\frac{1}{560}, \frac{8}{315}, -\frac{1}{5}, \frac{8}{5}, -\frac{205}{72}, \frac{8}{5}, -\frac{1}{5}, \frac{8}{315}, -\frac{1}{560}\}$$

$$i = 11$$
$$\{\frac{1}{3150}, -\frac{5}{1008}, \frac{5}{126}, -\frac{5}{21}, \frac{5}{3}, -\frac{5269}{1800}, \frac{5}{3}, -\frac{5}{21}, \frac{5}{126}, -\frac{5}{1008}, \frac{1}{3150}\}$$

Here are the weights for even-order approximations for the first and second derivative around $x_0 = 0$ and node distance 1 [1014], [1001].

```
In[204]:= Table[weights[der1, k, {(-k + 1)/2, 1}], {k, 2, 8, 2}]
```

$$Out[204]= \{\{-1, 1\}, \{\frac{1}{24}, -\frac{9}{8}, \frac{9}{8}, -\frac{1}{24}\}, \{-\frac{3}{640}, \frac{25}{384}, -\frac{75}{64}, \frac{75}{64}, -\frac{25}{384}, \frac{3}{640}\},$$
$$\{\frac{5}{7168}, -\frac{49}{5120}, \frac{245}{3072}, -\frac{1225}{1024}, \frac{1225}{1024}, -\frac{245}{3072}, \frac{49}{5120}, -\frac{5}{7168}\}\}$$

```
In[205]:= Table[weights[der2, k, {(-k + 1)/2, 1}], {k, 4, 8, 2}]
```

$$Out[205]= \{\{\frac{1}{2}, -\frac{1}{2}, -\frac{1}{2}, \frac{1}{2}\}, \{-\frac{5}{48}, \frac{13}{16}, -\frac{17}{24}, -\frac{17}{24}, \frac{13}{16}, -\frac{5}{48}\},$$
$$\{\frac{259}{11520}, -\frac{499}{2304}, \frac{1299}{1280}, -\frac{1891}{2304}, -\frac{1891}{2304}, \frac{1299}{1280}, -\frac{499}{2304}, \frac{259}{11520}\}\}$$

We can also find the weights for a functional that is not given explicitly. To improve readability, we replace the functional by the corresponding moments.

```
In[206]:= weights[f, 4, {x0, h}] //. {f[var_^exp_., var_] -> m[exp],
                                       f[1, var_] -> m[0]} // Simplify
```

$$Out[206]= \{\frac{1}{6 h^3}(6 h^3 m[0] + x0^3 m[0] + 11 h^2 (x0 m[0] - m[1]) -$$
$$3 x0^2 m[1] + 3 x0 m[2] + 6 h (x0^2 m[0] - 2 x0 m[1] + m[2]) - m[3]),$$
$$\frac{1}{2 h^3}(-x0^3 m[0] + 3 x0^2 m[1] + h^2 (-6 x0 m[0] + 6 m[1]) - 3 x0 m[2] -$$
$$5 h (x0^2 m[0] - 2 x0 m[1] + m[2]) + m[3]), \frac{1}{2 h^3}(x0^3 m[0] + 3 h^2 (x0 m[0] - m[1]) -$$
$$3 x0^2 m[1] + 3 x0 m[2] + 4 h (x0^2 m[0] - 2 x0 m[1] + m[2]) - m[3]),$$
$$\frac{1}{6 h^3}(-x0^3 m[0] + 3 x0^2 m[1] + h^2 (-2 x0 m[0] + 2 m[1]) -$$
$$3 x0 m[2] - 3 h (x0^2 m[0] - 2 x0 m[1] + m[2]) + m[3])\}$$

Next, we examine a few integration Newton–Cotes-type formulas. We consider only integrating from $-n$ to n. To be able to compare with the references [10], we factor out the greatest common multiple of the denominators.

```
In[207]:= integral[f_, var_] := Integrate[f, {var, -k, k}] /. var -> 0;

          Do[k = (i - 1)/2;
```

```
wts = weights[integral, i, {-k, 1}];
(* bring result in nicer form *)
denominator = LCM @@ (Denominator /@ wts);
numerator = GCD @@ (denominator wts);
Print["i = ", i, "  ", numerator/denominator, wts denominator/numerator],
{i, 3, 13, 2}]
```

$$i = 3 \quad \frac{1}{3}\{1, 4, 1\}$$

$$i = 5 \quad \frac{2}{45}\{7, 32, 12, 32, 7\}$$

$$i = 7 \quad \frac{1}{140}\{41, 216, 27, 272, 27, 216, 41\}$$

$$i = 9 \quad \frac{4}{14175}\{989, 5888, -928, 10496, -4540, 10496, -928, 5888, 989\}$$

$$i = 11 \quad \frac{5}{299376}\{16067, 106300, -48525, 272400,$$
$$-260550, 427368, -260550, 272400, -48525, 106300, 16067\}$$

$$i = 13 \quad \frac{1}{5255250}\{1364651, 9903168, -7587864, 35725120, -51491295, 87516288,$$
$$-87797136, 87516288, -51491295, 35725120, -7587864, 9903168, 1364651\}$$

This agrees with the formulas from the mentioned reference [10].

1.6.5 Residues

The residue is the coefficient of the term $(z - z_0)^{-1}$ in the Laurent expansion of a function about the point z_0. (For a very detailed survey of applications of residues, see [1278] and [1279].) In *Mathematica*, we get the residue of a function at a given point via Residue.

Residue[*function*, {*var*, *expansionPoint*}]

    computes the residue of the function *function* with respect to the variable *var* at the point *expansionPoint*.

Such residues arise frequently in the solution of ordinary differential equations via Fourier transforms. Here are two examples.

In[1]:= **Residue[Exp[I k x]/(k^2 - ki^2), {k, ki}]**

Out[1]= $\dfrac{e^{i\,ki\,x}}{2\,ki}$

In[2]:= **Residue[Exp[I k x]/(k^4 - ki^4), {k, ki}]**

Out[2]= $\dfrac{e^{i\,ki\,x}}{4\,ki^3}$

The residue can also be computed at the point Infinity.

In[3]:= **Residue[1/z, {z, Infinity}]**

Out[3]= -1

Residues do not exist at branch points.

In[4]:= **Residue[z^(-1/2), {z, 0}]**

Out[4]= Residue$\left[\dfrac{1}{\sqrt{z}}, \{z, 0\}\right]$

Typically, residues are only defined for poles. Here, we give *Mathematica* an essential singularity.

In[5]:= **Residue[Exp[-1/z], {z, 0}]**

Out[5]= Residue[$e^{-1/z}$, {z, 0}]

Carrying out the defining contour integral for the residue shows that the last result is sensible.

In[6]:= **NIntegrate[Exp[-1/z], {z, 1, I, -1, -I, 1}]/(2Pi I) // Chop**

Out[6]= -1.

In a similar way, the following residues could be defined [1740]. But they stay currently unevaluated.

In[7]:= **{Residue[z^3 Exp[1/z], {z, 0}], Residue[Exp[Exp[1/z]], {z, 0}],
        Residue[Exp[Exp[Exp[1/z]]], {z, 0}]}**

Out[7]= $\left\{ \text{Residue}\left[e^{\frac{1}{z}} z^3, \{z, 0\} \right], \text{Residue}\left[e^{e^{\frac{1}{z}}}, \{z, 0\} \right], \text{Residue}\left[e^{e^{e^{\frac{1}{z}}}}, \{z, 0\} \right] \right\}$

Residues can also be calculated at the poles of special functions, such as at the following pole of order 2 of the product of two Γ functions.

In[8]:= **Residue[Gamma[z] Gamma[z + 3], {z, -5}]**

Out[8]= $\dfrac{-227 + 120 \text{ EulerGamma}}{14400}$

Unfortunately, the command `Residue` does not have an `Assumptions` option in the moment. So the following input does not work.

In[9]:= **Residue[1/Sin[x/Pi], {x, k}, Assumptions -> Element[k, Integers]]**

  Residue::argrx : Residue called with 3 arguments; 2 arguments are expected. More…

Out[9]= Residue$\left[\text{Csc}\left[\dfrac{x}{\pi} \right], \{x, k\}, \text{Assumptions} \rightarrow k \in \text{Integers} \right]$

1.6.6 Sums

We already encountered the function `Sum`. We used it for carrying out finite sums in a procedural way. The symbol `Sum` has a double meaning in *Mathematica*. It does the sum by evaluating one summand after the other, but it is also used to carry out symbolic summations.

In[1]:= **?Sum**

            Sum[f, {i, imax}] evaluates the sum of the expressions
             f as evaluated for each i from 1 to imax. Sum[f, {i,
             imin, imax}] starts with i = imin. Sum[f, {i, imin, imax,
             di}] uses steps di. Sum[f, {i, imin, imax}, {j, jmin,
             jmax}, ...] evaluates a sum over multiple indices. More…

Sum[*summand*, {*var*, *lowerLimit*, *upperLimit*}]
  finds (if possible) the sum $\sum_{var=lowerLimit}^{upperLimit}$ *summand*.

Similar to integration, a variety of special algorithms exist to do symbolic summation [1631], [1046].

Consider the following sums: $s_{j,n} = \sum_{i=i}^{n} i^j$.

In[2]:= **Do[CellPrint[Cell[BoxData[FormBox[RowBox[{ (* marker *) " ∘ ",
         MakeBoxes[#, TraditionalForm] & @@**

```
                    {Subscript[s, j, n] == Sum[i^j, {i, n}]}}],
                    TraditionalForm]], "Output"]], {j, 4}]
```

$\circ \ s_{1,n} = \dfrac{1}{2}\, n\,(n+1)$

$\circ \ s_{2,n} = \dfrac{1}{6}\, n\,(n+1)\,(2\,n+1)$

$\circ \ s_{3,n} = \dfrac{1}{4}\, n^2\,(n+1)^2$

$\circ \ s_{4,n} = \dfrac{1}{30}\, n\,(n+1)\,(2\,n+1)\,(3\,n^2+3\,n-1)$

Note that a commonly given closed formula for this sum in terms of Bernoulli numbers is $(j+1)^{-1}\sum_{i=0}^{j}\binom{j+1}{i}B_i\,(n+1)^{j-i+1}$ (for arbitrary positive integers n and j) ([365], [951], [1038], [813], [788], [1175], [1146], [1066]). *Mathematica* computes the following sum as a difference of two harmonic numbers (Zeta functions).

In[3]:= **Sum[i^j, {i, n}]**

Out[3]= HarmonicNumber[n, -j]

Here are a few examples of the capabilities of Sum. (We assume that the limits of summation are integers.) Here is a slight generalization of the harmonic series.

In[4]:= **Sum[1/(i + a), {i, n}]**

Out[4]= -PolyGamma[0, 1 + a] + PolyGamma[0, 1 + a + n]

The result contains a Polygamma function, which we will discuss in Chapter 3.

In the next input, we compute the first five moments of the binomial distribution.

In[5]:= **Table[Sum[Binomial[n, k] p^k (1 - p)^(n - k) k^i, {k, 0, n}] //**
PowerExpand // Factor, {i, 0, 5}]

Out[5]= $\{(-1)^n\,(1-p)^n\,(-1+p)^{-n},$

$(-1)^n\,n\,(1-p)^n\,(-1+p)^{-n}\,p,\ n\,(1-p)^n\,p\,(1-p+n\,p)\left(1+\dfrac{p}{1-p}\right)^n,$

$n\,(1-p)^n\,p\left(1+\dfrac{p}{1-p}\right)^n\,(1-3\,p+3\,n\,p+2\,p^2-3\,n\,p^2+n^2\,p^2),\ n\,(1-p)^n\,p$

$\left(1+\dfrac{p}{1-p}\right)^n\,(1-7\,p+7\,n\,p+12\,p^2-18\,n\,p^2+6\,n^2\,p^2-6\,p^3+11\,n\,p^3-6\,n^2\,p^3+n^3\,p^3),$

$n\,(1-p)^n\,p\left(1+\dfrac{p}{1-p}\right)^n\,(1-15\,p+15\,n\,p+50\,p^2-75\,n\,p^2+25\,n^2\,p^2-60\,p^3+$

$110\,n\,p^3-60\,n^2\,p^3+10\,n^3\,p^3+24\,p^4-50\,n\,p^4+35\,n^2\,p^4-10\,n^3\,p^4+n^4\,p^4)\}$

For symbolic limits, it is implicitly assumed that the limits are integer-valued.

In[6]:= **Sum[Exp[I k x], {k, n}]**

Out[6]= $\dfrac{e^{i\,x}\,(-1+e^{i\,n\,x})}{-1+e^{i\,x}}$

Similar to integration results, the results of Sum will often contain special functions and RootSum-objects.

In[7]:= **Sum[Binomial[n, k] k^3/(n - k), {k, n}]**

$$\text{Out[7]= } -2^{-1+n} n^2 - (-1 + 2^n) n^2 - \frac{n^2}{(-1+n)^2} + \frac{2^n n^2}{(-1+n)^2} +$$

$$\frac{n^3}{(-1+n)^2} - \frac{2^n n^3}{(-1+n)^2} - \frac{n^3}{-1+n} + \frac{2^n n^3}{-1+n} - 2^{-2+n} n (1+n) +$$

$$\frac{n^3 \pi \, \text{Csc}[n\,\pi] \, \text{HypergeometricPFQRegularized}[\{1-n,\ 1-n\},\ \{2-n\},\ -1]}{(-1+n)^2 \, \text{Gamma}[-1+n]} -$$

$$\frac{n^4 \pi \, \text{Csc}[n\,\pi] \, \text{HypergeometricPFQRegularized}[\{1-n,\ 1-n\},\ \{2-n\},\ -1]}{(-1+n)^2 \, \text{Gamma}[-1+n]}$$

In[8]:= `Sum[Log[k^2 + 2k - 1], {k, n}]`

$$\text{Out[8]= } \text{Log}[\text{Gamma}[2 - \sqrt{2} + n]] + \text{Log}[\text{Gamma}[2 + \sqrt{2} + n]] +$$

$$\frac{1}{2} \, (-\text{Log}[2] - 2\,\text{LogGamma}[2 - \sqrt{2}] - 2\,\text{LogGamma}[-1 + \sqrt{2}])$$

In[9]:= `Sum[ (k^2 - 3)/(k^3 + 4), {k, n}]`

$$\text{Out[9]= } \frac{1}{3} \Bigg(-\text{RootSum}[5 + 3\,\#1 + 3\,\#1^2 + \#1^3\ \&,\ \text{PolyGamma}[0,\ -\#1]\ \&] +$$

$$3\,\text{RootSum}\Big[5 + 3\,\#1 + 3\,\#1^2 + \#1^3\ \&,\ \frac{\text{PolyGamma}[0,\ -\#1]}{1 + 2\,\#1 + \#1^2}\ \&\Big] -$$

$$2\,\text{RootSum}\Big[5 + 3\,n + 3\,n^2 + n^3 + 3\,\#1 + 6\,n\,\#1 + 3\,n^2\,\#1 + 3\,\#1^2 + 3\,n\,\#1^2 + \#1^3\ \&,$$

$$\frac{\text{PolyGamma}[0,\ -\#1]}{1 + 2\,n + n^2 + 2\,\#1 + 2\,n\,\#1 + \#1^2}\ \&\Big] +$$

$$2\,n\,\text{RootSum}\Big[5 + 3\,n + 3\,n^2 + n^3 + 3\,\#1 + 6\,n\,\#1 + 3\,n^2\,\#1 + 3\,\#1^2 + 3\,n\,\#1^2 + \#1^3\ \&,$$

$$\frac{\text{PolyGamma}[0,\ -\#1]}{1 + 2\,n + n^2 + 2\,\#1 + 2\,n\,\#1 + \#1^2}\ \&\Big] +$$

$$n^2\,\text{RootSum}\Big[5 + 3\,n + 3\,n^2 + n^3 + 3\,\#1 + 6\,n\,\#1 + 3\,n^2\,\#1 + 3\,\#1^2 + 3\,n\,\#1^2 + \#1^3\ \&,$$

$$\frac{\text{PolyGamma}[0,\ -\#1]}{1 + 2\,n + n^2 + 2\,\#1 + 2\,n\,\#1 + \#1^2}\ \&\Big] +$$

$$2\,\text{RootSum}\Big[5 + 3\,n + 3\,n^2 + n^3 + 3\,\#1 + 6\,n\,\#1 + 3\,n^2\,\#1 + 3\,\#1^2 + 3\,n\,\#1^2 + \#1^3\ \&,$$

$$\frac{\text{PolyGamma}[0,\ -\#1]\ \#1}{1 + 2\,n + n^2 + 2\,\#1 + 2\,n\,\#1 + \#1^2}\ \&\Big] +$$

$$2\,n\,\text{RootSum}\Big[5 + 3\,n + 3\,n^2 + n^3 + 3\,\#1 + 6\,n\,\#1 + 3\,n^2\,\#1 + 3\,\#1^2 + 3\,n\,\#1^2 + \#1^3\ \&,$$

$$\frac{\text{PolyGamma}[0,\ -\#1]\ \#1}{1 + 2\,n + n^2 + 2\,\#1 + 2\,n\,\#1 + \#1^2}\ \&\Big] +$$

$$\text{RootSum}\Big[5 + 3\,n + 3\,n^2 + n^3 + 3\,\#1 + 6\,n\,\#1 + 3\,n^2\,\#1 + 3\,\#1^2 + 3\,n\,\#1^2 + \#1^3\ \&,$$

$$\frac{\text{PolyGamma}[0,\ -\#1]\ \#1^2}{1 + 2\,n + n^2 + 2\,\#1 + 2\,n\,\#1 + \#1^2}\ \&\Big]\Bigg)$$

In[10]:= `Sum[ (-1)^k/(α + β k), {k, n}]`

$$\text{Out[10]= } \frac{1}{2\,\beta} \Big(\text{PolyGamma}\Big[0,\ \frac{1}{2} + \frac{\alpha}{2\,\beta}\Big] - \text{PolyGamma}\Big[0,\ 1 + \frac{\alpha}{2\,\beta}\Big] -$$

$$(-1)^n \, \text{PolyGamma}\Big[0,\ \frac{1}{2} + \frac{n}{2} + \frac{\alpha}{2\,\beta}\Big] + (-1)^n \, \text{PolyGamma}\Big[0,\ 1 + \frac{n}{2} + \frac{\alpha}{2\,\beta}\Big]\Big)$$

Here is the absolute deviation of the binomial distribution. The result contains hypergeometric functions [660].

```
In[11]:= Module[{p, mean, S = Simplify[#, 0 <= p <= 1]&},
          (* binomial distribution *)
          p[r_, n_, p_] := Binomial[n, r] p^r (1 - p)^(n - r);
          (* mean value *)
          mean = Sum[Evaluate[r p[r, n, p]], {r, 0, n}] // S;
          (* absolute deviation *)
          Sum[Evaluate[(mean - r) p[r, n, p]], {r, 0, m}] +
          Sum[Evaluate[(r - mean) p[r, n, p]], {r, m + 1, n}] // S] //
                                                     TraditionalForm
```

Out[11]//TraditionalForm=

$$\left(2\left(\frac{1}{p}-1\right)\right)^{-m}(1-p)^{n-2}\,p\,\Gamma(n+1)$$

$$\left((p-1)(-m+n\,p-1)\,\Gamma(m+3)\,\Gamma(-m+n-1)\,{}_2F_1\!\left(1,\,m-n+1;\,m+2;\,\frac{p}{p-1}\right)+\right.$$

$$\left.p\,\Gamma(m+2)\,\Gamma(n-m)\,{}_2F_1\!\left(2,\,m-n+2;\,m+3;\,\frac{p}{p-1}\right)\right)\Big/\big(\Gamma(m+2)\,\Gamma(m+3)\,\Gamma(-m+n-1)\,\Gamma(n-m)\big)$$

Here are some more infinite sums.

In[12]:= `Sum[1/(k^2 2^k), {k, Infinity}]`

Out[12]= $\dfrac{1}{12}\,(\pi^2 - 6\,\text{Log}[2]^2)$

In[13]:= `Sum[Sin[k a]/k^2, {k, Infinity}]`

Out[13]= $\dfrac{1}{2}\,i\,(\text{PolyLog}[2,\,e^{-ia}] - \text{PolyLog}[2,\,e^{ia}])$

In[14]:= `Sum[z^k/k!/k, {k, Infinity}]`

Out[14]= $-\text{EulerGamma} - \text{Gamma}[0,\,-z] - \text{Log}[-z]$

In[15]:= `Sum[(k - 1)^2/(k + 2)^5, {k, Infinity}]`

Out[15]= $\dfrac{1}{480}\,(-1935 - 32\,\pi^4 + 480\,\text{Zeta}[3] + 4320\,\text{Zeta}[5])$

In[16]:= `Sum[(2n + 1)!/(2^(3n) (2n - 1) n!^2), {n, Infinity}]`

Out[16]= 1

And here is a sum whose summands contain the Riemann Zeta function.

In[17]:= `Sum[(Zeta[n] - 1)/n/2^n, {n, 2, Infinity}]`

Out[17]= $\dfrac{1}{2}\,(1 - \text{EulerGamma} - 2\,\text{Log}[2] + \text{Log}[\pi])$

As with `Integrate`, results of `Sum` are not automatically simplified. Applying `Simplify` and other functions often allows to get a simpler result.

In[18]:= `Assuming[z > 1,`
         `    Sum[2/(2k + 1) ((z - 1)/(z + 1))^(2k + 1), {k, 0, Infinity}]]`

Out[18]= $2\,\text{ArcTanh}\!\left[\dfrac{-1 + z}{1 + z}\right]$

In[19]:= `% // TrigToExp // Simplify[#, z > 1]&`

Out[19]= $\text{Log}[z]$

Here are two forms of the binomial theorem [1139]. The two sums contain symbolic products as arguments.

In[20]:= `1 + Sum[Product[α - k, {k, 0, k - 1}] t^k/k!, {k, Infinity}]`

Out[20]= $(1 + t)^\alpha$

In[21]:= `1/(-h)^α Sum[(-1)^k Product[α - k, {k, 0, k - 1}] (1 + h + h t)^k/k!,`
         `    {k, 0, Infinity}] // Simplify // PowerExpand`

Out[21]= $(1 + t)^\alpha$

Here is a much more complicated sum [1302]:

$$\sum_{k=1}^{\infty}\ln\!\left(\frac{1}{\varepsilon}\left(1+\frac{1}{\alpha\,k}\right)^{k\,\alpha+\beta}\right)=\ln\!\left(\prod_{k=1}^{\infty}\left(\frac{1}{\varepsilon}\left(1+\frac{1}{\alpha\,k}\right)^{k\,\alpha+\beta}\right)\right)$$

The necessary conditions for the convergence of the sum are $\varepsilon = e$ and $\beta = 1/2$.

In[22]:= **Series[Log[1/ε (1 + 1/(α k))^(α k + β)], {k, Infinity, 2}] // Simplify**

Out[22]= $\left(1 + \text{Log}\left[\frac{1}{\varepsilon}\right]\right) + \frac{-1 + 2\beta}{2\,\alpha\,k} + \frac{(2 - 3\beta)\,\left(\frac{1}{k}\right)^2}{6\,\alpha^2} + O\left[\frac{1}{k}\right]^3$

Because Sum currently does not have the Assumptions option, we carry out the summation with a "random" positive number.

In[23]:= **Sum[Log[1/E (1 + 1/(GoldenRatio k))^(GoldenRatio k + 1/2)],**
         {k, 1, Infinity}] /. GoldenRatio -> a // Simplify

Out[23]= $\frac{1}{12}\left(6 + \frac{6}{a} + a - 6\,\text{Log}[2] + 6\,\text{Log}[a] - \right.$

$\left. 12\,a\,\text{Log}[\text{Glaisher}] - 6\,\text{Log}[\pi] + 6\,\text{LogGamma}\left[\frac{1}{a}\right] - 12\,a\,\text{Zeta}^{(1,0)}\left[-1,\frac{1}{a}\right]\right)$

Here is a quick check of the result for another random number.

In[24]:= **Block[{a = Pi/E + EulerGamma + 1},**
         {N[%, 8],
          NSum[Log[1/E (1 + 1/(a k))^(a k + 1/2)], {k, 1, Infinity},
              VerifyConvergence -> False, Method -> Fit, NSumTerms -> 500,
              NSumExtraTerms -> 500, WorkingPrecision -> 500,
              PrecisionGoal -> 8] // N[#, 8]&}]

Out[24]= {0.014556901, 0.014556901}

There is a small extension in the definition for sums extending over all integers: The intuitive notation {k, -Infinity, Infinity} works, and the iterator variable k starts at the "integer" $-\infty$.

In[25]:= **Sum[1/(1 + k^2), {k, -Infinity, Infinity}]**

Out[25]= $\pi\,\text{Coth}[\pi]$

Multidimensional sums use iterators similar to multidimensional integrals. (For sums of noninteger dimensions, see [401].) Here is a simple example.

In[26]:= **{Sum[j, {k, n}, {j, k}], Sum[Sum[j, {j, k}], {k, n}]}**

Out[26]= $\left\{\frac{1}{6}\,n\,(1 + n)\,(2 + n),\ \frac{1}{6}\,n\,(1 + n)\,(2 + n)\right\}$

Of course, there are also many sums that *Mathematica* cannot sum.

In[27]:= **unevaluatedSum = Sum[n^-n, {n, Infinity}]**

Out[27]= $\sum_{n=1}^{\infty} n^{-n}$

Applying N yields results in a call to NSum and we get a numerical approximation of the sum.

In[28]:= **N[%]**

Out[28]= 1.29129

Be aware that the automatic comparisons carried out by Equal, Less, and so on, do not work immediately for unevaluated sums and integrals.

In[29]:= **unevaluatedSum < Pi/2**

Out[29]= $\sum_{n=1}^{\infty} n^{-n} < \frac{\pi}{2}$

For comparisons to be carried out by numericalization, the terms involved must be numeric. But Sum does not have the NumericFunction attribute. Nevertheless, we can declare the whole sum to be a numeric expression.

In[30]:= **NumericQ[unevaluatedSum] = True;**

Now, the comparison results in True.

In[31]:= **unevaluatedSum < Pi/2**

Out[31]= True

Because of Sum's double life, as procedural summation routine and as symbolic summation engine, it must decide what to do for finite limits. It will switch between a procedural and a symbolic approach. This explains the timings in the following example.

In[32]:= **ListPlot[Table[{k, Timing[Sum[j, {j, Round[10^k]}]][[1, 1]]},**
          {k, 1, 12, 1/4}],
          PlotJoined -> True, PlotRange -> All]

We mentioned above that there are two ways to "do" a sum—evaluating the summands one after another in a procedural way or carrying out the symbolic summation. The first possibility can be further subdivided into two possibilities. We can either form the cumulative sum after the calculation of each summand or first calculate all summands and then apply Plus to the resulting list. It depends on the nature of the summands which way is more appropriate. Let us take, for instance, the case of summing 100000 approximative real numbers. First, we use Sum.

In[33]:= **Sum[k + Sin[k] + Log[k]^3,**
          {k, (* approximate numbers in limits *) 1., 2 10.^5}] // Timing
Out[33]= {1.24 Second, 2.02879×10^{10}}

Next, we carry out a Do loop and form cumulative sums.

In[34]:= **(sum = 0; Do[sum = sum + k + Sin[k] + Log[k]^3, {k, 1., 2 10.^5}];**
          sum) // Timing
Out[34]= {1.47 Second, 2.02879×10^{10}}

The last method to calculate the sum is to first form a list of 100000 elements and then to change the list head to head Plus.

In[35]:= **Apply[Plus, Table[k + Sin[k] + Log[k]^3, {k, 1., 2 10.^5}]] // Timing**
Out[35]= {0.29 Second, 2.02879×10^{10}}

The last method was clearly the fastest, the reason being the autocompilation of Table (which was discussed in Chapter 1 of the Numerics volume [1737]). (It is the compilation of the whole process that makes the last method fast. Just making the calculation of the elements fast accounts for about 30% of the speed-up.)

Doing a compiled evaluation of the whole sum is of course the fastest method.

In[36]:= **Compile[{}, Module[{sum = 0.},**
                   Do[sum = sum + k + Sin[k] + Log[k]^3, {k, 1, 2 10.^5}];
                   sum]][] // Timing

Out[36]= {0.2 Second, 2.02879 × 10^10}

Both approaches, forming all cumulative sums and forming the sum at once, have sometimes advantages and disadvantages. Forming all partial sums clearly saves memory. In the last examples, forming the compiled sum cumulative was fastest.

Now, let us consider the case in which the individual summands are symbolic quantities that, after adding up, cancel partially. Here is the definition for the summands.

```
In[37]:= summand[k_] := Sum[Random[Integer, {-100, 100}] *
                               c[Random[Integer, {1, 10000}]], {j, 200}]
```

We carry out some timing as a function of the number of summed terms.

```
In[38]:= (* sum term by term *)
         doTimes = Table[{n, Timing[(sum = 0;
                                        Do[sum = sum + summand[k], {k, 1, n}];
                                        sum)][[1, 1]]},
                        {n, 1, 101, 5}];

         (* use Sum *)
         sumTimes = Table[{n, Timing[Sum[summand[k], {k, 1, n}]][[1, 1]]},
                        {n, 1, 101, 5}];

         (* form list and apply Times *)
         applyTimes = Table[{n, Timing[Apply[Plus,
                                        Table[summand[k], {k, 1, n}]]][[1, 1]]},
                        {n, 1, 101, 5}];
```

Here are the exponents of fits of the form t^α of the measured timings.

```
In[46]:= (* assume n ~ t^α and determine α *)
         Coefficient[Fit[DeleteCases[Log[#], {_, Indeterminate}], {1, t}, t], t]& /@
                                        {doTimes, sumTimes, applyTimes}
Out[47]= {1.96022, 1.28951, 1.06914}
```

We see that Sum and Apply[Plus[Table[...], {1}] are basically identical. We also see a much more pronounced $O(n^2)$ contribution in the complexity of the cumulative sum version from doTimes. The reason for this term is because of the sorting that takes place each time a new sum is formed.

Sum has the attribute HoldAll. As a result of this, the body of Sum is not evaluated until the summation variables are localized. Be aware of this difference to Integrate, which does not have the HoldAll attribute. Here this difference is demonstrated.

```
In[48]:= Clear[α, k, n]; α = k;
         {Sum[α, {k, 1, n}], Sum[Evaluate[α], {k, 1, n}]}
```
$$Out[49]= \left\{ k\,n,\ \frac{1}{2}\,n\,(1+n) \right\}$$

```
In[50]:= Clear[α, k, n]; α = k;
         {Integrate[α, {k, 1, n}], Integrate[Evaluate[α], {k, 1, n}]}
```
$$Out[51]= \left\{ -\frac{1}{2} + \frac{n^2}{2},\ -\frac{1}{2} + \frac{n^2}{2} \right\}$$

As a slightly larger example for the use of Sum for symbolic summation, let us construct one sheet of the Riemann surface of $(1 + z)^{1/2}$ by Weierstrass's method.

Mathematical Remark: Weierstrass's Method of Analytic Continuation

Let the analytic function $f(z)$ have the following Taylor expansion around $z = z_0$:

$$f(z_0 + \delta) = \sum_{i=0}^{\infty} a_n^{(0)} \delta^n.$$

Now, let us look for the series expansion of $f(z)$ around a (regular) point $z = z_1$ (inside the disk of convergence of the above Taylor expansion). Formally, we can expand $f(z)$ in the same manner around $z = z_1$:

$$f(z_1 + \delta) = \sum_{i=0}^{\infty} a_n^{(1)} \delta^n.$$

Using elementary arithmetic, it can be shown [483], [828], [829], [947], [118], [1626], [1668], [287] that the coefficients $a_n^{(1)}$ can be expressed through the coefficients $a_n^{(0)}$ in the following way:

$$a_n^{(1)} = \sum_{i=n}^{\infty} \binom{i}{n} a_i^{(0)} (z_1 - z_0)^{i-n}.$$

We will use this formula repeatedly by encircling the branch point $z = -1$ of $(1 + z)^{1/2}$. The nth coefficient of $(1 + z)^{1/2}$ is given by $(-1)^i (-3/2 + i)! / i! / (-2\pi^{1/2})$.

```
In[52]:= Sum[ (-1)^i (-3/2 + i)!/i! z^i, {i, 0, Infinity}]
```

```
Out[52]= -2 √π √1 + z
```

The function `analyticContinuation` implements the above formula for calculating the new coefficients of the Taylor series from the old ones (this function makes heavy use of the symbolic part of `Sum`).

```
In[53]:= analyticContinuation[startTerm_, expansionPointList_] :=
    FoldList[Sum[Binomial[n, i] (#1 /. i -> n) #2^(n - i),
        {n, i, Infinity}]&, startTerm,
        Apply[-Subtract[##]&, Partition[expansionPointList, 2, 1], {1}]]
```

Here is the procedure of calculating new series terms iterated eight times. We denote the expansion points by `eP[i]` and display them in the next output as p_i for brevity

```
In[54]:= (seriesTerms = analyticContinuation[(-1)^i (-3/2 + i)!/i!,
                            Table[eP[j], {j, 0, 8}]]) /.
    (* for a more compact display *) {eP[i_] :> Subscript[p, i], Gamma -> Γ}
```

$$\text{Out[54]= } \left\{ \frac{(-1)^i \left(-\frac{3}{2} + i\right)!}{i!} , \quad \frac{(p_0 - p_1)^i (-p_0 + p_1)^{-i} (1 - p_0 + p_1)^{\frac{1}{2} - i} \Gamma\left[-\frac{1}{2} + i\right]}{\Gamma[1 + i]} , \right.$$

$$\frac{\sqrt{1 - p_0 + p_1} \left(\frac{-1 + p_0 - p_2}{-1 + p_0 - p_1}\right)^{\frac{1}{2} - i} (-p_1 + p_2)^{-i} \left(\frac{(p_0 - p_1)(-p_1 + p_2)}{(-p_0 + p_1)(1 - p_0 + p_1)}\right)^i \Gamma\left[-\frac{1}{2} + i\right]}{\Gamma[1 + i]} ,$$

$$\frac{\sqrt{1 - p_0 + p_1} \sqrt{\frac{-1 + p_0 - p_2}{-1 + p_0 - p_1}} \left(\frac{-1 + p_0 - p_3}{-1 + p_0 - p_2}\right)^{\frac{1}{2} - i} (-p_2 + p_3)^{-i} \left(\frac{(-1 + p_0 - p_1)(p_0 - p_1)(-p_2 + p_3)}{(-p_0 + p_1)(1 - p_0 + p_1)(-1 + p_0 - p_2)}\right)^i \Gamma\left[-\frac{1}{2} + i\right]}{\Gamma[1 + i]} ,$$

$$\frac{1}{\Gamma[1 + i]} \left(\sqrt{1 - p_0 + p_1} \sqrt{\frac{-1 + p_0 - p_2}{-1 + p_0 - p_1}} \sqrt{\frac{-1 + p_0 - p_3}{-1 + p_0 - p_2}} \left(\frac{-1 + p_0 - p_4}{-1 + p_0 - p_3}\right)^{\frac{1}{2} - i} \right.$$

$$(-p_3 + p_4)^{-i} \left(\frac{(-1 + p_0 - p_1)\,(p_0 - p_1)\,(-p_3 + p_4)}{(-p_0 + p_1)\,(1 - p_0 + p_1)\,(-1 + p_0 - p_3)} \right)^{i} \Gamma\left[-\frac{1}{2} + i\right], \quad \frac{1}{\Gamma[1 + i]}$$

$$\left(\sqrt{1 - p_0 + p_1}\; \sqrt{\frac{-1 + p_0 - p_2}{-1 + p_0 - p_1}}\; \sqrt{\frac{-1 + p_0 - p_3}{-1 + p_0 - p_2}}\; \sqrt{\frac{-1 + p_0 - p_4}{-1 + p_0 - p_3}}\; \left(\frac{-1 + p_0 - p_5}{-1 + p_0 - p_4}\right)^{\frac{1}{2} - i} \right.$$

$$(-p_4 + p_5)^{-i} \left(\frac{(-1 + p_0 - p_1)\,(p_0 - p_1)\,(-p_4 + p_5)}{(-p_0 + p_1)\,(1 - p_0 + p_1)\,(-1 + p_0 - p_4)} \right)^{i} \Gamma\left[-\frac{1}{2} + i\right], \quad \frac{1}{\Gamma[1 + i]}$$

$$\left(\sqrt{1 - p_0 + p_1}\; \sqrt{\frac{-1 + p_0 - p_2}{-1 + p_0 - p_1}}\; \sqrt{\frac{-1 + p_0 - p_3}{-1 + p_0 - p_2}}\; \sqrt{\frac{-1 + p_0 - p_4}{-1 + p_0 - p_3}}\; \sqrt{\frac{-1 + p_0 - p_5}{-1 + p_0 - p_4}}\; \left(\frac{-1 + p_0 - p_6}{-1 + p_0 - p_5}\right)^{\frac{1}{2} - i} \right.$$

$$(-p_5 + p_6)^{-i} \left(\frac{(-1 + p_0 - p_1)\,(p_0 - p_1)\,(-p_5 + p_6)}{(-p_0 + p_1)\,(1 - p_0 + p_1)\,(-1 + p_0 - p_5)} \right)^{i} \Gamma\left[-\frac{1}{2} + i\right], \quad \frac{1}{\Gamma[1 + i]}$$

$$\left(\sqrt{1 - p_0 + p_1}\; \sqrt{\frac{-1 + p_0 - p_2}{-1 + p_0 - p_1}}\; \sqrt{\frac{-1 + p_0 - p_3}{-1 + p_0 - p_2}}\; \sqrt{\frac{-1 + p_0 - p_4}{-1 + p_0 - p_3}}\; \sqrt{\frac{-1 + p_0 - p_5}{-1 + p_0 - p_4}}\; \sqrt{\frac{-1 + p_0 - p_6}{-1 + p_0 - p_5}} \right.$$

$$\left(\frac{-1 + p_0 - p_7}{-1 + p_0 - p_6} \right)^{\frac{1}{2} - i} (-p_6 + p_7)^{-i} \left(\frac{(-1 + p_0 - p_1)\,(p_0 - p_1)\,(-p_6 + p_7)}{(-p_0 + p_1)\,(1 - p_0 + p_1)\,(-1 + p_0 - p_6)} \right)^{i} \Gamma\left[-\frac{1}{2} + i\right],$$

$$\frac{1}{\Gamma[1 + i]} \left(\sqrt{1 - p_0 + p_1}\; \sqrt{\frac{-1 + p_0 - p_2}{-1 + p_0 - p_1}}\; \sqrt{\frac{-1 + p_0 - p_3}{-1 + p_0 - p_2}}\; \sqrt{\frac{-1 + p_0 - p_4}{-1 + p_0 - p_3}} \right.$$

$$\sqrt{\frac{-1 + p_0 - p_5}{-1 + p_0 - p_4}}\; \sqrt{\frac{-1 + p_0 - p_6}{-1 + p_0 - p_5}}\; \sqrt{\frac{-1 + p_0 - p_7}{-1 + p_0 - p_6}}\; \left(\frac{-1 + p_0 - p_8}{-1 + p_0 - p_7}\right)^{\frac{1}{2} - i}$$

$$(-p_7 + p_8)^{-i} \left(\frac{(-1 + p_0 - p_1)\,(p_0 - p_1)\,(-p_7 + p_8)}{(-p_0 + p_1)\,(1 - p_0 + p_1)\,(-1 + p_0 - p_7)} \right)^{i} \Gamma\left[-\frac{1}{2} + i\right] \right\}$$

We can use `SymbolicSum` to get closed-form expressions for all of the series of interest here.

```
In[55]:= (summedSeries =  Sum[# z^i, {i, 0, Infinity}] & /@ seriesTerms) //
                                                                TableForm
```

Out[55]//TableForm=

$-2\sqrt{\pi}\ \sqrt{1+z}$							
$-2\sqrt{\pi}\ \sqrt{\frac{-1-z+eP[0]-eP[1]}{-1+eP[0]-eP[1]}}$	$\sqrt{1-eP[0]+eP[1]}$						
$-2\sqrt{\pi}\ \sqrt{1-eP[0]+eP[1]}$	$\sqrt{\frac{-1+eP[0]-eP[2]}{-1+eP[0]-eP[1]}}$	$\sqrt{\frac{-1-z+eP[0]-eP[2]}{-1+eP[0]-eP[2]}}$					
$-2\sqrt{\pi}\ \sqrt{1-eP[0]+eP[1]}$	$\sqrt{\frac{-1+eP[0]-eP[2]}{-1+eP[0]-eP[1]}}$	$\sqrt{\frac{-1+eP[0]-eP[3]}{-1+eP[0]-eP[2]}}$	$\sqrt{\frac{-1-z+eP[0]-eP[3]}{-1+eP[0]-eP[3]}}$				
$-2\sqrt{\pi}\ \sqrt{1-eP[0]+eP[1]}$	$\sqrt{\frac{-1+eP[0]-eP[2]}{-1+eP[0]-eP[1]}}$	$\sqrt{\frac{-1+eP[0]-eP[3]}{-1+eP[0]-eP[2]}}$	$\sqrt{\frac{-1+eP[0]-eP[4]}{-1+eP[0]-eP[3]}}$	$\sqrt{\frac{-1-z+eP[0]-eP[4]}{-1+eP[0]-eP[4]}}$			
$-2\sqrt{\pi}\ \sqrt{1-eP[0]+eP[1]}$	$\sqrt{\frac{-1+eP[0]-eP[2]}{-1+eP[0]-eP[1]}}$	$\sqrt{\frac{-1+eP[0]-eP[4]}{-1+eP[0]-eP[2]}}$	$\sqrt{\frac{-1+eP[0]-eP[3]}{-1+eP[0]-eP[3]}}$	$\sqrt{\frac{-1+eP[0]-eP[5]}{-1+eP[0]-eP[4]}}$	$\sqrt{\frac{-1-z+eP[0]-eP[5]}{-1+eP[0]-eP[5]}}$		
$-2\sqrt{\pi}\ \sqrt{1-eP[0]+eP[1]}$	$\sqrt{\frac{-1+eP[0]-eP[2]}{-1+eP[0]-eP[1]}}$	$\sqrt{\frac{-1+eP[0]-eP[3]}{-1+eP[0]-eP[2]}}$	$\sqrt{\frac{-1+eP[0]-eP[4]}{-1+eP[0]-eP[3]}}$	$\sqrt{\frac{-1+eP[0]-eP[5]}{-1+eP[0]-eP[4]}}$	$\sqrt{\frac{-1+eP[0]-eP[6]}{-1+eP[0]-eP[5]}}$	$\sqrt{\frac{-1-z+eP[0]-eP[6]}{-1+eP[0]-eP[6]}}$	
$-2\sqrt{\pi}\ \sqrt{1-eP[0]+eP[1]}$	$\sqrt{\frac{-1+eP[0]-eP[2]}{-1+eP[0]-eP[1]}}$	$\sqrt{\frac{-1+eP[0]-eP[3]}{-1+eP[0]-eP[2]}}$	$\sqrt{\frac{-1+eP[0]-eP[4]}{-1+eP[0]-eP[3]}}$	$\sqrt{\frac{-1+eP[0]-eP[5]}{-1+eP[0]-eP[4]}}$	$\sqrt{\frac{-1+eP[0]-eP[6]}{-1+eP[0]-eP[5]}}$	$\sqrt{\frac{-1+eP[0]-eP[7]}{-1+eP[0]-eP[6]}}$	$\sqrt{\frac{-1-z+eP[0]-eP[7]}{-1+eP[0]-eP[7]}}$
$-2\sqrt{\pi}\ \sqrt{1-eP[0]+eP[1]}$	$\sqrt{\frac{-1+eP[0]-eP[2]}{-1+eP[0]-eP[1]}}$	$\sqrt{\frac{-1+eP[0]-eP[3]}{-1+eP[0]-eP[2]}}$	$\sqrt{\frac{-1+eP[0]-eP[4]}{-1+eP[0]-eP[3]}}$	$\sqrt{\frac{-1+eP[0]-eP[5]}{-1+eP[0]-eP[4]}}$	$\sqrt{\frac{-1+eP[0]-eP[6]}{-1+eP[0]-eP[5]}}$	$\sqrt{\frac{-1+eP[0]-eP[7]}{-1+eP[0]-eP[6]}}$	$\sqrt{\frac{-1+eP[0]-eP[8]}{-1+eP[0]-eP[7]}}$ $\sqrt{\frac{-1-z+eP[0]-eP[8]}{-1+eP[0]-eP[8]}}$

The basic structure of all of the elements of the last list is just $\sqrt{1 + z}$. The large factors are correcting for branch cuts. Using `PowerExpand`, this becomes visible (the next result is of course no longer mathematically equivalent to the last one).

```
In[56]:= summedSeries // PowerExpand // Simplify
```

Out[56]= $\left\{-2\sqrt{\pi}\sqrt{1+z}, \; -\dfrac{2\sqrt{\pi}\sqrt{-1-z+eP[0]-eP[1]}\sqrt{1-eP[0]+eP[1]}}{\sqrt{-1+eP[0]-eP[1]}}, \right.$

$-\dfrac{2\sqrt{\pi}\sqrt{1-eP[0]+eP[1]}\sqrt{-1-z+eP[0]-eP[2]}}{\sqrt{-1+eP[0]-eP[1]}},$

$-\dfrac{2\sqrt{\pi}\sqrt{1-eP[0]+eP[1]}\sqrt{-1-z+eP[0]-eP[3]}}{\sqrt{-1+eP[0]-eP[1]}},$

$-\dfrac{2\sqrt{\pi}\sqrt{1-eP[0]+eP[1]}\sqrt{-1-z+eP[0]-eP[4]}}{\sqrt{-1+eP[0]-eP[1]}},$

$-\dfrac{2\sqrt{\pi}\sqrt{1-eP[0]+eP[1]}\sqrt{-1-z+eP[0]-eP[5]}}{\sqrt{-1+eP[0]-eP[1]}},$

$-\dfrac{2\sqrt{\pi}\sqrt{1-eP[0]+eP[1]}\sqrt{-1-z+eP[0]-eP[6]}}{\sqrt{-1+eP[0]-eP[1]}},$

$-\dfrac{2\sqrt{\pi}\sqrt{1-eP[0]+eP[1]}\sqrt{-1-z+eP[0]-eP[7]}}{\sqrt{-1+eP[0]-eP[1]}},$

$\left.-\dfrac{2\sqrt{\pi}\sqrt{1-eP[0]+eP[1]}\sqrt{-1-z+eP[0]-eP[8]}}{\sqrt{-1+eP[0]-eP[1]}}\right\}$

We see mainly the `Sqrt[1+z]` expression, and the prefactors determine the phase.

We now specialize the expansion points to lie on a circle of radius 1 around the point -1. (By using nine points on this circle, we make sure that the *j*th expansion point lies inside the disk of convergence of the $(j-1)$th series expansion.)

In[57]:= `eP[i_] = -1 + Exp[i I 2Pi/8];`

Let us piecewise define a function `sqrt` that is represented by the summed forms of the various series.

In[58]:= `Do[sqrt[i - 1, z_] = Evaluate[summedSeries[[i]] /.`
    `{z -> z - eP[i - 1]}], {i, 1, Length[summedSeries]}]`

We look at the resulting Riemann surface by showing the values of the various `sqrt[i, z]` inside their disks of convergence.

In[59]:= `Do[points[i] = Table[{Re[#], Im[#], Im[N[sqrt[i, #]]]}&[`
                               `N[eP[i] + r Exp[I φ]]],`
                `{r, 0, 0.99, 0.99/10}, {φ, 0, N[2Pi], N[2Pi]/16}],`
    `{i, 0, 8}]`

In[60]:= `Show[Graphics3D[{`
    `{Thickness[0.002], Table[(* the disks *)`
    `{Hue[i/8 0.76], Line /@ points[i],`
                    `Line /@ Transpose[points[i]]}, {i, 0, 8}]},`
    `{Thickness[0.01], GrayLevel[0.3], Line[{{-1, 0, -5}, {-1, 0, 2}}]},`
    `{Thickness[0.01], (* the continuation path *)`
    `Line[N[Append[#, First[#]]]& @`
    `Table[{Re[eP[i]], Im[eP[i]], Im[N[sqrt[i, eP[i]]]]}, {i, 1, 8}]]}}],`
        `PlotRange -> All, BoxRatios -> {1, 1, 1.5},`
        `ViewPoint -> {-2, -1, 1.1}, Axes -> True,`
        `AxesLabel -> (StyleForm[#, TraditionalForm]& /@`
                    `{"x", "y", "Sqrt[1 + x + I y]"})]`

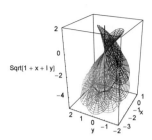

Because of the two-valuedness of $(1+z)^{1/2}$, the first function `sqrt[0, z]` (in red) and the last function `sqrt[8, z]` (in blue) do not coincide, and the branch cut of `Sqrt[1+z]` along the negative real axis is—because of the analytic continuation—missing.

As another application of Sum, let us look at the Hölder summation method [1744], [1054], [200], [656].

Given a divergent sum (divergent in the limit $n \to \infty$) $S_0^{(n)} = \sum_{j=1}^n a_j$ one recursively forms the (partial) sums $S_k^{(n)} = n^{-1} \sum_{j=1}^n S_{k-1}^{(n)}$ until $S_k^{(n)}$ converges (if this happens).

Let us take an example, the series of $-(x+1)^{-2}$ for $x = 1$. The nth term of the series is given by $a_j = (-1)^j j x^{j-1}$.

```
In[61]:= Series[-1/(1 + x)^2, {x, 0, 8}]
```
$$\text{Out[61]}= -1 + 2\,x - 3\,x^2 + 4\,x^3 - 5\,x^4 + 6\,x^5 - 7\,x^6 + 8\,x^7 - 9\,x^8 + O[x]^9$$

The first partial sums are formed.

```
In[62]:= sum1 = Sum[(-1)^j j x^(j - 1), {j, n}]
```
$$\text{Out[62]}= \frac{-1 + (-x)^n + n\,(-x)^n + n\,(-x)^n\,x}{(1 + x)^2}$$

The first partial sum does not converge for $n \to \infty$.

```
In[63]:= Table[sum1 /. x -> 1, {n, 12}]
```
$$\text{Out[63]}= \{-1, 1, -2, 2, -3, 3, -4, 4, -5, 5, -6, 6\}$$

So, let us form the second iteration.

```
In[64]:= sum2 = Sum[Evaluate[sum1 /. n -> j], {j, n}]/n // Together
```
$$\text{Out[64]}= \frac{-n - 2\,x - n\,x + 2\,(-x)^n\,x + n\,(-x)^n\,x + n\,(-x)^n\,x^2}{n\,(1 + x)^3}$$

This partial sum still does not converge.

```
In[65]:= Table[sum2 /. x -> 1, {n, 12}]
```
$$\text{Out[65]}= \left\{-1, 0, -\frac{2}{3}, 0, -\frac{3}{5}, 0, -\frac{4}{7}, 0, -\frac{5}{9}, 0, -\frac{6}{11}, 0\right\}$$

```
In[66]:= {sum2 /. x -> 1 /. (-1)^n -> -1, sum2 /. x -> 1 /. (-1)^n -> +1}
```
$$\text{Out[66]}= \left\{\frac{-4 - 4\,n}{8\,n}, 0\right\}$$

So, let us do one more iteration.

```
In[67]:= sum3 = Sum[Evaluate[sum2 /. n -> j], {j, n}]/n // Together
```

Out[67]= $-\dfrac{1}{n\,(1+n)\,(1+x)^3}\,(n+n^2+n\,x+n^2\,x+x^2+n\,x^2-(-x)^n\,x^2-$

$n\,(-x)^n\,x^2+2\,x\,\mathrm{HarmonicNumber}[n]+2\,n\,x\,\mathrm{HarmonicNumber}[n]-$

$2\,(-x)^n\,x^2\,\mathrm{Hypergeometric2F1}[1+n,\,1,\,2+n,\,-x]+2\,x\,\mathrm{Log}[1+x]+2\,n\,x\,\mathrm{Log}[1+x])$

Now, we finally have a convergent sum.

In[68]:= `Table[Expand[sum3 /. x -> 1], {n, 12}] // N`

Out[68]= {-1., -0.5, -0.555556, -0.416667, -0.453333, -0.377778,
       -0.405442, -0.354762, -0.377072, -0.339365, -0.3581, -0.328259}

The limit is -1/4, which is exactly what one obtains by substituting $x = 1$ into $-(x+1)^{-2}$.

In[69]:= `Table[Abs[1/4 + sum3 /. x -> 1 /. n -> N[10^k, 22]], {k, 20}] // N[#, 2]&`

Out[69]= $\{0.089,\ 0.015,\ 0.0020,\ 0.00026,\ 0.000032,\ 3.8\times10^{-6},\ 4.3\times10^{-7},\ 4.9\times10^{-8},$
$5.5\times10^{-9},\ 6.1\times10^{-10},\ 6.6\times10^{-11},\ 7.2\times10^{-12},\ 7.8\times10^{-13},\ 8.4\times10^{-14},$
$9.0\times10^{-15},\ 9.5\times10^{-16},\ 1.0\times10^{-16},\ 1.1\times10^{-17},\ 1.1\times10^{-18},\ 1.2\times10^{-19}\}$

The following picture visualizes the big smoothing property of forming recursively partial sums. We start with largely random numbers, and after a very few steps, we have a very smooth sequence.

In[70]:=
```
recursivePartialSumList[l_, n_] :=
   NestList[MapIndexed[#1/#2[[1]]&, Rest[FoldList[Plus, 0, #]]]&, l, n]
```

In[71]:=
```
Show[GraphicsArray[
   Block[{$DisplayFunction = Identity},
     {(* density plot of a sequence with increasing variations *)
      ListDensityPlot[recursivePartialSumList[
                      Table[Random[Real, {-i^2, i^2}/i^2], {i, 1200}], 10],
               Mesh -> False, ColorFunction -> (Hue[0.78#]&)],
      (* 3D plot for uniform random variables *)
      ListPlot3D[Log @ Abs @ recursivePartialSumList[
             Table[Random[Real, {-1, 1}], {i, 120}], 20],
             Mesh -> False]}]]]
```

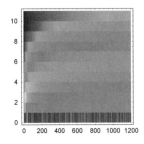
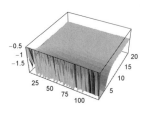

The next inputs use the Cesàro summation method [216] to establish the value $-1/4$.

In[72]:=
```
partialSums = Simplify[#, x > 0]& @
                 Sum[(-1)^(j + 1) (j + 1) x^j, {j, 0, k}]
```

Out[72]= $\dfrac{-1+(2+k)\,(-x)^{1+k}-(1+k)\,(-x)^{2+k}}{(1+x)^2}$

In[73]:=
```
(* multiply partial sums with a binomial and sum again *)
cesaroSum =
Sum[Evaluate[partialSums Binomial[n - k + p - 1, n - k]],
   {k, 0, n}]/Binomial[n + p, n] /. x -> 1 // Simplify
```

Out[74]= $-\dfrac{\frac{\mathrm{Gamma}[1+n+p]}{\mathrm{Gamma}[1+p]}+\frac{3\,\mathrm{Gamma}[n+p]\,\mathrm{Hypergeometric2F1}[1,-n,1-n-p,-1]}{\mathrm{Gamma}[p]}}{\mathrm{Gamma}[1+n]}+\dfrac{2\,\mathrm{Gamma}[-1+n+p]\,\mathrm{Hypergeometric2F1}[2,1-n,2-n-p,-1]}{\mathrm{Gamma}[n]\,\mathrm{Gamma}[p]}$

$$\overline{\qquad\qquad\qquad\qquad\qquad\qquad 4\,\mathrm{Binomial}[n+p,\,n]\qquad\qquad\qquad\qquad\qquad\qquad}$$

In[75]:= (* limits for different values for the parameter p *)
         Table[Limit[FullSimplify[cesaroSum], n -> Infinity], {p, 4}]

Out[76]= $\left\{-\dfrac{1}{4},\ -\dfrac{1}{4},\ -\dfrac{1}{4},\ -\dfrac{1}{4}\right\}$

The function `Integrate` gives finite results for (some) divergent integrals when using the option setting `GenerateConditions -> False`. `Sum` does not have the option `GenerateConditions`. But the function `SymbolicSum`SymbolicSum` does. The next input calculates a finite result for the divergent sum $\sum_{k=1}^{\infty}(-1)^{k}\ln(k)$.

In[77]:= **SymbolicSum`SymbolicSum[(-1)^k Log[k], {k, Infinity},**
                                         GenerateConditions -> False] // Simplify

Out[77]= $\dfrac{1}{2}\,\mathrm{Log}\left[\dfrac{\pi}{2}\right]$

Taking into account that $\partial\,k^{\varepsilon}/\partial\,\varepsilon = k^{\varepsilon}\ln(k)$, the last result can be understood in the following way (zeta regularization).

In[78]:= **Normal[Series[D[Sum[(-1)^k k^ε, {k, Infinity}], ε],**
                     {ε, 0, 0}]] // Simplify

Out[78]= $\dfrac{1}{2}\,\mathrm{Log}\left[\dfrac{\pi}{2}\right]$

We end this subsection by remarking that the symbolic analog of the function `NProduct`, namely the function `Product` should be mentioned here. Because its syntax and functionality is largely identical to the one of `Sum`, we just give three simple examples here.

In[79]:= **{Product[Sin[z + k Pi/v], {k, 0, v - 1}],**
         Product[1 - k^-4, {k, 2, Infinity}],
         Product[(1 - Prime[k]^-2)^4, {k, Infinity}]}

Out[79]= $\left\{2^{1-v}\,\mathrm{Sin}[z\,v],\ \dfrac{\mathrm{Sinh}[\pi]}{4\,\pi},\ \dfrac{1296}{\pi^{8}}\right\}$

We end with an infinite sum over finite products.

In[80]:= **Sum[n!/Product[x + k, {k, n}], {n, Infinity}]**

Out[80]= $\dfrac{1}{-1 + x}$

1.7 Differential and Difference Equations

1.7.0 Remarks

In this section, we discuss another of the very useful *Mathematica* commands for symbolic computations: `DSolve`, the function for the symbolic solution of ordinary differential equations (ODEs), systems of ODEs [1667], partial differential equations, and differential-algebraic equations. The function `DSolve` is quite powerful and will find closed-form solutions to many differential equations. Here we present examples for the most popular classes of differential equations. This listing is far from exhaustive.

1.7.1 Ordinary Differential Equations

The syntax for solving an ordinary differential equation is straightforward.

DSolve [*listOfODEsAndInitialValues*, *listOfFunctions*, *independentVariable*]

    tries to solve the ODE(s) with potential initial conditions given by *listOfODEsAndInitialValues* for the functions in *listOfFunctions*. The independent variable is *independentVariable*. In the case of a single differential equation without initial conditions with only one unknown function, the first and second arguments can appear without the braces.

We first look at a simple example. The result of a successfully solved differential equation is a list of lists of rules—structurally, like the result of Solve.

In[1]:= **y1 = DSolve[y''[x] == x^2, y[x], x]**

Out[1]= $\left\{\left\{y[x] \to \dfrac{x^4}{12} + C[1] + x\, C[2]\right\}\right\}$

Here is a more complicated example. Similar to the results of Integrate and Sum, DSolve-results often contain special functions, Root-objects, and RootSum-objects.

In[2]:= **DSolve[y'[x] == y[x]^2 - x, y[x], x]**

Out[2]= $\left\{\left\{y[x] \to \left(-\text{BesselJ}\left[-\dfrac{1}{3}, \dfrac{2}{3} \text{ i } x^{3/2}\right] C[1] + \text{i } x^{3/2} \left(-2\, \text{BesselJ}\left[-\dfrac{2}{3}, \dfrac{2}{3} \text{ i } x^{3/2}\right] - \right.\right.\right.\right.$
$\left.\text{BesselJ}\left[-\dfrac{4}{3}, \dfrac{2}{3} \text{ i } x^{3/2}\right] C[1] + \text{BesselJ}\left[\dfrac{2}{3}, \dfrac{2}{3} \text{ i } x^{3/2}\right] C[1]\right)\right) /$
$\left(2 x \left(\text{BesselJ}\left[\dfrac{1}{3}, \dfrac{2}{3} \text{ i } x^{3/2}\right] + \text{BesselJ}\left[-\dfrac{1}{3}, \dfrac{2}{3} \text{ i } x^{3/2}\right] C[1]\right)\right)\right\}\right\}$

The next picture shows a visualization of the last solution curves generated by choosing real values from the interval $[-4, 4]$ for the integration constant C[1].

In[3]:= **Show[Graphics[{Thickness[0.002],**
    Table[With[{c = Random[Real, {-3, 3}]},
    Line /@ DeleteCases[Partition[Table[{x,
      -(AiryAiPrime[x] + AiryBiPrime[x] c)/(AiryAi[x] + AiryBi[x] c)},
        {x, -4., 4., 1/50.}], 2, 1],
        (* delete steep vertical parts *)
        _?(#.#&[Subtract @@ #] > 5&)]], {50}]}], Frame -> True]

The specification of the functions in the second argument of DSolve is analogous to that for NDSolve; that is, if no argument is specified for the function to be found, DSolve returns a pure function (with the dummy variable typically being the independent variable from the input equations). Here this is demonstrated using the simple differential equation $y''(x) = -y(x)$ [1199], [1863].

In[4]:= **y2 = DSolve[{y''[x] == -y[x], y[0] == 0}, y, x]**

Out[4]= **{{y → Function[{x}, C[2] Sin[x]]}}**

As with NDSolve, the difference between using *func* or *func(var)* (or lists of this form) in the second argument is in the use of the result in later replacements. The form y[x] can only replace y[x] itself. The form with a pure function can contain other arguments, or purely y (e.g., in derivatives). So, in the following the expression, y2 can be used for substituting y into every y[*something*] or just for y as it appears in constructions such as Derivative[*i*][y][*arg*].

In[5]:= **{y /. y2, Integrate[(x y[z]) /. y2, z]}**

Out[5]= **{{Function[{x}, C[2] Sin[x]]}, {-x C[2] Cos[z]}}**

The solution y1 cannot be used in the same fashion.

In[6]:= **{y /. y1, Integrate[(x y[z]) /. y1, z]}**

Out[6]= $\left\{\{y\}, \left\{x \int y[z] \, dz\right\}\right\}$

Here is the difference for the third derivative.

In[7]:= **{y'''[x] /. y1, y'''[x] /. y2}**

Out[7]= **{{y$^{(3)}$[x]}, {-C[2] Cos[x]}}**

Because of the pure function form of the solution, arbitrary derivatives Derivative[*n*][*solution*] can also be immediately replaced.

Here is a very simple differential equation, for which the solution in quadratures is obvious and results in a special function, the function Erfi (which we discuss in Chapter 3).

In[8]:= **DSolve[{w'[z] == Exp[(I - 1/15) z^2], w[0] == 0}, w, z]**

Out[8]= $\left\{\left\{w \to \text{Function}\left[\{z\}, \dfrac{\text{Erfi}\left[\sqrt{-\frac{1}{15}+i}\ z\right]}{2\sqrt{-\frac{\frac{1}{15}-i}{\pi}}}\right]\right\}\right\}$

The solution of the next example contains the ProductLog function.

In[9]:= **DSolve[y'[x] == x/(y[x] - x^2), y, x]**

```
InverseFunction::ifun : Inverse functions are
    being used. Values may be lost for multivalued inverses. More…

Solve::ifun : Inverse functions are being used by Solve, so some solutions
    may not be found; use Reduce for complete solution information. More…
```

Out[9]= $\left\{\left\{y \to \text{Function}\left[\{x\}, x^2 + \dfrac{1}{2}\left(1 + \text{ProductLog}\left[-e^{-1-2\,x^2+C[1]}\right]\right)\right]\right\}\right\}$

Often, the built-in DSolve can integrate an ODE, but the internally called Solve cannot solve the resulting equation for the dependent variables and the solution is then given in implicit form or using InverseFunction[*complicatedFunction*]. The differential equation $x^n\,y'(x) - y(x)^2$ is an example where Solve can find a solution that is the inverse of a given function.

In[10]:= **DSolve[{x^n y'[x] - y[x]^2 + 1 == 0}, y[x], x]**

```
Solve::tdep : The equations appear to involve the variables
    to be solved for in an essentially non-algebraic way. More…
```

Out[10]= $\left\{\left\{y[x] \to \text{InverseFunction}\left[\dfrac{1}{2}\text{Log}[-1+\#1] - \dfrac{1}{2}\text{Log}[1+\#1]\ \&\right]\left[\dfrac{x^{1-n}}{1-n}+C[1]\right]\right\}\right\}$

Similarly to NDSolve, DSolve also solves some boundary-value problems (but it typically does not find all possible solutions). In the following example, these are $y_k(x) = c \sin(k\,x)$, $k \in \mathbb{Z}$. The $k = 1$ solution is returned in the following example.

In[11]:= **DSolve[{y''[x] == -y[x], y[0] == 0, y[Pi] == 0}, y[x], x]**

Out[11]= $\{\{y[x] \to C[2] \sin[x]\}\}$

Also, systems of coupled differential equations can be solved.

In[12]:= **DSolve[{y'[x] == z[x] + 2 x, z'[x] == y[x] + 2 z[x]},**
{y[x], z[x]}, x] // Short[#, 6]&

Out[12]//Short=

$$\{\{y[x] \to \tfrac{1}{4}\, e^{-(1+\sqrt{2})\,x} \left(e^{(1-\sqrt{2})\,x} - e^{(1+\sqrt{2})\,x}\right)$$
$$\left(3 - 2\sqrt{2} + (-1+\sqrt{2})\,x + e^{2\sqrt{2}\,x}\,(-3 - 2\sqrt{2} + x + \sqrt{2}\,x)\right) +$$
$$\tfrac{1}{8}\, e^{-(1+\sqrt{2})\,x} \left(2\, e^{(1-\sqrt{2})\,x} + \sqrt{2}\, e^{(1-\sqrt{2})\,x} + 2\, e^{(1+\sqrt{2})\,x} - \sqrt{2}\, e^{(1+\sqrt{2})\,x}\right)$$
$$\left(-10 + 7\sqrt{2} + (4 - 3\sqrt{2})\,x + e^{2\sqrt{2}\,x}\,(-10 - 7\sqrt{2} + (4 + 3\sqrt{2})\,x)\right) +$$
$$\tfrac{1}{4} \left(2\, e^{(1-\sqrt{2})\,x} + \sqrt{2}\, e^{(1-\sqrt{2})\,x} + 2\, e^{(1+\sqrt{2})\,x} - \sqrt{2}\, e^{(1+\sqrt{2})\,x}\right) C[1] -$$
$$\frac{\left(e^{(1-\sqrt{2})\,x} - e^{(1+\sqrt{2})\,x}\right) C[2]}{2\sqrt{2}},$$
$$z[x] \to -\frac{1}{4\sqrt{2}} \left(e^{-(1+\sqrt{2})\,x} \left(2\, e^{(1-\sqrt{2})\,x} - \sqrt{2}\, e^{\ll 1\gg} + 2 \ll 1\gg + \sqrt{2}\, e^{(\ll 1\gg)\,x}\right)\right.$$
$$\left. \left(3 - 2\sqrt{2} + (-1+\sqrt{2})\,x + e^{2\sqrt{2}\,x}\,(-3 - 2\sqrt{2} + x + \sqrt{2}\,x)\right)\right) -$$
$$\frac{e^{\ll 1\gg} \ll 1\gg (-10 + \ll 3\gg)}{4\sqrt{2}} - \frac{(\ll 1\gg) \ll 1\gg}{2\sqrt{2}} +$$
$$\tfrac{1}{4} \left(2\, e^{(1-\sqrt{2})\,x} - \sqrt{2}\, e^{(1-\sqrt{2})\,x} + 2\, e^{(1+\sqrt{2})\,x} + \sqrt{2}\, e^{(1+\sqrt{2})\,x}\right) C[2]\}\}$$

Not all functions appearing in the second argument of DSolve must occur differentiated in the first argument. So the following system of algebraic-differential equations is solved too. No derivative of $z1$ exists in the list of equations.

In[13]:= **DSolve[{z2'[x] == 1, z1[x] == z0'[x], z2[x] == z0'[x]},**
{z0[x], z1[x], z2[x]}, x]

Out[13]= $\left\{\left\{z0[x] \to -3x + \tfrac{x^2}{2} + C[1] + x\,C[2] + 3\,(x + C[2]),\ z1[x] \to x + C[2],\ z2[x] \to x + C[2]\right\}\right\}$

The following mathematically equivalent form works too.

In[14]:= **DSolve[{z2'[x] == 1, z1[x] == z0'[x], z1'[x] == z2[x]},**
{z0[x], z1[x], z2[x]}, x]

Out[14]= $\left\{\left\{z2[x] \to x + C[1],\ z0[x] \to \tfrac{x^3}{6} + \tfrac{1}{2}x^2\,C[1] + C[2] + x\,C[3],\ z1[x] \to \tfrac{x^2}{2} + x\,C[1] + C[3]\right\}\right\}$

Here is the degenerate case of a single "differential equation" of order 0.

In[15]:= **DSolve[{x[t]^3 + 2 x[t] + 3 == 0}, x, t]**

Out[15]= $\left\{\{x \to \text{Function}[\{t\}, -1]\},\ \left\{x \to \text{Function}\left[\{t\}, \tfrac{1}{2}\,(1 - i\sqrt{11})\right]\right\},\right.$
$$\left\{x \to \text{Function}\left[\{t\}, \tfrac{1}{2}\,(1 + i\sqrt{11})\right]\right\}\right\}$$

For linear ODEs with constant coefficients, the characteristic equation can, of course, also contain multiple zeros, resulting in the typical structure $x^n\, e^{a\,x}$.

In[16]:= **DSolve[{y'''[x] - 3y''[x] + 3y'[x] - y[x] == 0}, y, x]**

Out[16]= $\{\{y \rightarrow \text{Function}[\{x\}, e^x \, C[1] + e^x \, x \, C[2] + e^x \, x^2 \, C[3]]\}\}$

Here is an example of a linear ODE with symbolic constant coefficients.

In[17]:= **DSolve[{y'''[x] - (a + b + c) y''[x] +**
(a b + a c + b c) y'[x] - a b c y[x] == 0,
y[0] == 1, y'[0] == 0, y''[0] == 0}, y[x], x]

Out[17]= $\left\{\left\{y[x] \rightarrow \dfrac{b^2 \, c \, e^{a \, x} - b \, c^2 \, e^{a \, x} - a^2 \, c \, e^{b \, x} + a \, c^2 \, e^{b \, x} + a^2 \, b \, e^{c \, x} - a \, b^2 \, e^{c \, x}}{(b - c) \, (a^2 - a \, b - a \, c + b \, c)}\right\}\right\}$

In the case a==b==c, the form of the solution changes from exponential functions to products of exponential functions with powers.

In[18]:= **DSolve[{y'''[x] - (a + b + c) y''[x] +**
(a b + a c + b c) y'[x] - a b c y[x] == 0,
y[0] == 1, y'[0] == 0, y''[0] == 0} //. {a -> c, b -> c}, y[x], x]

Out[18]= $\left\{\left\{y[x] \rightarrow \dfrac{1}{2} \, e^{c \, x} \, (2 - 2 \, c \, x + c^2 \, x^2)\right\}\right\}$

This solution can also be calculated using `Limit`, starting with the nondegenerate case [604].

In[19]:= **Expand[Limit[Limit[Together[%%[[1, 1, 2]]], a -> c], b -> c]]**

Out[19]= $e^{c \, x} - c \, e^{c \, x} \, x + \dfrac{1}{2} \, c^2 \, e^{c \, x} \, x^2$

Here is a linear system with the same property—the characteristic polynomial has multiple roots. (These are Newton's equations for a free-falling body, including Coriolis's force; for a closed-form solution, see [691], [1012], [114], [394], and [1176].)

In[20]:= **ffBEqs = With[{r = {x[t], y[t], z[t]}, ω = {ωx, ωy, ωz}},**
Thread[D[r, t, t] == -2 Cross[ω, D[r, t]] + {0, 0, -g}]]

Out[20]= $\{x''[t] == -2 \, (-\omega z \, y'[t] + \omega y \, z'[t]),$
$y''[t] == -2 \, (\omega z \, x'[t] - \omega x \, z'[t]), z''[t] == -g - 2 \, (-\omega y \, x'[t] + \omega x \, y'[t])\}$

In[21]:= **dsol = DSolve[ffBEqs, {x[t], y[t], z[t]}, t];**

The solution returned from `DSolve` is quite large; so, we do not display it here.

In[22]:= **{ByteCount[dsol], LeafCount[dsol]}**

Out[22]= $\{116016, 5840\}$

Here is a check of the solution.

In[23]:= **Simplify[ffBEqs //.** (* make a pure function from solution dsol *)
(dsol[[1]] /. ((xyz:(x | y | z))[t] -> s_) :>
(xyz -> (Function @@ {t, s})))]

Out[23]= $\{\text{True, True, True}\}$

This is an often-occurring "problem" when doing symbolic calculations. Functions such as `Integrate`, `DSolve`, `Sum`, ... will return an answer, but either a large one or in a form "not wanted". Functions such as `TrigExpand`, `Simplify`, and so on are intentionally not applied inside `Integrate`, `DSolve`, `Sum`, and so on. Their automatic application would slow down these functions unacceptably and unpredictably. But *Mathematica* provides all tools so the users can "massage" the result in the form they want. Let us do some simplification with the last result. As expected, the absolute value of the angular velocity vector appears frequently in the result.

In[24]:= **Count[dsol, Expand[# (+ωx^2 + ωy^2 + ωz^2)], Infinity]& /@ {+1, -1}**

Out[24]= $\{27, 204\}$

Replacing the absolute value of the angular velocity vector with the symbol ω shrinks the size of the solution dramatically.

In[25]:= `ωRules = {+ωx^2 + ωy^2 + ωz^2 -> +ω^2, -ωx^2 - ωy^2 - ωz^2 -> -ω^2}`

```
dsol1 = dsol //. ωRules // PowerExpand;
```
Out[25]= $\{\omega x^2 + \omega y^2 + \omega z^2 \to \omega^2, \ -\omega x^2 - \omega y^2 - \omega z^2 \to -\omega^2\}$

In[27]:= `{ByteCount[dsol1], LeafCount[dsol1]}`

Out[27]= $\{37384, 2168\}$

Simplifying the result, we get a manageable expression. (For brevity, we display only the x-component of the solution.)

In[28]:= `((dsol1[[1, 1]] //. ωRules) // Simplify) //. ωRules`

Out[28]= $x[t] \to \dfrac{1}{16\,\omega^7}\ (e^{-4\,i\,t\,\omega}\ (4\,i\,e^{2\,i\,t\,\omega}\,\omega^4$

$\qquad (\omega y^2\,C[1] - \omega y\,(\omega x\,C[2] + i\,\omega\,C[3]) + \omega z\,(\omega z\,C[1] + i\,\omega\,C[2] - \omega x\,C[3])) + 4\,i\,e^{6\,i\,t\,\omega}$

$\qquad \omega^4\,(-\omega y^2\,C[1] + \omega y\,(\omega x\,C[2] - i\,\omega\,C[3]) + \omega z\,(-\omega z\,C[1] + i\,\omega\,C[2] + \omega x\,C[3])) +$

$\qquad 4\,e^{4\,i\,t\,\omega}\,\omega\,(g\,(2\,t\,\omega^4\,\omega y + \omega^2\,\omega x\,\omega z - 2\,t^2\,\omega^4\,\omega x\,\omega z) +$

$\qquad 2\,\omega^4\,(\omega z\,C[2] - \omega y\,C[3] + 2\,t\,\omega x\,(\omega x\,C[1] + \omega y\,C[2] + \omega z\,C[3]) + 2\,\omega^2\,C[4]))))$

As expected, the result contains six integration constants—the three initial positions and the three initial velocities of the particle. The last example shows one frequently encountered issue in DSolve. Using `ExpTo` `Trig //@ % /. _Complex -> 0`, we could recast the last result in a purely real form, which is more appropriate for the physical meaning of this example. Doing this results in an expression of reasonable size.

In[29]:=
```
Module[{num, den, im, aux, trigs},
    (* numerator and denominator *)
    {num, den} = {Numerator[#], Denominator[#]}&[
                        Together[ExpToTrig[%[[2]]]]];
    (* delete imaginary part *)
    im = Factor[Plus @@ Cases[Expand[num], _Complex _]];
    (* rewrite in minimal set of trigonometric expressions *)
    aux = TrigFactor[Expand[num - (Plus @@ Cases[Expand[num], _Complex _])]];
    (* cancel factors and
        collect with respect to trigonometric expressions *)
    trigs = Union[Cases[aux, _Sin | _Cos, Infinity]];
    Collect[Factor[Collect[aux, trigs, Simplify] //. ωRules]/(2 ω^3),
            trigs, Simplify] * (2 ω^3/den) //. ωRules]
```

Out[29]= $\dfrac{1}{2\,\omega}\ \Big(\dfrac{1}{2\,\omega^3}\ (g\,(2\,t\,\omega^2\,\omega y + \omega x\,\omega z - 2\,t^2\,\omega^2\,\omega x\,\omega z) +$

$\qquad 2\,\omega^2\,(\omega z\,C[2] - \omega y\,C[3] + 2\,t\,\omega x\,(\omega x\,C[1] + \omega y\,C[2] + \omega z\,C[3]) + 2\,\omega^2\,C[4])) +$

$\qquad \dfrac{(-\omega z\,C[2] + \omega y\,C[3])\,\text{Cos}[2\,t\,\omega]}{\omega}\ +$

$\qquad \dfrac{(\omega y^2\,C[1] + \omega z^2\,C[1] - \omega x\,\omega y\,C[2] - \omega x\,\omega z\,C[3])\,\text{Sin}[2\,t\,\omega]}{\omega^2}\ \Big)$

Here is an ODE containing an unspecified function $g(x)$. The solutions of the differential equations are all functions $f[g(x)]\,(x)$, such that the pair of functions $f(x)$, $g(x)$ obey the differentiation rule $(f(x)\,g(x))' = f'(x)\,g'(x)$ [1181].

In[30]:= `DSolve[f'[x] g'[x] == D[f[x] g[x], x], f[x], x]`

Out[30]= $\left\{\left\{f[x] \to e^{\int_1^x -\frac{g'[K\$715]}{g[K\$715] - g'[K\$715]}\,dK\$715}\,C[1]\right\}\right\}$

Here are the similar solutions for $(f(x)/g(x))' = f'(x)/g'(x)$.

In[31]:= **DSolve[f'[x]/g'[x] == D[f[x]/g[x], x], f[x], x]**

Out[31]= $\left\{\left\{f[x] \rightarrow e^{\int_1^x -\frac{g'[K\$749]^2}{g[K\$749]\,(g[K\$749]-g'[K\$749])}\,dK\$749}\,C[1]\right\}\right\}$

Before looking at some typical ODEs, we make a remark about the integration constants that usually appear in the integration of differential equations. The notation C[*i*] used in DSolve can be changed using GeneratedᐧᐧParameters.

In[32]:= **Options[DSolve]**

Out[32]= {GeneratedParameters → C}

GeneratedParameters

is an option for DSolve, and it specifies the notation to be used for the integration constants that arise in the solution of the corresponding differential equation.

Default:

C

Admissible:

arbitrary symbol

In the following example, the default C is employed.

In[33]:= **DSolve[(x^2 + 1)y''[x] - x y'[x] + y[x] == 0, y[x], x]**

Out[33]= $\left\{\left\{y[x] \rightarrow x\,C[1] + \left(-\sqrt{1+x^2} + x\,\text{ArcSinh}[x]\right)C[2]\right\}\right\}$

Here, we use the modified form of C.

In[34]:= **DSolve[y'[x] == y[x] + 1, y[x], x, GeneratedParameters :> C]**

Out[34]= $\left\{\left\{y[x] \rightarrow -1 + e^x\,C[1]\right\}\right\}$

The right-hand side of the GeneratedParameters option can also be a pure function. For each integration constant, the pure function is called with an increasing integer argument starting from 1.

In[35]:= **DSolve[y''''[x] == y[x] + 1, y[x], x,**
GeneratedParameters :> (Subscript[c, #]&)]

Out[35]= $\left\{\left\{y[x] \rightarrow -1 + e^x\,c_1 + \text{Cos}[x]\,c_2 + e^{-x}\,c_3 + \text{Sin}[x]\,c_4\right\}\right\}$

We now look at a few more or less difficult examples to get an idea of what *Mathematica* can do. (The examples are assembled randomly from [243], [1653], [1537], [971], and [1425]. (If *Mathematica* cannot find the solution to a differential equation, the reader should have a look at these references.) We begin with differential equations of first order.

■ **Linear Inhomogeneous ODE with Constant Coefficients** $(\sum_{k=0}^n a_k\,y^{(k)}(x) = g(x))$

Here is a simple example (we have seen this type of equation already above).

In[36]:= **DSolve[f'[x] + f[x] == 0, f[x], x]**

Out[36]= $\left\{\left\{f[x] \rightarrow e^{-x}\,C[1]\right\}\right\}$

This slightly more general example has multiple zeros of the characteristic equation and an inhomogeneous term on the right-hand side.

In[37]:= **DSolve[y''''[x] - 4y'''[x] + 6y''[x] - 4y'[x] + y[x] == x Sin[x],**
y[x], x] // Simplify

Out[37]= $\left\{\left\{y[x] \rightarrow e^x\,(C[1] + x\,(C[2] + x\,(C[3] + x\,C[4]))) - \frac{\text{Cos}[x]}{2} - \frac{1}{4}\,(2+x)\,\text{Sin}[x]\right\}\right\}$

Because *Mathematica* does not introduce auxiliary variables and expresses all results explicitly in terms of the input variables, the explicit solution of the characteristic equation can lead to quite large expressions (even in the solution of simple linear homogeneous and/or inhomogeneous ODEs with constant coefficients). To avoid this, *Mathematica* uses Root-objects whenever possible in the solution of such type differential equations.

In[38]:= `DSolve[a[4] y''''[x] + a[3] y'''[x] + a[2] y''[x] +`
            `a[1] y'[x] + a[0] y[x] == x^3, y[x], x]`

Out[38]= $\left\{\left\{y[x] \to \frac{1}{a[0]^4} (x^3 a[0]^3 - 3 x^2 a[0]^2 a[1] + 6 x a[0] a[1]^2 - 6 a[1]^3 - 6 x a[0]^2 a[2] + \right.\right.$

$12 a[0] a[1] a[2] - 6 a[0]^2 a[3]) + e^{x \, \text{Root}[a[0]+a[1] \, \#1+a[2] \, \#1^2+a[3] \, \#1^3+a[4] \, \#1^4 \, \&, \, 1]} C[1] +$

$e^{x \, \text{Root}[a[0]+a[1] \, \#1+a[2] \, \#1^2+a[3] \, \#1^3+a[4] \, \#1^4 \, \&, \, 2]} C[2] + e^{x \, \text{Root}[a[0]+a[1] \, \#1+a[2] \, \#1^2+a[3] \, \#1^3+a[4] \, \#1^4 \, \&, \, 3]}$

$\left.\left. C[3] + e^{x \, \text{Root}[a[0]+a[1] \, \#1+a[2] \, \#1^2+a[3] \, \#1^3+a[4] \, \#1^4 \, \&, \, 4]} C[4]\right\}\right\}$

If we rewrite the Root-objects in radicals, the solution would be much larger.

In[39]:= `{ByteCount[%], ByteCount[ToRadicals[%]]}`

Out[39]= `{5088, 86592}`

If *Mathematica* solves a differential equation except for quadratures, functions not explicitly given can also appear in the given differential equation.

In[40]:= `DSolve[f''[x] + f'[x] + f[x] == g[x], f[x], x]`

Out[40]= $\left\{\left\{f[x] \to e^{-x/2} C[2] \cos\left[\frac{\sqrt{3} \, x}{2}\right] + e^{-x/2} C[1] \sin\left[\frac{\sqrt{3} \, x}{2}\right] + \right.\right.$

$e^{-x/2} \left(\cos\left[\frac{\sqrt{3} \, x}{2}\right] \int_1^x -\frac{2 \, e^{K\$900/2} \, g[K\$900] \sin\left[\frac{\sqrt{3} \, K\$900}{2}\right]}{\sqrt{3}} \, dK\$900 + \right.$

$\left.\left.\left.\left(\int_1^x \frac{2 \, e^{K\$875/2} \cos\left[\frac{\sqrt{3} \, K\$875}{2}\right] g[K\$875]}{\sqrt{3}} \, dK\$875\right) \sin\left[\frac{\sqrt{3} \, x}{2}\right]\right)\right\}\right\}$

Note the appearance of the variables K$*integer* as the dummy integration variable. For the following linear inhomogeneous differential equation (with variable coefficients), we obtain nested unevaluated integrals.

In[41]:= `DSolve[Log[x] y'[x] + y[x] g[x] + Exp[x] == 0, y[x], x]`

Out[41]= $\left\{\left\{y[x] \to e^{\int_1^x -\frac{g[K\$932]}{\text{Log}[K\$932]} \, dK\$932} C[1] + e^{\int_1^x -\frac{g[K\$932]}{\text{Log}[K\$932]} \, dK\$932} \int_1^x -\frac{e^{K\$951-\int_1^{K\$951} -\frac{g[K\$932]}{\text{Log}[K\$932]} \, dK\$932}}{\text{Log}[K\$951]} \, dK\$951\right\}\right\}$

Many differential equations are converted to the form of an unevaluated integral. In this form, the built-in Integrate cannot integrate the function explicitly. Here is a trivial higher-order example of this kind.

In[42]:= `DSolve[y''''[x] == unknownFunction[x], y[x], x]`

Out[42]= $\left\{\left\{y[x] \to C[1] + x \, C[2] + x^2 \, C[3] + x^3 \, C[4] + \right.\right.$

$\int_1^x \left(\int_1^{K\$1053} \left(\int_1^{K\$1044} \left(\int_1^{K\$1032} \text{unknownFunction}[K\$1016] \, dK\$1016\right) dK\$1032\right) dK\$1044\right)$

$\left.\left. dK\$1053\right\}\right\}$

Now let us investigate some classes of first-order of differential equations [1063] in more detail.

■ **ODE with Separated Variables** $(y'(x) = g(x)/f(y(x)))$

The following example can be integrated in closed form. (The warning message again comes from Solve when generating the ProductLog function; see Chapter 3.)

In[43]:= `DSolve[y'[x] == Exp[x]/Log[y[x]], y[x], x]`

InverseFunction::ifun : Inverse functions are
   being used. Values may be lost for multivalued inverses. More…

Solve::ifun : Inverse functions are being used by Solve, so some solutions
   may not be found; use Reduce for complete solution information. More…

Out[43]= $\left\{\left\{y[x] \to \dfrac{e^x + C[1]}{ProductLog[\frac{e^{x+C[1]}}{e}]}\right\}\right\}$

- **Homogeneous ODE** $(y'(x) = f(x, y(x)), \; f(t\,\xi, t\,\psi) = f(\xi, \psi))$

Here, *Mathematica* finds a first integral of this homogeneous ODE. However, because it is a polynomial of fifth degree, we again obtain Root-objects in the solution.

In[44]:= DSolve[y'[x] == (4x + 6y[x])/(6x + 4y[x]), y[x], x]

Out[44]= $\{\{y[x] \to Root[-e^{2\,C[1]}\,x - x^5 - e^{2\,C[1]}\,\#1 + 5\,x^4\,\#1 - 10\,x^3\,\#1^2 + 10\,x^2\,\#1^3 - 5\,x\,\#1^4 + \#1^5\,\&,\,1]\},$
$\{y[x] \to Root[-e^{2\,C[1]}\,x - x^5 - e^{2\,C[1]}\,\#1 + 5\,x^4\,\#1 - 10\,x^3\,\#1^2 + 10\,x^2\,\#1^3 - 5\,x\,\#1^4 + \#1^5\,\&,\,2]\},$
$\{y[x] \to Root[-e^{2\,C[1]}\,x - x^5 - e^{2\,C[1]}\,\#1 + 5\,x^4\,\#1 - 10\,x^3\,\#1^2 + 10\,x^2\,\#1^3 - 5\,x\,\#1^4 + \#1^5\,\&,\,3]\},$
$\{y[x] \to Root[-e^{2\,C[1]}\,x - x^5 - e^{2\,C[1]}\,\#1 + 5\,x^4\,\#1 - 10\,x^3\,\#1^2 + 10\,x^2\,\#1^3 - 5\,x\,\#1^4 + \#1^5\,\&,\,4]\},$
$\{y[x] \to Root[-e^{2\,C[1]}\,x - x^5 - e^{2\,C[1]}\,\#1 + 5\,x^4\,\#1 - 10\,x^3\,\#1^2 + 10\,x^2\,\#1^3 - 5\,x\,\#1^4 + \#1^5\,\&,\,5]\}\}$

- **Exact ODE** $(y'(x) = P(x, y(x))/Q(x, y(x))$ with $\partial P(\xi, \psi)/\partial \psi = -\partial Q(\xi, \psi)/\partial \xi)$

The following ODE is also solvable with the DSolve function.

In[45]:= solExactODE = DSolve[y'[x] == (y[x] - x)/(y[x]^-2 - x), y[x], x]

Out[45]= $\left\{\left\{y[x] \to \dfrac{x^2 + 2\,C[1] - \sqrt{-16\,x + x^4 + 4\,x^2\,C[1] + 4\,C[1]^2}}{4\,x}\right\},\right.$

$\left.\left\{y[x] \to \dfrac{x^2 + 2\,C[1] + \sqrt{-16\,x + x^4 + 4\,x^2\,C[1] + 4\,C[1]^2}}{4\,x}\right\}\right\}$

Let us make a short graphical check of the solution by plotting the solution curves together with the vector field generated by the right-hand side of the differential equation.

In[46]:= Needs["Graphics`PlotField`"]

In[47]:= Module[{sols, vectorField, statusPlotplnr, solutionCurves,
         ε = 10^-4},
  (* the solution curves calculated by DSolve *)
  sols = Union[Last /@ Flatten[solExactODE]];
  (* the vector field, calculated at randomly chosen points *)
  vectorField = ListPlotVectorField[
  Table[Function[{x, y},
   {{x, y}, 1/10 {Cos[#], Sin[#]}&[ArcTan[(y - x)/(y^-2 - x)]]}
     ][Random[Real, {-2, 2}], Random[Real, {-2, 2}]], {2000}],
     DisplayFunction -> Identity];
  (* solution curves for randomly chosen integration constants *)
  solutionCurves =
  (* turn of message of Plot triggered because of nonreal values *)
  Internal`DeactivateMessages @ Table[
  {Plot[#, {x, -2, -ε}, DisplayFunction -> Identity],
   Plot[#, {x, ε, 2},   DisplayFunction -> Identity]}&[
         sols /. {C[1] -> Random[Real, {-12, 12}]}], {100}];
  (* restore old status of Plot message *)
  If[Head[statusPlotplnr] === String, On[Plot::plnr]];
  (* show vector field and solution curves *)
  Show[{vectorField, Graphics[{GrayLevel[1/2],
          Cases[solutionCurves, _Line, {0, Infinity}]}]},
```
```

```
PlotRange -> {{-2, 2}, {-2, 2}}, Axes -> False,
Frame -> True, DisplayFunction -> $DisplayFunction]]
```

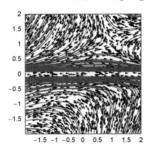

■ **Bernoulli ODE** $(y'(x) = f(x) y(x) + g(x) y(x)^n)$

A nontrivial solution is again found.

In[48]:= `DSolve[x^2 (x - 1) y'[x] == y[x]^2 + x(x - 2) y[x], y[x], x]`

Out[48]= $\left\{\left\{ y[x] \to \dfrac{x^2}{1 - C[1] + x\, C[1]} \right\}\right\}$

We insert the (nontrivial) solution into the ODE to check the result.

In[49]:= `(x^2 (x - 1) D[#, x] - (#^2 + x(x - 2) #))&[%[[1, 1, 2]]] // Simplify`

Out[49]= 0

■ **Jacobi ODE** $\left((ax + b\,y(x) + c) + (\tilde{a}x + \tilde{b}\,y(x) + \tilde{c})\,y'(x) + (\hat{a}x + \hat{b}\,y + \hat{c})(x\,y'(x) - y(x)) = 0\right)$

The following reduced example can also be solved, even though it cannot be solved explicitly for y[x]. This is an example for the situation mentioned above where the returned solution is mathematically correct, but is not explicitly solved for $y(x)$. While above, we used ToRules to convert a solution returned from Roots to a list of rules, a function ToRules is not needed here because Solve would return such a list of rules if it were be able to solve the equation.

In[50]:= `DSolve[(14x) + (4x + 5y[x]) y'[x] + (-4) (x y'[x] - y[x]) == 0,`
`          y[x], x]`

        Solve::tdep : The equations appear to involve the variables
             to be solved for in an essentially non-algebraic way. More…

Out[50]= $\mathtt{Solve}\Big[$

$$-\frac{1}{5}\sqrt{\frac{2}{33}}\,\mathtt{ArcTan}\Big[\frac{2 + \frac{5\,y[x]}{x}}{\sqrt{66}}\Big] + \frac{1}{10}\,\mathtt{Log}\Big[14 + \frac{4\,y[x]}{x} + \frac{5\,y[x]^2}{x^2}\Big] == C[1] - \frac{\mathtt{Log}[x]}{5},\ y[x]\Big]$$

Here is a quick check of the solution.

In[51]:= `Simplify[D[%[[1, 1]], x]]`

Out[51]= $\dfrac{y[x]\,(-y[x] + x\,y'[x])}{x\,(14\,x^2 + 4\,x\,y[x] + 5\,y[x]^2)}$

■ **Special Riccati ODE** $(y'(x) + a\,y(x)^2 = b\,x^\alpha)$

This special Riccati ODE is also solved.

In[52]:= `DSolve[y'[x] - y[x]^2 == x^v, y[x], x]`

Out[52]= $\left\{\left\{y[x] \rightarrow \left(-\text{BesselJ}\left[-\frac{1}{2+v}, \frac{2 x^{\frac{2+v}{2}}}{2+v}\right] C[1] + x^{\frac{2+v}{2}} \left(-2 \text{ BesselJ}\left[-1+\frac{1}{2+v}, \frac{2 x^{\frac{2+v}{2}}}{2+v}\right] + \right.\right.\right.\right.$

$\left.\left.\left. \text{BesselJ}\left[\frac{1+v}{2+v}, \frac{2 x^{1+\frac{v}{2}}}{2+v}\right] C[1] - \text{BesselJ}\left[-\frac{3+v}{2+v}, \frac{2 x^{\frac{2+v}{2}}}{2+v}\right] C[1]\right)\right)\right/$

$\left.\left.\left(2 x \left(\text{BesselJ}\left[\frac{1}{2+v}, \frac{2 x^{\frac{2+v}{2}}}{2+v}\right] + \text{BesselJ}\left[-\frac{1}{2+v}, \frac{2 x^{\frac{2+v}{2}}}{2+v}\right] C[1]\right)\right)\right\}\right\}$

■ **Abel ODE** of the first kind $(y'(x) = \sum_{k=0}^{3} f_k(x)\, y(x)^k)$

This special Abel ODE is solved in implicit form.

In[53]:= `DSolve[y'[x] == 1 + 2 y[x] + 3 y[x]^2 + 4 y[x]^3, y[x], x]`

> Solve::tdep : The equations appear to involve the variables
> to be solved for in an essentially non-algebraic way. More…

> Solve::tdep : The equations appear to involve the variables
> to be solved for in an essentially non-algebraic way. More…

Out[53]= $\text{Solve}\left[\frac{1}{2} \text{RootSum}\left[1 + 2 \#1 + 3 \#1^2 + 4 \#1^3 \&, \frac{\text{Log}[-\#1 + y[x]]}{1 + 3 \#1 + 6 \#1^2} \&\right] == x + C[1], y[x]\right]$

■ **Abel ODE** of the second kind $(y'(x)\,(y(x) + g(x)) = \sum_{k=0}^{2} f_k(x)\, y(x)^k)$

This special Abel ODE is also solved in implicit form. It contains a complicated integral.

In[54]:= `DSolve[(y[x] + x) y'[x] == 1 + 2 y[x] + 3 y[x]^2, y[x], x]`

> Solve::tdep : The equations appear to involve the variables
> to be solved for in an essentially non-algebraic way. More…

> Solve::tdep : The equations appear to involve the variables
> to be solved for in an essentially non-algebraic way. More…

> Solve::tdep : The equations appear to involve the variables
> to be solved for in an essentially non-algebraic way. More…

> General::stop :
> Further output of Solve::tdep will be suppressed during this calculation. More…

Out[54]= $\text{Solve}\Bigg[$

$x == e^{\frac{\text{ArcTan}\left[\frac{1+3 y[x]}{\sqrt{2}}\right]}{\sqrt{2}}} C[1] + e^{\frac{\text{ArcTan}\left[\frac{1+3 y[x]}{\sqrt{2}}\right]}{\sqrt{2}}} \int_1^{y[x]} \frac{e^{\frac{\text{ArcTan}\left[\frac{1+3 K\$8949}{\sqrt{2}}\right]}{\sqrt{2}}}}{1 + 2 K\$8949 + 3 K\$8949^2} \, dK\$8949, y[x]\Bigg]$

■ **Chini ODE** $(y'(x) = f(x)\, y(x)^n + g(x)\, y(x) + h(x))$

This special Chini ODE has two solutions.

In[55]:= `DSolve[y'[x] == Exp[x] y[x]^3 - Log[x] y[x], y[x], x]`

Out[55]= $\left\{\left\{y[x] \rightarrow -\frac{i\, e^x}{\sqrt{-x^2 x\, C[1] + 2 x^2 x \int_1^x e^{K\$99264+2\,(K\$99264 - K\$99264\, \text{Log}[K\$99264])}\, dK\$99264}}\right\},\right.$

$\left.\left\{y[x] \rightarrow \frac{i\, e^x}{\sqrt{-x^2 x\, C[1] + 2 x^2 x \int_1^x e^{K\$99264+2\,(K\$99264 - K\$99264\, \text{Log}[K\$99264])}\, dK\$99264}}\right\}\right\}$

■ **Lagrange ODE** $(x f(y'(x)) + g(y'(x)) = C\, y(x))$

The following Lagrange ODE with not explicitly defined functions f and g returns an implicit solution for $y(x)$.

In[56]:= `DSolve[y[x] == f[y'[x]] x + g[y'[x]], y[x], x]`

```
InverseFunction::ifun : Inverse functions are
     being used. Values may be lost for multivalued inverses. More…

InverseFunction::ifun : Inverse functions are
     being used. Values may be lost for multivalued inverses. More…

Solve::tdep : The equations appear to involve the variables
     to be solved for in an essentially non-algebraic way. More…

Solve::tdep : The equations appear to involve the variables
     to be solved for in an essentially non-algebraic way. More…

Solve::tdep : The equations appear to involve the variables
     to be solved for in an essentially non-algebraic way. More…

General::stop :
     Further output of Solve::tdep will be suppressed during this calculation. More…
```

Out[56]= $\text{Solve}\Big[\Big\{x == e^{\int_1^{K\$100033} \frac{f'[K\$100038]}{K\$100038-f[K\$100038]} dK\$100038} \, C[1] + e^{\int_1^{K\$100033} \frac{f'[K\$100038]}{K\$100038-f[K\$100038]} dK\$100038}$

$$\int \frac{e^{-\int_1^{K\$100033} \frac{f'[K\$100038]}{K\$100038-f[K\$100038]} dK\$100038} \, g'[K\$100033]}{K\$100033 - f[K\$100033]} dK\$100033,$$

$y[x] == x \, f[K\$100033] + g[K\$100033]\Big\}, \{y[x], K\$100033\}\Big]$

■ Clairaut ODE (a special case of a Lagrange ODE) $(y(x) = y'(x)\,x + \psi(y'(x)))$

The two solutions produced for the Clairaut ODE are correct in different parts of the complex x-plane.

In[57]:= **DSolve[y[x] == y'[x] x + 2 Sqrt[-y'[x]], y, x]**

Out[57]= $\Big\{\Big\{y \to \text{Function}\Big[\{x\}, \frac{4\,(-x + C[1])}{C[1]^2}\Big]\Big\}, \Big\{y \to \text{Function}\Big[\{x\}, -\frac{4\,(x + C[1])}{C[1]^2}\Big]\Big\}\Big\}$

To see that the individual solutions are not correct in the whole complex x-plane, we substitute the solutions in the original differential equation.

In[58]:= **(y[x] - (D[y[x], x] x + 2 Sqrt[-D[y[x], x]])) /. % // Simplify**

Out[58]= $\Big\{-4\,\sqrt{\frac{1}{C[1]^2}} + \frac{4}{C[1]}, \, -4\,\sqrt{\frac{1}{C[1]^2}} - \frac{4}{C[1]}\Big\}$

In[59]:= **{% /. C[1] -> 2, % /. C[1] -> 2I} // N // Chop**

Out[59]= $\{\{0, -4.\}, \{-4. \, \mathrm{i}, 0\}\}$

All solutions returned for this Clairaut ODE are correct in the whole complex x-plane.

In[60]:= **DSolve[y[x] == y'[x] x + 2 (y'[x])^2, y[x], x]**

Out[60]= $\Big\{\Big\{y[x] \to \frac{1}{2}\,\Big(16 - e^{C[1]} - 4\,x - \sqrt{-64\,e^{C[1]} + 16\,e^{C[1]}\,x - e^{C[1]}\,x^2}\Big)\Big\},$

$\Big\{y[x] \to \frac{1}{2}\,\Big(16 - e^{C[1]} - 4\,x + \sqrt{-64\,e^{C[1]} + 16\,e^{C[1]}\,x - e^{C[1]}\,x^2}\Big)\Big\},$

$\Big\{y[x] \to 2\,\Big(4 + 4\,e^{C[1]} - x - \sqrt{64\,e^{C[1]} - 16\,e^{C[1]}\,x + e^{C[1]}\,x^2}\Big)\Big\},$

$\Big\{y[x] \to 2\,\Big(4 + 4\,e^{C[1]} - x + \sqrt{64\,e^{C[1]} - 16\,e^{C[1]}\,x + e^{C[1]}\,x^2}\Big)\Big\}\Big\}$

In[61]:= **(y[x] == y'[x] x + 2 (y'[x])^2 /.**
               y -> Function[x, Evaluate[#[[1, 2]]]]& /@ %) // Simplify

Out[61]= {True, True, True, True}

■ ODE with Shifted Argument $(y'(x) = f(y(x) + a\,x))$

For the following ODE with shifted argument, we get an explicit result for $y(x)$.

In[62]:= **DSolve[y'[x] == Sin[y[x] + 3x], y[x], x] // Simplify**

> Solve::ifun : Inverse functions are being used by Solve, so some solutions
>   may not be found; use Reduce for complete solution information. More…

Out[62]= $\left\{\left\{y[x] \to -3x - 2\,\text{ArcTan}\left[\frac{1}{3}\left(1 - 2\sqrt{2}\,\text{Tan}\left[\frac{1}{3}\sqrt{2}\,(3x + C[1])\right]\right)\right]\right\},\right.$

$\left.\left\{y[x] \to -3x + 2\,\text{ArcTan}\left[\frac{1}{3}\left(-1 + 2\sqrt{2}\,\text{Tan}\left[\frac{1}{3}\sqrt{2}\,(3x + C[1])\right]\right)\right]\right\}\right\}$

Closed solutions can also be found for the following two special cases of the Riccati ODE with purely symbolic coefficients f[x], g[x], and h[x].

In[63]:= `h[x] = -g[x] - f[x];`
`DSolve[y'[x] == h[x] y[x]^2 + g[x] y[x] + f[x], y[x], x]`

Out[64]= $\left\{\left\{y[x] \to 1 + e^{\int_1^x (2\,(-f[K\$100494] - g[K\$100494]) + g[K\$100494])\,dK\$100494} \middle/ \right.\right.$

$\left(C[1] - \int_1^x e^{\int_1^{K\$100586}(2\,(-f[K\$100494] - g[K\$100494]) + g[K\$100494])\,dK\$100494}\right.$

$\left.\left.\left.(-f[K\$100586] - g[K\$100586])\,dK\$100586\right)\right\}\right\}$

In[65]:= `(y'[x] == h[x] y[x]^2 + g[x] y[x] + f[x] /.`
`        y -> Function[x, Evaluate[%[[1, 1, 2]]]]) // Simplify`

Out[65]= `True`

For the following Riccati ODE, a closed-form solution can also be found.

In[66]:= `Clear[h];`
`f[x] = 1/4 g[x]^2/h[x] - D[g[x]/h[x], x]/2;`

`DSolve[y'[x] == h[x] y[x]^2 + g[x] y[x] + f[x], y[x], x]`

Out[68]= $\left\{\left\{y[x] \to -\frac{g[x]}{2\,h[x]} + \frac{1}{C[1] - \int_1^x h[K\$101505]\,dK\$101505}\right\}\right\}$

In[69]:= `(y'[x] == (h[x] y[x]^2 + g[x] y[x] + f[x]) /.`
`        y -> Function[x, Evaluate[%[[1, 1, 2]]]]) // Simplify`

Out[69]= `True`

We end here with giving examples of the classical types of first-order ODEs. *Mathematica* can solve many more types of differential equations. Here is a nonlinear first-order ODE attributed to Cayley [314]. The result is quite a complicated function containing Root-objects. To better see the structure of the solution, we abbreviate the roots by $r[counter, degree]$.

In[70]:= `Module[{c = 0},`
`    DSolve[Q[k]^2 - Q[k] (k + 1/k) - 3 == 3(1 - k^2) Q'[k],`
`          Q, k] /. Root[p_, __] :>`
`          With[{c = c++, d = Exponent[p[C], C]}, r[c, d] /; True]]`

Out[70]= $\left\{\left\{Q \to \text{Function}\left[\{k\},\; 3\,(-1 + k^2)\right.\right.\right.$

$\left(\frac{e^{-\int_1^k r[3,4]\,dK\$101530}}{k^{1/6}\,(-1 + k^2)^{1/3}} - \frac{2\,e^{\int_1^k r[4,4]\,dK\$101530}\,k^{5/6}\,\int_1^k e^{-2\int_1^{K\$101602} r[5,4]\,dK\$101530}\,dK\$101602}{3\,(-1 + k^2)^{4/3}} - \right.$

$$\frac{e^{\int_1^k r[6,4]\, dK\$101530} \int_1^k e^{-2\int_1^{K\$101602} r[7,4]\, dK\$101530}\, dK\$101602}{6\, k^{7/6}\, (-1+k^2)^{1/3}} +$$

$$\frac{e^{\int_1^k r[8,4]\, dK\$101530} \left(\int_1^k e^{-2\int_1^{K\$101602} r[9,4]\, dK\$101530}\, dK\$101602\right) r[10,4]}{k^{1/6}\, (-1+k^2)^{1/3}} + C[1]$$

$$\left(-\frac{2\, e^{\int_1^k r[11,4]\, dK\$101530}\, k^{5/6}}{3\, (-1+k^2)^{4/3}} - \frac{e^{\int_1^k r[12,4]\, dK\$101530}}{6\, k^{7/6}\, (-1+k^2)^{1/3}} + \frac{e^{\int_1^k r[13,4]\, dK\$101530}\, r[14,4]}{k^{1/6}\, (-1+k^2)^{1/3}}\right)\right) \Bigg/$$

$$\left(\frac{e^{\int_1^k r[0,4]\, dK\$101530}\, C[1]}{k^{1/6}\, (-1+k^2)^{1/3}} + \frac{e^{\int_1^k r[1,4]\, dK\$101530} \int_1^k e^{-2\int_1^{K\$101602} r[2,4]\, dK\$101530}\, dK\$101602}{k^{1/6}\, (-1+k^2)^{1/3}}\right)\Bigg]\Bigg\}\Bigg\}$$

Similar to integrals and sums, there exist many first-order (as well as higher-order) differential equations for which *Mathematica* is unable to find a solution.

In[71]:= **DSolve[y'[x] + y[x]^2 Log[x] == 2, y[x], x]**

Out[71]= DSolve[Log[x] y[x]^2 + y'[x] == 2, y[x], x]

Differential equations of second order describe many physical processes [1871], but at the same time, they are much more difficult to solve. Only a few types have been given names. Many of the ones appearing more often in practice belong to the Painlevé type [397], [399], [377], [1700]: $y''(x) = f(x, y(x), y'(x))$.

Here, f is a rational function of $y(x)$ and $y'(x)$ with coefficients that are analytic functions of x. Unfortunately, at present, relatively few properties (when compared to the known properties of the named special functions of Chapter 3) of these equations are known, and fast numerical values for all parameters are also not available. For the state of the art, see [777], [1132], [398], and [1749]. For some of the modern techniques and algorithms for solving differential equations in general, see [1173], [1352], [890], [891], [1657], [892], and [1916]. A quite readable introduction can be found in [1091] and [1597]. For some of the original research on this subject, see [1143]. *Mathematica* does not recognize a Painlevé-type differential equation and does not bring any differential equation into a Painlevé normal form. In the next input, we try to solve the simplest Painlevé differential equation.

In[72]:= **DSolve[y''[x] == y[x]^2 + x, y[x], x]**

Out[72]= DSolve[y''[x] == x + y[x]^2, y[x], x]

We now give a few examples of relatively simple differential equations of second order. Here is a homogeneous differential equation that *Mathematica* solves.

In[73]:= **DSolve[x^2 y[x] y''[x] == y[x]^2 - 2 x y[x] y'[x] + x^2 y'[x]^2, y, x]**

Out[73]= $\left\{\left\{y \to \text{Function}\left[\{x\}, e^{-\frac{C[1]}{x}} x\, C[2]\right]\right\}\right\}$

Here is a check if the solution is correct.

In[74]:= **x^2 y[x] y''[x] == y[x]^2 - 2 x y[x] y'[x] + x^2 y'[x]^2 /.**
                        %[[1]] // Simplify

Out[74]= True

Here is a complicated-looking nonlinear second-order differential equation with a relatively simple solution.

In[75]:= **DSolve[w''[z] + w[z]^3 w'[z] ==**
          w[z] w'[z] (w[z]^4 + 4 w'[z])^(1/2), w[z], z]

Out[75]= $\left\{\left\{w[z] \rightarrow \sqrt{2} \ e^{\frac{C[1]}{4}} \ Tan\left[2\sqrt{2} \ e^{\frac{3 C[1]}{4}} \ (z + C[2])\right]\right\}\right\}$

Here is an equation in which the independent variables do not appear explicitly. It too is solved.

In[76]:= `DSolve[y[x] y''[x] - y'[x]^2 == y[x]^2 Log[y[x]], y[x], x]`

> Solve::tdep : The equations appear to involve the variables
> to be solved for in an essentially non-algebraic way. More...

> Solve::tdep : The equations appear to involve the variables
> to be solved for in an essentially non-algebraic way. More...

Out[76]= $\Big\{\{y[x] \rightarrow \text{InverseFunction}\Big[$

$-\dfrac{Log\left[Log[\#1] + \sqrt{C[1] + Log[\#1]^2}\right] \sqrt{C[1] + Log[\#1]^2} \ \#1}{\sqrt{(C[1] + Log[\#1]^2)} \ \#1^2} \ \&\Big] [x + C[2]]\},$

$\{y[x] \rightarrow \text{InverseFunction}\Big[\dfrac{Log\left[Log[\#1] + \sqrt{C[1] + Log[\#1]^2}\right] \sqrt{C[1] + Log[\#1]^2} \ \#1}{\sqrt{(C[1] + Log[\#1]^2)} \ \#1^2} \ \&\Big] [$

$x + C[2]]\}\Big\}$

The solution to the following interesting differential equations contains a singular point. The location of the singular point depends on the integration constants.

In[77]:= `DSolve[{y''[x] == y'[x]^2 (2 y[x] - 1)/(y[x]^2 + 1)}, y[x], x]`

> Solve::ifun : Inverse functions are being used by Solve, so some solutions
> may not be found; use Reduce for complete solution information. More...

Out[77]= $\{\{y[x] \rightarrow Tan[Log[-C[1] (-x - C[2])]]\}\}$

This is another example for solving a nontrivial ODE of second order.

In[78]:= `DSolve[x^5 y''[x] + 2x^4 y'[x] - 1 == 0, y[x], x]`

Out[78]= $\left\{\left\{y[x] \rightarrow \dfrac{1}{6 x^3} - \dfrac{C[1]}{x} + C[2]\right\}\right\}$

The following two second-order ODEs can also be solved explicitly.

In[79]:= `DSolve[y''[x] - y[x]/x^4 == 0, y[x], x]`

Out[79]= $\left\{\left\{y[x] \rightarrow e^{\frac{1}{x}} x C[1] + \dfrac{1}{2} e^{-1/x} x C[2]\right\}\right\}$

In[80]:= `DSolve[{(x - a)^2 (x - b)^2 y''[x] == c y[x]}, y[x], x] // Simplify`

Out[80]= $\Big\{\{y[x] \rightarrow \dfrac{1}{\sqrt{a^2 - 2 a b + b^2 + 4 c}} \Big((-a + x)^{\frac{-a+b+\sqrt{a^2-2 a b+b^2+4 c}}{2 (a-b)}} (-b + x)^{\frac{1}{2} - \frac{\sqrt{a^2-2 a b+b^2+4 c}}{2 (a-b)}}$

$\Big(\sqrt{a^2 - 2 a b + b^2 + 4 c} \ (-a + x)^{\frac{\sqrt{a^2-2 a b+b^2+4 c}}{a-b}} C[1] - (-b + x)^{\frac{\sqrt{a^2-2 a b+b^2+4 c}}{a-b}} C[2]\Big)\Big)\}\Big\}$

Here is a nonlinear second-order ODE that has a rational solution.

In[81]:= `DSolve[y''[x] + 3 y[x] y'[x] + y[x]^3 == 0, y, x]`

Out[81]= $\left\{\left\{y \rightarrow Function\left[\{x\}, \dfrac{2 e^{\frac{2 C[1]}{9}} (x + C[2])}{3 + e^{\frac{2 C[1]}{9}} x^2 + 2 e^{\frac{2 C[1]}{9}} x C[2] + e^{\frac{2 C[1]}{9}} C[2]^2}\right]\right\}\right\}$

Here is another nonlinear second-order ODE. This time an implicit solution is returned.

In[82]:= `DSolve[x''[t] + Exp[-x[t]^2] x[t] == c, x[t], t]`

> Solve::tdep : The equations appear to involve the variables
> to be solved for in an essentially non-algebraic way. More...

> Solve::tdep : The equations appear to involve the variables
> to be solved for in an essentially non-algebraic way. More...

Out[82]= $\mathtt{Solve}\left[\left(\int_{1}^{x[t]} \dfrac{1}{\sqrt{2\left(\dfrac{e^{-K\$102367^2}}{2} + c\,K\$102367\right) + C[1]}}\; dK\$102367\right)^2 == (t + C[2])^2,\, x[t]\right]$

The next differential equation can be solved in quadratures for a symbolic function $f(x)$.

In[83]:= `DSolve[y''[x] == f[x] y'[x], y[x], x]`

Out[83]= $\left\{\left\{y[x] \rightarrow\right.\right.$

$\qquad C[2] + \displaystyle\int_{1}^{x} e^{\int_{1}^{K\$103098} \frac{g[K\$103051]^2\, h[K\$103051] - 2\,h[K\$103051]\, g'[K\$103051] + 2\,g[K\$103051]\, h'[K\$103051]}{4\,h[K\$103051]^2}\; dK\$103051}\; C[1]$

$\qquad \left.\left. dK\$103098\right\}\right\}$

And the following differential equation can be solved in implicit form for any function f.

In[84]:= `DSolve[y''[x] == f[y[x]], y[x], x]`

> Solve::tdep : The equations appear to involve the variables
> to be solved for in an essentially non-algebraic way. More...

> Solve::tdep : The equations appear to involve the variables
> to be solved for in an essentially non-algebraic way. More...

Out[84]= $\mathtt{Solve}\left[\left(\int_{1}^{y[x]} \dfrac{1}{\sqrt{C[1] + 2\int_{1}^{K\$103251} f[K\$103235]\; dK\$103235}}\; dK\$103251\right)^2 == (x + C[2])^2,\, y[x]\right]$

Here is a relatively complicated nonlinear second-order differential equation. DSolve succeeds in solving for $\Upsilon[\xi]$ explicitly.

In[85]:= `DSolve[ξ D[ξ Υ'[ξ], ξ] == (ξ Υ'[ξ])^2 + α Exp[Υ[ξ]] + β, Υ, ξ]`

> Solve::ifun : Inverse functions are being used by Solve, so some solutions
> may not be found; use Reduce for complete solution information. More...

Out[85]= $\left\{\left\{\Upsilon \rightarrow \mathtt{Function}\left[\{\xi\},\, -\mathtt{Log}\left[-\beta\,\xi^{\sqrt{-\beta}}\, C[1]\left(-\dfrac{\sqrt{-\beta}\,\xi^{-2\sqrt{-\beta}}}{2\,\beta^2} + \dfrac{\alpha\,\xi^{-\sqrt{-\beta}}}{\beta^2\, C[1]} - C[2]\right)\right]\right]\right\}\right\}$

Here is a quick check for the correctness of the returned solution.

In[86]:= `ξ D[ξ Υ'[ξ], ξ] == (ξ Υ'[ξ])^2 + α Exp[Υ[ξ]] + β /. %[[1]] //`
`                                                       Simplify`

Out[86]= `True`

In addition to implicit solutions of differential equations, the modified DSolve can solve some differential equations whose solutions are special functions. Here are examples whose solutions are confluent hypergeometric, Bessel, and Legendre functions. The latter solution is also expressed in terms of hypergeometric functions.

In[87]:= `(* Bessel ODE *)`
`     DSolve[x^2 y''[x] + x y'[x] + (x^2 - v^2) y[x] == 0, y[x], x]`

Out[88]= $\left\{\left\{y[x] \rightarrow \mathtt{BesselJ}[v, x]\, C[1] + \mathtt{BesselY}[v, x]\, C[2]\right\}\right\}$

In[89]:= `(* Bessel-type ODE *)`
`     DSolve[x y''[x] + y'[x] + (-x/4 - m^2/(4x)) y[x] == 0, y[x], x]`

Out[90]= $\left\{\left\{y[x] \rightarrow \mathtt{BesselJ}\left[\dfrac{m}{2}, -\dfrac{i\,x}{2}\right] C[1] + \mathtt{BesselY}\left[\dfrac{m}{2}, -\dfrac{i\,x}{2}\right] C[2]\right\}\right\}$

In[91]:= `(* ODE that has solution x^α BesselZ[n, β x^γ] *)`
`     DSolve[y''[x] + (1 - 2 α)/x y'[x] + ((β γ x^(γ - 1))^2 +`

$$(\alpha\text{\textasciicircum}2 - n\text{\textasciicircum}2\ \gamma\text{\textasciicircum}2)/x\text{\textasciicircum}2)\ y[x] == 0,\ y[x],\ x] //$$
$$\text{Simplify // PowerExpand}$$

Out[92]= $\{\{y[x] \to 2^{-\frac{\alpha}{\gamma}} x^{\alpha}\ \beta^{\alpha/\gamma}$

$(\text{BesselJ}[-n,\ x^{\gamma}\ \beta]\ C[1]\ \text{Gamma}[1-n] + \text{BesselJ}[n,\ x^{\gamma}\ \beta]\ C[2]\ \text{Gamma}[1+n])\}\}$

In[93]:= (* simple ODE that can be solved in Bessel function *)
      DSolve[y''[x] == x^a y[x], y[x], x]

Out[94]= $\{\{y[x] \to (-1)^{\frac{1}{2+a}}\ (2+a)^{-\frac{1}{2+a}}\ x^{1-\frac{1+\frac{a}{2}}{2+a}}\ \text{BesselI}\Big[\frac{1}{2+a},\ \frac{2\ x^{\frac{2+a}{2}}}{2+a}\Big]\ C[2]\ \text{Gamma}\Big[1+\frac{1}{2+a}\Big] +$

$(2+a)^{-\frac{1}{2+a}}\ x^{\frac{1+\frac{a}{2}}{2+a}}\ \text{BesselI}\Big[\frac{1}{-2-a},\ \frac{2\ x^{\frac{2+a}{2}}}{2+a}\Big]\ C[1]\ \text{Gamma}\Big[\frac{1}{2+a}+\frac{a}{2+a}\Big]\}\}$

In[95]:= (* Legendre ODE *)
      DSolve[(1 - x^2) y''[x] - 2 x y'[x] + v (v + 1) y[x] == 0, y[x], x]
Out[96]= $\{\{y[x] \to C[1]\ \text{LegendreP}[v,\ x] + C[2]\ \text{LegendreQ}[v,\ x]\}\}$

Here are some Schrödinger equations for various potentials.

In[97]:= (* Calogero *)
      DSolve[-ψ''[z] + z^-2 ψ[z] == ε ψ[z], ψ[z], z]

Out[98]= $\{\{\psi[z] \to \sqrt{z}\ \text{BesselJ}\Big[\frac{\sqrt{5}}{2},\ z\ \sqrt{\varepsilon}\Big]\ C[1] + \sqrt{z}\ \text{BesselY}\Big[\frac{\sqrt{5}}{2},\ z\ \sqrt{\varepsilon}\Big]\ C[2]\}\}$

In[99]:= (* 1D hydrogen (or DiracDelta[z]) *)
      DSolve[-ψ''[z] + z^-1 ψ[z] == ε ψ[z], ψ[z], z]
Out[100]= $\{\{\psi[z] \to$

$e^{z\ (-\sqrt{-\varepsilon}+\sqrt{\varepsilon})-z\ \sqrt{\varepsilon}}\ z\ C[2]\ \text{Hypergeometric1F1}\Big[1+\frac{-2\ (-1-2\ \sqrt{\varepsilon})-4\ \sqrt{\varepsilon}}{4\ \sqrt{-\varepsilon}},\ 2,\ 2\ z\ \sqrt{-\varepsilon}\Big] +$

$e^{z\ (-\sqrt{-\varepsilon}+\sqrt{\varepsilon})-z\ \sqrt{\varepsilon}}\ z\ C[1]\ \text{HypergeometricU}\Big[1+\frac{-2\ (-1-2\ \sqrt{\varepsilon})-4\ \sqrt{\varepsilon}}{4\ \sqrt{-\varepsilon}},\ 2,\ 2\ z\ \sqrt{-\varepsilon}\Big]\}\}$

In[101]:= (* free particle *)
      DSolve[-ψ''[z] + z^0 ψ[z] == ε ψ[z], ψ[z], z]
Out[102]= $\{\{\psi[z] \to e^{z\ \sqrt{1-\varepsilon}}\ C[1] + e^{-z\ \sqrt{1-\varepsilon}}\ C[2]\}\}$

In[103]:= (* constant field *)
      DSolve[-ψ''[z] + z^1 ψ[z] == ε ψ[z], ψ[z], z]
Out[104]= $\{\{\psi[z] \to \text{AiryAi}[z - \varepsilon]\ C[1] + \text{AiryBi}[z - \varepsilon]\ C[2]\}\}$

In[105]:= (* harmonic oscillator *)
      DSolve[-ψ''[z] + z^2 ψ[z] == ε ψ[z], ψ[z], z]
Out[106]= $\{\{\psi[z] \to$

$e^{-\frac{z^2}{2}}\ C[1]\ \text{HermiteH}\Big[\frac{1}{2}\ (-1+\varepsilon),\ z\Big] + e^{-\frac{z^2}{2}}\ C[2]\ \text{Hypergeometric1F1}\Big[\frac{1-\varepsilon}{4},\ \frac{1}{2},\ z^2\Big]\}\}$

In[107]:= (* Calogero *)
      DSolve[-ψ''[z] + (α/z^2 + β z^2) ψ[z] == ε ψ[z], ψ[z], z]
Out[108]= $\{\{\psi[z] \to \frac{1}{\sqrt{z}}\ \Big(2^{\frac{1}{4}\ (2+\sqrt{1+4\alpha})}\ e^{-\frac{1}{2}\ z^2\ \sqrt{\beta}}\ (z^2)^{\frac{1}{4}\ (2+\sqrt{1+4\alpha})}\Big)\ C[1]$

$\text{HypergeometricU}\Big[\frac{2\beta+\sqrt{1+4\alpha}\ \beta-\sqrt{\beta}\ \varepsilon}{4\beta},\ \frac{1}{2}\ (2+\sqrt{1+4\alpha}),\ z^2\ \sqrt{\beta}\Big] +$

$\frac{1}{\sqrt{z}}\ \Big(2^{\frac{1}{4}\ (2+\sqrt{1+4\alpha})}\ e^{-\frac{1}{2}\ z^2\ \sqrt{\beta}}\ (z^2)^{\frac{1}{4}\ (2+\sqrt{1+4\alpha})}\Big)\ C[2]$

$\text{LaguerreL}\Big[-\frac{2\beta+\sqrt{1+4\alpha}\ \beta-\sqrt{\beta}\ \varepsilon}{4\beta},\ -1+\frac{1}{2}\ (2+\sqrt{1+4\alpha}),\ z^2\ \sqrt{\beta}\Big]\}\}$

```
In[109]:= (* nD Coulomb *)
         DSolve[-ψ''[z] + (α/z + β/z^2) ψ[z] == ε ψ[z], ψ[z], z]
```

$$\text{Out[110]}= \left\{\left\{\psi[z] \to e^{\frac{1}{2}\left(-2\,i\,z\sqrt{\varepsilon}+\left(1+\sqrt{1+4\beta}\right)\text{Log}[z]\right)}\, C[1]\, \text{HypergeometricU}\left[\right.\right.\right.$$

$$-\frac{i\,\alpha - \sqrt{\varepsilon} - \sqrt{1+4\,\beta}\,\sqrt{\varepsilon}}{2\,\sqrt{\varepsilon}},\ 1+\sqrt{1+4\,\beta},\ 2\,i\,z\,\sqrt{\varepsilon}\left.\right] + e^{\frac{1}{2}\left(-2\,i\,z\sqrt{\varepsilon}+\left(1+\sqrt{1+4\beta}\right)\text{Log}[z]\right)}$$

$$C[2]\, \text{LaguerreL}\left[\frac{i\,\alpha - \sqrt{\varepsilon} - \sqrt{1+4\,\beta}\,\sqrt{\varepsilon}}{2\,\sqrt{\varepsilon}},\ \sqrt{1+4\,\beta},\ 2\,i\,z\,\sqrt{\varepsilon}\right]\left.\right\}\right\}$$

```
In[111]:= (* Pöschl-Teller *)
         DSolve[-ψ''[z] + α/Cosh[z]^2 ψ[z] == ε ψ[z], ψ[z], z]
```

$$\text{Out[112]}= \left\{\left\{\psi[z] \to C[1]\, \text{LegendreP}\left[\frac{1}{2}\left(-1+\sqrt{1-4\,\alpha}\right),\ i\,\sqrt{\varepsilon},\ \text{Tanh}[z]\right] +\right.\right.$$

$$C[2]\, \text{LegendreQ}\left[\frac{1}{2}\left(-1+\sqrt{1-4\,\alpha}\right),\ i\,\sqrt{\varepsilon},\ \text{Tanh}[z]\right]\left.\right\}\right\}$$

```
In[113]:= (* modified Pöschl-Teller *)
         DSolve[-ψ''[z] + α/Cos[z]^2 ψ[z] == ε ψ[z], ψ[z], z]
```

$$\text{Out[114]}= \left\{\left\{\psi[z] \to (-1)^{\frac{1}{4}\left(1-\sqrt{1+4\alpha}\right)}\, C[1]\, \text{Cos}[z]^{\frac{1}{2}\left(1-\sqrt{1+4\alpha}\right)}\, \text{Hypergeometric2F1}\left[\right.\right.\right.$$

$$\frac{1}{4} - \frac{1}{4}\sqrt{1+4\,\alpha} - \frac{\sqrt{\varepsilon}}{2},\ \frac{1}{4} - \frac{1}{4}\sqrt{1+4\,\alpha} + \frac{\sqrt{\varepsilon}}{2},\ 1 - \frac{1}{2}\sqrt{1+4\,\alpha},\ \text{Cos}[z]^2\left.\right] +$$

$$(-1)^{\frac{1}{4}\left(1+\sqrt{1+4\alpha}\right)}\, C[2]\, \text{Cos}[z]^{\frac{1}{2}\left(1+\sqrt{1+4\alpha}\right)}\, \text{Hypergeometric2F1}\left[\right.$$

$$\frac{1}{4} + \frac{1}{4}\sqrt{1+4\,\alpha} - \frac{\sqrt{\varepsilon}}{2},\ \frac{1}{4} + \frac{1}{4}\sqrt{1+4\,\alpha} + \frac{\sqrt{\varepsilon}}{2},\ 1 + \frac{1}{2}\sqrt{1+4\,\alpha},\ \text{Cos}[z]^2\left.\right]\left.\right\}\right\}$$

```
In[115]:= (* Liouville *)
         DSolve[-ψ''[z] + Exp[z] ψ[z] == ε ψ[z], ψ[z], z]
```

$$\text{Out[116]}= \left\{\left\{\psi[z] \to (-1)^{-i\sqrt{\varepsilon}}\, \text{BesselI}\left[-2\,i\,\sqrt{\varepsilon},\ 2\,\sqrt{e^z}\right]\, C[1]\, \text{Gamma}\left[1 - 2\,i\,\sqrt{\varepsilon}\right] +\right.\right.$$

$$(-1)^{i\sqrt{\varepsilon}}\, \text{BesselI}\left[2\,i\,\sqrt{\varepsilon},\ 2\,\sqrt{e^z}\right]\, C[2]\, \text{Gamma}\left[1 + 2\,i\,\sqrt{\varepsilon}\right]\left.\right\}\right\}$$

```
In[117]:= (* Morse *)
         DSolve[-ψ''[z] + (α Exp[z] + β Exp[2 z]) ψ[z] == ε ψ[z], ψ[z], z]
```

$$\text{Out[118]}= \left\{\left\{\psi[z] \to\right.\right.$$

$$e^{-e^z\sqrt{\beta}+i\sqrt{\varepsilon}\,\text{Log}[e^z]}\, C[1]\, \text{HypergeometricU}\left[-\frac{-\alpha - \sqrt{\beta} - 2\,i\,\sqrt{\beta}\,\sqrt{\varepsilon}}{2\,\sqrt{\beta}},\ 1 + 2\,i\,\sqrt{\varepsilon},\ 2\,e^z\,\sqrt{\beta}\right] +$$

$$e^{-e^z\sqrt{\beta}+i\sqrt{\varepsilon}\,\text{Log}[e^z]}\, C[2]\, \text{LaguerreL}\left[\frac{-\alpha - \sqrt{\beta} - 2\,i\,\sqrt{\beta}\,\sqrt{\varepsilon}}{2\,\sqrt{\beta}},\ 2\,i\,\sqrt{\varepsilon},\ 2\,e^z\,\sqrt{\beta}\right]\left.\right\}\right\}$$

```
In[119]:= (* smooth step *)
         DSolve[-ψ''[z] + α/(1 + Exp[z]) ψ[z] == ε ψ[z], ψ[z], z]
```

$$\text{Out[120]}= \left\{\left\{\psi[z] \to (e^z)^{-\sqrt{\alpha-\varepsilon}}\, C[1]\right.\right.$$

$$\text{Hypergeometric2F1}\left[-\sqrt{\alpha-\varepsilon} - i\,\sqrt{\varepsilon},\ -\sqrt{\alpha-\varepsilon} + i\,\sqrt{\varepsilon},\ 1 - 2\,\sqrt{\alpha-\varepsilon},\ -e^z\right] +$$

$$(e^z)^{\sqrt{\alpha-\varepsilon}}\, C[2]\, \text{Hypergeometric2F1}\left[\sqrt{\alpha-\varepsilon} - i\,\sqrt{\varepsilon},\ \sqrt{\alpha-\varepsilon} + i\,\sqrt{\varepsilon},\ 1 + 2\,\sqrt{\alpha-\varepsilon},\ -e^z\right]\left.\right\}\right\}$$

```
In[121]:= (* periodic Mathieu *)
         DSolve[-ψ''[z] + Cos[z] ψ[z] == ε ψ[z], ψ[z], z]
```

$$\text{Out[122]}= \left\{\left\{\psi[z] \to C[1]\, \text{MathieuC}\left[4\,\varepsilon,\ 2,\ \frac{z}{2}\right] + C[2]\, \text{MathieuS}\left[4\,\varepsilon,\ 2,\ \frac{z}{2}\right]\right\}\right\}$$

In the next input for the harmonic oscillator, we give an additional boundary value. The resulting solution depends on one constant of integration.

```
In[123]:= DSolve[{-ψ''[z] + z^2 ψ[z] == ε ψ[z], ψ[-L] == 0}, ψ[z], z]
```

Out[123]= $\left\{\left\{\psi[z] \rightarrow \dfrac{1}{\text{HermiteH}\left[\frac{1}{2}(-1+\varepsilon), -L\right]}\right.\right.$

$\left(e^{-\frac{z^2}{2}} C[2] \left(-\text{HermiteH}\left[\frac{1}{2}(-1+\varepsilon), z\right] \text{Hypergeometric1F1}\left[\frac{1-\varepsilon}{4}, \frac{1}{2}, L^2\right] + \right.\right.$

$\left.\left.\left.\left.\text{HermiteH}\left[\frac{1}{2}(-1+\varepsilon), -L\right] \text{Hypergeometric1F1}\left[\frac{1-\varepsilon}{4}, \frac{1}{2}, z^2\right]\right)\right)\right\}\right\}$

For two constant we obtain only the trivial solution (which is one among the countable many possible solutions).

In[124]:= `DSolve[{-ψ''[z] + z^2 ψ[z] == ε ψ[z], ψ[-L] == 0, ψ[L] == 0}, ψ[z], z]`

Out[124]= $\{\{\psi[z] \rightarrow 0\}\}$

The next input solves the differential equation for a generalized harmonic oscillator [306].

In[125]:= `DSolve[-(1 + λ^2 x^2) y''[x] - λ^2 x y'[x] +`
        `α^2 x^2/(1 + λ^2 x^2) == ε y[x], y[x], x] // Simplify[#, λ > 0]&`

Out[125]= $\left\{\left\{y[x] \rightarrow\right.\right.$

$\dfrac{1}{2 \varepsilon \lambda^3 (\varepsilon + 4 \lambda^2)} \left(e^{-\frac{i\sqrt{\varepsilon} \text{ArcSinh}[x\lambda]}{\lambda}} \left(\cos\left[\frac{\sqrt{\varepsilon} \text{ArcSinh}[x\lambda]}{\lambda}\right] + i \sin\left[\frac{\sqrt{\varepsilon} \text{ArcSinh}[x\lambda]}{\lambda}\right]\right)\right.$

$\left(-i\, e^{2\,\text{ArcSinh}[x\lambda]} \alpha^2 \varepsilon (\sqrt{\varepsilon} - 2 i \lambda)\right.$

$\text{Hypergeometric2F1}\left[1 - \frac{i\sqrt{\varepsilon}}{2\lambda}, 1, 2 - \frac{i\sqrt{\varepsilon}}{2\lambda}, -e^{2\,\text{ArcSinh}[x\lambda]}\right] + (\sqrt{\varepsilon} + 2 i \lambda)$

$\left(i\, e^{2\,\text{ArcSinh}[x\lambda]} \alpha^2 \varepsilon\, \text{Hypergeometric2F1}\left[1 + \frac{i\sqrt{\varepsilon}}{2\lambda}, 1, 2 + \frac{i\sqrt{\varepsilon}}{2\lambda}, -e^{2\,\text{ArcSinh}[x\lambda]}\right] + \right.$

$(\sqrt{\varepsilon} - 2 i \lambda) \left(i\, \alpha^2 \sqrt{\varepsilon}\, \text{Hypergeometric2F1}\left[-\frac{i\sqrt{\varepsilon}}{2\lambda}, 1, 1 - \frac{i\sqrt{\varepsilon}}{2\lambda}, \right.\right.$

$\left. -e^{2\,\text{ArcSinh}[x\lambda]}\right] - i\, \alpha^2 \sqrt{\varepsilon}\, \text{Hypergeometric2F1}\left[\frac{i\sqrt{\varepsilon}}{2\lambda}, 1, 1 + \frac{i\sqrt{\varepsilon}}{2\lambda}, \right.$

$\left. -e^{2\,\text{ArcSinh}[x\lambda]}\right] + 2 \lambda \left(\alpha^2 + \varepsilon \lambda^2 C[1] \cos\left[\frac{\sqrt{\varepsilon} \text{ArcSinh}[x\lambda]}{\lambda}\right] + \right.$

$\left.\left.\left.\left.\left.\left.\varepsilon \lambda^2 C[2] \sin\left[\frac{\sqrt{\varepsilon} \text{ArcSinh}[x\lambda]}{\lambda}\right]\right)\right)\right)\right)\right)\right\}\right\}$

But the following Schrödinger equation [1552], [1553] for a simple $\mathcal{PT}$-invariant potential remains still unsolved.

In[126]:= `DSolve[{-ψ''[z] -6/(Cos[z] + 2 I Sin[z])^2 ψ[z] == ε ψ[z]}, ψ[z], z]`

Out[126]= $\text{DSolve}\left[\left\{-\dfrac{6\,\psi[z]}{(\cos[z] + 2 i \sin[z])^2} - \psi''[z] == \varepsilon\, \psi[z]\right\}, \psi[z], z\right]$

In[127]:= `(-ψ''[z] -6/(Cos[z] + 2 I Sin[z])^2 ψ[z] == ε ψ[z]) /.`
        `(* the solutions *)`
        `{{ψ -> (Exp[+I Sqrt[ε] #] (2 I - Sqrt[ε] I + 3/(2 I + Cot[#]))&)},`
        `{ψ -> (Exp[-I Sqrt[ε] #] (2 I + Sqrt[ε] I + 3/(2 I + Cot[#]))&)}} //`
        `Simplify`

Out[127]= `{True, True}`

In many quantum-mechanical model calculations, one uses piecewise-defined potentials in the following Schrödinger equation. The following Schrödinger equation with a piecewise-constant potential also gives a closed-form solution as a piecewise function (head `Piecewise`). Because the resulting solution is quite large, we display an abbreviated version of the result.

In[128]:= `𝒱[z_] = Piecewise[{{0, -2 < z < -1}, {1, -1 < z < 1}, {0, 1 < z < 2}}, 4];`

        `dsol1 = DSolve[{-ψ''[z] + 𝒱[z] ψ[z] == ε ψ[z], ψ[0] == 0, ψ'[0] == ψ1},`

```
                        ψ[z], z];
```

```
        LeafCount[dsol1]
Out[131]=  1972
```

```
In[132]:= (* display abbreviated result *)
        dsol1 //. {Sqrt[ε] -> s0, Sqrt[1 - ε] -> s1, Sqrt[4 - ε] -> s4} /.
                HoldPattern[Piecewise[l_, r___]] :> Piecewise[Take[l, 2]] //
                                                                    InputForm
```

```
Out[133]//InputForm=
        {{ψ[z] -> Piecewise[
            {{(E^(-s1 - 2*s4 - z*s4)*(s0*s4*ψ1*Cos[s0] -
              E^(2*s1)*s0*s4*ψ1*Cos[s0] + E^(4*s4 + 2*z*s4)*s0*s4*ψ1*Cos[s0] -
              E^(2*s1 + 4*s4 + 2*z*s4)*s0*s4*ψ1*Cos[s0] - Sqrt[ε - ε^2]*ψ1*Cos[s0] -
              E^(2*s1)*Sqrt[ε - ε^2]*ψ1*Cos[s0] + E^(4*s4 + 2*z*s4)*Sqrt[ε - ε^2]*ψ1*
              Cos[s0] + E^(2*s1 + 4*s4 + 2*z*s4)*Sqrt[ε - ε^2]*ψ1*Cos[s0] -
              s1*s4*ψ1*Sin[s0] - E^(2*s1)*s1*s4*ψ1*Sin[s0] - E^(4*s4 + 2*z*s4)*s1*
              s4*ψ1*Sin[s0] - E^(2*s1 + 4*s4 + 2*z*s4)*s1*s4*ψ1*Sin[s0] -
              ε*ψ1*Sin[s0] + E^(2*s1)*ε*ψ1*Sin[s0] + E^(4*s4 + 2*z*s4)*ε*ψ1*
              Sin[s0] - E^(2*s1 + 4*s4 + 2*z*s4)*ε*ψ1*Sin[s0]))/
              (4*Sqrt[4 - ε]*Sqrt[ε - ε^2]),
              Inequality[-Infinity, Less, z, LessEqual, -2]},
            {(ψ1*(s0*Cos[s0]*Cos[z*s0] - E^(2*s1)*s0*Cos[s0]*Cos[z*s0] +
                s1*Cos[z*s0]*Sin[s0] + E^(2*s1)*s1*Cos[z*s0]*Sin[s0] +
                s1*Cos[s0]*Sin[z*s0] + E^(2*s1)*s1*Cos[s0]*Sin[z*s0] -
                s0*Sin[s0]*Sin[z*s0] + E^(2*s1)*s0*Sin[s0]*Sin[z*s0]))/
                (2*E^s1*Sqrt[1 - ε]*Sqrt[ε]),
              Inequality[-2, Less, z, LessEqual, -1]}}, 0]}}
```

And here is the solution for the potential $V(z) = |z|$. It contains Airy functions.

```
In[134]:= V[z_] = Piecewise[{{z, z > 0}, {-z, z < 0}}];
```

```
        (dsol2 = DSolve[{-ψ''[z] + V[z] ψ[z] == ε ψ[z], ψ[0] == 1, ψ'[0] == 0},
                ψ[z], z]) // TraditionalForm
```

Out[135]//TraditionalForm=

$$\left\{\left\{\psi(z) \to \begin{cases} \dfrac{Ai'\left(\sqrt[3]{-1}\,\varepsilon\right)Bi\left(-\sqrt[3]{-1}\,(-z-\varepsilon)\right)-Ai\left(-\sqrt[3]{-1}\,(-z-\varepsilon)\right)Bi'\left(\sqrt[3]{-1}\,\varepsilon\right)}{Ai'\left(\sqrt[3]{-1}\,\varepsilon\right)Bi\left(\sqrt[3]{-1}\,\varepsilon\right)-Ai\left(\sqrt[3]{-1}\,\varepsilon\right)Bi'\left(\sqrt[3]{-1}\,\varepsilon\right)} & -\infty < z \leq 0 \\[4pt] \dfrac{Ai'(-\varepsilon)Bi(z-\varepsilon)-Ai(z-\varepsilon)Bi'(-\varepsilon)}{Ai'\left(\sqrt[3]{-1}\,\varepsilon\right)Bi\left(\sqrt[3]{-1}\,\varepsilon\right)-Ai\left(\sqrt[3]{-1}\,\varepsilon\right)Bi'\left(\sqrt[3]{-1}\,\varepsilon\right)} & \text{True} \end{cases}\right\}\right\}$$

Displaying the last two solutions over the z,ε-plane, shows characteristic "grooves". They correspond to the eigenvalues with respect to ε, so that solution vanishes quickly asymptotically as $z \to \infty$.

```
In[136]:= Show[GraphicsArray[
        Block[{$DisplayFunction = Identity},
                (* show Log[Abs[ψ[z, ε]]] *)
                Plot3D[Evaluate[Log[10, Abs[ψ[z] /. #[[1]]]] /. ψ1 -> 1],
                        {z, -12, 12}, {ε, -1, 6}, PlotPoints -> 90,
                        Mesh -> False]& /@ {dsol1, dsol2}]]]
```

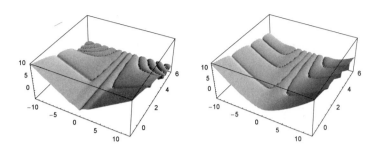

A restricted class of differential equations of higher order and systems of nonlinear differential equations are solved, too. Here are four examples.

In[137]:= **DSolve[x y''''[x] + 5y'''[x] == 24, y[x], x]**

Out[137]= $\left\{\left\{y[x] \to \dfrac{4\,x^3}{5} - \dfrac{C[1]}{24\,x^2} + C[2] + x\,C[3] + x^2\,C[4]\right\}\right\}$

In[138]:= **DSolve[z^8 w''''[z] == 2 w[z], w[z], z]**

Out[138]= $\left\{\left\{w[z] \to e^{\frac{2^{1/4}}{z}}\,z^3\,C[1] + e^{\frac{i\,2^{1/4}}{z}}\,z^3\,C[2] + e^{-\frac{i\,2^{1/4}}{z}}\,z^3\,C[3] + e^{-\frac{2^{1/4}}{z}}\,z^3\,C[4]\right\}\right\}$

In[139]:= **DSolve[y'''[x] y'[x] == y''[x]^3, y[x], x]**

    InverseFunction::ifun : Inverse functions are
       being used. Values may be lost for multivalued inverses. More…

    Solve::ifun : Inverse functions are being used by Solve, so some solutions
       may not be found; use Reduce for complete solution information. More…

Out[139]= $\left\{\left\{y[x] \to C[3] + \displaystyle\int_1^x \dfrac{-K\$106214 - C[2]}{\text{ProductLog}[-e^{-1+C[1]}\,(K\$106214 + C[2])]}\ dK\$106214\right\}\right\}$

In[140]:= **DSolve[{u'[x] == u[x] v[x], v'[x] == u[x]/v[x]}, {u, v}, x] //**
               (* for brevity, show only one solution *) #[[1]]&

    Solve::tdep : The equations appear to involve the variables
       to be solved for in an essentially non-algebraic way. More…

    Solve::tdep : The equations appear to involve the variables
       to be solved for in an essentially non-algebraic way. More…

    Solve::tdep : The equations appear to involve the variables
       to be solved for in an essentially non-algebraic way. More…

    General::stop :
    Further output of Solve::tdep will be suppressed during this calculation. More…

Out[140]= $\Big\{ v \to \mathrm{Function}\Big[\{x\},$

$$-(-3)^{1/3}\Big(C[1] + \mathrm{InverseFunction}\Big[\frac{1}{2\,C[1]^{1/3}}\Big(2\sqrt{3}\,\mathrm{ArcTan}\Big[\frac{1 + \frac{2\,(C[1]+\#1)^{1/3}}{C[1]^{1/3}}}{\sqrt{3}}\Big] +$$

$$2\,\mathrm{Log}[C[1]^{1/3} - (C[1]+\#1)^{1/3}] - \mathrm{Log}[C[1]^{2/3} + C[1]^{1/3}\,(C[1]+\#1)^{1/3} +$$

$$(C[1]+\#1)^{2/3}]\Big)\,\&\Big][-(-3)^{1/3}\,x + C[2]]\Big]^{1/3}\Big] ,$$

$$u \to \mathrm{Function}\Big[\{x\}, \mathrm{InverseFunction}\Big[\frac{1}{2\,C[1]^{1/3}}\Big(2\sqrt{3}\,\mathrm{ArcTan}\Big[\frac{1 + \frac{2\,(C[1]+\#1)^{1/3}}{C[1]^{1/3}}}{\sqrt{3}}\Big] +$$

$$2\,\mathrm{Log}[C[1]^{1/3} - (C[1]+\#1)^{1/3}] -$$

$$\mathrm{Log}[C[1]^{2/3} + C[1]^{1/3}\,(C[1]+\#1)^{1/3} + (C[1]+\#1)^{2/3}]\Big)\,\&\Big][-(-3)^{1/3}\,x + C[2]]\Big]\Big\}$$

As already demonstrated above in the Coriolis example, *Mathematica* can also solve systems of differential equations. Here is a more complicated system of three coupled nonlinear differential equations.

In[141]:= **DSolve[{x'[t] == y[t]   - z[t],**
          y'[t] == x[t]^2 + y[t],
          z'[t] == x[t]^2 + z[t]}, {x, y, z}, t]

Out[141]= $\Big\{ \{x \to \mathrm{Function}[\{t\}, e^{-C[3]}\,(e^t + e^{C[3]}\,C[1])],$

$$y \to \mathrm{Function}\Big[\{t\}, (-C[1] + e^{-C[3]}\,(e^t + e^{C[3]}\,C[1]))\,C[2] +$$

$$(-C[1] + e^{-C[3]}\,(e^t + e^{C[3]}\,C[1]))\,\Big(e^{-C[3]}\,(e^t + e^{C[3]}\,C[1]) -$$

$$\frac{C[1]^2}{-C[1] + e^{-C[3]}\,(e^t + e^{C[3]}\,C[1])} + 2\,C[1]\,\mathrm{Log}[-C[1] + e^{-C[3]}\,(e^t + e^{C[3]}\,C[1])]\Big)\Big],$$

$$z \to \mathrm{Function}\Big[\{t\}, C[1] - e^{-C[3]}\,(e^t + e^{C[3]}\,C[1]) + (-C[1] + e^{-C[3]}\,(e^t + e^{C[3]}\,C[1]))\,C[2] +$$

$$(-C[1] + e^{-C[3]}\,(e^t + e^{C[3]}\,C[1]))\,\Big(e^{-C[3]}\,(e^t + e^{C[3]}\,C[1]) -$$

$$\frac{C[1]^2}{-C[1] + e^{-C[3]}\,(e^t + e^{C[3]}\,C[1])} + 2\,C[1]\,\mathrm{Log}[-C[1] + e^{-C[3]}\,(e^t + e^{C[3]}\,C[1])]\Big)\Big]\Big\}\Big\}$$

Here is a check that the solution is correct.

In[142]:= **({x'[t] == y[t] - z[t], y'[t] == x[t]^2 + y[t],**
      z'[t] == x[t]^2 + z[t]} /. %) // Simplify

Out[142]= **{{True, True, True}}**

We chose random values for the constants C[1], C[2], and C[3] and display the solutions graphically.

In[143]:= **Show[Graphics3D[**
    Table[{Hue[Random[]], Thickness[0.002],
      ParametricPlot3D[Evaluate[Re[{x[t], y[t], z[t]}] /. %%[[1]] /.
        (* substitute random values for the C[i] *)
      Table[C[j] -> Random[Complex, {-3, 3} (1 + I)], {j, 3}]], {t, -3, 3},
        PlotPoints -> 500, DisplayFunction -> Identity][[1]]}, {100}]],
      PlotRange -> {{-10, 10}, {-20, 20}, {-30, 30}},
      BoxRatios -> {1, 1, 1}]

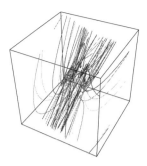

Mathematica always tries to explicitly solve for *y(x)*. As a result, we sometimes get some `Solve::ifun` messages from `Solve` in the following example.

In[144]:= `DSolve[y''[t] == Exp[y[t]], y, t]`

> `Solve::ifun : Inverse functions are being used by Solve, so some solutions`
> `may not be found; use Reduce for complete solution information. More…`

Out[144]= $\left\{\left\{y \rightarrow \text{Function}\left[\{t\}, \text{Log}\left[\frac{1}{2} C[1] \left(-1 + \text{Tanh}\left[\frac{1}{2} \sqrt{C[1]} (t + C[2])^2\right]^2\right)\right]\right]\right\}\right\}$

An implicit solution can be found for the following Abel-type differential equations. [334] Again, we get messages from `Solve`.

In[145]:= `DSolve[y'[x] == 1/(g[y[x]] x + f[y[x]]), y[x], x]`

> `Solve::tdep : The equations appear to involve the variables`
> `to be solved for in an essentially non-algebraic way. More…`
>
> `Solve::tdep : The equations appear to involve the variables`
> `to be solved for in an essentially non-algebraic way. More…`
>
> `Solve::tdep : The equations appear to involve the variables`
> `to be solved for in an essentially non-algebraic way. More…`
>
> `General::stop :`
> `Further output of Solve::tdep will be suppressed during this calculation. More…`

Out[145]= $\text{Solve}\left[x == e^{\int_1^{y[x]} g[K\$109499] \, dK\$109499} C[1] + \right.$

$\left. e^{\int_1^{y[x]} g[K\$109499] \, dK\$109499} \int_1^{y[x]} e^{-\int_1^{K\$109515} g[K\$109499] \, dK\$109499} f[K\$109515] \, dK\$109515, y[x]\right]$

Introducing a printing side effect in a user definition of `Solve` allows us to monitor which (implicit) equations are given to `Solve`. This is useful, for instance, in cases in which `Solve` succeeds in solving for *y*[x], but the result is too "messy" to be useful when compared with the implicit solution.

In[146]:= `Unprotect[Solve];`
`Solve[eqs_, vars_] := Null /;`
`        (Print["Calling: ", HoldForm[Solve[eqs, vars]]]; False)`

The following equation gives an implicit answer. As a side effect, we get the equations given to `Solve` printed.

In[148]:= `DSolve[{y'[x] == (y[x] - x)/(y[x] + x)}, y[x], x]`

> `Calling: Solve[x - y[x] + x y'[x] + y[x] y'[x] == 0, y'[x]]`
>
> `Calling: Solve`$\left[\text{ArcTan}\left[\frac{y[x]}{x}\right] + \frac{1}{2} \text{Log}\left[1 + \frac{y[x]^2}{x^2}\right] == C[1] - \text{Log}[x], y[x]\right]$
>
> `Solve::tdep : The equations appear to involve the variables`
> `to be solved for in an essentially non-algebraic way. More…`

Out[148]= $\text{Solve}\left[\text{ArcTan}\left[\frac{y[x]}{x}\right] + \frac{1}{2}\text{Log}\left[1 + \frac{y[x]^2}{x^2}\right] == C[1] - \text{Log}[x], \ y[x]\right]$

The above example $y''(t) == e^{y(t)}$ results in two carried out `Print` commands: One from the equation preprocessing step and one from the solution of the implicit equation found by DSolve for `y[x]`.

In[149]:= `DSolve[y''[t] == Exp[y[t]], y, t];`

Calling: $\text{Solve}[e^{y[t]} - y''[t] == 0, \ y''[t]]$

Calling: $\text{Solve}\left[\frac{4 \ \text{ArcTanh}\left[\frac{\sqrt{2 \ e^{y[t]} + C[1]}}{\sqrt{C[1]}}\right]^2}{C[1]} == (t + C[2])^2, \ y[t]\right]$

> Solve::ifun : Inverse functions are being used by Solve, so some solutions
>   may not be found; use Reduce for complete solution information. More…

We restore the original state of `Solve`.

In[150]:= `Solve[eqs_, vars_] =.`
`Protect[Solve];`

Now, let us deal with a slightly more complicated use of DSolve, the δ-expansion.

Mathematical Remark: δ-Expansion

Many nonlinear problems (equations, differential equations, and so on) cannot be solved in closed form. The idea of the so-called δ-expansion is to parametrize the nonlinearity with a parameter δ and then make a perturbation expansion around such a value δ_0 such that the original problem becomes a linear one for $\delta = \delta_0$. The various orders of perturbation expansion in $(\delta - \delta_0)^n$ typically require the solution of only linear problems. (For more details and applications, see [144], [1761], [1034], [822], [823], [437], [50], [52] and references therein.)

Here, we will deal with a very simple case (from [144]) of the application of the δ-expansion. The nonlinear differential equation $y'(x) = y(x)^\epsilon$, $y(0) = 1$ can be solved in closed form.

In[152]:= `DSolve[{y'[x] == y[x]^ε, y[0] == 1}, y[x], x] // Simplify`

> Solve::ifun : Inverse functions are being used by Solve, so some solutions
>   may not be found; use Reduce for complete solution information. More…
> Solve::ifun : Inverse functions are being used by Solve, so some solutions
>   may not be found; use Reduce for complete solution information. More…

Out[152]= $\left\{\left\{y[x] \to (1 + x - x \ \epsilon)^{\frac{1}{1-\epsilon}}\right\}\right\}$

For the parametrization $y'(x) = y(x)^{1+\delta}$, we get for $\delta = 0$ a linear problem. Making the ansatz $y(x) = \sum_{k=0}^{\infty} y_k(x) \, \delta^k$ and carrying out a series expansion of $y'(x) - y(x)^{1+\delta}$ around $\delta = 0$ yields a system of coupled equations for the $y_k(x)$.

In[153]:= `ord = 12;`
`eqs = CoefficientList[Normal[`
`Series[y'[x] - y[x]^(1 + δ) /. y -> Function[x,`
`Evaluate[Sum[y[i][x] δ^i, {i, 0, ord}]]], {δ, 0, ord}]], δ];`

Here are the first three equations for the $y_k(x)$. Although nonlinear by itself, the kth equation is a linear differential equation in $y_{k-1}(x)$.

In[155]:= `Take[eqs, 3]`

Out[155]= $\left\{-y[0][x] + y[0]'[x], -y[0][x] \left(\text{Log}[y[0][x]] + \dfrac{y[1][x]}{y[0][x]}\right) + y[1]'[x],\right.$

$\qquad -\dfrac{1}{2} y[0][x] \left(\left(\text{Log}[y[0][x]] + \dfrac{y[1][x]}{y[0][x]}\right)^2 + \right.$

$\qquad \left. 2\left(\dfrac{y[1][x]}{y[0][x]} + \dfrac{1}{2}\left(-\dfrac{y[1][x]^2}{y[0][x]^2} + \dfrac{2\,y[2][x]}{y[0][x]}\right)\right)\right) + y[2]'[x]\bigg\}$

It is now straightforward to implement the recursive solution of the $y_k(x)$. We take the inhomogeneous boundary condition $y(0) = 1$ into account for $y_0(x)$ and then use homogeneous boundary conditions for the $y_k(x)$, $k \geq 1$. Here, we calculate the first 12 orders using DSolve.

```
In[156]:= yiList = {};
       Do[AppendTo[yiList, MapAt[Factor,
            DSolve[{(eqs[[k + 1]] /. yiList /. Log[Exp[x]] -> x) == 0,
                y[k][0] == If[k == 0, 1, 0]},
                y[k][x], x][[1, 1]], 2]], {k, 0, ord - 2}]
```

Here are the first terms of the resulting expansion.

```
In[158]:= Sum[y[i][x] δ^i, {i, 0, 4}] /. yiList
```

Out[158]= $e^x + \dfrac{1}{2} e^x x^2 \delta + \dfrac{1}{24} e^x x^3 (8 + 3 x) \delta^2 +$

$\qquad \dfrac{1}{48} e^x x^4 (2 + x)(6 + x) \delta^3 + \dfrac{e^x x^5 (1152 + 1040 x + 240 x^2 + 15 x^3) \delta^4}{5760}$

To get an idea about the quality of the solution, we compare the various orders of the perturbation expansion with the exact result for $x = 1/7$ and $1/5 \leq \delta \leq 6$. We display the logarithm of the absolute value of the difference between the exact and the series solution.

```
In[159]:= diffList[δ_, x_] = Rest[FoldList[Plus, 0, (* all orders *)
                            Table[y[i][x] δ^i, {i, 0, ord - 2}] /. yiList]] -
                            (1 + x - (1 + δ) x)^(1/(1 - (1 + δ)));
```

```
In[160]:= ListPlot3D[Table[(* logarithm of the difference *)
                    N[Log[10, Abs[diffList[δ, 1/7]]], 22],
                    {δ, 1/5, 6, 1/5},
                    PlotRange -> All, MeshRange -> {{0, 10}, {1/5, 6}}]
```

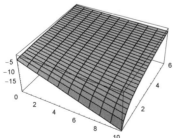

1.7.2 Partial Differential Equations

Partial differential equations of first order arise, for example, in the solution of the Hamilton-Jacobi differential equations of classical theoretical mechanics. (See, for example, [740], [265], [1110], [810], [1836], [473], [1531], and [1382].) The reduced Hamilton-Jacobi differential equation for a slanted trajectory has the following form:

$$\frac{1}{2m}\left(\left(\frac{\partial S(x,\,y)}{\partial x}\right)^2 + \left(\frac{\partial S(x,\,y)}{\partial y}\right)^2\right) + m\,g\,y = \varepsilon$$

(ε is the total energy, g the gravitational acceleration).

We do not need the most general solution of this differential equation, but only a so-called complete integral (see [972] and [1653]). This can be found using *Mathematica*. Let us just try it.

In[1]:= **DSolve[1/2m (D[S[x, y], x]^2 + D[S[x, y], y]^2) + m g y == ε,**
S[x, y], {x, y}]

> DSolve::nlpde : Solution requested to nonlinear partial
> differential equation. Trying to build a complete integral.

Out[1]= $\left\{\left\{S[x,\,y] \to -\dfrac{(-2\,g\,m\,y - C[1])^{3/2}}{3\,g\,m^{3/2}} - \dfrac{x\,\sqrt{2\,\varepsilon + C[1]}}{\sqrt{m}} + C[2]\right\},\right.$

$\left\{S[x,\,y] \to \dfrac{(-2\,g\,m\,y - C[1])^{3/2}}{3\,g\,m^{3/2}} - \dfrac{x\,\sqrt{2\,\varepsilon + C[1]}}{\sqrt{m}} + C[2]\right\},$

$\left\{S[x,\,y] \to -\dfrac{(-2\,g\,m\,y - C[1])^{3/2}}{3\,g\,m^{3/2}} + \dfrac{x\,\sqrt{2\,\varepsilon + C[1]}}{\sqrt{m}} + C[2]\right\},$

$\left.\left\{S[x,\,y] \to \dfrac{(-2\,g\,m\,y - C[1])^{3/2}}{3\,g\,m^{3/2}} + \dfrac{x\,\sqrt{2\,\varepsilon + C[1]}}{\sqrt{m}} + C[2]\right\}\right\}$

A parametric formula for the path of a point mass can be obtained as follows (we look only at the *x*-component and see the linear dependence of *x* on *t*). The two constants A[2] and C[1] can be adjusted to make the solution satisfy given initial conditions. For details of this approach, see the references cited above.

In[2]:= **Solve[D[%[[1, 1, 2]], ε] == t - A[2], x]**

Out[2]= $\left\{\left\{x \to -\sqrt{m}\,(t - A[2])\,\sqrt{2\,\varepsilon + C[1]}\,\right\}\right\}$

And, similar to the case for ordinary differential equations, some solutions are not found at all. The Hamilton–Jacobi differential equation [283] for the Kepler (Coulomb) problem in spherical coordinates remains unsolved in *Mathematica* Version 5.1.

In[3]:= **DSolve[1/(2m) (D[S[r, ϑ, φ], r]^2 + α/r 1/r^2 D[S[r, ϑ, φ], ϑ]^2 +**
1/r^2/Sin[ϑ]^2 D[S[r, ϑ, φ], φ]^2) == ε,
S[r, ϑ, φ], {r, ϑ, φ}]

> DSolve::nlpde : Solution requested to nonlinear partial
> differential equation. Trying to build a complete integral.

Out[3]= $\text{DSolve}\Big[\dfrac{\frac{\text{Csc}[\vartheta]^2\,S^{(0,0,1)}[r,\vartheta,\varphi]^2}{r^2} + \frac{\alpha\,S^{(0,1,0)}[r,\vartheta,\varphi]^2}{r^3} + S^{(1,0,0)}[r,\vartheta,\varphi]^2}{2\,m} == \varepsilon,$

$S[r,\,\vartheta,\,\varphi],\,\{r,\,\vartheta,\,\varphi\}\Big]$

Next, we solve a first-order partial differential equation.

In[4]:= **DSolve[{D[u[x, y], x]^a == 1/D[u[x, y], y]^b + γ}, u[x, y], {x, y}]**

> DSolve::nlpde : Solution requested to nonlinear partial
> differential equation. Trying to build a complete integral.

> Solve::ifun : Inverse functions are being used by Solve, so some solutions
> may not be found; use Reduce for complete solution information. More…

Out[4]= $\left\{\left\{u[x,\,y] \to C[1] + y\,C[2] + x\,(-C[2]^{-b}\,(-1 - \gamma\,C[2]^b))^{\frac{1}{a}}\right\}\right\}$

The general solution of a linear or quasilinear partial differential equation often depends not on arbitrary constants, but on arbitrary functions. (For nonlinear first-order PDEs, *Mathematica* returns complete integrals with arbitrary constants and a warning message).

In[5]:= **DSolve[D[u[x, y], x] == D[u[x, y], y], u[x, y], {x, y}]**

Out[5]= {{u[x, y] → C[1][x + y]}}

The general solution of the 1D wave equation also contains arbitrary functions.

In[6]:= **DSolve[D[φ[x, t], x, x] == α^2 D[φ[x, t], t, t], φ[x, t], {x, t}]**

Out[6]= $\{\{\phi[x, t] \rightarrow C[1]\left[t - x\sqrt{\alpha^2}\right] + C[2]\left[t + x\sqrt{\alpha^2}\right]\}\}$

Here is another example of a partial differential equation having a nice solution which is not found [478].

In[7]:= **pde[Φ_, {α_, β_}, {x_, t_}] :=**
      D[-Φ, t, t] + D[Φ, x, x] == α Exp[β Φ]

In[8]:= **DSolve[pde[Φ[x, t], {α, β}, {x, t}], Φ[x, t], {x, t}]**

Out[8]= DSolve$[-\Phi^{(0,2)}[x, t] + \Phi^{(2,0)}[x, t] == e^{\beta\Phi[x,t]}\alpha, \Phi[x, t], \{x, t\}]$

In[9]:= **(* the solution *)**
      Φ = 1/β Log[C[1]'[t + x] C[2]'[t - x]/
                  (1 + α β/8 C[1][t + x] C[2][t - x])^2];

      (* check solution *) pde[Φ, {α, β}, {x, t}] // Simplify

Out[12]= True

The general implicit solution $t\,c_1(\varphi(x, t)) = x\,c_2(\varphi(x, t))$ of the Bateman equation [601] is not found.

In[13]:= **DSolve[D[φ[x, t], x]^2 D[φ[x, t], t, t] + D[φ[x, t], t]^2 D[φ[x, t], x, x]**
      ==
                  2 D[φ[x, t], x] D[φ[x, t], t] D[φ[x, t], t, x],
                  φ[x, t], {x, t}]

Out[13]= DSolve$\left[\phi^{(0,2)}[x, t]\,\phi^{(1,0)}[x, t]^2 + \phi^{(0,1)}[x, t]^2\,\phi^{(2,0)}[x, t] ==\right.$
                  $\left.2\,\phi^{(0,1)}[x, t]\,\phi^{(1,0)}[x, t]\,\phi^{(1,1)}[x, t], \phi[x, t], \{x, t\}\right]$

Here is a quick check of the general solution.

In[14]:= **%[[1]] /. With[{ds = {{t}, {x}, {t, t}, {x, x}, {t, x}}},**
                  Solve[(* or, shorter solution: C[1][φ[x, t]] == x/t *)
                      (D[t C[1][φ[x, t]] + x C[2][φ[x, t]], ##]& @@@ ds) == 0,
                      D[φ[x, t], ##]& @@@ ds][[1]]] // Simplify

Out[14]= True

Let us consider another, less-known example of a partial differential equation from classical mechanics. Given a family of orbits in 2D in the form $f(x, y) = parameter$ and a relation between the total energy ε and a parameter *parameter* in the form $\varepsilon = \varepsilon(parameter)$, the Szebehely equation is a first-order linear partial differential equation for the (unknown) potential $V(x, y)$, giving rise to the orbits $f(x, y) = parameter$ [246], [775], [231], [230], [402], [747], [65], [1353], [232], [206]. This is the Szebehely equation:

$$f_x V_x + f_y V_y = 2 W(\varepsilon - V)$$

$$W = \frac{f_{xx} f_y^2 - 2 f_{xy} f_x f_y + f_{yy} f_x^2}{f_x^2 + f_y^2}.$$

Given $f(x, y)$ and $\varepsilon = \varepsilon(parameter)$, makeSzebehelyPDE generates the Szebehely equation.

In[15]:= **makeSzebehelyPDE[f_, ε_, parameter_, {x_, y_}, V_] :=**
      Module[{sol = Solve[f == 0, parameter], e, W},
        (* express ε in x and y *)
        e = ε /. sol[[1]];

```
W = (D[f, x, x] D[f, y]^2 - 2 D[f, x, y] D[f, x] D[f, y] +
        D[f, y, y] D[f, x]^2)/(D[f, x]^2 + D[f, y]^2);
(* the Szebehely pde *)
D[V[x, y], x] D[f, x] + D[V[x, y], y] D[f, y] - 2 (e - V[x, y]) W]
```

Here, we solve the Szebehely equation for circular orbits and $x^2 + y^2 = R^2$ and $\varepsilon = R^2$.

In[16]:= `makeSzebehelyPDE[x^2 + y^2 - R2, R2, R2, {x, y}, V]`

Out[16]= $-\dfrac{2\,(8\,x^2 + 8\,y^2)\,(x^2 + y^2 - V[x, y])}{4\,x^2 + 4\,y^2} + 2\,y\,V^{(0,1)}[x, y] + 2\,x\,V^{(1,0)}[x, y]$

Setting the arbitrary function $C[1][y/x]$ in the solution to 0 results in the potential of a 2D harmonic oscillator.

In[17]:= `DSolve[% == 0, V[x, y], {x, y}]`

Out[17]= $\left\{\left\{V[x, y] \to \dfrac{x^4 + x^2\,y^2 + 2\,C[1]\left[\frac{y}{x}\right]}{2\,x^2}\right\}\right\}$

Here is a quick check for the correctness of the solution.

In[18]:= `Simplify[%% /. V -> Function[{x, y}, Evaluate[%[[1, 1, 2]]]]]`

Out[18]= 0

For a listing of solutions of linear PDEs, see [1426].

1.7.3 Difference Equations

The discrete analog of differential equations are difference (recurrence) equations [7], [20], [1770], [1251], [845], [598], [26] (for the unification of differential and difference equations through time scales, see [30]). The *Mathematica* function to solve difference equations is `RSolve`.

`RSolve[`*listOfDEsAndInitialValues*`, `*listOfFunctions*`, `*independentVariable*`]`

    tries to solve the difference equation(s) with potential initial conditions given by *listOfDEsAndInitialValues* for the functions in *listOfFunctions*. The independent variable is *independentVariable*. In the case of a single difference equation without initial conditions with only one unknown function, the first and second arguments can appear without the braces.

We do not give a detailed listing of various classes of difference equations and their solutions here. We just discuss some examples and make some remarks concerning similarities and differences to the function `DSolve`.

We start with a simple linear difference equation.

In[1]:= `RSolve[u[n + 1] == u[n] - u[n - 1], u[n], n]`

Out[1]= $\left\{\left\{u[n] \to C[1]\,\mathrm{Cos}\left[\dfrac{n\,\pi}{3}\right] + C[2]\,\mathrm{Sin}\left[\dfrac{n\,\pi}{3}\right]\right\}\right\}$

Here is a fourth-order linear difference equation with constant coefficients. The powers of the roots of the characteristic equation are clearly visible in the solution.

In[2]:= `RSolve[Sum[α[k] u[n + k], {k, 0, 4}] == 0, u[n], n]`

Out[2]= $\{\{u[n] \to C[1]\,\mathrm{Root}[\alpha[0] + \#1\,\alpha[1] + \#1^2\,\alpha[2] + \#1^3\,\alpha[3] + \#1^4\,\alpha[4]\ \&,\ 1]^n +$
$\quad\quad C[2]\,\mathrm{Root}[\alpha[0] + \#1\,\alpha[1] + \#1^2\,\alpha[2] + \#1^3\,\alpha[3] + \#1^4\,\alpha[4]\ \&,\ 2]^n +$
$\quad\quad C[3]\,\mathrm{Root}[\alpha[0] + \#1\,\alpha[1] + \#1^2\,\alpha[2] + \#1^3\,\alpha[3] + \#1^4\,\alpha[4]\ \&,\ 3]^n +$
$\quad\quad C[4]\,\mathrm{Root}[\alpha[0] + \#1\,\alpha[1] + \#1^2\,\alpha[2] + \#1^3\,\alpha[3] + \#1^4\,\alpha[4]\ \&,\ 4]^n\}\}$

The analog of the Wronskian for difference equation is the Casoratian.

```
In[3]:= Casoratian[fs_, n_] :=
         Det[Table[fs /. n -> n + k, {k, 0, Length[fs] - 1}]]
```

Here is a third-order linear difference equation.

```
In[4]:= RSolve[Sum[(k + 1) u[n + k], {k, 0, 2}] == 0, u[n], n]
```

$$Out[4]= \ \{\{u[n] \to 3^{-n/2}\, C[2]\, Cos[n\, (\pi - ArcTan[\sqrt{2}\,])] - 3^{-n/2}\, C[1]\, Sin[n\, (-\pi + ArcTan[\sqrt{2}\,])]\}\}$$

This is its Casoratian.

```
In[5]:= Cs[n_] = Casoratian[Coefficient[u[n] /. %[[1]],
                    Union[Cases[%, _C, Infinity]]], n] // Simplify
```

$$Out[5]= \ -\sqrt{2}\ 3^{-1-n}$$

The Casoratian Cs_n of the solutions of a linear homogeneous kth-order difference equation $p(u_n, u_{n+1}, \ldots, u_{n+k}) = 0$ in u_n fulfills a first-order difference equation of the form $Cs_{n+1} = (-1)^{k+1}[u_n](p)/[u_{n+k}](p)$.

```
In[6]:= Cs[n]/Cs[n - 1] // Simplify
```

$$Out[6]= \ \frac{1}{3}$$

(The Casoratian can be used to construct solutions of some differential-difference equations [1170].)

We continue with the simplest possible linear difference equation with nonconstant coefficients. The result as a Pochhammer symbol.

```
In[7]:= RSolve[Γ[n + 1] == n Γ[n], Γ[n], n]
```

$$Out[7]= \ \{\{\Gamma[n] \to C[1]\, Pochhammer[1, -1 + n]\}\}$$

After adding an initial condition, we recover the familiar Gamma function.

```
In[8]:= RSolve[{Γ[n + 1] == n Γ[n], Γ[1] == 1}, Γ[n], n] // FullSimplify
```

$$Out[8]= \ \{\{\Gamma[n] \to Gamma[n]\}\}$$

And here is a corresponding inhomogeneous equation with a constant right-hand side term.

```
In[9]:= RSolve[{Γ[n + 1] - n Γ[n] == A, Γ[1] == 1}, Γ[n], n] // FullSimplify
```

$$Out[9]= \ \{\{\Gamma[n] \to Gamma[n] - A\, Gamma[n] + A\, e\, Gamma[n, 1]\}\}$$

If the right-hand side contains a function of n, then the returned solution contains a summation. The K\$$n$ are automatically generated dummy product variables.

```
In[10]:= RSolve[{Γ[n + 1] - n Γ[n] == A[n], Γ[1] == 1}, Γ[n], n] // FullSimplify
```

$$Out[10]= \ \left\{\left\{\Gamma[n] \to Gamma[n] \left(1 - \sum_{K\$258=K\$259}^{0} \frac{A[K\$258]}{Gamma[1 + K\$258]} + \sum_{K\$258=K\$259}^{-1+n} \frac{A[K\$258]}{Gamma[1 + K\$258]}\right)\right\}\right\}$$

The result of the following linear difference equation with nonconstant, not explicitly specified coefficients is a product. The C[1] indicates the linear independent solutions.

```
In[11]:= RSolve[{Ψ[n + 1] == α[n] Ψ[n]}, Ψ[n], n]
```

$$Out[11]= \ \left\{\left\{\Psi[n] \to C[1] \prod_{K\$489=K\$490}^{-1+n} \alpha[K\$489]\right\}\right\}$$

Here is the simplest possible nonlinear difference equation.

```
In[12]:= RSolve[u[n + 1] == u[n]^2, u[n], n]
```

$$Out[12]= \ \{\{u[n] \to e^{2^n\, C[1]}\}\}$$

Another class of recurrence equations that is solved by RSolve are q-difference equations. Instead of $n + k$, one has the multiplicative structure $q\,n$. Here are two simple examples of such a q-difference equation.

In[13]:= `RSolve[u[q n] == u[n] - u[n/q], u[n], n]`

Out[13]= $\left\{\left\{u[n] \rightarrow C[1]\, Cos\left[\frac{\pi\, Log[n]}{3\, Log[q]}\right] + C[2]\, Sin\left[\frac{\pi\, Log[n]}{3\, Log[q]}\right]\right\}\right\}$

In[14]:= `RSolve[u[q n] == u[n]^2, u[n], n]`

Out[14]= $\left\{\left\{u[n] \rightarrow e^{2^{\frac{Log[n]}{Log[q]}}\, C[1]}\right\}\right\}$

The shifts in difference equations must be explicit integers, and additive and multiplicative differences cannot be mixed.

In[15]:= `RSolve[u[n + 1/2] == u[n] - u[n - 1], u[n], n]`

> RSolve::piarg : All arguments in position 1 of
>
> $u\left[\frac{1}{2} + n\right] == -u[-1 + n] + u[n]$ should be either of the form n + Integer
>
> or q^Integer * n. Mixtures of these forms are not allowed. More…

Out[15]= $RSolve\left[u\left[\frac{1}{2} + n\right] == -u[-1 + n] + u[n], u[n], n\right]$

In[16]:= `RSolve[u[n + 1] == u[q n], u[n], n]`

> RSolve::piarg :
> All arguments in position 1 of u[1 + n] == u[n q] should be either of the form n +
> Integer or q^Integer * n. Mixtures of these forms are not allowed. More…

Out[16]= $RSolve[u[1 + n] == u[n\, q], u[n], n]$

While the (closed form) solutions of differential equations are fulfilled for all values of the independent variables, this is not always the case for difference equations. Here is a difference equation, whose solution is correct for all complex values of n.

In[17]:= `deq = {u[n + 2] == u[n + 1] + u[n], u[0] == 1, u[1] == 1};`

      `RSolve[deq, u, n] // Simplify`

Out[18]= $\left\{\left\{u \rightarrow Function\left[\{n\},\right.\right.\right.$

$$\left.\left.\left.\frac{1}{10}\left(5\left(\frac{1}{2} - \frac{\sqrt{5}}{2}\right)^n - \sqrt{5}\left(\frac{1}{2} - \frac{\sqrt{5}}{2}\right)^n + 5\left(\frac{1}{2} + \frac{\sqrt{5}}{2}\right)^n + \sqrt{5}\left(\frac{1}{2} + \frac{\sqrt{5}}{2}\right)^n\right)\right]\right\}\right\}$$

In[19]:= `deq /. %[[1]] // Simplify`

> N::meprec :
> Internal precision limit $MaxExtraPrecision = 50.` reached while evaluating
>
> $-1 + \frac{1}{10}\left(5\left(\frac{1}{2} - \frac{\sqrt{5}}{2}\right) - \sqrt{5}\left(\frac{1}{2} - \frac{\sqrt{5}}{2}\right) + 5\left(\frac{1}{2} + \frac{\sqrt{5}}{2}\right) + \sqrt{5}\left(\frac{1}{2} + \frac{\sqrt{5}}{2}\right)\right).$ More…

Out[19]= `{True, True, True}`

Here is a more complicated solution. It is expressed as a product with upper limit n. This means, the solutions returned from RSolve are not always valid for generic values of n, but only for (positive) integers.

In[20]:= `deq = {(-1 - 3 n + 3 n^2) u[n] == 3 n (1 + n) u[2 + n]};`

      `rsol = RSolve[deq, u, n] /. (* for smaller solution *)`
          `C[2] -> 0 // (Evaluate //@ #) & // InputForm`

Out[21]//InputForm=
```
{
  {u -> Function[{n},
      C[1]*Product[E^(((-1)^K$12378*((-2*I)*Pi + Log[4] + Log[Gamma[(3 -
```

```
            Sqrt[21])/12]^(-2)] +
                  2*Log[Gamma[(9 - Sqrt[21])/12]] - 2*Log[Gamma[(3 + Sqrt[21])/12]] +
                  2*Log[Gamma[(9 + Sqrt[21])/12]] +
            2*(-1)^K$12378*(Log[Gamma[K$12378/2]] -
                  Log[Gamma[(2 + K$12378)/2]] - Log[Gamma[(-3 - Sqrt[21] +
            6*K$12378)/12]] +
                  Log[Gamma[(3 - Sqrt[21] + 6*K$12378)/12]] - Log[Gamma[(-3 +
            Sqrt[21] + 6*K$12378)/12]] +
                  Log[Gamma[(3 + Sqrt[21] + 6*K$12378)/12]])))/2), {K$12378,
            K$12379, -1 + n}]]}}
```

In[22]:= (* extract dummy summation and product variables from an expression *)
```
         extractDummySummationVariables[expr_] :=
            Union[Cases[Cases[expr, HoldPattern[(Sum | Product)[_, {_, r__}]] :> {r},
                  Infinity],
                  dummySummationIndex_Symbol?(StringMatchQ[ToString[#], "K$*"]&),
                  Infinity]]
```

In[24]:= (deq /. rsol[[1]] /. C[1] -> 1 /. n -> # /.
```
            Thread[extractDummySummationVariables[rsol[[1]]] -> {3}])& /@
                              {-3, 3, 1/2} // N[#, 22]&
```
      N::meprec : Internal precision limit $MaxExtraPrecision = 50.` reached

      while evaluating $17 - 36\,e^{\frac{1}{2}\,(\ll 10\gg +2\,\mathrm{Plus}[\ll 6\gg])+\frac{1}{2}\,(-2\,i\,\pi +\ll 6\gg +2\,\mathrm{Plus}[\ll 5\gg])}$. More...

Out[24]= {{{False}}, {{True}}, {{False}}}

In general, for most of the exactly solvable differential equations, there is an exactly solvable difference equation obtainable by replacing the derivatives by finite differences (for the opposite transformation, see [1133]).

In[25]:= *ODEToDE*[eq_, {y_, x_}, {u_, n_}] := eq //.
```
            Derivative[k_][y][x] :> Sum[(-1)^j Binomial[k, j] u[n + k - j], {j, 0, k}]
         //.
            y[x] -> u[n] //. x -> n
```

Here are a few difference analogs of the differential equations solved in the last subsection.

In[26]:= {#, RSolve[#, u, n]}& @
```
            ODEToDE[y'[x] g'[x] == D[y[x] g[x], x] , {y, x}, {u, n}]
```
Out[26]= $\left\{(-u[n] + u[1 + n])\, g'[n] == g[n]\,(-u[n] + u[1 + n]) + u[n]\, g'[n],\right.$

$$\left\{\left\{u \to \mathrm{Function}\left[\{n\}, C[1] \prod_{K\$22259=K\$22260}^{-1+n} \frac{g[K\$22259] - 2\, g'[K\$22259]}{g[K\$22259] - g'[K\$22259]}\right]\right\}\right\}\right\}$$

In[27]:= {#, RSolve[#, u, n]}& @
```
            ODEToDE[Log[x] y'[x] + y[x] Cos[x] + Exp[x] == 0, {y, x}, {u, n}]
```
Out[27]= $\left\{e^n + \mathrm{Cos}[n]\, u[n] + \mathrm{Log}[n]\,(-u[n] + u[1+n]) == 0,\right.$

$$\left\{\left\{u \to \mathrm{Function}\left[\{n\}, C[1] \prod_{K\$22279=K\$22280}^{-1+n} \frac{-\mathrm{Cos}[K\$22279] + \mathrm{Log}[K\$22279]}{\mathrm{Log}[K\$22279]} + \right.\right.\right.$$

$$\left(\prod_{K\$22279=K\$22280}^{-1+n} \frac{-\mathrm{Cos}[K\$22279] + \mathrm{Log}[K\$22279]}{\mathrm{Log}[K\$22279]}\right)$$

$$\left.\left.\left.\sum_{K\$22288=K\$22289}^{-1+n} -\frac{e^{K\$22288}}{\mathrm{Log}[K\$22288]\prod_{K\$22279=K\$22280}^{K\$22288} \frac{-\mathrm{Cos}[K\$22279]+\mathrm{Log}[K\$22279]}{\mathrm{Log}[K\$22279]}}\right]\right\}\right\}\right\}$$

In[28]:= {#, RSolve[#, u, n]}& @
```
            ODEToDE[y''[x] == f[x] y'[x], {y, x}, {u, n}]
```

Out[28]= $\{u[n] - 2u[1+n] + u[2+n] == f[n] (-u[n] + u[1+n]),$

$$\{\{u \to \text{Function}[\{n\}, C[1] + C[2] \sum_{K\$22311=K\$22312}^{-1+n} \prod_{K\$22305=K\$22306}^{-1+K\$22311} (1 + f[K\$22305])]\}\}\}$$

In[29]:= {#, RSolve[#, u, n]}& @
         $O\mathcal{D}\mathcal{E}\text{To}\mathcal{D}\mathcal{E}[x \ y''''[x] + 5y'''[x] == 24, \{y, x\}, \{u, n\}]$

Out[29]= $\{5 (-u[n] + 3u[1+n] - 3u[2+n] + u[3+n]) +$

$n (u[n] - 4u[1+n] + 6u[2+n] - 4u[3+n] + u[4+n]) == 24,$

$\{\{u \to \text{Function}[\{n\}, \dfrac{126n - 255n^2 + 170n^3 - 45n^4 + 4n^5}{5(20 - 9n + n^2)} + \dfrac{C[1]}{20 - 9n + n^2} +$

$\dfrac{1}{24}(78 - 17n + n^2)C[2] + \dfrac{1}{24}(-66 + 17n - n^2)C[3] + \dfrac{1}{24}(42 - 13n + n^2)C[4]]\}\}\}$

In[30]:= {#, RSolve[#, u, n]}& @
         $O\mathcal{D}\mathcal{E}\text{To}\mathcal{D}\mathcal{E}[y[x] == y'[x] x + 2 (y'[x])^2, \{y, x\}, \{u, n\}]$

Out[30]= $\{u[n] == n (-u[n] + u[1+n]) + 2 (-u[n] + u[1+n])^2,$

$\{\{u \to \text{Function}[\{n\}, \dfrac{1}{32}(-4n^2 + 64(-\dfrac{n}{4} + C[1])^2)]\},$

$\{u \to \text{Function}[\{n\}, \dfrac{1}{32}(-4n^2 + 64(\dfrac{n}{4} + C[1])^2)]\}\}\}$

Here is a simple system of two coupled linear difference equations. We calculate the ratio of the two terms.

In[31]:= ratio = a[n]/b[n] /.
         RSolve[{a[n] == a[n - 1] + 2 b[n - 1],
                b[n] == a[n - 1] + 1 b[n - 1],
                a[0] == 1, b[0] == 2}, {a, b}, n][[1]]

Out[31]= $\dfrac{2\left((1-\sqrt{2})^n - 2\sqrt{2}(1-\sqrt{2})^n + (1+\sqrt{2})^n + 2\sqrt{2}(1+\sqrt{2})^n\right)}{4(1-\sqrt{2})^n - \sqrt{2}(1-\sqrt{2})^n + 4(1+\sqrt{2})^n + \sqrt{2}(1+\sqrt{2})^n}$

In the limit the ratio approaches $\sqrt{2}$ [563].

In[32]:= Limit[%, n -> Infinity] // RootReduce
Out[32]= $\sqrt{2}$

We end with a selection of various lower-order difference equations. The solution of the following difference equation is expressed in Pochhammer symbols.

In[33]:= RSolve[(2 + 2 n + 3 n^2)*u[n] + (-2 + 3 n + 2 n^2) u[1 + n] == 0, u[n], n] //
                                                                          Simplify

Out[33]= $\left\{\left\{u[n] \to \dfrac{(-\frac{3}{2})^{-1+n} C[1] \text{Pochhammer}[\frac{4}{3} - \frac{i\sqrt{5}}{3}, -1+n] \text{Pochhammer}[\frac{4}{3} + \frac{i\sqrt{5}}{3}, -1+n]}{\text{Pochhammer}[\frac{1}{2}, -1+n] \text{Pochhammer}[3, -1+n]}\right\}\right\}$

The solution of the next difference equation is expressed in Legendre functions.

In[34]:= RSolve[3 (1 + n + n^2) u[n] == 2 (1 + n) u[1 + n] + 3 u[2 + n], u[n], n]

Out[34]= $\left\{\left\{u[n] \to C[1] \text{LegendreP}[\frac{1}{2} i (i + \sqrt{3}), n, 2, \frac{1}{\sqrt{10}}] + \right.\right.$

$\left.\left. C[2] \text{LegendreQ}[\frac{1}{2} i (i + \sqrt{3}), n, 2, \frac{1}{\sqrt{10}}]\right\}\right\}$

The next solution contains hypergeometric functions.

In[35]:= RSolve[(-3 + 2n) n u[n] + (3 + n + n^2) u[2 + n] ==
               (2 + 3 n^2) u[1 + n], u[n], n] // Simplify

Out[35]= $\left\{\left\{u[n] \rightarrow \left(2^{-1-n} \; (-3 \; i + \sqrt{11}\,) \; (3 \; i + \sqrt{11}\,) \; (-1 + n) \; (3 - n + n^2) \; \text{Gamma}[-3 + 2 \, n]\right.\right.\right.$

$\left(-3 \; 2^n \; C[2] \; \text{Gamma}\left[-\frac{1}{2} - \frac{i \sqrt{11}}{2} + n\right] \; \text{Gamma}\left[-\frac{1}{2} + \frac{i \sqrt{11}}{2} + n\right] \; \text{HypergeometricPFQ}\Big[$

$\left\{1, \; -\frac{1}{2} - \frac{i \sqrt{11}}{2} + n, \; -\frac{1}{2} + \frac{i \sqrt{11}}{2} + n\right\}, \; \left\{-\frac{1}{2} + n, \; 1 + n\right\}, \; \frac{1}{2}\Big] +$

$n \; \text{Gamma}[-1 + 2 \, n] \; \left(6 \, \pi \, C[1] \; \text{Sech}\left[\frac{\sqrt{11} \; \pi}{2}\right] - C[2] \; \text{Gamma}\left[\frac{1}{2} - \frac{i \sqrt{11}}{2}\right]\right.$

$\text{Gamma}\left[\frac{1}{2} + \frac{i \sqrt{11}}{2}\right] \; \left(-2 + \sqrt{2} \; \text{Cosh}\left[\frac{\sqrt{11} \; \pi}{4}\right] - \sqrt{22} \; \text{Sinh}\left[\frac{\sqrt{11} \; \pi}{4}\right]\right)\Big)\Big)\Big) \Big/$

$\left(3 \, n \; \text{Gamma}\left[\frac{1}{2} - \frac{i \sqrt{11}}{2} + n\right] \; \text{Gamma}\left[\frac{1}{2} + \frac{i \sqrt{11}}{2} + n\right] \; \text{Gamma}[-1 + 2 \, n]\right)\Big\}\Big\}$

Quite simple-looking difference equations can give quite complicated solutions.

In[36]:= `RSolve[3 n^2 u[1 + n] == (-1 + n) (2 n u[n] + (n - 1) u[2 + n]),`
`        u[n], n] // Simplify // Short[#, 10]&`

Out[36]//Short=

$\left\{\left\{u[n] \rightarrow \frac{1}{3} \; 2^{-8+n} \; (48 + 148 \, n - 132 \, n^2 - 5 \, n^3 + 12 \, n^4 + n^5)\right.\right.$

$\left(C[1] + \left(8 \, C[2] \; \ll 3 \gg \; \ll 1 \gg \; ! \; \Big((\text{Gamma}[1 - \text{Root}[24 + 206 \; \#1 + 131 \; \#1^2 + 22 \; \#1^3 + \#1^4 \; \&,\right.\right.$

$\qquad 1]] \; \text{Gamma}[1 - \text{Root}[24 + 206 \; \#1 + 131 \; \#1^2 + 22 \; \#1^3 + \#1^4 \; \&, \; 2]]$

$\qquad \text{Gamma}[1 - \; \ll 1 \gg \;] \; \text{Gamma}[1 - \text{Root}[24 + 206 \; \#1 + 131 \; \#1^2 + 22 \; \#1^3 + \#1^4 \; \&, \; 4]]$

$\qquad \text{HypergeometricPFQ}[\ll 1 \gg]) \; / \; (8 \; \text{Gamma}[1 - \text{Root}[1152 + 1026 \; \#1 +$

$\qquad\qquad 287 \; \#1^2 + 30 \; \#1^3 + \#1^4 \; \&, \; 1]] \; \text{Gamma}[1 - \text{Root}[\ll 1 \gg \; \&, \; 2]]$

$\qquad \text{Gamma}[1 - \text{Root}[1152 + 1026 \; \#1 + 287 \; \#1^2 + 30 \; \#1^3 + \#1^4 \; \&, \; 3]]$

$\qquad \text{Gamma}[1 - \text{Root}[1152 + 1026 \; \#1 + 287 \; \#1^2 + 30 \; \#1^3 + \#1^4 \; \&, \; 4]]) - \frac{\ll 1 \gg}{\ll 1 \gg}\right)\right) \Big/$

$\qquad ((-\text{Root}[24 + 206 \; \#1 + 131 \; \#1^2 + 22 \; \#1^3 + \#1^4 \; \&, \; 1]) \, !$

$\qquad (-\text{Root}[24 + 206 \; \#1 + 131 \; \#1^2 + 22 \; \#1^3 + \#1^4 \; \&, \; 2]) \, !$

$\qquad (-\text{Root}[24 + 206 \; \#1 + 131 \; \#1^2 + 22 \; \#1^3 + \#1^4 \; \&, \; 3]) \, !$

$\qquad (-\text{Root}[24 + 206 \; \#1 + 131 \; \#1^2 + 22 \; \#1^3 + \#1^4 \; \&, \; 4]) \, !)\Big)\Big\}\Big\}$

1.8 Integral Transforms and Generalized Functions

Until now, in all sections of this chapter, we dealt with polynomials (where basically every possible operation can be done on algorithmically safe ground), rational functions, or at least analytic functions (like all of the elementary functions, all special functions of mathematical physics.) In many instances, when describing idealized situations, this set of smooth functions is not enough and a natural generalization is to deal with certain limits of such functions, or what is technically more suited, with linear, continuous functionals of (a subclass of) such functions. This leads to the world of generalized functions, also called distributions (and ultradistributions). Obviously, there are many such functionals; distributions of finite order can be represented as integrals over smooth functions. For the most common-named distributions, one typically chooses as the space of test functions the space of all functions with continuous nth derivative and compact support (such functions exists) or the space of all functions with continuous nth derivative that vanish faster than any power at $\pm\infty$—which generates the tempered distributions. Let Ξ be such a linear functional. One typically writes the functional as $\langle\Xi, \varphi\rangle = \Xi[\varphi]$, where φ is the test function and the scalar product notation reminds us of the Riesz–Fischer-theorem based L_2-case. In a slight abuse of notation, one often just refers then to the generalized function Ξ, but its sole existence is only inside $\langle\Xi, \varphi\rangle$ and must always be kept in mind. We remark that some, but not all, "ordinary" functions are distributions [597] and that at the same time some, but not all, distributions are "ordinary" functions (for a representation of all Schwartz distributions as smooth nonstandard functions, see [1369], [840]). For the many interesting mathematical properties of distributions, see, for example, [707], [1727], [1790], [876], [1789], [975], [1168], [170], [67], [389], [607], [666], [390], [1101], [1481], [893], [460], and [177]. Beyond being a convenient mathematical idealization, distributions such as the Dirac delta functions are unavoidable ingredients for a proper mathematical formulation of the solutions of partial differential equations [1463] and of many physical theories, such as electrodynamics [1519], [1520], [23], and quantum mechanics [195], [725], [194], [462], [1085], [1771], [196], [463], [461], [1300], [530] and engineering applications, such as the calculation of propeller noise in areoacoustics [602], [603], [605].

Currently, there are two distributions built-into *Mathematica*: `DiracDelta` and `UnitStep`.

    UnitStep[*x*]

         represents the Heaviside distribution $\theta(x)$.

The Heaviside distribution $\theta(x)$ is, like all generalized functions, defined for real x only (no nontrivial, nonproduct-like extensions into the complex plane are possible). It is 1 for positive x and 0 for negative x (in the world of distributions, it does not matter which finite value one chooses for $x = 0$).

In[1]:= **Plot[UnitStep[x], {x, -2, 2}, Frame -> True, Axes -> False,**
          PlotStyle -> {Hue[0], Thickness[0.01]}]

The picture clearly shows a step of unit size and justifies the *Mathematica* naming of the Heaviside distribution. Here is a piecewise defined function (not using `Piecewise`).

```
In[2]:= manyPeriodFunction[x_] =
         With[{θ = UnitStep},
           (* left and right tails *) θ[-x] + θ[x - 10 Pi] +
           (* the middle oscillations *)
           Sum[θ[x - 2 k Pi] θ[2 (k + 1) Pi - x] Cos[(k + 1) x], {k, 0, 4}]];
```

This function can be plotted.

```
In[3]:= Plot[Evaluate[manyPeriodFunction[x]], {x, -2 Pi, 12 Pi},
         Frame -> True, Axes -> False]
```

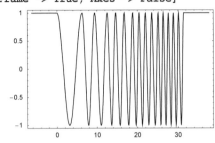

This function can be used inside `Integrate`.

```
In[4]:= Integrate[manyPeriodFunction[x], {x, 0, 10 Pi}]
Out[4]= 0
```

Here, the function inside the `UnitStep` is more complicated.

```
In[5]:= Integrate[UnitStep[Exp[x] - 2], {x, 0, Pi}]
Out[5]= π - Log[2]
```

This result agrees with the corresponding numerical integration.

```
In[6]:= {N[%], NIntegrate[UnitStep[Exp[x] - 2], {x, 0, Log[2], Pi}]}
Out[6]= {2.44845, 2.44845}
```

According to the definition of $\theta(x)$, we have the identity $\theta(x) = \theta(x)^2$. This definition is not used automatically. To get it into effect, one has to use the function `FunctionExpand` (to be discussed in Chapter 3).

```
In[7]:= UnitStep[x]^2
Out[7]= UnitStep[x]²
```

```
In[8]:= FunctionExpand[%]
Out[8]= UnitStep[x]
```

In addition to rewriting expressions in "simpler" functions, the function `FunctionExpand` will carry out a second class of transformations on the `UnitStep` function; it will rewrite $\theta(f(x))$ in the form $\sum_k c_k\,\theta(x - x_k)$, $c_k = \pm 1$. Here is an example.

```
In[9]:= FunctionExpand[UnitStep[(x - 1) (x - 3) (x - 5)]]
Out[9]= UnitStep[-5 + x] - UnitStep[-3 + x] + UnitStep[-1 + x]
```

In the next example, there is no real x such that $-x^4 - 1$ becomes positive. Contrary to *Mathematica*'s default assumption that every variable represents a generic complex value, inside distributions inside `FunctionEx` pand variables (as long as they can be uniquely found) are considered to be real, because, as stated, no distributions with complex arguments exist.

In[10]:= **FunctionExpand[UnitStep[-x^4 - 1]]**

Out[10]= 0

UnitStep has an obvious generalization to more than one variable.

UnitStep $[x_1,...,x_n]$
    represents the multidimensional Heaviside distribution $\theta(x_1, ..., x_n)$.

The definition for $\theta(x_1, ..., x_n)$ is simply $\theta(x_1, ..., x_n) = \theta(x_1) \cdots \theta(x_n)$. Here is a visualization of $\theta(x_1, x_2)$.

In[11]:= **Plot3D[UnitStep[x, y], {x, -2, 2}, {y, -2, 2},**
        PlotPoints -> 80, Mesh -> False]

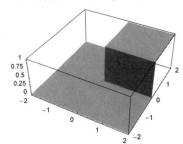

One of the big advantages of distributions is that they possess derivatives of any order. Given the linear functional $\langle \Xi, \varphi \rangle = \int_{-\infty}^{\infty} \varphi(t) \, \Xi(t) \, dt$ with the test function $\varphi(t)$ and the distribution $\Xi(t)$, one just *defines* the derivative $\Xi'(t)$ by $\langle \Xi', \varphi \rangle = -\langle \Xi, \varphi' \rangle$, this means by "partial integration $\int_{-\infty}^{\infty} \varphi(t) \, \Xi'(t) \, dt = -\int_{-\infty}^{\infty} \varphi'(t) \, \Xi(t) \, dt$". This definition takes into account that the test function $\varphi(t)$ is arbitrarily smooth and vanishes at infinity. We can define the derivative of any order for $\Xi(t)$ by using this procedure recursively. The fact that distributions can be differentiated arbitrarily often allows expanding any distribution $\Xi(x)$ in derivatives of the Dirac delta function in the form $\Xi(x) = \sum_{k=0}^{\infty} \mu_k[\Xi] \, (-1)^k / k! \, \delta^{(k)}(x)$ where the (x-independent) moments are $\mu_k[\Xi] = \int_{-\infty}^{\infty} \xi^k \, \Xi(\xi) \, d\xi$. (This is the so-called dual Taylor expansion) [978], [435], [596], [366], [367], [728]; for some of its applications, see [520], [366], [1719], [18], [731], [66], [1316].)

The differentiation of the distribution θ yields the Dirac δ function. It is often used as an idealization of a point-like object with a finite characteristic and is a computable function [1813].

In[12]:= **UnitStep'[x]**

Out[12]= DiracDelta[x]

DiracDelta $[x]$
    represents the Dirac δ distribution $\delta(x)$.

Similar to UnitStep, multidimensional versions of the Dirac δ distribution exist.

DiracDelta $[x_1,...,x_n]$
    represents the multidimensional Dirac δ distribution $\delta(x_1, ..., x_n) = \delta(x_1) \cdots \delta(x_n)$.

Because for all positive and negative x the distribution $\theta(x)$ is constant, a picture of $\delta(x)$ will just show a straight line.

In[13]:= **Plot[DiracDelta[x], {x, -2, 2}, Frame -> True, Axes -> False,**
          PlotStyle -> {Hue[0], Thickness[0.01]}]

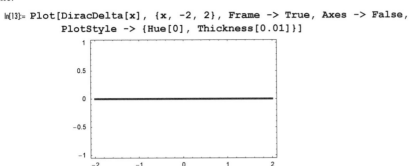

(For convenience and following common practice, for arguments that are not in the support of a generalized function *Mathematica* will numerically evaluate this functions. By doing evaluations like $\theta(2)$ or $\delta(2)$ one must always keep in mind that generalized functions are not functions, but rather equivalence classes of functions defined through integrals with smooth test functions. This means that adding pointwise finite values to distributions does not change them.)

Because Plot uses not exactly regularly spaced points the nonnumerical value at $x = 0$ was not even detected.

In[14]:= **% [[1, 3, 1]] /. Line[{___, a:{_?Negative, _}, b:{_?NonNegative, _}, ___}] :>**
          Line[{…, a, b, …}]

Out[14]= Line[{…, {-0.00412735, 0.}, {0.000593836, 0.}, …}]

The Dirac δ distribution $\delta(x)$ is a so-called singular distribution. All of its support is concentrated at $x = 0$ (a set of Lebesgue measure zero). Loosely speaking, it has the "value" infinity there. And the "size of the infinity" is such that $\int_{-\infty}^{\infty} \delta(x)\,dx = 1$. At $x = 0$, DiracDelta does not return Infinity, but stays unevaluated.

In[15]:= **DiracDelta[0]**

Out[15]= DiracDelta[0]

Be aware that the distribution UnitStep is not just "a function that is zero for $x < 0$ and 1 for $x > 0$" [344]. Many such functions exist, but their derivative is just 0, not a Dirac δ function. Here are two examples of such functions (they are discontinuous functions, but not generalized functions within *Mathematica*).

In[16]:= **myUnitStepFunction[1][x_] := (1 + Sqrt[x^2]/x)/2**
          myUnitStepFunction[2][x_] := 1 - (Log[x] + Log[1/x])/(2 I Pi)

In[18]:= **Show[GraphicsArray[**
          Plot[Chop[myUnitStepFunction[#][x]], {x, -2, 2}, Frame -> True,
              DisplayFunction -> Identity, Axes -> False, PlotRange -> All,
              PlotStyle -> {Hue[0], Thickness[0.01]}]& /@ {1, 2}]]

In[19]:= **Table[D[myUnitStepFunction[k][x], x] // Simplify, {k, 2}]**

Out[19]= {0, 0}

As mentioned, for complex arguments, distributions are undefined [70].

In[20]:= **{UnitStep[I], DiracDelta[1 + I]}**

Out[20]= {UnitStep[i], DiracDelta[1 + i]}

Because distributions are linear functionals, their natural life space is under an integral. (Writing $\theta(x)$ and $\delta(x)$ is already kind of an abuse of proper notation.) So it comes as no surprise that currently the function Integrate (and some integral transforms—see below) knows best how to deal with the functions UnitStep and Dirac‑Delta. Let us look at some examples. The integral $\int_0^\infty \log(x)\,(x^3 - 1)^{-1}\,dx$ can be written as an integral over the domain $\{-\infty, \infty\}$ as $\int_{-\infty}^\infty \theta(x) \log(x)\,(x^3 - 1)^{-1}\,dx$. Integrate knows how to deal with this form.

In[21]:= **Integrate[Log[x]/(x^3 + 1), {x, 0, Infinity}]**

Out[21]= $-\dfrac{2\,\pi^2}{27}$

In[22]:= **Integrate[Log[x]/(x^3 + 1) UnitStep[x], {x, -Infinity, Infinity}]**

Out[22]= $-\dfrac{2\,\pi^2}{27}$

Here is an integral containing the parameter a. Inside distributions, variables are considered to be real; so, we do not need Assumptions -> Im[a] == 0.

In[23]:= **Integrate[UnitStep[x - a], {x, 0, 1}]**

Out[23]= $-(-1 + a)\,\text{UnitStep}[1 - a] + a\,\text{UnitStep}[-a]$

In[24]:= **Plot[%, {a, -2, 2}, PlotRange -> All, Frame -> True, Axes -> False]**

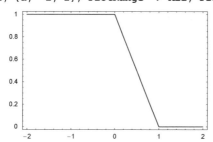

Here is an integral containing the symbolic function f. The result basically says that the value of the integral is just f[c] when $-a<c<a$.

In[25]:= **Integrate[DiracDelta[x - c] f[x], {x, -a, a}]**

Out[25]= f[c] UnitStep[-c - a UnitStep[-a] + a UnitStep[a], c - a UnitStep[-a] + a UnitStep[a]]

Next, we deal with a small physics example. The time dilation of a particle whose path starts and ends at the same point and undergoes an acceleration $a(\tau)$ can be calculated through the following formula [1271]. (T is the total duration of the movement in the inertial frame.)

In[26]:= **timeDilation[a_, T_] :=**
   Assuming[T > 0 && (* speed of light *) c > 0, Sqrt[
       Integrate[Exp[+Integrate[a[τ], {τ, 0, t}]], {t, 0, T}]*
       Integrate[Exp[-Integrate[a[τ], {τ, 0, t}]], {t, 0, T}]]]

For a movement of the form (in the inertial frame) $v(t) = v_0\,\theta(T/2 - t) - \theta(t - T/2)$ (forth and back with constant speed), the acceleration has the form of a Dirac δ function. As a result we obtain the well-known formula for the time dilation.

```
In[27]:= (* use a[τ] == D[ArcTanh[v[τ]], τ] with proper time τ *)
         aForthAndBackWithConstantSpeed[τ_] = -2 ArcTanh[v0/c] DiracDelta[τ - T/2];
```

```
In[29]:= Simplify[timeDilation[aForthAndBackWithConstantSpeed, T],
                  (* parameter conditions *) T > 0 && 0 < v0 < c,
                  TransformationFunctions -> {Automatic, ExpToTrig}] /.
                  (* write in well-known form *)
                  c/Sqrt[c^2 - v0^2] :> 1/Sqrt[1 - v0^2/c^2]
```

$$Out[29]= \frac{T}{\sqrt{1 - \frac{v0^2}{c^2}}}$$

Integrals of the form $\int_{-\infty}^{\infty} \delta(f(x))\,g(x)\,dx$ can be transformed to the form $\sum_k \int_{-\infty}^{\infty} \delta(x - x_k)/|f'(x_k)|\,g(x)\,dx$, where the x_k are the simple zeros of $f(x)$ [427].

```
In[30]:= (* because of symmetry the integral vanishes *)
         Integrate[DiracDelta[x^4 - 2] Sin[x], {x, -Infinity, Infinity}]
```

```
Out[31]= 0
```

Inhomogeneous linear ordinary differential equations can be solved through quadratures in case the homogeneous solution is known. The next input solves the equation of motion for a repeatedly kicked harmonic oscillator [984]. After integration of the Dirac δ distribution on the right-hand side we obtain Heaviside θ distribution in the result.

```
In[32]:= DSolve[{x''[t] + x[t] == Sum[(-1)^k DiracDelta[t - k], {k, 6}],
                 x[0] == 1, x'[0] == 0}, x[t], t] // Simplify
Out[32]= {{x[t] → Cos[t] - Sin[6 - t] UnitStep[-6 + t] + Sin[5 - t] UnitStep[-5 + t] -
            Sin[4 - t] UnitStep[-4 + t] + Sin[3 - t] UnitStep[-3 + t] -
            Sin[2 - t] UnitStep[-2 + t] + Sin[1 - t] UnitStep[-1 + t]}}
```

The Dirac δ distribution can be differentiated an arbitrary number of times. The distributions arising from differentiating $\delta(x)$ do not have special names. Here are the first five derivatives of the Dirac δ distribution.

```
In[33]:= Table[D[DiracDelta[x], {x, k}], {k, 3}]
Out[33]= {DiracDelta'[x], DiracDelta''[x], DiracDelta^(3)[x]}
```

The first derivative of the Dirac δ functions is often used as an ideal dipole [1309]. According to the definition of derivatives of distributions from above, we have the identity

$$\int_{-\infty}^{\infty} \delta^{(n)}(x)\,f(x)\,dx = (-1)^n \int_{-\infty}^{\infty} \delta(x)\,f^{(n)}(x)\,dx.$$

Here, we calculate the definite integral $\int_{-1}^{\xi} \delta^{(n)}(x)\,f(x)\,dx$.

```
In[34]:= Integrate[# f[x], {x, -1, ξ}]& /@ %
Out[34]= {DiracDelta[ξ] f[0] - UnitStep[
           -ξ UnitStep[-1 - ξ] + UnitStep[1 + ξ], -UnitStep[-1 - ξ] + ξ UnitStep[1 + ξ]] f'[0],
          f[0] DiracDelta'[ξ, ξ] - 2 DiracDelta[ξ] f'[0] + UnitStep[
           -ξ UnitStep[-1 - ξ] + UnitStep[1 + ξ], -UnitStep[-1 - ξ] + ξ UnitStep[1 + ξ]] f''[0],
          -3 DiracDelta'[ξ, ξ] f'[0] + f[0] DiracDelta''[ξ, ξ] + 3 DiracDelta[ξ] f''[0] -
           UnitStep[-ξ UnitStep[-1 - ξ] + UnitStep[1 + ξ],
           -UnitStep[-1 - ξ] + ξ UnitStep[1 + ξ]] f^(3)[0]}
```

If one views the property $\theta'(x) = \delta(x)$ as an algebraic property, it makes sense to also allow the opposite transformation, meaning $\int^x \delta(\xi)\,d\xi = \theta(x)$ (be aware that this indefinite integral has no direct relation with the integral forming the defining linear functional). *Mathematica* allows generalized functions in indefinite integrals. Here are some examples of such integrals.

In[35]:= **Integrate[DiracDelta[x], x]**

Out[35]= UnitStep[x]

In[36]:= **NestList[Integrate[#, x]&, %, 3]**

Out[36]= $\left\{\text{UnitStep}[x],\ x\,\text{UnitStep}[x],\ \frac{1}{2}\,x^2\,\text{UnitStep}[x],\ \frac{1}{6}\,x^3\,\text{UnitStep}[x]\right\}$

By partial integration of $\int_{-\infty}^{\infty} \delta^{(n)}(x)\,x^m\,dx$, it follows that $\delta^{(n)}(x)\,x^m = 0$ in the case of $m \geq n$. Similar to the transformation of expressions containing the UnitStep functions, such transformations do not happen automatically, but they can be carried out using FunctionExpand.

In[37]:= **x^2 DiracDelta[x]**

Out[37]= x^2 DiracDelta[x]

In[38]:= **FunctionExpand[%]**

Out[38]= 0

Here is a more complicated integral involving a fourth derivative of a Dirac delta function with a nonlinear argument.

In[39]:= **Integrate[DiracDelta''''[x^11 - 1], {x, -2, 2}]**

Out[39]= $\dfrac{288960}{161051}$

Mathematica can manipulate derivatives of the Dirac delta functions with arguments that have numeric roots in a variety of ways. Here is an expression containing the zeroth to fourth derivative of the Dirac delta function.

In[40]:= **δDeriv4 = D[UnitStep[x^11 - 1], {x, 5}]**

Out[40]= $55440\,x^6$ DiracDelta$[-1 + x^{11}]\ +$
$\qquad 1524600\,x^{17}$ DiracDelta$'[-1 + x^{11}] + 3194400\,x^{28}$ DiracDelta$''[-1 + x^{11}]\ +$
$\qquad 1464100\,x^{39}$ DiracDelta$^{(3)}[-1 + x^{11}] + 161051\,x^{50}$ DiracDelta$^{(4)}[-1 + x^{11}]$

Applying the specialized simplifier Simplify`SimplifyPseudoFunctions (be aware that this special simplifier lives in the context Simplify`—together with other specialized simplifiers) yields an expression that contains only Dirac delta functions with linear arguments. Because the only real root of $x^{11} - 1 = 0$ is $x = 1$, all Dirac delta functions have now the argument $x - 1$.

In[41]:= **δDeriv4 = Expand[δDeriv4];**

In[42]:= **Simplify`SimplifyPseudoFunctions[δDeriv4]**

Out[42]= $5040 \, \text{DiracDelta}[-1 + x] +$

$1524600 \, x^{17} \left(\dfrac{10}{121} \, \text{DiracDelta}[-1 + x] + \dfrac{1}{121} \, \text{DiracDelta}'[-1 + x] \right) +$

$3194400 \, x^{28} \left(\dfrac{210 \, \text{DiracDelta}[-1 + x]}{1331} + \dfrac{30 \, \text{DiracDelta}'[-1 + x]}{1331} + \dfrac{\text{DiracDelta}''[-1 + x]}{1331} \right) +$

$1464100 \, x^{39} \left(\dfrac{6720 \, \text{DiracDelta}[-1 + x]}{14641} + \dfrac{1140 \, \text{DiracDelta}'[-1 + x]}{14641} + \right.$

$\left. \dfrac{60 \, \text{DiracDelta}''[-1 + x]}{14641} + \dfrac{\text{DiracDelta}^{(3)}[-1 + x]}{14641} \right) +$

$161051 \, x^{50} \left(\dfrac{288960 \, \text{DiracDelta}[-1 + x]}{161051} + \dfrac{54600 \, \text{DiracDelta}'[-1 + x]}{161051} + \right.$

$\left. \dfrac{3600 \, \text{DiracDelta}''[-1 + x]}{161051} + \dfrac{100 \, \text{DiracDelta}^{(3)}[-1 + x]}{161051} + \dfrac{\text{DiracDelta}^{(4)}[-1 + x]}{161051} \right)$

The last result becomes obvious after applying `Simplify`SimplifyPseudoFunctions` to the original function $\theta(x^{11} - 1)$.

In[43]:= `Simplify`SimplifyPseudoFunctions[UnitStep[x^11 - 1]]`

Out[43]= `UnitStep[-1 + x]`

We repeat a remark from the above subsection on integration: Because differentiation of discontinuous functions is allowed within distribution theory, it is important to stay away from piecewise-defined functions (in the sense of `Piecewise`). `PiecewiseExpand` will convert `UnitStep` functions into `Piecewise` functions. But by doing so the important differentiation property of `UnitStep` is lost. The next input demonstrates this.

In[44]:= `With[{f = UnitStep[x + 1] - UnitStep[x - 1]},`
         `(* differentiate and integrate; then form definite integral *)`
         `Integrate[Integrate[D[# @ f, x], x], {x, -Infinity, Infinity}]& /@`
         `(* functions to apply to f *) {Identity, PiecewiseExpand}]`

Out[44]= `{2, 0}`

Here is an integral that does not evaluate. Because the theory of distributions is based on linear functionals, squares are not allowed in this theory [559] (within a neutrix approach products, powers, and compositions of distributions with singular support can be defined [1766], [1050], [1051], [1051], [1362], [881], [1454]).

In[45]:= `Integrate[DiracDelta[x]^2, {x, -Infinity, Infinity}]`

Out[45]= $\displaystyle\int_{-\infty}^{\infty} \text{DiracDelta}[x]^2 \, dx$

Many other integrals are not defined for distributions. When the integrand contains distributions with singular support (such as the Heaviside $\theta(x)$ distribution or the Dirac δ distribution), products of distributions with the same singular support are typically not unique [396], [1405] (for unique products, see [429]). So the following integrals stay unevaluated.

In[46]:= `Integrate[DiracDelta''[x] Sqrt[DiracDelta[x]^2], {x, -1, 1}]`

Out[46]= $\displaystyle\int_{-1}^{1} \sqrt{\text{DiracDelta}[x]^2} \; \text{DiracDelta}''[x] \, dx$

A `UnitStep` distribution can also appear in a numerical integration.

In[47]:= `NIntegrate[UnitStep[Tan[x]], {x, -2, -Pi/2, Pi/2, 2}]`

Out[47]= `2.`

As already used above, a `DiracDelta` distribution will typically go undetected in a numerical integration. `NIntegrate` is mostly a purely numerical routine that samples its first argument like a black box and does not symbolically analyze the content of the first argument. The chances to hit the x, such that $\sin(x - 1/2)$ vanishes, are virtually zero.

```
In[48]:= NIntegrate[1 + DiracDelta[Sin[x] - 1/2], {x, -1, 1}]
Out[48]= 2.
```

```
In[49]:= Integrate[1 + DiracDelta[Sin[x] - 1/2], {x, -1, 1}]
```
$$Out[49]= \ 2 + \frac{2}{\sqrt{3}}$$

Distributions like the Dirac δ distribution or the Heaviside θ distribution are often viewed as limits from continuous functions. Limits typically encountered in textbooks are [1826], [22], [655] (for statistics applications of such sequences, see [1804]; for applications to matrix arguments, see [1378]; for representations suitable for numerical calculations involving Dirac δ functions, see [1722]):

$$\theta(x) = \frac{1}{2}\left(\lim_{\varepsilon\to 0}\frac{2}{\pi}\arctan\left(\frac{x}{\varepsilon}\right) + 1\right) = \frac{1}{2}\left(\lim_{\varepsilon\to 0}\operatorname{erf}\left(\frac{x}{\varepsilon}\right) + 1\right)$$

$$\delta(x) = \frac{1}{\pi}\lim_{\varepsilon\to 0}\frac{\varepsilon}{x^2 + \varepsilon^2} = \lim_{\varepsilon\to 0}\frac{1}{\sqrt{\pi\varepsilon}}e^{-\frac{x^2}{\varepsilon}} = \lim_{\varepsilon\to 0}\frac{1}{\pi x}\sin\left(\frac{x}{\varepsilon}\right)$$

Be aware that *Mathematica* 5.1 will not recognize such limits.

```
In[50]:= Limit[ε/(x^2 + ε^2), ε -> 0]
Out[50]= 0
```

Here is a similar limit [653], [597].

$$\delta(x) = \lim_{\varepsilon\to 0}\frac{1}{2}\left(\coth'(x + i\varepsilon) - \coth'(x)\right).$$

```
In[51]:= Limit[D[ComplexExpand[Re[Coth[x + I ε]]], x] - D[Coth[x], x], ε -> 0]
Out[51]= 0
```

`Limit` will search for a limit inside the space of continuous functions, and according to the *Mathematica* philosophy, lower-dimensional exceptional values (here, just the one point $x = 0$) will be ignored.

In a similar way, the following sum evaluates to 0. Obviously, for $x = 0$, this is not the case (in a distributional sense we get $2\pi\delta(x)$).

```
In[52]:= Sum[Exp[I k x], {k, -Infinity, Infinity}]
Out[52]= 0
```

Let us come back to the evaluation of $\int_{-\infty}^{\infty}\delta^{(4)}(x^{11} - 1)\,dx$ from above. Instead of $\delta(x)$ we will use the sequence $\delta_\varepsilon(x) = (\pi\varepsilon)^{-1/2}\exp(-x^2/\varepsilon)$ with $\varepsilon \to 0$.

```
In[53]:= δ[ε_][x_] := 1/Sqrt[Pi ε]Exp[-x^2/ε]
```

Evaluating the integral $\int_{-\infty}^{\infty}\delta_\varepsilon^{(4)}(x^{11} - 1)\,dx$ yields a closed form result in hypergeometric functions.

```
In[54]:= (int[ε_] = Integrate[δ[ε]''''[x^11 - 1], {x, -Infinity, Infinity},
                    Assumptions -> ε > 0]) // TraditionalForm
```

Out[54]//TraditionalForm=

$$\frac{1}{11\sqrt{\pi}\ \varepsilon^{49/11}}$$

$$\left(8\,e^{-1/\varepsilon}\left(11\,(3\,\varepsilon^2-12\,\varepsilon+4)\,\Gamma\!\left(\frac{23}{22}\right){}_1F_1\!\left(\frac{1}{22};\frac{1}{2};\frac{1}{\varepsilon}\right)+2\left(-3\,(\varepsilon-2)\,\varepsilon\,\Gamma\!\left(\frac{23}{22}\right){}_1F_1\!\left(\frac{23}{22};\frac{1}{2};\frac{1}{\varepsilon}\right)+4\,(3\,\varepsilon-2)\right.\right.$$

$$\left.\left.\Gamma\!\left(\frac{23}{22}\right){}_1F_1\!\left(\frac{23}{22};\frac{3}{2};\frac{1}{\varepsilon}\right)+\varepsilon\,\Gamma\!\left(\frac{45}{22}\right)\!\left(\varepsilon\,{}_1F_1\!\left(\frac{45}{22};\frac{1}{2};\frac{1}{\varepsilon}\right)-8\,{}_1F_1\!\left(\frac{45}{22};\frac{3}{2};\frac{1}{\varepsilon}\right)\right)\right)\right)\right)$$

Using the asymptotic expansion of the function ${}_1F_1(a;b;z)$ yields the result from above, namely $288960/161051$.

```
In[55]:= (* Limit[int[ε], ε -> 0] needs some manual help *)
        (* we know that we expand around an essential singularity *)
        hypRule =  Internal`DeactivateMessages[Hypergeometric1F1[a_, b_, z_] ->
              Normal[Series[Hypergeometric1F1[α, β, ζ], {ζ, Infinity, 5}]] /.
                                {α -> a, β -> b, ζ -> z}];
```

```
In[58]:= (* avoid Series::"esss" messages *) Internal`DeactivateMessages @
        Limit[(* use 1F1 asymptotics and simplify integral *)
              (int[ε] /. hypRule // Together // ExpandAll // PowerExpand) /.
                        Exp[-1/ε] -> 0, ε -> 0] // FullSimplify
```

$$\text{Out[58]=}\quad \frac{288960}{161051}$$

Let us now use the representation $\delta_\varepsilon(x)=1/(2\,\varepsilon)\,(\theta(\varepsilon+x)\,\theta(\varepsilon-x))$. As $\varepsilon\to 0$, we have $\lim_{\varepsilon\to 0}\delta_\varepsilon(x)=\delta(x)$. A symmetric approximation of the nth derivative of a function $f(x)$ is

$$f^{(n)}(x)=\lim_{\varepsilon\to 0}\varepsilon^{-n}\sum_{k=0}^{n}(-1)^{n-k}\binom{n}{k}f(k\,\epsilon+(x-n\,\epsilon/2)).$$

```
In[59]:= (* Dirac δ approximation for ε -> 0 *)
        δ1[ε_][x_] := 1/(2 ε) UnitStep[ε + x] UnitStep[ε - x]

        (* Dirac δ derivative approximation *)
        δApproxDeriv[n_, x_, ε_, ε_] :=
        Sum[(-1)^(n - k) Binomial[n, k] δ1[ε][(x - n/2 ε) + k ε], {k, 0, n}]/ε^n
```

Here are the piecewise constant approximations of $\delta_\varepsilon^{(n)}(x)$ visualized.

```
In[63]:= With[{nMax = 60},
        Show[Graphics3D[{Thickness[0.002],
              Table[{Hue[0.8 n/nMax],
                Line[Table[{x, δApproxDeriv[n, x, 1/8, 1/12]},
                        {x, -3/2, 3/2, 1/100}]] /.
                    (* scale maximum to 1 *)
                      Line[l_] :> With[{m = Max[l]}, Line[{#1, n, #2/m}& @@@ l]]},
                  {n, 0, nMax}]}],
              Axes -> {True, True, False}, PlotRange -> All,
              BoxRatios -> {2, 1, 1}, ViewPoint -> {0, -3, 2}]]
```

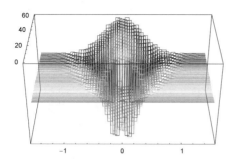

Using the Taylor expansion of a testfunction $f(x)$, we have $\int_{-\infty}^{\infty} \delta_\varepsilon^{(n)} f(x)\, dx = \sum_{k=0}^{\infty} (f^{(k)}(0) \int_{-\infty}^{\infty} \delta_\varepsilon^{(n)} x^k\, dx)$. The next inputs calculate $\int_{-\infty}^{\infty} \delta_\varepsilon^{(4)} x^k\, dx$ for $0 \le n \le 8$.

```
In[64]:= (* adapted integration routine *)
        integrateθ[expr_, {x_, -Infinity, Infinity}] :=
        With[{I = Integrate[PiecewiseExpand[#], {x, -Infinity, Infinity},
                        Assumptions -> Element[{ε, ϵ}, Reals]]&},
              Simplify[If[Head[#] === Plus, I /@ #, I[#]]]&[Expand[expr]]]
```

```
In[66]:= Table[{k, integrateθ[1/k! x^k δApproxDeriv[4, x, ε, ϵ],
                        {x, -Infinity, Infinity}]}, {k, 0, 8}]
```

Out[66]= $\{\{0, 0\}, \{1, 0\}, \{2, 0\}, \{3, 0\}, \{4, \{ 1 \quad \varepsilon > 0\}, \{5, 0\},$
$\{6, \{ \frac{1}{6} (\epsilon^2 + \varepsilon^2) \quad \varepsilon > 0\}, \{7, 0\}, \{8, \{ \frac{1}{720} (9\,\epsilon^4 + 20\,\epsilon^2\,\varepsilon^2 + 6\,\varepsilon^4) \quad \varepsilon > 0\}\}$

In the limit all terms with $k \ne n$ vanish [229].

```
In[67]:= Limit[Limit[%, ε -> 0], ϵ -> 0]
```

Out[67]= $\{\{0, 0\}, \{1, 0\}, \{2, 0\}, \{3, 0\}, \{4, 1\}, \{5, 0\}, \{6, 0\}, \{7, 0\}, \{8, 0\}\}$

In the general case, we find $\lim_{\varepsilon \to 0} \int_{-\infty}^{\infty} \delta_\varepsilon^{(k)} x^k\, dx = k!\,(-1)^k$.

```
In[68]:= (* now take double limit *)
        Limit[Limit[Table[{k,
          integrateθ[1/k! x^k δApproxDeriv[k, x, ε, ϵ], {x, -Infinity, Infinity}]},
                        {k, 0, 8}], ε -> 0], ϵ -> 0]
```

Out[69]= $\{\{0, 1\}, \{1, -1\}, \{2, 1\}, \{3, -1\}, \{4, 1\}, \{5, -1\}, \{6, 1\}, \{7, -1\}, \{8, 1\}\}$

Here is an example that involves a 2D integration: $\int_{-\infty}^{\infty} \int_{-\infty}^{\infty} \delta(\mathcal{E} - (q^2 + p^2))\, dq\, dp$. We use two different approximations of the Dirac δ distribution and carry out the integral in polar coordinates. In the limit of $\varepsilon \to 0$, we obtain the value π for this integral (the phase space volume of a harmonic oscillator with energy $\mathcal{E}$ [1584]).

```
In[70]:= (* another Dirac δ approximation for ε -> 0 *)
        δ2[ε_][x_] := (1/Pi) ε/(x^2 + ε^2)
```

```
In[72]:= Series[Integrate[2Pi δ[ε][ℰ - r^2] r, {r, 0, Infinity},
                        Assumptions -> ε > 0 && ℰ > 0],
              {ε, 0, 1}, Assumptions -> ε > 0 && ℰ > 0] // Expand //
                                    Internal`DeactivateMessages
```

Out[72]= $\pi + e^{-\frac{\mathcal{E}^2}{\varepsilon}} \left(-\frac{\sqrt{\pi}\,\sqrt{\varepsilon}}{2\,\mathcal{E}} + \frac{\sqrt{\pi}\,\varepsilon^{3/2}}{4\,\mathcal{E}^3} + O[\varepsilon]^2 \right)$

```
In[73]:= Series[Integrate[2Pi δ2[ε][ℰ - r^2] r, {r, 0, Infinity},
                        GenerateConditions -> False],
              {ε, 0, 3}, Assumptions -> ε > 0 && ℰ > 0]
```

Out[73]= $\pi - \frac{\varepsilon}{\mathcal{E}} + \frac{\varepsilon^3}{3\,\mathcal{E}^3} + O[\varepsilon]^4$

Be also aware that currently only the integral transforms will return distributions in their output for inputs that do not contain distributions. Functions like D and `Derivative` know how to deal with certain distributions, but their output will only contain them when their input does. So in the following example (Green's function of the Laplace operator in 3D), we just get 0 and not $-4\pi\delta(x, y, z)$. (In such cases one should use the proper pseudofunction $\theta(r)/r$ [593], [178], [659], [1706], [22], [1812], [1072], [1832], [176], [857], [858].)

```
In[74]:= r = Sqrt[x^2 + y^2 + z^2];

      Together[D[1/r, x, x] + D[1/r, y, y] + D[1/r, z, z]]
Out[75]= 0
```

In a similar sense one cannot conclude from the following input that the function $K_0\!\left((x^2 + y^2)^{1/2}\right)$ fulfills the 2D Helmholtz equation everywhere. At the origin we have again a term containing $\delta(x, y)$ [293], [376], [1100], [79], [122], [1875], [404], [592], [294]. (We discuss the Bessel function $K_\nu(z)$ and the function `FullSim` `plify` in Chapter 3.) For the occurrence of the Dirac δ function for the irregular solution of the radial Schrödinger equation, see [1211].

```
In[76]:= Function[ψ, D[ψ, x, x] + D[ψ, y, y] - k^2 ψ][
              BesselK[0, k Sqrt[x^2 + y^2]]] // FullSimplify
Out[76]= 0
```

The Dirac δ distribution plays an important role for the solution of a differential equation. Given the linear differential operator $\hat{L}_x$ (acting on the variable x), the solution $y(x)$ of the differential equation $\hat{L}_x\, y(x) = f(x)$ can be obtained easily from the fundamental solution (or Green's function) $G(x, y)$ by $y(x) = \int_{-\infty}^{\infty} G(x, y)\, f(y)\, dy$. The fundamental solution obeys the differential equation $\hat{L}_x\, G(x, y) = \delta(x - y)$. (The fundamental solution obeys the same boundary condition as the solution $y(x)$.) Here, we calculate the fundamental solution of the differential operator $\hat{L}_x = \partial_{xx} + 1$.

```
In[77]:= DSolve[{y''[x] + y[x] == DiracDelta[x]}, y[x], x]
Out[77]= {{y[x] → C[1] Cos[x] + C[2] Sin[x] + Sin[x] UnitStep[x]}}
```

A direct substitution of the last result in the differential equation will not immediately confirm the result.

```
In[78]:= y''[x] + y[x] /. y -> Function[x, Evaluate[%[[1, 1, 2]]]]
Out[78]= 2 Cos[x] DiracDelta[x] + Sin[x] DiracDelta'[x]
```

We have to use the properties $\delta'(x)\, f(x) = -\delta(x)\, f'(x)$ and $f(x)\, \delta(x) = f(0)\, \delta(x)$ "by hand".

```
In[79]:= % /. DiracDelta'[x] f_[x] :> -f'[x] DiracDelta[x] /.
             DiracDelta[x] f_[x] :> f[0] DiracDelta[x]
Out[79]= DiracDelta[x]
```

The solution of the initial value problem is obtained using the fundamental solution $G(x)$ (Green's function) [1704] for arbitrary initial values and adding the initial conditions $y^{(n)}(0)$ in the form $\sum_{k=1}^{n} \partial^{k-1} G(x)/\partial x^n\, y^{(n-k)}(0)$ to the right-hand side as an inhomogeneous term. Here is a simple example—the differential equation $y''(x) + y(x) = e^{-x}$ with initial conditions $y(0) = y_0$ and $y'(0) = y_p$. We use DSolve to solve the initial value problem.

```
In[80]:= sol = DSolve[{y''[x] + y[x] == Exp[-x],
                y[0] == y0, y'[0] == yp}, y[x], x][[1, 1, 2]] // Expand
Out[80]= - Cos[x]/2 + y0 Cos[x] + 1/2 e^-x Cos[x]^2 + Sin[x]/2 + yp Sin[x] + 1/2 e^-x Sin[x]^2
```

This is a fundamental solution for this problem.

```
In[81]:= gf[x_] = Limit[DSolve[{y''[x] + y[x] == DiracDelta[x],
                (* right sided initial conditions; after δ kicked *)
```

```
                        y[ε] == 0, y'[ε] == 1}, y[x], x][[1, 1, 2]] /.
                        DiracDelta[c_] Sin[c_] :> 0 // Simplify,
                        ε -> 0, Direction -> -1]
```
Out[81]= Sin[x] UnitStep[x]

Now, we use the fundamental solution to build the solution of the inhomogeneous equation and to fulfill the initial conditions.

In[82]:= soll = Integrate[Expand[gf[x - ξ] Exp[-ξ]], {ξ, 0, Infinity},
                        GenerateConditions -> False] +
        (* the initial conditions as part of the inhomogeneous part *)
        (gf[x - ξ] yp /. ξ -> 0) + (D[gf[x - ξ], x] y0 /. ξ -> 0) /.
                        DiracDelta[c_] Sin[c_] :> 0

Out[82]= y0 Cos[x] UnitStep[x] + yp Sin[x] UnitStep[x] + $\frac{1}{2}$ (e⁻ˣ - Cos[x] + Sin[x]) UnitStep[x]

For $x > 0$ (the region under consideration), the solution so-obtained agrees with the one from DSolve.

In[83]:= Expand[sol - %] // Simplify[#, x > 0]&

Out[83]= 0

Within the realm of distributions, differential equations get more solutions than just the classical ones. Let us look at the first-order differential equation $\xi^2 u'(\xi) = 1$.

In[84]:= ode = ξ^2 u'[ξ] - 1;

In the space of ordinary functions, we have the solution $u(\xi) = c_1 - 1/\xi$.

In[85]:= DSolve[ode == 0, u[ξ], ξ]

Out[85]= $\left\{\left\{u[\xi] \to -\frac{1}{\xi} + C[1]\right\}\right\}$

In the space of generalized functions we have the solution $u_{GF}(\xi) = c_1 + c_2\, \theta(\xi) + c_3\, \delta(\xi) - 1/\xi$. Let us check this.

In[86]:= uGF[ξ_] = c[1] + c[2] UnitStep[ξ] + c[3] DiracDelta[ξ] - 1/ξ

Out[86]= $-\frac{1}{\xi}$ + c[1] + c[3] DiracDelta[ξ] + c[2] UnitStep[ξ]

Directly substituting the solution into *Mathematica* does not give zero.

In[87]:= ξ^2 uGF'[ξ] - 1 // Expand

Out[87]= ξ^2 c[2] DiracDelta[ξ] + ξ^2 c[3] DiracDelta'[ξ]

Using Simplify, we can get zero.

In[88]:= Simplify[%]

Out[88]= 0

To get the last zero, we have to add the two rules $x^n\, \delta(x) = 0$ and $x^n\, \delta^{(v)}(x) = (-1)^n\, v!\, /(v-n)!\, \delta^{(v-n)}(x)$.

In[89]:= δSimplify[expr_, x_] :=
        With[{rules = {x^n_. Derivative[v_][DiracDelta][x] :>
                        (-1)^n v!/(v - n)! Derivative[v - n][DiracDelta][x],
                        x^n_. DiracDelta[x] :> 0}},
            FixedPoint[Expand[#] //. rules&, expr]]

Now it is straightforward to see that $u_{GF}(\xi)$ is indeed a solution of the differential equation $\xi^2 u(\xi) = 1$.

In[90]:= δSimplify[%%, ξ]

Out[90]= 0

No option of `DSolve` is currently available to generate solutions of differential equations that are distributions. Let us deal with a slightly more complicated example, the hypergeometric differential equation

$$x(1-x)\,y''(x) + (\gamma - (\alpha + \beta + 1)\,x)\,y'(x) - \alpha\,\beta\,y(x) = 0.$$

Classically, the solutions are hypergeometric functions (see Chapter 3). These become rational functions for integer parameters. Here is an example.

```
In[91]:= ode2F1[x_, y_, {α_, β_, γ_}] =
          x (1 - x) y''[x] + (γ - (α + β +1) x) y'[x] - α β y[x];
```

```
In[92]:= With[{α = 12, β = 7, γ = 10},
          DSolve[ode2F1[x, y, {α, β, γ}] == 0, y, x]]
```

$$\text{Out[92]= } \left\{\left\{y \to \text{Function}\left[\{x\}, \frac{(28 + 3\,x\,(7 + 2\,x))\,C[1]}{6\,x^9} + \right.\right.\right.$$
$$\left.\left.\left. \frac{(-28 + 231\,x - 825\,x^2 + 1650\,x^3 - 1980\,x^4 + 1386\,x^5 - 462\,x^6)\,C[2]}{2310\,(-1 + x)^9\,x^9}\right]\right\}\right\}$$

This solution fulfills the differential equation.

```
In[93]:= With[{α = 12, β = 7, γ = 10},
          ode2F1[x, %[[1, 1, 2]], {α, β, γ}]] // Together
```

Out[93]= 0

But the following sum of derivatives of Dirac δ functions is also a (weak) solution [978].

```
In[94]:= yGF[x_, {α_, β_, γ_}] :=
          Sum[(-1)^k Pochhammer[β - γ + 1, k]/k!/Pochhammer[β - α + 1, k]*
              Derivative[β + k - 1][DiracDelta][x], {k, 0, γ - β - 1}] /;
                  IntegerQ[α] && IntegerQ[β] && IntegerQ[γ] && α >= γ > β
```

Here is a distributional solution of our special case of the hypergeometric differential equation.

```
In[95]:= yGF[x, {12, 7, 10}]
```

$$\text{Out[95]= } \text{DiracDelta}^{(6)}[x] - \frac{1}{2}\,\text{DiracDelta}^{(7)}[x] + \frac{1}{12}\,\text{DiracDelta}^{(8)}[x]$$

Substituting this solution into the differential equation and applying our δ`Simplify` shows that this is indeed a solution.

```
In[96]:= With[{α = 12, β = 7, γ = 10, y = Function[x, Evaluate[%]]},
          ode2F1[x, y, {α, β, γ}]] // Expand
```

$$\text{Out[96]= } -84\,\text{DiracDelta}^{(6)}[x] + 52\,\text{DiracDelta}^{(7)}[x] -$$
$$20\,x\,\text{DiracDelta}^{(7)}[x] - 12\,\text{DiracDelta}^{(8)}[x] + 11\,x\,\text{DiracDelta}^{(8)}[x] -$$
$$x^2\,\text{DiracDelta}^{(8)}[x] + \frac{5}{6}\,\text{DiracDelta}^{(9)}[x] - \frac{13}{6}\,x\,\text{DiracDelta}^{(9)}[x] +$$
$$\frac{1}{2}\,x^2\,\text{DiracDelta}^{(9)}[x] + \frac{1}{12}\,x\,\text{DiracDelta}^{(10)}[x] - \frac{1}{12}\,x^2\,\text{DiracDelta}^{(10)}[x]$$

```
In[97]:= δSimplify[%, x]
```

Out[97]= 0

For some more uses of series of Dirac δ distributions, see [289], [1728], [969], [1729]; for a spectacular weak solution of the Euler PDEs, see [1615]; for distributional solutions of functional equations, see [454], [456], [1569], and [372].

As a little application of how to deal with the `UnitStep` and the `DiracDelta` function in *Mathematica*, let us check that $\psi(x, t) = \theta(2\,(x - k\,t)\,\gamma + \pi)\,\theta(\pi - 2\,\gamma\,(x - k\,t))\,\cos^{\delta+1}(\gamma\,(x - k\,t))\,e^{i\,(k\,x - \omega\,t)}$ is a "finite length solitonic" solution (also called compacton [1509], [1093], [1165], [1360], [1475], [1166], [406], [1817], [560], [1818], [1873], [1819]) of the following nonlinear Schrödinger equation [300]:

$$i\,\frac{\partial \psi(x,\,t)}{\partial t} = -\frac{1}{2}\,\frac{\partial^2 \psi(x,\,t)}{\partial x^2} + \frac{1}{8}\,\xi\left(\frac{1}{\rho(x,\,t)}\,\frac{\partial \rho(x,\,t)}{\partial x}\right)^2 \psi(x,\,t)$$

where $\rho(x,\,t) = \psi(x,\,t)\,\overline{\psi(x,\,t)}$, $0 < \xi < 1$, $\delta = \xi/(1-\xi)$, and $\omega = (k^2 + \gamma^2\,(\delta+1))/2$. (For arbitrarily narrow solitons, see [434].)

Here, we implement the equations from above.

```
In[98]:= δ = ξ/(1 - ξ);
        ω = 1/2 (k^2 + γ^2 (1 + δ));
```

```
In[100]:= Ω[ψ_] := Module[{ψc = ψ /. c_Complex :> Conjugate[c], ρ, j},
             ρ = ψ ψc; ξ/8 (D[ρ, x]/ρ)^2]
```

Without the finite length restriction (the terms $\theta(2\,(x-k\,t)\,\gamma + \pi)\,\theta(\pi - 2\,\gamma\,(x-k\,t))$ in $\psi(x,\,t)$, it is straightforward that $\psi(x,\,t)$ is a solution of the equation.

```
In[101]:= ψ[x_, t_] = Cos[γ (x - k t)]^(1 + δ) Exp[I (k x - ω t)];
```

```
In[102]:= Factor[I D[ψ[x, t], t] + 1/2 D[ψ[x, t], {x, 2}] - Ω[ψ[x, t]] ψ[x, t]]
Out[102]= 0
```

Including the finite length condition makes things a bit more tricky. Here is the finite length solution.

```
In[103]:= ψ1[x_, t_] =
          ψ[x, t] UnitStep[2 γ (x - k t) + Pi] UnitStep[Pi - 2 γ (x - k t)];
```

Just plainly redoing the calculation above will not give the desired result.

```
In[104]:= Simplify[Factor[I D[ψ1[x, t], t] + 1/2 D[ψ1[x, t], {x, 2}] -
                    Ω[ψ1[x, t]] ψ[x, t]]] === 0
Out[104]= False
```

So let us do the calculation step by step. First, we form the first time derivative with respect to t.

```
In[105]:= D[ψ1[x, t], x]
```

$$\text{Out[105]= } 2\,e^{\mathbf{i}\,(k\,x - \frac{1}{2}\,t\,(k^2 + \gamma^2\,(1 + \frac{\xi}{1-\xi})))}\,\gamma\,\text{Cos}[(-k\,t + x)\,\gamma]^{1 + \frac{\xi}{1-\xi}}$$
$$\text{DiracDelta}[\pi + 2\,(-k\,t + x)\,\gamma]\,\text{UnitStep}[\pi - 2\,(-k\,t + x)\,\gamma] -$$
$$2\,e^{\mathbf{i}\,(k\,x - \frac{1}{2}\,t\,(k^2 + \gamma^2\,(1 + \frac{\xi}{1-\xi})))}\,\gamma\,\text{Cos}[(-k\,t + x)\,\gamma]^{1 + \frac{\xi}{1-\xi}}\,\text{DiracDelta}[\pi - 2\,(-k\,t + x)\,\gamma]$$
$$\text{UnitStep}[\pi + 2\,(-k\,t + x)\,\gamma] + \mathbf{i}\,e^{\mathbf{i}\,(k\,x - \frac{1}{2}\,t\,(k^2 + \gamma^2\,(1 + \frac{\xi}{1-\xi})))}\,k\,\text{Cos}[(-k\,t + x)\,\gamma]^{1 + \frac{\xi}{1-\xi}}$$
$$\text{UnitStep}[\pi - 2\,(-k\,t + x)\,\gamma]\,\text{UnitStep}[\pi + 2\,(-k\,t + x)\,\gamma] -$$
$$e^{\mathbf{i}\,(k\,x - \frac{1}{2}\,t\,(k^2 + \gamma^2\,(1 + \frac{\xi}{1-\xi})))}\,\gamma\,\left(1 + \frac{\xi}{1-\xi}\right)\,\text{Cos}[(-k\,t + x)\,\gamma]^{\frac{\xi}{1-\xi}}\,\text{Sin}[(-k\,t + x)\,\gamma]$$
$$\text{UnitStep}[\pi - 2\,(-k\,t + x)\,\gamma]\,\text{UnitStep}[\pi + 2\,(-k\,t + x)\,\gamma]$$

We implement a generalization of $x\,\delta(x) = 0$ for the form $f(t)\,\delta(g(t))$ to simplify the expression above.

```
In[106]:= δrule = Times[factors__, DiracDelta[y_]] :>
               Module[{t0, factor1},
                   (* the t such that y vanishes *)
                   t0 = t /. Solve[y == 0, t][[1]];
                   (* the value of factor at t0 *)
                   factor1 = Times[factors] //. _UnitStep -> 1 /. t -> t0;
                   (* the zero result *) 0 /;
                   ((Together //@ factor1) /. 0^_ -> 0) === 0];
```

Applying δrule to the first time derivative gives a better result—no Dirac δ functions appear anymore.

In[107]:= **timeDeriv1 = Expand[D[ψ1[x, t], t]] /. δrule**

Out[107]= $-\dfrac{1}{2}\,\mathbb{i}\,e^{\mathbb{i}\,(k\,x-\frac{1}{2}\,t\,(k^2+\gamma^2\,(1+\frac{\xi}{1-\xi})))}\,k^2\,\text{Cos}[(-k\,t+x)\,\gamma]^{1+\frac{\xi}{1-\xi}}$
  $\text{UnitStep}[\pi-2\,(-k\,t+x)\,\gamma]\,\text{UnitStep}[\pi+2\,(-k\,t+x)\,\gamma]\,-$
  $\dfrac{1}{2}\,\mathbb{i}\,e^{\mathbb{i}\,(k\,x-\frac{1}{2}\,t\,(k^2+\gamma^2\,(1+\frac{\xi}{1-\xi})))}\,\gamma^2\,\text{Cos}[(-k\,t+x)\,\gamma]^{1+\frac{\xi}{1-\xi}}\,\text{UnitStep}[\pi-2\,(-k\,t+x)\,\gamma]$
  $\text{UnitStep}[\pi+2\,(-k\,t+x)\,\gamma]\,-\dfrac{1}{2\,(1-\xi)}\Big(\mathbb{i}\,e^{\mathbb{i}\,(k\,x-\frac{1}{2}\,t\,(k^2+\gamma^2\,(1+\frac{\xi}{1-\xi})))}\,\gamma^2\,\xi$
  $\text{Cos}[(-k\,t+x)\,\gamma]^{1+\frac{\xi}{1-\xi}}\,\text{UnitStep}[\pi-2\,(-k\,t+x)\,\gamma]\,\text{UnitStep}[\pi+2\,(-k\,t+x)\,\gamma]\Big)\,+$
  $e^{\mathbb{i}\,(k\,x-\frac{1}{2}\,t\,(k^2+\gamma^2\,(1+\frac{\xi}{1-\xi})))}\,k\,\gamma\,\text{Cos}[(-k\,t+x)\,\gamma]^{\frac{\xi}{1-\xi}}\,\text{Sin}[(-k\,t+x)\,\gamma]$
  $\text{UnitStep}[\pi-2\,(-k\,t+x)\,\gamma]\,\text{UnitStep}[\pi+2\,(-k\,t+x)\,\gamma]\,+$
  $\dfrac{1}{1-\xi}\Big(e^{\mathbb{i}\,(k\,x-\frac{1}{2}\,t\,(k^2+\gamma^2\,(1+\frac{\xi}{1-\xi})))}\,k\,\gamma\,\xi\,\text{Cos}[(-k\,t+x)\,\gamma]^{\frac{\xi}{1-\xi}}\,\text{Sin}[(-k\,t+x)\,\gamma]$
  $\text{UnitStep}[\pi-2\,(-k\,t+x)\,\gamma]\,\text{UnitStep}[\pi+2\,(-k\,t+x)\,\gamma]\Big)$

In a similar way, we deal with the first and second space derivative.

In[108]:= **spaceDeriv1 = Expand[D[ψ1[x, t], x]] /. δrule**

Out[108]= $\mathbb{i}\,e^{\mathbb{i}\,(k\,x-\frac{1}{2}\,t\,(k^2+\gamma^2\,(1+\frac{\xi}{1-\xi})))}\,k\,\text{Cos}[(-k\,t+x)\,\gamma]^{1+\frac{\xi}{1-\xi}}\,\text{UnitStep}[\pi-2\,(-k\,t+x)\,\gamma]$
  $\text{UnitStep}[\pi+2\,(-k\,t+x)\,\gamma]\,-e^{\mathbb{i}\,(k\,x-\frac{1}{2}\,t\,(k^2+\gamma^2\,(1+\frac{\xi}{1-\xi})))}\,\gamma\,\text{Cos}[(-k\,t+x)\,\gamma]^{\frac{\xi}{1-\xi}}$
  $\text{Sin}[(-k\,t+x)\,\gamma]\,\text{UnitStep}[\pi-2\,(-k\,t+x)\,\gamma]\,\text{UnitStep}[\pi+2\,(-k\,t+x)\,\gamma]\,-$
  $\dfrac{1}{1-\xi}\Big(e^{\mathbb{i}\,(k\,x-\frac{1}{2}\,t\,(k^2+\gamma^2\,(1+\frac{\xi}{1-\xi})))}\,\gamma\,\xi\,\text{Cos}[(-k\,t+x)\,\gamma]^{\frac{\xi}{1-\xi}}\,\text{Sin}[(-k\,t+x)\,\gamma]$
  $\text{UnitStep}[\pi-2\,(-k\,t+x)\,\gamma]\,\text{UnitStep}[\pi+2\,(-k\,t+x)\,\gamma]\Big)$

In[109]:= **spaceDeriv2 = Expand[D[spaceDeriv1, x]] /. δrule;**

The nonlinear term still needs to be dealt with.

In[110]:= **ψ1c = ψ1[x, t] /. c_Complex :> Conjugate[c];**
    ρ1 = ψ1[x, t] ψ1c

Out[111]= $\text{Cos}[(-k\,t+x)\,\gamma]^{2+\frac{2\,\xi}{1-\xi}}\,\text{UnitStep}[\pi-2\,(-k\,t+x)\,\gamma]^2\,\text{UnitStep}[\pi+2\,(-k\,t+x)\,\gamma]^2$

The rule ruleθ simplifies powers of Heaviside distributions.

In[112]:= **ruleθ = u_UnitStep^e_ :> u**

Out[112]= $u\_\text{UnitStep}^{e\_}\,\rightsquigarrow\,u$

In[113]:= **ρ1 = ρ1 /. ruleθ**

Out[113]= $\text{Cos}[(-k\,t+x)\,\gamma]^{2+\frac{2\,\xi}{1-\xi}}\,\text{UnitStep}[\pi-2\,(-k\,t+x)\,\gamma]\,\text{UnitStep}[\pi+2\,(-k\,t+x)\,\gamma]$

After carrying out the spatial differentiation, we again apply our rule δrule.

In[114]:= **ρDeriv1 = Expand[D[ρ1, x]] /. δrule**

Out[114]= $-2\,\gamma\,\text{Cos}[(-k\,t+x)\,\gamma]^{1+\frac{2\,\xi}{1-\xi}}\,\text{Sin}[(-k\,t+x)\,\gamma]\,\text{UnitStep}[\pi-2\,(-k\,t+x)\,\gamma]$
  $\text{UnitStep}[\pi+2\,(-k\,t+x)\,\gamma]\,-\dfrac{1}{1-\xi}\Big(2\,\gamma\,\xi\,\text{Cos}[(-k\,t+x)\,\gamma]^{1+\frac{2\,\xi}{1-\xi}}$
  $\text{Sin}[(-k\,t+x)\,\gamma]\,\text{UnitStep}[\pi-2\,(-k\,t+x)\,\gamma]\,\text{UnitStep}[\pi+2\,(-k\,t+x)\,\gamma]\Big)$

In the process of forming the expression $1/\rho(x,t)\,\partial\rho(x,t)/\partial x$, we must be especially careful. Formally, the terms $\theta(\pi-2\,(x-k\,t)\,\gamma)\,\theta(2\,(x-k\,t)\,\gamma+\pi)$ cancel because inside Times they are treated like a commutative, associative quantity.

In[115]:= **ξ/8 (ρDeriv1/ρ1)^2 // Expand**

Out[115]= $\frac{1}{2} \gamma^2 \xi \, \text{Tan}[(-k\,t+x)\,\gamma]^2 + \frac{\gamma^2 \xi^2 \, \text{Tan}[(-k\,t+x)\,\gamma]^2}{1-\xi} + \frac{\gamma^2 \xi^3 \, \text{Tan}[(-k\,t+x)\,\gamma]^2}{2\,(1-\xi)^2}$

We restore the finite length conditions "by hand".

In[116]:= **Ω[ψ1] = % UnitStep[2 γ (x - k t) + Pi] UnitStep[Pi - 2 γ (x - k t)]**

Out[116]= $\left(\frac{1}{2} \gamma^2 \xi \, \text{Tan}[(-k\,t+x)\,\gamma]^2 + \frac{\gamma^2 \xi^2 \, \text{Tan}[(-k\,t+x)\,\gamma]^2}{1-\xi} + \frac{\gamma^2 \xi^3 \, \text{Tan}[(-k\,t+x)\,\gamma]^2}{2\,(1-\xi)^2} \right)$
$\text{UnitStep}[\pi - 2\,(-k\,t+x)\,\gamma] \, \text{UnitStep}[\pi + 2\,(-k\,t+x)\,\gamma]$

Putting everything together, we arrive at the zero we were hoping for. This indeed shows that $\psi1[x, t]$ describes a finite length soliton of the above nonlinear Schrödinger equation.

In[117]:= **Factor[Expand[I timeDeriv1 + 1/2 spaceDeriv2 - Ω[ψ1] ψ1[x, t]] /. ruleθ]**

Out[117]= 0

Here is a space-time picture of the absolute value of the finite length soliton for certain parameters. It is really a localized, moving, shape-invariant solution of a nonlinear wave equation that is concentrated at every time on a compact space domain. For a fixed time (right graphic), one sees that the transition between the zero-elongation and the nonzero-elongation domain is smooth (which is needed to fulfill the second-order differential equation).

In[118]:= **Φ = With[{k = 2, γ = 1/2, ξ = 1/2}, Evaluate[ψ1[x, t]]]**

Out[118]= $e^{i \left(-\frac{9t}{4} + 2x\right)} \, \text{Cos}\left[\frac{1}{2}\,(-2\,t+x)\right]^2 \, \text{UnitStep}[\pi + 2\,t - x] \, \text{UnitStep}[\pi - 2\,t + x]$

In[119]:= **Show[GraphicsArray[**
```
        Block[{$DisplayFunction = Identity},
          {(* 3D plot of the compacton *)
           Plot3D[Evaluate[Abs[Φ]], {x, -12, 12}, {t, -4, 4},
                Mesh -> False, PlotPoints -> 140, PlotRange -> All],
           (* plot of the compacton at a fixed time *)
           Plot[Evaluate[Abs[Φ] /. t -> 2], {x, -0, 8}, PlotRange -> All,
                AspectRatio -> 1/3, Frame -> True, Axes -> False]}]]]
```

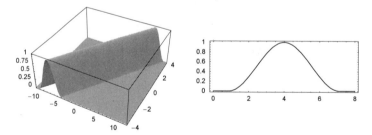

Until now, we encountered only one possibility that *Mathematica* would return a Dirac δ function if we did not input one; this was by differentiation of the UnitStep function. More functions generate generalized functions, also in case one does not explicitly input the UnitStep or the DiracDelta distribution. The most important one is the Fourier transform [1387]. The Fourier transform $\mathcal{F}_t[f(t)](\omega)$ of a function $f(t)$ is defined as $\mathcal{F}_t[f(t)](\omega) = (2\pi)^{-1/2} \int_{-\infty}^{\infty} e^{i\omega t} f(t)\,dt$ [920]. (The square brackets in the traditional form notation $\mathcal{F}_t[f(t)](\omega)$ indicate the fact that the Fourier transform of $f(t)$ is a linear functional of $f(t)$ and a function of ω.)

> `FourierTransform[f(t) , t, ω]`
>
>   represents the Fourier transform of the function $f(t)$ with respect to the variable t and the kernel $e^{i\omega t}$.

Here is the Fourier transform of an "ordinary" function.

In[120]:= `Clear[t, ω, x, y, s, Ω, a, b, term]`
        `FourierTransform[Exp[-x^2] x^3, x, y]`

Out[121]= $-\dfrac{i\, e^{-\frac{y^2}{4}}\, y\, (-6 + y^2)}{8\sqrt{2}}$

The Fourier transformation is a linear operation.

In[122]:= `FourierTransform[α Sin[x^2] + β Exp[-x^2], x, y]`

Out[122]= $\dfrac{1}{2}\left(\sqrt{2}\, e^{-\frac{y^2}{4}}\,\beta + \alpha\,\text{Cos}\left[\dfrac{y^2}{4}\right] - \alpha\,\text{Sin}\left[\dfrac{y^2}{4}\right]\right)$

Derivative operators transform under a Fourier transformation into multiplication operators. This property makes them useful for solving ordinary and partial differential equations [532], [750], [443], [183], [879].

In[123]:= `FourierTransform[y''[x], x, ξ]`

Out[123]= $-\xi^2\,\text{FourierTransform}[y[x], x, \xi]$

The Fourier transform of the function 1 is essentially a Dirac δ distribution [257].

In[124]:= `FourierTransform[1, t, ω]`

Out[124]= $\sqrt{2\pi}\,\text{DiracDelta}[\omega]$

The following Fourier transform of cos(t) and sin(t) too gives a result that contains Dirac δ distribution.

In[125]:= `FourierTransform[α Cos[t] + β Sin[t], t, ω]`

Out[125]= $\sqrt{\dfrac{\pi}{2}}\,\alpha\,\text{DiracDelta}[-1 + \omega] + i\sqrt{\dfrac{\pi}{2}}\,\beta\,\text{DiracDelta}[-1 + \omega] +$

$\sqrt{\dfrac{\pi}{2}}\,\alpha\,\text{DiracDelta}[1 + \omega] - i\sqrt{\dfrac{\pi}{2}}\,\beta\,\text{DiracDelta}[1 + \omega]$

Be aware that carrying out the "integral" (using `Integrate`) will not result in a Dirac δ distribution.

In[126]:= `Integrate[Exp[I k t] Exp[I ω t], {t, -Infinity, Infinity},`
        `            Assumptions -> Im[k] == 0]/(2 Pi)`

          Integrate::idiv : Integral of $e^{it\,(k+\omega)}$ does not converge on $\{-\infty, \infty\}$. More…

Out[126]= $\dfrac{\text{Integrate}[e^{i\,k\,t + i\,t\,\omega},\ \{t, -\infty, \infty\},\ \text{Assumptions} \to \text{Im}[k] == 0]}{2\pi}$

The asymptotic behavior of a Fourier transform measures the smoothness of a function. Here is a Fourier transform of a function that oscillates rapidly near the origin. In the Fourier transform, these oscillations result in a discontinuity at the origin (a jump of size $\sqrt{2\pi}$) [569].

In[127]:= `(* cry of a bat *)`
        `FourierTransform[Exp[-I/x] - 1, x, y] //` `(* simplify result *)`
        `                Expand // FullSimplify // FunctionExpand`

Out[128]= $\dfrac{\sqrt{\frac{\pi}{2}}\,(y + \text{Abs}[y])\,\text{BesselI}[1, 2\sqrt{\text{Abs}[y]}\,]}{\text{Abs}[y]^{3/2}}$

In[129]:= `FourierTransform[%, y, x] // TrigToExp`

Out[129]= $-1 + e^{\frac{i}{x}}$

The following function $\mathsf{sF}[g][x]$ is equal to its Fourier transform [299], [1155], [1849], [409]. (Here g is an arbitrary function).

```
In[130]:= sF[g_][x_] := g[x] + g[-x] + FourierTransform[g[p], p, +x] +
                              FourierTransform[g[p], p, -x]
```

For an unspecified function g, no transformations are carried out. (*Mathematica* does not know anything about asymptotics of the function g, so even the existence of the Fourier transform is not guaranteed.)

```
In[131]:= sF[g][x]  -  FourierTransform[sF[g][x], x, p] /. p -> x // Simplify
```

Out[131]= FourierTransform[g[x], x, -x] + FourierTransform[g[x], x, x] -
        FourierTransform[FourierTransform[g[x], x, -x] +
        FourierTransform[g[x], x, x] + g[-x] + g[x], x, x] + g[-x] + g[x]

For functions g, such that `FourierTransform` evaluates nontrivially, we can show the identity.

```
In[132]:= Function[g, Simplify[sF[g][x] -
                        FourierTransform[sF[g][x], x, p] /. p -> x]] /@
          {(* classical self-Fourier functions *)
           Function[x, Sqrt[2 Pi] DiracDelta[x - Pi]],
           Function[x, Sqrt[2 Pi] DiracDelta[x + Pi]],
           Function[x, Exp[I Pi x] + Exp[-I Pi x]],
           Function[x, Exp[-x^2/2]],
           (* Function[x, Sech[SqrtPi2 x]], *)
           (* some functions *) Function[x, Exp[-x^4]],
           Function[x,  1/(1 + x^4)]} /. (* use symmetry of DiracDelta *)
                            DiracDelta[Pi - x] -> DiracDelta[x - Pi]
```

Out[132]= {0, 0, 0, 0, 0, 0}

(For the self-dual functions of the Fourier sin transform, see [159].)

Fourier transforms can be defined for distributions too. Here is the Fourier transform of the Heaviside θ function.

```
In[133]:= FourierTransform[UnitStep[t], t, ω]
```

Out[133]= $\dfrac{i}{\sqrt{2\pi}\,\omega} + \sqrt{\dfrac{\pi}{2}}\;\mathrm{DiracDelta}[\omega]$

The term $\frac{1}{\omega}$ is to be understood as the principal value distribution $\mathcal{P}(\frac{1}{\omega})$, defined by

$$\left\langle \mathcal{P}\!\left(\frac{1}{\omega}\right),\ \varphi \right\rangle = \lim_{\varepsilon \to 0^+} \left(\int_{-\infty}^{-\varepsilon} \frac{\varphi(\omega)}{\omega}\,d\omega - \int_{\varepsilon}^{\infty} \frac{\varphi(\omega)}{\omega}\,d\omega \right).$$

For sufficiently nice functions the last identity can be integrated by parts to yield $\langle \mathcal{P}(1/\omega),\ \varphi \rangle = \int_{-\infty}^{\infty} \varphi'(\omega) \ln(|\omega|)\,d\omega$.

Be aware that for distributions with the same singular support, the law of associativity does not hold generically. A typical example is the following: $((\delta(x)\,x)\,\mathcal{P}(\frac{1}{x}) = 0) \neq ((\delta(x)\,(x\,\mathcal{P}(\frac{1}{x}))) = \delta(x))$. As result the convolution theorem does not hold in the following example.

```
In[134]:= g1[t_] := DiracDelta[t] t + 1
          g2[t_] := 1/t
          (* Fourier transform of product *)
          FourierTransform[g1[t] g2[t], t, ω]
```

Out[137]= $\dfrac{1}{\sqrt{2\pi}} + i\,\sqrt{\dfrac{\pi}{2}}\;\mathrm{Sign}[\omega]$

```
In[138]:= f1[ω_] = FourierTransform[g1[t], t, ω];
         f2[ω_] = FourierTransform[g2[t], t, ω];
         (* convolution of Fourier transforms *)
         Integrate[f1[w - ω] f2[w], {w, -Infinity, Infinity},
                   Assumptions -> Im[ω] == 0]/Sqrt[2 Pi]
```

$$Out[141]= \quad i \sqrt{\frac{\pi}{2}} \ \text{Sign}[\omega]$$

The Fourier transform of $\sqrt{2\pi}\ (\delta(\omega) - \frac{i}{\pi\omega})$ is basically the UnitStep function. *Mathematica* returns the result in the form of the Sign function.

```
In[142]:= FourierTransform[Sqrt[2Pi] (DiracDelta[ω] - I/(Pi ω)), ω, t]
```

$$Out[142]= \quad 1 + \text{Sign}[t]$$

Here are two Fourier transforms that result in more singular results (for various possibilities to give sense to such strongly divergent integrals, see [1550]).

```
In[143]:= FourierTransform[t^+3, t, ω]
```

$$Out[143]= \quad i \sqrt{2\pi} \ \text{DiracDelta}^{(3)}[\omega]$$

```
In[144]:= FourierTransform[1/t^2, t, ω]
```

$$Out[144]= \quad -\sqrt{\frac{\pi}{2}} \ \omega \ \text{Sign}[\omega]$$

The inverse Fourier transformation recovers the original function.

```
In[145]:= InverseFourierTransform[%, ω, t] // Expand
```

$$Out[145]= \quad \frac{1}{t^2}$$

On the last transformation, we encountered the singularity $\frac{1}{\omega^2}$. This distribution occurs frequently in Fourier analysis [702], [178]. This singularity is not of Cauchy type. It was treated in the Hadamard sense Pf($\frac{1}{\omega^2}$), which means to discard any integer power and logarithmic singular part via $\langle Pf(f(x)), \varphi \rangle = fp \int_{-\infty}^{\infty} \varphi(x) f(x)\, dx$, where fp indicates the finite part in the Hadamard sense. For the special case Pf($\frac{1}{\omega^2}$), we have

$$\left\langle Pf\left(\frac{1}{\omega^2}\right), \varphi \right\rangle = \lim_{\varepsilon \to 0^+} \left(\int_{\varepsilon}^{\infty} \frac{\varphi(\omega) - \varphi(-\omega) - 2\varphi(0)}{\omega^2} \, d\omega \right).$$

Because the Fourier transform is an isometry in L_1, we can also "backtransform" and even extend the backtransformation to distributions. The inverse Fourier transform $\mathcal{F}_\omega^{-1}[f(\omega)](t)$ of a function $f(\omega)$ is defined as $\mathcal{F}_\omega^{-1}[f(\omega)](t) = (2\pi)^{-1/2} \int_{-\infty}^{\infty} e^{-i\omega t} f(\omega)\, d\omega$.

InverseFourierTransform[$f(\omega)$, ω, t]

    represents the inverse Fourier transform of the function $f(\omega)$ with respect to the variable ω and the kernel $e^{-i\omega t}$.

Interpreting the $\frac{1}{\omega^2}$ from the last output as Pf($\frac{1}{\omega^2}$), the inverse Fourier transformation of $|t|$ recovers $|t|$ (in a slightly rewritten form).

```
In[146]:= FourierTransform[Abs[t], t, ω]
```

$$Out[146]= \quad -\frac{\sqrt{\frac{2}{\pi}}}{\omega^2}$$

```
In[147]:= InverseFourierTransform[%, ω, t]
```

$$Out[147]= \quad t \ \text{Sign}[t]$$

We have the important identity: $\mathcal{F}_\omega^{-1}[\mathcal{F}_t[f(t)](\omega)](t) = f(t)$. Here is an example in which both transformations are explicitly carried out.

In[148]:= `InverseFourierTransform[FourierTransform[Exp[-x^2], x, y], y, x]`

Out[148]= e^{-x^2}

Mathematica also uses this identity for "symbolic" functions.

In[149]:= `InverseFourierTransform[FourierTransform[g[x], x, y], y, x]`

Out[149]= `g[x]`

Be aware that this rule gets applied automatically, independent of the existence of the individual transforms.

In[150]:= `InverseFourierTransform[` (* a function that grows exponentially *)
            `FourierTransform[Exp[x], x, y], y, x]`

Out[150]= e^x

Mathematica (in most cases) assumes that generically every expression is a generic complex number. But, currently, it does not make any assumptions about the function space where, say, `f[x]` is coming from (`f[x]` is just an expression, not a function in *Mathematica*). This means `f'[x]` is always considered a valid expression despite the fact that in a (Wiener-)measure sense, most continuous functions are nowhere differentiable. In a similar way, the convergence of integrals is for symbolic expressions not taken into account, but rather treated in a formal way.

We have the following important relation (Sokhotsky–Plemelj formula) [1272], [945], [444]:

$$\lim_{\varepsilon \to 0^+} \frac{1}{x \pm i\varepsilon} = \mathcal{P}\left(\frac{1}{x}\right) \mp i\pi\,\delta(x).$$

Let us check the relation for an example case.

In[151]:= `Limit[Integrate[Cos[x + 1]/(x + I ε), {x, -Infinity, Infinity},`
            `Assumptions -> Re[ε] > 0 && Im[ε] == 0] //`
            `Expand // PowerExpand // Simplify, ε -> 0] // Expand`

Out[151]= $-i\,\pi\,\text{Cos}[1] - \pi\,\text{Sin}[1]$

The δ integral immediately yields cos(1), and the principal value integral yields $-\pi\sin(1)$.

In[152]:= `Integrate[-I Pi DiracDelta[x] Cos[x + 1], {x, -Infinity, Infinity},`
            `PrincipalValue -> True] +`
            `Integrate[Cos[x + 1]/x, {x, -Infinity, Infinity},`
            `PrincipalValue -> True]`

Out[152]= $-i\,\pi\,\text{Cos}[1] - \pi\,\text{Sin}[1]$

We encountered already two of the options of `FourierTransform` in the discussion of the options of `Integrate` (but be aware that the option setting for the `GenerateConditions` option is now `False`, for `Integrate` the default setting was `True`), but a third one is new.

In[153]:= `Options[FourierTransform]`

Out[153]= `{Assumptions :→ $Assumptions,`
            `GenerateConditions → False, FourierParameters → {0, 1}}`

The meaning of the option `FourierParameters` is best seen by forming the Fourier transform of a constant function and comparing it with the default behavior.

In[154]:= `{FourierTransform[1, x, y],`
            `FourierTransform[1, x, y, FourierParameters -> {a, b}]}`

Out[154]= $\left\{\sqrt{2\pi}\ \text{DiracDelta[y]},\ \dfrac{\sqrt{2\pi}\ \sqrt{(2\pi)^a}\ \text{DiracDelta[y]}}{\sqrt{\text{Abs[b]}}}\right\}$

The setting `FourierTransform[`$f(t)$`, ` t`, ` ω`, FourierParameters -> {`$a,\ b$`}]` represents the integral

$$^{(a,\,b)}\mathcal{F}_t[f(t)]\,(\omega) = \sqrt{\dfrac{|b|}{(2\pi)^{1-a}}}\ \int_{-\infty}^{\infty} e^{ib\omega t}\, f(t)\,dt.$$

Be aware that `FourierTransform` does not have the `PrincipalValue` option that `Integrate` has. $1/x$ is automatically interpreted in the principal value sense. Be also aware that $\mathcal{P}(\frac{1}{x})$ does not exist as a named function in *Mathematica*.

The next input demonstrates the Poincaré–Bertrand identity [1327], [769], [445], [164], [1830]

$$\mathcal{P}\!\left(\frac{1}{x}\right)\mathcal{P}\!\left(\frac{1}{y}\right) = \mathcal{P}\!\left(\frac{1}{x-y}\right)\!\left(\mathcal{P}\!\left(\frac{1}{y}\right) - \mathcal{P}\!\left(\frac{1}{x}\right)\right) + \pi^2\,\delta(x)\,\delta(y)$$

by using the option setting `PrincipalValue -> True` for a constant test function.

In[155]:= `Module[{`(* some abbreviations for a shorter input *)
```
             I = Integrate[##, PrincipalValue -> True]&, A = Assumptions},
      I[1/x, {x, -1, 1}] I[1/y, {y, -1, 1}] ==
      (* calculate integrals with assumptions arising from outer limits *)
      (I[I[1/(x - y) 1/y, {x, -1, 1}, A -> -1 < y < 0], {y, -1, 0}] +
       I[I[1/(x - y) 1/y, {x, -1, 1}, A -> 0 < y < +1], {y, 0, +1}]) -
      (I[I[1/(x - y) 1/x, {y, -1, 1}, A -> -1 < x < 0], {x, -1, 0}] +
       I[I[1/(x - y) 1/x, {y, -1, 1}, A -> 0 < x < +1], {x, 0, +1}]) +
      Pi^2 I[DiracDelta[x] DiracDelta[y], {x, -1, 1}, {y, -1, 1}]]
```
Out[155]= `True`

Applying an operation to a function (such as differentiation, shifting the argument, …) often possesses an equivalent for the Fourier transform of the function. The following input shows differentiation as an example.

In[156]:= `FourierTransform[Derivative[2][x][t], t, ω]`

Out[156]= $-\omega^2$ `FourierTransform[x[t], t, ω]`

But for many other cases `FourierTransform` does not do the corresponding operation for the Fourier-transformed functions. The reason that *Mathematica* does not carry out these operations is the following: Depending on the circumstances, the direction in which the identities should be applied varies. (In addition, one could argue that for an abstract function $f(t)$ the existence of the Fourier transform is not guaranteed; $f(t) \in L^1_{\text{loc}}(\mathbb{R},\,d\mu)$ is needed.) Here are two examples of Fourier transforms that do not get rewritten.

In[157]:= `FourierTransform[x[t + α], t, ω]`

Out[157]= `FourierTransform[x[t + α], t, ω]`

In[158]:= `FourierTransform[Integrate[x[t + τ] y[τ],`
                   `{τ, -Infinity, Infinity}], t, ω]`

Out[158]= `FourierTransform`$\left[\displaystyle\int_{-\infty}^{\infty} \text{x}[t + τ]\ \text{y}[τ]\ dτ,\ t,\ ω\right]$

Be aware of the following consequence of the option setting `GenerateConditions -> False` in `Fourier⁀Transform`: `FourierTransform` could correctly (within the properties that follow from its definition) return 0 when the Fourier transform is not identically zero or not defined mathematically. The latter case could happen do to a Hadamard regularization (exponentially divergent integrals would result in distributions of infinite order) and the former case could happen because not all possible values of the variable will be considered. Here is an example where the default option setting of FourierTransform returns 0. Using Generate Conditions -> True, we get the appropriate restrictions on the independent variable. (For applications of such-type integrals, see [1633].)

In[159]:= `ft = FourierTransform[Sqrt[α^2 - x^2] UnitStep[x + α] UnitStep[α - x], x, y]`

Out[159]= $\dfrac{\sqrt{\frac{\pi}{2}}\ \sqrt{\alpha^2}\ \text{BesselJ}\left[1, \frac{y}{\sqrt{\frac{1}{\alpha^2}}}\right]\text{UnitStep}[2\,\alpha]}{y}$

In[160]:= `(* default option for GenerateConditions, with assumptions, and *)`
```
{FourierTransform[ft, y, x],
 FourierTransform[ft, y, x, Assumptions -> Im[α] == 0 && α/x > 1],
 FourierTransform[ft, y, x, GenerateConditions -> True]}
```

Out[161]= $\left\{0,\ \dfrac{\sqrt{\alpha^2}\ \sqrt{-x^2+\alpha^2}\ \text{UnitStep}[\alpha]}{\text{Abs}[\alpha]},\ \text{If}\left[\dfrac{\alpha^2}{x^2} < 1, 0, \text{FourierTransform}\left[\right.\right.$

$\dfrac{\sqrt{\frac{\pi}{2}}\ \sqrt{\alpha^2}\ \text{BesselJ}\left[1, \frac{y}{\sqrt{\frac{1}{\alpha^2}}}\right]\text{UnitStep}[2\,\alpha]}{y},\ y, x, \text{GenerateConditions} \to \text{True}\left.\left.\left.\right]\right]\right\}$

Another important integral transform implemented in *Mathematica* is the Laplace transform. The Laplace transform $\mathcal{L}_t[f(t)](s)$ of an ordinary function $f(t)$ is defined as $\mathcal{L}_t[f(t)](s) = \int_0^\infty e^{-st} f(t)\,dt$ [1444].

`LaplaceTransform[f(t), t, s]`

represents the Laplace transform of the function $f(t)$ with respect to the variable t and the kernel e^{-st}.

Here is the Laplace transform of an "ordinary" function.

In[162]:= `LaplaceTransform[Exp[-x^2] x^3, x, y]`

Out[162]= $\dfrac{1}{16}\left(2\,(4 + y^2) - e^{\frac{y^2}{4}}\sqrt{\pi}\ y\,(6 + y^2)\,\text{Erfc}\left[\frac{y}{2}\right]\right)$

Here is a more complicated Laplace transform. The result contains Bessel and Struve functions (see Chapter 3).

In[163]:= `LaplaceTransform[1/(1 + s^2)^v, s, t]`

Out[163]= $\dfrac{1}{\text{Gamma}[v]}$

$\left(2^{-\frac{1}{2}-v}\,\pi^{3/2}\,t^{-\frac{1}{2}+v}\left(2\,\text{BesselJ}\left[-\frac{1}{2} + v, t\right]\text{Csc}[2\,\pi\,v] - \text{BesselJ}\left[\frac{1}{2} - v, t\right]\text{Sec}[\pi\,v] +\right.\right.$

$\text{Csc}[\pi\,v]\,\text{StruveH}\left[\frac{1}{2} - v, t\right]\left.\left.\right)\right)$

Like the Fourier transform, the Laplace transform is a linear operation. Here, we use two abstract (meaning no explicit definitions are given) functions $a(x)$ and $b(x)$.

In[164]:= `LaplaceTransform[a[x] + b[x], x, y]`

Out[164]= `LaplaceTransform[a[x], x, y] + LaplaceTransform[b[x], x, y]`

Similar to the Fourier transform, differential operators transform into multiplication operators (plus some boundary terms).

In[165]:= `LaplaceTransform[Ω''[x], x, y]`

Out[165]= y^2 `LaplaceTransform[Ω[x], x, y]` $- y\,\Omega[0] - \Omega'[0]$

Derivative$[-n]$ $(n \in \mathbb{N})$ is interpreted as n-fold integration. This interpretation is in accord with the Laplace transform of an integral corresponding to multiplication of s^{-1} [532], [750].

In[166]:= `LaplaceTransform[Integrate[Ω[ξ], {ξ, 0, x}], x, y]`

Out[166]= $\dfrac{\text{LaplaceTransform}[\Omega[x], x, y]}{y}$

Laplace transforms are understood as right-sided Laplace transforms for distributions. This means that the classical definition is generalized in such a way that the test functions for the linear functional are 1 on the support of the distribution. This definition allows for a nontrivial Laplace transform of the Dirac δ function.

In[167]:= `LaplaceTransform[DiracDelta[t], t, s]`

Out[167]= 1

In[168]:= `InverseLaplaceTransform[%, t, s]`

Out[168]= `DiracDelta[s]`

The Laplace transform can also be inverted [174]. The inverse Fourier transform $\mathcal{L}_s^{-1}[f(s)]\,(t)$ of a function $f(s)$ is defined as $\mathcal{L}_s^{-1}[f(s)]\,(t) = (2\pi i)^{-1} \int_{c-i\infty}^{c+i\infty} e^{st} f(s)\,ds$ for an appropriately chosen c.

InverseLaplaceTransform$[f(s),s,t]$

  represents the inverse Laplace transform of the function $f(s)$ with respect to the variable s and the kernel e^{st}.

Again, we have the property $\mathcal{L}_s^{-1}[\mathcal{L}_t[f(t)]\,(s)]\,(t) = f(t)$.

In[169]:= `InverseLaplaceTransform[1/(1 + s^2 + s^4), s, t]`

Out[169]= $\dfrac{1}{6}\,e^{-t/2}\left(-3\,(-1+e^t)\,\text{Cos}\left[\dfrac{\sqrt{3}\,t}{2}\right] + \sqrt{3}\,(1+e^t)\,\text{Sin}\left[\dfrac{\sqrt{3}\,t}{2}\right]\right)$

In[170]:= `LaplaceTransform[%, t, s] // Simplify`

Out[170]= $\dfrac{1}{1 + s^2 + s^4}$

For a detailed account of calculating direct and inverse Laplace transforms in *Mathematica*, see [760].

Using the Laplace transformation, we can understand some results returned by Sum better. Let us take the following sum [1229]: $\sum_{n=0}^{\infty}(-1)^n\,n!\,z^n$. It obviously has a zero convergence radius. But nevertheless Symbol $\backslash$ icSum`SymbolicSum returns an answer, containing the incomplete Gamma function—to be discussed in Chapter 3.

In[171]:= `SymbolicSum`SymbolicSum[(-1)^n n! z^n, {n, 0, Infinity},`
                              `GenerateConditions -> False]`

Out[171]= $\dfrac{e^{\frac{1}{z}}\,\text{Gamma}[0, \frac{1}{z}]}{z}$

And the answer returned by Sum is sensible [1490]. A divergent sum of the form $\sum_{k=0}^{\infty} c_k\,z^k$ is called Borel summable [1658], [147], [1437], [1384] if $\mathcal{L}_t[\phi(z\,t)]\,(1)$ exists [594], [538], [934], [641], [539], [1020], [285], [997], [1664] ([1383] in case poles are present). Here, $\phi(z)$ is defined by $\phi(z) = \sum_{k=0}^{\infty} c_k/k!\,z^k$. Using these definitions, we get exactly the above result returned by Sum.

In[172]:= `ϕ[z_] = Sum[(-1)^n n!/n! z^n, {n, 0, Infinity}]`

Out[172]= $\dfrac{1}{1 + z}$

In[173]:= `LaplaceTransform[ϕ[z t], t, 1] // Simplify[#, z > 0]&`

Out[173]= $\dfrac{e^{\frac{1}{z}}\,\text{Gamma}[0, \frac{1}{z}]}{z}$

Looking at the Laplace transform of the single terms of the series, it becomes obvious that the trick is effectively to cancel $n!$ terms from the numerators [791].

In[174]:= **LaplaceTransform[Sum[c[k]/k! (z t)^k, {k, 0, 6}], t, 1]**

Out[174]= $c[0] + z\,c[1] + z^2\,c[2] + z^3\,c[3] + z^4\,c[4] + z^5\,c[5] + z^6\,c[6]$

The summed form $e^{1/z}\,\Gamma(0, 1/z)/z$ for the divergent series $\sum_{n=0}^{\infty}(-1)^n\,n!\,z^n$ is actually very sensible. If we calculate the partial sums for small z and truncate at the point where the series starts to grow (excluding the smallest element), we obtain a very close numerical agreement between the sum and its closed form. For $z = 10^{-2}$, we obtain more than 30 agreeing digits!

In[175]:= **With[{z = 10^-2},**
   ListPlot[Log[10, Abs[(* difference *)
       N[(Exp[1/z] Gamma[0, 1/z])/z -
       Rest[FoldList[Plus, 0, Table[(-1)^n n! z^n,
       {n, 0, 270}]]], 22]/(Exp[1/z] Gamma[0, 1/z])/z]]]]

By summing up to a finite upper bound, we see explicitly the "size" of the diverging part.

In[176]:= **Sum[(-1)^n n! z^n, {n, 0, o}] // Expand**

Out[176]= $\dfrac{e^{\frac{1}{z}}\,\text{Gamma}[0, \frac{1}{z}]}{z} + e^{\frac{1}{z}}\left(\dfrac{1}{z}\right)^{1+o}(-z)^o\,\text{Gamma}[2+o]\,\text{Gamma}\!\left[-1-o, \dfrac{1}{z}\right]$

(The inverse operation, meaning multiplying the terms of sum by $k!$, is carried out by the Sumudu transform [1814], [1825], [95].)

Using the function `SymbolicSum`SymbolicSum`, we can also sum a near relative of the last sum [1229].

In[177]:= **SymbolicSum`SymbolicSum[ζ^k k!, {k, Infinity},**
                    GenerateConditions -> False]

Out[177]= $-1 - \dfrac{e^{-1/\zeta}\,\text{Gamma}\!\left[0, -\frac{1}{\zeta}\right]}{\zeta}$

Here is another, slightly more complicated example. This time, the Borel summed form contains a Bessel function (see Chapter 3).

In[178]:= **s2[z_] = SymbolicSum`SymbolicSum[Evaluate[term = z^n (3n)!/(3^n n!^2)],**
                    {n, 0, Infinity},
                    GenerateConditions -> False]

Out[178]= $\dfrac{e^{-\frac{1}{18\,z}}\sqrt{-\frac{1}{z}}\,\text{BesselK}[\frac{1}{6}, -\frac{1}{18\,z}]}{3\sqrt{\pi}}$

In[179]:= **φ[z_] = Sum[Evaluate[term/n!], {n, 0, Infinity}]**

Out[179]= $\text{HypergeometricPFQ}\!\left[\left\{\dfrac{1}{3}, \dfrac{2}{3}\right\}, \{1\}, 9\,z\right]$

In[180]:= **LaplaceTransform[φ[z t], t, 1] // Simplify**

Out[180]= $\dfrac{e^{-\frac{1}{18z}}\,\text{BesselK}[\frac{1}{6},\,-\frac{1}{18z}]}{3\,\sqrt{\pi}\,\sqrt{-z}}$

Borel summation is an important ingredient in the resummation for divergent series arising from perturbation theory. For the Borel treatment of the anharmonic oscillator see, [1071], [1440], [761], and [726]. Here is an example from quantum electrodynamics in external fields [1350], [541], [1759], [1760], [540], [291].

In[181]:= `term[n_] := (2 (-1)^(-1 + n) e^(2n) (-3 + 2n)!)/Pi^(2n)`

In[182]:= `φ[x_] = SymbolicSum`SymbolicSum[Evaluate[term[n] x^n/n!],`
`                    {n, 2, Infinity}, GenerateConditions -> False]`

Out[182]= $-\dfrac{e^4\,x^2\,\text{HypergeometricPFQ}\left[\{1,\,1,\,\frac{3}{2}\},\,\{3\},\,-\frac{4\,e^2\,x}{\pi^2}\right]}{\pi^4}$

In[183]:= `LaplaceTransform[φ[x t], t, 1] /. x -> 1 // PowerExpand`

Out[183]= $-\dfrac{e^2\,(-2\,\text{Cos}[\frac{\pi}{e}]\,\text{CosIntegral}[\frac{\pi}{e}] + \text{Sin}[\frac{\pi}{e}]\,(\pi - 2\,\text{SinIntegral}[\frac{\pi}{e}]))}{\pi^2}$

Sometimes a Borel-summed divergent sum can have opposite sign than the original summands or even have an unexpected imaginary part [1618], [1619], [1214]. Here is a divergent sum of purely positive summands.

$$\sum_{k=0}^{\infty}\left(\prod_{j=0}^{k-1}\left(\frac{1}{2}+j\right)z^k\right)=\sum_{k=0}^{\infty}\left(\frac{1}{2}\right)_k z^k$$

(Here $(a)_k$ is the Pochhammer symbol and formally the last sum can be expressed as the hypergeometric function $_2F_0(1,\,1/2;\,;z)$.) The sum calculated by *Mathematica* contains the special function `Erfi` (which is real-valued for real arguments).

In[184]:= `SymbolicSum`SymbolicSum[Product[1/2 + j, {j, 0, k - 1}] z^k,`
`                    {k, 0, Infinity}, GenerateConditions -> False] //`
`                                Simplify[#, z > 0]&`

Out[184]= $\dfrac{e^{-1/z}\,\sqrt{\pi}\,\left(i + \text{Erfi}\left[\frac{1}{\sqrt{z}}\right]\right)}{\sqrt{z}}$

Such "spurious" contributions are often unavoidable. They arise when the modified sum (made convergent with the additional $1/k!$ term) has branch cuts along the integration path of the Laplace transform to be carried out [1620], [661]. This is the case for the current example.

In[185]:= `φ[z_] = Sum[Pochhammer[1/2, k] z^k/k!, {k, 0, Infinity}]`

Out[185]= $\dfrac{1}{\sqrt{1-z}}$

In[186]:= `LaplaceTransform[φ[z t], t, 1] // Simplify[#, z > 0]&`

Out[186]= $\dfrac{e^{-1/z}\,\sqrt{\pi}\,\left(-i + \text{Erfi}\left[\frac{1}{\sqrt{z}}\right]\right)}{\sqrt{z}}$

Ignoring the imaginary contribution shows that the sum agrees favorably with a numerical summation of the divergent sum that takes into account all decreasing in absolute value summands [1664]. For $z = 0.005$ the two results agree to about 87 digits. (For an optimal truncation, one should sum until the smallest element occurs, but not taking the smallest element itself into account.) (By subtracting an appropriate integral term, one can achieve a good agreement for all values of z [1464].)

In[187]:= `sumUntilGrowing[summand_, {k_, k0_, k1_}] :=`
`        Block[{sum = 0, oldSummand = Infinity, k = k0},`
`              While[newSummand = summand;`
`                    Abs[newSummand] < Abs[oldSummand],`

```
                    oldSummand = newSummand;
                    sum = sum + newSummand; k++];
            sum]
```

In[188]:= `With[{z = N[5/1000, 100]},`
`        Abs[# - sumUntilGrowing[Pochhammer[1/2, k] z^k, {k, 0, Infinity}]]/#&[`
`        Sqrt[Pi] Exp[-1/z] Erfi[1/Sqrt[z]]/Sqrt[z]]]`

Out[188]= $1.6266184768 \times 10^{-87}$

A direct call to the function $_2F_0$ too returns an imaginary part.

In[189]:= `HypergeometricPFQ[{1, 1/2}, {}, 1/2.]`

Out[189]= $1.27998 + 0.339235 \, i$

To make sure a sum is summed in Borel sense, one can use `SymbolicSum`SymbolicSum[body, iterator,`
`GenerateConditions -> False]`.

We could now go on and use `Sum` (and `Series` or `Limit`) to invert a Laplace transform via the following
formula [1840], [1743], but we end here.

$$\mathcal{L}_s^{-1}[f(s)](t) = \lim_{t \to \infty} t \sum_{k=1}^{\infty} \frac{(-1)^{k-1}}{(k-1)!} e^{kst} f(kt).$$

The Laplace and the inverse Laplace transforms are frequently used to solve linear differential equations. In the
following, we will use the Laplace and the inverse Laplace transform to find approximate solutions to nonlinear
initial value problems [1019], [1285]. To linearize the nonlinear problem, we use the Adomian decomposition
[17], [680], [1545], [1137].

Mathematical Remark: Adomian decomposition

Given a nonlinear function of appropriate smoothness $f(x)$ and making the ansatz $x = \sum_{k=0}^{\infty} \lambda^k \xi_k$ yields the
following expansion $f(x) = \sum_{k=0}^{\infty} A_k(\xi_0, \ldots, \xi_{k-1})$ where the Adomian polynomials $A_k(\xi_0, \ldots, \xi_{k-1})$ are

$$A_k(\xi_0, \ldots, \xi_{k-1}) = \frac{1}{k!} \frac{d^k}{d\lambda^k} f\left(\sum_{k=0}^{\infty} \lambda^k \xi_k\right)\bigg|_{\lambda=0}.$$

The function `AdomianPolynomials` calculates the first o Adomian polynomials [1], [4], [356]. The usefulness of the Adomian decomposition stems from the fact that Adomian polynomial $A_k(\xi_0, \ldots, \xi_{k-1})$ is a linear
function in ξ_{k-1}.

In[190]:= `AdomianPolynomials[f_, ξ_, o_] :=`
`        Module[{x, λ}, CoefficientList[Series[f[Sum[λ^k x[k],`
`                    {k, 0, o}]], {λ, 0, o}], λ] /. x -> ξ]`

In[191]:= `Clear[f, f, x, y, η, ξ, a, b, s];`
`        AdomianPolynomials[f, x, 3] /. x[k_] :> Subscript[x, k]`

Out[192]= $\left\{f[x_0], \; x_1 f'[x_0], \; x_2 f'[x_0] + \frac{1}{2} x_1^2 f''[x_0], \; x_3 f'[x_0] + x_1 x_2 f''[x_0] + \frac{1}{6} x_1^3 f^{(3)}[x_0]\right\}$

We consider the second-order nonlinear initial value problem $y''(x) + a(x) y'(x) + b(x) y(x) = f(y(x))$ with initial
conditions $y(0) = y_0$, $y'(0) = y_1$. Laplace transforming the differential equation yields the following equation.

In[193]:= `Solve[LaplaceTransform[y''[x] + a[x] y'[x] + b[x] y[x] -`
                     `f[y[x]], x, s] == 0,`
              `LaplaceTransform[y[x], x, s]]`

Out[193]= $\{\{$LaplaceTransform$[$y$[$x$]$, x, s$] \to$

   $\frac{1}{s^2}$ (LaplaceTransform$[$f$[$y$[$x$]]$, x, s$] -$ LaplaceTransform$[$b$[$x$]$ y$[$x$]$, x, s$] -$

   LaplaceTransform$[$a$[$x$]$ y$'[$x$]$, x, s$] + $ s y$[0] + $ y$'[0])\}\}$

Expanding now $y(x)$ in an Adomian way $y(x) = \sum_{k=0}^{\infty} \eta_k(x)$, the following iteration procedure arises naturally.

$$\mathcal{L}_x[\eta_k(x)](s) = -\frac{1}{s^2} \left(\mathcal{L}_x[a(x) \eta'_{k-1}(x)](s) + \mathcal{L}_x[b(x) \eta_{k-1}(x)](s) + \mathcal{L}_x[A_{k-1}(\eta_0(x), \ldots, \eta_{k-2}(x))](s)\right)$$

Let us calculate the first few $\eta_k(x)$ for the differential equation $y''(x) + \cos(x) y(x) = y(x)^3$ under the initial conditions $y(0) = 1$, $y'(0) = 0$. A direct implementation of the last formula allows calculating the $\eta_k(x)$ through inverse Laplace transform $\mathcal{L}_s^{-1}[\mathcal{L}_x[\eta_k(x)](s)](x)$.

In[194]:= `a[x_] := 0; b[x_] := Cos[x]; f[y_] := y^3;`

In[195]:= `η[0][x_] = 1;`
         `η[n_][x_] := η[n][x] = InverseLaplaceTransform[`
             `Together[-s^-2 TrigExpand[LaplaceTransform[`
             (* presimplify integrand for faster transform *)
             `Expand[TrigToExp[a[x] D[η[n - 1][x], x] + b[x] η[n - 1][x] -`
             `AdomianPolynomials[f, H, n][[n]] /. H -> (η[#][x]&)]], x, s]]],`
                                                       `s, x] // Expand`

Here are the first three partial sums of the $\eta_k(x)$ shown.

In[197]:= `yApproxList[x_] = Rest[FoldList[Plus, 0, Table[η[k][x], {k, 0, 5}]]];`
         `Take[yApproxList[x], 3]`

Out[198]= $\left\{1, \frac{x^2}{2} + \cos[x], \frac{55}{8} - \frac{5 x^2}{4} + \frac{x^4}{8} - 6 \cos[x] + \frac{1}{2} x^2 \cos[x] + \frac{1}{8} \cos[2 x] - 2 x \sin[x]\right\}$

We compare the approximate solutions with a high-precision numerical solution `ndsol`. The following graphics show, that with each $\eta_k(x)$ the solution becomes substantially better and the fifth approximation has an error less than 10^{-10} for $0 \le x \le 0.8$.

In[199]:= (* high-precision numerical solution *)
         `ndsol = NDSolve[{yN''[x] + a[x] yN'[x] + b[x] yN[x] == f[yN[x]],`
                 `yN[0] == 1, yN'[0] == 0}, yN, {x, 0, 5/2},`
                 `WorkingPrecision -> 50, MaxSteps -> 10^5,`
                 `PrecisionGoal -> 30, AccuracyGoal -> 30];`

In[201]:= `Show[GraphicsArray[`
         `Block[{$DisplayFunction = Identity,`
                 (* order increases from red to blue *)
                 `t = Table[Hue[k/7], {k, 0, 5}]},`
             {(* absolute differences *)
             `Plot[Evaluate[Join[yApproxList[x], yN[x] /. ndsol]], {x, 0, 5/2},`
                 `PlotRange -> All, PlotStyle -> Prepend[t, GrayLevel[0]]],`
             (* logarithms of the differences *)
             `MapIndexed[(δN[#2[[1]]][x_?NumberQ] :=`
                 `Log[10, Abs[SetPrecision[#1, 60] - yN[SetPrecision[x, 60]] /.`
                                         `ndsol[[1]]]])&, yApproxList[x]];`
             (* show logarithms of the differences *)
             `Plot[Evaluate[Table[δN[k][x], {k, 6}]], {x, 0, 5/2},`
                 `PlotRange -> {All, {-10, 2}}, PlotStyle -> t]}]]]`

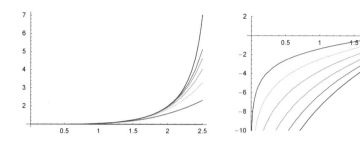

For the application of the Adomian decomposition to boundary value problems, see [455], [1816].

1.9 Additional Symbolics Functions

Now, we are nearly at the end of our chapter about symbolic computations. Many features of *Mathematica* have been discussed, but as many have not been discussed. The next section will deal with some applications of the discussed functions. In addition to the functionality built into the *Mathematica* kernel, a number of important packages in the standard package directory of *Mathematica* are useful for symbolic calculations, and they enhance the power of the corresponding built-in functions and offer new functionality. In addition to Calculus`Limit`, Calculus`PDSolve1`, and Calculus`DSolve`, which were already mentioned above, the following packages are often very useful: Calculus`VectorAnalysis`, DiscreteMath` RSolve`, and Calculus`VariationalMethods`. The functions contained in these packages can be deduced immediately from their names. Because of space and time limitations, we look only briefly at what these packages can accomplish.

The package Calculus`VariationalMethods` implements the calculation of variational derivatives of integrals and the associated Euler-Lagrange equation (for an introduction to variational calculations, see, e.g., [240], [664], or for somewhat more detail, see [439] and [1806]).

In[1]:= **Needs ["Calculus`VariationalMethods`"]**

In[2]:= **?VariationalD**

> VariationalD[f, u[x], x] or VariationalD[f, u[x,y,...]],
> {x,y,...}] returns the first variational derivative of the
> functional defined by the integrand f which is a function
> of u, its derivatives, and x,y,.... VariationalD[f, {u[x,
> y,...], v[x,y,...],...}, {x,y,...}] gives a list of first
> variational derivatives with respect to u, v, ... More...

Here is a well-known example.

In[3]:= **VariationalD[Sqrt[1 - y'[x]^2], y[x], x]**

Out[3]= $\dfrac{y''[x]}{\left(1 - y'[x]^2\right)^{3/2}}$

Here are the Newton equations of motion for a 2D anharmonic oscillator.

In[4]:= **EulerEquations[m/2 (x'[t]^2 + y'[t]^2) + (x[t]^4 + x[t] y[t] + y[t]^4),**
{x[t], y[t]}, t]

Out[4]= $\{4\, x[t]^3 + y[t] - m\, x''[t] == 0,\ x[t] + 4\, y[t]^3 - m\, y''[t] == 0\}$

The next input gives the equation of motion for a damped nonlinear 1D oscillator [1800].

In[5]:= **EulerEquations[Exp[μ t] (x'[t]^2/2 + a/2 x[t]^2 + b/(n + 1) x[t]^n),**
    {x[t]}, t]

Out[5]= $\left\{ e^{t\,\mu} \left(a\, x[t] + \dfrac{b\, n\, x[t]^{-1+n}}{1+n} - \mu\, x'[t] - x''[t] \right) == 0 \right\}$

Another sometimes useful functionality is provided by the package `DiscreteMath`RSolve`` is the possibility to calculate symbolic terms of power series for the nth term. The function `SeriesTerm` does this. Here is a simple example—the nth term of $\sin(x^2)$ around $x = 0$.

In[6]:= **Needs["DiscreteMath`RSolve`"]**

In[7]:= **?SeriesTerm**

> SeriesTerm[f, {z, z0, n}, opts] gives the coefficient of the term (z-z0)^
> n in the Laurent series expansion for f about the point z = z0. **More...**

In[8]:= **SeriesTerm[Sin[x^2], {x, 0, n}]**

Out[8]= $-\dfrac{i\,(-1)^{n/4}\,\text{KroneckerDelta}[\text{Mod}[-2+n, 4]]\,\text{UnitStep}[-2+n]}{\text{Gamma}[1+\frac{n}{2}]}$

Here, we use the last result and compare it with the explicit terms returned by `Series`.

In[9]:= **Plus @@ MapIndexed[#1 x^(#2[[1]] - 1)&, Table[%, {n, 0, 36}]]**

Out[9]= $x^2 - \dfrac{x^6}{6} + \dfrac{x^{10}}{120} - \dfrac{x^{14}}{5040} + \dfrac{x^{18}}{362880} - \dfrac{x^{22}}{39916800} +$
    $\dfrac{x^{26}}{6227020800} - \dfrac{x^{30}}{1307674368000} + \dfrac{x^{34}}{355687428096000}$

In[10]:= **Series[Sin[x^2], {x, 0, 36}] // Normal**

Out[10]= $x^2 - \dfrac{x^6}{6} + \dfrac{x^{10}}{120} - \dfrac{x^{14}}{5040} + \dfrac{x^{18}}{362880} - \dfrac{x^{22}}{39916800} +$
    $\dfrac{x^{26}}{6227020800} - \dfrac{x^{30}}{1307674368000} + \dfrac{x^{34}}{355687428096000}$

Let us have a look at the integral $\int_0^\infty e^{-x^2}(1-\alpha x)^{-2}\,dx$. Assuming that $\alpha < 0$, the integrand is free of poles, and we can directly carry out the integral.

In[11]:= **int = Integrate[Exp[-x^2]/(1 - α x)^2, {x, 0, Infinity},**
    Assumptions -> α < 0]

Out[11]= $-\dfrac{e^{-\frac{1}{\alpha^2}}\left(e^{\frac{1}{\alpha^2}}\sqrt{\pi}\,\alpha + e^{\frac{1}{\alpha^2}}\alpha^2 - \pi\,\text{Erfi}[\frac{1}{\alpha}] + \text{Gamma}[0, -\frac{1}{\alpha^2}] + \text{Log}[-\frac{1}{\alpha^2}] + 2\,\text{Log}[-\alpha] \right)}{\alpha^3}$

Expanding the denominator in a series around $x = 0$ results in the following series term.

In[12]:= **SeriesTerm[1/(1 - α x)^2, {x, 0, k}]**

Out[12]= $(1 + k)\,\alpha^k$

For $\alpha x > 1$ the series is divergent (corresponding to the singularity of $1/(1 - \alpha x^2)$ at $x = 1/\alpha$).

Exchanging summation and integration yields a divergent sum. But due to the automatic Borel summation of `SymbolicSum`SymbolicSum` for such type sums, we get the a closed-form result as for the integral.

$$\int_0^\infty \frac{e^{-x^2}}{(1 - \alpha x)^2}\, dx \doteq \int_0^\infty e^{-x^2}\left(\sum_{k=0}^\infty (1 + k)\, \alpha^k\, x^k\right) dx \doteq$$

$$\sum_{k=0}^\infty (1 + k)\, \alpha^k \left(\int_0^\infty e^{-x^2}\, x^k\, dx\right) \doteq \sum_{k=0}^\infty (1 + k)\, \alpha^k\, \frac{\Gamma((1 + k)/2)}{2}$$

In[13]:= `Integrate[x^k Exp[-x^2], {x, 0, Infinity}, Assumptions -> k >= 0]`

Out[13]= $\frac{1}{2}$ `Gamma`$\left[\frac{1 + k}{2}\right]$

In[14]:= `sum = SymbolicSum`SymbolicSum[α^k (1 + k) Gamma[(1 + k)/2]/2,`
            `{k, 0, Infinity},`
            `GenerateConditions -> False]`

Out[14]= $\frac{1}{2}\left(-\dfrac{2\, e^{-\frac{1}{\alpha^2}}\left(e^{\frac{1}{\alpha^2}}\sqrt{\pi} - \pi\sqrt{-\frac{1}{\alpha^2}} + e^{\frac{1}{\alpha^2}}\alpha - \pi\sqrt{\frac{1}{\alpha^2}}\,\text{Erfi}\!\left[\sqrt{\frac{1}{\alpha^2}}\right]\right)}{\alpha^2} - \dfrac{2\, e^{-\frac{1}{\alpha^2}}\,\text{Gamma}[0, -\frac{1}{\alpha^2}]}{\alpha^3}\right)$

Using the functions `FullSimplify` we can show that the sum and the integral are identical. (`FullSim`ⁱ
`plify` simplifies identities with special functions, we will discuss it in Chapter 3.)

In[15]:= `FullSimplify[int - sum, α < 0]`

Out[15]= 0

Here is another example. We first sum a series [734] and then recover the *n*th term.

In[16]:= `Sum[(ℱ[x] - ℱ[y])^n/(x - y)^(n + 1) λ^n, {n, Infinity}] // Simplify`

Out[16]= $\dfrac{\lambda\,(\mathcal{F}[x] - \mathcal{F}[y])}{(x - y)\,(x - y - \lambda\,\mathcal{F}[x] + \lambda\,\mathcal{F}[y])}$

In[17]:= `SeriesTerm[%, {λ, 0, n}] // Simplify[#, n > 1]&`

Out[17]= $\dfrac{\left(\frac{\mathcal{F}[x] - \mathcal{F}[y]}{x - y}\right)^n}{x - y}$

As a small application of the function `SeriesTerm`, we will prove the following identity (due to Ramanujan)
about the Taylor series coefficients of three rational functions [856], [570].

$$\left([x^k]\left(\frac{9\,x^2 + 53\,x + 1}{x^3 - 82\,x^2 - 82\,x + 1}\right)\right)^3 + \left([x^k]\left(\frac{-12\,x^2 - 26\,x + 2}{x^3 - 82\,x^2 - 82\,x + 1}\right)\right)^3 =$$

$$\left([x^k]\left(\frac{-10\,x^2 + 8\,x + 2}{x^3 - 82\,x^2 - 82\,x + 1}\right)\right)^3 + (-1)^k$$

Using the function `Series`, we can easily explicitly verify the identity for the first few coefficients.

In[18]:= `abc = {1 + 53 x + 9 x^2, 2 - 26 x - 12 x^2, 2 + 8 x - 10 x^2}/`
            `(1 - 82 x - 82 x^2 + x^3);`

         `(#1 + #2 - #3)& @@@ Transpose[#[[3]]^3& /@ Series[abc, {x, 0, 12}]]`

Out[20]= `{1, -1, 1, -1, 1, -1, 1, -1, 1, -1, 1, -1, 1}`

`SeriesTerm[f, {z, z₀, k}]` gives us a closed form for the *k*th term of the Taylor series of the function *f*
expanded around $z = z_0$.

In[21]:= `{aS[k_], bS[k_], cS[k_]} = SeriesTerm[abc, {x, 0, k}] // Simplify`

Out[21]= $\{\dfrac{1}{85\,(6887+747\,\sqrt{85}\,)}\,\Big((83+9\,\sqrt{85}\,)^{-1-k}$

$\qquad \Big(2^{5-k}\,\big(-4^{k}\,(348508+37801\,\sqrt{85}\,)+(83+9\,\sqrt{85}\,)^{2\,k}\,(4920812+533737\,\sqrt{85}\,)\big)-$

$\qquad 172\,(83+9\,\sqrt{85}\,)^{k}\,(285769+30996\,\sqrt{85}\,)\,e^{i\,k\,\pi}\big)\Big),$

$\dfrac{1}{85\,(6887+747\,\sqrt{85}\,)}\,\big(2^{2-k}\,(83+9\,\sqrt{85}\,)^{-1-k}$

$\qquad \big(7\,\big(4^{k}\,(508799+55187\,\sqrt{85}\,)+(83+9\,\sqrt{85}\,)^{2\,k}\,(5778119+626725\,\sqrt{85}\,)\big)+$

$\qquad 2^{4+k}\,(83+9\,\sqrt{85}\,)^{k}\,(285769+30996\,\sqrt{85}\,)\,e^{i\,k\,\pi}\big)\big),$

$\dfrac{1}{85\,(6887+747\,\sqrt{85}\,)}\,\big(2^{2-k}\,(83+9\,\sqrt{85}\,)^{-1-k}\,\big(3\,4^{k}\,(954859+103569\,\sqrt{85}\,)+(83+9\,\sqrt{85}\,)^{2\,k}$

$\qquad (50288457+5454549\,\sqrt{85}\,)-2^{4+k}\,(83+9\,\sqrt{85}\,)^{k}\,(285769+30996\,\sqrt{85}\,)\,e^{i\,k\,\pi}\big)\big)\}$

Simplifying the resulting expressions proves the above identity.

In[22]:= `aS[k]^3 + bS[k]^3 - cS[k]^3 // RootReduce //`
`        Simplify[#, Element[k, Integers] && k > 0]&`

Out[22]= $(-1)^{k}$

We end with another application of the series terms also due to Ramanujan: Calculating integrals through series terms. For a sufficiently nice function $f(x)$, the kth moment $\mu_k[f(x)] = \int_0^\infty x^k\,f(x)\,dx$ can be calculated through the analytic continuation of the series coefficient $c(k) = [x^k]\,(f(x))$ to negative integer k by $\mu_k = -(-1)^{-k}\,k!\,(-k-1)!\,c_{-k-1}$ (Ramanujan's master theorem [157]).

Here is a simple example.

In[23]:= `f[x_] = x^2 Exp[-x] Sin[x]^2;`

In[24]:= `c[k_] = SeriesTerm[f[x], {x, 0, k}];`
`        intc[k_] = k! (-1)^(-k - 1) (-k - 1)! c[-k - 1]`

Out[25]= $\dfrac{1}{2\,\pi}\,\Big((-1)^{-4-2\,k}\,\big(-1+5^{\frac{1}{2}\,(-3-k)}\,\mathrm{Cos}[(-3-k)\,\mathrm{ArcTan}[2]]\big)$

$\qquad (-1-k)!\,k!\,\mathrm{Gamma}[3+k]\,\mathrm{Sin}[(-3-k)\,\pi]\Big)$

This is the result of the direct integration.

In[26]:= `intI[k_] = Integrate[x^k f[x], {x, 0, Infinity}, Assumptions -> k > 0]`

Out[26]= $\dfrac{1}{2}\,\big(1-5^{-\frac{3}{2}-\frac{k}{2}}\,\mathrm{Cos}[(3+k)\,\mathrm{ArcTan}[2]]\big)\,\mathrm{Gamma}[3+k]$

For negative integer k, `intc[k]` is indeterminate. For concrete k we could use `Limit` or `Series` to obtain a value. For generic k, we simplify first the Gamma functions using `FullSimplify`.

In[27]:= `intI[k]/intc[k] // FullSimplify // Simplify[#, Element[k, Integers]]&`

Out[27]= 1

(For calculating series terms of arbitrary order, see [1040], [1041], and [1043].)

1.10 Three Applications

1.10.0 Remarks

In this section, we will discuss three larger calculations. Here, "larger" mainly refers to the necessary amount of operations to calculate the result and not so much to the number of lines of *Mathematica* programs to carry it out. The first two are "classical" problems. Historically, the first one was solved in an ingenious method. Here we will implement a straightforward calculation. Carrying out the calculation of an extension of the second one $(\cos(2\pi/65537))$ took more than 10 years at the end of the nineteenth century. The third problem is a natural continuation from the visualizations discussed in Section 3.3 of the Graphics volume [1736]. The code is adapted to *Mathematica* Version 5.1. As mentioned in the Introduction, later versions of *Mathematica* may allow for a shorter implementation and more efficient implementation.

1.10.1 Area of a Random Triangle in a Square

In the middle of the last century, J. J. Sylvester proposed calculating the expectation value of the convex hull of n randomly chosen points in a plane square. For $n = 1$, the problem is trivial, and for $n = 2$, the question is relatively easy to answer. For $n \geq 3$, the straightforward formulation of the problem turns out to be technically quite difficult because of the multiple integrals to be evaluated. In 1885, M. W. Crofton came up with an ingenious trick to solve special cases of this problem. (His formulae are today called Crofton's theorem.) At the same time, he remarked:

The intricacy and difficulty to be encountered in dealing with such multiple integrals and their limits is so great that little success could be expected in attacking such questions directly by this method [direct integration]; and most of what has been done in the matter consists in turning the difficulty by various considerations, and arriving at the result by evading or simplifying the integration. [1031]

The general setting of the problem is to calculate the expectation value of the $\min(n-1, d)$-dimensional volume of the convex hull of n points in d dimensions, for instance, the volume of a random tetrahedron formed by four randomly chosen points in $\mathbb{R}^3$. For details about what is known, the Crofton theorem and related matters, see [35], [262], [562], [263], [1217], [832], [1031], [1238], [277], and [1410]. For an ingenious elementary derivation for the $n = 3$ case, see [1596]; for a tetrahedron in a cube, see [1910]. For the case of a tetrahedron inside a tetrahedron, see [1196].)

In this subsection, we will show that using the integration capabilities of *Mathematica* it is possible to tackle such problems directly—this means by carrying out the integrations. (This subsection is based on [1733].)

In the following, let the plane polygon be a unit square. We will calculate the expectation value of the area of a random triangle within this unit square (by an affine coordinate transformation, the problem in an arbitrary convex quadrilateral can be reduced to this case).

Here is a sketch of the situation.

```
In[1]:= With[{P1 = {0.2, 0.3}, P2 = {0.8, 0.2}, P3 = {0.4, 0.78}},
         Show[Graphics[
              {{Thickness[0.01],
                Line[{{0, 0}, {1, 0}, {1, 1}, {0, 1}, {0, 0}}]},
               {Thickness[0.002], Hue[0], Line[{P1, P2, P3, P1}]}},
```

```
{Text["P1", {0.16, 0.26}], Text["P2", {0.84, 0.16}],
    Text["P3", {0.40, 0.82}]}}],
AspectRatio -> Automatic]]
```

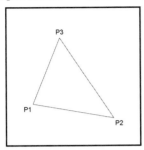

Let $\{x_1, y_1\}$, $\{x_2, y_2\}$, and $\{x_3, y_3\}$ be the vertices of the triangle. Then, the area A_T of the triangle is given by the absolute value of the determinant:

$$A_T = \frac{1}{2} \begin{Vmatrix} x_1 & y_1 & 1 \\ x_2 & y_2 & 1 \\ x_3 & y_3 & 1 \end{Vmatrix} = \frac{1}{2} |x_3 y_1 - x_2 y_1 + x_1 y_2 - x_3 y_2 - x_1 y_3 + x_2 y_3|.$$

This means the integral to be calculated is

$$A = \frac{1}{2} \int_0^1 \int_0^1 \int_0^1 \int_0^1 \int_0^1 \int_0^1 |x_3 y_1 - x_2 y_1 + x_1 y_2 - x_3 y_2 + x_2 y_3 - x_1 y_3| \, dy_3 \, dx_3 \, dy_2 \, dx_2 \, dy_1 \, dx_1.$$

Depending on the orientation of the triangle, the determinant changes sign, and there are two obvious possibilities to take care of this. The first would be to use $((x_3 y_1 + x_1 y_2 - x_3 y_2 - x_1 y_3 + x_2 y_3 - x_2 y_1)^2)^{1/2}$ as the integrand, and the second would be to divide the six-dimensional integration region into such subregions, so that inside each the expression $x_3 y_1 + x_1 y_2 - x_3 y_2 - x_1 y_3 + x_2 y_3 - x_2 y_1$ does not change sign.

We will follow this second approach here, because it makes interesting use of various *Mathematica* functions.

A subdivision of the six-dimensional unit cube into regions where the expression $x_3 y_1 - x_2 y_1 + \cdots$ has a constant sign can be most easily obtained using the cylindrical algebraic decomposition. Because for integration purposes we are only interested in subregions that have a nonvanishing six-dimensional Lebesgue measure, we use the function GenericCylindricalAlgebraicDecomposition. Here is the cylindrical algebraic description of the areas where the area of the triangle is positive; it can be calculated quickly.

In[2]:= **AppendTo[$ContextPath, "Experimental`"];**

In[3]:= **signedTriangleArea = -x2 y1 + x3 y1 + x1 y2 - x3 y2 - x1 y3 + x2 y3 > 0;**
unitCube6D = 0 < x1 < 1 && 0 < y1 < 1 && 0 < x2 < 1 && 0 < y2 < 1 &&
0 < x3 < 1 && 0 < y3 < 1

Out[4]= 0 < x1 < 1 && 0 < y1 < 1 && 0 < x2 < 1 && 0 < y2 < 1 && 0 < x3 < 1 && 0 < y3 < 1

In[5]:= **Timing[cad = GenericCylindricalAlgebraicDecomposition[**
(* the triangle area *) signedTriangleArea &&
(* the six-dimensional unit cube *) unitCube6D,
{x1, y1, x2, y2, x3, y3}];]

Out[5]= {0.61 Second, Null}

We convert the resulting nested expression to a list of single six-dimensional regions.

In[6]:= `l1 = cad[[1]] //. a_ && (b_ || c_) :> a && b || a && c;`

This is one of the resulting regions.

In[7]:= `l1[[99]]`

Out[7]= $0 < x1 < \dfrac{1}{2}$ && $1 - x1 < y1 < 1$ && $x1 < x2 < \dfrac{x1}{y1}$ && $\dfrac{-y1 + x2\,y1}{-1 + x1} < y2 < \dfrac{x1 - x2 + x2\,y1}{x1}$ &&

$\dfrac{x1 - x2 + x2\,y1 - x1\,y2}{y1 - y2} < x3 < 1$ && $\dfrac{x2\,y1 - x3\,y1 - x1\,y2 + x3\,y2}{-x1 + x2} < y3 < 1$

Because we want to integrate the area function over the regions, we will convert from $\xi_l < \xi < \xi_u$ to $\{\xi_l, \xi, \xi_u\}$, a form that can be easily used inside `Integrate` as the integration regions.

In[8]:= `regions = Apply[List, Apply[{#3, #1, #5} &, l1, {2}], {0, 2}];`

We have 216 individual regions.

In[9]:= `Length[regions]`

Out[9]= `216`

There is no need to calculate the regions where the area of the triangle has a negative sign. The resulting integrals will give exactly the same contribution by symmetry of the problem. Here is the resulting description of one of the regions.

In[10]:= `regions[[111]]`

Out[10]= $\left\{\left\{x1, \dfrac{1}{2}, 1\right\}, \{y1, 0, 1 - x1\}, \left\{x2, 0, \dfrac{-x1 + y1}{-1 + y1}\right\}, \left\{y2, \dfrac{-y1 + x2\,y1}{-1 + x1}, \dfrac{x1 - x2 + x2\,y1}{x1}\right\},\right.$

$\left.\left\{x3, 0, \dfrac{x2\,y1 - x1\,y2}{y1 - y2}\right\}, \left\{y3, 0, \dfrac{x2\,y1 - x3\,y1 - x1\,y2 + x3\,y2}{-x1 + x2}\right\}\right\}$

Now, inside each of the 216 six-dimensional regions, we have to integrate the area as a function of x_1, y_1, x_2, y_2, x_3, y_3. The area is only a multilinear polynomial in each of the six integration variables. Because the limits of the integrations are rational functions of the outer integration variables, the integrals we deal with will be typically quite complicated and contain rational functions, logarithms, and polylogarithms in the result after carrying out some integrations.

We will not use *Mathematica*'s definite integration here. As mentioned the following in Subsection 1.6.2: If a definite integral contains parameters, *Mathematica* tries to determine if the integral exists for all parameters and if the integrand contains singular points. Because the integrals involved possess up to five parameters (the outer integration variables), this turns out to be quite a complicated integration. *Mathematica* can carry out the integrations, but some for some regions, it will be a time-consuming task. Here is a typical result for one of the regions. The result contains the transcendental numbers π and $\ln(2)$.

In[11]:= `area = Det[{{x1, y1, 1}, {x2, y2, 1}, {x3, y3, 1}}]/2;`

In[12]:= `Integrate[area, Sequence @@ regions[[2]]]`

Out[12]= $\dfrac{1}{288}\,(19 - 5\,\pi^2 + 23\,\mathrm{Log}[2] + 30\,\mathrm{Log}[2]^2)$

On the other hand, by the construction of the integration regions and by the nature of the area function, we do not have to worry about any of these problems—inside each of the integration regions the integrand is a smooth, not sign-changing function. So, we use the much faster indefinite integration capabilities of *Mathematica* and calculate the definite integrals from the indefinite ones. If we are lucky, we could just substitute the integration limits into the indefinite integral to obtain the definite integral. But sometimes this might give indeterminate expressions. In such a situation, we must be more careful and must calculate the needed limits by using a series expansion around the integration endpoints.

Now, let us implement the integration described above: The function `multiDimensionalIntegrate` is our own multidimensional, definite integration function. It recursively calls the function `fastIntegrate`, defined below.

```
In[13]:= multiDimensionalIntegrate[int_, iters__, iter_] :=
         multiDimensionalIntegrate[fastIntegrate[int, iter], iters];

         multiDimensionalIntegrate[int_, iter_] := fastIntegrate[int, iter]
```

The function `fastIntegrate` is our 1D integrator. It carries out the indefinite integral first. We use `Limit` to determine the values of the integral at the integration limits. This will take care of potentially indeterminate expressions that would arise from blind substitution of the limit points into the indefinite integral.

```
In[15]:= fastIntegrate[integrand_, {x_, l_, u_}] :=
         Module[{indefiniteIntegral, lValue, uValue},
             (* the indefinite integral *)
             indefiniteIntegral = Integrate[LogExpand[integrand //. x -> ξ], ξ];
             (* the lower limit *)
             lValue = Limit[indefiniteIntegral, ξ -> l, Direction -> -1];
             (* the upper limit *)
             uValue = Limit[indefiniteIntegral, ξ -> u, Direction -> +1];
             Factor[Together[uValue - lValue]]]
```

To speed up the indefinite integration and the calculation of the limits, we apply some transformation rules implemented in `LogExpand` to the expressions. `LogExpand` splits all `Log[expr]` into as many subparts as possible to simplify the integrands. Because we know that the integrals we are dealing with are real quantities, we do not have to worry about branch cut problems associated with the logarithm function, and so drop all imaginary parts at the end.

```
In[16]:= LogExpand[expr_] := PowerExpand //@ Together //@ expr
```

Now, we have all functions together and can actually carry out the integration. To get an idea about the form of the expressions appearing in the six integrations, let us have a look at the individual integration results of the first region. (The indefinite integrals are typically quite a bit larger than the definite ones, as shown in the following results.)

This is the description of the first six-dimensional region.

```
In[17]:= regions[[1]]
```

$$\text{Out[17]}= \left\{ \left\{ x1, 0, \frac{1}{2} \right\}, \{y1, 0, x1\}, \left\{ x2, 0, \frac{-x1 + y1}{-1 + y1} \right\}, \left\{ y2, 0, \frac{x2\ y1}{x1} \right\},$$
$$\left\{ x3, \frac{x2\ y1 - x1\ y2}{y1 - y2}, 1 \right\}, \left\{ y3, 0, \frac{x2\ y1 - x3\ y1 - x1\ y2 + x3\ y2}{-x1 + x2} \right\} \right\}$$

After the integration with respect to y3, we get a rational function in the remaining five variables x3, y2, x2, y1, and x1.

```
In[18]:= fastIntegrate[area, regions[[1, -1]]]
```

$$\text{Out[18]}= \frac{(-x2\ y1 + x3\ y1 + x1\ y2 - x3\ y2)^2}{4\ (x1 - x2)}$$

After the second integration with respect to x3, we still have a rational function in five variables y2, x2, y1, and x1.

```
In[19]:= fastIntegrate[%, regions[[1, -2]]]
```

$$\text{Out[19]}= \frac{(y1 - x2\ y1 - y2 + x1\ y2)^3}{12\ (x1 - x2)\ (y1 - y2)}$$

After the third integration with respect to y2, logarithmic terms emerge.

In[20]:= `fastIntegrate[%, regions[[1, -3]]]`

Out[20]= $\dfrac{1}{72\,x1^3\,(x1-x2)}$

$\Big(y1^3\,\big(6\,x1^2\,x2 - 6\,x1^5\,x2 - 6\,x1\,x2^2 - 9\,x1^2\,x2^2 + 15\,x1^4\,x2^2 + 2\,x2^3 + 3\,x1\,x2^3 + 6\,x1^2\,x2^3 -$

$\qquad 11\,x1^3\,x2^3 + 6\,x1^6\,Log[-y1] - 18\,x1^5\,x2\,Log[-y1] + 18\,x1^4\,x2^2\,Log[-y1] -$

$\qquad 6\,x1^3\,x2^3\,Log[-y1] - 6\,x1^6\,Log\big[\big(-1+\tfrac{x2}{x1}\big)\,y1\big] + 18\,x1^5\,x2\,Log\big[\big(-1+\tfrac{x2}{x1}\big)\,y1\big] -$

$\qquad 18\,x1^4\,x2^2\,Log\big[\big(-1+\tfrac{x2}{x1}\big)\,y1\big] + 6\,x1^3\,x2^3\,Log\big[\big(-1+\tfrac{x2}{x1}\big)\,y1\big]\big)\Big)$

After the fourth integration with respect to x2, we still have a mixture of rational and logarithmic terms. The size of the result increases.

In[21]:= `fastIntegrate[%, regions[[1, -4]]] // Simplify`

Out[21]= $-\dfrac{1}{216\,x1^3\,(-1+y1)^3}$

$\Big(y1^3\,\big(-2\,x1^3 + 24\,x1^4 - 33\,x1^5 + 13\,x1^6 - 21\,x1^3\,y1 + 9\,x1^4\,y1 + 18\,x1^5\,y1 - 12\,x1^6\,y1 +$

$\qquad 21\,x1^3\,y1^2 - 9\,x1^4\,y1^2 - 18\,x1^5\,y1^2 + 12\,x1^6\,y1^2 + 2\,y1^3 - 3\,x1\,y1^3 + 3\,x1^2\,y1^3 -$

$\qquad 22\,x1^3\,y1^3 + 30\,x1^4\,y1^3 - 12\,x1^5\,y1^3 + 6\,x1^3\,(-1+3\,x1-3\,x1^2+2\,x1^3)\,(-1+y1)^3$

$\qquad Log[-x1] + 6\,x1^3\,(-y1^3+3\,x1\,y1^3-3\,x1^2\,y1^3+x1^3\,(1-3\,y1+3\,y1^2))\,Log[x1] -$

$\qquad 6\,x1^3\,Log\big[-\tfrac{(-1+x1)\,y1}{-1+y1}\big] + 18\,x1^4\,Log\big[-\tfrac{(-1+x1)\,y1}{-1+y1}\big] -$

$\qquad 18\,x1^5\,Log\big[-\tfrac{(-1+x1)\,y1}{-1+y1}\big] + 12\,x1^6\,Log\big[-\tfrac{(-1+x1)\,y1}{-1+y1}\big] +$

$\qquad 18\,x1^3\,y1\,Log\big[-\tfrac{(-1+x1)\,y1}{-1+y1}\big] - 54\,x1^4\,y1\,Log\big[-\tfrac{(-1+x1)\,y1}{-1+y1}\big] +$

$\qquad 54\,x1^5\,y1\,Log\big[-\tfrac{(-1+x1)\,y1}{-1+y1}\big] - 36\,x1^6\,y1\,Log\big[-\tfrac{(-1+x1)\,y1}{-1+y1}\big] -$

$\qquad 18\,x1^3\,y1^2\,Log\big[-\tfrac{(-1+x1)\,y1}{-1+y1}\big] + 54\,x1^4\,y1^2\,Log\big[-\tfrac{(-1+x1)\,y1}{-1+y1}\big] -$

$\qquad 54\,x1^5\,y1^2\,Log\big[-\tfrac{(-1+x1)\,y1}{-1+y1}\big] + 36\,x1^6\,y1^2\,Log\big[-\tfrac{(-1+x1)\,y1}{-1+y1}\big] +$

$\qquad 6\,x1^3\,y1^3\,Log\big[-\tfrac{(-1+x1)\,y1}{-1+y1}\big] - 18\,x1^4\,y1^3\,Log\big[-\tfrac{(-1+x1)\,y1}{-1+y1}\big] +$

$\qquad 18\,x1^5\,y1^3\,Log\big[-\tfrac{(-1+x1)\,y1}{-1+y1}\big] - 12\,x1^6\,y1^3\,Log\big[-\tfrac{(-1+x1)\,y1}{-1+y1}\big] -$

$\qquad 6\,x1^6\,Log\big[\tfrac{(-1+x1)\,y1}{-1+y1}\big] + 18\,x1^6\,y1\,Log\big[-\tfrac{(-1+x1)\,y1}{-1+y1}\big] -$

$\qquad 18\,x1^6\,y1^2\,Log\big[\tfrac{(-1+x1)\,y1}{-1+y1}\big] + 6\,x1^3\,y1^3\,Log\big[\tfrac{(-1+x1)\,y1}{-1+y1}\big] -$

$\qquad 18\,x1^4\,y1^3\,Log\big[\tfrac{(-1+x1)\,y1}{-1+y1}\big] + 18\,x1^5\,y1^3\,Log\big[\tfrac{(-1+x1)\,y1}{-1+y1}\big]\big)\Big)$

After the fifth integration with respect to y1, we get a special function, the dilogarithm $Li_2(x)$.

In[22]:= `fastIntegrate[%, regions[[1, -5]]] // Simplify`

Out[22]= $\dfrac{1}{864\,x1^3}$

$(120\,i\,\pi - 120\,x1 - 180\,i\,\pi\,x1 + 120\,x1^2 + 180\,i\,\pi\,x1^2 - 130\,x1^3 - 660\,i\,\pi\,x1^3 + 180\,\pi^2\,x1^3 +$

$\quad 960\,x1^4 + 900\,i\,\pi\,x1^4 - 540\,\pi^2\,x1^4 - 1419\,x1^5 - 360\,i\,\pi\,x1^5 + 540\,\pi^2\,x1^5 + 591\,x1^6 -$

$\quad 180\,\pi^2\,x1^6 + 180\,(-1+x1)^3\,x1^3\,Log[-1+x1]^2 - 360\,i\,\pi\,x1^3\,Log[x1] +$

$\quad 1080\,i\,\pi\,x1^4\,Log[x1] - 1080\,i\,\pi\,x1^5\,Log[x1] + 360\,i\,\pi\,x1^6\,Log[x1] -$

$\quad 360\,x1^3\,Log[1-x1]\,Log[x1] + 1080\,x1^4\,Log[1-x1]\,Log[x1] -$

$$1080 \, x1^5 \, \text{Log}[1 - x1] \, \text{Log}[x1] + 360 \, x1^6 \, \text{Log}[1 - x1] \, \text{Log}[x1] +$$
$$60 \, (-1 + x1)^3 \, \text{Log}[-1 + x1] \, (2 + 3 \, x1 + 6 \, x1^2 - 6 \, i \, \pi \, x1^3 - 6 \, x1^3 \, \text{Log}[x1]) +$$
$$360 \, (-1 + x1)^3 \, x1^3 \, \text{PolyLog}[2, \, x1])$$

The sixth integration with respect to $x1$ results in the final result for the first region, a numeric quantity containing π, $\ln(2)$, and $\ln^2(2)$.

`In[23]:=` **fastIntegrate[%, regions[[1, -6]]]**

`Out[23]=` $\dfrac{-1771 + 170 \, \pi^2 + 800 \, \text{Log}[2] - 960 \, \text{Log}[2]^2}{18432}$

All of the last six integrations get carried out recursively using our above-defined function `multiDimension`‌ `alIntegrate`.

`In[24]:=` **multiDimensionalIntegrate[area, Sequence @@ regions[[1]]]**

`Out[24]=` $\dfrac{-1771 + 170 \, \pi^2 + 800 \, \text{Log}[2] - 960 \, \text{Log}[2]^2}{18432}$

Now, we carry out all of the 216 multidimensional integrations. We take the real part because we expanded logarithm in intermediate steps. This could potentially introduce imaginary parts in the form $i \, k \, \pi$. This calculation will take a few minutes.

`In[25]:=` **Simplify[#, TransformationFunctions -> {Automatic,**
   (# /. Log[x_] :> Log[2, x]/Log[2])&}]& @
(Re[Together[Plus @@
   Apply[multiDimensionalIntegrate[area, ##]&, regions, {1}]]] // Timing)

`Out[25]=` $\left\{1133.01 \, \text{Second}, \, \dfrac{11}{288}\right\}$

All π and $\log(2)$ terms cancelled, and we got (taking into account the triangles with negative orientation) for the expectation value, the simple result $A = 11/144$.

The degree of difficulty to do multidimensional integrals is often depending sensitively from the order of the integration. As a check of the last result and for comparison, we now first evaluate the three integrations over the y_i and then the three integration over the x_i. For this situation, we have only 62 six-dimensional regions.

`In[26]:=` **cad2 = GenericCylindricalAlgebraicDecomposition[**
   signedTriangleArea && unitCube6D,
   {x1, x2, x3, y1, y2, y3}];

regions2 = Apply[List, Apply[{{#3, #1, #5} &,
   cad2[[1]] //. a_ && (b_ || c_) :> a && b || a && c, {2}], {0,
   2}];

Length[regions2]
`Out[30]=` 62

And doing the integrations and simplifying the result takes now only a few seconds. Again, we obtain the result 11/288.

`In[31]:=` **Simplify[Together[Re[Plus @@**
   Apply[multiDimensionalIntegrate[area, ##]&, regions2, {1}]]],
   TransformationFunctions ->
   {(# /. Log[k_Integer] :> (Plus @@ ((#2 Log[#1])& @@@
      FactorInteger[k])))&}] // Timing

`Out[31]=` $\left\{32.51 \, \text{Second}, \, \dfrac{11}{288}\right\}$

Using numerical integration, we can calculate an approximative value of this integral to support the result 11/144.

```
In[32]:= (SeedRandom[111];
         NIntegrate[Evaluate[Abs[area]], {x1, 0, 1}, {y1, 0, 1},
                    {x2, 0, 1}, {y2, 0, 1}, {x3, 0, 1}, {y3, 0, 1},
                    Method -> QuasiMonteCarlo, MaxPoints -> 10^6,
                    PrecisionGoal -> 3])
Out[32]= 0.0763889
```

This result confirms the above result.

```
In[33]:= N[2 %%[[2]]]
Out[33]= 0.0763889
```

We could now go on and calculate the probability distribution for the areas. The six-dimensional integral to be calculated is now

$$p(A) \sim \int_0^1 \int_0^1 \int_0^1 \int_0^1 \int_0^1 \int_0^1 \delta(A - \mathcal{A}(x_1, x_2, x_3, y_1, y_2, y_3)) \, dy_3 \, dx_3 \, dy_2 \, dx_2 \, dy_1 \, dx_1,$$

$$\mathcal{A}(x_1, x_2, x_3, y_1, y_2, y_3) = |x_3 \, y_1 - x_2 \, y_1 + x_1 \, y_2 - x_3 \, y_2 + x_2 \, y_3 - x_1 \, y_3|.$$

(Here we temporarily changed $A \to 2A$ so that all variables involved range over the interval [0, 1].)

This time, before subdividing the integration variable space into subregions, we carry out the integral over y_3 to eliminate the Dirac δ function. To do this, we use the identity

$$\int_a^b \delta(y - f(x)) \, dx = \int_a^b \sum_k \frac{\delta(y - x_{0,k})}{|f'(x_{0,k})|} \, dx$$

where the $x_{0,k}$ are the zeros of $f(x)$ in [a, b].

Expressing y_3 through x_1, x_2, x_3, y_1, y_2, and A yields the following expression.

```
In[34]:= soly3 = Solve[A == (* or - *) (-x2 y1 + x3 y1 + x1 y2 - x3 y2 -
                                        x1 y3 + x2 y3), y3][[1, 1, 2]]
```

$$Out[34]= \frac{-A - x2\ y1 + x3\ y1 + x1\ y2 - x3\ y2}{x1 - x2}$$

And the derivative from the denominator becomes $|x_1 - x_2|$.

```
In[35]:= D[-x2 y1 + x3 y1 + x1 y2 - x3 y2 - x1 y3 + x2 y3, y3]
Out[35]= -x1 + x2
```

Now is a good time to obtain a decomposition of the space into subregions. In addition to the constraints following from the geometric constraints of the integration variables being from the unit square, we add three more inequalities: 1) $0 < y_3(x_1, x_2, x_3, y_1, y_2; A) < 1$ to ensure the existence of a zero inside the Dirac δ function argument; 2) $A > 0$ for positive oriented areas; and 3) $x_1 > x_2$ to avoid the absolute value in the denominator (the case $x_1 < x_2$ follows from symmetry).

```
In[36]:= cad = Experimental`GenericCylindricalAlgebraicDecomposition[
                0 < soly3 < 1 && A > 0 && x1 > (* or < *) x2 &&
                0 < x1 < 1 && 0 < x2 < 1 &&
                0 < x3 < 1 && 0 < y2 < 1 && 0 < y1 < 1,
                {A, x1, x2, x3, y1, y2}];
```

This time, we get a total of 1282 subregions.

```
In[37]:= (l1 = cad[[1]] //. a_ && (b_ || c_) :> (a && b) || (a && c)) // Length
Out[37]= 1282
```

One expects the probability distribution $p(A)$ to be a piecewise smooth function of λ. Six λ-interval arise naturally from the decomposition.

```
In[38]:= Union[First /@ l1]
```

$$Out[38]= \ 0 < A < \frac{1}{6} \ || \ \frac{1}{6} < A < \frac{1}{5} \ || \ \frac{1}{5} < A < \frac{1}{4} \ || \ \frac{1}{4} < A < \frac{1}{3} \ || \ \frac{1}{3} < A < \frac{1}{2} \ || \ \frac{1}{2} < A < 1$$

```
In[39]:= ASortedRegions = {#[[1, 1, 2]] < A < #[[1, 1, 3]] , Rest /@ #}& /@
            Split[Sort[(# /. Inequality[a_, Less, b_, Less, c_] :>
                {b, a, c} /. And -> List)& /@ (List @@ l1)], #1[[1]] === #2[[1]]&];
```

Here is the number of regions for the six λ-intervals.

```
In[40]:= {#1, Length[#2] "subregions"}& @@@ ASortedRegions
```

$$Out[40]= \ \left\{ \left\{ 0 < A < \frac{1}{6}, \ 317 \text{ subregions} \right\}, \ \left\{ \frac{1}{6} < A < \frac{1}{5}, \ 324 \text{ subregions} \right\}, \right.$$
$$\left\{ \frac{1}{5} < A < \frac{1}{4}, \ 310 \text{ subregions} \right\}, \ \left\{ \frac{1}{4} < A < \frac{1}{3}, \ 216 \text{ subregions} \right\},$$
$$\left. \left\{ \frac{1}{3} < A < \frac{1}{2}, \ 99 \text{ subregions} \right\}, \ \left\{ \frac{1}{2} < A < 1, \ 16 \text{ subregions} \right\} \right\}$$

The regions themselves look quite similar to the above ones.

```
In[41]:= {#1, #2[[1]]}& @@@ ASortedRegions
```

$$Out[41]= \ \left\{ \left\{ 0 < A < \frac{1}{6}, \ \{x1, 0, A\}, \{x2, 0, x1\}, \{x3, A + x1, 1\}, \right. \right.$$
$$\left. \left\{ y1, -\frac{A}{x2 - x3}, \frac{-A - x1 + x2}{x2 - x3} \right\}, \left\{ y2, 0, \frac{-A - x2\,y1 + x3\,y1}{-x1 + x3} \right\} \right\} \right\},$$
$$\left\{ \frac{1}{6} < A < \frac{1}{5}, \ \{x1, 0, A\}, \{x2, 0, x1\}, \{x3, A + x1, 1\}, \right.$$
$$\left. \left\{ y1, -\frac{A}{x2 - x3}, \frac{-A - x1 + x2}{x2 - x3} \right\}, \left\{ y2, 0, \frac{-A - x2\,y1 + x3\,y1}{-x1 + x3} \right\} \right\} \right\},$$
$$\left\{ \frac{1}{5} < A < \frac{1}{4}, \ \{x1, 0, A\}, \{x2, 0, x1\}, \{x3, A + x1, 1\}, \right.$$
$$\left. \left\{ y1, -\frac{A}{x2 - x3}, \frac{-A - x1 + x2}{x2 - x3} \right\}, \left\{ y2, 0, \frac{-A - x2\,y1 + x3\,y1}{-x1 + x3} \right\} \right\} \right\},$$
$$\left\{ \frac{1}{4} < A < \frac{1}{3}, \ \{x1, 0, A\}, \{x2, 0, x1\}, \{x3, A + x1, 1\}, \right.$$
$$\left. \left\{ y1, -\frac{A}{x2 - x3}, \frac{-A - x1 + x2}{x2 - x3} \right\}, \left\{ y2, 0, \frac{-A - x2\,y1 + x3\,y1}{-x1 + x3} \right\} \right\} \right\},$$
$$\left\{ \frac{1}{3} < A < \frac{1}{2}, \ \{x1, 0, A\}, \{x2, 0, x1\}, \{x3, A + x1, 1\}, \right.$$
$$\left. \left\{ y1, -\frac{A}{x2 - x3}, \frac{-A - x1 + x2}{x2 - x3} \right\}, \left\{ y2, 0, \frac{-A - x2\,y1 + x3\,y1}{-x1 + x3} \right\} \right\} \right\},$$
$$\left\{ \frac{1}{2} < A < 1, \ \{x1, 0, 1 - A\}, \{x2, 0, x1\}, \{x3, A + x1, 1\}, \right.$$
$$\left. \left. \left\{ y1, -\frac{A}{x2 - x3}, \frac{-A - x1 + x2}{x2 - x3} \right\}, \left\{ y2, 0, \frac{-A - x2\,y1 + x3\,y1}{-x1 + x3} \right\} \right\} \right\}$$

To carry out the remaining five-dimensional integration, we could use a function like `x1x2x3y1y2Integra`-
`tion`. The optional functions S and $\mathcal{T}$ could carry out simplifications of the integrands to ease the work of
`Integrate` and the results of `Integrate` to ease the work of `Limit`.

```
In[42]:= x1x2x3y1y2Integration[region_, startIntegrand:i0_,
                    S_:Identity, T_:Identity] :=
         Module[{DL, y2y1x3x2x1Regions},
```

```
(* form difference of integral at upper limit minus
   integral at lower limit *)
𝒟ℒ = (Limit[#1, #2[[1]] -> #2[[3]], Direction -> +1] -
       Limit[#1, #2[[1]] -> #2[[2]], Direction -> -1])&;
(* integration regions *)
y2y1x3x2x1Regions = Reverse[region];
(* carry out the five integrations *)
Fold[𝒟ℒ[𝒯[Integrate[𝒮[#1], #2[[1]]]]], #2]&, i0,
    y2y1x3x2x1Regions]]
```

Here is a typical result for one of the 1282 five-dimensional integrals.

```
In[43]:= (* no special simplifiers needed for the subregion {5, 40} *)
         x1x2x3y1y2Integration[ASortedRegions[[5, 2, 40]], 1/(x1 - x2)] //
                                                     Simplify[#, A > 0]&
```

$$Out[44]= \frac{1}{8} \, (4 \, A^2 \, (-3 + \text{Log}[16]) + (-1 + A) \, (-5 + 9 \, A + 2 \, (1 + A) \, \text{Log}[1 - A] - 2 \, (1 + A) \, \text{Log}[A]))$$

Naturally, we will start carrying out the integrations for the area interval $1/2 < A < 1$, which has the smallest number of regions.

```
In[45]:= Module[{j = 6, sumA = 0, res},
         Do[res = x1x2x3y1y2Integration[ASortedRegions[[j, 2, k]], 1/(x1 - x2)];
            sumA = sumA + res;
            If[MemberQ[res, _Integrate, {0, Infinity}], Print[{k, res}]],
            {k, Length[ASortedRegions[[j, 2]]]}];
         sumA];
```

The result is a nice short expression in the area A. Because it contains polylogarithms, we simplify it using FullSimplify (which simplifies special functions).

```
In[46]:= pA = % // FullSimplify[#, ASortedRegions[[6, 1]]]&
```

$$Out[46]= \frac{1}{12} \left(18 - A \, (18 + (12 + 7 \, A) \, \pi^2) + \right.$$
$$15 \, A^2 \, (-6 + \text{Log}[A]) \, \text{Log}[A] + 6 \, \text{Log}[1 - A] \, (3 \, (-1 + A) \, (1 + 5 \, A) + A^2 \, \text{Log}[A]) -$$
$$\left. 6 \, A^2 \, \text{PolyLog}\left[2, \frac{-1 + A}{A}\right] + 6 \, A \, (12 + 7 \, A) \, \text{PolyLog}[2, A]\right)$$

Applying an identity for the dilogarithm allows obtaining the following concise result for the probability of the area (in this step, we restore the original units of A).

$$p(A) = -4 \, (-12 \, (\ln(2 \, A) - 5) \, \ln(2 \, A) \, A^2 -$$
$$24 \, (A + 1) \, \text{Li}_2(2 \, A) \, A + 4 \, (A + 1) \, \pi^2 \, A + 6 \, A + (12 \, (2 - 5 \, A) \, A + 3) \, \ln(1 - 2 \, A) - 3)$$

```
In[47]:= polyLogRule = PolyLog[2, (A - 1)/A] ->
                       PolyLog[2, A] + Log[1 - A] Log[A] - Log[A]^2/2 - Pi^2/6;
```

```
In[48]:= p[A_] = 8 (pA /. polyLogRule) /. A -> 2A // FullSimplify[#, 0 < A < 1]&
```

$$Out[48]= -4 \, (-3 + 6 \, A + 4 \, A \, (1 + A) \, \pi^2 + (3 + 12 \, (2 - 5 \, A) \, A) \, \text{Log}[1 - 2 \, A] -$$
$$12 \, A^2 \, (-5 + \text{Log}[2 \, A]) \, \text{Log}[2 \, A] - 24 \, A \, (1 + A) \, \text{PolyLog}[2, 2 \, A])$$

Using the other five intervals for A would give the same result, but only after more work because of the substantially larger number of regions.

Here is a quick check of the normalization and the average area. Now, it is also straightforward to calculate the variance.

```
In[49]:= {Integrate[p[A], {A, 0, 1/2}], Integrate[A p[A], {A, 0, 1/2}],
         Integrate[(A - 11/144)^2 p[A], {A, 0, 1/2}]}
```

Out[49]= $\left\{1, \dfrac{11}{144}, \dfrac{95}{20736}\right\}$

Interestingly, the probability distribution is the solution of a simple linear forth-order differential equation with polynomial coefficients.

```
In[50]:= (* eliminate logarithms and polylogarithms *)
        GroebnerBasis[Numerator[Together[
                Table[dp[k] - D[p[A], {A, k}], {k, 0, 4}]]], {},
                Reverse @ Union[Cases[p[A], _Log | _PolyLog, Infinity]],
                Sort -> True] // Factor
```

Out[51]= {A (768 dp[1] - 384 dp[2] - 768 A dp[2] - 384 A dp[3] +
         768 A^2 dp[3] + 4 A^2 dp[3]^2 - 8 A^3 dp[3]^2 + 4 A^4 dp[3]^2 + 4 A^3 dp[3] dp[4] -
         12 A^4 dp[3] dp[4] + 8 A^5 dp[3] dp[4] + A^4 dp[4]^2 - 4 A^5 dp[4]^2 + 4 A^6 dp[4]^2),
         9216 - 18432 A - 768 dp[0] + 384 A dp[2] + 384 A^2 dp[2] + 480 A^2 dp[3] - 768 A^3 dp[3] -
         4 A^3 dp[3]^2 + 8 A^4 dp[3]^2 - 4 A^5 dp[3]^2 + 96 A^3 dp[4] - 192 A^4 dp[4] - 4 A^4 dp[3] dp[4] +
         12 A^5 dp[3] dp[4] - 8 A^6 dp[3] dp[4] - A^5 dp[4]^2 + 4 A^6 dp[4]^2 - 4 A^7 dp[4]^2}

```
In[52]:= Sort[(* find short differential equation *)
                GroebnerBasis[%, {dp[4], dp[3], dp[2], dp[1], dp[0]}],
                (Length[#1] < Length[#2])&][[1]]
```

Out[52]= -96 + 192 A + 8 dp[0] - 8 A dp[1] + 4 A^2 dp[2] - A^2 dp[3] - A^3 dp[4] + 2 A^4 dp[4]

Here is a quick check of this differential equation.

```
In[53]:= A^3 (1 - 2 A) p''''[A] + A^2 p'''[A] - 4 A^2 p''[A] + 8 A p'[A] -
                8 p[A] - 96 (2 A - 1) // Simplify
```

Out[53]= 0

We end by a comparing the calculated density with a modeled one. The function `modelData` generates *o* random triangles in the unit square and returns the histogram data for *bins* bins.

```
In[54]:= modelData[o_, bins_] :=
        MapIndexed[(* scale *) {#2[[1]]/bins/2, 2 bins #1/o}&, #]& @
            Compile[{}, Module[{T = Table[0, {bins}], A,
                                p1, p2, p3, d1x, d1y, d2x, d2y, cA},
                        Do[(* the random points *)
                            {p1, p2, p3} = Table[Random[], {3}, {2}];
                            {d1x, d1y} = p3 - p1; {d2x, d2y} = p2 - p1;
                            (* area of the triangle *)
                            A = Abs[d1x d2y - d1y d2x]/2;
                            (* discretized area *)
                            cA = Ceiling[bins A/(1/2)];
                            (* increase area counters *)
                            T[[cA]] = T[[cA]] + 1, {i, o}];
                        (* return counts *) T]][]
```

Using 10 million random triangles and 100 bins, we obtain an excellent agreement between the modeled data and the theoretical probability density. The right graphic shows the weighted density *A p(A)*.

```
In[55]:= Show[GraphicsArray[
        Block[{$DisplayFunction = Identity},
            {(* probability density *)
            Plot[p[A], {A, 0, 1/2}, PlotRange -> All, PlotStyle -> {Hue[0]},
                (* modeled probabilities *)
                Prolog -> {PointSize[0.01], GrayLevel[0],
                        Point /@ modelData[10^6, 100]}, PlotRange -> All],
```

```
(* weighted probability density *)
Plot[A p[A], {A, 0, 1/2}, PlotRange -> All]}]]]
```

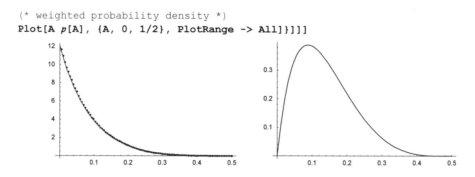

Having calculated and modeled the probability distribution of the area, a natural problem arising is the form of the probability density $p(x, y)$ of a point $\{x, y\}$ being inside a randomly chosen triangle. Geometrically "inside" means that the point $\{x, y\}$ lies in the three left (or right) half planes generated through the three triangle edges (assuming the three triangle edges are uniformly oriented). This means the integral to be evaluated for the probability density is

$$p(x, y) = 2 \int_0^1 \int_0^1 \int_0^1 \int_0^1 \int_0^1 \int_0^1 \theta_{2,1}(x, y)\, \theta_{3,3}(x, y)\, \theta_{1,3}(x, y)\, dy_3\, dx_3\, dy_2\, dx_2\, dy_1\, dx_1,$$

$$\theta_{i,j}(x, y) = \theta((p_i - p_j).(p - p_j))$$

where $p_i = \{x_i, y_i\}$ are the triangle vertices and $p = \{x, y\}$. (The factor 2 accounts for the orientation of the triangles).

```
In[56]:= θθθ = Module[{p1 = {x1, y1}, p2 = {x2, y2}, p3 = {x3, y3},
                p = {x, y}, normal = (Reverse[#]{1, -1})&},
               (* point p is to the left of each edge *)
               normal[p2 - p1].(p - p1) > 0 &&
               normal[p3 - p2].(p - p2) > 0 &&
               normal[p1 - p3].(p - p3) > 0]
Out[56]= (x1 - x2) (y - y1) + (x - x1) (-y1 + y2) > 0 &&
         (x2 - x3) (y - y2) + (x - x2) (-y2 + y3) > 0 && (-x1 + x3) (y - y3) + (x - x3) (y1 - y3) > 0
```

To calculate the sixfold integral, we will follow the already twice successfully-used strategy to first calculate a decomposition of the integration domain. Because of the obvious fourfold rotational symmetry of $p(x, y)$ around the square center $\{1/2, 1/2\}$, we use the restrictions $1/2 < x < 1$, $1/2 < y < x$ to reduce the number of cells returned. We obtain 327 cells (without using the symmetry, we would obtain 3738 cells).

```
In[57]:= cadxy = Experimental`GenericCylindricalAlgebraicDecomposition[
                θθθ && 0 < x1 < 1 && 0 < x2 < 1 && 0 < x3 < 1 &&
                0 < y1 < 1 && 0 < y2 < 1 && 0 < y3 < 1 &&
                1/2 < x < 1 && 1/2 < y < x,
               {x, y, x1, x2, x3, y1, y2, y3}];

In[58]:= l1 = cadxy[[1]] //. a_ && (b_ || c_) :> a && b || a && c;

In[59]:= Length[l1]
Out[59]= 327
```

All cells span the specified x,y-domain. This means, the density $p(x, y)$ is continuous within this domain.

```
In[60]:= Union[Take[#, 2]& /@ l1]
Out[60]=  1/2 < x < 1 && 1/2 < y < x
```

The cells of the 6D integration domain have similar-looking boundaries as the cells from the above calculations.

In[61]:= **xyRegions = (# /. Inequality[a_, Less, b_, Less, c_] :>**
                    {b, a, c} /. And -> List}& /@
              ((* remove x and y parts *) List @@ Drop[#, 2]& /@ 11);

In[62]:= **xyRegions[[1]]**

Out[62]= $\left\{\left\{x1, 0, \frac{-x+y}{-1+y}\right\}, \{x2, 0, x1\}, \{x3, x, 1\}, \left\{y1, 0, \frac{-x1\,y+x2\,y}{-x+x2}\right\},\right.$

$\left.\{y2, 0, 1\}, \left\{y3, \frac{x2\,y-x3\,y-x\,y2+x3\,y2}{-x+x2}, \frac{x1\,y-x3\,y-x\,y1+x3\,y1}{-x+x1}\right\}\right\}$

The values of the integrals are slightly simpler looking than in the last calculation—no polylogarithms occur in the result.

In[63]:= **x1x2x3y1y2Integration[xyRegions[[-1]], 1]**

Out[63]= $-\frac{(-1+x)^3\,(-1+y)^4\,\text{Log}[1-x]}{12\,y^2} + \frac{1}{72\,y^8}\left((-1+x)^3\,(-1+y)^3\right.$

$\left(-1+4\,y-3\,y^2-y^3-2\,y^4-6\,y^5+12\,y^6+6\,(-1+y)\,y^6\,\text{Log}\left[-\frac{(-1+x)\,(-1+y)^2}{y^2}\right]\right)\right)$

We now carry out the 327 6D integrals. This takes a few minutes.

In[64]:= **(integrals =**
          Table[res = x1x2x3y1y2Integration[xyRegions[[k]], 2];
               If[MemberQ[res, _Integrate, {0, Infinity}], Print[{k, res}]];
               res, {k, Length[xyRegions]}]); // Timing

Out[64]= {1057.76 Second, Null}

The unsimplified form of the result is a large expression.

In[65]:= **sum1 = Expand[Plus @@ integrals];**
          Length[sum1]

Out[66]= 7520

Here are the first and the last terms of sum1.

In[67]:= **{Take[sum1, 6], Take[sum1, -6]}**

Out[67]= $\left\{\frac{1609}{12} + \frac{163}{12\,(1-x)} + \frac{427}{8\,(-1+x)} + \frac{1}{4\,x} + \frac{5}{12\,(-1+x)\,x} - \frac{4733\,x}{12},\right.$

$\frac{x\,y^4\,\text{Log}\left[\frac{y\,(-1+x+2\,y-2\,x\,y-y^2+x\,y^2)}{(-1+y)^3}\right]}{2\,(-1+y)} - \frac{x^2\,y^4\,\text{Log}\left[\frac{y\,(-1+x+2\,y-2\,x\,y-y^2+x\,y^2)}{(-1+y)^3}\right]}{-1+y} +$

$\frac{x^3\,y^4\,\text{Log}\left[\frac{y\,(-1+x+2\,y-2\,x\,y-y^2+x\,y^2)}{(-1+y)^3}\right]}{2\,(-1+y)} - \frac{3\,x\,y^5\,\text{Log}\left[\frac{y\,(-1+x+2\,y-2\,x\,y-y^2+x\,y^2)}{(-1+y)^3}\right]}{2\,(-1+y)^2} +$

$\left.\frac{3\,x^2\,y^5\,\text{Log}\left[\frac{y\,(-1+x+2\,y-2\,x\,y-y^2+x\,y^2)}{(-1+y)^3}\right]}{(-1+y)^2} - \frac{3\,x^3\,y^5\,\text{Log}\left[\frac{y\,(-1+x+2\,y-2\,x\,y-y^2+x\,y^2)}{(-1+y)^3}\right]}{2\,(-1+y)^2}\right\}$

We see various logarithms with relatively complicated arguments. In sum1 more than 60 different logarithms occur.

In[68]:= **Cases[sum1, _Log, Infinity] // Union // Length**

Out[68]= 63

We simplify the logarithms by power expanding them. This will result in some spurious imaginary parts which we remove by taking the real part.

In[69]:= **sum2 = sum1 //. Log[ξ_] :> PowerExpand[Log[Factor[ξ]]];**

Now, we have only a handful of different logarithms left.

In[70]:= `Cases[sum2, _Log, Infinity] // Union`

Out[70]= `{Log[1 - x], Log[-1 + x], Log[x], Log[-1 + y], Log[y]}`

In[71]:= `sum3 = ComplexExpand[Re[sum2]];`

Simplifying `sum3` gives the following final result for the probability density of the point $\{x, y\}$ being inside a randomly chose triangle.

$$p(x, y) =$$

$$\frac{1}{2} \frac{(x-1)^2}{xy(y-1)} (12 x^2 y^4 - 30 x y^4 + 3 y^4 - 24 x^2 y^3 + 60 x y^3 - 6 y^3 + 25 x^2 y^2 - 43 x y^2 + 4 y^2 - $$

$$13 x^2 y + 13 x y - y + x^2 - x) - \frac{3}{2} (x-1)^2 (6 x y^2 - 4 y^2 - 6 x y + 4 y + 2 x - 1) \log\left(\frac{1}{x} - 1\right) +$$

$$\frac{3}{2} (x-1)^2 (2 y - 1) (2 x y^2 - 2 y^2 - 2 x y + 2 y + 2 x - 1) \log\left(\frac{1}{y} - 1\right)$$

In[72]:= `p[x_, y_] = FullSimplify[sum3, 1/2 < x < 1 && 1/2 < y < x] //`
          `Collect[#, _Log, Factor] &`

Out[72]= $\dfrac{1}{2 x (-1 + y) y}$

$((-1 + x)^2 (-x + x^2 - y + 13 x y - 13 x^2 y + 4 y^2 - 43 x y^2 + 25 x^2 y^2 - 6 y^3 + 60 x y^3 - $
$\quad 24 x^2 y^3 + 3 y^4 - 30 x y^4 + 12 x^2 y^4)) - $

$\dfrac{3}{2} (-1 + x)^2 (-1 + 2 x + 4 y - 6 x y - 4 y^2 + 6 x y^2) \, \text{Log}\left[-1 + \dfrac{1}{x}\right] +$

$\dfrac{3}{2} (-1 + x)^2 (-1 + 2 y) (-1 + 2 x + 2 y - 2 x y - 2 y^2 + 2 x y^2) \, \text{Log}\left[-1 + \dfrac{1}{y}\right]$

Here is a quick check that the power expanding did indeed not change the numerical value of the density.

In[73]:= `{sum1, p[x, y]} /. x -> 2/E /. y -> 2/Pi // N[#, 10] &`

Out[73]= `{0.13621111723 + 0. × 10⁻¹² i, 0.1362111172}`

The probability that the center of the square is inside a randomly chosen triangle is exactly 25%.

In[74]:= `p[1/2, 1/2]`

Out[74]= $\dfrac{1}{4}$

The next graphic shows the calculated probability density on the left and a modeled one on the right. The two densities look identical.

In[75]:=
```
Show[GraphicsArray[
  Block[{$DisplayFunction = Identity},
  {(* exact probability *)
  Module[{ε = 10^-12, pp = 20, polys},
  (* plot in the domain 1/2 < x < 1 ∧ 1/2 < y < x *)
  polys = Map[(# - {1/2, 1/2, 0}) &, Cases[
    ParametricPlot3D[Evaluate @ {x, 1/2 + η (x - 1/2),
                                 p[x, 1/2 + η (x - 1/2)]},
                    {x, 1/2, 1 - ε}, {η, 0, 1},
                    PlotPoints -> pp], _Polygon, Infinity], {-2}];
  Graphics3D[{EdgeForm[], (* generate seven other parts *)
    Map[(# + {1/2, 1/2, 0}) &,
      {Apply[{#2, #1, #3} &, #, {-2}], #} &[
```

```
            {Map[(# {-1, 1, 1})&, #, {-2}], #}&[
                {Map[(# {1, -1, 1})&, #, {-2}], #}&[polys]]], {-2}]},
         BoxRatios -> {1, 1, 1/2}, Axes -> True]],
(* modeled probability *)
Module[{d = 60, o = 10^4, data, if},
data = Compile[{},
Module[{T = Table[0, {d}, {d}], p1, p2, p3, xc, yc, mp, σ},
  Do[{p1, p2, p3} = Table[Random[], {3}, {2}]; mp = (p1 + p2 + p3)/3;
     (* orientation of the normals *)
     σ = Sign[(Reverse[p2 - p1]{1, -1}).(mp - p1)];
     (* are discretized square points inside triangle? *)
     Do[If[σ (Reverse[p2 - p1]{1, -1}).({x, y} - p1) > 0 &&
         σ (Reverse[p3 - p2]{1, -1}).({x, y} - p2) > 0 &&
         σ (Reverse[p1 - p3]{1, -1}).({x, y} - p3) > 0,
         (* increase counters *)
         {xc, yc} = Round[{x, y} (d - 1)] + 1;
         T[[xc, yc]] =  T[[xc, yc]] + 1],
       {x, 0, 1, 1/(d - 1)}, {y, 0, 1, 1/(d - 1)}], {o}]; T]][];
(* interpolated scaled counts *)
if = Interpolation[Flatten[MapIndexed[Flatten[
              {(#2 - {1, 1})/(d - 1), #1}]&, data/o, {2}], 1]];
(* interpolated observed frequencies *)
Plot3D[if[x, y], {x, 0, 1}, {y, 0, 1}, Mesh -> False]}]]]
```

We end by integrating the calculated probability density $p(x, y)$ over the unit square. $p(x, y)$ is the probability that the point $\{x, y\}$ is inside a randomly chosen triangle. This means the average of $p(x, y)$ is again the area of a randomly chosen triangle, namely $11/144$.

In[76]:= 8 Integrate[p[x, y], {x, 1/2, 1}, {y, 1/2, x}]

Out[76]= $\dfrac{11}{144}$

For a similar probabilistic problem, the Heilbronn triangle problem, see [936].

1.10.2 $\cos\left(\frac{2\pi}{257}\right)$ à la Gauss

In the early morning of March 29 in 1796, Carl Friedrich Gauss (while still in bed) recognized how it is possible to construct a regular 17-gon by ruler and compass; or more arithmetically and less geometrically speaking, he expressed $\cos(\frac{2\pi}{17})$ in terms of square roots and the four basic arithmetic operations of addition, subtraction, multiplication, and division only. (This discovery was the reason why he decided to become a mathematician [1472], [704], [1792].) His method works immediately for all primes of the form $2^{2^j}+1$, so-called Fermat numbers F_j [1080]. For $j=0$ to 4, we get the numbers 3, 5, 17, 257, and 65537. ($j=5$, ..., 14 do not give primes; we return to this at the end of this section.) The problem to be solved is to express the roots of $z^p=1$, where p is a Fermat prime in square roots. One obvious solution of this equation is $z=1$. After dividing $z^p=1$ by this solution, we get as the new equation to be solved:

$$z^{p-1}+z^{p-2}+\cdots+z+1=0.$$

It can be shown that there are no further rational zeros; so this equation cannot be simplified further in an easy way. Let us denote (by following Gauss's notation here and in the following) the solution $\exp\left(\frac{\lambda 2\pi i}{p}\right)$, λ integer, $1\le\lambda\le p-1$ of this equation by $\overline{\lambda}$ (which is, of course, a solution, but which contains a pth root). Gauss's idea, which solves the above equation exclusively in square roots, is to group the roots of the above equation in a recursive way such that the explicit values of the sums of these roots can be expressed in numbers and square roots. Each step then rearranges these roots until finally only groups of length two remain. These last groups are then just of the form $\cos\left(\frac{j2\pi}{p}\right)$.

Let us describe this idea in more detail. First, we need the number-theoretic notion of a primitive root: the number g is called a primitive root of p if the set of numbers $\{g^i \bmod p\}_{i=0,\ldots,p-2}$ are (up to their order) identical with the numbers 1, ..., $p-1$. The following input shows that the number 3 is a primitive root of 5, 17, 257, and 65357. (To avoid the calculation of huge numbers caused by exponentiation, we use `PowerMod[g, i, n]`, which is mathematically equivalent to `Mod[g^i, n]`, but much faster.)

```
In[1]:= Function[p, Sort[Array[PowerMod[3, #, p]&, p - 1, 0]] ==
                        Range[p - 1]] /@ {5, 17, 257, 65537}
Out[1]= {True, True, True, True}
```

(Only integers n of the form 2, 4, p^j, $2\,p^j$, where p is an odd prime and $j>0$ have primitive roots; they have $\phi(\phi(n))$ different ones; $\phi(n)$ is Euler's totient function.) The order of the integers in `Array[PowerMod[base, #, prime]&, p-1, 0]` exhibits some interesting symmetry, as can be visualized for the *prime* = 257 case with the following input.

```
In[2]:= primitiveRootsGraphics[b_] := Graphics[
          {Thickness[0.002], Line[Append[#, First[#]]&[
          (* connect numbers in their permuted order *)
          {Cos[#], Sin[#]}& /@ N[2Pi Array[
              PowerMod[b, #, 257]&, 256, 0]/257]]]},
          PlotRange -> All, AspectRatio -> Automatic]

In[3]:= (* reduced residue system exists for the following 128 numbers *)
        rssNumbers = Flatten[Position[Table[Sort[Array[
                  PowerMod[i, #, 257]&, 256, 0]] ==
                  Range[256], {i, 256}], True]];

In[5]:= (* visualizations of the powermod sequences *)
        Function[bs, Show[GraphicsArray[Function[b,
```

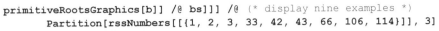

```
        primitiveRootsGraphics[b]] /@ bs]]] /@ (* display nine examples *)
            Partition[rssNumbers[[{1, 2, 3, 33, 42, 43, 66, 106, 114}]], 3]
```

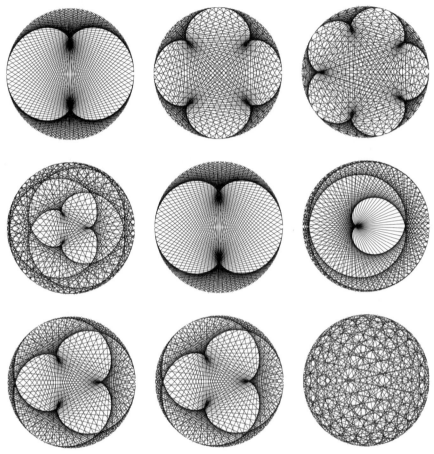

```
        Show[primitiveRootsGraphics[#]]& /@ rssNumbers
```

(For some interesting discussions about the number of crossings and the number of regions in such pictures, see [1428].)

The next concept we need is that of the so-called periods. A period $\overset{f}{\overline{\lambda}}$ to the primitive root g, containing the root $\overline{\lambda}$ and having length f, is defined by the expression below. (The dependence on the fixed quantities p and g is suppressed.)

$$\overset{f}{\overline{\lambda}} = \sum_{j=0}^{f-1} \overline{\lambda\, g^{\frac{j(p-1)}{f}}}$$

Because the root $\overline{\lambda + p}$ is equivalent to the root $\overline{\lambda}$, we implement the construction of the periods in the following way. (We again use `PowerMod` because of speed and denote $\overline{\lambda}$ by R[λ].)

```
In[7]:= period[λ_, f_, p_, g_] := Plus @@ (R /@ Mod[Mod[λ, p] Array[
                                   PowerMod[g^((p - 1)/f), #, p]&, f, 0], p])
```

Let us look at two examples for the prime 17 and the primitive root 3.

```
In[8]:= period[1, 8, 17, 3]
```

Out[8]= R[1] + R[2] + R[4] + R[8] + R[9] + R[13] + R[15] + R[16]

```
In[9]:= period[3, 8, 17, 3]
```

Out[9]= R[3] + R[5] + R[6] + R[7] + R[10] + R[11] + R[12] + R[14]

We see that their sum just gives the sum of all roots. This is always the case if p is a Fermat prime; here, the case $p = 257$ is checked.

```
In[10]:= period[1, 128, 257, 3] + period[3, 128, 257, 3] == Plus @@ Array[R, 256]
```

Out[10]= True

```
In[11]:= period[5, 128, 257, 3] + period[9, 128, 257, 3] == Plus @@ Array[R, 256]
```

Out[11]= True

Dividing the last period again into subperiods by using the above definition for the periods, we find that the period period[3, 8, 17, 3] can be expressed as the sum of the following periods.

```
In[12]:= period[3, 4, 17, 3]
```

Out[12]= R[3] + R[5] + R[12] + R[14]

```
In[13]:= period[11, 4, 17, 3]
```

Out[13]= R[6] + R[7] + R[10] + R[11]

It can be shown that one can always represent a period in this way: One root of one of the new periods is identical to the old one ($\overline{\lambda}$), whereas one root of the other period is generated by the root $\overline{\left(\lambda g^{\frac{p-1}{f}}\right)} \bmod(p-1)$. The other roots of the two periods under consideration follow immediately from the above definition of the periods. In our example, we have this for one root of the second period: $\left(3 3^{\frac{17-1}{8}}\right) \bmod 16 = 11$. In doing this division process for the periods repeatedly, we end up in periods of length two. These periods are of the form $\overline{1 + p - 1}, \overline{2 + p - 2}, ...$, which give immediately $2\cos\left(\frac{2\pi}{p}\right), 2\cos\left(2 \times 2\frac{\pi}{p}\right), ...$ To explicitly calculate the values of the periods in square roots, we need the following theorem: The (numerical) values Λ_1, Λ_2 of two periods λ_1, λ_2 (which contain no higher roots than square roots and are to be discriminated from the periods λ_1, λ_2 themselves) obtained by splitting one period are the solutions of a quadratic equation. If Λ_1 and Λ_2 are the solutions of $\Lambda^2 + a_1 \Lambda + a_2 = 0$; by Vieta's theorem, we have $\Lambda_1 \Lambda_2 = -a_1$ and $\Lambda_1 + \Lambda_2 = a_2$. The sum of the two periods is just the period before splitting, and the (numerical) value of the starting period is -1. It is important to observe that the product of two periods of length f, obtained by splitting a period $\overline{\lambda}^{2f}$, can always be expressed as a linear combination of periods of length $2 f$. The explicit formula for carrying out this multiplication of two periods is given by

$$\overline{\lambda\mu}^{f} = \overline{\lambda_1 \mu_1}^{f} + \overline{\lambda_2 \mu_1}^{f} + \cdots + \overline{\lambda_f \mu_1}^{f}$$

where

$$\overline{\lambda}^{f} = \overline{\lambda_1} + \overline{\lambda_2} + \cdots + \overline{\lambda_f} \quad \text{and} \quad \overline{\mu}^{f} = \overline{\mu_1} + \overline{\mu_2} + \cdots + \overline{\mu_f}.$$

After this multiplication, the periods on the right-hand side can then be expressed as periods of length $2f$ or as pure numbers. (For $f = \frac{p-1}{2}$, they can always be expressed as pure numbers, which ensures that we have appropriate starting values for the recursive calculation.) Here, the above two periods of length 8 of $p = 17$ (`period[1, 8, 17, 3]` and `period[3, 8, 17, 3]`; $\mu_1 = 1$, $\lambda_1 = 3$, $\lambda_2 = 5$, $\lambda_3 = 6$, $\lambda_4 = 7$, $\lambda_5 = 10$, $\lambda_6 = 11, \lambda_7 = 12, \lambda_8 = 14$) are multiplied in this manner.

```
In[14]:= period[ 3 + 1, 8, 17, 3] + period[ 5 + 1, 8, 17, 3] +
        period[ 6 + 1, 8, 17, 3] + period[ 7 + 1, 8, 17, 3] +
        period[10 + 1, 8, 17, 3] + period[11 + 1, 8, 17, 3] +
        period[12 + 1, 8, 17, 3] + period[14 + 1, 8, 17, 3] // Factor
Out[14]= 4 (R[1] + R[2] + R[3] + R[4] + R[5] + R[6] + R[7] +
           R[8] + R[9] + R[10] + R[11] + R[12] + R[13] + R[14] + R[15] + R[16])
```

By taking into account the original equation this obviously simplifies to -4. (The value of the period of length 16 was -1.)

The two values for the periods of length $\frac{p-1}{2}$ can be given in closed form.

```
In[15]:= {1/2 (-1 + I^(((p - 1)/2)^2) Sqrt[p]),
         1/2 (-1 - I^(((p - 1)/2)^2) Sqrt[p])};
```

This agrees with the direct numerical calculation, as shown here for $p = 17$.

```
In[16]:= % /. p -> 17 // N
Out[16]= {1.56155, -2.56155}
```

```
In[17]:= {period[1, 8, 17, 3] /. (R -> (Exp[2Pi I #/17.]&)),
         period[3, 8, 17, 3] /. (R -> (Exp[2Pi I #/17.]&))} // N // Chop
Out[17]= {1.56155, -2.56155}
```

In the general case, the values of the periods with $f < \frac{p-1}{2}$ and their products have to be determined by the above-described recursive procedures.

By calculating the values of the periods Λ_1 and Λ_2 in this manner (up to the correspondence between periods and their numerical values), we can recursively continue splitting periods until we end at the periods of length 2.

Now, we implement the calculation of $\cos(\frac{2\pi}{257})$ via the routine `GaussSolve[p, Λ]`. Its arguments are the prime p and the name of the periods Λ. The eight functions introduced inside `GaussSolve`, λ, `newλs`, `Timesλ`, `allλs`, `rules1`, `rules`, `Simplifyλ`, and `solStep`, have the following meanings. (In every case, we use the primitive root $g = 3$, which is possible as we have proved above by explicit calculation for the primes p of interest here.)

$\lambda[t, f]$ gives the numbers of the roots of the period of length f containing the root t (without using the head R as above).

`newλs[t, f]` returns a list with two roots that generate the two periods of length $f/2$ by splitting the period $\overset{f}{t}$ of length f, which contains the root $\bar{t}$.

`Timesλ[t, u, f]` carries out the above-described multiplication of the two periods $\overset{f}{t}$ and $\overset{f}{u}$.

`allλs[f]` gives a list containing one root from every period of length f with $2 \leq f \leq \frac{p-1}{2}$.

`rules1[f]` gives a list of rules that replace all of the $\overset{f}{\lambda_2}, \overset{f}{\lambda_3}, ..., \overset{f}{\lambda_f}$ by $\overset{f}{\lambda_1}$ if the $\overset{f}{\lambda_i}$ arise from $\overset{f}{\lambda} = \overset{f}{\lambda_1} + \overset{f}{\lambda_2} + \cdots + \overset{f}{\lambda_f}$. (This is possible because obviously the relation $\overset{f}{\lambda_1} = \overset{f}{\lambda_2} = \cdots = \overset{f}{\lambda_f}$ holds as a result of

$\overset{f}{\lambda} = \overline{\lambda_1} + \overline{\lambda_2} + \cdots + \overline{\lambda_f}$ (a period can be generated by any of its roots) and ensures that we have only the value of one root from every period to calculate.)

rules2 [*f*] replaces the period containing the root $\overline{\left(\lambda g^{\frac{p-1}{f}}\right)} \mod(p-1)$ by $\overset{2f}{\lambda} - \overset{f}{\lambda}$ to have only one new period introduced explicitly in solList at each subdivision.

Simplifyλ[*t*, *u*, *f*] simplifies the product of two periods calculated with Times1[*t*, *u*, *f*] using the rules rules1 and rules2.

solStep[*t*, *f*] carries out one subdivision of the period $\overset{f}{t}$, and calculates the values of the periods $\overset{f/2}{u}$ and $\overset{f/2}{v}$ after splitting. For the determination of which root belongs to which period, we compare the numerical values of the periods. (This comparison could be done with much more effort symbolically, but because this has no influence on the result, we use this quick way.) Carrying out solStep[*t*, *f*] has no result, but only a side effect: The symbolic values for the new periods obtained by splitting $\overset{f}{t}$ are added to the list solList.

The algorithm carried out by GaussSolve is quite simple: After the definition of the just-described functions and the calculation of all numerical values of the periods (which are stored in solNList), we collect in solArgs the ordered list of all periods $\overset{f}{\lambda}$ that have to be subdivided in the form $\{\lambda, f\}$. Then, solStep is applied to all of the elements of solArgs. The result of GaussSolve[*p*, Λ] gives a list of replacement rules for the symbolic values of the periods $\Lambda[\lambda, f]$. We collect not only the result, but also all values for the periods, using this list, we can later also easily get cos(j2π/*p*) (*j* other than 1). Because for *p* > 17 the various lists of rules that are in use inside GaussSolve are quite big, we use Dispatch to accelerate their application (with the exception of the list solList, which is not used actively internally, but only serves as a container for the results).

```
In[18]:= GaussSolve[p: (3 | 5 | 17 | 257 | 65537), Λ_Symbol] :=
    Module[{g = 3, λ, newλs, Timesλ, allλs, rules1, rules2, Simplifyλ,
            solStep, solArgs, solNList, solList = {Λ[1, p - 1] -> - 1}},
        (* the λ's *)
        λ[t_, f_] := λ[t, f] = Function[γ, Mod[Mod[t, p] Array[
                        PowerMod[γ, #, p]&, f, 0], p]][g^((p - 1)/f)];
        (* newλs function definition with remembering *)
        newλs[t_, f_] := newλs[t, f] =
            {t, Mod[Mod[t, p] PowerMod[g, (p - 1)/f, p], p]};
        (* Timesλ function for λ multiplication *)
        Timesλ[t_, u_, f_] := Plus @@ (Λ[#, f]& /@ Mod[λ[u, f] + t, p]);
        (* allλs lists *)
        allλs[p - 1] = {1};
        allλs[f_] := allλs[f] = Flatten[Map[newλs[#, 2f]&, allλs[2f], {-1}]];
        (* rules1 for λ canonicalization *)
        rules1[f_] := rules1[f] =
        Dispatch[Map[Λ[#, f]&, Flatten[Function[a,
          Apply[Rule, Transpose[{Rest[a], Table[#,
          {Length[Rest[a]]}]}&[First[a]]}], {1}]] /@
                    (λ[#, f]& /@ allλs[f])], {-1}]];
        (* rules2 for λ eliminating one λ *)
        rules2[(p - 1)/2] = Λ[g, (p - 1)/2] -> - 1 - Λ[1, (p - 1)/2];
        rules2[f_] := rules2[f] = Dispatch[
          Λ[#[[2, 2]], f] -> Λ[#[[1]], 2f] - Λ[#[[2, 1]], f]& /@
                Map[{#, newλs[#, 2f]}&, allλs[2f], {-1}]];
        (* Simplifyλ for simplifying products of λs *)
        Simplifyλ[t_, u_, f_] := Fold[Expand[#1 //. #2]&,
```

```
        Expand[Timesλ[t, u, f] //. rules1[f]],
            rules2 /@ (f 2^Range[0, Log[2, (p - 1)/f] - 1])];
(* solStep for period subdivision *)
solStep[t_, f_] :=
Module[{u, v, x1Px2, x1Tx2, sol1, sol2, sol1N, sol2N, numSol1},
{u, v} = newλs[t, f];
 x1Px2 = Λ[t, f]; x1Tx2 = Simplifyλ[u, v, f/2];
 {sol1, sol2} = # + Sqrt[#^2 - x1Tx2]{1, -1}&[x1Px2/2];
 numSol1 = Λ[u, f/2] //. solNList;
 {sol1N, sol2N} = N[{sol1, sol2} //. solNList];
 solList = Flatten[{solList,
 If[Abs[sol1N - numSol1] < Abs[sol2N - numSol1],
     {Λ[u, f/2] -> sol1, Λ[v, f/2] -> sol2},
     {Λ[u, f/2] -> sol2, Λ[v, f/2] -> sol1}}]}]; ];
(* solNList for numerical values of the periods *)
solNList = Dispatch[Apply[(Λ @ ##) ->
    (Plus @@ Exp[N[2Pi I λ[##]/p]])&,
       Flatten[Function[i, {#, i}& /@ allλs[i]] /@
               (2^Range[Log[2, p - 1], 1, -1]), 1], {1}]];
(* stepArgs for period arguments *)
stepArgs = Flatten[Function[i, {#, i}& /@
    allλs[i]] /@ (2^Range[Log[2, p - 1], 2, -1]), 1];
(* do the work *) solStep @@ #& /@ stepArgs; solList]
```

Now, let us calculate the two simple cases $p = 3$ and $p = 5$ as a warm up.

In[19]:= **(Λ[1, 2] //. GaussSolve[3, Λ])/2**

Out[19]= $-\dfrac{1}{2}$

In[20]:= **(Λ[1, 2] //. GaussSolve[5, Λ])/2 // Expand**

Out[20]= $-\dfrac{1}{4} + \dfrac{\sqrt{5}}{4}$

The results agree with the well-known expressions for $\cos(2\pi/3)$ and $\cos(2\pi/5)$. Here is the list of the values of the periods for $p = 17$.

In[21]:= **(list17 = GaussSolve[17, Λ]) // InputForm**

Out[21]//InputForm=
```
{Λ[1, 16] -> -1, Λ[1, 8] -> Λ[1, 16]/2 + Sqrt[4 + Λ[1, 16]^2/4],
 Λ[3, 8] -> Λ[1, 16]/2 - Sqrt[4 + Λ[1, 16]^2/4],
 Λ[1, 4] -> Λ[1, 8]/2 + Sqrt[1 + Λ[1, 8]^2/4],
 Λ[9, 4] -> Λ[1, 8]/2 - Sqrt[1 + Λ[1, 8]^2/4],
 Λ[3, 4] -> Λ[3, 8]/2 + Sqrt[1 + Λ[3, 8]^2/4],
 Λ[10, 4] -> Λ[3, 8]/2 - Sqrt[1 + Λ[3, 8]^2/4],
 Λ[1, 2] -> Λ[1, 4]/2 + Sqrt[Λ[1, 4]^2/4 - Λ[3, 4]],
 Λ[13, 2] -> Λ[1, 4]/2 - Sqrt[Λ[1, 4]^2/4 - Λ[3, 4]],
 Λ[9, 2] -> Λ[9, 4]/2 - Sqrt[1 + Λ[1, 8] + Λ[3, 4] + Λ[9, 4]^2/4],
 Λ[15, 2] -> Λ[9, 4]/2 + Sqrt[1 + Λ[1, 8] + Λ[3, 4] + Λ[9, 4]^2/4],
 Λ[3, 2] -> Λ[3, 4]/2 + Sqrt[Λ[1, 4] - Λ[1, 8] + Λ[3, 4]^2/4],
 Λ[5, 2] -> Λ[3, 4]/2 - Sqrt[Λ[1, 4] - Λ[1, 8] + Λ[3, 4]^2/4],
 Λ[10, 2] -> Λ[10, 4]/2 - Sqrt[-Λ[1, 4] + Λ[10, 4]^2/4],
 Λ[11, 2] -> Λ[10, 4]/2 + Sqrt[-Λ[1, 4] + Λ[10, 4]^2/4]}
```

Here is the final expression for $\cos(\frac{2\pi}{17})$.

In[22]:= **(Λ[1, 2] //. list17)/2 // Expand // Factor**

Out[22]= $\dfrac{1}{16}\left(-1+\sqrt{17}+\sqrt{2\,(17-\sqrt{17}\,)}\;+\right.$

$$\left.\sqrt{2\left(34+6\sqrt{17}-\sqrt{2\,(17-\sqrt{17}\,)}+\sqrt{34\,(17-\sqrt{17}\,)}-8\sqrt{2\,(17+\sqrt{17}\,)}\,\right)}\,\right)$$

We numerically check this result. Because the result *is* 0, we cannot get any significant digit, and so the `N::meprec` message is issued.

In[23]:= `(% - Cos[2Pi/17]) // SetPrecision[#, 1000]&`

Out[23]= $0.\times 10^{-1000}$

Next is the result for $\cos(2\times2\pi/17)$. (Because we have eliminated most of the $L[j,\,2]$'s with even j, we make use of $\cos(2\,j\pi/p)=\cos(2\,(p-j)\,\pi/p)$ and use $L[15,\,2]$.)

In[24]:= `(A[15, 2] //. list17)/2 // Expand // Factor`

Out[24]= $\dfrac{1}{16}\left(-1+\sqrt{17}-\sqrt{2\,(17-\sqrt{17}\,)}\;+\right.$

$$\left.\sqrt{2\left(34+6\sqrt{17}+\sqrt{2\,(17-\sqrt{17}\,)}-\sqrt{34\,(17-\sqrt{17}\,)}+8\sqrt{2\,(17+\sqrt{17}\,)}\,\right)}\,\right)$$

In[25]:= `(% - Cos[2 2Pi/17]) // SetPrecision[#, 1000]&`

Out[25]= $0.\times 10^{-1000}$

Using the powerful function `RootReduce` we could also prove the last equality symbolically.

In[26]:= `(%% // Simplify // RootReduce) -`
`(Together[TrigToExp[Cos[2 2Pi/17]]] // RootReduce)`

Out[26]= 0

In[27]:= `Together[TrigToExp[Cos[2 2Pi/17]]]`

Out[27]= $-\dfrac{1}{2}\,(-1)^{13/17}\,(1+(-1)^{8/17})$

The last value of interest here is $\cos(8\times2\pi/17)$.

In[28]:= `(A[9, 2] //. list17)/2 // Expand // Factor`

Out[28]= $\dfrac{1}{16}\left(-1+\sqrt{17}-\sqrt{2\,(17-\sqrt{17}\,)}\;-\right.$

$$\left.\sqrt{2\left(34+6\sqrt{17}+\sqrt{2\,(17-\sqrt{17}\,)}-\sqrt{34\,(17-\sqrt{17}\,)}+8\sqrt{2\,(17+\sqrt{17}\,)}\,\right)}\,\right)$$

Here is again a quick numerical check for the last result.

In[29]:= `(% - Cos[8 2Pi/17]) // SetPrecision[#, 1000]&`

Out[29]= $0.\times 10^{-1000}$

Now, as promised in the title of this subsection, we calculate $\cos(2\pi/257)$ [1487], [638].

In[30]:= `list257 = GaussSolve[257, A];`

We select only those parts that are explicitly needed for the evaluation of $\cos(2\pi/257)$.

In[31]:= `Flatten[Function[{lhs, rhs},`
`(* until we have all needed A's *)`

```
FixedPoint[{#, Complement[Union[Cases[
(* what is in the rhs *)
rhs[[#]]& /@ Flatten[Position[lhs, #]& /@
    Last[#]], _Λ, {0, Infinity}]], Flatten[#]]}&,
(* this we need of course *) {{Λ[1, 2]}},
SameTest -> (Last[#2] === {}&)]][
            (* all lhs and rhs from list257 *)
            First /@ list257, Last /@ list257]];

solListPiD257 = (list257[[#]]& /@ Flatten[Function[lhs,
        Position[lhs, #]& /@ %][First /@ list257]]);
```

Here is a shortened version of this list of replacement rules necessary to express $\cos(\frac{2\pi}{257})$.

In[33]:= `solListPiD257 // Short[#, 6]&`

Out[33]//Short=

$$\{\Lambda[1, 2] \to \frac{1}{2} \Lambda[1, 4] + \sqrt{\frac{1}{4} \Lambda[1, 4]^2 + \Lambda[1, 16] - \Lambda[1, 32] + \Lambda[136, 8] + \Lambda[197, 4]},$$

$$\Lambda[1, 4] \to \frac{1}{2} \Lambda[1, 8] + \sqrt{\frac{1}{4} \Lambda[1, 8]^2 - \Lambda[3, 8] + \Lambda[131, 8] - \Lambda[131, 16]},$$

$$\Lambda[1, 16] \to \frac{1}{2} \Lambda[1, 32] +$$

$$\sqrt{-\Lambda[1, 32] + \frac{1}{4} \Lambda[1, 32]^2 - \Lambda[1, 128] + 2 \Lambda[3, 32] - 2 \Lambda[3, 64] - \Lambda[9, 32]},$$

$$\Lambda[1, 32] \to \frac{1}{2} \Lambda[1, 64] + \sqrt{5 + 2 \Lambda[1, 64] + \frac{1}{4} \Lambda[1, 64]^2 + \Lambda[1, 128]}, \ll 28 \gg,$$

$$\Lambda[243, 32] \to \frac{1}{2} \Lambda[3, 64] + \sqrt{4 - \Lambda[1, 128] + 2 \Lambda[3, 64] + \frac{1}{4} \Lambda[3, 64]^2},$$

$$\Lambda[27, 64] \to \frac{1}{2} \Lambda[3, 128] - \sqrt{16 + \frac{1}{4} \Lambda[3, 128]^2},$$

$$\Lambda[81, 32] \to \frac{1}{2} \Lambda[1, 64] - \sqrt{5 + 2 \Lambda[1, 64] + \frac{1}{4} \Lambda[1, 64]^2 + \Lambda[1, 128]},$$

$$\Lambda[215, 32] \to \frac{1}{2} \Lambda[9, 64] - \sqrt{5 - 2 \Lambda[1, 64] + 3 \Lambda[1, 128] + \frac{1}{4} \Lambda[9, 64]^2}\}$$

The value for $\cos(\frac{2\pi}{257})$ is now easily obtained, but because of its size, we do not display it here.

In[34]:= `(cos2PiD257 = (Λ[1, 2] //. Dispatch[solListPiD257])/2) // ByteCount`

Out[34]= 1822680

It contains only square roots, but it contains a lot of them.

In[35]:= `Cases[cos2PiD257, Power[_, 1/2],`
`            {0, Infinity}, Heads -> True] // Length`

Out[35]= 5133

If the reader wants to see all of them, the following code opens a new notebook with the typeset formula for the square root version of $\cos(2\pi/257)$.

```
NotebookPut[Notebook[{Cell[BoxData[
                FormBox[MakeBoxes[#, TraditionalForm]&[cos2PiD257],
                        TraditionalForm]], "Output",
        ShowCellBracket -> False, CellMargins -> {{0, 0}, {5, 5}},
        PageWidth -> Infinity, FontColor -> GrayLevel[1],
        (* allow to see all square roots *) CellHorizontalScrolling -> True]},
        WindowSize -> {Automatic, Fit},
        Background -> RGBColor[0.31, 0., 0.51],
        ScrollingOptions   -> {"HorizontalScrollRange" -> 500000},
        WindowMargins -> {{0, 0}, {Automatic, 10}},
        WindowElements -> {"HorizontalScrollBar"},
        WindowFrameElements -> {"CloseBox"}]]
```

Here is a numerical check of the result.

In[36]:= **(cos2PiD257 - Cos[2Pi/257]) // SetPrecision[#, 1000]&**

Out[36]= $0. \times 10^{-996}$

One could now go on and calculate the following quite large calculation for the denominator 65537.

```
    165537 = GaussSolve[65537, L]
```

It will take around one day on a modern workstation. Here are the first lines of the result (of size 55 MB).

```
{Λ[  1, 65536] -> -1,
 Λ[  1, 32768] -> Λ[1, 65536]/2 + Sqrt[16384 + Λ[1, 65536]^2/4],
 Λ[  3, 32768] -> Λ[1, 65536]/2 - Sqrt[16384 + Λ[1, 65536]^2/4],
 Λ[  1, 16384] -> Λ[1, 32768]/2 - Sqrt[ 4096 + Λ[1, 32768]^2/4],
 Λ[  9, 16384] -> Λ[1, 32768]/2 + Sqrt[ 4096 + Λ[1, 32768]^2/4],
 Λ[  3, 16384] -> Λ[3, 32768]/2 - Sqrt[ 4096 + Λ[3, 32768]^2/4],
 Λ[ 27, 16384] -> Λ[3, 32768]/2 + Sqrt[ 4096 + Λ[3, 32768]^2/4],
 Λ[  1,  8192] -> Λ[1, 16384]/2 - Sqrt[ 1040 + 32 Λ[1, 16384] +
                    Λ[1, 16384]^2/4 + 16 Λ[1, 32768]],
 Λ[ 81,  8192] -> Λ[1, 16384]/2 + Sqrt[1040 + 32 Λ[1, 16384] +
                    Λ[1, 16384]^2/4 + 16 Λ[1, 32768]],
 Λ[  9,  8192] -> Λ[9, 16384]/2 - Sqrt[1040 - 32 Λ[1, 16384] +
                    48 Λ[1, 32768] + Λ[9, 16384]^2/4],
 Λ[729,  8192] -> Λ[9, 16384]/2 - Sqrt[1040 - 32 Λ[1, 16384] +
                    48 Λ[1, 32768] + Λ[9, 16384]^2/4],
 Λ[  3,  8192] -> Λ[3, 16384]/2 + Sqrt[1024 - 16 Λ[1, 32768] +
                    32 Λ[3, 16384] + Λ[3, 16384]^2/4]}
```

(Although the above implementation strictly follows Gauss's original work, we could have used more efficient procedures. See [835].)

Let us briefly discuss the numbers n for which the value $\cos(2\pi/n)$ can be expressed in square roots (or geometrically speaking, which n-gons can be constructed by ruler and compass [466], [825]?).

The above-mentioned number $2^{2^5} - 1 = 4294967295$ is not a prime number; the factors are all Fermat numbers F_j with $j = 0, \ldots, 4$.

In[37]:= **FactorInteger[2^(2^5) - 1]**

Out[37]= {{3, 1}, {5, 1}, {17, 1}, {257, 1}, {65537, 1}}

(It is not by accident that we have the relation $F_n = 2 + F_{n-1} F_{n-2} \cdots F_0$; for factorizations of Fermat numbers, see [1079] and [757].) By solving some simple linear diophantine equations (how to do this in *Mathematica* is discussed in [1801]), we easily find that $1 / (2^{2^5} - 1)$ can be expressed as the following sum.

In[38]:= 1/(3 5 17 257 65537) -
          (29129863/3 - 58259726/5 + 33291272/17 - 4194176/257 + 32768/65537)

Out[38]= 0

In general, it is possible to calculate $\cos(2\pi/n)$ in square roots if n is a product q of prime Fermat numbers that contains no factor more than once or if $n = q\,2^j$ (j is a nonnegative integer).

For an explicit ruler and compass construction of the regular 257-gon, see [758].

It should be noted that the above approach could be easily generalized to the case of splitting one period into more than two periods. Then, the resulting equations obeyed by the periods are no longer quadratic, but of the order of the number of newly generated periods. For more details on this subject, see [109], [480], [508], [1080], [1290], [1688], [1280], [883], [375], [1660], [1717], [953], [663], [1329], [1821], and [1820]. See [131], [944], [311], [1778], [1058], and [1670] for the explicit geometrical construction of the *n*-gon.

1.10.3 Implicitization of a Trefoil Knot

In this subsection, we will deal with a larger polynomial calculation, the implicitization of a trefoil knot. Implicitization of a parametrically defined surface is, in general, a very useful operation because both the parametric and implicit representations have advantages (see [115], [1593], [1067], and [1411]). The parametric version is easy to plot, and the implicit one allows an easy decision on which side of the surface a given point lies or if a point lies on the surface. In Chapter 3 of the Graphics volume [1736], we discussed and constructed several polynomial surfaces, but not a knotted one. Using the technique described there of gluing together simple implicitly defined objects (like ellipsoids, spheres, tori, cones, tubes) appropriately spaced and oriented, we could generate knotted objects.

But we consider this approach cheating, and we will construct a clean "knot tube" here.

Using the function `TubeFunctionalShort` from Chapter 2 of the Graphics volume [1736], here is a picture of the trefoil knot under consideration.

```
In[1]:= TubeFunctionalShort[
                curve_List        (* parametric representation      *),
                cp_               (* curve parameter                *),
                rr_               (* tube radius                    *),
                cpInt_List        (* domain of the curve parameter  *),
                ppq_              (* PlotPoints cross section        *),
                ppl_              (* PlotPoints transversal          *),
                opts___           (* possible options for the plot  *)] :=
        Show[Graphics3D[Drop[MapThread[Polygon[{#1, #2, #3, #4}]&,
          {#[[1]], #[[2]], RotateRight[#[[3]]], #[[3]]}&[
           {#, RotateRight[#], Transpose[RotateRight[
            Transpose[#]]]}&[ Function[uln,
          MapThread[Map[Function[temp, temp + #1], #2]&,
            {Function[cp, Evaluate[N[curve]]] /@ uln,
             Partition[(Plus @@ #)& /@ Distribute[{
              Transpose[Function[{d1n, d2n},
               {(#/Sqrt[#.#])& /@ MapThread[(#2/(#1.#1) -
```

```
               (#1.#2) #1/(#1.#1)^2)&, {d1n, d2n}],
        MapThread[(Function[{t1, t2, t3, k1, k2, k3},
                  {t2 k3 - t3 k2, t3 k1 - t1 k3, t1 k2 - t2 k1}] @@
               Flatten[{##}]&), {(#/Sqrt[#.#])& /@
          d1n, (#/Sqrt[#.#])& /@ MapThread[(#2/(#1.#1) -
            (#1.#2) #1/(#1.#1)^2)&, {d1n, d2n}]}]}][
      Function[cp, Evaluate[N[D[curve, cp]]]] /@ u1n,
      Function[cp, Evaluate[N[D[curve, {cp, 2}]]]] /@ u1n]],
        rr N[{Cos[#], Sin[#]}& /@
      N[Range[0, 2Pi(1 - 1/ppq), 2Pi/ppq]]]},
            List, List, List, Times], {ppq}]}]][
         N[Range[cpInt[[1]], cpInt[[2]],
          -Subtract @@ cpInt/ppl]]]]], 2], 1]], opts]
```

```
In[2]:= TubeFunctionalShort[
            {Sin[t] + 2 Sin[2 t], Cos[t] - 2 Cos[2 t], Sin[3 t]}
                                (* parametric representation   *),
            t                   (* curve parameter             *),
            1/2                 (* tube radius                 *),
            {0, 2Pi}            (* domain of the curve parameter *),
            12                  (* PlotPoints cross section     *),
            100                 (* PlotPoints transversal       *)]
```

Our trefoil knot is a tube of radius 1/2 along the space curve parametrized by

$$\{c_x(t),\ c_y(t),\ c_z(t)\} = \{\sin(t) + 2\sin(2\,t),\ \cos(t) - 2\cos(2\,t),\ \sin(3\,t)\}$$

(a so-called canal surface [193]). Our goal is to find a function $f(x,\ y,\ z)$, such that the tuboidal surface shown last is given by $f(x,\ y,\ z) = 0$. To make use of *Mathematica*'s built-in function to construct such an implicit representation, we calculate a polynomial $p(x,\ y,\ z)$ describing the above curve.

A first attempt could be to explicitly write down the parametric representation $\{x(s,\ t),\ y(s,\ t),\ z(s,\ t)\}$ of the trefoil knot and then try to eliminate the two parametric variables s and t. Because of the normalization conditions on the normal and the binormal vectors to ensure the tube radius of $1/2$, this parametrization is quite a complicated function, involving square roots of polynomials in trigonometric functions of s and t. Therefore, there seems little chance to eliminate s and t from these expressions. So, we choose another approach based on the following two equations [1295]:

$$|\{x,\ y,\ z\} - \{c_x(t),\ c_y(t),\ c_z(t)\}|^2 - \left(\frac{1}{2}\right)^2 = 0$$

$$\left\{\frac{dc_x(t)}{dt},\ \frac{dc_y(t)}{dt},\ \frac{dc_z(t)}{dt}\right\} \cdot (\{x,\ y,\ z\} - \{c_x(t),\ c_y(t),\ c_z(t)\}) = 0.$$

The meaning of these two equations is the following: The first equation expresses the condition that every point $\{x, y, z\}$ of the tube is an element of a sphere of radius $1/2$ around a point $\{c_x(t), c_y(t), c_z(t)\}$ of the space curve. The second equation ensures that the vector from the space curve point $\{c_x(t), c_y(t), c_z(t)\}$ to the point $\{x, y, z\}$ of the tube is always perpendicular to the tangent at the space curve, so that the tube's radius is constant in the perpendicular cross section. This means, these two equations describe exactly the tube we want to implicitize.

The advantage of this description is that now we have only one parameter, the curve parameter t, to eliminate. We do this by using rational parametrization of the trigonometric functions to transform the problem into a polynomial one. If this is done, we can use known elimination algorithms to eliminate the curve parameter. The resulting polynomial then contains the implicit representation we are looking for as some factor. (For the inverse problem—the derivation of a parametrized description from an implicit description, see [1570].)

There are basically two ways to implicitize equations of the above type [1069], [1244]. The first possibility is to add the equation $\sin(t)^2 + \cos(t)^2 - 1 = 0$ to the equations under consideration and then to eliminate $\sin(x)$ and $\cos(x)$ from the resulting set of polynomial equations. The second possibility is to use the new variable $s = \tan(t/2)$. In this case, we keep the number of variables, but introduce fractions of polynomials. After clearing denominators (we will not care here about the zeros of the denominators; for how to take these single points into account properly, see [414], [432]), we eliminate s from the resulting equations. In the following, we choose the second way. Here, we demonstrate this method for calculating the implicit representation of a torus, using the built-in function `Eliminate` to eliminate the one variable s.

```
In[3]:= With[{c = {R Sin[t], R Cos[t], 0} (* the spine curve of a torus *)},
    Factor @ (Subtract @@ (Eliminate[# == 0& /@
    (({#.#&[{x, y, z} - c] - r^2, D[c, t].({x, y, z} - c)} /.
    {Cos[t] -> (1 - s^2)/(1 + s^2), Sin[t] -> 2 s/(1 + s^2)})), {s}]))]
```
$$Out[3]= r^4 - 2\,r^2\,R^2 + R^4 - 2\,r^2\,x^2 - 2\,R^2\,x^2 + x^4 - 2\,r^2\,y^2 -$$
$$2\,R^2\,y^2 + 2\,x^2\,y^2 + y^4 - 2\,r^2\,z^2 + 2\,R^2\,z^2 + 2\,x^2\,z^2 + 2\,y^2\,z^2 + z^4$$

The last output is the implicit representation of the torus.

For our trefoil knot, the two equations derived above are given by the following inputs.

```
In[4]:= c = {Sin[t] + 2 Sin[2t], Cos[t] - 2 Cos[2t], Sin[3t]}
```
$$Out[4]= \{Sin[t] + 2\,Sin[2\,t],\ Cos[t] - 2\,Cos[2\,t],\ Sin[3\,t]\}$$

```
In[5]:= {#.#&[{x, y, z} - c] - (1/2)^2, D[c, t].({x, y, z} - c)}
```
$$Out[5]= \{-\frac{1}{4} + (y - Cos[t] + 2\,Cos[2\,t])^2 + (x - Sin[t] - 2\,Sin[2\,t])^2 + (z - Sin[3\,t])^2,$$
$$(Cos[t] + 4\,Cos[2\,t])\,(x - Sin[t] - 2\,Sin[2\,t]) +$$
$$(y - Cos[t] + 2\,Cos[2\,t])\,(-Sin[t] + 4\,Sin[2\,t]) + 3\,Cos[3\,t]\,(z - Sin[3\,t])\}$$

We express all trigonometric functions of multiple arguments in $\sin(t)$ and $\cos(t)$.

```
In[6]:= % // TrigExpand
```
$$Out[6]= \{\frac{21}{4} + x^2 + y^2 + z^2 - 2\,y\,Cos[t] + 4\,y\,Cos[t]^2 - 4\,Cos[t]^3 - \frac{Cos[t]^6}{2} - 2\,x\,Sin[t] -$$
$$8\,x\,Cos[t]\,Sin[t] - 6\,z\,Cos[t]^2\,Sin[t] - 4\,y\,Sin[t]^2 + 12\,Cos[t]\,Sin[t]^2 +$$
$$\frac{15}{2}\,Cos[t]^4\,Sin[t]^2 + 2\,z\,Sin[t]^3 - \frac{15}{2}\,Cos[t]^2\,Sin[t]^4 + \frac{Sin[t]^6}{2},$$
$$x\,Cos[t] + 4\,x\,Cos[t]^2 + 3\,z\,Cos[t]^3 - y\,Sin[t] + 8\,y\,Cos[t]\,Sin[t] -$$
$$18\,Cos[t]^2\,Sin[t] - 9\,Cos[t]^5\,Sin[t] - 4\,x\,Sin[t]^2 -$$
$$9\,z\,Cos[t]\,Sin[t]^2 + 6\,Sin[t]^3 + 30\,Cos[t]^3\,Sin[t]^3 - 9\,Cos[t]\,Sin[t]^5\}$$

Using rational parametrizations for $\sin(t)$ and $\cos(t)$, we get the following two polynomials in s after writing both equations over a common denominator.

In[7]:= % /. {Cos[t] -> (1 - s^2)/(1 + s^2),
             Sin[t] -> 2 s/(1 + s^2)}

Out[7]= $\{ \dfrac{21}{4} + \dfrac{32\,s^6}{(1+s^2)^6} - \dfrac{120\,s^4\,(1-s^2)^2}{(1+s^2)^6} + \dfrac{30\,s^2\,(1-s^2)^4}{(1+s^2)^6} - \dfrac{(1-s^2)^6}{2\,(1+s^2)^6} +$

$\dfrac{48\,s^2\,(1-s^2)}{(1+s^2)^3} - \dfrac{4\,(1-s^2)^3}{(1+s^2)^3} - \dfrac{16\,s\,(1-s^2)\,x}{(1+s^2)^2} - \dfrac{4\,s\,x}{1+s^2} + x^2 - \dfrac{16\,s^2\,y}{(1+s^2)^2} +$

$\dfrac{4\,(1-s^2)^2\,y}{(1+s^2)^2} - \dfrac{2\,(1-s^2)\,y}{1+s^2} + y^2 + \dfrac{16\,s^3\,z}{(1+s^2)^3} - \dfrac{12\,s\,(1-s^2)^2\,z}{(1+s^2)^3} + z^2,$

$-\dfrac{288\,s^5\,(1-s^2)}{(1+s^2)^6} + \dfrac{240\,s^3\,(1-s^2)^3}{(1+s^2)^6} - \dfrac{18\,s\,(1-s^2)^5}{(1+s^2)^6} + \dfrac{48\,s^3}{(1+s^2)^3} -$

$\dfrac{36\,s\,(1-s^2)^2}{(1+s^2)^3} - \dfrac{16\,s^2\,x}{(1+s^2)^2} + \dfrac{4\,(1-s^2)^2\,x}{(1+s^2)^2} + \dfrac{(1-s^2)\,x}{1+s^2} +$

$\dfrac{16\,s\,(1-s^2)\,y}{(1+s^2)^2} - \dfrac{2\,s\,y}{1+s^2} - \dfrac{36\,s^2\,(1-s^2)\,z}{(1+s^2)^3} + \dfrac{3\,(1-s^2)^3\,z}{(1+s^2)^3} \}$

In[8]:= {p1, p2} = Numerator /@ Together /@ %

Out[8]= $\{3 + 450\,s^2 - 243\,s^4 + 2268\,s^6 - 1107\,s^8 + 66\,s^{10} + 35\,s^{12} - 80\,s\,x - 272\,s^3\,x - 288\,s^5\,x - 32\,s^7\,x +$
$112\,s^9\,x + 48\,s^{11}\,x + 4\,x^2 + 24\,s^2\,x^2 + 60\,s^4\,x^2 + 80\,s^6\,x^2 + 60\,s^8\,x^2 + 24\,s^{10}\,x^2 + 4\,s^{12}\,x^2 +$
$8\,y - 64\,s^2\,y - 312\,s^4\,y - 448\,s^6\,y - 232\,s^8\,y + 24\,s^{12}\,y + 4\,y^2 + 24\,s^2\,y^2 + 60\,s^4\,y^2 +$
$80\,s^6\,y^2 + 60\,s^8\,y^2 + 24\,s^{10}\,y^2 + 4\,s^{12}\,y^2 - 48\,s\,z + 16\,s^3\,z + 288\,s^5\,z + 288\,s^7\,z +$
$16\,s^9\,z - 48\,s^{11}\,z + 4\,z^2 + 24\,s^2\,z^2 + 60\,s^4\,z^2 + 80\,s^6\,z^2 + 60\,s^8\,z^2 + 24\,s^{10}\,z^2 + 4\,s^{12}\,z^2,$
$-54\,s + 342\,s^3 - 972\,s^5 + 1404\,s^7 - 318\,s^9 - 18\,s^{11} + 5\,x - 4\,s^2\,x - 63\,s^4\,x -$
$112\,s^6\,x - 73\,s^8\,x - 12\,s^{10}\,x + 3\,s^{12}\,x + 14\,s\,y + 38\,s^3\,y + 12\,s^5\,y - 52\,s^7\,y -$
$58\,s^9\,y - 18\,s^{11}\,y + 3\,z - 36\,s^2\,z - 81\,s^4\,z + 81\,s^8\,z + 36\,s^{10}\,z - 3\,s^{12}\,z\}$

We now have to calculate the resultant of two polynomials of degree 12 with symbolic coefficients. This is potentially quite a large calculation. There are various possibilities, for instance, the subresultant method [1255], [251], [1274], [386], [387], and [1153] (which requires calculating greatest common divisors of multivariate polynomials, which we are not going to do here to avoid very complicated polynomial work), using Sylvester's determinant formulation (which, as discussed in Section 1.2, results in calculating the determinant of a not sparse 24×24 matrix with symbolic entries, a very stunning task, too), or making an ansatz for the resulting polynomial in x, y, and z and determining the unknown coefficients (for instance, via interpolation [1197], [1198], [699]) or Cayley's method [980], [1823], [1824], [773], and [708] for the resultant calculation (and use Laplace's expansion [1109], [1777] recursively of the determinant). Here we use the Method->Modular option setting of the built-in function Resultant.

In[9]:= (treFoilKnotPoly[x_, y_, z_] =
          Factor[Resultant[p1, p2, s, Method -> Modular]]); // Timing
Out[9]= {58.55 Second, Null}

In[10]:= treFoilKnotPoly[x, y, z] // Short[#, 16]&
Out[10]//Short=

     1329227995784915872903807060280344576
       $(2164764194414099357625 - 7014002132627225648640\,x^2 + 2557575896813793543168\,x^4 +$
       $163721455988819921280\,x^6 - 35920473838281441792\,x^8 + 11101452668745744384\,x^{10} -$
       $4896337446674124800\,x^{12} - 836052734165745664\,x^{14} + 292894441143336960\,x^{16} +$
       $2947572217413632\,x^{18} - 5608909354041344\,x^{20} + 379637528002560\,x^{22} +$
       $8916100448256\,x^{24} - 17950247748010258851456\,x^2\,y + 16014563930164085155584\,x^4\,y -$
       $4092306021947893469184\,x^6\,y + 107486645236502704128\,x^8\,y +$
       $101077487797146107904\,x^{10}\,y - 39569792051317112832\,x^{12}\,y +$
       $2121291115181899776\,x^{14}\,y + 1455106505387802624\,x^{16}\,y - 213048086451191808\,x^{18}\,y +$

$5501130696032256 x^{20}\,y + \ll 1638 \gg +\,427972821516288\,y^7\,z^{14} +$
$60867245726760960\,y^8\,z^{14} - 570630428688384\,y^9\,z^{14} + 959377691847426048\,x^3\,z^{15} +$
$248224236479447040\,x^5\,z^{15} + 6847565144260608\,x^5\,y\,z^{15} - 2878133075542278144\,x\,y^2\,z^{15} -$
$496448472958894080\,x^3\,y^2\,z^{15} - 22825217147535360\,x^3\,y^3\,z^{15} -$
$744672709438341120\,x\,y^4\,z^{15} + 6847565144260608\,x\,y^5\,z^{15} + 1236413481360556032\,z^{16} +$
$252979490051850240\,x^2\,z^{16} + 9130086859014144\,x^6\,z^{16} - 337813213783523328\,x^2\,y\,z^{16} +$
$252979490051850240\,y^2\,z^{16} + 27390260577042432\,x^4\,y^2\,z^{16} +$
$112604404594507776\,y^3\,z^{16} + 27390260577042432\,x^2\,y^4\,z^{16} + 9130086859014144\,y^6\,z^{16} +$
$36520347436056576\,x^3\,z^{17} - 109561042308169728\,x\,y^2\,z^{17} + 36520347436056576\,z^{18})$

The result is a large polynomial. It is of total degree 24, has 1212 terms and coefficients with up to 23 digits.

In[11]:= (* remove large factored-out integer prefactor *)
```
treFoilKnotPoly[x_, y_, z_] = treFoilKnotPoly[x, y, z] [[-1]];
```

In[13]:= { (* number of terms *) **Length** @ treFoilKnotPoly[x, y, z],
      (* total degree *) **Exponent**[treFoilKnotPoly[τ, τ, τ], τ],
      (* largest coefficient *)
      Max[**Abs**[**Cases**[treFoilKnotPoly[x, y, z], _**Integer**, {-1}]]] // **N**}

Out[13]= {1212, 24, 3.97887×10²²}

We now use the just-calculated implicit formula to make a graphic of the knot.

In[14]:= **Needs**["Graphics`ContourPlot3D`"]

```
ContourPlot3D[Evaluate[treFoilKnotPoly[x, y, z]],
             {x, -4, 4}, {y, -4, 4}, {z, -3/2, 3/2},
             MaxRecursion -> 1, Boxed -> False,
             PlotPoints -> {{21, 5}, {21, 5}, {16, 5}}]
```

We could now continue and slowly morph the knot into a ball (similar to Section 3.3 of the Graphics volume) [1745]. Here are the coordinate plane cross sections of such a morphing. We leave it to the reader to implement the corresponding 3D animation.

In[16]:= **Function**[sign, (* add and subtract constant *)
    Show[**GraphicsArray**[**Show** /@ **Transpose**[**Table**[
    ContourPlot[**Evaluate**[treFoilKnotPoly[#1, #2, #3] +
                    (* build and cut connections *) sign 10^ε],
              {#4, -5, 5}, {#5, -4, 4}, PlotPoints -> 120,
              Contours -> {0}, ContourShading -> False,
              ContourStyle -> {Hue[0.8 (ε - 55)/6]},
              DisplayFunction -> Identity, Frame -> False,
              PlotLabel -> #6 <> "-plane"]& @@@
        (* the 3 coordinate plane data *)

```
{{x, y, 0, x, y, "x,y"}, {x, 0, z, x, z, "x,z"},
 {0, y, z, y, z, "y,z"}}, {ε, 18, 26, 1/3}]]]]] /@ {+1, -1}
```

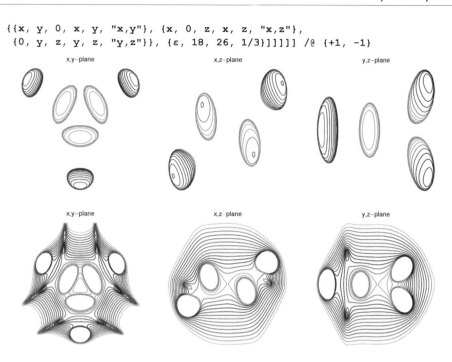

And here are two 3D plots of the resulting surfaces. By adding a constant to the polynomial, we squeeze the tube and by subtracting a constant, we thicken the tube.

```
In[17]:= Show[GraphicsArray[
       (* show squeezed and fattened version *)
       Graphics3D[{EdgeForm[], Cases[
        ContourPlot3D[Evaluate[treFoilKnotPoly[x, y, z] + #],
                     {x, -5, 5}, {y, -5, 5}, {z, -2, 2}, Boxed -> False,
                     MaxRecursion -> 1, DisplayFunction -> Identity,
                     PlotPoints -> {{21, 6}, {21, 6}, {13, 6}}],
                             _Polygon, Infinity] /. (* cut vertices off *)
                   Polygon[l_] :> Polygon[Plus @@@
                                       Partition[Append[l, l[[1]]], 2, 1]/2]},
        Boxed -> False]& /@ (* two constant values *) {8 10^21, -10^23}]]
```

In a similar manner, one can implicitize many other surfaces, when their parametrization is in terms of trigonometric or hyperbolic functions, for instance, the Klein bottle from Section 2.2.1 of the Graphics volume [1736].

Here is their implicit form together with the code for making a picture of the resulting polynomial. (For the implicitization of a "realistic looking" Klein bottle, see [1734].)

```
Needs["Graphics`ContourPlot3D`"]

Clear[x, y, z, r, φ]

Show[Graphics3D[
(* convert back from polar coordinates to Cartesian coordinates *)
Apply[{#1 Cos[#2], #1 Sin[#2], #3}&,
Cases[ContourPlot3D[Evaluate[
768 x^4 - 1024 x^5 - 128 x^6 + 512 x^7 - 80 x^8 - 64 x^9 + 16 x^10 +
144 x^2 y^2 - 768 x^3 y^2 - 136 x^4 y^2 + 896 x^5 y^2 - 183 x^6 y^2 -
176 x^7 y^2 + 52 x^8 y^2 + 400 y^4 + 256 x y^4 - 912 x^2 y^4 +
256 x^3 y^4 + 315 x^4 y^4 - 144 x^5 y^4 - 16 x^6 y^4 + 4 x^8 y^4 -
904 y^6 - 128 x y^6 + 859 x^2 y^6 - 16 x^3 y^6 - 200 x^4 y^6 +
16 x^6 y^6 + 441 y^8 + 16 x y^8 - 224 x^2 y^8 + 24 x^4 y^8 - 76 y^10 +
16 x^2 y^10 + 4 y^12 - 2784 x^3 y z + 4112 x^4 y z - 968 x^5 y z -
836 x^6 y z + 416 x^7 y z - 48 x^8 y z + 1312 x y^3 z + 2976 x^2 y^3 z -
5008 x^3 y^3 z - 12 x^4 y^3 z + 2016 x^5 y^3 z - 616 x^6 y^3 z -
64 x^7 y^3 z + 32 x^8 y^3 z - 1136 y^5 z - 4040 x y^5 z +
2484 x^2 y^5 z + 2784 x^3 y^5 z - 1560 x^4 y^5 z - 192 x^5 y^5 z +
128 x^6 y^5 z + 1660 y^7 z + 1184 x y^7 z - 1464 x^2 y^7 z -
192 x^3 y^7 z + 192 x^4 y^7 z - 472 y^9 z - 64 x y^9 z + 128 x^2 y^9 z +
32 y^11 z - 752 x^4 z^2 + 1808 x^5 z^2 - 1468 x^6 z^2 + 512 x^7 z^2 -
64 x^8 z^2 + 6280 x^2 y^2 z^2 - 5728 x^3 y^2 z^2 - 4066 x^4 y^2 z^2 +
5088 x^5 y^2 z^2 - 820 x^6 y^2 z^2 - 384 x^7 y^2 z^2 + 96 x^8 y^2 z^2 -
136 y^4 z^2 - 7536 x y^4 z^2 + 112 x^2 y^4 z^2 + 8640 x^3 y^4 z^2 -
2652 x^4 y^4 z^2 - 1152 x^5 y^4 z^2 + 400 x^6 y^4 z^2 + 2710 y^6 z^2 +
4064 x y^6 z^2 - 3100 x^2 y^6 z^2 - 1152 x^3 y^6 z^2 + 624 x^4 y^6 z^2 -
1204 y^8 z^2 - 384 x y^8 z^2 + 432 x^2 y^8 z^2 + 112 y^10 z^2 +
3896 x^3 y z^3 - 7108 x^4 y z^3 + 3072 x^5 y z^3 + 768 x^6 y z^3 -
768 x^7 y z^3 + 128 x^8 y z^3 - 3272 x y^3 z^3 - 4936 x^2 y^3 z^3 +
8704 x^3 y^3 z^3 - 80 x^4 y^3 z^3 - 2496 x^5 y^3 z^3 + 608 x^6 y^3 z^3 +
2172 y^5 z^3 + 5632 x y^5 z^3 - 2464 x^2 y^5 z^3 - 2688 x^3 y^5 z^3 +
1056 x^4 y^5 z^3 - 1616 y^7 z^3 - 960 x y^7 z^3 + 800 x^2 y^7 z^3 +
224 y^9 z^3 + 752 x^4 z^4 - 1792 x^5 z^4 + 1472 x^6 z^4 - 512 x^7 z^4 +
64 x^8 z^4 - 3031 x^2 y^2 z^4 + 1936 x^3 y^2 z^4 + 2700 x^4 y^2 z^4 -
2304 x^5 y^2 z^4 + 448 x^6 y^2 z^4 + 697 y^4 z^4 + 3728 x y^4 z^4 -
24 x^2 y^4 z^4 - 3072 x^3 y^4 z^4 + 984 x^4 y^4 z^4 - 1204 y^6 z^4 -
1280 x y^6 z^4 + 880 x^2 y^6 z^4 + 280 y^8 z^4 - 800 x^3 y z^5 +
1488 x^4 y z^5 - 768 x^5 y z^5 + 128 x^6 y z^5 + 992 x y^3 z^5 +
1016 x^2 y^3 z^5 - 1728 x^3 y^3 z^5 + 480 x^4 y^3 z^5 - 472 y^5 z^5 -
960 x y^5 z^5 + 576 x^2 y^5 z^5 + 224 y^7 z^5 + 16 x^4 z^6 +
388 x^2 y^2 z^6 - 384 x^3 y^2 z^6 + 96 x^4 y^2 z^6 - 76 y^4 z^6 -
384 x y^4 z^6 + 208 x^2 y^4 z^6 + 112 y^6 z^6 - 64 x y^3 z^7 +
32 x^2 y^3 z^7 + 32 y^5 z^7 + 4 y^4 z^8 /.
(* to polar coordinates *) {x -> r Cos[φ], y -> r Sin[φ]}],
{r, 0.6, 3.3}, {φ, 0, 2Pi}, {z, -1.3, 1.3}, PlotPoints -> {18, 40, 24},
MaxRecursion -> 0, DisplayFunction -> Identity], _Polygon, Infinity],
{-2}]]]
```

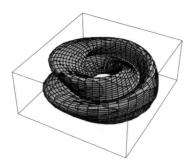

For more on the subject of implicitization of surfaces, see [1197], [351], and [1591] and references cited therein.

We end with another implicit surface originating from a trefoil knot. Starting with a parametrized space curve $c(t)$, we construct the parametrized surface $(c(t + \alpha/2) + c(t + \alpha/2))/\alpha$ (the average of two symmetrically located points with respect to t). The following code calculates the implicit form of this surface for the trefoil knot. We use the function Resultant to eliminate the parametrization variables. For brevity, we express the resulting surface in cylindrical coordinates.

```
(* a function to convert from trigonometric to polynomial variables *)
F[expr_] := Numerator[Together[TrigToExp[expr] /.
                           {t -> Log[T]/I, α -> Log[A]/I}]]
(* make algebraic form of average *)
cAv = ((c /. t -> t + α) + (c /. t -> t - α))/2
cAvAlg = F[{x, y, z} - cAv]/{I, 1, I}
(* eliminate parametrization variables *)
res1 = Resultant[cAvAlg[[1]], cAvAlg[[2]], A] // Factor
res2 = Resultant[cAvAlg[[1]], cAvAlg[[3]], A] // Factor
res3 = Resultant[res1[[-1]] /. T -> Sqrt[T2],
                 res2[[-1, 1]] /. T -> Sqrt[T2], T2,
                 Method -> SylvesterMatrix];
(* express implicit form of surface in cylindrical coordinates *)
cAvImpl  = Factor[res3][[3, 1]] /.
                  {x -> r Cos[φ], y -> r Sin[φ]} // FullSimplify
```

In[18]:= cAvImpl = r^6 (2 + r) (r - 2) (1 - 44 r^2 + 64 r^4) +
    24 r^4 (-12 - 3 r^2 + 80 r^4) z^2 -
    128 r^2 (-123 + 36 r^2 + 64 r^4) z^4 - 8192 z^6 +
    r^3 (2 z (993 r^4 - 80 r^6 - 4144 z^2 + 8192 z^4 +
    r^2 (84 - 5760 z^2)) Cos[3 φ] + r^3 (-4 + 177 r^2 - 300 r^4 + 64 r^6 -
                32 (-109 + 48 r^2) z^2) Cos[6 φ] - 64 r^6 z Cos[9 φ] -
    16 (3 r^6 (-4 + r^2) + 2 r^2 (69 - 114 r^2 + 64 r^4) z^2 -
                                256 (-2 + 3 r^2) z^4) Sin[3 φ] +
    4 r^3 z (157 - 174 r^2 + 512 z^2) Sin[6 φ] - 48 r^6 (-4 + r^2) Sin[9 φ]);

In[19]:= Needs["Graphics`ContourPlot3D`"]

In[20]:= (* a function for making a hole in a polygon *)
    makeHole[Polygon[l_], f_] :=
    Module[{mp = Plus @@ l/Length[l], l, ℓ},
      l = Append[l, First[l]]; ℓ = (mp + f (# - mp))& /@ l;
      {(* new polygons *) MapThread[Polygon[Join[#1, Reverse[#2]]]&,
                          Partition[#, 2, 1]& /@ {l, ℓ}]}]

The next pair of graphics shows the parametric and the implicit version of this surface. We make use of the threefold rotational symmetry of the surface in the generation of the implicit plot.

```
In[22]:= Show[GraphicsArray[
     Block[{$DisplayFunction = Identity, polysCart,
          R = {{-1, Sqrt[3], 0}, {-Sqrt[3], -1, 0}, {0, 0, 2}}/2.},
     {(* the parametrized 3D plot *)
     ParametricPlot3D[Evaluate[Append[
        ((c /. t -> t + α) + (c /. t -> t - α))/2,
        {EdgeForm[], SurfaceColor[#, #, 3]&[Hue[(t + Pi)/(2Pi)]]}]],
                         {t, -Pi, Pi}, {α, 0, Pi/2}, Axes -> False,
                         PlotPoints -> {64, 32}, BoxRatios -> {1, 1, 0.6},
                         PlotRange -> {{-3, 3}, {-3, 3}, {-1, 1}}] /.
                p_Polygon :> makeHole[p, 0.76],
     (* the implicit 3D plot; use symmetry *)
     polysCart = Apply[{#1 Cos[#2], #1 Sin[#2], #3}&,
       Cases[(* contour plot in cylindrical coordinates *)
       ContourPlot3D[cAvImpl, {r, 0, 3}, {φ, -Pi/3, Pi/3}, {z, -1, 1},
                     PlotPoints -> {28, 24, 32}, MaxRecursion -> 0],
                       _Polygon, Infinity], {-2}];
     Graphics3D[{EdgeForm[], (* generate all three parts of the surface *)
        {polysCart, Map[ R.#&, polysCart, {-2}],
                Map[R.R.#&, polysCart, {-2}]}} /.
       p_Polygon :> {SurfaceColor[#, #, 2.4]&[
                     Hue[Sqrt[#.#]&[0.24 Plus @@ p[[1]]/Length[p[[1]]]]]],
                     makeHole[p, 0.72]}, BoxRatios -> {1, 1, 0.6}}]}]]]
```

For the volume of such tubes, see [309].

Exercises

1.¹² The 2 in the Factorization of $x^i - 1$, Heron's Formula, Volume of Tetrahedron, Circles of Apollonius, Circle ODE, Modular Transformations, Two-Point Taylor Expansion, Quotential Derivatives

a) Program a function which finds all i for which numbers other than 0 or ± 1 appear as coefficients of x^j ($0 \le j \le i$) in the factorized decomposition of $x^j - 1$ ($1 \le i \le 500$) [586]. Do not use temporary variables (no `Block` or `Module` constructions).

b) Let P_1, P_2, and P_3 be three points in the plane. Starting from the formula $A = |(P_2 - P_1) \times (P_3 - P_1)|/2$ for the area A of the triangle formed by P_1, P_2, and P_3, derive a formula for the area which only contains the lengths of the three sides of the triangle (Heron's area formula).

c) Let P_1, P_2, P_3, and P_4 be four points in $\mathbb{R}^3$. Starting from the formula $V = (areaOfOneFace\ height\ /3)$ for the volume V of the tetrahedron formed by P_1, P_2, P_3, and P_4, derive a formula for the volume which only contains the lengths of the six edges of the tetrahedron [841].

d) Given are three circles in the plane that touch each other pairwise. In the "middle" between these three circles now put a fourth circle that touches each of the three others. Calculate the radius of this circle as an explicit function of the radius of the three other circles (see [1630], [416], [155], [1680], [839], and [695]).

e) Calculate the differential equation that governs all circles in the x,y-plane (from I.I.5.6 of [896]).

f) Show that the three equations

$$u^4 - v(u)^4 - 2\,u\,v(u)\,(1 - u^2\,v(u)^2) = 0$$
$$u^6 - v(u)^6 + 5\,v(u)^2\,u^2\,(u^2 - v(u)^2) - 4\,u\,v(u)\,(1 - u^4\,v(u)^4) = 0$$
$$(1 - u^8)\,(1 - v(u)^8) - (1 - u\,v(u))^8 = 0$$

are solutions of the (so-called modular) differential equation [1438]

$$\left(\left(\frac{1 + k^2}{k - k^3} \right)^2 - \left(\frac{1 + l}{l - l^3} \right)^2 l'(k)^2 \right) l'(k)^2 + 3\,l''(k)^2 - 2\,l'(k)\,l'''(k) = 0.$$

The change of variables between $\{k,\ l\}$ and $\{u,\ v\}$ is given by $k^{\frac{1}{4}} = u$ and $l^{\frac{1}{4}} = v$.

g) The function

$$w(x) = c_1\,e^{-\int \frac{f(x)}{1 - h(x)}\,dx} \left(c_2 + \int \frac{e^{\int g(x)\,dx + \int \frac{f(x)}{1 - h(x)}\,dx}}{1 - h(x)}\,dx \right)$$

fulfills a linear second-order differential equation [700]. Derive this differential equation.

h) Prove the following two identities (from [1838] and [897]):

$$\tan\left(\frac{1}{4}\,\tan^{-1}(4) \right) = 2\left(\cos\left(\frac{6\pi}{17} \right) + \cos\left(\frac{10\pi}{17} \right) \right)$$

$$\cos\left(\frac{\pi}{7} \right) = \frac{1}{6} + \frac{\sqrt{7}}{6}\left(\cos\left(\frac{1}{3}\,\cos^{-1}\left(\frac{1}{2\sqrt{7}} \right) \right) + \sqrt{3}\,\sin\left(\frac{1}{3}\,\cos^{-1}\left(\frac{1}{2\sqrt{7}} \right) \right) \right)$$

i) Given a rectangular box of size $w_1 \times h_1 \times d_1$. Is it possible to put a second box of size $w_2 \times h_2 \times d_2$ in the first one such that $1/w_2 + 1/h_2 + 1/d_2$ is equal to, less than, or greater than $w_1 + h_1 + d_1$?

j) What geometric object is described by the following three inequalities?

$$|\phi x| + |y| < 1 \ \wedge \ |\phi y| + |z| < 1 \ \wedge \ |x| + |\phi z| < 1$$

(ϕ is the Golden ratio.)

k) Check the following integral identity [1062]:

$$\int_0^\xi \left(\int_x^\infty \frac{f(t)}{t} dt \right)^2 dx = \int_0^\xi \left(\frac{1}{x} \int_0^x f(t) dt \right)^2 dx + \left(\sqrt{\xi} \int_\xi^\infty \frac{f(t)}{t} dt + \frac{1}{\sqrt{\xi}} \int_0^\xi f(t) dt \right)^2.$$

l) Check the following identity [1900] for small integer n and r:

$$\sum_{k=1}^n \frac{p(a_k)}{(x - a_k)^{r+1} \prod_{\substack{l=1 \\ l \ne k}}^n (a_k - a_l)} =$$

$$\frac{(-1)^r}{r! \prod_{k=1}^n (x - a_k)} \left[p^{(r)}(x) + \sum_{j=1}^r (-1)^j \binom{r}{j} p^{(r-j)}(x) A\left(\sum_{i=1}^n (x - a_i)^{-1}, \sum_{i=1}^n (x - a_i)^{-2}, \ldots, \sum_{i=1}^n (x - a_i)^{-j} \right) \right]$$

Here $p(z)$ is a polynomial of degree equal to or less than n; the a_k are arbitrary complex numbers and the multivariate polynomials $A(\tau_1, \ldots, \tau_j)$ are defined through

$$A(\tau_1, \tau_2, \ldots, \tau_j) = \sum_{\substack{k_1, k_2, \ldots, k_j \\ k_1 + 2k_2 + \cdots + jk_j = j}} \frac{j!}{k_1! \, k_2! \cdots k_j!} \left(\frac{\tau_1}{1} \right)^{k_1} \left(\frac{\tau_2}{2} \right)^{k_2} \cdots \left(\frac{\tau_j}{j} \right)^{k_j}.$$

m) Given five points in $\mathbb{R}^2$, find all relations between the oriented areas (calculated, say, with the determinantal formula from Subsection 1.9.2) of the nine triangles that one can form using the points.

n) Is it possible to position six points $P_1, \ldots, P_6$ in the plane in such a way that they have the following integer distances between them [814]?

	P_1	P_2	P_3	P_4	P_5	P_6
P_1	0	87	158	170	127	68
P_2	87	0	85	127	136	131
P_3	158	85	0	68	131	174
P_4	170	127	68	0	87	158
P_5	127	136	131	87	0	85
P_6	68	131	174	158	85	0

o) Show that there are no 3×3 Hadamard matrices [78], [1866], [681]. (An $n \times n$ Hadamard matrix $\mathbf{H}_n$ is a matrix with elements ± 1 that fulfills $\mathbf{H}_n.\mathbf{H}_n^T = n\,\mathbf{1}_n$.)

p) The two-point Taylor series of order for a function $f(z)$ analytic in z_1, z_2 is defined through [1158]

$$f(z) = \sum_{n=0}^{o} (c_n(z_1, z_2)(z - z_1) + c_n(z_2, z_1)(z - z_2))(z - z_1)^n (z - z_2)^n + R_{o+1}(z, z_1, z_2).$$

Here $R_{o+1}(z, z_1, z_2)$ is the remainder term and the coefficients $c_n(z_1, z_2)$ are given as

$$c_0(z_1, z_2) = \frac{f(z_2)}{z_2 - z_1}$$

$$c_n(z_1, z_2) = \sum_{k=0}^{n} \frac{(k+n-1)!}{k!\, n!\, (n-k)!} \; \frac{(-1)^k k \, f^{(n-k)}(z_1) + (-1)^{n+1} n \, f^{(n-k)}(z_2)}{(z_1 - z_2)^{k+n+1}}.$$

Calculate the two-point Taylor series $T_{0,2\pi}^{(20)}[\sin](z)$ of order 20 for $f(z) = \sin(z)$, $z_1 = 0$, and $z_2 = 2\pi$. Find $\max_{z_1 \le z \le z_2} |f(z) - T_{20}(z)|$.

q) While for a smooth function $y(x)$, the relation $dy(x)/dx = 1/(dx(y)/dy)$ holds; the generalization $d^n y(x)/dx^n = 1/(d^n x(y)/dy^n)$ for $n \ge 2$ in general does not hold. Find functions $y(x)$ such that the generalization holds for $n = 2$ [245]. Can you find one for $n = 3$?

r) Define a function (similar to the built-in function D) that implements the quotential derivatives $q^n./qx^n$ of a function $f(x)$ defined recursively by [1297]

$$\frac{q^n f(x)}{dx^n} = \frac{q}{qx}\left(\frac{q^{n-1} f(x)}{dx^{n-1}}\right)$$

with the first quotential derivatives $q./qx$ defined as

$$\frac{q^1 f(x)}{qx^1} = \frac{qf(x)}{qx} = \lim_{q\to1} \ln\!\left(\frac{f(q\,x)}{f(x)}\right).$$

Show that $qf(y(x))/qx = qf(y(x))/qy \; qy(x)/qx$.

Define the multivariate quotential derivative recursively starting with the rightmost ones, meaning

$$\frac{q^2 f(x, y)}{qx\,qy} = \frac{q}{qx}\left(\frac{qf(x, y)}{qy}\right).$$

Show by explicit calculation that

$$\frac{qf(x, y)}{qy}\,\frac{q^2 f(x, y)}{qx\,qy} = \frac{qf(x, y)}{qx}\,\frac{q^2 f(x, y)}{qy\,qx}.$$

s) Conjecture the value of the following sum: $\sum_{k=1}^{\infty} (\prod_{j=1}^{k} a_{j-1}/(x + a_j))$. Here $a_0 = 1$, $a_k \in \mathbb{C}$, $a_k \ne 0$, $x \ne 0$ [1648].

2.11 Horner's Form, Bernoulli Polynomials, Squared Zeros, Polynomialized Radicals, Zeros of Icosahedral Equation, Iterated Exponentials, Matrix Sign Function, Appell–Nielsen Polynomials

a) Given a polynomial $p(x)$, rewrite it in Horner's form.

b) Bernoulli polynomials $B_n(x)$ are uniquely characterized by the property $\int_x^{x+1} B_n(t)\,dt = x^n$. Use this method to implement the calculation of Bernoulli polynomials $B_n(x)$. Try to use only built-in variables (with the exception of x and n, of course).

c) Given the polynomial $x^4 + a_3 x^3 + a_2 x^2 + a_1 x + a_0$ with zeros x_1, x_2, x_3, and x_4, calculate the coefficients (as functions of a_0, a_1, a_2, and a_3) of a polynomial that has the zeros x_1^2, x_2^2, x_3^2, and x_4^2.

d) Express the real zeros of

$$-1 + x + 2\sqrt{1+x^2} - 3\sqrt[3]{1+x^3} + 5\sqrt[5]{1+x^5} - 4 = 0$$

as the zeros of a polynomial.

e) Show that all nontrivial solutions of $x^{10} + 11\,x^5 - 1 = 0$ stay invariant under the following 60 substitutions:

$$x \longrightarrow \epsilon^i\, x$$

$$x \longrightarrow -\frac{\epsilon^i}{x}$$

$$x \longrightarrow \frac{\epsilon^j\,(\epsilon^i + x\,(\epsilon^4 + \epsilon))}{x - \epsilon^i\,(\epsilon^4 + \epsilon)}$$

$$x \longrightarrow -\frac{\epsilon^j\,(x - \epsilon^i\,(\epsilon^4 + \epsilon))}{\epsilon^i + x\,(\epsilon^4 + \epsilon)}$$

f) Iterated exponentials $\exp(c_1\, z \exp(c_2\, z \exp(c_3\, z \cdots)))$ can be used to approximate functions [966], [1881], [1882], [47]. Find values for c_1, c_2, ..., c_{10} such that $\exp(c_1\, z \exp(c_2\, z \exp(c_3\, z \cdots)))$ approximates the function $1 + \ln(1 + z)$ around $z = 0$ as best as possible.

g) Motivate symbolically the result of the following input.

```
m = Table[1/(i+j+1), {i, 5}, {j, 5}];
```

```
FixedPoint[(# + Inverse[#])/2&, N[m], 100]
```

h) Efficiently calculate the list of coefficients of the polynomial

$$(x^4 + x^3 + x^2 + x + 1)^{500}\,(x^2 + x + 1)^{1000}\,(x + 1)^{2000}$$

without making use of any polynomial function like `Expand`, `Coefficient`, `CoefficientList`, ...

i) What is the minimal distance between the roots of $z^3 + c^2\,z + 1 = 0$ for real c?

j) Let $f^{(k)}(z) = f(f^{(k-1)}(z))$, $f^{(1)}(z) = f(z) = z^2 - c$. Then the following remarkable identity holds [1135], [119]:

$$\exp\left(-\sum_{k=1}^{\infty} c_k\,\frac{z^k}{k}\right) = 1 + \sum_{k=1}^{\infty} \frac{(\frac{z}{2})^k}{\prod_{j=1}^{k} f^{(j)}(0)}$$

where

$$c_k = \sum_{j=1}^{2^k} \frac{1}{f^{(k)\prime}(z_j)\,(f^{(k)\prime}(z_j) - 1)}.$$

The sum appearing in the definition of the c_k extends over all 2^k roots of $f^{(k)}(z) = z$. Expand both sides of the identity in a series around $z = 0$ and check the equality of the terms up to order z^4 explicitly.

k) Write a one-liner that, for a given integer m, quickly calculates the matrix of values

$$c_{e,d} = \lim_{x \to 0} \frac{\partial^d \left(\frac{x}{\sin(x)} \right)^e}{\partial x^d}$$

for $1 \le e \le m,\, 0 \le d \le m$.

l) The Appell–Nielsen polynomials $p_n(z)$ are defined through the recursion $p_n'(z) = p_{n-1}(z)$, the symmetry constraint $p_n(z) = (-1)^n\, p_n(-z-1)$, and the initial condition $p_0(z) = 1$ [324], [1341]. Write a one-liner that calculates the first n Appell–Nielsen polynomials. Visualize the polynomials.

m) Write a one-liner that uses `Integrate` (instead of the typically used `D`) to derive the first n terms of the Taylor expansion of a function f around x that is based on the following identity [729], [549]

$$f(x + h) = \sum_{k=0}^{n-1} \frac{h^k}{k!}\, f^{(k)}(x) + \int_0^h \int_0^{h_1} \cdots \int_0^{h_{n-1}} f^{(n)}(x + h_n)\, dh_n \ldots dh_2\, dh_1.$$

n) A generalization of the classical Taylor expansion of a function $f(x)$ around a point x_0 into functions $\varphi_k(x)$, $k = 0, 1, \ldots, n$ (where the $\varphi_k(x)$ might be other functions that the monomials x^k) can be written as [1839]

$$f(x) \approx -\frac{1}{W(\varphi_0(x_0), \ldots, \varphi_n(x_0))}
\begin{vmatrix}
0 & \varphi_0(x) & \varphi_1(x) & \cdots & \varphi_n(x) \\
f(x_0) & \varphi_0(x_0) & \varphi_1(x_0) & \cdots & \varphi_n(x_0) \\
f'(x_0) & \varphi_0'(x_0) & \varphi_1'(x_0) & \cdots & \varphi_n'(x_0) \\
\vdots & \vdots & \vdots & \ddots & \vdots \\
f^{(n)}(x_0) & \varphi_0^{(n)}(x_0) & \varphi_1^{(n)}(x_0) & \cdots & \varphi_n^{(n)}(x_0)
\end{vmatrix}.$$

Here the $W(\varphi_0(\xi), \ldots, \varphi_n(\xi))$ is the Wronskian of the $\varphi_0(\xi), \ldots, \varphi_n(\xi)$ and it is assumed not to vanish at x_0. Implement this approximation and approximate $f(x) = \cos(x)$ around $x_0 = 0$ through $\exp(x)$, $\exp(x/2)$, $\ldots$, $\exp(x/m)$. Can this formula be used for $m = 25$?

o) Show that the function [1222]

$$w(z) = \frac{((u(z) + 2\,z\,u'(z))^2 - 4\,z\,u'(z)^2)^2}{8\,(u(z)\,u'(z)\,(u(z) + 2\,(z - 1)\,u'(z))\,(u(z) + 2\,z\,u'(z)))}$$

where $u(z) = c_1\, f_1(z) + c_2\, f_2(z)$ and $f_{1,2}(z)$ are solutions of $(1 - z)\,z\,f''(z) + (1 - 2z)\,f'(z) - f(z)/4 = 0$ fulfills the following special case of the Painlevé VI equation:

$$w''(z) = \frac{1}{2}\left(\frac{1}{w(z)} + \frac{1}{w(z) - 1} + \frac{1}{w(z) - z}\right)w'(z)^2 - \left(\frac{1}{z} + \frac{1}{z - 1} + \frac{1}{w(z) - z}\right)w'(z) +$$

$$\frac{w(z)\,(w(z) - 1)\,(w(z) - z)}{2\,(z - 1)^2\,z^2}\left(\frac{z(z - 1)}{(w(z) - z)^2} + 4\right).$$

3.^{l1} Nested Integration, `Derivative[-n]`, `PowerFactor`, Rational Painlevé II Solutions

a) Given that the following definition is plugged into *Mathematica*, what will be the result of `f[2][x]`?

```
f[n_][x_] := Integrate[f[n - 1][x - z], {z, 0, x}]

f[0][x_] = Exp[-x];
```

Consider the evaluation process. How would one change the first two inputs to get the "correct" result as if from

```
Nest[Integrate[#/.{x->x-z}, {z, 0, x}]&, Exp[-x], 2]
```

b) Find two (univariate) functions f and g, such that `Integrate[f, x] + Integrate[g, x]` gives a different result than does `Integrate[f+g, x]`. Find a (univariate) functions f and integration limits x_l, x_m, and x_u, such that `Integrate[f, {x, x_l, x_u}]` gives a different result than does `Integrate[f, {x, x_l, x_m}] + Integrate[f, {x, x_m, x_u}]`.

c) What does the following code do?

```
Derivative[i_Integer?Negative][f_] :=
With[{pI = Integrate[f[C], C]},
      derivative[i + 1][Function[pI] /. C -> #] /;
                  FreeQ[pI, Integrate, {0, Infinity}]]
```

Predict the results of `Derivative[+4][Exp[1 #]&]` and `Derivative[-4][Exp[1 #]&]`.

d) Is it possible to find a function $f(x, y)$ such that `D[Integrate[f(x, y), x], y]` is different from `Integrate[D[f(x, y), y], x]`?

e) Write a function `PowerFactor` that does the "reverse" of the function `PowerExpand`. It should convert products of radicals into one radical with the base having integer powers. It should also convert sums of logarithms into one logarithm and $s\log(a)$ into $\log(a^s)$.

f) The rational solutions of $w''(z) = 2\,w(z)^3 - 4\,z\,w(z) + 4\,k$, $k \in \mathbb{N}^+$ (a special Painlevé II equation) can be expressed in the following way [967], [910], [1215], [952]:
Let the polynomials $q_k(z)$ be defined by the generating function $\sum_{k=0}^{\infty} q_k(z)\,\xi^k = \exp(z\xi + \xi^3/3)$ (for $k < 0$, let $q_k(z) = 0$).
Let the determinants $\sigma_k(z)$ be defined by matrices $(a_{ij})_{0 \le i, j \le k-1}$ with $a_{ij} = q_{k+i-2j}(z)$ (for $k = 0$, let $\sigma_0(z) = 1$).
Then, $w_k(z)$ is given as $w_n(z) = \partial\log(\sigma_{k+1}(z)/\sigma_k(z))/\partial z$. Calculate the first few $w_k(z)$ explicitly.

4.^{l1} Differential Equations for the Product, Quotient of Solutions of Linear Second-Order Differential Equations

Let $y_1(z)$ and $y_2(z)$ be two linear independent solutions of

$$y''(z) + f(z)\,y'(z) + g(z)\,y(z) = 0$$

The product $u(z) = y_1(z)\,y_2(z)$ obeys a linear third-order differential equation

$$u'''(z) + a_p[f(z), g(z)]\,u''(z) + b_p[f(z), g(z)]\,u'(z) + c_p[f(z), g(z)]\,u(z) = 0$$

The quotient $w(z) = y_1(z)/y_2(z)$ obeys (Schwarz's differential operator; see, for instance, [847] and [1906])

$$w'''(z)\, w'(z) + a_q[f(z),\, g(z)]\, w''(z)^2 + b_q[f(z),\, g(z)]\, w'(z)^2 = 0$$

Calculate a_p, b_p, c_p and a_q, b_q. (For analogous equations for the solutions of higher-order differential equations, see [1024].)

5.^{L1} Singular Points of ODEs, Integral Equation

a) First-order ordinary differential equations of the form $y'(x) = P(x, y)/Q(x, y)$ possess singular points $\{x_i^*, y_i^*\}$ [215], [1403], [1746], [1105], [467], [949]. These are defined by $P(x_i^*, y_i^*) = Q(x_i^*, y_i^*) = 0$. It is possible to trace the typical form of the solution curves in the neighborhood of a singular point by solving $y'(x) = (a\,x + b\,y)/(c\,x + d\,y)$. Some typical forms include the following examples:

a knot point $y'(x) = \dfrac{2\,y(x)}{x}, \quad y'(x) = \dfrac{y(x) + x}{x}, \quad y'(x) = \dfrac{y(x)}{x}$

a vortex point $y'(x) = -\dfrac{y(x)}{x}$

an eddy point $y'(x) = \dfrac{y(x) - x}{y(x) + x}.$

Investigate which of the given differential equations can be solved analytically by *Mathematica*, and plot the behavior of the solution curves in a neighborhood of the singular point $\{0, 0\}$.

b) Consider the general case (meaning λ is a regular value) of the Fredholm integral equation of the second kind

$$y(x) - \lambda \int_a^b \mathcal{K}(x, \xi)\, y(\xi)\, d\xi = f(x)$$

with a separable kernel (meaning $\mathcal{K}(x, \xi)$ can be written in the form $\mathcal{K}(x, \xi) = \sum_{j=1}^n g_j(x)\, h_j(\xi)$). Implement a function that solves such Fredholm integral equations in the classical manner, by forming and solving a linear system in the $r_j = \int_a^b h_j(\xi)\, y(\xi) = \int_a^b h_j(x)\, y(x)$.

The solution $y(x)$ can be written in the form $y(x) = f(x) + \lambda \int_a^b \Gamma(x, \xi)\, f(\xi)\, d\xi$ with the resolvent kernel $\Gamma(x, \xi)$.

Implement a function that calculates the truncated form of the Fredholm resolvent [1628], [1427], [1455], [1884], [1102], [623]

$$\Gamma^F(x, \xi) = \frac{D(x, \xi; \lambda)}{D(\lambda)}$$

$$D(x, \xi; \lambda) = \sum_{k=0}^{\infty} \frac{(-1)^k}{k!}\, \lambda^k\, d_k(x, \xi), \quad D(\lambda) = \sum_{k=0}^{\infty} \frac{(-1)^k}{k!}\, \lambda^k\, c_k.$$

where the c_k and $d_k(x, \xi)$ are recursively defined through

$$c_0 = 1, \quad c_k = \int_a^b d_{k-1}(\xi, \xi) \, d\xi$$

$$d_0(\xi, \xi) = \mathcal{K}(\xi, \xi), \quad d_k(x, \xi) = c_k \, \mathcal{K}(x, \xi) - k \int_a^b \mathcal{K}(x, \rho) \, d_{k-1}(\rho, \xi) \, d\rho.$$

Finally, implement a function that calculates a truncated form of the Neumann form of the resolvent (valid for a limited set of λ) that utilizes iterated kernels.

$$\Gamma^N(x, \xi) = \sum_{k=1}^{\infty} \lambda^k \, \mathcal{K}_n(x, \xi)$$

$$\mathcal{K}_1(x, \xi) = \mathcal{K}(x, \xi), \quad \mathcal{K}_n(x, \xi) = \int_a^b \mathcal{K}(x, \rho) \, \mathcal{K}_{n-1}(\rho, \xi) \, d\rho.$$

Use the implemented functions to solve the equation $y(x) - \lambda \int_0^1 \sin(x + \xi) \, y(\xi) \, d\xi = \cos(x)$ exactly and up to $O(\lambda^5)$.

6.¹¹ Inverse Sturm–Liouville Problems, Graeffe Method

a) In recent years, so-called "inverse problems" have assumed an increasingly important role (see, e.g., [323], [776], [718], [1886], [1523], [92], and [161]). Very roughly speaking, they are as follows: For a given result, find the associated problem that led to the result. For Sturm–Liouville eigenvalue problems

$$-y''(x) + v(x) \, y(x) = \lambda \, y(x)$$

with given eigenvalues λ_i, the solutions are completely known. We consider the following special case: Given eigenvalues $\lambda_1, \ldots, \lambda_n$ of the problem

$$-y_i''(x) + v(x) \, y_i(x) = \lambda_i \, y_i(x), \quad y_i(0) = 0, \quad 0 \le x \le \infty$$

find one (eigenvalue-independent) potential that leads to these eigenvalues (given only finitely many eigenvalues, of course, infinitely many solutions exist). The following method solves this problem constructively:

$$\varphi_0(x, \lambda) = \frac{\sin(\lambda \, x)}{\lambda}$$

$$u(x)_{ij} = \delta_{ij} + \int_0^x \varphi_0\left(s, \sqrt{\lambda_j}\right) \varphi_0\left(s, \sqrt{\lambda_i}\right) ds \quad i, j = 1, 2, \ldots, n$$

$$v(x) = -2 \, \frac{d^2 \ln(\det U(x))}{dx^2}.$$

Here, $U(x)$ is a matrix with matrix elements $u_{i\,j}(x)$. Then, the corresponding (normalizable) eigenfunctions $\psi_i(x)$ are given by

$$\psi_i(x) = \sum_{j=1}^{n} u_{ij}^{-1}(x) \, \varphi_0\left(x, \sqrt{\lambda_j}\right)$$

Implement these formulas, and for $n = 1$, verify that the resulting $v(x)$ and $\psi_1(x)$ satisfy the equation

$$-\psi_1''(x) + v(x)\,\psi_1(x) = \lambda_1\,\psi_1(x)$$

Plot the result for $n = 2$. For details see [1662], [1250], [1885], [322], [1753], [852], [548], [873], [33], [1579], [1551], [1375], [1794], [405], [321], and [1446]. For the vector-valued version, see [342].

a) Implement the Graeffe method [829], [878], [1186], [1187], [1655] to calculate the roots of the polynomial $p = z^5 + 5\,z^4 - 10\,z^3 - 10\,z^2 + 5\,z + 2 = 0$ to 100 digits. The Graeffe method calculates the polynomials

$$p_{n+1}(z^2) = p_n(z)\,p_n(-z)$$
$$p_0(z) = p.$$

The distinct real roots z_k of the polynomial p of degree d are then given by

$$|z_k| = \lim_{n \to \infty} \left| \frac{a_{n,k-1}}{a_{n,k}} \right|^{2^{-n}}$$

where $p_{n+1}(z) = \sum_{k=0}^{d} a_{n,k}\,z^k$.

7.^{13} Finite Element Method: Lagrange and Hermite Interpolation

This problem is meant for readers who are familiar with the finite element method (FEM). This is not the right place to give a detailed account of the method itself. Interested readers should consult any typical FEM book, such as [1103], [373], [276], [1587], [82], [83], [84], [546], [134], [799], [1764], [31], and [140] for some computer algebra applications. For applications in physics, see, for example, [783], [1314], [1055], [1467], [239], [837], [14], [1037], [1128], [1589], [13], and [724], and the references therein.

a) Suppose we are given an elliptic partial differential equation of second order in two space dimensions

$$\frac{\partial^2 u(x,\,y)}{\partial x^2} + \frac{\partial^2 u(x,\,y)}{\partial y^2} + \lambda\,u(x,\,y) = f(x,\,y)$$

For most finite-element calculations, it is common to use Lagrange basis functions of the first or second order on triangles. Higher-order elements may be needed in some applications to get higher accuracy (say up to eighth order in some applications [798]). To carry out the calculation, we need to compute the corresponding stiffness matrices, element vectors, and, for eigenvalue problems, the corresponding mass matrices. These should be computed for the case of linear isoparametric maps of the unit triangle onto arbitrary triangles. (If this mapping is nonlinear, this will not be possible, in general, without using numerical integration.) To accomplish this, solve the following subproblems:

■ Compute and plot the position of the data points for Lagrange interpolation [469], [1149]. Usually, the data points are numbered as follows: The three outer vertices are given the numbers 1, 2, and 3. Now starting with vertex 1, continue in a spiral winding in the same sense as 1, 2, 3 to the center.

■ Compute the Lagrange shape functions of nth-order for the unit triangle.

■ Program a corresponding integration routine to efficiently compute integrals of the form

$$\int_0^1 \int_0^{1-\xi} \xi^p\,\eta^q\,d\eta\,d\xi = \frac{p!\,q!}{(p+q+2)}$$

■ Compute the element vectors, the mass matrices, and the stiffness matrices for various orders.

b) Now, consider the 1D case. The Hermite element functions $\chi_{k,l}^{(p,d)}(\xi)$ are defined through their smooth interpolation property

$$\frac{\partial^n \chi_{k,l}^{(p,d)}(\xi)}{\partial \xi^n}\bigg|_{\xi=\xi_m} = \delta_{k,m}\,\delta_{l,n}.$$

Here $p+1$ indicates the number of nodes in one cell and the index d indicates the order of smoothness. The two integer labels k and l range over $0 \le k \le p$ and $0 \le l \le d$, and ξ runs from 0 to 1. (In most cases, one uses equidistantly spaced ξ_m, meaning $\xi_m = m/p$.)

Consider the eigenvalue problem for a harmonic oscillator [1681], [561], [1456], [1457], [1754]

$$-\psi_j''(x) + x^2\,\psi_j(x) = \varepsilon_j\,\psi_j(x)$$
$$\psi(-\infty) = \psi(\infty) = 0$$

with the exact eigenvalues $\varepsilon_j = 2\,j + 1$, $j \in \mathbb{N}$. Discretize the problem in the interval $[-L, L]$ into e elements and use the ansatz

$$\psi_j(x) = \sum_{j=0}^{e-1}\sum_{k=0}^{p}\sum_{l=0}^{d} c_{k,l}^{(j,p,d)}\,\chi_{k,l}^{(p,d)}(\xi(x))\,\theta(x-x_j)\,\theta(x_{j+1}-x).$$

Here $\xi(x) = (x-x_j)/\delta$, $x_j = -L + j/e\,\delta$, $\delta = 2L/e$ and the Dirichlet boundary conditions enforce $c_{0,0}^{(0,p,d)} = c_{0,0}^{(e-1,p,d)} = 0$. Continuity and smoothness at the leftmost and rightmost nodes enforces $c_{p,l}^{(j,p,d)} = c_{0,l}^{(j+1,p,d)}$ for $1 \le j \le j-2$.

Construct the algebraic eigenvalue problem corresponding to the minimization problem

$$\int_{-L}^{L}\left(\left(\frac{\partial\psi_j(x)}{\partial x}\right)^2 + (x^2 - \varepsilon_j)\,\psi_j(x)\right)dx$$

and calculate the lowest eigenvalues.

Is it possible to get the lowest eigenvalue correct to 20 digits with resulting matrices of dimension 64×64 or less?

8.¹² Helium Atom, Improved Variational Method

a) In 1933, Hylleraas and Undheim ([887], see also [167], [1282], [647], [1372], and [854]) published the results of a variational calculation [1827] of the ground-state ($_2S$ state) energy $\lambda_0(k_{\min}, c_{\min})$ of an (ortho) Helium atom [991]. Without going into the physical background (see [1249], or for more depth, see [168], [1061], [1449], [1025], [1398], [1558], [888], [821], [77], [1052], [1049], [1713], [1286], [737], [1335], [1373], [1904], [1527], and [1702], and the references therein), we present the underlying mathematical problem.

Find the smallest zero $\lambda_0(k, c)$ of the determinant of the 6×6 matrix D with elements D_{ij} defined below. (The minimum is around $k_{\min} \approx 0.5$, $c_{\min} \approx 0.5$.) The experimental value of the ground-state energy is -1.08762. How well does the calculation perform?

$$D_{ij} = \lambda N_{ij}(c) + k\left(L_{ij}(c) + L'_{ij}(c)\right) - k^2 M_{ij}(c)$$

$$N_{ij}(c) = \int_0^\infty ds \int_0^s du \int_0^u dt \, \frac{u(s^2 - t^2)}{8} \, \varphi_i \varphi_j$$

$$L_{ij}(c) = \int_0^\infty ds \int_0^s du \int_0^u dt \, 2s \, u \, \varphi_i \varphi_j$$

$$L'_{ij}(c) = \int_0^\infty ds \int_0^s du \int_0^u dt \, \frac{(s^2 - t^2)}{4} \, \varphi_i \varphi_j$$

$$M_{ij}(c) = \int_0^\infty ds \int_0^s du \int_0^u dt \, (u(s^2 - t^2)(\varphi_{i,s}\varphi_{j,s} + \varphi_{i,t}\varphi_{j,t} + \varphi_{i,u}\varphi_{j,u}) +$$

$$s(u^2 - t^2)(\varphi_{i,s}\varphi_{j,u} + \varphi_{i,u}\varphi_{j,s}) + t(s^2 - u^2)(\varphi_{i,t}\varphi_{j,u} + \varphi_{i,u}\varphi_{j,t})).$$

Here are the $\varphi_i(u, s, t)$:

$$\varphi_1(u, s, t) = e^{-s/2}\sinh(ct/2)$$
$$\varphi_2(u, s, t) = s\,\varphi_1(u, s, t)$$
$$\varphi_3(u, s, t) = t\,e^{-s/2}\cosh(ct/2)$$
$$\varphi_4(u, s, t) = u\,\varphi_1(u, s, t)$$
$$\varphi_5(u, s, t) = u\,\varphi_2(u, s, t)$$
$$\varphi_6(u, s, t) = u\,\varphi_3(u, s, t).$$

$\varphi_{i,s}$ is the partial derivative $\partial\varphi_i(u, s, t)/\partial s$, and the analogous formula holds for $\varphi_{i,t}$ and $\varphi_{i,u}$.

b) A possible scheme for improving variational calculations with one variational parameter is the following (see [1675] and [1905]): Given a (real) trial function $\psi_0(\beta; x)$ in which β is the variational parameter, one constructs additional functions (which we take here orthonormalized) $\psi_1(\beta; x)$, $\psi_2(\beta; x)$, and so on (all depending on the same parameter β) via

$$\psi_i(\beta; x) = \frac{\partial\psi_{i-1}(\beta; x)}{\partial\beta}$$

$$\int_L \psi_i^2(\beta; x)\,dx = 1.$$

(Here, the integration extends over the range where the $\psi_0(\beta; x)$ are defined.) As the trial function, it uses

$$\Psi(\beta; x) = \sum_{i=0}^n c_i(\beta)\,\psi_i(\beta; x).$$

For the eigenvalue problem for a hermitian operator H, $H(x)\psi(x) = \varepsilon\,\psi_i(x)$, the condition for $\varepsilon(\beta)$ to be the best approximation to the lowest eigenvalue becomes

$$\det |H_{ij}(\beta) - \varepsilon(\beta)\,M_{ij}(\beta)| = 0$$

Here, $H_{ij}(\beta)$ and $M_{ij}(\beta)$ (which reflects the nonorthogonality of the $\psi_i(\beta; x)$) are defined as

$$H_{ij}(\beta) = \int_L \psi_i(\beta; x) H(x) \psi_j(\beta; x) dx$$

$$M_{ij}(\beta) = \int_L \psi_i(\beta; x) \psi_j(\beta; x) dx.$$

Carry out the above calculations for the following particular realizations of the described scheme [1905] in $L = (-\infty, \infty)$:

$$H = -\frac{1}{2}\frac{d^2}{dx^2} + \frac{1}{2}x^2 + x^4$$

$$\psi_0(\beta; x) = \sqrt{\frac{\beta}{\sqrt{\pi}}}\, e^{-\beta^2 \frac{x^2}{2}}.$$

Use different techniques to calculate explicit numerical values for the lowest eigenvalue for $n \le 6$. The so-defined $\varepsilon(\beta)$ has to be minimized with respect to β.

9.$^{\text{L2}}$ Hyperspherical Coordinates, Constant Negative Curvature Surface

a) The standard n-dimensional spherical coordinates are defined by the following relations [1742], [187] (for nonstandard n-dimensional spherical coordinates, see [1339]):

$$
\begin{aligned}
x_1 &= r\cos\vartheta_1 \\
x_2 &= r\sin\vartheta_1\cos\vartheta_2 \\
x_3 &= r\sin\vartheta_1\sin\vartheta_2\cos\vartheta_3 \\
&\vdots \\
x_{n-1} &= r\sin\vartheta_1\sin\vartheta_2\cdots\sin\vartheta_{n-2}\cos\vartheta_{n-1} \\
x_n &= r\sin\vartheta_1\sin\vartheta_2\cdots\sin\vartheta_{n-2}\sin\vartheta_{n-1}.
\end{aligned}
$$

Here, the x_i are Cartesian coordinates. Compute and simplify the Jacobian of this coordinate transformation for the first n. Compare the times needed by `Simplify` with the times needed by manual simplification.

b) The following gives a parametric description of a family of constant negative curvature [1748] surfaces [1230] ($0 < c < 1$ is the family parameter) [1502], [1571], [1572], [1503], [1573].

$$
\mathbf{r}(u, v) = \{0, 0, x\} + \frac{2\,d\cosh(c\,x)\sin(d\,y)}{c\,\mathcal{D}}\{\sin(y), -\cos(y), 0\} +
$$

$$
\frac{2\,d^2\cosh(c\,x)\,(\cosh^2(c\,x) - \sin^2(d\,y))}{c\,\mathcal{D}\tilde{\mathcal{D}}}\{\cos(y)\cos(d\,y), \sin(y)\cos(d\,y), -\sinh(c\,x)\}
$$

$$\mathcal{D} = d^2\cosh^2(c\,x) + c^2\sin^2(d\,y)$$

$$\tilde{\mathcal{D}} = \cos^2(d\,y)\cosh^2(c\,x) + \sin^2(d\,y)\sinh^2(c\,x)$$

$$c = \sqrt{1 - d^2}$$

$$x = u + v,\quad y = u - v$$

Show that $\mathbf{r}(u, v)$ has a constant curvature $\mathcal{K}$. The curvature $\mathcal{K}$ can be calculated through the following formulas.

$$\mathcal{K} = \frac{1}{|\mathbf{n}|^2}\frac{eg - f^2}{\mathcal{E}\mathcal{G} - \mathcal{F}^2}$$

$$\mathcal{E} = \frac{\partial \mathbf{r}(u,\, v)}{\partial u} \cdot \frac{\partial \mathbf{r}(u,\, v)}{\partial u}, \quad \mathcal{G} = \frac{\partial \mathbf{r}(u,\, v)}{\partial v} \cdot \frac{\partial \mathbf{r}(u,\, v)}{\partial v}, \quad \mathcal{F} = \frac{\partial \mathbf{r}(u,\, v)}{\partial u} \cdot \frac{\partial \mathbf{r}(u,\, v)}{\partial v}$$

$$e = \frac{\partial^2 \mathbf{r}(u,\, v)}{\partial u^2} \cdot \mathbf{n}, \quad f = \frac{\partial^2 \mathbf{r}(u,\, v)}{\partial u \partial v} \cdot \mathbf{n}, \quad g = \frac{\partial^2 \mathbf{r}(u,\, v)}{\partial v^2} \cdot \mathbf{n}$$

$$\mathbf{n} = \frac{\partial \mathbf{r}(u,\, v)}{\partial u} \times \frac{\partial \mathbf{r}(u,\, v)}{\partial v}$$

Visualize some of the surfaces of the family.

10.^{L2} Throw from a Finite Height, Pendulum Throw, Spring System

a) Determine the optimal throw angle (optimal with respect of achieving the furthest possible throw distance) for a throw starting from a finite height h. Visualize the optimality of this angle. Calculate the envelope of all throw curves for a variable throw angle [505], [662], [108], [76], [506], [631].

b) Consider a mathematical pendulum with mass m, length l, and minimum height h above the ground. Suppose during the movement of the pendulum that the string is suddenly cut. The subsequent trajectory of the mass is a parabola. Determine the angle of release φ for which the point of contact $x_s = w$ of the pendulum with the horizontal floor is farthest. Try to find an analytic solution. The horizontal distance is measured from the abscissa of the point from which the pendulum is hung, and the angle φ is measured from the vertical. The maximum angle should be less than $\pi/2$. Plot the solution.

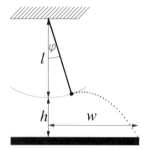

c) Take a Platonic solid and replace all edges by springs of identical stiffness and the vertices with point masses. For small elongations, analyze the "breathing mode" (meaning all vertices move radial and in phase) of the resulting structure. How many normal modes contribute to the "breathing mode" and how does its frequency compare with other eigenmodes.

11.^{L1} Sturm–Liouville Problems in Normal Form

The general eigenvalue problem for a linear differential equation of second order takes the form

$$f_2(x)\, y''(x) + f_1(x)\, y'(x) + (f_0(x) + \lambda\, g(x))\, y(x) = 0.$$

Often, it is useful to rewrite this eigenvalue problem in the Liouville normal form

$$\eta''(\xi) + (\lambda + v(\xi))\, \eta(\xi) = 0.$$

In this form, we can better determine the $v(\xi)$ for which there is an analytical solution of the problem (see [112], [1522], [1291], [27], [648], [1265], [127], [1322], [1169], [136], and [1843]; for multidimensional analytic solutions, see [1781]). This transformation is given by (we assume sufficient smoothness for the corresponding functions)

$$\eta(\xi) = \Phi(x)\, y(x)$$

$$\xi = \xi(x) = \int^x \sqrt{\frac{g(x)}{f_2(x)}}\, dx$$

$$\Phi(x) = \sqrt[4]{\frac{g(x)}{f_2(x)}}\, \exp\!\left(\frac{1}{2}\int^x \frac{f_1(x)}{f_2(x)}\, dx\right).$$

Determine the resulting $v(\eta)$, and implement this transformation. Try to generate error messages when the function is called with inappropriate arguments.

12.[11] Noncentral Collision, Bernstein Polynomials, Bernstein Operator

a) A sphere of mass M is at rest at the point $\{0, 0\}$. Suppose a second sphere (of the same diameter) with velocity $\{v_x, v_y\}$ collides with the first one in an elastic noncentral collision. Suppose the coordinates of the center of the second sphere at the time of collision are $\{x, y\}$. Compute the velocity of both spheres after of the collision, and plot their motion. Suppose the momentum, energy, and angular momentum conservation laws hold (the angular momentum law implies that no tangential forces are present).

b) Calculate a closed form of the envelope of the scaled maxima $n^{1/2} x_m$ of the Bernstein polynomials $b_{n,k}(x) = \sum_{k=0}^{n} \binom{n}{k} x^k (1-x)^{n-k}$, $0 \le k \le n$ as $n \to \infty$ [1172].

c) The eigenvectors of the Bernstein operator $\hat{B}_n$

$$\hat{B}_n(f(x)) = \sum_{k=0}^{n} \binom{n}{k} x^k (1-x)^{n-k}\, f\!\left(\frac{k}{n}\right)$$

are polynomials of degree less than or equal to n. Find the eigenvectors for $n = 36$ and visualize them in the interval $0 \le x \le 1$.

13.[12] A Sensitive Linear System of Equations, Bisector Surface, Surface Connecting Three Cylinders, Double Torus Surface

a) Suppose we are given the following linear inhomogeneous system of equations [1530]:

$$
\begin{aligned}
- 367296\,t - 43199\,u + 519436\,v - 954302\,w &= 1 \\
259718\,t - 477151\,u - 367295\,v - 1043199\,w &= 1 \\
886731\,t + 88897\,u - 1254026\,v - 1132096\,w &= 1 \\
627013\,t + 566048\,u - 886732\,v + 911103\,w &= 0.
\end{aligned}
$$

How many digits are needed to be sure that the solution is correct to 10 digits? How sensitive to a small change of the coefficients is the solution of this system? (See also [742] and [1480].)

b) The bisector surface of two geometrical objects (points, lines, surfaces) is the set of points that have the same distance to each of the two geometrical objects [611], [567], [566], [610], [1022], [162], [568], [1406], [819]. Calculate the implicit form of the bisector surface of a torus and a point in generic position. Visualize the bisector surface for the point at the center of the torus.

c) Construct a polynomial surface $p(x, y, z) = 0$ that in a smooth way connects the three half-infinite cylinders [116], [1858], [629], [626], [1811], [336], [1809] given by their parametric representations

$$\{x, \cos(\varphi), \sin(\varphi)\} \quad 0 \le \varphi \le 2\pi, 3 \le x < \infty$$
$$\{\cos(\varphi), y, \sin(\varphi)\} \quad 0 \le \varphi \le 2\pi, 3 \le y < \infty$$
$$\{\cos(\varphi), \sin(\varphi), z\} \quad 0 \le \varphi \le 2\pi, 3 \le z < \infty.$$

Make a picture of that part of the polynomial $p(x, y, z) = 0$ that connects the three cylinders and the three cylinders themselves.

d) Calculate the implicit form of a polynomial surface of a double torus, such that the two holes of the double torus tightly enclose the hole of the given torus with radii $r = 1$ and $R = 3$, and the given torus tightly encloses the "middle bridge" of the double torus.

14.^{L2} Transformation of Variables in a PDE, det(e^A) = e^{TrA}, Matrix Derivative, Lewis–Carroll Identities

a) Program a function `DerivativeVariableTransformation` that carries out transformations of variables in expressions that contain partial derivatives. Starting with Cartesian coordinates, use it to transform $\Delta f(x, y, z)$ into spherical coordinates.

b) Given the result of evaluating the following determinant.

```
A=Array[a, {3, 3}]; Det[MatrixExp[A]]
```

By virtue of $\det(e^A) = e^{TrA}$ this simplifies to `Exp[a[1, 1] + a[2, 2] + a[3, 3]]`. Use *Mathematica* to carry out this simplification.

c) Construct a one-liner that proves the matrix identity

$$\frac{1}{1 - \lambda \mathbf{A}} = \sum_{k=0}^{\infty} \lambda^k \mathbf{A}^k$$

for a generic symbolic 3×3 matrix $\mathbf{A}$.

d) The differential quotient $d_{\mathbb{H}}(t, \mathbb{A})$ of the matrix exponential $\exp(t\,\mathbb{A})$ in (matrix) direction $\mathbb{H}$ is defined as [1313], [1488], [1738], [1690]

$$d_{\mathbb{H}}(t, \mathbb{A}) = \lim_{\varepsilon \to 0} \frac{e^{t(\mathbb{A} + \varepsilon \mathbb{H})} - e^{t\,\mathbb{A}}}{\varepsilon}.$$

$d_{\mathbb{H}}(t, \mathbb{A})$ can be expressed in the following way:

$$d_{\mathbb{H}}(t, \mathbb{A}) = \int_0^t e^{(t - \tau)\,\mathbb{A}} \mathbb{H} e^{\tau\,\mathbb{A}} \, d\tau.$$

For $n = 2$, show the equivalence of the two expressions for $d_{\mathbb{H}}(t, \mathbb{A})$ by explicit calculation.

f) Take a generic 3×3 matrix $\mathcal{A}$ and the resulting 81 matrices of size 2×2 that arise from extracting two columns and two rows from $\mathcal{A}$ and the 9 matrices of size 1×1 that arise from using one columns and one row from $\mathcal{A}$. Find algebraic relations that are obeyed by the determinants of these 91 matrices.

15.^{L2} 1 + 2 + 3 + ⋯ = −1/12, Casimir Effect

a) Often, in physics books and journals, one sees the following sum: $1 + 2 + 3 + \cdots = -1/12$ [1130], [572], [184], [1200]. Motivate this finite result for this divergent sum.

b) Calculate the following expression (the order of the limits matter):

$$\lim_{\lambda \to 0} \lim_{m \to \infty} \left(\sum_{n=0}^{m} n^3 e^{-\lambda n} - \int_0^m x^3 e^{-\lambda x} \, dx \right)$$

c) Consider the following formally divergent expression [1566], [1903], [228] (arising from the classical limit of the Casimir effect)

$$\mathcal{A}(l, L) = \int_{-\infty}^{\infty} \int_{-\infty}^{\infty} \sum_{n=-\infty}^{\infty} \left(\ln\left(\sqrt{\frac{\pi^2 n^2}{l^2} + k_x^2 + k_y^2} \Big/ \sqrt{\frac{\pi^2 n^2}{(L/2)^2} + k_x^2 + k_y^2} \right) + \right.$$

$$\left. \ln\left(\sqrt{\frac{\pi^2 n^2}{(L-l)^2} + k_x^2 + k_y^2} \Big/ \sqrt{\frac{\pi^2 n^2}{(L/2)^2} + k_x^2 + k_y^2} \right) \right) dk_x \, dk_y.$$

Find a finite result for $\mathcal{A}(l, L)$.

16.^{L2} Random Functions, Use of Numerical Techniques

a) Implement a function (of one variable) `RandomFunction` that produces a random function (of a given depths n) sometimes involving trigonometric functions, sometimes power functions, sometimes trigonometric functions of power functions, ... Use such a random function to find an integral that could be done, but *Mathematica* is not able to do it.

b) Find at least ten functions discussed in this chapter that, when called with purely symbolic input (this means input with infinite precision), potentially make internal use of numerical techniques. Show convincingly that these functions really use approximative numbers internally.

c) Find all built-in functions for which `Derivative[2]` [*function*] evaluates to a nontrivial result.

17.^{L3} Thomas–Fermi Equation, Yoccoz Function, $y(x) = x/\ln(x)$ Inversion, Lagrange–Bürmann Theorem, Divisor Sums, Multiple Differentiation

a) Iteratively compute the first few terms of a series expansion of a solution $f(x)$ of the differential equation

$$f''(x) = \frac{\sqrt{f(x)^3}}{\sqrt{x}}$$

satisfying the initial conditions

$$f(0) = f_0, \quad f'(0) = f_{s0}$$

(This is the differential equation for the radial electron density distribution of atoms in the Thomas–Fermi approximation. For details on the Thomas–Fermi approximation, see [743], [117], [1645], [668], [1380], [519], [489], [1714], [1715], [582], [1203], [1202], [1027], [1120], [144], [316], [954], [1292], [584], [622], and [861].)

b) Carrying out the Majorana substitution $\{x, \varphi(x)\} \to \{t, u(t)\}$ defined through [588], [589], [590]

$$t = \frac{1}{\sqrt[6]{144}} \sqrt{x} \sqrt[6]{\varphi(x)}$$

$$u(t) = -\sqrt[3]{\frac{16}{3}} \frac{\varphi'(x)}{\sqrt[3]{\varphi(x)^4}}$$

in the Thomas–Fermi equation $\varphi''(x) = x^{-1/2} \varphi(x)^{3/2}$ yields a first-order differential equation for $u(t)$. Derive this differential equation for $u(t)$.

c) The Yoccoz function [1207], [1877] $u(z)$ is the limit $u(z) = \lim_{n \to \infty} u_n(z)$ of the polynomials

$$u_0(z) = 1$$

$$u_n(z) = u_{n-1}(z) - \frac{z^{n-1}}{2} u_{n-1}^2(z).$$

Calculate the first 100 terms of the expansion of $u(z)$ around $z = 0$:

$$u(z) = \sum_{k=0}^{\infty} c_k z^k.$$

Is the calculation of the first 1000 terms feasible? Make a graphic of $\arg(u(0.999\,e^{i\varphi}))$ and conjecture some properties of $\arg(u(r\,e^{i\varphi}))$ as $r \to 1^-$.

d) The inverse function $x(y)$ to $y(x) = x/\ln(x)$ can be asymptotically for $x \to \infty$, expressed in the form (see [191], [393], [644], [851], [924], and [124])

$$x(y) = y\ln(y) + y\ln(\ln(y)) + y \sum_{i=1}^{\infty} \sum_{j=1}^{i} a_{ij} \frac{(\ln(\ln(y)))^j}{\ln^i(y)}.$$

Determine the coefficients a_{ij} for $i,\,j < 5$.

e) Calculate the series expansion up to order three of $g(z) = \tan(z)$ if $z = z(w)$ is defined as the inverse function of $\exp(z) + \ln(z + 1)$ around $z_0 = 1$. Calculate the result one time directly and one time using the Lagrange–Bürmann theorem [1837], [1489], [332], [886], [303], [1616], [1647]. It states that if $w = f(z)$, $w_0 = w(z_0)$, $w'(z_0) \neq 0$, the series expansion of $g(f^{-1}(w))$ around w_0 is given by

$$g(f^{-1}(w)) = g(z_0) + \sum_{i=1}^{\infty} \frac{1}{i!} \left[\frac{d^{i-1}}{d\xi^{i-1}} \left(g'(\xi) \left(\frac{\xi - z_0}{f(\xi) - w_0} \right)^i \right) \right]_{\xi = z_0} (w - w_0)^i$$

Determine the coefficients a_{ij} for $i,\,j < 5$.

f) Using the three functions

$$P(q) = 1 - 24 \sum_{k=1}^{\infty} \frac{k\, q^k}{1 - q^k}$$

$$Q(q) = 1 + 240 \sum_{k=1}^{\infty} \frac{k^3\, q^k}{1 - q^k}$$

$$Q(q) = 1 - 504 \sum_{k=1}^{\infty} \frac{k^5\, q^k}{1 - q^k},$$

it is possible to express many sums of the form [1458], [1779]

$$S_{\alpha,o}(q) = \sum_{k=1}^{\infty} k^{\alpha}\, \sigma_o(k)\, q^k$$

($\sigma_o(k)$ being the divisor sum function) as polynomials in $P(q)$, $Q(q)$, and $R(q)$. Find at least ten such polynomials.

g) The Eisenstein series $\mathcal{E}_n(q)$, defined by

$$\mathcal{E}_n(q) = 1 - \frac{2n}{B_n} \sum_{k=1}^{\infty} \sigma_{n-1}(k)\, q^k$$

(where B_n are the Bernoulli numbers and $\sigma_n(k)$ are the divisor sums) fulfill identities of the form [721], [705]

$$\sum_{m=1}^{o} \prod_{k=1}^{l^{(m)}} c_{m,k}\, \mathcal{E}_{n_k^{(m)}}(q)^{e_k^{(m)}} = 0$$

with integer c_m, k. In the last identity, the indices $n_k^{(m)}$ and the exponents $e_k^{(m)}$ fulfill the constraint

$$\sum_{k=1}^{l^{(m)}} n_k^{(m)} e_k^{(m)} = d, \quad m = 1, \ldots, o$$

where d is a given positive even integer. Examples of such relations are $\mathcal{E}_4(q)^2 - \mathcal{E}_8(q) = 0$ and $250\,\mathcal{E}_6(q)^2 + 441\,\mathcal{E}_4(q)\,\mathcal{E}_8(q) - 691\,\mathcal{E}_{12}(q) = 0$. Based on comparing coefficients of powers of q, implement a search for such identities and find at least ten.

h) The polynomials $S_k(x)$ defined by

$$S_k(x) = \frac{1}{k+1} \sum_{j=0}^{k} (-1)^j \binom{k+1}{j} B_j\, x^{k+1-j}$$

(here B_j are the Bernoulli numbers) have the property $S_k(n) = \sum_{j=1}^{n} j^k$. In addition, they fulfill identities of the form [1479]

$$S_k(x)^2 = \sum_{j=0}^{2o+1} c_j \, S_j(x).$$

Implement a one-liner that calculates this identity for a given k.

i) Consider the following infinite product representation of the function exp [1741]:

$$e^z = \prod_{k=1}^{\infty} \frac{1 - a_k \, z^k}{1 + a_k \, z^k}.$$

Calculate the first 100 of the a_k. How good an approximation of e (meaning $z = 1$) results from this? What is the radius of convergence of this infinite product?

j) Given a vector-valued differential equation $x_i'(s) = f_i(x_1(s), \ldots, x_n(s))$, the higher derivatives $\frac{\partial^2 x_i(s)}{\partial s^2}$, $\frac{\partial^3 x_i(s)}{\partial s^3}$, $\ldots$ can be expressed as functions of partial derivatives of the f_i with respect to the x_js. Using the convention that a sum is understood over all doubly occurring subscripts and that $f_{i;j,k,\ldots}$ stands as a shortcut for $\frac{\partial^n f_i(s)}{\partial x_j \partial x_k \cdots}$, the derivatives can be concisely written as $\frac{\partial x_i(s)}{\partial s} = f_i$, $\frac{\partial^2 x_i(s)}{\partial s^2} = f_{i;j} \, f_j$, $\frac{\partial^3 x_i(s)}{\partial s^3} = f_{i;j,k} \, f_j \, f_k + f_{i;j} \, f_{j;k} \, f_k$, ... [658]. Calculate explicit expressions for the first six $\frac{\partial^n x_i(s)}{\partial s^n}$. Write the result in the most concise form by renaming dummy summation indices.

18.¹³ Trig Values in Radicals, Ramanujan Identities, Modular Transformation

a) If possible, compute exact values (containing only square and cube roots and no exponentials and logarithms) for $\cos(j \pi / n)$ and $\sin(j \pi / n)$ (let $2 < n < 10$ and $j < n/2$).

b) Write a program that expresses every trigonometric function with argument $p/q\,\pi$ (p, q integer) that can be expressed as a function containing only addition, subtraction, multiplication, division, and square rooting. (Do not use the function `FunctionExpand`.)

c) Prove the following three identities that are from Ramanujan [158]:

$$\frac{\sin(2\pi/7)}{\sin^2(3\pi/7)} - \frac{\sin(\pi/7)}{\sin^2(2\pi/7)} + \frac{\sin(3\pi/7)}{\sin^2(\pi/7)} = 2\sqrt{7}$$

$$\frac{\sin^2(2\pi/7)}{\sin^4(3\pi/7)} + \frac{\sin^2(\pi/7)}{\sin^4(2\pi/7)} + \frac{\sin^2(3\pi/7)}{\sin^4(\pi/7)} = 28$$

$$\frac{\sin^2(3\pi/7)}{\sin^4(\pi/7)} \left(2 \frac{\sin(2\pi/7)}{\sin(3\pi/7)} + 4 \frac{\sin(3\pi/7)}{\sin(\pi/7)} \right) +$$

$$\frac{\sin^2(\pi/7)}{\sin^4(2\pi/7)} \left(-2 \frac{\sin(3\pi/7)}{\sin(\pi/7)} + 4 \frac{\sin(\pi/7)}{\sin(2\pi/7)} \right) -$$

$$\frac{\sin^2(2\pi/7)}{\sin^4(3\pi/7)} \left(2 \frac{\sin(\pi/7)}{\sin(2\pi/7)} + 4 \frac{\sin(2\pi/7)}{\sin(3\pi/7)} \right) = 280.$$

After making the following substitutions, for which n do these identities remain true?

$$\sin(\pi/7) \longrightarrow \sin(n\pi/7)$$
$$\sin(2\pi/7) \longrightarrow \sin(2n\pi/7)$$
$$\sin(3\pi/7) \longrightarrow \sin(3n\pi/7).$$

d) Determine all transformations of the form (modular transformations [1791], [1822])

$$y(x, k) = \frac{\alpha(k) + \beta(k)\,x}{\alpha'(k) + \beta'(k)\,x}$$

such that the following equation holds:

$$\left(\frac{\partial y(x, k)}{\partial x}\right)^2 = \frac{1}{M(k)}\,\frac{(1 - y(x, k)^2)(1 - l(k)^4\,y(x, k)^2)}{(1 - x^2)(1 - k^4\,x^2)}$$

Calculate all corresponding $l(k)^4$ and $M(k)$. Represent the results in a " nice" form.

19.^{12} Forced Damped Oscillation, $(1 + z/n)^n$-Series, e-Series, q-Logarithm

a) Suppose we are given the differential equation of a 1D oscillation subject to a harmonic force [63]

$$m\,x''(t) + \gamma\,x'(t) + k\,x(t) = F_0\,\cos(\omega t)$$

Compute the series expansion (in γ) of the maximum of the nth derivative (with respect to ω) of the amplitude of the resulting oscillation in the forced state.

If in addition to the external force $F_0\cos(\omega t)$, a stochastic force $\mathcal{F}(t)$ with the properties $\langle\mathcal{F}(t)\rangle = 0$ and $\langle\mathcal{F}(t)\mathcal{F}(t')\rangle = 2\gamma\,k_B\,T\,\delta(t - t')$ ($\langle.\rangle$ indicates time averaging) acts on the oscillator, the motion of the oscillator becomes different for each realization of the force $\mathcal{F}(t)$. Averaging over all realizations yields the probability distribution [329], [1289], [129]

$$W(x,\ t;\ x_0,\ v_0) = \frac{1}{\sqrt{4\pi\,q\,I_{\psi\psi}(t)}}\,\exp\left(-\frac{(x - x_c(t;\ x_0,\ v_0))^2}{4\pi\,q\,I_{\psi\psi}(t)}\right)$$

for the probability to find the oscillator as position x at time t if the oscillator started at time 0 at position x_0 with velocity v_0.

Here $q = \gamma\,k_B\,T/m$, $I_{\psi\psi}(t) = \int_0^t \psi(\tau)^2\,d\tau$, $\psi(\tau) = (e^{\mu_1(t-\tau)} - e^{\mu_2(t-\tau)})/(\mu_1 - \mu_2)$, $\mu_{1/2} = -\gamma/2 \pm(\gamma^2/4 - \omega_0^2)^{1/2}$, $\omega_0 = (k/m)^{1/2}$, and $x_c(x,\ t;\ x_0,\ v_0)$ is the solution of the oscillator ODE with initial conditions $x(0) = x_0$, $x'(0) = v_0$. Calculate $\langle x(t)\rangle_P$ and $\langle(x - \langle x(t)\rangle_P)^2\rangle_P$ where $\langle f(x)\rangle_P = \int_{-\infty}^{\infty} f(x)\,W(x,\ t;\ x_0,\ v_0)$.

b) Calculate the first 10 terms of the series of $(1 + z/n)^n$ around $n = \infty$.

c) To improve the well-known formula $\lim_{n\to\infty}(1 + 1/n)^n = e$, we write [151]

$$e = \left(1 + \frac{1}{n}\right)^{n + \sum_{k=0}^{\infty}\alpha_k\,n^{-k}}.$$

Calculate the first 100 of the α_k. How precise is the approximation for $n = 1$? Find the optimal α_0, α_1, β_2, and β_3 in the following formula:

$$e = \lim_{n\to\infty}\left(1 + \frac{1}{n} + \frac{\beta_2}{n^2} + \frac{\beta_3}{n^3}\right)^{n+\alpha_0+\frac{\alpha_1}{n}}.$$

d) The q-logarithm $\ln_q(x)$ can be defined through $\ln_q(x) = (x^{1-q} - 1)/(1 - q)$ [1323] and fulfills the following identity for positive q and x_i [1872]

$$\ln_q\left(\prod_{k=1}^{n} x_k\right) = \sum_{k=0}^{n} (1-q)^{k-1}\left(\sum_{l_1=1}^{n}\sum_{l_2=l_1+1}^{n}\cdots\sum_{l_k=l_{k-1}+1}^{n}\prod_{j=1}^{k}\ln_q(x_{l_j})\right).$$

Verify this identity for $2 \le n \le 10$ by explicit calculation of the left and right sides.

20.¹³ S_{mn}, Symmetric Determinant, Fermat Test

Implement an efficient computation for [1365], [1599], [959], [186]

$$S_{mn} = \Psi(m\,n + 1) - \frac{\int_0^\infty dq_1 \int_0^\infty dq_2 \cdots \int_0^\infty dq_m\, Q_{m\,n}(q_1, q_2, \ldots, q_m)}{m\,n \int_0^\infty dq_1 \int_0^\infty dq_2 \cdots \int_0^\infty dq_m\, P_{mn}(q_1, q_2, \ldots, q_m)}, \quad m \le n$$

$$Q_{mn}(q_1, q_2, \ldots, q_m) = \prod_{\substack{i,j=1 \\ i<j}}^{m}(q_i - q_j)^2 \prod_{k=1}^{m} e^{-q_k}\, q_k^{n-m}$$

$$P_{mn}(q_1, q_2, \ldots, q_m) = Q_{mn}(q_1, q_2, \ldots, q_m)\sum_{k=1}^{m} q_k \ln(q_k).$$

where $\Psi(z)$ is the Digamma function. Compute the results for S_{ij}, $1 \le i, j \le 4$. Can one compute S_{55} by brute force?

b) Let the symmetrized determinant of a matrix $\mathbf{X}$ of dimension $n \times n$ and elements x_{ij} be defined as [126]

$$\text{sdet}\,\mathbf{X} = \frac{1}{n!}\sum_{\sigma\in S_n}\sum_{\tau\in S_n}\text{signature}(\sigma)\,\text{signature}(\tau)\,x_{\sigma(1)\,\tau(1)}\cdots x_{\sigma(n)\,\tau(n)}.$$

The summations run over all elements of the permutations S_n of $\{1, 2, \ldots, n\}$.

Does there exist a 4×4 matrix $\mathbf{X}$ with elements

$$x_{ij} = c_{ij}^{(a)} a + c_{ij}^{(b)} b + c_{ij}^{(c)} c + c_{ij}^{(e)} e$$

where the $c_{ij}^{()}$ are ± 1 or 0 and $a, b, c,$ and e are noncommuting quantities with the multiplication table

	a	*b*	*c*	*e*
a	e	c	b	a
b	c	e	a	b
c	b	a	e	c
e	a	b	c	e

such that $\text{sdet}\,\mathbf{X} = a + b + c + e$?

b) Show that the following expression R (from [1136] and http://www.bway.net/~lewis/fermat/FerTest1) vanishes identically. Use at least three different methods to show that $R = 0$. Try to minimize memory usage.

$$R = q_2^2 q_3^2 (l_1 - 1)(l_2 - 1)(g^2 - n_{1,1} n_{2,2}) + q_3^2 l_1 l_2 (g^2 + n_{1,1} + n_{2,2} - n_{1,1} n_{2,2} - 1) +$$
$$q_2 q_3^2 (g^2 l_1 + g^2 l_2 - 2 g^2 l_1 l_2 + l_2 n_{1,1} - l_1 l_2 n_{1,1} + l_1 n_{2,2} - l_1 l_2 n_{2,2} - l_1 n_{1,1} n_{2,2} -$$
$$l_2 n_{1,1} n_{2,2} + 2 l_1 l_2 n_{1,1} n_{2,2}) - q_3 q_4 (g^2 + n_{1,1} + n_{2,2} - n_{1,1} n_{2,2} - 1)(l_2 p_{1,1} + l_1 p_{2,2}) -$$
$$q_2^2 q_3 (g^2 - n_{1,1} n_{2,2})(2 - l_1 - l_2 - p_{1,1} + l_2 p_{1,1} - p_{2,2} + l_1 p_{2,2}) +$$
$$q_1 q_2 q_3 q_4 (l_1 - 1)(l_2 - 1)(n_{2,2} p_{1,1} - g p_{1,2} - g p_{2,1} + n_{1,1} p_{2,2}) +$$
$$q_1 q_3 l_1 l_2 (2 - n_{1,1} - n_{2,2} - p_{1,1} + n_{2,2} p_{1,1} - g p_{1,2} - g p_{2,1} - p_{2,2} + n_{1,1} p_{2,2}) +$$
$$q_1 q_3 q_4 (-l_2 p_{1,1} + l_1 l_2 p_{1,1} + l_2 n_{2,2} p_{1,1} - l_1 l_2 n_{2,2} p_{1,1} - g l_2 p_{1,2} + g l_1 l_2 p_{1,2} -$$
$$g l_1 p_{2,1} + g l_1 l_2 p_{2,1} - l_1 p_{2,2} + l_1 l_2 p_{2,2} + l_1 n_{1,1} p_{2,2} - l_1 l_2 n_{1,1} p_{2,2}) +$$
$$q_1 q_2 q_3 (-l_2 n_{1,1} + l_1 l_2 n_{1,1} - l_1 n_{2,2} + l_1 l_2 n_{2,2} + l_1 n_{2,2} p_{1,1} - l_1 l_2 n_{2,2} p_{1,1} -$$
$$g l_1 p_{1,2} + g l_1 l_2 p_{1,2} - g l_2 p_{2,1} + g l_1 l_2 p_{2,1} + l_2 n_{1,1} p_{2,2} - l_1 l_2 n_{1,1} p_{2,2}) +$$
$$q_2 q_3 (-g^2 l_1 - g^2 l_2 - l_2 n_{1,1} - l_1 n_{2,2} + l_1 n_{1,1} n_{2,2} + l_2 n_{1,1} n_{2,2} + g^2 l_2 p_{1,1} + l_2 n_{1,1} p_{1,1} -$$
$$l_2 n_{1,1} n_{2,2} p_{1,1} + g l_1 p_{1,2} + g l_2 p_{2,1} + g^2 l_1 p_{2,2} + l_1 n_{2,2} p_{2,2} - l_1 n_{1,1} n_{2,2} p_{2,2}) +$$
$$q_2 q_3 q_4 (-g^2 p_{1,1} + g^2 l_2 p_{1,1} - n_{2,2} p_{1,1} + l_2 n_{2,2} p_{1,1} + n_{1,1} n_{2,2} p_{1,1} - l_2 n_{1,1} n_{2,2} p_{1,1} +$$
$$g p_{1,2} - g l_1 p_{1,2} + g p_{2,1} - g l_2 p_{2,1} - g^2 p_{2,2} + g^2 l_1 p_{2,2} - n_{1,1} p_{2,2} + l_1 n_{1,1} p_{2,2} +$$
$$n_{1,1} n_{2,2} p_{2,2} - l_1 n_{1,1} n_{2,2} p_{2,2}) + q_1^2 q_4^2 (l_1 - 1)(l_2 - 1)(p_{1,2} p_{2,1} - p_{1,1} p_{2,2}) -$$
$$q_1 q_4^2 (2 - l_1 - l_2 - n_{1,1} + l_2 n_{1,1} - n_{2,2} + l_1 n_{2,2})(p_{1,2} p_{2,1} - p_{1,1} p_{2,2}) -$$
$$q_4^2 (g^2 + n_{1,1} + n_{2,2} - n_{1,1} n_{2,2} - 1)(p_{1,2} p_{2,1} - p_{1,1} p_{2,2}) -$$
$$q_1^2 l_1 l_2 (1 - p_{1,1} - p_{1,2} p_{2,1} - p_{2,2} + p_{1,1} p_{2,2}) +$$
$$q_1 q_2 (l_2 n_{1,1} + l_1 n_{2,2})(1 - p_{1,1} - p_{1,2} p_{2,1} - p_{2,2} + p_{1,1} p_{2,2}) +$$
$$q_2^2 (g^2 - n_{1,1} n_{2,2})(1 - p_{1,1} - p_{1,2} p_{2,1} - p_{2,2} + p_{1,1} p_{2,2}) +$$
$$q_1^2 q_4 (l_2 p_{1,1} - l_1 l_2 p_{1,1} + l_1 p_{1,2} p_{2,1} + l_2 p_{1,2} p_{2,1} - 2 l_1 l_2 p_{1,2} p_{2,1} +$$
$$l_1 p_{2,2} - l_1 l_2 p_{2,2} - l_1 p_{1,1} p_{2,2} - l_2 p_{1,1} p_{2,2} + 2 l_1 l_2 p_{1,1} p_{2,2}) +$$
$$q_1 q_4 (-l_2 p_{1,1} + l_2 n_{1,1} p_{1,1} + g l_2 p_{1,2} + g l_1 p_{2,1} - l_1 p_{1,2} p_{2,1} - l_2 p_{1,2} p_{2,1} + l_2 n_{1,1} p_{1,2} p_{2,1} + l_1 n_{2,2}$$
$$p_{1,2} p_{2,1} - l_1 p_{2,2} + l_1 n_{2,2} p_{2,2} + l_1 p_{1,1} p_{2,2} + l_2 p_{1,1} p_{2,2} - l_2 n_{1,1} p_{1,1} p_{2,2} - l_1 n_{2,2} p_{1,1} p_{2,2}) +$$
$$q_1 q_2 q_4 (-n_{2,2} p_{1,1} + l_1 n_{2,2} p_{1,1} + g p_{1,2} - g l_2 p_{1,2} + g p_{2,1} - g l_1 p_{2,1} - n_{1,1} p_{1,2} p_{2,1} +$$
$$l_2 n_{1,1} p_{1,2} p_{2,1} - n_{2,2} p_{1,2} p_{2,1} + l_1 n_{2,2} p_{1,2} p_{2,1} - n_{1,1} p_{2,2} + l_2 n_{1,1} p_{2,2} +$$
$$n_{1,1} p_{1,1} p_{2,2} - l_2 n_{1,1} p_{1,1} p_{2,2} + n_{2,2} p_{1,1} p_{2,2} - l_1 n_{2,2} p_{1,1} p_{2,2}) +$$
$$q_2 q_4 (g^2 p_{1,1} + n_{2,2} p_{1,1} - n_{1,1} n_{2,2} p_{1,1} - g p_{1,2} - g p_{2,1} + 2 g^2 p_{1,2} p_{2,1} + n_{1,1} p_{1,2} p_{2,1} +$$
$$n_{2,2} p_{1,2} p_{2,1} - 2 n_{1,1} n_{2,2} p_{1,2} p_{2,1} + g^2 p_{2,2} + n_{1,1} p_{2,2} - n_{1,1} n_{2,2} p_{2,2} -$$
$$2 g^2 p_{1,1} p_{2,2} - n_{1,1} p_{1,1} p_{2,2} - n_{2,2} p_{1,1} p_{2,2} + 2 n_{1,1} n_{2,2} p_{1,1} p_{2,2}).$$

The q_i in the last polynomial are given by the following rational functions:

$$q_1 = -\frac{c_2 c_6}{c_3 c_5}, \quad q_2 = \frac{c_2 c_4}{c_1 c_5}, \quad q_3 = \frac{c_2 c_9}{c_3 c_8}, \quad q_4 = \frac{c_2 c_7}{c_1 c_8},$$

$$c_1 = (l_1 - 1) a_{1,2} a_{2,1} - (l_2 - 1) a_{2,2}$$
$$c_2 = l_1 a_{1,2} (1 + a_{2,1}) - l_2 (a_{1,2} + a_{2,2})$$
$$c_3 = l_1 (1 + a_{2,1}) a_{2,2} + l_2 ((l_1 - 1) a_{1,2} a_{2,1} - (l_1 + a_{2,1}) a_{2,2})$$
$$c_4 = -a_{2,1} (g - a_{1,2} (n_{1,1} - 1)) + a_{2,2} (1 + g a_{1,2} - n_{2,2})$$
$$c_5 = -g + g a_{1,2}^2 - g a_{2,1} + a_{1,2} (g a_{2,2} + (1 + a_{2,1}) n_{1,1} - n_{2,2}) - a_{2,2} n_{2,2}$$

$$c_6 = g\,a_{2,1}^2 - a_{2,2}\,(g^2 + g\,(a_{1,2} + a_{2,2}) - n_{1,1}\,(n_{2,2} - 1)) +$$
$$a_{2,1}\,(g + a_{2,2}\,(n_{2,2} - n_{1,1}) + a_{1,2}\,(g^2 - (n_{1,1} - 1)\,n_{2,2}))$$
$$c_7 = -a_{2,1}\,p_{1,2} + a_{1,2}\,(a_{2,1}\,(p_{1,1} - 1) + a_{2,2}\,p_{2,1}) - a_{2,2}\,(-1 + p_{2,2})$$
$$c_8 = -(1 + a_{2,1})\,p_{1,2} + a_{1,2}^2\,p_{2,1} + a_{1,2}\,((1 + a_{2,1})\,p_{1,1} + a_{2,2}\,p_{2,1} - p_{2,2}) - a_{2,2}\,p_{2,2}$$
$$c_9 = a_{2,2}^2\,p_{2,1} + a_{2,1}\,(-p_{1,2}\,(1 + a_{2,1} + a_{1,2}\,p_{2,1}) + a_{1,2}\,(p_{1,1} - 1)\,p_{2,2}) +$$
$$a_{2,2}\,((1 + a_{2,1})\,p_{1,1} + (a_{1,2} + p_{1,2})\,p_{2,1} - (a_{2,1} + p_{1,1})\,p_{2,2})$$

21.¹³ WKB Approximations of Higher Order, GHZ State, Entropic Uncertainty Relation

a) This problem is meant for those who are already familiar with WKB approximations in quantum mechanics [1600], [889]—an approximate solution for the eigenvalue of the 1D Schrödinger equation. (In the following, $m = 1$ and $\hbar = h/(2\pi)$ explicitly stand for the derivation of the relevant equations determined by a "small" parameter, the so-called WKB approximation.)

$$\left(-\frac{\hbar^2}{2}\frac{d^2}{dx^2} + V(x) - E\right)\Psi(x) = 0$$

We restrict ourselves here to the case in which the potential has just one minimum. Assuming a solution of the form

$$\Psi(x) = \exp\!\left(i\,\frac{1}{\hbar}\int^x S(x)\,dx\right)$$

$$S(x) = \sum_{k=0}^{\infty}\left(\frac{\hbar}{i}\right)^k S_k(x)$$

for the Schrödinger equation, we get the following recurrence formula for $S_n(x)$ [1115], [802], [1499], [727], [491]:

$$S_0(x) = \sqrt{E - V(x)}$$

$$S'_{n-1}(x) = \sum_{m=0}^{n} S_{n-m}(x)\,S_m(x).$$

Now, interpreting x as a complex variable and $V(x)$ as a complex function, we are led to the following condition for the eigenvalue E [1739]:

$$\oint S_0(x)\,dx + \sum_{k=1}^{\infty}\left(\frac{\hbar}{i}\right)^k \oint S_k(x)\,dx = \left(n + \frac{1}{2}\right)$$

Thus, E_j ($j = 0, 1, \ldots$) is an eigenvalue for the above differential equation if and only if this equation is satisfied. The x-integrals are taken along a path in the complex x-plane that contains both of the classical turning points x_1 and x_2. Both x_1 and x_2 are solutions of the equation $V(x) = E$.

Frequently, the potential $V(x)$ is given only for real numbers x (perhaps from a numerical calculation). Then, the above integral formula is of little use. Even if we put the integration path on the real axis, for $n > 1$, in general, $S_n(x)$ has a nonintegrable singularity. Moreover, the integrands that arise can be greatly simplified by integration by parts.

One way out of these difficulties is to use integration by parts with respect to x, and to reduce the strongly singular integrands to integrable integrands using integration by parts over E [654], [1327]. (For details, see [1077], [1076], [536], [51], [1266], [142], [71], [16], [1345], [1856], [160], [1006], [1007], [641], [1497], [345], [333], [1008], [1009], [44], [544], [620], [594], [272], [1876], [1195], and [1796].)

Implement the computation of these reduced forms of the integrals over the $S_n(x)$ for $n = 1$ to 10. To this end, first find $S_n(x)$, and then integrate the parts containing $V'(x)$ as often as possible by parts. Write the resulting expressions over all denominators that appear, and integrate the terms so obtained as often as possible by parts, without increasing the maximum singularity in the current denominator. Finally, integrate the expressions thus obtained by parts with respect to E, and write the results in an appropriate form.

For $V(x) = x^4$ [1498], [1797], calculate the correction terms explicitly. Compare for the value $\hbar = 1/10$ and $\hbar = 1/100$ the convergence behavior of $n = 3$ and $n = 100$.

b) This is another problem for readers who are with quantum mechanics. For the three-particle Greenberger–Horne–Zeilinger state [1248], [449], [139], [1154], [1898], [288], [11], [786], and [1720] $|\Psi\rangle = (|\uparrow\rangle_1 |\uparrow\rangle_2 |\uparrow\rangle_3 - |\downarrow\rangle_1 |\downarrow\rangle_2 |\downarrow\rangle_3)/2^{1/2}$ find the most general observable $\hat{\Xi}_\pm = (\vec{n}_1 . \vec{\sigma})_1 \otimes 1 \otimes 1 + 1 \otimes (\vec{n}_2 . \vec{\sigma})_2 \otimes 1 + 1 \otimes 1 \otimes (\vec{n}_3 . \vec{\sigma})_3$ (with normalized $|\vec{n}_k| = 1$) such that $\hat{\Xi}_\pm |\Psi\rangle = \pm 1 |\Psi\rangle$. ($\vec{\sigma}$ is the vector of the three Pauli matrices.)

c) And one more exercise for the friends of quantum mechanics: Does there exists a pure (entangled) state of four (different) spin $1/2$ particles [1539], [320], [1654], [565], [844], [542], [339], [1388] such that all two-particle reduced density matrices are multiples of the identity?

d) The entropic uncertainty relation is $\delta x \, \delta p \geq \pi e \, \hbar$ [1554], [880], [1180]. Here the δa is the exponential of the differential entropy $\delta a = \exp(-\int_{-\infty}^\infty w(a) \ln(w(a)) \, da)$ of the normalized to $\int_{-\infty}^\infty w(a) \, da = 1$ probability density $p(a)$ [880]. For the simplest case of a 1D scattering on a slit of length $2L$ with the position probability density $w(x) = \theta(L + x) \theta(L - x) (2L)^{-1/2}$, calculate δp whose probability density is $w(p) = |\int_{-\infty}^\infty \exp(i \, p \, x) w(x)^{1/2} \, dx|^2$ [1554].

22.^{13} QES Condition, Integral of Rational Function, Integrals of Roots, Triangle Roots

a) Calculate the symbolic value of α such that $i = \int_{-x_0}^{x_0} \sqrt{\alpha - x^6 + 4 \, x^2} \, dx$, where x_0 is the real positive root of a $\alpha - x^6 + 4 \, x^2 = 0$ takes the value $i = 2\pi$. (The answer for α is a "nice" algebraic number [145], [1753].)

b) Calculate the value of the integral $\int_{-\infty}^\infty (z^8 + 7 \, z^6 + 5 \, z^4 + 3 \, z^2 + 1)^{-4} \, dz$. Express the result in a nice form [210], [1284].

c) Calculate the following integral $(x > 1)$ [1594]:

$$\lim_{\gamma \to 0} \int_1^\infty \frac{z^\gamma - 1}{\gamma \, z^2} (1 + x (z^\gamma - 1))^{-\frac{1}{\gamma} - 1} \, dz.$$

d) Let $\mathcal{R}_k(\xi, a_0 + a_1 \xi + \cdots + a_n \xi^n)$ stand for the root that is represented by the Root-object $\texttt{Root}[a_0 + a_1 \# + \cdots + a_n \#^n, \ k]$. Calculate the following integrals symbolically (express the results using Root-objects):

$$\int \ln(x^2) \, \mathcal{R}_1(\xi, \ -x - x\xi + \xi^6) \, dx$$

$$\int \exp(\mathcal{R}_3(\xi, \ -x - \xi^2 + \xi^7)) \ln(\mathcal{R}_3(\xi, \ -x - \xi^2 + \xi^7)) \, \mathcal{R}_3(\xi, \ -x - \xi^2 + \xi^7) \, dx$$

$$\int \sqrt{\frac{R_2(\xi, -x - \xi + \xi^3)}{R_3(\xi, -x - \xi + \xi^3)}}\, dx$$

$$\int \sqrt[3]{\frac{R_2(\xi, -x - x\xi + \xi^3)}{R_3(\xi, -x - x\xi + \xi^3)}}\, dx$$

$$\int_0^1 \frac{R_2(\xi, -x - \xi + \xi^3)}{R_2(\xi, x - \xi + \xi^3) - 1}\, dx$$

$$\int_1^\infty \left(\frac{1}{R_1(\xi, -x + \xi + \xi^5)} - \frac{1}{5x} - \frac{1}{\sqrt[5]{x}}\right) dx$$

e) Under which conditions on a_1, a_2, a_3 can the three roots of the cubic $x^3 + a_1 x^2 + a_2 x + a_3 = 0$ be interpreted as the side length of a nondegenerate triangle [1338]? Visualize the volume in a_1, a_2, a_3-space for which this happens. For random a_1, a_2, a_3 from the interval $[-1, 1]$, what is the probability that the roots are the side length of a nondegenerate triangle?

23.^{12} Riemann Surface of Cubic

Visualize the Riemann surface of $x(a)$, where $x = x(a)$ is implicitly given by $x^3 + x^2 + ax - 1/2 = 0$. Do not use `ContourPlot3D`.

24.^{12} Celestial Mechanics, Lagrange Points

a) For the so-called Kepler equation (see [1332], [1699], [1521], [781], [1236], [303], [343], and [388]) $L = M + \epsilon \sin(L)$ find a series solution for small ϵ in the form

$$L \approx M + \sum_{i=1}^{n}\left(\sum_{j=i}^{n\ \mathrm{or}\ n-1} a_{ij}\, \epsilon^j\right)\sin(i\, M)$$

with n around 10.

b) Find a short time-series solution (power series in t up to order 10, for example) for the equation of motion for a body in a spherical symmetric gravitational field (to avoid unnecessary constants, appropriate units are chosen)

$$\mathbf{r}''(t) = \frac{\mathbf{r}(t)}{r(t)^3}$$

with the initial conditions $\mathbf{r}(0) = \mathbf{r}_0$, $\mathbf{r}'(0) = \mathbf{v}_0$. Here, $\mathbf{r}(t)$ is the time-dependent position vector of the body and $r(t) = |\mathbf{r}(t)|$. To shorten the result, introduce the abbreviations

$$s = \frac{\mathbf{r}_0.\mathbf{v}_0}{r_0^2} \qquad w = \frac{\mathbf{v}_0.\mathbf{v}_0}{r_0^2} \qquad u = \frac{1}{r_0^3}$$

(Do not use explicit lists as vectors, first because this is explicitly dependent on the dimension, and second because it slows down the calculation considerably. It is better to implement an abstract vector type for $\mathbf{r}(t)$ and define appropriate rules for it.)

c) The Lagrange points $\{x(\mu), y(\mu)\}$ of the restricted three-body problem are the solutions of the following system of equations [430], [1421], [1755], [801], [1433], [745], [137]:

$$-\frac{\partial V(x, y)}{\partial x} = -\frac{\partial V(x, y)}{\partial y} = 0.$$

The potential $V(x, y)$ is given by the following expression:

$$V(x, y) = -\frac{1}{2}(x^2 + y^2) - \frac{1-\mu}{r_1} - \frac{\mu}{r_2}$$

$$r_1 = \sqrt{(x - x_1)^2 + y^2}$$

$$r_2 = \sqrt{(x - x_2)^2 + y^2}$$

$$x_1 = -\mu$$

$$x_2 = 1 - \mu$$

Calculate explicit symbolic solutions for the Lagrange points. For the parameter value $\mu = \frac{1}{10}$, calculate all real solutions (do not do this by a direct call to `Solve`).

25.^{12} Algebraic Lissajous Curves, Light Ray Reflection Inside a Closed Region

a) Derive an implicit representation $f(x, y)$ of the Lissajous curves $\{x(t), y(t)\} = \{\cos(t), \sin(2t)\}$, $\{x(t), y(t)\} = \{\cos(3t), \sin(2t)\}$ and some higher-order Lissajous curves.

b) Describe the region that is given implicitly by

$$(9x^2 - 24x^4 + 16x^6 - 4y^2 + 4y^4)^2 - \frac{1}{2} < 0$$

explicitly in terms of a conjunction of 2D sets. Each set described by a pair of inequalities of the form

$$x_l \le x \le x_u, \quad y_l(x) \le y \le y_u(x)$$

Use this description to generate a picture of this region (the name `lissa` stems from the fact that $9x^2 - 24x^4 + 16x^6 - 4y^2 + 4y^4$ is the implicit description of a Lissajous curve). (Do not use `CylindricalDe` `composition`, `Reduce` or similar functions.)

c) Derive the implicit representation $f(x, y)$ of the evolute of the Lissajous curves $\{x(t), y(t)\} = \{\cos(t), \sin(2t)\}$. For an implicitly defined curve $g(x, y)$, the points $\{\xi, \eta\}$ of the evolute are given by

$$\xi = x - \frac{g_x(g_x^2 + g_y^2)}{g_y^2 g_{xx} - 2g_x g_y g_{xy} + g_x^2 g_{yy}} \qquad \eta = y - \frac{g_y(g_x^2 + g_y^2)}{g_y^2 g_{xx} - 2g_x g_y g_{xy} + g_x^2 g_{yy}}$$

where $g_x = \partial g(x, y)/\partial x$, $g_y = \partial g(x, y)/\partial y$, and so on.

d) Make an animation showing how the orthopodic locus [850], [1865] of the Lissajous curves $\{x(t), y(t)\} = \{\cos(t), \sin(2t)\}$ is generated. The orthopodic locus of a curve is the set of all points where two tangents on the given point intersect perpendicularly. Derive the implicit representation $f(x, y)$ of the orthopodic locus of the Lissajous curves under consideration. Do not solve nonpolynomial equations numerically here.

e) Calculate the implicit equation of the cissoid [850] of the Lissajous curves $\{x(t), y(t)\} = \{\cos(t), \sin(2t)\}$. The cissoid of a curve and a given point P is the set of all points that lie on a line through P and are at the same time the midpoint of the line segment formed by two points of the given curve. For $P = \{1/2, 1/2\}$, make a picture of the implicit representation and compare it with a straightforward calculation of the cissoid.

f) Make a picture with the implicit equation of the $\{\cos(3\,\varphi) - \sin(\varphi)/5, \sin(2\,\varphi) + \cos(\varphi)/3\}$, $0 \leq \varphi \leq 2\pi$ written along the curve itself.

g) Follow a light ray that is ideally reflected multiple times inside the two-dimensional region defined by $x^4 + y^4 - 2x^3 + 2y + 2x - 1 \leq 0$ (a billiard problem [955]). Determine some (and if possible all) stationary light rays.

h) The hedgehog of a function $h(t)$ is the envelope of the family of lines $x \cos(t) + y \sin(t) = h(t)$ [1212]. Calculate and visualize the hedgehog of $h(t) = -\cos(t) + 2 \cos(6\,t)$.

i) Make an animation of the following function inside the region $x^4 + y^4 < 1$ as k varies from 0 to 150:

$$I_k(x, y) = \int_0^L \exp(i\,k\,d(\{x, y\}, \{\xi(s), \eta(s)\}))\, \delta(g(\{x, y\}, \{\xi(s), \eta(s)\}))\, ds.$$

Here s is the arclength of the curve C implicitly described through $x^4 + y^4 = 1$, L its length, $g(\{x, y\}, \{\xi(s), \eta(s)\}) = a(s)\,y + b(s)\,x + c(s)$ is the implicit equation for the lines through the point $\{x, y\}$ that are perpendicular to C at $\{\xi(s), \eta(s)\}$, and $d(\{x, y\}, \{\xi(s), \eta(s)\})$ is the distance of the point $\{x, y\}$ to the point $\{\xi(s), \eta(s)\}$. For some selected k, compare the resulting graphics with a corresponding graphic of

26.^{L2} Change of Variables in a Differential Equation, Functional Equation

a) Given the following initial value problem [621]:

$$y''(x) = x^3 + y(x)^3 + y'(x)^3$$
$$y(0) = 0$$
$$y'(0) = 0$$

Derive a new differential equation for $u(x_1)$ when using the following change of variables on the above differential equation:

$$x_1 = y'(x)$$
$$u(x_1) = y(x)$$

Calculate 10 terms of the series of $u(x_1)$ around x_1 and compare the quality of the series solution with a numerical solution.

b) Let the function $f_q(z)$ be defined through the functional equation $f_q(q\,z) = (q\,f_q(z)^2 - q + 2)/2$ [763], [703], [1721], [412], [249], [1182], [1183]. Make an animation of $f_{1.6\exp(i\,\varphi)}(z)$ as a function of real φ.

27.^{L2} Discriminant Surface, Multivalued Surface, 27 Lines on the Clebsch Surface

a) Make a picture of the surface [1776], [1852], [993] defined implicitly by

$$256\,z^3 + 128\,x^2\,z^2 + 16\,x^4\,z - 144\,x\,z\,y^2 - 4\,x^3\,y^2 + 27\,y^4 = 0$$

in the x,y-region $(-1, 2) \times (-3/2, 3/2)$ and in the corresponding z-region. Using `ContourPlot3D` will not show all details with sufficient quality.

b) Make a picture of the following implicitly defined surface

$$\sin(x)\cos(y) + \sin(y)\cos(z) + \sin(z)\cos(x) = 0$$

in the cube $0 \le x \le 2\pi, 0 \le y \le 2\pi, 0 \le z \le 2\pi$. Do not use `ContourPlot3D` for generating the picture.

c) The Clebsch surface (see also Exercise 7 of Chapter 1 of the Numerics volume [1737]), defined by the implicit equation

$$32 - 216\,x^2 + 648\,x^2\,y - 216\,y^2 - 216\,y^3 - 150\,z + 216\,x^2\,z + 216\,y^2\,z + 231\,z^2 - 113\,z^3 = 0$$

has the remarkable property that 27 straight lines lie on it [550], [1501], [830], [1033], [639], [1595], [1590], [28], [855]. Calculate these 27 lines explicitly. Visualize these lines that lie on the surface.

d) Calculate the implicit representations of "generalized" Clebsch surfaces, which, instead of a threefold rotation symmetry, have an n-fold rotation symmetry ($n = 4, 5, 6, \dots$) and has a similar shape. Visualize some of them.

28.^{L2} 28 Bitangents on a Real Plane Quartic, Curve Intersections, Pentaellipse

a) In [555], the following implicit equation for a plane quartic with 28 real bitangents is given. (A bitangent is a line that is tangent to two points, including the degenerate case in which these two points coincide) and to a curve at the same time.

$$x^4 + y^4 - 6\,(x^2 + y^2) + 10 = 0$$

Calculate the explicit form of the bitangents, and make a picture that illustrates that these lines are bitangents.

b) Consider the parametric curve $\{\mathrm{Re}(z(\varphi)), \mathrm{Im}(z(\varphi))\}$ where $z(\varphi)$ is implicitly defined by $2\,z(\varphi)^6 = q(\varphi) + q(\varphi)^3$ and $q = R\exp(i\,\varphi)$, $R = 9/10$. Find exact values for the points where the curve self-intersects.

c) An ellipse can be defined as the set of points whose sum of distances from two points is a constant. A natural generalization of this definition is the set of points whose sum of distances from n points is a constant c [613], [1598], [1407]. For the case of $n = 5$ and the five points being the vertices of a regular pentagon, derive a polynomial $p_c(x, y)$ (with integer coefficients) such that $p_c(x, y) = 0$ contains the pentaellipse. Visualize $p_c(x, y) = 0$.

29.^{L2} Maxwell's Equations *Are* Galilei Invariant, X-Waves,
Fields in a Moving Media, Thomas Precession, Retarded Potential Expansion

a) Show that the vacuum Maxwell equations [929] (the fields $\vec{E}, \vec{H}$ are functions of the coordinates x, y, z.)

$$\frac{\partial \vec{E}}{\partial t} = \mathrm{curl}\,\vec{H}$$

$$\frac{\partial \vec{H}}{\partial t} = -\mathrm{curl}\,\vec{E}$$

$$\mathrm{div}\,\vec{E} = 0$$

$$\mathrm{div}\,\vec{H} = 0$$

are invariant under the Galilei transformation

$$\vec{r}' = \vec{r} + \vec{v}\,t$$

$$t' = t$$

if the fields are transformed in the following way:

$$\vec{E}' = \vec{E} - \vec{v} \times \vec{H} - \left(\vec{v}.\vec{r} + \frac{t}{2}\,\vec{v}.\vec{v}\right)\mathrm{rot}\,\vec{H} + \frac{1}{2}\left(v^2\vec{E} - \vec{v}\vec{v}.\vec{E} + \vec{r}.\vec{v}\left(\nabla.\vec{v}\,\vec{E} - 2\,\vec{v}\times\mathrm{rot}\,\vec{E}\right) + (\vec{r}.\vec{v})^2\,\Delta\,\vec{E}\right) + O(v^3)$$

$$\vec{H}' =$$

$$\vec{H} + \vec{v} \times \vec{E} - \left(\vec{v}.\vec{r} + \frac{t}{2}\,\vec{v}.\vec{v}\right)\mathrm{rot}\,\vec{E} + \frac{1}{2}\left(v^2\vec{H} - \vec{v}\vec{v}.\vec{H} + \vec{r}.\vec{v}\left(\nabla.\vec{v}\,\vec{H} - 2\,\vec{v}\times\mathrm{rot}\,\vec{H}\right) + (\vec{r}.\vec{v})^2\,\Delta\,\vec{H}\right) + O(v^3)$$

$$v =$$

$$|\vec{v}|$$

For the derivation of these formulae, see [679] and [1064]. (For related discussions see [1125], [735], [1065].)

b) The relativistic relations between the electric and magnetic vacuum fields $\mathcal{E}$, $\mathcal{H}$ and the electric and magnetic fields in media $\mathcal{D}$, $\mathcal{B}$ [1500] are given by [834], [1899], [1234]

$$\mathcal{D} + \frac{1}{c}\,v \times \mathcal{H} = \varepsilon\left(\mathcal{E} + \frac{1}{c}\,v \times \mathcal{B}\right)$$

$$\mathcal{B} - \frac{1}{c}\,v \times \mathcal{E} = \mu\left(\mathcal{H} - \frac{1}{c}\,v \times \mathcal{D}\right)$$

Here, v is the velocity of the moving media, ε and μ are the relative permittivity and permeability. Express the fields $\mathcal{D}$, $\mathcal{B}$ as a function of $\mathcal{E}$, $\mathcal{H}$ in compact vector form. (Do not guess the form; derive it.)

c) Show that $\mathrm{Re}(\varphi_v(\rho, z; t))$ where [1889], [1603], [165]

$$\varphi_v(\rho, z; t) = \frac{z - vt + i\alpha}{\left((z - vt + i\alpha)^2 + \left(1 - \frac{v^2}{c^2}\right)\rho^2\right)^{3/2}}$$

is a solution of the 3D wave equation (ρ, and z are cylindrical coordinates, c is the group velocity, α and v are arbitrary parameters). Visualize the solution for $v \lesseqgtr c$.

d) Solve the following set of equations for $\psi = \psi(\xi, \eta, \alpha)$, eliminating the variables ρ, and θ [352].

$$e^{-\xi\sigma_3}\,e^{\eta\,(\sin(\alpha)\,\sigma_1 + \cos(\alpha)\,\sigma_3)} = e^{\rho(\sin(\theta)\,\sigma_1 + \cos(\theta)\,\sigma_3)}\,e^{i\psi\sigma_2}$$

Here e^{matrix} denotes the matrix exponential functions and the three σ_k are the Pauli matrices

$$\sigma_1 = \begin{pmatrix} 0 & 1 \\ 1 & 0 \end{pmatrix}, \quad \sigma_2 = \begin{pmatrix} 0 & -i \\ i & 0 \end{pmatrix}, \quad \sigma_3 = \begin{pmatrix} 1 & 0 \\ 0 & -1 \end{pmatrix}.$$

($\psi(\xi, \eta, \alpha)$ is the Thomas precession angle [411], [1546], [1751], [156], [1005], [1750], [1303], [838], [353], [922], [128], [1478] that arises from two successive Lorentz transformations in different directions as one Lorentz transformation and a rotation.)

e) Calculate the first few terms of the expansion in c^{-1} [458], [755], [1868], [756], [853], [688], [1798], [998], [1799], [179], [222], [792], [1237] of the Lienárd–Wiechert potential $\varphi(\mathbf{r}, t)$

$$\varphi(\mathbf{r}, t) = \frac{e}{r(\tau) - \frac{1}{c}\,\mathbf{r}(\tau).\mathbf{v}(\tau)}.$$

Here e is the electric charge, c is the speed of light, $\mathbf{r}(\tau) = \mathbb{R} - \mathbb{R}_e(\tau)$, where $\mathbb{R}$ is the observation point and $\mathbb{R}_e(\tau)$ the position of the charge. The scalar distance is $r(\tau) = |\mathbf{r}(\tau)|$ and $\mathbf{v}(\tau) = \partial \mathbb{R}_e(\tau)/\partial \tau$ is the velocity of the charge. The retarded time τ is defined through $t = \tau + \mathbf{r}(\tau)/c$.

f) Find a nontrivial "spherical", standing wave-like solutions of the homogeneous Maxwell equations.

$$\operatorname{div} \mathcal{E}(\mathbf{r},\, t) = 0$$
$$\operatorname{div} \mathcal{B}(\mathbf{r},\, t) = 0$$
$$\frac{\partial \mathcal{E}(\mathbf{r},\, t)}{\partial t} = \operatorname{curl} \mathcal{B}(\mathbf{r},\, t)$$
$$\frac{\partial \mathcal{B}(\mathbf{r},\, t)}{\partial t} = -\operatorname{curl} \mathcal{E}(\mathbf{r},\, t)$$

Here by "spherical" we mean a solution that does not depend on the azimuthal coordinate φ, but only on r and ϑ. By "standing wave-like", we mean $\mathcal{E}(\mathbf{r}).\mathcal{B}(\mathbf{r}) = 0$. For a wave-like solution, assume the time dependence of the form $\mathcal{E}(\mathbf{r}) \sim \sin(t)$ and $\mathcal{B}(\mathbf{r}) \sim \cos(t)$. Use for the spatial part of the components $e_{\mathrm{a}} \in \{\mathcal{E}_r,\, \mathcal{E}_\vartheta,\, \mathcal{E}_\varphi,\, \mathcal{B}_r,\, \mathcal{B}_\vartheta,\, \mathcal{B}_\varphi\}$ of the fields in spherical coordinates $\mathcal{E}(\mathbf{r}) = \mathcal{E}_r\, \mathbf{e}_r + \mathcal{E}_\vartheta\, \mathbf{e}_\vartheta + \mathcal{E}_\varphi\, \mathbf{e}_\varphi$, $\mathcal{B}(\mathbf{r}) = \mathcal{B}_r\, \mathbf{e}_r + \mathcal{B}_\vartheta\, \mathbf{e}_\vartheta + \mathcal{B}_\varphi\, \mathbf{e}_\varphi$ an ansatz of the form [370]

$$e_{\mathrm{a}} \sim \sum_{\rho_r=-3}^{1} \sum_{\rho_c,\rho_s=0}^{1} \sum_{k,l=0}^{o} \sum_{\alpha_{c,k},\alpha_{s,k}=0}^{1} c^{(e,\mathrm{a})}_{\rho_r,\rho_c,\rho_s,k,l,\alpha_{c,k},\alpha_{s,k}}\, r^{\rho_r} \cos(r)^{\rho_c} \sin(r)^{\rho_s} \cos(k\,\vartheta)^{\alpha_{c,k}} \sin(l\,\vartheta)^{\alpha_{s,l}}$$

for some (small integer) o. Here the $c^{(e,\mathrm{a})}_{\rho_r,\rho_c,\rho_s,k,l,\alpha_{c,k},\alpha_{s,k}}$ are constants independent of r and ϑ. Visualize the motion of a charged particle in such fields.

30.[12] Asymptotic Series for $n!$, q-Series to q-Product, q-Binomial, gcd-Free Partitions

a) The series of $n!$ is around $n = \infty$ asymptotic. So to get the best numerical result from such a series in dependence of n, the series should be truncated in an n-dependent manner. Calculate after how many terms it is best to truncate the series for $n = 1!, \ldots, 15!$ (For a detailed exposition on the optimal truncation of divergent series, see [121] and [487].)

b) Ramanujan gave the following expansion for the factorial function for large n [157], [46]:

$$n! = \sqrt{\pi}\, \left(\frac{n}{e}\right)^n \left(8\,n^3 + 4\,n^2 + n + \frac{1}{30} + R(n)\right)^{1/6}.$$

Assuming that $R(n)$ is of the form $\sum_{k=1}^{\infty} c_k\, n^{-k}$, calculate the first few c_k. How many c_k are needed to calculate $100!$ to 100 correct digits?

c) Given a q-series $\Sigma = 1 + \sum_{k=1}^{n} c_k\, q^k$, $c_k \in \mathbb{Z}$, write a one-liner `qSeriesToqProduct` that expresses the series Σ as a product

$$\Pi = \prod_{k=1}^{m} (1 - q^k)^{\alpha_k},\ \alpha_k \in \mathbb{Z},$$

such that $\Pi - \Sigma = O(q)^{n+1}$ [698], [211].

d) The q-binomial coefficients [600], [96], [1651], [60], [1059], [141], [1574], [1094], [1026], [1538], [256]

$$\begin{bmatrix} n \\ k \end{bmatrix}_q = \frac{\prod_{j=1}^{n-k+1} (1-q^j)}{\prod_{j=1}^{k} (1-q^j)}$$

that are seemingly rational functions in q, are actually polynomials in q. For $1 \le k, n \le 250$, visualize the number of factors in the factored form of the q-binomial coefficient s. (The coefficient of q^k of $\begin{bmatrix} m+n \\ n \end{bmatrix}_q$ is the number of partitions of k elements into at most n parts each part having at most m elements.)

e) A Taylor series $c_0 + \sum_{j=1}^{\infty} c_j q^k$ is called multiplicative if $c_{k \times l} = c_k c_l$ for all pairs k, l, such that $\gcd(k, l) = 1$. Products of the form

$$\frac{\prod_{j=1}^{m} \prod_{k=1}^{\infty} (1-q^{\alpha_j k})^{\mu_j}}{\prod_{j=1}^{n} \prod_{k=1}^{\infty} (1-q^{\beta_j k})^{\nu_j}}$$

where $\sum_{j=1}^{m} \alpha_j \mu_j = \sum_{j=1}^{n} \beta_j \nu_j = s$ and the α_j, β_j are all divisors of $n = \max(\alpha_j, \beta_j)$ are sometimes multiplicative series [636], [1461]. Find all such type multiplicative series for $1 \le n, s \le 24$.

f) Given a set of distinct primes $\{p_1, \ldots, p_r\}$, the number of partitions $P_{p_1, \ldots, p_k}(n)$ of a positive integer n into integers $\{m_1, \ldots, m_j\}$ (meaning $\sum_{l=1}^{j} m_l = n$) such that $\gcd(m_l, p_q) = 1$, $1 \le l \le j$, $1 \le q \le r$ is given by [1496]

$$P_{p_1, \ldots, p_k}(n) = [z^n] \left(\prod_{o=1}^{\infty} C_s(z^o)^{(-1)^{r-1}} \right)$$

where $s = \prod_{q=1}^{r} p_q$ and $C_s(z)$ is the sth cyclotomic polynomial (in *Mathematica* Cyclotomic[s, z]).

Use this formula to calculate $P_{2,3,5,7,11}(111)$. Compare the result with a direct calculation of the partitions themselves.

31.¹² *One ODE for the Kepler Problem, Euler Equations, $u(x) = \alpha \exp(\beta (z - z_0)^{-1/2})$, Lattice Green's Function ODE*

a) Construct one differential equation for the x-component $x(t)$ of the radius vector of the 2D Kepler problem:

$$x''(t) = \frac{x(t)}{\sqrt{x(t)^2 + y(t)^2}}$$

$$y''(t) = \frac{y(t)}{\sqrt{x(t)^2 + y(t)^2}}$$

b) Construct one differential equation for the $x(t)$ of the radius vector of equations of motion for a freely rotating body:

$$x'(t) = A\, y(t)\, z(t)$$
$$y'(t) = B\, x(t)\, z(t)$$
$$z'(t) = C\, x(t)\, y(t)$$

c) Find a polynomial differential equation $P(u(x), u'(x), u''(x), u'''(x)) = 0$ for the function $u(x) = \alpha \exp(\beta / \sqrt{z - z_0})$ [1088].

d) Given the Darboux–Halphen system of differential equations [8]

$$w_1'(z) = w_1(z)(w_2(z) + w_3(z)) - w_2(z)w_3(z)$$
$$w_2'(z) = w_2(z)(w_1(z) + w_3(z)) - w_1(z)w_3(z)$$
$$w_3'(z) = w_3(z)(w_1(z) + w_2(z)) - w_1(z)w_2(z)$$

find an (ordinary nonlinear) differential equation for $W(z) = 2(w_1(z) + w_2(z) + w_3(z))$.

e) Derive *x*-free, polynomial differential equations for the two functions $f(x) = \exp(\exp(\exp(e^x)))$ and $g(x) = x^x$.

f) The following function $G_\alpha(w)$ [1610], [1287], [1542], [551], [992], [957], [958] fulfills a linear homogeneous differential equation with respect to w with polynomial coefficients in α and w [911], [468], [956]:

$$G_\alpha(w) = \frac{1}{\pi^3} \int_0^\pi \int_0^\pi \int_0^\pi \frac{1}{w - (\cos(\vartheta_1) + \cos(\vartheta_2) + \alpha \cos(\vartheta_3))} \, d\vartheta_1 \, d\vartheta_2 \, d\vartheta_3$$

Use a series expansion in w to find such a differential equation.

32.^{12} Puzzles

a) Can one find an example in which `DSolve[eqs, vars]` remains unsolved, but applying `ExpandAll[` `DSolve[eqs, vars]]` gives a result?

b) What will be the result of the following inputs?

`Integrate[Integrate[x^x, x]^3, Integrate[x^x, x]]`

`D[D[1[1], 1]^2, D[1[1], 1]]`

`Sum[Sum[k, {1, 4}]^2, {Sum[k, {1, 4}], 1, 4}]`

`Unprotect[Sum]; Sum[Sum[k, {1, 4}]^2, {Sum[k, {1, 4}], 1, 4}]`

`Solve[Positive[x]^2 == 1, Positive[x]]`

`Integrate[x[1], {x, 0, 2}, {x[1], -1, 1}]`

`Sum[x[1], {x, 0, 2}, {x[1], -1, 1}]`

c) What will be the result of the following input?

`Unprotect[Message];`

`Message /: Message[___] := Null /; (Abort[]; False);`

`Integrate[1/(Sqrt[C^2] - C), {C, -1, 1}]`

d) Find a short (surely less than 20 characters) symbolic input that results in a mathematically wrong result. Do not use any variables or approximative numbers.

e) Why does the following input (mathematically equal to 1) generate messages, instead of returning just 0?

```
t[x_] := ((x+1)^2 - x^2 - x - 1)/x
t'[0]
```

f) What will be the result of the following input?

```
Block[{E}, D[E[x], x] /. head_[arg_] :> arg[head]]
```

g) Find a rational function $r(x, y, z)$ (with rational coefficients) such that `Together` does not act idempotent on it (meaning `Together[r(x, y, z)]` and `Together[Together[r(x, y, z)]]` are different).

h) Predict the result of the following input:

```
f1 = Factor[(x[2] - x[1])^3];
f2 = Factor[(x[2] - x[3])^3];
x[1] = x[3];
f1 - f2 === 0
```

i) Is there a *Mathematica* function (from the `System`` context) that, independently from any current and further progress in *Mathematica*, one will always be able to fool to give wrong result?

j) The function `Developer`ZeroQ` tries quickly to determine if an expression is identical to zero. Often it will correctly determine if an expression is zero, but it might sometimes give wrong results. (In physics terminology, one would call `Developer`ZeroQ` a FAPP-function [138], [74]; for PEF see [1328].) For exact, numeric input this function will, among other techniques, use high-precision numericalization to find this out. Find a short (surely less than 10 characters) expression for which `Developer`ZeroQ` erroneously asserts that it is zero.

k) Find a function $f(x, y)$, such that *Mathematica* can evaluate `Integrate[f(x, y), {x, 0, 1}, {y, 0, 1}]` in closed form, but cannot evaluate `Integrate[f(x, y), {y, 0, 1}, {x, 0, 1}]` in closed form.

l) Predict the result of the following inputs:

```
makeDef1[d_, f_, A_] :=
  f /: d[1][f] /; (clearDef1[d, f, A]; makeDef2[d, f, A]; False) := Null

clearDef1[d_, f_, A_] :=
  f /: d[1][f] /; (clearDef1[d, f, A]; makeDef2[d, f, A]; False) =.

makeDef2[d_, f_, A_] := d /: HoldPattern[d[1][f][0]] :=
(clearDef2[d, f]; makeDef1[d, f, A]; A)

clearDef2[d_, f_] := d /: HoldPattern[d[1][f][0]] =.

makeDef1[Derivative, f, f0];
D[f[x], x] /. x -> 0
```

m) Write a function that tries to find functions $f(x)$ that are compositions of integer powers, exponential, and logarithmic functions, such that `Limit[f(x), x -> 0]` stays unevaluated.

n) Can you find a rational function in trigonometric functions for which the built-in `Integrate` cannot find the indefinite integral, but can after a substitution of variables?

o) Find an example of a quadratic polynomial p in the variable x with exact, numeric coefficients, such that `Solve[p == 0, x]` gives a mathematically wrong result.

p) What is wrong, from a scoping point of view, with the following naive definition of the Fourier transform $\mathcal{F}_t[f(t)](\omega) = (2\pi)^{-1/2} \int_{-\infty}^{\infty} \exp(i\,\omega\,t)\,f(t)\,dt$?

```
myFourierTransform[f_, t_, ω_, opts___] :=
  1/Sqrt[2 Pi] Integrate[f Exp[I t ω], {t, -Infinity, Infinity}, opts]
```

q) Find pairs of univariate functions f and their inverses f^{-1}, such that

- $f(f^{-1}(z)) = f(f^{-1}(z)) = z$ holds for all complex z

- $f(f^{-1}(z)) = f(f^{-1}(z)) = z$ holds for almost all complex

- $f(f^{-1}(z)) = z$ holds for all complex z, but $f(f^{-1}(z)) = z$ does not

- neither $f(f^{-1}(z)) = z$ nor $f(f^{-1}(z)) = z$ holds for all complex z.

r) Why does the following input give a mathematically wrong result?

```
Simplify[x DiracDelta`[x], x == 0]
```

s) Motivate symbolically the result of the input

```
NIntegrate[Sin[x] x^(3/2), {x, 0, Infinity}, Method -> Oscillatory].
```

What is the exact value of this result?

33.^{12} 2D Newton–Leibniz Example, 1D Integral, Square Root of Differential Operator

a) Consider the following double integral:

$$\int_0^1 \int_0^1 \sqrt{(x-y)^2}\ dx\,dy.$$

Substituting limits in the (double) indefinite integral does not give the right result, despite the fact that the integrand and the (double) indefinite integral are continuous functions. How can one "fix" the double indefinite integral by doing the indefinite integrals?

b) Calculate the value of the integral $\int_0^{\infty} \exp(-2x)\,(\coth(x) - 1/x)\,dx$ through its indefinite integral.

c) Compositions of the differential operator $\partial/\partial x$ can be formally defined through the two symbolic identities [1913], [709], [1845], [789], [107], [790], [1325], [678], [808], [1348]

$$\frac{\partial^i}{\partial x^i}\left(\frac{\partial^j}{\partial x^j}.\right) = \frac{\partial^{i+j}}{\partial x^{i+j}}.$$

$$\frac{\partial^i}{\partial x^i}(f(x).) = \sum_{k=0}^{\infty}\binom{i}{k}f^{(k)}(x)\frac{\partial^i}{\partial x^i}.$$

representing order independence and the Leibniz rule.

Here $\partial^i/\partial x^i$ for negative i are understood as the inverse positive i, meaning $\partial^i/\partial x^i(\partial^{-i}./\partial x^{-i}) = \text{identity}(.)$. (Practically this means that $\partial^{-1}/\partial x^{-1}$ represents one-time indefinite integration [$\partial^{-1}./\partial x^{-1} = \int_{-\infty}^{x}.dx$] [678], [1555], [357] and the Leibniz rule represents partial integration in this formalism.)

Use these definitions to calculate the first ten α_k of the square root [436], [479], [1577], [1901], [1788]

$$\sqrt{\mathcal{L}} = \frac{\partial}{\partial x}\cdot + \sum_{k=0}^{\infty} \alpha_{-k}(x)\frac{\partial^{-k}}{\partial x^{-k}}.$$

where $\mathcal{L}$ is the differential operator $\mathcal{L} = \partial^2./\partial x^2 + u(x)$. containing an unspecified function $u(x)$.

34.^{L2} Coefficients = Roots of a Univariate Polynomial, Amoebas, Tiling

a) Find all polynomials $x^3 + a x^2 + b x + c$ that have a, b, c as their roots. Find the corresponding coefficients/roots for higher-order polynomials and visualize the coefficients/roots in the complex plane.

b) The amoeba of a system of polynomial equations $p_j(z_1, z_2, \ldots, z_n) = 0$, $j = 1, \ldots, m$ is the image of all solutions $\{z_1, z_2, \ldots, z_n\}$ under the map $\{z_1, z_2, \ldots, z_n\} \to \{\log(|z_1|), \log(|z_2|), \ldots, \log(|z_2|)\}$ [1785], [1711], [1260], [1259], [1261], [1505], [708]. Describe and visualize the amoeba of the equation $w = 1 + 2z - z^3$.

c) A bivariate function $f(x, y)$ with $\lim_{|x|+|y|\to\infty} f(x, y) = \infty$ induces a tiling of the plane with a tile defined as the set of all points $\{x, y\}$, such that $|f(x, y)| \le |f(x + i, y + j)|$ for all points $\{i, j\}$, $i, j \in \mathbb{Z}$ of a square lattice the plane [451]. Find an exact parametric description for the tile in case $f(x, y) = (3x + y)(y + 3x)$ (this example comes from [451]). Calculate the volume of the tile through integration and visualize the resulting tiling.

35.^{L2} Cartesian Leaf Area, Triple Integral, Average Distance

a) Calculate the 2D area given by $x^3 - 3yx + y^3 < 0$ in the first quadrant [509].

b) Find the value of the following integral [803]:

$$\int_0^1 \int_0^1 \int_0^1 \frac{1}{(1 + x^2 + y^2 + z^2)^2} \, dx\,dy\,dz$$

Use a spherical coordinate system to calculate the integral.

c) Find the value of the following integral [553]:

$$\int_0^\infty \int_0^\infty (x^2 + x y + y^2)^y \, e^{-(x+y)} \, dx\,dy$$

d) Determine analytically the average distance between two randomly selected points in the unit square [534]. Use the independence of the components of the Cartesian coordinates to determine the probability that two randomly selected points in the interval $(0, 1)$ are separated by a distance l. Find the average distance of two randomly chosen points in a 3D unit cube [1705], [1217].

e) The April 2003 issue of the *American Mathematical Monthly* contains the problem to calculate [554]

$$\mathcal{I}(a, b, c) = \int_0^\infty \int_0^\infty \int_0^\infty \frac{e^{-x-y-z}\sin(a x + b y + c z)}{\sqrt{x + y + z}} \, dz\,dy\,dx.$$

Write a one-liner that calculates this integral. Simplify the result for real a, b, c and calculate the exact value of $\mathcal{I}(1, 1, 1)$.

f) Calculate the following triple integral:

$$\int_0^\infty \int_0^x \int_0^y \frac{2\cos(x) - \cos(x - y - z) - \cos(x - y + z)}{y^2 (x - z)^2} \, dz \, dy \, dx.$$

36.[12] Duffing Equation, Secular Terms

a) Given the differential equation (Duffing equation) [950], [56], [1611], [1011], [965], [53]

$$y''(t) + y(t) + \epsilon y(t)^3 = 0$$
$$y(0) = 0, \ y'(0) = 0$$

(ϵ is a small parameter) using an ansatz of the form $y(t) = \sum_{i=0}^n \epsilon^i y_i(t)$ calculate the first six terms $y_i(t)$ [1613]. Compare with the exact solution for $\epsilon = 1/10$.

Derive the first three orders for general initial conditions [1385], [1191].

b) Frequently, one is interested in periodic solutions of differential equation. Consider, for instance, the following nonlinear differential equation [1893]

$$\Omega^2 \frac{\partial^2 X(t)}{\partial t^2} + \omega_0^2 X(t) = \sum_{k=2}^\infty J_k X(t)^k.$$

Here Ω, ω_0, and the J_k are given parameter. Assume we can expand Ω and $X(t)$ naturally in a power series in ϵ:

$$\Omega = \sum_{k=0}^\infty \epsilon^k \omega_k$$

$$X(t) = \epsilon \sum_{k=0}^\infty \epsilon^k x_k(t).$$

For the lowest-order solution, we take $x_0(t) = c_{0,c} \cos(t)$. The differential equations for the $x_k(t)$ are inhomogeneous harmonic oscillator differential equations with solutions $x_k(t) = c_{k,c} \cos(t) + c_{k,s} \sin(t) + x_k^{(\text{inh})}(t)$. Calculate conditions for the first 10 of the ω_k, such that the $x_k(t)$ become periodic functions.

37.[12] Implicitization of Various Surfaces

Calculate polynomial implicit representations $p(x, y, z) = 0$ for the following five surfaces:

a) A "spindle" (come up with your own parametric description of a spindle).

b) A cube-rooted sphere (this means a surface obtained by taking the third root (in case of negative values ξ, take $-(-\xi)^{1/3}$ of the coordinate values of a sphere) [1907].

c) A cubed sphere (this means a surface obtained by taking the third power of the coordinate values of a sphere).

d) A torus with a cross section of a cubed circle, in which the cubed circle is rotated when moved along the "outer" circle.

e) A circle that is rotated around its diameter and its radius changes periodically.

Use ContourPlot3D to make pictures of all calculated implicit surfaces.

38.[12] Riemann Surface of Kronig–Penney Dispersion Relation

Make a picture of the imaginary part of $K = K(e)$, where $K(e)$ is defined implicitly by [1659]

$$\cos(K) = \cos\left(\sqrt{e}\right) + \frac{4}{\sqrt{e}}\sin\left(\sqrt{e}\right).$$

Include all sheets of the Riemann surface in the picture, and let the region of the complex e-plane be

$$-10 \le \operatorname{Re}(e) \le 60,\ -10 \le \operatorname{Im}(e) \le 10.$$

(This equation comes up in the quantum-mechanical treatment of a particle in a periodic potential; and it is the eigenvalue equation of the Kronig–Penney model [1053], [38].)

39.[12] Envelopes of Secants in an Ellipse, Lines Intersecting Four Lines

a) Given an ellipse $x^2 + 4y^2 - 4 = 0$, imagine all possible line segments with a start point and an end point on this ellipse, such that the length of the line segments is exactly 1. These line segments envelope a closed region inside the ellipse. Calculate the implicit representation of this region. Visualize how this region becomes formed by the enveloping line segments.

b) Given four generic lines in $\mathbb{R}^3$, how many lines (generically) exist that cross all for lines? (See [1448], [859], [1639], and [1710]; for similar problems see [1638] and [1640].)

40.[12] Shortest Path in a Triangle Billiard

Given the triangle $P_1P_2P_3$ with vertices $P_1 = \{0, 0\}$, $P_2 = \{1, 0\}$, $P_3 = \{1, 2\}$. Now, imagine a ball that starts at a point Q inside the triangle bounces against wall $P_1\,P_2$, is reflected there, bounces against wall P_2P_3, is reflected there, bounces against $P_3\,P_1$ and comes back to the point Q. Calculate the exact value of the smallest length of such a path [509].

41.[12] Differential Equation Singularities, Weak Measurement Identity, Logarithmic Residue

a) Find a numerical approximation for the distance of the nearest singularity of

$$y'(x) = 1 - x\,y(x)^2$$
$$y(0) = 0$$

from the origin [143].

b) In [1756], [24], [25], [1504], [221], [1634], [1635] (see also [529]) the following peculiar identity was given (η is a real variable):

$$\sum_{k=0}^{n} c_k^{(n)}(\eta)\, f\left(x - d_k^{(n)}\right) \approx f(x - \eta)$$

Here the shifts $d_k^{(n)}$ and the weights $c_k^{(n)}(\eta)$ are given by

$$d_k^{(n)} = \frac{k}{n}$$

$$c_k^{(n)}(\eta) = \binom{n}{k} \eta^k (1 - \eta)^{n-k}.$$

(The identity is peculiar because the superposition of shifted copies of f with, for $|\eta| > 1$ large coefficients approximates another shifted copy of f.)

Assuming a Fourier expansion of $f(x)$ exists, show that for $n \to \infty$, exact equality holds. Under which conditions does the identity hold approximately for finite n?

c) Let $f(z) = \sum_{k=-n}^{\infty} c_k z^k$ for some positive integer n. Then we have the residue $\text{res}_{z=0}(f(z)) = c_1$. For $1 \le n \le 10$, express the residue as a function of $F(0)$ and $\Phi^{(k)}(0)$ where $F(z) = z^n f(z)$ and $\Phi(z) = F'(z)/F(z)$ [1152].

Let ABC be a triangle. The sides $\overline{AB}$ and $\overline{AC}$ have the same length. A line starting from C intersects the line AB at the point D, and a line starting from B intersects the line AC at the point E. The following angles are given: ∢EBC=60°, ∢BCD = 70°, and ∢ABE = 20°. Determine the angle ∢CDE. Here is a sketch of the problem:

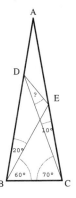

Determine the angle ∢CDE. From [1116], see also http://www.dcs.st-andrews.ac.uk/~ad/mathrecs/advent/advent.html.

43.^{12} Differential Equation for Polynomial, Graph Eigenvalues

a) Given a polynomial $p(x, y)$ in two variables, the y in $p(x, y) = 0$ can be viewed as an implicitly defined function of x. Write a program, in which given an explicit bivariate polynomial $p(x, y)$, it derives a linear differential equation for $y(x)$ [1042], [1043], [1045], [379], [380].

b) Calculate the eigenvalues of the Laplace operator for the graph formed by the vertices (nodes) and edges (connections between nodes) of a stellated icosahedron. The action of Laplace Δ operator on a function f at node q of a graph is given by [733], [171], [275], [918], [1477], [214]

$$\Delta f(q) = numberOfNeighborsOfq\, f(q) - \sum_{\text{neighbor nodes of } q} f(p).$$

c) Conjecture exact values for the smallest and the largest eigenvalues of the Laplace operator for the graph formed by the vertices (nodes) and edges (connections between nodes) of the 120-cell (see Exercise 17 of Chapter 2 of the Graphics volume [1736]).

44.^{12} Fourier Transform Eigenvalues, $\int_{-\infty}^{\infty} \theta(x)\,\delta(x)\,dx$, $\delta^{(n)}\,(f(x))$, PDF for Sums and Determinants, Fourier Transform and Series, Functional Differentiation

a) The eigenfunctions of the Fourier transform are of the form $\exp(-x^2/2)\sum_{k=0}^{n} c_k\, x^k$. Find the eigenvalues and eigenfunctions corresponding to $0 \le n \le 12$.

b) Find explicit sequences $\delta_\varepsilon(x)$ and $\theta_\epsilon(x)$ such that $\lim_{\varepsilon \to 0}\delta_\varepsilon(x) = \delta(x)$ and $\lim_{\epsilon \to 0}\theta_\epsilon(x) = \theta(x)$ and $\int_{-\infty}^{\infty}\delta_\varepsilon(x)\,\theta_\epsilon(x)\,dx$ takes on different values depending on how fast ε and ϵ approach 0. (This shows that $\int_{-\infty}^{\infty}\delta(x)\,\theta(x)\,dx$ is not a well-defined integral.)

c) Let ξ be the only simple zero of $f(x) = 0$. Using the following three formulas [978]:

$$g(x)\,\delta^{(n)}(x - \xi) = \sum_{k=0}^{n} (-1)^k \binom{n}{k} g^{(k)}(\xi)\,\delta^{(n-k)}(x - \xi),$$

$$\delta(f(x)) = \frac{\delta(x - \xi)}{|f'(\xi)|},$$

$$\delta^{(n)}(f(x)) = \frac{1}{f'(x)}\frac{\partial}{\partial x}\,\delta^{(n-1)}(f(x)),$$

find expansions for $\delta^{(n)}(f(x))$ of the form $\delta^{(n)}(f(x)) = \sum_{k=0}^{n} c_k(\xi)\,\delta^{(k)}(x - \xi)$ for $2 \le n \le 10$. Check some of the resulting formulas numerically.

d) The probability density $p_z(\xi)$ for a sum $z = x + y$ of two random variables x and y with distribution functions $p_x(\xi)$ and $p_y(\xi)$ is given by [213], [516], [1909], [1216], [1617], [477]

$$p_z(\xi) = \int_{-\infty}^{\infty}\int_{-\infty}^{\infty} p_x(s)\,p_y(t)\,\delta(\xi - s - t)\,ds\,dt.$$

(This can also be viewed as a rewritten definition for B-splines.) Use this formula to calculate the distribution functions for n ($n = 2, \ldots, 5$) uniformly in $[0, 1]$ distributed random numbers [1844].

e) Calculate the probability distribution of the determinant of a 2×2 matrix with random, uniform in $[0, 1]$ distributed elements. Carry out a numerical simulation, and compare it with the theoretical distribution.

f) The divided differences $[x_1\, x_2 \ldots x_n]$ of a function f at the points $x_1, \ldots, x_n$ have the integral representation [964], [1264], [452]

$$[x_1\, x_2 \ldots x_n] = \int_{0}^{\infty}\int_{0}^{\infty}\cdots\int_{0}^{\infty}\delta(1 - (x_1 + \cdots + x_n))\,f^{(n-1)}(\tau_1\, x_1 + \tau_2\, x_2 + \cdots + \tau_n\, x_n)\,dx_n \ldots dx_2\,dx_1.$$

Use this formula to calculate $[x_1\, x_2 \ldots x_8]$.

g) Calculate the limits $\lim_{\varepsilon \to 0} f_\varepsilon^{(\xi)}(x)$ and $\lim_{\varepsilon \to 1} f_\varepsilon^{(\xi)}(x)$ where [464], [1023], [1293]

$$f_\varepsilon^{(\xi)}(x) = \frac{1}{2\sqrt{\pi}\,\sqrt{1 - \varepsilon}}\,e^{\frac{i\pi}{4}}\,e^{-\frac{i}{2}\frac{\varepsilon x^2 - 2\xi x + \varepsilon \xi^2}{1 - \varepsilon}}.$$

Derive a first-order and an x-free differential equation for $f_\varepsilon^{(\xi)}(x)$.

h) Given the Fourier transform $F^{(L)}(k) = (2\pi)^{-1/2} \int_{-L}^{L} \exp(i\,k\,x)\,f(x)\,dx$ of a function $f(x)$, visualize how the Fourier transform $F_o^{(L)}(k)$ of $2\,o + 1$ periodic continuations of $f(x)$

$$F_o^{(L)}(k) = \frac{1}{\sqrt{2\pi}} \int_{-\infty}^{\infty} \exp(i\,k\,x)\,f_o(x)\,dx$$

$$f_o(x) = \sum_{j=-o}^{o} \theta(x - (2\,j\,L - L))\,\theta((2\,j\,L + L) - x)\,f(x + 2\,j\,L)$$

approaches the Fourier series coefficients $c_n^{(L)} = 1/(2\,L) \int_{-L}^{L} \exp(i\,n\,x\,\pi/L)\,f(x)\,dx$ for $o \to \infty$ [1470]. Use $f(x) = \exp(-x^2)$ as the example function.

i) The functional derivative of a functional $F[f(x)]$ is defined through (see for instance [770], [173], [1021], [732], [1641], [898], [1793], [905])

$$\frac{\delta F[f(x)]}{\delta f(y)} = \lim_{\epsilon \to 0} \frac{F[f(x) + \epsilon\,\delta(x - y)] - F[f(x)]}{\epsilon}.$$

Implement a function that attempts the functional differentiation for a given functional $F[f(x)]$. Use this function to calculate

$$\frac{\delta}{\delta f(y)} \int_{-\infty}^{\infty} f(x)^3\,f'(x)^2\,f''(x)\,dx \; .$$

Defining the Korteweg–de Vries bracket $\{F, G\} = \int_{-\infty}^{\infty} \delta F(y)/\delta u(y)\,\partial(\delta F(y)/\delta u(y))/\partial y\,dy$ [1208], show that $\{Q_i, Q_j\} = 0$ where

$$Q_0 = \int_{-\infty}^{\infty} u(x)\,dx$$

$$Q_1 = \int_{-\infty}^{\infty} \frac{1}{2}\,u(x)^2\,dx$$

$$Q_2 = \int_{-\infty}^{\infty} \left(\frac{1}{2}\,u'(x)^2 - u(x)^3\right)dx$$

$$Q_3 = \int_{-\infty}^{\infty} \left(\frac{5}{2}\,u(x)^4 - 5\,u(x)\,u'(x)^2 + \frac{1}{2}\,u''(x)^2\right)dx.$$

Implement multiple functional differentiation defined through

$$\frac{\delta^n F[f(x)]}{\delta f(y_1) \cdots \delta f(y_n)} = \frac{\delta}{\delta f(y_1)} \left(\frac{\delta^{n-1} F[f(x)]}{\delta f(y_2) \cdots \delta f(y_n)}\right)$$

and calculate

$$\frac{\delta^3 \exp(\int_{-\infty}^{\infty} f(x)\,h(x)\,dx)}{\delta f(y_1)\,\delta f(y_2)\,\delta f(y_3)} \bigg|_{f(y)=0} \; .$$

45.^{12} Operator Splitting of Order 5

For many applications, it is useful to decompose exponentials of sums of operators into a product of exponentials of only one operator. (Examples are the solution of the time-dependent Schrödinger equation and the calculation of higher-order Runge–Kutta formulas for the solution of ordinary differential equations.) Calculate the real-valued coefficients ω_1, ω_2, ω_3, ω_4, ω_5, ω_6, and ω_7 of the fifth-order splitting formula [512], [657], [348], [1231], [1355], [1878], [1908], [1557], [1691], [182], [1304], [1258], [995], [996], [1257], [349], [1692], [1460], [1354], [988], [350], [739]:

$$e^{t(\hat{A}+\hat{B})} = e^{t\,\omega_1\,\hat{A}}\, e^{t\,\omega_2\,\hat{B}}\, e^{t\,\omega_3\,\hat{A}}\, e^{t\,\omega_4\,\hat{B}}\, e^{t\,\omega_5\,\hat{A}}\, e^{t\,\omega_6\,\hat{B}}\, e^{t\,\omega_7\,\hat{A}} + O(t^5).$$

Here, $\hat{A}$ and $\hat{B}$ are the two (in general noncommuting) operators.

46.^{12} Maximal Triangle Area and Tetrahedron of Maximal Volume

a) Given three concentric circles of radius r_1, r_2, and r_3, calculate the maximal area of the triangle that has one vertex at each of the circles. For $r_1 = 1$, $r_2 = 2$, $r_3 = 3$, calculate the explicit value of the area [640].

b) Given the areas of the four faces of a tetrahedron, find the maximal volume that the tetrahedron can have. Express the result in a compact way. For the two sets of four areas, $\{\frac{9}{10}, \frac{10}{10}, \frac{11}{10}, \frac{12}{10}\}$ and $\{1, 1, 1, 1\}$, calculate a closed-form symbolic result for the maximal volume. [712], [42], [842]

Solutions

1. The 2 in the Factorization of $x^i - 1$, Heron's Formula, Volume of Tetrahedron, Circles of Apollonius, Circle ODE, Modular Transformations, Two-Point Taylor Expansion, Quotential Derivatives

a) Note the multiple, completely independent, appearances of #. We use C and n as variables. (To see some progress when the program is running, we print every occurrence that is not -1, 1, or 0, rather than producing a list as the result.)

```
In[1]:= Array[Function[n,
            If[# != {}, (* use Print here to see progress *)
               CellPrint[Cell[TextData[{"○ Degree: ",  Cell[BoxData[
                           FormBox[ToString[n], TraditionalForm]]],
                        ". Coefficients other than 0 and ±1 are: ",
                        Cell[BoxData[FormBox[MakeBoxes[#, TraditionalForm],
                                          TraditionalForm]]]}], "PrintText"]]]&[
              Cases[Union @@ (* investigate all coefficients *)
                        (CoefficientList[#, C]& /@
              If[Head[#] === Times, List @@ #, {#}]&[Factor[C^n - 1]]),
                        (* anything other than + 1, 0  and - 1 *)
                        _?(((# != 1) && (# != -1) && (# != 0))&)]]], 500, 0];
```

○ Degree: 105. Coefficients other than 0 and ±1 are: {−2}

○ Degree: 165. Coefficients other than 0 and ±1 are: {2}

○ Degree: 195. Coefficients other than 0 and ±1 are: {−2}

○ Degree: 210. Coefficients other than 0 and ±1 are: {−2, 2}

○ Degree: 255. Coefficients other than 0 and ±1 are: {2}

○ Degree: 273. Coefficients other than 0 and ±1 are: {2}

○ Degree: 285. Coefficients other than 0 and ±1 are: {−2}

○ Degree: 315. Coefficients other than 0 and ±1 are: {−2}

○ Degree: 330. Coefficients other than 0 and ±1 are: {−2, 2}

○ Degree: 345. Coefficients other than 0 and ±1 are: {2}

○ Degree: 357. Coefficients other than 0 and ±1 are: {−2}

○ Degree: 385. Coefficients other than 0 and ±1 are: {−3, −2, 2}

○ Degree: 390. Coefficients other than 0 and ±1 are: {−2}

○ Degree: 420. Coefficients other than 0 and ±1 are: {−2, 2}

○ Degree: 429. Coefficients other than 0 and ±1 are: {−2}

○ Degree: 455. Coefficients other than 0 and ±1 are: {−2, 2}

○ Degree: 495. Coefficients other than 0 and ±1 are: {2}

Also, larger integers can appear in the factorizations.

```
In[2]:= (* return {degree, coefficient} *)
        Function[d, {d, Max[Abs[(CoefficientList[#, C]& /@
                   If[Head[#] === Times, List @@ #, {#}]&[Factor[C^d - 1]])]]}] /@
                   (* lowest degrees that have maximal coefficient ±k *)
                        {1, 105, 385, 1365, 1785, 2805, 3135, 10353, 6545,
                        12155, 21385, 11165, 21505, 10465, 16555, 19285}
```

```
Out[3]= {{1, 1}, {105, 2}, {385, 3}, {1365, 4}, {1785, 5}, {2805, 6},
         {3135, 7}, {10353, 8}, {6545, 9}, {12155, 10}, {21385, 11},
         {11165, 12}, {21505, 13}, {10465, 14}, {16555, 15}, {19285, 16}}
```

b) Here, we choose an approach in the sense of geometrical theorem-proving ([359] and [360]) by reducing the geometrical problem to a polynomial problem. Without loss of generality, we can choose the following coordinates for the three points.

In[1]:= `p1 = {0, 0, 0}; p2 = {0, p2y, 0}; p3 = {p3x, p3y, 0};`

This is the square of the area.

In[2]:= `ASquare = (Cross[p2 - p1, p3 - p1].Cross[p2 - p1, p3 - p1])/4`

Out[2]= $\dfrac{p2y^2\, p3x^2}{4}$

We start with a step-by-step method to derive the result. Because the area has dimension *length²*, it should be expressed as a homogeneous polynomial of degree 2 in the squares of the lengths of the three edges. So, we make the following ansatz for the square of the area.

In[3]:= `ASquareAnsatz = c1 lS12^2 + c2 lS13^2 + c3 lS23^2 +`
`                c5 lS12 lS13 + c5 lS12 lS23 + c6 lS13 lS23`

Out[3]= $c1\, lS12^2 + c5\, lS12\, lS13 + c2\, lS13^2 + c5\, lS12\, lS23 + c6\, lS13\, lS23 + c3\, lS23^2$

Expanding the expression `ASquareAnsatz - ASquare` and comparing the coefficients, we can derive a linear system of equations in the c_i.

In[4]:= `eq1 = 4 (ASquareAnsatz /.`
`           {lS12 -> (p2 - p1).(p2 - p1), lS13 -> (p3 - p1).(p3 - p1),`
`            lS23 -> (p3 - p2).(p3 - p2)}) - 4 ASquare`

Out[4]= $-p2y^2\, p3x^2 + 4\,\big(c1\, p2y^4 + c5\, p2y^2\,(p3x^2 + p3y^2) + c2\,(p3x^2 + p3y^2)^2 + c5\, p2y^2\,(p3x^2 + (-p2y + p3y)^2) +$
$\qquad c6\,(p3x^2 + p3y^2)\,(p3x^2 + (-p2y + p3y)^2) + c3\,(p3x^2 + (-p2y + p3y)^2)^2\big)$

In[5]:= `(eq2 = Expand[eq1]) // Length`

Out[5]= 25

These are all monomials in the polynomial `eq2`.

In[6]:= `monomials = Union[((List @@ eq2) //. {i_Integer f_ -> f}) //.`
`                    {c1 -> 1, c2 -> 1, c3 -> 1, c4 -> 1, c5 -> 1, c6 -> 1}];`
`           Length[monomials]`

Out[7]= 9

We extract the equations.

In[8]:= `eqns = Coefficient[eq2, #] == 0& /@ monomials;`

Then we solve them.

In[9]:= `sol = Solve[eqns, {c1, c2, c3, c4, c5, c6}]`

    `Solve::svars : Equations may not give solutions for all "solve" variables. More…`

Out[9]= $\left\{\left\{c1 \to -\dfrac{1}{16},\ c2 \to -\dfrac{1}{16},\ c3 \to -\dfrac{1}{16},\ c5 \to \dfrac{1}{8},\ c6 \to \dfrac{1}{8}\right\}\right\}$

We replace the symbolic squares of the side length by explicit squares of side length.

In[10]:= `ASquare = Factor[((ASquareAnsatz /. sol)[[1]]) /.`
`                  {lS12 -> 112^2, lS13 -> 113^2, lS23 -> 123^2}]`

Out[10]= $-\dfrac{1}{16}\,(112 - 113 - 123)\,(112 + 113 - 123)\,(112 - 113 + 123)\,(112 + 113 + 123)$

Factoring the result yields the square of the area (Heron's formula).

In[11]:= `Factor[ASquare]`

Out[11]= $-\dfrac{1}{16}\,(112 - 113 - 123)\,(112 + 113 - 123)\,(112 - 113 + 123)\,(112 + 113 + 123)$

Instead of using the previous step-by-step way to derive the result, we can use more powerful commands to carry out the whole calculation at once. In the next input, `GroebnerBasis` is used to derive the result.

In[12]:= `GroebnerBasis[`
`             {(* area^2 expressed in vertices coordinates *)`
`              areaSquared - p2y^2 p3x^2/4,`

```
                     (* edges expressed in vertices coordinates *)
                     lS12 - (p2 - p1).(p2 - p1),
                     lS13 - (p3 - p1).(p3 - p1),
                     lS23 - (p3 - p2).(p3 - p2)},
                     (* variables to keep *) {areaSquared, lS12, lS13, lS23},
                     (* variables to eliminate *)
                     {p2y, p3x, p3y}, MonomialOrder -> EliminationOrder]
```

Out[12]= $\{16 \text{ areaSquared} + lS12^2 - 2 \, lS12 \, lS13 + lS13^2 - 2 \, lS12 \, lS23 - 2 \, lS13 \, lS23 + lS23^2\}$

c) We essentially choose the same approach as in problem a). Again, without loss of generality, we can choose the following coordinates for the four points of a general tetrahedron.

In[1]:= `p1 = {0, 0, 0}; p2 = {p2x, 0, 0}; p3 = {p3x, p3y, 0}; p4 = {p4x, p4y, p4z};`

The square of the area of the triangle in the x,y-plane and the corresponding height are given by *areaOfOneFace heightAboveThisFace* $/3$ ([768], [1652], [354]).

In[2]:= `cr = Cross[p2 - p1, p3 - p1]`

Out[2]= `{0, 0, p2x p3y}`

In[3]:= `ASquare = (cr.cr)/4 // Expand`

Out[3]= $\dfrac{p2x^2 \, p3y^2}{4}$

In[4]:= `hSquare = Together[(cr.(p4 - p1))^2/(cr.cr)] // Expand`

Out[4]= $p4z^2$

We make an ansatz for the volume V in the form $V^2 = homogeneousPolynomialOfDegree3InTheEdgeLengths$. These are all degree-three monomials that we can form with the squares of the four side lengths.

In[5]:= `lList = {lS12, lS13, lS14, lS23, lS24, lS34}`

Out[5]= `{lS12, lS13, lS14, lS23, lS24, lS34}`

In[6]:= `tab1 = Table[lList[[i]]^3, {i, 6}]`

Out[6]= $\{lS12^3, lS13^3, lS14^3, lS23^3, lS24^3, lS34^3\}$

In[7]:= `tab2 = Complement[Flatten[Table[lList[[i]]^2 lList[[j]],`
`                                {i, 6}, {j, 6}]], tab1];`

In[8]:= `tab3 = Complement[Flatten[Table[lList[[i]] lList[[j]] lList[[k]],`
`                                {i, 6}, {j, 6}, {k, 6}]], tab1, tab2];`

This is the corresponding ansatz for the volume V. (We use `Unique` to generate the coefficients to be determined because there are 56 of them now, and we do not want to write them all explicitly down.)

In[9]:= `(VSquareAnsatz = Plus @@ MapIndexed[Unique[c] #&,`
`                Join[tab1, tab2, tab3]]) // Short[#, 6]&`

Out[9]//Short= $c\$16 \, lS12^3 + c\$22 \, lS12^2 \, lS13 + c\$23 \, lS12 \, lS13^2 + c\$17 \, lS13^3 + c\$24 \, lS12^2 \, lS14 +$
$c\$52 \, lS12 \, lS13 \, lS14 + c\$25 \, lS13^2 \, lS14 + c\$26 \, lS12 \, lS14^2 + c\$27 \, lS13 \, lS14^2 + c\$18 \, lS14^3 +$
$c\$28 \, lS12^2 \, lS23 + c\$53 \, lS12 \, lS13 \, lS23 + c\$29 \, lS13^2 \, lS23 + c\$54 \, lS12 \, lS14 \, lS23 + \ll 29 \gg +$
$c\$67 \, lS14 \, lS23 \, lS34 + c\$45 \, lS23^2 \, lS34 + c\$68 \, lS12 \, lS24 \, lS34 + c\$69 \, lS13 \, lS24 \, lS34 +$
$c\$70 \, lS14 \, lS24 \, lS34 + c\$71 \, lS23 \, lS24 \, lS34 + c\$46 \, lS24^2 \, lS34 + c\$47 \, lS12 \, lS34^2 +$
$c\$48 \, lS13 \, lS34^2 + c\$49 \, lS14 \, lS34^2 + c\$50 \, lS23 \, lS34^2 + c\$51 \, lS24 \, lS34^2 + c\$21 \, lS34^3$

In[10]:= `(* the variables *)`
`allVars = List @@ VSquareAnsatz //.`
`         {lS12 -> 1, lS13 -> 1, lS14 -> 1, lS23 -> 1, lS24 -> 1, lS34 -> 1}`

Out[11]= `{c$16, c$22, c$23, c$17, c$24, c$52, c$25, c$26, c$27, c$18, c$28, c$53, c$29, c$54,`
`c$55, c$30, c$31, c$32, c$33, c$19, c$34, c$56, c$35, c$57, c$58, c$36, c$59, c$60,`
`c$61, c$37, c$38, c$39, c$40, c$41, c$20, c$42, c$62, c$43, c$63, c$64, c$44, c$65,`
`c$66, c$67, c$45, c$68, c$69, c$70, c$71, c$46, c$47, c$48, c$49, c$50, c$51, c$21}`

Expanding the expression `VSquareAnsatz - ASquare×hSquare/9` and comparing the coefficients, we can derive a linear system of equations in the $c\$i$.

In[12]:= `eq1 = 9 4 (VSquareAnsatz /.`
`             {lS12 -> (p2 - p1).(p2 - p1), lS13 -> (p3 - p1).(p3 - p1),`

```
                    1S14 -> (p4 - p1).(p4 - p1), 1S23 -> (p3 - p2).(p3 - p2),
                    1S24 -> (p4 - p2).(p4 - p2), 1S34 -> (p4 - p3).(p4 - p3)}) *
                  Denominator[hSquare] - 4 ASquare Numerator[hSquare];
```

In[13]:= `(eq2 = Expand[eq1]) // Length`

Out[13]= 1672

In[14]:= `monomials = Function[{localVars},`
        `(* locally inside Block set all the c$i to 1 *)`
        `Block[localVars,`
            `Union[((List @@ eq2) //. {i_Integer f_ -> f})]], {HoldAll}] @@`
            `(* setting all c$i to 1, enclosed in Hold *)`
            `DeleteCases[Hold @@ {Map[Hold[Set[#, 1]]&, allVars]},`
                        `Hold, {3}, Heads -> True];`

In[15]:= `Length[monomials]`

Out[15]= 156

The polynomial is now a bit larger in comparison to the 2D case of the last subexercise, but we can proceed in an analogous way.

In[16]:= `eqns = Coefficient[eq2, #] == 0& /@ monomials;`

In[17]:= `(sol = Solve[eqns, allVars]) // Short[#, 6]&`

Out[17]//Short= $\{\{c\$16 \to 0,\ c\$22 \to 0,\ c\$23 \to 0,\ c\$17 \to 0,\ c\$24 \to 0,\ c\$52 \to 0,\ c\$25 \to 0,\ c\$26 \to 0,$

$c\$27 \to 0,\ c\$18 \to 0,\ c\$28 \to 0,\ c\$53 \to -\frac{1}{144},\ c\$29 \to 0,\ c\$54 \to \frac{1}{144},\ c\$55 \to \frac{1}{144},$

$c\$30 \to -\frac{1}{144},\ c\$31 \to 0,\ c\$32 \to 0,\ c\$33 \to -\frac{1}{144},\ c\$19 \to 0,\ c\$34 \to 0,\ c\$56 \to \frac{1}{144},$

$c\$35 \to -\frac{1}{144},\ c\$57 \to -\frac{1}{144},\ c\$58 \to \frac{1}{144},\ c\$36 \to 0,\ c\$59 \to 0,\ \ll1\gg,\ \ll1\gg,$

$c\$37 \to 0,\ c\$38 \to 0,\ c\$39 \to -\frac{1}{144},\ c\$40 \to 0,\ c\$41 \to 0,\ c\$20 \to 0,\ c\$42 \to -\frac{1}{144},$

$c\$62 \to \frac{1}{144},\ c\$43 \to 0,\ c\$63 \to \frac{1}{144},\ c\$64 \to -\frac{1}{144},\ c\$44 \to 0,\ c\$65 \to \frac{1}{144},$

$c\$66 \to 0,\ c\$67 \to \frac{1}{144},\ c\$45 \to 0,\ c\$68 \to \frac{1}{144},\ c\$69 \to \frac{1}{144},\ c\$70 \to 0,\ c\$71 \to -\frac{1}{144},$

$c\$46 \to 0,\ c\$47 \to -\frac{1}{144},\ c\$48 \to 0,\ c\$49 \to 0,\ c\$50 \to 0,\ c\$51 \to 0,\ c\$21 \to 0\}\}$

We obtain the following.

In[18]:= `Factor[((VSquareAnsatz /. sol)[[1]]) /.`
        `{1S12 -> 112^2, 1S13 -> 113^2, 1S14 -> 114^2,`
        `1S23 -> 123^2, 1S24 -> 124^2, 1S34 -> 134^2}]`

Out[18]= $\frac{1}{144}$ $(-112^2\ 113^2\ 123^2 + 112^2\ 114^2\ 123^2 + 113^2\ 114^2\ 123^2 - 114^4\ 123^2 - 114^2\ 123^4 + 112^2\ 113^2\ 124^2 -$
        $113^4\ 124^2 - 112^2\ 114^2\ 124^2 + 113^2\ 114^2\ 124^2 + 113^2\ 123^2\ 124^2 + 114^2\ 123^2\ 124^2 - 113^2\ 124^4 -$
        $112^4\ 134^2 + 112^2\ 113^2\ 134^2 + 112^2\ 114^2\ 134^2 - 113^2\ 114^2\ 134^2 + 112^2\ 123^2\ 134^2 +$
        $114^2\ 123^2\ 134^2 + 112^2\ 124^2\ 134^2 + 113^2\ 124^2\ 134^2 - 123^2\ 124^2\ 134^2 - 112^2\ 134^4)$

This result for the square of the volume agrees with the textbook version written in the form of a determinant ([243], [355]).

In[19]:= `(1/288 Det[{{0, 123^2, 124^2, 112^2, 1}, {123^2, 0, 134^2, 113^2, 1},`
        `{124^2, 134^2, 0, 114^2, 1}, {112^2, 113^2, 114^2, 0, 1},`
        `{1, 1, 1, 1, 0}}] // Factor) == %`

Out[19]= True

Instead of carrying out many relatively simple steps (as one would do it with paper and pencil), as done in the above calculation, we could again use the very powerful GroebnerBasis command to solve this problem. We express the volume as a function of the coordinates and then eliminate the coordinates in favor of the length of the six edges. This approach leads, of course, to the same result.

In[20]:= `GroebnerBasis[`
        `{(* volume^2 expressed in vertices coordinates *)`
        `volumeSquared - p2x^2 p3y^2 p4z^2/36,`
        `(* edges expressed in vertices coordinates *)`
        `1S12 - (p2 - p1).(p2 - p1), 1S13 - (p3 - p1).(p3 - p1),`

```
          1S14 - (p4 - p1).(p4 - p1), 1S23 - (p3 - p2).(p3 - p2),
          1S24 - (p4 - p2).(p4 - p2), 1S34 - (p4 - p3).(p4 - p3)},
          (* variables to keep *)
          {volumeSquared, 1S12, 1S13, 1S14, 1S23, 1S24, 1S34},
          (* variables to eliminate *) {p2x, p3x, p3y, p4x, p4y, p4z},
          MonomialOrder -> EliminationOrder]
```

Out[20]= $\{1S12\ 1S13\ 1S23 - 1S12\ 1S14\ 1S23 - 1S13\ 1S14\ 1S23 + 1S14^2\ 1S23 + 1S14\ 1S23^2 - 1S12\ 1S13\ 1S24 +$
$1S13^2\ 1S24 + 1S12\ 1S14\ 1S24 - 1S13\ 1S14\ 1S24 - 1S13\ 1S23\ 1S24 - 1S14\ 1S23\ 1S24 + 1S13\ 1S24^2 +$
$1S12^2\ 1S34 - 1S12\ 1S13\ 1S34 - 1S12\ 1S14\ 1S34 + 1S13\ 1S14\ 1S34 - 1S12\ 1S23\ 1S34 - 1S14\ 1S23\ 1S34 -$
$1S12\ 1S24\ 1S34 - 1S13\ 1S24\ 1S34 + 1S23\ 1S24\ 1S34 + 1S12\ 1S34^2 + 144\ \text{volumeSquared}\}$

For generalizations of such formulas, see [1533], [1534], [1535], [1366], [686], [1536], and [1174]. For related problems, see [1151], [1492], and [1491].

d) One might immediately think of calling `Solve` with the set of equations under interest. Let $\{x1, y1\}$, $\{x2, y2\}$, and $\{x3, y3\}$ be the midpoints of the three given circles and r1, r2, and r3, their radii, respectively. The midpoint of the fourth circle is denoted by $\{x, y\}$, and its radius is r.

```
In[1]:= eqs1 = (* the three circles touch the fourth one *)
           {(x - x1)^2 + (y - y1)^2 == (r1 + r )^2,
            (x - x2)^2 + (y - y2)^2 == (r2 + r )^2,
            (x - x3)^2 + (y - y3)^2 == (r3 + r )^2};

       eqs2 = (* the three circles touch each other *)
           {(x1 - x2)^2 + (y1 - y2)^2 == (r1 + r2)^2,
            (x1 - x3)^2 + (y1 - y3)^2 == (r1 + r3)^2,
            (x2 - x3)^2 + (y2 - y3)^2 == (r2 + r3)^2};

In[4]:= (* try for a direct solution using built-in solvers *)
       TimeConstrained[
          #[Join[eqs1, eqs2], {r}, {x, y, x1, y1, x2, y2, x3, y3}],
                120]& /@ {Solve, Reduce}
```

Out[5]= $\left\{\left\{\left\{r \to \dfrac{-r1\ r2\ r3\ (-2\ r1\ r2 - 2\ r1\ r3 - 2\ r2\ r3) - 4\ \sqrt{r1^4\ r2^3\ r3^3 + r1^3\ r2^4\ r3^3 + r1^3\ r2^3\ r3^4}}{2\ (r1^2\ r2^2 - 2\ r1^2\ r2\ r3 - 2\ r1\ r2^2\ r3 + r1^2\ r3^2 - 2\ r1\ r2\ r3^2 + r2^2\ r3^2)}\right\},\right.\right.$

$\left.\left\{r \to \dfrac{-r1\ r2\ r3\ (-2\ r1\ r2 - 2\ r1\ r3 - 2\ r2\ r3) + 4\ \sqrt{r1^4\ r2^3\ r3^3 + r1^3\ r2^4\ r3^3 + r1^3\ r2^3\ r3^4}}{2\ (r1^2\ r2^2 - 2\ r1^2\ r2\ r3 - 2\ r1\ r2^2\ r3 + r1^2\ r3^2 - 2\ r1\ r2\ r3^2 + r2^2\ r3^2)}\right\}\right\},$

$\$Aborted\}$

We give a second solution. Without restricting generality, we can assume the midpoint of one of the three circles to be the origin and the midpoint of another one of the three circles to be on the *x*-axis. This decreases the number of variables involved by 3, and so reduces the complexity of the problem.

```
In[6]:= x1 = 0; y1 = 0; y2 = 0;
```

As a first step, we eliminate the midpoint coordinates of the fourth circle from the describing equations of the fourth circle touching the three others, not taking into account the fact that the three circles touch each other pairwise.

```
In[7]:= fa = Factor[Subtract @@ Eliminate[eqs1, {x, y}]];
       Short[fa, 4]
```

Out[8]//Short= $4\ r^2\ r1^2\ x2^2 + 4\ r\ r1^3\ x2^2 + r1^4\ x2^2 - 8\ r^2\ r1\ r3\ x2^2 - 4\ r\ r1^2\ r3\ x2^2 + 4\ r^2\ r3^2\ x2^2 -$
$4\ r\ r1\ r3^2\ x2^2 - 2\ r1^2\ r3^2\ x2^2 + 4\ r\ r3^3\ x2^2 + \ll 87 \gg + x2^4\ y3^2 - 4\ r\ r1\ x2\ x3\ y3^2 -$
$2\ r1^2\ x2\ x3\ y3^2 + 4\ r\ r2\ x2\ x3\ y3^2 + 2\ r2^2\ x2\ x3\ y3^2 - 2\ x2^3\ x3\ y3^2 + 2\ x2^2\ x3^2\ y3^2 + x2^2\ y3^4$

Because the triangle formed by the midpoints of the three circles determines a triangle that at the same time is determined uniquely by the radii of the three circles, we eliminate the explicit coordinates of the midpoints and express all quantities as functions of the three radii.

```
In[9]:= GroebnerBasis[Flatten[{fa, Subtract @@@ eqs2}], {r1, r2, r3, r},
                       (* variables to eliminate *) {x2, x3, y3},
                       MonomialOrder -> EliminationOrder]
```

Out[9]= $\{r^2\ r1^2\ r2^2 - 2\ r^2\ r1^2\ r2\ r3 - 2\ r^2\ r1\ r2^2\ r3 + r^2\ r1^2\ r3^2 -$
$2\ r^2\ r1\ r2\ r3^2 - 2\ r\ r1^2\ r2\ r3^2 + r^2\ r2^2\ r3^2 - 2\ r\ r1\ r2^2\ r3^2 + r1^2\ r2^2\ r3^2\}$

The result is a simple quadratic equation for r.

```
In[10]:= Solve[% == 0, r] // Simplify
```

Out[10]= $\{\{r \rightarrow \dfrac{r1\ r2^2\ r3^2 + r1^2\ r2\ r3\ (r2 + r3) - 2\sqrt{r1^3\ r2^3\ r3^3\ (r1 + r2 + r3)}}{r1^2\ (r2 - r3)^2 + r2^2\ r3^2 - 2\ r1\ r2\ r3\ (r2 + r3)}\},$

$\{r \rightarrow \dfrac{r1\ r2^2\ r3^2 + r1^2\ r2\ r3\ (r2 + r3) + 2\sqrt{r1^3\ r2^3\ r3^3\ (r1 + r2 + r3)}}{r1^2\ (r2 - r3)^2 + r2^2\ r3^2 - 2\ r1\ r2\ r3\ (r2 + r3)}\}\}$

To see which of the two solutions is the correct one, we just substitute numerical values for the radii of the three circles.

In[11]= % /. {r1 -> 1, r2 -> 1, r3 -> 1} // N

Out[11]= {{r → 0.154701}, {r → -2.1547}}

This means, that the first of the above solutions is the one of interest.

In[12]= %%[[1, 1, 2]]

Out[12]= $\dfrac{r1\ r2^2\ r3^2 + r1^2\ r2\ r3\ (r2 + r3) - 2\sqrt{r1^3\ r2^3\ r3^3\ (r1 + r2 + r3)}}{r1^2\ (r2 - r3)^2 + r2^2\ r3^2 - 2\ r1\ r2\ r3\ (r2 + r3)}$

Taking into account that r1, r2, and r3 are positive real numbers, we can simplify the result.

In[13]= Factor //@ (
          % /. {Sqrt[r1^3 r2^3 r3^3 (r1 + r2 + r3)] ->
                r1 r2 r3 Sqrt[r1 r2 r3 (r1 + r2 + r3)],
                Sqrt[r1] Sqrt[r2] Sqrt[r3] Sqrt[r1 + r2 + r3] ->
                Sqrt[r1 r2 r3 (r1 + r2 + r3)]})

Out[13]= $\dfrac{r1\ r2\ r3\ (r1\ r2 + r1\ r3 + r2\ r3 - 2\sqrt{r1\ r2\ r3\ (r1 + r2 + r3)}\ )}{r1^2\ r2^2 - 2\ r1^2\ r2\ r3 - 2\ r1\ r2^2\ r3 + r1^2\ r3^2 - 2\ r1\ r2\ r3^2 + r2^2\ r3^2}$

Multiplying numerator and denominator of the last expression by the numerator with reversed sign of the square root, the expression simplifies further.

In[14]= Simplify[Factor[Expand[
          Numerator[%] (Numerator[%] /. p : Power[_, 1/2] :> -p)]] /
          (Denominator[%] (Numerator[%] /. p : Power[_, 1/2] :> -p))]

Out[14]= $\dfrac{r1\ r2\ r3}{r2\ r3 + r1\ (r2 + r3) + 2\sqrt{r1\ r2\ r3\ (r1 + r2 + r3)}}$

The coordinates of the midpoint of the circle are straightforward to calculate once the radius is determined. For the corresponding problem of placing a fifth sphere between four given spheres (when possible), see [1731], [518].

e) This is the implicit equation for a circle.

In[1]= circle = (y[x] - y0)^2 + (x - x0)^2 - r^2

Out[1]= $-r^2 + (x - x0)^2 + (-y0 + y[x])^2$

We differentiate the implicit equation three times (we want to eliminate the three parameters x_0, y_0, and r) with respect to x to obtain a set of algebraic equations.

In[2]= eqs = Table[D[circle, {x, i}], {i, 0, 3}]

Out[2]= $\{-r^2 + (x - x0)^2 + (-y0 + y[x])^2,\ 2\,(x - x0) + 2\,(-y0 + y[x])\,y'[x],$
$2 + 2\,y'[x]^2 + 2\,(-y0 + y[x])\,y''[x],\ 6\,y'[x]\,y''[x] + 2\,(-y0 + y[x])\,y^{(3)}[x]\}$

Using GroebnerBasis with EliminationOrder, we get the nonlinear differential equation of third order obeyed by all circles in the x,y-plane.

In[3]= GroebnerBasis[eqs, {x, y[x], y'[x], y''[x], y'''[x]},
                      {x0, y0, r}, MonomialOrder -> EliminationOrder]

Out[3]= $\{3\,y'[x]\,y''[x]^2 - y^{(3)}[x] - y'[x]^2\,y^{(3)}[x]\}$

f) We start by rewriting the given differential equation as a differential equation for $v(u)$. Derivatives transform according to the following rule: $\dfrac{d}{dk} = \dfrac{du}{dk}\dfrac{d}{du} = 1 \Big/ \Big(\dfrac{dk}{du}\Big) \dfrac{d}{du} = \dfrac{1}{4u^3}\dfrac{d}{du}$. Let $1i$ the ith rewritten derivative.

In[1]= l1 = 1/(4u^3) D[v[u]^4, u]

Out[1]= $\dfrac{v[u]^3\,v'[u]}{u^3}$

In[2]= l2 = Together[1/(4u^3) D[l1, u]]

Out[2]= $\dfrac{-3\,v[u]^3\,v'[u] + 3\,u\,v[u]^2\,v'[u]^2 + u\,v[u]^3\,v''[u]}{4\,u^7}$

In[3]:= **l3 = Together[1/(4u^3) D[l2, u]]**

Out[3]= $\frac{1}{16\,u^{11}}$ (21 v[u]³ v'[u] - 27 u v[u]² v'[u]² +
      6 u² v[u] v'[u]³ - 9 u v[u]³ v''[u] + 9 u² v[u]² v'[u] v''[u] + u² v[u]³ v^{(3)}[u])

Using the three quantities $l1$, $l2$, and $l3$, we obtain the following differential equation for $v(u)$.

In[4]:= **ode = 3 l2^2 - 2 l1 l3 + l1^2 (((1 + u^8)/(u^4 - u^12))^2 -**
              ((1 + v[u]^8)/(v[u]^4 - v[u]^12))^2 l1^2)

Out[4]= $\dfrac{v[u]^6\,v'[u]^2\left(\frac{(1+u^8)^2}{(u^4-u^{12})^2}-\frac{v[u]^6\,(1+v[u]^8)^2\,v'[u]^2}{u^6\,(v[u]^4-v[u]^{12})^2}\right)}{u^6}$ +

      $\dfrac{3\,(-3\,v[u]^3\,v'[u]+3\,u\,v[u]^2\,v'[u]^2+u\,v[u]^3\,v''[u])^2}{16\,u^{14}}$ -

      $\dfrac{1}{8\,u^{14}}$ (v[u]³ v'[u] (21 v[u]³ v'[u] - 27 u v[u]² v'[u]² +
          6 u² v[u] v'[u]³ - 9 u v[u]³ v''[u] + 9 u² v[u]² v'[u] v''[u] + u² v[u]³ v^{(3)}[u]))

Now, we use the modular equations under consideration to derive expressions for $v'(u)$, $v''(u)$, and $v'''(u)$.

In[5]:= **derivatives[poly_] := Reverse[Flatten[Table[**
              Solve[D[poly, {u, i}] == 0, Derivative[i][v][u]], {i, 3}]]]

The first equation, supposedly a solution of ode, is the following.

In[6]:= **modularEquation3 = u^4 - v^4 - 2 u v (1 - u^2 v^2) /. v -> v[u];**

We calculate explicit forms for $v'(u)$, $v''(u)$, and $v'''(u)$ from the last equation.

In[7]:= **derivatives3 = derivatives[modularEquation3]**

Out[7]= {v^{(3)}[u] →

      $-\dfrac{1}{-u+3\,u^3\,v[u]^2-2\,v[u]^3}$ (3 (4 u + 2 v[u]³ + 18 u v[u]² v'[u] + 18 u² v[u] v'[u]² + 2 u³ v'[u]³ -
          4 v[u] v'[u]³ - v''[u] + 9 u² v[u]² v''[u] + 6 u³ v[u] v'[u] v''[u] - 6 v[u]² v'[u] v''[u])),

      v''[u] → $-\dfrac{2\,(3\,u^2+3\,u\,v[u]^3-v'[u]+9\,u^2\,v[u]^2\,v'[u]+3\,u^3\,v[u]\,v'[u]^2-3\,v[u]^2\,v'[u]^2)}{-u+3\,u^3\,v[u]^2-2\,v[u]^3}$,

      v'[u] → $\dfrac{-2\,u^3+v[u]-3\,u^2\,v[u]^3}{-u+3\,u^3\,v[u]^2-2\,v[u]^3}$ }

Inserting these derivatives into ode and clearing denominators gives the following result.

In[8]:= **(modularInserted3 =**
          Factor[Numerator[Together[ode //. derivatives3]]]) // Short[#, 6]&

Out[8]//Short= -v[u]⁴ (≪1≫) ≪1≫
      (u⁸ - 2 u¹⁶ + u²⁴ + 2 u² v[u]² - 14 u¹⁰ v[u]² - 10 u¹⁸ v[u]² + 6 u²⁶ v[u]² + 4 u⁷ v[u]³ -
      40 u¹⁵ v[u]³ + 20 u²³ v[u]³ + 2 u⁴ v[u]⁴ + 197 u¹² v[u]⁴ - 124 u²⁰ v[u]⁴ + 81 u²⁸ v[u]⁴ -
      312 u⁹ v[u]⁵ + 504 u¹⁷ v[u]⁵ - 216 u²⁵ v[u]⁵ + v[u]²⁴ - 676 u⁸ v[u]²⁴ -
      4283 u¹⁶ v[u]²⁴ - 702 u²⁴ v[u]²⁴ + 216 u⁵ v[u]²⁵ + 1872 u¹³ v[u]²⁵ - 1224 u²¹ v[u]²⁵ +
      6 u² v[u]²⁶ + 90 u¹⁰ v[u]²⁶ + 2202 u¹⁸ v[u]²⁶ + 486 u²⁶ v[u]²⁶ - 324 u⁷ v[u]²⁷ -
      1656 u¹⁵ v[u]²⁷ - 324 u²³ v[u]²⁷ + 81 u⁴ v[u]²⁸ + 414 u¹² v[u]²⁸ + 81 u²⁰ v[u]²⁸)

One of the factors of the last equation is simply the starting polynomial modularInserted3.

In[10]:= **Cancel[PolynomialGCD[modularInserted3, modularEquation3]/**
                  modularEquation3]

Out[10]= 1

This means that modularInserted3 satisfies the differential equation ode.

In a similar way, we can deal with the second and third equations.

In[11]:= **modularEquation5 =**
          u^6 - v^6 + 5 u^2 v^2 (u^2 - v^2) - 4 u v (1 - u^4 v^4) /. v -> v[u];
          (* repeat steps from above *)
          derivatives5 = derivatives[modularEquation5];
          modularInserted5 = Factor[Numerator[Together[ode //. derivatives5]]];

```
        Cancel[PolynomialGCD[modularInserted5, modularEquation5]/
                    modularEquation5]
```

Out[15]= 1

```
In[16]:= modularEquation7 = (1 - u^8)(1 - v^8) - (1 - u v)^8 /. v -> v[u];
        (* repeat steps from above *)
        derivatives7 = derivatives[modularEquation7];
        modularInserted7 = Factor[Numerator[Together[ode //. derivatives7]]];
        Cancel[PolynomialGCD[modularInserted7, modularEquation7]/
                    modularEquation7]
```

Out[20]= -1

g) This is the function under consideration.

```
In[1]:= W[x_] := (c1 E^-Integrate[f[x]/(1 - h[x]), x] *
            (c2 +  Integrate[E^(Integrate[g[x], x] +
                        Integrate[f[x]/(1 - h[x]), x])/(1 - h[x]), x]))
```

We will differentiate the function twice and replace the exponentials of the undone integrals with new variables.

```
In[2]:= rules = {Exp[-Integrate[f[x]/(1 - h[x]), x]] -> int1,
            Integrate[1/(1 - h[x]) Exp[Integrate[f[x]/(1 - h[x]), x]]*
                Exp[Integrate[g[x], x]], x] -> int2,
            Exp[Integrate[g[x], x]] -> int3};
```

```
In[3]:= w0 = W[x] /. rules
```

Out[3]= c1 int1 (c2 + int2)

```
In[4]:= w1 = D[W[x], x] /. rules
```

Out[4]= $\dfrac{c1\ int3}{1 - h[x]} - \dfrac{c1\ int1\ (c2 + int2)\ f[x]}{1 - h[x]}$

```
In[5]:= w2 = D[W[x], x, x] /. rules
```

Out[5]= $-\dfrac{c1\ int3\ f[x]}{(1 - h[x])^2} + \dfrac{c1\ int1\ (c2 + int2)\ f[x]^2}{(1 - h[x])^2} + \dfrac{c1\ int3\ g[x]}{1 - h[x]} -$

$\dfrac{c1\ int1\ (c2 + int2)\ f'[x]}{1 - h[x]} + \dfrac{c1\ int3\ h'[x]}{(1 - h[x])^2} - \dfrac{c1\ int1\ (c2 + int2)\ f[x]\ h'[x]}{(1 - h[x])^2}$

A polynomial system, formed from the above three results, allows us to eliminate the three auxiliary variables int1, int2, and int3.

```
In[6]:= Factor[GroebnerBasis[Numerator[
            Together[{w[x] - w0, w'[x] - w1, w''[x] - w2}]], {},
            {int1, int2, int3}, MonomialOrder -> EliminationOrder]]
```

Out[6]= { (-1 + h[x]) (f[x] g[x] w[x] - w[x] f'[x] - f[x] w'[x] +
        g[x] w'[x] - g[x] h[x] w'[x] + h'[x] w'[x] - w''[x] + h[x] w''[x])}

The second factor of the last result is the differential equation we were looking for.

```
In[7]:= Collect[%[[1, -1]], {w[x], w'[x], w''[x]}]
```

Out[7]= w[x] (f[x] g[x] - f'[x]) + (-f[x] + g[x] - g[x] h[x] + h'[x]) w'[x] + (-1 + h[x]) w''[x]

```
In[8]:= % /. w -> W // Together
```

Out[8]= 0

h) None of the built-in functions is currently able to prove this identity directly.

```
In[1]:= Tan[1/4 ArcTan[4]] - 2 (Cos[6Pi/17] + Cos[10 Pi/17]) //  Simplify
```

Out[1]= $-2 \left(\text{Cos}\left[\dfrac{6\pi}{17} \right] + \text{Cos}\left[\dfrac{10\pi}{17} \right] \right) + \text{Tan}\left[\dfrac{\text{ArcTan}[4]}{4} \right]$

Also, the function FullSimplify—to be discussed in Chapter 3— is not able to establish the identity.

```
In[2]:= FullSimplify[%]
```

Out[2]= $-2 \left(\text{Cos}\left[\dfrac{6\pi}{17} \right] + \text{Cos}\left[\dfrac{10\pi}{17} \right] \right) + \text{Tan}\left[\dfrac{\text{ArcTan}[4]}{4} \right]$

However, it is nevertheless straightforward to prove this identity: we transform both sides of the equation into explicit algebraic numbers (using `TrigToExp` to get rid of the trigonometric functions and `Together` and `ExpandAll` to eliminate the exponential functions) by applying various functions that change the form, but not the mathematical content, of an expression. Then a call to `RootReduce` will transform this algebraic number into a canonical form.

In[3]:= `RootReduce[Together[ExpandAll[TrigToExp[Tan[1/4 ArcTan[4]]]]]]`

Out[3]= $\text{Root}[1 - \#1 - 6 \,\#1^2 + \#1^3 + \#1^4 \,\&, \, 3]$

In[4]:= `RootReduce[Expand[Together[TrigToExp[2 (Cos[6Pi/17] + Cos[10 Pi/17])]]]]`

Out[4]= $\text{Root}[1 - \#1 - 6 \,\#1^2 + \#1^3 + \#1^4 \,\&, \, 3]$

Also, the radical form of this expression looks kind of neat.

In[5]:= `ToRadicals[%]`

Out[5]= $\dfrac{1}{4}\left(-1 - \sqrt{17} + 2\sqrt{\dfrac{17}{2} + \dfrac{\sqrt{17}}{2}}\,\right)$

Now let us prove the second identity. This is the right-hand side of the identity to be proved.

In[6]:= `rhs = (1/6 + 1/6 Sqrt[7] (Cos[1/3 ArcCos[1/(2 Sqrt[7])]] +`
           `Sqrt[3] Sin[1/3 ArcCos[1/(2 Sqrt[7])]]))`

Out[6]= $\dfrac{1}{6} + \dfrac{1}{6}\,\sqrt{7}\left(\text{Cos}\Big[\dfrac{1}{3}\,\text{ArcCos}\Big[\dfrac{1}{2\sqrt{7}}\Big]\Big] + \sqrt{3}\,\text{Sin}\Big[\dfrac{1}{3}\,\text{ArcCos}\Big[\dfrac{1}{2\sqrt{7}}\Big]\Big]\right)$

Numerically, the identity holds to a high precision.

In[7]:= `Cos[Pi/7`100] - rhs`

Out[7]= $0. \times 10^{-101}$

Making all numbers algebraic and applying `RootReduce` again proves the identity.

In[8]:= `(* apply RootReduce to parts and then to whole expression *)`
         `RootReduce[RootReduce /@ Expand[Simplify[Together //@ TrigToExp[rhs]]]]`
Out[9]= $\text{Root}[1 - 4\,\#1 - 4\,\#1^2 + 8\,\#1^3 \,\&, \, 3]$

In[10]:= `Together[TrigToExp[Cos[Pi/7]]] // RootReduce`

Out[10]= $\text{Root}[1 - 4\,\#1 - 4\,\#1^2 + 8\,\#1^3 \,\&, \, 3]$

i) We assume the one corner of the boxes is at the origin, and the edges of the boxes are oriented along the coordinate axes. Using quantifiers, it is straightforward to express the problem at hand.

In[1]:= `existsSecondBox[relation_] :=`
          `Exists[{w1, h1, d1},`
                `Element[{w1, h1, d1}, Reals] &&`
                      `(* first box has positive size *)`
                      `w1 > 0 && h1 > 0 && d1 > 0,`
                `Exists[{w2, h2, d2}, Element[{w2, h2, d2}, Reals] &&`
                            `(* second box is inside the first box *)`
                            `w1 > w2 > 0  && h1 > h2 > 0 && d1 > d2 > 0,`
                `(* the relation to be tested *)`
                `(1/w2 + 1/h2 + 1/d2) ~ relation ~ (w1 + h1 + d1)]]`

`Resolve` shows that a second box for all three cases exists.

In[2]:= `AppendTo[$ContextPath, "Experimental`"];`

         `Resolve[existsSecondBox[#]]& /@ {Equal, Less, Greater}`
Out[3]= `{True, True, True}`

j) We will start by visualizing the geometric region described by the three inequalities. To do this, we use the function `CylindricalDecomposition` to generate a set of cells forming the region.

In[1]:= `(cad = CylindricalDecomposition[`
                `Abs[φ x] + Abs[y] < 1 &&  Abs[φ y] + Abs[z] < 1 &&`
                `Abs[φ z] + Abs[x] < 1 &&`

```
(* algebraic definition of GoldenRatio *)
(2 φ - 1)^2 == 5 && φ > 0, {φ, x, y, z}]) // Short[#, 6]&
```

Out[1]//Short= $\phi = \frac{1}{2}\left(1 + \sqrt{5}\right)$ &&

$$\left(\left(\left(\frac{1 - \phi - \phi^2}{1 + \phi^3} < x \le -\frac{1}{1 + \phi} \;\&\&\; -1 - x\,\phi < y < 1 + x\,\phi \;\&\&\; \frac{-1 - x}{\phi} < z < \frac{1 + x}{\phi}\right) \;||\; \left(-\frac{1}{1 + \phi} < x \le \frac{1 - \phi}{\phi^2} \;\&\&\right.\right.\right.$$

$$\left(\left(-1 - x\,\phi < y < \frac{1 + x - \phi}{\phi^2} \;\&\&\; -1 - y\,\phi < z < 1 + y\,\phi\right) \;||\; \left(\frac{1 + x - \phi}{\phi^2} \le y \le \frac{-1 - x + \phi}{\phi^2} \;\&\&\right.\right.$$

$$\left.\frac{-1 - x}{\phi} < z < \frac{1 + x}{\phi}\right) \;||\; \left(\frac{-1 - x + \phi}{\phi^2} < y < 1 + x\,\phi \;\&\&\; -1 + y\,\phi < z < 1 - y\,\phi\right)\right)\right) \;||$$

$$\left(\frac{1 - \phi}{\phi^2} < x \le 0 \;\&\&\; \left(\left(-\frac{1}{\phi} < y < \frac{1 + x - \phi}{\phi^2} \;\&\&\; -1 - y\,\phi < z < 1 + y\,\phi\right) \;||\; \left(\frac{1 + x - \phi}{\phi^2} \le y \le \frac{-1 - x + \phi}{\phi^2} \;\&\&\right.\right.\right.$$

$$\left.\frac{-1 - x}{\phi} < z < \frac{1 + x}{\phi}\right) \;||\; \left(\frac{-1 - x + \phi}{\phi^2} < y < \frac{1}{\phi} \;\&\&\; -1 + y\,\phi < z < 1 - y\,\phi\right)\right)\right) \;||$$

$$\left(0 < x < \frac{\ll 1 \gg}{\ll 1 \gg} \;\&\&\; (\ll 1 \gg)\right) \;||\; \ll 1 \gg \;||\; \left(\frac{-1 + \phi}{\phi^2} < x < \frac{1}{1 + \phi} \;\&\&\; (\ll 1 \gg)\right) \;||$$

$$\left(x = \frac{1}{2}\left(3 - \sqrt{5}\right) \;\&\&\; \frac{1 - x - \phi}{\phi^2} < y < \frac{-1 + x + \phi}{\phi^2} \;\&\&\; \frac{-1 + x}{\phi} < z < \frac{1 - x}{\phi}\right) \;||$$

$$\left(\frac{1}{1 + \phi} < x < \frac{1 - \phi + \phi^2}{-1 + \phi^3} \;\&\&\; -1 + x\,\phi < y < 1 - x\,\phi \;\&\&\; \frac{-1 + x}{\phi} < z < \frac{1 - x}{\phi}\right)\right)$$

The result cad contains the numerical value of ϕ and the regions in the x, y, z-space.

```
In[2]:= {cad[[1]], cad[[2]] // Length, cad[[2]] // Head}
```

Out[2]= $\left\{\phi = \frac{1}{2}\left(1 + \sqrt{5}\right),\; 8,\; \text{Or}\right\}$

We expand the Or and obtain 16 single subregions.

```
In[3]:= (regions = cad[[2]] /. (xi_ && rest_Or) :> (And[xi, #]& /@ rest) /.
                     LessEqual -> Less /. Or -> List) // Length
```

Out[3]= 18

They are three- and two-dimensional.

```
In[4]:= Take[regions, 5]
```

Out[4]= $\left\{\frac{1 - \phi - \phi^2}{1 + \phi^3} < x < -\frac{1}{1 + \phi} \;\&\&\; -1 - x\,\phi < y < 1 + x\,\phi \;\&\&\; \frac{-1 - x}{\phi} < z < \frac{1 + x}{\phi},\right.$

$$-\frac{1}{1 + \phi} < x < \frac{1 - \phi}{\phi^2} \;\&\&\; -1 - x\,\phi < y < \frac{1 + x - \phi}{\phi^2} \;\&\&\; -1 - y\,\phi < z < 1 + y\,\phi,$$

$$-\frac{1}{1 + \phi} < x < \frac{1 - \phi}{\phi^2} \;\&\&\; \frac{1 + x - \phi}{\phi^2} < y < \frac{-1 - x + \phi}{\phi^2} \;\&\&\; \frac{-1 - x}{\phi} < z < \frac{1 + x}{\phi},$$

$$-\frac{1}{1 + \phi} < x < \frac{1 - \phi}{\phi^2} \;\&\&\; \frac{-1 - x + \phi}{\phi^2} < y < 1 + x\,\phi \;\&\&\; -1 + y\,\phi < z < 1 - y\,\phi,$$

$$\left.\frac{1 - \phi}{\phi^2} < x < 0 \;\&\&\; -\frac{1}{\phi} < y < \frac{1 + x - \phi}{\phi^2} \;\&\&\; -1 - y\,\phi < z < 1 + y\,\phi\right\}$$

The function toPolygon forms the six polygons bounding the 3D and forms the one polygon forming the 2D regions.

```
In[5]:= f = {#1, #2 /. x -> #1, #3 /. y -> #2 /. x -> #1}&;

In[6]:= toPolygon[Inequality[x1_, Less, x, Less, x2_] &&
                  Inequality[y1_, Less, y, Less, y2_] &&
                  Inequality[z1_, Less, z, Less, z2_]] :=
        Module[(* points *)
        {p111 = f @@ {x1, y1, z1}, p112 = f @@ {x1, y1, z2},
         p121 = f @@ {x1, y2, z1}, p122 = f @@ {x1, y2, z2},
         p211 = f @@ {x2, y1, z1}, p212 = f @@ {x2, y1, z2},
         p221 = f @@ {x2, y2, z1}, p222 = f @@ {x2, y2, z2}},
         (* the six bounding polygons *) Polygon /@
        {{p111, p211, p221, p121}, {p112, p212, p222, p122},
         {p111, p211, p212, p112}, {p121, p221, p222, p122},
         {p111, p121, p122, p112}, {p211, p221, p222, p212}} /.
                           φ -> GoldenRatio]

In[7]:= (* make polygon from Boolean expression *)
        toPolygon[(x == x1_) && Inequality[y1_, Less, y, Less, y2_] &&
                  Inequality[z1_, Less, z, Less, z2_]] :=
```

```
Module[{p1 = f @@ {x1, y1, z1}, p2 = f @@ {x1, y2, z1},
       p3 = f @@ {x1, y2, z2}, p4 = f @@ {x1, y1, z2}},
      (* the single polygon *)
   Polygon[{p1, p2, p3, p4} /. φ -> GoldenRatio]]
```

Here are the 16 regions shown individually.

```
In[9]:= SeedRandom[123];
      Show[GraphicsArray[#]]& /@ Partition[
        (* the individual 3D graphics *) tab =
      Table[Graphics3D[{SurfaceColor[Hue[Random[]], Hue[Random[]], 2.1],
                         toPolygon[regions[[k]]]},
                 PlotRange -> {{-1, 1}, {-1, 1}, {-1, 1}} 0.6],
          {k, Length[regions]}] // N, 4]
```

Displaying them together, let us conjecture that the body is a dodecahedron. The next picture shows more clearly how the 16 subregions form the dodecahedron.

```
In[11]:= (* contract all polygons of the graphics gr *)
      contract[gr_, α_] :=
      Module[{vertices, center},
       vertices = Level[N[Cases[gr, _Polygon, Infinity]], {-2}];
```

```
              center = Plus @@ vertices/Length[vertices];
              gr /. Polygon[l_] :> Polygon[(center + α (# - center))& /@ l]]
In[13]:= (* cut a hole in a polygon *)
          holePolygon[Polygon[l_], factor_] :=
          Module[{mp = Plus @@ l/4, g = Partition[Append[#, First[#]], 2, 1]&},
          MapThread[Polygon[Join[#1, Reverse[#2]]]&,
                    {g[l], g[(mp + factor(# - mp))& /@ l]}]]
In[15]:= Show[GraphicsArray[
          Block[{$DisplayFunction = Identity},
           {(* the individual pieces as solids *)
            Show[tab, PlotRange -> All, Axes -> True],
            (* the individual pieces with holes *)
            Show[(contract[#, 0.8]& /@ tab) /. p_Polygon :> holePolygon[p, 0.7],
                 Boxed -> False]}]]]
```

The resulting polyhedron looks like a dodecahedron. Now, we will show that it actually is a dodecahedron. These are all the vertices of the subregions.

```
In[16]:= vertices = Union[RootReduce[Union[Together //@ (
                        Level[Cases[tab, _Polygon, Infinity], {3}] /.
                        GoldenRatio -> (1 + Sqrt[5])/2)]]];

In[17]:= Length[vertices]

Out[17]= 92
```

Twenty of these vertices are the vertices of the dodecahedron.

```
In[18]:= squaredDistances = RootReduce[#.#]& /@ vertices;

In[19]:= (dodeVertices = Select[vertices,
                RootReduce[#.#] == Max[squaredDistances]&]) // Length
Out[19]= 66
```

Groups of five of the dodecahedron vertices form the 12 faces of the dodecahedron. Each of the 12 vertices has three nearest neighbors.

```
In[20]:= vertexVertexDistances =
          Table[RootReduce[#.#&[dodeVertices[[i]] - dodeVertices[[j]]]],
                {i, 20}, {j, 20}];
In[21]:= With[{edgeLength = Min[DeleteCases[vertexVertexDistances, 0, {2}]]},
              Count[#, edgeLength]& /@ vertexVertexDistances]
Out[21]= {1, 1, 1, 1, 1, 1, 1, 1, 1, 1, 1, 1, 1, 1, 1, 1, 1, 1, 1, 1}
```

For the general descriptions of polyhedra through inequalities, see [220], [779].

k) We will check the identity by viewing the left-hand and the right-hand sides as functions of ξ. Differentiation of the integral is a purely formal process. All integrals are assumed to exist, so that the $\partial\, identity/\partial\xi$ holds.

```
In[1]:= Integrate[Integrate[f[t]/t, {t, x, Infinity}]^2, {x, 0, ξ}] -
          (Integrate[(1/x Integrate[f[t], {t, 0, x}])^2, {x, 0, ξ}] +
           (Sqrt[ξ] Integrate[f[t]/t, {t, ξ, Infinity}] +
            1/Sqrt[ξ] Integrate[f[t], {t, 0, ξ}])^2)
```

$$Out[1]= -\left(\frac{\int_0^\xi f[t]\,dt}{\sqrt{\xi}} + \sqrt{\xi}\int_\xi^\infty \frac{f[t]}{t}\,dt\right)^2 - \int_0^\xi \frac{\left(\int_0^x f[t]\,dt\right)^2}{x^2}\,dx + \int_0^\xi \left(\int_x^\infty \frac{f[t]}{t}\,dt\right)^2 dx$$

```
In[2]:= D[%, ξ] // Simplify

Out[2]= 0
```

In addition, we have to check an initial value; we choose $\xi = 0$.

```
In[3]:= Series[%%, {ξ, 0, 1}]

Out[3]= O[ξ]²
```

l) We start by implementing the A-polynomials. To generate the terms of the j-dimensional sums that fulfill the side condition $k_1 + 2 k_2 + \cdots + j k_j = j$, we will use the function `Experimental`IntegerPartitions`.

```
In[1]:= (* add the context of IntegerPartitions to context path *)
        AppendTo[$ContextPath, "Experimental`"];

In[3]:= (* make result in condensed form *)
        myIntegerPartitions[n_Integer?Positive] := Reverse /@
            (({First[#], Length[#]}& /@ Split[#])& /@ IntegerPartitions[n])

In[5]:= partitions[n_] := (Last /@ Sort[Join[#, {#, 0}& /@
            Complement[Range[n], First /@ #]]])& /@ myIntegerPartitions[n]

In[6]:= A[varList_] :=
        Module[{n = Length[varList],
                v = MapIndexed[#1/#2[[1]]&, varList]},
                Plus @@ ((n!/(Times @@ Factorial[#])(Times @@ (v^#)))& /@
                partitions[n])]
```

Here are the first few A-polynomials.

```
In[7]:=
        Table[A[Table[Subscript[τ, k], {k, n}]], {n, 5}]
Out[7]= {τ₁, τ₁² + τ₂, τ₁³ + 3 τ₁ τ₂ + 2 τ₃, τ₁⁴ + 6 τ₁² τ₂ + 3 τ₂² + 8 τ₁ τ₃ + 6 τ₄,
        τ₁⁵ + 10 τ₁³ τ₂ + 15 τ₁ τ₂² + 20 τ₁² τ₃ + 20 τ₂ τ₃ + 30 τ₁ τ₄ + 24 τ₅}
```

The implementation of the relation itself is straightforward.

```
In[8]:= relation[as_List, r_] :=
        With[{n = Length[as], (* left-hand side *)
              p = Function[Evaluate[Sum[c[i] #^i, {i, 0, Length[as] - 1}]]]},
              Sum[p[as[[k]]]/(x - as[[k]])^(r + 1)/
                Product[If[k == 1, 1, as[[k]] - as[[l]]], {l, 1, n}], {k, n}] ==
                (* right-hand side *)
              (-1)^r/(r! Product[x - as[[k]], {k, n}])*
              (Derivative[r][p][x] +
              Sum[(-1)^j Binomial[r, j] Derivative[r - j][p][x] *
                A[Table[Sum[(x - as[[i]])^-m, {i, n}], {m, j}]], {j, r}])]
```

The relation either autoevaluates to `True` or is a rational function. This means that `Together` is the appropriate function to apply. Here we check the first 16 relations of this type.

```
In[9]:= checkRelation[n_, r_] :=
        With[{expr = relation[Table[Subscript[τ, k], {k, n}], r]},
            If[expr === True, True,
                Factor[Together[Subtract @@ expr]] === 0]]

In[10]:= Table[checkRelation[n, r], {n, 4}, {r, 4}]

Out[10]= {{True, True, True, True}, {True, True, True, True},
         {True, True, True, True}, {True, True, True, True}}
```

m) We start by generating all possible triples `allTriangles` out of five elements.

```
In[1]:= n = 5;
        allTriangles = Subsets[Table[P[j], {j, n}], {3}]
Out[2]= {{P[1], P[2], P[3]}, {P[1], P[2], P[4]}, {P[1], P[2], P[5]},
        {P[1], P[3], P[4]}, {P[1], P[3], P[5]}, {P[1], P[4], P[5]},
        {P[2], P[3], P[4]}, {P[2], P[3], P[5]}, {P[2], P[4], P[5]}, {P[3], P[4], P[5]}}
```

Without loss of generality, we can assume that the first point is at the origin and that the second one lies on the x-axis.

```
In[3]:= P[1][1] = 0; P[1][2] = 0;
```

To obtain relations between the areas of the triangles, we express the areas through the vertex coordinates and then eliminate the vertex coordinates. We use the oriented areas $\mathbb{A}_{i,j,k}$.

In[4]:= A[{{x1_, y1_}, {x2_, y2_}, {x3_, y3_}}] :=
          Det[{{x1, y1, 1}, {x2, y2, 1}, {x3, y3, 1}}]/2;

In[5]:= eqs = Map[Subscript[A, Sequence @@ (First /@ #)] -
          A[Table[#[k], {k, 2}]& /@ #]&, allTriangles]

Out[5]= $\{\mathbb{A}_{1,2,3} + \frac{1}{2}$ (P[2][2] P[3][1] - P[2][1] P[3][2]),

          $\mathbb{A}_{1,2,4} + \frac{1}{2}$ (P[2][2] P[4][1] - P[2][1] P[4][2]),

          $\mathbb{A}_{1,2,5} + \frac{1}{2}$ (P[2][2] P[5][1] - P[2][1] P[5][2]),

          $\mathbb{A}_{1,3,4} + \frac{1}{2}$ (P[3][2] P[4][1] - P[3][1] P[4][2]),

          $\mathbb{A}_{1,3,5} + \frac{1}{2}$ (P[3][2] P[5][1] - P[3][1] P[5][2]),

          $\mathbb{A}_{1,4,5} + \frac{1}{2}$ (P[4][2] P[5][1] - P[4][1] P[5][2]),

          $\mathbb{A}_{2,3,4} + \frac{1}{2}$ (P[2][2] P[3][1] - P[2][1] P[3][2] -
              P[2][2] P[4][1] + P[3][2] P[4][1] + P[2][1] P[4][2] - P[3][1] P[4][2]),

          $\mathbb{A}_{2,3,5} + \frac{1}{2}$ (P[2][2] P[3][1] - P[2][1] P[3][2] - P[2][2] P[5][1] +
              P[3][2] P[5][1] + P[2][1] P[5][2] - P[3][1] P[5][2]),

          $\mathbb{A}_{2,4,5} + \frac{1}{2}$ (P[2][2] P[4][1] - P[2][1] P[4][2] - P[2][2] P[5][1] +
              P[4][2] P[5][1] + P[2][1] P[5][2] - P[4][1] P[5][2]),

          $\mathbb{A}_{3,4,5} + \frac{1}{2}$ (P[3][2] P[4][1] - P[3][1] P[4][2] - P[3][2] P[5][1] +
              P[4][2] P[5][1] + P[3][1] P[5][2] - P[4][1] P[5][2])$\}$

We have ten equations, and we have to eliminate eight vertex coordinates.

In[6]:= pointCoords = Union[Cases[eqs, _P[_], Infinity]]

Out[6]= {P[2][1], P[2][2], P[3][1], P[3][2], P[4][1], P[4][2], P[5][1], P[5][2]}

In[7]:= {Length[eqs], Length[pointCoords]}

Out[7]= {10, 8}

These are the resulting relations between the oriented areas. The last one, quadratic in the areas, may be unexpected.

In[8]:= gb1 = GroebnerBasis[eqs, {}, pointCoords,
                          MonomialOrder -> EliminationOrder]

Out[8]= $\{\mathbb{A}_{2,3,4} - \mathbb{A}_{2,3,5} + \mathbb{A}_{2,4,5} - \mathbb{A}_{3,4,5}, \mathbb{A}_{1,3,4} - \mathbb{A}_{1,3,5} + \mathbb{A}_{1,4,5} - \mathbb{A}_{3,4,5}, \mathbb{A}_{1,2,4} - \mathbb{A}_{1,2,5} + \mathbb{A}_{1,4,5} - \mathbb{A}_{2,4,5},$
          $\mathbb{A}_{1,2,3} - \mathbb{A}_{1,2,5} + \mathbb{A}_{1,3,5} - \mathbb{A}_{2,3,5}, \mathbb{A}_{1,4,5} \mathbb{A}_{2,3,5} - \mathbb{A}_{1,3,5} \mathbb{A}_{2,4,5} + \mathbb{A}_{1,2,5} \mathbb{A}_{3,4,5}\}$

For similar relations between the volumes of five tetrahedra formed by five points in $\mathbb{R}^3$, see [1213].

n) It is straightforward to calculate the explicit positions of the six points $P_1, \ldots, P_6$. We just form all equations that are implied by the distances and solve these equations.

In[1]:= (* pairwise distances *)
          MapIndexed[Set[d[#2[[1]], #2[[1]] + #2[[2]]], #1]&,
          {{87, 158, 170, 127, 68}, {85, 127, 136, 131},
          {68, 131, 174}, {87, 158}, {85}}, {2}];

In[3]:= (* fix position of two points *)
          p[1][1] = 0; p[1][2] = 0; p[2][1] = 87; p[2][2] = 0;

In[5]:= (* Euclidean distance between two points *)
          d2[i_, j_] := (p[i][1] - p[j][1])^2 + (p[i][2] - p[j][2])^2;
          (* 15 equations describing the distances *)
          eqs = Flatten[Table[d2[i, j] - d[i, j]^2, {i, 5}, {j, i + 1, 6}]];

In[9]:= (* undetermined point coordinates *)
          ps = Cases[eqs, p[_][_], Infinity] // Union;
          sol = Solve[# == 0& /@ eqs, ps]

Out[11]= $\{\{p[3][2] \to -\dfrac{40\sqrt{2002}}{29}, p[4][2] \to -\dfrac{80\sqrt{2002}}{29},$

$p[5][2] \to -\dfrac{80\sqrt{2002}}{29}, p[6][2] \to -\dfrac{40\sqrt{2002}}{29}, p[3][1] \to \dfrac{4218}{29},$

$p[4][1] \to \dfrac{3390}{29}, p[5][1] \to \dfrac{867}{29}, p[6][1] \to -\dfrac{828}{29}\},$

$\{p[3][2] \to \dfrac{40\sqrt{2002}}{29}, p[4][2] \to \dfrac{80\sqrt{2002}}{29}, p[5][2] \to \dfrac{80\sqrt{2002}}{29}, p[6][2] \to \dfrac{40\sqrt{2002}}{29},$

$p[3][1] \to \dfrac{4218}{29}, p[4][1] \to \dfrac{3390}{29}, p[5][1] \to \dfrac{867}{29}, p[6][1] \to -\dfrac{828}{29}\}\}$

The following graphic visualizes the relative positions of the six points.

In[12]:= `Show[Graphics[{PointSize[0.02],`
`Table[Point[Table[p[i][j], {j, 2}]], {i, 6}]}] /. sol[[1]],`
`Frame -> True, AspectRatio -> Automatic, PlotRange -> All]`

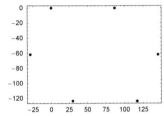

o) This defines an nD Hadamard matrix.

In[1]:= `H[n_] := Array[h, {n, n}];`

Forming the defining equations and imposing the condition that the matrix elements can only take the values ± 1 through $h_{ij}^2 = 1$ yields ten equations for the nine unknowns h_{ij}. The inconsistency of this set of equations can be easily shown through `GroebnerBasis`.

In[2]:= `With[{n = 3},`
`GroebnerBasis[Flatten[{H.Transpose[H] - n IdentityMatrix[n], H^2 - 1}],`
`Flatten[H[n]], MonomialOrder -> DegreeReverseLexicographic]]`

Out[2]= `{1}`

p) The implementation of the formulas for the $c_n(z_1, z_2)$ is straightforward.

In[1]:= `TwoPointTaylorSeries[f_, z_, {z1_, z2_}, o_] :=`
`Module[{c},`
`   c[0][ζ1_, ζ2_] := f[ζ2]/(ζ2 - ζ1);`
`   c[n_][ζ1_, ζ2_] := c[n][ζ1, ζ2] =`
`     (* simplify coefficients *) Together @ Factor @`
`       Sum[(n + k - 1)!/(k! n! (n - k)!)*`
`         ((-1)^(n + 1) n Derivative[n - k][f][ζ2] +`
`         (-1)^(k) k Derivative[n - k][f][ζ1])/`
`         (ζ1 - ζ2)^(n + k + 1), {k, 0, n}];`
`   (* the series *)`
`   Sum[(c[n][z1, z2] (z - z1) + c[n][z1, z2] (z - z2))*`
`     (z - z1)^n (z - z2)^n, {n, 0, o}]]`

The two-point Taylor series $\mathcal{T}_{0,2\pi}^{(20)}[\sin](z)$ of order 20 for $f(z) = \sin(z)$ can be calculated in a fraction of a second.

In[2]:= `(tps20[z_] = TwoPointTaylorSeries[Sin, z, {0, 2Pi}, 20];) // Timing`

Out[2]= `{0.07 Second, Null}`

Here are the first two and the last term of the just-calculated series.

In[3]:= `(List @@ tps20[z])[[{2, 1, -1}]]`

Out[3]= $\left\{z(-2\pi + z)\left(\dfrac{z}{4\pi^2} + \dfrac{-2\pi + z}{4\pi^2}\right), z^2(-2\pi + z)^2\left(-\dfrac{3z}{16\pi^4} - \dfrac{3(-2\pi + z)}{16\pi^4}\right), z^{20}(-2\pi + z)^{20}\right.$

$(((-30460093978369311506250 + 468616830436450946250\pi^2 - 19138705387194393000\pi^4 +$

$321658914070494000\pi^6 - 2639877451336125\pi^8 + 11354311618650\pi^{10} - 25815032100\pi^{12} +$

$$29418840\, \pi^{14} - 14421\, \pi^{16} + 2\, \pi^{18})\, z) / (48592050592872923136000\, \pi^{40}) +$$
$$((-3046009397836931150625 + 46861683043645094 6250\, \pi^2 - 19138705387194393000\, \pi^4 +$$
$$321658914070494000\, \pi^6 - 2639877451336125\, \pi^8 + 11354311618650\, \pi^{10} - 25815032100\, \pi^{12} +$$
$$29418840\, \pi^{14} - 14421\, \pi^{16} + 2\, \pi^{18})\,(-2\, \pi + z)) / (48592050592872923136000\, \pi^{40}))\}$$

At $z = z_1$ and $z = z_2$ the approximation $\texttt{tps20[z]}$ reproduces the local Taylor expansion.

`In[4]:= Normal[Series[tps20[z], {z, #, 20}]& /@ {0, 2Pi} // Simplify]`

$$\text{Out[4]=} \quad \{z - \frac{z^3}{6} + \frac{z^5}{120} - \frac{z^7}{5040} + \frac{z^9}{362880} - \frac{z^{11}}{39916800} +$$
$$\frac{z^{13}}{6227020800} - \frac{z^{15}}{1307674368000} + \frac{z^{17}}{355687428096000} - \frac{z^{19}}{121645100408832000},$$
$$-2\,\pi + z - \frac{1}{6}\,(-2\,\pi + z)^3 + \frac{1}{120}\,(-2\,\pi + z)^5 - \frac{(-2\,\pi + z)^7}{5040} + \frac{(-2\,\pi + z)^9}{362880} - \frac{(-2\,\pi + z)^{11}}{39916800} +$$
$$\frac{(-2\,\pi + z)^{13}}{6227020800} - \frac{(-2\,\pi + z)^{15}}{1307674368000} + \frac{(-2\,\pi + z)^{17}}{355687428096000} - \frac{(-2\,\pi + z)^{19}}{121645100408832000}\}$$

The two-point Taylor series is just the Hermite interpolating polynomial at z_1 and z_2 [226], [447]. Using $\texttt{Interpolating}$ $\texttt{Polynomial}$ and specifying the derivatives at z_1 and z_2 confirms this.

```
In[5]:= With[{o = 10}, Together @
        (TwoPointTaylorSeries[Sin, z, {0, 2Pi}, o] -
        InterpolatingPolynomial[
        Table[{j 2Pi, Table[Derivative[k][Sin][j 2Pi],
                           {k, 0, o}]}, {j, 0, 1}], z])]
Out[5]= 0
```

Now let us determine the precision of the approximation in the interval $[0, 2\pi]$. A visualization shows that the maximum error occurs near $z \approx 2.7$ and $z = 3.3$.

```
In[6]:= $MaxExtraPrecision = 200;
        δ[z_?NumericQ] := N[tps20[SetPrecision[z, Infinity]] -
                        Sin[SetPrecision[z, Infinity]], 20]

In[8]:= data = Table[{z, δ[z]}, {z, 0 - 1, 2Pi + 1, (2Pi - 0 + 2)/555}];

In[9]:= Show[GraphicsArray[
        ListPlot[#, PlotJoined -> True, DisplayFunction -> Identity]& /@
             N[{data, {#1, Log[10, Abs[#2]]}& @@@ data}]]]
```

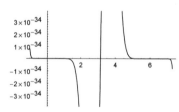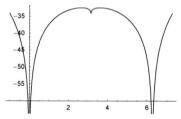

A numerical minimization yields $\max_{z_1 \le z \le z_2} |\sin(z) - \mathcal{T}_{0,2\pi}^{(20)}[\sin](z)| \approx 3.3 \times 10^{-33}$.

```
In[10]:= FindMinimum[δ[z], {z, 2, 2.1}, WorkingPrecision -> 100]
Out[10]= {-3.30638 × 10^-33, {z → 2.66256}}
```

For the *n*-point Taylor expansion, see [1159].

q) Because we have the identity $dy(x)/dx = 1/(dx(y)/dy)$, we use $p(y) = dy(x)/dx$ as the function for which we will find a differential equation.

```
In[1]:= reciprocalOdE[p_, y_, o_] := Cancel[Numerator[Together[
                        Nest[p[y] D[#, y]&, p[y], o - 1] -
                        1/D[1/p[y], {y, o - 1}]]]/p[y]] == 0
```

The complexity of the differential equations increases quickly with *n*.

```
In[2]:= Table[reciprocalOdE[p, y, o], {o, 2, 4}] // Simplify
```

Out[2]= $\{p[y] + p'[y]^2 = 0, 2 p'[y]^4 + p[y] p'[y]^2 p''[y] = p[y]^2 (1 + p''[y]^2),$
$6 p'[y]^6 + 18 p[y] p'[y]^4 p''[y] + p[y]^4 p^{(3)}[y]^2 +$
$p[y]^2 p'[y]^2 (-24 p''[y]^2 + 7 p'[y] p^{(3)}[y]) + p[y]^3 (1 - 2 p'[y] p''[y] p^{(3)}[y]) = 0\}$

For $n = 2$, we obtain the simple solution $(x - c_1)(y - c_2) = 4$.

In[3]:= DSolve[{x'[y] == 1/p[y], reciprocalOdE[p, y, 2]}, {x[y], p[y]}, y]

Out[3]= $\{\{p[y] \rightarrow \frac{1}{4} (-y^2 - 2 i y C[1] + C[1]^2), x[y] \rightarrow \frac{4}{y + i C[1]} + C[2]\},$
$\{p[y] \rightarrow \frac{1}{4} (-y^2 + 2 i y C[1] + C[1]^2), x[y] \rightarrow \frac{4}{y - i C[1]} + C[2]\}\}$

The general solution for $n = 3$ would contain integrals over elliptic integrals.

In[4]:= TimeConstrained[DSolve[{x'[y] == 1/p[y], reciprocalOdE[p, y, 3]},
{x[y], p[y]}, y], 100]

Solve::tdep : The equations appear to involve the variables
to be solved for in an essentially non-algebraic way. More…

Solve::tdep : The equations appear to involve the variables
to be solved for in an essentially non-algebraic way. More…

Solve::tdep : The equations appear to involve the variables
to be solved for in an essentially non-algebraic way. More…

General::stop :
Further output of Solve::tdep will be suppressed during this calculation. More…

Out[4]= $Aborted

Guided by the solution for $n = 2$, we try a similar ansatz for $n = 3$.

In[5]:= eq = (x - C[1]) (y - C[2]) - C[3];
{X = x /. Solve[eq == 0, x][[1]], Y = y /. Solve[eq == 0, y][[1]]}

Out[6]= $\{\frac{y C[1] - C[1] C[2] + C[3]}{y - C[2]}, \frac{x C[2] - C[1] C[2] + C[3]}{x - C[1]}\}$

Demanding that $d^n y(x)/dx^n = 1/(d^n x(y)/dy^n)$ holds yields immediately $c_3 = 6$.

In[7]:= Together[D[Y, {x, 3}] D[X, {y, 3}] /. y -> Y] == 1

Out[7]= $\frac{36}{C[3]^2} = 1$

Trying now the same ansatz for $n = 4, 5, \ldots$ yields the conjecture $(x - c_1)(y - c_2) = n!^{2/(n-1)}$ [245].

In[8]:= Table[Together[D[Y, {x, k}] D[X, {y, k}] /. y -> Y] == 1,
{k, 2, 6}]

Out[8]= $\{\frac{4}{C[3]} = 1, \frac{36}{C[3]^2} = 1, \frac{576}{C[3]^3} = 1, \frac{14400}{C[3]^4} = 1, \frac{518400}{C[3]^5} = 1\}$

r) We start by having a look at the expression $f(q x)/f(x)$ around $q = 1$. The following result shows that $f(q x)/f(x) \approx x f'(x)/f(x)$.

In[1]:= Series[Log[q, f[q x]/f[x]], {q, 1, 2}] // Simplify

Out[1]= $\frac{x f'[x]}{f[x]} + \frac{x (-x f'[x]^2 + f[x] (f'[x] + x f''[x])) (q - 1)}{2 f[x]^2} + \frac{1}{12 f[x]^3}$
$(x (4 x^2 f'[x]^3 - 3 x f[x] f'[x] (f'[x] + 2 x f''[x]) + f[x]^2 (-f'[x] + x (3 f''[x] + 2 x f^{(3)}[x]))))$
$(q - 1)^2) + O[q - 1]^3$

The function q implements the quotential derivative in straightforward way, according to the definition.

In[2]:= q[f_, {x_, n_Integer}] := Module[{q},
Nest[Together[Limit[Log[q, (# /. x -> q x)/#], q -> 1,
Analytic -> True]]&, f, n]]

q[f_, x_] := q[f, {x, 1}]

Here are the first few quotential derivatives of the univariate functions $f(x)$.

In[5]:= Table[q[f[x], {x, n}], {n, 0, 3}]

Out[5]= $\{ f[x], \dfrac{x \, f'[x]}{f[x]}, \dfrac{f[x] \, f'[x] - x \, f'[x]^2 + x \, f[x] \, f''[x]}{f[x] \, f'[x]},$

$(x \, (-f[x] \, f'[x]^3 + x \, f'[x]^4 + f[x]^2 \, f'[x] \, f''[x] - x \, f[x] \, f'[x]^2 \, f''[x] - x \, f[x]^2 \, f''[x]^2 +$
$x \, f[x]^2 \, f'[x] \, f^{(3)}[x])) \, / \, (f[x] \, f'[x] \, (f[x] \, f'[x] - x \, f'[x]^2 + x \, f[x] \, f''[x])) \}$

The first of the above identities is easily proved.

In[6]:= `q[f[g[x]], x] - q[f[g[x]], g[x]] q[g[x], x]`

Out[6]= 0

Next, we extend the definition of q to the multivariate case. We follow closely the syntax of D.

In[7]:= `q[f_, vars__] := Fold[q, f, Reverse @ {vars}]`

Here are the two simplest second quotential derivatives.

In[8]:= `{q[f[x, y], x, y], q[f[x, y], y, x]}`

Out[8]= $\{ \dfrac{-x \, f^{(0,1)}[x, y] \, f^{(1,0)}[x, y] + x \, f[x, y] \, f^{(1,1)}[x, y]}{f[x, y] \, f^{(0,1)}[x, y]},$

$\dfrac{-y \, f^{(0,1)}[x, y] \, f^{(1,0)}[x, y] + y \, f[x, y] \, f^{(1,1)}[x, y]}{f[x, y] \, f^{(1,0)}[x, y]} \}$

The last two quotential derivatives differ from each other. Multiplying with the first quotential derivatives makes them equal.

In[9]:= `q[f[x, y], x] q[f[x, y], y, x] - q[f[x, y], y] q[f[x, y], x, y] //`
                                                                          `Together`

Out[9]= 0

s) We start by forming the first few partial sums.

In[1]:= `summands = Rest[FoldList[Times, 1,`
                `Table[If[k == 1, 1, a[k - 1]]/(x + a[k]), {k, 6}]]];`

In[2]:= `partialSums = Together @ Rest[FoldList[Plus, 0, summands]];`

The numerators and denominators have simple forms when their coefficients in x are expressed as elementary symmetric polynomials in the a_k.

In[3]:= `Needs["Algebra`SymmetricPolynomials`"]`

In[4]:= `With[{n = 6}, Function[numDen,`
                `(Collect[numDen[#], x]& @ partialSums[[n]]) //.`
                `Table[SymmetricPolynomial[Table[a[j], {j, n}], k] ->`
                        `Subscript[S, k], {k, n}]] /@ {Numerator, Denominator}]`

Out[4]= $\{ x^5 + x^4 \, S_1 + x^3 \, S_2 + x^2 \, S_3 + x \, S_4 + S_5, \; x^6 + x^5 \, S_1 + x^4 \, S_2 + x^3 \, S_3 + x^2 \, S_4 + x \, S_5 + S_6 \}$

For each n, the ratio of numerator and denominator has the form $1/x + O(1/x)^n$. From this follows that the conjectured value of the sum is $1/x$.

In[5]:= `Table[ExpandAll @`
                `Series[(x^n + Sum[x^(n - j) Subscript[S, j], {j, n}])/`
                        `(x^(n + 1) + Sum[x^(n + 1 - j) Subscript[S, j], {j, n + 1}]),`
                        `{x, Infinity, n}], {n, 6}]`

Out[5]= $\{ \frac{1}{x} + O[\frac{1}{x}]^2, \; \frac{1}{x} + O[\frac{1}{x}]^3, \; \frac{1}{x} + O[\frac{1}{x}]^4, \; \frac{1}{x} + O[\frac{1}{x}]^5, \; \frac{1}{x} + O[\frac{1}{x}]^6, \; \frac{1}{x} + O[\frac{1}{x}]^7 \}$

2. Horner's Form, Bernoulli Polynomials, Squared Zeros, Polynomialized Radicals, Zeros of Icosahedral Equation, Iterated Exponentials, Matrix Sign Function, Appell–Nielsen Polynomials

a) The implementation is straightforward. We first extract all coefficients and then build the nested form.

In[1]:= `ToHornerForm[poly_, x_] :=`
                `Fold[(#1 x + #2)&, 0, Reverse[CoefficientList[poly, x]]]`

Here is an example.

In[2]:= `ToHornerForm[Sum[a[i] x^i, {i, 0, 18}], x]`

```
Out[2]= a[0] +
       x (a[1] + x (a[2] + x (a[3] + x (a[4] + x (a[5] + x (a[6] + x (a[7] + x (a[8] + x (a[9] + x (a[10] +
                  x (a[11] + x (a[12] + x (a[13] + x (a[14] + x (a[15] +
                     x (a[16] + x (a[17] + x a[18])))))))))))))))))))
```

```
In[3]:= Expand[%] == Sum[a[i] x^i, {i, 0, 18}]
```

```
Out[3]= True
```

For a more complete implementation to rewrite polynomials in Horner form, see the package Algebra`Horner`.

```
In[4]:= Needs["Algebra`Horner`"]
```

```
In[5]:= ??Horner
```

> Horner[p] gives the polynomial or rational function of polynomials p in Horner or
> nested form with respect to the default variables Variables[p]. Horner[p, v]
> gives p in Horner form with respect to the variable or variable list v. Horner[
> `p1/p2, v1, v2] gives the rational function of polynomials p1/p2 in Horner form
> using the variables or variable lists v1 and v2 for p1 and p2 respectively. More...
> Attributes[Horner] = {Protected, ReadProtected}

For multivariate polynomials see [1399]; for intervals, see [315].

b) Here, the method is directly implemented. For the polynomials, we use the form Sum[C[i] C^i, {i, 0, n}]. (C is a built-in variable, and we use it for the coefficients as well as for the polynomial variable.) The integral on the left-hand side of the above equation can easily be carried out. Using CoefficientList, we determine the coefficients of C. Equating these coefficients to 0 or 1, we get a system of linear equations for the C[i] that are solved by Solve.

```
In[1]:= myBernoulli[n_Integer?(# >= 0&), x_] :=
        Plus @@ (Array[C[#] x^# &, n + 1, 0] /.
        Solve[(* generate and solve the equations for the coefficients *)
              Thread[(* the coefficients of the powers of C *)
                 CoefficientList[Plus @@ Array[
                 (* the integral carried out *)
                 C[#] ((C + 1)^(1 + #) -
                        C^(1 + #))/(1 + #)&, n + 1, 0], C] ==
                 (* the right-hand side *)
                 Array[If[# === n, 1, 0]&, n + 1, 0]],
              (* the coefficients *)Array[C, n + 1, 0]][[1]])
```

The calculated and built-in polynomials agree. Here is a quick check for the first few polynomials.

```
In[2]:= Union @ Table[BernoulliB[n, z] === myBernoulli[n, z], {n, 0, 12}]
```

```
Out[2]= {True}
```

c) Let the zeros of the original polynomial be xa[i] and of the new polynomial be xb[i]. Then the following relations hold.

```
In[1]:= n = 4;
        (eqs = Join[Thread[Table[a[i], {i, 0, n}] ==
                    CoefficientList[Product[x - xa[i], {i, n}], x]],
                    Thread[Table[b[i], {i, 0, n}] ==
                    CoefficientList[Product[x - xb[i], {i, n}], x]],
                    Table[xb[i] == xa[i]^2, {i, n}],
                    {Sum[b[i] x^i, {i, 0, n}] == 0}]) // Short[#, 6]&
```

```
Out[2]//Short= {a[0] == xa[1] xa[2] xa[3] xa[4],
       a[1] == -xa[1] xa[2] xa[3] - xa[1] xa[2] xa[4] - xa[1] xa[3] xa[4] - xa[2] xa[3] xa[4],
       a[2] == xa[1] xa[2] + xa[1] xa[3] + xa[2] xa[3] + xa[1] xa[4] + xa[2] xa[4] + xa[3] xa[4],
       a[3] == -xa[1] - xa[2] - xa[3] - xa[4], a[4] == 1, b[0] == xb[1] xb[2] xb[3] xb[4],
       b[1] == -xb[1] xb[2] xb[3] - xb[1] xb[2] xb[4] - xb[1] xb[3] xb[4] - xb[2] xb[3] xb[4],
       b[2] == <<1>>, b[3] == <<1>>, b[4] == 1, xb[1] == xa[1]^2, xb[2] == xa[2]^2,
       xb[3] == xa[3]^2, xb[4] == xa[4]^2, b[0] + x b[1] + x^2 b[2] + x^3 b[3] + x^4 b[4] == 0}
```

Using Eliminate, we express all xb[i] through the old coefficients xa[i].

```
In[3]:= Eliminate[eqs, Join[Table[xa[i], {i, n}], Table[xb[i], {i, n}],
                       Table[b[i], {i, 0, n}]]]
```

Out[3]= $2 x^2 a[1] a[3] + x^3 a[3]^2 ==$
$x^4 + 2 x^2 a[0] + a[0]^2 - x a[1]^2 + 2 x^3 a[2] + 2 x a[0] a[2] + x^2 a[2]^2 \&\& a[4] == 1$

So the new polynomial is given by the following expression.

In[4]:= `Collect[-Subtract @@ (%[[1]]), x]`

Out[4]= $x^4 + a[0]^2 + x (-a[1]^2 + 2 a[0] a[2]) + x^2 (2 a[0] + a[2]^2 - 2 a[1] a[3]) + x^3 (2 a[2] - a[3]^2)$

Here is a short numerical check for random coefficients.

In[5]:= `SeedRandom[999];`
`randomCoeffs = Table[a[i] -> Random[Complex, {-1 - I, 1 + I}],`
`                     {i, 0, 3}];`

`(* the squared roots of the original polynomial *)`
`Sort[x^2 /. {ToRules[`
`        NRoots[(Sum[a[i] x^i, {i, 0, 3}] + x^4 == 0) /.`
`                                 randomCoeffs, x]]}]`

Out[8]= $\{-0.747577 - 1.99359 i, -0.0935436 + 0.309006 i,$
$-0.00842973 + 0.777293 i, 1.29986 - 0.752807 i\}$

In[9]:= `(* the roots of the calculated polynomial *)`
`Sort[x /. {ToRules[NRoots[(x^4 + x^3 (2a[2] - a[3]^2) +`
`             x^2 (2a[0] + a[2]^2 - 2a[1] a[3]) +`
`         x (2a[0] a[2] - a[1]^2) + a[0]^2 == 0) /. randomCoeffs, x]]}]`
Out[10]= $\{-0.747577 - 1.99359 i, -0.0935436 + 0.309006 i,$
$-0.00842973 + 0.777293 i, 1.29986 - 0.752807 i\}$

d) This is a sketch of the function under consideration.

In[1]:= `f[x_] = -(1 + x) + 2 (1 + x^2)^(1/2) - 3 (1 + x^3)^(1/3) +`
`          5 (1 + x^5)^(1/5) - 4`
Out[1]= $-5 - x + 2 \sqrt{1 + x^2} - 3 (1 + x^3)^{1/3} + 5 (1 + x^5)^{1/5}$

In[2]:= `Plot[Evaluate[f[x]], {x, -1, 5}, AxesOrigin -> {0, 0}]`

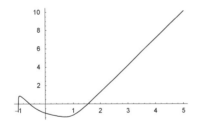

A direct call to `Solve` will not yield a result in a reasonable amount of time. (Of course, `Solve` will calculate all, not only the real, solutions.)

In[3]:= `TimeConstrained[Solve[f[x] == 0, x], 100]`

Out[3]= `$Aborted`

Also, `Reduce` cannot find the roots quickly.

In[4]:= `TimeConstrained[Reduce[Element[x, Reals] && f[x] == 0, x], 100]`

Out[4]= `$Aborted`

Using numerical techniques, such as `FindRoot`, we can find a numerical approximation for the real zeros.

In[5]:= `FindRoot[Evaluate[f[x]], {x, #}]& /@ {-0.99, -0.5, 2.}`

Out[5]= $\{\{x \to -0.999478\}, \{x \to -0.566712\}, \{x \to 1.54208\}\}$

To find a polynomial in x, which includes among its zeros the zeros of the expression under consideration, we rewrite the roots in the form in a polynomial form *squareRoot*$^2 = (1 + x^2)$, *cubeRoot*$^3 = (1 + x^3)$, and *quintRoot*$^5 = (1 + x^5)$ and eliminate the variables *squareRoot*, *cubeRoot*, and *quintRoot* from the resulting system of polynomial equations. Using `Resultant`, we eliminate these eliminations step by step.

```
In[6]:= Resultant[-(1 + x) + 2 squareRoot - 3 cubeRoot + 5 quintRoot - 4,
                   squareRoot^2 - (1 + x^2), squareRoot]
```

$\text{Out[6]}= 21 + 30 \text{ cubeRoot} + 9 \text{ cubeRoot}^2 - 50 \text{ quintRoot} - 30 \text{ cubeRoot quintRoot} + 25 \text{ quintRoot}^2 + 10 x + 6 \text{ cubeRoot } x - 10 \text{ quintRoot } x - 3 x^2$

```
In[7]:= Resultant[%, cubeRoot^3 - (1 + x^3), cubeRoot]
```

$\text{Out[7]}= 19980 - 89640 \text{ quintRoot} + 210825 \text{ quintRoot}^2 - 289250 \text{ quintRoot}^3 + 226875 \text{ quintRoot}^4 - 93750 \text{ quintRoot}^5 + 15625 \text{ quintRoot}^6 + 17928 x - 84330 \text{ quintRoot } x + 173550 \text{ quintRoot}^2 x - 181500 \text{ quintRoot}^3 x + 93750 \text{ quintRoot}^4 x - 18750 \text{ quintRoot}^5 x + 6381 x^2 - 12750 \text{ quintRoot } x^2 + 11850 \text{ quintRoot}^2 x^2 - 7500 \text{ quintRoot}^3 x^2 + 1875 \text{ quintRoot}^4 x^2 + 9370 x^3 - 13710 \text{ quintRoot } x^3 + 9750 \text{ quintRoot}^2 x^3 - 3250 \text{ quintRoot}^3 x^3 + 4365 x^4 - 7650 \text{ quintRoot } x^4 + 3825 \text{ quintRoot}^2 x^4 + 4320 x^5 - 4320 \text{ quintRoot } x^5 + 1404 x^6$

```
In[8]:= Resultant[%, quintRoot^5 - (1 + x^5), quintRoot]
```

$\text{Out[8]}= 44934823396753480354062 5 + 138037607596004585977500 0 x + 200022675682551157523625 0 x^2 + 31292103521667411829092 50 x^3 + 386330699582964000975427 5 x^4 + 497249639279067721687981 8 x^5 + 650932799579114233890865 5 x^6 + 553922134037161002246837 0 x^7 + 724805141568090695971396 5 x^8 + 852365294270416511681840 x^9 + 8194069297089616714539996 x^{10} + 9275509092199201030302210 x^{11} + 6411322724844182787916630 x^{12} + 6741386638088728450185540 x^{13} + 6950397711983936711623020 x^{14} + 3922955300222612332507320 x^{15} + 4124282347731958883301420 x^{16} + 2779454624016891927119340 x^{17} + 2690481887448230206743710 x^{18} + 2518482531261881452939650 x^{19} + 873650701758416216304 60 x^{20} - 1968098076267760883043 60 x^{21} - 1949741058666680705590 35 x^{22} + 1072676662945205011714 50 x^{23} + 2148696496900244923887 75 x^{24} - 1515727620539259059483 10 x^{25} - 1503655817767865990874 45 x^{26} - 766704016711095127761 50 x^{27} - 388887091910105111905 50 x^{28} - 106035074967194514054 00 x^{29} - 100611296033342286635 1 x^{30}$

These are the real roots greater than -1 of the resulting polynomial of degree 30.

```
In[9]:= (roots = Cases[{ToRules[Roots[% == 0, x]]},
                 {x -> _?(Im[N[#]] == 0 && # > -1&)}]) // Short[#, 12]&
```

$\text{Out[9]//Short}= \{\{x \to \text{Root}[\ll 41 \gg + 1968098076267760883043 60 \#1^{21} + 1949741058666680705590 35 \#1^{22} - 1072676662945205011714 50 \#1^{23} - 2148696496900244923887 75 \#1^{24} + 1515727620539259059483 10 \#1^{25} + 1503655817767865990874 45 \#1^{26} + 766704016711095127761 50 \#1^{27} + 388887091910105111905 50 \#1^{28} + 106035074967194514054 00 \#1^{29} + 100611296033342286635 1 \#1^{30} \&, 2]\},$
$\{x \to \text{Root}[\ll 48 \gg + 1503655817767865990874 45 \#1^{26} + 766704016711095127761 50 \#1^{27} + 388887091910105111905 50 \#1^{28} + 106035074967194514054 00 \#1^{29} + 100611296033342286635 1 \#1^{30} \&, 3]\},$
$\{x \to \text{Root}[\ll 42 \gg + 1949741058666680705590 35 \#1^{22} - 1072676662945205011714 50 \#1^{23} - 2148696496900244923887 75 \#1^{24} + 1515727620539259059483 10 \#1^{25} + 1503655817767865990874 45 \#1^{26} + 766704016711095127761 50 \#1^{27} + 388887091910105111905 50 \#1^{28} + 106035074967194514054 00 \#1^{29} + 100611296033342286635 1 \#1^{30} \&, 4]\}\}$

```
In[10]:= N[roots, 30]
```

$\text{Out[10]}= \{\{x \to -0.999478342973600816405486392030\},$
$\{x \to -0.566711964200604453002972160669\}, \{x \to 1.54208432959745310883104021458\}\}$

The roots obey the original equation. Here is a quick numerical check.

```
In[11]:= f[x /. #]& /@ %
```

$\text{Out[11]}= \{0. \times 10^{-28}, 0. \times 10^{-30}, 0. \times 10^{-29}\}$

We end with a look at the Riemann surface of the function $w = f(z)$.

```
In[12]:= (* continuation of f to other sheets *)
         fC[x_, {k2_, k3_, k5_}] =
```

```
               -(1 + x) + 2 Exp[k2 I 2Pi/2] (1 + x^2)^(1/2) -
                    3 Exp[k3 I 2Pi/3] (1 + x^3)^(1/3) +
                    5 Exp[k5 I 2Pi/3] (1 + x^5)^(1/5) - 4;
```

```
In[14]:= With[{ε = 10^-6},
         Show[GraphicsArray[
         Show[Table[ParametricPlot3D[Evaluate[{r Cos[φ], r Sin[φ],
                       Re[fC[r Exp[I φ], {k2, k3, k5}]]}]},
                     {r, 0, 2},
                     Evaluate[{φ, j 2Pi/60 + ε, (j + 1) 2Pi/60 - ε}],
                     PlotPoints -> {20, 2},
                     DisplayFunction -> Identity],
              (* show all sheets *)
              {k2, 2}, {k3, 3}, {k5, 5},
              (* show three quarters; avoid crossing branch cuts *)
              {j, 0, 44}] /. (* cut off polygon corners *) Polygon[l_] :>
              Polygon[Plus @@@ Partition[Append[l, First[l]], 2, 1]/4],
              BoxRatios -> {1, 1, 3/2}, Axes -> False]& /@
              (* show real and imaginary part *) {Re, Im}]]]
```

e) The following is a list of the 60 substitutions.

```
In[1]:= allTransformations = (x -> #)& /@ Together /@
       Flatten[{Table[ε^i x, {i, 5}], Table[-ε^i/x, {i, 5}],
                Table[ ε^j ((ε + ε^4) x + ε^i)/
                      (x - ε^i (ε + ε^4)), {i, 5}, {j, 5}],
                Table[-ε^j (x - ε^i (ε + ε^4))/
                      ((ε + ε^4) x + ε^i), {i, 5}, {j, 5}]}];
```

The original polynomial factors into three factors of degree 2, 4, and 4.

```
In[2]:= factors0 = Factor[-1 + 11 x^5 + x^10]
```
```
Out[2]= (-1 + x + x^2) (1 - 2 x + 4 x^2 - 3 x^3 + x^4) (1 + 3 x + 4 x^2 + 2 x^3 + x^4)
```

To show that the transformed equations have the same solutions as the original one, we first carry out the substitution. We do not use the explicit value of ϵ at this stage. Because $\epsilon^5 = 1$, we then reduce all terms of the form ϵ^n to $\epsilon^{n \bmod 5}$. From the resulting polynomial in x and ϵ, we eliminate ϵ by using its defining equation $\epsilon^5 - 1 = 0$. The resulting polynomial must then contain the three factors $(x^2 + x - 1)$, $(x^4 + 4 x^2 - 2 x - 3 x^3 + 1)$, and $(x^4 + 2 x^3 + 4 x^2 + 3 x + 1)$. Here, this is implemented.

```
In[3]:= Table[PolynomialGCD[Factor[Resultant[
               Numerator[Together[(-1 + 11 x^5 + x^10) /.
                        allTransformations[[i]]]] /.
                   ε^n_ :> ε^Mod[n, 5], ε^5 - 1, ε]],
             factors0] === factors0,
             {i, Length[allTransformations]}] // Union
```
```
Out[3]= {True}
```

f) The function `iteratedExponential` implements the iterated exponential.

```
In[1]:= iteratedExponential[n_] :=
          Fold[Exp[#2 z #1]&, Exp[c[n] z], Reverse[Table[c[i], {i, n - 1}]]]
```

Here is an example.

```
In[2]:= iteratedExponential[3]
```
```
Out[2]= e^{e^{e^{e^{z c[3]} z c[2]} z c[1]}}
```

To achieve the best possible matching between $1 + \ln(1 + z)$ and $\exp(c_1 z \exp(c_2 z \exp(c_3 z \cdots)))$, we will adjust the constants $c_1, c_2, \ldots, c_{10}$ in such a way that their Taylor series will agree around $z = 0$. Here are the first terms of its series expansion. The term of z^k is a polynomial in $c_1, c_2, \ldots, c_k$.

```
In[3]:= Factor //@ Series[iteratedExponential[3], {z, 0, 4}]
```

$$Out[3]= \ 1 + c[1] \ z + \frac{1}{2} \ c[1] \ (c[1] + 2 \ c[2]) \ z^2 +$$

$$\frac{1}{6} \ c[1] \ (c[1]^2 + 6 \ c[1] \ c[2] + 3 \ c[2]^2 + 6 \ c[2] \ c[3]) \ z^3 + \frac{1}{24} \ c[1] \ (c[1]^3 + 12 \ c[1]^2 \ c[2] +$$

$$24 \ c[1] \ c[2]^2 + 4 \ c[2]^3 + 24 \ c[1] \ c[2] \ c[3] + 24 \ c[2]^2 \ c[3] + 12 \ c[2] \ c[3]^2) \ z^4 + O[z]^5$$

The ith series term of $1 + \ln(1 + z)$ is given as $(-1)^{i+1}/i$.

```
In[4]:= logSeries = Series[1 + Log[1 + z], {z, 0, 10}]
```

$$Out[4]= \ 1 + z - \frac{z^2}{2} + \frac{z^3}{3} - \frac{z^4}{4} + \frac{z^5}{5} - \frac{z^6}{6} + \frac{z^7}{7} - \frac{z^8}{8} + \frac{z^9}{9} - \frac{z^{10}}{10} + O[z]^{11}$$

```
In[5]:= 1 + Sum[(-1)^(i + 1)/i z^i, {i, 1, 10}]
```

$$Out[5]= \ 1 + z - \frac{z^2}{2} + \frac{z^3}{3} - \frac{z^4}{4} + \frac{z^5}{5} - \frac{z^6}{6} + \frac{z^7}{7} - \frac{z^8}{8} + \frac{z^9}{9} - \frac{z^{10}}{10}$$

Taking into account that the term of z^k of the iterated polynomial is a polynomial in $c_1, c_2, \ldots, c_k$, it is straightforward to calculate recursively the first 10 of the c_k. Doing it recursively by backsubstituting the previously calculated values for the c_k is much faster than making a series expansion of the iterated exponential with 10 symbolic coefficients and then solving the coupled nonlinear system of equations for the c_k.

```
In[6]:= cList[n_] :=
        Module[{cRules, iExp, ser, coeff, sol},
          cRules = {c[1] -> 1};
          Do[(* iterated exponential with backsubstituted cs *)
             iExp = iteratedExponential[i] /. cRules;
             (* series to order i of the iterated exponential *)
             ser = Series[iExp, {z, 0, i}];
             (* coefficients of the power series *)
             coeff = SeriesCoefficient[ser, i];
             (* solve for the new coefficient *)
             sol = Solve[coeff == (-1)^(i + 1)/i, c[i]][[1]];
             (* collect the value of c[i] *)
             AppendTo[cRules, sol[[1]]], {i, 2, n}];
             (* return list of all c[i], {i, 1, n} *)
          cRules]
```

Here are the first 10 c_k.

```
In[7]:= cL = cList[10]
```

$$Out[7]= \ \left\{c[1] \to 1, \ c[2] \to -1, \ c[3] \to -\frac{2}{3}, \ c[4] \to -\frac{29}{48}, \right.$$

$$c[5] \to -\frac{24697}{41760}, \ c[6] \to -\frac{1220847911}{2062693440}, \ c[7] \to -\frac{12552829066927958983}{21153173808962192256},$$

$$c[8] \to -\frac{7865030956644535351853906364276597875903}{1327660875234599079505918525536641178240 0}, \ c[9] \to$$

$$-\frac{36926419075288597576903378839705784490074085320283337241694728970368715785606958 9}{62652563301875399351550911907462274845264773336660547744240415947496033437043200}$$

$$, \ c[10] \to$$

$$-5425656785819317199971611482267085601932492541176219189734044183054877004407537 0 \text{\}.}$$
$$8369975492655302281060963038439560892600017511782282334954401249855231838423737 9/$$
$$9254139234504392431376236082961962231747499376153730907912962439277552813783780 3 \text{\}.}$$
$$3432927898602630879696070751291107385876946569112066517805346571271048918396979 \text{\}.}$$
$$20\}$$

It seems that their numerical value approaches a constant.

```
In[8]:= Map[MapAt[N[#, 22]&, #, -1]&, cL]
```

$$Out[8]= \ \{c[1] \to 1.000000000000000000000, \ c[2] \to -1.000000000000000000000,$$
$$c[3] \to -0.6666666666666666666667, \ c[4] \to -0.6041666666666666666667,$$

c[5] → -0.5914032567049808429119, c[6] → -0.591870748859316680621 2,
c[7] → -0.593425326161200789659 5, c[8] → -0.592397584605690521034 1,
c[9] → -0.589384011271303678398 0, c[10] → -0.586295132192258344050 4}

Here is a quick check that the result has the series terms of $1 + \ln(1 + z)$ around $z = 0$.

```
In[9]:= lnApproximation = iteratedExponential[10] /. cL;
        Series[lnApproximation, {z, 0, 10}]
```

Out[10]= $1 + z - \dfrac{z^2}{2} + \dfrac{z^3}{3} - \dfrac{z^4}{4} + \dfrac{z^5}{5} - \dfrac{z^6}{6} + \dfrac{z^7}{7} - \dfrac{z^8}{8} + \dfrac{z^9}{9} - \dfrac{z^{10}}{10} + O[z]^{11}$

Lastly, let us look at the quality of the approximation calculated. The next picture shows the logarithm of the relative difference along the real axis (we turn off messages that would be generated due to the singularity of $\log(\xi)$ at $\xi = 0$). The right picture shows the logarithm of the absolute difference for complex values of z.

```
In[11]:= Show[GraphicsArray[
         Block[{$DisplayFunction = Identity},
         (* suppress messages from numericalization *)
         Internal`DeactivateMessages[
         {(* along the real axis *)
         Plot[Log[10, Abs[1 + Log[1 + z] - lnApproximation]/
                                  Abs[1 + Log[1 + z]]], {z, -1, 2}],
         (* over the complex plane *)
         Plot3D[Evaluate[Log[10, Abs[1 + Log[1 + z] - lnApproximation]] /.
                   z -> x + I y],
                {x, -1, 2}, {y, -2, 2}, PlotPoints -> 30,
                Mesh -> False, PlotRange -> All]}]]]]
```

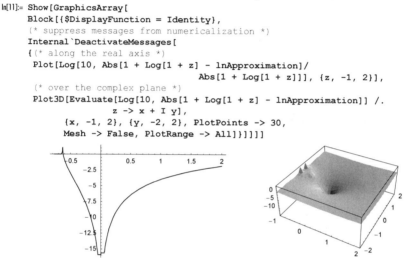

g) Running the input under consideration results in the five-dimensional identity matrix.

```
In[1]:= m = Table[1/(i + j + 1), {i, 5}, {j, 5}];
        FixedPoint[(# + Inverse[#])/2&, N[m], 100]
Out[2]= {{1., 0., 0., 0., 0.}, {0., 1., 0., 0., 0.},
         {0., 0., 1., 0., 0.}, {0., 0., 0., 1., 0.}, {0., 0., 0., 0., 1.}}
```

The fixed points of the map $x \to (x + x^{-1})/2$ are the values $x \to \pm 1$. This means that $x \to (x + x^{-1})/2$ represents a matrix version of the Sign function [472], [843]. To calculate the matrix version of a function, we decompose the matrix into its spectral components.

```
In[3]:= {ε, 𝒱} = Eigensystem[m];
```

We introduce the abbreviation $\mathcal{R}$ for a root object and write the kth eigenvalues and kth eigenvector in a form that makes the dependence on k explicit.

```
In[4]:= {ε, 𝒱} =
        With[{𝑅 = Root[(-1 + 5603255 # - 246181488000 #^2 + 191979048240000 #^3 -
                   4545067645440000 #^4 + 5175372787200000 #^5)&, #]&},
        {(* eigenvalues *)
        Table[𝑅[k], {k, 5}],
        (* eigenvectors *)
        Table[{7/66 (1 - 1206272 𝑅[k] + 8667711360 𝑅[k]^2 - 767331532800 𝑅[k]^3 +
                   1408264704000 𝑅[k]^4),
                 28/33 (-1 + 643230 𝑅[k] - 1974974400 𝑅[k]^2 + 44008272000 𝑅[k]^3),
                 42/55 (+3 - 682640 𝑅[k] - 397252800 𝑅[k]^2 + 39118464000 𝑅[k]^3),
                 28/11 (-1 - 92260 𝑅[k] + 158760000 𝑅[k]^2 + 9779616000 𝑅[k]^3),
                 1 + 346080 𝑅[k] + 789566400 𝑅[k]^2 + 21337344000 𝑅[k]^3}/
                 (1 + 346080 𝑅[k] + 789566400 𝑅[k]^2 + 21337344000 𝑅[k]^3), {k, 5}]}];
```

The symbolic spectral decomposition contains many Root-objects.

```
In[5]:= Count [{ε, ν}, _Root, Infinity]

Out[5]= 130
```

The eigenvalues of m give rise to the above-shown identity matrix.

```
In[6]:= Sign [DiagonalMatrix[ε]]

Out[6]= {{1, 0, 0, 0, 0}, {0, 1, 0, 0, 0}, {0, 0, 1, 0, 0}, {0, 0, 0, 1, 0}, {0, 0, 0, 0, 1}}
```

For a hermitian matrix, we know that the eigenvectors are orthogonal and the decomposition $\mathcal{V}^T.\mathcal{E}.\mathcal{V}$ holds. Here, $\mathcal{V}$ is the matrix of the normalized eigenvectors and $\mathcal{E}$ is the diagonal matrix of the eigenvalues. Normalizing the eigenvectors we can use $f(m) = \mathcal{V}^T.f(\mathcal{D}(\mathcal{E})).\mathcal{V}$ to verify that the above fixed point ($\mathcal{D}(\mathcal{E})$ is the diagonal matrix of the eigenvalues $\mathcal{E}$).

```
In[7]:= νn = RootReduce [#/RootReduce [Sqrt [#.#]]]& /@
                  (RootReduce [Simplify [#]]& /@ ν);

In[8]:= Transpose [νn] . Sign [DiagonalMatrix[ε]] .νn;

In[9]:= Table [RootReduce [%[[i, j]]], (* use symmetry *) {i, 5}, {j, i, 5}]

Out[9]= {{1, 0, 0, 0, 0}, {1, 0, 0, 0}, {1, 0, 0}, {1, 0}, {1}}
```

h) This is the polynomial to be expanded.

```
In[1]:= poly = (x + 1)^2000 (x^2 + x + 1)^1000 (x^4 + x^3 + x^2 + x + 1)^500;
```

To obtain the coefficients of the polynomial we could use a rule-based approach for expansion. But this would be relatively slow. A fast, nonalgebraic way to get the list of coefficients is to transform the polynomial into a Taylor series. This uses the built-in functions Series and SeriesData.

```
In[2]:= (cl = (poly + O[x]^6001) [[3]]); // Timing

Out[2]= {153.83 Second, Null}
```

The next graphic visualizes the magnitude of the coefficients.

```
In[3]:= ListPlot [Log [10, N[cl]] // N, PlotRange -> All]
```

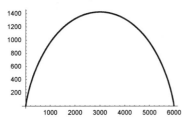

i) The following graphic shows the three roots as a function of c (because c appears in the polynomial quadratically we can restrict to the case $c > 0$).

```
In[1]:= Show [Graphics3D [{PointSize[0.01],
         Table [roots = {Re[#], Im[#]}& /@  (z /. NSolve [z^3 + c^2 z + 1 == 0, z]);
         Point [Append [#, c]]& /@ roots, {c, 0, 6, 6/200}]}],
              BoxRatios -> {1, 1, 1}, Axes -> True, ViewPoint -> {2, 2, 1}]
```

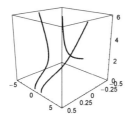

A graphic of the numerically calculated minimal distance $d(c)$ between the three roots shows a clear minimum around $c \approx \pm 1.1$.

```
In[2]:= δMin[c_] :=
        With[{roots = Last /@ (List @@ NRoots[x^3 + c^2 x + 1 == 0, x])},
            Min[Abs[Flatten[Table[roots[[i]] - roots[[j]],
                                {i, 3}, {j, i - 1}]]]]]
```

```
In[3]:= Plot[δMin[c], {c, -5, 5}]
```

To calculate an exact value for the minimal distance we express the distance between the roots in a polynomial way. To do this we separate real and imaginary parts and find a polynomial $p(c, d) = 0$.

```
In[4]:= z = x + I y;
        p = Expand[z^3 + c^2 z + 1];
```

```
In[6]:= {im = Plus @@ Cases[p, _Complex _] // Factor, re = p - im // Factor}
```

Out[6]= $\{i\,y\,(c^2 + 3\,x^2 - y^2),\ 1 + c^2\,x + x^3 - 3\,x\,y^2\}$

Now, we eliminate the remaining variables x1, x2, y1, y2, and Z and obtain one polynomial equation in c and d.

```
In[7]:= gb = Factor[GroebnerBasis[Flatten[
            {(* the two roots *)
            {re, im} /. {x -> x1, y -> y1}, {re, im} /. {x -> x2, y -> y2},
            (* the roots are different *) 1 - Z (x1 - x2),
            (* distance between the two roots *)
            d^2 - (x1 - x2)^2 - (y1 - y2)^2}],
            {c, d}, {x1, x2, y1, y2, Z}, MonomialOrder -> EliminationOrder]][[1]]
```

Out[7]= $-d^2\,(27 + 4\,c^6 - 3\,c^2\,d^4 + d^6)\,(-27 - 4\,c^6 + 3\,c^2\,d^4 + d^6)\,(27 + 4\,c^6 + 9\,c^4\,d^2 + 6\,c^2\,d^4 + d^6)$

Viewing d as a function of c gives the two equations $p(c, d) = 0$, $\partial p(c, d(c))/\partial c = 0$.

```
In[8]:= gbD = Factor[D[gb/d^2 /. d -> d[c], c] /. d'[c] -> 0 /. d[c] -> d]
```

Out[8]= $12\,c\,(2\,c^2 + d^2)\,(2187\,c^2 + 648\,c^8 + 48\,c^{14} + 486\,c^6\,d^2 + 72\,c^{12}\,d^2 -$
$243\,c^4\,d^4 - 36\,c^{10}\,d^4 - 189\,c^2\,d^6 - 82\,c^8\,d^6 - 27\,d^8 - 4\,c^6\,d^8 + 21\,c^4\,d^{10} + 2\,c^2\,d^{12} - d^{14})$

Using the numerical solution of the system allows us to extract the corresponding symbolic solution.

```
In[9]:= FindRoot[Evaluate[{gb == 0, gbD == 0}], {c, 1}, {d, 1.5}]
```

Out[9]= $\{c \to 1.09112,\ d \to 1.54308\}$

```
In[10]:= {#, N[#]}& @ Select[{c, d} /. Solve[{gb == 0, gbD == 0}, {c, d}],
                        # == {1.09112`5, 1.54308`5}&]
```

Out[10]= $\left\{\left\{\left\{\dfrac{\sqrt{3}}{2^{2/3}},\ \dfrac{\sqrt{3}}{2^{1/6}}\right\}\right\},\ \{\{1.09112,\ 1.54308\}\}\right\}$

Analytical bounds on the minimal distance between roots can be found in [1256], [1262] and [1563].

j) This is the definition of the functions of the $f^{(k)}(z)$.

```
In[1]:= f[1, z_] := z^2 + c
        f[n_, z_] := f[1, f[n - 1, z]]
```

The most difficult calculation of this exercise is the calculation of the c_k. Because c_k is a rational function of all roots of a polynomial, it will be a rational function of the coefficients of the original polynomial. This means that it will be a rational function in c. To avoid the explicit calculation of any expressions containing radicals or/and roots of c, we use the function RootReduce in the following definition.

```
In[3]:= c[k_] := RootSum[Function[x, Evaluate[f[k, x] - x]],
                Function[x, Evaluate[1/(#^2 - #)&[
                        D[f[k, x], x]]]]]] // Simplify //
                                Together // Factor
```

The remaining steps are straightforward and show the identity of the first four nontrivial terms.

```
In[4]:= o = 4;
        Sum[z^k/k c[k], {k, o}]
```

$$Out[5]= -\frac{z}{2c} + \frac{(-1+c)\,z^2}{8\,c^2\,(1+c)} - \frac{(1-2c-c^2+c^3)\,z^3}{24\,c^3\,(1+c+2\,c^2+c^3)} + \frac{(-1+2c+3\,c^2+c^3-3\,c^4-6\,c^5-4\,c^6+c^7+c^8)\,z^4}{64\,c^4\,(1+c)^2\,(1+2\,c^2+3\,c^3+3\,c^4+3\,c^5+c^6)}$$

```
In[6]:= Exp[-% + O[z]^(o + 1)] // Simplify
```

$$Out[6]= 1 + \frac{z}{2c} + \frac{z^2}{4\,c^2+4\,c^3} + \frac{z^3}{8\,c^3\,(1+c)\,(1+c+2\,c^2+c^3)} +$$

$$\frac{z^4}{16\,c^4\,(1+c)^2\,(1+c+2\,c^2+c^3)\,(1+2\,c^2+3\,c^3+3\,c^4+3\,c^5+c^6)} + O[z]^5$$

```
In[7]:= 1 + Sum[(z/2)^k/Product[f[j, 0], {j, k}], {k, o}] + O[z]^(o + 1) //
                        MapAll[Together[Factor[#]]&, #]&
```

$$Out[7]= 1 + \frac{z}{2c} + \frac{z^2}{4\,c^2\,(1+c)} + \frac{z^3}{8\,c^3\,(1+2\,c+3\,c^2+3\,c^3+c^4)} +$$

$$\frac{z^4}{16\,c^4\,(1+c)^2\,(1+c+4\,c^2+6\,c^3+10\,c^4+14\,c^5+13\,c^6+10\,c^7+5\,c^8+c^9)} + O[z]^5$$

k) While a straightforward implementation of the $c_{e,d}$ is possible, it suffers from being very slow.

```
In[1]:= derivativeValuesL[m_] :=
        Table[Limit[D[(x/Sin[x])^e, {x, d}], x -> 0], {e, m}, {d, 0, m}]
In[2]:= derivativeValuesL[6] // Timing
```

$$Out[2]= \{15.09\ \text{Second},$$

$$\{\{1, 0, \frac{1}{3}, 0, \frac{7}{15}, 0, \frac{31}{21}\}, \{1, 0, \frac{2}{3}, 0, \frac{8}{5}, 0, \frac{160}{21}\}, \{1, 0, 1, 0, \frac{17}{5}, 0, \frac{457}{21}\},$$

$$\{1, 0, \frac{4}{3}, 0, \frac{88}{15}, 0, \frac{992}{21}\}, \{1, 0, \frac{5}{3}, 0, 9, 0, \frac{1835}{21}\}, \{1, 0, 2, 0, \frac{64}{5}, 0, \frac{3056}{21}\}\}\}$$

To avoid repeating any work, and using the fact that $x/\sin(x)$ has a Taylor series starting with $1 + x^2/6 + \cdots$, we can implement the following much faster function `derivativeValuesS`. We first form all series of the $x/\sin(x)$ recursively and then calculate the value of the oth derivative at $x = 0$ from the coefficient of x^o in the Taylor series.

```
In[3]:= derivativeValuesS[m_] :=
        Module[{x},
        Table[Table[d! #[[d + 1]], {d, 0, m}]&[
                CoefficientList[#[[e]], x]], {e, m}]&[
                        Function[s, NestList[s #&, s, m - 1]][
                                Series[x/Sin[x], {x, 0, m}]]]]
```

This resulted in a thousand times faster calculation of the $c_{e,o}$.

```
In[4]:= derivativeValuesS[6] // Timing
```

$$Out[4]= \{0.01\ \text{Second},$$

$$\{\{1, 0, \frac{1}{3}, 0, \frac{7}{15}, 0, \frac{31}{21}\}, \{1, 0, \frac{2}{3}, 0, \frac{8}{5}, 0, \frac{160}{21}\}, \{1, 0, 1, 0, \frac{17}{5}, 0, \frac{457}{21}\},$$

$$\{1, 0, \frac{4}{3}, 0, \frac{88}{15}, 0, \frac{992}{21}\}, \{1, 0, \frac{5}{3}, 0, 9, 0, \frac{1835}{21}\}, \{1, 0, 2, 0, \frac{64}{5}, 0, \frac{3056}{21}\}\}\}$$

The calculation for $m = 100$ now takes less than a minute.

```
In[5]:= (cs = derivativeValuesS[100]); // Timing
```

$$Out[5]= \{3.4\ \text{Second}, \text{Null}\}$$

The following graphic shows the logarithm of the $c_{e,d}$ for odd d.

```
In[6]:= ListPlot3D[Log[10, MapIndexed[If[EvenQ[#2[[1]]],
                                #1]&, #1]& /@ cs],
                MeshRange -> {{1, 100}, {0, 100}}, Mesh -> False]
```

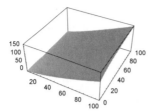

For a semi-closed form of the $c_{e,d}$, see [1336].

l) It is straightforward to write a one-liner that calculates the first n Appell–Nielsen polynomials. To avoid an explicit counter n, we operate recursively on a two-element list $\{\pm 1, poly\}$.

```
In[1]:= AppellNielsenPolynomialList[n_, z_] :=
          (* extract polynomials *) Last /@
          NestList[Function[{s, p}, {-s, Function[I, I /. (* use symmetry *)
                    Solve[-s I == (I /. z -> -(z + 1)), C][[1]] /. C -> 0][
                    (* use recursion *) Integrate[p, z] + C]}] @@ #&, {1, 1}, n]
```

The next plot shows the logarithm of the absolute value of the first 36 Appell–Nielsen polynomials. The steep vertical cusps are the zeros of the polynomials. While the majority of the roots seem to have identical numerical values, most of them are actually slightly different.

```
In[2]:= With[{o = 35},
          Plot[Evaluate[Log @ Abs @ AppellNielsenPolynomialList[o, z]],
               {z, -6, 6}, PlotRange -> {-45, 5}, PlotPoints -> 200,
               Frame -> True, Axes -> False,
               PlotStyle -> Table[{Thickness[0.002], Hue[0.8 k/o]},
                              {k, o + 1}]]]
```

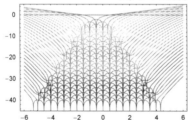

m) Evaluating the multiple integral of $f^{(n)}(x + h_n)$ for a given integer n yields (up to sign and the term $f(x + h)$) the first $n - 1$ terms of the Taylor series. Because the integrand at each integration stage is a complete differential, the n iterated integrations can all be carried out completely. The following function `TaylorTerms` implements the multiple integral (we use the same integration variable h for each integration).

```
In[1]:= TaylorTerms[n_, {f_, x_, h_}] := Expand[f[x + h] -
                    Nest[Integrate[#, {h, 0, h}]&, Derivative[n][f][x + h], n]]
```

And here are the first ten Taylor terms obtained from integration for a not explicitly specified function f.

```
In[2]:= TaylorTerms[10, {f, x, h}]
```

$$Out[2]= f[x] + h \, f'[x] + \frac{1}{2} h^2 \, f''[x] + \frac{1}{6} h^3 \, f^{(3)}[x] + \frac{1}{24} h^4 \, f^{(4)}[x] +$$
$$\frac{1}{120} h^5 \, f^{(5)}[x] + \frac{1}{720} h^6 \, f^{(6)}[x] + \frac{h^7 \, f^{(7)}[x]}{5040} + \frac{h^8 \, f^{(8)}[x]}{40320} + \frac{h^9 \, f^{(9)}[x]}{362880}$$

The next input expands $f(x) = \cos(x)$ around $x = \pi$.

```
In[3]:= TaylorTerms[10, {Cos, Pi, h}]
```

$$Out[3]= -1 + \frac{h^2}{2} - \frac{h^4}{24} + \frac{h^6}{720} - \frac{h^8}{40320}$$

For direct integral analogues of the Taylor formula, see [1358].

n) Here is the above formula implemented as the function `GeneralizedTaylorExpansion`.

```
In[1]:= GeneralizedTaylorExpansion[f_, φs_, x_, x0_] :=
    Module[{n = Length[φs], W, 𝔰},
      (* the Wronskian *)
      W = Det[Table[D[φs, {x, k}], {k, 0, n - 1}]] /. x -> x0;
      (* the second determinant *)
      𝔰 = Det[Join[{Prepend[φs, 0]},
            Table[D[Prepend[φs, f], {x, k}], {k, 0, n - 1}] /. x -> x0]];
      (* the approximation *) -Expand[𝔰/W]]
```

For small *m* (say $m \le 10$), it works fine.

```
In[2]:= ExpBasis[m_] := Table[Exp[x/k], {k, m}]
    GeneralizedTaylorExpansion[Cos[x], ExpBasis[8], x, 0] // Timing
```

$$\text{Out[3]= } \left\{ 0.27 \text{ Second}, \ -\frac{173539328\,e^{x/8}}{63} + \frac{631657481\,e^{x/7}}{72} - 10614240\,e^{x/6} + \right.$$
$$\left. \frac{435546875\,e^{x/5}}{72} - \frac{14643200\,e^{x/4}}{9} + \frac{1431027\,e^{x/3}}{8} - \frac{47840\,e^{x/2}}{9} + \frac{5135\,e^x}{504} \right\}$$

Due to the calculation of a determinant with symbolic entries, this form is not suited for larger *m*. The Taylor-like approximation carried out by `GeneralizedTaylorExpansion` cancels the leading monomial terms in a classical Taylor expansion of $f(x) - \sum_{k=0}^{n} c_k\,\varphi_k(x)$ around $x = x_0$. We can so reformulate the problem to the determination of the c_k. Assuming no additional degeneracy and the presence of all monomials, the following function `GeneralizedTaylor⁝ Expansion1` solves for the c_k and returns the resulting sum.

```
In[4]:= GeneralizedTaylorExpansion1[f_, φs_, x_, x0_] :=
    Module[{n = Length[φs], vars = Table[C[k], {k, Length[φs]}], sol},
        sol = Solve[# == 0& /@ CoefficientList[Series[f - vars.φs,
                                    {x, x0, n - 1}], x], vars];
        vars.φs /. sol[[1]]]
```

If all the c_k are numbers, the resulting linear system can be solved much more quickly then the above symbolic determinant.

```
In[5]:= GeneralizedTaylorExpansion1[Cos[x], ExpBasis[8], x, 0] // Timing
```

$$\text{Out[5]= } \left\{ 0.05 \text{ Second}, \ -\frac{173539328\,e^{x/8}}{63} + \frac{631657481\,e^{x/7}}{72} - 10614240\,e^{x/6} + \right.$$
$$\left. \frac{435546875\,e^{x/5}}{72} - \frac{14643200\,e^{x/4}}{9} + \frac{1431027\,e^{x/3}}{8} - \frac{47840\,e^{x/2}}{9} + \frac{5135\,e^x}{504} \right\}$$

Now, we can also deal with $m = 25$.

```
In[6]:= fApprox[x_] = GeneralizedTaylorExpansion1[Cos[x], ExpBasis[25], x, 0];
```

Some of the resulting numbers have up to 50 digits.

```
In[7]:= {#, N[#]}& @ Max[Abs[{Numerator[#], Denominator[#]}& /@
        Cases[fApprox[x], _Integer | _Rational, {0, Infinity}]]]
```

$$\text{Out[7]= } \{2308198100282732312393818574491888284683227539062}5, \ 2.3082 \times 10^{49} \}$$

The left graphic shows the function $f(x) = \cos(x)$ and its approximation $f\text{Approx}[x]$. The right graphic shows the logarithm of the absolute error between $f(x)$ and its approximation.

```
In[8]:= dataApprox = Table[{x, fApprox[N[x, 100]]} // N, {x, -5Pi, 5Pi, 10Pi/201}];

In[9]:= (* avoid messages from points where the approximation is basically exact *)
    Off[Graphics::gptn];

    Show[GraphicsArray[
    Block[{$DisplayFunction = Identity},
      {(* Cos[x] and approximation *)
      Show[{Plot[Cos[x], {x, -5Pi, 5Pi}, PlotStyle -> {Hue[0]}],
        ListPlot[dataApprox // N, PlotRange -> {-2, 2}, PlotJoined -> True]}],
      (* logarithm of absolute error *)
      ListPlot[N[{#1, Log[10, Abs[#2 - Cos[#1]]]}]& @@@ dataApprox,
                PlotRange -> All, PlotJoined -> True]}]]]
```

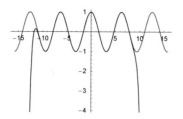
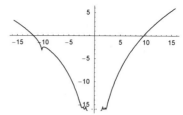

o) `PainlevéODEVIS` is the differential operator for the special Painlevé VI equation under consideration.

```
In[1]:= PainlevéODEVIS[w_, z_] := D[w, z, z] -
        (1/2 (1/w + 1/(w - 1) + 1/(w - z)) D[w, z]^2 -
        (1/z + 1/(z - 1) + 1/(w - z)) D[w, z] +
        1/2 w (w - 1)(w - z)/(z^2 (z - 1)^2)(4 + z (z - 1)/(w - z)^2))
```

The function `yChazy` is the proposed solution.

```
In[2]:= yChazy[z_] = With[{u = (c1 f[1][#] + c2 f[2][#])&},
        1/8 ((u[z] + 2z u'[z])^2 - 4z u'[z]^2)^2/
        (u[z] u'[z] (2(z - 1) u'[z] + u[z])(u[z] + 2z u'[z]))];
```

Substituting now `yChazy[z]` into `PainlevéODEVIS`, replacing the second and third derivatives of $f[k]$ by using its defining differential equation, and simplifying the result shows that `yChazy[z]` is a solution.

```
In[3]:= Together[PainlevéODEVIS[yChazy[z], z] //.
        {f[k_]'''[z] :> (8 (1 - 2z) f[k]''[z] - 9 f[k]'[z])/(4 z(z - 1)),
         f[k_]''[z]  :> (4 (1 - 2z) f[k]'[z] - f[k][z])/(4 z(z - 1))}]
Out[3]= 0
```

We end by remarking that the explicit solution of $u(z)$ is $u(z) = c_1 K(z) + c_2 K(1 - z)$ where K is the complete elliptic integral of the first kind.

```
In[4]:= Together[z (1 - z) f''[z] + (1 - 2z) f'[z] - f[z]/4 /.
        f -> Function[z, c[1] EllipticK[z] + c[2] EllipticK[1 - z]]]
Out[4]= 0
```

3. Nested Integration, `Derivative[-n]`, `PowerFactor`, Rational Painlevé II Solutions

a) First, we look at the actual result.

```
In[1]:= f[n_][x_] := Integrate[f[n - 1][x - z], {z, 0, x}]

        f[0][x_] = Exp[-x];

        f[2][x]
Out[3]= 1/2 (-x Cosh[x] + Sinh[x] + x Sinh[x])
```

We compare it to the following.

```
In[4]:= fn[2][x] = Nest[Integrate[# /. {x -> x - z}, {z, 0, x}]&, Exp[-x], 2]
Out[4]= -1 + e^{-x} + x
```

```
In[5]:= Expand[TrigToExp[fn[2][x] - f[2][x]]]
Out[5]= -1 + (5 e^{-x})/4 - e^x/4 + x + (e^{-x} x)/2
```

The reason for this in the first moment unexpected result is that `Integrate` does not localize its integration variable. (It is impossible for `Integrate` to do this because it has no `HoldAll` attribute, and so it cannot avoid the evaluation of all its arguments before `Integrate` can go to work.) So the integration variables are not screened from each other in nested integrations. Here is what happens in detail by calculating `f[2][x]`.

```
f[2][x] → Integrate[f[2 - 1][x - z], {z, 0, x}]
```

Now the two variables (from a mathematical point of view—dummy variables) z interfere.

```
f[1][x - z] → Integrate[f[0][(x - z) - z], {z, 0, (x - z)}]

Integrate[f[0][(x - z) - z], {z, 0, x}] → Integrate[Exp[-((x - z) - z)], {z, 0, x}]

Integrate[Exp[-((x - z) - z)], {z, 0, x}] → Exp[x - 2 z]/2 - Exp[-x]/2

Integrate[Exp[x - 2z]/2 - Exp[-x]/2, {z, 0, x}] → Exp[x]/4 - (1 + 2 x)/4 Exp[-x]
```

By using On[], we could follow all of the above steps in more detail, but because of the extensive output, we do not show it here.

To screen the integration variables in nested integrations, we could, for instance, use the following construction for the function definiteIntegrate. (We implement it here only for 1D integrals—the generalization to multidimensional integrals is obvious.)

```
In[6]:= SetAttributes[definiteIntegrate, HoldAll]

       definiteIntegrate[integrand_, {iVar_, lowerLimit_, upperLimit_}] :=
       Function[x, Integrate[#, {x, lowerLimit, upperLimit}]]& @@
       (* avoid evaluation of integrand; substitute new integration variable *)
               (Hold[integrand] //. iVar -> x)][
       (* create a unique integration variable *) Unique[x]]
```

(Note that definiteIntegrate has the attribute HoldAll and that an additional Hold on the right-hand side is necessary to avoid any evaluation. A unique integration variable is created via Unique[x].)

Using the function definiteIntegrate in the recursive definition of f now gives the "expected" result.

```
In[8]:= f1[n_][x_] := definiteIntegrate[f1[n - 1][x - z], {z, 0, x}]

       f1[0][x_] = Exp[-x];
```

Now, we get from f[2][x] the expected result.

```
In[10]:= f1[2][x]
```

Out[10]= $-1 + e^{-x} + x$

```
In[11]:= f1[2][x] - fn[2][x] // TrigToExp // Expand
```

Out[11]= 0

For the simple example under consideration, we could use a simpler way of creating different dummy integration variables. Here is an example.

```
In[12]:= f2[n_][x_] := Integrate[f2[n - 1][x - z[x]], {z[x], 0, x}]

       f2[0][x_] = Exp[-x];

       f2[2][x]
```

Out[14]= $-1 + e^{-x} + x$

b) Obviously, Integrate[f, x] + Integrate[g, x] and Integrate[$f + g$, x] can only differ by an x-independent constant. It turns out that finding a pair of functions f and g is not difficult; low-degree polynomials and powers already do the job.

```
In[1]:= Integrate[(1 + x)^2, x] + Integrate[x^α, x]
```

Out[1]= $x + x^2 + \dfrac{x^3}{3} + \dfrac{x^{1+\alpha}}{1 + \alpha}$

```
In[2]:= Integrate[(1 + x)^2 + x^α, x]
```

Out[2]= $\dfrac{1}{3}(1 + x)^3 + \dfrac{x^{1+\alpha}}{1 + \alpha}$

```
In[3]:= % - %% // Expand
```

Out[3]= $\dfrac{1}{3}$

Now, let us deal with the definite integrals. The function f should have a discontinuity at x_m. We choose the branch cut of the square root function as the discontinuity. We take x_l and x_u on opposite sides of the branch cut and x_m directly on the branch cut.

```
In[4]:= Integrate[Sqrt[z], {z, -1 - I, -1 + I}]
```

$$\text{Out[4]}= \frac{4\,\mathrm{i}}{3} - \frac{2}{3}\,(-1 - \mathrm{i})^{3/2} + \frac{2}{3}\,(-1 + \mathrm{i})^{3/2}$$

```
In[5]:= Integrate[Sqrt[z], {z, -1 - I, 0}] + Integrate[Sqrt[z], {z, 0, -1 + I}]
```

$$\text{Out[5]}= -\frac{2}{3}\,(-1 - \mathrm{i})^{3/2} + \frac{2}{3}\,(-1 + \mathrm{i})^{3/2}$$

```
In[6]:= % - %% // Expand
```

$$\text{Out[6]}= -\frac{4\,\mathrm{i}}{3}$$

c) First, the input adds a new rule to `Derivative` (which does not have the attribute `Protected`) for a negative integer argument for an arbitrary function. Now, we look at the actual code. `With` evaluates its first argument, which means the local variable `pI`, and sets the value to `Integrate[f[C], C]`. In the case that the result does not contain `Integrate`, `pI` becomes a pure function by substituting `Slot[1]` for `C` and adding the head `Function`. The whole expression so constructed again has a `Derivative` wrapped around it, but with the order incremented by one. In summary, this means that taking a `Derivative` of negative order n is interpreted as an iterated n-fold integration. Let us look at some examples.

```
In[1]:= Derivative[i_Integer?Negative][f_] :=
          (* because the test is the whole calculation, use With and
             then use pI as test and as the result *)
          With[{pI = Integrate[f[C], C]},
               (* test if Integrate appears in result *)
               Derivative[i + 1][Function[pI] /. C -> #] /;
               FreeQ[pI, Integrate, {0, Infinity}]]
```

```
In[2]:= Derivative[-3][Exp]
```

$$\text{Out[2]}= e^{\#1}\,\&$$

```
In[3]:= Derivative[-3][#^3 + Sin[#]&]
```

$$\text{Out[3]}= \frac{\#1^6}{120} + \text{Cos}[\#1]\,\&$$

Here are the two derivatives `Derivative[+4][Exp[1 #]&]` and `Derivative[-4][Exp[1 #]&]`.

```
In[4]:= {Derivative[+4][Exp[1 #]&], Derivative[-4][Exp[1 #]&]}
```

$$\text{Out[4]}= \{e^{\#1}\,\&,\ e^{\#1}\,\&\}$$

Actually, the rule we added to `Derivative` is already present in *Mathematica*. `Derivative[`*negativeInterger*`][`*f*`]` is treated as the n-fold integral of the pure function f.

```
In[5]:= Derivative[i_Integer?Negative][f_] =.
```

```
In[6]:= Derivative[-3][#^3 + Sin[#]&]
```

$$\text{Out[6]}= \text{Cos}[\#1] + \frac{\#1^6}{120}\,\&$$

d) In the case that $f(x, y)$ factors in the form $g(x)\,h(y)$, this is not possible. For a generic $f(x, y)$, the difference between `D[Integrate[f(x,y), x], y]` and `Integrate[D[f(x,y), y], x]` is a constant with respect to the integration variable x; this means it is a function of y. It remains to find a function such that the two integrals are doable, meaning that they both can be done in closed form. There are many such functions. Here are two examples.

```
In[1]:= f[x_, y_] := Log[x + Exp[y]];
          {D[Integrate[f[x, y], x], y], Integrate[D[f[x, y], y], x]}
```

$$\text{Out[2]}= \{e^y + e^y\,\text{Log}[e^y + x],\ e^y\,\text{Log}[e^y + x]\}$$

```
In[3]:= %[[1]] - %[[2]] // Together // Factor
```

$$\text{Out[3]}= e^y$$

```
In[4]:= f[x_, y_] := Log[x + y];
          (D[Integrate[f[x, y], x], y] - Integrate[D[f[x, y], y], x]) //
                                              Together // Factor
```

Out[5]= 1

e) The following replacement rule `rulePower` converts products of (explicit) radicals into one radical, with the base having only integer powers.

```
In[1]:= rulePower = t:_Times? (MemberQ[#, Power[_, _Rational]]&) :>
        Module[{product = List @@ t, rads, rest},
                 (* select the radicals *)
                 rads = Cases[product, Power[_, _Rational], {1}];
                 rest = Complement[product, rads];
                 (* the new exponent *)
                 exp = LCM @@ Denominator[Last /@ rads];
                 (Times @@ rads^exp)^(1/exp) (Times @@ rest)];
```

Here is an example showing `rulePower` at work.

```
In[2]:= a^(2/3) b^(3/4) c^(4/5) (d + e)^(5/6) f^(1/n) g /. rulePower
```

$$Out[2]= \left(a^{40} \, b^{45} \, c^{48} \, (d+e)^{50}\right)^{1/60} f^{\frac{1}{n}} \, g$$

The rule `ruleLogSum` rewrites sums of logarithms as one logarithm.

```
In[3]:= ruleLogSum = p:_Plus :>
        Module[{sum = List @@ p, logs, rest},
                 (* select the logarithms *)
                 logs = Cases[sum, _Log, {1}];
                 rest = Complement[sum, logs];
                 Plus[Sequence @@ rest, Log[Times @@ (First /@ logs)]]];
```

Here is an example. The term −Log[c] has the head `Times` and is not matched by the rule `ruleLogSum`.

```
In[4]:= Log[a] + Log[b] - Log[c] /. ruleLogSum
```

$$Out[4]= Log[a\,b] - Log[c]$$

The rule `ruleLogProduct` rewrites products involving logarithms.

```
In[5]:= ruleLogProduct = c_ Log[a_] :> Log[a^c];
```

Now terms of the form −Log[c] are rewritten too.

```
In[6]:= 1 - Log[a] + Log[b] Log[c] /. ruleLogProduct
```

$$Out[6]= 1 + Log\left[\frac{1}{a}\right] + Log[b^{Log[c]}]$$

The rule `ruleLogProduct` rewrites products involving logarithms.

```
In[7]:= ruleLogPower = Log[a_]^e_ :> Log[a^(Log[a]^(e - 1))];
```

Here is an example.

```
In[8]:= Log[a]^3 /. ruleLogPower
```

$$Out[8]= Log\left[a^{Log[a]^2}\right]$$

Now, we put all rules together in the function `PowerFactor`. To make sure that every rule gets applied whenever possible, we use `ReplaceRepeated` and `MapAll`.

```
In[9]:= PowerFactor[expr_] := MapAll[(# //. rulePower //. ruleLogSum //.
                              ruleLogProduct //. ruleLogPower)&, expr]
```

Here is `PowerFactor` applied to a more complicated input.

```
In[10]:= 1 + a^(1/3) b^(2/3) c /d^(5/3) (z^3)^(1/2) + Log[s^2] +
         ((Log[x] + Log[z^2])^2 + 1)^(1/2) +
         3(Log[a] - Log[b] Log[c]) + Log[x]^3 Log[y]^3
```

$$Out[10]= 1 + \frac{a^{1/3} \, b^{2/3} \, c \, \sqrt{z^3}}{d^{5/3}} + 3\,(Log[a] - Log[b]\,Log[c]) +$$
$$Log[s^2] + Log[x]^3 \, Log[y]^3 + \sqrt{1 + (Log[x] + Log[z^2])^2}$$

```
In[11]:= PowerFactor[%]
```

Out[11]= $1 + c \left(\dfrac{a^2 \, b^4 \, z^9}{d^{10}}\right)^{1/6} + \text{Log}\left[a^3 \, b^{\text{Log}\left[\frac{1}{c^3}\right]} \, s^2 \, \left(x^{\text{Log}[x^{\text{Log}[x]}]}\right)^{\text{Log}\left[y^{\text{Log}[y^{\text{Log}[y]}]}\right]}\right] + \sqrt{1 + \text{Log}\left[(x \, z^2)^{\text{Log}[x \, z^2]}\right]}$

PowerExpand rewrites the expression in the opposite direction.

In[12]:= **PowerExpand[%]**

Out[12]= $1 + \dfrac{a^{1/3} \, b^{2/3} \, c \, z^{3/2}}{d^{5/3}} + 3 \, \text{Log}[a] - 3 \, \text{Log}[b] \, \text{Log}[c] +$

$2 \, \text{Log}[s] + \text{Log}[x]^3 \, \text{Log}[y]^3 + \sqrt{1 + (\text{Log}[x] + 2 \, \text{Log}[z])^2}$

PowerFactor recovers the above expression.

In[13]:= **PowerFactor[%]**

Out[13]= $1 + c \left(\dfrac{a^2 \, b^4 \, z^9}{d^{10}}\right)^{1/6} + \text{Log}\left[a^3 \, b^{\text{Log}\left[\frac{1}{c^3}\right]} \, s^2 \, \left(x^{\text{Log}[x^{\text{Log}[x]}]}\right)^{\text{Log}\left[y^{\text{Log}[y^{\text{Log}[y]}]}\right]}\right] + \sqrt{1 + \text{Log}\left[(x \, z^2)^{\text{Log}[x \, z^2]}\right]}$

We could now continue and extend the rulePower to complex powers. The above rule rulePower was designed to work with rational powers. For complex powers it will not work.

In[14]:= **PowerFactor[x^I y^I (1/x)^I (1/y)^I**
        (1 - I z)^((1 - I)/2) (1 + I z)^((I - 1)/2)]

Out[14]= $\left(\dfrac{1}{x}\right)^{\text{i}} x^{\text{i}} \left(\dfrac{1}{y}\right)^{\text{i}} y^{\text{i}} (1 - \text{i} \, z)^{\frac{1}{2} - \frac{\text{i}}{2}} (1 + \text{i} \, z)^{-\frac{1}{2} + \frac{\text{i}}{2}}$

Now, we have to deal with exponents e and $-e$ appropriately.

In[15]:= **rulePower = t:_Times?(MemberQ[#, Power[_, _Complex]]&) :>**
        Module[{product = List @@ t, crads, rest, exp, cradsN},
                (* select the radicals *)
                crads = Cases[product, Power[_, _Complex], {1}];
                rest = Complement[product, crads];
                (* the new exponent *)
                exp = LCM @@ Denominator[Last /@ crads];
                cradsN = crads^exp;
                If[exp =!= 1, (Times @@ cradsN)^(1/exp),
                (* complementary powers *)
                    Times @@ (Function[l, Times @@
                    (#[[2, 1]]^(l[[1, 2, 2]]/
                    #[[2, 2]])& /@ l)^l[[1, 2, 2]]] /@
                    Split[{Sort[{#[[2]], -#[[2]]}], #}& /@ cradsN,
                        #1[[1]] === #2[[1]]&)]] (Times @@ rest)];

In[16]:= **PowerFactor[x^I y^I (1/x)^I (1/y)^I**
        (1 - I z)^((1 - I)/2) (1 + I z)^((I - 1)/2)]

Out[16]= $\sqrt{\left(\dfrac{1 - \text{i} \, z}{1 + \text{i} \, z}\right)^{1 - \text{i}}}$

f) We use the series of the generating function to define the $q_k(z)$ for the first k.

In[1]:= **q[_, z_] = 0;**

        (* make definitions for the q *)
        MapIndexed[(q[#2[[1]], z_] = #1)&,
            CoefficientList[Series[Exp[z ξ + ξ^3/3],
                {ξ, 0, 20}], ξ] // Expand];

Given the $q_k(z)$, the definition of the $\sigma_k(z)$ is straightforward.

In[4]:= **σ[0, z_] := 1;**
        σ[k_, z_] := σ[k, z] =
            Det[Table[q[k + i - 2 j, z], {i, 0, k - 1}, {j, 0, k - 1}]]

Now, we can calculate the first few $w_k(z)$.

In[6]:= **kMax = 10;**

        Do[w[k, z_] = D[Log[σ[k + 1, z]/σ[k, z]], z] // Together // Factor,
            {k, kMax}]

Here are the first four $w_k(z)$.

```
In[8]:= Table[w[k, z], {k, 4}]
```

$$Out[8]= \{\frac{1}{z}, \frac{1 + 2 z^3}{(-1 + z) z (1 + z + z^2)}, \frac{3 z^2 (10 - 2 z^3 + z^6)}{(-1 + z) (1 + z + z^2) (-5 - 5 z^3 + z^6)},$$

$$\frac{875 - 1750 z^3 + 1400 z^6 + 250 z^9 - 50 z^{12} + 4 z^{15}}{z (-5 - 5 z^3 + z^6) (-175 - 15 z^6 + z^9)}\}$$

Here is a quick check for the correctness of the calculated functions.

```
In[9]:= Table[Together[D[w[k, z], {z, 2}] - (2w[k, z]^3 - 4 z w[k, z] + 4 k)],
          {k, kMax}]
Out[9]= {0, 0, 0, 0, 0, 0, 0, 0, 0, 0}
```

The degree of the numerator and the denominator is a growing function of k.

```
In[10]:= Table[Exponent[{Numerator[w[k, z]], Denominator[w[k, z]]}, z],
           {k, kMax}]
Out[10]= {{0, 1}, {3, 4}, {8, 9}, {15, 16}, {24, 25},
          {35, 36}, {48, 49}, {63, 64}, {80, 81}, {99, 100}}
```

The zeros of the denominator of the $w[k, z]$ form an interesting pattern in the complex z-plane [970], [908]. A contour plot of the lines $\mathrm{Im}(w_k(z)) = 0$ also shows an interesting pattern.

```
In[11]:=
      Show[GraphicsArray[
        {Graphics[{PointSize[0.01], Point[{Re[#], Im[#]}]}& /@
          (z /. NSolve[Denominator[w[kMax, z]] == 0, z])},
            PlotRange -> All, Frame -> True, AspectRatio -> Automatic],
         (* the contour plot of Im[w[kMax, z]] *)
         ContourPlot[Evaluate[Im[w[kMax, z] /. z -> x + I y]],
                      {x, -4, 6}, {y, -5, 5}, AspectRatio -> Automatic,
                      PlotPoints -> 400, Contours -> {0},
                      DisplayFunction -> Identity]}]]
```

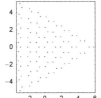 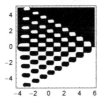

Here are some related polynomials whose zeros form pentagonal patterns [378].

```
In[12]:= (* bilinear Hirota derivative of order n *)
         HirotaD[f_[z_], g_[z_], {z_, n_}] :=
         Module[{d},
                 d[p_Plus] := d /@ p;
                 d[p_Times] := (p/(#1 #2) D[#1, z] #2 - p/(#1 #2)  #1 D[#2, z])& @@
                                 (Function[fg, Select[p, MemberQ[#, fg, {0, Infinity},
                                                    Heads -> True]&]] /@ {f, g});
                 Nest[d, f[z] g[z], n]]
In[14]:= (* symmetric fourth order Hirota derivative for f == g *)
         D4[f_, z_] = (HirotaD[f[z], g[z], {z, 4}] /. g -> f);
In[16]:= (* recursive definition for the polynomials Q2[n] *)
         Off[RuleDelayed::rhs];
         Q2[0][z_] = 1; Q2[1][z_] = z;
         Q2[n_][z_] := Q2[n][z_] = Factor[Cancel[Together[
                         -(2 D4[Q2[n - 1], z] - z Q2[n - 1][z]^2)/Q2[n - 2][z]]]]
In[20]:= Table[Subscript[Q, k][z] == Q2[k][z], {k, 6}] // TableForm // TraditionalForm
```

Out[20]//TraditionalForm=

$$Q_1(z) = z$$

$$Q_2(z) = z^3$$

$$Q_3(z) = z(z^5 - 144)$$

$$Q_4(z) = z^{10} - 1008\,z^5 - 48384$$

$$Q_5(z) = z^{15} - 4032\,z^{10} - 3048192\,z^5 + 146313216$$

$$Q_6(z) = z(z^{20} - 12096\,z^{15} - 21337344\,z^{10} - 33798352896\,z^5 - 4866962817024)$$

The coefficients of the $Q_k(z)$ are quickly increasing functions of the degree.

```
In[21]:= Table[{k, Ceiling[Log[10, Max[Abs[CoefficientList[Q2[k][z], z]]]]]},
            {k, 16}]
```

```
Out[21]= {{1, 0}, {2, 0}, {3, 3}, {4, 5}, {5, 9}, {6, 13}, {7, 18}, {8, 26}, {9, 34},
         {10, 43}, {11, 54}, {12, 65}, {13, 79}, {14, 94}, {15, 110}, {16, 127}}
```

The following graphic shows the roots for the first 21 polynomials $Q_k(z)$. The right graphic shows the curves $\mathrm{Re}(Q_{14}(z)) = 0$ in red and $\mathrm{Im}(Q_{14}(z)) = 0$ in blue.

```
In[22]:= Show[GraphicsArray[
            {(* zeros of the first polynomials *)
             Graphics[{PointSize[0.01],
                Table[{Hue[k/20], Point[{Re[#], Im[#]}]}& /@
                   N[z /. {ToRules[Roots[Q2[k][z] == 0, z]]}]}, {k, 16}]}],
                PlotRange -> All, Frame -> True, AspectRatio -> Automatic],
             (* curves of vanishing real and imaginary parts of Q2[14, z] *)
             Show[ContourPlot[Evaluate[#1[Q2[14][z]] /. z -> x + I y],
                   {x, -20, 20}, {y, -20, 20}, DisplayFunction -> Identity,
                   PlotPoints -> 400, Contours -> {0}, ContourShading -> False,
                   ContourStyle -> {{Thickness[0.001], #2}}]& @@@
                {{Re, RGBColor[1, 0, 0]}, {Im, RGBColor[0, 0, 1]}}]}]]
```

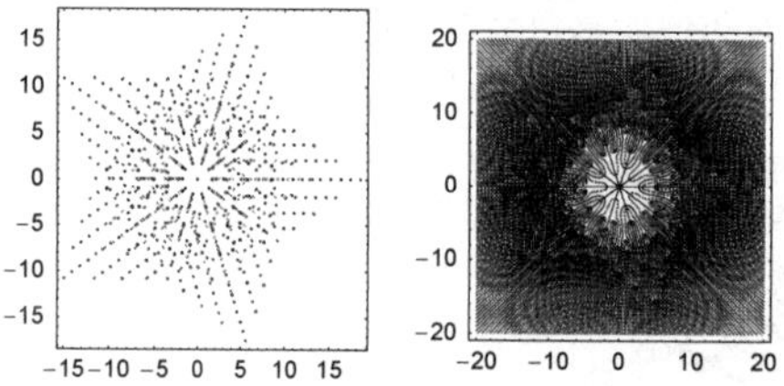

4. Differential Equations for the Product, Quotient of Solutions of Second-Order Differential Equations

The idea for the determination of a_p, b_p, c_p and a_q, b_q is to use $u(z) = y_1(z)\,y_2(z)$ and $w(z) = y_1(z)/y_2\,(z)$ in the third-order differential equations and to replace all second and third derivatives of $y_1(z)$ and $y_2(z)$ by using the differential equation, $y''(z) + f(z)\,y'(z) + g(z)\,y(z) = 0$. In the resulting equation, we set all coefficients of $y_1^p(z)\,y_1{}'^m\,y_2^p(z)\,y_2{}'^p$ to zero.

Here are the rules for the replacement of $y_1'''(z)$, $y_2'''(z)$, $y_1''(z)$, and $y_2''(z)$, which directly follow from the differential equation.

```
In[1]:= derRepRule =
         (* from differentiating the differential equation *)
         {y1''[z]   -> -f[z] y1'[z] - g[z] y1[z],
          y2''[z]   -> -f[z] y2'[z] - g[z] y2[z],
          y1'''[z] ->  f[z] g[z] y1[z] - y1[z] g'[z] + f[z]^2 y1'[z] -
                       g[z] y1'[z] - f'[z] y1'[z],
          y2'''[z] ->  f[z] g[z] y2[z] - y2[z] g'[z] + f[z]^2 y2'[z] -
                       g[z] y2'[z] - f'[z] y2'[z]};
```

Now, we carry out this replacement in the linear third-order differential equation.

```
In[2]:= expr = ((   D[y1[z] y2[z], {z, 3}] + ap D[y1[z] y2[z], {z, 2}] +
                 bp D[y1[z] y2[z], z]      + cp    y1[z] y2[z]) //.
                                          derRepRule) // Expand
```

```
Out[2]= cp y1[z] y2[z] - 2 ap g[z] y1[z] y2[z] + 2 f[z] g[z] y1[z] y2[z] - 2 y1[z] y2[z] g'[z] +
       bp y2[z] y1'[z] - ap f[z] y2[z] y1'[z] + f[z]² y2[z] y1'[z] - 4 g[z] y2[z] y1'[z] -
       y2[z] f'[z] y1'[z] + bp y1[z] y2'[z] - ap f[z] y1[z] y2'[z] + f[z]² y1[z] y2'[z] -
       4 g[z] y1[z] y2'[z] - y1[z] f'[z] y2'[z] + 2 ap y1'[z] y2'[z] - 6 f[z] y1'[z] y2'[z]
```

Next, we equate the coefficients of y1[z] y2[z], y1'[z] y2[z], y1[z] y2'[z], and y1'[z] y2'[z] to zero.

```
In[3]:= Coefficient[expr, #] == 0& /@
           {y1[z]  y2[z], y1'[z]  y2[z], y1[z]  y2'[z], y1'[z]  y2'[z]}
Out[3]= {cp - 2 ap g[z] + 2 f[z] g[z] - 2 g'[z] == 0, bp - ap f[z] + f[z]² - 4 g[z] - f'[z] == 0,
        bp - ap f[z] + f[z]² - 4 g[z] - f'[z] == 0, 2 ap - 6 f[z] == 0}
```

Finally, we solve the resulting system of linear equations for a, b, and c.

```
In[4]:= Solve[%, {ap, bp, cp}] // Simplify
Out[4]= {{cp → 2 (2 f[z] g[z] + g'[z]), bp → 2 f[z]² + 4 g[z] + f'[z], ap → 3 f[z]}}
```

So the differential equation obeyed by $u(z) = y_1(z)\, y_2(z)$ is

$$u'''(z) + 3\, f(z)\, u''(z) + (2\, f^2(z) + 4\, g(z) + f'(z))\, u'(z) + (4\, f(z)\, g(z) + 2\, g'(z))\, u(z) = 0$$

We follow the same technique for the quotient $w(z) = y_1(z)/y_2(z)$.

```
In[5]:= expr = ((Numerator[Together[
              D[y1[z]/y2[z], {z, 3}] D[y1[z]/y2[z], {z, 1}] +
              aq D[y1[z]/y2[z], {z, 2}]^2 +
              bq D[y1[z]/y2[z], {z, 1}]^2]]) //. derRepRule) // Expand
Out[5]= bq y2[z]⁴ y1'[z]² + f[z]² y2[z]⁴ y1'[z]² + aq f[z]² y2[z]⁴ y1'[z]² +
       2 g[z] y2[z]⁴ y1'[z]² - y2[z]⁴ f'[z] y1'[z]² - 2 bq y1[z] y2[z]³ y1'[z] y2'[z] -
       2 f[z]² y1[z] y2[z]³ y1'[z] y2'[z] - 2 aq f[z]² y1[z] y2[z]³ y1'[z] y2'[z] -
       4 g[z] y1[z] y2[z]³ y1'[z] y2'[z] + 2 y1[z] y2[z]³ f'[z] y1'[z] y2'[z] -
       6 f[z] y2[z]³ y1'[z]² y2'[z] + 4 aq f[z] y2[z]³ y1'[z]² y2'[z] + bq y1[z]² y2[z]² y2'[z]² +
       f[z]² y1[z]² y2[z]² y2'[z]² + aq f[z]² y1[z]² y2[z]² y2'[z]² + 2 g[z] y1[z]² y2[z]² y2'[z]² -
       y1[z]² y2[z]² f'[z] y2'[z]² - 12 f[z] y1[z] y2[z]² y1'[z] y2'[z]² -
       8 aq f[z] y1[z] y2[z]² y1'[z] y2'[z]² + 6 y2[z]² y1'[z]² y2'[z]² + 4 aq y2[z]² y1'[z]² y2'[z]² +
       6 f[z] y1[z]² y2[z] y2'[z]³ + 4 aq f[z] y1[z]² y2[z] y2'[z]³ - 12 y1[z] y2[z] y1'[z] y2'[z]³ -
       8 aq y1[z] y2[z] y1'[z] y2'[z]³ + 6 y1[z]² y2'[z]⁴ + 4 aq y1[z]² y2'[z]⁴
```

Now, we have many more products of the form $y_1^n(z)\, y_1'^m(z)\, y_2^o(z)\, y_2'^p(z)$. Next, we collect them.

```
In[6]:= Union[(List @@ %) //.
           {f'[z] -> 1, g'[z] -> 1, f[z] -> 1, g[z] -> 1, aq -> 1, bq -> 1,
            i_Integer ws_ -> ws}]
Out[6]= {y2[z]⁴ y1'[z]², y1[z] y2[z]³ y1'[z] y2'[z], y2[z]³ y1'[z]² y2'[z],
        y1[z]² y2[z]² y2'[z]², y1[z] y2[z]² y1'[z] y2'[z]², y2[z]² y1'[z]² y2'[z]²,
        y1[z]² y2[z] y2'[z]³, y1[z] y2[z] y1'[z] y2'[z]³, y1[z]² y2'[z]⁴}
```

We calculate the values of a_q, b_q.

```
In[7]:= Coefficient[expr, #] == 0& /@ %
Out[7]= {bq + f[z]² + aq f[z]² + 2 g[z] - f'[z] == 0, -2 bq - 2 f[z]² - 2 aq f[z]² - 4 g[z] + 2 f'[z] == 0,
        6 f[z] + 4 aq f[z] == 0, bq + f[z]² + aq f[z]² + 2 g[z] - f'[z] == 0,
        -12 f[z] - 8 aq f[z] == 0, 6 + 4 aq == 0, 6 f[z] + 4 aq f[z] == 0, -12 - 8 aq == 0, 6 + 4 aq == 0}
```

```
In[8]:= Solve[%, {aq, bq}]
Out[8]= {{bq → ½ (f[z]² - 4 g[z] + 2 f'[z]), aq → -³⁄₂}}
```

So for $u(z)$, we have the differential equation

$$w'''(z)\, w'(z) - \frac{3}{2}\, w''(z)^2 + \left(\frac{f^2(z)}{2} - 2\, g(z) + f'(z) \right) w'(z)^2 = 0$$

Using the function `GroebnerBasis`, we can easily calculate the differential equations under interest by generating a closed set of polynomial equations and then eliminating all derivatives of $y_1(z)$ and $y_2(z)$. Here, this is done for the product $w(z) = y_1(x)\, y_2(z)$.

In[9]:=
```
GroebnerBasis[(* make polynomial system *)
Numerator[Together[Flatten[{
(* differentiate first equation *)
Table[D[y1''[z] + f[z] y1'[z] + g[z] y1[z], {z, k}], {k, 0, 1}],
(* differentiate second equation *)
Table[D[y2''[z] + f[z] y2'[z] + g[z] y2[z], {z, k}], {k, 0, 1}],
(* differentiate ratio *)
Table[D[w[z] - y1[z] y2[z], {z, k}], {k, 0, 3}]}]]], {},
(* variables to eliminate *)
  {y1[z], y1'[z], y1''[z], y1'''[z],
   y2[z], y2'[z], y2''[z], y2'''[z]},
            MonomialOrder -> EliminationOrder,
            CoefficientDomain -> RationalFunctions]
```

Out[9]= $\{4\,f[z]\,g[z]\,w[z] + 2\,w[z]\,g'[z] + 2\,f[z]^2\,w'[z] + 4\,g[z]\,w'[z] + f'[z]\,w'[z] + 3\,f[z]\,w''[z] + w^{(3)}[z]\}$

5. Singular Points of ODEs, Integral Equation

a) We solve the differential equation $y'(x) = 2\,y(x)/x$ symbolically.

In[1]:= `DSolve[{y'[x] == 2y[x]/x, y[x0] == y0}, y[x], x][[1]]`

Out[1]= $\left\{ y[x] \rightarrow \dfrac{x^2\,y0}{x0^2} \right\}$

In[2]:= `sol1[x_, {x0_, y0_}] = x^2 y0 /x0^2;`

Here is a plot of the solution.

In[3]:=
```
opts[label_] =
    Sequence[Axes -> False, AspectRatio -> 1,
            PlotLabel -> StyleForm[label, FontWeight -> "Bold",
                                    FontSize -> 7]];

Plot[Evaluate[Table[sol1[x, {Cos[φ], Sin[φ]}], {φ, Pi/30, 2Pi, 2Pi/12}]],
    {x, -1.1, 1}, Evaluate[opts["Knot point"]]]
```

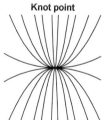
Knot point

Now, we solve $y'(x) = (y(x) + x)/x$.

In[5]:= `DSolve[{y'[x] == (x + y[x])/x, y[x0] == y0}, y[x], x][[1]]`

Out[5]= $\left\{ y[x] \rightarrow \dfrac{x\,y0 + x\,x0\,\text{Log}[x] - x\,x0\,\text{Log}[x0]}{x0} \right\}$

In[6]:= `sol2[x_, {x0_, y0_}] = x Log[x] + x (y0/x0 - Log[x0]);`

The `Off[Plot::plnr]` serves to suppress any error message resulting from computation of complex values.

In[7]:=
```
Off[Plot::plnr];
Plot[Evaluate[Table[sol2[x, {Cos[φ], Sin[φ]}], {φ, Pi/10, 2Pi, 2Pi/22}]],
    {x, -1.1, 1}, Evaluate[opts["Knot point"]]]
```

Knot point

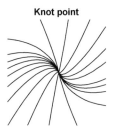

Next, we solve $y'(x) = y(x)/x$.

In[9]:= `DSolve[{y'[x] == y[x]/x, y[x0] == y0}, y[x], x][[1]]`

Out[9]= $\left\{ y[x] \rightarrow \dfrac{x\,y0}{x0} \right\}$

In[10]:= `sol3[x_, {x0_, y0_}] = x y0/x0;`

Here again is a plot.

In[11]:= `Plot[Evaluate[Table[sol3[x, {Cos[φ], Sin[φ]}], {φ, Pi/10, 2Pi, 2Pi/22}]],`
`{x, -1.1, 1}, Evaluate[opts["Knot point"]]]`

Knot point

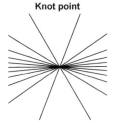

Now, we solve $y'(x) = -y(x)/x$.

In[12]:= `DSolve[{y'[x] == -y[x]/x, y[x0] == y0}, y[x], x][[1]]`

Out[12]= $\left\{ y[x] \rightarrow \dfrac{x0\,y0}{x} \right\}$

In[13]:= `sol4[x_, {x0_, y0_}] = x0 y0/x;`

The associated plot is as follows.

In[14]:= `Plot[Evaluate[Table[sol4[x, {Cos[φ], Sin[φ]}], {φ, 0, 2Pi, 2Pi/13}]],`
`{x, -1.1, 1}, PlotRange -> {{-2, 2}, {-2, 2}},`
`Evaluate[opts["Saddle point"]]]`

Saddle point

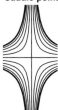

For $y'(x) = -x/y(x)$, we get two solutions from `DSolve`.

In[15]:= `DSolve[{y'[x] == -x/y[x], y[x0] == y0}, y[x], x]`

Out[15]= $\left\{ \left\{ y[x] \rightarrow -\sqrt{-x^2 + x0^2 + y0^2} \right\}, \left\{ y[x] \rightarrow \sqrt{-x^2 + x0^2 + y0^2} \right\} \right\}$

In[16]:= `sol5[x_, {x0_, y0_}] = {-Sqrt[-x^2 + x0^2 + y0^2], Sqrt[-x^2 + x0^2 + y0^2]};`

Now, we plot both solutions.

```
In[17]:= Show[Table[Plot[Evaluate[sol5[x, {x0, 0}]], {x, -x0, x0},
                      DisplayFunction -> Identity],
                  {x0, 0.1, 1, 0.1}], DisplayFunction -> $DisplayFunction,
              Evaluate[opts["Vortex point"]]]
```

Vortex point

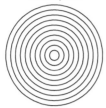

Remaining is the differential equation that gives eddy points. Again, we can find a solution, although not explicitly for $y(x)$.

```
In[18]:= DSolve[{y'[x] == (y[x] - x)/(y[x] + x)}, y[x], x]
```

Solve::tdep : The equations appear to involve the variables
to be solved for in an essentially non-algebraic way. More...

$$Out[18]= \ \text{Solve}\left[\text{ArcTan}\left[\frac{y[x]}{x}\right] + \frac{1}{2}\text{Log}\left[1 + \frac{y[x]^2}{x^2}\right] == C[1] - \text{Log}[x], y[x]\right]$$

```
In[19]:= sol6[{x_, y_}] = -2 ArcTan[y/x] + Log[1/(x^2 (1 + y^2/x^2))];
```

We now plot this result. Unfortunately, for this transcendental equation, ImplicitPlot is of little use because it can only plot polynomial equations. Also, ContourPlot does not give a very good result because of the branch cut of Log.

```
In[20]:= ContourPlot[Evaluate[Re[sol6[{x, y}]]],
                   {x, -2, 2}, {y, -2, 2}, PlotPoints -> 100,
                   Contours -> 20, ContourShading -> False]
```

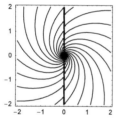

Therefore, we now create a special implementation. We could try a numerical implementation using FindRoot, for example, of the following form. However, it is difficult to represent larger pieces like this. The form of the solution suggests the use of polar coordinates.

```
In[21]:= sol6[{r Cos[φ], r Sin[φ]}] // Simplify
```

$$Out[21]= \ -2\,\text{ArcTan}[\text{Tan}[\varphi]] + \text{Log}\left[\frac{1}{r^2}\right]$$

```
In[22]:= Solve[% == c, r]
```

$$Out[22]= \ \left\{\left\{r \to -e^{\frac{1}{2}(-c-2\,\text{ArcTan}[\text{Tan}[\varphi]])}\right\}, \left\{r \to e^{\frac{1}{2}(-c-2\,\text{ArcTan}[\text{Tan}[\varphi]])}\right\}\right\}$$

We arrive at the following formula.

```
In[23]:= % // PowerExpand
```

$$Out[23]= \ \left\{\left\{r \to -e^{\frac{1}{2}(-c-2\,\varphi)}\right\}, \left\{r \to e^{\frac{1}{2}(-c-2\,\varphi)}\right\}\right\}$$

The final graphics of the integral curves is the following.

```
In[24]:= Show[Table[
          ParametricPlot[Evaluate[Exp[c - φ]{Cos[φ], Sin[φ]}],
                       {φ, c + 0.1, 3Pi + c}, DisplayFunction -> Identity],
                   {c, 0, 2Pi 24/25, 2Pi/25}],
```

```
                DisplayFunction -> $DisplayFunction,
                PlotRange -> All, Evaluate[opts["Eddy point"]]]]
                    Eddy point
```

We demonstrate the appearance of various singular points in the following example. For a "random" bivariate function $\psi(x, y)$, we will integrate the equations $x'(s) = \partial\psi(x(s), y(s))/\partial x(s)$, $y'(t) = -\partial\psi(x(s), y(s))/\partial x(s)$. We see saddle points, vortex points, and knot points [85], [448].

```
In[25]:= Module[{L = 4, pp = 41, T = 5, o = 90, ms = 100, ℓ = 3/2,
            ps = 21, λ = 2, ipo, ipoX, ipoY, eqs, pathList, nsol},
        SeedRandom[123];
        (* a streamfunction *)
        ipo = (* smooth interpolation *) Interpolation[
        (* random data *)
        Flatten[Table[{x, y, Random[Real, {-1, 1}]},
              {x, -L, L, 2L/pp}, {y, -L, L, 2L/pp}], 1],
              InterpolationOrder -> 8];
        (* derivatives *)
        {ipoX, ipoY} = D[ipo[x[t], y[t]], #]& /@ {x[t], y[t]};
        (* differential equations for flow lines *)
        eqs = Thread[{x'[t], y'[t]} == #/Sqrt[#.#]&[{ipoX, -ipoY}]];
        (* calculate flow lines *)
        pathList = Table[
        ((* solve for flow lines *)
        Internal`DeactivateMessages[
         nsol = NDSolve[Join[eqs, {x[0] == x0, y[0] == y0}],
           {x, y}, {t, 0, #}, MaxSteps -> ms, PrecisionGoal -> 3,
                     AccuracyGoal -> 3]];
          (* visualize flow lines *)
        ParametricPlot[Evaluate[{x[t], y[t]} /. nsol],
                    {t, 0, DeleteCases[nsol[[1, 1, 2, 1, 1]], 0.][[1]]},
                    (* color flow lines differently *)
                    PlotStyle -> {{Thickness[0.002], RGBColor[
                                 (x0 + ℓ)/(2ℓ), 0.2, (y0 + ℓ)/(2ℓ)]}},
                    DisplayFunction -> Identity,
                    PlotPoints -> 200]& /@ {T, -T},
                    (* grid of initial conditions *)
                    {x0, -ℓ, ℓ, 2ℓ/ps}, {y0, -ℓ, ℓ, 2ℓ/ps}];
        (* display flow lines and stream function *)
        Show[(* contour plot of the stream function *)
          {ContourPlot[Evaluate[ipo[x, y]], {x, -L/2, L/2}, {y, -L/2, L/2},
                     PlotPoints -> 400, Contours -> 60,
                     ContourLines -> False, PlotRange -> All,
                     DisplayFunction -> Identity],
           Show[pathList]}, DisplayFunction -> $DisplayFunction,
              Frame -> True, Axes -> False, FrameTicks -> False,
              PlotRange -> {{-λ, λ}, {-λ, λ}}, AspectRatio -> Automatic]]
```

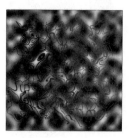

For higher-order singularities, see [1726].

b) We start with implementing the exact solution for separable kernels. The function `iSolve` (named in analogy to `DSolve`) attempts this. Because a kernel might be separable, but structurally not in separated form, we allow for an optional function S that attempts to separate the kernel. While we could be more elaborate with respect to matching the pattern of a Fredholm integral equation of the second kind, we require here the canonical form. The step-by-step implementation of `iSolve` is self-explanatory.

```
In[1]:= iSolve[eq:(y_[x_] + λ_ Integrate[𝒦_ y[ξ_], {ξ_, a_, b_}] == f_),
             y_, x_, S_:Identity] :=
        Module[{I = Integrate, intExpand, eq1, integrals, cRules,
                functions, eq2, eqs, cs, separableQ},
           (* thread integrals over sums and
              pull integration variable-independent out *)
           intExpand = Function[int, (int //. I[p_Plus, i_] :> (I[#, i]& /@ p) //.
                         HoldPattern[Integrate[c_?(FreeQ[#, ξ, Infinity]&) rest_,
                                      {ξ, a, b}]] :> c Integrate[rest, {ξ, a, b}])];
           (* separate kernel *)
           eq1 = intExpand[ExpandAll[S //@ (Subtract @@ eq)]];
           (* integrals over y[ζ] and kernel functions *)
           integrals = Union[Cases[eq1, _I, Infinity]];
           (* replace integrals by variables c[i] *)
           cRules = Rule @@@ Transpose[{integrals /. ξ -> ζ_,
                                        cs = Array[c, Length[integrals]]}];
           (* was the kernel 𝒦 separable? *)
           separableQ = FreeQ[cRules, x, Infinity];
           (* kernel functions h_j[.] *)
           functions = ((First /@ integrals)/y[ξ]) /. ξ -> x;
           (* replace integrals by variables *)
           eq2 = eq1 /. cRules;
           (* make linear system in the c[i] *)
           eqs  = intExpand[ExpandAll[I[eq2 #, {x, a, b}]& /@
                            functions]] //. cRules;
           (* solve linear system, backsubstitute into eq2
              and solve for y[x] *)
           Solve[(eq2 /. Solve[(# == 0)& /@ eqs, cs][[1]]) == 0, y[x]] /;
                             (* was iSolve applicable? *) separableQ]
```

The next inputs solve the example equation $y(x) - \lambda \int_0^1 \sin(x + \xi)\, y(\xi)\, d\xi = \cos(x)$.

```
In[2]:= 𝒦[x_, ξ_] := Sin[x + ξ]
        f[x_] := Cos[x]

        IEq = y[x] - λ Integrate[𝒦[x, ξ] y[ξ], {ξ, 0, 1}] == f[x];

        IEqSol = iSolve[y[x] - λ Integrate[𝒦[x, ξ] y[ξ], {ξ, 0, 1}] ==
                    f[x], y, x, TrigExpand] // Simplify
```

$$Out[5]= \left\{\left\{y[x] \rightarrow -\frac{2\,(\lambda\,\mathrm{Cos}[2-x] - (-4+\lambda)\,\mathrm{Cos}[x] + 2\,\lambda\,\mathrm{Sin}[x])}{-8 + \lambda^2\,(1 + \mathrm{Cos}[2]) + 8\,\lambda\,\mathrm{Sin}[1]^2}\right\}\right\}$$

Here is a quick check for the correctness of the result.

```
In[6]:= yExact = IEqSol[[1, 1, 2]];
        IEq /. y -> Function[x, Evaluate[yExact]] // Simplify
```

In[7]:= True

In the calculation of the truncated Fredholm and Neumann resolvents, we have to carry out many definite integrals. Because we do not worry about convergence and hope to carry out all integrals successfully term by term, we do not use the built-in function `Integrate` directly, but rather implement a function `integrate`, that expands products and powers.

```
In[8]:= integrate[l_List, i_] := Integrate[#, i]& /@ l
        integrate[p_Plus, i_] := Integrate[#, i]& /@ p
        integrate[p:Times[___, _Plus] | p:Power[_Plus, _Integer], i_] :=
                                        integrate[Expand[p], i]
        integrate[e_, i_] := Integrate[e, i]
```

The function `FredholmResolventList` calculates a list of the successive resolvent approximations arising from truncating the Fredholm minor and the Fredholm determinant at λ^{o+1}.

```
In[12]:= FredholmResolventList[𝒦_, {ξ_, a_, b_}, {x_, ξ_}, o_, S_:Identity] :=
         Module[{c, d, k, x, y, z},
            (* make recursive definitions for Fredholm minor and determinant *)
            (* avoid variable interference by applying Set and SetDelayed *)
            Set @@ {k[x_, y_], 𝒦 /. {x -> x, ξ -> y}};
            Set @@ {d[0][x_, y_], k[x, y]};
            SetDelayed @@ {d[k_][x_, y_], Unevaluated @
             With[{px = Pattern[#, _]& @@ {x}, py = Pattern[#, _]& @@ {y}},
                d[k][px, py] = S @ (c[k] k[x, y] -
                               k integrate[k[x, z] d[k - 1][z, y], {z, a, b}])]};
            c[0] := 1;
            c[k_] := c[k] = S @ integrate[d[k - 1][y, y], {y, a, b}];
            (* calculate c[k] and d[k] recursively and form
               successive resolvent approximations *)
            Divide @@ Transpose[Rest[FoldList[Plus, 0,
                  Table[(-1)^k/k! λ^k {d[k][x, ξ], c[k]}, {k, 0, o}]]]]]
```

For the example integral equation, all higher c_k and $d_k(x, \xi)$ vanish identically and we obtain the exact solution.

```
In[13]:= FSerKernels = FredholmResolventList[𝒦[x, ξ], {ξ, 0, 1}, {x, ξ}, 3,
                                             Simplify]
```

$$Out[13]= \left\{ Sin[x + ξ], \frac{-\frac{1}{2} λ (-Cos[x - ξ] + Cos[1 - x - ξ] Sin[1]) + Sin[x + ξ]}{1 - λ Sin[1]^2} , \right.$$

$$\frac{-\frac{1}{2} λ (-Cos[x - ξ] + Cos[1 - x - ξ] Sin[1]) + Sin[x + ξ]}{1 - \frac{1}{4} λ^2 Cos[1]^2 - λ Sin[1]^2} ,$$

$$\left. \frac{-\frac{1}{2} λ (-Cos[x - ξ] + Cos[1 - x - ξ] Sin[1]) + Sin[x + ξ]}{1 - \frac{1}{4} λ^2 Cos[1]^2 - λ Sin[1]^2} \right\}$$

In[14]:= yFSerSols[x_] = f[x] + λ integrate[FSerKernels f[ξ], {ξ, 0, 1}];

In[15]:= yFSerSols[x][[-1]] == yExact // Simplify

Out[15]= True

We end with the implementation of the iterated kernels. The function `NeumannResolventList` calculates the resolvent arising from $o + 1$ iterated kernels.

```
In[16]:= NeumannResolventList[𝒦_, {ξ_, a_, b_}, {x_, ξ_}, o_, S_:Identity] :=
         Module[{k, x, y, z, kernels},
            Set @@ {k[x_, y_], 𝒦 /. {x -> x, ξ -> y}};
            kernels = NestList[S[integrate[k[x, z] (# /. x -> z), {z, 0, 1}]]&,
                            k[x, y], o] /. {x -> x, y -> ξ};
            Rest[FoldList[Plus, 0, MapIndexed[λ^(#2[[1]] - 1) #1&, kernels]]]]
```

The iterated kernels become increasingly complicated functions.

```
In[17]:= NSerKernels = NeumannResolventList[𝒦[x, ξ], {ξ, 0, 1},
                                            {x, ξ}, 5, Simplify];
```

```
        {LeafCount /@ NSerKernels, Short[NSerKernels, 12]}
```

$$Out[19]= \left\{ \{4, 26, 78, 168, 292, 454\}, \right.$$

$$\left\{ Sin[x + ξ], \frac{1}{2} λ (Cos[x - ξ] - Cos[1 + x + ξ] Sin[1]) + Sin[x + ξ], \ll1\gg, \ll1\gg, \right.$$

$$\frac{1}{2} \lambda \left(\text{Cos}[x - \xi] - \text{Cos}[1 + x + \xi]\, \text{Sin}[1]\right) + \text{Sin}[x + \xi] + \frac{1}{16} \lambda^2 \left(\ll 1 \gg\right) + \frac{1}{64} \lambda^3 \left(\ll 1 \gg\right) + \frac{1}{256} \lambda^4$$
$$\left(48\, \text{Cos}[x - \xi] - 30\, \text{Cos}[2 + x - \xi] + \ll 14 \gg + 18\, \text{Sin}[4 + x + \xi] - 4\, \text{Sin}[6 + x + \xi] + \text{Sin}[8 + x + \xi]\right),$$
$$\frac{1}{2} \lambda \left(\text{Cos}[x - \xi] - \text{Cos}[1 + x + \xi]\, \text{Sin}[1]\right) + \text{Sin}[x + \xi] + \frac{1}{16} \lambda^2 \left(4\, \text{Cos}[x - \xi] - \right.$$
$$2\, \text{Cos}[2 + x - \xi] - 2\, \text{Cos}[2 - x + \xi] + 5\, \text{Sin}[x + \xi] - 2\, \text{Sin}[2 + x + \xi] + \text{Sin}[4 + x + \xi]) +$$
$$\frac{1}{64} \lambda^3 \left(16\, \text{Cos}[x - \xi] - 6\, \text{Cos}[2 + x - \xi] + 2\, \text{Cos}[4 + x - \xi] - 6\, \text{Cos}[2 - x + \xi] + 2\, \text{Cos}[4 - x + \xi] + \right.$$
$$4\, \text{Sin}[2 - x - \xi] + 13\, \text{Sin}[x + \xi] - 11\, \text{Sin}[2 + x + \xi] + 3\, \text{Sin}[4 + x + \xi] - \text{Sin}[6 + x + \xi]) +$$
$$\frac{1}{256} \lambda^4 \left(48\, \text{Cos}[x - \xi] - 30\, \text{Cos}[2 + x - \xi] + 8\, \text{Cos}[4 + x - \xi] - 2\, \text{Cos}[6 + x - \xi] - \right.$$
$$30\, \text{Cos}[2 - x + \xi] + 8\, \text{Cos}[4 - x + \xi] - 2\, \text{Cos}[6 - x + \xi] + 16\, \text{Sin}[2 - x - \xi] - 4\, \text{Sin}[4 - x - \xi] +$$
$$49\, \text{Sin}[x + \xi] - 36\, \text{Sin}[2 + x + \xi] + 18\, \text{Sin}[4 + x + \xi] - 4\, \text{Sin}[6 + x + \xi] + \text{Sin}[8 + x + \xi]) +$$
$$\frac{1}{1024} \left(\lambda^5 \left(176\, \text{Cos}[x - \xi] - 110\, \text{Cos}[2 + x - \xi] + 46\, \text{Cos}[4 + x - \xi] - 10\, \text{Cos}[6 + x - \xi] + \right.\right.$$
$$2\, \text{Cos}[8 + x - \xi] - 110\, \text{Cos}[2 - x + \xi] + 46\, \text{Cos}[4 - x + \xi] - 10\, \text{Cos}[6 - x + \xi] + 2\, \text{Cos}[8 - x + \xi] +$$
$$80\, \text{Sin}[2 - x - \xi] - 20\, \text{Sin}[4 - x - \xi] + 4\, \text{Sin}[6 - x - \xi] + 161\, \text{Sin}[x + \xi] - 145\, \text{Sin}[2 + x + \xi] +$$
$$70\, \text{Sin}[4 + x + \xi] - 26\, \text{Sin}[6 + x + \xi] + 5\, \text{Sin}[8 + x + \xi] - \text{Sin}[10 + x + \xi])))\}\}$$

In[20]:= `yNSerSols[x_] = f[x] + λ integrate[NSerKernels f[ξ], {ξ, 0, 1}];`

As expected, the terms of the series expansion of the exact solution in λ around $\lambda = 0$ agree with the result obtained using the Neumann resolvent.

In[21]:= `Series[yExact - yNSerSols[x][[-1]], {λ, 0, 6}] // Simplify`

Out[21]= $O[\lambda]^7$

For Adomian polynomial-based solutions of Fredholm integral equations, see [457].

6. Inverse Sturm–Liouville Problems, Graeffe Method

a) Here, we first implement the formula. The integrals involved are integrated immediately to avoid repeated calculations.

In[1]:=
```
InverseSturmLiouville[listOfEigenvalues_List, var_Symbol] :=
InverseSturmLiouville[listOfEigenvalues, var] =
Module[{φ0, u, uMatrix, v, ef, s, lde},
    lde = Sqrt /@ listOfEigenvalues;
    (* direct implementation of above formulas *)
    φ0[s_, k_] = Sin[k s]/k;
    u[ki_, kj_] = (Cos[kj var] Sin[ki var])/(ki (+ki^2 - kj^2)) +
                  (Cos[ki var] Sin[kj var])/(kj (-ki^2 + kj^2));
    u[ki_, ki_] = 1 + var/(2 ki^2) - Sin[2 ki var]/(4 ki^3);
    uMatrix = Outer[u, lde, lde];
    v = (-2D[Log[#], {var, 2}])&[Det[uMatrix]];
    ef = Inverse[uMatrix].(φ0[var, #]& /@ lde);
    (* make no simplification here; it takes too long on average *)
    {v, ef}]
```

Here are $v(x)$ and the eigenfunction for a given (energy) eigenvalue.

In[2]:= `{v, ef} = InverseSturmLiouville[{k}, x] // Simplify`

Out[2]= $\left\{-\dfrac{32 k \left(\sqrt{k}\, (2 k + x)\, \text{Cos}[\sqrt{k}\, x] - \text{Sin}[\sqrt{k}\, x]\right)\, \text{Sin}[\sqrt{k}\, x]}{\left(-2 \sqrt{k}\, (2 k + x) + \text{Sin}[2 \sqrt{k}\, x]\right)^2},\ \left\{\dfrac{4 k\, \text{Sin}[\sqrt{k}\, x]}{2 \sqrt{k}\, (2 k + x) - \text{Sin}[2 \sqrt{k}\, x]}\right\}\right\}$

We now check whether the differential equation is satisfied.

In[3]:= `- D[ef, {x, 2}] + v ef - k ef // Simplify`

Out[3]= `{0}`

We now prescribe two eigenvalues. We do not write out the result or simplify it because of its size.

In[4]:= `{pot, {ef1, ef2}} = InverseSturmLiouville[{k1, k2}, x];`

Here is a plot of the result. The isolated appearance of a point spectrum inside the continuous spectrum is remarkable.

In[5]:=
```
Plot[Evaluate[
    (* to be shown: the potential in thick and the eigenfunctions at their
       eigenvalues (adding the eigenvalue to their function
```

```
                   value for better visibility) *)
                 {pot, ef1 + k1^2, ef2 + k2^2, k1^2, k2^2} /.
                 {k1 -> 1, k2 -> 3/2}],
            {x, 0, 20},
         (* setting options to get a pretty picture *)
            PlotRange -> All, PlotPoints -> 200,
            PlotStyle -> {Thickness[0.007], Thickness[0.002], Thickness[0.002],
                         {Thickness[0.002], Dashing[{0.02, 0.02}]},
                         {Thickness[0.002], Dashing[{0.02, 0.02}]}},
            Frame -> True, FrameLabel -> ({#["r"], #["V"], None, "ε"}&[
                 StyleForm["r", FontWeight -> "Bold", FontSize -> 6]&)]
```

For the practical importance of such conditions, see [301], [1474], [1887], [1888], [113], and [1663]. For a nontrivial background potential, see [1367]; for bound states in gaps, see [1508].

b) The function `GraeffeSolve` implements the calculation of the polynomials $p_k(z)$ and the root $z_{n,k}$. After the $|z_k|$ are calculated as precisely as possible given the initial precision *prec*, $\pm z_k$ is formed and the appropriate sign is selected.

```
In[1]:= Off[RuleDelayed::rhs];
        GraeffeSolve[poly_, z_Symbol, prec_] :=
        Module[{k = 1, oldRoots = {0, 0}, newRoots}, Clear[p];
          p[0, ζ_] = N[poly /. z -> ζ, prec];
          (* polynomial recursion *)
          p[k_, ζ_] := p[k, ζ_] = Expand[p[k - 1, z] p[k - 1, -z]] /.
                (* avoid 0. z^o *) {_?(# == 0&) -> 0, z^n_ :> ζ^(n/2)};
          While[FreeQ[p[k, ζ], Overflow[] | Underflow[], Infinity] &&
                (coeffs = CoefficientList[p[k, ζ], ζ];
                (* next polynomial; normalized *)
                p[k, ζ_] = Expand[p[k, ζ]/Max[Abs[coeffs]]];
                (* new root approximations *)
                newRoots = Abs[Divide @@@ Partition[coeffs, 2, 1]]^(2^-k);
                (* are roots still changing? *)
                newRoots =!= oldRoots), oldRoots = newRoots; k++];
          {z -> #}& /@ (* add sign *)
          Select[Join[newRoots, -newRoots], (poly /. z -> #) == 0&]]
```

Here the function `GraeffeSolve` is used to solve $p = z^5 + 5\,z^4 - 10\,z^3 - 10\,z^2 + 5\,z + 2 = 0$. We start with 100 digits.

```
In[3]:= poly[z_] := 2 + 5 z - 10 z^2 - 10 z^3 + 5 z^4 + z^5;
        (* display shortened result *)
        (grs = GraeffeSolve[poly[z], z, 110]) // N[#, 10]&
Out[5]= {{z -> 0.5973232647}, {z -> 1.949083754},
         {z -> -0.2882370815}, {z -> -0.9438772110}, {z -> -6.314292726}}
```

Substituting the so-calculated roots in the original equation shows that they fulfill the equation.

```
In[6]:= {Precision /@ grs, poly[z] /. grs}
Out[6]= {{106.176, 106.177, 106.177, 106.176, 106.177},
         {0. × 10^-105, 0. × 10^-104, 0. × 10^-106, 0. × 10^-105, 0. × 10^-102}}
```

7. Finite Element Method: Lagrange and Hermite Interpolation

a) We begin with the numbering of the nodes. For Lagrange interpolation of nth order, we need $(n + 1)(n + 2)/2$ points in a triangle. Let the unit triangle have coordinates $P_1^{(e)} = \{0, 0\}$, $P_2^{(e)} = \{1, 0\}$, and $P_3^{(e)} = \{0, 1\}$.

The function `Numeration[n]` computes the data points `PD[n, i]`, which will be globally available for further use after the call of `Numeration`. (Be aware that `PD` is not in the list of local variables of `Block`; it is a global variable and will be used often in the following.)

```
In[1]:= numKnots[n_] = (n + 1) (n + 2)/2;

      Numeration[n_Integer?Positive] := Numeration[n] =
      Block[{count, steps, ord, step, zs, σ, x, y},
      (* the three given points *)
            PD[n, 1] = {0, 0};  PD[n, 2] = {1, 0};  PD[n, 3] = {0, 1};
            count = 4; steps = Quotient[n, 3] + 1; ord = n;
      (* go in spiral manner, along lines parallel to
         the edges of the triangle to the center *)
      (* outside *)
      step = 1; v = n - 1;
        Clear[x, y]; y = 0; x[t_] = t/n;
        Do[PD[n, count] = {x[i], y}; count = count + 1, {i, v}];
        Clear[x, y]; y[t_] = t/n; x[t_] = 1 - t/n;
        Do[PD[n, count] = {x[i], y[i]}; count = count + 1, {i, v}];
        Clear[x, y]; x = 0; y[t_] = 1 - t/n;
        Do[PD[n, count] = {x, y[i]}; count = count + 1, {i, v}];
      (* spiral in *)
      If[steps >= 2,
      Do[ord = ord - 3;
        Which[ord >= 1, v = ord; σ = step - 1;
                Clear[x, y]; y = σ/n; x[κ_] = σ/n + (κ - 1)/n;
                Do[PD[n, count] = {x[i], y}; count = count + 1, {i, v}];
                Clear[x, y]; y[κ_] = σ/n + (κ - 1)/n;
                          x[κ_] = 1 - 2 σ/n - (κ - 1)/n;
                Do[PD[n, count] = {x[i], y[i]}; count = count + 1, {i, v}];
                Clear[x, y]; x = σ/n; y[κ_] = 1 - 2 σ/n - (κ - 1)/n;
                Do[PD[n, count] = {x, y[i]}; count = count + 1, {i, v}],
                ord == 0, PD[n, count] = {(step - 1)/n, (step - 1)/n}],
            {step, 2, steps}]]]
```

We look at a few examples.

```
In[3]:= Do[Numeration[i], {i, 3}]
       Short[DownValues[PD], 12]
```

Out[4]//Short=
$\{$ HoldPattern[PD[1, 1]] :→ {0, 0},

HoldPattern[PD[1, 2]] :→ {1, 0}, HoldPattern[PD[1, 3]] :→ {0, 1},
HoldPattern[PD[2, 1]] :→ {0, 0}, HoldPattern[PD[2, 2]] :→ {1, 0},

HoldPattern[PD[2, 3]] :→ {0, 1}, HoldPattern[PD[2, 4]] :→ $\left\{\frac{1}{2}, 0\right\}$,

HoldPattern[PD[2, 5]] :→ $\left\{\frac{1}{2}, \frac{1}{2}\right\}$, HoldPattern[PD[2, 6]] :→ $\left\{0, \frac{1}{2}\right\}$,

HoldPattern[PD[3, 1]] :→ {0, 0}, HoldPattern[PD[3, 2]] :→ {1, 0},

HoldPattern[PD[3, 3]] :→ {0, 1}, HoldPattern[PD[3, 4]] :→ $\left\{\frac{1}{3}, 0\right\}$,

HoldPattern[PD[3, 5]] :→ $\left\{\frac{2}{3}, 0\right\}$, HoldPattern[PD[3, 6]] :→ $\left\{\frac{2}{3}, \frac{1}{3}\right\}$,

HoldPattern[PD[3, 7]] :→ $\left\{\frac{1}{3}, \frac{2}{3}\right\}$, HoldPattern[PD[3, 8]] :→ $\left\{0, \frac{2}{3}\right\}$,

HoldPattern[PD[3, 9]] :→ $\left\{0, \frac{1}{3}\right\}$, HoldPattern[PD[3, 10]] :→ $\left\{\frac{1}{3}, \frac{1}{3}\right\}\}$

Here is a plot; the idea of the numbering should become somewhat clearer.

```
In[5]:= NumerationPlot[n_Integer?Positive, opts___Rule] :=
       (Numeration[n];
       Show[Graphics[{{GrayLevel[0], (* the nodes *)
                      Table[Text["P" <> ToString[i], PD[n, i]],
                           {i, numKnots[n]}]},
                    (* the plot label *)
                     {GrayLevel[0], Text[StyleForm["Order " <> ToString[n],
                       FontFamily -> "Times", FontWeight -> "Bold",
                                         FontSize -> 5], {0.5, 0.9}]}}],
```

```
          opts, Axes -> False, PlotRange -> All, AspectRatio -> 1,
          Prolog -> {Hue[0.17], Polygon[{{0, 0}, {1, 0}, {0, 1}}]},
          TextStyle -> {FontSize -> 4}])
In[6]:= Show[GraphicsArray[#]]& /@
          Table[NumerationPlot[3i + j, DisplayFunction -> Identity],
              {i, 0, 2}, {j, 3}]
```

Order 1 · Order 2 · Order 3 · Order 4 · Order 5 · Order 6 · Order 7 · Order 8 · Order 9

Now, we discuss the computation of the Lagrange-shape functions. (We restrict ourselves to the Lagrange case [1562], although in present-day FEM calculations, Hermite-shape functions are being used more and more; see [1162] and [1010]. For the construction of shape functions with given interior nodes, see [1189].) The ith shape function $\varphi_i^{(e)}(\xi, \eta)$ is determined by the condition $\varphi_i^{(e)}(\xi_j, \eta_j) = \delta_{ij}$, where (ξ_j, η_j) are the coordinates of the jth data point $P_j^{(e)}$. Unfortunately, `Interpolat` `ingPolynomial` works only for one variable; and so, we have to program the computation of the interpolating two-dimensional polynomials.

```
In[7]:= ShapeFunction[n_Integer?Positive, i_Integer?Positive, ξ_, η_] :=
        (ShapeFunction[n, i, ξ, η] =
        Module[{x, y, tablePoly, tableAi, eqns},
            (* do the numeration *) Numeration[n];
            (* the monomials of the polynomial *)
            tablePoly = Flatten[Table[Table[x^i y^(j - i), {i, 0, j}], {j, 0, n}]];
            (* the coefficients to be determined *)
            tableAi = Table[a[i], {i, numKnots[n]}];
            (* make the left-hand side of the equations
               shapeFunction[i][Node[j]] == Kronecker[i, j] *)
            eqns = tablePoly.tableAi;
            sol = Simplify[eqns /. (* solve the equations *)
              Solve[ (* replace actual node coordinates *)
                Table[(eqns /. {x -> PD[n, j][[1]],
                       y -> PD[n, j][[2]]}) == KroneckerDelta[i, j],
                    {j, numKnots[n]}], tableAi]];
              (sol //. {x -> ξ, y -> η})[[1]]]) /; (i <= numKnots[n])
```

We now look at a few shape functions.

```
In[8]:= Do[Do[ (* make a reasonably nice looking print *)
          Print["SF[" <> ToString[j] <> ", " <> ToString[
                    i] <> "], ", ξ, η] = ",
                ShapeFunction[j, i, ξ, η]], {i, numKnots[j]}], {j, 2}]
```

$$SF[1, 1, \xi, \eta] = 1 - \eta - \xi$$

$$SF[1, 2, \xi, \eta] = \xi$$

$$SF[1, 3, \xi, \eta] = \eta$$

$$SF[2, 1, \xi, \eta] = 1 - 3\eta + 2\eta^2 + (-3 + 4\eta)\xi + 2\xi^2$$

$$SF[2, 2, \xi, \eta] = \xi(-1 + 2\xi)$$

$$SF[2, 3, \xi, \eta] = \eta(-1 + 2\eta)$$

$$SF[2, 4, \xi, \eta] = -4\xi(-1 + \eta + \xi)$$

$$SF[2, 5, \xi, \eta] = 4\eta\xi$$

$$SF[2, 6, \xi, \eta] = -4\eta(-1 + \eta + \xi)$$

Next, we check the property $\varphi_i^{(e)}(\xi_j, \eta_j) = \delta_{ij}$ explicitly for the first two sets of shape functions.

```
In[9]:= Table[ShapeFunction[1, i, ξ, η] /.
                {ξ -> PD[1, j][[1]], η -> PD[1, j][[2]]},
                {i, numKnots[1]}, {j, numKnots[1]}] // TableForm
```

Out[9]//TableForm=

1	0	0
0	1	0
0	0	1

```
In[10]:= Table[ShapeFunction[2, i, ξ, η] /.
                {ξ -> PD[2, j][[1]], η -> PD[2, j][[2]]},
                {i, numKnots[2]}, {j, numKnots[2]}] // TableForm
```

Out[10]//TableForm=

1	0	0	0	0	0
0	1	0	0	0	0
0	0	1	0	0	0
0	0	0	1	0	0
0	0	0	0	1	0
0	0	0	0	0	1

Now, we want to plot the shape functions. Because they are defined on a triangle, the built-in commands `Plot3D` and `ParametricPlot3D` are not immediately applicable (although one could proceed with `ParametricPlot3D`). We thus go to a little more trouble and decompose the unit triangle into subtriangles that we can use to plot the shape functions.

```
In[11]:= ShapeFunctionPlot[n_Integer?Positive, k_Integer?Positive,
                           pp_Integer?Positive, opts___Rule] :=
        Module[{x, y, s, t, tabv, tabd},
               (* calculate all z-values *)
               auxFu[ξ_, η_] = ShapeFunction[n, k, ξ, η];
               t = Table[{x, y}, {y, 0, 1, 1/pp}, {x, 0, 1 - y, 1/pp}];
               s = Table[auxFu[x, y], {y, 0, 1, 1/pp}, {x, 0, 1 - y, 1/pp}];
               (* triangulate the basic triangle {{0, 0}, {1, 0}, {0, 1}} *)
               tabv = Table[Polygon[
                     {{t[[i, j, 1]], t[[i, j, 2]], s[[i, j]]},
                      {t[[i + 1, j, 1]], t[[i + 1, j, 2]], s[[i + 1, j]]},
                      {t[[i + 1, j + 1, 1]], t[[i + 1, j + 1, 2]], s[[i + 1, j + 1]]},
                      {t[[i, j + 1, 1]], t[[i, j + 1, 2]], s[[i, j + 1]]}},
                           {j, 1, pp - 1}, {i, 1, pp - j}];
               (* boundary *)
               tabd = Table[Polygon[
                     {{t[[j, - 2, 1]], t[[j, -2, 2]], s[[j, -2]]},
                      {t[[j, - 1, 1]], t[[j, -1, 2]], s[[j, -1]]},
                      {t[[j + 1, -1, 1]], t[[j + 1, -1, 2]], s[[j + 1, -1]]}},
                           {j, pp}];
               (* display shape functions *)
               Show[Graphics3D[{{GrayLevel[0.6], tabv, tabd}},
                    opts, Axes -> False, AxesEdge -> {{-1, -1}, {+1, -1}, {-1, -1}},
```

```
            Lighting -> False, PlotRange -> All, BoxRatios -> {1, 1, 0.7},
            AxesLabel -> {x, y, None}, Boxed -> True,
            TextStyle -> {FontFamily -> "Times", FontSize -> 6},
            PlotLabel -> "SF[" <> ToString[n] <> ", " <>
                              ToString[k] <> "]"]] /; (k <= numKnots[n])
```

We can look at the shape functions for the first three orders.

In[12]:= Show[GraphicsArray[Table[ShapeFunctionPlot[1, i, 5,
                          DisplayFunction -> Identity],
                          {i, numKnots[1]}]]]

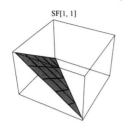
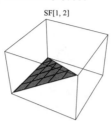
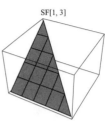

In[13]:= Show[GraphicsArray[#]]& /@
              Table[ShapeFunctionPlot[2, 3i + j, 12,
                      DisplayFunction -> Identity],
                  {i, 0, 1}, {j, 3}]

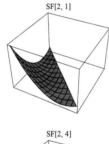
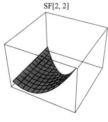
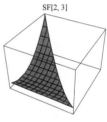

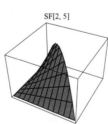
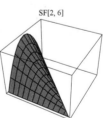

In[14]:= (* suppress message for only one picture in the last row *)
         Show[GraphicsArray[#]]& /@
              Table[ShapeFunctionPlot[3, 3i + j, 12,
                                  DisplayFunction -> Identity],
                  {i, 0, 2}, {j, 3}]

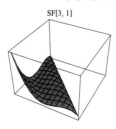
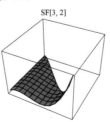
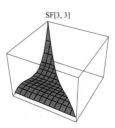

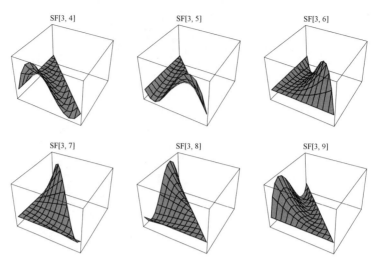

In[16]:= ShapeFunctionPlot[3, 10, 12]

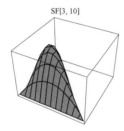

SF[3, 10]

We turn now to the computation of the integrals of the element vector and to the entries in the stiffness and mass matrices. Because these involve integrals of the shape functions $\psi_i(x, y)$ over the triangle with vertices $P_1 = \{x_1, y_1\}$, $P_2 = \{x_2, y_2\}$, and $P_3 = \{x_3, y_3\}$, we first find the corresponding Jacobian determinant. Here is the transformation from the $\{\xi, \eta\}$ coordinates to the $\{x, y\}$ coordinates:

$$x(\xi, v) = a_x \xi + b_x \eta + c_x,$$
$$y(\xi, \eta) = a_y \xi + b_y \eta + c_y.$$

In[17]:= Clear[x, y, x1, x2, x3, y1, y2, y3, ax, ay, bx, by, cx, cy, ξ, η];

```
Numeration[1];

Solve[ (* the equations made from the nodes *)
    Flatten[{Table[ax PD[1, i][[1]] + bx PD[1, i][[2]] + cx ==
                ToExpression["x" <> ToString[i]], {i, 1, 3}],
            Table[ay PD[1, i][[1]] + by PD[1, i][[2]] + cy ==
                ToExpression["y" <> ToString[i]], {i, 1, 3}]}],
                (* the coefficients to be determined *)
        {ax, ay, bx, by, cx, cy}] // TableForm[#[[1]]]&
```

Out[20]//TableForm=
```
ax → -x1 + x2
ay → -y1 + y2
bx → -x1 + x3
by → -y1 + y3
cx → x1
cy → y1
```

Thus, we express the coordinates x, y through ξ and η.

```
In[21]:= x[ξ_, η_] = (ax ξ + bx η + cx) /.
           {ax -> - x1 + x2, bx -> - x1 + x3, cx -> x1} // Simplify
Out[21]= x3 η + x2 ξ - x1 (-1 + η + ξ)
```

```
In[22]:= y[ξ_, η_] = (ay ξ + by η + cy) /.
           {ay -> - y1 + y2, by -> - y1 + y3, cy -> y1} // Simplify
Out[22]= y3 η + y2 ξ - y1 (-1 + η + ξ)
```

We then get the following Jacobian determinant.

```
In[23]:= Simplify[Det[Outer[D, {x[ξ, η], y[ξ, η]}, {ξ, η}]]]
Out[23]= x3 (y1 - y2) + x1 (y2 - y3) + x2 (-y1 + y3)
```

Next, we implement the relation $\int_0^1 \int_0^{1-\xi} \xi^p \, \eta^q \, d\eta \, d\xi = p!\,q!/(p+q+2)$ (for our applications p and q are positive integers).

The function `TriangularIntegration` implements the integration of polynomials over the unit triangle.

```
In[24]:= (* Additivity of the integration *)
         TriangularIntegration[p_Plus, {x_, y_}] :=
         TriangularIntegration[#, {x, y}]& /@ p;

         (* Factors that do not depend on the integration
            variables are moved in front of the integral *)
         TriangularIntegration[c_ z_, {x_, y_}] :=
         c TriangularIntegration[z, {x, y}] /; FreeQ[c, x] && FreeQ[c, y];

         (* let q be 0 *)
         TriangularIntegration[x_^p_., {x_, y_}] :=
                TriangularIntegration[x^p, {x, y}] = p!/(p + 2)!;

         (* let p be 0 *)
         TriangularIntegration[y_^q_., {x_, y_}] :=
                TriangularIntegration[x^q, {x, y}] = q!/(q + 2)!;

         (* the actual integration formula *)
         TriangularIntegration[x_^p_. y_^q_., {x_, y_}] :=
                TriangularIntegration[x^p y^q, {x, y}] = (p! q!) /(p + q + 2)!;

         (* integration of a constant *)
         TriangularIntegration[c_, {x_, y_}] :=
                (c/2) /; FreeQ[c, x] && FreeQ[c, y];
```

(For the efficient integration of analytic functions over triangles, see [446].) By comparing our triangular integration with the built-in command `Integrate`, we see that our work was justified.

```
In[36]:= Timing[TriangularIntegration[a + b x + c y^2 + d x^3 y^6 +
                           e x^12 y^ 16, {x, y}]]
```
$$Out[36]= \left\{0.\text{ Second}, \frac{a}{2} + \frac{b}{6} + \frac{c}{12} + \frac{d}{9240} + \frac{e}{26466926850}\right\}$$

```
In[37]:= Timing[Integrate[a + b x + c y^2 + d x^3 y^6 + e x^12 y^ 16,
                           {x, 0, 1}, {y, 0, 1 - x}]] // Simplify
```
$$Out[37]= \left\{1.35 \text{ Second}, \frac{a}{2} + \frac{b}{6} + \frac{c}{12} + \frac{d}{9240} + \frac{e}{26466926850}\right\}$$

Now to the heart of this problem: the computation of the element vector and the mass and stiffness matrices. For the element vector, we have

$$f_i = \int_{RT} \psi_i(x, y) \, dxdy = J \int_{UT} \varphi_i^{(e)}(\xi, \eta) \, d\xi \, d\eta = J \, f_i^{(e)}.$$

Here, RT denotes the real triangle, whereas UT denotes the unit triangle. J is the Jacobian determinant $|\partial(x, y)/\partial(\xi, \eta)|$. We get this relationship by means of the relations

$$x(\xi, \eta) = \sum_j x_j \, \varphi_j^{(e)}(\xi, \eta), \quad y(\xi, \eta) = \sum_j y_j \, \varphi_j^{(e)}(\xi, \eta)$$

(where x_j, y_j are the coordinates of the point P_j in the actual triangle RT), which hold for the isoparametric mappings $\psi_i(x(\xi, \eta), y(\xi, \eta)) = \varphi_i^{(e)}(\xi, \eta)$.

Thus, we compute only the element vector in the unit triangle $f_i^{(e)}$ (i.e., we do not explicitly write the Jacobian determinant).

```
In[38]:= ElementVectorElement[n_Integer?Positive, i_Integer?Positive] :=
    (ElementVectorElement[n, i] =
        TriangularIntegration[Expand[ShapeFunction[n, i, ξ, η]], {ξ, η}]) /;
                                                    (i <= numKnots[n])


    ElementVector[n_Integer?Positive] := ElementVector[n] =
            Table[ElementVectorElement[n, i], {i, numKnots[n]}]
```

Here are the element vectors of orders 1, 2, and 3.

```
In[40]:= ElementVector[1]
```

$$Out[40]= \left\{\frac{1}{6}, \frac{1}{6}, \frac{1}{6}\right\}$$

```
In[41]:= ElementVector[2]
```

$$Out[41]= \left\{0, 0, 0, \frac{1}{6}, \frac{1}{6}, \frac{1}{6}\right\}$$

```
In[42]:= ElementVector[3]
```

$$Out[42]= \left\{\frac{1}{60}, \frac{1}{60}, \frac{1}{60}, \frac{3}{80}, \frac{3}{80}, \frac{3}{80}, \frac{3}{80}, \frac{3}{80}, \frac{3}{80}, \frac{9}{40}\right\}$$

Here is the element vector for order 12 visualized.

```
In[43]:= Module[{n = 12, evec}, evec = ElementVector[n];
    Show[GraphicsArray[
        {(* values in consecutive order *)
        ListPlot[evec, PlotRange -> All, Frame -> True,
                    Axes -> False, PlotStyle -> {PointSize[0.008]},
                    DisplayFunction -> Identity],
        (* values over the base points; coloring according to size *)
        Graphics3D[{Hue[0.76 #[[2]]/Max[evec]],
            Line[{Append[#[[1]], 0], Append[#[[1]], Abs[#[[2]]]]}]}& /@
                Transpose[{Table[PD[n, k], {k, numKnots[n]}], evec}],
                BoxRatios -> {1, 1, 0.5}, PlotRange -> All, Axes -> True]}]]]
```

The computation of the mass matrix is essentially analogous to that for the eigenvector. Using similar notation as in the element vector case, we have

$$m_{ij} = \int_{RT} \psi_i(x, y)\, \psi_j(x, y)\, dxdy = J \int_{UT} \varphi_i^{(e)}(\xi, \eta)\, \varphi_j^{(e)}(\xi, \eta)\, d\xi\, d\eta = J\, m_{ij}^{(e)}.$$

Again, we find only the coordinate-free part.

```
In[44]:= MassMatrixElement[n_Integer?Positive,
                        i_Integer?Positive, j_Integer?Positive] :=
    (MassMatrixElement[n, i, j] = (* because of symmetry *)
    MassMatrixElement[n, j, i] =
        TriangularIntegration[Expand[ShapeFunction[n, i, ξ, η] *
                            ShapeFunction[n, j, ξ, η]], {ξ, η}]) /;
                ((i <= numKnots[n]) && (j <= numKnots[n]))
```

```
MassMatrix[n_Integer?Positive] := MassMatrix[n] =
    Table[MassMatrixElement[n, i, j],
        {i, numKnots[n]}, {j, numKnots[n]}]
```

Here are the mass matrices for the two lowest orders.

In[46]:= **MassMatrix[1] // TableForm**

Out[46]//TableForm=

$$\begin{pmatrix} \frac{1}{12} & \frac{1}{24} & \frac{1}{24} \\ \frac{1}{24} & \frac{1}{12} & \frac{1}{24} \\ \frac{1}{24} & \frac{1}{24} & \frac{1}{12} \end{pmatrix}$$

In[47]:= **MassMatrix[2] // TableForm[#, TableAlignments -> Center]&**

Out[47]//TableForm=

$$\begin{pmatrix} \frac{1}{60} & -\frac{1}{360} & -\frac{1}{360} & 0 & -\frac{1}{90} & 0 \\ -\frac{1}{360} & \frac{1}{60} & -\frac{1}{360} & 0 & 0 & -\frac{1}{90} \\ -\frac{1}{360} & -\frac{1}{360} & \frac{1}{60} & -\frac{1}{90} & 0 & 0 \\ 0 & 0 & -\frac{1}{90} & \frac{4}{45} & \frac{2}{45} & \frac{2}{45} \\ -\frac{1}{90} & 0 & 0 & \frac{2}{45} & \frac{4}{45} & \frac{2}{45} \\ 0 & -\frac{1}{90} & 0 & \frac{2}{45} & \frac{2}{45} & \frac{4}{45} \end{pmatrix}$$

The computation of the stiffness matrix is somewhat more complicated. Here, we need derivatives with respect to the variables x and y of the actual triangle RT, but our integration of $\varphi_i^{(e)}(\xi, \eta)$ takes over the unit triangle UT:

$$s_{ij} = \int_{RT} \left(\frac{\partial}{\partial x} \varphi_i(x, y) \frac{\partial}{\partial x} \varphi_j(x, y) + \frac{\partial}{\partial y} \varphi_i(x, y) \frac{\partial}{\partial y} \varphi_j(x, y) \right) dx\, dy.$$

Now with $\psi_i(x(\xi, \eta), y(\xi, \eta)) = \varphi_i^{(e)}(\xi, \eta)$, we have

$$\frac{\partial}{\partial x} \psi_i(x, y) = \frac{\partial}{\partial x} \psi_i(x(\xi, \eta), y(\xi, \eta)) = \frac{\partial}{\partial x} \varphi_i^{(e)}(\xi(x, y), \eta(x, y))$$

$$= \frac{\partial}{\partial \xi} \varphi_i^{(e)}(\xi, \eta) \cdot \frac{\partial}{\partial x} \xi(x, y) + \frac{\partial}{\partial \eta} \varphi_i^{(e)}(\xi, \eta) \cdot \frac{\partial}{\partial x} \eta(x, y).$$

Similarly,

$$\frac{\partial}{\partial y} \psi_i(x, y) = \frac{\partial}{\partial y} \psi_i(x(\xi, \eta), y(\xi, \eta)) = \frac{\partial}{\partial y} \varphi_i^{(e)}(\xi(x, y), \eta(x, y))$$

$$= \frac{\partial}{\partial \xi} \varphi_i^{(e)}(\xi, \eta) \cdot \frac{\partial}{\partial y} \xi(x, y) + \frac{\partial}{\partial \eta} \varphi_i^{(e)}(\xi, \eta) \cdot \frac{\partial}{\partial y} \eta(x, y).$$

Thus, by inversion of the above $x(\xi, \eta)$, $y(\xi, \eta)$, we get $\xi = \xi(x, y)$ and $\eta = \eta(x, y)$. Here, it is computed.

In[48]:= **((Solve[{x1 - η x1 - ξ x1 + ξ x2 + η x3 == x,**
          y1 - η y1 - ξ y1 + ξ y2 + η y3 == y}, {ξ, η}] // Simplify) //.
        { -(x2 y1) + x3 y1 + x1 y2 - x3 y2 - x1 y3 + x2 y3 -> J,
        -(-(x2 y1) + x3 y1 + x1 y2 - x3 y2 - x1 y3 + x2 y3) -> -J}) // Simplify

Out[48]= $\left\{ \left\{ \xi \to \dfrac{x3\,(-y + y1) + x1\,(y - y3) + x\,(-y1 + y3)}{x3\,(y1 - y2) + x1\,(y2 - y3) + x2\,(-y1 + y3)},\ \eta \to \dfrac{x2\,(y - y1) + x\,(y1 - y2) + x1\,(-y + y2)}{J} \right\} \right\}$

We now can calculate the following four quantities: $\frac{\partial}{\partial x} \xi(x, y)$, $\frac{\partial}{\partial y} \eta(x, y)$, $\frac{\partial}{\partial y} \xi(x, y)$, $\frac{\partial}{\partial y} \eta(x, y)$.

In[49]:= **Ξ[x_, y_] = (x1 y - x3 y - x y1 + x3 y1 + x y3 - x1 y3)/J;**
       H[x_, y_] = (x2 y - x1 y + x y1 - x2 y1 - x y2 + x1 y2)/J;

In[51]:= **{dξdx = D[Ξ[x, y], x], dηdx = D[H[x, y], x],**
       dξdy = D[Ξ[x, y], y], dηdy = D[H[x, y], y]}

Out[51]= $\left\{ \dfrac{-y1 + y3}{J},\ \dfrac{y1 - y2}{J},\ \dfrac{x1 - x3}{J},\ \dfrac{-x1 + x2}{J} \right\}$

We now rewrite $\frac{\partial}{\partial x} \psi_i(x, y) \frac{\partial}{\partial x} \psi_j(x, y) + \frac{\partial}{\partial y} \psi_i(x, y) \frac{\partial}{\partial y} \psi_j(x, y)$ in the form

$$
\left(\frac{\partial}{\partial\xi}\,\varphi_i^{(e)}(\xi,\eta)\cdot\frac{\partial}{\partial\xi}\,\varphi_j^{(e)}(\xi,\eta)\right)\cdot\left(\left(\frac{\partial}{\partial x}\,\xi(x,y)\right)^2+\left(\frac{\partial}{\partial y}\,\xi(x,y)\right)^2\right)+
$$

$$
\left(\frac{\partial}{\partial\eta}\,\varphi_i^{(e)}(\xi,\eta)\cdot\frac{\partial}{\partial\eta}\,\varphi_j^{(e)}(\xi,\eta)\right)\cdot\left(\left(\frac{\partial}{\partial x}\,\eta(x,y)\right)^2+\left(\frac{\partial}{\partial y}\,\eta(x,y)\right)^2\right)+
$$

$$
\left(\frac{\partial}{\partial\xi}\,\varphi_i^{(e)}(\xi,\eta)\cdot\frac{\partial}{\partial\eta}\,\varphi_j^{(e)}(\xi,\eta)+\frac{\partial}{\partial\eta}\,\varphi_i^{(e)}(\xi,\eta)\cdot\frac{\partial}{\partial\xi}\,\varphi_j^{(e)}(\xi,\eta)\right)\times
$$

$$
\left(\frac{\partial}{\partial x}\,\xi(x,y)\cdot\frac{\partial}{\partial x}\,\eta(x,y)+\frac{\partial}{\partial y}\,\xi(x,y)\cdot\frac{\partial}{\partial y}\,\eta(x,y)\right)
$$

and introduce

$$
A\;=\;\left(\left(\frac{\partial}{\partial x}\,\xi(x,y)\right)^2+\left(\frac{\partial}{\partial y}\,\xi(x,y)\right)^2\right)J=\frac{(x_3-x_1)^2+(y_3-y_1)^2}{J}
$$

$$
C\;=\;\left(\left(\frac{\partial}{\partial x}\,\eta(x,y)\right)^2+\left(\frac{\partial}{\partial y}\,\eta(x,y)\right)^2\right)J=\frac{(x_2-x_1)^2+(y_2-y_1)^2}{J}
$$

$$
B\;=\;\left(\frac{\partial}{\partial x}\,\xi(x,y)\,\frac{\partial}{\partial x}\,\eta(x,y)+\frac{\partial}{\partial y}\,\xi(x,y)\,\frac{\partial}{\partial y}\,\eta(x,y)\right)J=-\frac{(y_3-y_1)(y_2-y_1)+(x_3-x_1)(x_2-x_1)}{J}.
$$

This leads to the following result:

$$
s_{ij}\;=\;\int_{RT}\left(\frac{\partial}{\partial x}\,\psi_i(x,y)\,\frac{\partial}{\partial x}\,\psi_j(x,y)+\frac{\partial}{\partial y}\,\psi_i(x,y)\,\frac{\partial}{\partial y}\,\psi_j(x,y)\right)d\xi d\eta
$$

$$
=\;A\int_{UT}\frac{\partial}{\partial\xi}\,\varphi_i^{(e)}(\xi,\eta)\,\frac{\partial}{\partial\xi}\,\varphi_j^{(e)}(\xi,\eta)\,d\xi d\eta+
$$

$$
C\int_{UT}\frac{\partial}{\partial\eta}\,\varphi_i^{(e)}(\xi,\eta)\,\frac{\partial}{\partial\eta}\,\varphi_j^{(e)}(\xi,\eta)\,d\xi d\eta+
$$

$$
B\int_{UT}\left(\frac{\partial}{\partial\xi}\,\varphi_i^{(e)}(\xi,\eta)\,\frac{\partial}{\partial\eta}\,\varphi_j^{(e)}(\xi,\eta)+\frac{\partial}{\partial\eta}\,\varphi_i^{(e)}(\xi,\eta)\,\frac{\partial}{\partial\xi}\,\varphi_j^{(e)}(\xi,\eta)\right)d\xi d\eta.
$$

```mathematica
In[52]:= StiffnessMatrixElement[n_Integer?Positive,
                                i_Integer?Positive, j_Integer?Positive] :=
        (StiffnessMatrixElement[n, i, j] = (* because of symmetry *)
         StiffnessMatrixElement[n, j, i] =
        With[{SF = ShapeFunction},
        (* sum of the three terms *)
        A TriangularIntegration[
            Expand[D[SF[n, i, ξ, η], ξ] D[SF[n, j, ξ, η], ξ]], {ξ, η}] +
        C TriangularIntegration[
            Expand[D[SF[n, i, ξ, η], η] D[SF[n, j, ξ, η], η]], {ξ, η}] +
        B TriangularIntegration[
            Expand[D[SF[n, i, ξ, η], ξ] D[SF[n, j, ξ, η], η] +
                    D[SF[n, i, ξ, η], η] D[SF[n, j, ξ, η], ξ]], {ξ, η}]]) /;
                    ((i <= numKnots[n]) && (j <= numKnots[n]))


        StiffnessMatrix[n_Integer?Positive] := StiffnessMatrix[n] =
            Table[StiffnessMatrixElement[n, i, j],
                    {i, numKnots[n]}, {j, numKnots[n]}]
```

As an example, we again look at the two lowest orders.

```mathematica
In[54]:= StiffnessMatrix[1] // TableForm
```

$$
Out[54]//TableForm=\quad
\begin{array}{ccc}
\frac{A}{2}+B+\frac{C}{2} & -\frac{A}{2}-\frac{B}{2} & -\frac{B}{2}-\frac{C}{2}\\[4pt]
-\frac{A}{2}-\frac{B}{2} & \frac{A}{2} & \frac{B}{2}\\[4pt]
-\frac{B}{2}-\frac{C}{2} & \frac{B}{2} & \frac{C}{2}
\end{array}
$$

In[55]:= StiffnessMatrix[2] //
    TableForm[#, TableAlignments -> Center, TableSpacing -> {1, 1}]&

Out[55]//TableForm=

$$
\begin{array}{cccccc}
\frac{A}{2} + B + \frac{C}{2} & \frac{A}{6} + \frac{B}{6} & \frac{B}{6} + \frac{C}{6} & -\frac{2A}{3} - \frac{2B}{3} & 0 & -\frac{2B}{3} - \frac{2C}{3} \\[6pt]
\frac{A}{6} + \frac{B}{6} & \frac{A}{2} & -\frac{B}{6} & -\frac{2A}{3} - \frac{2B}{3} & \frac{2B}{3} & 0 \\[6pt]
\frac{B}{6} + \frac{C}{6} & -\frac{B}{6} & \frac{C}{2} & 0 & \frac{2B}{3} & -\frac{2B}{3} - \frac{2C}{3} \\[6pt]
-\frac{2A}{3} - \frac{2B}{3} & -\frac{2A}{3} - \frac{2B}{3} & 0 & \frac{4A}{3} + \frac{4B}{3} + \frac{4C}{3} & -\frac{4B}{3} - \frac{4C}{3} & \frac{4B}{3} \\[6pt]
0 & \frac{2B}{3} & \frac{2B}{3} & -\frac{4B}{3} - \frac{4C}{3} & \frac{4A}{3} + \frac{4B}{3} + \frac{4C}{3} & -\frac{4A}{3} - \frac{4B}{3} \\[6pt]
-\frac{2B}{3} - \frac{2C}{3} & 0 & -\frac{2B}{3} - \frac{2C}{3} & \frac{4B}{3} & -\frac{4A}{3} - \frac{4B}{3} & \frac{4A}{3} + \frac{4B}{3} + \frac{4C}{3}
\end{array}
$$

For a larger order, we will visualize the resulting mass and stiffness matrices. Here are these two matrices shown for $n = 10$ for the unit triangle.

In[56]:= With[{n = 10},
    Show[GraphicsArray[
        ListDensityPlot[(* scale *) ArcTan[#], PlotRange -> All, Mesh -> False,
                DisplayFunction -> Identity]& /@
        (* calculate exact mass and stiffness matrices *)
        {MassMatrix[n], StiffnessMatrix[n] /. {A -> 1, C -> 1, B -> 0}}]]]

The subject of finite elements contains many other opportunities for programming with *Mathematica*. For example, we mention algorithms for minimizing the bandwidth of sparse matrices (following, e.g., Cuthill–McKee [424], Gibbs–Poole–Stockmeyer ([723] and [711]), or Sloan [1625]). Because of their special nature, we do not go any further into the explicit implementation of these finite-element computations.

b) We start by implementing the interpolating functions $\chi_{k,l}^{(p,d)}(\xi)$. Using the function InterpolatingPolynomial, their construction is straightforward for explicitly given integers e, p, d, k, and l. While the unexpanded form has a better stability for numerical evaluation, we expand the functions here to speed-up the integrations to be carried out later.

In[1]:= χ[p_, d_][k_, l_, ξ_] :=
    Expand[InterpolatingPolynomial[
    Table[{j/p, Table[KroneckerDelta[j, k]*
                KroneckerDelta[l, i], {i, 0, d}]}, {j, 0, p}], ξ]]

Here are two examples:

In[2]:= {χ[3, 0][0, 0, ξ], χ[2, 2][1, 1, ξ]}

Out[2]= $\left\{ 1 - \frac{11\,\xi}{2} + 9\,\xi^2 - \frac{9\,\xi^3}{2},\ -32\,\xi^3 + 160\,\xi^4 - 288\,\xi^5 + 224\,\xi^6 - 64\,\xi^7 \right\}$

We sidestep a moment and visualize some of the $\chi_{k,l}^{(p,d)}(\xi)$. The function maxAbs[p, d][k, l] calculates the maximum of the absolute value of the $\chi_{k,l}^{(p,d)}(\xi)$ over the ξ-interval [0, 1].

In[3]:= maxAbs[p_, d_][k_, l_] :=
    Module[{f = χ[p, d][k, l, ξ], extξs},
    (* solve for extrema *)
    extξs = Select[N[{ToRules[Roots[D[f, ξ, ξ] == 0, ξ,
                    Cubics -> False, Quartics -> False]]}, 50],
                (Im[ξ /. #] == 0 && 0 <= (ξ /. #) <= 1)&];
    Max[Abs[f /. Join[extξs, {{ξ -> 0}, {ξ -> 1}}]]]]

The magnitude of the functions decreases quickly with higher-order continuity.

```
In[4]:= With[{p = 4, d = 4},
          Table[{j, Max[Table[maxAbs[p, d][i, j], {i, 0, p}]] // N},
            {j, 0, d}]]
```
Out[4]= {{0, 5.69702}, {1, 0.369672}, {2, 0.0478262}, {3, 0.000512063}, {4, 0.000037622}}

The function hpPlot implements a numerically reliable plotting function that we will use to safely plot the (for larger p and d high-order) polynomials.

```
In[5]:= hpPlot[f_, {ξ_, ξ0_, ξ1_}, prec_, opts___] :=
         Module[{f = Function @@ {ξ, f}, f},
           (* use high-precision arithmetic *)
           f[x_?NumericQ] := f[SetPrecision[x, prec]];
           Plot[f[ξ], {ξ, ξ0, ξ1}, opts, PlotRange -> All,
             Frame -> True, DisplayFunction -> Identity]]
```

The next four graphics show some functions from the family $\chi_{k,l}^{(4,5)}$. Be aware of the greatly different ordinate values on the four graphics.

```
In[6]:= With[{p = 4, d = 5},
         Show[GraphicsArray[Table[Show[
           Table[hpPlot[χ[p, d][k, 1, ξ], {ξ, 0, 1}, 100,
                 PlotStyle -> {{Thickness[0.002], Hue[k/6]}}],
             {k, 0, p}], {1, 0, 3}]]]]]
```

For a fixed l in $\chi_{k,l}^{(p,d)}$, the functions are roughly of equal magnitude. The following graphics show this for $p = d = 1$, $p = d = 2$, and $p = d = 3$.

```
In[7]:= graph[p_, d_][k_, l_] :=
         Plot[Evaluate[χ[p, d][k, 1, ξ]], {ξ, 0, 1},
             DisplayFunction -> Identity, FrameTicks -> None,
             PlotRange -> All, Frame -> True, Axes -> False]

         Show[GraphicsArray[#]]& /@ Table[
             Table[graph[μ, μ][k, 1], {1, 0, μ}, {k, 0, μ}], {μ, 3}]
```

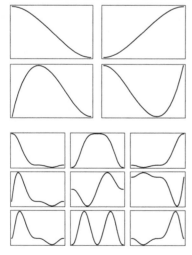

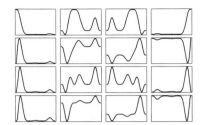

Before starting the implementation of the functions to solve the eigenvalue problem, we will renumber the $\chi_{k,l}^{(p,d)}$. For fixed p and d, we want to number the functions $\chi_{k,l}^{(p,d)}(\xi) = \chi_h^{(p,d)}(\xi)$ using one index to easily assemble the global finite element matrices. We number them consecutively with increasing k, and within each k with increasing l. The function reducesIn dices does the inverse: given the linear numbering h, it generates the pairs (k, l).

```
In[10]:= reducesIndices[p_, d_][h_] :=
            Sequence[Floor[h/(d + 1)], h - (d + 1) Floor[h/(d + 1)]]
```

Here are the sixteen pairs corresponding to $\chi_h^{(3,3)}(\xi)$.

```
In[11]:= Table[{k, {reducesIndices[3, 3][k]}}, {k, 0, 15}]

Out[11]= {{0, {0, 0}}, {1, {0, 1}}, {2, {0, 2}}, {3, {0, 3}}, {4, {1, 0}},
          {5, {1, 1}}, {6, {1, 2}}, {7, {1, 3}}, {8, {2, 0}}, {9, {2, 1}}, {10, {2, 2}},
          {11, {2, 3}}, {12, {3, 0}}, {13, {3, 1}}, {14, {3, 2}}, {15, {3, 3}}}
```

Here is the definition for the $\chi_{k,l}^{(p,d)}(\xi) = \chi_h^{(p,d)}(\xi)$. The local element matrices of the shape functions $\chi_i^{(p,d)}(\xi)$ are defined in the interval $[0, 1]$.

```
In[12]:= χ[p_, d_][k_, ξ_] := (* remember *) χ[p, d][k, ξ] =
              χ[p, d][reducesIndices[p, d][k], ξ]
```

For the second-order eigenvalue problem, we now must calculate the following element stiffness and mass matrices $t_{i,j}^{(p,d)}$, $m_{i,j}^{(p,d)}$, and $v_{i,j}^{(\sigma,p,d)}$ (primes denote differentiation with respect to ξ):

$$t_{i,j}^{(p,d)} = \int_0^1 \chi_i^{(p,d)\prime}(\xi)\, \chi_j^{(p,d)\prime}(\xi)\, d\xi$$

$$m_{i,j}^{(p,d)} = \int_0^1 \chi_i^{(p,d)}(\xi)\, \chi_j^{(p,d)}(\xi)\, d\xi$$

$$v_{i,j}^{(\sigma,p,d)} = \int_0^1 \chi_i^{(p,d)}(\xi)\, \xi^\sigma\, \chi_j^{(p,d)}(\xi)\, d\xi$$

In the definitions of the $v_{i,j}^{(\sigma,p,d)}$, we used a power function ξ^σ because of the x^2 potential of the harmonic oscillator.

These definitions for the matrices $t_{i,j}^{(p,d)}$, $m_{i,j}^{(p,d)}$, and $v_{i,j}^{(\sigma,p,d)}$ can straightforwardly be implemented. We store the calculated values for the matrix elements because they are identical for all elements and will be used repeatedly later. We use the symmetries in the indices i and j to avoid doubling the work.

```
In[13]:= t[p_, d_][i_, j_] := (t[p, d][i, j] = t[p, d][j, i] =
              Integrate[Expand[D[χ[p, d][i, ξ], ξ] D[χ[p, d][j, ξ], ξ]], {ξ, 0, 1}])

          m[p_, d_][i_, j_] := (m[p, d][i, j] = m[p, d][j, i] =
              Integrate[Expand[χ[p, d][i, ξ] χ[p, d][j, ξ]], {ξ, 0, 1}])

          v[p_, d_][e_, i_, j_] := (v[p, d][e, i, j] = v[p, d][e, j, i] =
              Integrate[Expand[ξ^e χ[p, d][i, ξ] χ[p, d][j, ξ]], {ξ, 0, 1}])
```

The next three density plots show the element matrices for $p = d = 4$. The periodic structure results from the 1D ordering scheme. The matrix elements decrease rapidly for increasing l for each k.

```
In[17]:= Module[{p = 4, d = 4, o},
              (* element matrix size *) o = (p + 1) (d + 1) - 1;
              Show[GraphicsArray[
              ListDensityPlot[ArcTan[#],
                  Mesh -> False, DisplayFunction -> Identity,
```

```
          FrameTicks -> False, PlotRange -> All,
          ColorFunction -> (Hue[0.78 #]&)]& /@
       (* the three matrices t, m, and v *)
       {Table[t[p, d][i, j], {i, 0, o}, {j, 0, o}],
        Table[m[p, d][i, j], {i, 0, o}, {j, 0, o}],
        Table[v[p, d][2, i, j], {i, 0, o}, {j, 0, o}]}]]]]
```

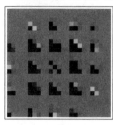  

The matrices $t_{i,j}^{(p,d)}$, $m_{i,j}^{(p,d)}$, and $v_{i,j}^{(e,p,d)}$ refer to the normalized element $(0, 1)$. We now construct the corresponding matrices $\mathcal{T}_{i,j}^{(p,d)}$, $\mathcal{M}_{i,j}^{(p,d)}$, and $\mathcal{V}_{i,j}^{(e,p,d)}$ for the interval $[x_j, x_j + \delta]$. The matrix $\mathcal{M}_{i,j}^{(p,d)}$ gets just multiplied by the length δ of the interval. Similarly, $\mathcal{T}_{i,j}^{(p,d)}$ scales with $1/\delta$. $\mathcal{V}_{i,j}^{(e,p,d)}$ contains information about the explicitly x-dependent function x^2 and contains information about the element length δ and the element position x_j.

```
In[18]:= T[p_, d_, δ_][i_, j_, ___]    := 1/δ t[p, d][i, j]
         V[p_, d_, δ_][i_, j_, xj_] := δ (xj^2 v[p, d][0, i, j] +
                                           2 xj δ v[p, d][1, i, j] +
                                           δ^2 v[p, d][2, i, j])
         M[p_, d_, δ_][i_, j_, ___]    := δ m[p, d][i, j]
```

The next step is to assemble the local element matrices $\mathcal{T}_{i,j}^{(p,d)}$, $\mathcal{V}_{i,j}^{(p,d)}$, and $\mathcal{M}_{i,j}^{(e,p,d)}$ into global matrices $\mathbb{T}_{i,j}$, $\mathbb{V}_{i,j}$ and $\mathbb{M}_{i,j}$. The coefficients $c_{k,l}^{(j,p,d)}$ multiplied by the element functions form the resulting function $\psi(x)$. For $1 < k < p - 1$, only element functions inside the jth element contribute. The two coefficients $c_{p,l}^{(j,p,d)}$ and $c_{0,l}^{(j+1,p,d)}$ are both attached to the same x-node (being the functions associated with the leftmost node of the element $[x_j, x_j + \delta]$ and the functions associated with the rightmost node from the interval $[x_{j+1}, x_{j+1} + \delta]$) and are so identical. This means that the resulting global stiffness and mass matrices have dimension $(e\, p + 1)(d + 1) \times (e\, p + 1)(d + 1)$. (Because all elements have the same length δ, we do not have to worry about prefactors in the mapping of the from $[0, 1]$ to the intervals $[x_j, x_j + \delta]$ for the $\chi_{k,l}^{(p,d)}$ with $l > 0$.)

The function $\mathtt{TVM}$ assembles the three matrices $\mathbb{T}_{i,j}$, $\mathbb{V}_{i,j}$ and $\mathbb{M}_{i,j}$.

```
In[21]:= TVM[{elements_, L_}, elementNodes_, diffOrder_] :=
         Module[{e = elements, p = elementNodes,
                 d = diffOrder, δ = 2L/elements, o, m},
           (* element matrix size *) o = (p + 1) (d + 1);
           ((* blank matrix to be filled *)
           m = Table[0, {(e p + 1) (d + 1)}, {(e p + 1) (d + 1)}];
           Do[(* matrices for all elements *)
             Do[m[[p (d + 1) k + i + 1, p (d + 1) k + j + 1]] =
               m[[p (d + 1) k + i + 1, p (d + 1) k + j + 1]] +
                                    #[i, j, -L + k/e 2L],
               {i, 0, o - 1}, {j, 0, o - 1}], {k, 0, e - 1}]; m)& /@
         {T[p, d, δ], V[p, d, δ], M[p, d, δ]}]
```

The resulting matrices $\mathbb{T}_{i,j}$, $\mathbb{V}_{i,j}$ and $\mathbb{M}_{i,j}$ have a block structure due to the individual elements assembled. Here the structure of the resulting matrices is shown for $e = 5$, $p = d = 3$.

```
In[22]:= Show[GraphicsArray[
         ListDensityPlot[Sign[#], Mesh -> False,
                         PlotRange -> All, FrameTicks -> None,
                         DisplayFunction -> Identity]& /@ TVM[{5, 1}, 3, 3]]]
```

  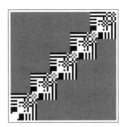

Now the boundary conditions $\psi(-L) = \psi(L) = 0$ must be applied. To do this, the function `dirichletize` sets the corresponding values of the node values to zero, meaning it removes the corresponding two equations.

```
In[23]:= dirichletize[{t_, v_, m_}, diffOrder_] :=
    Map[Delete[#, {{1}, {-(diffOrder + 1)}}]&, {t, v, m}, {1, 2}]
```

The three matrices $\mathbb{T}_{i,j}$, $\mathbb{V}_{i,j}$ and $\mathbb{M}_{i,j}$ (we use the same variables even after imposing the boundary conditions) have now to satisfy the following algebraic eigenvalue problem:

$$\sum_{j=1}^{(e\,p+1)(d+1)} (\mathbb{T}_{i,j} + \mathbb{V}_{i,j})\,\psi_j = \varepsilon \sum_{j=1}^{(e\,p+1)(d+1)} \mathbb{M}_{i,j}\,\psi_j.$$

(For a finite difference approach that yields a generalized eigenvalue problem, see [537].) Now that we have the matrices forming the generalized eigenvalue problem, we can deal with solving it. In absence of a dedicated generalized eigenvalue solver, we have to transform the problem to a classical eigenvalue problem $\mathbb{A}.\psi = \varepsilon\,\psi$ for a matrix $\mathbb{A}$. To do this, we multiply from the left with the inverse of $\mathbb{M}_{i,j}$. To ensure real eigenvalues, an explicitly symmetric matrix $\mathbb{A}$ is desirable. So instead of $\mathbb{M}_{i,j}^{-1}(\mathbb{T}_{i,j} + \mathbb{V}_{i,j})$ we make use of the normality of $\mathbb{A}$ and form $(\mathbb{M}_{i,j}^{-1}(\mathbb{T}_{i,j}^{\mathsf{T}} + \mathbb{V}_{i,j}^{\mathsf{T}}) + \mathbb{M}_{i,j}^{-1\,\mathsf{T}}(\mathbb{T}_{i,j} + \mathbb{V}_{i,j}))/2$ [875].

```
In[24]:= symmetricLhs[{t_, v_, m_}] :=
    With[{i = Inverse[m]},
        (i.Transpose[t + v] + Transpose[i].(t + v))/2];
```

This yields the following function `lowestEigenvalues[{e, L}, p, d, m, WorkingPrecision -> wp]` for the calculation of the *m* lowest eigenvalues. We give this function a `WorkingPrecision` option to allow the use of high-precision arithmetic. (For the (relatively fast) matrix inversion, we use a higher precision than for the eigenvalue calculation.)

```
In[25]:= lowestEigenvalues[{elements_, L_}, elementNodes_, diffOrder_,
                numberOfEigenvalues_, WorkingPrecision -> wp_] :=
    Take[Eigenvalues[SetPrecision[symmetricLhs @
        (* calculate global matrices *)
        N[dirichletize[TVM[{elements, L},
            elementNodes, diffOrder], diffOrder], 2 wp], wp]],
        -numberOfEigenvalues]
```

Here are two examples. The first example uses 32 elements and linear Lagrange interpolation. The resulting matrices are of size 51. The agreement with the exact values $2\,j + 1$ is moderate.

```
In[26]:= lowestEigenvalues[{50, 6}, 1, 0, 5,
                WorkingPrecision -> $MachinePrecision]
Out[26]= {9.14521689982773, 7.088879599427062,
    5.046391357619056, 3.017911029232807, 1.003596313258432}
```

Now, we must optimize *L*. For $p = 1, d = 0$, $p = 2, d = 0$, and $p = 2, d = 2$ and resulting matrix size of about 64, we calculate and display the difference between the calculated and the exact eigenvalues. Because we minimize a variational expression, all our lowest eigenvalue estimates will always be above the correct values.

```
In[27]:= {evs10, evs20, evs22} =
    Table[{L, #}& /@ (lowestEigenvalues[{#1, L}, #2, #3, 5,
                WorkingPrecision -> $MachinePrecision] -
            (* exact values *) {9, 7, 5, 3, 1}),
        (* L-values *) {L, 2, 12, 1/2}]& @@@
    (* L, p, and d *) {{50, 1, 0}, {25, 2, 0}, {12, 2, 2}};
```

In[28]:=

```
Show[GraphicsArray[Graphics[Line /@ Transpose[
    Apply[{{#1, Log[10, #2]}&, #, {-2}]], Frame -> True]& /@
                                    {evs10, evs20, evs22}]]
```

For $d = 0$, we get $L_{opt} \approx 4$; and for larger d, the value of L_{opt} increases. The last graphics suggest that it is possible to calculate the ground-state to 10 correct digits with a resulting matrix of dimension 50×50. Let us now try to get 20 correct digits.

With eight elements and $p = d = 2$, we get a 61×61 matrix and 13 correct digits.

In[29]:= `lowestEigenvalues[{10, 6}, 2, 2, 1, WorkingPrecision -> 20] - 1`

Out[29]= $\{6.715385 \times 10^{-13}\}$

We will now increase the smoothness of the eigenfunctions. To use similar resulting matrix sizes, we have to reduce the number of elements accordingly. With four elements and $p = 3$ and $d = 4$, we get a 63×63 matrix and 20 correct digits. Higher-order smoothness gives considerably better results than just continuity [787].

In[30]:= `lowestEigenvalues[{4, 7}, 3, 4, 1, WorkingPrecision -> 60] - 1`

Out[30]= $\{2.9710069844126506438596840428671413299971 \times 10^{-20}\}$

With only two elements and $p = 2$ and $d = 12$ (meaning high-order continuity), we get a 63×63 matrix and 25 correct digits.

In[31]:= `lowestEigenvalues[{2, 8}, 2, 12, 1, WorkingPrecision -> 60] - 1`

Out[31]= $\{1.4440561957059222478901951763095147 \times 10^{-25}\}$

Even with only one single element and $p = 12$, $d = 4$, we can get 24 correct digits with a resulting 63×63 matrix.

In[32]:= `lowestEigenvalues[{1, 15/2}, 12, 4, 1, WorkingPrecision -> 60] - 1`

Out[32]= $\{6.28804898816808241805856957757574148377 \times 10^{-24}\}$

(For some exact solutions to low-degree finite element approximations of the harmonic oscillator, see [392]; for multidimensional oscillators, see [1429]. For a bra-ket formulation of the finite element method, see [1147].)

Using the known symmetry of the ground-state, we could solve the eigenvalue problem in the interval $[0, L]$ with Neumann boundary conditions on the left interval end. With the same matrix size limitation of 64, this reduction would allow us to get nearly 50 correct digits. We could further optimize by using elements of different length and nonuniform distributions of nodes inside the elements [201] or using special element basis functions [665]. We leave further experiments to the reader.

8. Helium Atom, Improved Variational Method

a) This problem is a perfect example of the effective use of computer algebra and a combination of symbolic and numerical calculations. Although appropriate commands can be found in the package `Calculus`VariationalMethods``, we implement everything ourselves to better see which part of the computation is symbolic and which part is numerical.

We will compute the matrices M_{ij}, N_{ij}, L_{ij}, and L'_{ij}, symbolically, and then compute the values of the determinants for prescribed numerical values of c and k.

We begin with the implementation of the wave functions $\varphi_i(u, s, t)$. We make use of the representation of hyperbolic functions in terms of exponential functions to speed up the integration.

In[1]:=
```
ψ[1] =   Exp[-s/2] (Exp[c t/2] - Exp[-c t/2])/2;
ψ[2] = s ψ[1];
ψ[3] = t Exp[-s/2] (Exp[c t/2] + Exp[-c t/2])/2;
ψ[4] = u ψ[1];
```

```
ψ[5] = u ψ[2];
ψ[6] = u ψ[3];
```

A direct implementation of the triple integrals is possible, but is extremely time-inefficient as *Mathematica* requires quite some time for just one single (out of the many) integral.

```
In[7]:= i = 1; j = 2;
        TimeConstrained[Simplify[Integrate[
        u (s^2 - t^2) (D[ψ[i], s] D[ψ[j], s] + D[ψ[i], t] D[ψ[j], t] +
                       D[ψ[i], u] D[ψ[j], u]) +
        s (u^2 - t^2) (D[ψ[i], s] D[ψ[j], u] + D[ψ[i], u] D[ψ[j], s]) +
        t (s^2 - u^2) (D[ψ[i], t] D[ψ[j], u] + D[ψ[i], u] D[ψ[j], t]),
                           {s, 0, Infinity}, {u, 0, s}, {t, 0, u}]], 10]
Out[8]= $Aborted
```

We thus accelerate the integration process somewhat by hand. Here is the expression to be integrated.

```
In[9]:= temp =
        u (s^2 - t^2) (D[ψ[i], s] D[ψ[j], s] + D[ψ[i], t] D[ψ[j], t] +
                       D[ψ[i], u] D[ψ[j], u]) +
        s (u^2 - t^2) (D[ψ[i], s] D[ψ[j], u] + D[ψ[i], u] D[ψ[j], s]) +
        t (s^2 - u^2) (D[ψ[i], t] D[ψ[j], u] + D[ψ[i], u] D[ψ[j], t]);
```

First, we note that indefinite integration is typically many times faster than is definite integration (because it deals with convergence, possible singularities along the integration path, ...), and for the integrals of interest here, it gives the same results as carrying out the indefinite integration and simply substituting the limits. This can be easily checked.

```
In[10]:= Timing[Integrate[temp, {t, 0, u}]; ]
Out[10]= {1.64 Second, Null}

In[11]:= Timing[(((# /. {t -> u}) - (# /. {t -> 0}))&[Integrate[temp, t]]);]
Out[11]= {0.05 Second, Null}
```

The time-critical problem here, however, is the *s*-integration. First, we carry out the other two integrations.

```
In[12]:= tInt[x_] := ((# /. {t -> u}) - (# /. {t -> 0}))&[
                                          Together[Integrate[x, t]]]

         uInt[x_] := ((# /. {u -> s}) - (# /. {u -> 0}))&[Integrate[x, u]]

         Short[temp1 = uInt[tInt[temp]], 8]
```

$$
\text{Out[14]//Short=} \quad \frac{e^{-s} (15 (-2 + s + c^2 s) (-12 + c^2 s^2) - 15 (24 - 12 (1 + c^2) s - 2 c^2 s^2 + c^2 (1 + c^2) s^3))}{240 c^5} -
$$

$$
\frac{1}{240 c^5} (e^{-s-cs} (-8 c^5 e^{cs} s^5 (2 + (-1 + c^2) s) +
$$

$$
15 e^{2cs} (-2 + s + c^2 s) (-12 + 12 cs - 4 c^2 s^2) - 15 (-2 c^2 s^2 (1 + cs) + c^2 (1 + c^2) s^3 (1 + cs) +
$$

$$
2 (12 + 12 cs + 5 c^2 s^2 + c^3 s^3) - (1 + c^2) s (12 + 12 cs + 5 c^2 s^2 + c^3 s^3))))
$$

```
In[15]:= Integrate[temp1, {s, 0, Infinity}] // Timing
Out[15]= {39.77 Second,
```

$$
\frac{1}{60 c^5} \text{If}\Big[-1 < \text{Re}[c] < 1, \ \frac{480 c^7 (15 - 24 c^2 + 26 c^4 - 14 c^6 + 3 c^8)}{(-1 + c^2)^4}, \ \text{Integrate}[e^{-(1+c) s}
$$

$$
(2 c^5 e^{cs} s^5 (2 + (-1 + c^2) s) - 15 (-6 + 3 (-1 + c)^2 s + c (3 - 2 c + 3 c^2) s^2 + (c^2 + c^4) s^3) +
$$

$$
15 e^{2cs} (-6 + 3 (1 + c)^2 s - c (3 + 2 c + 3 c^2) s^2 + (c^2 + c^4) s^3)),
$$

$$
\{s, 0, \infty\}, \text{Assumptions} \to \text{Re}[c] \le -1 \ || \ \text{Re}[c] \ge 1]\Big]\}
$$

We can do better "manually". Looking at the integrand, we note that the resulting integrand of the *s*-integral has the form $\sum_{k=1}^{m} (f_k(c) s^{n_k} e^{-s} + g_k(c) s^{n_k} e^{cs} e^{-s})$.

Integrals of this form are easy to integrate.

```
In[16]:= Integrate[Exp[α s] s^n, {s, 0, Infinity},
                   Assumptions -> Re[n] > -1 && Re[α] < 0]
Out[16]= (-α)^{-1-n} Gamma[1 + n]
```

The convergence of the integral is ensured on physical grounds (all n in s^n are nonnegative and $0 < c < 1$). Taking account of the structure, we thus implement the following integration over s: we first factor the exponents and then do the integration by pattern matching.

```
In[17]:= sInt[x_] := Together[(Expand[x] //. {E^y_ :> E^Factor[y]}) //.
            (* do the integration *)
            {s^e_. Exp[s f_.] -> e! (-f)^(-e - 1), Exp[s f_.] -> - 1/f}]
```

Now, the remaining third integral can be quickly computed.

```
In[18]:= Together[sInt[temp1]] // Timing
```

$$Out[18]= \left\{0.02 \text{ Second}, \frac{8 c^2 (15 - 24 c^2 + 26 c^4 - 14 c^6 + 3 c^8)}{(-1 + c^2)^4}\right\}$$

Thus, we introduce myIntegration for carrying out these three-dimensional integrals.

```
In[19]:= myIntegration[x_] := Together[sInt[uInt[tInt[x]]]]
```

Note that another possible way of calculating the integrals is by using so-called perimetric coordinates. For these coordinates, the above triple integrals can be factored into three single integrals. For details, see [403], [1397], and [2].

Here are the four 6×6 matrices with their 144 triple integrals. Because of space limitations, we do not print them out.

Here is the matrix M.

```
In[20]:= mij[i_, j_] :=
           u (s^2 - t^2) (D[ψ[i], s] D[ψ[j], s] + D[ψ[i], t] D[ψ[j], t] +
                          D[ψ[i], u] D[ψ[j], u]) +
           s (u^2 - t^2) (D[ψ[i], s] D[ψ[j], u] + D[ψ[i], u] D[ψ[j], s]) +
           t (s^2 - u^2) (D[ψ[i], t] D[ψ[j], u] + D[ψ[i], u] D[ψ[j], t])

         mMat = Table[myIntegration[mij[i, j]], {i, 6}, {j, 6}];
```

Here is the matrix L.

```
In[22]:= lij[i_, j_] := 2 s u ψ[i] ψ[j]
         lMat = Table[myIntegration[lij[i, j]], {i, 6}, {j, 6}];
```

Here is the matrix L'.

```
In[24]:= lsij[i_, j_] := 1/4 (s^2 - t^2) ψ[i] ψ[j]
         lsMat = Table[myIntegration[lsij[i, j]], {i, 6}, {j, 6}];
```

Here is the matrix N.

```
In[26]:= norm[i_, j_] := 1/8 u (s^2 - t^2) ψ[i] ψ[j]
         normMat = Table[myIntegration[norm[i, j]], {i, 6}, {j, 6}];
```

The matrices require about 70 kB of memory.

```
In[28]:= ByteCount[{mMat, lMat, lsMat, normMat}]
```

```
Out[28]= 87432
```

After some time, all integrals are computed, and we can proceed with the numerical part. (The symbolic computation of the determinants takes some time.) Here is the matrix D_{ij}.

```
In[29]:= dMat[k_, c_] = 1 normMat + k (lMat - lsMat) - k^2 mMat;
```

Because D_{ij} is a symmetric matrix, the λ_i are real, and the smallest value is in first place if we sort the zeros using Sort[NSolve[...]]. We first substitute parameters and then calculate the determinant to avoid the very time-consuming symbolic calculation of the determinant. Using FindRoot, we now determine the minimum as a function of k and c. We apply InputForm to get enough digits written out.

```
In[30]:= E0[k_Real, c_Real] := Sort[1 /. NSolve[Det[dMat[k, c]] == 0, 1]][[1]]

         FindMinimum[E0[k, c], {k, 0.65, 0.67}, {c, 0.54, 0.56},
                     PrecisionGoal -> 8] // InputForm
```

```
Out[31]//InputForm= {-1.0876016547794582, {k -> 0.6766749130540749, c -> 0.552822387840412}}
```

We check the quality of the last solution by carrying out the analogous high-precision minimum finding.

```
In[32]:= EOHP[k_Real, c_Real, p_] := E0[SetPrecision[k, p], SetPrecision[c, p]]
```

```
In[33]:= FindMinimum[EOHP[k, c, 40], {k, 676/1000, 677/1000},
                                       {c, 552/1000, 553/1000},
                    WorkingPrecision -> 40,
                    AccuracyGoal -> 30, PrecisionGoal -> 12] // InputForm
```

```
Out[33]//InputForm= {-1.0876016547830254915450861587018273848625644829649632900073`40.000000000000014,
        {k -> 0.6766760643548078457583170318586682382950652654933703877719`40.,
         c -> 0.5528364423528775157085408833263768816668407810365569721 82`40.}}
```

One could go a few steps more on the symbolic side by producing a polynomial equation for 1 in the following manner.

```
(* generate polynomial *)
det = Factor[Numerator[Together[Det[dMat[k, c]]]]]/c^16;
(* the two equations *)
{eqk, eqc} = Factor[{D[det, k], D[det, c]}]
(* degrees of the two polynomials in c, k, and l *)
Print[Exponent[#, {c, k, l}]& /@ {det, eqk, eqc}];
(* find solution *)
FindRoot[Evaluate[# == 0& /@ {det, eqk, eqc}],
         {c, 55/100}, {k, 76/100}, {l, -108/100},
         WorkingPrecision -> 22, MaxIterations -> 40]
```

```
{{48, 12, 6}, {48, 11, 5}, {47, 12, 6}}
```

```
{c → 0.5528364400016718638307,
 k → 0.676676064714730 1423004, l → -1.08760165478302549 1819}
```

The theoretical result is in excellent agreement with the experimental result; it agrees to almost six digits. Note that this calculation was first carried out in 1930.

For this model in *d*D, see [1684]. For the minimization of the helium ground-state energy using piecewise linear wave functions, see [1226], and for high precision results, see [1585], [513], [1171], [517], [987], and [523].

For the interesting problem of two electrons in a box and in a sphere and other potentials, see [961], [1772], [330], [36], [1716], [1389], [1390], [1543], and [1192]; for polygonal domains, see [417], [1547].

For the general problem of bound states of three charges, see [1209], [1210], [87], [88], [772].

For the use of functionals of a set of functions that contain the variational parameters in variational calculations, see [1374].

b) Now the various trial functions are not independent of each other [1676] and we will calculate exact expressions for all integrals. To speed up integration, we do not let *Mathematica* carry out all integrations involved, but define our own integration function myInt, which integrates products of a power of x and Exp[-b^2 x^2]. The generic integral of this type is calculated here.

```
In[1]:= Simplify[Integrate[x^n Exp[-b^2 x^2], {x, -Infinity, Infinity},
                   GenerateConditions -> False]]
```

$$\text{Out[1]}= \quad \frac{1}{2}\,(1 + (-1)^n)\,(b^2)^{-\frac{1}{2}-\frac{n}{2}}\,\text{Gamma}\left[\frac{1+n}{2}\right]$$

Using this result, we implement myInt. The power of x is determined by the position in the corresponding Coefficient : List, which we use here to handle arbitrary polynomials in x.

```
In[2]:= myInt[int_] :=
        Plus @@ MapIndexed[#1 1/2 ((1 + (-1)^(#2[[1]] - 1))*
                         b^(-#2[[1]]) Gamma[#2[[1]]/2])&,
                    CoefficientList[Cancel[int/Exp[-b^2 x^2]], x]]
```

So, we can calculate the first six normalized $\psi_i(\beta; x)$.

```
In[3]:= ψ[0] = #/Sqrt[myInt[#^2]]&[Exp[-b^2 x^2/2]]
```

$$\text{Out[3]}= \quad \frac{e^{-\frac{1}{2}b^2 x^2}}{\sqrt{\frac{1}{b}}\,\pi^{1/4}}$$

In[4]:= `Do[ψ[i] = Factor[#/Sqrt[myInt[#^2]]&[D[ψ[i - 1], x]]], {i, 6}]`

In[5]:= `??ψ`

Global`ψ

$$\psi[0] = \frac{e^{-\frac{1}{2}b^2 x^2}}{\sqrt{\frac{1}{b}}\ \pi^{1/4}}$$

$$\psi[1] = -\frac{\sqrt{2}\ e^{-\frac{1}{2}b^2 x^2}\ x}{\left(\frac{1}{b}\right)^{5/2}\sqrt{b^2}\ \pi^{1/4}}$$

$$\psi[2] = \frac{2\ e^{-\frac{1}{2}b^2 x^2}\left(-1+\sqrt{b^2}\ x\right)\left(1+\sqrt{b^2}\ x\right)}{\sqrt{3}\ \sqrt{\frac{1}{b}}\ \pi^{1/4}}$$

$$\psi[3] = \frac{2\sqrt{\frac{2}{15}}\ e^{-\frac{1}{2}b^2 x^2}\ x\ (3-b^2 x^2)}{\left(\frac{1}{b}\right)^{5/2}\sqrt{b^2}\ \pi^{1/4}}$$

$$\psi[4] = \frac{4\ e^{-\frac{1}{2}b^2 x^2}\left(3-6 b^2 x^2 + b^4 x^4\right)}{\sqrt{105}\ \sqrt{\frac{1}{b}}\ \pi^{1/4}}$$

$$\psi[5] = \frac{4\sqrt{\frac{2}{105}}\ e^{-\frac{1}{2}b^2 x^2}\ x\ \left(-15+10 b^2 x^2 - b^4 x^4\right)}{3\left(\frac{1}{b}\right)^{5/2}\sqrt{b^2}\ \pi^{1/4}}$$

$$\psi[6] = \frac{8\ e^{-\frac{1}{2}b^2 x^2}\left(-15+45 b^2 x^2 - 15 b^4 x^4 + b^6 x^6\right)}{3\sqrt{1155}\ \sqrt{\frac{1}{b}}\ \pi^{1/4}}$$

Due to the symmetry of $H_{ij}(\beta)$ and $M_{ij}(\beta)$ in i and j, we implement the following calculation of the matrix elements.

In[6]:= `mI[i_, j_] := mI[i, j] = mI[j, i] = Factor[myInt[ψ[i] ψ[j]]]`

In[7]:= `hI[i_, j_] := hI[i, j] = hI[j, i] = Factor[myInt[ψ[i] *`
`                    (-1/2 D[ψ[j], {x, 2}] + (1/2 x^2 + x^4) ψ[j])]]`

The determinants in a factored form (dropping the irrelevant denominator) are now given by the following.

In[8]:= `evEq[n_] := evEq[n] = Factor[Numerator[Together[Det[PowerExpand[`
`                    Table[hI[i, j] - e mI[i, j], {i, 0, n}, {j, 0, n}]]]]]]`

In[9]:= `evEq[0]`

Out[9]= $3 + b^2 + b^6 - 4 b^4 e$

In[10]:= `evEq[1]`

Out[10]= $(3 + b^2 + b^6 - 4 b^4 e)\ (15 + 3 b^2 + 3 b^6 - 4 b^4 e)$

In[11]:= `evEq[2]`

Out[11]= $(15 + 3 b^2 + 3 b^6 - 4 b^4 e)$
$(45 + 30 b^2 + 3 b^4 + 78 b^6 + 14 b^8 + 3 b^{12} - 168 b^4 e - 24 b^6 e - 24 b^{10} e + 16 b^8 e^2)$

In[12]:= `evEq[3]`

Out[12]= $(525 + 210 b^2 + 15 b^4 + 450 b^6 + 54 b^8 + 15 b^{12} - 360 b^4 e - 40 b^6 e - 40 b^{10} e + 16 b^8 e^2)$
$(45 + 30 b^2 + 3 b^4 + 78 b^6 + 14 b^8 + 3 b^{12} - 168 b^4 e - 24 b^6 e - 24 b^{10} e + 16 b^8 e^2)$

In[13]:= `evEq[4]`

Out[13]= $(525 + 210 b^2 + 15 b^4 + 450 b^6 + 54 b^8 + 15 b^{12} - 360 b^4 e - 40 b^6 e - 40 b^{10} e + 16 b^8 e^2)$
$(1575 + 1575 b^2 + 315 b^4 + 7590 b^6 + 2670 b^8 + 165 b^{10} + 1419 b^{12} + 165 b^{14} + 15 b^{18} - 11340 b^4 e - 3240$
$b^6 e - 180 b^8 e - 6120 b^{10} e - 584 b^{12} e - 180 b^{16} e + 2640 b^8 e^2 + 240 b^{10} e^2 + 240 b^{14} e^2 - 64 b^{12} e^3)$

In[14]:= `evEq[5]`

Out[14]= $(1575 + 1575\,b^2 + 315\,b^4 + 7590\,b^6 + 2670\,b^8 + 165\,b^{10} +$
  $1419\,b^{12} + 165\,b^{14} + 15\,b^{18} - 11340\,b^4\,e - 3240\,b^6\,e - 180\,b^8\,e - 6120\,b^{10}\,e -$
  $584\,b^{12}\,e - 180\,b^{16}\,e + 2640\,b^8\,e^2 + 240\,b^{10}\,e^2 + 240\,b^{14}\,e^2 - 64\,b^{12}\,e^3)$
  $(33075 + 19845\,b^2 + 2835\,b^4 + 72030\,b^6 + 17262\,b^8 + 819\,b^{10} + 9795\,b^{12} +$
  $819\,b^{14} + 105\,b^{18} - 41580\,b^4\,e - 9240\,b^6\,e - 420\,b^8\,e - 15960\,b^{10}\,e -$
  $1256\,b^{12}\,e - 420\,b^{16}\,e + 4368\,b^8\,e^2 + 336\,b^{10}\,e^2 + 336\,b^{14}\,e^2 - 64\,b^{12}\,e^3)$

In[15]:= **evEq[6]**

Out[15]= $(33075 + 19845\,b^2 + 2835\,b^4 + 72030\,b^6 + 17262\,b^8 + 819\,b^{10} +$
  $9795\,b^{12} + 819\,b^{14} + 105\,b^{18} - 41580\,b^4\,e - 9240\,b^6\,e - 420\,b^8\,e - 15960\,b^{10}\,e -$
  $1256\,b^{12}\,e - 420\,b^{16}\,e + 4368\,b^8\,e^2 + 336\,b^{10}\,e^2 + 336\,b^{14}\,e^2 - 64\,b^{12}\,e^3)$
  $(99225 + 132300\,b^2 + 39690\,b^4 + 1013040\,b^6 + 528045\,b^8 + 65100\,b^{10} + 492030\,b^{12} +$
  $112812\,b^{14} + 4950\,b^{16} + 24612\,b^{18} + 2100\,b^{20} + 105\,b^{24} - 1164240\,b^4\,e - 498960\,b^6\,e -$
  $55440\,b^8\,e - 1495200\,b^{10}\,e - 285600\,b^{12}\,e - 11312\,b^{14}\,e - 160656\,b^{16}\,e -$
  $11312\,b^{18}\,e - 1680\,b^{22}\,e + 480480\,b^8\,e^2 + 87360\,b^{10}\,e^2 + 3360\,b^{12}\,e^2 + 141120\,b^{14}\,e^2 +$
  $9536\,b^{16}\,e^2 + 3360\,b^{20}\,e^2 - 26880\,b^{12}\,e^3 - 1792\,b^{14}\,e^3 - 1792\,b^{18}\,e^3 + 256\,b^{16}\,e^4)$

To get an estimation of the value of β for the minimum of ε, we solve the case with $\Psi(\beta; x) = \psi_0(\beta; x)$ exactly.

In[16]:= **Select[Solve[{ (D[evEq[0] /. e -> e[b], b] /. {e'[b] -> 0,**
           e[b] -> e}) == 0, evEq[0] == 0}, {b, e}],
         (Im[e] == 0 && b > 0 /. N[#])&]

Out[16]= $\left\{\left\{e \to \dfrac{13}{16}, b \to \sqrt{2}\right\}\right\}$

Here is a sketch of the behavior of the $\psi(\beta; x)$, including more terms.

In[17]:= **Show[GraphicsArray[#]]& /@**
        Partition[Table[
         ListPlot[Table[{b, NRoots[evEq[i] == 0, e][[1, 2]]},
                  {b, 0.8, 2.5, 0.025}],
            PlotRange -> {0.8, 0.82}, PlotJoined -> True,
            AxesOrigin -> {0.8, 0.8}, DisplayFunction -> Identity,
            PlotLabel -> StyleForm["evEq[" <> ToString[i] <> "]", "MR"]],
                  {i, 6}], 3]

Let us now numerically compute the minimizing values for β. We compare three different methods for the case of evEq[3]. One method is to use FindMinimum for the lowest value of e, which we calculate by solving the polynomial in e with NRoots.

In[18]:= **oFevEq3[$b\_$?NumericQ] :=**
         Block[{b = b}, NRoots[evEq[3] == 0, e, 20][[1, 2]]]

        Timing[FindMinimum[oFevEq3[b], {b, ##}, WorkingPrecision -> 25,
              PrecisionGoal -> 12, Compiled -> False]& @@@
            (* two initial intervals *) {{11/10, 12/10}, {17/10, 18/10}}]

Out[19]= {0.02 Second, {{0.80741457234272701782504888, {b → 1.203732086388522417922241}},
        {0.804174817459669786692035, {b → 1.7220450774800853432000884}}}}

A second method would be to eliminate ε from the two polynomial equations (after using the condition $\varepsilon'(\beta) = 0$ for the lowest eigenvalue to be extremal)

$$\frac{\partial}{\partial \beta} \det(H_{ij}(\beta) - \varepsilon(\beta) M_{ij}(\beta)) = 0$$

$$\det(H_{ij}(\beta) - \varepsilon(\beta) M_{ij}(\beta)) = 0$$

and solve the resulting polynomial in β with NRoots and after backsubstitution of all roots search for the lowest value of ε. The advantage of this approach is that we do not need an initial value.

```
In[20]:= Function[eq,
        N[Sort[{NRoots[(evEq[3] /. b -> #) == 0, e][[1, 2]], #}& /@
            Cases[NRoots[eq == 0, b], _Real?Positive, {-1}],
            #1[[1]] < #2[[1]]&], 12]][polyInb =
        Numerator[Factor[Resultant[(D[evEq[3] /. e -> e[b], b] /.
            e'[b] -> 0) /. e[b] -> e, evEq[3], e]]]] // Timing
Out[20]= {0.06 Second,
        {{0.804175, 1.72205}, {0.805235, 1.65348}, {0.805269, 1.78449}, {0.807359, 1.59397},
         {0.807415, 1.20373}, {0.81036, 1.51951}, {0.810867, 1.32443}, {0.8125, 1.41421}}}
```

The disadvantage is that for larger values of n, the polynomial in β is typically of high degree.

```
In[21]:= polyInb
```

$$\text{Out[21]= } 3039929748475084800\, b^{60}\, (-2 + b^2)\, (3 + 2 b^2 + b^4)\, (-10 - b^2 + b^6)$$
$$(-2100 - 546 b^2 - 42 b^4 + 689 b^6 + 92 b^8 + 3 b^{10} - 50 b^{12} - 3 b^{14} + b^{18})$$
$$(-660 - 270 b^2 - 30 b^4 + 381 b^6 + 68 b^8 + 3 b^{10} - 38 b^{12} - 3 b^{14} + b^{18})$$
$$(19800 + 10080 b^2 + 1710 b^4 + 13080 b^6 +$$
$$3375 b^8 + 288 b^{10} + 566 b^{12} + 72 b^{14} + 2 b^{16} + 96 b^{18} + 8 b^{20} + 3 b^{24})^2$$

Furthermore, an accurate numerical solution is time-consuming. We could, of course, avoid the explicit intermediate calculation of polyInb by direct use of NSolve for the two equations { (((D[evEq[3] /. e -> e[b], b] /. e'[b] -> 0) /. e[b] -> e) == 0, evEq[3] == 0}. This would not have resulted in a faster solution. Actually, the quality of the solution is not guaranteed.

```
In[22]:= Sort[Select[NSolve[{((D[evEq[3] /. e -> e[b], b] /. e'[b] -> 0) /.
            e[b] -> e) == 0, evEq[3] == 0}, {e, b}],
            Im[e] == 0 && Im[b] == 0 && Re[b] > 0 /. #&],
            #1[[1, 2]] < #2[[1, 2]]&] // Timing
Out[22]= {2.38 Second, {{e -> 0.804175, b -> 1.72205}, {e -> 0.807415, b -> 1.20373},
        {e -> 0.8125, b -> 1.41421}, {e -> 2.73917, b -> 1.78449}, {e -> 2.74605, b -> 1.32443},
        {e -> 2.75994, b -> 1.51951}, {e -> 5.22069, b -> 1.59397}, {e -> 8.00984, b -> 1.65348}}}
```

A combined use of GroebnerBasis for preprocessing the equations and then Solve with an appropriately Working Precision option setting will work fine for the smaller examples, but its running time will be prohibitive for larger examples.

```
In[23]:= gbSolve[order_] := Sort[Select[Solve[# == 0& /@ GroebnerBasis[
            {((D[evEq[order] /. e -> e[b], b] /. e'[b] -> 0) /.
            e[b] -> e), evEq[2]}, {e, b}], {e, b}, WorkingPrecision -> 50],
            Im[e] == 0 && Im[b] == 0 && Re[b] > 0 /. #&], #1[[1, 2]] < #2[[1, 2]]&]

In[24]:= gbSolve[2] // N // Timing
Out[24]= {0.12 Second, {{e -> 0.804175, b -> 1.72205}, {e -> 0.807415, b -> 1.20373},
        {e -> 0.8125, b -> 1.41421}, {e -> 2.75994, b -> 1.51951}, {e -> 5.22069, b -> 1.59397}}}

In[25]:= gbSolve[3] // N // Timing
Out[25]= {0.3 Second,
        {{e -> 0.804175, b -> 1.72205}, {e -> 0.807415, b -> 1.20373}, {e -> 0.8125, b -> 1.41421},
         {e -> 2.75994, b -> 1.51951}, {e -> 2.76423, b -> 1.48811}, {e -> 5.22069, b -> 1.59397}}}

In[26]:= gbSolve[4] // N // Timing
```

Out[26]= {0.53 Second,
    {{e → 0.804175, b → 1.72205}, {e → 0.804221, b → 1.70854}, {e → 0.807415, b → 1.20373},
    {e → 0.808069, b → 1.16986}, {e → 2.75994, b → 1.51951}, {e → 2.76274, b → 1.49408},
    {e → 5.22069, b → 1.59397}, {e → 5.228, b → 1.56519}, {e → 9.74293, b → 0.824128}}}

In[27]:= gbSolve[5] // N // Timing

Out[27]= {1.35 Second, {{e → 0.804175, b → 1.72205}, {e → 0.804209, b → 1.71047},
    {e → 0.807415, b → 1.20373}, {e → 0.80788, b → 1.17482}, {e → 5.22069, b → 1.59397},
    {e → 5.22575, b → 1.56996}, {e → 7.16626, b → 0.903199}, {e → 9.44823, b → 1.09849}}}

A further related solution method is to solve evEq[3] == 0 explicitly for e (which is possible if the evEq[i] is a polynomial of degree less than five in radicals), minimizing the resulting expressions with respect to b. Here, this is done.

```
In[28]:= Sort[{e -> #[[1]], b -> #[[2]]}& /@ Flatten[
            {{#, b} /. FindRoot[Evaluate[D[#, b] == 0], {b, 11/10}],
             {#, b} /. FindRoot[Evaluate[D[#, b] == 0], {b, 18/10}]}& /@
                (#[[1, 2]]& /@ Solve[evEq[3] == 0, e]), 1],
            #1[[1, 2]] < #2[[1, 2]]&] // Timing
```

Out[28]= {0.06 Second, {{e → 0.804175, b → 1.72205}, {e → 0.807415, b → 1.20373},
    {e → 2.73917, b → 1.78449}, {e → 2.74605, b → 1.32443}, {e → 5.22069, b → 1.59397},
    {e → 5.22069, b → 1.59397}, {e → 8.00984, b → 1.65348}, {e → 8.00984, b → 1.65348}}}

A third method is to solve the above two equations in ε and β with FindRoot.

```
In[29]:= frSolve[n_, startbs_] :=
         Function[eq,
         FindRoot[eq, {e, 8/10}, {b, #},
                 AccuracyGoal -> 20, WorkingPrecision -> 25]& /@ startbs][
         {(D[evEq[n] /. e -> e[b], b] /. e'[b] -> 0) /. e[b] -> e, evEq[n]}]

In[30]:= Timing[frSolve[3, {12/10, 18/10}]]
```

Out[30]= {0.04 Second, {{e → 0.80741457234272701782504777, b → 1.2037320863888404096736660},
    {e → 0.8041748174596697866920035, b → 1.720450774801061760653876}}}

Because this method seems most appropriate for the problem and parameters under consideration here, we finally use it to calculate the other values of n.

```
In[31]:= TableForm[Table[Flatten[{i, N[{e, b} /.
         Function[res, res[[Position[#, Min[#]]&[#[[2, 1]]& /@
                 res]][[1, 1]], 1]][
         {#, {e, b} /. #}& /@ frSolve[i, {11/10, 14/10, 19/10}]], 20}]},
         {i, 1, 6}], TableHeadings -> {None,
         StyleForm[#, FontWeight -> "Bold", FontFamily -> "Times"]& /@
                 {"order", "e", "b"}}]
```

order	e	b
1	0.81250000000000000000	1.4142135623730950488
2	0.80417481745966978669	1.7220450774801061761
3	0.80417481745966978669	1.7220450774801061761
4	0.80380028862649106880	1.9072606757405776050
5	0.80380028862649106880	1.9072606757405776050
6	0.80379477116055950769	1.4099008797774434924

Out[31]//TableForm= (rows 3 shown at left)

9. Hyperspherical Coordinates, Constant Negative Curvature Surface

a) Here is the computation of the Jacobian for the change of coordinate systems. The use of Outer permits a short, elegant, and fast implementation.

```
In[1]:= x[dim_][n_] := x[dim][n] = r Product[Sin[θ[i]], {i, n - 1}] Cos[θ[n]];

         (* the last coordinate *)
         x[dim_][dim_] := x[dim][dim] = r Product[Sin[θ[i]], {i, dim - 1}]
         var[n_] := var[n] = Union[Array[θ, n - 1], {r}]
```

Now, we apply Simplify to simplify the resulting sums of products of trigonometric functions.

```
In[5]:= NaivJacobiDeterminant[dim_] :=
        Simplify[Det[Outer[D, Array[x[dim], dim], Union[Array[θ, dim - 1], {r}]]]]
```

For comparison, consider the following `FastJacobiDeterminant`, which works only with the identity $\sin(x)^2 + \cos(x)^2 = 1$ (in two slightly different versions at two stages).

```
In[6]:= FastJacobiDeterminant[dim_] := Factor[
          (Det[Outer[D, Array[x[dim], dim], Union[Array[ϑ, dim - 1], {r}]]] //.
              {a_ Cos[x_]^2 + a_ Sin[x_]^2 -> a}) /.
                    {Cos[x_]^2 + Sin[x_]^2 -> 1}
```

Here are the computations of some Jacobian determinants with the times required.

```
In[7]:= timings[k_Integer] := {k, {Timing[NaivJacobiDeterminant[k]],
                                    Timing[FastJacobiDeterminant[k]]}}
```

```
In[8]:= Table[timings[k], {k, 2, 7}]
```

```
Out[8]= {{2, {{0.01 Second, r}, {0. Second, r}}},
         {3, {{0. Second, r² Sin[ϑ[1]]}, {0. Second, r² Sin[ϑ[1]]}}},
         {4, {{0.03 Second, r³ Sin[ϑ[1]]² Sin[ϑ[2]]}, {0. Second, r³ Sin[ϑ[1]]² Sin[ϑ[2]]}}},
         {5, {{0.05 Second, r⁴ Sin[ϑ[1]]³ Sin[ϑ[2]]² Sin[ϑ[3]]},
           {0.02 Second, r⁴ Sin[ϑ[1]]³ Sin[ϑ[2]]² Sin[ϑ[3]]}}},
         {6, {{0.14 Second, r⁵ Sin[ϑ[1]]⁴ Sin[ϑ[2]]³ Sin[ϑ[3]]² Sin[ϑ[4]]},
           {0.14 Second, r⁵ Sin[ϑ[1]]⁴ Sin[ϑ[2]]³ Sin[ϑ[3]]² Sin[ϑ[4]]}}},
         {7, {{30.96 Second, r⁶ Sin[ϑ[1]]⁵ Sin[ϑ[2]]⁴ Sin[ϑ[3]]³ Sin[ϑ[4]]² Sin[ϑ[5]]},
           {0.31 Second, r⁶ Sin[ϑ[1]]⁵ Sin[ϑ[2]]⁴ Sin[ϑ[3]]³ Sin[ϑ[4]]² Sin[ϑ[5]]}}}}
```

The enormous savings of time using an appropriate transformation rule is obvious.

b) This is the family of curves under consideration. To avoid complicated formulas we do not use $d = d(c)$ here.

```
In[1]:= r[u_, v_] =
        With[{x = u + v, y = u - v},
             {0, 0, x} + 2d/c Sin[d y] Cosh[c x]/
             (d^2 Cosh[c x]^2 + c^2 Sin[d y]^2) {Sin[y], -Cos[y], 0} +
             2d^2/c Cosh[c x] (Cosh[c x]^2 - Sin[d y]^2)/
             (d^2 Cosh[c x]^2 + c^2 Sin[d y]^2)/
             (Cos[d y]^2 Cosh[c x]^2 + Sinh[c x]^2 Sin[d y]^2)*
             {Cos[y] Cos[d y], Sin[y] Cos[d y], -Sinh[c x]}];
```

It is straightforward to calculate the quantities $\mathcal{E}, \mathcal{F}, \mathcal{G}, e, f,$ and g.

```
In[2]:= ℰ = D[r[u, v], u].D[r[u, v], u];
        ℱ = D[r[u, v], u].D[r[u, v], v];
        𝒢 = D[r[u, v], v].D[r[u, v], v];
```

```
In[5]:= n = Cross[D[r[u, v], u], D[r[u, v], v]];
        e = D[r[u, v], u, u].n;
        f = D[r[u, v], u, v].n;
        g = D[r[u, v], v, v].n;
```

The resulting expression for $\mathcal{K}$ is quite large (and so very probably cannot be simplified in a reasonable time using `Simplify[𝒦]`).

```
In[9]:= (𝒦 = (e g - f^2)/n.n/(ℰ 𝒢 - ℱ^2)) // ByteCount
```

```
Out[9]= 3002480
```

Random numerical substitutions indicate that we have $\mathcal{K} = -1$.

```
In[10]:= Table[Block[{c = Random[], d, u = Random[], v = Random[]},
                      d = Sqrt[1 - c^2]; 𝒦], {6}]
```

```
Out[10]= {-1., -1., -1., -1., -1., -1.}
```

To symbolically show $\mathcal{K} = -1$ we transform the expression $\mathcal{K}$ that contains trigonometric and hyperbolic functions into a purely polynomial one. We first rewrite the trigonometric and hyperbolic functions in exponential form and then use new variables for the various exponential terms.

```
In[11]:= makePoly[expr_] := expr /.
         (trig:(Cos | Cosh | Sin | Sinh))[x_] :> TrigToExp[trig[Expand[x]]]
```

```
In[12]:= expOfVarToVars[E^l_] := Times @@ (Function[x,
         Which[(* one polynomial variable per exponential term *)
              (* Exp[+I ...] case *)
```

```
          x === -  u,  1/u1, x === -  v,  1/v1,
          x === -c u,  1/cu, x === -c v,  1/cv,
          x ===  c u,   cu, x ===  c v,   cv,
          x === -d u,  1/du, x === -d v,  1/dv,
          x ===  d u,   du, x ===  d v,   dv,
          (* Exp[-I ...] case *)
          x === -I   u, 1/iu1, x === -I   v, 1/iv1,
          x ===  I   u,  iu1, x ===  I   v,  iv1,
          x === -I c u, 1/icu, x === -I c v, 1/icv,
          x ===  I c u,  icu, x ===  I c v,  icv,
          x === -I d u, 1/idu, x === -I d v, 1/idv,
          x ===  I d u,  idu, x ===  I d v,  idv]] /@
                 If[Head[l] === Plus, List @@ l, {l}])
In[13]:= transform[expr_] :=
         Together[makePoly[expr] /. e:Power[E, l_] :> expOfVarToVars[e]]
In[14]:= {&1, F1, G1} = transform /@ {&, F, G};
         {e1, f1, g1} = transform /@ {e, f, g};
         nn1 = transform[n.n];
```

Writing all expressions involved over a common denominator results in an expression whose size is about 1% of the above one.

```
In[17]:= &GF = Together[&1 g1 - F1^2];
         egf = Together[e1 g1 - f1^2];
         K = Together[egf/(nn1 &GF)];
In[20]:= ByteCount[K]
Out[20]= 33264
```

Now it is straightforward to use the relation $c = c(d)$ and to obtain the result $K = -1$. c only appears in even powers in K, so that all square roots disappear.

```
In[21]:= Cases[K, c^_, Infinity] // Union
Out[21]= {c^2, c^4}

In[22]:= Together[K /. c^n_?EvenQ :> (1 - d^2)^(n/2)]
Out[22]= -1
```

Now let us visualize some of the surfaces for various d.

```
In[23]:= r[{u_, v_}, d_] = r[u, v] /. c -> Sqrt[1 - d^2];

In[24]:= picture[d_, pp_, opts___] :=
         With[{L = 6},
          Show[Graphics3D[(* remove some polygons to better see inside *)
              MapIndexed[If[EvenQ[#2[[1]]], #, {}]&,
               ParametricPlot3D[Evaluate[Append[r[{u, v}, d],
                {EdgeForm[], SurfaceColor[Hue[u/L], Hue[u/L], 2.6]}]],
                {u, -L, L}, {v, -L, L}, opts, PlotPoints -> pp,
                DisplayFunction -> Identity][[1]]]],
               BoxRatios -> {1, 1, 1.8}, PlotRange -> {All, All, {-4, 4}},
               DisplayFunction -> $DisplayFunction, ViewPoint -> {3, 0, 1}]];
In[25]:= Show[GraphicsArray[
          Block[{$DisplayFunction = Identity},
               {picture[0.25, 101], picture[0.45, 100], picture[0.77, 101]}]]]]
```

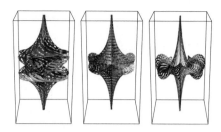

For similar parametrizations of constant positive curvature surfaces, see [1805].

10. Throw from a Finite Height, Pendulum Throw, Spring System

a) The horizontal and the vertical components of the motion satisfy (using standard notation)

$$x(t) = v_{x0} \, t + x_0,$$
$$y(t) = -\frac{g}{2} \, t^2 + v_{y0} \, t + h.$$

Introducing the throw angle α via $v_{x0} = v_0 \cos(\alpha)$, $v_{y0} = v_0 \sin(\alpha)$ and eliminating the time parameter from the parametric description of the throw, we obtain the following equation.

```
In[1]:= Eliminate[{x == v0 Cos[α] t,
                   y == -g/2 t^2 + v0 Sin[α] t + h}, t]
Out[1]= g x² - 2 v0² x Cos[α] Sin[α] == v0² (2 h - 2 y) Cos[α]²
```

The maximum distance is obtained by setting y equal to 0 (hitting the ground). The throw distance x_0 is a function of the throw angle α, this means $x_0 = x_0(\alpha)$.

```
In[2]:= (Subtract @@ %) /. {y -> 0, x -> x[α]}
Out[2]= -2 h v0² Cos[α]² - 2 v0² Cos[α] Sin[α] x[α] + g x[α]²
```

Differentiating the last equation with respect to α and applying the condition for an extremum, we obtain the following equation for α.

```
In[3]:= Eliminate[{% == 0, (D[%, α] /. x'[α] -> 0) == 0}, x[α]]
Out[3]= h v0⁴ Cos[α]² (Cos[α]⁴ - Sin[α]⁴) == 2 g h² v0² Cos[α]² Sin[α]²
```

After eliminating all Sin[α] from the last equation, *Mathematica* can solve for the optimal throw angle.

```
In[4]:= Solve[% /. Sin[α]^n_ -> (1 - Cos[α]^2)^(n/2), α]
```

> Solve::ifun : Inverse functions are being used by Solve, so some solutions
> may not be found; use Reduce for complete solution information. More…

$$Out[4]= \left\{ \left\{ \alpha \to -\frac{\pi}{2} \right\}, \left\{ \alpha \to \frac{\pi}{2} \right\}, \left\{ \alpha \to -\text{ArcCos}\left[-\frac{\sqrt{2 g h + v0^2}}{\sqrt{2 g h + 2 v0^2}} \right] \right\}, \left\{ \alpha \to \text{ArcCos}\left[-\frac{\sqrt{2 g h + v0^2}}{\sqrt{2 g h + 2 v0^2}} \right] \right\},$$
$$\left\{ \alpha \to -\text{ArcCos}\left[\frac{\sqrt{2 g h + v0^2}}{\sqrt{2 g h + 2 v0^2}} \right] \right\}, \left\{ \alpha \to \text{ArcCos}\left[\frac{\sqrt{2 g h + v0^2}}{\sqrt{2 g h + 2 v0^2}} \right] \right\} \right\}$$

Obviously, the last solution is the physically sensible one. For small heights h, we have the following series solution for the optimal throw angle. For positive heights, the optimal throw angle is smaller than 45°.

```
In[5]:= Series[%[[-1, 1, 2]], {h, 0, 2}]
```

$$Out[5]= \frac{\pi}{4} - \frac{g h}{2 v0^2} + \frac{g^2 h^2}{2 v0^4} + O[h]^3$$

Let us now visualize the optimality of the throw angle just calculated. We do this by showing the optimal throw in comparison to other throws. Here is the description of a general throw.

```
In[6]:= throw[α_, h_, v0_, g_] := Line[
          Table[{v0 Cos[α] t, -g/2 t^2 + v0 Sin[α] t + h},
                {t, 0, #, #/50}]&[ v0/g Sin[α] + Sqrt[2h/g + v0^2/g^2 Sin[α]^2]]]
```

This shows two examples for different starting heights. The violet curve is for an angle of $\pi/4$, the optimal throw angle for $h = 0$ is in red.

```
In[7]:= throwPicture[h_, v0_, l_, {α1_, α2_}, opts___] := With[{g = 10},
          Graphics[{{(* the ground *) GrayLevel[0.8],
                          Polygon[{{0, 0}, {1, 0}, {1, -3}, {0, -3}}]},
                          (* various throws *)
                          Table[{GrayLevel[0], Thickness[0.002],
                                  throw[α, h, v0, g]}, {α, α1, α2, 5 Degree}],
                          (* the optimal throw *)
                          {Hue[0], Thickness[0.01],
                           throw[ArcCos[Sqrt[(2g h + v0^2)/(2g h + 2v0^2)]], h, v0, g]},
                          {Hue[0.78], Thickness[0.002], throw[Pi/4, h, v0, g]}},
                          opts, Frame -> True, PlotRange -> All, AspectRatio -> Automatic]]

In[8]:= Show[GraphicsArray[
          {throwPicture[10, 10, 20, {-20, 80} Degree] // (* name it *) (pic1 = #)&,
           throwPicture[-2, 10, 10, { 40, 80} Degree]}] // N]
```

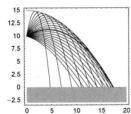

We obtain the envelope by eliminating α from the following two equations $C(x, y, \alpha) = 0$ and $\partial C(x, y, \alpha)/\partial\alpha = 0$ where $C(x, y, \alpha)$ is the implicit description of the throw curve.

To obtain polynomial equations, we use a rational parametrization for the trigonometric functions.

```
In[10]:= Eliminate[{# == 0, D[#, σ] == 0}&[
          g x^2 - 2 v0^2 x Cos[α] Sin[α] - v0^2 (2 h - 2 y) Cos[α]^2 /.
          {Cos[α] -> (1 - σ^2)/(1 + σ^2), Sin[α] -> 2 σ/(1 + σ^2)}], σ]
Out[10]= g² x⁴ + g v0² x² (-2 h + 2 y) = v0⁴ x²
```

```
In[11]:= Solve[%, y] // Simplify
```

$$Out[11]= \left\{\left\{y \to h + \frac{v0^2}{2\,g} - \frac{g\,x^2}{2\,v0^2}\right\}\right\}$$

Here are the above throw curves together with their envelope [199], [295], [282].

```
In[12]:= With[{g = 10, h = 10, v0 = 10},
          Show[{pic1, Graphics[{RGBColor[0, 1, 0], Thickness[0.006],
                          (* envelope *) Line[N[
                          Table[{x, -((-2 g h v0^2 - v0^4 + g^2 x^2)/
                                      (2 g v0^2))}, {x, 0, 10, 1/10}]]]}]}] // N]]
```

For the also interesting question of the maximum length of the arc length of the trajectory, see [510]; for the inclusion of friction, see [371], [1283]; for the throw process itself, see [418].

b) Again, the two components of the motion satisfy (using standard notation) are

$$x(t) = v_{x0}\, t + x_0,$$
$$y(t) = -\frac{g}{2}\, t^2 + v_{y0}\, t + y_0.$$

Solving the first equation for $t = (x - x_0)/v_{x0}$ and substituting in the second equation, we get $y(x)$. At this point, the initial position $\{x_0, y_0\}$ and the components of the initial velocity v_{x0} and v_{y0} are still unspecified.

```
In[1]:= y = -g/2 t^2 + vy0 t + y0 /. {t -> (x - x0)/vx0}
```

$$Out[1]=\quad \frac{vy0\,(x - x0)}{vx0} - \frac{g\,(x - x0)^2}{2\,vx0^2} + y0$$

The components of the initial velocity v_0 satisfy $v_{x0} = v_0 \cos(\varphi)$, $v_{y0} = v_0 \sin(\varphi)$. Here, v_0 is the absolute value of the initial velocity v_0.

```
In[2]:= y = y /. {vx0 -> v0 Cos[φ], vy0 -> v0 Sin[φ]}
```

$$Out[2]=\quad y0 - \frac{g\,(x - x0)^2\,Sec[\varphi]^2}{2\,v0^2} + (x - x0)\,Tan[\varphi]$$

v_0^2 follows from the conservation of energy law $m\,g\,(l - l\cos(\varphi_m)) = m\,g\,(l - l\cos(\varphi)) + m/2\,v_0^2$, which gives $v_0^2 = 2\,g\,l\,(\cos(\varphi) - \cos(\varphi_m))$. Here, φ_m is the maximum angle, φ is the instantaneous angle, and l is the length of the pendulum. (We use v0q for v_0^2 to avoid confusion with the vector $\mathbf{v_0}$.)

```
In[3]:= Solve[(* energy conservation *)
            m g (1 - l Cos[φMax]) ==
            m g (1 - l Cos[φ]) + m/2 v0q, v0q] // Simplify
```

$$Out[3]=\quad \{\{v0q \rightarrow 2\,g\,l\,(Cos[\varphi] - Cos[\varphi Max])\}\}$$

This yields the following equation.

```
In[4]:= y = y /. {v0^(-2) -> 1/(2 l g (Cos[φ] - Cos[φMax]))}
```

$$Out[4]=\quad y0 - \frac{(x - x0)^2\,Sec[\varphi]^2}{4\,l\,(Cos[\varphi] - Cos[\varphi Max])} + (x - x0)\,Tan[\varphi]$$

Note that the constant of gravitational acceleration g now no longer appears!

The components of initial position $\{x_0, y_0\}$ of the trajectory can be found directly from the following geometry: $x_0 = l\sin(\varphi)$, $y_0 = h + (l - l\cos(\varphi))$. Here h is the height of the bob of the pendulum at $\varphi = 0$.

```
In[5]:= y = y /. {x0 -> l Sin[φ], y0 -> h + (l - l Cos[φ])}
```

$$Out[5]=\quad h + l - l\,Cos[\varphi] - \frac{Sec[\varphi]^2\,(x - l\,Sin[\varphi])^2}{4\,l\,(Cos[\varphi] - Cos[\varphi Max])} + (x - l\,Sin[\varphi])\,Tan[\varphi]$$

Because $y = 0$ holds at the point of collision with the earth, we can simplify by multiplying by $4\,l(\cos(\varphi) - \cos(\varphi_m))\cos(\varphi)^2$.

```
In[6]:= y = Together[y] (4 l (Cos[φ] - Cos[φMax]) Cos[φ]^2 (-1)) // Expand
```

$$Out[6]=\quad x^2 - 4\,h\,l\,Cos[\varphi]^3 - 4\,l^2\,Cos[\varphi]^3 + 4\,l^2\,Cos[\varphi]^4 + 4\,h\,l\,Cos[\varphi]^2\,Cos[\varphi Max] +$$
$$4\,l^2\,Cos[\varphi]^2\,Cos[\varphi Max] - 4\,l^2\,Cos[\varphi]^3\,Cos[\varphi Max] - 2\,l\,x\,Sin[\varphi] - 4\,l\,x\,Cos[\varphi]^2\,Sin[\varphi] +$$
$$4\,l\,x\,Cos[\varphi]\,Cos[\varphi Max]\,Sin[\varphi] + l^2\,Sin[\varphi]^2 + 4\,l^2\,Cos[\varphi]^2\,Sin[\varphi]^2 - 4\,l^2\,Cos[\varphi]\,Cos[\varphi Max]\,Sin[\varphi]^2$$

To determine the angle that results in the maximum distance, we must solve $y(x_s(\varphi)) = 0$ and $\partial x_s(\varphi)\,\partial\varphi = 0$.

A direct implementation of the form

```
{eq1, eq2} = y /. Solve[y == 0, x];
{Solve[D[eq1, φ] == 0, φ], Solve[D[eq2, φ] == 0, φ]}
```

fails to find a result even after several hours. We thus intervene by hand. y is a polynomial of second degree in x. We write y in the form $x^2 + p\,x + q = 0$.

```
In[7]:= {q, p} = Take[CoefficientList[y, x], 2];
```

With $a = -p/2$, $b = p^2/4 - q$, x_s becomes $a \pm \sqrt{b}$, and from (prime denotes differentiation with respect to φ) $\partial x(\varphi)/\partial\varphi = a'(\varphi) \pm 1/\sqrt{2\,b(\varphi)}\;b'(\varphi) = 0$, we get $4\,b(\varphi)^2\,a'(\varphi)^2 - b'(\varphi)^2 = 0$.

```
In[8]:= {a, b} = {-p/2, p^2/4 - q} // Expand;
        eq = 4 D[a, φ]^2 b - D[b, φ]^2 // Expand;
```

We simplify this a little more by multiplying by l^2.

In[10]:= `(eq = eq/l^2 // Expand) // Short[#, 6]&`

Out[10]//Short= $16\,h\,l\,\mathrm{Cos}[\varphi]^5 + 16\,l^2\,\mathrm{Cos}[\varphi]^5 - 16\,l^2\,\mathrm{Cos}[\varphi]^6 + 64\,h\,l\,\mathrm{Cos}[\varphi]^7 + 64\,l^2\,\mathrm{Cos}[\varphi]^7 - $
$64\,l^2\,\mathrm{Cos}[\varphi]^8 + 64\,h\,l\,\mathrm{Cos}[\varphi]^9 + \ll78\gg + 320\,h\,l\,\mathrm{Cos}[\varphi]^4\,\mathrm{Cos}[\varphi\mathrm{Max}]\,\mathrm{Sin}[\varphi]^4 + $
$320\,l^2\,\mathrm{Cos}[\varphi]^4\,\mathrm{Cos}[\varphi\mathrm{Max}]\,\mathrm{Sin}[\varphi]^4 - 320\,l^2\,\mathrm{Cos}[\varphi]^5\,\mathrm{Cos}[\varphi\mathrm{Max}]\,\mathrm{Sin}[\varphi]^4 - $
$256\,h\,l\,\mathrm{Cos}[\varphi]^3\,\mathrm{Cos}[\varphi\mathrm{Max}]^2\,\mathrm{Sin}[\varphi]^4 - 256\,l^2\,\mathrm{Cos}[\varphi]^3\,\mathrm{Cos}[\varphi\mathrm{Max}]^2\,\mathrm{Sin}[\varphi]^4 + $
$256\,l^2\,\mathrm{Cos}[\varphi]^4\,\mathrm{Cos}[\varphi\mathrm{Max}]^2\,\mathrm{Sin}[\varphi]^4 + 64\,h\,l\,\mathrm{Cos}[\varphi]^2\,\mathrm{Cos}[\varphi\mathrm{Max}]^3\,\mathrm{Sin}[\varphi]^4 + $
$64\,l^2\,\mathrm{Cos}[\varphi]^2\,\mathrm{Cos}[\varphi\mathrm{Max}]^3\,\mathrm{Sin}[\varphi]^4 - 64\,l^2\,\mathrm{Cos}[\varphi]^3\,\mathrm{Cos}[\varphi\mathrm{Max}]^3\,\mathrm{Sin}[\varphi]^4$

We then replace all appearances of $\sin(\varphi)$ by $\cos(\varphi)$.

In[11]:= `eq = eq /. {Sin[φ]^n_?EvenQ -> (1 - Cos[φ]^2)^(n/2)} // Expand;`

Finally, introducing $c = \cos(\varphi)$, $c_m = \cos(\varphi_m)$, we write the resulting expression in the following somewhat shorter form.

In[12]:= `eq = Expand[eq l h] //. {Cos[φ] -> c, Cos[φMax] -> cm}`

Out[12]= $-144\,c^4\,h^3\,l + 144\,c^6\,h^3\,l + 192\,c^3\,cm\,h^3\,l - 192\,c^5\,cm\,h^3\,l - 64\,c^2\,cm^2\,h^3\,l + $
$64\,c^4\,cm^2\,h^3\,l - 288\,c^4\,h^2\,l^2 + 144\,c^5\,h^2\,l^2 + 288\,c^6\,h^2\,l^2 + 384\,c^3\,cm\,h^2\,l^2 - 48\,c^4\,cm\,h^2\,l^2 - $
$384\,c^5\,cm\,h^2\,l^2 - 288\,c^6\,cm\,h^2\,l^2 - 128\,c^2\,cm^2\,h^2\,l^2 - 128\,c^3\,cm^2\,h^2\,l^2 + 128\,c^4\,cm^2\,h^2\,l^2 + $
$384\,c^5\,cm^2\,h^2\,l^2 + 64\,c^2\,cm^3\,h^2\,l^2 - 128\,c^4\,cm^3\,h^2\,l^2 - 144\,c^4\,h\,l^3 + 144\,c^5\,h\,l^3 + $
$144\,c^6\,h\,l^3 - 144\,c^8\,h\,l^3 + 192\,c^3\,cm\,h\,l^3 - 48\,c^4\,cm\,h\,l^3 - 336\,c^5\,cm\,h\,l^3 - 288\,c^6\,cm\,h\,l^3 + $
$480\,c^7\,cm\,h\,l^3 - 64\,c^2\,cm^2\,h\,l^3 - 128\,c^3\,cm^2\,h\,l^3 + 256\,c^4\,cm^2\,h\,l^3 + 384\,c^5\,cm^2\,h\,l^3 - $
$448\,c^6\,cm^2\,h\,l^3 + 64\,c^2\,cm^3\,h\,l^3 - 64\,c^3\,cm^3\,h\,l^3 - 128\,c^4\,cm^3\,h\,l^3 + 128\,c^5\,cm^3\,h\,l^3$

This polynomial of eighth degree in $c = \cos(\varphi)$ can be solved explicitly.

In[13]:= `Factor[eq]`

Out[13]= $-16\,c^2\,(3\,c - 2\,cm)^2\,h\,l\,(-h - l + c\,l)\,(-h + c^2\,h - l + c^2\,l + c^3\,l + cm\,l - 2\,c^2\,cm\,l)$

It is worth noting that a one-line input to arrive at eq starting from the above y would be the following GroebnerBasis containing input.

In[14]:= `gb = GroebnerBasis[{y, D[y /. x -> x[φ], φ] /. {x'[φ] -> 0, x[φ] -> x},`
         `(* algebraic relation between Sin and Cos *)`
         `Sin[φ]^2 + Cos[φ]^2 - 1},`
         `{Cos[φ], Cos[φMax], h, l}, {Sin[φ], x},`
         `MonomialOrder -> EliminationOrder} /.`
         `{Cos[φ] -> c, Cos[φMax] -> cm} // Factor`

Out[14]= $\{c^2\,(3\,c - 2\,cm)\,l\,(-h - l + c\,l)\,(-h + c^2\,h - l + c^2\,l + c^3\,l + cm\,l - 2\,c^2\,cm\,l)\}$

Now, we look for the relevant roots of eq (we select from the last product the factor that contains the parameters of the problem).

In[15]:= `(sol = Solve[gb[[-1, -1]] == 0, c]) // Short[#, 12]&`

Out[15]//Short= $\left\{\left\{c \to -\dfrac{h + l - 2\,cm\,l}{3\,l} + (2^{1/3}\,(h + l - 2\,cm\,l)^2)\,\Big/\right.\right.$

$\left(3\,l\,\Big(-2\,h^3 - 6\,h^2\,l + 12\,cm\,h^2\,l + 21\,h\,l^2 + 24\,cm\,h\,l^2 - 24\,cm^2\,h\,l^2 + 25\,l^3 - 15\,cm\,l^3 - 24\,cm^2\,l^3 + \right.$

$\left.16\,cm^3\,l^3 + \sqrt{-4\,(h + l - 2\,cm\,l)^6 + (-2\,h^3 - 6\,h^2\,l + 12\,cm\,h^2\,l + \ll9\gg + 16\,cm^3\,l^3)^2}\Big)^{1/3}\right) +$

$\dfrac{1}{3\,2^{1/3}\,l}\Big(\big(-2\,h^3 - 6\,h^2\,l + 12\,cm\,h^2\,l + 21\,h\,l^2 + 24\,cm\,h\,l^2 - 24\,cm^2\,h\,l^2 + 25\,l^3 - 15\,cm\,l^3 - $

$24\,cm^2\,l^3 + 16\,cm^3\,l^3 + \sqrt{(-4\,(h + l - 2\,cm\,l)^6 + (-2\,h^3 - 6\,h^2\,l + 12\,cm\,h^2\,l + 21\,h\,l^2 + }$

$\left.\overline{24\,cm\,h\,l^2 - \ll1\gg + 25\,l^3 - 15\,cm\,l^3 - 24\,cm^2\,l^3 + 16\,cm^3\,l^3)^2}\big)^{1/3}\Big)\right\},$

$\left\{c \to -\dfrac{h + l - 2\,cm\,l}{3\,l} - \dfrac{(1 + i\sqrt{3})\,(h + l - 2\,cm\,l)^2}{3\,2^{2/3}\,l\,\big(-2\,h^3 - 6\,h^2\,l + \ll10\gg + \ll1\gg + \sqrt{-4\,(\ll1\gg)^6 + (\ll1\gg)^2}\big)^{1/3}} - \right.$

$\left.\dfrac{(1 - i\sqrt{3})\,\big(-2\,h^3 - 6\,h^2\,l + \ll10\gg + 16\,\ll2\gg^3\,l^3 + \sqrt{-4\,(h + l - 2\,cm\,l)^6 + (\ll1\gg)^2}\big)^{1/3}}{6\,2^{1/3}\,l}\right\},$

$\left\{c \to -\dfrac{h + l - 2\,cm\,l}{3\,l} - \big((1 - i\sqrt{3})\,(h + l - 2\,cm\,l)^2\big)\,\Big/\right.$

$$\left(3\ 2^{2/3}\ 1\ \left(-2\,h^3 - 6\,h^2\,1 + 12\ cm\,h^2\,1 + \ll 9\gg + 16\ cm^3\,1^3 + \right.\right.$$

$$\left.\sqrt{-4\,(h + 1 - 2\ cm\ 1)^6 + (\ll 13\gg + 16\ cm^3\,1^3)^2}\,\right)^{1/3} -$$

$$\frac{1}{6\ 2^{1/3}\ 1}\ \left(\left(1 + i\ \sqrt{3}\right)\left(-2\,h^3 - 6\,h^2\,1 + 12\ cm\,h^2\,1 + 21\,h\,1^2 + \ll 8\gg + 16\ cm^3\,1^3 + \right.\right.$$

$$\left.\left.\left.\sqrt{-4\,(h + 1 - 2\ cm\ 1)^6 + (-2\,h^3 - 6\,h^2\,1 + \ll 10\gg + 16\ cm^3\,1^3)^2}\,\right)^{1/3}\right)\right\}\right\}$$

(Another way to solve the above two equations is given in [1730].) To find the solution of interest to us, we insert some physically meaningful values for l, h, and c_m.

In[16]:= `sol /. {cm -> Cos[80. Degree], l -> 1, h -> 1} // N`

Out[16]= `{{c → 0.853635}, {c → -1.25317 + 0.754364 i}, {c → -1.25317 - 0.754364 i}}`

In[17]:= `sol[[Position[%, {c -> _?(Im[#] === 0 && 0 < # < 1&)}][[1, 1]]]];`

The relevant solution is thus the sixth one $(-1 < (c = \cos(\varphi)) < 1)$, which we now incorporate into `OptimalThrowAngle`.

In[18]:= `aux[h_, l_, φMax_] =`
    `Block[{cm}, cm = Cos[φMax];`
               `(* just the long result from above *) %[[1, 2]]];`

    `optimalThrowAngle[h_, l_, φMax_] := ArcCos[aux[h, l, φMax]]`

The answer does not depend explicitly on l and h, but only on l/h. This can be easily seen by replacing l with $h\,x$ and testing to see if the result is independent of h.

In[20]:= `Cancel[PowerExpand[PowerExpand[(aux[h, l, φMax] //. l -> h x) //.`
        `Sqrt[s_Plus] :> Sqrt[Collect[s, h]]] //.`
        `Power[s_Plus, n_] :> Power[Collect[s, h], n]]]`

Out[20]= $\dfrac{-1 - x + 2\,x\,Cos[\varphi Max]}{3\,x} +$

    $(2^{1/3}\,(-1 - x + 2\,x\,Cos[\varphi Max])^2)\,/\,\left(3\,x\,\left(-2 - 6\,x + 21\,x^2 + 25\,x^3 + 12\,x\,Cos[\varphi Max] + 24\,x^2\,Cos[\varphi Max] - \right.\right.$
        $15\,x^3\,Cos[\varphi Max] - 24\,x^2\,Cos[\varphi Max]^2 - 24\,x^3\,Cos[\varphi Max]^2 + 16\,x^3\,Cos[\varphi Max]^3 +$
        $3\,\sqrt{3}\,\sqrt{(-4\,x^2 - 16\,x^3 + 3\,x^4 + 38\,x^5 + 23\,x^6 + 28\,x^3\,Cos[\varphi Max] + 84\,x^4\,Cos[\varphi Max] +}$
        $30\,x^5\,Cos[\varphi Max] - 26\,x^6\,Cos[\varphi Max] - 72\,x^4\,Cos[\varphi Max]^2 - 144\,x^5\,Cos[\varphi Max]^2 -$
        $\left.\left.45\,x^6\,Cos[\varphi Max]^2 + 80\,x^5\,Cos[\varphi Max]^3 + 80\,x^6\,Cos[\varphi Max]^3 - 32\,x^6\,Cos[\varphi Max]^4)\right)^{1/3}\right) +$

    $\dfrac{1}{3\,2^{1/3}\,x}\,\left(\left(-2 - 6\,x + 21\,x^2 + 25\,x^3 + 12\,x\,Cos[\varphi Max] + 24\,x^2\,Cos[\varphi Max] - 15\,x^3\,Cos[\varphi Max] - \right.\right.$
        $24\,x^2\,Cos[\varphi Max]^2 - 24\,x^3\,Cos[\varphi Max]^2 + 16\,x^3\,Cos[\varphi Max]^3 +$
        $3\,\sqrt{3}\,\sqrt{(-4\,x^2 - 16\,x^3 + 3\,x^4 + 38\,x^5 + 23\,x^6 + 28\,x^3\,Cos[\varphi Max] + 84\,x^4\,Cos[\varphi Max] +}$
        $30\,x^5\,Cos[\varphi Max] - 26\,x^6\,Cos[\varphi Max] - 72\,x^4\,Cos[\varphi Max]^2 - 144\,x^5\,Cos[\varphi Max]^2 -$
        $\left.\left.45\,x^6\,Cos[\varphi Max]^2 + 80\,x^5\,Cos[\varphi Max]^3 + 80\,x^6\,Cos[\varphi Max]^3 - 32\,x^6\,Cos[\varphi Max]^4)\right)^{1/3}\right)$

At this point, we leave the solution in the above form. We now look at some typical values for the throw angle.

In[21]:= `Plot3D[optimalThrowAngle[1, l, φMax],`
    `{l, 1/3, 3}, {φMax, Pi/10, Pi/2}, PlotPoints -> 12]`

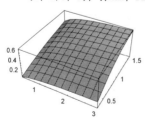

The average result is around $30°$ measured from the vertical for a large parameter range. This is considerably smaller than the $45°$ angle that is the known optimal throw angle starting from a plane. Now, we plot the trajectory. For an arbitrary throw angle, the horizontal throw distance corresponds to $y(x) = 0$.

```
In[22]:= Solve[y == 0, x][[2, 1, 2]]
```

```
Out[22]= l Sin[φ] + 2 l Cos[φ]² Sin[φ] - 2 l Cos[φ] Cos[φMax] Sin[φ] +
         2 √ (h l Cos[φ]³ + l² Cos[φ]³ - l² Cos[φ]⁴ - h l Cos[φ]² Cos[φMax] -
             l² Cos[φ]² Cos[φMax] + l² Cos[φ]³ Cos[φMax] + l² Cos[φ]⁴ Sin[φ]² -
             2 l² Cos[φ]³ Cos[φMax] Sin[φ]² + l² Cos[φ]² Cos[φMax]² Sin[φ]²)
```

```
In[23]:= throwDistance[h_, l_, φMax_, φ_] :=
         N[(-(l (-2 - 4 Cos[φ]^2 + 4 Cos[φ] Cos[φMax])*
              Sin[φ]) - 4 l^(1/2) Cos[φ] (-Cos[φ] + Cos[φMax])^(1/2)*
              (-h - l + l Cos[φ] - l Cos[φ] Sin[φ]^2 +
              l Cos[φMax] Sin[φ]^2)^(1/2))/2] // Chop
```

The time of flight is given by $t = (x - x_0)/v_{x0}$.

```
In[24]:= flightTime[h_, l_, φMax_, φ_] :=
         N[(throwDistance[h, l, φMax, φ] - l Sin[φ])/
           (Cos[φ] Sqrt[2 l g (Cos[φ] - Cos[φMax])])] // Chop
```

The trajectory $\{x(t), y(t)\}$ is given by the following.

```
In[25]:= path[t_, h_, l_, φMax_, φ_] :=
         (* superposition of x- and y- motion *)
         {Cos[φ] Sqrt[2 l g (Cos[φ] - Cos[φMax])] t + l Sin[φ],
          -g/2 t^2 + Sin[φ] Sqrt[2 l g(Cos[φ] - Cos[φMax])] t +
          h + (l - l Cos[φ])} // N;
```

With these functions, we can construct the following schematic plot of the trajectory. The path corresponding to the largest horizontal throw distance is shown with a heavy line. For comparison, the other paths are depicted with thin lines.

```
In[26]:= flightTime[0.7, 1.8, 110 Degree, 30 Degree]
```

```
Out[26]= 2.76595
         ────────
           √g
```

```
In[27]:= throwVisualization[h_, l_, φMax_, opts___] :=
         Module[{(* for φ > 90 Degree eliminate small imaginary part *)
                 opt = Re @ N[optimalThrowAngle[h, l, φMax]]},
         Block[{g = 9.81},
         Show[(* a few possible trajectories *)
          {Table[ParametricPlot[Evaluate[path[t, h, l, φMax, φ]],
                               Evaluate[{t, 0, flightTime[h, l, φMax, φ]}],
                               DisplayFunction -> Identity,
                               PlotStyle -> {Thickness[0.002], Hue[0.78 φ/φMax],
                                             Dashing[{1/80, 1/80}]}],
                 {φ, φMax/15, φMax 14/15, φMax/15}],
           (* the optimal trajectory *)
           ParametricPlot[Evaluate[path[t, h, l, φMax, opt]],
                          Evaluate[{t, 0, flightTime[h, l, φMax, opt]}],
                          DisplayFunction -> Identity,
                          PlotStyle -> {GrayLevel[0], Thickness[0.006]}],
           (* the pendulum *)
           Graphics[{Thickness[0.012], GrayLevel[1/2],
                     Circle[{0, l + h}, l, {3Pi/2, 3Pi/2 + φMax}]}],
           (* the ground *)
           Graphics[{GrayLevel[3/4], Rectangle[{0, 0},
                     {1.1 throwDistance[h, l, φMax, opt], -(l + h)/12}]}]},
            opts, PlotRange -> All, AspectRatio -> Automatic,
            Axes -> None, PlotLabel -> StyleForm[
            "optimal angle = " <> ToString[N[opt/Pi 180, 3]] <> " Degree",
            FontFamily -> "Courier", FontWeight -> "Bold", FontSize -> 9]]]]
```

Here are two examples.

```
In[28]:= Show[GraphicsArray[{throwVisualization[0.7, 1.8,  75 Degree],
                             throwVisualization[1.8, 0.7, 110 Degree]}]]
```

optimal angle = 35.1221 Degree   optimal angle = 29.1316 Degree

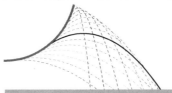   

An important application of this exercise is to the problem of maximizing the distance that one can jump from a moving swing. In this application, for solving the further very important problem of how to get maximum impetus, see [652], [312], [1144], [1629], [422], [40], [34], [1601], [1122], [1846], and [274]. For the related problem of when to throw something from a rotating wheel to achieve the maximal height, see [1333] and [753]; for jumping from a bungee chord, see [1712]; for jumping from a springboard, see [670].

c) It is straightforward to write down the equations of motions for the point masses under the forces of the springs. Expanding the forces into a series for small elongations yields a linear system of homogeneous equations. Its solutions are formed by the eigenvectors of the corresponding coefficient matrix. To determine the "breathing mode", we form the scalar product of a symmetrically elongated initial state with the eigenvectors. The function breathingData does all this and returns a two-argument list. Its first argument contains all eigenmodes and its second is the squared eigenfrequency of the "breathing mode" and the number of contributing modes.

```
In[1]:= Needs["Graphics`Polyhedra`"]

In[2]:= breathingData[plato_] :=
    Module[{dim = Length[Vertices[plato]], edges, neighbors, ℱ, force,
            ℱLinearized, coefficientRules, M, evals, evecs,
            allEigenvalues, iVec, iCoeffs, breatingModeEigenvalue},
        (* the edges *)
        edges = Union[Sort /@
            Flatten[Partition[Append[#, First[#]], 2, 1]& /@ Faces[plato], 1]];
        (* neighbors of the vertex k *)
        neighbors[k_] :=
        Union[Flatten[DeleteCases[Select[edges, MemberQ[#, k]&], k, Infinity]]];
        (* resulting force on a vertex *)
        ℱ[k_] := Plus @@ (force[{k, #}]& /@ neighbors[k]);
        (* force on vertex i due to vertex j *)
        force[{i_, j_}] :=
        Module[{p0 = N[Vertices[plato][[#]], 22]&, p},
            p = (p0[#] + {x[#], y[#], z[#]})&;
            #/Sqrt[#.#]&[p[j] - p[i]]*
            (Sqrt[#.#]&[p[j] - p[i]] - Sqrt[#.#]&[p0[j] - p0[i]])];
        (* force for small elongations *)
        ℱLinearized[k_] := Expand[Normal[(* keep linear terms only *)
            Series[ℱ[k] /. xyz:(_x | _y | _z) :> ε xyz, {ε, 0, 1}]]/ε];
        (* make rules of all nonvanishing matrix elements *)
        coefficientRules = Module[{υ, xC, yC, zC}, υ[0] := {};
        υ[p_Plus] := υ /@ (List @@ p); υ[l_List] := υ /@ l;
        υ[f_. x[j_]] := {j, f};
        υ[f_. y[j_]] := {dim + j, f}; υ[f_. z[j_]] := {2dim + j, f};
        (* nonvanishing matrix elements *)
        Flatten[Table[{xC, yC, zC} = Chop[ℱLinearized[k]];
                    {({k, #1} -> #2)& @@@ υ[xC],
                    ({dim + k, #1} -> #2)& @@@ υ[yC],
                    ({2dim + k, #1} -> #2)& @@@ υ[zC]}, {k, dim}]]];
        (* form matrix *)
        M = Module[{m = Table[0, {3dim}, {3dim}],
                  cR = coefficientRules},
                Do[m[[cR[[j, 1, 1]], cR[[j, 1, 2]]]] =
                    m[[cR[[j, 1, 1]], cR[[j, 1, 2]]]] + cR[[j, 2]],
                  {j, Length[coefficientRules]}]; m];
        (* eigenvalues and eigenvectors *)
```

```
{evals, evecs} = Eigensystem[M];
(* count eigenvalues *)
allEigenvalues = {SetPrecision[First[#], 5], Length[#]}& /@
Split[Sort[Chop[SetAccuracy[SetPrecision[
            evals, 11], 11], 10^-10]]];
(* eigenvalues forming the breathing mode *)
iVec = Flatten[Transpose[N[Vertices[plato], 22]]];
iCoeffs = Chop[iVec.#& /@ evecs, 10^-10];
breatingModeEigenvalue = {First[#], Length[#]}& /@
Split[Sort[SetPrecision[evals[[Flatten[
            Position[iCoeffs, _?(# != 0&)]]]], 5]]];
(* return all and breathing eigenvalues *)
Apply[{Sqrt[Abs[#1]], #2}&,
        {allEigenvalues, breatingModeEigenvalue}, {-2}]]
```

For the tetrahedron and the octahedron, the "breathing mode" is the highest mode.

In[3]:= **breathingData[Tetrahedron]**

Out[3]= {{{2.0000, 1}, {1.41421, 3}, {1.00000, 2}, {0, 6}}, {{2.0000, 1}}}

In[4]:= **breathingData[Octahedron]**

Out[4]= {{{2.0000, 1}, {1.73205, 3}, {1.41421, 3}, {1.00000, 5}, {0, 6}}, {{2.0000, 1}}}

Due to the orthogonality of the faces of the cube, linear coupling terms are largely missing and 12 modes contribute to the "breathing mode".

In[5]:= **breathingData[Hexahedron]**

Out[5]= {{{1.41421, 12}, {0, 12}}, {{1.41421, 12}}}

For the dodecahedron and the icosahedron, the "breathing mode" is again unique, but it is not the highest frequency mode.

In[6]:= **breathingData[Dodecahedron]**

Out[6]= {{{1.61803, 10}, {1.54336, 3}, {1.51954, 4}, {1.30038, 4},
        {1.17557, 5}, {1.00000, 3}, {0.87403, 1}, {0, 30}}, {{0.87403, 1}}}

In[7]:= **breathingData[Icosahedron]**

Out[7]= {{{1.85123, 4}, {1.84776, 5}, {1.73205, 3}, {1.66251, 1}, {1.34500, 4},
        {1.00000, 5}, {0.87403, 3}, {0.76537, 5}, {0, 6}}, {{1.66251, 1}}}

For a group-theoretical discussion of the eigenmodes, see [1114], [547].

11. Sturm–Liouville Problems in Normal Form

We begin with the computation of $v(\xi)$. We implement the left side of the differential equation and insert the given transformation of the dependent variables.

In[1]:= **ODE[y_, x_] := f2[x] D[y, {x, 2}] + f1[x] D[y, x] + (f0[x] + λ g[x]) y**

    aux1 = ODE[η[ξ[x]]/Φ[x], x] // Together // (# Φ[x]^3)&

Out[2]= f0[x] η[ξ[x]] Φ[x]² + λ g[x] η[ξ[x]] Φ[x]² + f1[x] Φ[x]² η'[ξ[x]] ξ'[x] −
    f1[x] η[ξ[x]] Φ[x] Φ'[x] − 2 f2[x] Φ[x] η'[ξ[x]] ξ'[x] Φ'[x] + 2 f2[x] η[ξ[x]] Φ'[x]² +
    f2[x] Φ[x]² ξ'[x]² η''[ξ[x]] + f2[x] Φ[x]² η'[ξ[x]] ξ''[x] − f2[x] η[ξ[x]] Φ[x] Φ''[x]

The quantities $\xi[x]$, $\eta[\xi]$ and their derivatives can now be given explicitly.

In[3]:= **ξv[x_] = Integrate[Sqrt[g[x]/f2[x]], x];**
    ξvs = D[ξv[x], x];
    ξvss = D[ξv[x], {x, 2}] // Together;

    Φv[x_] = (g[x]/f2[x])^(1/4) Exp[1/2 Integrate[f1[x]/f2[x], x]];
    Φvs = D[Φv[x], x] // Together;
    Φvss = D[Φv[x], {x, 2}] // Together;

We insert this in aux1, simplify the resulting expression, and "backsubstitute" (the backsubstituted quantities carry an additional v in their names).

In[9]:= **Numerator**[(* carry out all backsubstitutions *)
        aux1 //. {ξ[x] -> ξv[x], ξ'[x] -> ξvs, ξ''[x] -> ξvss,
                  Φ[x] -> Φv[x], Φ'[x] -> Φvs,
                  Φ''[x] -> Φvss} // Together] //.
                {Integrate[(g[x]/f2[x])^(1/2), x] -> ξ,
                 Exp[Integrate[f1[x]/f2[x], x]] -> Φ^2 Sqrt[f2[x]/g[x]]}

Out[9]= $-\Phi^2 \sqrt{\dfrac{f2[x]}{g[x]}}$ (4 f1[x]² g[x]² η[ξ] - 16 f0[x] f2[x] g[x]² η[ξ] -

        16 λ f2[x] g[x]³ η[ξ] + 8 f2[x] g[x]² η[ξ] f1'[x] - 8 f1[x] g[x]² η[ξ] f2'[x] +
        3 g[x]² η[ξ] f2'[x]² + 2 f2[x] g[x] η[ξ] f2'[x] g'[x] - 5 f2[x]² η[ξ] g'[x]² -
        4 f2[x] g[x]² η[ξ] f2''[x] + 4 f2[x]² g[x] η[ξ] g''[x] - 16 f2[x] g[x]³ η''[ξ])

We now cancel common factors.

In[10]:= **Factor**[%]/(Φ^2 Sqrt[f2[x]/g[x]])

Out[10]= -4 f1[x]² g[x]² η[ξ] + 16 f0[x] f2[x] g[x]² η[ξ] +
        16 λ f2[x] g[x]³ η[ξ] - 8 f2[x] g[x]² η[ξ] f1'[x] + 8 f1[x] g[x]² η[ξ] f2'[x] -
        3 g[x]² η[ξ] f2'[x]² - 2 f2[x] g[x] η[ξ] f2'[x] g'[x] + 5 f2[x]² η[ξ] g'[x]² +
        4 f2[x] g[x]² η[ξ] f2''[x] - 4 f2[x]² g[x] η[ξ] g''[x] + 16 f2[x] g[x]³ η''[ξ]

We rearrange the result in an appropriate form.

In[11]:= **Collect**[%, {η[ξ], η''[ξ]}]

Out[11]= η[ξ] (-4 f1[x]² g[x]² + 16 f0[x] f2[x] g[x]² + 16 λ f2[x] g[x]³ - 8 f2[x] g[x]² f1'[x] +
        8 f1[x] g[x]² f2'[x] - 3 g[x]² f2'[x]² - 2 f2[x] g[x] f2'[x] g'[x] + 5 f2[x]² g'[x]² +
        4 f2[x] g[x]² f2''[x] - 4 f2[x]² g[x] g''[x]) + 16 f2[x] g[x]³ η''[ξ]

Now, we divide through by the coefficient of $\eta''(\xi)$.

In[12]:= **#/(16 f2[x] g[x]^3)& /@ %**

Out[12]= $\dfrac{1}{16\,\text{f2}[x]\,g[x]^3}$
        (η[ξ] (-4 f1[x]² g[x]² + 16 f0[x] f2[x] g[x]² + 16 λ f2[x] g[x]³ - 8 f2[x] g[x]² f1'[x] +
        8 f1[x] g[x]² f2'[x] - 3 g[x]² f2'[x]² - 2 f2[x] g[x] f2'[x] g'[x] +
        5 f2[x]² g'[x]² + 4 f2[x] g[x]² f2''[x] - 4 f2[x]² g[x] g''[x])) + η''[ξ]

This leads to the following result for $v(\xi)$.

In[13]:= **Expand**[%[[1]]/η[ξ]] - λ

Out[13]= $\dfrac{f0[x]}{g[x]} - \dfrac{f1[x]^2}{4\,f2[x]\,g[x]} - \dfrac{f1'[x]}{2\,g[x]} + \dfrac{f1[x]\,f2'[x]}{2\,f2[x]\,g[x]} -$

        $\dfrac{3\,f2'[x]^2}{16\,f2[x]\,g[x]} - \dfrac{f2'[x]\,g'[x]}{8\,g[x]^2} + \dfrac{5\,f2[x]\,g'[x]^2}{16\,g[x]^3} + \dfrac{f2''[x]}{4\,g[x]} - \dfrac{f2[x]\,g''[x]}{4\,g[x]^2}$

The following formula can be found in the references given above:

$$\frac{f_0(x)}{g(x)} + \frac{1}{2}\left(\frac{g(x)}{f_2(x)}\right)'\left(\frac{f_2(x)}{g(x)}\right)^2 \frac{\Phi'(x)}{\Phi(x)} - \frac{f_2(x)}{g(x)}\frac{\Phi''(x)}{\Phi(x)}.$$

This is identical to the one above.

In[14]:= **f0[x]/g[x] + 1/2D[g[x]/f2[x], x] (f2[x]/g[x])^2 ***
        D[Φv[x], x]/Φv[x] - f2[x]/g[x] D[Φv[x], {x, 2}]/Φv[x];

        Expand[MapAll[Cancel, %]] === %%
Out[15]= True

Now, we implement the transformation to normal form. Most error messages implemented in the following deal with the case when the differential equations are nonlinear, homogeneous, and of second order. We try to analyze the input carefully to generate helpful warning messages. This is the reason for the test of these properties in the ODE. In addition, it may not be possible to compute $\xi(x)$ explicitly (for example, we cannot integrate explicitly, or afterwards, we cannot solve for x). In the case in which we get several solutions for x (as a default), we look only at the first solution.

The application of Solve permits only generic transformations. Using Eliminate, we can sometimes do more. We prefer to use Solve to avoid too many solutions. The optional function func serves to help with the integration; PowerExpand is often useful here. Now, we give the implementation. If the argument is inappropriate or *Mathematica* cannot do the necessary integration, ToNormalForm[*arguments*] remains unevaluated. We achieve this with a construction of the form With[{*res* = *allCalculations*}, *res* /; Return[$Failed]]. If any problems are found in *allCalculations*, then $Failed is returned. This is tested in the outer With, and only in the case in which $Failed was not returned is the result of *allCalculations* returned.

```
In[16]:= (* various error messages *)
         (* _..._ points out the problem at hand *)
         ToNormalForm::cantint   = "Cannot explicitly integrate.";
         ToNormalForm::cantsolve = "Cannot solve the equation.";
         ToNormalForm::notlinode2 = "ODE is not a linear ODE of _second_ order.";
         ToNormalForm::notLINode2 = "ODE is not a _linear_ ODE 2. order.";
         ToNormalForm::nothomode2 = "ODE is not a _homogeneous_ ODE 2. order.";
         ToNormalForm::notevpode2 = "ODE is not an _eigenvalue_ problem.";
         ToNormalForm::multisol  = "Solution for the new variable is not unique.";

         ToNormalForm[ode_Equal, y_, x_, λ_,
                      yNew_, xNew_, func_:Identity, i_Integer:1] :=
         With[{res =
         Module[{right, f2, f1, f0, g, ξ1, pot, η, ξ},
         Check[(* is it a linear ODE of degree 2? *)
             If[Max[#[[0, 0, 1]]& /@
               Cases[ode, Derivative[_][y][x], {0, Infinity}]] > 2,
               Message[ToNormalForm::notlinode2]], Return[$Failed]];
           right = Expand[ode[[1]] - ode[[2]]];
         Check[(* is it not an eigenvalue problem? *)
             If[FreeQ[Expand[right], λ],
           Message[ToNormalForm::notevpode2]], Return[$Failed]];
           f2 = Coefficient[right, y''[x]];
         Check[ (* it is not of second order? *)
               If[f2 === 0, Message[ToNormalForm::notlinode2]],
               Return[$Failed]];
           f1 = Coefficient[right, y'[x]]; g = Coefficient[right, λ]/y[x];
           f0 = Coefficient[right, y[x]] - λ g;
           {f2, f1, f0, g} = If[# === {}, 0, #]& /@ {f2, f1, f0, g};
         Check[(* is it linear? *)
           If[MemberQ[#, y''[x] | y'[x] | y[x], {0, Infinity}],
               Message[ToNormalForm::notLINode2]]& /@ {f2, f1, f0, g},
               Return[$Failed]];
         Check[(* is it homogeneous? *)
               If[Simplify[Expand[right -
                       (f2 y''[x] + f1 y'[x] + (f0 + λ g)y[x])]] =!= 0,
             Message[ToNormalForm::nothomode2]], Return[$Failed]];
           ξ1 = Integrate[func[Sqrt[g/f2]], x];
         Check[(* can it explicitly integrated? *)
         If[FreeQ[ξ1, Integrate],
             Check[sol = Solve[ξ1 == ξ, x],
                   Message[ToNormalForm::cantsolve]; Return[$Failed]],
               Message[ToNormalForm::cantint]], Return[$Failed]];
         If[Length[sol] > 1, Message[ToNormalForm::multisol]];
           (* write the result in a nice form *)
           pot = Together[f0/g - f1^2/(4 f2 g) - D[f1, x]/(2 g) +
                         (f1 D[f2, x])/(2 f2 g) - (3 D[f2, x]^2)/(16 f2 g) -
                         (D[f2, x] D[g, x])/(8 g^2) + (5 f2 D[g, x]^2)/(16 g^3) +
                         D[f2, x]/(4 g) - (f2 D[g, {x, 2}])/(4 g^2)];
           (* form new equation *)
           ((η''[ξ] + λ η[ξ] + # η[ξ] == 0 & /@ (* go to new variables *)
                   (pot /. sol // Simplify))[[i]]) //.
                   {η -> yNew, ξ -> xNew}]}, res /; res =!= Return[$Failed]]
```

Here are a few examples. First, we look at a differential equation of third order, an inhomogeneous differential equation, a nonlinear differential equation, and a differential equation not containing an eigenvalue. These are all unsuitable inputs.

`In[26]:= ToNormalForm[c'''[r] + la c'[r] == 0, c, r, la, y, x]`

> ToNormalForm::notlinode2 : ODE is not a linear ODE of _second_ order.

`Out[26]=` ToNormalForm[la c′[r] + c$^{(3)}$[r] = 0, c, r, la, y, x]

`In[27]:= ToNormalForm[c''[r] + 34 c'[r] == e c[r] + 56, c, r, e, y, x]`

> ToNormalForm::nothomode2 : ODE is not a _homogeneous_ ODE 2. order.

`Out[27]=` ToNormalForm[34 c′[r] + c″[r] = 56 + e c[r], c, r, e, y, x]

`In[28]:= ToNormalForm[c'[r] + la c[r] c'[r] == 0, c, r, la, y, x]`

> ToNormalForm::notlinode2 : ODE is not a linear ODE of _second_ order.

`Out[28]=` ToNormalForm[c′[r] + la c[r] c′[r] = 0, c, r, la, y, x]

`In[29]:= ToNormalForm[c'[r] + la c[r] c'[r] == 0, c, r, al, y, x]`

> ToNormalForm::notevpode2 : ODE is not an _eigenvalue_ problem.

`Out[29]=` ToNormalForm[c′[r] + la c[r] c′[r] = 0, c, r, al, y, x]

If the differential equation is already in normal form, the most that happens is that the variables are renamed.

`In[30]:= ToNormalForm[z''[s] + a s z[s] + e z[s] == 0, z, s, e, z, s]`

`Out[30]=` a s z[s] + e z[s] + z″[s] = 0

`In[31]:= ToNormalForm[z''[s] + a s z[s] + e z[s] == 0, z, s, e, ψ, t]`

`Out[31]=` a t ψ[t] + e ψ[t] + ψ″[t] = 0

Now, we give a few nontrivial examples.

`In[32]:= ToNormalForm[D[(a s + b) z'[s], s] + a s z[s] + e z[s] == 0,`
`          z, s, e, ψ, t]`

`Out[32]=` $\frac{1}{4}\left(a - 4b + \frac{1}{t^2} + a^2 t^2\right)\psi[t] + e\,\psi[t] + \psi''[t] = 0$

`In[33]:= ToNormalForm[D[D[ψ[r]/r, r] r^2, r]/r^2 + v0/r ψ[r]/r +`
`          1^2/r^2 ψ[r]/r + e ψ[r]/r == 0, ψ, r, e, φ, x]`

`Out[33]=` $-\frac{(2 - 4\,1^2 + x - 4\,v0\,x)\,\varphi[x]}{4\,x^2} + e\,\varphi[x] + \varphi''[x] = 0$

`In[34]:= ToNormalForm[D[D[ψ[r]/r, r] r^2, r]/r^2 + v0/r ψ[r]/r +`
`          eF r ψ[r]/r + 1^2/r^2 ψ[r]/r + e ψ[r]/r == 0, ψ, r, e, φ, x]`

`Out[34]=` $-\frac{(2 - 4\,1^2 + x - 4\,v0\,x - 4\,eF\,x^3)\,\varphi[x]}{4\,x^2} + e\,\varphi[x] + \varphi''[x] = 0$

In the following example, multiple choices for the new variable exist.

`In[35]:= ToNormalForm[D[1/(a z + b) p'[z], z] + (a z + b1) p[z] + e p[z] == 0,`
`          p, z, e, ψ, z]`

> ToNormalForm::multisol : Solution for the new variable is not unique.

`Out[35]=` $\left(-b + b1 - \frac{5}{12\,z^2} - \frac{1}{3\,2^{2/3}\,3^{1/3}\,a^{1/3}\,z^{4/3}} + \left(\frac{3}{2}\right)^{2/3}a^{2/3}\,z^{2/3}\right)\psi[z] + e\,\psi[z] + \psi''[z] = 0$

The next input fails because the resulting equation cannot be solved.

`In[36]:= ToNormalForm[x^2 y''[x] + x y'[x] + (x^2 - n^2) y[x] == 0,`
`          y, x, n^2, y, x]`

> Solve::tdep : The equations appear to involve the variables
>   to be solved for in an essentially non-algebraic way. More…

> ToNormalForm::cantsolve : Cannot solve the equation.

`Out[36]=` ToNormalForm[$(-n^2 + x^2)$ y[x] + x y′[x] + x² y″[x] = 0, y, x, n², y, x]

Next, we use `ToNormalForm` with a nontrivial seventh argument on the differential equation from the last input.

```
In[37]:= ToNormalForm[x^2 y''[x] + x y'[x] + (x^2 - n^2) y[x] == 0,
                       y, x, n^2, y, x, PowerExpand]
```

$$Out[37]= \frac{1}{2} (1 - e^{-ix} - 2 e^{-2ix}) y[x] + n^2 y[x] + y''[x] == 0$$

12. Noncentral Collision, Bernstein Polynomials, Bernstein Operator

a) Here are the velocities $\{v'_x, v'_y\}$ and $\{V'_x, V'_y\}$ of the two spheres after the collision. $\{v'_x, v'_y\}$ and $\{V'_x, V'_y\}$ can be computed directly from the three physical laws of conservation of energy, momentum, and angular momentum.)

```
In[1]:= sol = Solve[{ (* x - component of impulse *)
                    m vx == m vxs + m Vxs,
                     (* y - component of impulse *)
                    m vy == m vys + M Vys,
                     (* energy conservation *)
                    m/2 (vx^2  +  vy^2) ==
                    m/2 (vxs^2 + vys^2) + M/2 (Vxs^2 + Vys^2),
                     (* angular momentum of the center of gravity
                        with respect to (0, 0) *)
                    x m vy - y m vx ==  x m vys - y m vxs},
                    {vxs, vys, Vxs, Vys}] // Simplify
```

$$Out[1]= \left\{\left\{Vxs \to 0,\ Vys \to 0,\ vxs \to vx,\ vys \to vy\right\},\ \left\{Vxs \to \frac{2 m M x\ (vx\, x + vy\, y)}{(m + M)\ (M\, x^2 + m\, y^2)},\right.\right.$$

$$Vys \to \frac{2 m^2 y\ (vx\, x + vy\, y)}{(m + M)\ (M\, x^2 + m\, y^2)},\ vxs \to \frac{M^2\, vx\, x^2 + m^2\, vx\, y^2 + m M\ (-2\, vy\, x\, y + vx\, (-x^2 + y^2))}{(m + M)\ (M\, x^2 + m\, y^2)},$$

$$\left.\left.vys \to \frac{M^2\, vy\, x^2 + m^2\, vy\, y^2 + m M\ (-2\, vx\, x\, y + vy\, (x^2 - y^2))}{(m + M)\ (M\, x^2 + m\, y^2)}\right\}\right\}$$

The first solution is the trivial one after the collision. We are interested in the second solution.

```
In[2]:= sol = #[[2]]& /@ sol[[2]]
```

$$Out[2]= \left\{\frac{2 m M x\ (vx\, x + vy\, y)}{(m + M)\ (M\, x^2 + m\, y^2)},\ \frac{2 m^2 y\ (vx\, x + vy\, y)}{(m + M)\ (M\, x^2 + m\, y^2)},\ \frac{M^2\, vx\, x^2 + m^2\, vx\, y^2 + m M\ (-2\, vy\, x\, y + vx\, (-x^2 + y^2))}{(m + M)\ (M\, x^2 + m\, y^2)},\right.$$

$$\left.\frac{M^2\, vy\, x^2 + m^2\, vy\, y^2 + m M\ (-2\, vx\, x\, y + vy\, (x^2 - y^2))}{(m + M)\ (M\, x^2 + m\, y^2)}\right\}$$

Here is a plot [1039]: after scaling all quantities involved, we show the positions of the two spheres before the collision, at the point of collision (twice), and after the collision. The time scale was chosen by scaling with the radius of the sphere.

```
In[3]:= collisionGraphics[m_, M_, vx_, vy_, x_, y_] :=
        Module[{vxs, vys, Vxs, Vys, v, R},
        (* the velocity components after the collision *)
        Vxs = (2 m M x (vx x + vy y))/((m + M) (M x^2 + m y^2));
        Vys = (2 m^2 y (vx x + vy y))/((m + M) (M x^2 + m y^2));
        vxs = (-(m M vx x^2) + M^2 vx x^2 -
               2 m M vy x y + m^2 vx y^2 + m M vx y^2)/
               ((m + M) (M x^2 + m y^2));
        vys = (m M vy x^2 + M^2 vy x^2 -
               2 m M vx x y +  m^2 vy y^2 - m M vy y^2)/
               ((m + M) (M x^2 + m y^2));
        (* scaled time *)
        R = Sqrt[x^2 + y^2]/2;
        tm = 3 R/Max[Sqrt[vx^2 + vy^2], Sqrt[vxs^2 + vys^2],
                    Sqrt[Vxs^2 + Vys^2]];
        (* make array of graphics *)
        Show[GraphicsArray[#]]& /@ Join[ (* before the collision *)
        Partition[Table[(* make a couple of pictures at various times *)
        Graphics[{Line[{{-2, 0}, {2, 0}}], Line[{{0, -2}, {0, 2}}],
                 {GrayLevel[0], Disk[{0, 0}, R]},
                 {GrayLevel[1/2], Disk[{x, y} + t {vx, vy}, R]}},
                 AspectRatio -> Automatic, PlotRange -> {{-3, 3}, {-3, 3}},
                 PlotLabel -> "t = " <> ToString[N[t, 3]]],
              {t, -tm, 0, tm/7}], 4],
                             (* after the collision *)
        Partition[Table[(* make a couple of pictures at various times *)
```

```
Graphics[{Line[{{-2, 0}, {2, 0}}], Line[{{0, -2}, {0, 2}}],
        {GrayLevel[0], Disk[{0, 0} + t {Vxs, Vys}, R]},
        {GrayLevel[1/2], Disk[{x, y} + t {vxs, vys}, R]}},
        AspectRatio -> Automatic, PlotRange -> {{-3, 3}, {-3, 3}},
        PlotLabel -> StyleForm["t = " <> ToString[N[t, 3]]]],
  {t, 0, tm, tm/7}], 4]]]
```

Here is the result of a central collision ($|x/y| = |v_x/v_y|$) of two spheres with the same masses.

In[4]:= collisionGraphics[1, 1, -1, 0, 1, 0]

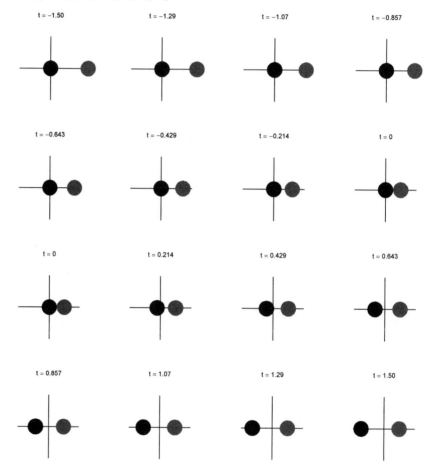

Here is a noncentral collision ($|x/y| \neq |v_x/v_y|$) for two different masses.

In[5]:= collisionGraphics[1, 2, -1, -1, 1, 1/2]

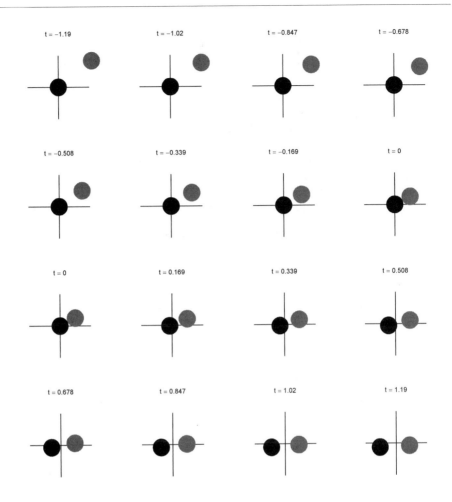

| t = -1.19 | t = -1.02 | t = -0.847 | t = -0.678 |

| t = -0.508 | t = -0.339 | t = -0.169 | t = 0 |

| t = 0 | t = 0.169 | t = 0.339 | t = 0.508 |

| t = 0.678 | t = 0.847 | t = 1.02 | t = 1.19 |

Collision can still be refined somewhat by checking whether the input arguments would result in a collision at all, and the plot could be animated. For a detailed discussion of noncentral collisions, also with friction, see [738], [974], [233], [1003], and [496].

b) These are the Bernstein polynomials.

```
In[1]:= b[n_, k_, x_] := Binomial[n, k] x^k (1 - x)^(n - k)
```

Their maximum occurs at $x_m = k/n$.

```
In[2]:= Solve[Together[D[x^k (1 - x)^(n - k), x]] == 0, x]
```

$$\text{Out[2]= } \left\{\left\{x \to \frac{k}{n}\right\}\right\}$$

The following graphic shows the 33 Bernstein polynomials for $n = 32$ and marks their maxima.

```
In[3]:= With[{n = 32},
          Plot[Evaluate[Table[b[n, k, x], {k, 0, n}]], {x, 0, 1},
              PlotRange -> {0, 0.4}, Epilog -> {Hue[0], PointSize[0.01],
                  Table[Point[{k/n, b[n, k, k/n]}],
                      {k, 1, n - 1}]}]]
```

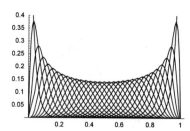

In the limit of large n, we get the following expression for the scaled maxima.

```
In[4]:= (* direct attempt *)
        Limit[Sqrt[n] b[n, x n, x], n -> ∞]
```

$$\text{Out[5]= } \text{Limit}\left[\sqrt{n}\ (1-x)^{n-n\,x}\,x^{n\,x}\ \text{Binomial}[n,\ n\,x],\ n \to \infty\right]$$

```
In[6]:= Off[Series::esss];
        Series[Sqrt[n] b[n, x n, x], {n, ∞, 0},
              Assumptions -> 0 < x < 1] // Normal // Simplify[#, 0 < x < 1]&
```

$$\text{Out[7]= } \frac{1}{\sqrt{2\,\pi}\ \sqrt{-(-1+x)\,x}}$$

```
In[8]:= (* make series expansion for large n, then take limit *)
        Off[Series::esss];
        Series[Sqrt[n] b[n, x n, x], {n, ∞, 0},
              Assumptions -> 0 < x < 1] // Normal // Simplify[#, 0 < x < 1]&
```

$$\text{Out[10]= } \frac{1}{\sqrt{2\,\pi}\ \sqrt{-(-1+x)\,x}}$$

In the limit $n \to \infty$, this expression is $(2\,\pi\,x(1-x))^{-1/2}$.

```
In[11]:= θ[x_] = Limit[%, n -> Infinity];
```

The next graphic shows the limit curve of the scaled maxima together with the scaled maxima for $n = 256$.

```
In[12]:= Plot[θ[x], {x, 0, 1}, PlotStyle -> Hue[0], PlotRange -> {0, 3},
              Prolog -> With[{n = 256}, {PointSize[0.006],
                      Table[Point[{k/n, Sqrt[n] b[n, k, k/n]}],
                            {k, 1, n - 1}]}]]
```

c) `BernsteinB` is the Bernstein operator.

```
In[1]:= BernsteinB[n_, f_, x_] := Sum[Binomial[n, k] x^k (1 - x)^(n - k) *
                                  (f /. x -> k/n), {k, 0, n}]
```

To obtain the eigenfunctions of $\hat{B}_n$, we make the ansatz $f_k(x) = \sum_{j=0}^{n} c_j^{(k)}\,x^j$. Calculating $\hat{B}_n(f_k(x))$ and extracting the coefficients of the x^j yields a matrix M, whose eigensystem yields the $f_k(x)$.

```
In[2]:= n = 12; f = Sum[c[k] x^k, {k, 0, n}];
        bs = BernsteinB[n, f, x];
In[4]:= (* coefficients of the x^j *)
        cl = CoefficientList[bs, x] // Expand;
        (* c[k] coefficients inside the coefficients of the x^j *)
        M = Table[Coefficient[cl[[k]], c[j]], {k, n + 1}, {j, 0, n}];
```

Interestingly, all eigenvalues are rational and no algebraic numbers appear.

```
In[8]:= Take[evals = Eigenvalues[M], -5]
```

$$\text{Out[8]= } \left\{ \frac{1925}{41472}, \frac{1925}{124416}, \frac{1925}{497664}, \frac{1925}{2985984}, \frac{1925}{35831808} \right\}$$

```
In[9]:= Union[Head /@ evals]
```

Out[9]= {Integer, Rational}

The eigenvalues can actually be expressed in closed form.

```
In[10]:= Eigenvalues[M] == Table[n!/(n - k)!/n^k, {k, n, 0, - 1}]
```

Out[10]= False

Now, we calculate the eigenvectors of M.

```
In[11]:= evecs = Eigenvectors[M];
         xPowers = Array[x^#&, n + 1, 0];
         evecsx = xPowers.#& /@ evecs;
```

The resulting eigenvectors have rational x-coefficients with hundreds of digits in the numerators and denominators.

```
In[14]:= Round[Log[10, #]]& @
             {Max[Numerator[#]], Max[Denominator[#]]}&[
                 Abs[Cases[evecs, _Rational, Infinity]]]
```

Out[14]= {23, 25}

To visualize the eigenvectors, we scale them to the same maximal height over the x-interval $[0, 1]$. In addition, to avoid numerical errors when calculating the eigenfunctions, we define a function evecsxHP that uses high-precision arithmetic for their numerical evaluation.

```
In[15]:= scale[k_] := 1/Max[Abs[evecsx[[k]]] /.
                            {ToRules[NRoots[D[evecsx[[k]], x] == 0, x]]}]]
         evecsxHP[k_, ξ_?NumberQ] := evecsx[[k]] /. x -> SetPrecision[ξ, 200]
```

Here are the 36 eigenfunctions. For a nicer picture, we use the same overall sign for the linear coefficient of the polynomials.

```
In[17]:= Plot[Evaluate[Table[Sign[Coefficient[evecsx[[k]], x, 1]] scale[k]*
                       evecsxHP[k, x], {k, n + 1}]], {x, 0, 1},
                       PlotRange -> All, Frame -> True,
                       PlotStyle -> {Thickness[0.002]},
                       Axes -> False, PlotRange -> All]
```

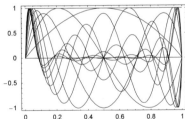

For closed form expressions of the eigenvalues and eigenvectors, see [1802], [408].

13. A Sensitive Linear System of Equations, Bisector Surface, Surface Connecting Three Cylinders, Double Torus Surface

a) A first look at the condition number indicates that the system is (quite) ill-conditioned.

```
In[1]:= M = {{-367296,  -43199, +0519436, -0954302},
             {+259718, -477151, -0367295, -1043199},
             {+886731, +088897, -1254026, -1132096},
             {+627013, +566048, -0886732, +0911103}};
```

```
In[2]:= Max[#]/Min[#]&[Abs[Eigenvalues[N[M]]]]
```

Out[2]= 8.1582×10^{14}

Here are the equations and the exact solution of the system.

```
In[3]:= eqs = Thread[M.{t, u, v, w} == {1, 1, 1, 0}]
```

Out[3]= {-367296 t - 43199 u + 519436 v - 954302 w == 1, 259718 t - 477151 u - 367295 v - 1043199 w == 1,
886731 t + 88897 u - 1254026 v - 1132096 w == 1, 627013 t + 566048 u - 886732 v + 911103 w == 0}

In[4]:= Solve[eqs, {t, u, v, w}]

Out[4]= {{t → 886731088897000000, u → 886731088897, v → 627013566048000000, w → 627013566048}}

In[5]:= N[%]

Out[5]= {{t → 8.86731×10^17, u → 8.86731×10^11, v → 6.27014×10^17, w → 6.27014×10^11}}

In[6]:= exactSolution = ({t, u, v, w} /. %%)[[1]];

An attempt to find a numerical solution with machine accuracy fails completely.

In[7]:= Solve[N[eqs], {t, u, v, w}]

RowReduce::luc : Result for RowReduce of badly conditioned matrix
{{-367296., -43199., 519436., -954302., -1.}, {259718., -477151., -367295.,
-1.0432×10^6, -1.}, {≪1≫}, {627013., 566048., -886732., 911103., 0.}}
may contain significant numerical errors. More…

Out[7]= {{t → 1.90642×10^10, u → 19064.2, v → 1.34804×10^10, w → 13480.4}}

If we increase the number of digits used in the calculation step by step and monitor the accuracy of the results obtained, we can get the at least one certified digit. We have to work with the following number of digits.

In[8]:= prec = 10;
While[sol = Check[Solve[SetPrecision[eqs, prec], {t, u, v, w}],
$Failed];
sol === {} || sol === $Failed || Precision[sol] < 2,
prec = prec + 1]

RowReduce::luc : Result for RowReduce of badly conditioned
matrix ≪1≫ may contain significant numerical errors. More…

RowReduce::luc : Result for RowReduce of badly conditioned matrix
{{-367296.00000000000000, -43199.000000000000, 519436.00000000000,
-954302.00000000000, -1.0000000000000000000}, {≪1≫}, {≪1≫},
{627013.00000000000, 566048.00000000000, - ≪26≫, ≪26≫, 0.×10^-20}}
may contain significant numerical errors. More…

RowReduce::luc : Result for RowReduce of badly conditioned
matrix {{-367296.000000000000, -43199.0000000000000,
519436.000000000000, -954302.000000000000, -1.0000000000000000000000},
{259718.000000000000, ≪3≫, -1.00000000000000000000},
≪1≫, {627013.000000000000, 566048.000000000000,
-886732.000000000000, 911103.000000000000, 0.×10^-21}}
may contain significant numerical errors. More…

General::stop :
Further output of RowReduce::luc will be suppressed during this calculation. More…

We had to use more than 25 digits to get a certified solution.

In[10]:= prec

Out[10]= 26

Even with 27 digits, the solution gives a residual correct to only two digits.

In[11]:= Solve[SetPrecision[eqs, prec], {t, u, v, w}]

Out[11]= {{t → 8.86731088896999939783937617×10^17, u → 8.86731088896999939783937617×10^11,
v → 6.27013566047999957420813953×10^17, w → 6.27013566047999957420813953×10^11}}

In[12]:= (Subtract @@@ eqs) /. %

Out[12]= {{0.×10^-3, 0.×10^-3, 0.×10^-2, 0.×10^-2}}

To find the sensitivity of the solution to small changes, we find the Taylor expansion for small a[i, j] for the solutions of the following generalization of the above system of equations.

In[13]:= `eqn1 = Thread[(M (1 + Array[a, {4, 4}])).{t, u, v, w} == {1, 1, 1, 0}]`

Out[13]= {-367296 t (1 + a[1, 1]) − 43199 u (1 + a[1, 2]) + 519436 v (1 + a[1, 3]) − 954302 w (1 + a[1, 4]) =
1, 259718 t (1 + a[2, 1]) − 477151 u (1 + a[2, 2]) −
367295 v (1 + a[2, 3]) − 1043199 w (1 + a[2, 4]) == 1,
886731 t (1 + a[3, 1]) + 88897 u (1 + a[3, 2]) − 1254026 v (1 + a[3, 3]) −
1132096 w (1 + a[3, 4]) == 1, 627013 t (1 + a[4, 1]) +
566048 u (1 + a[4, 2]) − 886732 v (1 + a[4, 3]) + 911103 w (1 + a[4, 4]) == 0}

In[14]:= `coeff = Flatten[Table[a[i, j], {i, 4}, {j, 4}]];`
`coeffRules = Flatten[Table[a[i, j] -> 0, {i, 4}, {j, 4}]];`

We now compute the sensitivity. The quantities being computed below are the following:

$$\left.\frac{\frac{\partial}{\partial a[i,\,j]} \text{ exact solution with } a[i,\,j]}{\text{exact solution with } a[i,\,j]}\right|_{a[i,j]=0}.$$

To find these values, we compute the exact solution, develop it in a series, and extract the desired relation using `(N[Di : vide @@ #[[3]]]) &`.

In[16]:= `tab = Table[`
`(* the exact solutions *)`
`soltemp = Solve[eqn1 /. Delete[coeffRules, i], {t, u, v, w}];`
`(* calculate relative change *)`
`res = (N[Divide @@ Reverse[#[[3]]]]) & /@`
`(* make series of exact solution wrt a[i, j] *)`
`(Series[#, {coeff[[i]], 0, 1}] & /@ (({t, u, v, w} /. soltemp)[[1]]));`
`NumberForm[#, 2, NumberSigns -> {"-", "+"}] & /@ res, {i, 16}] //`
`(* format as a table *) TableForm[#, TableHeadings ->`
`{coeff, StyleForm[#, FontWeight -> "Bold"] & /@ {"t", "u", "v", "w"}}] &`

	t	u	v	w
a[1, 1]	$+2.3 \times 10^{23}$	$+2.3 \times 10^{23}$	$+2.3 \times 10^{23}$	$+2.3 \times 10^{23}$
a[1, 2]	$+2.7 \times 10^{16}$	$+2.7 \times 10^{16}$	$+2.7 \times 10^{16}$	$+2.7 \times 10^{16}$
a[1, 3]	-2.3×10^{23}	-2.3×10^{23}	-2.3×10^{23}	-2.3×10^{23}
a[1, 4]	$+4.3 \times 10^{17}$	$+4.3 \times 10^{17}$	$+4.3 \times 10^{17}$	$+4.3 \times 10^{17}$
a[2, 1]	-2.3×10^{23}	-2.3×10^{23}	-2.3×10^{23}	-2.3×10^{23}
a[2, 2]	$+4.2 \times 10^{17}$	$+4.2 \times 10^{17}$	$+4.2 \times 10^{17}$	$+4.2 \times 10^{17}$
a[2, 3]	$+2.3 \times 10^{23}$	$+2.3 \times 10^{23}$	$+2.3 \times 10^{23}$	$+2.3 \times 10^{23}$
Out[16]//TableForm= a[2, 4]	$+6.5 \times 10^{17}$	$+6.5 \times 10^{17}$	$+6.5 \times 10^{17}$	$+6.5 \times 10^{17}$
a[3, 1]	$+5.6 \times 10^{23}$	$+5.6 \times 10^{23}$	$+5.6 \times 10^{23}$	$+5.6 \times 10^{23}$
a[3, 2]	$+5.6 \times 10^{16}$	$+5.6 \times 10^{16}$	$+5.6 \times 10^{16}$	$+5.6 \times 10^{16}$
a[3, 3]	-5.6×10^{23}	-5.6×10^{23}	-5.6×10^{23}	-5.6×10^{23}
a[3, 4]	$-5. \times 10^{17}$	$-5. \times 10^{17}$	$-5. \times 10^{17}$	$-5. \times 10^{17}$
a[4, 1]	-5.6×10^{23}	-5.6×10^{23}	-5.6×10^{23}	-5.6×10^{23}
a[4, 2]	-5.1×10^{17}	-5.1×10^{17}	-5.1×10^{17}	-5.1×10^{17}
a[4, 3]	$+5.6 \times 10^{23}$	$+5.6 \times 10^{23}$	$+5.6 \times 10^{23}$	$+5.6 \times 10^{23}$
a[4, 4]	-5.7×10^{17}	-5.7×10^{17}	-5.7×10^{17}	-5.7×10^{17}

Thus, a relative change of the first coefficient of size $2 \times 10^{-23} = 4 \times 10^{-24}$ already leads to a change in the first digit of the result!

In[17]:= `With[{ε = 4 10^-24},`
`Solve[{-367296 t (1 + ε) − 43199 u + 519436 v − 954302 w == 1,`
`259718 t − 477151 u − 367295 v − 1043199 w == 1,`
`886731 t + 88897 u − 1254026 v − 1132096 w == 1,`
`627013 t + 566048 u − 886732 v + 911103 w == 0},`
`{t, u, v, w}] // N]`

Out[17]= `{{t → 1.21591×10¹⁹, u → 1.21591×10¹³, v → 8.59776×10¹⁸, w → 8.59776×10¹²}}`

For comparison, this was the exact result without perturbation:
`{{t → 8.867×10¹⁷, u → 8.867×10¹¹, v → 6.270×10¹⁷, w → 6.270×10¹²}}`.

Another way to look at the sensitivity is the following [1747] (now all perturbations enter with the same parameter t): Let the original linear system be $A.x = b$. Now, we change A to $A + t\,\delta A$ and b to $b + t\,\delta b$. Then, for $\dot{x}(t)$, we get

$$\mathbf{x}(t) \approx \mathbf{x}(0) + t\,\dot{\mathbf{x}}(0) = \mathbf{x}(0) + t\,\mathbf{A}^{-1}(\delta \mathbf{b} - \delta \mathbf{A}\,.\mathbf{x}(0))$$

with $\mathbf{x}(0) = \mathbf{A}^{-1}.\mathbf{b}$.

Here, the term $\dot{x}(0)$ describes the sensitivity of the system to small perturbations. For our example, we get the following by taking into account only changes in the matrix A. (The following are a different measure of sensitivity than is the above one; so, we get different numbers now.)

```
In[18]:= (* avoid numericalization the arguments of δa *)
         SetAttributes[δa, NHoldAll]

         -Inverse[M].Array[δa, {4, 4}].
                 LinearSolve[M, {1, 1, 1, 0}] // Expand // N[#, 2]&
```

$$
\begin{aligned}
\text{Out[20]= } \{ &-5.6\times10^{35}\,\delta a[1, 1] - 5.6\times10^{29}\,\delta a[1, 2] - 4.0\times10^{35}\,\delta a[1, 3] - 4.0\times10^{29}\,\delta a[1, 4] - \\
& 7.8\times10^{35}\,\delta a[2, 1] - 7.8\times10^{29}\,\delta a[2, 2] - 1\,\delta a[2, 3] - 5.5\times10^{29}\,\delta a[2, 4] + \\
& 5.6\times10^{35}\,\delta a[3, 1] + 5.6\times10^{29}\,\delta a[3, 2] + 3.9\times10^{35}\,\delta a[3, 3] + 3.9\times10^{29}\,\delta a[3, 4] - \\
& 7.9\times10^{35}\,\delta a[4, 1] - 7.9\times10^{29}\,\delta a[4, 2] - 5.6\times10^{35}\,\delta a[4, 3] - 5.6\times10^{29}\,\delta a[4, 4], \\
& -5.6\times10^{29}\,\delta a[1, 1] - 5.6\times10^{23}\,\delta a[1, 2] - 4.0\times10^{29}\,\delta a[1, 3] - 4.0\times10^{23}\,\delta a[1, 4] - \\
& 7.8\times10^{29}\,\delta a[2, 1] - 7.8\times10^{23}\,\delta a[2, 2] - 5.5\times10^{29}\,\delta a[2, 3] - 5.5\times10^{23}\,\delta a[2, 4] + \\
& 5.6\times10^{29}\,\delta a[3, 1] + 5.6\times10^{23}\,\delta a[3, 2] + 3.9\times10^{29}\,\delta a[3, 3] + 3.9\times10^{23}\,\delta a[3, 4] - \\
& 7.9\times10^{29}\,\delta a[4, 1] - 7.9\times10^{23}\,\delta a[4, 2] - 5.6\times10^{29}\,\delta a[4, 3] - 5.6\times10^{23}\,\delta a[4, 4], \\
& -4.0\times10^{35}\,\delta a[1, 1] - 4.0\times10^{29}\,\delta a[1, 2] - 2.8\times10^{35}\,\delta a[1, 3] - 2.8\times10^{29}\,\delta a[1, 4] - \\
& 5.5\times10^{35}\,\delta a[2, 1] - 5.5\times10^{29}\,\delta a[2, 2] - 3.9\times10^{35}\,\delta a[2, 3] - 3.9\times10^{29}\,\delta a[2, 4] + \\
& 3.9\times10^{35}\,\delta a[3, 1] + 3.9\times10^{29}\,\delta a[3, 2] + 2.8\times10^{35}\,\delta a[3, 3] + 2.8\times10^{29}\,\delta a[3, 4] - \\
& 5.6\times10^{35}\,\delta a[4, 1] - 5.6\times10^{29}\,\delta a[4, 2] - 4.0\times10^{35}\,\delta a[4, 3] - 4.0\times10^{29}\,\delta a[4, 4], \\
& -4.0\times10^{29}\,\delta a[1, 1] - 4.0\times10^{23}\,\delta a[1, 2] - 2.8\times10^{29}\,\delta a[1, 3] - 2.8\times10^{23}\,\delta a[1, 4] - \\
& 5.5\times10^{29}\,\delta a[2, 1] - 5.5\times10^{23}\,\delta a[2, 2] - 3.9\times10^{29}\,\delta a[2, 3] - 3.9\times10^{23}\,\delta a[2, 4] + \\
& 3.9\times10^{29}\,\delta a[3, 1] + 3.9\times10^{23}\,\delta a[3, 2] + 2.8\times10^{29}\,\delta a[3, 3] + 2.8\times10^{23}\,\delta a[3, 4] - \\
& 5.6\times10^{29}\,\delta a[4, 1] - 5.6\times10^{23}\,\delta a[4, 2] - 4.0\times10^{29}\,\delta a[4, 3] - 4.0\times10^{23}\,\delta a[4, 4] \}
\end{aligned}
$$

We could also use `Intervals` to measure the sensitivity of the system. In the next input, we use an `Interval`-object for one of the integer coefficients of the first equation. We calculate the relative uncertainty of the result.

```
In[21]:= Table[{n,
         X = Interval[43199 + {-10^-n, 10^-n}];
         (* second coefficient of first equation is now an interval *)
         eqn = {-367296 t -        X u +  519436 v -   954302 w  == 1,
                 259718 t -  477151 u -  367295 v - 1043199 w  == 1,
                 886731 t +   88897 u - 1254026 v - 1132096 w  == 1,
                 627013 t + 566048 u -  886732 v +  911103 w  == 0};
         (* (length of the resulting solution interval)/
            (average of the resulting solution interval) *)
         Abs[(Subtract @@ #[[1]])/(Plus @@ #[[1]]/2)]& /@
            ({t, u, v, w} /. Solve[eqn, {t, u, v, w}][[1]] /.
            Interval[{0, 0}] -> 0)] // N}, {n, 15, 30}]
```

$$
\begin{aligned}
\text{Out[21]= } \{ &\{15, \{0.00126203, 0.0073015, 0.00126202, 0.00126202\}\}, \\
& \{16, \{0.000126203, 0.00073015, 0.000126202, 0.000126202\}\}, \\
& \{17, \{0.0000126203, 0.000073015, 0.0000126202, 0.0000126202\}\}, \\
& \{18, \{1.26203\times10^{-6}, 7.3015\times10^{-6}, 1.26202\times10^{-6}, 1.26202\times10^{-6}\}\}, \\
& \{19, \{1.26203\times10^{-7}, 7.3015\times10^{-7}, 1.26202\times10^{-7}, 1.26202\times10^{-7}\}\}, \\
& \{20, \{1.26203\times10^{-8}, 7.3015\times10^{-8}, 1.26202\times10^{-8}, 1.26202\times10^{-8}\}\}, \\
& \{21, \{1.26203\times10^{-9}, 7.3015\times10^{-9}, 1.26202\times10^{-9}, 1.26202\times10^{-9}\}\}, \\
& \{22, \{1.26203\times10^{-10}, 7.3015\times10^{-10}, 1.26202\times10^{-10}, 1.26202\times10^{-10}\}\}, \\
& \{23, \{1.26203\times10^{-11}, 7.3015\times10^{-11}, 1.26202\times10^{-11}, 1.26202\times10^{-11}\}\}, \\
& \{24, \{1.26203\times10^{-12}, 7.3015\times10^{-12}, 1.26202\times10^{-12}, 1.26202\times10^{-12}\}\}, \\
& \{25, \{1.26203\times10^{-13}, 7.3015\times10^{-13}, 1.26202\times10^{-13}, 1.26202\times10^{-13}\}\}, \\
& \{26, \{1.26203\times10^{-14}, 7.3015\times10^{-14}, 1.26202\times10^{-14}, 1.26202\times10^{-14}\}\}, \\
& \{27, \{1.26203\times10^{-15}, 7.3015\times10^{-15}, 1.26202\times10^{-15}, 1.26202\times10^{-15}\}\},
\end{aligned}
$$

$\{28, \{1.26203 \times 10^{-16}, 7.3015 \times 10^{-16}, 1.26202 \times 10^{-16}, 1.26202 \times 10^{-16}\}\},$
$\{29, \{1.26203 \times 10^{-17}, 7.3015 \times 10^{-17}, 1.26202 \times 10^{-17}, 1.26202 \times 10^{-17}\}\},$
$\{30, \{1.26203 \times 10^{-18}, 7.3015 \times 10^{-18}, 1.26202 \times 10^{-18}, 1.26202 \times 10^{-18}\}\}\}$

b) This is the parametric representation of the torus.

```
In[1]:= torus[φ1_, φ2_, r1_, r2_] = {r1 Cos[φ1] + r2 Cos[φ1] Cos[φ2],
                                       r1 Sin[φ1] + r2 Sin[φ1] Cos[φ2],
                                       r2              Sin[φ2]};
```

Let the point be at $\{\xi, \eta, \zeta\}$. The bisector surface has coordinates $\{X, Y, Z\}$.

```
In[2]:= T = torus[φ1, φ2, r1, r2];
        eqs = {({X, Y, Z} - T).D[T, φ1], ({X, Y, Z} - T).D[T, φ2],
               #.#&[{ξ, η, ζ} - {X, Y, Z}] - #.#&[{X, Y, Z} - T]} // Factor
Out[3]= {-(r1 + r2 Cos[φ2]) (-Y Cos[φ1] + X Sin[φ1]),
        r2 (Z Cos[φ2] - X Cos[φ1] Sin[φ2] + r1 Cos[φ1]² Sin[φ2] - r2 Cos[φ2] Sin[φ2] + r2 Cos[φ1]²
           Cos[φ2] Sin[φ2] - Y Sin[φ1] Sin[φ2] + r1 Sin[φ1]² Sin[φ2] + r2 Cos[φ2] Sin[φ1]² Sin[φ2]),
        -2 Z ζ + ζ² - 2 Y η + η² - 2 X ξ + ξ² + 2 r1 X Cos[φ1] - r1² Cos[φ1]² + 2 r2 X Cos[φ1] Cos[φ2] -
         2 r1 r2 Cos[φ1]² Cos[φ2] - r2² Cos[φ1]² Cos[φ2]² + 2 r1 Y Sin[φ1] + 2 r2 Y Cos[φ2] Sin[φ1] -
         r1² Sin[φ1]² - 2 r1 r2 Cos[φ2] Sin[φ1]² - r2² Cos[φ2]² Sin[φ1]² + 2 r2 Z Sin[φ2] - r2² Sin[φ2]²}
```

To obtain an implicit representation of the bisector surface, we use `GroebnerBasis` to eliminate the parametric variables $\varphi1$ and $\varphi2$.

```
In[4]:= gb = GroebnerBasis[
          Join[Cancel[eqs/{-r1 - r2 Cos[φ2], r2, 1}],
               {Cos[φ1]^2 + Sin[φ1]^2 - 1, Cos[φ2]^2 + Sin[φ2]^2 - 1}],
               {}, {Cos[φ1], Sin[φ1], Cos[φ2], Sin[φ2]},
                    MonomialOrder -> EliminationOrder];
```

The resulting basis is large. It has 470 terms and total degree 4.

```
In[5]:= {Length[gb[[1]]], Max[(Plus @@ Exponent[#, {X, Y, Z}]) & /@
          ((List @@ gb[[1]]) /. {r1 -> 1, r2 -> 1, ξ -> 1, η -> 1, ζ -> 1})]}
Out[5]= {470, 4}
```

Writing for brevity R^2 for $\xi^2 + \eta^2 + \zeta^2$ and collecting with respect to X, Y, and Z, the implicit equation for the bisector takes the following form.

```
In[6]:= bisectorImplicit = Plus @@ Flatten[
          MapIndexed[(Times @@ (#1 {X, Y, Z}^(#2 - 1))) &,
          Factor[CoefficientList[gb[[1]], {X, Y, Z}] /. ξ^n_ :>
                  (R^2 - η^2 - ζ^2)^Floor[n/2] ξ^Mod[n, 2]], {3}]];

        Short[bisectorImplicit, 12]
Out[8]//Short= (-R + r1 - r2)⁶ (R + r1 - r2)⁶ (-R + r1 + r2)⁶ (R + r1 + r2)⁶ +
        512 (-R + r1 - r2)³ (R + r1 - r2)³ (-R + r1 + r2)³ (R + r1 + r2)³ (-R² + r1² + r2²)³ Z ζ³ -
        32768 (-R² + r1² + r2²)³ Z³ (r2 - ζ)³ ζ³ (r2 + ζ)³ + 4096 Z⁴ (r2 - ζ)⁶ (r2 + ζ)⁶ -
        512 Z² (R⁴ r2² - 2 R² r1² r2² + r1⁴ r2² - 2 R² r2⁴ - 2 r1² r2⁴ + r2⁶ -
           3 R⁴ ζ² + 6 R² r1² ζ² - 3 r1⁴ ζ² + 6 R² r2² ζ² - 2 r1² r2² ζ² - 3 r2⁴ ζ²)³ +
        512 (-R + r1 - r2)³ (R + r1 - r2)³ (-R + r1 + r2)³ (R + r1 + r2)³ (-R² + r1² + r2²)³ Y η³ +
        ≪35≫ + 7077888 (-R² + r1² + r2²)³ X Y Z ζ³ η³ ξ³ - 262144 X Y Z² (r2² - 3 ζ²)³ η³ ξ³ -
        262144 X Y² Z ζ³ (r1² + r2² - 3 η²)³ ξ³ - 262144 X Y³ η³ (r1² + r2² - η²)³ ξ³ -
        262144 X³ Z ζ³ (-R² + r1² + r2² + ζ² + η²)³ ξ³ - 262144 X³ Y η³ (-R² + r1² + r2² + ζ² + η²)³ ξ³ -
        32768 X Y² (-R² r1² + r1⁴ - R² r2² - 2 r1² r2² + r2⁴ + 3 R² η² - 3 r1² η² - 3 r2² η²)³ ξ³ - 32768 X³
        (R⁴ - 2 R² r1² + r1⁴ - 2 R² r2² - 2 r1² r2² + r2⁴ - R² ζ² + r1² ζ² + r2² ζ² - R² η² + r1² η² + r2² η²)³ ξ³
```

Now let us visualize the bisector surface for the case $\{\xi, \eta, \zeta\} = \{0, 0, 0\}$. Obviously, the bisector surface has rotational symmetry around the z-axis.

```
In[9]:= bisector1 = gb[[1]] /. {r1 -> 3, r2 -> 1, ξ -> 0, η -> 0, ζ -> 0}
Out[9]= 4096 - 5120 X² + 1024 X⁴ - 5120 Y² + 2048 X² Y² + 1024 Y⁴ - 512 Z² - 256 X² Z² - 256 Y² Z² + 16 Z⁴
```

The bisector surface splits into two disconnected parts. We use the representation $z = z(x)$.

In[10]:= Factor[bisector1 /. Y -> 0]

Out[10]= $16 (16 - 24 X + 8 X^2 - Z^2) (16 + 24 X + 8 X^2 - Z^2)$

In[11]:= X[1][Z_] = Abs[6 - Sqrt[2] Sqrt[2 + Z^2]]/4;
        X[2][Z_] = (6 + Sqrt[2] Sqrt[2 + Z^2])/4;

Now, we have everything together and can create the polygons of the torus and the bisector surface.

In[13]:= (* cut a hole in a polygon *)
        makeHole[Polygon[l_], factor_] :=
        Module[{mp = Plus @@ l/Length[l], ℓ}, ℓ = (mp + factor(# - mp))& /@ l;
          {MapThread[Polygon[Join[#1, Reverse[#2]]]&,
                    Partition[Append[#, First[#]], 2, 1]& /@ {l, ℓ}]}]

In[15]:= (* the polygons of the original torus *)
        torusPolys = makeHole[#, 0.7]& /@ Cases[
        ParametricPlot3D[Evaluate[torus[φ1, φ2, 3, 1]],
                    {φ1, 0, 2Pi}, {φ2, 0, 2Pi},
                    PlotPoints -> {60, 16}, DisplayFunction -> Identity],
                        _Polygon, Infinity];

For the bisector surface, we rotate the line forming the cross section around the *z*-axis.

In[17]:= (* bisector surface *)
        bisectorPolys = makeHole[#, 0.9]& /@ Cases[
        Table[ParametricPlot3D[Evaluate[{X[j][Z] Cos[φ], X[j][Z] Sin[φ], Z}],
                    {φ, 0, 2Pi}, {Z, -6, 6},
                PlotPoints -> {60, 49}, DisplayFunction -> Identity], {j, 2}],
                    _Polygon, Infinity];

The torus is shown in red and the bisector surface in green.

In[19]:= Show[Graphics3D[{EdgeForm[],
            {SurfaceColor[Hue[0], Hue[0], 2.8], torusPolys},
            {SurfaceColor[Hue[0.22], Hue[0.22], 2.9], bisectorPolys}}],
            PlotRange -> All]

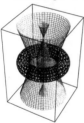

c) We choose a heuristic way to solve the problem, but other, more algorithmically oriented derivations, are possible too. We start by making the following ansatz for the general polynomial $p(x, y, z)$ of degree 4 in three variables.

In[1]:= poly[x_, y_, z_] = Sum[c[i, j, k] x^i y^j z^k,
                        {i, 0, 4}, {j, 0, 4 - i}, {k, 0, 4 - i - j}]

Out[1]= $c[0, 0, 0] + z c[0, 0, 1] + z^2 c[0, 0, 2] + z^3 c[0, 0, 3] + z^4 c[0, 0, 4] +$
        $y c[0, 1, 0] + y z c[0, 1, 1] + y z^2 c[0, 1, 2] + y z^3 c[0, 1, 3] + y^2 c[0, 2, 0] +$
        $y^2 z c[0, 2, 1] + y^2 z^2 c[0, 2, 2] + y^3 c[0, 3, 0] + y^3 z c[0, 3, 1] + y^4 c[0, 4, 0] +$
        $x c[1, 0, 0] + x z c[1, 0, 1] + x z^2 c[1, 0, 2] + x z^3 c[1, 0, 3] + x y c[1, 1, 0] +$
        $x y z c[1, 1, 1] + x y z^2 c[1, 1, 2] + x y^2 c[1, 2, 0] + x y^2 z c[1, 2, 1] + x y^3 c[1, 3, 0] +$
        $x^2 c[2, 0, 0] + x^2 z c[2, 0, 1] + x^2 z^2 c[2, 0, 2] + x^2 y c[2, 1, 0] + x^2 y z c[2, 1, 1] +$
        $x^2 y^2 c[2, 2, 0] + x^3 c[3, 0, 0] + x^3 z c[3, 0, 1] + x^3 y c[3, 1, 0] + x^4 c[4, 0, 0]$

Due to the symmetry of the three cylinders, the polynomial $p(x, y, z)$ should have a threefold rotational symmetry, along the direction {1, 1, 1}. By applying this symmetry, some coefficients can be eliminated. (The message in the following input is generated because it is not possible to solve for all of the coefficients.)

In[2]:= poly1[x_, y_, z_] = poly[x, y, z] /. Solve[
        Flatten[(* equations from applying rotational symmetry *)
        {# == 0& /@ Flatten[
          CoefficientList[poly[x, y, z] - poly[y, z, x], {x, y, z}]],

```
        # == 0& /@ Flatten[
    CoefficientList[poly[y, z, x] - poly[z, x, y], {x, y, z}]]}],
            (* all coefficients *)
            Cases[poly[x, y, z], _c, {0, Infinity}]][[1]]
      Solve::svars : Equations may not give solutions for all "solve" variables. More…
```

Out[2]= $c[0, 0, 0] + x\,c[1, 0, 0] + y\,c[1, 0, 0] + z\,c[1, 0, 0] + x\,y\,c[1, 1, 0] +$
$x\,z\,c[1, 1, 0] + y\,z\,c[1, 1, 0] + x\,y\,z\,c[1, 1, 1] + x^2\,c[2, 0, 0] + y^2\,c[2, 0, 0] +$
$z^2\,c[2, 0, 0] + x\,y^2\,c[2, 0, 1] + x^2\,z\,c[2, 0, 1] + y\,z^2\,c[2, 0, 1] + x^2\,y\,c[2, 1, 0] +$
$y^2\,z\,c[2, 1, 0] + x\,z^2\,c[2, 1, 0] + x^2\,y\,z\,c[2, 1, 1] + x\,y^2\,z\,c[2, 1, 1] + x\,y\,z^2\,c[2, 1, 1] +$
$x^2\,y^2\,c[2, 2, 0] + x^2\,z^2\,c[2, 2, 0] + y^2\,z^2\,c[2, 2, 0] + x^3\,c[3, 0, 0] + y^3\,c[3, 0, 0] +$
$z^3\,c[3, 0, 0] + x\,y^3\,c[3, 0, 1] + x^3\,z\,c[3, 0, 1] + y\,z^3\,c[3, 0, 1] + x^3\,y\,c[3, 1, 0] +$
$y^3\,z\,c[3, 1, 0] + x\,z^3\,c[3, 1, 0] + x^4\,c[4, 0, 0] + y^4\,c[4, 0, 0] + z^4\,c[4, 0, 0]$

Next, we demand that (at least) eight points from each of the three cylinders should lie on the surface defined implicitly by the polynomial at $x = l$, $y = l$, and $z = l$ (later we set l equal to 3). This again eliminates some coefficients.

```
In[3]:= sol1 = Map[Together, Solve[Flatten[{
        (* eight points on every cylinder *)
        Table[poly1[Cos[φ], Sin[φ], 1] == 0, {φ, 0, 2Pi, 2Pi/8}],
        Table[poly1[Cos[φ], 1, Sin[φ]] == 0, {φ, 0, 2Pi, 2Pi/8}],
        Table[poly1[1, Cos[φ], Sin[φ]] == 0, {φ, 0, 2Pi, 2Pi/8}]}],
        Cases[poly1[x, y, z], _c, {0, Infinity}] // Union], {3}];
      Solve::svars : Equations may not give solutions for all "solve" variables. More…
```

In[4]:= poly2[x_, y_, z_] = Numerator[Factor[(poly1[x, y, z] /. sol1)[[1]]]]

Out[4]= $l^3\,c[1, 1, 0] - l^2\,x\,c[1, 1, 0] - l^2\,y\,c[1, 1, 0] + l\,x\,y\,c[1, 1, 0] - l^2\,z\,c[1, 1, 0] +$
$l\,x\,z\,c[1, 1, 0] + l\,y\,z\,c[1, 1, 0] - x\,y\,z\,c[1, 1, 0] - l\,c[2, 0, 0] - l^3\,c[2, 0, 0] +$
$l\,x^2\,c[2, 0, 0] + l\,y^2\,c[2, 0, 0] + l\,z^2\,c[2, 0, 0] - l^2\,c[2, 1, 0] + l^4\,c[2, 1, 0] -$
$l^3\,x\,c[2, 1, 0] - l^3\,y\,c[2, 1, 0] + l\,x^2\,y\,c[2, 1, 0] - l^3\,z\,c[2, 1, 0] +$
$l\,x^2\,z\,c[2, 1, 0] + l\,x\,y\,z\,c[2, 1, 0] - x^2\,y\,z\,c[2, 1, 0] + l\,y^2\,z\,c[2, 1, 0] - x\,y^2\,z\,c[2, 1, 0] +$
$l\,x\,z^2\,c[2, 1, 0] + l\,y\,z^2\,c[2, 1, 0] - x\,y\,z^2\,c[2, 1, 0] + l^2\,c[3, 0, 0] - l^4\,c[3, 0, 0] -$
$l\,x\,c[3, 0, 0] + l\,x^3\,c[3, 0, 0] - l\,y\,c[3, 0, 0] + l\,y^3\,c[3, 0, 0] - l\,z\,c[3, 0, 0] -$
$l\,x\,y\,z\,c[3, 0, 0] + x^2\,y\,z\,c[3, 0, 0] + x\,y^2\,z\,c[3, 0, 0] + x\,y\,z^2\,c[3, 0, 0] + l\,z^3\,c[3, 0, 0] +$
$l^3\,c[3, 1, 0] + l^5\,c[3, 1, 0] - l^2\,x\,c[3, 1, 0] - l^4\,x\,c[3, 1, 0] - l^2\,y\,c[3, 1, 0] -$
$l^4\,y\,c[3, 1, 0] + l\,x^3\,y\,c[3, 1, 0] + l\,x\,y^3\,c[3, 1, 0] - l^2\,z\,c[3, 1, 0] - l^4\,z\,c[3, 1, 0] +$
$l\,x^3\,z\,c[3, 1, 0] - x\,y\,z\,c[3, 1, 0] - l^2\,x\,y\,z\,c[3, 1, 0] + l\,x^2\,y\,z\,c[3, 1, 0] +$
$l\,x\,y^2\,z\,c[3, 1, 0] + l\,y^3\,z\,c[3, 1, 0] + l\,x\,y\,z^2\,c[3, 1, 0] + l\,x\,z^3\,c[3, 1, 0] +$
$l\,y\,z^3\,c[3, 1, 0] - l\,c[4, 0, 0] - 2\,l^3\,c[4, 0, 0] - l^5\,c[4, 0, 0] + l\,x^4\,c[4, 0, 0] +$
$2\,l\,x^2\,y^2\,c[4, 0, 0] + l\,y^4\,c[4, 0, 0] + 2\,l\,x^2\,z^2\,c[4, 0, 0] + 2\,l\,y^2\,z^2\,c[4, 0, 0] + l\,z^4\,c[4, 0, 0]$

To have a smooth transition between the cylinders and the surface to be determined, we also demand that the normals of the cylinders are parallel to the normals of our surface at $x = l$, $y = l$, and $z = l$. Due to the rotational symmetry of each cylinder, this means that the normal should have no component in the direction of the cylinder axes at this cross section. This constraint fixes some more of the $c[i, j, k]$.

```
In[5]:= sol2 = Map[Together, Solve[Flatten[{
        (* parallel components of eight normals on every cylinder *)
        Table[Evaluate[D[poly2[x, y, z], z] == 0 /.
                       {x -> Cos[φ], y -> Sin[φ], z -> 1}],
              {φ, 0, 2Pi, 2Pi/8}],
        Table[Evaluate[D[poly2[x, y, z], y] == 0 /.
                       {x -> Cos[φ], y -> 1, z -> Sin[φ]}],
              {φ, 0, 2Pi, 2Pi/8}],
        Table[Evaluate[D[poly2[x, y, z], x] == 0 /.
                       {x -> 1, y -> Cos[φ], z -> Sin[φ]}],
              {φ, 0, 2Pi, 2Pi/8}]}],
        Cases[poly2[x, y, z], _c, {0, Infinity}] // Union], {3}]
      Solve::svars : Equations may not give solutions for all "solve" variables. More…
```

Out[5]= $\{\{c[1, 1, 0] \to (-1 + l^2)\,c[3, 1, 0], c[2, 0, 0] \to 3\,l^2\,c[3, 1, 0] - 2\,c[4, 0, 0] - 2\,l^2\,c[4, 0, 0],$
$c[2, 1, 0] \to -2\,l\,c[3, 1, 0], c[3, 0, 0] \to -2\,l\,c[3, 1, 0]\}\}$

In[6]:= poly3[x_, y_, z_] = Numerator[Factor[poly2[x, y, z] /. sol2[[1]]]]

Out[6]= -1 (3 l² c[3, 1, 0] + l⁴ c[3, 1, 0] - 2 l x c[3, 1, 0] - 3 l² x² c[3, 1, 0] + 2 l x³ c[3, 1, 0] -
    2 l y c[3, 1, 0] + x y c[3, 1, 0] - l² x y c[3, 1, 0] + 2 l x² y c[3, 1, 0] - x³ y c[3, 1, 0] -
    3 l² y² c[3, 1, 0] + 2 l x y² c[3, 1, 0] + 2 l y³ c[3, 1, 0] - x y³ c[3, 1, 0] - 2 l z c[3, 1, 0] +
    x z c[3, 1, 0] - l² x z c[3, 1, 0] + 2 l x² z c[3, 1, 0] - x³ z c[3, 1, 0] + y z c[3, 1, 0] -
    l² y z c[3, 1, 0] + 2 l x y z c[3, 1, 0] - x² y z c[3, 1, 0] + 2 l y² z c[3, 1, 0] -
    x y² z c[3, 1, 0] - y³ z c[3, 1, 0] - 3 l² z² c[3, 1, 0] + 2 l x z² c[3, 1, 0] +
    2 l y z² c[3, 1, 0] - x y z² c[3, 1, 0] + 2 l z³ c[3, 1, 0] - x z³ c[3, 1, 0] - y z³ c[3, 1, 0] -
    c[4, 0, 0] - 2 l² c[4, 0, 0] - l⁴ c[4, 0, 0] + 2 x² c[4, 0, 0] + 2 l² x² c[4, 0, 0] -
    x⁴ c[4, 0, 0] + 2 y² c[4, 0, 0] + 2 l² y² c[4, 0, 0] - 2 x² y² c[4, 0, 0] - y⁴ c[4, 0, 0] +
    2 z² c[4, 0, 0] + 2 l² z² c[4, 0, 0] - 2 x² z² c[4, 0, 0] - 2 y² z² c[4, 0, 0] - z⁴ c[4, 0, 0])

Now let us see what we have so far. poly3[x, y, z] still contains two parameters.

In[7]:= cs = Cases[poly3[x, y, z], _c, {0, Infinity}] // Union

Out[7]= {c[3, 1, 0], c[4, 0, 0]}

But all properties we wanted are already satisfied for all points of the boundary of the three cylinders and not only at the eight points used above. Let us check this explicitly. This shows that the boundaries of the three cylinders agree with the surface at $x = l$, $y = l$, and $z = l$.

In[8]:= {poly3[l, Cos[φ], Sin[φ]], poly3[Cos[φ], l, Sin[φ]],
    poly3[Cos[φ], Sin[φ], l]} // Simplify

Out[8]= {0, 0, 0}

Here, we show that the parallel components of the normals vanish along the cylinder boundaries at $x = l$, $y = l$, and $z = l$.

In[9]:= (Simplify /@ Flatten[CoefficientList[D[poly3[x, y, z], x] /.
                {x -> l, y -> Cos[φ], z -> Sin[φ]}, cs]]) & /@ {x, y, z}

Out[9]= {{0, 0, 0, 0}, {0, 0, 0, 0}, {0, 0, 0, 0}}

We fix the two free parameters by looking for values such that the surface of our polynomial is simply connected and looks nice. Here are some cross sections of $p(x, y, z) = 0$ in the y,z- plane.

After renaming the undetermined coefficients, we have the following polynomial surface connecting the three cylinders.

In[10]:= finalPoly[x_, y_, z_, {l_, α_, β_}] =
    α (-3 l^2 - l^4 + 2 l x + 3 l^2 x^2 - 2 l x^3 + 2 l y - x y +
    l^2 x y - 2 l x^2 y + x^3 y + 3 l^2 y^2 - 2 l x y^2 - 2 l y^3 +
    x y^3 + 2 l z - x z + l^2 x z - 2 l x^2 z + x^3 z - y z +
    l^2 y z - 2 l x y z + x^2 y z - 2 l y^2 z + x y^2 z + y^3 z +
    3 l^2 z^2 - 2 l x z^2 - 2 l y z^2 + x y z^2 - 2 l z^3 +
    x z^3 + y z^3) - β 2 l (-l^2 + x^2 + y^2 + z^2 - 1)^2;

We look at some of the possible shapes for the connector by using various values for α and β.

In[11]:= Show[Flatten[Table[
    ContourPlot[Evaluate[finalPoly[0, y, z, {3, α, β}]],
                {y, -2, 6}, {z, -2, 6},
                Contours -> {0}, ContourShading -> False,
                DisplayFunction -> Identity, PlotPoints -> 60,
                ContourStyle -> {Hue[Random[]], Thickness[0.002]}],
        {α, 50, 60, 1}, {β, -6, -12, -1}]],
    DisplayFunction -> $DisplayFunction]

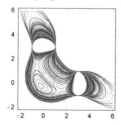

Choosing the parameters $\beta \to -7$, $\alpha \to 55$, we can make a picture of our polynomial. The right graphic sorts out the polygons needed to connect the three cylinders.

```
In[12]:= Needs["Graphics`ContourPlot3D`"]

In[13]:= (* 3D contour plot of the surface *)
        cp = ContourPlot3D[Evaluate[finalPoly[x, y, z, {3, 55, -7}]],
                    {x, -2, 5}, {y, -2, 5}, {z, -2, 5},
                    MaxRecursion -> 1, PlotPoints -> {15, 3},
                    PlotRange -> All, Axes -> True,
                    DisplayFunction -> Identity]

In[15]:= (* extract polygons within corner *)
        connector = Select[cp[[1]], Max[Flatten[#[[1]]]] <= 3&];

In[17]:= Show[GraphicsArray[{cp, Graphics3D[connector]}]]
```

We add the three cylinders to the end. To get a pleasing picture, we arrange the mesh lines of the cylinders to fit the mesh lines of `connector`. We use `ListSurfacePlot3D` from the package `Graphics`Graphics3D`` to generate the cylinders.

```
In[18]:= Needs["Graphics`Graphics3D`"]

In[19]:= (* along the x-axis *)
        (* boundary polygons of the polynomial part *)
        tubePolysx = Select[connector, MemberQ[#[[1]]& /@ #[[1]], 3.] &&
                                Max[#[[1]]& /@ #[[1]]] <= 3&];
        (* list of angles of points along the circular boundary *)
        pxList = Append[#, First[#]]&[Sort[Apply[ArcTan,
                Delete[#, 1]& /@ Union[Cases[Flatten[Last /@ tubePolysx, 1],
                                    {3., _, _}]], {1}]]];
        (* points of the cylinder *)
        tabx = Table[{x, 0, 0} + {0, Cos[#], Sin[#]}& /@ pxList,
                    {x, 3 + 1/4, 5, 1/4}];
        (* boundary points of polynomial *)
        lastPointsx = Append[#, First[#]]&[
        Sort[Union[Cases[Flatten[Last /@ tubePolysx, 1], {3., _, _}]],
            ArcTan @@ Delete[#1, 1] < ArcTan @@ Delete[#2, 1]&]];
        (* the cylinder along the x - axis *)
        tubex = ListSurfacePlot3D[Prepend[tabx, lastPointsx],
                            DisplayFunction -> Identity][[1]];

In[30]:= (* along the y-axis *)
        (* boundary polygons of the polynomial part *)
        tubePolysy = Select[connector, MemberQ[#[[2]]& /@ #[[1]], 3.] &&
                Max[#[[2]]& /@ #[[1]]] <= 3&];
        (* list of angles of points along the circular boundary *)
        pyList = Append[#, First[#]]&[Sort[Apply[ArcTan,
                Delete[#, 2]& /@ Union[Cases[Flatten[Last /@ tubePolysy, 1],
                                    {_, 3., _}]], {1}]]];
        (* points of the cylinder *)
        taby = Table[{0, y, 0} + {Cos[#], 0, Sin[#]}& /@ pyList,
                    {y, 3 + 1/4, 5, 1/4}];
        (* boundary points of polynomial *)
        lastPointsy = Append[#, First[#]]&[
        Sort[Union[Cases[Flatten[Last /@ tubePolysy, 1], {_, 3., _}]],
            ArcTan @@ Delete[#1, 2] < ArcTan @@ Delete[#2, 2]&]];
```

```
        (* the cylinder along the x - axis *)
        tubey = ListSurfacePlot3D[Prepend[taby, lastPointsy],
                         DisplayFunction -> Identity][[1]];
In[41]:= (* along the z-axis *)
        (* boundary polygons of the polynomial part *)
        tubePolysz = Select[connector, MemberQ[#[[3]]& /@ #[[1]], 3.] &&
             Max[#[[3]]& /@ #[[1]]] <= 3&];
        (* list of angles of points along the circular boundary *)
        pzList = Append[#, First[#]]&[Sort[Apply[ArcTan,
                 Delete[#, 3]& /@ Union[Cases[Flatten[Last /@ tubePolysz, 1],
                                       {_, _, 3.}]], {1}]]]];
        (* points of the cylinder *)
        tabz = Table[{0, 0, z} + {Cos[#], Sin[#], 0}& /@ pzList,
                 {z, 3 + 1/4, 5, 1/4}];
        (* boundary points of polynomial *)
        lastPointsz = Append[#, First[#]]&[
        Sort[Union[Cases[Flatten[Last /@ tubePolysz, 1], {_, _, 3.}]],
             ArcTan @@ Delete[#1, 3] < ArcTan @@ Delete[#2, 3]&]];
        (* the cylinder along the x - axis *)
        tubez = ListSurfacePlot3D[Prepend[tabz, lastPointsz],
                         DisplayFunction -> Identity][[1]];
```

Here the three cylinder and the inner part of the polynomial are shown.

```
In[52]:= connectedTubes =
        Graphics3D[{EdgeForm[Thickness[0.002]],
                       (* the three tube pieces and the connecting quartic *)
                       connector, tubex, tubey, tubez}, Boxed -> False];
In[53]:= Show[GraphicsArray[
              {(* show all polygons *)
               Append[connectedTubes, PlotRange -> All],
               (* remove polygons in front *)
               Join[connectedTubes, Graphics3D[PlotRange -> {All, {0, 4}, All},
                                       ViewPoint -> {1, -3, 1}]]}]]
```

For blending using PDEs, see [1879].

d) We will first make an ansatz for the implicit representation of the double torus, then determine the unknowns in the ansatz, and then prove that the so obtained formula fulfills the requirements to enclose the original torus. Because of the symmetry of the problem at hand, we use a polynomial with only even powers in x, y, and z. A torus is described by a polynomial of degree 4; so, we try to model the double torus by a slightly more complicated polynomial of total degree 6.

```
In[1]:= degree = 6;
```

```
        ansatz[x_, y_, z_] =
        Sum[If[i + j + k > degree, 0, c[i, j, k] x^i y^j z^k],
            {i, 0, degree, 2}, {j, 0, degree, 2}, {k, 0, degree, 2}];
```

These are the to-be-determined coefficients.

```
In[3]:= cs = Cases[ansatz[x, y, z], _c, Infinity]
```

```
Out[3]= {c[0, 0, 0], c[0, 0, 2], c[0, 0, 4], c[0, 0, 6], c[0, 2, 0], c[0, 2, 2],
         c[0, 2, 4], c[0, 4, 0], c[0, 4, 2], c[0, 6, 0], c[2, 0, 0], c[2, 0, 2], c[2, 0, 4],
         c[2, 2, 0], c[2, 2, 2], c[2, 4, 0], c[4, 0, 0], c[4, 0, 2], c[4, 2, 0], c[6, 0, 0]}
```

To determine some of the coefficients cs, we require the double torus to touch the torus at 12 points at each of its holes. We also require the double torus's "middle bridge" to fit exactly into the torus's hole. In addition, we make the cross section of the leftmost and rightmost arm of the double torus also circles of radius r. All of these conditions we enforce at 12 equidistant points on the circles. This leads to the following set of equations. (We take the original torus in the x,z-plane.)

```
In[4]:= n = 12;
      eqs1 = Flatten[{(* right hole of the double torus *)
                 Table[ansatz[+R + r Cos[φ], r Sin[φ], 0],
                       {φ, 0, 2Pi, 2Pi/n}],
                 (* middle bridge of the double torus *)
                 Table[ansatz[(R - r) Cos[φ], 0, (R - r) Sin[φ]],
                       {φ, 0, 2Pi, 2Pi/n}],
                 (* outer arm of the double torus *)
                 Table[ansatz[+R + 2r + r Cos[φ], 0, r Sin[φ]],
                       {φ, 0, 2Pi, 2Pi/n}],
                 (* extension of the double torus in y-direction *)
                 ansatz[0, 3r, 0]}];
```

The solution is a large expression. We fix the coefficient $c[6, 0, 0]$ to be 1.

```
In[6]:= sol = Solve[# == 0& /@ eqs1, cs] /. c[6, 0, 0] -> 1;

      Solve::svars : Equations may not give solutions for all "solve" variables. More…
```

```
In[7]:= ByteCount[sol]

Out[7]= 660480
```

Fortunately, the expression can be simplified considerably by writing it over common denominators.

```
In[8]:= sol1 = {MapAt[Together, #, 2]& /@ sol[[1]]};
      ByteCount[sol1]
Out[9]= 8088
```

The coefficients are all rational functions in R and r.

```
In[10]:= sol1
```

$$Out[10]= \left\{\left\{c[0, 2, 0] \to \frac{171 r^6 + 114 r^5 R - 161 r^4 R^2 - 120 r^3 R^3 - 11 r^2 R^4 + 6 r R^5 + R^6}{9 r^2}, \right.\right.$$

$$c[0, 4, 0] \to \frac{-99 r^4 - 66 r^3 R + 7 r^2 R^2 + 12 r R^3 + 2 R^4}{9 r^2}, \quad c[0, 6, 0] \to \frac{9 r^2 + 6 r R + R^2}{9 r^2},$$

$$c[2, 2, 0] \to -\frac{2 (99 r^4 + 60 r^3 R + 10 r^2 R^2 + 6 r R^3 + R^4)}{9 r^2}, \quad c[2, 4, 0] \to \frac{27 r^2 + 12 r R + 2 R^2}{9 r^2},$$

$$c[4, 2, 0] \to \frac{27 r^2 + 6 r R + R^2}{9 r^2}, \quad c[0, 0, 2] \to 3 r^4 + 28 r^3 R + 30 r^2 R^2 + 4 r R^3 - R^4,$$

$$c[0, 0, 4] \to 5 r^2 + 10 r R + R^2, \quad c[2, 0, 2] \to -2 (3 r^2 - 2 r R + R^2), \quad c[0, 0, 6] \to 1, \quad c[2, 0, 4] \to 3,$$

$$c[4, 0, 2] \to 3, \quad c[0, 0, 0] \to -9 r^6 - 6 r^5 R + 17 r^4 R^2 + 12 r^3 R^3 - 7 r^2 R^4 - 6 r R^5 - R^6,$$

$$\left.\left. c[2, 0, 0] \to 19 r^4 + 12 r^3 R + 18 r^2 R^2 + 12 r R^3 + 3 R^4, \quad c[4, 0, 0] \to -11 r^2 - 6 r R - 3 R^2\right\}\right\}$$

We substitute the solution sol1 and obtain the following polynomial.

```
In[11]:= ansatz1[x_, y_, z_] = Numerator[(ansatz[x, y, z] /. sol1[[1]] /.
                 c[6, 0, 0] -> 1) // Together // Factor]
```

$$Out[11]= -81 r^8 - 54 r^7 R + 153 r^6 R^2 + 108 r^5 R^3 - 63 r^4 R^4 - 54 r^3 R^5 - 9 r^2 R^6 + 171 r^6 x^2 + 108 r^5 R x^2 +$$
$$162 r^4 R^2 x^2 + 108 r^3 R^3 x^2 + 27 r^2 R^4 x^2 - 99 r^4 x^4 - 54 r^3 R x^4 - 27 r^2 R^2 x^4 + 9 r^2 x^6 + 171 r^6 y^2 +$$
$$114 r^5 R y^2 - 161 r^4 R^2 y^2 - 120 r^3 R^3 y^2 - 11 r^2 R^4 y^2 + 6 r R^5 y^2 + R^6 y^2 - 198 r^4 x^2 y^2 -$$
$$120 r^3 R x^2 y^2 - 20 r^2 R^2 x^2 y^2 - 12 r R^3 x^2 y^2 - 2 R^4 x^2 y^2 + 27 r^2 x^4 y^2 + 6 r R x^4 y^2 + R^2 x^4 y^2 -$$
$$99 r^4 y^4 - 66 r^3 R y^4 + 7 r^2 R^2 y^4 + 12 r R^3 y^4 + 2 R^4 y^4 + 27 r^2 x^2 y^4 + 12 r R x^2 y^4 + 2 R^2 x^2 y^4 +$$
$$9 r^2 y^6 + 6 r R y^6 + R^2 y^6 + 27 r^6 z^2 + 252 r^5 R z^2 + 270 r^4 R^2 z^2 + 36 r^3 R^3 z^2 - 9 r^2 R^4 z^2 - 54 r^4 x^2 z^2 +$$
$$36 r^3 R x^2 z^2 - 18 r^2 R^2 x^2 z^2 + 27 r^2 x^4 z^2 + 45 r^4 z^4 + 90 r^3 R z^4 + 9 r^2 R^2 z^4 + 27 r^2 x^2 z^4 + 9 r^2 z^6 +$$
$$9 r^2 y^2 z^2 c[0, 2, 2] + 9 r^2 y^2 z^4 c[0, 2, 4] + 9 r^2 y^4 z^2 c[0, 4, 2] + 9 r^2 x^2 y^2 z^2 c[2, 2, 2]$$

The following coefficients are still undetermined.

```
In[12]:= cs = Cases[ansatz1[x, y, z], _c, Infinity] // Union
Out[12]= {c[0, 2, 2], c[0, 2, 4], c[0, 4, 2], c[2, 2, 2]}
```

We will try the simplest possible choice for them—we set them all to 0. This results in the following homogeneous polynomial.

In[13]:= `doubleTorus[x_, y_, z_, {R_, r_}] = Factor[ansatz1[x, y, z] /. _c -> 0]`

Out[13]= $-81\,r^8 - 54\,r^7\,R + 153\,r^6\,R^2 + 108\,r^5\,R^3 - 63\,r^4\,R^4 - 54\,r^3\,R^5 - 9\,r^2\,R^6 + 171\,r^6\,x^2 + 108\,r^5\,R\,x^2 +$
$162\,r^4\,R^2\,x^2 + 108\,r^3\,R^3\,x^2 + 27\,r^2\,R^4\,x^2 - 99\,r^4\,x^4 - 54\,r^3\,R\,x^4 - 27\,r^2\,R^2\,x^4 + 9\,r^2\,x^6 + 171\,r^6\,y^2 +$
$114\,r^5\,R\,y^2 - 161\,r^4\,R^2\,y^2 - 120\,r^3\,R^3\,y^2 - 11\,r^2\,R^4\,y^2 + 6\,r\,R^5\,y^2 + R^6\,y^2 - 198\,r^4\,x^2\,y^2 -$
$120\,r^3\,R\,x^2\,y^2 - 20\,r^2\,R^2\,x^2\,y^2 - 12\,r\,R^3\,x^2\,y^2 - 2\,R^4\,x^2\,y^2 + 27\,r^2\,x^4\,y^2 + 6\,r\,R\,x^4\,y^2 + R^2\,x^4\,y^2 -$
$99\,r^4\,y^4 - 66\,r^3\,R\,y^4 + 7\,r^2\,R^2\,y^4 + 12\,r\,R^3\,y^4 + 2\,R^4\,y^4 + 27\,r^2\,x^2\,y^4 + 12\,r\,R\,x^2\,y^4 + 2\,R^2\,x^2\,y^4 +$
$9\,r^2\,y^6 + 6\,r\,R\,y^6 + R^2\,y^6 + 27\,r^6\,z^2 + 252\,r^5\,R\,z^2 + 270\,r^4\,R^2\,z^2 + 36\,r^3\,R^3\,z^2 - 9\,r^2\,R^4\,z^2 - 54\,r^4\,x^2\,z^2 +$
$36\,r^3\,R\,x^2\,z^2 - 18\,r^2\,R^2\,x^2\,z^2 + 27\,r^2\,x^4\,z^2 + 45\,r^4\,z^4 + 90\,r^3\,R\,z^4 + 9\,r^2\,R^2\,z^4 + 27\,r^2\,x^2\,z^4 + 9\,r^2\,z^6$

The polynomial `doubleTorus` really describes a closed surface. This can be seen from the asymptotic behavior far away from the origin. In every direction $\{x, y, z\}$, the polynomial `doubleTorus[x, y, z, {R, r}]` is positive definite.

In[14]:= `Series[doubleTorus[t x, t y, t z, {R, r}], {t, Infinity, 1}]`

Out[14]= $\dfrac{1}{(\frac{1}{t})^6}\,(9\,r^2\,x^6 + 27\,r^2\,x^4\,y^2 + 6\,r\,R\,x^4\,y^2 + R^2\,x^4\,y^2 + 27\,r^2\,x^2\,y^4 +$

$12\,r\,R\,x^2\,y^4 + 2\,R^2\,x^2\,y^4 + 9\,r^2\,y^6 + 6\,r\,R\,y^6 + R^2\,y^6 + 27\,r^2\,x^4\,z^2 + 27\,r^2\,x^2\,z^4 + 9\,r^2\,z^6) +$

$\dfrac{1}{(\frac{1}{t})^4}\,(-99\,r^4\,x^4 - 54\,r^3\,R\,x^4 - 27\,r^2\,R^2\,x^4 - 198\,r^4\,x^2\,y^2 - 120\,r^3\,R\,x^2\,y^2 - 20\,r^2\,R^2\,x^2\,y^2 -$

$12\,r\,R^3\,x^2\,y^2 - 2\,R^4\,x^2\,y^2 - 99\,r^4\,y^4 - 66\,r^3\,R\,y^4 + 7\,r^2\,R^2\,y^4 + 12\,r\,R^3\,y^4 + 2\,R^4\,y^4 -$
$54\,r^4\,x^2\,z^2 + 36\,r^3\,R\,x^2\,z^2 - 18\,r^2\,R^2\,x^2\,z^2 + 45\,r^4\,z^4 + 90\,r^3\,R\,z^4 + 9\,r^2\,R^2\,z^4) +$

$\dfrac{1}{(\frac{1}{t})^2}\,(171\,r^6\,x^2 + 108\,r^5\,R\,x^2 + 162\,r^4\,R^2\,x^2 + 108\,r^3\,R^3\,x^2 + 27\,r^2\,R^4\,x^2 + 171\,r^6\,y^2 +$

$114\,r^5\,R\,y^2 - 161\,r^4\,R^2\,y^2 - 120\,r^3\,R^3\,y^2 - 11\,r^2\,R^4\,y^2 + 6\,r\,R^5\,y^2 +$
$R^6\,y^2 + 27\,r^6\,z^2 + 252\,r^5\,R\,z^2 + 270\,r^4\,R^2\,z^2 + 36\,r^3\,R^3\,z^2 - 9\,r^2\,R^4\,z^2) +$

$(-81\,r^8 - 54\,r^7\,R + 153\,r^6\,R^2 + 108\,r^5\,R^3 - 63\,r^4\,R^4 - 54\,r^3\,R^5 - 9\,r^2\,R^6) + O\!\left[\dfrac{1}{t}\right]^2$

For $z = 0$, the polynomial `doubleTorus` contains the two factors $(r^2 - R^2 - 2\,R\,x - x^2 - y^2)$ and $(r^2 - R^2 + 2\,R\,x - x^2 - y^2)$ that represent the circles from the torus's cross section. And for $y = 0$, `doubleTorus` contains the factor $(r^2 - 2\,r\,R + R^2 - x^2 - z^2)$ that represents the middle hole of the torus, and the factors $(3\,r^2 + 4\,r\,R + R^2 - 4\,r\,x - 2\,R\,x + x^2 + z^2)$ and $(3\,r^2 + 4\,r\,R + R^2 + 4\,r\,x + 2\,R\,x + x^2 + z^2)$ represent the leftmost and rightmost arms of the double torus. This means that not only the 12 points, but also all points from the cross section fulfill the required conditions.

In[15]:= `Factor[doubleTorus[x, y, 0, {R, r}]]`

Out[15]= $-(r^2 - R^2 - 2\,R\,x - x^2 - y^2)\,(r^2 - R^2 + 2\,R\,x - x^2 - y^2)$
$(81\,r^4 + 54\,r^3\,R + 9\,r^2\,R^2 - 9\,r^2\,x^2 - 9\,r^2\,y^2 - 6\,r\,R\,y^2 - R^2\,y^2)$

For $y = 0$, `ansatz1` contains the factor $(r^2 - 2\,r\,R + R^2 - x^2 - z^2)$ that represents the middle hole of thee torus and the factors $(3\,r^2 + 4\,r\,R + R^2 - 4\,r\,x - 2\,R\,x + x^2 + z^2)$ and $(3\,r^2 + 4\,r\,R + R^2 + 4\,r\,x + 2\,R\,x + x^2 + z^2)$ that represent the left- and rightmost arm of the double torus. This means that the polynomial `doubleTorus` fulfills the required conditions.

In[16]:= `Factor[doubleTorus[x, 0, z, {R, r}]]`

Out[16]= $-9\,r^2\,(r^2 - 2\,r\,R + R^2 - x^2 - z^2)$
$(3\,r^2 + 4\,r\,R + R^2 - 4\,r\,x - 2\,R\,x + x^2 + z^2)\,(3\,r^2 + 4\,r\,R + R^2 + 4\,r\,x + 2\,R\,x + x^2 + z^2)$

Here is a cross section of the family of double tori for various R (we fix r to be 1). We see that for the R-values displayed, `doubleTorus` has the intended shape.

In[17]:=
```
Show[Table[
   ContourPlot[Evaluate[doubleTorus[x, y, 0, {R, 1}]],
              {x, -R - 4, R + 4}, {y, -4, 4},
              Contours -> {0}, ContourShading -> False,
              AspectRatio -> Automatic, PlotPoints -> 200,
              ContourStyle -> {Hue[0.8 (R - 1)/2], Thickness[0.002]},
              DisplayFunction -> Identity], (* use 20 R *)
     {R, 1, 3, 1/10}], DisplayFunction -> $DisplayFunction]
```

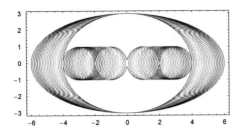

We see that for the *R* values displayed doubleTorus has the intended shape.

For too large values of *R*, our sextic ansatz breaks down. The following contour plot demonstrates this.

```
In[18]:= With[{R = 100},
        ContourPlot[Evaluate @ doubleTorus[x, y, 0, {R, 1}],
                {x, - R - 2, - R + 2}, {y, -2, 2},
                PlotPoints -> 100, Contours -> {0}]]
```

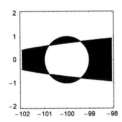

Now, we will visualize the resulting double torus. This is a parametric representation of the torus.

```
In[19]:= torus[φ1_, φ2_, {R_, r_}] := {R Cos[φ1] + r Cos[φ1] Cos[φ2], Sin[φ2],
                                        R Sin[φ1] + r Sin[φ1] Cos[φ2]};
```

Here is the corresponding implicit representation.

```
In[20]:= torusPoly[x_, y_, z_, {R_, r_}] = Factor @
        GroebnerBasis[Flatten[{{x, y, z} - torus[φ1, φ2, {R, r}],
                            (* algebraic relation between Cos and Sin *)
                            Sin[φ1]^2 + Cos[φ1]^2 - 1,
                            Sin[φ2]^2 + Cos[φ2]^2 - 1}], {x, y, z},
                            {Sin[φ1], Cos[φ1], Sin[φ2], Cos[φ2]}][[1]]
```

$$Out[20]= r^4 - 2 r^2 R^2 + R^4 - 2 r^2 x^2 - 2 R^2 x^2 + x^4 - 2 r^4 y^2 +$$
$$2 r^2 R^2 y^2 + 2 r^2 x^2 y^2 + r^4 y^4 - 2 r^2 z^2 - 2 R^2 z^2 + 2 x^2 z^2 + 2 r^2 y^2 z^2 + z^4$$

The two polynomials torusPoly and doubleTorusPoly assume negative values inside the corresponding bodies. (For the torus only, values $R > r$ are sensible.)

```
In[21]:= torusPoly[R, 0, 0, {R, r}] // Factor
```

$$Out[21]= r^2 (r - 2 R) (r + 2 R)$$

```
In[22]:= doubleTorus[0, 0, 0, {R, r}] // Factor
```

$$Out[22]= - 9 r^2 (r - R)^2 (r + R)^2 (3 r + R)^2$$

After having constructed the double torus, let us visualize it and see how it fits together with the original torus. For the graphic of the double torus, we use ContourPlot3D. Because of the symmetry of doubleTorusPoly, we construct only one-eight of its surface. We will choose the parameter values $R = 3$ and $r = 1$. (As mentioned, not for all r, R do we get a sensible solution.) The right graphic shows the torus and the double torus together.

```
In[23]:= Needs["Graphics`ContourPlot3D`"]

In[24]:= cp = ContourPlot3D[Evaluate[doubleTorus[x, y, z, {3, 1}]],
                            {x, 0, 7}, {y, 0, 4}, {z, 0, 4},
                            MaxRecursion -> 1, PlotPoints -> {15, 4},
                            DisplayFunction -> Identity];
```

```
In[25]:= doubleTorusPolys = (* reflect on coordinate planes *)
                    {#, Map[# {1, 1, -1}&, #, {-2}]}&[
                      {#, Map[# {1, -1, 1}&, #, {-2}]}&[
                        {#, Map[# {-1, 1, 1}&, #, {-2}]}&[
                          Cases[cp, _Polygon, Infinity]]]];
```

```
In[26]:= (* the torus *) torusPolys =
          Cases[ParametricPlot3D[Evaluate[torus[φ1, φ2, {3, 1}]],
                {φ1, 0, 2Pi}, {φ2, 0, 2Pi}, PlotPoints -> {120, 24},
                DisplayFunction -> Identity], _Polygon, Infinity];
```

```
In[27]:= Show[GraphicsArray[
            {(* only the double torus *)
            Graphics3D[{EdgeForm[], SurfaceColor[Hue[0], Hue[0], 2.2],
                        doubleTorusPolys}, ViewPoint -> {2, -1, 2.5}],
            (* double torus an ordinary torus *)
            Graphics3D[{EdgeForm[],
              {SurfaceColor[Hue[0.00], Hue[0.00], 2.2], doubleTorusPolys},
              {SurfaceColor[Hue[0.12], Hue[0.12], 2.2], torusPolys}}]}]]
```

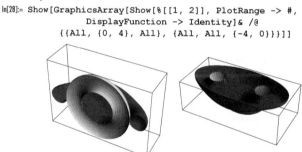

The following two cross sections show nicely that the two bodies fit perfectly together and penetrate each other in a symmetrical way.

```
In[28]:= Show[GraphicsArray[Show[%[[1, 2]], PlotRange -> #,
                DisplayFunction -> Identity]& /@
            {{All, {0, 4}, All}, {All, All, {-4, 0}}}]]
```

Here is a short symbolic proof that the two surfaces have no interior points in common. The function `Resolve` allows us to prove this as a one-liner.

```
In[29]:= AppendTo[$ContextPath, "Experimental`"]
```

```
Out[29]= {Graphics`ContourPlot3D`, Utilities`FilterOptions`, Global`, System`, Experimental`}
```

```
In[30]:= Resolve[
          Exists[{x, y, z}, Element[{x, y, z}, Reals],
                  (* because of symmetry, use one-eight *)
                  x > 0 && y > 0 && z > 0 && torusPoly[x, y, z, {3, 1}] < 0 &&
                  doubleTorus[x, y, z, {3, 1}] < 0]]
```

```
Out[30]= False
```

If the reader liked this exercise, he can continue and construct a triple torus that perfectly fits around the just-constructed double torus and then ....

14. Transformation of Variables in a PDE, $\det(e^A) = e^{\text{Tr}\,A}$, Matrix Derivative

a) Here is a suggestion for how to program this problem, which can be greatly extended. The notation is obvious. First, we identify the interesting variables (derivatives that are not altered are essentially not influenced) and replace them everywhere, except in partial derivatives. The replacement of the partial derivatives of the old variables x_{old_j} through the new variables x_{new_j} (these relations are input in the form of a list of equations with head Equal) is accomplished using

$$\frac{\partial}{\partial x_{old_1}} = \frac{\partial x_{new_1}}{\partial x_{old_1}} \frac{\partial}{\partial x_{new_1}} + \cdots + \frac{\partial x_{new_n}}{\partial x_{old_1}} \frac{\partial}{\partial x_{new_n}}.$$

Here, the corresponding Jacobian matrix is obtained by inverting the inverse Jacobian matrix $\partial x_{old_i}/\partial x_{new_j}$ one time, which is immediately accessible. The new derivatives are computed in the form of pure functions d[*oldVariable*] by applying them to the not-renamed (original) dependent functions. The number of derivatives with respect to the old variables determines how often d[*oldVariable*] is applied. The result is written as a polynomial in the derived functions. The optional function simp carries out additional simplifications on the resulting expression.

If we are interested in transforming higher derivatives, we should further optimize the code; in particular, we should try to factor the differential expression before the change of variables. Further on, we should check that the transformation is nonsingular, ... (In the following routine DerivativeVariablesTransformation, we built a few expressions in the following by joining strings together to avoid the evaluation of such expressions as {D[#1, var1], D[#1, var3], D[#1, var3], ... }; we could have done this also by using appropriate stringfree constructions, which avoid evaluation.)

```
In[1]:= DerivativeVariablesTransformation[expression_,
               oldVarsOfNewVars:{__Equal}, newVars_, simp_:Identity] :=
        Module[{oldVars, allDerivatives, allVars, remainingVars, jacobi,
                dRules, remainingRules, operators, variableReplacements,
                rules, d, auxTab, allNewDerivatives, exprNew, aux,
                exprTemp1, exprTemp2, exprTemp3, D, dot},
        (* the old independent involved variables *)
        oldVars = First /@ oldVarsOfNewVars;
        (* all derivatives *)
        allDerivatives = Cases[expression, Derivative[___][_][___],
                               {0, Infinity}] // Union;
        (* local renaming *)
        auxTab = MapIndexed[(#1 -> aux[#2[[1]]])&, allDerivatives];
        exprTemp1 = expression //. auxTab;
        exprTemp2 = exprTemp1 //. ((Rule @@ #)& /@ oldVarsOfNewVars);
        exprTemp3 = exprTemp2 //. (Reverse /@ auxTab);
        (* variables with respect to which is differentiated *)
        allVars = Union[Flatten[List @@ # & /@ allDerivatives]];
        (* remaining variables *)
        remainingVars = Complement[allVars, oldVars];
        (* the Jacobi matrix; simplified - takes time !! *)
        jacobi = Transpose @ Simplify @ Inverse @
                Outer[D, #[[2]]& /@ oldVarsOfNewVars, newVars];
        (* how to transform the derivatives in form of applying
           pure Functions; the d's are differential operators *)
        dRules = Table[d[oldVars[[i]]] = Function @@ {
                   dot[jacobi[[i]], Function[nV, D[#, nV]] /@ newVars]} //.
                       {D -> D, dot -> Dot}, {i, Length[oldVars]}];
        (* not to change *)
        remainingRules = Function[rV, d[rV] = ((Function @@ {D[#, rV]}) /.
                                   {D -> D})] /@ remainingVars;
        (* the pure functions *)
        operators = Function[expr, Composition @@ Flatten[
           Function[p, Table[d[p], {expr[[0, 0, Position[
                       expr, p][[1, 1]]]}]]]] /@ (List @@ expr)]];
        (* replace variables *)
        variableReplacements = MapThread[Rule, {oldVars, newVars}];
        rules = (# -> operators[#][#[[0, 1]] @@
                ((List @@ #) /. variableReplacements)])& /@ allDerivatives;
        (* variables not changed, but which are involved in D *)
```

```
        allNewDerivatives = Union @ Cases[#[[2]]& /@ rules,
            Derivative[___][_?(MemberQ[#, oldVars]&)][___],
                {0, Infinity}];
        exprNew = exprTemp3 /. rules;
        (* apply a little bit of make up *)
        simp[Which[Head[expression] === Plus,
                Collect[exprNew, allNewDerivatives],
                Head[expression] === Equal,
                Collect[#, allNewDerivatives]& /@ exprNew,
                True, exprNew]]]
```

As a first application of `DerivativeVariablesTransformation`, we transform the 3D wave equation from the Cartesian to the cylindrical coordinate system. The derivatives with respect to t remain unchanged. (We set $z = = Z$ to clearly differentiate between the old and the new coordinates.)

In[2]:=
```
DerivativeVariablesTransformation[
    D[u[x, y, z, t], {x, 2}] + D[u[x, y, z, t], {y, 2}] +
    D[u[x, y, z, t], {z, 2}] -
    (* this differentiation is not to be touched *)
    1/c^2 D[u[x, y, z, t], {t, 2}] == 0,
                {x == r Cos[φ], y == r Sin[φ], z == Z}, {r, φ, Z}]
```

Out[2]= $-\dfrac{u^{(0,0,0,2)}[r, \varphi, Z, t]}{c^2} + u^{(0,0,2,0)}[r, \varphi, Z, t] -$

$\dfrac{1}{r}\left(\text{Sin}[\varphi]\left(-\dfrac{\text{Cos}[\varphi]\,u^{(0,1,0,0)}[r, \varphi, Z, t]}{r} - \dfrac{\text{Sin}[\varphi]\,u^{(0,2,0,0)}[r, \varphi, Z, t]}{r} - \right.\right.$

$\left.\left.\text{Sin}[\varphi]\,u^{(1,0,0,0)}[r, \varphi, Z, t] + \text{Cos}[\varphi]\,u^{(1,1,0,0)}[r, \varphi, Z, t]\right)\right) +$

$\dfrac{1}{r}\left(\text{Cos}[\varphi]\left(-\dfrac{\text{Sin}[\varphi]\,u^{(0,1,0,0)}[r, \varphi, Z, t]}{r} + \dfrac{\text{Cos}[\varphi]\,u^{(0,2,0,0)}[r, \varphi, Z, t]}{r} + \right.\right.$

$\left.\left.\text{Cos}[\varphi]\,u^{(1,0,0,0)}[r, \varphi, Z, t] + \text{Sin}[\varphi]\,u^{(1,1,0,0)}[r, \varphi, Z, t]\right)\right) +$

$\text{Cos}[\varphi]\left(\dfrac{\text{Sin}[\varphi]\,u^{(0,1,0,0)}[r, \varphi, Z, t]}{r^2} - \dfrac{\text{Sin}[\varphi]\,u^{(1,1,0,0)}[r, \varphi, Z, t]}{r} + \right.$

$\left.\text{Cos}[\varphi]\,u^{(2,0,0,0)}[r, \varphi, Z, t]\right) + \text{Sin}[\varphi]\left(-\dfrac{\text{Cos}[\varphi]\,u^{(0,1,0,0)}[r, \varphi, Z, t]}{r^2} + \right.$

$\left.\dfrac{\text{Cos}[\varphi]\,u^{(1,1,0,0)}[r, \varphi, Z, t]}{r} + \text{Sin}[\varphi]\,u^{(2,0,0,0)}[r, \varphi, Z, t]\right) = 0$

Simplified, we get the well-known result.

In[3]:= `Simplify[%]`

Out[3]= $u^{(0,0,2,0)}[r, \varphi, Z, t] + \dfrac{u^{(0,2,0,0)}[r, \varphi, Z, t]}{r^2} + \dfrac{u^{(1,0,0,0)}[r, \varphi, Z, t]}{r} + u^{(2,0,0,0)}[r, \varphi, Z, t] ==$

$\dfrac{u^{(0,0,0,2)}[r, \varphi, Z, t]}{c^2}$

Next, we express the 4D Laplace operator in hyperspherical coordinates [105], [1301], [1394].

In[4]:=
```
(DerivativeVariablesTransformation[
    (D[#, {x, 2}] + D[#, {y, 2}] + D[#, {z, 2}] + D[#, {w, 2}])&[
                                        u[x, y, z, w]],
    (* 4D Cartesian ⟹ 4D spherical coordinates *)
    {x == r Cos[𝒪[1]],
     y == r Cos[𝒪[2]] Sin[𝒪[1]],
     z == r Cos[𝒪[3]] Sin[𝒪[1]] Sin[𝒪[2]],
     w == r Sin[𝒪[1]] Sin[𝒪[2]] Sin[𝒪[3]]},
    {r, 𝒪[1], 𝒪[2], 𝒪[3]}, (* could use more specific simplifyer here *)
        (Simplify /@ #)&] // Expand //
            Collect[#, Derivative[___][u][___], Simplify]&) /.
                    (* shorten output *) 𝒪[k_] :> Subscript[𝒪, k]
```

Out[4]= $\dfrac{Csc[\vartheta_1]^2\, Csc[\vartheta_2]^2\, u^{(0,0,0,2)}\,[r,\,\vartheta_1,\,\vartheta_2,\,\vartheta_3]}{r^2}\; +\; \dfrac{Cot[\vartheta_2]\, Csc[\vartheta_1]^2\, u^{(0,0,1,0)}\,[r,\,\vartheta_1,\,\vartheta_2,\,\vartheta_3]}{r^2}\; +$

$\dfrac{Csc[\vartheta_1]^2\, u^{(0,0,2,0)}\,[r,\,\vartheta_1,\,\vartheta_2,\,\vartheta_3]}{r^2}\; +\; \dfrac{2\, Cot[\vartheta_1]\, u^{(0,1,0,0)}\,[r,\,\vartheta_1,\,\vartheta_2,\,\vartheta_3]}{r^2}\; +$

$\dfrac{u^{(0,2,0,0)}\,[r,\,\vartheta_1,\,\vartheta_2,\,\vartheta_3]}{r^2}\; +\; \dfrac{3\, u^{(1,0,0,0)}\,[r,\,\vartheta_1,\,\vartheta_2,\,\vartheta_3]}{r}\; +\, u^{(2,0,0,0)}\,[r,\,\vartheta_1,\,\vartheta_2,\,\vartheta_3]$

In quantum mechanics, the operator $x\,\partial/\partial y - y\,\partial/\partial x$ turns out to be essentially the angular momentum in the z-direction. Here, we transform the variables of this differential operator, applied to $u(x, y)$, to polar coordinates.

```
In[5]:= DerivativeVariablesTransformation[
          y D[u[x, y], x] - D[u[x, y], y] x,
          {x == r Cos[φ], y == r Sin[φ]}, {r, φ}, Simplify]
```
Out[5]= $-u^{(0,1)}\,[r,\,\varphi]$

Here is the Laplace operator in spherical coordinates [1420].

```
In[6]:= DerivativeVariablesTransformation[
          (D[#, {x, 2}] + D[#, {y, 2}] + D[#, {z, 2}])&[u[x, y, z, t]],
          {x == r Cos[φ] Sin[θ], y == r Sin[φ] Sin[θ], z == r Cos[θ]},
          {r, φ, θ}, (* could use more specific simplifier here *)
                      (Simplify /@ #)&] // Expand //
                      Collect[#, Derivative[__][u][__], Simplify]&
```
Out[6]= $\dfrac{Cot[\theta]\, u^{(0,0,1,0)}\,[r,\,\varphi,\,\theta,\,t]}{r^2}\; +\; \dfrac{u^{(0,0,2,0)}\,[r,\,\varphi,\,\theta,\,t]}{r^2}\; +$

$\dfrac{Csc[\theta]^2\, u^{(0,2,0,0)}\,[r,\,\varphi,\,\theta,\,t]}{r^2}\; +\; \dfrac{2\, u^{(1,0,0,0)}\,[r,\,\varphi,\,\theta,\,t]}{r}\; +\, u^{(2,0,0,0)}\,[r,\,\varphi,\,\theta,\,t]$

This is the Levi–Civita transformation that transforms a 2D Kepler problem to a harmonic oscillator problem [1134], [1098].

```
In[7]:= DerivativeVariablesTransformation[
          D[u[x, y], x, x] + D[u[x, y], y, y],
          {x == u1^2 - u2^2, y == 2 u1 u2}, {u1, u2}, Simplify]
```
Out[7]= $\dfrac{u^{(0,2)}\,[u1,\,u2] + u^{(2,0)}\,[u1,\,u2]}{4\,(u1^2 + u2^2)}$

We can also treat purely symbolic relations.

```
In[8]:= DerivativeVariablesTransformation[D[u[x, y], x] - D[v[x, y], y],
          {x == x[ξ, η], y == y[ξ, η]}, {ξ, η}, Simplify]
```
Out[8]= $(y^{(0,1)}\,[ξ,\,η]\, u^{(1,0)}\,[ξ,\,η] + x^{(0,1)}\,[ξ,\,η]\, v^{(1,0)}\,[ξ,\,η] - v^{(0,1)}\,[ξ,\,η]\, x^{(1,0)}\,[ξ,\,η]\, -$

$u^{(0,1)}\,[ξ,\,η]\, y^{(1,0)}\,[ξ,\,η])\, /\, (y^{(0,1)}\,[ξ,\,η]\, x^{(1,0)}\,[ξ,\,η] - x^{(0,1)}\,[ξ,\,η]\, y^{(1,0)}\,[ξ,\,η])$

For another implementation, see [3].

b) This is the determinant to be simplified.

```
In[1]:= A = Array[a, {3, 3}];
        det = Det[MatrixExp[A]];
```

The determinant `det` is a relatively large symbolic expression.

```
In[3]:= {ByteCount[det], LeafCount[det]}
```
Out[3]= {93016, 4098}

A direct call to `Simplify` will take a long time and not give the desired result.

```
In[4]:= TimeConstrained[Simplify[det - Tr[A]], 100] === 0
```
Out[4]= False

Here is the matrix exponential of $\mathbb{A}$ written in a concise form.

```
In[5]:= ExpA = With[{a11 = a[1, 1], a12 = a[1, 2], a13 = a[1, 3],
                      a21 = a[2, 1], a22 = a[2, 2], a23 = a[2, 3],
                      a31 = a[3, 1], a32 = a[3, 2], a33 = a[3, 3], e = Exp[#]},
        Map[Function[r,
        RootSum[#^3 - #^2 a[1, 1] - # a12 a21 - #^2 a22 + # a11 a22 - # a13 a31 +
                a13 a22 a31 - a12 a23 a31 - a13 a21 a32 - # a23 a32 + a11 a23 a32 -
                #^2 a33 + # a11 a33 + a12 a21 a33 + # a22 a33 - a11 a22 a33 & ,
```

```
              Function[r]]],
        {{e #^2 - e # a22 - e a23 a32 - e # a33 + e a22 a33,
          e # a12 + e a13 a32 - e a12 a33, e # a13 - e a13 a22 + e a12 a23},
         {e # a21 + e a23 a31 - e a21 a33, e #^2 - e # a11 - e a13 a31 -
          e # a33 + e a11 a33, -e a13 a21 - e # a23 + e a11 a23},
         {e # a31 - e a22 a31 + e a21 a32, -e a12 a31 - e # a32 + e a11 a32,
          e #^2 - e # a11 - e a12 a21 - e # a22 + e a11 a22}}/
        (3 #^2 - 2 # a11 - a12 a21 - 2 # a22 + a11 a22 - a13 a31 - a23 a32 -
         2 # a33 + a11 a33 + a22 a33), {2}] {{1, 1, 1}, {1, 1, -1}, {1, -1, 1}}];
```

In[6]:= `det = Det[ExpA];`

So, we will carry out various transformations "by hand". `det` is a sum of six terms, each being a `RootSum`.

In[7]:= `Module[{c = 0}, det /. _RootSum :> RS[c++]]`

Out[7]= `(RS[0] RS[1] + RS[2] RS[3]) RS[4] +`
`RS[5] (RS[6] RS[7] + RS[8] RS[9]) + RS[10] (-RS[11] RS[12] + RS[13] RS[14])`

We start by rewriting the `RootSum`s as `Root`s and factor the resulting polynomials. For brevity, we denote the three roots by `R[1]`, `R[2]`, and `R[3]`.

In[8]:= `det1 = Map[Factor, #, {2}]& /@ (* RootSum → sum of Root-objects *)`
`(Normal[Expand[det]] /. Root[_, k_, ___] :> R[k]);`

These are the three `Root`-objects that appear in the determinant.

In[9]:= `(Cases[Normal[det], _Root, Infinity] // Union) /.`
`a[ij__] :> Subscript[a, ij]`

Out[9]= $\{\text{Root}[\#1^3 - \#1^2 a_{1,1} - \#1 a_{1,2} a_{2,1} - \#1^2 a_{2,2} + \#1 a_{1,1} a_{2,2} - \#1 a_{1,3} a_{3,1} +$
$a_{1,3} a_{2,2} a_{3,1} - a_{1,2} a_{2,3} a_{3,1} - a_{1,3} a_{2,1} a_{3,2} - \#1 a_{2,3} a_{3,2} + a_{1,1} a_{2,3} a_{3,2} -$
$\#1^2 a_{3,3} + \#1 a_{1,1} a_{3,3} + a_{1,2} a_{2,1} a_{3,3} + \#1 a_{2,2} a_{3,3} - a_{1,1} a_{2,2} a_{3,3} \&, 1],$
$\text{Root}[\#1^3 - \#1^2 a_{1,1} - \#1 a_{1,2} a_{2,1} - \#1^2 a_{2,2} + \#1 a_{1,1} a_{2,2} - \#1 a_{1,3} a_{3,1} +$
$a_{1,3} a_{2,2} a_{3,1} - a_{1,2} a_{2,3} a_{3,1} - a_{1,3} a_{2,1} a_{3,2} - \#1 a_{2,3} a_{3,2} + a_{1,1} a_{2,3} a_{3,2} -$
$\#1^2 a_{3,3} + \#1 a_{1,1} a_{3,3} + a_{1,2} a_{2,1} a_{3,3} + \#1 a_{2,2} a_{3,3} - a_{1,1} a_{2,2} a_{3,3} \&, 2],$
$\text{Root}[\#1^3 - \#1^2 a_{1,1} - \#1 a_{1,2} a_{2,1} - \#1^2 a_{2,2} + \#1 a_{1,1} a_{2,2} - \#1 a_{1,3} a_{3,1} +$
$a_{1,3} a_{2,2} a_{3,1} - a_{1,2} a_{2,3} a_{3,1} - a_{1,3} a_{2,1} a_{3,2} - \#1 a_{2,3} a_{3,2} + a_{1,1} a_{2,3} a_{3,2} -$
$\#1^2 a_{3,3} + \#1 a_{1,1} a_{3,3} + a_{1,2} a_{2,1} a_{3,3} + \#1 a_{2,2} a_{3,3} - a_{1,1} a_{2,2} a_{3,3} \&, 3]\}$

The whole determinant is formed by the following 10 polynomials (with different arguments). Here we identify all `Root`-objects.

In[10]:= `allPolys = Cases[det1, _Plus?(FreeQ[#, _Plus, {1, Infinity}]&),`
`Infinity] /. _R -> R // Union`

Out[10]= $\{R^2 - R\,a[1, 1] - a[1, 2]\,a[2, 1] - R\,a[2, 2] + a[1, 1]\,a[2, 2],$
$a[1, 3]\,a[2, 1] + R\,a[2, 3] - a[1, 1]\,a[2, 3], R\,a[1, 3] - a[1, 3]\,a[2, 2] + a[1, 2]\,a[2, 3],$
$a[1, 2]\,a[3, 1] + R\,a[3, 2] - a[1, 1]\,a[3, 2], R\,a[3, 1] - a[2, 2]\,a[3, 1] + a[2, 1]\,a[3, 2],$
$R^2 - R\,a[1, 1] - a[1, 3]\,a[3, 1] - R\,a[3, 3] + a[1, 1]\,a[3, 3],$
$R\,a[1, 2] + a[1, 3]\,a[3, 2] - a[1, 2]\,a[3, 3], R\,a[2, 1] + a[2, 3]\,a[3, 1] - a[2, 1]\,a[3, 3],$
$R^2 - R\,a[2, 2] - a[2, 3]\,a[3, 2] - R\,a[3, 3] + a[2, 2]\,a[3, 3],$
$3\,R^2 - 2\,R\,a[1, 1] - a[1, 2]\,a[2, 1] - 2\,R\,a[2, 2] + a[1, 1]\,a[2, 2] -$
$a[1, 3]\,a[3, 1] - a[2, 3]\,a[3, 2] - 2\,R\,a[3, 3] + a[1, 1]\,a[3, 3] + a[2, 2]\,a[3, 3]\}$

The two expressions `collapsePolyRules` and `expandPolyRules` are list of rules that rewrite the polynomials of `det1`.

In[11]:= `collapsePolyRules = Flatten[Table[`
`MapIndexed[((# /. R -> R[j]) -> P[#2[[1]]], j])&, allPolys], {j, 3}]];`

`expandPolyRules = Reverse /@ collapsePolyRules;`

In abbreviated form, the first term of the determinant is the following.

In[14]:= `det2 = det1 //. collapsePolyRules;`

In[15]:= `det2[[1]]`

Out[15]= $\left(\dfrac{e^{R[1]}\,P[3,\,1]}{P[10,\,1]} + \dfrac{e^{R[2]}\,P[3,\,2]}{P[10,\,2]} + \dfrac{e^{R[3]}\,P[3,\,3]}{P[10,\,3]}\right) \left(\dfrac{e^{R[1]}\,P[5,\,1]}{P[10,\,1]} + \dfrac{e^{R[2]}\,P[5,\,2]}{P[10,\,2]} + \dfrac{e^{R[3]}\,P[5,\,3]}{P[10,\,3]}\right)$

$\left(-\dfrac{e^{R[1]}\,P[6,\,1]}{P[10,\,1]} - \dfrac{e^{R[2]}\,P[6,\,2]}{P[10,\,2]} - \dfrac{e^{R[3]}\,P[6,\,3]}{P[10,\,3]}\right)$

By Schanuel's conjecture [362], [1483], we assume the exponents of the various sums of the roots are independent quantities. So, we collect with respect to these exponents and simplify their coefficients.

In[16]:= det3 = Expand[det2];

In[17]:= exps = Cases[det3, Power[E, _], Infinity] // Union

Out[17]= $\{e^{3\,R[1]},\ e^{3\,R[2]},\ e^{2\,R[1]+R[2]},\ e^{R[1]+2\,R[2]},\ e^{3\,R[3]},$
$e^{2\,R[1]+R[3]},\ e^{R[1]+R[2]+R[3]},\ e^{2\,R[2]+R[3]},\ e^{R[1]+2\,R[3]},\ e^{R[2]+2\,R[3]}\}$

For the first exponent Exp[3 R[1]] it is easily shown that its prefactor vanishes. This is the expression.

In[18]:= fac[1] = Factor[Together[Plus @@
                    Cases[det3, exps[[1]] _, Infinity]]]

Out[18]= $-\dfrac{1}{P[10,\,1]^3}\,(e^{3\,R[1]}\,(P[3,\,1]\,P[5,\,1]\,P[6,\,1] - P[2,\,1]\,P[5,\,1]\,P[7,\,1] - P[3,\,1]\,P[4,\,1]\,P[8,\,1] +$
$P[1,\,1]\,P[7,\,1]\,P[8,\,1] + P[2,\,1]\,P[4,\,1]\,P[9,\,1] - P[1,\,1]\,P[6,\,1]\,P[9,\,1]))$

In[19]:= fac1[1] = Expand[Cases[fac[1], _Plus][[1]] /. expandPolyRules];

Its factored form resembles the defining polynomial for the three roots R[1], R[2], and R[3].

In[20]:= fac1[1] // Factor

Out[20]= $-(R[1]^3 - R[1]^2\,a[1,\,1] - R[1]\,a[1,\,2]\,a[2,\,1] - R[1]^2\,a[2,\,2] +$
$R[1]\,a[1,\,1]\,a[2,\,2] - R[1]\,a[1,\,3]\,a[3,\,1] + a[1,\,3]\,a[2,\,2]\,a[3,\,1] -$
$a[1,\,2]\,a[2,\,3]\,a[3,\,1] - a[1,\,3]\,a[2,\,1]\,a[3,\,2] - R[1]\,a[2,\,3]\,a[3,\,2] +$
$a[1,\,1]\,a[2,\,3]\,a[3,\,2] - R[1]^2\,a[3,\,3] + R[1]\,a[1,\,1]\,a[3,\,3] +$
$a[1,\,2]\,a[2,\,1]\,a[3,\,3] + R[1]\,a[2,\,2]\,a[3,\,3] - a[1,\,1]\,a[2,\,2]\,a[3,\,3])^2$

In[21]:= (* the polynomial fulfilled by the roots *)
        rootPoly = (Cases[Normal[det], _Root, Infinity] // Union)[[1, 1]]

Out[22]= $\#1^3 - \#1^2\,a[1,\,1] - \#1\,a[1,\,2]\,a[2,\,1] - \#1^2\,a[2,\,2] + \#1\,a[1,\,1]\,a[2,\,2] - \#1\,a[1,\,3]\,a[3,\,1] +$
$a[1,\,3]\,a[2,\,2]\,a[3,\,1] - a[1,\,2]\,a[2,\,3]\,a[3,\,1] - a[1,\,3]\,a[2,\,1]\,a[3,\,2] -$
$\#1\,a[2,\,3]\,a[3,\,2] + a[1,\,1]\,a[2,\,3]\,a[3,\,2] - \#1^2\,a[3,\,3] + \#1\,a[1,\,1]\,a[3,\,3] +$
$a[1,\,2]\,a[2,\,1]\,a[3,\,3] + \#1\,a[2,\,2]\,a[3,\,3] - a[1,\,1]\,a[2,\,2]\,a[3,\,3]\ \&$

In[23]:= PolynomialReduce[fac1[1], rootPoly[R[1]]][[2]]

Out[23]= 0

We automate the process of showing that the prefactors of all the other eight exponents of exps1 vanish too.

In[24]:= exps1 = DeleteCases[exps, Exp[R[1] + R[2] + R[3]]]

Out[24]= $\{e^{3\,R[1]},\ e^{3\,R[2]},\ e^{2\,R[1]+R[2]},\ e^{R[1]+2\,R[2]},\ e^{3\,R[3]},\ e^{2\,R[1]+R[3]},\ e^{2\,R[2]+R[3]},\ e^{R[1]+2\,R[3]},\ e^{R[2]+2\,R[3]}\}$

In[25]:= gb = GroebnerBasis[
        (* R[1], R[2], and R[3] are the three roots of rootPoly *)
        CoefficientList[rootPoly[ξ] - (ξ - R[1])(ξ - R[2])(ξ - R[3]), ξ],
                        {R[1], R[2], R[3]}];

In[26]:= Table[fac[j] = Factor[Together[Plus @@
                        Cases[det3, exps1[[1]] _, Infinity]]];
        fac1[j] = Expand[Cases[fac[j], _Plus][[1]] /. expandPolyRules];
        (* reduce the factors *)
        PolynomialReduce[fac1[j], gb, {R[1], R[2], R[3]}][[2]],
        {j, Length[exps1]}]

Out[26]= {0, 0, 0, 0, 0, 0, 0, 0, 0}

The prefactor of the remaining exponent Exp[R[1] + R[2] + R[3]] simplifies to 1.

In[27]:= fac[10] = 1/Exp[R[1] + R[2] + R[3]] Factor[Together[Plus @@
                    Cases[det3, Exp[R[1] + R[2] + R[3]] _, Infinity]]]

Out[27]= $-\dfrac{1}{P[10, 1] P[10, 2] P[10, 3]}$

(P[3, 3] P[5, 2] P[6, 1] + P[3, 2] P[5, 3] P[6, 1] + P[3, 3] P[5, 1] P[6, 2] +
    P[3, 1] P[5, 3] P[6, 2] + P[3, 2] P[5, 1] P[6, 3] + P[3, 1] P[5, 2] P[6, 3] -
    P[2, 3] P[5, 2] P[7, 1] - P[2, 2] P[5, 3] P[7, 1] - P[2, 3] P[5, 1] P[7, 2] -
    P[2, 1] P[5, 3] P[7, 2] - P[2, 2] P[5, 1] P[7, 3] - P[2, 1] P[5, 2] P[7, 3] -
    P[3, 3] P[4, 2] P[8, 1] - P[3, 2] P[4, 3] P[8, 1] + P[1, 3] P[7, 2] P[8, 1] +
    P[1, 2] P[7, 3] P[8, 1] - P[3, 3] P[4, 1] P[8, 2] - P[3, 1] P[4, 3] P[8, 2] +
    P[1, 3] P[7, 1] P[8, 2] + P[1, 1] P[7, 3] P[8, 2] - P[3, 2] P[4, 1] P[8, 3] -
    P[3, 1] P[4, 2] P[8, 3] + P[1, 2] P[7, 1] P[8, 3] + P[1, 1] P[7, 2] P[8, 3] +
    P[2, 3] P[4, 2] P[9, 1] + P[2, 2] P[4, 3] P[9, 1] - P[1, 3] P[6, 2] P[9, 1] -
    P[1, 2] P[6, 3] P[9, 1] + P[2, 3] P[4, 1] P[9, 2] + P[2, 1] P[4, 3] P[9, 2] -
    P[1, 3] P[6, 1] P[9, 2] - P[1, 1] P[6, 3] P[9, 2] + P[2, 2] P[4, 1] P[9, 3] +
    P[2, 1] P[4, 2] P[9, 3] - P[1, 2] P[6, 1] P[9, 3] - P[1, 1] P[6, 2] P[9, 3])

In[28]:= **PolynomialReduce[Numerator[Together[fac[10] - (* the 1 *) 1]] /.
                    expandPolyRules, gb, {R[1], R[2], R[3]}][[2]]**

Out[28]= 0

It remains to simplify the argument of Exp, namely $R[1] + R[2] + R[3]$ to $a[1,1] + a[2,2] + a[3,3]$.

In[29]:= **gb[[-1]]**

Out[29]= R[1] + R[2] + R[3] - a[1, 1] - a[2, 2] - a[3, 3]

In[30]:= **PolynomialReduce[R[1] + R[2] + R[3], gb, {R[1], R[2], R[3]}][[2]]**

Out[30]= a[1, 1] + a[2, 2] + a[3, 3]

c) Here is a generic symbolic 3×3 matrix **A**.

In[1]:= **A = {{a11, a12, a13}, {a21, a22, a23}, {a31, a32, a33}};**

The implementation of the identity is straightforward by using the functions Inverse, MatrixPower, and Sum. To speed up the calculation we sprinkle Together and Simplify throughout the code.

In[2]:= **Inverse[IdentityMatrix[3] - λ A] ==
            Map[Simplify[Together[Sum[λ^k #, {k, 0, Infinity}]]]&,
                Together[MatrixPower[A, k]], {2}] // Simplify**

Out[2]= True

For some more complicated matrix identities, see [1415].

d) We start with two generic 2×2 matrices $\mathbb{A}$ and $\mathbb{H}$.

In[1]:= **n = 2;
        A = Table[Subscript[a, i, j], {i, n}, {j, n}];
        H = Table[Subscript[h, i, j], {i, n}, {j, n}];**

d1 and d2 are the two expressions for the matrix differential quotients. d1Pre is the argument of the limit definition.

In[4]:= **d1Pre = (MatrixExp[t (A + ε H)] - MatrixExp[t A])/ε;**

In[5]:= **d2 = Map[Integrate[#, {τ, 0, t}, GenerateConditions -> False]&,
                MatrixExp[(t - τ) A].H.MatrixExp[τ A], {2}];**

d1Pre and d2 are relatively large expressions.

In[6]:= **{LeafCount[d1Pre], LeafCount[d2]}**

Out[6]= {4363, 6868}

To calculate d1 from d1Pre we have to carry out the limit. A direct call to Limit gives a relatively short result.

In[7]:= **Limit[d1Pre[[1, 1]], ε -> 0] // Head**

Out[7]= Plus

To calculate the limit a series expansion will often be carried out.

In[8]:= **ser = Series[d1Pre[[1, 1]], {ε, 0, 0}];**

But the identity cannot be shown immediately.

```
In[9]:= Union[Flatten[Limit[d1Pre[[1, 1]], ε -> 0] - d2 // Together // Factor]] === {0}

Out[9]= False
```

Automatic internal simplification cannot always reduce terms to zero.

```
In[10]:= Short[Together //@ ser[[3]][[1]], 6]
```

$$Out[10]//Short= \left(e^{-\frac{1}{2} t \left(-a_{1,1} - a_{2,2} + \sqrt{a_{1,1}^2 + 4 a_{1,2} a_{2,1} - 2 a_{1,1} a_{2,2} + a_{2,2}^2} \right)} \right.$$

$$\left(-t a_{1,1}^3 h_{1,1} + \ll 81 \gg + 2 e^{\frac{1}{2} t \left(-a_{1,1} - a_{2,2} + \sqrt{a_{1,1}^2 + 4 a_{1,2} a_{2,1} - 2 a_{1,1} a_{2,2} + a_{2,2}^2} \right)} \right) \ll 1 \gg \frac{1}{2} \ll 1 \gg \ll 1 \gg (\ll 1 \gg)$$

$$\left. t a_{1,2} a_{2,1} \sqrt{a_{1,1}^2 + 4 a_{1,2} a_{2,1} - 2 a_{1,1} a_{2,2} + a_{2,2}^2} h_{2,2} \right) \Big/$$

$$\left(2 \left(a_{1,1}^2 + 4 a_{1,2} a_{2,1} - 2 a_{1,1} a_{2,2} + a_{2,2}^2 \right)^{3/2} \right)$$

d1 is the simplified limit version of d1Pre. No ε is present anymore.

```
In[11]:= d1 = Map[Normal[Together //@ Series[#, {ε, 0, 0}]]&, d1Pre, {2}];

In[12]:= Position[d1, ε]

Out[12]= {}
```

Now it remains to show that diff is zero.

```
In[13]:= diff = d1 - d2;
```

To show that an expression is identically zero, it is always a good idea to transform the problem to a rational function. This means we have to replace the following square roots.

```
In[14]:= Cases[diff, Power[_, _Rational], Infinity] // Union
```

$$Out[14]= \left\{ \frac{1}{\left(a_{1,1}^2 + 4 a_{1,2} a_{2,1} - 2 a_{1,1} a_{2,2} + a_{2,2}^2 \right)^{3/2}}, \sqrt{a_{1,1}^2 + 4 a_{1,2} a_{2,1} - 2 a_{1,1} a_{2,2} + a_{2,2}^2} \right\}$$

```
In[15]:= aux = With[{s = Subscript},
            s[a, 1, 1]^2 + s[a, 2, 2]^2 + 4 s[a, 1, 2] s[a, 2, 1] -
            2 s[a, 1, 1] s[a, 2, 2]];
```

Replacing the square roots by functions of $\mathcal{R}$, "togethering" and factoring the expression, and finally using the definition for $\mathcal{R}$ gives the result $\{\{0,0\}, \{0,0\}\}$ we were looking for.

```
In[16]:= diff1 = diff //. {1/aux^(3/2) :> 1/(aux R),
                          1/aux^(1/2) :> 1/R, aux^(1/2) :> R};

In[17]:= Factor[ExpandAll[Numerator[Together[diff1]]]] /. R^2 -> aux

Out[17]= {{0, 0}, {0, 0}}
```

For details of the derivative with respect to a symmetric matrix, see [907]; for derivatives with respect to infinite matrices, see [1689].

e) We start by constructing the 81 matrices. A[rows, columns] is the matrix formed by using the rows *rows* and the columns *columns* from a generic 3×3 matrix with symbolic elements $a_{i,j}$.

```
In[1]:= d = 3;
        A[] = Table[a[i, j], {i, d}, {j, d}];
        A[rows_, colums_] := #[[colums]]& /@ A[][[rows]];

        (* generate all equations *)
        eqs = (A[##] - Det[A[##]])& @@@
        Join[{{}}, Flatten[Table[{{i}, {j}}, {i, d}, {j, d}], 1],
                Flatten[Table[{{i1, i2}, {j1, j2}}, {i1, d}, {i2, d},
                              {j1, d}, {j2, d}], 3]];
```

Using GroebnerBasis to eliminate the variables $a[i, j]$ gives us a set of 124 polynomials that have to vanish.

```
In[6]:= (gb = GroebnerBasis[eqs, Union[Cases[eqs, _a, Infinity]],
                           Flatten[A[]]]) // Length

Out[6]= 124
```

These 124 polynomials can be naturally grouped according to the number of monomials in them. We have four different groups.

```
In[7]:= groups = Map[Last, #, {2}]& @ Split[Sort[{Union[Cases[#, _𝒜, Infinity] /.
                   {𝒜[l_, _] :> Length[l]}], #}& /@ gb], #1[[1]] === #2[[1]]&];

In[8]:= Length /@ groups

Out[8]= {45, 27, 51, 1}
```

To represent the results in a concise form, we introduce the notation $\mathcal{A}_{rows}^{columns}$ to denote the determinant of the matrix that was formed by using the rows *rows* and the columns *columns* from $\mathcal{A}$.

```
In[9]:= MakeBoxes[𝒜[rs_, cs_], fmt_] :=
         SubsuperscriptBox["𝒜", #[rs], #[cs]]&[StringJoin[ToString /@ #]&]

        MakeBoxes[𝒜[], fmt_] := "𝒜"
```

The first group contains the identities of determinants that arise from 2×2 matrices with identical rows and/or columns.

```
In[12]:= groups[[1]]
```

Out[12]= $\{\mathcal{A}_{11}^{11}, \mathcal{A}_{11}^{12}, \mathcal{A}_{11}^{13}, \mathcal{A}_{11}^{21}, \mathcal{A}_{11}^{22}, \mathcal{A}_{11}^{23}, \mathcal{A}_{11}^{31}, \mathcal{A}_{11}^{32}, \mathcal{A}_{11}^{33}, \mathcal{A}_{12}^{11}, \mathcal{A}_{12}^{22}, \mathcal{A}_{12}^{33}, \mathcal{A}_{13}^{11}, \mathcal{A}_{13}^{22},$
$\mathcal{A}_{13}^{33}, \mathcal{A}_{21}^{11}, \mathcal{A}_{21}^{22}, \mathcal{A}_{21}^{33}, \mathcal{A}_{22}^{11}, \mathcal{A}_{22}^{12}, \mathcal{A}_{22}^{13}, \mathcal{A}_{22}^{21}, \mathcal{A}_{22}^{22}, \mathcal{A}_{22}^{23}, \mathcal{A}_{22}^{31}, \mathcal{A}_{22}^{32}, \mathcal{A}_{22}^{33}, \mathcal{A}_{23}^{11}, \mathcal{A}_{23}^{22},$
$\mathcal{A}_{23}^{33}, \mathcal{A}_{31}^{11}, \mathcal{A}_{31}^{22}, \mathcal{A}_{31}^{33}, \mathcal{A}_{32}^{11}, \mathcal{A}_{32}^{22}, \mathcal{A}_{32}^{33}, \mathcal{A}_{33}^{11}, \mathcal{A}_{33}^{12}, \mathcal{A}_{33}^{13}, \mathcal{A}_{33}^{21}, \mathcal{A}_{33}^{22}, \mathcal{A}_{33}^{23}, \mathcal{A}_{33}^{31}, \mathcal{A}_{33}^{32}, \mathcal{A}_{33}^{33}\}$

The second group contains again trivial identities, namely the ones arising from exchanging rows and columns in the 2×2 matrices.

```
In[13]:= Partition[groups[[2]], 3] // TableForm
```

Out[13]//TableForm=

$\mathcal{A}_{12}^{12} - \mathcal{A}_{21}^{21}$	$\mathcal{A}_{12}^{21} + \mathcal{A}_{21}^{21}$	$\mathcal{A}_{21}^{12} + \mathcal{A}_{21}^{21}$
$\mathcal{A}_{12}^{13} - \mathcal{A}_{21}^{31}$	$\mathcal{A}_{12}^{31} + \mathcal{A}_{21}^{31}$	$\mathcal{A}_{21}^{13} + \mathcal{A}_{21}^{31}$
$\mathcal{A}_{12}^{23} - \mathcal{A}_{21}^{32}$	$\mathcal{A}_{12}^{32} + \mathcal{A}_{21}^{32}$	$\mathcal{A}_{21}^{23} + \mathcal{A}_{21}^{32}$
$\mathcal{A}_{13}^{12} - \mathcal{A}_{31}^{21}$	$\mathcal{A}_{13}^{21} + \mathcal{A}_{31}^{21}$	$\mathcal{A}_{31}^{12} + \mathcal{A}_{31}^{21}$
$\mathcal{A}_{13}^{13} - \mathcal{A}_{31}^{31}$	$\mathcal{A}_{13}^{31} + \mathcal{A}_{31}^{31}$	$\mathcal{A}_{31}^{13} + \mathcal{A}_{31}^{31}$
$\mathcal{A}_{13}^{23} - \mathcal{A}_{31}^{32}$	$\mathcal{A}_{13}^{32} + \mathcal{A}_{31}^{32}$	$\mathcal{A}_{31}^{23} + \mathcal{A}_{31}^{32}$
$\mathcal{A}_{23}^{12} - \mathcal{A}_{32}^{21}$	$\mathcal{A}_{23}^{21} + \mathcal{A}_{32}^{21}$	$\mathcal{A}_{32}^{12} + \mathcal{A}_{32}^{21}$
$\mathcal{A}_{23}^{13} - \mathcal{A}_{32}^{31}$	$\mathcal{A}_{23}^{31} + \mathcal{A}_{32}^{31}$	$\mathcal{A}_{32}^{13} + \mathcal{A}_{32}^{31}$
$\mathcal{A}_{23}^{23} - \mathcal{A}_{32}^{32}$	$\mathcal{A}_{23}^{32} + \mathcal{A}_{32}^{32}$	$\mathcal{A}_{32}^{23} + \mathcal{A}_{32}^{32}$

The third group finally contains some more interesting identities. We sort the identities from this group according to the number of determinants appearing in the polynomials. `𝒜Count` counts the number of determinants in an expression.

```
In[14]:= (* count 𝒜's in a polynomial recursively *)
         𝒜Count[expr_Plus] := Plus @@ (𝒜Count /@ expr)
         𝒜Count[expr_Times] := Plus @@ (𝒜Count /@ (List @@ expr))
         𝒜Count[expr_] := 0 /; FreeQ[expr, _𝒜, {0, Infinity}]
         𝒜Count[Power[_𝒜, e_]] := e
         𝒜Count[_𝒜] := 1
```

The third group splits naturally into ten subgroups. The first subgroup contains the expansion of 2×2 formulas. The second subgroup contains the symmetric Laplace expansion-like identities.

```
In[20]:= subGroups3 = Split[Sort[groups[[3]], 𝒜Count[#1] < 𝒜Count[#2]&],
                                   𝒜Count[#1] === 𝒜Count[#2]&];

         (* number of groups and their length *)
         {𝒜Count[#[[1]]], Length[#]}& /@ subGroups3

Out[23]= {{5, 9}, {6, 12}, {8, 3}, {11, 2}, {12, 9}, {13, 5}, {15, 2}, {21, 1}, {22, 7}, {42, 1}}

In[24]:= Print[TableForm[#]]& /@ Take[subGroups3, {1, 2}];
```

$$-\mathcal{A}_3^3\,\mathcal{A}_3^2 + \mathcal{A}_2^2\,\mathcal{A}_3^3 - \mathcal{A}_{32}^{32}$$
$$-\mathcal{A}_3^3\,\mathcal{A}_3^1 + \mathcal{A}_1^1\,\mathcal{A}_3^3 - \mathcal{A}_{32}^{31}$$
$$-\mathcal{A}_3^2\,\mathcal{A}_3^1 + \mathcal{A}_1^1\,\mathcal{A}_3^2 - \mathcal{A}_{32}^{21}$$
$$-\mathcal{A}_1^3\,\mathcal{A}_3^2 + \mathcal{A}_1^2\,\mathcal{A}_3^3 - \mathcal{A}_{31}^{32}$$
$$-\mathcal{A}_1^3\,\mathcal{A}_3^1 + \mathcal{A}_1^1\,\mathcal{A}_3^3 - \mathcal{A}_{31}^{31}$$
$$-\mathcal{A}_1^2\,\mathcal{A}_3^1 + \mathcal{A}_1^1\,\mathcal{A}_3^2 - \mathcal{A}_{31}^{21}$$
$$-\mathcal{A}_1^3\,\mathcal{A}_2^2 + \mathcal{A}_1^2\,\mathcal{A}_2^3 - \mathcal{A}_{21}^{32}$$
$$-\mathcal{A}_1^3\,\mathcal{A}_2^1 + \mathcal{A}_1^1\,\mathcal{A}_2^3 - \mathcal{A}_{21}^{31}$$
$$-\mathcal{A}_1^2\,\mathcal{A}_2^1 + \mathcal{A}_1^1\,\mathcal{A}_2^2 - \mathcal{A}_{21}^{21}$$
$$\mathcal{A}_3^3\,\mathcal{A}_{32}^{21} - \mathcal{A}_3^2\,\mathcal{A}_{32}^{31} + \mathcal{A}_3^1\,\mathcal{A}_{32}^{32}$$
$$\mathcal{A}_2^3\,\mathcal{A}_{32}^{21} - \mathcal{A}_2^2\,\mathcal{A}_{32}^{31} + \mathcal{A}_2^1\,\mathcal{A}_{32}^{32}$$
$$\mathcal{A}_3^3\,\mathcal{A}_{21}^{32} - \mathcal{A}_3^2\,\mathcal{A}_{31}^{32} + \mathcal{A}_1^3\,\mathcal{A}_{32}^{32}$$
$$\mathcal{A}_3^2\,\mathcal{A}_{21}^{32} - \mathcal{A}_2^2\,\mathcal{A}_{31}^{32} + \mathcal{A}_1^2\,\mathcal{A}_{32}^{32}$$
$$\mathcal{A}_3^3\,\mathcal{A}_{21}^{31} - \mathcal{A}_3^2\,\mathcal{A}_{31}^{31} + \mathcal{A}_1^3\,\mathcal{A}_{32}^{31}$$
$$\mathcal{A}_1^1\,\mathcal{A}_{21}^{31} - \mathcal{A}_2^1\,\mathcal{A}_{31}^{31} + \mathcal{A}_1^1\,\mathcal{A}_{32}^{31}$$
$$\mathcal{A}_3^2\,\mathcal{A}_{21}^{31} - \mathcal{A}_2^2\,\mathcal{A}_{21}^{31} + \mathcal{A}_1^2\,\mathcal{A}_{32}^{21}$$
$$\mathcal{A}_1^3\,\mathcal{A}_{21}^{31} - \mathcal{A}_2^1\,\mathcal{A}_{21}^{31} + \mathcal{A}_1^1\,\mathcal{A}_{32}^{21}$$
$$\mathcal{A}_3^3\,\mathcal{A}_{31}^{21} - \mathcal{A}_3^2\,\mathcal{A}_{31}^{31} + \mathcal{A}_3^1\,\mathcal{A}_{31}^{31}$$
$$\mathcal{A}_1^3\,\mathcal{A}_{31}^{21} - \mathcal{A}_1^2\,\mathcal{A}_{31}^{31} + \mathcal{A}_1^1\,\mathcal{A}_{31}^{32}$$
$$\mathcal{A}_2^3\,\mathcal{A}_{21}^{21} - \mathcal{A}_2^2\,\mathcal{A}_{21}^{31} + \mathcal{A}_2^1\,\mathcal{A}_{21}^{32}$$
$$\mathcal{A}_1^3\,\mathcal{A}_{21}^{21} - \mathcal{A}_1^2\,\mathcal{A}_{21}^{31} + \mathcal{A}_1^1\,\mathcal{A}_{21}^{32}$$

The third to ninths subgroups contains a variety of symmetry-based identities.

The tenth subgroup contains a relatively complicated identity that has nine summands.

```
In[25]:= subGroups3[[-1, 1]]
```
Out[25]= $-\left(\mathcal{A}_3^3\right)^2 \mathcal{A}_{21}^{32}\,\mathcal{A}_{31}^{31}\,\mathcal{A}_{21}^{21} + \left(\mathcal{A}_3^3\right)^2 \mathcal{A}_{21}^{31}\,\mathcal{A}_{31}^{32}\,\mathcal{A}_{32}^{21} + \left(\mathcal{A}_3^3\right)^2 \mathcal{A}_{21}^{32}\,\mathcal{A}_{21}^{21}\,\mathcal{A}_{31}^{31} - \left(\mathcal{A}_3^3\right)^2 \mathcal{A}_{21}^{21}\,\mathcal{A}_{31}^{32}\,\mathcal{A}_{32}^{31} -$
$\left(\mathcal{A}_{31}^{32}\right)^2 \left(\mathcal{A}_{32}^{31}\right)^2 - \left(\mathcal{A}_3^3\right)^2 \mathcal{A}_{21}^{31}\,\mathcal{A}_{31}^{21}\,\mathcal{A}_{32}^{32} + \left(\mathcal{A}_3^3\right)^2 \mathcal{A}_{21}^{21}\,\mathcal{A}_{31}^{31}\,\mathcal{A}_{32}^{32} + 2\,\mathcal{A}_{31}^{31}\,\mathcal{A}_{31}^{32}\,\mathcal{A}_{32}^{31}\,\mathcal{A}_{32}^{32} - \left(\mathcal{A}_{31}^{31}\right)^2 \left(\mathcal{A}_{32}^{32}\right)^2$

The fourth and last group contains the Laplace-expansion of the 3×3 determinant through three 2×2 determinants.

```
In[26]:= groups[[4]]
```
Out[26]= $\{\mathcal{A} - \mathcal{A}_3^3\,\mathcal{A}_{21}^{21} + \mathcal{A}_3^2\,\mathcal{A}_{21}^{31} - \mathcal{A}_3^1\,\mathcal{A}_{21}^{32}\}$

We could now continue to generate more identities. If we eliminate all determinants of the 2×2 matrices, we get the expansion of the 3×3 determinant through the nine matrix elements.

```
In[27]:= GroebnerBasis[gb, {}, Union[Cases[gb, 𝒜[{_, _}, {_, _}], Infinity]]]
```
Out[27]= $\{-\mathcal{A} - \mathcal{A}_1^3\,\mathcal{A}_2^2\,\mathcal{A}_3^1 + \mathcal{A}_1^2\,\mathcal{A}_2^3\,\mathcal{A}_3^1 + \mathcal{A}_1^3\,\mathcal{A}_2^1\,\mathcal{A}_3^2 - \mathcal{A}_1^1\,\mathcal{A}_2^3\,\mathcal{A}_3^2 - \mathcal{A}_1^2\,\mathcal{A}_2^1\,\mathcal{A}_3^3 + \mathcal{A}_1^1\,\mathcal{A}_2^2\,\mathcal{A}_3^3\}$

If we eliminate all determinants of the 1×1 matrices (meaning the matrix elements themselves), we get an expansion of the 3×3 determinant through then determinants of 2×2 matrices.

```
In[28]:= GroebnerBasis[gb, {}, Union[Cases[gb, 𝒜[{_}, {_}], Infinity]]] //
              Select[#, (Head[#] === Plus && Length[#] > 2)&]&
```
Out[28]= $\{-\mathcal{A}^2 - \mathcal{A}_{12}^{23}\,\mathcal{A}_{13}^{13}\,\mathcal{A}_{23}^{12} + \mathcal{A}_{13}^{13}\,\mathcal{A}_{13}^{23}\,\mathcal{A}_{23}^{12} + \mathcal{A}_{13}^{23}\,\mathcal{A}_{13}^{12}\,\mathcal{A}_{23}^{13} - \mathcal{A}_{13}^{12}\,\mathcal{A}_{13}^{23}\,\mathcal{A}_{23}^{13} - \mathcal{A}_{13}^{13}\,\mathcal{A}_{13}^{12}\,\mathcal{A}_{23}^{23} + \mathcal{A}_{13}^{12}\,\mathcal{A}_{13}^{13}\,\mathcal{A}_{23}^{23}\}$

By selecting various 1×1 and 2×2 matrices, we can generate many more identities. Here is a Lewis–Carroll identity [904].

```
In[29]:= all𝒜s = Cases[eqs, _𝒜, Infinity] // Union;
```

```
In[30]:= GroebnerBasis[gb, #, Complement[all𝒜s, #]]& @
            {𝒜[{1, 3}, {1, 3}], 𝒜[{1, 2}, {1, 2}],
             𝒜[{1, 3}, {1, 2}], 𝒜[{1, 2}, {1, 3}], 𝒜[{1}, {1}], 𝒜[]}
```
Out[30]= $\{-\mathcal{A}\,\mathcal{A}_1^1 - \mathcal{A}_{12}^{13}\,\mathcal{A}_{13}^{12} + \mathcal{A}_{12}^{12}\,\mathcal{A}_{13}^{13}\}$

By selecting a random combination of determinants, we obtain a variety of representations for the 3×3 determinant.

```
In[31]:= selectGoodIdentity[l_List] :=
            Select[l, (Head[#] === Plus && MemberQ[#, 𝒜[], Infinity])&]
```

```
In[32]:= SeedRandom[111];
        Table[While[(gI = selectGoodIdentity[
                GroebnerBasis[gb, #, Complement[all𝒜s, #]]& @
                  all𝒜s[[Join[{1}, Table[Random[Integer, {2, 91}],
                    {12}]]]]]) === {}]; gI, {5}]
```
$$\text{Out[33]= } \{\{\mathcal{A}\,\mathcal{A}_2^3 + \mathcal{A}_{12}^{32}\,\mathcal{A}_{23}^{31} + \mathcal{A}_{12}^{13}\,\mathcal{A}_{23}^{32}\}, \{\mathcal{A}\,\mathcal{A}_3^1 - \mathcal{A}_{13}^{31}\,\mathcal{A}_{23}^{12} + \mathcal{A}_{13}^{21}\,\mathcal{A}_{32}^{31}\},$$
$$\{\mathcal{A}\,\mathcal{A}_3^1 - \mathcal{A}_{13}^{31}\,\mathcal{A}_{23}^{12} - \mathcal{A}_{23}^{31}\,\mathcal{A}_{31}^{12}\}, \{\mathcal{A}\,\mathcal{A}_2^3 - \mathcal{A}_{12}^{32}\,\mathcal{A}_{23}^{12} + \mathcal{A}_{12}^{12}\,\mathcal{A}_{23}^{32}\}, \{\mathcal{A}\,\mathcal{A}_2^3 + \mathcal{A}_{12}^{32}\,\mathcal{A}_{23}^{31} - \mathcal{A}_{12}^{31}\,\mathcal{A}_{32}^{23}\}\}$$

15. $1 + 2 + 3 + \cdots = -1/12$, Casimir Effect

a) We first try to use Abel and Hölder summation [61]. Because of the relatively strong singularity, these methods fail. (For the generalized Cesaro summation of this sum, see [1666].)

```
In[1]:= Sum[k x^k, {k, Infinity}] // Apart
```
$$\text{Out[1]= } \frac{1}{(-1 + x)^2} + \frac{1}{-1 + x}$$

```
In[2]:= b[n_] = Sum[k, {k, n}];
        Sum[b[k], {k, n}]/(n + 1)
```
$$\text{Out[3]= } \frac{1}{6}\, n\,(2 + n)$$

A standard way to deal with divergent sums and integrals is to add exponential terms that give rise to convergence.

```
In[4]:= Sum[k Exp[-ε k], {k, Infinity}]
```
$$\text{Out[4]= } \frac{e^\varepsilon}{(-1 + e^\varepsilon)^2}$$

Letting now ε go to 0, we have the following behavior.

```
In[5]:= Series[%, {ε, 0, 2}]
```
$$\text{Out[5]= } \frac{1}{\varepsilon^2} - \frac{1}{12} + \frac{\varepsilon^2}{240} + O[\varepsilon]^3$$

We subtract the integer order quadratic divergence and obtain $-1/12$, a result first obtained by Euler [976]. Another way to obtain this result is to get convergence by summing n^{-s} and then letting $s \to -1$ [1556]. Again, we obtain the result $-1/12$ [1028], [595], [573].

```
In[6]:= Sum[k^-s, {k, Infinity}] /. s -> -1
```
$$\text{Out[6]= } -\frac{1}{12}$$

For a detailed discussion why the last two approaches yield the same result, see [1693], [1694], [1695]. For corresponding multidimensional sums, see [552]. For Ramanujan's derivation of this result, see [157], [459].

Unfortunately, *Mathematica*'s Sum function does not carry out this removal of infinities automatically, and currently there is also no option to allow for this.

b) We start by calculating the sum and the integral with finite upper limits [1188].

```
In[1]:= sum = Sum[n^3 Exp[-λ n], {n, 0, m}]
```
$$\text{Out[1]= } \frac{e^{-3\lambda} + 4\,e^{-2\lambda} + e^{-\lambda}}{(1 - e^{-\lambda})^4} - $$
$$(e^{-\lambda})^{1+m} \left(\frac{3\,e^{-\lambda}}{(1 - e^{-\lambda})^2} + \frac{3\,(e^{-2\lambda} + e^{-\lambda})}{(1 - e^{-\lambda})^3} + \frac{e^{-3\lambda} + 4\,e^{-2\lambda} + e^{-\lambda}}{(1 - e^{-\lambda})^4} + \frac{e^\lambda}{-1 + e^\lambda} + \frac{6\,e^\lambda\,m}{(-1 + e^\lambda)^2} + \right.$$
$$\left. \frac{3\,e^\lambda\,m}{-1 + e^\lambda} + \frac{3\,e^\lambda\,(1 + e^\lambda)\,m}{(-1 + e^\lambda)^3} + \frac{3\,e^\lambda\,m^2}{(-1 + e^\lambda)^2} + \frac{3\,e^\lambda\,m^2}{-1 + e^\lambda} + \frac{e^\lambda\,m^3}{-1 + e^\lambda} \right)$$

```
In[2]:= integral = Integrate[n^3 Exp[-λ n], {n, 0, m}]
```
$$\text{Out[2]= } \frac{6 - e^{-m\lambda}\,(6 + 6\,m\,\lambda + 3\,m^2\,\lambda^2 + m^3\,\lambda^3)}{\lambda^4}$$

We simplify the result.

```
In[3]:= Short[difference = sum - integral // Together // PowerExpand, 6]
```

Out[3]//Short= $\dfrac{1}{(-1 + e^\lambda)^4\ \lambda^4}$

$(e^{-m\lambda}\ (6 - 24\ e^\lambda + 36\ e^{2\lambda} - 24\ e^{3\lambda} + 6\ e^{4\lambda} - 6\ e^{m\lambda} + 24\ e^{\lambda+m\lambda} - 36\ e^{2\lambda+m\lambda} + 24\ e^{3\lambda+m\lambda} - 6\ e^{4\lambda+m\lambda} +$

$\ll 29\gg\ + e^{3\lambda+m\lambda}\ \lambda^4 + 3\ e^\lambda\ m\ \lambda^4 - 3\ e^{3\lambda}\ m\ \lambda^4 - 3\ e^\lambda\ m^2\ \lambda^4 + 6\ e^{2\lambda}\ m^2\ \lambda^4 -$

$3\ e^{3\lambda}\ m^2\ \lambda^4 + m^3\ \lambda^4 - 3\ e^\lambda\ m^3\ \lambda^4 + 3\ e^{2\lambda}\ m^3\ \lambda^4 - e^{3\lambda}\ m^3\ \lambda^4))$

All we now have to do is apply taking limit twice, and we are done (the order matters in this example). The first limit is trivial to do by hand (the terms e^{-ml} vanish faster than does any power of m for $m \to \infty$).

In[4]:= **Limit[difference, m -> Infinity, Assumptions -> λ > 0]**

Out[4]= $\dfrac{-6 - 6\ e^{4\lambda} + 4\ e^{2\lambda}\ (-9 + \lambda^4) + e^\lambda\ (24 + \lambda^4) + e^{3\lambda}\ (24 + \lambda^4)}{(-1 + e^\lambda)^4\ \lambda^4}$

And the second limit yields the result we are looking for: $1/120$.

In[5]:= **Series[%, {λ, 0, 2}]**

Out[5]= $\dfrac{1}{120} - \dfrac{\lambda^2}{504} + O[\lambda]^3$

Taking the limits in the opposite order would have resulted in a nonconvergent result.

In[6]:= **Series[difference, {λ, 0, 0}] // Simplify**

Out[6]= $\dfrac{1}{4}\ m^2\ (1 + 2\ m) + O[\lambda]^1$

Another possibility would have been to use the Abel–Plana formula [1541], [1506], [894] for the first summation/integration step.

$$\sum_{n=0}^{\infty} f(n) - \int_0^\infty f(x)\,dx = \frac{f(0)}{2} + i \int_0^\infty \frac{f(it) - f(-it)}{e^{2\pi t} - 1}\,dt$$

In[7]:= **f[ξ_] := ξ^3 Exp[-λ ξ]**

        1/2 f[0] + I Integrate[(f[I t] - f[-I t])/(Exp[2 Pi t] - 1),
                                 {t, 0, Infinity},
                                 Assumptions -> (Im[λ] == 0)] // Simplify

Out[8]= $\dfrac{-48 + \lambda^4\ (2 + \text{Cosh}[\lambda])\ \text{Csch}[\frac{\lambda}{2}]^4}{8\ \lambda^4}$

In[9]:= **Series[%, {λ, 0, 3}]**

Out[9]= $\dfrac{1}{120} - \dfrac{\lambda^2}{504} + O[\lambda]^4$

Still another possibility is the use of the extended Poisson summation formula [543]. (Here $\mathcal{F}_n[f(n)]\,(v)$ denotes the Fourier transform of $f(n)$)

$$\sum_{n=-\infty}^{\infty} f(n) - \int_{-\infty}^\infty f(x)\,dx = -\sqrt{2\pi}\ \mathcal{F}_n[f(n)]\,(0) + \sum_{v=-\infty}^{\infty} \sqrt{2\pi}\ \mathcal{F}_n[f(n)]\,(v).$$

To obtain a sum running from $-\infty$ to ∞, we extend the summands symmetrically for negative n. Proceeding in a similar way for the integral, we obtain the following function f.

In[10]:= **f[ξ_] := 1/2 Sqrt[ξ^2]^3 Exp[-λ Sqrt[ξ^2]]**

It is straightforward to implement the above Poisson formula.

In[11]:= **F[v_] = Sqrt[2 Pi] FourierTransform[f[n], n, v]**

Out[11]= $\dfrac{6\ (\lambda^4 - 6\ \lambda^2\ v^2 + v^4)}{(\lambda^2 + v^2)^4}$

In[12]:= **Sum[F[2 Pi v], {v, -Infinity, Infinity}] - F[0]**

Out[12]= $-\dfrac{6}{\lambda^4} + \dfrac{1}{8}\ \left(2\ \text{Coth}\left[\frac{\lambda}{2}\right]^2\ \text{Csch}\left[\frac{\lambda}{2}\right]^2 + \text{Csch}\left[\frac{\lambda}{2}\right]^4\right)$

In the limit $\lambda \to 0$, we again obtain the value $1/120$.

In[13]:= **Series[%, {λ, 0, 1}]**

Out[13]= $\frac{1}{120} + O[\lambda]^2$

For a Zeta function approach to calculate such differences, see [576]. For a review of methods to calculate $\int_0^\infty f(x)\,dx - \sum_{k=0}^\infty f(n)$, see [125]; for multidimensional examples, see [1605], [1606];

Note that the above-calculated difference between the sum and the integral plays an important role in the quantum electrodynamical explanation of the Casimir effect; see, for instance, [1108], [471], [203], [833], [1200], [1419], [1270], [606], [511], [1694], [575], [1263], [1117], [202], [1096], and [1299].

In[14]:=

c) This is the expression to be summed and integrated.

In[1]:= **Σ = Log[Sqrt[Pi^2 n^2/1^2   + kx^2 + ky^2]/**
               Sqrt[Pi^2 n^2/(L/2)^2   + kx^2 + ky^2]] +
         Log[Sqrt[Pi^2 n^2/(L - 1)^2 + kx^2 + ky^2]/
               Sqrt[Pi^2 n^2/(L/2)^2   + kx^2 + ky^2]];

Not unexpectedly, *Mathematica* cannot evaluate the sum and the double integrals directly.

In[2]:= **Integrate[Sum[Evaluate[Σ], {n, -Infinity, Infinity}],**
                 {kx, -Infinity, Infinity}, {kx, -Infinity, Infinity},
                 GenerateConditions -> False]

     Sum::div : Sum does not converge. More…

     Sum::div : Sum does not converge. More…

Out[2]= $\text{Integrate}\left[\sum_{n=-\infty}^{\infty} \left(\text{Log}\left[\frac{\sqrt{kx^2 + ky^2 + \frac{n^2 \pi^2}{1^2}}}{\sqrt{kx^2 + ky^2 + \frac{4 n^2 \pi^2}{L^2}}} \right] + \text{Log}\left[\frac{\sqrt{kx^2 + ky^2 + \frac{n^2 \pi^2}{(-1+L)^2}}}{\sqrt{kx^2 + ky^2 + \frac{4 n^2 \pi^2}{L^2}}} \right] \right), \right.$

$\left. \{kx, -\infty, \infty\}, \{kx, -\infty, \infty\}, \text{GenerateConditions} \rightarrow \text{False} \right]$

Because we are dealing with divergent integrals, we take the liberty to exchange the integrals and the sum. To simplify the integrand we first simplify the arguments of the logarithms taking into account they are always positive. Performing the resulting integration over the k_x, k_y-plane in polar coordinates using the GenerateConditions -> False option setting to get regularized results for divergent integrals gives a nice result containing only integer powers and logarithms of n.

In[3]:= **int = Integrate[Pi (PowerExpand[Σ] /. kx^2 + ky^2 -> κ),**
                 {κ, 0, Infinity}, GenerateConditions -> False]

Out[3]= $\frac{1}{2} n^2 \pi^3 \left(\frac{1 + \text{Log}\left[\frac{1^2}{n^2 \pi^2}\right]}{1^2} - \frac{8 \left(1 + \text{Log}\left[\frac{L^2}{4 n^2 \pi^2}\right]\right)}{L^2} + \frac{1 + \text{Log}\left[\frac{(1-L)^2}{n^2}\right]}{(1 - L)^2} - 2\,\text{Log}[\pi] \right)$

We can understand the finite result for this divergent integral by considering $\partial(a+\kappa)^\alpha / \partial\alpha = (a+\kappa)^\alpha \ln((a+\kappa))$ for $\alpha = 0$.

In[4]:= **Series[D[Integrate[(κ + a)^α, {κ, 0, ∞},**
                 GenerateConditions -> False], α], {α, 0, 2}]

Out[4]= $(a - a\,\text{Log}[a]) + (-2 a + 2 a\,\text{Log}[a] - a\,\text{Log}[a]^2)\,\alpha +$

$\left(3 a - 3 a\,\text{Log}[a] + \frac{3}{2} a\,\text{Log}[a]^2 - \frac{1}{2} a\,\text{Log}[a]^3\right) \alpha^2 + O[\alpha]^3$

This means that we use $\int_0^\infty \ln(\kappa + a)\,d\kappa = a(1 - \ln(a))$.

In[5]:= **intH = Pi (PowerExpand[Σ] /. kx^2 + ky^2 -> κ) /.**
                       Log[a_ + κ] :> a (1 + Log[1/a]);

In[6]:= **Simplify[int - intH]**

Out[6]= 0

(Using a convergence-generating factor such as $\exp(-\varepsilon\,\kappa)$, then make a series expansion around $\kappa = 0$ and ignore algebraically and logarithmically diverging terms gives a similar result. The difference $a\gamma$ will disappear after the following summation.)

In[7]:= **ser = Integrate[Log[a + κ] Exp[-ε κ], {κ, 0, Infinity},**
                 Assumptions -> Re[a] > 0 && Re[ε] > 0]

Out[7]= $\dfrac{e^{a\,\varepsilon}\,\text{Gamma}[0,\,a\,\varepsilon] - \text{Log}[\varepsilon] + \text{Log}[a\,\varepsilon]}{\varepsilon}$

In[8]:= `Assuming[a > 0 && ε > 0, Series[%, {ε, 0, 1}]] /. _Arg -> 0 // Simplify`

Out[8]= $\dfrac{-\text{EulerGamma} - \text{Log}[\varepsilon]}{\varepsilon} - a\,(-1 + \text{EulerGamma} + \text{Log}[a] + \text{Log}[\varepsilon]) + O[\varepsilon]^1$

Using the last result in the sum over n yields a divergent sum. We change the double infinite sum to a single infinite sum by observing that int depends only on n^2 and the $n = 0$ term does not contribute to the finite result. We carry out the summation in a zeta-regularization sense. The sums of pure integer powers of n vanish and the product term $n^2 \ln(n)$ gives a nonvanishing contribution.

In[9]:= `Series[PowerExpand[int], {n, 0, 1}]`

Out[9]= $O[n]^2$

In[10]:= `res = Expand[PowerExpand[int]] /.`
`        n^2 Log[n] -> Zeta[3]/(4 Pi^2) /. n -> 0`

Out[10]= $-\dfrac{\pi\,\text{Zeta}[3]}{4\,1^2} - \dfrac{\pi\,\text{Zeta}[3]}{4\,(1-L)^2} + \dfrac{2\,\pi\,\text{Zeta}[3]}{L^2}$

We end by writing the last result in a symmetric form.

In[11]:= `-4 Pi Zeta[3] Apart[-1 /(Pi Zeta[3]/4) Factor[res], l]`

Out[11]= $-4\left(\dfrac{1}{1^2} + \dfrac{1}{(1-L)^2} - \dfrac{8}{L^2}\right)\pi\,\text{Zeta}[3]$

As a quick explanation for the result $\sum_{k=1}^{\infty} k^2 \ln(k) = \zeta(3)/(4\,\pi^2)$, we proceed in a way similar to part a) of this exercise. Taking into account that $\partial n^{\alpha}/\partial\alpha = \ln(n)\,n^{\alpha}$, we immediately obtain the desired result.

In[12]:= `D[Sum[n^α, {n, Infinity}], α] /. α -> 2`

Out[12]= $\dfrac{\text{Zeta}[3]}{4\,\pi^2}$

16. Random Functions, Use of Numerical Techniques

a) One difficulty with this problem is that it is not so easy to define what constitutes a "random function". Here is a possible heuristic definition: RandomFunction[n] should be a function of depth "about" n, which makes use of randomly selected functions of a given class of functions. For our random univariate functions, we assume the independent variable is always x.

We now give a possible implementation: the functions involved can have several arguments. Here, we restrict ourselves to functions with one, two, or three arguments. The lists are arbitrarily extendable in number and size (in particular, for more than three arguments, Times and Plus are appropriate). Here, we avoid special functions and instead just use elementary functions.

```
In[1]:= functionWithiArguments[1] := #[[Random[Integer, {1, Length[#]}]]]&[
                    (* we use only elementary functions here *)
                    {Sin, Cos, Tan, Cot, ArcSin, ArcCos, Exp, Log,
                     Sqrt, Log, Minus, Sinh, Tanh, Power[#, 1/3]& (* ... *)}]

        functionWithiArguments[2] := #[[Random[Integer, {1, Length[#]}]]]&[
                    {Power, Plus, Times, Log, Divide
                     (* plus many special functions *)}]

        functionWithiArguments[3] :=
          #[[Random[Integer, {1, Length[#]}]]]&[{Plus, Times}]

        (* expandable:
             functionWithiArguments[4] := ... *)
```

The list of random functions should also include constants. To be able to combine these more easily later, we do not use the built-in command Unique directly, but instead employ the following modified version.

In[5]:= `myUnique := c[Unique[my]]`

The expression myUnique produces the previously unused constant of the form c[my$*number*]. Here are some examples.

In[6]:= `Table[myUnique, {20}]`

Out[6]= {c[my$16], c[my$17], c[my$18], c[my$19], c[my$20], c[my$21],
    c[my$22], c[my$23], c[my$24], c[my$25], c[my$26], c[my$27], c[my$28],
    c[my$29], c[my$30], c[my$31], c[my$32], c[my$33], c[my$34], c[my$35]}

In order to get an arbitrary ordering of the arguments of the various types and x, we permute all possible arguments. Because we only need one permutation, we implement `aPermutation` instead of the (for many arguments much, much slower) built-in `Permutations`.

```
In[7]:= aPermutation[l_List] :=
        (* take random elements from a list until this list is empty *)
        Nest[Function[a, {Append[a[[1]], a[[2, #]]],
        Drop[a[[2]], {#}]}&[Random[Integer, {1, Length[a[[2]]]}]]],
                          {{}, l}, Length[l]][[1]]
```

Here are two examples.

```
In[8]:= aPermutation[{1, 2, 3, 4, 5, 6, 7, 8, 9, 10, 11, 12}]
```

Out[8]= {5, 3, 7, 6, 12, 10, 9, 2, 8, 11, 4, 1}

```
In[9]:= aPermutation[{x, x, x, x, c[1], c[2], c[3], c[4]}]
```

Out[9]= {c[3], c[4], x, x, c[1], c[2], x, x}

We now proceed to the actual implementation: a random function of depth 2 is just one of the above functions in `function WithiArguments` with appropriate arguments. We must be careful to make sure that at least one x appears in the random function.

```
In[10]:= randomFunction[2] :=
         Function[m, functionWithiArguments[m][
         Sequence @@ (* mix *) aPermutation[Flatten[(* the elements *)
         {Table[x, {#}], Table[myUnique, {m - #}]}]]]&[
                      Random[Integer, {1, m}]]][Random[Integer, {1, 3}]]
```

Here is again an example.

```
In[11]:= Table[randomFunction[2], {12}]
```

Out[11]= $\left\{x^2\, c[my\$36],\ x^{1/3},\ \dfrac{Log[x]}{Log[c[my\$37]]},\ 3\,x,\ x\,c[my\$38]\,c[my\$39],\right.$
$\left.Tan[x],\ Log[x],\ x^2,\ 1,\ ArcSin[x],\ Log[x],\ x^{c[my\$40]}\right\}$

We define a random function of depth n recursively as a random function that has at least one argument of depth $n - 1$.

```
In[12]:= randomFunction[n_Integer?(# >= 3&)] :=
         Function[m, functionWithiArguments[m][
         Sequence @@ (* mix *) aPermutation[Flatten[
         (* the elements *)
         {Table[(* call _one_ level lower *)
                  randomFunction[n - 1], {#}],
           Table[(* call _any_ lower level *)
                  randomFunction[Random[Integer, {2, n - 1}]],
                  {m - #}]}]]&[Random[Integer, {1, m}]]][
                             Random[Integer, {1, 3}]]
```

In principle, this definition should already work.

```
In[13]:= Table[randomFunction[3], {6}]
```

Out[13]= $\left\{x\,c[my\$41] + Cos[x] + Cot[x],\right.$
$x^{c[my\$42]} + \dfrac{c[my\$43]}{x} + x^2\,c[my\$44],\ x^3\,c[my\$45]\,c[my\$46]\,c[my\$47]\,Sin[x],$
$\left.1 + 2\,x + x^x + c[my\$48],\ 2\,ArcCos[x] + Sinh[x],\ \dfrac{Log[Tan[x]]}{Log[x + c[my\$49]]}\right\}$

We now refine it somewhat. First, functions of constants are again constants (here, we make use of the structure of constants generated by the above `myUnique` function). The following simplification rules for x independent parts cover most but not all cases. On the other hand, we do not want to wait too long for a result.

```
In[14]:= (* f[constants] -> newConstant *)
         simpRule = {f_[__c] :> myUnique, _?NumericQ*(_c)^_. :> myUnique};
```

Secondly, in this form, the constants do not look especially nice; c1, c2, ... , looks better than c[my$*number*]. So, we can implement the function `randomFunctionPre`.

```
In[16]:= randomFunctionPre[n_] :=
         Function[p, p //. ((Rule @@ #)& /@ Transpose[
           (* replace c[...] by ci for shortness *)
           {#, Table[ToExpression["c" <> ToString[i]],
               {i, Length[#]}]}]]&[Cases[p, c[_], {-2}]]][
             randomFunction[n] //. simpRule]

         Table[randomFunctionPre[3], {6}]
```

$$\text{Out[17]}= \left\{2\,x\,\text{Tan}[x], \frac{\text{Log}[\frac{\text{Log}[x]}{c1}]}{\text{Log}[\text{Cot}[x]]}, \frac{c1+x}{2\,x}, 3\,x\,(c1+2\,x),\right.$$
$$\left.\text{ArcCos}[\text{Tanh}[x]], (c1+c2+x)\,(c3+c4+x)\,\text{ArcCos}[x]\right\}$$

Complications can and still do arise. Because of the completely independent choice of functions, the result can turn out to be *x*-independent, we could be dividing by 0, and so on. We check for the most common of these problems, and keep calling RandomFunctionH until a satisfactory result is obtained. We do not insist on an exact depth of *n*, because not all functions collected in `functionWithiArguments` have the same depth because some of them are composite. This leads us finally the following definition.

```
In[18]:= RandomFunction[n_Integer?(# >= 2&)] :=
         Module[{aux, test1, test2, test3, test4},
             While[Check[ (* until we have a "good" result *)
                 aux = TimeConstrained[randomFunctionPre[n],
                     (* possibly increase for large n *) 10 n], " error "];
                 test1 = !(And @@ (* might be 0/0 or related
                                    objects were generated *)
                     (FreeQ[aux, #, {0, Infinity}]& /@
                     {Indeterminate, DirectedInfinity}));
                     (* the variable x should be present *)
                 test2 = !MemberQ[aux, x, {0, Infinity}];
                     (* may be do here: test3 = Depth[aux] =!= n; *)
                 test4 = aux === $Aborted;
                 test5 = aux === " error ";
                 !(test1 == False && test2 == False &&
                     (* test3 == False && *)  test4 == False &&
                     test5 == False), Null];
             aux]
```

We now look at a few examples of random functions of various depths.

```
In[19]:= Table[RandomFunction[2], {12}]
```

$$\text{Out[19]}= \left\{\frac{c1}{x}, e^x, c1+c2+x, 2\,x, c1^x, \frac{\text{Log}[x]}{c1}, x^2, \frac{c1}{\text{Log}[x]}, \text{Tan}[x], \text{ArcSin}[x], c1\,x^2, x^2\right\}$$

```
In[20]:= Table[RandomFunction[3], {12}]
```

$$\text{Out[20]}= \left\{\text{Tan}[x^x], c1\,x^2, \sqrt{\text{Cot}[x]}, \frac{c1\,\text{ArcSin}[x]}{x}, \frac{\text{Log}[3\,x]}{\text{Log}[x^3]}, e^{x^3},\right.$$
$$\left.c1\,x\,\text{Cos}[x], \text{ArcCos}[2\,x], \text{Cot}[x^3], c1\,c2\,x^2, \text{Tan}[\text{Sin}[x]], \text{Tanh}[x]^{1/3}\right\}$$

```
In[21]:= Table[RandomFunction[4], {12}]
```

$$\text{Out[21]}= \left\{1+\text{Log}[\text{Tan}[x]]+\text{Sin}[x]^{c1\,x^2}, \text{Cos}[c1\,x]^{1/3}, 1-\sqrt{x}, \text{ArcSin}[\text{Log}[\text{Cot}[x]]],\right.$$
$$\frac{x^{1/3}\,(x^{c1}+\text{Log}[x]+\frac{\text{Log}[x]}{c2})\,(e^x+\text{Sin}[x]+\text{Tanh}[x])}{\text{Log}[x]}, \text{Sinh}\left[x^{\frac{1}{3}+c1}\,(c2+2\,x)\right],$$
$$\frac{c2\,\text{Sin}[c3+x]}{c1\,x^2}, \frac{x^2}{3}+\frac{\text{ArcSin}[x]}{c1+2\,x}+\frac{\text{Log}[x]}{c2}, x^3\,\text{Cot}[1]\,\text{Tan}\left[\frac{\text{Log}[x]}{c1}\right],$$
$$\left.\sqrt{x^2}\,\text{Log}[x]^{1/3}, (x^x)^{c1\,c2\,x^3}, c1\,c2\,c3\,x^4\,\text{ArcCos}[x]^2\,\text{Cos}[3\,x]\,\text{Cot}[x]\right\}$$

```
In[22]:= Table[RandomFunction[5], {6}]
```

Out[22]= $\left\{ \sqrt{x^{1/3} + c1\ x^2}\ \text{Log}[2\ x^{2/3}]\ \sqrt{\dfrac{\text{Log}[x^{1/3}]}{\text{Log}[\text{ArcCos}[x]]}},\ \left(\dfrac{1 + x^{1/3} + \text{Cot}[x]}{(c1 + 2\ x)^{1/3}}\right)^{1/3},\right.$

$\quad \text{Coth}[\text{Log}[x]\ \text{Tan}[x]]\ \text{Sin}[x]^{c1+x},\ -\text{Tanh}\left[\sqrt{x} + c1\ c2\ x^{1+x}\ \text{ArcCos}[x] - \dfrac{\text{Log}[x^{1/3}]}{\text{Log}[c3 + c4 + x]}\right],$

$\quad \text{Cot}[(c1\ c2\ x)^{-x}]\left(\text{ArcCos}[x^x] + \dfrac{\text{Sin}[x]}{\text{ArcSin}[x]} + \text{Tan}[x]\right),$

$\quad \left. (c1^x + c2 + 2\ x + c3\ c4\ c5\ x^3)\ \text{Log}\left[\dfrac{c7 + c8 + x}{c6 + 2\ x}\right]\ \text{Sin}[x]^{3x}\ \text{Sinh}[x^x]\right\}$

In[23]:= RandomFunction[6]

Out[23]= $2 + c1 + c2 + c3 + 5\ x + x^3 + \text{ArcCos}[\text{Log}[x]] + \text{Cos}[x] + \text{Cos}[x^2\ (c4 + c5 + x)\ \text{Log}[x]] +$

$\quad (c6 + c7 + x)\ (3\ x + \text{Cot}[x]^{1/3} + \text{Log}[x^2]) + \text{Log}[(c8 + c9 + x)\ (c10 + 2\ x)] + \dfrac{c11\ x^3\ \text{Tanh}[x]}{\text{Log}[x]}$

We can use RandomFunction to test *Mathematica*'s integration capabilities by first integrating the random functions, and then differentiating them and comparing with the originals.

```
In[24]:= IntegrateTest[n_] :=
    Module[{v = RandomFunction[n], int},
            int = Integrate[v, x];
            If[FreeQ[int, Integrate, {0, Infinity}, Heads -> True],
                If[(* hopefully, Simplify can do it *)
                    Simplify[D[int, x] - v] =!= 0, v, Null], Null]]
```

```
In[25]:=
    SeedRandom[111];
    Do[If[# =!= Null, Print[#]]&[IntegrateTest[3]], {10}] // Timing
Out[26]= {1.25 Second, Null}
```

Of course, we could first differentiate the random functions and then integrate. (Here, additive constants may be lost, as, for instance, in the case Integrate[D[Log[2x], x], x] – Log[2x].)

```
In[27]:= IntegrateTest2[n_] :=
    Module[{v = RandomFunction[n], d, int},
            d = D[v, x];
            int = Integrate[d, x];
            If[MemberQ[int, Integrate, {0, Infinity}, Heads -> True],
                (* Risch should do it *)
                Print["Antiderivative of the complete differential",
                        d, " not found"],
                (* tricky case or a bug? *)
                If[MemberQ[Simplify[int - v], x, {0, Infinity}],
                    Print["Equality cannot be established: ", v, " ⟷ ", int],
                    Null]]]
```

Because indefinite integration is correct modulo a differential-algebraic constant, the branch cut structure of some functions $f(x)$ might change after carrying out $\frac{\partial}{\partial x} \int^x f(x)\,dx$. Here, we find such an example.

```
In[28]:= SeedRandom[175068687549];
    Do[IntegrateTest2[3], {1}]
    Equality cannot be established: Log[x³] ⟷ 3 Log[x]
```

To get "positive" results with these tests, we have to look at several thousand examples for $n = 3$. (We found an example quickly because we used a "lucky" seed for the random number generator.)

Now let us use $n = 5$ to find an example for a doable integral that will not be done by *Mathematica*. We count the number of tries needed to find such an integrand.

```
In[30]:= SeedRandom[7777777];
    Module[{counter, v, d, int},
            counter = 0;
            While[counter = counter + 1;
                    v = RandomFunction[4];
                    (* differentiation *)
                    d = D[v, x];
                    (* integration *)
                    int = Integrate[d, x];
```

$$\texttt{FreeQ[int, Integrate, \{0, Infinity\}, Heads -> True], Null];}$$
$$\texttt{\{counter, v\}]}$$

Out[31]= $\left\{ 6, \ 2^{\texttt{c1} \texttt{x}^2 \texttt{ ArcSin[x] Tanh[x]}} \ \texttt{x}^{\texttt{c1} \texttt{x}^2 \texttt{ ArcSin[x] Tanh[x]}} \right\}$

Often such type integrands contain radicals.

In[32]:= $\texttt{D[\% [[2]], x]}$

Out[32]= $2^{\texttt{c1} \texttt{x}^2 \texttt{ ArcSin[x] Tanh[x]}} \ \texttt{x}^{\texttt{c1} \texttt{x}^2 \texttt{ ArcSin[x] Tanh[x]}} \ \texttt{Log[2]}$

$$\left(\texttt{c1 } \texttt{x}^2 \texttt{ ArcSin[x] Sech[x]}^2 + \frac{\texttt{c1 } \texttt{x}^2 \texttt{ Tanh[x]}}{\sqrt{1 - \texttt{x}^2}} + 2 \texttt{ c1 x ArcSin[x] Tanh[x]} \right) +$$

$$2^{\texttt{c1} \texttt{x}^2 \texttt{ ArcSin[x] Tanh[x]}} \ \texttt{x}^{\texttt{c1} \texttt{x}^2 \texttt{ ArcSin[x] Tanh[x]}} \ \left(\texttt{c1 x ArcSin[x] Tanh[x]} + \right.$$

$$\left. \texttt{Log[x]} \left(\texttt{c1 } \texttt{x}^2 \texttt{ ArcSin[x] Sech[x]}^2 + \frac{\texttt{c1 } \texttt{x}^2 \texttt{ Tanh[x]}}{\sqrt{1 - \texttt{x}^2}} + 2 \texttt{ c1 x ArcSin[x] Tanh[x]} \right) \right)$$

In[33]:= $\texttt{Integrate[\% [[2]], x]}$

Out[33]= $\displaystyle\int 2^{\texttt{c1} \texttt{x}^2 \texttt{ ArcSin[x] Tanh[x]}} \ \texttt{x}^{\texttt{c1} \texttt{x}^2 \texttt{ ArcSin[x] Tanh[x]}} \ \left(\texttt{c1 x ArcSin[x] Tanh[x]} + \right.$

$$\left. \texttt{Log[x]} \left(\texttt{c1 } \texttt{x}^2 \texttt{ ArcSin[x] Sech[x]}^2 + \frac{\texttt{c1 } \texttt{x}^2 \texttt{ Tanh[x]}}{\sqrt{1 - \texttt{x}^2}} + 2 \texttt{ c1 x ArcSin[x] Tanh[x]} \right) \right) \texttt{dx}$$

The function RandomFunction can also be used to discover functions for which *Mathematica* is not able to find series expansions or limit values, and so on. With an analogous implementation, we can also construct random functions of several variables.

Such techniques are typically very useful for automatic testing for larger pieces of code that need to be robust in all cases.

Using the function randomFunction, we can easily implement the generation of random numeric expressions (say for investigating the uniformity conjecture [1485], [1486]). randomNumericExpression[n, *iMax*] generates a random numeric expression by (about) n-fold composition that has no integers larger than *iMax* (to avoid expressions with huge integers that might arise from repeatedly forming powers of smaller integers).

In[34]:= ```
randomNumericExpressionPre[n_, nMax_] :=
    randomFunction[n] //. (* substitute integers for symbols *)
                    (x | _c) :> Random[Integer, {-nMax, nMax}]
```

In[35]:= ```
randomNumericExpression[n_, iMax_] :=
    Module[{rNe},
    While[(* better reproducibility *) Developer`ClearCache[];
        rNe = MemoryConstrained[TimeConstrained[
                    randomNumericExpressionPre[n, iMax], 100], 10^6];
        (* is it a "good" expression? *)
        MatchQ[rNe, _Integer] || rNe === $Aborted ||
        rNe === Overflow[] || rNe === Underflow[] ||
        Max[Abs[Cases[rNe, _Integer, {-1}]]] > iMax ||
        Max[{Numerator[#], Denominator[#]}& /@
            Abs[Cases[rNe, _Rational, {-1}]]] > iMax ||
        Not[NumericQ[rNe]], Null]; rNe]
```

Here are four examples of random numeric expressions.

In[36]:= ```
Off[General::dbyz]; Off[Infinity::indet]; Off[Power::infy];
Off[Power::indet]; Off[General::ovfl]; Off[General::unfl];
Off[N::meprec];
```

```
Table[randomNumericExpression[8, 10], {4}]
```

Out[39]= $\left\{ \texttt{Cos[Cos[Sinh[Tan[Cot[2]]]]]}, \ \dfrac{\sqrt{-1 - \texttt{Cot[Cot[7]]}} + 2 \texttt{ ArcCos[10] Cos[6] Log[7] + Tan[7]}}{\texttt{ArcCos[-10]}} \right.,$

$$\left. \texttt{Cos[Tanh[ArcSin[ArcSin[Log[-Tan[8]] - Sinh[9]]]]]}, \ \frac{\pi}{2} + \texttt{ArcCos}\left[\frac{1}{\texttt{ArcSin[4]}} \right] \right\}$$

b) To find out if a function uses inexact numbers when evaluating its input, we have to monitor internal evaluation steps.

We could use Trace or On, but these functions produce large outputs and slow down calculations substantially. So, we add additional rules to Real and Complex (with inexact arguments) as well as to N. In case any of these heads is used, we
```

record this and immediately after this, we remove the rule to avoid the slowdown of any further computation. The argument `HoldAll` is needed to avoid the evaluation of the input before `Real`, `Complex`, and `N` are overloaded. `inexactNumber` `UseChecker` returns a list with three elements. The first element is the result of the calculation, the second indicates if inexact numbers were used in this calculation, and the third indicates if explicit calls to `N` were carried out.

```
In[1]:= SetAttributes[inexactNumberUseChecker, HoldAll];

    inexactNumberUseChecker[f_] :=
    Module[{realsWereUsed = False, complexesWereUsed = False,
            NWasCalled = False},
      Unprotect[{N, Real, Complex}];
      (* intercept calls to N *)
      DownValues[N] = {HoldPattern[N[x__]] :> Null /;
                        (NWasCalled = True; Clear[N]; False)};
      (* intercept calls to a function with an approximative argument *)
      Real /: _[___, _Real, ___] := Null /;
                      (realsWereUsed = True; Clear[Real]; False);
      Complex /: _?(# =!= Precision&)[___, c_Complex, ___] := Null /;
            (If[Precision[c] < Infinity,
                complexesWereUsed = True; Clear[Complex]]; False);
      (* return result and if approximative numbers were used
         and if N was called explicitly *)
      res = {f, realsWereUsed || complexesWereUsed, NWasCalled};
      Protect[{N, Real, Complex}]; (* return result *) res]
```

Here are three sample inputs demonstrating that `inexactNumberUseChecker` works correctly.

```
In[3]:= inexactNumberUseChecker[Sin[3.]]
```

Out[3]= {0.14112, True, False}

```
In[4]:= inexactNumberUseChecker[Sin[N @ Pi]]
```

Out[4]= {1.22465×10⁻¹⁶, True, True}

$$\text{Out[4]= } \{1.22465 \times 10^{-16},\ \text{True, True}\}$$

```
In[5]:= inexactNumberUseChecker[Sin[1. + I]]
```

Out[5]= {1.29846 + 0.634964 i, True, False}

Not every possible input to the functions of this chapter will cause the use of inexact numbers. Expressions with "hidden zeros" will more frequently cause numericalization to make certain decisions. So, we will use the zero `zero` in the next examples. Because "hidden zeros" will frequently trigger `N::meprec` messages, we will turn them off.

```
In[6]:= (* the hidden zero *)
    zero = Sqrt[2] + Sqrt[3] - Sqrt[5 + 2 Sqrt[6]];
    Off[N::meprec]; Off[Clear::wrsym];
```

Many symbolic functions are actually using numerical techniques internally. In the following, we will see that at least the `Solve`, `Integrate`, `Factor`, `Roots`, `Root`, `RootSum`, `Sum`, `Series`, `DSolve`, `FourierTransform`, `Laplace` `Transform`, `CylindricalDecomposition` are of such type.

`Solve` uses numerical techniques for verification.

```
In[9]:= inexactNumberUseChecker[Solve[Cos[x] + Sin[x] == Pi, {x}]]
```

> Solve::ifun : Inverse functions are being used by Solve, so some solutions
> may not be found; use Reduce for complete solution information. More…

$$\text{Out[9]= } \left\{\left\{\left\{x \to \text{ArcCos}\left[\frac{\pi}{2} - \frac{1}{2} i \sqrt{-2 + \pi^2}\right]\right\}, \left\{x \to \text{ArcCos}\left[\frac{\pi}{2} + \frac{1}{2} i \sqrt{-2 + \pi^2}\right]\right\}\right\}, \text{True, True}\right\}$$

Avoiding the verification of the solutions does not avoid the use of inexact numbers.

```
In[10]:= inexactNumberUseChecker[Solve[Cos[x] + Sin[x] == Pi, {x},
                            VerifySolutions -> False]]
```

> Solve::ifun : Inverse functions are being used by Solve, so some solutions
> may not be found; use Reduce for complete solution information. More…

$$\text{Out[10]= } \left\{\left\{\left\{x \to -\text{ArcCos}\left[\frac{\pi}{2} - \frac{1}{2} i \sqrt{-2 + \pi^2}\right]\right\}, \left\{x \to \text{ArcCos}\left[\frac{\pi}{2} - \frac{1}{2} i \sqrt{-2 + \pi^2}\right]\right\},\right.\right.$$
$$\left.\left.\left\{x \to -\text{ArcCos}\left[\frac{\pi}{2} + \frac{1}{2} i \sqrt{-2 + \pi^2}\right]\right\}, \left\{x \to \text{ArcCos}\left[\frac{\pi}{2} + \frac{1}{2} i \sqrt{-2 + \pi^2}\right]\right\}\right\}, \text{True, True}\right\}$$

For definite integrals, `Integrate` will use numerical techniques to find singularities along the integration path.

In[11]:= `inexactNumberUseChecker[Integrate[1/(x - zero), {x, -1, 1}]]`

Integrate::idiv :

Integral of $\dfrac{1}{-\sqrt{2}-\sqrt{3}+\sqrt{5+2\sqrt{6}}+x}$ does not converge on {-1, 1}. More…

Out[11]= $\left\{\displaystyle\int_{-1}^{1}\dfrac{1}{-\sqrt{2}-\sqrt{3}+\sqrt{5+2\sqrt{6}}+x}\,dx,\ \text{True, True}\right\}$

Also factoring over extension fields uses numerical techniques.

In[12]:= `inexactNumberUseChecker[Factor[3 x^3 + 7 x^2 - 9,`
             `Extension -> {(1501/2 - (27 Sqrt[2445])/2)^(1/3)}]] // Simplify`
Out[12]= $\{-9 + 7 x^2 + 3 x^3,\ \text{True, True}\}$

Most functions related to `Root` need numerical techniques to isolate the roots.

In[13]:= `inexactNumberUseChecker[Roots[zero x^7 + 5 x - 1 == 0, x]]`

Out[13]= $\left\{x = \dfrac{1}{5},\ \text{True, False}\right\}$

In[14]:= `roots = Root[#^5 + 3 # - 1&, 3] + Root[#^3 + 5 # - 3&, 1];`
         `inexactNumberUseChecker[RootReduce[roots]]`
Out[15]= $\{\text{Root}[-13809 + 13411\ \#1 - 13772\ \#1^2 + 21527\ \#1^3 - 14495\ \#1^4 + 13712\ \#1^5 - 5588\ \#1^6 + 4232\ \#1^7 -$
         $1770\ \#1^8 + 1265\ \#1^9 - 303\ \#1^{10} + 259\ \#1^{11} - 15\ \#1^{12} + 25\ \#1^{13} + \#1^{15}\ \&,\ 7],\ \text{True, False}\}$

The results of `CylindricalDecomposition` often contain `Root`s. As a result, it uses numerical techniques.

In[16]:= `inexactNumberUseChecker[`
           `CylindricalDecomposition[x^2 + x^6 + y^8 == 1, {y, x}]]`
Out[16]= $\left\{-1 \le y \le 1\ \&\&\ \left(x == -\sqrt{\text{Root}[-1 + y^8 + \#1 + \#1^3\ \&,\ 1]}\ ||\ x == \sqrt{\text{Root}[-1 + y^8 + \#1 + \#1^3\ \&,\ 1]}\right),\right.$
         $\left.\text{True, False}\right\}$

In the next example, numerical techniques are used to obtain the number of steps from the iterator of `Sum`.

In[17]:= `inexactNumberUseChecker[Sum[1/k, {k, 2, 3 + zero}]]`

Sum::itflrw :

Warning: In evaluating $\text{Floor}\left[2 + \sqrt{2} + \sqrt{3} - \sqrt{5 + 2\sqrt{6}}\right]$ to find the number of iterations to use for Sum, \$MaxExtraPrecision = 50.` was encountered. An upper estimate will be used for the number of iterations. More…

Out[17]= $\left\{\dfrac{5}{6},\ \text{True, False}\right\}$

Also, `Series` might numericalize certain expressions internally.

In[18]:= `inexactNumberUseChecker[Series[ArcSin[x + zero], {x, 1, 0}]]`

Out[18]= $\left\{\dfrac{\pi}{2} + (-1)^{\text{Floor}\left[-\frac{\text{Arg}[-1+x]}{2\pi}\right]} \left(-i\sqrt{2}\sqrt{x-1} + O[x-1]^1\right),\ \text{True, True}\right\}$

`DSolve` heavily depends on `Solve`. As a result, it too uses inexact numbers at some steps.

In[19]:= `inexactNumberUseChecker[DSolve[{y''[x] - y'[x] + y[x] == 0,`
                               `y[0] == 1, y'[0] == 0}, y[x], x]]`
Out[19]= $\left\{\left\{y[x] \to \dfrac{1}{3} e^{x/2} \left(3 \cos\left[\dfrac{\sqrt{3}\,x}{2}\right] - \sqrt{3}\,\sin\left[\dfrac{\sqrt{3}\,x}{2}\right]\right)\right\}\right\},\ \text{True, True}\right\}$

Integral transforms use numerical techniques for verifying convergence.

In[20]:= `inexactNumberUseChecker[FourierTransform[Exp[-x^2], x, k]]`

Out[20]= $\left\{\dfrac{e^{-\frac{k^2}{4}}}{\sqrt{2}},\ \text{True, True}\right\}$

In[21]:= `inexactNumberUseChecker[LaplaceTransform[Exp[-x^2], x, k]]`

Out[21]= $\left\{\dfrac{1}{2} e^{\frac{k^2}{4}} \sqrt{\pi}\ \text{Erfc}\left[\dfrac{k}{2}\right],\ \text{True, True}\right\}$

c) These are all built-in functions.

```
In[1]:= allFunctions = DeleteCases[Names["*"], "allFunctions"];
```

We eliminate all values, meaning the commands that evaluated to something other than themselves.

```
In[2]:= allFunctions1 = Select[allFunctions, ToString[ToExpression[#]] === #&];
```

Because many of the following inputs would generate large amounts of messages, we shut off all messages.

```
In[3]:= Get[ToFileName[{$TopDirectory, "SystemFiles", "Kernel",
                "TextResources", $Language}, "Messages.m"]];

In[4]:= (* list of all messages *)
        allMessages = Flatten[(Messages @@ #)& /@ (ToHeldExpression[#]& /@
                                        DeleteCases[allFunctions, "I"])];

In[6]:= (Off @@ First[#])& /@ allMessages;
```

Often argument-counting and -checking messages are associated with General. So, we shut off all these messages too.

```
In[7]:= generalMessageNames = #[[1, 1, 2]]& /@ Messages[General];

In[8]:= Function[c, Function[n, Off[c::n], {HoldAll}] /@ c] /@ allFunctions2;
```

Further, we delete some "dangerous" (meaning they would exit the program) functions from allFunctions1.

```
In[9]:= allFunctions2  = ToExpression /@ DeleteCases[allFunctions1,
                (* "dangerous" functions *)
                "Abort" | "Break" | "Continue" | "Exit" | "Quit" |
                "Remove" | "Goto" | "Throw" | "Install" | "On[]" |
                "ConsoleMessage" | "Interrupt" | "$Inspector" |
                "CellEvaluationDuplicate" | "CoefficientArrays" |
                "$SyntaxHandler" | "$PreRead" |
                _?(Function[c, Or @@ (StringMatchQ[c, #]& /@
                {"*Link*", "*Print*", "*Write*", "*Edit*", "*File*",
                "*Input*", "*Dialog*", "*Convert*", "*Notebook*"})])]];
```

Selecting now all functions for which Derivative[2] [*function*] is different from function yields about 280 functions.

```
In[10]:= l1 =  Select[allFunctions2,
            ((Derivative[2][#] /. Derivative -> d) =!= d[2][#])&];

In[11]:= l1 // Length

Out[11]= 282
```

About 70 of these functions are elementary and special functions like Cos and ArcTan. Calculating the second derivative of these functions yields often a different function. These functions carry the attribute NumericFunction. Or they are mathematical constants.

```
In[12]:= l2 = DeleteCases[l1,
                _?((MemberQ[Attributes[#], NumericFunction] ||
                MemberQ[Attributes[#], Constant] ||
                NumberQ[#] || PolyGamma === # || ProductLog === #)&)];

In[13]:= l2 // Length

Out[13]= 203
```

We group the remaining 180 or so functions according to their result for Derivative[2] [*function*].

```
In[14]:= ({#[[1, 1]] /. _?((LeafCount[#] > 50)&) :> tooBig,
        (Last /@ #) }& /@ Split[Sort[{Derivative[2][#], #}& /@ l2,
                            OrderedQ[{#1[[1]], #2[[1]]}]&],
                        #1[[1]] === #2[[1]]&])

Out[14]= {{0 &, {AbortProtect, Accuracy, And, Apart, ApartSquareFree, ArrayDepth, ArrayQ, AtomQ,
            Boole, ByteCount, Cancel, Catch, Chop, Clear, ClearAll, Complement, ComplexExpand,
            Compose, Composition, CompoundExpression, D, Debug, Denominator, Depth,
            Dispatch, Distribute, Do, Dot, Drop, Dt, Equal, Evaluate, EvenQ, ExactNumberQ,
            Expand, ExpandAll, ExpandDenominator, ExpandNumerator, ExpToTrig, Factor,
            FactorSquareFree, FactorTerms, Find, FindList, First, Flatten, FrontEndExecute,
```

FrontEndTokenExecute, FullSimplify, FunctionExpand, Greater, GreaterEqual,
Hash, Head, HeadCompose, Identity, Inequality, InexactNumberQ, IntegerQ,
InterpolatingFunction, Intersection, IntervalIntersection, IntervalUnion, Join,
Last, LeafCount, Length, Less, LessEqual, ListQ, LogicalExpand, MachineNumberQ,
MakeBoxes, MatrixQ, Most, N, Normal, NProduct, NSum, NumberQ, Numerator, NumericQ,
OddQ, Off, On, OptionQ, Or, OrderedQ, Part, PiecewiseExpand, PolynomialGCD,
PolynomialLCM, PolynomialQ, PowerExpand, Precedence, Precision, PrimeQ, Product,
Rationalize, Refine, Refresh, Reinstall, Release, ReleaseHold, ResetDirectory,
Resolve, Rest, Reverse, RootReduce, RotateLeft, RotateRight, RuleCondition,
Run, SameQ, Set, SetDelayed, Share, Signature, Simplify, Sort, Sow, StackBegin,
StackComplete, StackInhibit, StringQ, StringToStream, Sum, Table, Take, TensorQ,
TensorRank, Thread, TimeUsed, ToBoxes, ToExpression, Together, ToRadicals,
ToString, Total, TrigExpand, TrigFactor, TrigReduce, TrigToExp, TrueQ, Unequal,
Union, UnsameQ, Unset, UpSet, UpSetDelayed, ValueQ, VectorQ, While, Xor}},
$\{\partial_{\{\#1,2\}}$ Hold[#1] &, {HeldPart}}, $\{\partial_{\{\#1,2\}}$ Integrate[#1] &, {Integrate}},
$\{\partial_{\{\#1,2\}}$ Limit[#1] &, {Limit}},
$\{\partial_{\{\#1,2\}}$ NIntegrate[#1] &, {NIntegrate}},
$\{\partial_{\{\#1,2\}}$ (! #1) &, {Nand, Nor}},
$\{\partial_{\{\#1,2\}}$ Piecewise[#1] &, {Piecewise}},
$\{\partial_{\{\#1,2\}}$ Root[#1] &, {Root}},
$\{\partial_{\{\#1,2\}}$ RootSum[#1] &, {RootSum}},
$\{\partial_{\{\#1,2\}}$ SeriesData[#1] &, {SeriesData}},
$\{\partial_{\{\#1,2\}}$ SparseArray[#1] &, {SparseArray}},
{0 & &, {Function}}, {If[#1] &, {If}},
{{} &, {Contexts, DirectoryStack, Options, Stack, Trace}},
{{0} &, {Decompose, Dimensions, List, Ordering, Permutations, Variables}},
$\{\{\{0 \to 0\}\}$ &, {NSolve}},
{{0, 0} &, {AbsoluteTiming, FactorTermsList, Subsets, Timing}}, {{0, {}} &, {Reap}},
{{{0, 0}, {0, 0}} &, {BoxRegion, FactorList, FactorSquareFreeList, TrigFactorList}},
{NonCommutativeMultiply[0] &, {NonCommutativeMultiply}},
{Rule[0] &, {Rule}}, {RuleDelayed[0] &, {RuleDelayed}},
$\{O[\#1]^0$ &, {0}}, {Switch[#1] &, {Switch}},
{Which[#1] &, {Which}}, {BoxData''[#1] &, {StripBoxes}},
{*tooBig*, {ListContourPlot}}, {*tooBig*, {ListDensityPlot}},
{ErrorBox'[#1]^2 ErrorBox''[ErrorBox[#1]] + ErrorBox'[ErrorBox[#1]] ErrorBox''[#1] &,
  {MakeExpression}}, {Slot''[#1] &, {Split}}, {*tooBig*, {ListPlot3D}},
$\left\{\text{Export}^{(1,0,0)\ (1,0,0)}[\#1, \text{NotebookObject}[\ll 4\_\text{Symbolics\_1a.nb}\gg], \text{HTML}] \&, \{\text{HTMLSave}\}\right\}$,
$\left\{\text{Export}^{(1,0,0)\ (1,0,0)}[\#1, \text{NotebookObject}[\ll 4\_\text{Symbolics\_1a.nb}\gg], \text{TeX}] \&, \{\text{TeXSave}\}\right\}$,
{Derivative'', {Derivative}}}}

Most of the results of Derivative[2][*function*] are of the form 0& or some list version of this. The reason for these results is the sequence of evaluations carried out when evaluating Derivative[2][*function*]. Derivative[2][*function*] evaluates *function*[#] and differentiates the result. *function*[#] often evaluates to # and differentiating this twice with respect to # yields 0. Here is the sequence of evaluations for N and Decompose.

In[15]:= **Trace[Derivative[2][N]]**

Out[15]= {N'', {N[#1], #1}, 0 &}

In[16]:= **Trace[Derivative[2][Decompose]]**

Out[16]= {Decompose'', {Decompose[#1], {#1}}, {0} &}

17. Thomas–Fermi Equation, Yoccoz Function, $y(x) = x/\ln(x)$ Inversion, Lagrange–Bürmann Theorem, Divisor Sums, Multiple Differentiation

a) Because the right-hand side of the differential equation contains fractional powers of the independent as well of the dependent variable, we cannot use a series as an ansatz in Solve directly. Thus, we rewrite the differential equation as an integral equation [99] $f(x) = f_0 + f_{s0}\, x + \int_0^x \int_0^x \xi^{-1/2}\, f(\xi)^{3/2}\, d\xi\, dx$.

We now iterate this several times. Here is the corresponding recursive *Mathematica* implementation (the Function part serves to sum up the coefficients of the powers of x). For efficiency, we do not use the built-in Integrate for the definite integrations to be carried out, but define our own `integrate`, which is based on the (much faster) indefinite integration capabilities of *Mathematica*.

```
In[1]:= seriesSol[0, x_] = f0 + fs0 x;

       seriesSol[ord_, x_] := seriesSol[ord, x] =
       Module[{ξ, x},
       Function[res, Plus @@ (* write result in nice form *)
       (Simplify[Plus @@ Cases[res, _ x^#]]& /@
       Union[{{1}, #[[2]]& /@ Cases[res, x^_, {0, Infinity}]])][
       (* starting values *)
       f0 + fs0 x + (* Integrate existing series *)
           integrate[integrate[Normal[Series[ (* rename variables
                                               for definite integration *)
               (seriesSol[ord - 1, x] //. x -> x)^(3/2)/x^(1/2), {x, 0, ord}]],
               {x, 0, ξ}], {ξ, 0, x}] // Expand] +
                               (* lost by Cases for powers of x *) f0]
In[3]:= (* simple definite integration *)
       integrate[f_, {x_, x0_, x1_}] :=
           ((# /. x -> x1) - (# /. x -> x0))&[Integrate[f, x]]

       integrate[f_Plus, {x_, x0_, x1_}] := integrate[#, {x, x0, x1}]& /@ f
```

Here is the result of the first two iterations.

```
In[7]:= seriesSol[0, x]
```

Out[7]= $f0 + fs0\, x$

```
In[8]:= seriesSol[1, x]
```

Out[8]= $f0 + fs0\, x + \frac{4}{3}\, f0^{3/2}\, x^{3/2} + \frac{2}{5}\, \sqrt{f0}\, fs0\, x^{5/2}$

```
In[9]:= seriesSol[2, x]
```

Out[9]= $f0 + fs0\, x + \frac{4}{3}\, f0^{3/2}\, x^{3/2} + \frac{2}{5}\, \sqrt{f0}\, fs0\, x^{5/2} +$

$\frac{f0^2\, x^3}{3} + \frac{3\, fs0^2\, x^{7/2}}{70\, \sqrt{f0}} + \frac{2}{15}\, f0\, fs0\, x^4 + \frac{(32\, f0^4 - 3\, fs0^3)\, x^{9/2}}{756\, f0^{3/2}}$

```
In[10]:= seriesSol[3, x]
```

Out[10]= $f0 + fs0\, x + \frac{4}{3}\, f0^{3/2}\, x^{3/2} + \frac{2}{5}\, \sqrt{f0}\, fs0\, x^{5/2} + \frac{f0^2\, x^3}{3} + \frac{3\, fs0^2\, x^{7/2}}{70\, \sqrt{f0}} +$

$\frac{2}{15}\, f0\, fs0\, x^4 + \frac{(56\, f0^4 - 3\, fs0^3)\, x^{9/2}}{756\, f0^{3/2}} + \frac{fs0^2\, x^5}{175} + \frac{fs0\, (992\, f0^4 + 45\, fs0^3)\, x^{11/2}}{47520\, f0^{5/2}}$

Here is a later stage of the iteration.

```
In[11]:= seriesSol[10, x] // Simplify // Short[#, 6]&
```

Out[11]//Short= $f0 + fs0\, x + \frac{4}{3}\, f0^{3/2}\, x^{3/2} + \ll 21 \gg +$

$\frac{(-113117235\, f0^{12} + 3608878924\, f0^8\, fs0^3 - 3762973290\, f0^4\, fs0^6 + 365783040\, fs0^9)\, x^{12}}{3176090742825\, f0^7} +$

$(fs0^2\, (32277293455912960\, f0^{12} - 93370159717661952\, f0^8\, fs0^3 + 25825890288785400\, f0^4\, fs0^6 -$

$501368610117375\, fs0^9)\, x^{25/2}) / (56992982228582400000\, f0^{19/2})$

For an Adomian decomposition-based series solution of the Thomas–Fermi equation, see [1815].

b) These are the three given equations. We consider *x* as a function of the new independent variable *t*.

```
In[1]:= U = u[t] - (-(16/3)^(1/3) φ[x[t]]^(-4/3) φ'[x[t]]);
       T = t - (144^(-1/6) x[t]^(1/2) φ[x[t]]^(1/6));
       TF = φ''[x[t]] - φ[x[t]]^(3/2)/Sqrt[x];
```

Differentiating U and T with respect to t gives us two further equations. We start by clearing denominators.

```
In[4]:= eqs1 = Numerator[Together[{TF, D[U, t], U, D[T, t], T}]] /. x[t] -> x;
```

Next, we form a polynomial system by introducing new variables for the symbolic quantities appearing in fractional powers.

In[5]:= `eqs2 = PowerExpand[eqs1 /. φ[x] -> Φ^6 /. φ'[x] -> φ1 /. φ''[x] -> φ2 /.`
   `x'[t] -> xs/(2^(1/3) 3^(2/3)) /. x -> y^2]`

Out[5]= $\{-\Phi^9 + y\,\varphi2,\ -8\,xs\,\varphi1^2 + 6\,xs\,\Phi^6\,\varphi2 + 9\,\Phi^{14}\,u'[t],$
   $2\,2^{1/3}\,3^{2/3}\,\varphi1 + 3\,\Phi^8\,u[t],\ 36\,y\,\Phi^5 - 3\,xs\,\Phi^6 - xs\,y^2\,\varphi1,\ 6\,t - 2^{1/3}\,3^{2/3}\,y\,\Phi\}$

Now, we have five equations and we have to eliminate five variables: Φ, $\varphi2$, $\varphi1$, y, and xs. This means we cannot use `Eliminate` or `GroebnerBasis[eqs2, {}, {Φ, φ2, φ1, y, xs}]`. We eliminate one variable after another and look for factors containing only t, $u[t]$, and $u'[t]$. The function `eliminate` eliminates the variable *var* from the equations *eqs* and factors the result.

In[6]:= `eliminate[eqs_, var_] := Cases[Factor[`
   `Resultant[First[eqs], #, var]]& /@ Rest[eqs], _Plus, Infinity]`

Eliminating the variables in the order $\varphi2$, $\varphi1$, y, and finally xs yields the differential equation $u'(t) \to -8\,(t\,u(t)^2 - 1)/(t^2\,u(t) - 1)$.

In[7]:= `eliminate[eqs2, φ2]`

Out[7]= $\{6\,xs\,\Phi^{15} - 8\,xs\,y\,\varphi1^2 + 9\,y\,\Phi^{14}\,u'[t],$
   $2\,2^{1/3}\,3^{2/3}\,\varphi1 + 3\,\Phi^8\,u[t],\ 36\,y\,\Phi^5 - 3\,xs\,\Phi^6 - xs\,y^2\,\varphi1,\ 6\,t - 2^{1/3}\,3^{2/3}\,y\,\Phi\}$

In[8]:= `eliminate[%, φ1]`

Out[8]= $\{2\,2^{2/3}\,3^{1/3}\,xs\,\Phi - 2\,xs\,y\,\Phi^2\,u[t]^2 + 3\,2^{2/3}\,3^{1/3}\,y\,u'[t],$
   $-3456\,y^2 + 576\,xs\,y\,\Phi - 24\,xs^2\,\Phi^2 + 2\,xs^2\,y^3\,\Phi^5 + 3\,xs\,y^4\,\Phi^4\,u'[t],\ 6\,t - 2^{1/3}\,3^{2/3}\,y\,\Phi\}$

In[9]:= `eliminate[%, y]`

Out[9]= $\{-4\,xs^4\,\Phi^8\,u[t]^2 + 1152\,2^{1/3}\,3^{2/3}\,xs^2\,\Phi^4\,u[t]^4 - 96\,2^{2/3}\,3^{1/3}\,xs^3\,\Phi^6\,u[t]^6 + 4\,xs^4\,\Phi^8\,u[t]^8 -$
   $20736\,xs\,\Phi^2\,u[t]^2\,u'[t] + 864\,2^{1/3}\,3^{2/3}\,xs^2\,\Phi^4\,u[t]^4\,u'[t] - 24\,2^{2/3}\,3^{1/3}\,xs^3\,\Phi^6\,u[t]^6\,u'[t] +$
   $15552\,2^{2/3}\,3^{1/3}\,u'[t]^2 - 7776\,xs\,\Phi^2\,u[t]^2\,u'[t]^2 + 108\,2^{1/3}\,3^{2/3}\,xs^2\,\Phi^4\,u[t]^4\,u'[t]^2 +$
   $3888\,2^{2/3}\,3^{1/3}\,u'[t]^3 - 648\,xs\,\Phi^2\,u[t]^2\,u'[t]^3 + 243\,2^{2/3}\,3^{1/3}\,u'[t]^4,$
   $-2\,xs\,\Phi^2 + 2\,t\,xs\,\Phi^2\,u[t]^2 - 3\,2^{2/3}\,3^{1/3}\,t\,u'[t]\}$

In[10]:= `eliminate[%, xs]`

Out[10]= $\{-8 + 8\,t\,u[t]^2 - u'[t] + t^2\,u[t]\,u'[t],\ 8 - 8\,t\,u[t]^2 + u'[t] + t^2\,u[t]\,u'[t]\}$

In[11]:= `Solve[First[%] == 0, u'[t]] // Simplify`

Out[11]= $\left\{\left\{u'[t] \to \dfrac{8 - 8\,t\,u[t]^2}{-1 + t^2\,u[t]}\right\}\right\}$

c) We start with a direct implementation of the recursion relation.

In[1]:= `uE[0][z_] = 1;`
   `uE[n_][z_] := uE[n][z] = Expand[uE[n - 1][z] - z^(n - 1)/2 uE[n - 1][z]^2]`

The direct implementation allows us easily to calculate the first ten terms of the series expansion.

In[3]:= `Take[uE[10][z], 10]`

Out[3]= $\dfrac{1}{2} - \dfrac{z}{8} - \dfrac{z^2}{8} - \dfrac{z^3}{16} - \dfrac{9\,z^4}{128} - \dfrac{z^5}{128} - \dfrac{7\,z^6}{128} + \dfrac{3\,z^7}{256} - \dfrac{29\,z^8}{1024} - \dfrac{z^9}{256}$

Due to the exponentially increasing number of terms of $u_n(z)$, the direct calculation of 100 terms is impossible. Already $u_{10}(z)$ is a polynomial with 1014 terms.

In[4]:= `uE[10][z] // Length`

Out[4]= 1014

To keep the number of needed terms reasonable, we will use a `SeriesData`-object instead. Making sure that we have enough series terms, we have the following implementation.

In[5]:= `u[0, o_][z_] = 1;`
   `u[n_, o_][z_] :=  u[n, o][z] =`
   `Normal[Series[u[n - 1, o][z] - 1/2 z^(n - 1) u[n - 1, o][z]^2,`
   `            {z, 0, o}]];`

   `u[n_][z_] := Normal[u[n + 1, n + 2][z] + O[z]^(n + 1)]`

Using the function u it is possible to calculate the first 100 series terms quickly.

In[9]:= u[100][x]; // Timing

Out[9]= {5.64 Second, Null}

In[10]:= Short[u[100][x] /. (* nicer form of denominators *)
            Rational[p_, q_] :> (p "2"^-#&[Log[2, q]]), 12]

Out[10]//Short= $\dfrac{1}{2} - \dfrac{x}{2^3} - \dfrac{x^2}{2^3} - \dfrac{x^3}{2^4} - \dfrac{9\,x^4}{2^7} - \dfrac{x^5}{2^7} - \dfrac{7\,x^6}{2^7} + \dfrac{3\,x^7}{2^8} - \dfrac{29\,x^8}{2^{10}} - \dfrac{x^9}{2^8} - \dfrac{25\,x^{10}}{2^{11}} +$

$\dfrac{559\,x^{11}}{2^{15}} - \dfrac{1233\,x^{12}}{2^{15}} + \dfrac{649\,x^{13}}{2^{15}} - \dfrac{9\,x^{14}}{2^{13}} - \dfrac{657\,x^{15}}{2^{16}} - \dfrac{2095\,x^{16}}{2^{18}} + \dfrac{1049\,x^{17}}{2^{16}} - \dfrac{8437\,x^{18}}{2^{19}} +$

$\ll 92 \gg + \dfrac{7690524842088915723 5099\,x^{83}}{2^{85}} - \dfrac{744339392829979569 80403\,x^{84}}{2^{83}} -$

$\dfrac{28148738472862860602151\,x^{85}}{2^{86}} + \dfrac{17786871659465997057436 05\,x^{86}}{2^{88}} +$

$\dfrac{51780709826190065919 5371\,x^{87}}{2^{88}} - \dfrac{195151033334673882067 8695\,x^{88}}{2^{89}} +$

$\dfrac{40832974676042873768291455\,x^{89}}{2^{94}} - \dfrac{37034628192076675685904489\,x^{90}}{2^{92}} -$

$\dfrac{14661832617054818360228079\,x^{91}}{2^{94}} + \dfrac{109140455151164357231 02875\,x^{92}}{2^{91}} +$

$\dfrac{49352326245213302865420967\,x^{93}}{2^{94}} + \dfrac{153711162894716551516 7481631\,x^{94}}{2^{95}} -$

$\dfrac{39479410097544119514 4202855\,x^{95}}{2^{97}} - \dfrac{227315887382772870225366455\,x^{96}}{2^{95}} +$

$\dfrac{89079969916497966203 5894285\,x^{97}}{2^{98}} + \dfrac{100146352635569747874 2953913\,x^{98}}{2^{101}} -$

$\dfrac{27153353436069896914 6070383\,x^{99}}{2^{97}} - \dfrac{398556489605471797800 4815265\,x^{100}}{2^{101}}$

Doubling the number of terms multiplies the time by about a factor of $8 = 2^3$.

In[11]:= u[200][x]; // Timing

Out[11]= {54.92 Second, Null}

This means that the first 1000 series terms using u[1000][x] can be calculated in a few hours. The result for c_{1000} is:

$$c_{1000} = -2^{-999} \times$$

20596597206547221775348128198667711072845591129164645600499852338845498385101242006105 39˙.
71772320476161765015023715101909820200164095551924529438045514977473064917451164552235˙.
86487666192915420935506951282756039329662387964917626104681167956079615434274785056328˙.
8750785661107936946594213271986032 1823

Now, let us visualize $\arg(u(0.999\,e^{i\,\varphi}))$. For approximative numerical values of z, we can calculate $u(z)$ by iterating the recursion relation. For an efficient calculation, we compile the function.

In[12]:= uN = Compile[{{x, _Complex}},
            Module[{c = 0}, FixedPoint[(# - x^(c++)/2. #^2)&, 1. + 0. I]]];

This leads to the following graphic. The graphic immediately suggest that $\arg(u(0.999\,e^{i\,\varphi}))$ is discontinuous at rational values of $\alpha = p/q$ and the discontinuity height is $2\,q$. For comparison, we also show $u_{200}(0.999\,e^{i\,\alpha\,2\pi})$ [266], [307].

In[13]:= Show[GraphicsArray[
            Block[{$DisplayFunction = Identity},
            {(* argument along the unit circle *)
            With[{ε = 10^-12},
                Plot[Arg[uN[0.999 Exp[I φ]]], {φ, ε, 2Pi - ε},
                    Frame -> True, Axes -> False, Compiled -> False]],
            (* real and imaginary part along the unit circle *)
            Module[{pp = ParametricPlot[Evaluate[{Im[#], Re[#]}&[u[200][x] /.
                                            x -> 0.999 Exp[I φ]]],
                            {φ, 0, 2Pi}, PlotPoints -> 10000, PlotRange -> All,
                            AspectRatio -> Automatic, Axes -> False], λ},
                λ   = Length[pp[[1, 1, 1, 1]]];
                (* color curve from red to red *)
                Graphics[{Thickness[0.002], MapIndexed[{Hue[#2[[1]]]/λ],

```
Line[#1]}&, Partition[pp[[1, 1, 1, 1]], 2, 1]]},
   PlotRange -> All, AspectRatio -> Automatic]}]]]
```

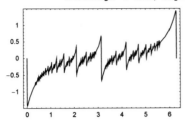

d) To get a system of equations for the coefficients a_{ij}, we write $x(y)$ in the form $x(y) = y \ln(y) + rest(y)$ and plug this $x(y)$ into the equation $y \ln(x(y)) = x(y)$ [766]. The term $\ln(x(y))$ can be written as

$$\ln(x(y)) = \ln(y \ln(y) + rest(y)) = \ln\!\left(y \ln(y)\left(1 + \frac{rest(y)}{y \ln(y)}\right)\right)$$

$$= \ln(y) + \ln(\ln(y)) + \sum_{k=1}^{\infty} \frac{(-1)^{k+1}}{k}\left(\frac{rest(y)}{y \ln(y)}\right)^{k}.$$

Comparing equal powers of $(\ln(\ln(y)))^j / \ln^i(y)$ then gives the required equations for the a_{ij}.

The implementation is nearly straightforward. The only difficulty is that a direct brute force calculation of $(rest(y)/(y \ln(y)))^k$ with $rest$ containing 25 terms and $k = 5$ generates nearly 10 million terms that are too much to handle in a reasonable time. So, we implement a function $\texttt{rest[n, pow]}$, which calculates only the required terms of the pow power of $rest(y)/(y \ln(y))$ up to order n.

```
In[1]:= rest[n_, pow_] :=
   Module[{iterVars},
      (* list of iterator variables *)
      iterVars = Table[a[i], {i, pow}];
      (* list of iterator limits *)
      iterUpperLimits = Drop[n - FoldList[Plus, 0, iterVars], -1];
      (* the sum *)
      Sum[y^pow Log[y]^(-Plus @@ iterVars) Times @@ (b /@ iterVars), ##]& @@
         (* generating the iterator *)
         Thread[{iterVars, Table[0, {pow}], iterUpperLimits}] //.
         (* inner sum *) {b[i_] :> If[i === 0, Log[Log[y]],
                          Sum[Log[Log[y]]^j a[i, j], {j, i}]]}]
```

$\texttt{rest[5, 1]}$ just gives $rest$ to order 5.

```
In[2]:= rest[5, 1]
```

$$\text{Out[2]= } y \, \text{Log}[\text{Log}[y]] + \frac{y \, a[1, 1] \, \text{Log}[\text{Log}[y]]}{\text{Log}[y]} + \frac{y \, (a[2, 1] \, \text{Log}[\text{Log}[y]] + a[2, 2] \, \text{Log}[\text{Log}[y]]^2)}{\text{Log}[y]^2} +$$

$$\frac{y \, (a[3, 1] \, \text{Log}[\text{Log}[y]] + a[3, 2] \, \text{Log}[\text{Log}[y]]^2 + a[3, 3] \, \text{Log}[\text{Log}[y]]^3)}{\text{Log}[y]^3} +$$

$$\frac{y \, (a[4, 1] \, \text{Log}[\text{Log}[y]] + a[4, 2] \, \text{Log}[\text{Log}[y]]^2 + a[4, 3] \, \text{Log}[\text{Log}[y]]^3 + a[4, 4] \, \text{Log}[\text{Log}[y]]^4)}{\text{Log}[y]^4} +$$

$$+ \frac{1}{\text{Log}[y]^5} \, (y \, (a[5, 1] \, \text{Log}[\text{Log}[y]] + a[5, 2] \, \text{Log}[\text{Log}[y]]^2 +$$

$$a[5, 3] \, \text{Log}[\text{Log}[y]]^3 + a[5, 4] \, \text{Log}[\text{Log}[y]]^4 + a[5, 5] \, \text{Log}[\text{Log}[y]]^5))$$

These are the possible expressions of the form $(\ln(\ln(y)))^j / \ln^i(y)$.

```
In[3]:= scales = Flatten[Table[y Log[Log[y]]^j/Log[y]^i, {i, 0, 5}, {j, i}]]
```

Out[3]= $\left\{\dfrac{y\,\text{Log}[\text{Log}[y]]}{\text{Log}[y]}, \dfrac{y\,\text{Log}[\text{Log}[y]]}{\text{Log}[y]^2}, \dfrac{y\,\text{Log}[\text{Log}[y]]^2}{\text{Log}[y]^2}, \dfrac{y\,\text{Log}[\text{Log}[y]]}{\text{Log}[y]^3}, \dfrac{y\,\text{Log}[\text{Log}[y]]^2}{\text{Log}[y]^3}, \right.$

$\dfrac{y\,\text{Log}[\text{Log}[y]]^3}{\text{Log}[y]^3}, \dfrac{y\,\text{Log}[\text{Log}[y]]}{\text{Log}[y]^4}, \dfrac{y\,\text{Log}[\text{Log}[y]]^2}{\text{Log}[y]^4}, \dfrac{y\,\text{Log}[\text{Log}[y]]^3}{\text{Log}[y]^4}, \dfrac{y\,\text{Log}[\text{Log}[y]]^4}{\text{Log}[y]^4},$

$\left.\dfrac{y\,\text{Log}[\text{Log}[y]]}{\text{Log}[y]^5}, \dfrac{y\,\text{Log}[\text{Log}[y]]^2}{\text{Log}[y]^5}, \dfrac{y\,\text{Log}[\text{Log}[y]]^3}{\text{Log}[y]^5}, \dfrac{y\,\text{Log}[\text{Log}[y]]^4}{\text{Log}[y]^5}, \dfrac{y\,\text{Log}[\text{Log}[y]]^5}{\text{Log}[y]^5}\right\}$

These are the coefficients to be determined.

In[4]:= `coeffs = Table[a[i, j], {i, 5}, {j, i}] // Flatten`

Out[4]= `{a[1, 1], a[2, 1], a[2, 2], a[3, 1], a[3, 2], a[3, 3], a[4, 1],`
`a[4, 2], a[4, 3], a[4, 4], a[5, 1], a[5, 2], a[5, 3], a[5, 4], a[5, 5]}`

So, we can calculate the expression $y\ln(x(y)) - x(y)$, which should be 0.

In[5]:= `order = 6;`

```
zero = y (Log[y] + Log[Log[y]]) +
       Sum[(-1)^(i + 1) (rest[order, i]/(y Log[y])^i)/i, {i, 1, order}]) -
       (y Log[y] + rest[order, 1]) // Expand;
```

Equating the coefficients of $(\ln(\ln(y)))^j / \ln^i(y)$ to zero gives the following set of equations.

In[7]:= `Plus @@ # == 0& /@`
`Function[lnTerm, Cases[Cancel[Cases[zero, _. lnTerm]/lnTerm],`
`                          _?(FreeQ[#, y]&)]] /@ scales`

Out[7]= $\left\{1 - a[1, 1] == 0, a[1, 1] - a[2, 1] == 0, -\dfrac{1}{2} - a[2, 2] == 0, a[2, 1] - a[3, 1] == 0,\right.$

$-a[1, 1] + a[2, 2] - a[3, 2] == 0, \dfrac{1}{3} - a[3, 3] == 0, a[3, 1] - a[4, 1] == 0,$

$-\dfrac{1}{2}a[1, 1]^2 - a[2, 1] + a[3, 2] - a[4, 2] == 0, a[1, 1] - a[2, 2] + a[3, 3] - a[4, 3] == 0,$

$-\dfrac{1}{4} - a[4, 4] == 0, a[4, 1] - a[5, 1] == 0, -a[1, 1] a[2, 1] - a[3, 1] + a[4, 2] - a[5, 2] == 0,$

$a[1, 1]^2 + a[2, 1] - a[1, 1] a[2, 2] - a[3, 2] + a[4, 3] - a[5, 3] == 0,$

$\left.-a[1, 1] + a[2, 2] - a[3, 3] + a[4, 4] - a[5, 4] == 0, \dfrac{1}{5} - a[5, 5] == 0\right\}$

These can be easily solved.

In[8]:= `Solve[%, coeffs]`

Out[8]= $\left\{\left\{a[5, 1] \to 1, a[5, 2] \to -5, a[5, 3] \to \dfrac{35}{6}, a[5, 4] \to -\dfrac{25}{12}, a[5, 5] \to \dfrac{1}{5},\right.\right.$

$a[4, 1] \to 1, a[4, 2] \to -3, a[4, 3] \to \dfrac{11}{6}, a[4, 4] \to -\dfrac{1}{4}, a[3, 1] \to 1,$

$\left.\left.a[3, 2] \to -\dfrac{3}{2}, a[3, 3] \to \dfrac{1}{3}, a[2, 1] \to 1, a[2, 2] \to -\dfrac{1}{2}, a[1, 1] \to 1\right\}\right\}$

We finally use this result in `rest[5, 1]`.

In[9]:= `rest[5, 1] /. %`

Out[9]= $\left\{y\,\text{Log}[\text{Log}[y]] + \dfrac{y\,\text{Log}[\text{Log}[y]]}{\text{Log}[y]} + \dfrac{y\,(\text{Log}[\text{Log}[y]] - \frac{1}{2}\,\text{Log}[\text{Log}[y]]^2)}{\text{Log}[y]^2} + \right.$

$\dfrac{y\,(\text{Log}[\text{Log}[y]] - \frac{3}{2}\,\text{Log}[\text{Log}[y]]^2 + \frac{1}{3}\,\text{Log}[\text{Log}[y]]^3)}{\text{Log}[y]^3} +$

$\dfrac{y\,(\text{Log}[\text{Log}[y]] - 3\,\text{Log}[\text{Log}[y]]^2 + \frac{11}{6}\,\text{Log}[\text{Log}[y]]^3 - \frac{1}{4}\,\text{Log}[\text{Log}[y]]^4)}{\text{Log}[y]^4} +$

$\left.\dfrac{y\,(\text{Log}[\text{Log}[y]] - 5\,\text{Log}[\text{Log}[y]]^2 + \frac{35}{6}\,\text{Log}[\text{Log}[y]]^3 - \frac{25}{12}\,\text{Log}[\text{Log}[y]]^4 + \frac{1}{5}\,\text{Log}[\text{Log}[y]]^5)}{\text{Log}[y]^5}\right\}$

Let us look at a numerical example: $10^6 = x/\ln(x)$. The single calculated summands of the series for $x(y)$ have the following values.

In[10]:= `Prepend[Apply[List, %[[1]]], (* add main term *) y Log[y]] //.`
`                {a_ b_Plus :> a List @@ b}`

Out[10]= $\{y \text{Log}[y], y \text{Log}[\text{Log}[y]], \dfrac{y \text{Log}[\text{Log}[y]]}{\text{Log}[y]}, \{\dfrac{y \text{Log}[\text{Log}[y]]}{\text{Log}[y]^2}, -\dfrac{y \text{Log}[\text{Log}[y]]^2}{2 \text{Log}[y]^2}\},$

$\{\dfrac{y \text{Log}[\text{Log}[y]]}{\text{Log}[y]^3}, -\dfrac{3 y \text{Log}[\text{Log}[y]]^2}{2 \text{Log}[y]^3}, \dfrac{y \text{Log}[\text{Log}[y]]^3}{3 \text{Log}[y]^3}\},$

$\{\dfrac{y \text{Log}[\text{Log}[y]]}{\text{Log}[y]^4}, -\dfrac{3 y \text{Log}[\text{Log}[y]]^2}{\text{Log}[y]^4}, \dfrac{11 y \text{Log}[\text{Log}[y]]^3}{6 \text{Log}[y]^4}, -\dfrac{y \text{Log}[\text{Log}[y]]^4}{4 \text{Log}[y]^4}\},$

$\{\dfrac{y \text{Log}[\text{Log}[y]]}{\text{Log}[y]^5}, -\dfrac{5 y \text{Log}[\text{Log}[y]]^2}{\text{Log}[y]^5},$

$\dfrac{35 y \text{Log}[\text{Log}[y]]^3}{6 \text{Log}[y]^5}, -\dfrac{25 y \text{Log}[\text{Log}[y]]^4}{12 \text{Log}[y]^5}, \dfrac{y \text{Log}[\text{Log}[y]]^5}{5 \text{Log}[y]^5}\}\}$

In[11]:= `% /. y -> 10.^6`

Out[11]= $\{1.38155 \times 10^7, 2.62579 \times 10^6, 190061., \{13757.1, -18061.6\},$
$\{995.771, -3922.03, 2288.54\}, \{72.0763, -567.772, 911.076, -326.222\},$
$\{5.21706, -68.4945, 209.828, -196.773, 49.6017\}\}$

Here is their sum.

In[12]:= `N[Plus @@ Flatten[%], 12]`

Out[12]= 1.66265×10^7

This is the direct numerical solution. The asymptotic result agrees well with the numeric one.

In[13]:= `N[FindRoot[10^6 == x/Log[x], {x, 10^6},`
`            AccuracyGoal -> 60, WorkingPrecision -> 40], 10]`

Out[13]= $\{x \to 1.662650890 \times 10^7\}$

The asymptotic inversion of $y = x/\ln x$ can also be obtained by iteration of $x_{k+1}(y) = y \ln x_k(y)$ [453] and [154], or in a closed form, [393]. Here is the *Mathematica* implementation for this solution.

In[14]:= `Clear[s, a, i, j, k, n];`

`s[0, 0] = 1; s[0, k_] = 0; s[n_, 0] = 0;`
`s[n_, k_] := s[n, k] = s[n - 1, k - 1] - (n - 1)s[n - 1, k]`
`a[i_, j_] := s[i, i - j + 1]/j!`

The so-obtained coefficients a_{ij} are in agreement with the above-calculated ones.

In[18]:= `Table[a[i, j], {i, 1, 5}, {j, 1, i}]`

Out[18]= $\{\{1\}, \{1, -\frac{1}{2}\}, \{1, -\frac{3}{2}, \frac{1}{3}\}, \{1, -3, \frac{11}{6}, -\frac{1}{4}\}, \{1, -5, \frac{35}{6}, -\frac{25}{12}, \frac{1}{5}\}\}$

Still another way to tackle the problem of the series expansion of the $x(y)$ defined by $y \ln(x(y)) = x(y)$ is to solve this equation explicitly in terms of the `ProductLog` function $W(z)$ (see Chapter 3). The solution is $x(y) = y W_{-1}(-1/y)$, and then we could carry out a series expansion at $y = \infty$.

e) Let us start with the direct calculation. We use `InverseSeries` to calculate $z = (f^{-1}(w))$, change variables from z to w, and apply the function g.

In[1]:= `directCalculation[f_, g_, w_Symbol, z0_, ord_Integer?Positive] :=`
`    Module[{z}, Normal[g[InverseSeries[Series[f[z], {z, z0, ord}], w]]]]`

Here is an example.

In[2]:= `f[z_] = Exp[z] + Log[z + 1];`
`g[w_] := Tan[w];`

`dcs = directCalculation[f, g, w, 1, 3] // Simplify`

Out[4]= $-\dfrac{2 (e - w + \text{Log}[2]) \text{Sec}[1]^2}{1 + 2 e} -$

$\dfrac{(e - w + \text{Log}[2])^2 \text{Sec}[1]^3 (-\text{Cos}[1] + 4 e (\text{Cos}[1] - 2 \text{Sin}[1]) - 4 \text{Sin}[1])}{(1 + 2 e)^3} +$

$\dfrac{1}{6 (1 + 2 e)^5} ((e - w + \text{Log}[2])^3 \text{Sec}[1]^4 (3 (-11 + 5 \text{Cos}[2] - 4 \text{Sin}[2]) +$

$32 e^2 (-5 + \text{Cos}[2] + 3 \text{Sin}[2]) + 4 e (-23 + 25 \text{Cos}[2] + 6 \text{Sin}[2]))) + \text{Tan}[1]$

Implementing the Lagrange–Bürmann formula is also straightforward. (To avoid expressions of the form $0/0$, we use `Normal[Series[..., z, z0, 0]]` instead of the naive $\xi \to z0$ substitution.)

```
In[5]:= LagrangeBürmannSeries[f_, g_, w_Symbol, z0_, ord_Integer?Positive] :=
    Module[{ξ},
    Module[{gp = D[g[ξ], ξ], w0 = f[z0]},
        g[z0] + Sum[(1/i! Normal[Series[D[gp ((ξ - z0)/(f[ξ] - w0))^i,
            {ξ, i - 1}], {ξ, z0, 0}]]) (w - w0)^i, {i, ord}]]]
```

Here is the result.

```
In[6]:= lbs = LagrangeBürmannSeries[f, g, w, 1, 3] // Simplify
```

$$Out[6]= -\frac{2(e - w + \text{Log}[2])\,\text{Sec}[1]^2}{1 + 2\,e} -$$

$$\frac{(e - w + \text{Log}[2])^2\,\text{Sec}[1]^3\,(-\text{Cos}[1] + 4\,e\,(\text{Cos}[1] - 2\,\text{Sin}[1]) - 4\,\text{Sin}[1])}{(1 + 2\,e)^3} +$$

$$\frac{1}{6\,(1 + 2\,e)^5}\,((e - w + \text{Log}[2])^3\,\text{Sec}[1]^4\,(3\,(-11 + 5\,\text{Cos}[2] - 4\,\text{Sin}[2]) +$$

$$32\,e^2\,(-5 + \text{Cos}[2] + 3\,\text{Sin}[2]) + 4\,e\,(-23 + 25\,\text{Cos}[2] + 6\,\text{Sin}[2]))) + \text{Tan}[1]$$

The two expressions calculated are equal.

```
In[7]:= dcs == lbs
```

```
Out[7]= True
```

For a multivariate version of the Lagrange–Bürmann formula, see [751], [6], [716], [831], [1073], [1902], [1118], and [1855].

f) To express $S_{\alpha,o}(q)$ polynomially in $P(q)$, $Q(q)$, and $R(q)$ we must find the coefficients $c_{i,j,k}^{(\alpha,o)}$ in

$$S_{\alpha,o}(q) = \sum_{i,j,k=1}^{d(\alpha,o)} c_{i,j,k}^{(\alpha,o)}\,P(q)^i\,Q(q)^j\,R(q)^k.$$

Using a truncated series representation for $P(q)$, $Q(q)$, and $R(q)$ and collecting terms with respect to powers of q gives a set of equations (by carrying out the series expansion to a high order we get as many terms as needed) to determine $c_{i,j,k}^{(\alpha,o)}$. Next, we implement the truncated series.

```
In[1]:= PS[n_] := PS[n] = Series[1 - 24 Sum[k q^k/(1 - q^k), {k, n}], {q, 0, n}]
    QS[n_] := QS[n] = Series[1 + 240 Sum[k^3 q^k/(1 - q^k), {k, n}], {q, 0, n}]
    RS[n_] := RS[n] = Series[1 - 504 Sum[k^5 q^k/(1 - q^k), {k, n}], {q, 0, n}]
```

Correspondingly, we also use a truncated series for $S_{\alpha,o}(q)$.

```
In[4]:= S[α_, o_, n_] := Sum[k^α DivisorSigma[o, k] q^k, {k, n}] + O[q]^(n + 1)
```

`PQRPoly` generates all monomials of the form $P(q)^i\,Q(q)^j\,R(q)^k$ of total degree less than or equal to d.

```
In[5]:= PQRPoly[d_] := PQRPoly[d] = Sum[c[i, j, k] P^i Q^j R^k,
    {i, 0, d}, {j, 0, d - i}, {k, 0, d - i - j}]
```

The function `makeIdentity[α, o, d]` tries to find polynomial representation of degree less than or equal to d. We take about 50% more equations than unknowns $c_{i,j,k}^{(\alpha,o)}$ to find unique expressions for the coefficients.

```
In[6]:= makeIdentity[α_, o_, d_] :=
    Module[{ansatz = PQRPoly[d], δ, sbz, sol, id, ΣTerm},
        δ = Ceiling[3/2 Length[ansatz]];
        (* the series *)
        sbz = S[α, o, δ] - (ansatz /. {P -> PS[δ], Q -> QS[δ], R -> RS[δ]});
        (* solve for coefficients *)
        Off[Solve::svars];
        sol = Solve[# == 0& /@ CoefficientList[sbz, q],
                Cases[ansatz, _c, Infinity]];
        (* simplify result *)
        If[sol === {}, {}, id = Numerator[Together[
                Subscript[S, α, o] - ansatz /. sol[[1]] /.
        (* fix undetermined coefficients *) _c -> 0]] // Expand;
        ΣTerm = Select[id, MemberQ[#, S, Infinity]&];
```

```
(* format result *)
STerm == -id + STerm]]
```

For odd *o*, we can now easily find the polynomials we were looking for. (Because of relations between $P(q)$, $Q(q)$, and $R(q)$, the results are not unique.)

```
In[7]:= Do[If[# =!= {}, Print[#]]&[makeIdentity[o, k, o + 2]],
          {o, 1, 3}, {k, 9}]
```

$$288\, S_{1,1} == -P^2 + Q$$

$$720\, S_{1,3} == P\,Q - R$$

$$1008\, S_{1,5} == Q^2 - P\,R$$

$$720\, S_{1,7} == P\,Q^2 - Q\,R$$

$$1584\, S_{1,9} == 3\,Q^3 - 5\,P\,Q\,R + 2\,R^2$$

$$1728\, S_{2,1} == -P^3 + 3\,P\,Q - 2\,R$$

$$1728\, S_{2,3} == P^2\,Q + Q^2 - 2\,P\,R$$

$$1728\, S_{2,5} == 2\,P\,Q^2 - P^2\,R - Q\,R$$

$$8640\, S_{2,7} == 9\,P^2\,Q^2 + 5\,Q^3 - 18\,P\,Q\,R + 4\,R^2$$

$$1728\, S_{2,9} == 6\,P\,Q^3 - 5\,P^2\,Q\,R - 5\,Q^2\,R + 4\,P\,R^2$$

$$6912\, S_{3,1} == -P^4 + 6\,P^2\,Q + 3\,Q^2 - 8\,P\,R$$

$$3456\, S_{3,3} == P^3\,Q + 3\,P\,Q^2 - 3\,P^2\,R - Q\,R$$

$$5184\, S_{3,5} == 6\,P^2\,Q^2 + Q^3 - 2\,P^3\,R - 6\,P\,Q\,R + R^2$$

$$3456\, S_{3,7} == 3\,P^3\,Q^2 + 5\,P\,Q^3 - 9\,P^2\,Q\,R - 3\,Q^2\,R + 4\,P\,R^2$$

$$1728\, S_{3,9} == 9\,P^2\,Q^3 + 2\,Q^4 - 5\,P^3\,Q\,R - 15\,P\,Q^2\,R + 6\,P^2\,R^2 + 3\,Q\,R^2$$

g) We start by implementing order *o* power series for the Eisenstein functions.

```
In[1]:= makeℰDefinitions[] :=
        (Clear[ℰ]; ℰ[k_, q_, o_] := ℰ[k, q, o] =
          1 - 2k/BernoulliB[k] Sum[DivisorSigma[k - 1, j] q^j, {j, o}] +
          O[q]^(o + 1))
```

The search for identities of the prescribed kind is straightforward for a given *d* and suppresses identities of lower order. We use the function `Experimental`IntegerPartitions` for the explicit partitions of an integer.

```
In[2]:= AppendTo[$ContextPath, "Experimental`"];
```

```
In[3]:= (* make result in condensed form *)
        myIntegerPartitions[n_Integer?Positive] := Reverse /@
          (({First[#], Length[#]}& /@ Split[#])& /@ IntegerPartitions[n])
```

```
In[5]:= makeℰRelations[d_] :=
        ((* the possible indices and powers of the E's *)
         ips = Select[myIntegerPartitions[d],
                       (And @@ (EvenQ[First[#]]& /@ #))&];
         (* order of q-series *)
         λ = Length[ips] + 20;
         (* the q-series of the monomials *)
         prods = Times @@@ Apply[ℰ[#1, q, λ]^#2&, ips, {2}];
         (* the coefficient lists *)
         cls = If[Length[#] < λ + 1, Join[#,
                   Table[0, {λ - Length[#]}]], #]& /@ (#[[3]]& /@ prods);
         (* quick modular null space *)
         nsm = NullSpace[Transpose[cls], Modulus -> Prime[10^6]];
         (* loop over all null space spanning vectors *)
         tab = Table[posis = Flatten[Position[nsm[[k]], _?(# =!= 0&),
                                      {1}, Heads -> False]];
         (* full null space *)
         ns = NullSpace[Transpose[cls[[posis]]]];
         If[ns =!= {}, Factor[Numerator[Together[(* extract monomials *)
                 ns[[1]].(Times @@@ Apply[Subscript[ℰ, #1]^#2&,
                     ips[[posis]], {2}])]]]], {k, Length[nsm]}];
```

```
(* extract sums *)
res = Union[Cases[tab, _Plus, Infinity]])
```

We get the first identity of the kind described for $d = 8$.

```
In[6]:= Table[{d, makeδDefinitions[]; makeδRelations[d]}, {d, 2, 8, 2}]
```

Out[6]= $\{\{2, \{\}\}, \{4, \{\}\}, \{6, \{\}\}, \{8, \{\varepsilon_4^2 - \varepsilon_8\}\}\}$

For larger d, we obtain more than one identity. To canonicalize the identities found, we use their Gröbner basis. The function reduceδRelations does this canonicalization.

```
In[7]:= reduceδRelations[eqs_] :=
       GroebnerBasis[eqs, Union[Cases[eqs, Subscript[δ, _], Infinity]],
                MonomialOrder -> DegreeReverseLexicographic]
```

Here are the identities for $d = 10$, $d = 12$, $d = 14$, and $d = 16$. We display only the relations containing the actual d.

```
In[8]:= Do[makeδDefinitions[];
         Print @ Select[reduceδRelations[makeδRelations[d]],
                  MemberQ[#, Subscript[δ, d], Infinity]&],
         {d, 10, 16, 2}]
```

$\{\varepsilon_6\,\varepsilon_8 - \varepsilon_4\,\varepsilon_{10}, \ \varepsilon_4\,\varepsilon_6 - \varepsilon_{10}\}$

$\{441\,\varepsilon_8^2 + 250\,\varepsilon_6\,\varepsilon_{10} - 691\,\varepsilon_4\,\varepsilon_{12}, \ 250\,\varepsilon_6^2 + 441\,\varepsilon_4\,\varepsilon_8 - 691\,\varepsilon_{12}\}$

$\{\varepsilon_{10}^2 - \varepsilon_6\,\varepsilon_{14}, \ \varepsilon_8\,\varepsilon_{10} - \varepsilon_4\,\varepsilon_{14}, \ \varepsilon_4\,\varepsilon_{10} - \varepsilon_{14}, \ \varepsilon_6\,\varepsilon_8 - \varepsilon_{14}\}$

$\{5528\,\varepsilon_{10}\,\varepsilon_{12} - 1911\,\varepsilon_8\,\varepsilon_{14} - 3617\,\varepsilon_6\,\varepsilon_{16}, \ 7601\,\varepsilon_8\,\varepsilon_{12} + 3250\,\varepsilon_6\,\varepsilon_{14} - 10851\,\varepsilon_4\,\varepsilon_{16}, $
$3250\,\varepsilon_6\,\varepsilon_{10} + 7601\,\varepsilon_4\,\varepsilon_{12} - 10851\,\varepsilon_{16}, \ 1911\,\varepsilon_8^2 - 5528\,\varepsilon_4\,\varepsilon_{12} + 3617\,\varepsilon_{16}, $
$336146624\,\varepsilon_{12}^2\,\varepsilon_{14} + 49686000\,\varepsilon_{10}\,\varepsilon_{14}^2 - 241923045\,\varepsilon_8\,\varepsilon_{14}\,\varepsilon_{16} - 143909579\,\varepsilon_6\,\varepsilon_{16}^2, $
$17966000\,\varepsilon_6\,\varepsilon_{12}\,\varepsilon_{14} - 6210750\,\varepsilon_4\,\varepsilon_{14}^2 + 27492817\,\varepsilon_4\,\varepsilon_{12}\,\varepsilon_{16} - 39248067\,\varepsilon_{16}^2, $
$1911\,\varepsilon_4\,\varepsilon_8\,\varepsilon_{14} - 5528\,\varepsilon_{12}\,\varepsilon_{14} + 3617\,\varepsilon_{10}\,\varepsilon_{16}, $
$42018328\,\varepsilon_4\,\varepsilon_{12}^2 + 6210750\,\varepsilon_{14}^2 - 20736261\,\varepsilon_4\,\varepsilon_8\,\varepsilon_{16} - 27492817\,\varepsilon_{12}\,\varepsilon_{16}\}$

As visible in the last results, the coefficients occurring in the relations are quickly growing. Luckily, these integers factor nicely. So, for $d = 18$ we display the factored form of the coefficients.

```
In[9]:= f[i_?Negative] = -f[-i];
       f[i_?(# === 1 || PrimeQ[#]&)] = i;
       f[i_] := (HoldForm @@ {C @@ FactorInteger[i]}) /. {List -> Power, C -> Times}

       factorCoefficients[expr_] := expr /. i_Integer r_ :> f[i] r
```

```
In[14]:= Function[d, factorCoefficients[reduceδRelations[Select[makeδRelations[d],
                      MemberQ[#, Subscript[δ, d], Infinity]&]]]] /@ {18}
```

Out[14]= $\{\{(2^1\,11^1\,691^1)\,\varepsilon_6\,\varepsilon_{12} + (3^2\,5^1\,7^2\,13^1)\,\varepsilon_4\,\varepsilon_{14} - 43867\,\varepsilon_{18}, $
$(2^2\,5^3\,11^1)\,\varepsilon_8^3 + (3^3\,7^2\,29^1)\,\varepsilon_4\,\varepsilon_{14} - 43867\,\varepsilon_{18}, \ (2^1\,3^2\,5^4\,7^2\,13^1)\,\varepsilon_4\,\varepsilon_6^2\,\varepsilon_{14} - $
$(3^3\,7^2\,29^1\,691^1)\,\varepsilon_4\,\varepsilon_{12}\,\varepsilon_{14} - (2^1\,5^3\,43867^1)\,\varepsilon_6^2\,\varepsilon_{18} + (691^1\,43867^1)\,\varepsilon_{12}\,\varepsilon_{18}, $
$(3^3\,7^2\,11^1\,29^1\,691^2)\,\varepsilon_4\,\varepsilon_{12}^2\,\varepsilon_{14} + (3^4\,5^5\,7^4\,13^2)\,\varepsilon_4^2\,\varepsilon_6\,\varepsilon_{14}^2 - (11^1\,691^2\,43867^1)\,\varepsilon_{12}^2\,\varepsilon_{18} - $
$(2^1\,3^2\,5^4\,7^2\,13^1\,43867^1)\,\varepsilon_4\,\varepsilon_6\,\varepsilon_{14}\,\varepsilon_{18} + (5^3\,43867^2)\,\varepsilon_6\,\varepsilon_{18}^2, $
$- (2^1\,3^3\,7^2\,11^2\,29^1\,691^3)\,\varepsilon_4\,\varepsilon_{12}^2\,\varepsilon_{14} + (3^6\,5^6\,7^6\,13^3)\,\varepsilon_4^3\,\varepsilon_{14}^3 + (2^1\,11^2\,691^3\,43867^1)\,\varepsilon_{12}^3\,\varepsilon_{18} - $
$(3^5\,5^5\,7^4\,13^2\,43867^1)\,\varepsilon_4^3\,\varepsilon_{14}^2\,\varepsilon_{18} + (3^3\,5^4\,7^2\,13^1\,43867^2)\,\varepsilon_4\,\varepsilon_{14}\,\varepsilon_{18}^2 - (5^3\,43867^3)\,\varepsilon_{18}^3\}\}$

Also, the number relations increase with d.

```
In[15]:= Table[{d, Length[Select[makeδRelations[d], MemberQ[#,
                      Subscript[δ, d], Infinity]&]]}, {d, 20, 26, 2}]
```

Out[15]= $\{\{20, 10\}, \{22, 12\}, \{24, 18\}, \{26, 22\}\}$

h) These are the functions $S_k(x)$.

```
In[1]:= S[k_, x_] := Sum[(-1)^j Binomial[k + 1, j] BernoulliB[j] x^(k + 1 - j),
                {j, 0, k}]/(k + 1)
```

To find relations of the given form we make a generic ansatz, compare coefficients in ξ, and solve for the c_j. This is done in the following one-liner.

```
In[2]:= findQuadraticIdentity[o_, ξ_] := Subscript[S, o][ξ]^2 ==
                Sum[c[k] Subscript[S, k][ξ], {k, 0, 2o + 1}] /.
```

```
Solve[# == 0& /@ CoefficientList[S[o, ξ]^2 - Sum[c[k] S[k, ξ],
                {k, 0, 2o + 1}], ξ], Table[c[k], {k, 0, 2o + 1}]]
```

Here are the first 12 of these identities.

In[3]:= `Flatten[Table[findQuadraticIdentity[o, ξ], {o, 12}]] // TraditionalForm`

Out[3]//TraditionalForm= $\Big\{S_1(\xi)^2 = S_3(\xi), S_2(\xi)^2 = \dfrac{S_3(\xi)}{3} + \dfrac{2 S_5(\xi)}{3}, S_3(\xi)^2 = \dfrac{S_5(\xi)}{2} + \dfrac{S_7(\xi)}{2}, S_4(\xi)^2 = -\dfrac{1}{15} S_5(\xi) + \dfrac{2 S_7(\xi)}{3} + \dfrac{2 S_9(\xi)}{5},$

$S_5(\xi)^2 = -\dfrac{1}{6} S_7(\xi) + \dfrac{5 S_9(\xi)}{6} + \dfrac{S_{11}(\xi)}{3}, S_6(\xi)^2 = \dfrac{S_7(\xi)}{21} - \dfrac{S_9(\xi)}{3} + S_{11}(\xi) + \dfrac{2 S_{13}(\xi)}{7},$

$S_7(\xi)^2 = \dfrac{S_9(\xi)}{6} - \dfrac{7 S_{11}(\xi)}{12} + \dfrac{7 S_{13}(\xi)}{6} + \dfrac{S_{15}(\xi)}{4}, S_8(\xi)^2 = -\dfrac{1}{15} S_9(\xi) + \dfrac{4 S_{11}(\xi)}{9} - \dfrac{14 S_{13}(\xi)}{15} + \dfrac{4 S_{15}(\xi)}{3} + \dfrac{2 S_{17}(\xi)}{9},$

$S_9(\xi)^2 = -\dfrac{3}{10} S_{11}(\xi) + S_{13}(\xi) - \dfrac{7 S_{15}(\xi)}{5} + \dfrac{3 S_{17}(\xi)}{2} + \dfrac{S_{19}(\xi)}{5},$

$S_{10}(\xi)^2 = \dfrac{5 S_{11}(\xi)}{33} - S_{13}(\xi) + 2 S_{15}(\xi) - 2 S_{17}(\xi) + \dfrac{5 S_{19}(\xi)}{3} + \dfrac{2 S_{21}(\xi)}{11},$

$S_{11}(\xi)^2 = \dfrac{5 S_{13}(\xi)}{6} - \dfrac{11 S_{15}(\xi)}{4} + \dfrac{11 S_{17}(\xi)}{3} - \dfrac{11 S_{19}(\xi)}{4} + \dfrac{11 S_{21}(\xi)}{6} + \dfrac{S_{23}(\xi)}{6},$

$S_{12}(\xi)^2 = -\dfrac{691 S_{13}(\xi)}{1365} + \dfrac{10 S_{15}(\xi)}{3} - \dfrac{33 S_{17}(\xi)}{5} + \dfrac{44 S_{19}(\xi)}{7} - \dfrac{11 S_{21}(\xi)}{3} + 2 S_{23}(\xi) + \dfrac{2 S_{25}(\xi)}{13}\Big\}$

Using `GroebnerBasis`, we could find many more relations between the $S_k(x)$.

In[4]:= `GroebnerBasis[Table[Subscript[S, k][x] - S[k, x], {k, 6}],`
        `{}, {x}, MonomialOrder -> EliminationOrder] //`
                                    `TraditionalForm`

Out[4]//TraditionalForm= $\{S_2(x) + 5 S_4(x) + 540 S_4(x) S_5(x) - 42 S_1(x) S_6(x) - 504 S_3(x) S_6(x),$

$432 S_5(x)^2 + 24 S_3(x) S_5(x) + 5 S_5(x) + S_3(x) - 42 S_2(x) S_6(x) - 420 S_4(x) S_6(x),$

$108 S_5(x) S_2(x) + 2 S_2(x) - 5 S_4(x) - 84 S_1(x) S_6(x) - 21 S_6(x),$

$576 S_5(x)^2 + 4 S_1(x) S_5(x) + 7 S_5(x) + S_3(x) - 28 S_2(x) S_6(x) - 560 S_4(x) S_6(x),$

$150 S_4(x)^2 - 3360 S_6(x) S_4(x) + 3456 S_5(x)^2 + 7 S_3(x) + 41 S_5(x) - 294 S_2(x) S_6(x),$

$S_2(x) + 60 S_3(x) S_4(x) - 5 S_4(x) - 42 S_1(x) S_6(x) - 14 S_6(x),$

$3456 S_5(x)^2 + 37 S_5(x) + 5 S_3(x) + 30 S_2(x) S_4(x) - 168 S_2(x) S_6(x) - 3360 S_4(x) S_6(x),$

$S_2(x) + 30 S_1(x) S_4(x) - 10 S_4(x) - 21 S_6(x), 8 S_3(x)^2 + S_3(x) + 864 S_5(x)^2 + 9 S_5(x) - 42 S_2(x) S_6(x) - 840 S_4(x) S_6(x),$

$12 S_2(x) S_3(x) - 5 S_4(x) - 7 S_6(x), 4 S_1(x) S_3(x) - S_3(x) - 3 S_5(x), 3 S_2(x)^2 - S_3(x) - 2 S_5(x),$

$6 S_1(x) S_2(x) - S_2(x) - 5 S_4(x), S_1(x)^2 - S_3(x), 248832 S_5(x)^3 - 1512 S_5(x)^2 - 53 S_5(x) -$

$28224 S_1(x) S_6(x)^2 - 225792 S_3(x) S_6(x)^2 - 1176 S_6(x)^2 - 13 S_3(x) + 1218 S_2(x) S_6(x) + 6720 S_4(x) S_6(x)\}$

In[5]:= `% /. (* check *) Subscript[S, k_][x_] -> S[k, x] // Simplify`

Out[5]= {0, 0, 0, 0, 0, 0, 0, 0, 0, 0, 0, 0, 0, 0, 0}

i) It is straightforward to implement a recursive procedure that calculates the coefficients a_k.

In[1]:= `o = 100;`
        `as = {};`

        `Do[ser = Exp[z] - (Product[(1 - a[k] z^k)/(1 + a[k] z^k),`
                                {k, n}] /. as) + O[z]^(n + 1);`
            `coeff = SeriesCoefficient[ser, n];`
            `(* solve for last coefficient a[n] *)`
            `sol = Solve[coeff == 0, a[n]];`
            `as = AppendTo[as, sol[[1, 1]]], {n, o}]`

The coefficients a_k decrease quickly.

In[4]:= `ListPlot[N[Log[10, Abs[(Last /@ as) /. 0 -> 1]]]]`

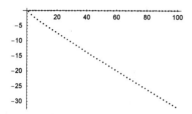

The function $\delta[z]$ is the difference between the exponential function and the truncated product.

```
In[5]:= δ[z_] = (Exp[z] - Product[(1 - a[k] z^k)/(1 + a[k] z^k), {k, 1, o}] /. as);
```

Taking into account the first 50 nonvanishing factors approximates *e* to 32 digits.

```
In[6]:= δ[1] // N[#, 22]&
```

```
Out[6]= -2.812620072471284307359×10^-32
```

Here are the first few factors explicitly. The linear term in the denominator shows that the radius of convergence is 2.

```
In[7]:= δ[z] - Exp[z] /. z^(_?(# > 12&)) -> 0
```

$$Out[7]= -\frac{\left(1+\frac{z}{2}\right)\left(1-\frac{z^3}{24}\right)\left(1-\frac{z^5}{160}\right)\left(1-\frac{z^7}{896}\right)\left(1-\frac{z^9}{5184}\right)\left(1-\frac{z^{11}}{22528}\right)}{\left(1-\frac{z}{2}\right)\left(1+\frac{z^3}{24}\right)\left(1+\frac{z^5}{160}\right)\left(1+\frac{z^7}{896}\right)\left(1+\frac{z^9}{5184}\right)\left(1+\frac{z^{11}}{22528}\right)}$$

The following plot also clearly shows that the radius of convergence is 2.

```
In[8]:= Show[GraphicsArray[
          Block[{$DisplayFunction = Identity, ε = 10^-3},
          (* avoid messages *) Internal`DeactivateMessages[
          {(* along the real axis *)
          Plot[Log[10, Abs[δ[SetPrecision[z, 200]]]], {z, 0, 3},
                  Compiled -> False],
          (* over the complex plane *)
          ParametricPlot3D[{r Cos[φ], r Sin[φ],
                            Log[10, Abs[δ[SetPrecision[r Exp[I φ], 200]]]]},
                            {r, ε, 3}, {φ, 0, 2Pi},
                            Compiled -> False, BoxRatios -> {1, 1, 1}]}]]]]
```

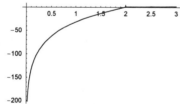
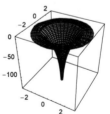

j) In the following, we use a straightforward method to calculate the $\partial^n x_i(s)/\partial s^n$. Faster implementations, based on rooted tree enumerations, are possible. We will use the *Mathematica* expression $f[i][j,k,...]$ for $f_{i;j,k,...}$. Using the formula $\partial g/\partial s = g_{;k} f_k$, we can implement a derivative operator d. Its second argument is the new dummy summation index *newIndex*. Inside d, we add a variable s to all functions, then use D to carry out the differentiations, and then rewrite the result again in terms of $f[i][j,k,...]$s. $f[i][]$ represents the component $f[i]$ without any differentiation.

```
In[1]:= d[expr_, newIndex_] :=
          Module[{aux1, aux2, aux3, s},
          (* append temporary differentiation variable s *)
          aux1 = expr /. f[i_][derivs___] -> f[i][derivs][s];
          (* carry out differentiation *)
          aux2 = D[aux1, s];
          (* reformat partial derivatives *)
          aux3 = aux2 /. {Derivative[1][f[i_][derivs___]][s] :>
```

```
                                          f[i][derivs, newIndex],
                        f[i_][derivs___][s] :> f[i][derivs]};
        (* multiply by D[f, newIndex] *) Expand[aux3 f[newIndex][]]]
```

Here are the first few derivatives. We use j, k, and l as the dummy summation variables.

```
In[2]:= d2 = d[f[i][], j]
```

```
Out[2]= f[i][j] f[j][]
```

```
In[3]:= d3 = d[d2, k]
```

```
Out[3]= f[i][j, k] f[j][] f[k][] + f[i][j] f[j][k] f[k][]
```

```
In[4]:= d4 = d[d3, l]
```

```
Out[4]= f[i][j, k, l] f[j][] f[k][] f[l][] + f[i][j, l] f[j][k] f[k][] f[l][] +
        f[i][j, k] f[j][l] f[k][] f[l][] + f[i][j] f[j][k, l] f[k][] f[l][] +
        f[i][j, k] f[j][] f[k][l] f[l][] + f[i][j] f[j][k] f[k][l] f[l][]
```

The number of terms of the resulting expressions grows factorially with increasing number of differentiations.

```
In[5]:= Length /@ FoldList[d, f[i][], Table[δ[j], {j, 8}]]
```

```
Out[5]= {0, 2, 2, 6, 24, 120, 720, 5040, 40320}
```

For a more readable and more concise version of the multiple-differentiated expressions, we implement a function `formatNicely` that uses subscripts.

```
In[6]:= formatNicely[expr_] := expr //. (* use subscripts *)
        {f[i_][derivs__] :> Subscript[f,
         StringReplace[ToString[i] <> ";" <> ToString[{derivs}],
                       {" " -> "", "{" -> "", "}" -> ""}]],
        f[i_][] :> Subscript[f, ToString[i]]};
```

```
In[7]:= formatNicely[d4]
```

```
Out[7]= f_{i;j,k,l} f_j f_k f_l + f_{i;j,l} f_{j;k} f_k f_l + f_{i;j,k} f_{j;l} f_k f_l +
        f_{i;j,k} f_{j;l} f_k f_l + f_{i;j,k} f_j f_{k;l} f_l + f_{i;j} f_{j;k} f_{k;l} f_l
```

In the last expressions, the two terms $f_{i;j,l} f_{j;k} f_k f_l$ and $f_{i;j,k} f_{j;l} f_k f_l$ are equivalent. By renaming the dummy summations indices $k \leftrightarrow l$ in the last expressions, the second expression becomes identical to the first. The function `canonicalize` identifies expressions which are identical with respect to the renaming of summation variables by comparing with all possible permutations. This is a straightforward brute-force approach for canonicalization.

```
In[8]:= (* pure f-term and the prefactor *)
        fPair[expr_] := {#, expr/#}&[DeleteCases[expr, _Integer, Infinity]];
        (* sort partial derivative variables -- their order does not matter *)
        sortDerivs[expr_] := expr //. f[i_][derivs___] :>
                                    f[i][Sequence @@ Sort[{derivs}]]
```

```
In[12]:= canonicalize[expr_] :=
        Module[{vars, perms, rules, newList, oldList, newPair, permElems, pos},
            (* all permutations of the variables *)
            vars = DeleteCases[Union[Level[expr, {-1}]], i | _Integer];
            perms  = Permutations[vars];
            rules = Apply[Rule, Transpose[{vars, #}], {1}]& /@ perms;
            newList = {};
            (* list of f-terms and how many of them *)
            oldList = sortDerivs[fPair /@ (List @@ expr)];
            (* check if elements are equivalent *)
        While[oldList =!= {},
                newPair = oldList[[1]];
                oldList = Rest[oldList];
                permElems = Union[sortDerivs[newPair[[1]] /. rules]];
                pos = Position[First /@ newList,
                            _?(MemberQ[permElems, #, Infinity]&), {1}, 1];
                If[pos === {}, (* new term *) AppendTo[newList, newPair],
                    (* increase counter *) pos = pos[[1, 1]];
                    newList[[pos, 2]] = newList[[pos, 2]] + newPair[[2]]]];
            (* return simplified result *) Plus @@ Apply[Times, newList, {1}]]
```

Using the function `canonicalize` three terms in d4 turn out to be identical.

In[13]:= `formatNicely[canonicalize[d4]]`

Out[13]= $f_{i;j,k,1}\ f_j\ f_k\ f_1 + 3\ f_{i;j,1}\ f_{j;k}\ f_k\ f_1 + f_{i;j}\ f_{j;k,1}\ f_k\ f_1 + f_{i;j}\ f_{j;k}\ f_{k;1}\ f_1$

Now it is straightforward to calculate the fifth and sixth derivative. We use m, n, and o as the dummy summation variables.

In[14]:= `formatNicely[d5 = canonicalize[d[d4, m]]]`

Out[14]= $f_{i;j,k,1,m}\ f_j\ f_k\ f_1\ f_m + 6\ f_{i;j,1,m}\ f_{j;k}\ f_k\ f_1\ f_m + 4\ f_{i;j,m}\ f_{j;k,1}\ f_k\ f_1\ f_m +$
$f_{i;j}\ f_{j;k,1,m}\ f_k\ f_1\ f_m + 4\ f_{i;j,m}\ f_{j;k}\ f_{k;1}\ f_1\ f_m + 3\ f_{i;j}\ f_{j;k,m}\ f_{k;1}\ f_1\ f_m +$
$3\ f_{i;j,k}\ f_{j;m}\ f_{k;1}\ f_1\ f_m + f_{i;j}\ f_{j;k}\ f_{k;1,m}\ f_1\ f_m + f_{i;j}\ f_{j;k}\ f_{k;1}\ f_{1;m}\ f_m$

In[15]:= `Short[formatNicely[d6 = canonicalize[d[d5, n]]], 12]`

Out[15]//Short= $f_{i;j,k,1,m,n}\ f_j\ f_k\ f_1\ f_m\ f_n + 10\ f_{i;j,1,m,n}\ f_{j;k}\ f_k\ f_1\ f_m\ f_n +$
$10\ f_{i;j,m,n}\ f_{j;k,1}\ f_k\ f_1\ f_m\ f_n + 5\ f_{i;j,n}\ f_{j;k,1,m}\ f_k\ f_1\ f_m\ f_n + f_{i;j}\ f_{j;k,1,m,n}\ f_k\ f_1\ f_m\ f_n +$
$10\ f_{i;j,m,n}\ f_{j;k}\ f_{k;1}\ f_1\ f_m\ f_n + 15\ f_{i;j,n}\ f_{j;k,m}\ f_{k;1}\ f_1\ f_m\ f_n + 6\ f_{i;j}\ f_{j;k,m,n}\ f_{k;1}\ f_1\ f_m\ f_n +$
$15\ f_{i;j,k,n}\ f_{j;m}\ f_{k;1}\ f_1\ f_m\ f_n + 10\ f_{i;j,k}\ f_{j;m,n}\ f_{k;1}\ f_1\ f_m\ f_n + 5\ f_{i;j,n}\ f_{j;k}\ f_{k;1,m}\ f_1\ f_m\ f_n +$
$4\ f_{i;j}\ f_{j;k,n}\ f_{k;1,m}\ f_1\ f_m\ f_n + f_{i;j}\ f_{j;k}\ f_{k;1,m,n}\ f_1\ f_m\ f_n + 5\ f_{i;j,n}\ f_{j;k}\ f_{k;1}\ f_{1;m}\ f_m\ f_n +$
$4\ f_{i;j}\ f_{j;k,n}\ f_{k;1}\ f_{1;m}\ f_m\ f_n + 3\ f_{i;j}\ f_{j;k}\ f_{k;1,n}\ f_{1;m}\ f_m\ f_n + f_{i;j}\ f_{j;k}\ f_{k;1}\ f_{1;m,n}\ f_m\ f_n +$
$10\ f_{i;j,k}\ f_{j;m}\ f_{k;1}\ f_{1;n}\ f_m\ f_n + 3\ f_{i;j}\ f_{j;k,m}\ f_{k;1}\ f_1\ f_{m;n}\ f_n + f_{i;j}\ f_{j;k}\ f_{k;1}\ f_{1;m}\ f_{m;n}\ f_n$

The seventh derivative has 48 terms and takes a little while to canonicalize.

In[16]:= `Length[d7 = canonicalize[d[d6, o]]] // Timing`

Out[16]= $\{22.31\ \text{Second},\ 48\}$

In[17]:= `formatNicely[d7]`

Out[17]= $f_{i;j,k,1,m,n,o}\ f_j\ f_k\ f_1\ f_m\ f_n\ f_o + 15\ f_{i;j,1,m,n,o}\ f_{j;k}\ f_k\ f_1\ f_m\ f_n\ f_o + 20\ f_{i;j,m,n,o}\ f_{j;k,1}\ f_k\ f_1\ f_m\ f_n\ f_o +$
$15\ f_{i;j,n,o}\ f_{j;k,1,m}\ f_k\ f_1\ f_m\ f_n\ f_o + 6\ f_{i;j,o}\ f_{j;k,1,m,n}\ f_k\ f_1\ f_m\ f_n\ f_o + f_{i;j}\ f_{j;k,1,m,n,o}\ f_k\ f_1\ f_m\ f_n\ f_o +$
$20\ f_{i;j,m,n,o}\ f_{j;k}\ f_{k;1}\ f_1\ f_m\ f_n\ f_o + 45\ f_{i;j,n,o}\ f_{j;k,m}\ f_{k;1}\ f_1\ f_m\ f_n\ f_o + 36\ f_{i;j,o}\ f_{j;k,m,n}\ f_{k;1}\ f_1\ f_m\ f_n\ f_o +$
$10\ f_{i;j}\ f_{j;k,m,n,o}\ f_{k;1}\ f_1\ f_m\ f_n\ f_o + 45\ f_{i;j,k,n,o}\ f_{j;m}\ f_{k;1}\ f_1\ f_m\ f_n\ f_o + 60\ f_{i;j,k,o}\ f_{j;m,n}\ f_{k;1}\ f_1\ f_m\ f_n\ f_o +$
$15\ f_{i;j,k}\ f_{j;m,n,o}\ f_{k;1}\ f_1\ f_m\ f_n\ f_o + 15\ f_{i;j,n,o}\ f_{j;k}\ f_{k;1,m}\ f_1\ f_m\ f_n\ f_o + 24\ f_{i;j,o}\ f_{j;k,n}\ f_{k;1,m}\ f_1\ f_m\ f_n\ f_o +$
$10\ f_{i;j}\ f_{j;k,n,o}\ f_{k;1,m}\ f_1\ f_m\ f_n\ f_o + 6\ f_{i;j,o}\ f_{j;k}\ f_{k;1,m,n}\ f_1\ f_m\ f_n\ f_o + 5\ f_{i;j}\ f_{j;k,o}\ f_{k;1,m,n}\ f_1\ f_m\ f_n\ f_o +$
$f_{i;j}\ f_{j;k}\ f_{k;1,m,n,o}\ f_1\ f_m\ f_n\ f_o + 10\ f_{i;j,k}\ f_{j;m,n}\ f_{k;1,o}\ f_1\ f_m\ f_n\ f_o + 15\ f_{i;j,n,o}\ f_{j;k}\ f_{k;1}\ f_{1;m}\ f_m\ f_n\ f_o +$
$24\ f_{i;j,o}\ f_{j;k,n}\ f_{k;1}\ f_{1;m}\ f_m\ f_n\ f_o + 10\ f_{i;j}\ f_{j;k,n,o}\ f_{k;1}\ f_{1;m}\ f_m\ f_n\ f_o + 18\ f_{i;j,o}\ f_{j;k}\ f_{k;1,n}\ f_{1;m}\ f_m\ f_n\ f_o +$
$15\ f_{i;j}\ f_{j;k,o}\ f_{k;1,n}\ f_{1;m}\ f_m\ f_n\ f_o + 6\ f_{i;j}\ f_{j;k}\ f_{k;1,n,o}\ f_{1;m}\ f_m\ f_n\ f_o + 6\ f_{i;j,o}\ f_{j;k}\ f_{k;1}\ f_{1;m,n}\ f_m\ f_n\ f_o +$
$5\ f_{i;j}\ f_{j;k,o}\ f_{k;1}\ f_{1;m,n}\ f_m\ f_n\ f_o + 4\ f_{i;j}\ f_{j;k}\ f_{k;1,o}\ f_{1;m,n}\ f_m\ f_n\ f_o + f_{i;j}\ f_{j;k}\ f_{k;1}\ f_{1;m,n,o}\ f_m\ f_n\ f_o +$
$60\ f_{i;j,k,o}\ f_{j;m}\ f_{k;1}\ f_{1;n}\ f_m\ f_n\ f_o + 20\ f_{i;j,k}\ f_{j;m,o}\ f_{k;1}\ f_{1;n}\ f_m\ f_n\ f_o + 45\ f_{i;j,k}\ f_{j;m}\ f_{k;1,o}\ f_{1;n}\ f_m\ f_n\ f_o +$
$15\ f_{i;j,k}\ f_{j;m}\ f_{k;1}\ f_{1;n,o}\ f_m\ f_n\ f_o + 18\ f_{i;j,o}\ f_{j;k,m}\ f_{k;1}\ f_1\ f_{m;n}\ f_n\ f_o + 15\ f_{i;j}\ f_{j;k,m,o}\ f_{k;1}\ f_1\ f_{m;n}\ f_n\ f_o +$
$10\ f_{i;j}\ f_{j;k,m}\ f_{k;1,o}\ f_1\ f_{m;n}\ f_n\ f_o + 6\ f_{i;j,o}\ f_{j;k}\ f_{k;1}\ f_{1;m}\ f_{m;n}\ f_n\ f_o + 5\ f_{i;j}\ f_{j;k,o}\ f_{k;1}\ f_{1;m}\ f_{m;n}\ f_n\ f_o +$
$4\ f_{i;j}\ f_{j;k}\ f_{k;1,o}\ f_{1;m}\ f_{m;n}\ f_n\ f_o + 3\ f_{i;j}\ f_{j;k}\ f_{k;1}\ f_{1;m,o}\ f_{m;n}\ f_n\ f_o + 10\ f_{i;j}\ f_{j;k,n}\ f_{k;1}\ f_{1;o}\ f_{m;n}\ f_n\ f_o +$
$f_{i;j}\ f_{j;k}\ f_{k;1}\ f_{1;m}\ f_{m;n,o}\ f_n\ f_o + 10\ f_{i;j,k}\ f_{j;m}\ f_{k;1}\ f_{1;n}\ f_{m;o}\ f_n\ f_o + 15\ f_{i;j,k,n}\ f_{j;m}\ f_{k;1}\ f_1\ f_m\ f_{n;o}\ f_o +$
$15\ f_{i;j,n}\ f_{j;k}\ f_{k;1,m}\ f_1\ f_m\ f_{n;o}\ f_o + 3\ f_{i;j}\ f_{j;k}\ f_{k;1,n}\ f_{1;m}\ f_m\ f_{n;o}\ f_o + f_{i;j}\ f_{j;k}\ f_{k;1}\ f_{1;m}\ f_{m;n}\ f_{n;o}\ f_o$

Although `canonicalize` worked, it is of course a very slow function with a worst-case complexity $n!$. In general, the algorithmic simplification of tensor-like expressions is a very difficult problem [817], [1430], [1431], [1432]. For the simple case under consideration, it is relatively straightforward to implement a more efficient way to canonicalize the derivatives. As a first step towards canonicalization, we sort the products of f terms in such a way that the highest derivatives come first. Equal orders of differentiation are sorted lexicographically. To avoid automatic reordering inside the products, we change the head Times to List.

In[18]:= `productList[dk_] :=`
`        Apply[List, (List @@ Expand[dk]) //. i_Integer r__ :> r, {1}]`

In[19]:= `factorList[dk_] :=`
`        (List @@ Expand[dk])/Apply[Times, productList[dk], {1}]`

In[20]:= `sortFProducts[fProductList_] :=`
`        Last /@ Sort[R = {Length[#], #}& /@ fProductList,`
`                Which[#1[[1]] > #2[[1]], True,`
`                    #1[[1]] < #2[[1]], False,`
`                    (* lex *) #1[[1]] == #2[[1]],`
`                        OrderedQ[{#1[[2]], #2[[2]]}]]&]`

Here are the sorted terms of d4.

```
In[21]:= sortFProducts /@ productList[d4]

Out[21]= {{f[i][j, k, l], f[j][], f[k][], f[l][]}, {f[i][j, l], f[j][k], f[k][], f[l][]},
         {f[i][j, k], f[j][l], f[k][], f[l][]}, {f[j][k, l], f[i][j], f[k][], f[l][]},
         {f[i][j, k], f[k][l], f[j][], f[l][]}, {f[i][j], f[j][k], f[k][l], f[l][]}}
```

As a next step, we reindex the expressions in such a way that the indices are lexicograhically increasing. As a subsidiary function, we need UnsortedUnion. Given a list *l* with multiple elements, UnsortedUnion[*l*] returns a list where the multiple elements are dropped, but the order of the elements is not changed.

```
In[22]:= UnsortedUnion[l_] :=
         Module[{alreadyFound, bag = {}},
             (* step through and collect elements that are new *)
             If[Head[alreadyFound[#]] === alreadyFound,
                     alreadyFound[#] = True; bag = {bag, #}]& /@ l;
             (* return elements as they were found *)
             Flatten[bag]]

In[23]:= reindex[fProductList_] :=
         Module[{sortedIndices, indicesInOrder},
             (* sorted list of all indices *)
             sortedIndices = fProductList /. f[i_][jk___] :> {i, jk};
             (* remove doubles *)
             indicesInOrder = UnsortedUnion[DeleteCases[Flatten[sortedIndices], i]];
             (* replace by sorted list of indices *)
             reindexRules = Rule @@@ Transpose[{indicesInOrder, Sort[indicesInOrder]}];
             fProductList /. reindexRules]
```

For our simplest nontrivial expression, d4 we get now the following.

```
In[24]:= sortFProducts /@ reindex /@ sortFProducts /@ productList[d4]

Out[24]= {{f[i][j, k, l], f[j][], f[k][], f[l][]}, {f[i][j, k], f[j][l], f[k][], f[l][]},
         {f[i][j, k], f[j][l], f[k][], f[l][]}, {f[j][k, l], f[i][j], f[k][], f[l][]},
         {f[i][j, k], f[k][l], f[j][], f[l][]}, {f[i][j], f[j][k], f[k][l], f[l][]}}
```

Now comes the main step: reindexing the dummy variables such that the expressions become canonical. To do this we sequentially go through the product list and deal with each group of "identical" indices. The first time a dummy index is needed we use the first (in lexicographic order) one that is still available from the indices that are in the same group as the original index. Here is an example of how this works. Say we have the product f[*i*] [*j, k, l, m*] f[*m*] [*l*], and we encounter, say, index *m*. If *m* has not been seen so far, then make an assignment of the form indicesGroup[*m*]. If we encounter the index *m* the second time (meaning in f[*m*] [*l*]), then evaluate indicesGroup[*m*]. Take the first element of the returned result and use it as the new dummy variable (here *j*). Remove assignments of the form indicesGroup[*m*] and update other assignments for dummies from the same group as *m*. Add *j*→*m* and *m*→*j* to the renameList of rules collected. Then go to the next index. At the end apply renameRules in a recursive way.

```
In[25]:= canonicalizeFProduct[fProductList_] :=
         Module[{indicesGroup, renameRules = {}, reIndex, activeIndicesGroup,
                 indicesUsedWithinGroup, activeIndex, usableIndices, indexToBeUsed},
             (* recursively reindex *)
             reIndex[l_] := Fold[(# /. #2)&, l, Partition[renameRules, 2]];
             (* loop over all groups of variables *)
             Do[activeIndicesGroup = reIndex[fProductList[[k1]] /.
                     f -> List /. List[ii_][jk___] :> DeleteCases[{ii, jk}, i]];
                 indicesUsedWithinGroup = {};
                 (* loop over all variables from one group *)
                 Do[activeIndex = activeIndicesGroup[[k2]];
                     If[(* index appears for the first time *)
                        Head[indicesGroup[activeIndex]] === indicesGroup,
                            indicesGroup[activeIndex] =
                              DeleteCases[activeIndicesGroup,
                                          Alternatives @@ indicesUsedWithinGroup],
                     (* use the earliest available index *)
                     usableIndices = indicesGroup[activeIndex];
                     indexToBeUsed = First[usableIndices];
```

```
             AppendTo[indicesUsedWithinGroup, activeIndex];
             (* update the reindex rules *)
             renameRules = Flatten[{renameRules,
                               activeIndex -> indexToBeUsed,
                               indexToBeUsed -> activeIndex}];
             indicesGroup[indexToBeUsed] =. ;
             (* update definitions for indicesGroup *)
             DownValues[indicesGroup] = (RuleDelayed @@ {#[[1]],
             DeleteCases[Union[#[[2]]], indexToBeUsed, {-1}]})& /@
               (DownValues[indicesGroup] /. indexToBeUsed -> activeIndex)],
             {k2, Length[activeIndicesGroup]}],
             {k1, Length[fProductList]}];
        (* the reindexed result *)
        reIndex[fProductList] /. f[i_][l__] :> f[i] @@ Sort[{l}]]
```

Here are some examples demonstrating `canonicalizeFProducs` action.

```
In[26]:= canonicalizeFProduct[{f[i][j, k, 1, m], f[m][1], f[k][]}]

Out[26]= {f[i][j, k, 1, m], f[j][k], f[1][]}
```

```
In[27]:= canonicalizeFProduct[{f[i][j, k], f[k][1], f[j][], f[1][]}]

Out[27]= {f[i][j, k], f[j][1], f[k][], f[1][]}
```

```
In[28]:= canonicalizeFProduct[{f[i][j, k, 1], f[1][m], f[j][], f[k][], f[m][]}]

Out[28]= {f[i][j, k, 1], f[j][m], f[k][], f[1][], f[m][]}
```

```
In[29]:= canonicalize2[expr_] :=
         Module[{factors = factorList[expr],
                 fTerms = canonicalizeFProduct /@ sortFProducts /@
                   reindex /@ sortFProducts /@ productList[expr]},
         factors.(Times @@@ fTerms)]
```

`canonicalize2` works much faster than `canonicalize` and produces equivalent results (with differently ordered dummy indices).

```
In[30]:= canonicalize2[d4] // formatNicely

Out[30]= f_{i;j,k,1} f_j f_k f_1 + f_{i;j} f_{j;k,1} f_k f_1 + 3 f_{i;j,k} f_{j;1} f_k f_1 + f_{i;j} f_{j;k} f_{k;1} f_1
```

```
In[31]:= canonicalize2[d5 = d[d4, m]] // formatNicely

Out[31]= f_{i;j,k,1,m} f_j f_k f_1 f_m + f_{i;j} f_{j;k,1,m} f_k f_1 f_m + 4 f_{i;j,k} f_{j;1,m} f_k f_1 f_m +
         6 f_{i;j,k,1} f_{j;m} f_k f_1 f_m + 3 f_{i;j} f_{j;k,1} f_{k;m} f_1 f_m + 3 f_{i;j,k} f_{j;1} f_{k;m} f_1 f_m +
         4 f_{i;j,k} f_{j;1} f_k f_{1;m} f_m + f_{i;j} f_{j;k} f_{k;1} f_{1;m} f_m + f_{i;j} f_{j;k,1} f_k f_1 f_{m;j}
```

```
In[32]:= canonicalize2[d6 = d[d5, n]] // formatNicely

Out[32]= f_{i;j,k,1,m,n} f_j f_k f_1 f_m f_n + 5 f_{i;j,n} f_{j;k,1,m} f_k f_1 f_m f_n +
         f_{i;j} f_{j;k,1,m,n} f_k f_1 f_m f_n + 10 f_{i;j,k,1} f_{j;m,n} f_k f_1 f_m f_n + 10 f_{i;j,k,1,m} f_{j;n} f_k f_1 f_m f_n +
         4 f_{i;k} f_{j;m,n} f_{k;j,1} f_1 f_m f_n + 6 f_{i;j} f_{j;k,1,m} f_{k;n} f_1 f_m f_n + 10 f_{i;j,k} f_{j;1,m} f_{k;n} f_1 f_m f_n +
         15 f_{i;j,k,1} f_{j;m} f_{k;n} f_1 f_m f_n + 3 f_{i;j,k} f_{j;1} f_{k;n} f_{1;m} f_m f_n +
         5 f_{i;j,k} f_{j;1} f_k f_{1;m,n} f_m f_n + 15 f_{i;j,k,1} f_{j;1,m} f_k f_{1;n} f_m f_n + 3 f_{i;j} f_{j;k,1} f_{k;m} f_{1;n} f_m f_n +
         3 f_{i;j,k} f_{j;1} f_{k;m} f_{1;n} f_m f_n + 3 f_{i;m} f_{j;n} f_{k;j,1} f_1 f_{m;k} f_n + 10 f_{i;j,k,1} f_{j;m} f_k f_1 f_{m;n} f_n +
         4 f_{i;j} f_{j;k,1} f_{k;m} f_1 f_{m;n} f_n + 4 f_{i;j,k} f_{j;1} f_{k;m} f_1 f_{m;n} f_n + 5 f_{i;j,k} f_{j;1} f_k f_{1;m} f_{m;n} f_n +
         f_{i;j} f_{j;k} f_{k;1} f_{1;m} f_{m;n} f_n + f_{i;n} f_{j;k,1,m} f_k f_1 f_m f_{n;j} + f_{i;m} f_{j;k,1} f_k f_1 f_{m;n} f_{n;j}
```

Now the canonicalization of d7 is much faster. Canonicalization of all 5040 terms using `canonicalize2` is about ten times faster then the above canonicalization of the 120 terms of the differentiated precanonicalized d6 from above using `canonicalize`.

```
In[33]:= canonicalize2[d7 = d[d6, o]]; // Timing

Out[33]= {1.6 Second, Null}
```

For a tree-oriented approach to calculate such expressions, see [281], [1669], [1349], [247], [248], [280], [208], [331], [902], [903], [900], [1632], [1512], [475], [181], and [901]. For a Faà di Bruno-like formula for vector arguments, see [395], [1273], [580], [646], [5], and [1247]. For expressions through Bell polynomials, see [1346].

18. Trig Values in Radicals, Ramanujan Identities, Modular Transformation

a) The following procedural implementation is almost self-explanatory. The idea is to use `Solve` to solve the equations for $\cos(j\pi/n)$ and $\sin(j\pi/n)$, which arise from $\exp(i\,j\pi/n)^n = \exp(i\,j\pi) = (\cos(j\pi/n) + i\sin(j\pi/n))^n$.

Because `Solve` produces several solutions, we subsequently have to pick out the relevant one. We do this using numerical techniques. (We could also only look for a primitive root and generate other ones by exponentiation of this one.)

```
In[1]:= SinCosExact[arg_?((Head[Simplify[#/Pi]] === Rational &&
                    0 < N[#] < N[Pi/2])&)] :=
           Module[{denominator, numerator, exponentDenominator, re, im, poly,
                   sol, equations, exSol, nSol, dirSol, pre, pos, c, s},
             If[FreeQ[Cos[arg], Cos, {0, Infinity}], {Cos[arg], Sin[arg]},
               {denominator, numerator} = {Denominator[#],
                               Numerator[#]}&[Simplify[arg/Pi]];
               exponentDenominator = If[EvenQ[denominator], denominator/2, denominator];
               (* make two coupled polynomial equations for
                  Cos[arg] and for Sin[arg] *)
               poly = Expand[(c + I s)^exponentDenominator];
               im = Plus @@ Cases[poly, Complex[_, _] _];
               re = poly - im;
               equations = {re == Re[Exp[I arg exponentDenominator]],
                            Cancel[im/I] == Im[Exp[I arg exponentDenominator]]};
               (* solve the two equations *)
               sol = Solve[equations, {c, s}];
               (* analyze solution if they are all _solved_ *)
               If[sol == {{}} ||
                  MemberQ[sol, Roots, {0, Infinity}, Heads -> True],
                  Print["No analytic solution computable"],
                  exSol = {c, s} /. sol; nSol = N[exSol];
               (* compare with numerical value to choose the correct root *)
                  dirSol = {Cos[arg], Sin[arg]} // N;
                  pre = (Plus @@ #)& /@ Map[Abs, (# - dirSol& /@ nSol), {2}];
                  pos = Position[pre, Min[pre]];
                  Simplify[{c, s} /. sol[[pos[[1, 1]]]]]]]];
```

Here are a few examples (and a comparison with the numerical results).

```
In[2]:= arg = Pi/4; SinCosExact[arg]
```

$$\text{Out[2]= } \left\{ \frac{1}{\sqrt{2}}, \frac{1}{\sqrt{2}} \right\}$$

```
In[3]:= % == {Cos[arg], Sin[arg]}
```

Out[3]= True

```
In[4]:= arg = Pi/5; SinCosExact[arg]
```

$$\text{Out[4]= } \left\{ \frac{1}{4}\left(1 + \sqrt{5}\right), \frac{1}{2}\sqrt{\frac{1}{2}\left(5 - \sqrt{5}\right)} \right\}$$

```
In[5]:= % == {Cos[arg], Sin[arg]}
```

Out[5]= True

```
In[6]:= arg = 3Pi/8; SinCosExact[arg]
```

$$\text{Out[6]= } \left\{ \frac{1}{2}\left(-1 + \sqrt{2}\right)\sqrt{2 + \sqrt{2}}, \frac{\sqrt{2 + \sqrt{2}}}{2} \right\}$$

```
In[7]:= N[%] == {Cos[arg], Sin[arg]}
```

Out[7]= True

Unfortunately, the result of `Solve` frequently contains an explicit `I` (by Hölder's theorem, this is unavoidable [867]), and `Chop` only gives a real numerical solution. Further on, the following result contains more than square and cube roots.

```
In[8]:= arg = Pi/7; SinCosExact[arg]
```

Out[8]= $\{ \frac{1}{2} \left(19 + \frac{14\ 7^{2/3}}{(\frac{3}{2}(9+i\sqrt{3}))^{1/3}} + 7 \left(\frac{2}{3}\right)^{2/3} (63 + 7\ i\ \sqrt{3})^{1/3} \right)^{1/7}$,

$\frac{1}{2\ 3^{10/21}} \left(\sqrt{ \left(\frac{1}{-9\ i + \sqrt{3}} \left(\left(\frac{1}{-9\ i + \sqrt{3}} (-63\ i\ 2^{2/3}\ (21\ (9 + i\ \sqrt{3}))^{1/3} - 14\ i\ 2^{1/3}\ (21\ (9 + i\ \sqrt{3}))^{2/3} + \right. \right. \right. \right.$

$7\ 2^{2/3}\ 3^{5/6}\ (63 + 7\ i\ \sqrt{3})^{1/3} + 57\ (-9\ i + \sqrt{3})))^{2/7}\ (-2\ 6^{1/3}\ (63 + 7\ i\ \sqrt{3})^{2/3}$

$\left. \left. \left. \left. (-i + \sqrt{3}) + 4\ 2^{2/3}\ (63 + 7\ i\ \sqrt{3})^{1/3}\ (3\ i + 2\ \sqrt{3}) + 21\ (3^{1/6} - 3\ i\ 3^{2/3})) \right) \right) \right) \right\}$

In[9]:= **N[%] == {Cos[arg], Sin[arg]}**

Out[9]= False

Due to space limitations, we do not carry out the computation of all sin and cos of possible numerator–denominator arrangements, as suggested in the exercise.

In[10]:= **Sort @ Union @ Flatten[Range[1, # - 1]/#& /@ Range[1, 10]]**

Out[10]= $\{ \frac{1}{10}, \frac{1}{9}, \frac{1}{8}, \frac{1}{7}, \frac{1}{6}, \frac{1}{5}, \frac{2}{9}, \frac{1}{4}, \frac{2}{7}, \frac{3}{10}, \frac{1}{3}, \frac{3}{8}, \frac{2}{5}, \frac{3}{7},$
$\frac{4}{9}, \frac{1}{2}, \frac{5}{9}, \frac{4}{7}, \frac{3}{5}, \frac{5}{8}, \frac{2}{3}, \frac{7}{10}, \frac{5}{7}, \frac{3}{4}, \frac{7}{9}, \frac{4}{5}, \frac{5}{6}, \frac{6}{7}, \frac{7}{8}, \frac{8}{9}, \frac{9}{10} \}$

We can also find exact solutions for larger denominators. Here is an example.

In[11]:= **arg = 5 Pi/14; SinCosExact[arg]**

Out[11]= $\{ \frac{1}{2\ 3^{10/21}}$

$\left(\sqrt{ \left(\frac{1}{-9\ i + \sqrt{3}} \left(\left(\frac{1}{-9\ i + \sqrt{3}} (-63\ i\ 2^{2/3}\ (21\ (9 + i\ \sqrt{3}))^{1/3} - 14\ i\ 2^{1/3}\ (21\ (9 + i\ \sqrt{3}))^{2/3} + 7\ 2^{2/3} \right. \right. \right.$

$3^{5/6}\ (63 + 7\ i\ \sqrt{3})^{1/3} + 57\ (-9\ i + \sqrt{3})))^{2/7}\ (-2\ 6^{1/3}\ (63 + 7\ i\ \sqrt{3})^{2/3}\ (-i + \sqrt{3}) +$

$\left. \left. \left. 4\ 2^{2/3}\ (63 + 7\ i\ \sqrt{3})^{1/3}\ (3\ i + 2\ \sqrt{3}) + 21\ (3^{1/6} - 3\ i\ 3^{2/3})) \right) \right) \right),$

$\frac{1}{2} \left(19 + \frac{14\ 7^{2/3}}{(\frac{3}{2}(9 + i\ \sqrt{3}))^{1/3}} + 7 \left(\frac{2}{3}\right)^{2/3} (63 + 7\ i\ \sqrt{3})^{1/3} \right)^{1/7} \}$

In[12]:= **{%, {Cos[arg], Sin[arg]}} // N**

Out[12]= $\{\{0.433884 + 1.21938 \times 10^{-15}\ i,\ 0.900969 + 4.63312 \times 10^{-19}\ i\},\ \{0.433884, 0.900969\}\}$

In some sense, the solution of this exercise could be written straightforwardly, without using *Mathematica* at all. It depends on which roots one allows in the result [375], [1821]. If we also allow any order, then we can immediately write the following expressions for cos($j 2\pi/n$) and sin($j 2\pi/n$):

$$\cos(j\ 2\ \pi/n) = \frac{1}{2}\ ((-1)^{2\ j/n} + (-1)^{-2\ j/n})$$

$$\sin(j\ 2\ \pi/n) = \frac{1}{2\ i}\ ((-1)^{2\ j/n} - (-1)^{-2\ j/n}).$$

Last, but not least, we could use the function FunctionExpand (to be discussed in Chapter 3) to express $\cos\left(j\ 2\ \frac{\pi}{n}\right)$ in radicals. Here are two examples.

In[13]:= **FunctionExpand[Cos[Pi/9]]**

Out[13]= $\frac{1}{2} \left(-\frac{(-1 - i\ \sqrt{3})^{4/3}}{2\ 2^{1/3}} - \frac{(-1 + i\ \sqrt{3})^{4/3}}{2\ 2^{1/3}} \right)$

In[14]:= **FunctionExpand[Sin[Pi/14]]**

$$\text{Out[14]= } \frac{1}{2}\left(\frac{1}{3}\left(\frac{1}{2}\left(1-i\sqrt{7}\right)-\frac{\frac{1}{2}\left(-1+i\sqrt{7}\right)+\frac{1}{2}\left(-1-i\sqrt{7}\right)\left(\frac{1}{2}\left(-1+i\sqrt{3}\right)+\frac{1}{4}\left(-1+i\sqrt{3}\right)^2\right)}{\left(6+\frac{3}{4}\left(-1+i\sqrt{3}\right)\left(-1+i\sqrt{7}\right)+\frac{1}{2}\left(-1-i\sqrt{7}\right)\left(1+\frac{3}{4}\left(-1+i\sqrt{3}\right)^2\right)\right)^{1/3}}-\right.$$

$$\left(6+\frac{3}{4}\left(-1+i\sqrt{3}\right)\left(-1+i\sqrt{7}\right)+\frac{1}{2}\left(-1-i\sqrt{7}\right)\left(1+\frac{3}{4}\left(-1+i\sqrt{3}\right)^2\right)\right)^{1/3}\right)+\frac{1}{3}\left(\frac{1}{2}\right.$$

$$\left(1+i\sqrt{7}\right)-$$

$$\frac{\left(-1+i\sqrt{3}\right)\left(\frac{1}{2}\left(-1-i\sqrt{7}\right)+\frac{1}{2}\left(-1+i\sqrt{7}\right)\left(\frac{1}{2}\left(-1+i\sqrt{3}\right)+\frac{1}{4}\left(-1+i\sqrt{3}\right)^2\right)\right)}{2\left(6+\frac{3}{4}\left(-1+i\sqrt{3}\right)\left(-1-i\sqrt{7}\right)+\frac{1}{2}\left(-1+i\sqrt{7}\right)\left(1+\frac{3}{4}\left(-1+i\sqrt{3}\right)^2\right)\right)^{1/3}}-$$

$$\left.\left.\frac{1}{4}\left(-1+i\sqrt{3}\right)^2\left(6+\frac{3}{4}\left(-1+i\sqrt{3}\right)\left(-1-i\sqrt{7}\right)+\frac{1}{2}\left(-1+i\sqrt{7}\right)\left(1+\frac{3}{4}\left(-1+i\sqrt{3}\right)^2\right)\right)^{1/3}\right)\right)$$

b) As discussed in Subsection 1.10.2, the denominator q of a fraction p/q must have the following form to allow sin and cos to be expressed in the required manner $q = 2^n\, 3^\alpha\, 5^\beta\, 17^\gamma\, furtherFermatPromesToPower0r1$, where n is a nonnegative integer and α, β, and γ are either 0 or 1.

Because we know explicit expressions involving only square roots for the fractions $p_3/3$, $p_5/5$, and $p_{17}/17$, and because by using half-angle formulas and multiple-angle formulas we can produce expressions involving only square roots for arguments of the form $\bar{p}/2^n\,\pi$, the first step is to express p/q in the following form (where the denominators actually present are given by the prime factorization of $q = 2^i\, 3^{0Or1}\, 5^{0Or1}\, 17^{0Or1}$): $p/q = p_2/2^i + p_3/3 + p_5/5 + p_{17}/17$.

Due to the size of the expressions calculated in Section 1.8.2 for $\cos(2\pi/257)$, here and in the following, we assume that no larger Fermat prime than 17 is present in the denominator; the inclusion of the case that larger ones are present is straightforward.

Multiplying the above equation by q, we get a linear Diophantine equation in several unknowns (p_2, p_3, p_5, and p_{17}). There is a straightforward way to solve a Diophantine equation of the following form: $a_1 x_1 + a_2 x_2 + \cdots + a_n x_n = d$.

Here, the a_i and d are given integer coefficients and the x_i are the unknowns. We search for the smallest coefficient a_j among the a_i, and solve for the corresponding x_j. In the resulting expression $x_j = \sum_{\substack{k=1 \\ k\neq j}}^{n} a_k/a_j\, x_k$ the integer part of the a_k/a_j is explicitly separated (because a_j is the smallest coefficient, there is always a nonzero integer part)

$$x_j = \sum_{\substack{k=1 \\ k\neq j}}^{n}\left(\frac{a_k}{a_j}\right)_{(int)} x_k + \sum_{\substack{k=1 \\ k\neq j}}^{n}\left(\frac{a_k}{a_j}\right)_{(fract)} x_k.$$

Because we look for integer solutions x_j only, the second sum again must be an (still undetermined) integer b_j. So, we have a new equation of a form similar to the starting equation with one less unknown variable $\sum_{\substack{k=1 \\ k\neq j}}^{n}(a_k/a_j)_{(fract)}\, x_k = b_j$.

Now selecting again the smallest coefficient among the coefficients of this sum, the above process is repeated until it ends with one equation connecting the last two coefficients. Then the calculated values for the parameters b_j, ... are substituted back. Here, this algorithm is implemented. We use C[i] as the undetermined parameters of the solution. (There are of course more efficient, number-theory-based algorithms to solve linear Diophantine equations. But because of the obvious working principle of the algorithm described above, we use this one.)

```
In[1]:= eliminateSmallestPart[eq_, i_, vars_] :=
         Module[{coeffTimesVarList, coeffList, positionSmallestCoeff,
                 varToSolveFor, eqSolved, integerPart, fractionalPart},
           (* find smallest coefficient *)
           coeffTimesVarList =
               Cases[eq, Alternatives @@ Join[_ #& /@ vars, {_ _C}]];
           coeffList = coeffTimesVarList //. Join[# -> 1& /@ vars, {_C -> 1}];
           positionSmallestCoeff = If[Length[coeffTimesVarList] > 1,
                   Position[Abs[coeffList], Min[Abs[coeffList]]][[1, 1]], 1];
           (* variable corresponding to smallest coefficient *)
           varToSolveFor = Cases[coeffTimesVarList[[positionSmallestCoeff]],
                   Alternatives @@ Join[#& /@ vars, {_C}] ];
           (* solve for this variable *)
           eqSolved = Solve[eq == 0, varToSolveFor][[1, 1, 2]] // Expand;
```

```
                (* write all fractions in the form (integerPart + fractionalPart) *)
                integerPart = eqSolved //. r_Rational :> Round[r];
                fractionalPart = eqSolved - integerPart;
                {varToSolveFor[[1]] -> integerPart + C[i], C[i] - fractionalPart}]
In[2]:= linearDiophantineSolve[equation_Equal, varList_] :=
        Module[{equation1, coeffs, rhs, currentEquation, aux, integerPartBag,
                n, substitutedSolutions, substitutedSolutionswithOutTempVars,
                rhss, remainingFreeVars, additionalParameterVars, rules,
                expandAll = If[Head[#] === Rule, ExpandAll /@ #, ExpandAll @ #]&},
          (* write equation in form Sum[coeff[i] var[i], {i, m}] = integer *)
          equation1 = Subtract @@ equation;
          coeffs = Coefficient[equation1, #]& /@ varList;
          rhs = Expand[coeffs.varList - equation1];
          (* solvability condition *)
          If[Mod[rhs, GCD @@ coeffs] =!= 0, {},
          (* recursively solve for var with smallest
             coefficient until process terminates *)
            integerPartBag = {};
            currentEquation = equation1; i = 0;
            While[i = i + 1; aux = eliminateSmallestPart[
                                          currentEquation, i, varList];
                  currentEquation = aux[[2]];
                  If[i > 1,
                      (* termination conditions *)
                      currentEquation =!= C[i] &&
                      (MemberQ[currentEquation,
                             Alternatives @@ varList, {0, Infinity}] ||
                       MemberQ[Solve[currentEquation == 0, C[i - 1]][[1, 1, 2]],
                             _Rational, {0, Infinity}]), True],
                  AppendTo[integerPartBag, aux]];
            AppendTo[integerPartBag, aux];
            n = Length[integerPartBag];
            sol = If[MemberQ[integerPartBag[[-1, 2]], C[n - 1], {0, Infinity}],
                     Solve[integerPartBag[[-1, 2]] == 0, C[n - 1]], {{C[n] -> 0}}];
            (* substitute solutions back *)
            substitutedSolutions =
              Fold[{Flatten[{expandAll[#2 //. Flatten[#1]], #1}]}&,
                   {expandAll[integerPartBag[[-1, 1]] //. sol[[1, 1]]]},
                        Flatten[(First /@ Rest[Reverse[integerPartBag]]) //. sol]];
            substitutedSolutionswithOutTempVars =
                   DeleteCases[Flatten[substitutedSolutions], _C -> _];
            rhss = Last /@ substitutedSolutionswithOutTempVars;
            (* introduce parameter variables in
               case a variable from varList is used *)
            remainingFreeVars = Union[Cases[rhss, Alternatives @@ varList,
                                            {0, Infinity}]];
            additionalParameterVars =
                   Table[C[j], {j, i + 1, i + 1 + Length[remainingFreeVars] - 1}];
            (* make rules for the result *)
            rules = Apply[Rule, Transpose[{remainingFreeVars,
                                          additionalParameterVars}], {1}];
            (* return result, recursively backsubstituted *)
            Flatten[{substitutedSolutionswithOutTempVars //. rules, rules}]]] /;
          (* test if equation is a linear diophantine equation *)
          Function[eq1, Function[coeffs, Not[And @@ (NumberQ /@ N[varList])] &&
            Union[Exponent[eq1, varList]] === {1} && And @@ (IntegerQ /@ coeffs) &&
            IntegerQ[Expand[eq1 - coeffs.varList]]][
                  Coefficient[eq1, #]& /@ varList]][Subtract @@ equation]
```

Let us test `linearDiophantineSolve` using a few examples.

```
In[3]:= linearDiophantineSolve[16 x[1] + 17 x[2] == 2, {x[1], x[2]}]

Out[3]= {x[1] -> -2 + 17 C[1], x[2] -> 2 - 16 C[1]}
```

```
In[4]:= 16 x[1] + 17 x[2] == 2 //. % // ExpandAll

Out[4]= True

In[5]:= linearDiophantineSolve[17 x[1] + 5 x[2] == 15, {x[1], x[2]}]

Out[5]= {x[2] → 3 - 17 C[2], x[1] → 5 C[2]}

In[6]:= 17 x[1] + 5 x[2] == 15 //. % // ExpandAll

Out[6]= True

In[7]:= linearDiophantineSolve[17 x + 5 y + 8 z == 7, {x, y, z}]

Out[7]= {y → -2 - 17 C[2] - 5 C[3], x → 1 + 5 C[2] + C[3], z → C[3]}

In[8]:= 17 x + 5 y + 8 z == 7 //. % // ExpandAll

Out[8]= True
```

While it is instructive to implement a solution of a class of special equations, we could, of course, also just have used Reduce with the domain of integers specified.

```
In[9]:= linearDiophantineSolveR[eqs_, vars_] :=
          {ToRules[DeleteCases[Reduce[eqs, vars, Integers], _Element]]}

In[10]:= linearDiophantineSolveR[17 x + 5 y + 8 z == 7, {x, y, z}]

Out[10]= {{x → C[1], y → 3 + 3 C[1] + 8 C[2], z → -1 - 4 C[1] - 5 C[2]}}

In[11]:= 17 x + 5 y + 8 z == 7 //. % // ExpandAll

Out[11]= {True}
```

Now, we use the routine linearDiophantineSolve for determining the numerators in the above decomposition of p/q. Actually, we do not need the full parametrized solution, but only one special one. Because we do not need explicit variable names for the unknowns later, the following function lDS takes as its arguments a list of the coefficients a_i and the right-hand side d, and it returns a list of a special solution set x_i. (One special solution can be obtained in a shorter way; see [269], [1507], and [1356].)

```
In[12]:= lDS[abList_, d_] :=
          Module[{vars, sol, res, paramVars},
            vars = Table[Unique[x], {Length[abList]}];
            sol = linearDiophantineSolve[vars.abList == d, vars];
            res = vars /. sol;
            paramVars = Union[Cases[res, _C, {0, Infinity}]];
            res /. Apply[Rule, (* just take all parameters equal to 0 *)
                  Transpose[{paramVars, Table[0, {Length[paramVars]}]}], {1}]]
```

Let us test lDS again using a few examples.

```
In[13]:= lDS[{123, 73, 90}, 3]

Out[13]= {2, 9, -10}

In[14]:= %.{123, 73, 90}

Out[14]= 3

In[15]:= lDS[{15, 40, 24}, 1]

Out[15]= {-1, 1, -1}

In[16]:= %.{15, 40, 24}

Out[16]= 1

In[17]:= lDS[{11, -29, 17}, -3]

Out[17]= {-4, -2, -1}

In[18]:= %.{11, -29, 17}

Out[18]= -3
```

Now, we can implement a function `sumOfFractions` that gives the summands of the partial fraction decomposition of a fraction p/q with the denominators being given by the prime factorization of q.

```
In[19]:= sumOfFractions[frac:Rational[p_, q_]] :=
         Module[{fa0 = FactorInteger[q], fa, lds},
             If[Length[fa0] > 1,
                 fa = Apply[Power, fa0, {1}];
                 lds = lDS @@ {1/fa q, Numerator[frac]};
                 tmp = lds/fa; If[(Plus @@ tmp) > 1, Append[tmp, -1], tmp], frac]]
```

Here are two examples.

```
In[20]:= sumOfFractions[1/120]
```

$$\text{Out[20]}= \left\{-\frac{1}{8}, \frac{1}{3}, -\frac{1}{5}\right\}$$

```
In[21]:= Plus @@ %
```

$$\text{Out[21]}= \frac{1}{120}$$

```
In[22]:= sumOfFractions[13/(2^3 5 17)]
```

$$\text{Out[22]}= \left\{\frac{9}{8}, -\frac{2}{5}, -\frac{12}{17}\right\}$$

```
In[23]:= Plus @@ %
```

$$\text{Out[23]}= \frac{13}{680}$$

To test if a given fraction belongs to the class that represents arguments so that the corresponding trigonometric expressions can be expressed in square roots, we implement a function `squareRootableQ[`p/q`]` that returns `True` or `False` depending on whether it is possible to express trigonometric functions of $p/q\,\pi$ by using only addition, subtraction, multiplication, division, and square rooting.

```
In[24]:= squareRootableQ[frac:Rational[p_, q_]] := squareRootableQ[frac] =
         Module[{fa0 = FactorInteger[q], oddPrimeParts},
             If[(* are the only primes 2 and/or 3 and/or 5 and/or 17? *)
                 Complement[First /@ fa0, {2, 3, 5, 17}] === {},
                 (* are the primes and/or 3 and/or 5 and/or 17 with power 0 or 1? *)
                 oddPrimeParts = Cases[fa0, {3, _} | {5, _} | {17, _}];
                 If[Length[oddPrimeParts] > 0,
                     If[Union[Last /@ oddPrimeParts] =!= {1}, False, True], True],
                 False]]
```

Taking into account the above-stated conditions when this can be done, we easily verify the following examples.

```
In[25]:= squareRootableQ[1/(2^3 5^3)]
```

Out[25]= False

```
In[26]:= squareRootableQ[1/(2^4 3 5 17)]
```

Out[26]= True

```
In[27]:= squareRootableQ[131/(2^12 17)]
```

Out[27]= True

To actually express a trigonometric function of $p/q\,\pi$ we need explicit expressions for basic fractions of the form $p_2/2^n\,\pi$, $p_3/3\,\pi$, $p_5/5\,\pi$, $p_{17}/17\,\pi$. For fractions less than $1/2$, we implement these here. We use our own functions `sin` and `cos` here, to avoid any autosimplification of built-in functions. (The other trigonometric functions we express later through `sin` and `cos`.)

```
In[28]:= sin[0] = 0;
         cos[0] = 1;

         sin[Pi/3] = Sqrt[3]/2;
         cos[Pi/3] = 1/2;

         sin[Pi/5] = Sqrt[(5 - Sqrt[5])/2]/2;
         cos[Pi/5] = (1 + Sqrt[5])/4;
```

```
        sin[2 Pi/5] = Sqrt[(5 + Sqrt[5])/2]/2;
        cos[2 Pi/5] = (-1 + Sqrt[5])/4;
In[36]:= (* use some abbreviations to shorten definitions *)
        With[{s = Sqrt}, Module[{α = s[17 - s[17]], β = s[17 + s[17]]},
        cos[1 Pi/17] = s[15 + s[17] + s[2] α + s[2] s[34 + 6 s[17] - s[2] α +
                       s[34] α - 8 s[2] β]]/(4 s[2]);
        sin[1 Pi/17] = s[8 - s[2] s[15 + s[17] - s[2] α +
                       s[2] s[34 + 6 s[17] + s[2] α - s[34] α + 8 s[2] β]]]/4;
        cos[2 Pi/17] = s[15 + s[17] - s[2] α + s[2] s[34 + 6 s[17] + s[2] α -
                       s[34] α + 8 s[2] β]]/(4 s[2]);
        sin[2 Pi/17] = s[8 - s[2] s[15 + s[17] + s[2] α - s[2] s[34 + 6 s[17] -
                       s[2] α + s[34] α - 8 s[2] β]]]/4;
        cos[3 Pi/17] = s[15 - s[17] + s[2] β + s[2] s[34 - 6 s[17] + 8 s[2] α -
                       s[2] β - s[34] β]]/(4 s[2]);
        sin[3 Pi/17] = s[8 - s[2] s[15 - s[17] - s[2] β + s[2] s[34 - 6 s[17] -
                       8 s[2] α + s[2] β + s[34] β]]]/4;
        cos[4 Pi/17] = s[15 + s[17] + s[2] α - s[2] s[34 + 6 s[17] - s[2] α +
                       s[34] α - 8 s[2] β]]/(4 s[2]);
        sin[4 Pi/17] = s[8 - s[2] s[15 + s[17] - s[2] α - s[2] s[34 + 6 s[17] +
                       s[2] α - s[34] α + 8 s[2] β]]]/4;
        cos[5 Pi/17] = s[15 - s[17] + s[2] β - s[2] s[34 - 6 s[17] + 8 s[2] α -
                       s[2] β - s[34] β]]/(4 s[2]);
        sin[5 Pi/17] = s[17 + s[17] - s[2] β + s[2] s[34 - 6 s[17] + 8 s[2] α -
                       s[2] β - s[34] β]]/(4 s[2]);
        cos[6 Pi/17] = s[15 - s[17] - s[2] β + s[2] s[34 - 6 s[17] - 8 s[2] α +
                       s[2] β + s[34] β]]/(4 s[2]);
        sin[6 Pi/17] = s[17 + s[17] + s[2] β - s[2] s[34 - 6 s[17] - 8 s[2] α +
                       s[2] β + s[34] β]]/(4 s[2]);
        cos[7 Pi/17] = s[15 - s[17] - s[2] β - s[2] s[34 - 6 s[17] - 8 s[2] α +
                       s[2] β + s[34] β]]/(4 s[2]);
        sin[7 Pi/17] = s[17 + s[17] + s[2] β + s[2] s[34 - 6 s[17] - 8 s[2] α +
                       s[2] β + s[34] β]]/(4 s[2]);
        cos[8 Pi/17] = s[15 + s[17] - s[2] α - s[2] s[34 + 6 s[17] + s[2] α -
                       s[34] α + 8 s[2] β]]/(4 s[2]);
        sin[8 Pi/17] = s[17 - s[17] + s[2] α + s[2] s[34 + 6 s[17] + s[2] α -
                       s[34] α + 8 s[2] β]]/(4 s[2])]];
```

Now remaining is the implementation of the reduction of every "square rootable" trigonometric function to the previously-defined cases.

All trigonometric functions are periodic with period 2π.

```
In[38]:= cos[(fract_Rational | fract_Integer) Pi] :=
            (cos[fract Pi] = cos[Mod[fract, 2] Pi]) /; fract >= 2

        sin[(fract_Rational | fract_Integer) Pi] :=
            (sin[fract Pi] = sin[Mod[fract, 2] Pi]) /; fract >= 2
```

Trigonometric functions with arguments $\pi < p/q\,\pi < 2\pi$ can be reduced to arguments in the range $(0, \pi)$.

```
In[40]:= cos[(fract_Rational | fract_Integer) Pi] :=
            (cos[fract Pi] = +cos[(2 - fract) Pi]) /; fract >= 1

        sin[(fract_Rational | fract_Integer) Pi] :=
            (sin[fract Pi] = -sin[(2 - fract) Pi]) /; fract >= 1
```

We can actually go one step further: trigonometric functions with arguments $\pi/2 < p/q\,\pi < \pi$ can be reduced to arguments in the range $(0, \pi/2)$.

```
In[42]:= cos[fract_Rational Pi] :=
            (cos[fract Pi] = -cos[(1 - fract) Pi]) /; fract > 1/2

        sin[fract_Rational Pi] :=
            (sin[fract Pi] = +sin[(1 - fract) Pi]) /; fract > 1/2
```

And in the case $q = 2^n$, we can go still one step further and reduce all arguments to the range $(0, \pi/4)$.

```
In[44]:= cos[fract_Rational Pi] :=
            (cos[fract Pi] = sin[(1/2 - fract) Pi]) /; fract > 1/4

         sin[fract_Rational Pi] :=
            (sin[fract Pi] = cos[(1/2 - fract) Pi]) /; fract > 1/4
```

Taking again into account the periodicity of trigonometric functions, negative arguments can always be transformed to positive ones.

```
In[46]:= cos[fract_Rational Pi] :=
            (cos[fract Pi] = +cos[-fract Pi]) /; fract < 0

         sin[fract_Rational Pi] :=
            (sin[fract Pi] = -sin[-fract Pi]) /; fract < 0
```

The next rule governs the case when the numerator is 1 and the denominator is a power of 2 by recursively using half-angle formulas.

```
In[48]:= cos[fract_Rational Pi] :=
            (cos[fract Pi] = Sqrt[Together[(1 + cos[2 fract Pi])/2]]) /;
               fract > 0 && MatchQ[FactorInteger[1/fract], {{2, _?(# > 1&)}}] &&
               Numerator[fract] === 1

         sin[fract_Rational Pi] :=
            (sin[fract Pi] = Sqrt[Together[(1 - cos[2 fract Pi])/2]]) /;
               fract > 0 && MatchQ[FactorInteger[1/fract], {{2, _?(# > 1&)}}] &&
               Numerator[fract] === 1
```

Using multiple-angle formulas, arguments $p/q\,\pi$ can be reduced to a sum of products of trigonometric functions with argument $1/q\,\pi$.

```
In[50]:= cos[fract_Rational Pi] := (cos[fract Pi]  =
         With[{z = 1/Denominator[fract] Pi, n = Numerator[fract]},
           Factor @ Together @
           (* this might be a bit dangerous from a timing point of view
              for large n *)
           Sum[(-1)^k Binomial[n, 2k] cos[z]^(n - 2k) (1 - cos[z]^2)^k,
             {k, 0, n/2}]]) /;
               Numerator[fract] > 1 && IntegerQ[Log[2, Denominator[fract]]]

         sin[fract_Rational Pi] := (sin[fract Pi]  =
         With[{z = 1/Denominator[fract] Pi, n = Numerator[fract]},
           Factor @ Together @
           (* this might be a bit dangerous from a timing point of view *)
           Sum[(-1)^(k) Binomial[n, 2k + 1] sin[z]^(2k + 1) cos[z]^(n - 2k - 1),
             {k, 0, (n - 1)/2}]]) /;
               Numerator[fract] > 1 && IntegerQ[Log[2, Denominator[fract]]]
```

The last rules govern the generic case $p/q\,\pi$. If the argument is such that the trigonometric function can be expressed in square roots, the decomposition of p/q is actually carried out and the resulting sum of products of trigonometric functions contains the already implemented cases.

```
In[52]:= cos[fract_Rational Pi] := (cos[fract Pi]  =
         Module[{singles, auxVars, expa},
           (* write fract as a sum of fractions with
              denominators 3 or 5 or 17 or powers of 2 *)
           singles = sumOfFractions[fract];
           (* introduce dummy variable for expanding the sum of fractions *)
           auxVars = Table[cc[i], {i, Length[singles]}];
           expa = cos[Plus @@ auxVars]  //.
                       {sin[x_ + y_] :> sin[x] cos[y] + cos[x] sin[y],
                        cos[x_ + y_] :> cos[x] cos[y] - sin[x] sin[y]};
           (* substitute fractions for dummy variables *)
           expa //. Apply[Rule, Transpose[{auxVars, singles Pi}], {1}]]) /;
               squareRootableQ[fract]

         sin[fract_Rational Pi]  := (sin[fract Pi]  =
         Module[{singles, auxVars, expa},
```

```
(* write fract as a sum of fractions with
   denominators 3 or 5 or 17 or powers of 2 *)
singles = sumOfFractions[fract];
(* introduce dummy variable for expanding the sum of fractions *)
auxVars = Table[cc[i], {i, Length[singles]}];
expa = sin[Plus @@ auxVars]  //.
          {sin[x_ + y_] :> sin[x] cos[y] + cos[x] sin[y],
           cos[x_ + y_] :> cos[x] cos[y] - sin[x] sin[y]};
(* substitute fractions for dummy variables *)
expa //. Apply[Rule, Transpose[{auxVars, singles Pi}], {1}]]) /;
squareRootableQ[fract]
```

We reduce the trigonometric functions tan, cot, sec, and csc to rational functions of sin and cos with the same arguments.

```
In[54]:= tan[z:(fract_Rational Pi)] := (tan[z] = sin[z]/cos[z]) /;
                                            squareRootableQ[fract]

         cot[z:(fract_Rational Pi)] := (cot[z] = cos[z]/sin[z]) /;
                                            squareRootableQ[fract]

         sec[z:(fract_Rational Pi)] := (sec[z] = 1/cos[z]) /;
                                            squareRootableQ[fract]

         csc[z:(fract_Rational Pi)] := (csc[z] = 1/sin[z]) /;
                                            squareRootableQ[fract]
```

Now, we have everything together to define a function `TrigToSquareRoots[expr]` that expresses every trigonometric function in *expr*, which can be expressed in square roots.

```
In[58]:= TrigToSquareRoots[expr_] :=
         (expr //. {Sin -> sin, Cos -> cos, Tan -> tan, Cot -> cot,
                    Sec -> sec, Csc -> csc}) //.
            (* restore nontransformed expressions *)
                {sin -> Sin, cos -> Cos, tan -> Tan, cot -> Cot,
                 sec -> Sec, csc -> Csc}
```

Here are a few examples. First are some fractions with denominators being fractions of 2.

```
In[59]:= sin[1/16Pi]
```

$$\text{Out[59]}= \frac{1}{2}\sqrt{2 - \sqrt{2 + \sqrt{2}}}$$

```
In[60]:= sin[Pi/256]
```

$$\text{Out[60]}= \frac{1}{2}\sqrt{2 - \sqrt{2 + \sqrt{2 + \sqrt{2 + \sqrt{2 + \sqrt{2 + \sqrt{2}}}}}}}$$

Here are two compound denominators.

```
In[61]:= cos[1/(2 3 5) Pi]
```

$$\text{Out[61]}= -\frac{1}{8}\sqrt{3}\left(-1-\sqrt{5}\right) + \frac{1}{4}\sqrt{\frac{1}{2}\left(5-\sqrt{5}\right)}$$

```
In[62]:= cos[1/272 Pi]
```

Out[62]= $\frac{1}{8} \sqrt{\left| \frac{1}{2} \left(2 + \sqrt{2 + \sqrt{2}}\right) \left(15 + \sqrt{17} + \sqrt{2\left(17 - \sqrt{17}\right)} + \right.\right.}$

$$\left.\left.\sqrt{2\left(34 + 6\sqrt{17} - \sqrt{2\left(17 - \sqrt{17}\right)} + \sqrt{34\left(17 - \sqrt{17}\right)} - 8\sqrt{2\left(17 + \sqrt{17}\right)}\right)}\right)\right|} +$$

$$\frac{1}{8} \sqrt{\left| \left(2 - \sqrt{2 + \sqrt{2}}\right) \left(8 - \sqrt{\frac{1}{2}\left(2\left(15 + \sqrt{17} - \sqrt{2\left(17 - \sqrt{17}\right)} + \right.\right.}\right.\right.}$$

$$\left.\left.\left.\left.\sqrt{2\left(34 + 6\sqrt{17} + \sqrt{2\left(17 - \sqrt{17}\right)} - \sqrt{34\left(17 - \sqrt{17}\right)} + 8\sqrt{2\left(17 + \sqrt{17}\right)}\right)}\right)\right)}\right)\right|}$$

These are all fractions $1/q$ with numerator 1, which allow the corresponding trigonometric functions with argument π/q to be expressed in square roots.

In[63]:= `Select[1/Range[1000], squareRootableQ]`

Out[63]= $\left\{ \frac{1}{2}, \frac{1}{3}, \frac{1}{4}, \frac{1}{5}, \frac{1}{6}, \frac{1}{8}, \frac{1}{10}, \frac{1}{12}, \frac{1}{15}, \frac{1}{16}, \frac{1}{17}, \frac{1}{20}, \frac{1}{24}, \frac{1}{30}, \frac{1}{32}, \frac{1}{34}, \frac{1}{40}, \frac{1}{48}, \frac{1}{51}, \right.$
$\frac{1}{60}, \frac{1}{64}, \frac{1}{68}, \frac{1}{80}, \frac{1}{85}, \frac{1}{96}, \frac{1}{102}, \frac{1}{120}, \frac{1}{128}, \frac{1}{136}, \frac{1}{160}, \frac{1}{170}, \frac{1}{192}, \frac{1}{204}, \frac{1}{240}, \frac{1}{255},$
$\left. \frac{1}{256}, \frac{1}{272}, \frac{1}{320}, \frac{1}{340}, \frac{1}{384}, \frac{1}{408}, \frac{1}{480}, \frac{1}{510}, \frac{1}{512}, \frac{1}{544}, \frac{1}{640}, \frac{1}{680}, \frac{1}{768}, \frac{1}{816}, \frac{1}{960} \right\}$

We test all of these fractions using the `Cos` function.

In[64]:= `Chop[N[TrigToSquareRoots[Cos[# Pi]] - Cos[# Pi]]& /@ %] // Union`

Out[64]= `{0}`

Here is a more complicated expression, which contains various trigonometric functions and arguments.

In[65]:= `largeExpr = Sin[Pi/15] + Cos[Pi/544] + Tan[Pi/17] + Sec[5 Pi/256] //`
$\qquad\qquad$ `TrigToSquareRoots;`

This expression contains quite a few square roots.

In[66]:= `Count[largeExpr, Power[_, 1/2] | Power[_, -1/2]]`

Out[66]= 0

This is a quick numerical test using 100 digits that the applied transformation was correct.

In[67]:= `Off[N::meprec];`
$\qquad\quad$ `N[Sin[Pi/15] + Cos[Pi/544] + Tan[Pi/17] + Sec[5 Pi/256] -`
$\qquad\quad$ `largeExpr, 1000]`
Out[68]= $0. \times 10^{-1049}$

The resulting square roots can be sometimes simplified; see [209] and [1111].

All of the above was an interesting exercise for programming. If one is just interested in the result, the built-in function `FunctionExpand` (to be discussed in Chapter 3) will rewrite trigonometric functions of arguments of the form *rational* π in radicals, if possible in square roots.

In[69]:= `FunctionExpand[Sin[Pi/(2^4 5)]]`

Out[69]= $\frac{1}{8}\left(1 + \sqrt{5}\right)\left(-\sqrt{2 - \sqrt{2 + \sqrt{2}}} - \sqrt{\left(2 + \sqrt{2}\right)\left(2 - \sqrt{2 + \sqrt{2}}\right)}\right) +$

$$\frac{1}{4}\sqrt{\frac{1}{2}\left(5 - \sqrt{5}\right)}\left(-\sqrt{2 + \sqrt{2 + \sqrt{2}}} + \sqrt{\left(2 + \sqrt{2}\right)\left(2 + \sqrt{2 + \sqrt{2}}\right)}\right)$$

`FunctionExpand` can also express $\sec(\pi/25)$ in radicals, although not in square roots.

In[70]:= `FunctionExpand[Sec[Pi/25]]`

Out[70]= $-\dfrac{2}{\left(\dfrac{\frac{1}{2}\left(-1-\sqrt{5}\right) - i\sqrt{2 + \frac{1}{2}\left(1 - \sqrt{5}\right)}}{2 \cdot 2^{1/5}}\right)^{6/5} - \left(\dfrac{\frac{1}{2}\left(-1-\sqrt{5}\right) + i\sqrt{2 + \frac{1}{2}\left(1 - \sqrt{5}\right)}}{2 \cdot 2^{1/5}}\right)^{6/5}}$

c) As a warm-up exercise, we will guide the following calculation by hand; at the end, we will give a more automatic answer. To solve this problem, we look for a simplified version of the above identities, which we are able to recognize as an equation fulfilled by $\sin(\pi/7)$, $\sin(2\pi/7)$, and $\sin(3\pi/7)$. (We could in principle calculate exact values for $\sin(\pi/7)$, $\sin(2\pi/7)$, and $\sin(3\pi/7)$ with part a) of this exercise, plug them into the identities, and try to simplify them. Because the resulting expressions are quite big, this approach would need days on a good workstation; so, we do not go this way.)

From part a) of this exercise, we have the following two equations for $c1 = \cos(\pi/7)$ and $s1 = \sin(\pi/7)$.

`In[1]:= {TrigExpand[Cos[7 V]] == -1, TrigExpand[Sin[7 V]] == 0} /.`
$$\qquad\qquad\qquad \texttt{\{Cos[V] -> c1, Sin[V] -> s1\}}$$

`Out[1]=` $\{c1^7 - 21\,c1^5\,s1^2 + 35\,c1^3\,s1^4 - 7\,c1\,s1^6 == -1,\ 7\,c1^6\,s1 - 35\,c1^4\,s1^3 + 21\,c1^2\,s1^5 - s1^7 == 0\}$

Elimination of $s1$ yields the following.

`In[2]:= d1 = Eliminate[%, s1]`

`Out[2]=` $-36991\,c1^7 - 970880\,c1^{14} + 1163264\,c1^{21} + 2097152\,c1^{28} == -1$

This equation factors into simpler pieces, which we will use later.

`In[3]:= Factor[Subtract @@ d1]`

`Out[3]=` $(1 + c1)\ (1 - 4\,c1 - 4\,c1^2 + 8\,c1^3)\ (1 - c1 + c1^2 - c1^3 + c1^4 - c1^5 + c1^6)$
$\qquad (1 - 8\,c1 + 36\,c1^2 - 64\,c1^3 + 64\,c1^4 - 64\,c1^5 + 64\,c1^6)$
$\qquad (1 + 6\,c1 + 8\,c1^2 - 8\,c1^3 + 64\,c1^4 - 64\,c1^5 + 64\,c1^6)$
$\qquad (1 + 6\,c1 + 36\,c1^2 + 104\,c1^3 + 176\,c1^4 + 160\,c1^5 + 64\,c1^6)$

All roots of $d1$ are well separated.

`In[4]:= Min[Flatten[MapIndexed[Drop[#1, #2]&,`
$\qquad\qquad\qquad$ `Outer[Abs[Sqrt[#1^2 - #2^2]]&, #, #]&[`
$\qquad\qquad\qquad\qquad\qquad$ `c1 /. NSolve[d1, c1]]]]]`

`Out[4]=` 0.207287

Therefore, we are sure to have `Cos[Pi/7]` as a solution of $1 - 4c1 - 4c1\wedge2 + 8c1\wedge3 == 0$.

`In[5]:= {NRoots[1 - 4 c1 - 4 c1^2 + 8 c1^3 == 0, c1], Cos[Pi/7] // N}`

`Out[5]=` $\{c1 == -0.62349\ ||\ c1 == 0.222521\ ||\ c1 == 0.900969,\ 0.900969\}$

Now let us simplify the three given identities. First, we rewrite in terms of `Sin[2Pi/7]` and `Sin[3Pi/7]` in `Sin[` `Pi/7]` and `Cos[Pi/7]`.

`In[6]:= TrigExpand[{Sin[2 a], Sin[3 a]}]`

`Out[6]=` $\{2\,\mathrm{Cos}[a]\,\mathrm{Sin}[a],\ 3\,\mathrm{Cos}[a]^2\,\mathrm{Sin}[a] - \mathrm{Sin}[a]^3\}$

`In[7]:= repRules = {s2 -> 2 c1 s1, s3 -> 3 s1 - 4 s1^3};`

So, we have the first identity.

`In[8]:= id1 = s2/s3^2 - s1/s2^2 + s3/s1^2 == 2 Sqrt[7]`

`Out[8]=` $-\dfrac{s1}{s2^2} + \dfrac{s2}{s3^2} + \dfrac{s3}{s1^2} == 2\sqrt{7}$

This gives the following equation.

`In[9]:= (Numerator[#]^2 - (id1[[2]] Denominator[#])^2)&[Together[id1[[1]]]]`

`Out[9]=` $-28\,s1^4\,s2^4\,s3^4 + (s1^2\,s2^3 - s1^3\,s3^2 + s2^2\,s3^3)^2$

Elimination of $s1$ results in a polynomial.

`In[10]:= Eliminate[{(% /. repRules) == 0, c1^2 + s1^2 == 1}, s1]`

`Out[10]=` $13\,c1^2 + 16\,c1^3 + 446\,c1^4 - 144\,c1^5 - 10710\,c1^6 - 32\,c1^7 + 97675\,c1^8 + 4832\,c1^9 - 491575\,c1^{10} -$
$\qquad 23856\,c1^{11} + 1511080\,c1^{12} + 55984\,c1^{13} - 2925680\,c1^{14} - 74176\,c1^{15} + 3516544\,c1^{16} + 56832\,c1^{17} -$
$\qquad 2385920\,c1^{18} - 23552\,c1^{19} + 536576\,c1^{20} + 4096\,c1^{21} + 397312\,c1^{22} - 311296\,c1^{24} + 65536\,c1^{26} == 1$

This has solutions similar to the equation $d1$.

`In[11]:= PolynomialGCD[Subtract @@ %, Subtract @@ d1] // Factor`

In[11]= $(1 + c1) \, (1 - 4 \, c1 - 4 \, c1^2 + 8 \, c1^3)$

Here is the same calculation (in an abbreviated form) for the second identity.

```
In[12]:= id2 = s2^2/s3^4 + s1^2/s2^4 + s3^2/s1^4 == 28;
```

```
In[13]:= PolynomialGCD[Subtract @@ Eliminate[{(
              (Numerator[#] - id2[[2]] Denominator[#])&[
                  Together[id2[[1]]]] /. repRules) == 0,
                  c1^2 + s1^2 == 1}, s1], Subtract @@ d1] // Factor
```

Out[13]= $(1 + c1) \, (1 - 4 \, c1 - 4 \, c1^2 + 8 \, c1^3)$

Here it is for the third identity.

```
In[14]:= id3 = s3^2/s1^4 ( 2 s2/s3 + 4 s3/s1) +
              s1^2/s2^4 (-2 s3/s1 + 4 s1/s2) -
              s2^2/s3^4 ( 2 s1/s2 + 4 s2/s3) == 280;
```

```
In[15]:= PolynomialGCD[Subtract @@ Eliminate[{(
              (Numerator[#] - id3[[2]] Denominator[#])&[
                  Together[id3[[1]]]] /. repRules) == 0,
                  c1^2 + s1^2 == 1}, s1], Subtract @@ d1] // Factor
```

Out[15]= $(1 + c1) \, (1 - 4 \, c1 - 4 \, c1^2 + 8 \, c1^3)$

This gives the same result. Because we have done all calculations so far for $\cos(n\pi/7)$, we must determine which of the possible $\pm\sin(n\pi/7) = \sin(\pm n\pi/7)$ do not fulfill the identities. This is easily done by plugging in the numerical values in all possible combinations. (We exclude the solution $n = 0$ because this would give zeros in the denominators.) The testing must of course give N::meprec messages.

```
In[16]:= (* avoid messages from true identities *) Off[N::meprec];
         TableForm[
         Outer[N[#1 //. {s1 -> Sin[1 #2 Pi/7], s2 -> Sin[2 #2 Pi/7],
                         s3 -> Sin[3 #2 Pi/7]}]&,
             {id1, id2, id3}, Range[6]],
             TableHeadings -> {{"id1", "id2", "id3"},
                               {" 1\n", " 2", " 3", " 4", " 5", " 6"}}]
```

	1	2	3	4	5	6
Out[17]//TableForm= id1	True	False	False	False	False	False
id2	True	True	True	True	True	True
id3	True	False	True	False	True	False

This also answers the question for which integers n there are similar identities.

Now, we will give a straightforward solution using RootReduce. We first transform the trigonometric functions into exponentials using TrigToExp. Then we transform the exponentials into explicit algebraic numbers of the form $(-1)^{rational}$ using Together, and finally we use RootReduce to canonicalize the resulting algebraic number. These three operations are put together in the function simplify.

```
In[18]:= simplify[expr_] :=
         RootReduce[RootReduce[Together[TrigToExp[#]]]& /@ expr]
```

Here are the three identities checked.

```
In[19]:= simplify @ (Sin[2Pi/7]/Sin[3Pi/7]^2 - Sin[1Pi/7]/Sin[2Pi/7]^2 +
                     Sin[3Pi/7]/Sin[1Pi/7]^2)
```

Out[19]= $2\sqrt{7}$

```
In[20]:= simplify @ (Sin[2Pi/7]^2/Sin[3Pi/7]^4 + Sin[1Pi/7]^2/Sin[2Pi/7]^4 +
                     Sin[3Pi/7]^2/Sin[1Pi/7]^4)
```

Out[20]= 28

```
In[21]:= simplify @
         (Sin[3Pi/7]^2/Sin[1Pi/7]^4 *
                 ( 2 Sin[2Pi/7]/Sin[3Pi/7] + 4 Sin[3Pi/7]/Sin[1Pi/7]) +
             Sin[1Pi/7]^2/Sin[2Pi/7]^4 *
                 (-2 Sin[3Pi/7]/Sin[1Pi/7] + 4 Sin[1Pi/7]/Sin[2Pi/7]) -
```

```
        Sin[2Pi/7]^2/Sin[3Pi/7]^4 *
             ( 2 Sin[1Pi/7]/Sin[2Pi/7] + 4 Sin[2Pi/7]/Sin[3Pi/7]))
```

Out[21]= 280

In a similar way, we can now check the $sin(expr) \longrightarrow sin(n\ expr)$ part of the exercise.

```
In[22]:= Table[{k, simplify @ (Sin[k 2Pi/7]/Sin[k 3Pi/7]^2 -
                                Sin[k 1Pi/7]/Sin[k 2Pi/7]^2 +
                                Sin[k 3Pi/7]/Sin[k 1Pi/7]^2)}, {k, 1, 6}]
Out[22]= {{1, 2 √7 }, {2, Root[-107584 + 14672 #1² - 588 #1⁴ + 7 #1⁶ &, 5]},
          {3, -2 √7 }, {4, Root[-107584 + 14672 #1² - 588 #1⁴ + 7 #1⁶ &, 1]},
          {5, -2 √7 }, {6, Root[-107584 + 14672 #1² - 588 #1⁴ + 7 #1⁶ &, 4]}}
```

```
In[23]:= Table[{k, simplify @ (Sin[k 2Pi/7]^2/Sin[k 3Pi/7]^4 +
                                Sin[k 1Pi/7]^2/Sin[k 2Pi/7]^4 +
                                Sin[k 3Pi/7]^2/Sin[k 1Pi/7]^4)}, {k, 1, 6}]
Out[23]= {{1, 28}, {2, 28}, {3, 28}, {4, 28}, {5, 28}, {6, 28}}
```

```
In[24]:= Table[{k, simplify @
          (Sin[k 3Pi/7]^2/Sin[k 1Pi/7]^4 *
              ( 2 Sin[k 2Pi/7]/Sin[k 3Pi/7] + 4 Sin[k 3Pi/7]/Sin[k 1Pi/7]) +
          Sin[k 1Pi/7]^2/Sin[k 2Pi/7]^4 *
              (-2 Sin[k 3Pi/7]/Sin[k 1Pi/7] + 4 Sin[k 1Pi/7]/Sin[k 2Pi/7]) -
          Sin[k 2Pi/7]^2/Sin[k 3Pi/7]^4 *
              ( 2 Sin[k 1Pi/7]/Sin[k 2Pi/7] + 4 Sin[k 2Pi/7]/Sin[k 3Pi/7]))},
              {k, 1, 6}]
Out[24]= {{1, 280}, {2, Root[-10941952 - 38976 #1 + 280 #1² + #1³ &, 1]},
          {3, 280}, {4, Root[-10941952 - 38976 #1 + 280 #1² + #1³ &, 2]},
          {5, 280}, {6, Root[-10941952 - 38976 #1 + 280 #1² + #1³ &, 3]}}
```

Additional interesting trigonometric identities can be derived, for example, [1414], [368], [132], [715], [485], [916], and [1810]. For instance, the following identities hold [133]:

```
DedekindCotIdentity[a_Integer?Positive, c_Integer?Positive] :=
4c Sum[#[n/c] #[a n/c], {n, 1, c - 1}]&[# - Floor[#] - 1/2&] ==
    Sum[Cot[Pi n/c] Cot[Pi a n/c], {n, 1, c - 1}] /; GCD[a, c] === 1
```

```
BenczeCotIdentity[m_Integer?Positive][n_Integer?Positive] :=
Module[{k}, Sum[Product[Cot[k[r] Pi/(2n + 1)]^2, {r, m}], ##]& @@
            Table[{k[j], If[j === 1, 1, k[j - 1] + 1], n}, {j, m}]] ==
            (2n)!/((2m + 1)!(2n - 2m)!)
```

d) This is the transformation of interest.

In[1]:= y = (α + β x)/(αs + βs x)

Out[1]= $\dfrac{\alpha + x\,\beta}{\alpha s + x\,\beta s}$

We can express the value of $M(k)$ in terms of $\alpha(k)$, $\beta(k)$, $\alpha'(k)$, $\beta'(k)$ by substituting the value $x = 0$ in the given equation.

```
In[2]:= M = ((1 - y^2) (1 - 1^4 y^2)/((1 - x^2) (1 - k^4 x^2))/D[y, x]^2 /.
          x -> 0) // Together
```

Out[2]= $\dfrac{(\alpha^2 - \alpha s^2)\,(1^4\,\alpha^2 - \alpha s^2)}{(\alpha s\,\beta - \alpha\,\beta s)^2}$

One could think that a possibility to solve the problem would be to construct

$$y'(x)^2\,M(k)\,(1-x^2)(1-k^4\,x^2) - (1-y(x,\,k)^2)(1-l(k)^4\,y(x,\,k)^2) = 0$$

and equate all coefficients of the various powers of x to zero and solve the resulting system of equations.

```
In[3]:= TimeConstrained[
    #[# == 0& /@ CoefficientList[Numerator[Together[
        D[y, x]^2 M (1 - x^2)(1 - k^4 x^2) - (1 - y^2)(1 - 1^4 y^2)]], x],
        {α, αs, β, βs, 1}], 120]& /@ {Solve, Reduce}
Out[3]= {$Aborted, $Aborted}
```

Unfortunately, `Solve` will never return a solution of the last code. So, we must help a bit manually. In the first step, we rewrite the part $(1 - x^2)(1 - k^4 x^2)$ as follows.

```
In[4]:= yPart = (1 - y^2)(1 - 1^4 y^2) 1/D[y, x]^2 1/ M // Together // Factor
```

$$Out[4]= \frac{(\alpha - \alpha s + x\,\beta - x\,\beta s)\,(1^2\,\alpha - \alpha s + 1^2\,x\,\beta - x\,\beta s)\,(\alpha + \alpha s + x\,\beta + x\,\beta s)\,(1^2\,\alpha + \alpha s + 1^2\,x\,\beta + x\,\beta s)}{(\alpha - \alpha s)\,(1^2\,\alpha - \alpha s)\,(\alpha + \alpha s)\,(1^2\,\alpha + \alpha s)}$$

We apply some cosmetic manipulations to this expression.

```
In[5]:= yPartFactored = Times @@ ({1, x}.#& /@
    ((((CoefficientList[#, x]& /@ (List @@
        Numerator[yPart]))/(List @@ Denominator[yPart]) //
                        Cancel)) // Cancel))
```

$$Out[5]= \left(1 + \frac{x\,(\beta - \beta s)}{\alpha - \alpha s}\right)\left(1 + \frac{x\,(1^2\,\beta - \beta s)}{1^2\,\alpha - \alpha s}\right)\left(1 + \frac{x\,(\beta + \beta s)}{\alpha + \alpha s}\right)\left(1 + \frac{x\,(1^2\,\beta + \beta s)}{1^2\,\alpha + \alpha s}\right)$$

If the product of these four factors $f_1\,f_2\,f_3\,f_4$ with $f_1 = 1 + (\beta - \beta')x/(\alpha - \alpha')$, $f_2 = 1 + (\beta + \beta')x/(\alpha + \alpha')$, $f_3 = 1 + (\beta' - \beta\,l^2)x/(\alpha' - \alpha\,l^2)$, and $f_4 = 1 + (\beta' + \beta\,l^2)x/(\alpha' + \alpha\,l^2)$ is equal to $(1 - x^2)(1 - k^4\,x^2)$, this means that either

$$f_1\,f_2 = (1 - x^2) \quad \text{and} \quad f_3\,f_4 = (1 - k^4\,x^2) \quad \text{or}$$
$$f_1\,f_3 = (1 - x^2) \quad \text{and} \quad f_2\,f_4 = (1 - k^4\,x^2) \quad \text{or}$$
$$f_1\,f_4 = (1 - x^2) \quad \text{and} \quad f_2\,f_3 = (1 - k^4\,x^2) \quad \text{or}$$
$$\cdots$$

Here, the six possible combinations of two factors from `yPartFactored` are calculated.

```
In[6]:= yPartFactoredList =
    Flatten[Table[{yPartFactored[[i]] yPartFactored[[j]],
        yPartFactored/(yPartFactored[[i]] yPartFactored[[j]])},
                    {i, 4}, {j, i + 1, 4}], 1];
```

We now generate all possible equations listed above and extract the coefficients of the powers of x.

```
In[7]:= allEquations = DeleteCases[Join[
    Apply[Join, Map[CoefficientList[#, x]&,
        # - {1 - x^2, 1 - k^4 x^2}& /@ yPartFactoredList, {2}], {1}],
    Apply[Join, Map[CoefficientList[#, x]&,
        # - {1 - k^4 x^2, 1 - x^2}& /@ yPartFactoredList, {2}], {1}]],
            0, {2}];
```

```
Short[allEquations, 8]
```

$$Out[8]//Short= \left\{\left\{\frac{\beta - \beta s}{\alpha - \alpha s} + \frac{1^2\,\beta - \beta s}{1^2\,\alpha - \alpha s},\ 1 + \frac{(\beta - \beta s)\,(1^2\,\beta - \beta s)}{(\alpha - \alpha s)\,(1^2\,\alpha - \alpha s)},\ \frac{\beta + \beta s}{\alpha + \alpha s} + \frac{1^2\,\beta + \beta s}{1^2\,\alpha + \alpha s},\ k^4 + \frac{(\beta + \beta s)\,(1^2\,\beta + \beta s)}{(\alpha + \alpha s)\,(1^2\,\alpha + \alpha s)}\right\},\right.$$

$$\left\{\frac{\beta - \beta s}{\alpha - \alpha s} + \frac{\beta + \beta s}{\alpha + \alpha s},\ 1 + \frac{(\beta - \beta s)\,(\beta + \beta s)}{(\alpha - \alpha s)\,(\alpha + \alpha s)},\right.$$

$$\left.\frac{1^2\,\beta - \beta s}{1^2\,\alpha - \alpha s} + \frac{1^2\,\beta + \beta s}{1^2\,\alpha + \alpha s},\ k^4 + \frac{(1^2\,\beta - \beta s)\,(1^2\,\beta + \beta s)}{(1^2\,\alpha - \alpha s)\,(1^2\,\alpha + \alpha s)}\right\},\ \ll 8\gg,$$

$$\left\{\frac{1^2\,\beta - \beta s}{1^2\,\alpha - \alpha s} + \frac{1^2\,\beta + \beta s}{1^2\,\alpha + \alpha s},\ k^4 + \frac{(1^2\,\beta - \beta s)\,(1^2\,\beta + \beta s)}{(1^2\,\alpha - \alpha s)\,(1^2\,\alpha + \alpha s)},\ \frac{\beta - \beta s}{\alpha - \alpha s} + \frac{\beta + \beta s}{\alpha + \alpha s},\ 1 + \frac{(\beta - \beta s)\,(\beta + \beta s)}{(\alpha - \alpha s)\,(\alpha + \alpha s)}\right\},$$

$$\left.\left\{\frac{\beta + \beta s}{\alpha + \alpha s} + \frac{1^2\,\beta + \beta s}{1^2\,\alpha + \alpha s},\ k^4 + \frac{(\beta + \beta s)\,(1^2\,\beta + \beta s)}{(\alpha + \alpha s)\,(1^2\,\alpha + \alpha s)},\ \frac{\beta - \beta s}{\alpha - \alpha s} + \frac{1^2\,\beta - \beta s}{1^2\,\alpha - \alpha s},\ 1 + \frac{(\beta - \beta s)\,(1^2\,\beta - \beta s)}{(\alpha - \alpha s)\,(1^2\,\alpha - \alpha s)}\right\}\right\}$$

This system has no generic solutions.

```
In[9]:= Solve[Thread[Flatten[allEquations] == 0], {α, αs, β, βs, 1}]
Out[9]= {}
```

Looking at the left-hand sides of the equations, contained in `allEquations`, we see that the variables $\alpha(k)$, $\beta(k)$, $\alpha'(k)$, $\beta'(k)$, and $l(k)$ appear only in certain combinations. So, we introduce some auxiliary variables, solve the resulting equations, and finally solve for the variables $\alpha(k)$, $\beta(k)$, $\alpha'(k)$, $\beta'(k)$, and $l(k)$. The solutions so obtained are sometimes of the form `Sqrt[`*someExpression*`^2]`. They can be somewhat simplified. Because the result could be ±*something*, we generate both possibilities with the help of the following function.

```
In[10]:= getRidOfSquareRoots[sol_] :=
        Module[{aux1, squareRoots, powerExpandedSquareRoots,
                allPowerExpandedSquareRoots, repRules},
          If[(* no square root present *)
            FreeQ[sol, Power[_, 1/2] | Power[_, -1/2]],
            Factor //@ Together //@ sol,
            (* square roots present *)
            (* eliminate all square roots by using PowerExpand
               and generate all possible solutions *)
          For[aux1 = (Factor //@ Together //@ sol) /.
                       Power[b_, 1/2] :> Power[Factor[b], 1/2],
              aux1 = (Factor //@ Together //@ aux1) /.
                       Power[b_, 1/2] :> Power[Factor[b], 1/2];
          (* all square roots on lowest levels *)
          squareRoots =
          Union[Cases[aux1, Power[_?(FreeQ[#, Power[_,  1/2] |
                                              Power[_, -1/2]]&),  1/2] |
                             Power[_?(FreeQ[#, Power[_,  1/2] |
                                              Power[_, -1/2]]&), -1/2],
                      {0, Infinity}]];
          squareRoots =!= {}, Null,
          (* the list of square roots after applying PowerExpand *)
          powerExpandedSquareRoots = PowerExpand[squareRoots];
          (* all possible combinations of +1, -1 of all square roots *)
          factorList = If[Length[(squareRoots)] === 1, {{-1}, {1}},
                Flatten[Outer[(-1)^{##}&, ##]& @@ Table[{0, 1},
                    {Length[squareRoots]}], Length[squareRoots] - 1]];
          allPowerExpandedSquareRoots =
                powerExpandedSquareRoots # & /@ factorList;
          (* replacement rules for replacing the square roots *)
          repRules = MapThread[Rule, {Table[squareRoots,
                {Length[allPowerExpandedSquareRoots]}],
                            allPowerExpandedSquareRoots}, 2];
          (* all possible combinations generated *)
          aux1 = aux1 //. repRules]; Sequence @@ aux1]]
```

Here, the solution process is implemented and carried out for all 12 sets of equations.

```
In[11]:= Off[Solve::svars];

        possibleSolutions = Union[Flatten[Table[
        (* introduce "obvious" auxiliary variables *)
        eqn1 = allEquations[[i]] //.
                {β - βs -> βm, -β + βs -> -βm,
                 α - αs -> αm, -α + αs -> -αm,
                 β + βs -> βp,  α + αs -> αp,
                 βs - β l^2 -> βlm, -βs + β l^2 -> -βlm,
                 αs - α l^2 -> αlm, -αs + α l^2 -> -αlm,
                 βs + β l^2 -> βlp, αs + α l^2 -> αlp};
        (* eliminate a few of the auxiliary variables *)
        sol1 = Solve[# == 0& /@ eqn1, {αm, αp, βm, βp, αlm, αlp, βlm, βlp}];
        (* the so simplified equations *)
        eqn2 = {β - βs - βm, α - αs - αm, β + βs - βp, α + αs - αp,
                βs - β l^2 - βlm, αs - α l^2 - αlm,
                βs + β l^2 - βlp, αs + α l^2 - αlp} //. sol1;
        (* eliminate all still-present auxiliary variables
           and solve for a, as, b, bs, l *)
        preSolution0 = Flatten[
        Solve[Eliminate[# == 0& /@ #, {αm, αp, βm, βp, αlm, αlp, βlm, βlp}],
              {α, αs, β, βs, l}]& /@ eqn2, 1];
              (* get rid of square roots if present *)
        preSolution1 = Union[getRidOfSquareRoots /@ preSolution0];
```

```
(* get rid of trivial solution *)
preSolution2 = DeleteCases[Sort[#, OrderedQ[{#1[[1]], #2[[1]]}]&]& /@
                 MapAll[Together, preSolution1],
                 {α -> 0, αs -> 0, β -> 0, βs -> 0, ___}];
(* y[x], M and l^4 *)
Union[Factor //@ Together //@ ({(α + β x)/(αs + βs x),
         (αs^2 - α^2)(αs^2 - α^2 l^4)/(αs β - α βs)^2, l^4} //.
              preSolution2)], {i, 12}], 1]];
```

```
Short[possibleSolutions, 8]
```

Out[13]//Short= $\left\{\left\{-\dfrac{1}{x}, \dfrac{1}{k^4}, \dfrac{1}{k^4}\right\}, \left\{\dfrac{1}{x}, \dfrac{1}{k^4}, \dfrac{1}{k^4}\right\}, \left\{-\dfrac{1}{k^2 x}, 1, k^4\right\}, \left\{\dfrac{1}{k^2 x}, 1, k^4\right\},\right.$

$\left\{-x, 1, k^4\right\}, \{x, 1, k^4\}, \left\{-k^2 x, \dfrac{1}{k^4}, \dfrac{1}{k^4}\right\}, \left\{k^2 x, \dfrac{1}{k^4}, \dfrac{1}{k^4}\right\}, \ll 8\gg,$

$\left\{-\dfrac{(-1+k)(-1+kx)}{(1+k)(1+kx)}, -\dfrac{4}{(-1+k)^4}, \dfrac{(1+k)^4}{(-1+k)^4}\right\}, \left\{\dfrac{(-1+k)(-1+kx)}{(1+k)(1+kx)}, -\dfrac{4}{(-1+k)^4}, \dfrac{(1+k)^4}{(-1+k)^4}\right\},$

$\left\{-\dfrac{(1+k)(-1+kx)}{(-1+k)(1+kx)}, -\dfrac{4}{(1+k)^4}, \dfrac{(-1+k)^4}{(1+k)^4}\right\}, \left\{\dfrac{(1+k)(-1+kx)}{(-1+k)(1+kx)}, -\dfrac{4}{(1+k)^4}, \dfrac{(-1+k)^4}{(1+k)^4}\right\},$

$\left\{-\dfrac{(-1+k)(1+kx)}{(1+k)(-1+kx)}, -\dfrac{4}{(-1+k)^4}, \dfrac{(1+k)^4}{(-1+k)^4}\right\}, \left\{\dfrac{(-1+k)(1+kx)}{(1+k)(-1+kx)}, -\dfrac{4}{(-1+k)^4}, \dfrac{(1+k)^4}{(-1+k)^4}\right\},$

$\left.\left\{-\dfrac{(1+k)(1+kx)}{(-1+k)(-1+kx)}, -\dfrac{4}{(1+k)^4}, \dfrac{(-1+k)^4}{(1+k)^4}\right\}, \left\{\dfrac{(1+k)(1+kx)}{(-1+k)(-1+kx)}, -\dfrac{4}{(1+k)^4}, \dfrac{(-1+k)^4}{(1+k)^4}\right\}\right\}$

Because of the use of PowerExpand, it might be that we have generated some fictitious solutions that do not satisfy the original equation. So, we test the solutions possibleSolutions and keep only the correct ones.

In[14]:= **Clear[y, M];**

```
test[sol_] :=
If[Factor[Together[D[sol[[1]], x]^2 sol[[2]] (1 - x^2) (1 - k^4 x^2) -
          (1 - sol[[1]]^2) (1 - sol[[3]] sol[[1]]^2)]] === 0, sol,
    Sequence @@ {}]
```

In[16]:= **testedSolutions = test /@ possibleSolutions;**

So, we finally arrive at the following 24 solutions.

In[17]:= **TableForm[testedSolutions, TableHeadings -> {None, {"y[x] \n \n",**
            "M[k] \n \n", "l[k]^4 \n"}}, TableAlignments -> Center]

y[x]	M[k]	l[k]^4
$-\dfrac{1}{x}$	$\dfrac{1}{k^4}$	$\dfrac{1}{k^4}$
$\dfrac{1}{x}$	$\dfrac{1}{k^4}$	$\dfrac{1}{k^4}$
$-\dfrac{1}{k^2 x}$	1	k^4
$\dfrac{1}{k^2 x}$	1	k^4
$-x$	1	k^4
x	1	k^4
$-k^2 x$	$\dfrac{1}{k^4}$	$\dfrac{1}{k^4}$
$k^2 x$	$\dfrac{1}{k^4}$	$\dfrac{1}{k^4}$
$-\dfrac{(-i+k)(-i+kx)}{(i+k)(i+kx)}$	$-\dfrac{4}{(-i+k)^4}$	$\dfrac{(i+k)^4}{(-i+k)^4}$
$\dfrac{(-i+k)(-i+kx)}{(i+k)(i+kx)}$	$-\dfrac{4}{(-i+k)^4}$	$\dfrac{(i+k)^4}{(-i+k)^4}$
$-\dfrac{(i+k)(-i+kx)}{(-i+k)(i+kx)}$	$-\dfrac{4}{(i+k)^4}$	$\dfrac{(-i+k)^4}{(i+k)^4}$
$\dfrac{(i+k)(-i+kx)}{(-i+k)(i+kx)}$	$-\dfrac{4}{(i+k)^4}$	$\dfrac{(-i+k)^4}{(i+k)^4}$
$-\dfrac{(-i+k)(i+kx)}{(i+k)(-i+kx)}$	$-\dfrac{4}{(-i+k)^4}$	$\dfrac{(i+k)^4}{(-i+k)^4}$
$\dfrac{(-i+k)(i+kx)}{(i+k)(-i+kx)}$	$-\dfrac{4}{(-i+k)^4}$	$\dfrac{(i+k)^4}{(-i+k)^4}$
$-\dfrac{(i+k)(i+kx)}{(-i+k)(-i+kx)}$	$-\dfrac{4}{(i+k)^4}$	$\dfrac{(-i+k)^4}{(i+k)^4}$

Out[17]//TableForm= (label positioned at left of table)

$$\frac{(i+k)(i+k\,x)}{(-i+k)(-i+k\,x)} \qquad -\frac{4}{(i+k)^4} \qquad \frac{(-i+k)^4}{(i+k)^4}$$

$$-\frac{(-1+k)(-1+k\,x)}{(1+k)(1+k\,x)} \qquad -\frac{4}{(-1+k)^4} \qquad \frac{(1+k)^4}{(-1+k)^4}$$

$$\frac{(-1+k)(-1+k\,x)}{(1+k)(1+k\,x)} \qquad -\frac{4}{(-1+k)^4} \qquad \frac{(1+k)^4}{(-1+k)^4}$$

$$\frac{(1+k)(-1+k\,x)}{(-1+k)(1+k\,x)} \qquad -\frac{4}{(1+k)^4} \qquad \frac{(-1+k)^4}{(1+k)^4}$$

$$\frac{(1+k)(-1+k\,x)}{(-1+k)(1+k\,x)} \qquad -\frac{4}{(1+k)^4} \qquad \frac{(-1+k)^4}{(1+k)^4}$$

$$-\frac{(-1+k)(1+k\,x)}{(1+k)(-1+k\,x)} \qquad -\frac{4}{(-1+k)^4} \qquad \frac{(1+k)^4}{(-1+k)^4}$$

$$\frac{(-1+k)(1+k\,x)}{(1+k)(-1+k\,x)} \qquad -\frac{4}{(-1+k)^4} \qquad \frac{(1+k)^4}{(-1+k)^4}$$

$$-\frac{(1+k)(1+k\,x)}{(-1+k)(-1+k\,x)} \qquad -\frac{4}{(1+k)^4} \qquad \frac{(-1+k)^4}{(1+k)^4}$$

$$\frac{(1+k)(1+k\,x)}{(-1+k)(-1+k\,x)} \qquad -\frac{4}{(1+k)^4} \qquad \frac{(-1+k)^4}{(1+k)^4}$$

These solutions play an important role in the so-called first-order modular transformation of elliptic functions; see [583], [217], and [545].

Now, we recommend potentially restarting *Mathematica* here. Here is another (more complicated, but at the same time more straightforward) way to calculate all solutions. The following calculation (not counting the first call to Reduce) will take a few minutes on a year-2005 computer. This is the system of equations under consideration.

```
In[18]:= eqs = {2 a as^2 b + 2 a^2 as bs - 4 as^3 bs - 4 a^3 b l^4 +
          2 a as^2 b l^4 + 2 a^2 as bs l^4,
          a^2 as^2 - as^4 + as^2 b^2 + 4 a as b bs + a^2 bs^2 -
          6 as^2 bs^2 + a^2 as^2 k^4 - as^4 k^4 - a^4 l^4 +
          a^2 as^2 l^4 - 6 a^2 b^2 l^4 + as^2 b^2 l^4 +
          4 a as b bs l^4 + a^2 bs^2 l^4 -
          a^4 k^4 l^4 + a^2 as^2 k^4 l^4,
          2 as b^2 bs + 2 a b bs^2 - 4 as bs^3 - 4 a b^3 l^4 +
          2 as b^2 bs l^4 + 2 a b bs^2 l^4,
          b^2 bs^2 - bs^4 - a^2 as^2 k^4 + as^4 k^4 - b^4 l^4 +
          b^2 bs^2 l^4 + a^4 k^4 l^4 - a^2 as^2 k^4 l^4};
```

Trying to solve this system directly fails within a reasonable time.

```
In[19]:= TimeConstrained[Reduce[eqs == 0, {l, a, as, b, bs}], 1000]

Out[19]= $Aborted
```

We have five equations in four variables.

```
In[20]:= Exponent[#, {k, l, a, as, b, bs}]& /@ eqs

Out[20]= {{0, 4, 3, 3, 1, 1}, {4, 4, 4, 4, 2, 2}, {0, 4, 1, 1, 3, 3}, {4, 4, 4, 4, 4, 4}}
```

Unfortunately, it is not possible to eliminate a, b, as, and bs from the equations to get one equation in just l and k.

```
In[21]:= GroebnerBasis[eqs, {l, k}, {a, as, b, bs},
                        MonomialOrder -> EliminationOrder] // Timing

Out[21]= {0.36 Second, {}}
```

So, we will split the problem. In the first part, we will calculate all possible $l = l(k)$ values. Because we are looking for rational functions $a(k)$, $a'(k)$, $b(k)$, $b'(k)$, $l(k)$ we need the resulting equations to be linear (over the Gaussian integers) in l. We find all such factors by calculating a GroebnerBasis with l as the highest variable.

```
In[22]:= (gb = GroebnerBasis[eqs, {a, as, b, bs, l}, ParameterVariables -> {k},
                        CoefficientDomain -> RationalFunctions]); // Timing

Out[22]= {4.33 Second, Null}

In[23]:= Length[gb]

Out[23]= 66
```

Now, we first factor this GroebnerBasis over the integers.

```
In[24]:= gb1 = Factor /@ gb;
```

We split the equations into their factors and keep (in case of powers) only the bases.

```
In[25]:= allFactors1 = Union[Flatten[If[Head[#] === Times,
                                  List @@ #, {#}]& /@ gb1]];

         allFactors2 = If[Head[#] === Power, #[[1]], #]& /@ allFactors1;
```

Now, we factor gain, this time over the Gaussian integers.

```
In[27]:= allFactors3 = Factor[#, GaussianIntegers -> True]& /@ allFactors2;
         allFactors4 = Union[Flatten[If[Head[#] === Times,
                                  List @@ #, {#}]& /@ allFactors3]];

In[29]:= allFactors5 = If[Head[#] === Power, #[[1]], #]& /@ allFactors4;
```

We now have 107 factors.

```
In[30]:= Length[allFactors5]

Out[30]= 107
```

Twenty-eight of them contain the variable l linearly.

```
In[31]:= s1 = Select[allFactors5,
              Variables[#] === {k, l} || Variables[#] === {l}&]

Out[31]= {k - l, k - i l, k + i l, -1 + l, -i + l, i + l, 1 + l, k + l, -1 + k l, -i + k l, i + k l,
          1 + k l, -1 - k - l + k l, -i - i k - l + k l, i + i k - l + k l, 1 + k - l + k l, -i - k - i l + k l,
          1 - i k - i l + k l, -1 + i k - i l + k l, i + k - i l + k l, i - k + i l + k l, -1 - i k + i l + k l,
          1 + i k + i l + k l, -i + k + i l + k l, 1 - k + l + k l, i - i k + l + k l, -i + i k + l + k l, -1 + k + l + k l}
```

So, we have the following potential values for l.

```
In[32]:= s2 = Union[Flatten[(l /. Solve[# == 0, l])& /@ s1]]
```

$$Out[32]= \left\{-1, -i, i, 1, \frac{i(-1-k)}{-1+k}, \frac{-1-k}{-1+k}, \frac{1}{k}, -\frac{1}{k}, \frac{i}{k}, \frac{i}{k}, \frac{1}{k}, -k, -ik, ik, k, \frac{i(1-k)}{1+k}, \frac{1-k}{1+k}, \right.$$

$$\frac{i(-1+k)}{1+k}, \frac{-1+k}{1+k}, \frac{i(1+k)}{-1+k}, \frac{1+k}{-1+k}, \frac{1-2ik-k^2}{1+k^2}, \frac{1+2ik-k^2}{1+k^2}, \frac{i-2k-ik^2}{1+k^2},$$

$$\left. \frac{i+2k-ik^2}{1+k^2}, \frac{-i-2k+ik^2}{1+k^2}, \frac{-i+2k+ik^2}{1+k^2}, \frac{-1-2ik+k^2}{1+k^2}, \frac{-1+2ik+k^2}{1+k^2} \right\}$$

```
In[33]:= Length[s2]

Out[33]= 28
```

Concerning the variables a, as and b, bs, we go on in a similar way: We calculate Gröbner bases with a and as (b and bs, respectively) as the highest variables, factor them over the Gaussian integers, and take out the factors that are linear.

Here the variables b and bs are eliminated.

```
In[34]:= (gbA = GroebnerBasis[eqs, {a, as, l}, {b, bs},
                    ParameterVariables -> {k},
                    CoefficientDomain -> RationalFunctions]); // Timing

Out[34]= {3.22 Second, Null}

In[35]:= Exponent[#, {a, as, b, bs, l}]& /@ gbA

Out[35]= {{4, 9, 0, 0, 28}, {5, 7, 0, 0, 28}, {5, 8, 0, 0, 20}, {6, 8, 0, 0, 20},
          {6, 9, 0, 0, 24}, {7, 7, 0, 0, 24}, {8, 8, 0, 0, 16}, {9, 8, 0, 0, 16}}

In[36]:= gbA1 = Factor /@ gbA;
```

These are the linear occurrences of as and a.

```
In[37]:= gbA2 = If[Head[#] === Times,
         Factor[#, GaussianIntegers -> True]& /@ #,
                    Factor[#, GaussianIntegers -> True]]& /@ gbA1;

In[38]:= (gbA3 = gbA2 /. p_Plus? (Variables[#] === {k, l}&) :> 1);

In[39]:= resAS = Select[Union[DeleteCases[
            If[Head[#] === Power, #[[1]], #]& /@
              Flatten[If[Head[#] === Times,
                      List @@ #, {#}]& /@ gbA3], _?NumberQ, {1}]],
            Exponent[#, as] === 1 || Exponent[#, a] === 1&]
```

```
Out[39]= {a, a - as, as, -a + as, a + as, as - i a l,
         as + i a l, -as + a l, as + a l, as - a l², -as + a l², as + a l²}
```

Here, the variables a and as are eliminated.

```
In[40]:= (gbB = GroebnerBasis[eqs, {b, bs, l}, {a, as},
                   ParameterVariables -> {k},
                   CoefficientDomain -> RationalFunctions]); // Timing
Out[40]= {3.48 Second, Null}
```

The polynomials we now get have still a high degree.

```
In[41]:= Exponent[#, {a, as, b, bs, l}]& /@ gbB
Out[41]= {{0, 0, 4, 9, 28}, {0, 0, 5, 7, 28}, {0, 0, 5, 8, 20}, {0, 0, 6, 8, 20},
         {0, 0, 6, 9, 24}, {0, 0, 7, 7, 24}, {0, 0, 8, 8, 16}, {0, 0, 9, 8, 16}}

In[42]:= gbB1 = Factor /@ gbB;
```

These are the linear occurrences of bs and b.

```
In[43]:= gbB2 = If[Head[#] === Times,
           Factor[#, GaussianIntegers -> True]& /@ #,
                     Factor[#, GaussianIntegers -> True]]& /@ gbB1;
In[44]:= (gbB3 = gbB2 /. p_Plus?(Variables[#] === {k, l}&) :> 1);

In[45]:= resBS = Select[Union[DeleteCases[
                     If[Head[#] === Power, #[[1]], #]& /@
                     Flatten[If[Head[#] === Times, List @@ #, {#}]& /@ gbB3],
                     _?NumberQ, {1}]], Exponent[#, bs] === 1 ||
                     Exponent[#, b] === 1&]
Out[45]= {b, b - bs, bs, -b + bs, b + bs, bs - i b l,
         bs + i b l, -bs + b l, bs + b l, bs - b l², -bs + b l², bs + b l²}
```

prep preprocesses expressions for Solve. Basically, it removes all multiplicities, because we are not interested in them and so can speed up things considerably.

```
In[46]:= prep[p_Plus] = p;
         prep[p_Power] := p[[1]];
         prep[p_Times] := If[Head[#] === Power, #[[1]], #]& /@ p;
```

test[bruch] tests if a, as, b, bs, l fulfill the original equations.

```
In[49]:= test[rational_, i_] :=
         ((* numerator and denominator *)
         {num, den} = {Numerator[#], Denominator[#]}&[Together[rational]];
         (* a, b, as, and bs *)
         {{a1, b1}, {as1, bs1}} = If[FreeQ[#, x, {0, Infinity}], {#, 0},
                               CoefficientList[#, x]]& /@ {num, den};
         (* the check *)
         Together[eqs //. {l -> s2[[i]],
                   a -> a1, b -> b1, as -> as1, bs -> bs1}])
```

For every value of $l(k)$, we now calculate all possible values of a, as and b, bs.

We do this by solving (when possible) for as and bs and substitute the solutions into the original equations.

Then we calculate a GroebnerBasis with respect to the renamed variables and solve for them. Finally, we substitute all results in the original equations and test if they are fulfilled.

```
In[50]:= (* avoid messages from infinite expressions *)
         Off[Solve::svars]; Off[Power::infy]; Off[Infinity::indet]
In[52]:= Do[bag = {};
           Do[(* solve for as or a *)
               vara = DeleteCases[Variables[resAS[[i]]], l];
               sola = Solve[resAS[[i]] == 0, vara[[-1]]];
               (* solve for bs or b *)
               varb = DeleteCases[Variables[resBS[[j]]], l];
               solb = Solve[resBS[[j]] == 0, varb[[-1]]];
```

```
(* generate reduced equation for only two variables *)
eqs1 = Numerator[Together[#]]& /@
        Flatten[(eqs //. sola //. solb) //. 1 -> s2[[h]]];
(* still present variables in the equations *)
remVars = DeleteCases[Variables[eqs1], k];
(* groebnerize remaining variables *)
fgb = Factor /@ GroebnerBasis[eqs1, remVars,
                               ParameterVariables -> {k}];
fgb2 = Union[prep /@ fgb];
(* solve for remaining variables *)
sol = Union[Solve[# == 0& /@ fgb2, remVars]];
res = Cancel[Flatten[((a + b x)/(as + bs x)  //.
              sola //. solb //. sol) //. 1 -> s2[[h]]]];
(* test if solutions fulfill original equations *)
If[MemberQ[#, x, {0, Infinity}],
    If[test[#, h] === {0, 0, 0, 0}, AppendTo[bag, Factor //@ #]];
        (* printing some results while one is waiting
           in case waiting is too boring:
           Print[test[#, i]; InputForm[#]] *)]& /@ res,
    {i, Length[resAS]}, {j, Length[resBS]}];
union = Union[bag];
(* keep the good solutions *)
result[h] = {s2[[h]], union}, {h, 1, Length[s2]}]
```

And here are the 24 different transformations.

In[53]:= **Flatten[Last /@ Table[result[i], {i, 1, 28}]] // Union**

Out[53]= $\{-\dfrac{1}{x}, \dfrac{1}{x}, -\dfrac{1}{k^2 x}, \dfrac{1}{k^2 x}, -x, x, -k^2 x, k^2 x, -\dfrac{(-i+k)(-i+kx)}{(i+k)(i+kx)},$

$\dfrac{(-i+k)(-i+kx)}{(i+k)(i+kx)}, -\dfrac{(i+k)(-i+kx)}{(-i+k)(i+kx)}, \dfrac{(i+k)(-i+kx)}{(-i+k)(i+kx)}, -\dfrac{(-i+k)(i+kx)}{(i+k)(-i+kx)},$

$\dfrac{(-i+k)(i+kx)}{(i+k)(-i+kx)}, -\dfrac{(i+k)(i+kx)}{(-i+k)(-i+kx)}, \dfrac{(i+k)(i+kx)}{(-i+k)(-i+kx)},$

$-\dfrac{(-1+k)(-1+kx)}{(1+k)(1+kx)}, \dfrac{(-1+k)(-1+kx)}{(1+k)(1+kx)}, -\dfrac{(1+k)(-1+kx)}{(-1+k)(1+kx)}, \dfrac{(1+k)(-1+kx)}{(-1+k)(1+kx)},$

$-\dfrac{(-1+k)(1+kx)}{(1+k)(-1+kx)}, \dfrac{(-1+k)(1+kx)}{(1+k)(-1+kx)}, -\dfrac{(1+k)(1+kx)}{(-1+k)(-1+kx)}, \dfrac{(1+k)(1+kx)}{(-1+k)(-1+kx)}\}$

We could have also started by determining relations between the functions $a(k)$, $a'(k)$, $b(k)$, $b'(k)$. The first polynomial of the following Gröbner basis gives such relations.

In[54]:= **GroebnerBasis[eqs, {1, k, a, as, b, bs}] // First // Factor**

Out[54]= $-(as - b)^3 (as + b)^3 (b - bs)^2 bs (b + bs)^2 (as b - a bs)$
$(as b + as bs + b bs - bs^2)(as b + as bs - b bs + bs^2)(as b - as bs + b bs + bs^2)$

19. Forced Damped Oscillation, $(1 + z/n)^n$-Series, e-Series, q-Logarithm

a) We first compute the solution of the differential equation and transform it into a suitable form.

In[1]:= **(sol = DSolve[m x''[t] + γ x'[t] + k x[t] ==
              F0 Cos[ω t], x[t], t]) // Short[#, 12]&**

Out[1]//Short= $\{\{x[t] \to e^{\frac{t(-\gamma-\sqrt{-4 k m+\gamma^2})}{2 m}} C[1] + e^{\frac{t(-\gamma+\sqrt{-4 k m+\gamma^2})}{2 m}} C[2] -$

$\dfrac{4 (F0\, k\, m^2 \cos[t\,\omega] - F0\, m^3\, \omega^2 \cos[t\,\omega] + F0\, m^2\, \gamma\, \omega \sin[t\,\omega])}{(2 k m - \gamma^2 + \gamma\sqrt{-4 k m + \gamma^2} - 2 m^2\, \omega^2)(-2 k m + \gamma^2 + \gamma\sqrt{-4 k m + \gamma^2} + 2 m^2\, \omega^2)}\}\}$

Here is the long-time behavior ($t \to \infty$).

In[2]:= **((x[t] /. sol) /. {C[1] -> 0, C[2] -> 0}) // Simplify**

Out[2]= $\left\{\dfrac{F0 ((k - m\,\omega^2) \cos[t\,\omega] + \gamma\,\omega \sin[t\,\omega])}{k^2 - 2 k m\,\omega^2 + \gamma^2\,\omega^2 + m^2\,\omega^4}\right\}$

In[3]:= **Collect[%[[1]], {Cos[ω t], Sin[ω t]}]**

Out[3]= $\dfrac{F0 (k - m\,\omega^2) \cos[t\,\omega]}{k^2 - 2 k m\,\omega^2 + \gamma^2\,\omega^2 + m^2\,\omega^4} + \dfrac{F0\,\gamma\,\omega \sin[t\,\omega]}{k^2 - 2 k m\,\omega^2 + \gamma^2\,\omega^2 + m^2\,\omega^4}$

In[4]:= **Together /@ %**

Out[4]= $\dfrac{F0\,(k-m\,\omega^2)\,\text{Cos}[t\,\omega]}{k^2-2\,k\,m\,\omega^2+\gamma^2\,\omega^2+m^2\,\omega^4} + \dfrac{F0\,\gamma\,\omega\,\text{Sin}[t\,\omega]}{k^2-2\,k\,m\,\omega^2+\gamma^2\,\omega^2+m^2\,\omega^4}$

We make a comparison with the following.

In[5]:= **TrigExpand[x0 Cos[ω t - φ]]**

Out[5]= x0 Cos[φ] Cos[t ω] + x0 Sin[φ] Sin[t ω]

We get the following for the square of the amplitude x0.

In[6]:= **((F0 k - F0 m w^2)/(k^2 + γ^2 w^2 - 2 k m w^2 + m^2 w^4))^2 +**
     ((F0 γ w)/(k^2 + γ^2 w^2 - 2 k m w^2 + m^2 w^4))^2 // Simplify

Out[6]= $\dfrac{F0^2}{k^2-2\,k\,m\,\omega^2+\gamma^2\,\omega^2+m^2\,\omega^4}$

For convenience, we introduce the notation $\omega_0^2 = k/m$ and $g = \gamma/m$. Thus, the solution for $t \to \infty$ (forcing condition) can be written as $x(t) = x_0 \cos(\omega t + \varphi)$ where $x_0 = F_0/m\left((\omega_0^2 - \omega^2)^2 + g^2\,\omega^2\right)^{-1/2}$.

The amplitude of the nth derivative is then $\max(x^{(n)}(t)) = \omega^n F_0/m\left((\omega_0^2 - \omega^2)^2 + g^2\,\omega^2\right)^{-1/2}$.

Now, we find the relevant minimum, simplify the result somewhat, develop everything in a series about $\gamma = 0$, and simplify again.

In[7]:= **ω0 Map[(* mainly cosmetics *)**
        #/ω0&, Map[Together, Collect[Expand[
          MapAll[PowerExpand, Simplify[MapAll[PowerExpand,
                (* make series around g = 0 *)
            Normal[Series[#[[1]][[Position[#[[2]],
                        {ω -> ω0}][[1, 1]], 1, 2]], {γ, 0, 6}]&[
          (* all is positive here; so, we use PowerExpand *)
              {#, MapAll[PowerExpand, # /. {γ -> 0, n -> 0}]}&[
        Solve[(* prepare equations for Solve *)
              (Factor[Numerator[Together[
                D[w^n/Sqrt[(w0^2 - w^2)^2 + γ^2 w^2], w]]]]/w^n) == 0,
          w]]]]]]]], γ]]]

Out[7]= $\left(1 + \dfrac{(-1+9\,n-25\,n^2+31\,n^3-18\,n^4+4\,n^5)\,\gamma^6}{128\,\omega0^6} + \dfrac{(-1+4\,n-5\,n^2+2\,n^3)\,\gamma^4}{32\,\omega0^4} + \dfrac{(-1+n)\,\gamma^2}{4\,\omega0^2}\right)\omega0$

Here is a plot of the results (the deviation of ω_0): the maximum of the amplitude (as a function of g) moves to smaller frequencies (in comparison to ω_0), the maximum of the velocity remains at the same place as that of $g = 0$, and the maximum of the acceleration moves toward larger frequencies.

In[8]:= **Needs["Graphics`Legend`"]**

```
Plot[Evaluate[Table[
((γ^6 (-1 + 9n - 25n^2 + 31n^3 - 18n^4 + 4n^5))/(128) +
 (γ^4 (-1 + 4n - 5n^2 + 2n^3))/(32) + (γ^2 (-1 + n))/(4)),
            {n, 0, 3}]], {γ, 0, 0.14},
        (* color curves differently *)
    PlotStyle -> {{AbsoluteThickness[1], GrayLevel[0]},
                {AbsoluteThickness[1], RGBColor[0, 1, 0]},
                {AbsoluteThickness[1], RGBColor[0, 0, 1]},
                {AbsoluteThickness[1], RGBColor[1, 0, 0]}},
    PlotRange -> {{0, 0.22}, {-0.005, 0.01}}, AxesOrigin -> {0, -0.005},
    AxesLabel -> {StyleForm[HoldForm[γ], "Input"], None},
    PlotLegend -> {"x0", "v0", "a0", "x0'''"},
    LegendPosition -> {0.44, -0.36}, LegendSize -> {0.42, 0.9}]
```

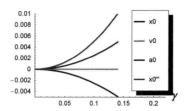

For the relativistic case, see [1276]. For supersymmetric partner solutions, see [1514]. For a special property of the solutions of the damped harmonic oscillator, see [1580].

We continue with the solution of the Fokker–Planck equation for the forced damped oscillator under the influence of a random force. We calculate the quantity $x - x_c(t; x_0, v_0)$.

```
In[10]:= x - (x[t] /. DSolve[{x''[t] + γ x'[t] + ω0^2 x[t] == F0 Cos[ω t],
                   x[0] == x0, x'[0] == v0}, {x[t]}, {t}][[1]]]);
```

For shorter outputs, we use the abbreviations $\mu_{1/2}$ instead of their explicit values expressed in γ and ω_0.

```
In[11]:= xMxC1 = x + 1/(μ1 - μ2)*
            (x0 (μ2 Exp[μ1 t]) - μ1 Exp[μ2 t]) + v0 (-Exp[μ1 t] + Exp[μ2 t]) +
            F0 (-μ1 Exp[μ1 t]/(μ1^2 + ω^2) + μ2 Exp[μ2 t]/(μ2^2 + ω^2)) +
            F0 (μ1/(μ1^2 + ω^2) - μ2/(μ2^2 + ω^2)) Cos[ω t] +
            F0 ω (-1/(μ1^2 + ω^2) + 1/(μ2^2 + ω^2)) Sin[ω t]);
```

```
In[12]:= Block[{μ1 = -γ/2 + Sqrt[γ^2/4 - ω0^2],
                μ2 = -γ/2 - Sqrt[γ^2/4 - ω0^2]}, xMxC1 - %%] // Simplify
Out[12]= 0
```

Next, we calculate the integral $I_{\psi\psi}(t)$.

```
In[13]:= ψ[τ_] = (Exp[μ1 (t - τ)] - Exp[μ2 (t - τ)])/(μ1 - μ2);
```

```
In[14]:= Iψψ = Integrate[ψ[τ] ψ[τ], {τ, 0, t}] // Together
```

$$Out[14]= \frac{-\mu1^2 + e^{2\,t\,\mu2}\,\mu1^2 + 2\,\mu1\,\mu2 + e^{2\,t\,\mu1}\,\mu1\,\mu2 + e^{2\,t\,\mu2}\,\mu1\,\mu2 - 4\,e^{t\,(\mu1+\mu2)}\,\mu1\,\mu2 - \mu2^2 + e^{2\,t\,\mu1}\,\mu2^2}{2\,\mu1\,(\mu1 - \mu2)^2\,\mu2\,(\mu1 + \mu2)}$$

So, we have the following probability distribution for the oscillator's position.

```
In[15]:= W[x_, t_] =
            With[{q = γ kB T/m}, 1/(4 Pi q Iψψ)^(1/2) Exp[-xMxC1^2/(4 q Iψψ)]];
```

To calculate the averages $\langle f(x)\rangle_P$, we will not directly use Integrate on the expressions containing $W(x, t; x_0, v_0)$. Due to the complex functions in the exponents, convergence testing on such expression could take a long time. Instead, we carry out the needed integrals using pattern matching.

```
In[16]:= intRules = Table[Exp[α_ (x + r_)^2] ->
            (Integrate[x^k Exp[α (x - r)^2], {x, -Infinity, Infinity},
               Assumptions -> Re[α] < 0 && Re[r α] < 0] /. r -> -r),
            {k, 0, 2}]
```

$$Out[16]= \left\{ e^{(x+r\_)^2\,α\_} \to \frac{\sqrt{\pi}}{\sqrt{-α}},\ e^{(x+r\_)^2\,α\_} \to -\frac{\sqrt{\pi}\ r}{\sqrt{-α}},\ e^{(x+r\_)^2\,α\_} \to \frac{\sqrt{\pi}\ (1 - 2\,r^2\,α)}{2\,(-α)^{3/2}} \right\}$$

The probability density is normalized.

```
In[17]:= (W[x, t] /. intRules[[1]]) // PowerExpand // Simplify
Out[17]= 1
```

Because of its Gaussian shape, the average position agrees with the position of maximal probability which in turn is the solution of the deterministic trajectory.

```
In[18]:= xAv = W[x, t] /. intRules[[2]] // PowerExpand // Simplify
```

Out[18]= $-\dfrac{1}{\mu1 - \mu2}\left(-e^{t\,\mu1}\,v0 + e^{t\,\mu2}\,v0 - e^{t\,\mu2}\,x0\,\mu1 + e^{t\,\mu1}\,x0\,\mu2 - \dfrac{e^{t\,\mu1}\,F0\,\mu1}{\mu1^2 + \omega^2} + \dfrac{e^{t\,\mu2}\,F0\,\mu2}{\mu2^2 + \omega^2} + \right.$

$\qquad\left. F0\,\left(\dfrac{\mu1}{\mu1^2 + \omega^2} - \dfrac{\mu2}{\mu2^2 + \omega^2}\right)\,Cos[t\,\omega] + F0\,\omega\,\left(-\dfrac{1}{\mu1^2 + \omega^2} + \dfrac{1}{\mu2^2 + \omega^2}\right)\,Sin[t\,\omega]\right)$

In[19]:= `- (xMxCl - x) == xAv // FullSimplify`

Out[19]= True

For short times, the deterministic solution is the following.

In[20]:= `(Series[xAv, {t, 0, 2}] //`
`        Simplify[#, kB T > 0 && γ > 0 && m > 0]&) /.`
`        {μ1 + μ2 -> -γ, μ1 μ2 -> ω0^2}`

Out[20]= $x0 + v0\,t + \dfrac{1}{2}\,(F0 - v0\,\gamma - x0\,\omega0^2)\,t^2 + O[t]^3$

The average of the squared position deviation from its average $\langle(x - \langle x(t)\rangle_P)^2\rangle_P$ is of the form $k_B\,T/m\,\zeta(t;\gamma)$.

In[21]:= `δxS = Simplify[#, kB T > 0 && γ > 0 && m > 0]& @ PowerExpand @`
`        (Plus @@ {W[x, t] /. intRules[[3]],`
`        -2 xAv W[x, t] /. intRules[[2]], xAv^2 /. intRules[[1]]})`

Out[21]= $\dfrac{kB\,T\,\gamma\,((-1 + e^{2\,t\,\mu2})\,\mu1^2 + (2 + e^{2\,t\,\mu1} + e^{2\,t\,\mu2} - 4\,e^{t\,(\mu1+\mu2)})\,\mu1\,\mu2 + (-1 + e^{2\,t\,\mu1})\,\mu2^2)}{m\,\mu1\,(\mu1 - \mu2)^2\,\mu2\,(\mu1 + \mu2)}$

In[22]:= `Series[%, {t, 0, 4}] // Simplify`

Out[22]= $\dfrac{2\,kB\,T\,\gamma\,t^3}{3\,m} + \dfrac{kB\,T\,\gamma\,(\mu1 + \mu2)\,t^4}{2\,m} + O[t]^5$

We end with a graphic showing the probability density $W(x, t; x_0, v_0)$ for some concrete initial conditions over the t,x,-plane. We clearly see the oscillations of the deterministic forced damped oscillator and the broadening of the distribution with time.

In[23]:= `Block[{m = 1, γ = 0.3, F0 = 6, ω0 = Pi, ω = 2,`
`        x0 = 3, v0 = 2, kB = 1, T = 10, μ1, μ2},`
`        {μ1, μ2} = -γ/2 + Sqrt[γ^2/4 - ω0^2] {+1, -1};`
`        Plot3D[Evaluate[W[x, t]], {t, 1/3, 24}, {x, -8, 8},`
`        PlotPoints -> {240, 800}, PlotRange -> All,`
`        Mesh -> False, BoxRatios -> {3, 1, 0.6}]]`

b) The built-in `Series` command does not give a result for $(1 + z/n)^n$ around $n = \infty$.

In[1]:= `Series[(1 + z/n)^n, {n, Infinity, 3}]`

Out[1]= $e^z - \dfrac{e^z\,z^2}{2\,n} + \left(\dfrac{e^z\,z^3}{3} + \dfrac{e^z\,z^4}{8}\right)\left(\dfrac{1}{n}\right)^2 + \left(-\dfrac{1}{4}\,e^z\,z^4 - \dfrac{e^z\,z^5}{6} - \dfrac{e^z\,z^6}{48}\right)\left(\dfrac{1}{n}\right)^3 + O\left[\dfrac{1}{n}\right]^4$

Using $(1 + z/n)^n = \exp(\ln((1 + z/n)^n)) = \exp(n\,\ln((1 + z/n)))$ it is straightforward to calculate the first 10 terms of the series.

In[2]:= `Factor //@ Exp[n Series[Log[1 + z/n], {n, Infinity, 10}]]`

Out[2]= $e^z - \dfrac{e^z\,z^2}{2\,n} + \dfrac{1}{24}\,e^z\,z^3\,(8 + 3\,z)\,\left(\dfrac{1}{n}\right)^2 - \dfrac{1}{48}\,(e^z\,z^4\,(2 + z)\,(6 + z))\,\left(\dfrac{1}{n}\right)^3 +$

$\qquad \dfrac{e^z\,z^5\,(1152 + 1040\,z + 240\,z^2 + 15\,z^3)\,\left(\frac{1}{n}\right)^4}{5760} - \dfrac{(e^z\,z^6\,(4 + z)\,(480 + 408\,z + 68\,z^2 + 3\,z^3))\,\left(\frac{1}{n}\right)^5}{11520} +$

$\qquad \dfrac{e^z\,z^7\,(414720 + 526176\,z + 211456\,z^2 + 35280\,z^3 + 2520\,z^4 + 63\,z^5)\,\left(\frac{1}{n}\right)^6}{2903040} -$

$$\frac{(e^z \, z^8 \, (6 + z) \, (120960 + 151104 \, z + 55792 \, z^2 + 7800 \, z^3 + 450 \, z^4 + 9 \, z^5)) \, (\frac{1}{n})^7}{5806080} +$$

$$\frac{1}{1393459200} \Big(e^z \, z^9 \, (154828800 + 239376384 \, z + 128424960 \, z^2 +$$

$$32287360 \, z^3 + 4220160 \, z^4 + 292320 \, z^5 + 10080 \, z^6 + 135 \, z^7) \, \Big(\frac{1}{n}\Big)^8 \Big) -$$

$$\frac{1}{2786918400} \Big((e^z \, z^{10} \, (8 + z) \, (34836480 + 53568000 \, z + 27721728 \, z^2 + 6455424 \, z^3 +$$

$$749312 \, z^4 + 44880 \, z^5 + 1320 \, z^6 + 15 \, z^7)) \, \Big(\frac{1}{n}\Big)^9 \Big) + O\Big[\frac{1}{n}\Big]^{10}$$

For a similar series, see [795].

c) Taking the logarithm of the equation, we get the following identity: $n + \sum_{k=0}^{\infty} \alpha_k \, n^{-k} = 1/\ln(1 + 1/n)$.
This allows for a direct calculation of the α_k using `Series`. Here are the first 11 α_k.

```
In[1]:= 1/Series[Log[1 + 1/n], {n, Infinity, 10}]
```

$$\text{Out[1]= } \frac{1}{\frac{1}{n}} + \frac{1}{2} - \frac{1}{12\,n} + \frac{1}{24} \Big(\frac{1}{n}\Big)^2 - \frac{19}{720} \Big(\frac{1}{n}\Big)^3 + \frac{3}{160} \Big(\frac{1}{n}\Big)^4 -$$

$$\frac{863 \, (\frac{1}{n})^5}{60480} + \frac{275 \, (\frac{1}{n})^6}{24192} - \frac{33953 \, (\frac{1}{n})^7}{3628800} + \frac{8183 \, (\frac{1}{n})^8}{1036800} + O\Big[\frac{1}{n}\Big]^9$$

The first 100 terms can be calculated in less than a second and yield an approximation for $n = 1$ correct to about 0.01%.

```
In[2]:= (ser = 1/Series[Log[1 + 1/n], {n, Infinity, 101}]); // Timing
```

```
Out[2]= {0.13 Second, Null}
```

```
In[3]:= {ser[[5]], ser[[3, -1]]} // Short[#, 3]&
```

```
Out[3]//Short= {100, - (52211582251198720757103406246580767681044649 2532143
           ≪58≫ 677333569678144960964574199571230941106066784044 6019) /
           (175515158329790183633032053055797594164828801391742 ≪62≫
           5882288752804939740133457920000000000000000000000000000) }
```

```
In[4]:= N[E - ((1 + 1/n)^Normal[ser] /. n -> 1), $MachinePrecision + 1]
```

```
Out[4]= 0.00027840472848168629
```

Now let us deal with the case where the base and the exponent have more than one term. Proceeding as above, we get the following system of equations.

```
In[5]:= ansatz = (1 + 1/n + β[2]/n^2 + β[3]/n^3)^(n + α[0] + α[1]/n)
```

$$\text{Out[5]= } \Big(1 + \frac{1}{n} + \frac{\beta[2]}{n^2} + \frac{\beta[3]}{n^3}\Big)^{n + \alpha[0] + \frac{\alpha[1]}{n}}$$

```
In[6]:= ser = Exp[Series[Log[ansatz] //
               PowerExpand, {n, Infinity, 5}]] // Simplify;
```

```
In[7]:= eqs = Table[SeriesCoefficient[ser, k], {k, 1, 4}];
```

We solve the resulting polynomial system and select the best solution.

```
In[8]:= sol = Solve[# == 0& /@ eqs, {α[0], α[1], β[2], β[3]}];
        Length[sol]
```

```
Out[9]= 6
```

```
In[10]:= fsol = sol[[Position[#, Min[#]]&[Abs[N[E -
                (ansatz /. sol /. n -> 1), 22]]][[1, 1]]]] // RootReduce
```

```
Out[10]= {α[1] → Root[1 + 360 #1 + 23280 #1² + 572800 #1³ + 2937600 #1⁴ + 5760000 #1⁵ + 4608000 #1⁶ &, 2],
         β[3] → Root[
             1751 - 16200 #1 + 92880 #1² - 13305600 #1³ + 4665600 #1⁴ + 93312000 #1⁵ + 373248000 #1⁶ &, 1],
         β[2] → Root[-5 + 324 #1² - 1440 #1³ + 1440 #1⁴ + 2880 #1⁶ &, 2],
         α[0] → Root[31 - 504 #1 + 3024 #1² - 8640 #1³ + 12240 #1⁴ - 8640 #1⁵ + 2880 #1⁶ &, 1]}
```

For $n = 1$, we obtain the following approximation for e.

```
In[11]:= (ansatz /. fsol /. n -> 1) // RootReduce
```

Out[11]= Root [21445272851 - 64519746480 #1 + 81867505680 #1^2 -
55767225600 #1^3 + 21419769600 #1^4 - 4385664000 #1^5 + 373248000 #1^6 &,
2]$^{\text{Root} [107635123-439120320 \text{ #1}+748639440 \text{ #1}^2-688560000 \text{ #1}^3+364780800 \text{ #1}^4-107136000 \text{ #1}^5+13824000 \text{ #1}^6 \&, 1]}$

It is correct to about 0.002%.

In[12]:= N[1 - E/%]

Out[12]= -0.0000205567

c) It is straightforward to implement the left side. We generate the nested iterators on the right side using Table.

In[1]:= rhs[q_, x_, n_] := Logq[q, Product[x[i], {i, n}]]

```
lhs[q_, x_, n_] := Sum[(1 - q)^(k - 1)*
  (Sum[Product[Logq[q, x[i[j]]], {j, k}], ##]& @@
    Table[{i[j], If[j == 1, 0, i[j - 1]] + 1, n}, {j, k}]),
  {k, n}]
```

Here is the definition of the q-logarithm $\ln_q(x)$.

In[3]:= Logq[q_, x_] := (x^(1 - q) - 1)/(1 - q)

Here are the left and right sides for the simplest nontrivial case, $n = 2$.

In[4]:= {rhs[q, x, 2], lhs[q, x, 2]}

Out[4]= $\left\{ \dfrac{-1 + (x[1] \, x[2])^{1-q}}{1-q}, \dfrac{-1 + x[1]^{1-q}}{1-q} + \dfrac{-1 + x[2]^{1-q}}{1-q} + \dfrac{(-1 + x[1]^{1-q})\,(-1 + x[2]^{1-q})}{1-q} \right\}$

By using Simplify with the assumptions about positivity of the x_i in the second argument, we establish the identities under consideration without problem for small n.

In[5]:= Table[Simplify[rhs[q, x, n] - lhs[q, x, n],
                And @@ Table[x[j] > 0, {j, n}]],
          {n, 6}]

Out[5]= {0, 0, 0, 0, 0, 0}

For larger n, we presimplify the right side using Together to speed the calculations.

In[6]:= Table[Simplify[rhs[q, x, n] - Together[lhs[q, x, n]],
                And @@ Table[x[j] > 0, {j, n}]], {n, 7, 10}]

Out[6]= {0, 0, 0, 0}

For other deformations of the logarithm, see [977], [205].

20. S_{mn}, Symmetrized Determinant, Fermat Test

a) First, we implement the two functions $P_{mn}(q_1, q_2, \ldots, q_m)$ and $Q_{mn}(q_1, q_2, \ldots, q_m)$ without the exponential factors. (Because P_{mn} contains the function Q_{mn}, we store the Q_{mn}.)

In[1]:= Qs[m_, n_] := Qs[m, n] = Product[q[k]^(n - m), {k, m}]*
          Product[(q[i] - q[j])^2, {i, m}, {j, i + 1, m}];

          Ps[m_, n_] := Sum[q[i] Log[q[i]], {i, m}] Qs[m, n]

Here are two examples.

In[3]:= Qs[3, 3]

Out[3]= $(q[1] - q[2])^2 \, (q[1] - q[3])^2 \, (q[2] - q[3])^2$

In[4]:= Ps[4, 4]

Out[4]= $(q[1] - q[2])^2 \, (q[1] - q[3])^2 \, (q[2] - q[3])^2 \, (q[1] - q[4])^2 \, (q[2] - q[4])^2$
$(q[3] - q[4])^2 \, (Log[q[1]] \, q[1] + Log[q[2]] \, q[2] + Log[q[3]] \, q[3] + Log[q[4]] \, q[4])$

The real problem is the computation of the multiple integrals. With the following, we have to wait "forever": Integrate[..., {q[1], 0, Infinity}, ..., {q[n], 0, Infinity}]

If we look at the resulting expressions, we note that only the following two basic types of integrals appear (with different integers n).

```
In[5]:= Integrate[Exp[-q[i]] q[i]^n, {q[i], 0, Infinity}, Assumptions -> n >= 0]
Out[5]= Gamma[1 + n]
```

Then the second class of needed integrals $\int_0^\infty \log(q_i) e^{-q_i} q_i^n \, dq_i$ follows by differentiation with respect to n.

```
In[6]:= D[%, n]
Out[6]= Gamma[1 + n] PolyGamma[0, 1 + n]
```

This suggests that the corresponding integrations should be programmed. For convenience, we drop the term `Exp[-q[i]]`, which arises in every integration. We have used the two constructions `myIntegrate /@ a` in expressions with the head `Times` or `Plus` to avoid their slower recursive evaluation. The only remaining difficulty is in the `Log` terms; here, we must avoid an infinite recursive call.

```
In[7]:= (* additivity *)
        myIntegrate[a_Plus] := myIntegrate /@ a;

        (* integrand is a constant *)
        myIntegrate[fac_?NumberQ] = fac;

        (* q^n Exp[-q] *)
        myIntegrate[q[_]^n_.] = n!;

        (* every variable treated separately *)
        myIntegrate[a_Times?(FreeQ[#, Log]&)] := myIntegrate /@ a;

        (* separate variables in mixed case *)
        myIntegrate[Log[q[i_]] rest_] :=
          (myIntegrate[Log[q[i]] #]*
           myIntegrate[rest/#])&[Cases[rest, q[i]^_.][[1]]] /;
             Union[Cases[rest, q[_], {0, Infinity}]] != {q[i]}

        (* Log[q] q^n Exp[-q] *)
        myIntegrate[Log[q[i_]] q[i_]^n_.] = n! PolyGamma[0, n + 1];
```

For the Digamma function, we could also use the somewhat more explicit expressions ("with storage" via a `SetDelayed[`⋮`Set[...]]` construction) $\Psi(0) = -\gamma$, $\Psi(n) = \sum_{k=1}^{n-1} 1/k - \gamma$.

We now compare the required times for a simple example.

```
In[19]:= Integrate[Expand[Exp[-q[1]] Exp[-q[2]] Qs[2, 2]],
                   {q[1], 0, Infinity}, {q[2], 0, Infinity}] // Timing
Out[19]= {0.77 Second, 2}

In[20]:= myIntegrate[Expand[Qs[2, 2]]]
Out[20]= 2

In[21]:= Do[myIntegrate[Expand[Qs[2, 2]]], {100}] // Timing
Out[21]= {0.01 Second, Null}
```

Because the application of `myIntegrate` can be highly recursive, we set the following.

```
In[22]:= $RecursionLimit = Infinity;
```

Putting all of this together, we have the following.

```
In[23]:= S[m_, n_] :=
         PolyGamma[0, 1 + n m] - myIntegrate[Expand[Ps[m, n]]]/
         (m n myIntegrate[Expand[Qs[m, n]]]) // Together
```

We now look at a few values.

```
In[24]:= Table[S[m, n], {n, 2, 4}, {m, n}]
```

$$Out[24]= \left\{\left\{0, \frac{1}{3}\right\}, \left\{0, \frac{9}{20}, \frac{1669}{2520}\right\}, \left\{0, \frac{107}{210}, \frac{21341}{27720}, \frac{664789}{720720}\right\}\right\}$$

One can actually write S_{mn} in the shorter form: $S_{mn} = (\sum_{k=1}^{mn} 1/k) - (m-1)/(2n)$.

```
In[25]:= s[m_, n_] = Sum[1/k, {k, n + 1, n m}] - (m - 1)/(2n);

        Table[Table[s[m, k], {m, k}], {k, 2, 4}]
```
$$\text{Out[26]}= \{\{0, \tfrac{1}{3}\}, \{0, \tfrac{9}{20}, \tfrac{1669}{2520}\}, \{0, \tfrac{107}{210}, \tfrac{21341}{27720}, \tfrac{664789}{720720}\}\}$$

For the use of this function, see [1365], [1599], and [959]. Qs[5, 5] cannot be computed so easily (without the use of a lot of memory).

```
In[27]:= Qs[5, 5]
```

$$\text{Out[27]}= (q[1] - q[2])^2 (q[1] - q[3])^2 (q[2] - q[3])^2 (q[1] - q[4])^2 (q[2] - q[4])^2$$
$$(q[3] - q[4])^2 (q[1] - q[5])^2 (q[2] - q[5])^2 (q[3] - q[5])^2 (q[4] - q[5])^2$$

After the application of Expand, it would temporarily have 3^Length[%] = 59049 summands (and in the final form, it would still have 2961 terms).

In this case, it may be better not to multiply out, but to find every term in the sum by hand, and then integrate afterward. The following routine constructs the summands arising from multiplying out an expression of the form

$$(a_{11} + a_{12} + \cdots)^{e_{11}} (a_{21} + a_{22} + \cdots)^{e_{21}} (a_{31} + a_{32} + \cdots)^{e_{31}}$$

and then integrates them. The interesting thing about bigInt is the automatic construction of iterators and of the expressions on which these iterators operate (we suppress the error messages that are generated by Part), which we have also used in Chapter 5 of the Programming volume [1735].

```
In[28]:= bigInt[arg_] :=
        Module[{pluses, res, plus1, iteras, taker, a, sum},
            (* look for all powers of a sum *)
            pluses = Cases[arg, _Plus^_.];
            (* just a factor *)
            res = arg/(Times @@ pluses);
            (* analyze powers, build Lists *)
            plus1 = Flatten[If[Head[#1] === Power, Table[#[[1]], {#[[2]]}],
                            #1]& /@ pluses];
            (* build iterators *)
            iteras = MapIndexed[{a[#2[[1]]], #1}&, Length /@ plus1];
            (* call Part with nonnumeric indices gives error message *)
            Off[Part::pspec];
            taker = MapIndexed[Part[#1, a[#2[[1]]]]&, plus1];
            On[Part::pspec];
            sum = 0; (* simple Do loop over the iterators *)
            Do[sum = sum + Expand[myIntegrate[res Times @@ taker]]; Null,
                (* evaluate the iterators *)
                Evaluate[Sequence @@ iteras]]; sum]
```

For explicit computations, it is convenient to multiply out the squares in Qs and Ps, because by the binomial formula, we then have "only" the above 59049 elements to integrate. With a longer (but requiring less memory) calculation, we get the value 9953280.

```
In[29]:= bigInt[Expand /@ Qs[5, 5]]

Out[29]= 9953280
```

We get the value for Ps[5, 5].

```
In[30]:= bigInt[Expand /@ Ps[5, 5]]

Out[30]= 667699200 - 248832000 EulerGamma
```

Thus, we get this for S_{55}. It agrees with the direct formula for S_{55}.

```
In[31]:= {PolyGamma[0, 1 + 25] -
            (667699200 - 248832000 EulerGamma)/(25 9953280) // Together,
            (* direct formula *) s[5, 5]}
```
$$\text{Out[31]}= \left\{ \frac{10107221087}{8923714800}, \frac{10107221087}{8923714800} \right\}$$

b) We start by implementing the noncommutative multiplication CT for the symbols a, b, c, and e. We also implement linearity over coefficients $\alpha[i, j, k]$.

```
In[1]:= SetAttributes[CT, Flat]
```

```
CT[e, a] = a; CT[a, e] = a; CT[e, b] = b; CT[b, e] = b;
CT[e, c] = c; CT[c, e] = c; CT[a, b] = c; CT[b, a] = c;
CT[a, c] = b; CT[c, a] = b; CT[b, c] = a; CT[c, b] = a;
CT[a, a] = e; CT[e, e] = e; CT[b, b] = e; CT[c, c] = e;
```

```
(* linearity *)
CT[a___, b_ + c_, d_, e___] := CT[a, b, d, e] + CT[a, c, d, e]
CT[a___, b_, c_ + d_, e___] := CT[a, b, c, e] + CT[a, b, d, e]
CT[a___, (A_ α | A_Integer) b_, c___] := A CT[a, b, c]
CT[a___, 0, b___] := 0
```

Next, we calculate the symmetric determinant of a 4×4 matrix with generic elements of the form $c_{ij}^{(a)} a + c_{ij}^{(b)} b + c_{ij}^{(c)} c + c_{ij}^{(e)} e$.

```
In[11]:= o = 4;
         perms = Permutations[Range[o]];
In[13]:= m = Table[Sum[α[i, j, k] {a, b, c, e}[[k]], {k, o}],
                {i, o}, {j, o}];
In[14]:= (* general definition of the symmetric determinant *)
         sDet[m_] := 1/o! Sum[Signature[perms[[i]]] Signature[perms[[j]]]*
                     CT @@ (m[[##]])& @@@ Transpose[{perms[[i]], perms[[j]]}]),
                 {i, Length[perms]}, {j, Length[perms]}];
In[16]:= sum = sDet[m];
```

In expanded form, the symmetric determinant under consideration has 6144 terms.

```
In[17]:= Length[Expand[sum]]
```

```
Out[17]= 6144
```

We extract the coefficients of a, b, c, and e and equate all of them to 1.

```
In[18]:= cs = Coefficient[Expand[sum], {a, b, c, e}];
         αs = Cases[cs, _α, Infinity] // Union;
```

To find actual values for the α's that make the four equations `cs` equal to 1, we could carry out a random search. This is relatively time-consuming. Instead, we choose random values for some variables, simplify the resulting equations and then try various assignments for the remaining variables.

```
In[20]:= o = 56;
         SeedRandom[1000];
         (* random assignment until we succeed *)
         While[r = Random[Integer, {1, 10^20}]; SeedRandom[r];
                αsT = Take[αs, o]; αsR = Take[αs, {o + 1, 64}];
                Rs = (Rule @@@ Transpose[{αsT,
                            Table[Random[Integer, {-1, 1}], {o}]}]);
                gb = GroebnerBasis[(cs - {1, 1, 1, 1}) /. Rs, αsR,
                            MonomialOrder -> Lexicographic];
                tfs = Table[Union[gb] == {0}, Evaluate[Sequence @@
                            (Table[{#, -1, 1}]& /@ αsR)]] //
                                            Flatten // Union;
         FreeQ[tfs, True], Null]
```

We add the successful assignments to the list of rules Rs.

```
In[24]:= cas = Cases[Flatten[Table[C[αsR, gb == Table[0, {Length[gb]}]]],
                Evaluate[Sequence @@ (Table[{#, -1, 1}]& /@ αsR)]]],
                C[_, True]]
Out[24]= {C[{-1, -1, 1, 1, 0, 0, -1, -1}, True],
          C[{0, 0, 0, 0, 0, 0, -1, -1}, True], C[{0, 0, 1, 1, 1, 1, 0, 0}, True],
          C[{1, 1, -1, -1, 0, 0, -1, -1}, True], C[{1, 1, 0, 0, 1, 1, 0, 0}, True]}
```

```
In[25]:= αsRules = Join[Rs, Rule @@@ Transpose[{αsR, cas[[1, 1]]}]];
```

Here is one possible form of the matrix we were looking for.

In[26]:= **(M = m /. αsRules) // MatrixForm**

Out[26]//MatrixForm=
$$
\begin{pmatrix}
a+b-c-e & a+b-e & a-c+e & -a+b+c \\
c & -a-b-e & -a-c+e & a+c \\
-c & b+e & a+b & c \\
0 & b-e & -a-b+c+e & -c-e
\end{pmatrix}
$$

In[27]:= **sDet[M] // Expand**

Out[27]= **a + b + c + e**

c) We start by implementing all relevant definitions. To avoid evaluation of the q_i, we define the four variables $q1$, $q2$, $q3$, and $q4$.

```
In[1]:= c1 = a22 + a12 a21 (11 - 1) - a22 12;
        c2 = a12 (1 + a21) 11 - (a12 + a22) 12;
        c3 = (1 + a21) a22 11 + a12 a21 (11 - 1) 12 -
             a22 (a21 + 11) 12;
        c4 = a22 - a21 g + a12 (a22 g + a21 (n11 - 1)) - a22 n22;
        c5 = (-1 - a21 + a12 (a12 + a22)) g + a12 (1 + a21) n11 -
             (a12 + a22) n22;
        c6 = a21^2 g - a22 (g (a12 + a22 + g) + n11 - n11 n22) +
             a21 (g + a12 g^2 - a22 n11 + (a12 + a22 - a12 n11) n22);
        c7 = a22 + a12 a21 (p11 - 1) - a21 p12 + a12 a22 p21 - a22 p22;
        c8 = a12 p11 + a12 a21 p11 - p12 - a21 p12 + a12^2 p21 +
             a12 a22 p21 - (a12 + a22) p22;
        c9 = a22^2 p21 - a21 p12 (1 + a21 + a12 p21) + a12 a21 (p11 - 1) p22 +
             a22 (p11 + a21 p11 + a12 p21 + p12 p21 - (a21 + p11) p22);
In[10]:= q1 = -c2 c6/(c3 c5);
        q2 = c2 c4/(c1 c5);
        q3 = c2 c9/(c3 c8);
        q4 = c2 c7/(c1 c8);
In[14]:= qToq = {q1 -> q1, q2 -> q2, q3 -> q3, q4 -> q4};
In[15]:= R =
        -q1^2 11 12 (1 - p11 - p12 p21 - p22 + p11 p22) +
        q1 q2 (12 n11 + 11 n22) (1 - p11 - p12 p21 - p22 + p11 p22) +
        q2^2 (g^2 - n11 n22) (1 - p11 - p12 p21 - p22 + p11 p22) +
        q1 q3 11 12 (2 - n11 - n22 - p11 + n22 p11 - g p12 - g p21 -
                     p22 + n11 p22) + q2 q3 (-g^2 11 - g^2 12 - 12 n11 - 11 n22 +
                     11 n11 n22 + 12 n11 n22 + g^2 12 p11 + 12 n11 p11 -
                     12 n11 n22 p11 + g 11 p12 + g 12 p21 + g^2 11 p22 +
                     11 n22 p22 - 11 n11 n22 p22) +
        q1 q2 q3 (-12 n11 + 11 12 n11 - 11 n22 + 11 12 n22 + 11 n22 p11 -
                     11 12 n22 p11 - g 11 p12 + g 11 12 p12 - g 12 p21 +
                     g 11 12 p21 + 12 n11 p22 - 11 12 n11 p22) -
        q2^2 q3 (g^2 - n11 n22) (2 - 11 - 12 - p11 + 12 p11 - p22 + 11 p22) +
        q3^2 11 12 (-1 + g^2 + n11 + n22 - n11 n22) +
        q2 q3^2 (g^2 11 + g^2 12 - 2 g^2 11 12 - 12 n11 - 11 12 n11 + 11 n22 -
                     11 12 n22 - 11 n11 n22 - 12 n11 n22 + 2 11 12 n11 n22) +
        q2^2 q3^2 (11 - 1) (12 - 1) (g^2 - n11 n22) +
        q1 q4 (-12 p11 + 12 n11 p11 + g 12 p12 + g 11 p21 - 11 p12 p21 -
                     12 p12 p21 + 12 n11 p12 p21 + 11 n22 p12 p21 - 11 p22 +
                     11 n22 p22 + 11 p11 p22 + 12 p11 p22 - 12 n11 p11 p22 -
                     11 n22 p11 p22) +
        q1^2 q4 (12 p11 - 11 12 p11 + 11 p12 p21 + 12 p12 p21 -
                     2 11 12 p12 p21 + 11 p22 - 11 12 p22 - 11 p11 p22 -
                     12 p11 p22 + 2 11 12 p11 p22) +
        q2 q4 (g^2 p11 + n22 p11 - n11 n22 p11 - g p12 - g p21 +
                     2 g^2 p12 p21 + n11 p12 p21 + n22 p12 p21 - 2 n11 n22 p12 p21 +
                     g^2 p22 + n11 p22 - n11 n22 p22 - 2 g^2 p11 p22 - n11 p11 p22 -
                     n22 p11 p22 + 2 n11 n22 p11 p22) +
        q1 q2 q4 (-n22 p11 + 11 n22 p11 + g p12 - g 12 p12 + g p21 -
                     g 11 p21 - n11 p12 p21 + 12 n11 p12 p21 - n22 p12 p21 +
```

```
                      11 n22 p12 p21 - n11 p22 + 12 n11 p22 + n11 p11 p22 -
                      12 n11 p11 p22 + n22 p11 p22 - 11 n22 p11 p22) -
       q3 q4 (g^2 + n11 + n22 - n11 n22 - 1) (12 p11 + 11 p22) +
       q1 q3 q4 (-12 p11 + 11 12 p11 + 12 n22 p11 - 11 12 n22 p11 -
                 g 12 p12 + g 11 12 p12 - g 11 p21 + g 11 12 p21 -
                 11 p22 + 11 12 p22 + 11 n11 p22 - 11 12 n11 p22) +
       q2 q3 q4 (-g^2 p11 + g^2 12 p11 - n22 p11 + 12 n22 p11 +
                 n11 n22 p11 - 12 n11 n22 p11 + g p12 - g 11 p12 +
                 g p21 - g 12 p21 - g^2 p22 + g^2 11 p22 - n11 p22 +
                 11 n11 p22 + n11 n22 p22 - 11 n11 n22 p22) +
       q1 q2 q3 q4 (11 - 1) (12 - 1) (n22 p11 - g p12 - g p21 + n11 p22) -
       q4^2 (g^2 + n11 + n22 - n11 n22 - 1) (p12 p21 - p11 p22) -
       q1 q4^2 (2 - 11 - 12 - n11 + 12 n11 - n22 + 11 n22)*
                (p12 p21 - p11 p22) +
       q1^2 q4^2 (11 - 1) (12 - 1) (p12 p21 - p11 p22);
```

The most direct way to show that R vanishes would be `Together[R/.qToq]`. Because of the complexity of the expression under consideration, this will probably take some time. After inspecting the form of R for a moment, one recognizes that the common denominator is $c_1^2 c_3^2 c_5^2 c_8^2$. Because this denominator does not identically vanish, we can reduce the problem to the expansion of a polynomial. `Expand[((c1 c3 c5 c8)^2 #) & /@ (R/.qToq)]` should than give the result 0. If we expand one term after the other and sum the intermediate results, we can quickly show that R vanishes.

```
In[16]:= R1 = ((c1 c3 c5 c8)^2 #)& /@ (R /. qToq);
         Length[R1]
Out[17]= 21
```

The next input compares the three methods and the maximal amount of used memory. The built-in function `Expand` was the fastest and most memory-efficient of the three methods.

```
In[18]:= (* use Expand *)
         {(ex = Expand[R1]); // Timing, {ex, MaxMemoryUsed[]}}
Out[19]= {{648.62 Second, Null}, {0, 453929064}}
```

```
In[20]:= (* "manual" expansion *)
         {Timing[sum = 0; Do[sum = sum + Expand[R1[[k]]], {k, 21}]],
          {sum, MaxMemoryUsed[]}}
Out[21]= {{1348.34 Second, Null}, {0, 453929064}}
```

```
In[22]:= (* use Together *)
         {(tog = Together[R1]); // Timing, {tog, MaxMemoryUsed[]}}
Out[23]= {{902.12 Second, Null}, {0, 454312304}}
```

Because the 21 terms of the last inputs after expansion had a couple of hundred thousand terms each, the memory usage of the last inputs was considerable. We could reduce the memory usage by expanding R before carrying out the $q_i \to q_i$ substitution. This reduces the memory usage by a factor 3, but needs about three to four times as long.

```
R2 = ((c1 c3 c5 c8)^2 #)& /@ (Expand[R] /. qToq);
Length[R2]

(* rerun this subexercise without evaluating the above Timing[...] *)
Timing[sum = 0; Do[sum = sum + Expand[R2[[k]]], {k, 239}]]
{sum, MaxMemoryUsed[]}
```

21. WKB Approximations of Higher Order, GHZ State, Entropic Uncertainty Relation

a) Because arbitrary complete differentials can be added to the expressions of different orders, the expressions to be found are not uniquely defined. Here, we implement an adaptation of the method from [142]. We begin with the computation of $S_n(x)$. We separate the definition into even and odd indices, because for even indices, the term $S_k^2(x)$ does not appear. Because the higher terms of $S_n(x)$ are required several times in the definition, we use a `SetDelayed[Set[...]]` construction.

```
In[1]:= S[0] = Sqrt[ε - v[x]];
```

```
S[k_?EvenQ] := S[k] = Expand[(-1/2 D[S[k - 1], x] -
    Sum[S[m] S[k - m], {m, 1, k/2}] + 1/2 S[k/2]^2)/S[0]]

S[k_?OddQ] := S[k] = Expand[(-1/2 D[S[k - 1], x] -
    Sum[S[m] S[k - m], {m, 1, (k - 1)/2}])/S[0]]
```

Here is an example.

In[4]:= `S[8]`

Out[4]=
$$
-\frac{1282031525\, v'[x]^8}{8388608\,(\varepsilon - v[x])^{23/2}} - \frac{256406305\, v'[x]^6\, v''[x]}{524288\,(\varepsilon - v[x])^{21/2}} - \frac{121782417\, v'[x]^4\, v''[x]^2}{262144\,(\varepsilon - v[x])^{19/2}} -
$$
$$
\frac{4321753\, v'[x]^2\, v''[x]^3}{32768\,(\varepsilon - v[x])^{17/2}} - \frac{174317\, v''[x]^4}{32768\,(\varepsilon - v[x])^{15/2}} - \frac{8905935\, v'[x]^5\, v^{(3)}[x]}{65536\,(\varepsilon - v[x])^{19/2}} -
$$
$$
\frac{3164229\, v'[x]^3\, v''[x]\, v^{(3)}[x]}{16384\,(\varepsilon - v[x])^{17/2}} - \frac{47919\, v'[x]\, v''[x]^2\, v^{(3)}[x]}{1024\,(\varepsilon - v[x])^{15/2}} - \frac{281237\, v'[x]^2\, v^{(3)}[x]^2}{16384\,(\varepsilon - v[x])^{15/2}} -
$$
$$
\frac{13161\, v''[x]\, v^{(3)}[x]^2}{4096\,(\varepsilon - v[x])^{13/2}} - \frac{1841055\, v'[x]^4\, v^{(4)}[x]}{65536\,(\varepsilon - v[x])^{17/2}} - \frac{223431\, v'[x]^2\, v''[x]\, v^{(4)}[x]}{8192\,(\varepsilon - v[x])^{15/2}} -
$$
$$
\frac{10461\, v''[x]^2\, v^{(4)}[x]}{4096\,(\varepsilon - v[x])^{13/2}} - \frac{3847\, v'[x]\, v^{(3)}[x]\, v^{(4)}[x]}{1024\,(\varepsilon - v[x])^{13/2}} - \frac{251\, v^{(4)}[x]^2}{2048\,(\varepsilon - v[x])^{11/2}} -
$$
$$
\frac{36195\, v'[x]^3\, v^{(5)}[x]}{8192\,(\varepsilon - v[x])^{15/2}} - \frac{2547\, v'[x]\, v''[x]\, v^{(5)}[x]}{1024\,(\varepsilon - v[x])^{13/2}} - \frac{209\, v^{(3)}[x]\, v^{(5)}[x]}{1024\,(\varepsilon - v[x])^{11/2}} -
$$
$$
\frac{2135\, v'[x]^2\, v^{(6)}[x]}{4096\,(\varepsilon - v[x])^{13/2}} - \frac{119\, v''[x]\, v^{(6)}[x]}{1024\,(\varepsilon - v[x])^{11/2}} - \frac{11\, v'[x]\, v^{(7)}[x]}{256\,(\varepsilon - v[x])^{11/2}} - \frac{v^{(8)}[x]}{512\,(\varepsilon - v[x])^{9/2}}
$$

Now, we integrate the terms containing $V'(x)$. A direct use of the built-in `Integrate` function does not lead to a transformation to the desired form with integrated $(\varepsilon - v[x])\,\hat{}\,(\cdots)$ and differentiated $v'[x]\ldots$.

In[5]:= `Integrate[v'[x] v''[x] (ε - v[x])^(-3/2), x]`

Out[5]=
$$
\int \frac{v'[x]\, v''[x]}{(\varepsilon - v[x])^{3/2}}\, dx
$$

Hence, we suggest the following program to carry out the integration. (The rules follow from a direct calculation after changing the integration variable.)

In[6]:=
```
(* additivity *)
IntegrationOverV[f_. a_Plus * (ε - v[x])^n_?(# < - 1&)] :=
    IntegrationOverV /@ Expand[f a (ε - v[x])^n]

(* integration by parts *)
IntegrationOverV[a_. v'[x]^v_. (ε - v[x])^n_?(# < - 1&)] :=
    Expand[D[a v'[x]^(v - 1), x] 1/(n + 1) (ε - v[x])^(n + 1)];

(* zero case *)
IntegrationOverV[a_. v'[x] (ε - v[x])^-1] := 0

(* additivity *)
IntegrationOverV[x_Plus] := IntegrationOverV /@ x

(* nothing to do *)
IntegrationOverV[x_?(FreeQ[Numerator[#], v'[x], {0, Infinity}]&)] = x;
```

Now the above expression is integrated.

In[16]:= `IntegrationOverV[v'[x] v''[x] (ε - v[x])^(-3/2)]`

Out[16]=
$$
-\frac{2\, v^{(3)}[x]}{\sqrt{\varepsilon - v[x]}}
$$

Taking our example `S[4]` expression as an example, we get the following result.

In[17]:= `S[4]`

Out[17]=
$$
-\frac{1105\, v'[x]^4}{2048\,(\varepsilon - v[x])^{11/2}} - \frac{221\, v'[x]^2\, v''[x]}{256\,(\varepsilon - v[x])^{9/2}} - \frac{19\, v''[x]^2}{128\,(\varepsilon - v[x])^{7/2}} - \frac{7\, v'[x]\, v^{(3)}[x]}{32\,(\varepsilon - v[x])^{7/2}} - \frac{v^{(4)}[x]}{32\,(\varepsilon - v[x])^{5/2}}
$$

In[18]:= `IntegrationOverV[S[4]]`

Out[18]= $\dfrac{1105\ v'[x]^2\ v''[x]}{3072\ (\varepsilon - v[x])^{9/2}} + \dfrac{11\ v''[x]^2}{112\ (\varepsilon - v[x])^{7/2}} + \dfrac{221\ v'[x]\ v^{(3)}[x]}{896\ (\varepsilon - v[x])^{7/2}} + \dfrac{9\ v^{(4)}[x]}{160\ (\varepsilon - v[x])^{5/2}}$

We drop the part consisting of a complete differential because it does not contribute to the integral along a closed path.

The integration by parts is more difficult to realize. The actual integration is carried out with the function IntegrationBy Parts.

In[19]:= `IntegrationByParts[x_Plus] := IntegrationByParts /@ x`

```
IntegrationByParts[a_. (ε - v[x])^n_] :=
Module[{int, intWith, intFree, intWith1, intWith2},
   (* Integrate the factor a *)
   int = Integrate[a, x];
   (* analyze the usefulness of the result *)
   If[Head[int] === Plus,
      intFree = Select[int, FreeQ[#, Integrate,
                       {0, Infinity}, Heads -> True]&];
      (* still contains undone Integrate *)
      intWith = int - intFree,
      If[FreeQ[int, Integrate, {0, Infinity}, Heads -> True],
         intFree = int; intWith = 0, intFree = 0; intWith = int]];
   (* rewrite part that contains Integrate *)
   If[intWith =!= 0, intWith1 =
      If[Head[intWith] === Plus,
         Cases[intWith, HoldPattern[_. Integrate[_, _]]],
      intWith1 = {intWith}];
      (* undone *)
      intWith2 = Plus @@ (# //. {α_. Integrate[β_, x] -> α β}& /@ intWith1)];
   If[intWith =!= 0, intWith2, 0] (ε - v[x])^n +
            Expand[-intFree (-n (ε - v[x])^(n - 1) v'[x])]]
```

IntegrationByPartsTogether carries out the integration of the collected terms with the same denominators. myTogether collects these terms.

In[21]:=
```
IntegrationByPartsTogether[y_Plus] :=
Module[{tog, tog1, max, toIntegrate, res},
       tog = myTogether[y];
       If[Head[tog] =!= Times,
          (* largest divergent power *)
          tog1 = (Denominator /@ Apply[List, tog]) //.
                                {n_. Power[_, m_] -> m};
          max = Max[tog1];
          toInt = Cases[tog, _. Power[ε - v[x], pot_?(# > -max&)]];
          (* integrate by parts *)
          (tog - (Plus @@ toIntegrate)) +
                Plus @@ (IntegrationByParts /@ toIntegrate),
          tog]]

IntegrationByPartsTogether[x_] = x;

myTogether[x_Plus] := (* concentrate on denominators only *)
Plus @@ ((Together[Plus @@ Cases[x, _. #^-1]])& /@
   Union[Flatten[Cases[#, Power[_, _], {0, Infinity}]& /@
                      (Denominator /@ List @@ x)]])

myTogether[x_Times] = x;
```

We apply IntegrationByParts to S[4].

In[25]:= `IntegrationByParts[S[4]]`

Out[25]= $\dfrac{1547\ v'[x]^4}{2048\ (\varepsilon - v[x])^{11/2}} - \dfrac{19\ v''[x]^2}{128\ (\varepsilon - v[x])^{7/2}} - \dfrac{9\ v'[x]\ v^{(3)}[x]}{64\ (\varepsilon - v[x])^{7/2}}$

Here is the result of applying IntegrationOverV and IntegrationByParts to S[4].

In[26]:= `IntegrationByParts[IntegrationOverV[S[4]]]`

$$\text{Out[26]= } -\frac{1105\, v'[x]^4}{2048\,(\varepsilon - v[x])^{11/2}} + \frac{11\, v''[x]^2}{112\,(\varepsilon - v[x])^{7/2}} + \frac{95\, v'[x]\, v^{(3)}[x]}{896\,(\varepsilon - v[x])^{7/2}}$$

Next, we give an explicit representation for the result in the form of derivatives with respect to E of definite integrals over x. To avoid the differentiation, we apply `HoldForm`. We avoid higher powers in the denominators to get integrable singularities. (For another method to deal with the singularities of the form $(\varepsilon - v(x))^{-k/2}$, see [490].)

```
In[27]:= finalForm[y_Plus] :=
         Module[{pre, lcm, u, expr, v},
                 pre = finalFormAux /@ List @@ y;
                 (* pull out biggest denominator *)
                 lcm = LCM @@ ((Denominator[#[[2]]] //.
                           {n_. Power[ε - v[x], _] -> n})& /@ pre);
                 1/lcm Plus @@ Table[u = lcm pre[[i, 2]];
                 (* use HoldForm to avoid the differentiation
                    to be carried out *)
                 HoldForm[D[Integrate[expr, {x, x1, x2}], {ε, v}]] /.
                         {expr -> u, v -> pre[[i, 1]]}, {i, Length[pre]}]];


         finalForm[y_Times] :=
         Module[{pre = finalFormAux[y], expr, v, denominator},
                 (* pull out denominator *)
                 denominator = Denominator[pre[[2, 1]]];
                 1/denominator HoldForm[D[Integrate[HoldForm[expr],
                                        {x, x1, x2}], {ε, v}]] /.
                     {expr -> pre[[2]] denominator, v -> pre[[1]]}]


         finalForm[0] := HoldForm[0];

         (* subsidiary function for preparing output in nice form *)
         finalFormAux[0] = 0;

         finalFormAux[a_. (ε - v[x])^n_] :=
         {-(n + 1/2), ExpandNumerator[a * (* from the differentiations wrt e *)
         Product[1/(i + 1), {i, n, -3/2}]] (ε - v[x])^(-1/2)}
```

Here are the correction terms of order 1 to 10. The function `WKBCorrection` applies both of the functions `Integration OverV` and `IntegrationByPartsTogether` as often as possible.

```
In[33]:= WKBCorrection[ord_Integer?Positive] :=
             finalForm[FixedPoint[IntegrationByPartsTogether,
                           FixedPoint[IntegrationOverV, S[ord]]]]
```

All odd terms are complete differentials, and they disappear identically.

```
In[34]:= (* odd orders *)
         WKBCorrection /@ {1, 3, 5, 7, 9}
Out[35]= {0, 0, 0, 0, 0}
```

The even orders become increasingly complicated integrals.

```
In[36]:= WKBCorrection[2]
```

$$\text{Out[36]= } \frac{1}{24}\, \partial_{\{\varepsilon, 1\}}\left(\int_{x1}^{x2} \frac{v''[x]}{\sqrt{\varepsilon - v[x]}}\, dx\right)$$

```
In[37]:= WKBCorrection[4]
```

$$\text{Out[37]= } \frac{\partial_{\{\varepsilon, 3\}}\left(\int_{x1}^{x2} \frac{7\, v''[x]^2}{\sqrt{\varepsilon - v[x]}}\, dx\right) + \partial_{\{\varepsilon, 2\}}\left(\int_{x1}^{x2} -\frac{5\, v^{(4)}[x]}{\sqrt{\varepsilon - v[x]}}\, dx\right)}{2880}$$

```
In[38]:= WKBCorrection[6]
```

$$\text{Out[38]= } \frac{\partial_{\{\varepsilon, 5\}}\left(\int_{x1}^{x2} \frac{93\, v''[x]^3}{\sqrt{\varepsilon - v[x]}}\, dx\right) + \partial_{\{\varepsilon, 4\}}\left(\int_{x1}^{x2} \frac{-8\, v^{(3)}[x]^2 - 189\, v''[x]\, v^{(4)}[x]}{\sqrt{\varepsilon - v[x]}}\, dx\right) + \partial_{\{\varepsilon, 3\}}\left(\int_{x1}^{x2} \frac{35\, v^{(6)}[x]}{\sqrt{\varepsilon - v[x]}}\, dx\right)}{725760}$$

```
In[39]:= WKBCorrection[8]
```

Out[39]= $\dfrac{1}{174182400}$

$$\left(\partial_{\{\epsilon,7\}} \left(\int_{x1}^{x2} \frac{1143\ v''[x]^4}{\sqrt{\epsilon - v[x]}}\ dx \right) + \partial_{\{\epsilon,6\}} \left(\int_{x1}^{x2} \frac{2\ \left(-176\ v''[x]\ v^{(3)}[x]^2 - 2223\ v''[x]^2\ v^{(4)}[x]\right)}{\sqrt{\epsilon - v[x]}}\ dx \right) + \right.$$

$$\partial_{\{\epsilon,5\}} \left(\int_{x1}^{x2} \frac{5\ \left(297\ v^{(4)}[x]^2 + 64\ v^{(3)}[x]\ v^{(5)}[x] + 308\ v''[x]\ v^{(6)}[x]\right)}{\sqrt{\epsilon - v[x]}}\ dx \right) +$$

$$\left. \partial_{\{\epsilon,4\}} \left(\int_{x1}^{x2} -\frac{175\ v^{(8)}[x]}{\sqrt{\epsilon - v[x]}}\ dx \right) \right)$$

In[40]:= **WKBCorrection[10]**

Out[40]= $\dfrac{1}{22992076800}$

$$\left(\partial_{\{\epsilon,9\}} \left(\int_{x1}^{x2} \frac{7665\ v''[x]^5}{\sqrt{\epsilon - v[x]}}\ dx \right) + \partial_{\{\epsilon,8\}} \left(\int_{x1}^{x2} \frac{2\ \left(-2632\ v''[x]^2\ v^{(3)}[x]^2 - 23991\ v''[x]^3\ v^{(4)}[x]\right)}{\sqrt{\epsilon - v[x]}}\ dx \right) + \right.$$

$$\partial_{\{\epsilon,7\}} \left(\int_{x1}^{x2} \frac{1}{\sqrt{\epsilon - v[x]}} \left(3760\ v^{(3)}[x]^2\ v^{(4)}[x] + 46215\ v''[x]\ v^{(4)}[x]^2 + \right. \right.$$

$$\left. 9152\ v''[x]\ v^{(3)}[x]\ v^{(5)}[x] + 23606\ v''[x]^2\ v^{(6)}[x]\right)\ dx \right) +$$

$$\partial_{\{\epsilon,6\}} \left(\int_{x1}^{x2} \frac{5\ \left(-416\ v^{(5)}[x]^2 - 3146\ v^{(4)}[x]\ v^{(6)}[x] - 352\ v^{(3)}[x]\ v^{(7)}[x] - 1001\ v''[x]\ v^{(8)}[x]\right)}{\sqrt{\epsilon - v[x]}} \right.$$

$$\left. \left. dx \right) + \partial_{\{\epsilon,5\}} \left(\int_{x1}^{x2} \frac{385\ v^{(10)}[x]}{\sqrt{\epsilon - v[x]}}\ dx \right) \right)$$

Here, the explicit correction terms for a quartic oscillator are calculated. The turning points are $x_{1,2} = \pm \epsilon^{1/4}$.

In[41]:= **tab = Table[Expand[DeleteCases[WKBCorrection[i] /.**
   {x1 -> -ε^(1/4), x2 -> ε^(1/4),
   Derivative[n_][v][x] :> D[x^4, {x, n}], v[x] -> x^4} /.
   Integrate[f_, range_] :> Integrate[f, range,
   Assumptions -> ε > 0 && ε^(1/4) > 0],
   HoldForm, Infinity, Heads -> True]], {i, 10}]

Out[41]= $\left\{ 0,\ \dfrac{\sqrt{\pi}\ \text{Gamma}\left[\frac{3}{4}\right]}{4\ \epsilon^{3/4}\ \text{Gamma}\left[\frac{1}{4}\right]},\ 0,\ -\dfrac{5\ \sqrt{\pi}\ \text{Gamma}\left[\frac{5}{4}\right]}{192\ \epsilon^{9/4}\ \text{Gamma}\left[\frac{3}{4}\right]} + \dfrac{21\ \sqrt{\pi}\ \text{Gamma}\left[\frac{5}{4}\right]}{512\ \epsilon^{9/4}\ \text{Gamma}\left[\frac{7}{4}\right]}, \right.$

$$0,\ \dfrac{451\ \sqrt{\pi}\ \text{Gamma}\left[\frac{3}{4}\right]}{3072\ \epsilon^{15/4}\ \text{Gamma}\left[\frac{1}{4}\right]} - \dfrac{1023\ \sqrt{\pi}\ \text{Gamma}\left[\frac{7}{4}\right]}{8192\ \epsilon^{15/4}\ \text{Gamma}\left[\frac{9}{4}\right]},\ 0,$$

$$-\dfrac{21879\ \sqrt{\pi}\ \text{Gamma}\left[\frac{5}{4}\right]}{229376\ \epsilon^{21/4}\ \text{Gamma}\left[\frac{3}{4}\right]} + \dfrac{341445\ \sqrt{\pi}\ \text{Gamma}\left[\frac{5}{4}\right]}{917504\ \epsilon^{21/4}\ \text{Gamma}\left[\frac{7}{4}\right]} - \dfrac{2273427\ \sqrt{\pi}\ \text{Gamma}\left[\frac{9}{4}\right]}{2621440\ \epsilon^{21/4}\ \text{Gamma}\left[\frac{11}{4}\right]},\ 0,$$

$$\left. \dfrac{4696439\ \sqrt{\pi}\ \text{Gamma}\left[\frac{3}{4}\right]}{1572864\ \epsilon^{27/4}\ \text{Gamma}\left[\frac{1}{4}\right]} - \dfrac{12784435\ \sqrt{\pi}\ \text{Gamma}\left[\frac{7}{4}\right]}{2097152\ \epsilon^{27/4}\ \text{Gamma}\left[\frac{9}{4}\right]} + \dfrac{90439335\ \sqrt{\pi}\ \text{Gamma}\left[\frac{11}{4}\right]}{8388608\ \epsilon^{27/4}\ \text{Gamma}\left[\frac{13}{4}\right]} \right\}$$

int0 is the main term in the quantization condition.

In[42]:= **int0 = Integrate[Sqrt[ε - x^4], {x, -ε^(1/4), ε^(1/4)},**
   Assumptions -> ε > 0 && ε^(1/4) > 0]

Out[42]= $\dfrac{\sqrt{\pi}\ \epsilon^{3/4}\ \text{Gamma}\left[\frac{5}{4}\right]}{\text{Gamma}\left[\frac{7}{4}\right]}$

The correction terms to order *m* are included in quantizationCondition.

In[43]:= **quantizationCondition[ε_, n_, ℏ_, m_] :=**
   int0 + Sum[(ℏ/I)^k tab[[k]], {k, 1, m}] - (n + 1/2) ℏ

For $\hbar = 1/10$ and $n = 3$, here is the quantization condition as a function of ϵ.

In[44]:= **Plot[Evaluate[Table[quantizationCondition[ε, 3, 1/10, m], {m, 10}]],**
   {ε, 0, 0.2}, PlotRange -> {-1, 1},
   PlotStyle -> Table[Hue[i 0.7], {i, 10}]]

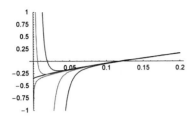

We use `FindRoot` to calculate numerical values for the energies.

```
In[45]:= Table[FindRoot[Evaluate[quantizationCondition[ε, n, 1/10, m] == 0],
              {ε, n/10}, WorkingPrecision -> 30,
              AccuracyGoal -> 30], {n, 3, 6}, {m, 0, 10, 2}] //
                                      Chop // SetPrecision[#, 12]&
Out[45]= {{{ε → 0.117135745040}, {ε → 0.120416463738}, {ε → 0.120222216706},
          {ε → 0.120107781681}, {ε → 0.120274460120}, {ε → 0.120629148891}},
         {{ε → 0.163762855886}, {ε → 0.166556102541}, {ε → 0.166453993552},
          {ε → 0.166417245284}, {ε → 0.166450176728}, {ε → 0.166494255257}},
         {{ε → 0.214000918687}, {ε → 0.216452839098}, {ε → 0.216392177550},
          {ε → 0.216377467323}, {ε → 0.216386360994}, {ε → 0.216394410720}},
         {{ε → 0.267392893358}, {ε → 0.269590798970}, {ε → 0.269551612597},
          {ε → 0.269544780576}, {ε → 0.269547751437}, {ε → 0.269549686197}}}
```

The higher *n*, the better is the convergence.

```
In[46]:= Table[FindRoot[Evaluate[quantizationCondition[ε, 100, 1/10, m] == 0],
              {ε, 1/10}, WorkingPrecision -> 30,
              AccuracyGoal -> 30], {m, 0, 10, 2}] //
                                      Chop // SetPrecision[#, 12]&
Out[46]= {{ε → 10.2995946763}, {ε → 10.2999506238}, {ε → 10.2999505968},
         {ε → 10.2999505968}, {ε → 10.2999505968}, {ε → 10.2999505968}}
```

Here, the same for ℏ = 1/100 is done.

```
In[47]:= Plot[Evaluate[Table[quantizationCondition[ε, 3, 1/100, m], {m, 10}]],
             {ε, 0, 0.02}, PlotRange -> {-1/10, 1/10},
             PlotStyle -> Table[Hue[i 0.7], {i, 10}]]
```

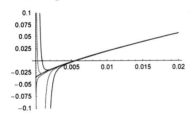

```
In[48]:= Table[FindRoot[
             Evaluate[quantizationCondition[ε, n, 1/100, m] == 0],
                 {ε, 5n/100}, WorkingPrecision -> 30,
                 AccuracyGoal -> 30], {n, 3, 6}, {m, 0, 10, 2}] //
                                      Chop // SetPrecision[#, 12]&
Out[48]= {{{ε → 0.00543695966194}, {ε → 0.00558923713468}, {ε → 0.00558022098614},
          {ε → 0.00557490938281}, {ε → 0.00558264591060}, {ε → 0.00559910910500}},
         {{ε → 0.00760119843241}, {ε → 0.00773084945723}, {ε → 0.00772610997782},
          {ε → 0.00772440427430}, {ε → 0.00772593281652}, {ε → 0.00772797876060}},
         {{ε → 0.00993304274561}, {ε → 0.0100468508096}, {ε → 0.0100440351500},
          {ε → 0.0100433523617}, {ε → 0.0100437651693}, {ε → 0.0100441388045}},
         {{ε → 0.0124112786800}, {ε → 0.0125132964214}, {ε → 0.0125114775511},
          {ε → 0.0125111604368}, {ε → 0.0125112983320}, {ε → 0.0125113881355}}}
```

```
In[49]:= Table[FindRoot[Evaluate[quantizationCondition[ε, 100, 1/10, m] == 0],
                {ε, 10}, WorkingPrecision -> 30,
                AccuracyGoal -> 30], {m, 0, 10, 2}] //
                                       Chop // SetPrecision[#, 12]&
Out[49]= {{ε → 10.2995946763}, {ε → 10.2999506238}, {ε → 10.2999505968},
          {ε → 10.2999505968}, {ε → 10.2999505968}, {ε → 10.2999505968}}
```

Using a Padé approximation of the truncated series, we can further improve the accuracy of the results [45], [1608].

For fractional power potentials, see [1321]. For other ways to derive the WKB coefficients, see [1318] and [1319]. For the related quasi-linearization method, see [1084], [1193], [1082], [1194], [1083], [1450], [919].

For summing the corrections to all orders, see [1497] and [1544]. For a discussion when the WKB approximations become exact, see [1511]. For an investigation about the accuracy of the higher orders, see [1451]. For an improved momentum definition, see [1320]

b) Let $\sigma[k, j]$ stand for σ_j acting on particle k, and let $\psi[l, "\updownarrow"]$ stand for $|\updownarrow\rangle_l$. The following implements the action of $\sigma[k, j]$ on the two basis states $|\uparrow\rangle_l$, $|\downarrow\rangle_l$. By explicitly carrying the particle number in the ket as well as in the operators, we can use ordinary multiplication instead of a noncommutative one.

```
In[1]:= δ = KroneckerDelta;
        (* Pauli matrices *)
        σ[1] = {{0,  1}, {1,  0}};
        σ[2] = {{0, -I}, {I,  0}};
        σ[3] = {{1,  0}, {0, -1}};
        (* act on a ket of a given particle *)
        σ[k_, j_][ψ[l_, "↑"]] := δ[k, 1] σ[j].{1, 0}.{ψ[1, "↑"], ψ[1, "↓"]} +
                                  (1 - δ[k, 1])ψ[1, "↑"];
        σ[k_, j_][ψ[l_, "↓"]] := δ[k, 1] σ[j].{0, 1}.{ψ[1, "↑"], ψ[1, "↓"]} +
                                  (1 - δ[k, 1])ψ[1, "↓"];
        (* pull out constant factors *)
        σ[k_, j_][c_?(FreeQ[#, _ψ, {0, ∞}]&) t_] := c σ[k, j][t]
        (* sums and products *)
        σ[k_, j_][pt_Plus | pt_Times] := σ[k, j] /@ pt;
```

This is the Greenberger–Horne–Zeilinger state under consideration. (We omit the normalization constant.)

```
In[13]:= Ψ = (ψ[1, "↑"]ψ[2, "↑"]ψ[3, "↑"] - ψ[1, "↓"]ψ[2, "↓"]ψ[3, "↓"]);
```

Here are two examples of operators $\hat{\Xi}_\pm$.

```
In[14]:= {(σ[1, 1] @ σ[2, 2] @ σ[3, 2] @ Ψ)/Ψ,
          (σ[1, 1] @ σ[2, 1] @ σ[3, 1] @ Ψ)/Ψ} // Cancel
Out[14]= {1, -1}
```

For the $\vec{n}_k$, we can use the following obvious parametrization.

```
In[15]:= n[k_, 1] = Cos[φ[k]] Sin[θ[k]];
         n[k_, 2] = Sin[φ[k]] Sin[θ[k]];
         n[k_, 3] = Cos[θ[k]];
```

Acting now with $(\vec{n}_1.\vec{\sigma})_1 \otimes (\vec{n}_2.\vec{\sigma})_2 \otimes (\vec{n}_3.\vec{\sigma})_3$ on $|\Psi\rangle$, we get the following result sum3 for arbitrary $\vec{n}_k$.

```
In[18]:= sum1 = Sum[n[1, j] σ[1, j][Ψ], {j, 3}];
         sum2 = Sum[n[2, j] σ[2, j][sum1], {j, 3}];
         sum3 = Sum[n[3, j] σ[3, j][sum2], {j, 3}];
```

This gives the following condition for $\hat{\Xi}_\pm |\Psi\rangle = \pm 1 |\Psi\rangle$.

```
In[21]:= sum = Expand[sum3 - ±1 Ψ];
         Short[sum, 4]
Out[22]//Short= Cos[θ[1]] Cos[θ[2]] Cos[θ[3]] ψ[1, ↓] ψ[2, ↓] ψ[3, ↓] + ±1 ψ[1, ↓] ψ[2, ↓] ψ[3, ↓] + ≪73≫ +
                Cos[φ[1]] Sin[θ[1]] Sin[θ[2]] Sin[θ[3]] Sin[φ[2]] Sin[φ[3]] ψ[1, ↑] ψ[2, ↑] ψ[3, ↑] -
                i Sin[θ[1]] Sin[θ[2]] Sin[θ[3]] Sin[φ[1]] Sin[φ[2]] Sin[φ[3]] ψ[1, ↑] ψ[2, ↑] ψ[3, ↑]
```

The Hilbert space of three two-state particles is spanned by the following eight states.

```
In[23]:= ψΠs = {ψ[1, "↑"] ψ[2, "↑"] ψ[3, "↑"], ψ[1, "↑"] ψ[2, "↑"] ψ[3, "↓"],
               ψ[1, "↑"] ψ[2, "↓"] ψ[3, "↑"], ψ[1, "↑"] ψ[2, "↓"] ψ[3, "↓"],
```

$$\psi[1, \text{"↓"}] \; \psi[2, \text{"↑"}] \; \psi[3, \text{"↑"}], \; \psi[1, \text{"↓"}] \; \psi[2, \text{"↑"}] \; \psi[3, \text{"↓"}],$$
$$\psi[1, \text{"↓"}] \; \psi[2, \text{"↓"}] \; \psi[3, \text{"↑"}], \; \psi[1, \text{"↓"}] \; \psi[2, \text{"↓"}] \; \psi[3, \text{"↓"}]\};$$

eqs1 is the list of the coefficients of the $\psi\Pi$s in sum.

In[24]:= `eqs1 = Coefficient[sum, #]& /@ ψΠs;`

Because eqs1 is polynomial in $\cos(\varphi_k)$, $\sin(\varphi_k)$, $\cos(\vartheta_k)$, $\sin(\vartheta_k)$, we will use these four quantities as polynomial variables.

In[25]:= `vars = Flatten[Table[{Cos[φ[k]], Sin[φ[k]],`
`                        Cos[ϑ[k]], Sin[ϑ[k]]}, {k, 3}]]`

Out[25]= `{Cos[φ[1]], Sin[φ[1]], Cos[ϑ[1]], Sin[ϑ[1]], Cos[φ[2]], Sin[φ[2]],`
`         Cos[ϑ[2]], Sin[ϑ[2]], Cos[φ[3]], Sin[φ[3]], Cos[ϑ[3]], Sin[ϑ[3]]}`

Supplementing eqs1 with the usual trigonometric constraints allows us to solve for the $\cos(\varphi_k)$, $\sin(\varphi_k)$, $\cos(\vartheta_k)$, $\sin(\vartheta_k)$.

In[26]:= `eqs2 = Table[{Cos[φ[k]]^2 + Sin[φ[k]]^2 - 1,`
`                Cos[ϑ[k]]^2 + Sin[ϑ[k]]^2 - 1}, {k, 3}] // Flatten;`
`       eqs = Flatten[{eqs1, eqs2}];`

In[28]:= `gbp1 = Simplify[GroebnerBasis[eqs /. ±1 -> 1, vars]]`

Out[28]= `{-Cos[ϑ[3]]², Cos[ϑ[3]], 0, -Cos[ϑ[2]]², Cos[ϑ[2]], 0, -Cos[ϑ[1]]²,`
`         Cos[ϑ[1]], Sin[φ[1]] - Sin[ϑ[1]] Sin[ϑ[2]] Sin[ϑ[3]] Sin[φ[2] + φ[3]],`
`         Cos[φ[1]] + Cos[φ[2] + φ[3]] Sin[ϑ[1]] Sin[ϑ[2]] Sin[ϑ[3]]}`

In[29]:= `solp1 = Solve[# == 0& /@ gbp1, vars] // Union`

`       Solve::svars : Equations may not give solutions for all "solve" variables. More…`

Out[29]= `{{Cos[φ[1]] → -Cos[φ[2] + φ[3]] Sin[ϑ[1]] Sin[ϑ[2]] Sin[ϑ[3]],`
`         Sin[φ[1]] → Sin[ϑ[1]] Sin[ϑ[2]] Sin[ϑ[3]] Sin[φ[2] + φ[3]],`
`         Cos[ϑ[1]] → 0, Cos[ϑ[2]] → 0, Cos[ϑ[3]] → 0}}`

Making use of the results for the ϑ_k, we finally obtain the following condition for the φ_k.

In[30]:= `Drop[solp1[[1]] /. {_ϑ -> Pi/2}, -3]`

Out[30]= `{Cos[φ[1]] → -Cos[φ[2] + φ[3]], Sin[φ[1]] → Sin[φ[2] + φ[3]]}`

This means the most general $\hat{\Xi}_+$ is given by

$$\hat{\Xi}_+ = (\cos(\pi - \varphi_2 - \varphi_3)\,\sigma_1 + \sin(\pi - \varphi_2 - \varphi_3)\,\sigma_2)_1 \otimes$$
$$(\cos(\varphi_2)\,\sigma_1 + \sin(\varphi_2)\,\sigma_2)_2 \otimes (\cos(\varphi_3)\,\sigma_1 + \sin(\varphi_3)\,\sigma_2)_3.$$

In a similar manner, we obtain for $\hat{\Xi}_-$ the result

$$\hat{\Xi}_- = (\cos(-\varphi_2 - \varphi_3)\,\sigma_1 + \sin(-\varphi_2 - \varphi_3)\,\sigma_2)_1 \otimes$$
$$(\cos(\varphi_2)\,\sigma_1 + \sin(\varphi_2)\,\sigma_2)_2 \otimes (\cos(\varphi_3)\,\sigma_1 + \sin(\varphi_3)\,\sigma_2)_3.$$

In[31]:= `gbm1 = Simplify[GroebnerBasis[eqs /. ±1 -> -1, vars]];`
`       solm1 = Solve[# == 0& /@ gbm1, vars] // Union`
`       Solve::svars : Equations may not give solutions for all "solve" variables. More…`

Out[32]= `{{Cos[φ[1]] → Cos[φ[2] + φ[3]] Sin[ϑ[1]] Sin[ϑ[2]] Sin[ϑ[3]],`
`         Sin[φ[1]] → -Sin[ϑ[1]] Sin[ϑ[2]] Sin[ϑ[3]] Sin[φ[2] + φ[3]],`
`         Cos[ϑ[1]] → 0, Cos[ϑ[2]] → 0, Cos[ϑ[3]] → 0}}`

A moment of reflection shows that these are the to-be-expected results for $\hat{\Xi}_\pm$.

c) The most general pure state of four (different) spin $1/2$ particles is

$$|\Psi\rangle = \sum_{i_1,i_2,i_3,i_4=0}^{1} c_{i_1,i_2,i_3,i_4} \, |i_1\rangle_1 \, |i_2\rangle_2 \, |i_3\rangle_3 \, |i_4\rangle_4$$

where $|0\rangle = |\downarrow\rangle$, and $|1\rangle = |\uparrow\rangle$. Without loss of generality we can always choose a coordinate system so that $c_{1,0,0,0} = c_{0,1,0,0} = c_{0,0,1,0} = c_{0,0,0,1} = 0$ [1854], [844], [12]. The corresponding density matrix is $\rho 4$. In the following

implementation, we write `KetBra[{i1, i2, i3, i4}, {j1, j2, j3, j4}]`
for $|i1\rangle_1 |i2\rangle_2 |i3\rangle_3 |i4\rangle_4 \langle i\,4|_4 \langle i\,3|_3 \langle i\,2|_2 \langle i\,1|_1$. cn and cc stand for c and its conjugate $\overline{c}$.

```
In[1]:= p4 = Sum[cn[i1, i2, i3, i4] cc[j1, j2, j3, j4] *
            KetBra[{i1, i2, i3, i4}, {j1, j2, j3, j4}],
            {i1, 0, 1}, {i2, 0, 1}, {i3, 0, 1}, {i4, 0, 1},
            {j1, 0, 1}, {j2, 0, 1}, {j3, 0, 1}, {j4, 0, 1}] //.
            ((cn[s__] | cc[s__]) /; Count[{s}, 1] === 1) -> 0;
```

The function `makeReducedDensityMatrix[i, j]` generates the reduced two-particle density matrix $\rho2_{i,j} = \mathrm{Tr}_{k,l}\, \rho4$.

```
In[2]:= makeReducedDensityMatrix[i_, j_] :=
        Module[{lhs, rhs},
          (* the particles to be traced out *)
          {k, l} = Complement[{1, 2, 3, 4}, {i, j}];
          (* the ket-bras to be contracted *)
          lhs = f_ KetBra[{i1_, i2_, i3_, i4_},
            MapAt[{i1_, i2_, i3_, i4_}[[l]]&,
              MapAt[{i1_, i2_, i3_, i4_}[[k]]&, {j1_, j2_, j3_, j4_}, k], l]];
          (* the surviving ket-bras *)
          rhs = f KetBra[{i1, i2, i3, i4}[[{i, j}]],
                          {j1, j2, j3, j4}[[{i, j}]]];
          (* collect wrt the surviving ket-bras *)
          collect[Plus @@ Cases[p4, RuleDelayed @@ {lhs, rhs}]]]
In[3]:= collect[p_] := Collect[p, Cases[p, _KetBra, Infinity] // Union];
```

Here are the six reduced two-particle density matrices $\rho2_{1,2}$, $\rho2_{1,3}$, $\rho2_{1,4}$, $\rho2_{2,3}$, $\rho2_{2,4}$, and $\rho2_{3,4}$.

```
In[4]:= {p2[1, 2], p2[1, 3], p2[1, 4], p2[2, 3], p2[2, 4], p2[3, 4]} =
        Apply[makeReducedDensityMatrix,
              {{1, 2}, {1, 3}, {1, 4}, {2, 3}, {2, 4}, {3, 4}}, {1}];
```

Equating all diagonal elements of the reduced two-particle density matrices to 1 and the nondiagonal elements to 0 yields 96 equations in 24 variables.

```
In[5]:= eqs = Flatten[Apply[List,
            {p2[1, 2], p2[1, 3], p2[1, 4], p2[2, 3],
             p2[2, 4], p2[3, 4]}, {1}]] /. (* diagonal elements *)
                {f_ KetBra[x_, x_] :> f - 1} /. _KetBra :> 1];
In[6]:= vars = Cases[eqs, _cc | _cn, Infinity] // Union;
In[7]:= {Length[vars], Length[eqs]}
Out[7]= {24, 96}
```

Using `GroebnerBasis`, we can quickly show that the 96 polynomial equations in cn and cc have no solutions [844]. (Because there does not exist a solution to the system of equations eqs, we use the monomial order `DegreeReverseLex icographic`. Using this order, a Gröbner basis is typically computed quite quickly.) For the reconstruction of a state from its reduced density matrices, see [948].

```
In[8]:= GroebnerBasis[eqs, vars, MonomialOrder -> DegreeReverseLexicographic]
Out[8]= {1}
```

d) The entropic uncertainty δx is the expected value $2L$.

```
In[1]:= δx = Exp[-Integrate[1/(2L) Log[1/(2L)], {x, -L, L}]]
Out[1]= 2 L
```

The corresponding $w(p)$ is $w(p) = \sin^2(L\,p)/(L\,p^2\,\pi)$.

```
In[2]:= w[p_] = (1/Sqrt[2 Pi] Integrate[1/Sqrt[2 L] Exp[I p x],
                  {x, -L, L}])^2 //
                  ExpToTrig // Simplify
```

$$Out[2]= \frac{\mathrm{Sin}[L\,p]^2}{L\,p^2\,\pi}$$

The direct evaluation of the entropic uncertainty of the momentum fails.

In[3]:= `Integrate[w[p] Log[w[p]], {p, -Infinity, Infinity}]`

> ∞::indet : Indeterminate expression 0 (-∞) encountered. More…

> ∞::indet : Indeterminate expression 0 (-∞) encountered. More…

Out[3]= $\displaystyle\int_{-\infty}^{\infty} \frac{Log\left[\frac{Sin[Lp]^2}{Lp^2\,\pi}\right]Sin[Lp]^2}{L\,p^2\,\pi}\,dp$

The infinite number of integrable singularities at $p_k = k\,\pi/L$ are a problem for the `Integrate` function. Expanding the argument of the logarithm gives one simple and one complicated integral [476].

In[4]:= `int1 = -1/(L Pi) Integrate[Sin[L p]^2/p^2 Log[L Pi p^2], {p, -Infinity, Infinity},`
`Assumptions -> L > 0]`

Out[4]= $-2 + 2\,EulerGamma - Log\left[\frac{\pi}{4\,L}\right]$

The complicated integral with the countable number of integrable singularities is not done by *Mathematica*.

In[5]:= `1/(L Pi) Integrate[Sin[L p]^2/p^2 Log[Sin[L p]^2], {p, -Infinity, Infinity}]`

> ∞::indet : Indeterminate expression 0 (-∞) encountered. More…

> ∞::indet : Indeterminate expression 0 (-∞) encountered. More…

Out[5]= $\displaystyle\frac{\int_{-\infty}^{\infty}\frac{Log[Sin[Lp]^2]Sin[Lp]^2}{p^2}\,dp}{L\,\pi}$

We use the periodicity of parts of the integrand to obtain an integral over a finite range having only two singularities at the end points.

$$\int_{-\infty}^{\infty}\frac{\sin^2(pL)}{p^2}\ln(\sin^2(pL))\,dp = \sum_{k=-\infty}^{\infty}\int_0^{\pi/L}\frac{\sin^2((p-k\pi/L)L)}{(p-k\pi/L)^2}\ln(\sin^2((p-k\pi/L)L))\,dp$$

$$= \int_0^{\pi/L}\left(\sum_{k=-\infty}^{\infty}(p-k\pi/L)^{-2}\right)\sin^2(pL)\ln(\sin^2(pL))\,dp = L^2\int_0^{\pi/L}\ln(\sin^2(pL))\,dp$$

In[6]:= `Sum[1/(p - k Pi/L)^2, {k, -Infinity, Infinity}]`

Out[6]= $L^2\,Csc[Lp]^2$

In[7]:= `int2 = 1/(L Pi) L^2 1/L Integrate[2 Log[Sin[p]], {p, 0, Pi}]`

Out[7]= $-Log[4]$

So, we obtain the value $\pi\exp(2\,(1-\gamma))/L$ for the entropic momentum uncertainty.

In[8]:= `δp = Exp[-(int1 + int2)] // Together // Simplify`

Out[8]= $\dfrac{e^{2-2\,EulerGamma}\,\pi}{L}$

In[9]:= `δx δp`

Out[9]= $2\,e^{2-2\,EulerGamma}\,\pi$

This is slightly larger than the possible minimal value.

In[10]:= `N[{%, %/(E Pi)}]`

Out[10]= `{14.6354, 1.7138}`

22. QES Condition, Integral of Rational Function, Integrals of Roots, Triangle Roots

a) Here, the value of the integral is calculated numerically.

In[1]:= `poly[a_, x_] = a - x^6 + 4 x^2;`

In[2]:= `int[a_?NumberQ] :=`
`Module[{x0 = Cases[x /. NSolve[poly[a, x] == 0, x],`

```
                    _Real?Positive][[1]]},
        NIntegrate[Evaluate[Sqrt[poly[a, x]]], {x, -x0, x0}]]
```

The following plot shows that the value of *a* is about 3.

```
In[3]:= Plot[int[a] - 2Pi, {a, 1/2, 12}]
```

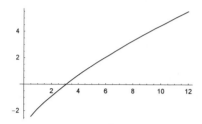

We calculate a high-precision value of *a*.

```
In[4]:= int[a_?NumberQ] :=
        With[{α = SetPrecision[a, 40]},
        Module[{(* the root *)
            x0 = Cases[x /. NSolve[poly[α, x] == 0, x,
                WorkingPrecision -> 40], _Real?Positive][[1]]},
            (* the integral to high precision *)
            NIntegrate[Evaluate[SetPrecision[Sqrt[poly[α, x]], 40]],
                {x, -x0, x0}, WorkingPrecision -> 40]]]

In[5]:= FindRoot[int[a] - 2 Pi, {a, 2, 3},
            WorkingPrecision -> 30, AccuracyGoal -> 30, Compiled -> False]

Out[5]= {a -> 3.07920143567800407738212682934}
```

Using the package `NumberTheory`Recognize``, we find a low-order algebraic number that has the same value.

```
In[6]:= Needs["NumberTheory`Recognize`"]

In[7]:= Recognize[%%[[1, 2]], 2, t]

Out[7]= -256 + 27 t²
```

$$Out[7]= -256 + 27\,t^2$$

```
In[8]:= Solve[% == 0, t]
```

$$Out[8]= \left\{\left\{t \to -\frac{16}{3\sqrt{3}}\right\}, \left\{t \to \frac{16}{3\sqrt{3}}\right\}\right\}$$

For this value of a, the polynomial actually factors.

```
In[9]:= Factor[16/3/Sqrt[3] - x^6 + 4 x^2, Extension -> Automatic]
```

$$Out[9]= \frac{1}{27}\left(4\sqrt{3} - 3x^2\right)\left(2\sqrt{3} + 3x^2\right)^2$$

Calculating the exact value for x0, we can symbolically integrate the integral and obtain the value $2\,\pi$.

```
In[10]:= Solve[16/3/Sqrt[3] - x^6 + 4 x^2 == 0, x]
```

$$Out[10]= \left\{\left\{x \to -\frac{2}{3^{1/4}}\right\}, \left\{x \to \frac{2}{3^{1/4}}\right\}, \left\{x \to -\frac{i\sqrt{2}}{3^{1/4}}\right\}, \left\{x \to -\frac{i\sqrt{2}}{3^{1/4}}\right\}, \left\{x \to \frac{i\sqrt{2}}{3^{1/4}}\right\}, \left\{x \to \frac{i\sqrt{2}}{3^{1/4}}\right\}\right\}$$

```
In[11]:= Integrate[Sqrt[%%], {x, -2/3^(1/4), 2/3^(1/4)}]

Out[11]= 2 π
```

b) The integrand is a rational function with no singularities on the real axis that decays as z^{-32} as $z \to \infty$. *Mathematica* applies straightforwardly the residue theorem and gives the following, slightly lengthy answer.

```
In[1]:= int = Integrate[1/(z^8 + 5 z^6 + 14 z^4 + 5 z^2 +1)^4,
                {z, -Infinity, Infinity}];

In[2]:= ByteCount[int]

Out[2]= 123048
```

Here is the approximative value of the integral.

In[3]:= `N[int, 50]`

Out[3]= `0.3974924065625265618865802143700865387175865360363 6+ 0. × 10⁻⁵¹ i`

The result is a product of $2\pi i$ and a sum that contains four terms.

In[4]:= `int1 = int/(2 Pi I);`
`Head[int1]`

Out[5]= `Plus`

Each of the four terms has the structure of a rational function of some radicals. Here is the structure of the first term.

In[6]:= `Short[int1[[1]], 2]`

Out[6]//Short=
$$
\left(\frac{36\,(\ll 1\gg)\,\left(5 + 84\,\text{Root}[1 + \ll 3\gg + \#1^4\,\&,\,1] + 75\,\ll 1\gg + 28\,\text{Root}[1 + \ll 3\gg + \ll 2\gg^4\,\&,\,1]^3\right)}{(\ll 1\gg)^2} - \right.
$$
$$
\left.\frac{12\,(\ll 1\gg)}{\ll 1\gg} - \frac{3\,(\ll 1\gg)\,\left(\frac{20\,\ll 1\gg}{\ll 1\gg^2} - \frac{\ll 1\gg}{\ll 1\gg}\right)}{\ll 1\gg}\right) \Big/
$$
$$
\left(3\,\left(-10\,\sqrt{\text{Root}[1 + 5\,\#1 + 14\,\#1^2 + 5\,\#1^3 + \#1^4\,\&,\,1]} - 56\,\ll 1\gg^{\ll 1\gg} - \ll 1\gg - 8\,\ll 1\gg^{7/2}\right)^4\right)
$$

To simplify `int`, we could apply `RootReduce` to first find a canonical form for `int`. Although conceptually this is a useful operation, practically, it would take a long time because of the high-degree radicals. So, we simplify the numerators and denominators of the four summands by applying `Together` and `Simplify`.

In[7]:= `{nums, dens} = {Simplify[Numerator[#]]& /@ (List @@ int1),`
`Simplify[Expand[Denominator[#]]]& /@ (List @@ int1)};`

The last operations shrank the expressions considerably.

In[8]:= `ByteCount[{nums, dens}]`

Out[8]= `31280`

Now, we first unite complex conjugate roots and the apply `RootReduce` to the result. Numericalization shows which terms are complex conjugate to each other.

In[9]:= `N[nums/dens]`

Out[9]= `{1.32273 × 10⁻⁷ + 5.44783 × 10⁻⁸ i, −1.32273 × 10⁻⁷ + 5.44783 × 10⁻⁸ i,`
`0.552571 − 0.0316315 i, −0.552571 − 0.0316315 i}`

In[10]:= `ord = Last /@ Sort[MapIndexed[{Abs[#1], #2[[1]]}&, N[nums/dens]]]`

Out[10]= `{1, 2, 3, 4}`

In[11]:= `RootReduce[`
`RootReduce[Simplify[nums[[ord[[1]]]]]/dens[[ord[[1]]]] +`
`nums[[ord[[2]]]]]/dens[[ord[[2]]]]] +`
`RootReduce[Simplify[nums[[ord[[3]]]]]/dens[[ord[[3]]]] +`
`nums[[ord[[4]]]]]/dens[[ord[[4]]]]]]]`

Out[11]= `Root[−1125785801146975639794752509671025 − 28037052748639128099069111936666 8288 #1² +`
`23031468885814795508979931416834 8672 #1⁴ &, 3]`

The last result is a root of a biquadratic polynomial. It can be easily written as a nested radical.

In[12]:= `2 Pi I ToRadicals[%] // Simplify`

Out[12]=
$$
\frac{\sqrt{\frac{1}{598}\left(-17044684650701397835331 + 3364717264299039577344\,\sqrt{26}\right)}\,\pi}{3421555072}
$$

In[13]:= `N[%, 50]`

Out[13]= `0.3974924065625265618865802143700865387175865360363 6`

c) Not unexpectedly, *Mathematica* cannot calculate this complicated integral directly.

In[1]:= `int = (z^γ − 1)/γ/z^2 (1 + (z^γ − 1) x)^(−γ^(−1) − 1);`

In[2]:= `Integrate[int, {z, 1, Infinity}, Assumptions -> x > 1]`

Out[2]= Integrate$\left[\dfrac{(-1 + z^Y)\ (1 + x\ (-1 + z^Y))^{-1-\frac{1}{Y}}}{z^2\ Y}, \{z, 1, \infty\}, \text{Assumptions} \rightarrow x > 1\right]$

In[3]:= TimeConstrained[Limit[%, Y -> 0], 10]

Out[3]= $Aborted

We will try a term-by-term integration after expanding the integrand in a power series around $\gamma = 0$. The terms might not be integrable, so potentially we might have to regularize the result.

In[4]:= Off[Power::infy];
        Limit[int, Y -> 0] + Y Limit[D[int, Y], Y -> 0]

Out[5]= $z^{-2-x} \text{Log}[z] + \dfrac{1}{2} z^{-2-x} Y \text{Log}[z]^2 (1 - 2x + (-1+x) x \text{Log}[z])$

Now a direct call to Integrate succeeds.

In[6]:= Integrate[%, {z, 1, Infinity}, Assumptions -> x > 1,
                  GenerateConditions -> False] /. Y -> 0

Out[6]= $\dfrac{1 + x\ (2 + x)}{(1 + x)^4}$

Here is a quick check for a random value of x and a small value of γ.

In[7]:= With[{Y = 10^-20, x = E Pi},
            {(* numerical value for small Y *)
             NIntegrate[(z^Y - 1)/Y (1 + (z^Y - 1)x)^(-1/Y - 1)/z^2,
                 {z, 1, 2, Infinity},
                 PrecisionGoal -> 20, WorkingPrecision -> 22],
             (* exact result *)
             (1 + 2x + x^2)/(1 + x)^4 // N[#, 20]&}]

Out[7]= {0.010988222609486848337, 0.010988222609486848337}

d) *Mathematica* cannot carry out the first integral directly.

In[1]:= root = Root[#^6 - x # - x &, 1];
        integrand = Log[x^2] root;
        Integrate[integrand, x]

Out[3]= $\displaystyle\int \text{Log}[x^2]\ \text{Root}[-x - x\ \#1 + \#1^6\ \&, 1]\ dx$

It seems natural to introduce a new integration variable X=root.

In[4]:= newIntegrand = integrand /. root -> X

Out[4]= $X \text{Log}[x^2]$

Expressing the original integration variable x through X is straightforward.

In[5]:= sol = Solve[root[[1]][X] == 0, x]

Out[5]= $\left\{\left\{x \rightarrow \dfrac{X^6}{1 + X}\right\}\right\}$

So, we have the following new integrand.

In[6]:= newIntegrand2 = newIntegrand /. sol[[1]]

Out[6]= $X \text{Log}\left[\dfrac{X^{12}}{(1 + X)^2}\right]$

The change of variables $\int f(x)\,dx = \int f(x(X))\,(dx/dX)\,dX$ requires the calculation of the derivative dx/dX as a function of X.

In[7]:= gb = GroebnerBasis[{#, D[#, X]}&[root[[1]][X] /. x -> x[X]] /. x[X] -> x,
                {}, {x}, MonomialOrder -> EliminationOrder];

In[8]:= dxdX = x'[X] /. Solve[gb[[1]] == 0, x'[X]][[1]]

Out[8]= $\dfrac{6 X^5 + 5 X^6}{(1 + X)^2}$

So, we arrive at the following result for the first integral.

In[9]:= `int = (Integrate[newIntegrand2 dxdX, X] // Simplify) /. X -> root`

Out[9]= $-\Bigg(-720 + 360\,\mathrm{Log}\Big[\dfrac{\mathrm{Root}[-x - x\,\#1 + \#1^6\,\&,\,1]^{12}}{(1 + \mathrm{Root}[-x - x\,\#1 + \#1^6\,\&,\,1])^2}\Big] + 1764\,\mathrm{Log}[1 + \mathrm{Root}[-x - x\,\#1 + \#1^6\,\&,\,1]]\,+$

$360\,\mathrm{Log}\Big[\dfrac{\mathrm{Root}[-x - x\,\#1 + \#1^6\,\&,\,1]^{12}}{(1 + \mathrm{Root}[-x - x\,\#1 + \#1^6\,\&,\,1])^2}\Big]\,\mathrm{Log}[1 + \mathrm{Root}[-x - x\,\#1 + \#1^6\,\&,\,1]]\,+$

$360\,\mathrm{Log}[1 + \mathrm{Root}[-x - x\,\#1 + \#1^6\,\&,\,1]]^2 + 2556\,\mathrm{Root}[-x - x\,\#1 + \#1^6\,\&,\,1]\,+$

$1764\,\mathrm{Log}[1 + \mathrm{Root}[-x - x\,\#1 + \#1^6\,\&,\,1]]\,\mathrm{Root}[-x - x\,\#1 + \#1^6\,\&,\,1]\,+$

$360\,\mathrm{Log}\Big[\dfrac{\mathrm{Root}[-x - x\,\#1 + \#1^6\,\&,\,1]^{12}}{(1 + \mathrm{Root}[-x - x\,\#1 + \#1^6\,\&,\,1])^2}\Big]$

$\mathrm{Log}[1 + \mathrm{Root}[-x - x\,\#1 + \#1^6\,\&,\,1]]\,\mathrm{Root}[-x - x\,\#1 + \#1^6\,\&,\,1]\,+$

$360\,\mathrm{Log}[1 + \mathrm{Root}[-x - x\,\#1 + \#1^6\,\&,\,1]]^2\,\mathrm{Root}[-x - x\,\#1 + \#1^6\,\&,\,1]\,+$

$2358\,\mathrm{Root}[-x - x\,\#1 + \#1^6\,\&,\,1]^2 - 180\,\mathrm{Log}\Big[\dfrac{\mathrm{Root}[-x - x\,\#1 + \#1^6\,\&,\,1]^{12}}{(1 + \mathrm{Root}[-x - x\,\#1 + \#1^6\,\&,\,1])^2}\Big]$

$\mathrm{Root}[-x - x\,\#1 + \#1^6\,\&,\,1]^2 - 426\,\mathrm{Root}[-x - x\,\#1 + \#1^6\,\&,\,1]\,+$

$60\,\mathrm{Log}\Big[\dfrac{\mathrm{Root}[-x - x\,\#1 + \#1^6\,\&,\,1]^{12}}{(1 + \mathrm{Root}[-x - x\,\#1 + \#1^6\,\&,\,1])^2}\Big]\,\mathrm{Root}[-x - x\,\#1 + \#1^6\,\&,\,1]^3\,+$

$153\,\mathrm{Root}[-x - x\,\#1 + \#1^6\,\&,\,1]^4 - 30\,\mathrm{Log}\Big[\dfrac{\mathrm{Root}[-x - x\,\#1 + \#1^6\,\&,\,1]^{12}}{(1 + \mathrm{Root}[-x - x\,\#1 + \#1^6\,\&,\,1])^2}\Big]$

$\mathrm{Root}[-x - x\,\#1 + \#1^6\,\&,\,1]^4 - 75\,\mathrm{Root}[-x - x\,\#1 + \#1^6\,\&,\,1]^5\,+$

$18\,\mathrm{Log}\Big[\dfrac{\mathrm{Root}[-x - x\,\#1 + \#1^6\,\&,\,1]^{12}}{(1 + \mathrm{Root}[-x - x\,\#1 + \#1^6\,\&,\,1])^2}\Big]\,\mathrm{Root}[-x - x\,\#1 + \#1^6\,\&,\,1]^5\,+$

$44\,\mathrm{Root}[-x - x\,\#1 + \#1^6\,\&,\,1]^6 - 12\,\mathrm{Log}\Big[\dfrac{\mathrm{Root}[-x - x\,\#1 + \#1^6\,\&,\,1]^{12}}{(1 + \mathrm{Root}[-x - x\,\#1 + \#1^6\,\&,\,1])^2}\Big]$

$\mathrm{Root}[-x - x\,\#1 + \#1^6\,\&,\,1]^6 + 500\,\mathrm{Root}[-x - x\,\#1 + \#1^6\,\&,\,1]^7\,-$

$300\,\mathrm{Log}\Big[\dfrac{\mathrm{Root}[-x - x\,\#1 + \#1^6\,\&,\,1]^{12}}{(1 + \mathrm{Root}[-x - x\,\#1 + \#1^6\,\&,\,1])^2}\Big]\,\mathrm{Root}[-x - x\,\#1 + \#1^6\,\&,\,1]^7\,-$

$4320\,\mathrm{Log}[\mathrm{Root}[-x - x\,\#1 + \#1^6\,\&,\,1]]\,(1 + \mathrm{Root}[-x - x\,\#1 + \#1^6\,\&,\,1])\,+$

$4320\,\mathrm{PolyLog}[2,\,-\mathrm{Root}[-x - x\,\#1 + \#1^6\,\&,\,1]]\,(1 + \mathrm{Root}[-x - x\,\#1 + \#1^6\,\&,\,1])\Bigg)\Bigg/$

$(360\,(1 + \mathrm{Root}[-x - x\,\#1 + \#1^6\,\&,\,1]))$

Differentiating the last result allows us straightforwardly to check the integral.

In[10]:= `D[int, x] - integrand // Simplify`

Out[10]= `0`

Mathematica also cannot carry out the second integral directly.

In[11]:= `root = Root[#^7 - #^2 - x &, 3];`
`integrand = Exp[root] Log[root] root;`
`Integrate[integrand, x]`

Out[13]= $\int e^{\mathrm{Root}[-x - \#1^2 + \#1^7\,\&,\,3]}\,\mathrm{Log}[\mathrm{Root}[-x - \#1^2 + \#1^7\,\&,\,3]]\,\mathrm{Root}[-x - \#1^2 + \#1^7\,\&,\,3]\,dx$

Again, we introduce a new integration variable X=root. This time we have not to solve for x as a function of X.

In[14]:= `newIntegrand = integrand /. root -> X`

Out[14]= $e^X\,X\,\mathrm{Log}[X]$

We proceed as above and express the integrand and dx/dX as a function of X. Then, we carry out the resulting integral and substitute `root` for X.

In[15]:= `sol = Solve[root[[1]][X] == 0, x]`

Out[15]= $\{\{x \to -X^2 + X^7\}\}$

In[16]:= `newIntegrand2 = newIntegrand /. sol[[1]]`

Out[16]= $e^X\,X\,\mathrm{Log}[X]$

In[17]:= `gb = GroebnerBasis[{#, D[#, X]}&[root[[1]][X] /. x -> x[X]] /. x[X] -> x,`
`                    {}, {x}, MonomialOrder -> EliminationOrder];`

In[18]:= `dxdX = x'[X] /. Solve[gb[[1]] == 0, x'[X]][[1]]`

Out[18]= $-2\,X + 7\,X^6$

In[19]:= `int = (Integrate[newIntegrand2 dxdX, X] // Simplify) /. X -> root`

Out[19]= $35284\ \text{ExpIntegralEi}[\text{Root}[-x - \#1^2 + \#1^7 \&, 3]] +$
$e^{\text{Root}[-x-\#1^2+\#1^7\&,3]}\ \big(-91482 + 56198\,\text{Root}[-x - \#1^2 + \#1^7 \&, 3] - 19278\,\text{Root}[-x - \#1^2 + \#1^7 \&, 3]^2 +$
$4466\,\text{Root}[-x - \#1^2 + \#1^7 \&, 3]^3 - 749\,\text{Root}[-x - \#1^2 + \#1^7 \&, 3]^4 +$
$91\,\text{Root}[-x - \#1^2 + \#1^7 \&, 3]^5 - 7\,\text{Root}[-x - \#1^2 + \#1^7 \&, 3]^6 + \text{Log}[\text{Root}[-x - \#1^2 + \#1^7 \&, 3]]$
$(-35284 + 35284\,\text{Root}[-x - \#1^2 + \#1^7 \&, 3] - 17642\,\text{Root}[-x - \#1^2 + \#1^7 \&, 3]^2 +$
$5880\,\text{Root}[-x - \#1^2 + \#1^7 \&, 3]^3 - 1470\,\text{Root}[-x - \#1^2 + \#1^7 \&, 3]^4 + 294$
$\text{Root}[-x - \#1^2 + \#1^7 \&, 3]^5 - 49\,\text{Root}[-x - \#1^2 + \#1^7 \&, 3]^6 + 7\,\text{Root}[-x - \#1^2 + \#1^7 \&, 3]^7\big)\big)$

Again, we can use differentiation to check the result.

In[20]:= `D[int, x] - integrand // Simplify`

Out[20]= 0

The third integrand is a square root of a ratio of roots.

In[21]:= `Clear[x, integrand];`
`        poly[x_] = #^3 - # - x&;`

In[23]:= `ratio[x_] = Root[poly[x], 2]/Root[poly[x], 3];`
`        integrand[x_] = (ratio[x]^2)^(1/2);`

A direct integration fails again.

In[25]:= `Integrate[integrand[x], x]`

Out[25]= $\displaystyle\int \sqrt{\frac{\text{Root}[-x - \#1 + \#1^3 \&, 2]^2}{\text{Root}[-x - \#1 + \#1^3 \&, 3]^2}}\ dx$

In this example, some discontinuities come from the `Root`-objects and some from the square root. The following graphic shows the imaginary part of the integrand and the argument of the square root.

In[26]:= `Plot[{Im[ratio[x]], Im[integrand[x]]}, {x, -10, 10},`
`        PlotStyle -> {{Thickness[0.02], GrayLevel[0.4]}, {Hue[0]}},`
`        Frame -> True, Axes -> False]`

Along the real line, the discontinuities arising from the roots are at $x \approx \pm 0.385$.

In[27]:= `NSolve[{poly[x][ξ] == 0, D[poly[x][ξ], ξ] == 0}, {x}, {ξ}]`

Out[27]= $\{\{x \to -0.3849\}, \{x \to 0.3849\}\}$

We proceed as above and introduce the new integration variable $X = \mathcal{R}_2(\xi,\, -x - \xi + \xi^3)/\mathcal{R}_3(\xi,\, -x - \xi + \xi^3)$.

In[28]:= `xXEq = Factor[GroebnerBasis[Numerator[Together[`
`                {poly[x][r2], poly[x][r3],   (* definition for variable X *)`
`                X^2 - (r2/r3)^2}]], {X, x}, {r2, r3},`
`                MonomialOrder -> EliminationOrder][[1]]]`

Out[28]= $x^2\,(-1 + X)\,(1 + X)\,(x^2 - 3\,x^2\,X - X^2 + 6\,x^2\,X^2 + 2\,X^3 - 7\,x^2\,X^3 - X^4 + 6\,x^2\,X^4 - 3\,x^2\,X^5 + x^2\,X^6)$
$(x^2 + 3\,x^2\,X - X^2 + 6\,x^2\,X^2 - 2\,X^3 + 7\,x^2\,X^3 - X^4 + 6\,x^2\,X^4 + 3\,x^2\,X^5 + x^2\,X^6)$

It turns out that the last factor is the one of interest for us.

```
In[29]:= sol = x'[X] /. Solve[{xXEq[[-1]] == 0, D[xXEq[[-1]] /.
                            x -> x[X], X] == 0} /. x[X] -> x, {x'[X]}, {x}]
```

$$Out[29]= \left\{ -\left(\sqrt{4 + 20\,X + 25\,X^2 - 20\,X^3 - 58\,X^4 - 20\,X^5 + 25\,X^6 + 20\,X^7 + 4\,X^8}\right) \middle/ \right.$$
$$\left(\sqrt{(4 + 28\,X + 104\,X^2 + 260\,X^3 + 480\,X^4 + 684\,X^5 +}\right.$$
$$\left.768\,X^6 + 684\,X^7 + 480\,X^8 + 260\,X^9 + 104\,X^{10} + 28\,X^{11} + 4\,X^{12})\right),$$
$$\left(\sqrt{4 + 20\,X + 25\,X^2 - 20\,X^3 - 58\,X^4 - 20\,X^5 + 25\,X^6 + 20\,X^7 + 4\,X^8}\right) \middle/$$
$$\left(\sqrt{(4 + 28\,X + 104\,X^2 + 260\,X^3 + 480\,X^4 + 684\,X^5 +}\right.$$
$$\left.\left.768\,X^6 + 684\,X^7 + 480\,X^8 + 260\,X^9 + 104\,X^{10} + 28\,X^{11} + 4\,X^{12})\right)\right\}$$

In the new integration variable X, the integral is relatively simple.

```
In[30]:= intPreX = Integrate[Sqrt[X^2] sol[[1]], X] // Simplify
```

$$Out[30]= \frac{\sqrt{X^2}\,(1 + X + X^2)\,\sqrt{(2 + 5\,X - 5\,X^3 - 2\,X^4)^2}\,\left(2 + 6\,X + 9\,X^2 + 7\,X^3 - 3\,(1 + X + X^2)^{3/2}\,\text{ArcSinh}\left[\frac{1+2\,X}{\sqrt{3}}\right]\right)}{3\,X\,\sqrt{(1 + X)^2}\,(1 + X + X^2)^5\,(-2 - 3\,X + 3\,X^2 + 2\,X^3)}$$

Unfortunately, due to the outer square root, the integral is not correct everywhere. The ratio of the derivative of `intPre[x]` and the original integrand `integrand[x]` is a differential algebraic constant of value ± 1. The following graphic shows this clearly.

```
In[31]:= intPre[x_] = intPreX /. X -> ratio[x];

In[32]:= aux1[R2_, R3_, x_] = Simplify[D[intPre[x], x] /. {Root[poly[x], 2] -> R2,
                                                            Root[poly[x], 3] -> R3}];

In[33]:= der[x_] := Module[{R2 = Root[poly[x], 2], R3 = Root[poly[x], 3]},
                        aux1[R2, R3, x]]

In[34]:= Plot3D[Re[der[x + I y]/integrand[x + I y]], {x, -3, 3}, {y, -3, 3},
              PlotPoints -> 60, Mesh -> False]
```

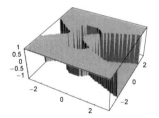

Let $\alpha(x)$ be this differential algebraic constant. Let $I_P(x)$ be the current integral `intPre[x]` and $I(x)$ the correct value. Differentiating the relation $I(x) = \alpha(x)\,I_P(x)$ and taking into account $\alpha'(x) = 0$ yields $I(x) = integrand(x)/I'_P(x)\,I_P(x)$. Calculating $I(x)$ according to this formula yields the following result. (We abbreviate the two roots for brevity.)

```
In[35]:= int = integrand[x]/D[intPre[x], x] intPre[x];

In[36]:= Simplify[int /. {Root[poly[x], 2] -> R2, Root[poly[x], 3] -> R3}]
```

$$Out[36]= \left(2\,(-1 + 3\,R2^2)\,\sqrt{\frac{R2^2}{R3^2}}\,R3\,(R2^2 + R2\,R3 + R3^2)\,(-1 + 3\,R3^2)\right.$$

$$\left.\left(7\,R2^3 + 9\,R2^2\,R3 + 6\,R2\,R3^2 + 2\,R3^3 - 3\,R3^3\,\left(\frac{R2^2 + R2\,R3 + R3^2}{R3^2}\right)^{3/2}\,\text{ArcSinh}\left[\frac{2\,R2 + R3}{\sqrt{3}\,R3}\right]\right)\right) \middle/$$

$$(3\,R2\,(R2 - R3)^2\,(6\,R2^4 + 21\,R2^3\,R3 - 2\,R3^2 + 6\,R3^4 + R2\,R3\,(-5 + 21\,R3^2) + R2^2\,(-2 + 27\,R3^2)))$$

Using the defining polynomial relations for R2 and R3, we could eliminate all powers of degree ≥ 3, but we do not need this here. A quick numerical check (because of the large size of the derivative we use a numerical check here) shows that `int` is the correct indefinite integral for all complex x.

```
In[37]:= aux2[R2_, R3_, x_] = Simplify[D[int, x] /.
                {Root[poly[x], 2] -> R2, Root[poly[x], 3] -> R3}];

(* compiled version for faster numerical evaluation *)
```

```
        aux2c = Compile[{{R2, _Complex}, {R3, _Complex}, {x, _Complex}},
                   Evaluate[aux2[R2, R3, x]]];
```

In[40]:= `der[x_] := Module[{R2 = Root[poly[x], 2], R3 = Root[poly[x], 3]},`
`                   aux2c[R2, R3, x]]`

In[41]:= `Plot3D[Re[der[x + I y]/integrand[x + I y]] - 1,`
`              {x, -3, 3}, {y, -3, 3}, PlotPoints -> 60, Mesh -> False]`

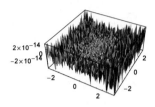

Although the fourth integrand looks more complicated than the third one, it is actually easier to integrate.

In[42]:= `poly[x_] = #^3 - x # - x&;`

In[43]:= `ratio[x_] = Root[poly[x], 2]/Root[poly[x], 3];`
`            integrand[x_] = (ratio[x]^3)^(1/3);`

A direct integration fails again.

In[45]:= `TimeConstrained[Integrate[integrand[x], x], 100]`

Out[45]= $\displaystyle \int \left(\frac{\text{Root}[-x - x\,\#1 + \#1^3 \,\&, \, 2]^3}{\text{Root}[-x - x\,\#1 + \#1^3 \,\&, \, 3]^3} \right)^{1/3} dx$

In[46]:= `TimeConstrained[Integrate[ToRadicals[integrand[x]], x], 100]`

Out[46]= `$Aborted`

Again, we let the new integration variable be the ratio of the roots $X = \mathcal{R}_2(\xi, \, -x - \xi + \xi^3)/\mathcal{R}_3(\xi, \, -x - \xi + \xi^3)$.

In[47]:= `xXEq = Factor[GroebnerBasis[Numerator[Together[`
`                   {poly[x][r2], poly[x][r3],`
`                   X^3 - (r2/r3)^3}]], {X, x}, {r2, r3},`
`                   MonomialOrder -> EliminationOrder][[1]]]`

Out[47]= $-x\,(-1+X)\,(1+X+X^2)\,(-1-3X-6X^2+x\,X^2-7X^3+2\,x\,X^3-6X^4+x\,X^4-3X^5-X^6)$
$(1-3X+3X^2+x\,X^2-4X^3-x\,X^3+9X^4-5\,x\,X^4+x^2\,X^4-9X^5+13\,x\,X^5-2\,x^2\,X^5+6X^6-16\,x\,X^6+$
$3\,x^2\,X^6-9X^7+13\,x\,X^7-2\,x^2\,X^7+9X^8-5\,x\,X^8+x^2\,X^8-4X^9-x\,X^9+3X^{10}+x\,X^{10}-3X^{11}+X^{12})$

This time, the second factor is the one of relevance.

In[48]:= `sol = x'[X] /. Solve[{xXEq[[-2]] == 0, D[xXEq[[-2]] /.`
`                   x -> x[X], X] == 0} /. x[X] -> x, {x'[X]}, {x}]`

Out[48]= $\left\{ -\dfrac{2 + 7X + 9X^2 + 5X^3 - 5X^4 - 9X^5 - 7X^6 - 2X^7}{X^3\,(1+X)^3} \right\}$

The resulting indefinite integral is quite simple.

In[49]:= `intX = Integrate[(X^3)^(1/3) sol[[1]], X] // Simplify`

Out[49]= $\dfrac{(X^3)^{1/3}\,(12 + 36X + 30X^2 + 3X^3 + 10X^4 + 11X^5 + 4X^6 - 6X\,(1+X)^2\,\text{Log}[X] + 6X\,(1+X)^2\,\text{Log}[1+X])}{6X^2\,(1+X)^2}$

A numerical check shows that int is the correct indefinite integral.

In[50]:= `int = intX /. X -> ratio[x];`

In[51]:= `Clear[diff];`
`            diff[x_] = D[int, x] - integrand[x] // Simplify;`

In[53]:= `Plot3D[Re[diff[x + I y]], {x, -3, 3}, {y, -3, 3},`
`              PlotPoints -> 60, Mesh -> False]`

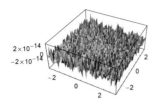

In[54]:= **diff[x]**

Out[54]= $\left(8\, x^3\, \left(x\, \left(7\, x + (3 + 9\, x)\, \text{Root}[-x - x\, \#1 + \#1^3\, \&,\, 3] + 2\, (4 + x)\, \text{Root}[-x - x\, \#1 + \#1^3\, \&,\, 3]^2\right) + \right.\right.$

$\qquad x\, \text{Root}[-x - x\, \#1 + \#1^3\, \&,\, 2]$

$\qquad \left(3 + 9\, x + 11\, (1 + x)\, \text{Root}[-x - x\, \#1 + \#1^3\, \&,\, 3] + (15 + 2\, x)\, \text{Root}[-x - x\, \#1 + \#1^3\, \&,\, 3]^2\right) +$

$\qquad \text{Root}[-x - x\, \#1 + \#1^3\, \&,\, 2]^2\, \left(2\, x\, (4 + x) + x\, (15 + 2\, x)\, \text{Root}[-x - x\, \#1 + \#1^3\, \&,\, 3] + \right.$

$\qquad \left.\left.\left.(3 + 9\, x)\, \text{Root}[-x - x\, \#1 + \#1^3\, \&,\, 3]^2\right)\right)\right) \Big/$

$\qquad \left(\text{Root}[-x - x\, \#1 + \#1^3\, \&,\, 1]^3\, \left(x - 3\, \text{Root}[-x - x\, \#1 + \#1^3\, \&,\, 2]^2\right)\right.$

$\qquad \left(\dfrac{\text{Root}[-x - x\, \#1 + \#1^3\, \&,\, 2]^3}{\text{Root}[-x - x\, \#1 + \#1^3\, \&,\, 3]^3}\right)^{2/3}\, \text{Root}[-x - x\, \#1 + \#1^3\, \&,\, 3]^6$

$\qquad \left.\left(x - 3\, \text{Root}[-x - x\, \#1 + \#1^3\, \&,\, 3]^2\right)\right)$

Mathematica cannot carry out the fifth integral directly.

In[55]:= **integrand = Root[#^3 - # - x&, 2]/(Root[#^3 - # + x&, 2] - 1);**
Integrate[integrand, {x, 0, 1}]

Out[56]= $\displaystyle\int_0^1 \dfrac{\text{Root}[-x - \#1 + \#1^3\, \&,\, 2]}{-1 + \text{Root}[x - \#1 + \#1^3\, \&,\, 2]}\, dx$

Also, the indefinite integral cannot be calculated.

In[57]:= **Integrate[integrand, x]**

Out[57]= $\displaystyle\int \dfrac{\text{Root}[-x - \#1 + \#1^3\, \&,\, 2]}{-1 + \text{Root}[x - \#1 + \#1^3\, \&,\, 2]}\, dx$

Of course, we can calculate a numerical approximation of this integral.

In[58]:= **NIntegrate[integrand, {x, 0, 2/(3 Sqrt[3]), 1}]**

Out[58]= $0.927235 - 0.186968\, i$

We introduce a new integration variable X=integrand.

In[59]:= **xXEq = Eliminate[{r1^3 - r1 - x == 0, r2^3 - r2 + x == 0,**
**   X - r1/(r2 - 1) == 0},**
**   {r1, r2}] // Subtract @@ #& // Factor**

Out[59]= $(x - X + 3\, x\, X - 2\, x^2 + 3\, x\, X^2 + x\, X^3)$

$\qquad (x - 2\, X - 3\, x\, X + 5\, X^2 + 6\, x\, X^2 - 7\, x\, X^3 - 5\, X^4 + 6\, x\, X^4 + 2\, X^5 - 3\, x\, X^5 + x\, X^6)$

This time, we have two different possibilities for $\frac{dx}{dX}$.

In[60]:= **sol = Solve[{xXEq == 0, D[xXEq /. x -> x[X], X] == 0} /.**
**   x[X] -> x, {x'[X]}, {x}]**

Out[60]= $\left\{\left\{x'[X] \to \dfrac{1 + 2\, X - 2\, X^2}{(1 + X)^4}\right\},\, \left\{x'[X] \to \dfrac{2 - 6\, X - 15\, X^2 + 40\, X^3 - 15\, X^4 - 6\, X^5 + 2\, X^6}{(1 - X + X^2)^4}\right\}\right\}$

The first one is correct in certain parts of the complex *x*-plane.

In[61]:= **int1 = Integrate[X sol[[1, 1, 2]], X] /. X -> integrand**

Out[61]= $-2 \text{Log}\left[1 + \dfrac{\text{Root}\left[-x - \#1 + \#1^3 \&, 2\right]}{-1 + \text{Root}\left[x - \#1 + \#1^3 \&, 2\right]}\right] - \dfrac{9 + \dfrac{16 \text{Root}\left[-x-\#1+\#1^3\&,2\right]^2}{(-1+\text{Root}\left[x-\#1+\#1^3\&,2\right])^2} + \dfrac{23 \text{Root}\left[-x-\#1+\#1^3\&,2\right]}{-1+\text{Root}\left[x-\#1+\#1^3\&,2\right]}}{2\left(1 + \dfrac{\text{Root}\left[-x-\#1+\#1^3\&,2\right]}{-1+\text{Root}\left[x-\#1+\#1^3\&,2\right]}\right)^3}$

In[62]:= `Plot[Evaluate[Abs[D[int1, x] - integrand]], {x, -1, 1}, PlotRange -> All]`

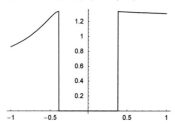

The boundaries of the interval where `int1` is correct coincide with the points of discontinuity of `Root[#^3 - # - x&, 2]`.

In[63]:= `Show[GraphicsArray[Plot[Evaluate[#[Root[#^3 - # - x&, 2]]], {x, -1, 1}, PlotRange -> All, DisplayFunction -> Identity]& /@ {Re, Im}]]`

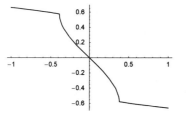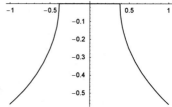

The two points are branch points of $r(x)^3 - r(x) + x = 0$.

In[64]:= `Solve[{r1^3 - r1 + x == 0, D[r1^3 - r1 - x, r1] == 0}, {x}, {r1}]`

Out[64]= $\left\{\left\{x \to -\dfrac{2}{3\sqrt{3}}\right\}, \left\{x \to \dfrac{2}{3\sqrt{3}}\right\}\right\}$

The integral `int1` is a continuous function of x at the branch points.

In[65]:= `Show[GraphicsArray[`
`    Plot[Evaluate[#[int1]], {x, -1, 1},`
`        PlotRange -> All, Frame -> True, Axes -> False,`
`        DisplayFunction -> Identity]& /@ {Re, Im}]]`

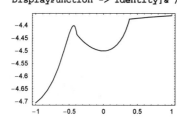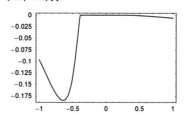

In a similar manner, we find that `int2` is the correct indefinite integral outside of the interval $\left(-2/(3\sqrt{3}), 2/(3\sqrt{3})\right)$.

In[66]:= `int2 = Integrate[X sol[[2, 1, 2]], X] /. X -> integrand`

Out[66]= $\text{Log}\left[1 + \dfrac{\text{Root}\left[-x - \#1 + \#1^3 \&, 2\right]^2}{(-1 + \text{Root}\left[x - \#1 + \#1^3 \&, 2\right])^2} - \dfrac{\text{Root}\left[-x - \#1 + \#1^3 \&, 2\right]}{-1 + \text{Root}\left[x - \#1 + \#1^3 \&, 2\right]}\right] +$

$\left(9 - \dfrac{2 \text{Root}\left[-x - \#1 + \#1^3 \&, 2\right]^5}{(-1 + \text{Root}\left[x - \#1 + \#1^3 \&, 2\right])^5} +\right.$

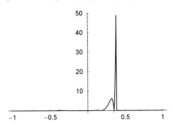

$$\frac{32\,\text{Root}\,[-x - \#1 + \#1^3\, \&,\, 2]^4}{(-1 + \text{Root}\,[x - \#1 + \#1^3\, \&,\, 2])^4} - \frac{54\,\text{Root}\,[-x - \#1 + \#1^3\, \&,\, 2]^3}{(-1 + \text{Root}\,[x - \#1 + \#1^3\, \&,\, 2])^3} +$$

$$\left.\frac{49\,\text{Root}\,[-x - \#1 + \#1^3\, \&,\, 2]^2}{(-1 + \text{Root}\,[x - \#1 + \#1^3\, \&,\, 2])^2} - \frac{25\,\text{Root}\,[-x - \#1 + \#1^3\, \&,\, 2]}{-1 + \text{Root}\,[x - \#1 + \#1^3\, \&,\, 2]}\right)\Bigg/$$

$$\left(2\left(1 + \frac{\text{Root}\,[-x - \#1 + \#1^3\, \&,\, 2]^2}{(-1 + \text{Root}\,[x - \#1 + \#1^3\, \&,\, 2])^2} - \frac{\text{Root}\,[-x - \#1 + \#1^3\, \&,\, 2]}{-1 + \text{Root}\,[x - \#1 + \#1^3\, \&,\, 2]}\right)^3\right)$$

In[67]:= **Plot[Evaluate[Abs[D[int2, x] - integrand]], {x, -1, 1}, PlotRange -> All]**

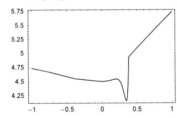

int2 too is a continuous function at the branch points.

In[68]:= **Show[GraphicsArray[**
          Plot[Evaluate[#[int2]], {x, -1, 1},
              PlotRange -> All, Frame -> True, Axes -> False,
              DisplayFunction -> Identity]& /@ {Re, Im}]]

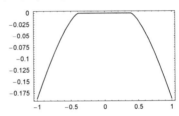

Using now the two indefinite integrals int1 and int2 allows us to calculate the definite integral.

In[69]:= **(int1 /. x -> 2/(3 Sqrt[3])) - (int1 /. x -> 0) +**
          (int2 /. x -> 1) - (int2 /. x -> 2/(3 Sqrt[3])) // RootReduce

Out[69]= $-\text{Log}\left[\frac{3}{2}\right] - 2\,\text{Log}\left[\frac{1}{2}\,\left(3 + \sqrt{3}\,\right)\right] +$

   Log[Root[27 - 108 #1 + 162 #1² - 112 #1³ + 35 #1⁴ - 4 #1⁵ + #1⁶ &, 5]] +
   Root[362560396897 - 1532000472480 #1 + 8733315168288 #1² - 24291598615200 #1³ +
      33284161513440 #1⁴ - 15784654934016 #1⁵ + 750008916864 #1⁶ - 18351484416 #1⁷ +
      690216360192 #1⁸ - 101220378624 #1⁹ - 41721661440 #1¹⁰ + 2176782336 #1¹² &, 12]

Combining the logarithms, we have the following final result.

In[70]:= **% //. Log[x_] + a_. Log[y_] :> Log[x y^a] // RootReduce**

Out[70]= Log[Root[16777216 - 603979776 #1 + 8493465600 #1² - 59567505408 #1³ + 223494930432 #1⁴ -
      460940378112 #1⁵ + 742572417024 #1⁶ - 857061015552 #1⁷ + 588411475200 #1⁸ -
      217134038016 #1⁹ + 38722917024 #1¹⁰ - 2754990144 #1¹¹ + 387420489 #1¹² &, 5]] +
   Root[362560396897 - 1532000472480 #1 + 8733315168288 #1² - 24291598615200 #1³ +
      33284161513440 #1⁴ - 15784654934016 #1⁵ + 750008916864 #1⁶ - 18351484416 #1⁷ +
      690216360192 #1⁸ - 101220378624 #1⁹ - 41721661440 #1¹⁰ + 2176782336 #1¹² &, 12]

The numerical value of the last expression agrees with the above result from NIntegrate.

In[71]:= **N[%]**

Out[71]= 0.927235 - 0.186968 i

Mathematica also cannot carry out the last integral directly.

```
In[72]:= integrand = 1/Root[#^5 + # - x&, 1] - (x^(-1/5) + 1/5 x^-1);
        Integrate[integrand, {x, 1, Infinity}]
```

$$Out[73]= \int_1^\infty \left(-\frac{1}{5\,x} - \frac{1}{x^{1/5}} + \frac{1}{\text{Root}[-x + \#1 + \#1^5\ \&,\ 1]}\right) dx$$

We can use NIntegrate to get a numerical approximation of the integral.

```
In[74]:= NIntegrate[Evaluate[integrand], {x, 1, Infinity},
        Method -> DoubleExponential,
        WorkingPrecision -> 50, PrecisionGoal -> 5]
```

```
Out[74]= 0.1253021
```

We introduce a new integration variable X=Root[#^5+#-x&, 1] to integrate the first summand $\int_1^\infty 1/\mathcal{R}_1(\xi, -x+\xi+\xi^5)\,d\xi$.

```
In[75]:= Solve[{X^5 + X - x == 0, D[X^5 + X - x[X], X] == 0} /.
        x[X] -> x, {x'[X]}, {x}]
```

$$Out[75]= \{\{x'[X] \to 1 + 5\,X^4\}\}$$

The integral is now easily doable and verifiable.

```
In[76]:= int1 = Integrate[1/X (1 + 5 X^4), X] /. X -> Root[#^5 + # - x&, 1]
```

$$Out[76]= \text{Log}[\text{Root}[-x + \#1 + \#1^5\ \&,\ 1]] + \frac{5}{4}\,\text{Root}[-x + \#1 + \#1^5\ \&,\ 1]^4$$

```
In[77]:= D[int1, x] - 1/Root[#^5 + # - x&, 1] // Simplify
```

```
Out[77]= 0
```

The integration over the second and third summand is trivial.

```
In[78]:= int2 = Integrate[(x^(-1/5) + 1/5 x^-1), x] // PowerExpand
```

$$Out[78]= \frac{5\,x^{4/5}}{4} + \frac{\text{Log}[x]}{5}$$

The series expansion of $1/\mathcal{R}_1(\xi, -x+\xi+\xi^5)\,d\xi$ around $x = \infty$ shows that the second and third summand of the integral are simply the leading terms of the inverse root.

```
In[79]:= rootSer = InverseSeries[Series[y^5 + y, {y, Infinity, 20}], x]
```

$$Out[79]= \frac{1}{\left(\frac{1}{x}\right)^{1/5}} - \frac{1}{5}\left(\frac{1}{x}\right)^{3/5} - \frac{1}{25}\left(\frac{1}{x}\right)^{7/5} - \frac{1}{125}\left(\frac{1}{x}\right)^{11/5} + \frac{21\left(\frac{1}{x}\right)^{19/5}}{15625} + \frac{78\left(\frac{1}{x}\right)^{23/5}}{78125} + O\left[\frac{1}{x}\right]^5$$

```
In[80]:= 1/rootSer
```

$$Out[80]= \left(\frac{1}{x}\right)^{1/5} + \frac{1}{5\,x} + \frac{2}{25}\left(\frac{1}{x}\right)^{9/5} + \frac{4}{125}\left(\frac{1}{x}\right)^{13/5} + \frac{7}{625}\left(\frac{1}{x}\right)^{17/5} + \frac{44\left(\frac{1}{x}\right)^{21/5}}{15625} + O\left[\frac{1}{x}\right]^{27/5}$$

The remaining series terms are convergent at $x = \infty$. At $x = \infty$, int1 differs from int2 by 1.

```
In[81]:= Series[int1 /. _Root -> Normal[rootSer], {x, Infinity, 2}]
```

$$Out[81]= \frac{5}{4\left(\frac{1}{x}\right)^{4/5}} + \left(-1 - \frac{1}{5}\,\text{Log}\left[\frac{1}{x}\right]\right) - \frac{1}{10}\left(\frac{1}{x}\right)^{4/5} - \frac{1}{50}\left(\frac{1}{x}\right)^{8/5} + O\left[\frac{1}{x}\right]^{11/5}$$

Putting all subresults together, we arrive at the following short result for the integral.

```
In[82]:= int2 - int1 - 1 /. x -> 1 // RootReduce
```

$$Out[82]= -\text{Log}[\text{Root}[-1 + \#1^2 + \#1^3\ \&,\ 1]] + \text{Root}[59 + 332\,\#1 - 288\,\#1^2 + 64\,\#1^3\ \&,\ 1]$$

The numerical value of the last result agrees with the above value from NIntegrate.

```
In[83]:= N[%]
```

```
Out[83]= 0.125302
```

For an algorithmic treatment of the integration of algebraic functions, see [440].

e) Using the function Resolve, it is straightforward to obtain a cylindrical description of the region in a_1, a_2, a_3-space such that the roots are the side length of a nondegenerate triangle. To obtain a canonical description of the volume, we apply the function CylindricalDecomposition.

```
In[1]:= conds = CylindricalDecomposition[#, {a1, a2, a3}]& @
            Resolve[Exists[{x1, x2, x3},
                Element[{a1, a2, a3, x1, x2, x3}, Reals],
                (* x1, x2, x3 are roots of cubic *)
                x1 x2 x3 == a3 && x1 x2 + x1 x3 + x2 x3 == a2 &&
                x1 + x2 + x3 == -a1 &&
                (* triangle conditions *)
                x1 > 0 && x2 > 0 && x3 > 0 &&
                x1 + x2 > x3 && x2 + x3 > x1 && x1 + x3 > x2]]
```

$$\text{Out[1]= } a1 < 0 \&\& \left(\left(\frac{a1^2}{4} < a2 \le \frac{5\,a1^2}{16}\,\&\&\right.\right.$$

$$\frac{1}{27}\,(2\,a1^3 - 9\,a1\,a2) - \frac{2}{27}\,\sqrt{a1^6 - 9\,a1^4\,a2 + 27\,a1^2\,a2^2 - 27\,a2^3} \le a3 < \frac{1}{8}\,(a1^3 - 4\,a1\,a2)\right) \;\|\|$$

$$\left(\frac{5\,a1^2}{16} < a2 < \frac{a1^2}{3}\,\&\&\,\frac{1}{27}\,(2\,a1^3 - 9\,a1\,a2) - \frac{2}{27}\,\sqrt{a1^6 - 9\,a1^4\,a2 + 27\,a1^2\,a2^2 - 27\,a2^3} \le\right.$$

$$\left.a3 \le \frac{1}{27}\,(2\,a1^3 - 9\,a1\,a2) + \frac{2}{27}\,\sqrt{a1^6 - 9\,a1^4\,a2 + 27\,a1^2\,a2^2 - 27\,a2^3}\right) \;\|\|$$

$$\left.\left(a2 = \frac{a1^2}{3}\,\&\&\,a3 = \frac{1}{27}\,(2\,a1^3 - 9\,a1\,a2) - \frac{2}{27}\,\sqrt{a1^6 - 9\,a1^4\,a2 + 27\,a1^2\,a2^2 - 27\,a2^3}\right)\right)$$

For the visualization and probability calculation, we can ignore lower-dimensional parts of conds.

```
In[2]:= conds2 = DeleteCases[conds //. a_ && (b_ || c_) :> a && b || a && c,
            _?(MemberQ[#, Equal, Infinity, Heads -> True]&)] /.
                                                LessEqual -> Less;
```

conds2 yields immediately a piecewise parametrization of the region in a_1, a_2, a_3-space. The right graphic shows the cross section of the 3D region for $a_1 = -1$.

```
In[3]:= makePoints[a1 < 0 && Inequality[a2l_, Less, a2, Less, a2u_] &&
                Inequality[a3l_, Less, a3, Less, a3u_], {ε_, pp_}] :=
            (* points along boundaries;
               avoid endpoints for numerical evaluation *)
            Table[{{a1, a2, Re[#]}, {a1, -2 + ε, Re[a2l + ε], Re[a2u - ε],
                {a2, Re[a2l + ε], Re[a2u - ε],
                    Re[(a2u - a2l - 2ε)/pp]}}]& /@ {a3l, a3u};
```

```
In[4]:= Needs["Graphics`Graphics3D`"]
```

```
In[5]:= Show[GraphicsArray[{
            Graphics3D[{EdgeForm[], Cases[(* the region in a1,a2,a3-space *)
                Map[ListSurfacePlot3D[#, DisplayFunction -> Identity]&,
                    makePoints[#, {10^-12, 24}]& /@ conds2, {2}],
                _Polygon, Infinity]}, BoxRatios -> {1, 1, 1},
                Axes -> True, PlotRange -> All] // N,
            (* zoomed in outline plot in the a1 == -1 plane *)
            ContourPlot[Evaluate[If[conds2 /. a1 -> -1, 1, -1]],
                {a2, 0.25, 0.35}, {a3, 0.01, 0.04},
                PlotPoints -> 250, ContourShading -> False,
                Contours -> {0}, DisplayFunction -> Identity]}]]
```

conds2 also yields immediately the boundaries for the 3D region that allows for a triangle interpretation of the roots. A1 is the smallest value of a1.

```
In[6]:= toIntegral[a1 < 0 && Inequality[a2l_, Less, a2, Less, a2u_] &&
                Inequality[a3l_, Less, a3, Less, a3u_], A1_] :=
```

```
           Integrate[1, {a1, A1, 0}, {a2, a21, a2u}, {a3, a31, a3u},
                      GenerateConditions -> False]
```

Calculating the volume of the region for $-1 < a_1, a_2, a_3 < 1$ gives a 0.00014% probability that for random a_1, a_2, a_3 from the interval $[-1, 1]$, the roots are the side length of a nondegenerate triangle.

```
In[7]:= Simplify[toIntegral[conds2[[1]], A1] +
               toIntegral[conds2[[2]], A1], A1 < 0]
```

$$Out[7]= \frac{A1^6}{92160}$$

```
In[8]:= {#, N[#]}&[%/2^3 /. A1 -> 1]
```

$$Out[8]= \left\{ \frac{1}{737280},\ 1.35634 \times 10^{-6} \right\}$$

A "random" check shows that it takes on average about three-quarters of a million trials until we find a cubic whose roots can be interpreted as triangle side lengths.

```
In[9]:= (* do the three numbers x1, x2, and x3 form the edges of a triangle? *)
        triangleQ[{x1_, x2_, x3_}] :=
        If[Union[Head /@ {x1, x2, x3}] === {Real},
           If[x1 + x2 > x3 && x2 + x3 > x1 && x1 + x3 > x2,
              True, False], False]

In[11]:= SeedRandom[18];
         counter = 0;
         While[{a1, a2, a3} = Table[Random[Real, {-1, 1}], {3}];
               {x1, x2, x3} = x /. {ToRules[NRoots[
                                     x^3 + a1 x^2 + a2 x + a3 == 0, x]]};
               Not[triangleQ[{x1, x2, x3}]], counter++];
         counter

Out[14]= 728655
```

23. Riemann Surface of Cubic

Because the branch points and the branch cuts are unknown in this example, it is not so obvious how to generate a picture like those in the corresponding examples treated in Chapter 2 of the Graphics volume [1736]. Of course, we could use the general numerical algorithm implemented in Subsection 1.12.1 of the Numerics volume [1737], but here we will use some more symbolic calculations to generate the Riemann surface under consideration.

We start by looking at the branch points of the Riemann surface. Because they connect the various sheets, they occur at the points a_{bp} where the polynomial $x^3 + x^2 + a\,x - 1/2 = 0$ has multiple roots.

```
In[1]:= Eliminate[{# == 0, D[#, x] == 0}, x]&[eq = x^3 + x^2 + a x - 1/2]

Out[1]= 36 a - 4 a^2 + 16 a^3 == -19
```

That means we have three branch points bps that have the following coordinates.

```
In[2]:= bps = #[[1, 2]]& /@ NSolve[%, a]

Out[2]= {-0.460727, 0.355363 - 1.56562 i, 0.355363 + 1.56562 i}
```

Solving the above cubic equation eq for x yields three solutions corresponding to three sheets of the Riemann surface.

```
In[3]:= sol = Solve[eq == 0, x]
```

Out[3]= $\{\{x \to -\dfrac{1}{3} - \dfrac{-4 + 12\,a}{6\,2^{1/3}\,\left(23 + 18\,a + 3\sqrt{3}\,\sqrt{19 + 36\,a - 4\,a^2 + 16\,a^3}\,\right)^{1/3}} +$

$\dfrac{\left(23 + 18\,a + 3\sqrt{3}\,\sqrt{19 + 36\,a - 4\,a^2 + 16\,a^3}\,\right)^{1/3}}{3\,2^{2/3}}\Big\},$

$\{x \to -\dfrac{1}{3} + \dfrac{(1 + i\sqrt{3}\,)\,(-4 + 12\,a)}{12\,2^{1/3}\,\left(23 + 18\,a + 3\sqrt{3}\,\sqrt{19 + 36\,a - 4\,a^2 + 16\,a^3}\,\right)^{1/3}} -$

$\dfrac{(1 - i\sqrt{3}\,)\,\left(23 + 18\,a + 3\sqrt{3}\,\sqrt{19 + 36\,a - 4\,a^2 + 16\,a^3}\,\right)^{1/3}}{6\,2^{2/3}}\Big\},$

$\{x \to -\dfrac{1}{3} + \dfrac{(1 - i\sqrt{3}\,)\,(-4 + 12\,a)}{12\,2^{1/3}\,\left(23 + 18\,a + 3\sqrt{3}\,\sqrt{19 + 36\,a - 4\,a^2 + 16\,a^3}\,\right)^{1/3}} -$

$\dfrac{(1 + i\sqrt{3}\,)\,\left(23 + 18\,a + 3\sqrt{3}\,\sqrt{19 + 36\,a - 4\,a^2 + 16\,a^3}\,\right)^{1/3}}{6\,2^{2/3}}\Big\}\Big\}$

Unfortunately, a direct visualization of the real or imaginary part of $x(a)$ by putting together three `Plot3D`-generated pictures of the above solution `sol` does not give a satisfactory result. (We use the imaginary part here; the real part results in very similar pictures.) This is because branch cuts that come out from the branch points give nearly vertical walls in the picture.

```
In[4]:= Show[GraphicsArray[ (* the three pictures *)
         Plot3D[Evaluate[Im[#[[1, 2]] /. a -> ax + I ay]], {ax, -3, 3},
             {ay, -3, 3}, PlotRange -> All, PlotPoints -> 45,
             DisplayFunction -> Identity]& /@ sol]]
```

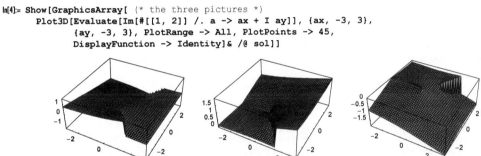

The use of `NSolve` or `NRoots` also does not avoid this problem, because in their output, the roots are ordered lexicographically, which, of course, does not give better pictures. To avoid these discontinuities in the graphic, we have to be a little bit more careful. First, we determine where the above branch cuts are located. Looking at the solution `sol`, we recognize that they are caused by negative values of the expression under the square root, this means by $(19 + 36\,a - 4\,a^2 + 16\,a^3)$, which is simply the left-hand side of the defining equation for the branch points. Here, the branch cuts are calculated and graphically represented. The underlying `ContourGraphics` shows the real part of the first solution of `sol`.

```
In[5]:= ContourPlot[Evaluate[Im[sol[[1, 1, 2]] /. a -> ax + I ay]],
             {ax, -3, 3}, {ay, -3, 3}, PlotPoints -> 120,
             ContourShading -> False, Contours -> 50,
             (* the branch cuts implicitly parametrized *)
             Epilog -> (* use ParametricPlot for smooth curves *)
             ParametricPlot[Evaluate[{Re[#[[1, 2]]], Im[#[[1, 2]]]}& /@
                 Solve[19 + 36 a - 4 a^2 + 16 a^3 == -t, a]], {t, 0, 600},
                 PlotStyle -> {{Thickness[0.02], Dashing[{0.01, 0.04}]}},
                 PlotRange -> {{-3, 3}, {-3, 3}}, AspectRatio -> Automatic,
                 DisplayFunction -> Identity][[1]]]
```

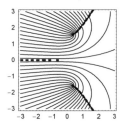

By using $a = r e^{i\varphi}$ in $19 + 36\,a - 4\,a^2 + 16\,a^3$, we can obtain an implicit equation for the branch cuts.

```
In[6]:= Module[{y1, realPart, imaginaryPart},
          (* change to polar coordinates *)
          y1 = (19 + 36a - 4 a^2 + 16 a^3 /.
                {a -> r (Cos[φ] + I Sin[φ])}) // Expand;
          (* take out real and imaginary parts *)
          imaginaryPart = Plus @@ Cases[y1, _ _Complex];
          realPart = y1 - ii;
          (* make nice equations *) Factor[Simplify[Subtract @@
          Eliminate[{realPart == -t, imaginaryPart == 0}, {t}]]]]
```
```
Out[6]=  -r (9 + 4 r² - 2 r Cos[φ] + 8 r² Cos[2 φ]) Sin[φ]
```

For a given value of r, we can determine the corresponding φ. (We do this here numerically to make sure that we pick the "correct" one in the first quadrant.)

```
In[7]:= pOfr[r_] := pOfr[r] =
         FindRoot[9 Sin[φ] - r Sin[2φ] + 4 r^2 Sin[3φ], {φ, 1.3}][[1, 2]]
```

Now, we have to put all of these together to make a picture of the Riemann surface of $x(a)$.

To avoid the vertical walls due to the branch cuts, we subdivide the complex a-plane in various regions. For $|a|<r1$ (where r1 is the distance of the nearest branch point to the origin), no problems arise; all three sheets are disconnected in this region. For $r1<|a|<r2$ (r2 being the distance of the two branch points further away from the origin) we avoid the line $\varphi = \pi$, where two sheets intersect each other, and in the region $|a|>r2$, we model the surface by three pieces, consisting of the various solutions, arranged in such a manner that they fit together continuously over the branch point. (Because of the complicated function structure inside `ParametricPlot3D`, we precompile the above solutions to speed up the numerical calculations. The chosen values for the plot points make a homogeneously divided surface. The φ division is homogeneous; this could be refined to fit exactly within the branch cuts.)

```
In[8]:= {r1, r2} = {Abs[bps[[1]]], Abs[bps[[2]]]}
```
```
Out[8]= {0.460727, 1.60544}
```

```
In[9]:= Do[f[i] = Compile[{{a, _Complex}}, Evaluate[sol[[i, 1, 2]]]], {i, 3}]
```

```
In[10]:= rsfP[reim_] := rsfP[reim] =
         Module[{ε = 10^-6 (* not visible in picture;
                             but avoids touching the branch cuts *),
                 np = 41, nr = 2, opts = Sequence[Compiled -> False,
                                          DisplayFunction -> Identity]},
         Flatten[{(* the various parts, avoiding crossing any branch cut *)
         (* the inner parts wrt r *)
         Table[ParametricPlot3D[{r Cos[φ], r Sin[φ], reim @ f[i][N[r Exp[I φ]]]},
                               {r, 0, r1}, {φ, 0, 2Pi}, PlotPoints -> {2nr, np},
                               Evaluate[opts]][[1]], {i, 3}],
         (* the middle part wrt r *)
         Table[ParametricPlot3D[{r Cos[φ], r Sin[φ], reim @ f[i][N[r Exp[I φ]]]},
                               {r, r1, r2}, {φ, -Pi + ε, Pi - ε},
                               PlotPoints -> {5nr, np}, Evaluate[opts]][[1]], {i, 3}],
         (* the outer part wrt r *)
         ParametricPlot3D[{r Cos[φ], r Sin[φ],
         Which[φ < pOfr[r],       reim @ f[1][N[r Exp[I φ]]],
               φ < Pi + ε,        reim @ f[3][N[r Exp[I φ]]],
               φ < 2Pi - pOfr[r], reim @ f[2][N[r Exp[I φ]]],
```

```
          True,                  reim @ f[1][N[r Exp[I φ]]]]},
                {r, r2, 3.2}, {φ, 0, 2Pi},
                   PlotPoints -> {7nr, np}, Evaluate[opts]][[1]],
   ParametricPlot3D[{r Cos[φ], r Sin[φ],
   Which[φ < N[Pi] + ε,        reim @ f[2][N[r Exp[I φ]]],
          True,                reim @ f[3][N[r Exp[I φ]]]]},
                {r, r2, 3}, {φ, 0, 2Pi}, PlotPoints -> {7nr, np},
                Evaluate[opts]][[1]],
   ParametricPlot3D[{r Cos[φ], r Sin[φ],
   Which[φ < pOfr[r],          reim @ f[3][N[r Exp[I φ]]],
          φ < 2Pi - pOfr[r],   reim @ f[1][N[r Exp[I φ]]],
          True,                reim @ f[2][N[r Exp[I φ]]]]},
                {r, r2, 3}, {φ, 0, 2Pi}, PlotPoints -> {7nr, np},
                Evaluate[opts]][[1]]}]];
```

So, we can display the whole surface. If all the pieces are put together now, no spurious vertical walls will be there. For better visibility, we make holes in the polygons.

```
In[11]:= Show[GraphicsArray[
   Function[reIm,
   Block[{$DisplayFunction = Identity, opts = Sequence[
          Boxed -> False, BoxRatios -> {1, 1, 2},
          ViewPoint -> {1.53, -1.9, 1.15},
          PlotRange -> {{-3.25, 3.25}, {-3.25, 3.25}, {-2.3, 1.45}}]},
   Show[Graphics3D[{EdgeForm[], SurfaceColor[Hue[0.22], Hue[0.12], 2.1],
   (* make holes in the polygons to better see that it is one surface *)
   Function[l, Module[{mp = Mean[l], λ, nOld, nNew},
       λ = (mp + 0.74 (# - mp))& /@ l;
       {nOld, nNew} = Partition[Append[#, First[#]], 2, 1]& /@ {l, λ};
       {MapThread[Polygon[Join[#1, Reverse[#2]]]&, {nOld, nNew}]}]] /@
       (* the polygons of the last picture *)
       (First /@ Cases[rsfP[reIm], _Polygon, Infinity])}], opts]]] /@
       (* show real part and imaginary part *) {Re, Im}]]
```

24. Celestial Mechanics, Lagrange Points

a) We solve the equation by iteration. To avoid unnecessary big terms in the first stages, we take care of the increase of the order in every step. At the end, we use `TrigFactor /@ ...` to reduce all trigonometric functions to sin functions.

```
In[1]:= Function[ord, (* rewrite at the end *)
   Collect[#, Cases[#, _Sin, {0, Infinity}]]&[
   (* simplify the sin functions *)
   TrigFactor /@ Expand[Fold[Normal[M + e Sin[#1 +
       (* this makes automatically the expansion *) O[e]^#2]]&,
          M, Range[ord]]]]][10]
```

$$Out[1]= M + (e + e^2 \cos[M] + e^3 \cos[M]^2 + e^4 \cos[M]^3 + e^5 \cos[M]^4 +$$
$$e^6 \cos[M]^5 + e^7 \cos[M]^6 + e^8 \cos[M]^7 + e^9 \cos[M]^8 + e^{10} \cos[M]^9) \sin[M] +$$
$$\left(-\frac{e^3}{2} - \frac{5}{3} e^4 \cos[M] - \frac{11}{3} e^5 \cos[M]^2 - \frac{20}{3} e^6 \cos[M]^3 - \frac{65}{6} e^7 \cos[M]^4 -$$
$$\frac{49}{3} e^8 \cos[M]^5 - \frac{70}{3} e^9 \cos[M]^6 - 32 e^{10} \cos[M]^7\right) \sin[M]^3 +$$
$$\left(\frac{13 e^5}{24} + \frac{47}{15} e^6 \cos[M] + \frac{1291}{120} e^7 \cos[M]^2 + \frac{427}{15} e^8 \cos[M]^3 +$$

$$\frac{1281}{20} \, e^9 \, Cos[M]^4 + \frac{644}{5} \, e^{10} \, Cos[M]^5\Big) \, Sin[M]^5 +$$

$$\left(-\frac{541 \, e^7}{720} - \frac{1957}{315} \, e^8 \, Cos[M] - \frac{36619 \, e^9 \, Cos[M]^2}{1260} - \frac{6368}{63} \, e^{10} \, Cos[M]^3\right) \, Sin[M]^7 +$$

$$\left(\frac{9509 \, e^9}{8064} + \frac{5141}{405} \, e^{10} \, Cos[M]\right) \, Sin[M]^9$$

Note that by using the Lagrange-Bürmann theorem, it is possible to give this result in closed form [829].

b) In the following, we implement a straightforward realization of the calculation of the required series. (A more elegant, recursive possibility can be found in [1588], [192], and [254]; an application of the series can be found in [426], [317].) We do not directly use a list of the form {x[t], y[t], z[t]} for the position vector, because first, this is very memory-consuming, and second, it makes it quite difficult to introduce the given abbreviations. So, we define our own rules for the time derivative d[..., t], and for the position vector R[t] and its length r[t]. (We use capital and small letters for the vector and its length.) Because R[t] is not an explicit list (meaning it does not have the head List), we use dot instead of Dot for the scalar product. Here are all relevant definitions for the time evolution of R[t].

```
In[1]:= (* scalar product is commutative *)
        SetAttributes[dot, Orderless]

        (* linearity of d *)
        d[sum_Plus, t] := d[#, t]& /@ sum

        (* Leibniz rule for differentiation of products *)
        d[product_Times, t] := Plus @@
        MapIndexed[Function[{arg, pos}, MapAt[d[#, t]&, arg, pos[[1]]]],
                   Array[product&, Length[product]]]

        (* time derivative of a constant vanishes *)
        d[_?(FreeQ[#, r[t], {0, Infinity}] &&
            FreeQ[#, R[t], {0, Infinity}] &), t] = 0;

        (* time derivative of r[t] via explicit differentiation *)
        d[r[t]^n_., t] := n r[t]^(n - 1) dot[R[t], d[R[t], t]]/r[t]

        (* time derivative of scalar product *)
        d[dot[a_, b_], t] := dot[d[a, t], b] + dot[a, d[b, t]]

        (* time derivative of powers of scalar products *)
        d[dot[R[t], d[R[t], t]]^n_, t] :=
          n dot[R[t], d[R[t], t]]^(n - 1) d[dot[R[t], d[R[t], t]], t]

        (* equation of motion for d[d[R[t], t], t] *)
        d[d[R[t], t], t] = -R[t]/r[t]^3;

        (* pulling out nonvectors from the scalar product *)
        dot[a_?(FreeQ[#, R[t]] && FreeQ[#, d[R[t], t]]&) b_, c_] := a dot[b, c]

        (* definition of r[t] through scalar product of R[t] *)
        dot[R[t], R[t]] = r[t]^2;
```

The most important ingredient in the last definition is the use of the equation of motion. This allows us to reduce all higher-order time derivatives as linear combinations of the two vectors R[0] and R'[0].

The derivatives needed in $r(t) = r(0) + r'(0) + 1/2! \, r''(0) \, t^2 + 1/3! \, r'''(0) \, t^3 + \cdots$ are calculated recursively.

```
In[21]:= term[0] = R[t];
         term[n_] := term[n] = Expand[d[term[n - 1], t]];
```

The command shorten introduces the abbreviations given above. shorten[term[i]] gives a list with the coefficients of r_0 and v_0 in term[i]. (Because the lowest-order terms are not sums, we give their short forms separately.)

```
In[23]:= (* initial terms *)
         shorten[term[0]] = { 1, 0};
         shorten[term[1]] = { 0, 1};
         shorten[term[2]] = {-u, 0};
```

```
shorten[term_] := shorten[term] =
(* collect wrt R[t] and d[R[t], t]; factor coefficients *)
{Factor[Plus @@ (Cases[#, R[t] _.]/R[t])],
 Factor[Plus @@ (Cases[#, d[R[t], t] _.]/d[R[t], t])]}&[
(* introduce shortcuts *)
(term //. {dot[d[R[t], t], R[t]] -> s r[t]^2,
           dot[d[R[t], t], d[R[t], t]] -> w r[t]^2}) //.
                {r[t]^n_ -> u^(-n/3)}]
```

Now, we have everything together to define and to test the resulting series.

```
In[28]:= newtonSeries[ord_] := {r0, v0}.
            Table[Sum[shorten[term[i]][[1]] t^i/i!, {i, 0, ord}], {i, 2}]
```

Here are the first few terms of the resulting series expansion.

```
In[29]:= newtonSeries[5]
```

$$\text{Out[29]= } r0 \left(1 - \frac{t^2 u}{2} + \frac{1}{2} s t^3 u + \frac{1}{8} s t^5 u (7 s^2 + 2 u - 3 w) - \frac{1}{24} t^4 u (15 s^2 + 2 u - 3 w)\right) +$$
$$v0 \left(1 - \frac{t^2 u}{2} + \frac{1}{2} s t^3 u + \frac{1}{8} s t^5 u (7 s^2 + 2 u - 3 w) - \frac{1}{24} t^4 u (15 s^2 + 2 u - 3 w)\right)$$

Much larger expansions can be calculated quite quickly too.

```
In[30]:= Short[newtonSeries[15], 8] // Timing
```

$$\text{Out[30]= } \left\{1.48 \text{ Second, } r0 \left(1 - \frac{t^2 u}{2} + \frac{1}{2} s t^3 u + \ll11\gg + \right.\right.$$
$$\frac{s \ll2\gg (\ll1\gg)}{13305600} - \frac{t^{12} u (13749310575 s^{10} + \ll35\gg + 9450 u^{\ll1\gg} w^4 r[t])}{479001600} +$$
$$\frac{s t^{13} u (35137127025 s^{10} + 56015709750 s^8 u + \ll35\gg + 548100 u^{4/3} w^4 r[t])}{691891200} + \frac{1}{435891456000}$$
$$(s t^{15} u (71152682225625 s^{12} + 137034795397500 s^{10} u + 87316737725700 s^8 u^2 + \ll47\gg +$$
$$13314596400 s^2 u^{4/3} w^4 r[t] + 1387649880 u^{7/3} w^4 r[t] - 390152700 u^{4/3} w^5 r[t])) -$$
$$\frac{1}{87178291200} (t^{14} u (7905853580625 s^{12} + 13914302301900 s^{10} u + 8047182795750 s^8 u^2 +$$
$$1864391484960 s^6 u^3 + 192016033056 s^4 u^4 + 10842907680 s^2 u^5 + \ll43\gg +$$
$$531789300 s^2 u^{4/3} w^4 r[t] + 8138340 u^{7/3} w^4 r[t] - 3231900 u^{4/3} w^5 r[t]))\right) +$$
$$\left.\left. v0 \left(1 - \frac{\ll1\gg u}{2} + \ll16\gg + \frac{\ll1\gg}{\ll11\gg 0} - \frac{t^{14} u (\ll1\gg)}{87178291200}\right)\right)\right\}$$

For a similar expansion for a quantum-mechanical problem, see [784].

c) This is the potential function $V(x, y)$.

```
In[1]:= x1 = -μ; x2 = 1 - μ;
```

```
In[2]:= V[x, y] =
        With[{r1 = Sqrt[(x - x1)^2 + y^2], r2 = Sqrt[(x - x2)^2 + y^2]},
             -(1/2 (x^2 + y^2) + (1 - μ)/r1 + μ/r2)];
```

The two equations to be solved (meaning equilibrium points) are the following.

```
In[3]:= eqs1 = {D[V[x, y], x], D[V[x, y], y]}
```

$$\text{Out[3]= } \left\{-x + \frac{μ (-1 + x + μ)}{(y^2 + (-1 + x + μ)^2)^{3/2}} + \frac{(1 - μ) (x + μ)}{(y^2 + (x + μ)^2)^{3/2}}, -y + \frac{y μ}{(y^2 + (-1 + x + μ)^2)^{3/2}} + \frac{y (1 - μ)}{(y^2 + (x + μ)^2)^{3/2}}\right\}$$

Solving the last two equations directly fails in a reasonable amount of time. So, we will solve the two equations "manually".

```
In[4]:= TimeConstrained[#[Thread[eqs1 == {0, 0}], {x, y}], 100]& /@
                                    {Solve, Reduce}
```

```
Out[4]= {$Aborted, $Aborted}
```

To avoid the presence of radicals, we first rewrite the two equations in a polynomial form by introducing two auxiliary variables R1 and R2.

```
In[5]:= eqs2 = Factor[Numerator[Together[eqs1]]] /.
               {Sqrt[(x - x2)^2 + y^2] -> R2, Sqrt[(x - x1)^2 + y^2] -> R1}
```

Out[5]= $\{R2\,x - 2\,R2\,x^2 + R2\,x^3 - R1\,R2\,x^3 + 2\,R1\,R2\,x^4 - R1\,R2\,x^5 + R2\,x\,y^2 - R1\,R2\,x\,y^2 +$
$2\,R1\,R2\,x^2\,y^2 - 2\,R1\,R2\,x^3\,y^2 - R1\,R2\,x\,y^4 + R2\,\mu - 5\,R2\,x\,\mu - R1\,x^2\,\mu + 5\,R2\,x^2\,\mu - 2\,R1\,R2\,x^2\,\mu +$
$R1\,x^3\,\mu - R2\,x^3\,\mu + 6\,R1\,R2\,x^3\,\mu - 4\,R1\,R2\,x^4\,\mu - R1\,y^2\,\mu + R2\,y^2\,\mu + R1\,x\,y^2\,\mu - R2\,x\,y^2\,\mu +$
$2\,R1\,R2\,x\,y^2\,\mu - 4\,R1\,R2\,x^2\,y^2\,\mu - 3\,R2\,\mu^2 - 2\,R1\,x\,\mu^2 + 7\,R2\,x\,\mu^2 - R1\,R2\,x\,\mu^2 + 3\,R1\,x^2\,\mu^2 -$
$3\,R2\,x^2\,\mu^2 + 6\,R1\,R2\,x^2\,\mu^2 - 6\,R1\,R2\,x^3\,\mu^2 + R1\,y^2\,\mu^2 - R2\,y^2\,\mu^2 - 2\,R1\,R2\,x\,y^2\,\mu^2 - R1\,\mu^3 +$
$3\,R2\,\mu^3 + 3\,R1\,x\,\mu^3 - 3\,R2\,x\,\mu^3 + 2\,R1\,R2\,x\,\mu^3 - 4\,R1\,R2\,x^2\,\mu^3 + R1\,\mu^4 - R2\,\mu^4 - R1\,R2\,x\,\mu^4,\ -y$
$(-R2 + 2\,R2\,x - R2\,x^2 + R1\,R2\,x^2 - 2\,R1\,R2\,x^3 + R1\,R2\,x^4 - R2\,y^2 + R1\,R2\,y^2 - 2\,R1\,R2\,x\,y^2 + 2\,R1\,R2\,x^2\,y^2 +$
$R1\,R2\,y^4 + 3\,R2\,\mu - 4\,R2\,x\,\mu + 2\,R1\,R2\,x\,\mu - R1\,x^2\,\mu + R2\,x^2\,\mu - 6\,R1\,R2\,x^2\,\mu + 4\,R1\,R2\,x^3\,\mu -$
$R1\,y^2\,\mu + R2\,y^2\,\mu - 2\,R1\,R2\,y^2\,\mu + 4\,R1\,R2\,x\,y^2\,\mu - 3\,R2\,\mu^2 + R1\,R2\,\mu^2 - 2\,R1\,x\,\mu^2 + 2\,R2\,x\,\mu^2 -$
$6\,R1\,R2\,x\,\mu^2 + 6\,R1\,R2\,x^2\,\mu^2 + 2\,R1\,R2\,y^2\,\mu^2 - R1\,\mu^3 + R2\,\mu^3 - 2\,R1\,R2\,\mu^3 + 4\,R1\,R2\,x\,\mu^3 + R1\,R2\,\mu^4)\}$

The second equation of eqs2 shows that we can split the problem into the case $y = 0$, and into the case in which the more complicated term in the second element of eqs2 is equal to zero. Let us start with the $y = 0$ case. Eliminating R1 and R2 gives the following polynomial system.

```
In[6]:= gb1 = GroebnerBasis[{eqs2[[1]], y, (x - x2)^2 + y^2 - R2^2,
                                            (x - x1)^2 + y^2 - R1^2},
                            {x, y}, {R1, R2},
                            MonomialOrder -> EliminationOrder,
                            ParameterVariables -> {μ},
                            CoefficientDomain -> Rationals] // Factor
```

Out[6]= $\{y,\ (-1 + x + \mu)\,(x + \mu)\,(1 - x + x^2 - \mu + 2\,x\,\mu + \mu^2)$
$(1 - x - x^2 + x^3 - 2\,\mu - x\,\mu + 2\,x^2\,\mu + x\,\mu^2)\,(-1 + 2\,x - x^2 + x^3 - 2\,x^4 + x^5 + 3\,\mu - 4\,x\,\mu +$
$2\,x^2\,\mu - 6\,x^3\,\mu + 4\,x^4\,\mu - 3\,\mu^2 + x\,\mu^2 - 6\,x^2\,\mu^2 + 6\,x^3\,\mu^2 - 2\,x\,\mu^3 + 4\,x^2\,\mu^3 + x\,\mu^4)$
$(1 - 2\,x + x^2 + x^3 - 2\,x^4 + x^5 - 3\,\mu + 4\,x\,\mu + 2\,x^2\,\mu - 6\,x^3\,\mu + 4\,x^4\,\mu + 3\,\mu^2 + x\,\mu^2 - 6\,x^2\,\mu^2 +$
$6\,x^3\,\mu^2 - 2\,x\,\mu^3 + 4\,x^2\,\mu^3 + x\,\mu^4)\,(-1 + 2\,x - x^2 + x^3 - 2\,x^4 + x^5 + 3\,\mu - 4\,x\,\mu + 4\,x^2\,\mu -$
$6\,x^3\,\mu + 4\,x^4\,\mu - 3\,\mu^2 + 5\,x\,\mu^2 - 6\,x^2\,\mu^2 + 6\,x^3\,\mu^2 + 2\,\mu^3 - 2\,x\,\mu^3 + 4\,x^2\,\mu^3 + x\,\mu^4)\}$

Solving this system results in x-coordinates that are solutions of fifth-order equations.

```
In[7]:= (* do not give radicals in the solution—
         Root-objects are shorter *)
        SetOptions[Roots, Cubics -> False, Quartics -> False];

        sol1 = Union[Append[#, y -> 0]& /@
                        (ToRules[Roots[# == 0, x]]& /@ Rest[gb1])];

        Short[sol1, 6]
```

Out[11]//Short= $\{\{x \to \frac{1}{2}\,(1 - i\,\sqrt{3} - 2\,\mu),\ y \to 0\},\ \{x \to \frac{1}{2}\,(1 + i\,\sqrt{3} - 2\,\mu),\ y \to 0\},\ \{x \to 1 - \mu,\ y \to 0\},\ \ll17\gg,$
$\{x \to \text{Root}[-1 + 3\,\mu - 3\,\mu^2 + 2\,\mu^3 + 2\,\#1 - 4\,\mu\,\#1 + 5\,\mu^2\,\#1 - 2\,\mu^3\,\#1 + \mu^4\,\#1 - \#1^2 + 4\,\mu\,\#1^2 -$
$6\,\mu^2\,\#1^2 + 4\,\mu^3\,\#1^2 + \#1^3 - 6\,\mu\,\#1^3 + 6\,\mu^2\,\#1^3 - 2\,\#1^4 + 4\,\mu\,\#1^4 + \#1^5\ \&,\ 4],\ y \to 0\},$
$\{x \to \text{Root}[-1 + 3\,\mu - 3\,\mu^2 + 2\,\mu^3 + 2\,\#1 - 4\,\mu\,\#1 + 5\,\mu^2\,\#1 - 2\,\mu^3\,\#1 + \mu^4\,\#1 - \#1^2 + 4\,\mu\,\#1^2 -$
$6\,\mu^2\,\#1^2 + 4\,\mu^3\,\#1^2 + \#1^3 - 6\,\mu\,\#1^3 + 6\,\mu^2\,\#1^3 - 2\,\#1^4 + 4\,\mu\,\#1^4 + \#1^5\ \&,\ 5],\ y \to 0\}\}$

The second case results in a more complicated set of equations after the elimination of R1 and R2.

```
In[12]:= gb2 = GroebnerBasis[{eqs2[[1]], eqs2[[2, -1]],
                             (x - x2)^2 + y^2 - R2^2, (x - x1)^2 + y^2 - R1^2},
                             {x, y}, {R1, R2}, MonomialOrder -> EliminationOrder,
                             ParameterVariables -> {μ},
                             CoefficientDomain -> Rationals] // Factor;

         Short[gb2, 8]
```

Out[14]//Short= $\{(-1 + \mu)\,\mu\,(-1 + 2\,x + 2\,\mu)$
$(-7 + 7\,x - 7\,x^2 - 3\,y^2 - 48\,x\,y^2 + 48\,x^2\,y^2 + 24\,y^4 + 7\,\mu - 14\,x\,\mu - 48\,y^2\,\mu + 96\,x\,y^2\,\mu - 7\,\mu^2 + 48\,y^2\,\mu^2),$
$(-1 + \mu)\,\mu\,(-1 + 2\,x + 2\,\mu)\,(5 - 13\,x + 21\,x^2 - 16\,x^3 + 8\,x^4 + 9\,y^2 - 13\,\mu +$
$42\,x\,\mu - 48\,x^2\,\mu + 32\,x^3\,\mu + 21\,\mu^2 - 48\,x\,\mu^2 + 48\,x^2\,\mu^2 - 16\,\mu^3 + 32\,x\,\mu^3 + 8\,\mu^4),$
$(-1 + \mu)\,\mu\,(\ll1\gg)\,(\ll1\gg),\ -4608 + 18432\,x - 27648\,x^2 + \ll169\gg + 9031680\,x\,\mu^9 + 1999872\,\mu^{10},$
$-(-1 + \mu)\,\mu\,(1708 - 7784\,x + 13860\,x^2 - 12152\,x^3 + 6076\,x^4 + 3939\,y^2 - 11328\,x\,y^2 + 11328\,x^2\,y^2 +$
$3504\,y^4 - 6240\,x\,y^4 + 6240\,x^2\,y^4 - 1440\,y^6 - 2880\,y^8 - 2304\,y^{10} - 7784\,\mu + 27720\,x\,\mu -$
$36456\,x^2\,\mu + 24304\,x^3\,\mu - 11328\,y^2\,\mu + 22656\,x\,y^2\,\mu - 6240\,y^4\,\mu + 12480\,x\,y^4\,\mu + 13860\,\mu^2 -$
$36456\,x\,\mu^2 + 36456\,x^2\,\mu^2 + 11328\,y^2\,\mu^2 + 6240\,y^4\,\mu^2 - 12152\,\mu^3 + 24304\,x\,\mu^3 + 6076\,\mu^4)\}$

Rewriting gb2 in lexicographic order shows more clearly the nature of the solutions.

In[15]:= `gb3 = GroebnerBasis[gb2, {x, y}, ParameterVariables -> {μ}] // Factor;`

`Short[gb3, 8]`

Out[16]//Short= $\{y^2 (-3 + 4 y^2) (1 + 4 y^2)^3 (21 + 12 y^2 + 16 y^4) (21 + 24 y^2 + 16 y^4) (-1 + μ) μ,$

$y^2 (1 + 4 y^2) (21 + 12 \ll 1 \gg + 16 y^4) \ll 1 \gg μ (-1 + 2 x + 2 μ), \ll 1 \gg,$

$-16777662720 + 67110650880 x - 100665976320 x^2 + 67110650880 x^3 - 16777662720 x^4 +$

$16777662720 x^6 - 67110650880 x^7 + 100665976320 x^8 - 67110650880 x^9 + 16777662720 x^{10} -$

$33555325440 y^2 + 67110650880 x y^2 - 33555325440 x^2 y^2 + 50332988160 x^4 y^2 -$

$201331952640 x^5 y^2 + \ll 176 \gg + 4680967898880 x μ^8 - 8030845467141 y^2 μ^8 -$

$133040302378416 y^4 μ^8 - 206881083710496 y^6 μ^8 + 6186983622336 y^8 μ^8 +$

$284442471251712 y^{10} μ^8 + 474522386614272 y^{12} μ^8 + 315544814714880 y^{14} μ^8 +$

$138124091179008 y^{16} μ^8 + 4613857248000 μ^9 - 1040215088640 x μ^9 - 922771449600 μ^{10}\}$

The first two equations determine the possible y-values.

In[17]:= `Prepend[Union[{(ToRules[Roots[# == 0, y]] /@`
          `Union[Cases[gb3, _Plus?(FreeQ[#, x]&), Infinity]])], {y -> 0}]`

Out[17]= $\{\{y \to 0\}, \{y \to -\frac{i}{2}\}, \{y \to \frac{i}{2}\}, \{y \to -\frac{\sqrt{3}}{2}\}, \{y \to \frac{\sqrt{3}}{2}\}, \{y \to -\sqrt{-\frac{3}{4} - \frac{i\sqrt{3}}{2}}\},$

$\{y \to \sqrt{-\frac{3}{4} - \frac{i\sqrt{3}}{2}}\}, \{y \to -\sqrt{-\frac{3}{4} + \frac{i\sqrt{3}}{2}}\}, \{y \to \sqrt{-\frac{3}{4} + \frac{i\sqrt{3}}{2}}\}, \{y \to -\sqrt{-\frac{3}{8} - \frac{5i\sqrt{3}}{8}}\},$

$\{y \to \sqrt{-\frac{3}{8} - \frac{5i\sqrt{3}}{8}}\}, \{y \to -\sqrt{-\frac{3}{8} + \frac{5i\sqrt{3}}{8}}\}, \{y \to \sqrt{-\frac{3}{8} + \frac{5i\sqrt{3}}{8}}\}\}$

Here is the whole solution set of gb3.

In[18]:= `sol2 = Solve[# == 0& /@ gb3, {x, y}] // Union;`

`Short[sol2, 6]`

Out[19]//Short= $\{\{x \to \frac{1}{4} (5 - i\sqrt{3} - 4μ), y \to -\sqrt{-\frac{3}{8} + \frac{5i\sqrt{3}}{8}}\}, \{x \to \frac{1}{4} (5 - i\sqrt{3} - 4μ), y \to \sqrt{-\frac{3}{8} + \frac{5i\sqrt{3}}{8}}\},$

$\{x \to \frac{1}{4} (5 + i\sqrt{3} - 4μ), y \to -\sqrt{-\frac{3}{8} - \frac{5i\sqrt{3}}{8}}\}, \{x \to \frac{1}{4} (5 + i\sqrt{3} - 4μ), y \to \sqrt{-\frac{3}{8} - \frac{5i\sqrt{3}}{8}}\},$

$\{x \to \frac{1}{2} (1 - 2μ), y \to -\frac{i}{2}\}, \{x \to \frac{1}{2} (1 - 2μ), y \to \frac{i}{2}\}, \{x \to \frac{1}{2} (1 - 2μ), y \to -\frac{\sqrt{3}}{2}\},$

$\{x \to \frac{1}{2} (1 - 2μ), y \to \frac{\sqrt{3}}{2}\}, \{\ll 1 \gg\}, \ll 1 \gg, \{x \to \frac{1}{2} (1 - 2μ), y \to -\sqrt{\ll 1 \gg}\},$

$\{x \to \frac{1}{2} (1 - 2μ), y \to \sqrt{-\frac{3}{4} + \frac{i\sqrt{3}}{2}}\}, \{x \to \frac{1}{2} (1 - i\sqrt{3} - 2μ), y \to 0\},$

$\{x \to \frac{1}{2} (1 + i\sqrt{3} - 2μ), y \to 0\}, \{x \to -\frac{1}{4} i (-i + \sqrt{3} - 4iμ), y \to -\sqrt{-\frac{3}{8} - \frac{5i\sqrt{3}}{8}}\},$

$\{x \to -\frac{1}{4} i (-i + \sqrt{3} - 4iμ), y \to \sqrt{-\frac{3}{8} - \frac{5i\sqrt{3}}{8}}\},$

$\{x \to \frac{1}{4} i (i + \sqrt{3} + 4iμ), y \to -\sqrt{-\frac{3}{8} + \frac{5i\sqrt{3}}{8}}\},$

$\{x \to \frac{1}{4} i (i + \sqrt{3} + 4iμ), y \to \sqrt{-\frac{3}{8} + \frac{5i\sqrt{3}}{8}}\}\}$

Now let us deal with the special case $μ = 1/10$. We first select all real solutions.

In[20]:= `solAll1 = Select[RootReduce[Join[sol1, sol2] /. μ -> 1/10],`
          `(Im[x] == 0 && Im[y] == 0 /. #)&]`

Out[20]= $\left\{\left\{x \to \dfrac{9}{10},\ y \to 0\right\},\ \left\{x \to -\dfrac{1}{10},\ y \to 0\right\},\ \{x \to \text{Root}[80 - 109\ \#1 - 80\ \#1^2 + 100\ \#1^3\ \&,\ 1],\ y \to 0\},\right.$

$\{x \to \text{Root}[80 - 109\ \#1 - 80\ \#1^2 + 100\ \#1^3\ \&,\ 2],\ y \to 0\},$

$\{x \to \text{Root}[80 - 109\ \#1 - 80\ \#1^2 + 100\ \#1^3\ \&,\ 3],\ y \to 0\},$

$\{x \to \text{Root}[-7300 + 16081\ \#1 - 8560\ \#1^2 + 4600\ \#1^3 - 16000\ \#1^4 + 10000\ \#1^5\ \&,\ 1],\ y \to 0\},$

$\{x \to \text{Root}[7300 - 15919\ \#1 + 11440\ \#1^2 + 4600\ \#1^3 - 16000\ \#1^4 + 10000\ \#1^5\ \&,\ 1],\ y \to 0\},$

$\{x \to \text{Root}[-7280 + 16481\ \#1 - 6560\ \#1^2 + 4600\ \#1^3 - 16000\ \#1^4 + 10000\ \#1^5\ \&,\ 1],\ y \to 0\},$

$\left.\left\{x \to \dfrac{2}{5},\ y \to -\dfrac{\sqrt{3}}{2}\right\},\ \left\{x \to \dfrac{2}{5},\ y \to \dfrac{\sqrt{3}}{2}\right\}\right\}$

Because in the process of rewriting a system of equations that contains radicals into a polynomial system some solutions of the polynomial system may not be solutions of the original system containing radicals, we backsubstitute the solutions solAll1 and check if they satisfy the original equations eqs1.

In[21]:= (* to avoid messages due to vanishing denominators *)
Off[Power::infy]; Off[Infinity::indet];

solAll2 = Select[solAll1, (RootReduce[eqs1 /. μ -> 1/10 /. #] === {0, 0})&]

Out[23]= $\left\{\{x \to \text{Root}[-7300 + 16081\ \#1 - 8560\ \#1^2 + 4600\ \#1^3 - 16000\ \#1^4 + 10000\ \#1^5\ \&,\ 1],\ y \to 0\},\right.$

$\{x \to \text{Root}[7300 - 15919\ \#1 + 11440\ \#1^2 + 4600\ \#1^3 - 16000\ \#1^4 + 10000\ \#1^5\ \&,\ 1],\ y \to 0\},$

$\{x \to \text{Root}[-7280 + 16481\ \#1 - 6560\ \#1^2 + 4600\ \#1^3 - 16000\ \#1^4 + 10000\ \#1^5\ \&,\ 1],\ y \to 0\},$

$\left.\left\{x \to \dfrac{2}{5},\ y \to -\dfrac{\sqrt{3}}{2}\right\},\ \left\{x \to \dfrac{2}{5},\ y \to \dfrac{\sqrt{3}}{2}\right\}\right\}$

The so-obtained solutions are exactly the same as the ones produced by a direct call to Solve with specified parameters. (Using Solve on eqs1 for an explicitly specified μ is quite fast.)

In[24]:= Solve[# == 0& /@ (eqs1 /. μ -> 1/10), {x, y}]

Out[24]= $\left\{\{y \to 0,\ x \to \text{Root}[-7300 + 16081\ \#1 - 8560\ \#1^2 + 4600\ \#1^3 - 16000\ \#1^4 + 10000\ \#1^5\ \&,\ 1]\},\right.$

$\{y \to 0,\ x \to \text{Root}[-7280 + 16481\ \#1 - 6560\ \#1^2 + 4600\ \#1^3 - 16000\ \#1^4 + 10000\ \#1^5\ \&,\ 1]\},$

$\{y \to 0,\ x \to \text{Root}[7300 - 15919\ \#1 + 11440\ \#1^2 + 4600\ \#1^3 - 16000\ \#1^4 + 10000\ \#1^5\ \&,\ 1]\},$

$\left.\left\{y \to -\dfrac{\sqrt{3}}{2},\ x \to \dfrac{2}{5}\right\},\ \left\{y \to \dfrac{\sqrt{3}}{2},\ x \to \dfrac{2}{5}\right\}\right\}$

The following picture shows the location of the five Lagrange points on the surface $V(x, y)$.

In[25]:= Show[Graphics3D[{
        {EdgeForm[], SurfaceColor[Hue[0.3]],
        Cases[Graphics3D[Plot3D[Evaluate[V[x, y] /. μ -> 1/10],
                            {x, -3/2, 3/2}, {y, -1.2, 1.2},
                            PlotPoints -> {120, 80},
                            DisplayFunction -> Identity]],
                            _Polygon, Infinity]},
        {PointSize[0.02], Hue[0],
        Point[{x, y, 0.05 + V[x, y] /. μ -> 1/20} /. #]& /@ N[%]}}],
        BoxRatios -> {3/2, 1.2, 1/2}]

25. Algebraic Lissajous Curves, Light Ray Reflection Inside a Closed Region

a) In analogy with the circle, we look for an algebraic function of the following form: $\sum_{i,j=0}^{n} a_{i,j}\, x^i\, y^j$.

For the curves of interest here, we calculate the coefficients a_{ij} in a straightforward brute force way. By converting the trigonometric functions into exponentials and expanding the resulting expressions, we get a system of linear equations for the coefficients of exp($i\,k\,t$), which we solve and extract the interesting part from it. Here, this method is implemented. (For the general case, this method should (and can) be refined to shorten the calculation time.)

```
In[1]:= algebraizise[{x_, y_}, t_, maxPower_, {X_, Y_}] :=
         Module[{repRules, x1, y1, sum, terms, poly, vars, c1, c2, sol, facs, a},
           (* trigs to exponentials *)
           repRules = {Sin[a_] -> (((Exp[I a] - Exp[-I a])/(2I))),
                       Cos[a_] -> (((Exp[I a] + Exp[-I a])/(2 )))};
           {x1, y1} = {x, y} //. repRules;
           (* the ansatz *)
           sum = Sum[a[i, j] x1^i y1^j, {i, 0, maxPower}, {j, 0, maxPower}];
           (* write as polynomial in Exp[I t] *)
           terms = Collect[sum, Exp[I t]];
           (* change Exp[i t] to new variable *)
           poly = terms //. Exp[u_] -> C[u];
           (* all powers present *)
           vars = Cases[poly, _C, {0, Infinity}];
           (* determine the coefficients *)
           c1 = Coefficient[poly, #]& /@ vars;
           (* take care of the term Exp[0 I t] *)
           c2 = Append[c1, Expand[poly - c1.vars]];
           (* solve the linear system *)
           sol = Solve[# == 0& /@ c2, Flatten[
                       Table[a[i, j], {i, 0, maxPower}, {j, 0, maxPower}]]];
           (* factor the result after backsubstitution *)
           facs = Factor[Sum[a[i, j] X^i Y^j,
                             {i, 0, maxPower}, {j, 0, maxPower}] //. sol][[1]];
           (* extract the interesting piece *)
           Times @@ Cases[facs, _Plus?(FreeQ[#, _a, {0, Infinity}]&)]]
```

Here is an example.

```
In[2]:= algebraizise[{Sin[3θ], Cos[2θ]}, 0, 4, {x, y}]

        Solve::svars : Equations may not give solutions for all "solve" variables. More…

Out[2]= -1 + 2 x^2 - 3 y + 4 y^3
```

The disadvantage of the last implementation is the fact that we explicitly have to specify the maximal degree of the linear system. A better way is to polynomialize the equations and then eliminate the auxiliary variable using Resultant.

```
In[3]:= algebraizise[{x_, y_}, t_, {X_, Y_}] :=
         Module[{aux1, aux2, aux3, aux4, s},
           (* transform trigs of multiple arguments to simple arguments *)
           aux1 = {x, y} /.
           {Cos[n_Integer t] :> Sum[(-1)^k Binomial[n, 2k] *
                                    Cos[t]^(n - 2k) Sin[t]^(2k), {k, 0, n/2}],
            Sin[n_Integer t] :> Sum[(-1)^(k) Binomial[n, 2 k + 1] *
                                    Sin[t]^(2k + 1) Cos[t]^(n - 2k - 1),
                                    {k, 0, (n - 1)/2}]};
           (* use rational parametrization for trigs *)
           aux2 = ({X, Y} - aux1) /. {Cos[t] -> (1 - s^2)/(1 + s^2),
                                      Sin[t] -> 2s/(1 + s^2)};
           (* make polynomial equations *)
           aux3 = Numerator[Together[aux2]];
           (* eliminate parametrization variable s *)
           aux4 = Factor[Resultant[##, s]& @@ aux3];
           (* if present - delete X and Y independent prefactor *)
           If[Head[aux4] === Times, DeleteCases[aux4, _?IntegerQ], aux4]]
```

As a first test, we treat the circle.

```
In[4]:= algebraizise[{Cos[t], Sin[t]}, t, {x, y}]

Out[4]= -1 + x^2 + y^2
```

Here are the two Lissajous curves of interest.

```
In[5]:= algebraizise[{Cos[t], Sin[2t]}, t, {x, y}]
```

Out[5]= $-4 x^2 + 4 x^4 + y^2$

In[6]:= `algebraizise[{Cos[3t], Sin[2t]}, t, {x, y}]`

Out[6]= $-4 x^2 + 4 x^4 + 9 y^2 - 24 y^4 + 16 y^6$

For the second one, we compare the contour plot of the implicit representation with the plot of the parametric representation.

In[7]:= `Show[GraphicsArray[{`
`      (* the parametrized form *)`
`      ParametricPlot[{Cos[3t], Sin[2t]}, {t, 0, 2Pi}, Frame -> True,`
`                    Axes -> False, DisplayFunction -> Identity,`
`                    AspectRatio -> 1],`
`      (* the implicit form *)`
`      ContourPlot[4 x^4 - 4 x^2 + 9 y^2 - 24 y^4 + 16 y^6,`
`                  {x, -1.1, 1.1}, {y, -1.1, 1.1},`
`                  Contours -> {0}, PlotPoints -> 160,`
`                  DisplayFunction -> Identity, AspectRatio -> 1]}]]`

For some other, more effective methods of how to implicitize parametrically given curves, see [414], [872], [1246], and [748].

b) This is the polynomial under consideration.

In[1]:= `lissa = (9 x^2 - 24 x^4 + 16 x^6 - 4 y^2 + 4 y^4)^2 - 1/2;`

To get an idea about the shape of the region under consideration, we use `ContourPlot`. The black part is the region to be described. Because of the obvious symmetry of `lissa` with respect to the x-axis and y-axis, we restrict our investigation to the region $x \geq 0$, $y \geq 0$.

In[2]:= `sketch = ContourPlot[Evaluate[lissa], {x, 0, 1.1}, {y, 0, 1.1},`
`                    PlotPoints -> 60, ContourShading -> True,`
`                    Contours -> {0}]`

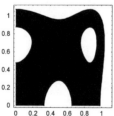

We divide the region under interest in vertical stripes $x_l < x < x_u$, in such a manner that in each stripe, $y_l(x)$ and $y_u(x)$ are real analytic functions. In principle, the $y_l(x)$ and $y_u(x)$ may change their form (form = the branch of the roots of `lissa`) if a beginning or end of a hole or extremum is traversed. Analytically, this means $y'(x) = \infty$. The points x_l and x_u are so described by the following equation.

In[3]:= `res = Resultant[lissa, D[lissa, y], y] // Factor`

Out[3]= $562949953421312 (1 - 36 x^2 + 258 x^4 - 928 x^6 + 1728 x^8 - 1536 x^{10} + 512 x^{12})^2$
$(-1 + 162 x^4 - 864 x^6 + 1728 x^8 - 1536 x^{10} + 512 x^{12})$

Here, these points are explicitly calculated.

In[4]:= **xPoints = Simplify @ Sort[Cases[Union[x /. Solve[res == 0, x]],**
(* or use Roots with Root-solutions to
avoid numerical selection *)
_?((Im[Chop[N[#]]] == 0 && Re[Chop[N[#]]] > 0)&)],
Re[N[#1]] < Re[N[#2]]&]

Out[4]= $\left\{ \dfrac{1}{2} \sqrt{\dfrac{-1 - i\sqrt{3} + 4\left(1 - \sqrt{2} + i\sqrt{2}\left(-1+\sqrt{2}\right)\right)^{1/3} + i\left(i+\sqrt{3}\right)\left(1-\sqrt{2}+i\sqrt{2}\left(-1+\sqrt{2}\right)\right)^{2/3}}{2\left(1-\sqrt{2}+i\sqrt{2}\left(-1+\sqrt{2}\right)\right)^{1/3}}} \right.,$

$\dfrac{1}{2}\sqrt{-\dfrac{\left(-1+\left(1-\sqrt{2}+i\sqrt{2}\left(-1+\sqrt{2}\right)\right)^{1/3}\right)^2}{\left(1-\sqrt{2}+i\sqrt{2}\left(-1+\sqrt{2}\right)\right)^{1/3}}},$

$\dfrac{1}{2}\sqrt{\dfrac{1 - i\sqrt{3} + 4\left(1 - \sqrt{2} + i\sqrt{2}\left(-1+\sqrt{2}\right)\right)^{1/3} + \left(1+i\sqrt{3}\right)\left(1-\sqrt{2}+i\sqrt{2}\left(-1+\sqrt{2}\right)\right)^{2/3}}{2\left(1-\sqrt{2}+i\sqrt{2}\left(-1+\sqrt{2}\right)\right)^{1/3}}},$

$\dfrac{1}{2}\sqrt{\dfrac{-1 + i\sqrt{3} + 4\left(1 - \sqrt{2} + i\sqrt{2}\left(-1+\sqrt{2}\right)\right)^{1/3} + \left(-1-i\sqrt{3}\right)\left(1-\sqrt{2}+i\sqrt{2}\left(-1+\sqrt{2}\right)\right)^{2/3}}{2\left(1-\sqrt{2}+i\sqrt{2}\left(-1+\sqrt{2}\right)\right)^{1/3}}},$

$\dfrac{1}{2}\sqrt{2 + \dfrac{1}{\left(1-\sqrt{2}+i\sqrt{2}\left(-1+\sqrt{2}\right)\right)^{1/3}} + \left(1-\sqrt{2}+i\sqrt{2}\left(-1+\sqrt{2}\right)\right)^{1/3}},$

$\dfrac{1}{2}\sqrt{\dfrac{1 + i\sqrt{3} + 4\left(1 - \sqrt{2} + i\sqrt{2}\left(-1+\sqrt{2}\right)\right)^{1/3} + \left(1-i\sqrt{3}\right)\left(1-\sqrt{2}+i\sqrt{2}\left(-1+\sqrt{2}\right)\right)^{2/3}}{2\left(1-\sqrt{2}+i\sqrt{2}\left(-1+\sqrt{2}\right)\right)^{1/3}}},$

$\left.\dfrac{1}{2}\sqrt{2 + \dfrac{1}{\left(1+\sqrt{2}-\sqrt{2}\left(1+\sqrt{2}\right)\right)^{1/3}} + \left(1+\sqrt{2}-\sqrt{2}\left(1+\sqrt{2}\right)\right)^{1/3}} \right\}$

Adding the vertical lines corresponding to xPoints to sketch, we clearly see the six resulting stripes.

In[5]:= **Show[{sketch, Graphics[{Hue[0], Line[{{#, 0}, {#, 1.1}}]& /@ xPoints}]}]**

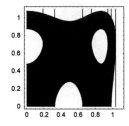

Inside each of the stripes, we need the solutions of lissa with respect to *y*. Because lissa is a quartic polynomial in *y*, we can explicitly solve for *y(x)* in radicals.

In[6]:= **sol = Simplify[Union[Solve[lissa == 0, y]]]**

Out[6]= $\left\{ \left\{y \to -\dfrac{1}{2}\sqrt{2 + \sqrt{2}\sqrt{2-\sqrt{2}} - 18x^2 + 48x^4 - 32x^6}\right\}, \right.$

$\left\{y \to \dfrac{1}{2}\sqrt{2 + \sqrt{2}\sqrt{2-\sqrt{2}} - 18x^2 + 48x^4 - 32x^6}\right\},$

$\left\{y \to -\dfrac{1}{2}\sqrt{2 - \sqrt{2}\sqrt{2-\sqrt{2}} - 18x^2 + 48x^4 - 32x^6}\right\},$

$\left\{y \to \dfrac{1}{2}\sqrt{2 - \sqrt{2}\sqrt{2-\sqrt{2}} - 18x^2 + 48x^4 - 32x^6}\right\},$

$$\left\{ y \to -\frac{1}{2} \sqrt{2 + \sqrt{2} \sqrt{2 + \sqrt{2} - 18\, x^2 + 48\, x^4 - 32\, x^6}} \right\},$$

$$\left\{ y \to \frac{1}{2} \sqrt{2 + \sqrt{2} \sqrt{2 + \sqrt{2} - 18\, x^2 + 48\, x^4 - 32\, x^6}} \right\},$$

$$\left\{ y \to -\frac{1}{2} \sqrt{2 - \sqrt{2} \sqrt{2 + \sqrt{2} - 18\, x^2 + 48\, x^4 - 32\, x^6}} \right\},$$

$$\left\{ y \to \frac{1}{2} \sqrt{2 - \sqrt{2} \sqrt{2 + \sqrt{2} - 18\, x^2 + 48\, x^4 - 32\, x^6}} \right\} \right\}$$

To see which of the last solutions is describing which part of the boundary inside each stripe, we look at the curves described by the solutions.

```
In[7]:= (* avoid messages from spurious numericalization *)
        Off[Plot::plnr];

        Show[GraphicsArray[#]]& /@ Partition[
        Show[{sketch,
              Graphics[{Hue[0], Line[{{#, 0}, {#, 1.1}}]& /@ xPoints}],
              Plot[Evaluate[#[[1, 2]]], {x, 0, 1.2},
                   PlotStyle -> {{RGBColor[0, 1, 0], Thickness[0.02]}},
                   DisplayFunction -> Identity, PlotRange -> {0, 1.1}]},
              DisplayFunction -> Identity, PlotRange -> {0, 1.1}]& /@ sol, 4]
```

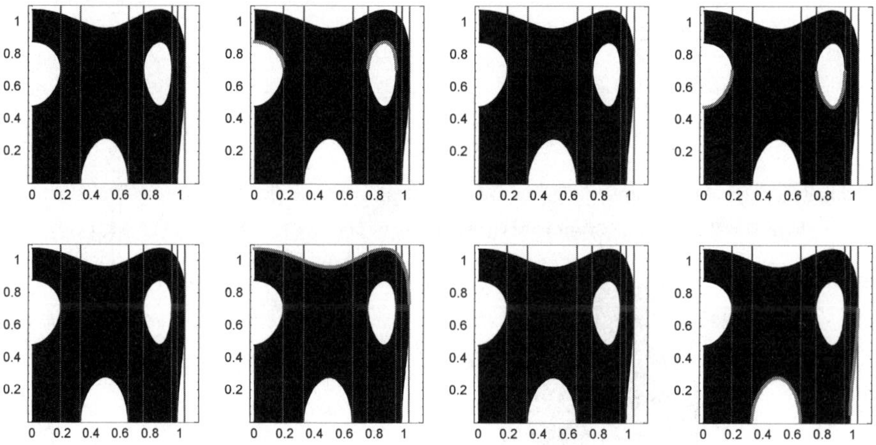

So, we finally have the following x_l, x_u, $y_l(x)$, and $y_u(x)$.

```
In[10]:= u = (1 - Sqrt[2] + I Sqrt[2 Sqrt[2] - 2]);
         v = 2 - 18 x^2 + 48 x^4 - 32 x^6;

         x1 = 0;
         x2 = Sqrt[4 - u^(-1/3) - I Sqrt[3]/u^(1/3) -
                   u^(1/3) + I Sqrt[3] u^(1/3)]/(2 Sqrt[2]);
         x3 = Sqrt[2 - u^(-1/3) - u^(1/3)]/2;
         x4 = Sqrt[4 + u^(-1/3) - I Sqrt[3]/u^(1/3) +
                   u^(1/3) + I Sqrt[3] u^(1/3)]/(2 Sqrt[2]);
         x5 = Sqrt[4 - u^(-1/3) + I Sqrt[3]/u^(1/3) -
                   u^(1/3) - I Sqrt[3] u^(1/3)]/(2 Sqrt[2]);
         x6 = Sqrt[2 + u^(-1/3) + u^(1/3)]/2;
         x7 = Sqrt[4 + u^(-1/3) + I Sqrt[3]/u^(1/3) +
                   u^(1/3) - I Sqrt[3] u^(1/3)]/(2 Sqrt[2]);
         x8 = ((1 + Sqrt[2] + Sqrt[2 + 2 Sqrt[2]])^(-1/6) +
               (1 + Sqrt[2] + Sqrt[2 + 2 Sqrt[2]])^( 1/6))/2;
```

```
y1 = 0;
y2 = Sqrt[2 - Sqrt[2] Sqrt[v - Sqrt[2]]]/2;
y3 = Sqrt[2 + Sqrt[2] Sqrt[v - Sqrt[2]]]/2;
y4 = Sqrt[2 - Sqrt[2] Sqrt[v + Sqrt[2]]]/2;
y5 = Sqrt[2 + Sqrt[2] Sqrt[v + Sqrt[2]]]/2;
```

In terms of $x0$, $x1$, $x2$, $x3$, $x4$, $x5$, $x6$, $x7$, $x8$, $y0$, $y1$, $y2$, $y3$, $y4$, and $y5$, the region `lissa < 0` can be described in the following manner.

```
In[25]:= (* avoid messages from spurious numericalization *)
         Off[Less::"nord"];

In[27]:= negativeRegion =
         x1 < x < x2 && y1 < y < y2 || x1 < x < x2 && y3 < y < y5 ||
         x2 < x < x3 && y1 < y < y5 || x3 < x < x4 && y4 < y < y5 ||
         x4 < x < x5 && y1 < y < y5 || x5 < x < x6 && y1 < y < y2 ||
         x5 < x < x6 && y3 < y < y5 || x6 < x < x7 && y1 < y < y5 ||
         x7 < x < x8 && y4 < y < y5;
```

We suppressed messages from `Less` in the last input. They would have been issued because `Less` uses numerical techniques to establish if a comparison is correct. The numericalization of the nested radical expressions for the xi and yi will yield an imaginary term of the form $0. \times 10^{-exp}\,i$.

```
In[28]:= N[x2, 22]

Out[28]= 0.1894673314863441628283 + 0. × 10⁻²⁴ i
```
$$\text{Out[28]} = 0.1894673314863441628283 + 0.\times 10^{-24}\,i$$

Because *Mathematica* cannot exclude the possibility that the imaginary part is not identical to zero, it issues a message. For the variables of interest xi and yi, we know that they are purely real; so, we can safely suppress this message.

For every `And` of `negativeRegion`, we now form a polygon by discretizing $y_l(x)$ and $y_u(x)$.

```
In[29]:= makePoly[xl_ < x < xu_ && yl_ < y < yu_, pp_] :=
         Polygon[Re @ N @ Join[
                 Table[Evaluate[{x, yl}], {x, xl, xu, (xu - xl)/pp}],
         Reverse[Table[Evaluate[{x, yu}], {x, xl, xu, (xu - xl)/pp}]]]]
```

The next picture shows the nine pieces from `negativeRegion` in different colors.

```
In[30]:= Show[Graphics[{Hue[Random[]], makePoly[#, 30]}& /@
                        (List @@ negativeRegion)],
             AspectRatio -> Automatic, PlotRange -> All]
```

Here, the region `lissa < 0` as a whole is shown.

```
In[31]:= Show[Graphics[{#, (* mirror in other three quadrants *)
                        Map[{+1, -1}#&, #, {-2}],
                        Map[{-1, +1}#&, #, {-2}],
                        Map[{-1, -1}#&, #, {-2}]}&[
             (* original polygons *) Cases[%, _Polygon, Infinity]]],
             AspectRatio -> Automatic, PlotRange -> All]
```

For the solution of similar, but more general problems, see the literature for cylindrical algebraic decomposition, e.g., [260], [1081], [59], [1206], [1225], [90], [91], [242], [89], [1275], [1123], and [941]. For an efficient way to determine the topology of a plane algebraic curve, see [749].

c) There are (at least) two possible ways to solve this problem.

One would be to derive the parametric representation of the evolute and then try to implicitize it. Because this is a difficult problem, we start with the already implicit representation of the Lissajous curve derived in a) and derive the implicit representation of the evolute from it. We carry out this program by eliminating the variables x and y from the equation of the evolute to get an implicit equation for ξ and η.

This is the equation of the Lissajous curve under consideration. (In the following, we write only the left-hand side of the equation, and the right-hand side is always zero.)

In[1]:= f[x_, y_] = 4 x^4 - 4 x^2 + y^2;

The x-coordinate of the evolute is given by the following equation.

In[2]:= eqx = ξ - (x - D[f[x, y], x] (D[f[x, y], x]^2 + D[f[x, y], y]^2)/
       (D[f[x, y], y]^2 D[f[x, y], {x, 2}] -
       2 D[f[x, y], x] D[f[x, y], y] D[D[f[x, y], x], y] +
       D[f[x, y], x]^2 D[f[x, y], {y, 2}]))

Out[2]= $-x + \dfrac{(-8x + 16x^3)\left((-8x + 16x^3)^2 + 4y^2\right)}{2(-8x + 16x^3)^2 + 4(-8 + 48x^2)y^2} + \xi$

Here is the y-component.

In[3]:= eqy = η - (y - D[f[x, y], y] (D[f[x, y], x]^2 + D[f[x, y], y]^2)/
       (D[f[x, y], y]^2 D[f[x, y], {x, 2}] -
       2 D[f[x, y], x] D[f[x, y], y] D[D[f[x, y], x], y] +
       D[f[x, y], x]^2 D[f[x, y], {y, 2}]))

Out[3]= $-y + \dfrac{2y\left((-8x + 16x^3)^2 + 4y^2\right)}{2(-8x + 16x^3)^2 + 4(-8 + 48x^2)y^2} + \eta$

By taking the numerator of the last two equations, we get two polynomials in the four variables x, y, ξ, and η.

In[4]:= numx = Numerator[Together[eqx]] // Factor

Out[4]= $-20 x^3 + 112 x^5 - 208 x^7 + 128 x^9 - 4 x^3 y^2 + 4 x^2 \xi - 16 x^4 \xi + 16 x^6 \xi - y^2 \xi + 6 x^2 y^2 \xi$

In[5]:= numy = Numerator[Together[eqy]] // Factor

Out[5]= $5 y^3 - 24 x^2 y^3 + 16 x^2 \eta - 64 x^4 \eta + 64 x^6 \eta - 4 y^2 \eta + 24 x^2 y^2 \eta$

Now, we eliminate the variable y from the last two equations.

In[6]:= Resultant[numx, f[x, y], y] // Factor

Out[6]= $16 x^6 \left(-5 + 24 x^2 - 48 x^4 + 32 x^6 + 3 x \xi - 2 x^3 \xi\right)^2$

In[7]:= Resultant[numy, f[x, y], y] // Factor

Out[7]= $64 x^6 \left(-25 + 315 x^2 - 1371 x^4 + 2473 x^6 - 1968 x^8 + 576 x^{10} + 36 x^2 \eta^2 - 48 x^4 \eta^2 + 16 x^6 \eta^2\right)$

Finally, from the resulting two equations, we eliminate the variable x and end up with a polynomial in ξ and η.

In[8]:= evolute[ξ_, η_] =
       (Resultant[First[Last[%%]], Last[%], x] // Factor)/24209548902400

Out[8]= $11390625 - 119556000\,\eta^2 + 391474944\,\eta^4 - 435732480\,\eta^6 + 277807104\,\eta^8 - 94371840\,\eta^{10} +$
$16777216\,\eta^{12} - 13638375\,\xi^2 + 14221008\,\eta^2\,\xi^2 - 120144384\,\eta^4\,\xi^2 - 182624256\,\eta^6\,\xi^2 + 120127488\,\eta^8\,\xi^2 -$
$3145728\,\eta^{10}\,\xi^2 + 5828139\,\xi^4 + 44999028\,\eta^2\,\xi^4 + 146914752\,\eta^4\,\xi^4 - 3154944\,\eta^6\,\xi^4 - 245760\,\eta^8\,\xi^4 -$
$1088713\,\xi^6 - 13107936\,\eta^2\,\xi^6 - 3703296\,\eta^4\,\xi^6 - 4096\,\eta^6\,\xi^6 + 92232\,\xi^8 - 200880\,\eta^2\,\xi^8 - 2916\,\xi^{10}$

Here is the plot of the parametric representation and the contour plot of the implicit representation.

```
In[9]:= Show[GraphicsArray[{
        Graphics[{Thickness[0.002], With[{ε = 10^-8},
        (* the parametric plot *)
        Apply[ParametricPlot[Evaluate[
        With[{x = Cos[t], y = Sin[2t]},
        {x - D[y, t] (D[x, t]^2 + D[y, t]^2)/
                     (D[x, t] D[y, {t, 2}] - D[y, t] D[x, {t, 2}]),
         y + D[x, t] (D[x, t]^2 + D[y, t]^2)/
                     (D[x, t] D[y, {t, 2}] - D[y, t] D[x, {t, 2}])}]],
                {t, ##}, DisplayFunction -> Identity][[1]]&, # + {ε, -ε}& /@
                Partition[Table[φ, {φ, 0, 2Pi, Pi/2}], 2, 1], 1]]},
            PlotRange -> {{-3, 3}, {-3/2, 3/2}},
            Frame -> True, AspectRatio -> 1],
        (* the implicit plot *)
        ContourPlot[evolute[ξ, η], {ξ, -3, 3}, {η, -3/2, 3/2},
                    PlotPoints -> 160, Contours -> {0}, AspectRatio -> 1,
                    DisplayFunction -> Identity]}]]
```

d) The main issue in the construction of the orthopodic locus is to determine at which point of the curve the second tangent starts for a given first tangent. So our goal is to derive an equation $h(x_1, x_2)$ that connects the two points $\{x_1, y_1\}$, $\{x_2, y_2\}$ that touch the quartic with each other. We can get such an equation by writing the equations of the curve and the tangents and eliminating all variables except x_1 and x_2.

This is the implicit representation of the curve of interest here. (Again, we only write the left-hand side of the equations; the right-hand side is always zero.)

```
In[1]:= f[x_, y_] = 4 x^4 - 4 x^2 + y^2
```
Out[1]= $-4\,x^2 + 4\,x^4 + y^2$

These are the equations to be fulfilled by the point of the curve where the first and second tangents are attached.

```
In[2]:= eq[1] = f[x1, y1];
        eq[2] = f[x2, y2];
```

The function ys represents the derivative $y'(x)$, derived by implicit differentiation.

```
In[4]:= ys[x_, y_] = -D[f[x, y], x]/D[f[x, y], y]
```
Out[4]= $-\dfrac{-8\,x + 16\,x^3}{2\,y}$

Using ys, the condition that the two tangents intersect perpendicularly can be written as follows.

```
In[5]:= eq[3] = 1 + ys[x1, y1] ys[x2, y2]
```
Out[5]= $1 + \dfrac{(-8\,x1 + 16\,x1^3)\,(-8\,x2 + 16\,x2^3)}{4\,y1\,y2}$

These are the parametric forms of $x_1(t)$, $y_1(t)$, $x_2(s)$ and $y_2(s)$.

```
In[6]:= eq[4] = x - (x1 + t);
        eq[5] = y - (y1 + t ys[x1, y1]);
        eq[6] = x - (x2 + s);
        eq[7] = y - (y2 + s ys[x2, y2])
```

Out[8]= $-s + x - x2$

Out[9]= $y + \dfrac{s\,(-8\,x2 + 16\,x2^3)}{2\,y2} - y2$

Having written down all equations of interest here, we convert them to pure polynomials by taking the numerators.

```
In[10]:= Table[eq1[i] = Numerator[Together[eq[i]]], {i, 7}]
```

Out[10]= $\{-4\,x1^2 + 4\,x1^4 + y1^2,\ -4\,x2^2 + 4\,x2^4 + y2^2,\ 16\,x1\,x2 - 32\,x1^3\,x2 - 32\,x1\,x2^3 + 64\,x1^3\,x2^3 + y1\,y2,$
$-t + x - x1,\ -4\,t\,x1 + 8\,t\,x1^3 + y\,y1 - y1^2,\ -s + x - x2,\ -4\,s\,x2 + 8\,s\,x2^3 + y\,y2 - y2^2\}$

Now, we start to eliminate the variables we are not interested in. This eliminates the parameter t from the parametric representation of the first tangent.

```
In[11]:= eq1[4] = Resultant[eq1[4], eq1[5], t]
```

Out[11]= $-(x - x1)\,(-4\,x1 + 8\,x1^3) - y\,y1 + y1^2$

This eliminates the parameter s from the parametric representation of the second tangent.

```
In[12]:= eq1[5] = Resultant[eq1[6], eq1[7], s]
```

Out[12]= $-(x - x2)\,(-4\,x2 + 8\,x2^3) - y\,y2 + y2^2$

Using the implicit representation of the curve, we eliminate y_1 from the equation describing the condition that the two tangents cross perpendicularly.

```
In[13]:= eq2[1] = (Resultant[eq1[1], eq1[3], y1] /. y2^2 -> 4 x2^2 - 4 x2^4) //
                                                       Expand // Factor
```

Out[13]= $16\,x1^2\,x2^2$
$(15 - 63\,x1^2 + 64\,x1^4 - 63\,x2^2 + 255\,x1^2\,x2^2 - 256\,x1^4\,x2^2 + 64\,x2^4 - 256\,x1^2\,x2^4 + 256\,x1^4\,x2^4)$

This is the equation $h(x_1, x_2)$ we were looking for.

The same process for the second tangent leads to the same equation.

```
In[14]:= eq2[2] = (Resultant[eq1[2], eq1[3], y2] /. y1^2 -> 4 x1^2 - 4 x1^4) //
                                                       Expand // Factor
```

Out[14]= $16\,x1^2\,x2^2$
$(15 - 63\,x1^2 + 64\,x1^4 - 63\,x2^2 + 255\,x1^2\,x2^2 - 256\,x1^4\,x2^2 + 64\,x2^4 - 256\,x1^2\,x2^4 + 256\,x1^4\,x2^4)$

We eliminate y_1 from the implicit representation of the first tangent.

```
In[15]:= eq2[3] = Resultant[eq1[1], eq1[4], y1] // Factor
```

Out[15]= $4\,x1^2\,(4\,x^2 - 16\,x^2\,x1^2 + 8\,x\,x1^3 + 16\,x^2\,x1^4 - 16\,x\,x1^5 + 4\,x1^6 - y^2 + x1^2\,y^2)$

We also eliminate y_2 from the implicit representation of the second tangent.

```
In[16]:= eq2[4] = Resultant[eq1[2], eq1[5], y2] // Factor
```

Out[16]= $4\,x2^2\,(4\,x^2 - 16\,x^2\,x2^2 + 8\,x\,x2^3 + 16\,x^2\,x2^4 - 16\,x\,x2^5 + 4\,x2^6 - y^2 + x2^2\,y^2)$

Thus, we have to eliminate the variables x_1 and x_2 to get the implicit representation of the orthopodic locus. We do not carry out the two missing steps here because they take a long time and the result is an expression that is not of interest here. `eq2[5] = Resultant[Last[eq2[1]], Last[eq2[3]], x1]` results in a polynomial with 175 terms, and the last step `Resultant[eq2[5], Last[eq2[4]], x2]` yields a polynomial of degree 64 in x and degree 96 in y with 1073 terms.

Let us now concentrate on the animation. The equation $h(x_1, x_2) = $ `eq2[1]` can be explicitly solved with respect to x_2.

```
In[17]:= x2 /. Solve[15 - 63 x1^2 + 64 x1^4 - 63 x2^2 + 255 x1^2 x2^2 -
                256 x1^4 x2^2 + 64 x2^4 - 256 x1^2 x2^4 +
                256 x1^4 x2^4 == 0, x2]
```

Out[17]= $\{-\sqrt{\left(\dfrac{63}{2\,(64-256\,x1^2+256\,x1^4)} - \dfrac{255\,x1^2}{2\,(64-256\,x1^2+256\,x1^4)} + \right.}$

$\dfrac{128\,x1^4}{64-256\,x1^2+256\,x1^4} - \dfrac{\sqrt{129-642\,x1^2+1025\,x1^4-512\,x1^6}}{2\,(64-256\,x1^2+256\,x1^4)}$,

$\sqrt{\left(\dfrac{63}{2\,(64-256\,x1^2+256\,x1^4)} - \dfrac{255\,x1^2}{2\,(64-256\,x1^2+256\,x1^4)} + \dfrac{128\,x1^4}{64-256\,x1^2+256\,x1^4} - \right.}$

$\dfrac{\sqrt{129-642\,x1^2+1025\,x1^4-512\,x1^6}}{2\,(64-256\,x1^2+256\,x1^4)}$,

$-\sqrt{\left(\dfrac{63}{2\,(64-256\,x1^2+256\,x1^4)} - \dfrac{255\,x1^2}{2\,(64-256\,x1^2+256\,x1^4)} + \right.}$

$\dfrac{128\,x1^4}{64-256\,x1^2+256\,x1^4} + \dfrac{\sqrt{129-642\,x1^2+1025\,x1^4-512\,x1^6}}{2\,(64-256\,x1^2+256\,x1^4)}$,

$\sqrt{\left(\dfrac{63}{2\,(64-256\,x1^2+256\,x1^4)} - \dfrac{255\,x1^2}{2\,(64-256\,x1^2+256\,x1^4)} + \dfrac{128\,x1^4}{64-256\,x1^2+256\,x1^4} + \right.}$

$\dfrac{\sqrt{129-642\,x1^2+1025\,x1^4-512\,x1^6}}{2\,(64-256\,x1^2+256\,x1^4)}$ $\}$

We inspect the four solutions.

```
In[18]:= Show[GraphicsArray[
            Plot[Evaluate[#[%]], {x1, -1, 1}, Frame -> True,
            Axes -> False, DisplayFunction -> Identity]& /@ {Re, Im}]]
```

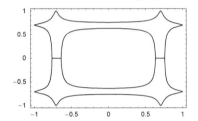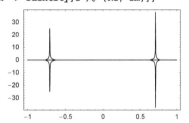

We see that the four solutions are not real for every $-1 \le x_1 \le 1$. We find the regions where the first and second solution for x_2 are not real with the following.

```
In[19]:= Function[data, Abs[data[[#]]]& /@ {Position[#, Min[#]][[1, 1]],
                Position[#, Max[#]][[1, 1]]}&[Abs[N[data]]]][
        x1 /. Solve[(63 - 255 x1^2 + 256 x1^4)^2 -
                (129 - 642 x1^2 + 1025 x1^4 - 512 x1^6) == 0, x1]]
```

Out[19]= $\{\dfrac{1}{8}\sqrt{\dfrac{1}{2}\left(63-\sqrt{129}\right)}, \sqrt{\dfrac{63}{128}+\dfrac{\sqrt{129}}{128}}\}$

So, we can implement the following solution of the equation $h(x_1, x_2) = \text{eq2}[1]$.

```
In[20]:= x2Roots[x1_] :=
        With[{denom = Sqrt[2] Sqrt[64 - 256 x1^2 + 256 x1^4],
            num1 = Sqrt[129 - 642 x1^2 + 1025 x1^4 - 512 x1^6],
            num2 = 63 - 255 x1^2 + 256 x1^4},
        If[N[Sqrt[63 - Sqrt[129]]/(8 Sqrt[2])] < Abs[x1] <
          N[Sqrt[63 + Sqrt[129]]/(8 Sqrt[2])],
          Drop[#, 2], #]&[{-Sqrt[num2 - num1]/denom, Sqrt[num2 - num1]/denom,
                -Sqrt[num2 + num1]/denom, Sqrt[num2 + num1]/denom}]]
```

For the actual construction of the two tangents, we need the two y-values corresponding to every x-value.

```
In[21]:= Clear[y];

        {y[1][x_], y[2][x_]} = {+Sqrt[4 x^2 - 4 x^4], -Sqrt[4 x^2 - 4 x^4]};
```

We also need the crossing point of the two tangents originating from x_1, y_1 and x_2, y_2 and having slope y_1' and y_2'.

```
In[23]:= cp[{x1_, y1_}, {x2_, y2_}, ys1_, ys2_] =
    {0, y1} + {x2 ys2 - x1 ys1 + y1 - y2,
              ys1(x2 ys2 - x1 ys2 + y1 - y2)}/(ys2 - ys1);
```

So, we can program for the points of the orthopodic locus.

```
In[24]:= orthopodicLocus[x1_] := orthopodicLocus[x1] =
    Module[{x2p = x2Roots[x1] // N, ε = 10^-6, x2real, snips},
    Table[(* y[1] and y[2] *) ysa = ys[x1, y[i][x1]];
    (* select perpendicular crossings *)
    x2real = Select[x2p, Abs[N[(1 + ysa ys[#, y[j][#]])]] < εε];
    snips = cp[{x1, y[i][x1]},
              {#, y[j][#]}, ysa, ys[#, y[j][#]]]& /@ x2real;
      snips, {i, 2}, {j, 2}]]
```

This is the original curve.

```
In[25]:= curve = ParametricPlot[{Cos[1t], Sin[2t]}, {t, 0, 2Pi},
                    PlotStyle -> {{Thickness[0.01], Hue[0]}},
                    DisplayFunction -> Identity];
```

This shows the original curve and its orthopodic locus.

```
In[26]:= Module[{a = ArcCos @ N[Sqrt[63 - Sqrt[129]]/(8 Sqrt[2])],
              b = ArcCos @ N[Sqrt[63 + Sqrt[129]]/(8 Sqrt[2])],
              ε = 10.^-6, pp = 60},
    Show[{curve, (* the orthopodic locus *)
    Graphics[{#, Map[# {-1, 1}&, #, {-2}]}&[
    Map[Line, Transpose[#, {4, 2, 3, 1, 5}], {3}]& /@
    {Table[orthopodicLocus[Cos[t]], {t, ε, b - ε, (b - 2ε)/pp}],
    Table[orthopodicLocus[Cos[t]], {t, a + ε, N[Pi/2 - ε],
                        N[(Pi/2b - 2ε)]/pp}]}]]},
        DisplayFunction -> $DisplayFunction, Axes -> False]]
```

For the animation, we should also display the tangents. We vary x_1 between 1 and -1 and mark the points of the orthopodic locus traversed.

```
In[27]:= orthopodicLocusAndTangents[x1_] := orthopodicLocusAndTangents[x1] =
    Module[{x2p = x2Roots[x1] // N, ε = 10^-6, x2real, snips},
    Table[(* y[1] and y[2] *) ysa = ys[x1, y[i][x1]];
    (* select perpendicular crossings *)
    x2real = Select[x2p, Abs[N[(1 + ysa ys[#, y[j][#]])]] < εε];
    snips = cp[{x1, y[i][x1]}, {#, y[j][#]}, ysa, ys[#, y[j][#]]]& /@ x2real;
    {MapThread[Line[{##}]&, {Table[Evaluate[{x1, y[i][x1]}], {Length[snips]}],
                  snips, {#, y[j][#]}& /@ x2real}],
      Point /@ snips}, {i, 2}, {j, 2}]]
```

Here are a few configurations for various x_1.

```
In[28]:= Show[GraphicsArray[#]]& /@ Map[
    Show[{curve, Graphics[{Thickness[0.002], PointSize[0.015],
          (* the tangents *)
          orthopodicLocusAndTangents[#]}]}, Frame -> True, Axes -> False,
          PlotRange -> {{-2, 2}, {-2, 2}}, AspectRatio -> Automatic,
          FrameTicks -> None, DisplayFunction -> Identity]&,
          Partition[Table[N[t], {t, 1/10, 9/10, 1/10}], 3], {2}]
```

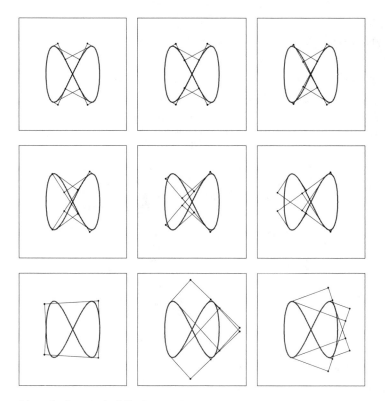

The animation could now be done via the following.

```
(* points of the orthopodic locus already traversed *)
rememberPoints[x1_] :=
Cases[DownValues[orthopodicLocusAndTangents], _Point, {-3}];

orthopodicLocusAnimation[n_] := With[{ε = 10^-6},
Do[Show[{curve, Graphics[
        {Thickness[0.002], PointSize[0.01],
        (* the current x1 value *) orthopodicLocusAndTangents[N[Cos[t]]],
        (* already traversed points of the orthopodic locus *) rememberPoints[N[t]]}]},
        Frame -> True, Axes -> False, PlotRange -> {{-2, 2}, {-2, 2}},
        AspectRatio -> Automatic, FrameTicks -> None,
        DisplayFunction -> $DisplayFunction],
    {t, ε, Pi - ε, (Pi - 2ε)/n}]];

(* 250 frames give a smooth animation *)
orthopodicLocusAnimation[250]
```

e) This is the implicit representation of the curve under investigation.

In[1]:= `f[x_, y_] = 4 x^4 - 4 x^2 + y^2;`

Because the cissoid is the midpoint of two points on the curve, these two points {x1, y1} and {x2, y2} fulfill the implicit representation.

In[2]:= `{eq[1] = f[x1, y1], eq[2] = f[x2, y2]}`

Out[2]= $\{-4\,x1^2 + 4\,x1^4 + y1^2,\ -4\,x2^2 + 4\,x2^4 + y2^2\}$

The points of the cissoid lie on straight lines through the point {px, py}.

```
In[3]:= eq[3] = Resultant[x - (px + t(x1 - px)), y - (py + t(y1 - py)), t]
```

$$\text{Out[3]}= \ (px - x1)\,(-py + y) - (-px + x)\,(py - y1)$$

```
In[4]:= eq[4] = Resultant[x - (px + s(x2 - px)), y - (py + s(y2 - py)), s]
```

$$\text{Out[4]}= \ (px - x2)\,(-py + y) - (-px + x)\,(py - y2)$$

The following two equations are the explicit form of the left-hand side of the equation that states the following: the points of the cissoid are midpoints of the line segments formed by two points from the given curve.

```
In[5]:= eq[5] = x - (x1 + x2)/2; eq[6] = y - (y1 + y2)/2;
```

Putting all equations together and eliminating the coordinates of the two points {x1, y1} and {x2, y2}, we arrive at the implicit representation of the cissoid of the Lissajous curve under consideration here.

```
In[6]:= res = (Subtract @@ Eliminate[# == 0& /@ Array[eq, 6],
                                      {x1, y1, x2, y2}]) // Factor
```

$$\text{Out[6]}= \ (-4\,x^2 + 4\,x^4 + y^2)$$
$$(16\,px^3\,x^2 - 8\,px\,py^2\,x^2 - 48\,px^2\,x^3 + 8\,py^2\,x^3 + 48\,px\,x^4 - 128\,px^3\,x^4 + 16\,px\,py^2\,x^4 - 16\,x^5 +$$
$$384\,px^2\,x^5 - 16\,py^2\,x^5 - 384\,px\,x^6 + 256\,px^3\,x^6 + 128\,x^7 - 768\,px^2\,x^7 + 768\,px\,x^8 -$$
$$256\,x^9 + 2\,py^3\,x\,y + 16\,px\,py\,x^2\,y - 16\,py\,x^3\,y + 32\,px^2\,py\,x^3\,y - 96\,px\,py\,x^4\,y +$$
$$64\,py\,x^5\,y - px\,py^2\,y^2 - 5\,py^2\,x\,y^2 - 8\,px\,x^2\,y^2 - 16\,px^3\,x^2\,y^2 + 8\,x^3\,y^2 +$$
$$16\,px^2\,x^3\,y^2 + 32\,px\,x^4\,y^2 - 32\,x^5\,y^2 + 2\,px\,py\,y^3 + 4\,py\,x\,y^3 - px\,y^4 - x\,y^4)$$

The first factor of the last result is just the implicit representation of the curve. It appears here because of the possibility that {x1, y1} and {x2, y2} are identical. The second part gives the nontrivial solution. Using ContourPlot, let us now have a look at the cissoid.

```
In[7]:= (* special point for this picture *) px = 1/2; py = 1/2;

Show[Block[{$DisplayFunction = Identity},
     (* the original curve in gray *)
     {lissa = ParametricPlot[{Cos[t], Sin[2t]}, {t, 0, 2Pi},
          PlotStyle -> {{Thickness[0.01], GrayLevel[1/2]}}],
      (* the cissoid *)
      ContourPlot[Evaluate[res[[2]]], {x, -1, 1}, {y, -1, 1},
                  Contours -> {0}, ContourShading -> False,
                  PlotPoints -> 200, ContourStyle -> {Thickness[0.02]}]}],
     Frame -> True, Axes -> False, PlotRange -> All,
     AspectRatio -> Automatic]
```

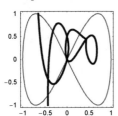

Note that due to the implicit form, points outside of the original curve are also contained in the above representation of the cissoid.

For comparison, let us construct the cissoid for the last example explicitly. The first problem is the calculation of all points of the Lissajous curve that go through a given line. To avoid the numerical solution of transcendental equations, we calculate a polynomial equation for the *x*-values of these points.

```
In[9]:= x1Eq = Resultant[eq[1], dx py + dy x1 - dx y1 - dy px, y1] // Factor
```

$$\text{Out[9]}= \ \frac{1}{4}\,(dx^2 - 2\,dx\,dy + dy^2 + 4\,dx\,dy\,x1 - 4\,dy^2\,x1 - 16\,dx^2\,x1^2 + 4\,dy^2\,x1^2 + 16\,dx^2\,x1^4)$$

Next, rotating a line through {px, py}, we calculate for each configuration all x-values on the Lissajous curve, all corresponding y-values, and select all points which actually lie on the given line. Then, forming all pairs of different points, we get all lines segments, the midpoints of which form the cissoid. The next input carries out this construction.

```
In[10]:= Show[{lissa, Graphics[{PointSize[0.02],
        Table[(* direction of line *) dx = Cos[t]; dy = Sin[t];
        MapIndexed[(* only different points *) Drop,
        (* all point pairs *) Outer[Point[(#1 + #2)/2]&, #, #, 1]]&[
        Flatten[(Function[x, Select[{x[[1]], #}& /@ x[[2]],
                 (* select all points on line *)
                 (Abs[# - t] < 10^-5 || Abs[# - N[(t - Pi)]] < 10^-5)&[
                         (ArcTan @@ (# - {px, py}))]&]] /@
        ({#, Cases[(* all y-values *) NRoots[f[#, y] == 0, y], _Real, {-1}]}& /@
                 (* all x-values *)
                 Cases[NRoots[x1Eq == 0, x1], _Real, {-1}])), 1]],
        (* a bunch of lines *) {t, N[0], N[Pi], N[Pi]/5555}]}]},
        PlotRange -> All, AspectRatio -> Automatic,
        Frame -> True, Axes -> False]
```

f) We first calculate the implicit form from the parametric representation given, then transform the equation to a list of the letters of the individual characters, and place them along the curve.

For the calculation of the implicit form, we rewrite all trigonometric functions of a multiple argument into powers of $\sin(\varphi)$ and $\cos(\varphi)$ and then call Eliminate with the given equations and the identity $\cos(\varphi)^2 + \sin(\varphi)^2 = 1$.

```
In[1]:= eq = Factor[(Subtract @@ Eliminate[Append[TrigExpand //@
        {x == Cos[3φ] - Sin[φ]/5, y == Sin[2φ] + Cos[φ]/3},
        (* add algebraic relation between Cos and Sin *)
        Sin[φ]^2 + Cos[φ]^2 == 1], {Cos[φ], Sin[φ]}])] == 0
```

$$Out[1]= 55919 + 4860\,x - 1470875\,x^2 - 121500\,x^3 + 1822500\,x^4 + 846180\,y +$$
$$902250\,x\,y - 648000\,x^2\,y - 3645000\,x^3\,y + 2779254\,y^2 + 1714500\,x\,y^2 + 1822500\,x^2\,y^2 -$$
$$4101840\,y^3 + 3510000\,x\,y^3 - 11712600\,y^4 + 2916000\,y^5 + 7290000\,y^6 == 0$$

Now, we transform the FullForm of the *Mathematica* expression eq into a string (for compactness, we eliminate all spaces).

```
In[2]:= string = StringReplace[ToString[FullForm[eq]], " " -> ""]
```

```
Out[2]= Equal[Plus[55919,Times[4860,x],Times[-1470875,Power[x,2]],Times[-121500,Power[
        x,3]],Times[1822500,Power[x,4]],Times[846180,y],Times[902250,x,y],Times[
        -648000,Power[x,2],y],Times[-3645000,Power[x,3],y],Times[2779254,Power[
        y,2]],Times[1714500,x,Power[y,2]],Times[1822500,Power[x,2],Power[y,2]],
        Times[-4101840,Power[y,3]],Times[3510000,x,Power[y,3]],Times[-11712600,
        Power[y,4]],Times[2916000,Power[y,5]],Times[7290000,Power[y,6]]],0]
```

To adjust the equation properly, we treat every letter individually. Here are all letters from string.

```
In[3]:= chars = Characters[string];
```

To have the same spacing between all letters, we now calculate the arc length of the given curve (for later work, we do not use NIntegrate here, but NDSolve, because the resulting InterpolatingFunction will be reused later).

```
In[4]:= sol = NDSolve[{arcLength'[φ] == Sqrt[D[Cos[3φ] - Sin[φ]/5, φ]^2 +
                                      D[Sin[2φ] + Cos[φ]/3, φ]^2],
                arcLength[0] == 0}, arcLength, {φ, 0, 2Pi}]
```

```
Out[4]= {{arcLength → InterpolatingFunction[{{0., 6.28319}}, <>]}}
```

The length of the whole curve is about 15.35.

```
In[5]:= curveLength = (arcLength[N[2Pi]] /. sol)[[1]]
Out[5]= 15.3488
```

Now, we divide the curve into pieces of equal length using the interpolating function of sol.

```
In[6]:= φTab = Table[φ /. FindRoot[sol[[1, 1, 2]][φ] ==
                        i/(Length[chars] + 2) curveLength,
                    {φ, 0, N[2Pi]}, Compiled -> False], {i, Length[chars]}];
```

The letters should change their orientation along the curve, so they are always parallel to the tangent of the curve.

```
In[7]:= letterDir[φ_] = {D[Cos[3φ] - Sin[φ]/5, φ], D[Sin[2φ] + Cos[φ]/3, φ]}
```

$$Out[7]= \left\{ -\frac{Cos[\varphi]}{5} - 3\,Sin[3\,\varphi],\ 2\,Cos[2\,\varphi] - \frac{Sin[\varphi]}{3} \right\}$$

As a guide for the eye, we underlay a thick version of the curve under the equation.

```
In[8]:= curve = ParametricPlot[{Cos[3φ] - Sin[φ]/5, Sin[2φ] + Cos[φ]/3},
                    {φ, 0, N[2Pi]}, DisplayFunction -> Identity];
```

Finally, we can display the curve and its implicit representation.

```
In[9]:= Show[Graphics[{
            {GrayLevel[0], Thickness[0.04], curve[[1, 1, 1]]},
            {GrayLevel[1], Table[(* the characters *)
            Text[StyleForm[chars[[i]], FontFamily -> "Courier", FontSize -> 8],
                {Cos[3#] - Sin[#]/5, Sin[2#] + Cos[#]/3}, {0, 0},
                (* their direction *)
                letterDir[#]]&[φTab[[i]]], {i, Length[chars]}]}}],
                AspectRatio -> Automatic, FrameTicks -> None,
                Frame -> True, PlotRange -> {{-1.5, 1.5}, {-1.5, 1.5}}]]
```

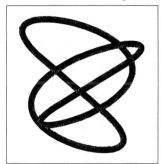

g) The polynomial system to be solved in this exercise is relatively complicated. We will use three different approaches to solve it.

Let us start and have a look at the region under consideration.

```
In[1]:= poly0[x_, y_] = x^4 + y^4 - 2 x^3 + 2 y - 1 + 2 x;

In[2]:= cp = Graphics[ContourPlot[Evaluate[poly0[x, y]], {x, -2, 2}, {y, -2, 1},
                    PlotPoints -> 60, Contours -> {0}]];
```

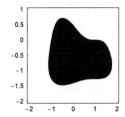

After calculating the *x*-values enclosing the region, we could have also used Plot to get a picture of the region as follows.

```
Module[{domain, sol, ε = 10^-6},
(* boundaries with respect to x *)
domain = Cases[NRoots[poly0[x,
Cases[NRoots[Numerator[D[poly0[x, y], y]/D[poly0[x, y], x]] == 0, y],
      _Real, {-1}][[1]]] == 0, x], _Real, {-1}] + {ε, -ε};
(* get y = y(x) *) sol = Solve[poly0[x, y] == 0, y];
(* form polygons from discretized y(x) curves *)
Show[Graphics[Polygon[Join[#1, Reverse[#2]]]]& @@
(Plot[Evaluate[sol[[#, 1, 2]]], Evaluate[{x, Sequence @@ domain},
      DisplayFunction -> Identity]][[1, 1, 1, 1]]& /@ {3, 4})],
      AspectRatio -> 1, PlotRange -> {{-2, 2}, {-2, 1}}, Frame -> True]]
```

To follow the light ray, we must calculate where a given ray hits the boundary of our region and its new direction after the reflection. Taking a ray originating from {x0, y0} and pointing in the direction {dx, dx}, the implicit equation for the *x*-value of the intersection with the region is given by the following.

In[3]:= **polyInx[dx_, dy_, x0_, y0_] =**
       Factor[Resultant[dy x0 + dx y - dx y0 - dy x, poly0[x, y], y]]

Out[3]= $-dx^4 + 2 dx^4 x + 2 dx^3 dy x - 2 dx^4 x^3 + dx^4 x^4 + dy^4 x^4 - 2 dx^3 dy x0 -$
       $4 dy^4 x^3 x0 + 6 dy^4 x^2 x0^2 - 4 dy^4 x x0^3 + dy^4 x0^4 + 2 dx^4 y0 + 4 dx dy^3 x^3 y0 -$
       $12 dx dy^3 x^2 x0 y0 + 12 dx dy^3 x x0^2 y0 - 4 dx dy^3 x0^3 y0 + 6 dx^2 dy^2 x^2 y0^2 -$
       $12 dx^2 dy^2 x x0 y0^2 + 6 dx^2 dy^2 x0^2 y0^2 + 4 dx^3 dy x y0^3 - 4 dx^3 dy x0 y0^3 + dx^4 y0^4$

We can now use this result to straightforwardly implement the following routine that calculates the point of intersection. We first calculate the roots of polyInx[dx, dy, x0, y0]. We select the real ones and calculate the corresponding *y*-values as the roots of poly0. Again, we select the real solutions. Then we discard the starting point (which is, of course, also a solution), check if the directions of the so-calculated points are the same as the direction of {dx, dx}, and finally, take the nearest point to the starting point. (Due to the fact that the region is concave, there might be more than one intersection in a given direction.) Here is the corresponding *Mathematica* implementation.

In[4]:= **nextPoint[{x0_, y0_}, {dx_, dy_}] :=**
```
    Module[{s1, s2, s3, s4, s5, ε = 10^-7},
       (* maybe add here a check if point is on curve and if normal points
          toward the interior; but we start always inside;
          so, we do not check this here *)
       (* all solutions of polyInx *)
       s1 = Chop[Last /@ List @@ NRoots[polyInx[dx, dy, x0, y0] == 0, x]];
       (* discard solutions with imaginary part *)
       s2 = Union[Re @ Cases[s1, _?((Abs[Im[#]] < ε)&)],
                  SameTest -> (Abs[#1 - #2] < ε&)];
       (* calculate corresponding y-values and discard them if they have
          a "nonvanishing" imaginary part *)
       s3 = Flatten[Function[xv, {xv, #}& /@
        Union[Re @ Cases[Last /@ List @@ NRoots[poly0[xv, y] == 0, y],
               _?((Abs[Im[#]] < ε)&)],
                  SameTest -> (Abs[#1 - #2] < ε&)]] /@ s2, 1];
       (* discard the starting point {x0, y0} *)
       s4 = Delete[s3, Position[#, Min[#]]&[(#.#&[{x0, y0} - #])& /@ s3][[1, 1]]];
        (* which solutions are on the ray in the specified direction;
          ε is chosen somewhat arbitrary *)
       s5 = Select[s4, Chop[Abs[(#[[1]] - x0) dy - (#[[2]] - y0)dx], ε] === 0&];
       If[s5 === {}, s5 = {s4[[Position[#, Min[#]]&[
                        Abs[(#[[1]] - x0) dy - (#[[2]] - y0)dx]& /@
                                             s4][[1, 1]]]]}];
       (* choose the nearest point
          (if boundary is concave, s5 may have more than one element);
          for the general case this test needs refinement! *)
       If[Length[s5] === 1, s5[[1]],
          s5[[Position[#, Min[#]]&[#.#&[{x0, y0} - #]& /@ s5][[1, 1]]]]]]]
```

For calculating the new direction of the light ray after it has undergone one reflection, we need the normal to the boundary of our region.

```
In[5]:= normal[{x_, y_}] = -{D[poly0[x, y], x], D[poly0[x, y], y]};
```

Now, we have everything together to implement `newPointAndDirection`, that calculates the point of intersection of a ray originating from `{x0, y0}` and pointing in the direction `{dx, dx}` and its direction after reflection.

```
In[6]:= newPointAndDirection[{{x0_, y0_}, {dx_, dy_}}] :=
        Module[{p1, dir1, n},
                (* the next point *)
                p1 = nextPoint[{x0, y0}, {dx, dy}];
                n = #/Sqrt[#.#]&[normal[p1]];
                (* the reflection law *)
                dir1 = #/Sqrt[#.#]&[-{dx, dy} + 2 n({dx, dy}.n)];
                {p1, dir1}]
```

Let us have a look at some rays that are reflected multiple times. (To see better what happens after multiple reflections, we use rays which all originate from `{x0, y0}`, but start in the direction `{dx, dx} + {0, r div}`, $-rays \le r \le rays$.)

```
In[7]:= reflectionPicture[{x0_, y0_}, {dx_, dy_},
                          rays_, reflections_, div_, opts___Rule] :=
        Show[{cp, Graphics[{Thickness[0.004],
        (* a small arrow at the starting point *)
        {GrayLevel[0.2],
        Function[{d, n}, {Line[{{x0, y0} - 0.1 d, {x0, y0} - 0.35 d}],
         Polygon[{{x0, y0} - 0.1 d, {x0, y0} - 0.2 d - 0.03 n,
                  {x0, y0} - 0.2 d + 0.03 n}]}][#/Sqrt[#.#]&[{dx, dy}],
             Reverse[#/Sqrt[#.#]&[{dx, dy}]] {-1, 1}]},
        Transpose[Table[(* the path *)
          {Hue[(r + rays) 0.78/(2 rays + 1)], Line[#]}& /@
          (Partition[First /@ NestList[(* construct the path *)
            newPointAndDirection, {{x0, y0},
                          {dx, dy} + {0, r div}}, reflections], 2, 1]),
                           {r, -rays, rays}]]}]},
          opts, PlotRange -> All, AspectRatio -> Automatic,
          Frame -> False, Axes -> False]
```

Here are three examples (the little arrow indicates the starting point and the starting direction of the light ray). The spreading of the ray after multiple reflections is clearly visible.

```
In[8]:= Show[GraphicsArray[
        Module[{x0, y0, dp = DisplayFunction -> Identity},
        {x0 = 0; y0 = Min[Cases[NRoots[poly0[x0, y] == 0, y],
                        _?(Im[#] == 0&), {-1}]];
        reflectionPicture[{x0, y0}, {1, 1}, 12, 20, 10^-7, dp],
        x0 = -0.55; y0 = Max[Cases[NRoots[poly0[x0, y] == 0, y],
                        _?(Im[#] == 0&), {-1}]];
        reflectionPicture[{x0, y0}, {1, -0.39}, 12, 17, 10^-6, dp],
        x0 = 1.7; y0 = Min[Cases[NRoots[poly0[x0, y] == 0, y],
                        _?(Im[#] == 0&), {-1}]];
        reflectionPicture[{x0, y0}, {-1, 0.11}, 12, 12, 10^-6, dp]}]]]
```

To find some stationary light rays, that means light rays that are each reflected into itself, we need to find points P_1 and P_2 on the boundary of our region, so that their normal aims in the opposite direction and that the direction given by $\overline{P_1 P_2}$ also points in the same direction.

We start by forming an equation that expresses that the normal in the point P_1 lies on the same line as $\overline{P_1 P_2}$.

```
In[9]:= r1 = Resultant[Sequence @@
            Apply[Subtract, Thread[normal[{x1, y1}] ==
                      a1({x2, y2} - {x1, y1})], {1}], a1]
```

$Out[9]= (x1 - x2)(-2 - 4 y1^3) - (-2 + 6 x1^2 - 4 x1^3)(y1 - y2)$

The following equation expresses that the normal in the point P_2 lies on the same line as $\overline{P_1 P_2}$.

```
In[10]:= r2 = Resultant[Sequence @@
            Apply[Subtract, Thread[normal[{x2, y2}] ==
                      a2({x2, y2} - {x1, y1})], {1}], a2]
```

$Out[10]= -(-2 + 6 x2^2 - 4 x2^3)(y1 - y2) + (x1 - x2)(-2 - 4 y2^3)$

These two equations, together with the equations for the boundary of our region applied to P_1 and P_2, form four equations for the four unknowns $x1$, $x2$, $y1$, and $y2$.

```
In[11]:= eqs[{x1_, x2_, y1_, y2_}] =
            {r1, r2, poly0[x1, y1], poly0[x2, y2]} // Expand
```

$Out[11]= \{-2 x1 + 2 x2 + 2 y1 - 6 x1^2 y1 + 4 x1^3 y1 - 4 x1 y1^3 + 4 x2 y1^3 - 2 y2 + 6 x1^2 y2 - 4 x1^3 y2,$
$\quad -2 x1 + 2 x2 + 2 y1 - 6 x2^2 y1 + 4 x2^3 y1 - 2 y2 + 6 x2^2 y2 - 4 x2^3 y2 - 4 x1 y2^3 + 4 x2 y2^3,$
$\quad -1 + 2 x1 - 2 x1^3 + x1^4 + 2 y1 + y1^4, -1 + 2 x2 - 2 x2^3 + x2^4 + 2 y2 + y2^4\}$

A direct try to solve these equations unfortunately fails.

```
In[12]:= TimeConstrained[Solve[eqs[{x1, x2, y1, y2}] == 0,
                      {x1, x2, y1, y2}], 120]
```

$Out[12]= \$Aborted$

```
In[13]:= TimeConstrained[Reduce[Element[{x1, x2, y1, y2}, Reals] &&
                      eqs[{x1, x2, y1, y2}] == 0,
                      {x1, x2, y1, y2}], 120]
```

$Out[13]= \$Aborted$

The equations can also be found using Fermat's principle of minimizing the length of the path.

```
In[14]:= With[{l = 2 Sqrt[#.#]&[{x1, y1[x1]} - {x2, y2[x2]}]},
            Expand[Numerator[Together[#]]]& /@ ({D[l, x1], D[l, x2]} /.
                  {y1'[x1] -> -((1 - 3 x1^2 + 2 x1^3)/(1 + 2 y1^3)),
                   y2'[x2] -> -((1 - 3 x2^2 + 2 x2^3)/(1 + 2 y2^3))} //.
                  {y1[x1] -> y1, y2[x2] -> y2})]
```

$Out[14]= \{2 x1 - 2 x2 - 2 y1 + 6 x1^2 y1 - 4 x1^3 y1 + 4 x1 y1^3 - 4 x2 y1^3 + 2 y2 - 6 x1^2 y2 + 4 x1^3 y2,$
$\quad -2 x1 + 2 x2 + 2 y1 - 6 x2^2 y1 + 4 x2^3 y1 - 2 y2 + 6 x2^2 y2 - 4 x2^3 y2 - 4 x1 y2^3 + 4 x2 y2^3\}$

We could now try to solve this system symbolically, but here we are only interested in a few solutions, which we search numerically. To avoid FindRoot spending too much time in cases when the process does not converge (and to prevent all these messages from printing), we catch the generated messages via $MessagePrePrint and abort immediately the FindRoot calculation via $MessagePrePrint = (Abort[]) &. To prevent the whole While loop searching until five solutions are found is aborted, we catch the generated abort via CheckAbort and start a new numerical search with FindRoot for new randomly selected starting values.

```
In[15]:= rayBag = {};
```

```
        ($MessagePrePrint = (Abort[])&);
In[17]:= SeedRandom[999]
```

```
        While[Length[rayBag] < 5,
        If[# =!= Null, (* startpoint and endpoint should be different *)
            If[((({x1, y1} - {x2, y2}).({x1, y1} - {x2, y2}) /. #) > 0.01,
                      AppendTo[rayBag, #], Null]]&[
        CheckAbort[
        FindRoot[Evaluate[# == 0& /@ eqs[{x1, x2, y1, y2}]],
                  {x1, Random[Real, {-1, 1}]}, {x2, Random[Real, {-1, 1}]},
```

```
                        {y1, Random[Real, {-1, 1}]}, {y2, Random[Real, {-1, 1}]},
                        MaxIterations -> 50], Null]]]
```

In[19]:= **$MessagePrePrint = Short;**

Here are the so-generated stationary rays.

In[20]:= **rayBag**

Out[20]= {{x1 → 1.74831, x2 → -1.08927, y1 → -0.89113, y2 → -0.423101},
          {x1 → 1.74831, x2 → -1.08927, y1 → -0.89113, y2 → -0.423101},
          {x1 → 1., x2 → 1., y1 → -5.55112×10⁻¹⁷, y2 → -1.25992},
          {x1 → -1.08927, x2 → 1.74831, y1 → -0.423101, y2 → -0.89113},
          {x1 → -0.5, x2 → -0.5, y1 → -1.46606, y2 → 0.713887}}

Finally, let us have a look at some stationary points over 12 reflections to make sure that they are reflected into themselves. (Be aware that following the light ray over many more reflections requires the use of much more accurate starting values and a higher working precision in all of the above-defined routines.)

In[21]:= **Show[GraphicsArray[#]]& /@**
        Map[reflectionPicture[#, normal[#], 0, 12, 0,
                                DisplayFunction -> Identity]&,
          {Take[#, {1, 4}], Take[#, {5, 7}]}]&[
            (* the data found after a long numerical search *)
            {{ 1.0000000000000000, 0.0000000000000000},
            {-0.1635363018502669, -1.4295866669246624},
            {-0.9118376919491141, -1.3493071403532364},
            { 1.6904691540714618, -1.0495609404364403},
            { 1.7483143392872937, -0.8911301339349427},
            {-0.5000000000000000, -1.4660592721926984},
            { 1.7559486066393841, -0.7937005259840998}}], {2}]
```

It is also possible to determine all rays that are reflected into themselves symbolically. This is the system of equations to be solved.

In[22]:= **Clear[x1, y1, x2, y2]**

```
        eqs = {2 (-x1 + x2 + y1 - 3 x1^2 y1 + 2 x1^3 y1 -
                 2 x1 y1^3 + 2 x2 y1^3 - y2 + 3 x1^2 y2 - 2 x1^3 y2),
               -2 (x1 - x2 - y1 + 3 x2^2 y1 - 2 x2^3 y1 + y2 -
                  3 x2^2 y2 + 2 x2^3 y2 + 2 x1 y2^3 - 2 x2 y2^3),
                -1 + 2 x1 - 2 x1^3 + x1^4 + 2 y1 + y1^4,
                -1 + 2 x2 - 2 x2^3 + x2^4 + 2 y2 + y2^4};
```

There are two equations that are linear.

In[24]:= **Exponent[#, {x1, x2, y1, y2}]& /@ eqs**

Out[24]= {{3, 1, 3, 1}, {1, 3, 1, 3}, {4, 0, 4, 0}, {0, 4, 0, 4}}

We use the linearity in x2 to eliminate x2.
```

In[25]:= s1 = Factor //@ Solve[eqs[[1]] == 0, x2]

Out[25]= $\left\{\left\{x2 \to \dfrac{x1 - y1 + 3 x1^2 y1 - 2 x1^3 y1 + 2 x1 y1^3 + y2 - 3 x1^2 y2 + 2 x1^3 y2}{1 + 2 y1^3}\right\}\right\}$

These are the three remaining equations.

In[26]:= eqs1 = Factor[Numerator[Together[#]]]& /@
                    (Rest[eqs] /. s1[[1]]);

We only use the last factor of the first polynomial in the following.

In[27]:= Head /@ eqs1

Out[27]= {Times, Plus, Plus}

The small factors of the last result shows that we must specially treat the cases $x1 == 1$, $x1 == -1/2$ and $y1 == y2$. So, we have the following set of new equations.

In[28]:= eqs2 = {eqs1[[1, -1]], eqs1[[2]], eqs1[[3]]};

We eliminate $y2$ from the first and the last equations (because their degree is not too high).

In[29]:= Exponent[#, {x1, x2, y1, y2}]& /@ eqs2

Out[29]= {{6, 0, 8, 2}, {4, 0, 4, 0}, {12, 0, 12, 4}}

This gives the following equation in $x1$ and $y1$.

In[30]:= Short[res1 = Factor[Resultant[eqs2[[1]], eqs2[[3]], y2]], 6]

Out[30]//Short= $(1 + 2 y1^3)^8 (-162 + 324 x1 + 4860 x1^2 - 12852 x1^3 - 24084 x1^4 + 92772 x1^5 +$
$47304 x1^6 - 307692 x1^7 - 175797 x1^8 + 903702 x1^9 + 849684 x1^{10} - 3541878 x1^{11} -$
$618653 x1^{12} + 10091588 x1^{13} - 7593428 x1^{14} - 11183404 x1^{15} + 19436983 x1^{16} + \ll711\gg +$
$282624 x1^2 y1^{26} + 585728 x1^3 y1^{26} - 2469888 x1^4 y1^{26} + 2555904 x1^5 y1^{26} - 851968 x1^6 y1^{26} +$
$319488 x1 y1^{27} - 319488 x1^2 y1^{27} - 8192 y1^{28} + 16384 x1 y1^{28} + 258048 x1^2 y1^{28} -$
$532480 x1^3 y1^{28} + 266240 x1^4 y1^{28} + 32768 y1^{29} + 49152 x1 y1^{30} - 49152 x1^2 y1^{30} + 4096 y1^{32})$

We will later treat the case $(1 + 2y1^3) == 0$; for the moment, we keep only the last factor of res1.

In[31]:= Exponent[res1[[-1]], {x1, y1}]

Out[31]= {32, 32}

Now, we use the defining equation for the region in $x1$ and $y1$ to eliminate $x1$ from res1.

In[32]:= Short[res2 = Factor[Resultant[res1[[-1]], eqs2[[2]], x1]], 6]

Out[32]//Short= $20736 y1 \ll1\gg^2 (-2 + \ll8\gg + 4 y1^7)$
$(-303022080 + 13795246080 y1 - 264576761856 y1^2 + 3029976640512 y1^3 - 9822182549760 y1^4 -$
$239145217464960 y1^5 + 2278318437774864 y1^6 - 12718345283083584 y1^7 -$
$40395269463620544 y1^8 + \ll144\gg + 8121052977661063725056 y1^{105} -$
$8121381740433861771264 y1^{106} + 1435819618799752052736 y1^{107} +$
$1238958271888132931584 y1^{108} - 688294138250287644672 y1^{109} + 51881467707308113920 y1^{110} +$
$1106804644442257309696 y1^{111} - 27670116110564327424 y1^{112} + 4611686018427387904 y1^{114})$

Now, we have a univariate polynomial. The above-mentioned factor $(1 + 2y1^3)$ appears; so it is already taken care of.

Let us have a look at the number of real solutions of res2.

In[33]:= Needs["Algebra`RootIsolation`"]

In[34]:= CountRoots[res2, {y1, -Infinity, Infinity}]

Out[34]= 14

This means there are 13 different ones (because one factor in res2 appears in power 2). We treat the four factors of res2 individually and calculate numerical values of the real solutions.

In[35]:= l1 = Select[List @@ res2, MemberQ[#, y1, {0, Infinity}]&];
         Length[l1]

Out[36]= 4

```
In[37]:= y1Sols1 = {0};
         y1Sols2 = Select[Last /@ List @@ Union[Roots[ll[[2]] == 0, y1]],
                          Im[#] === 0&]
Out[38]= {-2^{1/3}}
```

```
In[39]:= y1Sols3 = Select[Last /@ List @@ Union[Roots[ll[[3]] == 0, y1]],
                          Im[#] === 0&]
Out[39]= {Root[-2 + 4 #1 - 4 #1^2 - #1^3 + 8 #1^4 - 4 #1^5 + 4 #1^7 &, 1],
          Root[-2 + 4 #1 - 4 #1^2 - #1^3 + 8 #1^4 - 4 #1^5 + 4 #1^7 &, 2],
          Root[-2 + 4 #1 - 4 #1^2 - #1^3 + 8 #1^4 - 4 #1^5 + 4 #1^7 &, 3]}
```

```
In[40]:= N[y1Sols4 = Select[Last /@ List @@ Union[Roots[ll[[4]] == 0,
                         y1]], Im[N[#]] === 0&] // N[#, 60]&, 22]
Out[40]= {-1.429586666924662374836, -1.349307140353236296777,
          -1.049560940436440356309, -0.891130133934942653374, -0.423101214228750219190Q,
          0.223256853851107094349Q, 0.52262495509000088239228, 0.707919396489508500801Q}
```

Here are all 13 of them.

```
In[41]:= N[y1Solutions = SetPrecision[Sort[{y1 -> #}& /@
                 N[Union[Join[y1Sols1, y1Sols2, y1Sols3, y1Sols4]], 60]], 60], 22]
Out[41]= {{y1 -> -1.429586666924662374836}, {y1 -> -1.349307140353236296777},
          {y1 -> -1.336866549304587795266}, {y1 -> -1.259921049894873164767},
          {y1 -> -1.067282287436716936973}, {y1 -> -1.049560940436440356309},
          {y1 -> -0.891130133934942653374}, {y1 -> -0.423101214228750219190Q},
          {y1 -> 0}, {y1 -> 0.223256853851107094349Q}, {y1 -> 0.52262495509000088239228},
          {y1 -> 0.687888935627353015147Q}, {y1 -> 0.707919396489508500801Q}}
```

```
In[42]:= Length[y1Solutions]
Out[42]= 13
```

Now, we use the defining equation for the region to calculate corresponding y1 values. If there are multiple roots, we only use one.

```
In[43]:= Short[N[x1y1Solutions = Flatten[
         Function[y1Solution, Prepend[y1Solution, x1 -> #1]& /@
           Select[Union[Cases[NRoots[(eqs[[3]] /. y1Solution) == 0, x1, 60],
                          _?NumberQ, {-1}],
                        SameTest -> (Abs[N[#1] - N[#2]] < 10^-6&)],
                Abs[Im[#]] < 10^-6&] /@ y1Solutions, 1], 22], 6]
Out[43]//Short= {{x1 -> -0.757128703682847040037Q, y1 -> -1.429586666924662374836},
          {x1 -> ±0.163536301850266935078Q, y1 -> -1.429586666924662374836},
          <<22>>, {x1 -> -0.565595603360497805520Q, y1 -> 0.707919396489508500801Q},
          {x1 -> -0.430336051883199932589Q, y1 -> 0.707919396489508500801Q}}
```

Now, we need values for x2. We first derive an equation for x2 that involves x1, y1 (which we already know), and x2.

```
In[44]:= Factor[Resultant[eqs[[1]], eqs[[2]], y2]]
Out[44]= 16 (x1 - x2)^2 (1 + 2 y1^3)
          (-3 x1 + 18 x1^3 - 24 x1^4 - 19 x1^5 + 54 x1^6 - 36 x1^7 + 8 x1^8 - 3 x2 + 6 x1 x2 + 18 x1^2 x2 -
          24 x1^3 x2 - 19 x1^4 x2 + 54 x1^5 x2 - 36 x1^6 x2 + 8 x1^7 x2 - 12 x1^2 x2^2 + 8 x1^3 x2^2 +
          18 x1^4 x2^2 - 24 x1^5 x2^2 + 8 x1^6 x2^2 + 6 x1 y1 - 18 x1^3 y1 + 12 x1^4 y1 - 6 x2 y1 +
          18 x1^2 x2 y1 - 12 x1^3 x2 y1 - 6 y1^2 + 36 x1^2 y1^2 - 24 x1^3 y1^2 - 54 x1^4 y1^2 + 72 x1^5 y1^2 -
          24 x1^6 y1^2 - 8 x1^2 y1^3 + 16 x1 x2 y1^3 - 8 x2^2 y1^3 + 12 x1 y1^4 - 36 x1^3 y1^4 + 24 x1^4 y1^4 -
          12 x2 y1^4 + 36 x1^2 x2 y1^4 - 24 x1^3 x2 y1^4 - 8 x1^2 y1^6 + 16 x1 x2 y1^6 - 8 x2^2 y1^6)
```

The case (1 + 2 y1^3) == 0 is already taken care of, and with the case x1 == x2, we will deal later.

```
In[45]:= res3 = %[[-1]];
```

Now, we use the defining equation for the region to calculate corresponding y1-values. In the case of multiple roots, we only use one. We also take care of the case x1 == x2 here.

```
In[46]:= Short[N[x1y1x2Solutions = Join[Flatten[
         Function[x1y1Solution, Append[x1y1Solution, x2 -> #]& /@
             Select[Union[Cases[NRoots[(res3 /. x1y1Solution) == 0, x2, 60],
```

```
                        _?NumberQ, {-1}],
           SameTest -> (Abs[N[#1] - N[#2]] < 10^-6&)],
                   Abs[Im[#]] < 10^-6&]] /@ x1y1Solutions, 1],
               Append[#, x2 -> (x1 /. #)]& /@ x1y1Solutions], 22], 6]
Out[46]//Short=  {{x1 → -0.7571287036828470400372,
           y1 → -1.429586666924662374836, x2 → -0.4343795050968671692383},
           {x1 → -0.7571287036828470400372, y1 → -1.429586666924662374836,
           x2 → 0.04126593644302874747043}, ≪65≫, {x1 → -0.4303360518831999325893,
           y1 → 0.7079193964895085008010, x2 → -0.4303360518831999325893}}
```

Now, we need only the remaining values for y2. We use the defining equation for the region to calculate corresponding y1-values. If there are multiple roots, we only use one. We also take care of the case y1 == y2 here.

```
In[47]:= Short[N[x1y1x2y2Solutions = Join[Flatten[
           Function[x1y1x2Solution, Append[x1y1x2Solution, y2 -> #]& /@
                   Select[Union[Cases[NRoots[(eqs[[4]] /. x1y1x2Solution) == 0,
           y2, 60], _?NumberQ, {-1}],
             SameTest -> (Abs[N[#1] - N[#2]] < 10^-6&)],
                     Abs[Im[#]] < 10^-6&]] /@ x1y1x2Solutions, 1],
                 Append[#, y2 -> (y1 /. #)]& /@ x1y1x2Solutions], 22], 6]
Out[47]//Short=  {{x1 → -0.7571287036828470400372, y1 → -1.429586666924662374836,
           x2 → -0.4343795050968671692383, y2 → -1.464333170108635120131},
           ≪188≫, {x1 → -0.4303360518831999325893, y1 → 0.7079193964895085008010,
           x2 → -0.4303360518831999325893, y2 → 0.7079193964895085008010}}

In[48]:= Length[x1y1x2y2Solutions]

Out[48]=  190
```

We now sort out all solutions that do not satisfy the original equations.

```
In[49]:= Short[N[finalSolutions1 =
           Select[x1y1x2y2Solutions /. xy_?NumberQ :> Re[xy],
               Chop[eqs /. #, 10^-6] === {0, 0, 0, 0}&], 22], 12]
Out[49]//Short=  {{x1 → -0.163536301850266935078, y1 → -1.429586666924662374836,
           x2 → -0.5655956033604978055208, y2 → 0.7079193964895085008010},
           {x1 → -0.9118376919491139816145, y1 → -1.349307140353236296777,
           x2 → 0.2980272976828714194084, y2 → 0.223256853851107094349},
           ≪67≫, {x1 → -0.5655956033604978055208, y1 → 0.7079193964895085008010,
           x2 → -0.5655956033604978055208, y2 → 0.7079193964895085008010},
           {x1 → -0.4303360518831999325893, y1 → 0.7079193964895085008010,
           x2 → -0.4303360518831999325893, y2 → 0.7079193964895085008010}}
```

And we sort out all solutions in which the point {x1, y1} coincides with the point {x2, y2}.

```
In[50]:= Short[N[finalSolutions2 = Select[finalSolutions1,
                     #.#&[N[({x1, y1} /. #) - ({x2, y2} /. #)]] > 10^-6&], 22], 6]
Out[50]//Short=  {{x1 → -0.163536301850266935078, y1 → -1.429586666924662374836,
           x2 → -0.5655956033604978055208, y2 → 0.7079193964895085008010},
           ≪9≫, {x1 → 1.0000000000000000000, y1 → 0,
           x2 → 1.0000000000000000000, y2 → -1.2599210498948731648}}

In[51]:= Length[finalSolutions2]

Out[51]=  11
```

Now, we must take care of the remaining special cases. The case x==1 is already in the solution set finalSolutions3. It remains to treat the cases y1==y2 and x1==-1/2. Let us start with the last one. We calculate all possible solutions.

```
In[52]:= gb1 = GroebnerBasis[eqs /. x1 -> -1/2, {x2, y1, y2}]

Out[52]=  {-27 + 32 y2 + 16 y2^4, -27 + 32 y1 + 16 y1^4, 1 + 2 x2}

In[53]:= SetOptions[Roots, Quartics -> False, Cubics -> False];

In[54]:= Short[N[additionalSolutionsA1 = N[Prepend[#, x1 -> -1/2]& /@
               Union[Select[Solve[# == 0& /@ gb1, {y1, x2, y2}],
               (And @@ (Abs[Im[N[#]]] < 10^-8& /@ ({y1, x2, y2} /. #)))&]],
               60], 22], 6]
```

Out[54]//Short= `{{x1 → -0.500000000000000000000, x2 → -0.500000000000000000000,`
`y1 → -1.466059272192698319512, y2 → -1.466059272192698319512},`
`{x1 → -0.500000000000000000000, x2 → -0.500000000000000000000,`
`y1 → -1.466059272192698319512, y2 → 0.713886573052310827470},`
`{x1 → -0.500000000000000000000, x2 → -0.500000000000000000000,`
`y1 → «44», y2 → -1.466059272192698319512},`
`{x1 → -0.500000000000000000000, x2 → -0.500000000000000000000,`
`y1 → 0.713886573052310827470, y2 → 0.713886573052310827470}}`

Again, we drop coinciding points.

In[55]:= `N[additionalSolutionsA2 = Select[additionalSolutionsA1,`
`                  #.#&[N[({x1, y1} /. #) - ({x2, y2} /. #)]] > 10^-6&], 22]`

Out[55]= `{{x1 → -0.500000000000000000000, x2 → -0.500000000000000000000,`
`y1 → -1.466059272192698319512, y2 → 0.713886573052310827470},`
`{x1 → -0.500000000000000000000, x2 → -0.500000000000000000000,`
`y1 → 0.713886573052310827470, y2 → -1.466059272192698319512}}`

Now the condition y1 == y2 is treated.

In[56]:= `gb2 = GroebnerBasis[eqs /. y2 -> y1, {x1, x2, y1}]`

Out[56]= $\{-1 + 2\,x2 - 2\,x2^3 + x2^4 + 2\,y1 + y1^4,\ x1 - x2 + 2\,x1\,y1^3 - 2\,x2\,y1^3,\ -1 + 2\,x1 - 2\,x1^3 + x1^4 + 2\,y1 + y1^4\}$

In[57]:= `N[additionalSolutionsB1 = N[Prepend[#, y2 -> (y1 /. #)]& /@`
`Union[Select[N[Solve[# == 0& /@ gb2, {x1, x2, y1}], 60],`
`(And @@ (Abs[Im[N[#]]] < 10^-8& /@ ({x1, x2, y1} /. #)))&]], 60], 22]`
`    Solve::svars : Equations may not give solutions for all "solve" variables. More…`

Out[57]= `{{y2 → -0.793700525984099737375, x1 → -1.124209761148022281501,`
`x2 → -1.124209761148022281501, y1 → -0.793700525984099737375},`
`{y2 → -0.793700525984099737375, x1 → -1.124209761148022281501,`
`x2 → 1.755948606639384186096, y1 → -0.793700525984099737375},`
`{y2 → -0.793700525984099737375, x1 → 1.755948606639384186096,`
`x2 → -1.124209761148022281501, y1 → -0.793700525984099737375},`
`{y2 → -0.793700525984099737375, x1 → 1.755948606639384186096,`
`x2 → 1.755948606639384186096, y1 → -0.793700525984099737375}}`

Again, we drop coinciding points.

In[58]:= `N[additionalSolutionsB2 = Select[additionalSolutionsB1,`
`            #.#&[N[({x1, y1} /. #) - ({x2, y2} /. #)]] > 10^-6&], 22]`

Out[58]= `{{y2 → -0.793700525984099737375, x1 → -1.124209761148022281501,`
`x2 → 1.755948606639384186096, y1 → -0.793700525984099737375},`
`{y2 → -0.793700525984099737375, x1 → 1.755948606639384186096,`
`x2 → -1.124209761148022281501, y1 → -0.793700525984099737375}}`

So, we finally have the following solutions.

In[59]:= `Short[N[finalSolutions4 = Join[finalSolutions2, additionalSolutionsA2,`
`                  additionalSolutionsB2], 22], 6]`

Out[59]//Short= `{{x1 → -0.163536301850266935078, y1 → -1.429586666924662374836,`
`x2 → -0.565595603360497805520, y2 → 0.707919396489508500801},`
`«13», {y2 → -0.793700525984099737375, x1 → 1.755948606639384186096,`
`x2 → -1.124209761148022281501, y1 → -0.793700525984099737375}}`

All of them fulfill the original system of equations. We see that within the precision of the roots the original equations are fulfilled.

In[60]:= `Abs[Max[eqs /. finalSolutions4]]`

Out[60]= $0.\times 10^{-56}$

Finally, we drop coinciding points and solutions that are the same up to an exchange of x1↔x2 and y1↔y2.

In[61]:= `(finalSolutions5 = Union[Union[finalSolutions4,`
`        SameTest -> (#.#&[({x1, y1, x2, y2} /. #1) -`
`                          ({x1, y1, x2, y2} /. #2)] < 10^-6&)],`

```
              SameTest -> (#.#&[({x1, y1, x2, y2} /. #1) -
                                 ({x2, y2, x1, y1} /. #2)] < 10^-6&)]) // N
Out[61]=  {{x1 → -1.08927, y1 → -0.423101, x2 → 1.74831, y2 → -0.89113},
          {x1 → -0.911838, y1 → -1.34931, x2 → 0.298027, y2 → 0.223257},
          {x1 → -0.811667, y1 → 0.522625, x2 → 1.69047, y2 → -1.04956},
          {x1 → -0.565596, y1 → 0.707919, x2 → -0.163536, y2 → -1.42959},
          {x1 → -0.5, x2 → -0.5, y1 → -1.46606, y2 → 0.713887},
          {x1 → 1., y1 → -1.25992, x2 → 1., y2 → 0.},
          {y2 → -0.793701, x1 → -1.12421, x2 → 1.75595, y1 → -0.793701}}

In[62]:= Length[%]

Out[62]= 7

In[63]:= Show[GraphicsArray[#]]& /@
         Map[Show[{cp, Graphics[{Hue[0], Thickness[0.002],
                          Line[{{x1, y1} /. #, {x2, y2} /. #}]}]},
              DisplayFunction -> Identity,
              Frame -> False]&, (* the seven solutions found *)
           {Take[finalSolutions5, {1, 4}], Take[finalSolutions5, {5, 7}]}, {2}]
```

We give one more way to solve the polynomial system under consideration. First, we add an equation that guarantees $x1 \neq x2$ and $y1 \neq y2$.

```
In[64]:= eqs = AppendTo[eqs, 1 - z ((x1 - x2)^2 + (y1 - y2)^2)];
```

Next, we eliminate the auxiliary variable z that was introduced in the last input.

```
In[65]:= (gbElim = GroebnerBasis[eqs, {x1, y1, x2, y2}, {z},
                         MonomialOrder -> EliminationOrder];) // Timing

Out[65]= {24. Second, Null}
```

`gbElim` is quite a large polynomial system and its direct solution is not easy. So in a first step we convert it to a degree reverse lexicographic basis.

```
In[66]:= {Length[gbElim], ByteCount[gbElim]}

Out[66]= {29, 280904}

In[67]:= (gbDegRevLex =  GroebnerBasis[gbElim, {x1, y1, x2, y2},
         MonomialOrder -> DegreeReverseLexicographic];) // Timing
Out[67]= {0.62 Second, Null}

In[68]:= {Length[gbDegRevLex], ByteCount[gbDegRevLex]}

Out[68]= {29, 280904}
```

Using the internal function `Internal`GroebnerWalk` allows us to convert this basis to a lexicographic basis.

```
In[69]:= (gbLex = Internal`GroebnerWalk[gbDegRevLex,
                        {x1, y1, x2, y2}];) // Timing
Out[69]= {40.48 Second, Null}
```

Again, this is a large polynomial system with large coefficients.

```
In[70]:= {Length[gbLex], ByteCount[gbLex],
          Cases[gbLex, _Integer, {-1}] // Abs // Max // N}
Out[70]= {14, 786552, 1.393077647037545×10^926}
```

The univariate equation in `y2` in `gbLex` has degree 125.

```
In[71]:= pos = Position[Variables /@ gbLex, {y2}][[1, 1]];
         Exponent[gbLex[[pos]], y2]
Out[72]= 125
```

And the real roots of this univariate equation are all the values of `y1` and `y2` we found above. (Because of the symmetry `y1↔y2`, we now have 13 possible values for `y2`.)

```
In[73]:= roots = Roots[gbLex[[pos]] == 0, y2];
         y2Rules = {ToRules[Select[N[roots], Im[#[[2]]] == 0&]]}
Out[74]= {{y2 → 0.}, {y2 → -1.42959}, {y2 → -1.34931}, {y2 → -1.04956}, {y2 → -0.89113},
          {y2 → -0.423101}, {y2 → 0.223257}, {y2 → 0.522625}, {y2 → 0.707919},
          {y2 → -1.46606}, {y2 → 0.713887}, {y2 → -0.793701}, {y2 → -1.25992}}
```

If the reader liked this kind of exercise, one could go on and calculate the "mirror cabinet map" [1087] of this curve or calculate multiple reflections using more reflecting obstacles [474] or in 3D [1696].

h) We start by displaying the family of lines that defines the hedgehog.

```
In[1]:= h[t_] := -Cos[t] + 2 Cos[6t]
```

```
In[2]:= makeLine[t_] := Line[{{-16, (h[t] - (-16) Cos[t])/Sin[t]},
                              {+16, (h[t] - (+16) Cos[t])/Sin[t]}}]
```

The hedgehog is clearly visible.

```
In[3]:= With[{ε = 10^-10},
         Show[Graphics[{Thickness[0.002], Table[makeLine[t] // N,
                        {t, ε, 2Pi - ε, (2Pi - 2 ε)/200}]}],
              AspectRatio -> Automatic, PlotRange -> {{-16, 15}, {-15, 15}}]]
```

A parametric description of the hedgehog can be easily derived by differentiating *h*(*t*).

```
In[4]:= sol = Solve[{x Cos[t] + y Sin[t] == h[t],
                     D[x Cos[t] + y Sin[t] == h[t], t]}, {x, y}] // Simplify
Out[4]= {{x → -1 + 7 Cos[5 t] - 5 Cos[7 t], y → -7 Sin[5 t] - 5 Sin[7 t]}}
```

An implicit description of the hedgehog can be derived using `GroebnerBasis`.

```
In[5]:= p = x Cos[t] + y Sin[t] - TrigExpand[h[t]]
```

$$Out[5]= \ Cos[t] + x\,Cos[t] - 2\,Cos[t]^6 + y\,Sin[t] + 30\,Cos[t]^4\,Sin[t]^2 - 30\,Cos[t]^2\,Sin[t]^4 + 2\,Sin[t]^6$$

```
In[6]:= gb = GroebnerBasis[{p, D[p, t], Cos[t]^2 + Sin[t]^2 - 1}, {},
                           {Cos[t], Sin[t]}, MonomialOrder -> EliminationOrder]
```

Out[6]= $\{-16251426263790507 + 29386064477201390\,x - 4644791923465925\,x^2 -$
$13745865871591140\,x^3 + 2886388151433605\,x^4 + 2883803620261858\,x^5 - 241214062307085\,x^6 -$
$272359270453560\,x^7 - 10246432071225\,x^8 + 8457102353490\,x^9 + 1110082997849\,x^{10} +$
$73523537500\,x^{11} + 6719409375\,x^{12} + 136718750\,x^{13} + 9765625\,x^{14} + 14692661644250695\,y^2 -$
$19341530105566620\,x\,y^2 - 1297651391070090\,x^2\,y^2 + 7009279626187300\,x^3\,y^2 +$
$77933288967345\,x^4\,y^2 - 972790065588600\,x^5\,y^2 - 151681970698860\,x^6\,y^2 - 11599893086040\,x^7\,y^2 -$
$11491214073255\,x^8\,y^2 - 3341060187500\,x^9\,y^2 - 331098206250\,x^{10}\,y^2 + 820312500\,x^{11}\,y^2 +$
$68359375\,x^{12}\,y^2 - 4831985411516655\,y^4 + 4213733749856610\,x\,y^4 + 1165522105283865\,x^2\,y^4 -$
$802231166816520\,x^3\,y^4 + 14280059080770\,x^4\,y^4 + 186233459120940\,x^5\,y^4 + 78824377478490\,x^6\,y^4 +$
$20489329875000\,x^7\,y^4 + 2567318578125\,x^8\,y^4 + 2050781250\,x^9\,y^4 + 205078125\,x^{10}\,y^4 +$
$694135882609435\,y^6 - 362700044081480\,x\,y^6 - 209630129026780\,x^2\,y^6 - 66207481086040\,x^3\,y^6 -$
$63955619146510\,x^4\,y^6 - 28434593075000\,x^5\,y^6 - 4732718637500\,x^6\,y^6 + 2734375000\,x^7\,y^6 +$
$341796875\,x^8\,y^6 - 39222698735185\,y^8 + 17389200353490\,x\,y^8 + 18931061989245\,x^2\,y^8 +$
$10239196187500\,x^3\,y^8 + 2563217015625\,x^4\,y^8 + 2050781250\,x^5\,y^8 + 341796875\,x^6\,y^8 +$
$380074085349\,y^{10} - 670399537500\,x\,y^{10} - 334379456250\,x^2\,y^{10} + 820312500\,x^3\,y^{10} +$
$205078125\,x^4\,y^{10} + 5899096875\,y^{12} + 136718750\,x\,y^{12} + 68359375\,x^2\,y^{12} + 9765625\,y^{14}\}$

The next graphic shows the parametric and implicit representations.

```
In[7]:= pp = 300;
       cp = ContourPlot[Evaluate[N[gb[[1]]]], {x, -13, 12}, {y, -12, 12},
                       PlotPoints -> pp, Contours -> {0},
                       Epilog -> (* parametric representation *)
                       ParametricPlot[Evaluate[{x, y} /. sol[[1]]], {t, 0, 2Pi},
                       PlotPoints -> 200, DisplayFunction -> Identity,
                       PlotStyle -> {Hue[0]}][[1]]]
```

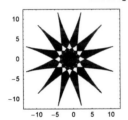

Here is a more general contour plot of the polynomial of the implicit representation of the hedgehog.

```
In[9]:= pp = 300; c = 60;
       (* preliminary contour plot *)
       cp = ContourPlot[Evaluate[Log[Abs[N[gb[[1]]]]]],
                       {x, -6, 4}, {y, -5, 5}, PlotPoints -> pp,
                       DisplayFunction -> Identity];
       (* equally spaced contour values *)
       cls = #[[Round[pp^2/c/2]]]& /@
                       Partition[Sort[Flatten[cp[[1]]]], Round[pp^2/c]];
       (* pure (color) function for alternating coloring *)
       color[γ_] = Which @@ Flatten[Append[Prepend[
       MapIndexed[{#1[[1]] <= γ < #1[[2]],
                       GrayLevel @ If[OddQ[#2[[1]]], 1, 0]}&,
                                       Partition[cls, 2, 1]],
                       {γ < cls[[1]], GrayLevel @ 0}],
                       {γ >= cls[[1]], GrayLevel @ If[EvenQ[Length[cls]], 0, 1]}]];
       (* final contour plot *)
       ListContourPlot[cp[[1]], Contours -> cls, ColorFunction -> color,
                       ContourLines -> False, ColorFunctionScaling -> False,
                       FrameTicks -> None]
```

i) We start with a purely numerical calculation of $I_k(x, y)$ for each value of k and discretized arclength. We first discretize C in equal arclength pieces through φValues$[pp\varphi]$ and then add the contribution of the perpendicular rays on a dense set of $\{x, y\}$-values using the function addRaysCF. Its arguments are the parameter k, the discretization dimensions $(2\,d + 1)^2$ for the rays, and a list of polar angles φ. A parametric description of the curve $x^4 + y^4 = 1$ is $R(\varphi)\{\cos(\varphi), \sin(\varphi)\}, 0 \le \varphi \le 2\,\pi$.

```
In[1]:= R[φ_] = (Cos[φ]^4 + Sin[φ]^4)^(-1/4);
```

```
In[2]:= (* angles values φ for equal arclength *)
        φValues[ppφ_] := Module[{ndsol, f, 1, ipo, s},
          (* calculate arclength as a function of φ *)
          ndsol = NDSolve[{s'[φ] == Sqrt[#.#]&[D[R[φ] {Cos[φ], Sin[φ]}, φ]],
                    s[0] == 0}, s, {φ, 0, 2Pi}, PrecisionGoal -> 8];
          f = ndsol[[1, 1, 2]]; L = f[2Pi];
          (* interpolate arclength(φ) *)
          ipo = Interpolation[{f[#], #}& /@ ndsol[[1, 1, 2]][[3, 1]]];
          (* equally arclength spaced φ values *)
          Table[ipo[j/ppφ L], {j, 0, ppφ - 1}]];
```

```
In[4]:= (* discretize normals and add their contributions *)
        addRaysCF = Compile[{k, {d, _Integer}, {pp, _Integer}, {φs, _Real, 1}},
        Module[{m = Table[0. + 0. I, {2d}, {2d}], r, p0, normal, ξ, η, s, i, j, φ},
        (* loop over all φ-values *)
        Do[φ = φs[[kφ]]; r = (Cos[φ]^4 + Sin[φ]^4)^(-1/4); p0 = r {Cos[φ], Sin[φ]};
            normal = -{Cos[φ]^3, Sin[φ]^3}/Sqrt[Cos[φ]^6 + Sin[φ]^6];
            Do[{ξ, η} = p0 + s normal;
                If[(* inside only *) ξ^4 + η^4 < 1.,
                    (* add ray contribution *)
                    {i, j} = Round[{ξ, η} d] + {d, d};
                    If[1 <= i <= 2d && 1 <= j <= 2d,
                        m[[i, j]] = m[[i, j]] + Exp[I k s]]], {s, 0., 3., 3/pp}],
            {kφ, Length[φs]}]; m]];
```

Here is the resulting graphic for $k = 120$ and $d = 360$ using 3600 rays. Each ray is discretized at 3600 points.

```
In[6]:= ListDensityPlot[Log[Abs[addRaysCF[120, 360, 3600, φValues[3600]]] + 1],
                Mesh -> False, PlotRange -> All, FrameTicks -> False]
```

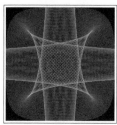

While the last approach works, it will be inefficient for a multi-frame animation. Rewriting $I_k(x, y)$ as

$$I_k(x, y) = \int_0^{2\pi} \exp(i\,k\,d(\{x, y\}, \{\xi(s(\varphi)), \eta(s(\varphi))\})) \, \delta(g(\{x, y\}, \{\xi(s(\varphi)), \eta(s(\varphi))\})) \, ds$$

$$= \sum_{j=1}^{n} \exp\!\left(i\,k\,((x - \xi(\varphi_j))^2 + (x - \eta(\varphi_j))^2)^{1/2}\right) \frac{1}{|a'(\varphi_j)\,y + b'(\varphi_j)\,x + c'(\varphi_j)|} \, s'(\varphi_j)$$

(where we used $\delta(f(x)) = \sum \delta(x - x_k)/|f'(x_k)|$ for the simple zeros x_k of $f(x)$ and φ is the polar angle), we see that at each point $\{x, y\}$ only a finite number of (k-independent) points $\{\xi(\varphi_j), \eta(\varphi_j)\}$ contribute to $I_k(x, y)$. So, we will calculate these points $\{\xi(\varphi_j), \eta(\varphi_j)\}$ of C once and then it will be possible to quickly add their contribution for each k-value.

The equation for a normal through $\{x, y\}$ is given by $x\,\eta(s(\varphi))^3 - \xi(s(\varphi))\,\eta(s(\varphi))^3 + \xi(s(\varphi))^3\,\eta(s(\varphi)) - y\,\xi(s(\varphi))^3 = 0$. The next inputs derive this equation.

```
In[7]:= superCircle[{ξ_, η_}] = ξ^4 + η^4 - 1;
```

```
In[8]:= normalLine[{ξ_, η_}, {x_, y_}] =
          Resultant[x - (ξ + s/4 D[superCircle[{ξ, η}], ξ]),
                    y - (η + s/4 D[superCircle[{ξ, η}], η]), s]
Out[8]= η^3 (x - ξ) - (y - η) ξ^3
```

By eliminating η using the implicit equation for C, we obtain a degree 16 polynomial in ξ.

```
In[9]:= ξRootPoly[ξ_, {x_, y_}] = Resultant[normalLine[{ξ, η}, {x, y}],
                                             superCircle[{ξ, η}], η]
Out[9]= -x^4 + 4 x^3 ξ - 6 x^2 ξ^2 + 4 x ξ^3 - ξ^4 + 3 x^4 ξ^4 - 12 x^3 ξ^5 + 20 x^2 ξ^6 -
        16 x ξ^7 + 5 ξ^8 - 3 x^4 ξ^8 + 12 x^3 ξ^9 - 4 x y^2 ξ^9 - 22 x^2 ξ^10 + 4 y^2 ξ^10 + 20 x ξ^11 -
        8 ξ^12 + x^4 ξ^12 + y^4 ξ^12 - 4 x^3 ξ^13 + 4 x y^2 ξ^13 + 8 x^2 ξ^14 - 4 y^2 ξ^14 - 8 x ξ^15 + 4 ξ^16
```

Using the last polynomial, the function $\xi\eta$Points$[\{x, y\}]$ calculates all points $\{\xi(x, y), \eta(x, y)\}$ through the point $\{x, y\}$.

```
In[10]:= ξηPoints[{x_, y_}] :=
           Module[{ξRoots, ξηRoots, prec = 30},
             ξRoots = Union[Select[Last /@ (List @@ N[Union[Roots[
                         ξRootPoly[ξ, {x, y}] == 0, ξ]], prec]), Im[#] == 0&]];
             ξηRoots = Function[ξ, Union[{{ξ, +(1 - ξ^4)^(1/4)},
                                          {ξ, -(1 - ξ^4)^(1/4)}}]] /@ ξRoots;
             Select[Flatten[ξηRoots, 1],
                    (#[[1]]^3 (y - #[[2]]) == #[[2]]^3 (x - #[[1]]))&]]
```

Generically either two, four, six, or eight normals go through a point $\{x, y\}$.

```
In[11]:= Length[ξηPoints[#]]& /@
           {{0, 0}, {0, 1/3}, {1/Pi, 1/E}, {1/2, 1/2}, {9/10, 1/10}}
Out[11]= {8, 8, 6, 4, 4}
```

We visualize the normals for some $\{x, y\}$-points. For a nicer graphic, and for later use, we also display the curve C and the wavefront caustics.

```
In[12]:= (* implicit equation of the wavefront caustics *)
         causticsImplicit[{x_, y_}] =
         Module[{det, res1, res2},
           (* determinant that must vanish; caustics are envelopes *)
           det = η^3/4 Factor[Together[Det @
           Outer[D, {ξ + s D[superCircle[{ξ, η}], ξ],
                     η + s D[superCircle[{ξ, η}], η]} /. η -> η[ξ], {ξ, s}] /.
             Solve[D[superCircle[{ξ, η[ξ]}], ξ] == 0, η'[ξ]][[1]] /. η[ξ] -> η]];
           (* eliminate line parameter s *)
           res1 = Resultant[x - ξ - 4 s ξ^3, det, s]/(-4 ξ^2) // Factor;
           res2 = Resultant[y - η - 4 s η^3, det, s]/(-4 η^2) // Factor;
           (* eliminate boundary values ξ and η *)
           GroebnerBasis[{res1, res2, superCircle[{ξ, η}]}, {x, y}, {ξ, η},
                         MonomialOrder -> EliminationOrder][[1]] // Factor]
Out[13]= -16384 + 1645824 x^4 - 41190729 x^8 - 7093520 x^12 - 510432 x^16 - 17664 x^20 - 256 x^24 - 41472 x^2 y^2 +
         24378756 x^6 y^2 - 95136168 x^10 y^2 - 4402800 x^14 y^2 + 47616 x^18 y^2 - 768 x^22 y^2 + 1645824 y^4 -
         165417846 x^4 y^4 + 4231638144 x^8 y^4 + 668513790 x^12 y^4 + 56248737 x^16 y^4 + 1873896 x^20 y^4 +
```

$27216\,x^{24}\,y^4 + 24378756\,x^2\,y^6 - 4479640944\,x^6\,y^6 + 60597220152\,x^{10}\,y^6 + 9209786484\,x^{14}\,y^6 +$
$623186732\,x^{18}\,y^6 + 21816540\,x^{22}\,y^6 + 314928\,x^{26}\,y^6 - 41190729\,y^8 + 4231638144\,x^4\,y^8 -$
$146454777036\,x^8\,y^8 + 189418755687\,x^{12}\,y^8 + 18322312311\,x^{16}\,y^8 + 945824769\,x^{20}\,y^8 +$
$36807210\,x^{24}\,y^8 + 531441\,x^{28}\,y^8 - 95136168\,x^2\,y^{10} + 60597220152\,x^6\,y^{10} - 404710248120\,x^{10}\,y^{10} +$
$243005113656\,x^{14}\,y^{10} + 3634060140\,x^{18}\,y^{10} - 206907696\,x^{22}\,y^{10} - 7093520\,y^{12} + 668513790\,x^4\,y^{12} +$
$189418755687\,x^8\,y^{12} - 197747121606\,x^{12}\,y^{12} + 147813819975\,x^{16}\,y^{12} - 3881389185\,x^{20}\,y^{12} +$
$2657205\,x^{24}\,y^{12} - 4402800\,x^2\,y^{14} + 9209786484\,x^6\,y^{14} + 243005113656\,x^{10}\,y^{14} +$
$126355591008\,x^{14}\,y^{14} + 706304772\,x^{18}\,y^{14} - 510432\,y^{16} + 56248737\,x^4\,y^{16} + 18322312311\,x^8\,y^{16} +$
$147813819975\,x^{12}\,y^{16} + 12092644710\,x^{16}\,y^{16} + 5314410\,x^{20}\,y^{16} + 47616\,x^2\,y^{18} + 623186732\,x^6\,y^{18} +$
$3634060140\,x^{10}\,y^{18} + 706304772\,x^{14}\,y^{18} - 17664\,y^{20} + 1873896\,x^4\,y^{20} + 945824769\,x^8\,y^{20} -$
$3881389185\,x^{12}\,y^{20} + 5314410\,x^{16}\,y^{20} - 768\,x^2\,y^{22} + 21816540\,x^6\,y^{22} - 206907696\,x^{10}\,y^{22} -$
$256\,y^{24} + 27216\,x^4\,y^{24} + 36807210\,x^8\,y^{24} + 2657205\,x^{12}\,y^{24} + 314928\,x^6\,y^{26} + 531441\,x^8\,y^{28}$

```
In[14]:= (* graphic of the supercircle *)
         superCircleGraphics =
         Graphics[{Thickness[0.005], RGBColor[0, 0, 1],
                   Line[Table[R[φ] {Cos[φ], Sin[φ]}, {φ, 0., 2.Pi, 2Pi/240}]]}];

         (* graphic of the wavefront caustics of the supercircle *)
         (* domain to visualize *) L = 11/10;

         causticsGraphics = Graphics[
         ContourPlot[Evaluate[causticsImplicit[{x, y}]], {x, -L, L}, {y, -L, L},
                     PlotPoints -> 600, Contours -> {0},
                     ContourShading -> False, DisplayFunction -> Identity,
                     ContourStyle -> {{Thickness[0.008], Hue[0]}}]];

In[20]:= Show[GraphicsArray[Function[xys,
         Show[{(* supercircle and its wave caustics *)
               superCircleGraphics, causticsGraphics,
               (* normals through point {x,y} *)
               Graphics[{Thickness[0.002], GrayLevel[1/2],
                         Function[xy, Line[{xy, #}]& /@ ξηPoints[xy]] /@ xys}],
            AspectRatio -> Automatic, DisplayFunction -> Identity]] /@
         (* two points {x, y} *)
         {{{1/6, 1/5}},
          {{0, Root[(-128 + 6429 #^4 + 552 #^8 + 16 #^12)& , 2]}}}]]
```

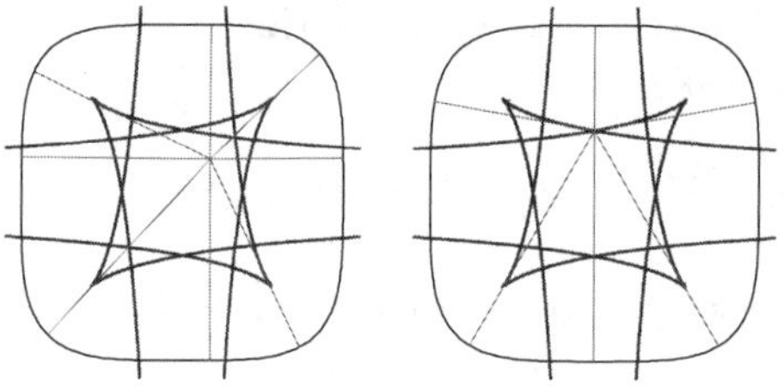

Now, we calculate the intersection points of the normals for a dense set of {x, y}-points. Because of the symmetry of C, we restrict the {x, y}-points to the first quadrant.

```
In[21]:= ξηData = With[{pp = 200},
                  Table[{{x, y}, ξηPoints[{x, y}]}, {y, 0, L, L/pp}, {x, 0, L, L/pp}]];
```

The following graphic shows the number of normals as a function of x and y. The wavefront caustics divide the regions of identical number of normals.

```
In[22]:= ListDensityPlot[Map[Length[Last[#]]&, ξηData, {2}], Mesh -> False,
                  MeshRange -> {{0, 1}, {0, 1}}, PlotRange -> All]
```

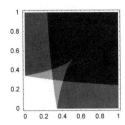

For the above expression for $I_k(x, y)$, we can easily calculate the expressions $|a'(\varphi) y + b'(\varphi) x + c'(\varphi)|$ and $s'(\varphi)$.

```
In[23]:= δDen[{x_, y_}, φ_] =
          Abs[Simplify[D[normalLine[{R[φ] Cos[φ], R[φ] Sin[φ]}, {x, y}], φ]]]
```

$$
Out[23]= \text{Abs}\left[\frac{1}{(\text{Cos}[\varphi]^4 + \text{Sin}[\varphi]^4)^2} \right.
$$
$$
\left(\text{Cos}[\varphi]^6 - 3\,\text{Cos}[\varphi]^4\,\text{Sin}[\varphi]^2 + \text{Sin}[\varphi]^6 + 3\,x\,\text{Cos}[\varphi]^3\,\text{Sin}[\varphi]^2\,(\text{Cos}[\varphi]^4 + \text{Sin}[\varphi]^4)^{1/4} -\right.
$$
$$
\left.\left. 3\,\text{Cos}[\varphi]^2\,\text{Sin}[\varphi]^3\,(\text{Sin}[\varphi] - y\,(\text{Cos}[\varphi]^4 + \text{Sin}[\varphi]^4)^{1/4})\right)\right]
$$

```
In[24]:= s[φ_] = Sqrt[D[R[φ] Cos[φ], φ]^2 + D[R[φ] Sin[φ], φ]^2] // Simplify
```

$$
Out[24]= 2\sqrt{\frac{5 + 3\,\text{Cos}[4\,\varphi]}{(3 + \text{Cos}[4\,\varphi])^{5/2}}}
$$

Now, we have everything together to generate the animation. ΦCF is a compiled function that calculates the sum, for a given value of k, a point xy, and the intersection points $\xi\eta$s of the normals with C.

```
In[25]:= ICF = Compile[{k, {xy, _Real, 1}, {ξηs, _Real, 2}},
          Module[{x = xy[[1]], y = xy[[2]], ξ, η, φ, l, Φ = 0.I, ψ},
              (* sum over all normals *)
              Sum[{ξ, η} = ξηs[[j]]; φ = ArcTan[ξ, η];
                  l = Sqrt[(x - ξ)^2 + (y - η)^2];
                  #1/#2 Exp[I k l], {j, Length[ξηs]}]]&[s[φ], δDen[{x, y}, φ]];
```

The function `fullMatrix` uses the reflection symmetry of C to generate an array of $I_k(x, y)$-values for all points inside C (meaning it adds the points from the second, third, and fourth quadrants).

```
In[26]:= fullMatrix[m_] := Join[Reverse[Rest[#]], #]& /@ Join[Reverse[Rest[m]], m]
```

Here is finally the resulting animation.

```
In[27]:= (* show sum of ray contributions, supercircle, and caustics *)
          superCircleDataGraphics[m_] :=
          Show[{ListDensityPlot[fullMatrix[m], Mesh -> False,
                      MeshRange -> L {{-1, 1}, {-1, 1}},
                      PlotRange -> Automatic, FrameTicks -> None,
                      DisplayFunction -> Identity],
                  superCircleGraphics, causticsGraphics},
              DisplayFunction -> $DisplayFunction, Frame -> False]
```

```
In[29]:= animationGraphicsI[k_, f_] :=
          superCircleDataGraphics[f @ Apply[ICF[k, ##]&, N[ξηData], {2}]]
```

```
In[30]:= Show[GraphicsArray[Block[{$DisplayFunction = Identity},
              animationGraphicsI[#, Log[Abs[Re[#]] + 1]&]& /@ #]]& /@
                  (* selected frames *) {{0, 10, 30}, {60, 90, 120}}
```

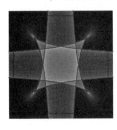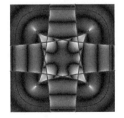

 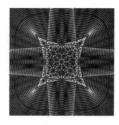 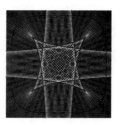

```
Do[animationGraphicsI[k, Log[Abs[Re[#]] + 1]&], {k, 1, 150}]
```

Now, we will calculate and visualize $\tilde{I}_k(x, y)$. This time we have to carry out a numerical integration for each point $\{x, y\}$. Because of the periodic nature of the integrand, we use the `Trapezoidal` method. And, for visualization purposes, the precision goal value 3 is sufficient.

```
In[31]:= IT[k_, {x_, y_}, opts___] := NIntegrate[Evaluate[
          Exp[I k Sqrt[(x - R[φ] Cos[φ])^2 + (y - R[φ] Sin[φ])^2]] s[φ]] ,
                                        {φ, 0, 2Pi}, opts]

In[32]:= animationGraphicsIT[k_, f_] :=
          Module[{pp = 200, dataIT, m},
          (* calculate values for 0 <= x <= y *)
          dataIT = Table[Table[IT[k, {x, y}, PrecisionGoal -> 3, AccuracyGoal -> 2,
                                        MaxRecursion -> 16, Method -> Trapezoidal],
                          {x, 0, y, 1/pp}], {y, 0, L, 1/pp}];
          (* mirror to obtain remaining values for first quadrant *)
          m = Table[0.I, {pp}, {pp}];
          Do[m[[i, j]] = dataIT[[i, j]]; m[[j, i]] = dataIT[[i, j]], {i, pp}, {j, i}];
          (* add supercircle and caustics *)
          superCircleDataGraphics[f @ m]]
```

The next graphics show $I_k(x, y)$ and $\tilde{I}_k(x, y)$, the absolute value and the argument for two k-values. No direct relation between the two graphic exists. $I_k(x, y)$ has caustics (we display the logarithm), while $\tilde{I}_k(x, y)$ has no singularities.

```
In[33]:= (* absolute value for k == 120; use logarithmic scale for I *)
          Show[GraphicsArray[
          Block[{$DisplayFunction = Identity},
                {animationGraphicsI[120, Log[Abs[#] + 1]&],
                 animationGraphicsIT[120, Abs]}]]]
```

 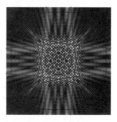

```
In[35]:= (* argument value for k == 30 *)
          Show[GraphicsArray[
          Block[{$DisplayFunction = Identity},
                {animationGraphicsI[30, Arg], animationGraphicsIT[30, Arg]}]]]
```

26. Change of Variables in a Differential Equation, Functional Equation

a) We use the chain rule for rewriting the quantities y'[x] and y''[x] as functions of u'[x1] and u''[x1]:

```
y'[x] == D[y[x], x]
   == D[u[x1], x]
   == D[x1, x] D[u[x1], x1]
   == y''[x] u'[x1]
```

```
y''[x] == D[y'[x], x]
   == D[y''[x] u'[x1], x]
   == D[y''[x], x] u'[x1] + y''[x] D[u'[x1], x]
   == y'''[x] u'[x1] + D[x1, x] D[u'[x1], x1]
   == y'''[x] u'[x1] + y''[x]^2 u''[x1]
```

So, we arrive at the following set of equations. (Be aware that we added the differentiated form of the original equation to get an expression for y'''[x].)

```
In[1]:= eqs = {x1 == y'[x],
            u[x1] == y[x],
            y'[x] == y''[x] u'[x1],
            y''[x] == y'''[x] u'[x1] + y''[x]^2 u''[x1],
            y''[x] == x^3 + y[x]^3 + y'[x]^3,
            y'''[x] == 3 x^2 + 3 y[x]^2 y'[x] + 3 y'[x]^2 y''[x]}
```

Out[1]= $\{x1 = y'[x], u[x1] = y[x], y'[x] = u'[x1] y''[x], y''[x] = u''[x1] y''[x]^2 + u'[x1] y^{(3)}[x],$
$y''[x] = x^3 + y[x]^3 + y'[x]^3, y^{(3)}[x] = 3 x^2 + 3 y[x]^2 y'[x] + 3 y'[x]^2 y''[x]\}$

Now, we eliminate the five variables x, y[x], y'[x], y''[x], and y'''[x] from the above six equations to get the desired differential equation for u[x1].

(We temporarily use the variables y, ys, yss, ysss, u, us, uss for GroebnerBasis.)

```
In[2]:= eqs1 = Apply[Subtract,
      eqs /. y[x] -> y /. y'[x] -> ys /. y''[x] -> yss /. y'''[x] -> ysss /.
         u[x1] -> u /. u'[x1] -> us /. u''[x1] -> uss, {1}]
```

Out[2]= $\{x1 - ys, u - y, ys - us yss, yss - uss yss^2 - us ysss,$
$-x^3 - y^3 - ys^3 + yss, -3 x^2 - 3 y^2 ys - 3 ys^2 yss + ysss\}$

```
In[3]:= gb = GroebnerBasis[eqs1, {x1, u, us, uss}, {x, y, ys, yss, ysss},
                    MonomialOrder -> EliminationOrder][[1]]
```

Out[3]= $27 u^6 us^9 - 54 u^3 us^8 x1 + 27 us^7 x1^2 - us^3 x1^3 + 9 u^2 us^5 x1^3 - 27 u^4 us^7 x1^3 + 54 u^3 us^9 x1^3 +$
$27 u^6 us^9 x1^3 - 54 us^8 x1^4 + 3 us^2 uss x1^4 - 18 u^2 us^4 uss x1^4 + 27 u^4 us^6 uss x1^4 + 9 us^4 x1^5 -$
$54 u^2 us^6 x1^5 + 81 u^4 us^8 x1^5 - 3 us uss^2 x1^5 + 9 u^2 us^3 uss^2 x1^5 + 27 us^9 x1^6 - 18 us^3 uss x1^6 +$
$54 u^2 us^5 uss x1^6 + uss^3 x1^6 - 27 us^5 x1^7 + 81 u^2 us^7 x1^7 + 9 us^2 uss^2 x1^7 + 27 us^4 uss x1^8 + 27 us^6 x1^9$

```
In[4]:= newOde = gb /. u -> u[x1] /. us -> u'[x1] /. uss -> u''[x1]
```

Out[4]= $-x1^3 u'[x1]^3 + 9 x1^5 u'[x1]^4 - 27 x1^7 u'[x1]^5 + 9 x1^3 u[x1]^2 u'[x1]^5 + 27 x1^9 u'[x1]^6 -$
$54 x1^5 u[x1]^2 u'[x1]^6 + 27 x1^2 u'[x1]^7 + 81 x1^7 u[x1]^2 u'[x1]^7 - 27 x1^3 u[x1]^4 u'[x1]^7 -$
$54 x1^4 u'[x1]^8 - 54 x1 u[x1]^3 u'[x1]^8 + 81 x1^5 u[x1]^4 u'[x1]^8 + 27 x1^6 u'[x1]^9 +$
$54 x1^3 u[x1]^3 u'[x1]^9 + 27 u[x1]^6 u'[x1]^9 + 27 x1^3 u[x1]^6 u'[x1]^9 + 3 x1^4 u[x1]^2 u'[x1]^2 u''[x1] -$
$18 x1^6 u'[x1]^3 u''[x1] + 27 x1^8 u'[x1]^4 u''[x1] - 18 x1^4 u[x1]^2 u'[x1]^4 u''[x1] +$
$54 x1^6 u[x1]^2 u'[x1]^5 u''[x1] + 27 x1^4 u[x1]^4 u'[x1]^6 u''[x1] - 3 x1^5 u'[x1] u''[x1]^2 +$
$9 x1^7 u'[x1]^2 u''[x1]^2 + 9 x1^5 u[x1]^2 u'[x1]^3 u''[x1]^2 + x1^6 u''[x1]^3$

Now let us calculate the series solution for u[x1]. Because we do not know the general form of the series terms of u[x1], we calculate the series for y[x], which is a pure power series. Then we convert this series into a series for u[x1].

Here, the first 13 terms of the series for y[x] are calculated.

```
In[5]:= cBag = {};
        Do[(* substitute already-calculated terms *)
           y[x_] = 0 + 0 x + Sum[c[i] x^i, {i, 2, n}] + O[x]^(n + 1) /. cBag;
           (* plug in y[x] in the differential equation *)
           odeRest = y''[x] - (x^3 + y[x]^3 + y'[x]^3);
           (* the rest must vanish *)
           mustBeZero = SeriesCoefficient[odeRest, n - 2];
           (* solve for the next coefficient c[n] *)
           sol = Solve[mustBeZero == 0, c[n]][[1, 1]];
           AppendTo[cBag, sol], {n, 2, 50}]
```

```
In[7]:= seriesSolutionY = 0 + 0 x + Sum[c[i] x^i, {i, 2, 50}] +
                          O[x]^(51 + 1) /. cBag
```

$$\text{Out[7]}= \frac{x^5}{20} + \frac{x^{14}}{11648} + \frac{x^{17}}{2176000} + \frac{3\,x^{23}}{6735872} + \frac{393\,x^{26}}{121139200000} + \frac{3\,x^{29}}{706764800000} +$$
$$\frac{183\,x^{32}}{60428386304} + \frac{58948767\,x^{35}}{2042397220864000000} + \frac{692109\,x^{38}}{9265501306880000000} +$$
$$\frac{552680721620117\,x^{41}}{2361957612806471680000000000} + \frac{2792961751271\,x^{44}}{9966584725603195289600000000} +$$
$$\frac{3669093790017\,x^{47}}{3398294084344532172800000000} + \frac{2383434617247962925203\,x^{50}}{1222419755980981230632960000000000000} + O[x]^{52}$$

Substituting the series into the differential equation for y[x] verifies that the solution is correct.

```
In[8]:= y1''[x] - (x^3 + y1[x]^3 + y1'[x]^3) /.
                   y1 -> Function[x, Evaluate[seriesSolutionY]]
```

$$\text{Out[8]}= O[x]^{50}$$

Now, we calculate x as a function of x1 by inverting the relation x1 = y[x].

```
In[9]:= xOfx1 = InverseSeries[D[seriesSolutionY, x], x1]
```

$$\text{Out[9]}= \sqrt{2}\,x1^{1/4} - \frac{x1^{5/2}}{26} - \frac{x1^{13/4}}{1000\sqrt{2}} + \frac{97\,x1^{19/4}}{14872\sqrt{2}} + \frac{1489\,x1^{11/2}}{4550000} +$$
$$\frac{2971\,x1^{25/4}}{476000000\sqrt{2}} - \frac{81\,x1^7}{136214} - \frac{2206812117\,x1^{31/4}}{14246632400000\sqrt{2}} - \frac{299578749\,x1^{17/2}}{74410700000000} +$$
$$\frac{933275794381453793\,x1^{37/4}}{10056210063099200000000\sqrt{2}} + \frac{6544294647507\,x1^{10}}{216019906252150000} +$$
$$\frac{348018511894286811\,x1^{43/4}}{11032734596884000000000\sqrt{2}} - \frac{576983666084852929 79397\,x1^{23/2}}{15761485195982805479200000000} + O[x1]^{12}$$

Substituting this series into the series solution for y[x] gives the series solution for u[x1] we are looking for.

```
In[10]:= seriesSolutionU = (seriesSolutionY /. x -> xOfx1) // Simplify
```

$$\text{Out[10]}= \frac{1}{5}\sqrt{2}\,x1^{5/4} - \frac{5\,x1^{7/2}}{182} - \frac{13\,x1^{17/4}}{17000\sqrt{2}} + \frac{1843\,x1^{23/4}}{342056\sqrt{2}} +$$
$$\frac{16379\,x1^{13/2}}{59150000} + \frac{2971\,x1^{29/4}}{552160000\sqrt{2}} - \frac{567\,x1^8}{1089712} - \frac{68411175627\,x1^{35/4}}{498632134000000\sqrt{2}} -$$
$$\frac{299578749\,x1^{19/2}}{83164900000000} + \frac{34531204392113790341\,x1^{41/4}}{412304612587067200000000\sqrt{2}} + \frac{6544294647507\,x1^{11}}{237621896877365000} +$$
$$\frac{1496479601145 4332873\,x1^{47/4}}{5185385260535480000000000\sqrt{2}} - \frac{13270624319951617385 26131\,x1^{25/2}}{39403712989957013698000000000} + O[x1]^{13}$$

Here is a quick check for the correctness of seriesSolutionU.

```
In[11]:= newOde /. u -> Function[x1, Evaluate[seriesSolutionU]]
```

$$\text{Out[11]}= O[x1]^{31/2}$$

Now, we see the structure of the series; it is a series in x1^(1/4). So, we could equivalently have done the following.

```
cBag = {};
Do [(* substitute already calculated terms *)
    u[x1_] = Sqrt[2]/5 x1^(5/4) + Sum[c[i/4] x1^(i/4), {i, 6, n}] +
                                   O[x1]^((n + 1)/4) /. cBag;
    (* plug in y[x] in the differential equation *) odeRest = newOde;
    (* the rest must vanish *) mustBeZero = odeRest[[3, 1]];
    (* solve for the next coefficient c[n] *)
    sol = Solve[mustBeZero == 0, c[n/4]][[1, 1]];
    AppendTo[cBag, sol], {n, 6, 60}]
```

Here is a plot of the series solution. Plotting each partial sum of the series solution gives a rough idea about the range where the series solution is correct. This is done in the right picture.

```
In[12]:= seriesSolutionU1 = Normal[seriesSolutionU];
```

```
In[13]:= Show[GraphicsArray[
        Block[{$DisplayFunction = Identity},
          {Plot[Evaluate[seriesSolutionU1], {x1, 0, 3}],
           Plot[Evaluate[Rest[FoldList[Plus, 0,
                 List @@ seriesSolutionU1]]], {x1, 0, 3},
             PlotStyle -> Table[{Thickness[0.002], Hue[i/18]}, {i, 13}]]}]]]
```

Now let us compare the series solution with the numerical solution for u[x1]. We solve newOde with respect to u''[x].

```
In[14]:= s1 = (Solve[newOde == 0, u''[x1]] // Simplify) /.
              u[x1] -> u /. u'[x1] -> us
```

$$Out[14]= \left\{\left\{u''[x1] \to \frac{us\, x1^5 - 3\, u^2\, us^3\, x1^5 - 3\, us^2\, x1^7 + 3\left(-us^7\, x1^{12}\left(x1 - us\left(u^3 + x1^3\right)\right)^2\right)^{1/3}}{x1^6}\right\},\right.$$

$$\left\{u''[x1] \to \frac{2\, us\, x1^5 - 6\, u^2\, us^3\, x1^5 - 6\, us^2\, x1^7 + 3\, i\, (i + \sqrt{3})\left(-us^7\, x1^{12}\left(x1 - us\left(u^3 + x1^3\right)\right)^2\right)^{1/3}}{2\, x1^6}\right\},$$

$$\left.\left\{u''[x1] \to -\frac{-2\, us\, x1^5 + 6\, u^2\, us^3\, x1^5 + 6\, us^2\, x1^7 + 3\, (1 + i\, \sqrt{3})\left(-us^7\, x1^{12}\left(x1 - us\left(u^3 + x1^3\right)\right)^2\right)^{1/3}}{2\, x1^6}\right\}\right\}$$

It turns out that the second solution is the one that corresponds to the above series solution.

We now use NDSolve to solve the resulting differential equation. We cannot start with the numerical solution at x==0 because the right-hand side does not evaluate to a number. Instead, we use our above series solution to start at a small distance away from the origin.)

```
In[15]:= ε = 10^-2; (* start away from the origin *)
         nsol = NDSolve[(* the differential equation *)
           {u''[x1] ==  1/(2 x1^6) *
            (2 x1^5 u'[x1] - 6 x1^7 u'[x1]^2 -
             6 x1^5 u[x1]^2 u'[x1]^3 - 3 (-x1^12 u'[x1]^7
               (x1 - (x1^3 + u[x1]^3) u'[x1])^2)^(1/3) +
             3 I Sqrt[3] (-x1^12 u'[x1]^7 *
               (x1 - (x1^3 + u[x1]^3)  u'[x1])^2)^(1/3)),
             (* initial values from series solution *)
             u[ε] == seriesSolutionU1 /. x1 -> ε,
             u'[ε] == D[seriesSolutionU1, x1] /. x1 -> ε}, u[x1], {x1, ε, 3},
             MaxSteps -> 10000, WorkingPrecision -> 20,
             PrecisionGoal -> 15, AccuracyGoal -> 12];
```

Here, the numerically calculated solution shown (we use Re because due to numerical inaccuracies, we got a small imaginary part from extracting the cube root). The right plot shows the difference of the partial sums of the above-calculated series solution to the numerical solution.

```
In[17]:= Show[GraphicsArray[
         Block[{$DisplayFunction = Identity},
           {Plot[Evaluate[Re[u[x1]] /. nsol], {x1, ε, 3}, PlotRange -> All],
            Plot[Evaluate[Re[u[x1] /. nsol[[1]]] - (* list of partial sums *)
                        Rest[FoldList[Plus, 0, List @@ seriesSolutionU1]]],
                {x1, ε, 3}, PlotRange -> All, PlotStyle ->
                        Table[{Thickness[0.002], Hue[i/18]}, {i, 13}]]}]]]
```

Because for $x < 2$ the difference between the series solution and the numerical solution is quite small, let us look at the logarithm of the difference.

```
In[18]:= Plot[Evaluate[Log[10, Abs[Re[u[x1] /. nsol[[1]]] -
                        (* list of partial sums *)
                        Rest[FoldList[Plus, 0, List @@ seriesSolutionU1]]]]],
              {x1, ε, 3}, PlotRange -> All, PlotStyle ->
                        Table[{Thickness[0.002], Hue[i/18]}, {i, 13}]]
```

The large cusps in the above curve are due to the different signs of the last term of the partial sums.

b) Because $|q| > 1$, it is straightforward to use the functional equation to iteratively make the argument smaller.

```
In[1]:= fContinued[q_, z_] := (q/2 (fContinued[q, z/q]^2 - 1) + 1) /; Abs[z] > ε
```

For a small enough argument z, we will use the first few terms of the series of $f_q(z)$. Using a generic ansatz of the form $f_q(z) = c_0(q) + c_1(q) z + c_2(q) z^2 + \cdots$, it is straightforward to calculate as many series terms as wanted. Here we calculate the first four.

```
In[2]:= eqs = ((f[q x] - (2 - q + q f[x]^2)/2) /.
             f[x_] :> Sum[c[k] x^k, {k, 0, 4}]) + O[x]^5
```

$$\text{Out[2]=} \left(-1 + \frac{q}{2} + c[0] - \frac{1}{2} q\,c[0]^2\right) + (q\,c[1] - q\,c[0]\,c[1])\,x +$$
$$\left(-\frac{1}{2} q\,c[1]^2 + q^2\,c[2] - q\,c[0]\,c[2]\right) x^2 + (-q\,c[1]\,c[2] + q^3\,c[3] - q\,c[0]\,c[3])\,x^3 +$$
$$\left(-\frac{1}{2} q\,c[2]^2 - q\,c[1]\,c[3] + q^4\,c[4] - q\,c[0]\,c[4]\right) x^4 + O[x]^5$$

Inspecting the equations shows that, for generic q, we must have $c_0(q) = 1$ and without loss of generality, we choose $c_1(q) = 1$.

```
In[3]:= cl = CoefficientList[eqs, x] // Factor
```

Out[3]= $\left\{-\frac{1}{2}\,(-1+c[0])\,(-2+q+q\,c[0]),\, -q\,(-1+c[0])\,c[1],\, \frac{1}{2}\,q\,(-c[1]^2+2\,q\,c[2]-2\,c[0]\,c[2]),\right.$

$\left. q\,(-c[1]\,c[2]+q^2\,c[3]-c[0]\,c[3]),\, \frac{1}{2}\,q\,(-c[2]^2-2\,c[1]\,c[3]+2\,q^3\,c[4]-2\,c[0]\,c[4])\right\}$

In[4]:= `sol = Solve[# == 0& /@ (cl /. {c[0] -> 1, c[1] -> 1}),`
`{c[2], c[3], c[4]}] // Simplify`

Out[4]= $\left\{\left\{c[3]\to\frac{1}{2\,(-1+q)^2\,(1+q)},\, c[4]\to\frac{5+q}{8\,(-1+q)^3\,(1+2\,q+2\,q^2+q^3)},\, c[2]\to\frac{1}{2\,(-1+q)}\right\}\right\}$

In[5]:= `fSeries[q_, z_] = 1 + z + Sum[c[k] z^k, {k, 2, 4}] /. sol[[1]]`

Out[5]= $1+z+\frac{z^2}{2\,(-1+q)}+\frac{z^3}{2\,(-1+q)^2\,(1+q)}+\frac{(5+q)\,z^4}{8\,(-1+q)^3\,(1+2\,q+2\,q^2+q^3)}$

In[6]:= `fContinued[q_, z_] := fSeries[q, z] /; Abs[z] <= ε`

Here is a quick high-precision check that the so-calculated values for $f_q(z)$ fulfill the original functional equation.

In[7]:= `Block[{ε = 10^-10, q = Pi/2, z = N[I E, 30], f = fContinued},`
`f[q, q z] - (2 - q + q f[q, z]^2)/2]`

Out[7]= $0.\times10^{-28}+0.\times10^{-29}\,i$

For an animation, we want to calculate the function values quickly. So, we implement a compiled version of the above formulas.

In[8]:= `cf = Compile[{{a, _Complex}, {z, _Complex}},`
`Module[{ε = 10.^-10, ζ = z, c = 0, z0, f0},`
`While[Abs[ζ] > ε, ζ = ζ/a; c++];`
`Nest[(a/2 (#^2 - 1) + 1)&,`
`1 + ζ + ζ^2/(2(a - 1)) + ζ^3/(2(a - 1)^2(a + 1)) +`
`ζ^4 (5 + a)/(8(a - 1)^3(1 + a)(a^2 + a + 1)), c]]];`

In[9]:= `(* check if compilation was successful *)`
`Union[Head /@ Flatten[cf[[4]]]]`

Out[10]= `{Integer, Real}`

For the visualization of $f_q(z)$, we will choose a contour plot of $\text{Im}(f_q(z)) \leq 0$ and simultaneously display the contour lines of $\text{Re}(f_q(z)) = 0$.

In[11]:= `graphic[{r_:1.65, φ_}, L_:25, pp_:400, opts___] :=`
`Module[{(* generate data *)`
`data = Table[cf[r Exp[I φ], x + I y],`
`{y, -L, L, 2L/pp}, {x, -L, L, 2L/pp}],`
`(* set graphics options *)`
`opts1 = Sequence[Contours -> {0}, DisplayFunction -> Identity]},`
`Show[{ListContourPlot[Im[data], opts1],`
`ListContourPlot[Re[data], opts1, ContourShading -> False,`
`ContourStyle -> {Hue[0]}]},`
`opts, Frame -> False]]`

In[12]:= `With[{o = 16, L = 25, pp = 200},`
`Show[GraphicsArray[#]]& /@ Partition[`
`Table[graphic[{φ}, L, pp, DisplayFunction -> Identity],`
`{φ, 0, 2Pi(1 - 1/o), 2Pi/o}], 4]]`

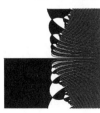

Here is the corresponding animation.

```
With[{o = 128, L = 25, pp = 240},
     Do[graphic[{φ}, L, pp, DisplayFunction -> $DisplayFunction],
        {φ, 0, 2Pi(1 - 1/o), 2Pi/o}]]
```

For some theoretical considerations of related functional equations, see [1526].

27. Discriminant Surface, Multivalued Surface, 27 Lines on the Clebsch Surface

a) Here, the implicit definition is implemented.

In[1]:= disc = 256 z^3 + 128 x^2 z^2 + 16 x^4 z - 144 x z y^2 - 4 x^3 y^2 + 27 y^4;

This determines the lower values for the z-coordinate for the picture. (This can be easily verified with the next picture.)

In[2]:= {#, N[#]}&[Solve[disc == 0 /. {x -> 2, y -> 3/2}, z]]

Out[2]= $\left\{\left\{\left\{z \to \frac{5}{16}\right\}, \left\{z \to \frac{1}{32}\left(-37 - 13\sqrt{13}\right)\right\}, \left\{z \to \frac{1}{32}\left(-37 + 13\sqrt{13}\right)\right\}\right\},\right.$
$\left.\{\{z \to 0.3125\}, \{z \to -2.62101\}, \{z \to 0.308505\}\}\right\}$

To get a rough overview of the surface, we use ContourPlot3D.

In[3]:= Needs["Graphics`ContourPlot3D`"]

In[4]:= ContourPlot3D[disc, {x, -1, 2}, {y, -3/2, 3/2},
                    (* to fit good *) {z, -(592 + 208Sqrt[13])/512, 1},
                    MaxRecursion -> 1, PlotPoints -> {12, 3},
                    PlotRange -> All, Axes -> True, AxesLabel -> {x, y, z}]

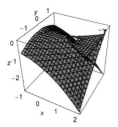

Now, we see the problem. This surface has some sharp edges, and due to the regularly spaced tensor product grid used by `ContourPlot3D`, the sharp edges cannot be modeled with high quality. Some of the partial derivatives must vanish at the sharp edges. We can use this fact to get a parametric representation $\{x, y(x), z(x)\}$ of the sharp edges.

In[5]:= `Solve[{D[disc, y] == 0, D[disc, z] == 0, disc == 0}, {y, z}]`

Out[5]= $\left\{\left\{z \to -\dfrac{x^2}{4},\ y \to 0\right\},\ \left\{z \to \dfrac{x^2}{12},\ y \to -\dfrac{2}{3}\sqrt{\dfrac{2}{3}}\,x^{3/2}\right\},\right.$

$\left.\left\{z \to \dfrac{x^2}{12},\ y \to -\dfrac{2}{3}\sqrt{\dfrac{2}{3}}\,x^{3/2}\right\},\ \left\{z \to \dfrac{x^2}{12},\ y \to \dfrac{2}{3}\sqrt{\dfrac{2}{3}}\,x^{3/2}\right\},\ \left\{z \to \dfrac{x^2}{12},\ y \to \dfrac{2}{3}\sqrt{\dfrac{2}{3}}\,x^{3/2}\right\}\right\}$

Here is the above picture together with the parametrized edges.

In[6]:= `Show[{%%, Graphics3D[{`
`    {Thickness[0.02], (* the sharp edges *) Line /@ {`
`    Table[{x, -(2x/3)^(3/2), x^2/12}, {x, 0, 2, 0.01}],`
`    Table[{x, +(2x/3)^(3/2), x^2/12}, {x, 0, 2, 0.01}],`
`    Table[{x, 0, -x^2/4}, {x, 0, 2, 0.01}]}}}]},`
`    PlotRange -> All, Axes -> True]`

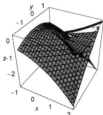

To get a better resolution of the sharp edges, we calculate all parts of the surface bounded by these edges separately. From the above crude picture (and as well from the equation), we see that $z = z(x, y)$ is in general a three-valued function, and by sorting the three real roots of `disc == 0`, we get all parts of the surface separately. `prePoints` are the x,y-coordinates of the points used to calculate the corresponding z-values numerically.

In[7]:= `pp = 12;`
`    prePoints = Table[{x, i/pp 2/3 Sqrt[2/3] x^(3/2)},`
`                      {x, 0, 2, 2/pp}, {i, 0, pp}] // N;`

This is the grid formed by these x,y-coordinates. (Because of symmetry, we calculate only points with $y > 0$.)

In[9]:= `Show[Graphics[{Line /@ prePoints, Line /@ Transpose[prePoints]}],`
`    AxesLabel -> {"x", "y"}, Axes -> True]`

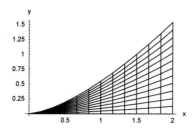

We calculate and sort the roots $z(x, y)$.

```
In[10]:= roots = Map[Sort[List @@ (Last (* the last is of interest *) /@
                NRoots[(* the polynomial *) disc == 0 /.
                    Apply[Rule, Transpose[{{x, y}, #}], {1}], z])]&,
                prePoints, {2}];
```

It gives the following two parts of the surface, which are bounded by the sharp edges.

```
In[11]:= {s2, s3} = Function[p, Map[#[[p]]&, roots, {2}]] /@ {2, 3};
         {f2, f3} = MapThread[Append, {prePoints, #}, 2]& /@ {s2, s3};
```

For the graphical visualization, we again use `ListSurfacePlot3D` from the package `Graphics`Graphics3D``.

```
In[13]:= Needs["Graphics`Graphics3D`"]
```

```
In[14]:= pic1 = Show[ListSurfacePlot3D[#, DisplayFunction -> Identity]& /@
                {f2, f3}, DisplayFunction -> $DisplayFunction,
                PlotRange -> All, Axes -> True]
```

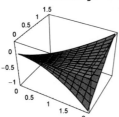

The missing third part of the surface is now easily calculated and displayed.

```
In[15]:= f1 = Table[{ξ, η, If[η === 0 && ξ <= 0, 0,
                (* choose correct root *) Min[Cases[Chop[Last /@
                NRoots[disc == 0 /. {x -> ξ, y -> η}, z]], _Real]]]},
                {ξ, -1, 2, 2/pp}, {η, 0, 3/2, (3/2)/pp}] /.
                            {0, 0, Infinity} -> {0, 0, 0};
In[16]:= pic2 = ListSurfacePlot3D[f1, PlotRange -> All, Axes -> True]
```

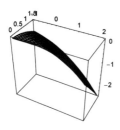

After mirroring the surface constructed so far to get the second half, we finally have the following picture, which is much better than the earlier `ContourPlot3D`.

```
In[17]:= Show[Graphics3D[{(* mirror on x,z-plane *)
                pic1[[1]], Map[{1, -1, 1}#&, pic1[[1]], {-2}],
```

```
        pic2[[1]], Map[{1, -1, 1}#&, pic2[[1]], {-2}]}],
        PlotRange -> All, ViewPoint -> {1.8, -2.2, 2.2}, Boxed -> False]
```

b) To get an impression of the form of the surface under consideration, let us use `ContourPlot3D` for a first sketch. (We define the variable `pi` as the numerical value of `Pi`, which we will use very often in the following.)

```
In[1]:= Needs["Graphics`ContourPlot3D`"]
        pi = N[Pi];

In[3]:= ContourPlot3D[Sin[x] Cos[y] + Sin[y] Cos[z] + Sin[z] Cos[x],
                {x, 0, 2pi}, {y, 0, 2pi}, {z, 0, 2pi}, Axes -> True,
                PlotPoints -> 18, MaxRecursion -> 0]
```

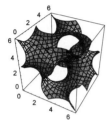

This is the implicit form of the surface.

```
In[4]:= F[x_, y_, z_] = Sin[x] Cos[y] + Sin[y] Cos[z] + Sin[z] Cos[x];
```

To avoid the use of `ContourPlot3D` for the generation of the picture, we need a parametric description of the surface (at least locally). The x, y, and z appear in a symmetrical way in `F`; so, we pick z as the dependent variable and we try to construct $z = z(x, y)$. We have four different solutions for z.

```
In[5]:= Solve[F[x, y, z] == 0, z] // Simplify
```

```
        Solve::ifun : Inverse functions are being used by Solve, so some solutions
                may not be found; use Reduce for complete solution information. More…
```

$$Out[5]= \left\{ \left\{ z \to -\text{ArcCos}\left[-\frac{1}{\text{Cos}[x]^2 + \text{Sin}[y]^2} \left(\text{Cos}[y] \, \text{Sin}[x] \, \text{Sin}[y] + \right. \right. \right. \right.$$
$$\left. \left. \text{Cos}[x] \, \text{Csc}[y] \sqrt{\text{Sin}[y]^2 \, (\text{Cos}[x]^2 - \text{Cos}[y]^2 \, \text{Sin}[x]^2 + \text{Sin}[y]^2)} \right) \right] \right\},$$
$$\left\{ z \to \text{ArcCos}\left[-\frac{1}{\text{Cos}[x]^2 + \text{Sin}[y]^2} \left(\text{Cos}[y] \, \text{Sin}[x] \, \text{Sin}[y] + \right. \right. \right.$$
$$\left. \left. \text{Cos}[x] \, \text{Csc}[y] \sqrt{\text{Sin}[y]^2 \, (\text{Cos}[x]^2 - \text{Cos}[y]^2 \, \text{Sin}[x]^2 + \text{Sin}[y]^2)} \right) \right] \right\},$$
$$\left\{ z \to -\text{ArcCos}\left[\frac{1}{\text{Cos}[x]^2 + \text{Sin}[y]^2} \left(-\text{Cos}[y] \, \text{Sin}[x] \, \text{Sin}[y] + \right. \right. \right.$$
$$\left. \left. \text{Cos}[x] \, \text{Csc}[y] \sqrt{\text{Sin}[y]^2 \, (\text{Cos}[x]^2 - \text{Cos}[y]^2 \, \text{Sin}[x]^2 + \text{Sin}[y]^2)} \right) \right] \right\},$$
$$\left\{ z \to \text{ArcCos}\left[\frac{1}{\text{Cos}[x]^2 + \text{Sin}[y]^2} \left(-\text{Cos}[y] \, \text{Sin}[x] \, \text{Sin}[y] + \right. \right. \right.$$
$$\left. \left. \text{Cos}[x] \, \text{Csc}[y] \sqrt{\text{Sin}[y]^2 \, (\text{Cos}[x]^2 - \text{Cos}[y]^2 \, \text{Sin}[x]^2 + \text{Sin}[y]^2)} \right) \right] \right\} \right\}$$

Taking into account that *Mathematica* chooses the values of arccos in the range $[0, \pi]$, we define the following solutions for $z = z(x, y)$.

```
In[6]:= {sol1[x_, y_], sol2[x_, y_]} =
        (-Cos[y] Sin[x] Sin[y] - # Sqrt[Cos[x]^2 *
```

$$(Cos[x]\verb|^|2 - Cos[y]\verb|^|2 \; Sin[x]\verb|^|2 + Sin[y]\verb|^|2)])/$$
$$(Cos[x]\verb|^|2 + Sin[y]\verb|^|2)\& \; /@ \; \{+1, \; -1\}$$

Out[6]= $\left\{ \dfrac{-Cos[y]\,Sin[x]\,Sin[y] - \sqrt{Cos[x]^2\,(Cos[x]^2 - Cos[y]^2\,Sin[x]^2 + Sin[y]^2)}}{Cos[x]^2 + Sin[y]^2}, \right.$

$\left. \dfrac{-Cos[y]\,Sin[x]\,Sin[y] + \sqrt{Cos[x]^2\,(Cos[x]^2 - Cos[y]^2\,Sin[x]^2 + Sin[y]^2)}}{Cos[x]^2 + Sin[y]^2} \right\}$

In[7]:= `f1[x_, y_] = +ArcCos[sol1[x, y]]; f2[x_, y_] = +ArcCos[sol2[x, y]];`
`f3[x_, y_] = -ArcCos[sol1[x, y]]; f4[x_, y_] = -ArcCos[sol2[x, y]];`

Now, we must look for which regions of the x,y-plane where the four functions f1, f2, f3, and f4 are really solutions. The solution should satisfy the following two conditions: a) $z(x, y)$ must be real, and b) $z(x, y)$ must be a solution of the implicit equation. For better visibility of the various regions (shown in black if the two conditions are satisfied) that are related to each other by translations and reflections, we add some grid lines.

In[9]:= `grid = {GrayLevel[1/2], Line[{{0, #}, {2pi, #}}]& /@ (pi/2 Range[0, 4]),`
`Line[{{#, 0}, {#, 2pi}}]& /@ (pi/2 Range[0, 4])};`

In[10]:= `Show[GraphicsArray[#]]& /@ Map[ContourPlot[`
`  If[(* does the "solution" fulfill the equation? *)`
`    Abs[F[x, y, #1]] < 10^-10 &&`
`    (* is the solution real-valued? *) Im[#] === 0, -1, 1],`
`    {x, 0, N[2Pi]}, {y, 0, N[2Pi]}, Contours -> {0},`
`    PlotPoints -> 100, AspectRatio -> Automatic,`
`    Epilog -> grid, DisplayFunction -> Identity,`
`    Evaluate[FrameTicks -> {#, #}&[Table[i Pi/2, {i, 0, 4}]]]]&,`
`  (* the four solutions *)`
`  {{f1[x, y], f2[x, y]}, {f3[x, y], f4[x, y]}}, {2}]`

We see (modulo translation and reflection symmetry) three types of regions where our function F can be locally inverted by our f1 or f2 or f3 or f4. Let us analytically calculate the boundaries of these three regions. Because of the underlying symmetry, we restrict the calculation to the square of $\pi/2 \le x \le \pi, 0 \le y \le \pi/2$. The first restriction is given by the condition of real-valuedness of the f1, f2, f3, and f4. This means that arguments of the square root appearing in the fs must be nonnegative. Solving *argumentOfTheSquareRoot* $== 0$ for y thus gives the boundary curve in the form $y = y(x)$.

In[11]:= `Solve[{Cos[x]^4 - Cos[x]^2 Cos[y]^2 Sin[x]^2 +`
`  Cos[x]^2 Sin[y]^2 == 0 /. Cos[y]^2 -> 1 - Sin[y]^2},`
`  {Sin[y]}] // Simplify`

Out[11]= $\left\{ \left\{ Sin[y] \to -\dfrac{\sqrt{-Cos[2x]}}{\sqrt{1 + Sin[x]^2}} \right\}, \; \left\{ Sin[y] \to \dfrac{\sqrt{-Cos[2x]}}{\sqrt{1 + Sin[x]^2}} \right\} \right\}$

The second condition is that F[x, y, fi[x, y]] == 0 holds. A little bit of inspection shows that this means that f1[x, y] for a fixed x, x > x_{min} is a solution of F[x, y, fi[x, y]] == 0 for increasing values of y as long as soll[x, y] ≤ −1. A direct solution of soll[x, y] == −1 works just fine.

In[12]:= **Solve[soll[x, y] == -1, y]**

> Solve::ifun : Inverse functions are being used by Solve, so some solutions
> may not be found; use Reduce for complete solution information. More…

Out[12]= $\left\{\left\{y \to -\text{ArcCos}\left[-\dfrac{1}{\sqrt{1 + \text{Sin}[x]^2}}\right]\right\}, \left\{y \to \text{ArcCos}\left[-\dfrac{1}{\sqrt{1 + \text{Sin}[x]^2}}\right]\right\},\right.$
$\left.\left\{y \to -\text{ArcCos}\left[\dfrac{1}{\sqrt{1 + \text{Sin}[x]^2}}\right]\right\}, \left\{y \to \text{ArcCos}\left[\dfrac{1}{\sqrt{1 + \text{Sin}[x]^2}}\right]\right\}\right\}$

We rewrite this solution for sin(y) instead for y.

In[13]:= **Sin[y /. %] // Simplify // PowerExpand // Union**

Out[13]= $\left\{-\dfrac{\text{Sin}[x]}{\sqrt{1 + \text{Sin}[x]^2}}, \dfrac{\text{Sin}[x]}{\sqrt{1 + \text{Sin}[x]^2}}\right\}$

So, we have analytical descriptions of the two arcs bounding the region of validity of F[x, y, fi[x, y]] == 0. Our next step is to generate mesh for the three (up to translation and reflection symmetry) different regions. To get a nice-looking mesh, we subdivide the region in a polar coordinate system-like fashion generated by radial rays going outward from the point $(\pi/2, 0)$ and divide the rays into equal parts. To calculate the starting length for the rays, we have to solve the equation $y(x) = y_{ray}(x)$.

Here, the above two equations are rewritten as an equation in r for the two arcs. (We take $(\pi/2, 0)$ as the origin.)

In[14]:= **y1[r_, φ_] = ArcSin[Sqrt[-Cos[2(Pi/2 + r Cos[φ])]]/**
              Sqrt[1 + Sin[(Pi/2 + r Cos[φ])]^2]] ==
                            r Sin[φ] // ExpandAll

Out[14]= $\text{ArcSin}\left[\dfrac{\sqrt{\text{Cos}[2\,r\,\text{Cos}[φ]]}}{\sqrt{1 + \text{Cos}[r\,\text{Cos}[φ]]^2}}\right] == r\,\text{Sin}[φ]$

In[15]:= **y2[r_, φ_] = ArcSin[Sin[Pi/2 + r Cos[φ]]/**
              Sqrt[1 + Sin[Pi/2 + r Cos[φ]]^2]] ==
                            r Sin[φ] // ExpandAll

Out[15]= $\text{ArcSin}\left[\dfrac{\text{Cos}[r\,\text{Cos}[φ]]}{\sqrt{1 + \text{Cos}[r\,\text{Cos}[φ]]^2}}\right] == r\,\text{Sin}[φ]$

We solve the two equations numerically using FindRoot.

In[16]:= **radius1[φ_] := FindRoot[Evaluate[y1[r, φ]], {r, Pi/4}][[1, 2]]**
       radius2[φ_] := FindRoot[Evaluate[y2[r, φ]], {r, pi/4}][[1, 2]]

Now, we have all ingredients together to generate the meshes for the three regions of interest.

In[18]:= **ppr = 12; ppl = 12;**

In[19]:= **tab[1] = Table[#[[1]] + i/ppr(#[[2]] - #[[1]]), {i, 0, ppr}]& /@**
       Table[{radius1[φ] {Cos[φ], Sin[φ]},
                       (* the endpoint along the sides of the square *)
                       Which[φ <= pi/4, {pi/2, pi/2 Tan[φ]},
                             pi/4 <= φ <= 3pi/4, {pi/2 Cot[φ], pi/2},
                             φ > 3pi/4, {-pi/2, -pi/2 Tan[φ]}]},
                     {φ, 0, pi/2, pi/4/ppl}];

In[20]:= **tab[2] = Table[#[[1]] + i/ppr(#[[2]] - #[[1]]), {i, 0, ppr}]& /@**
       Table[{radius2[φ] {Cos[φ], Sin[φ]},
                       (* the endpoint along the sides of the square *)
                       Which[φ <= pi/4, {pi/2, pi/2 Tan[φ]},
                             pi/4 <= φ <= 3pi/4, {pi/2 Cot[φ], pi/2},
                             φ > 3pi/4, {-pi/2, -pi/2 Tan[φ]}]},
                     {φ, 0, pi/2, pi/4/ppl}];

       tab[3] = Table[#[[1]] + i/ppr(#[[2]] - #[[1]]), {i, 0, ppr}]& /@
       Table[{radius1[φ] {Cos[φ], Sin[φ]}, radius2[φ] {Cos[φ], Sin[φ]}},
                     {φ, 0, pi/2, pi/4/ppl}];

This shows the three meshes.

```
In[22]:= Show[GraphicsArray[
          Graphics[{Line /@ tab[#], Line /@ Transpose[tab[#]]},
            Frame -> True, AspectRatio -> Automatic]& /@ {1, 2, 3}]]
```

By translation and reflection, we generate the corresponding meshes in the other 15 pieces of the above-shown grid of the region $0 \le x < 2\pi, 0 \le y < 2\pi$.

```
In[23]:= Do[(* the first row *)
            x[i, 1, 1] = Map[{-1,  1}# + {pi/2, 0}&, tab[i], {-2}];
            x[i, 1, 2] = Map[# + {pi/2, 0}&, tab[i], {-2}];
            x[i, 1, 3] = Map[{-1,  1}# + {3pi/2, 0}&, tab[i], {-2}];
            x[i, 1, 4] = Map[# + {3pi/2, 0}&, tab[i], {-2}];
            (* the second row *)
            x[i, 2, 1] = Map[{-1, -1}# + {pi/2, pi}&, tab[i], {-2}];
            x[i, 2, 2] = Map[{ 1, -1}# + {pi/2, pi}&, tab[i], {-2}];
            x[i, 2, 3] = Map[{-1, -1}# + {3pi/2, pi}&, tab[i], {-2}];
            x[i, 2, 4] = Map[{ 1, -1}# + {3pi/2, pi}&, tab[i], {-2}];
            (* the third row *)
            x[i, 3, 1] = Map[{-1,  1}# + {pi/2, pi}&, tab[i], {-2}];
            x[i, 3, 2] = Map[# + {pi/2, pi}&, tab[i], {-2}];
            x[i, 3, 3] = Map[{-1,  1}# + {3pi/2, pi}&, tab[i], {-2}];
            x[i, 3, 4] = Map[# + {3pi/2, pi}&, tab[i], {-2}];
            (* the fourth row *)
            x[i, 4, 1] = Map[{-1, -1}# + {pi/2, 2pi}&, tab[i], {-2}];
            x[i, 4, 2] = Map[{ 1, -1}# + {pi/2, 2pi}&, tab[i], {-2}];
            x[i, 4, 3] = Map[{-1, -1}# + {3pi/2, 2pi}&, tab[i], {-2}];
            x[i, 4, 4] = Map[{ 1, -1}# + {3pi/2, 2pi}&, tab[i], {-2}],
            (* the three meshes *) {i, 3}]
```

Looking at the above contour plot that shows the validity, we can easily determine which mesh describes the region of validity of the four solutions f1, f2, f3, and f4 in each of the 16 pieces. (The validity of the various solutions in the various regions could also be established analytically with a bit more effort.)

```
In[24]:= mesh1 = {x[1, 1, 4], x[1, 2, 1], x[1, 3, 3], x[1, 4, 2],
                  x[2, 1, 1], x[2, 2, 4], x[2, 3, 2], x[2, 4, 3],
                  x[3, 1, 2], x[3, 2, 3], x[3, 3, 1], x[3, 4, 4]};

         mesh2 = {x[1, 1, 2], x[1, 2, 3], x[1, 3, 1], x[1, 4, 4],
                  x[2, 1, 3], x[2, 2, 2], x[2, 3, 4], x[2, 4, 1],
                  x[3, 1, 4], x[3, 2, 1], x[3, 3, 3], x[3, 4, 2]};

         mesh3 = {x[1, 1, 3], x[1, 2, 2], x[1, 3, 4], x[1, 4, 1],
                  x[2, 1, 2], x[2, 2, 3], x[2, 3, 1], x[2, 4, 4],
                  x[3, 1, 1], x[3, 2, 4], x[3, 3, 2], x[3, 4, 3]};

         mesh4 = {x[1, 1, 1], x[1, 2, 4], x[1, 3, 2], x[1, 4, 3],
                  x[2, 1, 4], x[2, 2, 1], x[2, 3, 3], x[2, 4, 2],
                  x[3, 1, 3], x[3, 2, 2], x[3, 3, 4], x[3, 4, 1]};
```

For comparison with the above contour graphic, here are the four meshed regions for f*i* to f4.

```
In[28]:= Show[GraphicsArray[#]]& /@
          Map[Graphics[{Thickness[0.002], (* make lines from mesh *)
          Map[{Line /@ #, Line /@ Transpose[#]}&, #]},
```

```
                AspectRatio -> Automatic, PlotRange -> All,
                Epilog -> grid]&, {mesh1, mesh2, mesh3, mesh4}, {1}]
```

To generate the three-dimensional graphic, we use the command `ListSurfacePlot3D` from the package `Graphics`
`Graphics3D``.

In[29]:= **Needs["Graphics`Graphics3D`"]**

Here, each of the four solutions is shown.

In[30]:= **Show[GraphicsArray[#]]& /@**
      MapThread[Function[{func, mesh}, Show[
      ListSurfacePlot3D[(* generate z-values on the mesh *)
              Map[Flatten[{#, Re[func @@ #]}]&, #, {-2}],
                      DisplayFunction -> Identity]& /@ mesh]],
          (* the solutions and corresponding meshes *)
          {{{f1, f2}, {f3, f4}}, {{mesh1, mesh2}, {mesh3, mesh4}}}, 2]

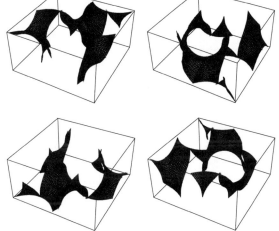

Together, they form the complete surface described by $\sin(x)\cos(y) + \sin(y)\cos(z) + \sin(z)\cos(x) = 0$.

In[31]:= **Show[GraphicsArray[**
          Graphics3D[{EdgeForm[#], (* extract polygons from last graphic *)
                  allPolys = Cases[%, _Polygon, {0, Infinity}]},
                  Axes -> True, PlotRange -> All]& /@
                  {{Thickness[0.001]}, {}}]]

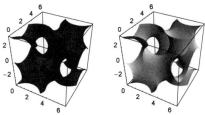

Due to the 2π periodicity of the surface along the x, y, and z-axes, we can continue this surface. With enough memory, one could, for instance, display the surface continued to seven adjacent cubes.

```
(* the Gauss curvature of an implicitly given surface;
   see Chapter 3 of the Graphics volume *)
κ[{x_, y_, z_}] = Together @
Module[{d},
 d[ds__] := Derivative[ds][Function[{x, y, z},
          Sin[x] Cos[y] + Sin[y] Cos[z] + Sin[z] Cos[x]]][x, y, z];
 (-(d[0, 0, 2] d[0, 1, 0] d[1, 0, 0] +
    d[0, 0, 1] (-d[0, 1, 1] d[1, 0, 0] - d[0, 1, 0] d[1, 0, 1] +
    d[0, 0, 1] d[1, 1, 0]))^2 +
  (d[0, 0, 2] d[0, 1, 0]^2 + d[0, 0, 1] (-2 d[0, 1, 0] d[0, 1, 1] +
    d[0, 0, 1] d[0, 2, 0])) (d[0, 0, 2] d[1, 0, 0]^2 +
    d[0, 0, 1] (-2 d[1, 0, 0] d[1, 0, 1] + d[0, 0, 1] d[2, 0, 0])))/
  (d[0, 0, 1]^2 (d[0, 0, 1]^2 + d[0, 1, 0]^2 + d[1, 0, 0]^2)^2)];

(* color polygons according to Gauss curvature *)
colorPolygon[p:Polygon[l_]] :=
Module[{c = Hue[-Plus @@ (κ /@ l)/4]}, {SurfaceColor[c, c, 2.2], p}]

Show[Graphics3D[{EdgeForm[{}],
 Table[(colorPolygon /@ allPolys) /. (* use periodicity of the surface *)
        Polygon[l_] :> Polygon[(# + 2Pi {i, j, k})& /@ l],
   {i, 0, 1}, {j, 0, 1}, {k, 0, 1}]}],
  Boxed -> False, Axes -> False]
```

For similar surfaces, see [689] and [690].

c) This is the implicit equation of the surface under consideration.

```
In[1]:= clebsch = 32 - 216 x^2 + 648 x^2 y - 216 y^2 - 216 y^3 - 150 z +
        216 x^2 z + 216 y^2 z + 231 z^2 - 113 z^3;
```

Let us make the following ansatz for the (18, as will be shown in the following) lines which are not in a plane parallel to the x,y-plane: $\{x, y, z\} = \{x_0, y_0, 0\} + t\,\{x_1, y_1, 1\}$.

Here, t is the parameter of the line. The coefficients x_0, x_1, y_0, and y_1 are to be determined. (Every line not parallel to the x,y-plane will cross the x,y-plane; so, we have taken $z_0 = 0$, and $z_1 = 1$ can always be achieved by a reparametrization.)

We insert this parametrization for the lines into `clebsch`.

```
In[2]:= clebsch /. {x -> x0 + x1 t, y -> y0 + y1 t, z -> t}
```

```
Out[2]= 32 - 150 t + 231 t^2 - 113 t^3 - 216 (x0 + t x1)^2 + 216 t (x0 + t x1)^2 +
        648 (x0 + t x1)^2 (y0 + t y1) - 216 (y0 + t y1)^2 + 216 t (y0 + t y1)^2 - 216 (y0 + t y1)^3
```

Extracting the coefficients with respect to t gives us a set of polynomial equations for x_0, x_1, y_0, and y_1.

```
In[3]:= cl1 = CoefficientList[%, t]
```

```
Out[3]= {32 - 216 x0^2 + 648 x0^2 y0 - 216 y0^2 - 216 y0^3,
         -150 + 216 x0^2 - 432 x0 x1 + 1296 x0 x1 y0 + 216 y0^2 + 648 x0^2 y1 - 432 y0 y1 - 648 y0^2 y1,
         231 + 432 x0 x1 - 216 x1^2 + 648 x1^2 y0 + 1296 x0 x1 y1 + 432 y0 y1 - 216 y1^2 - 648 y0 y1^2,
         -113 + 216 x1^2 + 648 x1^2 y1 + 216 y1^2 - 216 y1^3}
```

We can now straightforwardly solve these equations.

```
In[4]:= sol2 = Solve[# == 0& /@ GroebnerBasis[cl1, {x0, y0, x1, y1}],
                {x0, y0, x1, y1}] // RootReduce // ToRadicals // Union
```

$$\text{Out[4]= } \left\{\left\{x0 \to 0, y0 \to \frac{1}{3}, x1 \to -\frac{1}{4\sqrt{3}}, y1 \to -\frac{7}{12}\right\}, \left\{x0 \to 0, y0 \to \frac{1}{3}, x1 \to \frac{1}{4\sqrt{3}}, y1 \to -\frac{7}{12}\right\},\right.$$

$$\left\{x0 \to -\frac{1}{2\sqrt{3}}, y0 \to -\frac{1}{6}, x1 \to \frac{1}{\sqrt{3}}, y1 \to \frac{1}{6}\right\}, \left\{x0 \to -\frac{1}{2\sqrt{3}}, y0 \to -\frac{1}{6}, x1 \to \frac{\sqrt{3}}{4}, y1 \to \frac{5}{12}\right\},$$

$$\left\{x0 \to \frac{1}{2\sqrt{3}}, y0 \to -\frac{1}{6}, x1 \to -\frac{1}{\sqrt{3}}, y1 \to \frac{1}{6}\right\}, \left\{x0 \to \frac{1}{2\sqrt{3}}, y0 \to -\frac{1}{6}, x1 \to -\frac{\sqrt{3}}{4}, y1 \to \frac{5}{12}\right\},$$

$$\left\{x0 \to -\frac{1}{2}\sqrt{\frac{1}{6}\left(7-3\sqrt{5}\right)}, y0 \to \frac{1}{12}\left(1-3\sqrt{5}\right), x1 \to \frac{1}{4}\sqrt{\frac{1}{6}\left(27-7\sqrt{5}\right)}, y1 \to \frac{1}{24}\left(1+9\sqrt{5}\right)\right\},$$

$$\left\{x0 \to \frac{1}{2}\sqrt{\frac{1}{6}\left(7-3\sqrt{5}\right)}, y0 \to \frac{1}{12}\left(1-3\sqrt{5}\right), x1 \to -\frac{1}{4}\sqrt{\frac{1}{6}\left(27-7\sqrt{5}\right)}, y1 \to \frac{1}{24}\left(1+9\sqrt{5}\right)\right\},$$

$$\left\{x0 \to -\sqrt{\frac{1}{6}\left(3-\sqrt{5}\right)}, y0 \to \frac{1}{3}, x1 \to \frac{1}{4}\sqrt{\frac{1}{6}\left(67-15\sqrt{5}\right)}, y1 \to \frac{1}{24}\left(-11-3\sqrt{5}\right)\right\},$$

$$\left\{x0 \to -\frac{1}{2}\sqrt{\frac{1}{6}\left(3-\sqrt{5}\right)}, y0 \to \frac{1}{12}\left(-5-3\sqrt{5}\right), x1 \to \sqrt{\frac{1}{6}\left(3-\sqrt{5}\right)}, y1 \to \frac{1}{12}\left(5+3\sqrt{5}\right)\right\},$$

$$\left\{x0 \to \frac{1}{2}\sqrt{\frac{1}{6}\left(3-\sqrt{5}\right)}, y0 \to \frac{1}{12}\left(-5-3\sqrt{5}\right), x1 \to -\sqrt{\frac{1}{6}\left(3-\sqrt{5}\right)}, y1 \to \frac{1}{12}\left(5+3\sqrt{5}\right)\right\},$$

$$\left\{x0 \to \sqrt{\frac{1}{6}\left(3-\sqrt{5}\right)}, y0 \to \frac{1}{3}, x1 \to -\frac{1}{4}\sqrt{\frac{1}{6}\left(67-15\sqrt{5}\right)}, y1 \to \frac{1}{24}\left(-11-3\sqrt{5}\right)\right\},$$

$$\left\{x0 \to -\sqrt{\frac{1}{8}+\frac{\sqrt{5}}{24}}, y0 \to \frac{1}{12}\left(-5+3\sqrt{5}\right), x1 \to \sqrt{\frac{1}{2}+\frac{\sqrt{5}}{6}}, y1 \to \frac{1}{12}\left(5-3\sqrt{5}\right)\right\},$$

$$\left\{x0 \to \sqrt{\frac{1}{8}+\frac{\sqrt{5}}{24}}, y0 \to \frac{1}{12}\left(-5+3\sqrt{5}\right), x1 \to -\sqrt{\frac{1}{2}+\frac{\sqrt{5}}{6}}, y1 \to \frac{1}{12}\left(5-3\sqrt{5}\right)\right\},$$

$$\left\{x0 \to -\sqrt{\frac{7}{24}+\frac{\sqrt{5}}{8}}, y0 \to \frac{1}{12}\left(1+3\sqrt{5}\right), x1 \to \sqrt{\frac{9}{32}+\frac{7\sqrt{5}}{96}}, y1 \to \frac{1}{24}\left(1-9\sqrt{5}\right)\right\},$$

$$\left\{x0 \to \sqrt{\frac{7}{24}+\frac{\sqrt{5}}{8}}, y0 \to \frac{1}{12}\left(1+3\sqrt{5}\right), x1 \to -\sqrt{\frac{9}{32}+\frac{7\sqrt{5}}{96}}, y1 \to \frac{1}{24}\left(1-9\sqrt{5}\right)\right\},$$

$$\left\{x0 \to -\sqrt{\frac{1}{2}+\frac{\sqrt{5}}{6}}, y0 \to \frac{1}{3}, x1 \to \sqrt{\frac{67}{96}+\frac{5\sqrt{5}}{32}}, y1 \to \frac{1}{24}\left(-11+3\sqrt{5}\right)\right\},$$

$$\left.\left\{x0 \to \sqrt{\frac{1}{2}+\frac{\sqrt{5}}{6}}, y0 \to \frac{1}{3}, x1 \to -\sqrt{\frac{67}{96}+\frac{5\sqrt{5}}{32}}, y1 \to \frac{1}{24}\left(-11+3\sqrt{5}\right)\right\}\right\}$$

As previously mentioned, there are 18 lines.

```
In[5]:= Length[sol2]
```

```
Out[5]= 18
```

For the picture to be constructed later, we numericalize the data for x_0, x_1, y_0, and y_1.

```
In[6]:= sol3 = N[sol2];
        sol4 = {x0, x1, y0, y1} /. sol3;
```

We will make a picture for the surface for all values of z, which arise when x and y vary inside a circle of radius R. To make the lines fit inside the region where the surface is drawn, we determine the relevant parameter values.

```
In[8]:= tRange[{x0_, x1_, y0_, y1_}, R_] =
        t /. Solve[(x0 + x1 t)^2 + (y0 + y1 t)^2 == R^2, {t}]
```

Out[8]= $\{ \dfrac{-2\,x0\,x1 - 2\,y0\,y1 - \sqrt{(2\,x0\,x1 + 2\,y0\,y1)^2 - 4\,(-R^2 + x0^2 + y0^2)\,(x1^2 + y1^2)}}{2\,(x1^2 + y1^2)}$,

$\dfrac{-2\,x0\,x1 - 2\,y0\,y1 + \sqrt{(2\,x0\,x1 + 2\,y0\,y1)^2 - 4\,(-R^2 + x0^2 + y0^2)\,(x1^2 + y1^2)}}{2\,(x1^2 + y1^2)} \}$

So, we get the following lines. (We take $R = 0.9$ as an appropriate parameter for a neat picture here.)

```
In[9]:= lines1 = Apply[Function[{t0, t1, x0, x1, y0, y1},
                        Line[{{x0, y0, 0} + t0 {x1, y1, 1},
                              {x0, y0, 0} + t1 {x1, y1, 1}}]],
                  Flatten[{tRange[#, 0.9], #}]& /@ sol4, {1}];
```

Now, we must treat the case that there are lines that are parallel to the x,y-plane. We make a similar ansatz as above: $\{x, y, z\} = \{x_0, y_0, z_0\} + t\,\{1, y_1, 0\}$.

```
In[10]:= clebsch /. {x -> x0 + t, y -> y0 + y1 t, z -> z0}
```

Out[10]= $32 - 216\,(t + x0)^2 + 648\,(t + x0)^2\,(y0 + t\,y1) - 216\,(y0 + t\,y1)^2 -$
$216\,(y0 + t\,y1)^3 - 150\,z0 + 216\,(t + x0)^2\,z0 + 216\,(y0 + t\,y1)^2\,z0 + 231\,z0^2 - 113\,z0^3$

We again extract the coefficients with respect to t.

```
In[11]:= cl1 = CoefficientList[%, t]
```

Out[11]= $\{32 - 216\,x0^2 + 648\,x0^2\,y0 - 216\,y0^2 - 216\,y0^3 - 150\,z0 + 216\,x0^2\,z0 + 216\,y0^2\,z0 + 231\,z0^2 - 113\,z0^3,$
$-432\,x0 + 1296\,x0\,y0 + 648\,x0^2\,y1 - 432\,y0\,y1 - 648\,y0^2\,y1 + 432\,x0\,z0 + 432\,y0\,y1\,z0,$
$-216 + 648\,y0 + 1296\,x0\,y1 - 216\,y1^2 - 648\,y0\,y1^2 + 216\,z0 + 216\,y1^2\,z0,\ 648\,y1 - 216\,y1^3\}$

The resulting equations can be easily solved for x_0, y_0, y_1, and z_0.

```
In[12]:= sol1 = Solve[# == 0& /@ cl1, {x0, y0, y1, z0}]
```

        Solve::svars : Equations may not give solutions for all "solve" variables. More…

Out[12]= $\{\{x0 \to -\dfrac{y0}{\sqrt{3}},\ z0 \to 1,\ y1 \to -\sqrt{3}\},\ \{x0 \to \dfrac{y0}{\sqrt{3}},\ z0 \to 1,\ y1 \to \sqrt{3}\},$

$\{x0 \to \dfrac{1}{27}\,(-2\sqrt{3} - 9\sqrt{3}\,y0),\ z0 \to \dfrac{2}{3},\ y1 \to -\sqrt{3}\},$

$\{x0 \to \dfrac{1}{9}\,(-2\sqrt{3} - 3\sqrt{3}\,y0),\ z0 \to 0,\ y1 \to -\sqrt{3}\},\ \{x0 \to \dfrac{1}{9}\,(2\sqrt{3} + 3\sqrt{3}\,y0),\ z0 \to 0,\ y1 \to \sqrt{3}\},$

$\{x0 \to \dfrac{1}{27}\,(2\sqrt{3} + 9\sqrt{3}\,y0),\ z0 \to \dfrac{2}{3},\ y1 \to \sqrt{3}\},\ \{z0 \to 0,\ y1 \to 0,\ y0 \to \dfrac{1}{3}\},$

$\{z0 \to \dfrac{2}{3},\ y1 \to 0,\ y0 \to \dfrac{1}{9}\},\ \{z0 \to 1,\ y1 \to 0,\ y0 \to 0\}\}$

There are nine remaining solutions.

```
In[13]:= Length[sol1]
```

Out[13]= 9

Replacing the undetermined values of variables appearing in an parametric way in the solution sol1 by 0, we can again calculate explicit representations for these nine lines.

```
In[14]:= sol2 = # /. (# -> 0& /@ Cases[Last /@ #, _Symbol, {-1}])& /@ sol1
```

Out[14]= $\{\{x0 \to 0,\ z0 \to 1,\ y1 \to -\sqrt{3}\},\ \{x0 \to 0,\ z0 \to 1,\ y1 \to \sqrt{3}\},$

$\{x0 \to -\dfrac{2}{9\sqrt{3}},\ z0 \to \dfrac{2}{3},\ y1 \to -\sqrt{3}\},\ \{x0 \to -\dfrac{2}{3\sqrt{3}},\ z0 \to 0,\ y1 \to -\sqrt{3}\},$

$\{x0 \to \dfrac{2}{3\sqrt{3}},\ z0 \to 0,\ y1 \to \sqrt{3}\},\ \{x0 \to \dfrac{2}{9\sqrt{3}},\ z0 \to \dfrac{2}{3},\ y1 \to \sqrt{3}\},$

$\{z0 \to 0,\ y1 \to 0,\ y0 \to \dfrac{1}{3}\},\ \{z0 \to \dfrac{2}{3},\ y1 \to 0,\ y0 \to \dfrac{1}{9}\},\ \{z0 \to 1,\ y1 \to 0,\ y0 \to 0\}\}$

```
In[15]:= sol3 = Join[#, # -> 0& /@ Complement[{x0, y0, y1, z0}, First /@ #]]& /@ sol2;
         sol4 = N[{x0, 1, y0, y1, z0} /. sol3];
In[17]:= lines2 = Apply[Function[{t0, t1, x0, x1, y0, y1, z0},
                        Line[{{x0, y0, z0} + t0 {x1, y1, 0},
                              {x0, y0, z0} + t1 {x1, y1, 0}}]],
                  Flatten[{tRange[Drop[#, -1], 0.9], #}]& /@ sol4, {1}];
```

So, we finally have the following 27 lines.

In[18]:= **twentySevenLines = Join[lines1, lines2];**

Let us have a look at them. Viewing from far above (right picture), we see how the calculated lines reflect the threefold rotational symmetry of the Clebsch surface.

In[19]:= **Show[GraphicsArray[**
   Graphics3D[twentySevenLines, PlotRange -> All, ViewPoint -> #]& /@
                         {{1.3, -2.4, 2}, {0, 0, 100}}]]

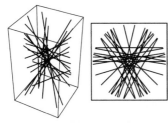

Now let us make a picture of the Clebsch surface and the 27 lines. Here is a picture of the Clebsch surface. As in Chapter 1 of the Numerics volume [1737], we use ContourPlot3D to calculate this picture and exploit the symmetry of the surface.

In[20]:= **Needs["Graphics`ContourPlot3D`"]**

In[21]:= **clebschSurface = Graphics3D[{**
       {#, Function[m, Map[m.#&, #, {-2}]][
       (* make other parts of the surface *)
         N[{{-1, Sqrt[3], 0}, {-Sqrt[3], -1, 0}, {0, 0, 2}}/2]],
       Function[m, Map[m.#&, #, {-2}]][
         N[{{-1, -Sqrt[3], 0}, {Sqrt[3], -1, 0}, {0, 0, 2}}/2]]}&[
       Union[#, Map[{-1, 1, 1}#&, #, {-2}]]&[
         (* change to Cartesian coordinates *)
         Apply[{#1 Cos[#2], #1 Sin[#2], #3}&,
       (* make 1/6 of the surface with ContourPlot3D
         in cylindrical coordinates *)
       ContourPlot3D[32 - 216 r^2 - 150 z + 216 r^2 z + 231 z^2 -
                       113 z^3 + 216 r^3 Sin[3φ],
                       {r, 0, 0.9}, {φ, N[-Pi/2], N[-Pi/6]},
       (* carefully selected z-boundaries to fit exactly *)
                       {z, -0.975, 2.173}, MaxRecursion -> 1,
                       PlotPoints -> {{8, 4}, {8, 3}, {22, 3}},
                       DisplayFunction -> Identity][[1]], {-2}]]]}}]

Just displaying this surface and the lines together in one picture does not yield a reasonable quality. Due to the approximative nature of the polygons from the surface, they intersect in a complicated manner with the lines, and so some parts of the lines will be "below" and some parts "above" the surface. To avoid this, we "lift" the lines slightly in the direction of the local normal on the surface. Here is the normal vector for our cubic.

In[22]:= **normalVector[{x_, y_, z_}] =**
   #/Sqrt[#.#]&[{D[#, x], D[#, y], D[#, z]}&[clebsch]];

The routine liftLines lifts its first argument (a line) to both sides of the surface to avoid intersections with the polygons by a height height.

In[23]:= **liftLines[Line[{p1_, p2_}], height_, pp_] :=**
   Function[{l, normal}, (* move slightly up and down *)
   {Line[MapThread[Plus, {l, +height normal}]],
    Line[MapThread[Plus, {l, -height normal}]]}] @@
     {#, normalVector /@ #}&[
       Evaluate[p1 + # (p2 - p1)]& /@ (Range[0, pp]/pp)];

Now, we can display the lines and the surface together.

In[24]:= **Show[Graphics3D[{**
       {(* the 27 lines *) GrayLevel[1], Thickness[0.004],
        liftLines[#, 0.005, 200]& /@ twentySevenLines},
       (* the Clebsch surface *)
       {EdgeForm[{GrayLevel[0], Thickness[0.002]}],

```
Flatten[clebschSurface[[1]]]}}],
    PlotRange -> All, Axes -> False, Boxed -> False,
    ViewPoint -> {1.87, -1.6, 0.9}, BoxRatios -> {1, 1, 2.45},
    Background -> GrayLevel[0.9]]
```

d) This is the original implicit formula for the Clebsch surfaces used here.

In[1]:= `clebsch3 = 32 - 216 x^2 + 648 x^2 y - 216 y^2 - 216 y^3 - 150 z +`
        `216 x^2 z + 216 y^2 z + 231 z^2 - 113 z^3;`

The threefold rotational symmetry becomes obvious in cylindrical coordinates.

In[2]:= `Simplify[clebsch3 /. {x -> r Sin[φ], y -> r Cos[φ]}]`

Out[2]= $(-1 + z) (-32 + 216 r^2 + 118 z - 113 z^2) - 216 r^3 \cos[3 φ]$

A straightforward generalization from the threefold rotational symmetry to an *n*-fold rotational symmetry is achieved by using $\cos[n φ]$ instead of $\cos[3 φ]$.

In[3]:= `clebsch[n_] := (z - 1) (-32 + 216 r^2 + 118 z - 113 z^2) - 216 r^3 Cos[n φ]`

To obtain the corresponding implicit representation, we eliminate the coordinates r and $φ$ and use the Cartesian x and y instead. Using, say, the elimination capabilities of `GroebnerBasis`, this is straightforward.

In[4]:= `generalizedClebsch[n_] :=`
        `GroebnerBasis[{TrigExpand[clebsch[n]], x - r Sin[φ], y - r Cos[φ],`
        `        Sin[φ]^2 + Cos[φ]^2 - 1}, {x, y, z}, {r, Sin[φ], Cos[φ]},`
        `        MonomialOrder -> EliminationOrder][[1]] // Factor`

Another possibility to obtain an implicit representation would be the direct substitution of r and $φ$ as functions of x and y and ongoing simplification of the resulting formula.

In[5]:= `generalizedClebsch1[n_] :=`
        `Module[{aux, pre},`
        `    (* remove r and φ *)`
        `    aux = clebsch[n] /. {r -> Sqrt[x^2 + y^2],`
        `                         φ -> ArcCos[y/Sqrt[x^2 + y^2]]};`
        `    (* remove trigs *)`
        `    pre = Numerator[Factor[Together[TrigExpand[aux]]]];`
        `    (* remove square roots if present *)`
        `    If[MemberQ[pre, Sqrt[x^2 + y^2], Infinity],`
        `            Factor[Expand[(pre - #)^2 - #^2]]&[`
        `                Plus @@ Cases[pre, Sqrt[x^2 + y^2] _]], pre]]`

For $n = 3$, we recover the original Clebsch surface.

In[6]:= `generalizedClebsch[3]`

Out[6]= $32 - 216 x^2 + 648 x^2 y - 216 y^2 - 216 y^3 - 150 z + 216 x^2 z + 216 y^2 z + 231 z^2 - 113 z^3$

In[7]:= `generalizedClebsch[3] == generalizedClebsch1[3]`

Out[7]= `True`

For $n = 5$, $n = 7$, we get the following implicit representations of Clebsch surfaces.

In[8]:= `generalizedClebsch[5]`

Out[8]= $-32 x^2 + 216 x^4 + 1080 x^4 y - 32 y^2 + 432 x^2 y^2 - 2160 x^2 y^3 + 216 y^4 + 216 y^5 + 150 x^2 z -$
        $216 x^4 z + 150 y^2 z - 432 x^2 y^2 z - 216 y^4 z - 231 x^2 z^2 - 231 y^2 z^2 + 113 x^2 z^3 + 113 y^2 z^3$

In[9]:= `generalizedClebsch[7]`

Out[9]= $32 x^4 - 216 x^6 + 1512 x^6 y + 64 x^2 y^2 - 648 x^4 y^2 - 7560 x^4 y^3 + 32 y^4 - 648 x^2 y^4 + 4536 x^2 y^5 - 216 y^6 - 216 y^7 - 150 x^4 z + 216 x^6 z - 300 x^2 y^2 z + 648 x^4 y^2 z - 150 y^4 z + 648 x^2 y^4 z + 216 y^6 z + 231 x^4 z^2 + 462 x^2 y^2 z^2 + 231 y^4 z^2 - 113 x^4 z^3 - 226 x^2 y^2 z^3 - 113 y^4 z^3$

For an even number *n*, we get the following implicit representations of Clebsch surfaces.

In[10]:= `generalizedClebsch[2]`

Out[10]= $-1024 + 13824 x^2 - 46656 x^4 + 46656 x^6 + 13824 y^2 - 93312 x^2 y^2 - 46656 x^4 y^2 - 46656 y^4 - 46656 x^2 y^4 + 46656 y^6 + 9600 z - 78624 x^2 z + 93312 x^4 z - 78624 y^2 z + 186624 x^2 y^2 z + 93312 y^4 z - 37284 z^2 + 164592 x^2 z^2 - 46656 x^4 z^2 + 164592 y^2 z^2 - 93312 x^2 y^2 z^2 - 46656 y^4 z^2 + 76532 z^3 - 148608 x^2 z^3 - 148608 y^2 z^3 - 87261 z^4 + 48816 x^2 z^4 + 48816 y^2 z^4 + 52206 z^5 - 12769 z^6$

In[11]:= `generalizedClebsch[4]`

Out[11]= $-1024 x^2 + 13824 x^4 - 46656 x^6 + 46656 x^8 - 1024 y^2 + 27648 x^2 y^2 - 139968 x^4 y^2 - 559872 x^6 y^2 + 13824 y^4 - 139968 x^2 y^4 + 1772928 x^4 y^4 - 46656 y^6 - 559872 x^2 y^6 + 46656 y^8 + 9600 x^2 z - 78624 x^4 z + 93312 x^6 z + 9600 y^2 z - 157248 x^2 y^2 z + 279936 x^4 y^2 z - 78624 y^4 z + 279936 x^2 y^4 z + 93312 y^6 z - 37284 x^2 z^2 + 164592 x^4 z^2 - 46656 x^6 z^2 - 37284 y^2 z^2 + 329184 x^2 y^2 z^2 - 139968 x^4 y^2 z^2 + 164592 y^4 z^2 - 139968 x^2 y^4 z^2 - 46656 y^6 z^2 + 76532 x^2 z^3 - 148608 x^4 z^3 + 76532 y^2 z^3 - 297216 x^2 y^2 z^3 - 148608 y^4 z^3 - 87261 x^2 z^4 + 48816 x^4 z^4 - 87261 y^2 z^4 + 97632 x^2 y^2 z^4 + 48816 y^4 z^4 + 52206 x^2 z^5 + 52206 y^2 z^5 - 12769 x^2 z^6 - 12769 y^2 z^6$

The expression `generalizedClebsch` is more advantageous than is `generalizedClebsch1` because it does not contain additional factors.

In[12]:= `generalizedClebsch1[2]`

Out[12]= $(x^2 + y^2) (-1024 + 13824 x^2 - 46656 x^4 + 46656 x^6 + 13824 y^2 - 93312 x^2 y^2 - 46656 x^4 y^2 - 46656 y^4 - 46656 x^2 y^4 + 46656 y^6 + 9600 z - 78624 x^2 z + 93312 x^4 z - 78624 y^2 z + 186624 x^2 y^2 z + 93312 y^4 z - 37284 z^2 + 164592 x^2 z^2 - 46656 x^4 z^2 + 164592 y^2 z^2 - 93312 x^2 y^2 z^2 - 46656 y^4 z^2 + 76532 z^3 - 148608 x^2 z^3 - 148608 y^2 z^3 - 87261 z^4 + 48816 x^2 z^4 + 48816 y^2 z^4 + 52206 z^5 - 12769 z^6)$

Now let us visualize some of the generalized Clebsch surfaces. For efficiency, we will make use of the symmetry of the surface and generate a contour plot of a part of the surface and then rotate this part to generate the whole surface.

In[13]:= `Needs["Graphics`ContourPlot3D`"]`

In[14]:=
```
clebschPicture[n_, opts___] :=
  Module[{cpPolys, CartesianPolys, R},
         (* contour plot in cylindrical coordinates *)
         cpPolys = ContourPlot3D[Evaluate[clebsch[n]],
                        {r, 0, 0.9}, {φ, 0, N[2Pi/n]},
                        {z, -0.975, 2.173}, MaxRecursion -> 1,
                        PlotPoints -> {{8, 3}, {12, 3}, {22, 3}},
                        DisplayFunction -> Identity][[1]];
         (* polygons in Cartesian coordinates *)
         CartesianPolys = Apply[{#1 Cos[#2], #1 Sin[#2], #3}&,
                        cpPolys, {-2}];
         (* rotation matrices *)
         Do[R[i] = N[{{ Cos[i 2Pi/n], Sin[i 2Pi/n], 0},
                     {-Sin[i 2Pi/n], Cos[i 2Pi/n], 0},
                     {0, 0, 1}}], {i, 0, n - 1}];
         (* show generalized Clebsch surface *)
         Show[Graphics3D[{EdgeForm[],
            SurfaceColor[Hue[Random[]], Hue[Random[]], 2.6],
            Table[Map[R[i].#&, CartesianPolys, {-2}],
                     {i, 0, n - 1}]}], opts, Boxed -> False]]
```

In[15]:= `Show[GraphicsArray[#]]& /@ Partition[clebschPicture[#,`
`           DisplayFunction -> Identity]& /@ Range[2, 7], 3]`

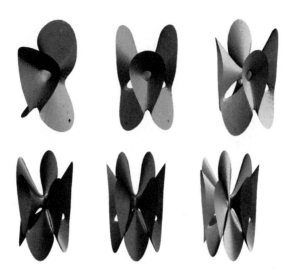

We could now continue and calculate the straight lines that could be found on these surfaces.

28. 28 Bitangents on a Real Plane Quartic, Curve Intersections, Pentaellipse

a) This is the implicit equation of the quartic under consideration.

    In[1]:= F[x_, y_] = x^4 + y^4 - 6 (x^2 + y^2) + 10;

Let us have a look at the form of this curve.

    In[2]:= sketch = ContourPlot[Evaluate[F[x, y]], {x, -3, 3}, {y, -3, 3},
                        Contours -> {0}, PlotPoints -> 100]

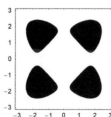

We see four equal components in symmetrical position. Between each of the six possible pairs of these regions, there are obviously four (two outer and two crossing) bitangents. As it will turn out later, the remaining four bitangents touch the regions in the point nearest to $\{0, 0\}$ where a degenerate bitangent occurs.

Let us now explicitly calculate the bitangents. We will derive an equation for the slope m of the possible bitangents, and then calculate the constants b in the equations $y = m\,x + b$ of the bitangents (assuming $m \neq \infty$ and treating the case $m = \infty$ later on).

We could also try to calculate the points where the bitangents touch the quartic and then calculate m and b for each line. Let us attach a tangent at the point $\{x_1, y_1\}$. These tangents will generically intersect the quartic at two additional points or nowhere else. We will derive an equation that forces the tangent to have just one additional intersection; this means to become a bitangent.

The following four equations describe a tangent attached to point $\{x_1, y_1\}$. The point $\{x_2, y_2\}$ represents the additional point(s) of intersection.

    In[3]:= {eq1, eq2, eq3, eq4} = Numerator[Together[#]]& /@
               {(* definition of slope m *)

```
        y2 - y1 - m (x2 - x1),
        (* condition for slope m at point P1 *)
        m + D[F[x1, y1], x1]/D[F[x1, y1], y1],
        (* point P1 lies on the curve *) F[x1, y1],
        (* point P2 lies on the curve *) F[x2, y2]}
Out[3]=  {m x1 - m x2 - y1 + y2, -3 x1 + x1³ - 3 m y1 + m y1³,
          10 - 6 x1² + x1⁴ - 6 y1² + y1⁴, 10 - 6 x2² + x2⁴ - 6 y2² + y2⁴}
```

One possible solution to this equation is $\{x_1, y_1\} = \{x_2, y_2\}$. Here, we are interested only in the case of degeneracy. So, we will try to get rid of this solution. Let us first derive an equation describing the additional points of intersection that depend on x_1. To achieve this, we eliminate m, y_1, and y_2 from the four equations.

```
In[4]:= res1 = Resultant[eq1, eq2, m] // Factor
Out[4]=  -3 x1² + x1⁴ + 3 x1 x2 - x1³ x2 - 3 y1² + y1⁴ + 3 y1 y2 - y1³ y2
```

```
In[5]:= res2 = Resultant[res1, eq3, y1] // Factor
Out[5]=  100 - 1200 x1² + 3960 x1⁴ - 2160 x1⁶ + 324 x1⁸ - 120 x1 x2 + 1120 x1³ x2 - 2736 x1⁵ x2 + 1440 x1⁷ x2 -
          216 x1⁹ x2 + 216 x1² x2² - 1440 x1⁴ x2² + 1536 x1⁶ x2² - 576 x1⁸ x2² + 72 x1¹⁰ x2² - 108 x1³ x2³ +
          432 x1⁵ x2³ - 360 x1⁷ x2³ + 112 x1⁹ x2³ - 12 x1¹¹ x2³ + 81 x1⁴ x2⁴ - 108 x1⁶ x2⁴ + 54 x1⁸ x2⁴ -
          12 x1¹⁰ x2⁴ + x1¹² x2⁴ - 60 y2² + 756 x1² y2² - 2664 x1⁴ y2² + 1800 x1⁶ y2² - 444 x1⁸ y2² +
          36 x1¹⁰ y2² - 108 x1³ x2 y2² + 720 x1⁵ x2 y2² - 552 x1⁷ x2 y2² + 144 x1⁹ x2 y2² - 12 x1¹¹ x2 y2² +
          54 x1² x2² y2² - 360 x1⁴ x2² y2² + 276 x1⁶ x2² y2² - 72 x1⁸ x2² y2² + 6 x1¹⁰ x2² y2² +
          10 y2⁴ - 126 x1² y2⁴ + 453 x1⁴ y2⁴ - 360 x1⁶ y2⁴ + 120 x1⁸ y2⁴ - 18 x1¹⁰ y2⁴ + x1¹² y2⁴
```

```
In[6]:= res3 = Resultant[res2, eq4, y2] // Factor
Out[6]=  16 (x1 - x2)⁸
          (900 - 12240 x1² + 53145 x1⁴ - 84600 x1⁶ + 80964 x1⁸ - 52680 x1¹⁰ + 24630 x1¹² - 8712 x1¹⁴ +
          2256 x1¹⁶ - 360 x1¹⁸ + 25 x1²⁰ - 600 x1 x2 + 8400 x1³ x2 - 42336 x1⁵ x2 + 94368 x1⁷ x2 -
          89232 x1⁹ x2 + 34816 x1¹¹ x2 - 864 x1¹³ x2 - 3360 x1¹⁵ x2 + 936 x1¹⁷ x2 - 80 x1¹⁹ x2 -
          300 x2² + 5920 x1² x2² - 44478 x1⁴ x2² + 172992 x1⁶ x2² - 386208 x1⁸ x2² +
          410272 x1¹⁰ x2² - 228852 x1¹² x2² + 71040 x1¹⁴ x2² - 12132 x1¹⁶ x2² + 1024 x1¹⁸ x2² -
          30 x1²⁰ x2² + 100 x1 x2³ - 1980 x1³ x2³ + 14352 x1⁵ x2³ - 49680 x1⁷ x2³ + 86008 x1⁹ x2³ -
          72168 x1¹¹ x2³ + 32592 x1¹³ x2³ - 8208 x1¹⁵ x2³ + 1092 x1¹⁷ x2³ - 60 x1¹⁹ x2³ + 25 x2⁴ -
          630 x1² x2⁴ + 6558 x1⁴ x2⁴ - 34416 x1⁶ x2⁴ + 90654 x1⁸ x2⁴ - 102684 x1¹⁰ x2⁴ +
          60696 x1¹² x2⁴ - 20448 x1¹⁴ x2⁴ + 3969 x1¹⁶ x2⁴ - 414 x1¹⁸ x2⁴ + 18 x1²⁰ x2⁴)²
```

The last result shows the solution $\{x_1, y_1\} = \{x_2, y_2\}$, which we are not interested in. We will ignore this term in the following. The condition that at $\{x_2, y_2\}$ we again have a tangency condition means that x_2 has a double root (x_1 viewed as a parameter).

```
In[7]:= res4 = Resultant[res3[[-1, 1]], D[res3[[-1, 1]], x2], x2] // Factor
Out[7]=  41472 (-1 + x1)² x1⁸ (1 + x1)² (-3 + x1²)¹⁰ (-1 - 2 x1 + x1²)⁷
          (-1 + 2 x1 + x1²)⁷ (10 - 6 x1² + x1⁴) (1 - 4 x1² + x1⁴) (5 - 22 x1² + 5 x1⁴)
          (5 - 14 x1² + 5 x1⁴) (5 - 30 x1² + 9 x1⁴) (25 - 630 x1² + 6558 x1⁴ - 34416 x1⁶ +
            90654 x1⁸ - 102684 x1¹⁰ + 60696 x1¹² - 20448 x1¹⁴ + 3969 x1¹⁶ - 414 x1¹⁸ + 18 x1²⁰)
          (25 - 5310 x1² + 67781 x1⁴ - 309960 x1⁶ + 431458 x1⁸ - 332820 x1¹⁰ +
            169378 x1¹² - 55944 x1¹⁴ + 10757 x1¹⁶ - 990 x1¹⁸ + 25 x1²⁰)²
```

The various possibilities to make the last expression zero are collected in the list terms.

```
In[8]:= terms = Cases[res4, _Plus, Infinity]
Out[8]=  {-1 + x1, 1 + x1, -3 + x1², -1 - 2 x1 + x1², -1 + 2 x1 + x1², 10 - 6 x1² + x1⁴,
          1 - 4 x1² + x1⁴, 5 - 22 x1² + 5 x1⁴, 5 - 14 x1² + 5 x1⁴, 5 - 30 x1² + 9 x1⁴,
          25 - 630 x1² + 6558 x1⁴ - 34416 x1⁶ + 90654 x1⁸ - 102684 x1¹⁰ + 60696 x1¹² -
          20448 x1¹⁴ + 3969 x1¹⁶ - 414 x1¹⁸ + 18 x1²⁰, 25 - 5310 x1² + 67781 x1⁴ - 309960 x1⁶ +
          431458 x1⁸ - 332820 x1¹⁰ + 169378 x1¹² - 55944 x1¹⁴ + 10757 x1¹⁶ - 990 x1¹⁸ + 25 x1²⁰}
```

Using these possibilities, we can derive equations for the potential slopes m to be obeyed. To do this, we first derive an equation between x_1 and m.

```
In[9]:= res5 = Resultant[eq2, eq3, y1]
```

Out[9]= $10\,m^4 + 54\,m^2\,x1^2 - 126\,m^4\,x1^2 + 81\,x1^4 - 360\,m^2\,x1^4 + 453\,m^4\,x1^4 - 108\,x1^6 + 276\,m^2\,x1^6 -$
$360\,m^4\,x1^6 + 54\,x1^8 - 72\,m^2\,x1^8 + 120\,m^4\,x1^8 - 12\,x1^{10} + 6\,m^2\,x1^{10} - 18\,m^4\,x1^{10} + x1^{12} + m^4\,x1^{12}$

We do not use the last two high-order polynomials in `terms` in the following because, as it turns out, they do not give any solution of interest to us.

In[10]:= `mEquations = Factor[Resultant[#, res5, x1]]& /@ Drop[terms, -2]`

Out[10]= $\{16\,(-1+m)\,(1+m)\,(-1+5\,m^2),\ 16\,(-1+m)\,(1+m)\,(-1+5\,m^2),\ 4096\,m^8,\ 4096,$
$\quad 4096,\ 400\,(5+162\,m^2+1458\,m^4)^2,\ 256\,(-1+m)^4\,(1+m)^4\,(1-22\,m^2+13\,m^4)^2,$
$\quad 419430400\,(-2+m^2)^4\,(20-148\,m^2+101\,m^4)^2,\ 419430400\,(-1+2\,m^2)^4\,(5-244\,m^2+212\,m^4)^2,$
$\quad 1638400\,(-1-4\,m+m^2)^2\,(-1+4\,m+m^2)^2\,(5-2970\,m^2+149\,m^4)^2\}$

Because some of the solutions for *m* may be complex and because the multiple use of resultants might have introduced spurious solutions, we now select the correct ones. In a first step, we solve explicitly for all possible slopes resulting from `mEquations` and select the real ones.

In[11]:= `mPreSols1 = Union[Flatten[Solve[# == 0, m]& /@ mEquations]] // Simplify;`
`mPreSols2 = Simplify[Select[Last /@ mPreSols1, Im[Chop[N[#]]] == 0&]]`

Out[12]= $\left\{-1,\ 0,\ 1,\ -\dfrac{1}{\sqrt{2}},\ \dfrac{1}{\sqrt{2}},\ -\sqrt{2},\ \sqrt{2},\ -\dfrac{1}{\sqrt{5}},\ \dfrac{1}{\sqrt{5}},\ -\sqrt{\dfrac{1}{13}\left(11-6\sqrt{3}\right)},\ \sqrt{\dfrac{1}{13}\left(11-6\sqrt{3}\right)},\right.$

$\quad -\sqrt{\dfrac{1}{13}\left(11+6\sqrt{3}\right)},\ \sqrt{\dfrac{1}{13}\left(11+6\sqrt{3}\right)},\ -2-\sqrt{5},\ 2-\sqrt{5},\ -2+\sqrt{5},\ 2+\sqrt{5},$

$\quad -\sqrt{\dfrac{1}{149}\left(1485+664\sqrt{5}\right)},\ \sqrt{\dfrac{1}{149}\left(1485+664\sqrt{5}\right)},\ -\sqrt{\dfrac{5}{1485+664\sqrt{5}}},\ \sqrt{\dfrac{5}{1485+664\sqrt{5}}},$

$\quad -\sqrt{\dfrac{1}{106}\left(61-24\sqrt{6}\right)},\ \sqrt{\dfrac{1}{106}\left(61-24\sqrt{6}\right)},\ -\sqrt{\dfrac{2}{101}\left(37-12\sqrt{6}\right)},\ \sqrt{\dfrac{2}{101}\left(37-12\sqrt{6}\right)},$

$\quad \left.-\sqrt{\dfrac{1}{106}\left(61+24\sqrt{6}\right)},\ \sqrt{\dfrac{1}{106}\left(61+24\sqrt{6}\right)},\ -\sqrt{\dfrac{2}{101}\left(37+12\sqrt{6}\right)},\ \sqrt{\dfrac{2}{101}\left(37+12\sqrt{6}\right)}\right\}$

Let us now calculate the constants *b*. First, we derive an equation connecting *m* and *b*.

In[13]:= `res6 = Numerator[Together[#]]& /@`
`Resultant[F[x1, m x1 + b], m + D[F[x1, y1], x1]/D[F[x1, y1], y1] /.`
`y1 -> m x1 + b, x1] // Factor`

Out[13]= $(1+m^4)$
$(10 - 126\,b^2 + 453\,b^4 - 360\,b^6 + 120\,b^8 - 18\,b^{10} + b^{12} - 360\,m^2 - 1638\,b^2\,m^2 + 1764\,b^4\,m^2 - 792\,b^6\,m^2 +$
$180\,b^8\,m^2 - 18\,b^{10}\,m^2 + 3270\,m^4 - 2844\,b^2\,m^4 + 2094\,b^4\,m^4 - 792\,b^6\,m^4 + 120\,b^8\,m^4 - 720\,m^6 - 2844\,b^2\,m^6 +$
$1764\,b^4\,m^6 - 360\,b^6\,m^6 + 3270\,m^8 - 1638\,b^2\,m^8 + 453\,b^4\,m^8 - 360\,m^{10} - 126\,b^2\,m^{10} + 10\,m^{12})$

Now for every value *m* from `mPreSols2`, we calculate the possible values of *b*. Then for the resulting line $y = m\,x + b$, we have to check if it is really a bitangent. We do this by calculating its intersection with the quartic and then see if the quantity $(y_2 - y_1)/(x_2 - x_1)$ is identical to the *m* we started with. The function `goodSlopeQ` does exactly this check for a given pair $\{m, b\}$. Here, we take into account the case of degeneracy, in which the two points where the bitangent touches the quartic coincide.

In[14]:= `slopeFromTwoPoints[{{x1_, y1_}, {x2_, y2_}}] :=`
`If[Abs[x2 - x1] < 10^-6,`
`If[Abs[y2 - y1], Indeterminate, Infinity], (y2 - y1)/(x2 - x1)]`

In[15]:= `goodSlopeQ[{m1_, b1_}] :=`
`Module[{x1Roots, countRoots, touchPoints, x1y1Pairs, twoPointPairs},`
`x1Roots = Re @ Cases[NRoots[F[x1, m1 x1 + b1] == 0, x1],`
`_?(Im[Chop[#]] === 0&), {-1}];`
`If[Length[x1Roots] =!= 4, False,`
`(* counting coinciding roots *)`
`countRoots = Count[Chop[Flatten[Table[x1Roots[[i]] - x1Roots[[j]],`
`{i, 3}, {j, i + 1, 4}]], 10^-6], 0];`
`Which[(* double tangent at one point *) countRoots === 6, True,`
`(* only one tangent *)                countRoots < 2, False,`
`(* two tangents at two different points *)  True,`
`(* x-values of the two points *)`

```
        touchPoints = Union[x1Roots, SameTest -> (Abs[#1 - #2] < 10^-6&)];
        (* coordinates of the points *)
        x1y1Pairs = Chop[N[{#, m1 # + b1}]]& /@ touchPoints;
        (* all possible pairs *)
        twoPointPairs = Flatten[Table[{x1y1Pairs[[i]], x1y1Pairs[[j]]},
              {i, Length[x1y1Pairs] - 1},
              {j, i + 1, Length[x1y1Pairs]}], 1];
        (* is the slope calculated via y2 - y1 - m*(x2 - x1) equal to m1 ? *)
        Min[Abs[N[(slopeFromTwoPoints /@ twoPointPairs) - m1]]] < 10^-6]]
```

The function `selectGoodShifts` calculates the values of b for a given m such that we have a bitangent and returns a list of all possible $\{m, b\}$ pairs.

```
In[16]:= selectGoodShifts[μ_] :=
        Module[{bEquation, bSymbSolsPre, bSymbSols},
        (* the resulting equation for b *)
        bEquation = Factor[(res6[[-1]] /. m -> μ)];
        (* possible values for b in y == m x + b *)
        bSymbSolsPre = Union[Solve[bEquation == 0, b]];
        (* the real solutions for b *)
        bSymbSols = Select[bSymbSolsPre, Chop[Im[N[#[[1, 2]]]]] == 0&];
        If[FreeQ[bSymbSols, ToRules, Infinity],
            (* all pairs of m and b *)
            mbPairs = {μ, #}& /@ (Last /@ Flatten[bSymbSols]);
            (* select the real bitangents *)
            Select[mbPairs, goodSlopeQ] // Simplify]]
```

Finally, we have the following solutions for our bitangents.

```
In[17]:= mbPairList = Flatten[selectGoodShifts /@ mPreSols2, 1]
```

$$Out[17]= \{\{-1, -2\}, \{-1, 2\}, \{-1, -\sqrt{2}\}, \{-1, \sqrt{2}\}, \{0, -1 - \sqrt{2}\}, \{0, 1 - \sqrt{2}\},$$

$$\{0, -1 + \sqrt{2}\}, \{0, 1 + \sqrt{2}\}, \{1, -2\}, \{1, 2\}, \{1, -\sqrt{2}\}, \{1, \sqrt{2}\}, \{-\frac{1}{\sqrt{2}}, -\sqrt{\frac{5}{2}}\},$$

$$\{-\frac{1}{\sqrt{2}}, \sqrt{\frac{5}{2}}\}, \{\frac{1}{\sqrt{2}}, -\sqrt{\frac{5}{2}}\}, \{\frac{1}{\sqrt{2}}, \sqrt{\frac{5}{2}}\}, \{-\sqrt{2}, -\sqrt{5}\}, \{-\sqrt{2}, \sqrt{5}\},$$

$$\{\sqrt{2}, -\sqrt{5}\}, \{\sqrt{2}, \sqrt{5}\}, \{-2 - \sqrt{5}, 0\}, \{2 - \sqrt{5}, 0\}, \{-2 + \sqrt{5}, 0\}, \{2 + \sqrt{5}, 0\}\}$$

There are now 24 solutions.

```
In[18]:= Length[mbPairList]

Out[18]= 24
```

The remaining four solutions result from the case $m = \infty$. Because of the symmetry of the quartic under consideration, there is no need to calculate this case explicitly; we can easily get the four bitangents that are parallel to the y-axis from the pairs with $m = 0$, which are parallel to the x-axis.

Because we used numerical approximations in `goodSlopeQ`, let us make sure that all solutions from `mbPairList` are really bitangents by symbolically calculating the quantity $(y_2 - y_1)/(y_2 - y_1) - m$.

```
In[19]:= test[{m_, b_}] :=
        Module[{sol, rules, x, y, x1, x2, y1, y2},
        (* intersection (better touch) points *)
        sol = Union[Solve[{F[x, y] == 0, y == m x + b}, {y, x}]];
        Which[(* degenerate case *) Length[sol] === 1, True,
            (* too many touch points *)
            Length[sol] === 3 || Length[sol] === 4, False,
            (* two touch points *) Length[sol] === 2,
            (* make replacement rules for the two points *)
            rules = Flatten[{sol[[1]] /. {x -> x1, y -> y1},
                            sol[[2]] /. {x -> x2, y -> y2}}];
            (* calculate slope from intersection points
               and compare with m *)
            Together[((y2 - y1)/(x2 - x1) /. rules) - m] === 0]]
```

```
In[20]:= (test /@ mbPairList) // Union
```

```
Out[20]= {True}
```

So, we have the following 28 equations describing the bitangents.

```
In[21]:= (# == 0& /@
           Join[Apply[Numerator[Together[y - #1 x - #2]]&, mbPairList, {1}],
              (* the m == Infinity equations *)
              Apply[Numerator[Together[x - #2]]&, Cases[mbPairList, {0, _}], {1}]])
```

$$Out[21]= \{2 + x + y = 0, \ -2 + x + y = 0, \ \sqrt{2} + x + y = 0, \ -\sqrt{2} + x + y = 0, \ 1 + \sqrt{2} + y = 0, \ -1 + \sqrt{2} + y = 0,$$
$$1 - \sqrt{2} + y = 0, \ -1 - \sqrt{2} + y = 0, \ 2 - x + y = 0, \ -2 - x + y = 0, \ \sqrt{2} - x + y = 0,$$
$$-\sqrt{2} - x + y = 0, \ \sqrt{10} + \sqrt{2} \ x + 2 \ y = 0, \ -\sqrt{10} + \sqrt{2} \ x + 2 \ y = 0, \ \sqrt{10} - \sqrt{2} \ x + 2 \ y = 0,$$
$$-\sqrt{10} - \sqrt{2} \ x + 2 \ y = 0, \ \sqrt{5} + \sqrt{2} \ x + y = 0, \ -\sqrt{5} + \sqrt{2} \ x + y = 0, \ \sqrt{5} - \sqrt{2} \ x + y = 0,$$
$$-\sqrt{5} - \sqrt{2} \ x + y = 0, \ 2 \ x + \sqrt{5} \ x + y = 0, \ -2 \ x + \sqrt{5} \ x + y = 0, \ 2 \ x - \sqrt{5} \ x + y = 0,$$
$$-2 \ x - \sqrt{5} \ x + y = 0, \ 1 + \sqrt{2} + x = 0, \ -1 + \sqrt{2} + x = 0, \ 1 - \sqrt{2} + x = 0, \ -1 - \sqrt{2} + x = 0\}$$

Let us now add the lines representing the bitangents to the above picture sketch. The function makeLine generates graphic lines with Line, which fit into the above plot range.

```
In[22]:= makeLine[{0, b_}] := {Line[{{-3, b}, {3, b}}], Line[{{b, -3}, {b, 3}}]}
```

```
     makeLine[{m_, b_}] := Line[{{-3, -3m + b}, {3, 3 m + b}}]
```

To better see the points of tangency, we also calculate the points where the bitangents touch the quartic.

```
In[24]:= makePoint[{0, b_}] := {Point[{#, b}], Point[{b, #}]}& /@
              Re[Cases[NRoots[F[x, b] == 0, x], _?NumberQ, {-1}]]
```

```
     makePoint[{m_, b_}] := Point[{#, m # + b}]& /@
              Re[Cases[NRoots[F[x, m x + b] == 0, x], _?NumberQ, {-1}]]
```

We finally have the following picture.

```
In[26]:= Show[{Graphics[sketch] /. GrayLevel[0.] -> GrayLevel[1/2],
              (* the lines and points of the 28 tangents *)
              Graphics[{{Thickness[0.002], makeLine /@ mbPairList},
                        {PointSize[0.0125], makePoint /@ mbPairList}}]},
              PlotRange -> {{-3, 3}, {-3, 3}}, AspectRatio -> Automatic,
              Frame -> False]
```

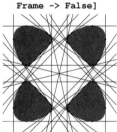

For other explicit equations of quartics with real bitangents, see [185], [1731], and [1612]. For a GröbnerBasis use for calculating all 28 bitangents, see [1731]. For the inverse problem, namely determining the curve from its 28 bitangents, see [302].

b) Using NRoots, we can calculate a dense set of curve points. We recognize 12 self-intersections.

```
In[1]:= eq[q_, z_] = 2 z^6 - q^3 - q;
```

```
     pic = With[{R = 0.9, pp = 1000},
                Show[Graphics[{PointSize[0.002],
                        Table[Point[{Re[#], Im[#]}]& /@
                              Cases[NRoots[eq[R Exp[I φ], z] == 0, z],
                        _Real | _Complex, {-1}], {φ, 0, 2Pi, 2Pi/pp}]}],
                PlotRange -> All, Frame -> True, AspectRatio -> Automatic]]
```

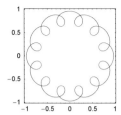

To calculate the intersection points, we find the solutions of $\{Re(z(\varphi)),\ Im(z(\varphi))\} = \{Re(z(\phi)),\ Im(z(\phi))\}$ for $\varphi \neq \phi$ [1392].

```
In[3]:= eqs0 = {eq[r (Cos[φ] + I Sin[φ]), zr + I zi],
              eq[r (Cos[φ] + I Sin[φ]), zr + I zi]} // Expand
Out[3]= {-2 zi⁶ + 12 i zi⁵ zr + 30 zi⁴ zr² - 40 i zi³ zr³ - 30 zi² zr⁴ + 12 i zi zr⁵ + 2 zr⁶ - r Cos[φ] -
        r³ Cos[φ]³ - i r Sin[φ] - 3 i r³ Cos[φ]² Sin[φ] + 3 r³ Cos[φ] Sin[φ]² + i r³ Sin[φ]³,
        -2 zi⁶ + 12 i zi⁵ zr + 30 zi⁴ zr² - 40 i zi³ zr³ - 30 zi² zr⁴ + 12 i zi zr⁵ + 2 zr⁶ - r Cos[φ] -
        r³ Cos[φ]³ - i r Sin[φ] - 3 i r³ Cos[φ]² Sin[φ] + 3 r³ Cos[φ] Sin[φ]² + i r³ Sin[φ]³}
```

We separate real and imaginary parts and obtain a coupled set of four equations.

```
In[4]:= eqs1 = Flatten[Function[expr, Flatten[Expand[{expr - #, #/I}]]&[
                      Plus @@ Cases[expr, _Complex ]]]]] /@ eqs0]
Out[4]= {-2 zi⁶ + 30 zi⁴ zr² - 30 zi² zr⁴ + 2 zr⁶ - r Cos[φ] - r³ Cos[φ]³ + 3 r³ Cos[φ] Sin[φ]²,
        12 zi⁵ zr - 40 zi³ zr³ + 12 zi zr⁵ - r Sin[φ] - 3 r³ Cos[φ]² Sin[φ] + r³ Sin[φ]³,
        -2 zi⁶ + 30 zi⁴ zr² - 30 zi² zr⁴ + 2 zr⁶ - r Cos[φ] - r³ Cos[φ]³ + 3 r³ Cos[φ] Sin[φ]²,
        12 zi⁵ zr - 40 zi³ zr³ + 12 zi zr⁵ - r Sin[φ] - 3 r³ Cos[φ]² Sin[φ] + r³ Sin[φ]³}
```

Eliminating the two sin functions yields a set of 18 equations.

```
In[5]:= gb1 = GroebnerBasis[Join[eqs1, {Cos[φ]^2 + Sin[φ]^2 - 1,
                      Cos[φ]^2 + Sin[φ]^2 - 1}], {}, {Sin[φ], Sin[φ]},
                      MonomialOrder -> EliminationOrder] // Factor;
```

Not unexpectedly, many of the equations of gb1 have factors of the form $\cos(\varphi) - \cos(\phi)$. (These factors represent the trivial solution that the two points are identical. The situation $\varphi = -\phi$ does not occur here.)

```
In[6]:= Short[gb1, 12]

Out[6]//Short= {-r (Cos[φ] - Cos[φ]) (Cos[φ] + Cos[φ]), -r (1 - 3 r² + 4 r² Cos[φ]²) (Cos[φ] - Cos[φ]),
        ≪14≫, -zr¹¹ (1 - 3 r² + 4 r² Cos[φ]²) (Cos[φ] - Cos[φ]),
        -819 r² zi⁴ + 1638 r⁴ zi⁴ - 819 r⁶ zi⁴ + 12246 r² zi² zr² - 24492 r⁴ zi² zr² + 12246 r⁶ zi² zr² -
        11683 r² zr⁴ + 23366 r⁴ zr⁴ - 11683 r⁶ zr⁴ + 65536 zr¹⁶ - 7281792 r zi⁴ zr⁶ Cos[φ] +
        21845376 r³ zi⁴ zr⁶ Cos[φ] + 7703808 r zi² zr⁸ Cos[φ] - 23111424 r³ zi² zr⁸ Cos[φ] -
        526976 r zr¹⁰ Cos[φ] + 1580928 r³ zr¹⁰ Cos[φ] + 819 r² zi⁴ Cos[φ]² - 8190 r⁴ zi⁴ Cos[φ]² +
        7371 r⁶ zi⁴ Cos[φ]² + 17238 r² zi² zr² Cos[φ]² - 54444 r⁴ zi² zr² Cos[φ]² +
        155142 r⁶ zi² zr² Cos[φ]² + 258787 r² zr⁴ Cos[φ]² - 1599454 r⁴ zr⁴ Cos[φ]² +
        2329083 r⁶ zr⁴ Cos[φ]² - 29127168 r³ zi⁴ zr⁶ Cos[φ]³ + 30815232 r³ zi² zr⁸ Cos[φ]³ -
        2107904 r³ zr¹⁰ Cos[φ]³ + 6552 r⁴ zi⁴ Cos[φ]⁴ - 19656 r⁶ zi⁴ Cos[φ]⁴ + 137904 r⁴ zi² zr² Cos[φ]⁴ -
        413712 r⁶ zi² zr² Cos[φ]⁴ + 2070296 r⁴ zr⁴ Cos[φ]⁴ - 6210888 r⁶ zr⁴ Cos[φ]⁴ +
        13104 r⁶ zi⁴ Cos[φ]⁶ + 275808 r⁶ zi² zr² Cos[φ]⁶ + 4140592 r⁶ zr⁴ Cos[φ]⁶}
```

We eliminate these factors and then eliminate the two cos functions. This gives two equations for $Re(z(\varphi))$ and $Im(z(\varphi))$.

```
In[7]:= gb1a = DeleteCases[gb1, Cos[φ] - Cos[φ], Infinity];
       gb2 = GroebnerBasis[gb1a, {}, {Cos[φ], Cos[φ]}]
Out[8]= {zi⁶ - 15 zi⁴ zr² + 15 zi² zr⁴ - zr⁶, r⁴ + r⁶ - 247104 zi⁴ zr⁸ + 264576 zi² zr¹⁰ - 17728 zr¹²}
```

We solve this equation and extract the $Re(z(\varphi))$ and $Im(z(\varphi))$ that are real. This results in 12 points.

```
In[9]:= sol1 = Solve[# == 0& /@ gb2, {zi, zr}] // Union;

In[10]:= iPoints = {zr, zi} /. Select[sol1 /. r -> 9/10,
                      (Im[N[{zr, zi} /. #, 22]] == {0., 0.})&];

In[11]:= Length[iPoints]
```

Out[11]= 12

Because of the symmetry of the curve, it is enough to show the intersection points in the first quadrant.

In[12]:= **Select[iPoints, (#[[1]] >= 0 && #[[2]] >= 0)&]**

Out[12]= $\left\{\left\{\dfrac{3^{2/3}\,181^{1/12}}{2\,2^{1/6}\,\sqrt{5}},\ \dfrac{3^{2/3}\,181^{1/12}}{2\,2^{1/6}\,\sqrt{5}}\right\},\ \left\{\left(\dfrac{1604367891}{1000000}-\dfrac{46314099\,\sqrt{3}}{50000}\right)^{1/12}\dfrac{1}{2\,2^{1/6}},\right.\right.$

$\left.\dfrac{50000\left(\dfrac{46314099\left(\dfrac{1604367891}{1000000}-\dfrac{46314099\,\sqrt{3}}{50000}\right)^{1/12}}{50000\,2^{1/6}}+\dfrac{46314099\,\sqrt{3}\left(\dfrac{1604367891}{1000000}-\dfrac{46314099\,\sqrt{3}}{50000}\right)^{1/12}}{100000\,2^{1/6}}\right)}{46314099}\right\},$

$\left\{\left(\dfrac{1604367891}{16384000000}+\dfrac{46314099\,\sqrt{3}}{819200000}\right)^{1/12},\right.$

$\left.\dfrac{50000\left(\dfrac{46314099\left(\dfrac{1604367891}{16384000000}+\dfrac{46314099\,\sqrt{3}}{819200000}\right)^{1/12}}{25000}-\dfrac{46314099\,\sqrt{3}\left(\dfrac{1604367891}{16384000000}+\dfrac{46314099\,\sqrt{3}}{819200000}\right)^{1/12}}{50000}\right)}{46314099}\right\}\right\}$

Here is the curve together with the just-calculated intersection points shown.

In[13]:= **Show[{pic, Graphics[{**
    {PointSize[0.02], Hue[0.0], Point[#]& /@ iPoints}}]},
    PlotRange -> All, Frame -> True, AspectRatio -> Automatic]

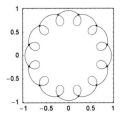

c) Without loss of generality, we assume the pentagon vertices to lie on the unit circle. So the pentaellipse is described by the following formula: $\sum_{k=1}^{5} |\{\sin(k\,2\pi/5),\ \cos(k\,2\pi/5)\} - \{x,\ y\}| = c$.

To derive a polynomial equation from this description we have to eliminate the five square roots. We could, for instance, introduce auxiliary variables for the square roots and then use GroebnerBasis with an elimination order. Instead, we will follow a more straightforward approach. We eliminate one square root on one side the equation and then square the equation. Doing this repeatedly eliminates all square roots. The function step eliminates the linearly in *eq* occurring variable *v* by squaring and replaces v^2 by V.

In[1]:= **step[eq_, v_, V_] :=**
    Module[{rhs, lhs},
        (* isolate v *)
        lhs = Plus @@ Cases[Expand[eq], f_. v, {1}];
        rhs = Expand[eq - lhs];
        Expand[Factor[lhs]^2 - rhs^2 /. v^2 -> V] /.
            {r[k_]^e_ :> (* use squares if possible *)
                If[EvenQ[e], R[k]^(e/2), R[k]^((e - 1)/2) r[k]]}]

Writing r[k] for $|\{\sin(k\,2\pi/5),\ \cos(k\,2\pi/5)\} - \{x,\ y\}|$, the following expression s0 is our pentaellipse.

In[2]:= **s0 = r[1] + r[2] + r[3] + r[4] + r[5] - c;**

The next call to step eliminates r[1].

In[3]:= **s1 = step[s0, r[1], R[1]]**

Out[3]= $-c^2 + 2\,c\,r[2] + 2\,c\,r[3] - 2\,r[2]\,r[3] + 2\,c\,r[4] - 2\,r[2]\,r[4] - 2\,r[3]\,r[4] +$
    $2\,c\,r[5] - 2\,r[2]\,r[5] - 2\,r[3]\,r[5] - 2\,r[4]\,r[5] + R[1] - R[2] - R[3] - R[4] - R[5]$

The next four calls to step eliminate all square roots containing *x* and *y*.

```
In[4]:= s2 = step[s1, r[2], R[2]];
        s3 = step[s2, r[3], R[3]];
        s4 = step[s3, r[4], R[4]];
        s5 = step[s4, r[5], R[5]];
```

The resulting polynomial s5 is quite large. Expanded it has 20349 terms.

```
In[8]:= {Length[#], ByteCount[#], LeafCount[#]}&[Expand @ s5]

Out[8]= {20349, 7944376, 358531}
```

s5 is of degree 32 in c.

```
In[9]:= Exponent[s5, c]

Out[9]= 32
```

To get an explicit polynomial, we have to expand the squared distances $R[k]$. Let $d[p1, p2]$ denote the distance between the points $p1$ and $p2$.

```
In[10]:= p[5][j_] = {Sin[j 2Pi/5], Cos[j 2Pi/5]};
         d[p1_, p2_] := Sqrt[(p1 - p2).(p1 - p2)]
```

The symmetry of the pentaellipse suggests that the coefficients of the powers of c in s5 are most naturally expressed as symmetric functions in the five $R[k]$. Using the package `Algebra`SymmetricPolynomials`` we eliminate all the $R[k]$.

```
In[12]:= Needs["Algebra`SymmetricPolynomials`"]

In[13]:= rules = Append[(* general case *)
         (# -> Expand[# /. R[j_] :> d[p[5][j], {x, y}]^2])& /@
         Table[SymmetricPolynomial[{R[1], R[2], R[3], R[4], R[5]}, k], {k, 5}],
             (* pure product case *)
             R[1]^e_ R[2]^e_ R[3]^e_ R[4]^e_ R[5]^e_ ->
             ExpandAll[((R[1] R[2] R[3] R[4] R[5])^e /.
                        R[j_] :> d[p[5][j], {x, y}]^2)]]

Out[13]= {R[1] + R[2] + R[3] + R[4] + R[5] -> 5 + 5 x^2 + 5 y^2,
          R[1] R[2] + R[1] R[3] + R[2] R[3] + R[1] R[4] + R[2] R[4] + R[3] R[4] + R[1] R[5] +
            R[2] R[5] + R[3] R[5] + R[4] R[5] -> 10 + 15 x^2 + 10 x^4 + 15 y^2 + 20 x^2 y^2 + 10 y^4,
          R[1] R[2] R[3] + R[1] R[2] R[4] + R[1] R[3] R[4] + R[2] R[3] R[4] + R[1] R[2] R[5] +
            R[1] R[3] R[5] + R[2] R[3] R[5] + R[1] R[4] R[5] + R[2] R[4] R[5] + R[3] R[4] R[5] ->
            10 + 15 x^2 + 15 x^4 + 10 x^6 + 15 y^2 + 30 x^2 y^2 + 30 x^4 y^2 + 15 y^4 + 30 x^2 y^4 + 10 y^6,
          R[1] R[2] R[3] R[4] + R[1] R[2] R[3] R[5] + R[1] R[2] R[4] R[5] + R[1] R[3] R[4] R[5] +
            R[2] R[3] R[4] R[5] -> 5 + 5 x^2 + 5 x^4 + 5 x^6 + 5 x^8 + 5 y^2 + 10 x^2 y^2 + 15 x^4 y^2 +
            20 x^6 y^2 + 5 y^4 + 15 x^2 y^4 + 30 x^4 y^4 + 5 y^6 + 20 x^2 y^6 + 5 y^8, R[1] R[2] R[3] R[4] R[5] ->
            1 + x^10 - 10 x^4 y + 5 x^8 y^2 + 20 x^2 y^3 + 10 x^6 y^4 - 2 y^5 + 10 x^4 y^6 + 5 x^2 y^8 + y^10,
          R[1]^e_ R[2]^e_ R[3]^e_ R[4]^e_ R[5]^e_ ->
          (1 + x^10 - 10 x^4 y + 5 x^8 y^2 + 20 x^2 y^3 + 10 x^6 y^4 - 2 y^5 + 10 x^4 y^6 + 5 x^2 y^8 + y^10)^e_}

In[14]:= eliminateRs[expr_] := Factor[Plus @@
         SymmetricReduction[expr, {R[1], R[2], R[3], R[4], R[5]}] //.
             (* use explicit values for the distances *) rules]

In[15]:= poly = Collect[s5, c, eliminateRs];
```

Luckily, not only did we eliminate all $R[k]$, but also all square roots resulting from $\sin(k\, 2\pi/5)$ and $\cos(k\, 2\pi/5)$.

```
In[16]:= Union[Head /@ ((List @@ poly) /. {x -> 1, y -> 1, c -> 1})]

Out[16]= {Integer}
```

poly is still quite large, but it is considerably smaller than s5.

```
In[17]:= {Length[#], ByteCount[#], LeafCount[#]}&[Expand @ poly]

Out[17]= {1477, 282048, 14180}

In[18]:= Max[Plus @@@ (First /@ Internal`DistributedTermsList[Expand[poly],
                                            {x, y, c}][[1]])]

Out[18]= 32
```

Switching finally to a polar coordinate system results in polynomial `polarPoly` of manageable size.

```
In[19]:= polarPoly = Collect[poly, c,
                   Simplify[# /. {x -> r Cos[φ], y -> r Sin[φ]}]&];
```

```
In[20]:= {Length[#], ByteCount[#], LeafCount[#]}&[Expand @ polarPoly]
```

```
Out[20]= {260, 46920, 2363}
```

```
In[21]:= Short[polarPoly, 12]
```

$$Out[21]//Short= -c^{32} + 80\ c^{30}\ (1 + r^2) - 40\ c^{28}\ (67 + 138\ r^2 + 67\ r^4) + \ll 20 \gg +$$
$$16\ c^2\ (977589 - 14744025\ r^2 + 86704425\ r^4 - 196046325\ r^6 + 164746225\ r^8 +$$
$$242872651\ r^{10} - 858832475\ r^{12} + 1245410575\ r^{14} + 1245410575\ r^{16} - 858832475\ r^{18} +$$
$$242872651\ r^{20} + 164746225\ r^{22} - 196046325\ r^{24} + 86704425\ r^{26} - 14744025\ r^{28} +$$
$$977589\ r^{30} - 12582912\ r^{10}\ (7 - 25\ r^2 + 50\ r^4 + 50\ r^6 - 25\ r^8 + 7\ r^{10})\ \text{Cos}[10\ φ] +$$
$$512\ r^5\ (27243 + 202050\ r^2 - 1474225\ r^4 + 2213400\ r^6 + 500550\ r^8 - 6870196\ r^{10} + 500550\ r^{12} +$$
$$2213400\ r^{14} - 1474225\ r^{16} + 202050\ r^{18} + 27243\ r^{20})\ \text{Sin}[5\ φ] + 134217728\ r^{15}\ \text{Sin}[15\ φ])$$

Now let us visualize $p_c(x, y)$. We start by generating an efficient routine to calculate the numerical version of `polarPoly`.

```
In[22]:= optimize = (* use function from Experimental` context *)
             Experimental`OptimizeExpression[#,
                OptimizationLevel -> 1, ExcludedForms -> {}]&;
```

```
In[23]:= optPolarPoly = optimize[Evaluate[N[polarPoly]]];
```

```
In[24]:= cfpolarPoly = Compile[{r, φ, c}, Evaluate[optPolarPoly]];
```

We will make explicit use of the symmetry and calculate the contours $p_c(x, y) = 0$ for only 1/10th of the whole graphic and generate the remaining 9/10th by reflection and rotation. `R[i]` are the rotation matrices to do this and `makeAll` generates the other nine graphic parts.

```
In[25]:= Do[R[i] = {{ Cos[2Pi/5 i], Sin[2Pi/5 i]},
                    {-Sin[2Pi/5 i], Cos[2Pi/5 i]}}, {i, 0, 4}]
```

```
In[26]:= makeAll[gr_] :=
             With[{gr = {gr, gr /. (pl:(Polygon | Line))[l_] :> pl[{-1, 1}#& /@ l]}},
                Table[gr /. (pl:(Polygon | Line))[l_] :> pl[R[i].#& /@ l], {i, 0, 4}]]
```

We will use `ContourPlot` in a polar coordinate system. The function `addPoints` refines polygons generated by `Graphics[ContourGraphics[...]]`.

```
In[27]:= addPoints[points_, δε_] :=
             Module[{n, l}, Join @@ (
                Function[s, If[(l = Sqrt[#. #]&[Subtract @@ s]) < δε, s,
                   n = Floor[l/δε] + 1; (* form segments *)
                   Table[# + i/n (#2 - #1), {i, 0, n - 1}]& @@ s]] /@
                Partition[Append[points, First[points]], 2, 1])]
```

```
In[28]:= p[5][j_] = {Sin[j 2Pi/5], Cos[j 2Pi/5]};
         d[p1_, p2_] := Sqrt[(p1 - p2).(p1 - p2)]
```

We will calculate the contour lines of the original expression `s0` using `p`.

```
In[30]:= p[x_, y_, c_] = s0 /. r[j_] :> d[p[5][j], {x, y}];
```

The function `cGraphics` finally calculates all contour lines for a given value of c. The contour lines of the original expression will be shown in red.

```
In[31]:= cGraphics[c_, opts___] :=
             Module[{R = 7, ppr = 300, ppφ = 120, ε = 0.1,
                   cp, grC, L1, L2, grD, cpO, outerCircle},
                (* contour plot in r,φ-coordinates *)
                cp = ContourPlot[cfpolarPoly[r, φ, c],
                   {r, 0, R}, {φ, 3Pi/10, Pi/2}, Compiled -> False,
                   ContourLines -> False, DisplayFunction -> Identity,
                   PlotPoints -> {ppr, ppφ}, Contours -> {0}];
                (* convert contour graphics into Cartesian polygons *)
                grC = Graphics[cp] /. Polygon[l_] :> Polygon[addPoints[l, 0.1]] /.
                   Polygon[l_] :> Polygon[Apply[#1 {Cos[#2], Sin[#2]}&, l, {1}]];
                (* deal especially with boundaries of the Pi/5 wedge *)
```

```
L1 = {GrayLevel[Sign[Plus @@ (Last /@ #)] /. -1 -> 0],
     Line[{0, #}& @@@ #]}& /@
       Split[Table[{r, cfpolarPoly[r, Pi/2, c]}, {r, 0, R, R/(ppr)}],
             Sign[#1] == Sign[#2]&];
L2 = {GrayLevel[Sign[Plus @@ (Last /@ #)] /. -1 -> 0],
     Line[# {Cos[3Pi/10], Sin[3Pi/10]}& @@@ #]}& /@
       Split[Table[{N @ r, cfpolarPoly[r, 3Pi/10, c]}, {r, 0, R, R/(2ppr)}],
             Sign[#1] == Sign[#2]&];
grD = {Graphics[{Thickness[0.004], L1, L2 }]};
 (* contour plot of the radical version *)
cpO = ContourPlot[Evaluate[p[r Cos[φ], r Sin[φ], c]],
    {r, 0, R}, {φ, 3Pi/10, Pi/2},
    ContourShading -> False, DisplayFunction -> Identity,
    PlotPoints -> {ppr, ppφ}, Contours -> {0}];
(* convert contour graphics into polygons *)
grO = Graphics[cpO] /. Line[l_] :> {Hue[0], Thickness[0.002],
                  Line[Apply[#1 {Cos[#2], Sin[#2]}&, 1, {1}]]};
(* outer white boundary *)
outerCircle = Graphics[{GrayLevel[1],
                 Polygon[Join[0.99 #, 1.01 Reverse[#]]]&[
                 Table[R{Cos[φ], Sin[φ]}, {φ, 0, 2Pi, 2Pi/360}]]]}];
(* display all elements *)
Show[{makeAll[grC], makeAll[grD], makeAll[grO], outerCircle},
     opts, Frame -> False, DisplayFunction -> $DisplayFunction] // N]
```

The following array of graphics shows the contour lines of the polynomial and the original version for various values of the parameter *c*. As expected, all graphics show a fivefold symmetry.

```
In[32]:= Show[GraphicsArray[
        cGraphics[#, DisplayFunction -> Identity]& /@ #]]& /@
                   Partition[Table[c, {c, 0, 11, 11/14}], 3]
```

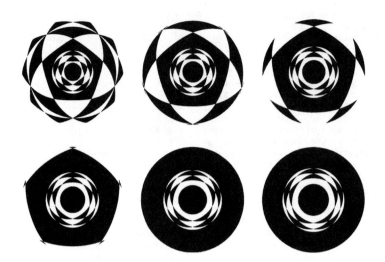

Here is the corresponding animation.

```
Do[cGraphics[c], {c, 0, 11, 1/8}]
```

For other generalizations of ellipses, see [617].

29. Maxwell's Equations *Are* Galilei Invariant, X-Waves, Fields in a Moving Media, Thomas Precession, Retarded Potential Expansion

a) Here, the four-vector analysis operations div, curl, Δ, and $\nabla.\vec{v}$ are implemented.

```
In[1]:= curl[{ax_, ay_, az_}] :=
          {D[az, y] - D[ay, z], D[ax, z] - D[az, x], D[ay, x] - D[ax, y]}
In[2]:= div[{ax_, ay_, az_}] := D[ax, x] + D[ay, y] + D[az, z]

In[3]:= laplace[a_] := D[a, x, x] + D[a, y, y] + D[a, z, z]

In[4]:= nablaV[{vx_, vy_, vz_}, ψ_] := vx D[ψ, x] + vy D[ψ, y] + vz D[ψ, z]
```

The unprimed fields fulfill the Maxwell equations in the unprimed coordinates. We write substitution rules that will replace all time derivatives by the corresponding space derivatives.

```
In[5]:= maxwellRules1 =
          {Derivative[1, dx_, dy_, dz_][Ex][t_, x_, y_, z_] ->
              -Derivative[0, dx, dy, dz + 1][Hy][t, x, y, z] +
              Derivative[0, dx, dy + 1, dz][Hz][t, x, y, z],
           Derivative[1, dx_, dy_, dz_][Ey][t_, x_, y_, z_] ->
              Derivative[0, dx, dy, dz + 1][Hx][t, x, y, z] -
              Derivative[0, dx + 1, dy, dz][Hz][t, x, y, z],
           Derivative[1, dx_, dy_, dz_][Ez][t_, x_, y_, z_] ->
              -Derivative[0, dx, dy + 1, dz][Hx][t, x, y, z] +
              Derivative[0, dx + 1, dy, dz][Hy][t, x, y, z],
           Derivative[1, dx_, dy_, dz_][Hx][t_, x_, y_, z_] ->
              Derivative[0, dx, dy, dz + 1][Ey][t, x, y, z] -
              Derivative[0, dx, dy + 1, dz][Ez][t, x, y, z],
           Derivative[1, dx_, dy_, dz_][Hy][t_, x_, y_, z_] ->
              -Derivative[0, dx, dy, dz + 1][Ex][t, x, y, z] +
              Derivative[0, dx + 1, dy, dz][Ez][t, x, y, z],
           Derivative[1, dx_, dy_, dz_][Hz][t_, x_, y_, z_] ->
              Derivative[0, dx, dy + 1, dz][Ex][t, x, y, z] -
              Derivative[0, dx + 1, dy, dz][Ey][t, x, y, z]};
```

In a similar way, we solve the last two Maxwell equations with respect to the E_x and H_x.

```
In[6]:= maxwellRules2 =
        {Derivative[0, dx_?(# >= 1&), dy_, dz_][Ex][t_, x_, y_, z_] ->
                -Derivative[0, dx - 1, dy + 1, dz][Ey][t, x, y, z] +
                -Derivative[0, dx - 1, dy, dz + 1][Ez][t, x, y, z],
         Derivative[0, dx_?(# >= 1&), dy_, dz_][Hx][t_, x_, y_, z_] ->
                -Derivative[0, dx - 1, dy + 1, dz][Hy][t, x, y, z] +
                -Derivative[0, dx - 1, dy, dz + 1][Hz][t, x, y, z]};
```

The following implements some abbreviations for the fields, the position, and the velocity vector.

```
In[7]:= e = {Ex[t, x, y, z], Ey[t, x, y, z], Ez[t, x, y, z]};
       h = {Hx[t, x, y, z], Hy[t, x, y, z], Hz[t, x, y, z]};
       v = {vx, vy, vz};
       r = {x, y, z};
```

These are the transformed fields.

```
In[11]:= eP = e - Cross[v, h] - (v.r + 1/2 v.v t) curl[h] +
              1/2 (v.v e - v v.e + r.v (nablaV[v, e] - 2 Cross[v, curl[e]]) +
              (r.v)^2 laplace[e]);

In[12]:= hP = h + Cross[v, e] + (v.r + 1/2 v.v t) curl[e] +
              1/2 (v.v h - v v.h + r.v (nablaV[v, h] - 2 Cross[v, curl[h]]) +
              (r.v)^2 laplace[h]);
```

The function Υ checks if eq is zero up to order 3 in $\tilde{v}$.

```
In[13]:= Υ[eq_] := Expand[Expand[eq] /. maxwellRules1 /. maxwellRules2] /.
                  {vx -> λ vx, vy -> λ vy, vz -> λ vz} /.
                  λ^(n_)?(# >= 3&) -> 0 /. λ -> 1
```

According to the chain rule and using the Galilei transformation, the derivatives transform as $\nabla' = \nabla$ and $\partial_{t'} = \partial_t - \nabla.\tilde{v}$. Here are the transformed fields used in the Maxwell equations.

```
In[14]:= Υ[D[hP, t] - nablaV[v, hP] + curl[eP]]
Out[14]= {0, 0, 0}

In[15]:= Υ[D[eP, t] - nablaV[v, eP] - curl[hP]]
Out[15]= {0, 0, 0}

In[16]:= Υ[div[eP]]
Out[16]= 0

In[17]:= Υ[div[hP]]
Out[17]= 0
```

This means that the transformed fields fulfill the Maxwell equations to the given order. Only the relation between the fields in a media determines if the symmetry is the one of the Galilei group or of the Lorentz group.

For the transition from the Lorentz invariance to the Galilei invariance in the Maxwell equations, see [868]. To calculate the maximal symmetry group of a free nonrelativistic point particle, see [917]. For the invariance of the Maxwell equations with respect to intertial transformations, see [270].

b) We start by defining the five vector-valued functions $\mathcal{E}$, $\mathcal{H}$, $\mathcal{D}$, $\mathcal{B}$, v. In the following, *symbolc*[i] denotes the ith component of the three-vector *symbol*.

```
In[1]:= {𝓔, 𝓗, 𝓓, 𝓑, v} = Map[(# /@ Range[3])&, {𝓔c, 𝓗c, 𝓓c, 𝓑c, vc}];
```

Componentwise, it is straightforward to solve the constitutive equations for $\mathcal{D}$ and $\mathcal{B}$.

```
In[2]:= eqs = Flatten[Thread /@
              {𝓓 + 1/c Cross[v, 𝓗] == ε (𝓔 + 1/c Cross[v, 𝓑]),
               𝓑 - 1/c Cross[v, 𝓔] == μ (𝓗 - 1/c Cross[v, 𝓓])}]
```

$$\text{Out[2]= } \left\{ 𝓓c[1] + \frac{-vc[3]\,𝓗c[2] + vc[2]\,𝓗c[3]}{c} == ε\left(\frac{-vc[3]\,𝓑c[2] + vc[2]\,𝓑c[3]}{c} + 𝓔c[1]\right), \right.$$

$$𝓓c[2] + \frac{vc[3]\,𝓗c[1] - vc[1]\,𝓗c[3]}{c} == ε\left(\frac{vc[3]\,𝓑c[1] - vc[1]\,𝓑c[3]}{c} + 𝓔c[2]\right),$$

$$\mathcal{D}c[3] + \frac{-vc[2]\,\mathcal{H}c[1] + vc[1]\,\mathcal{H}c[2]}{c} = \varepsilon\left(\frac{-vc[2]\,\mathcal{B}c[1] + vc[1]\,\mathcal{B}c[2]}{c} + \mathcal{E}c[3]\right),$$

$$\mathcal{B}c[1] - \frac{-vc[3]\,\mathcal{E}c[2] + vc[2]\,\mathcal{E}c[3]}{c} = \mu\left(-\frac{-vc[3]\,\mathcal{D}c[2] + vc[2]\,\mathcal{D}c[3]}{c} + \mathcal{H}c[1]\right),$$

$$\mathcal{B}c[2] - \frac{vc[3]\,\mathcal{E}c[1] - vc[1]\,\mathcal{E}c[3]}{c} = \mu\left(-\frac{vc[3]\,\mathcal{D}c[1] - vc[1]\,\mathcal{D}c[3]}{c} + \mathcal{H}c[2]\right),$$

$$\mathcal{B}c[3] - \frac{-vc[2]\,\mathcal{E}c[1] + vc[1]\,\mathcal{E}c[2]}{c} = \mu\left(-\frac{-vc[2]\,\mathcal{D}c[1] + vc[1]\,\mathcal{D}c[2]}{c} + \mathcal{H}c[3]\right)\}$$

In[3]:= `sol = Simplify[Solve[eqs, Join[D, B]]]`

Out[3]= `{{Dc[1] →`
$$(c^2\,\varepsilon\,\mathcal{E}c[1] - \varepsilon\,(vc[2]^2\,\mathcal{E}c[1] + vc[3]^2\,\mathcal{E}c[1] - vc[1]\,vc[2]\,\mathcal{E}c[2] - vc[1]\,vc[3]\,\mathcal{E}c[3]) + \varepsilon\,\mu$$
$$vc[1]\,(vc[1]\,\mathcal{E}c[1] + vc[2]\,\mathcal{E}c[2] + vc[3]\,\mathcal{E}c[3])) -$$
$$c\,(-1 + \varepsilon\,\mu)\,(vc[3]\,\mathcal{H}c[2] - vc[2]\,\mathcal{H}c[3]))\,/\,(c^2 - \varepsilon\,\mu\,(vc[1]^2 + vc[2]^2 + vc[3]^2)),$$
$$\mathcal{B}c[2] \to (-c\,(-1 + \varepsilon\,\mu)\,(vc[3]\,\mathcal{E}c[1] - vc[1]\,\mathcal{E}c[3]) + c^2\,\mu\,\mathcal{H}c[2] -$$
$$\mu\,((-1 + \varepsilon\,\mu)\,vc[1]\,vc[2]\,\mathcal{H}c[1] + vc[1]^2\,\mathcal{H}c[2] + vc[3]\,(vc[3]\,\mathcal{H}c[2] - vc[2]\,\mathcal{H}c[3]) +$$
$$\varepsilon\,\mu\,vc[2]\,(vc[2]\,\mathcal{H}c[2] + vc[3]\,\mathcal{H}c[3])))\,/\,(c^2 - \varepsilon\,\mu\,(vc[1]^2 + vc[2]^2 + vc[3]^2)),\ \mathcal{D}c[3] \to$$
$$(-\varepsilon^2\,\mu\,vc[3]\,(vc[1]\,\mathcal{E}c[1] + vc[2]\,\mathcal{E}c[2] + vc[3]\,\mathcal{E}c[3]) + c\,(vc[2]\,\mathcal{H}c[1] - vc[1]\,\mathcal{H}c[2]) +$$
$$\varepsilon\,(vc[2]\,vc[3]\,\mathcal{E}c[2] + c^2\,\mathcal{E}c[3] - vc[1]^2\,\mathcal{E}c[3] - vc[2]^2\,\mathcal{E}c[3] - c\,\mu\,vc[2]\,\mathcal{H}c[1] +$$
$$vc[1]\,(vc[3]\,\mathcal{E}c[1] + c\,\mu\,\mathcal{H}c[2])))\,/\,(c^2 - \varepsilon\,\mu\,(vc[1]^2 + vc[2]^2 + vc[3]^2)),\ \mathcal{D}c[2] \to$$
$$(-\varepsilon^2\,\mu\,vc[2]\,(vc[1]\,\mathcal{E}c[1] + vc[2]\,\mathcal{E}c[2] + vc[3]\,\mathcal{E}c[3]) + c\,(-vc[3]\,\mathcal{H}c[1] + vc[1]\,\mathcal{H}c[3]) +$$
$$\varepsilon\,(c^2\,\mathcal{E}c[2] - vc[1]^2\,\mathcal{E}c[2] - vc[3]^2\,\mathcal{E}c[2] + vc[2]\,vc[3]\,\mathcal{E}c[3] + c\,\mu\,vc[3]\,\mathcal{H}c[1] +$$
$$vc[1]\,(vc[2]\,\mathcal{E}c[1] - c\,\mu\,\mathcal{H}c[3])))\,/\,(c^2 - \varepsilon\,\mu\,(vc[1]^2 + vc[2]^2 + vc[3]^2)),$$
$$\mathcal{B}c[1] \to (c\,(-1 + \varepsilon\,\mu)\,(vc[3]\,\mathcal{E}c[2] - vc[2]\,\mathcal{E}c[3]) + c^2\,\mu\,\mathcal{H}c[1] -$$
$$\mu\,(vc[2]^2\,\mathcal{H}c[1] + vc[3]^2\,\mathcal{H}c[1] - vc[1]\,vc[2]\,\mathcal{H}c[2] - vc[1]\,vc[3]\,\mathcal{H}c[3] + \varepsilon\,\mu\,vc[1]$$
$$(vc[1]\,\mathcal{H}c[1] + vc[2]\,\mathcal{H}c[2] + vc[3]\,\mathcal{H}c[3])))\,/\,(c^2 - \varepsilon\,\mu\,(vc[1]^2 + vc[2]^2 + vc[3]^2)),$$
$$\mathcal{B}c[3] \to (c\,(-1 + \varepsilon\,\mu)\,(vc[2]\,\mathcal{E}c[1] - vc[1]\,\mathcal{E}c[2]) + c^2\,\mu\,\mathcal{H}c[3] -$$
$$\mu\,((-1 + \varepsilon\,\mu)\,vc[1]\,vc[3]\,\mathcal{H}c[1] + (-1 + \varepsilon\,\mu)\,vc[2]\,vc[3]\,\mathcal{H}c[2] + vc[1]^2\,\mathcal{H}c[3] +$$
$$vc[2]^2\,\mathcal{H}c[3] + \varepsilon\,\mu\,vc[3]^2\,\mathcal{H}c[3]))\,/\,(c^2 - \varepsilon\,\mu\,(vc[1]^2 + vc[2]^2 + vc[3]^2)))\}\}$$

Inspecting these solutions for a while would allow one to guess and then to verify the correct vector version of the solution. (For some four versions of electrodynamic equations that are more difficult to translate into three-vector equations, see, for instance, equation (8) of [1644], [714], or [81].) Instead of following this heuristic approach, we will choose a more mechanical one. The scalar quantities appearing in the solutions are either vector components (the possible vectors are $\mathcal{E}$, $\mathcal{H}$, v, $\mathcal{E}\times\mathcal{H}$, $v\times\mathcal{E}$, or $v\times\mathcal{H}$) or scalars (the possible "pure" scalars are $\mathcal{E}.\mathcal{E}$, $\mathcal{H}.\mathcal{H}$, $v.v$, $v.\mathcal{E}$, or $v.\mathcal{H}$).

In[4]:= `vars = {vc[2], vc[3], Ec[2], Ec[3], Hc[2], Hc[2],`
`        vxEc[1], vxHc[1], ExHc[1], vE, vH, EH, EE, HH, vv};`

The goal is to eliminate in the solutions for the ith component all occurrences of vector components other than i. Such an elimination can be performed using `PolynomialReduce`. We preprocess the defining set of equations by forming a Gröbner basis. The variables to eliminate come first in the list of variables. Here, this done for the first component of $\mathcal{D}$.

In[5]:= `gb = GroebnerBasis[`
`        Flatten[{vE - v.E, vH - v.H, EH - E.H, EE - E.E, HH - H.H, vv - v.v,`
`                vxE[1] - Cross[v, E][[1]], vxH[1] - Cross[v, H][[1]],`
`                ExH[1] - Cross[E, H][[1]]}], vars];`

In[6]:= `{num, den} = {Numerator[#], Denominator[#]}&[Together[Dc[1] /. First[sol]]];`

In[7]:= `Simplify[PolynomialReduce[#, gb, vars,`
`                MonomialOrder -> Lexicographic][[-1]]]& /@ {num, den}`

Out[7]= $\{\dfrac{1}{2\,vx\mathcal{H}[1]\,\mathcal{E}x\mathcal{H}[1]}$
$$(\mathcal{H}\mathcal{H}\,\varepsilon\,(-1 + \varepsilon\,\mu)\,vc[1]\,vx\mathcal{E}[1]^2 + \mathcal{E}\mathcal{E}\,\varepsilon\,(1 - \varepsilon\,\mu)\,vc[1]\,vx\mathcal{H}[1]^2 - \varepsilon\,vc[1]\,vx\mathcal{H}[1]^2\,\mathcal{E}c[1]^2 +$$
$$\varepsilon^2\,\mu\,vc[1]\,vx\mathcal{H}[1]^2\,\mathcal{E}c[1]^2 - 2\,c\,vx\mathcal{H}[1]^2\,\mathcal{E}x\mathcal{H}[1] + 2\,c\,\varepsilon\,\mu\,vx\mathcal{H}[1]^2\,\mathcal{E}x\mathcal{H}[1] +$$
$$2\,c^2\,\varepsilon\,vx\mathcal{H}[1]\,\mathcal{E}c[1]\,\mathcal{E}x\mathcal{H}[1] - 2\,vv\,\varepsilon\,vx\mathcal{H}[1]\,\mathcal{E}c[1]\,\mathcal{E}x\mathcal{H}[1] +$$
$$2\,\varepsilon\,vc[1]^2\,vx\mathcal{H}[1]\,\mathcal{E}c[1]\,\mathcal{E}x\mathcal{H}[1] - 2\,\varepsilon^2\,\mu\,vc[1]^2\,vx\mathcal{H}[1]\,\mathcal{E}c[1]\,\mathcal{E}x\mathcal{H}[1] +$$
$$vv\,\varepsilon\,vc[1]\,\mathcal{E}x\mathcal{H}[1]^2 - vv\,\varepsilon^2\,\mu\,vc[1]\,\mathcal{E}x\mathcal{H}[1]^2 - \varepsilon\,vc[1]^3\,\mathcal{E}x\mathcal{H}[1]^2 + \varepsilon^2\,\mu\,vc[1]^3\,\mathcal{E}x\mathcal{H}[1]^2 +$$
$$\varepsilon\,vc[1]\,vx\mathcal{E}[1]^2\,\mathcal{H}c[1]^2 - \varepsilon^2\,\mu\,vc[1]\,vx\mathcal{E}[1]^2\,\mathcal{H}c[1]^2),\ c^2 - vv\,\varepsilon\,\mu\}$$

In a completely analogous way, we deal with the first component of $\mathcal{B}$.

In[8]:= `{num, den} = {Numerator[#], Denominator[#]}&[Together[Bc[1] /. First[sol]]];`

```
In[9]:= Simplify[PolynomialReduce[#, gb, vars,
                    MonomialOrder -> Lexicographic][[-1]]]& /@ {num, den}
```

Out[9]= $\Big\{ \dfrac{1}{2 \, vx\mathcal{E}[1] \, \mathcal{E}x\mathcal{H}[1]}$

$(\mathcal{H}\mathcal{H} \mu \, (-1 + \varepsilon \mu) \, vc[1] \, vx\mathcal{E}[1]^2 + \mathcal{E}\mathcal{E} \mu \, (1 - \varepsilon \mu) \, vc[1] \, vx\mathcal{H}[1]^2 - \mu \, vc[1] \, vx\mathcal{H}[1]^2 \, \mathcal{E}c[1]^2 +$
$\varepsilon \mu^2 \, vc[1] \, vx\mathcal{H}[1]^2 \, \mathcal{E}c[1]^2 + 2 \, c \, vx\mathcal{E}[1]^2 \, \mathcal{E}x\mathcal{H}[1] - 2 \, c \, \varepsilon \mu \, vx\mathcal{E}[1]^2 \, \mathcal{E}x\mathcal{H}[1] -$
$vv \, \mu \, vc[1] \, \mathcal{E}x\mathcal{H}[1]^2 + vv \, \varepsilon \mu^2 \, vc[1] \, \mathcal{E}x\mathcal{H}[1]^2 + \mu \, vc[1]^3 \, \mathcal{E}x\mathcal{H}[1]^2 -$
$\varepsilon \mu^2 \, vc[1]^3 \, \mathcal{E}x\mathcal{H}[1]^2 + 2 \, c^2 \, vx\mathcal{E}[1] \, \mathcal{E}x\mathcal{H}[1] \, \mathcal{H}c[1] - 2 \, vv \, \mu \, vx\mathcal{E}[1] \, \mathcal{E}x\mathcal{H}[1] \, \mathcal{H}c[1] +$
$2 \, \mu \, vc[1]^2 \, vx\mathcal{E}[1] \, \mathcal{E}x\mathcal{H}[1] \, \mathcal{H}c[1] - 2 \, \varepsilon \mu \, vc[1]^2 \, vx\mathcal{E}[1] \, \mathcal{E}x\mathcal{H}[1] \, \mathcal{H}c[1] +$
$\mu \, vc[1] \, vx\mathcal{E}[1]^2 \, \mathcal{H}c[1]^2 - \varepsilon \mu^2 \, vc[1] \, vx\mathcal{E}[1]^2 \, \mathcal{H}c[1]^2), \ c^2 - vv \, \varepsilon \mu \Big\}$

The last two equations show that the first component has been appropriately expressed in other quantities that are either scalars or contain only the first components of the fields and the velocity. Proceeding in a similar way shows that the second and third components give a similar result. This means that the corresponding vector equation can be obtained by simply dropping the component index, and we arrive at:

$$\mathcal{D} = \frac{1}{c^2 - \varepsilon \mu \, v.v} \, (\varepsilon (1 - \varepsilon \mu) \, v.\mathcal{E} \, v + c \, (\varepsilon \mu - 1) \, v \times \mathcal{H} + \varepsilon c^2 \, \mathcal{E} - \varepsilon \, v.v \, \mathcal{E}),$$

$$\mathcal{B} = \frac{1}{c^2 - \varepsilon \mu \, v.v} \, (\mu (1 - \varepsilon \mu) \, v.\mathcal{H} \, v + c \, (1 - \varepsilon \mu) \, v \times \mathcal{E} + \mu c^2 \, \mathcal{H} - \mu \, v.v \, \mathcal{H}).$$

c) This is the complex potential $\varphi_v(\rho, z; t)$.

```
In[1]:= φ[{ρ_, z_}, t_, {α_, v_}] :=
           (z - t v + I α)/Sqrt[(z - t v + I α)^2 + (1 - (v/c)^2) ρ^2]^3
```

It is straightforward to show that the real part and the imaginary part of $\varphi_v(\rho, z; t)$ fulfill the 3D wave equation.

```
In[2]:= With[{(* X-wave *) φ = φ[{ρ, z}, t, {α, v}]},
           (* 1D wave equation *)
           1/c^2 D[φ, t, t] - (1/ρ D[ρ D[φ, ρ], ρ] + D[φ, z, z])] //
                                                          Together
```

Out[2]= 0

Here is the real part written in an explicitly real way.

```
In[3]:= ComplexExpand[Re[φ[{ρ, z}, t, {α, v}]],
                TargetFunctions -> {Re, Im}] // Simplify
```

Out[3]= $\Big((-t \, v + z) \, \text{Cos}\Big[\dfrac{3}{2} \, \text{ArcTan}\Big[(-t \, v + z)^2 - \alpha^2 + \Big(1 - \dfrac{v^2}{c^2}\Big) \rho^2, \, 2 \, (-t \, v + z) \, \alpha \Big]\Big] +$

$\alpha \, \text{Sin}\Big[\dfrac{3}{2} \, \text{ArcTan}\Big[(-t \, v + z)^2 - \alpha^2 + \Big(1 - \dfrac{v^2}{c^2}\Big) \rho^2, \, 2 \, (-t \, v + z) \, \alpha \Big]\Big] \Big) \Big/$

$\Big(4 \, (-t \, v + z)^2 \, \alpha^2 + \Big((-t \, v + z)^2 - \alpha^2 + \Big(1 - \dfrac{v^2}{c^2}\Big) \rho^2 \Big)^2 \Big)^{3/4}$

$\varphi_v(\rho, z; t)$ is a solution of the wave equation for any value of v. Because $\varphi_v(\rho, z; t) = \varphi_v(\rho, z - v \, t; 0)$, it is a (unphysical) solution that propagates with any possible speed v. The following graphics show the $\text{Re}(\varphi_v(\rho, z; 0))$ for $v = 1/2$ and $v = 2$ in the ρ, z-plane.

```
In[4]:= Show[GraphicsArray[
           Plot3D[Evaluate[Re[φ[{ρ, z}, 0, {1, # c}]] /. c -> 1],
                {ρ, 0, 5}, {z, -5, 5}, PlotPoints -> 120,
                DisplayFunction -> Identity,
                Mesh -> False, PlotRange -> All]& /@ {1/2, 1, 2}]]
```

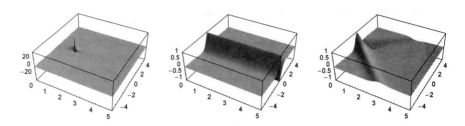

For similar superluminal solutions of the wave equations, see [1532], [1466], [1471], [1892], [1127], [166], [1656], [1604], [849], [41], [1268], [207], [1911], [1890], [1891], [1465], [507], and [1347]; for localized, finite energy solutions, see [1787].

d) We start by forming the products of the matrix exponentials.

```
In[1]:= σ[1] = {{0, 1}, {1, 0}};
        σ[2] = {{0, -I}, {I, 0}};
        σ[3] = {{1, 0}, {0, -1}};
```

```
In[4]:= m1 = MatrixExp[-ξ σ[3]].MatrixExp[η (Sin[α] σ[1] + Cos[α] σ[3])] //
                                                        Simplify
```

$$Out[4]= \left\{\left\{\frac{1}{2} e^{-\eta-\xi} (1 + e^{2\eta} + (-1 + e^{2\eta}) \cos[\alpha]), \frac{1}{2} e^{-\eta-\xi} (-1 + e^{2\eta}) \sin[\alpha]\right\},\right.$$
$$\left.\left\{\frac{1}{2} e^{-\eta+\xi} (-1 + e^{2\eta}) \sin[\alpha], \frac{1}{2} e^{-\eta+\xi} (1 + e^{2\eta} + \cos[\alpha] - e^{2\eta} \cos[\alpha])\right\}\right\}$$

```
In[5]:= m2 = MatrixExp[ρ (Sin[θ] σ[1] + Cos[θ] σ[3])].MatrixExp[I ψ σ[2]] //
          Expand // Simplify // TrigExpand // Expand // Simplify
```

$$Out[5]= \left\{\left\{\frac{1}{2} e^{-\rho} ((1 + e^{2\rho}) \cos[\psi] + (-1 + e^{2\rho}) \cos[\theta + \psi]),\right.\right.$$
$$\left.\frac{1}{2} e^{-\rho} ((1 + e^{2\rho}) \sin[\psi] + (-1 + e^{2\rho}) \sin[\theta + \psi])\right\},$$
$$\left\{\frac{1}{2} e^{-\rho} (-(1 + e^{2\rho}) \sin[\psi] + (-1 + e^{2\rho}) \sin[\theta + \psi]),\right.$$
$$\left.\left.\frac{1}{2} e^{-\rho} ((1 + e^{2\rho}) \cos[\psi] - (-1 + e^{2\rho}) \cos[\theta + \psi])\right\}\right\}$$

After threading m1 == m2, we have a set of four pseudopolynomial equations to be solved. Due to the complexity of this problem, a direct call to Solve will not succeed. To be on safe ground for equation solving, we rewrite the equations m1 == m2 in polynomial form by replacing the exponents with variables and clearing denominators.

```
In[6]:= eqs1 = Expand[Numerator[Together[(Flatten[m1 - m2] // Expand) /.
                    {Exp[ρ] -> eρ, Exp[-ρ] -> 1/eρ}]]] // TrigExpand
```

$$Out[6]= \{e\rho + e^{2\eta} e\rho - e\rho \cos[\alpha] + e^{2\eta} e\rho \cos[\alpha] - e^{\eta+\xi} \cos[\psi] - e^{\eta+\xi} e\rho^2 \cos[\psi] +$$
$$e^{\eta+\xi} \cos[\theta] \cos[\psi] - e^{\eta+\xi} e\rho^2 \cos[\theta] \cos[\psi] - e^{\eta+\xi} \sin[\theta] \sin[\psi] + e^{\eta+\xi} e\rho^2 \sin[\theta] \sin[\psi],$$
$$-e\rho \sin[\alpha] + e^{2\eta} e\rho \sin[\alpha] + e^{\eta+\xi} \cos[\psi] \sin[\theta] - e^{\eta+\xi} e\rho^2 \cos[\psi] \sin[\theta] -$$
$$e^{\eta+\xi} \sin[\psi] - e^{\eta+\xi} e\rho^2 \sin[\psi] + e^{\eta+\xi} \cos[\theta] \sin[\psi] - e^{\eta+\xi} e\rho^2 \cos[\theta] \sin[\psi],$$
$$-e^{\xi} e\rho \sin[\alpha] + e^{2\eta+\xi} e\rho \sin[\alpha] + e^{\eta} \cos[\psi] \sin[\theta] - e^{\eta} e\rho^2 \cos[\psi] \sin[\theta] +$$
$$e^{\eta} \sin[\psi] + e^{\eta} e\rho^2 \sin[\psi] + e^{\eta} \cos[\theta] \sin[\psi] - e^{\eta} e\rho^2 \cos[\theta] \sin[\psi],$$
$$e^{\xi} e\rho + e^{2\eta+\xi} e\rho + e^{\xi} e\rho \cos[\alpha] - e^{2\eta+\xi} e\rho \cos[\alpha] - e^{\eta} \cos[\psi] - e^{\eta} e\rho^2 \cos[\psi] -$$
$$e^{\eta} \cos[\theta] \cos[\psi] + e^{\eta} e\rho^2 \cos[\theta] \cos[\psi] + e^{\eta} \sin[\theta] \sin[\psi] - e^{\eta} e\rho^2 \sin[\theta] \sin[\psi]\}$$

We now introduce auxiliary variables for the nonpolynomially occurring variables. This gives the following equations

```
In[7]:= eqs2 = {eρ + eρ ηP^2 - eρ Cos[α] + eρ ηP^2 Cos[α] - ηP ξP Cos[ψ] -
        eρ^2 ηP ξP Cos[ψ] + ηP ξP Cos[θ] Cos[ψ] - eρ^2 ηP ξP Cos[θ] Cos[ψ] -
        ηP ξP Sin[θ] Sin[ψ] + eρ^2 ηP ξP Sin[θ] Sin[ψ],
        -eρ Sin[α] + eρ ηP^2 Sin[α] + ηP ξP Cos[ψ] Sin[θ] -
        eρ^2 ηP ξP Cos[ψ] Sin[θ] - ηP ξP Sin[ψ] - eρ^2 ηP ξP Sin[ψ] +
        ηP ξP Cos[θ] Sin[ψ] - eρ^2 ηP ξP Cos[θ] Sin[ψ], -eρ ξP Sin[α] +
        eρ ηP^2 ξP Sin[α] + ηP Cos[ψ] Sin[θ] - eρ^2 ηP Cos[ψ] Sin[θ] +
        ηP Sin[ψ] + eρ^2 ηP Sin[ψ] + ηP Cos[θ] Sin[ψ] - eρ^2 ηP Cos[θ] Sin[ψ],
        eρ ξP + eρ ηP^2 ξP + eρ ξP Cos[α] - eρ ηP^2 ξP Cos[α] - ηP Cos[ψ] -
        eρ^2 ηP Cos[ψ] - ηP Cos[θ] Cos[ψ] + eρ^2 ηP Cos[θ] Cos[ψ] +
        ηP Sin[θ] Sin[ψ] - eρ^2 ηP Sin[θ] Sin[ψ]};
```

We have to eliminate the three variables $e\rho$, $\mathtt{Sin[\theta]}$, and $\mathtt{Cos[\theta]}$ from four the equations $\mathtt{eqs1}$. Normally we would supplement our system with $\mathtt{Sin[\theta]2+Cos[\theta]^2-1}$, but because of the excess equation, this is not needed and we can eliminate $\mathtt{Sin[\theta]}$ and $\mathtt{Cos[\theta]}$ immediately.

```
In[8]:= GroebnerBasis[eqs2, {ξP, ηP, Sin[α], Cos[α], Sin[ψ], Cos[ψ]},
                {Sin[θ], Cos[θ], eρ}, Sort -> False,
                MonomialOrder -> EliminationOrder] // Factor
```

$\mathtt{Out[8]}=$ $\{-\eta P\,\xi P\,(\mathrm{Cos}[\psi]\,\mathrm{Sin}[\alpha] - \eta P^2\,\mathrm{Cos}[\psi]\,\mathrm{Sin}[\alpha] - \xi P^2\,\mathrm{Cos}[\psi]\,\mathrm{Sin}[\alpha] +$
$\eta P^2\,\xi P^2\,\mathrm{Cos}[\psi]\,\mathrm{Sin}[\alpha] + \mathrm{Sin}[\psi] + \eta P^2\,\mathrm{Sin}[\psi] + \xi P^2\,\mathrm{Sin}[\psi] + \eta P^2\,\xi P^2\,\mathrm{Sin}[\psi] -$
$\mathrm{Cos}[\alpha]\,\mathrm{Sin}[\psi] + \eta P^2\,\mathrm{Cos}[\alpha]\,\mathrm{Sin}[\psi] + \xi P^2\,\mathrm{Cos}[\alpha]\,\mathrm{Sin}[\psi] - \eta P^2\,\xi P^2\,\mathrm{Cos}[\alpha]\,\mathrm{Sin}[\psi])\}$

We could now evaluate $\mathtt{Solve[gb[[1]] == 0, \psi]}$. This gives the correct, but slightly clumsy, result. Instead, we solve for $\mathtt{Tan[\psi]}$ and simplify the result (to avoid autoevaluation we use $\mathtt{tan}$ instead of $\mathtt{Tan}$ temporarily). This gives a nice, concise answer.

```
In[9]:= GroebnerBasis[{%, Cos[ψ]^2 + Sin[ψ]^2 - 1,
                tan[ψ] Cos[ψ] - Sin[ψ]}, {}, {Sin[ψ], Cos[ψ]},
                MonomialOrder -> EliminationOrder] /. tan -> Tan
```

$\mathtt{Out[9]}=$ $\{-\eta P\,\xi P\,\mathrm{Sin}[\alpha] + \eta P^3\,\xi P\,\mathrm{Sin}[\alpha] + \eta P\,\xi P^3\,\mathrm{Sin}[\alpha] - \eta P^3\,\xi P^3\,\mathrm{Sin}[\alpha] -$
$\eta P\,\xi P\,\mathrm{Tan}[\psi] - \eta P^3\,\xi P\,\mathrm{Tan}[\psi] - \eta P\,\xi P^3\,\mathrm{Tan}[\psi] - \eta P^3\,\xi P^3\,\mathrm{Tan}[\psi] + \eta P\,\xi P\,\mathrm{Cos}[\alpha]\,\mathrm{Tan}[\psi] -$
$\eta P^3\,\xi P\,\mathrm{Cos}[\alpha]\,\mathrm{Tan}[\psi] - \eta P\,\xi P^3\,\mathrm{Cos}[\alpha]\,\mathrm{Tan}[\psi] + \eta P^3\,\xi P^3\,\mathrm{Cos}[\alpha]\,\mathrm{Tan}[\psi]\}$

```
In[10]:= Solve[%[[1]] == 0, Tan[ψ]] /. {ξP -> Exp[ξ], ηP -> Exp[η]} //
                FullSimplify
```

$\mathtt{Out[10]}=$ $\left\{\left\{\mathrm{Tan}[\psi] \to \dfrac{1}{\mathrm{Cot}[\alpha] - \mathrm{Coth}[\eta]\,\mathrm{Coth}[\xi]\,\mathrm{Csc}[\alpha]}\right\}\right\}$

e) Because we will work with time-dependent vectors, we will use the built-in function $\mathtt{Dot}$ in the following. We start adding the $\mathtt{Orderless}$ attribute to $\mathtt{Dot}$ because we will only deal with products of vectors and not matrices.

```
In[1]:= Unprotect[Dot];
        SetAttributes[Dot, Orderless];
```

Next, because we work with symbolic vectors, we teach $\mathtt{Dot}$ linearity and distributivity over nonvector quantities. We will use the doublestruck letters $\mathbb{r}$ and $\mathbb{v}$ for vectors.

```
In[3]:= (* distribute over sums *)
        Dot[a_Plus, b_] := Dot[#, b]& /@ a

        (* extract nonvector prefactors *)
        Dot[f_ a_, b_] := f Dot[a, b] /;
          Not[MatchQ[f, _. (r[_] | v[_] | r | v |
                      Derivative[_][r | v][_] | Derivative[_][r | v])]]
```

The next definition implements the scalar product for two Taylor series, both having vector-valued coefficients.

```
In[7]:= Dot[s_SeriesData, σ_SeriesData] :=
        (* calculate components *)
        ExpandAll[Normal[s]].ExpandAll[Normal[σ]] +
        (* make series again *)
        O[s[[1]]]^Min[s[[-2]], σ[[-2]]] /; First[s] === First[σ]
```

For uniformity, we write higher-order derivatives of $\mathbb{r}$ as derivatives of $\mathbb{v}$.

```
In[8]:= Derivative[k_][r][τ_] := -Derivative[k - 1][v][τ]
```

The basic identity to be used repeatedly in the following is the τ-derivative of $r(\tau)$. It is easily derived from the defining equations [1624], [204].

```
In[9]:= r'[τ_] := -r[τ].v[τ]/r[τ]
```

To obtain an expansion for the potential φ in inverse powers of c, we start by expressing the implicitly defined $\tau(t)$ (parametrically depending on γ) in inverse powers of c. Using $\gamma = 1/c$, we have $\tau(t) = \sum_{k=0}^{\infty} \gamma^k / k!\, \partial^k \tau(t)/\partial\gamma^k|_{\gamma=0}$.

The 0th-order time represents an infinite speed of light, meaning $\tau = t$. By $\mathtt{sol\gamma[k]}$, we denote the replacement rule for $\partial^k \tau(t)/\partial\gamma^k|_{\gamma=0}$.

```
In[10]:= solγ[0] = {τ[t][0] -> t};
```

The higher derivatives of $\tau(t)$ with respect to γ can be obtained through implicit differentiation. Here this is done for the first order.

```
In[11]:= eq[0] = τ[t][γ] == (t - γ r[τ[t][γ]]);
```

```
In[12]:= eq[1] = D[eq[0], γ]
```

$$\text{Out[12]= } \tau[t]'[\gamma] == -r[\tau[t][\gamma]] + \frac{\gamma\, r[\tau[t][\gamma]].v[\tau[t][\gamma]]\, \tau[t]'[\gamma]}{r[\tau[t][\gamma]]}$$

```
In[13]:= sol[1] = Solve[eq[1], τ[t]'[γ]][[1]]
```

$$\text{Out[13]= } \left\{ \tau[t]'[\gamma] \to \frac{r[\tau[t][\gamma]]^2}{\gamma\, r[\tau[t][\gamma]].v[\tau[t][\gamma]] - r[\tau[t][\gamma]]} \right\}$$

```
In[14]:= solγ[1] = sol[1] /. γ -> 0 /. solγ[0]
```

$$\text{Out[14]= } \{ \tau[t]'[0] \to -r[t] \}$$

It is straightforward to automate the last steps for the higher derivatives.

```
In[15]:= o = 5;

       Do[eq[k] = D[eq[k - 1], γ];
          sol[k] = Solve[eq[k], Derivative[k][τ[t]][γ]][[1]];
          solγ[k] = Fold[ReplaceAll, sol[k] /. γ -> 0,
                         Table[solγ[j], {j, k - 1, 0, -1}]] // Simplify,
       {k, 2, o}]
```

Here are the first few terms of the resulting power series for τ. Because now all quantities depend only on the unretarded time t, we suppress all time dependencies for brevity.

```
In[17]:= τS = Sum[γ^k/k! solγ[k][[1, 2]], {k, 0, o}] + O[γ]^(o + 1) /. x_[t] :> x
```

$$\text{Out[17]= } t - r\gamma - r.v\,\gamma^2 + \left(-\frac{(r.v)^2}{2\,r} + \frac{1}{2}\, r\,r.v' - \frac{r\,v.v}{2} \right) \gamma^3 +$$

$$\left(r.v\,r.v' - \frac{1}{6}\, r^2\, r.v'' - r.v\,v.v + \frac{1}{2}\, r^2\, v.v' \right) \gamma^4 +$$

$$\left(\frac{(r.v)^4}{8\,r^3} + \frac{3\,(r.v)^2\,r.v'}{4\,r} - \frac{3}{8}\, r\,(r.v')^2 - \frac{1}{2}\, r\,r.v\,r.v'' + \frac{1}{24}\, r^3\, r.v^{(3)} - \frac{3\,(r.v)^2\,v.v}{4\,r} + \right.$$

$$\left. \frac{3}{4}\, r\,r.v'\,v.v - \frac{3}{8}\, r\,(v.v)^2 + \frac{3}{2}\, r\,r.v\,v.v' - \frac{1}{6}\, r^3\, v.v'' - \frac{1}{8}\, r^3\, v'.v' \right) \gamma^5 + O[\gamma]^6$$

Having a series expansion for τ, we can easily calculate series expansions for $r(\tau)$, $v(\tau)$, and $r(\tau)$.

```
In[18]:= rS = (t - τS)/γ;
         {rS, vS} = {r[τS], v[τS]} + O[γ]^(o + 1) /. x_[t] :> x // Simplify;
         (* show first two terms *) {rS, vS} + O[γ]^3
```

$$\text{Out[20]= } \left\{ r + r\,v\,\gamma + \left(v\,r.v - \frac{r^2\,v'}{2} \right) \gamma^2 + O[\gamma]^3,\ v - r\,v'\,\gamma + \left(-r.v\,v' + \frac{r^2\,v''}{2} \right) \gamma^2 + O[\gamma]^3 \right\}$$

So, we finally arrive at the following first terms for in $\gamma = 1/c$. The 0th-order term is just the Newtonian distance; a first order term is absent. The higher-order terms are relatively complicated expressions, involving scalar products between the various vectors.

```
In[21]:= φS = 1/(rS - γ rS.vS) // Simplify
```

$$\text{Out[21]= } \frac{1}{r} - \frac{((r.v)^2 + r^2\,(r.v' - v.v))\,\gamma^2}{2\,r^3} + \frac{1}{3}\,(r.v'' - 3\,v.v')\,\gamma^3 +$$

$$\frac{1}{8\,r^5}\,((3\,(r.v)^4 + 6\,r^2\,(r.v)^2\,(r.v' - v.v) + 4\,r^4\,r.v\,(r.v'' - 3\,v.v') +$$

$$r^4\,(3\,(r.v')^2 - r^2\,r.v^{(3)} - 6\,r.v'\,v.v + 3\,(v.v)^2 + 4\,r^2\,v.v'' + 3\,r^2\,v'.v'))\,\gamma^4) + O[\gamma]^5$$

For the non-Abelian case, see [1559].

f) Because we have to test various possible nonvanishing field component arrangements (such as $\mathcal{E}(r) = \mathcal{E}_r\,e_r$, $\mathcal{B}(r) = \mathcal{B}_\vartheta\,e_\vartheta + \mathcal{B}_\varphi\,e_\varphi$ or $\mathcal{E}(r) = \mathcal{E}_\vartheta\,e_\vartheta$, $\mathcal{B}(r) = \mathcal{B}_r\,e_r + \mathcal{B}_\varphi\,e_\varphi$) where the $\mathcal{B}_{r,\vartheta,\varphi}$ and $\mathcal{E}_{r,\vartheta,\varphi}$ are functions of r and ϑ), we first define a set of functions and then deal with the six possible field sets at once. We start by expressing the unit vectors of a spherical coordinate system through Cartesian coordinates.

```
In[1]:= er := {Cos[φ] Sin[θ], Sin[φ] Sin[θ], Cos[θ]}
        eφ := {-Sin[φ], Cos[φ], 0}
        eθ := {Cos[φ] Cos[θ], Sin[φ] Cos[θ], -Sin[θ]}
```

The function `makeMaxwellEquations` constructs the above given ansatz (with the coefficients $c[a, ir, irc, irs, irse, j\theta c, j\theta s, j\theta se]$) for given field components, and plugs the ansatz into the four Maxwell equations [710]. We use some vector analysis functions for spherical coordinates from the package `Calculus`VectorAnalysis``.

```
In[4]:= Needs["Calculus`VectorAnalysis`"]
```

```
In[5]:= makeMaxwellEquations[ℰℬComponents_, o_] :=
       Module[{ℰA, ℬA},
         (* the ansatz; ℰℬComponentsAnsatz is globally visible *)
         ℰℬComponentsAnsatz =
         Sum[c[#, ir, irc, irs, irse, jθc, jθs, jθse] r^ir*
           Cos[irc r] If[irs === irse === 0, 0, Sin[irs r]^irse]*
           Cos[jθc θ] If[jθs === jθse === 0, 0, Sin[jθs θ]^jθse],
           {ir, -3, -1}, {irc, 0, 1}, {irs, 0, 1}, {irse, 0, 1},
           {jθc, 0, o}, {jθs, 0, o}, {jθse, 0, 1}]& /@ ℰℬComponents;
         (* field definitions *)
         {ℰA, ℬA} = {#[[1]] Sin[t], #[[2]] Cos[t]}& @ Partition[
         If[# === {}, 0, ℰℬComponentsAnsatz[[#[[1, 1]]]]]&[
           Position[ℰℬComponents, #]]& /@ {ℰr, ℰθ, ℰφ, ℬr, ℬθ, ℬφ}, 3];
         (* use ansatz in Maxwell equations *)
         {Div[ℰA, Spherical[r, θ, φ]],
         Sin[θ] Div[ℬA, Spherical[r, θ, φ]],
         Sin[θ] (Curl[ℰA, Spherical[r, θ, φ]] + D[ℬA, t]),
         Sin[θ] (Curl[ℬA, Spherical[r, θ, φ]] - D[ℰA, t])}]
```

The function `makeMaxwellEquations` returns a list of four (possibly quite large) unsimplified Poisson polynomials. To obtain a linear system of equations for the coefficients $c[...]$, we write all trigonometric functions in exponential form and extract their coefficients. `extractCoefficients` takes the result of `makeMaxwellEquations` as its argument and returns a list with two elements—a list of rules for the coefficients that are obviously zero and a list of the nontrivial equations.

```
In[6]:= extractCoefficients[mwes_] :=
         (* trivial and nontrivial equations in the c[...] *)
         {zeroCoefficientsRules[#], nontrivialRelations[#]}&[
         (* write equations in canonical form *)
           Join @@ (makecEquations /@ Expand[Map[TrigToExp,
             Cases[Expand[#], _Plus, {0, Infinity}], {2}]])]& /@ mwes;

       (* separate Poisson monomial and coefficient *)
       monomialsAndCoefficients[expr_] :=
       With[{m = Times @@ DeleteCases[List @@ expr, _c | _?NumericQ,
                         {1}]}, {m, expr/m}]

       (* collect with respect to Poisson monomials *)
       makecEquations[p_] := If[Head[p] =!= Plus, 0,
       Plus @@@ Map[Last, Split[Sort[monomialsAndCoefficients /@
                         (List @@ p)], #1[[1]] == #2[[1]]&], {2}]]

       (* make rules . -> 0 for coefficients that do not appear in sums *)
       zeroCoefficientsRules[l_] :=
       With[{h = Cases[l, _?(Head[#] =!= Plus&)]},
           (# -> 0)& /@ Union[(h/(h /. _c -> 1))]]

       (* extract equations that contain only a single coefficient *)
       nontrivialRelations[l_] := Cases[l, _?(Head[#] === Plus&)]
```

Next, we have to solve the linear system of equations that we obtained from the second argument of `extractCoeffi`⟶ `cients`. The function `solveCoefficientEquations` does this. It returns the solutions plugged into the ansatz `ℰℬComponentsAnsatz`.

```
In[18]:= solveCoefficientEquations[ceqs_] :=
         Module[{dpr, allEqs, allVars, sol},
                 (* use trivial zeros of the c[...] *)
                 dpr = Dispatch[Join @@ (First /@ ceqs)];
                 (* all equations *)
                 allEqs = Flatten[(Last /@ ceqs) //. dpr];
```

```
(* all variables c[...] *)
allVars = Union[Cases[allEqs, _c, Infinity]];
(* print number of equations and variables *)
Print[Length[allEqs], " equations in ",
      Length[allVars], " variables."];
(* solve linear system of the c[...] *)
sol = Solve[Take[allEqs, All] == 0, allVars];
(* substitute solution into ansatz *)
EBComponentsAnsatz //. dpr //. sol[[1]]]
```

The solutions we obtain at this point might not be realizable due to diverging field values at the origin. makeFiniteVal ·
ueAtOrigin avoids divergent terms by making a series expansion around $r = 0$ and guaranteeing divergent terms to
vanish. We start with $o = 2$ and, if needed, will use larger values of o later.

```
In[19]:= makeFiniteValueAtOrigin[EBComponentsR_] :=
        Module[{zeroEqs, zeroCs, solFinite},
        If[EBComponentsR =!= {0, 0, 0},
           (* series terms of the fields ~ r^-n *)
           zeroEqs = makecEquations /@
           Flatten[ExpandAll[TrigToExp[finiteOriginValue /@
                       EBComponentsR]]];
           zeroCs = Union[Cases[zeroEqs, _c, Infinity]];
           (* solve equations for the c[...] *)
           solFinite = Solve[zeroEqs == 0, zeroCs];
             (* resulting field components; finite at the origin *)
           EBComponentsR //. solFinite[[1]] // Simplify]

In[20]:= (* series terms that must vanish *)
        finiteOriginValue[expr_] :=
        With[{ser = Series[expr, {r, 0, 0}]},
             If[(* expr is a constant *) Head[ser] =!= SeriesData, {},
                Table[SeriesCoefficient[ser, k], {k, ser[[4]], -1}]]]
```

Finally, the function EBResult forms the resulting time-dependent vectors of the $\mathcal{E}(\mathbf{r})$ and $\mathcal{B}(\mathbf{r})$ fields in spherical
coordinates.

```
In[22]:= EBResult[fieldComponents_, EBres_] :=
          Function[r, (* substitute subscripted c's for c[...] *)
            r //. Table[#[[k]] -> Subscript[c, k],
            {k, Length[#]}]&[Union[Cases[r, _c, Infinity]]]][
        (* add time dependence *)
        {Sin[t] #[[1]], Cos[t] #[[2]]}& @ (* make E and B field *)
          ((If[MemberQ[fieldComponents, #], EBres[[
              Position[fieldComponents, #][[1, 1]]]], 0]& /@ #)& /@
                {{Er, E0, Eφ}, {Br, B0, Bφ}})]
```

Now, we will search for nontrivial solutions for the six possible component assignments that guarantee $\mathcal{E}(\mathbf{r}).\mathcal{B}(\mathbf{r}) = 0$. As a
side effect, we print the number of linear equations and the number of variables.

```
In[23]:= (* turn off message telling that not all coefficients can be
            uniquely determined *)
        Off[Solve::svars]

        Function[fieldComponents,
          (* use ansatz in Maxwell equations *)
          mwes = makeMaxwellEquations[fieldComponents, 2];
          (* form linear equations for the coefficients *)
          ceqs = extractCoefficients[mwes];
          (* solve equations and form fields *)
          EBComponentsR = solveCoefficientEquations[ceqs];
          (* make finite field values at origin *)
          EBres = makeFiniteValueAtOrigin[EBComponentsR] //
                        TrigReduce // Simplify;
          (* the resulting field components *)
          {E, B} = EBResult[fieldComponents, EBres]] /@
            (* the six field component assignments *)
            {{Er, B0, Bφ}, {E0, Br, Bφ}, {Eφ, Br, B0},
              {Er, E0, Bφ}, {Er, Eφ, B0}, {E0, Eφ, Br}}
```

```
2400 equations in 416 variables.
2530 equations in 402 variables.
1666 equations in 425 variables.
1848 equations in 414 variables.
2248 equations in 414 variables.
2682 equations in 404 variables.
```

Out[25]= $\left\{\{\{0, 0, 0\}, \{0, 0, 0\}\}, \{\{0, 0, 0\}, \{0, 0, 0\}\},\right.$

$\left\{\left\{0, 0, \dfrac{(r \cos[r] - \sin[r]) \sin[t] \sin[\vartheta] (c_1 - c_2)}{r^2}\right\},\right.$

$\left\{\dfrac{2 \cos[t] \cos[\vartheta] (r \cos[r] - \sin[r]) (c_1 - c_2)}{r^3},\right.$

$\left.\dfrac{\cos[t] (r \cos[r] + (-1 + r^2) \sin[r]) \sin[\vartheta] (c_1 - c_2)}{r^3}, 0\right\}\right\},$

$\left\{\left\{\dfrac{2 \cos[\vartheta] (r \cos[r] - \sin[r]) \sin[t] c_1}{r^3}, \dfrac{(r \cos[r] + (-1 + r^2) \sin[r]) \sin[t] \sin[\vartheta] c_1}{r^3}, 0\right\},\right.$

$\left.\left\{0, 0, \dfrac{\cos[t] (r \cos[r] - \sin[r]) \sin[\vartheta] c_1}{r^2}\right\}\right\},$

$\left.\{\{0, 0, 0\}, \{0, 0, 0\}\}, \{\{0, 0, 0\}, \{0, 0, 0\}\}\right\}$

We indeed found two nontrivial "spherical", standing wave-like solutions of the homogeneous Maxwell equations. Here is a quick check.

In[26]:= **(With[{cs = Spherical[r, ϑ, φ], 𝓔 = #[[1]], 𝓑 = #[[2]]},**
{Div[𝓔, cs], Div[𝓑, cs],
Curl[𝓔, cs] + D[𝓑, t], Curl[𝓑, cs] - D[𝓔, t]} //
TrigToExp // ExpandAll // Together]& /@ %) // Union

Out[26]= **{{0, 0, {0, 0, 0}, {0, 0, 0}}}**

For visualizations of the first of these fields, and potential applications for ball lightnings, see [370]. The second solution is the dual solution to the first.

Now, we want to solve the equations of motions of a test particle in one of these fields. The equations of motion of a test particle are $m\,\ddot{\mathbf{r}}(t) = e\,(\mathcal{E}(\mathbf{r}(t)) + \dot{\mathbf{r}}(t) \times \mathcal{B}(\mathbf{r}(t)))$.

In[27]:= **(* the time-dependent field components *)**
{𝓔φt, 𝓑rt, 𝓑ϑt} =
{Sin[ϑ] (Sin[r] - r Cos[r]) Sin[t]/r^2,
2 Cos[t] Cos[ϑ] (Sin[r] - r Cos[r])/r^3,
-Cos[t] Sin[ϑ](r Cos[r] + (r^2 - 1) Sin[r])/r^3} /.
{r -> r[t], φ -> φ[t], ϑ -> ϑ[t]};
(* the time-dependent unit vectors *)
{ert, eϑt, eφt} = {er, eϑ, eφ} /. {r -> r[t], φ -> φ[t], ϑ -> ϑ[t]};

(* equations of motion *)
eqsNewton[α_] = D[r[t] ert, t, t] == 𝓔φt eφt -
Cross[D[r[t] ert, t], 𝓑rt ert + 𝓑ϑt eϑt];

Here are some trajectories for selected initial conditions.

In[34]:= **nsol[α_, {r0_, ϑ0_, φ0_}, T_] :=**
NDSolve[Join[Thread[eqsNewton[α]],
{r[0] == r0, ϑ[0] == ϑ0, φ[0] == φ0},
{r'[0] == 0, ϑ'[0] == 0, φ'[0] == 0}],
{r, ϑ, φ}, {t, 0, T}, MaxSteps -> 10^5,
MaxStepSize -> 0.1];

In[35]:= **Show[GraphicsArray[**
Function[{α, inits, T},
(* solve odes and plot curve *)
ParametricPlot3D[Evaluate[
{r[t] Cos[φ[t]] Sin[ϑ[t]], r[t] Sin[φ[t]] Sin[ϑ[t]],
r[t] Cos[ϑ[t]], {Thickness[0.002], Hue[0]}} /.
nsol[α, inits, T]], {t, 0, T}, PlotPoints -> 4 T,
PlotRange -> All, BoxRatios -> {1, 1, 1},
DisplayFunction -> Identity, Axes -> False]] @@@

```
(* some parameter  sets *)
{{-0.509325, {6.558630, 3.133570, 2.263920},  200},
 { 1.000000, {9.736281, 1.447010, 4.663740}, 4000},
 { 1.903598, {3.668112, 1.193021, 1.937031}, 2000}}]]
```

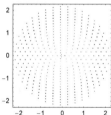

30. Asymptotic Series for *n!*, *q*-Series to *q*-Product, *q*-Binomial, gcd-Free Partitions

a) Here, the first few terms of the series are calculated.

In[1]:= (* turn of message *) **Off[Series::esss];**

    Normal[Series[n!, {n, Infinity, 5}]]

Out[2]= $e^{-1-n} \, n^{1+n}$

$$\left(\frac{1}{6} \, e \left(\frac{1}{n}\right)^{3/2} \sqrt{\frac{\pi}{2}} + \frac{1}{144} \, e \left(\frac{1}{n}\right)^{5/2} \sqrt{\frac{\pi}{2}} - \frac{139 \, e \left(\frac{1}{n}\right)^{7/2} \sqrt{\frac{\pi}{2}}}{25920} - \frac{571 \, e \left(\frac{1}{n}\right)^{9/2} \sqrt{\frac{\pi}{2}}}{1244160} + e \sqrt{\frac{1}{n}} \, \sqrt{2\pi} \right)$$

We separate the purely Laurent part ser of the series for the first 150 series terms.

In[3]:= **fac[n_] = Exp[-n] n^(n + 1) Sqrt[Pi/2] Sqrt[n];**

In[4]:= **ser[n_] = Sort[List @@ PowerExpand[Expand[**
                  1/fac[n] Normal[Series[n!, {n, Infinity, 150}]]]]],
                  (* increasing negative powers of n *)
                  Exponent[#1, n] > Exponent[#2, n]&];

Here is a graphics of the zeros of the first 50 partial sums.

In[5]:= **Show[Graphics[Table[{Hue[k/50], Point[{Re[#], Im[#]}]}& /@**
    (n /. NSolve[(Plus @@ Take[ser[n], k]) == 0, n])}, {k, 2, 40}]],
    PlotRange -> All, Frame -> True, AspectRatio -> Automatic]

The function pos gives the number of terms and the difference to the exact value of *n*! for a given *n*.

In[6]:= **pos[n_] :=**
    Module[{data = fac[N[n, 250]] FoldList[Plus, 0, ser[N[n, 500]]] - n!, p},
        (* where is the difference minimal? *)
        p = Position[Abs[data], Min[Abs[data]]][[1, 1]];
        {n, p, N @ data[[p]]}]

Here, the number of terms and the differences to the exact result for the first 150 factorials are calculated.

In[7]:= **data = Table[pos[k], {k, 150}];**

The next graphic shows how the relative precision of the approximation is increasing, the absolute difference has a minimum, and the position of the smallest element is increasing.

```
In[8]:= Show [GraphicsArray [
       Block [{$DisplayFunction = Identity},
          {(* absolute difference *)
           ListPlot[{#1, Log[10, Abs[#3]]}& @@@ data],
           (* relative difference *)
           ListPlot[{#1, Log[10, Abs[#3]/#1!]}& @@@ data],
           (* position of smallest term *)
           ListPlot[{#1, #2}& @@@ data]}]]]
```

Here is the convergence to a complex value of n shown. After approaching and encircling the exact value, including further terms of the formally divergent series results in a divergent path of the partial sums.

```
In[9]:= Show [Graphics [
       {{Hue[0], PointSize[0.03],
         Point[{Re[N[(5 + I)!]], Im[N[(5 + I)!]]}]},
        {Thickness[0.002], GrayLevel[0],
         Line[{Re[#], Im[#]}& /@ N[N[fac[5 + I]*
                 FoldList[Plus, 0, ser[5 + I]], 250]]]}}],
       Frame -> True, PlotRange ->
       (* appropriate rectangle around the exact value (5 + I)! *)
         {{-15.38094275, -15.38094276}, {108.56846148, 108.5684615}}]
```

b) We start by comparing the Ramanujan expansion with the asymptotic expansion of the factorial.

```
In[1]:= (* suppress messages *) Off[Series::esss];
       Series[n!, {n, Infinity, 4}]
```

$$Out[2]= e^{-1-n} n^{1+n} \left(e \sqrt{2\pi} \sqrt{\frac{1}{n}} + \frac{1}{6} e \sqrt{\frac{\pi}{2}} \left(\frac{1}{n}\right)^{3/2} + \frac{1}{144} e \sqrt{\frac{\pi}{2}} \left(\frac{1}{n}\right)^{5/2} - \frac{139 \left(e \sqrt{\frac{\pi}{2}}\right) \left(\frac{1}{n}\right)^{7/2}}{25920} + O\left[\frac{1}{n}\right]^{9/2} \right)$$

Both have the same prefactors $\sqrt{\pi}\,(n/e)^n$. Equating the remaining terms yields the following form for $R(n)$.

```
In[3]:= sol = Solve[(8 n^3 + 4 n^2 + n + 1/30 + R)^(1/6) == S, R]
```

$$Out[3]= \left\{\left\{R \to \frac{1}{30}\left(-1 - 30 n - 120 n^2 - 240 n^3 + 30 S^6\right)\right\}\right\}$$

Taking now 110 series terms of the classical expansion allows us to calculate 100 of the c_k.

```
In[4]:= facSer = Series[n!, {n, Infinity, 110}]/(Sqrt[Pi] (n/E)^n);
       rSer = sol[[1, 1, 2]] /. S -> facSer;

In[6]:= terms[n_] = Reverse[List @@ Normal[rSer]];
```

Here are the first few terms of the resulting series.

```
In[7]:= RamanujanFactorialSeries[x_, o_] :=
       Sqrt[Pi] (x/E)^x (8 x^3 + 4 x^2 + x + 1/30 + (Plus @@ Take[terms[x], o]))^(1/6)

In[8]:= RamanujanFactorialSeries[n, 6]
```

Out[8]= $e^{-n} n^n \left(\dfrac{1}{30} + \dfrac{233934691}{6386688000\, n^6} - \dfrac{10051}{716800\, n^5} - \dfrac{9511}{403200\, n^4} + \dfrac{3539}{201600\, n^3} + \dfrac{79}{3360\, n^2} - \dfrac{11}{240\, n} + n + 4\, n^2 + 8\, n^3 \right)^{1/6} \sqrt{\pi}$

The following two graphics show the logarithm of the relative error as a function of the number of terms for $n = 10$ and $n = 100$. The left graphic clearly shows that the series is asymptotic (as is the original series for $n!$).

```
In[9]:= $MaxExtraPrecision = 1000;
       Show[GraphicsArray[Function[v,
       ListPlot[Table[{k, Log[10,
                       Abs[(1 - RamanujanFactorialSeries[v, k]/v!) // N[#, 22]&]]},
               {k, 100}], DisplayFunction -> Identity]] /@ {10, 100}]]
```

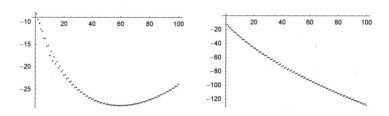

From the right graphic, we infer that we need 68 terms to obtain 100 correct digits of 100!.

```
In[11]:= With[{v = 100},
               (1 - RamanujanFactorialSeries[v, 68]/v!) // N[#, 22]&]
Out[11]= 6.336585081488976309222 × 10^{-102}
```

For a closed form for the c_k, see [983].

The interested reader could now go on and find similar expansion of the form

$$ n\,! = \sqrt{\pi} \left(\frac{n}{e} \right)^n \left(2^o\, n^o + \left(\sum_{k=0}^{o-1} c_k\, n^k \right) + R(n) \right)^{\frac{1}{2o}}. $$

c) For a given q-series Σ, there is a unique α_k, that will annihilate the smallest nonzero c_k. The simple relation $\alpha_k = c_k$ holds. Here is an example.

```
In[1]:= ser = 1 + ck q^8 + O[q]^9
Out[1]= 1 + ck q^8 + O[q]^9

In[2]:= ser (1 - q^8)^αk
Out[2]= 1 + (ck - αk) q^8 + O[q]^9
```

The function factorOutqProduct implements the calculation of $(1 - q^k)^{\alpha_k}$.

```
In[3]:= factorOutqProduct[HoldPattern[SeriesData[q_, 0,
               {1, zeros:(0 ...), c_?(# =!= 0&), ___}, 0, n_, 1]]] :=
               (1 - q^(Length[{zeros}] + 1))^c
In[4]:= factorOutqProduct[HoldPattern[SeriesData[q_, 0, {1}, 0, n_, 1]]] := 1
```

After factoring out a product $(1 - q^k)^{\alpha_k}$, we multiply the series with this product and continue the process. For the example series Σ, this is done in the following inputs.

```
In[5]:= Σ = 1 + q + q^2 + q^3 + 2 q^4 + 2 q^5 + 3 q^6 + O[q]^7
Out[5]= 1 + q + q^2 + q^3 + 2 q^4 + 2 q^5 + 3 q^6 + O[q]^7

In[6]:= factorOutqProduct[Σ]
Out[6]= 1 - q

In[7]:= Σ = Σ %
```

Out[7]= $1 + q^4 + q^6 + O[q]^7$

In[8]:= `factorOutqProduct[Σ]`

Out[8]= $1 - q^4$

In[9]:= `Σ = Σ %`

Out[9]= $1 + q^6 + O[q]^7$

In[10]:= `factorOutqProduct[Σ]`

Out[10]= $1 - q^6$

In[11]:= `Σ = Σ %`

Out[11]= $1 + O[q]^7$

We automate the repeated factor extraction and series calculation in the function `toqProduct`.

```
In[12]:= toqProduct[s_] := 1/FixedPoint[Function[x,
            {First[x] #, Last[x] #}&[factorOutqProduct[First[x]]]],
            {s, 1}, 20, SameTest -> (First[#1] === First[#2]&)][[2]]
```

A random q-series shows `toqProduct` at work.

```
In[13]:= SeedRandom[111];
         randomqSeries[q_, degree_, maxCoeff_] :=
           1 + Sum[Random[Integer, maxCoeff{-1, 1}] q^k, {k, degree}] +
                                                  O[q]^(degree + 1);

         Σ = randomqSeries[q, 8, 8]
```

Out[15]= $1 - 4 q + 6 q^2 + 3 q^3 - q^4 - 4 q^5 + 3 q^6 - q^7 + 2 q^8 + O[q]^9$

In[16]:= `toqProduct[Σ]`

Out[16]= $\dfrac{(1 - q)^4 (1 - q^7)^6 (1 - q^8)^{487}}{(1 - q^3)^7 (1 - q^4)^{26} (1 - q^5)^{58} (1 - q^6)^{79}}$

In[17]:= `Σ - %`

Out[17]= $O[q]^9$

Now it is straightforward to unite the two functions `factorOutqProduct` and `toqProduct` in one function `qSeries` `ToqProduct`.

```
In[18]:= qSeriesToqProduct[qs:qSeries_SeriesData] :=
           With[{q = First[qSeries], H = HoldPattern, S = SeriesData},
             FixedPoint[Function[x, {First[x] Series[#, {qs[[1]], 0, qs[[5]]}],
                         Last[x] #}&[First[x] /.
               {H[S[q_, 0, {1, ξ:(0 ...), c_?(# != 0&),        }, 0, n_, 1]] :>
                                              (1 - q^(Length[{ξ}] + 1))^c,
                H[S[q_, 0, {1}, 0, n_, 1]] :> 1}]], {qSeries, 1},
                     SameTest -> (First[#1] === First[#2]&)][[2]]^-1 /;
             MatchQ[qSeries, H[S[_, 0, {1, _Integer ...}, 0, _, 1]]]
```

Here are six examples of q-series and the corresponding q-products together with a quick check of their correctness.

```
In[19]:= Table[{#, qSeriesToqProduct[#]}&[randomqSeries[q, 12, 10]], {6}]
```

Out[19]= $\{\{1 + 8 q - 2 q^2 - 8 q^3 + q^4 - 2 q^5 + 7 q^6 + 7 q^7 + 6 q^8 - q^9 - 3 q^{10} - q^{11} + 8 q^{12} + O[q]^{13},$

$\dfrac{(1 - q^2)^{38} (1 - q^4)^{1072} (1 - q^6)^{47937} (1 - q^8)^{2367636} (1 - q^{10})^{124995187} (1 - q^{12})^{6871862430}}{(1 - q)^8 (1 - q^3)^{176} (1 - q^5)^{7070} (1 - q^7)^{333211} (1 - q^9)^{17098143} (1 - q^{11})^{922946336}}\},$

$\{1 - 5 q - 5 q^2 + q^3 - 8 q^4 + 7 q^5 - 8 q^6 + 8 q^7 + 9 q^8 - 7 q^9 + 6 q^{10} - 6 q^{11} - 8 q^{12} + O[q]^{13},$

$(1 - q)^5 (1 - q^2)^{15} (1 - q^3)^{64} (1 - q^4)^{288} (1 - q^5)^{1377} (1 - q^6)^{6709} (1 - q^7)^{33854}$
$(1 - q^8)^{173441} (1 - q^9)^{903993} (1 - q^{10})^{4766354} (1 - q^{11})^{25390989} (1 - q^{12})^{136365761}\},$

$\{1 - 10 q - 2 q^2 - 8 q^3 + 7 q^4 + 2 q^5 + 3 q^6 - 5 q^7 + 7 q^8 + 6 q^9 - 5 q^{10} - 5 q^{11} + q^{12} + O[q]^{13},$

$(1 - q)^{10} (1 - q^2)^{47} (1 - q^3)^{358} (1 - q^4)^{2749} (1 - q^5)^{22782} (1 - q^6)^{194688} (1 - q^7)^{1714557}$
$(1 - q^8)^{15397222} (1 - q^9)^{140491494} (1 - q^{10})^{1297808271} (1 - q^{11})^{12109970996} (1 - q^{12})^{113939839261}\},$

$\{1 - 6 q + 8 q^2 + 7 q^3 + 4 q^4 - q^5 + 3 q^6 + 2 q^7 + 7 q^8 - 7 q^9 + 4 q^{10} + 6 q^{11} + q^{12} + O[q]^{13},$

$$\Big((1-q)^6 (1-q^2)^7 (1-q^3)^{15} (1-q^4)^{17}\Big) \Big/ \Big((1-q^5)^9 (1-q^6)^{242} (1-q^7)^{1264}$$
$$(1-q^8)^{5275} (1-q^9)^{19134} (1-q^{10})^{62911} (1-q^{11})^{188107} (1-q^{12})^{501007}\Big)\Big\},$$
$$\{1 - 7q - 5q^2 - 5q^3 - 7q^4 - 2q^5 - 10q^7 + 7q^8 - q^9 + q^{10} - 2q^{11} + 9q^{12} + O[q]^{13},$$
$$(1-q)^7 (1-q^2)^{26} (1-q^3)^{152} (1-q^4)^{885} (1-q^5)^{5571} (1-q^6)^{35876} (1-q^7)^{238732}$$
$$(1-q^8)^{1617327} (1-q^9)^{11136926} (1-q^{10})^{77623873} (1-q^{11})^{546536713} (1-q^{12})^{3879983029}\Big\},$$
$$\{1 + 10q - 5q^2 + 2q^3 + 7q^4 - 6q^5 + 3q^6 - 6q^7 - 4q^8 - 3q^9 + 2q^{10} + 3q^{12} + O[q]^{13},$$
$$\frac{(1-q^2)^{60} (1-q^4)^{2998} (1-q^6)^{222035} (1-q^8)^{18302850} (1-q^{10})^{1610865017} (1-q^{12})^{147667629297}}{(1-q)^{10} (1-q^3)^{382} (1-q^5)^{25382} (1-q^7)^{1994530} (1-q^9)^{170650405} (1-q^{11})^{15359200847}}\Big\}\Big\}$$

In[20]:= **Subtract @@@ %**

Out[20]= $\{O[q]^{13}, O[q]^{13}, O[q]^{13}, O[q]^{13}, O[q]^{13}, O[q]^{13}\}$

Here are three well-known series.

In[21]:= **qSeriesToqProduct[Sum[q^k, {k, 0, 20}] + O[q]^21]**

Out[21]= $\dfrac{1}{1-q}$

In[22]:= **qSeriesToqProduct[Sum[(k + 1) q^k, {k, 0, 20}] + O[q]^21]**

Out[22]= $\dfrac{1}{(1-q)^2}$

In[23]:= **qSeriesToqProduct[Sum[(-1)^k q^((6k + 1)^2), {k, -360, 360}]/q + O[q]^361]**

Out[23]= $(1-q^{24}) (1-q^{48}) (1-q^{72}) (1-q^{96}) (1-q^{120}) (1-q^{144}) (1-q^{168})$
$(1-q^{192}) (1-q^{216}) (1-q^{240}) (1-q^{264}) (1-q^{288}) (1-q^{312}) (1-q^{336}) (1-q^{360})$

Here are two very nice products using the function PartitionsP and PartitionsQ from Chapter 2 of the Numerics volume [1737].

In[24]:= **Sum[PartitionsP[k] q^k, {k, 0, 20}] + O[q]^21**

Out[24]= $1 + q + 2q^2 + 3q^3 + 5q^4 + 7q^5 + 11q^6 + 15q^7 + 22q^8 + 30q^9 + 42q^{10} + 56q^{11} + 77q^{12} +$
$101q^{13} + 135q^{14} + 176q^{15} + 231q^{16} + 297q^{17} + 385q^{18} + 490q^{19} + 627q^{20} + O[q]^{21}$

In[25]:= **qSeriesToqProduct[%]**

Out[25]= $1 / ((1-q) (1-q^2) (1-q^3) (1-q^4) (1-q^5) (1-q^6) (1-q^7) (1-q^8) (1-q^9) (1-q^{10}) (1-q^{11})$
$(1-q^{12}) (1-q^{13}) (1-q^{14}) (1-q^{15}) (1-q^{16}) (1-q^{17}) (1-q^{18}) (1-q^{19}) (1-q^{20}))$

In[26]:= **Sum[PartitionsQ[k] q^k, {k, 0, 20}] + O[q]^21**

Out[26]= $1 + q + q^2 + 2q^3 + 2q^4 + 3q^5 + 4q^6 + 5q^7 + 6q^8 + 8q^9 + 10q^{10} + 12q^{11} +$
$15q^{12} + 18q^{13} + 22q^{14} + 27q^{15} + 32q^{16} + 38q^{17} + 46q^{18} + 54q^{19} + 64q^{20} + O[q]^{21}$

In[27]:= **qSeriesToqProduct[%]**

Out[27]= $\dfrac{1}{(1-q) (1-q^3) (1-q^5) (1-q^7) (1-q^9) (1-q^{11}) (1-q^{13}) (1-q^{15}) (1-q^{17}) (1-q^{19})}$

The next inputs recover the product $\prod_{k=2}^{n} (1 - q^{F_k})$ [1495], [80].

In[28]:= **Product[(1 - q^Fibonacci[k]), {k, 2, 30}] + O[q]^31**

Out[28]= $1 - q - q^2 + q^4 + q^7 - q^8 + q^{11} - q^{12} - q^{13} + q^{14} + q^{18} - q^{19} - q^{20} + q^{22} + q^{23} - q^{24} + q^{29} - q^{30} + O[q]^{31}$

In[29]:= **qSeriesToqProduct[%]**

Out[29]= $(1-q) (1-q^2) (1-q^3) (1-q^5) (1-q^8) (1-q^{13}) (1-q^{21})$

For many more q-series that have nice q-product forms, see [407], [1914].

d) While a direct implementation to calculate the number of factors is possible, it is prohibitively slow.

In[1]:= **qBinomialPolynomial[n_, k_, q_] :=**
   Product[1 - q^j, {j, n - k + 1, n}]/Product[1 - q^j, {j, 1, k}]

In[2]:= **qBinomialPolynomial[150, 60, q] // Cancel // Factor // Length // Timing**

Out[2]= $\{5.09\,\text{Second}, 94\}$

We observe that we must repeatedly factor terms of the form $1 - q^j$. We will do this once, abbreviate the factors, and remember the factors.

```
In[3]:= o = 250;
        factoredqPolys = Table[-Factor[1 - q^k], {k, 2, o}];
In[5]:= (* all single factors; ordered by degree *)
        allFactors = Sort[Union[Flatten[List @@@ factoredqPolys]],
                     (Exponent[#1, q] <= Exponent[#2, q]&)];
In[7]:= (* replace factors by variables *)
        rules = MapIndexed[(#1 -> C[#2[[1]] + 1])&, allFactors] // Reverse;
In[9]:= (* simplified form of factored form of 1 - q^j *)
        F[1] = C[2];
        MapIndexed[(F[#2[[1]]] + 1] = (#1 /. rules))&, factoredqPolys, {1}];
```

Using the symbolic factors $\mathcal{F}[j]$, the number of factors of the q-binomial coefficients can now be calculated hundreds of times faster.

```
In[12]:= qBL[n_, k_] := If[k > n, 0, Length[
         Product[F[j], {j, n - k + 1, n}]/Product[F[j], {j, 1, k}]]]
In[13]:= qBL[150, 60] // Timing
Out[13]= {0. Second, 94}
```

The following graphic shows the number of factors as a function of k and n.

```
In[14]:= tab = Table[Table[qBL[n, k], {k, o}], {n, o}];
```

```
ListPlot3D[tab, Mesh -> False]
```

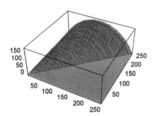

For interesting identities of the q-binomial coefficients, see [676].

e) We start by calculating the series of $\prod_{k=1}^{\infty} (1 - q^{\alpha k})$ to order o.

```
In[1]:= ΠSeries[k_, o_] := ΠSeries[k, o] =
        Series[Product[1 - q^(j k), {j, Ceiling[o/k] + 1}], {q, 0, o}]
```

Now let us implement a function `multiplicativeIdentities` that generates all identities that must be satisfied by the coefficients c_k for $k \le n$.

```
In[2]:= multiplicativeIdentities[c_, n_] := multiplicativeIdentities[c, n] =
        And @@ ((c[#1] c[#2] == c[#1 #2])& @@@
        Select[Union[Sort /@ (* pairs of divisors *)
        Flatten[Function[k, {#, k/#}& /@ Divisors[k]] /@ Range[2, n], 1]],
        (FreeQ[#, 1] && (GCD @@ #) === 1)&])
```

Here are all 15 identities that the coefficients up to c_{30} must fulfill.

```
In[3]:= multiplicativeIdentities[c, 30]
Out[3]= c[2] c[3] == c[6] && c[2] c[5] == c[10] && c[2] c[7] == c[14] && c[2] c[9] == c[18] &&
        c[2] c[11] == c[22] && c[2] c[13] == c[26] && c[2] c[15] == c[30] &&
        c[3] c[4] == c[12] && c[3] c[5] == c[15] && c[3] c[7] == c[21] && c[3] c[8] == c[24] &&
        c[3] c[10] == c[30] && c[4] c[5] == c[20] && c[4] c[7] == c[28] && c[5] c[6] == c[30]
```

The first 1000 of the c_{30} must fulfill 1494 relations.

```
In[4]:= multiplicativeIdentities[c, 1000] // Length
Out[4]= 1494
```

The function `multiplicativeSeriesQ` checks if a given *ser* is a multiplicative series.

```
In[5]:= multiplicativeSeriesQ[ser_SeriesData] :=
         With[{cl = Rest[CoefficientList[ser, ser[[1]]]]},
             multiplicativeIdentities[c, Length[cl]] /. c[j_] :> cl[[j]]]
```

$q \prod_{k=1}^{\infty} (1 - q^k)^{24}$ is a well-known multiplicative series.

```
In[6]:= multiplicativeSeriesQ[q ΠSeries[24, 20]]
```

```
Out[6]= True
```

Now, we have all functions that create and analyze a series. Next, we must create the factors of the products. The function `divisorGroups` generates a list of all pairs of lists of divisors of the integer *n*. The two pairs of numbers will be used as the α_j and β_j in the numerator and denominator powers of *q*.

```
In[7]:= divisorGroups[n_] :=
         Module[{d = Divisors[n], λ}, λ = Length[d];
             Select[Union[{Sort[#], Sort[Complement[d, #]]}& /@ Level[
                 (* table of all possible divisors *)
                 Table[Table[If[α[k] === 0, Sequence @@ {}, d[[k]]], {k, λ}],
                 Evaluate[Sequence @@ Table[{α[k], 0, 1}, {k, λ}]]], {-2}]],
                 FreeQ[#, {}]&]]
```

Here are the 14 possibilities to split the divisors of 15.

```
In[8]:= divisorGroups[15]
```

```
Out[8]= {{{1}, {3, 5, 15}}, {{3}, {1, 5, 15}}, {{5}, {1, 3, 15}},
         {{15}, {1, 3, 5}}, {{1, 3}, {5, 15}}, {{1, 5}, {3, 15}},
         {{1, 15}, {3, 5}}, {{3, 5}, {1, 15}}, {{3, 15}, {1, 5}}, {{5, 15}, {1, 3}},
         {{1, 3, 5}, {15}}, {{1, 3, 15}, {5}}, {{1, 5, 15}, {3}}, {{3, 5, 15}, {1}}}
```

The number of such pairs is a highly fluctuating, but in average increasing function of *n*.

```
In[9]:= ListPlot[Table[{k, Log[10, 1 + Length[divisorGroups[k]]]}, {k, 100}],
             PlotRange -> All, Frame -> True]
```

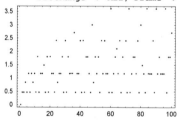

Given the α_js (or β_js) and an integer *s*, the function `ΠProducts` generates all products $\prod_{j=1}^{m} \Pi_{\alpha_j}^{\mu_j}$ where $\Pi_{\alpha_j} = \prod_{k=1}^{\infty} (1 - q^{\alpha_j k})$ and the μ_j fulfill the following sum condition: $\sum_{j=1}^{m} \alpha_j \mu_j = s$.

```
In[10]:= ΠProducts[divs_, s_] :=
          Module[{d = Reverse[Sort[divs]], t1},
              (* table of all divisors; exclude smallest *)
              t1 = Table[Evaluate[Table[{d[[k]], α[k]}, {k, Length[d] - 1}]],
              Evaluate[Sequence @@ Table[{α[k], 0, Floor[(s - Sum[α[j] d[[j]],
                  {j, 1, k - 1}])/d[[k]]]}, {k, Length[d] - 1}]]];
              (* append smallest divisor *)
              t2 = Append[#, {d[[-1]], (s - (Plus @@ (Times @@@ #)))/d[[-1]]}]& /@
                          Flatten[t1, Max[{0, Length[d] - 2}]]];
              Times @@@ Apply[Subscript[Π, #1]^#2&, (* select integer pairs *)
              Select[DeleteCases[t2, {_, 0}, Infinity],
                          FreeQ[#, _Rational, Infinity]&], {2}]]

          ΠProducts[{}, s_] := {}

          ΠProducts[{d_}, s_] := If[IntegerQ[s/d], {Subscript[Π, d]^(s/d)}, {}]
```

Here is an example.

```
In[14]:= ΠProducts[{2, 3, 6}, 18]
```

$$\text{Out[14]= } \{\Pi_2^9,\ \Pi_2^6\,\Pi_3^2,\ \Pi_2^3\,\Pi_3^4,\ \Pi_3^6,\ \Pi_2^6\,\Pi_6,\ \Pi_2^3\,\Pi_3^2\,\Pi_6,\ \Pi_3^4\,\Pi_6,\ \Pi_2^3\,\Pi_6^2,\ \Pi_3^2\,\Pi_6^2,\ \Pi_6^3\}$$

For a given list of the α_k and β_k and a given s, forming now all possible combinations of potential numerators and denominators gives all possible quotients. The function ΠQuotients forms these quotients.

```
In[15]:= ΠQuotients[{n_List, d_List}, s_, maxTerms_:Infinity] :=
           With[{nums = ΠProducts[n, s], dens = ΠProducts[d, s]},
               If[Length[nums] Length[dens] > maxTerms, {},
                   Flatten[Outer[Divide, ΠProducts[n, s], ΠProducts[d, s]]]]]
```

Finally, the function allΠQuotients generates all quotients for a given n and a given s.

```
In[16]:= allΠQuotients[n_, s_, maxTerms_:Infinity] :=
           Union[Flatten[ΠQuotients[#, s, maxTerms]& /@ divisorGroups[n]]]
```

Here is an explicit example. For $n = s = 8$, we get 34 possible fractions.

```
In[17]:= allΠQuotients[8, 8]
```

$$\text{Out[17]= } \left\{ \frac{\Pi_1^8}{\Pi_2^4},\ \frac{\Pi_1^4}{\Pi_2^8},\ \frac{\Pi_1^8}{\Pi_4^2},\ \frac{\Pi_1^6\,\Pi_2}{\Pi_4^2},\ \frac{\Pi_1^4\,\Pi_2^2}{\Pi_4^2},\ \frac{\Pi_1^2\,\Pi_2^3}{\Pi_4^2},\ \frac{\Pi_2^4}{\Pi_4^2},\ \frac{\Pi_1^8}{\Pi_2^2\,\Pi_4},\ \frac{\Pi_1^4}{\Pi_1^4\,\Pi_4},\ \frac{\Pi_1^4\,\Pi_4}{\Pi_2^4},\ \frac{\Pi_2^2\,\Pi_4}{\Pi_1^8},\ \frac{\Pi_2^2}{\Pi_1^4}, \right.$$

$$\frac{\Pi_2^4}{\Pi_4^2},\ \frac{\Pi_4^2}{\Pi_1^4\,\Pi_2^3},\ \frac{\Pi_2^2}{\Pi_1^4\,\Pi_4^2},\ \frac{\Pi_4^2}{\Pi_1^6\,\Pi_2},\ \frac{\Pi_1^8}{\Pi_8},\ \frac{\Pi_1^6\,\Pi_2}{\Pi_8},\ \frac{\Pi_1^4\,\Pi_2^2}{\Pi_8},\ \frac{\Pi_1^2\,\Pi_2^3}{\Pi_8},\ \frac{\Pi_2^4}{\Pi_8},\ \frac{\Pi_1^4\,\Pi_4}{\Pi_8},\ \frac{\Pi_1^2\,\Pi_2\,\Pi_4}{\Pi_8},$$

$$\left. \frac{\Pi_2^2\,\Pi_4}{\Pi_8},\ \frac{\Pi_4^2}{\Pi_8},\ \frac{\Pi_8}{\Pi_1^8},\ \frac{\Pi_8}{\Pi_2^4},\ \frac{\Pi_8}{\Pi_1^2\,\Pi_2^3},\ \frac{\Pi_8}{\Pi_1^4\,\Pi_2^2},\ \frac{\Pi_8}{\Pi_1^6\,\Pi_2},\ \frac{\Pi_8}{\Pi_4^2},\ \frac{\Pi_8}{\Pi_1^4\,\Pi_4},\ \frac{\Pi_8}{\Pi_2^2\,\Pi_4},\ \frac{\Pi_8}{\Pi_1^2\,\Pi_2\,\Pi_4} \right\}$$

Now, we have everything together to check the 35312 possible series for mutiplicativeness. We do a quick check for 20 series terms and if this test succeeds, we test with 200 terms.

```
In[18]:= o1 = 20; o2 = 100; bag = {};

         counter = 0;
         Do[(* all quotients for given n and s *)
             quots = allΠQuotients[n, s, Infinity];
             (* cumulative number of quotients checked *)
             counter = counter + Length[quots];
             (* newly found multiplicative series *)
             sel = Select[quots, multiplicativeSeriesQ[# /.
                           Subscript[Π, x_] :> ΠSeries[x, o1]]&];
             If[sel =!= {},
                 (* keep newly found multiplicative series *)
                 sel1 = Select[sel, multiplicativeSeriesQ[# /.
                               Subscript[Π, x_] :> ΠSeries[x, o2]]&];
                 If[sel1 =!= {}, bag = Union[Join[bag, sel1]]]],
             {n, 2, 24}, {s, n, 24}]
```

Seven multiplicative series were found.

```
In[21]:= {counter, Length[bag]}
```

```
Out[21]= {35312, 7}
```

A check with 800 series terms confirms that these seven series are all multiplicative.

```
In[22]:= multiplicativeSeriesQ[# /. Subscript[Π, x_] :> ΠSeries[x, 800]]& /@ bag
```

```
Out[22]= {True, True, True, True, True, True, True}
```

Here are these seven quotients and the first terms of their series.

```
In[23]:= bag
```

$$\text{Out[23]= } \left\{ \frac{\Pi_2^5\,\Pi_3^3}{\Pi_1\,\Pi_6^2},\ \frac{\Pi_2^2\,\Pi_3^3}{\Pi_1\,\Pi_6^2},\ \frac{\Pi_2^2\,\Pi_3}{\Pi_1\,\Pi_6},\ \frac{\Pi_2^2\,\Pi_5}{\Pi_1\,\Pi_{10}},\ \frac{\Pi_3^3\,\Pi_4^3}{\Pi_1\,\Pi_2\,\Pi_6\,\Pi_{12}},\ \frac{\Pi_3^2\,\Pi_5^2}{\Pi_1\,\Pi_{15}},\ \frac{\Pi_2\,\Pi_4\,\Pi_5\,\Pi_{10}}{\Pi_1\,\Pi_{20}} \right\}$$

```
In[24]:= bag /. Subscript[Π, x_] :> ΠSeries[x, 12] // TableForm
```

$$1 + q - 3\,q^2 - 5\,q^3 - 3\,q^4 + 6\,q^5 + 15\,q^6 + 8\,q^7 - 3\,q^8 - 23\,q^9 - 18\,q^{10} + 12\,q^{11} + 15\,q^{12} + O[q]^{13}$$

$$1 + q - 2\,q^3 - 3\,q^4 + 2\,q^7 - 2\,q^9 + 6\,q^{12} + O[q]^{13}$$

Out[24]//TableForm=
$$1 + q - q^4 - 2\,q^9 + O[q]^{13}$$
$$1 + q - q^2 - q^4 - 3\,q^5 - q^8 + q^9 + 3\,q^{10} + O[q]^{13}$$
$$1 + q + 3\,q^2 + q^3 + 3\,q^4 + 3\,q^6 + 2\,q^7 + 3\,q^8 + q^9 + 3\,q^{12} + O[q]^{13}$$
$$1 + q + 2\,q^2 + q^3 + 3\,q^4 + q^5 + 2\,q^6 + 4\,q^8 + q^9 + 2\,q^{10} + 3\,q^{12} + O[q]^{13}$$
$$1 + q + q^2 + 2\,q^3 + q^4 + q^5 + 2\,q^6 + 2\,q^7 + q^8 + 3\,q^9 + q^{10} + 2\,q^{12} + O[q]^{13}$$

f) Here is an implementation of the formula given. We do not have to calculate an infinite product of cyclotomic polynomials, but we can stop when the term z^n of the resulting formal series in z no longer changes. To avoid building polynomials of large degree, we truncate the cyclotomic polynomials using `SeriesData`.

```
In[1]:= GCDFreePartition[primes_?(VectorQ[#, PrimeQ]&),
                         n_Integer?Positive] :=
        Module[{m, r, s, fp, z, j = 1},
               {m, r} = {Times @@ primes, Length[primes]};
               (* truncated polynomial *)
               s[j_] = Cyclotomic[m, z^j]^((-1)^(r - 1)) + O[z]^(n + 1);
               (* multiply as long as needed *)
               fp = FixedPoint[(j++; s[j] #)&, s[j]];
               (* extract coefficient *)
               SeriesCoefficient[fp, n]]
```

For $P_{2,3,5,7,11}(111)$, the function `GCDFreePartition` yields the answer 1036.

```
In[2]:= n = 111; primes = {2, 3, 5, 7, 11};
        GCDFreePartition[primes, n]
Out[3]= 1036
```

Now, we will calculate the partitions explicitly using the function `GCDFreePartitions`. We first select the possible m_j and then recursively subtract them form the initial integer n. The result is a list of lists, each sublist being of the form $\{\{m_1, \rho_1\}, \dots, \{m_j, \rho_j\}\}$ where all the m_i are now different and ρ_l represents the number of times m_l appears in the partition.

```
In[4]:= GCDFreePartitions[primes_?(VectorQ[#, PrimeQ]&),
                          n_Integer?Positive] :=
        Module[{m = Times @@ primes, pps, l, aux},
               (* possible m's *)
               pps = Select[Reverse @ Range[n], GCD[#, m] === 1&];
               l = Length[pps];
               aux = Select[DeleteCases[Level[
                  (* recursively subtract m's from n *)
                  Table[(* occurrences of the m's *)
                        Evaluate[Table[j[i], {i, l}]],
                        (* iterators *)
                        Evaluate[Sequence @@
                        Table[{j[i], 0, (n - Sum[pps[[k]] j[k],
                              {k, 1, i - 1}])/pps[[i]]}, {i, l}]]],
                  {-2}], {}], (pps.# === n)&];
               (* delete nonappearing m's *)
               DeleteCases[#, {_, 0}, {2}]& @ (Reverse /@
               MapIndexed[{pps[[#2[[2]]]], #1}&, aux, {2}])]
```

For $P_{2,3,5,7,11}(111)$, the function `GCDFreePartitions` yields 1036 different partitions.

```
In[5]:= gcffp111 = GCDFreePartitions[primes, n];
```

```
In[6]:= {Length[gcffp111], Length[Union[gcffp111]],
         Union[Apply[Plus, Apply[Times, gcffp111, {2}], {1}]]}
Out[6]= {1036, 1036, {111}}
```

Here are the first and the last few calculated partitions.

```
In[7]:= Short[gcffp111, 6]
```

Out[7]//Short= {{{1, 111}}, {{1, 98}, {13, 1}}, {{1, 85}, {13, 2}}, {{1, 72}, {13, 3}}, {{1, 59}, {13, 4}},
       {{1, 46}, {13, 5}}, {{1, 33}, {13, 6}}, {{1, 20}, {13, 7}}, {{1, 7}, {13, 8}},
       {{1, 94}, {17, 1}}, {{1, 81}, {13, 1}, {17, 1}}, {{1, 68}, {13, 2}, {17, 1}},
       ≪1012≫, {{1, 9}, {19, 1}, {83, 1}}, {{1, 5}, {23, 1}, {83, 1}},

{{1, 22}, {89, 1}}, {{1, 9}, {13, 1}, {89, 1}}, {{1, 5}, {17, 1}, {89, 1}},
{{1, 3}, {19, 1}, {89, 1}}, {{1, 14}, {97, 1}}, {{1, 1}, {13, 1}, {97, 1}},
{{1, 10}, {101, 1}}, {{1, 8}, {103, 1}}, {{1, 4}, {107, 1}}, {{1, 2}, {109, 1}}}

31. *One ODE for the Kepler Problem, Euler Equations, $v(x) = \alpha \, \exp(\beta\,(z - z_0)^{-1/2})$, Lattice Green's Function ODE*

a) This is the original differential equation (written in a polynomial form).

In[1]:= `eq1 = x''[t]^2 (x[t]^2 + y[t]^2) - x[t]^2`

Out[1]= $-x[t]^2 + (x[t]^2 + y[t]^2)\, x''[t]^2$

We use the second differential equation of the original problem (the one for the y-component) to eliminate $y''(t)$ in the calculations below.

In[2]:= `yEq = y''[t]^2 (x[t]^2 + y[t]^2) - y[t]^2;`

Differentiating the first equation two times yields a differential equation which contains $y''(t)$ and $x'''(t)$ and $x^{(4)}(t)$.

In[3]:= `eq2 = D[eq1, t]`

Out[3]= $-2\,x[t]\,x'[t] + (2\,x[t]\,x'[t] + 2\,y[t]\,y'[t])\,x''[t]^2 + 2\,(x[t]^2 + y[t]^2)\,x''[t]\,x^{(3)}[t]$

In[4]:= `eq3 = D[eq2, t]`

Out[4]= $-2\,x'[t]^2 - 2\,x[t]\,x''[t] + x''[t]^2\,(2\,x'[t]^2 + 2\,y'[t]^2 + 2\,x[t]\,x''[t] + 2\,y[t]\,y''[t]) +$
$\quad 4\,(2\,x[t]\,x'[t] + 2\,y[t]\,y'[t])\,x''[t]\,x^{(3)}[t] +$
$\quad 2\,(x[t]^2 + y[t]^2)\,x^{(3)}[t]^2 + 2\,(x[t]^2 + y[t]^2)\,x''[t]\,x^{(4)}[t]$

Now, we have four equations (eq1, eq2, eq3, and yEq), so that we can eliminate $y(t)$, $y'(t)$, and $y''(t)$ to get one nonlinear differential equation of order 4 for $x(t)$. This differential equation factors into three factors.

In[5]:= `GroebnerBasis[{eq1, eq2, eq3, yEq},`
`                {x[t], x'[t], x''[t], x'''[t], x''''[t]}, {y[t], y'[t], y''[t]},`
`                MonomialOrder -> EliminationOrder] // Factor`

Out[5]= $\{x[t]^3\,(-2\,x'[t]\,x''[t]\,x^{(3)}[t] + 2\,x'[t]\,x''[t]^3\,x^{(3)}[t] +$
$\quad 2\,x[t]\,x^{(3)}[t]^2 - 3\,x[t]\,x''[t]^2\,x^{(3)}[t]^2 - x[t]\,x''[t]\,x^{(4)}[t] + x[t]\,x''[t]^3\,x^{(4)}[t])$
$\quad (2\,x''[t]^3 - 4\,x''[t]^5 + 2\,x''[t]^7 - 2\,x'[t]\,x''[t]\,x^{(3)}[t] + 2\,x'[t]\,x''[t]^3\,x^{(3)}[t] +$
$\quad 2\,x[t]\,x^{(3)}[t]^2 - 3\,x[t]\,x''[t]^2\,x^{(3)}[t]^2 - x[t]\,x''[t]\,x^{(4)}[t] + x[t]\,x''[t]^3\,x^{(4)}[t])\}$

For the elimination of the time variable t, see [989], [1004], [391].

b) This is the original set of differential equations.

In[1]:= `eqs = {x'[t] - a y[t] z[t], y'[t] - b x[t] z[t], z'[t] - c x[t] y[t]}`

Out[1]= $\{-a\,y[t]\,z[t] + x'[t],\ -b\,x[t]\,z[t] + y'[t],\ -c\,x[t]\,y[t] + z'[t]\}$

We differentiate these three equations two times with respect to t and introduce some new variables.

In[2]:= `gl1 = Flatten[{eqs, D[eqs, t], D[eqs, t, t]}] /.`
`        {x[t] -> x,   x'[t] -> x1,   x''[t] -> x2,   x'''[t] -> x3,`
`         y[t] -> y,   y'[t] -> y1,   y''[t] -> y2,   y'''[t] -> y3,`
`         z[t] -> z,   z'[t] -> z1,   z''[t] -> z2,   z'''[t] -> z3}`

Out[2]= $\{x1 - a\,y\,z,\ y1 - b\,x\,z,\ -c\,x\,y + z1,\ x2 - a\,y1\,z - a\,y\,z1,\ y2 - b\,x1\,z - b\,x\,z1,\ -c\,x1\,y - c\,x\,y1 + z2,$
$\quad x3 - a\,y2\,z - 2\,a\,y1\,z1 - a\,y\,z2,\ y3 - b\,x2\,z - 2\,b\,x1\,z1 - b\,x\,z2,\ -c\,x2\,y - 2\,c\,x1\,y1 - c\,x\,y2 + z3\}$

Now, we have nine equations and we want to eliminate eight variables y, y1, y2, y3, z, z1, z2, and z3. Because y1 and z1 appear linear, it is straightforward to eliminate them.

In[3]:= `gl2 = Delete[gl1, {{2}, {3}}] /. {y1 -> b x z, z1 -> c x y}`

Out[3]= $\{x1 - a\,y\,z,\ x2 - a\,c\,x\,y^2 - a\,b\,x\,z^2,\ -b\,c\,x^2\,y + y2 - b\,x1\,z,$
$\quad -c\,x1\,y - b\,c\,x^2\,z + z2,\ x3 - 2\,a\,b\,c\,x^2\,y\,z - a\,y2\,z - a\,y\,z2,$
$\quad -2\,b\,c\,x\,x1\,y + y3 - b\,x2\,z - b\,x\,z2,\ -c\,x2\,y - c\,x\,y2 - 2\,b\,c\,x\,x1\,z + z3\}$

The variable z3 also appears linearly; so, we can eliminate it.

In[4]:= `Exponent[#, {y, y2, y3, z, z2, z3}]& /@ gl2`

Out[4]= {{1, 0, 0, 1, 0, 0}, {2, 0, 0, 2, 0, 0}, {1, 1, 0, 1, 0, 0},
{1, 0, 0, 1, 1, 0}, {1, 1, 0, 1, 1, 0}, {1, 0, 1, 1, 1, 0}, {1, 1, 0, 1, 0, 1}}

In[5]:= sol = Solve[gl2[[-1]] == 0, z3][[1]]

Out[5]= {z3 → c x2 y + c x y2 + 2 b c x x1 z}

In[6]:= gl3 = Drop[gl2, -1] /. sol

Out[6]= {x1 - a y z, x2 - a c x y² - a b x z², -b c x² y + y2 - b x1 z, -c x1 y - b c x² z + z2,
x3 - 2 a b c x² y z - a y2 z - a y z2, -2 b c x x1 y + y3 - b x2 z - b x z2}

The same we do for y3.

In[7]:= Exponent[#, {y, y2, y3, z, z2}]& /@ gl3

Out[7]= {{1, 0, 0, 1, 0}, {2, 0, 0, 2, 0}, {1, 1, 0, 1, 0},
{1, 0, 0, 1, 1}, {1, 1, 0, 1, 1}, {1, 0, 1, 1, 1}}

In[8]:= sol = Solve[gl3[[-1]] == 0, y3][[1]]

Out[8]= {y3 → 2 b c x x1 y + b x2 z + b x z2}

In[9]:= gl4 = Drop[gl3, -1] /. sol

Out[9]= {x1 - a y z, x2 - a c x y² - a b x z², -b c x² y + y2 - b x1 z,
-c x1 y - b c x² z + z2, x3 - 2 a b c x² y z - a y2 z - a y z2}

The same we do for z2.

In[10]:= Exponent[#, {y, y2, z, z2}]& /@ gl4

Out[10]= {{1, 0, 1, 0}, {2, 0, 2, 0}, {1, 1, 1, 0}, {1, 0, 1, 1}, {1, 1, 1, 1}}

In[11]:= sol = Solve[gl4[[-1]] == 0, z2][[1]]

Out[11]= $\left\{ z2 \to \dfrac{x3 - 2 a b c x² y z - a y2 z}{a y} \right\}$

In[12]:= gl5 = Numerator[Together[#]]& /@ (Drop[gl4, -1] /. sol)

Out[12]= {x1 - a y z, x2 - a c x y² - a b x z², -b c x² y + y2 - b x1 z, x3 - a c x1 y² - 3 a b c x² y z - a y2 z}

The same we do for z.

In[13]:= Exponent[#, {y, y2, z}]& /@ gl5

Out[13]= {{1, 0, 1}, {2, 0, 2}, {1, 1, 1}, {2, 1, 1}}

In[14]:= sol = Solve[gl5[[-1]] == 0, z][[1]]

Out[14]= $\left\{ z \to \dfrac{x3 - a c x1 y²}{a (3 b c x² y + y2)} \right\}$

In[15]:= gl6 = Numerator[Together[#]]& /@ (Drop[gl5, -1] /. sol)

Out[15]= {3 b c x² x1 y - x3 y + a c x1 y³ + x1 y2, -b x x3² + 9 a b² c² x⁴ x2 y² + 2 a b c x x1 x3 y² -
9 a² b² c³ x⁵ y⁴ - a² b c² x x1² y⁴ + 6 a b c x² x2 y y2 - 6 a² b c² x³ y³ y2 + a x2 y2² - a² c x y² y2²,
-b x1 x3 - 3 a b² c² x⁴ y² + a b c x1² y² + 2 a b c x² y y2 + a y2²}

The same we do for y2.

In[16]:= Exponent[#, {y, y2}]& /@ gl6

Out[16]= {{3, 1}, {4, 2}, {2, 2}}

In[17]:= sol = Solve[gl6[[1]] == 0, y2][[1]]

Out[17]= $\left\{ y2 \to \dfrac{-3 b c x² x1 y + x3 y - a c x1 y³}{x1} \right\}$

In[18]:= gl7 = Numerator[Together[#]]& /@ (Drop[gl6, 1] /. sol)

Out[18]= {-b x x1² x3² + 2 a b c x x1³ x3 y² + a x2 x3² y² - a² b c² x x1⁴ y⁴ - 2 a² c x1 x2 x3 y⁴ -
a² c x x3² y⁴ + a³ c² x1² x2 y⁶ + 2 a³ c² x x1 x3 y⁶ - a⁴ c³ x x1² y⁸, -b x1³ x3 + a b c x1⁴ y² -
4 a b c x² x1 x3 y² + a x3² y² + 4 a² b c² x² x1² y⁴ - 2 a² c x1 x3 y⁴ + a³ c² x1² y⁶}

Let us factor the two remaining equations.

In[19]:= **g17a = Factor[g17]**

Out[19]= $\{-(-x3 + a\,c\,x1\,y^2)^2\,(b\,x\,x1^2 - a\,x2\,y^2 + a^2\,c\,x\,y^4),$
$(-x3 + a\,c\,x1\,y^2)\,(b\,x1^3 + 4\,a\,b\,c\,x^2\,x1\,y^2 - a\,x3\,y^2 + a^2\,c\,x1\,y^4)\}$

Eliminating y from the nonequal factors, we arrive at the following equation for $x(t)$.

In[20]:= **Factor[Resultant[g17a[[1, -1]], g17a[[2, -1]], y]& @@ g17] /.**
        {x -> x[t], x1 -> x'[t], x2 -> x''[t], x3 -> x'''[t]}

Out[20]= $a^8\,b^2\,c^2\,x'[t]^4\,(4\,b\,c\,x[t]^3\,x'[t] + x'[t]\,x''[t] - x[t]\,x^{(3)}[t])^4$

c) The function $u(x)$ has three parameters α, β, and z_0. This means a corresponding third-order differential equation exists.

In[1]:= **u = α Exp[β/Sqrt[z - z0]]**

Out[1]= $e^{\frac{\beta}{\sqrt{z-z0}}}\,\alpha$

We start by explicitly calculating the first three derivatives of $u(x)$. up[i] is the ith derivative of $u(x)$.

In[2]:= **eqs1 = Numerator[Together[Table[up[i] - D[u, {z, i}], {i, 3}]]]**

Out[2]= $\{e^{\frac{\beta}{\sqrt{z-z0}}}\,\alpha\,\beta + 2\,z\,\sqrt{z-z0}\,up[1] - 2\,\sqrt{z-z0}\,z0\,up[1],$
$-3\,e^{\frac{\beta}{\sqrt{z-z0}}}\,z\,\alpha\,\beta + 3\,e^{\frac{\beta}{\sqrt{z-z0}}}\,z0\,\alpha\,\beta - e^{\frac{\beta}{\sqrt{z-z0}}}\,\sqrt{z-z0}\,\alpha\,\beta^2 + 4\,z^3\,\sqrt{z-z0}\,up[2] -$
$12\,z^2\,\sqrt{z-z0}\,z0\,up[2] + 12\,z\,\sqrt{z-z0}\,z0^2\,up[2] - 4\,\sqrt{z-z0}\,z0^3\,up[2],$
$15\,e^{\frac{\beta}{\sqrt{z-z0}}}\,z\,\alpha\,\beta - 15\,e^{\frac{\beta}{\sqrt{z-z0}}}\,z0\,\alpha\,\beta + 9\,e^{\frac{\beta}{\sqrt{z-z0}}}\,\sqrt{z-z0}\,\alpha\,\beta^2 + e^{\frac{\beta}{\sqrt{z-z0}}}\,\alpha\,\beta^3 + 8\,z^4\,\sqrt{z-z0}\,up[3] -$
$32\,z^3\,\sqrt{z-z0}\,z0\,up[3] + 48\,z^2\,\sqrt{z-z0}\,z0^2\,up[3] - 32\,z\,\sqrt{z-z0}\,z0^3\,up[3] + 8\,\sqrt{z-z0}\,z0^4\,up[3]\}$

The parameter α and the exponential function are easily eliminated by introducing the original function $u(x)$.

In[3]:= **eqs2 = eqs1 /. Exp[β/Sqrt[z - z0]] -> up[0]/α**

Out[3]= $\{\beta\,up[0] + 2\,z\,\sqrt{z-z0}\,up[1] - 2\,\sqrt{z-z0}\,z0\,up[1],$
$-3\,z\,\beta\,up[0] + 3\,z0\,\beta\,up[0] - \sqrt{z-z0}\,\beta^2\,up[0] + 4\,z^3\,\sqrt{z-z0}\,up[2] -$
$12\,z^2\,\sqrt{z-z0}\,z0\,up[2] + 12\,z\,\sqrt{z-z0}\,z0^2\,up[2] - 4\,\sqrt{z-z0}\,z0^3\,up[2],$
$15\,z\,\beta\,up[0] - 15\,z0\,\beta\,up[0] + 9\,\sqrt{z-z0}\,\beta^2\,up[0] + \beta^3\,up[0] + 8\,z^4\,\sqrt{z-z0}\,up[3] -$
$32\,z^3\,\sqrt{z-z0}\,z0\,up[3] + 48\,z^2\,\sqrt{z-z0}\,z0^2\,up[3] - 32\,z\,\sqrt{z-z0}\,z0^3\,up[3] + 8\,\sqrt{z-z0}\,z0^4\,up[3]\}$

Because we are looking for a polynomial differential equation, we eliminate the square roots $\sqrt{z-z_0}$ by squaring the equations.

In[4]:= **eliminateSquareRoot[expr_Plus] := Expand[#^2 - (expr - #)^2]&[**
        Select[expr, MemberQ[#1, Sqrt[z - z0]] &]]

Factoring eqs2, we see one unwanted factor: $z - z_0$.

In[5]:= **(eqs3 = Factor[eliminateSquareRoot /@ eqs2]) // Short[#, 4]&**

Out[5]//Short= $\{-\beta^2\,up[0]^2 + 4\,z^3\,up[1]^2 - 12\,z^2\,z0\,up[1]^2 + 12\,z\,z0^2\,up[1]^2 - 4\,z0^3\,up[1]^2,$
$(z - z0)\,(\ll 1\gg),\ -225\,z^2\,\beta^2\,up[0]^2 + 450\,z\,z0\,\beta^2\,up[0]^2 - 225\,z0^2\,\beta^2\,up[0]^2 + 51\,z\,\beta^4\,up[0]^2 -$
$51\,z0\,\beta^4\,up[0]^2 - \beta^6\,up[0]^2 + \ll15\gg + 8064\,z^5\,z0^4\,up[3]^2 - 8064\,z^4\,z0^5\,up[3]^2 +$
$5376\,z^3\,z0^6\,up[3]^2 - 2304\,z^2\,z0^7\,up[3]^2 + 576\,z\,z0^8\,up[3]^2 - 64\,z0^9\,up[3]^2\}$

After eliminating this factor, we arrive at the following three polynomial equations.

In[6]:= **(eqs4 = MapAt[Rest, eqs3, 2]) // Short[#, 4]&**

Out[6]//Short= $\{-\beta^2\,up[0]^2 + 4\,z^3\,up[1]^2 - 12\,z^2\,z0\,up[1]^2 + 12\,z\,z0^2\,up[1]^2 - 4\,z0^3\,up[1]^2,$
$\ll1\gg,\ -225\,z^2\,\beta^2\,up[0]^2 + 450\,z\,z0\,\beta^2\,up[0]^2 - 225\,z0^2\,\beta^2\,up[0]^2 + 51\,z\,\beta^4\,up[0]^2 -$
$51\,z0\,\beta^4\,up[0]^2 - \beta^6\,up[0]^2 + \ll15\gg + 8064\,z^5\,z0^4\,up[3]^2 - 8064\,z^4\,z0^5\,up[3]^2 +$
$5376\,z^3\,z0^6\,up[3]^2 - 2304\,z^2\,z0^7\,up[3]^2 + 576\,z\,z0^8\,up[3]^2 - 64\,z0^9\,up[3]^2\}$

We now try to eliminate the variables z, z_0, and β.

In[7]:= **GroebnerBasis[eqs4, {up[0], up[1], up[2], up[3]},**
        {β, z, z0}, MonomialOrder -> EliminationOrder]

Out[7]= $\{\}$

Unfortunately, the last try did not work. It is not too unexpected since we tried to eliminate three parameters from three equations. So let us be less ambitious and only eliminate two variables. We try all three possibilities.

```
In[8]:= GroebnerBasis[eqs4, {up[0], up[1], up[2], up[3]},
                  {β, z}, MonomialOrder -> EliminationOrder]

Out[8]= {}
```

```
In[9]:= GroebnerBasis[eqs4, {up[0], up[1], up[2], up[3]},
                  {β, z0}, MonomialOrder -> EliminationOrder]

Out[9]= {}
```

```
In[10]:= gb = GroebnerBasis[eqs4, {up[0], up[1], up[2], up[3]},
                   {z, z0 }, MonomialOrder -> EliminationOrder];
```

The last Gröbner basis is a pretty large one.

```
In[11]:= {ByteCount[gb], LeafCount[gb]}

Out[11]= {452104, 20260}
```

We select all nontrivial expressions (no single factors like up[0] are of interest for us).

```
In[12]:= candidates = Union[Cases[Factor[gb], _Plus, Infinity]];
```

We have 50 such terms.

```
In[13]:= Length[candidates]

Out[13]= 50
```

Fortunately, among the 50 candidates, we find some that do not even contain β.

```
In[14]:= Select[candidates, FreeQ[#, β]&]

Out[14]= {up[1]^4 + up[0] up[1]^2 up[2] - 5 up[0]^2 up[2]^2 + 3 up[0]^2 up[1] up[3],
          -17 up[1]^4 + 19 up[0] up[1]^2 up[2] - 5 up[0]^2 up[2]^2 + 3 up[0]^2 up[1] up[3],
          17 up[1]^4 - 19 up[0] up[1]^2 up[2] + 5 up[0]^2 up[2]^2 + 3 up[0]^2 up[1] up[3],
          -up[1]^4 - up[0] up[1]^2 up[2] + 5 up[0]^2 up[2]^2 + 3 up[0]^2 up[1] up[3]}
```

And among the last four differential equations, we find one such that the original function is a solution.

```
In[15]:= Select[%, (Together[# /. up[i_] :> D[u, {z, i}]] === 0)&]

Out[15]= {up[1]^4 + up[0] up[1]^2 up[2] - 5 up[0]^2 up[2]^2 + 3 up[0]^2 up[1] up[3]}
```

So the differential equation we were looking for is:

$$u'(x)^4 + u(x) u'(x)^2 u''(x) - 5 u(x)^2 u''(x)^2 + 3 u(x)^2 u'(x) u'''(x) = 0.$$

Here is a quick final check.

```
In[16]:= Together[x'[x]^4 + x[x] x'[x]^2 x''[x] - 5 x[x]^2 x''[x]^2 +
                3 x[x]^2 x'[x] x'''[x] /.
                   x -> Function[x, α Exp[β/Sqrt[x - x0]]]]

Out[16]= 0
```

For algorithms to derive minimal differential equations from algebraic and differential equations, see [1469].

d) This is the Darboux–Halphen system.

```
In[1]:= eqs = {w1'[z] - (w1[z] (w2[z] + w3[z]) - w2[z] w3[z]),
               w2'[z] - (w2[z] (w1[z] + w3[z]) - w1[z] w3[z]),
               w3'[z] - (w3[z] (w1[z] + w2[z]) - w1[z] w2[z]),
               W[z] - 2 (w1[z] + w2[z] + w3[z])};
```

The function makeOde tries to find a single oth-order ordinary differential equation for $W(z)$. It does this by differentiating the Darboux–Halphen system and eliminating $w_1(z)$, $w_2(z)$, $w_3(z)$, and their derivatives.

```
In[2]:= makeOde[o_] :=
        Module[{eqs1, funcs, ws},
          (* the equations *)
```

```
eqs1 = Append[Flatten[Table[D[eqs, {z, k}], {k, 0, o - 1}]],
           D[Last[eqs], {z, o}]];
(* the functions *)
funs = Flatten[Table[D[{w1[z], w2[z], w3[z]}, {z, k}], {k, 0, o}]];
(* the derivatives of W *)
Ws = Flatten[Table[D[W[z], {z, k}], {k, 0, o}]];
(* eliminate w1, w2, and w3 *)
GroebnerBasis[eqs1, Ws, funs, MonomialOrder -> EliminationOrder]]
```

There are no first or second-order differential equations for $W(z)$.

In[3]:= **{makeOde[1], makeOde[2]}**

Out[3]= {{}, {}}

But there is a third-order nonlinear differential equation for $W(z)$—the Chazy equation [815].

In[4]:= **makeOde[3]**

Out[4]= {3 W'[z]^2 - 2 W[z] W''[z] + W$^{(3)}$[z]}

makeOde[4] gives again the Chazy equation, its derivative, and an additional fourth order differential equation.

In[5]:= **makeOde[4]**

Out[5]= {4 W'[z] W''[z] - 2 W[z] W$^{(3)}$[z] + W$^{(4)}$[z], 3 W'[z]^2 - 2 W[z] W''[z] + W$^{(3)}$[z],
          8 W[z] W''[z]^2 - 6 W[z] W'[z] W$^{(3)}$[z] - 4 W''[z] W$^{(3)}$[z] + 3 W'[z] W$^{(4)}$[z]}

e) To derive x-free differential equations, we have to differentiate $f(x)$ repeatedly and to eliminate the resulting nested exponential functions. These are the first four derivatives of $\exp(\exp(\exp(e^x)))$.

In[1]:= **f[x_] = Exp[Exp[Exp[Exp[x]]]];**
        eqs1 = Table[F[i] - D[f[x], {x, i}], {i, 0, 4}] // ExpandAll;

We convert the terms in eqs1 to polynomial form by replacing the n-nested exponential function by en.

In[3]:= **eqs2 = eqs1 //. {Exp[Exp[Exp[x]]] +**
        **   a_. Exp[Exp[x]] + b_. Exp[x] + c_. x] :> e4 e3^a e2^b e1^c,**
        **   Exp[Exp[Exp[Exp[x]]]] -> e4}**

Out[3]= {-e4 + F[0], -e1 e2 e3 e4 + F[1], -e1 e2 e3 e4 - e1^2 e2 e3 e4 - e1^2 e2^2 e3 e4 - e1^2 e2^2 e3^2 e4 + F[2],
          -e1 e2 e3 e4 - 3 e1^2 e2 e3 e4 - e1^3 e2 e3 e4 - 3 e1^2 e2^2 e3 e4 - 3 e1^3 e2^2 e3 e4 -
          e1^3 e2^3 e3 e4 - 3 e1^2 e2^2 e3^2 e4 - 3 e1^3 e2^2 e3^2 e4 - 3 e1^3 e2^3 e3^2 e4 - e1^3 e2^3 e3^3 e4 + F[3],
          -e1 e2 e3 e4 - 7 e1^2 e2 e3 e4 - 6 e1^3 e2 e3 e4 - e1^4 e2 e3 e4 - 7 e1^2 e2^2 e3 e4 - 18 e1^3 e2^2 e3 e4 -
          7 e1^4 e2^2 e3 e4 - 6 e1^3 e2^3 e3 e4 - 6 e1^4 e2^3 e3 e4 - e1^4 e2^4 e3 e4 - 7 e1^2 e2^2 e3^2 e4 -
          18 e1^3 e2^2 e3^2 e4 - 7 e1^4 e2^2 e3^2 e4 - 18 e1^3 e2^3 e3^2 e4 - 18 e1^4 e2^3 e3^2 e4 - 7 e1^4 e2^4 e3^2 e4 -
          6 e1^3 e2^3 e3^3 e4 - 6 e1^4 e2^3 e3^3 e4 - 6 e1^4 e2^4 e3^3 e4 - e1^4 e2^4 e3^4 e4 + F[4]}

Eliminating e1, e2, and e4 is straightforward—these three variables appear linearly.

In[4]:= **eqs3 = Rest[eqs2] /. e4 -> F[0];**
        sol1 = Solve[eqs3[[1]] == 0, e1];
        eqs4 = Numerator[Together[Rest[eqs3] /. sol1[[1]]]] // Factor;
        sol2 = Solve[eqs4[[1]] == 0, e2];
        eqs5 = Numerator[Together[Rest[eqs4] /. sol2[[1]]]] // Factor;
        (* ignore multiplicative factors *)
        eqs6 = Cases[eqs5, _Plus, Infinity];

Using Resultant, we eliminate the remaining variable e3 from the two equations that are left.

In[11]:= **res = Cases[Factor[Resultant[##, e3]& @@ eqs6], _Plus, Infinity];**

In[12]:= **res /.   (* abbreviate result *)**
         **   F[k_] :> Derivative[k][f] // TraditionalForm**

Out[12]//TraditionalForm=

$\{3 (f')^{12} + 8 f (f')^{11} + 11 f^2 (f')^{10} - 8 f f'' (f')^{10} + 13 f^3 (f')^9 - 18 f^2 f'' (f')^9 + 4 f^2 f^{(3)} (f')^9 + 11 f^4 (f')^8 +$
$3 f^2 (f'')^2 (f')^8 - 22 f^3 f'' (f')^8 + 14 f^3 f^{(3)} (f')^8 - 3 f^3 f^{(4)} (f')^8 + 5 f^5 (f')^7 - 3 f^3 (f'')^2 (f')^7 - 22 f^4 f'' (f')^7 +$
$18 f^4 f^{(3)} (f')^7 + 4 f^3 f'' f^{(3)} (f')^7 - 4 f^4 f^{(4)} (f')^7 + 3 f^6 (f')^6 - 3 f^3 (f'')^3 (f')^6 - 7 f^4 (f'')^2 (f')^6 + 3 f^4 (f^{(3)})^2 (f')^6 -$
$12 f^5 f'' (f')^6 + 11 f^5 f^{(3)} (f')^6 - 6 f^4 f'' f^{(3)} (f')^6 - 2 f^5 f^{(4)} (f')^6 + 4 f^4 f'' f^{(4)} (f')^6 + 10 f^4 (f'')^3 (f')^5 -$
$3 f^5 (f'')^2 (f')^5 + 5 f^5 (f^{(3)})^2 (f')^5 - 5 f^6 f'' (f')^5 + 10 f^6 f^{(3)} (f')^5 - 18 f^4 (f'')^2 f^{(3)} (f')^5 - 10 f^5 f'' f^{(3)} (f')^5 -$
$3 f^6 f^{(4)} (f')^5 + 3 f^5 f'' f^{(4)} (f')^5 - f^5 f^{(3)} f^{(4)} (f')^5 + 11 f^4 (f'')^4 (f')^4 + 12 f^5 (f'')^3 (f')^4 - 9 f^6 (f'')^2 (f')^4 +$

$$9 f^6 (f^{(3)})^2 (f')^4 + f^5 f'' (f^{(3)})^2 (f')^4 + f^6 (f^{(4)})^2 (f')^4 - 15 f^5 (f'')^2 f^{(3)} (f')^4 + f^6 f'' f^{(3)} (f')^4 + 2 f^6 f'' f^{(4)} (f')^4 -$$
$$5 f^6 f^{(3)} f^{(4)} (f')^4 + 6 f^5 (f'')^4 (f')^3 + 3 f^6 (f'')^3 (f')^3 + f^6 (f^{(3)})^3 (f')^3 + 15 f^6 f'' (f^{(3)})^2 (f')^3 + 3 f^5 (f'')^3 f^{(3)} (f')^3 -$$
$$26 f^6 (f'')^2 f^{(3)} (f')^3 + 6 f^6 (f'')^2 f^{(4)} (f')^3 - 7 f^6 f'' f^{(3)} f^{(4)} (f')^3 - 3 f^5 (f'')^5 (f')^2 + 15 f^6 (f'')^4 (f')^2 +$$
$$9 f^6 (f'')^2 (f^{(3)})^2 (f')^2 - 28 f^6 (f'')^3 f^{(3)} (f')^2 + 5 f^6 (f'')^3 f^{(4)} (f')^2 + 12 f^6 (f'')^5 f' - 14 f^6 (f'')^4 f^{(3)} f' + 5 f^6 (f'')^6\}$$

Here is a quick check of the result.

```
In[13]:= res /. F[k_] :> D[f[x], {x, k}] // Expand
Out[13]= {0}
```

We proceed in a similar manner for the function $g(x)$.

```
In[14]:= g[x_] = x^x;
eqs1 = Table[G[k] - D[g[x], {x, k}], {k, 0, 3}] // ExpandAll
Out[15]= {-x^x + G[0], -x^x + G[1] - x^x Log[x], -x^{-1+x} - x^x + G[2] - 2 x^x Log[x] - x^x Log[x]^2,
         x^{-2+x} - 3 x^{-1+x} - x^x + G[3] - 3 x^{-1+x} Log[x] - 3 x^x Log[x] - 3 x^x Log[x]^2 - x^x Log[x]^3}
```

The first three derivatives of $g(x)$ contain $\ln(x)$, x^x, and x. Assuming that these three quantities are linear independent we replace them with the polynomial variables log, xx, and ξ.

```
In[16]:= eqs2 = eqs1 //. {Log[x] -> log, x^(x + c_) -> xx ξ^c, x^x -> xx}
```
$$Out[16]= \left\{-xx + G[0], -xx - \log xx + G[1], -xx - 2 \log xx - \log^2 xx - \frac{xx}{\xi} + G[2],\right.$$
$$\left.-xx - 3 \log xx - 3 \log^2 xx - \log^3 xx + \frac{xx}{\xi^2} - \frac{3 xx}{\xi} - \frac{3 \log xx}{\xi} + G[3]\right\}$$

```
In[17]:= eqs3 = Numerator[Together[eqs2]];
gb = GroebnerBasis[eqs3, {}, {log, ξ, xx}] // Factor;
In[19]:= gb /. G[k_] :> Derivative[k][g] // TraditionalForm
Out[19]//TraditionalForm=
```
$$\{(g')^4 + 2 g (g')^3 - 2 g g'' (g')^2 - 3 g^2 g'' g' + g^2 (g'')^2 + g^3 g^{(3)}\}$$

Here is a quick check of the result.

```
In[20]:= gb /. G[k_] :> D[g[x], {x, k}] // Expand
Out[20]= {0}
```

f) A direct calculation of the integral is basically hopeless (for a closed-form, see [468], [899]). So, to obtain the differential equation, we will use the first terms of a series representation of $G_\alpha(w)$. The form of the integrand suggests $w = \infty$ as the most natural expansion points ($w = 0$ would give singularities due to $w = \cos(\vartheta_1) + \cos(\vartheta_2) + \alpha \cos(\vartheta_3)$).

$$G_\alpha(w) = \frac{1}{\pi^3} \sum_{k=0}^{\infty} \frac{1}{w^{k+1}} \left(\int_0^\pi \int_0^\pi \int_0^\pi (\cos(\vartheta_1) + \cos(\vartheta_2) + \alpha \cos(\vartheta_3))^k \, d\vartheta_1 \, d\vartheta_2 \, d\vartheta_3 \right)$$

The triple integral over the power of $\cos(\vartheta_1) + \cos(\vartheta_2) + \alpha \cos(\vartheta_3)$ can be conveniently calculated for small k by the multinomial theorem. The terms of the resulting integrands factor and we only need the following basic integral

$$\frac{1}{\pi} \int_0^\pi \cos^n(\vartheta) \, d\vartheta = \frac{(1 + (-1)^n)(\frac{n}{2} - \frac{1}{2})!}{n \sqrt{\pi} (\frac{n}{2} - 1)!}.$$

The function $I[n, \alpha]$ calculates the triple integral over $(\cos(\vartheta_1) + \cos(\vartheta_2) + \alpha \cos(\vartheta_3))^n$.

```
In[1]:= h[0] = 1;
h[n_] = Integrate[Cos[θ]^n, {θ, 0, Pi}]/Pi // Simplify;
int[n_] := int[n] = h[n]
In[4]:= I[n_, α_] := I[n, α] =
Sum[Multinomial[n1, n2, n - n1 - n2]*
    If[n - n1 - n2 >= 0,
        int[n1] int[n2] α^(n - n1 - n2) int[n - n1 - n2], 0],
    {n1, 0, n}, {n2, 0, n}]
```

The integral vanishes for odd n because of symmetry.

```
In[5]:= {I[7, α], I[8, α]}
```

$$Out[5]= \left\{0, \; \frac{1225}{64} + \frac{175\,\alpha^2}{2} + \frac{945\,\alpha^4}{16} + \frac{35\,\alpha^6}{4} + \frac{35\,\alpha^8}{128}\right\}$$

```
In[6]:= (* compare with direct (much slower) calculation *)
        Integrate[(Cos[φ1] + Cos[φ2] + α Cos[φ3])^#,
                  {φ1, 0, Pi}, {φ2, 0, Pi},
                  {φ3, 0, Pi}]/Pi^3& /@ {7, 8} // Expand
```

$$Out[7]= \left\{0, \; \frac{1225}{64} + \frac{175\,\alpha^2}{2} + \frac{945\,\alpha^4}{16} + \frac{35\,\alpha^6}{4} + \frac{35\,\alpha^8}{128}\right\}$$

```
In[8]:= I[n_?OddQ, α_] := 0
```

Let us next find the order of the differential equation we are looking for. We start by assuming the following general form of a differential equation.

$$\sum_{j=0}^{o}\left(\left(\sum_{l=0}^{d} c_{j,l}(\alpha)\,w^l\right)\frac{\partial^j G_\alpha(w)}{\partial w^j}\right) = 0.$$

Here the $c_{j,l}(\alpha)$ are yet-to-be-determined polynomials in α. We also do not know the values of d and o in the moment. So, we will increase the value of the product $d \times o$ and for each $d \times o$ try all possible realizations for d and o until we find candidates for d and o. For symbolic α, even this is virtually hopeless at this point. To obtain a manageable problem we will a) use a concrete value for α (17 in the following; any other number will do too) and b) only look for modular solutions now. A nontrivial solution for the $c_{j,l}(\alpha)$ for a concrete α is necessary (but not sufficient) for a nontrivial solution for general α.

The following input searches for the first two candidates of d and o.

```
In[9]:= αp = 17;
        founddoPairs = 0;
        dTo = 2;
        While[founddoPairs < 2,
               (* all values for d and o for a given d × o *)
               doPairs = {#, dTo/#}& /@ Divisors[dTo];
               λ = 1;
             While[λ < Length[doPairs] && founddoPairs < 2,
               {d, o} = doPairs[[λ]];
               (* sufficient high order *)
               do = 2 d o + 8;
               (* the series of the function G(w) for concrete α *)
               G = Sum[w^(-k - 1) I[k, αp], {k, 0, do + 4}] //
                                          Series[#, {w, Infinity, do}]&;
               (* differential equation of series *)
               eq = Sum[Sum[c[j, l] w^l, {l, 0, d}] D[G, {w, j}], {j, 0, o}];
               (* coefficients in w *)
               eqs = DeleteCases[eq[[3]], 0];
               (* all occurring cs *)
               vars = Cases[eqs, _c, Infinity] // Union;
               (* modular solution *)
               sol = Solve[Append[# == 0& /@ eqs, Modulus == Prime[10000]],
                          vars, Mode -> Modular];
               (* was a nontrivial solution found? *)
               If[Union[Last /@ Rest[sol[[1]]]] =!= {0},
                  (* print candidate values for d and o *)
                  Print[{d, o}]; founddoPairs++]; λ++];
               (* increase d × o *) dTo++]
        Solve::svars : Equations may not give solutions for all "solve" variables. More…

        {7, 5}

        Solve::svars : Equations may not give solutions for all "solve" variables. More…

        {9, 4}
```

Let us proceed with the found candidate differential equation of order 4. We have $d = 9$. We now generate the coefficients of w with symbolic α and calculate a full (not only a modular) solution.

```
In[13]:= d = 9; o = 4; do = 2 d o + 8;
        (* the series of the function G(w) for symbolic α *)
        G = Sum[w^(-k - 1) I[k, α], {k, 0, do + 4}] //
                                        Series[#, {w, ∞, do}]&;
        eq = Sum[Sum[c[j, l] w^l, {l, 0, d}] D[G, {w, j}], {j, 0, o}];
        (* specialize α *)
        eqs = DeleteCases[eq[[3]], 0];
        (* use "random" prime for α *)
        eqsN = Block[{α = 17}, eqs];
        vars = Cases[eqsN, _c, Infinity] // Union;
        sol = Solve[# == 0& /@ eqsN, vars];

        Solve::svars : Equations may not give solutions for all "solve" variables. More…
```

To reduce the complexity of the problem, we eliminate the vanishing $c_{j,l}(\alpha)$.

```
In[23]:= zerocs = Select[sol[[1]], Last[#] === 0&]
```

```
Out[23]= {c[0, 0] → 0, c[0, 1] → 0, c[0, 2] → 0, c[0, 4] → 0, c[0, 6] → 0, c[0, 7] → 0,
          c[0, 8] → 0, c[0, 9] → 0, c[1, 1] → 0, c[1, 3] → 0, c[1, 5] → 0, c[1, 7] → 0,
          c[1, 8] → 0, c[1, 9] → 0, c[2, 0] → 0, c[2, 2] → 0, c[2, 4] → 0, c[2, 6] → 0,
          c[2, 8] → 0, c[2, 9] → 0, c[3, 1] → 0, c[3, 3] → 0, c[3, 5] → 0, c[3, 7] → 0,
          c[3, 9] → 0, c[4, 0] → 0, c[4, 2] → 0, c[4, 4] → 0, c[4, 6] → 0, c[4, 8] → 0}
```

```
In[24]:= eqsA = DeleteCases[eqs //. zerocs, 0];
```

Now 38 equations remain.

```
In[25]:= Length[eqsA]
```

```
Out[25]= 38
```

Now, we make an ansatz for the $c_{j,l}(\alpha)$ in the form $c_{j,l}(\alpha) = \sum_{k=0}^{e} c_{j,l,k} \alpha^k$. We now loop over e to find its smallest possible value for a nontrivial solution.

```
In[26]:= e = 2;
        While[
         (* substitute polynomial in α for the c[i, 1] *)
         eqsB = eqsA //. c[j_, l_] :> Sum[c[j, 1, k] α^k, {k, 0, e}];
         (* extract c[j, 1, k]'s *)
         varsB = Cases[eqsB, _c, Infinity] // Union;
         (* the coefficients of the c[j, 1, k]'s *)
         tab = Table[Coefficient[eqsB[[k]], #]& /@ varsB, {k, Length[eqsB]}];
         (* substitute prime values for α *)
         eqsC = Join @@ Table[tab /. α -> Prime[j],
                              {j, 2, Round[3 Length[varsB]/Length[tab]]}];
         (* calculate modular null space *)
         nsp = NullSpace[eqsC, Modulus -> Prime[10^6]];
         Length[nsp] === 0, (* increase e *) e++];
         e
```

```
Out[28]= 8
```

So, we have $e = 8$. We extract the vanishing $c_{j,l,k}$ and eliminate these $c_{j,l,k}$ in the list of equations eqsB. Now, we are left with 38 equations (with symbolic α) in 51 variables.

```
In[29]:= zeroRules = Rule @@@ Cases[Transpose[{varsB, nsp[[1]]}], {_, 0}];
        eqsC = eqsB //. zeroRules;
        varsC = Cases[eqsC, _c, Infinity] // Union;
        {Length[eqsC], Length[varsC]}
```

```
Out[32]= {38, 51}
```

Specializing α yields now a set of equations that can be solved exactly.

```
In[33]:= eqsD = Join @@ Table[eqsC /. α -> Prime[j], {j, 2, 6}];
        solD = Solve[# == 0& /@ eqsD, varsC];
        Short[solD, 6]
```

Solve::svars : Equations may not give solutions for all "solve" variables. More…

Out[35]//Short= $\{\{c[0, 3, 0] \to 36\,c[4, 9, 0], c[0, 3, 2] \to -9\,c[4, 9, 0], c[0, 5, 0] \to 3\,c[4, 9, 0],$

$c[1, 0, 0] \to -\dfrac{320}{3}\,c[4, 9, 0], c[1, 0, 2] \to 80\,c[4, 9, 0], c[1, 0, 4] \to -40\,c[4, 9, 0],$

$c[1, 0, 6] \to \dfrac{20}{3}\,c[4, 9, 0], c[1, 2, 0] \to -\dfrac{944}{3}\,c[4, 9, 0], c[1, 2, 2] \to -\dfrac{88}{3}\,c[4, 9, 0],$

$c[1, 2, 4] \to 28\,c[4, 9, 0], \ll30\gg, c[4, 1, 8] \to \dfrac{5}{3}\,c[4, 9, 0],$

$c[4, 3, 0] \to \dfrac{320}{3}\,c[4, 9, 0], c[4, 3, 2] \to -\dfrac{128}{3}\,c[4, 9, 0], c[4, 3, 4] \to 28\,c[4, 9, 0],$

$c[4, 3, 6] \to -6\,c[4, 9, 0], c[4, 5, 0] \to -\dfrac{112}{3}\,c[4, 9, 0], c[4, 5, 2] \to -\dfrac{20}{3}\,c[4, 9, 0],$

$c[4, 5, 4] \to 8\,c[4, 9, 0], c[4, 7, 0] \to -\dfrac{4}{3}\,c[4, 9, 0], c[4, 7, 2] \to -\dfrac{14}{3}\,c[4, 9, 0]\}\}$

One of the $c_{j,l,k}$ remained undetermined (as it has to be for a linear homogeneous differential equation and the ansatz used).

In[36]:= ucs = Union[Flatten[Cases[Last[#], _c, Infinity]& /@ solD[[1]]]]

Out[36]= {c[4, 9, 0]}

Now, we collect all results, substitute them in the full ansatz for the differential equation, and simplify the coefficients of the resulting differential equation.

In[37]:= (* substitute results in ansatz *)
    ode1 = Sum[Sum[c[j, l] w^l, {l, 0, 9}] D[$\mathcal{G}$[w], {w, j}], {j, 0, 4}] /.
    zerocs /. c[j_, l_] :> Sum[c[j, l, k] α^k, {k, 0, 8}] /.
    zeroRules /. solD /. ucs[[1]] -> 1;

In[39]:= (* simplify coefficients *)
    ode2 = Collect[Numerator[Together[ode1]],
              Table[D[$\mathcal{G}$[w], {w, j}], {j, 0, o}],
                 Collect[#, w, Factor]&][[1]]

Out[40]= $(9\,w^5 - 27\,w^3\,(-2+\alpha)\,(2+\alpha))\,\mathcal{G}[w] +$
$(87\,w^6 + w^4\,(836 - 245\,\alpha^2) + 20\,(-2+\alpha)\,(2+\alpha)\,(4 - 2\,\alpha^2 + \alpha^4) + 4\,w^2\,(-236 - 22\,\alpha^2 + 21\,\alpha^4))\,\mathcal{G}'[w] +$
$(111\,w^7 - 68\,w^5\,(-11 + 5\,\alpha^2) - 20\,w\,(-2+\alpha)\,(2+\alpha)\,(4 - 2\,\alpha^2 + \alpha^4) + w^3\,(-2800 - 308\,\alpha^2 + 249\,\alpha^4))$
$\mathcal{G}''[w] + (36\,w^8 - 5\,(-2+\alpha)^3\,\alpha^2\,(2+\alpha)^3 - 7\,w^6\,(-16 + 19\,\alpha^2) +$
$3\,w^4\,(-448 - 60\,\alpha^2 + 51\,\alpha^4) - w^2\,(-2+\alpha)\,(2+\alpha)\,(320 + 36\,\alpha^2 + 51\,\alpha^4))\,\mathcal{G}^{(3)}[w] +$
$(3\,w^9 + 5\,w\,(-2+\alpha)^3\,\alpha^2\,(2+\alpha)^3 - 2\,w^7\,(2 + 7\,\alpha^2) + 4\,w^5\,(-28 - 5\,\alpha^2 + 6\,\alpha^4) -$
$2\,w^3\,(-2+\alpha)\,(2+\alpha)\,(40 - 6\,\alpha^2 + 9\,\alpha^4))\,\mathcal{G}^{(4)}[w]$

We end with a check of the derived differential equation by calculating the first 100 nonvanishing series terms.

In[41]:= do = 200;
    (ode2 /. $\mathcal{G}$ -> Function[w,
              Evaluate[Sum[w^(-k - 1) I[k, α], {k, 0, do + 4}]]]) //
                     Series[#, {w, Infinity, do}]&

Out[42]= $O\left[\dfrac{1}{w}\right]^{201}$

For other approaches to derive this differential equation, see [911] and [468]. For other complicated integrals that can be calculated through differential equations, see [425].

32. Puzzles

a) Yes, just make the order of Derivative something that is not explicitly an integer.

In[1]:= DSolve[Derivative[1 + (Pi + 1)^2 - Pi^2 - 2 Pi][y][x] == x^2, y, x]

DSolve::ndord : Derivative order $1 - 2\,\pi - \pi^2 + (1+\pi)^2$ in term
    $y^{(1-2\,\pi-\pi^2+(1+\pi)^2)}[x]$ should be a non-negative machine-sized integer. More…

Out[1]= DSolve$\left[y^{(1-2\,\pi-\pi^2+(1+\pi)^2)}[x] = x^2, y, x\right]$

In[2]:= ExpandAll[%]

DSolve::ndord : Derivative order $1 - 2\,\pi - \pi^2 + (1+\pi)^2$ in term
    $y^{(1-2\,\pi-\pi^2+(1+\pi)^2)}[x]$ should be a non-negative machine-sized integer. More…

Out[2]= $\left\{\left\{y \rightarrow \text{Function}\left[\{x\}, \frac{x^4}{12} + C[1] + x\, C[2]\right]\right\}\right\}$

b) Let us evaluate the integral under consideration.

In[1]:= `Integrate[Integrate[x^x, x]^3, Integrate[x^x, x]]`

Out[1]= $\dfrac{1}{4}\left(\displaystyle\int x^x\, dx\right)^4$

Mathematica tries to evaluate the two occurrences of `Integrate[x^x, x]`. Both integrals remain undone. Then, the `Integrate[x^x, x]` is viewed as a variable and we have an integral of the form `Integrate[y^3, y]` which gives `y^4/4` with `y = Integrate[x^x, x]`.

In `D[D[1[1], 1]^2, D[1[1], 1]]`, the arguments evaluate first. This yields the messages `General::ivar` because `1[1]` cannot be differentiated with respect to 1. Then, `D[1[1], 1]` becomes the differentiation variable, and the result after differentiation is `2D[1[1], 1]`.

In[2]:= `D[D[1[1], 1]^2, D[1[1], 1]]`

         General::ivar : 1 is not a valid variable. More…

         General::ivar : 1 is not a valid variable. More…

Out[2]= $2\, \partial_1 1[1]$

Now let us look at the nested `Sum`. `Sum` has the `HoldAll` attribute. To localize its iterator variables in a first try, it tries to make `Sum[k, {1, 4}]` the local iterator variable. `Sum` locally creates a `Block`-like environment, and because of the `Protected` attribute, assignments cannot be carried out. This yields the `Sum::write` message. In a next step, `Sum` evaluates the iterator variable and the body; this means in the example under consideration, the expression `Sum[k, {1, 4}]`. Here, 1 would be the iterator variable and we get the `Sum::itraw` messages. Because all tries to localize iterator variables fail, we end up with `Sum[Sum[k, {1, 4}]^2, {Sum[k, {1, 4}], 1, 4}]`.

In[3]:= `Sum[Sum[k, {1, 4}]^2, {Sum[k, {1, 4}], 1, 4}]`

         Sum::write : Tag Sum in $\displaystyle\sum_{1=1}^{4} k$ is Protected. More…

         Sum::itraw : Raw object 1 cannot be used as an iterator. More…

         Sum::itraw : Raw object 1 cannot be used as an iterator. More…

Out[3]= $\displaystyle\sum_{\Sigma_{1=1}^{4} k=1}^{4}\left(\sum_{1=1}^{4} k\right)^2$

Repeating the last evaluation, but unprotecting `Sum`, enables the use of `Sum[k, {1, 4}]` as an iterator variable. We get the result 30.

In[4]:= `Unprotect[Sum];`
     `Sum[Sum[k, {1, 4}]^2, {Sum[k, {1, 4}], 1, 4}]`
Out[5]= 30

In the next example, the second argument of `Solve`, the expression `Positive[x]` is the variable to be solved for. So, we get the result `{{Positive[x]→-1}, {Positive[x]→1}}`. (Only expressions with head `List`, `Plus`, `Times`, `Power`, and `Equal` are not allowed as "solve variables".)

In[6]:= `Solve[Positive[x]^2 == 1, Positive[x]]`

Out[6]= $\{\{\text{Positive}[x] \rightarrow -1\}, \{\text{Positive}[x] \rightarrow 1\}\}$

Now let us deal with the double integral and the double sum. The integration example is easy. `Integrate` does not have any options. The inner integral $\int_{-1}^{1} x_1\, dx_1$ gives zero; the outer integral does too.

In[7]:= `Integrate[x[1], {x, 0, 2}, {x[1], -1, 1}]`

Out[7]= 0

In distinction to `Integrate`, the function `Sum` has the attribute `HoldAll`. `Sum` uses this attribute to scope the summation variables in a `Block`-like manner. So the outer sum localizes the symbol x. The local variable of the inner sum is not a symbol, but the expression x [1]. So it cannot be localized exactly the same way as a symbol is in Block. When the inner sum is carried out, the values of the iterator are assigned to x [1]. But the head is the variable of the outer iterator and has an integer value. So the assignment 0[1] =-1 gets evaluated. Because the head Integer is protected, we get Set::write messages and the sum evaluates to 3 0[1] + 3 1[1] + 3 2[1].

```
In[8]:= Sum[x[1], {x, 0, 2}, {x[1], -1, 1}]

        Set::write : Tag Integer in 0[1] is Protected. More…

        Set::write : Tag Integer in 0[1] is Protected. More…

        Set::write : Tag Integer in 0[1] is Protected. More…

        General::stop :
          Further output of Set::write will be suppressed during this calculation. More…

Out[8]= 3 0[1] + 3 1[1] + 3 2[1]
```

If we unprotect the symbol Integer, we get no messages and the result of the sum is 0 too.

```
In[9]:= Unprotect[Integer];
        Sum[x[1], {x, 0, 2}, {x[1], -1, 1}]
Out[10]= 0
```

c) Let us evaluate the input under consideration.

```
In[1]:= Unprotect[Message]
        Message /: Message[___] := Null /; (Abort[]; False);
        Integrate[1/(Sqrt[C^2] - C), {C, -1, 1}]
Out[1]= {Message}

Out[3]= $Aborted
```

So the result of the input under consideration is $Aborted. This is not surprising. As a side effect, an abort is happening whenever a message is issued. And a message is issued because the integral can be evaluated, but substituting limits produces Power::infy messages.

```
In[4]:= HoldPattern[Message[___]] =.

In[5]:= Integrate[1/(Sqrt[C^2] - C), {C, -1, 1}]

        Integrate::idiv : Integral of  \frac{1}{-C + \sqrt{C^2}}  does not converge on {-1, 1}. More…

Out[5]= \int_{-1}^{1} \frac{1}{-C + \sqrt{C^2}} \, dC
```

d) To produce a mathematically wrong result (that is not obviously a bug) we must search for exceptions from general rules that are used by design.

One rule, mentioned already in Chapter 2 of the Programming volume [1735], is that Times [0, *somethingThatDoesNot-EvaluateToZero*] evaluates to zero. So, we should look for *somethingThatDoesNotEvaluateToZero* that is zero. Then *Mathematica* would give 0 as the result of 0/0. There are an infinite number of (theoretically impossible to detect) zeros, but already togethered (meaning written over a common denominator) versus nontogethered rational functions or expanded versus factored polynomials will do. To avoid variables we use a constant instead, say E. So, we are led to the following 17-character input (without counting white space).

```
In[1]:= 0/(1/E + 1 - (1 + E)/E)
Out[1]= 0
```

Here is a 19 character (again without counting white space) factored-expanded version of a similar input.

```
In[2]:= 0/((E + 1)^2 - E^2 - 2 E - 1)
Out[2]= 0
```

We had to rely on Times; Power for instance will numericalize its arguments.

```
In[3]:= {(1/E + 1 - (1 + E)/E)^0, 0^(1/E + 1 - (1 + E)/E)}
```

N::meprec : Internal precision limit

$MaxExtraPrecision = 50.` reached while evaluating $1 + \frac{1}{e} - \frac{1+e}{e}$. More...

N::meprec : Internal precision limit

$MaxExtraPrecision = 50.` reached while evaluating $1 + \frac{1}{e} - \frac{1+e}{e}$. More...

Out[3]= $\left\{ \left(1 + \frac{1}{e} - \frac{1+e}{e}\right)^0, \ 0^{1+\frac{1}{e}-\frac{1+e}{e}} \right\}$

e) This is the function definition for t.

In[1]:= **t[x_] := ((x + 1)^2 - x^2 - x - 1)/x**

Evaluating t'[0] leads to messages, and the result is Indeterminate.

In[2]:= **t'[0]**

Power::infy : Infinite expression $\frac{1}{0}$ encountered. More...

Power::infy : Infinite expression $\frac{1}{0^2}$ encountered. More...

∞::indet : Indeterminate expression 0 ComplexInfinity encountered. More...

Out[2]= Indeterminate

Evaluating t'[0] starts with evaluating t' as a pure function and then supplies the argument 0 to this pure function.

In[3]:= **t'**

Out[3]= $\frac{-1 - 2\, \#1 + 2\,(1 + \#1)}{\#1} - \frac{-1 - \#1 - \#1^2 + (1 + \#1)^2}{\#1^2}$ &

But evaluating t' forms a derivative without any ongoing simplification. The resulting expression has ξ in the denominator and substituting 0 for ξ leads to messages.

In[4]:= **D[t[ξ], ξ]**

Out[4]= $\frac{-1 - 2\,\xi + 2\,(1 + \xi)}{\xi} - \frac{-1 - \xi - \xi^2 + (1 + \xi)^2}{\xi^2}$

Be aware that, because of the HoldAll attribute of Function, applying Simplify to the pure function has no effect.

In[5]:= **{Simplify[t'], Simplify[t'[[1]]]}**

Out[5]= $\left\{ \frac{-1 - 2\, \#1 + 2\,(1 + \#1)}{\#1} - \frac{-1 - \#1 - \#1^2 + (1 + \#1)^2}{\#1^2}\, \&, \ 0 \right\}$

f) E has the attribute Constant and so Derivative[*positiveInteger*][E] evaluates to 0&. Inside Block, the attribute of E is temporarily nonexistent. As a result, D[E[x], x] evaluates to E'[x].

In[1]:= **Block[{E}, Print @ D[E[x], x]]**

E'[x]

The replacement rule *head_*[*arg_*] :> *arg*[*head*] gets applied to the resulting expression Derivative[1][E][x]. Derivative[1][E] matches *head* and E matches *arg*. The result is x[E'].

In[2]:= **Block[{E}, Print[D[E[x], x] /. head_[arg_] :> arg[head]]]**

x[E']

After leaving the Block, the constant E has again the attribute Constant and E' evaluates to 0&.

In[3]:= **Block[{E}, D[E[x], x] /. head_[arg_] :> arg[head]]**

Out[3]= x[0 &]

g) The function randomRationalFunction implements the generation of random rational functions in *x, y, z*.

In[1]:= (* a random variable *)
   randomVar := {x, y, z}[[Random[Integer, {1, 3}]]];
   (* a random product of powers *)
   randomTerm := Product[randomVar^Random[Integer, {-5, 5}],
           {Random[Integer, {1, 3}]}];

```
(* a random Laurent polynomial *)
randomPoly := Sum[Random[Integer, {-5, 5}] randomTerm,
                {Random[Integer, {1, 3}]}];
(* a random rational function *)
randomRationalFunction := randomPoly/randomPoly
```

Now, we search for a function such that `Together` does not act idempotent on it.

```
In[9]:= SeedRandom[444];
        While[r2 = Together[r1 = Together[
              r0 = Sum[randomRationalFunction, {3}]]];
          r2 === r1, Null];

        {r1, r2}
```

$$Out[12]= \{ \frac{1}{15 \, (-4 + 5 \, x^3 - 4 \, y) \, y^5} \, (-100 \, x^3 + 125 \, x^6 - 100 \, x^3 \, y + 80 \, x^3 \, y^6 - 100 \, x^6 \, y^6 + 80 \, x^3 \, y^7 +$$
$$45 \, x^3 \, y^9 - 12 \, y^6 \, z + 15 \, x^3 \, y^6 \, z - 12 \, y^7 \, z - 48 \, x^4 \, y^6 \, z^5 + 60 \, x^7 \, y^6 \, z^5 - 48 \, x^4 \, y^7 \, z^5),$$
$$-\frac{1}{15 \, y^5 \, (4 - 5 \, x^3 + 4 \, y)} \, (-100 \, x^3 + 125 \, x^6 - 100 \, x^3 \, y + 80 \, x^3 \, y^6 - 100 \, x^6 \, y^6 + 80 \, x^3 \, y^7 +$$
$$45 \, x^3 \, y^9 - 12 \, y^6 \, z + 15 \, x^3 \, y^6 \, z - 12 \, y^7 \, z - 48 \, x^4 \, y^6 \, z^5 + 60 \, x^7 \, y^6 \, z^5 - 48 \, x^4 \, y^7 \, z^5)\}$$

The nonuniqueness in the form of `Together[r(x, y, z)]` results often from the factorization of the denominator.

h) Because of the last `SameQ` statement, the question is if the result will be either `True` or `False`. The result is `False`.

```
In[1]:= f1 = Factor[(x[2] - x[1])^3];
        f2 = Factor[(x[2] - x[3])^3];
        x[1] = x[3];
        f1 - f2 === 0
Out[4]= False
```

Here is what happens in detail: Because of the `Orderless` attribute of `Plus`, the summands of $(x[2] - x[1])^3$ become reordered.

```
In[5]:= Clear[x];
        {(x[2] - x[1])^3, (x[2] - x[3])^3}
Out[6]= {(-x[1] + x[2])^3, (x[2] - x[3])^3}
```

`Factor` operates on multivariate polynomials without an explicit variable specification. This means that the result is unique only up to a factor ± 1 and a corresponding change in one of the factors. The coefficient -1 of $x[2]$ in $(-x[1] + x[2])^3$ gets pulled out, but no -1 is pulled out from $(x[2] - x[3])^3$.

```
In[7]:= Factor[%]
Out[7]= {-(x[1] - x[2])^3, (x[2] - x[3])^3}
```

Identifying now $x[3]$ with $x[1]$ yields two mathematically identical, but structurally different, expressions.

```
In[8]:= x[1] = x[3]; %
Out[8]= {-(-x[2] + x[3])^3, (x[2] - x[3])^3}
```

So `SameQ` returns `False`.

```
In[9]:= SameQ @@ %
Out[9]= False
```

i) *Mathematica* functions can be differentiated into two classes: The ones that evaluate always nontrivially (if given the expected number of arguments) and the ones that may return unchanged. Structural functions (such as `Level`, `Position`, ...) will always evaluate. But there is hardly any chance to fool them. Mathematical functions (such as `Sum`, `Integrate`, ...) might give a wrong result in one version of *Mathematica*, but are less likely to be fooled in the following version. They probably either evaluate correctly or do return "unevaluated" (`Solve` could return unevaluated in case it is not possible to determine if a leading coefficient of a polynomial vanishes). So the question is: Is there a set of functions that have to do nontrivial mathematical work and have to evaluate? A promising set of such functions are the ones ending with Q. They have to evaluate (the ones that should return `True` or `False` when called with the correct number of arguments) nontrivially and they have to carry out a mathematical, not structural, task. Here are all functions ending with Q.

In[1]:= **Names ["*Q"]**

Out[1]= {ArgumentCountQ, ArrayQ, AtomQ, DigitQ, EllipticNomeQ, EvenQ, ExactNumberQ, FreeQ,
HypergeometricPFQ, InexactNumberQ, IntegerQ, IntervalMemberQ, InverseEllipticNomeQ,
LegendreQ, LetterQ, LinkConnectedQ, LinkReadyQ, ListQ, LowerCaseQ, MachineNumberQ,
MatchLocalNameQ, MatchQ, MatrixQ, MemberQ, NameQ, NumberQ, NumericQ, OddQ, OptionQ,
OrderedQ, PartitionsQ, PolynomialQ, PrimeQ, SameQ, StringFreeQ, StringMatchQ,
StringQ, SyntaxQ, TensorQ, TrueQ, UnsameQ, UpperCaseQ, ValueQ, VectorQ}

A moment of reflection shows that `IntervalMemberQ` is such a function we are looking for. Because it is impossible to recognize all transcendental zeros algorithmically, inputs such as `IntervalMemberQ[Interval[{ a, b}], (a+b)/2 + i hiddenZero]` or `IntervalMemberQ[Interval[{ a, b}], a±hiddenZero]` for complicated enough *hiddenZero* might give a wrong answer. Similar to the functions `Equal`, `Less`, ... the function `IntervalMemberQ` must make use of numerical techniques to determine if an exact numeric expression lies within certain bounds.

Here is an example that uses one of the real solutions of an irreducible cubic with three real roots. The unavoidable imaginary part after numericalization fools `IntervalMemberQ` into giving the wrong result.

In[2]:= **r = Solve[1 - 3x - 2x^2 + x^3 == 0, x][[2, 1, 2]]**

Out[2]= $\frac{2}{3} - \frac{1}{6} \left(1 + i \sqrt{3}\right) \left(\frac{1}{2} \left(43 + 3 i \sqrt{771}\right)\right)^{1/3} - \frac{13 \left(1 - i \sqrt{3}\right)}{3 \cdot 2^{2/3} \left(43 + 3 i \sqrt{771}\right)^{1/3}}$

The quantity `r` is mathematically a real number, but numericalization to *any* precision will always yield a number with the head `Complex`.

In[3]:= **N[r, 30]**

Out[3]= $0.286462065031600498058212760431 + 0. \times 10^{-31} i$

In[4]:= **IntervalMemberQ[Interval[{0, 1}], r]**

Out[4]= False

For algebraic numbers, `RootReduce` can, in principle, be used to give the correct result.

In[5]:= **IntervalMemberQ[Interval[{0, 1}], RootReduce @ r]**

Out[5]= True

Here is a transcendental zero. We again get a wrong result, but this time with a message that warns us.

In[6]:= **hZ = Sin[Pi/256] - Sqrt[2 - Sqrt[2 + Sqrt[2 + Sqrt[2 +**
                Sqrt[2 + Sqrt[2 + Sqrt[2]]]]]]]]/2

Out[6]= $-\frac{1}{2} \sqrt{2 - \sqrt{2 + \sqrt{2 + \sqrt{2 + \sqrt{2 + \sqrt{2 + \sqrt{2}}}}}}} + \text{Sin}\left[\frac{\pi}{256}\right]$

In[7]:= **IntervalMemberQ[Interval[{1, 2}], 1 + hZ + Exp[-1000]]**

         N::meprec : Internal precision limit $MaxExtraPrecision = 50.` reached
         while evaluating $\frac{1}{2} \sqrt{2 - \sqrt{2 + \text{Power}[\ll 2\gg]}} - \frac{1}{e^{1000}} - \text{Sin}\left[\frac{\pi}{256}\right]$. More…

Out[7]= False

Of course, raising the value of $MaxExtraPrecision to a sufficiently large value helps here.

In[8]:= **$MaxExtraPrecision = 10^4;**
      IntervalMemberQ[Interval[{1, 2}], 1 + hZ + Exp[-1000]]
Out[9]= True

Currently *Mathematica* evaluates functions with iterator constructs like `Table`, `Sum`, `Product`, etc. if the bounds are real numeric. In case the number of steps cannot be uniquely decided currently, a message is issued and the numericalized number of steps is carried out. This behavior could also be used to generate a wrong result. Here is an example (no $MaxExtraPrecision messages are generated).

In[10]:= **Table[k, {k, 1, 2 + hZ - 10^-1000}]**

Out[10]= {1}

Of course, one could imagine such `Table`, `Sum`, `Product`-constructs, etc. to stay unevaluated. But because of the Q-ending, `IntervalMemberQ` will always evaluate to `True` or `False`.

For a general discussion, why some calculations sometimes must give wrong results, see [1851].

j) First, we must decide if to look for an expression that contains indeterminates or to look for an expression that is numeric. Surely, `Developer`ZeroQ` will handle polynomials and rational functions correctly. To fool it, we at least need radicals or transcendental function. Although it is surely possible to find a transcendental identity that is not recognized by `Developer`ZeroQ` (for instance, `JacobiSN[z, m]^2 + JacobiCN[z, m]^2 - 1` will do), it will not be so straightforward to find a short one. To fool numerical techniques we have to find an expression involving more than transcendental arithmetic and that is actually very small. An example would be, for instance, the difference between a Root-object and its radical form.

```
In[1]:= Developer`ZeroQ[(# - ToRadicals[#] - 10^-1000)&[
                Roots[x^3 - 2x + 5 == 0, x, Cubics -> False][[1, 2]]]]
```

Out[1]= True

Using the `Algebraics` option of `ZeroQ` the last wrong result can be avoided.

```
In[2]:= Developer`ZeroQ[(# - ToRadicals[#] - 10^-1000)&[
                Roots[x^3 - 2x + 5 == 0, x, Cubics -> False][[1, 2]]],
                Algebraics -> True]
```

Out[2]= False

Here is another example. It uses the small difference between π and a high-order rational approximation to π.

```
In[3]:= Developer`ZeroQ[E^(Pi - FromContinuedFraction[
                ContinuedFraction[Pi, 200]]) - 1]
```

Out[3]= True

Because in the last example the transcendental number e^π occurred, the `Algebraics` option cannot cure the last wrong result.

```
In[4]:= Developer`ZeroQ[E^(Pi - FromContinuedFraction[
                ContinuedFraction[Pi, 200]]) - 1,
                Algebraics -> True]
```

Out[4]= True

But again, this is a relatively large input. So let us look for a short transcendental function. The following search gives some candidates.

```
In[5]:= Select[Names["*"], (StringLength[#] <= 3 &&
                MemberQ[Attributes[#], NumericFunction])&]
```

Out[5]= {Abs, Arg, Cos, Cot, Csc, Erf, Exp, Im, Log, Max, Min, Mod, Re, Sec, Sin, Tan}

Inspecting the last list makes `Erf` a prime candidate (we discuss the special functions in detail in Chapter 3). It is a transcendental function and approaches 1 exponentially fast as $z \to \infty$ ($\mathrm{erf}(z) \approx 1 - \exp(z^2)/(\pi^{1/2} z)$). This means erf(99) is a number with more than 4000 9's. Any quick numerical check will not detect the difference to 1. So the eight-character input `Erf@99-1` is an example what we were looking for.

```
In[6]:= Developer`ZeroQ[Erf@99-1]
```

Out[6]= True

In addition to giving sometimes mathematically incorrect answers, the function `Developer`ZeroQ` can be very slow on relatively simple inputs. Here is an example where a high-precision numericalization is used to determine if the expression vanishes.

```
In[7]:= Developer`ZeroQ[Nest[Tan, E, 8]] // Timing
```

Out[7]= {3.18 Second, False}

k) It is easy to come up with such examples. We just have to find a function $f(x, y)$ where $\int f(x, y)\, dy$ is easy to do and gives a simple result, while $\int f(x, y)\, dx$ is more difficult to do and gives a more complicated result. One example is $f(x, y) = y^x$. Multiplying then y^x by an x-dependent function results in a function we are looking for.

```
In[1]:= f[x_, y_] := y^x x^(-1/3)
```

```
       {Integrate[f[x, y], {x, 0, 1}, {y, 0, 1}],
        Integrate[f[x, y], {y, 0, 1}, {x, 0, 1}]}
```

$$Out[2]= \left\{ \frac{\pi}{\sqrt{3}} - Log[2], \int_0^1 \frac{Gamma[\frac{2}{3}] - Gamma[\frac{2}{3}, -Log[y]]}{(-Log[y])^{2/3}} \, dy \right\}$$

While the last returned integral can in principle be carried out, it requires a clever change of variables. Numerically the two results agree, of course.

```
In[3]:= {N[%], {NIntegrate[f[x, y], {x, 0, 1}, {y, 0, 1}],
                NIntegrate[f[x, y], {y, 0, 1}, {x, 0, 1}]}}
Out[3]= {{1.12065, 1.12065}, {1.12065, 1.12065}}
```

l) We start by evaluating the inputs under consideration.

```
In[1]:= makeDef1[d_, f_, A_] :=
        f /: d[1][f] /; (clearDef1[d, f, A]; makeDef2[d, f, A]; False) := Null

       clearDef1[d_, f_, A_] :=
        f /: d[1][f] /; (clearDef1[d, f, A]; makeDef2[d, f, A]; False) =.

       makeDef2[d_, f_, A_] :=
        d /: HoldPattern[d[1][f][0]] := (clearDef2[d, f]; makeDef1[d, f, A]; A)

       clearDef2[d_, f_] := d /: HoldPattern[d[1][f][0]] =.
In[5]:= makeDef1[Derivative, f, f0];
       D[f[x], x] /. x -> 0
Out[6]= f0
```

So the result was f0. Here is what happened: First definitions for the functions makeDef1, clearDef1, makeDef2, and clearDef2 were made. Then makeDef1[Derivative, f, f0] was evaluated. This made an upvalue for f.

```
In[7]:= makeDef1[Derivative, f, f0]
       ??f
       Global`f
```

```
       f' /; (clearDef1[Derivative, f, f0]; makeDef2[Derivative, f, f0]; False) ^:= Null
```

Then D[f[x], x] evaluated to Derivative[1][f][x]. The head of the last expression, Derivative[1][f], fires the upvalue for f. While the upvalue rule itself does not match (the condition gives always False), as a side effect of evaluating the condition, the two statements clearDef1[Derivative, f, f0] and makeDef2[Derivative, f, f0] are evaluated. The first one removes the upvalue for f and the second one creates a subvalue for Derivative.

```
In[9]:= clearDef1[Derivative, f, f0]
       makeDef2[Derivative, f, f0]
       SubValues[Derivative] // (* select relevant definition *)
              Select[#, MemberQ[#, f, Infinity, Heads -> True]&]&
Out[11]= {HoldPattern[HoldPattern[f'[0]]] :>
          (clearDef2[Derivative, f]; makeDef1[Derivative, f, f0]; f0)}
```

Then the replacement x -> 0 gets evaluated. As a result, the expression Derivative[1][f][0] is created. The subvalue definition of Derivative matches now and the right-hand side of the last rule is evaluated. By doing so the subvalue definition itself is destroyed, the original upvalue definition for f is restored and the expression f0 is returned.

Using On[], the previously described steps can be seen.

```
In[12]:= Clear[f, Derivative];
        makeDef1[Derivative, f, f0];
        On[]; $MessagePrePrint = Identity;
        D[f[x], x] /. x -> 0
           On::trace : On[] --> Null. More...

           Set::trace : $MessagePrePrint = Identity --> Identity. More...
```

```
CompoundExpression::trace : On[]; $MessagePrePrint = Identity; --> Null. More...

D::trace : ∂ₓf[x] --> Derivative[1][f][x]. More...

clearDef1::trace : clearDef1[Derivative, f, f0] --> f /: Derivative[1][f] /;
    (clearDef1[Derivative, f, f0]; makeDef2[Derivative, f, f0]; False) =.. More...

TagUnset::trace :
  f /: Derivative[1][f] /; (clearDef1[Derivative, f, f0]; makeDef2[Derivative, f, f0];
    False) =. --> Null. More...

makeDef2::trace :
  makeDef2[Derivative, f, f0] --> Derivative /: HoldPattern[Derivative[1][f][0]] :=
    (clearDef2[Derivative, f]; makeDef1[Derivative, f, f0]; f0). More...

TagSetDelayed::trace : Derivative /: HoldPattern[Derivative[1][f][0]] :=
    (clearDef2[Derivative, f]; makeDef1[Derivative, f, f0]; f0) --> Null. More...

CompoundExpression::trace :
  clearDef1[Derivative, f, f0]; makeDef2[Derivative, f, f0]; False --> False. More...

ReplaceAll::trace : ∂ₓf[x] /. x → 0 --> Derivative[1][f][x] /. x → 0. More...

ReplaceAll::trace : Derivative[1][f][x] /. x → 0 --> Derivative[1][f][0]. More...

Derivative::trace : Derivative[1][f][0] -->
    clearDef2[Derivative, f]; makeDef1[Derivative, f, f0]; f0. More...

clearDef2::trace : clearDef2[Derivative, f] -->
    Derivative /: HoldPattern[Derivative[1][f][0]] =.. More...

TagUnset::trace : Derivative /: HoldPattern[Derivative[1][f][0]] =. --> Null. More...

makeDef1::trace : makeDef1[Derivative, f, f0] --> f /: Derivative[1][f] /;
    (clearDef1[Derivative, f, f0]; makeDef2[Derivative, f, f0]; False) := Null. More...

TagSetDelayed::trace :
  f /: Derivative[1][f] /; (clearDef1[Derivative, f, f0]; makeDef2[Derivative, f, f0];
    False) := Null --> Null. More...

CompoundExpression::trace :
  clearDef2[Derivative, f]; makeDef1[Derivative, f, f0]; f0 --> f0. More...
```

Out[15]= f0

After evaluating D[f[x], x] /. x->0 we have again the same state of affairs for the definitions as before—the upvalue for f and no definitions for Derivative.

```
In[16]:= Off[];
        ??f
        Global`f

        Derivative[1][f] /;
          (clearDef1[Derivative, f, f0]; makeDef2[Derivative, f, f0]; False) ^:= Null

In[18]:= ??Derivative

        f' represents the derivative of a function f of one argument.
            Derivative[n1, n2, ... ][f] is the general form, representing a function
            obtained from f by differentiating n1 times with respect to the first
            argument, n2 times with respect to the second argument, and so on. More...
        Attributes[Derivative] = {NHoldAll, ReadProtected}
```

m) It is straightforward to find such examples. We just randomly nest the functions $\exp(x)$, $\ln(x)$, $-x$, $1/x$, and $x \pm 1$ (trigonometric and inverse trigonometric functions could be easily added to the list of building blocks). The function notLimitableFunction returns a function such that Limit $[f(x), x->0]$ stays unevaluated.

```
In[1]:= rf := #[[Random[Integer, {1, Length[#]}]]]& @
                {Exp, Log, x #&, -#&, #^2&, 1/#&, (# + 1)&, (# - 1)&}

In[2]:= rF[n_] := Nest[rf[#]&, x, n]
```

Here are some examples of such functions.

```
In[3]:= SeedRandom[1];
        Table[rF[n], {n, 2, 16}]
```

$$Out[4]= \left\{x, -1 + e^{x^2}, 1 + x^2 \text{Log}[x]^2, 2 + \text{Log}\left[\frac{1}{1+x}\right], e^{\frac{2}{x^2(1+x)}}, x \text{Log}\left[1 + e^{-x^3}\right], -1 - \frac{1}{-1+x^3},\right.$$

$$1 + (-2 + x^2 \text{Log}[x])^2, \text{Log}\left[(1 + \text{Log}[\text{Log}[e^{-1-x}]])^2\right], 1 + \text{Log}\left[\text{Log}\left[\text{Log}\left[-\text{Log}\left[e^{e^x}\right]\right]^2\right]\right],$$

$$(-2 + x^2 \text{Log}[\text{Log}[x^x]]^2)^2, e \text{Log}\left[\text{Log}\left[x\left(-1 + x\left(1 + \frac{1}{\text{Log}[x]}\right)\right)\right]\right],$$

$$\left.\frac{1}{\text{Log}\left[1 - e^{e^{\frac{1}{1+\text{Log}\left[\text{Log}\left[\text{Log}\left[\frac{1}{\text{Log}[x]}\right]\right]\right]}}}x\right]}, 1 + e^{-\frac{1}{\text{Log}[x \text{Log}[e^{-x^5}]]}}, e^{\left(-1+\text{Log}\left[e^{e^{e^{\frac{2}{\text{Log}[1+\text{Log}[x]]}}}}\right]^2\right)^2}\right\}$$

```
In[5]:= (* avoid messages *)
        Off[N::meprec]; Off[General::ovfl]; Off[Power::infy];

        notLimitableFunction[nestDepth_:20] :=
        Module[{f},
               While[f = Nest[rf[#]&, x, Random[Integer, {nestDepth/2, nestDepth}]];
                     TimeConstrained[Head[Limit[f, x -> 0]], 10] =!= Limit, Null];
               f]
```

And here is a try to find an example of a function returned by notLimitableFunction.

```
In[8]:= SeedRandom[123];

        TimeConstrained[notLimitableFunction[], 1000]
```
        ∞::indet : Indeterminate expression $e^{\text{ComplexInfinity}}$ encountered. More…

$$Out[9]= x \text{Log}\left[\text{Log}\left[1 - e^{-4+2\text{Log}\left[\frac{1}{x^2}\right]^4}\right]\right]$$

Here is an example of a function whose limit cannot be determined.

```
In[10]:= Limit[E^((-1 + E^E^(1/(E^2 x^2 Log[Log[-x]^2]^2))) x), x -> 0]
```

$$Out[10]= \text{Limit}\left[e^{\left(-1+e^{e^{\frac{1}{e^2 x^2 \text{Log}[\text{Log}[-x]^2]^2}}}\right)x}, x \to 0\right]$$

n) If it is at all possible to find such functions, then the "right" choice of variables will surely be one that converts everything into rational functions. Assuming the trigonometric functions have integer arguments, we implement the following substitution of variables.

```
In[1]:= myTrigIntegrate[int_, x_] :=
        Module[{ξ}, intTξ = TrigToExp[int] //. Exp[c_ x] :> ξ^(c/I);
               Integrate[-I intTξ/ξ, ξ] /. ξ -> Exp[I x]]
```

The function myTrigIntegrate works as expected.

```
In[2]:= myTrigIntegrate[(Cos[x]^2 + 2)/(Tan[x] + Sin[2x]), x]
```

$$Out[2]= -\frac{1}{2}\text{Log}[e^{ix}] + \text{Log}[-1 + e^{2ix}] - \frac{1}{4}\text{Log}[1 + 4e^{2ix} + e^{4ix}]$$

```
In[3]:= D[%, x] - (Cos[x]^2 + 2)/(Tan[x] + Sin[2x]) // Simplify
```

```
Out[3]= 0
```

Next, we define a function randomTrigSum that returns a random sum of trigonometric function. The function randomTrigRational returns the quotient of two random sum of trigonometric function—a rational trigonometric function.

```
In[4]:= (* random sum of trigonometric functions *)
        randomTrigSum[c_, n_, o_, p_] :=
        Sum[Random[Integer, {-c, c}] {Cos, Sin, Csc, Sec, Tan, Cot}[[
                                      Random[Integer, {1, 6}]]][
               (2 Random[Integer] - 1)*
               Random[Integer, {1, c}] x]^Random[Integer, {-p, p}], {n}]

        (* random rational function of trigonometric functions *)
```

```
randomTrigRational[c_, n_, o_, p_] :=
Module[{num, den},
        While[den = randomTrigSum[c, n, o, p];
              FreeQ[Simplify[den], x, Infinity], Null];
        num = randomTrigSum[c, n, o, p]; num/den]
```

Here are some examples of random rational trigonometric functions.

In[9]:= `Table[randomTrigRational[3, 3, 3, 3], {3}]`

Out[9]= $\left\{\dfrac{\text{Sin}[x] - 3\,\text{Sin}[x]^3 + 2\,\text{Sin}[3\,x]^3}{-2\,\text{Cos}[x]^3 - 3\,\text{Tan}[3\,x]^2},\right.$

$\dfrac{-2 - \text{Cos}[2\,x]^3 + \text{Tan}[3\,x]^3}{3\,\text{Cot}[x]^3 + 3\,\text{Csc}[2\,x] + 2\,\text{Tan}[2\,x]}, \left. \dfrac{3\,\text{Tan}[x]}{-\text{Cot}[x] + 3\,\text{Cot}[2\,x]^2 - \text{Sec}[3\,x]}\right\}$

Now, we generate random rational trigonometric functions and try to find an integrand that the built-in `Integrate` cannot integrate. The function `findFunction`, beginning with seeding the random number generator, does this.

In[10]:= `findFunction[seed_, {c_, n_, o_, p_}, maxTries_] :=`
```
Module[{integrand, builtInIntegral, myIntegral, counter = 0},
   (* seed random number generator *) SeedRandom[seed];
   While[(counter++) < maxTries &&
         (((* built-in integration *)
           integrand = randomTrigRational[c, n, o, p];
           builtInIntegral = Integrate[integrand, x];
           FreeQ[builtInIntegral, Integrate, {0, Infinity},
                 Heads -> True]) ||
          ((* manual substitution of variables *)
           myIntegral = myTrigIntegrate[integrand, x];
           MemberQ[ToRadicals[Normal[myIntegral]],
                   Root, {0, Infinity}, Heads -> True])),
         Null]; Print[counter];
   If[counter == maxTries + 1, (* no integrand found *) $Failed,
      (* found integrand *) integrand]]
```

Here is try to find a rational trigonometric function with the property we were looking for. We try 1000 functions.

In[11]:= `findFunction[1, {2, 2, 2, 2}, 1000]`

        1001

Out[11]= `$Failed`

In[12]:= `If[% =!= $Failed, {Integrate[#, x], myTrigIntegrate[#, x],`
              `(* quick check of the result of myTrigIntegrate *)`
              `D[myTrigIntegrate[#, x], x] - % // Simplify}&[%], Null]`

o) *Mathematica* will always find the correct solution for rational coefficients. If we allow nonrational coefficients, then we might find such an example. Because it is algorithmically undecidable if an expression is identical zero and because for a polynomial `Solve` will never stay unevaluated, we use such a "nearly zero" expression as the coefficient of the leading power. For a sufficient complex "nearly zero" expression *Mathematica* cannot determine if it is zero. When the coefficient is zero we might get more solutions than exist and when the leading coefficient is zero it might get found to be zero, and *Mathematica* will miss one solution.

We do not have to search too hard to find an expression that is zero, but is not recognized as such by *Mathematica*. The expression `zero` is an example.

In[1]:= `zero = (# - Together[TrigToExp[#]])&[Sin[Pi/13]]`

Out[1]= $\dfrac{1}{2}\,(-1)^{11/26}\,(-1 + (-1)^{2/13}) + \text{Sin}\!\left[\dfrac{\pi}{13}\right]$

In[2]:= `Simplify[zero]`

Out[2]= $\dfrac{1}{2}\,(-1)^{11/26}\,(-1 + (-1)^{2/13}) + \text{Sin}\!\left[\dfrac{\pi}{13}\right]$

Using `zero` as the leading coefficient in a quadratic yields correctly one solution.

In[3]:= `Solve[zero x^2 + x - 1 == 0, x]`

Out[3]= $\{\{x \to 1\}\}$

Perturbing the leading coefficient now slightly, yields an expression that *Mathematica* cannot symbolically determine to be nonvanishing. So it uses numerical techniques and within the default setting for $MaxExtraPrecision, it finds that the leading coefficient vanishes.

```
In[4]:= Solve[(zero + 10^-100) x^2 + x - 1 == 0, x]
```
```
Out[4]= {{x → 1}}
```

p) On a first glance, myFourierTransform seems to work as intended.

```
In[1]:= myFourierTransform[f_, t_, ω_, opts___] :=
          1/Sqrt[2 Pi] Integrate[f Exp[I t ω], {t, -Infinity, Infinity}, opts]
```
```
In[2]:= myFourierTransform[Exp[-t^2], t, ω, Assumptions -> Im[k] == 0]
```
$$Out[2]= \frac{e^{-\frac{\omega^2}{4}}}{\sqrt{2}}$$

A problem arises in the case myFourierTransform[$f(\omega)$, ω, ω].

```
In[3]:= myFourierTransform[Exp[-ω^2], ω, ω, Assumptions -> Im[ω] == 0]
```
$$Out[3]= \frac{1}{\sqrt{2 - 2 i}}$$

$\mathcal{F}_t[f(t)](\omega)$ is a functional of f, depending on the variable ω. The variable t itself is of no relevance, only the map $t \to f(t)$. This means the "integration variable" has to be a truly dummy variable and the result has to depend on the third argument of myFourierTransform. The function myBetterFourierTransform uses a dummy integration variable and only later substitutes the third argument. It also guards this substitution against the situation of an unevaluated integral.

```
In[4]:= myBetterFourierTransform[f_, t_, ω_, opts___] :=
          Module[{κ, res},
          res = 1/Sqrt[2 Pi] Integrate[f Exp[I t κ], {t, -Infinity, Infinity},
                                      Sequence @@ ({opts} /. ω -> κ)] /. κ -> ω;
          res /; FreeQ[res, _Integrate, {0, Infinity}]];
```

Now, we get the expected result for $\mathcal{F}_\omega[f(\omega)](\omega)$.

```
In[5]:= myBetterFourierTransform[Exp[-ω^2], ω, ω, Assumptions -> Im[ω] == 0]
```
$$Out[5]= \frac{e^{-\frac{\omega^2}{4}}}{\sqrt{2}}$$

q) It is easy to find a function such that $f(f^{-1}(z)) = f(f^{-1}(z)) = z$.

```
In[1]:= f[z_] := α z + β; fInv[z_] := (z - β)/α
```
```
          {f[fInv[z]], fInv[f[z]]}
```
```
Out[2]= {z, z}
```

It is also easy to find a function such that $f(f^{-1}(z)) = f(f^{-1}(z)) = z$ almost everywhere. We just add a function to $f(z)$ that vanishes almost everywhere, but not everywhere.

```
In[3]:= f[z_] := z + (Log[z] + Log[1/z])/(2 I Pi)
          fInv[z_] := z
```

Along the negative real axis, the $f(f^{-1}(z)) = f(f^{-1}(z)) = z$ does not hold.

```
In[5]:= {f[fInv[z]] - z, fInv[f[z]] - z} /.
          {{z -> Random[Complex, {-1 - I, 1 + I}, 20]},
          {z -> Random[Real, {0, -4}, 20]}}
```
```
Out[5]= {{0. × 10^-21 + 0. × 10^-21 i, 0. × 10^-21 + 0. × 10^-21 i},
          {1.00000000000000000000 + 0. × 10^-21 i, 1.00000000000000000000 + 0. × 10^-21 i}}
```

The third class, the pairs f, f^{-1} such that $f(f^{-1}(z)) = z$ everywhere, but $f(f^{-1}(z)) = z$ not everywhere is the standard situation of the named functions of *Mathematica*. Here are two typical examples.

```
In[6]:= f[z_] := Exp[z]; fInv[z_] := Log[z]
```
```
          {f[fInv[z]], fInv[f[z]]}
```
```
Out[7]= {z, Log[e^z]}
```

In[8]:= **f[z_] := Sin[z]; fInv[z_] := ArcSin[z]**

{f[fInv[z]], fInv[f[z]]}
Out[9]= {z, ArcSin[Sin[z]]}

(The pair EllipticNomeQ and InverseEllipticNomeQ are special in the sense that InverseEllipticNomeQ plays, despite its name, the role of f.)

The last type of functions is easily generated by starting with a function that has a branch cut. Here is an example.

In[10]:= **f[z_] = z^2 + Sqrt[z];**
fInv[z_] = (z /. Solve[f[z] == w, z][[2]]) /.
w -> z // Simplify

Out[11]= $-\dfrac{\sqrt{8z + \dfrac{32\,2^{1/3}\,z^2}{\left(128\,z^3+3\left(9+\sqrt{81+768\,z^3}\right)\right)^{1/3}} + 2^{2/3}\left(128\,z^3+3\left(9+\sqrt{81+768\,z^3}\right)\right)^{1/3}}}{2\sqrt{6}} +$

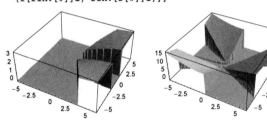

$$\frac{2\sqrt{6}}{\sqrt{8z + \dfrac{32\,2^{1/3}\,z^2}{\left(128\,z^3+3\left(9+\sqrt{81+768\,z^3}\right)\right)^{1/3}} + 2^{2/3}\left(128\,z^3+3\left(9+\sqrt{81+768\,z^3}\right)\right)^{1/3}}}$$

The following plot shows the region where the two identities $f(f^{-1}(z)) = z$ and $f(f^{-1}(z)) = z$ hold and not hold.

In[12]:= **Show[GraphicsArray[**
Plot3D[Chop[Abs[#[(x + I y)] - (x + I y)]],
{x, -6, 6}, {y, -6, 6},
DisplayFunction -> Identity, Mesh -> False,
PlotPoints -> 120, PlotRange -> All]& /@
{f[fInv[#]]&, fInv[f[#]]&}]]

r) The two-argument form of Simplify with an assumption expressing equality of a variable to an expression can do nothing else then substituting the value for the variable in the expression (it cannot take any limit). Here is a simple example.

In[1]:= **Simplify[Sin[x]/x, x == 0]**

Simplify::infd : Expression $\dfrac{Sin[x]}{x}$ simplified to Indeterminate. More...

Out[1]= Indeterminate

And because of the rule for Times that $0 \times$*something* evaluates to 0, we get in the example under consideration 0.

In[2]:= **Simplify[x DiracDelta'[x], x == 0]**

Out[2]= 0

Using the function Simplify`SimplifyPseudoFunctions instead, we get the mathematically correct result.

In[3]:= **Simplify`SimplifyPseudoFunctions[x DiracDelta'[x]]**

Out[3]= -DiracDelta[x]

s) The numerical integration gives a finite result for this diverging integral. Taking the steps of the method option setting Oscillatory into account, such a finite result is to be expected. For such-type integrals, the function is integrated between the zeros of the oscillatory function (here sin) and the resulting series of partial integrals is extrapolated. And, as discussed in the Numerics volume, the Wynn extrapolation method will give finite results for diverging series.

In[1]:= NIntegrate[Sin[x^2] x^(3/2), {x, 0, Infinity}, Method -> Oscillatory]

    SequenceLimit::seqlim : The general form of the sequence
        could not be determined, and the result may be incorrect. More…

Out[1]= 0.418703

Using a larger setting for the WorkingPrecision option yields more digits.

In[2]:= NIntegrate[Sin[x] x^(3/2), {x, 0, Infinity}, Method -> Oscillatory,
            WorkingPrecision -> 20, PrecisionGoal -> 12]

    SequenceLimit::seqlim : The general form of the sequence
        could not be determined, and the result may be incorrect. More…

Out[2]= -0.9399856030

Carrying out the symbolic integral and ignoring the given convergence condition gives for the exact value of this integral $-3/4\,(\pi/2)^{1/2}$.

In[3]:= int1 = Integrate[Sin[x] x^α, {x, 0, Infinity}]

Out[3]= If$\left[-2 < \text{Re}[α] < 0, \text{Cos}\left[\frac{\pi\,α}{2}\right] \text{Gamma}[1+α, 0],\right.$

    $\left.\text{Integrate}[x^α \text{Sin}[x], \{x, 0, ∞\}, \text{Assumptions} \to \text{Re}[α] \le -2 \,||\, \text{Re}[α] \ge 0]\right]$

In[4]:= {int1[[2]] /. α -> 3/2, N[int1[[2]] /. α -> 3/2, 12]}

Out[4]= $\left\{-\dfrac{3\sqrt{\frac{\pi}{2}}}{4}, -0.939985602987\right\}$

We obtain the same result by integrating up to a large finite upper limit X and ignoring terms of the form $\cos(X)\,X^α$ and $\sin(X)\,X^α$.

In[5]:= int2 = Integrate[Sin[x] x^(3/2), {x, 0, X}]

Out[5]= $\dfrac{1}{4}\left(-3\sqrt{2\,\pi}\ \text{FresnelS}\left[\sqrt{\frac{2}{\pi}}\ \sqrt{X}\,\right] + 2\sqrt{X}\ (-2\,X\,\text{Cos}[X] + 3\,\text{Sin}[X])\right)$

In[6]:= Off[Series::esss];
    Series[Series[int2, {X, Infinity, 2}], {X, Infinity, 2}]   // Normal // Expand

Out[7]= $-\dfrac{3\sqrt{\frac{\pi}{2}}}{4} - \dfrac{\text{Cos}[X]}{\left(\frac{1}{X}\right)^{3/2}} + \dfrac{3}{4}\sqrt{\dfrac{1}{X}}\,\text{Cos}[X] - \dfrac{9}{16}\left(\dfrac{1}{X}\right)^{5/2}\text{Cos}[X] + \dfrac{3\,\text{Sin}[X]}{2\sqrt{\frac{1}{X}}} + \dfrac{3}{8}\left(\dfrac{1}{X}\right)^{3/2}\text{Sin}[X]$

The exact form of the definite integral between consecutive zeros of the sin function is given by the following expression.

In[8]:= intP[k_] = Integrate[Sin[x] x^(3/2), {x, k Pi, (k + 1) Pi},
                    Assumptions -> k > 0] // Simplify[#, Element[k, Integers]]&

Out[8]= $\dfrac{1}{4}\sqrt{\pi}$

    $\left(4\,(-1)^k\left(k^{3/2} + \sqrt{1+k} + k\sqrt{1+k}\right)\pi + 3\sqrt{2}\ \text{FresnelS}[\sqrt{2}\ \sqrt{k}\,] - 3\sqrt{2}\ \text{FresnelS}[\sqrt{2}\ \sqrt{1+k}\,]\right)$

As mentioned above, the sequence limit of this series with alternating increasing in magnitude elements gives a finite result (the same as the above NIntegrate calculation).

In[9]:= SequenceLimit[FoldList[Plus, 0, N[#, 30]& @ Table[intP[k], {k, 0, 30}]],
                    Method -> Fit]

    SequenceLimit::seqlim : The general form of the sequence
        could not be determined, and the result may be incorrect. More…

Out[9]= -0.9399856029866251884

Another way to obtain the result is the summation of the series with a convergence achieving factor $\exp(-\varepsilon k)$ and taking the limit $\varepsilon \to 0$. We sum the tail through its asymptotic expansion and take the first term exactly into account.

In[10]:= `(ser = Series[intP[k], {k, Infinity, 8}] // Normal // PowerExpand //`
`Simplify[#, Element[k, Integers]]&) //`
`Short[#, 3]&`

Out[10]//Short= $\frac{1}{4}\sqrt{\pi}\left(\frac{(-1)^k\,(99-144\,k+224\,k^2-384\,k^3+768\,k^4-2048\,k^5+12288\,k^6+49152\,k^7+65536\,k^8)\,\pi}{8192\,k^{13/2}}+\right.$

$\frac{3\,(\ll 1\gg)}{256\,k^{17}\ll 1\gg\ll 1\gg\pi^9}-$

$\left.3\sqrt{2}\left(\frac{1}{2}+\frac{(-1)^k\,(\ll 1\gg)\,(2027025-41580\,(21-6\,k+k^2)\,\pi^2+\ll 1\gg-\ll 1\gg+256\,k^8\,\pi^8)}{8388608\sqrt{2}\,k^{33/2}\,\pi^9}\right)\right)$

In[11]:= `(sum = Sum[Evaluate[ser Exp[-ε k]], {k, 15, Infinity}]) // Short[#, 3]&`

Out[11]//Short= $\frac{1}{33554432\,\pi^{17/2}}$

$\left(67108864\,\pi^{10}\left(-14\sqrt{14}\,e^{-14\,\varepsilon}+13\sqrt{13}\,e^{-13\,\varepsilon}-24\sqrt{3}\,e^{-12\,\varepsilon}+\ll 15\gg+e^{-\varepsilon}+\text{PolyLog}\left[-\frac{3}{2},-e^{-\varepsilon}\right]\right)+\right.$

$\left.\ll 96\gg+25945920\,\pi^6\,(\ll 1\gg)\right)$

In[12]:= `Series[sum, {ε, 0, 0}] + Sum[intP[k], {k, 0, 14}] // N[#, 10]&`

Out[12]= $-0.9399856030+O[\varepsilon]^1$

33. 2D Newton–Leibniz Example, 1D Integral, Square Root of Differential Operator

a) *Mathematica* can do the definite double integral immediately.

In[1]:= `Integrate[Sqrt[(x - y)^2], {x, 0, 1}, {y, 0, 1}]`

Out[1]= $\frac{1}{3}$

This is the indefinite integral.

In[2]:= `int = Integrate[Sqrt[(x - y)^2], x, y]`

Out[2]= $\frac{1}{2}x\sqrt{(x-y)^2}\,y$

The integral is "correct".

In[3]:= `D[%, x, y] // Simplify`

Out[3]= $\sqrt{(x-y)^2}$

The 2D indefinite integral is a continuous (but not everywhere differentiable) function of x and y.

In[4]:= `Plot3D[int, {x, 0, 1}, {y, 0, 1}, PlotPoints -> 61]`

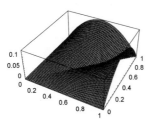

But the substitution of limits gives the obviously wrong (a positive integrand should give a positive integral) result 0.

In[5]:= `((# /. x -> 1) - (# /. x -> 0))&[(int /. y -> 1) - (int /. y -> 0)]`

Out[5]= 0

Carrying out the definite integration using `Integrate` too gives a wrong result.

In[6]:= `Integrate[Sqrt[(x - y)^2], {x, 0, 1}, {y, 0, 1}]`

Out[6]= $\dfrac{1}{3}$

This means that the Newton–Leibniz theorem was not applicable because the indefinite integral did not fulfill all suppositions needed. Now, we do the two integrations step-by-step.

```
In[7]:= int = Integrate[Sqrt[(x - y)^2], x] // Factor
```

Out[7]= $\dfrac{x\,(x - 2\,y)\,\sqrt{(x - y)^2}}{2\,(x - y)}$

The integral has a jump for a generic value of

```
In[8]:= (* suppress messages due to division by 0 *)
        Off[Power::"infy"]; Off[Infinity::"indet"];
        Off[Plot3D::"plnc"]; Off[Plot3D::"gval"]

        indefIntGraphics[int_] :=
        Show[GraphicsArray[
          Block[{$DisplayFunction = Identity},
                {Plot[Evaluate[int /. y -> 0.3], {x, 0, 1}],
                 Plot3D[Evaluate[int], {x, 0, 1}, {y, 0, 1},
                        PlotPoints -> {61, 61}, Mesh -> False]}]]];
```

```
In[12]:= indefIntGraphics[int];
```

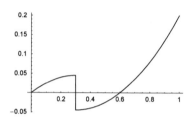 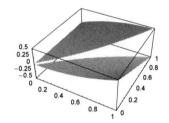

Let us fix this jump by adding a function that makes everything together smooth.

```
In[13]:= δ = (-x (x - 2y) /. x -> y) (Sqrt[(x - y)^2]/(x - y) + 1)/2
```

Out[13]= $\dfrac{1}{2}\left(1 + \dfrac{\sqrt{(x - y)^2}}{x - y}\right) y^2$

```
In[14]:= int1 = int + δ // Together // Factor
```

Out[14]= $\dfrac{1}{2}\left(x\,\sqrt{(x - y)^2} - \sqrt{(x - y)^2}\,y + y^2\right)$

```
In[15]:= indefIntGraphics[int1]
```

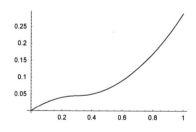 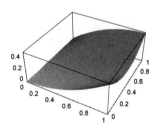

Now let us deal with the second integration.

```
In[16]:= int2 = Integrate[int1, y] // Together // Factor
```

Out[16]= $\dfrac{y\left(3\,x^2\,\sqrt{(x - y)^2} - 3\,x\,\sqrt{(x - y)^2}\,y + x\,y^2 + \sqrt{(x - y)^2}\,y^2 - y^3\right)}{6\,(x - y)}$

Again, we have a jump.

In[17]:= indefIntGraphics[int2]

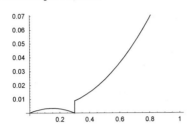 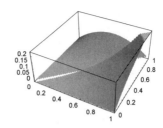

We again add a function $\delta2$ to obtain a smooth function of two variables.

In[18]:= int2a = int2 // Expand

Out[18]= $\dfrac{x^2 \sqrt{(x-y)^2}\, y}{2\,(x-y)} - \dfrac{x \sqrt{(x-y)^2}\, y^2}{2\,(x-y)} + \dfrac{x\, y^3}{6\,(x-y)} + \dfrac{\sqrt{(x-y)^2}\, y^3}{6\,(x-y)} - \dfrac{y^4}{6\,(x-y)}$

In[19]:= int2a /. Sqrt[(x - y)^2]/(x - y) -> 1

Out[19]= $\dfrac{x^2\, y}{2} - \dfrac{x\, y^2}{2} + \dfrac{y^3}{6} + \dfrac{x\, y^3}{6\,(x-y)} - \dfrac{y^4}{6\,(x-y)}$

In[20]:= Together[%] /. y -> x

Out[20]= $\dfrac{x^3}{3}$

In[21]:= δ2 = x^3/3 (Sqrt[(x - y)^2]/(x - y) + 1)/2

Out[21]= $\dfrac{1}{6}\, x^3 \left(1 + \dfrac{\sqrt{(x-y)^2}}{x-y}\right)$

In[22]:= int3 = int2 - δ2 // Together // Factor

Out[22]= $-\dfrac{1}{6}\,(x-y)\left(x^2 + x\sqrt{(x-y)^2} + x\,y - \sqrt{(x-y)^2}\, y + y^2\right)$

In[23]:= Collect[int3, Sqrt[(x - y)^2]] // Factor

Out[23]= $\dfrac{1}{6}\left(-x^3 - ((x-y)^2)^{3/2} + y^3\right)$

Now, we have a nice, smooth, correct double indefinite result.

In[24]:= indefIntGraphics[int3]

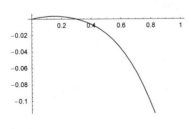 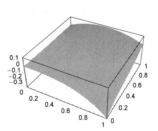

Differentiation yields $((x - y)^2)^{1/2}$.

In[25]:= D[int3, x, y] // Together // Factor

Out[25]= $\sqrt{(x-y)^2}$

In[26]:= (int3 /. y -> 1) - (int3 /. y -> 0)

Out[26]= $-\dfrac{1}{6}\,(-1+x)\left(1 - \sqrt{(-1+x)^2} + x + \sqrt{(-1+x)^2}\, x + x^2\right) + \dfrac{1}{6}\, x\left(x^2 + x\sqrt{x^2}\right)$

In[27]:= (% /. x -> 1) - (% /. x -> 0)

Out[27]= $\dfrac{1}{3}$

This is the result we also get from numerical integration.

```
In[28]:= NIntegrate[Sqrt[(x - y)^2], {x, 0, 1}, {y, 0, 1}]
Out[28]= 0.333333
```

b) A direct evaluation of the definite integral gives the result $\gamma - 1/2$.

```
In[1]:= Off[Series::esss];
        Integrate[integrand = Exp[-2 x] (Coth[x] - 1/x), {x, 0, Infinity}]
```

$$\text{Out[2]=}\; -\frac{1}{2} + \text{EulerGamma}$$

But we can also calculate the indefinite integral.

```
In[3]:= indefInt = Integrate[integrand, x] // Simplify
```

$$\text{Out[3]=}\; \frac{e^{-2x}}{2} - \text{ExpIntegralEi}[-2x] + \text{Log}[-1 + e^{-2x}]$$

The indefinite integral is a continuous function along the positive real line. Its value at $x = 0$ is $\gamma - 1/2$.

```
In[4]:= Series[indefInt, {x, 0, 0}, Assumptions -> x < 0]
```

$$\text{Out[4]=}\; \left(\frac{1}{2} - \text{EulerGamma}\right) + O[x]^1$$

After making the indefinite integral real-valued along the positive real axis and expanding the function `ExpIntegralEi`, we see that the upper limit contribution to the integral vanishes.

```
In[5]:= indefInt /. {Log[Exp[-2 x] - 1] -> -I Pi,
        ExpIntegralEi[-2 x] :>
        (Normal[Series[ExpIntegralEi[ξ], {ξ, Infinity, 1}]] /. ξ -> -2x)}
```

$$\text{Out[5]=}\; \frac{e^{-2x}}{2} + \frac{e^{-2x}}{2x}$$

Another possibility to calculate the integral is the following. We split the one convergent integral into two divergent integrals and use the option setting `GenerateConditions -> False`. This allows evaluating one of the integrals. We also multiply each term with a symbolic factor of the form x^μ to have a nontrivial parameter dependence. This again yields again the result $\gamma - 1/2$.

Multiplying each term with a symbolic factor of the form x^μ yields again the result $\gamma - 1/2$.

```
In[6]:= Integrate[x^μ #, {x, 0, Infinity},
                  GenerateConditions -> False]& /@  Expand[integrand]
```

$$\text{Out[6]=}\; -2^{-\mu}\,\text{Gamma}[\mu] + 2^{-1-\mu}\,\text{Gamma}[1+\mu]\;(-1 + \text{PolyLog}[1+\mu,\,1] + \text{Zeta}[1+\mu])$$

```
In[7]:= Series[%, {μ, 0, 0}] // Simplify // Normal
```

$$\text{Out[7]=}\; -\frac{1}{2} + \text{EulerGamma}$$

The value $\gamma - 1/2$ agrees with the result of a numerical integration.

```
In[8]:= Off[General::unfl]; Off[General::ovfl];
        {NIntegrate[Evaluate[integrand], {x, 0, Infinity},
                    WorkingPrecision -> 30], N[EulerGamma - 1/2, 30]}
Out[9]= {0.0772156649015328606 07, 0.077215664901532860606512090082 4}
```

c) We will denote $\partial^i/\partial x^i$ by `d[i]` in the following. Because the order of x-dependent functions with respect to `d[i]` matters, we will introduce our own noncommutative operator product `I`. The following input defines the basic properties of `d` and `I` needed in the following.

```
In[1]:= (* unite adjacent differential operators *)
        I[a___, d[i_], d[j_], b___] := I[a, d[i + j], b]

        (* bring all d's to the right of x-dependent functions *)
        I[start___, d[i_], f_, rest___] :=
          Sum[Binomial[i, k] I[start, D[f, {x, k}], d[i - k], rest],
              {k, 0, o}] /; Head[f] =!= d && Head[f] =!= I

        (* pull out numeric factors *)
        I[a___, f_?NumericQ b_, c___] := f I[a, b, c] // Expand
```

```
(* distribute d's over sums *)
I[a___, p_Plus, b___] := I[a, #, b]& /@ p
```

```
(* flatten out nested I's *)
I[a___, I[b___], c___] := I[a, b, c]
```

Now let us calculate α_0. Truncating the ansatz for $\sqrt{\mathcal{L}}$ at $k = 0$ and truncating the infinite binomial sum yields the following expression for $\sqrt{\mathcal{L}} \left(\sqrt{\mathcal{L}} \, (.) \right) = \mathcal{L}(.)$.

```
In[12]:= o = 3;
        δ = I[d[1] + I[α[0][x], d[0]]];
        I[δ, δ]
Out[14]= I[d[2]] + 2 I[α[0][x], d[1]] + I[α[0]'[x], d[0]] + I[α[0][x], α[0][x], d[0]]
```

Because $\mathcal{L}$ does not contain any term of the form $\partial / \partial x$, it follows immediately that $\alpha_0 = 0$.

Next, let us repeat this procedure to calculate $\alpha_1(x)$.

```
In[15]:= δ = I[d[1] + I[α[-1][x], d[-1]]];
        I[δ, δ]
Out[16]= I[d[2]] + 2 I[α[-1][x], d[0]] + I[α[-1]'[x], d[-1]] +
        I[α[-1][x], α[-1][x], d[-2]] - I[α[-1][x], α[-1]'[x], d[-3]] +
        I[α[-1][x], α[-1]''[x], d[-4]] - I[α[-1][x], α[-1]^(3)[x], d[-5]]
```

Equating the terms of the form $\ldots(x)$. with $u(x)$. yields the result $\alpha_{-1}(x) = u(x)/2$.

```
In[17]:= Plus @@ (Cases[%, _. I[___, d[0]]])
Out[17]= 2 I[α[-1][x], d[0]]
```

```
In[18]:= Solve[2 α[-1][x] == u[x], α[-1][x]]
Out[18]= {{α[-1][x] → u[x]/2}}
```

Now it is straightforward to automate the process of the calculation of $\alpha_{-k}(x)$ after $\alpha_0(x), \ldots, \alpha_{-k+1}(x)$ are already calculated.

```
In[19]:= (* current approximation of sqrtL *)
        sqrtL = d[1] + I[u[x]/2, d[-1]];
In[21]:= Do[(* take enough terms *) o = Abs[k] + 3;
        (* new ansatz approximation for sqrtL *)
        δ = I[sqrtL + I[α[k][x], d[k]]];
        (* calculate sqrtL @ sqrtL *)
        δI = I[δ, δ];
        (* original L does not contain D[., {x, k}] *)
        zero = (Plus @@ (Cases[δI, _. I[___, d[k + 1]]])) /.
                        I[ts__, _] :> Times[ts];

        (* solve for α[d] *)
        sol = Solve[zero == 0, α[k][x]];
        (* new approximation for sqrtL *)
        sqrtL = sqrtL + I[α[k][x] /. sol[[1]], d[k]], {k, -2, -10, -1}]
```

We implement some formatting rules to display the result in a concise form.

```
In[22]:= With[{tf = TraditionalForm},
        (* format x-dependent parts to the left of ∂'s *)
        MakeBoxes[d[k_], tf] := SuperscriptBox["∂", MakeBoxes[k, tf]];
        (* format d's as ∂'s *)
        MakeBoxes[C[us_, d[k_]], tf] :=
                RowBox[{MakeBoxes[us, tf], MakeBoxes[d[k], tf]}];
        (* sort d's in decreasing order *)
        MakeBoxes[LList[args__], tf] := RowBox[Drop[Flatten[
                Transpose[{MakeBoxes[#, tf]& /@ {args},
                        Table["+", {Length[{args}]}]}], 1], -1]]]
```

We now simplify and format the calculated first terms of the factorization of $\mathcal{L} = \partial^2 . / \partial x^2 + u(x)$..

```
In[23]:= res1 = Simplify[(* unite prefactors of equal d[k] *)
                (sqrtL /. I -> C) //. α_. C[a_, d_] :> C[α a, d] //.
```

```
              C[a_, d_] + C[b_, d_] :> C[a + b, d]] /. {u[x] -> u,
                  Derivative[j_][u][x_] :> Derivative[j][u]};
In[24]:= res2 = Sort[ℒList @@ res1, Function[f, f[#1] > f[#2]][
                  Cases[#, d[_], {0, Infinity}, Heads -> True][[1, 1]]&]&];
```

```
In[25]:= res2 // TraditionalForm
```

Out[25]//TraditionalForm=

$$
\partial^1 + \frac{u}{2}\partial^{-1} + -\frac{u'}{4}\partial^{-2} + \frac{1}{8}(u'' - u^2)\partial^{-3} + \frac{3\,u\,u'}{8} - \frac{u^{(3)}}{16}\partial^{-4} +
$$

$$
\frac{1}{32}(2\,u^3 - 14\,u''\,u - 11\,(u')^2 + u^{(4)})\partial^{-5} + \frac{1}{64}(-30\,u'\,u^2 + 30\,u^{(3)}\,u + 60\,u'\,u'' - u^{(5)})\partial^{-6} +
$$

$$
\frac{1}{128}(-5\,u^4 + 110\,u''\,u^2 + 170\,(u')^2\,u - 62\,u^{(4)}\,u - 91\,(u'')^2 - 148\,u'\,u^{(3)} + u^{(6)})\partial^{-7} +
$$

$$
\frac{1}{256}(140\,u'\,u^3 - 350\,u^{(3)}\,u^2 - 1400\,u'\,u''\,u + 126\,u^{(5)}\,u - 350\,(u')^3 + 490\,u''\,u^{(3)} + 350\,u'\,u^{(4)} - u^{(7)})\partial^{-8} +
$$

$$
\frac{1}{512}(14\,u^5 - 700\,u''\,u^3 - 14\,(115\,(u')^2 - 73\,u^{(4)})\,u^2 + (3178\,(u'')^2 + 4984\,u'\,u^{(3)} - 254\,u^{(6)})\,u -
$$
$$
699\,(u^{(3)})^2 + 4718\,(u')^2\,u'' - 1262\,u''\,u^{(4)} - 810\,u'\,u^{(5)} + u^{(8)})\partial^{-9} +
$$

$$
\frac{1}{1024}(-630\,u'\,u^4 + 2940\,u^{(3)}\,u^3 + 42\,(420\,u'\,u'' - 67\,u^{(5)})\,u^2 + 6\,(1470\,(u')^3 - 2702\,u^{(4)}\,u' - 4130\,u''\,u^{(3)} + 85\,u^{(7)})\,u -
$$
$$
18438\,(u')^2\,u^{(3)} + 3780\,u^{(3)}\,u^{(4)} + 3192\,u''\,u^{(5)} - 42\,u'\,(543\,(u'')^2 - 44\,u^{(6)}) - u^{(9)})\partial^{-10}
$$

With some handwork one can, for the special differential operator $\mathcal{L} = \partial^2./\partial x^2 + u(x)$, even derive a closed-form nonlinear recursion relation for the $\alpha_{-k}(x) = \alpha_{-k}(u(x), u'(x), \ldots, u^{(k)}(x))$ [1577]

$$
\alpha_k(x) = -\frac{1}{2}\left(\alpha'_{k+1}(x) + \sum_{j=1}^{-k-2}\sum_{i=1}^{-j-k-1}\binom{-j}{-i-j-k-1}\alpha_{-j}(x)\,\alpha_{-i}^{(-i-j-k-1)}(x)\right)
$$

where the recursion starts with $\alpha_{-1}(x) = u(x)/2$.

```
In[26]:= α[-1] = u[x]/2;
```

```
      α[k_] := α[k] = Factor[-(D[α[k + 1], x] +
              Sum[Binomial[-j, -k - 1 - j - i]*
              α[-j] D[α[-i], {x, -k - 1 - j - i}],
                  {j, -k - 2}, {i, -k - 1 - j}])/2]
```

This recursion yields the just-calculated result.

```
In[28]:= (* above calculated coefficients *)
      Expand[Table[Cases[res1, C[_, d[k]]][[1, 1]], {k, -1, -10, -1}] -
      (* coefficients from the recursion relation *)
      (Table[α[k], {k, -1, -10, -1}] /.    {u[x] -> u,
                  Derivative[j_][u][x_] :> Derivative[j][u]})]
```

Out[29]= {0, 0, 0, 0, 0, 0, 0, 0, 0, 0}

34. Coefficients = Roots of a Univariate Polynomial, Amoebas, Tiling

a) We write the polynomial in fully factored form: $x^3 + a\,x^2 + b\,x + c$ as $(x - x_1)(x - x_2)(x - x_3)$. The following expresses a, b, and c, as functions of the roots x_1, x_2, and x_3.

```
In[1]:= coeffs = Drop[CoefficientList[(x - x1)(x - x2)(x - x3), x], -1]
```

Out[1]= {-x1 x2 x3, x1 x2 + x1 x3 + x2 x3, -x1 - x2 - x3}

Forming all possible combinations of coefficients and roots, we have the following equations.

```
In[2]:= eqs = (coeffs - #)& /@ Permutations[{x1, x2, x3}]
```

```
Out[2]= {{-x1 - x1 x2 x3, -x2 + x1 x2 + x1 x3 + x2 x3, -x1 - x2 - 2 x3},
         {-x1 - x1 x2 x3, x1 x2 - x3 + x1 x3 + x2 x3, -x1 - 2 x2 - x3},
         {-x2 - x1 x2 x3, -x1 + x1 x2 + x1 x3 + x2 x3, -x1 - x2 - 2 x3},
         {-x2 - x1 x2 x3, x1 x2 - x3 + x1 x3 + x2 x3, -2 x1 - x2 - x3},
         {-x3 - x1 x2 x3, -x1 + x1 x2 + x1 x3 + x2 x3, -x1 - 2 x2 - x3},
         {-x3 - x1 x2 x3, -x2 + x1 x2 + x1 x3 + x2 x3, -2 x1 - x2 - x3}}
```

Solving these equations gives the following 28 solutions.

```
In[3]:= (* avoid large outputs containing radicals *)
        SetOptions[Roots, Cubics -> False, Quartics -> False];

In[5]:= sols = Union[Sort /@ Flatten[RootReduce[
            Solve[# == 0& /@ #, {x1, x2, x3}]& /@ eqs], 1]]

Out[5]= {{x1 → -2, x2 → 0, x3 → 1}, {x1 → -2, x2 → 1, x3 → 0},
          {x1 → -1, x2 → 1, x3 → -1}, {x1 → 0, x2 → -2, x3 → 1}, {x1 → 0, x2 → 0, x3 → 0},
          {x1 → 0, x2 → 1, x3 → -2}, {x1 → 1, x2 → -2, x3 → 0}, {x1 → 1, x2 → -1, x3 → -1},
          {x1 → 1, x2 → 0, x3 → -2}, {x1 → Root[2 - 2 #1 + #1³ &, 1], x2 → Root[-2 + 4 #1 - 2 #1² + #1³ &, 1],
            x3 → Root[-1 + 2 #1² + 2 #1³ &, 1]}, {x1 → Root[2 - 2 #1 + #1³ &, 1],
            x2 → Root[-1 + 2 #1² + 2 #1³ &, 1], x3 → Root[-2 + 4 #1 - 2 #1² + #1³ &, 1]},
          {x1 → Root[2 - 2 #1 + #1³ &, 2], x2 → Root[-2 + 4 #1 - 2 #1² + #1³ &, 3],
            x3 → Root[-1 + 2 #1² + 2 #1³ &, 2]}, {x1 → Root[2 - 2 #1 + #1³ &, 2],
            x2 → Root[-1 + 2 #1² + 2 #1³ &, 2], x3 → Root[-2 + 4 #1 - 2 #1² + #1³ &, 3]},
          {x1 → Root[2 - 2 #1 + #1³ &, 3], x2 → Root[-2 + 4 #1 - 2 #1² + #1³ &, 2],
            x3 → Root[-1 + 2 #1² + 2 #1³ &, 3]}, {x1 → Root[2 - 2 #1 + #1³ &, 3],
            x2 → Root[-1 + 2 #1² + 2 #1³ &, 3], x3 → Root[-2 + 4 #1 - 2 #1² + #1³ &, 2]},
          {x1 → Root[-2 + 4 #1 - 2 #1² + #1³ &, 1], x2 → Root[2 - 2 #1 + #1³ &, 1],
            x3 → Root[-1 + 2 #1² + 2 #1³ &, 1]}, {x1 → Root[-2 + 4 #1 - 2 #1² + #1³ &, 1],
            x2 → Root[-1 + 2 #1² + 2 #1³ &, 1], x3 → Root[2 - 2 #1 + #1³ &, 1]},
          {x1 → Root[-2 + 4 #1 - 2 #1² + #1³ &, 2], x2 → Root[2 - 2 #1 + #1³ &, 3],
            x3 → Root[-1 + 2 #1² + 2 #1³ &, 3]}, {x1 → Root[-2 + 4 #1 - 2 #1² + #1³ &, 2],
            x2 → Root[-1 + 2 #1² + 2 #1³ &, 3], x3 → Root[2 - 2 #1 + #1³ &, 3]},
          {x1 → Root[-2 + 4 #1 - 2 #1² + #1³ &, 3], x2 → Root[2 - 2 #1 + #1³ &, 2],
            x3 → Root[-1 + 2 #1² + 2 #1³ &, 2]}, {x1 → Root[-2 + 4 #1 - 2 #1² + #1³ &, 3],
            x2 → Root[-1 + 2 #1² + 2 #1³ &, 2], x3 → Root[2 - 2 #1 + #1³ &, 2]},
          {x1 → Root[-1 + 2 #1² + 2 #1³ &, 1], x2 → Root[2 - 2 #1 + #1³ &, 1],
            x3 → Root[-2 + 4 #1 - 2 #1² + #1³ &, 1]}, {x1 → Root[-1 + 2 #1² + 2 #1³ &, 1],
            x2 → Root[-2 + 4 #1 - 2 #1² + #1³ &, 1], x3 → Root[2 - 2 #1 + #1³ &, 1]},
          {x1 → Root[-1 + 2 #1² + 2 #1³ &, 2], x2 → Root[2 - 2 #1 + #1³ &, 2],
            x3 → Root[-2 + 4 #1 - 2 #1² + #1³ &, 3]}, {x1 → Root[-1 + 2 #1² + 2 #1³ &, 2],
            x2 → Root[-2 + 4 #1 - 2 #1² + #1³ &, 3], x3 → Root[2 - 2 #1 + #1³ &, 2]},
          {x1 → Root[-1 + 2 #1² + 2 #1³ &, 3], x2 → Root[2 - 2 #1 + #1³ &, 3],
            x3 → Root[-2 + 4 #1 - 2 #1² + #1³ &, 2]}, {x1 → Root[-1 + 2 #1² + 2 #1³ &, 3],
            x2 → Root[-2 + 4 #1 - 2 #1² + #1³ &, 2], x3 → Root[2 - 2 #1 + #1³ &, 3]}}
```

Here are the corresponding numerical values.

```
In[6]:= {x1, x2, x3} /. N[sols]

Out[6]= {{-2., 0., 1.}, {-2., 1., 0.}, {-1., -1., 1.}, {-1., 1., -1.},
          {0., -2., 1.}, {0., 0., 0.}, {0., 1., -2.}, {1., -2., 0.}, {1., -1., -1.},
          {1., 0., -2.}, {-1.76929, 0.638897, 0.565198}, {-1.76929, 0.565198, 0.638897},
          {0.884646 - 0.589743 i, 0.680552 + 1.63317 i, -0.782599 - 0.521714 i},
          {0.884646 - 0.589743 i, -0.782599 - 0.521714 i, 0.680552 + 1.63317 i},
          {0.884646 + 0.589743 i, 0.680552 - 1.63317 i, -0.782599 + 0.521714 i},
          {0.884646 + 0.589743 i, -0.782599 + 0.521714 i, 0.680552 - 1.63317 i},
          {0.638897, -1.76929, 0.565198}, {0.638897, 0.565198, -1.76929},
          {0.680552 - 1.63317 i, 0.884646 + 0.589743 i, -0.782599 + 0.521714 i},
          {0.680552 - 1.63317 i, -0.782599 + 0.521714 i, 0.884646 + 0.589743 i},
          {0.680552 + 1.63317 i, 0.884646 - 0.589743 i, -0.782599 - 0.521714 i},
          {0.680552 + 1.63317 i, -0.782599 - 0.521714 i, 0.884646 - 0.589743 i},
          {0.565198, -1.76929, 0.638897}, {0.565198, 0.638897, -1.76929},
          {-0.782599 - 0.521714 i, 0.884646 - 0.589743 i, 0.680552 + 1.63317 i},
          {-0.782599 - 0.521714 i, 0.680552 + 1.63317 i, 0.884646 - 0.589743 i},
          {-0.782599 + 0.521714 i, 0.884646 + 0.589743 i, 0.680552 - 1.63317 i},
          {-0.782599 + 0.521714 i, 0.680552 - 1.63317 i, 0.884646 + 0.589743 i}}
```

Here is a quick numerical check of the correctness of the solutions.

```
In[7]:= poly = {-(x1 x2 x3), x1 x2 + x1 x3 + x2 x3,
               -x1 - x2 - x3}.{1, x, x^2} + x^3
```

$$Out[7]= \ x^3 + x^2 \ (-x1 - x2 - x3) - x1 \ x2 \ x3 + x \ (x1 \ x2 + x1 \ x3 + x2 \ x3)$$

```
In[8]:= ((Sort[N[{x1, x2, x3} /. #]] ==
          Sort[x /. NSolve[(poly /. #) == 0, x]])& /@
                        DeleteCases[sols, {x1 -> 0, x2 -> 0, x3 -> 0}]) // Union
```

Out[8]= {True}

We continue with the calculation and visualizations of such solutions for various degrees of the polynomial. This time we solve the resulting equations numerically using NSolve.

```
In[9]:= Show[Graphics[
          (* color solutions as a function of the degree of the polynomial *)
          Table[{Hue[(n - 2)/5], PointSize[0.016 - n 0.002], Point[{Re[#], Im[#]}]}& /@
          Cases[(* construct and solve equations for root == coefficients *)
                NSolve[(Drop[CoefficientList[Product[x - ξ[k], {k, n}], x], -1] -
                Table[ξ[k], {k, n}]) == 0, Table[ξ[k], {k, n}]],
                _?NumberQ, {-1}]}, {n, 2, 6}]], Frame -> True, PlotRange -> All]
```

For polynomials of degree seven or higher NSolve will run a very long time. Instead of calculating all solutions at once, we use FindRoot with different starting values to calculate many individual solutions. Here are some of the solutions for degree eight shown. (Because we use random complex starting values, we will surely miss some special real solutions.)

```
In[10]:= Module[{n = 8, o = 2 10^4, zs, eqs, jac, ξ},
          (* seed random number generator *)
          SeedRandom[222];
          (* coefficients/roots *)
          zs = Table[ξ[k], {k, n}];
          (* equations *)
          eqs = Drop[CoefficientList[Product[x - ξ[k], {k, n}], x], -1] - zs;
          (* jacobian (calculate only once) *)
          jac = Outer[D, eqs, zs];
          (* show coefficients/roots in the complex plane *)
          Show[Graphics[{PointSize[0.003], (* use symmetry *)
                {Point[{Re[#], +Im[#]}], Point[{Re[#], -Im[#]}]}& /@
          Flatten[Table[Check[Table[ξ[k], {k, n}] /.
             FindRoot[Evaluate[eqs], (* random starting values *)
                        Evaluate[Table[{ξ[k],
                        6 Random[] Exp[2 Pi I Random[]]}, {k, n}]],
                        Method -> "Newton", Jacobian -> jac], {}], {o}]]}],
             PlotRange -> All, Frame -> True]] // Timing
```

b) The polynomial $w = 1 + 2z - z^3$ is already in "solved" form. The description is a parametric representation of the solution, z can assume any complex value. The next graphic sketches the part of the amoeba that results from all $|z| < 4$ by displaying 480×120 points. We see long thin spikes emerge.

```
In[1]:= f[z_] := 1 + 2z - z^3;
        log[z_] := Log[Abs[z]];
```

```
In[3]:= (* parameter settings *) R = 4; ppr = 480; ppφ = 120;
        (* avoid messages *) Off[Graphics::gptn];
        gr = Show[Graphics[{PointSize[0.003],
        Table[{Hue[0.78 r/R], Point[{log[N[r Exp[I φ]]], log[f[N[r Exp[I φ]]]]}]},
            {r, 0, R, R/ppr}, {φ, 0, 2Pi, 2Pi/ppφ}]} // N],
            Frame -> True, PlotRange -> All]
```

For an exact description of the amoeba, we first observe that applying log to $|z_k|$ just stretches a figure in $\mathbb{R}^n$ [1784]. So it is enough to get a description of the map $\{z_1, z_2, \ldots, z_n\} \to \{|z_1|, |z_2|, \ldots, |z_n|\}$. Writing $z_k = \mathrm{Re}(z_k) + i\,\mathrm{Im}(z_k)$ yields a quantified expression that can be resolved using `Resolve`. The z parametrizes the solution and is the quantified (complex) variable to be eliminated. (We use the notations $z_1 = x$, and $z_2 = y$.)

```
In[6]:= ComplexExpand[(Re[#]^2 + Im[#]^2)&[1 + 2z - z^3 /.
            z -> xr + I xi]] // Simplify
```

$$Out[6]= xi^6 + xi^4\,(4 + 3\,xr^2) + (-1 - 2\,xr + xr^3)^2 + xi^2\,(4 + 6\,xr + 3\,xr^4)$$

```
In[7]:= AppendTo[$ContextPath, "Experimental`"]
```

$$Out[7]= \{Global`, System`, Experimental`\}$$

```
In[8]:= amoeba =
        Exists[{xr, xi}, Element[{x, y, xr, xi}, Reals],
            (* x parametrizes the solution *)
            x^2 == xr^2 + xi^2 &&
            (* y is the solution *)
            y^2 == xi^6 + xi^4 (4 + 3 xr^2) + (1 + 2 xr - xr^3)^2 +
                xi^2 (4 + 6 xr + 3 xr^4) &&
            (* Abs implies positive x and y *)
            x > 0 && y > 0] // Resolve
```

$$Out[8]= \left(0 < x \le \mathrm{Root}[-2 + 9\,\#1^2 + 8\,\#1^3\,\&,\,1]\,\&\&\,1 - 2x + x^3 \le y \le 1 + 2x - x^3\right) \,||$$

$$\left(\mathrm{Root}[-2 + 9\,\#1^2 + 8\,\#1^3\,\&,\,1] < x < \frac{1}{2}\left(-1 + \sqrt{5}\right)\,\&\&\,1 - 2x + x^3 \le y \le \mathrm{Root}[-5 + 20\,x^4 + 10\,x^6 - \right.$$
$$\left. 20\,x^8 - 20\,x^{10} - 5\,x^{12} - 54\,\#1^2 - 144\,x^2\,\#1^2 - 108\,x^4\,\#1^2 - 22\,x^6\,\#1^2 + 27\,\#1^4\,\&,\,2]\right)\,||$$

$$\left(x == \frac{1}{2}\left(-1 + \sqrt{5}\right)\,\&\&\,1 - 2x + x^3 < y \le \mathrm{Root}[-5 + 20\,x^4 + 10\,x^6 - 20\,x^8 - 20\,x^{10} - \right.$$
$$\left. 5\,x^{12} - 54\,\#1^2 - 144\,x^2\,\#1^2 - 108\,x^4\,\#1^2 - 22\,x^6\,\#1^2 + 27\,\#1^4\,\&,\,2]\right)\,||$$

$$\left(\frac{1}{2}\left(-1 + \sqrt{5}\right) < x < \mathrm{Root}[-1 + \#1^2 + \#1^4\,\&,\,2]\,\&\&\,-1 + 2x - x^3 \le y \le \mathrm{Root}[-5 + 20\,x^4 + 10\,x^6 - \right.$$
$$\left. 20\,x^8 - 20\,x^{10} - 5\,x^{12} - 54\,\#1^2 - 144\,x^2\,\#1^2 - 108\,x^4\,\#1^2 - 22\,x^6\,\#1^2 + 27\,\#1^4\,\&,\,2]\right)\,||$$

$$(x == \mathrm{Root}[-1 + \#1^2 + \#1^4\,\&,\,2]\,\&\&\,-1 + 2x - x^3 \le y \le \mathrm{Root}[-5 + 20\,x^4 + 10\,x^6 - 20\,x^8 -$$
$$20\,x^{10} - 5\,x^{12} - 54\,\#1^2 - 144\,x^2\,\#1^2 - 108\,x^4\,\#1^2 - 22\,x^6\,\#1^2 + 27\,\#1^4\,\&,\,4])\,||$$

$$(\mathrm{Root}[-1 + \#1^2 + \#1^4\,\&,\,2] < x < 1\,\&\&\,-1 + 2x - x^3 \le y \le \mathrm{Root}[-5 + 20\,x^4 + 10\,x^6 - 20\,x^8 -$$
$$20\,x^{10} - 5\,x^{12} - 54\,\#1^2 - 144\,x^2\,\#1^2 - 108\,x^4\,\#1^2 - 22\,x^6\,\#1^2 + 27\,\#1^4\,\&,\,2])\,||$$

$$(x == 1\,\&\&\,0 < y \le \mathrm{Root}[-20 - 328\,\#1^2 + 27\,\#1^4\,\&,\,2])\,||$$

$$\left(1 < x \le \sqrt{2}\,\&\&\,1 - 2x + x^3 \le y \le \mathrm{Root}[-5 + 20\,x^4 + 10\,x^6 - 20\,x^8 - \right.$$

$$20\,x^{10} - 5\,x^{12} - 54\,\#1^2 - 144\,x^2\,\#1^2 - 108\,x^4\,\#1^2 - 22\,x^6\,\#1^2 + 27\,\#1^4\ \&,\ 2]\Big)\ ||$$

$$\Big(\sqrt{2} < x < \tfrac{1}{2}\left(1 + \sqrt{5}\right) \&\&\ 1 + 2\,x - x^3 \le y \le \mathrm{Root}\,[-5 + 20\,x^4 + 10\,x^6 - 20\,x^8 - 20\,x^{10} -$$

$$5\,x^{12} - 54\,\#1^2 - 144\,x^2\,\#1^2 - 108\,x^4\,\#1^2 - 22\,x^6\,\#1^2 + 27\,\#1^4\ \&,\ 2]\Big)\ ||$$

$$\Big(x == \tfrac{1}{2}\left(1 + \sqrt{5}\right) \&\&\ 1 + 2\,x - x^3 < y \le \mathrm{Root}\,[-5 + 20\,x^4 + 10\,x^6 - 20\,x^8 - 20\,x^{10} -$$

$$5\,x^{12} - 54\,\#1^2 - 144\,x^2\,\#1^2 - 108\,x^4\,\#1^2 - 22\,x^6\,\#1^2 + 27\,\#1^4\ \&,\ 2]\Big)\ ||$$

$$\Big(x > \tfrac{1}{2}\left(1 + \sqrt{5}\right) \&\&\ -1 - 2\,x + x^3 \le y \le \mathrm{Root}\,[-5 + 20\,x^4 + 10\,x^6 - 20\,x^8 - 20\,x^{10} -$$

$$5\,x^{12} - 54\,\#1^2 - 144\,x^2\,\#1^2 - 108\,x^4\,\#1^2 - 22\,x^6\,\#1^2 + 27\,\#1^4\ \&,\ 2]\Big)$$

The following contour plot shows the resulting region in the x,y-plane.

```
In[9]:= cp = ContourPlot[Evaluate[If[amoeba, -1, 1]], {x, 0, 3}, {y, 0, 10},
        PlotPoints -> 200, Contours -> {0}, ContourLines -> False]
```

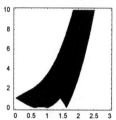

Using the cylindrical description amoeba, we can discretize the boundary and apply the (nonfolding) transformation log.

```
In[10]:= plot[Inequality[xl_, Less, x, Less, xu_] &&
         Inequality[yl_, Less, y, Less, yu_], pp_] :=
       With[{ε = N[10^-40, 60]}, (* make points along boundaries *)
       Polygon[((#2[Table[Log @ {x, #1},
                   {x, xl + ε, xu - ε, (xu - xl - 2ε)/pp}]]) & @@@
           {{yl, Identity}, {yu, Reverse}}) // (Join @@ #)&]];
```

Here is the resulting picture of the amoeba.

```
In[11]:= Show[{Graphics[plot[#, 120]& /@
         (List @@ (DeleteCases[amoeba, x == _ && _] /. LessEqual -> Less /.
                   x > xl_ :> Inequality[xl, Less, x, Less, 4xl]))], gr} // N,
         Frame -> True, PlotRange -> {{-5, 2}, {-5, 5}},
         AspectRatio -> Automatic]
```

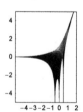

c) We start by sketching the tile to be constructed. For any concrete f, only a finite amount of lattice points have to be taken into account to bound the tile completely. The following graphics suggest that taking 48 neighboring lattice points into account is sufficient.

```
In[1]:= f[{x_, y_}] = (3 x + y) (x + 3 y);
```

```
In[2]:= (* take (2 n + 1)^2 - 1 neighbor lattice points into account *)
       conds[n_] :=
       And @@ Flatten[Table[If[i == j == 0, {},
                       Abs[f[{x, y}]] < Abs[f[{x + i, y + j}]]],
                       {i, -n, n}, {j, -n, n}]];
```

```
In[4]:= Show[GraphicsArray[Table[
          ContourPlot[Evaluate[If[conds[k], -1, 1]], {x, -4, 4}, {y, -4, 4},
              Contours -> {0}, PlotPoints -> 240,
              DisplayFunction -> Identity], {k, 2, 4}]]]
```

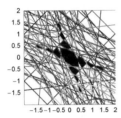

The next graphic shows the tile from conds[3] together with the 48 corresponding contours $|f(x, y)| = |f(x + i, y + j)|$.

```
In[5]:= Show[{%[[1, 2]], (* the 48 bounding curves *) Table[
          ContourPlot[Evaluate[Subtract @@ conds[3][[k]]], {x, -2, 2}, {y, -2, 2},
              Contours -> {0}, PlotPoints -> 240,
              DisplayFunction -> Identity, ContourShading -> False,
              ContourStyle -> {{Hue[Random[]]}}], {k, Length[conds[3]]}]},
          DisplayFunction -> $DisplayFunction,
          PlotRange -> {{-2, 2}, {-2, 2}}]
```

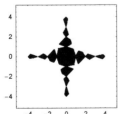

The symmetry $f(x, y) = f(y, x)$ suggests changing the coordinate system to simplify the problem of the construction of an exact description of the tile. The obvious change is $\xi = 3x + y$, and $\eta = x + 3y$.

```
In[6]:= Solve[{3x + y == ξ, x + 3y == η}, {x, y}]
```

$$Out[6]= \left\{\left\{x \to \frac{1}{8}\,(-\eta + 3\,\xi),\; y \to \frac{1}{8}\,(3\,\eta - \xi)\right\}\right\}$$

```
In[7]:= condsξη[n_] := condsξη[n] =
              conds[n] //. {x -> (3ξ - η)/8, y -> (3η - ξ)/8};
```

Here is the resulting tile. The twofold mirror symmetry of the resulting figure suggests finding an exact description of the tile for the region $\xi \geq 0$, $-\xi \leq \eta \leq \xi$. The other three congruent parts can than easily obtained by mirroring and rotating.

```
In[8]:= ContourPlot[Evaluate[If[condsξη[3], -1, 1]], {ξ, -5, 5}, {η, -5, 5},
          Contours -> {0}, PlotPoints -> 240]
```

To find an exact description of the tile in the ξ,η-coordinates means nothing else other than carrying out a decomposition of $\mathbb{R}^2$ with respect to the 48 inequalities condsξη[3]. Because we are not interested in lower-dimensional subparts, we will use the function CylindricalDecomposition.

```
In[9]:= (* initial two inequalities and symmetry restrictions *)
        gcad[2] = CylindricalDecomposition[
          And @@ Flatten[{{ξ > 0, -ξ < η < ξ, Take[condsξη[3], 2]}}], {ξ, η}];
In[11]:= (* use one inequality after the other *)
        Do[gcad[k] = CylindricalDecomposition[
                        And @@ {gcad[k - 1], condsξη[3][[k]]}, {ξ, η}],
          {k, 3, Length[condsξη[3]]}]
```

The following plot of the sizes of the intermediate decompositions shows that about 15 of the inequalities border the tile.

```
In[13]:= ListPlot[Table[ByteCount[gcad[k]], {k, 2, 48}],
                  PlotRange -> All, PlotJoined -> True]
```

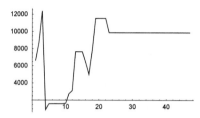

Here is the final decomposition.

```
In[14]:= gcad[48]
```

$$
\text{Out[14]= } \left(0 < \xi \le \frac{3}{4} \text{ && } (-\xi < \eta < 0 \,||\, 0 < \eta < \xi)\right) \,||\, \left(\frac{3}{4} < \xi \le \frac{1}{2} \left(-1 + \sqrt{7}\right) \text{ && } (-\xi < \eta < 0 \,||\, 0 < \eta < 3 - 3\,\xi)\right) \,||\,
$$

$$
\left(\frac{1}{2} \left(-1 + \sqrt{7}\right) < \xi < 1 \text{ && } \left(\frac{-3 + 3\,\xi}{-1 + 2\,\xi} < \eta < 0 \,||\, 0 < \eta < 3 - 3\,\xi\right)\right) \,||\,
$$

$$
\left(1 < \xi \le \frac{5}{4} \text{ && } \left(3 - 3\,\xi < \eta < 0 \,||\, 0 < \eta < \frac{-3 + 3\,\xi}{-1 + 2\,\xi}\right)\right) \,||\,
$$

$$
\left(\frac{5}{4} < \xi \le \frac{1}{2} \left(-1 + \sqrt{13}\right) \text{ && } \left(-2 + \xi < \eta < 0 \,||\, 0 < \eta < \frac{-3 + 3\,\xi}{-1 + 2\,\xi}\right)\right) \,||\,
$$

$$
\left(\frac{1}{2} \left(-1 + \sqrt{13}\right) < \xi \le \frac{1}{2} \left(7 - \sqrt{13}\right) \text{ && } \left(-2 + \xi < \eta < 0 \,||\, 0 < \eta < \frac{3 - \xi}{3}\right)\right) \,||\,
$$

$$
\left(\frac{1}{2} \left(7 - \sqrt{13}\right) < \xi < 2 \text{ && } \left(-2 + \xi < \eta < 0 \,||\, 0 < \eta < \frac{2 - \xi}{-1 + \xi}\right)\right) \,||\,
$$

$$
\left(2 < \xi \le \frac{9}{4} \text{ && } \left(\frac{2 - \xi}{-1 + \xi} < \eta < 0 \,||\, 0 < \eta < -2 + \xi\right)\right) \,||\,
$$

$$
\left(\frac{9}{4} < \xi \le \frac{1}{6} \left(11 + \sqrt{13}\right) \text{ && } \left(\frac{2 - \xi}{-1 + \xi} < \eta < 0 \,||\, 0 < \eta < \frac{3 - \xi}{3}\right)\right) \,||\,
$$

$$
\left(\frac{1}{6} \left(11 + \sqrt{13}\right) < \xi < 3 \text{ && } \left(\frac{-3 + \xi}{-3 + 2\,\xi} < \eta < 0 \,||\, 0 < \eta < \frac{3 - \xi}{3}\right)\right) \,||\,
$$

$$
\left(3 < \xi \le \frac{15}{4} \text{ && } \left(\frac{3 - \xi}{3} < \eta < 0 \,||\, 0 < \eta < \frac{-3 + \xi}{-3 + 2\,\xi}\right)\right) \,||\,
$$

$$
\left(\frac{15}{4} < \xi \le \frac{1}{2} \left(5 + \sqrt{7}\right) \text{ && } \left(-4 + \xi < \eta < 0 \,||\, 0 < \eta < \frac{-3 + \xi}{-3 + 2\,\xi}\right)\right) \,||\,
$$

$$
\left(\frac{1}{2} \left(5 + \sqrt{7}\right) < \xi < 4 \text{ && } (-4 + \xi < \eta < 0 \,||\, 0 < \eta < 4 - \xi)\right)
$$

For more convenient further manipulations, we expand the obtained decomposition.

```
In[15]:= tileD1 = DeleteCases[LogicalExpand[gcad[48]], (ξ == _) && _ ] /.
                                                      LessEqual -> Less;
```

Rewriting the bounds on the variables ξ and η in tileD1 in a form appropriate for Integrate allows obtaining an exact result for the area of the tile.

```
In[16]:= toIntegrationDomain[l_] :=
           {{Cases[l, _?(FreeQ[#, η]&) < ξ], Cases[l, ξ < _]},
            {Cases[l, _ < η], Cases[l, η < _]}}
In[17]:= iRanges = {{{ξ, #[[1, 1, 1, 1]], #[[1, 2, 1, 2]]},
                      {η, #[[2, 1, 1, 1]], #[[2, 2, 1, 2]]}}}& /@
                    toIntegrationDomain /@ (List @@ tileD1);
```

Here are the first few integration domains.

In[18]:= **Take[iRanges, 6]**

Out[18]= $\{\{\{\xi, 0, \frac{3}{4}\}, \{\eta, 0, \xi\}\}, \{\{\xi, \frac{3}{4}, \frac{1}{2}(-1+\sqrt{7})\}, \{\eta, 0, 3-3\xi\}\},$
$\{\{\xi, 1, \frac{5}{4}\}, \{\eta, 0, \frac{-3+3\xi}{-1+2\xi}\}\}, \{\{\xi, \frac{5}{4}, \frac{1}{2}(-1+\sqrt{13})\}, \{\eta, 0, \frac{-3+3\xi}{-1+2\xi}\}\},$
$\{\{\xi, 2, \frac{9}{4}\}, \{\eta, 0, -2+\xi\}\}, \{\{\xi, \frac{9}{4}, \frac{1}{6}(11+\sqrt{13})\}, \{\eta, 0, \frac{3-\xi}{3}\}\}\}$

And here is the area of the whole tile in ξ,η-coordinates.

In[19]:= **Expand[4 Plus @@ (Integrate[1, ##]& @@@ iRanges)]**

Out[19]= $8 - 4 \, \text{Log}\left[\frac{5}{4}\right] - \frac{3}{2} \, \text{Log}\left[\frac{9}{4}\right] + \frac{1}{2} \, \text{Log}\left[\frac{531441}{64}\right] + \frac{\text{Log}[387420489]}{3} - 3 \, \text{Log}[-2+\sqrt{7}] - $
$3 \, \text{Log}[2+\sqrt{7}] + 4 \, \text{Log}\left[-\frac{2}{-5+\sqrt{13}}\right] - 3 \, \text{Log}[-2+\sqrt{13}] - 3 \, \text{Log}[2+\sqrt{13}] + 4 \, \text{Log}\left[\frac{15}{2 \, (5+\sqrt{13})}\right]$

Simplifying the result yields the area 8 in the ξ,η-plane.

In[20]:= **Simplify[% //. a_. Log[α_] + b_. Log[β_] :> Log[α^a β^b]]**

Out[20]= 8

Taking into account the value $1/8$ of the functional determinant from changing variables from x,y to ξ,η shows that the tile in the x,y-plane has area 1—as it has to be for a tile of a square lattice.

In[21]:= **Det @ Outer[D, {(3ξ - η)/8, (3η - ξ)/8}, {ξ, η}]**

Out[21]= $\frac{1}{8}$

For a graphical representation of the tile, we sort the integration regions with respect to ξ and we separate the lower and upper boundaries.

In[22]:= **ξηRegions = Sort[iRanges, (#1[[1, 2]] < #2[[1, 2]])&];**

The two functions upperCurve[ξ] and lowerCurve[ξ] are the upper and lower bounding curves of the quarter of the tile. quarterPoly is a polygon representing the quarter of the tile.

In[23]:= **{upperCurve[ξ_], lowerCurve[ξ_]} = Function[{r, p},**
  Which @@ Flatten[(* ranges and ξ-dependent functions *)
    {#1[[1, 2]] <= #1[[1, 1]] <= #1[[1, 3]], #1[[2, p]]}& /@
      Cases[ξηRegions, {_, r}]]] @@@ {{{η, 0, _}, 3}, {{η, _, 0}, 2}};

In[24]:= **quarterPoly = With[{pp = 12}, N @**
  Polygon[Join[Flatten[#1, 1], Reverse[Flatten[#2, 1]]]]& @@
  (Function[{r, p}, (* ranges and ξ-dependent functions *)
    Table[{ξ, #[[2, p]]} /. ξ -> #[[1, 2]] + k/pp (#[[1, 3]] - #[[1, 2]]),
      {k, 0, pp}]& /@ Cases[ξηRegions, {_, r}]] @@@
        {{{η, 0, _}, 3}, {{η, _, 0}, 2}})];

Here is the quarter of the tile and its boundary.

In[25]:= **Show[{Graphics[quarterPoly],**
  Plot[{upperCurve[ξ], lowerCurve[ξ]}, {ξ, 0, 4},
    PlotStyle -> {{Hue[0], Thickness[0.01]}},
    DisplayFunction -> Identity]},
  Frame -> True, PlotRange -> All]

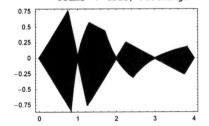

Mapping now back to the x,y-plane yields the following polygonal description of the tile.

```
In[26]:= fullPoly = {quarterPoly, Map[{-1, -1}#&, quarterPoly, {-2}],
                Map[Reverse[#]&, quarterPoly, {-2}],
                Map[{-1, -1}Reverse[#]&, quarterPoly, {-2}]};

In[27]:= tile = Apply[{3#1 - #2, -#1 + 3 #2}/8&, fullPoly, {-2}];

In[28]:= Show[Graphics[tile], Frame -> True, PlotRange -> All,
            AspectRatio -> Automatic]
```

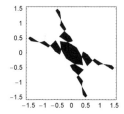

We end with a graphic of a randomly colored array of the tile. Each individual tile is centered at the lattice points $\{i, j\}$ of the square lattice.

```
In[29]:= With[{o = 5},
        Show[Graphics[
        Table[{{Hue[Random[]], T = Map[({i, j} + #)&, tile, {-2}]},
            {GrayLevel[0], Thickness[0.002], T /. Polygon -> Line}},
            {i, -o, o}, {j, -o, o}]],
            Frame -> False, PlotRange -> (o - 2) {{-1, 1}, {-1, 1}},
            AspectRatio -> Automatic]]
```

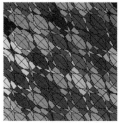

35. Cartesian Leaf Area, Triple Integral, Average Distance

a) Here, the area under consideration is shown.

```
In[1]:= ContourPlot[x^3 + y^3 - 3 x y, {x, 0, 2}, {y, 0, 2},
            Contours -> {0}, PlotPoints -> 40]
```

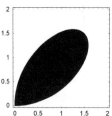

Trying to calculate the integral directly as a double integral using `Boole` to specify the region, fails.

```
In[2]:= Integrate[Boole[x^3 + y^3 - 3 x y < 0],
            {x, 0, Infinity}, {y, 0, Infinity}]
```

$$Out[2]= \int_0^{\text{Root}[-4+\#1^3\&,1]} (\text{Root}[x^3 - 3 x \#1 + \#1^3 \&, 1] + 2\,\text{Root}[x^3 - 3 x \#1 + \#1^3 \&, 3])\,dx$$

Piecewise, we can uniquely solve for $y = y(x)$.

In[3]:= **sol = Solve[x^3 + y^3 - 3 x y == 0, y]**

Out[3]= $\left\{\left\{y \to \dfrac{2^{1/3} x}{\left(-x^3 + \sqrt{-4 x^3 + x^6}\right)^{1/3}} + \dfrac{\left(-x^3 + \sqrt{-4 x^3 + x^6}\right)^{1/3}}{2^{1/3}}\right\},\right.$

$\left\{y \to -\dfrac{(1 + i \sqrt{3}) x}{2^{2/3} \left(-x^3 + \sqrt{-4 x^3 + x^6}\right)^{1/3}} - \dfrac{(1 - i \sqrt{3}) \left(-x^3 + \sqrt{-4 x^3 + x^6}\right)^{1/3}}{2 \cdot 2^{1/3}}\right\},$

$\left.\left\{y \to -\dfrac{(1 - i \sqrt{3}) x}{2^{2/3} \left(-x^3 + \sqrt{-4 x^3 + x^6}\right)^{1/3}} - \dfrac{(1 + i \sqrt{3}) \left(-x^3 + \sqrt{-4 x^3 + x^6}\right)^{1/3}}{2 \cdot 2^{1/3}}\right\}\right\}$

Here, the three solutions are shown.

In[4]:= **(* avoid messages from Plot due to non-real values *)**
Off[Plot::plnr];

Show[GraphicsArray[Plot[Evaluate[y /. #], {x, 0, 2},
  PlotRange -> All, DisplayFunction -> Identity]& /@ sol]]

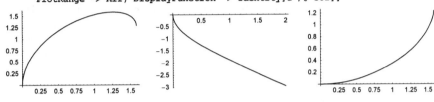

We calculate the "rightmost point" (meaning where $1/y'(x)$ vanishes).

In[7]:= **Solve[D[x^3 + y[x]^3 - 3 x y[x], x] == 0, y'[x]]**

Out[7]= $\left\{\left\{y'[x] \to \dfrac{x^2 - y[x]}{x - y[x]^2}\right\}\right\}$

In[8]:= **Select[Solve[{x == y^2, x^3 + y^3 - 3 x y == 0}, {x, y}],**
  Im[x /. #] == 0 && Im[y /. #] == 0 &&
  (x /. #1) > 0 && (y /. #1) > 0&]

Out[8]= $\{\{x \to 2^{2/3}, y \to 2^{1/3}\}\}$

We could now continue and carefully integrate between the solutions 1 and 3 to obtain the area.

Another possibility is to calculate the integral in polar coordinates:

$$\int_0^\infty \int_0^\infty \theta(x^3 + y^3 - 3 x y)\, dx\, dy = \int_0^{\pi/2} \int_0^{r(\varphi)} r\, dr\, d\varphi = \int_0^{\pi/2} \frac{r(\varphi)^2}{2}\, d\varphi.$$

The next input calculates $r(\varphi)$.

In[9]:= **Solve[((x^3 + y^3 - 3 x y /.**
  {x -> r Cos[φ], y -> r Sin[φ]}) == 0, r]

Out[9]= $\left\{\{r \to 0\}, \{r \to 0\}, \left\{r \to \dfrac{3 \cos[\varphi] \sin[\varphi]}{\cos[\varphi]^3 + \sin[\varphi]^3}\right\}\right\}$

The resulting 1D integral over φ is easily carried out. We obtain the area $3/2$.

In[10]:= **Integrate[((3 Cos[φ] Sin[φ])/(Cos[φ]^3 + Sin[φ]^3))^2/2,**
  {φ, 0, Pi/2}]

Out[10]= $\dfrac{3}{2}$

b) In Cartesian coordinates, *Mathematica* cannot carry out the three definite integrations symbolically.

In[1]:= **Integrate[(1 + u^2 + v^2 + w^2)^-2, {u, 0, 1}, {v, 0, 1}, {w, 0, 1}]**

Out[1]= $\displaystyle\int_0^1 \dfrac{\text{ArcCot}\left[\sqrt{2 + u^2}\right]}{(1 + u^2) \sqrt{2 + u^2}}\, du$

Of course, we can calculate a numerical approximation of the remaining 1D integral.

In[2]:= **N[%]**

Out[2]= 0.308425

In[3]:= **NIntegrate[(1 + x^2 + y^2 + z^2)^-2, {x, 0, 1}, {y, 0, 1}, {z, 0, 1},**
PrecisionGoal -> 8]

Out[3]= 0.308425

Integrals of this form often are rational multiples (of small denominator) of powers of π [1242]. This seems to be true also for this integral. The next input suggests the conjectured value $\pi^2/32$ for the integral.

In[4]:= **Rationalize[%/Pi^Range[4], 10^-12]**

Out[4]= $\{ \dfrac{58607}{596966}, \dfrac{1}{32}, \dfrac{11400}{1146053}, \dfrac{2670}{843259} \}$

To confirm our conjecture $\pi^2/32$, we use a spherical coordinate system to calculate the triple integral. Making also use of the symmetry of the integrand, we have the following identity (see Exercise 9 of Chapter 3 of the Graphics volume [1736]). The limits for the r-integration follow from the plane $x = 1$, and the limit for the ϑ-integration follows from the line $z = y$ in the plane $x = 1$.

$$\int_0^1 \int_0^1 \int_0^1 f(x, y, z)\, dx\, dy\, dz =$$
$$6 \int_0^{\frac{\pi}{4}} \int_{\arctan(1/\sin(\varphi))}^{\frac{\pi}{2}} \int_0^{1/(\cos(\varphi)\sin(\vartheta))} f(r\cos(\varphi)\sin(\vartheta),\ r\sin(\varphi)\sin(\vartheta),\ r\cos(\vartheta))\, dr\, d\vartheta\, d\varphi$$

Here is the numerical integration carried out in spherical coordinates.

In[5]:= **NIntegrate[(1 + r^2)^-2 r^2 Sin[θ],**
{φ, 0, Pi/4}, {θ, ArcTan[1/Sin[φ]], Pi/2},
{r, 0, 1/(Cos[φ] Sin[θ])}] 6

Out[5]= 0.308425

The integration over r results in a relatively simple expression.

In[6]:= **int1 = Integrate[6 (1 + r^2)^-2 r^2 Sin[θ],**
{r, 0, 1/(Cos[φ] Sin[θ])}] // Simplify

Out[6]= $\dfrac{3\,\mathrm{Sec}[\varphi]\,(-1 + \mathrm{ArcTan}[\mathrm{Csc}[\theta]\,\mathrm{Sec}[\varphi]]\,(\mathrm{Csc}[\theta]\,\mathrm{Sec}[\varphi] + \mathrm{Cos}[\varphi]\,\mathrm{Sin}[\theta]))}{1 + \mathrm{Csc}[\theta]^2\,\mathrm{Sec}[\varphi]^2}$

The definite integration over ϑ can be carried out directly. Due to the presence of the symbolic parameter φ, *Mathematica* the result is quite lengthy. Because we will have to integrate the resulting expression over φ, we will not use this result.

In[7]:= **(int2 = Integrate[int1, {θ, ArcTan[1/Sin[φ]], Pi/2},**
GenerateConditions -> False]) // Short[#, 2]&

Out[7]//Short= $3\ (\ll 1\gg)\ \ll 1\gg$

$$\left| \dfrac{i\sqrt{2}\ \ll 1\gg\ \ll 1\gg\ \sqrt{\ll 1\gg}\ \left(\dfrac{\mathrm{ArcTan}\left[\tfrac{1}{\sqrt{\ll 1\gg}}\right]\,(-2-\ll 1\gg + \ll 1\gg)}{\sqrt{4+2\,\ll 1\gg -2\sqrt{2}\,\sqrt{\ll 1\gg^2\,(3+\ll 1\gg)}}} + \dfrac{\ll 1\gg}{\sqrt{\ll 1\gg}} \right)}{(\pi - 2\,\mathrm{ArcTan}[\mathrm{Csc}[\varphi]])\,\left(1 + 2\,i\,\sqrt{\ll 1\gg^2} + \mathrm{Cos}[2\,\varphi]\right)\,(\sqrt{2} + i\,\sqrt{1 + \mathrm{Cos}[2\,\varphi]})} - \dfrac{\ll 1\gg}{\ll 1\gg} \right|$$

But the indefinite integral can be carried out too.

In[8]:= **int2Indef = Integrate[int1, θ] // Simplify[#, 0 < φ < Pi/4]&**

$$\text{Out[8]= } 3\left(-\text{ArcTan}[\text{Csc}[\vartheta]\,\text{Sec}[\varphi]]\,\text{Cos}[\vartheta]-\frac{1}{\sqrt{2}\,\sqrt{3+\text{Cos}[2\,\varphi]}}\right.$$

$$\left(\text{Cos}[\varphi]\left(\frac{\text{ArcTan}\left[\frac{\text{Tan}\left[\frac{\vartheta}{2}\right]}{\sqrt{2+\text{Cos}[2\,\varphi]-\sqrt{2}\,\text{Cos}[\varphi]\,\sqrt{3+\text{Cos}[2\,\varphi]}}}\right](-2\,\text{Cos}[\varphi]+\sqrt{2}\,\sqrt{3+\text{Cos}[2\,\varphi]})}{\sqrt{2+\text{Cos}[2\,\varphi]}-\sqrt{2}\,\text{Cos}[\varphi]\,\sqrt{3+\text{Cos}[2\,\varphi]}}+\right.\right.$$

$$\left.\left.\left.\frac{\text{ArcTan}\left[\frac{\text{Tan}\left[\frac{\vartheta}{2}\right]}{\sqrt{2+\text{Cos}[2\,\varphi]+\sqrt{2}\,\text{Cos}[\varphi]\,\sqrt{3+\text{Cos}[2\,\varphi]}}}\right](2\,\text{Cos}[\varphi]+\sqrt{2}\,\sqrt{3+\text{Cos}[2\,\varphi]})}{\sqrt{2+\text{Cos}[2\,\varphi]}+\sqrt{2}\,\text{Cos}[\varphi]\,\sqrt{3+\text{Cos}[2\,\varphi]}}\right)\right)\right)$$

To get shorter expressions, we rewrite the last result by using a stripped down version of the addition theorem for the arctan function (stripped down because we ignore piecewise present multiples of π—for the definite integral, they do not matter as long as we do not pass a branch cut with the integration path). We could also continue with the above form of int2Indef and carry out all further calculations, but some intermediate expressions would be much larger.

In[9]:= int2Indef = int2Indef /. a_ ArcTan[x_] - a_ ArcTan[y_] :>
                     a ArcTan[(x - y)/(1 + x y)] // Simplify

$$\text{Out[9]= } 3\left(-\text{ArcTan}[\text{Csc}[\vartheta]\,\text{Sec}[\varphi]]\,\text{Cos}[\vartheta]-\frac{1}{\sqrt{2}\,\sqrt{3+\text{Cos}[2\,\varphi]}}\right.$$

$$\left(\text{Cos}[\varphi]\left(\frac{\text{ArcTan}\left[\frac{\text{Tan}\left[\frac{\vartheta}{2}\right]}{\sqrt{2+\text{Cos}[2\,\varphi]-\sqrt{2}\,\text{Cos}[\varphi]\,\sqrt{3+\text{Cos}[2\,\varphi]}}}\right](-2\,\text{Cos}[\varphi]+\sqrt{2}\,\sqrt{3+\text{Cos}[2\,\varphi]})}{\sqrt{2+\text{Cos}[2\,\varphi]}-\sqrt{2}\,\text{Cos}[\varphi]\,\sqrt{3+\text{Cos}[2\,\varphi]}}+\right.\right.$$

$$\left.\left.\left.\frac{\text{ArcTan}\left[\frac{\text{Tan}\left[\frac{\vartheta}{2}\right]}{\sqrt{2+\text{Cos}[2\,\varphi]+\sqrt{2}\,\text{Cos}[\varphi]\,\sqrt{3+\text{Cos}[2\,\varphi]}}}\right](2\,\text{Cos}[\varphi]+\sqrt{2}\,\sqrt{3+\text{Cos}[2\,\varphi]})}{\sqrt{2+\text{Cos}[2\,\varphi]}+\sqrt{2}\,\text{Cos}[\varphi]\,\sqrt{3+\text{Cos}[2\,\varphi]}}\right)\right)\right)$$

We rewrite the last result in a slightly more condensed form.

In[10]:= int2Indef = -3 ArcTan[Csc[θ] Sec[φ]] Cos[θ] -
              (3 Sqrt[2] ArcTanh[(Sqrt[-3 - Cos[2 φ]] Tan[θ])/Sqrt[2]] Cos[φ])/
                     Sqrt[-3 - Cos[2 φ]];

In[11]:= D[int2Indef, θ] - int1 // Simplify

Out[11]= 0

This indefinite integral int2Indef is a continuous function of φ and ϑ over the integration domain.

In[12]:= With[{ε = 10^-6}, Show[GraphicsArray[
            ParametricPlot3D[{φ, (* map from rectangular plotparameter domains *)
                    τ ArcTan[1/Sin[φ]], #[int2Indef] /.
                      θ -> τ ArcTan[1/Sin[φ]]},
                 {φ, ε, 2Pi/4 - ε}, {τ, ε, 1 - ε},
                 DisplayFunction -> Identity,
                 PlotRange -> All, PlotLabel -> #]& /@
                 (* show real and imaginary part *) {Re, Im}]]]

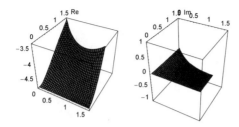

The value of int2Indef at $\vartheta = \pi/2$ cannot be determined by substitution; we need to take the limit "by hand".

In[13]:= **int2Upper = Assuming[0 < φ < Pi/4,**
                      Limit[int2Indef, ϑ -> Pi/2, Direction -> +1]]

Out[13]= $-\dfrac{3\,\pi\,\text{Cos}[\varphi]}{\sqrt{2}\,\sqrt{3 + \text{Cos}[2\,\varphi]}}$

The lower limit is easy to calculate.

In[14]:= **int2Lower = int2Indef /. ϑ -> ArcTan[1/Sin[φ]] //**
                              Simplify[#, 0 < φ < Pi/4]&

Out[14]= $-\dfrac{3\,\sqrt{2}\,\text{ArcTan}\left[\frac{\sqrt{3+\text{Cos}[2\,\varphi]}\,\text{Csc}[\varphi]}{\sqrt{2}}\right]\,\text{Cos}[\varphi]}{\sqrt{3 + \text{Cos}[2\,\varphi]}} - \dfrac{3\,\text{ArcTan}\left[\sqrt{1 + \text{Csc}[\varphi]^2}\,\text{Tan}[\varphi]\right]}{\sqrt{1 + \text{Csc}[\varphi]^2}}$

Carrying out the remaining definite integral over φ confirms our conjecture that the value of the integral is $\pi^2/32$.

In[15]:= **FullSimplify[int2Upper - int2Lower, 0 < φ < Pi/4]**

Out[15]= $-\dfrac{3\,\pi\,\text{Cos}[\varphi]}{\sqrt{2}\,\sqrt{3 + \text{Cos}[2\,\varphi]}} + \dfrac{3\,\sqrt{2}\,\text{ArcTan}\left[\frac{\sqrt{3+\text{Cos}[2\,\varphi]}\,\text{Csc}[\varphi]}{\sqrt{2}}\right]\,\text{Cos}[\varphi]}{\sqrt{3 + \text{Cos}[2\,\varphi]}} + \dfrac{3\,\text{ArcTan}\left[\sqrt{1 + \text{Csc}[\varphi]^2}\,\text{Tan}[\varphi]\right]}{\sqrt{1 + \text{Csc}[\varphi]^2}}$

In[16]:= **Integrate[(* slightly simplified form of the integrand *)**
                  (3 Cos[φ] ((2 ArcTanh[(Sqrt[-3 - Cos[2 φ]] Csc[φ])/Sqrt[2]])/
                  Sqrt[-3 - Cos[2 φ]] - Pi/Sqrt[3 + Cos[2 φ]]))/Sqrt[2] +
                  (3 ArcTan[Sqrt[1 + Csc[φ]^2] Tan[φ]])/Sqrt[1 + Csc[φ]^2],
                  {φ, 0, Pi/4}]

Out[16]= $\dfrac{\pi^2}{32}$

c) A direct attempt to evaluate the integral fails (not unexpectedly, else it would not be an exercise in the *GuideBooks*).

In[1]:= **integrand = (x^2 + x y + y^2)^v Exp[-(x + y)];**

       Integrate[integrand, {x, 0, Infinity}, {y, 0, Infinity}]

Out[2]= $\displaystyle\int_0^\infty \int_0^\infty e^{-x-y}\,(x^2 + x\,y + y^2)^v\,dy\,dx$

An attempt to evaluate the integral in polar coordinates works.

In[3]:= **intPolar = integrand /. {x -> r Cos[φ], y -> r Sin[φ]} // Simplify**

Out[3]= $e^{-r\,(\text{Cos}[\varphi]+\text{Sin}[\varphi])}\,(r^2\,(1 + \text{Cos}[\varphi]\,\text{Sin}[\varphi]))^v$

In[4]:= **Integrate[intPolar r, {r, 0, Infinity}, {φ, 0, Pi/2},**
                  Assumptions :> Re[v] > -1]

Out[4]= $\dfrac{1}{(-i+\sqrt{3})\,(1+2\,v)}\left(\left(-i-\sqrt{3}\right)^{-v}\left(-\dfrac{i}{i+\sqrt{3}}\right)^{-v}\text{Gamma}[2\,(1+v)]\right.$

$\left(\left(-\dfrac{1+i\,\sqrt{3}}{i+\sqrt{3}}\right)^v\,(i+\sqrt{3})\,\text{Hypergeometric2F1}\left[-1-2\,v,\,-v,\,-2\,v,\,\dfrac{2\,\sqrt{3}}{-i+\sqrt{3}}\right]-\right.$

$\left.\left.2\,i\left(\dfrac{1-i\,\sqrt{3}}{-i+\sqrt{3}}\right)^v\,\text{Hypergeometric2F1}\left[-1-2\,v,\,-v,\,-2\,v,\,\dfrac{2\,\sqrt{3}}{i+\sqrt{3}}\right]\right)\right)$

The result for this manifestly real integral contains i. We try another approach to obtain an explicitly real result. The structure of the integrand strongly suggests using the new variable $u = x + y$. A natural companion of this variable is $v = x - y$. This is the integrand after the change of variables.

```
In[5]:= intpm = (integrand /. Solve[{u == x + y, v == x - y}, {x, y}][[1]])/
            (* functional determinant from change of variables *)
            (Outer[D, {x + y, x - y}, {x, y}] // Det // Abs) //
                                                                Factor //@ #&
```

$$Out[5]= 2^{-1-2\,v}\, e^{-u}\, (3\, u^2 + v^2)^{v}$$

A moment of reflection shows that the new limits are $u : 0 \ldots \infty$, $v : -u \ldots u$. Now, we obtain an explicit result for the integral in terms of a product of a Gamma function and a Gauss hypergeometric function.

```
In[6]:= Integrate[intpm, {u, 0, Infinity}, {v, -u, u}]
```

$$Out[6]= \left(\frac{3}{4}\right)^{v} \text{Hypergeometric2F1}\left[\frac{1}{2}, -v, \frac{3}{2}, -\frac{1}{3}\right]$$
$$\text{If}[\text{Re}[v] > -1, \text{Gamma}[2 + 2\,v], \text{Integrate}[e^{-u}\, u^{1+2\,v}, \{u, 0, \infty\}, \text{Assumptions} \to \text{Re}[v] \le -1]]$$

```
In[7]:= I[v_] = 3^v 2^(-2v) Gamma[2v + 2] Hypergeometric2F1[1/2, -v, 3/2, -1/3]
```

$$Out[7]= 2^{-2\,v}\, 3^{v}\, \text{Gamma}[2 + 2\,v]\, \text{Hypergeometric2F1}\left[\frac{1}{2}, -v, \frac{3}{2}, -\frac{1}{3}\right]$$

Here is a quick check of the result for integer v. In this case, the direct calculation of the double integral in Cartesian coordinates is straightforward.

```
In[8]:= Table[Integrate[integrand, {x, 0, Infinity}, {y, 0, Infinity}] -
            I[v], {v, 0, 12}]
```

$$Out[8]= \{0, 0, 0, 0, 0, 0, 0, 0, 0, 0, 0, 0, 0\}$$

d) The mean distance between two points in a d-dimensional unit cube is

$$\int_0^1 dx_1^{(1)} \int_0^1 dx_2^{(1)} \cdots \int_0^1 dx_1^{(d)} \int_0^1 dx_2^{(d)} \left((x_1^{(1)} - x_2^{(1)})^2 + \cdots + (x_1^{(d)} - x_2^{(d)})^2\right)^{1/2}.$$

In general, the $2 \times d$-dimensional integral can be reduced to a d-dimensional one. The squares of the differences can be converted to a new variable with probability density $2(1 - l)$. (The density $2(1 - l)$ follows immediately from elementary geometric considerations.) This reduces by half the number of integrations that have to be carried out. For the mean distance in the unit square, we have the following by using the probability densities for the projections on the coordinate axes.

```
In[1]:= Integrate[Sqrt[x^2 + y^2] 2(1 - x) 2(1 - y), {x, 0, 1}, {y, 0, 1}]
```

$$Out[1]= \frac{1}{15}\left(2 + \sqrt{2} + 5\, \text{ArcSinh}[1]\right)$$

The result can still be simplified somewhat.

```
In[2]:= % // FullSimplify
```

$$Out[2]= \frac{1}{15}\left(2 + \sqrt{2} + 5\, \text{ArcSinh}[1]\right)$$

Here is the numerical result.

```
In[3]:= N[%]
```

$$Out[3]= 0.521405$$

For comparison, we now conduct a numerical test for 100000 pairs of points.

```
In[4]:= With[{num = 100000},
            (Plus @@ Table[Sqrt[(Random[] - Random[])^2 +
                           (Random[] - Random[])^2], {num}])/num]
```

$$Out[4]= 0.521571$$

The result for the 3D cube is given by the following expression:

$$\bar{x} = \frac{1}{420}\left(-210 \log(2) + 42 \log\left(1 + \sqrt{2}\right) + \right.$$
$$\left. 42 \log\left(2 + \sqrt{2}\right) + 378 \log\left(1 + \sqrt{3}\right) - 21 \log\left(2 + \sqrt{3}\right) + 68 \sqrt{2} - 24 \sqrt{3} - 28\,\pi + 16\right).$$

In *Mathematica*, this can be calculated in the following way: Instead of trying to use definite integration, we use indefinite integration and substitute the limits by hand [1311]. The advantage is that we avoid that *Mathematica* has to check for singularities in intermediate expressions that still contain outer integration variables.

```
In[5]:= (* rewrite Logs with complicated arguments
            in Logs with simpler arguments *)
        logRewrite[expr_] := FixedPoint[PowerExpand[# //.
                Log[a_] :> Log[Factor[Together[a]]]]&, TrigToExp[expr]]
```

```
In[7]:= (* calculate the indefinite integral wrt x, y, and z *)
        Short[indefiniteIntegral =
        Integrate[# 2(1 - z), z]& /@ Expand[logRewrite[
            Integrate[Integrate[Sqrt[x^2 + y^2 + z^2] *
                        2(1 - x), x] 2(1 - y), y]]], 8]
```

Out[8]//Short=
$$\frac{1024\,z}{405} - \frac{4\,y^3\,z}{9} + \frac{y^4\,z}{4} - \frac{1024\,\ll1\gg\,\ll1\gg}{1215} + \ll92\gg +$$

$$\frac{1}{4}\,x^4\,\left(4\,z - z^2 + \ll8\gg + x\,(2\,i + x)\,\text{Log}\left[-\frac{4\,\left(x^2 + i\,x\,z + y\,\left(y + \sqrt{x^2 + y^2 + z^2}\,\right)\right)}{(2 - i\,x)\,x\,y^2\,(x + i\,z)}\right]\right) +$$

$$\frac{1}{450}\,i\,\left(-30\,(4\,x^4 + y^3\,(-5\,i + 4\,y))\,z - 15\,(5\,x^2 + (5 + 4\,i\,y)\,y^2)\,z^2 +\right.$$

$$10\,(4\,x^2 + y\,(-5\,i + 4\,y))\,z^3 + 15\,(5 + 2\,i\,y)\,z^4 - 48\,z^5 + \ll5\gg + 15\,(5 + 4\,i\,y)\,y^4\,\text{Log}[y^2 + z^2] +$$

$$15\,x\,(3\,x^4 + 10\,x^2\,y\,(-2\,i + y) + 5\,y^3\,(-4\,i + 3\,y))\,\text{Log}\left[z + \sqrt{x^2 + y^2 + z^2}\,\right] +$$

$$30\,(5 + 4\,i\,y)\,y^4\,\text{Log}\left[\frac{40\,\left(x^2 + y\,(y - i\,z) + x\,\sqrt{x^2 + y^2 + z^2}\,\right)}{x^2\,y^4\,(-5\,i + 4\,y)\,(y - i\,z)}\right] +$$

$$30\,z^4\,(-5 + 4\,z)\,\text{Log}\left[x^2 + i\,y\,z + z^2 + x\,\sqrt{x^2 + y^2 + z^2}\,\right] +$$

$$15\,(5 + 4\,i\,x)\,x^4\,\text{Log}\left[\frac{40\,\left(x^2 - i\,x\,z + y\,\left(y + \sqrt{x^2 + y^2 + z^2}\,\right)\right)}{x^4\,(-5\,i + 4\,x)\,y^2\,(x - i\,z)}\right] +$$

$$\left.15\,i\,x^4\,(5\,i + 4\,x)\,\text{Log}\left[-\frac{40\,\left(x^2 + i\,x\,z + y\,\left(y + \sqrt{x^2 + y^2 + z^2}\,\right)\right)}{x^4\,(5\,i + 4\,x)\,y^2\,(x + i\,z)}\right]\right) - \frac{1}{450}\,i\,(\ll1\gg)$$

```
In[9]:= (* replace limit procedure *)
        replaceLimits[expr_, xyz_] :=
        Module[{exprNew = Expand[logRewrite[expr]] /.
                (* get rid of terms that would produce 0 Infinity and that
                   in the limit give 0 *) xyz^n_?Positive Log[xyz] -> 0, δ},
            δ = (exprNew /. xyz -> 1) - (exprNew /. xyz -> 0);
                (* in case direct substitution fails, use series expansion *)
            If[δ === Indeterminate, Normal[Series[exprNew, {xyz, 1, 0}]] -
                Normal[Series[exprNew, {xyz, 0, 0}]], δ]]
```

```
In[11]:= (* substitute all limits *)
        Fold[replaceLimits, indefiniteIntegral, {z, y, x}] // Together
```

Out[12]= $\frac{1}{420}\,\left(16 + 68\,\sqrt{2} - 24\,\sqrt{3} + (21 - 84\,i)\,\text{Log}[1 - i] +\right.$

$(21 + 84\,i)\,\text{Log}[1 + i] - 189\,\text{Log}[2] + 42\,\text{Log}[1 + \sqrt{2}\,] + 42\,\text{Log}[2 + \sqrt{2}\,] +$

$\left.378\,\text{Log}[1 + \sqrt{3}\,] - (21 - 84\,i)\,\text{Log}[(2 - i) + \sqrt{3}\,] - (21 + 84\,i)\,\text{Log}[(2 + i) + \sqrt{3}\,]\right)$

```
In[13]:= FullSimplify[%]
```

Out[13]= $\frac{1}{420}\,\left(16 + 68\,\sqrt{2} - 24\,\sqrt{3} - 28\,\pi + 42\,\text{ArcSinh}[1] -\right.$

$\left.168\,\text{Log}[2] + 42\,\text{Log}[2 + \sqrt{2}\,] + 378\,\text{Log}[1 + \sqrt{3}\,] - 21\,\text{Log}[4\,\left(2 + \sqrt{3}\,\right)]\right)$

```
In[14]:= N[%]
```

Out[14]= 0.661707

Collecting all logarithms, the last result can be further simplified to $\left(4 + 17\,\sqrt{2} - 6\,\sqrt{3} - 7\,\pi\right)/105 +$ $\log\left(4\,\left(3\,\left(3 + 2\,\sqrt{2}\,\right)\right)^{1/2} + 7\,\sqrt{2} + 7\right)/5$.

Here is the comparison with a numerical simulation.

```
In[15]:= Compile[{{n, _Integer}},
                Module[{sum = 0.},
                    Do[sum = sum + Sqrt[(Random[] - Random[])^2 +
```

```
                              (Random[] - Random[])^2 +
                              (Random[] - Random[])^2],
                        {n}]; sum/n]][10^6]
```

Out[15]= 0.661383

Carrying out the 3D integration numerically also yields a similar result, namely 0.661707.

In[16]:= NIntegrate[2(1 - x) 2(1 - y) 2(1 - z) Sqrt[x^2 + y^2 + z^2],
                   {x, 0, 1}, {y, 0, 1}, {z, 0, 1}]

Out[16]= 0.661707

In a similar way, we could calculate the average line length in a triangle. For the triangle with vertices $\{0, 0\}$, $\{1, 0\}$, $\{0, 1\}$, we obtain the value $(1 + 4\,2^{1/2} + \mathrm{arcsinh}(1)(4 + 2^{1/2}))/30 = 0.41429....$

We outline another approach in which we borrow some ideas from one of the next subexercise. We represent the square root as a Laplace integral.

In[17]:= LaplaceTransform[-1/Sqrt[Pi]/2 p^(-3/2), p, x]

Out[17]= $\sqrt{x}$

This means that now we have to consider the integral

$$\bar{x}_d = -\frac{1}{2\sqrt{\pi}} \int\limits_0^\infty dp\, \frac{1}{p^{3/2}} \int_0^1 d\xi_1 \int_0^1 d\xi_2 \cdots \int_0^1 d\xi_d\, 2^d\,(1-\xi_1)(1-\xi_2)...(1-\xi_d)\exp(-p(\xi_1^2 + \xi_2^2 + ... + \xi_d^2))$$

We start with the inner integrals and observe that we can factorize the multiple integral. A single factor integral has then the following form.

In[18]:= int1D[p_] = Integrate[Exp[-p ξ^2] 2 (1 - ξ), {ξ, 0, 1}]

Out[18]= $\dfrac{-1 + e^{-p} + \sqrt{p}\,\sqrt{\pi}\,\mathrm{Erf}[\sqrt{p}\,]}{p}$

Because the last expression behaves like $1 - \pi/6$ for small values of p, the outermost p-integral diverges at the origin. So we use a Hadamard-regularized version of the last integral:

$$\bar{x}_d = -\frac{1}{2\sqrt{\pi}} \int\limits_0^\infty \frac{1}{p^{3/2}} \left(\left(2 \int_0^1 (1-\xi) e^{-p\xi^2}\, d\xi \right)^d - 1 \right) dp.$$

Here the corresponding integral is carried out for the above Laplace transform.

In[19]:= Integrate[-1/Sqrt[Pi]/2 p^(-3/2) (Exp[-p Σ] - 1), {p, 0, Infinity},
                    Assumptions -> Re[Σ] > 0]

Out[19]= $\sqrt{\Sigma}$

This is the resulting integrand. We remove the square root argument inside the erf function through a change of variables.

In[20]:= integrand[dim_, ζ_] = -1/Sqrt[Pi]/2 (p^(-3/2) (int1D[p]^dim - 1) /.
                   (* change of integration variable *)
                                        p -> ζ^2 // PowerExpand) 2 ζ

Out[20]= $-\dfrac{-1 + \zeta^{-2\,\mathrm{dim}}\left(-1 + e^{-\zeta^2} + \sqrt{\pi}\,\zeta\,\mathrm{Erf}[\zeta]\right)^{\mathrm{dim}}}{\sqrt{\pi}\,\zeta^2}$

We canonicalize the integrand by reducing powers of ζ in the denominators through partial integration. We do not keep the integrated parts explicitly. From the form of the integrand, we see that at infinity they do not contribute and the contributions from the origin cancel the singular contributions from the remaining integrals. Instead, we drop all terms that at are singular at the origin at the end.

In[21]:= (* use linearity and extract prefactors *)
         partialIntegrate[p_Plus]   := partialIntegrate /@ p;
         partialIntegrate[a_?NumericQ b_] := a partialIntegrate[b]

         (* the main partial integration rule; drop endpoint contributions *)

```
partialIntegrateRule = partialIntegrate[a_ ζ^n_?(# < -1&)] :>
                      -partialIntegrate[(1/(n + 1)  ζ^(n + 1)) D[a, ζ]];
```

In[26]:= `dropSingularTerms[expr_, ζ_] := expr - Normal[Series[expr, {ζ, 0, -1}]]`

In[27]:= `makeReducedIntegrand[dim_, ζ_] := dropSingularTerms[#, ζ]& @`
         `(FixedPoint[ExpandAll[# /. partialIntegrateRule]&,`
                    `partialIntegrate[ExpandAll[integrand[dim, ζ]]]]] /.`
                                    `partialIntegrate -> Identity)`

Here are the resulting 1D integrals for the average distances for the first four dimensions.

In[28]:= `Table[makeReducedIntegrand[d, ζ], {d, 4}] // Simplify // TraditionalForm`

Out[28]//TraditionalForm=

$$\left\{ \frac{2e^{-\zeta^2}}{3\sqrt{\pi}}, \; \frac{e^{-2\zeta^2}\left(4\left(1+e^{\zeta^2}\right)\zeta + 5\,e^{\zeta^2}\sqrt{\pi}\,\mathrm{erf}(\zeta)\right)}{15\sqrt{\pi}\,\zeta}, \right.$$

$$\frac{e^{-3\zeta^2}\left(-42\,e^{2\zeta^2}\pi\,\zeta\,\mathrm{erf}(\zeta)^2 + 21\,e^{\zeta^2}\left(4+e^{\zeta^2}\right)\sqrt{\pi}\,\mathrm{erf}(\zeta) + 4\left(-9+17\,e^{\zeta^2}+2\,e^{2\zeta^2}\right)\zeta\right)}{105\sqrt{\pi}\,\zeta},$$

$$\frac{1}{1890\sqrt{\pi}\,\zeta}\left(e^{-4\zeta^2}\left(-252\,e^{3\zeta^2}\pi^{3/2}\,\mathrm{erf}(\zeta)^3 - 144\,e^{2\zeta^2}\left(17+4\,e^{\zeta^2}\right)\pi\,\zeta\,\mathrm{erf}(\zeta)^2 + \right.\right.$$

$$\left.\left.9\,e^{\zeta^2}\left(197 + 208\,e^{\zeta^2} + 15\,e^{2\zeta^2}\right)\sqrt{\pi}\,\mathrm{erf}(\zeta) + 4\left(-338 + 216\,e^{\zeta^2} + 219\,e^{2\zeta^2} + 8\,e^{3\zeta^2}\right)\zeta\right)\right)\right\}$$

And for dimensions 1, 2, 3, we recover the above-calculated results.

In[29]:= `Table[Integrate[makeReducedIntegrand[d, ζ], {ζ, 0, Infinity}], {d, 3}] //`
                                                        `Simplify`

Out[29]= $\left\{ \frac{1}{3}, \; \frac{1}{15}\left(2 + \sqrt{2} + 5\,\mathrm{ArcSinh}[1]\right), \right.$

$\left. \frac{1}{105}\left(4 + 17\sqrt{2} - 6\sqrt{3} - 7\pi + 21\,\mathrm{ArcSinh}[1] + 84\,\mathrm{ArcSinh}\left[\frac{1}{\sqrt{2}}\right]\right)\right\}$

The average distances for $d = 4$ can with some effort be integrated in closed form, but the resulting 1D integral can easily be integrated numerically.

For an asymptotic expansion for the k-dimensional case, see [57]; for the shape that minimizes the average value of the distance, see [146].

e) The "unpleasant" term is the square root in the denominator. It makes the integrand nonfactorizable. By using the inverse Laplace transform $s(\xi)$ of the square root function, we can make it factorizable.

In[1]:= `InverseLaplaceTransform[1/Sqrt[x], x, p]`

Out[1]= $\dfrac{1}{\sqrt{p}\,\sqrt{\pi}}$

Here is the resulting integral after exchanging the integration order.

$$I(a, b, c) = \frac{1}{\sqrt{\pi}} \int\limits_0^\infty \frac{1}{\sqrt{s}}\left(\int\limits_0^\infty\int\limits_0^\infty\int\limits_0^\infty e^{-s(x+y+z)}\,e^{-x-y-z}\sin(a\,x + b\,y + c\,z)\,dz\,dy\,dx\right)ds.$$

This yields the following input for the calculation of $I(a, b, c)$. *Mathematica* can calculate the resulting fourfold integral.

In[2]:= `1/Sqrt[Pi] Integrate[`
         `Integrate[1/Sqrt[s] Exp[-(1 + s) (x + y + z)] Sin[a x + b y + c z],`
                  `{x, 0, Infinity}, {y, 0, Infinity}, {z, 0, Infinity},`
                  `Assumptions -> Re[s] > 0, GenerateConditions -> False],`
                  `{s, 0, Infinity}] // (* abbreviate *) Short[#, 6]&`

Out[2]//Short= $\dfrac{1}{b\,c\,\sqrt{\pi}}$

         `If[(-1 ≤ Im[a] ≤ 1 || Re[a] ≠ 0) && (-1 ≤ Im[b] ≤ 1 || Re[b] ≠ 0) && (-1 ≤ Im[c] ≤ 1 || Re[c] ≠ 0),`

$$-\frac{\ll 1\gg}{\ll 1\gg} + \ll 33\gg + \frac{\ll 1\gg}{\ll 1\gg},$$

$$\texttt{Integrate}\Big[\frac{b\,(1+s)^2\,\big(\sqrt{c^2}\,\texttt{Abs}[c]+\sqrt{a^2}\,c\,\texttt{Sign}[a]\big)+\sqrt{b^2}\,\texttt{Abs}[b]\,(c\,(\ll 1\gg)^2-\ll 1\gg)}{\sqrt{s}\,(a^2+(1+s)^2)\,(b^2+(1+s)^2)\,(c^2+(1+s)^2)},$$

$$\{s,\,0,\,\infty\},\,\texttt{Assumptions}\rightarrow !\,(\ll 1\gg)\Big]\Big]$$

Here is a concise way of writing the last integral. It nicely shows its structure and its symmetry.

```
In[3]:= T[a_, b_, c_] =
    With[{s = Sqrt, s = Sign}, I Sqrt[Pi]*
       ((c - b) s[-I + b] s[I + b] s[-I + c] s[I + c]
          (s[ I + a] s[-I/s[-I + a]] s[s[-I + a]] -
           s[-I + a] s[ I/s[ I + a]] s[s[ I + a]]) +
        (a - c) s[-I + a] s[I + a] s[-I + c] s[I + c]
          (s[ I + b] s[-I/s[-I + b]] s[s[-I + b]] -
           s[-I + b] s[ I/s[ I + b]] s[s[ I + b]]) +
        (b - a) s[-I + a] s[I + a] s[-I + b] s[I + b]
          (s[ I + c] s[-I/s[-I + c]] s[s[-I + c]] -
           s[-I + c] s[ I/s[ I + c]] s[s[ I + c]]))/
       (2 (a - b) (a - c) (b - c) s[-I + a]*
         s[I + a] s[-I + b] s[I + b] s[-I + c] s[I + c])]
```

$$\text{Out[3]=}\ \Big(i\,\sqrt{\pi}\,\Big(\sqrt{-i+b}\,\sqrt{i+b}\,\sqrt{-i+c}\,\sqrt{i+c}\,(-b+c)$$

$$\Big(\sqrt{i+a}\,\sqrt{-\frac{i}{\text{Sign}[-i+a]}}\,\sqrt{\text{Sign}[-i+a]}-\sqrt{-i+a}\,\sqrt{\frac{i}{\text{Sign}[i+a]}}\,\sqrt{\text{Sign}[i+a]}\Big)+$$

$$\sqrt{-i+a}\,\sqrt{i+a}\,(a-c)\,\sqrt{-i+c}\,\sqrt{i+c}\,\Big(\sqrt{i+b}\,\sqrt{-\frac{i}{\text{Sign}[-i+b]}}\,\sqrt{\text{Sign}[-i+b]}-$$

$$\sqrt{-i+b}\,\sqrt{\frac{i}{\text{Sign}[i+b]}}\,\sqrt{\text{Sign}[i+b]}\Big)+\sqrt{-i+a}\,\sqrt{i+a}\,\sqrt{-i+b}\,\sqrt{i+b}\,(-a+b)$$

$$\Big(\sqrt{i+c}\,\sqrt{-\frac{i}{\text{Sign}[-i+c]}}\,\sqrt{\text{Sign}[-i+c]}-\sqrt{-i+c}\,\sqrt{\frac{i}{\text{Sign}[i+c]}}\,\sqrt{\text{Sign}[i+c]}\Big)\Big)\Big)\Big/$$

$$\Big(2\,\sqrt{-i+a}\,\sqrt{i+a}\,(a-b)\,\sqrt{-i+b}\,\sqrt{i+b}\,(a-c)\,(b-c)\,\sqrt{-i+c}\,\sqrt{i+c}\Big)$$

A numerical check at random complex parameter values confirms the last result.

```
In[4]:= With[{a = 1 + 1/4 I, b = Pi - 2/11 I, c = E + 1/Pi I, L = (* truncate *) 12},
    {(* numericalized symbolic result *) T[a, b, c] // N,
     (* numerical integration result *)
     NIntegrate[Exp[-(x + y + z)] Sin[a x + b y + c z]/Sqrt[x + y + z],
         {x, 0, L}, {y, 0, L}, {z, 0, L}, Method -> GaussKronrod]}]
Out[4]= {0.0263132 - 0.0115489 i, 0.0263161 - 0.0115477 i}
```

The value $I(1, 1, 1)$ cannot be obtained by substitution because indeterminate expressions arise. Expanding $I(a, b, c)$ in a series around $\{a, b, c\} = \{1, 1, 1\}$ gives the result $\pi^{1/2}\,3/32\,(1+2^{1/2})^{1/2}$.

```
In[5]:= T[1, 1, 1]
```

$$\texttt{Power::infy : Infinite expression } \frac{1}{0} \texttt{ encountered. More...}$$

$$\texttt{Power::infy : Infinite expression } \frac{1}{0} \texttt{ encountered. More...}$$

$$\texttt{Power::infy : Infinite expression } \frac{1}{0} \texttt{ encountered. More...}$$

General::stop :
 Further output of Power::infy will be suppressed during this calculation. More...

∞::indet : Indeterminate expression

$$\frac{0\,\sqrt{\pi}\,\texttt{ComplexInfinity}\,\texttt{ComplexInfinity}\,\ll 1\gg}{\sqrt{1-i}\,\sqrt{1-i}\,\sqrt{1-i}\,\sqrt{1+i}\,\sqrt{1+i}\,\sqrt{1+i}}\ \texttt{encountered. More...}$$

Out[5]= Indeterminate

In[6]:= `Sqrt[Pi] FullSimplify[Simplify[ExpandAll[Normal[Simplify[Series[`
    `Limit[Limit[FunctionExpand[`$\mathcal{T}$`[a, b, c]], a -> b], b -> c], {c, 1, 0}]]]]/Sqrt[Pi]]]`

Out[6]= $\frac{3}{32} \sqrt{\left(1 + \sqrt{2}\right)} \pi$

For real a, b, c, we would prefer a manifestly real result. By using the following two simplifications, we arrive at the expression $\mathcal{I}$Real$[a, b, c]$.

In[7]:= `FullSimplify[Sqrt[abc - I] Sqrt[abc + I], abc > 0]`

Out[7]= $\sqrt{1 + abc^2}$

In[8]:= `ComplexExpand[FunctionExpand[Sqrt[-I/Sign[abc - I]] Sqrt[Sign[abc - I]]],`
                  `TargetFunctions -> {Re, Im}] // Simplify[#, abc > 0]&`

Out[8]= $\cos\left[\frac{1}{2}\left(-\text{ArcTan}[abc] + \text{ArcTan}[abc, -1]\right)\right] + i \sin\left[\frac{1}{2}\left(-\text{ArcTan}[abc] + \text{ArcTan}[abc, -1]\right)\right]$

In[9]:= `f[a_, b_, c_] = -Sqrt[2] (a - b) Sqrt[1 + a^2] Sqrt[1 + b^2]*`
                `Sqrt[c^2]/c Sqrt[Sqrt[1 + c^2] - 1];`

    `(* integral is a symmetric function in a, b, c *)`
    $\mathcal{I}$`Real[a_, b_, c_] = Sqrt[Pi] (f[a, b, c] + f[b, c, a] + f[c, a, b])/`
                `(2 (a - c) (b - c) (a - b)*`
                `Sqrt[1 + a^2] Sqrt[1 + b^2] Sqrt[1 + c^2])`

Out[11]= $\left(\left(-\dfrac{\sqrt{2} \sqrt{a^2} \sqrt{-1 + \sqrt{1 + a^2}} \sqrt{1 + b^2} (b - c) \sqrt{1 + c^2}}{a} - \right.\right.$

$\dfrac{\sqrt{2} \sqrt{1 + a^2} \sqrt{b^2} \sqrt{-1 + \sqrt{1 + b^2}} (-a + c) \sqrt{1 + c^2}}{b} - $

$\left.\dfrac{\sqrt{2} \sqrt{1 + a^2} (a - b) \sqrt{1 + b^2} \sqrt{c^2} \sqrt{-1 + \sqrt{1 + c^2}}}{c}\right) \sqrt{\pi}\right) /$

$\left(2 \sqrt{1 + a^2} (a - b) \sqrt{1 + b^2} (a - c) (b - c) \sqrt{1 + c^2}\right)$

For $a = b = c = 1$, we again obtain the result $3/32 (1 + 2^{1/2})^{1/2}$.

In[12]:= `Series[Series[`$\mathcal{I}$`Real[a, b, c], {b, a, 0}], {c, a, 0}] // Normal // FullSimplify`

Out[12]= $-\dfrac{3\sqrt{a^2}\left(a^6 + a^2\left(3 - 5\sqrt{1 + a^2}\right) + a^4\left(8 - 5\sqrt{1 + a^2}\right) + 4\left(-1 + \sqrt{1 + a^2}\right)\right)\sqrt{\frac{\pi}{2}}}{8 a (1 + a^2)^3 \left(-1 + \sqrt{1 + a^2}\right)^{5/2}}$

In[13]:= `% /. a -> 1 // FullSimplify`

Out[13]= $\frac{3}{32} \sqrt{\left(1 + \sqrt{2}\right)} \pi$

For similar integrals, see [1361]. For some more challenging multidimensional integrals, see [675], [816], [1363].

f) A direct call to `Integrate` does not give a result.

In[1]:= `integrand = (2 Cos[x] - Cos[x - y - z] - Cos[x - y + z])/ (y^2 (x - z)^2);`

In[2]:= `Integrate[integrand, {x, 0, Infinity}, {y, 0, x}, {z, 0, y}]`

Out[2]= $\displaystyle\int_0^\infty \int_0^x \dfrac{1}{x\, y^2\, (-x + y)}\, (x \cos[x] - 2 y \cos[x] + x \cos[x - 2 y] -$

$2 x \cos[x - y] + 2 y \cos[x - y] + 2 x (x - y) \cos[x - y] \text{CosIntegral}[-x] \sin[x] -$

$2 x (x - y) \cos[x - y] \text{CosIntegral}[-x + y] \sin[x] - x^2 \cos[2 x - y] \text{SinIntegral}[x] +$

$x y \cos[2 x - y] \text{SinIntegral}[x] - x^2 \cos[y] \text{SinIntegral}[x] + x y \cos[y] \text{SinIntegral}[x] +$

$x^2 \cos[2 x - y] \text{SinIntegral}[x - y] - x y \cos[2 x - y] \text{SinIntegral}[x - y] +$

$x^2 \cos[y] \text{SinIntegral}[x - y] - x y \cos[y] \text{SinIntegral}[x - y])\, dy\, dx$

Carrying out a properly seeded numerical Monte–Carlo integration yields a value that is suspiciously near to -2π.

In[3]:= `NIntegrate[Evaluate[integrand], {x, 0, Infinity}, {y, 0, x}, {z, 0, y},`
                `Method -> MonteCarlo[Floor[2Pi^Pi]], MaxPoints -> 10^6]`

Out[3]= `-6.28195`

Frequently integrals ranging from 0 to ∞ are easier to carry out than integrals with finite limits; so, we try to carry out the x-integration first. But the inner integral limit limits depend on the variable x. So, we change the inner integration variables to have parameter-free limits.

In[4]:= `newIntegrand = x (y integrand /. z -> Z y) /. y -> Y x`

Out[4]= $\dfrac{2\,Cos[x] - Cos[x - x\,Y - x\,Y\,Z] - Cos[x - x\,Y + x\,Y\,Z]}{Y\,(x - x\,Y\,Z)^2}$

Now, it is possible to carry out the outer x-integral.

In[5]:= `Integrate[newIntegrand, {x, 0, Infinity},`
         `Assumptions -> Element[{Y, Z}, Reals]]`

Out[5]= $\dfrac{\pi\,(-2 + Abs[1 + Y\,(-1 + Z)] + Abs[-1 + Y + Y\,Z])}{2\,Y\,(-1 + Y\,Z)^2}$

And the inner two integrals, now over the unit square, are also easily done now. As conjectured, the result of the triple integral is -2π.

In[6]:= `Integrate[%, {Y, 0, 1}, {Z, 0, 1}]`

Out[6]= $-2\,\pi$

36. Duffing Equation, Secular Terms

a) Here, a series for `y[t]` is implemented.

In[1]:= `Y = Sum[ε^i y[i][t], {i, 0, 6}] + O[ε]^7`

Out[1]= $y[0][t] + y[1][t]\,\varepsilon + y[2][t]\,\varepsilon^2 + y[3][t]\,\varepsilon^3 + y[4][t]\,\varepsilon^4 + y[5][t]\,\varepsilon^5 + y[6][t]\,\varepsilon^6 + O[\varepsilon]^7$

We substitute the series into the differential equation.

In[2]:= `eqs = Factor[Numerator[Together[#]]]& /@`
         `CoefficientList[Normal[D[Y, {t, 2}] + Y + ε Y^3], ε];`

The first equation becomes the following.

In[3]:= `eqs[[1]]`

Out[3]= $y[0][t] + y[0]''[t]$

Taking the initial conditions into account, the solution is simply cos.

In[4]:= `solList = {y[0] -> (Cos[#]&)}`

Out[4]= $\{y[0] \to (Cos[\#1]\,\&)\}$

Now, we recursively solve for the higher orders with homogeneous boundary conditions.

In[5]:= `Do[(* substitute already calculated solutions *)`
         `eq[i] = eqs[[i + 1]] //. solList;`
         `(* solve for next solutions *)`
         `sol[i] = DSolve[{eq[i] == 0, y[i][0] == 0, y[i]'[0] == 0},`
             `y[i], t][[1, 1]];`
         `AppendTo[solList, sol[i]], {i, 5}] // Timing`

Out[5]= $\{5.81\,Second,\,Null\}$

These are the first terms of the series.

In[6]:= `Sum[ε^i y[i][t], {i, 0, 3}] /. solList`

Out[6]= $Cos[t] + \dfrac{1}{32}\,\varepsilon\,(5\,Cos[t] - 4\,Cos[t]\,Cos[2\,t] -$

$Cos[t]\,Cos[4\,t] - 12\,t\,Sin[t] - 8\,Sin[t]\,Sin[2\,t] - Sin[t]\,Sin[4\,t]) +$

$\dfrac{1}{1024}\,(\varepsilon^2\,(-43\,Cos[t] - 72\,t^2\,Cos[t] + 36\,Cos[t]^3 + 9\,Cos[t]\,Cos[4\,t] - 2\,Cos[t]\,Cos[6\,t] +$

$24\,t\,Sin[t] - 144\,t\,Cos[2\,t]\,Sin[t] - 36\,t\,Cos[4\,t]\,Sin[t] + 78\,Sin[t]\,Sin[2\,t] +$

$36\,t\,Cos[t]\,Sin[4\,t] + 3\,Sin[t]\,Sin[4\,t] - 2\,Sin[t]\,Sin[6\,t])) +$

$\dfrac{1}{32768}\,(\varepsilon^3\,(425\,Cos[t] + 648\,t^2\,Cos[t] - 300\,Cos[t]^3 - 48\,Cos[t]\,Cos[2\,t] -$

$864\,t^2\,Cos[t]\,Cos[2\,t] + 12\,Cos[t]\,Cos[2\,t]^2 - 156\,Cos[t]\,Cos[4\,t] +$

$648\,t^2\,Cos[t]\,Cos[4\,t] + 70\,Cos[t]\,Cos[6\,t] - 3\,Cos[t]\,Cos[8\,t] - 576\,t\,Sin[t] +$

$$288\, t^3\, \text{Sin}[t] + 2664\, t\, \text{Cos}[2\, t]\, \text{Sin}[t] + 288\, t\, \text{Cos}[4\, t]\, \text{Sin}[t] - 120\, t\, \text{Cos}[6\, t]\, \text{Sin}[t] +$$
$$504\, t\, \text{Cos}[t]\, \text{Sin}[2\, t] - 1458\, \text{Sin}[t]\, \text{Sin}[2\, t] + 1728\, t^2\, \text{Sin}[t]\, \text{Sin}[2\, t] -$$
$$648\, t\, \text{Cos}[t]\, \text{Sin}[4\, t] + 78\, \text{Sin}[t]\, \text{Sin}[4\, t] + 648\, t^2\, \text{Sin}[t]\, \text{Sin}[4\, t] +$$
$$120\, t\, \text{Cos}[t]\, \text{Sin}[6\, t] + 62\, \text{Sin}[t]\, \text{Sin}[6\, t] - 3\, \text{Sin}[t]\, \text{Sin}[8\, t]))$$

We calculate a numerical solution of the differential equation for $\varepsilon = 1/10$.

```
In[7]:= ε = 10^-1;
    nSol = NDSolve[{y''[t] + y[t] + ε y[t]^3 == 0,
                    y[0] == 1, y'[0] == 0}, y[t], {t, 0, 8/ε}];
```

The numerical solution together with the series solutions to various orders in ϵ are shown next. For easier comparison, we calculate and plot (right graphic) the differences between the numerical and the series solutions in the L_2-norm. We calculate the differences by solving their differential equations numerically.

```
In[9]:= (* calculate interpolating functions for squared differences *)
    Do[d[j] = (* solve differential equation for differences *)
        NDSolve[{diff'[t] == ((y[t] /. First[nSol]) -
            (Sum[ε^i y[i][t], {i, 0, j}] /. solList))^2,
            diff[0] == 0}, diff[t], {t, 0, 8/ε}], {j, 0, 5}];
In[11]:= Show[GraphicsArray[
    Block[{$DisplayFunction = Identity},
    {(* show all solutions *)
    Show[{(* the numerical solution *)
        Plot[Evaluate[y[t] /. nSol], {t, 0, 8/ε},
            PlotPoints -> 250, DisplayFunction -> Identity,
            PlotStyle -> {{GrayLevel[0], Thickness[0.005]}}],
        Table[(* partial sums of the symbolic solution *)
            Plot[Evaluate[Sum[ε^i y[i][t], {i, 0, j}] /. solList],
                {t, 0, 8/ε}, PlotPoints -> 250,
                DisplayFunction -> Identity,
                PlotStyle -> {{Hue[0.8 j/5], Thickness[0.002]}}],
            {j, 0, 5}]}, DisplayFunction -> $DisplayFunction,
        Frame -> True, Axes -> False],
    (* show all solutions *)
    Show[Table[
        Plot[(* display logarithm of squared error *)
            Evaluate[Log[10, Abs[diff[t] /. d[j]]]], {t, 0, 8/ε},
            PlotPoints -> 250, DisplayFunction -> Identity,
            PlotStyle -> {{Hue[0.8 j/5], Thickness[0.002]}}],
        {j, 0, 5}], DisplayFunction -> $DisplayFunction,
            Frame -> True, Axes -> False]}]]]
```

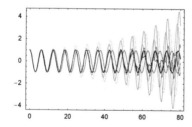 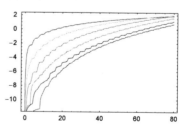

For more general perturbation methods of oscillators, see [106], [556], [1870], and [1386].

For the case of general initial conditions, we proceed similarly as above. We will impose general initial conditions to $y_0(t)$ and vanishing boundary conditions to all other $y_k(t)$.

```
In[13]:= ε =.;
    Y = Sum[ε^i y[i][t], {i, 0, 4}] + O[ε]^5;
    eqs = Factor[Numerator[Together[#]]]& /@
        CoefficientList[Normal[D[Y, {t, 2}] + Y + ε Y^3], ε];
```

For a more concisely written result, we define a function `writeNice`. It collects terms with respect to the initial conditions.

```
In[16]:= writeNice := (a_ -> b_) :> (a -> Internal`FromDistributedTermsList[
              Factor @ Internal`DistributedTermsList[b, {a, b}]])
```

Now the solution for the zeroth order is $y_0(t) = a\cos(t) + b\sin(t)$.

```
In[17]:= solList = DSolve[{eqs[[1]] == 0, y[0][0] == a, y[0]'[0] == b},
                     y[0][t], t][[1]]
Out[17]= {y[0][t] → a Cos[t] + b Sin[t]}
```

Iterating backsubstitution and differential equation solving, we obtain the following result.

```
In[18]:= Do[eq[k] = eqs[[k]] //. solList;
           solList = Append[solList,
             DSolve[{eq[k] == 0, y[k - 1][0] == 0, y[k - 1]'[0] == 0},
                 y[k - 1][t], t][[1, 1]] /. writeNice], {k, 2, 4}];

In[19]:= solList // Short[#, 16]& // TraditionalForm
```

Out[19]//TraditionalForm=

$$\Big\{y(0)(t) \to a\cos(t) + b\sin(t), \; y(1)(t) \to \ll 1 \gg, \; \ll 1 \gg,$$

$$y(3)(t) \to \frac{1}{32768}\big((288\sin(t)\,t^3 + 648\cos(t)\,t^2 - 864\cos(t)\cos(2t)\,t^2 + 648\cos(t)\cos(4t)\,t^2 + 1728\sin(t)\sin(2t)\,t^2 + 648\sin(t)$$

$$\sin(4t)\,t^2 + 2664\cos(2t)\sin(t)\,t + 288\cos(4t)\sin(t)\,t - 120\cos(6t)\sin(t)\,t - 576\sin(t)\,t + 504\cos(t)\sin(2t)\,t -$$

$$648\cos(t)\sin(4t)\,t + 120\cos(t)\sin(6t)\,t - 900\cos^3(t) + 731\cos(t) + 252\cos(t)\cos(2t) - 150\cos(t)\cos(4t) +$$

$$70\cos(t)\cos(6t) - 3\cos(t)\cos(8t) - 1458\sin(t)\sin(2t) + 78\sin(t)\sin(4t) + 62\sin(t)\sin(6t) - 3\sin(t)\sin(8t))\,a^7\big) +$$

$$\frac{1}{32768}\,\big(b\,(-288\cos(t)\,t^3 - 4320\cos(2t)\sin(t)\,t^2 - 1944\cos(4t)\sin(t)\,t^2 + 2088\sin(t)\,t^2 -$$

$$3456\cos(t)\sin(2t)\,t^2 + 1944\cos(t)\sin(4t)\,t^2 + \ll14\gg + 21\cos(8t)\sin(t) - 12935\sin(t) +$$

$$1902\cos(t)\sin(2t) - 3006\cos(t)\sin(4t) + 558\cos(t)\sin(6t) - 21\cos(t)\sin(8t))\,a^6\big) -$$

$$\frac{3\,b^2\,(\ll29\gg + \ll1\gg - 21\sin(t)\sin(8t))\,a^5}{32768} + \frac{b^3\,(\ll1\gg)\,a^4}{32768} + \frac{b^4\,(\ll1\gg)\,a^3}{32768} -$$

$$\frac{3\,b^5\,(288\cos(t)\,t^3 + \ll30\gg + 21\cos(t)\sin(8t))\,a^2}{32768} +$$

$$\frac{b^6\,(288\sin(t)\,t^3 + 2664\cos(t)\,t^2 + \ll28\gg + 2502\sin(t)\sin(4t) + 638\sin(t)\sin(6t) + 21\sin(t)\sin(8t))\,a}{32768} +$$

$$\frac{1}{32768}\,(b^7\,(-288\cos(t)\,t^3 + 864\cos(2t)\sin(t)\,t^2 + 648\cos(4t)\sin(t)\,t^2 + 3528\sin(t)\,t^2 + 1728\cos(t)\sin(2t)\,t^2 -$$

$$648\cos(t)\sin(4t)\,t^2 + 11904\cos(t)\sin(t)\,t + \ll22\gg + 906\cos(t)\sin(4t) + 142\cos(t)\sin(6t) + 3\cos(t)\sin(8t)))\Big\}$$

Here is a quick check of the so-obtained solution.

```
In[20]:= y[t_] = Y + O[ε]^4 /. solList;
         {y[0], y'[0], y''[t] + y[t] + ε y[t]^3}
Out[21]= {a + O[ε]^4, b + O[ε]^4, O[ε]^4}
```

b) Defining Ω and $X(t)$ as truncated power series and substituting these series into the differential equations yields the resulting differential equations for the $x_k(t)$.

```
In[1]:= o = 12;

       Ω = Sum[ε^k Subscript[ω, k], {k, 0, o}];
       X = ε Sum[ε^k Subscript[x, k][t], {k, 0, o}];

       εSer = Ω^2 D[X, t, t] + Subscript[ω, 0]^2 X -
               Sum[Subscript[J, k]/k! X^k, {k, 2, o}] + O[ε]^o;

In[5]:= εTerms = εSer[[3]];
```

Here are the first three resulting differential equations.

```
In[6]:= Table[εTerms[[j]], {j, 3}]
```

$$Out[6]= \Big\{\omega_0^2\,x_0[t] + \omega_0^2\,x_0''[t], \; -\frac{1}{2}\,J_2\,x_0[t]^2 + \omega_0^2\,x_1[t] + 2\,\omega_0\,\omega_1\,x_0''[t] + \omega_0^2\,x_1''[t],$$

$$-\frac{1}{6}\,J_3\,x_0[t]^3 - J_2\,x_0[t]\,x_1[t] + \omega_0^2\,x_2[t] + (\omega_1^2 + 2\,\omega_0\,\omega_2)\,x_0''[t] + 2\,\omega_0\,\omega_1\,x_1''[t] + \omega_0^2\,x_2''[t]\Big\}$$

We see that the inhomogeneous part of the differential equation for $x_k(t)$ depends only the earlier $x_j(t)$:

$$\omega_0^2 \frac{\partial^2 x_k(t)}{\partial t^2} + \omega_0^2 x_k(t) = p_k(x_0(t), \ldots, x''_{k-1}(t), \omega_0, \ldots, \omega_{k-1}, J_2, \ldots, J_{k+1}).$$

We will store the already obtained results for the $x_k(t)$ and the ω_k in the two lists xRules and ωRules. For the zeroth order, we only have a nontrivial result for $x_0(t)$.

```
In[7]:= xRules = {Subscript[x, 0] ->
                Function[t, Subscript[c, "s", 0] Cos[t]]};
        ωRules = {};
```

Now, we consider the first order. The resulting differential equation for $x_1(t)$ is the following.

```
In[9]:= μ = 1;
        ode = εTerms[[μ + 1]] //. xRules //. ωRules
```

$$\text{Out[10]= } -2 \cos[t] \, \omega_0 \, \omega_1 \, c_{s,0} - \frac{1}{2} \cos[t]^2 \, J_2 \, c_{s,0}^2 + \omega_0^2 \, x_1[t] + \omega_0^2 \, x_1''[t]$$

Solving this differential equation is straightforward.

```
In[11]:= dsol = ExpandAll[TrigFactor //@
                DSolve[ode == 0, Subscript[x, μ][t], t]] /.
               (* rename constants *)
               {C[_] Cos[t] :> Subscript[c, "c", μ] Cos[t],
                C[_] Sin[t] :> Subscript[c, "s", μ] Sin[t]}
```

$$\text{Out[11]= } \left\{\left\{x_1[t] \rightarrow \right.\right.$$
$$\left.\left. \cos[t] \, c_{c,1} + \frac{\cos[t] \, \omega_1 \, c_{s,0}}{2 \, \omega_0} + \frac{t \, \sin[t] \, \omega_1 \, c_{s,0}}{\omega_0} + \frac{J_2 \, c_{s,0}^2}{4 \, \omega_0^2} - \frac{\cos[2 \, t] \, J_2 \, c_{s,0}^2}{12 \, \omega_0^2} + \sin[t] \, c_{s,1}\right\}\right\}$$

To get a periodic solution for $x_1(t)$, we cannot afford the term $c_{s,0} \, t \sin(t) \, \omega_1 / \omega_0$. In general, the secular terms (meaning linearly growing in t) of the form $t \, periodicFunctionInt$ spoil the periodicity and must be eliminated [1185].

```
In[12]:= secularTerms = Plus @@ Cases[dsol[[1, 1, 2]], t^_. _]
```

$$\text{Out[12]= } \frac{t \, \sin[t] \, \omega_1 \, c_{s,0}}{\omega_0}$$

With the ω_k, $k \geq 1$ to be determined, we get the condition $\omega_1 = 0$ for a periodic $x_1(t)$.

```
In[13]:= ωCondition = Factor //@ Solve[secularTerms == 0, Subscript[ω, μ]]
```

$$\text{Out[13]= } \{\{\omega_1 \rightarrow 0\}\}$$

We now use the condition $\omega_1 = 0$ to obtain the final periodic solution for $x_1(t)$.

```
In[14]:= ωRules = AppendTo[ωRules, ωCondition[[1, 1]]];
```

```
In[15]:= dsolG = ExpandAll[dsol //. ωRules];
```

```
In[16]:= xRules = AppendTo[xRules, Subscript[x, μ] ->
                Function[t, Evaluate[dsolG[[1, 1, 2]]]]]
```

$$\text{Out[16]= } \left\{x_0 \rightarrow \text{Function}[t, \, c_{s,0} \cos[t]], \right.$$
$$\left. x_1 \rightarrow \text{Function}\left[t, \, \cos[t] \, c_{c,1} + \frac{J_2 \, c_{s,0}^2}{4 \, \omega_0^2} - \frac{\cos[2 \, t] \, J_2 \, c_{s,0}^2}{12 \, \omega_0^2} + \sin[t] \, c_{s,1}\right]\right\}$$

It seems now straightforward to wrap up the last inputs in a Do-loop to calculate $x_2(t)$ to $x_{10}(t)$. We just expect all results to become a bit larger for the higher orders. Unfortunately they do not just become a bit larger, but considerable larger. So before calculating the higher orders of the $x_k(t)$ we will pause a moment and ponder about some optimizations. For the higher orders, the inhomogeneous term in $\omega_0^2 \, x_k''(t) + \omega_0^2 \, x_k(t) = p_k$ will be a large sum. When DSolve calculates the inhomogeneous solution, DSolve will carry out integrals of the form $\int \sin(t) \, p_k \, dt$ and $\int \cos(t) \, p_k \, dt$. These integrals will take a while to evaluate for large expressions p_k with head Plus. So, we implement an optimized function $\mathcal{DSolve}$ for this purpose. It carefully constructs the solution of $\omega_0^2 \, x_k''(t) + \omega_0^2 \, x_k(t) = p_k$. The solution is

```
In[17]:= (* general solution *)
         sol = c1 Cos[t] + c2 Sin[t] -
               Cos[t] Integrate[pk[t] Sin[t], t] +
               Sin[t] Integrate[pk[t] Cos[t], t];
```

```
                 (* check *)
                 D[sol, t, t] + sol - pk[t] // Simplify
    Out[20]= 0

    In[21]:= DSolve[lhs_ == 0, Subscript[x_, μ_][t_], t_] :=
             Module[{f = Expand[lhs - Subscript[ω, 0]^2 *
                          (Subscript[x, μ]''[t] + Subscript[x, μ][t])]},
               {{Subscript[x, μ][t] -> Expand[
                 (* homogeneous solution *)
                 Subscript[c, "c", μ] Cos[t] + Subscript[c, "s", μ] Sin[t] -
                 (* inhomogeneous solution *)
                 (-Cos[t] Integrate[f Sin[t], t] +
                  Sin[t] Integrate[f Cos[t], t])/Subscript[ω, 0]^2]}}]
```

We handle the integrals $\int \sin(t)\, p_k\, dt$ and $\int \cos(t)\, p_k\, dt$ with the specialized function *IntegrateCosSin*. This function first collects the terms of p_k with respect to $\sin^a(t)\cos^b(t)$, and then uses a look-up for these integrals.

```
    In[22]:= Integrate[f_, t_] := With[{f = Expand[f]},
               If[Head[f] === Plus, IntegrateCosSin[f, t], Integrate[f, t]]]

    In[23]:= IntegrateCosSin[f_, t_] :=
             Module[{cosSinPoly, coeffs},
               cosSinPoly = Expand[Plus @@ (TrigExpand /@ (List @@ Expand[f]))];
               (* collect wrt powers of Sin[t] and Cos[t] *)
               coeffs = Internal`DistributedTermsList[cosSinPoly,
                                      {Cos[t], Sin[t]}][[1]];
               (* integrate and expand *)
               Expand[Plus @@ ((CosSinIntegral[#1] #2)& @@@ coeffs)]]

    In[24]:= (* cached integrals *)
             CosSinIntegral[{m_, n_}] := CosSinIntegral[{m, n}] =
                         TrigExpand[Integrate[Cos[t]^m Sin[t]^n, t]]
```

By adding some other optimizations (like preprocessing the differential equation using `Together` before substituting the already calculated results for the $x_k(t)$), we have the following modification of the above code for $x_1(t)$. Now it takes a couple of minutes to calculate $x_2(t)$ to $x_{10}(t)$.

```
    In[26]:= Do[(* the ODE for the current order *)
               ode = Together[eTerms[[μ + 1]]] //. xRules //. ωRules;
               (* solve ODE *)
               dsol = DSolve[ode == 0, Subscript[x, μ][t], t];
               (* secular terms *)
               secularTerms = Factor[Plus @@ Cases[dsol[[1, 1, 2]], t^_. _]];
               (* condition for the current ω *)
               ωCondition = Factor //@ Solve[secularTerms == 0, Subscript[ω, μ]];
               (* update ωRules and ωRules *)
               ωRules = AppendTo[ωRules, ωCondition[[1, 1]]];
               dsolG = ExpandAll[dsol //. ωRules];
               xRules = AppendTo[xRules, Subscript[x, μ] ->
                               Function[t, Evaluate[dsolG[[1, 1, 2]]]]],
             {μ, 2, o - 2}]
```

The resulting conditions on the ω_k are of the form $\omega_k = f(\omega_0, J_2, \ldots, J_{k+1}, c_{c,0}, \ldots, c_{c,k-2}, c_{s,1}, \ldots, c_{s,k-2})$. As a function of k, they quickly grow in size.

```
    In[27]:= {ByteCount /@ ωRules, LeafCount /@ ωRules,
             (* number of terms in the sum *)
             Length[Cases[#, _Plus, Infinity][[1]]]& /@ Rest[ωRules]}
    Out[27]= {{96, 768, 792, 4184, 4664, 15344, 21432, 58936, 84952, 196648},
             {5, 37, 39, 207, 235, 763, 1079, 2882, 4238, 9555}, {2, 2, 11, 11, 36, 46, 124, 166, 372}}
```

Here are the first three of the resulting conditions on the ω_k.

```
    In[28]:= Take[ωRules, 3] // TraditionalForm
```

Out[28]//TraditionalForm=

$$\left\{\omega_1 \to 0,\ \omega_2 \to -\frac{(5\,J_2^2 + 3\,J_3\,\omega_0^2)\,c_{s,0}^2}{48\,\omega_0^3},\ \omega_3 \to -\frac{(5\,J_2^2 + 3\,J_3\,\omega_0^2)\,c_{c,1}\,c_{s,0}}{24\,\omega_0^3}\right\}$$

Here is an abbreviated version for the condition on ω_{10}.

In[29]:= ωRules[[-1]] // Short[#, 12]& // TraditionalForm

Out[29]//TraditionalForm=

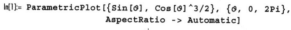

$$\omega_{10} \to -\frac{1}{1155790798848000\,\omega_0^{19}}$$

$$(6531840\,J_{11}\,c_{s,0}^{10}\,\omega_0^{18} + 6270566400\,J_9\,c_{c,2}\,c_{s,0}^7\,\omega_0^{18} + 21946982400\,J_9\,c_{c,1}^2\,c_{s,0}^6\,\omega_0^{18} + 376233984000\,J_7\,c_{c,4}\,c_{s,0}^5\,\omega_0^{18} +$$
$$940584960000\,J_7\,c_{c,2}^2\,c_{s,0}^4\,\omega_0^{18} + 1881169920000\,J_7\,c_{c,1}\,c_{c,3}\,c_{s,0}^4\,\omega_0^{18} + \ll 398 \gg + 1084527234000\,J_2^7\,J_4\,c_{s,0}^{10}\,\omega_0^4 +$$
$$8730857520000\,J_2^8\,c_{c,2}\,c_{s,0}^7\,\omega_0^4 + 23512232520000\,J_2^8\,c_{c,1}^2\,c_{s,0}^6\,\omega_0^4 + 5240271960000\,J_2^8\,c_{s,0}^6\,c_{s,1}^2\,\omega_0^4 +$$
$$2295577811625\,J_2^8\,J_3\,c_{s,0}^{10}\,\omega_0^2 + 491692665625\,J_2^{10}\,c_{s,0}^{10})$$

37. Implicitization of Various Surfaces

a) Here is a possible cross section of a "spindle".

In[1]:= ParametricPlot[{Sin[θ], Cos[θ]^3/2}, {θ, 0, 2Pi},
                    AspectRatio -> Automatic]

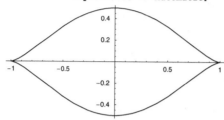

This curve has the following implicit representation.

In[2]:= Numerator[Together[#]]& /@ ({x, y} - {Sin[θ], Cos[θ]^3/2} /.
     {Sin[θ] -> 2s/(1 + s^2), Cos[θ] -> (1 - s^2)/(1 + s^2)})

Out[2]= {-2 s + x + s² x, -1 + 3 s² - 3 s⁴ + s⁶ + 2 y + 6 s² y + 6 s⁴ y + 2 s⁶ y}

In[3]:= Factor[Resultant[##, s]& @@ %]

Out[3]= 64 (-1 + 3 x² - 3 x⁴ + x⁶ + 4 y²)

Here is a visualization of the implicit representation. The right graphic shows this curve rotated around the *x*-axis. The resulting formula describes a spindle.

In[4]:= Show[GraphicsArray[
    Block[{$DisplayFunction = Identity},
     {(* cross section *)
     ContourPlot[-1 + 3 x^2 - 3 x^4 + x^6 + 4 y^2,
            {x, -3/2, 3/2}, {y, -1, 1},
            PlotPoints -> 80, Contours -> {0}],
     (* parametrized rotated curve *)
     ParametricPlot3D[Evaluate[
     {{1, 0, 0}, {0, Cos[φ], Sin[φ]}, {0, -Sin[φ], Cos[φ]}}.
     {Sin[θ], 1/2 Cos[θ]^3, 0}], {φ, 0, Pi}, {θ, 0, 2Pi},
              PlotPoints -> {20, 21}]}]]]

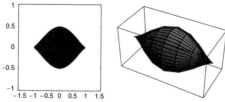

First, we generate a rational parametrization for the spindle.

In[5]:= {x, y, z} - {Sin[θ], Cos[φ] Cos[θ]^3/2, -Cos[θ]^3 Sin[φ]/2} /.
        {Sin[θ] -> 2 s/(1 + s^2), Cos[θ] -> (1 - s^2)/(1 + s^2),
         Sin[φ] -> 2 u/(1 + u^2), Cos[φ] -> (1 - u^2)/(1 + u^2)}

Out[5]= $\left\{-\dfrac{2\,s}{1 + s^2} + x,\ -\dfrac{(1 - s^2)^3\,(1 - u^2)}{2\,(1 + s^2)^3\,(1 + u^2)} + y,\ \dfrac{(1 - s^2)^3\,u}{(1 + s^2)^3\,(1 + u^2)} + z\right\}$

Then, we generate an equivalent system of polynomials.

In[6]:= Numerator[Together[#]]& /@ %

Out[6]= $\{-2\,s + x + s^2\,x,\ -1 + 3\,s^2 - 3\,s^4 + s^6 + u^2 - 3\,s^2\,u^2 + 3\,s^4\,u^2 -$
        $s^6\,u^2 + 2\,y + 6\,s^2\,y + 6\,s^4\,y + 2\,s^6\,y + 2\,u^2\,y + 6\,s^2\,u^2\,y + 6\,s^4\,u^2\,y + 2\,s^6\,u^2\,y,$
        $u - 3\,s^2\,u + 3\,s^4\,u - s^6\,u + z + 3\,s^2\,z + 3\,s^4\,z + s^6\,z + u^2\,z + 3\,s^2\,u^2\,z + 3\,s^4\,u^2\,z + s^6\,u^2\,z\}$

From the polynomials, we eliminate the parameterization variables s and u and get rid of unimportant factors.

In[7]:= Factor /@ GroebnerBasis[%, {x, y, z}, {s, u},
                    MonomialOrder -> EliminationOrder]

Out[7]= $\{(-1 + x)^2\,(1 + x)^2\,(-1 + 3\,x^2 - 3\,x^4 + x^6 + 4\,y^2 + 4\,z^2)\}$

Here is a graphics check for the correctness of the so-obtained result—we display the surface defined by the implicit formula.

In[8]:= Needs["Graphics`ContourPlot3D`"]

In[9]:= ContourPlot3D[-1 + 3x^2 - 3x^4 + x^6 + 4y^2 + 4z^2,
              {x, -1, 1}, {y, -1/2, 1/2}, {z, -1/2, 1/2},
              PlotPoints -> {30, 20, 20}, MaxRecursion -> 0,
              PlotRange -> All]

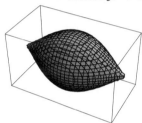

In this case, we could have obtained the implicit equation easier. Because the implicit equation of the cross section had the form $p_1(x) + p_2(y)$, because of symmetry, the 3D polynomial must have the form $p_1(x) + p_2(y) + p_2(z)$.

b) Instead of taking x== (Cos[φ] Sin[θ])^(1/3), we use x^3==Cos[φ] Sin[θ]. This has the advantage that everything is polynomial in x, y, and z. Here, rational parametrizations for the trigonometric expressions that parametrically define a sphere are constructed.

In[1]:= {x^3, y^3, z^3} - {Cos[φ] Sin[θ], Sin[φ] Sin[θ], Cos[θ]} /.
        {Sin[θ] -> 2 s/(1 + s^2), Cos[θ] -> (1 - s^2)/(1 + s^2),
         Sin[φ] -> 2 u/(1 + u^2), Cos[φ] -> (1 - u^2)/(1 + u^2)}

Out[1]= $\left\{-\dfrac{2\,s\,(1 - u^2)}{(1 + s^2)\,(1 + u^2)} + x^3,\ -\dfrac{4\,s\,u}{(1 + s^2)\,(1 + u^2)} + y^3,\ -\dfrac{1 - s^2}{1 + s^2} + z^3\right\}$

In[2]:= Numerator[Together[#]]& /@ %

Out[2]= $\{-2\,s + 2\,s\,u^2 + x^3 + s^2\,x^3 + u^2\,x^3 + s^2\,u^2\,x^3,\ -4\,s\,u + y^3 + s^2\,y^3 + u^2\,y^3 + s^2\,u^2\,y^3,\ -1 + s^2 + z^3 + s^2\,z^3\}$

Eliminating the parametrization variables s and u gives the following result.

In[3]:= Factor[GroebnerBasis[%, {x, y, z}, {s, u},
                    MonomialOrder -> EliminationOrder]]

Out[3]= $\{(-1 + z)\,(1 + z + z^2)\,(-1 + x^6 + y^6 + z^6)\}$

Obviously, the last factor is the equation needed.

Here are pictures of the parametric cube rooted sphere and the implicit one.

In[4]:= Needs["Graphics`ContourPlot3D`"]

```
Show[GraphicsArray[
Block[{$DisplayFunction = Identity},
{(* parametric form *)
 Module[{cubeRoot}, (* use real root *)
     cubeRoot[x_] := If[x > 0, x^(1/3), -(-x)^(1/3)];
     ParametricPlot3D[Evaluate[cubeRoot /@
                     {Cos[φ] Sin[θ], Sin[φ] Sin[θ], Cos[θ]}],
                     {φ, 0, 2 Pi}, {θ, 0, Pi}]],
 (* implicit form *)
 ContourPlot3D[x^6 + y^6 + z^6 - 1, {x, -1, 1}, {y, -1, 1}, {z, -1, 1},
               PlotPoints -> 20, MaxRecursion -> 0,
               PlotRange -> All, Axes -> True]}]]]
```

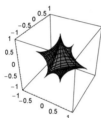

c) Here is a picture of the cubed sphere.

In[1]:= `ParametricPlot3D[Evaluate[{Cos[φ] Sin[θ], Sin[φ] Sin[θ], Cos[θ]}^3],`
                   `{φ, 0, 2Pi}, {θ, 0, Pi},`
                   `PlotRange -> All, PlotPoints -> 30]`

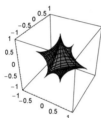

In a similar way as the last example, we replace the trigonometric functions by corresponding rational parametrizations and polynomialize everything.

In[2]:= `{x, y, z} - {Cos[φ] Sin[θ], Sin[φ] Sin[θ], Cos[θ]}^3 /.`
        `{Sin[θ] -> 2 s/(1 + s^2), Cos[θ] -> (1 - s^2)/(1 + s^2),`
        `Sin[φ] -> 2 u/(1 + u^2), Cos[φ] -> (1 - u^2)/(1 + u^2)}`

Out[2]= $\left\{-\dfrac{8\,s^3\,(1-u^2)^3}{(1+s^2)^3\,(1+u^2)^3} + x, -\dfrac{64\,s^3\,u^3}{(1+s^2)^3\,(1+u^2)^3} + y, -\dfrac{(1-s^2)^3}{(1+s^2)^3} + z\right\}$

In[3]:= `eqs = Numerator[Together[#]]& /@ %`

Out[3]= $\{-8\,s^3 + 24\,s^3\,u^2 - 24\,s^3\,u^4 + 8\,s^3\,u^6 + x + 3\,s^2\,x + 3\,s^4\,x + s^6\,x + 3\,u^2\,x + 9\,s^2\,u^2\,x + 9\,s^4\,u^2\,x +$
$3\,s^6\,u^2\,x + 3\,u^4\,x + 9\,s^2\,u^4\,x + 9\,s^4\,u^4\,x + 3\,s^6\,u^4\,x + u^6\,x + 3\,s^2\,u^6\,x + 3\,s^4\,u^6\,x + s^6\,u^6\,x,$
$-64\,s^3\,u^3 + y + 3\,s^2\,y + 3\,s^4\,y + s^6\,y + 3\,u^2\,y + 9\,s^2\,u^2\,y + 9\,s^4\,u^2\,y + 3\,s^6\,u^2\,y +$
$3\,u^4\,y + 9\,s^2\,u^4\,y + 9\,s^4\,u^4\,y + 3\,s^6\,u^4\,y + u^6\,y + 3\,s^2\,u^6\,y + 3\,s^4\,u^6\,y + s^6\,u^6\,y,$
$-1 + 3\,s^2 - 3\,s^4 + s^6 + z + 3\,s^2\,z + 3\,s^4\,z + s^6\,z\}$

Unfortunately, this time a call to `GroebnerBasis` will not return in a reasonable time.

In[4]:= `TimeConstrained[`
            `Factor[GroebnerBasis[eqs, {x, y, z}, {s, u},`
                `MonomialOrder -> EliminationOrder]], 100]`

Out[4]= `$Aborted`

So, we guide the calculation by hand. The last equation does not contain the variable *u*; so, we eliminate u from the first two equations.

In[5]:= `Short[polyIns = Factor[Resultant[eqs[[1]], eqs[[2]], u]], 12]`

Out[5]//Short= $68719476736\ s^{18}$

$(-262144\ s^{18} + 12288\ s^{12}\ x^2 + 73728\ s^{14}\ x^2 + 184320\ s^{16}\ x^2 + 245760\ s^{18}\ x^2 + 184320\ s^{20}\ x^2 +$
$73728\ s^{22}\ x^2 + 12288\ s^{24}\ x^2 - 192\ s^6\ x^4 - 2304\ s^8\ x^4 - 12672\ s^{10}\ x^4 - 42240\ s^{12}\ x^4 - 95040\ s^{14}\ x^4 -$
$152064\ s^{16}\ x^4 - 177408\ s^{18}\ x^4 - 152064\ s^{20}\ x^4 - 95040\ s^{22}\ x^4 - 42240\ s^{24}\ x^4 - 12672\ s^{26}\ x^4 -$
$2304\ s^{28}\ x^4 - 192\ s^{30}\ x^4 + x^6 + 18\ s^2\ x^6 + 153\ s^4\ x^6 + 816\ s^6\ x^6 + 3060\ s^8\ x^6 + 8568\ s^{10}\ x^6 +$
$18564\ s^{12}\ x^6 + 31824\ s^{14}\ x^6 + 43758\ s^{16}\ x^6 + 48620\ s^{18}\ x^6 + 43758\ s^{20}\ x^6 + 31824\ s^{22}\ x^6 +$
$18564\ s^{24}\ x^6 + 8568\ s^{26}\ x^6 + \ll73\gg\ + 2448\ s^6\ x^2\ y^4 + 9180\ s^8\ x^2\ y^4 + 25704\ s^{10}\ x^2\ y^4 +$
$55692\ s^{12}\ x^2\ y^4 + 95472\ s^{14}\ x^2\ y^4 + 131274\ s^{16}\ x^2\ y^4 + 145860\ s^{18}\ x^2\ y^4 + 131274\ s^{20}\ x^2\ y^4 +$
$95472\ s^{22}\ x^2\ y^4 + 55692\ s^{24}\ x^2\ y^4 + 25704\ s^{26}\ x^2\ y^4 + 9180\ s^{28}\ x^2\ y^4 + 2448\ s^{30}\ x^2\ y^4 + 459\ s^{32}\ x^2\ y^4 +$
$54\ s^{34}\ x^2\ y^4 + 3\ s^{36}\ x^2\ y^4 + y^6 + 18\ s^2\ y^6 + 153\ s^4\ y^6 + 816\ s^6\ y^6 + 3060\ s^8\ y^6 + 8568\ s^{10}\ y^6 +$
$18564\ s^{12}\ y^6 + 31824\ s^{14}\ y^6 + 43758\ s^{16}\ y^6 + 48620\ s^{18}\ y^6 + 43758\ s^{20}\ y^6 + 31824\ s^{22}\ y^6 +$
$18564\ s^{24}\ y^6 + 8568\ s^{26}\ y^6 + 3060\ s^{28}\ y^6 + 816\ s^{30}\ y^6 + 153\ s^{32}\ y^6 + 18\ s^{34}\ y^6 + s^{36}\ y^6)$

The polynomial `polyIns` contains only even powers of s, as the last equation of the polynomialized equation does.

In[6]:= `Cases[polyIns, s^_, Infinity] // Union`

Out[6]= $\{s^2,\ s^4,\ s^6,\ s^8,\ s^{10},\ s^{12},\ s^{14},\ s^{16},\ s^{18},\ s^{20},\ s^{22},\ s^{24},\ s^{26},\ s^{28},\ s^{30},\ s^{32},\ s^{34},\ s^{36}\}$

Thinking of the formula `Resultant[p[x^n], q[x^n], x]==Resultant[p[y], q[y], y]^n`, where *n* is integer, we introduce a new parameter S==s^2. We use the nontrivial factor of `polyIns` only.

In[7]:= `(res = Factor[Resultant[polyIns[[-1]] /. s^(n_) :> S^(n/2),`
         `eqs[[-1]] /. s^(n_) :> S^(n/2), S]]) // Short[#, 8]&`

Out[7]//Short= 18014398509481984

$(-1 + 9\ x^2 - 36\ x^4 + 84\ x^6 - 126\ x^8 + 126\ x^{10} - 84\ x^{12} + 36\ x^{14} - 9\ x^{16} + x^{18} + 9\ y^2 + 9\ x^2\ y^2 - 234\ x^4\ y^2 +$
$711\ x^6\ y^2 - 990\ x^8\ y^2 + 711\ x^{10}\ y^2 - 234\ x^{12}\ y^2 + 9\ x^{14}\ y^2 + 9\ x^{16}\ y^2 - 36\ y^4 - 234\ x^2\ y^4 -$
$513\ x^4\ y^4 + 2961\ x^6\ y^4 - 2961\ x^8\ y^4 + 513\ x^{10}\ y^4 + 234\ x^{12}\ y^4 + 36\ x^{14}\ y^4 + 84\ y^6 + 711\ x^2\ y^6 +$
$\ll219\gg\ + 1512\ x^2\ y^2\ z^{10} - 1512\ x^4\ y^2\ z^{10} - 711\ x^6\ y^2\ z^{10} + 513\ y^4\ z^{10} - 1512\ x^2\ y^4\ z^{10} +$
$513\ x^4\ y^4\ z^{10} + 711\ y^6\ z^{10} - 711\ x^2\ y^6\ z^{10} + 126\ y^8\ z^{10} - 84\ z^{12} - 234\ x^2\ z^{12} + 234\ x^4\ z^{12} +$
$84\ x^6\ z^{12} - 234\ y^2\ z^{12} - 3339\ x^2\ y^2\ z^{12} - 234\ x^4\ y^2\ z^{12} + 234\ y^4\ z^{12} - 234\ x^2\ y^4\ z^{12} + 84\ y^6\ z^{12} +$
$36\ z^{14} + 9\ x^2\ z^{14} + 36\ x^4\ z^{14} + 9\ y^2\ z^{14} - 9\ x^2\ y^2\ z^{14} + 36\ y^4\ z^{14} - 9\ z^{16} + 9\ x^2\ z^{16} + 9\ y^2\ z^{16} + z^{18})$

Here, the cubed sphere is shown using the implicit representation. We show the contour surfaces for various values of the calculated polynomial.

In[8]:= `Needs["Graphics`ContourPlot3D`"]`

```
Show[GraphicsArray[Function[c,
(* one-eight of the surface *)
cp3D = ContourPlot3D[Evaluate[N[res[[-1]]]],
                     {x, 0, 3/2}, {y, 0, 3/2}, {z, 0, 3/2},
                     Contours -> {c}, PlotPoints -> {20, 4},
                     MaxRecursion -> 1, PlotRange -> All,
                     DisplayFunction -> Identity];
Graphics3D[{EdgeForm[], (* form other surface patches *)
       {#, Map[{-1, 1, 1}#&, #, {-2}]}&[
        {#, Map[{1, -1, 1}#&, #, {-2}]}&[
         {#, Map[{1, 1, -1}#&, #, {-2}]}&[
          Cases[cp3D, _Polygon, Infinity]]]]}]] /@
                (* contour values *) {0, 1, -1/4}]]
```

d) The following yields a parametrization of the surface under consideration.

```
In[1]:= parametrization =
        {3 Cos[φ1] + Cos[φ1]( Cos[φ1] Cos[φ2]^3 + Sin[φ1] Sin[φ2]^3),
         3 Sin[φ1] + Sin[φ1]( Cos[φ1] Cos[φ2]^3 + Sin[φ1] Sin[φ2]^3),
                    (-Sin[φ1] Cos[φ2]^3 + Cos[φ1] Sin[φ2]^3)};
```

Here, a picture of the twisted torus is shown.

```
In[2]:= ParametricPlot3D[Evaluate[parametrization], {φ1, 0, 2 Pi}, {φ2, 0, 2 Pi},
                         PlotPoints -> {72, 23}]
```

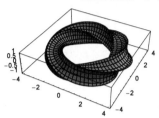

Using `GroebnerBasis`, we eliminate the parameter variables.

```
In[3]:= Short[res = GroebnerBasis[Join[parametrization -
                {x, y, z}, {Sin[φ1]^2 + Cos[φ1]^2 - 1,
                            Sin[φ2]^2 + Cos[φ2]^2 - 1}], {x, y, z},
                {Cos[φ1], Sin[φ1], Cos[φ2], Sin[φ2]},
                MonomialOrder -> EliminationOrder], 6]
```

$$Out[3]//Short= \quad \{262144\, x^8 - 245760\, x^{10} + 89088\, x^{12} - 15680\, x^{14} + 1392\, x^{16} - 60\, x^{18} + x^{20} + 3288064\, x^6\, y^2 -$$
$$1835328\, x^8\, y^2 + 533016\, x^{10}\, y^2 - 95018\, x^{12}\, y^2 + 9516\, x^{14}\, y^2 - 486\, x^{16}\, y^2 + 10\, x^{18}\, y^2 + 10834809\, x^4\, y^4 -$$
$$6402948\, x^6\, y^4 + 1684566\, x^8\, y^4 - 281814\, x^{10}\, y^4 + \ll322\gg + 324\, x^5\, y^3\, z^9 - 324\, x^3\, y^5\, z^9 - 324\, x\, y^7\, z^9 +$$
$$48\, x^8\, z^{10} + 6\, x^{10}\, z^{10} + 246\, x^6\, y^2\, z^{10} + 30\, x^8\, y^2\, z^{10} + 396\, x^4\, y^4\, z^{10} + 60\, x^6\, y^4\, z^{10} + 246\, x^2\, y^6\, z^{10} +$$
$$60\, x^4\, y^6\, z^{10} + 48\, y^8\, z^{10} + 30\, x^2\, y^8\, z^{10} + 6\, y^{10}\, z^{10} + x^8\, z^{12} + 4\, x^6\, y^2\, z^{12} + 6\, x^4\, y^4\, z^{12} + 4\, x^2\, y^6\, z^{12} + y^8\, z^{12}\}$$

Due to the sharp edges, a picture of the implicit form needs many plot points. We display the equicontour surfaces of the resulting polynomial for three contour values. The surface of the contour value 0 represents the original parametrized surface.

```
In[4]:= Needs["Graphics`ContourPlot3D`"]

        Show[GraphicsArray[
        ContourPlot3D[Evaluate[N[res[[-1]]]],
                      {x, -4.5, 4.5}, {y, -4.5, 4.5}, {z, -1.5, 1.5},
                      PlotPoints -> {{20, 4}, {20, 4}, {10, 4}},
                      MaxRecursion -> 1, PlotRange -> All,
                      DisplayFunction -> Identity, Contours -> {#}]& /@
                      (* contour values *) {-10^8, 0, 10^8}]]
```

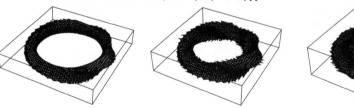

e) The following represents a parametric version of the rotated disk.

```
In[1]:= parametrization = {{Cos[φ], Sin[φ], 0}, {-Sin[φ], Cos[φ], 0},
                           {0, 0, 1}}.(Cos[φ/4]^2 {Cos[s], 0, Sin[s]});
```

Here is a graphic showing this surface. To get a better view of the inner parts, we cut some holes in the polygons of this surface (right graphic).

```
In[2]:= makeHole[Polygon[l_], factor_] :=
        Module[{mp = Plus @@ l/Length[l], newPoints, nOld, nNew},
```

```
                    newPoints = (mp + factor(# - mp))& /@ l;
                    nOld = Partition[Append[#, First[#]]&[l], 2, 1];
                    nNew = Partition[Append[#, First[#]]&[newPoints], 2, 1];
                    {MapThread[Polygon[Join[#1, Reverse[#2]]]&, {nOld, nNew}],
                    Line[Append[#, First[#]]&[newPoints]}]
```

In[3]:=
```
        Show[GraphicsArray[
        Block[{$DisplayFunction = Identity},
         {(* parametrized surface *)
          ParametricPlot3D[Evaluate[parametrization],
                          {s, -Pi, Pi}, {φ, 0, 4Pi}, PlotRange -> All],
          (* lower half of this surfaces with holed polygons *)
          Show[Graphics3D[
          {EdgeForm[], Thickness[0.002], (* add some color *)
           SurfaceColor[RGBColor[1, 0, 0], RGBColor[0, 0, 1], 2.12],
           Cases[ParametricPlot3D[Evaluate[
                 parametrization], {s, -Pi, Pi}, {φ, 0, 4Pi},
              PlotRange -> All, DisplayFunction -> Identity,
              PlotPoints -> {20, 40}], _Polygon, Infinity] /.
                  (* cut holes into polygons *)
                  p_Polygon :> makeHole[p, 0.72]}],
                      PlotRange -> {{-1, 1}, {-1, 1}, {-1, 0}}]}]]]
```

Again, using the GroebnerBasis command with EliminationOrder easily yields the implicit representation we are looking for.

In[4]:= res = Factor /@ GroebnerBasis[Join[TrigExpand[
            parametrization /. {φ -> 4φ}] - {x, y, z},
            {Cos[φ]^2 + Sin[φ]^2 - 1, Cos[s]^2 + Sin[s]^2 - 1}],
            {x, y, z}, {Cos[φ], Sin[φ], Cos[s], Sin[s]}]

Out[4]= $\{-256\,x^6 + 2304\,x^8 - 6144\,x^{10} + 4096\,x^{12} - 608\,x^4\,y^2 + 9344\,x^6\,y^2 - 30720\,x^8\,y^2 + 24576\,x^{10}\,y^2 + y^4 -$
$448\,x^2\,y^4 + 14208\,x^4\,y^4 - 61440\,x^6\,y^4 + 61440\,x^8\,y^4 - 96\,y^6 + 9600\,x^2\,y^6 - 61440\,x^4\,y^6 + 81920\,x^6\,y^6 +$
$2432\,y^8 - 30720\,x^2\,y^8 + 61440\,x^4\,y^8 - 6144\,y^{10} + 24576\,x^2\,y^{10} + 4096\,y^{12} - 256\,x^4\,z^2 + 4608\,x^6\,z^2 -$
$18432\,x^8\,z^2 + 16384\,x^{10}\,z^2 - 352\,x^2\,y^2\,z^2 + 14080\,x^4\,y^2\,z^2 - 73728\,x^6\,y^2\,z^2 + 81920\,x^8\,y^2\,z^2 - 96\,y^4\,z^2 +$
$14336\,x^2\,y^4\,z^2 - 110592\,x^4\,y^4\,z^2 + 163840\,x^6\,y^4\,z^2 + 4864\,y^6\,z^2 - 73728\,x^2\,y^6\,z^2 + 163840\,x^4\,y^6\,z^2 -$
$18432\,y^8\,z^2 + 81920\,x^2\,y^8\,z^2 + 16384\,y^{10}\,z^2 + 2304\,x^4\,z^4 - 18432\,x^6\,z^4 + 24576\,x^8\,z^4 +$
$4736\,x^2\,y^2\,z^4 - 55296\,x^4\,y^2\,z^4 + 98304\,x^6\,y^2\,z^4 + 2432\,y^4\,z^4 - 55296\,x^2\,y^4\,z^4 + 147456\,x^4\,y^4\,z^4 -$
$18432\,y^6\,z^4 + 98304\,x^2\,y^6\,z^4 + 24576\,y^8\,z^4 - 6144\,x^4\,z^6 + 16384\,x^6\,z^6 - 12288\,x^2\,y^2\,z^6 +$
$49152\,x^4\,y^2\,z^6 - 6144\,y^4\,z^6 + 49152\,x^2\,y^4\,z^6 + 16384\,y^6\,z^6 + 4096\,x^4\,z^8 + 8192\,x^2\,y^2\,z^8 + 4096\,y^4\,z^8\}$

The ContourPlot3D verifies the correctness of the calculated implicit form. The middle graphic shows again a picture with holes in the polygons. And the right graphic shows contour lines in the plane $z = 0$.

In[5]:= Needs["Graphics`ContourPlot3D`"]

```
        Show[GraphicsArray[
        Block[{$DisplayFunction = Identity, cp2D, cp3D, cls},
         {(* implicit surface *)
          cp3D = ContourPlot3D[Evaluate[N[res[[1]]]],
                      {x, -1.15, 1.15}, {y, -1.15, 1.15}, {z, -1.15, 0},
                      PlotPoints -> {25, 25, 18}, MaxRecursion -> 0],
          (* lower half of this surfaces with holed polygons *)
          Graphics3D[{EdgeForm[], Thickness[0.002],
                      SurfaceColor[Hue[0.45], Hue[0.34], 2.12],
```

```
                Cases[cp3D, _Polygon, Infinity] /.
            p_Polygon :> makeHole[p, 0.75]}, PlotRange -> All],
      (* 2D contour lines in cross section plane z == 0*)
      cp2D = ContourPlot[Evaluate[res[[1]] /. z -> 0],
                      {x, -1.15, 1.15}, {y, -1.15, 1.15}, PlotPoints -> 240];
      cls = #[[30]]& /@ Partition[Sort[Flatten[cp2D[[1]]]], 960];
      ListContourPlot[cp2D[[1]], ContourShading -> False,
                    Contours -> cls,  FrameTicks -> None,
                    ContourStyle -> {Thickness[0.003]}]}]]]]
```

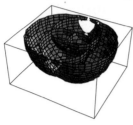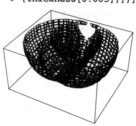

38. Riemann Surface of Kronig–Penney Dispersion Relation

This is the function to be drawn.

```
In[1]:= f[e_] = ArcCos[Cos[Sqrt[e]] + 4/Sqrt[e] Sin[Sqrt[e]]];
```

Here are the real and the imaginary parts slightly above and below the imaginary axis.

```
In[2]:= (* small number to avoid discontinuities *) ε = 10^-8;
      Show[GraphicsArray[
      Function[pm, Plot[Evaluate[#[f[er  + pm I ε]]], {er, -50, 200},
                      DisplayFunction -> Identity, Frame -> True,
                      Axes -> False]& /@ {Re, Im}] /@ {+1, -1}]]
```

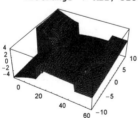

Trying to use `Plot3D[Im[...]]` results in a graphic with some discontinuities.

```
In[4]:= Plot3D[Evaluate[Im[f[er + I ei]]], {er, -10, 60}, {ei, -10, 10},
            PlotRange -> All, PlotPoints -> 80]
```

Most of the discontinuities are all of the following type. The discontinuities seen in the picture are due to the branch cuts of `ArcCos` (right graphic).

```
In[5]:= Show[GraphicsArray[
        Block[{$DisplayFunction = Identity},
          {(* shape around branch cuts *)
```

```
Plot3D[Im[ArcCos[3/2 - (er + I ei)^2]], {er, -3/2, 3/2}, {ei, -1, 1},
    PlotRange -> All, PlotPoints -> 40],
(* branch cuts of ArcCos *)
Plot3D[Im[-ArcCos[(er + I ei)]], {er, -2, 2}, {ei, -1, 1},
    PlotRange -> All, PlotPoints -> 40]}]]]
```

The discontinuities along the real axis of $\text{Im}[f[e]]$ start and end when the expression $\cos(\sqrt{e}) + 4/\sqrt{e} \sin(\sqrt{e})$ equals ± 1. Here are these values calculated.

```
In[6]:= bandEdges =
    Apply[(e /. FindRoot[Cos[#1] == Cos[Sqrt[e]] + 4/Sqrt[e] Sin[Sqrt[e]],
            {e, #2}, AccuracyGoal -> 25, WorkingPrecision -> 30])&,
        {{0, 4}, {Pi, 10}, {Pi, 20}, {0, 39}, {0, 51},
        {Pi, 88}, {Pi, 103}, {0, 156}, {0, 173}, {Pi, 250}}, {1}]
Out[6]= {4.6386303295802987739506764064, 9.8696044010893586188344909998,
    20.9567972007820985332003083282, 39.47841760435743447533796399995,
    53.10320127388161213227974555030, 88.82643960980422756951041899989,
    103.50966939047474099766319566, 157.91367041742973790135185598,
    173.09789879629047205316861507070, 246.74011002723965470862274997}
```

Between the bandEdges, the function $f[e]$ is purely real.

```
In[7]:= pic = Show[Apply[
    Plot[Evaluate[f[e]], {e, #1 + ε, #2 - ε},
        DisplayFunction -> Identity]&, Partition[bandEdges, 2], {1}],
        DisplayFunction -> $DisplayFunction, PlotRange -> All]
```

Adding multiples of π, we can redraw the last picture in the following way. Slightly more common is the right picture, which is obtained by interchanging K and e.

```
In[8]:= Show[GraphicsArray[
    Block[{$DisplayFunction = Identity},
    {(* show 𝒦(ε) *)
     Graphics[{{Thickness[0.01], GrayLevel[0.7],
      Line[Table[{e, Sqrt[e]}, {e, 0, 250}]]},
      MapIndexed[If[OddQ[#2[[1]]],
                    {#[[1]], +#1[[2]] + (#2[[1]] - 1) Pi},
                    {#[[1]], -#1[[2]] + #2[[1]] Pi}]&,
                        Cases[#, _Line, Infinity]& /@ pic[[1]],
                {-2}]} // N, PlotRange -> All, Frame -> True],
    (* show ε(𝒦) *)
     Graphics[{#, Map[{-1, 1}#&, N[#], {-2}]}& @
      MapIndexed[If[OddQ[#2[[1]]],
                    {+#1[[2]] + (#2[[1]] - 1) Pi, #1[[1]]},
                    {-#1[[2]] + #2[[1]] Pi, #1[[1]]}]&,
```

```
Cases[#, _Line, Infinity]& /@ pic[[1]], {-2}],
  PlotRange -> All, Frame -> True,
  FrameTicks -> {Table[{i Pi, i Pi}, {i, -5, 5}],
                 Automatic, None, None}}]]]]
```

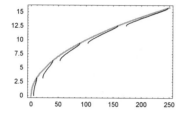

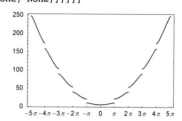

Now, we must deal with the lines of singularities that originate between two band edges and go into the complex plane. This kind of singularity is caused by having the imaginary part of f[e] change its sign. Here are the lines where this happens.

In[9]:= ContourPlot[Im[f[er + I ei]], {er, -100, 250}, {ei, -30, 30},
          PlotPoints -> 100, Contours -> {0}, ContourShading -> False]

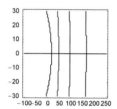

We can get an implicit equation describing these lines by taking the real part of f[e].

In[10]:= Imf = Plus @@ (Cases[Expand[ComplexExpand[Cos[Sqrt[er + I ei]] +
                4/(Sqrt[er + I ei]) Sin[Sqrt[er + I ei]]] /.
           {Abs[I ei + er] -> Sqrt[er^2 + ei^2],
            Arg[I ei + er] -> ArcTan[er, ei]}], _Complex _]/I)

$$
\text{Out[10]=} \quad -\frac{1}{(ei^2 + er^2)^{1/4}}\left(4 \operatorname{Cosh}\left[(ei^2 + er^2)^{1/4} \operatorname{Sin}\left[\frac{1}{2} \operatorname{ArcTan}[er, ei]\right]\right]\right.
$$
$$
\left. \operatorname{Sin}\left[\frac{1}{2} \operatorname{ArcTan}[er, ei]\right] \operatorname{Sin}\left[(ei^2 + er^2)^{1/4} \operatorname{Cos}\left[\frac{1}{2} \operatorname{ArcTan}[er, ei]\right]\right]\right) +
$$
$$
\frac{1}{(ei^2 + er^2)^{1/4}}\left(4 \operatorname{Cos}\left[\frac{1}{2} \operatorname{ArcTan}[er, ei]\right] \operatorname{Cos}\left[(ei^2 + er^2)^{1/4} \operatorname{Cos}\left[\frac{1}{2} \operatorname{ArcTan}[er, ei]\right]\right]\right.
$$
$$
\left. \operatorname{Sinh}\left[(ei^2 + er^2)^{1/4} \operatorname{Sin}\left[\frac{1}{2} \operatorname{ArcTan}[er, ei]\right]\right]\right) -
$$
$$
\operatorname{Sin}\left[(ei^2 + er^2)^{1/4} \operatorname{Cos}\left[\frac{1}{2} \operatorname{ArcTan}[er, ei]\right]\right] \operatorname{Sinh}\left[(ei^2 + er^2)^{1/4} \operatorname{Sin}\left[\frac{1}{2} \operatorname{ArcTan}[er, ei]\right]\right]
$$

By viewing er as a function of ei in the last equation, we see that the following expression must be zero.

In[11]:= Imf1 = Imf /. er -> er[ei]

$$
\text{Out[11]=} \quad -\frac{1}{(ei^2 + er[ei]^2)^{1/4}}
$$
$$
\left(4 \operatorname{Cosh}\left[(ei^2 + er[ei]^2)^{1/4} \operatorname{Sin}\left[\frac{1}{2} \operatorname{ArcTan}[er[ei], ei]\right]\right] \operatorname{Sin}\left[\frac{1}{2} \operatorname{ArcTan}[er[ei], ei]\right]\right.
$$
$$
\left. \operatorname{Sin}\left[\operatorname{Cos}\left[\frac{1}{2} \operatorname{ArcTan}[er[ei], ei]\right] (ei^2 + er[ei]^2)^{1/4}\right]\right) + \frac{1}{(ei^2 + er[ei]^2)^{1/4}}
$$
$$
\left(4 \operatorname{Cos}\left[\frac{1}{2} \operatorname{ArcTan}[er[ei], ei]\right] \operatorname{Cos}\left[\operatorname{Cos}\left[\frac{1}{2} \operatorname{ArcTan}[er[ei], ei]\right] (ei^2 + er[ei]^2)^{1/4}\right]\right.
$$
$$
\left. \operatorname{Sinh}\left[(ei^2 + er[ei]^2)^{1/4} \operatorname{Sin}\left[\frac{1}{2} \operatorname{ArcTan}[er[ei], ei]\right]\right]\right) -
$$
$$
\operatorname{Sin}\left[\operatorname{Cos}\left[\frac{1}{2} \operatorname{ArcTan}[er[ei], ei]\right] (ei^2 + er[ei]^2)^{1/4}\right]
$$
$$
\operatorname{Sinh}\left[(ei^2 + er[ei]^2)^{1/4} \operatorname{Sin}\left[\frac{1}{2} \operatorname{ArcTan}[er[ei], ei]\right]\right]
$$

Differentiating this expression with respect to `ei` and solving for `er'[ei]` gives us a differential equation for these lines.

```
In[12]:= Short[ode = Equal @@ (Solve[D[Imf1, ei] == 0, er'[ei]][[1, 1]]), 12]
```

$$
\text{Out[12]//Short= } er'[ei] = \Big(4\,\text{Cos}\big[\tfrac{1}{2}\,\text{ArcTan}[er[ei], ei]\big]^2\,\text{Cos}\big[\text{Cos}\big[\tfrac{1}{2}\,\text{ArcTan}[er[ei], ei]\big]\,(ei^2 + er[ei]^2)^{1/4}\big]
$$

$$
\text{Cosh}\big[(ei^2 + er[ei]^2)^{1/4}\,\text{Sin}\big[\tfrac{1}{2}\,\text{ArcTan}[er[ei], ei]\big]\big]\,er[ei]\,(ei^2 + er[ei]^2)^{1/4} + \ll 19 \gg\Big) \Big/
$$

$$
\Big(4\,ei\,\text{Cos}\big[\tfrac{1}{2}\,\text{ArcTan}[er[ei], ei]\big]^2\,\text{Cos}\big[\text{Cos}\big[\tfrac{1}{2}\,\text{ArcTan}[er[ei], ei]\big]\,(ei^2 + er[ei]^2)^{1/4}\big]
$$

$$
\text{Cosh}\big[(ei^2 + er[ei]^2)^{1/4}\,\text{Sin}\big[\tfrac{1}{2}\,\text{ArcTan}[er[ei], ei]\big]\big]\,(ei^2 + er[ei]^2)^{1/4} +
$$

$$
\ll 14 \gg + 4\,er[ei]\,(ei^2 + er[ei]^2)^{1/4}\,\text{Sin}\big[\tfrac{1}{2}\,\text{ArcTan}[er[ei], ei]\big]^2
$$

$$
\text{Sin}\big[\text{Cos}\big[\tfrac{1}{2}\,\text{ArcTan}[er[ei], ei]\big]\,(ei^2 + er[ei]^2)^{1/4}\big]
$$

$$
\text{Sinh}\big[(ei^2 + er[ei]^2)^{1/4}\,\text{Sin}\big[\tfrac{1}{2}\,\text{ArcTan}[er[ei], ei]\big]\big]\Big)
$$

Let us solve the resulting differential equation numerically. The `fri` are the initial conditions.

```
In[13]:= fr1 = er /. FindRoot[Evaluate[Imf == 0 /. ei -> 1], {er, 15},
                               WorkingPrecision -> 22, AccuracyGoal -> 20]
Out[13]= 14.63566800020600747569
```

```
In[14]:= ε = 10^-9; (* starting slightly away from 0 *)
        nsol1 = NDSolve[{ode, er[1] == fr1}, er, {ei, ε, 10},
                        PrecisionGoal -> 12, MaxSteps -> 5000][[1, 1, 2]]
Out[15]= InterpolatingFunction[{{1.×10⁻⁹, 10.}}, <>]
```

```
In[16]:= fr2 = er /. FindRoot[Evaluate[Imf == 0 /. ei -> 1], {er, 45},
                               WorkingPrecision -> 22, AccuracyGoal -> 20]
Out[16]= 45.96726058494253542007
```

```
In[17]:= nsol2 = NDSolve[{ode, er[1] == fr2}, er, {ei, ε, 10},
                        PrecisionGoal -> 12, MaxSteps -> 5000][[1, 1, 2]]
Out[17]= InterpolatingFunction[{{1.×10⁻⁹, 10.}}, <>]
```

Here are the resulting curves.

```
In[18]:= ParametricPlot[Evaluate[{{nsol1[ei], ei}, {nsol2[ei], ei}}],
                        {ei, ε, 10}]
```

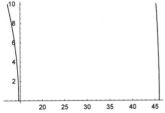

Now, we have the location of all discontinuities and we can subdivide the upper-half-plane into regions in such a way that inside these regions, `f[e]` is a smooth function.

```
In[19]:= makePart[reim_, {left_, right_}, {ppr_, ppi_}] :=
        Module[{erLeft, erRight, ε = 10^-3},
        (* list of data points *)
        Table[{erLeft, erRight} = {left[ei] + ε, right[ei] - ε};
              Table[{er, ei, reim[f[er + I ei]]},
                    {er, erLeft, erRight, (erRight - erLeft)/ppr}],
              {ei, ε, 10, (10 - ε)/ppi}]];
```

```
In[20]:= (* the regions in the parameter plane; cut along branch cuts *)
        regions = {{{-10.&, bandEdges[[1]]&}, {15, 40}},
                   {{bandEdges[[1]]&, bandEdges[[2]]&}, {5, 40}},
                   {{bandEdges[[2]]&, nsol1}, {4, 40}},
```

```
            {{nsol1, bandEdges[[3]]&}, {6, 40}},
            {{bandEdges[[3]]&, bandEdges[[4]]&}, {20, 40}},
            {{bandEdges[[4]]&, nsol2}, {6, 40}},
            {{nsol2, 60.&}, {15, 40}}};
```
In[22]:= ImSurfaceParts = Apply[makePart[Im, ##]&, regions, {1}];

Here, the resulting surface is shown.

In[23]:= Needs["Graphics`Graphics3D`"];

In[24]:= ImPolys1 = Cases[ListSurfacePlot3D[#,
            DisplayFunction -> Identity]& /@ ImSurfaceParts, _Polygon, Infinity];

In[25]:= Show[Graphics3D[ImPolys1]]

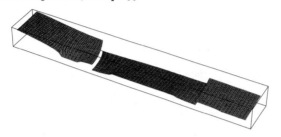

Now, we must calculate all other parts and sheets of the surface. Potential other sheets either come from the other sheet of the square root or the arccos function. Switching to the other sheet of the square root does not change f[e].

In[26]:= f[e] == (f[e] /. Power[e, pmOneHalf_] -> -Power[e, pmOneHalf])

Out[26]= True

Now let us make use of the other sheets of arccos. Rewriting ArcCos in logarithms and square roots shows that we must look at the other sheets of the square root and the logarithm function.

In[27]:= ArcCos[x] // TrigToExp

Out[27]= $\frac{\pi}{2} + i\,\text{Log}\left[i\,x + \sqrt{1 - x^2}\right]$

The other sheets of the logarithm do not contribute here because they are different from the principal sheet by a purely imaginary quantity. The second sheet of the square root produces a second sheet in our picture. The value of f[e] on this sheet is simply the negative of the value from the first sheet. Here, the two sheets for ArcCos are shown.

In[28]:= Show[GraphicsArray[Function[pm,
            Plot3D[Im[Pi/2 + I Log[I (xr + I xi) + pm Sqrt[1 - (xr + I xi)^2]]],
                {xr, -2, 2}, {xi, -1, 1}, PlotPoints -> 20,
                DisplayFunction -> Identity]] /@ {+1, -1}]]

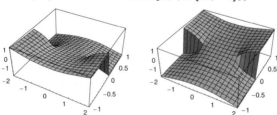

Now, we generate all other parts of the surface and display them together.

In[29]:= ImPolys2 = {ImPolys1, Map[{1, 1, -1} #&, ImPolys1, {-2}]};
         Show[Graphics3D[ImPolys2], BoxRatios -> {1, 1, 1}]

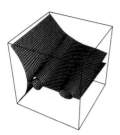

For the more complicated problem of determining the Riemann surface of $e(V)$ (defined implicitly via $\cos(K)=$ $\cos(e^{1/2}) + V\, e^{-1/2} \sin(e^{1/2}))$ as a function of the potential strength V, see [1752], [637].

39. Envelopes of Secants in an Ellipse, Lines Intersecting Four Lines

a) This is the equation of the ellipse under consideration.

In[1]:= `ellipse[x_, y_] := x^2 + 4 y^2 - 4`

This is the implicit representation of a line that joins the two points $\{x1, y1\}$ and $\{x2, y2\}$.

In[2]:= `line[{x_, y_}, {{x1_, y1_}, {x2_, y2_}}] =`
      `    Numerator[Together[y - (y2 - y1)/(x2 - x1) (x - x1) - y1]]`

Out[2]= `x1 y - x2 y - x y1 + x2 y1 + x y2 - x1 y2`

The distance between the two points $\{x1, y1\}$ and $\{x2, y2\}$ is given by the following expression.

In[3]:= `dist[{x1_, y1_}, {x2_, y2_}] = (x1 - x2)^2 + (y1 - y2)^2 - 1`

Out[3]= `-1 + (x1 - x2)² + (y1 - y2)²`

Now, we must calculate the envelope. If $f(x, y, c)$ is a family of curves in the x,y-plane, parametrized by c, then the envelope (an implicit function of x and y) formed by this family is obtained by eliminating c from the two equations $f(x, y, c) = 0$, $\partial f(x, y, c)/\partial c = 0$.

In our case, let the x-coordinate of the first point x_1 be the parameter to be eliminated. y_1 due to the defining equation of the ellipse as well as x_2 and y_2 (due to the distance restriction) are then implicitly depending on x_1.

So, we have the following set of equations (it is important to take the derivatives of the individual equations into account, too, because they are algebraically independent from the equations).

In[4]:= `{(* the first point lies on the ellipse *)`
      `  ellipse[x1, y1[x1]], D[ellipse[x1, y1[x1]], x1],`
      `  (* the second point lies on the ellipse *)`
      `  ellipse[x2[x1], y2[x1]], D[ellipse[x2[x1], y2[x1]], x1],`
      `  (* the distance between the two points is 1 *)`
      `  dist[{x1, y1[x1]}, {x2[x1], y2[x1]}],`
      `  D[dist[{x1, y1[x1]}, {x2[x1], y2[x1]}], x1],`
      `  (* the line between the two points *)`
      `  line[{x, y}, {{x1, y1[x1]}, {x2[x1], y2[x1]}}],`
      `  D[line[{x, y}, {{x1, y1[x1]}, {x2[x1], y2[x1]}}], x1]}`

Out[4]= `{-4 + x1² + 4 y1[x1]², 2 x1 + 8 y1[x1] y1'[x1], -4 + x2[x1]² + 4 y2[x1]²,`
      `  2 x2[x1] x2'[x1] + 8 y2[x1] y2'[x1], -1 + (x1 - x2[x1])² + (y1[x1] - y2[x1])²,`
      `  2 (x1 - x2[x1]) (1 - x2'[x1]) + 2 (y1[x1] - y2[x1]) (y1'[x1] - y2'[x1]),`
      `  x1 y - y x2[x1] - x y1[x1] + x2[x1] y1[x1] + x y2[x1] - x1 y2[x1],`
      `  y - y2[x1] - y x2'[x1] + y1[x1] x2'[x1] - x y1'[x1] + x2[x1] y1'[x1] + x y2'[x1] - x1 y2'[x1]}`

Replacing the nested variables, we get the following polynomial system.

In[5]:= `% /. {x2'[x1] -> x2s, y1'[x1] -> y1s, y2'[x1] -> y2s,`
      `      y1[x1] -> y1, y2[x1] -> y2, x2[x1] -> x2}`

Out[5]= `{-4 + x1² + 4 y1², 2 x1 + 8 y1 y1s, -4 + x2² + 4 y2², 2 x2 x2s + 8 y2 y2s,`
      `  -1 + (x1 - x2)² + (y1 - y2)², 2 (x1 - x2) (1 - x2s) + 2 (y1 - y2) (y1s - y2s),`
      `  x1 y - x2 y - x y1 + x2 y1 + x y2 - x1 y2, y - x2s y + x2s y1 - x y1s + x2 y1s - y2 + x y2s - x1 y2s}`

From these equations, we must eliminate the variables x2s, y1s, y2s, x1, y1, x2, and y2. This is easily done using GroebnerBasis.

```
In[6]:= gb1 = GroebnerBasis[%, {x, y}, {x2s, y1s, y2s, x1, y1, x2, y2},
                    MonomialOrder -> EliminationOrder]
```

Out[6]= $\{-19440 + 29160\,x^2 - 18495\,x^4 + 6165\,x^6 - 1080\,x^8 + 80\,x^{10} + 7776\,y^2 - 3456\,x^2\,y^2 + 6390\,x^4\,y^2 - 3876\,x^6\,y^2 + 624\,x^8\,y^2 + 7344\,y^4 - 22320\,x^2\,y^4 + 1077\,x^4\,y^4 + 1648\,x^6\,y^4 + 5472\,y^6 + 4440\,x^2\,y^6 + 2000\,x^4\,y^6 + 1296\,y^8 + 1152\,x^2\,y^8 + 256\,y^{10}\}$

Here, this region is shown.

```
In[7]:= cp = ContourPlot[Evaluate[%[[1]]], {x, -2, 2}, {y, -1, 1},
                    PlotPoints -> 120, Contours -> {0},
                    ContourStyle -> {Thickness[0.005], RGBColor[0, 0, 1]},
                    ContourShading -> False, AspectRatio -> Automatic]
```

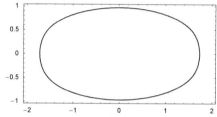

The routine makeLines calculates all line segments such that one point has the *x*-coordinate x.

```
In[8]:= makeLines[x_] :=
        Module[{x1y1Sol = Prepend[#, x1 -> x]& /@
                  (* calculate y coordinate of the first point *)
                  NSolve[x^2 + 4 y1^2 - 4 == 0, y1], x2y2Sol},
                  x2y2Sol = Function[x1y1, Join[x1y1, #]& /@
                  Select[(* calculate the second point *)
                  NSolve[{x2^2 + 4 y2^2 - 4 == 0,
                          (x1 - x2)^2 + (y1 - y2)^2 - 1 == 0} /. x1y1, {x2, y2}],
                  And @@ (Im[#] == 0& /@ ({x2, y2} /. #))&]] /@ x1y1Sol;
                  (* form the line *)
                  Line[{{x1, y1}, {x2, y2}}] /. Flatten[x2y2Sol, 1]]
```

Here are some of these line segments together with the previously calculated envelope.

```
In[9]:= Show[{Graphics[{
            {Thickness[0.002], Hue[0], (* the ellipse itself *)
            Line[Table[{x, Sqrt[4 - x^2]/2}, {x, -2, 2, 1/100}]],
            Line[Table[{x, -Sqrt[4 - x^2]/2}, {x, -2, 2, 1/100}]]},
            {Thickness[0.002], GrayLevel[1/2], (* the line segments *)
            Table[makeLines[x], {x, -2, 2, 1/10}]}}],
            (* the envelope *) cp} // N,
            PlotRange -> All, Frame -> True, AspectRatio -> Automatic]
```

The contour surfaces of the resulting polynomial looks much more complicated as a function of two complex variables.

```
In[10]:= Show[GraphicsArray[Show @ Table[
            (* show nodal lines for changing imaginary part of x and y *)
            ContourPlot[Evaluate[ComplexExpand[#[gb1[[1]]] /.
```

```
                              {x -> xr + I im, y -> yr + I im}]]],
            {xr, -3, 3}, {yr, -2, 2},
            PlotPoints -> 120, Contours -> {0},
            FrameTicks -> None, DisplayFunction -> Identity,
            ContourStyle -> {Thickness[0.002], Hue[im/2.4]},
            ContourShading -> False, AspectRatio -> Automatic],
            {im, 2, 0, -2/60}]& /@
            (* show nodal lines of real and imaginary part *) {Re, Im}]]
```

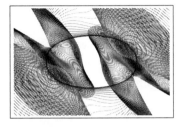 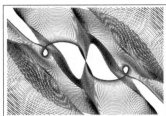

b) Without loss of generality, we will take the first line along the *x*-axis, and the second line in the *x,y*-plane intersecting the first line at origin. Each of the four lines will be parametrized in such a way that *x*-constant is 0 and *x*-parameter is 1. $\xi[i, j]$ is the *j*th component of the *i*th line.

```
In[1]:= fourLines = Flatten[
          Table[a[i, j] + b[i, j] t[i] - ξ[i, j], {i, 4}, {j, 3}]] /.
              {a[_, 1] -> 0, b[_, 1] -> 1} /.
              {a[1, 2] -> 0, a[1, 3] -> 0, b[1, 2] -> 0, b[1, 3] -> 0,
               a[2, 1] -> 0, a[2, 3] -> 0, a[2, 2] -> 0, b[2, 3] -> 0}
Out[1]= {t[1] - ξ[1, 1], -ξ[1, 2], -ξ[1, 3], t[2] - ξ[2, 1], b[2, 2] t[2] - ξ[2, 2], -ξ[2, 3],
         t[3] - ξ[3, 1], a[3, 2] + b[3, 2] t[3] - ξ[3, 2], a[3, 3] + b[3, 3] t[3] - ξ[3, 3],
         t[4] - ξ[4, 1], a[4, 2] + b[4, 2] t[4] - ξ[4, 2], a[4, 3] + b[4, 3] t[4] - ξ[4, 3]}
```

The unknown intersecting line(s) we parametrize in a similar manner. These lines intersect four lines `fourLines` at $\xi[i, j]$.

```
In[2]:= intersectingLines =
          Flatten[Table[A[j] + B[j] T[i] - ξ[i, j], {i, 4}, {j, 3}]] /.
                 {A[1] -> 0, B[1] -> 1};

In[3]:= eqs = Join[fourLines, intersectingLines];
```

Now, we have 24 equations with 4 unknowns A[2], A[3], B[2], and B[3], 9 parameters, and 20 variables to eliminate.

```
In[4]:= {Length[eqs], Length[Union[Cases[eqs, _A | _B, Infinity]]]& /@
                     (* the variables *) {_A | _B, _a | _b, _ξ | _t | _T}}
Out[4]= {24, {4, 4, 4}}
```

In a first step, we eliminate all ξ, t, and T variables.

```
In[5]:= GroebnerBasis[eqs, {},
                 Join[Union[Cases[eqs, _ξ, Infinity]],
                     Table[T[i], {i, 4}], Table[t[i], {i, 4}]],
                 MonomialOrder -> EliminationOrder]
Out[5]= {A[3] B[2] - A[2] B[3], A[3] b[2, 2],
         a[4, 3] b[4, 2] - A[3] b[4, 2] - a[4, 2] b[4, 3] + A[2] b[4, 3] - a[4, 3] B[2] + a[4, 2] B[3],
         a[3, 3] b[3, 2] - A[3] b[3, 2] - a[3, 2] b[3, 3] + A[2] b[3, 3] - a[3, 3] B[2] + a[3, 2] B[3],
         A[2] b[2, 2] B[3]}
```

In a second step, we solve for the unknowns A[2], A[3], B[2], and B[3].

```
In[6]:= GroebnerBasis[%, {A[2], A[3], B[2], B[3]},
                 CoefficientDomain -> RationalFunctions]
Out[6]= {(a[3, 3] a[4, 3] b[3, 2] - a[3, 2] a[4, 3] b[3, 3] - a[3, 3] a[4, 3] b[4, 2] +
          a[3, 3] a[4, 2] b[4, 3]) B[3] + (-a[3, 3] a[4, 2] + a[3, 2] a[4, 3]) B[3]^2,
         a[4, 3] b[3, 3] b[4, 2] - a[3, 3] b[3, 2] b[4, 3] + a[3, 2] b[3, 3] b[4, 3] -
         a[4, 2] b[3, 3] b[4, 3] + (-a[4, 3] b[3, 3] + a[3, 3] b[4, 3]) B[2] +
```

```
      (a[4, 2] b[3, 3] - a[3, 2] b[4, 3]) B[3], A[3],
      -a[3, 3] a[4, 3] b[3, 2] + a[3, 2] a[4, 3] b[3, 3] + a[3, 3] a[4, 3] b[4, 2] -
      a[3, 3] a[4, 2] b[4, 3] + A[2] (-a[4, 3] b[3, 3] + a[3, 3] b[4, 3]) +
      (a[3, 3] a[4, 2] - a[3, 2] a[4, 3]) B[3]}
```

The last Gröbner basis is pseudotriangular. The equation in B[3] is quadratic. This means that generically two lines exist that intersect four (generic) lines.

```
In[7]:= Exponent[#, {A[2], A[3], B[2], B[3]}]& /@ %
```

```
Out[7]= {{0, 0, 0, 2}, {0, 0, 1, 1}, {0, 1, 0, 0}, {1, 0, 0, 1}}
```

40. Shortest Path in a Triangle Billiard

These are the vertices of the triangle under consideration.

```
In[1]:= P1 = {0, 0}; P2 = {1, 0}; P3 = {1, 2}; Q = {qx, qy};
```

The normals of the three triangle sides are the following.

```
In[2]:= normal12 = {-1, 1} Reverse[P2 - P1];
      normal23 = {-1, 1} Reverse[P3 - P2];
      normal31 = {-1, 1} Reverse[P1 - P3];
```

These are the points where the ball is reflected.

```
In[5]:= P12 = P1 + r (P2 - P1);
      P23 = P2 + s (P3 - P2);
      P31 = P3 + t (P1 - P3);
```

The conditions for the reflections result in the following set of equations.

```
In[8]:= n = #/Sqrt[#.#]&;
      eqs1 = Factor[Numerator[Together[#]]]& /@
        {(n[Q   - P12].normal12)^2 - (n[P23 - P12].normal12)^2,
         (n[P12 - P23].normal23)^2 - (n[P31 - P23].normal23)^2,
         (n[Q   - P31].normal31)^2 - (n[P23 - P31].normal31)^2}
```

```
Out[9]= {(-qy + qy r + 2 qx s - 2 r s) (-qy + qy r - 2 qx s + 2 r s),
      16 (1 - r - s + r s - t + r t - s t) (1 - r - s + r s - t + r t + s t),
      (-10 - 6 qx + 8 qy + 10 s + 6 qx s - 8 qy s + 10 t + 10 qx t - 5 qy t - 10 s t)
      (2 - 2 qx - 2 s + 2 qx s - 2 t + 2 qx t - qy t + 2 s t)}
```

Next, we solve for the reflection points.

```
In[10]:= sol = Union[Solve[# == 0& /@ eqs1, {r, s, t}]] // Simplify
```

```
      Solve::svars : Equations may not give solutions for all "solve" variables. More...
```

$$Out[10]= \left\{\left\{r \to \frac{qy - 2\,qx\,s}{qy - 2\,s}, \; t \to \frac{2\,(-1 + qx + s - qx\,s)}{-2 + 2\,qx - qy + 2\,s}\right\},\right.$$

$$\left\{r \to 1, \; t \to \frac{2 - 2\,qx}{2 - 2\,qx + qy}, \; s \to 0\right\}, \; \left\{r \to 1, \; t \to \frac{10 + 6\,qx - 8\,qy}{10 + 10\,qx - 5\,qy}, \; s \to 0\right\},$$

$$\left\{r \to \frac{-2\,qx + qy}{-2 + qy}, \; t \to 0, \; s \to 1\right\}, \; \left\{r \to \frac{2\,qx + qy}{2 + qy}, \; t \to 0, \; s \to 1\right\},$$

$$\left\{r \to qx + 2\,qy, \; t \to \frac{1}{5}\,(5 - qx - 2\,qy), \; s \to \frac{1}{4}\,(-1 + qx + 2\,qy)\right\}, \; \left\{r \to \frac{2\,qx^2 + 5\,qy + 3\,qx\,qy - 2\,qy^2}{2\,qx + 4\,qy},\right.$$

$$t \to \frac{1}{10}\,(10 + 2\,qx^2 - 3\,qy - 2\,qy^2 + qx\,(-4 + 3\,qy)), \; s \to \frac{2\,qx^2 + qy - 2\,qy^2 + qx\,(-2 + 3\,qy)}{2\,(-5 + qx + 2\,qy)}\right\},$$

$$\left\{r \to \frac{qx^2 + qy^2}{qx + 2\,qy}, \; t \to \frac{-5\,qx + qx^2 + qy^2}{5\,qx}, \; s \to \frac{-qx + qx^2 + (-2 + qy)\,qy}{4\,qx - 2\,qy}\right\},$$

$$\left\{r \to \frac{2\,qx^2 + qy\,(-5 + 2\,qy)}{2\,qx - qy}, \; t \to \frac{2\,(5 - 2\,qx + qx^2 - 4\,qy + qy^2)}{5\,(-2 + qy)}, \; s \to \frac{-qx + qx^2 + (-2 + qy)\,qy}{-5 + qx + 2\,qy}\right\}\right\}$$

We have nine possible solutions.

```
In[11]:= Length[sol]
```

```
Out[11]= 9
```

To see which of the solutions of sol describes reflection points inside the triangle, let us look at a "typical" configuration.

```
In[12]:= Block[{qx = 0.4, qy = 0.1, P12, P23, P31},
    Show[GraphicsArray[
    MapIndexed[Graphics[
    {Thickness[0.02], Line[{P1, P2, P3, P1}]},
     Thickness[0.002], Hue[0],  (* billiard path *)
     {P12, P23, P31} = {P1 + r (P2 - P1), P2 + s (P3 - P2),
                        P3 + t (P1 - P3)} /. N[#];
     Line[{{qx, qy}, P12, P23, P31, {qx, qy}}]},
        PlotRange -> All, AspectRatio -> Automatic, FrameTicks -> None,
        Frame -> True, PlotLabel -> #2[[1]] + 1]&, Rest[sol]]]]]
```

So it is the seventh solution that is of interest to us.

```
In[13]:= goodSol =
    {r -> (2 qx^2 + 5 qy + 3 qx qy - 2 qy^2)/(2 qx + 4 qy),
     t -> 1/10*(10 + 2 qx^2 - 3 qy - 2 qy^2 + qx (-4 + 3 qy)),
     s -> (2 qx^2 + qy - 2 qy^2 + qx*(-2 + 3 qy))/(2 (-5 + qx + 2 qy))};
```

Here are 15 different paths with randomly chosen starting points for this case.

```
In[14]:= Block[{qx, qy, P12, P23, P31},
    Show[Graphics[
    {{Thickness[0.02], GrayLevel[0], Line[{P1, P2, P3, P1}]},
     (* random start points *)
     Table[qx = Random[Real, {0.2, 0.4}]; qy = Random[Real, {0, qx}];
     {Thickness[0.002], Hue[Random[]],  (* billiard path *)
     {P12, P23, P31} = {P1 + r (P2 - P1), P2 + s (P3 - P2),
                        P3 + t (P1 - P3)} /. N[goodSol];
     Line[{{qx, qy}, P12, P23, P31, {qx, qy}}]}, {15}]},
        PlotRange -> All, Frame -> True, FrameTicks -> None,
        AspectRatio -> Automatic]]]
```

Here is the square of the length of the balls flight.

```
In[15]:= pL = ((Q    - P12).(Q    - P12) + (P12 - P23).(P12 - P23) +
        (P23 - P31).(P23 - P31) + (Q    - P31).(Q    - P31))
```

$$Out[15]= \ qy^2 + (qx - r)^2 + (-1 + r)^2 + 4 s^2 + t^2 + (-1 + qx + t)^2 + (-2 + qy + 2 t)^2 + (-2 + 2 s + 2 t)^2$$

Let us graphically look if there is a minimum somewhere. The right graphic shows a magnification of the interesting area.

```
In[16]:=
    Show[GraphicsArray[
    Block[{$DisplayFunction = Identity, ε = 10^-6},
    {(* show function over full region *)
    ContourPlot[Evaluate[Sqrt[pL /. sol[[7]]]],
                {qx, ε, 1 - ε}, {qy, ε, 2 - ε}, PlotPoints -> 60,
```

```
                     Contours -> 30, ColorFunction -> (Hue[0.7 (1 - #)]&)],
       (* show function near minima *)
       ContourPlot[Evaluate[Sqrt[pL /. sol[[7]]]],
                     {qx, 0.2, 0.6}, {qy, 0, 0.2}, PlotPoints -> 60,
                     Contours -> 30, ColorFunction -> (Hue[0.7 (1 - #)]&)]}]]]
```

So it seems there is a minimum near qx==0.46, qy==0.08.

In[17]:= **FindMinimum[Evaluate[Sqrt[pL /. sol[[7]]]], {qx, 0.46}, {qy, 0.08}]**

Out[17]= {1.03603, {qx → 0.46341, qy → 0.0842235}}

As a warm-up for the following, let us calculate the exact coordinates of Q. We differentiate the square path length with respect to qx and qy to obtain two equations for qx and qy.

In[18]:= **eqsxy = Factor[Numerator[Together[#]]]& /@**
                 ({D[pL /. goodSol, qx], D[pL /. goodSol, qy]})

Out[18]= {1750 qx^3 – 4650 qx^4 + 4020 qx^5 – 2308 qx^6 + 654 qx^7 – 114 qx^8 + 8 qx^9 – 3125 qx qy +
         12375 qx^2 qy – 29950 qx^3 qy + 30600 qx^4 qy – 20721 qx^5 qy + 6991 qx^6 qy – 1404 qx^7 qy +
         114 qx^8 qy + 9375 qy^2 + 9750 qx qy^2 – 60600 qx^2 qy^2 + 85625 qx^3 qy^2 – 72705 qx^4 qy^2 +
         30456 qx^5 qy^2 – 7238 qx^6 qy^2 + 697 qx^7 qy^2 – 16000 qy^3 – 31800 qx qy^3 + 100950 qx^2 qy^3 –
         122080 qx^3 qy^3 + 68220 qx^4 qy^3 – 19992 qx^5 qy^3 + 2366 qx^6 qy^3 + 13600 qy^4 + 33900 qx qy^4 –
         91320 qx^2 qy^4 + 79840 qx^3 qy^4 – 31080 qx^4 qy^4 + 4788 qx^5 qy^4 – 11960 qy^5 – 12336 qx qy^5 +
         39888 qx^2 qy^5 – 25088 qx^3 qy^5 + 5656 qx^4 qy^5 + 11888 qy^6 – 2688 qx qy^6 – 6048 qx^2 qy^6 +
         3248 qx^3 qy^6 – 6848 qy^7 + 4224 qx qy^7 – 96 qx^2 qy^7 + 2176 qy^8 – 1088 qx qy^8 – 384 qy^9,
         3125 qx^2 – 5250 qx^3 + 2825 qx^4 + 90 qx^5 – 1121 qx^6 + 418 qx^7 – 93 qx^8 + 6 qx^9 –
         9375 qx qy + 375 qx^2 qy + 5225 qx^3 qy + 1075 qx^4 qy – 7677 qx^5 qy + 3597 qx^6 qy –
         973 qx^7 qy + 73 qx^8 qy + 27000 qx qy^2 – 21450 qx^2 qy^2 + 8250 qx^3 qy^2 – 17460 qx^4 qy^2 +
         11052 qx^5 qy^2 – 3906 qx^6 qy^2 + 354 qx^7 qy^2 + 10500 qy^3 – 58100 qx qy^3 + 30900 qx^2 qy^3 –
         11960 qx^3 qy^3 + 11740 qx^4 qy^3 – 6804 qx^5 qy^3 + 812 qx^6 qy^3 – 33800 qy^4 + 51800 qx qy^4 +
         2160 qx^2 qy^4 – 6720 qx^3 qy^4 – 1960 qx^4 qy^4 + 616 qx^5 qy^4 + 31680 qy^5 – 7632 qx qy^5 –
         20304 qx^2 qy^5 + 9744 qx^3 qy^5 – 1008 qx^4 qy^5 – 14144 qy^6 – 5696 qx qy^6 + 11424 qx^2 qy^6 –
         2464 qx^3 qy^6 + 5184 qy^7 + 1088 qx qy^7 – 1472 qx^2 qy^7 – 2688 qy^8 + 384 qx qy^8 + 512 qy^9}

We solve these two equations by eliminating qx and qy.

In[19]:= **Factor[Resultant[##, qx]& @@ eqsxy]**

Out[19]= -1318359375000000000000000000000000000000000000000000
         (-2 + qy)^5 qy^5 (36 – 72 qy + 144 qy^2 – 60 qy^3 + 25 qy^4)
         (-73432 + 1158680 qy – 4054004 qy^2 + 8580275 qy^3 – 11251347 qy^4 +
         10225920 qy^5 – 6050050 qy^6 + 2530125 qy^7 – 783125 qy^8 + 153125 qy^9)

The smallest root of the last polynomial is the one of interest for us.

In[20]:= **Roots[%[[-1]] == 0, qy][[1]]**

Out[20]= qy == Root[-73432 + 1158680 #1 – 4054004 #1^2 + 8580275 #1^3 – 11251347 #1^4 +
         10225920 #1^5 – 6050050 #1^6 + 2530125 #1^7 – 783125 #1^8 + 153125 #1^9 &, 1]

In[21]:= **N[%]**

Out[21]= qy == 0.0842235

In[22]:= **Factor[Resultant[##, qy]& @@ eqsxy]**

Out[22]= $2374945115996160000000000000000000000000000000000000000000000 (-1 + qx)^5 qx^5$
$(121 - 44 qx + 86 qx^2 + 20 qx^3 + 25 qx^4) (-57586 + 205195 qx - 135918 qx^2 - 365590 qx^3 +$
$809654 qx^4 - 365680 qx^5 - 409850 qx^6 + 801750 qx^7 - 577500 qx^8 + 153125 qx^9)$

Again, the smallest root of the last polynomial is the one of interest for us.

In[23]:= `Roots[%[[-1]] == 0, qx][[1]]`

Out[23]= $qx == \text{Root}[-57586 + 205195 \#1 - 135918 \#1^2 - 365590 \#1^3 +$
$809654 \#1^4 - 365680 \#1^5 - 409850 \#1^6 + 801750 \#1^7 - 577500 \#1^8 + 153125 \#1^9 \&, 1]$

In[24]:= `N[%]`

Out[24]= $qx == 0.46341$

The two numerical values agree with our expectations from the above contour graphics.

Now let us calculate the exact value of the path length. We differentiate now (where LL is the square of the path length) with respect to qx and qy to obtain two polynomial equations for qx and qy after togethering.

In[25]:= `pL1 = Factor[Numerator[Together[#]]]&[(LL - pL) /. goodSol]`

Out[25]= $-500 qx^2 + 250 LL qx^2 + 900 qx^3 - 100 LL qx^3 - 1020 qx^4 + 10 LL qx^4 + 536 qx^5 -$
$204 qx^6 + 36 qx^7 - 4 qx^8 - 750 qx\, qy + 1000 LL qx\, qy + 2150 qx^2 qy - 600 LL qx^2 qy -$
$3960 qx^3 qy + 80 LL qx^3 qy + 3020 qx^4 qy - 1518 qx^5 qy + 334 qx^6 qy - 44 qx^7 qy - 2625 qy^2 +$
$1000 LL qy^2 + 2800 qx\, qy^2 - 1200 LL qx\, qy^2 - 4655 qx^2 qy^2 + 240 LL qx^2 qy^2 + 5520 qx^3 qy^2 -$
$4015 qx^4 qy^2 + 1184 qx^5 qy^2 - 193 qx^6 qy^2 + 4200 qy^3 - 800 LL qy^3 - 3740 qx\, qy^3 +$
$320 LL qx\, qy^3 + 3520 qx^2 qy^3 - 4040 qx^3 qy^3 + 1880 qx^4 qy^3 - 412 qx^5 qy^3 - 4220 qy^4 +$
$160 LL qy^4 + 1600 qx\, qy^4 - 360 qx^2 qy^4 + 960 qx^3 qy^4 - 380 qx^4 qy^4 + 2112 qy^5 + 832 qx\, qy^5 -$
$608 qx^2 qy^5 + 32 qx^3 qy^5 - 496 qy^6 - 512 qx\, qy^6 + 272 qx^2 qy^6 + 128 qy^7 + 64 qx\, qy^7 - 64 qy^8$

In[26]:= `eqs = Factor[Numerator[Together[#]]]& /@ {D[pL1, qx], D[pL1, qy]}`

Out[26]= $\{-2 (500 qx - 250 LL qx - 1350 qx^2 + 150 LL qx^2 + 2040 qx^3 - 20 LL qx^3 - 1340 qx^4 + 612 qx^5 - 126 qx^6 +$
$16 qx^7 + 375 qy - 500 LL qy - 2150 qx\, qy + 600 LL qx\, qy + 5940 qx^2 qy - 120 LL qx^2 qy -$
$6040 qx^3 qy + 3795 qx^4 qy - 1002 qx^5 qy + 154 qx^6 qy - 1400 qy^2 + 600 LL qy^2 + 4655 qx\, qy^2 -$
$240 LL qx\, qy^2 - 8280 qx^2 qy^2 + 8030 qx^3 qy^2 - 2960 qx^4 qy^2 + 579 qx^5 qy^2 + 1870 qy^3 -$
$160 LL qy^3 - 3520 qx\, qy^3 + 6060 qx^2 qy^3 - 3760 qx^3 qy^3 + 1030 qx^4 qy^3 - 800 qy^4 + 360 qx\, qy^4 -$
$1440 qx^2 qy^4 + 760 qx^3 qy^4 - 416 qy^5 + 608 qx\, qy^5 - 48 qx^2 qy^5 + 256 qy^6 - 272 qx\, qy^6 - 32 qy^7),$
$-2 (375 qx - 500 LL qx - 1075 qx^2 + 300 LL qx^2 + 1980 qx^3 - 40 LL qx^3 - 1510 qx^4 + 759 qx^5 -$
$167 qx^6 + 22 qx^7 + 2625 qy - 1000 LL qy - 2800 qx\, qy + 1200 LL qx\, qy + 4655 qx^2 qy - 240 LL qx^2 qy -$
$5520 qx^3 qy + 4015 qx^4 qy - 1184 qx^5 qy + 193 qx^6 qy - 6300 qy^2 + 1200 LL qy^2 + 5610 qx\, qy^2 -$
$480 LL qx\, qy^2 - 5280 qx^2 qy^2 + 6060 qx^3 qy^2 - 2820 qx^4 qy^2 + 618 qx^5 qy^2 + 8440 qy^3 -$
$320 LL qy^3 - 3200 qx\, qy^3 + 720 qx^2 qy^3 - 1920 qx^3 qy^3 + 760 qx^4 qy^3 - 5280 qy^4 - 2080 qx\, qy^4 +$
$1520 qx^2 qy^4 - 80 qx^3 qy^4 + 1488 qy^5 + 1536 qx\, qy^5 - 816 qx^2 qy^5 - 448 qy^6 - 224 qx\, qy^6 + 256 qy^7)\}$

These equations, together with the defining equation for LL, allow us to eliminate qx and qy.

In[27]:= `eqs2 = Flatten[{eqs, pL1}];`

`Exponent[#, {qx, qy, LL}]& /@ eqs2`

Out[28]= $\{\{7, 7, 1\}, \{7, 7, 1\}, \{8, 8, 1\}\}$

First, we eliminate qx to obtain two equations in qy and LL.

In[29]:= `(* calculate and name the to resultants *)`
`res1 = Factor @ Resultant[eqs2[[1]], eqs2[[2]], qx];`
`res2 = Factor @ Resultant[eqs2[[1]], eqs2[[3]], qx];`

`Length /@ {res1, res2}`

Out[32]= $\{5, 4\}$

Now, we eliminate qy. res3 is a large polynomial.

In[33]:= `res3 = Resultant[res1[[5]], res2[[4]], qy];`

`Exponent[res3, LL]`

In[34]:= 109

It turns out the last factor is the relevant one.

In[35]:= `Roots[Factor[res3][[4]] == 0, LL][[1]]`

Out[35]= $LL == Root[-5159780352 + 14691041280 \#1 - 17490898944 \#1^2 + 11371207680 \#1^3 - 4210382592 \#1^4 + 767433600 \#1^5 - 1305600 \#1^6 - 14726500 \#1^7 + 713125 \#1^8 + 306250 \#1^9 \&, 1]$

So, we arrive at the following exact expression for the path length.

In[36]:= `Sqrt[%[[2]]] // RootReduce`

Out[36]= $Root[-5159780352 + 14691041280 \#1^2 - 17490898944 \#1^4 + 11371207680 \#1^6 - 4210382592 \#1^8 + 767433600 \#1^{10} - 1305600 \#1^{12} - 14726500 \#1^{14} + 713125 \#1^{16} + 306250 \#1^{18} \&, 2]$

The numerical value is about 1.03603, which is in agreement with the value found above using `FindMinimum`.

In[37]:= `N[%]`

Out[37]= 1.03603

41. Differential Equation Singularities, Weak Measurement Identity, Logarithmic Residue

a) *Mathematica* can solve the differential equation directly.

In[1]:= `sol = DSolve[{y'[x] == 1 - x y[x]^2, y[0] == 1}, y[x], x]`

Out[1]= $\{\{y[x] \rightarrow \left(-\sqrt{3} \text{ AiryAi}[x] \text{ Gamma}\left[\frac{1}{3}\right] + \right.$
$\text{AiryBi}[x] \text{ Gamma}\left[\frac{1}{3}\right] + 3^{5/6} \text{ AiryAi}[x] \text{ Gamma}\left[\frac{2}{3}\right] + 3^{1/3} \text{ AiryBi}[x] \text{ Gamma}\left[\frac{2}{3}\right]\right) /$
$\left(-\sqrt{3} \text{ AiryAiPrime}[x] \text{ Gamma}\left[\frac{1}{3}\right] + \text{AiryBiPrime}[x] \text{ Gamma}\left[\frac{1}{3}\right] + \right.$
$\left.3^{5/6} \text{ AiryAiPrime}[x] \text{ Gamma}\left[\frac{2}{3}\right] + 3^{1/3} \text{ AiryBiPrime}[x] \text{ Gamma}\left[\frac{2}{3}\right]\right)\}\}$

A graphical look at the solution shows that the singularity nearest to the origin is around $0.3 + 1.2\,i$.

In[2]:= `(* 3D plot; suppress displaying the graphic *)`
`pl3D = Plot3D[Evaluate[Re[N[y[x] /. sol[[1]]] /. x -> x + I y]],`
`            {x, -2, 2}, {y, -2, 2}, PlotPoints -> 90,`
`            Mesh -> False, DisplayFunction -> Identity];`

In[4]:= `(* show 3D graphic and corresponding contour plot *)`
`Show[GraphicsArray[`
`  {pl3D, ListContourPlot[pl3D[[1]], Contours -> 30,`
`                    DisplayFunction -> Identity,`
`                    MeshRange -> {{-2, 2}, {-2, 2}}}]]]`

We search numerically for the zero of the inverse to find the precise location of the nearest singularity.

In[6]:= `FindRoot[Evaluate[1/y[x] /. sol[[1]]], {x, 0.3 + 1.2 I}]`

Out[6]= $\{x \rightarrow 0.313409 + 1.18753\,i\}$

In[7]:= `Abs[x] /. %`

Out[7]= 1.22819

b) These are the shifts and weights.

In[1]:= d[k_, n_] := k/n

c[k_, n_, η_] := n!/((n - k)! k!) η^k (1 - η)^(n - k)

Based on the assumed existence of a Fourier expansion, we have to show the above identity for $f(x) = \exp(i\,l\,x)$. This can be done straightforwardly.

In[3]:= Sum[c[k, n, η] Exp[I (l x - d[k, n])], {k, 0, n}]

$$\text{Out[3]=} \quad \frac{e^{i\,l\,x}\,(1 - \eta)^n\,\left(\dfrac{e^{-\frac{i}{n}}\left(-e^{\frac{i}{n}} - \eta + e^{\frac{i}{n}}\,\eta\right)}{-1 + \eta}\right)^n\,n!}{\text{Gamma}[1 + n]}$$

In[4]:= % /. Gamma[n + 1] :> n! //
        Simplify[#, Element[n, Integers] && n > 0]&

$$\text{Out[4]=} \quad e^{i\,l\,x}\left(1 + \left(-1 + e^{-\frac{i}{n}}\right)\eta\right)^n$$

In[5]:= Limit[%, n -> ∞]

$$\text{Out[5]=} \quad e^{i\,(l\,x - \eta)}$$

Expanding the term $(1 + (\exp(-i/n) - 1)\,\eta)^n$ for large n, we obtain the following.

In[6]:= Exp[n Series[Log[1 + η (Exp[-I/n] - 1)], {n, Infinity, 4}]] //
                                                                Simplify

$$\text{Out[6]=} \quad e^{-i\,\eta} + \frac{e^{-i\,\eta}\,(-1 + \eta)\,\eta}{2\,n} + \frac{1}{24}\,e^{-i\,\eta}\,\eta\,(4\,i + (3 - 12\,i)\,\eta - (6 - 8\,i)\,\eta^2 + 3\,\eta^3)\left(\frac{1}{n}\right)^2 +$$
$$\frac{1}{48}\,e^{-i\,\eta}\,\eta\,(2 - (14 + 4\,i)\,\eta + (23 + 16\,i)\,\eta^2 - (9 + 20\,i)\,\eta^3 - (3 - 8\,i)\,\eta^4 + \eta^5)\left(\frac{1}{n}\right)^3 + O\left[\frac{1}{n}\right]^4$$

This means for the identity to hold we must have $n \gg \eta^2$.

Let us end by visualizing the identity under consideration. The left graphic shows the difference between $e^{i(x-\eta)}$ and $\sum_{k=0}^{n} c_k^{(n)}(\eta)\exp\!\big(i\big(x - d_k^{(n)}\big)\big)$ over the η,n-plane. The right graphic shows the $c_k^{(n)}(\eta)\exp\!\big(i\big(x - d_k^{(n)}\big)\big)$ for $n = 50$ and the partial sums $\sum_{k=0}^{o} c_k^{(n)}(\eta)\exp\!\big(i\big(x - d_k^{(n)}\big)\big)$ that approach $\exp(i(x - \eta))$.

In[7]:= (* difference between approximation and shifted exponential *)
        δ[x_, n_, η_] := Sum[c[k, n, η] Exp[I (x - d[k, n])],
                                {k, 0, n}] - Exp[I (x - η)]

In[9]:= (* sum terms *)
        curves = With[{n = 50},
                     Table[c[k, n, 0.9] Exp[I(x - d[k, n])], {k, 0, n}]];

        (* cumulative sums *)
        sumCurves = Rest[FoldList[Plus, 0, curves]];

In[14]:= Show[GraphicsArray[
         Block[{$DisplayFunction = Identity},
         {(* difference *)
         With[{x = 1, o = 20}, ListPlot3D[
             Table[Abs[δ[N[1, 100], n, η]], {n, o}, {η, 1/100, 8, 8/o}],
             PlotRange -> All, MeshRange -> {{0, 6}, {1, o}}]],
         (* cumulative sums *)
         Plot[Evaluate[Re @ Join[{curves, sumCurves}]],
             {x, 0, 2Pi}, PlotRange -> All,
             PlotStyle -> Join[Table[{GrayLevel[1/2]}, {51}],
                             Table[{Hue[k/60]}, {k, 51}]]]}]]]

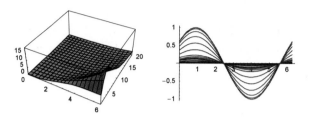

c) We will use a purely algebraic technique in the following. For a given n, analyzing the coefficients of the negative powers of z shows that the residue will be a function of the $\Phi^{(k)}(0)$ for $k \le n - 2$. So, we first generate all $\Phi(0)$, $\Phi'(0)$, ..., $\Phi^{(n-2)}(0)$ and then eliminate all the coefficients c_k. The following function `makeLogarithmicResidue` implements this procedure.

```
In[1]:= makeLogarithmicResidue[n_] :=
        Module[{f, c, z, Fs, Фs, δs, eqs, gb, sol},
          (* starting series; keep only relevant negative powers *)
          f = Sum[c[k] z^k, {k, -n, -1}];
          (* series of F *) Fs = Expand[z^n f];
          (* series of Φ *) Фs = D[Fs, z]/Fs;
          (* defining equations for derivatives of Φ *)
          δs = Table[Derivative[j][Ф][0] -
                Normal[Series[D[Фs, {z, j}], {z, 0, 0}]], {j, 0, n - 2}] //.
                c[-n] -> F[0];
          (* make polynomial equations *)
          eqs = Flatten[{Subscript[res, n] - c[-1],
                        Numerator[Together[δs]]}] /. c[-n] -> F[0];
          (* eliminate c's *)
          gb = GroebnerBasis[eqs, {}, Table[c[k], {k, -1, -n + 1, -1}],
                            MonomialOrder -> EliminationOrder];
          (* solve with respect to res *)
          sol = Solve[gb[[1]] == 0, Subscript[res, n]];
          (* simplify result *)
          (Equal @@ sol[[1, 1]]) // (Factor //@ #)&]
```

Here are the resulting logarithmic residues.

```
In[2]:= Table[makeLogarithmicResidue[k], {k, 10}] // TraditionalForm
```

Out[2]//TraditionalForm= $\Big\{ res_1 = F(0),\ res_2 = F(0)\,\Phi(0),\ res_3 = \dfrac{1}{2}\,F(0)\,(\Phi(0)^2 + \Phi'(0)),\ res_4 = \dfrac{1}{6}\,F(0)\,(\Phi(0)^3 + 3\,\Phi'(0)\,\Phi(0) + \Phi''(0)),$

$res_5 = \dfrac{1}{24}\,F(0)\,(\Phi(0)^4 + 6\,\Phi'(0)\,\Phi(0)^2 + 4\,\Phi''(0)\,\Phi(0) + 3\,\Phi'(0)^2 + \Phi^{(3)}(0)),$

$res_6 = \dfrac{1}{120}\,F(0)\,(\Phi(0)^5 + 10\,\Phi'(0)\,\Phi(0)^3 + 10\,\Phi''(0)\,\Phi(0)^2 + 15\,\Phi'(0)^2\,\Phi(0) + 5\,\Phi^{(3)}(0)\,\Phi(0) + 10\,\Phi'(0)\,\Phi''(0) + \Phi^{(4)}(0)),$

$res_7 = \dfrac{1}{720}\,F(0)\,(\Phi(0)^6 + 15\,\Phi'(0)\,\Phi(0)^4 + 20\,\Phi''(0)\,\Phi(0)^3 + 45\,\Phi'(0)^2\,\Phi(0)^2 + 15\,\Phi^{(3)}(0)\,\Phi(0)^2 +$
$\qquad 60\,\Phi'(0)\,\Phi''(0)\,\Phi(0) + 6\,\Phi^{(4)}(0)\,\Phi(0) + 15\,\Phi'(0)^3 + 10\,\Phi''(0)^2 + 15\,\Phi'(0)\,\Phi^{(3)}(0) + \Phi^{(5)}(0)),$

$res_8 = \dfrac{1}{5040}\,(F(0)\,(\Phi(0)^7 + 21\,\Phi'(0)\,\Phi(0)^5 + 35\,\Phi''(0)\,\Phi(0)^4 + 105\,\Phi'(0)^2\,\Phi(0)^3 + 35\,\Phi^{(3)}(0)\,\Phi(0)^3 +$
$\qquad 210\,\Phi'(0)\,\Phi''(0)\,\Phi(0)^2 + 21\,\Phi^{(4)}(0)\,\Phi(0)^2 + 105\,\Phi'(0)^3\,\Phi(0) + 70\,\Phi''(0)^2\,\Phi(0) + 105\,\Phi'(0)\,\Phi^{(3)}(0)\,\Phi(0) +$
$\qquad 7\,\Phi^{(5)}(0)\,\Phi(0) + 105\,\Phi'(0)^2\,\Phi''(0) + 35\,\Phi''(0)\,\Phi^{(3)}(0) + 21\,\Phi'(0)\,\Phi^{(4)}(0) + \Phi^{(6)}(0))),$

$res_9 = \dfrac{1}{40320}\,\big(F(0)\,\big(\Phi(0)^8 + 28\,\Phi'(0)\,\Phi(0)^6 + 56\,\Phi''(0)\,\Phi(0)^5 + 210\,\Phi'(0)^2\,\Phi(0)^4 + 70\,\Phi^{(3)}(0)\,\Phi(0)^4 +$
$\qquad 560\,\Phi'(0)\,\Phi''(0)\,\Phi(0)^3 + 56\,\Phi^{(4)}(0)\,\Phi(0)^3 + 420\,\Phi'(0)^3\,\Phi(0)^2 + 280\,\Phi''(0)^2\,\Phi(0)^2 + 420\,\Phi'(0)\,\Phi^{(3)}(0)\,\Phi(0)^2 +$
$\qquad 28\,\Phi^{(5)}(0)\,\Phi(0)^2 + 840\,\Phi'(0)^2\,\Phi''(0)\,\Phi(0) + 280\,\Phi''(0)\,\Phi^{(3)}(0)\,\Phi(0) + 168\,\Phi'(0)\,\Phi^{(4)}(0)\,\Phi(0) + 8\,\Phi^{(6)}(0)\,\Phi(0) +$
$\qquad 105\,\Phi'(0)^4 + 280\,\Phi'(0)\,\Phi''(0)^2 + 35\,\Phi^{(3)}(0)^2 + 210\,\Phi'(0)^2\,\Phi^{(3)}(0) + 56\,\Phi''(0)\,\Phi^{(4)}(0) + 28\,\Phi'(0)\,\Phi^{(5)}(0) + \Phi^{(7)}(0)\big)\big),$

$res_{10} = \dfrac{1}{362880}\,\big(F(0)\,\big(\Phi(0)^9 + 36\,\Phi'(0)\,\Phi(0)^7 + 84\,\Phi''(0)\,\Phi(0)^6 + 378\,\Phi'(0)^2\,\Phi(0)^5 + 126\,\Phi^{(3)}(0)\,\Phi(0)^5 +$
$\qquad 1260\,\Phi'(0)\,\Phi''(0)\,\Phi(0)^4 + 126\,\Phi^{(4)}(0)\,\Phi(0)^4 + 1260\,\Phi'(0)^3\,\Phi(0)^3 + 840\,\Phi''(0)^2\,\Phi(0)^3 + 1260\,\Phi'(0)\,\Phi^{(3)}(0)\,\Phi(0)^3 +$
$\qquad 84\,\Phi^{(5)}(0)\,\Phi(0)^3 + 3780\,\Phi'(0)^2\,\Phi''(0)\,\Phi(0)^2 + 1260\,\Phi''(0)\,\Phi^{(3)}(0)\,\Phi(0)^2 + 756\,\Phi'(0)\,\Phi^{(4)}(0)\,\Phi(0)^2 +$

$$36\,\Phi^{(6)}(0)\,\Phi(0)^2 + 945\,\Phi'(0)^4\,\Phi(0) + 2520\,\Phi'(0)\,\Phi''(0)^2\,\Phi(0) + 315\,\Phi^{(3)}(0)^2\,\Phi(0) + 1890\,\Phi'(0)^2\,\Phi^{(3)}(0)\,\Phi(0) +$$
$$504\,\Phi''(0)\,\Phi^{(4)}(0)\,\Phi(0) + 252\,\Phi'(0)\,\Phi^{(5)}(0)\,\Phi(0) + 9\,\Phi^{(7)}(0)\,\Phi(0) + 280\,\Phi''(0)^3 + 1260\,\Phi'(0)^3\,\Phi''(0) +$$
$$1260\,\Phi'(0)\,\Phi''(0)\,\Phi^{(3)}(0) + 378\,\Phi'(0)^2\,\Phi^{(4)}(0) + 126\,\Phi^{(3)}(0)\,\Phi^{(4)}(0) + 84\,\Phi''(0)\,\Phi^{(5)}(0) + 36\,\Phi'(0)\,\Phi^{(6)}(0) + \Phi^{(8)}(0)\big)\big)\big\}$$

42. Geometry Puzzle

Using *Mathematica*'s symbolic capabilities, it is straightforward to calculate the angle <CDE. To do this, we calculate the points D and E. This allows us to express the cosine of <CDE and we also get <CDE itself.

There are the three vertices of the triangle.

```
In[1]:= b = {-1, 0};
        c = {+1, 0};
        a = {0, Tan[80 Degree]};
```

We define a function `intersectionPoint` that calculates the point of intersection between two lines. Every line is given in the form of one point on this line and a direction.

```
In[4]:= intersectionPoint[{p1_, dir1_}, {p2_, dir2_}] :=
        First[p1 + s dir1 /. Solve[Thread[p1 + s dir1 = p2 + t dir2], {s, t}]]
```

So, we get the following expression for the coordinates of the point D.

```
In[5]:= d = intersectionPoint[{a, b - a}, {c, {Cos[110 Degree], Sin[110 Degree]}}]
```

$$\text{Out[5]= } \Big\{1 + \frac{2\,\mathrm{Cos}[110\,°]\,\mathrm{Tan}[80\,°]}{\mathrm{Sin}[110\,°] - \mathrm{Cos}[110\,°]\,\mathrm{Tan}[80\,°]},$$
$$\mathrm{Tan}[80\,°] - \mathrm{Tan}[80\,°]\left(-1 - \frac{2\,\mathrm{Cos}[110\,°]\,\mathrm{Tan}[80\,°]}{\mathrm{Sin}[110\,°] - \mathrm{Cos}[110\,°]\,\mathrm{Tan}[80\,°]}\right)\Big\}$$

It is best to transform every number into an algebraic number. For trigonometric functions with arguments of the form *rational* π, this can always be done. The function `toAlgebraicNumber` implements this conversion.

```
In[6]:= toAlgebraicNumber[y_] :=
        TrigToExp[y] /. Degree -> Pi/180 /. E^α_ -> (-1)^(α/(I Pi))
```

Now the point D looks the following way.

```
In[7]:= toAlgebraicNumber[d]
```

$$\text{Out[7]= } \Big\{1 + \frac{\mathbbm{i}\,(-(-1)^{4/9} - (-1)^{5/9})\,(-(-1)^{7/18} + (-1)^{11/18})}{((-1)^{4/9} - (-1)^{5/9})\left(\frac{1}{2}\mathbbm{i}\,(-(-1)^{7/18} - (-1)^{11/18}) - \frac{\mathbbm{i}\,(-(-1)^{4/9}-(-1)^{5/9})\,(-(-1)^{7/18}+(-1)^{11/18})}{2\,((-1)^{4/9}-(-1)^{5/9})}\right)},$$

$$\frac{\mathbbm{i}\,(-(-1)^{4/9} - (-1)^{5/9})}{(-1)^{4/9} - (-1)^{5/9}} - \frac{1}{(-1)^{4/9} - (-1)^{5/9}}\left(\mathbbm{i}\,(-(-1)^{4/9} - (-1)^{5/9})\left(-1 - \right.\right.$$

$$\left.\left.\frac{\mathbbm{i}\,(-(-1)^{4/9} - (-1)^{5/9})\,(-(-1)^{7/18} + (-1)^{11/18})}{((-1)^{4/9} - (-1)^{5/9})\left(\frac{1}{2}\mathbbm{i}\,(-(-1)^{7/18} - (-1)^{11/18}) - \frac{\mathbbm{i}\,(-(-1)^{4/9}-(-1)^{5/9})\,(-(-1)^{7/18}+(-1)^{11/18})}{2\,((-1)^{4/9}-(-1)^{5/9})}\right)}\right)\right)$$

$$\Big\}$$

The function `RootReduce` generates this normal form of the above algebraic number.

```
In[8]:= d = RootReduce[Simplify[%]]
```

Out[8]= {Root[-1 - 3 #1 + #1³ &, 2], Root[-3 + 18 #1² - 15 #1⁴ + #1⁶ &, 6]}

Doing the same for the point E, we get its coordinates as an algebraic number, too.

```
In[9]:= e = RootReduce[Simplify[toAlgebraicNumber[
            intersectionPoint[{a, c - a},
                {b, {Cos[60 Degree], Sin[60 Degree]}}]]]]
```

Out[9]= {Root[-1 + 3 #1² + #1³ &, 3], Root[-27 + 81 #1² - 18 #1⁴ + #1⁶ &, 5]}

Now that we have explicit coordinates of the points D and E, the cosine of <CDE can be calculated by the following elementary geometry formula.

```
In[10]:= cos<cde = RootReduce[RootReduce[(c - d).(e - d)]/
            RootReduce[Sqrt[RootReduce[(c - d).(c - d)]*
                RootReduce[(e - d).(e - d)]]]]
```

Out[10]= Root[-1 - 6 #1 + 8 #1³ &, 3]

Taking the arccos of this expression and simplifying it, we get the final answer.

```
In[11]:= ToRadicals[RootReduce[TrigToExp[ArcCos[cos<cde]]]]
```

$$Out[11]= \frac{\pi}{9}$$

Knowing the solution, it is straightforward to verify it.

43. Differential Equation for Polynomial, Graph Eigenvalues

a) A polynomial of degree n in y has n solutions. A linear differential equation of degree n too has n solutions. This means that the differential equation we are looking for must be of order n. The following code calculates the differential equation by making an ansatz of the form $\sum_{k=0}^{n} a_k(x) \, y^{(k)}(x) = 0$ for the differential equation and then determining the coefficients $a_k(x)$. To determine the coefficients $a_k(x)$, we differentiate the original polynomial k times and substitute the resulting expressions recursively into the ansatz for the differential equation. Then we reduce the resulting polynomial modulo the original polynomial. This gives us a new polynomial of degree less than or equal n. The coefficients of this polynomial are linear functions of the $a_k(x)$. We solve for the $a_k(x)$, set remaining undetermined $a_k(x)$ to 1, and finally substitute the $a_k(x)$ in the original ansatz.

```
In[1]:= polynomialToDifferentialEquation[polyEq_, y_, x_] :=
        Module[{poly, n, ode, derivatives, odePoly, oldMessageState,
                leadingTerm, α, reducedOdePoly, CEqs, CSols, commonCs, res},
         (* make a polynomial from the equation *)
         poly = (Subtract @@ polyEq) /. y -> y[x];
         (* the degree of the polynomial *)
         n = Exponent[poly, y[x]];
         (* ansatz for the differential equation;
            the C[i] are to be determined coefficients depending on x *)
         ode = Sum[C[i] D[y[x], {x, i}], {i, 0, n}];
         (* the derivatives expressed in lower derivatives *)
         derivatives = Flatten[Table[Solve[D[poly, {x, i}] == 0,
                             D[y[x], {x, i}]], {i, n}]];
         (* substitute all derivatives in the polynomial *)
         odePoly = Numerator[Together[ode //. derivatives]];
         (* turn off a Solve message *)
         oldMessageState = Solve::svars; Off[Solve::svars];
         (* express the leading term of the polynomial *)
         lT = Solve[(poly /. y[x]^n -> α) == 0, α][[1, 1, 2]];
         (* reduce odePoly by using the original polynomial *)
         reducedOdePoly = FixedPoint[Expand[# /. y[x]^m_ :>
                         lT^Quotient[m, n] y[x]^Mod[m, n]]&, odePoly];
         (* make linear equations for the C[i] *)
         CEqs = (# == 0&) /@ CoefficientList[reducedOdePoly, y[x]];
         (* solve the equations CEqs for the C[i] *)
         CSols = Solve[CEqs, Table[C[i], {i, 0, n}]][[1]];
         (* the C[i] appearing on the rhs of CSols *)
         commonCs = Cases[Last /@ CSols, _C, Infinity];
         (* the resulting differential equation *)
         res = Numerator[Together[(ode /. CSols /. ((# -> 1)& /@ commonCs))]];
         (* some (optional) cosmetics to get a nicer looking solution *)
         res = Collect[res, Table[D[y[x], {x, i}], {i, 0, n}], Factor];
         (* turn on Solve message again *)
         If[Head[oldMessageState] === String, On[Solve::svars]];
         (* return the differential equation *) res == 0]
```

We start with a simple quadratic polynomial.

```
In[2]:= ode = polynomialToDifferentialEquation[y^2 + x == 0, y, x]
```

$$Out[2]= (1 - 2x) \, y[x] + 4x^2 \, y'[x] + 4x^2 \, y''[x] = 0$$

The solution of the polynomial $y = \sqrt{x}$ satisfies the derived differential equation.

```
In[3]:= Expand /@ (ode /. y -> (Function[x, x^(1/2)]))
```

```
Out[3]= True
```

Here is a slightly more complicated quadratic polynomial.

In[4]:= **ode = polynomialToDifferentialEquation[y^2 + a y + b x == 0, y, x]**

Out[4]= $2\, b\, y'[x] + (-a^2 + 4\, b\, x)\, y''[x] = 0$

A call to DSolve solves the derived differential equation.

In[5]:= **DSolve[ode, y[x], x]**

Out[5]= $\left\{\left\{y[x] \to -\dfrac{\sqrt{a^2 - 4\, b\, x}\ C[1]}{2\, b} + C[2]\right\}\right\}$

By fixing the integration constants, we could reproduce the solutions of the original quadratic polynomial.

In[6]:= **Solve[y^2 + a y + b x == 0, y]**

Out[6]= $\left\{\left\{y \to \dfrac{1}{2}\left(-a - \sqrt{a^2 - 4\, b\, x}\right)\right\}, \left\{y \to \dfrac{1}{2}\left(-a + \sqrt{a^2 - 4\, b\, x}\right)\right\}\right\}$

This is a simple cubic polynomial.

In[7]:= **ode = polynomialToDifferentialEquation[y^3 + x == 0, y, x]**

Out[7]= $(-10 + 6\, x - 9\, x^2)\, y[x] + 27\, x^3\, y'[x] + 27\, x^3\, y''[x] + 27\, x^3\, y^{(3)}[x] = 0$

The root $y = \sqrt[3]{x}$ of the polynomial is also a solution of the differential equation.

In[8]:= **Expand /@ (ode[[1]] /. y -> (Function[x, x^(1/3)]))**

Out[8]= 0

And here is a slightly more complicated cubic.

In[9]:= **ode = polynomialToDifferentialEquation[y^3 + y + x == 0, y, x]**

Out[9]= $-9\,(-8 + 27\, x)\, y[x] + 3\,(-32 + 513\, x^2)\, y'[x] +$
$\qquad 3\,(-32 + 513\, x^2)\, y''[x] - (4 + 27\, x^2)\,(4 - 27\, x + 27\, x^2)\, y^{(3)}[x] = 0$

Here is a quick check for the correctness of the differential equation.

In[10]:= **Together //@ (ode /. MapAt[Function[x, #]&, #, {1, 2}]& /@**
Solve[y^3 + y + x == 0, y])

Out[10]= {True, True, True}

Here is a more complicated example: a quartic polynomial.

In[11]:= **ode = polynomialToDifferentialEquation[y^4 + y + x^9 == 0, y, x]**

Out[11]= $-29160\, x^{26}\,(-1296 - 27\, x - 4992\, x^{27} + 256\, x^{28})\, y[x] +$
$\qquad 8\,(272646 + 12393\, x + 4187997\, x^{27} - 143154\, x^{28} - 5432320\, x^{54} + 243200\, x^{55})\, y'[x] +$
$\qquad 8\, x\,(-37908 - 2187\, x - 1778706\, x^{27} + 6480\, x^{28} - 2122240\, x^{54} + 135168\, x^{55})\, y''[x] +$
$\qquad x^3\,(-27 + 256\, x^{27})^2\, y^{(3)}[x] + x^3\,(-27 + 256\, x^{27})^2\, y^{(4)}[x] = 0$

The last example is a quintic. The above implementation allows for additional parameters, here a and b.

In[12]:= **polynomialToDifferentialEquation[y^5 + a x + b y == 0, y, x]**

Out[12]= $-721875\, a^8\, x^2\,(-267 + 70\, x)\, y[x] + 15\, a^4\,(-524288\, b^5 - 290609375\, a^4\, x^3 + 86568750\, a^4\, x^4)\, y'[x] +$
$\qquad 9375\, a^4\, x\,(-4864\, b^5 - 707875\, a^4\, x^3 + 281875\, a^4\, x^4)\, y''[x] +$
$\qquad 625\, a^4\, x^2\,(-68352\, b^5 + 1353125\, a^4\, x^4)\, y^{(3)}[x] + 625\, a^4\, x^2\,(-68352\, b^5 + 1353125\, a^4\, x^4)\, y^{(4)}[x] -$
$\qquad (256\, b^5 + 3125\, a^4\, x^4)\,(256\, b^5 - 31250\, a^4\, x^3 + 3125\, a^4\, x^4)\, y^{(5)}[x] = 0$

We check the last differential equation for the first root (the Root-object with root number 1). As mentioned, for the simplification of Root-objects we need the function FullSimplify.

In[13]:= **% /. Derivative[k_][y][x] :> D[Root[a*x + b*#1 + #1^5 &, 1], {x, k}] /.**
y[x] -> Root[a*x + b*#1 + #1^5 &, 1] // FullSimplify

Out[13]= True

For related differential equations, see [410].

b) We use the package Graphics`Polyhedra` to construct a list of neighbors for each vertex of a stellated icosahedron. We number the 20 vertices from the stellation process as 13, ..., 32.

```
In[1]:= Needs["Graphics`Polyhedra`"]

In[2]:= neighbors = {#[[1, 1]],(* union *) Union[Flatten[Last /@ #]]}& /@
           Split[Sort[Join[Sequence @@
               (* neighbors formed by the stellation process *)
           Apply[{{#1, {#2, #3}}, {#2, {#1, #3}}, {#3, {#1, #2}}}&,
               Faces[Icosahedron], {1}], (* original neighbors *)
           Sequence @@ MapIndexed[{{12 + #2[[1]], #1},
                           {#1[[1]], {12 + #2[[1]]}}},
                           {#1[[2]], {12 + #2[[1]]}},
                           {#1[[3]], {12 + #2[[1]]}}}&, Faces[Icosahedron]]],
                       #1[[1]] < #2[[1]]&], #1[[1]] === #2[[1]]&];
```

Here are the first and last elements of the list `neighbors`.

```
In[3]:= Take[neighbors, 2]

Out[3]= {{1, {2, 3, 4, 5, 6, 13, 14, 15, 16, 17}}, {2, {1, 3, 6, 7, 11, 13, 17, 18, 22, 27}}}
```

```
In[4]:= Take[neighbors, -2]

Out[4]= {{31, {10, 11, 12}}, {32, {7, 11, 12}}}
```

Here is a wireframe of the stellated icosahedron.

```
In[5]:= vertices = Join[Vertices[Icosahedron],
               Apply[Plus, Map[Vertices[Icosahedron][[#]]&,
                       Faces[Icosahedron], {-1}], {1}]];

In[6]:= Show[Graphics3D[N @ Map[vertices[[#]]&,
           Function[x, Line[{x[[1]], #}]& /@ x[[2]]] /@ neighbors, {-1}]]]
```

We construct the matrix representation of the Laplace operator. It acts on the vector of nodes.

```
In[7]:= n = Length[neighbors];
        mat = Table[0, {n}, {n}];
        Do[mat[[k, k]] = Length[neighbors[[k, 2]]];
           (mat[[k, #]] = -1)& /@ neighbors[[k, 2]], {k, n}]
```

Here are the resulting eigenvalues, together with their multiplicity.

```
In[10]:= {#[[1]], Length[#]}& /@
            Split[Sort[Eigenvalues[mat] // Simplify, #1 < #2&]]

Out[10]= {{0, 1}, {Root[500 - 600 #1 + 214 #1² - 26 #1³ + #1⁴ &, 1], 3},
         {7 - √19, 5}, {Root[500 - 600 #1 + 214 #1² - 26 #1³ + #1⁴ &, 2], 3},
         {3, 8}, {8, 1}, {Root[500 - 600 #1 + 214 #1² - 26 #1³ + #1⁴ &, 3], 3},
         {7 + √19, 5}, {Root[500 - 600 #1 + 214 #1² - 26 #1³ + #1⁴ &, 4], 3}}
```

c) To calculate the vertices and the edges, we repeat the inputs from the mentioned exercise from Chapter 2 of the Graphics volume [1736].

```
In[1]:= vertices =
        With[{τ = GoldenRatio, g = Function[{f, 1}, Flatten[
            f /@ Flatten[Outer[List, Sequence @@ 1], 3], 1]],
            evenPermutations = Function[1, 1[[#]]& /@ Select[
            Permutations[{1, 2, 3, 4}], Signature[#] === 1&]]},
        Union[Join @@ Flatten[{g[Permutations, #]& /@
        {{{+2, -2}, {+2, -2}, {0}, {0}},
        {{+Sqrt[5], -Sqrt[5]}, {+1, -1}, {+1, -1}, {+1, -1}},
```

```
          {{+τ, -τ}, {+τ, -τ}, {+τ, -τ}, {+τ^-2, -τ^-2}}},
     g[evenPermutations, #]& /@
       {{{+τ^2, -τ^2}, {+1/τ, -1/τ}, {+1/τ, -1/τ}},
        {{+τ^2, -τ^2}, {+1/τ^2, -1/τ^2}, {+1, -1}, {0}},
        {{+Sqrt[5], -Sqrt[5]}, {+1/τ, -1/τ}, {+τ, -τ}, {0}},
        {{+2, -2}, {+1, -1}, {+τ, -τ}, {+1/τ, -1/τ}}}}, 1]]];
```

In[2]:= verticesN = N[vertices];
     (edgeList = DeleteCases[Union[Sort /@ Flatten[
      Table[s = verticesN[[1]];
         res = {s, #}& /@ Select[verticesN, #.#&[s - #] < 1&];
         verticesN = Rest[verticesN]; res, {599}], 1]], {a_, a_}]);

The expression edgeConnectivity is a list of pairs of connected vertices.

In[4]:= edgeConnectivity = edgeList /. (Rule @@@ Transpose[
        {Union[Flatten[edgeList, 1]], Table[k, {k, 600}]}]);

Each vertex is connected to four others.

In[5]:= (Function[e, Length[Select[edgeConnectivity,
                              MemberQ[#, e]&]]] /@ Range[600]) // Union

Out[5]= {4}

mat is the matrix corresponding to the Laplace operator for the 120-cell.

In[6]:= mat = 4 IdentityMatrix[600] - ReplacePart[Table[0, {600}, {600}], 1,
        Join[edgeConnectivity, Reverse /@ edgeConnectivity]];

mat is a very sparse matrix.

In[7]:= ListDensityPlot[Sign[Abs[mat]], Mesh -> False,
                  ColorFunction -> (GrayLevel[1 - #]&)]

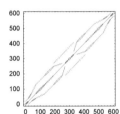

A machine-precision calculation of the eigenvalues of mat is relatively fast.

In[8]:= (evs = Sort[Re[Eigenvalues[N[mat]]]]); // Timing

Out[8]= {0.48 Second, Null}

The eigenvalues have a high degeneracy.

In[9]:= ListPlot[evs, Frame -> True, Axes -> False]

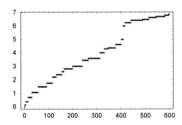

Here are the 10 smallest and the 10 largest eigenvalues.

In[10]:= Take[evs, 10]

Out[10]= {6.21725 × 10^-15, 0.145898, 0.145898, 0.145898,
       0.145898, 0.381966, 0.381966, 0.381966, 0.381966, 0.381966}

In[11]:= **Take[evs, -10]**

Out[11]= {6.79129, 6.79129, 6.79129, 6.79129, 6.79129, 6.79129, 6.8541, 6.8541, 6.8541, 6.8541}

An exact diagonalization of a 600×600 matrix would be extremely time-consuming. So, we will not carry it out here. The above values suggest that 0 is the smallest eigenvalue. It is possible to symbolically confirm that 0 is an eigenvalue.

In[12]:= **Det[mat]**

Out[12]= 0

Using the function `Recognize` from the package `NumberTheory`Recognize`` leads us to the conjecture that the largest eigenvalue is $(7 + 3\,5^{1/2})/4$.

In[13]:= **Needs["NumberTheory`Recognize`"]**

    Recognize[evs[[-1]], 10, x]

Out[14]= $1 - 7\,x + x^2$

In[15]:= **evMax = Max[x /. Solve[% == 0, x]]**

Out[15]= $\frac{1}{2}\left(7 + 3\sqrt{5}\right)$

A machine-precision check does not give an encouraging result.

In[16]:= **Det[mat - DiagonalMatrix[Table[N[evMax], {600}]]]**

Out[16]= 6.54443×10^{75}

A low-precision high-precision check gives a high-precision zero with negative accuracy.

In[17]:= **prec = $MachinePrecision + 1;**
    Det[mat - DiagonalMatrix[Table[N[evMax, prec], {600}]]]

Out[18]= $0. \times 10^{42}$

A high-precision high-precision check yields hundreds of canceling digits. This confirms the conjecture.

In[19]:= **prec = 200;**
    Det[mat - DiagonalMatrix[Table[N[evMax, prec], {600}]]]

Out[20]= $0. \times 10^{-690}$

Now, we could continue with the second highest eigenvalue.

In[21]:= **Recognize[evs[[-5]], 10, x]**

Out[21]= $15 - 9\,x + x^2$

44. Fourier Transform Eigenvalues, $\int_{-\infty}^{\infty} \theta(x)\,\delta(x)\,dx$, $\delta^{(n)}(f(x))$, PDF for Sums and Determinants, Fourier Transform and Series, Functional Differentiation

a) The eigenvalue for $n = 0$ is obviously 1.

In[1]:= **FourierTransform[Exp[-x^2/2], x, y]**

Out[1]= $e^{-\frac{y^2}{2}}$

Fourier transforming the function $\exp(-x^2/2)$ allows us to immediately read the eigenvalue as i and the eigenfunctions as y.

In[2]:= **FourierTransform[x Exp[-x^2/2], x, y]**

Out[2]= $i\,e^{-\frac{y^2}{2}}\,y$

For $n \geq 2$, we must work more programmatically. We Fourier transform the general ansatz $\exp(-x^2/2)\sum_{k=0}^{n} c_k\,x^k$ and determine the c_k. This is easily implemented.

In[3]:= **Off[Solve::"svars"]**

```
Table[(* the ansatz *)
    f = Sum[c[i] x^i, {i, 0, n}];
    (* the Fourier transform of the ansatz *)
    ft = FourierTransform[f Exp[-x^2/2], x, y]/Exp[-y^2/2] /. y -> x;
```

```
(* conditions for being eigenfunctions *)
cl = CoefficientList[f - λ ft, x];
(* make sure the c[n] term does not vanish *)
gb = GroebnerBasis[Append[cl, 1 - z c[n]],
                   Append[Table[c[i], {i, 0, n}], λ], z];
(* solve for the coefficients *)
sol = Solve[# == 0& /@ gb, Append[Table[c[i], {i, 0, n}], λ]];
(* return eigenvalue and eigenfunctions *)
{λ, f} /. sol[[1]] /. _c -> 1, {n, 3, 12}]
```

Out[4]= $\left\{\left\{i, -\dfrac{3x}{2} + x^3\right\}, \{1, 1 - 3x^2 + x^4\}, \{-i, x - 5x^3 + x^5\}\right.$,

$\left\{-1, \dfrac{13}{4} + x^2 - \dfrac{15x^4}{2} + x^6\right\}, \left\{i, \dfrac{99x}{4} + x^3 - \dfrac{21x^5}{2} + x^7\right\}, \{1, 1 + 102x^2 + x^4 - 14x^6 + x^8\}$,

$\{-i, x + 310x^3 + x^5 - 18x^7 + x^9\}, \left\{-1, -\dfrac{1877}{4} + x^2 + 780x^4 + x^6 - \dfrac{45x^8}{2} + x^{10}\right\}$,

$\left.\left\{i, -\dfrac{20691x}{4} + x^3 + 1722x^5 + x^7 - \dfrac{55x^9}{2} + x^{11}\right\}, \{1, 1 - 31083x^2 + x^4 + 3451x^6 + x^8 - 33x^{10} + x^{12}\}\right\}$

The eigenfunctions are just the Hermite polynomials and the eigenvalues are ±1 and ±i [1317], [1223], [97], [1092], [1850]. (In some cases, the above procedure could produce superpositions of various eigenfunctions belonging to the same eigenvalue.)

In[5]:= `Table[(FourierTransform[HermiteH[k, x] Exp[-x^2/2], x, y]/`
`         Exp[-y^2/2]/HermiteH[k, y]) // Cancel, {k, 12}]`

Out[5]= `{i, -1, -i, 1, i, -1, -i, 1, i, -1, -i, 1}`

For the general form of functions that are their equal to their Fourier transform, see [1245] and [973]. For the eigenvalues and eigenfunctions of the Fourier transform over a finite interval, see [667], [1288], [1016], [227]. For the eigenvalues and eigenvectors of the discrete Fourier transform, see [1220], [482], [785], [1241], [1418], and [103].

b) Here is a simple choice for $\theta_\epsilon(x)$.

In[1]:= `θ[x_, ε_] = 1/(Exp[-x/ε] + 1);`

The following picture shows a 3D plot of $\theta_\epsilon(x)$.

In[2]:= `Plot3D[θ[x, ε], {x, -3, 3}, {ε, 10^-6, 1},`
`          PlotPoints -> 60, Mesh -> False, ViewPoint -> {-1, -2, 1}]`

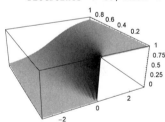

For $\delta_\epsilon(x)$, we make a related choice. The additional parameter ϵs is a small shift that (later) vanishes together with ϵ.

In[3]:= `δ[x_, ε_, εs_] = D[1/(Exp[-x/ε - εs] + 1), x]`

Out[3]= $\dfrac{e^{-\frac{x}{\varepsilon} - \varepsilon s}}{\left(1 + e^{-\frac{x}{\varepsilon} - \varepsilon s}\right)^2 \varepsilon}$

Here is a picture that shows the convergence to a peak of area 1 concentrated at $x = 1$.

In[4]:= `Plot3D[δ[x, ε, 3 ε], {x, -3, 3}, {ε, 0.01, 1},`
`          PlotPoints -> 60, Mesh -> False,`
`          ViewPoint -> {-1, -2, 1}, PlotRange -> All]`

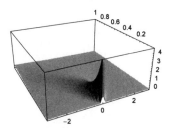

For a fixed ε and εs a multiple of ε we see how the maximum of δ shifts with respect to Θ. For εs being a negative multiple of ε, the peak of δ is within the area where Θ is 1. For εs being a positive multiple of ε, the peak of δ is within the area where θ is 0.

```
In[5]:= With[{ε = 0.1},
       Show[Table[
       Plot[Evaluate[{δ[x, ε, k ε], Θ[x, ε]}],
            {x, -1, 1}, PlotRange -> All, DisplayFunction -> Identity,
            Frame -> True, Axes -> False,
            PlotStyle -> {{Thickness[0.002], Hue[(k + 50)/100 0.2]},
                          {Thickness[0.002], Hue[0.7]}}],
            {k, -50, 50, 5}], DisplayFunction -> $DisplayFunction]]
```

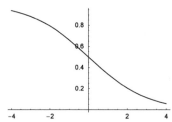

Choosing $\epsilon = \varepsilon = e$, we can easily evaluate the integral $\int_{-\infty}^{\infty} \delta_\varepsilon(x)\,\theta_\epsilon(x)\,dx$. We let εs be es.

```
In[6]:= overlap[e_, es_] =
          Integrate[θ[y, 1] δ[y, 1, es/e], {y, -Infinity, Infinity},
                    Assumptions -> es > 0 && e > 0]
```

$$Out[6]= \frac{e + e^{\frac{es}{e}}\,(-e + es)}{e\left(-1 + e^{\frac{es}{e}}\right)^2}$$

If we let es = k e, we obtain the following e-independent result for the overlap.

```
In[7]:= overlap[e, k e] // PowerExpand
```

$$Out[7]= \frac{e + e^k\,(-e + e\,k)}{e\,(-1 + e^k)^2}$$

The overlap integral can take on any value between 0 and 1.

```
In[8]:= Plot[Evaluate[%], {k, -4, 4}]
```

For some similar, not well-defined expressions of the form $\int_{-\infty}^{\infty} \delta(x) f(x) \, dx$, see [774]; for a well-defined limit of this type, see [1281].

c) This is the well-known formula for $\delta(f(x))$. We associate all definitions with δ instead of with the built-in function `DiracDelta`.

```
In[1]:= Derivative[0][δ][f] = δ[x - ξ]/Abs[f'[ξ]]
```

$$Out[1]= \frac{\delta[x - \xi]}{\text{Abs}[f'[\xi]]}$$

The first derivative then takes the following form.

```
In[2]:= d1 = Derivative[1][δ][f] = 1/f'[x] D[Derivative[0][δ][f], x]
```

$$Out[2]= \frac{\delta'[x - \xi]}{\text{Abs}[f'[\xi]] \, f'[x]}$$

To eliminate the x-dependence of the factor $1/f'(x)$, we implement the identity

$$g(x)\,\delta^{(n)}(x - \xi) = \sum_{k=0}^{n} (-1)^k \binom{n}{k} g^{(k)}(\xi)\,\delta^{(n-k)}(x - \xi)$$

which transforms the x-dependence of the prefactor to a ξ-dependence.

```
In[3]:= Derivative /:
        g_?(MemberQ[#, x, {0, Infinity}]&)*Derivative[n_][δ][x - ξ] :=
          (Sum[(-1)^k Binomial[n, k] (D[g, {x, k}] /. x -> ξ) *
            Derivative[n - k][δ][x - ξ], {k, 0, n}])
```

Using this rule, $\delta'(f(x))$ takes on the following form.

```
In[4]:= d1
```

$$Out[4]= \frac{\delta'[x - \xi]}{\text{Abs}[f'[\xi]] \, f'[\xi]} + \frac{\delta[x - \xi] \, f''[\xi]}{\text{Abs}[f'[\xi]] \, f'[\xi]^2}$$

Now let us look at the second derivative.

```
In[5]:= Derivative[2][δ][f] = 1/f'[x] D[Derivative[1][δ][f], x]
```

$$Out[5]= \frac{\frac{\delta'[x-\xi] \, f''[\xi]}{\text{Abs}[f'[\xi]] \, f'[\xi]^2} + \frac{\delta''[x-\xi]}{\text{Abs}[f'[\xi]] \, f'[\xi]}}{f'[x]}$$

The above rule for `Derivative` did not go into effect because the expression was not of the form of a product of something x-dependent with $\delta^{(k)}(x - \xi)$. To get the rule working, we must expand the expression.

```
In[6]:= Expand[%]
```

$$Out[6]= \frac{3 \, \delta'[x - \xi] \, f''[\xi]}{\text{Abs}[f'[\xi]] \, f'[\xi]^3} + \frac{3 \, \delta[x - \xi] \, f''[\xi]^2}{\text{Abs}[f'[\xi]] \, f'[\xi]^4} + \frac{\delta''[x - \xi]}{\text{Abs}[f'[\xi]] \, f'[\xi]^2} - \frac{\delta[x - \xi] \, f^{(3)}[\xi]}{\text{Abs}[f'[\xi]] \, f'[\xi]^3}$$

```
In[7]:= d2 = Simplify[%]
```

$$Out[7]= \frac{3 \, \delta[x - \xi] \, f''[\xi]^2 + f'[\xi]^2 \, \delta''[x - \xi] + f'[\xi] \, (3 \, \delta'[x - \xi] \, f''[\xi] - \delta[x - \xi] \, f^{(3)}[\xi])}{\text{Abs}[f'[\xi]] \, f'[\xi]^4}$$

For $f(x) = x$, we get an identity.

```
In[8]:= d2 /. f -> Function[x, x]
```

$$Out[8]= \delta''[x - \xi]$$

To "numerically" check the formula for d2, we use a smoothed version of the Dirac δ distribution.

```
In[9]:= smoothδ[x_, δ_] := 1/Sqrt[Pi δ]Exp[-x^2/δ]
```

This smoothed version can be easily integrated by `NIntegrate`.

```
In[10]:= With[{δ = 0.001},
           NIntegrate[Evaluate[Derivative[2, 0][smoothδ][x^3 - 1, δ]],
             {x, 0, 3}, MaxRecursion -> 10]]
```

$$Out[10]= 0.371279$$

Comparing with the symbolic result, we see a good agreement.

```
In[11]:= {#, N[#]}&[Integrate[d2 /. f -> Function[x, x^3 - 1] /.
                ξ -> 1 /. δ -> DiracDelta, {x, 0, 3}]]
```

$$\text{Out[11]}= \left\{\frac{10}{27}, 0.37037\right\}$$

Here is another function used for the check.

```
In[12]:= {With[{δ = 0.001},
            NIntegrate[Evaluate[Derivative[2, 0][smoothδ][Cos[x], δ]],
                {x, 1, 2}, MaxRecursion -> 10]],
        Integrate[d2 /. f -> Function[x, Cos[x]] /.
                ξ -> Pi/2 /. δ -> DiracDelta, {x, 0, 3}]}
```

Out[12]= {1.00226, 1}

Now let us continue and calculate higher orders.

```
In[13]:= Derivative[2][δ][f] = d2
```

$$\text{Out[13]}= \frac{3\,\delta[x-\xi]\,f''[\xi]^2 + f'[\xi]^2\,\delta'[x-\xi] + f'[\xi]\,(3\,\delta'[x-\xi]\,f''[\xi] - \delta[x-\xi]\,f^{(3)}[\xi])}{\text{Abs}[f'[\xi]]\,f'[\xi]^4}$$

```
In[14]:= Derivative[3][δ][f] = Together[Expand[1/f'[x] D[Derivative[2][δ][f], x]]]
```

$$\text{Out[14]}= \frac{1}{\text{Abs}[f'[\xi]]\,f'[\xi]^6}$$
$$(15\,f'[\xi]\,\delta'[x-\xi]\,f''[\xi]^2 + 15\,\delta[x-\xi]\,f''[\xi]^3 + 6\,f'[\xi]^2\,f''[\xi]\,f''[\xi] - 4\,f'[\xi]^2\,\delta'[x-\xi]$$
$$f^{(3)}[\xi] - 10\,\delta[x-\xi]\,f'[\xi]\,f''[\xi]\,f^{(3)}[\xi] + f'[\xi]^3\,\delta^{(3)}[x-\xi] + \delta[x-\xi]\,f'[\xi]^2\,f^{(4)}[\xi])$$

For a nicer looking output, we implement a function collect that determines the c_k in $\delta^{(n)}(f(x)) = \sum_{k=0}^{n} c_k(\xi)\,\delta^{(k)}(x-\xi)$

explicitly.

```
In[15]:= collect[expr_] :=
        With[{aux = Together[expr]},
        Collect[Numerator[expr], (* the derivatives of δ *)
                Table[Derivative[i][δ][x - ξ],
                {i, 0, Max[#[[0, 0, 1]]& /@ Cases[aux,
                        Derivative[_][δ][x - ξ], Infinity]]}],
                Factor]/Denominator[aux]]
```

```
In[16]:= collect[Derivative[4][δ][f] =
        Together[Expand[1/f'[x] D[Derivative[3][δ][f], x]]]]
```

$$\text{Out[16]}= \frac{1}{\text{Abs}[f'[\xi]]\,f'[\xi]^8}\,(-5\,f'[\xi]^2\,\delta''[x-\xi]\,(-9\,f''[\xi]^2 + 2\,f'[\xi]\,f^{(3)}[\xi]) + 10\,f'[\xi]^3\,f''[\xi]\,\delta^{(3)}[x-\xi] +$$
$$5\,f'[\xi]\,\delta'[x-\xi]\,(21\,f''[\xi]^3 - 12\,f'[\xi]\,f''[\xi]\,f^{(3)}[\xi] + f'[\xi]^2\,f^{(4)}[\xi]) +$$
$$f'[\xi]^4\,\delta^{(4)}[x-\xi] + \delta[x-\xi]\,(105\,f''[\xi]^4 - 105\,f'[\xi]\,f''[\xi]^2\,f^{(3)}[\xi] +$$
$$10\,f'[\xi]^2\,f^{(3)}[\xi]^2 + 15\,f'[\xi]^2\,f''[\xi]\,f^{(4)}[\xi] - f'[\xi]^3\,f^{(5)}[\xi]))$$

Continuing in this way, we can calculate expressions for any n in $\delta^{(n)}(f(x))$ the following expression.

```
In[17]:= Do[Derivative[k + 1][δ][f] =
            Together[Expand[1/f'[x] D[Derivative[k][δ][f], x]]],
        {k, 4, 9}];
```

Here is the (already large) result for $\delta^{(10)}(f(x))$. For shortness, we write f_k for $f^{(k)}(\xi)$.

```
In[18]:= collect[Derivative[10][δ][f]] /.
        Derivative[n_][f_][_] -> Subscript[f, n]
```

$$\text{Out[18]}= \frac{1}{\text{Abs}[f_1]\,f_1^{20}}$$
$$(11\,f_1\,(59520825\,f_2^9 - 150368400\,f_1\,f_2^7\,f_3 + 116953200\,f_1^2\,f_2^5\,f_3^2 - 30576000\,f_1^3\,f_2^3\,f_3^3 + 1911000\,f_1^4\,f_2$$
$$f_3^4 + 29238300\,f_1^2\,f_2^6\,f_4 - 34398000\,f_1^3\,f_2^4\,f_3\,f_4 + 8599500\,f_1^4\,f_2^2\,f_3^2\,f_4 - 254800\,f_1^5\,f_3^3\,f_4 +$$
$$2149875\,f_1^4\,f_2^3\,f_4^2 - 573300\,f_1^5\,f_2\,f_3\,f_4^2 + 6825\,f_1^6\,f_4^3 - 4127760\,f_1^3\,f_2^5\,f_5 + 3439800\,f_1^4\,f_2^3\,f_3\,f_5 -$$
$$458640\,f_1^5\,f_2\,f_3^2\,f_5 - 343980\,f_1^5\,f_2^2\,f_4\,f_5 + 32760\,f_1^6\,f_3\,f_4\,f_5 + 9828\,f_1^6\,f_2\,f_5^2 + 429975\,f_1^4\,f_2^4\,f_6 -$$
$$229320\,f_1^5\,f_2^2\,f_3\,f_6 + 10920\,f_1^6\,f_3^2\,f_6 + 16380\,f_1^6\,f_2\,f_4\,f_6 - 504\,f_1^7\,f_5\,f_6 - 32760\,f_1^5\,f_2^2\,f_7 +$$
$$9360\,f_1^6\,f_2\,f_3\,f_7 - 360\,f_1^7\,f_4\,f_7 + 1755\,f_1^6\,f_2^2\,f_8 - 180\,f_1^7\,f_3\,f_8 - 60\,f_1^7\,f_2\,f_9 + f_1^8\,f_{10})\,\delta_1 -$$
$$11\,f_1^2\,(-28194075\,f_2^8 + 58476600\,f_1\,f_2^6\,f_3 - 34398000\,f_1^2\,f_2^4\,f_3^2 + 5733000\,f_1^3\,f_2^2\,f_3^3 -$$

$$127400\, f_1^4\, f_3^4 - 10319400\, f_1^2\, f_2^5\, f_4 + 8599500\, f_1^3\, f_2^3\, f_3\, f_4 - 1146600\, f_1^4\, f_2\, f_3^2\, f_4 -$$
$$429975\, f_1^4\, f_2^2\, f_4^2 + 40950\, f_1^5\, f_3\, f_4^2 + 1289925\, f_1^3\, f_2^4\, f_5 - 687960\, f_1^4\, f_2^2\, f_3\, f_5 +$$
$$32760\, f_1^5\, f_3^2\, f_5 + 49140\, f_1^5\, f_2\, f_4\, f_5 - 756\, f_1^6\, f_5^2 - 114660\, f_1^4\, f_2^3\, f_6 + 32760\, f_1^5\, f_2\, f_3\, f_6 -$$
$$1260\, f_1^6\, f_4\, f_6 + 7020\, f_1^5\, f_2^2\, f_7 - 720\, f_1^6\, f_3\, f_7 - 270\, f_1^6\, f_2\, f_8 + 5\, f_1^7\, f_9)\, \delta_2 +$$
$$55\, f_1^3\, (1670760\, f_2^7 - 2751840\, f_1\, f_2^5\, f_3 + 1146600\, f_1^2\, f_2^3\, f_3^2 - 101920\, f_1^3\, f_2\, f_3^3 +$$
$$429975\, f_1^2\, f_2^4\, f_4 - 229320\, f_1^3\, f_2^2\, f_3\, f_4 + 10920\, f_1^4\, f_3^2\, f_4 + 8190\, f_1^4\, f_2\, f_4^2 - 45864\, f_1^3\, f_2^3\, f_5 +$$
$$13104\, f_1^4\, f_2\, f_3\, f_5 - 504\, f_1^5\, f_4\, f_5 + 3276\, f_1^4\, f_2^2\, f_6 - 336\, f_1^5\, f_3\, f_6 - 144\, f_1^5\, f_2\, f_7 + 3\, f_1^6\, f_8)\, \delta_3 -$$
$$55\, f_1^4\, (-343980\, f_2^6 + 429975\, f_1\, f_2^4\, f_3 - 114660\, f_1^2\, f_2^2\, f_3^2 + 3640\, f_1^3\, f_3^3 - 57330\, f_1^2\, f_2^3\, f_4 -$$
$$16380\, f_1^3\, f_2\, f_3\, f_4 - 315\, f_1^4\, f_4^2 + 4914\, f_1^3\, f_2^2\, f_5 - 504\, f_1^4\, f_3\, f_5 - 252\, f_1^4\, f_2\, f_6 + 6\, f_1^5\, f_7)\, \delta_4 + 231\, f_1^5$$
$$(12285\, f_2^5 - 10920\, f_1\, f_2^3\, f_3 + 1560\, f_1^2\, f_2\, f_3^2 + 1170\, f_1^2\, f_2^2\, f_4 - 120\, f_1^3\, f_3\, f_4 - 72\, f_1^3\, f_2\, f_5 + 2\, f_1^4\, f_6)\, \delta_5 -$$
$$231\, f_1^6\, (-1365\, f_2^4 + 780\, f_1\, f_2^2\, f_3 - 40\, f_1^2\, f_3^2 - 60\, f_1^2\, f_2\, f_4 + 2\, f_1^3\, f_5)\, \delta_6 +$$
$$330\, f_1^7\, (78\, f_2^3 - 24\, f_1\, f_2\, f_3 + f_1^2\, f_4)\, \delta_7 -$$
$$165\, f_1^8\, (-9\, f_2^2 + f_1\, f_3)\, \delta_8 + 55\, f_1^9\, f_2\, \delta_9 + f_1^{10}\, \delta_{10} +$$
$$(654729075\, f_2^{10} - 1964187225\, f_1\, f_2^8\, f_3 + 1929727800\, f_1^2\, f_2^6\, f_3^2 - 714714000\, f_1^3\, f_2^4\, f_3^3 + 84084000\, f_1^4\, f_2^2$$
$$f_3^4 - 1401400\, f_1^5\, f_3^4 + 413513100\, f_1^2\, f_2^7\, f_4 - 643242600\, f_1^3\, f_2^5\, f_3\, f_4 + 252252000\, f_1^4\, f_2^3\, f_3^2\, f_4 -$$
$$21021000\, f_1^5\, f_2\, f_3^3\, f_4 + 47297250\, f_1^4\, f_2^4\, f_4^2 - 23648625\, f_1^5\, f_2^2\, f_3\, f_4^2 + 1051050\, f_1^6\, f_3^2\, f_4^2 +$$
$$525525\, f_1^6\, f_2\, f_4^3 - 64324260\, f_1^3\, f_2^6\, f_5 + 75675600\, f_1^4\, f_2^4\, f_3\, f_5 - 18918900\, f_1^5\, f_2^2\, f_3^2\, f_5 +$$
$$560560\, f_1^6\, f_3^3\, f_5 - 9459450\, f_1^5\, f_2^3\, f_4\, f_5 + 2522520\, f_1^6\, f_2\, f_3\, f_4\, f_5 - 45045\, f_1^7\, f_4^2\, f_5 +$$
$$378378\, f_1^6\, f_2^2\, f_5^2 - 36036\, f_1^7\, f_3\, f_5^2 + 7567560\, f_1^4\, f_2^5\, f_6 - 6306300\, f_1^5\, f_2^3\, f_3\, f_6 + 840840\, f_1^6\, f_2\, f_3^2\, f_6 +$$
$$630630\, f_1^6\, f_2^2\, f_4\, f_6 - 60060\, f_1^7\, f_3\, f_4\, f_6 - 36036\, f_1^7\, f_2\, f_5\, f_6 + 462\, f_1^8\, f_6^2 - 675675\, f_1^5\, f_2^4\, f_7 +$$
$$360360\, f_1^6\, f_2^2\, f_3\, f_7 - 17160\, f_1^7\, f_3^2\, f_7 - 25740\, f_1^7\, f_2\, f_4\, f_7 + 792\, f_1^8\, f_5\, f_7 + 45045\, f_1^6\, f_2^3\, f_8 -$$
$$12870\, f_1^7\, f_2\, f_3\, f_8 + 495\, f_1^8\, f_4\, f_8 - 2145\, f_1^7\, f_2^2\, f_9 + 220\, f_1^8\, f_3\, f_9 + 66\, f_1^8\, f_2\, f_{10} - f_1^9\, f_{11})\, \delta[x - \xi])$$

For f being the identity function, the last expression simplifies to $\delta^{(10)}(x - \xi)$.

```
In[19]:= Derivative[10][δ][f] /. f -> Function[x, x]
```

```
Out[19]= δ^(10) [x - ξ]
```

Let us check a second example numerically, $\delta^{(6)}(x)$. Here is the exact result for $f(x) = x^3 - 1$.

```
In[20]:= Integrate[Derivative[6][δ][f] /. f -> Function[x, x^3 - 1] /.
              ξ -> 1 /. δ -> DiracDelta, {x, 0, 3}] // N
```

```
Out[20]= 95.7659
```

Here is a plot of the smoothed function. The function values are in the order 10^{19}; so to find an integral of the order 10^2 we cannot use machine arithmetic to carry out the integration.

```
In[21]:= With[{δ = 10^-5},
            Plot[Evaluate[Derivative[6, 0][smoothδ][x^3 - 1, δ]],
                {x, 0.995, 1.005}, PlotRange -> All,
                Frame -> True, Axes -> False]]
```

Using high-precision arithmetic, we obtain a good agreement with the exact value.

```
In[22]:= With[{δ = 10^-5},
            NIntegrate[Evaluate[Derivative[6, 0][smoothδ][x^3 - 1, δ]],
                {x, 0, 2}, MaxRecursion -> 15,
                PrecisionGoal -> 6, WorkingPrecision -> 30]]
```

```
Out[22]= 95.7781
```

d) We start by implementing a function ΘSimplify for simplifying products of UnitStep functions.

```
In[1]:= ΘSimplify[expr_] :=
          With[{Θ = UnitStep}, Expand[expr] /. Θ[x_, y_] :> Θ[x] Θ[y] /.
```

```
θ[x_]^n_ :> θ[x] //.
θ[z + n1_.] θ[z + n2_.] :> θ[z + Min[n1, n2]] //.
θ[n1_. - z] θ[n2_. - z] :> θ[Min[n1, n2] - z] //.
(θ[z + n1_.] θ[n2_. - z] :> 0 /; n2 <= -n1)]
```

Now, we implement the double integration that yields the probability distribution for a sum. This is nearly straightforward. The probability distribution $p[n][z]$ will be a piecewise-defined function. The function Integrate would return an answer containing an If statement and would give only one branch explicitly. To get all branches, we explicitly sum over all pieces and give the corresponding assumptions in Integrate.

In[2]:= `Off[RuleDelayed::rhs];`

```
(* uniform probability distribution in [0, 1] *)
p[1][z_] := UnitStep[z] UnitStep[1 - z];

(* probability distribution for sum of n variables *)
p[n_][z_] := p[n][z_] =
Module[{res},
    (* carry out the double integration *)
    res = θSimplify[Sum[UnitStep[z - ξ] UnitStep[(ξ + 1) - z] *
    Integrate[p[1][y] p[n - 1][z - y]
    (* p[n - 1][z - y] arises from
        Integrate[DiracDelta[z - x - y] p[n - 1][x],
                  {x, -Infinity, Infinity}] *),
            {y, -Infinity, Infinity},
            Assumptions -> ξ < z < ξ + 1], {ξ, 0, n - 1}]];
    (* write result in nice form *)
    Collect[res, Union[#/(# /. _UnitStep -> 1)&[List @@ res]], Factor]]
```

Finally, for better readability of the result, we introduce the characteristic function $\chi_{a,b}$ that is 1 inside the interval $[a, b]$ and zero everywhere else.

In[7]:= `toχRule = UnitStep[n1_ - z] UnitStep[z + n2_.] :> Subscript[χ, -n2, n1];`

So, we arrive at the following expressions for the probability densities for n random variables with uniform distribution.

In[8]:= `p[2][z] /. toχRule`

Out[8]= $z \chi_{0,1} + (2 - z) \chi_{1,2}$

In[9]:= `p[3][z] /. toχRule`

Out[9]= $\frac{1}{2} z^2 \chi_{0,1} + \frac{1}{2} (-3 + 6 z - 2 z^2) \chi_{1,2} + \frac{1}{2} (-3 + z)^2 \chi_{2,3}$

In[10]:= `p[4][z] /. toχRule`

Out[10]= $\frac{1}{6} z^3 \chi_{0,1} + \frac{1}{6} (4 - 12 z + 12 z^2 - 3 z^3) \chi_{1,2} + \frac{1}{6} (-44 + 60 z - 24 z^2 + 3 z^3) \chi_{2,3} - \frac{1}{6} (-4 + z)^3 \chi_{3,4}$

In[11]:= `p[5][z] /. toχRule`

Out[11]= $\frac{1}{24} z^4 \chi_{0,1} + \frac{1}{24} (-5 + 20 z - 30 z^2 + 20 z^3 - 4 z^4) \chi_{1,2} + \frac{1}{24} (155 - 300 z + 210 z^2 - 60 z^3 + 6 z^4) \chi_{2,3} + \frac{1}{24} (-655 + 780 z - 330 z^2 + 60 z^3 - 4 z^4) \chi_{3,4} + \frac{1}{24} (-5 + z)^4 \chi_{4,5}$

All probability densities are properly normalized.

In[12]:= `Table[Integrate[p[k][z], {z, 0, k}], {k, 1, 5}]`

Out[12]= {1, 1, 1, 1, 1}

Here are plots of the calculated probability densities.

In[13]:= `Show[GraphicsArray[Table[`
`Plot[Evaluate[p[k][z]], {z, -1, k + 1}, Frame -> True, Axes -> False,`
`PlotRange -> All, DisplayFunction -> Identity], {k, 1, 5}]]]`

We could, of course, also express the functions using Piecewise instead of UnitStep. Here this is done.

```
In[14]:= q[1][z_] := Piecewise[{{1, 0 < z < 1}}]

        q[n_][z_] := q[n][z_] =
            Integrate[q[1][y] q[n - 1][z - y], {y, -Infinity, Infinity}] //.
            (* do not care about individual points *)
            HoldPattern[Piecewise[{a___, {_, z == _}, b___}, c_]] :>
                Piecewise[{a, b}, c]

In[16]:= Do[q[k][z], {k, 6}]
        q[6][z]
```

$$
\text{Out[17]}= \begin{cases} -\frac{1}{120}\,(-6+z)^5 & 5 < z < 6 \\[4pt] \frac{z^5}{120} & 0 < z \le 1 \\[4pt] \frac{1}{120}\,(6 - 30\,z + 60\,z^2 - 60\,z^3 + 30\,z^4 - 5\,z^5) & 1 < z < 2 \\[4pt] \frac{1}{60}\,(2193 - 3465\,z + 2130\,z^2 - 630\,z^3 + 90\,z^4 - 5\,z^5) & 3 < z < 4 \\[4pt] \frac{1}{120}\,(-10974 + 12270\,z - 5340\,z^2 + 1140\,z^3 - 120\,z^4 + 5\,z^5) & 4 < z < 5 \\[4pt] \frac{1}{60}\,(-237 + 585\,z - 570\,z^2 + 270\,z^3 - 60\,z^4 + 5\,z^5) & 2 < z < 3 \end{cases}
$$

Another possibility to generate the probability distributions for sums consists of using the characteristic function of the starting probability distribution [1308], [1725], [752]. The next input uses InverseFourierTransform to generate the centered versions of the first 21 distribution functions.

```
In[18]:= With[{o = 20},
        (* center all curves at z == 0 *)
        Plot[Evaluate[Table[Sqrt[2/Pi] InverseFourierTransform[(Sin[ω]/ω)^j, ω, t],
                {j, 0, o}]],
            {t, -o/2, o/2}, PlotRange -> All, Frame -> True, Axes -> False,
            PlotStyle -> Table[Hue[0.8 j/o], {j, 0, o}]]]
```

In the limit $n \to \infty$, the $p[n][z]$ become Gaussian [1057]. After rescaling, they approach $(2\pi)^{-1/2} \exp(-x^2/2)$. Here this is demonstrated.

```
In[19]:= (* scale height and horizontal width *)
        scaledp[n_, x_] := Sqrt[n/(6 Pi)] *
                InverseFourierTransform[(Sin[ω]/ω)^n, ω, x] /.
                            {x -> x Sqrt[n/3]}

In[21]:= With[{o = 20},
        (* show relative error compared to Gaussian limit *)
        Plot[Evaluate[Table[scaledp[j, x] /(1/Sqrt[2 Pi] Exp[-1/2 x^2]),
                {j, 2, o}]],
            {x, -o/4, o/4}, PlotRange -> {1/4, 3/2}, Frame -> True,
            Axes -> False, PlotStyle -> Table[Hue[0.8 j/o], {j, 2, o}]]]
```

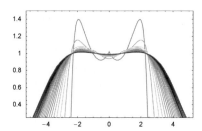

For the distribution of products of random variables, see [1581]; For the distribution of sums of nonidentically uniformly distributed random numbers, see [235].

e) The determinant of a 2×2 matrix is $a_{11} a_{22} - a_{12} a_{21}$. Assuming that the two elements a_{11}, and a_{22} are uniformly distributed in [0, 1], the probability density for the product is $p(x) = -\theta(x)\,\theta(1-x)\ln(x)$. Here this result is derived [683].

In[1]:= `Integrate[DiracDelta[x - a11 a22], {a11, 0, 1}]`

Out[1]= $\dfrac{\text{UnitStep}[-a22, -x, -a22 + x] + \text{UnitStep}[a22, a22 - x, x]}{\text{Abs}[a22]}$

In[2]:= `Integrate[%, {a22, 0, 1}]`

Out[2]= $\begin{cases} -\text{Log}[x] & 0 < x < 1 \\ 0 & x \geq 1 \,||\, \\ \text{Integrate}\left[\frac{\text{UnitStep}[-a22,-x,-a22+x]+\text{UnitStep}[a22,a22-x,x]}{\text{Abs}[a22]}, \{a22, 0, 1\}, \text{Assumptions} \to x = 0\right] & \text{True} \end{cases}$

Obviously, the probability density for the product $a_{12}\,a_{21}$ also is $-\theta(x)\,\theta(1-x)\ln(x)$.

In[3]:= `f[aa_] = -UnitStep[aa] UnitStep[1 - aa] Log[aa]`

Out[3]= `-Log[aa] UnitStep[1 - aa] UnitStep[aa]`

Using this result we can calculate a closed form expression for the probability distribution of the determinant, this means of the difference of the two product terms.

In[4]:= `Integrate[(* for a better result *) PiecewiseExpand @`
`                Integrate[DiracDelta[x - a11a22 + a12a21] *`
`                Log[a11a22] Log[a12a21], {a11a22, 0, 1}], {a12a21, 0, 1}] //`
`                                                         InputForm`

Out[4]//InputForm= `Piecewise[{{2, x == 0},`
`                {(12 - 12*x - Pi^2*x - 6*Log[1 - x] +`
`                    6*x*Log[1 - x] + 6*x*Log[x] - 3*x*Log[x]^2 +`
`                    6*x*PolyLog[2, x])/6, Inequality[0, Less, x,`
`                  Less, 1]}, {0, x >= 1 || x <= -1}},`
`               Integrate[Piecewise[{{Log[a12a21]*Log[a12a21 + x],`
`                   Inequality[0, LessEqual, a12a21 + x, LessEqual,`
`                   1]}}, 0], {a12a21, 0, 1}, Assumptions ->`
`               Inequality[-1, Less, x, Less, 0]]]`

Mathematica was able to carry out the integral in closed form. It contains a special function, the dilogarithm function Li_2. The last result is only correct for negative x. Because of the symmetry of the problem, we add the mirrored part for positive x.

In[5]:= `pdf[x_] =`
`         (2 - 2x + (Pi^2 x)/6 + (-1 + x) Log[1 - x] + x Log[x] (1 + Log[-1/x]) -`
`             x PolyLog[2, 1/x]) UnitStep[1 - x] UnitStep[x] +`
`         (2 + 2x - (Pi^2 x)/6 - (1 + x) Log[1 + x] - x Log[-x] (1 + Log[1/x]) +`
`             x PolyLog[2, -1/x]) UnitStep[-x] UnitStep[1 + x]`

Out[5]= $\left(2 - 2x + \dfrac{\pi^2 x}{6} + (-1 + x)\,\text{Log}[1 - x] + x\left(1 + \text{Log}\left[-\dfrac{1}{x}\right]\right)\text{Log}[x] - x\,\text{PolyLog}\left[2, \dfrac{1}{x}\right]\right)\text{UnitStep}[1 - x]$

$\text{UnitStep}[x] + \left(2 + 2x - \dfrac{\pi^2 x}{6} - x\left(1 + \text{Log}\left[\dfrac{1}{x}\right]\right)\text{Log}[-x] - (1 + x)\,\text{Log}[1 + x] + x\,\text{PolyLog}\left[2, -\dfrac{1}{x}\right]\right)$

$\text{UnitStep}[-x]\,\text{UnitStep}[1 + x]$

Here is a graphic of the probability distribution.

```
In[6]:= exactPdfGraphic =
    Plot[pdf[x], {x, -3/2, 3/2}, Frame -> True, Axes -> False,
        PlotStyle -> {Thickness[0.006], Hue[0]}]
```

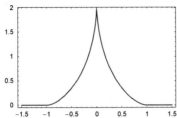

We carry out a numerical simulation using 10^6 random 2×2 matrices. The data agree very well with the theoretical distribution.

```
In[7]:= Module[{data = Table[Det[Table[Random[], {2}, {2}]], {10^6}],
            binSize = (* optimal bin size *) 0.0084},
    Show[{Graphics[(* make histogram *)
    Apply[Polygon[{{#1, 0}, {#1, #3}, {#2, #3}, {#2, 0}}]&,
            (* bin the data *)
            {First[#] binSize, (First[#] + 1) binSize,
            Length[#]/Length[data]/binSize}& /@
                    Split[Quotient[Sort[data], binSize]], {1}],
            Frame -> True], (* exact distribution *) exactPdfGraphic},
        PlotRange -> {-0.2, 2.2}]]
```

For probability densities of determinants of matrices with Gaussian distributions, see [374].

f) After carrying out the integration over x_n using the Dirac delta function, we are left with an $n - 1$ dimensional integral. It is straightforward to implement the remaining integrations. Unfortunately, this is relatively slow.

```
In[1]:= iDomains[n_] := Table[{t[k + 1], 0, 1 - Sum[t[i], {i, k}]},
                        {k, n - 2, 0, -1}]

In[2]:= startIntegrand[n_] :=
    Derivative[n - 1][f][Array[t, n].Array[x, n]] /.
                t[n] -> 1 - Sum[t[i], {i, n - 1}]

In[3]:= ddInt1[x_, n_] := Fold[Integrate, startIntegrand[n], iDomains[n]]

In[4]:= ((ddInt1[x, 5] // Timing) /. x[k_] :> Subscript[x, k]) // Short[#, 6]&
```

$$Out[4]//Short= \left\{ 6.58 \text{ Second}, \frac{1}{(x_3 - x_4)(x_3 - x_5)(x_4 - x_5)} \left((x_3 - x_4) \left(\frac{f[x_1]-f[x_2]}{x_1-x_2} + \frac{-f[x_1]+f[x_5]}{x_1-x_5} \right) \frac{}{x_2 - x_5} + \right. \right.$$

$$\frac{\left(\frac{f[x_1]-f[x_2]}{x_1-x_2} + \frac{-f[\ll 1 \gg]+\ll 1 \gg}{x_1-x_3} \right)(x_4 - x_5)}{x_2 - x_3} + \frac{\left(\frac{f[x_1]-f[x_2]}{x_1-x_2} + \frac{-f[x_1]+f[x_4]}{x_1-x_4} \right)(-x_3 + x_5)}{x_2 - x_4} \left. \left. \right) \right\}$$

We can speed up the integration by a substantial factor by calculating only the indefinite integral and substituting the limits.

```
In[5]:= integrate[f_, {x_, xl_, xu_}] :=
                ((# /. x -> xu) - (# /. x -> xl))&[Integrate[f, x]]

In[6]:= ddInt2[x_, n_] := Fold[integrate, startIntegrand[n], iDomains[n]]
```

```
In[7]:= ddInt2[x, 5]; // Timing
Out[7]= {0.61 Second, Null}
```

Simplifying the functions after each integration step can speed up the calculation even more.

```
In[8]:= ddInt3[x_, n_] :=
            Fold[Together[integrate[##]]&, startIntegrand[n], iDomains[n]]
In[9]:= ddInt3[x, 5]; // Timing
Out[9]= {0.4 Second, Null}
```

Pulling out factors that do not depend on the integration variables and integrating sums termwise, yields a further speed-up.

```
In[10]:= iStep[{_, f_}, {x_, xl_, xu_}] :=
           Module[{f1 = Together[f], num, den},
                 {num, den} = {Expand[Numerator[f1]], Denominator[f1]};
                 (* thread over sums *)
                 {den, If[Head[num] === Plus,
                          integrate[#, {x, xl, xu}]& /@ num,
                          integrate[num, {x, xl, xu}]]}]
In[11]:= ddInt4[x_, n_] := (1/(Times @@ (First /@ #)) #[[-1, 2]])&[
                 FoldList[iStep, {1, startIntegrand[n]}, iDomains[n]]]
In[12]:= ddInt4[x, 5]; // Timing
Out[12]= {0.1 Second, Null}
```

The results of the various *dd*Int*i* agree.

```
In[13]:= dd[{x0_, x1_}, f_] := (f[x0] - f[x1])/(x0 - x1)

         dd[xl_List, f_] := (* direct definition *)
             (dd[Drop[xl, -1], f] - dd[Drop[xl, 1], f])/(First[xl] - Last[xl])
In[15]:= dd[Array[x, 5], f]; // Timing
Out[15]= {0. Second, Null}
```

```
In[16]:= dd[Array[x, 5], f] - ddInt4[x, 5] // Together
Out[16]= 0
```

```
In[17]:= Together[dd[Array[x, 5], f] - #]& /@
                 {ddInt4[x, 5], ddInt3[x, 5], ddInt2[x, 5], ddInt1[x, 5]}
Out[17]= {0, 0, 0, 0}
```

For computational efficient methods to calculate divided differences, see [1476].

g) This is the function under consideration.

```
In[1]:= f[ε_, ξ_][x_] := Exp[I Pi/4]/(2 Sqrt[Pi] Sqrt[1 - ε])*
                          Exp[-I/2 (ε x^2 - 2 ξ x + ε ξ^2)/(1 - ε)].
```

The first limit $\lim_{\varepsilon\to 0} f_\varepsilon^{(\xi)}(x)$ is straightforward to carry out; the result is an exponential function.

```
In[2]:= Limit[f[ε, ξ][x], ε -> 0]
```

$$Out[2]= \frac{\left(\frac{1}{2} + \frac{i}{2}\right) e^{i x \xi}}{\sqrt{2 \pi}}$$

The built-in Limit cannot find the second limit.

```
In[3]:= Limit[f[ε, ξ][x], ε -> 1]
Out[3]= Indeterminate
```

The last result indicates that the limit is not a function. So, we will search the limit as a generalized function. We start by investigating the limiting behavior of the argument of the exponential function.

```
In[4]:= Series[-I/2 (ε x^2 - 2 ξ x + ε ξ^2)/(1 - ε), {ε, 1, 1}] // Simplify
```

$$Out[4]= \frac{i (x - \xi)^2}{2 (\varepsilon - 1)} + \frac{1}{2} i (x^2 + \xi^2) + O[\varepsilon - 1]^2$$

The function $g(x) = e^{i\pi/4} 2^{-1} \pi^{-1/2}(1 - \varepsilon)^{1/2} \exp(i\zeta^2/(2(1 - \varepsilon)))$ is a highly oscillating function. In the limit $\varepsilon \to 1$, the oscillations increase indefinitely. But the saddle point $\zeta = 0$ is exceptional. Independently of ε, around $\zeta \approx 0$ the function is stationary and grows as $\varepsilon \to 0$. $\int_{-\infty}^{\infty} g(\zeta)\, d\zeta$ tends to 1 as ε tends to 1. This means $\int_{-\infty}^{\infty} f_\varepsilon^{(\xi)}(x) h(x)\, dx$ tends to $h(\xi)$ as $\varepsilon \to 1$. So, we have $\lim_{\varepsilon \to 1} f_\varepsilon^{(\xi)}(x) = \delta(x - \xi)$.

```
In[5]:= Integrate[Exp[I Pi/4]/(2 Sqrt[Pi] Sqrt[1 - ε])*
            Exp[I/2 (x - ξ)^2/(2 (ε - 1))],
            {x, -Infinity, Infinity},
            (* or later: Assumptions -> 0 < ε < 1, *)
            GenerateConditions -> False] // FullSimplify[#, 0 < ε < 1]&
```

```
Out[5]= 1
```

This means $\int_{-\infty}^{\infty} f_\varepsilon^{(\xi)}(x) h(x)\, dx$ tends to $h(\xi)$ as $\varepsilon \to 1$. So, we have $\lim_{\varepsilon \to 1} f_\varepsilon^{(\xi)}(x) = \delta(x - \xi)$.

Now let us look for differential equations for $f_\varepsilon^{(\xi)}(x)$. The first-order differential equation

$$f_\varepsilon^{(\xi)'}(x) = -\frac{i(2x\varepsilon - 2\xi)}{2(1 - \varepsilon)} f_\varepsilon^{(\xi)}(x)$$

follows immediately after differentiation.

```
In[6]:= D[f[ε, ξ][x], x]/f[ε, ξ][x]
```

```
Out[6]=   i (2 x ε - 2 ξ)
        - ─────────────
           2 (1 - ε)
```

To derive an x-free differential equation, we differentiate $f_\varepsilon^{(\xi)}(x)$ multiple times. Then we algebraize the resulting equations by replacing exponentials involving x by new variables $\in 1$ and $\in 2$.

```
In[7]:= eqs = Table[Numerator[Together[Derivative[j][f[ε, ξ]] - D[f[ε, ξ][x], {x, j}]]],
            {j, 0, 3}] /. (* make algebraic system *)
            {Exp[I x ξ/(1 - ε)] :> ε1,
             Exp[I x^2 ε/(2 (1 - ε)) + r_] :> ε2 Exp[r]}
```

$$Out[7]= \left\{ -(-1)^{1/4}\, \in 1 + 2\, e^{\frac{i\,\varepsilon\,\xi^2}{2\,(1-\varepsilon)}} \sqrt{\pi}\, \sqrt{1 - \varepsilon}\, \in 2\, f[\varepsilon, \xi], \right.$$

$$-(-1)^{3/4}\, x\, \varepsilon\, \in 1 + (-1)^{3/4}\, \in 1\, \xi - 2\, e^{\frac{i\,\varepsilon\,\xi^2}{2\,(1-\varepsilon)}} \sqrt{\pi}\, \sqrt{1 - \varepsilon}\, \in 2\, f[\varepsilon, \xi]' +$$

$$2\, e^{\frac{i\,\varepsilon\,\xi^2}{2\,(1-\varepsilon)}} \sqrt{\pi}\, \sqrt{1 - \varepsilon}\, \varepsilon\, \in 2\, f[\varepsilon, \xi]',\ (-1)^{3/4}\, \varepsilon\, \in 1 - (-1)^{3/4}\, \varepsilon^2\, \in 1 + (-1)^{1/4}\, x^2\, \varepsilon^2\, \in 1 -$$

$$2\, (-1)^{1/4}\, x\, \varepsilon\, \in 1\, \xi + (-1)^{1/4}\, \in 1\, \xi^2 + 2\, e^{\frac{i\,\varepsilon\,\xi^2}{2\,(1-\varepsilon)}} \sqrt{\pi}\, \sqrt{1 - \varepsilon}\, \in 2\, f[\varepsilon, \xi]'' -$$

$$4\, e^{\frac{i\,\varepsilon\,\xi^2}{2\,(1-\varepsilon)}} \sqrt{\pi}\, \sqrt{1 - \varepsilon}\, \varepsilon\, \in 2\, f[\varepsilon, \xi]'' + 2\, e^{\frac{i\,\varepsilon\,\xi^2}{2\,(1-\varepsilon)}} \sqrt{\pi}\, \sqrt{1 - \varepsilon}\, \varepsilon^2\, \in 2\, f[\varepsilon, \xi]'',$$

$$-3\,(-1)^{1/4}\, x\, \varepsilon^2\, \in 1 + 3\,(-1)^{1/4}\, x\, \varepsilon^3\, \in 1 + (-1)^{3/4}\, x^3\, \varepsilon^3\, \in 1 + 3\,(-1)^{1/4}\, \varepsilon\, \in 1\, \xi -$$

$$3\,(-1)^{1/4}\, \varepsilon^2\, \in 1\, \xi - 3\,(-1)^{3/4}\, x^2\, \varepsilon^2\, \in 1\, \xi + 3\,(-1)^{3/4}\, x\, \varepsilon\, \in 1\, \xi^2 - (-1)^{3/4}\, \in 1\, \xi^3 -$$

$$2\, e^{\frac{i\,\varepsilon\,\xi^2}{2\,(1-\varepsilon)}} \sqrt{\pi}\, \sqrt{1 - \varepsilon}\, \in 2\, f[\varepsilon, \xi]^{(3)} + 6\, e^{\frac{i\,\varepsilon\,\xi^2}{2\,(1-\varepsilon)}} \sqrt{\pi}\, \sqrt{1 - \varepsilon}\, \varepsilon\, \in 2\, f[\varepsilon, \xi]^{(3)} -$$

$$\left. 6\, e^{\frac{i\,\varepsilon\,\xi^2}{2\,(1-\varepsilon)}} \sqrt{\pi}\, \sqrt{1 - \varepsilon}\, \varepsilon^2\, \in 2\, f[\varepsilon, \xi]^{(3)} + 2\, e^{\frac{i\,\varepsilon\,\xi^2}{2\,(1-\varepsilon)}} \sqrt{\pi}\, \sqrt{1 - \varepsilon}\, \varepsilon^3\, \in 2\, f[\varepsilon, \xi]^{(3)} \right\}$$

Eliminating $\in 1$, $\in 2$, and x yields the following polynomials.

```
In[8]:= gb = GroebnerBasis[(* eliminate ε1 *)
            Cases[Factor[Resultant[First[eqs], #, ε1]]& /@ Rest[eqs]],
            _Plus, {2}], {}, (* eliminate ε2 and x *){ε2, x},
            MonomialOrder -> EliminationOrder] // Factor;
```

Here are two of the possible differential equations, a second-order and a third-order one.

```
In[9]:= {gb[[1, -1]], gb[[3, -1]]} // Simplify
```

$$Out[9]= \{ \varepsilon\, f[\varepsilon, \xi]^2 - i\,(-1 + \varepsilon)\, (f[\varepsilon, \xi]')^2 + i\,(-1 + \varepsilon)\, f[\varepsilon, \xi]\, f[\varepsilon, \xi]'',$$
$$2\,\varepsilon\, (f[\varepsilon, \xi]')^2 + f[\varepsilon, \xi]''\, (-\varepsilon\, f[\varepsilon, \xi] - i\,(-1 + \varepsilon)\, f[\varepsilon, \xi]'') + i\,(-1 + \varepsilon)\, f[\varepsilon, \xi]'\, f[\varepsilon, \xi]^{(3)} \}$$

We end with a check of the correctness of the derived differential equations.

```
In[10]:= % /. Derivative[j_][f[ε, ξ]] :> D[f[ε, ξ][x], {x, j}] /.
            f[ε, ξ] :> f[ε, ξ][x] // Simplify
```

Out[10]= {0, 0}

h) Using the definition of $F_o^{(L)}(k)$, we can derive the following relation between $F_o^{(L)}(k)$ and $F^{(L)}(k)$ [290].

$$F_o^{(L)}(k) = \frac{1}{\sqrt{2\pi}} \int_{-\infty}^{\infty} \exp(i\,k\,x)\, f_o(x)\, dx = \frac{1}{\sqrt{2\pi}} \sum_{j=-o}^{o} \int_{j2L-L}^{j2L+L} \exp(i\,k\,x)\, f(x+2\,j\,L)\, dx$$

$$= \frac{1}{\sqrt{2\pi}} \sum_{j=-o}^{o} \int_{-L}^{L} \exp(i\,k(\xi-2\,j\,L))\, f(\xi)\, d\xi = \left(\sum_{j=-o}^{o} \exp(-2\,i\,k\,j\,L) \right) \times \frac{1}{\sqrt{2\pi}} \int_{-L}^{L} \exp(i\,k\,\xi)\, f(\xi)\, d\xi$$

$$= \csc(k\,L)\,\sin(k\,L\,(2\,o+1))\, F^{(L)}(k).$$

(For other representation of Fourier transforms as infinite sums, see [49], [188].)

We implement the definitions of the $c_n^{(L)}$. The evaluated integrals contain the error function erf(ξ).

In[1]:= `f[x_] := Exp[-x^2];`

```
{c[L_][n_] = 1/(2L) Integrate[f[x] Exp[I n x/L Pi], {x, -L, L},
                    Assumptions -> L > 0] // Simplify,

 c[L_][0]  = 1/(2L) Integrate[f[x] Exp[I 0 x/L Pi], {x, -L, L},
                    Assumptions -> L > 0] // Simplify}
```

Out[2]= $\left\{ \left(e^{-\frac{n^2\,\pi^2}{4\,L^2}}\,\sqrt{\pi} \left((2\,L^2 - i\,n\,\pi)\,\sqrt{(2\,L^2 + i\,n\,\pi)^2}\; \text{Erf}\left[\sqrt{\left(L - \frac{i\,n\,\pi}{2\,L}\right)^2}\,\right] + \right. \right.$

$$\left. \left. \sqrt{(2\,L^2 - i\,n\,\pi)^2}\; (2\,L^2 + i\,n\,\pi)\; \text{Erf}\left[\sqrt{\left(L + \frac{i\,n\,\pi}{2\,L}\right)^2}\,\right] \right) \right) \Bigg/$$

$$\left(4\,L\,\sqrt{(2\,L^2 - i\,n\,\pi)^2}\,\sqrt{(2\,L^2 + i\,n\,\pi)^2} \right),\; \frac{\sqrt{\pi}\;\text{Erf}[L]}{2\,L} \right\}$$

$F^{(L)}(k)$ is defined similarly. Formally, we have $c_n^{(L)} = \sqrt{\pi/2}\,/\,L\,F^{(L)}(n\,\pi/L)$. (Here $c_n^{(L)}$ is defined for integer n only.)

In[3]:= `F[L_][k_]  = 1/Sqrt[2 Pi] Integrate[f[x] Exp[I k x], {x, -L, L},`
                      `Assumptions -> L > 0] // Simplify;`

The formula for $F_o^{(L)}(k)$ contains the oscillating function csc(ξ) sin($\xi\,\eta$).

In[4]:= `δ[ξ_, η_] := Csc[ξ] Sin[ξ η]`
        `F[L_][o_, k_] = δ[k L, 2 o + 1] F[L][k];`

The left graphic shows the original functions $f(x)$ together with its truncated Fourier sum. The right graphic shows the logarithm of the relative error. The error is quite small in the interval $[-5, 5]$ and increases outside because the Fourier sum defines a periodic function versus the single-bump function $f(x)$.

In[6]:= `Show[GraphicsArray[`
        `    Block[{L = 4, m = 12, $DisplayFunction = Identity},`
        `       {(* original function and Fourier sum *)`
        `        Plot[Evaluate[{f[x], (* Fourier sum in red *)`
        `             Sum[c[L][n] Exp[-I n x/L Pi], {n, -m, m}]}],`
        `                {x, -2L, 2L}, PlotStyle -> {GrayLevel[0], Hue[0]},`
        `                PlotRange -> All, Frame -> True, Axes -> False],`
        `         (* relative error plot *)`
        `         Plot[Evaluate[Log[Abs[f[x] -`
        `              Sum[c[L][n] Exp[-I n x/L Pi], {n, -m, m}]]]], {x, -2L, 2L},`
        `                PlotRange -> All, Frame -> True, Axes -> False]}]]]`

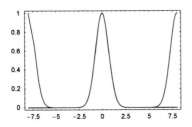
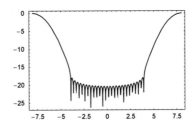

The next two graphics show 25 Fourier series coefficients $c_n^{(L)}$ and $|F_o^{(L)}(k)|$ for $L = 4$ and $o = 4$. A qualitative similarity between the two graphics exists, but the scales in horizontal and vertical direction disagree.

```
In[7]:= Show[GraphicsArray[
          Block[{L = 4, m = 12, o = 4, $DisplayFunction = Identity},
            {(* the Fourier series coefficients *)
             fourierSeriesGraphic = Graphics[{Hue[0],
                            Table[Line[{{n, 0}, {n, Abs[c[L][n]]}}],
                                  {n, -m, m}]},
                      PlotRange -> All, Frame -> True, Axes -> False],
             (* the Fourier transform *)
             Plot[Abs[F[L][o, k]], {k, -o, o},
                  PlotPoints -> 200, PlotRange -> All,
                  Frame -> True, Axes -> False]}]]]
```

Now, we will consider the limit $o \to \infty$. The function $\delta(\xi, \eta) = \csc(\xi)\sin(\xi\eta)$ converges to a sequence of delta functions in the limit $\eta \to \infty$. This can be most easily seen by observing that $\lim_{\eta\to\infty} \delta(0, \eta) = \infty$ and $\lim_{\eta\to\infty} \int_{-\pi/2}^{\pi/2} \delta(\xi, \eta)\,d\xi = \pi$. The following two graphics demonstrate this.

```
In[8]:= Show[GraphicsArray[
          Block[{$DisplayFunction = Identity},
            {(* δ[ξ, η] gets large spikes at ξ - j π *)
             Plot3D[δ[ξ, η], {ξ, -3, 3}, {η, 0, 12},
                    PlotPoints -> 200, Mesh -> False,
                    ClipFill -> None, PlotRange -> {-5, 12}],
             (* the integral converges to π *)
             Plot[Evaluate[Integrate[δ[ξ, η], {ξ, -Pi/2, Pi/2}]],
                  {η, 10, 200}, PlotPoints -> 200]}]]]
```

Using $\sum_{j=-\infty}^{\infty} \exp(-2\,i\,k\,j\,L) = \pi \sum_{n=-\infty}^{\infty} \delta(k\,L - n\,\pi)$, we see that in the limit $o \to \infty$ the following identity holds:

$$F_\infty^{(L)}(k) = \pi \sum_{n=-\infty}^{\infty} \delta(k\,L - n\,\pi)\,F^{(L)}(k) = \sqrt{2\,\pi} \sum_{n=-\infty}^{\infty} \delta\left(k - \frac{n\,\pi}{L}\right) c_n^{(L)}.$$

This means that in the large o limit $F_o^{(L)}(k)$ converges to a series of delta functions. These delta functions are located at $n\,\pi/L$ and their strength is exactly the Fourier series coefficient. Using the inverse Fourier transform and integrating out the Dirac delta functions shows that we recover the Fourier series representation:

$$f(x) = \frac{1}{\sqrt{2\,\pi}} \int_{-\infty}^{\infty} e^{-ikx}\,F_\infty^{(L)}(k) = \sum_{n=-\infty}^{\infty} \exp\left(-i\,n\,\pi\,\frac{x}{L}\right) c_n^{(L)}.$$

For small ξ we have $\delta(\xi, \eta) \approx \eta$. This enables us to normalize $F_o^{(L)}(k)$ in such a way that the maxima of $F_o^{(L)}(n\,\pi/L)/(2\,o + 1)$ are located at integer n and the height of the maxima approaches the values $c_n^{(L)}$.

```
In[9]:= Block[{L = 4, m = 6, o = 4},
          Show[{Plot[Abs[Sqrt[Pi/2] F[L][o, n Pi/L]/L/(2o + 1)], {n, -m, m},
                  PlotPoints -> 500, DisplayFunction -> Identity],
                (* discrete Fourier transform *) fourierSeriesGraphic},
            DisplayFunction -> $DisplayFunction,
            PlotRange -> {{-6, 6}, All}, Frame -> True,
            Axes -> False, AspectRatio -> 1/4]]
```

i) The implementation of the definition is straightforward. We substitute the ϵ-dependent term $f(x) + \epsilon\,\delta(x - y)$ as a pure function to avoid any possible variable conflict with respect to the (dummy) variable x. The function $\mathcal{FD}[F[f(x)], f(y)]$ tries to calculate $\delta\,F[f(x)]\,/\,\delta\,f(y)$. We add an optional simplifying function $\mathcal{S}$.

```
In[1]:= FD[functional_, f_[y_], SimplifyFunction -> S_] :=
          Module[{x, s1, s2, s3},
            Assuming[(* for integration purposes *) Element[y, Reals],
            (* definition *)
            s1 = ((functional /. f :> Function[x, f[x] +
                          ε DiracDelta[x - y]]) - functional)/ε;
            (* carry out and simplify integrals *)
            s2 = s1 //. Integrate[a_, b_List] :>
                  Integrate[a, b, Assumptions -> Element[y, Reals]] //.
              α_. Integrate[a_, b_, as_] + β_. Integrate[c_, b_, as_] :>
                            Integrate[Expand[α a + β c], b, as] //.
                    Integrate[a_ (ε d_ DiracDelta + c_), b_, as_] :>
                    Integrate[Expand[a ε d], b, as] + Integrate[a c, b, as];
            (* expand around ε == 0 *)
            s3 = Limit[Normal[Series[#, {ε, 0, 1}]]& //@ s2, ε -> 0] ;
            (* (maybe) simplify result *) S[s3]]]
```

By default, we will not carry out any simplification.

```
In[2]:= (* no simplification by default *)
          FD[functional_, f_[y_]] :=
          FD[functional, f[y], SimplifyFunction -> Identity]
```

Here is a first example, the functional derivative of $\int_{-\infty}^{\infty} f(x)^2\,dx$ with respect to $f(y)$.

```
In[4]:= FD[Integrate[f[y]^2, {y, -Infinity, Infinity}], f[x]]

Out[4]= 2 f[x]
```

For the identity functional $F[f(x)] = f(x)$ we get $\delta f(x)/\delta f(y) = \delta(x - y)$.

In[5]:= `FD[f[y], f[x]]`

Out[5]= `DiracDelta[-x + y]`

For a functional of the form $F[f(x)] = \int_{-\infty}^{\infty} L(f(x), f'(x)) \, dx$, the functional derivative is, of course, just the classical Euler–Lagrange equation $\partial L/\partial f(x) - \partial(\partial L/\partial f'(x))/\partial x$ [1381].

In[6]:= `FD[Integrate[F[f[y], f'[y]], {y, -Infinity, Infinity}], f[x]] -`
        `(* the Lagrange equations *)`
        `(D[F[f[x], f'[x]], f[x]] - D[D[F[f[x], f'[x]], f'[x]], x])`

Out[6]= `0`

Here is a special case—the equation for the density of 1D quantum system [1401], [1548], [1549].

In[7]:= `FD[Integrate[1/4 n'[y]^2/n[y] + 1/4/n[y] - v[y] n[y],`
        `                {y, -Infinity, Infinity}], n[x]] // Expand`

Out[7]= $-\dfrac{1}{4\,n[x]^2} - v[x] + \dfrac{n'[x]^2}{4\,n[x]^2} - \dfrac{n''[x]}{2\,n[x]}$

For the functional $F[f(x)] = \int_a^b f(x)^2 \, dx$, the functional derivative is nonvanishing only in the interval $[a, b]$; this corresponds to the two Heaviside functions in the next result.

In[8]:= `FD[Integrate[f[x]^2, {x, a, b}], f[y]]`

Out[8]= `2 f[y] UnitStep[y - b UnitStep[a - b] - a UnitStep[-a + b]]`
        `UnitStep[-y + a UnitStep[a - b] + b UnitStep[-a + b]]`

The Lagrange equation for a point mass in a potential $V(x)$ follow from the functional $F[x(t)] = T[x(t)] - V(x(t))$.

In[9]:= `FD[Integrate[1/2 x'[t]^2 - V[x[t]],`
        `                {t, -Infinity, Infinity}], x[t]]`

Out[9]= $-V'[x[t]] - x''[t]$

Here are some inequivalent Lagrangians for the harmonic oscillator [1160], [722], [1161].

In[10]:= `Table[FD[Integrate[1/2 Sum[Binomial[n, k] (ω^2 x[t]^2)^(n - k) x'[t]^(2k)/(2k - 1),`
         `                {k, 0, n}],`
         `                {t, -Infinity, Infinity}], x[t]] // Factor, {n, 5}]`

Out[10]= $\{-\omega^2\,x[t] - x''[t], \; -2\,(\omega^2\,x[t]^2 + x'[t]^2)\,(\omega^2\,x[t] + x''[t]),$
         $\quad -3\,(\omega^2\,x[t]^2 + x'[t]^2)^2\,(\omega^2\,x[t] + x''[t]),$
         $\quad -4\,(\omega^2\,x[t]^2 + x'[t]^2)^3\,(\omega^2\,x[t] + x''[t]), \; -5\,(\omega^2\,x[t]^2 + x'[t]^2)^4\,(\omega^2\,x[t] + x''[t])\}$

Now let us deal with $F[f(x)] = f(x)^3\,f'(x)^2\,f''(x)$. Without explicit simplification, we get an integral containing derivatives of the Dirac delta function.

In[11]:= `FD[Integrate[f[x]^3 f'[x]^2 f''[x],`
         `                {x, -Infinity, Infinity}], f[ξ]]`

Out[11]= $6\,f[\xi]\,f'[\xi]^2\,(f'[\xi]^2 + 2\,f[\xi]\,f''[\xi])$

Because such-type integrals can be evaluated via partial integration, we implement a corresponding rule: $\int_{-\infty}^{\infty} f(x)\,\delta^{(n)}(x - \xi)\,dx = (-1)^n\,f^{(n)}(\xi)$.

In[12]:= `δIntegrateSimplify = (ExpandAll[#] //.`
         `        {Integrate[p_Plus, d_] :> (Integrate[#, d]& /@ p),`
         `         Integrate[f_ Derivative[n_][DiracDelta][x_ + ξ_.],`
         `                {x_, -Infinity, Infinity}] :>`
         `                ((-1)^n D[f, {x, n}] /. x -> -ξ)})&;`

Now, we get a fully integrated result.

In[13]:= `FD[Integrate[f[x]^3 f'[x]^2 f''[x],`
         `                {x, -Infinity, Infinity}], f[ξ],`
         `                SimplifyFunction -> δIntegrateSimplify] // Expand`

Out[13]= $6\,f[\xi]\,f'[\xi]^4 + 12\,f[\xi]^2\,f'[\xi]^2\,f''[\xi]$

Here is the Euler–Lagrange equation for a functional of the form $F[f(x)] = \int_{-\infty}^{\infty} L(f(x), f'(x), f''(x))\,dx$.

```
In[14]:= (el2 = FD[Integrate[F[f[y], f'[y], f''[y]],
                    {y, -Infinity, Infinity}], f[x],
                    SimplifyFunction -> δIntegrateSimplify]) /.
         (* shorten result *)
         (h: (Derivative[__][F]))[__] :> h // Simplify
```

$$
\text{Out[14]= } F^{(1,0,0)} - F^{(1,1,0)} f'[x] + F^{(2,0,1)} f'[x]^2 - F^{(0,2,0)} f''[x] +
$$
$$
F^{(1,0,1)} f''[x] + 2 F^{(1,1,1)} f'[x] f''[x] + F^{(0,2,1)} f''[x]^2 + 2 F^{(1,0,2)} f'[x] f^{(3)}[x] +
$$
$$
2 F^{(0,1,2)} f''[x] f^{(3)}[x] + F^{(0,0,3)} f^{(3)}[x]^2 + F^{(0,0,2)} f^{(4)}[x]
$$

The function `VariationalD` from the package `Calculus`VariationalMethods`` calculates the Euler–Lagrange equation from the above integrand.

```
In[15]:= Needs["Calculus`VariationalMethods`"]
         Expand[VariationalD[F[f[x], f'[x], f''[x]], f[x], x] - el2]
Out[16]= 0
```

Now, we show that the Korteweg–de Vries bracket vanishes. These are the Q_i [1540].

```
In[17]:= Q[0] := Integrate[u[x], {x, -Infinity, Infinity}]
         Q[1] := Integrate[u[x]^2/2, {x, -Infinity, Infinity}]
         Q[2] := Integrate[-u[x]^3 + u'[x]^2/2, {x, -Infinity, Infinity}]
         Q[3] := Integrate[5/2 u[x]^4 - 5 u[x] u'[x]^2 + u''[x]^2/2,
                    {x, -Infinity, Infinity}]
```

The function `KdVBracket` implements the Korteweg–de Vries bracket.

```
In[21]:= KdVBracket[cq1_, cq2_] :=
         Module[{q1, q2, integrand, yIndefIntegral, yDefIntegral},
           {q1, q2} = FD[#, u[y], SimplifyFunction -> δIntegrateSimplify]& /@
                       {cq1, cq2};
           integrand = q1 D[q2, y];
           yIndefIntegral = Integrate[integrand, y];
           If[FreeQ[yIndefIntegral, Integrate, {0, Infinity}, Heads -> True],
              (* substitute for completeness; integral must vanish now *)
              yDefIntegral = (yIndefIntegral /. y -> +Infinity) -
                             (yIndefIntegral /. y -> -Infinity);
              (* u and its derivatives vanish at ±∞ *)
              yDefIntegral /. {u[+Infinity] -> 0, Derivative[k_][u][+Infinity] -> 0,
                              u[-Infinity] -> 0, Derivative[k_][u][-Infinity] -> 0},
              yIndefIntegral]]
```

All of the $\{Q_i, Q_j\}$ vanish.

```
In[22]:= Table[KdVBracket[Q[i], Q[j]], {i, 0, 3}, {j, 0, 3}] // Flatten
Out[22]= {0, 0, 0, 0, 0, 0, 0, 0, 0, 0, 0, 0, 0, 0, 0, 0}
```

It is straightforward to implement repeated functional differentiation.

```
In[23]:= (* multiple differentiation *)
         FD[functional_, f1_, fs__, Simplifyer -> S_] :=
         FD[FD[functional, f1, SimplifyFunction -> S],
            fs, SimplifyFunction -> S]

         (* multiple differentiation; no simplification *)
         FD[functional_, f1_, fs__] := FD[FD[functional, f1], fs] /;
                              FreeQ[{fs}, Rule | RuleDelayed, {1}]
```

Here are again two examples.

```
In[27]:= FD[Integrate[f[y]^2, {y, -Infinity, Infinity}], f[x], f[z]]
Out[27]= 2 DiracDelta[x - z]
```

```
In[28]:= FD[Integrate[f[y] f'[y]^2, {y, -Infinity, Infinity}], f[x], f[z]]
Out[28]= -2 (DiracDelta'[x - z] f'[x] + f[x] DiracDelta''[x - z] + DiracDelta[x - z] f''[x])
```

The second functional derivative of the above-used point particle action is $-(\partial^2./\partial\tau_2^2 + V(x(\tau_1))) \delta(\tau_1 - \tau_2)$..

```
In[29]:= FD[Integrate[1/2 x'[t]^2 - V[x[t]],
                {t, -Infinity, Infinity}], x[τ1], x[τ2]]

Out[29]= -DiracDelta''[τ1 - τ2] - DiracDelta[τ1 - τ2] V''[x[τ1]]
```

For $F[f(x)] = \int_{-\infty}^{\infty} f(x) h(x)\, dx$, we have the general result $\delta^n F / (\delta f(y_1) \cdots \delta f(y_n)) = F[f] \, h(y_1) \cdots h(y_n)$. Here is the case $n = 3$ is explicitly calculated.

```
In[30]:= FD[Exp[Integrate[f[y] h[y], {y, -Infinity, Infinity}]],
                f[x], f[z], f[w]]

Out[30]= e^Integrate[f[y] h[y],{y,-∞,∞},Assumptions→True] h[w] h[x] h[z]
```

For functional differentiation with nontrivial constraints, see [682].

45. Operator Splitting of Order 5

For a more concise and more easily to read output in the following, all operators will carry a hat and operator products will be indicated by $\otimes$.

We start by defining some properties for the operator product $\otimes$. It should self-denest, be linear, expand out sums and products, and be linear over nonoperator quantities. We will write operators with a hat, implementing these quantities in the form `OverHat[operator]`. The functions `operatorQ` and `notOperatorQ` determine if an expression is considered to be an operator. (This implementation is similar to the ones from the solution of Exercise 8 of Chapter 5 of the Programming volume [1735].)

```
In[1]:= a___ ⊗ (⊗b___) ⊗ c___      := a ⊗ b ⊗ c;
        a___ ⊗ (b_ + c_) ⊗ d___    := a ⊗ b ⊗ d + a ⊗ c ⊗ d;
        ⊗ x_?notOperatorQ := x;
        a___ ⊗ x_?notOperatorQ ⊗ b___   := x (a ⊗ b);
        a___ ⊗ (x_?notOperatorQ y_) ⊗ b___   := x(a ⊗ y ⊗ b);
        a___ ⊗ b_Plus^n_ ⊗ c___   := (a ⊗ ## ⊗ c)& @@ Table[b, {n}];
        a___ ⊗ OverHat[A_]^n_. ⊗ OverHat[A_]^(m_.) ⊗ c___ :=
                a ⊗ OverHat[A]^(n + m) ⊗ c;


        notOperatorQ[φ_] := FreeQ[φ, _OverHat];


        operatorQ[φ_] := !notOperatorQ[φ];

In[11]:= (* avoid large sizes of prefix ⊗ *)
        MakeBoxes[CircleTimes[x_], fmt_] := MakeBoxes[CircleTimes["", x], fmt]
```

Next, we define an operator $\mathcal{T}$ acting on the exponential of an expression Ψ (containing operators); it gives the Taylor series expansion to order o around $t = 0$.

```
In[13]:= T[E^Φ_?operatorQ, o_:4] := Sum[⊗ (Φ^i/i!), {i, 0, o}]
```

Using $\mathcal{T}$, we calculate the first few terms in the series expansion of $\exp(t(\hat{A} + \hat{B}))$. Here, `OverHat[A]` and `OverHat[B]` are the two (in general noncommuting) operators.

```
In[14]:= lhs = T[E^(t (OverHat[A] + OverHat[B]))]
```

$$
\begin{aligned}
\text{Out[14]= } & 1 + t\left(\otimes\hat{A} + \otimes\hat{B}\right) + \frac{1}{2}t^2\left(\otimes\hat{A}^2 + \otimes\hat{B}^2 + \hat{A}\otimes\hat{B} + \hat{B}\otimes\hat{A}\right) + \\
& \frac{1}{6}t^3\left(\otimes\hat{A}^3 + \otimes\hat{B}^3 + \hat{A}\otimes\hat{B}^2 + \hat{A}^2\otimes\hat{B} + \hat{B}\otimes\hat{A}^2 + \hat{B}^2\otimes\hat{A} + \hat{A}\otimes\hat{B}\otimes\hat{A} + \hat{B}\otimes\hat{A}\otimes\hat{B}\right) + \\
& \frac{1}{24}t^4\left(\otimes\hat{A}^4 + \otimes\hat{B}^4 + \hat{A}\otimes\hat{B}^3 + \hat{A}^2\otimes\hat{B}^2 + \hat{A}^3\otimes\hat{B} + \hat{B}\otimes\hat{A}^3 + \hat{B}^2\otimes\hat{A}^2 + \hat{B}^3\otimes\hat{A} + \hat{A}\otimes\hat{B}\otimes\hat{A}^2 + \right. \\
& \left. \hat{A}\otimes\hat{B}^2\otimes\hat{A} + \hat{A}^2\otimes\hat{B}\otimes\hat{A} + \hat{B}\otimes\hat{A}\otimes\hat{B}^2 + \hat{B}\otimes\hat{A}^2\otimes\hat{B} + \hat{B}^2\otimes\hat{A}\otimes\hat{B} + \hat{A}\otimes\hat{B}\otimes\hat{A}\otimes\hat{B} + \hat{B}\otimes\hat{A}\otimes\hat{B}\otimes\hat{A}\right)
\end{aligned}
$$

To calculate the first few terms of $\exp(t(\hat{A} + \hat{B}))$, we define a function P that gives the series of a product of terms that come from $\mathcal{T}$. To avoid accumulating unnecessary high-order terms, after every multiplication, we delete the terms of order greater than o in t.

```
In[15]:= SetAttributes[P, HoldFirst]


        P[⊗ Φ___, t_, o_:4] := Fold[DeleteCases[⊗ ##, t^i_?(#1 > o&) _] &,
                First[{Φ}], Rest[{Φ}]]
```

So, we can calculate the first terms of $\exp(t(\hat{A} + \hat{B}))$.

`In[17]:= (rhs = P[T[E^(t Subscript[ω, 1] OverHat[A])] ⊗`
`                 T[E^(t Subscript[ω, 2] OverHat[B])] ⊗`
`                 T[E^(t Subscript[ω, 3] OverHat[A])] ⊗`
`                 T[E^(t Subscript[ω, 4] OverHat[B])] ⊗`
`                 T[E^(t Subscript[ω, 5] OverHat[A])] ⊗`
`                 T[E^(t Subscript[ω, 6] OverHat[B])] ⊗`
`                 T[E^(t Subscript[ω, 7] OverHat[A])], t]) // Short[#, 6]&`

$$Out[17]//Short= \; 1 + t \otimes \hat{A}\, \omega_1 + \frac{1}{2}\, t^2 \otimes \hat{A}^2\, \omega_1^2 + \frac{1}{6}\, t^3 \otimes \hat{A}^3\, \omega_1^3 + \frac{1}{24}\, t^4 \otimes \hat{A}^4\, \omega_1^4 + t \otimes \hat{B}\, \omega_2 +$$

$$t^2\, \hat{A} \otimes \hat{B}\, \omega_1\, \omega_2 + \frac{1}{2}\, t^3\, \hat{A}^2 \otimes \hat{B}\, \omega_1^2\, \omega_2 + \frac{1}{6}\, t^4\, \hat{A}^3 \otimes \hat{B}\, \omega_1^3\, \omega_2 + \frac{1}{2}\, t^2 \otimes \hat{B}^2\, \omega_2^2 + \frac{1}{2}\, t^3\, \hat{A} \otimes \hat{B}^2\, \omega_1\, \omega_2^2 +$$

$$\frac{1}{4}\, t^4\, \hat{A}^2 \otimes \hat{B}^2\, \omega_1^2\, \omega_2^2 + \ll 307 \gg + \frac{1}{2}\, t^4\, \hat{B}^2 \otimes \hat{A}^2\, \omega_4\, \omega_6\, \omega_7^2 + \frac{1}{2}\, t^4\, \hat{A} \otimes \hat{B} \otimes \hat{A}^2\, \omega_5\, \omega_6\, \omega_7^2 +$$

$$\frac{1}{4}\, t^4\, \hat{B}^2 \otimes \hat{A}^2\, \omega_6^2\, \omega_7^2 + \frac{1}{6}\, t^3 \otimes \hat{A}^3\, \omega_7^3 + \frac{1}{6}\, t^4 \otimes \hat{A}^4\, \omega_1\, \omega_7^3 + \frac{1}{6}\, t^4\, \hat{B} \otimes \hat{A}^3\, \omega_2\, \omega_7^3 +$$

$$\frac{1}{6}\, t^4 \otimes \hat{A}^4\, \omega_3\, \omega_7^3 + \frac{1}{6}\, t^4\, \hat{B} \otimes \hat{A}^3\, \omega_4\, \omega_7^3 + \frac{1}{6}\, t^4 \otimes \hat{A}^4\, \omega_5\, \omega_7^3 + \frac{1}{6}\, t^4\, \hat{B} \otimes \hat{A}^3\, \omega_6\, \omega_7^3 + \frac{1}{24}\, t^4 \otimes \hat{A}^4\, \omega_7^4$$

Up to order 4 in t, the difference between right-hand side and left-hand side should vanish identically.

`In[18]:= (shouldBeZero = rhs - lhs) // Short[#, 12]&`

$$Out[18]//Short= \; -t \left(\otimes \hat{A} + \otimes \hat{B} \right) - \frac{1}{2}\, t^2 \left(\otimes \hat{A}^2 + \otimes \hat{B}^2 + \hat{A} \otimes \hat{B} + \hat{B} \otimes \hat{A} \right) -$$

$$\frac{1}{6}\, t^3 \left(\otimes \hat{A}^3 + \otimes \hat{B}^3 + \hat{A} \otimes \hat{B}^2 + \hat{A}^2 \otimes \hat{B} + \hat{B} \otimes \hat{A}^2 + \hat{B}^2 \otimes \hat{A} + \hat{A} \otimes \hat{B} \otimes \hat{A} + \hat{B} \otimes \hat{A} \otimes \hat{B} \right) -$$

$$\frac{1}{24}\, t^4 \left(\otimes \hat{A}^4 + \otimes \hat{B}^4 + \hat{A} \otimes \hat{B}^3 + \hat{A}^2 \otimes \hat{B}^2 + \hat{A}^3 \otimes \hat{B} + \hat{B} \otimes \hat{A}^3 + \hat{B}^2 \otimes \hat{A}^2 + \hat{B}^3 \otimes \hat{A} + \hat{A} \otimes \hat{B} \otimes \hat{A}^2 + \right.$$

$$\left. \hat{A} \otimes \hat{B}^2 \otimes \hat{A} + \hat{A}^2 \otimes \hat{B} \otimes \hat{A} + \hat{B} \otimes \hat{A} \otimes \hat{B}^2 + \hat{B} \otimes \hat{A}^2 \otimes \hat{B} + \hat{B}^2 \otimes \hat{A} \otimes \hat{B} + \hat{A} \otimes \hat{B} \otimes \hat{A} \otimes \hat{B} + \hat{B} \otimes \hat{A} \otimes \hat{B} \otimes \hat{A} \right) +$$

$$t \otimes \hat{A}\, \omega_1 + \frac{1}{2}\, t^2 \otimes \hat{A}^2\, \omega_1^2 + \frac{1}{6}\, t^3 \otimes \hat{A}^3\, \omega_1^3 + \frac{1}{24}\, t^4 \otimes \hat{A}^4\, \omega_1^4 + t \otimes \hat{B}\, \omega_2 + t^2\, \hat{A} \otimes \hat{B}\, \omega_1\, \omega_2 +$$

$$\frac{1}{2}\, t^3\, \hat{A}^2 \otimes \hat{B}\, \omega_1^2\, \omega_2 + \frac{1}{6}\, t^4\, \hat{A}^3 \otimes \hat{B}\, \omega_1^3\, \omega_2 + \frac{1}{2}\, t^2 \otimes \hat{B}^2\, \omega_2^2 + \frac{1}{2}\, t^3\, \hat{A} \otimes \hat{B}^2\, \omega_1\, \omega_2^2 + \frac{1}{4}\, t^4\, \hat{A}^2 \otimes \hat{B}^2\, \omega_1^2\, \omega_2^2 +$$

$$\frac{1}{6}\, t^3 \otimes \hat{B}^3\, \omega_2^3 + \frac{1}{6}\, t^4\, \hat{A} \otimes \hat{B}^3\, \omega_1\, \omega_2^3 + \frac{1}{24}\, t^4 \otimes \hat{B}^4\, \omega_2^4 + \ll 298 \gg + \frac{1}{2}\, t^4\, \hat{B} \otimes \hat{A}^3\, \omega_4\, \omega_5\, \omega_7^2 +$$

$$\frac{1}{4}\, t^4 \otimes \hat{A}^4\, \omega_5^2\, \omega_7^2 + \frac{1}{2}\, t^3\, \hat{B} \otimes \hat{A}^2\, \omega_6\, \omega_7^2 + \frac{1}{2}\, t^4\, \hat{A} \otimes \hat{B} \otimes \hat{A}^2\, \omega_1\, \omega_6\, \omega_7^2 + \frac{1}{2}\, t^4\, \hat{B}^2 \otimes \hat{A}^2\, \omega_2\, \omega_6\, \omega_7^2 +$$

$$\frac{1}{2}\, t^4\, \hat{A} \otimes \hat{B} \otimes \hat{A}^2\, \omega_3\, \omega_6\, \omega_7^2 + \frac{1}{2}\, t^4\, \hat{B}^2 \otimes \hat{A}^2\, \omega_4\, \omega_6\, \omega_7^2 + \frac{1}{2}\, t^4\, \hat{A} \otimes \hat{B} \otimes \hat{A}^2\, \omega_5\, \omega_6\, \omega_7^2 +$$

$$\frac{1}{4}\, t^4\, \hat{B}^2 \otimes \hat{A}^2\, \omega_6^2\, \omega_7^2 + \frac{1}{6}\, t^3 \otimes \hat{A}^3\, \omega_7^3 + \frac{1}{6}\, t^4 \otimes \hat{A}^4\, \omega_1\, \omega_7^3 + \frac{1}{6}\, t^4\, \hat{B} \otimes \hat{A}^3\, \omega_2\, \omega_7^3 +$$

$$\frac{1}{6}\, t^4 \otimes \hat{A}^4\, \omega_3\, \omega_7^3 + \frac{1}{6}\, t^4\, \hat{B} \otimes \hat{A}^3\, \omega_4\, \omega_7^3 + \frac{1}{6}\, t^4 \otimes \hat{A}^4\, \omega_5\, \omega_7^3 + \frac{1}{6}\, t^4\, \hat{B} \otimes \hat{A}^3\, \omega_6\, \omega_7^3 + \frac{1}{24}\, t^4 \otimes \hat{A}^4\, \omega_7^4$$

For generic, noncommuting operators $\hat{A}$ and $\hat{B}$, we have the following independent operator products in `shouldBeZero`.

`In[19]:= allOperators = Union[Cases[shouldBeZero, _CircleTimes, {0, Infinity}]]`

$$Out[19]= \; \left\{ \otimes \hat{A},\; \otimes \hat{A}^2,\; \otimes \hat{A}^3,\; \otimes \hat{A}^4,\; \otimes \hat{B},\; \otimes \hat{B}^2,\; \otimes \hat{B}^3,\; \otimes \hat{B}^4,\; \hat{A} \otimes \hat{B},\; \hat{A} \otimes \hat{B}^2,\; \hat{A} \otimes \hat{B}^3,\; \hat{A}^2 \otimes \hat{B}, \right.$$

$$\hat{A}^2 \otimes \hat{B}^2,\; \hat{A}^3 \otimes \hat{B},\; \hat{B} \otimes \hat{A},\; \hat{B} \otimes \hat{A}^2,\; \hat{B} \otimes \hat{A}^3,\; \hat{B}^2 \otimes \hat{A},\; \hat{B}^2 \otimes \hat{A}^2,\; \hat{B}^3 \otimes \hat{A},\; \hat{A} \otimes \hat{B} \otimes \hat{A},\; \hat{A} \otimes \hat{B} \otimes \hat{A}^2,$$

$$\left. \hat{A} \otimes \hat{B}^2 \otimes \hat{A},\; \hat{A}^2 \otimes \hat{B} \otimes \hat{A},\; \hat{B} \otimes \hat{A} \otimes \hat{B},\; \hat{B} \otimes \hat{A} \otimes \hat{B}^2,\; \hat{B} \otimes \hat{A}^2 \otimes \hat{B},\; \hat{B}^2 \otimes \hat{A} \otimes \hat{B},\; \hat{A} \otimes \hat{B} \otimes \hat{A} \otimes \hat{B},\; \hat{B} \otimes \hat{A} \otimes \hat{B} \otimes \hat{A} \right\}$$

We take the c-number coefficients with respect to these operators—they should vanish.

`In[20]:= (operatorCoefficients =`
`          Coefficient[shouldBeZero, #]& /@ allOperators) // Short[#, 6]&`

$$Out[20]//Short= \; \left\{ -t + t\, \omega_1 + t\, \omega_3 + t\, \omega_5 + t\, \omega_7,\; -\frac{t^2}{2} + \frac{1}{2}\, t^2\, \omega_1^2 + t^2\, \omega_1\, \omega_3 + \frac{1}{2}\, t^2\, \omega_3^2 + \right.$$

$$t^2\, \omega_1\, \omega_5 + t^2\, \omega_3\, \omega_5 + \frac{1}{2}\, t^2\, \omega_5^2 + t^2\, \omega_1\, \omega_7 + t^2\, \omega_3\, \omega_7 + t^2\, \omega_5\, \omega_7 + \frac{1}{2}\, t^2\, \omega_7^2,\; \ll 26 \gg,$$

$$-\frac{t^4}{24} + t^4\, \omega_1\, \omega_2\, \omega_3\, \omega_4 + t^4\, \omega_1\, \omega_2\, \omega_3\, \omega_6 + t^4\, \omega_1\, \omega_2\, \omega_5\, \omega_6 + t^4\, \omega_1\, \omega_4\, \omega_5\, \omega_6 + t^4\, \omega_3\, \omega_4\, \omega_5\, \omega_6,$$

$$\left. -\frac{t^4}{24} + t^4\, \omega_2\, \omega_3\, \omega_4\, \omega_5 + t^4\, \omega_2\, \omega_3\, \omega_4\, \omega_7 + t^4\, \omega_2\, \omega_3\, \omega_6\, \omega_7 + t^4\, \omega_2\, \omega_5\, \omega_6\, \omega_7 + t^4\, \omega_4\, \omega_5\, \omega_6\, \omega_7 \right\}$$

For every power of t, we extract the corresponding coefficients. This leads to the following, quite complicated, polynomial system of nonlinear equations for $\omega_1, \omega_2, \omega_3, \omega_4, \omega_5, \omega_6$, and ω_7.

`In[21]:= (algebraicEquations =`
`          DeleteCases[Union[Flatten[Map[Numerator[Together[#]]&,`

```
                         CoefficientList[#, t]& /@ operatorCoefficients,
                                      {2}]]], 0]);
```

Short[algebraicEquations, 12]

Out[22]//Short= $\{-1 + \omega_2 + \omega_4 + \omega_6, \; -1 + 2\,\omega_1\,\omega_2 + 2\,\omega_1\,\omega_4 + 2\,\omega_3\,\omega_4 + 2\,\omega_1\,\omega_6 + 2\,\omega_3\,\omega_6 + 2\,\omega_5\,\omega_6,$

$-1 + 6\,\omega_2\,\omega_3\,\omega_4 + 6\,\omega_2\,\omega_3\,\omega_6 + 6\,\omega_2\,\omega_5\,\omega_6 + 6\,\omega_4\,\omega_5\,\omega_6,$

$-1 + 24\,\omega_1\,\omega_2\,\omega_3\,\omega_4 + 24\,\omega_1\,\omega_2\,\omega_3\,\omega_6 + 24\,\omega_1\,\omega_2\,\omega_5\,\omega_6 + 24\,\omega_1\,\omega_4\,\omega_5\,\omega_6 + 24\,\omega_3\,\omega_4\,\omega_5\,\omega_6, \; \ll 23 \gg,$

$-1 + \omega_1^3 + 3\,\omega_1^2\,\omega_3 + 3\,\omega_1\,\omega_3^2 + \omega_3^3 + 3\,\omega_1^2\,\omega_5 + 6\,\omega_1\,\omega_3\,\omega_5 + 3\,\omega_3^2\,\omega_5 + 3\,\omega_1\,\omega_5^2 + 3\,\omega_3\,\omega_5^2 + \omega_5^3 + 3\,\omega_1^2\,\omega_7 +$

$6\,\omega_1\,\omega_3\,\omega_7 + 3\,\omega_3^2\,\omega_7 + 6\,\omega_1\,\omega_5\,\omega_7 + 6\,\omega_3\,\omega_5\,\omega_7 + 3\,\omega_5^2\,\omega_7 + 3\,\omega_1\,\omega_7^2 + 3\,\omega_3\,\omega_7^2 + 3\,\omega_5\,\omega_7^2 + \omega_7^3,$

$-1 + 4\,\omega_2\,\omega_3^3 + 12\,\omega_2\,\omega_3^2\,\omega_5 + 12\,\omega_2\,\omega_3\,\omega_5^2 + 4\,\omega_2\,\omega_5^3 + 4\,\omega_4\,\omega_5^3 + 12\,\omega_2\,\omega_3^2\,\omega_7 + 24\,\omega_2\,\omega_3\,\omega_5\,\omega_7 +$

$12\,\omega_2\,\omega_5^2\,\omega_7 + 12\,\omega_4\,\omega_5^2\,\omega_7 + 12\,\omega_2\,\omega_3\,\omega_7^2 + 12\,\omega_2\,\omega_5\,\omega_7^2 + 12\,\omega_4\,\omega_5\,\omega_7^2 + 4\,\omega_2\,\omega_7^3 + 4\,\omega_4\,\omega_7^3 + 4\,\omega_6\,\omega_7^3,$

$-1 + \omega_1^4 + 4\,\omega_1^3\,\omega_3 + 6\,\omega_1^2\,\omega_3^2 + 4\,\omega_1\,\omega_3^3 + \omega_3^4 + 4\,\omega_1^3\,\omega_5 + 12\,\omega_1^2\,\omega_3\,\omega_5 + 12\,\omega_1\,\omega_3^2\,\omega_5 + 4\,\omega_3^3\,\omega_5 + 6\,\omega_1^2\,\omega_5^2 +$

$12\,\omega_1\,\omega_3\,\omega_5^2 + 6\,\omega_3^2\,\omega_5^2 + 4\,\omega_1\,\omega_5^3 + 4\,\omega_3\,\omega_5^3 + \omega_5^4 + 4\,\omega_1^3\,\omega_7 + 12\,\omega_1^2\,\omega_3\,\omega_7 + 12\,\omega_1\,\omega_3^2\,\omega_7 + 4\,\omega_3^3\,\omega_7 +$

$12\,\omega_1^2\,\omega_5\,\omega_7 + 24\,\omega_1\,\omega_3\,\omega_5\,\omega_7 + 12\,\omega_3^2\,\omega_5\,\omega_7 + 12\,\omega_1\,\omega_5^2\,\omega_7 + 12\,\omega_3\,\omega_5^2\,\omega_7 + 4\,\omega_5^3\,\omega_7 + 6\,\omega_1^2\,\omega_7^2 +$

$12\,\omega_1\,\omega_3\,\omega_7^2 + 6\,\omega_3^2\,\omega_7^2 + 12\,\omega_1\,\omega_5\,\omega_7^2 + 12\,\omega_3\,\omega_5\,\omega_7^2 + 6\,\omega_5^2\,\omega_7^2 + 4\,\omega_1\,\omega_7^3 + 4\,\omega_3\,\omega_7^3 + 4\,\omega_5\,\omega_7^3 + \omega_7^4 \}$

Using `GroebnerBasis`, we transform the last set of equations into a much nicer looking set of just seven equations in triangular form. On a vintage year-2005 computer, this main step in the whole computation takes only a fraction of a second.

In[23]:= **ωs = Table[Subscript[ω, i], {i, 7}];**

TableForm[gb = GroebnerBasis[algebraicEquations, ωs]]

Out[24]//TableForm=
$-1 + 18\,\omega_7 - 144\,\omega_7^2 + 624\,\omega_7^3 - 1440\,\omega_7^4 + 1152\,\omega_7^5$

$\omega_6 - 2\,\omega_7$

$-3 + 4\,\omega_5 + 28\,\omega_7 - 192\,\omega_7^2 + 624\,\omega_7^3 - 576\,\omega_7^4$

$-3 + 2\,\omega_4 + 32\,\omega_7 - 192\,\omega_7^2 + 624\,\omega_7^3 - 576\,\omega_7^4$

$-1 + 2\,\omega_3 + 2\,\omega_7$

$1 + 2\,\omega_2 - 28\,\omega_7 + 192\,\omega_7^2 - 624\,\omega_7^3 + 576\,\omega_7^4$

$1 + 4\,\omega_1 - 28\,\omega_7 + 192\,\omega_7^2 - 624\,\omega_7^3 + 576\,\omega_7^4$

Now it is straightforward to explicitly calculate all five (that there are five solutions is immediately visible from the first equation of gb) solutions for ω_1, ω_2, ω_3, ω_4, ω_5, ω_6, and ω_7.

In[25]:= **(sol = Solve[# == 0& /@ gb, ωs]) // Short[#, 6]&**

Out[25]//Short= $\Big\{\Big\{\omega_1 \to \frac{1}{3} + \frac{1}{3\,2^{2/3}} + \frac{1}{6\,2^{1/3}}, \; \omega_2 \to \frac{2}{3} + \frac{1}{3\,2^{1/3}} + \frac{2^{1/3}}{3},$

$\omega_3 \to \frac{1}{12}\,(2 - 2\,2^{1/3} - 2^{2/3}), \; \omega_4 \to \frac{1}{3}\,(-1 - 2\,2^{1/3} - 2^{2/3}), \; \omega_5 \to \frac{1}{12}\,(2 - 2\,2^{1/3} - 2^{2/3}),$

$\omega_6 \to \frac{2}{3} + \frac{1}{3\,2^{1/3}} + \frac{2^{1/3}}{3}, \; \omega_7 \to \frac{1}{3} + \frac{1}{3\,2^{2/3}} + \frac{1}{6\,2^{1/3}} \Big\}, \; \ll 3 \gg,$

$\Big\{\omega_1 \to \frac{1}{24}\,\big(3 + i\,\sqrt{15}\big), \; \omega_2 \to \frac{1}{12}\,\big(3 + i\,\sqrt{15}\big), \; \omega_3 \to \frac{1}{24}\,\big(9 + i\,\sqrt{15}\big), \; \omega_4 \to \frac{1}{2},$

$\omega_5 \to \frac{1}{24}\,\big(9 - i\,\sqrt{15}\big), \; \omega_6 \to \frac{1}{4} - \frac{1}{4}\,i\,\sqrt{\frac{5}{3}}, \; \omega_7 \to \frac{1}{24}\,\big(3 - i\,\sqrt{15}\big)\Big\}\Big\}$

Using `Root`-objects, the last result can be written in a more compact form as follows.

In[26]:= **RootReduce[%]**

Out[26]= $\big\{\{\omega_1 \to \text{Root}[-1 + 12\,\#1 - 48\,\#1^2 + 48\,\#1^3 \&, 1],$

$\omega_2 \to \text{Root}[-1 + 6\,\#1 - 12\,\#1^2 + 6\,\#1^3 \&, 1], \; \omega_3 \to \text{Root}[1 - 24\,\#1^2 + 48\,\#1^3 \&, 1],$

$\omega_4 \to \text{Root}[1 - 3\,\#1 + 3\,\#1^2 + 3\,\#1^3 \&, 1], \; \omega_5 \to \text{Root}[1 - 24\,\#1^2 + 48\,\#1^3 \&, 1],$

$\omega_6 \to \text{Root}[-1 + 6\,\#1 - 12\,\#1^2 + 6\,\#1^3 \&, 1], \; \omega_7 \to \text{Root}[-1 + 12\,\#1 - 48\,\#1^2 + 48\,\#1^3 \&, 1]\},$

$\{\omega_1 \to \text{Root}[-1 + 12\,\#1 - 48\,\#1^2 + 48\,\#1^3 \&, 3], \; \omega_2 \to \text{Root}[-1 + 6\,\#1 - 12\,\#1^2 + 6\,\#1^3 \&, 3],$

$\omega_3 \to \text{Root}[1 - 24\,\#1^2 + 48\,\#1^3 \&, 2], \; \omega_4 \to \text{Root}[1 - 3\,\#1 + 3\,\#1^2 + 3\,\#1^3 \&, 2],$

$\omega_5 \to \text{Root}[1 - 24\,\#1^2 + 48\,\#1^3 \&, 2], \; \omega_6 \to \text{Root}[-1 + 6\,\#1 - 12\,\#1^2 + 6\,\#1^3 \&, 3],$

$\omega_7 \to \text{Root}[-1 + 12\,\#1 - 48\,\#1^2 + 48\,\#1^3 \&, 3]\}, \; \{\omega_1 \to \text{Root}[-1 + 12\,\#1 - 48\,\#1^2 + 48\,\#1^3 \&, 2],$

$\omega_2 \to \text{Root}[-1 + 6\,\#1 - 12\,\#1^2 + 6\,\#1^3 \&, 2], \; \omega_3 \to \text{Root}[1 - 24\,\#1^2 + 48\,\#1^3 \&, 3],$

$\omega_4 \to \text{Root}[1 - 3\,\#1 + 3\,\#1^2 + 3\,\#1^3 \&, 3], \; \omega_5 \to \text{Root}[1 - 24\,\#1^2 + 48\,\#1^3 \&, 3],$

$\omega_6 \to \text{Root}[-1 + 6\,\#1 - 12\,\#1^2 + 6\,\#1^3 \&, 2], \; \omega_7 \to \text{Root}[-1 + 12\,\#1 - 48\,\#1^2 + 48\,\#1^3 \&, 2]\},$

$\big\{\omega_1 \to \frac{1}{24}\,\big(3 - i\,\sqrt{15}\big), \; \omega_2 \to \frac{1}{12}\,\big(3 - i\,\sqrt{15}\big), \; \omega_3 \to \frac{1}{24}\,\big(9 - i\,\sqrt{15}\big), \; \omega_4 \to \frac{1}{2},$

$$\omega_5 \to \frac{1}{24}\left(9 + i\sqrt{15}\right),\ \omega_6 \to \frac{1}{12}\left(3 + i\sqrt{15}\right),\ \omega_7 \to \frac{1}{24}\left(3 + i\sqrt{15}\right)\},$$

$$\{\omega_1 \to \frac{1}{24}\left(3 + i\sqrt{15}\right),\ \omega_2 \to \frac{1}{12}\left(3 + i\sqrt{15}\right),\ \omega_3 \to \frac{1}{24}\left(9 + i\sqrt{15}\right),\ \omega_4 \to \frac{1}{2},$$

$$\omega_5 \to \frac{1}{24}\left(9 - i\sqrt{15}\right),\ \omega_6 \to \frac{1}{12}\left(3 - i\sqrt{15}\right),\ \omega_7 \to \frac{1}{24}\left(3 - i\sqrt{15}\right)\}\}$$

The solution with all the ω_i real is the most interesting one.

```
In[27]:= Simplify[Select[sol, And @@ (# == 0&) /@ Im[ωs /. #1] & ]]
```

$$Out[27]= \{\{\omega_1 \to \frac{1}{12}\left(4 + 2\ 2^{1/3} + 2^{2/3}\right),\ \omega_2 \to \frac{1}{6}\left(4 + 2\ 2^{1/3} + 2^{2/3}\right),$$

$$\omega_3 \to \frac{1}{12}\left(2 - 2\ 2^{1/3} - 2^{2/3}\right),\ \omega_4 \to -\frac{1}{3}\left(1 + 2^{1/3}\right)^2,\ \omega_5 \to \frac{1}{12}\left(2 - 2\ 2^{1/3} - 2^{2/3}\right),$$

$$\omega_6 \to \frac{1}{6}\left(4 + 2\ 2^{1/3} + 2^{2/3}\right),\ \omega_7 \to \frac{1}{12}\left(4 + 2\ 2^{1/3} + 2^{2/3}\right)\}\}$$

So, we finally arrive at the following result.

```
In[28]:= (CircleTimes @@ MapThread[E^(t Subscript[w, #1] OverHat[#2])&,
                                    {Range[7], {A, B, A, B, A, B, A}}]) /.
                                    First[%] // TraditionalForm
```

Out[28]//TraditionalForm=

$$e^{\frac{1}{12}\left(4+2\sqrt[3]{2}+2^{2/3}\right)t\,\hat{A}} \otimes e^{\frac{1}{6}\left(4+2\sqrt[3]{2}+2^{2/3}\right)t\,\hat{B}} \otimes e^{\frac{1}{12}\left(2-2\sqrt[3]{2}-2^{2/3}\right)t\,\hat{A}} \otimes$$

$$e^{-\frac{1}{3}\left(1+\sqrt[3]{2}\right)^2 t\,\hat{B}} \otimes e^{\frac{1}{12}\left(2-2\sqrt[3]{2}-2^{2/3}\right)t\,\hat{A}} \otimes e^{\frac{1}{6}\left(4+2\sqrt[3]{2}+2^{2/3}\right)t\,\hat{B}} \otimes e^{\frac{1}{12}\left(4+2\sqrt[3]{2}+2^{2/3}\right)t\,\hat{A}}$$

For similar formulas for evolution loops, see [1254].

46. Maximal Triangle Area and Tetrahedron of Maximal Volume

a) Let p1, p2, and p3 be the vertices of the triangle. Without loss of generality, we can assume the point p1 to lie on the first circle.

```
In[1]:= p1 = {r1, 0}; p2 = {x2, y2}; p3 = {x3, y3};
```

This yields the following expression for the area of the triangle as a function of the coordinates of the vertices p2 and p3.

```
In[2]:= areaSquared = Cross[{x2, y2, 0} - {r1, 0, 0},
                            {x3, y3, 0} - {r1, 0, 0}][[3]]^2/4
```

$$Out[2]= \frac{1}{4}\ (r1\ y2 - x3\ y2 - r1\ y3 + x2\ y3)^2$$

The constraint that the vertices p2 and p3 are lying on circles we take into account using Lagrange multipliers.

```
In[3]:= Λ = areaSquared + λ (p2.p2 - r2^2) + μ (p3.p3 - r3^2)
```

$$Out[3]= \frac{1}{4}\ (r1\ y2 - x3\ y2 - r1\ y3 + x2\ y3)^2 + (-r2^2 + x2^2 + y2^2)\ \lambda + (-r3^2 + x3^2 + y3^2)\ \mu$$

For the maximal area, we so get the following system of six equations.

```
In[4]:= eqs = Factor[D[Λ, #]]& /@ {x2, y2, x3, y3, λ, μ}
```

$$Out[4]= \{\frac{1}{2}\ (r1\ y2\ y3 - x3\ y2\ y3 - r1\ y3^2 + x2\ y3^2 + 4\ x2\ \lambda),$$

$$\frac{1}{2}\ (r1^2\ y2 - 2\ r1\ x3\ y2 + x3^2\ y2 - r1^2\ y3 + r1\ x2\ y3 + r1\ x3\ y3 - x2\ x3\ y3 + 4\ y2\ \lambda),$$

$$\frac{1}{2}\ (-r1\ y2^2 + x3\ y2^2 + r1\ y2\ y3 - x2\ y2\ y3 + 4\ x3\ \mu),$$

$$\frac{1}{2}\ (-r1^2\ y2 + r1\ x2\ y2 + r1\ x3\ y2 - x2\ x3\ y2 + r1^2\ y3 - 2\ r1\ x2\ y3 + x2^2\ y3 + 4\ y3\ \mu),$$

$$-r2^2 + x2^2 + y2^2,\ -r3^2 + x3^2 + y3^2\}$$

Denoting by A2 the square of the maximal triangle area, it is straightforward to eliminate all the not needed variables to obtain A2 as a function of r1, r2, and r3.

```
In[5]:= gb = GroebnerBasis[Flatten[{A2 - areaSquared, eqs}],
                           {A, r1, r2, r3}, {x2, y2, λ, μ, x3, y3},
                           MonomialOrder -> EliminationOrder]
```

$$Out[5]= \{256\ A2^4 + 16\ A2^3\ r1^4 - 160\ A2^3\ r1^2\ r2^2 - 8\ A2^2\ r1^6\ r2^2 + 16\ A2^3\ r2^4 + 32\ A2^2\ r1^4\ r2^4 +$$

$$A2\ r1^8\ r2^4 - 8\ A2^2\ r1^2\ r2^6 - 2\ A2\ r1^6\ r2^6 + A2\ r1^4\ r2^8 - 160\ A2^3\ r1^2\ r3^2 - 8\ A2^2\ r1^6\ r3^2 -$$

$$160\ A2^3\ r2^2\ r3^2 - 16\ A2^2\ r1^4\ r2^2\ r3^2 - 2\ A2\ r1^8\ r2^2\ r3^2 - 16\ A2^2\ r1^2\ r2^4\ r3^2 +$$

$$2\ A2\ r1^6\ r2^4\ r3^2 - 8\ A2^2\ r2^6\ r3^2 + 2\ A2\ r1^4\ r2^6\ r3^2 - 2\ A2\ r1^2\ r2^8\ r3^2 + 16\ A2^3\ r3^4 +$$
$$32\ A2^2\ r1^4\ r3^4 + A2\ r1^8\ r3^4 - 16\ A2^2\ r1^2\ r2^2\ r3^4 + 2\ A2\ r1^6\ r2^2\ r3^4 + 32\ A2^2\ r2^4\ r3^4 -$$
$$6\ A2\ r1^4\ r2^4\ r3^4 + 2\ A2\ r1^2\ r2^6\ r3^4 + A2\ r2^8\ r3^4 - 8\ A2^2\ r1^2\ r3^6 - 2\ A2\ r1^6\ r3^6 - 8\ A2^2\ r2^2\ r3^6 +$$
$$2\ A2\ r1^4\ r2^2\ r3^6 + 2\ A2\ r1^2\ r2^4\ r3^6 - 2\ A2\ r2^6\ r3^6 + A2\ r1^4\ r3^8 - 2\ A2\ r1^2\ r2^2\ r3^8 + A2\ r2^4\ r3^8\}$$

For the special radii given, the last formula leads to the following maximal area.

`In[6]:= rRules = {r1 -> 1, r2 -> 2, r3 -> 3};`

`In[7]:= Solve[(gb[[1]] /. rRules) == 0, A2] // RootReduce`

`Out[7]= {{A2 → 0}, {A2 → Root[900 + 133 #1 - 392 #1² + 16 #1³ &, 3]},`
`        {A2 → Root[900 + 133 #1 - 392 #1² + 16 #1³ &, 1]},`
`        {A2 → Root[900 + 133 #1 - 392 #1² + 16 #1³ &, 2]}}`

`In[8]:= N[%, 22]`

`Out[8]= {{A2 → 0}, {A2 → 24.05727870023588814247},`
`        {A2 → -1.323686350138371303274}, {A2 → 1.766407649902483160808}}`

Let us compare the last result with a direct maximization of the area. (We use `FindMinimum` on the negative of the area.) We consider $x2$ and $x3$ the independent variables and express $y2$ and $y3$ piecewise as functions of $x2$ and $x3$.

`In[9]:= Apply[Function[{pm1, pm2},`
`        FindMinimum[Evaluate[(* substitute y2 and y3; use ± *)`
`        -areaSquared /. {y2 -> pm1 Sqrt[r2^2 - x2^2],`
`                         y3 -> pm2 Sqrt[r3^2 - x3^2]} /. rRules],`
`            {x2, 1/2, -2, 2}, {x3, -1/2, -3, 3}]],`
`        {{1, 1}, {1, -1}, {-1, 1}, {-1, -1}}, {1}]`

```
Power::infy : Infinite expression  1     encountered. More…
                                   ───
                                   √0.

∞::indet : Indeterminate expression
    2.48599 + ComplexInfinity + ComplexInfinity encountered. More…

FindMinimum::ngnum :
    The gradient is not a vector of numbers at {x2, x3} = {2., -1.67924}. More…

Power::infy : Infinite expression  1     encountered. More…
                                   ───
                                   √0.

∞::indet : Indeterminate expression
    -2.48599 + ComplexInfinity + ComplexInfinity encountered. More…

FindMinimum::ngnum :
    The gradient is not a vector of numbers at {x2, x3} = {2., -1.67924}. More…
```

`Out[9]= {{-0.508168, {x2 → 0.5, x3 → -0.5}}, {-24.0573, {x2 → -1.45876, x3 → -1.45876}},`
`        {-24.0573, {x2 → -1.45876, x3 → -1.45876}}, {-0.508168, {x2 → 0.5, x3 → -0.5}}}`

This confirms the above result.

Because we expect A2 to be a symmetric function of r1, r2, and r3, we use the function `SymmetricReduction` (from the standard package `Algebra`SymmetricPolynomials``) to rewrite gb in a more symmetric way. Let s[i] be the *i*th symmetric polynomial in r1, r2, and r3.

`In[10]:= Needs["Algebra`SymmetricPolynomials`"]`

`In[11]:= simpl[coeff_] :=`
`        With[{sr = SymmetricReduction[coeff,`
`                     {r1, r2, r3}, {s[1], s[2], s[3]}]},`
`             If[sr[[2]] === 0, Simplify[sr[[1]]], $Failed]]`

`In[12]:= Collect[gb[[1]], A2, simpl]`

`Out[12]= 256 A2⁴ + 16 A2³ (s[1]⁴ - 4 s[1]² s[2] - 8 s[2]² + 24 s[1] s[3]) -`
`        A2 (-s[1] s[2] + s[3])² (-s[1]² s[2]² + 4 s[2]³ + 4 s[1]³ s[3] - 18 s[1] s[2] s[3] + 27 s[3]²) +`
`        8 A2² (-s[1]⁴ s[2]² + 2 s[1]⁵ s[3] - 8 s[1]³ s[2] s[3] -`
`        16 s[1] s[2]² s[3] + s[1]² (4 s[2]³ + 15 s[3]²) + 2 (s[2]⁴ + 9 s[2] s[3]²))`

b) Without loss of generality, we can take the following coordinates for the four vertices p1, p2, p3, and p4.

In[1]:= p1 = {0, 0, 0}; p2 = {p2x, 0, 0}; p3 = {p3x, p3y, 0};
      p4 = {p4x, p4y, p4z};

The oriented volume v of the tetrahedron and the square of the areas aSijk of the four faces can then be expressed through the coordinates by using formulas from elementary geometry. The expression aSijk is the square of the area of the face bounded by the vertices p[i], p[j], and p[k].

In[3]:= v = 1/6 Det[Append[#, 1]& /@ {p1, p2, p3, p4}]

Out[3]= $-\frac{1}{6}$ p2x p3y p4z

In[4]:= {aS123, aS124, aS134, aS234} =
      Simplify[#.#/4]&[Cross[##]]& @@@ {{p2 - p1, p3 - p1},
      {p2 - p1, p4 - p1}, {p3 - p1, p4 - p1}, {p3 - p2, p4 - p2}}

Out[4]= $\{\frac{p2x^2\,p3y^2}{4}$, $\frac{1}{4}$ p2x^2 (p4y^2 + p4z^2), $\frac{1}{4}$ ((p3y p4x - p3x p4y)^2 + p3x^2 p4z^2 + p3y^2 p4z^2),

$\frac{1}{4}$ ((p3y p4x - p3x p4y + p2x (-p3y + p4y))^2 + p3y^2 p4z^2 + (p2x p4z - p3x p4z)^2)}

The volume, via the determinantal formula, is not positive definite. That is why we will use its square in the following. To determine the maximal volume, we proceed in the most natural way—we calculate the extremal value of the square of the maximal volume V2 under the restriction of given values for aS123, aS124, aS134, and aS234. These conditions are incorporated into the function to be minimized using Lagrange multipliers $\lambda1$, $\lambda2$, $\lambda3$, and $\lambda4$. The four constraints are φi, where aSijk is the square of the area of a face expressed through the coordinates of the vertices and is the given value of the faces ASijk.

In[5]:= {φ1, φ2, φ3, φ4} =
      {AS123 - aS123, AS124 - aS124, AS134 - aS134, AS234 - aS234};

So, we arrive at the following function L to minimize.

In[6]:= L = v^2 + λ1 φ1 + λ2 φ2 + λ3 φ3 + λ4 φ4;

Technically, the problem of finding the maximal volume now reduces to the elimination of the variables $\lambda1$, $\lambda2$, $\lambda3$, $\lambda4$, p2x, p3x, p3y, p4x, p4y, and p4z from the following 11 equations.

In[7]:= eqs = {V2 - v^2, (* derivative wrt coordinates *)
      D[L, p2x], D[L, p3x], D[L, p3y],
      D[L, p4x], D[L, p4y], D[L, p4z],
      (* derivative wrt Lagrange multipliers *)
      D[L, λ1], D[L, λ2], D[L, λ3], D[L, λ4]}

Out[7]= $\{-\frac{1}{36}$ p2x^2 p3y^2 p4z^2 + V2, $\frac{1}{18}$ p2x p3y^2 p4z^2 - $\frac{1}{2}$ p2x p3y^2 λ1 - $\frac{1}{2}$ p2x (p4y^2 + p4z^2) λ2 +

$\frac{1}{4}$ (-2 (-p3y + p4y) (p3y p4x - p3x p4y + p2x (-p3y + p4y)) - 2 p4z (p2x p4z - p3x p4z)) λ4,

$\frac{1}{4}$ (2 p4y (p3y p4x - p3x p4y) - 2 p3x p4z^2) λ3 +

$\frac{1}{4}$ (2 p4y (p3y p4x - p3x p4y + p2x (-p3y + p4y)) + 2 p4z (p2x p4z - p3x p4z)) λ4,

$\frac{1}{18}$ p2x^2 p3y p4z^2 - $\frac{1}{2}$ p2x^2 p3y λ1 + $\frac{1}{4}$ (-2 p4x (p3y p4x - p3x p4y) - 2 p3y p4z^2) λ3 +

$\frac{1}{4}$ (-2 (-p2x + p4x) (p3y p4x - p3x p4y + p2x (-p3y + p4y)) - 2 p3y p4z^2) λ4,

$-\frac{1}{2}$ p3y (p3y p4x - p3x p4y) λ3 - $\frac{1}{2}$ p3y (p3y p4x - p3x p4y + p2x (-p3y + p4y)) λ4,

$-\frac{1}{2}$ p2x^2 p4y λ2 + $\frac{1}{2}$ p3x (p3y p4x - p3x p4y) λ3 -

$\frac{1}{2}$ (p2x - p3x) (p3y p4x - p3x p4y + p2x (-p3y + p4y)) λ4,

$\frac{1}{18}$ p2x^2 p3y^2 p4z - $\frac{1}{2}$ p2x^2 p4z λ2 + $\frac{1}{4}$ (-2 p3x^2 p4z - 2 p3y^2 p4z) λ3 +

$\frac{1}{4}$ (-2 p3y^2 p4z - 2 (p2x - p3x) (p2x p4z - p3x p4z)) λ4, AS123 - $\frac{p2x^2\,p3y^2}{4}$,

AS124 - $\frac{1}{4}$ p2x^2 (p4y^2 + p4z^2), AS134 + $\frac{1}{4}$ (-(p3y p4x - p3x p4y)^2 - p3x^2 p4z^2 - p3y^2 p4z^2),

AS234 + $\frac{1}{4}$ (-(p3y p4x - p3x p4y + p2x (-p3y + p4y))^2 - p3y^2 p4z^2 - (p2x p4z - p3x p4z)^2)}

A direct call to Solve or Eliminate will not give an answer in any reasonable amount of time. One of the reasons is that besides the maximum we are interested in, degenerate cases (meaning some of the coordinates are equal to zero) will appear.

So, we will simplify the problem a bit by "hand work". In solving a complicated system of polynomial equations, it is often a good idea to eliminate variables that appear linearly. All of the λis appear linearly in L. This is obvious from the construction of L, and later, we care about eventually vanishing denominators. Here are the degrees of the 11 equations in the variables λi.

```
In[8]:= Exponent[#, {λ1, λ2, λ3, λ4}]& /@ eqs
```

```
Out[8]= {{0, 0, 0, 0}, {1, 1, 0, 1}, {0, 0, 1, 1}, {1, 0, 1, 1}, {0, 0, 1, 1},
         {0, 1, 1, 1}, {0, 1, 1, 1}, {0, 0, 0, 0}, {0, 0, 0, 0}, {0, 0, 0, 0}, {0, 0, 0, 0}}
```

We start by eliminating $\lambda 1$.

```
In[9]:= sol1 = Solve[eqs[[2]] == 0, λ1]
```

$$Out[9]= \left\{\left\{\lambda 1 \to \frac{1}{9\,p2x\,p3y^2}\,(p2x\,p3y^2\,p4z^2 - 9\,p2x\,p4y^2\,\lambda 2 - 9\,p2x\,p4z^2\,\lambda 2 - \right.\right.$$
$$9\,p2x\,p3y^2\,\lambda 4 + 9\,p3y^2\,p4x\,\lambda 4 + 18\,p2x\,p3y\,p4y\,\lambda 4 - 9\,p3x\,p3y\,p4y\,\lambda 4 -$$
$$\left.\left.9\,p3y\,p4x\,p4y\,\lambda 4 - 9\,p2x\,p4y^2\,\lambda 4 + 9\,p3x\,p4y^2\,\lambda 4 - 9\,p2x\,p4z^2\,\lambda 4 + 9\,p3x\,p4z^2\,\lambda 4)\right\}\right\}$$

The coordinates p2x and p3y cannot vanish for a nondegenerate solution, and we arrive at the following 10 equations.

```
In[10]:= eqs1 = Factor[Numerator[Together[Delete[eqs, 2] /. sol1[[1]]]]]
```

$$Out[10]= \{-p2x^2\,p3y^2\,p4z^2 + 36\,V2,\ p3y\,p4x\,p4y\,\lambda 3 - p3x\,p4y^2\,\lambda 3 - p3x\,p4z^2\,\lambda 3 -$$
$$p2x\,p3y\,p4y\,\lambda 4 + p3y\,p4x\,p4y\,\lambda 4 + p2x\,p4y^2\,\lambda 4 - p3x\,p4y^2\,\lambda 4 + p2x\,p4z^2\,\lambda 4 - p3x\,p4z^2\,\lambda 4,$$
$$p2x^2\,p4y^2\,\lambda 2 + p2x^2\,p4z^2\,\lambda 2 - p3y^2\,p4x^2\,\lambda 3 + p3x\,p3y\,p4x\,p4y\,\lambda 3 - p3y^2\,p4z^2\,\lambda 3 +$$
$$p2x\,p3y^2\,p4x\,\lambda 4 - p3y^2\,p4x^2\,\lambda 4 - p2x^2\,p3y\,p4y\,\lambda 4 + p3x\,p3y\,p4x\,p4y\,\lambda 4 +$$
$$p2x^2\,p4y^2\,\lambda 4 - p2x\,p3x\,p4y^2\,\lambda 4 + p2x^2\,p4z^2\,\lambda 4 - p2x\,p3x\,p4z^2\,\lambda 4 - p3y^2\,p4z^2\,\lambda 4,$$
$$-p3y\,(p3y\,p4x\,\lambda 3 - p3x\,p4y\,\lambda 3 - p2x\,p3y\,\lambda 4 + p3y\,p4x\,\lambda 4 + p2x\,p4y\,\lambda 4 - p3x\,p4y\,\lambda 4),$$
$$-p2x^2\,p4y\,\lambda 2 + p3x\,p3y\,p4x\,\lambda 3 - p3x^2\,p4y\,\lambda 3 + p2x^2\,p3y\,\lambda 4 - p2x\,p3x\,p3y\,\lambda 4 -$$
$$p2x\,p3y\,p4x\,\lambda 4 + p3x\,p3y\,p4x\,\lambda 4 - p2x^2\,p4y\,\lambda 4 + 2\,p2x\,p3x\,p4y\,\lambda 4 - p3x^2\,p4y\,\lambda 4,$$
$$p4z\,(2\,p2x^2\,p3y^2 - 9\,p2x^2\,\lambda 2 - 9\,p3x^2\,\lambda 3 - 9\,p3y^2\,\lambda 3 - 9\,p2x^2\,\lambda 4 + 18\,p2x\,p3x\,\lambda 4 - 9\,p3x^2\,\lambda 4 - 9\,p3y^2\,\lambda 4),$$
$$4\,AS123 - p2x^2\,p3y^2,\ 4\,AS124 - p2x^2\,p4y^2 - p2x^2\,p4z^2,$$
$$4\,AS134 - p3y^2\,p4x^2 + 2\,p3x\,p3y\,p4x\,p4y - p3x^2\,p4y^2 - p3x^2\,p4z^2 - p3y^2\,p4z^2,$$
$$4\,AS234 - p2x^2\,p3y^2 + 2\,p2x\,p3y^2\,p4x - p3y^2\,p4x^2 + 2\,p2x^2\,p3y\,p4y -$$
$$2\,p2x\,p3x\,p3y\,p4y - 2\,p2x\,p3y\,p4x\,p4y + 2\,p3x\,p3y\,p4x\,p4y - p2x^2\,p4y^2 +$$
$$2\,p2x\,p3x\,p4y^2 - p3x^2\,p4y^2 - p2x^2\,p4z^2 + 2\,p2x\,p3x\,p4z^2 - p3x^2\,p4z^2 - p3y^2\,p4z^2\}$$

Two of the resulting equations are products.

```
In[11]:= Position[eqs1, _Times, {1}]
```

```
Out[11]= {{4}, {6}}
```

```
In[12]:= {eqs1[[4]], eqs1[[6]]}
```

$$Out[12]= \{-p3y\,(p3y\,p4x\,\lambda 3 - p3x\,p4y\,\lambda 3 - p2x\,p3y\,\lambda 4 + p3y\,p4x\,\lambda 4 + p2x\,p4y\,\lambda 4 - p3x\,p4y\,\lambda 4),$$
$$p4z\,(2\,p2x^2\,p3y^2 - 9\,p2x^2\,\lambda 2 - 9\,p3x^2\,\lambda 3 - 9\,p3y^2\,\lambda 3 - 9\,p2x^2\,\lambda 4 + 18\,p2x\,p3x\,\lambda 4 - 9\,p3x^2\,\lambda 4 - 9\,p3y^2\,\lambda 4)\}$$

Again, p3y and p4z cannot vanish for nondegenerate solutions, and this allows us to simplify eqs1.

```
In[13]:= eqs1a = If[Head[#] === Plus, #,
                    Cases[#, _Plus, {1}][[1]]]& /@ eqs1
```

$$Out[13]= \{-p2x^2\,p3y^2\,p4z^2 + 36\,V2,\ p3y\,p4x\,p4y\,\lambda 3 - p3x\,p4y^2\,\lambda 3 - p3x\,p4z^2\,\lambda 3 -$$
$$p2x\,p3y\,p4y\,\lambda 4 + p3y\,p4x\,p4y\,\lambda 4 + p2x\,p4y^2\,\lambda 4 - p3x\,p4y^2\,\lambda 4 + p2x\,p4z^2\,\lambda 4 - p3x\,p4z^2\,\lambda 4,$$
$$p2x^2\,p4y^2\,\lambda 2 + p2x^2\,p4z^2\,\lambda 2 - p3y^2\,p4x^2\,\lambda 3 + p3x\,p3y\,p4x\,p4y\,\lambda 3 - p3y^2\,p4z^2\,\lambda 3 +$$
$$p2x\,p3y^2\,p4x\,\lambda 4 - p3y^2\,p4x^2\,\lambda 4 - p2x^2\,p3y\,p4y\,\lambda 4 + p3x\,p3y\,p4x\,p4y\,\lambda 4 +$$
$$p2x^2\,p4y^2\,\lambda 4 - p2x\,p3x\,p4y^2\,\lambda 4 + p2x^2\,p4z^2\,\lambda 4 - p2x\,p3x\,p4z^2\,\lambda 4 - p3y^2\,p4z^2\,\lambda 4,$$
$$p3y\,p4x\,\lambda 3 - p3x\,p4y\,\lambda 3 - p2x\,p3y\,\lambda 4 + p3y\,p4x\,\lambda 4 + p2x\,p4y\,\lambda 4 - p3x\,p4y\,\lambda 4,$$
$$-p2x^2\,p4y\,\lambda 2 + p3x\,p3y\,p4x\,\lambda 3 - p3x^2\,p4y\,\lambda 3 + p2x^2\,p3y\,\lambda 4 - p2x\,p3x\,p3y\,\lambda 4 -$$
$$p2x\,p3y\,p4x\,\lambda 4 + p3x\,p3y\,p4x\,\lambda 4 - p2x^2\,p4y\,\lambda 4 + 2\,p2x\,p3x\,p4y\,\lambda 4 - p3x^2\,p4y\,\lambda 4,$$
$$p2x^2\,p3y^2 - 9\,p2x^2\,\lambda 2 - 9\,p3x^2\,\lambda 3 - 9\,p3y^2\,\lambda 3 - 9\,p2x^2\,\lambda 4 + 18\,p2x\,p3x\,\lambda 4 - 9\,p3x^2\,\lambda 4 - 9\,p3y^2\,\lambda 4,$$
$$4\,AS123 - p2x^2\,p3y^2,\ 4\,AS124 - p2x^2\,p4y^2 - p2x^2\,p4z^2,$$
$$4\,AS134 - p3y^2\,p4x^2 + 2\,p3x\,p3y\,p4x\,p4y - p3x^2\,p4y^2 - p3x^2\,p4z^2 - p3y^2\,p4z^2,$$
$$4\,AS234 - p2x^2\,p3y^2 + 2\,p2x\,p3y^2\,p4x - p3y^2\,p4x^2 + 2\,p2x^2\,p3y\,p4y -$$
$$2\,p2x\,p3x\,p3y\,p4y - 2\,p2x\,p3y\,p4x\,p4y + 2\,p3x\,p3y\,p4x\,p4y - p2x^2\,p4y^2 +$$
$$2\,p2x\,p3x\,p4y^2 - p3x^2\,p4y^2 - p2x^2\,p4z^2 + 2\,p2x\,p3x\,p4z^2 - p3x^2\,p4z^2 - p3y^2\,p4z^2\}$$

In `eqs1a`, the Lagrange multiplier $\lambda 2$ appears linearly in the third equation.

```
In[14]:= Exponent[#, {λ2, λ3, λ4}]& /@ eqs1a
```

```
Out[14]= {{0, 0, 0}, {0, 1, 1}, {1, 1, 1}, {0, 1, 1},
         {1, 1, 1}, {1, 1, 1}, {0, 0, 0}, {0, 0, 0}, {0, 0, 0}}
```

Under the previously stated nondegeneracy condition, we can solve for $\lambda 2$.

```
In[15]:= sol2 = Solve[eqs1a[[3]] == 0, λ2]
```

$$Out[15]= \left\{\left\{\lambda 2 \rightarrow \frac{1}{p2x^2 \ (p4y^2 + p4z^2)} \ (p3y^2 \ p4x^2 \ \lambda 3 - p3x \ p3y \ p4x \ p4y \ \lambda 3 + p3y^2 \ p4z^2 \ \lambda 3 - \right.\right.$$
$$p2x \ p3y^2 \ p4x \ \lambda 4 + p3y^2 \ p4x^2 \ \lambda 4 + p2x^2 \ p3y \ p4y \ \lambda 4 - p3x \ p3y \ p4x \ p4y \ \lambda 4 -$$
$$\left.\left. p2x^2 \ p4y^2 \ \lambda 4 + p2x \ p3x \ p4y^2 \ \lambda 4 - p2x^2 \ p4z^2 \ \lambda 4 + p2x \ p3x \ p4z^2 \ \lambda 4 + p3y^2 \ p4z^2 \ \lambda 4)\right\}\right\}$$

Backsubstitution of this equation gives the following system of nine equations.

```
In[16]:= eqs2 = Factor[Numerator[Together[Delete[eqs1a, 3] /. sol2[[1]]]]]
```

```
Out[16]= {-p2x² p3y² p4z² + 36 V2, p3y p4x p4y λ3 - p3x p4y² λ3 - p3x p4z² λ3 -
         p2x p3y p4y λ4 + p3y p4x p4y λ4 + p2x p4y² λ4 - p3x p4y² λ4 + p2x p4z² λ4 - p3x p4z² λ4,
         p3y p4x λ3 - p3x p4y λ3 - p2x p3y λ4 + p3y p4x λ4 + p2x p4y λ4 - p3x p4y λ4,
         -p3y² p4x² p4y λ3 + 2 p3x p3y p4x p4y² λ3 - p3x² p4y³ λ3 + p3x p3y p4x p4z² λ3 -
         p3x² p4y p4z² λ3 - p3y² p4y p4z² λ3 + p2x p3y² p4x p4y λ4 - p3y² p4x² p4y λ4 -
         p2x p3x p3y p4y² λ4 - p2x p3y p4x p4y² λ4 + 2 p3x p3y p4x p4y² λ4 + p2x p3x p4y³ λ4 -
         p3x² p4y³ λ4 + p2x² p3y p4z² λ4 - p2x p3x p3y p4z² λ4 - p2x p3y p4x p4z² λ4 +
         p3x p3y p4x p4z² λ4 + p2x p3x p4y p4z² λ4 - p3x² p4y p4z² λ4 - p3y² p4y p4z² λ4,
         p2x² p3y² p4y² + p2x² p3y² p4z² - 9 p3y² p4x² λ3 + 9 p3x p3y p4x p4y λ3 - 9 p3x² p4y² λ3 -
         9 p3y² p4y² λ3 - 9 p3x² p4z² λ3 - 18 p3y² p4z² λ3 + 9 p2x p3y² p4x λ4 -
         9 p3y² p4x² λ4 - 9 p2x² p3y p4y λ4 + 9 p3x p3y p4x p4y λ4 + 9 p2x p3x p4y² λ4 -
         9 p3x² p4y² λ4 - 9 p3y² p4y² λ4 + 9 p2x p3x p4z² λ4 - 9 p3x² p4z² λ4 - 18 p3y² p4z² λ4,
         4 AS123 - p2x² p3y², 4 AS124 - p2x² p4y² - p2x² p4z²,
         4 AS134 - p3y² p4x² + 2 p3x p3y p4x p4y - p3x² p4y² - p3x² p4z² - p3y² p4z²,
         4 AS234 - p2x² p3y² + 2 p2x p3y² p4x - p3y² p4x² + 2 p2x² p3y p4y -
         2 p2x p3x p3y p4y - 2 p2x p3y p4x p4y + 2 p3x p3y p4x p4y - p2x² p4y² +
         2 p2x p3x p4y² - p3x² p4y² - p2x² p4z² + 2 p2x p3x p4z² - p3x² p4z² - p3y² p4z²}
```

This time, we do not have any products.

```
In[17]:= Position[eqs2, _Times, {1}]
```

```
Out[17]= {}
```

```
In[18]:= Exponent[#, {λ3, λ4}]& /@ eqs2
```

```
Out[18]= {{0, 0}, {1, 1}, {1, 1}, {1, 1}, {1, 1}, {0, 0}, {0, 0}, {0, 0}, {0, 0}}
```

We proceed as above and eliminate $\lambda 3$.

```
In[19]:= sol3 = Solve[eqs2[[2]] == 0, λ3]
```

$$Out[19]= \left\{\left\{\lambda 3 \rightarrow \frac{p2x \ p3y \ p4y \ \lambda 4 - p3y \ p4x \ p4y \ \lambda 4 - p2x \ p4y^2 \ \lambda 4 + p3x \ p4y^2 \ \lambda 4 - p2x \ p4z^2 \ \lambda 4 + p3x \ p4z^2 \ \lambda 4}{p3y \ p4x \ p4y - p3x \ p4y^2 - p3x \ p4z^2}\right\}\right\}$$

```
In[20]:= eqs3 = Factor[Numerator[Together[Delete[eqs2, 2] /. sol3[[1]]]]]
```

```
Out[20]= {-p2x² p3y² p4z² + 36 V2, p2x p3y (p3x - p4x) p4z² λ4,
         p2x p3y p4z² (p2x p3y p4x p4y - p3x p3y p4x p4y - p2x p3x p4y² +
         p3x² p4y² - p3y² p4y² + p3y p4y³ - p2x p3x p4z² + p3x² p4z² + p3y p4y p4z²) λ4,
         p2x p3y (p2x p3y² p4x p4y³ - p2x p3x p3y p4y⁴ + p2x p3y² p4x p4y p4z² -
         2 p2x p3x p3y p4y² p4z² - p2x p3x p3y p4z⁴ - 9 p2x p3y p4x p4y² λ4 +
         9 p3y p4x² p4y² λ4 + 9 p2x p3x p4y³ λ4 - 9 p3x² p4y³ λ4 - 9 p3y² p4y³ λ4 +
         9 p3y p4y⁴ λ4 - 9 p3x p3y p4x p4z² λ4 + 9 p3x² p4y p4z² λ4 + 9 p2x p3x p4y p4z² λ4 -
         9 p3x² p4y p4z² λ4 - 18 p3y² p4y p4z² λ4 + 27 p3y p4y² p4z² λ4 + 18 p3y p4z⁴ λ4),
         4 AS123 - p2x² p3y², 4 AS124 - p2x² p4y² - p2x² p4z²,
         4 AS134 - p3y² p4x² + 2 p3x p3y p4x p4y - p3x² p4y² - p3x² p4z² - p3y² p4z²,
```

$$4\,AS234 - p2x^2\,p3y^2 + 2\,p2x\,p3y^2\,p4x - p3y^2\,p4x^2 + 2\,p2x^2\,p3y\,p4y -$$
$$2\,p2x\,p3x\,p3y\,p4y - 2\,p2x\,p3y\,p4x\,p4y + 2\,p3x\,p3y\,p4x\,p4y - p2x^2\,p4y^2 +$$
$$2\,p2x\,p3x\,p4y^2 - p3x^2\,p4y^2 - p2x^2\,p4z^2 + 2\,p2x\,p3x\,p4z^2 - p3x^2\,p4z^2 - p3y^2\,p4z^2\}$$

This time, three of the resulting equations factor. We eliminate the factors p2x, p3y, p4z that cannot vanish for a nondegenerate tetrahedron.

`In[21]:= Position[eqs3, _Times, {1}]`

`Out[21]= {{2}, {3}, {4}}`

`In[22]:= {eqs3[[2]], eqs3[[3]], eqs3[[4]]}`

$$\text{Out[22]= } \{p2x\,p3y\,(p3x - p4x)\,p4z^2\,\lambda4,$$
$$p2x\,p3y\,p4z^2\,(p2x\,p3y\,p4x\,p4y - p3x\,p3y\,p4x\,p4y - p2x\,p3x\,p4y^2 + p3x^2\,p4y^2 -$$
$$p3y^2\,p4y^2 + p3y\,p4y^3 - p2x\,p3x\,p4z^2 + p3x^2\,p4z^2 + p3y\,p4y\,p4z^2)\,\lambda4,$$
$$p2x\,p3y\,(p2x\,p3y^2\,p4x\,p4y^3 - p2x\,p3x\,p3y\,p4y^4 + p2x\,p3y^2\,p4x\,p4y\,p4z^2 -$$
$$2\,p2x\,p3x\,p3y\,p4y^2\,p4z^2 - p2x\,p3x\,p3y\,p4z^4 - 9\,p2x\,p3y\,p4x\,p4y^2\,\lambda4 +$$
$$9\,p3y\,p4x^2\,p4y^2\,\lambda4 + 9\,p2x\,p3x\,p4y^3\,\lambda4 - 9\,p3x^2\,p4y^3\,\lambda4 - 9\,p3y^2\,p4y^3\,\lambda4 +$$
$$9\,p3y\,p4y^4\,\lambda4 - 9\,p3x\,p3y\,p4x\,p4z^2\,\lambda4 + 9\,p3y\,p4x^2\,p4z^2\,\lambda4 + 9\,p2x\,p3x\,p4y\,p4z^2\,\lambda4 -$$
$$9\,p3x^2\,p4y\,p4z^2\,\lambda4 - 18\,p3y^2\,p4y\,p4z^2\,\lambda4 + 27\,p3y\,p4y^2\,p4z^2\,\lambda4 + 18\,p3y\,p4z^4\,\lambda4)\}$$

`In[23]:= eqs3a = If[Head[#] === Plus, #, Cases[#, _Plus, {1}][[1]]]& /@ eqs3`

$$\text{Out[23]= } \{-p2x^2\,p3y^2\,p4z^2 + 36\,V2,\ p3x - p4x,\ p2x\,p3y\,p4x\,p4y - p3x\,p3y\,p4x\,p4y -$$
$$p2x\,p3x\,p4y^2 + p3x^2\,p4y^2 - p3y^2\,p4y^2 + p3y\,p4y^3 - p2x\,p3x\,p4z^2 + p3x^2\,p4z^2 + p3y\,p4y\,p4z^2,$$
$$p2x\,p3y^2\,p4x\,p4y^3 - p2x\,p3x\,p3y\,p4y^4 + p2x\,p3y^2\,p4x\,p4y\,p4z^2 - 2\,p2x\,p3x\,p3y\,p4y^2\,p4z^2 -$$
$$p2x\,p3x\,p3y\,p4z^4 - 9\,p2x\,p3y\,p4x\,p4y^2\,\lambda4 + 9\,p3y\,p4x^2\,p4y^2\,\lambda4 + 9\,p2x\,p3x\,p4y^3\,\lambda4 - 9\,p3x^2\,p4y^3\,\lambda4 -$$
$$9\,p3y^2\,p4y^3\,\lambda4 + 9\,p3y\,p4y^4\,\lambda4 - 9\,p3x\,p3y\,p4x\,p4z^2\,\lambda4 + 9\,p3y\,p4x^2\,p4z^2\,\lambda4 + 9\,p2x\,p3x\,p4y\,p4z^2\,\lambda4 -$$
$$9\,p3x^2\,p4y\,p4z^2\,\lambda4 - 18\,p3y^2\,p4y\,p4z^2\,\lambda4 + 27\,p3y\,p4y^2\,p4z^2\,\lambda4 + 18\,p3y\,p4z^4\,\lambda4,$$
$$4\,AS123 - p2x^2\,p3y^2,\ 4\,AS124 - p2x^2\,p4y^2 - p2x^2\,p4z^2,$$
$$4\,AS134 - p3y^2\,p4x^2 + 2\,p3x\,p3y\,p4x\,p4y - p3x^2\,p4y^2 - p3x^2\,p4z^2 - p3y^2\,p4z^2,$$
$$4\,AS234 - p2x^2\,p3y^2 + 2\,p2x\,p3y^2\,p4x - p3y^2\,p4x^2 + 2\,p2x^2\,p3y\,p4y -$$
$$2\,p2x\,p3x\,p3y\,p4y - 2\,p2x\,p3y\,p4x\,p4y + 2\,p3x\,p3y\,p4x\,p4y - p2x^2\,p4y^2 +$$
$$2\,p2x\,p3x\,p4y^2 - p3x^2\,p4y^2 - p2x^2\,p4z^2 + 2\,p2x\,p3x\,p4z^2 - p3x^2\,p4z^2 - p3y^2\,p4z^2\}$$

From `eqs3[[2]]`, the condition p4x = p3x follows. (This result afterward justifies the assumption that the denominator in sol3 does not vanish.)

`In[24]:= eqs3b = Delete[Factor[eqs3a /. p4x -> p3x], 2]`

$$\text{Out[24]= } \{-p2x^2\,p3y^2\,p4z^2 + 36\,V2,\ (p2x\,p3x - p3x^2 - p3y\,p4y)\,(p3y\,p4y - p4y^2 - p4z^2),$$
$$(p3y\,p4y - p4y^2 - p4z^2)\,(p2x\,p3x\,p3y\,p4y^2 + p2x\,p3x\,p3y\,p4z^2 - 9\,p2x\,p3x\,p4y\,\lambda4 + 9\,p3x^2\,p4y\,\lambda4 -$$
$$9\,p3y\,p4y^2\,\lambda4 - 18\,p3y\,p4z^2\,\lambda4),\ 4\,AS123 - p2x^2\,p3y^2,\ 4\,AS124 - p2x^2\,p4y^2 - p2x^2\,p4z^2,$$
$$4\,AS134 - p3x^2\,p3y^2 + 2\,p3x^2\,p3y\,p4y - p3x^2\,p4y^2 - p3x^2\,p4z^2 - p3y^2\,p4z^2,$$
$$4\,AS234 - p2x^2\,p3y^2 + 2\,p2x\,p3x\,p3y^2 - p3x^2\,p3y^2 + 2\,p2x^2\,p3y\,p4y -$$
$$4\,p2x\,p3x\,p3y\,p4y + 2\,p3x^2\,p3y\,p4y - p2x^2\,p4y^2 + 2\,p2x\,p3x\,p4y^2 -$$
$$p3x^2\,p4y^2 - p2x^2\,p4z^2 + 2\,p2x\,p3x\,p4z^2 - p3x^2\,p4z^2 - p3y^2\,p4z^2\}$$

Now only one Lagrange multiplier is left to eliminate: $\lambda4$.

`In[25]:= Exponent[#, {λ4}]& /@ eqs3b`

`Out[25]= {{0}, {0}, {1}, {0}, {0}, {0}, {0}}`

`In[26]:= sol4 = Solve[eqs3b[[3]] == 0, λ4]`

$$\text{Out[26]= } \left\{\left\{\lambda4 \to \frac{p2x\,p3x\,p3y\,p4y^2 + p2x\,p3x\,p3y\,p4z^2}{9\,(p2x\,p3x\,p4y - p3x^2\,p4y + p3y\,p4y^2 + 2\,p3y\,p4z^2)}\right\}\right\}$$

`In[27]:= eqs4 = Factor[Numerator[Together[Delete[eqs3b, 3] /. sol4[[1]]]]]`

$$\text{Out[27]= } \{-p2x^2\,p3y^2\,p4z^2 + 36\,V2,\ (p2x\,p3x - p3x^2 - p3y\,p4y)\,(p3y\,p4y - p4y^2 - p4z^2),$$
$$4\,AS123 - p2x^2\,p3y^2,\ 4\,AS124 - p2x^2\,p4y^2 - p2x^2\,p4z^2,$$
$$4\,AS134 - p3x^2\,p3y^2 + 2\,p3x^2\,p3y\,p4y - p3x^2\,p4y^2 - p3x^2\,p4z^2 - p3y^2\,p4z^2,\ 4\,AS234 -$$
$$p2x^2\,p3y^2 + 2\,p2x\,p3x\,p3y^2 - p3x^2\,p3y^2 + 2\,p2x^2\,p3y\,p4y - 4\,p2x\,p3x\,p3y\,p4y + 2\,p3x^2\,p3y\,p4y -$$
$$p2x^2\,p4y^2 + 2\,p2x\,p3x\,p4y^2 - p3x^2\,p4y^2 - p2x^2\,p4z^2 + 2\,p2x\,p3x\,p4z^2 - p3x^2\,p4z^2 - p3y^2\,p4z^2\}$$

Now, after elimination of all four Lagrange multipliers, we have a nontrivial product.

```
In[28]:= Position[eqs4, _Times, {1}]
```
```
Out[28]= {{2}}
```
```
In[29]:= eqs4[[2]]
```
$$Out[29]= \ (p2x\ p3x - p3x^2 - p3y\ p4y)\ (p3y\ p4y - p4y^2 - p4z^2)$$

In agreement with the assumption about the nonvanishing of the denominator in sol3, we assume that p2x p3x − p3x^2 − p3y p4y must vanish.

```
In[30]:= eqs4a = MapAt[#/(p3y p4y - p4y^2 - p4z^2)&, eqs4, 2]
```
$$Out[30]= \ \{-p2x^2\ p3y^2\ p4z^2 + 36\ V2,\ p2x\ p3x - p3x^2 - p3y\ p4y,$$
$$4\ AS123 - p2x^2\ p3y^2,\ 4\ AS124 - p2x^2\ p4y^2 - p2x^2\ p4z^2,$$
$$4\ AS134 - p3x^2\ p3y^2 + 2\ p3x^2\ p3y\ p4y - p3x^2\ p4y^2 - p3x^2\ p4z^2 - p3y^2\ p4z^2,\ 4\ AS234 -$$
$$p2x^2\ p3y^2 + 2\ p2x\ p3x\ p3y^2 - p3x^2\ p3y^2 + 2\ p2x^2\ p3y\ p4y - 4\ p2x\ p3x\ p3y\ p4y + 2\ p3x^2\ p3y\ p4y -$$
$$p2x^2\ p4y^2 + 2\ p2x\ p3x\ p4y^2 - p3x^2\ p4y^2 - p2x^2\ p4z^2 + 2\ p2x\ p3x\ p4z^2 - p3x^2\ p4z^2 - p3y^2\ p4z^2\}$$

By now, we have reduced the number of variables to be eliminated from ten to five (we have eliminated the four Lagrange multipliers and the identity p3x=p4x came as a freebie this way), and we also have excluded some degenerate cases. The resulting system of six polynomials has the following structure.

```
In[31]:= Exponent[#, {p2x, p3x, p3y, p4y, p4z}]& /@ eqs4a
```
```
Out[31]= {{2, 0, 2, 0, 2}, {1, 2, 1, 1, 0}, {2, 0, 2, 0, 0},
          {2, 0, 0, 2, 2}, {0, 2, 2, 2, 2}, {2, 2, 2, 2, 2}}
```

GroebnerBasis is now able to eliminate the remaining variables, and we arrive at the following system of polynomials.

```
In[32]:= gb = GroebnerBasis[eqs4a, {}, {p2x, p3x, p3y, p4y, p4z},
                           MonomialOrder -> EliminationOrder];
```

As to be expected, the resulting Gröbner basis has just one element.

```
In[33]:= Length[gb]
```
```
Out[33]= 1
```

The polynomial gb[[1]] is quite large.

```
In[34]:= Length[poly = gb[[1]]]
```
```
Out[34]= 434
```

Here are its first terms.

```
In[35]:= Take[poly, 10]
```
$$Out[35]= \ AS123^9\ AS124\ AS134\ AS234 - 8\ AS123^8\ AS124^2\ AS134\ AS234 + 28\ AS123^7\ AS124^3\ AS134\ AS234 -$$
$$56\ AS123^6\ AS124^4\ AS134\ AS234 + 70\ AS123^5\ AS124^5\ AS134\ AS234 -$$
$$56\ AS123^4\ AS124^6\ AS134\ AS234 + 28\ AS123^3\ AS124^7\ AS134\ AS234 -$$
$$8\ AS123^2\ AS124^8\ AS134\ AS234 + AS123\ AS124^9\ AS134\ AS234 - 8\ AS123^8\ AS124\ AS134^2\ AS234$$

We rewrite the polynomial in the square of the volume v2 in a more symmetric form. The coefficients of v2 are symmetric in AS123, AS124, AS134, and AS234. Using this symmetry by introducing the elementary symmetric polynomials; we arrive at the following final result. We denote the kth elementary symmetric polynomial in AS123, AS124, AS134, and AS234 by S[k].

```
In[36]:= Needs["Algebra`SymmetricPolynomials`"]
```

```
In[37]:= writeSymmetric[f_] :=
         First[SymmetricReduction[f, {AS123, AS124, AS134, AS234}]] //.
         Append[Table[SymmetricPolynomial[{AS123, AS124, AS134, AS234}, k] ->
                S[k], {k, 3}],
             AS123^e_. AS124^e_. AS134^e_. AS234^e_. -> S[4]^e]
```
```
In[38]:= Collect[poly, v2, writeSymmetric]
```
$$Out[38]= \ -1162261467\ V2^8 + 531441\ V2^6\ S[1]^3 - 19131876\ V2^6\ S[1]\ S[2] + 6561\ V2^4\ S[1]^4\ S[2] -$$
$$52488\ V2^4\ S[1]^2\ S[2]^2 + 104976\ V2^4\ S[2]^3 + 114791256\ V2^6\ S[3] - 236196\ V2^4\ S[1]^3\ S[3] +$$

$$81 \, V2^2 \, S[1]^6 \, S[3] + 944784 \, V2^4 \, S[1] \, S[2] \, S[3] - 972 \, V2^2 \, S[1]^4 \, S[2] \, S[3] +$$
$$3888 \, V2^2 \, S[1]^2 \, S[2]^2 \, S[3] - 5184 \, V2^2 \, S[2]^3 \, S[3] - 2834352 \, V2^4 \, S[3]^2 + 2204496 \, V2^4 \, S[1]^2 \, S[4] -$$
$$3240 \, V2^2 \, S[1]^5 \, S[4] + S[1]^8 \, S[4] - 3779136 \, V2^4 \, S[2] \, S[4] + 25920 \, V2^2 \, S[1]^3 \, S[2] \, S[4] -$$
$$16 \, S[1]^6 \, S[2] \, S[4] - 51840 \, V2^2 \, S[1] \, S[2]^2 \, S[4] + 96 \, S[1]^4 \, S[2]^2 \, S[4] - 256 \, S[1]^2 \, S[2]^3 \, S[4] +$$
$$256 \, S[2]^4 \, S[4] - 46656 \, V2^2 \, S[1]^2 \, S[3] \, S[4] + 186624 \, V2^2 \, S[2] \, S[3] \, S[4] -$$
$$124416 \, V2^2 \, S[1] \, S[4]^2 - 128 \, S[1]^4 \, S[4]^2 + 1024 \, S[1]^2 \, S[2] \, S[4]^2 - 2048 \, S[2]^2 \, S[4]^2 + 4096 \, S[4]^3$$

Now, we substitute actual numerical values for the areas of the faces.

```
In[39]:= rules = {AS123 -> 9/10, AS124 -> 10/10,
              AS134 -> 11/10, AS234 -> 12/10};
```

```
In[40]:= poly /. rules
```

$$\text{Out[40]= } \frac{2376}{48828125} + \frac{80828118 \, V2^2}{1953125} + \frac{307435338 \, V2^4}{3125} + \frac{4419463356 \, V2^6}{125} - 1162261467 \, V2^8$$

The polynomial has a unique positive root for $v2$.

```
In[41]:= Select[V2 /. {ToRules[Roots[% == 0, V2, Quartics -> False]]}, Positive]
```

$$\text{Out[41]= } \left\{ \sqrt{\text{Root}\left[-88 - 74840850 \, \# - 177913968750 \, \#^2 - 63938995312500 \, \#^3 + 2101890673828125 \, \#^4 \, \&, \, 4 \right]} \right\}$$

```
In[42]:= N[%, 22]
```

```
Out[42]= {0.1817034821860167543051}
```

We compare the last result with a numerical solution using FindRoot. We select random start values and repeat the FindRoot call until it finds a solution within the specified restrictions. We calculate five values.

```
In[43]:= numSols = Table[
         While[Check[
             nsol = FindRoot[Evaluate[# == 0 & /@
                          (Rest[eqs] /. rules)], Evaluate[
                 Sequence @@ ({#, Random[Real, {0, 2}]}&) /@
                 {p2x, p3x, p3y, p4x, p4y, p4z, λ1, λ2, λ3, λ4}],
                 MaxIterations -> 30], $Failed] === $Failed, Null];
         nsol, {5}]
```

```
FindRoot::cvmit : Failed to converge to the
    requested accuracy or precision within 30 iterations. More…

FindRoot::cvmit : Failed to converge to the
    requested accuracy or precision within 30 iterations. More…

FindRoot::cvmit : Failed to converge to the
    requested accuracy or precision within 30 iterations. More…

General::stop : Further output of
    FindRoot::cvmit will be suppressed during this calculation. More…

FindRoot::lstol :
    The line search decreased the step size to within tolerance specified
        by AccuracyGoal and PrecisionGoal but was unable to find a sufficient
        decrease in the merit function.  You may need more than MachinePrecision
        digits of working precision to meet these tolerances. More…

FindRoot::lstol :
    The line search decreased the step size to within tolerance specified
        by AccuracyGoal and PrecisionGoal but was unable to find a sufficient
        decrease in the merit function.  You may need more than MachinePrecision
        digits of working precision to meet these tolerances. More…
```

```
Out[43]= {{p2x → 1.42766, p3x → 0.662231, p3y → 1.32901, p4x → 0.662231, p4y → 0.381406,
    p4z → 1.34797, λ1 → 0.0808675, λ2 → 0.0699647, λ3 → 0.0607095, λ4 → 0.0525245},
    {p2x → 1.42766, p3x → 0.662231, p3y → 1.32901, p4x → 0.662231, p4y → 0.381406,
    p4z → 1.34797, λ1 → 0.0808675, λ2 → 0.0699647, λ3 → 0.0607095, λ4 → 0.0525245},
    {p2x → 1.42766, p3x → 0.662231, p3y → 1.32901, p4x → 0.662231, p4y → 0.381406,
    p4z → 1.34797, λ1 → 0.0808675, λ2 → 0.0699647, λ3 → 0.0607095, λ4 → 0.0525245},
    {p2x → 1.42766, p3x → 0.662231, p3y → 1.32901, p4x → 0.662231, p4y → 0.381406,
```

$$p4z \to 1.34797, \ \lambda1 \to 0.0808675, \ \lambda2 \to 0.0699647, \ \lambda3 \to 0.0607095, \ \lambda4 \to 0.0525245\},$$
$$\{p2x \to 1.42766, \ p3x \to 0.662231, \ p3y \to 1.32901, \ p4x \to 0.662231, \ p4y \to 0.381406,$$
$$p4z \to 1.34797, \ \lambda1 \to 0.0808675, \ \lambda2 \to 0.0699647, \ \lambda3 \to 0.0607095, \ \lambda4 \to 0.0525245\}\}$$

This result agrees with the above one.

```
In[44]:= v^2 /. numSols
```

```
Out[44]= {0.181703, 0.181703, 0.181703, 0.181703, 0.181703}
```

When all four faces have the same area, the resulting volume is the same as that the regular polyhedron.

```
In[45]:= poly /. {AS123 -> 1, AS124 -> 1, AS134 -> 1, AS234 -> 1}
```

$$Out[45]= \ 34012224 \ V2^6 - 1162261467 \ V2^8$$

```
In[46]:= Select[V2 /. {ToRules[Roots[% == 0, V2, Quartics -> False]]},
              Positive]
```

$$Out[46]= \ \left\{ \frac{8}{27\sqrt{3}} \right\}$$

Here are the vertices of a regular tetrahedron with faces of area 1.

```
In[47]:= P1 = {0, 0, 0};
         P2 = {2/3^(1/4), 0, 0};
         P3 = {1/3^(1/4), 3^(1/4), 0};
         P4 = {1/3^(1/4), 1/3^(3/4), 2 2^(1/2)/3^(3/4)};
```

```
In[51]:= {#.#/4&[Cross[P2 - P1, P3 - P1]], #.#/4&[Cross[P2 - P1, P4 - P1]],
          #.#/4&[Cross[P3 - P1, P4 - P1]], #.#/4&[Cross[P3 - P2, P4 - P2]]}
```

```
Out[51]= {1, 1, 1, 1}
```

This is the square of its volume.

```
In[52]:= (1/6 Det[Append[#, 1]& /@ {P1, P2, P3, P4}])^2
```

$$Out[52]= \ \frac{8}{27\sqrt{3}}$$

References

1   K. Abbaoui, Y. Cherruault, V. Seng. *J. Phys.* A 31, 4301 (1998).

2   P. C. Abbott, E. N. Maslen. *J. Phys.* B 19, 1595 (1986).

3   P. Abbott. *The Mathematica Journal* 7, 255 (1999).

4   F. Abdelwahid. *Appl. Math. Comput.* 141, 447 (2003).

5   A. Abdesselam. *arXiv:math.CO*/0212121 (2002).

6   A. Abdesselam. *J. Phys.* A 36, 9471 (2003).

7   M. J. Ablowitz, R. Halburd, B. Herbst. *Nonlinearity* 13, 889 (2000).

8   M. J. Ablowitz, S. Chakravarty, R. Halburd. *Stud. Appl. Math.* 103, 75 (2001).

9   R. Abraham, J. E. Marsden, T. Ratiu. *Manifolds, Tensor Analysis and Applications*, Springer-Verlag, New York, 1988.

10 M. Abramowitz, A. I. Stegun. *Handbook of Mathematical Functions*, National Bureau of Standards, Washington, 1964.

11 A. Aćin, A. Andrianov, E. Jané, R. Tarrach. *arXiv:quant-ph*/0009107 (2000).

12 A. Acín, D. Bruß, M. Lewenstein, A. Sanpera. *arXiv:quant-ph*/0103025 (2001).

13 J. Ackermann, B. Erdmann, R. Roitzsch. *J. Chem. Phys.* 101, 7643 (1994).

14 J. Ackermann, R. Roitzsch. *Chem. Phys. Lett.* 214, 109 (1993).

15 W. Adams, P. Loustaunau. *An Introduction to Gröbner Bases*, American Mathematical Society, Providence, 1994.

16 R. Adhikari, R. Dutt, A. Khare, U. P. Sukhatme. *Phys. Rev.* A 38, 1679 (1988).

17 G. Adomian. *Solving Frontier Problems of Physics: The Decomposition Method* , Kluwer, Dordrecht, 1994.

18 I. V. Adrianov, J. Awrejcewicz. *Phys. Lett.* A 319, 53 (2003).

19 D. Aerts, S. Aerts, J. Broekaert, L. Gabora. *Found. Phys.* 30, 1387 (2000).

20 R. P. Agarwal. *Difference Equations and Inequalities*, Marcel Dekker, New York, 1992.

21 R. P. Agnew. *Proc. Am. Math. Soc.* 2, 237 (1951).

22 J. M. Aguirregabiria, A. Hernández, M. Rivas. *Am. J. Phys.* 70, 180 (2002).

23 J. M. Aguirregabiria, J. Llosa, A. Molina. *arXiv:physics*/0411032 (2004).

24 Y. Aharonov, L. Vaidman. *arXiv:quant-ph*/0105101 (2001).

25 Y. Aharonov, L. Vaidman in J. G. Muga, R. Sala Mayato, I. L. Egusquiza (eds.). *Time in Quantum Mechanics*, Springer-Verlag, Berlin, 2002.

26 C. D. Ahlbrandt, A. C. Peterson. *Discrete Hamiltonian Systems*, Kluwer, Dordrecht, 1996

27 Z. Ahmed. *Phys. Rev.* A 47, 4761 (1993).

28 R. W. Ahrens, G. Szekeres. *Austral. Math. J.* 10, 485 (1969).

29 I. A. Ajwa, Z. Liu, P. S. Wang (1995). http://symbolicnet.mcs.kent.edu /areas/groebner/grobner.ps

30 E. Akin–Bohner, M. Bohner. *Methods Appl. Anal.* 10, 11 (2003).

31 J. E. Akin. *Finite Elements for Analysis and Design*, Academic Press, London, 1994.

32 A. G. Akritas. *Fibon. Quart.* 31, 325 (1993).

33 T. Aktosun. *J. Math. Phys.* 34, 1619 (1993).

34 L. D. Akulenko. *J. Appl. Math. Mech.* 57, 301 (1993).

35 V. S. Alagar. *J. Appl. Prob.* 14, 284 (1977).

36   A. Alavi. *J. Chem. Phys.* 113, 7735 (2000).

37   S. I. Al'ber. *Commun. Pure Appl. Math.* 34, 259 (1981).

38   S. Albeverio, F. Gesztesy, R. Høegh–Krohn, H. Holden. *Solvable Models in Quantum Mechanics*, Springer-Verlag, New York, 1988.

39   D. S. Alexander. *A History of Complex Dynamics*, Vieweg, Braunschweig, 1994.

40   R. M. Alexander. *Mechanics of Animal Locomotion*, Springer-Verlag, Berlin, 1992.

41   I. Alexeev, K. Y. Kim, H. M. Milchberg. *Phys. Rev. Lett.* 88, 073901 (2002).

42   M. M. Ali. *Pac. J. Math.* 33, 1 (1970).

43   G. Almkvist. *arXiv:math.CA*/0208244 (2002).

44   G. Alv'arez, R. F. Alvrez. *J. Phys.* A 28, 5767 (1995).

45   G. Álvarez, V. Martin–Mayor, J. J. Ruiz–Lorenzo. *J. Phys.* A 33, 841 (2000).

46   H. Alzer. *Bull. Lond. Math. Soc.* 35, 601 (2003).

47   G. A. Ambartsumyan, A. V. Burobin. *Math. Notes* 73, 155 (2003).

48   A. Amengual. *Am. J. Phys.* 68, 175 (2000).

49   D. B. Ames. *Proc. Am. Math. Soc.* 1, 99 (1950).

50   P. Amore, R. A. Sáenz. *arXiv:math-ph*/0405030 (2004).

51   P. Amore, J. A. Lopez. *arXiv:quant-ph*/0405090 (2004).

52   P. Amore. *arXiv:math-ph*/0411049 (2004).

53   P. Amore, A. Raya. *arXiv:math-ph*/0412060 (2004).

54   B. Amrhein, O. Gloor, W. Kuechlin. *Theor. Comput. Sc.* 187, 179 (1997).

55   H. Anai in K. Nishizawa, T. Saito, T. Hilano (eds.). *NLA99 Computer Algebra*, Josai University, Saitama, 2000.

56   C. M. Andersen, J. F. Geer. *SIAM J. Appl. Math.* 42, 678 (1982).

57   R. S. Andersson, R. P. Brent, D. J. Daley, P. A. P. Morani. *SIAM J. Appl. Math.* 30, 22 (1976).

58   N. Andersson, S. Linnäus. *Phys. Rev.* D 46, 4179 (1992).

59   C. Andradas, L. Bröcker, J. M. Ruiz. *Constructible Sets in Real Geometry*, Springer-Verlag, New York, 1996.

60   G. E. Andrews. *SIAM Rev.* 16, 441 (1974).

61   G. E. Andrews, R. Askey, R. Roy. *Special Functions*, Cambridge University Press, Cambridge, 1999.

62   G. E. Andrews, B. C. Berndt, J. Sohn, A. J. Yee, A. Zaharescu. *Trans. Am. Math. Soc.* 355, 2379 (2003).

63   A. A. Andronov, A. A. Vitt, S. E. Khaikin. *Theory of Oscillations*, Pergamon Press, Oxford, 1966.

64   I. O. Angell. *Acta Cryst.* A 43, 244 (1987).

65   M.-C. Anisiu, C. Blaga, G. Bozis. *Celest. Mech. Dynam. Astron.* 88, 245 (2004).

66   I. Antoniou, B. Qiao, Z. Suchanecki. *Chaos, Solitons, Fractals* 8, 77 (1997).

67   P. Antosik, J. Misusinski, R. Sikorski. *Theory of Distributions*, Elsevier, Amsterdam, 1973.

68   A. R. Angel. *Coll. Math. J.* 8, 278 (1977).

69   J. Angeles, G. Hommel, P. Kovács (eds.). *Computational Kinematics*, Kluwer, Dordrecht, 1993.

70   I. Antoniou, Z. Suchanecki, S. Tasaki in I. Antoniou, G. Lumer (eds.). *Generalized Functions, Operator Theory, and Dynamical Systems*, Chapman & Hall, Boca Raton, 1999.

71   T. Aoki, T. Kawai, Y. Takei. *Sugaku Exp.* 8, 217 (1995).

72   F. Apéry. *Tôhoku Math. J.* 44, 103 (1992).

73   F. Apéry, G. Franzoni. *Rendiconti Sem. Fac. Sc. Univ. Cagliari* 69, 1 (1999).

74   D. M. Appleby. *arXiv:quant-ph*/0408058 (2004).

75   T. M. Apostol. *Am. Math. Monthly* 107, 738, (2000).

76   T. A. Apostolatos. *Am. J. Phys.* 71, 261 (2003).

77   N. Aquino, A. Flores–Riveros, J. F. Rivas–Silva. *Phys. Lett.* A 307, 326 (2003).

78   K. T. Arasu. *J. Alg. Combinat.* 14, 103 (2001).

79   V. S. Araujo, F. A. B. Coutinho, J. F. Perez. *Am. J. Phys.* 72, 203 (2004).

80   F. Ardila. *arXiv:math.CO*/0409418 (2004).

81   G. Ares de Parga, R. Mares. *J. Math. Phys.* 40, 4807 (1999).

82   J. Argyris, H.-P. Mlejnek. *Die Methode der Finiten Elemente*, v. I, Vieweg, Braunschweig, 1986.

83   J. Argyris, H.-P. Mlejnek. *Die Methode der Finiten Elemente*, v. II, Vieweg, Braunschweig, 1987.

84   J. Argyris, H.-P. Mlejnek. *Die Methode der Finiten Elemente*, v. III, Vieweg, Braunschweig, 1988.

85   J. Argyris, G. Faust, M. Haase. *An Exploration of Chaos*, North-Holland, Amsterdam, 1994.

86   J. Arias-De-Reyna. *Proc. Am. Math. Soc.* 109, 165 (1990).

87   E. A. G. Armour. *J. Phys.* B 11, 2803 (1983).

88   E. A. G. Armour. *J. Phys.* B 16, 1295 (1983).

89   S. Arnborg, H. Feng. *J. Symb. Comput.* 5, 131 (1988).

90   D. A. Arnon, G. E. Collins, S. McCallum. *SIAM J. Comput.* 13, 865 (1984).

91   D. S. Arnon, M. Mignotte. *J. Symb. Comput.* 5, 237 (1988).

92   D. J. Arrigo, F. Hickling. *J. Phys.* A 35, L389 (2002).

93   E. Artin. *Galoissche Theorie*, Teubner, Leipzig, 1959.

94   J. H. Asad, R. S. Hijjawi, A. Sakaji, J. M. Khalifeh. *Int. J. Theor. Phys.* 43, 2223 (2004).

95   M. A. Asiru. *Int. J. Math. Sci. Technol.* 33, 441 (2002).

96   R. Askey. *CRM Proc. Lecture Notes* 9, 13 (1996).

97   N. M. Atakishiyev, L. E. Vincent, K. B. Wolf. *J. Comput. Appl. Math.* 107, 73 (1999).

98   D. Atkinson, F. J. van Steenwijk. *Am. J. Phys.* 67, 486 (1999).

99   G. M. Attia, H. M. Nour. *Int. J. Appl. Math.* 10, 1 (2002).

100   C. Audet, P. Hansen, F. Messine. *J. Combinat. Th.* A 98, 46 (2002).

101   C. Audet, P. Hansen, F. Messine, S. Perron. *J. Combinat. Th.* A 108, 63 (2004).

102   G. Auletta. *Foundations and Interpretations of Quantum Mechanics*, World Scientific, Singapore, 2000.

103   L. Auslander, R. Tolimieri. *Bull. Am. Math. Soc.* 6, 847 (1979).

104   W. Auzinger, H. Stetter in R. P. Agarwal, Y. M. Chow, S. J. Wilson (eds.). *International Series of Numerical Mathematics 86*, Birkhäuser, Basel, 1988.

105   J. Avery. *Hyperspherical Harmonics*, Kluwer, Dordrecht, 1989.

106   J. Awrejcewicz, I. V. Andrianov, L. I. Manevitch. *Asymptotic Approaches in Nonlinear Dynamics*, Springer-Verlag, Berlin, 1998.

107   A. W. Babister. *Transcendental Functions Satisfying Nonhomogeneous Linear Differential Equations*, MacMillan, New York, 1967.

108   M. Baće, S. Ilijić, Z. Narančić, L. Bistričić. *Eur. J. Phys.* 23, 637 (2002).

109   P. Bachmann. *Die Lehre von der Kreistheilung*, Teubner, Leipzig, 1872.

110   P. Bachmann. *Math. Ann.* 18, 321 (1881).

111   B. Bagchi, F. Cannata, C. Quesne. *arXiv:quant-ph*/0003085 (2000).

112   V. G. Bagrov, D. M. Gitman. *Exact Solutions of Relativistic Wave Equations*, Kluwer, Dordrecht, 1990.

113   V. G. Bagrov, B. F. Samsonov, L. A. Shekoyan. *arXiv:quant-ph*/9804032 (1998).

114   L. Y. Bahar. *Am. J. Phys.* 59, 1103 (1991).

115   C. L. Bajaj in D. C. Handscomb (ed.). *The Mathematics of Surfaces*, Clarendon Press, Oxford, 1989.

116   C. L. Bajaj, I. Ihm. *ACM Trans. Graphics* 11, 61 (1992).

117   E. Baker. *Phys. Rev.* 36, 630 (1930).

118   G. A. Baker, Jr. *Quantitative Theory of Critical Phenomena*, Academic Press, Boston, 1990.

119   V. Baladi, Y. Jiang, H. H. Rugh. *mp_arc* 01-207 (2001).   http://rene.ma.utexas.edu/mp_arc/c/01/01-311.ps.gz

120   A. Balfagón, X. Jaén. *arXiv:gr-qc*/9912062 (1999).

121   W. Balser. *From Divergent Power Series to Analytic Functions*, Springer-Verlag, Berlin, 1994.

122   W. Band. *Found. Phys.* 18, 549 (1988).

123   S. Barnett, D. D. Siljak. *SIAM Rev.* 19, 472 (1977).

124   D. A. Barry, P. J. Culligan-Hensley, S. J. Barry. *ACM Trans. Math. Softw.* 21, 161 (1995).

125   G. Barton. *J. Phys.* A 14, 1009 (1981).

126   A. Barvinok. *arXiv:math.CO*/0007153 (2000).

127   M. V. Basilevsky, V. M. Ryaboy in I. Prigigine, S. A. Rice (eds.). *Advances in Chemical Physics* LXXXIV, Wiley, New York, 1993.

128   S. Baskal, Y. S. Kim. *arXiv:math-ph*/0401032 (2004).

129   G. Bassi, A. Bazzani, H. Mais, G. Truchetti. *Physica* A 347, 17 (2005).

130   S. Basu, R. Pollack, M.-F. Roy. *Algorithms in Real Algebraic Geometry*, Springer-Verlag, Berlin, 2003.

131   F. L. Bauer in P. Hilton, F. Hirzebruch, R. Remmert (eds.). *Miscellanea Mathematica*, Springer-Verlag, Berlin, 1991.

132   M. Beck. *arXiv:math.NT*/0112077 (2001).

133   M. Beck. *Acta Arithm.* 109, 109 (2003).

134   G. Beer, J. O. Watson. *Introduction to Finite and Boundary Element Methods for Engineers*, Wiley, West Sussex, 1992.

135   T. Becker, V. Weispfennig. *Gröbner Bases*, Springer-Verlag, New York, 1993.

136   H. Beker. *Found. Phys.* 23, 851 (1993).

137   V. V. Beletsky. *Essays on the Motion of Celestial Bodies*, Birkhäuser, Basel, 2001.

138   J. S. Bell. *Phys. World* 8, 33 (1990).

139   N. Belnap, L. E. Szabó. *arXiv:quant-ph*/9510002 (1995).

140   A. I. Beltzer. *Engineering Analysis with Maple/Mathematica*, Academic Press, London, 1995.

141   H. B. Benaoum. *J. Phys.* A 32, 2037 (1999).

142   C. M. Bender. *Phys. Rev.* D 16, 1740 (1977).

143   C. M. Bender, S. A. Orszag. *Advanced Mathematical Methods for Scientists and Engineers*, McGraw–Hill, New York, 1978.

144   C. M. Bender, K. A. Milton, S. S. Pinsky, L. M. Simmons. *J. Math. Phys.* 30, 1447 (1989).

145   C. M. Bender, G. V. Dunne, M. Moshe. *Phys. Rev.* A 55, 2625 (1997).

146   C. M. Bender, M. A. Bender, E. D. Demaine, S. P. Fekete. *J. Phys.* A 37, 147 (2004).

147   M. Beneke. *Phys. Rep.* 317, 1 (1999).

148   F. Benhamou, F. Goualard, E. Languénou, M. Christie. *arXiv:cs.AI*/0007002 (2000).

149   M. A. Bennett, B. Reznick. *arXiv:math.NT*/0209072 (2002).

150 F. Benoy, A. King, F. Mesnard. *arXiv:cs.PL/*0311002 (2003).

151 A. Benyi. *Fibon. Quart.* 40, 295 (2002).

152 A. Bérczes, L. Hajdu in K. Györy, A. Pethö, V. T. Sós (eds.). *Number Theory*, de Gruyter, Berlin, 1998.

153 C. Berg. *Ann. Prob.* 16, 910 (1988).

154 L. Berg. *Asymptotische Darstellungen und Entwicklungen*, Verlag der Wissenshaften, Berlin, 1968.

155 M. Berger, P. Pansu, J.-P. Berry, X. Saint-Raymond. *Problems in Geometry*, Springer-Verlag, New York, 1984.

156 H. L. Berk. *Am. J. Phys.* 69, 996 (2001).

157 B. C. Berndt. *Ramanujan's Notebooks* v.1, Springer-Verlag, New York, 1985.

158 B. C. Berndt, L.-C. Zhang. *J. Number Th.* 48, 224 (1994).

159 B. C. Berndt, Y.-S. Choi, S.-Y. Kang in B. C. Berndt, F. Gesztesy (eds.). *Continued Fractions: From Analytic Number Theory to Constructive Approximation*, American Mathematical Society, Providence, 1999.

160 M. V. Berry, K. E. Mount. *Rep. Progr. Phys.* 35, 315 (1972).

161 M. Bertero in P. W. Hawkes (ed.). *Advances in Electronics and Electron Physics* 75, Academic Press, New York, 1989.

162 E. Bertin, J.-M. Chassery in A. Le Méhauté, L. L. Schumaker (eds.). *Curves and Surfaces in Geometric Design*, A. K. Peters, Wellesley, 1994.

163 D. Bertrand, W. Zudilin. *arXiv:math.NT/*0006176 (2000).

164 G. Bertrand. *Compt. Rend.* 172, 1458 (1921).

165 I. Besieris, M. Abdel–Rahman. *Progr. Electromag. Res.* 19, 1 (1998).

166 I. M. Besieris, A. M. Shaarawi, L. P. Lightart. *J. Electromag. Waves Appl.* 14, 593 (2000).

167 H. Bethe in H. Geiger, K. Scheel (eds.). *Quantentheorie*, Handbuch der Physik 24.1, Springer-Verlag, Berlin, 1933.

168 H. A. Bethe, E. Salpeter. *Quantum Mechanics of One and Two Electron Atoms*, Springer-Verlag, Berlin, 1957.

169 G. Beylkin, M. J. Mohlenkamp. *Proc. Natl. Acad. Sci.* 99, 10246 (2002).

170 H. A. Biagoni. *A Nonlinear Theory of Generalized Functions*, Springer-Verlag, Berlin, 1990.

171 N. Biggs. *Algebraic Graph Theory*, Cambridge University Press, Cambridge, 1993.

172 P. Bikker, A. Y. Uteshev. *J. Symb. Comput.* 28, 45 (1999).

173 J. J. Binney, N. J. Dowrick, A. J. Fisher, M. E. J. Newman. *The Theory of Critical Phenomena*, Clarendon Press, Oxford, 1992.

174 T. Biswas, S. D. Joglekar. *J. Math. Phys.* 40, 369 (1999).

175 M. Bjelica. *FILOMAT* 9, 117 (1995).

176 B. Blaive, J. Metzger. *J. Math. Phys.* 25, 1721 (1984).

177 P. Blanchard, E. Brüning. *Mathematical Methods in Physics*, Birkhäuser, Boston, 2003.

178 L. Blanchet, G. Faye. *arXiv:gr-qc/*0004008 (2000).

179 L. Blanchet, G. Faye. *arXiv:gr-qc/*0006100 (2000).

180 L. Blanchet, T. Damour, G. Eposito–Farèse. *arXiv:gr-qc/*0311052 (2004).

181 S. Blanes, F. Casas, J. Ros. *BIT* 40, 434 (2000).

182 S. Blanes, P. C. Moan. *Phys. Lett.* A 265, 35 (2000).

183 D. Bleecker, G. Csordas. *Basic Partial Differential Equations*, International Press, Cambridge, 1998.

184 S. Bloch. *J. Algebra* 182, 476 (1996).

185 R. Blum, A. P. Guinand. *Can. Math. Bull.* 7, 399 (1964).

186 R. Blume–Kohout. *arXiv:quant-ph/*0408147 (2004).

187 L. E. Blumenson. *Am. Math. Monthly* 67, 63 (1960).

188 R. P. Boas. *Proc. Am. Math. Soc.* 3, 444 (1952).

189 R. P. Boas. *SIAM Rev.* 12, 116 (1970).

190 R. P. Boas, Jr., M. B. Marcus. *Am. Math. Monthly* 81, 760 (1974).

191 R. P. Boas, Jr. *Am. Math. Monthly* 84, 237 (1988).

192 D. Boccaletti, G. Pucacco. *Theory of Orbits*, Springer-Verlag, Heidelberg, 1996.

193 W. Boehm, H. Prautsch. *Geometric Concepts for Geometric Design*, A K Peters, Wellesley, 1993.

194 A. Böhm, M. Gadella. *Dirac Kets, Gamov Vectors and Gel'fand Triplets*, Springer-Verlag, Berlin, 1989.

195 A. Bohm, H.-D. Doebner, P. Kielanowski. *Irreversibility and Causality*, Springer-Verlag, Berlin, 1998.

196 A. Bohm, M. Gadella, S. Wickramasekara in I. Antoniou, G. Lumer (eds.). *Generalized Functions, Operator Theory, and Dynamical Systems*, Chapman & Hall, Boca Raton, 1999.

197 A. Bokulich. *Phil. Sci.* 70, 609 (2003).

198 L. A. Bokut', P. S.Kolesnikov. *J. Math. Sc.* 116, 2894 (2003).

199 V. C. Boltyanskii. *Envelopes*, Pergamon Press, New York, 1964.

200 J. Boos, P. Cass. *Classical and Modern Methods in Summability*, Oxford University Press, Oxford, 2000.

201 L. Borcea, V. Druskin. *Inverse Problems* 18, 979 (2002).

202 M. Bordag, G. L. Klimchitskaya, V. M. Mostepanenko. *Phys. Lett.* A 200, 95 (1995).

203 M. Bordag, U. Mohideen, V. M. Mostepanenko. *arXiv:quant-ph/*0106045 (2001).

204 V. A. Bordovitsyn, T. P. Pozdeeva. *Russ. Phys. J.* 46, 457

205 E. P. Borges, I. Roditi. *Phys. Lett.* A 246, 399 (1998).

206 F. Borghero, G. Bozis. *Meccanica* 37, 545 (2002).

207 V. V. Borisov, I. I. Simonenko. *Can. J. Phys.* 75, 573 (1997).

208 F. Bornemann. *arXiv:math.NA/*0211049 (2002).

209 A. Borodin, R. Fagin, J. Hopcroft, M. Tompa. *J. Symb. Comput.* 1, 169 (1985).

210 G. Boros, V. H. Moll. *J. Comput. Appl. Math.* 106, 361 (1999).

211 S. Borofsky. *Ann. Math.* 32, 23 (1931).

212 G. Boros, V. Moll. *Irresistible Integrals*, Cambridge University Press, Cambridge, 2004.

213 A. A. Borovkov. *Probability Theory*, Gordon and Breach, Amsterdam, 1998.

214 K. Borre. *Plane Networks and Their Applications*, Birkhäuser, Boston, 2001.

215 R. L. Borrelli, C. S. Coleman. *Differential Equations*, Wiley, New York, 1996.

216 D. Borwein, T. Markovich. *Proc. Am. Math. Soc.* 103, 1108 (1988).

217 J. M. Borwein, P. B. Borwein. *Pi and the AGM*, Wiley, New York, 1987.

218 P. B. Borwein, W. Chen, K. Dilcher. *Can. J. Math.* 47, 65 (1995).

219 M. Boshernitzan. *Ann. Math.* 124, 273 (1986).

220 H. Bosse, M. Groetschel, M. Henk. *arXiv:math.MG/*0307190 (2003).

221 A. Botero. *arXiv:quant-ph/*0306082 (2003).

222 S. Botrić. *Il. Nuov. Cim.* 111, 1161 (1996).

223 O. Bottema, Z. Djordjebić, R. R. Janić, D. S. Mitronović, P. M. Vasi. *Geometric Inequalities*, Wolters-Noordhoff Publishing, Groningen, 1969.

224 D. Bowman, J. Mc Laughlin. *arXiv:math.NT/*0403027 (2004).

225   D. Bowman, J. Mc Laughlin. *J. Comput. Appl. Anal.* 172, 363 (2004).

226   J. P. Boyd. *J. Comput. Phys.* 178, 118 (2002).

227   J. P. Boyd. *J. Comput. Phys.* 199, 688 (2004).

228   T. H. Boyer. *Am. J. Phys.* 71, 990 (2003).

229   T. B. Boykin. *Am. J. Phys.* 71, 462 (2003).

230   G. Bozis. *Celest. Mech.* 31, 129 (1983).

231   G. Bozis. *Inverse Problems* 11, 687 (2003).

232   G. Bozis, M.-C. Anisiu. *Inverse Problems* 21, 487 (2005).

233   R. M. Brach. *J. Appl. Mech.* 51, 164 (1984).

234   D. H. Bradley, D. J. Broadhurst. *arXiv:math.CA/*9906134 (1999).

235   D. M. Bradley, R. C. Gupta. *arXiv:math.ST/*0411298 (2004).

236   K. A. Brakke. *J. Geom. Anal.* 2, 11 (1992).

237   K. A. Brakke. *J. Geom. Anal.* 5, 455 (1996).

238   O. Bratteli, D. W. Robinson. *Operator Algebras and Quantum Statistical Mechanics*, Springer-Verlag, New York, 1987.

239   M. Braun, W. Schweizer, J. Linderberg. *Int. J. Quant. Chem. Symp.* 26, 717 (1992).

240   U. Brechtken-Manderscheid. *Introduction to the Calculus of Variations*, Chapman & Hall, London, 1991.

241   K. Briggs. *arXiv:math.CA/*0211142 (2002).

242   L. Bröcker. *Jahresber. DMV* 97, 130 (1995).

243   I. N. Bronshtein, K. A. Semandyayev. *Handbook of Mathematics*, Van Nostrand, New York, 1985.

244   M. Bronstein. *Symbolic Integration I*, Springer-Verlag, Berlin, 1997.

245   C. J. Brookes. *Am. Math. Monthly* 74, 578 (1967).

246   R. Broucke, H. Lass. *Celest. Mech.* 16, 215 (1977).

247   C. Brouder. *arXiv:hep-th/*9906111 (1999).

248   C. Brouder. *BIT* 44, 425 (2004).

249   B. A. Brown, A. R. Brown, F. Shlesinger. *J. Stat. Phys.* 110, 1087 (2003).

250   C. W. Brown. *J. Symb. Comput.* 31, 521 (2001).

251   W. S. Brown, J. F. Traub. *J. ACM* 18, 505 (1971).

252   R. W. Bruggeman. *Families of Automorphic Forms*, Birkhäuser, Basel, 1994.

253   Č. Brukner, M. Zukowski, A. Zeilinger. *arXiv:quant-ph/*0106119 (2001).

254   A. Brumberg. *Analytical Techniques of Celestial Mechanics*, Springer-Verlag, Heidelberg, 1995.

255   J. C. Brunelli. *Braz. J. Phys.* 30, 455 (2000).

256   S. Brunetti, A. Del Lungo. *Adv. Appl. Math.* 33, 487 (2004).

257   Y. A. Brychkov, A. P. Prudnikov. *Integral Transforms of Generalized Functions*, Gordon and Breach, New York, 1989.

258   B. Buchberger in N. K. Bose (ed.). *Multidimensional System Theory*, Reidel, Dordrecht, 1985.

259   B. Buchberger in J. R. Rice (ed.). *Mathematical Aspects of Scientific Software*, Springer-Verlag, New York, 1988.

260   B. Buchberger, G. E. Collins, B. Kutzler. *Ann. Rev. Comput. Sci.* 3, 85 (1988).

261   B. Buchberger, F. Winkler (eds.). *Gröbner Bases and Applications*, Cambridge University Press, Cambridge, 1998.

262   C. Buchta in A. Dold, B. Eckmann (eds.). *Zahlentheoretische Analysis*, Springer-Verlag, Berlin, 1983.

263   C. Buchta. *J. reine angew. Math.* 347, 212 (1984).

264   R. C. Buck. *J. Diff. Eq.* 41, 239 (1981).

265   A. Budó. *Theoretische Mechanik*, Verlag der Wissenschaften, Berlin, 1980.

266   X. Buff, C. Henriksen, J. H. Hubbard. *Exper. Math.* 10, 481 (2001).

267   P. S. Bullen. *A Dictionary of Inequalities*, Longman, Bath, 1998.

268   P. S. Bullen. *Handbook of Means and Their Inequalities* , Kluwer, Dordrecht, 2003.

269   P. Bundschuh. *Einführung in die Zahlentheorie*, Springer-Verlag, Berlin, 1988.

270   B. Buonaura. *Found. Phys.* 17, 627 (2004).

271   H. Burkhardt in H. Burkhardt, W. Wirtinger, R. Fricke (eds.). *Encyklopädie der mathematischen Wissenschaften*, Teubner, Leipzig, 1904–1908.

272   M. Burdick, H.-J. Schmidt. *J. Phys.* A 27, 579 (1994).

273   F. J. Bureau. *Commun. Pure Appl. Math.* 8, 143 (1955).

274   W. Bürger. *Bild der Wissenschaften* n10, 89 (1994).

275   R. Burioni, D. Cassi. *J. Phys.* A 38, R45 (2005).

276   D. S. Burnett. *Finite Element Analysis*, Addison-Wesley, Reading, 1987.

277   B. L. Burrows, R. F. Talbot. *Int. J. Math. Edu. Sci. Technol.* 27, 253 (1996).

278   P. Busch. *arXiv:quant-ph*/0110023 (2001).

279   W. Businger, P. A. Chevalier, N. Droux, W. Hett. *The Mathematica Journal* 4, n2, 70 (1994).

280   J. C. Butcher, G. Wanner. *Appl. Num. Math.* 22, 113 (1996).

281   J. C. Butcher. *Numerical Methods for Ordinary Differential Equations*, Wiley, Chichester, 2003.

282   E. I. Butikov. *Eur. J. Phys.* 24, L5 (2003).

283   J. Butterfield. *arXiv:quant-ph*/0210140 (2002).

284   P. L. Butzer, S. Jansche. *Fourier Anal. Appl.* 3, 325 (1997).

285   J. G. Byatt–Smith. *Stud. Appl. Math.* 105, 83 (2000).

286   V. Bykov, A. Kytmanov, M. Lazman, M. Passare (ed.). *Elimination Methods in Polynomial Computer Algebra*, Kluwer, Dordrecht, 1998.

287   F. W. Byron, R. W. Fuller. *Mathematics of Classical and Quantum Physics* v. 1, Addison Wesley, Reading, 1970.

288   A. Cabello. *arXiv:quant-ph*/0007065 (2000).

289   R. Caboz, J.-P. Codaccioni, F. Constantinescu. *Math. Meth. Appl. Sci.* 7, 416 (1985).

290   J.-L. Calais, W. Weyrich. *Int. J. Quant. Chem.* 63, 223 (1997).

291   E. Caliceti. *arXiv:math-ph*/0210029 (2002).

292   F. Calogero. *Commun. Appl. Math.* 3, 267 (1999).

293   H. E. Camblong, C. R. Ordóñez. *arXiv:hep-th*/0110176 (2001).

294   H. Camblong, C. R. Ordóñez. *arXiv:hep-th*/0305035 (2003).

295   W. B. Campbell. *Am. Math. Monthly* 44, 319 (1937).

296   J. A. Campball. *J. Comput. Phys.* 10, 308 (1972).

297   L. Cangelmi. *Rend. Sem. Mat. Univ. Pol. Torino* 53, 207 (1995).

298   J. F. Canny. *The Complexity of Robot Motion Planning*, MIT Press, Cambridge, 1988.

299   M. J. Caola. *J. Phys.* A 24, L1143 (1991).

300   E. C. Caparelli, V. V. Dodonov, S. S. Mizrahi. *arXiv:quant-ph*/9811016 (1998).

301   F. Capasso, C. Sirtori, J. Faist, D. L. Sivco, S.-N. G. Chu, A. Y. Cho. *Nature* 358, 565 (1992).

302   L. Caporaso, E. Sernesi. *arXiv:math.AG*/0008239 (2000).

303   C. Carathéodory. *Funktionentheorie*, Birkhäuser, Basel, 1950.

304   F. Caravaglios. *Nucl. Phys.* B 589, 475 (2000).

305   J. Carette in J. Gutierrez (ed.). *ISSAC 04*, ACM Press, New York, 2004.

306   J. F. Cariñena, M. F. Rañada, M. Santander. *arXiv:hep-th*/0501106 (2005).

307   T. Carletti. *mp_arc* 03-248 (2001).   http://rene.ma.utexas.edu/mp_arc/c/03/03-248.ps.gz

308   L. Carlitz. *Fibon. Quart.* 6, 193 (1968).

309   M. C. Carmen Domingo Juan, V. Miquel in N. Bokan, M. Djorić, A. T. Fomenko, Z. Rakić, J. Wess (eds.). *Contemporary Geometry and Related Topics*, World Scientific, New Jersey, 2004.

310   R. D. Carmichael. *Am. Math. Monthly* 15, 78 (1908).

311   J.-C. Carrega. *Théorie des corps*, Hermann, Paris, 1989.

312   W. B. Case. *Am. J. Phys.* 64, 215 (1996).

313   R. F. Caviness, J. R. Johnson (eds.). *Quantifier Elimination and Cylindrical Algebraic Decomposition*, Springer-Verlag, Wien, 1998.

314   A. Cayley. *Mess. Math.* 4, 69 (1875).

315   M. Ceberio, L. Granvilliers. *Computing* 69, 51 (2002).

316   A. Cedillo. *J. Math. Phys.* 34, 2713 (1993).

317   A. Celletti, G. Pinzari . *mp_arc* 05-15 http://dell5.ma.utexas.edu/mp_arc-bin/mpa?yn=05-15 (2005).

318   J. L. Cereceda. *arXiv:quant-ph*/0010191 (2000).

319   J. L. Cereceda. *arXiv:quant-ph*/0101143 (2001).

320   N. J. Cerf. *Phys. Rev. Lett.* 84, 4497 (2000).

321   V. M. Chabanov, B. N. Zakhariev, S. Brandt, H. D. Dahmen, T. Stroh. *Phys. Rev.* A 52, R 3389 (1995).

322   V. M. Chabanov. *J. Phys.* A 37, 9139 (2004).

323   K. Chadan, D. Colton, L. Päivärinta, W. Rundell. *An Introduction to Inverse and Inverse Spectral Problems*, SIAM, Philadelphia, 1997.

324   A. M. Chak. *Math. Comput.* 22, 673 (1968).

325   S. Chakrabarti, J. Pal, J. Shamanna, B. Taludar. *Czech. J. Phys.* 52, 853 (2002).

326   B. L. Chalmers, F. T. Metcalf. *SIAM J. Num. Anal.* 11, 950 (1974).

327   C. S. Chan, D. P. Robbins, D. S. Yuen. *arXiv:math.CO*/9810154 (1998).

328   A. Chandler, A. Heyworth. *arXiv:math.CO*/0002119 (2000).

329   S. Chandrasekhar. *Rev. Mod. Phys.* 15, 1 (1943).

330   M. E. Changa, A. V. S. Scherbinin, V. I. Pupyshev. *Int. J. Quant. Chem.* 96, 167 (2004).

331   F. Chapaton. *arXiv:math.QA*/0209104 (2002).

332   T. Chaundy. *The Differential Calculus*, Clarendon Press, Oxford, 1935.

333   L. V. Cheboratev. *J. Phys.* A 29, 7229 (1996).

334   E. S. Cheb-Terrab, A. D. Roche. *arXiv:math-ph*/0001037 (2000).

335   P. L. Chebyshev. *Collected Works*, v. 5, Academy of Science, Moscow, 1951.

336   C. Chen, F. Chen, Y. Feng. *Graphical Models* 63, 212 (2001).

337   F. Chen, D. Wang (eds.). *Geometric Computation*, World Scientific, Singapore, 2004.

338   G. Chen, Y. Mao, C. K. Chui. *Chaos, Solitons, Fractals* 21, 749 (2004).

339   H. Chen. *arXiv:quant-ph*/0110103 (2001).

340   R. Chen, Z. Xu, L. Sun. *Phys. Rev.* E 47, 3799 (1993).

341   Y. Chen. *SIAM J. Matrix Anal. Appl.* 22, 965 (2001).

342   H.-H. Chern. *arXiv:math.SP*/9902041 (1999).

343   C. Chicone. *Ordinary Differential Equations with Applications*, Springer-Verlag, New York, 1999.

344   E. Chicurel–Uziel. *Comput. Aided Geom. Design* 21, 23 (2004).

345   M. S. Child. *Semiclassical Mechanics with Molecular Applications*, Clarendon Press, Oxford, 1991.

346   M. S. Child, S.-H. Dong, X.-G. Wang. *J. Phys.* A 33, 5653 (2000).

347   L. Childs. *A Concrete Introduction to Higher Algebra*, Springer-Verlag, New York, 1992.

348   S. A. Chin, D. W. Kidwell. *arXiv:physics*/0006082 (2000).

349   S. A. Chin, C.-R. Chen. *arXiv:physics*/0012017 (2000).

350   S. A. Chin. *arXiv:physics*/0312005 (2003).

351   E.-W. Chionh, R. N. Goldman. *Visual Comput.* 8, 171 (1992).

352   S.-H. Chiu, T. K. Kuo. *arXiv:physics*/0010082 (2000).

353   S.-H. Chiu, T. K. Kuo. *J. Phys.* A 36, 555 (2003).

354   E. C. Cho. *Appl. Math. Lett.* 4, 51 (1991).

355   E. C. Cho. *Appl. Math. Lett.* 8, 71 (1995).

356   H.-W. Choi, J.-G. Shin. *Appl. Math. Comput.* 146, 257 (2003).

357   K.-M. Chong. *Coll. Math. J.* 13, 155 (1982).

358   Y. Choquet-Bruhat, C. DeWitt-Morette, M. Dillard-Bleick. *Analysis, Manifolds and Physics*, North Holland, Amsterdam, 1982.

359   S. C. Chou. *Mechanical Theorem Proving*, Reidel, Dordrecht, 1988.

360   S.-C. Chou, M. Rathi in D.-Z. Du, F. Hwang (eds.). *Computing in Euclidean Geometry*, World Scientific, Singapore, 1995.

361   A. Choudhry. *Lin. Alg. Appl.* 387, 183 (2004).

362   T. Y. Chow. *Am. Math. Monthly* 106, 440 (1999).

363   S.-C. Chou in D. Kueker, C. Smith (eds.). *Learning and Geometry: Computational Aspects*, Birkhäuser, Boston, 1996.

364   S. Christensen, L. Parker. *The Mathematica Journal* 1, n1, 51 (1990).

365   J. G. Christiano. *Am. Math. Monthly* 68, 149 (1961).

366   D. Chruscinski. *arXiv:math-ph*/0301024 (2003).

367   D. Chruscinski. *arXiv:math-ph*/0307047 (2003).

368   W. Chu, A. Marini. *Adv. Appl. Math.* 23, 115 (1999).

369   A. E. Chubykalo, R. A. Flores, J. A. Pérez. *arXiv:math.CA*/9906079 (1999).

370   A. E. Chubykalo, A. Espinoza. *J. Phys.* A 35, 8043 (2002).

371   P. S. Chudinov. *Eur. J. Phys.* 25, 73 (2004).

372   J. Chung, S. Lee. *Aequ. Math.* 65, 267 (2003).

373   P. G. Ciarlet, J. L. Lions. *Handbook of Numerical Analysis II, Finite Element Analysis* v. 1–3, Elsevier, Amsterdam, 1991.

374   G. M. Cicuta, M. L. Mehta. *arXiv:cond-mat*/0011136 (2000).

375   J. Cigler. *Körper, Ringe, Gleichungen*, Spektrum, Heidelberg, 1995.

376   M. A. Cirone, K. Rzazewski, W. P. Schleich, F. Straub, J. A. Wheeler. *arXiv:quant-ph*/0108069 (2001).

377   P. A. Clarkson. *J. Comput. Appl. Math.* 153, 127 (2003).

378   P. A. Clarkson, E. L. Mansfield. *Nonlinearity* 16, R1 (2003).

379   J. Cockle. *Phil. Mag.* 20, 145 (1860).

380   J. Cockle. *Phil. Mag.* 23, 135 (1862).

381   J. K. Cohen, D. R. DeBaun. *The Mathematica Journal* 2, n2, 62 (1992).

382   J. S. Cohen. *Computer Algebra and Symbolic Computation: Mathematical Methods*, A K Peters, Wellesley, 2003.

383   M. Colins. *Educ. Times* 13, 30 (1870).

384   S. M. Collart, S. M. Kalkbrenner, D. Mall. *J. Symb. Comput.* 24, 465 (1997).

385   D. Collins, N. Gisin, N. Linden, S. Massar, S. Popescu. *arXiv:quant-ph*/0106024 (2001).

386   G. E. Collins. *J. ACM* 14, 128 (1967).

387   G. E. Collins. *J. ACM* 18, 515 (1971).

388   P. Colwell. *Am. Math. Monthly* 99, 45 (1992).

389   J. F. Colombeau. *New Generalized Functions and Multiplication of Distributions*, North–Holland, Amsterdam, 1984.

390   J. F. Colombeau. *Multiplication of Distributions*, Springer-Verlag, Berlin, 1992.

391   G. Comenetz. *Am. J. Math.* 58, 225 (1936).

392   A. K. Common, F. Ebrahimi, S. T. Hafez. *J. Phys.* A 22, 3229 (1989).

393   L. Comtet. *C. R. Acad. Sci. Paris* A 270, 1085 (1970).

394   D. Condurache, M. H. Matcovschi. *Bul. Inst. Polit. Din Iasi* 46, n1/2, 14 (2000).

395   G. M. Constantine, T. H. Savits. *Trans. Am. Math. Soc.* 348, 503 (1996).

396   F. Constantinescu. *Distributions and Their Application in Physics*, Pergamon Press, Oxford, 1980.

397   J.-P. R. Conte in D. Benest, C. Froeschle (eds.). *An Introduction to Methods of Complex Analysis and Geometry for Classical Mechanics and Non-linear Waves*, Frontiers, France 1994.

398   R. Conte. *The Painlevé Property—One Century Later*, Springer-Verlag, New York, 1999.

399   R. Conte, M. Musette. *arXiv:nlin.SI*/0211048 (2002).

400   P. Conti, C. Traverso. *Buchberger's Algorithm and Integer Programming*, Springer-Verlag, Berlin, 1991.

401   R. Contino, A. Gambassi. *J. Math. Phys.* 44, 570 (2003).

402   G. Contopoulos, G. Bozis. *J. Inv. Ill-Posed Problems* 8, 147 (2000).

403   A. S. Coolidge, E. N. Maslen. *Phys. Rev.* 51, 855 (1937).

404   S. A. Coon, B. R. Holstein. *Am. J. Phys.* 70, 513 (2002).

405   F. Cooper, A. Khare, U. Sukhatme. *Phys. Rep.* 251, 267 (1995).

406   F. Cooper, J. M. Hymann, A. Khare. *Phys. Rev.* E 64, 026601 (2001).

407   S. Cooper, M. Hirschhorn. *Bull. Austral. Math. Soc.* 63, 353 (2001).

408   S. Cooper, S. Waldron. *J. Approx. Th.* 117, 103 (2002).

409   C. J. Corcoran. *J. Phys.* A 37, L461 (2004).

410   O. Cormier, M. F. Singer, B. M. Trager, F. Ulmer. *J. Symb. Comput.* 34, 355 (2002).

411   J. P. Costella, B. H. J. McKellar, A. A. Rawlinson. *arXiv:hep-ph*/0102244 (2001).

412   O. Costin, M. Kruskal. *Theor. Math. Phys.* 133, 1455 (2002).

413   W. E. Couch, R. J. Torrence. *J. Phys.* A 26, 5491 (1993).

414   D. Cox, J. Little, D. O'Shea. *Ideals, Varieties and Algorithms*, Springer-Verlag, New York, 1992.

415   D. A. Cox in D. A. Cox, B. Sturmfels (eds.). *Applications of Computational Algebraic Geometry,* American Mathematical Society, Providence, 1997.

416   H. S. M. Coxeter. *Am. Math. Monthly* 75, 5 (1968).

417   C. E. Creffield, W. Häusler, J. H. Jefferson, S. Sarkar. *Phys. Rev.* B 59, 10719 (1999).

418   R. Cross. *Am. J. Phys.* 72, 305 (2004).

419   J. Cserti. *arXiv:cond-mat*/9909120 (1999).

420   J. Cserti. *Am. J. Phys.* 68, 896 (2000).

421   B. Ćurgus, V. Mascioni. *arXiv:math.CA*/0502037 (2005).

422   S. M. Curry. *Am. J. Phys.* 44, 924 (1976).

423   A. Cusumano. *Coll. Math. J.* 34, 72 (2003).

424   E. Cuthill, J. McKee. in *Proceedings of the 24th National Conference of the ACM,* 157, Brandon Press, New York, 1969.

425   M. Czachor, H. Czyz. *arXiv:hep-th*/0110351 (2001).

426   J. P. Dahl, A. Wolf, W. P. Schleich. *Fortschr. Phys.* 52, 1118 (2004).

427   E. Dahlberg. *J. Phys.* A 6, 1800 (1973).

428   U. Dammert. *J. Phys.* A 26, 4785 (1993).

429   B. P. Damyanov. *arXiv:math.FA*/0203251 (2002).

430   J. M. A. Danby. *Celestial Mechanics* MacMillan, New York, 1988.

431   C. d'Andrea, A. Dickenstein. *arXiv:math.AG*/0007036 (2000).

432   C. d'Andrea. *arXiv:math.AG*/0101260 (2001).

433   C. D'Andrea, T. Krick, A. Szanto. *arXiv:math.AG*/0501281 (2005).

434   V. G. Danilov, V. M. Shelkovich. *arXiv:math-ph*/0012002 (2000).

435   V. G. Danilov. *arXiv:math-ph*/0105025 (2001).

436   A. Das. *arXiv:hep-th*/0110125 (2001).

437   B. K. Datta. *Indian J. Pure Appl. Math.* 34, 237 (2003).

438   G. Dattoli, C. Cesarano, D. Sacchetti. *Appl. Math. Comput.* 134, 595 (2003).

439   R. Dautray, J. L. Lions. *Mathematical Analysis and Numerical Methods for Science and Technology, v. 2 Functional and Variational Methods,* Springer-Verlag, Berlin, 1990.

440   J. H. Davenport. *On the Integration of Algebraic Functions,* Springer-Verlag, Berlin, 1981.

441   J. H. Davenport, Y. Siret, E. Tournier. *Computer-Algebra-Systems and Algorithms for Algebraic Computations,* Academic Press, London, 1993.

442   J. H. Davenport. *J. Symb. Comput.* 34, 259 (2002).

443   B. Davies. *Integral Transforms and Their Applications,* Springer-Verlag, New York, 1978.

444   K. T. R. Davies, R. W. Davies. *Can. J. Phys.* 67, 759 (1989).

445   K. T. R. Davies, R. W. Davies, G. D. White. *J. Math. Phys.* 31, 1356 (1990).

446   P. J. Davis. *Math. Comput.* 18, 569 (1964).

447   P. J. Davis. *Interpolation and Approximation,* Dover, New York, 1975.

448   A. A. Davydov, G. Ishikawa, S. Izumiya, W.-Z. Sun. *arXiv:math.DS*/0302134 (2003).

449   J. A. de Barros, P. Suppes. *arXiv:quant-ph*/0001034 (2000).

450   N. Debergh, J. Ndimubandi, B. Van den Bossche. *Ann. Phys.* 298, 361 (2002).

451   C. de Boor, K. Höllig. *Am. Math. Monthly* 98, 793 (1991).

452   C. de Boor. *arXiv:math.CA*/0502036 (2005).

453   N. G. De Bruijn. *Asymptotic Methods in Analysis*, North Holland, Amsterdam, 1961.

454   E. Y. Deeba, E. L. Koh. *Proc. Am. Math. Soc.* 116, 157 (1992).

455   E. Deeba, S. A. Khuri, S. Xie. *J. Comput. Phys.* 159, 125 (2000).

456   E. Deeba, S. Xie. *Appl. Math. Lett.* 16, 669 (2003).

457   E. Deeba, S. Xie. *Int. J. Math. Math. Sci.* 20, 1057 (2004).

458   S. R. de Groot. *The Maxwell Equations*, North-Holland, Amsterdam, 1969.

459   E. Delabaere in F. Chyzak (ed.). *Preprint INRIA* n 5003 (2003).   http://www.inria.fr/RRRT/RR-5003.html

460   R. de la Madrid, A. Bohm, M. Gadella. *arXiv:quant-ph*/0109154 (2001).

461   R. de la Madrid. *J. Phys.* A 35, 319 (2002).

462   R. de la Madrid. *arXiv:quant-ph*/0407195 (2004).

463   R. de la Madrid. *arXiv:quant-ph*/0502053 (2005).

464   A. C. de la Torre. *arXiv:quant-ph*/0109129 (2001).

465   A. C. de la Torre, D. Goyeneche. *Am. J. Phys.* 71, 49 (2002).

466   J. Delattre, R. Bkouche in Inter-IREM Commission (ed.). *History of Mathematics*, Ellipsis, Paris, 1997.

467   J. M. Délery. *Annu. Rev. Fluid Mech.* 33, 129 (2001).

468   R. T. Delves, G. S. Joyce. *Ann. Phys.* 291, 71 (2001).

469   S. De Marchi, M. M. Cecchi. *J. Comput. Appl. Math.* 57, 99 (1995).

470   C. Dembowski, B. Dietz, H.-D. Gräf, H. L. Harney, A. Heine, W. D. Weiss, A. Richter. *arXiv:nlin.CD*/0402015 (2004).

471   M. Demetrian. *arXiv:hep-th*/0204020 (2002).

472   E. D. Denman, A. N. Beavers. *Appl. Math. Comput.* 2, 63 (1976).

473   E. A. Desloge. *Classical Mechanics* v. 2, Wiley, New York, 1982.

474   C. P. Dettmann, E. G. D. Cohen. *arXiv:nlin.CD*/0001062 (2000).

475   P. Deuflhard, F. Bornemann. *Numerische Mathematik II*, de Gruyter, Berlin, 1994.

476   A. De Vos. *Open Sys. Inform. Dyn.* 9, 97 (2002).

477   L. Devroye. *Non-Uniform Random Variate Generation*, Springer-Verlag, New York, 1986

478   E. D. D'Hoker, R. Jackiw. *Phys. Rev.* D 26, 3517 (1982).

479   L. A. Dickey. *Soliton Equations and Hamiltonian Systems*, World Scientific, Singapore, 1991.

480   L. E. Dickson. *Algebraic Theories*, Dover, New York, 1926.

481   S. Dickson. *The Mathematica Journal* 1, n1, 38 (1990).

482   B. W. Dickinson, K. Steiglitz. *IEEE Trans. Acoust., Speech, Signal Proc.* 30, 25 (1982).

483   P. Dienes. *The Taylor Series* , Clarendon Press, Oxford, 1931.

484   U. Dierkes, S. Hildebrandt, A. Küster, O. Wohlrab. *Minimal Surfaces* v. I, II, Springer-Verlag, Berlin, 1991, 1992.

485   U. Dieter. *J. Number Th.* 18, 289 (1984).

486   K. Dilcher, L. A. Rubel. *J. Math. Anal. Appl.* 198, 98 (1996).

487   R. B. Dingle. *Asymptotic Expansions: Their Derivation and Interpretation*, Academic Press, London, 1973.

488   R. Z. Djordjević in T. M. Rassias, H. M. Srivastava (eds.). *Analytic and Geometric Inequalities and Applications*, Kluwer, Dordrecht, 1999.

489   I. K. Dmitrieva, G. I. Plindov. *Properties of Atoms and Ions in the Light of the Statistical Theory*, Nauka i Technika, Minsk, 1991.

490   G. A. Dobrovolsky, R. S. Tutik. *J. Phys.* A 33, 6593 (2000).

491   I. V. Dobrovolska, R. S. Tutik. *arXiv:quant-ph*/0108142 (2001).

492   M. P. do Carmo. *Differentialgeometrie von Kurven und Flächen*, Vieweg, Braunschweig, 1983.

493   A. Dolzmann, T. Sturm, V. Weispfenning in B. H. Matzat, G.-M. Greuel, G. Hiss (eds.). *Algorithmic Algebra and Number Theory*, Springer-Verlag, Berlin, 1998.

494   A. Dolzmann, T. Sturm, V. Weispfenning. *Technical Report MIP-9720*, FMI, University Passau (1999). http://www.fmi.uni-passau.de/~dolzmann/refs/MIP-9720.ps.Z

495   A Dolzmann. *Technical Report MIP-9903*, FMI, University Passau (1999). http://www.fmi.uni-passau.de/~dolzmann/refs/MIP-9903.ps.Z

496   A. Doménech, M. T. Doménech. *Eur. J. Phys.* 14, 177 (1993).

497   D. Dominici. *Int. J. Math. Math. Sci.* 58, 3699 (2003).

498   S.-H. Dong, X.-W. Hou, Z.-Q. Ma. *arXiv:quant-ph*/9808037 (1998).

499   S.-H. Dong, X.-W. Hou, Z.-Q. Ma. *arXiv:quant-ph*/9810056 (1998).

500   S.-H. Dong, Z.-Q. Ma. *arXiv:quant-ph*/9901036 (1999).

501   S.-H. Dong, Z.-Q. Ma. *arXiv:quant-ph*/9901037 (1999).

502   S.-H. Dong, Z.-Q. Ma, G. Esposito. *arXiv:quant-ph*/9902081 (1999).

503   S.-H. Dong. *Physica Scripta* 65, 289 (2002).

504   S.-H. Dong. *Found. Phys. Lett.* 15, 385 (2002).

505   D. Donnelly. *Am. J. Phys.* 60, 1149 (1992).

506   D. Donnelly. *Comput. Sc. Eng.* 4, n1, 92 (2002).

507   R. Donnelly, R. W. Ziolkowski. *Proc. R. Soc. Lond.* A 440, 541 (1993).

508   H. Dörrie. *Quadratische Gleichungen*, R. Oldenburg, München, 1943.

509   H. Dörrie. *Mathematische Miniaturen*, Sändin, Wiesbaden, 1969.

510   Z.-L. Dou, S. G. Staples. *Math. Mag.* 30, 44 (1999).

511   J. P. Dowling. *Math. Mag.* 62, 325 (1989).

512   A. J. Dragt. *Phys. Rev. Lett.* 75, 1946 (1995).

513   G. W. F. Drake in S. G. Karshenboim, F. S. Pavone, G. F. Bassani, M. Inguscio, T. W. Hänsch (eds.). *The Hydrogen Atom: Precision Physics of Simple Atomic Systems*, Springer-Verlag, Berlin, 2001.

514   V. Drakopoulos. *Chaos, Solitons, Fractals* 13, 233 (2002).

515   P. Dräxler, G. O. Michler, C. M. Ringel (eds.). *Computational Methods for Representations of Groups and Algebras* Birkhäuser, Basel, 2000.

516   A. Drăgulescu, V. M. Yakovenko. *arXiv:cond-mat*/0008305 (2000).

517   G. W. F. Drake. in J. D. Brown, M. T. Chu, D. C. Ellison, R. J. Plemmons (eds.). *Proc. Cornelius Lanczos International Centenary Conference*, SIAM, Philadelphia, 1994.

518   K. Drechsler, U. Sterz. *Acta Math. Univ. Comenianae* 68, 37 (1999).

519   R. M. Dreizler, E. K. U. Gross. *Density Functional Theory*, Springer-Verlag, Berlin, 1990.

520   D. J. Driebe, G. E. Ordóñez. *J. Stat. Phys.* 89, 1087 (1997).

521   M. Drmota, M. Skalba. *Acta Arith.* 71, 65 (1995).

522   Q.-K. Du. *Int. J. Numer. Meth. Eng.* 51, 1195 (2001).

523   B. Duan, X.-Y. Gu, Z.-Q. Ma. *Eur. J. Phys.* D 19, 9 (2002).

524   A. Dubickas. *Liet. Matem. Rink* 139, 310 (1999).

525   A. Dubickas in K. Györy, H. Iwaniec, J. Urbanowicz (eds.). *Number Theory in Progress*, de Gruyter, Berlin, 1999.

526   A. Dubickas, J. Steuding. *Elem. Math.* 59, 133 (2004).

527   D. A. Dubin, M. A. Hennings, T. B. Smith. *Mathematical Aspects of Weyl Quantization and Phase*, World Scientific, Singapore, 2000.

528   B. A. Dubrovin, A. T. Fomenko, S. P. Novikov. *The Geometry of Surfaces, Transformation Groups, and Fields*, Springer-Verlag, Berlin, 1992.

529   I. M. Duck, P. M. Stevenson, E. C. G. Sudarshan. *Phys. Rev.* D 40, 2112 (1989).

530   D. Dürr. *Bohmsche Mechanik als Grundlage der Quantenmechanik*, Springer-Verlag, Berlin, 2001.

531   R. J. Duffin. *Proc. Natl. Acad. Sci. USA* 78, 4661 (1981).

532   D. G. Duffy. *Transform Methods for Solving Partial Differential Equations*, CRC Press, Boca Raton, 1994.

533   D. S. Dummit. *Math. Comput.* 57, 387 (1991).

534   S. R. Dunbar. *Coll. J. Math.* 28, 187 (1997).

535   N. Dunford, J. T. Schwartz. *Linear Operators* v.2, Interscience, New York, 1963.

536   J. L. Dunham. *Phys. Rev.* 41, 721 (1932).

537   D. Dunn, B. Grieves. *J. Phys.* A 22, L1093 (1989).

538   G. V. Dunne, T. M. Hall. *arXiv:hep-th*/9902064 (1999).

539   G. V. Dunne, C. Schubert. *arXiv:hep-th*/9907290 (1999).

540   G. V. Dunne. *arXiv:hep-th*/0011036 (2000).

541   G. V. Dunne. *arXiv:hep-th*/0406216 (2004).

542   W. Dür. *arXiv:quant-ph*/0006105 (2000).

543   A. L. Durán, R. Estrada, R. P. Kanwal. *J. Math. Anal. Appl.* 218, 581 (1998).

544   R. Dutt, A. Gangopadhyaya, A. Khare, A. Pagnamenta, U. Sukhatme. *Phys. Rev.* A 48, 1845 (1993).

545   M. Dutta, L. Debnath. *Elements of the Theory of Elliptic and Associated Functions with Applications*, World Press Private, Calcutta, 1965.

546   D. L. Dwoyer, M. Y. Hussaini, R. G. Voigt (eds.) *Finite Elements: Theory and Applications*, Springer-Verlag, New York, 1988.

547   A. Eastaugh, P. van Nieuwenhuizen. *J. Math. Phys.* 30, 252 (1989).

548   M. S. P. Eastham, H. Kalf. *Schroedinger–Type Operators with Continuous Spectra*, Pitman, Boston, 1982.

549   W. H. Echols. *Ann. Math.* 10, 17 (1895).

550   F. E. Eckhardt. *Math. Ann.* 10, 227 (1876).

551   E. N. Economou. *Green Functions in Quantum Physics*, Springer-Verlag, Berlin, 1983.

552   A. Edery. *arXiv:math-ph*/0212059 (2002).

553   G. A. Edgar, D. Henslay, D. B. West (ed.). *Am. Math. Monthly* 109, 665 (2002).

554   G. A. Edgar, D. Hensley, D. B. West (eds.). *Am. Math. Monthly* 110, 340 (2003).

555   W. L. Edge. *Proc. R. Soc. Edinb.* 124 A, 729 (1994).

556   V. Edneral in V. G. Ganzha, E. W. Mayr, E. V. Vorozhtsov (eds.). *Computer Algebra in Scientific Computing*, Springer-Verlag, Berlin, 1999.

557   A. Edrei, E. B. Saff, R. S. Varga. *Zeros of Sections of Power Series*, Springer-Verlag, Berlin, 1983.

558   M. Edson, L. Q. Zamboni. *Theor. Comput. Sc.* 326, 241 (2004).

559   Y. V. Egorov. *Russ. Math. Surv.* 45, 1 (1990).

560   U. A. Eichmann, J. P. Draayer, A. Ludu. *arXiv:math-ph*/0201054 (2002).

561   R. Eid. *Int. J. Quant. Chem.* 71, 147 (1999).

562   B. Eisenberg, R. Sullivan. *Am. Math. Monthly* 107, 129 (2000).

563   T. Eisenberg. *Int. J. Math. Edu. Sci. Technol.* 34, 153 (2003).

564   D. Eisenbud. *Commutative Algebra with a View Towards Algebraic Geometry*, Springer-Verlag, New York, 1995.

565   J. Eisert, H. J. Briegel. *arXiv:quant-ph*/0007081 (2000).

566   G. Elber, M.-S. Kim. *ACM Trans. Graphics* 17, 32 (1998).

567   G. Elber, M.-S. Kim. *IEEE Comput. Graphics Appl.* 19, n6, 76 (1999).

568   G. Elber, M.-S. Kim in W. F. Bronsvoort, D. C. Anderson (ed.). *Proc. Fifth Symposium on Solid Modeling and Applications*, ACM Press, New York, 1999.

569   T. Elbouayachi. *IMI Research Reports* 00:04 (2000).   http://www.math.sc.edu/~imip/00papers/0004.ps

570   M. Elia in A. K. Agarwal, B. C. Berndt, C. F. Krattenthaler, G. L. Mullen, K. Ramachandra, M. Waldschmidt (eds.). *Number Theory and Discrete Mathematics*, Birkhäuser, Basel, 2002.

571   E. Elizalde, S. D. Odintsov, A. Romeo, A. A. Bytsenk, S. Zerbini. *Zeta Regularization Techniques with Applications*, World Scientific, Singapore, 1994.

572   E. Elizalde. *Ten Physical Applications of Spectral Zeta Functions*, Springer-Verlag, Berlin, 1995.

573   E. Elizalde. *arXiv:gr-qc*/0409076 (2004).

574   G. Elfving. *The History of Mathematics in Finland 1828-1918*, Frenckell, Helsinki, 1981.

575   E. Elizalde. *Nuov. Cim.* B 104, 685 (1989).

576   E. Elizalde. *arXiv:hep-th*/0309075 (2003).

577   C. Elsner. *J. Math. Anal. Appl.* 244, 533 (2000).

578   C. Elsner. *Comm. Appl. Nonlin. Anal.* 9, n3, 23 (2002).

579   B. C. Elsner. *Abh. Math. Seminar Univ. Hamburg.* 74, 33 (2004).

580   L. H. Encinas, J. M. Masque. *Appl. Math. Lett.* 16, 975 (2003).

581   A. Endler, J. A. C. Gallas. *Phys. Rev.* E 65, 036231 (2002).

582   B.-G. Englert. *Semiclassical Theory of Atoms*, Springer-Verlag, Berlin, 1988.

583   A. Enneper. *Elliptische Funktionen*, Verlag Louis Nebert, Halle, 1890.

584   L. N. Epele, H. Fanchiotti, C. A. G. Canal, J. A. Ponciano. *Phys. Rev.* A 60, 280 (1999).

585   A. Erdely. *Am. Math. Monthly* 58, 629 (1951).

586   P. Erdös. *Bull. Am. Math. Soc.* 52, 179 (1946).

587   T. Ernst. *Uppsala U.U.D.M. Report* 2000:16 (2000).

588   S. Esposito. *arXiv:physics*/0111167 (2001).

589   S. Esposito. *arXiv:math-ph*/0204040 (2002).

590   S. Esposito. *Int. J. Theor. Phys.* 41, 2417 (2002).

591   A. v. d. Essen. *Nieuw Archief Wiskunde* 11, 21 (1993).

592   J. G. Esteve. *Phys. Rev.* D 66, 125013 (2002).

593   R. Estrada, R. P. Kanwal. *J. Math. Anal. Appl.* 141, 195 (1989).

594   R. Estrada, R. P. Kanwal. *Asymptotic Analysis: A Distributional Approach*, Birkhäuser, Basel 1994.

595   R. Estrada, J. M. Garcia-Bondía, J. C. Várilly. *arXiv:funct-an*/9702001 (1997).

596   R. Estrada. *Proc. R. Soc. Lond.* A 454, 2425 (1998).

597   R. Estrada, S. A. Fulling. *J. Phys.* A 35, 3079 (2002).

598   G. Everest, A, van der Poorten, I. Shparlinski, T. Ward. *Recurrence Sequences*, American Mathematical Society, Providence, 2003.

599   J. Ewing, G. Schober. *Num. Math.* 61, 59 (1992).

600   H. Exton. *q-Hypergeometric Functions and Applications*, Ellis Horwood, Chichester, 1983.

601   D. B. Fairlie. *J. Phys.* A 37, 5375 (2004).

602   F. Farassat. *NASA Technical Paper* 3428 (1994).   http://techreports.larc.nasa.gov/ltrs/PDF/tp3428.pdf

603   F. Farassat. *NASA Technical Memorandum* 110285 (1996).   http://techreports.larc.nasa.gov/ltrs/PDF/tp3428.pdf

604   F. Farassat, M. K. Myers. *J. Sound Vibr.* 195, 340 (1996).

605   F. Farassat. *Theor. Comput. Fluid Dyn.* 10, 155 (1998).

606   C. Farina, J. J. Passos, A. C. Tort. *arXiv:hep-th*/0007201 (2000).

607   E. Farkas, M. Grosser, M. Kunzinger, R. Steinbauer. *arXiv:math.FA*/9912214 (1999).

608   H. M. Farkas, I. Kra. *Theta Constants, Riemann Surfaces and the Modular Group*, American Mathematical Society, Providence, 2001.

609   R. T. Farouki, C. A. Neff. *Comput. Aided Geom. Design* 7, 83 (1990).

610   R. T. Farouki, J. K. Johnstone. *Comput. Aided Geom. Design* 11, 117 (1994).

611   R. T. Farouki, R. Ramamurthy. *Int. J. Comput. Geom. Appl.* 8, 599 (1998).

612   R. T. Farouki, H. P. Moon and B. Ravani. *Adv. Comput. Math.* 13, 199 (2000).

613   R. T. Farouki, W. Gu, H. P. Moon in R. Martin, W. Wang (eds.). *Geometric Modeling and Processing 2000* IEEE Computer Society Press, New York, 2000.

614   R. T. Farouki, H. P. Moon in R. Cipolla, R. Martin (eds.). *The Mathematics of Surfaces IX*, Springer-Verlag, London, 2000.

615   R. T. Farouki. *Geom. Dedicata* 85, 283 (2001).

616   R. T. Farouki, H. P. Moon in H.-C. Hege, K. Polthier (eds.). *Visualizations and Mathematics III*, Springer-Verlag, Berlin, 2003.

617   R. T. Farouki, C. Y. Han. *Appl. Num. Math.* 51, 257 (2004).

618   R. T. Farouki, C. Y. Han. *Num. Alg.* 36, 13 (2004).

619   H. Fearn, R. H. Gibb. *arXiv:quant-ph*/0310059 (2003).

620   M. V. Fedoryuk. *Asymptotic Analysis*, Springer-Verlag, Berlin, 1993.

621   G.J. Fee, M. B. Managan. *SIGSAM Bull.* 31, n1, 22 (1997).

622   C. Fefferman, L. Seco. *Adv. Math.* 111, 88 (1995).

623   J. Feinberg. *arXiv:math-ph*/0402029 (2004).

624   N. I. Fel'dman, Y. V. Nesterenko in A. N. Parshin, I. R. Shafarevich (eds.). *Number Theory IV*, Springer-Verlag, Berlin, 1998.

625   A. Felikson, P. Tumarkin. *arXiv:math.MG*/0502167 (2005).

626   G. Feng, H. Ren, Y. Zhou in X.-S. Gao, D. Wang (eds.). *Mathematics Mechanization and Applications*, Academic Press, San Diego, 2000.

627   M. Feng. *Phys. Rev.* A 64, 034101 (2001).

628   X. Feng. *Int. J. Theor. Phys.* 39, 207 (2000).

629   Y. Feng, F. Chen, J. Deng, C. Chen, X. Tang in F. Chen, D. Wang (eds.). *Geometric Computation*, World Scientific, Singapore, 2004.

630   M. H. Fenrick. *Introduction to the Galois Correspondence*, Birkhäuser, Boston, 1998.

631   J. L. Fernández–Chapou, A. L. Salas–Brito, C. A. Vargas. *Am. J. Phys.* 72, 1109 (2004).

632   W. L. Ferrar. *A Text-Book of Convergence*, Clarendon Press, Oxford, 1969.

633   S. Filipp, K. Svozil. *arXiv:quant-ph*/0105083 (2001).

634 S. Finch. *Preprint* (2003).   http://people.bu.edu/srfinch/asym.pdf

635 B. Fine, G. Rosenberger. *The Fundamental Theorem of Algebra,* Springer-Verlag, New York, 1997.

636 N. Fine. *Basic Hypergeometric Series and Applications*, American Mathematical Society, Providence, 1988.

637 N. E. Firsova in A. A. Bytsenko, F. L. Williams (eds.). *Mathematical Methods in Physics*, World Scientific, Singapore, 2000.

638 A. Fischer. *J. reine angew. Math.* 11, 201 (1834).

639 G. Fischer (ed.). *Mathematical Models: Commentary*, Vieweg, Braunschweig, 1986.

640 I. Fischer. *Math. Mag.* 72, 148 (1999).

641 J. Fischer. *Int. J. Mod. Phys.* A 12, 3625 (1997).

642 W. Fischer, E. Koch. *Phil. Trans. R. Soc. Lond.* A 354, 2105 (1996).

643 A. D. Fitt, G.T.Q. Hoare. *Math. Gazette* 227, (1993).

644 P. Flajolett, P. J. Grabner, P. Kirschenhofer, H. Prodinger. *J. Comput. Appl. Math.* 58, 103 (1995).

645 H. Flanders. *J. Math. Anal. Appl.* 40, 30 (1972).

646 H. Flanders. *Am. Math. Monthly* 108, 559 (2001).

647 A. Flores–Riveros. *Int. J. Quant. Chem.* 66, 287 (1997).

648 S. Flügge, H. Marshall. *Practical Quantum Mechanics*, Springer-Verlag, Berlin, 1974.

649 A. Fodgen. *J. de Phys.* 51, Coll. C 7, 149 (1990).

650 A. T. Fomenko. *The Plateau Problem*, Gordon and Breach, New York, 1990.

651 S. Fomin. *Adv. Appl. Math.* 28, 119 (2002).

652 O. Föppl. *ZAMM* 2, 150 (1908).

653 G. W. Ford, R. F. O'Connel. *arXiv:quant-ph*/0301073 (2003).

654 L. H. Ford, N. F. Svaiter. *arXiv:quant-ph*/0204126 (2002).

655 W. B. Ford. *Am. J. Math.* 38, 397 (1916).

656 W. B. Ford. *Studies on Divergent Series and Summability and the Asymptotic Developments of Functions Defined by MacLaurin Series*, Chelsea, New York, 1960.

657 E. Forest. *Phys. Rev. Lett.* 75, 1946 (1995).

658 B. L. Foster. *Proc. R. Soc. Lond.* A 423, 443 (1989).

659 C. P. Frahm. *Am. J. Phys.* 51, 826 (1983).

660 J. S. Frame. *Am. Math. Monthly* 52, 377 (1945).

661 V. Franceschini, V. Grecchi, H. J. Silverstone. *Phys. Rev.* A 32, 1338 (1985).

662 A. P. French. *Am. J. Phys.* 61, 805 (1993).

663 R. Fricke. *Lehrbuch der Algebra*, 2 vol. Vieweg, Braunschweig, 1926.

664 H. M. Fried. *Functional Methods and Eikonal Models*, Editions Frontieres, Gif-sur-Yvette, 1990.

665 I. Fried, M. Chavez. *J. Sound Vibr.* 275, 415 (2004).

666 B. R. Frieden. *Physics from Fisher Information,* Cambridge University Press, Cambridge, 1998.

667 R. Frieden in E. Wolf (ed.). *Progress in Optics IX*, North Holland, Amsterdam, 1971

668 H. Friedrich. *Theoretical Atomic Physics*, Springer-Verlag, Berlin, 1998.

669 R. Fröberg. *An Introduction to Gröbner Bases*, Wiley, Chichester, 1997.

670 C. Frohlich. *Am. J. Phys.* 47, 583 (1979).

671 N. Fröman, P. O. Fröman. *Ann. Phys.* 83, 103 (1974).

672 N. Fröman, P. O. Fröman in S. A. Fulling, F. J. Narcowich (eds.). *Forty More Years of Ramifications: Spectral Asymptotics and Its Applications, Discourses in Mathematics and Its Applications*, n1, Department of Mathematics, Texas University, College Station, 1991.

673 N. Fröman, P. O. Fröman, N. Andersson, A. Hökback. *Phys. Rev.* D 45, 2609 (1992).

674 N. Fröman, P. O. Fröman. *Technique of the Comparison Equation Adapted to the Phase Integral Method*, Springer-Verlag, New York, 1994.

675 D. M. Fromm, R. N. Hill. *Phys. Rev.* A 36, 1013 (1987).

676 A. M. Fu, A. Lascoux. *arXiv:mat.CO*/0211406 (2002).

677 A. M. Fu, A. Lascoux. *arXiv:math.CO*/0404063 (2004).

678 B. Fuchssteiner. *Physica* D 4, 47 (1981).

679 W. I. Fushchich, W. M. Shtelen, N. I. Serov. *Symmetry Analysis and Exact Solutions of Equations of Nonlinear Mathematical Physics*, Kluwer, Dordrecht, 1993.

680 L. Gabet. *Comput. Math. Appl.* 27, 41 (1994).

681 H. G. Gadiyar, K. M. S. Maini, R. Padma, H. S. Sharatchanda. *arXiv:math-ph*/0206018 (2002).

682 T. Gál. *J. Phys.* A 35, 5899 (2002).

683 J. Galambos, I. Simonelli. *Products of Random Variables*, Marcel Dekker, New York, 2004.

684 E. A. Galapon. *Proc. R. Soc. Lond.* A 458, 451 (2002).

685 D. Gale. *Math. Intell.* 13, n1, 42 (1991).

686 R. V. Galiulin, S. N. Mikhalev, I. K. Sabitov. *Math. Notes* 76, 25 (2004).

687 S. Gallot, D. Hulin, J. Lafontaine. *Riemannian Geometry*, Springer-Verlag, Berlin, 1987.

688 D. V. Gal'tsov. *Phys. Rev.* D 66, 025016 (2002).

689 P. J. F. Gandy, S. Bardhan, A. L. Mackay, J. Klinowski. *Chem. Phys. Lett.* 336, 187 (2001).

690 P. J. F. Gandy, J. Klinowski. *J. Math. Chem.* 31, 1 (2002).

691 F. R. Gantmacher. *The Theory of Matrices*, Chelsea, New York, 1959.

692 F. R. Gantmacher. *Applications of the Theory of Matrices*, Interscience, New York, 1959.

693 X.-S. Gao, D.-K. Wang. *J. Geom.* 53, 79 (1995).

694 X.-S. Gao, D. Wang (eds.). *Mathematics Mechanization and Applications*, Academic Press, San Diego, 2000.

695 M. Gardner. *Fractal Music, Hypercards and More*, Freeman, New York, 1992.

696 G. W. Gardiner. *Handbook of Stochastic Methods*, Springer-Verlag, Berlin, 1985.

697 T. M. Garoni, N. E. Frankel. *J. Math. Phys.* 43, 2670 (2002).

698 F. Garvan. *Sém. Lothar. Combinat.* 42, B42d (1999).

699 M. Gasca in W. Dahmen, M. Gasca, C. A. Micchelli (eds.). *Computation of Curves and Surfaces*, Kluwer, Dordrecht, 1990.

700 Y. Gaspar. *arXiv:math.CA*/9810149 (1998).

701 V. Gasparian, G. Schoen, J. Ruiz, M. Ortuño. *Ann. Physik* 7, 756 (1998).

702 C. Gasquet, P. Witomski. *Fourier Analysis with Applications*, Springer-Verlag, New York, 1999.

703 L. Gatteschi in G. Allasia, L. Amerio, A. Conte, R. Conti, D. Galletto, L. Gatteschi, P. Germain, G. Grioli, E. Magenes, E. Marchi, C. Morawetz, S. Nocilla, O. Oleinik, R. Piva, G. Salvini, E. Vesentini (eds.). *Tricomi's Ideas and Contemporary Applied Mathematics*, Academia Nazionale dei Lincei, Rome, 1998.

704 C. F. Gauss. *Werke* v.10, Georg Olms Verlag, Hildesheim, 1981.

705 E.-U. Gekeler. *Arch. Math.* 77, 5 (2001).

706   B. R. Gelbaum, J.M.H. Olmstedt. *Theorems and Counterexamples in Mathematics*, Springer-Verlag, New York, 1990.

707   I. M. Gelfand, G. E. Schilow. *Generalized Functions*, v. I-V, Academic Press, New York, 1964.

708   I. M. Gelfand, M. Kapranov, A. Zelevinsky. *Discriminants, Resultants and Multidimensional Resultants*, Birkhäuser, Boston, 1994.

709   I. M. Gel'fand, L. Dikii. *Funct. Anal. Appl.* 10, 259 (1971).

710   N. George, A. Gamliel in H. N. Kritikos, D. L. Jaggard (eds.). *Recent Advances in Electromagnetic Theory*, Springer-Verlag, New York, 1990.

711   P. L. George. *Automatic Mesh Generation*, Wiley, Chichester, 1991.

712   L. Gerber. *Pac. J. Math.* 56, 97 (1975).

713   V. Gerdt, D. Yanovich, M. Znojil. *arXiv:math-ph*/0310012 (2003).

714   A. Gersten. *arXiv:physics*/9911028 (1999).

715   A. Gervois, M. L. Lehta. *J. Math. Phys.* 36, 5098 (1995).

716   I. Gessel. *J. Combinat. Th.* A 45, 178 (1987).

717   F. Gesztesy, R. Weikard in W. F. Ames, E. M. Harrell II, J. V. Herod (eds.). *Differential Equations with Applications to Mathematical Physics*, Academic Press, Boston, 1993.

718   F. Gesztesy, B. Simon. *Acta Math.* 176, 49 (1996).

719   F. Gesztesy, H. Holden. *Soliton Equations and Their Algebro-Geometric Solutions: (1+1)-Dimensional Continuous Models*, Cambridge University Press, Cambridge, 2003.

720   W. E. Gettys, J. R. Ray, E. Breitenberger. *Am. J. Phys.* 49, 162 (1981).

721   E. Ghate. *J. Ramanujan Math. Soc.* 15, 71 (2000).

722   S. Ghosh, J. Shamanna, B. Talukdar. *Can. J. Phys.* 82, 561 (2004).

723   N. E. Gibbs, W. G. Poole, Jr., P. K. Stockmeyer. *SIAM J. Num. Anal.* 13, 236 (1976).

724   W. R. Gibbs. *Computation in Modern Physics*, World Scientific, Singapore, 1994.

725   F. Gieres. *arXiv:quant-ph*/9907069 (1999).

726   S. Giller, P. Milczarski. *J. Phys.* A 32, 955 (1999).

727   S. Giller. *arXiv:quant-ph*/0107021 (2001).

728   D. T. Gillespie. *Am. J. Phys.* 49, 552 (1981).

729   J. C. Glashan. *Am. J. Math.* 4, 277 (1881).

730   M. L. Glasser. *Math. Comput.* 40, 561 (1983).

731   F. Gleisberg, W. Wonneberger. *arXiv:cond-mat*/0208376 (2002).

732   J. Glimm, A. Jaffe. *Quantum Physics, A Functional Integral Point of View*, Springer-Verlag, New York, 1981.

733   C. Godsil, G. Royle. *Algebraic Graph Theory*, Springer-Verlag, New York, 2001.

734   M. J. Goldberg, T. Hrycak, S. Kim. *J. Comput. Appl. Math.* 158, 473 (2003).

735   R. A. Goldin, V. Shtelen. *arXiv:quant-ph*/0006067 (2000).

736   J. R. Goldman. *The Queen of Mathematics*, A K Peters, Wellesley, 1998.

737   S. P. Goldman. *Phys. Rev. Lett.* 73, 2547 (1994).

738   W. Goldsmith. *Impact*, Edward Arnold, London, 1960.

739   G. Goldstein, D. Baye. *Phys. Rev.* E 70, 056703 (2004).

740   H. Goldstein. *Classical Mechanics*, Addison-Wesley, Reading, 1980.

741   O. Golinelli. *arXiv:cond-mat*/9707023 (1997).

742   G. H. Golub, C. F. Van Loan. *Matrix Computations*, Johns Hopkins University Press, Baltimore, 1989.

743   P. Gombas. *Die statistische Theorie des Atoms und ihre Anwendungen*, Springer-Verlag, Wien, 1949.

744   A. Gómez, J. D. Meiss. *arXiv:nlin.CD*/0304035 (2003).

745   G. Gómez, J. Llibre, R. Martínez, C. Simó. *Dynamics and Mission Design Near Libration Points*, World Scientific, Singapore, 2001.

746   B. Gönül, O. Özer, M. Kocak, D. Tutcu, Y. Cancelik. *arXiv:quant-ph*/0106142 (2001).

747   G. Gontopoulous, G. Bozis. *J. Inv. Ill-Posed Problems* 8, 147 (2000).

748   L. Gonzalez-Vega, G. Trujillo in A. H. M. Levelt (ed.). *ISSAC'95*, ACM Press, New York, 1995.

749   L. Gonzalez–Vega, I. Necula. *Comput. Aided Geom. Design* 19, 719 (2002).

750   E. A. Gonzáles-Velasco. *Fourier Analysis and Boundary Value Problems*, Academic Press, San Diego, 1995.

751   I. J. Good. *Proc. Cambr. Phil. Soc.* 56, 367 (1960).

752   I. J. Good. *Proc. R. Soc. Lond.* 307, 317 (1968).

753   F. O. Goodman. *Am. J. Phys.* 63, 82 (1995).

754   D. B. Goodner. *Am. Math. Monthly* 70, 303 (1963).

755   A. N. Gordeyev. *J. Phys.* A 8, 1048 (1975).

756   A. N. Gordeyev. *Teor. Mat. Fiz.* 105, 256 (1995).

757   G. B. Gostin. *Math. Comput.* 64, 393 (1995).

758   C. Gottlieb. *Math. Intell.* 21, n1, 31 (1999).

759   X. Gourdon, B. Salvy. *Discr. Math.* 153, 145 (1996).

760   U. Graf. *Applied Laplace Transforms and z-Transforms for Scientists and Engineeres*, Birkhäuser, Basel, 2004.

761   S. Graffi, V. Grecchi, B. Simon. *Phys. Lett.* B 32, 631 (1970).

762   R. L. Graham. *J. Combinat. Th.* A 18, 165 (1975).

763   P. Gralewicz, K. Kowalski. *arXiv:math-ph*/0002044 (2000).

764   A. Gray. *Modern Differential Geometry of Curves and Surfaces*, CRC Press, Boca Raton, 1993.

765   E. L. Green in K. G. Fischer, P. Loustaunau, J. Shapiro, E. L. Green, D. Farkas (eds.). *Computational Algebra*, Marcel Dekker, New York, 1994.

766   D. H. Greene, D. E. Knuth. *Mathematics for the Analysis of Algorithms*, Birkhäuser, Boston, 1990.

767   A. G. Greenhill. *Lond. Math. Soc. Proc.* 25, 195 (1894).

768   R. J. Gregorac. *Nieuw Archief Wiskunde* 9, 267 (1991).

769   W. Greiner, B. Müller, J. Rafelski. *Quantum Electrodynamic of Strong Fields*, Springer-Verlag, Berlin, 1985.

770   W. Greiner, J. Reinhardt. *Field Quantization*, Springer-Verlag, Berlin, 1996.

771   A. A. Grib, W. A. Rodrigues, Jr. *Nonlocality in Quantum Physics*, Kluwer, Dordrecht, 1999.

772   D. K. Gridnev, C. Greiner, W. Greiner. *arXiv:math-ph*/0502022 (2005).

773   H. B. Griffith. *Am. Math. Monthly* 88, 328 (1981).

774   D. Griffith, S. Walborn. *Am. J. Phys.* 67, 446 (1999).

775   S. Grigororiadou, G. Bozis, B. Elmabsout. *Celest. Mech. Dynam. Astron.* 74, 211 (1999).

776   C. W. Groetsch. *Inverse Problems in the Mathematical Sciences*, Vieweg, Braunschweig, 1993.

777   V. Gromak, I. Laine, S. Shimomura. *Painlevé Differential Equations in the Complex Plane*, de Gruyter, Berlin, 2002.

778   D. Gronau in W. Förg–Rob, D. Gronau, C. Mira, N. Netzter, G. Targonsky (eds.). *Iteration Theory*, World Scientific, Singapore, 1996.

779   M. Grötschel, M. Henk. *arXiv:math.MG*/0203268 (2002).

780   K. Große–Brauckmann, K. Polthier. *Exper. Math.* 6, 13 (1997).

781   N. Grossmann. *The Sheer Joy of Celestial Mechanics*, Birkhäuser, Basel, 1996.

782   L. C. Grove, J. M. McShane. *Expos. Math.* 12, 289 (1994).

783   R. Gruber (ed.). *Finite Elements in Physics*, North Holland, Amsterdam, 1987.

784   G. J. Grübl, C. Leubner. *Am. J. Phys.* 48, 484 (1980).

785   F. A. Grünbaum. *J. Math. Anal. Appl.* 88, 355 (1982).

786   J. Gruska, H. Imai in M. Margenstern, Y. Rogozhin (eds.). *Machines, Computations, and Universality*, Springer-Verlag, Berlin, 2001.

787   W. Gui, I. Babuška. *Num. Math.* 49, 577 (1986).

788   S.-L. Guo, F. Qi. *J. Anal. Appl.* 4, 1123 (1999).

789   N. Gurappa, P. K. Panighari, T. Shreecharan, S. R. Ranjani in B. G. Sidharth, M. V. Altaisky (eds.). *Frontiers of Fundamental Physics 4*, Kluwer, New York, 2001.

790   N. Gurappa, P. K. Panigrahi. *arXiv:quant-ph*/0204130 (2002).

791   V. Gurarii, V. Katsnelson. *Preprint NTZ* 3/2000 (2000).   http://www.uni-leipzig.de/~ntz/abs/abs0300.htm

792   M. Gürses, O. Sarioglu. *hep-th*/0303078 (2003).

793   J. Gutierrez, T. Recio. *Math. Comput. Simul.* 51, 441 (2000).

794   R. K. Guy. *Unsolved Problems in Number Theory*, Springer-Verlag, New York, 1994.

795   M. Gyllenberg, Y. Ping. *J. Math. Anal. Appl.* 264, 687 (2001).

796   J. M. Habeb, M. Hajja. *Expos. Math.* 21, 285 (2003).

797   W. Hackbusch (ed.). *Numerical Techniques for Boundary Element Methods*, Vieweg, Braunschweig, 1992.

798   S. Hackel, D. Heinemann, B. Fricke. *Z. Angew. Math. Mech.* 5, S 488 (1995).

799   H. Haf. *Höhere Mathematik für Ingenieure* v. V, Teubner, Stuttgart, 1993.

800   T. R. Hagedorn. *J. Algebra* 233, 704 (2000).

801   Y. Hagihara. *Celestial Mechanics* v. IV/1, MIT Press, Cambridge, 1972.

802   J. Hainz, H. Grabert. *arXiv:quant-ph*/9904103 (1999).

803   M. Hajja, P. Walker. *Am. Math. Monthly* 106, 963 (1999).

804   T. Hales. *arXiv:math.MG*/0205208 (2002).

805   T. Hales. *arXiv:math.RT*/0205207 (2002).

806   T. C. Hales in B. Aronov, S. Basu, J. Pach, M. Sharir (eds.). *Discrete and Computational Geometry*, Springer-Verlag, Berlin, 2003.

807   H. Halvorson. *arXiv:quant-ph*/0110102 (2001).

808   M. Hamanaka, K. Toda. *J. Phys.* A 36, 11981 (2003).

809   B. M. Hambly, J. Jordan. *Adv. Appl. Prob.* 36, 824 (2004).

810   G. Hamel. *Theoretische Mechanik*, Springer-Verlag, Berlin, 1967.

811   G. Hanrot, F. Morain in B. Mourrain (ed.). *ISSAC 2001*, ACM, Baltimore, 2001.

812   G. Hanrot, F. Morain. *Preprint INRIA* n 4109 (2001).   http://www.inria.fr/rrrt/rr-4109.html

813   E. R. Hansen. *A Table of Series and Products*, Prentice-Hall, Englewood Cliffs, 1975.

814   H. Harborth in P. L. Butzer, E. T. Jongen, W. Oberschelp (eds.). *Charlemagne and His Heritage—1200 Years of Civilization and Science in Europe*, Brepols, Turnhout, 1998.

815   J. Harnad. *arXiv:solv-int*/9902013 (1999).

816   F. E. Harris. *Phys. Rev.* A 55, 1820 (1997).

817 J. F. Harris. *Ph. D. Thesis*, University of Canterbury, 1999.

818 J. Harrison. *Theorem Proving with the Real Numbers*, Springer-Verlag, London, 1998.

819 E. Hartmann. *Visual Comput.* 17, 445 (2001).

820 P. Hartmann. *Am. Math. Monthly* 70, 255 (1963).

821 I. L. Hawk, D. L. Hardcastle. *J. Comput. Phys.* 21, 197 (1976).

822 J. He. *Appl. Math. Mech.* 23, 634 (2002).

823 J.-H. He. *Appl. Math. Comput.* 156, 591 (2004).

824 T.-X. He. *Dimensionality Reducing Expansion of Multivariate Integration*, Birkhäuser, Boston, 2001.

825 H. Heineken. *Ukrainian J. Math.* 54, 1212 (2002).

826 W. D. Heiss, W.-H. Steeb. *J. Math. Phys.* 32, 3003 (1991).

827 G. Helzer. *The Mathematica Journal* 5, n1, 67 (1995).

828 P. Henrici. *SIAM J. Num. Anal.* 3, 67 (1966).

829 P. Henrici. *Applied and Computational Complex Analysis*, v.1, Wiley, New York, 1974.

830 A. Henderson. *The Twenty Seven Lines upon the Cubic Surface*, Cambridge University Press, Cambridge, 1911.

831 P. Henrici. *Jahresber. DMV* 86, 115 (1984).

832 N. Henze. *J. Appl. Prob.* 720, 111 (1983).

833 A. Herdegen. *arXiv:hep-th/*0008207 (2000).

834 E. Herlt, N. Sali. *Spezielle Relativitätstheorie*, Akademie-Verlag, Berlin, 1978.

835 J. Hermes. *Nachrichten Königl. Gesell. Wiss. Göttingen* 170 (1894).

836 M. A. Hernández, M. A. Salanova. *Publ. Math. Debrecen* 53, 11 (1998).

837 H. Herold. *Phys. Rev.* A 48, 1916 (1993).

838 L. Herrera, A. Di Prisco. *Found. Phys. Lett.* 15, 373 (2002).

839 M. Herrmann. *Math. Ann.* 145, 256 (1962).

840 R. A. Herrmann. *arXiv:math.FA/*0403303 (2004).

841 R. Hersh. *Coll. Math. J.* 35, 112 (2004).

842 G. Hettner (eds.). *C. W. Borchardt's Gesammelte Werke*, Georg Reimer, Berlin, 1888.

843 N. J. Highham. *SIAM J. Sci. Stat. Comput.* 7, 1160 (1986).

844 A. Higuchi, A. Sudbery. *arXiv:quant-ph/*0005013 (2000).

845 F. B. Hildebrand. *Finite-Difference Equations and Simulations*, Prentice-Hall, Englewood Cliffs, 1968.

846 G. N. Hile, R. Z. Yeh. *Am. Math. Monthly* 95, 739 (1988).

847 E. Hille. *Ordinary Differential Equations in the Complex Domain*, Wiley, New York, 1976.

848 P. Hillion. *J. Phys.* A 26, 6021 (1993).

849 P. Hillion. *Eur. J. Phys.* B 30, 527 (2002).

850 H. Hilton. *Plane Algebraic Curves*, Oxford University Press, London, 1932.

851 E. J. Hinch. *Perturbation Methods*, Cambridge University Press, Cambridge, 1991.

852 D. B. Hinton, M. Klaus, J. K. Shaw. *Proc. Lond. Math. Soc.* 62, 607 (1991).

853 T. Hirayama, T. Hara. *Progr. Theor. Phys.* 103, 907 (2000).

854 J. E. Hirsch. *arXiv:cond-mat/*0109385 (2001).

855 J. W. P. Hirschfeld. *Rendiconti Mat.* 26, 115 (1967).

856   M. D. Hirschhorn. _Math. Mag._ 68, 199 (1995).

857   V. Hnizdo. _Eur. J. Phys._ 25, 351 (2004).

858   V. Hnizdo. _arXiv:physics_/0409072 (2004).

859   W. Hodge, D. Pedoe. _Methods of Algebraic Geometry_ II, Cambridge University Press, Cambridge, 1953.

860   D. Hoffman. _Math. Intell._ 9, n3, 8 (1987).

861   G. G. Hoffman. _J. Comput. Phys._ 116, 154 (1995).

862   C. M. Hoffman. _Geometric and Solid Modelling_, Morgan Kaufmann, San Mateo, 1989.

863   D. Hoffman, W. H. Meeks, III. _Am. Math. Monthly_ 97, 702 (1990).

864   D. Hoffman, F. Wei, H. Karcher. _Bull. Am. Math. Soc._ 29, 77 (1993).

865   D. Hoffman, H. Karcher in R. Osserman (ed.). _Geometry V_, Springer-Verlag, Berlin, 1997.

866   M. Hofri. _Analysis of Algorithms_, Oxford University Press, New York, 1995.

867   A. Hölder. _Math. Ann._ 38, 307 (1891).

868   P. Holland, H. R. Brown. _Stud. Hist. Phil. Mod. Phys._ 34, 161 (2003).

869   D. Home in A. M. Mitra (ed.). _Quantum Field Theory_, Hindustan Book Agency, New Dehli, 2000.

870   H. Hong, R. Liska, S. Steinberg. _Comput. J._ 36, 432 (1993).

871   H. Hong, R. Liska, S. Steinberg. _J. Symb. Comput._ 24, 161 (1997).

872   H. Hong, J. Schicho. _RISC preprints_ 97-09 (1997). ftp://ftp.risc.uni-linz.ac.at/pub/techreports/1997/97-09.ps.gz

873   K. I. Hopcraft, P. R. Smith. _An Introduction to Electromagnetic Inverse Scattering_, Kluwer, Dordrecht, 1992.

874   J. Horgan. _Scientific American_ n10, 92 (1993).

875   R. A. Horn, C. R. Johnson. _Matrix Analysis_, Cambridge University Press, Cambridge, 1985.

876   R. F. Hoskins, J. Sousa Pinto. _Distributions, Ultradistributions and Other Generalized Functions_, Ellis Horwood, New York, 1994.

877   X. Hou, H. Li, D. Wang, L. Yang. _Math. Intell._ 23, n1, 9 (2001).

878   A. S. Householder. _The Numerical Treatment of a Single Nonlinear Equation_, McGraw–Hill, New York, 1970.

879   K. B. Howell. _Principles of Fourier Analysis_, Chapman & Hall, Boca Raton, 2001.

880   Z. Hradil, J. Reháček. _arXiv:quant-ph_/0309184 (2003).

881   K. Huaizhong, B. Fisher. _Publ. Math. Debrecen_ 40, 279 (1992).

882   Q. Huang, T. A. Cruse. _Int. J. Numer. Meth. Eng._ 36, 2643 (1993).

883   H. P. Hudson in _Squaring the Circle and Other Monographs_, Chelsea, New York, 1953.

884   P. M. Hummel, C. L. Seebeck, Jr. _Am. Math. Monthly_ 56, 243 (1949).

885   D. J. Hurley, M. A. Vandyck. _J. Phys._ A 33, 6981 (2000).

886   A. Hurwitz, R. Courant. _Vorlesungen über allgemeine Funktionentheorie und elliptische Funktionen_, vol. 1, Springer-Verlag, Berlin, 1968.

887   E. A. Hylleraas, B. Undheim. _Z. Physik_ 65, 759 (1933).

888   E. A. Hylleraas. _Adv. Quant. Chem._ 1, 1 (1964).

889   T. Hyouguchi, R. Seto, M. Ueda, S. Adachi. _Ann. Phys._ 312, 177 (2004).

890   N. H. Ibragimov in W. F. Ames, E. M. Harrell II, J. V. Herod (eds.). _Differential Equations with Applications to Mathematical Physics_, Academic Press, Boston, 1993.

891   N. H. Ibragimov (ed.). _CRC Handbook of Lie Group Analysis of Differential Equations_, v. 1, CRC Press, Boca Raton, 1994.

892   N. H. Ibragimov. _Math. Intell._ 16, n1, 20 (1994).

893   I. Imai. *Applied Hyperfunction Theory*, Kluwer, Dordrecht, 1992.

894   N. Inui. *J. Phys. Soc. Jpn.* 72, 1035 (2003).

895   I. M. Isaacs. *Am. Math. Monthly* 92, 571 (1985).

896   E. L. Ince. *Ordinary Differential Equations*. Dover, New York, 1956.

897   D. Iannucci. *Am. Math. Monthly* 106, 778 (1999).

898   B. Z. Iliev. *arXiv:hep-th*/0102002 (2003).

899   S. Inawashiro, S. Katsura, Y. Abe. *J. Math. Phys.* 14, 560 (1973).

900   A. Iserles, A. Marthinsen, S. P. Nørsett. *BIT* 39, 281 (1999).

901   A. Iserles in V. B. Priezzhev, V. P. Spiridonov. *Self-Similar Systems*, Joint Institute for Nuclear Research, Dubna, 1999.

902   A. Iserles. *Found. Comput. Math.* 1, 129 (2001).

903   A. Iserles in R. A. DeVore, A. Iserles, E. Süli (eds.). *Foundations of Computational Mathematics*, Cambridge University Press, Cambridge, 2001.

904   M. Ishikawa, M. Wakayama. *Adv. Stud. Pure Appl. Math.* 28, 133 (2000).

905   J. N. Islam. *Found. Phys.* 24, 593 (1994).

906   J. Isralowitz. *arXiv:math.HO*/0411620 (2004).

907   M. Itskov. *ZAMM* 82, 535 (2002).

908   K. Iwasaki, H. Kimura, S. Shimomura, M. Yoshida. *From Gauss to Painlevé*, Vieweg, Braunschweig, 1991.

909   K. Iwasaki, H. Kawamuko. *Proc. Am. Math. Soc.* 127, 29 (1999).

910   K. Iwasaki, K. Kajiwara, T. Nakamura. *arXiv:nlin.SI*/0112043 (2001).

911   G. Iwata. *Natural Science Report, Ochanomizu University* 30, 17 (1979).

912   K. S. K. Iyengar. *Proc. Cambr. Phil. Soc.* 37, 9 (1941).

913   R. Jackiw, A. Shimony. *arXiv:physics*/0105046 (2001).

914   F. H. Jackson. *Mess. Math.* 38, 62 (1909).

915   J. D. Jackson in G. R. Screaton (ed.). *Dispersion Relations,* Oliver and Boyd, Edinburgh, 1961.

916   A. A. Jagers. *Elem. Math.* 35, 123 (1980).

917   O. Jahn, V. V. Sreedhar. *Am. J. Phys.* 69, 1039 (2001).

918   D. Jakobsson, S. D. Miller, I. Rivin, Z. Rudnick. in D. A. Hejhal, J. Friedman, M. C. Gutzwiller, A. M. Odlyzko (eds.). *Emerging Applications of Number Theory*, Springer-Verlag, New York, 1999.

919   M. Jameel. *J. Phys.* A 21, 1719 (1988).

920   J. F. James. *A Student's Guide to Fourier Transforms*, Cambridge University Press, Cambridge, 2002.

921   W. Janous. *Math. Bilten* 24, 57 (2000).

922   D. E. Jaramillo, N. Vanegas. *arXiv:physics*/0307106 (2003).

923   D. J. Jeffrey. *Math. Mag.* 67, 294 (1994).

924   D. J. Jeffrey, R. M. Corless, D. E. G. Hare, D. E. Knuth. *Compt. Rend.* 320, s1, 1449 (1995).

925   D. J. Jeffrey, A. D. Rich. *ACM Trans. Math. Softw.* 20, 124 (1994).

926   D. J. Jeffrey. *J. Symb. Comput.* 24, 563 (1997).

927   D. J. Jeffrey, A. D. Rich in M. J. Wester (ed.). *Computer Algebra Systems*, Wiley, Chichester, 1999.

928   D. Jeffrey, M. Giesbrecht, R. Corless. *ORCCA Technical Report* TR-00-19 (2000). http://www.orcca.on.ca/TechReports/2000/TR-00-19.html

929   O. D. Jefimenko. *Z. Naturf.* 54a, 637 (1999).

930   M. Jeng. *Am. J. Phys.* 68, 37 (2000).

931   R. D. Jenks, R. S. Sutor. *Axiom*, Springer-Verlag, Berlin, 1992.

932   C. U. Jensen, A. Ledet, N. Yui. *Generic Polynomials*, Cambridge University Press, Cambridge, 2002.

933   J. L. Jensen. *Acta Math.* 22, 359 (1899).

934   U. D. Jentschura. *arXiv:hep-ph*/0001135 (2000).

935   G. Jeronimo, J. Sabia. *J. Algebra* 227, 633 (2000).

936   T. Jiang, M. Li, P. Vitányi. *arXiv:math.CO*/9902043 (1999).

937   Y. Jin, B. Kalantari. *Adv. Appl. Math.* 34, 156 (2005).

938   C. R. Johnson, H. M. Shapiro. *SIAM J. Alg. Discr. Methods* 7, 627 (1986).

939   C. R. Johnson, K. Okubo, R. Reams. *Lin. Alg. Appl.* 323, 51 (2001).

940   D. W. Johnson. *J. Phys.* A 8, 490 (1975).

941   J. R. Johnson, R. Caviness (eds.). *Quantifier Elimination and Cylindrical Algebraic Decomposition*, Springer-Verlag, Wien, 1996.

942   W. P. Johnson. *Am. Math. Monthly* 109, 273 (2002).

943   E. H. Johnston (ed.). *Math. Mag.* 75, 317 (2002).

944   A. Jones, S. A. Morris, K. R. Pearson. *Abstract Algebra and Famous Impossibilities*, Springer-Verlag, New York, 1991.

945   D. S. Jones. *Proc. R. Soc. Lond.* A 371, 479 (1980).

946   D. S. Jones. *Math. Meth. Appl. Sci.* 19, 1017 (1996).

947   G. A. Jones, D. Singerman. *Complex Functions—An Algebraic and Geometric Viewpoint*, Cambridge University Press, Cambridge, 1987.

948   N. S. Jones, N. Linden. *Phys. Rev.* A 71, 012324 (2005).

949   D. M. Jordan, H. L. Porteous. *Am. Math. Monthly* 79, 587 (1972).

950   J. V. José, E. J. Saletan. *Classical Dynamics: A Contemporary Approach*, Cambridge University Press, Cambridge, 1998.

951   D. G. Joshi. *J. Indian Acad. Math.* 18, 199 (1996).

952   N. Joshi, K. Kajiwara, M. Mazzocco. *arXiv:nlin.SI*/0406035 (2004).

953   M. Jospehy in M. Behara, R. Fritsch, R. G. Lintz (eds.). *Symposia Gaussiana*, de Gruyter, Berlin, 1995.

954   J. J. Jou, C. M. Liu. *Chinese J. Math.* 21, 299 (1993).

955   J. Stat. Phys. 83 n1/2 (1996).

956   G. S. Joyce. *J. Phys.* A 36, 911 (2003).

957   G. S. Joyce, R. T. Delves, I. J. Zucker. *J. Phys.* A 36, 8661 (2003).

958   G. S. Joyce, R. T. Delves. *J. Phys.* A 37, 3645 (2004).

959   R. Jozsa, D. Robb, W. K. Wootters. *Phys. Rev.* A 49, 668 (1994).

960   *J. Symb. Comput.* 24, n2 (1997).

961   J. Jung, J. E. Alvarellos. *J. Chem. Phys.* 118, 10825 (2003).

962   V. Kac, P. Cheung. *Quantum Calculus*, Springer-Verlag, New York, 2002.

963   J. P. Kahane. *Bull. Lond. Math. Soc.* 29, 257 (1997).

964   W. Kahan, R. J. Fateman. *SIGSAM Bull.* 33, n2, 7 (1999).

965   P. B. Kahn, Y. Zarmi. *Am. J. Phys.* 72, 538 (2004).

966   S. W. Kahng. *Math. Comput.* 23, 621 (1969).

967   K. Kajiwara, T. Masuda. *arXiv:solv-int*/9903015 (1999).

968   C. M. Kalker-Kalkman. *Mech. Mach. Th.* 28, 523 (1993).

969   A. Kamanthai. *Southeast Asian Bull. Math.* 23, 627 (1999).

970   Y. Kametaka, M. Noda, Y. Fukui, S. Hirano. *Mem. Fac. Eng. Ehime Univ.* 9, 1 (1986).

971   E. Kamke. *Differentialgleichungssysteme: Lösungsmethoden und Lösungen* v. I, Geest & Portig, Leipzig, 1961.

972   E. Kamke. *Differentialgleichungen* v. 2, Geest & Portig, Leipzig, 1961.

973   D. W. Kammler. *A First Course in Fourier Analysis*, Prentice-Hall, Upper Saddle River, 2000.

974   T. R. Kane, D. A. Levinson. *Dynamics: Theory and Applications*, McGraw–Hill, New York, 1985.

975   A. Kaneko. *Introduction to Hyperfunctions*, Kluwer, Dordrecht, 1988.

976   M. Kaneko, N. Kurokawa, M. Wakayama. *arXiv:math.QA*/0206171 (2002).

977   G. Kaniadakis. *arXiv:cond-mat*/0111467 (2001).

978   R. P. Kanwal. *Generalized Functions*, Birkhäuser, Basel, 1998.

979   D. Kapur. *Artif. Intell.* 37, 61 (1988).

980   D. Kapur, Y. N. Lakshman in B. Donald, D. Kapur, J. Mundy (eds.). *Symbolic and Numerical Computation for Artificial Intelligence*, Academic Press, New York, 1992.

981   D. Kapur in V. Saraswat, P. Van Hentenryck (eds.). *Constraint Programming*, MIT Press, Cambridge, 1995.

982   D. Kapur in D. Wang (ed.). *Automated Deduction in Geometry*, Springer-Verlag, Berlin, 1997.

983   E. A. Karatsuba. *J. Comp. Appl. Math.* 135, 225 (2001).

984   E. A. Karatsuba. *J. Math. Phys.* 45, 4310 (2004).

985   H. Karcher, F. Wei, D. Hoffmann in K. Uhlenbeck (ed.). *Global Analysis in Modern Mathematics*, Publish or Perish, Houston, 1993.

986   H. Karcher, K. Polthier. *Phil. Trans. R. Soc. Lond.* A 354, 2077 (1996).

987   S. G. Karshenboim. *arXiv:hep-ph*/0007278 (2000).

988   T. Kashiwa, Y. Ohnuki, M. Suzuki. *Path Integral Methods*, Clarendon Press, Oxford, 1997.

989   E. Kasner. *Trans. Am. Math. Soc.* 8, 135 (1907).

990   T. Kasper. *Math. Mag.* 53, 195 (1980).

991   T. Kato. *Trans. Am. Math. Soc.* 70, 212 (1951).

992   S. Katsura, T. Morita, S. Inawashiro, T. Horiguchi, Y. Abe. *J. Math. Phys.* 12, 892 (1973).

993   G. Katz. *arXiv:math.AG*/0211281 (2002).

994   R. S. Kaushal. *Classical and Quantum Mechanics of Noncentral Potentials*, Narosa, New Delhi, 1998.

995   T. Kawarabayashi, T. Ohtsuki. *Phys. Rev.* B 51, 10897 (1995).

996   T. Kawarabayashi, T. Ohtsuki. *Phys. Rev.* B 53, 6975 (1996).

997   D. I. Kazakov, V. S. Popov. *JETP* 95, 581 (2002).

998   P. O. Kazinski, S. L. Lyakhovich, A. A. Sharapov. *Phys. Rev.* D 66, 025017 (2002).

999   F. Keck, H. J. Korsch, S. Mossmann. *J. Phys.* A 36, 2125 (2003).

1000   K. S. Kedlaya. *arXiv:math.CO*/9810127 (1998).

1001   H. B. Keller, V. Pereyra. *Math. Comput.* 32, 955 (1978).

1002   J. B. Keller. *J. Math. Phys.* 5, 548 (1964).

1003   J. B. Keller. *ASME J. Appl. Mech.* 53, 1 (1986).

1004   L. M. Kells. *Am. J. Math.* 46, 258 (1924).

1005   W. L. Kennedy. *Eur. J. Phys.* 23, 235 (2002).

1006   R. N. Kesarwani, Y. P. Varshni. *J. Math. Phys.* 21, 90 (1980).

1007   N. R. Kesarwani, Y. P. Varshni. *Can. J. Phys.* 58, 363 (1980).

1008   N. R. Kesarwani, Y. P. Varshni. *J. Math. Phys.* 22, 1983 (1981).

1009   N. R. Kesarwani, Y. P. Varshni. *J. Math. Phys.* 23, 92 (1982).

1010   J. Ketter, S. D. Riemenschneider, Z. Shen. *SIAM J. Math. Anal.* 25, 962 (1994).

1011   J. Kevorkian, J. D. Cole. *Perturbation Methods in Applied Mathematics*, Springer-Verlag, New York, 1981.

1012   O. Keyes. *Am. J. Phys.* 57, 378 (1989).

1013   I. M. Khalatnikov, A. Y. Kamenshchik. *arXiv:cond-mat*/0002181 (2000).

1014   I. R. Khan, R. Ohba. *J. Comput. Appl. Math.* 126, 269 (2000).

1015   A. B. Kharazishvili. *Strange Functions in Real Analysis*, Marcel Dekker, New York, 2000.

1016   K. Khare, N. George. *J. Opt. Soc. Am.* 21, 1179 (2004).

1017   A. G. Khovanskii. *Fields Inst. Commun.* 24, 325 (1999).

1018   A. G. Khovanskii. *Russ. Math. Surv.* 59, 661 (2004).

1019   S. A. Khuri. *J. Appl. Math.* 1, 141 (2001).

1020   J. Killingbeck. *Rep. Progr. Phys.* 40, 963 (1977).

1021   C. K. Kim, S. K. You. *arXiv:cond-mat*/0212557 (2002).

1022   M.-S. Kim, G. Elber in R. Cipolla, R. Martin (eds.). *The Mathematics of Surfaces IX*, Springer-Verlag, London, 2000.

1023   S. Kim, C. Lee, K. Lee. *arXiv:hep-th*/0110249 (2001).

1024   W. J. Kim. *Pac. J. Math.* 31, 717 (1969).

1025   T. Kinoshita. *Phys. Rev.* 105, 1490 (1957).

1026   L. M. Kirousis, Y. C. Stamatiou, M. Vamvakari. *Studies Appl. Math.* 107, 43 (2001).

1027   D. A. Kirschnitz, J. E. Losovik, G. V. Schpatakovski. *Usp. Fiz. Nauk* 117, 3 (1975).

1028   K. Kirsten. *arXiv:hep-th*/0005133 (2000).

1029   K. Kladko, P. Fulde. *Int. J. Quant. Chem.* 66, 377 (1998).

1030   G. Klambauer. *Am. Math. Monthly* 85, 668 (1978).

1031   V. Klee. *Am. Math. Monthly* 76, 286 (1969).

1032   A. Klein, M. Wessler. *arXiv:math.CO*/0212262 (2002).

1033   F. Klein. *Math. Ann.* 14, 551 (1873).

1034   P. Kleinert. *Proc. R. Soc. Lond.* A 435, 129 (1991).

1035   A. Kneser. *Math. Ann.* 41, 344 (1893).

1036   R. A. Knoebel. *Am. Math. Monthly* 88, 235 (1981).

1037   P. N. Knupp, S. Steinberg. *Fundamentals of Grid Generation*, CRC Press, Boca Raton, 1994.

1038   D. Knuth. *Math. Comput.* 61, 277 (1993).

1039   J. Kocik. *Am. J. Phys.* 67, 516 (1999).

1040   W. Koepf. *SIGSAM Bull.* 27, n1, 20 (1993).

1041   W. Koepf. *ZIB SC 93-27*, Berlin (1993). http://www.zib.de/PaperWeb/abstracts/SC-93-27

1042   W. Koepf. *Complex Variables* 25, 23 (1994).

1043   W. Koepf. *The Mathematica Journal* 4, n2, 62 (1994).

1044   W. Koepf. *AAECC* 7, 21 (1996).

1045 W. Koepf. *Math. Semesterber.* 44, 173 (1997).

1046 W. Koepf. *Hypergeometric Summation: An Algorithmic Approach to Summation and Special Function Identities*, Vieweg, Braunschweig, 1998.

1047 T. W. Körner. *Fourier Analysis*, Cambridge University Press, Cambridge, 1988.

1048 T. W. Körner. *A Companion to Analysis*, American Mathematical Society, Providence, 2003 .

1049 T. Koga, K. Matsui. *Z. Phys.* D 27, 97 (1993).

1050 E. K. Koh, C. K. Li. *Int. J. Math. Math. Sci.* 16, 749 (1993).

1051 E. K. Koh, L. C. Kuan. *Math. Nachr.* 157, 243 (1992).

1052 T. Koga. *J. Chem. Phys.* 92, 1276 (1992).

1053 W. Kohn. *Phys. Rev.* 115, 809 (1959).

1054 M. Kohno. *Global Analysis in Linear Differential Equations*, Kluwer, Dordrecht, 1999.

1055 K. Kojima, K. Mitsunaga, K. Kyuma. *Appl. Phys. Lett.* 55, 882 (1989).

1056 J. E. Kolassa. *Series Approximation Methods in Statistics*, Springer-Verlag, New York, 1994.

1057 J. A. C. Kolk. *Indag. Math.* 14, 445 (2003).

1058 K. Kommerell. *Das Grenzgebiet der elementaren und höheren Mathematik*, Koehler, Leipzig, 1936.

1059 T. H. Koornwinder. *arXiv:math.CA/9403216* (1994).

1060 R. H. J. Kop, P. de Vries, R. Sprik, A. Lagendijk. *Opt. Commun.* 138, 118 (1997).

1061 V. I. Korobov. *arXiv:physics/9912052* (1999).

1062 R. A. Kortram. *Technical Report* 9942/99 University of Nijmegen (1999). http://www-math.sci.kun.nl/math/onderzoek/reports/rep1999/rep9942.ps.gz

1063 Y. N. Kosovtsov. *arXiv:math-ph/0207032* (2002).

1064 G. A. Kotel'nikov. *arXiv:physics/9802038* (1998).

1065 G. A. Kotel'nikov. *arXiv:physics/0012018* (2000).

1066 T. C. T. Kotiah. *Int. J. Math. Edu. Sci. Technol.* 24, 863 (1993).

1067 I. S. Kotsireas in F. Chen, D. Wang (eds.). *Geometric Computation* World Scientific, Singapore, 2004.

1068 P. Kovács. *Rechnergestüzte symbolische Roboterkinematik*, Vieweg, Wiesbaden, 1993.

1069 P. Kovács, G. Hommel in J. Angeles, G. Hommel, P. Kovács (eds.). *Computational Kinematics*, Kluwer, Dordrecht, 1993.

1070 P. Kovács in J. Fleischer, J. Grabmeier, F. W. Hehl, W. Küchlin (eds.). *Computer Algebra in Science and Industry*, World Scientific, Singapore, 1995.

1071 V. Kowalenko. *J. Phys.* A 31, L 663 (1998).

1072 K. Kowalski, K. Podlaski, J. Rembielinksi. *arXiv:quant-ph/0206176* (2002).

1073 C. Krattenthaler. *Trans. Am. Math. Soc.* 305, 431 (1988).

1074 C. Krattenthaler. *arXiv:math.CO/9902004* (1999).

1075 M. Kreuzer and L. Robbiano. *Computational Commutative Algebra*, Springer-Verlag, Berlin 2000.

1076 J. B. Krieger, M. L. Lewis, C. Rosenzweig. *J. Chem. Phys.* 47, 2942 (1967).

1077 J. B. Krieger, C. Rosenzweig. *Phys. Rev.* 164, 171 (1967).

1078 A. Kriegl, M. Losik, P. W. Michor. *ESI Preprint* 1214 (2002).   http://www.esi.ac.at/Preprint-shadows/esi1214.html

1079 M. Krizek, J. Chleboun. *Math. Bohemica* 119, 437 (1994).

1080 M. Krízek, F. Luca, L. Somer. *17 Lectures on Fermat Numbers*, Springer-Verlag, New York, 2001.

1081 S. Krishnan, D. Manocha in A.H.M. Levelt (ed.). *ISSAC'95*, ACM Press, New York, 1995.

1082 R. Krivec, V. B. Mandelzweig. *Comp. Phys. Commun.* 138, 69 (2001).

1083 R. Krivec, V. B. Mandelzweig. *Comp. Phys. Commun.* 152, 165 (2003).

1084 R. Krivec, V. B. Mandelzweig. *arXiv:math-ph*/0406023 (2004).

1085 J. Król. *Found. Phys.* 34, 843 (2004).

1086 A. R. Krommer, C. W. Ueberhuber. *Computational Integration*, SIAM, Philadelphia, 1998.

1087 C. A. Kruelle, A. Kittel, J. Peinke, R. Richter. *Z. Naturf.* 52a, 581 (1997).

1088 M. D. Kruskal, N. Joshi, R. Halburd. *arXiv:solv-int*/9710023 (1997).

1089 G. Krylov, M. Robnik. *J. Phys.* A 33, 1233 (2000).

1090 D. Kueker, C. Smith (eds.). *Learning and Geometry*, Birkhäuser, Basel, 1996.

1091 M. Kuga. *Galois' Dream*, Birkhäuser, Basel, 1993.

1092 C. N. Kumar, P. K. Panigrahi. *arXiv:solv-int*/9904020 (1999).

1093 K. Kumar. *Eur. J. Phys.* 20, 501 (1999).

1094 B. A. Kupershmidt. *J. Nonlin. Math. Phys.* 7, 244, (2000).

1095 K. Kurokawa. *IEEE Trans. Antennas Prop.* 49, 1315 (2001).

1096 N. Kurokawa, M. Wakayama. *Indag. Math.* 13, 63 (2002).

1097 N. Kurokawa, H. Ochiai, M. Wakayama. *J. Ramanujan Math. Soc.* 17, 101 (2002).

1098 P. Kustaanheimo, E. Steifel. *J. reine angew. Math.* 218, 204 (1965).

1099 H. R. Kutt. *Num. Math.* 24, 205 (1974).

1100 G. F. Kventsel, J. Katriel. *Phys. Rev.* A 24, 2299 (1981).

1101 P. K. Kythe. *Fundamental Solutions for Differential Operators and Applications*, Birkhäuser, Basel, 1996.

1102 P. K. Kythe, P. Puri. *Computational Methods for Linear Integral Equations.*, Birkhäuser, Boston, 2002.

1103 P. K. Kythe, D. Wei. *An Introduction to Linear and Nonlinear Finite Element Analysis*, Birkhäuser, Boston, 2004

1104 M. L. Lakshmanan, P. Kaliappan, K. Larsson, F. Karlsson, P. O. Fröman. *Phys. Rev.* A 49, 3296 (1994).

1105 M. Lakshmanan, S. Rajasekar. *Nonlinear Dynamics*, Springer-Verlag, Berlin, 2003.

1106 T. V. Lakshminarasimhan. *Am. Math. Monthly* 72, 877 (1965).

1107 D. Laksov, A. Thorup. *Trans. Am. Math. Soc.* 351, 1293 (1999).

1108 S. K. Lamoreaux. *Am. J. Phys.* 67, 850 (1999).

1109 P. Lancaster, M. Tismenetsky. *The Theory of Matrices*, Academic Press, Orlando, 1985.

1110 L. D. Landau, E. M. Lifschitz. *Course of Theoretical Physics, v. I, Mechanics*, Pergamon Press, Oxford, 1982.

1111 S. Landau. *Math. Intell.* 16, n2, 49 (1994).

1112 S. Landau. *SIAM J. Comput.* 21, 85 (1992).

1113 S. K. Lando. *Lectures on Generating Functions*, American Mathematical Society, Providence, 2003.

1114 D. Langbein. *J. Phys.* A 10, 1031 (1977).

1115 R. E. Langer. *Bull. Am. Math. Soc.* 40, 545 (1934).

1116 E. M. Langley. *Math. Gaz.* 11, 173 (1922).

1117 A. Larraza, B. Denardo. *Phys. Lett.* A 248, 151 (1998).

1118 A. Lascoux. *Symmetric Functions and Combinatorial Operators on Polynomials*, American Mathematical Society, Providence, 2003.

1119 D. Laugwitz. *Differential and Riemannian Geometry*, Academic Press, New York, 1965.

1120 B. J. Laurenzi. *J. Math. Phys.* 31, 2535 (1990).

1121 N. Lauritzen. *Concrete Abstract Algebra*, Cambridge University Press, Cambridge, 2003.

1122 E. K. Lavrovskii, A. M. Formalskii. *J. Appl. Math. Mech.* 57, 311 (1993).

1123 D. Lazard. *J. Symb. Comput.* 5, 261 (1988).

1124 S. Leader. *Am. Math. Monthly* 93, 348 (1986).

1125 M. Le Bellac, J.-M. Levy-Leblond. *Nuov. Cim.* B 14, 217 (1973).

1126 M. H. Lee. *Can. J. Phys.* 73, 106 (1995).

1127 J. Lekner. *J. Opt.* A 6, 711 (2004).

1128 S. K. Lenczowski, R. J. M. van de Veerdonk, M. A. M. Gijs, J. B. Giesbers, H. H. J. M. Janssen. *J. Appl. Phys.* 75, 5154 (1994).

1129 J. Lenells. *J. Phys.* A 38, 869 (2005).

1130 J. Lepowski. *arXiv:math.QA*/9909178 (1999).

1131 J. Lesko. *Coll. Math. J.* 35, 171 (2004).

1132 D. Levi, P. Winternitz (eds.). *Painlevé Transcendents*, Plenum Press, New York, 1992.

1133 D. Levi, P. Tempesta, P. Winternitz. *J. Math. Phys.* 45, 4077 (2004).

1134 T. Levi–Civita. *Oper. Math.* 2, 411 (1956).

1135 G. Levin, M. Sodin, P. Yuditskii. *Commun. Math. Phys.* 141, 119 (1991).

1136 R. H. Lewis, P. F. Stiller. *Math. Comput. Simul.* 49, 205 (1999).

1137 S. Liao. *Beyond Perturbation*, Chapman & Hall, Boca Raton, 2004.

1138 H. Li, F. Van Oystaeyen. *A Primer of Algebraic Geometry*, Marcel Dekker, New York, 1999.

1139 S.-J. Liao. *J. Indian Math. Soc.* 66, 125 (1999).

1140 R. L. Liboff. *Kinetic Theory*, Prentice Hall, Englewood Cliffs, 1986.

1141 D. Lichtblau. *The Mathematica Journal* 6, n4, 81 (1996).

1142 D. Lichtblau. *Worldwide Mathematica Conference 1998*, Chicago, 1998.
http://library.wolfram.com/conferences/conference98/abstracts/various_ways_to_tackle_algebraic_equations.html

1143 S. Lie, E. Study, F. Engel. *Beiträge zur Theorie der Differentialinvarianten*, Teubner, Stuttgart, 1993.

1144 A. Liebetegger. *Math. Gaz.* 34, 84 (1950).

1145 I. K. Lifanov, L. N. Poltavskii, G. M. Vainikko. *Hypersingular Integral Equations and Their Applications*, Chapman & Hall, Boca Raton, 2004.

1146 C. Lin, L. Zhipeng. *arXiv:math.HO*/0408082 (2004).

1147 J. Linderberg. *Comput. Phys. Rep.* 6, 209 (1987).

1148 H. Lippmann. *Angewandte Tensorrechnung*, Springer-Verlag, Berlin, 1993.

1149 V. D. Liseikin. *Grid Generation Methods*, Springer-Verlag, Berlin, 1999.

1150 R. Liska, S. Steinberg. *Comput. J.* 36, 497 (1993).

1151 P. Lisoněk, R. B. Israel in C. Traverso (ed.). *Proc. 2000 ISSAC*, ACM Press, New York, 2000.

1152 Z.-G. Liu. *Ramanujan J.* 5, 129 (2001).

1153 L. Llovet, R. Martínez, J. A. Jaén. *J. Comput. Appl. Math.* 49, 145 (1993).

1154 S. Lloyd. *arXiv:quant-ph*/9704013 (1997).

1155 A. W. Lohmann in J. C. Dainty (ed.). *Current Trends in Optics*, Academic Press, London, 1994.

1156 Z. A. Lomnicki. *Adv. Appl. Prob.* 4, 109 (1972).

1157 M. Longuet–Higgins. *Proc. R. Soc. Lond.* A 428, 283 (1990).

1158 J. L. Lopéz, N. M. Temme. *arXiv:math.CA*/0205064 (2002).

1159   J. L. López, N. M. Temme. *arXiv:math.CA*/0410436 (2004).

1160   G. Lopez. *Rev. Mex. Fis.* 48, 10 (2002).

1161   J. Lopuszanski. *The Inverse Variational Problem In Classical Mechanics*, World Scientific, Singapore, 1999.

1162   R. A. Lorentz. *Multivariate Birkhoff Interpolation*, Springer-Verlag, New York, 1992.

1163   E. R. Loubenets. *Phys. Rev.* A 69, 042102 (2004).

1164   D. Lovett. *Demonstrating Science with Soap Films*, Institute of Physics, Bristol, 1994.

1165   A. Ludu, G. Stoitcheva, J. P. Draayer. *arXiv:math-ph*/0003030 (2000).

1166   A. Ludu, R. F. O'Connell, J. P. Draayer. *arXiv:nlin.PS*/0008026 (2000).

1167   J.-G. Luque, J.-Y. Thibon. *arXiv:math.CO*/0204026 (2002).

1168   J. Lützen. *The Prehistory of the Theory of Distributions*, Springer-Verlag, New York, 1982.

1169   G. Lvai, B. W. Williams. *J. Phys.* A 26, 3301 (1993).

1170   W.-X. Ma, K. Maruno. *Physica* A 343, 219 (2004).

1171   Z.-Q. Ma, A.-Y. Dai. *arXiv:physics*/9905051 (1999).

1172   R. Mabry. *Am. Math. Monthly* 110, 59 (2003).

1173   M. A. H. MacCallum in A. M. Cohen, L. van Gastel, S. Verduyn Lunel (eds.). *Computer Algebra for Industry 2: Problem Solving in Practice*, Wiley, Chichester, 1995.

1174   A. L. Mackay. *J. Math. Chem.* 21, 197 (1997).

1175   G. Mackiw. *Math. Mag.* 73, 44 (2000).

1176   W. D. MacMillan. *Am. J. Math.* 37, 95 (1915).

1177   F. S. Maculay. *Algebraic Theory of Modular Systems*, Cambridge University Press, Cambridge, 1916.

1178   R. Maeder. *The Mathematica Journal* 2, n2, 25 (1990).

1179   J. R. Magnus, H. Neudecker. *Matrix Differential Calculus with Applications in Statistics and Econometrics*, Wiley, Chichester, 1988

1180   V. Majerník, E. Majerníkova. *J. Phys.* A 35, 5751 (2002).

1181   L. G. Maharam, E. P. Shaugnessy. *Coll. Math. J.* 7, 38 (1976).

1182   K. Mahler. *Proc. R. Soc. Lond.* A 378, 155 (1981).

1183   K. Mahler. *Proc. R. Soc. Lond.* A 389, 1 (1983).

1184   A. A. Mailybaev, O. N. Kirillov, A. P. Seyranian. *arXiv:math-ph*/0411024 (2004).

1185   A. Makhlouf, R. Chemlal. *Int. J. Diff. Eq. Appl.* 5, 283 (2002).

1186   G. Malajovich, J. P. Zubelli. *J. Complexity* 17, 541 (2001).

1187   G. Malajovich, J. P. Zubelli. *Num. Math.* 89, 749 (2001).

1188   I. Malakhov, P. Silaev, K. Sveshnikov. *arXiv:hep-th*/0501187 (2005).

1189   E. A. Malsch, G. Dasgupta. *Int. J. Numer. Meth. Eng.* 61, 1153 (2004).

1190   V. A. Malyshev. *St. Petersburg Math. J.* 13, 893 (2002).

1191   S. Mandal. *Phys. Lett.* A 305, 37 (2002).

1192   S. Mandal, P. K. Mukherjee, G. H. F. Diercksen. *J. Phys.* B 36, 4483 (2003).

1193   V. B. Mandelzweig. *J. Math. Phys.* 40, 6266 (1999).

1194   V. B. Mandelzweig, F. Tabakin. *Comp. Phys. Commun.* 141, 268 (2001).

1195   V. B. Mandelzweig. *arXiv:quant-ph*/0409137 (2004).

1196   D. Mannion. *Adv. Appl. Prob.* 26, 577 (2000).

1197   D. Manocha, J. F. Canny. *Comput. Aided Geom. Design* 9, 25 (1992).

1198   D. Manocha. Ph. D. thesis, Berkeley, 1992.

1199   E. B. Manoukian, N. Yongram. *Int. J. Theor. Phys.* 41, 1327 (2002).

1200   L. A. Manzoni, W. F. Wreszinski. *arXiv:hep-th/*0101128 (2001).

1201   V. N. Marachevsky. *Physica Scripta* 64, 205 (2001).

1202   N. H. March. *Adv. Phys.* 6, 1 (1957).

1203   N. H. March in S. Lundquist, N. H. March (eds.). *Theory of the Inhomogeneous Electron Gas*, Plenum, New York, 1983.

1204   V. A. Marchenko. *Sturm-Liouville Operators and Applications*, Birkhäuser, Basel, 1986.

1205   E. Marchisotto, G.-A. Zakeri. *Coll. Math. J.* 25, 295 (1994).

1206   D. Marker. *Notices Am. Math. Soc.* 43, 753 (1996).

1207   S. Marmi. *arXiv:math.DS/*0009232 (2000).

1208   J. E. Marsden, T. S. Ratiu. *Introduction to Mechanics and Symmetry*, Springer-Verlag, New York, 1999.

1209   A. Martin, J.-M. Richard, T. T. Wu. *Phys. Rev.* A 46, 3697 (1992).

1210   A. Martin, J.-M. Richard, T. T. Wu. *Phys. Rev.* A 52, 2557 (1995).

1211   A. Martin. *arXiv:quant-ph/*0411196 (2004).

1212   Y. Martinez–Maure. *Demonstratio Math.* 34, 59 (2001).

1213   E. V. Martyushev. *arXiv:math.MG/*0011022 (2000).

1214   M. Marucho. *arXiv:hep-th/*0402209 (2004).

1215   T. Masuda. *arXiv:nlin.SI/*0302026 (2003).

1216   A. M. Mathai. *A Handbook of Generalized Special Functions for Statistical and Physical Sciences*, Clarendon Press, Oxford, 1993.

1217   A. M. Mathai. *An Introduction to Geometrical Probability*, Gordon and Breach, Amsterdam, 1999.

1218   Y. Matiyasevich. *Math. Comput. Simul.* 67, 125 (2004).

1219   L. Mattner. *Doc. Math. J. DMV* 4, 601 (1999).

1220   V. B. Matveev. *Inverse Problems* 17, 633 (2001).

1221   H. A. Mavromatis. *Am. J. Phys.* 68, 287 (2000).

1222   M. Mazzocco. *Math. Ann.* 321, 157 (2001).

1223   A. C. McBride, F. H. Kerr. *IMA J. Appl. Math.* 39, 159 (1987).

1224   J. McCleary. *Geometry from a Differentiable Viewpoint*, Cambridge University Press, Cambridge, 1994.

1225   S. McCallum in W. Bosma, A. van der Poorten (eds.). *Computational Algebra and Number Theory*, Kluwer, Amsterdam, 1995.

1226   M. McCartney. *Eur. J. Phys.* 18, 90 (1997).

1227   P. R. McCreary, T. J. Murphy, C. Carter. *The Mathematica Journal* 9, 564 (2004).

1228   C. McCune. *Quart. J. Math.* 52, 329 (2001).

1229   D. G. C. McKeon. *Irish Math. Soc. Bull.* 42, 45 (1999).

1230   R. McLachlan. *Math. Intell.* 16, n4, 31 (1994).

1231   R. McLachlan, G. R. W. Quispel. *Acta Numerica* 11, 341 (2002).

1232   T. C. McMillan. *Coll. Math. J.* 34, 11 (2003).

1233   J. M. McNamee. http://pigeon.elsevier.nl/mcs/section/cam/mcnamee/index.html.

1234   J. P. Mc Tavish. *Eur. J. Phys.* 21, 229 (2000).

1235   J. G. McWhirter, I. K. Proudler. *Mathematics in Signal Processing*, Clarendon Press, Oxford, 1998.

1236   K. Mätzel, K. Nehrkorn. *Formelmanipulation mit dem Computer*, Akademie-Verlag, Berlin, 1985.

1237   M. M. de Souza. *arXiv:hep-th*/9708096 (1997).

1238   M. W. Meckes. *arXiv:math.MG*/0305411 (2003).

1239   A. Meder. *Monatsh. Math. Phys.* 14, 349 (1903).

1240   E. Meeron. *J. Chem. Phys.* 27, 1238 (1957).

1241   M. L. Mehta. *J. Math. Phys.* 28, 781 (1987).

1242   M. L. Mehta. *Random Matrices*, Academic Press, Boston, 1991.

1243   A. Meir, A. Sharma. *SIAM J. Num. Anal.* 5, 488 (1968).

1244   H. Melenk in J. Fleischer, J. Grabmeier, F. W. Hehl, W. Küchlin (eds.). *Computer Algebra in Science and Industry*, World Scientific, Singapore, 1995.

1245   D. Mendlovic, H. M. Ozaktas, A. W. Lohmann. *Opt. Commun.* 105, 36 (1994).

1246   J. C. Merino. *Coll. Math. J.* 34, 122 (2003).

1247   J. Merker. *arXiv:math.DG*/0411650 (2004).

1248   N. D. Mermin. *Am. J. Phys.* 58, 731 (1990).

1249   A. Messiah. *Quantum Mechanics*, North Holland, Amsterdam, 1976.

1250   N. Meyer-Vernet. *Am. J. Phys.* 50, 354 (1982).

1251   R. E. Mickens. *Difference Equations*, Van Nostrand, New York, 1990.

1252   R. E. Mickens. *Application of Nonstandard Finite Difference Schemes*, World Scientific, Singapore, 2000.

1253   A. E. Middleditch, C. M. P. Reade, A. J. Gomes. *Int. J. Shape Model.* 6, 175 (2000).

1254   B. Mielnik. *J. Math. Phys.* 27, 2290 (1988).

1255   M. Mignotte in B. Buchberger, G. E. Collins, R. Loos, R. Albrecht (eds.). *Computer Algebra: Symbolic and Algebraic Computations*, Springer-Verlag, Wien, 1983.

1256   M. Mignotte, D. Stefănescu. *Polynomials*, Springer-Verlag, Singapore, 1999.

1257   T. Y. Mikhailova, V. I. Pupyshev. *J. Phys.* A 31, 4263 (1998).

1258   T. Y. Mikhailova, V. I. Pupyshev. *Phys. Lett.* A 257, 1 (1999).

1259   G. Mikhalkin. *Ann. Math.* 151, 309 (2000).

1260   G. Mikhalkin. *arXiv:math.AG*/0108225 (2001).

1261   G. Mikhalkin. *arXiv:math.AG*/0403015 (2004).

1262   G. V. Milanović, T. M. Rassias in T. M. Rassias (ed.). *Survey on Classical Inequalities*, Kluwer, Dordrecht, 2000.

1263   W. P. Milloni. *The Quantum Vacuum*, Academic Press, Boston, 1994.

1264   L. M. Milne-Thomson. *The Calculus of Finite Differences*, MacMillan, London, 1951.

1265   R. Milson *arXiv:solv-int*/9706007 (1997).

1266   P. Milczarski. *arXiv:quant-ph*/9807039 (1998).

1267   G. A. Miller. *Am. Math. Monthly* 14, 213 (1907).

1268   P. W. Milonni. *J. Phys.* B 35, R31 (2002).

1269   K. A. Milton, Y. J. Ng. *Phys. Rev.* E 57, 5504 (1998).

1270   K. A. Milton. *arXiv:hep-th*/9901011 (1999).

1271   E. Minguzzi. *arXiv:physics*/0411233 (2004).

1272   R. N. Miroshin. *Math. Notes* 70, 682 (2001).

1273 R. L. Mishkov. *Int. J. Math. Math. Sci.* 24, 481 (2000).

1274 B. Mishra. *Algorithmic Algebra*, Springer-Verlag, New York, 1993.

1275 B. Mishra in J. E. Goodman, J. O'Rourke (eds.). *Handbook of Discrete and Computational Geometry*, CRC Press, Boca Raton, 1997.

1276 T. P. Mitchell, D. L. Pope. *J. SIAM.* 10, 49 (1962).

1277 D. S. Mitronović, J. E. Pečarić, V. Volenec. *Recent Advances in Geometric Inequalities*, Kluwer, Dordrecht, 1989.

1278 D. S. Mitronović, J. D. Keckic. *The Cauchy Method of Residues*, v. 1 Reidel, Dordrecht, 1984.

1279 D. S. Mitronović, J. D. Keckic. *The Cauchy Method of Residues*, v. 2 Kluwer, Dordrecht, 1993.

1280 A. Mitzscherling. *Das Problem der Kreisteilung*, Teubner, Leipzig, 1913.

1281 S. S. Mizrahi, D. Galetti. *J. Phys.* A 35, 3535 (2002).

1282 B. L. Moiseiwitsch in D. R. Bates (ed.). *Quantum Theory* v.1, Academic Press, New York, 1961.

1283 P. Mohazzabi, J. C. Fields. *Can. J. Phys.* 82, 197 (2004).

1284 V. Moll. *Notices Am. Math. Soc.* 49, 311 (2002).

1285 S. Momami. *Int. J. Mod. Phys.* C 15, 967 (2004).

1286 D. Moncrieff, S. Wilson. *Chem. Phys. Lett.* 209, 423 (1993).

1287 E. W. Montroll in E. F. Beckenbach (ed.). *Applied Combinatorial Mathematics*, Wiley, New York, 1981.

1288 I. C. Moore, M. Cada. *Appl. Comput. Harmon. Anal.* 16, 208 (2004).

1289 W. M. Moore, J. H. Steffen, P. E. Boynton. *arXiv:physics*/0412102 (2004).

1290 T. Mora. *Solving Polynomial Equation Systems*, v. 1, Cambridge University Press, Cambridge, 2003.

1291 J. Morales, G. Arreaga, J. J. Pena, J. Lopez-Bonilla. *Int. J. Quant. Chem. Symp.* 26, 171 (1992).

1292 D. A. Morales. *J. Math. Phys.* 35, 3916 (1994).

1293 J. D. Morales–Guzmán, J. Morales, J. J. Peña. *Int. J. Quant. Chem.* 85, 239 (2001).

1294 P. Morandi. *Field and Galois Theory*, Springer-Verlag, New York, 1996.

1295 A. Morgan. *Solving Polynomial Systems Using Continuation for Engineering and Scientific Problems*, Prentice-Hall, Englewood Cliffs, 1987.

1296 H. Moritz, B. Hofmann-Wellenhof. *Geometry, Relativity, Geodesy*, Whichmann, Karlsruhe, 1993.

1297 R. E. Moritz. *Am. Math. Monthly* 14, 1 (1907).

1298 L. Moser, J. R. Pounder. *Can. Math. Bull.* 5, 70 (1962).

1299 V. M. Mostepanenko, N. N. Trunov. *The Casimir Effect and Its Applications*, Clarendon Press, Oxford, 1997.

1300 G. E. Moyano, J. L. Villaveces. *Int. J. Quant. Chem.* 71, 121 (1999).

1301 C. Müller. *Analysis of Spherical Symmetries in Euclidean Spaces*, Springer-Verlag, New York, 1998.

1302 M. Müller, D. Schleicher. *arXiv:math.GM*/0502109 (2005).

1303 R. A. Muller. *Am. J. Phys.* 60, 313 (1992).

1304 T. Munehisa, Y. Munehisa. *arXiv:cond-mat*/0001304 (2000).

1305 N. Munro (ed.). *Symbolic Methods in Control System Analysis and Design*, Institute of Electrical Engineers, London, 1999.

1306 V. D. Mur, V. S. Popov. *JETP* 77, 18 (1993).

1307 A. Muresan. *Rev. d'Anal. Num. l'Approx.* 26, 131 (1997).

1308 K. P. N. Murthy. *arXiv:cond-mat*/0104215 (2001).

1309 R. E. Musafir. *J. Sound Vibr.* 236, 904 (2000).

1310 O. Mustafa. *arXiv:math-ph*/0101030 (2001).

1311   U. Mutze. *mp_arc* 04-165 (2004).   http://www.ma.utexas.edu/mp_arc-bin/mpa?yn=04-165

1312   R. Myoka, K. Sato. *Arch. Math.* 63, 565 (1994).

1313   I. Najfeld, T. F. Havel. *Adv. Appl. Math.* 16, 321 (1995).

1314   K. Nakamura, A. Shimizu, M. Koshiba, K. Hayata. *IEEE Quant. Electron.* 25, 889 (1989).

1315   G. Nakos, N. Glinos. *The Mathematica Journal* 4, n3, 70 (1994).

1316   V. Namias. *Am. J. Phys.* 45, 624 (1977).

1317   V. Namias. *J. Inst. Math. Appl.* 25, 241 (1980).

1318   A. Nanayakkara. *Phys. Lett.* A 289, 39 (2001).

1319   A. Nanayakkara, I. Dasanayake. *Phys. Lett.* A 294, 158 (2002).

1320   A. Nanayakkara, N. Ranatunga. *Int. J. Theor. Phys.* 41, 1355 (2002).

1321   A. Nanayakkara. *Pramana* 61, 739 (2003).

1322   G. A. Natanzon. *Teor. Mat. Fiz.* 38, 146 (1979).

1323   J. Naudts. *arXiv:cond-mat*/0203489 (2002).

1324   N. D. Naumov. *Vyss. Ucheb. Zaved., Fizika* n2, 72 (1993).

1325   Z. Navickas. *Lithuanian Math. J.* 42, 387 (2002).

1326   G. Nedzhibov, M. Petkov in Z. Li, L. Vulkov, J. Wasniewski (ed.). *Numerical Analysis and Its Applications*, Springer-Verlag, Berlin, 2005.

1327   M. L. Nekrasov. *math-ph*/0303024 (2003).

1328   K. Nemoto, S. L. Braunstein. *arXiv:quant-ph*/0207135 (2002).

1329   E. Netto. *Vorlesungen über Algebra*, v.1, v.2, Teubner, Leipzig, 1896.

1330   A. Neubacher. Diploma thesis, RISC Linz Report 92-66 (1992).

1331   J. W. Neuberger. *Proc. Am. Math. Soc.* 101, 45 (1987).

1332   W. Neutsch, K. Scherer. *Celestial Mechanics*, BI, Mannheim, 1992.

1333   N. D. Newby. *Am. J. Phys.* 45, 1116 (1977).

1334   D. J. Newman, T. J. Rivlin. *J. Approx. Th.* 5, 405 (1972).

1335   F. T. Newman. *Int. J. Quant. Chem.* 63, 1066 (1997).

1336   C. V. Newsom. *Am. Math. Monthly* 38, 500 (1931).

1337   Y. J. Ng, H. van Dam. *arXiv:hep-th*/0502163 (2005).

1338   C. P. Niculescu. *J. Inequ. Pure Appl. Math.* 1, Art. 17 (2000).   http://jipam.vu.edu.au/v1n2/014_99.html

1339   A. F. Nikiforov, S. K. Suslov, V. B. Uvarov. *Classical Orthogonal Polynomials*, Springer-Verlag, Berlin, 1991.

1340   J. Nielsen, B. Roth in G. Giralt, G. Hirzinger (eds.). *Robotics Research*, Springer-Verlag, London, 1996.

1341   N. Nielsen. *Traité Elémentaire des Nombres de Bernoulli*, Gauther–Villars, Paris, 1923.

1342   P. Nikolov, T. Valchev. *arXiv:math-ph*/0410056 (2004).

1343   B. W. Ninham. *Num. Math.* 8, 444 (1966).

1344   J. C. C. Nitsche. *Lectures on Minimal Surfaces*, Cambridge University Press, Cambridge, 1989.

1345   Z. Niu, W. L. Wendland, X. Wang, H. Zhou. *Comput. Meth. Appl. Mech. Eng.* 194, 1057 (2005).

1346   S. Noschese, P. E. Ricci. *J. Comput. Anal. Appl.* 5, 333 (2003).

1347   Z. A. Nowacki. *arXiv:quant-ph*/0208048 (2002).

1348   W. Oevel, W. Schief in P. A. Clarkson (ed.). *Applications of Analytic and Geometric Methods to Nonlinear Differential Equations*, Plenum Press, New York, 1993.

1349 W. Oevel in P. A. Clarkson, F. W. Nijhoff (eds.). *Symmetries and Integrability of Difference Equations*, Cambridge University Press, Cambridge, 1999.

1350 V. I. Ogievetski. *Dokl. Akad. Nauk USSR* 109, 919 (1956).

1351 A. Oldknow. *Am. Math. Monthly* 103, 319 (1996).

1352 P. J. Olver. *Applications of Lie Groups to Differential Equations*, Springer-Verlag, New York, 1993.

1353 G. T. Omarova, T. B. Omarov. *Celest. Mech. Dynam. Astron.* 85, 25 (2003).

1354 I. P. Omelyan, I. M. Myrglod, R. Folk. *arXiv:cond-mat*/0111055 (2001).

1355 I. P. Omelyan, I. M. Mryglod, R. Folk. *Comput. Phys. Commun.* 151, 172 (2003).

1356 O. Ore. *Number Theory and Its History*, McGraw-Hill, New York, 1948.

1357 J. S. E. Ortiz, C. S. Rajapakse, G. H. Gunaratne. *Phys. Rev.* B 66, 144203 (2002).

1358 T. J. Osler. *Math. Comput.* 26, 449 (1972).

1359 R. Ossermann. *A Survey of Minimal Surfaces*, Van Nostrand, New York, 1969.

1360 M. Öster, Y. B. Gaididei, M. Johansson, P. L. Christiansen. *Physica* D 198, 29 (2004).

1361 S. Ouvry. *arXiv:cond-mat*/0502366 (2005).

1362 E. Özcag. *Appl. Math. Lett.* 14, 419 (2001).

1363 K. Pachucki, M. Pachucki, E. Remiddi. *arXiv:physics*/0405057 (2004).

1364 A. Padoa. *Period. Mat.* 5, 80 (1925).

1365 D. N. Page. *Phys. Rev. Lett.* 71, 1291 (1993).

1366 I. Pak. *arXiv:math.MG*/0408104 (2004).

1367 J. Pal, B. Talukdar, S. K. Adhikari. *Phys. Lett.* A 247, 198 (1998).

1368 J. L. Palacios. *Int. J. Quant. Chem.* 82, 135 (2001).

1369 E. Palmgren. *Indag. Math.* 11, 129 (2000).

1370 J. Palmore. *J. Dynamics Diff. Eq.* 6, 507 (1994).

1371 J. Palmore. *Complex Variables* 38, 21 (1999).

1372 X.-Y. Pan, V. Sahni, L. Massa. *arXiv:physics*/0310128 (2003).

1373 X.-Y. Pan, V. Sahni, L. Massa. *arXiv:physics*/0402066 (2004).

1374 X.-Y. Pan, V. Sahni, L. Massa. *Phys. Rev. Lett.* 93, 130401 (2004).

1375 J. Pappademos, U. Sukhatme, A. Pagnamenta. *Phys. Rev.* A 48, 3525 (1993).

1376 F. Pappalardi, A, J. van der Poorten. *arXiv:math.NT*/0403228 (2004).

1377 F. D. Parker. *Am. Math. Monthly* 62, 439 (1955).

1378 G. A. Parker, W. Zhu, Y. Huang, D. K. Hoffman, D. J. Kouri. *Comput. Phys. Commun.* 96, 27 (1996).

1379 L. Parker, S. M. Christensen. *MathTensor*, Addison-Wesley, Reading, 1994.

1380 R. G. Parr, W. Yang. *Density Functional Theory of Atoms and Molecules*, Oxford University Press, Oxford, 1989.

1381 L. A. Pars. *An Introduction to the Calculus of Variations*, Wiley, New York, 1985.

1382 L. A. Pars. *A Treatise on Analytical Mechanics*, Heinemann, London, 1965.

1383 R. R. Parwani. *arXiv:hep-th*/0010197 (2000).

1384 R. R. Parwani. *arXiv:math-ph*/0211064 (2002).

1385 A. Pathak, S. Mandal. *Phys. Lett.* A 286, 261 (2001).

1386 A. Pathak, S. Mandal. *arXiv:quant-ph*/0206011 (2002).

1387　R. S. Pathak. *Integral Transforms of Generalized Functions and Their Applications*, Gordon and Breach, Amsterdam, 1997.

1388　A. K. Pati. *arXiv:quant-ph*/0311031 (2003).

1389　S. H. Patil. *J. Phys.* B 34, 1049 (2001).

1390　S. H. Patil, Y. P. Varshni. *Can. J. Phys.* 82, 647 (2004).

1391　M. P. Pato. *Physica* A 312, 153 (2002).

1392　N. M. Patrikalakis, T. Maekawa in G. Farin, J. Hoschek, M.-S. Kim (eds.). *Handbook of Computer Aided Geometric Design*, Elsevier, Amsterdam, 2002.

1393　W. Paulsen. *Mathematica Educ. Res.* 8, n1, 5 (1998).

1394　G. Paz. *Eur. J. Phys.* 22, 337 (2001).

1395　P. W. Pedersen in P. M. A. Sloot, C. J. K. Tan, J. J. Dongarra, A. G. Hoekstra (eds.). *Computational Science—ICCS 2002*, v. 2, Springer-Verlag, Berlin, 2002.

1396　R. Peele, P. Stănică. *math.CO*/0010186 (2000).

1397　C. L. Pekeris. *Phys. Rev.* 112, 1649 (1958).

1398　C. L. Pekeris. *Phys. Rev.* 126, 1470 (1962).

1399　J. M. Peña, T. Sauer. *SIAM J. Num. Anal.* 37, 1186 (2000).

1400　J. J. Peña, G. Ovando, D. Morales–Guzmán, J. Morales. *Int. J. Quant. Chem.* 85, 244 (2001).

1401　J. K. Percus. *Int. J. Quant. Chem.* 69, 573 (1998).

1402　M. E. Perel'man, R. Englman. *Mod. Phys. Lett.* B 14, 907 (2000).

1403　L. Perko. *Differential Equations and Dynamical Systems*, Springer-Verlag, New York, 1996.

1404　E. Peschl. *Differentialgeometrie*, BI, Mannheim, 1973.

1405　A. Peterman. *arXiv:math-ph*/0001025 (2000).

1406　M. Peternell. *Graphical Models* 62, 202 (2000).

1407　I. Peterson. *Mathematical Treks*, Mathematical Association of America, Washington, 2002.

1408　M. S. Petkovic, S. Trickovic. *J. Comput. Appl. Math.* 64, 291 (1995).

1409　D. Petz. *An Invitation to the Algebra of Canonical Commutation Relations*, Leuven University Press, Leuven, 1990.

1410　R. Pfiefer. *Math. Mag.* 62, 309 (1989).

1411　L. Piegl, W. Tiller. *The NURBS Book*, Springer-Verlag, Berlin, 1995.

1412　U. Pinkall, K. Polthier. *Exper. Math.* 2, 15 (1993).

1413　C. G. Pinner, J. D. Vaaler. *Acta Arithm.* 78, 125 (1996).

1414　L. Pinzur. *J. Number Th.* 9, 361 (1977).

1415　J. Piskorski. *arXiv:hep-th*/0001058 (2000).

1416　I. Pitowsky. *QuantumProbability—Quantum Logic*, Springer-Verlag, Berlin, 1989.

1417　I. Pitowsky, K. Svozil. *arXiv:quant-ph*/0011060 (2000).

1418　M. Planat, H. C. Rosu, S. Perrine. *arXiv:math-ph*/0209002 (2002).

1419　G. Plunien, B. Müller, W. Greiner. *Phys. Rep.* 134, 87 (1986).

1420　B. Podolsky. *Phys. Rev.* 32, 812 (1928).

1421　H. Pollard. *Mathematical Introduction to Celestial Mechanics*, Prentice–Hall, Englewood Cliffs, 1966.

1422　H. Pollard, H. G. Diamond. *The Theory of Algebraic Numbers*, Dover, Mineola, 1999.

1423　K. Polthier in G. Dziuk, G. Huisken, J. Hutchinson (eds.). *Proceedings of the Centre for Mathematics and Its Applications*, Centre for Mathematics and Its Applications, Canberra, 1991.

1424 G. Pólya, S, Szegö. *Aufgaben und Lehrsaetze aus der Analysis, v. II.* Springer-Verlag, Berlin, 1964.

1425 A. D. Polyanin, V. F. Zaitsev. *Handbook of Exact Solutions of Ordinary Differential Equations*, CRC Press, Boca Raton, 1995.

1426 A. D. Polyanin. *Handbook of Linear Partial Differential Equations for Engineers and Scientists*, Chapman & Hall, Boca Raton, 2002.

1427 A. D. Polyanin, A. V. Manzhirov. *Handbook of Integral Equations*, CRC Press, Boca Raton, 1998.

1428 B. Poonen, M. Rubinstein. *SIAM J. Discr. Math.* 11, 135 (1998).

1429 V. Popescu. *Phys. Lett.* A 297, 338 (2002).

1430 R. Portugal. *J. Phys.* A 32, 7779 (1999).

1431 R. Portugal, B. F. Svaiter. *arXiv:math-ph*/0107031 (2001).

1432 R. Portugal, B. F. Svaiter. *arXiv:math-ph*/0107032 (2001).

1433 H. A. Posch, W. Thirring. *J. Math. Phys.* 41, 3430 (2000).

1434 N. Pottier, A. Mauger. *Physica* A 291, 327, (2001).

1435 H. Pottmann, M. Hofer in H.-C. Hege, K. Polthier (eds.). *Visualizations and Mathematics III*, Springer-Verlag, Berlin, 2003.

1436 R. B. Potts. *Utilitas Math.* 9, 73 (1976).

1437 R. E. Powell, S. M. Shah. *Summability Theory and Applications*, van Nostrand, New York, 1972.

1438 G. Prasad. *Great Mathematicians of the Nineteenth Century*, Mahamandal Press, Benares, 1933.

1439 V. V. Prasalov. *Polynomials*, Springer, Berlin, 2004.

1440 P. Prešnajder, P. Kubinec. *Acta Phys. Slov.* 41, 3 (1991).

1441 A. Pringsheim. *Math. Ann.* 22, 455 (1883).

1442 J.-P. Provost in D. Benest, C. Froeschle (eds.). *An Introduction to Methods of Complex Analysis and Geometry for Classical Mechanics and Nonlinear Waves*, Frontiers, France, 1994.

1443 A. P. Prudnikov, Y. A. Brychkov, O. I. Marichev. *Integrals and Series*, v. 1–3, Gordon and Breach, New York, 1990.

1444 A. P. Prudnikov, Y. A. Brychkov, O. I. Marichev. *Integrals and Series*, v. 4–5, Gordon and Breach, New York, 1992.

1445 M. V. Puiseux. *J. Math. Pures Appl.* 15, 365 (1850).

1446 D. L. Pursey, T. A. Weber. *Phys. Rev.* A 52, 4255 (1995).

1447 C. R. Putnam. *Commutation Properties of Hilbert Space Operators*, Springer-Verlag, Berlin, 1967.

1448 T. M. Putnam. *Am. Math. Monthly* 11, 86 (1904).

1449 H. M. A. Radi. *Phys. Rev.* A 12, 1137 (1975).

1450 K. Raghunathan, R. Vasudevan. *J. Phys.* A 20, 839 (1987).

1451 S. Rahav, O. Agam, S. Fishman. *arXiv:chao-dyn* /9905037 (1999).

1452 Q. I. Rahman, G. Schmeisser. *Analytic Theory of Polynomials*, Oxford University Press, Oxford, 2002.

1453 P. M. Rajković, M. S. Stanković, S. D. Marinković. *Mat. Vestnik* 54, 171 (2002).

1454 C. K. Raju. *J. Phys.* A 15, 381 (1981).

1455 L. B. Rall in M. Z. Nashed (ed.). *Generalized Inverses and Applications*, Academic Press, New York, 1976.

1456 L. R. Ram–Mohan, S. Saigal, D. Dossa, J. Shertzer. *Comput. Phys.* 4, n1, 50 (1990).

1457 L. R. Ram–Mohan. *Finite Element and Boundary Element Applications in Quantum Mechanics*, Oxford University Press, Oxford, 2002.

1458 S. Ramanujan. *Trans. Cambr. Phil. Soc.* 22, 159 (1916).

1459 S. Ramanujan. *The Lost Notebook and Other Unpublished Papers*, Narosa, New Dehli, 1988.

1460   G. Rangarajan. *arXiv:nlin.SI*/0106033 (2001).

1461   R. A. Rankin. *Modular Forms and Functions*, Cambridge University Pres, Cambridge, 1988.

1462   J. M. Rassias. *Counterexamples in Differential Equations and Related Topics*, World Scientific, Singapore, 1990.

1463   J. Rauch. *Partial Differential Equations*, Springer-Verlag, New York, 1997.

1464   A. A. Rawlinson. *J. Phys.* A 36, 10215 (2003).

1465   E. Recami, M. Zamboni–Rached, C. A. Dartora, K. Z. Nóbrega. *arXiv:physics*/0210047 (2002).

1466   E. Recami, M. Zamboni–Rached, , K. Z. Nóbrega, H. E. Hernández F.. *IEEE J. Sel. Topics Quant. Electronics* 9, 59 (2003).

1467   J. N. Reddy. *Applied Functional Analysis and Variational Methods in Engineering*, Krieger, Malabar, 1991.

1468   M. Reed, B. Simon. *Methods of Modern Mathematical Physics* v.2, Academic Press, New York, 1974.

1469   G. J. Reid, P. Lin, A. D. Wittkopf. *Studies Appl. Math.* 106, 1 (2001).

1470   W. P. Reid. *Am. Math. Monthly* 64, 663 (1957).

1471   K. Reivelt, P. Saari. *arXiv:physics*/0309079 (2003).

1472   H. Reinhardt (ed.). *C. F. Gauss:* Gedenkband anläßlich des 100. Todestages am 23. Februar 1955, Teubner, Leipzig, 1957.

1473   W. Reitberger. *Internat. Math. Nachr.* 184, 1 (2000).

1474   P. Rejto, M. Toboada. *J. Math. Anal. Appl.* 208, 85 (1997).

1475   M. Remoissenet in P. L. Christiansen, M. P. Sørensen, A. C. Scott (eds.). *Nonlinear Science at the Dawn of the 21st Century*, Springer-Verlag, Berlin, 2000.

1476   T. W. Reps, L. B. Rall. *Higher-Order Symb. Comput.* 16, 93 (2003).

1477   M. Requardt. *arXiv:math-ph*/0001026 (2000).

1478   J. A. Rhodes, M. D. Semon. *Am. J. Phys.* 72, 943 (2004).

1479   C. Ribenboim. *Classical Theory of Algebraic Numbers*, Springer-Verlag, New York, 2001.

1480   J. R. Rice. *Numerical Methods, Software and Analysis*, Academic Press, Boston, 1993.

1481   I. Richards, H. Youn. *Theory of Distributions: a Non-technical Introduction*, Cambridge University Press, Cambridge, 1990.

1482   D. Richardson. *J. Symb. Logic* 33, 514 (1968).

1483   D. Richardson in P. S. Wang (eds.). *Proceedings ISSAC 92*, ACM Press, New York, 1992.

1484   D. Richardson. *J. Symb. Comput.* 24, 627 (1997).

1485   D. Richardson in J. Blanck, V. Brattka, P. Hertling (eds.). *Computability and Complexity in Analysis*, Springer-Verlag, Berlin, 2000.

1486   D. Richardson, S. Langley in T. Mora (ed.). *ISSAC 2002*, ACM, New York, 2002.

1487   J. Richelot. *J. reine angew. Math.* 9, 1 (1832).

1488   R. F. Rinehart. *Proc. Am. Math. Soc.* 7, 2 (1956).

1489   J. Riordan. *Combinatorical Identities*, Wiley, New York, 1968.

1490   A. Rivero. *arXiv:hep-th*/0208180 (2002).

1491   I. Rivin. *arXiv:math.GM*/0211261 (2002).

1492   I. Rivin. *arXiv:math.MG*/0308239 (2003).

1493   E. Roanes-Macías, E. Roanes-Lozano in P. M. A. Sloot, C. J. K. Tan, J. J. Dongarra, A. G. Hoekstra (eds.). *Computational Science—ICCS 2002*, v. 2, Springer-Verlag, Berlin, 2002.

1494   D. P. Robbins. *arXiv:Math.CO*/9805108 (1998).

1495   N. Robbins. *Fibon. Quart.* 34, 306 (1996).

1496   N. Robbins. *Integers* 0, A6 (2000).   http://www.integers-ejcnt.org/vol0.html

1497   M. Robnik, L. Salasnich. *J. Phys.* A 30, 1711 (1997).

1498   M. Robnik, L. Salasnich, M. Vraničar. *arXiv:nlin.CD*/0003050 (2000).

1499   M. Robnik, V. G. Romanovski. *arXiv:nlin.CD*/0003069 (2000).

1500   J. J. Roche. *Am. J. Phys.* 68, 438 (2000).

1501   C. Rodenberg. *Math. Ann.* 14, 46 (1879).

1502   C. Rogers, W. K. Schief, M. E. Johnston in P. J. Vassiliou, I. G. Lisle (eds.). *Geometric Approaches to Differential Equations*, Cambridge University Press, Cambridge, 2000.

1503   C. Rogers, W. K. Schief. *Bäcklund and Darboux Transformations*, Cambridge University Press, Cambridge, 2002.

1504   D. Rohrlich, Y. Aharonov. *Phys. Rev.* A 66, 042102 (2002).

1505   M. Rojas in R. Goldman, R. Krasauskas (eds.). *Topics in Algebraic Geometry and Geometric Modeling*, American Mathematical Society, Providence, 2003.

1506   A. Romeo, A. A. Saharian. *J. Phys.* A 35, 1297 (2002).

1507   J. C. Rosales. *Int. J. Alg. Comput.* 7, 25 (1997).

1508   O. Rosas–Ortiz. *quant-ph*/0302189 (2003).

1509   P. Rosenau, J. M. Hymann. *Phys. Rev. Lett.* 70, 564 (1993).

1510   E. Rosenthal, B. Segev. *Phys. Rev.* A 66, 052110 (2002).

1511   C. Rosenzweig, J. B. Krieger. *J. Math. Phys.* 9, 849 (1968).

1512   S. Rossano, C. Brouder. *arXiv:nlin.PS*/0002012 (2000).

1513   W. Rossman in T. Kotake, S. Nishikawa, R. Schoen (eds.). *Geometry and Global Analysis*, Tohoku University, Sendai, 1993.

1514   H. C. Rosu, M. A. Reyes. *Phys. Rev.* E 57, 4850 (1998).

1515   J. Rotman. *Galois Theory*, Springer-Verlag, New York, 1998.

1516   I. Rotter. *arXiv:quant-ph*/0107018 (2001).

1517   I. Rotter. *Phys. Rev.* E 64, 036213 (2001).

1518   I. Rotter. *Phys. Rev.* E 65, 026217 (2002).

1519   E. G. P. Rowe. *Phys. Rev.* D 12, 1576 (1975).

1520   E. G. P. Rowe. *Phys. Rev.* D 18, 3639 (1978).

1521   A. E. Roy. *Orbital Motion*, Adam Hilger, Bristol, 1982.

1522   B. Roy, P. Roy, R. Roychoudhury. *Fortschr. Phys.* 39, 211 (1991).

1523   D. N. G. Roy, L. S. Couchman. *Inverse Problems and Inverse Scattering of Plane Waves*, Academic Press, San Diego, 2002.

1524   L. A. Rubel. *Bull. Am. Math. Soc.* 4, 345 (1981).

1525   L. A. Rubel. *Nieuw Archief Wiskunde* 6, 263 (1988).

1526   J. Rudnick in E. G. D. Cohen (eds.). *Fundamental Problems in Statistical Mechanics IV*, North-Holland, Amsterdam, 1985.

1527   M. B. Ruiz. *Int. J. Quant. Chem.* 101, 246 (2005).

1528   O. E. Ruiz S., P. M. Ferreira. *MapleTech* 3, n1, (1996).

1529   F. Rulf. *Monatsh. Math. Phys.* 29, 268 (1918).

1530   S. M. Rump in P. W. Gaffney, E. N. Houstis (eds.). *Programming Environments for High-Level Scientific Problem Solving*, North Holland, Amsterdam, 1992.

1531   H. Rund. *The Hamilton–Jacobi Theory in the Calculus of Variations*, van Nostrand, London, 1966.

1532   P. Saari, K. Reivelt. *Phys. Rev.* E 69, 036612 (2004).

1533   I. K. Sabitov. *Sbornik Math.* 189, 1533 (1998).

1534   I. K. Sabitov. *Dokl. Math.* 63, 170 (2001).

1535   I. K. Sabitov. *Siberian Adv. Math.* 13, 102 (2003).

1536   I. Sabitov. *Bull. Braz. Math. Soc.* 35, 199 (2004).

1537   P. L. Sachdev. *A Compendium on Ordinary Nonlinear Differential Equations*, Wiley, New York, 1997.

1538   V. N. Sachkov. *Combinatorial Methods in Discrete Mathematics*, Cambridge University Press, Cambridge, 1996.

1539   C. A. Sackett, D. Kielpinski, B. E. King, C. Langer, V. Meyer, C. J. Myatt, M.Rowe, Q. A. Turchette, W. M. Itano, D. J. Wineland, C. Monroe. *Nature* 404, 256 (2000).

1540   R. Z. Sagdeev, D. A. Usikov, G. M. Zaslavsky. *Nonlinear Physics*, Harwood, Chur, 1988.

1541   A. A. Saharian. *arXiv:hep-th*/0002239 (2000).

1542   A. Sakaji, R. S. Hijjawi, N. Shawagfeh, J. M. Khalifeh. *Int. J. Theor. Phys.* 41, 973 (2002).

1543   T. Sako, G. H. F. Diercksen. *J. Phys.* B 37, 1673 (2004).

1544   L. Salasnich, F. Sattin. *J. Phys.* A 30, 7597 (1997).

1545   M. M. Saleh, I. L. El-Kalla. *Int. J. Diff. Eq. Appl.* 7, 15 (2003).

1546   N. Salingaros. *J. Math. Phys.* 25, 706 (1984).

1547   E. A. Salter, A. Wierzbicki. *Int. J. Quant. Chem.* 89, 121 (2002).

1548   L. Šamaj, J. K. Percus. *J. Chem. Phys.* 111, 1809 (1999).

1549   L. Šamaj, J. K. Percus, P. Kalinay. *arXiv:math-ph*/0210004 (2002).

1550   S. G. Samko. *Hypersingular Integrals and Their Applications*, Gordon and Breach, New York, 2000.

1551   B. F. Samsonov. *J. Phys.* A 28, 6989 (1995).

1552   B. F. Samsonov. *arXiv:quant-ph*/0503040 (2005).

1553   B. F. Samsonov. *arXiv:quant-ph*/0503075 (2005).

1554   J. Sánchez–Ruiz. *Phys. Rev.* A 57, 1519 (1998).

1555   J. A. Sanders, J. P. Wang. *Physica* D 149, 1 (2001).

1556   E. M. Santangelo. *arXiv:hep-th*/0104025 (2001).

1557   J. M. Sanz-Serna, M. P. Calvo. *Numerical Hamiltonian Problems*, Chapman & Hall, London, 1994.

1558   J. Sapirstein. *Phys. Rev.* A 69, 042515 (2004).

1559   Ö. Sarioglu. *arXiv:hep-th*/0207227 (2002).

1560   R. J. Sasiela, J. D. Shelton. *J. Math. Phys.* 34, 2572 (1993).

1561   R. J. Sasiela. *Electromagnetic Wave Propagation in Turbulence*, Springer-Verlag, Berlin, 1994.

1562   T. Sauer, Y. Xu. *Math. Comput.* 64, 1147 (1995).

1563   N. V. Saveleva. *J. Vis. Mat. Fiz.* 39, 1603 (1999).

1564   H. Sawada, X.-T. Yan. *Math. Comput. Simul.* 67, 135 (2004).

1565   G. Schachner. *arXiv:quant-ph*/0312117 (2003).

1566   M. Schaden, L. Spruch. *Phys. Rev.* A 65, 034101 (2002).

1567   P. Schaefer. *Coll. Math. J.* 17, 66 (1986).

1568   M. Schechter. *Operator Methods in Quantum Mechanics*, North Holland, New York, 1981.

1569   V. Scheffer. *J. Geom. Anal.* 13, 343 (1993).

1570   J. Schicho. *Preprint* RISC 00-18 (2000). ftp://ftp.risc.uni-linz.ac.at/pub/techreports/2000/00-18.ps.gz

1571 W. K. Schief, B. G. Konopelchenko. *arXiv:nlin.SI*/0104036 (2001).

1572 W. K. Schief, B. G. Konopelchenko. *arXiv:nlin.SI*/0104037 (2001).

1573 W. K. Schief, B. G. Konopelchenko. *Proc. R. Soc. Lond.* A 459, 67 (2003).

1574 A. Schilling, S. O.Warnaar. *Ramanujan J.* 2, 459 (1998).

1575 R. Schimming. *Z. Anal. Anw.* 7, 203 (1988).

1576 R. Schimming. *Acta Appl. Math.* 39, 489 (1995).

1577 R. Schimming, S. Z. Rida. *Chaos, Solitons, Fractals* 14, 1007 (2002).

1578 A. Schinzel in P. Turán (ed.). *Topics in Number Theory*, North Holland, Amsterdam, 1976.

1579 W. A. Schnizer, H. Leeb. *J. Phys.* A 26, 5145 (1993).

1580 P. Schöpf, J. Schwaiger. *Aequ. Math.* 68, 282 (2004).

1581 R. Schulz–Arenstorff, J. C. Morelock. *Am. Math. Monthly* 66, 95 (1959).

1582 E. O. Schulz–DuBois. *Math. Comput.* 23, 845 (1969).

1583 A. Schulze–Halberg. *Found. Phys. Lett.* 13, 11 (2000).

1584 C. Schwartz. *J. Math. Phys.* 18, 110 (1977).

1585 C. Schwartz. *arXiv:physics*/0208004 (2002).

1586 H. A. Schwarz. *Gesammelte Mathematische Abhandlungen*, Chelsea, New York, 1972.

1587 H. R. Schwarz. *Die Methode der finiten Elemente*, Teubner, Stuttgart, 1991.

1588 P. Sconzo, A. R. LeSchack, R. Tobey. *Astron. J.* 70, 269 (1965).

1589 A. Scrinzi. *Comput. Phys. Commun.* 86, 67 (1985).

1590 T. W. Sederberg, J. P. Snively in R. R. Martin (ed.). *The Mathematics of Surfaces II*, Clarendon Press, Oxford, 1987.

1591 T. W. Sederberg, F. Chen. *Computer Graphics, Proceedings SIGGGRAPH 95* 119 (1995).

1592 T. W. Sederberg in D. A. Cox, B. Sturmfels (eds.). *Applications of Computational Algebraic Geometry*, American Mathematical Society, Providence, 1997.

1593 T. Sederberg, J. Zheng in G. Farin, J. Hoschek, M.-S. Kim (eds.). *Handbook of Computer Aided Geometric Design*, Elsevier, Amsterdam, 2002.

1594 J. Segers, J. Teugels. *Extremes* 3, 291 (2000).

1595 B. Segre. *The Nonsingular Cubic Surfaces*, Oxford University Press, Oxford, 1942.

1596 Z. F. Seidov. *arXiv:math.GM*/0002134 (2000).

1597 W. Seiler. *MathPAD* 7, 34 (1997).

1598 J. Sekino. *Am. Math. Monthly* 106, 193 (1999).

1599 S. Sen. *Phys. Rev. Lett.* 77, 1 (1996).

1600 M. N. Sergeenko. *arXiv:quant-ph*/0206179 (2002).

1601 A. P. Seyranian. *Dokl. Phys.* 49, 64 (2004).

1602 A. P. Seyranian, O. N. Kirillov, A. A. Mailybaev. *J. Phys.* A 38, 1723 (2005).

1603 A. Sezginer. *J. Appl. Phys.* 57, 678 (1985).

1604 A. M. Shaarawi, I. M. Besieris. *J. Phys.* 33, 7255 (2000).

1605 R. Shail. *J. Phys.* A 28, 6999 (1995).

1606 R. Shail. *Math. Comput.* 234, 789 (2000).

1607 J. Shallit. *Math. Intell.* 25, n2, 20 (2003).

1608 H. Shen, H. J. Silverstone. *Int. J. Quant. Chem.* 99, 336 (2004).

1609 J. Shen, G. Strang. *Proc. Am. Math. Soc.* 124, 3810 (1996).

1610 P. Sheng. *Introduction to Wave Scattering, Localization, and Mesoscopic Phenomena*, Academic Press, San Diego, 1995.

1611 A. Shidfar, A. A. Sadeghi. *Appl. Math. Lett.* 3, n4, 21 (1990).

1612 F. Shioda. *Comm. Math. Univ. Sancti Pauli* 44, 109 (1995).

1613 B. K. Shivamoggi. *Perturbation Methods for Differential Equations*, Birkhäuser, Boston, 2003.

1614 E. Shmerling, K. J. Hochberg. *Methol. Comput. Appl. Prob.* 6, 203 (2004).

1615 A. Shnirelman. *Comm. Pure Appl. Math.* 50, 1261 (1997).

1616 A. M. Shuravski. *Istor. Matem. Issled.* 22, 34 (1977).

1617 M. P. Silverman, W. Strange, T. C. Lipscombe. *Am. J. Phys.* 72, 1068 (2004).

1618 H. J. Silverstone, S. Nakai, J. G. Harris. *Phys. Rev.* A 32, 1341 (1985).

1619 H. J. Silverstone, J. G. Harris. *Phys. Rev.* A 32, 1965 (1985).

1620 H. J. Silverstone. *Int. J. Quant. Chem.* 29, 261 (1986).

1621 D. Singer. *Electr. J. Combinatorics* 8, R2 (2001).   http://www.combinatorics.org/Volume_8/Abstracts/v8i1r2.html

1622 M. Singh. *J. Phys.* A 23, 2307 (1990).

1623 A. A. Skorupski. *Rep. Math. Phys.* 17, 161 (1980).

1624 L. Škovrlj, T. Ivezić. *Int. J. Mod. Phys.* A 17, 2513 (2002).

1625 S. W. Sloan. *Int. J. Num. Methods Eng.* 28, 2651 (1989).

1626 R. D. Small. *Am. Math. Monthly* 88, 439 (1981).

1627 W. I. Smirnow. *Lehrgang der höheren Mathematik*, v. II, Verlag der Wissenschaften, Berlin, 1955.

1628 V. I. Smirnov. *A Course of Higher Mathematics* v. 4, Addison, New York, 1964.

1629 R. Snieder. *A Guided Tour of Mathematical Methods*, Cambridge University Press, Cambridge, 2004

1630 F. Söddy. *Nature* 137, 1021 (1936).

1631 A. Sofo. *Computational Techniques for the Summation of Series*, Kluwer, New York, 2003.

1632 M. Sofroniou. *J. Symb. Comput.* 18, 265 (1994).

1633 K. Sogo. *J. Phys. Soc. Jpn.* 68, 3469 (1999).

1634 D. Sokolovski. *arXiv:quant-ph*/0306153 (2003).

1635 D. Sokolovski, A. Z. Msezane, V. R. Shaginyan. *arXiv:quant-ph*/0401159 (2004).

1636 I. Soprounov. *arXiv:math.AG*/0203114 (2002).

1637 F. Sottile. *arXiv:math.AG*/0007142 (2000).

1638 F. Sottile, T. Theobald. *arXiv:math.AG*/0105180 (2001).

1639 F. Sottile. *arXiv:math.AG*/0107179 (2001).

1640 F. Sottile, T. Theobald. *arXiv:math.AG*/0206099 (2002).

1641 S. Souma, A. Suzuki. *Phys. Rev.* B 65, 115307 (2002).

1642 S. Spagnolo. *Ann. Univ. Ferrara–Sez.* 45, 327 (1999).

1643 M. Spivak. *A Comprehensive Introduction to Differential Geometry*, v. 1–3, Publish or Perish, Houston, 1979.

1644 H. Spohn. *arXiv:physics*/9911027 (1999).

1645 L. Spruch. *Rev. Mod. Phys.* 63, 151 (1991).

1646 W. Squire. *J. SIAM* 9, 94 (1961).

1647 H. M. Srivastava, H. L. Manocha. *A Treatise on Generating Functions*, Ellis Horwood, Chichester, 1984.

1648 H. M. Srivastava, J. Choi. *Series Associated with the Zeta and Related Functions*, Kluwer, Dordrecht, 2001.

1649 J. H. Staib. *Math. Mag.* 39, 223 (1966).

1650 J. Stalker. *Complex Analysis*, Birkhäuser, Boston, 1998.

1651 R. P. Stanley. *Enumerative Combinatorics* v.1, Cambridge University Press, Cambridge, 1997.

1652 R. Stärk. *Elem. Math.* 48, 107 (1994).

1653 W. W. Stepanov. *Lehrbuch der Differentialgleichungen*, Deutscher Verlag der Wissenschaften, Berlin, 1982.

1654 W.-H. Steeb, Y. Hardy. *Int. J. Mod. Phys.* C 11, 69 (2000).

1655 D. Stefănescu. *Math. Inequ. Appl.* 5, 335 (2002).

1656 P. R. Stepanishen. *J. Sound Vibr.* 222, 115 (1999).

1657 H. Stephani. *Differential Equations*, Cambridge University Press, Cambridge, 1989.

1658 B. Y. Sternin, V. E. Shatalov. *Borel–Laplace Transform and Asymptotic Theory*, CRC Press, Boca Raton, 1995.

1659 M. Steslicka, R. Kucharczyk, A. Akjouj, B. Djafari–Rouhani, L. Dobrzynski, S. G. Davison. *Surf. Sc. Rep.* 47, 93 (2002).

1660 I. Stewart. *Galois Theory*, Chapman and Hall, London, 1972.

1661 S. Stifter. *J. Intell. Robot. Syst.* 11, 79 (1994).

1662 F. H. Stillinger, D. R. Herrick. *Phys. Rev.* A 11, 446 (1975).

1663 H. Stillinger. *Physica* B 85, 270 (1977).

1664 M. Stingl. *arXiv:hep-ph*/0207349 (2002).

1665 G. Stoll. *Math. Ann.* 6, 613 (1873).

1666 R. Stone. *Pac. J. Math.* 217, 331 (2004).

1667 W. Strampp, V. Ganzha. *Differentialgleichungen mit Mathematica*, Vieweg, Braunschweig, 1995.

1668 F. Streetman, L. R. Ford. *Am. Math. Monthly* 38, 198 (1931).

1669 K. Strehmel, R. Weiner. *Numerik gewöhnlicher Differentialgleichungen*, Teubner, Stuttgart, 1999.

1670 J. Strommer. *Acta Math. Hungar.* 70, 259 (1996).

1671 A. Strzebonski. *The Mathematica Journal* 6, n4, 74 (1996).

1672 A. Strzebonski. *Mathematica Educ. Research* 6, n3, 30 (1996).

1673 A. Strzebonski. *J. Symb. Comput.* 24, 647 (1997).

1674 A. Strzebonski. *The Mathematica Journal* 4, 525 (2000).

1675 C. Stubbins, K. Das. *Phys. Rev.* A 47, 4506 (1993).

1676 C. Stubbins. *Progr. Theor. Phys.* Suppl. 138, 750 (2000).

1677 T. Sturm, V. Weispfenning. *Technical Report MIP-9708*, FMI, University Passau (1999). http://www.fmi.uni-passau.de/~redlog/paper/MIP-9708.ps.Z

1678 T. Sturm, V. Weispfenning. *Technical Report MIP-9804*, FMI, University Passau (1999). http://www.fmi.uni-passau.de/~redlog/paper/MIP-9804.ps.Z

1679 B. Sturmfels in D. A. Cox, B. Sturmfels (eds.). *Applications of Computational Algebraic Geometry*, American Mathematical Society, Providence, 1997.

1680 E. Study. *Math. Ann.* 49, 497 (1897).

1681 M. Sugawara. *Chem. Phys. Lett.* 295, 423 (1998).

1682 Z. Sui, G. Yu, Z. Yu. *Comm. Theor. Phys.* 30, 133 (1998).

1683 C. V. Sukumar in R. Bijker, O. Castaños, D. Fernández, H. Morales–Técotl, L. Urritia, C. Villarreal (eds.). *Supersymmetries in Physics and Its Applications* A 64, 034101 (2005).

1684 J. H. Summerfield, J. G. Loeser. *J. Math. Chem.* 25, 309 (1999).

1685 S. J. Summers. *Preprint mp-arc* 98-720 (1998).   http://rene.ma.utexas.edu/mp_arc-bin/mpa?yn=98-720

1686 S. J. Summers in M. Rédei, M. Stöltzner (eds.). *John von Neumann and the Foundations of Quantum Physics.*, Kluwer, Dordrecht, 2001.

1687 P. Suppes, J. A. de Barros, G. Oas. *arXiv:quant-ph*/9610010 (1996).

1688 D. Surowski, P. McCombs. *Missouri J. Math. Sc.* 15, 4 (2003).

1689 M. Suzuki. *Phys. Lett.* A 224, 337 (1997).

1690 M. Suzuki. *Int. J. Mod. Phys.* B 10, 1637 (1996).

1691 M. Suzuki. *Int. J. Mod. Phys.* C 10, 1385 (1999).

1692 M. Suzuki. *J. Stat. Phys.* 110, 945 (2003).

1693 N. F. Svaiter, B. F. Svaiter. *J. Math. Phys.* 32, 175 (1991).

1694 N. F. Svaiter, B. F. Svaiter. *J. Phys.* A 25, 979 (1992).

1695 N. F. Svaiter, B. F. Svaiter. *Phys. Rev.* D 47, 4581 (1993).

1696 D. Sweet, E. Ott, A. Yorke. *Nature* 399, 315 (1999).

1697 A. Szenes. *Duke Math. J.* 118, 189 (2002).

1698 B. Tabarrok, L. Tong in E. Onate, J. Periaux, A. Samuelson (eds.). *The Finite Element Method in the 1990s*, Springer-Verlag, Berlin, 1990.

1699 L. G. Taff. *Celestial Mechanics*, Wiley, New York, 1985.

1700 K. M. Tamizhmani, B. Grammaticos, T. Tamizhmani, A. Ramani in R. Sahadevan, M. Lakhsmaman (eds.). *Nonlinear Systems*, Narosa, New Dehli, 2002.

1701 M. Tanaka, V. Sladek, J. Sladek. *Appl. Mech. Rev.* 47, 457 (1994).

1702 G. Tanner, K. Richter, J.-M. Rost. *Rev. Mod. Phys.* 72, 497 (2000).

1703 A. Tarski. *A Decision Method for Elementary Algebra and Geometry*, University of California, Berkeley, 1951.

1704 R. Tazzioli. *Hist. Math.* 28, 232 (2001).

1705 A. F. F. Teixeira. *arXiv:math.GM*/0112296 (2001).

1706 M. A. A. Téllez. *Int. J. Math. Math. Sci.* 33, 789 (2003).

1707 H. Terrones. *Coll. de Phys.* C 7, n23, 345 (1990).

1708 H. Terrones, A. L. Mackay in J. M. Garcia-Ruiz, E. Louis, P. Meakin, L. M. Sander (eds.) *Growth Patterns in Physical Sciences and Biology*, Plenum Press, New York, 1993.

1709 E. C. Thayer. *Exper. Math.* 4, 19 (1995).

1710 T. Theobald. *Preprint* 12/99, TU München (1999).   http://www-lit.ma.tum.de/veroeff/html/990.14011.html

1711 T. Theobald. *Exper. Math.* 11, 513 (2002).

1712 W. F. D. Theron. *Eur. J. Phys.* 23, 643 (2002).

1713 J. M. Thijssen. *Computational Physics*, Cambridge University Press, Cambridge, 1999.

1714 W. Thirring. *A Course in Mathematical Physics, v. 4 Quantum Mechanics of Large Systems*, Springer-Verlag, New York, 1983.

1715 W. Thirring (ed.). *The Stability of Matter: From Atoms to Stars*, Springer-Verlag, Berlin, 1991.

1716 D. C. Thompson, A. Alavi. *Phys. Rev.* B 66, 235118 (2002).

1717 H. Tietze. *Famous Problems of Mathematics*, Graylock Press, New York, 1965.

1718 J.-P. Tignol. *Galois' Theory of Algebraic Equations*, Longman Scientific, Harlow, 1980.

1719 F. V. Tkachov. *Nucl. Instr. Meth. Phys. Res.* A 534, 274 (2004).

1720 Z.-Y. Tong, L.-M. Kuang. *arXiv:quant-ph*/0005070 (2000).

1721 T. Töpfer. *Analysis* 15, 25 (1995).

1722 A.-K. Tornberg, B. Engquist. *J. Comput. Phys.* 200, 462 (2004).

1723 G. Toth. *Finite Möbius Groups, Minimal Immersion of Spheres, and Moduli*, Springer-Verlag, Berlin, 2002.

1724 J. F. Traub. *Am. Math. Monthly* 69, 904 (1962).

1725 M. I. Tribelsky. *arXiv:math.PR*/0106037 (2001).

1726 X. Tricoche, G. Scheuermann, H. Hagen in R. Cipolla, R. Martin (eds.). *The Mathematics of Surfaces IX*, Springer-Verlag, London, 2000.

1727 H. Triebel. *Höhere Analysis*, Verlag der Wissenschaften, Berlin, 1972.

1728 H. Triebel. *Num. Funct. Anal. Optim.* 21, 307 (2000).

1729 H. Triebel. *The Structure of Functions*, Birkhäuser, Basel, 2001.

1730 M. Trott. *Mathematica Edu. Researc.* 5, n1, 21 (1996).

1731 M. Trott. *Mathematica Edu. Research* 6, n1, 15 (1997).

1732 M. Trott, H. Yu. *Mathematica Educ. Research* 6, n2, 47 (1997).

1733 M. Trott. *The Mathematica Journal* 7, n2, 189 (1998).

1734 M. Trott. *Mathematica Educ. Research* 8, n1, 24 (1999).

1735 M. Trott. *The Mathematica GuideBook for Programming*, Springer-Verlag, New York, 2004.

1736 M. Trott. *The Mathematica GuideBook for Graphics*, Springer-Verlag, New York, 2004.

1737 M. Trott. *The Mathematica GuideBook for Numerics*, Springer-Verlag, New York, 2005.

1738 C. Truesdell, W. Noll in S. Flügge (ed.). *Handbuch der Physik* III/2, Springer-Verlag, Berlin, (1965).

1739 N. N. Trunov. *Theor. Math. Phys.* 138, 407 (2004).

1740 L. Tsa-Tsien (ed.). *Problems and Solutions in Mathematics*, World Scientific, Singapore, 1998.

1741 M. E. Tshanga. *Vestn. Mosk. Univ. Mat.* n5, 46 (2001).

1742 S.-J. Tu, E. Fischbach. *arXiv:math-ph*/0004021 (2000).

1743 V. K. Tuan, D. T. Duc. *Fract. Calc. Appl. Anal.* 5, 387 (2002).

1744 J. Tucciarone. *Arch. Hist. Exact Sci.* 10, 1 (1973).

1745 G. Turk and J. O'Brien. *Proc. SIGGRAPH 99* 335, (1999).

1746 F. Twilt in A. van der Burgh, J. Simonis (eds.). *Topics in Engineering Mathematics*, Kluwer, Amsterdam, 1992.

1747 C. W. Ueberhuber. *Numerical Computations II*, Springer-Verlag, Berlin, 1997.

1748 K. Ueno, K. Shiga, S. Morita. *A Mathematical Gift, I*, American Mathematical Society, Providence, 2003.

1749 H. Umemura. *Sugaku Exp.* 11, 77 (1998).

1750 A. A. Ungar. *Found. Phys.* 27, 881 (1997).

1751 A. A. Ungar. *Beyond the Einstein Law and its Gyroscopic Thomas Precession*, Kluwer, Dordrecht, 2001.

1752 A. G. Ushveridze. *J. Phys. A* 21, 955 (1988).

1753 A. G. Ushveridze. *Quasi-Exactly Solvable Models in Quantum Mechanics*, Institute of Physics, Bristol, 1994.

1754 T. Utsumi, T. Yabe, J. Koga, T. Aoki, M. Sekine. *Comput. Phys. Commun.* 157, 121 (2004).

1755 T. Uzer, E. A. Lee, D. Farrelly, A. F. Brunello. *Contemp. Phys.* 41, 1 (2000).

1756 L. Vaidman. *Found. Phys.* 21, 947 (1991).

1757 G. M. Vainikko, I. K. Lifanov. *Diff. Uravn.* 38, 1233 (2002).

1758 G. M. Vainikko, I. K. Lifanov. *Sbornik Math.* 194, 1137 (2003).

1759   S. R. Valluri, W. J. Mielniczuk, D. R. Lamm. *Can. J. Phys.* 71, 389 (1993).

1760   S. R. Valluri, W. J. Mielniczuk, D. R. Lamm. *Can. J. Phys.* 72, 786 (1994).

1761   S. R. Valluri, R. Biggs, W. Harper, C. Wilson. *Can. J. Phys.* 77, 393 (1999).

1762   B. L. van der Waerden. *Algebra II*, Springer-Verlag, Berlin, 1967.

1763   N. G. van Kampen. *Physica* 74, 215, 239 (1973).

1764   J. J. I. M. van Kan in A. van der Burgh, J. Simonis (eds.). *Topics in Engineering Mathematics*, Kluwer, Dordrecht, 1992.

1765   I. van den Berg. *Nieuw Archief Wiskunde* 5, 340 (2001).

1766   J. G. van der Corput. *J. d'Anal. Math.* 7, 281 (1959/1960).

1767   A. J. van der Poorten. *Proc. Am. Math. Soc.* 29, 451 (1971).

1768   A. J. van der Poorten, X. C. Tran. *Monatsh. Math.* 131, 155, (2000).

1769   A. J. van der Poorten in M. Jutila, T. Metsänkylä (eds.). *Number Theory*, de Gruyter, Berlin, 2001.

1770   M. van der Put in B. L. J. Braaksma, G. K. Immink, M. van der Put (eds.). *The Stokes Phenomena and Hilbert's 16th Problem*, World Scientific, Singapore, 1996.

1771   S. J. L. van Eijndhoven, J. de Graaf. *A Mathematical Introduction to Dirac's Formalism*, North Holland, Amsterdam, 1986.

1772   K. Varga, P. Navratil, J. Usukura, Y. Suzuki. *Phys. Rev.* B 63, 205308 (2001).

1773   R. S. Varga. *Scientific Computation on Mathematical Problems and Conjectures*, Capital City Press, Montpelier, 1990.

1774   R. S. Varga, A. J. Carpenter. *Num. Math.* 68, 169 (1994).

1775   R. S. Varga, A. J. Carpenter. *Num. Alg.* 25, 363 (2000).

1776   V. V. Vassiliev in S. D. Chatterji (ed.). *Proceedings of the International Congress of Mathematicians*, Birkhäuser, Basel, 1995.

1777   R. Vein, P. Dale. *Determinants and Their Applications in Mathematical Physics*, Springer-Verlag, New York, 1999.

1778   O. L. Vélez. *Geom. Dedicata* 52, 205 (1994).

1779   K. Venkatachaliengar. *Development of Elliptic Functions According to Ramanujan*, Deptartment of Mathematics, Madurai Kamaraj University, 1988.

1780   R. P. Venkataram004. *arXiv:math.CA/0007011* (2000).

1781   A. P. Veselov, K. L. Styrkas, O. A. Chalykh. *Teor. Mat. Fiz.* 94, 253 (1993).

1782   G. Vilasi. *Hamiltonian Dynamics*, World Scientific, Singapore, 2001.

1783   M. B. Villarino. *arXiv:math.CA/0402354* (2004).

1784   O. Viro in C. Casacuberta, R. M. Miró–Roig, J. Verdera, S. Xambó–Descamps (eds.). *European Congress of Mathematics*, Birkhäuser, Basel, 2001.

1785   O. Viro. *Notices Am. Math. Soc.* 49, 916 (2002).

1786   A. Visconti. *Introductory Differential Geometry for Physicists*, World Scientific, Singapore, 1992.

1787   M. Visser. *arXiv:hep-th/0304081* (2003).

1788   A. A. Vladimirov. *arXiv:hep-th/0402097* (2004).

1789   V. S. Vladimirov. *Equations of Mathematical Physics*, Mir, Moscow, 1984.

1790   V. S. Vladimirov. *Methods of the Theory of Generalized Functions*, Taylor & Francis, London, 2003.

1791   S. G. Vladut. *Kronecker's Jugendtruam and Modular Functions*, Gordon and Breach, Amsterdam 1995.

1792   M. Volkerts. *Mitt. Math. Ges. Hamburg* 21, 5 (2002).

1793   V. Volterra. *Theory of Functionals*, Blackie, London, 1931.

1794 H. V. von Geramb (ed.). *Quantum Inversion: Theory and Applications*, Springer-Verlag, Berlin, 1994.

1795 J. von Neumann. *Math. Ann.* 104, 570 (1931).

1796 A. Voros. *J. Phys.* A 27, 4653 (1994).

1797 M. Vraničar, M. Robnik. *arXiv:nlin.CD*/0003051 (2000).

1798 C. Vrejoiu. *J. Phys.* A 35, 9911 (2002).

1799 C. Vrejoiu, D. Nicmorus. *arXiv:physics*/0307113 (2003).

1800 B. D. Vujanovic, S. E. Jones. *Variational Methods in Nonconservative Phenomena*, Academic Press, Boston, 1989.

1801 S. Wagon. *Mathematica in Action*, W. H. Freeman, New York, 1991.

1802 S. Waldron and S. Cooper. *J. Approx. Th.* 104, 133 (2000).

1803 P. Walker. *Am. Math. Monthly* 110, 337 (2003).

1804 G. Walter, J. Blum. *Ann. Stat.* 7, 328 (1979).

1805 R. Walter. *Geom. Dedicata* 23, 287 (1987).

1806 F. Y. Wan. *Introduction to the Calculus of Variations and Its Applications*, Chapman & Hall, London, 1994.

1807 D. Wang in B. Buchberger, F. Winkler (eds.). *Gröbner Bases and Applications*, Cambridge University Press, Cambridge 1998.

1808 D. Wang in E. B. Corrochano, G. Sobczyk (eds.). *Geometric Algebra with Applications in Science and Engineering*, Birkhäuser, Boston, 2001.

1809 D. Wang. *Elimination Practice*, Imperial College Press, London, 2004.

1810 K. Wang. *Proc. Am. Math. Soc.* 95, 11 (1985).

1811 W. Wang in G. Farin, J. Hoschek, M.-S. Kim (eds.). *Handbook of Computer Aided Geometric Design*, Elsevier, Amsterdam, 2002.

1812 B. B. Waphare. *Acta Cienca Indica* 26, 75 (2000).

1813 M. Washihara. *Math. Japonica* 50, 1 (1999).

1814 G. K. Watagula. *Int. J. Math. Sci. Technol.* 24, 35 (1993).

1815 A.-M. Wazwaz. *Appl. Math. Comput.* 105, 11 (1999).

1816 A.-M. Wazwaz. *Comput. Math. Appl.* 40, 679 (2000).

1817 A. M. Wazwaz. *Chaos, Solitons, Fractals* 13, 1053 (2001).

1818 A.-M. Wazwaz. *Appl. Math. Comput.* 132, 29 (2002).

1819 A.-M. Wazwaz. *Appl. Math. Comput.* 150, 399 (2004).

1820 H. Weber. *Lehrbuch der Algebra*, v. 1, v. 2., Chelsea, New York, 1962.

1821 A. Weber. *SIGSAM Bull.* 30, n3, 11 (1996).

1822 H. Weber. *Elliptische Funktionen und algebraische Zahlen*, Vieweg, Braunschweig, 1908.

1823 C. E. Wee, R. N. Goldman. *IEEE Comput. Graphics Appl.* n1, 69 (1995).

1824 C. E. Wee, R. N. Goldman. *IEEE Comput. Graphics Appl.* n3, 60 (1995).

1825 S. Weerakoon. *Int. J. Math. Sci. Technol.* 25, 277 (1994).

1826 G. W. Wei. *arXiv:math.SC*/0007113 (2000).

1827 A. Weinstein, W. Stenger. *Intermediate Problems for Eigenvalues*, Academic Press, New York, 1972.

1828 F. V. Weinstein. *arXiv:math.NT*/0307150 (2003).

1829 V. Weispfenning. *MIP 99-06* (1999). http://www.fmi.uni-passau.de/forschung/mip-berichte/MIP-9906.html

1830 H. A. Weldon. *Phys. Rev.* D 47, 594 (1993).

1831 H. A. Weldon. *Phys. Rev.* D 65, 076010 (2002).

1832   E. J. Weniger. *Int. J. Quant. Chem.* 90, 92 (2002).

1833   R. F. Werner, M. M. Wolf. *arXiv:quant-ph*/0102024 (2001).

1834   R. F. Werner, M. M. Wolf. *arXiv:quant-ph*/0107093 (2001).

1835   W. Whiteley in D. Kueker, C. Smith (eds.). *Learning and Geometry: Computational Aspects*, Birkhäuser, Boston, 1996.

1836   E. T. Whittaker. *A Treatise on the Analytical Dynamics of Particles and Rigid Bodies: With an Introduction to the Problem of Three Bodies*, Cambridge University Press, Cambridge, 1964.

1837   E. T. Whittaker, G. N. Watson. *A Course of Modern Analysis*, Cambridge University Press, Cambridge, 1952.

1838   J. Wickner. *Math. Mag.* 71, 390 (1998).

1839   D. V. Widder. *Trans. Am. Math. Soc.* 30, 126 (1930).

1840   D. V. Widder. *Am. Math. Monthly* 55, 489 (1948).

1841   H. Wilf. *arXiv:Math.CO*/9809121 (1998).

1842   J. B. Wilker. *Am. Math. Monthly* 94, 354 (1987).

1843   B. W. Williams, D. P. Poulis. *Eur. J. Phys.* 14, 222 (1993).

1844   R. C. Williamson, T. Downs. *Stat. Prob. Lett.* 7, 167 (1989).

1845   G. Wilson. *Proc. Cambr. Phil. Soc.* 86, 131 (1979).

1846   S. Wirkus, R. Rand, A. Ruina. *Coll. Math. J.* 29, 266 (1998).

1847   G. L. Wise, E. B. Hall. *Counterexamples in Probability and Real Analysis*, Oxford University Press, Oxford, 1993.

1848   I. Van de Woestijne in M. Boyom, J.-M. Morvan, C. Verstraelen (eds.). *Geometry and Topology of Submanifolds* II, World Scientific, Singapore, 1990.

1849   K. B. Wolf. *J. Math. Phys.* 18, 1046 (1977).

1850   K. B. Wolf. *Integral Transforms in Science and Engineering*, Plenum Press, New York, 1979.

1851   D. H. Wolpert. *Phys. Rev. E* 65, 016128 (2002).

1852   C. F. Woodcock, P. R. Graves-Morris. *Bull. Austral. Math. Soc.* 53, 149 (1996).

1853   N. M. J. Woodhouse. *Proc. R. Soc. Lond.* A 438, 197 (1992).

1854   W. K. Wootters. *Phil. Trans. R. Soc. Lond.* A 356, 1717 (2000).

1855   D. Wright. *J. Pure Appl. Alg.* 57, 191 (1989).

1856   C.-H. Wu, C.-I. Kuo, L. H. Ford. *arXiv:quant-ph*/0112056 (2001).

1857   F. Y. Wu. *arXiv:math-ph*/0402038 (2004).

1858   T. Wu, Y. Zhou. *Comput. Aided Geom. Design* 17, 759 (2000).

1859   W. Wu. *Mechanical Theorem Proving in Geometries*, Springer-Verlag, Wien, 1994.

1860   W. Wu. *Mathematics Mechanization*, Kluwer, Dordrecht, 2000.

1861   X. Wu. *Appl. Num. Math.* 44, 415 (2003).

1862   X. Wu. *Appl. Math. Comput.* 162, 539 (2005).

1863   C. E. Wulfman, B. G. Wybourne. *J. Phys.* A 9, 507 (1976).

1864   W. Wunderlich. *Elem. Math.* 17, 121 (1962).

1865   W. Wunderlich. *Aequ. Math.* 10, 71 (1971).

1866   Y. Y. Xian. *Theory and Applications of Higher-Dimensional Hadamard Matrices*, Science Press, Bejing, 2001.

1867   Z. Xian, Y. Zhongpeng, C. Chongguang. *SIAM J. Matrix Anal. Appl.* 21, 642 (1999).

1868   M. Xiaochun. *arXiv:physics*/0210112 (2002).

1869   G. Xin. *arXiv:math.CO*/0408377 (2002).

1870 K. Yagasaki. *Nonl. Dynam.* 18, 129 (1999).

1871 T. Yamakawa, V. Kreinovich. *Int. J. Theor. Phys.* 38, 1763 (1999).

1872 T. Yamano. *Physica* A 305, 486 (2002).

1873 Z. Yan, G. Bluman. *Comput. Phys. Commun.* 149, 11 (2002).

1874 L. Yang. *J. Symb. Comput.* 28, 225 (1999).

1875 A. Yelnikov. *arXiv:hep-th*/0112134 (2001).

1876 H. S. Yi, H. R. Lee, K. S. Sohn. *Phys. Rev.* A 49, 3277 (1994).

1877 J.-C. Yoccoz. *Astérisque* 231, 3 (1995).

1878 H. Yoshida. *Celest. Mech. Dynam. Astron.* 56, 27 (1993).

1879 L. You, J. J. Zhang, P. Comninos. *Visual Comput.* 20, 199 (2004).

1880 E. C. Young. *Vector and Tensor Analysis*, Marcel Dekker, New York, 1993.

1881 V. I. Yukalov, E. P. Yukalov. *Chaos, Solitons, Fractals* 14, 839 (2002).

1882 V. I. Yukalov, E. P. Yukalova. *arXiv:hep-ph*/0211349 (2002).

1883 S. B. Yuste, M. Sanchez. *Phys. Rev.* A 48, 3478 (1993).

1884 P. P. Zabreyko, A. I. Koshelev, M. A. Krasnosel'skii, S. G. Mikhlin, L. S. Rakovshchik, Y. Y. Stet'senko. *Integral Equations—A Reference Text*, Noordhoff, Leyden, 1975.

1885 B. N. Zakhariev, A. A. Suzko. *Direct and Inverse Problems: Potentials in Quantum Mechanics*, Springer-Verlag, Berlin, 1990.

1886 B. N. Zakhariev, V. M. Chabanov. *Inverse Problems* 13, R47 (1997).

1887 B. N. Zakhariev, N. A. Kostov, E. B. Plekhanov. *Sov. J. Part. Nucl.* 21, 384 (1990).

1888 B. N. Zakhariev. *Sov. J. Part. Nucl.* 23, 603 (1993).

1889 M. Zamboni–Rached, E. Recami, H. E. Hernández–Figueroa. *arXiv:physics*/0109062 (2001).

1890 M. Zamboni–Rached, K. Z. Nóbrega, H. E. Hernández–Figueroa, E. Recami. *arXiv:physics*/0209101 (2002).

1891 M. Zamboni–Rached, K. Z. Nóbrega, E. Recami, H. E. Hernández–Figueroa. *arXiv:physics*/0209104 (2002).

1892 M. Zamboni–Rached, F. Fontana, E. Recami. *Phys. Rev.* E 67, 036620 (2003).

1893 Y. Zarmi. *Am. J. Phys.* 70, 446 (2002).

1894 D. Zeilberger. *arXiv:math.CO*/9805137 (1998).

1895 D. Zeilberger. *Electr. Trans. Numer. Anal.* 9, 147 (1999).
http://etna.mcs.kent.edu/vol.9.1999/pp147-148.dir/pp147-148.html

1896 A. H. Zemanian. *IEEE Trans. Circuits Syst.* 35, 1346 (1988).

1897 A. H. Zemanian. *Transfiniteness for Graphs, Electrical Networks, and Random Walks*, Birkhäuser, Boston, 1996.

1898 H.-S. Zeng, L.-M. Kunag. *arXiv:quant-ph*/0005002 (2000).

1899 Y. Z. Zhang. *Special Relativity and Its Experimental Tests*, World Scientific, Singapore, 1997.

1900 Z. Zhang. *Math. Montisnigri* 11, 159 (1999).

1901 D. Zhang, D. Chen. *J. Phys. Soc. Jpn.* 72, 448 (2003).

1902 W. Zhao. *arXiv:math.CV*/0305162 (2003).

1903 T. Zheng. *Phys. Lett.* A 305, 337 (2002).

1904 W. Zheng, L.-A. Ying. *Int. J. Quant. Chem.* 97, 659 (2004).

1905 Y. Zhou, J. D. Mancini, P. F. Meier. *Phys. Rev.* A 51, 3337 (1995).

1906 B. Zhou, C.-J. Zho. *arXiv:hep-th*/9907193 (1999).

1907 L. Zhou, C. Kambhamettu. *Graphical Models* 63, 1 (2001).

1908   W. Zhu, X. Zhao. *J. Chem. Phys.* 105, 9536 (1996).

1909   R. K. P. Zia, B. Schmittmann. *arXiv:cond-mat*/0209044 (2002).

1910   A. Zinani. *Monatsh. Math.* 139, 341 (2003).

1911   R. W. Ziolkowski. *Phys. Rev.* A 39, 2005 (1989).

1912   M. Znojil. *arXiv:quant-ph*/9907054 (1999).

1913   J. P. Zubelli. *CRM Proc. Lecture Notes* 14, 139 (1998).

1914   I. J. Zucker. *J. Phys.* A 20, L13 (1987).

1915   W. Zudilin. *arXiv:math.NT*/0008237 (2000).

1916   D. Zwillinger. *Handbook of Differential Equations*, Academic Press, New York, 1992.

1917   D. Zwillinger. *Handbook of Integration*, Jones and Bartlett, Boston, 1992.

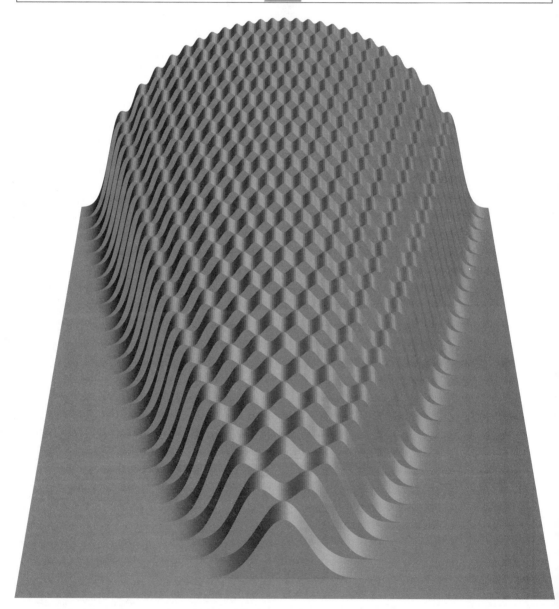

Classical Orthogonal Polynomials

2.0 Remarks

In this chapter, we discuss the classical orthogonal polynomials. Our main purpose is to provide formulas that uniquely define the polynomials (including normalization factors) and to demonstrate some of *Mathematica*'s integration, differentiation, and series expansion capabilities. Further, we will make use of graphics to visualize the orthogonal polynomials in a variety of ways. In this chapter, we sometimes use the word "check" not in the sense of a mathematical proof, but rather in the sense of checking formulas by means of special cases and computational examples. (For the use of *Mathematica* as a theorem prover, see [145], [330].) Occasionally we will encounter special functions (especially the Gamma function and the Gauss hypergeometric function) in this chapter when carrying out integrals involving orthogonal polynomials. We will discuss these functions in the next chapter.

2.1 General Properties of Orthogonal Polynomials

The classical orthogonal polynomials have been extensively studied in connection with eigenvalue problems for Sturm–Liouville differential operators (and this connection largely accounts for their practical importance) [148]:

$$-\frac{1}{r(z)}\left(p(x)\,\Psi'(z)\right)' + q(z)\,\Psi(z) \quad = \lambda\,\Psi(z)$$

$$c_1\,\Psi(a) + c_2\,\Psi'(a) \quad = \alpha$$

$$d_1\,\Psi(b) + d_2\,\Psi'(b) \quad = \beta$$

$$r(z),\ q(z) \in C^0(a,\,b),\ p(z) \in C^1(a,\,b),\ -\infty \le a < b \le \infty, \quad \lambda \in \mathbb{R}.$$

Sturm–Liouville operators typically arise in the solution of partial differential equations from physics (for instance, the heat equation, the wave equation, and the Schrödinger equation) by the method of separation of variables [530], [462], [466]. Using this method, a solution of these partial differential equations can be expressed as a multiple infinite sum (see [595], [563], and [150]).

The orthogonal polynomials to be discussed here include the following:

- Jacobi polynomials $P_n^{(a,b)}(z)$

- Legendre polynomials $P_n(z)$

- Associated Legendre polynomials (not true "polynomials") $P_n^m(z)$

- Hermite polynomials $H_n(z)$

- Laguerre polynomials $L_n^{(\alpha)}(z)$

- Gegenbauer polynomials $C_n^{(\alpha)}(z)$

- Chebyshev polynomials of the first type $T_n(z)$

- Chebyshev polynomials of the second type $U_n(z)$

This list does not include all known and named polynomials (or even all orthogonal polynomials), such as Courant–Hilbert, Appell, Adomian, Meixner, Pollaczek, Krawtchouk, Hahn, etc. [329]. Currently, these are not implemented in the *Mathematica* kernel.

The classical orthogonal polynomials $p_n(z)$ possess a number of important properties ($P_n^m(z)$ is somewhat of an exception; see below), including:

- $p_n(z)$ is a polynomial of degree n in z.

$$p_n(z) = \sum_{i=0}^{n} a_i \, z^i$$

(a_i is specific to the particular polynomials.)

- They satisfy a second-order linear differential equation.

$$g_2(z)\, p_n''(z) + g_1(z)\, p_n'(z) + g_0(z)\, p_n(z) = 0$$

($g_0(z)$, $g_1(z)$, $g_2(z)$ are specific to the particular polynomials.)

- They satisfy the following simple recurrence formula.

$$a_{1,n}\, p_{n+1}(z) = (a_{2,n} + z\, a_{3,n})\, p_n(z) - a_{4,n}\, p_{n-1}(z)$$

($a_{1,n}$, $a_{2,n}$, $a_{3,n}$, and $a_{4,n}$ depend on the particular polynomials.)

(A similar relation holds for the derivatives of the orthogonal polynomials; see [596].)

- The nth polynomial $p_n(z)$ can be expressed as the nth derivative of a simple expression (the Rodrigues' formula—the orthogonal polynomial systems having this property are exactly the classical orthogonal polynomials discussed in this chapter).

$$p_n(z) = \frac{1}{e_n\, w(z)}\, \frac{d^n \{w(z)\, g(z)^n\}}{dz^n}$$

(e_n, $g(z)$ and the weight function $w(z)$ are specific for the particular polynomials.)

- There are simple generating functions $g(z, t)$ [463].

$$g(z, t) = \sum_{n=0}^{\infty} d_n\, p_n(z)\, t^n$$

(d_n, $g(z, t)$ depend on the particular polynomials.) The polynomials $p_n(z)$ are (up to a constant factor d_n) the coefficients in the Taylor expansion of g about $t = 0$. This representation of the orthogonal polynomials is especially useful in the calculation of infinite sums of the form

$$g(z) = \sum_{n=0}^{\infty} \alpha(n) \, p_n(z)$$

which by some proper manipulations (differentiating, integrating, and multiplication of the generating function) can be expressed in closed forms (see, for instance, [393], [442], [5], and [1]).

■ The polynomials $p_n(z)$ ($n = 0, 1, \ldots$) form a complete function system with respect to the weight function $w(z)$. This means that an arbitrary function can be expanded in a generalized Fourier series (besides the general references on orthogonal polynomials, see for these expansion issues also [157], [632], [431], and [513]). In particular, if we set (including the square root of the weight functions into the basis [620])

$$\varphi_n(z) = \sqrt{w(z)} \, \frac{p_n(z)}{c_n}$$

(where c_n and the weight function $w(z)$ are specific for the particular polynomials), then (provided the integrals involved exist) any arbitrary function $y(z)$ can be written as

$$y(z) = \sum_{n=0}^{\infty} \alpha_n \, \varphi_n(z)$$

with

$$\alpha_n = \int_a^b y(z) \, \varphi_n(z) \, dz.$$

This means that the $\varphi_n(z)$ satisfy the following completeness and orthogonality relations:

$$\int_a^b \varphi_n(z) \, \varphi_m(z) \, dz = \delta_{nm}$$

$$\sum_{n=0}^{\infty} \varphi_n(z) \, \varphi_n(z') = \delta(z - z').$$

(These relations are often written in a somewhat different form without $\sqrt{w(z)}$. The form presented here is usually preferable for practical applications, however. And, not only the polynomials, but also their kth derivatives are orthogonal [535].)

For a fixed n, the following orthogonality holds (the sums extend over all the o simple zeros z_k of $p_o(z)$):

$$\sum_{k=1}^{o} \frac{b_{n-1}}{p_{o-1}(z_k) \, p'_o(z_k)} \, p_n(z_k) \, p_m(z_k) = \delta_{nm}$$

■ Derivatives of the orthogonal polynomials of order n can be expressed as linear combinations (with coefficients depending on z) of the same orthogonal polynomials of order n and $n - 1$. (This follows easily from the fact that the orthogonal polynomials presented above satisfy recurrence relations.)

■ The polynomials $p_n(z)$ can be expressed in terms of hypergeometric or confluent hypergeometric functions.

■ The $p_n(z)$ have more or less simple contour integral representations.

■ There are closed-form expressions for the following sums (which are important in various applications) [357]:

$$\sum_{i=0}^{n} \frac{p_i(z)\,p_i(z')}{c_i^2} = \frac{p_{n+1}(z)\,p_n(z') - p_n(z)\,p_{n+1}(z')}{(z - z')\,c_n^2}\,\frac{a_n}{a_{n+1}}.$$

(This is the Darboux–Christoffel formula.)

To avoid any potential disappointment let us mention the following fact.

> *Mathematica* cannot immediately use most of the above list of properties of orthogonal polynomials because, in *Mathematica*, all special functions are defined for arbitrary complex parameters (arguments), and not only for integer values of n. In these cases, these functions are not polynomials, but rather continuous functions of the (complex-valued) argument n.

For further discussion of the properties of the classical orthogonal polynomials, see [379], [2], [436], [405], [559], [445], [437], [187], [606], [503], [585], [341], [423], [565], [208], [55], [21], [117], [533], [59], [364], [88], [134], [101], [102], [443], [94], [433], [23], [211], [104], [340], [344], [469], [113], [205], [159], [384], [75], [312], [592], [603], [373], [106], [107], [89], and [151] as well as http://www.functions.wolfram.com and http://math.nist.gov/opsf. For path integral representations, see [322]. For some interesting generalizations of more general differential operators, see [549].

2.2 Hermite Polynomials

The Hermite polynomials play a major role in quantum-mechanical calculations (see Section 2.10 below). Multiplied by a Gaussian curve $\exp(-x^2/2)$, they are the eigenfunctions of the Hamiltonian (sum of kinetic and potential energy expressed in coordinates and momenta) of a harmonic oscillator. The Hermite polynomials arise as the solutions of the eigenvalue problem

$$\Psi''(z) - 2z\,\Psi'(z) + 2\lambda\,\Psi(z) = 0, \quad z \in (-\infty, \infty)$$

under the condition that the solution should grow not faster than a polynomial as $|z| \to \infty$. Here, the eigenvalues are $\lambda = n$ ($n \in \mathbb{N}$, $n \geq 0$). The *Mathematica* syntax for the Hermite polynomials follows.

HermiteH[*n*, *z*]

    ($n \in \mathbb{N}$, $n \geq 0$) gives the nth Hermite polynomial $H_n(z)$. If n is not a nonnegative integer, the value of the corresponding analytic continuation of $H_n(z)$ is understood.

We now look at the first few $H_n(z)$, both explicitly as polynomials and graphically.

```
In[1]:= With[{p = HermiteH},
          Table[HoldForm[p[v, z]] == p[n, z] /.
                v -> n, {n, 0, 6}]] // TableForm // TraditionalForm
```

Out[1]//TraditionalForm=

$$H_0(z) = 1$$
$$H_1(z) = 2\,z$$
$$H_2(z) = 4\,z^2 - 2$$
$$H_3(z) = 8\,z^3 - 12\,z$$
$$H_4(z) = 16\,z^4 - 48\,z^2 + 12$$
$$H_5(z) = 32\,z^5 - 160\,z^3 + 120\,z$$
$$H_6(z) = 64\,z^6 - 480\,z^4 + 720\,z^2 - 120$$

For nonnegative integers, orthogonal polynomials can be defined via the special functions of mathematical physics (see the following chapter). Here is an example for these Hermite "polynomials". Of course, *Mathematica* can always find a numerical approximation for this expression. It is the analytic continuation of $H_n(z)$ with respect to n.

In[2]:= `{HermiteH[1/3, 1], N[HermiteH[1/3, 1]]}`

Out[2]= $\left\{ \text{HermiteH}\left[\frac{1}{3}, 1\right], 1.31076\right\}$

Because in this chapter we are primarily interested in the case in which n is an integer (which corresponds to orthogonal *polynomials*), we do not present similar examples for noninteger n. We will encounter similar analytic continuations in the next chapter. An orthogonal polynomial of order n has exactly n real zeros. (For some quantitative discussions on the locations of these zeros, see [631] and [387].) Here are visualizations of the first few Hermite polynomials along the real axis.

In[3]:= `Show[GraphicsArray[##]] & /@`
           `Table[(* the individual plots *)`
               `Plot[HermiteH[3i + j, z], {z, -2, 2},`
                   `PlotLabel -> (StyleForm[TraditionalForm[`
                                   `HoldForm[HermiteH[#, z]]],`
                                   `FontSize -> 7]&[3i + j]),`
               `DisplayFunction -> Identity, AxesOrigin -> {0, 0}],`
               `{i, 0, 2}, {j, 0, 2}]`

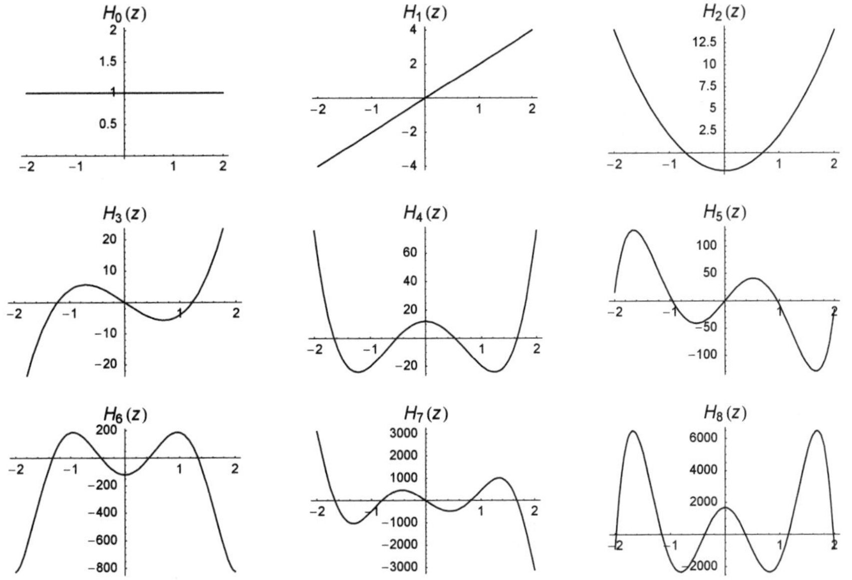

Let us look at the dependence of the zeros of the Hermite polynomials on the index n. It is well known that the zeros of $H_n(z)$ lie between the zeros of $H_{n+1}(z)$. The next picture clearly shows this behavior. We can use *Mathematica*'s capabilities to compute numerical values for noninteger first arguments of HermiteH for a visualization. In the right graphic, we show the surfaces $\mathrm{Re}(H_n(z)) = 0$ in red and $\mathrm{Im}(H_n(z)) = 0$ in blue in the $\mathrm{Re}(z),\mathrm{Im}(z),n$-space.

```
In[4]:= Needs["Graphics`ContourPlot3D`"]
```

```
In[5]:= (* values of the Hermite functions *)
        data = Table[HermiteH[k, N[x + I y]],
                {k, -1, 5, 1/5}, {y, -7/2, 7/2, 1/4}, {x, -7/2, 7/2, 1/4}];
```

```
In[7]:= (* cut a hole in a polygon *)
        makeHole[Polygon[l_], f_] :=
        Module[{mp = Plus @@ l/Length[l], ℓ}, ℓ = (mp + f*(# - mp))& /@ l;
            {MapThread[Polygon[Join[#1, Reverse[#2]]]&,
                Partition[Append[#, First[#]], 2, 1]& /@ {l, ℓ}],
            Line[Append[#, First[#]]]&[ℓ]}]
```

```
In[9]:= Show[GraphicsArray[
        {(* zeros in the z,n-plane *)
        Graphics[(* making small vertical lines at the zeros *)
        MapIndexed[Line[{{#1, #2[[1]] - 1/2}, {#1, #2[[1]] + 1/2}}]&,
        (* the zeros *)  Table[Cases[NRoots[HermiteH[n, z] == 0, z],
                                 _?NumberQ, {-1}], {n, 50}], {2}],
            PlotRange -> All, Frame -> True, AspectRatio -> 1,
            FrameLabel -> {"z", "n"}],
        (* zero surfaces in the Re[z],Im[z],n-space *)
        Graphics3D[{EdgeForm[], Thickness[0.001],
        {SurfaceColor[#2, #2, 2.2], makeHole[#, 0.72]& /@
          Cases[ListContourPlot3D[#1[data],
          DisplayFunction -> Identity, MeshRange ->
          {{-4, 4}, {-4, 4}, {-1, 5}}], _Polygon, Infinity]}& @@@
          {{Re, RGBColor[1, 0, 0]}, {Im, RGBColor[0, 0, 1]}}},
          Axes -> True, AxesLabel -> {"Re[z]", "Im[z]", "n"}]}]]
```

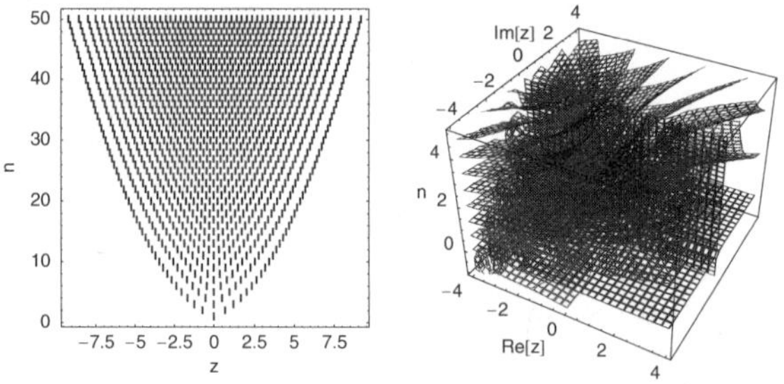

The next graphic shows how the new zeros arise when changing from $H_n(z)$ to $H_{n+1}(z)$.

```
In[10]:= DensityPlot[Sqrt[Abs[HermiteH[n, z]]], {z, -1.8, 1.8}, {n, 2, 4},
             PlotPoints -> {90, 60}, Mesh -> False, AspectRatio -> 1/2]
```

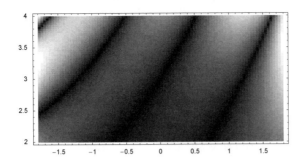

We now test whether the Hermite polynomials actually solve the differential equation given in the beginning of this section.

```
In[11]:= HermiteDifferentialOperator[w_, z_, n_] :=
              D[w, {z, 2}] - 2 z D[w, z] + 2 n w;

       Expand[HermiteDifferentialOperator[#[[1]], z, #[[2]]]]& /@
                 Table[{HermiteH[n, z], n}, {n, 0, 9}]
Out[12]= {0, 0, 0, 0, 0, 0, 0, 0, 0, 0}
```

Mathematica can also find the general solution of the Hermite differential operator. It contains hypergeometric $_1F_1$ functions, to be discussed in the next chapter.

```
In[13]:= DSolve[HermiteDifferentialOperator[w[ξ], ξ, v] == 0, w[ξ], ξ]
```

$$Out[13]= \left\{\left\{w[ξ] \to C[1] \ HermiteH[v, ξ] + C[2] \ Hypergeometric1F1\left[-\frac{v}{2}, \frac{1}{2}, ξ^2\right]\right\}\right\}$$

For v nonnegative integer and for appropriate values of the $C[k]$ we recover the Hermite polynomials.

```
In[14]:= Table[{k, DSolve[HermiteDifferentialOperator[w[ξ], ξ, k] == 0, w[ξ], ξ]},
              {k, 0, 4}] // FullSimplify
```

$$Out[14]= \left\{\left\{0, \left\{\left\{w[ξ] \to C[2] + \frac{1}{2} \sqrt{π} \ C[1] \ Erfi[ξ]\right\}\right\}\right\},\right.$$
$$\left\{1, \left\{\left\{w[ξ] \to 2 ξ C[1] + e^{ξ^2} C[2] - \sqrt{π} ξ C[2] \ Erfi[ξ]\right\}\right\}\right\},$$
$$\left\{2, \left\{\left\{w[ξ] \to -e^{ξ^2} ξ C[2] + \frac{1}{2} (-1 + 2 ξ^2) (C[1] + \sqrt{π} C[2] \ Erfi[ξ])\right\}\right\}\right\},$$
$$\left\{3, \left\{\left\{w[ξ] \to -e^{ξ^2} (-1 + ξ^2) C[2] + \frac{1}{2} ξ (-3 + 2 ξ^2) (8 C[1] + \sqrt{π} C[2] \ Erfi[ξ])\right\}\right\}\right\},$$
$$\left\{4, \left\{\left\{w[ξ] \to \right.\right.\right.$$
$$\left.\left.\left.\frac{1}{12} (-2 e^{ξ^2} ξ (-5 + 2 ξ^2) C[2] + (3 + 4 ξ^2 (-3 + ξ^2)) (3 C[1] + \sqrt{π} C[2] \ Erfi[ξ]))\right\}\right\}\right\}\right\}$$

```
In[15]:= Table[{k, DSolve[{HermiteDifferentialOperator[w[ξ], ξ, k] == 0,
                 (* initial conditions at the origin *)
                 w[0] == If[EvenQ[k], 2^k Sqrt[Pi]/Gamma[(1 - k)/2], 0],
                 w'[0] == If[EvenQ[k], 0, 2^k k Sqrt[Pi]/Gamma[(2 - k)/2]]},
                 w[ξ], ξ]}, {k, 0, 4}] // ExpandAll
Out[15]= {{0, {{w[ξ] → 1}}}, {1, {{w[ξ] → 2 ξ}}}, {2, {{w[ξ] → -2 + 4 ξ²}}},
         {3, {{w[ξ] → -12 ξ + 8 ξ³}}}, {4, {{w[ξ] → 12 - 48 ξ² + 16 ξ⁴}}}}
```

Next, we look at the orthogonality relation. The weight function for the Hermite polynomials is $\exp(-z^2)$, and the orthogonality interval is $(-\infty, \infty)$. We check this for the first four polynomials. (Here, we explicitly use Expand to speed up the integration.)

In[16]:= `Table[Integrate[Expand[HermiteH[n, z] HermiteH[m, z] Exp[-z^2]],`
`                  {z, -Infinity, Infinity}], {n, 0, 3}, {m, 0, 3}] //`
`                                                                      TableForm`

Out[16]//TableForm=

$\sqrt{\pi}$	0	0	0
0	$2\sqrt{\pi}$	0	0
0	0	$8\sqrt{\pi}$	0
0	0	0	$48\sqrt{\pi}$

It follows that the normalization constant c_n^2 must be $\sqrt{\pi}\, 2^n\, n!$.

In[17]:= `Table[Sqrt[Pi] 2^n n!, {n, 0, 3}]`

Out[17]= $\{\sqrt{\pi}, 2\sqrt{\pi}, 8\sqrt{\pi}, 48\sqrt{\pi}\}$

(For a 2D relative of this orthogonality property, see [583].) As mentioned, the various properties of orthogonal polynomials unfortunately cannot be verified with symbolic calculations. Here is an example of a failed attempt of this type.

In[18]:= `Function[int, TimeConstrained[int, 2], {HoldAll}] @`
`        Integrate[HermiteH[i, z] HermiteH[j, z] Exp[-z^2], {z, -Infinity, Infinity}]`

Out[18]= `$Aborted`

And the `Assumptions` option of `Integrate` currently does not allow us to specify a variable as being an integer.

In[19]:= `Function[int, TimeConstrained[int, 2], {HoldAll}] @`
`        Integrate[HermiteH[i, z] HermiteH[j, z] Exp[-z^2], {z, -Infinity, Infinity},`
`                  Assumptions -> Element[n, Integers]]`

Out[19]= `$Aborted`

For noninteger, real n, the weight function $\exp(-z^2)$ cannot compensate the growth of the Hermite functions (for negative z) and the integral diverges. Here is a numerical check for this statement.

In[20]:= `Table[NIntegrate[HermiteH[3/4, z] HermiteH[5/6, z] Exp[-z^2],`
`                  {z, -zM, zM}, Method -> DoubleExponential,`
`                  PrecisionGoal -> 2], {zM, 10, 30, 10}]`

Out[20]= $\{3.58516\times 10^{37}, 2.79915\times 10^{166}, 6.085059952500184\times 10^{382}\}$

The recurrence formula for the computation of the $(n+1)$-st Hermite polynomial from the previous two is

$$H_{n+1}(z) = 2\, z\, H_n(z) - 2\, n\, H_{n-1}(z).$$

We now check this.

In[21]:= `Table[HermiteH[n + 1, z] == Expand[`
`            2 z HermiteH[n, z] - 2 n HermiteH[n - 1, z]], {n, 1, 9}]`

Out[21]= `{True, True, True, True, True, True, True, True, True}`

For Hermite polynomials, Rodrigues' formula has a very simple form:

$$H_n(z) = (-1)^n \exp(z^2)\, \frac{d^n \exp(-z^2)}{dz^n}.$$

Here are the first few.

In[22]:= `Table[{HermiteH[n, z], Expand[ (-1)^n/Exp[-z^2] D[Exp[-z^2], {z, n}]]},`
`        {n, 0, 6}]`

Out[22]= {{1, 1}, {2 z, 2 z}, {-2 + 4 z², -2 + 4 z²}, {-12 z + 8 z³, -12 z + 8 z³},
      {12 - 48 z² + 16 z⁴, 12 - 48 z² + 16 z⁴}, {120 z - 160 z³ + 32 z⁵, 120 z - 160 z³ + 32 z⁵},
      {-120 + 720 z² - 480 z⁴ + 64 z⁶, -120 + 720 z² - 480 z⁴ + 64 z⁶}}

In[23]:= **Apply[Equal, %, {1}]**

Out[23]= {True, True, True, True, True, True, True}

The formula for the Hermite polynomials in terms of confluent hypergeometric functions is

$$H_{2n}(z) = (-1)^n \frac{(2n)!}{n!} \, _1F_1\left(-n, \frac{1}{2}, z^2\right)$$

$$H_{2n+1}(z) = (-1)^n \frac{(2n+1)!}{n!} \, 2z \, _1F_1\left(-n, \frac{3}{2}, z^2\right).$$

We also examine this formula.

In[24]:= **Table[Expand[(-1)^n (2n)!/n! Hypergeometric1F1[-n, 1/2, z^2]] ==**
         HermiteH[2n, z], {n, 0, 3}] // Union

Out[24]= {True}

In[25]:= **Table[Expand[(-1)^n (2n + 1)!/n! 2 z Hypergeometric1F1[-n, 3/2, z^2]] ==**
         HermiteH[2n + 1, z], {n, 0, 3}] // Union

Out[25]= {True}

Next, we turn to the generating function. One possible generating function is

$$e^{t(2z-t)} = \sum_{n=0}^{\infty} \frac{1}{n!} H_n(z) \, t^n.$$

We now check this.

In[26]:= **Expand /@ (Table[n!, {n, 0, 8}]***
         CoefficientList[Series[Exp[t(2z - t)], {t, 0, 8}], t])

Out[26]= {1, 2 z, -2 + 4 z², -12 z + 8 z³, 12 - 48 z² + 16 z⁴,
       120 z - 160 z³ + 32 z⁵, -120 + 720 z² - 480 z⁴ + 64 z⁶,
       -1680 z + 3360 z³ - 1344 z⁵ + 128 z⁷, 1680 - 13440 z² + 13440 z⁴ - 3584 z⁶ + 256 z⁸}

These are just the Hermite polynomials.

In[27]:= **% == Table[HermiteH[n, z], {n, 0, 8}]**

Out[27]= True

One possible integral formula for the Hermite polynomials is the following:

$$H_n(z) = \frac{2^{n+1}}{\sqrt{\pi}} \exp(z^2) \int_0^{\infty} \exp(-t^2) \, t^n \cos\left(2zt - \frac{n\pi}{2}\right) dt.$$

Here is the result of integration for an arbitrary *n*. Because, in general, *n* need not be a positive integer, the integral is not simplified in terms of Hermite polynomials, but instead is expressed in terms of hypergeometric functions.

In[28]:= **2^(n + 1) Exp[z^2]/Gamma[1/2]***
         Integrate[Exp[-t^2] t^n Cos[2 z t - n Pi/2], {t, 0, Infinity},
              GenerateConditions -> False] // Expand

Out[28]= $\dfrac{2^n\,e^{z^2}\,\text{Cos}\,[\frac{n\,\pi}{2}]\,\text{Gamma}\,[\frac{1+n}{2}]\,\text{Hypergeometric1F1}\,[\frac{1+n}{2},\,\frac{1}{2},\,-z^2]}{\sqrt{\pi}}\,+$

$\dfrac{2^{1+n}\,e^{z^2}\,\sqrt{z^2}\,\text{Gamma}\,[1+\frac{n}{2}]\,\text{Hypergeometric1F1}\,[\frac{2+n}{2},\,\frac{3}{2},\,-z^2]\,\text{Sign}\,[z]\,\text{Sin}\,[\frac{n\,\pi}{2}]}{\sqrt{\pi}}$

When n is a nonnegative integer, we again get the Hermite polynomials.

In[29]:= `{Expand[% /. n -> 4], HermiteH[4, z]}`

Out[29]= $\{12 - 48\,z^2 + 16\,z^4,\ 12 - 48\,z^2 + 16\,z^4\}$

Differentiation of a Hermite polynomial gives (modulo a prefactor) again a Hermite polynomial:

$$\frac{\partial^m H_n(z)}{\partial z^m} = \frac{2^m\,n!}{(n-m)!}\,H_{n-m}(z).$$

Here is a quick check for this relation.

In[30]:= `Table[Expand[D[HermiteH[n, z], {z, m}] - 2^m n!/(n - m)! HermiteH[n - m, z]],`
   `{n, 0, 10}, {m, 0, n}] // Flatten // Union`

Out[30]= $\{0\}$

As a result of the last differentiation formula, we have the following series expansion.

In[31]:= `Series[HermiteH[n, z], {z, z0, 5}]`

Out[31]= $\text{HermiteH}[n, z0] + 2\,n\,\text{HermiteH}[-1 + n, z0]\,(z - z0) +$
   $(-2\,n\,\text{HermiteH}[-2 + n, z0] + 2\,n^2\,\text{HermiteH}[-2 + n, z0])\,(z - z0)^2 +$
   $\dfrac{4}{3}\,(1 - n)\,(2 - n)\,n\,\text{HermiteH}[-3 + n, z0]\,(z - z0)^3 -$
   $\dfrac{2}{3}\,((1 - n)\,(2 - n)\,(3 - n)\,n\,\text{HermiteH}[-4 + n, z0])\,(z - z0)^4 +$
   $\dfrac{4}{15}\,(1 - n)\,(2 - n)\,(3 - n)\,(4 - n)\,n\,\text{HermiteH}[-5 + n, z0]\,(z - z0)^5 + O[z - z0]^6$

For a couple of interesting operational representations of Hermite polynomials, see [125] and [624].

After multiplying the Hermite polynomials by their weight functions, we obtain the functions

$$\phi_k(x) = \frac{1}{\sqrt{\sqrt{\pi}\,2^k\,k!}}\,e^{-x^2/2}\,H_k(x)$$

which are very important in quantum mechanics. $\phi_n^2(x)$ are just the probability densities for a quantum-mechanical particle in a harmonic oscillator potential in the stationary state labeled by n [569] (for the corresponding Sturm basis, see [16] and [553]; for oscillators with small frequency, see [502]). They fulfill the eigenvalue equation $-\phi_n''(x) + x^2\,\phi_n(x) = (2n + 1)\,\phi_n(x)$.

In[32]:= `φ[n_, x_] := 1/ Sqrt[Sqrt[Pi] 2^n n!] Exp[-x^2/2] HermiteH[n, x]`

The eigenfunctions fulfill the orthogonality relation $\int_{-\infty}^{\infty}\phi_n(x)\,\phi_m(x)\,dx = \delta_{n,m}$. In addition, they fulfill the following interesting identity: $\int_{-\infty}^{\infty}\phi_m(x)\,\phi_{2m+2}(x)/\phi_0(x)\,dx = 0$ [544]. Here is a quick check for the lowest states.

In[33]:= `Table[Integrate[φ[m, x]^2 φ[2m + 2, x]/φ[0, x], {x, -Infinity, Infinity}],`
   `{m, 0, 12}]`

Out[33]= $\{0, 0, 0, 0, 0, 0, 0, 0, 0, 0, 0, 0, 0\}$

Here are some of the eigenfunctions visualized.

```
In[34]:= Show[GraphicsArray[#]]& /@
        Table[(* the individual plots *)
            Plot[φ[3i + j, x], {x, -Sqrt[5 (3i + j) + 3], Sqrt[5 (3i + j) + 3]},
                PlotLabel -> (StyleForm[TraditionalForm[
                                HoldForm[HermiteH[#, x]]],
                                FontSize -> 6]&[3i + j]),
                AxesLabel -> {"x", "ψ(x)"},  TextStyle -> {FontSize -> 6},
                DisplayFunction -> Identity, PlotRange -> All,
                AxesOrigin -> {-Sqrt[6i + 2j + 3], 0}], {i, 0, 2}, {j, 0, 2}]
```

We now look at the probability of the higher state $n = 16$ in somewhat more detail. (With some imagination, one sees the square root singularities from the probability of a classical harmonic oscillator [477].) The curve is shown "filled" in the right graphic.

```
In[35]:= Show[GraphicsArray[{#, # /. Line -> Polygon}]]&[
        Plot[φ[16, x]^2, {x, -8, 8}, PlotPoints -> 200, PlotRange -> All,
            DisplayFunction -> Identity]]
```

Adding the normalized Hermite polynomials gives the integrated local density of states [371], [183], [385], [588], [418] of in a 1D quadratic potential that is embedded in three dimensions [499], [97], [98]. (For fast summation methods of Hermite functions, see [93].)

```
In[36]:= Module[{ε = 10^-10, pp = 124, eMax = 50, xMax, ψ, dos, δ},
         (* x-range between classical turning points *)
         xMax = 1.1 Sqrt[eMax];
         (* local density of states *)
         dos[e_, x_] := Sum[φ[n, x]^2 UnitStep[e - (2n + 1)], {n, 0, e}];
         (* make polygons *)
         δ = Flatten[MapIndexed[Function[{α, β},
                     {{#[[1]], 2β[[1]] - 1 + ε, #[[2]]}& /@ α,
                      {#[[1]], 2β[[1]] + 1 - ε, #[[2]]}& /@ α},
                     Table[{x, dos[n, N[x]]},
                            {n, 0, eMax, 2}, {x, -xMax, xMax, 2xMax/pp}]], 1];
         (* make graphics *)
         Show[Graphics3D[
         {EdgeForm[], SurfaceColor[Hue[0.05], Hue[0.22], 2.88],
          Table[Polygon[{δ[[i, j]], δ[[i + 1, j]],
                        δ[[i + 1, j + 1]], δ[[i, j + 1]]}],
               {i, Length[δ] - 1}, {j, Length[δ[[1]]] - 1}]}],
               PlotRange -> All, BoxRatios -> {2, 3, 0.6},
               Axes -> {True, True, False}, ViewPoint -> {0, -3, 1}]]
```

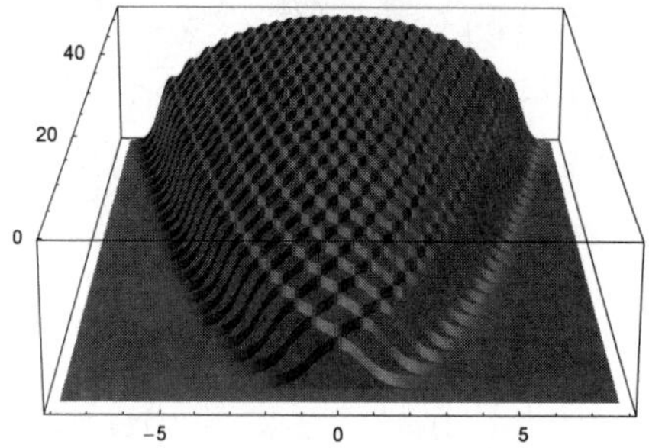

The time-dependent harmonic oscillator obeys the Schrödinger equation

$$i \frac{\partial \psi(x, t)}{\partial t} = -\frac{\partial^2 \psi(x, t)}{\partial x^2} + x^2 \psi(x, t).$$

It has stationary solutions $\phi_k(x, t) = \phi_k(x) \exp(-i(2k + 1)t)$. Using the special initial condition of a shifted eigenfunction $\psi(x, 0) = \phi_n(x - X)$, we will analyze the time-dependence of $\psi(x, t)$. The solution of the time-dependent Schrödinger equation can then be written as $\psi(x, t) = \sum_{k=0}^{\infty} c_k \phi_k(x, t)$, where $c_k = \int_{-\infty}^{\infty} \phi_n(x) \psi(x, 0) \, dx$ [278]. We use $n = 6$ and $X = 5/2$ in the following pictures. The left graphic shows $|\psi(x, t)|^2$ and the right graphic shows $\arg(\psi(x, t))$. To resolve the phase for larger x, we use high-precision values for the discretized $\phi_k(x)$.

```
In[37]:= Module[{ν = 6, X = 5/2, xMax = 8, T = 2Pi,
               o = 75, pp = 300, prec = 30},
         (* time-independent eigenfunctions *)
         φ[n_, x_] := 1/Sqrt[Sqrt[Pi] 2^n n!] Exp[-x^2/2] HermiteH[n, x];
         (* closed for of expansion coefficients *)
         c[n_, n_, ξ_] := Sqrt[n!/n!/2^(n - n)] Exp[-ξ^2/4] (-ξ)^(n - n) *
                          Hypergeometric1F1Regularized[-n, n - n + 1, ξ^2/2];
         Do[(* discretized time-independent eigenfunctions *)
            φN[n] = N[Table[Evaluate[φ[n, x]],
                           {x, -xMax, xMax, 2xMax/pp}], prec];
            (* expansion coefficients *)
```

```
        cN[n] = N[c[n, ν, X], prec], {n, 0, o}];
  (* construct array of time-dependent ψ-values *)
  ΨData = Table[Sum[cN[n] φN[n] Exp[-I (2n + 1) t], {n, 0, o}],
              {t, 0, T, T/pp}];
  (* make graphics *)
  Show[GraphicsArray[
  Block[{$DisplayFunction = Identity},
        {(* 3D plot of Abs[ψ[x, t]]^2 *)
          ListPlot3D[Abs[ΨData]^2, Mesh -> False, PlotRange -> All,
                  MeshRange -> {{-xMax, xMax}, {0, T}}],
          (* contour of Arg[ψ[x, t]]^2 *)
          ListContourPlot[Arg[ΨData]^2, PlotRange -> All,
            MeshRange -> {{-xMax, xMax}, {0, T}},
            ColorFunction -> (Hue[#/Pi^2]&), ContourLines -> False,
            ColorFunctionScaling -> False}]]]]
```

We give one more quantum-mechanical application of the Hermite polynomials. The time-independent solutions for a 2D harmonic oscillator are products of Hermite functions of the two coordinates. By forming a coherent state, the behavior of a classical particle distribution becomes visible [131]. Here are three examples of such coherent states.

```
In[38]:= (* 2D eigenfunctions of a harmonic oscillator *)
        φHO2D[{n_, m_}, {x_, y_}, {X_, Y_}] := 1/Sqrt[Pi 2^(n + m - 1) m! n! X Y]*
        HermiteH[n, Sqrt[2] x/X] HermiteH[m, Sqrt[2] y/Y] Exp[-(x/X)^2 - (y/Y)^2]

In[40]:= (* complete SU(2) coherent states *)
        φCSHO2D[n_, {p_, q_}, {x_, y_}, {X_, Y_}, A_] := (1 + Abs[A]^2)^(-n/2)*
        Sum[Sqrt[Binomial[n, k]] A^k φHO2D[{p k, q (n - k)}, {x, y}, {X, Y}],
            {k, 0, n}]

In[42]:= φCSHO2DContourPlot[n_, {p_, q_}, A_, {Lx_, Ly_}, pp_] :=
        ListContourPlot[#[[1]], PlotRange -> All, FrameTicks -> False,
                  (* use homogeneous contour spacing *)
                  Contours -> (#[[Round[pp^2/120]]]& /@
                      Partition[Sort[Flatten[#[[1]]]], Round[pp^2/60]]),
                        ColorFunction -> (Hue[0.8 #]&), ContourLines -> False]&[
        (* calculate square root of probability values *)
        DensityPlot[Evaluate @ (Abs[φCSHO2D[n, {p, q}, {x, y}, {1, 1}, A] -
                  φCSHO2D[n, {p, q}, {x, y}, {1, 1}, Conjugate[A]]]),
                  {x, -Lx, Lx},  {y, -Ly, Ly}, PlotPoints -> pp,
                  DisplayFunction -> Identity]]

In[43]:= Show[GraphicsArray[
        Block[{$DisplayFunction = Identity, pp = 400},
          {φCSHO2DContourPlot[40, {1, 1}, Exp[I Pi/2], {6, 6}, pp],
          (* main probability is along Lissajou curves *)
```

```
φCSHO2DContourPlot[16, {5, 3}, Exp[I Pi/2], {9, 7}, pp],
φCSHO2DContourPlot[12, {6, 6}, Exp[I Pi/16], {9, 9}, pp]}]]]
```

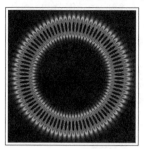 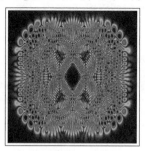 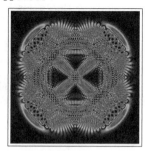

For exactly solvable potentials that are generated through Hermite polynomials at imaginary arguments, see [230]. For quantum field theoretical applications of the Hermite polynomials, see [626].

2.3 Jacobi Polynomials

The Jacobi polynomials arise as solutions of the eigenvalue problem

$$(1 - z^2)\,\Psi''(z) + (\beta - \alpha - (\alpha + \beta + 2)\,z)\,\Psi'(z) + \lambda(\lambda + \alpha + \beta + 1)\,\Psi(z) = 0$$

on the interval $(-1, 1)$, under the restriction that the solution should be finite at $z = \pm 1$. In this case, the eigenvalues are $\lambda = n\,(n \in \mathbb{N},\, n \geq 0)$. In *Mathematica*, the naming for the Jacobi polynomials is the following.

JacobiP[n, α, β, z]

   $(n \in \mathbb{N}, n \geq 0)$ gives the nth Jacobi polynomial $P_n^{(\alpha,\beta)}(z)$. If n is not a nonnegative integer, the corresponding analytic continuation of $P_n^{(\alpha,\beta)}(z)$ is understood.

We examine the first few $P_n^{(\alpha,\beta)}(z)$ explicitly.

In[1]:= **With[{p = JacobiP},**
 Table[HoldForm[p[v, α, β, z]] == p[n, α, β, z] /.
 v -> n, {n, 0, 2}]] // TableForm // TraditionalForm

Out[1]//TraditionalForm=

$P_0^{(\alpha,\beta)}(z) = 1$

$P_1^{(\alpha,\beta)}(z) = \frac{1}{2}\,(\alpha - \beta + z\,(\alpha + \beta + 2))$

$P_2^{(\alpha,\beta)}(z) = \frac{1}{8}\,(\alpha + \beta + 3)\,(\alpha + \beta + 4)\,(z - 1)^2 + \frac{1}{2}\,(\alpha + 2)\,(\alpha + \beta + 3)\,(z - 1) + \frac{1}{2}\,(\alpha + 1)\,(\alpha + 2)$

With three variables z, α, and β at our disposal, it is difficult to make the "appropriate" choices to get a good overview via a plot. Here are a few examples.

In[2]:= **Show[GraphicsArray[#]]& /@**
 Table[(* the individual plots *)
 Plot[JacobiP[3i + j, 0, 0, z], {z, -1, 1},
 PlotLabel -> (StyleForm[TraditionalForm[
 HoldForm[JacobiP[#, 0, 0, z]]],
 FontSize -> 7]&[3i + j]),

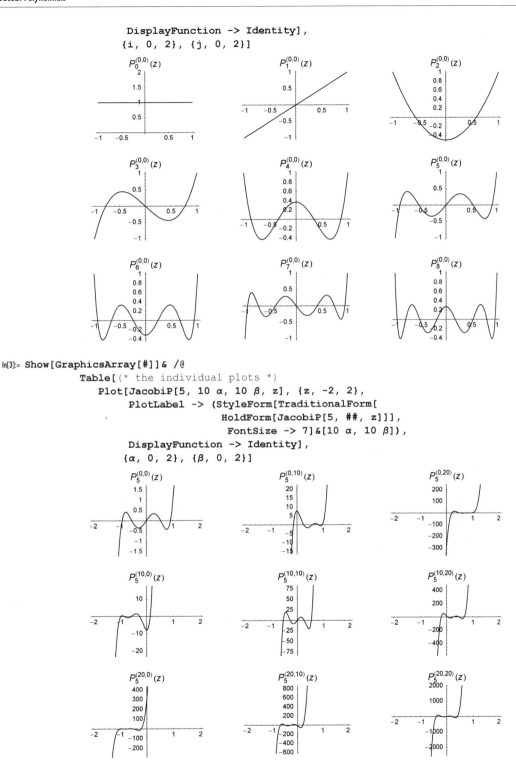

```
                        DisplayFunction -> Identity],
                        {i, 0, 2}, {j, 0, 2}]
```

```
In[3]:= Show[GraphicsArray[#]] & /@
           Table[(* the individual plots *)
             Plot[JacobiP[5, 10 α, 10 β, z], {z, -2, 2},
                 PlotLabel -> (StyleForm[TraditionalForm[
                                 HoldForm[JacobiP[5, ##, z]]],
                                 FontSize -> 7]&[10 α, 10 β]),
               DisplayFunction -> Identity],
             {α, 0, 2}, {β, 0, 2}]
```

Here, the dependence of the zeros of $P_n^{(\alpha,\beta)}(z)$ on α and β is shown for $n = 12$, $\beta = 12$ and for $n = 12$, $a = 12$.

```
In[4]:= Show[GraphicsArray[
          {Graphics[{PointSize[0.003], Table[(Point[{α, #}]& /@
                    (* calculate and extract roots *)
                    Last /@ (List @@ NRoots[JacobiP[12, α, 12, z] == 0, z])),
                    {α, 0, 80}]}, Frame -> True, FrameLabel -> {"z", "α"}],
           Graphics[{PointSize[0.003], Table[(Point[{β, #}]& /@
                    (* calculate and extract roots *)
                    Last /@ (List @@ NRoots[JacobiP[12, 12, β, z] == 0, z])),
                    {β, 0, 80}]}, Frame -> True, FrameLabel -> {"z", "β"}]}]]
```

Because the roots are real (and so are possible to order uniquely), we can also easily make 3D surfaces of the roots. We demonstrate this here for $P_8^{(\alpha,\beta)}(z)$.

```
In[5]:= Needs["Graphics`Graphics3D`"]

Show[Graphics3D[Function[poly,
Module[{mp = Plus @@ poly[[1]]/4,
        p1, p2, p3, p4, pn1, pn2, pn3, pn4, fac = 0.9},
(* this makes the holes in the polygons *)
{p1, p2, p3, p4} = poly[[1]];
{pn1, pn2, pn3, pn4} = (mp + fac (# - mp))& /@ {p1, p2, p3, p4};
{{EdgeForm[], {Polygon[{p1, p2, pn2, pn1}],
 Polygon[{p2, p3, pn3, pn2}], Polygon[{p3, p4, pn4, pn3}],
 Polygon[{p4, p1, pn1, pn4}]}}, {Thickness[0.002], GrayLevel[0],
 Line[{pn1, pn2, pn3, pn4, pn1}]}}]] /@
Flatten[(* collecting the surfaces of all 8 roots *)
ListSurfacePlot3D[#, DisplayFunction -> Identity][[1]]& /@
Transpose[(* the roots *) Table[{α, β, #}& /@
          Cases[NRoots[JacobiP[8, α, β, z] == 0, z], _?NumberQ, {-1}],
          {α, 10, 120, 10}, {β, 10, 120, 10}],
        (* transposing in appropriate form ListSurfacePlot3D *)
        {3, 2, 1, 4}]]], Axes -> True, PlotRange -> All,
        BoxRatios -> {1, 1, 1}, AxesLabel -> {"α", "β", "z"}]
```

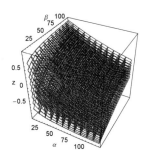

Here is a set of contour plots of $\mathrm{Re}(P_n^{(\alpha,\beta)}(z))$ over the complex z-plane. The contour plots are arranged in the n,α,β-space. This picture is best viewed in magnified form to better see the individual contour plot.

```
In[7]:= Show[Table[If[Random[] < 1/3, (* make 3D graphics *)
           Graphics3D[Graphics[ContourPlot[(* individual contour plots *)
                        Evaluate[Re[JacobiP[n, α, β, x + I y]]],
                        {x, -5, 5}, {y, -5, 5},
                        ContourShading -> False, Contours -> 12,
                        FrameTicks -> None, DisplayFunction -> Identity,
                        ContourStyle -> Table[{Thickness[0.002], Hue[k/15]},
                                              {k, 12}]]][[1]]] /.
                        (* position the contour plot *)
               Line[l_] :> Line[(# + {n, α, β})& /@
                 (Append[#, 0]& /@ (1/2(# + {5, 5})/10& /@ l))], {}],
               {n, 0, 5}, {α, 0, 5}, {β, 0, 5}],
           BoxRatios -> {3, 3, 1}, DisplayFunction -> $DisplayFunction]
```

We now test whether the Jacobi polynomials satisfy the above differential equation.

```
In[8]:= JacobiDifferentialOperator[w_, z_, n_] :=
           (1 - z^2) D[w, {z, 2}] + (β - α - (α + β + 2) z)  D[w, z] +
                                              n (n + α + β + 1) w;

        Expand[JacobiDifferentialOperator[#[[1]], z, #[[2]]]& /@
                    Table[{JacobiP[n, α, β, z], n}, {n, 0, 3}]]

Out[9]= {0, 0, 0, 0}
```

Now, we turn to the orthogonality relation. The weight function for the Jacobi polynomials is $(1-z)^\alpha (1+z)^\beta$, and the orthogonality interval is $(-1, 1)$. However, we cannot verify orthogonality in the generic case for integer order.

```
In[10]:= Table[Integrate[Expand[JacobiP[n, α, β, x] JacobiP[m, α, β, x] *
                         (1 - x)^α (1 + x)^β], {x, -1, 1},
```

Assumptions -> Re[α] > -1 && Re[β] > -1],
{n, 0, 1}, {m, 0, 1}] // FullSimplify

Out[10]= $\{\{$Gamma[1 + β] Hypergeometric2F1Regularized[1, -α, 2 + β, -1] +

Gamma[1 + α] Hypergeometric2F1Regularized[1, -β, 2 + α, -1],

$\dfrac{1}{4}\Big(-\dfrac{(2 + α + β)\ (2 + β + (α - β)\ \text{Hypergeometric2F1}[1, -α, 3 + β, -1])}{2 + 3 β + β^2}$ +

$\dfrac{(2 + α + β)\ (2 + α + (-α + β)\ \text{Hypergeometric2F1}[1, -β, 3 + α, -1])}{(1 + α)\ (2 + α)}$ +

2 (α - β) Gamma[1 + β] Hypergeometric2F1Regularized[1, -α, 2 + β, -1] +

2 (α - β) Gamma[1 + α] Hypergeometric2F1Regularized[1, -β, 2 + α, -1]$\Big)\}$,

$\{\dfrac{1}{4}\Big(-\dfrac{(2 + α + β)\ (2 + β + (α - β)\ \text{Hypergeometric2F1}[1, -α, 3 + β, -1])}{2 + 3 β + β^2}$ +

$\dfrac{(2 + α + β)\ (2 + α + (-α + β)\ \text{Hypergeometric2F1}[1, -β, 3 + α, -1])}{(1 + α)\ (2 + α)}$ +

2 (α - β) Gamma[1 + β] Hypergeometric2F1Regularized[1, -α, 2 + β, -1] +

2 (α - β) Gamma[1 + α] Hypergeometric2F1Regularized[1, -β, 2 + α, -1]$\Big)$,

$\dfrac{1}{2 + α + β}\left(2^{2+α+β}\left(\dfrac{2 + α + β}{1 + α}\right)^{-α}\left(\dfrac{2 + α + β}{1 + β}\right)^{-β}\right.$

$\left((1 + β)^3\ \text{Gamma}[1 + β]\ \text{Hypergeometric2F1Regularized}\Big[3, -α, 4 + β, -\dfrac{1 + β}{1 + α}\Big]\ +\right.$

$\left.\left.(1 + α)^3\ \text{Gamma}[1 + α]\ \text{Hypergeometric2F1Regularized}\Big[3, -β, 4 + α, -\dfrac{1 + α}{1 + β}\Big]\right)\right)\}\}\}$

In terms of Γ functions, the normalization factor c_n^2 becomes

$$\frac{2^{α+β+1}}{(2n+α+β+1)}\ \frac{\Gamma(n+α+1)\,\Gamma(n+β+1)}{n!\,\Gamma(n+α+β+1)}.$$

(The existence of the last integrals requires $α,\ β > -1$; for the orthogonality in case this condition is violated, see [337].)

The recurrence formula for the computation of the $(n+1)$-st Jacobi polynomial from the previous ones is

$$2\,(n+1)\,(n+α+β+1)\,(2n+α+β)\,P_{n+1}^{(α,β)}(z) =$$
$$((2n+α+β+1)\,(α^2 - β^2) + (2n+α+β)_3\,z)\,P_n^{(α,β)}(z) - 2\,(n+α)\,(n+β)\,(2n+α+β+2)\,P_{n-1}^{(α,β)}(z).$$

Here, $(a)_n$ is the Pochhammer symbol (see the next chapter). We now implement this recurrence formula.

In[11]:= **Table[Expand[(2 (n + 1) (n + α + β + 1) (2n + α + β))* **
                        **JacobiP[n + 1, α, β, z]] == **
        **Expand[((2n + α + β + 1) (α^2 - β^2) + **
                **Pochhammer[2n + α + β, 3] z) JacobiP[n, α, β, z] - **
            2 (n + α) (n + β) (2n + α + β + 2) JacobiP[n - 1, α, β, z]], {n, 1, 4}]

Out[11]= {True, True, True, True}

The Rodrigues' formula for Jacobi polynomials has the following form:

$$P_n^{(α,β)}(z) = \frac{1}{(-1)^n\,2^n\,n!\,(1-z)^α\,(1+z)^β}\ \frac{d^n\,\{(1-z)^α\,(1+z)^β\,(1-z^2)^n\}}{d\,z^n}.$$

Again, we look at the first few n.

```
In[12]:= Table[{JacobiP[n, α, β, z],
              Expand[1/((-1)^n 2^n n! (1 - z)^α (1 + z)^β)*
                     D[(1 - z)^α (1 + z)^β (1 - z^2)^n, {z, n}]]},
         {n, 0, 4}] // (Together[Subtract[##]]& @@@ #)&
Out[12]= {0, 0, 0, 0, 0}
```

A formula for the Jacobi polynomials in terms of hypergeometric functions is

$$P_n^{(\alpha,\beta)}(z) = \frac{(n+\alpha)!}{(n!\,\alpha!)} \,_2F_1\left(-n,\, n+\alpha+\beta+1,\, \alpha+1,\, \frac{1-z}{2}\right).$$

We also look at this formula.

```
In[13]:= Together /@ Simplify /@ (Table[(n + α)!/(n! α!) *
              Hypergeometric2F1[-n, n + α + β + 1, α + 1, (1 - z)/2],
         {n, 0, 2}] //. {(α + i_Integer)! -> α! Product[α + j, {j, 1, i}]})
```

$$Out[13]= \left\{1,\, \frac{1}{2}\,(2\,z + \alpha + z\,\alpha - \beta + z\,\beta),\right.$$

$$\frac{1}{8}\,(-4 + 12\,z^2 - \alpha + 6\,z\,\alpha + 7\,z^2\,\alpha + \alpha^2 + 2\,z\,\alpha^2 + z^2\,\alpha^2 - \beta - 6\,z\,\beta + 7\,z^2\,\beta -$$

$$\left. 2\,\alpha\,\beta + 2\,z^2\,\alpha\,\beta + \beta^2 - 2\,z\,\beta^2 + z^2\,\beta^2)\right\}$$

```
In[14]:= Table[Together[JacobiP[n, α, β, z]], {n, 0, 2}] == %
Out[14]= True
```

A generating function is given by

$$\frac{1}{R\,(1-t+R)^\alpha\,(1+t+R)^\beta} = \sum_{n=0}^{\infty} \frac{1}{2^{\alpha+\beta}}\, P_n^{(\alpha,\beta)}(z)\, t^n$$

where $R = \sqrt{1 - 2zt + t^2}$. Here is a quick check.

```
In[15]:= R = Sqrt[1 - 2 z t + t^2];
         Expand /@ (2^(α + β) CoefficientList[
                    Series[1/R 1/(1 - t + R)^α 1/(1 + t + R)^β,
                           {t, 0, 3}], t]) ==
         Table[Expand[JacobiP[n, α, β, z]], {n, 0, 3}]
Out[16]= True
```

(Integral representations for Jacobi polynomials are relatively complicated and, hence, are not presented here; see the references cited earlier.)

The zeros of the Jacobi polynomials have an interesting property. Assume n unit charges interacting with each other via a $-\log(distance)$ potential are located in the interval $(-1, 1)$. At the interval endpoints, we have charges of size q and p; then the equilibrium positions of the freely movable unit charges are at $x_1, x_2, \ldots, x_n$ are the zeros of the polynomial $P_n^{(2p-1,2q-1)}(x)$ (see [313], [580], [403], [207], [289], [517] and [288]).

Here is a quick check for this statement for $n = 3$.

```
In[17]:= equilibriumConditions[n_] :=
         Numerator /@ Together /@ Table[Sum[If[i =!= j, -1/(x[j] - x[i]), 0],
                           {i, 1, n}] + p/(x[j] - 1) - q/(x[j] + 1), {j, n}]
```

```
In[18]:= GroebnerBasis[Join[equilibriumConditions[3],
         {(* no two charge positions coincide *)
          1 - λ (x[1] - x[2])(x[1] - x[3])(x[2] - x[3])}],
```

```
      {x[1], p, q}, {x[2], x[3], λ},
      MonomialOrder -> EliminationOrder] /. x[1] -> x // Factor
```
Out[18]= $\{-5\,p + 3\,p^2 + 2\,p^3 - 5\,q + 6\,p^2\,q - 3\,q^2 + 6\,p\,q^2 + 2\,q^3 + 6\,x - 9\,p\,x - 3\,p^2\,x +$
$6\,p^3\,x + 9\,q\,x - 18\,p\,q\,x + 6\,p^2\,q\,x - 3\,q^2\,x - 6\,p\,q^2\,x - 6\,q^3\,x + 9\,p\,x^2 - 15\,p^2\,x^2 +$
$6\,p^3\,x^2 + 9\,q\,x^2 - 6\,p^2\,q\,x^2 + 15\,q^2\,x^2 - 6\,p\,q^2\,x^2 + 6\,q^3\,x^2 - 6\,x^3 + 13\,p\,x^3 -$
$9\,p^2\,x^3 + 2\,p^3\,x^3 - 13\,q\,x^3 + 18\,p\,q\,x^3 - 6\,p^2\,q\,x^3 - 9\,q^2\,x^3 + 6\,p\,q^2\,x^3 - 2\,q^3\,x^3\}$

For comparison, here is the factored form of $P_3^{(2p-1,2q-1)}(x)$.

In[19]:= `JacobiP[3, 2p - 1, 2q - 1, x] // Factor`

Out[19]= $\dfrac{1}{12}$ $(-5\,p - 3\,p^2 + 2\,p^3 + 5\,q - 6\,p^2\,q + 3\,q^2 + 6\,p\,q^2 - 2\,q^3 - 6\,x - 9\,p\,x + 3\,p^2\,x +$
$6\,p^3\,x - 9\,q\,x - 18\,p\,q\,x - 6\,p^2\,q\,x + 3\,q^2\,x - 6\,p\,q^2\,x + 6\,q^3\,x + 9\,p\,x^2 + 15\,p^2\,x^2 +$
$6\,p^3\,x^2 - 9\,q\,x^2 + 6\,p^2\,q\,x^2 - 15\,q^2\,x^2 - 6\,p\,q^2\,x^2 - 6\,q^3\,x^2 + 6\,x^3 + 13\,p\,x^3 +$
$9\,p^2\,x^3 + 2\,p^3\,x^3 + 13\,q\,x^3 + 18\,p\,q\,x^3 + 6\,p^2\,q\,x^3 + 9\,q^2\,x^3 + 6\,p\,q^2\,x^3 + 2\,q^3\,x^3)$

(For another interesting "physical" interpretation of the zeros of many orthogonal polynomials, see [276]; for the sums of powers of all zeros, see [4].)

One classical application of the Jacobi polynomials is the closed-form solution of the Pöschl–Teller potential $V(x) = (\alpha^2 - 1/4)\sin^{-2}(x) + (\beta^2 - 1/4)\cos^{-2}(x)$ [240], [154], [634], [635], [351], [321], [36], [220], [286], [625], [241], [147]. The differential equation, the eigenvalues, and the normalized solutions are given by the following expressions:

$$-\psi_n''(x) + \frac{1}{4}\left(\frac{4\alpha^2 - 1}{\sin^2(x)} + \frac{4\beta^2 - 1}{\cos^2(x)}\right)\psi_n(x) = \varepsilon_n\,\psi_n(x)$$

$$\varepsilon_n = (\alpha + \beta + 2n + 1)^2$$

$$\psi_n(x) = N_n\,\sin^{\alpha+\frac{1}{2}}(x)\,\cos^{\beta+\frac{1}{2}}(x)\,P_n^{(\alpha,\beta)}(\cos(2x))$$

$$N_n = \sqrt{\frac{2(2n + \alpha + \beta + 1)\,n!\,\Gamma(n + \alpha + \beta + 1)}{\Gamma(n + \alpha + 1)\,\Gamma(n + \beta + 1)}}.$$

Due to the inverse powers of the trigonometric functions, the natural boundary conditions are $\psi_n(0) = \psi_n(\frac{\pi}{2}) = 0$. (The square well type energy dependence $\varepsilon_n \sim n^2$ arises from supersymmetric partnership [212], [551], [319].) These formulas are straightforward to implement.

In[20]:= `V[{α_, β_}, x_] := (α^2 - 1/4)/Sin[x]^2 + (β^2 - 1/4)/Cos[x]^2`

In[ ]:= `ε[n_, {α_, β_}] := (α + β + 2n + 1)^2;`

In[ ]:= `N[n_, {α_, β_}] := Sqrt[2(α + β + 2n + 1) n! Gamma[α + β + n + 1]/`
`                      (Gamma[α + n + 1] Gamma[β + n + 1])]`

In[ ]:= `ψ[n_, {α_, β_}, x_] := N[n, {α, β}]*`
`        Sin[x]^(α + 1/2) Cos[x]^(β + 1/2) JacobiP[n, α, β, Cos[2x]]`

Similar to the harmonic oscillator case from above, let us visualize the 3D density of states $\mathcal{D}$ [371], [183] of a 1D Pöschl–Teller potential living in three dimensions [499]. (Because of the periodicity of the potential, we restrict the x-range to $0 \le x \le \pi/2$.)

In[24]:= `D[e_, {α_, β_}, x_] := Sum[ψ[n, {α, β}, x]^2 UnitStep[e - ε[n, {α, β}]],`
`                      {n, 0, Ceiling[(Sqrt[e] - α - β - 1)/2]}];`

In[25]:= `Needs["Graphics`Graphics3D`"]`

```
In[26]:= data =
    With[{εs = Table[ε[k, {1, 1}], {k, -1, 12}], pp = 400, ε = 10^-6},
    {(* before and after the discontinuities *)
    Table[{x, # - ε, D[# - ε, {1., 1.}, x]}, {x, 0, Pi/2, Pi/2/pp}],
    Table[{x, # + ε, D[# + ε, {1., 1.}, x]}, {x, 0, Pi/2, Pi/2/pp}]}& /@ εs];

In[27]:= Show[Graphics3D[{EdgeForm[], N @ Cases[
    ListSurfacePlot3D[Flatten[data, 1], DisplayFunction -> Identity],
    _Polygon, Infinity]}], BoxRatios -> {2, 3, 1}]
```

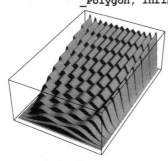

2.4 Gegenbauer Polynomials

Gegenbauer polynomials $C_n^{(\alpha)}(z)$ arise in the d-dimensional ($d > 3$) solution of the Laplace equation in spherical coordinates. They satisfy the following fundamental (generating function) relation [3]:

$$\frac{1}{|\mathbf{r} - \mathbf{r}'|^{d-2}} = \frac{1}{r_>^{d-2}} \sum_{n=0}^{\infty} \left(\frac{r_<}{r_>}\right)^n C_n^{(\alpha)}(\mathbf{e}.\mathbf{e}')$$

where $r_< = \min(|\mathbf{r}|, |\mathbf{r}'|)$, $r_> = \max(|\mathbf{r}|, |\mathbf{r}'|)$, $\mathbf{e} = \mathbf{r}/|\mathbf{r}|$, $\mathbf{e}' = \mathbf{r}'/|\mathbf{r}'|$, and $\alpha = (d - 2)/2$. Gegenbauer polynomials are eigenfunctions of the angular momentum operator $\hat{\Lambda}$, defined by [301], [31], [451], [171], [375], [293], [256]

$$\hat{\Lambda}^2 = -\sum_{i>j}^{d} \left(x_i \frac{\partial}{\partial x_j} - x_j \frac{\partial}{\partial x_i}\right)^2.$$

They satisfy

$$\left(\hat{\Lambda}^2 - \lambda(\lambda + d - 2)\right) C_\lambda^{(\alpha)}(\mathbf{e}\cdot\mathbf{e}') = 0$$

and are orthogonal over the d-dimensional unit sphere Ω_d

$$\int_{\Omega_d} C_n^{(\alpha)}(\mathbf{e}.\mathbf{e}')\, C_m^{(\alpha)}(\mathbf{e}'.\mathbf{e}'')\, d\Omega = \delta_{n,m} \frac{2\,\pi^{\alpha+1}}{\alpha!} \frac{\alpha}{n+\alpha} C_n^{(\alpha)}(\mathbf{e}.\mathbf{e}'').$$

(For the applications of the Gegenbauer polynomials as eigenfunctions of the angular momentum operator, see, in particular, [28], [419], [29], [333], [30], [191], [304], [112], [437], [581], and [424].)

Gegenbauer polynomials arise as the solution of the eigenvalue problem

$$(1 - z^2)\, \Psi''(z) - (2\,\alpha + 1)\, z\, \Psi'(z) + \lambda(\lambda + 2\,\alpha)\, \Psi(z) = 0\,, \quad z \in (-1, 1)$$

under the requirement that the solution is finite at ± 1. In this case, the eigenvalues are $\lambda = n$ ($n \in \mathbb{N}$, $n \geq 0$). The *Mathematica* notation for the Gegenbauer polynomials follows.

GegenbauerC[n, α, z]

> ($n \in \mathbb{N}$, $n \geq 0$) gives the nth Gegenbauer polynomial $C_n^{(\alpha)}(z)$. If n is not a nonnegative integer, the corresponding analytic continuation of $C_n^{(\alpha)}(z)$ is understood.

We look at the first few Gegenbauer polynomials explicitly.

In[1]:= **With[{p = GegenbauerC},**
Table[HoldForm[p[v, α, z]] == p[n, α, z] /.
v -> n, {n, 0, 4}]] // TableForm // TraditionalForm

Out[1]//TraditionalForm=

$$C_0^{(\alpha)}(z) = 1$$

$$C_1^{(\alpha)}(z) = 2\,z\,\alpha$$

$$C_2^{(\alpha)}(z) = 2\,z^2\,\alpha\,(\alpha + 1) - \alpha$$

$$C_3^{(\alpha)}(z) = \tfrac{4}{3}\,z^3\,\alpha\,(\alpha + 1)(\alpha + 2) - 2\,z\,\alpha\,(\alpha + 1)$$

$$C_4^{(\alpha)}(z) = \tfrac{2}{3}\,\alpha\,(\alpha + 1)(\alpha + 2)(\alpha + 3)\,z^4 - 2\,\alpha\,(\alpha + 1)(\alpha + 2)\,z^2 + \tfrac{1}{2}\,\alpha\,(\alpha + 1)$$

Again, it is difficult to choose appropriate parameters for a plot (look carefully at the scales at the vertical axes).

In[2]:= **Show[GraphicsArray[#]]& /@**
Table[(* the individual plots *)
Plot[GegenbauerC[3i + j, 1, z], {z, -1, 1},
PlotLabel -> (StyleForm[TraditionalForm[
HoldForm[GegenbauerC[#, 1, z]]],
FontSize -> 7]&[3i + j]),
DisplayFunction -> Identity], {i, 0, 2}, {j, 0, 2}]

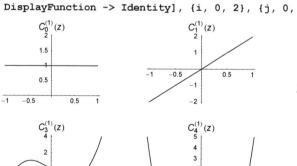

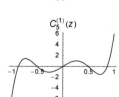

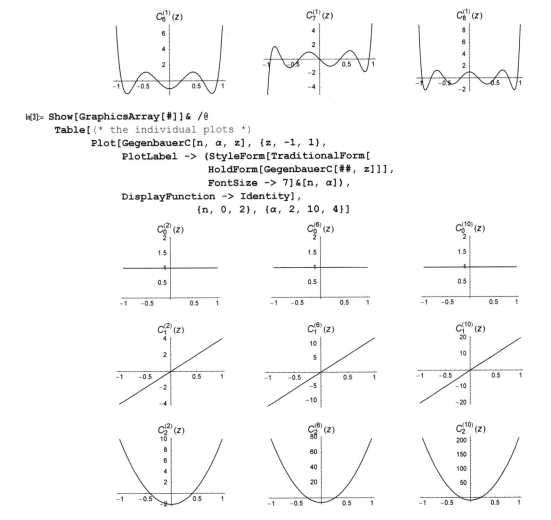

```
In[3]:= Show[GraphicsArray[#]]& /@
    Table[(* the individual plots *)
        Plot[GegenbauerC[n, α, z], {z, -1, 1},
            PlotLabel -> (StyleForm[TraditionalForm[
                            HoldForm[GegenbauerC[##, z]]],
                        FontSize -> 7]&[n, α]),
        DisplayFunction -> Identity],
            {n, 0, 2}, {α, 2, 10, 4}]
```

Here, we graphically determine for the regions $C_{18}^{(\alpha)}(z) > 0$ and $C_{18}^{(\alpha)}(z) < 0$ as a function of α. The left graphic is for real α and the other two graphic are for values of α with a fixed argument.

```
In[4]:= Show[GraphicsArray[
    Table[ContourPlot[Re[GegenbauerC[24, α Exp[I φ], z]],
                {z, -2, 2}, {α, -16, 16},
            Contours -> {0}, PlotPoints -> 360,
            DisplayFunction -> Identity],
        {φ, 0, Pi/2, Pi/4}]]]
```

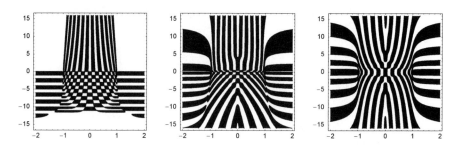

The following graphics shows $C_n^{(1)}(z)$ over the z,n-plane. It shows that (exclusively) for n being a nonnegative integer the function $C_n^{(1)}(-1)$ stays finite.

```
In[5]:= With[{ε = 10^-3},
    Plot3D[GegenbauerC[n, 1, z], {z, -1 + ε, 1 - ε}, {n, 0, 10},
        PlotPoints -> 120, Mesh -> False, PlotRange -> {-10, 10}]]
```

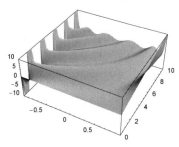

As a function of the parameter, the zeros of the Gegenbauer polynomials $C_{24}^{(\alpha)}(z)$ move in a complicated way in the complex plane. For $\alpha > -1/2$, all roots are in the interval $(-1, 1)$. The following graphic shows how the roots accumulate in this interval as α ranges from $-n$ to n [177]. The right graphic shows the location of the zeros for α varying over the unit circle.

```
In[6]:= With[{n = 24, o = 100},
    Show[GraphicsArray[
    Graphics[{PointSize[0.003],
        Table[{Hue[0.8 (α + n)/(2n)], Point[{Re[#], Im[#]}]}& /@
            Cases[NRoots[GegenbauerC[n, #[α], z] == 0, z],
            _?NumberQ, {-1}]}, {α, -n, n, 1/o}]},
        Frame -> True, PlotRange -> #2]& @@@
        (* function of α and plot range *)
        {{Identity, 8 {{-1, 1}, {-1, 1}}}, {Exp[I #]&, All}}]]]
```

We again test whether they actually solve the above differential equation.

```
In[7]:= GegenbauerDifferentialOperator[w_, z_, n_] :=
            (1 - z^2) D[w, {z, 2}] - (2 α + 1) z  D[w, z] + n (n + 2 α) w;

         Expand[GegenbauerDifferentialOperator[#[[1]], z, #[[2]]]& /@
                        Table[{GegenbauerC[n, α, z], n}, {n, 0, 3}]

Out[8]= {0, 0, 0, 0}
```

We now look at the orthogonality relation. The weight function for the Gegenbauer polynomials is $(1 - z^2)^{\alpha - 1/2}$, and the orthogonality interval is $(-1, 1)$. Again, there are difficulties with the verification of the orthogonality condition. The general integral remains unsolved.

```
In[9]:= Integrate[Expand[GegenbauerC[n, α, z] GegenbauerC[m, α, z]
                        (1 - z^2)^(α - 1/2)], {z, -1, 1}]
```

$$Out[9]= \int_{-1}^{1} (1 - z^2)^{-\frac{1}{2} + \alpha} \text{GegenbauerC}[m, \alpha, z] \text{GegenbauerC}[n, \alpha, z] \, dz$$

For the special choice $n = 1$, we can calculate a normalization.

```
In[10]:= int[α_] = Integrate[Expand[GegenbauerC[1, α, z] GegenbauerC[1, α, z]*
                        (1 - z^2)^(α - 1/2)], {z, -1, 1},
                    GenerateConditions -> False]
```

$$Out[10]= \frac{2 \sqrt{\pi} \, \alpha^2 \, \text{Gamma}[\frac{1}{2} + \alpha]}{\text{Gamma}[2 + \alpha]}$$

For general integer n, the normalization is $\pi \, 2^{1-2\alpha} \, \Gamma(2\alpha + n)/(n! \, (\alpha + n) \, \Gamma(\alpha)^2)$. For $n = 1$, this is just the last result. (Here we use the function `FullSimplify`, we will discuss it in more detail in the next chapter.)

```
In[11]:= ((Pi 2^(1 - 2 α) Gamma[2 α + n])/
             (n! (α + n) Gamma[α]^2) /. n -> 1) - % // FullSimplify
Out[11]= 0
```

The recurrence formula for the computation of the $(n + 1)$-st Gegenbauer polynomial from previous ones is

$$(n + 1) \, C_{n+1}^{(\alpha)}(z) = 2 \, (n + \alpha) \, z \, C_n^{(\alpha)}(z) - (n + 2\alpha - 1) \, C_{n-1}^{(\alpha)}(z).$$

```
In[12]:= Table[Expand[(n + 1)       GegenbauerC[n + 1, α, z]] ==
              Expand[2 (n + α) z     GegenbauerC[n, α, z] -
                     (n + 2α - 1) GegenbauerC[n - 1, α, z]], {n, 1, 4}]
Out[12]= {True, True, True, True}
```

For Gegenbauer polynomials, the Rodrigues' formula has the following form:

$$C_n^{(a)}(z) = \frac{\Gamma(\alpha + \frac{1}{2}) \Gamma(n + 2\alpha)}{(-1)^n \, 2^n \, n! \, \Gamma(2\alpha) \, \Gamma(n + \alpha + \frac{1}{2}) (1 + z^2)^{a - \frac{1}{2}}} \frac{d^n \left\{ (1 - z^2)^{\alpha - 1/2} (1 - z^2)^n \right\}}{d z^n}.$$

We now look at several examples using the following property of the Gamma function: $\Gamma(\alpha + 1) = \alpha \, \Gamma(\alpha)$.

```
In[13]:= Expand @ FullSimplify @
            Table[1/((-1)^n 2^n n! Gamma[2α] Gamma[α + n + 1/2]/
                   (Gamma[α + 1/2] Gamma[n + 2α]) (1 - z^2)^(α - 1/2)) *
                   D[(1 - z^2)^(α - 1/2) (1 - z^2)^n, {z, n}], {n, 0, 2}]
Out[13]= {1, 2 z α, -α + 2 z^2 α + 2 z^2 α^2}
```

```
In[14]:= Expand /@ Table[GegenbauerC[n, α, z], {n, 0, 2}]
```

Out[14]= $\{1,\ 2\ z\ \alpha,\ -\alpha + 2\ z^2\ \alpha + 2\ z^2\ \alpha^2\}$

The formula for the Gegenbauer polynomials in terms of the hypergeometric functions is

$$C_n^{(\alpha)}(z) = \frac{\Gamma(n+2\alpha)}{n!\,\Gamma(2\alpha)}\ {}_2F_1\left(-n,\, n+2\alpha,\, \alpha+\frac{1}{2},\, \frac{1-z}{2}\right).$$

We also examine it.

In[15]:= **Expand /@ Simplify /@**
      (Table[Gamma[n + 2α]/n!/Gamma[2α] *
              Hypergeometric2F1[-n, n + 2α, α + 1/2, (1 - z)/2],
              {n, 0, 3}] //. (* rewrite Gamma expressions ***)**
          {Gamma[2α + i_Integer] -> Gamma[2α] Product[(2α + j - 1), {j, i}]]})

Out[15]= $\left\{1,\ 2\ z\ \alpha,\ -\alpha + 2\ z^2\ \alpha + 2\ z^2\ \alpha^2,\ -2\ z\ \alpha + \dfrac{8\ z^3\ \alpha}{3} - 2\ z\ \alpha^2 + 4\ z^3\ \alpha^2 + \dfrac{4\ z^3\ \alpha^3}{3}\right\}$

In[16]:= **Expand /@ Table[GegenbauerC[n, α, z], {n, 0, 3}] == %**

Out[16]= True

Finally, here is the generating function:

$$\frac{1}{R^{2\alpha}} = \sum_{n=0}^{\infty} C_n^{(\alpha)}(z)\,t^n \quad \alpha \neq 0$$

$$-\ln(R^2) = \sum_{n=1}^{\infty} C_n(z)\,t^n \quad \alpha = 0$$

where $R = \sqrt{1 - 2zt + t^2}$. We check it as follows.

In[17]:= **R = Sqrt[1 - 2 z t + t^2];**
      Expand /@ CoefficientList[Series[R^(-2α), {t, 0, 4}], t]

Out[18]= $\left\{1,\ 2\ z\ \alpha,\ -\alpha + 2\ z^2\ \alpha + 2\ z^2\ \alpha^2,\ -2\ z\ \alpha + \dfrac{8\ z^3\ \alpha}{3} - 2\ z\ \alpha^2 + 4\ z^3\ \alpha^2 + \dfrac{4\ z^3\ \alpha^3}{3},\right.$

$\left.\dfrac{\alpha}{2} - 4\ z^2\ \alpha + 4\ z^4\ \alpha + \dfrac{\alpha^2}{2} - 6\ z^2\ \alpha^2 + \dfrac{22\ z^4\ \alpha^2}{3} - 2\ z^2\ \alpha^3 + 4\ z^4\ \alpha^3 + \dfrac{2\ z^4\ \alpha^4}{3}\right\}$

These are just the Gegenbauer polynomials.

In[19]:= **% == Table[Expand[GegenbauerC[n, α, z]], {n, 0, 4}]**

Out[19]= True

For the case $\alpha = 0$, we have the following generating function.

In[20]:= **R = Sqrt[1 - 2 z t + t^2];**

      Expand /@ Rest[CoefficientList[Series[-Log[R^2], {t, 0, 4}], t]]

Out[21]= $\left\{2\ z,\ -1 + 2\ z^2,\ -2\ z + \dfrac{8\ z^3}{3},\ \dfrac{1}{2} - 4\ z^2 + 4\ z^4\right\}$

In[22]:= **Table[Expand[GegenbauerC[n, z]], {n, 1, 4}]**

Out[22]= $\left\{2\ z,\ -1 + 2\ z^2,\ -2\ z + \dfrac{8\ z^3}{3},\ \dfrac{1}{2} - 4\ z^2 + 4\ z^4\right\}$

Integral representations for Gegenbauer polynomials are again relatively complicated; so, we do not give them here. Instead, we give a useful application of the Gegenbauer polynomials.

Mathematical Remark: Smoothing the Gibbs Phenomenon

An unexpected and interesting application of Gegenbauer polynomials is the removal of Gibbs oscillations [226], [298], [218], [302] (for the diagonal limit, see [96]).

Let $f(x)$ be analytic in $(-1,1)$ and $f(-1) \neq f(1)$. The Fourier coefficients $\hat{f}_k$ of $f(x)$ are given by $\hat{f}_k = 1/2 \int_{-1}^{1} f(x) \exp(-i k \pi x) \, dx$. The nth partial sum is then defined as $f_n(x) = \sum_{k=-n}^{n} \hat{f}_k \exp(i k \pi x)$.

The Gibbs phenomenon is the fact that $f_n(x)$ does not converge pointwise to $f(x)$, but only in L_2-norm (for smooth periodic functions, this problem does not occur).

Interestingly, the first n Fourier coefficients contain enough information to obtain an approximation $g_m(x)$ of $f(x)$ that is free of the Gibbs phenomenon and converges pointwise to $f(x)$. The new series is a series in Gegenbauer polynomials [225], [275] $\sum_{l=0}^{m} g_l(\lambda) C_l^{\lambda}(x)$, where $\lambda = m = \lfloor 2\pi e \, n/27 \rfloor$ and

$$g_l(\lambda) = \delta_{0l}\, \hat{f}_0 + \Gamma(\lambda)\, i^l\, (l+\lambda) \sum_{0<|k|\le n} J_{l+\lambda}(\pi k) \left(\frac{2}{k\pi}\right)^l \hat{f}_k.$$

Let us take $f(x) = x^3$ as an example for demonstrating the smoothing.

```
In[23]:= f[x_] := x^3
```

Then we have the following expressions for the Fourier coefficients.

```
In[24]:= fF[k_] = 1/2 Integrate[x^3 Exp[-I k Pi x], {x, -1, 1}]
```

$$\text{Out[24]}= \frac{i\,(k\pi\,(-6 + k^2\,\pi^2)\,\text{Cos}[k\pi] - 3\,(-2 + k^2\,\pi^2)\,\text{Sin}[k\pi])}{k^4\,\pi^4}$$

```
In[25]:= fF[0] = 1/2 Integrate[x^3 Exp[-I 0 Pi x], {x, -1, 1}]
```

Out[25]= 0

Let us have a look at the first partial sums.

```
In[26]:= f[n_, x_] := Sum[fF[k] Exp[I k Pi x], {k, -n, n}]

In[27]:= Plot[Evaluate[Re[N[Table[f[n, x], {n, 1, 20}]]]], {x, -1, 1},
            PlotStyle -> {{Thickness[0.002], GrayLevel[0]}},
            PlotRange -> All, Prolog -> {Thickness[0.01], Hue[0],
            Line[Table[{x, x^3}, {x, -1, 1, 0.01}]]}]
```

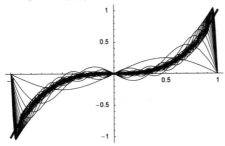

Here, the left graphic shows the pointwise difference between $f(x)$ and the 30th partial sum, and the Gibbs phenomenon is clearly visible. In the L_2-norm, the convergence is just fine (shown in the right graphic).

```
In[28]:= Show[GraphicsArray[
    Block[{$DisplayFunction = Identity},
      {(* pointwise difference *)
       Plot[Evaluate[Re[N[f[30, x]] - f[x]]], {x, -1, 1},
           PlotRange -> {-0.3, 0.3}, PlotPoints -> 200],
       (* L_2 difference *)
       ListPlot[Table[
         Log[10, NIntegrate[Evaluate[Abs[N[f[i, x]] - f[x]]^2], {x, -1, 1}]],
                {i, 30}], PlotRange -> All, AxesOrigin -> {0, 0}]}]]]
```

Now, let us implement the coefficients of the corresponding Gegenbauer series.

```
In[29]:= gG[n_, λ_, l_] := If[l == 0, fF[0], 0] +
    Gamma[λ] I^l (1 + λ)*
    (Sum[BesselJ[l + λ, Pi k] (2/(k Pi))^λ fF[k], {k, -n, -1}] +
     Sum[BesselJ[l + λ, Pi k] (2/(k Pi))^λ fF[k], {k, 1, n}])

In[30]:= fG[n_, x_] := Module[{λ = Floor[2/27 Pi E n]},
                     Sum[gG[n, λ, l] GegenbauerC[l, λ, x], {l, 0, λ}]]
```

The partial sums of the Gegenbauer series converge pointwise to $f(x)$:

```
In[31]:= p1 = Plot[Evaluate[Re[N[Table[fG[n, x], {n, 2, 15}]]]], {x, -1, 1},
           PlotStyle -> {{Thickness[0.002], GrayLevel[0]}},
           PlotRange -> All, Prolog -> {Thickness[0.01], Hue[0],
                         Line[Table[{x, x^3}, {x, -1, 1, 0.01}]]}}]
```

Here, the pointwise difference between $f(x)$ and the 30th partial sum of the Gegenbauer series is shown, the Gibbs phenomena disappeared, and the pointwise difference is about 100 times smaller (shown in the left graphic). The right graphic shows the logarithm of the maximum local deviation from $f(x)$—the Gibbs oscillations are dramatically suppressed.

```
In[32]:= Show[GraphicsArray[
    Block[{$DisplayFunction = Identity},
      {(* pointwise difference *)
       Plot[Evaluate[Re[N[fG[15, x]]] - f[x]], {x, -1, 1}, PlotRange -> All],
```

```
(* maximal local deviation *)
ListPlot[Log[10, Max /@ Apply[Abs[#1^3 - #2]&,
        (#[[3, 1, 1]]&) /@ pl[[1]], {-2}]],
    PlotRange -> All, AxesOrigin -> {0, 0}]]]]
```

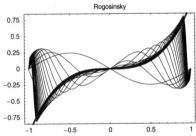

Another possibility to smooth the Gibbs oscillations is, of course, by using a Fejér sum [303], [465], [537] and Rogosinsky sums [430] (but finite Fejér and Rogosinsky sums will still deviate near the interval endpoints; to remove this one can divide by the overshoot [42]).

```
In[33]:= fFejer[n_, x_] := Sum[f[k, x], {k, 0, n - 1}]/n

    fRogosinsky[n_, x_] := (f[n, x + Pi/(2n + 1)] + f[n, x - Pi/(2n + 1)])/2

In[35]:= Show[GraphicsArray[
    Block[{$DisplayFunction = Identity},
    Plot[Evaluate[Re[N[Table[#[n, x], {n, 1, 25}]]]], {x, -1, 1},
        PlotStyle -> {{Thickness[0.002], GrayLevel[0]}}, PlotLabel -> #2,
        PlotRange -> All, Prolog -> {Thickness[0.01], Hue[0],
                    Line[Table[{x, x^3}, {x, -1, 1, 0.01}]]}]& @@@
    (* the two sums *) {{fFejer, "Fejer"}, {fRogosinsky, "Rogosinsky"}}]]]
```

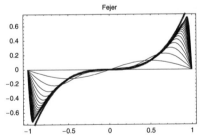

Still another method is the use of Padé approximation for the truncated Fourier series [176]. This dramatically suppresses the Gibbs phenomena. The following graphic shows the logarithm of the pointwise absolute error for 11 Fourier components.

```
In[36]:= Needs["Calculus`Pade`"]

In[37]:= fPadé[x_] =
    Module[{o = 10, pS, z, pade},
        (* form power series of nonnegative frequency part *)
        pS = #.(z^Range[0, Length[#] - 1])&[
            MapAt[#/2&, CoefficientList[f[o, x] /. Exp[a_ Pi x] :> z^(a/I) /.
                                z^_?Negative :> 0, z], 1]];
        (* Padé approximation of nonnegative frequency part *)
        pade = Pade[pS // N, {z, 0, Round[o/2], Round[o/2]}] /. z -> Exp[I Pi x];
```

```
(* add nonpositive frequency part *)
       pade + (pade /. c_Complex :> Conjugate[c])];
```

In[38]:= `Plot[Log[10, Abs[fPadé[x] - f[x]]], {x, -1, 1}]`

For further methods to improve the numerical evaluation of Fourier series, see [95], [217], [103], [633].

2.5 Laguerre Polynomials

The Laguerre polynomials arise as solutions of the eigenvalue problem

$$z \Psi''(z) + (\alpha + 1 - z) \Psi'(z) + \lambda \Psi(z) = 0 \quad z \in (0, \infty)$$

with the requirement that the solutions be finite at $z = 0$ and that at $+\infty$ they have at most polynomial growth. In this case, the eigenvalues are $\lambda = n$ ($n \in \mathbb{N}$, $n \geq 0$). The *Mathematica* function for the Laguerre polynomials is LaguerreL.

LaguerreL[n, α, z]

($n \in \mathbb{N}$, $n \geq 0$) gives the nth Laguerre polynomial $L_n^{(\alpha)}(z)$. If n is a nonnegative integer, the corresponding analytic continuation of $L_n^{(\alpha)}(z)$ is understood.

We now look at the first few Laguerre polynomials explicitly.

In[1]:= `With[{p = LaguerreL},`
`            Table[HoldForm[p[ν, α, z]] == p[n, α, z] /.`
`                        ν -> n, {n, 0, 3}]] // TableForm // TraditionalForm`

Out[1]//TraditionalForm=

$$L_0^\alpha(z) = 1$$
$$L_1^\alpha(z) = -z + \alpha + 1$$
$$L_2^\alpha(z) = \tfrac{1}{2}(z^2 - 2\alpha z - 4z + \alpha^2 + 3\alpha + 2)$$
$$L_3^\alpha(z) = \tfrac{1}{6}(-z^3 + 3\alpha z^2 + 9z^2 - 3\alpha^2 z - 15\alpha z - 18z + \alpha^3 + 6\alpha^2 + 11\alpha + 6)$$

The special case $L_n^{-n-1}(z)$, $n \in \mathbb{N}$ is just $(-1)^n$ times the partial sum of the exponential series. Here this is shown for $n = 10$.

In[2]:= `With[{n = 10}, LaguerreL[n, -n - 1, z] // Expand]`

Out[2]= $1 + z + \dfrac{z^2}{2} + \dfrac{z^3}{6} + \dfrac{z^4}{24} + \dfrac{z^5}{120} + \dfrac{z^6}{720} + \dfrac{z^7}{5040} + \dfrac{z^8}{40320} + \dfrac{z^9}{362880} + \dfrac{z^{10}}{3628800}$

Here are a few representative examples of these polynomials. Because there are more parameters (i.e., n, z, and α) than there are dimensions for visualization, it is hard to give enough visual information.

```
In[3]:= Show[GraphicsArray[#]]& /@
    Table[(* the individual plots *)
        Plot[LaguerreL[3i + j, 0, z], {z, 0, 8},
            PlotLabel -> (StyleForm[TraditionalForm[
                        HoldForm[LaguerreL[#, 0, z]]],
                    FontSize -> 7]&[3i + j]),
        DisplayFunction -> Identity], {i, 0, 2}, {j, 0, 2}]
```

$L_0^0(z)$ $L_1^0(z)$ $L_2^0(z)$

$L_3^0(z)$ $L_4^0(z)$ $L_5^0(z)$

$L_6^0(z)$ $L_7^0(z)$ $L_8^0(z)$

```
In[4]:= Show[GraphicsArray[#]]& /@
    Table[(* the individual plots *)
        Plot[LaguerreL[n, α, z], {z, 0, 8},
            PlotLabel -> (StyleForm[TraditionalForm[
                        HoldForm[LaguerreL[##, z]]],
                    FontSize -> 7]&[n, α]),
        DisplayFunction -> Identity],
                {n, 0, 2}, {α, 2, 10, 4}]
```

$L_0^2(z)$ $L_0^6(z)$ $L_0^{10}(z)$

$L_1^2(z)$ $L_1^6(z)$ $L_1^{10}(z)$

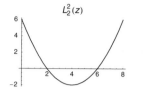

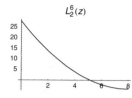

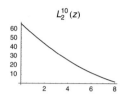

The next graphics are contour plots of $L_n^\alpha(z)$ over the complex z-plane for $n = 2, 6, 12$.

```
In[5]:= Show[GraphicsArray[#]]& /@
        (Function[n, Module[{cpD}, Table[Graphics[(* the individual contour plots *)
         (* make contour plot, then extract data to determine
            homogeneous contour levels *)
         ListContourPlot[cpD = ContourPlot[Evaluate[Re[LaguerreL[n, α, x + I y]]],
                {x, -5, 5}, {y, -5, 5}, PlotPoints -> 90,
                DisplayFunction -> Identity][[1]],
                ColorFunction -> (Hue[0.8 #]&), Contours ->
                (#[[15]]& /@ Partition[Sort[Flatten[cpD]], 270]),
                ContourStyle -> {Thickness[0.002]}, Frame -> True,
                FrameTicks -> None, DisplayFunction -> Identity] ] /.
                 (* position in larger plot *)
        (lp:(Line | Polygon))[l_] :>
          lp[(# + {n, α} - {1/2, 1/2})& /@ (0.9(# - {-5, -5})/10& /@ l)],
          {α, -3, 3}]]] /@ {2, 6, 12})
```

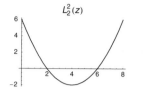

The orthogonal polynomials have all of their zeros inside their domain of orthogonality.

```
In[6]:= Show[Graphics[(* making small vertical lines at the zeros *)
        MapIndexed[Line[{{#1, #2[[1]] - 1/2}, {#1, #2[[1]] + 1/2}}]&,
        (* the zeros *)
        Table[Cases[NRoots[LaguerreL[n, z] == 0, z],
                _?NumberQ, {-1}], {n, 30}], {2}]],
          PlotRange -> All, Frame -> True, AspectRatio -> 1,
          FrameLabel -> {"z", "n"}]
```

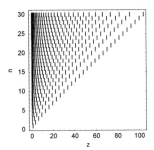

As a general trend, we observe that with growing (real) a, the zeros of $L_n^\alpha(z)$ spread and shift to the right. The right graphic shows the position of the roots of $L_{18}^{e^{i\varphi}}(z)$ as a function of φ.

```
In[7]:= Show[GraphicsArray[
        Block[{$DisplayFunction = Identity},
          {(* roots as a function of real α *)
           ContourPlot[LaguerreL[18, α, z], {z, 0., 125.}, {α, 0., 36.},
                       Contours -> {0}, PlotPoints -> 150,
                       ContourShading -> False],
          (* roots as a function of complex α *)
           Graphics[Table[{Hue[φ/(2Pi)], Point[{Re[#], Im[#]}]& /@
             (z /. NSolve[LaguerreL[18, Exp[I φ], z] == 0, z])},
             {φ, 0, 2Pi, 2Pi/120}], AspectRatio -> 1, Frame -> True]}]]]
```

This tests whether the Laguerre polynomials actually satisfy the differential equation.

```
In[8]:= LaguerreDifferentialOperator[w_, z_, n_] :=
           z D[w, {z, 2}] + (α + 1 - z) D[w, z] + n w;

        Expand[LaguerreDifferentialOperator[#[[1]], z, #[[2]]]]& /@
                 Table[{LaguerreL[n, α, z], n}, {n, 0, 3}]
Out[9]= {0, 0, 0, 0}
```

The second solution to the Laguerre differential equation is the hypergeometric function U.

```
In[10]:= DSolve[LaguerreDifferentialOperator[w[z], z, n] == 0, w[z], z]
Out[10]= {{w[z] → C[1] HypergeometricU[-n, 1 + α, z] + C[2] LaguerreL[n, α, z]}}
```

We turn now to the orthogonality relation. The weight function for the Laguerre polynomials is $e^{-z} z^\alpha$, and the orthogonality interval is $(0, \infty)$. To prevent this chapter from becoming too repetitive, we check the orthogonality relations ourselves rather than having *Mathematica* do it. This will save time and it is also very instructive. First, we look at a typical integral.

In[11]:= `Integrate[z^n z^α Exp[-z], {z, 0, Infinity}]`

Out[11]= If$[Re[n + \alpha] > -1$, Gamma$[1 + n + \alpha]$,
    Integrate$[e^{-z} z^{n+\alpha}$, {z, 0, ∞}, Assumptions $\to Re[n + \alpha] \le -1]]$

Now, we define our integration function `myInt`. We suppress the integration variable and will use always z.

In[12]:= `SetAttributes[myInt, Orderless];`

Integration is additive.

In[13]:= `myInt[s1_ + s2_] := myInt[s1] + myInt[s2]`

Coefficients that do not depend on the variable of integration can be brought out of the integral.

In[14]:= `myInt[coeff_?(FreeQ[#, z]&) z^n_ Exp[-z]] := coeff myInt[z^n Exp[-z]];`

Here is the basic integration formula to which everything reduces: $\int_0^\infty e^{-t} t^\alpha \, dt = \Gamma(1 + \alpha)$.

In[15]:= `myInt[z^exp_ Exp[-z]] = Gamma[1 + exp];`

Using `myInt`, we immediately get the following result.

In[16]:= `aux1 = Map[myInt, Table[Expand[LaguerreL[n, α, z] LaguerreL[m, α, z] *`
          `z^α Exp[-z]], {n, 0, 3}, {m, 0, 3}], {2}] // Simplify;`

In[17]:= `Take[aux1, 2] /. Gamma[α + k_] :> Subscript[Γ, k]`

Out[17]= $\{\{\Gamma_1, (1 + \alpha) \Gamma_1 - \Gamma_2, \frac{1}{2} ((2 + 3\alpha + \alpha^2) \Gamma_1 - 2 (2 + \alpha) \Gamma_2 + \Gamma_3),$

$\frac{1}{6} ((6 + 11\alpha + 6\alpha^2 + \alpha^3) \Gamma_1 - 3 (6 + 5\alpha + \alpha^2) \Gamma_2 + 9 \Gamma_3 + 3\alpha \Gamma_3 - \Gamma_4)\},$

$\{(1 + \alpha) \Gamma_1 - \Gamma_2, (1 + \alpha)^2 \Gamma_1 - 2 (1 + \alpha) \Gamma_2 + \Gamma_3,$

$\frac{1}{2} ((1 + \alpha)^2 (2 + \alpha) \Gamma_1 - 3 (2 + 3\alpha + \alpha^2) \Gamma_2 + 5 \Gamma_3 + 3\alpha \Gamma_3 - \Gamma_4), \frac{1}{6} ((1 + \alpha)^2 (6 + 5\alpha + \alpha^2) \Gamma_1 -$

$4 (6 + 11\alpha + 6\alpha^2 + \alpha^3) \Gamma_2 + 27 \Gamma_3 + 27\alpha \Gamma_3 + 6\alpha^2 \Gamma_3 - 10 \Gamma_4 - 4\alpha \Gamma_4 + \Gamma_5)\}\}$

We simplify the Gamma functions that appear in `aux1` by using $\Gamma(\alpha + i) = \Gamma(\alpha + 1) \prod_{j=1}^{i-1} (\alpha + j)$ for integers i greater than 1.

$$\Gamma(\alpha + i) = \Gamma(\alpha + 1) \prod_{j=1}^{i-1} (\alpha + j) \quad i \ge 2, i \in \mathbb{N}.$$

In[18]:= `aux2 = aux1 /. Gamma[α + i_?(# > 1&)] ->`
          `Gamma[α + 1] Product[α + j, {j, i - 1}];`

We can now factor by hand.

In[19]:= `Map[Gamma[α + 1] Expand[#/Gamma[α + 1]]&, aux2, {2}]`

Out[19]= $\{\{Gamma[1 + \alpha], 0, 0, 0\}, \{0, (1 + \alpha) Gamma[1 + \alpha], 0, 0\},$

$\{0, 0, (1 + \frac{3\alpha}{2} + \frac{\alpha^2}{2}) Gamma[1 + \alpha], 0\}, \{0, 0, 0, (1 + \frac{11\alpha}{6} + \alpha^2 + \frac{\alpha^3}{6}) Gamma[1 + \alpha]\}\}$

The normalization factors presented in the books mentioned at the start of this chapter are $\Gamma(n + \alpha + 1)/n!$. This is just what we obtained above.

In[20]:= `Table[Gamma[α + 1] Expand[Product[α + j, {j, n}]/n!], {n, 0, 3}]`

Out[20]= $\{Gamma[1 + \alpha], (1 + \alpha) Gamma[1 + \alpha],$

$(1 + \frac{3\alpha}{2} + \frac{\alpha^2}{2}) Gamma[1 + \alpha], (1 + \frac{11\alpha}{6} + \alpha^2 + \frac{\alpha^3}{6}) Gamma[1 + \alpha]\}$

The recurrence formula for the computation of the $(n + 1)$-st Laguerre polynomial in terms of previous ones is

$$(n + 1) L_{n+1}^{(\alpha)}(z) = (2 n + \alpha + 1 - z) L_n^{(\alpha)}(z) - (n + \alpha) L_{n-1}^{(\alpha)}(z).$$

```
In[21]:= Clear[α, n];
         Table[Expand[(n + 1)           LaguerreL[n + 1, α, z]] ==
               Expand[(2n + α + 1 - z) LaguerreL[n, α, z] -
                                        (n + α) LaguerreL[n - 1, α, z]], {n, 1, 4}]
Out[22]= {True, True, True, True}
```

For Laguerre polynomials, the formula of Rodrigues' has the following form:

$$L_n^{(\alpha)}(z) = \frac{1}{n!\, z^\alpha\, e^{-z}} \frac{d^n(z^\alpha\, e^{-z}\, z^n)}{dz^n}$$

Again, we look at the first few.

```
In[23]:= Expand /@ Table[1/(n! Exp[-z] z^α) *
               D[Exp[-z] z^α z^n, {z, n}], {n, 0, 3}]   ==
         Expand /@ Table[LaguerreL[n, α, z], {n, 0, 3}]
Out[23]= True
```

The formula for the Laguerre polynomials in terms of the hypergeometric functions is

$$L_n^{(\alpha)}(z) = \frac{(n + \alpha)!}{n!\, \alpha!} \, {}_1F_1(-n, \alpha + 1, z).$$

Here is an example.

```
In[24]:= With[{n = 4},
               Expand[Binomial[α + n, n] #& /@ Hypergeometric1F1[-n, α + 1, z]] ==
               Expand[LaguerreL[n, α, z]]]
Out[24]= True
```

The generating function is

$$\frac{e^{t z/(t-1)}}{(1 - t)^{\alpha+1}} = \sum_{n=0}^{\infty} L_n^{(\alpha)}(z)\, t^n.$$

We check this as follows.

```
In[25]:= Expand /@ CoefficientList[Series[(1 - t)^(-α - 1) Exp[t z/(t - 1)],
                                           {t, 0, 4}], t]
```

$$Out[25]= \left\{1,\ 1 - z + \alpha,\ 1 - 2 z + \frac{z^2}{2} + \frac{3\alpha}{2} - z\alpha + \frac{\alpha^2}{2},\right.$$

$$1 - 3 z + \frac{3 z^2}{2} - \frac{z^3}{6} + \frac{11\alpha}{6} - \frac{5 z\alpha}{2} + \frac{z^2\alpha}{2} + \alpha^2 - \frac{z\alpha^2}{2} + \frac{\alpha^3}{6},\ 1 - 4 z + 3 z^2 - \frac{2 z^3}{3} +$$

$$\left. \frac{z^4}{24} + \frac{25\alpha}{12} - \frac{13 z\alpha}{3} + \frac{7 z^2\alpha}{4} - \frac{z^3\alpha}{6} + \frac{35\alpha^2}{24} - \frac{3 z\alpha^2}{2} + \frac{z^2\alpha^2}{4} + \frac{5\alpha^3}{12} - \frac{z\alpha^3}{6} + \frac{\alpha^4}{24}\right\}$$

These are precisely the Laguerre polynomials.

```
In[26]:= % == Table[Expand[LaguerreL[n, α, z]], {n, 0, 4}]
Out[26]= True
```

An integral formula for the Laguerre polynomials is

$$L_n^{(\alpha)}(z) = \frac{z^{-\alpha/2} e^z}{n!} \int_0^\infty e^{-t} t^{n+\alpha/2} J_\alpha\left(2\sqrt{tx}\right) dt$$

where $J_\nu(z)$ is a Bessel function.

This nontrivial integral can be computed by *Mathematica*, although the result is not expressed as a Laguerre polynomial.

```
In[27]:= z^(-α/2) Exp[z] /n! *
         Integrate[Exp[-t] t^(n + α/2) BesselJ[α, 2Sqrt[t z]],
                   {t, 0, Infinity}, GenerateConditions -> False]
```
```
Out[27]=  e^z Gamma[1 + n + α] Hypergeometric1F1Regularized[1 + n + α, 1 + α, -z]
          ──────────────────────────────────────────────────────────────────────
                                           n!
```

Laguerre polynomials satisfy the following addition formula:

$$L_n^{(\alpha+\beta+1)}(x + y) = \sum_{m=0}^n L_m^{(\alpha)}(x)\, L_{n-m}^{(\beta)}(y).$$

We can easily check this for the first few Laguerre polynomials.

```
In[28]:= Table[Expand /@ (LaguerreL[n, α + β + 1, x + y] ==
               Sum[LaguerreL[m, α, x] LaguerreL[n - m, β, y], {m, 0, n}]),
               {n, 0, 5}]
```
```
Out[28]= {True, True, True, True, True, True}
```

Here, we make a picture of all partial sums of the terms of the Laguerre polynomial $L_6(z)$.

```
In[29]:= makeLines[x_] :=
         Function[coefficients, Line /@ Map[{Re[#], Im[#]}&,
             Union[Sort /@ Flatten[Partition[FoldList[Plus, 0, #], 2, 1]& /@
                 (* all possible orders of the summands *)
                 Permutations[coefficients *
                     (x^Range[0, Length[coefficients] - 1])], 1]], {2}]][
                           CoefficientList[LaguerreL[6, C], C]]
```
```
In[30]:= Show[Graphics[{Thickness[0.002],
         Table[{Hue[φ], makeLines[N[Exp[I φ 2Pi]]]}, {φ, 0, 23/24, 1/24}]}],
                   PlotRange -> All, Frame -> True, AspectRatio -> Automatic]
```

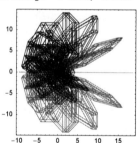

Orthogonal polynomials (although "only" polynomials) can exhibit a complicated behavior. The following graphics show a rescaled version of the image of the spiral $\varphi\, e^{i\varphi}$ under the map $z \to L_8^a(z)$ as a function of a.

```
In[31]:= frame[a_, opts___] :=
         Module[{pp = 5000},
```

```
      (* mapped points of the spiral *)
      data = Table[LaguerreL[8, a + 0.01, φ/30 Exp[I φ]],
                 {φ, 0., 20.Pi, 20.Pi/pp}];
      data = {Re[#], Im[#]}& /@ (data/Max[Abs[data]]);
      (* display mapped points *)
   Show[Graphics[MapIndexed[{Hue[0.8 #2[[1]]/pp], Line[#1]}&,
                             Partition[#, 2, 1]]& @ data],
       opts, AspectRatio -> Automatic, PlotRange -> {{-1, 1}, {-1, 1}}]];
```

```
In[32]:= Show[GraphicsArray[#]]& /@ Map[frame[#, DisplayFunction -> Identity]&,
            Partition[Table[a, {a, 1, -9, -10/15}] + 0.001, 4], {2}]
```

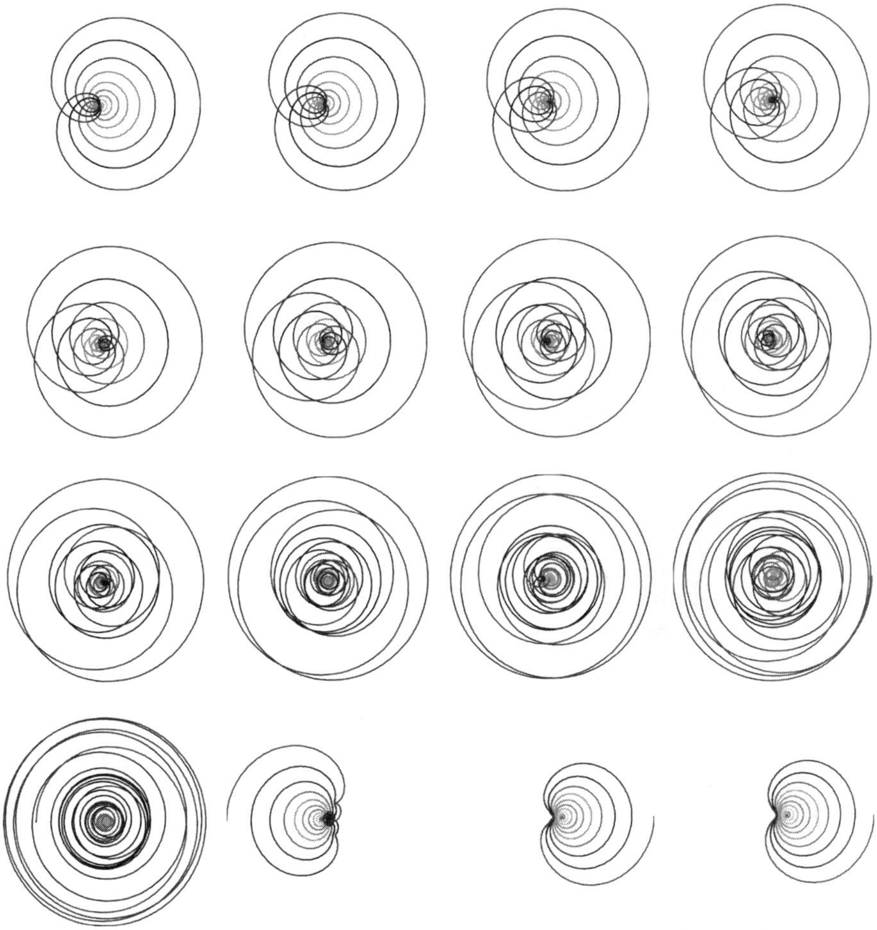

```
      With[{ε = 10^-3}, Do[frame[a + ε], {a, 1, -9, -10/120}]]
```

Let us give another possibility to visualize orthogonal polynomials. We fix $|z|$ and display $p_n(z)$ parametrized by $\arg(z)$ on the Riemann sphere of radius $|z|$. Varying $|z|$ sweeps out a surface.

```
In[33]:= toRiemannSphere[z_] := (* stereographic projection *)
         {Re[z], Im[z], Abs[z]^2/(1 + Abs[z]^2)}
```

```
In[34]:= toExpandingSphere[{r_, z_}] := {0, 0, 0.5} +
             ArcTan[r] (toRiemannSphere[z] - {0, 0, 0.5})

In[35]:= RiemannSpherePolynomialVisualization[
                                 func_, ppr_, ppφ_, rMax_, opts___] :=
         Module[{data, points, polys},
         (* function values along concentric circles *)
         data = Table[{r, func[N[r Exp[I φ]]]},
                     {r, 0, rMax, rMax/ppr}, {φ, 0, 2Pi, 2Pi/ppφ}];
         (* points on the expanded Riemann sphere *)
         points = Map[toExpandingSphere, data, {-2}];
         (* the polygons; use only every second *)
         polys = Table[If[(-1)^(i + j) === -1, {},
                         Polygon[{points[[i, j]], points[[i, j + 1]],
                                  points[[i + 1, j + 1]], points[[i + 1, j]]}]],
                         {i, 1, ppr}, {j, 1, ppφ}];
         (* display the picture *)
         Show[Graphics3D[{EdgeForm[{}],
              MapIndexed[{SurfaceColor[#, #, 3]&[Hue[#2[[1]]/ppr]], #}&,
                         polys]}], opts, BoxRatios -> {1, 1, 1},
              Boxed -> False, PlotRange -> All, SphericalRegion -> True]]
```

Here, we show an array of such "expanding Riemann spheres" for n in $L_n(z)$ ranging from -5 to 5, taking on also noninteger values.

```
In[36]:= Show[GraphicsArray[
         Function[n, RiemannSpherePolynomialVisualization[
                     LaguerreL[n, #]&, 60, 199, 5,
                     DisplayFunction -> Identity]] /@ #]]& /@
                     Partition[Table[n, {n, -6, 6, 12/11}], 4]
```

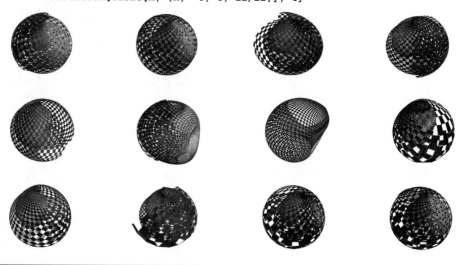

```
Do[RiemannSpherePolynomialVisualization[LaguerreL[n, #]&, 60, 199, 5],
   {n, -6, 6, 12/55}]
```

We end this section with a physics application of Laguerre polynomials: It is well known that the bound-state solutions $R_{n,l,m}(r)$ of the radial solution of the Schrödinger equation for the Coulomb problem contains associated Laguerre polynomials. Interestingly, the sum of the square of all orbitals $\psi_{n,l,m}(\mathbf{r})$ for a given n, meaning the normalized radial distribution [84], [133]

$$\mathcal{D}_n(r) = 4\pi r^2 \sum_{l=0}^{n-1} \sum_{m=-l}^{l} |\psi_{n,l,m}(\mathbf{r})|$$

depends on r only (by Unsöld's theorem) and can be expressed in a compact form through associated Laguerre polynomials [84]

$$\mathcal{D}_n(r) =$$
$$4\,n^{-7}\,Z^3\,r^2\,e^{-\frac{2rZ}{n}}\left(4\,Z^2\,r^2\,L_{n-2}^2\!\left(\frac{2Zr}{n}\right)^2 + L_{n-1}^1\!\left(\frac{2Zr}{n}\right)\!\left(n^2\,L_{n-1}^1\!\left(\frac{2Zr}{n}\right) - 4\,Z^2\,r^2\,L_{n-3}^3\!\left(\frac{2Zr}{n}\right)\right)\right).$$

```
In[37]:= M[n_][z_]  := z/n Exp[-z/2] LaguerreL[n - 1, 1, z]
        ρ[n_, Z_][r_] = Simplify[(Z/n)^3/Pi *
                        (M[n]'[2 Z/n r]^2 - M[n][2 Z/n r] M[n]''[2 Z/n r])];
        D[n_, Z_][r_] = 4Pi r^2 ρ[n, Z][r]
```

$$\text{Out[39]=}\quad \frac{1}{n^7}\left(4\,e^{-\frac{2\,r\,Z}{n}}\,r^2\,Z^3\left(4\,r^2\,Z^2\,\text{LaguerreL}\!\left[-2+n,\,2,\,\frac{2\,r\,Z}{n}\right]^2 + \text{LaguerreL}\!\left[-1+n,\,1,\,\frac{2\,r\,Z}{n}\right]\right.\right.$$
$$\left.\left.\left(-4\,r^2\,Z^2\,\text{LaguerreL}\!\left[-3+n,\,3,\,\frac{2\,r\,Z}{n}\right] + n^2\,\text{LaguerreL}\!\left[-1+n,\,1,\,\frac{2\,r\,Z}{n}\right]\right)\right)\right)$$

As expected, the *n*th shell contains n^2 electrons.

```
In[40]:= Table[Integrate[D[k, Z][r] /.
                         (* compensate symbolic differentiation for k = 1, 2 *)
                    _LaguerreL -> 0, {r, 0, Infinity},
                    Assumptions -> Z > 0], {k, 10}]
Out[40]= {1, 4, 9, 16, 25, 36, 49, 64, 81, 100}
```

Next, we visualize the scaled radial distribution $\mathcal{D}_n(z\,n^2\,r/2)$ for various n. The limiting curve as $n \to \infty$ [45] is shown in gray.

```
In[41]:= (* use high-precision calculation for larger n *)
        DHp[n_][ρ_?NumberQ]  :=
        With[{Z = 1}, D[n, Z][SetPrecision[Z n^2/2 ρ, 50]]]

In[43]:= Plot[Evaluate[Function[k, DHp[k][ρ]] /@ (* list of n-values *)
                    {1, 2, 3, 4, 5, 10, 50, 100, 500, 1000}],
               {ρ, 0, 6}, PlotRange -> All, Frame -> True, Axes -> False,
               PlotStyle -> Table[{Thickness[0.002], Hue[k/10]}, {k, 0, 10}],
               AxesLabel -> {"ρ", "D"}, (* limit n -> ∞ *)
               Prolog -> {{Thickness[0.01], GrayLevel[1/2],
                         Line[Table[{ρ, Re[ρ Sqrt[4 ρ - ρ^2]/(2 Pi)]},
                            {ρ, 0, 6, 1/20}] // N]}}]
```

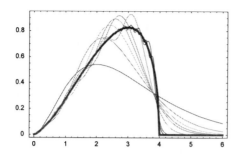

For the average electron distance in dD atoms, see [363].

2.6 Legendre Polynomials

The Legendre polynomials arise as the solutions of the eigenvalue problem

$$(1 - z^2)\, \Psi''(z) - 2\, z\, \Psi'(z) + \lambda(\lambda + 1)\, \Psi(z) = 0, \quad z \in (-1,\, 1)$$

with the requirement that the solution be finite at $z = \pm 1$. In this case, the eigenvalues are $\lambda = n$ ($n \in \mathbb{N}$, $n \geq 0$). The *Mathematica* formula for the Legendre polynomials follows.

LegendreP[n, z]

> ($n \in \mathbb{N}$, $n \geq 0$) gives the nth Legendre polynomial $P_n(z)$. If n is not a nonnegative integer, the value of the corresponding analytic continuation of $P_n(z)$ is understood.

We now look at the first few Legendre polynomials explicitly.

```
In[1]:= With[{p = LegendreP},
            Table[HoldForm[p[v, z]] == p[n, z] /.
                       v -> n, {n, 0, 6}]] // TableForm // TraditionalForm
```
Out[1]//TraditionalForm=

$P_0(z) = 1$

$P_1(z) = z$

$P_2(z) = \frac{3z^2}{2} - \frac{1}{2}$

$P_3(z) = \frac{5z^3}{2} - \frac{3z}{2}$

$P_4(z) = \frac{35z^4}{8} - \frac{15z^2}{4} + \frac{3}{8}$

$P_5(z) = \frac{63z^5}{8} - \frac{35z^3}{4} + \frac{15z}{8}$

$P_6(z) = \frac{231z^6}{16} - \frac{315z^4}{16} + \frac{105z^2}{16} - \frac{5}{16}$

Here are plots of these polynomials.

```
In[2]:= Show[GraphicsArray[#]]& /@
         Table[(* the individual plots *)
                Plot[LegendreP[3i + j, z], {z, -1, 1},
                    PlotLabel -> (StyleForm[TraditionalForm[
                                  HoldForm[LegendreP[#, z]]],
```

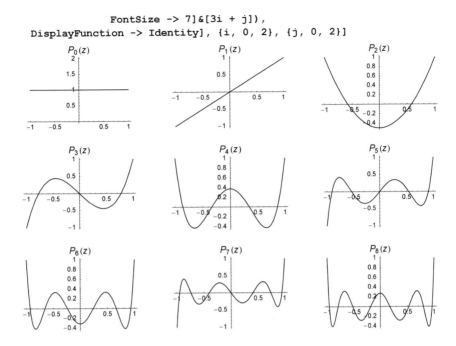

```
                    FontSize -> 7]&[3i + j]),
        DisplayFunction -> Identity], {i, 0, 2}, {j, 0, 2}]
```

This tests whether they satisfy the above differential equation.

```
In[3]:= LegendreDifferentialOperator[w_, z_, n_] :=
          (1 - z^2) D[w, {z, 2}] - 2 z D[w, z] + n (n + 1) w;

        Expand[LegendreDifferentialOperator[#[[1]], z, #[[2]]]]& /@
                        Table[{LegendreP[n, z], n}, {n, 0, 6}]
Out[4]= {0, 0, 0, 0, 0, 0, 0}
```

The solutions of the Legendre differential equation are Legendre functions.

```
In[5]:= DSolve[LegendreDifferentialOperator[w[z], z, n] == 0, w[z], z]
Out[5]= {{w[z] → C[1] LegendreP[n, z] + C[2] LegendreQ[n, z]}}
```

For the normalization, we note that the weight function is 1, and the interval of orthogonality is $(-1, 1)$.

```
In[6]:= Table[Integrate[Expand[LegendreP[i, z] LegendreP[j, z]], {z, -1, 1}],
          {i, 0, 3}, {j, 0, 3}] // TableForm
```

Out[6]//TableForm=

2	0	0	0
0	$\frac{2}{3}$	0	0
0	0	$\frac{2}{5}$	0
0	0	0	$\frac{2}{7}$

Thus, the normalization factor is $2/(2n + 1)$.

The recurrence formula for the computation of the $(n + 1)$th Legendre polynomial is

$$(n + 1) P_{n+1}(z) = (2n + 1) z P_n(z) - n P_{n-1}(z).$$

In[7]:= `Table[Expand[(n + 1)    LegendreP[n + 1, z]] ==`
        `      Expand[(2n + 1) z LegendreP[n, z] - n LegendreP[n - 1, z]],`
        `      {n, 1, 4}]`

Out[7]= {True, True, True, True}

The Rodrigues' formula for Legendre polynomials has the following form:

$$P_n(z) = \frac{1}{(-1)^n \, 2^n \, n!} \frac{d^n \, (1 - z^2)^n}{dz^n}.$$

Here are a few examples.

In[8]:= `Expand /@ Table[1/((-1)^n 2^n n!) *`
        `                D[(1 - z^2)^n, {z, n}], {n, 0, 3}] ==`
        `      Expand /@ Table[LegendreP[n, z], {n, 0, 3}]`

Out[8]= True

In terms of hypergeometric functions, the Legendre polynomials are given by

$$P_n(z) = {}_2F_1\left(-n, n + 1, 1, \frac{1 - z}{2}\right).$$

Here are the first few.

In[9]:= `Table[{Expand[Hypergeometric2F1[-n, n + 1, 1, (1 - z)/2]],`
        `             Expand[LegendreP[n, z]]}, {n, 0, 4}]`

Out[9]= $\left\{\{1, 1\}, \{z, z\}, \left\{-\frac{1}{2} + \frac{3\,z^2}{2}, -\frac{1}{2} + \frac{3\,z^2}{2}\right\},\right.$

$\left.\left\{-\frac{3\,z}{2} + \frac{5\,z^3}{2}, -\frac{3\,z}{2} + \frac{5\,z^3}{2}\right\}, \left\{\frac{3}{8} - \frac{15\,z^2}{4} + \frac{35\,z^4}{8}, \frac{3}{8} - \frac{15\,z^2}{4} + \frac{35\,z^4}{8}\right\}\right\}$

The generating function is

$$\frac{1}{R} = \sum_{n=0}^{\infty} P_n(z)\, t^n$$

where $R = \sqrt{1 - 2\,t\,z + t^2}$. We may check this as follows.

In[10]:= `Expand /@ CoefficientList[Series[1/Sqrt[1 - 2 t z + t^2],`
         `                                 {t, 0, 4}], t]`

Out[10]= $\left\{1, z, -\frac{1}{2} + \frac{3\,z^2}{2}, -\frac{3\,z}{2} + \frac{5\,z^3}{2}, \frac{3}{8} - \frac{15\,z^2}{4} + \frac{35\,z^4}{8}\right\}$

These are precisely the Legendre polynomials.

In[11]:= `% == Table[Expand[LegendreP[n, z]], {n, 0, 4}]`

Out[11]= True

Starting with the generating function, it is easy to verify the following representation for the Legendre polynomials:

$$P_n(z) = \operatorname*{res}_{t=0} t^{-n-1} \sqrt{1 - 2\,t\,z + t^2}.$$

First, we find the necessary residues.

In[12]:= `Table[Residue[1/(t^(1 + i) Sqrt[1 - 2x t + t^2]), {t, 0}], {i, 0, 4}]`

Out[12]= $\left\{1,\ x,\ \frac{1}{2}\ (-1 + 3\ x^2),\ \frac{1}{2}\ (-3\ x + 5\ x^3),\ \frac{1}{8}\ (3 - 30\ x^2 + 35\ x^4)\right\}$

We get Legendre polynomials.

In[13]:= **(Expand /@ %) == (Expand /@ Table[LegendreP[i, x], {i, 0, 4}])**

Out[13]= True

Like most orthogonal polynomials, Legendre polynomials fulfill a variety of interesting identities. One of them is the following determinantal one [219]. (Be aware that the calculation of the determinants in the next input is quite a task.)

In[14]:= **LegendrePolynomialDet[n_] := Factor[Det[**
           Table[LegendreP[j + i, z], {i, 0, n}, {j, 0, n}]]]

In[15]:= **Table[LegendrePolynomialDet[n], {n, 0, 6}]**

Out[15]= $\Big\{1,\ \frac{1}{2}\ (-1 + z)\ (1 + z),\ \frac{1}{16}\ (-1 + z)^3\ (1 + z)^3,\ \frac{1}{512}\ (-1 + z)^6\ (1 + z)^6,$

$\frac{(-1 + z)^{10}\ (1 + z)^{10}}{65536},\ \frac{(-1 + z)^{15}\ (1 + z)^{15}}{33554432},\ \frac{(-1 + z)^{21}\ (1 + z)^{21}}{68719476736}\Big\}$

In[16]:= **Table[2^(-(n - 1)^2) ((z - 1) (z + 1))^Binomial[n, 2], {n, 7}] === %**

Out[16]= True

The importance of the zeros of the Legendre polynomials for numerical integration is well known. Interestingly, the Legendre polynomial can also be used to numerically differentiate (through integration) [468].

In[17]:= **Table[Series[1/2 (2n + 1)!!/h^n/h Integrate[LegendreP[n, t/h] f[x + t],**
                                        {t, -h, h}],
                       {h, 0, n}] + O[h]^2, {n, 0, 6}]
Out[17]= $\{f[x] + O[h]^1,\ O[h]^2,\ f''[x] + O[h]^2,\ f^{(3)}[x] + O[h]^2,$
         $f^{(4)}[x] + O[h]^2,\ f^{(5)}[x] + O[h]^2,\ f^{(6)}[x] + O[h]^2\}$

In addition to the usual Legendre polynomials $P_n(z)$, the associated Legendre polynomials $P_n^m(z)$ are also of great importance. They satisfy the differential equation

$$(1 - z^2)\,w''(z) - 2\,z\,w'(z) + \left(\lambda(\lambda + 1) - \frac{m^2}{1 - z^2}\right)w(z) = 0,\quad z \in (-1,\ 1)$$

where $m \in \mathbb{N}$ and we have the requirement that the solution is finite at $z = \pm 1$. In this case, the eigenvalues are $\lambda = n$ (n, $m \in \mathbb{N}$, $n \geq 0$, $m = -n$, ..., n). The *Mathematica* formula for the associated Legendre polynomials follows.

LegendreP[n, m, z]

(n, $m \in \mathbb{N}$, $n \geq 0$, $m = -n$, $-n + 1$, ..., $n - 1$, n) gives the n, mth associated Legendre polynomial $P_n^m(z)$. If n, m do not lie in the prescribed domain, the corresponding analytic continuation of $P_n^m(z)$ is understood.

The connection between the Legendre polynomials and the associated Legendre polynomials is provided by the equation

$$P_n^m(z) = (1 - z^2)^{\frac{m}{2}}\ \frac{d^m\,P_n(z)}{dz^m}.$$

For fixed m, the $P_n^m(z)$ form a complete orthogonal system with the weight function 1 on $(-1, 1)$. Because of this close analogy with the Legendre polynomials, we do not go into as much detail here as we did for the other polynomials. Associated Legendre polynomials are not true polynomials because for odd m, they involve terms of the form $(1 - z^2)^{1/2}$.

```
In[18]:= With[{p = LegendreP},
            Table[HoldForm[p[v, m, z]] == FunctionExpand[p[n, m, 2, z]] /.
                v -> n, {n, 0, 4}]] // TableForm // TraditionalForm
```

Out[18]//TraditionalForm=

$$P_0^m(z) = \frac{(1-z)^{-m/2}\,(z+1)^{m/2}}{\Gamma(1-m)}$$

$$P_1^m(z) = \frac{(1-z)^{-m/2}\,(z+1)^{m/2}\,(z-m)}{\Gamma(2-m)}$$

$$P_2^m(z) = \frac{(1-z)^{-m/2}\,(z+1)^{m/2}\,(m^2 - 3\,z\,m + 3\,z^2 - 1)}{\Gamma(3-m)}$$

$$P_3^m(z) = \frac{(1-z)^{-m/2}\,(z+1)^{m/2}\,(-m^3 + 6\,z\,m^2 - 15\,z^2\,m + 4\,m + 15\,z^3 - 9\,z)}{\Gamma(4-m)}$$

$$P_4^m(z) = \frac{(1-z)^{-m/2}\,(z+1)^{m/2}\,(m^4 - 10\,z\,m^3 + 45\,z^2\,m^2 - 10\,m^2 - 105\,z^3\,m + 55\,z\,m + 105\,z^4 - 90\,z^2 + 9)}{\Gamma(5-m)}$$

For $P_n^m(z)$ with $m > n$, we have $P_n^m(z) = 0$.

```
In[19]:= {LegendreP[1, 2, z], LegendreP[23, 34, z]}
```

Out[19]= {0, 0}

We can look at the dependence on m in two examples. The left graphic shows $P_6^m(z)$ for integer m along the real z-axis and the right graphic shows $P_{12}^m(z)$ over the z,m-plane.

```
In[20]:= Show[GraphicsArray[
          Block[{$DisplayFunction = Identity, ε = 10^-6},
          {(* n = 6, integer m *)
          Plot[Evaluate[Table[LegendreP[6, m, z], {m, 0, 6}]],
              {z, -1, 1}, PlotRange -> All,
              PlotStyle -> Table[Hue[x], {x, 0, 7/10, 1/10}]],
          (* n = 12, real m *)
          Plot3D[Log @ Abs @ LegendreP[12, m, z],
              {z, -1 + ε, 1 - ε}, {m, -6, 6},  PlotPoints -> 90]}]]]
```

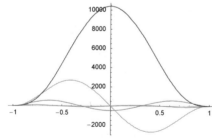 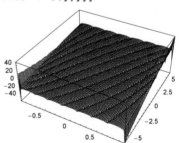

The zeros of $P_n^m(z)$ shift toward the right as m increases. Here is an example.

```
In[21]:= Show[GraphicsArray[
          Block[{$DisplayFunction = Identity},
          {(* roots as a function of real μ *)
          ContourPlot[Re[LegendreP[15, m, z]],
                  {z, -0.95, 0.95}, {m, -15., 15.},
                  Contours -> {0}, ContourShading -> False,
                  FrameLabel -> {"z", "m"}, PlotPoints -> 300],
          (* roots as a function of complex μ *)
          Graphics[Table[{Hue[φ/(2Pi)], Point[{Re[#], Im[#]}]}& /@
              (z /. NSolve[Sum[((Pochhammer[-18, k] Pochhammer[18 + 1, k])/
```

```
(Gamma[1 - Exp[I φ] + k] k!)) ((1 - z)/2)^k, {k, 0, 18}] == 0, z])},
    {φ, 0, 2Pi, 2Pi/120}], AspectRatio -> 1, Frame -> True]}]]]
```

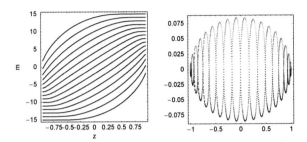

Now, we test whether the associated Legendre polynomials actually satisfy the above differential equation.

```
In[22]:= AssociatedLegendreDifferentialOperator[w_, z_, n_, m_] :=
    (1 - z^2) D[w, {z, 2}] - 2z D[w, z] + (n(n + 1) - m^2/(1 - z^2)) w;
```

```
    Map[Simplify[Expand[
        AssociatedLegendreDifferentialOperator[#[[1]], z, #[[2]], #[[3]]]]]&,
            Table[{LegendreP[n, m, z], n, m}, {n, 0, 4}, {m, 0, n}], {2}]
Out[23]= {{0}, {0, 0}, {0, 0, 0}, {0, 0, 0, 0}, {0, 0, 0, 0, 0}}
```

The normalization factor is

$$\frac{2}{(2n+1)} \frac{(n+m)!}{(n-m)!}.$$

The recurrence formula for computing the $(n + 1)$-st generalized Legendre polynomial is

$$(n - m + 1) P^m_{n+1}(z) = (2n + 1) z P^m_n(z) - (n + m) P^m_{n-1}(z).$$

For associated Legendre polynomials, the formula of Rodrigues' is

$$P^m_n(z) = \frac{1}{(-1)^n 2^n n!} (1 - z^2)^{m/2} d^{n+m} \frac{\{(1 - z^2)^n\}}{dz^{n+m}}.$$

The formula for associated Legendre polynomials in terms of hypergeometric functions is

$$P^m_n(z) = (-1)^n \frac{(n+m)!}{2^m m! (n-m)!} (1 - z^2)^{m/2} {}_2F_1\left(m - n, m + n + 1, m + 1, \frac{1 - z}{2}\right).$$

The generating function is

$$\frac{1}{R^{m+1/2}} = \sum_{n=0}^{\infty} \frac{2^m m! (1 - z^2)^{m/2}}{(2m)!} P^m_n(z) t^n$$

where $R = \sqrt{1 - 2tz + t^2}$.

The overlap integral $\int_{-1}^{1} P^m_n(z) P^{m'}_{n'}(z) dz$ is much more complicated than for the Legendre polynomials; see [392] and [621] for closed forms.

Associated Legendre functions satisfy the following interesting identity (Theorem of Unsöld [354]):

$$\sum_{m=-n}^{n} (-1)^m P_n^m(z) P_n^{-m}(z) = 1.$$

In[24]:= `Table[Sum[ (-1)^m LegendreP[n, +m, x] *`
                             `LegendreP[n, -m, x], {m, -n, n}] // Expand, {n, 0, 4}]`

Out[24]= `{1, 1, 1, 1, 1}`

For further properties of the associated Legendre polynomials, see [284], [311], and [378].

Using the associate Legendre polynomial, we can solve another quantum-mechanical eigenvalue problem, the modified Pöschl–Teller potential. For a given value of the parameter j the modified Pöschl–Teller potential [464], [170], [315], [105] has $\lfloor j \rfloor$ bound states. The differential equation, the eigenvalues, and the eigenfunctions are:

$$-\psi_n''(x) + \frac{j(j+1)}{\cosh^2(x)} \psi_n(x) = \varepsilon_n \psi_n(x)$$

$$\varepsilon_n = -(n-j)^2$$

$$\psi_n(x) = \sqrt{\frac{(j-n)\,n!}{(2\,j-n)!}} \; P_j^{j-n}(\tanh(x))$$

The eigenfunctions vanish as $x \to \pm\infty$ and are normalized to $\int_{-\infty}^{\infty} \psi_n(x)^2 = 1$. Here are the eigenfunctions defined.

In[25]:= `Clear[ψ, ε, n, j, x];`
    `ψ[j_][n_, x_] := Sqrt[(j - n) n!/(2j - n)!] LegendreP[j, j - n, Tanh[x]]`
    `ε[j_][n_] := -(n - j)^2`

In[28]:= `Table[-D[ψ[j][n, x], x, x] -`
               `j (j + 1)/Cosh[x]^2 ψ[j][n, x] - ε[j][n] ψ[j][n, x] //`
               `FullSimplify, {j, 0, 6, 1/2}, {n, 0, j - 1}] // Flatten // Union`

Out[28]= `{0}`

The next graphic shows a sketch of the potential and the eigenfunctions $\psi_n(x)$ for $j = 30$.

In[30]:= `With[{j = 30, ppx = 400, L = 4.4, f = 40},`
    `Module[{VLine, ψc},`
    `(* the curve of the potential *)`
    `VLine = {Thickness[0.01], GrayLevel[1/2],`
           `Line[Table[{x, -j (j + 1.)/Cosh[x]^2}, {x, -4, 4, 2L/ppx}]]};`
    `(* evaluate and compile the ψs *)`
    `Do[ψc[j, n] = Compile[z, Evaluate[Sqrt[(j - n) n!/(2j - n)!]*`
                          `Expand[LegendreP[j, j - n, z]]]],`
        `{n, j - 1, 0, -1}];`
    `(* the curves of the ψs *)`
    `ψs = Table[{Thickness[0.002], Hue[0.8 n/j],`
               `Line[Table[{x, f ψc[j, n][Tanh[x]] + ε[j][n]},`
               `{x, -4, 4, 2L/ppx}]]}, {n, 0, j - 1}];`
    `(* display potential and ψs *)`
    `Show[Graphics[{VLine, ψs}], Frame -> True]]]`

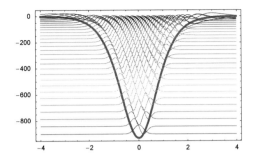

2.7 Chebyshev Polynomials of the First Kind

The Chebyshev polynomials $T_n(z)$ arise as the solutions of the eigenvalue problem

$$(1 - z^2)\,\Psi''(z) - z\,\Psi'(z) + \lambda^2\,\Psi(z) = 0, \quad z \in (-1,\,1)$$

with the requirement that the solution be finite at $z = \pm 1$. In this case, the eigenvalues are $\lambda = n$ ($n \in \mathbb{N}$, $n \geq 0$). The *Mathematica* formula for the Chebyshev polynomials $T_n(z)$ follows.

ChebyshevT $[n, z]$

    ($n \in \mathbb{N}$, $n \geq 0$) gives the nth Chebyshev polynomial $T_n(z)$. If n is not a nonnegative integer, the analytic continuation of $T_n(z)$ is understood.

Here are the first few Chebyshev polynomials $T_n(z)$.

```
In[1]:= With[{p = ChebyshevT},
           Table[HoldForm[p[v, z]] == p[n, z] /.
               v -> n, {n, 0, 6}]] // TableForm // TraditionalForm
```

Out[1]//TraditionalForm=

$$T_0(z) = 1$$
$$T_1(z) = z$$
$$T_2(z) = 2\,z^2 - 1$$
$$T_3(z) = 4\,z^3 - 3\,z$$
$$T_4(z) = 8\,z^4 - 8\,z^2 + 1$$
$$T_5(z) = 16\,z^5 - 20\,z^3 + 5\,z$$
$$T_6(z) = 32\,z^6 - 48\,z^4 + 18\,z^2 - 1$$

Here are their plots.

```
In[2]:= Show[GraphicsArray[#]]& /@
           Table[(* the individual plots *)
               Plot[ChebyshevT[3i + j, z], {z, -1, 1},
                   PlotLabel -> StyleForm[TraditionalForm[
                       HoldForm[ChebyshevT[3i + j, z]]], FontSize -> 7],
                   DisplayFunction -> Identity], {i, 0, 2}, {j, 0, 2}]
```

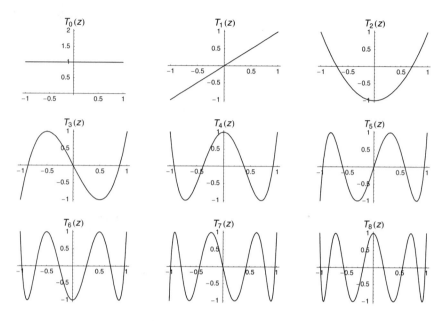

We can test whether they satisfy the above differential equation as follows.

```
In[3]:= ChebyshevTDifferentialOperator[w_, z_, n_] :=
          (1 - z^2) D[w, {z, 2}] - z D[w, z] + n^2 w;

        Expand[ChebyshevTDifferentialOperator[#[[1]], z, #[[2]]]]& /@
                         Table[{ChebyshevT[n, z], n}, {n, 0, 6}]
Out[4]= {0, 0, 0, 0, 0, 0, 0}
```

Now for the normalization. The weight function is $(1 - z^2)^{-1/2}$, and the interval of orthogonality is $(-1, 1)$.

```
In[5]:= Table[Integrate[Expand[ChebyshevT[i, x] *
                  ChebyshevT[j, x]]/Sqrt[1 - x^2], {x, -1, 1}],
              {i, 0, 3}, {j, 0, 3}] // TableForm
Out[5]//TableForm=
         π       0       0       0

         0       π/2     0       0

         0       0       π/2     0

         0       0       0       π/2
```

Thus, the normalization factor is $\pi/2$ for $n \neq 0$ and π for $n = 0$.

The recurrence formula for the computation of the $(n + 1)$-st Chebyshev polynomial $T_n(z)$ is

$$T_{n+1}(z) = 2 z T_n(z) - T_{n-1}(z).$$

```
In[6]:= Table[Expand[    ChebyshevT[n + 1, z]] ==
             Expand[2 z ChebyshevT[n, z] - ChebyshevT[n - 1, z]], {n, 1, 4}]
Out[6]= {True, True, True, True}
```

For Chebyshev polynomials $T_n(z)$, the formula of Rodrigues' is

$$T_n(z) = \frac{(-1)^n \sqrt{1-z^2} \sqrt{\pi}}{2^n \Gamma(n+\frac{1}{2})} \frac{d^n\left\{(1-z^2)^{-1/2}(1-z^2)^n\right\}}{dz^n}.$$

Here is the comparison of the first few polynomials.

```
In[7]:= Expand[Table[1/((-1)^n 2^n Gamma[n + 1/2]/Sqrt[Pi] (1 - z^2)^(-1/2)) *
                D[(1 - z^2)^(-1/2) (1 - z^2)^n, {z, n}], {n, 0, 3}]] ==
        Expand /@ Table[ChebyshevT[n, z], {n, 0, 3}]

Out[7]= True
```

In terms of hypergeometric functions, the Chebyshev polynomials $T_n(z)$ are given by

$$T_n(z) = {}_2F_1\left(-n, n, \frac{1}{2}, \frac{1-z}{2}\right).$$

Again, we look at the first few.

```
In[8]:= Table[{Expand[Hypergeometric2F1[-n, n, 1/2, (1 - z)/2]],
            Expand[ChebyshevT[n, z]]}, {n, 0, 4}]

Out[8]= {{1, 1}, {z, z}, {-1 + 2 z^2, -1 + 2 z^2},
         {-3 z + 4 z^3, -3 z + 4 z^3}, {1 - 8 z^2 + 8 z^4, 1 - 8 z^2 + 8 z^4}}
```

The generating function is

$$\frac{1-tz}{R^2} = \sum_{n=0}^{\infty} T_n(z)\, t^n$$

where $R = (1 - 2tz + t^2)^{1/2}$. To check this, we use the following input.

```
In[9]:= Expand /@ CoefficientList[Series[(1 - t z) /(1 - 2 t z + t^2),
                                        {t, 0, 4}], t]

Out[9]= {1, z, -1 + 2 z^2, -3 z + 4 z^3, 1 - 8 z^2 + 8 z^4}
```

It gives us exactly the Chebyshev polynomials $T_n(z)$.

```
In[10]:= % == Table[Expand[ChebyshevT[n, z]], {n, 0, 4}]

Out[10]= True
```

The Chebyshev polynomials $T_n(z)$ satisfy the following remarkable formula (this is a rewritten version of $T_n(\cos(z)) = \cos(n z)$):

$$T_n(z) = \frac{1}{2}\left(\left(z + i\sqrt{1-z^2}\right)^n + \left(z - i\sqrt{1-z^2}\right)^n\right).$$

We can check this as follows.

```
In[11]:= Table[Expand[1/2 ((z + I Sqrt[1 - z^2])^n + (z - I Sqrt[1 - z^2])^n)] ==
            ChebyshevT[n, z], {n, 0, 3}]

Out[11]= {True, True, True, True}
```

Chebyshev polynomials $T_n(z)$ have still another very remarkable symmetry property:

$$T_n(T_m(z)) = T_m(T_n(z)) = T_{nm}(z).$$

```
In[12]:= Table[Composition[ChebyshevT[i, #]&, ChebyshevT[j, #]&][x] -
                ChebyshevT[i j, x] // Expand,
            {i, 6}, {j, 6}] // Flatten // Union
Out[12]= {0}
```

The following simple formula for the Chebyshev polynomials $T_n(z)$ is also of interest: $T_n(z) = \cos(n \arccos(z))$. Using the function TrigExpand, we can easily check this formula.

```
In[13]:= {Table[Cos[i ArcCos[z]] // TrigExpand, {i, 0, 4}],
          Table[ChebyshevT[i, z], {i, 0, 4}]}
Out[13]= {{1, z, -1 + 2 z², -3 z + 4 z³, 1 - 8 z² + 8 z⁴}, {1, z, -1 + 2 z², -3 z + 4 z³, 1 - 8 z² + 8 z⁴}}
```

Using this formula, we can also explicitly calculate the zeros of $T_n(z)$. Here are the zeros for different values of n.

```
In[14]:= Show[Graphics[{PointSize[0.003],
            Table[If[m < n, Point[{N[Cos[(2m + 1) Pi/(2n)]], n}], {}],
                {n, 120}, {m, 120}]}], AspectRatio -> 1/3,
            Frame -> True, FrameLabel -> {"z0", "n"}]
```

Chebyshev polynomials $T_n(z)$ have the following interesting property, which makes them very useful in numerical analysis (see [229], [345], [247], and [86]): among all polynomials $p_n(z)$ of degree n with the property $|p_n(z)| \le 1$ for $-1 \le z \le 1$, they have the property that $|p_n(z)| \le |T_n(z)|$ for $z < -1$ and $z > 1$.

Because $|T_n(z)| \le 1$ for $-1 \le z \le 1$, we can conveniently take the function values $T_n(z)$ and wrap them around circles of radius $2n$; here, this done for all even n less than 82.

```
In[15]:= Show[GraphicsArray[
          Function[color, Graphics[
          Reverse[MapIndexed[Function[{li, pos},
          {(* color the shells *) color[pos],
          Polygon @@ (* wrap line around a circle *)
                  Map[N[(2 pos[[1]] + #[[2]])*
                      {Cos[Pi #[[1]]], Sin[#[[1]] Pi]}]&, li, {-2}]}],
              (* use Plot to make function graphs *)
              Flatten[(Function[chebys, Plot[chebys, {x, -1, 1},
                          PlotPoints -> 200, DisplayFunction -> Identity],
              {HoldAll}] @@ ((* avoid evaluation to explicit polynomials *)
          Apply[List, Hold @@ {Join @@ MapIndexed[# //. C -> 2 #2[[1]] - 2&,
                  (* the ChebyshevT to be displayed *)
              Table[Hold[ChebyshevT[C, x]], {C, 0, 82, 2}]]}, {1}])][[1]]], {1}]],
              PlotRange -> All, AspectRatio -> Automatic]] /@
              (* black or white and random coloring *)
              {GrayLevel[If[EvenQ[#[[1]]], 1, 0]]&, Hue[Random[]]&}]]
```

For more on Chebyshev polynomials, see, in particular, [478] and [389].

2.8 Chebyshev Polynomials of the Second Kind

The Chebyshev polynomials $U_n(z)$ arise as the solution of the eigenvalue problem

$$(1 - z^2)\,\Psi''(z) - 3\,z\,\Psi'(z) + \lambda(\lambda + 2)\,\Psi(z) = 0\,, \quad z \in (-1, 1)$$

with the requirement that the solution be finite at $z = \pm 1$. In this case, the eigenvalues are $\lambda = n$ ($n \in \mathbb{N}$, $n \geq 0$). The *Mathematica* formula for the Chebyshev polynomials $U_n(z)$ follows. (Note the spelling of Chebyshev; for other possibilities, see [270].)

ChebyshevU $[n, z]$

($n \in \mathbb{N}$, $n \geq 0$) gives the nth Chebyshev polynomial $U_n(z)$. If n is not a nonnegative integer, the analytic continuation of $U_n(z)$ is understood.

Here are the first few Chebyshev polynomials $U_n(z)$.

```
In[1]:= With[{p = ChebyshevU},
            Table[HoldForm[p[v, z]] == p[n, z] /.
                    v -> n, {n, 0, 6}]] // TableForm // TraditionalForm
```
Out[1]//TraditionalForm=

$U_0(z) = 1$

$U_1(z) = 2\,z$

$U_2(z) = 4\,z^2 - 1$

$U_3(z) = 8\,z^3 - 4\,z$

$U_4(z) = 16\,z^4 - 12\,z^2 + 1$

$U_5(z) = 32\,z^5 - 32\,z^3 + 6\,z$

$U_6(z) = 64\,z^6 - 80\,z^4 + 24\,z^2 - 1$

Here is a plot of these polynomials.

```
In[2]:= Show[GraphicsArray[#]]& /@
            Table[(* the individual plots *)
                Plot[ChebyshevU[3i + j, z], {z, -1, 1},
                    PlotLabel -> StyleForm[TraditionalForm[
                        HoldForm[ChebyshevU[3i + j, z]]],
```

```
                         FontSize -> 7],
       DisplayFunction -> Identity], {i, 0, 2}, {j, 0, 2}]
```

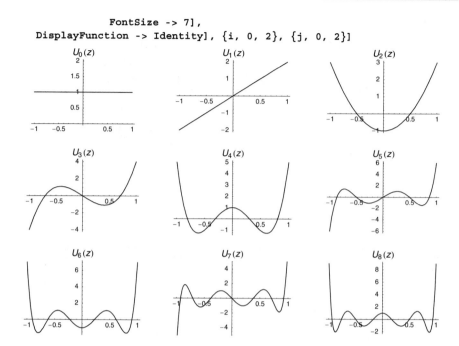

This tests whether they satisfy the differential equation.

```
In[3]:= ChebyshevUDifferentialOperator[w_, z_, n_] :=
          (1 - z^2) D[w, {z, 2}] - 3 z D[w, z] + n (n + 2) w;
```

```
        Expand[ChebyshevUDifferentialOperator[#[[1]], z, #[[2]]]]& /@
                        Table[{ChebyshevU[n, z], n}, {n, 0, 6}]
Out[4]= {0, 0, 0, 0, 0, 0, 0}
```

The weight function is $\sqrt{1-z^2}$, and the interval of orthogonality is $(-1, 1)$.

```
In[5]:= Table[Integrate[Expand[ChebyshevU[i, x] ChebyshevU[j, x]] Sqrt[1 - x^2],
                {x, -1, 1}], {i, 0, 3},
             {j, 0, 3}] - Pi/2 IdentityMatrix[4] // Flatten // Union
Out[5]= {0}
```

Thus, the normalization factor is $\pi/2$ for all n (including $n = 0$).

The recurrence formula for the computation of the $(n + 1)$-st Chebyshev polynomial $U_n(z)$ is

$$U_{n+1}(z) = 2 z U_n(z) - U_{n-1}(z).$$

```
In[6]:= Table[Expand[    ChebyshevU[n + 1, z]] ==
              Expand[2 z ChebyshevU[n, z] - ChebyshevU[n - 1, z]], {n, 1, 4}]
Out[6]= {True, True, True, True}
```

A Rodrigues' formula for Chebyshev polynomials $U_n(z)$ has the following form:

$$U_n(z) = \frac{(n+1)\sqrt{\pi}}{(-1)^n \, 2^{n+1} \, \Gamma(n+\frac{3}{2})\sqrt{1-z^2}} \frac{d^n\left\{\sqrt{1-z^2}\,(1-z^2)^n\right\}}{dz^n}.$$

We compare the first values of n.

```
In[7]:= Expand /@ Table[1/((-1)^n 2^(n + 1) *
              Gamma[n + 3/2]/Sqrt[Pi]/(n + 1) (1 - z^2)^(1/2)) *
            D[(1 - z^2)^(1/2) (1 - z^2)^n, {z, n}], {n, 0, 3}]  ==
        Expand /@ Table[ChebyshevU[n, z], {n, 0, 3}]
Out[7]= True
```

The formula for the Chebyshev polynomials $U_n(z)$ in terms of hypergeometric functions is

$$U_n(z) = (n+1) \, _2F_1\left(-n, \, n+2, \, \frac{3}{2}, \, \frac{1-z}{2}\right).$$

Here are the first few.

```
In[8]:= Table[{Expand[(n + 1) Hypergeometric2F1[-n, n + 2, 3/2, (1 - z)/2]],
             Expand[ChebyshevU[n, z]]}, {n, 0, 4}]
Out[8]= {{1, 1}, {2 z, 2 z}, {-1 + 4 z^2, -1 + 4 z^2},
         {-4 z + 8 z^3, -4 z + 8 z^3}, {1 - 12 z^2 + 16 z^4, 1 - 12 z^2 + 16 z^4}}
```

The generating function is

$$\frac{1}{R^2} = \sum_{n=0}^{\infty} U_n(z) \, t^n$$

where $R = \sqrt{1 - 2tz + t^2}$. We now check this.

```
In[9]:= Expand /@ CoefficientList[Series[1 /(1 - 2 t z + t^2), {t, 0, 4}], t]
Out[9]= {1, 2 z, -1 + 4 z^2, -4 z + 8 z^3, 1 - 12 z^2 + 16 z^4}
```

These are just the Chebyshev polynomials $U_n(z)$.

```
In[10]:= % == Table[Expand[ChebyshevU[n, z]], {n, 0, 4}]
Out[10]= True
```

The analog of the formula $T_n(z) = \cos(n \arccos(z))$ for the Chebyshev polynomials $T_n(z)$ for the Chebyshev polynomials $U_n(z)$ is

$$U_n(z) = \frac{\sin((n+1)\arccos(z))}{\sqrt{1-z^2}},$$

or, rewritten in trigonometric functions,

$$U_n(\cos(\vartheta)) = \frac{\sin((n+1)\,\vartheta)}{\sin(\vartheta)}.$$

```
In[11]:= {Table[(Sin[(i + 1) ArcCos[z]] // TrigExpand)/
            Sqrt[1 - z^2] // Expand, {i, 0, 4}],
        Table[ChebyshevU[i, z], {i, 0, 4}]}
Out[11]= {{1, 2 z, -1 + 4 z^2, -4 z + 8 z^3, 1 - 12 z^2 + 16 z^4},
         {1, 2 z, -1 + 4 z^2, -4 z + 8 z^3, 1 - 12 z^2 + 16 z^4}}
```

The last representation of $U_n(z)$ allows for an easy way to smoothly interpolate a given set of points [512], [479]. Let $p_0, \ldots, p_{n-1}$ be an odd number of points. Then the function

$$\mathcal{U}(x) = \sum_{k=0}^{n} \frac{p_k}{n} U_{n-1}\left(\sqrt{1-x^2} \sin\left(\frac{k}{n}\pi\right) + x \cos\left(\frac{k}{n}\pi\right)\right)$$

has the property $\mathcal{U}(-\cos(k/n\pi)) = p_k$ for $k = 0, \ldots, n-1$. Here is an example curve—the Peano curve used in Chapter 1 of the Graphics volume.

```
In[12]:= peanoLine = {{0, 0}, {0, -1}, {0, -2}, {1, -2}, {1, -1}, {1, 0}, {2, 0},
          {3, 0}, {3, -1}, {2, -1}, {2, -2}, {3, -2}, {3, -3}, {2, -3}, {1, -3},
          {0, -3}, {0, -4}, {0, -5}, {0, -6}, {1, -6}, {1, -5}, {1, -4}, {2, -4},
          {3, -4}, {3, -5}, {2, -5}, {2, -6}, {3, -6}, {3, -7}, {2, -7}, {1, -7},
          {0, -7}, {0, -8}, {0, -9}, {0, -10}, {0, -11}, {1, -11}, {1, -10},
          {2, -10}, {2, -11}, {3, -11}, {3, -10}, {3, -9}, {2, -9}, {1, -9},
          {1, -8}, {2, -8}, {3, -8}, {4, -8}, {4, -9}, {4, -10}, {4, -11},
          {5, -11}, {5, -10}, {6, -10}, {6, -11}, {7, -11}, {7, -10}, {7, -9},
          {6, -9}, {5, -9}, {5, -8}, {6, -8}, {7, -8}, {7, -7}, {7, -6}, {7, -5},
          {6, -5}, {6, -6}, {6, -7}, {5, -7}, {4, -7}, {4, -6}, {5, -6}, {5, -5},
          {4, -5}, {4, -4}, {5, -4}, {6, -4}, {7, -4}, {7, -3}, {7, -2}, {7, -1},
          {6, -1}, {6, -2}, {6, -3}, {5, -3}, {4, -3}, {4, -2}, {5, -2}, {5, -1},
          {4, -1}, {4, 0}, {5, 0}, {6, 0}, {7, 0}, {8, 0}, {8, -1}, {8, -2},
          {8, -3}, {9, -3}, {9, -2}, {10, -2}, {10, -3}, {11, -3}, {11, -2},
          {11, -1}, {10, -1}, {9, -1}, {9, 0}, {10, 0}, {11, 0}, {12, 0},
          {12, -1}, {12, -2}, {13, -2}, {13, -1}, {13, 0}, {14, 0}, {15, 0},
          {15, -1}, {14, -1}, {14, -2}, {15, -2}, {15, -3}, {14, -3}, {13, -3},
          {12, -3}, {12, -4}, {12, -5}, {12, -6}, {13, -6}, {13, -5}, {13, -4},
          {14, -4}, {15, -4}, {15, -5}, {14, -5}, {14, -6}, {15, -6}, {15, -7},
          {14, -7}, {13, -7}, {12, -7}, {11, -7}, {11, -6}, {11, -5}, {11, -4},
          {10, -4}, {10, -5}, {9, -5}, {9, -4}, {8, -4}, {8, -5}, {8, -6},
          {9, -6}, {10, -6}, {10, -7}, {9, -7}, {8, -7}, {8, -8}, {8, -9},
          {8, -10}, {8, -11}, {9, -11}, {9, -10}, {10, -10}, {10, -11}, {11, -11},
          {11, -10}, {11, -9}, {10, -9}, {9, -9}, {9, -8}, {10, -8}, {11, -8},
          {12, -8}, {12, -9}, {12, -10}, {13, -10}, {13, -9}, {13, -8}, {14, -8},
          {15, -8}, {15, -9}, {14, -9}, {14, -10}, {15, -10}, {15, -11},
          {14, -11}, {13, -11}, {12, -11}, {12, -12}, {12, -13}, {12, -14},
          {13, -14}, {13, -13}, {13, -12}, {14, -12}, {15, -12}, {15, -13},
          {14, -13}, {14, -14}, {15, -14}, {15, -15}, {14, -15}, {13, -15},
          {12, -15}, {11, -15}, {11, -14}, {11, -13}, {11, -12}, {10, -12},
          {10, -13}, {9, -13}, {9, -12}, {8, -12}, {8, -13}, {8, -14}, {9, -14},
          {10, -14}, {10, -15}, {9, -15}, {8, -15}, {7, -15}, {7, -14}, {7, -13},
          {7, -12}, {6, -12}, {6, -13}, {5, -13}, {5, -12}, {4, -12}, {4, -13},
          {4, -14}, {5, -14}, {6, -14}, {6, -15}, {5, -15}, {4, -15}, {3, -15},
          {3, -14}, {3, -13}, {3, -12}, {2, -12}, {2, -13}, {1, -13}, {1, -12},
          {0, -12}, {0, -13}, {0, -14}, {1, -14}, {2, -14}, {2, -15}, {1, -15},
          {0, -15}, {-1, -15}, {-1, -14}, {-1, -13}, {-1, -12}, {-1, -11},
          {-1, -10}, {-1, -9}, {-1, -8}, {-1, -7}, {-1, -6}, {-1, -5}, {-1, -4},
          {-1, -2}, {-1, -1}, {-1, 0}};
```

It is straightforward to implement the above formulas.

```
In[13]:= v = Length[peanoLine];
         Do[p[k] = peanoLine[[k + 1]], {k, 0, v - 1}]

In[15]:= (* the Lagrange-type interpolating polynomials *)
         σ[k_, n_][x_] := 1/n ChebyshevU[n - 1, Sin[Pi k/n] Sqrt[1 - x^2] -
                                                 Cos[Pi k/n] x]

         (* sum over all points *)
         U[x_] := Sum[p[k] σ[k, v][x], {k, 0, v - 1}];

In[19]:= (* take four points between original points *)
         interpolation = U /@ Table[N[-Cos[Pi j/v/5]], {j, 0, 5(v - 1)}];
```

The following graphic shows the original points and the interpolating curve.

```
In[21]:= Show[Graphics[{{PointSize[0.02], (* original points *)
                Table[{Hue[(k - 1)/v], Point[p[k]]}, {k, v - 1}]},
               {Thickness[0.002], (* interpolation *)
                Line[Append[interpolation, First[interpolation]]]}}],
          PlotRange -> All, AspectRatio -> Automatic]
```

The Chebyshev polynomials $U_n(z)$ are intimately related to the Fibonacci numbers. Here are two such relations shown [166].

```
In[22]:= Table[(-I)^n ChebyshevU[n, I/2], {n, 0, 12}]
```

```
Out[22]= {1, 1, 2, 3, 5, 8, 13, 21, 34, 55, 89, 144, 233}
```

```
In[23]:= With[{o = 12, U = ChebyshevU},
          Table[Resultant[U[n, x], U[n - 1, x] - I U[n - 2, x], x]/
              2^(n (n - 1))/(-1)^(n (n - 1)/2), {n, 0, 12}]]
```

```
Out[23]= {1, 1, 2, 3, 5, 8, 13, 21, 34, 55, 89, 144, 233}
```

A lot more interesting properties and pictures are hidden in the orthogonal polynomials As an example, let us view what happens with the zeros of $U_n(z)$ (for $n = 15$) if we replace the coefficients a_i in $a_i z^i$ by $a_i e^{i\varphi}$ and let φ run from 0 to 2π.

```
In[24]:= Module[{φ, z},
          Show[Graphics[{PointSize[0.003], Map[Point[{Re[#], Im[#]}]&,
          (* calculate zeros of polynomials *)
          Cases[Table[NRoots[# == 0, z], {φ, 0, 2Pi, 2Pi/500}],
              _?NumberQ, {-1}], {-1}]& /@ Drop[
          (* make new coefficients with factor Exp[I φ] *)
          FoldList[Plus[#1 /. φ -> 0, #2 Exp[I φ]]&, 0,
                List @@ Expand[ChebyshevU[15, z]]], 2]}],
          Frame -> True, PlotRange -> All]]
```

The coefficients of $U_n(x)$ modulo 3 form the "typical" hierarchical pattern that we repeatedly saw in Chapter 2 of the Numerics volume [567].

```
In[25]:= Show[Graphics[Table[
    MapIndexed[(* color the data array *)
              {(* better contrast for printed version *)
        If[Options[Graphics, ColorOutput][[1, 2]] === GrayLevel,
           Which[# == 0, GrayLevel[0.3], # == 1, GrayLevel[0.6],
                 # == 2, GrayLevel[0.9]],
           Which[# == 0, RGBColor[1, 0, 0], # == 1, RGBColor[0, 1, 0],
                 # == 2, RGBColor[0, 0, 1]]],
                Rectangle[{#2[[1]] - n/2, -n} - 1/2,
                          {#2[[1]] - n/2, -n} + 1/2]}&,
       (* array of data *)
       Mod[Round[CoefficientList[ChebyshevU[n, x], x]], 3]], {n, 300}]]]
```

Another example is to visualize how the single summands added together in all possible ways form the value of an orthogonal polynomial.

```
In[26]:= Show[GraphicsArray[
        Graphics3D[{Thickness[0.002],
        Table[{Hue[φ/(2Pi)], Map[{Re[#], r, Im[#]}&,
            Line[FoldList[Plus, 0, #]]& /@ (* all possible partial sums *)
              Permutations[(Function @@ {x, List @@ #[6, x]})[
                                                   r Exp[I φ]]], {-1}]},
          {r, 0, 3/2, 0.05}, {φ, 0, N[2Pi] - N[2Pi]/36, N[2Pi]/36}]},
        PlotRange -> All, Axes -> False, BoxRatios -> {1, 1, 1}]& /@
        (* use both Chebyshev polynomials *) {ChebyshevU, ChebyshevT}]]
```

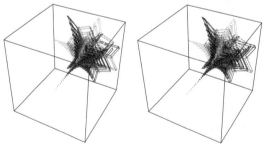

Next, we look at the regions in the z,n-plane where $U_n(z)$ is positive (in black).

```
In[27]:= Show[Graphics[
        Table[{GrayLevel[(* black or white? *)
        If[ChebyshevU[i, Plus @@ #/2] > 0, 1, 0]],
          (* divide positive from negative regions *)
           Rectangle[{#[[1]], i - 1/2}, {#[[2]], i + 1/2}]}& /@
```

```
Partition[Join[{-1}, Last /@ List @@ (* the zeros *)
    NRoots[ChebyshevU[i, z] == 0, z, 30], {1}], 2, 1], {i, 2, 50}]],
Frame -> True, PlotRange -> All, FrameLabel -> {"z", "n"}]
```

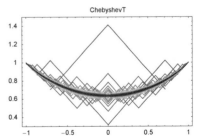

We end this section by visualizing how the interlacing of the zeros of $U_{n-1}(x)$ and $U_n(x)$ looks for larger n. By connecting the interlaced zeros with a piecewise straight curves of slope ± 1, the limit curve $2/\pi \left(x \arcsin(x) + (1 + x^2)^{1/2}\right)$ is approached [314]. (This limit curve is also obtained for rescaled zeros of other orthogonal polynomials.) The right graphic shows the curves for the zeros of $T_{n-1}(x)$ and $T_n(x)$. We use the first 100 polynomials that have zeros.

```
In[28]:= (* a curve with slope ±1 between the points minMaxList;
        rightmost line segment has slope -1 *)
    makeZigZagCurve[minMaxList_] :=
    FoldList[{#2[[1]], #1[[2]] + #2[[2]] (#2[[1]] - #1[[1]])}&,
            minMaxList[[-1]] {1, 1},
            MapIndexed[{#1, -(-1)^#2[[1]]}&, Reverse[minMaxList]]]
```

```
In[30]:= With[{o = 100},
    Show[GraphicsArray[
    Graphics[{{(* color curves *) MapIndexed[{Hue[0.8 #2[[1]]/o],
            (* interlace zeros and form zigzag curve *)
            Line[makeZigZagCurve[#1]]}&, Sort[Flatten[#]]& /@ Partition[
        Table[Chop[N[z /. {ToRules[Roots[#[k, z] == 0, z]]}]],
            {k, o}], 2, 1]]},
    {Thickness[0.002], GrayLevel[0], (* theoretical limit curve *)
    Line[Table[{x, 2/Pi (x ArcSin[x] + Sqrt[1 - x^2])},
        {x, -1, 1, 1/100}]]}} // N, PlotRange -> {0.3, 1.5},
    Frame -> True, PlotLabel -> #]& /@ {ChebyshevU, ChebyshevT}]]]
```

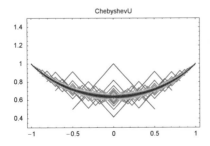
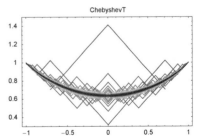

2.9 Relationships Among the Orthogonal Polynomials

There are a number of relationships among the various classical orthogonal polynomials. We do not discuss all of them here, but instead simply check some of them for special choices of parameter values.

- Gegenbauer polynomials are, up to the normalization constant, special Jacobi polynomials. More precisely,

$$C_n^\alpha(z) = \frac{\Gamma(\alpha + 1/2)\,\Gamma(2\,\alpha + n)}{(\Gamma(2\,\alpha)\,\Gamma(\alpha + n + 1/2))}\, P_n^{(\alpha - 1/2, \alpha - 1/2)}(z).$$

We can see this with an example.

```
In[1]:= JacobiGegenbauerRelation[n_, α_] :=
        Expand[Gamma[α + 1/2] Gamma[2α + n]/Gamma[2α]/Gamma[α + n + 1/2]*
            JacobiP[n, α - 1/2, α - 1/2, z]] ==
        Expand[GegenbauerC[n, α, z]]

        Apply[JacobiGegenbauerRelation, {{1, 1}, {2, 2}, {2, 1}, {1, 2}}, {1}]
Out[2]= {True, True, True, True}
```

- Gegenbauer polynomials $C_n^{(0)}(z)$ can be found from $C_n^{(\alpha)}(z)$ by

$$C_n^{(0)}(z) = \lim_{\alpha \to 0} \frac{C_n^{(\alpha)}(z)}{\alpha}, \quad n \neq 0.$$

Here are a few examples (note that $n \neq 0$).

```
In[3]:= Table[{Expand[GegenbauerC[n, z]], Limit[GegenbauerC[n, α, z]/α, α -> 0]},
            {n, 6}] // TableForm
Out[3]//TableForm=
```

$2\,z$	$2\,z$
$-1 + 2\,z^2$	$-1 + 2\,z^2$
$-2\,z + \frac{8\,z^3}{3}$	$-2\,z + \frac{8\,z^3}{3}$
$\frac{1}{2} - 4\,z^2 + 4\,z^4$	$\frac{1}{2} - 4\,z^2 + 4\,z^4$
$2\,z - 8\,z^3 + \frac{32\,z^5}{5}$	$2\,z - 8\,z^3 + \frac{32\,z^5}{5}$
$-\frac{1}{3} + 6\,z^2 - 16\,z^4 + \frac{32\,z^6}{3}$	$-\frac{1}{3} + 6\,z^2 - 16\,z^4 + \frac{32\,z^6}{3}$

- Hermite polynomials $H_n(z)$ can be found from Laguerre polynomials $L_n^{(\alpha)}(z)$ by

$$H_{2n}(z) = (-1)^n\, 2^{2n}\, n!\; L_n^{-\frac{1}{2}}(z^2)$$

$$H_{2n+1}(z) = (-1)^n\, 2^{2n+1}\, n!\; z\, L_n^{\frac{1}{2}}(z^2).$$

Here again are a few examples.

```
In[4]:= HermiteLaguerreRelation[n_Integer] := Apply[Equal,
        {{Expand[HermiteH[2n, z]],
          Expand[(-1)^n 2^(2n) n! LaguerreL[n, -1/2, z^2]]},
         {Expand[HermiteH[2n + 1, z]],
          Expand[(-1)^n 2^(2n + 1) n! z LaguerreL[n, 1/2, z^2]]}}, {1}]
```

```
In[5]:= Table[HermiteLaguerreRelation[n], {n, 0, 4}] // Flatten // Union

Out[5]= {True}
```

■ The Chebyshev polynomials T_n and U_n are closely connected to Gegenbauer polynomials:

$$U_n(z) = C_n^{(1)}(z)$$
$$T_n(z) = \frac{n}{2} C_n^{(0)}(z).$$

Here are six examples.

```
In[6]:= Table[Expand[ChebyshevU[n, z]] ==
            Expand[GegenbauerC[n, 1, z]], {n, 0, 5}]

Out[6]= {True, True, True, True, True, True}
```

```
In[7]:= Table[Expand[ChebyshevT[n, z]] ==
            Expand[n/2 GegenbauerC[n, z]], {n, 2, 7}]

Out[7]= {True, True, True, True, True, True}
```

The following two kinds of Chebyshev polynomials are closely related:

$$T_n(z) = U_n(z) - z\, U_{n-1}(z).$$

Here is a check.

```
In[8]:= Table[Expand[ChebyshevT[n, z]] ==
            Expand[ChebyshevU[n, z] - z ChebyshevU[n - 1, z]], {n, 0, 2}]

Out[8]= {True, True, True}
```

Using the trigonometric representations of the Chebychev polynomials, we can also easily prove this relation.

```
In[9]:= (Together @ ExpandAll[
          ChebyshevU[n, z] - z ChebyshevU[n - 1, z] /.
            ChebyshevU[n_, z_] -> Sin[(n + 1)ArcCos[z]]/Sin[ArcCos[z]]] /.
              Sin[a_ + b_] -> Sin[a] Cos[b] + Cos[a] Sin[b]) /.
                Cos[n ArcCos[z]] -> ChebyshevT[n, z]

Out[9]= ChebyshevT[n, z]
```

To complete this chapter, we discuss one more nontrivial relationship between orthogonal polynomials:

$$\frac{s^m}{m!} H_m(x - 2s) \exp(-s^2 + 2xs) = \sum_{k=0}^{\infty} L_m^{(k-m)}(2s^2) H_k(x) \frac{s^k}{k!}$$

for $m = 0, 1, \ldots$ ([399]).

We can examine the "correctness" of this assertion for some special cases. We fix m, and we compute both expressions up to the same order in s. In order to keep the powers on the right-hand side from becoming too large, we make use of the property of `SeriesData`-objects to collapse expressions of the form $f(x) + O(x)^n$ to order n.

```
In[10]:= hermiteLaguerreTest[m_Integer? (# >= 0&), n_Integer? (# >= 0&)] :=
           Cancel /@ Collect[Normal @
           (* make a series around s = 0 *)
           Series[s^m/m! HermiteH[m, x - 2s] Exp[-s^2 + 2s x], {s, 0, n}], s] ==
           Cancel /@ Collect[Normal[Sum[LaguerreL[m, k - m, 2s^2] *
```

```
                                    HermiteH[k, x]/k! s^k, {k, 0, n}] +
    (* make the expansion in a series *) O[s]^(n + 1)], s]
```

Here are a few special cases.

In[11]:= **Table[hermiteLaguerreTest[i, j], {i, 0, 6}, {j, 0, 6}] //**
                                                                        Flatten // Union

Out[11]= {True}

We could go on now and make some nice pictures involving various combinations of the orthogonal polynomials. Here is a contour plot that contains the two Chebyshev polynomials $T_7(z)$ and $U_8(z)$.

In[12]:= **ContourPlot[Re[ChebyshevT[7, I ChebyshevT[8, x + I y]]],**
                    {x, -3/2, 3/2}, {y, -0.35, 0.35},
                    Contours -> {0}, PlotPoints -> 222, FrameTicks -> None]

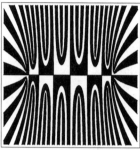

Expressions with more than one kind of orthogonal polynomial often arise as solutions of partial differential equations after separation of variables. A typical example is the quantum-mechanical treatment of the hydrogen atom. Here we give another, not so well known example. The (unnormalized) eigenfunctions of the A_{v-1} Calogero–Sutherland model [528], [245], [441], [231], [164]

$$-\frac{1}{2}\Delta \psi_{n,m}^{(\lambda)}(x_1, \ldots, x_v) + \left(\sum_{j=1}^{v} x_j^2\right)\psi_{n,m}^{(\lambda)}(x_1, \ldots, x_v) + \sum_{j=1}^{v}\sum_{k=j+1}^{v}\frac{\lambda(\lambda-1)}{(x_k-x_j)^2}\psi_{n,m}^{(\lambda)}(x_1, \ldots, x_v) =$$

$$\varepsilon_{n,m}\psi_{n,m}^{(\lambda)}(x_1, \ldots, x_v)$$

can be expressed through Hermite and Laguerre polynomials

$$\psi_{n,m}^{(\lambda)}(x_1, \ldots, x_v) = \left(\prod_{j=1}^{v}\prod_{k=j+1}^{v}(x_k-x_j)^\lambda\right)e^{-r^2/2}H_m(y)L_n^{(v-3+1(v-1)\lambda v)/2}(r^2-y^2)$$

where $r = (\sum_{j=1}^{v} x_j^2)^{1/2}$ and $y = v^{-1/2}(\sum_{j=1}^{v} x_j)^{1/2}$.

The eigenvalues $\varepsilon_{n,m}$ are $\varepsilon_{n,m}=(m+2n)/2+(v+\lambda v(v-1))$.

Here are the eigenfunctions $\psi_{n,m}^{(\lambda)}(x_1, \ldots, x_v)$, the Calogero–Sutherland Hamiltonian and the energy eigenvalues $\varepsilon_{n,m}$ defined.

In[13]:= (* wave functions *)
```
    ψCS[xs_List, λ_, n_Integer, m_Integer] :=
    Module[{v = Length[xs], y, r},
            y = Plus @@ xs/Sqrt[v]; r = Sqrt[xs.xs];
            Product[(xs[[k]] - xs[[j]])^λ,
```

```
                      {j, 1, v}, {k, j + 1, v}] Exp[-r^2/2] HermiteH[m, y]*
                      LaguerreL[n, (v - 3)/2 + λ v (v - 1)/2, r^2 - y^2]]
```

In[15]:= (* Hamiltonian *)
```
        HCS[xs_List, λ_] :=
        With[{v = Length[xs]},
          (Sum[-D[#, {xs[[k]], 2}] + xs[[k]]^2 #, {k, v}]/2 +
             λ (λ - 1) Sum[(xs[[k]] - xs[[j]])^-2, {j, 1, v}, {k, j + 1, v}] #)&]
```

In[17]:= εCS[v_, λ_, n_, m_] := (m + 2 n) + 1/2 (v + λ v (v - 1))

Verifying the eigenfunctions is already a large amount of work for small n and m. $\psi_{6,3}^{(\lambda)}(x_1, x_2, x_3)$ has 1396 terms when expanded. After differentiation and multiplication, we get 10273 terms.

In[18]:= Block[{ξ = {x1, x2, x3}, v = 3, n = 6, m = 4, Ψ},
```
          Ψ = ψCS[ξ, λ, n, m];
          HCS[ξ, λ][Ψ] - εCS[v, λ, n, m] Ψ // Together] // Timing
```
Out[18]= {2.74 Second, 0}

Here is a visualization of the state $\psi_{17,14}^{(16)}(x_1, x_2)$. (For similar graphics of multi-variable functions, see [286].) The right graphic shows the equisurfaces $\psi_{4,4}^{(4)}(x_1, x_2, x_3) = 10^4$.

In[19]:= Needs["Graphics`ContourPlot3D`"]

In[20]:= Show[GraphicsArray[
```
        Block[{$DisplayFunction = Identity},
        (* contour plot of ψCS[{x1, x2}, 16, 17, 14] *)
        {With[{L = 12},
          ContourPlot[Evaluate[N @ ψCS[{x1, x2}, 16, 17, 14]],
                      {x1, -L, L}, {x2, -L, L}, PlotPoints -> 240,
                      Contours -> 20, ContourLines -> False,
                      ColorFunction -> (Hue[0.8 #]&), PlotRange -> All]],
        (* 3D contour plot of ψCS[{x1, x2, x3}, 4, 4, 4] *)
        With[{L = 8},
          ContourPlot3D[Evaluate[N @ ψCS[{x1, x2, x3}, 4, 4, 4]],
                      {x1, -L, L}, {x2, -L, L}, {x3, -L, L},
                      Contours -> {10^4}, MaxRecursion -> 1,
                      ContourStyle -> {EdgeForm[]},
                      PlotPoints -> {36, 3}, ViewPoint -> {4, 1, 2}]]}]]]
```

Because we are at the end of our discussion of the orthogonal polynomials, let us examine how a value $p_n(z)$ (fixed n) evolves as the sum of terms of the form $c_i z^i$.

In[21]:= Module[{f, g, r, p, n = 8},
```
        (* make Function that gives a list of summands c[i] z^i *)
        f[poly_] := Function[C, #]& @ ((C^Range[0, Exponent[#, C]])*
                      N[CoefficientList[#, C]])&[poly[C]];
```

```
(* make a line from a list of summands c[i] z^i *)
g[poly_, y_] := Line[{Re[#], Im[#]}& /@
                     Rest[FoldList[Plus, 0, poly[y]]]];
Show[GraphicsArray[#]]& /@ Map[Function[poly,
   Graphics[{Thickness[0.002],
     Table[(* z varies inside unit disk *) g[poly, r Exp[I φ]],
        {r, 0, 1., 1/12.}, {φ, 0, 2Pi // N, 2Pi/12 // N}]},
        AspectRatio -> 1, Frame -> True, FrameTicks -> None,
        PlotRange -> All, DisplayFunction -> Identity]],
     (* the polynomials under investigation *) Table[f[#], {i, n}]& /@
   {HermiteH[i, #]&, LaguerreL[i, 2, #]&, JacobiP[i, 2 - I, 2 + I, #]&,
    LegendreP[i, 0, #]&, GegenbauerC[i, 1/2 + I/2, #]&,
    ChebyshevT[i, #]&, ChebyshevU[i, #]&}, {2}]]
```

There are a lot of interesting things to say and visualize about the classical orthogonal polynomials, for instance, finding and visualizing which coefficients and terms are the dominant ones.

```
In[22]:= coefficientDensityPlot[poly_, maxDegree_, opts___] :=
     Show[Graphics[{PointSize[0.003],
     (* make a grayed polygon from every coefficient *)
     MapIndexed[{GrayLevel[#], Rectangle[Reverse[#2] - {1/2, 1/2} ,
                                 Reverse[#2] + {1/2, 1/2}]}&,
     (* the rescaled list of coefficients *)
```

```
(# - Min[#])/(Max[#] - Min[#])& /@
(* list of all polynomials up to degree maxDegree *)
Table[CoefficientList[poly, x], {n, maxDegree}], {-1}]}],
            PlotRange -> All, AspectRatio -> 1, Frame -> True,
(* make label *)
FrameLabel -> {TraditionalForm[poly], "n"}, opts];

(* make a picture of every orthogonal polynomial *)
Show[GraphicsArray[#]]& /@
Map[coefficientDensityPlot[#, 50, DisplayFunction -> Identity]&,
(* the polynomials *)
  {{HermiteH[n, x], LaguerreL[n, x], ChebyshevT[n, x]},
   {ChebyshevU[n, x], GegenbauerC[n, 2, x], LegendreP[n, x]},
   {JacobiP[n, -4, 5, x], JacobiP[n, 2, 5, x], JacobiP[n, 30, 30, x]}}, {2}]
```

Because all orthogonal polynomials have simple real zeros only, we can construct a symmetric real tridiagonal companion matrix whose eigenvalues coincide with the zeros of the original polynomial [381], [508], [612]. The function SchmeisserCompanionMatrix generates this matrix for a given polynomial *poly*.

```
In[25]:= (* a message in case the polynomial roots are not all real *)
    SchmeisserCompanionMatrix::mhcr =
    "Polynomial `1` has complex roots in `2`.";

    SchmeisserCompanionMatrix[poly_, x_, S_:Expand] :=
```

```
Module[{n = Exponent[poly, x], v = 0, cond, LC, r, q, c, p},
(* leading coefficient of a polynomial *)
LC[p_] := If[p === 0, 1, Coefficient[p, x, Exponent[p, x]]];
(* initial conditions for the modified Euclidean algorithm *)
f[1] = S[poly/LC[poly]]; f[2] = S[1/n D[f[1], x]];
While[v == 0 || v <= n - 2 && c[v] != 0, v++];
If[f[v + 1] =!= 1,
   r[v] = -S[PolynomialRemainder[f[v], f[v + 1], x]];
   q[v] =  S[PolynomialQuotient[ f[v], f[v + 1], x]];
   c[v] = LC[r[v]]; If[c[v] != 0, f[v + 2] = S[r[v]/c[v]]];
   If[f[v + 2] === 0,
      c[v] = 0; f[v + 2] = S[D[f[v + 1], x]/LC[f[v + 1]]]],
   q[v] = S[f[v]]]]; q[n] = f[n];
(* for all zeros real no c should vanish *)
cond = If[#, #,  Message[SchmeisserCompanionMatrix::mhcr, poly, x];
                      False]&[FreeQ[Last /@ DownValues[c], 0, {1}]];
(* form the matrix *)
If[cond, Table[Which[i === j, -q[i] /. x -> 0,
                Abs[i - j] === 1, Sqrt[c[Min[i, j]]],
                True, 0], {i, n}, {j, n}]] /; cond] /; PolynomialQ[poly, x]
```

The next graphics show the structure of the eigenvectors of these companion matrices for various orthogonal polynomials.

```
In[29]:= With[{n = 100, prec = Sequence[]},
   Show[GraphicsArray[#]]& /@
   Map[Function[poly,
     ListDensityPlot[ArcTan @ (* the eigenvectors *) Eigenvectors[N[
                   SchmeisserCompanionMatrix[poly[n, x], x], prec]],
                   Mesh -> False, PlotRange -> All, FrameTicks -> False,
                   DisplayFunction -> Identity]],
         (* the polynomials *)
     {(* largest zero increases *)
      {HermiteH, LaguerreL, LaguerreL[#1, #1^2, #2]&,
       (* inverse argument *) Expand[#2^#1 ChebyshevU[#1, 1/#2]]&},
      (* zeros in fixed interval *)
      {ChebyshevT, ChebyshevU, LegendreP, JacobiP[#1, 3, 4, #2]&}}, {2}]]
```

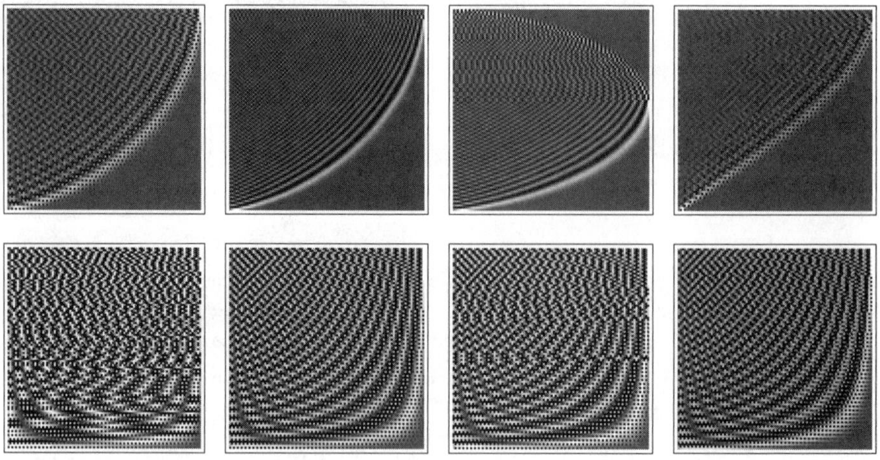

As a final example, we find the image of the iterated set of roots of the unit circle.

```
In[30]:= With[{roots = {7, 6, 5, 4, 3, 2}, n = 16},
    Show[GraphicsArray[#]]& /@
  Map[(* map onto the polynomials *)
    Function[poly,
    Graphics[{PointSize[0.003], Map[Point[{Re[#], Im[#]}]&,
    (* iterate the polynomial equation solving *)
    Rest[FoldList[Function[{x, y},
    (* the numerical values of the roots *)
    Cases[NRoots[poly[y, C] == #1, C]& /@ x, _?NumberQ, {-1}]],
        (* points on the unit circle *)
        Table[N[Exp[2Pi I i/n]], {i, 0, n - 1}], roots]], {-1}]],
        PlotRange -> All, Frame -> True, FrameTicks -> None,
        PlotLabel -> StyleForm[ToString[InputForm[
                         Head[poly[C, C]]]], "MR"]]],
        (* the polynomials *)
        {{LegendreP, ChebyshevT}, {ChebyshevU, LaguerreL},
        {HermiteH, JacobiP[#1, 5, -1, #2]&},
        {JacobiP[#1, -2 + I, I, #2]&,
         GegenbauerC[#1, -1/3, #2]&}}, {2}]]
```

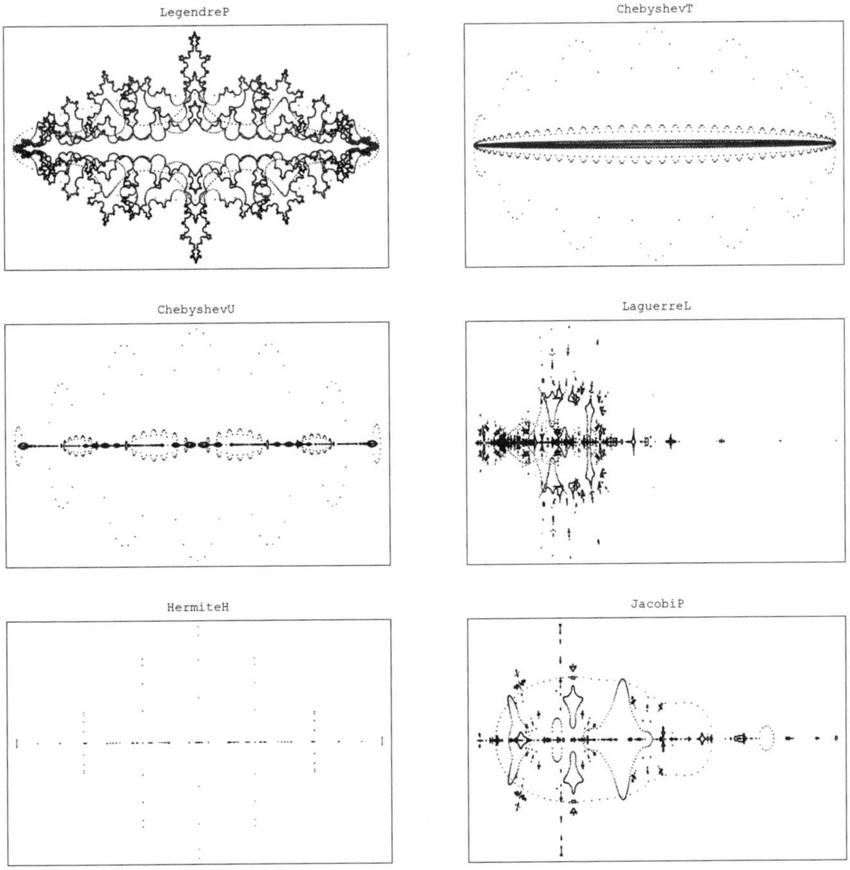

LegendreP

ChebyshevT

ChebyshevU

LaguerreL

HermiteH

JacobiP

JacobiP

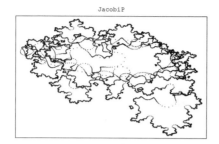

GegenbauerC

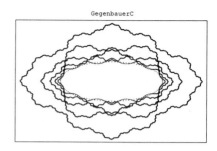

2.10 Ground-State of the Quartic Oscillator

Let us end this chapter with an application of the Hermite polynomials from quantum mechanics.

Mathematical Remark: Ground-State Energy of the Quartic Oscillator

The calculation of the ground-state energy ε_0 of the quartic oscillator in standard quantization [141]

$$-\psi_0''(z) + z^4\,\psi_0(z) = \varepsilon_0\,\psi_0(z)$$

is an important test ground for many algorithms (some were mentioned in exercises of Chapter 1 of the Numerics volume [567]; see also [377], [280], [281], [236], [522], [143], [514], [546], [126], [597], [258], [516], [269], [17], [617], [283], [144], [48], [598], [64], [421], [316], [629], [16], [268], [523], [398], [599], [262], [149], [32], [261], [600], [601], [470], [352], [40], [198], [161], [338], [232], [422], [331], [339], [174], [453], [575], [10], [11], [111], [571], [407], [243], [397], [18], [282], [12], [259] for various calculations about the anharmonic oscillator). For comparing the various methods, it is important to know the numerical value of ε_0 to a high precision (no explicit symbolic expression is known for ε_0). A very natural method is the expansion of $\psi_0(z)$ in harmonic oscillator eigenfunctions $\psi_0(z) = \sum_{k=0}^{\infty} \alpha_k\,\phi_k(z)$. This converts the original Sturm–Liouville problem into a Hill determinant problem [54], [204]. After truncating the resulting infinite matrix, the calculational task is the diagonalization of a matrix $\mathcal{H}$ with matrix elements

$$h_{m,n} = \int_{-\infty}^{\infty} \phi_m(z)\,(-\phi_n''(z) + z^4\,\phi_n(z))\,dz,$$

where the $\phi_n(z)$ are the eigenfunctions of the corresponding harmonic oscillator problem:

$$-\phi_n''(z) + z^2\,\phi_n(z) = (2\,n + 1)\,\phi_n(z)$$

with $\phi_n(z) = c_n \exp(-z^2/2)\,H_n(z)$ and $c_n = (\pi^{1/2}\,2^n\,n!)^{-1/2}$.

For the matrix diagonalization, we will use the first p (up to $p = 500$) harmonic oscillator states to calculate an approximate value for ε_0, which we will call $\varepsilon_0^{(p)}$. In addition, we will estimate the number of correct digits of ε_0. As we will see, for the p-range under consideration, the error $\delta_{\varepsilon_0}^{(p)} = |\varepsilon_0 - \varepsilon_0^{(p)}|$ obeys roughly the equation $\log_{10} \delta_{\varepsilon_0}^{(p)} \approx (0.1 \ldots 0.2)\,p$.

These are the exact, normalized wave functions for the harmonic oscillator.

```
In[1]:= φ[n_, z_] := Exp[-z^2/2] HermiteH[n, z]/c[n]
        c[n_] := Sqrt[Sqrt[Pi] 2^n n!]
```

Because orthogonal polynomials evaluate numerically also for noninteger first arguments we have a quick look at the behavior of the $\phi_n(z)$ for noninteger n. The following graphic shows that only for integer n, the functions $\phi_n(z)$ vanish as $z \to \pm\infty$.

```
In[3]:= φNHP[n_, z_] := φ[SetPrecision[n, 30], SetPrecision[z, 30]]
```

```
In[4]:= Plot3D[Log[10, Abs[φNHP[n, z]]], {z, -6, 6}, {n, -1/2, 2},
           PlotPoints -> 120, Mesh -> False]
```

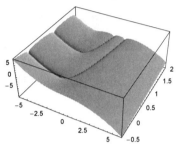

To avoid carrying out the time-consuming (symbolic or numeric) integration $\int_{-\infty}^{\infty}\phi_m(z)\,(-\phi_n''(z) + z^4\,\phi_n(z))\,dz$ explicitly, we will rewrite expressions of the form $z^m\,H_n(z)$ as linear combinations $\sum_{k=-m}^{m}\beta_k\,H_{n+k}(z)$, where the β_k are z-independent [367]. The function reduceProducts is implementing this. We use the three-term recursion relation for the Hermite polynomials here.

```
In[5]:= reduceProducts[expr_] :=
        FixedPoint[Expand[# /. z_^m_. HermiteH[n_, z_] :>
          (* recursion relation for Hermite polynomials *)
          z^(m - 1) (n HermiteH[n - 1, z] + HermiteH[n + 1, z]/2)]&, expr]
```

For our quartic oscillator, we get the following form of the matrix elements $\int_{-\infty}^{\infty}\phi_m(z)\,(-\phi_n''(z) + z^4\,\phi_n(z))\,dz$ [409], [447], [411], [179], [410], [56], [556], [368], [297], [256], [496] (for overlap integrals of shifted harmonic oscillator eigenfunctions, see [526], [309]). For the term $-\phi_n''(z)$, we use the differential equation of the harmonic oscillator and reexpress it as $(2n + 1)\,\phi_n(z) - z^2\,\phi_n(z)$.

```
In[6]:= makeh[V_, z_] := c[m]/c[n] Simplify[
        reduceProducts[((2n + 1) HermiteH[n, z] - z^2 HermiteH[n, z] +
                        (* the potential *) V HermiteH[n, z]] /.
          (* use orthogonality *) HermiteH[n_, z] -> KroneckerDelta[n, m]]
```

```
In[7]:= makeh[z^4, z] /. KroneckerDelta[a_, b_] :> Subscript[δ, a, b]
```

$$\text{Out[7]= } \frac{1}{\sqrt{2^n\,n!}}\left(\sqrt{2^m\,m!}\left(n\,(-6 + 11\,n - 6\,n^2 + n^3)\,\delta_{m,-4+n}\,+\right.\right.$$
$$\left.\left.\frac{1}{16}\,(32\,(-1 + n)^2\,n\,\delta_{m,-2+n} + 4\,(5 + 10\,n + 6\,n^2)\,\delta_{m,n} + 8\,\delta_{m,2+n} + 8\,n\,\delta_{m,2+n} + \delta_{m,4+n})\right)\right)$$

We rewrite the last expression slightly and add an outer If to avoid the unnecessary, but time-consuming, calculation of $2^{(m-n)/2-4}\,\sqrt{m!/n!}$ in the case $|n - m| > 4$.

```
In[8]:= h[n_, m_] := (h[n, m] = With[{δ = KroneckerDelta},
        If[n - m > 4, 0, (2^((m - n)/2 - 4) Sqrt[m!/n!])*
          (16 (n - 3) (n - 2) (n - 1) n δ[m, n - 4] +
          32 (n - 1)^2 n δ[m, n - 2] + 4 (5 + 2 n(5 + 3 n)) δ[m, n] +
          8 (1 + n) δ[m, n + 2] + δ[m, n + 4])]]) /; n >= m
```

```
(* use symmetry of H *)
h[n_, m_] := h[m, n] /; n < m
```

Here is a quick check for the correctness of $\mathcal{H}$ by comparing it with the result of the direct integration.

```
In[12]:= With[{m = 4},
        Table[h[m, n] == Integrate[φ[m, z]*
                (-D[φ[n, z], z, z] + z^4 φ[n, z]), {z, -Infinity, Infinity}],
            {n, 0, 8}]]
Out[12]= {True, True, True, True, True, True, True, True, True}
```

Now, let us calculate the lowest eigenvalue ε_0. We start by using the function `Eigenvalues` for doing this. `lowestEigenValue` gives the value of the lowest eigenvalue when using the first p harmonic oscillator states and numericalizing the matrix (before diagonalization) to precision *prec*.

```
In[13]:= lowestEigenValue[p_, prec_] :=
        Last[Eigenvalues[N[Table[h[n, m], {n, 0, p}, {m, 0, p}], prec]]]
```

For a 20×20 matrix, we obtain the value known from exercise 24 of Chapter 1 of the Numerics volume [567].

```
In[14]:= lowestEigenValue[20, $MachinePrecision - 1] // InputForm
Out[14]//InputForm=
        1.06036263818977739718596328988`14.954589770191003
```

For a 50×50 matrix, we get more correct digits.

```
In[15]:= lowestEigenValue[50, $MachinePrecision - 1] // InputForm
Out[15]//InputForm=
        1.0603620904844836941378815697`14.954589770191003
```

Comparing the last result with a high-precision evaluation shows that the last digits are not reliable from a numerical point of view.

```
In[16]:= lowestEigenValue[50, $MachinePrecision + 1]
Out[16]= 1.0603620904844837
```

To estimate the number of correct digits, we calculate and compare the lowest eigenvalue as a function of p.

```
In[17]:= ε0Data = Table[{k, lowestEigenValue[k, $MachinePrecision - 1]},
                {k, 5, 60, 5}];
```

The last picture suggests that we get around 10 digits per 50 harmonic oscillator states.

```
In[18]:= ListPlot[{#[[1]], Log[10, Abs[#[[2]]]]}& /@
                (# - {0, ε0Data[[-1, -1]]}& /@ Drop[ε0Data, -1]),
            Axes -> False, PlotRange -> All,
            Frame -> True, PlotJoined -> True]
```

It is interesting to also look at the eigenvectors. They describe the mixing of the harmonic oscillator states to build up the states of the anharmonic oscillators. We show a density plot of the absolute values of the eigenvector components. One sees that the lowest eigenfunctions are quite similar to the harmonic oscillator eigenfunctions. Higher states are complicated mixtures of harmonic oscillator states. The "checkerboard"-like overall structure results from the fact that the contribution of the antisymmetric (symmetric) harmonic oscillator states to the symmetric (antisymmetric) anharmonic oscillator states is identical zero [68]. The very high states are dominated by truncation effects and do not correctly mimic the anharmonic oscillator states.

```
In[19]:= With[{p = 100},
        With[{es = Eigensystem[N[Table[h[n, m], {n, 0, p}, {m, 0, p}]]]},
          ListDensityPlot[Abs[Reverse[es[[2]]]], Mesh -> False,
                ColorFunction -> (Hue[0.8 #]&)]]]
```

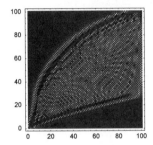

Now, let us calculate some more precise values for the ground-state energy. The call to the function Eigenvalues is quite expensive. It returns a list of *all* eigenvalues. On the other hand, we are only interested in the lowest one here. So, we use a numerical root-finding procedure for the characteristic polynomial. We do not calculate the characteristic polynomial symbolically, but rather for each λ numerically using Det. (Using the band-diagonality of the matrix $h_{m,n}$, we could also implement a five-term recursion relation for the determinant for a given λ.) Here is the 100×100 matrix with precision 500.

```
In[20]:= H100 = Table[N[h[n, m], 500], {n, 0, 100}, {m, 0, 100}];
```

We should use a high enough precision in all calculations. To make sure that we really have enough precision and to better watch the progress in the root finding, we will implement the following function *f*. *f* calculates the value of the characteristic polynomial and as a side effect prints the current value for λ, difference to the last value of λ and the value of the characteristic polynomial as well as the precision of these three numbers. The precision Precision[det] is especially important. We must be sure that the calculation of the determinant resulted in a value for det with valid digits.

```
In[21]:= f[λ_?NumberQ, mat_, prec_, info¿_] :=
        Module[{δ, det},
              (* data to be printed *)
              δ = λ - λOld; λOld = λ;
              (* the determinant value *)
              det = Det[mat - SetPrecision[λ, prec] *
                      IdentityMatrix[Length[mat]]];
              (* print progress? *)
              If[info¿ === True,
                 Print[{{Precision[λ]}, {N[δ, 3], Precision[δ]},
                      {N[det, 3], Precision[det]}}]];
              (* return result *) det]
```

The next input uses FindRoot on $f[\lambda, H100, 500, True]$ calculate $\lambda100$.

```
In[22]:= λ100 = λ /. FindRoot[f[λ, H100, 500, True], {λ, 106/100, 107/100},
                WorkingPrecision -> 500, Compiled -> False,
                AccuracyGoal -> 100]
```

$\{\{500.\}, \{0.\times 10^{-500}, 0.\}, \{5.98 \times 10^{296}, 490.841\}\}$

$\{\{500.\}, \{0.0100, 497.672\}, \{-1.57 \times 10^{298}, 492.241\}\}$

$\{\{500.\}, \{-0.00963, 497.655\}, \{-6.42 \times 10^{294}, 488.871\}\}$

$\{\{500.\}, \{-3.89 \times 10^{-6}, 494.263\}, \{-2.57 \times 10^{291}, 485.475\}\}$

$\{\{500.\}, \{-1.56 \times 10^{-9}, 490.867\}, \{7.00 \times 10^{282}, 476.909\}\}$

$\{\{500.\}, \{4.24 \times 10^{-18}, 482.301\}, \{-1.21 \times 10^{274}, 468.146\}\}$

$\{\{500.\}, \{-7.32 \times 10^{-27}, 473.538\}, \{-5.67 \times 10^{256}, 450.818\}\}$

$\{\{500.\}, \{-3.44 \times 10^{-44}, 456.21\}, \{3.40 \times 10^{213}, 407.595\}\}$

$\{\{500.\}, \{2.06 \times 10^{-87}, 412.987\}, \{-1.29 \times 10^{170}, 364.176\}\}$

$\{\{500.\}, \{-7.85 \times 10^{-131}, 369.568\}, \{-2.95 \times 10^{83}, 277.534\}\}$

$\{\{500.\}, \{-1.79 \times 10^{-217}, 282.926\}, \{9.22 \times 10^{-134}, 61.0286\}\}$

Out[22]= 1.0603620904841828996488338444895583827620302773177895252359268736814256297014 ⸜
         239352837864542715024952641768013830659609835272232149130928345583416857461 68 ⸜
         867200395213568684795160556783189231180979194915881113280634327338466441518 8 ⸜
         462595557682109111053666620304373760656647122796790138227551573989300830504 20 ⸜
         613362442948635851654024159517188885063366143924457753962057197844533602998 45 ⸜
         950640358891002448543889758452738534214591803417850930884080541952402525094 08 ⸜
         2571471411447373126986224292228545 0534

The last printed statement shows that we still had 61 digits left; so the result is reliable. For comparison, omitting the 100th harmonic oscillator state results in a change of ε_0 in the 22th digit, in agreement with the above-estimated precision gain of about 10 digits per 50 harmonic oscillator states. This means the dominating error is caused by the truncation, not the numerics.

```
In[23]:= H99 = Table[N[h[n, m], 500], {n, 0, 99}, {m, 0, 99}];
         λ99 = λ /. FindRoot[f[λ, H99, 500, False], {λ, 106/100, 107/100},
                WorkingPrecision -> 500, Compiled -> False,
                AccuracyGoal -> 100];
```

```
In[25]:= λ100 - λ99 // N
Out[25]= -6.63645 × 10^{-22}
```

We can save a lot of work by observing that ε_0 is a symmetric state and can only be composed by the symmetric harmonic oscillator states $\phi_{2n}(z)$ (one sees this in the arguments of the Kronecker delta functions in the definition of $h[m, n]$). This halves all matrix dimensions.

```
In[26]:= HS100 = Table[N[h[n, m], 100], {n, 0, 100, 2}, {m, 0, 100, 2}];
```

The last result agrees with $\lambda100$ within the precision of $\lambda100$.

```
In[27]:= (λ /. FindRoot[f[λ, HS100, 100, False], {λ, 106/100, 107/100},
                WorkingPrecision -> 500, Compiled -> False,
                AccuracyGoal -> 100]) - λ100 // N
Out[27]= -1.42613 × 10^{-99}
```

Now, let us take into account the first 250 symmetric states.

In[28]:= `HS500 = Table[N[h[n, m], 2000], {n, 0, 500, 2}, {m, 0, 500, 2}];`

We use the already-calculated value $\lambda 100$ for generating starting values for the root search.

In[29]:= `λ500 = N[λ /. FindRoot[f[λ, HS500, 2000, False],`
`                    {λ, N[λ100 (1 - 10^-20), 30], N[λ100 (1 + 10^-20), 30]},`
`                    WorkingPrecision -> 5000, Compiled -> False,`
`                    AccuracyGoal -> 1000], 100];`

  `N[λ500, 100]`

Out[31]= `1.0603620904841828996470460166926635455152087285289779332162452416959435639831`‥
`8469606419903390484787`4

This result is correct to about 73 digits (and agree with the 62 digits given in [594]; see also [576], [577]). To verify this, we use just one more state.

In[32]:= `HS502 = Table[N[h[n, m], 2000], {n, 0, 502, 2}, {m, 0, 502, 2}];`

In[33]:= `λ502 = λ /. FindRoot[f[λ, HS502, 2000, False],`
`                    {λ, N[λ500 (1 - 10^-50), 60], N[λ500 (1 + 10^-50), 60]},`
`                    WorkingPrecision -> 5000, Compiled -> False,`
`                    AccuracyGoal -> 1000];`

In[34]:= `N[λ500 - λ502]`
Out[34]= 2.71532×10^{-74}

Now, we could continue and use the first 500 symmetric states. The corresponding calculation takes a few minutes on a 2 GHz computer. The resulting ground-state energy has about 120 correct digits:

$$\varepsilon_0 = 1.060362090484182899647046016692663545515208728528977933216\dot{2}\cdot$$
$$4524169594356304434442112689629913467170351054624435858252558\ldots$$

In[35]:= `HS1000 = Table[N[h[n, m], 5000], {n, 0, 1000, 2}, {m, 0, 1000, 2}];`

  `λ1000 = λ /.`
  `FindRoot[f[λ, HS1000, 5000, False],`
`                    {λ, N[λ500 (1 - 10^-50), 60], N[λ500 (1 + 10^-50), 60]},`
`                    WorkingPrecision -> 5000, AccuracyGoal -> 2000, Compiled -> False];`

  `N[λ1000, 120]`
Out[39]= `1.0603620904841828996470460166926635455152087285289779332162452416959435630443`‥
`4442112689629913467170351054624435858252558`

Previously we calculated the kinetic and potential energy matrix elements all exactly. For the simple potential $V(z) = z^4$, the exact calculation of the matrix elements was easy. For a more complicated form of the potential, it is very unlikely that one will be able to carry out the integration exactly. But because of the oscillating nature of the higher eigenfunctions $\phi_n(z)$, a numerical integration will be expensive too. An alternative is the use of a Lagrange basis as the expansion basis [52], [53], [355], [49], [382], [515], [456], [475], [593], [51], [50], [142], [242], [356], [610], [510]:

We define Hermite polynomial-based Lagrange functions $\chi_k^{(n)}(z)$ through $\chi_k^{(n)}(z_l^{(n)}) = \delta_{k,l} \, w_k^{-1/2}$. This property can be most easily realized by $\chi_k^{(n)}(z) \sim H_n(z) / (z - z_k^{(n)})$. Adding the exponential weight function and needed normalizations (using $\partial H_n(z) / \partial z = 2 n \, H_{n-1}(z)$) we get [287], [559]

$$\chi_k^{(n)}(z) = \frac{1}{\sqrt{w_k^{(n)}}} \exp\!\left(\left(z^2 - \left(z_k^{(n)}\right)^2\right)\right) \frac{H_n(z)}{2\,n\,H_{n-1}\!\left(z_k^{(n)}\right)\left(z - z_k^{(n)}\right)}.$$

Here the $z_k^{(n)}$ ($k = 1, \ldots, n$) are the n real distinct zeros of the Hermite polynomial $H_n(z)$ and the $w_k^{(n)}$ are the weights in the corresponding Gauss-like integration approximation $\int_{-\infty}^{\infty} f(x)\,dx \approx \sum_{k=1}^{n} w_k^{(n)} f(x_k^{(n)})$. In the last formula, $f(x)$ is assumed to have the form $f(x) = g(x)\,e^{-x^2}$ where $g(x)$ is a smooth, polynomially at $\pm\infty$ increasing or decreasing function. Explicitly we have for the weights [2], [244], [611], [396]

$$w_k^{(n)} = \frac{\sqrt{\pi}\,2^{n-1}\,n!}{n^2\,H_{n-1}^2\!\left(z_k^{(n)}\right)} \exp\!\left(\left(z_k^{(n)}\right)^2\right).$$

We will use $n = 50$ in the following (to allow for a straightforward comparison with the above result). For larger n, we would get similarly encouraging results. The next inputs calculate the zeros $z_k^{(n)}$ and the weights $w_k^{(n)}$.

```
In[40]:= Clear[z0, w]; n = 50;
         Evaluate[Table[z0[n][k], {k, n}]] =
                 N[z /. {ToRules[Roots[HermiteH[n, z] == 0, z]]},
                   (* use sufficient precision *) 100];
```

```
In[42]:= (* the weights *)
         Do[w[n][k] = Exp[z0[n][k]^2] Sqrt[Pi] 2^(n - 1) n!/
                      (n HermiteH[n - 1, z0[n][k]])^2, {k, n}]
```

The weights are strongly centered around the origin.

```
In[44]:= ListPlot[Table[{z0[n][k], Abs[Log[10, w[n][k]]]}, {k, n}],
           PlotRange -> All, Frame -> True, Axes -> False]
```

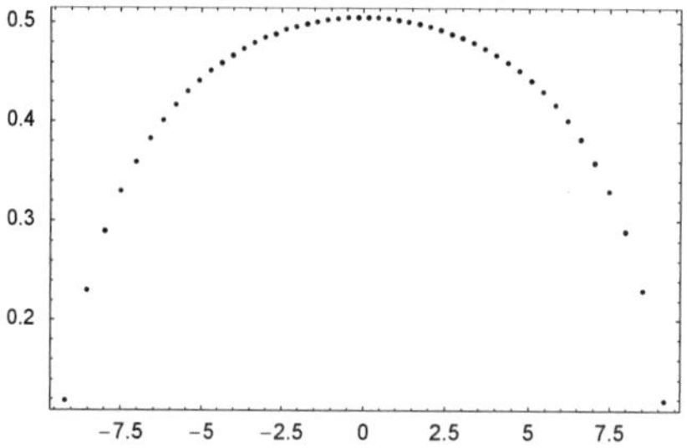

We continue by defining the Lagrange functions $\chi_k^{(n)}(z)$.

```
In[45]:= (* Lagrange interpolating functions *)
         χL[k_, z_] := Exp[-z^2/2] Exp[z0[n][k]^2/2]/Sqrt[w[n][k]] *
                       HermiteH[n, z]/((z - z0[n][k]) (2n HermiteH[n - 1, z0[n][k]]))

         (* last formula is of form 0/0 at z == z0[n][k] *)
         Do[χL[k, z0[n][k]] = 1/Sqrt[w[n][k]], {k, n}]
```

We also implement a typesetting rule for the zeros for nicer outputs.

```
In[50]:= (* for nicer formatted outputs *)
         MakeBoxes[z0[n_][k_], form_] := SubsuperscriptBox["z",
             MakeBoxes[k, form], RowBox[{"(", MakeBoxes[n, form], ")"}]]
```

The following graphic shows $\chi_1^{(n)}(z)$, $\chi_{n/2}^{(n)}(z)$, and $\chi_k^{(n)}(z)$. The $\chi_k^{(n)}(z)$ are localized around $z_k^{(n)}$.

```
In[52]:= Plot[{χL[1, z], χL[n/2, z], χL[n, z]}, {z, -12, 12},
            PlotStyle -> {Hue[0], Hue[0.2], Hue[0.8]}, PlotRange -> All,
            GridLines -> {Table[z0[n][k], {k, n}], None},
            Frame -> True, AspectRatio -> 1/6, Axes -> False]
```

Lagrange function are of degree $n - 1$, meaning because integration formulas are exact up to degree $2n - 1$, they are exactly orthogonal in the L^2 norm. Here is a quick check for this for $\chi_1^{(n)}(z)$ (using the factored form of the $H_n(z)$ we could even carry out the integral exactly, but a numerical quadrature is sufficient here).

```
In[53]:= Module[{d = 1, e = 1, nint},
           {Off[NIntegrate::ncvb]; (* exact numerical integration *)
            nint = NIntegrate[Evaluate[χL[d, z] χL[e, z]], {z, -Infinity, Infinity},
                     WorkingPrecision -> 20]; Off[NIntegrate::ncvb]; nint,
            (* Gauss approximation for integration *)
            Sum[w[n][k] χL[d, z0[n][k]] χL[e, z0[n][k]], {k, n}]} // N]
Out[53]= {1., 1.}
```

For the kinetic energy, we need the second derivative of the $\chi_1^{(n)}(z)$. The next input calculates the second derivative of the z-dependent part of the $\chi_k^{(n)}(z)$.

```
In[54]:= Block[{n}, HoldForm @@ {Together[D[Exp[-z^2/2] HermiteH[n, z]/
                              (z - z0[n][k]), z, z]] /. (* collect wrt HermiteH *)
                            p_Plus :> Collect[p, _HermiteH, Simplify]}] //
                                                             TraditionalForm
```

Out[54]//TraditionalForm=

$$
\frac{1}{\left(z - z_k^{(n)}\right)^3} \left(e^{-\frac{z^2}{2}} \left(4(n-1)n\, H_{n-2}(z) \left(z - z_k^{(n)}\right)^2 - \right.\right.
$$
$$
\left.\left. 4n\, H_{n-1}(z) \left(z^3 + \left(z_k^{(n)}\right)^2 z + z - (2z^2 + 1) z_k^{(n)}\right) + H_n(z) \left(z^4 - 2 z_k^{(n)} z^3 + z^2 + (z^2 - 1)\left(z_k^{(n)}\right)^2 + 2\right)\right)\right)
$$

We assign the last result to a function $\mathcal{D}2$.

```
In[55]:= Block[{n}, Function[rhs, SetDelayed @@ {D2[k_, z_], Exp[z0[n][k]^2/2]/
                              (2 n HermiteH[n - 1, z0[n][k]] Sqrt[w[n][k]]) rhs},
                            HoldAll] @@ %]
```

We also calculate the limiting case of the second derivative as z approaches $z_k^{(n)}$ (the only case needed later).

```
In[56]:= Block[{n, H}, HoldForm @@ {(Series[Cases[D2[k, z], _Plus][[1]] /.
                     HermiteH -> H, {z, z0[n][k], 3}] /. H -> HermiteH /.
                     HermiteH[n, z0[n][k]] -> 0 // Simplify // Normal) /.
                     p_Plus :> Collect[p, _HermiteH, Simplify]}] //
                                                             TraditionalForm
```

Out[56]//TraditionalForm=

$$
\frac{2}{3} n \left(z - z_k^{(n)}\right)^3 \left(4(n^2 - 3n + 2) H_{n-3}\left(z_k^{(n)}\right) - 6(n-1) H_{n-2}\left(z_k^{(n)}\right) z_k^{(n)} + 3 H_{n-1}\left(z_k^{(n)}\right) \left(\left(z_k^{(n)}\right)^2 - 1\right)\right)
$$

Using the identity $2(n-2) H_{n-3}(z) - 2z\, H_{n-2}(z) + H_{n-1}(z)$, the last result could be further simplified, but we will use it here as it is. So, we can make the following definitions for $\mathcal{D}2$ at $z = z_k^{(n)}$.

```
In[57]:= Do[D2[k, z0[n][k]] =
        Block[{n}, %[[1]]/(z - z0[n][k])^3/
            (2 n Sqrt[w[n][k]] HermiteH[n - 1, z0[n][k]])], {k, n}]
```

Now, we have all ingredients together to calculate the matrix elements $V_{i,j} = \int_{-\infty}^{\infty} \chi_i^{(n)}(z) V(z) \chi_j^{(n)}(z)\, dz$ and $T_{i,j} = -\int_{-\infty}^{\infty} \chi_i^{(n)}(z) \chi_j^{(n)}{}''\, dz$. It is crucial that we make use of the above-mentioned integration approximation $\int_{-\infty}^{\infty} f(z)\, dz \approx \sum_{k=1}^{n} w_k^{(n)} f(z_k^{(n)})$ consistently everywhere. Using $\chi_k^{(n)}(z_l^{(n)}) = \delta_{k,l}\, w_k^{-1/2}$, we arrive at the following simple approximations for $V_{i,j}$ and $T_{i,j}$.

$$V_{i,j} = \int_{-\infty}^{\infty} \chi_i^{(n)}(z) V(z) \chi_j^{(n)}(z)\, dz \approx \sum_{k=1}^{n} w_k^{(n)} \chi_i^{(n)}(z_k^{(n)}) V(z_k) \chi_j^{(n)}(z_k^{(n)})$$

$$= \sum_{k=1}^{n} w_k^{(n)} \frac{\delta_{i,k}}{\sqrt{w_k^{(n)}}} V(z_k^{(n)}) \frac{\delta_{j,k}}{\sqrt{w_k^{(n)}}} = \delta_{i,j} V(z_i^{(n)})$$

$$T_{i,j} = -\int_{-\infty}^{\infty} \chi_i^{(n)}(z) \frac{\partial^2 \chi_j^{(n)}(z)}{\partial z^2}\, dz \approx -\sum_{k=1}^{n} w_k^{(n)} \chi_i^{(n)}(z_k^{(n)}) \chi_j^{(n)}{}''(z_k^{(n)})$$

$$= -\sum_{k=1}^{n} w_k^{(n)} \frac{\delta_{i,k}}{\sqrt{w_k^{(n)}}} \chi_j^{(n)}{}''(z_k^{(n)}) = -\sqrt{w_i^{(n)}}\, \chi_i^{(n)}{}''(z_j^{(n)}).$$

The last equations mean that the potential energy matrix elements $V_{i,j}$ become a trial to calculate pointwise potential evaluation. Compared with a numerical calculation of these matrix elements, this is an enormous simplification. The kinetic energy matrix elements $T_{i,j}$ are basically the values of the second derivatives of the Lagrange functions at the zeros. Interestingly, the so-defined $T_{i,j}$ is symmetric (meaning $\chi_i^{(n)}{}''(z_j^{(n)}) = \chi_j^{(n)}{}''(z_i^{(n)})$), which is crucial for real eigenvalues of the resulting matrix eigenvalue problem. The following two short lines implement the last formulas.

```
In[58]:= (* the potential energy matrix elements are
             functionals of the potential V *)
        VχL[V_][{i_, j_}] := If[i == j, V[z0[n][i]], 0]
        (* the kinetic energy matrix elements are
             independent of the potential V *)
        TχL[{i_, j_}] := -Sqrt[w[n][i]] D2[j, z0[n][i]]
```

As a start, we will calculate the spectrum of the harmonic oscillator itself. The following calculation shows that all eigenvalues come out exactly. While the Lagrange functions were built from the exact solution of the harmonic oscillator eigenfunctions, this seems not surprising in the first moment. But, on the other hand, the potential energy matrix elements were calculated in the Gauss approximation and the integrands were polynomials of degree $2(n-1)+2$, meaning the integration rule was not exact. So, the exact result for all eigenvalues is somewhat surprising [550].

```
In[62]:= Block[{V, HLHOn, evsHOn}, (* harmonic oscillator *) V[x_] := x^2;
           HLHOn = Table[TχL[{i, j}] + VχL[V][{i, j}], {i, n}, {j, n}];
           evsHOn = Eigenvalues[HLHOn];
           evsHOn - Table[k, {k, 2n - 1, 1, -2}] // N // Abs // Max]

Out[62]= 0.
```

Now let us look at the quartic oscillator. This time the integrand of the potential energy is a polynomial of degree $2n+2$ and the integration Gauss approximation gives results that are sometimes very different from the exact integration result.

```
In[63]:= AHOVij[i_, j_] := Module[{nint}, {Off[NIntegrate::ncvb];
           (* potential energy matrix elements; use numerical integration *)
           nint = NIntegrate[Evaluate[χL[i, z] z^4 χL[j, z]],
                             {z, -Infinity, Infinity}, WorkingPrecision -> 20];
                             On[NIntegrate::ncvb]; nint,
              KroneckerDelta[i, j] z0[n][i]^4} // N]
```

```
In[64]:= {AHOVij[1, 1], AHOVij[n/2, n/2], AHOVij[n, n], AHOVij[1, n], AHOVij[1, n/2]}
Out[64]= {{25.2872, 0.000596849}, {309.853, 260.397},
          {7261.01, 7109.29}, {-66.7029, 0.}, {33.6446, 0.}}
```

The matrix elements for the kinetic energy are not exact either.

```
In[65]:= GenTij[i_, j_] := Module[{nint}, {Off[NIntegrate::ncvb];
           (* kinetic energy matrix elements; use numerical integration *)
           nint = NIntegrate[Evaluate[-χL[i, z] D[χL[j, z], z, z]],
                             {z, -Infinity, Infinity}, WorkingPrecision -> 20];
                             On[NIntegrate::ncvb]; nint,
                -Sqrt[w[n][i]] D2[j, z0[n][i]]} // N]
```

```
In[66]:= {GenTij[1, 1], GenTij[n/2, n/2], GenTij[n, n],
          GenTij[1, n], GenTij[1, n/2]}
Out[66]= {{33.1585, 33.6585}, {27.7877, 28.2877},
          {5.06113, 5.56113}, {0.477067, -0.0229328}, {-0.365821, 0.134179}}
```

But the consistent use of the Gauss approximation for all integrals cancels all errors magically [52] and we obtain the ground-state energy nearly as well as in the above exact approximation (the error in the above exact approximation was 4.4×10^{-13} and the error now is 6.3×10^{-13}).

```
In[67]:= Block[{V, HLAHOn, evsAHOn}, (* quartic oscillator *) V[x_] := x^4;
           (* discretized Hamiltonian *)
           HLAHOn = Table[TχL[{i, j}] + VχL[V][{i, j}], {i, n}, {j, n}];
           evsAHOn = Eigenvalues[HLAHOn];
           evsAHOn[[-1]] // N[#, 20]&]
Out[67]= 1.0603620904835530811
```

This excellent result suggests using the Lagrange function-based approach for other potentials too.

For the construction of Lagrange functions for the continuous spectrum, see [357], [358].

Instead of diagonalizing one large matrix, we could now continue many small matrices using the density matrix renormalization group approach [386], [343], [257], [123], [511] (see also [605]); but we will end here. (Unfortunately, the quartic oscillator is not suited for a Sturmian basis function approach [15], [16].)

Until now, we have dealt with a quantitative feature of the quartic oscillator—its ground-state energy. One could now continue to investigate more qualitative features of quartic oscillators. Without going into any details (see [60], [61], [521], [516]) let us quickly consider the following problem [60], [61], [521], [516]:

$$-\psi_n''(z; \alpha, \beta) + (\alpha z^2 + \beta z^4)\psi_n(z; \alpha, \beta) = \varepsilon_n(\alpha, \beta)\psi_n(z; \alpha, \beta)$$

Considering α and β as parameters we could investigate the dependence of $\varepsilon_n(\alpha, \beta)$ on these parameters. For a short visit into the fascinating arena of the analytic structure of $\varepsilon_n(\alpha, \beta)$ as a function of complex α and β and to encourage the reader to make further investigations, we will generate some graphics showing some aspects of the analytic structure. For speed, we restrict ourselves to machine-number calculations. So, we implement a compiled function $\mathcal{H}[p]$ that generates a compiled function that takes α and β as arguments and constructs a $(p/2) \times (p/2)$ Hill matrix. Because of the evenness of $(\alpha z^2 + \beta z^4)$, we again use only the even $\phi_m(z)$ in the expansion.

```
In[68]:= h[α_, β_][n_, m_] = makeh[α z^2 + β z^4, z];
```

```
In[69]:= H[p_] := H[p] = Compile[{{α, _Complex}, {β, _Complex}},
                  Evaluate[N[Table[If[Abs[n - m] > 4, 0.,
                      Expand[h[α, β][n, m]]],
                      {n, 0, p, 2}, {m, 0, p, 2}]]]]
```

The two functions `eigenvalues` and `lowestEigenvalue` return the k lowest eigenvalues and the lowest eigenvalue respectively.

```
In[70]:= eigenvalues[{α_, β_}, p_, k_] :=
              Reverse @ Take[Eigenvalues[H[p][α, β]], -k]
```

```
      lowestEigenvalue[{α_, β_}, p_] := Eigenvalues[H[p][α, β]][[-1]]
```

Let us consider the two limiting cases $\alpha = 1$ and β varying and $\beta = 1$ and α varying. The following two graphics show the five lowest eigenvalues as a function of β and α. In the case $\beta = 1$, the additional term αz^2 constitutes a small perturbation. For all α, the spectrum is discrete. In the case $\alpha = 1$, the additional term βz^4 does not always constitutes a small perturbation. For positive α, the spectrum stays discrete. But the smallest negative β gives a potential decreasing without bounds and the spectrum can no longer be discrete, but is continuous. This change in the nature of the spectrum is reflected in the right graphic in a clearly visible set of branch cuts for negative β. The $\varepsilon_n(1, \beta)$ are clearly the analytic continuations from each other. The graphic also correctly suggests that β is an accumulation point of branch points.

```
In[72]:= With[{o = 5, pp = 10^4},
        Show[GraphicsArray[
        Block[{$DisplayFunction = Identity},
          {(* α variable and β = 1 *)
          Show[Graphics[{PointSize[0.005],
              MapIndexed[{Hue[0.78 (#2[[1]] - 1)/o], Point /@ #1}&, Transpose[
                      Table[{α, #}& /@ eigenvalues[{α, 1}, 40, o],
                          {α, -1/2, 1/2, 1/pp}]]]}], Frame -> True,
                      AxesLabel -> {"α", None}],
          (* α = 1 and β variable *)
          Show[Graphics[{PointSize[0.005],
              MapIndexed[{Hue[0.78 (#2[[1]] - 1)/o], Point /@ #1}&, Transpose[
                      Table[{β, #}& /@ eigenvalues[{1, β}, 40, o],
                          {β, -1/2, 1/2, 1/pp}]]]}], Frame -> True,
                      AxesLabel -> {"β", None}]}]]]]
```

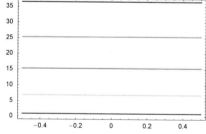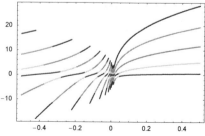

The last graphics show branch cuts along the negative real axis and so suggests to visualize $\varepsilon_n(1, \beta)$ over the complex β-plane. We will do this in the following. The left graphic shows $\mathrm{Im}(\varepsilon_n(1, \beta))$ as a 3D plot and the right graphic shows it as a contour plot. We should remark that for general complex β the boundary conditions $\psi_n(z; 1, \beta) \to 0$ as $z \to \infty$ (implied by using the harmonic oscillator eigenfunctions as the expansion basis) should be replaced by more general ones for complex z. But the qualitative feature of an (infinite) set arc-like branch cut around $\beta = 0$ is correctly represented.

```
In[73]:= (* convert a contour plot into one with homogeneous contour spacings *)
        homogeneousContourPlot[cp_ContourGraphics, n_, opts___] :=
        With[{pp2 = Length[Flatten[cp[[1]]]]},
        ListContourPlot[cp[[1]], Contours -> (#[[Round[pp2/n/2]]]& /@
                      Partition[Sort[Flatten[cp[[1]]]], Round[pp2/n]]),
                      PlotRange -> All, ColorFunction -> (Hue[0.8 #]&),
                      MeshRange -> (Options[cp, MeshRange][[1, -1]])]];

In[75]:= Show[GraphicsArray[
        Block[{$DisplayFunction = Identity},
        {(* 3D plot *)
        ParametricPlot3D[{r Cos[φ], r Sin[φ],
                        Im[lowestEigenvalue[{1, r Exp[I φ]}, 40]],
                        EdgeForm[]}, {r, 0, 1/2}, {φ, 0, 2Pi},
                        PlotPoints -> {160, 240}, Compiled -> False,
                        PlotRange -> All, BoxRatios -> {1, 1, 1/2}],
        (* contour plot *)
        homogeneousContourPlot[#, 25, AspectRatio -> 2/3]& @
        ContourPlot[Im[lowestEigenvalue[{1, r Exp[I φ]}, 60]],
                    {r, 0, 1/2}, {φ, Pi - 0.12, Pi + 0.12},
                    PlotPoints -> 400, Compiled -> False]}]]]
```

If, on the other hand, we first form the z^4-basis, then αz^2 will always be a small perturbation [516]. Unfortunately, the z^4-basis is not available in closed form. So, we form it by extracting the low-lying subspace from a large Hill matrix diagonalization. We take 400 harmonic oscillator states and calculate the first 40 quartic oscillator states (higher quartic oscillator states would show truncation errors).

```
In[76]:= quarticES =
        Module[{o = 800, H},
               H = N[Table[h[n, m], {n, 0, o, 2}, {m, 0, o, 2}]];
               Eigensystem[H]];

In[77]:= o = 40;
        {quarticEVals, quarticEVecs} = {Take[Reverse[quarticES[[1]]], o],
                        Take[#, o]& /@ Take[Reverse[quarticES[[2]]], o]};
```

Expressing the quartic oscillator eigenstates $\psi_n(z)$ through the harmonic oscillator states $\phi_k(z)$ (where we truncate the infinite sum for the numerical calculations) $\psi_n(z) = \sum_{k=0}^{\infty} c_k^{(n)} \phi_k(z)$, reduces the calculation of the matrix elements $\langle \psi_m(z) \,|\, z^2 \,|\, \psi_n(z) \rangle$ to the calculation of $\langle \phi_m(z) \,|\, z^2 \,|\, \phi_n(z) \rangle$:

$$\langle \psi_m(z) \,|\, z^2 \,|\, \psi_n(z) \rangle = \sum_{k=0}^{\infty} c_k^{(m)} \, c_k^{(n)} \, \langle \phi_m(z) \,|\, z^2 \,|\, \phi_n(z) \rangle.$$

```
In[79]:= makeMatrixElement[V_, z_] :=
         Sqrt[2^m m!]/Sqrt[2^n n!] Simplify[reduceProducts[V HermiteH[n, z]] /.
         (* use orthogonality *) HermiteH[n_, z] -> KroneckerDelta[n, m]]
```

```
In[80]:= m[m_, n_] = makeh[z^2, z] /. KroneckerDelta[a_, b_] :> If[a == b, 1, 0];
```

```
In[81]:= matElem[i_, j_] :=
         Module[{u = quarticEVecs[[i]], v = quarticEVecs[[j]]},
                 Sum[m[2(m - 1.), 2(n - 1.)] u[[m]] v[[n]], {m, o},
                     {n, Max[m - 2, 1], Min[o, m + 2]}]]]
```

Here is the resulting matrix $V_{ij} = \langle \psi_i(z) \,|\, z^2 \,|\, \psi_j(z) \rangle$ visualized.

```
In[82]:= VMat = Table[matElem[i, j], {i, o}, {j, o}];
```

```
In[83]:= Show[GraphicsArray[
         ListDensityPlot[#[VMat], Mesh -> False, PlotRange -> All,
                         ColorFunction -> (Hue[0.8 #]&),
                         DisplayFunction -> Identity]& /@ {Abs, Sign}]]
```

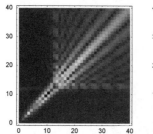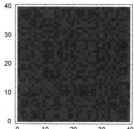

The Hill matrix for the potential αz^2 in the quartic oscillator basis can now be easily constructed. The pure quartic oscillator is now diagonal.

```
In[84]:= makeMat = Compile[{{α, _Complex}}, #1 + α #2]&[
                        (* exact quartic oscillator *) DiagonalMatrix[quarticEVals],
                        (* harmonic oscillator perturbation *) VMat];
```

We calculate the eigenvalues for a dense grid over the complex α-plane.

```
In[85]:= L = 12; pp = 300;
         data = Table[Eigenvalues[makeMat[x + I y]],
                      {y, -L, L, 2L/pp}, {x, -L, L, 2L/pp}];
```

The following graphics show the real part and the absolute value of the lowest eigenvalues [501]. We still see branch cuts, but they no longer accumulate around the origin.

```
In[87]:= Function[ξ, Show[GraphicsArray[Table[
         ListContourPlot[Map[ξ[#[[-j]]]&, data, {-2}], PlotRange -> All,
             ColorFunction -> (Hue[0.8 #]&), DisplayFunction -> Identity,
```

```
          Contours -> 20, ContourStyle -> {Thickness[0.002]},
          MeshRange -> {{-L, L}, {-L, L}}], {j, 3}]]]] /@ {Re, Abs}
```

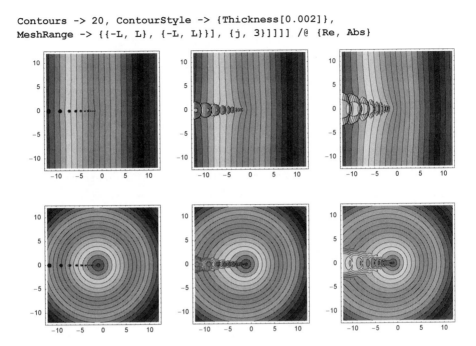

A detailed view at the eigenvalues along the negative real α-axis show how the various eigenvalues are all analytic continuations.

```
In[88]:= L = 12; pp = 5000;
         dataReal = Table[{x, #}& /@ Eigenvalues[makeMat[x]], {x, -L, 0, 2L/pp}];

In[90]:= Show[Graphics[Table[{Hue[(k - 1)/12], Point[{#1, Re[#2]}]& @@@
                                          Transpose[dataReal][[-k]]},
                       {k, 10}]], Frame -> True, PlotRange -> {-50, 50}]
```

Now that we have the Hill matrix for the quartic oscillator handy, let us use it one more time. Instead of solving the time-independent Schrödinger equation, we will use it to solve the time-dependent Schrödinger equation $i\,\partial\psi(x,t)/\partial t = \hat{\mathcal{H}}\,\psi(x,t)$.

Here $\hat{\mathcal{H}}$ is the differential operator $-\frac{\partial^2}{\partial x^2} + x^2$, or, in the basis of the harmonic oscillator eigenfunctions, it is just the matrix with elements h_{nm}. In the following, we will use $\psi(x, 0) = \phi_4(x)$ and visualize $\psi(x, t)$. This means we can restrict $\mathcal{H}$ to the symmetric subspace. Taking the first 100 eigenfunctions $\phi_n(x)$ will prove sufficient for a visualization.

```
In[91]:= o = 200;
        ℋ = N @ Table[ℏ[n, m], {n, 0, o, 2}, {m, 0, o, 2}];
```

For the conversion from the harmonic oscillator basis to explicit $\psi(x, t)$ we need explicit values for the $\phi_n(x)$.

```
In[93]:= L = 3.5; pp = 300;
        Do[ϕN[n] = Developer`ToPackedArray @ Table[ϕ[n, N[z]],
                      {z, -L, L, 2L/pp}], {n, 0, o, 2}];
```

Formally the time-dependent Schrödinger equation has the following solution (which can be viewed as a generalized Fourier transform [391], [194]) $\psi(x, t) = e^{-it\mathcal{H}}\psi(x, 0)$. For $\mathcal{H}$ in the harmonic oscillator basis and $\psi(x, t)$ in the truncated harmonic oscillator basis $\langle n | \psi(t)\rangle = \psi_n(t)$, this reads $\psi_n(t) = e^{-it(\hbar)_{n,m}}.\psi_n(0) = \mathcal{U}.\psi_n(0)$, where $\exp(-it(\hbar)_{n,m})$ is the matrix exponential of the matrix $(\hbar)_{n,m}$. The conversion from the $\psi_n(t)$ to $\psi(x, t)$ is given by $\psi(x, t) = \sum_{k=1}^{\infty} \psi_k(t)\,\phi_k(x)$. Here $\mathcal{U}$ is a time-independent matrix; so, we have to calculate it only once.

```
In[95]:= 𝒰 = With[{δt = 0.01}, MatrixExp[-I δt ℋ]];
```

The function cf carries out the calculation of the $\psi(x, 0)$, $\psi(x, \delta t)$, $\psi(x, 2\delta t)$, ...

```
In[96]:= ϕNAll = Table[ϕN[n], {n, 0, o, 2}];

        cf = Compile[{{initialcs, _Complex, 1}, {𝒰, _Complex, 2},
                      {ϕNs, _Real, 2}, {n, _Integer}},
                  Module[{cs = initialcs, λ = Length[initialcs]},
                       Table[(* time step *) cs = 𝒰.cs;
                          (* n -> x *)
                          Sum[cs[[k]] ϕNs[[k]], {k, λ}], {n}]]];
```

Now the calculation of 1000 values of $\psi(x, k\,\delta t)$ can be done quite quickly.

```
In[98]:= data = cf[Table[If[k == 4, 1, 0], {k, 0, o, 2}], 𝒰, ϕNAll, 1000];
```

The following graphics show $|\psi(x, t)|^2$ and $\arg(\psi(x, t))^2$. Time is running from left to right. The graphic shows a complicated, nonperiodic time-development.

```
In[99]:=   ListDensityPlot[# @ Transpose[data],
                      ColorFunction -> (Hue[0.8 #]&), Mesh -> False,
                      AspectRatio -> 1/4, FrameTicks -> None]& /@
                                     {Abs[#]^2&, Arg[#]^2&}
```

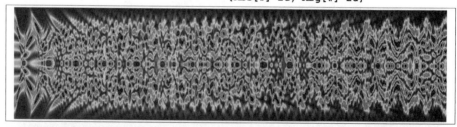

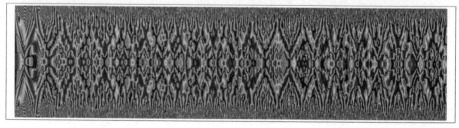

For other methods to effectively calculate the time-dependent $\psi(x, t)$ based on matrix powers, see [590], [591].

We could go on and investigate the trendy $\mathcal{PT}$-invariant oscillators [71], [173], [66], [67], [636], [295], [635], [162], [263], [264], [265], [266], [7], [34], [76], [172], [518], [156], [158], [69], [175]. Let us add the purely imaginary quantity $i\,\alpha\,z$ to the quartic term z^4 as the potential. Here are the matrix elements of the resulting Hill-matrix. (For the physical credibility of such Hamiltonians, see [70].)

```
In[100]:= (hPT[n_, m_] = makeh[x^4 + I α x, x]) /.
                              KroneckerDelta[a_, b_] :> Subscript[δ, a, b]
```

$$Out[100]= \frac{1}{16\sqrt{2^n\,n!}}$$
$$(\sqrt{2^m\,m!}\ (16\,n\,(-6 + 11\,n - 6\,n^2 + n^3)\,\delta_{m,-4+n} + 32\,(-1+n)^2\,n\,\delta_{m,-2+n} + 16\,i\,n\,\alpha\,\delta_{m,-1+n} +$$
$$20\,\delta_{m,n} + 40\,n\,\delta_{m,n} + 24\,n^2\,\delta_{m,n} + 8\,i\,\alpha\,\delta_{m,1+n} + 8\,\delta_{m,2+n} + 8\,n\,\delta_{m,2+n} + \delta_{m,4+n}))$$

Interestingly, for small values of α, the spectrum stays purely real. Here we show the real and imaginary parts of the first few eigenvalues. Within the numerical accuracy of the eigenvalue finder, the red points in the following graphic indicate purely real eigenvalues.

```
In[101]:= mPT[α_] = With[{o = 100}, Table[hPT[n, m], {m, 0, o}, {n, 0, o}]];
```

```
In[102]:= Show[Graphics[{PointSize[0.005], (* use symmetry *)
                    {#, # /. Point[{x_, y_}] :> Point[{-x, y}]}&[
          (* eigenvalues for positive α *)
          Table[evs = Select[Eigenvalues[mPT[N[α]]], Abs[#] < 100&];
                If[Abs[Im[#]] < 10^-6, {Hue[0.0], Point[{α, Re[#]}]},
                  {{Hue[0.22], Point[{α, Re[#]}]},
                    {Hue[0.78], Point[{α, Im[#]}]}}]& /@ evs,
                {α, 0, 60, 1/8}]]}], Frame -> True,
          PlotRange -> {{-30, 30}, {-20, 50}}]
```

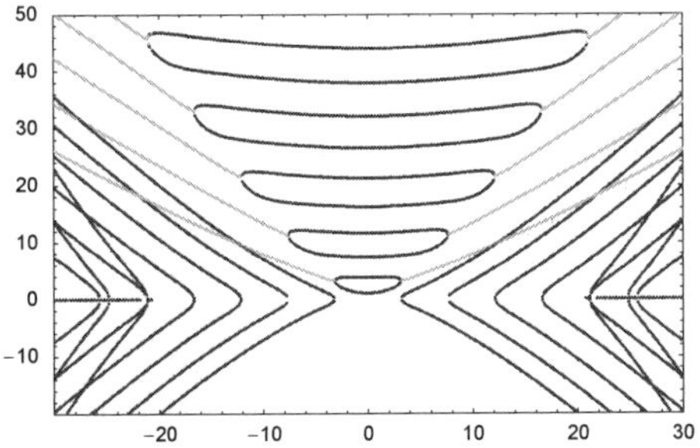

The next two graphic show the time-evolution of an initial Gaussian wave packet in a purely quartic potential and in a quartic potential with an additional $2\,i\,x$ term.

```
In[103]:= Module[{L = 3, T = 3, pp = 240, ppg = 320},
          (* solve time-dependent Schrödinger equation *)
          ndsols =
          NDSolve[{I D[ψ[x, t], t] == -D[ψ[x, t], x, x] + (x^4 + # x)ψ[x, t],
                  ψ[x, 0] == Exp[-8 x^2], ψ[-L, t] == ψ[L, t]},
                  {ψ}, {x, -L, L}, {t, 0, T},
                  PrecisionGoal -> 4, AccuracyGoal -> 4,
                  Method -> {"MethodOfLines", Method -> StiffnessSwitching,
                              "SpatialDiscretization" -> {"TensorProductGrid",
                              "DifferenceOrder" -> "Pseudospectral",
                              "MaxPoints" -> pp, "MinPoints" -> pp}}]& /@
              (* factor of term x. *) {0, 2 I};
```

```
(* show square of absolute value *)
Show[GraphicsArray[
        Plot3D[Evaluate[Abs[ψ[x, t]]^2 /. #[[1]]],
                {x, -L, L}, {t, 0, T}, PlotPoints -> ppg,
                Mesh -> False, PlotRange -> {0, 2},
                DisplayFunction -> Identity]& /@ ndsols]]]
```

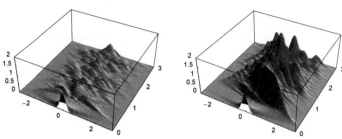

We leave it to the reader to continue and to investigate how the eigenfunctions behave as a function of α and for more general additional terms $(i\,z)^m$ [519], [65]. For general quartic oscillators, see [127]. For oscillators with potential $|z|^\alpha$, see [81].

Exercises

1.[11] Generating Function for $T_n(x)$, Mehler's Formula, Bauer–Rayleigh Expansion, and More

a) Verify the first few terms in the following series expansion:

$$e^{zx} \cosh\left(z \sqrt{x^2 - 1}\right) = \sum_{n=0}^{\infty} \frac{T_n(x)}{n!} z^n.$$

Using the trigonometric representations of the Chebychev polynomials, also prove this relation symbolically.

b) Check Mehler's formula for the lowest-order terms:

$$\frac{1}{\sqrt{1 - a^2}} \exp\left(-\frac{(z^2 + z'^2 - 2 z z' a)}{1 - a^2}\right) = \exp(-(z^2 + z'^2)) \sum_{n=0}^{\infty} \frac{a^n}{2^n n!} H_n(z) H_n(z').$$

c) Check the first few terms (in x) for the Bauer–Rayleigh expansion:

$$e^{ixy} = \sqrt{\frac{\pi}{2x}} \sum_{l=0}^{\infty} i^l (2l + 1) J_{l+1/2}(x) P_l(y).$$

Here, $J_\nu(z)$ is the Bessel function of order ν. (For a short derivation, see [561] and [215]; for generalizations, see [497], for pitfalls, see [380].)

d) Verify the first few terms of the following addition theorem for the Hermite polynomials:

$$\sum_{m=0}^{n} \frac{n!}{m! (n - m)!} H_{n-m}\left(\sqrt{2}\, z\right) H_n\left(\sqrt{2}\, z'\right) = 2^{n/2} H_n(z + z').$$

e) Verify the correctness of the following relations for the Chebyshev polynomials of the first kind for the first few n:

$$2\, T_m(x)\, T_n(x) = T_{m+n}(x) + T_{m-n}(x) \quad (m > n)$$
$$2\, T_n(x)^2 = 1 + T_{2n}(x).$$

Using the trigonometric representations of the Chebychev polynomials, prove this relation symbolically as well.

f) Verify the following formula for the products of Legendre polynomials [325], [459], [395] for some n and m:

$$P_m(x)\, P_n(x) = \sum_{l=|m-n|}^{m+n} b_{lmn}\, P_l(x).$$

The coefficients b_{lmn} are given by the following expression:

$$b_{lmn} = \begin{cases} (2l + 1)\, \dfrac{(m + n - l - 1)!!\, (l + n - m - 1)!!\, (m + n + l)!!\, (l + m - n - 1)!!}{(m + n - l)!!\, (l + n - m)!!\, (m + n + l + 1)!!\, (l + m - n)!!} & l + m + n \text{ even} \\ 0 & \text{otherwise.} \end{cases}$$

g) Check the following formula for Laguerre polynomials [472] for some n:

$$L_n(z) = (-1)^n \frac{e^z}{n!} \left(z \frac{d^2}{dz^2} + \frac{d}{dz} \right)^n e^z.$$

h) The zeros $x_k^{(n)}$ of the Hermite polynomials $H_n(x)$ fulfill the following nice relation [6], [118], [119]:

$$\sum_{k=1}^{n} \frac{1}{x_l^{(n)} - x_k^{(n)}} = x_l^{(n)}.$$

Check this relation for some n.

i) Define the following continued fraction $R_k(\epsilon, x)$ with k occurrences of x.

$$R_k(\epsilon, x) = \cfrac{1}{1 - \cfrac{x}{1 - \cfrac{\ddots}{1 - \cfrac{x}{1 - \epsilon x}}}}$$

Prove that $R_k(\epsilon, x)$ has the following closed-form representation [383]:

$$R_k(\epsilon, x) = \frac{1}{\sqrt{x}} \frac{U_{k-1}\left(\frac{1}{2\sqrt{x}}\right) - \epsilon \sqrt{x} \, U_{k-2}\left(\frac{1}{2\sqrt{x}}\right)}{U_k\left(\frac{1}{2\sqrt{x}}\right) - \epsilon \sqrt{x} \, U_{k-1}\left(\frac{1}{2\sqrt{x}}\right)}.$$

j) The 3D spherical harmonics $Y_l^m(\vartheta, \varphi) = ((2l+1)/(4\pi)(l-m)!/(l+m)!)^{1/2} P_l^m(\cos(\vartheta)) e^{im\varphi}$ (in *Mathematica* SphericalHarmonicY[l, m, ϑ, φ]) can, for nonnegative integers l and m, be written in the form $Y_l^m(\vartheta, \varphi) = c_{l,m}(x^2 + y^2 + z^2)^{-l} p(x, y, z)$ where $c_{l,m}$ is a numerical constant and $p(x, y, z)$ is a homogeneous polynomial of degree l over the Gaussian integers [586], [613], [300], [420]. Write a one-liner that writes a given Spherical HarmonicY[l, m, ϑ, φ] in this form.

k) Implement a one-liner that implements the calculation of the first n spherical harmonics

$$\mathbf{Y}_l(\vartheta, \varphi) = \{\{Y_{0,0}(\vartheta, \varphi)\}, \{Y_{1,-1}(\vartheta, \varphi), Y_{1,0}(\vartheta, \varphi), Y_{1,1}(\vartheta, \varphi)\}, \ldots, \{Y_{l,-l}(\vartheta, \varphi), \ldots, Y_{l,l}(\vartheta, \varphi)\}\}$$

that is based on the recursion [180], [536], [586]

$$\mathcal{Y}_{l,m}(\vartheta, \varphi) = \frac{1}{l-m} (\cos(\vartheta)\,\mathcal{Y}_{l-1,m} + (\cos(\varphi)\sin(\vartheta) - i\sin(\varphi)\sin(\vartheta))\,\mathcal{Y}_{l-1,m+1}) \text{ for } m \leq 0$$
$$\mathcal{Y}_{l,m}(\vartheta, \varphi) = (-1)^m \, \overline{\mathcal{Y}_{l,-m}(\vartheta, \varphi)}$$
$$\mathcal{Y}_{1,-1}(\vartheta, \varphi) = (\cos(\varphi)\sin(\vartheta) - i\sin(\varphi)\sin(\vartheta))/2$$
$$\mathcal{Y}_{1,-1}(\vartheta, \varphi) = \cos(\vartheta)$$
$$\mathcal{Y}_{1,1}(\vartheta, \varphi) = -(\cos(\varphi)\sin(\vartheta) + i\sin(\varphi)\sin(\vartheta))/2.$$

The normalized spherical harmonics are then given by

$$Y_{l,m}(\vartheta, \varphi) = \sqrt{\frac{(2l+1)(l+m)!(l-m)!}{4\pi}} \, \mathcal{Y}_{l,m}(\vartheta, \varphi).$$

For machine-precision ϑ and φ, implement a compiled version. To which precision can you calculate $Y_l(\vartheta, \varphi)$ using the compiled version?

2.¹² Generalized Fourier Series

Examine the convergence of the generalized Fourier series in terms of the classical orthogonal polynomials for the function $f(z) = \theta(z)(1 - \theta(z))$.

3.ᴸ¹ Transmission Through Layers, Sums of Zeros

a) Graphically examine the function

$$t(k) = \frac{1}{1 + k^2\, U_n(\cos(k) + \sin(k)/k)^2} \qquad n = 0, 1, 2\ldots, k \geq 0$$

where $U_n(z)$ are the Chebyshev polynomials.

b) Find a closed form for $\sigma_o^{(n)} = \sum_{k=1}^n \left(z_k^{(n)}\right)^o$ for positive integer n and o with $n > o/2$ where $z_k^{(n)}$ is the kth root of $T_n(z)$.

4.ᴸ¹ General Orthogonal Polynomials

Implement the following algorithm for the computation of the orthogonal polynomials corresponding to a weight function $w(z)$ on the interval (a, b). Let

$$c_{ij} = \int_a^b w(z)\, z^{i+j}\, dz$$

$$\Delta_n = \det \mathbf{C} \quad i, j = 0, 1, \ldots, n$$

$$d_{ij}^{(n)} = \begin{cases} c_{ij}, & j < n \\ z^j, & j = n \end{cases}$$

$$\Gamma_n = \det \mathbf{D}^n \quad i, j = 0, 1, \ldots, n.$$

Here, c_{ij} and $d_{ij}^{(n)}$ are the elements of the matrices $\mathbf{C}$ and $\mathbf{D}$.

Then, the nonnormalized form of the nth orthogonal polynomial $p_n(z)$ corresponding to the weight $w(z)$ is given by

$$p_n(z) = \frac{\Gamma_n}{\sqrt{\Delta_n \Delta_{n-1}}}$$

(See, e.g., [307], [9], [552], [155], [336], and [216]). Implement a normalization of the polynomials so that

$$\int_a^b p_n(z)\, p_m(z)\, w(z)dz = \delta_{nm}.$$

(One could also use the last equation for a direct recursive calculation of the nth orthogonal polynomial. Setting

$$p_n(z) = \sum_{i=0}^n a_i z^i$$

the orthogonality relations lead to a system of equations that can be solved for the a_i using `Solve`. However, since this system of equations is nonlinear (note the last equation), such an implementation would require a lot of unnecessary work. Another possibility would be to use Gram-Schmidt orthogonalization. A corresponding package is `LinearAlgebra`Orthogonalization`.)

5.^{11} Symmetric Polynomials

Symmetric polynomials, i.e., polynomials $f(x_1, x_2, ..., x_n)$ in several variables $x_1, x_2, ..., x_n$ for which $f(x_1, x_2, ..., x_n) = f(\pi(x_1, x_2, ..., x_n))$, where $\pi(x_1, x_2, ..., x_n)$ is an arbitrary permutation of the elements $x_1, x_2, ..., x_n$ holds, play a major role in algebra (see, e.g., [458], [328], [582], [562], [376], [140], [128], [539], [100]; for physics applications, see [39], [509]). The most important symmetric polynomials are the following:

▪ the power sums S_k

$$S_k(x_1, ..., x_n) = \sum_{i=1}^{n} x_i^k, \ k = 0, 1, ...$$

▪ the discriminant D

$$D(x_1, ..., x_n) = \prod_{\substack{i,j=1 \\ i>j}}^{n} (x_i - x_j)^2$$

▪ the elementary symmetric polynomials

$$C_k(x_1, ..., x_n) = \sum_{\substack{i_1,i_2,...,i_k=1 \\ i_1<i_2<\cdots<i_k \leq n}} x_{i_1} x_{i_2} \cdots x_{i_k}, \quad k = 1, 2, ...$$

▪ the Wronski polynomials

$$P_k(x_1, ..., x_n) = \sum_{\substack{i_1,i_2,...,i_n=0 \\ i_1+i_2+\cdots+i_n=k}}^{k} x_1^{i_1} x_2^{i_2} \cdots x_n^{i_n}, \ k = 1, 2, ...$$

a) For a given list of variables *varList*, length n, and given order k, implement the computation of these polynomials. Try not to use temporary variables (including iterator variables in `Table`, etc.). Do not use

```
ElementarySymmetricPolynomials[k_Integer?(#>=1&), varList_] :=
        Coefficient[Times @@ (C - #& /@ varList), C, k+1]
```
or similar implementations.

b) The following so-called Newton relations ([109], [235]) hold for the power sums and the elementary symmetric polynomials:

$$S_k(x_1, ..., x_n) + \sum_{j=1}^{k-1} (-1)^j C_j(x_1, ..., x_n) S_{k-j}(x_1, ..., x_n) + (-1)^k k C_k(x_1, ..., x_n) = 0, \ k \leq n$$

$$S_k(x_1, ..., x_n) + \sum_{j=1}^{n} (-1)^j C_j(x_1, ..., x_n) S_{k-j}(x_1, ..., x_n) = 0, \ k > n.$$

Verify these relationships for $n = 3$ and $k < 5$.

c) Show for some cases that the so-called Waring formula holds:

$$S_k(x_1, \ldots, x_n) = \sum_{\substack{i_1, i_2, \ldots, i_n = 0 \\ i_1 + 2i_2 + 3i_3 + \cdots + n\,i_n = k}} \frac{(-1)^{i_1 + i_2 + \cdots + i_n} (i_1 + i_2 + \cdots + i_n - 1)!}{i_1! \, i_2! \cdots i_n!} C_1^{i_1} C_2^{i_2} \cdots C_n^{i_n}.$$

d) Let $S_k^{(j)}(x_1, \ldots, x_n) = S_k^{(j)}(x_1, \ldots, x_{j-1}, x_{j+1}, \ldots, x_n)$ (meaning that the jth variable is removed from the n variables x_k). Then the following identity holds [124].

$$\sum_{j=1}^{n} S_k^{(j)}(x_1, \ldots, x_n) = (\alpha n + \beta k + \gamma) S_k(x_1, \ldots, x_n).$$

Determine the integers α, β, and γ.

e) The normalized elementary symmetric polynomials $\tilde{C}_k(x_1, \ldots, x_n) = n! \, (n - k)! \, C_k(x_1, \ldots, x_n)/n!$ fulfill for real x_k the inequalities [438], [394], [435]

$$\tilde{C}_{k-1}(x_1, \ldots, x_n) \, \tilde{C}_{k+1}(x_1, \ldots, x_n) \le \tilde{C}_k(x_1, \ldots, x_n)^2, \quad k = 1, \ldots, n - 1.$$

Verify these inequalities for $2 \le n \le 4$.

f) Using Vieta's relations from the above formulas, one can derive the following identity for the sum s_j of the jth powers

$$s_j = \sum_{i=1}^{n} x_i^j$$

of all roots x_i ($i = 1, \ldots, n$) of a polynomial equation $a_0 x^n + a_1 x^{n-1} a_2 x^{n-2} + \cdots + a_{n-1} x + a_n = 0$:

$$s_j = -j \frac{a_j}{a_0} - \sum_{k=1}^{j-1} s_k \frac{a_{k-j}}{a_0}.$$

For the s_j, a recursive implementation is obvious. In a functional style, implement a program that computes the first m of the s_j simultaneously. Try to avoid the use of any named variable. Use a polynomial and compare the result with a direct numerical calculation of the s_j.

6.¹² Generalized Lissajous Figures, Hyperspherical Harmonics, Hydrogen Orbitals

a) Construct "generalized Lissajous figures" by replacing the classical $\{sinOrCos(nt), sinOrCos(mt)\}$ by $\{p_n(t), p_m(t)\}$, $p_m(t)$ being an orthogonal polynomial. (The motivation for this choice of functions is the fact that $sin(n\,t)$ is an orthogonal function system, and the Sturm oscillation theorem holds (see, e.g., [460]), which accounts for the "oscillation up and down and from right to left" of the resulting curves.)

b) Check the formula $\left(\hat{\Lambda}^2 - \lambda(\lambda + d - 2)\right) C_\lambda^{(d-2)/2} (\mathbf{e} \cdot \mathbf{e}') = 0$ from the beginning of Section 2.4 for small d and λ.

For some small integers λ, show that the function

$$H_\lambda(\mathbf{r}, \mathbf{r}') = |\mathbf{r}|^\lambda \frac{C_\lambda^{(d-2)/2}\left(\frac{\mathbf{r}}{|\mathbf{r}|} \cdot \frac{\mathbf{r}'}{|\mathbf{r}'|}\right)}{C_\lambda^{(d-2)/2}(1)}$$

fulfills the Laplace equation and that $H_\lambda(\mathbf{e}', \mathbf{e}') = 1$ [47].

Multidimensional spherical harmonics $Y_{m_0}^{m_1,\dots,m_{d-2}}$ are defined as [47], [628], [362], [437], [192], [171], [82], [163]

$$Y_{m_0}^{m_1,\dots,m_{d-2}}(\vartheta_1, \vartheta_2, \dots, \vartheta_{d-2}, \phi) = e^{i\,m_{d-2}\,\phi} \prod_{k=0}^{d-3} \sin^{m_{k+1}}(\vartheta_{k+1}) C_{m_k-m_{k+1}}^{(d-k-2+2\,m_{k+1})/2}(\cos(\vartheta_{k+1})).$$

Here the $0 \le \vartheta_k \le \pi$ are the polar angles in a dD spherical coordinate system $\{r, \vartheta_1, \dots, \vartheta_{d-2}, \phi\}$. The m_j are nonnegative integers fulfilling $m_0 \ge m_1 \ge \cdots \ge m_{d-2} \ge 0$.

The angular part $\Delta_\circ\left(= \hat{\Lambda}^2\right)$ of the dD Laplace operator Δ in hyperspherical coordinates acts on a function $f(\vartheta_1, \dots, \vartheta_{d-2}, \phi)$ in the following way [290], [484], [153], [192], [437]:

$$\Delta_\circ f = \left(\sum_{k=0}^{d-3} \frac{\sin^{k-d+2}(\vartheta_{k+1})}{\prod_{j=1}^{k} \sin^2(\vartheta_j)} \frac{\partial}{\partial \vartheta_{k+1}} \left(\sin^{d-2-k}(\vartheta_{k+1}) \frac{\partial f}{\partial \vartheta_{k+1}}\right)\right) + \frac{1}{\prod_{j=1}^{d-2} \sin^2(\vartheta_j)} \frac{\partial^2 f}{\partial \phi^2}.$$

For small d, m_j, show that $Y_{m_0}^{m_1,\dots,m_{d-2}}(\vartheta_1, \vartheta_2, \dots, \vartheta_{d-2}, \phi)$ are eigenfunctions of the angular part of the dD Laplace operator

$$-\Delta_\circ\, Y_{m_0}^{m_1,\dots,m_{d-2}}(\vartheta_1, \vartheta_2, \dots, \vartheta_{d-2}, \phi) = m_0(m_0 + d - 2)\, Y_{m_0}^{m_1,\dots,m_{d-2}}(\vartheta_1, \vartheta_2, \dots, \vartheta_{d-2}, \phi).$$

(For the 1D case, see [108]; for $\exp(i\,\tau\,\Delta_\circ)$, see [369].)

c) The (not normalized) probability density for finding an electron in a hydrogen atom that the state $|n\,l\,m\rangle$ in spherical coordinates is $p_{n,l}(r, \vartheta) = |e^{-r}\,r^l\,L_{n-l-1}^{2\,l+1}(2\,r)\,P_n^l(\cos(\vartheta))|^2$ [619], [474]. Show $p_{12,6}(r, \vartheta) = c$ for various c.

7.12 Zeros of Hermite Polynomials, q-Hermite Polynomials, Pseudodifferential Operator, Moments of Hermite Polynomial Zeros

a) The nth Hermite polynomial $H_n(z)$ has n real zeros. Construct a visualization that shows what happens with the zeros of $H_n(z)$ if n changes continuously from one integer value to the next. (The command ND from the package NumericalMath`NLimit` is very useful here.)

b) The q-Hermite polynomials $H_n^{(q)}(z)$ [22], [24], [189], [87], [8], [73], [25], [527], [74], [190] can be defined in the following manner [279], [370]:

$$H_n^{(q)}(z) = 2\,z\,q^{-2\,n+3/2}\,H_{n-1}^{(q)}(z) - \frac{2}{q^2} \frac{\left(1 - \frac{1}{q^{2n-2}}\right)}{\left(1 - \frac{1}{q^2}\right)} H_{n-2}^{(q)}(z)$$

$$H_0^{(q)}(z) = 1$$
$$H_1^{(q)}(z) = 2\,q^{-1/2}\,z.$$

For small n calculate the positions and the order of the branch points of the function $z(q)$ implicitly defined by $H_n^{(q)}(z) = 0$ [637]. Make an animation that visualizes the dependence of the zeros $z_0(q)$ of $H_n^{(q)}(z_0(q))$. Vary r_q in $q = r_q \exp(i\,\varphi_q)$ from frame to frame.

c) By using expansion in orthogonal polynomials, calculate the lowest eigenvalue of the following pseudodifferential equation to at least correct digits for $m = 1$. (This describes ground-state energy of a relativistic harmonic oscillator with mass m in a half space [579], [213], [495], [251], [203], [252], [253], [428], [342], [487], [221], [222], [374], [146], [99]. But such-type Hamiltonians are not relativistically invariant [532], [541].)

$$\sqrt{m^2 + \frac{\partial^2}{\partial x^2}}\, \psi_\varepsilon(x) + x^2\, \psi_\varepsilon(x) = \varepsilon\, \psi_\varepsilon(x)$$

$$\psi_\varepsilon(0) = 0.$$

Sketch the dependence of the eigenvalue as a function of m.

c) The moments $\mu_m(n)$ of the n zeros $z_k^{(n)}$, $k = 1, \ldots, n$ of the Hermite polynomials $H_n(z)$ defined by

$$\mu_m(n) = \frac{1}{n} \sum_{k=1}^{n} \left(z_k^{(n)}\right)^m$$

can be expressed as $\mu_m(n) = \sum_{j=0}^{m/2} c_j^{(m)}\, n^j$ with rational $c_j^{(m)}$ [13]. Calculate the first few $\mu_m(n)$. Is it feasible to calculate $\mu_{100}(n)$?

8.^{l1} Iterated Polynomial Substitution

Starting with the polynomial $p(z) = \sum_{i=0}^{n} z^i$, repeatedly replace the powers z^j by initially randomly chosen and then fixed polynomials (for instance, orthogonal polynomials) of order j. Calculate the zeros of the resulting polynomials and display them graphically.

9.^{l2} Hermite Polynomials, Coherent States, Isospectral Potentials, Wave Packets

Let $\phi_n(x)$ be the normalized harmonic oscillator eigenfunctions

$$\phi_n(x) = \frac{1}{\sqrt{\sqrt{\pi}\, 2^n\, n!}}\, e^{-\frac{x^2}{2}}\, H_n(x).$$

a) Find the parameter α such that the integral $\int_{-\infty}^{\infty} \Psi(\alpha; x)\, \phi_k(x)\, dx$ becomes maximal (let $k = 0, 2$). Here, $\Psi(\alpha; x) \sim \exp(-\alpha\, x^4)$ and $\int_{-\infty}^{\infty} \Psi(\alpha; x)^2\, dx = 1$.

b) Find the parameters β_j, γ_i, and ξ_i such that the integral $\int_{-\infty}^{\infty} \Psi(x)\, \phi_k(x)\, dx$ ($k = 1, 3$) becomes maximal. Here,

$$\Psi(x) \sim \sum_{j=1}^{2} \beta_j\, e^{-\gamma_j(x - \xi_j)^2} \quad \text{and} \quad \int_{-\infty}^{\infty} \Psi(x)^2\, dx = 1.$$

c) The harmonic oscillator eigenfunctions $\phi_n(x)$ can be thought of as Fourier coefficients of time-dependent coherent states $\psi_C(\alpha; x, t)$ [273], [272], [274], [467], [169], [506].

$$\phi_n(x) = \frac{1}{2\pi}\, \exp(|\alpha|^2)\, \frac{\sqrt{n!}}{\alpha^n} \int_0^{2\pi} \psi_C(\alpha; x, t)\, e^{(2n+1)/2\, it}\, dt.$$

The time-dependent coherent states for a harmonic oscillator are given by [507], [324], [500], [292]:

$$\psi_C(\alpha; x, t) = -\frac{(-1)^{3/4}}{\sqrt[4]{\pi}} \frac{\exp\left(\left(-x^2 - \alpha^2 e^{-2it} + \overline{\alpha}^2 - 2\operatorname{Re}(\alpha)^2 + 2\sqrt{2}\, x\alpha e^{-it}\right)/2\right)}{\sqrt{1 - i\cot(t)}\,\sqrt{\sin(t)}}.$$

For a given n and appropriately chosen α, visualize how the harmonic oscillator eigenfunctions evolve from $\int_0^T \psi_C(\alpha; x, t)\, e^{(2n+1)/2\,it}\, dt$ as a function of T. (For the evolution of the harmonic oscillator eigenfunction from extended wave functions, see [414].)

d) $P_{qu}(x) = |\phi_n(x)|^2$ describes the probability density of finding a harmonic oscillator quantum particle with energy $2n + 1$ at position x. In the classical limit [482], [72], this probability tends to

$$P_{cl}(x) = \frac{\theta\left(x + \sqrt{2n+1}\right)\theta\left(\sqrt{2n+1} - x\right)}{\pi\sqrt{2n+1-x^2}}$$

[349], [481]. For a moderate large n, how good can an averaged $P_{qu}(x)$,

$$\tilde{P}_{qu}(x; \varepsilon) = \frac{1}{2\varepsilon} \int\limits_{x-\varepsilon}^{x+\varepsilon} P_{qu}(\xi)\, d\xi$$

[165], [480], [206], [209], [361], and [568] approximate $P_{cl}(x)$? What is the optimal ε?

e) The functions [210]

$$\Psi_n(\alpha, z) = \frac{1}{\sqrt{n!\, L_n\left(1 - \frac{1}{\eta}\right)}} \sum_{k=0}^{n} \left(\frac{1}{\eta} - 1\right)^{\frac{n-k}{2}} \frac{n!}{(n-k)!\,\sqrt{k!}}\, \phi_k(z)$$

interpolate smoothly between the harmonic oscillator eigenstates $\phi_k(z)$ and the time-independent coherent states

$$\Phi(z; \alpha) = e^{-\alpha^2/2} \sum_{k=0}^{\infty} \frac{\alpha^k}{\sqrt{k!}}\, \phi_k(z).$$

In the limit $\eta \to 1$, we have $\Psi_n(\alpha, z) = \phi_k(z)$, and in the limit $\eta \to 0$, $n \to \infty$, $n\eta^{1/2} = \alpha$, we have $\Psi_n(\alpha, z) = \Phi(z; \alpha)$. Visualize this transition between $\phi_{12}(z)$ and $\Phi(z; 2)$.

f) Given the eigenvalues ε_n and eigenfunctions $\phi_n(z)$ of the eigenvalue problem

$$-\phi_n''(z) + V(z)\,\phi_n(z) = \varepsilon_n\,\phi_n(z),$$

an isospectral (meaning having the same eigenvalues) eigenvalue problem (the notation $\psi_n[\phi_n](z; \lambda)$ indicates that ψ_n is a functional of ϕ_n)

$$-\psi_n[\phi_n]''(z; \lambda) + \mathcal{V}[\phi_n](z; \lambda)\,\psi_n[\phi_n](z; \lambda) = \varepsilon_n\,\psi_n[\phi_n](z; \lambda)$$

can be constructed by the following Darboux transformation [91], [488], [402], [372], [327], [401], [489], [317], [476], [449], [486], [318], [33], [412], [490], [323]:

$$\psi_n[\phi_n](z; \lambda) = \sqrt{\lambda(\lambda+1)}\, \frac{\phi_n(z)}{\int_{-\infty}^{z} \phi_n(z)^2\, dz + \lambda}$$

$$\mathcal{V}[\phi_n](z; \lambda) = V(z) - \frac{d}{dz^2} \ln\left(\int_{-\infty}^{z} \phi_n(z)^2 \, dz + \lambda\right)^2.$$

The new (normalized) eigenfunctions $\psi_n[\phi_n](z; \lambda)$ are functionals of the $\phi_n(z)$ and λ is a free parameter.

Use for the $\phi_n(x)$ the normalized harmonic oscillator eigenfunctions $\phi_n(x)$. Visualize some of the resulting $\psi_n[\phi_n](z; \lambda)$ and $\mathcal{V}[\phi_n](z; \lambda)$. Make conjectures for the exact symbolic values of the following two integrals:

$$\int_{-\infty}^{\infty} (\psi_n[\phi_n](z; \lambda) - \phi_n(z))^2 \, dz$$

$$\int_{-\infty}^{\infty} (\mathcal{V}[\phi_n](z; \lambda) - V(z)) \, dz.$$

g) "Wave packets" can be formed by superimposing energy eigenstates. For the harmonic oscillator, one possible superposition is [359], [360]

$$\Psi_n^{(m)}(z) = \sum_{k=-m}^{m} \phi_{n-k}(z).$$

For $n = 100$, which m gives the most localized wave packet? Visualize the t-dependence of the state

$$\Psi_n^{(m)}(z, t) = \sum_{k=-m}^{m} \phi_{n-k}(z) \, e^{-i(2k+1)t}.$$

h) A possible generalization of the harmonic oscillator eigenfunctions to d dimensions is given by the following functions ($d = 1$ yields the classical harmonic oscillator eigenfunctions $\phi_n(x)$) [390]:

$$\phi_n^{(d)}(x) = \frac{[n]_d!}{n!} \frac{1}{\sqrt{\sqrt{\pi} \, 2^n \, [n]_d!}} \, e^{-\frac{x^2}{2}} H_n^{(d)}(x).$$

Here $H_n^{(d)}(x)$ are d-dimensional generalizations of the Hermite polynomials

$$H_n^{(d)}(x) = \frac{n!}{[n]_d!} (-1)^n \, e^{\frac{x^2}{2}} \left(\frac{\partial_d}{\partial_d x}\right)^n e^{-\frac{x^2}{2}}$$

and the d-dimensional derivative $\partial_d . / \partial_d x$ and factorial $[n]_d!$ are defined by

$$\frac{\partial_d f(x)}{\partial_d x} = f'(x) + \frac{(d-1)}{2x} (f(x) - f(-x))$$

$$[n]_d! = \prod_{k=1}^{n} [n]_d = \prod_{k=1}^{n} \left(n + \frac{(d-1)}{2} (1 - (-1)^n)\right).$$

Calculate and visualize the first few $\phi_n^{(d)}(x)$ as a function of d. (For the multidimensional harmonic oscillator in integer dimensions, see for instance [555], [285], [115], [350], [604], [365], [538], [448].)

10.^{13} High-Order Perturbation Theory, Eigenvalue Differential Equation

a) The Rayleigh–Schrödinger perturbation theory treats the following problem: Given a (linear) eigenvalue problem

$$\left(\hat{L}_0 + g\,\hat{V}\right)\psi_n^{(g)}(z) = \varepsilon_n^{(g)}\,\psi_n^{(g)}(z)$$

the perturbed eigenvalues $\varepsilon_n^{(g)}$ and eigenfunctions $\psi_n^{(g)}(x)$ are to be found. The unperturbed problem is

$$\hat{L}_0\,\psi_n^{(0)}(z) = \varepsilon_n^{(0)}\,\psi_n^{(0)}(z)$$

and is assumed to be exactly solvable. $g\,\hat{V}$ is the perturbation. One starts with expanding the eigenvalues $\varepsilon_n^{(g)}$ and eigenfunctions $\psi_n^{(g)}(z)$ in powers of g using the expansion coefficients $^{(n)}\varepsilon_j$ and $^{(n)}c_{j,k}$.

$$\varepsilon_n^{(g)} = \sum_{j=0}^{\infty} g^j\, {}^{(n)}\varepsilon_j$$

$$\psi_n^{(g)}(x) = \sum_{j=0}^{\infty} g^j \left(\sum_{k=0}^{\infty} {}^{(n)}c_{j,k}\,\psi_k^{(0)}(z) \right).$$

Using the initial conditions $^{(n)}\varepsilon_0 = \varepsilon_n^{(0)}$, $^{(n)}c_{0,k} = \delta_{k,n}$, and $^{(k)}c_{j,k} = \delta_{j,k}$, it is straightforward to derive the following recursion relations for the $^{(n)}\varepsilon_j$ and $^{(n)}c_{j,k}$ [473], [260], [277], [199], [201], [110], [78]:

$$^{(n)}\varepsilon_j = \sum_{l=0}^{\infty} V_{m,l}\,\frac{N_l}{N_m}\,{}^{(n)}c_{j-1,l}$$

$$^{(n)}c_{j,k} = \frac{1}{\varepsilon_k^{(0)} - \varepsilon_n^{(0)}} \left(\sum_{l=1}^{j-1} {}^{(n)}\varepsilon_l\,{}^{(n)}c_{j-l,k} + \sum_{l=0}^{\infty} V_{k,l}\,\frac{N_l}{N_k}\,{}^{(n)}c_{j-1,l} \right).$$

Here $V_{k,l} = \left\langle \psi_k^{(0)} \middle| \hat{V} \middle| \psi_l^{(0)} \right\rangle$ and the perturbed eigenfunctions $\psi_n^{(g)}$ are normalized according to $\langle \psi_n^{(g)} | \psi_n^{(0)} \rangle = N_n$. (For matrix formulations, see [114], [79].)

For the eigenvalue problem $-\psi_n''(z) + (z^2 + z^4)\,\psi_n(z) = \varepsilon_n\,\psi_n(z)$ (mixed harmonic-quartic oscillator [60], [62], [61], [178]) one conveniently chooses the harmonic oscillator part as the unperturbed (and exactly solvable) part and $(g\,z^4)\,|_{g=1}$ as the perturbation.

$$\hat{L}_0 = -\frac{\partial^2}{\partial z^2}\cdot + z^2.$$

$$\hat{V} = z^4.$$

$$\left\langle \psi_k^{(0)} \middle| \hat{V} \middle| \psi_l^{(0)} \right\rangle = \int_{-\infty}^{\infty} \psi_k(z)\,z^4\,\psi_l(z)\,dz.$$

Because for each l, only finitely many of the $V_{k,l} = \langle \phi_k | z^4 | \phi_l \rangle$ (where $\phi_k(z)$ are the harmonic oscillator eigenfunctions) are nonvanishing, it is possible to calculate the sums in $^{(n)}\varepsilon_j$ and $^{(n)}c_{j,k}$ exactly for all n and j. Calculate the first few $^{(n)}\varepsilon_j$ and the first few hundred $^{(0)}\varepsilon_j$. Is the exact calculation of $^{(0)}\varepsilon_{1000}$ feasible?

Asymptotically, the $^{(0)}\varepsilon_j$ have the form [60], [524], [335], [120], [545], [331], [520]

$$^{(0)}\varepsilon_j = -(-1)^j \frac{2\sqrt{6}}{\pi^{3/2}} \left(\frac{3}{2}\right)^j \left(j - \frac{1}{2}\right)! \left(1 + \frac{\alpha_1}{j} + \frac{\alpha_2}{j^2} + \frac{\alpha_3}{j^3} + \cdots\right).$$

Conjecture exact values for α_1, α_2, and α_3.

For the mixed harmonic-quartic oscillator, the sum $\sum_{j=0}^{\infty} 1^j {}^{(0)}\varepsilon_j$ is divergent [181]. Given a (divergent) power series $\sum_{j=0}^{\infty} a_j x^j$, the diagonal Padé approximation of order M [521], [38], [514], [46], [37], [326], [543]

$$P_{(M,M)}(x) = \frac{\begin{vmatrix} a_1 & a_2 & \cdots & a_M & a_{M+1} \\ a_2 & a_3 & \cdots & a_{M+1} & a_{M+2} \\ \vdots & \vdots & \ddots & \vdots & \vdots \\ a_M & a_{M+1} & \cdots & a_{2M-1} & a_{2M} \\ \sum_{k=M}^{M} a_{k-M} x^k & \sum_{k=M-1}^{M} a_{k-M+1} x^k & \cdots & \sum_{k=2}^{M} a_k x^k & \sum_{k=0}^{M} a_k x^k \end{vmatrix}}{\begin{vmatrix} a_1 & a_2 & \cdots & a_M & a_{M+1} \\ a_2 & a_3 & \cdots & a_{M+1} & a_{M+2} \\ \vdots & \vdots & \ddots & \vdots & \vdots \\ a_M & a_{M+1} & \cdots & a_{2M-1} & a_{2M} \\ x^M & x^{M-1} & \cdots & x & 1 \end{vmatrix}}$$

allows calculating finite approximate values of such series by using the first M coefficients a_j. Use the diagonal Padé approximation to calculate the ground-state energy $\varepsilon_0^{(1)}$ to 10 correct digits.

b) Similar to the last subexercise, we consider again the problem $\left(\hat{L}_0 + g\,\hat{V}\right)\psi_n(g; z) = \varepsilon_n(g)\psi_n(g; z)$. The g-dependent quantities $\varepsilon_n(g)$ and $V_{k,l}(g) = \langle \psi_k(g) \mid \hat{V} \mid \psi_l(g) \rangle$ obey the following system of coupled differential equations [427], [261], [450], [534], [267]:

$$\frac{\partial \varepsilon_n(g)}{\partial g} = V_{n,n}^{(g)}$$

$$\frac{\partial V_{k,l}(g)}{\partial g} = \sum_{j \neq k} \frac{V_{k,j}(g)\, V_{j,l}(g)}{\varepsilon_k(g) - \varepsilon_j(g)} + \sum_{j \neq n} \frac{V_{k,j}(g)\, V_{j,l}(g)}{\varepsilon_l(g) - \varepsilon_j(g)}.$$

Use this system to calculate the ground-state energy of the quartic oscillator to 10 digits by using $\hat{L}_0 = -\frac{\partial^2}{\partial z^2} + z^2$. and $\hat{V} = (z^4 - z^2)..$

11.^{12} Sextic Oscillator, Time-Dependent Calogero Potential

a) Use expansion in eigenfunctions of the harmonic oscillator to make an animation of the time-development of the following initial value problem as a function of $\mathcal{K}$

$$i\frac{\partial \Psi(x, t)}{\partial t} = -\frac{\partial^2 \Psi(x, t)}{\partial x^2} + x^6 \Psi(x, t)$$

$$\Psi(x, 0) = e^{i\mathcal{K}x} \cos^2\left(\frac{\pi x}{4}\right) \theta(4 - x^2).$$

Let $\mathcal{K}$ range from 0 to 6 and $0 \le t \le 8$. How does this expansion compare with a direct numerical solution of the initial value problem using `NDSolve`?

b) Use expansion in eigenfunctions of the Calogero potential to make an animation of the time-development of the following initial value problem as a function of $\mathcal{K}$

$$i \frac{\partial \Psi(x, t)}{\partial t} = -\frac{\partial^2 \Psi(x, t)}{\partial x^2} + \left(x^2 + \frac{\gamma}{x^2} \right) \Psi(x, t)$$

$$\Psi(x, 0) = \theta(x - \pi) \, \theta(2\pi - x) \cos(\mathcal{K}(x - \pi)) \sqrt{\sin(x - \pi)} \ .$$

(For the special properties of the Calogero potential, see [122].)

Let $3 \le \mathcal{K} \le 12$, $\gamma = 2$, and $0 \le t \le 2$. The time-independent Schrödinger equation with the Calogero potential [214], [26], [570], [248], [195], [406], [425], [348], [137], [573], [572], [196]

$$-\frac{\partial^2 \psi_n(x)}{\partial x^2} + \left(x^2 + \frac{\gamma}{x^2} \right) \psi_n(x) = \varepsilon_n \, \psi_n(x)$$

has the following eigenvalues and (not normalized) eigenfunctions ($\alpha = (4\gamma + 1)^{1/2}/2$, $\gamma > -1/4$):

$$\varepsilon_n = 4n + 2\alpha + 2$$

$$\psi_n(x) = x^{\frac{1}{2}(2\alpha + 1)} e^{-\frac{x^2}{2}} L_n^\alpha(x^2).$$

(For $\gamma < -1/4$, see [233])

Solutions

1. Generating Function for $T_n(x)$, Mehler's Formula, Bauer–Rayleigh Expansion, and More

a) Here is a "straightforward" implementation.

```
In[1]:= MapIndexed[ (* coefficients * powers *)
                ((#2[[1]] - 1)! z^((-#2[[1]] + 1)) #1) &,
                List @@ Normal[ (* the series *)
                     Series[Exp[z x] Cosh[z Sqrt[x^2 - 1]], {z, 0, 5}]]];

In[2]:= Expand /@ %

Out[2]= {1, x, -1 + 2 x², -3 x + 4 x³, 1 - 8 x² + 8 x⁴, 5 x - 20 x³ + 16 x⁵}

In[3]:= % == Table[ChebyshevT[i, x], {i, 0, 5}]

Out[3]= True
```

For the symbolic proof, we use Sum. Using $\cos(n \arccos(x))$ for $T_n(x)$, we can compute the infinite sum.

```
In[4]:= ChebyshevT[i, x] // FunctionExpand

Out[4]= Cos[i ArcCos[x]]

In[5]:= Sum[z^n/n! Cos[n ArcCos[x]], {n, 0, Infinity}]

Out[5]= ½ (e^(e^(-i ArcCos[x]) z) + e^(e^(i ArcCos[x]) z))
```

To get rid of the exponentials, we use ComplexExpand.

```
In[6]:= ComplexExpand[% /. ArcCos[x] -> X] /. X -> ArcCos[x]

Out[6]= e^(x z) Cos[√(1 - x²) z]
```

To get the formula mentioned, we change the square root last.

```
In[7]:= % /. Sqrt[x_] -> I Sqrt[-x]

Out[7]= e^(x z) Cosh[√(-1 + x²) z]
```

b) This is another straightforward case.

```
In[1]:= ((Series[1/Sqrt[1 - a^2] Exp[-1/(1 - a^2)(z^2 + ζ^2 - 2z ζ a)],
              {a, 0, #}] // Normal // Expand)  ==
        (Exp[-z^2 - ζ^2] Sum[a^i/2^i/i! HermiteH[i, z] HermiteH[i, ζ],
              {i, 0, #}] // Expand)) & /@ Range[0, 6, 1]

Out[1]= {True, True, True, True, True, True, True}
```

c) Using the Bessel function BesselJ[l + 1/2, x], we can write the following program.

```
In[1]:= BauerRayleigh[n_, x_, y_] :=
        Sqrt[Pi/(2x)] Sum[(2l + 1) Exp[I Pi l/2]*
              BesselJ[l + 1/2, x] LegendreP[l, y], {l, 0, n}]

In[2]:= Normal[Series[BauerRayleigh[7, x, y], {x, 0, 6}]] // Expand

Out[2]= 1 + i x y - (x² y²)/2 - 1/6 i x³ y³ + (x⁴ y⁴)/24 + 1/120 i x⁵ y⁵ - (x⁶ y⁶)/720

In[3]:= Normal[Series[Exp[I x y], {x, 0, 6}]] // Expand

Out[3]= 1 + i x y - (x² y²)/2 - 1/6 i x³ y³ + (x⁴ y⁴)/24 + 1/120 i x⁵ y⁵ - (x⁶ y⁶)/720
```

d) This is again easy to check.

```
In[1]:= Table[Expand[Sum[
              HermiteH[m, x Sqrt[2]] HermiteH[n - m, y Sqrt[2]]*
              Binomial[n, m], {m, 0, n}]] ==
        Expand[Sqrt[2]^n HermiteH[n, x + y]], {n, 0, 2}]
```

```
Out[1]= {True, True, True}
```

e) This is also a simple exercise.

```
In[1]:= Table[Expand[2 ChebyshevT[n, z] ChebyshevT[m, z]] ==
            Expand[ ChebyshevT[m + n, z] + ChebyshevT[m - n, z]],
            {m, 0, 4}, {n, 0, m}]
Out[1]= {{True}, {True, True}, {True, True, True},
            {True, True, True, True}, {True, True, True, True, True}}

In[2]:= Table[Expand[2 ChebyshevT[n, z]^2] ==
            Expand[1 + ChebyshevT[2n, z]], {n, 0, 4}]
Out[2]= {True, True, True, True, True}
```

Using the trigonometric representation of the Chebychev polynomials and `TrigFactor`, these two relations are also easily proved symbolically.

```
In[3]:= TrigFactor[ExpandAll[
        2 ChebyshevT[m, x] ChebyshevT[n, x] -
        (ChebyshevT[m + n, x] + ChebyshevT[m - n, x]) /.
            ChebyshevT[n_, x_] -> Cos[n ArcCos[x]]]]
Out[3]= 0

In[4]:= TrigFactor[2 ChebyshevT[n, x]^2 - (1 + ChebyshevT[2n, x]) /.
                ChebyshevT[n_, x_] -> Cos[n ArcCos[x]]]
Out[4]= 0
```

f) Here is a direct implementation.

```
In[1]:= test[n_, m_] := Expand[LegendreP[n, x] LegendreP[m, x]] ==
        Expand[Sum[If[EvenQ[1 + m + n], (2l + 1)*
        (m + n - l - 1)!! (l + n - m - 1)!! (m + n + l)!! (1 + m - n - 1)!!/
        (m + n - 1)!!/(1 + n - m)!!/(m + n + l + 1)!!/(1 + m - n)!!, 0]*
        LegendreP[l, x], {l, Abs[m - n], m + n}]]
```

It immediately shows the correctness of the formulas for small n, m.

```
In[2]:= Array[test, {4, 4}, 0] // Flatten // Union
Out[2]= {True}
```

g) Here is a straightforward implementation.

```
In[1]:= myLaguerreL[n_, x_] := (-1)^n Exp[x]/n! *
                Nest[Together[x D[#, {x, 2}] + D[#, x]]&, Exp[-x], n]
```

It easily checks the first n.

```
In[2]:= Table[Together[myLaguerreL[i, z] - LaguerreL[i, z]], {i, 0, 10}]
Out[2]= {0, 0, 0, 0, 0, 0, 0, 0, 0, 0, 0}
```

h) Here is a numerical check.

```
In[1]:= hermiteRootTest[n_] :=
        Module[{roots, x},
                roots = x /. NSolve[HermiteH[n, x] == 0, x];
                Table[roots[[k]] - (Plus @@ (1/(roots[[k]] - Delete[roots, k]))),
                {k, n}]]

In[2]:= Table[Chop[hermiteRootTest[n]], {n, 12}] // Flatten // Union
Out[2]= {0}
```

Here is a symbolic check for $n = 5$.

```
In[3]:= roots = x /. RootReduce[{ToRules[Roots[HermiteH[5, x] == 0, x,
                        Cubics -> False, Quartics -> False]]}]
Out[3]= {0, Root[15 - 20 #1^2 + 4 #1^4 &, 3], Root[15 - 20 #1^2 + 4 #1^4 &, 2],
            Root[15 - 20 #1^2 + 4 #1^4 &, 4], Root[15 - 20 #1^2 + 4 #1^4 &, 1]}
```

```
In[4]:= Table[roots[[k]] - (Plus @@ (1/(roots[[k]] - Delete[roots, k]))),
           {k, 5}] // RootReduce
Out[4]= {0, 0, 0, 0, 0}
```

i) This is the closed-form representation of $R_k(\epsilon, x)$.

```
In[1]:= R[k_][e_, x_] :=
         With[{U = ChebyshevU},
           (U[k - 1, 1/(2 Sqrt[x])] - e Sqrt[x] U[k - 2, 1/(2 Sqrt[x])])/
           (U[k   , 1/(2 Sqrt[x])] - e Sqrt[x] U[k - 1, 1/(2 Sqrt[x])])/
           Sqrt[x]]
```

We will prove the correctness of $R_k(\epsilon, x)$ by induction. For small k, we can see the identity immediately.

```
In[2]:= Table[R[k][e, x], {k, 0, 3}] // Together
```
$$Out[2]= \left\{\epsilon, -\frac{1}{-1+x\,\epsilon}, \frac{-1+x\,\epsilon}{-1+x+x\,\epsilon}, \frac{1-x-x\,\epsilon}{1-2\,x-x\,\epsilon+x^2\,\epsilon}\right\}$$

```
In[3]:= {e, 1/(1 - e x), 1/(1 - x/(1 - e x)),
            1/(1 - x/(1 - x/(1 - e x)))} == % // Simplify
Out[3]= True
```

By construction, $R_k(\epsilon, x)$ fulfills the recursion relation $R_{k+1}(\epsilon, x) = 1/(1 - x R_k(\epsilon, x))$. Applying FullSimplify to this identity completes the proof.

```
In[4]:= R[k + 1][e, x] - 1/(1 - x R[k][e, x]) // FullSimplify
Out[4]= 0
```

j) To convert from the spherical coordinates ϑ, φ to Cartesian coordinates x, y, and z, we use the relations $\varphi = \arccos(z/r)$ and $\varphi = \arctan(x, y)$. Then we rewrite exponentials in trigonometric form and the function TrigExpand yields functions of x, y, and z. We treat remaining radicals containing rational functions of x, y, and z with a replacement rule. So, we can implement the following one-liner ExpandSphericalHarmonicY.

```
In[1]:= ExpandSphericalHarmonicY[{l_, m_}, {x_, y_, z_}] :=
         Factor[Together[TrigExpand[ExpToTrig[
           SphericalHarmonicY[l, m, ArcCos[z/Sqrt[x^2 + y^2 + z^2]],
                                    ArcTan[y/x]]]]]] /.
           Sqrt[1 - z^2/(x^2 + y^2 + z^2)]/Sqrt[1 + y^2/x^2] ->
           Sqrt[(x^2 + y^2)/(x^2 + y^2 + z^2)]/Sqrt[(x^2 + y^2)/x^2] /.
           Sqrt[(x^2 + y^2)/(x^2 + y^2 + z^2)]/Sqrt[(x^2 + y^2)/x^2] ->
           x/Sqrt[x^2 + y^2 + z^2]]
```

Here are some examples. We only display the polynomial part.

```
In[2]:= Table[ExpandSphericalHarmonicY[{l, m}, {x, y, z}]*
           (x^2 + y^2 + z^2)^(1/2) /. c_?NumericQ r_ :> r,
           {1, 5}, {m, -1, 1}]
Out[2]= {{x - i y, z, x + i y}, {(x - i y)^2, (x - i y) z, x^2 + y^2 - 2 z^2, (x + i y) z, (x + i y)^2},
         {(x - i y)^3, (x - i y)^2 z, (x - i y) (x^2 + y^2 - 4 z^2), z (3 x^2 + 3 y^2 - 2 z^2), (x + i y) (x^2 + y^2 - 4 z^2),
          (x + i y)^2 z, (x + i y)^3}, {(x - i y)^4, (x - i y)^3 z, (x - i y)^2 (x^2 + y^2 - 6 z^2),
          (x - i y) z (3 x^2 + 3 y^2 - 4 z^2), 3 x^4 + 6 x^2 y^2 + 3 y^4 - 24 x^2 z^2 - 24 y^2 z^2 + 8 z^4,
          (x + i y) z (3 x^2 + 3 y^2 - 4 z^2), (x + i y)^2 (x^2 + y^2 - 6 z^2), (x + i y)^3 z, (x + i y)^4},
         {(x - i y)^5, (x - i y)^4 z, (x - i y)^3 (x^2 + y^2 - 8 z^2), (x - i y)^2 z (x^2 + y^2 - 2 z^2),
          (x - i y) (x^4 + 2 x^2 y^2 + y^4 - 12 x^2 z^2 - 12 y^2 z^2 + 8 z^4),
          z (15 x^4 + 30 x^2 y^2 + 15 y^4 - 40 x^2 z^2 - 40 y^2 z^2 + 8 z^4),
          (x + i y) (x^4 + 2 x^2 y^2 + y^4 - 12 x^2 z^2 - 12 y^2 z^2 + 8 z^4),
          (x + i y)^2 z (x^2 + y^2 - 2 z^2), (x + i y)^3 (x^2 + y^2 - 8 z^2), (x + i y)^4 z, (x + i y)^5}}
```

The polynomial versions of the spherical harmonic agree with the built-in function SphericalHarmonicY. Here is a quick numerical check.

```
In[3]:= Module[{l, m, ϑ, φ, x, y, z, o = 10},
           Table[l = Random[Integer, {0, 5}];
                 m = Random[Integer, {-1, 1}];
                 ϑ = 2 Pi Random[]; φ = Pi Random[];
             (* the polynomial *)
             Y[{x_, y_, z_}] = ExpandSphericalHarmonicY[{l, m}, {x, y, z}];
```

```
(* from spherical coordinates *)
{x, y, z} = {Cos[φ] Sin[θ], Sin[φ] Sin[θ], Cos[θ]};
Chop[1 - Y[{x, y, z}]/SphericalHarmonicY[l, m, θ, φ]],
        {100}]] // Union
```

Out[3]= {0}

A direct polynomial representation of the $Y_l^m(x, y, z)$ is [586] (using $r = (x^2 + y^2 + z^2)^{1/2}$)

$$Y_l^m(x, y, z) = \sqrt{\frac{(2l+1)(l+m)!(l-m)!}{4\pi}} \frac{1}{r^l} \sum_{h,j,k=0}^{l} \frac{\delta_{h+j+k,l}\,\delta_{h-j,m}}{h!\,j!\,k!\,2^{h+j}} (-x - iy)^h (x - iy)^j z^k.$$

The results form `ExpandSphericalHarmonicY` agree with the results from `CartesianSphericalHarmonicY`.

```
In[4]:= CartesianSphericalHarmonicY[{l_, m_}, {x_, y_, z_}] :=
    Factor[Sqrt[x^2 + y^2 + z^2]^-1 Sqrt[(2l + 1)/(4Pi)(1 + m)!(1 - m)!]*
        Sum[1/(i! j! k! 2^i 2^j)*
            KroneckerDelta[i + j + k, l] KroneckerDelta[i - j, m]*
            (x - I y)^j (-x - I y)^i z^k,
            {i, 0, 1}, {j, 0, 1}, {k, 0, 1}]]

In[5]:= Table[Cancel[ExpandSphericalHarmonicY[{l, m}, {x, y, z}]/
                CartesianSphericalHarmonicY[{l, m}, {x, y, z}]],
            {l, 0, 10}, {m, -1, 1}] // Flatten // Union // Simplify

Out[5]= {1}
```

k) It is largely straightforward to implement this one-liner. We start with the unnormalized spherical harmonics for $l = 1$, $\{\mathcal{Y}_{1,-1}(\vartheta, \varphi), \mathcal{Y}_{1,0}(\vartheta, \varphi), \mathcal{Y}_{1,1}(\vartheta, \varphi)\}$ and then recursively calculate the $\mathcal{Y}_{l,m}(\vartheta, \varphi)$ for $l = \tilde{l} + 1$ and $m = -l, -l+1, ..., 0$ by the two-term recursion relation. Then we use the complex conjugation formula to obtain the unnormalized spherical harmonics for $l = \tilde{l} + 1$ and $m = 0, ..., l$. At the end, we change the normalization to obtain the spherical harmonics $Y_{l,m}(\vartheta, \varphi)$.

```
In[1]:= SphericalHarmonicLists[lMax_Integer?Positive, θ_, φ_] :=
    With[{x = Cos[φ] Sin[θ], y = Sin[φ] Sin[θ], z = Cos[θ]},
        MapIndexed[Function[{y, p}, (* change normalization *)
                Module[{l = p[[1]] - 1, m = p[[2]]},
                y Sqrt[(2l + 1) (1 + m)! (1 - m)!/(4Pi)]]],
            Prepend[Last /@ NestList[(* recurse over l *)
        Function[{l, L}, {l + 1, Expand @ Join[#,
        MapIndexed[(* l ≤ m ≤ l *)((-1)^#2[[1]]  #1)&,
            Rest[Reverse[ComplexExpand[Conjugate[#]]]]]]]&[
        MapIndexed[(#1/(2l + 3 - #2[[1]]))&, {z, x - I y}.#& /@
        Partition[Join[{0}, Take[L, l + 2]], 2, 1]]]]} @@ #&,
            (* l = 1 *) {1, {x - I y, 2z, -x - I y}/2}, lMax - 1],
            {SetPrecision[1, Precision[{θ, φ}]]}], {2}]]
```

Here are the first few spherical harmonics.

```
In[2]:= SphericalHarmonicLists[2, θ, φ] // Simplify
```

$$Out[2]= \left\{\left\{\frac{1}{2\sqrt{\pi}}\right\}, \left\{\frac{1}{2}\sqrt{\frac{3}{2\pi}} \,\mathrm{Sin}[\theta]\,(\mathrm{Cos}[\varphi] - i\,\mathrm{Sin}[\varphi]),\; \frac{1}{2}\sqrt{\frac{3}{\pi}}\,\mathrm{Cos}[\theta],\right.\right.$$

$$-\frac{1}{2}\sqrt{\frac{3}{2\pi}}\,\mathrm{Sin}[\theta]\,(\mathrm{Cos}[\varphi] + i\,\mathrm{Sin}[\varphi])\right\}, \left\{\frac{1}{4}\sqrt{\frac{15}{2\pi}}\,\mathrm{Sin}[\theta]^2\,(\mathrm{Cos}[2\varphi] - i\,\mathrm{Sin}[2\varphi]),\right.$$

$$\frac{1}{2}\sqrt{\frac{15}{2\pi}}\,\mathrm{Cos}[\theta]\,\mathrm{Sin}[\theta]\,(\mathrm{Cos}[\varphi] - i\,\mathrm{Sin}[\varphi]),\; \frac{1}{8}\sqrt{\frac{5}{\pi}}\,(1 + 3\,\mathrm{Cos}[2\theta]),$$

$$\left.\left.-\frac{1}{2}\sqrt{\frac{15}{2\pi}}\,\mathrm{Cos}[\theta]\,\mathrm{Sin}[\theta]\,(\mathrm{Cos}[\varphi] + i\,\mathrm{Sin}[\varphi]),\; \frac{1}{4}\sqrt{\frac{15}{2\pi}}\,\mathrm{Sin}[\theta]^2\,(\mathrm{Cos}[2\varphi] + i\,\mathrm{Sin}[2\varphi])\right\}\right\}$$

Here is a quick check showing that the spherical harmonics calculated by `SphericalHarmonicLists` agree with the built-in ones.

```
In[3]:= With[{λ = 10},
    SphericalHarmonicLists[λ, θ, φ] -
```

```
        Table[SphericalHarmonicY[1, m, ϑ, φ], {1, 0, λ}, {m, -1, 1}]] //
                ExpToTrig // TrigExpand // Expand // Flatten // Union
Out[3]= {0}
```

For machine-precision ϑ and φ, we will implement a compiled version. Because compiled code can only return tensor structures (meaning all sublists must have the same length), we first generate a flat list of length $(l+1)^2$ within compiled code that we later split into sublists of length 1, 3, 5, ... This time, we fold the normalization into the recursion relation to avoid the uncompilable calculation of the factorial function.

```
In[4]:= SphericalHarmonicListCF =
        Compile[{{lMax, _Integer}, ϑ, φ},
                Module[{x = Cos[φ] Sin[ϑ], y = Sin[φ] Sin[ϑ], z = Cos[ϑ],
                        YL = Table[0. I, {(lMax + 1)^2}], μ, λ, Y1m1},
                YL[[1]] = 1/(2 Sqrt[Pi + 0. I]);
                If[lMax != 0, Y1m1 = Sqrt[3/(2 Pi)] (x - I y)/2;
                    (* initial Y-values *)
                    YL[[2]] = Y1m1; YL[[3]] = Sqrt[3/Pi]/2 z + 0. I;
                    YL[[4]] = Sqrt[3/(2 Pi)] (-x - I y)/2;
                    Do[μ = 1^2 - (1 - 1); λ = (1 + 1)^2 - 1; (* next 1 *)
                        Do[YL[[λ + m]] =
                        (z If[m == -1, 0,
                                Sqrt[((21 + 1) (1 - m) (1 + m))/(21 - 1)] YL[[μ + m]]] +
                                2 Sqrt[2Pi/3] Y1m1 Sqrt[((21 + 1) (1 - m - 1) (1 - m))/
                                                        (21 - 1)] YL[[μ + m + 1]])/(1 - m),
                        {m, -1, 0}]; (* positive m *)
                        Do[YL[[λ + m]] = (-1)^m Conjugate[YL[[λ - m]]], {m, 1}],
                    {1, 2, lMax}]];
                (* return all Y-values *) YL]];
In[5]:= SphericalHarmonicListN[lMax_Integer?Positive, ϑ_, φ_] :=
        (* split result of SphericalHarmonicListCF into sublists for each 1 *)
        First /@ Rest[FoldList[{Take[#1[[2]], #2], Drop[#1[[2]], #2]}&,
                        {0, SphericalHarmonicListCF[lMax, ϑ, φ]},
                        Table[2 1 + 1, {1, 0, lMax}]]]
```

The function `SphericalHarmonicListCF` compiled successfully.

```
In[6]:= Union[Head /@ Flatten[SphericalHarmonicListCF[[4]]]]
Out[6]= {Integer, Real}
```

Here is again a quick check for the correctness of `SphericalHarmonicListN`.

```
In[7]:= With[{λ = 10, ϑ = 1, φ = 2},
        SphericalHarmonicListN[λ, ϑ, φ] -
        Table[SphericalHarmonicY[1, m, ϑ, φ], {1, 0, λ}, {m, -1, 1}]] //
                                                        Abs // Max
Out[7]= 1.83246×10^-15
```

Taking into account $|Y_{l,m}(\vartheta, \varphi)| \leq 1$ for all ϑ and φ, the error of `SphericalHarmonicListN` is maximal about 0.02%. Be aware that we must use high-precision arithmetic to obtain reliable values for $Y_{l,m}(\vartheta, \varphi)$. Here is an example for typical values of ϑ and φ.

```
In[8]:= With[{1 = 100, ϑ = 1, φ = 2},
        Module[{YLRecursion = SphericalHarmonicListN[1, ϑ, φ][[-1]],
                YLN   = Table[N[SphericalHarmonicY[1, m, ϑ, φ]],        {m, -1, 1}],
                YLN1  = Table[ SphericalHarmonicY[1, m, N[ϑ], N[φ]],   {m, -1, 1}],
                YLHP  = Table[N[SphericalHarmonicY[1, m, ϑ, φ], 50],   {m, -1, 1}]},
                Max[Abs[#]]& /@
                    {YLRecursion - YLN, YLRecursion - YLN1, YLRecursion - YLHP}]]
Out[8]= {169887., 169887., 0.000183444}
```

In addition to being more precise, calculating the values of the spherical harmonics recursively is also much faster.

```
In[9]:= With[{λ = 100, ϑ = 1., φ = 2.},
        {Timing[SphericalHarmonicListN[1, ϑ, φ];],
         Timing[Table[SphericalHarmonicY[1, m, ϑ, φ], {1, 0, λ}, {m, -1, 1}];]}]
```

```
Out[9]= {{0. Second, Null}, {41. Second, Null}}
```

2. Generalized Fourier Series

First, we define a few functions: `weight` computes the weight function corresponding to the polynomials, `normaliza`-
`tion` finds the corresponding normalization, `GeneralizedFourierCoefficient` computes the Fourier coefficients,
and `GeneralizedFourierSum` computes the partial sums of the products of the Fourier coefficients containing the
corresponding polynomials (*weightFunction normalization*)$^{1/2}$. The program is tailored somewhat to the special form of the
problem. For more general use, we would need to make the following improvements:

- a better check of whether optional arguments are admissible for the corresponding polynomials

- the better use of *Mathematica*'s ability to compute definite integrals numerically when needed

Here, `polynomial` is the type of the corresponding polynomials, *var* is the independent variable, n is the index of the
polynomials, *func* is the function to be analyzed, {*xl*, *xu*} are the limits of integration, and `para` is a list of additional
parameters of the polynomials (e.g., α and β in $P_n^{(\alpha,\beta)}$).

Here the weight functions and the normalization constants are implemented.

```
In[1]:= SetAttributes[weight, HoldAll];

     (* the weight functions *)
     weight[HermiteH[n_, z_]] = Exp[-z^2];
     weight[JacobiP[n_, α_, β_, z_]] = (1 - z)^α (1 + z)^β;
     weight[GegenbauerC[n_, z_]] = (1 - z^2)^(-1/2);
     weight[GegenbauerC[n_, α_, z_]] = (1 - z^2)^(α - 1/2);
     weight[LegendreP[n_, m_, z_]] = 1;
     weight[LegendreP[n_, z_]] = 1;
     weight[LaguerreL[n_, z_]] = Exp[-z];
     weight[LaguerreL[n_, α_, z_]] = Exp[-z] z^α;
     weight[ChebyshevT[n_, z_]] = 1/Sqrt[1 - z^2];
     weight[ChebyshevU[n_, z_]] = Sqrt[1 - z^2];

In[13]:= SetAttributes[normalization, HoldAll];

     (* the normalizations *)
     With[{Γ = Gamma},
     normalization[HermiteH[n_, z_]] = Sqrt[Pi] 2^n n!;
     normalization[JacobiP[n_, α_, β_, z_]] =
                       2^(α + β + 1)/(2n + α + β + 1)*
                       Γ[n + α + 1]Γ[n + β + 1]/n!/Γ[n + α + β + 1];
     (* separate definitions for the various cases *)
     normalization[GegenbauerC[0, z_]] = 1/2;
     normalization[GegenbauerC[n_, z_]] = 2Pi/n^2;
     normalization[GegenbauerC[n_, α_, z_]] =
                       Pi 2^(1 - 2α) Γ[n + 2α]/(n! (α + n) Γ[α]^2);
     normalization[LegendreP[n_, z_]] = 2/(2n + 1);
     normalization[LegendreP[n_, m_, z_]] = (n + m)!/(n - m)! 2/(2n + 1);
     normalization[LaguerreL[n_, z_]] = 1;
     normalization[LaguerreL[n_, α_, z_]] = Γ[n + 1 + α]/n!;
     normalization[ChebyshevT[0, z_]] = Pi;
     normalization[ChebyshevT[n_, z_]] = Pi/2;
     normalization[ChebyshevU[n_, z_]] = Pi/2];
```

The generalized Fourier coefficients are obtained by integration of the normalized orthogonal polynomials (including the
contribution of the weight function) with the function to be expanded.

```
In[16]:= GeneralizedFourierCoefficient[
           polynomial:(HermiteH  | JacobiP  | GegenbauerC |
                     LaguerreL | LegendreP | ChebyshevT  | ChebyshevU),
           n_Integer?(# >= 0&), func_, var_Symbol, {xl_, xu_}, para___] :=
         (* remember the value *)
         GeneralizedFourierCoefficient[
           polynomial, n, func, var, {xu, xo}, para] =
         Module[{poly, weight1, temp, fc, normFac, a, b, m},
```

```
                 (* make the polynomial *)
                 poly = polynomial[n, para, var];          \
                 (* get the weight function *)
                 weight1 = PowerExpand[Sqrt[weight[polynomial[n, para, var]]]];
                 (* do the integration *)
                 temp = Integrate[poly weight1 func, {var, xl, xu},
                                  GenerateConditions -> False];
                 If[FreeQ[temp, Integrate],
                    fc = temp, (* warn in case the integration failed *)
                    Print["! WARNING, not explicitly integrated:",
                          n, " - th ", polynomial, " - polynomial"]];
                 (* normalize *)
                 normFac = normalization[polynomial[n, para, var]];
                 fc/Sqrt[normFac]];
```

In[17]:= `GeneralizedFourierSum[`
         `    polynomial:(HermiteH    | JacobiP    | GegenbauerC |`
         `                LaguerreL   | LegendreP  | ChebyshevT  | ChebyshevU),`
         `    n_Integer?(# >= 0&), func_, var_Symbol, {xu_, xo_}, para___] :=`
         `GeneralizedFourierSum[polynomial, n, func, var, {xu, xo}, para] =`
         `(If[n =!= 0, (* all earlier terms *)`
         `    GeneralizedFourierSum[polynomial, n - 1, func,`
         `                          var, {xu, xo}, para], 0] +`
         `    (* the new term *)`
         `    GeneralizedFourierCoefficient[`
         `           polynomial, n, func, var, {xu, xo}, para]*`
         `    (* the polynomial and weight and normalization *)`
         `    polynomial[n, para, z]*`
         `    PowerExpand[Sqrt[weight[polynomial[n, para, z]]]]/`
         `    Sqrt[normalization[polynomial[n, para, z]]])`

We now look at three examples of the performance of `GeneralizedFourierCoefficient` and `GeneralizedFou` rierSum.

In[18]:= `Table[GeneralizedFourierCoefficient[HermiteH, i, 1, z, {0, 1}], {i, 0, 6}]`

Out[18]= $\left\{ \dfrac{\pi^{1/4} \text{Erf}\left[\frac{1}{\sqrt{2}}\right]}{\sqrt{2}}, \dfrac{2 - \frac{2}{\sqrt{e}}}{\sqrt{2}\ \pi^{1/4}}, \dfrac{-\frac{4}{\sqrt{e}} + \sqrt{2\pi}\ \text{Erf}\left[\frac{1}{\sqrt{2}}\right]}{2\sqrt{2}\ \pi^{1/4}}, \dfrac{4 - \frac{12}{\sqrt{e}}}{4\sqrt{3}\ \pi^{1/4}}, \right.$

$\left. \dfrac{-\frac{16}{\sqrt{e}} + 6\sqrt{2\pi}\ \text{Erf}\left[\frac{1}{\sqrt{2}}\right]}{8\sqrt{6}\ \pi^{1/4}}, \dfrac{56 - \frac{56}{\sqrt{e}}}{16\sqrt{15}\ \pi^{1/4}}, \dfrac{-\frac{144}{\sqrt{e}} + 60\sqrt{2\pi}\ \text{Erf}\left[\frac{1}{\sqrt{2}}\right]}{96\sqrt{5}\ \pi^{1/4}} \right\}$

In[19]:= `GeneralizedFourierSum[LaguerreL, 3, 1, z, {0, 1}, {2}]`

Out[19]= $\left\{ \dfrac{1}{2}\left(4 - \dfrac{6}{\sqrt{e}}\right) e^{-z/2}\ z + \dfrac{1}{6}\left(-4 + \dfrac{8}{\sqrt{e}}\right) e^{-z/2}\ (3 - z)\ z + \right.$

$\left. \dfrac{1}{24}\left(8 - \dfrac{11}{\sqrt{e}}\right) e^{-z/2}\ z\ (12 - 8 z + z^2) + \dfrac{1}{120}\left(-8 + \dfrac{16}{\sqrt{e}}\right) e^{-z/2}\ z\ (60 - 60 z + 15 z^2 - z^3) \right\}$

In[20]:= `GeneralizedFourierSum[ChebyshevU, 2, 1, z, {0, 1}]`

Out[20]= $\dfrac{16 z\ (1 - z^2)^{1/4}}{5\pi} + \dfrac{(1 - z^2)^{1/4}\ \text{Gamma}\left[\frac{1}{4}\right]}{3\sqrt{\pi}\ \text{Gamma}\left[\frac{3}{4}\right]} + \dfrac{(1 - z^2)^{1/4}\ (-1 + 4 z^2)\ \text{Gamma}\left[\frac{1}{4}\right]}{21\sqrt{\pi}\ \text{Gamma}\left[\frac{3}{4}\right]}$

We now present plots to illustrate the numerical values and convergence of the Parseval identity $\sum_{n=0}^{\infty} |\alpha_n|^2 = \int_0^1 |y(z)|^2\, dz$, where (z_l, z_u) is the orthogonality interval of the orthogonal polynomial under consideration. (Some of the integrals to be computed analytically are already complicated for simple constant functions.) We introduce an auxiliary function `gFAP` (short for `generalizedFourierApproximationPlot`) for the plot.

In[21]:= `gFAP[{polyData__}, max_, int_, opts___] :=`
         `    Plot[(* the various orders of the Fourier sums *)`
         `         Evaluate[Table[GeneralizedFourierSum[polyData], {n, 0, max}]],`
         `         Evaluate @ Join[{{polyData}[[4]]}, int],`
         `         (* setting options for a nice picture *)`
         `         PlotRange -> All, PlotLabel -> StyleForm[{polyData}[[1]],`
         `             FontFamily -> "Courier", FontWeight -> "Bold", FontSize -> 7],`
         `         Prolog -> {Thickness[0.01], GrayLevel[1/2],`
         `                    Line[{{0, 0}, {0, 1}, {1, 1}, {1, 0}}]},`

```
          PlotStyle -> Table[{Thickness[0.002], Hue[(i - 1)/max 0.7]},
                      {i, 0, max}], DisplayFunction -> Identity, opts]
```

Next, we collect the approximations for the various polynomials produced by `gFAP` in a `GraphicsArray`. Because the different orthogonal polynomials have different domains, we do not always use the same domain for the arguments. The gray curve in the background is the function being approximated, namely $\theta(z)\,\theta(1-z)$.

```
In[22]:= Off[Integrate::gener];
        Show[GraphicsArray[#]]& /@
         {{gFAP[{LaguerreL,    n, 1, z, {0, 1}, 1}, 10, {0, 4}],
           gFAP[{GegenbauerC,  n, 1, z, {0, 1}, 1}, 10, {-1, 1}]},
          {gFAP[{JacobiP,      n, 1, z, {0, 1}, 2, 2}, 10, {-1, 1}],
           gFAP[{HermiteH,     n, 1, z, {0, 1}}, 10, {-2, 3}]},
          {gFAP[{ChebyshevT,   n, 1, z, {0, 1}}, 10, {-0.99, 0.99}],
           gFAP[{ChebyshevU,   n, 1, z, {0, 1}}, 10, {-0.99, 0.99}]},
          {gFAP[{LegendreP,    n, 1, z, {0, 1}}, 10, {-1, 1}]}}
```

Now, we look at the Parseval identity. We again define an auxiliary function.

```
In[24]:= (* format numbers uniformly *)
        tokDigits[t_Real] := StringJoin[Flatten[{"0.",
                ToString /@ Take[RealDigits[t][[1]], 3]}]]
```

```
In[26]:= ParsevalSum[{polyData__}, max_] :=
            FoldList[Plus, 0, (* add up *)
                N[Table[ (* the single Fourier coefficients *)
                    GeneralizedFourierCoefficient[
                        polyData]^2, {n, 0, max}]]] //
                    Rest (* drop first 0 *)// (tokDigits /@ #)&
```

Then, we use it to create a numerical table of the sum of the absolute value squared of the Fourier coefficients.

```
In[27]:= parsevalData =
           {ParsevalSum[{LaguerreL,    n, 1, z, {0, 1}, 1}, 10],
            ParsevalSum[{GegenbauerC,  n, 1, z, {0, 1}, 1}, 10],
            ParsevalSum[{JacobiP,      n, 1, z, {0, 1}, 2, 2}, 10],
            ParsevalSum[{HermiteH,     n, 1, z, {0, 1}}, 10],
            ParsevalSum[{ChebyshevT,   n, 1, z, {0, 1}}, 10],
            ParsevalSum[{ChebyshevU,   n, 1, z, {0, 1}}, 10],
            ParsevalSum[{LegendreP,    n, 1, z, {0, 1}}, 10]} // Transpose;
```

```
In[28]:= TableForm[parsevalData, TableSpacing -> {1, 2},
           (* make table headings *)
           TableHeadings -> {Range[0, 10],  StyleForm[TraditionalForm[#],
                                                FontFamily -> "Times",
                                                FontWeight -> "Bold"]& /@
           {LaguerreL[n, z], GegenbauerC[n, α, z], LegendreP[n, z],
            HermiteH[n, z], ChebyshevT[n, z], ChebyshevU[n, z],
            JacobiP[n, α, β, z]}}]
```

	$L_n(z)$	$C_n^{(\alpha)}(z)$	$P_n(z)$	$H_n(z)$	$T_n(z)$	$U_n(z)$	$P_n^{(\alpha,\beta)}(z)$
0	0.248	0.486	0.416	0.413	0.456	0.486	0.500
1	0.503	0.893	0.826	0.587	0.739	0.893	0.875
2	0.689	0.903	0.876	0.623	0.776	0.903	0.875
3	0.801	0.923	0.876	0.750	0.920	0.923	0.929
4	0.857	0.925	0.892	0.750	0.924	0.925	0.929
5	0.880	0.954	0.937	0.821	0.938	0.954	0.951
6	0.886	0.955	0.944	0.824	0.939	0.955	0.951
7	0.887	0.961	0.944	0.859	0.960	0.961	0.962
8	0.887	0.961	0.948	0.865	0.960	0.961	0.962
9	0.888	0.970	0.963	0.880	0.965	0.970	0.969
10	0.891	0.971	0.965	0.887	0.965	0.971	0.969

Out[28]//TableForm=

3. Transmission through Layers, Sums of Zeros

a) Here is the definition of the function.

```
In[1]:= transmission[n_Integer?(# >= 0&), k_Symbol | k_?(Im[#] == 0&)] =
                1/(1 + (ChebyshevU[n, Cos[k] + Sin[k]/k]/k)^2)
```

$$Out[1]= \cfrac{1}{1 + \cfrac{ChebyshevU\left[n, Cos[k] + \frac{Sin[k]}{k}\right]^2}{k^2}}$$

Here is a plot of transmission.

```
In[2]:= Show[GraphicsArray[#]]& /@ Table[
           Plot[Evaluate[transmission[n + i, k]], {k, 10^- 10, 20},
               AxesOrigin -> {0, 0}, MaxBend -> 1, Ticks -> None,
               PlotRange -> All, DisplayFunction -> Identity,
               PlotPoints -> 20, Epilog -> {
                   {GrayLevel[0.9], Rectangle[{7, 0.45}, {13, 0.6}]},
                   Text["n = " <> ToString[n + i], {10, 1/2}]}],
                   {n, 0, 8, 4}, {i, 0, 3}]
```

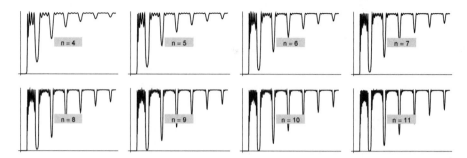

Treating the first argument of transmission as a continuous variable shows an interesting pattern in the n,k-plane [43], [455]. The emergence of "bands" is clearly visible for larger k [188].

```
In[3]:= transmission[v_, k_] = 1/(1 + (ChebyshevU[n, Cos[k] + Sin[k]/k]/k)^2);

In[4]:= With[{ε = 10^-8},
       ContourPlot[Evaluate[Abs[transmission[n, k]]], {k, ε, 20}, {n, ε, 12},
          PlotRange -> {0.8, 1.2}, PlotPoints -> 600,
          ColorFunction -> (Hue[5 #]&), ContourLines -> False,
          Contours -> 50, FrameTicks -> None, AspectRatio -> 0.6]]
```

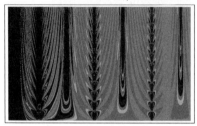

We call attention to the series of increasingly deeper minima and to the small oscillations on the wide plateaus. This function represents the transmission properties of layered structures (n is the number of layers). For some special applications, see [457], [193], [485], [239], [121], [454], [542], [186], [152], [587], [305], [334], [627], [505], [44], [238], [426], [228], [491], [623], [564], [227], [461], [548], [388], [531], [85], [558], [415], [35], [492], [498], [493], [602], [434], [139], [557], [132], [332], [504], [408], and [589].

b) We start by defining the function $\sigma[n, o]$ to calculate exact values for given integers n and o.

```
In[1]:= σ[n_, o_] := RootSum[Function[z, Evaluate[ChebyshevT[n, z]]], #^o&]
```

Because of the symmetry of $T_n(z)$, the $\sigma_o^{(n)}$ vanish for odd o.

```
In[2]:= Table[σ[n, 7], {o, 5}, {n, 1, 6}] // Flatten // Union

Out[2]= {0}
```

For even o, a plot suggests $\sigma_o^{(n)} \sim n$.

```
In[3]:= Show[Graphics[{PointSize[0.01],
            Table[{Hue[o/16], Table[Point[{n, σ[n, o]}], {n, 20}]}, {o, 2, 14, 2}]}],
         Frame -> True]
```

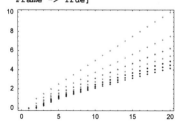

The following data confirm this conjecture.

```
In[4]:= Table[σ[n, 6]/n, {n, 12}]
```

$$\text{Out[4]= } \left\{0, \frac{1}{8}, \frac{9}{32}, \frac{5}{16}, \frac{5}{16}, \frac{5}{16}, \frac{5}{16}, \frac{5}{16}, \frac{5}{16}, \frac{5}{16}, \frac{5}{16}, \frac{5}{16}\right\}$$

```
In[5]:= Table[σ[n, 8]/n, {n, 12}]
```

$$\text{Out[5]= } \left\{0, \frac{1}{16}, \frac{27}{128}, \frac{17}{64}, \frac{35}{128}, \frac{35}{128}, \frac{35}{128}, \frac{35}{128}, \frac{35}{128}, \frac{35}{128}, \frac{35}{128}, \frac{35}{128}\right\}$$

Using the representation $T_n(z) = \cos(n \arccos(z))$, we can calculate $\sigma_o^{(n)}$ in closed form for a given o.

```
In[6]:= Table[{o, (* find large n value *)
       Simplify[#, And @@ Table[Not[Element[k/n, Integers]], {k, o/2}]]& @
       Sum[Cos[(Pi + k 2 Pi)/(2 n)]^o, {k, 0, n - 1}]]}, {o, 10}]
```

$$\text{Out[6]= } \left\{\{1, 0\}, \left\{2, \frac{n}{2}\right\}, \{3, 0\}, \left\{4, \frac{3n}{8}\right\}, \{5, 0\},\right.$$
$$\left.\left\{6, \frac{5n}{16}\right\}, \{7, 0\}, \left\{8, \frac{35n}{128}\right\}, \{9, 0\}, \left\{10, \frac{63n}{256}\right\}\right\}$$

Now, we need to find the $c(o)$ in $\sigma_o^{(n)} = c(o)\,n$. There are various heuristic ways to find a pattern in a sequence. Investigating the (sometimes simpler) differences or ratios of successive pairs is often a first step. Forming the ratio of successive pairs of the $\sigma_o^{(n)}$ shows already a simple pattern.

```
In[7]:= Divide @@@ Partition[Table[σ[o + 1, o]/(o + 1), {o, 2, 30, 2}], 2, 1]
```

$$\text{Out[7]= } \left\{\frac{4}{3}, \frac{6}{5}, \frac{8}{7}, \frac{10}{9}, \frac{12}{11}, \frac{14}{13}, \frac{16}{15}, \frac{18}{17}, \frac{20}{19}, \frac{22}{21}, \frac{24}{23}, \frac{26}{25}, \frac{28}{27}, \frac{30}{29}\right\}$$

Multiplying the ratios yields finally $\sigma_o^{(n)} = \prod_{j=1}^{o/2} (2j-1)/(2j) = 2^{-o/2}\,(o-1)!!/((o/2)!)\,n$ for $n > o/2$ [118].

```
In[8]:= Σ[n_, o_?EvenQ] := (o - 1)!!/(2^(o/2) (o/2)!) n
```

Here is a quick check for some larger values of n and o.

```
In[9]:= {{σ[14, 24], Σ[14, 24]}, {σ[56, 30], Σ[56, 30]}}
```

$$\text{Out[9]= } \left\{\left\{\frac{4732273}{2097152}, \frac{4732273}{2097152}\right\}, \left\{\frac{67863915}{8388608}, \frac{67863915}{8388608}\right\}\right\}$$

4. General Orthogonal Polynomials

Here is an almost direct implementation. Again, many possible refinements exist. For efficiency, we begin by finding the first n polynomials all at once, and save the results of the integration. We use these integrals later to find the normalizations of the polynomials using myInt. In order to better compare our results with known results, we take the square roots of all numerators.

```
In[1]:= GeneralOrthogonalPolynomials[n_, weight_, interval_, var_] :=
       Module[{oldesss, oldgener, vb, gram, spec, c, cc, myInt, poly, w, tab},
         (* state of messages *)
         oldesss = MessageName[Series, "esss"];
         oldgener = MessageName[Integrate, "gener"];
         Off[Series::esss]; Off[Integrate::gener];
         (* this gives the moments by integration; we reuse them *)
         cc[ipj_] := Integrate[weight var^ipj, Join[{var}, interval],
                        GenerateConditions -> False];
         c[ipj_] := c[ipj] = cc[ipj];
         (* the Gram determinant for the moments *)
         gram[-1] = 1;
         gram[m_] := gram[m] = Det[Array[c[#1 + #2]&, {m + 1, m + 1}, 0]];
         spec[m_] := Transpose[Append[
         Transpose[Array[c[#1 + #2]&, {m + 1, m}, 0]], var^Range[0, m]]] // Det;
         (* introducing local pattern-based integration of const var^i weight *)
         myInt[a_] := myInt /@ a /; (Length[a] > 1 && Head[a] === Plus);
         myInt[a_?(FreeQ[#, var]&) var^i_.] := a c[i];
         myInt[a_?(FreeQ[#, var]&)] = a c[0];
         myInt[var^i_.] := c[i];
         (* calculation of the polys *)
         tab = Table[poly = spec[i]/Sqrt[gram[i] gram[i - 1]];
           (* the polynomials *)po = Together @
```

```
Collect[poly/Sqrt[myInt[Expand[poly^2]]], var, Simplify];
w = Times @@ Union[Cases[Numerator[poly], Sqrt[_], {0, Infinity}]];
(* cosmetics to get a nicer result *)
Factor[Together[Collect[
Cancel[(Expand[w Numerator[poly]])/(w Denominator[poly])],
      var, Simplify]]], {i, 0, n}];
(* restore old message state *)
If[Head[oldesss] === String, On[Series::esss]];
If[Head[oldgener] === String, On[Integrate::gener]];
(* return result *) tab]
```

With $\exp(-x^2)$ as the weight function on the interval $(-\infty, \infty)$, we get the normalized Hermite polynomials $H_n(z)$.

In[2]:= `GeneralOrthogonalPolynomials[4, Exp[-x^2], {-Infinity, Infinity}, x]`

Out[2]= $\left\{ \dfrac{1}{\pi^{1/4}}, \ \dfrac{\sqrt{2}\,x}{\pi^{1/4}}, \ \dfrac{-1 + 2\,x^2}{\sqrt{2}\,\pi^{1/4}}, \ \dfrac{x\,(-3 + 2\,x^2)}{\sqrt{3}\,\pi^{1/4}}, \ \dfrac{3 - 12\,x^2 + 4\,x^4}{2\,\sqrt{6}\,\pi^{1/4}} \right\}$

In[3]:= `Cancel /@ Table[1/Sqrt[Sqrt[Pi] 2^i i!] HermiteH[i, x], {i, 0, 4}]`

Out[3]= $\left\{ \dfrac{1}{\pi^{1/4}}, \ \dfrac{\sqrt{2}\,x}{\pi^{1/4}}, \ \dfrac{-1 + 2\,x^2}{\sqrt{2}\,\pi^{1/4}}, \ \dfrac{-3\,x + 2\,x^3}{\sqrt{3}\,\pi^{1/4}}, \ \dfrac{3 - 12\,x^2 + 4\,x^4}{2\,\sqrt{6}\,\pi^{1/4}} \right\}$

For the weight function $x\,e^{-x}$ on the interval $(0, \infty)$, we get the Laguerre polynomials $L_n(1)$ (without a further restriction, these polynomials are only defined up to a factor ± 1).

In[4]:= `GeneralOrthogonalPolynomials[4, x Exp[-x], {0, Infinity}, x]`

Out[4]= $\left\{ 1, \ \dfrac{-2 + x}{\sqrt{2}}, \ \dfrac{6 - 6\,x + x^2}{2\,\sqrt{3}}, \ \dfrac{1}{12}\,(-24 + 36\,x - 12\,x^2 + x^3), \ \dfrac{120 - 240\,x + 120\,x^2 - 20\,x^3 + x^4}{24\,\sqrt{5}} \right\}$

In[5]:= `Table[1/Sqrt[Gamma[i + 1 + 1]/i!] LaguerreL[i, 1, x], {i, 0, 4}]`

Out[5]= $\left\{ 1, \ \dfrac{2 - x}{\sqrt{2}}, \ \dfrac{6 - 6\,x + x^2}{2\,\sqrt{3}}, \ \dfrac{1}{12}\,(24 - 36\,x + 12\,x^2 - x^3), \ \dfrac{120 - 240\,x + 120\,x^2 - 20\,x^3 + x^4}{24\,\sqrt{5}} \right\}$

With `GeneralOrthogonalPolynomials`, we can also construct nonclassical polynomials corresponding to arbitrary weight functions on arbitrary intervals (as long as the integrals that appear can be computed). Here are the first five orthogonal polynomials corresponding to the weight function $\sin(z)$ on the interval $(0, \pi)$.

In[6]:= `GeneralOrthogonalPolynomials[4, Sin[z], {0, Pi}, z]`

Out[6]= $\left\{ \dfrac{1}{\sqrt{2}}, \ -\dfrac{\pi - 2\,z}{\sqrt{2}\,(-8 + \pi^2)}, \ \dfrac{-2 + \pi\,z - z^2}{2\,\sqrt{10 - \pi^2}}, \ \dfrac{(\pi - 2\,z)\,(96 - 10\,\pi^2 - 8\,\pi\,z + \pi^3\,z + 8\,z^2 - \pi^2\,z^2)}{2\,\sqrt{-55296 + 20736\,\pi^2 - 2640\,\pi^4 + 122\,\pi^6 - \pi^8}}, \right.$

$\left. -\dfrac{-432 + 24\,\pi^2 + 2\,\pi^4 + 336\,\pi\,z - 34\,\pi^3\,z - 336\,z^2 + 24\,\pi^2\,z^2 + \pi^4\,z^2 + 20\,\pi\,z^3 - 2\,\pi^3\,z^3 - 10\,z^4 + \pi^2\,z^4}{2\,\sqrt{2}\,(429120 - 133632\,\pi^2 + 14352\,\pi^4 - 578\,\pi^6 + 5\,\pi^8)} \right\}$

In[7]:= `Outer[Integrate[#1 #2 Sin[z], {z, 0, Pi}]&, %, %]`

Out[7]= `{{1, 0, 0, 0, 0}, {0, 1, 0, 0, 0}, {0, 0, 1, 0, 0}, {0, 0, 0, 1, 0}, {0, 0, 0, 0, 1}}`

Here is a plot of these orthogonal polynomials.

In[8]:= `Plot[Evaluate[%], {z, 0, Pi}, PlotRange -> All,`
        `PlotStyle -> Table[{Thickness[0.01], GrayLevel[i/6]}, {i, 0, 4}]]`

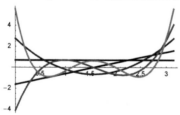

Here is one more example, for the Freud weight [432], [494], [19], [306], [77]. The polynomials orthogonal to the weight function $\exp(-x^4)$ contain many Gamma functions (for brevity, we write their argument as a subscript).

In[9]:= `GeneralOrthogonalPolynomials[5, Exp[-x^4],`
        `{-Infinity, Infinity}, x] /. Gamma[k_] :> Subscript[Γ, 4k]`

Out[9]= $\{\dfrac{1}{\sqrt{2}\,\sqrt{\Gamma_5}},\ \sqrt{2}\,x\,\sqrt{\dfrac{1}{\Gamma_3}},\ -\dfrac{\Gamma_3\,(\Gamma_3 - 4\,x^2\,\Gamma_5)}{\sqrt{2}\,\sqrt{-\Gamma_3^4\,\Gamma_5 + 4\,\Gamma_3^2\,\Gamma_5^3}},$

$\sqrt{2}\,x\,\sqrt{\dfrac{1}{-\Gamma_3\,\Gamma_5^2 + \Gamma_3^2\,\Gamma_7}}\left(-\Gamma_5 + x^2\,\sqrt{\dfrac{\Gamma_3\,(-\Gamma_3\,\Gamma_5^2 + \Gamma_3^2\,\Gamma_7)}{-\Gamma_5^2 + \Gamma_3\,\Gamma_7}}\right),$

$-\dfrac{\sqrt{2}\,(x^4\,\Gamma_3^2 - x^2\,\Gamma_3\,\Gamma_5 + \Gamma_5^2 - 4\,x^4\,\Gamma_5^2 - \Gamma_3\,\Gamma_7 + 4\,x^2\,\Gamma_5\,\Gamma_7)}{\sqrt{\Gamma_3^2\,\Gamma_5^2 - 4\,\Gamma_5^3 - 2\,\Gamma_3^3\,\Gamma_5\,\Gamma_7 + 8\,\Gamma_3\,\Gamma_5^2\,\Gamma_7 + 4\,\Gamma_3^2\,\Gamma_5\,\Gamma_7^2 - 16\,\Gamma_5^3\,\Gamma_7^2 + \Gamma_3^4\,\Gamma_9 - 8\,\Gamma_3^2\,\Gamma_5^2\,\Gamma_9 + 16\,\Gamma_3^4\,\Gamma_9}},$

$\Big(\sqrt{2}\,x\,(x^4\,\Gamma_3^2 - x^4\,\Gamma_3\,\Gamma_7 - x^2\,\Gamma_5\,\Gamma_7 + \Gamma_7^2 + x^2\,\Gamma_3\,\Gamma_9 - \Gamma_5\,\Gamma_9)\,(\Gamma_5^3 - 2\,\Gamma_3\,\Gamma_5\,\Gamma_7 + 4\,\Gamma_5\,\Gamma_7^2 + \Gamma_3^2\,\Gamma_9 - 4\,\Gamma_5^2\,\Gamma_9)\Big)\ /$

$\Big(\sqrt{(}\ \Gamma_5^8\,\Gamma_7^3 - 5\,\Gamma_3\,\Gamma_5^6\,\Gamma_7^4 + 8\,\Gamma_3^2\,\Gamma_5^4\,\Gamma_7^5 + 8\,\Gamma_5^5\,\Gamma_7^5 - 4\,\Gamma_3^3\,\Gamma_5^2\,\Gamma_7^6 - 24\,\Gamma_3\,\Gamma_5^3\,\Gamma_7^6 + 16\,\Gamma_3^2\,\Gamma_5^2\,\Gamma_7^7 + 16\,\Gamma_5^4\,\Gamma_7^7 -$
$16\,\Gamma_3\,\Gamma_5^2\,\Gamma_7^8 - 2\,\Gamma_5^9\,\Gamma_7\,\Gamma_9 + 10\,\Gamma_3\,\Gamma_5^7\,\Gamma_7^2\,\Gamma_9 - 14\,\Gamma_3^2\,\Gamma_5^5\,\Gamma_7^3\,\Gamma_9 - 24\,\Gamma_5^6\,\Gamma_7^3\,\Gamma_9 + 2\,\Gamma_3^3\,\Gamma_5^3\,\Gamma_7^4\,\Gamma_9 +$
$72\,\Gamma_3\,\Gamma_5^4\,\Gamma_7^4\,\Gamma_9 + 4\,\Gamma_3^4\,\Gamma_5\,\Gamma_7^5\,\Gamma_9 - 40\,\Gamma_3^2\,\Gamma_5^2\,\Gamma_7^5\,\Gamma_9 - 64\,\Gamma_5^3\,\Gamma_7^5\,\Gamma_9 - 8\,\Gamma_3^3\,\Gamma_5\,\Gamma_7^6\,\Gamma_9 + 64\,\Gamma_3\,\Gamma_5^2\,\Gamma_7^6\,\Gamma_9 +$
$\Gamma_3^5\,\Gamma_5^2\,\Gamma_7^2 - 9\,\Gamma_3^3\,\Gamma_5^6\,\Gamma_7\,\Gamma_9^2 + 16\,\Gamma_3^4\,\Gamma_5^7\,\Gamma_7^2\,\Gamma_9^2 + 20\,\Gamma_3^5\,\Gamma_5^6\,\Gamma_7\,\Gamma_9^2 - 40\,\Gamma_3\,\Gamma_5^5\,\Gamma_7^2\,\Gamma_9^2 - 11\,\Gamma_3^4\,\Gamma_5^5\,\Gamma_7^3\,\Gamma_9^2 -$
$16\,\Gamma_3^2\,\Gamma_5^4\,\Gamma_7^3\,\Gamma_9^2 + 80\,\Gamma_5^6\,\Gamma_5^3\,\Gamma_7\,\Gamma_9^2 - \Gamma_5^5\,\Gamma_5^4\,\Gamma_7^2\,\Gamma_9 + 40\,\Gamma_3^3\,\Gamma_5^5\,\Gamma_7^4\,\Gamma_9^2 - 64\,\Gamma_3\,\Gamma_5^4\,\Gamma_7^4\,\Gamma_9^2 - 16\,\Gamma_3^3\,\Gamma_5^2\,\Gamma_7^5\,\Gamma_9^2 +$
$2\,\Gamma_3^3\,\Gamma_5^5\,\Gamma_9^3 - 8\,\Gamma_3\,\Gamma_5^7\,\Gamma_7\,\Gamma_9^3 - 8\,\Gamma_3^4\,\Gamma_5^3\,\Gamma_7\,\Gamma_9^3 + 40\,\Gamma_3^2\,\Gamma_5^5\,\Gamma_7\,\Gamma_9^3 - 32\,\Gamma_3^7\,\Gamma_7\,\Gamma_9^3 + 6\,\Gamma_3^3\,\Gamma_5\,\Gamma_7^2\,\Gamma_9^3 - 24\,\Gamma_3^3\,\Gamma_5^3\,\Gamma_7^2\,\Gamma_9^3 -$
$8\,\Gamma_3^4\,\Gamma_5\,\Gamma_7^3\,\Gamma_9^3 + 32\,\Gamma_3^2\,\Gamma_5^3\,\Gamma_7^3\,\Gamma_9^3 + \Gamma_3^5\,\Gamma_5^2\,\Gamma_9^4 - 8\,\Gamma_3^3\,\Gamma_5^4\,\Gamma_9^4 + 16\,\Gamma_3\,\Gamma_5^6\,\Gamma_9^4 - \Gamma_3^5\,\Gamma_7\,\Gamma_9^4 + 8\,\Gamma_3^4\,\Gamma_5^2\,\Gamma_7\,\Gamma_9^4 -$
$16\,\Gamma_3^2\,\Gamma_5^4\,\Gamma_7\,\Gamma_9^4 + \Gamma_5^{10}\,\Gamma_{11} - 6\,\Gamma_3\,\Gamma_5^8\,\Gamma_7\,\Gamma_{11} + 13\,\Gamma_3^2\,\Gamma_5^6\,\Gamma_7^2\,\Gamma_{11} + 8\,\Gamma_5^8\,\Gamma_7^2\,\Gamma_{11} - 12\,\Gamma_3^3\,\Gamma_5^4\,\Gamma_7^3\,\Gamma_{11} -$
$32\,\Gamma_3\,\Gamma_5^6\,\Gamma_7^3\,\Gamma_{11} + 4\,\Gamma_3^4\,\Gamma_5^2\,\Gamma_7^4\,\Gamma_{11} + 40\,\Gamma_3^2\,\Gamma_5^4\,\Gamma_7^4\,\Gamma_{11} + 16\,\Gamma_5^6\,\Gamma_7^4\,\Gamma_{11} - 16\,\Gamma_3^3\,\Gamma_5^2\,\Gamma_7^5\,\Gamma_{11} - 32\,\Gamma_3\,\Gamma_5^4\,\Gamma_7^5\,\Gamma_{11} +$
$16\,\Gamma_3^2\,\Gamma_5^2\,\Gamma_7^6\,\Gamma_{11} + 2\,\Gamma_3^2\,\Gamma_5^7\,\Gamma_9\,\Gamma_{11} - 8\,\Gamma_5^9\,\Gamma_9\,\Gamma_{11} - 8\,\Gamma_3^3\,\Gamma_5^5\,\Gamma_7\,\Gamma_9\,\Gamma_{11} + 32\,\Gamma_3\,\Gamma_5^7\,\Gamma_7\,\Gamma_9\,\Gamma_{11} +$
$10\,\Gamma_3^4\,\Gamma_5^3\,\Gamma_7^2\,\Gamma_9\,\Gamma_{11} - 32\,\Gamma_3^2\,\Gamma_5^5\,\Gamma_7^2\,\Gamma_9\,\Gamma_{11} - 32\,\Gamma_3^7\,\Gamma_7^2\,\Gamma_9\,\Gamma_{11} - 4\,\Gamma_3^5\,\Gamma_5\,\Gamma_7^3\,\Gamma_9\,\Gamma_{11} + 64\,\Gamma_3\,\Gamma_5^5\,\Gamma_7^3\,\Gamma_9\,\Gamma_{11} +$
$8\,\Gamma_3^4\,\Gamma_5\,\Gamma_7^4\,\Gamma_9\,\Gamma_{11} - 32\,\Gamma_3^2\,\Gamma_5^3\,\Gamma_7^4\,\Gamma_9\,\Gamma_{11} + \Gamma_3^4\,\Gamma_5^4\,\Gamma_9^2\,\Gamma_{11} - 8\,\Gamma_3^2\,\Gamma_5^6\,\Gamma_9^2\,\Gamma_{11} + 16\,\Gamma_3^8\,\Gamma_9^2\,\Gamma_{11} - 2\,\Gamma_3^5\,\Gamma_5^2\,\Gamma_7\,\Gamma_9^2\,\Gamma_{11} +$
$16\,\Gamma_3^3\,\Gamma_5^4\,\Gamma_7\,\Gamma_9^2\,\Gamma_{11} - 32\,\Gamma_3\,\Gamma_5^6\,\Gamma_7\,\Gamma_9^2\,\Gamma_{11} + \Gamma_5^6\,\Gamma_7^2\,\Gamma_9^2\,\Gamma_{11} - 8\,\Gamma_3^4\,\Gamma_5^2\,\Gamma_7^2\,\Gamma_9^2\,\Gamma_{11} + 16\,\Gamma_3^2\,\Gamma_5^4\,\Gamma_7^2\,\Gamma_9^2\,\Gamma_{11})\Big)\}$

Next, we use a discontinuous weight function [129], [130]: $w(z) = \theta(x^2 - 1/9)$ in the interval $[-1, 1]$.

In[10]:= `gesθ = GeneralOrthogonalPolynomials[48, UnitStep[x^2 - 1/9], {-1, 1}, x];`
`Take[gesθ, 8]`

Out[11]= $\{\dfrac{\sqrt{3}}{2},\ \dfrac{9\,x}{2\,\sqrt{13}},\ \dfrac{1}{4}\,\sqrt{\dfrac{15}{61}}\,(-13 + 27\,x^2),$

$\dfrac{9}{4}\,\sqrt{\dfrac{21}{155233}}\,x\,(-121 + 195\,x^2),\ \dfrac{9}{16}\,\sqrt{\dfrac{3}{14680321}}\,(11941 - 60570\,x^2 + 57645\,x^4),$

$\dfrac{9}{16}\,\sqrt{\dfrac{33}{8763031729213}}\,x\,(5735105 - 23804130\,x^2 + 20311641\,x^4),$

$\dfrac{1}{32}\,\sqrt{\dfrac{39}{14740649659059661}}\,(-3669303965 + 30055366845\,x^2 - 65209017195\,x^4 + 40527071739\,x^6),$

$\dfrac{9}{32}\,\sqrt{\dfrac{5}{216326905573553286604393 7}}\,x$

$(-28286323822295 + 201363630065955\,x^2 - 395870479819353\,x^4 + 229508358263613\,x^6)\}$

In[12]:= `(* check orthonormality for the first few states *)`
`Table[Integrate[gesθ[[i]] gesθ[[j]] UnitStep[x^2 - 1/9], {x, -1, 1}],`
`     {i, 10}, {j, 10}] == IdentityMatrix[10]`

Out[13]= `True`

Here is a plot of the first few polynomials. The right graphic shows the zeros for the first 49 functions.

In[39]:= `Show[GraphicsArray[`
`     {(* plot of the first polynomials *)`
`       Plot[Evaluate[Take[gesθ, 10]], {x, -1, 1}, PlotRange -> All,`
`           DisplayFunction -> Identity, PlotStyle ->`
`           Table[{Hue[k/13], Thickness[0.002]}, {k, 0, 10}]],`
`       (* plot of the zeros *)`
`       Graphics[{(* positions of weight discontinuities *)`
`                 {GrayLevel[0.7], Thickness[0.01],`
`                  Line[{{-1/3, 0}, {-1/3, 48}}], Line[{{1/3, 0}, {1/3, 48}}]},`
`                 {PointSize[0.004],`
`                  Table[Point[{#, k}]& /@ (Re[x] /. N[Solve[gesθ[[k]] == 0, x]], 22]),`
`                     {k, 2, 49}]}},`
`                 PlotRange -> All, Frame -> True]}]]`

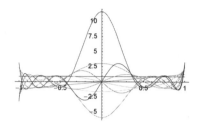 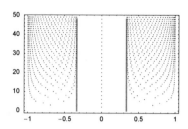

5. Symmetric Polynomials

a) We begin with the power sums. The implementation is immediate.

```
In[1]:= PowerSum[k_, varList_] := Plus @@ (varList^k)
```

We do not have to require that k be an integer. Here is an example.

```
In[2]:= PowerSum[k, {x, y, z}]
```

$$Out[2]= x^k + y^k + z^k$$

Now, we look at discriminants. First, we have to construct all pairs $\{i, j\}$ with $i > j$. We do this with the following construction.

```
In[3]:= n = 6;
        Flatten[MapThread[List, {Range[1, # - 1], Table[#, {# - 1}]}]& /@
                            Range[2, n], 1]
```

$$Out[4]= \{\{1, 2\}, \{1, 3\}, \{2, 3\}, \{1, 4\}, \{2, 4\}, \{3, 4\}, \{1, 5\},$$
$$\{2, 5\}, \{3, 5\}, \{4, 5\}, \{1, 6\}, \{2, 6\}, \{3, 6\}, \{4, 6\}, \{5, 6\}\}$$

Next, we extract the corresponding variables from the list varList using Part, subtract and square them, and convert the outer list to a product.

```
In[5]:= Discriminant[varList_] :=
        Times @@ ((Subtract @@ varList[[#]])^2& /@
        (* all ordered nonincreasing pairs *)
        Flatten[MapThread[List, {Range[1, # - 1], Table[#, {# - 1}]}]& /@
                            Range[2, Length[varList]], 1])
```

Here is an example.

```
In[6]:= Discriminant[{p1, p2, p3, p4, p5, p6}]
```

$$Out[6]= (p1 - p2)^2 (p1 - p3)^2 (p2 - p3)^2 (p1 - p4)^2 (p2 - p4)^2 (p3 - p4)^2 (p1 - p5)^2$$
$$(p2 - p5)^2 (p3 - p5)^2 (p4 - p5)^2 (p1 - p6)^2 (p2 - p6)^2 (p3 - p6)^2 (p4 - p6)^2 (p5 - p6)^2$$

The construction of the elementary symmetric polynomials is somewhat more complicated. Here, we have to compute all strictly ordered k-tuples $\{i_1, \ldots, i_k\}$ of integers less than or equal to n. The following construction does this recursively by computing one integer after another, where each new integer is smaller than the previous one.

```
In[7]:= k = 3; n = 6;
        Fold[Flatten[
        Apply[MapThread[Join,  (* fill out *)
                {List /@ Range[#1, First[#2] - 1],
                 Table[#2, {First[#2] - #1}]}]&, MapThread[List, {Table[#2,
                        {Length[#1]}], #1}], {1}], 1]&,
                (* the last number *) List /@ Range[k, n],
                (* the first numbers *) Range[k - 1, 1, -1]]
```

$$Out[8]= \{\{1, 2, 3\}, \{1, 2, 4\}, \{1, 3, 4\}, \{2, 3, 4\}, \{1, 2, 5\}, \{1, 3, 5\},$$
$$\{2, 3, 5\}, \{1, 4, 5\}, \{2, 4, 5\}, \{3, 4, 5\}, \{1, 2, 6\}, \{1, 3, 6\}, \{2, 3, 6\},$$
$$\{1, 4, 6\}, \{2, 4, 6\}, \{3, 4, 6\}, \{1, 5, 6\}, \{2, 5, 6\}, \{3, 5, 6\}, \{4, 5, 6\}\}$$

Proceeding in an analogous way and using Plus @@ (Times @@ varList[[#]]) & /@ ..., we get the following program.

```
In[9]:= ElementarySymmetricPolynomial[k_Integer?(# >= 1&), varList_List] :=
        Plus @@ ((* take the corresponding variables *)
                 (Times @@ varList[[#]])& /@
```

```
Fold[Flatten[Apply[MapThread[Join,
        {List /@ Range[#1, First[#2] - 1],
        Table[#2, {First[#2] - #1}]}]]&, MapThread[List, {Table[#2,
            {Length[#1]}], #1}], {1}], 1]&,
    List /@ Range[k, Length[varList]], Range[k - 1, 1, -1]]);
```

```
(* degenerate case *)
ElementarySymmetricPolynomial[0, varList_List] := 1
```

We now look at some examples.

```
In[13]:= Table[ElementarySymmetricPolynomial[k, {x1, x2, x3, x4, x5}], {k, 6}]
```

```
Out[13]= {x1 + x2 + x3 + x4 + x5, x1 x2 + x1 x3 + x2 x3 + x1 x4 + x2 x4 + x3 x4 + x1 x5 + x2 x5 + x3 x5 + x4 x5,
    x1 x2 x3 + x1 x2 x4 + x1 x3 x4 + x2 x3 x4 + x1 x2 x5 + x1 x3 x5 + x2 x3 x5 + x1 x4 x5 + x2 x4 x5 +
    x3 x4 x5, x1 x2 x3 x4 + x1 x2 x3 x5 + x1 x2 x4 x5 + x1 x3 x4 x5 + x2 x3 x4 x5, x1 x2 x3 x4 x5, 0}
```

Here is a recursive definition of the elementary symmetric polynomials.

```
In[14]:= ElementarySymmetricPolynomialRec[0, l_List] := 1
```

```
ElementarySymmetricPolynomialRec[_, {}] := 0
```

```
ElementarySymmetricPolynomialRec[k_Integer?(# >= 1&), l_List] :=
    Expand[  ElementarySymmetricPolynomialRec[k, Most[l]] +
    Last[l] ElementarySymmetricPolynomialRec[k - 1, Most[l]]]
```

```
In[17]:= (* check that recursive definition agrees with
        above definition *)
    Equal @@ (Table[#1[k, {x1, x2, x3, x4, x5}], {k, 6}]& /@
                    {ElementarySymmetricPolynomial,
                    ElementarySymmetricPolynomialRec})
```

```
Out[18]= True
```

We turn now to the implementation of the Wronski polynomials. The main difficulty is the construction of all partitions of a given k into integers that add up to k. The following piece of code finds all "bounds" (numbers appearing a maximum number of times when there are i numbers $\neq 0$) of such a partition.

```
In[19]:= k = 13; n = 6;
        (* all possible maximal partitions *)
        (Join[#, Table[0, {n - Length[#]}]]& /@
            ((Join[Table[#1 + 1, {#2}], Table[#1, {(k - (#1 + 1) #2)/#1}]]& @@
                {Quotient[k, #], Mod[k, #]})& /@
        (* number of nonzero elements *) Range[If[n > k, k, n]]]))
```

```
Out[21]= {{13, 0, 0, 0, 0, 0}, {7, 6, 0, 0, 0, 0}, {5, 4, 4, 0, 0, 0},
    {4, 3, 3, 3, 0, 0}, {3, 3, 3, 2, 2, 0}, {3, 2, 2, 2, 2, 2}}
```

For every such "bound", we now compute all corresponding suitable realizations.

```
In[22]:= auxFunction =
        (* auxFunction is a pure function *)
        (* fill out all possible elements within the limits *)
        (Join[{k - Plus @@ #}, #]& /@ Fold[Flatten[
        MapThread[Join, {If[#1 == 0, {{0}},
                        List /@ Range[Max[1, First[#2]], #1]],
                    If[#1 == 0, {#2}, Table[#2,
                        {#1 - Max[1, First[#2]] + 1}]]]}]& @@ #]& /@
            MapThread[List, {Table[#2, {Length[#1]}], #1}], 1]&,
            If[Last[#] == 0, {{0}}, List /@ Range[Last[#]]],
            Drop[Drop[Reverse[#], -1], 1]])&;
```

We now examine the performance of auxFunction using the above boundaries.

```
In[23]:= auxFunction[{13, 0, 0, 0, 0, 0}]
```

```
Out[23]= {{13, 0, 0, 0, 0, 0}}
```

```
In[24]:= auxFunction[{7, 6, 0, 0, 0, 0}]
```

```
Out[24]= {{12, 1, 0, 0, 0, 0}, {11, 2, 0, 0, 0, 0}, {10, 3, 0, 0, 0, 0},
         {9, 4, 0, 0, 0, 0}, {8, 5, 0, 0, 0, 0}, {7, 6, 0, 0, 0, 0}}

In[25]:= auxFunction[{5, 4, 4, 0, 0, 0}]

Out[25]= {{11, 1, 1, 0, 0, 0}, {10, 2, 1, 0, 0, 0}, {9, 3, 1, 0, 0, 0},
         {8, 4, 1, 0, 0, 0}, {9, 2, 2, 0, 0, 0}, {8, 3, 2, 0, 0, 0},
         {7, 4, 2, 0, 0, 0}, {7, 3, 3, 0, 0, 0}, {6, 4, 3, 0, 0, 0}, {5, 4, 4, 0, 0, 0}}

In[26]:= auxFunction[{4, 3, 3, 3, 0, 0}]

Out[26]= {{10, 1, 1, 1, 0, 0}, {9, 2, 1, 1, 0, 0}, {8, 3, 1, 1, 0, 0},
         {8, 2, 2, 1, 0, 0}, {7, 3, 2, 1, 0, 0}, {6, 3, 3, 1, 0, 0},
         {7, 2, 2, 2, 0, 0}, {6, 3, 2, 2, 0, 0}, {5, 3, 3, 2, 0, 0}, {4, 3, 3, 3, 0, 0}}

In[27]:= auxFunction[{3, 3, 3, 2, 2, 0}]

Out[27]= {{9, 1, 1, 1, 1, 0}, {8, 2, 1, 1, 1, 0}, {7, 3, 1, 1, 1, 0}, {7, 2, 2, 1, 1, 0},
         {6, 3, 2, 1, 1, 0}, {5, 3, 3, 1, 1, 0}, {6, 2, 2, 2, 1, 0}, {5, 3, 2, 2, 1, 0},
         {4, 3, 3, 2, 1, 0}, {5, 2, 2, 2, 2, 0}, {4, 3, 2, 2, 2, 0}, {3, 3, 3, 2, 2, 0}}

In[28]:= auxFunction[{3, 2, 2, 2, 2, 2}]

Out[28]= {{8, 1, 1, 1, 1, 1}, {7, 2, 1, 1, 1, 1}, {6, 2, 2, 1, 1, 1},
         {5, 2, 2, 2, 1, 1}, {4, 2, 2, 2, 2, 1}, {3, 2, 2, 2, 2, 2}}
```

We now form all permutations of the computed exponents, and again apply `Plus @@ ((Times @@ (varList^#)) & /@`
… .

Finally, we can sum the individual parts without using any temporary variables.

```
In[29]:= WronskiPolynomial[k_Integer?(# >= 1&), varList_] :=
         Plus @@ ((Times @@ (varList^#))& /@
         Flatten[Permutations /@ Flatten[
         (Join[{k - Plus @@ #}, #]& /@ Fold[Flatten[
         MapThread[Join, {If[#1 == 0, {{0}},
                     List /@ Range[Max[1, First[#2]], #1]],
                 If[#1 == 0, {#2}, Table[#2,
             {#1 - Max[1, First[#2]] + 1}]]]& @@ #]& /@
           MapThread[List, {Table[#2, {Length[#1]}], #1}], 1]&,
             If[Last[#] == 0, {{0}}, List /@ Range[Last[#]]],
             Drop[Drop[Reverse[#], -1], 1]]& /@
         (Join[#, Table[0, {Length[varList] - Length[#]}]]& /@
         ((Join[Table[#1 + 1, {#2}], Table[#1, {(k - (#1 + 1) #2)/#1}]]& @@
           {Quotient[k, #], Mod[k, #]})& /@
         Range[If[Length[varList] > k, k, Length[varList]]])), 1], 1]);

         (* degenerate case *)
         WronskiPolynomial[1, {a_}] := a
         WronskiPolynomial[0, varList_List] := 1
```

We can now give some examples.

```
In[34]:= WronskiPolynomial[1, {x1, x2, x3, x4, x5}]

Out[34]= x1 + x2 + x3 + x4 + x5

In[35]:= WronskiPolynomial[2, {x1, x2, x3, x4, x5}]

Out[35]= x1² + x1 x2 + x2² + x1 x3 + x2 x3 + x3² + x1 x4 + x2 x4 + x3 x4 + x4² + x1 x5 + x2 x5 + x3 x5 + x4 x5 + x5²
```

$$Out[35]= x1^2 + x1\,x2 + x2^2 + x1\,x3 + x2\,x3 + x3^2 + x1\,x4 + x2\,x4 + x3\,x4 + x4^2 + x1\,x5 + x2\,x5 + x3\,x5 + x4\,x5 + x5^2$$

```
In[36]:= WronskiPolynomial[3, {x1, x2, x3, x4, x5}]
```

$$Out[36]= x1^3 + x1^2\,x2 + x1\,x2^2 + x2^3 + x1^2\,x3 + x1\,x2\,x3 + x2^2\,x3 + x1\,x3^2 +$$
$$x2\,x3^2 + x3^3 + x1^2\,x4 + x1\,x2\,x4 + x2^2\,x4 + x1\,x3\,x4 + x2\,x3\,x4 + x3^2\,x4 + x1\,x4^2 +$$
$$x2\,x4^2 + x3\,x4^2 + x4^3 + x1^2\,x5 + x1\,x2\,x5 + x2^2\,x5 + x1\,x3\,x5 + x2\,x3\,x5 + x3^2\,x5 +$$
$$x1\,x4\,x5 + x2\,x4\,x5 + x3\,x4\,x5 + x4^2\,x5 + x1\,x5^2 + x2\,x5^2 + x3\,x5^2 + x4\,x5^2 + x5^3$$

```
In[37]:= WronskiPolynomial[4, {x1, x2, x3, x4, x5}] // Short[#, 4]&
```

Out[37]//Short= $x1^4 + x1^3 x2 + x1^2 x2^2 + x1 x2^3 + x2^4 + x1^3 x3 + x1^2 x2 x3 + x1 x2^2 x3 + x2^3 x3 + x1^2 x3^2 +$
$x1 x2 x3^2 + x2^2 x3^2 + x1 x3^3 + x2 x3^3 + x3^4 + x1^3 x4 + x1^2 x2 x4 + x1 x2^2 x4 + x2^3 x4 + \ll33\gg +$
$x2 x4^2 x5 + x3 x4^2 x5 + x4^3 x5 + x1^2 x5^2 + x1 x2 x5^2 + x2^2 x5^2 + x1 x3 x5^2 + x2 x3 x5^2 +$
$x3^2 x5^2 + x1 x4 x5^2 + x2 x4 x5^2 + x3 x4 x5^2 + x4^2 x5^2 + x1 x5^3 + x2 x5^3 + x3 x5^3 + x4 x5^3 + x5^4$

In[38]:= `WronskiPolynomial[5, {x1, x2, x3, x4, x5}] // Short[#, 4]&`

Out[38]//Short= $x1^5 + x1^4 x2 + x1^3 x2^2 + x1^2 x2^3 + x1 x2^4 + x2^5 + x1^4 x3 + x1^3 x2 x3 + x1^2 x2^2 x3 + x1 x2^3 x3 +$
$x2^4 x3 + x1^3 x3^2 + x1^2 x2 x3^2 + x1 x2^2 x3^2 + x2^3 x3^2 + x1^2 x3^3 + x1 x2 x3^3 + x2^2 x3^3 + \ll90\gg +$
$x2 x4^2 x5^2 + x3 x4^2 x5^2 + x4^3 x5^2 + x1^2 x5^3 + x1 x2 x5^3 + x2^2 x5^3 + x1 x3 x5^3 + x2 x3 x5^3 +$
$x3^2 x5^3 + x1 x4 x5^3 + x2 x4 x5^3 + x3 x4 x5^3 + x4^2 x5^3 + x1 x5^4 + x2 x5^4 + x3 x5^4 + x4 x5^4 + x5^5$

Here is an example for the case $k > n$.

In[39]:= `WronskiPolynomial[12, {x1, x2}]`

Out[39]= $x1^{12} + x1^{11} x2 + x1^{10} x2^2 + x1^9 x2^3 + x1^8 x2^4 + x1^7 x2^5 +$
$x1^6 x2^6 + x1^5 x2^7 + x1^4 x2^8 + x1^3 x2^9 + x1^2 x2^{10} + x1 x2^{11} + x2^{12}$

Here is a shorter but also much slower implementation that generates all possible summation indices tuples.

In[40]:= `WronskiPolynomialNaive[k_Integer?(# >= 1&), varList_] :=`
`    Plus @@ (Times @@@ (varList^#& /@ (* summation indices *)`
`        Select[Tuples @ Array[Range[0, k]&, Length[varList]], Total[#] == k&]))`

In[41]:= `(* timing comparison *)`
`    Timing[#1 = Table[#2[k, {x1, x2, x3, x4, x5}], {k, 8}];]& @@@`
`        {{wPs, WronskiPolynomial}, {wPNs, WronskiPolynomialNaive}}`

Out[42]= `{{0.03 Second, Null}, {0.79 Second, Null}}`

In[43]:= `(* check if results agree *)`
`    Table[WronskiPolynomialNaive[k, {x1, x2, x3, x4, x5}] ===`
`        WronskiPolynomial[k, {x1, x2, x3, x4, x5}], {k, 6}]`

Out[44]= `{True, True, True, True, True, True}`

We mention the possibility to calculate the Wronski polynomials through series expansions of a simple function of the polynomial whose zeros are *varList* [529].

In[45]:= `WronskiPolynomialSeriesBased[k_, varList_] :=`
`    Expand[SeriesCoefficient[`
`        Series[C^-Length[varList]/(Times @@ (1/C - varList)),`
`            {C, 0, k + 1}], k]]`

The Wronski polynomials calculated by `WronskiPolynomialsSeriesBased` agree with the above ones. Here is an example.

In[46]:= `WronskiPolynomialSeriesBased[4, {x1, x2, x3, x4, x5}] ===`
`    WronskiPolynomial[4, {x1, x2, x3, x4, x5}]`

Out[46]= `True`

Here is a similar series-based method to calculate the power sums.

In[47]:= `PowerSumSeriesBased[k_, varList_] :=`
`    Expand[SeriesCoefficient[`
`        -Series[C D[#, C]/#&[Times @@ (1/C - varList)],`
`            {C, 0, k + 1}], k]]`

In[48]:= `PowerSumSeriesBased[12, {x1, x2, x3, x4, x5}]`

Out[48]= $x1^{12} + x2^{12} + x3^{12} + x4^{12} + x5^{12}$

The Jacobian of the elementary symmetric polynomials and the Wronski polynomials with respect to the polynomial variables is the square root of the corresponding discriminant [347]. Here is a quick check for this result.

In[49]:= `Module[{ξ, ξs},`
`        Table[ξs = Array[ξ, d];`
`            (* divide Jacobian squared by discriminant *)`
`            Cancel[Det[Outer[D, Table[#[k, ξs], {k, d}], ξs]]^2/`
`                Discriminant[ξs]]& /@`
`            (* elementary symmetric and Wronski polynomials *)`

```
                         {ElementarySymmetricPolynomial, WronskiPolynomial},
                                   {d, 2, 6}]]
      Out[49]=  {{1, 1}, {1, 1}, {1, 1}, {1, 1}, {1, 1}}
```

We could continue and investigate symmetric (and antisymmetric) expressions in symmetric polynomials. Here is one example, the determinant of a $d \times d$ matrix with elements $p_j(x_1, x_2, \ldots, \check{x}_i, \ldots, x_d)$, $i = 1, \ldots, d$, $j = 0, \ldots, d-1$. In this expression, p is a symmetric polynomial and $\check{x}_i$ indicates the absence of x_i. For the elementary symmetric polynomials and the Wronski polynomials, we obtain discriminant-like polynomials [185], for the Wronski polynomials, most such determinants vanish.

```
      In[50]:= symmetricDet[d_, a_, poly_] := Expand @
               Det[Table[poly[j, Delete[Array[a, {d}], i]], {j, 0, d - 1}, {i, d}]]

      In[51]:= Table[Expand[symmetricDet[d, a, ElementarySymmetricPolynomial] -
                          Product[a[j] - a[i], {j, d - 1}, {i, j + 1, d}]],
                     {d, 1, 6}]
      Out[51]=  {0, 0, 0, 0, 0, 0}

      In[52]:= Table[symmetricDet[d, a, WronskiPolynomial], {d, 1, 8}]

      Out[52]=  {1, a[1] - a[2], 0, 0, 0, 0, 0, 0}

      In[53]:= Table[Expand[symmetricDet[d, a, PowerSum] + (-1)^Ceiling[d/2]*
                          (d - 1) Product[a[j] - a[i], {j, d - 1}, {i, j + 1, d}]],
                     {d, 1, 8}]
      Out[53]=  {0, 0, 0, 0, 0, 0, 0, 0}
```

b) We do not repeat definitions for the power sums and the elementary symmetric polynomials. (We use the definition of the elementary symmetric polynomials from the package `Algebra`SymmetricPolynomials`.)

```
      In[1]:= PowerSum[k_, varList_] := Plus @@ (varList^k)

      In[2]:= Needs["Algebra`SymmetricPolynomials`"]

      In[3]:= ElementarySymmetricPolynomials[k_, varList_] :=
                          SymmetricPolynomial[varList, k]
```

The Newton relations can be immediately implemented.

```
      In[4]:= NewtonRelation[k_Integer?(# >= 1&), varList_] := PowerSum[k, varList] +
               Sum[(-1)^i ElementarySymmetricPolynomials[i, varList]*
                  PowerSum[k - i, varList], {i, k - 1}]  +
               (-1)^k k ElementarySymmetricPolynomials[k, varList] /;
                                              k <= Length[varList];

              NewtonRelation[k_Integer, varList_] := PowerSum[k, varList] +
               Sum[(-1)^i ElementarySymmetricPolynomials[i, varList]*
                  PowerSum[k - i, varList], {i, Length[varList]}] /;
                                              k > Length[varList]
```

Next, we examine the test cases; they all satisfy the Newton relations.

```
      In[6]:= Table[NewtonRelation[k, {x, y, z}] // Expand, {k, 5}]
      Out[6]=  {0, 0, 0, 0, 0}
```

For symbolic methods to prove polynomial identities in n (n unspecified) variables, see [310].

c) We repeat definitions for the power sums and the elementary symmetric polynomials. (We again use the definition of the elementary symmetric polynomials from the package `Algebra`SymmetricPolynomials`.)

```
      In[1]:= PowerSum[k_, varList_] := Plus @@ (varList^k)

      In[2]:= Needs["Algebra`SymmetricPolynomials`"]

      In[3]:= ElementarySymmetricPolynomials[k_, varList_] :=
                          SymmetricPolynomial[varList, k]
```

First, we calculate the relevant k in the sum. To do this in an effective way, we modify the code from Exercise 20 in Chapter 6 of the Programming volume [566].

```
In[4]:= (* special case of only one variable *)
       allλs[sum_?(TrueQ[# > 0]&), 1] = {{sum}};

       allλs[sum_?(TrueQ[# > 0]&), numberOfVars_Integer?(TrueQ[# > 0]&)] :=
       Function[summands, (* make the difference fit *)
       Flatten[{sum - summands.#, #}]& /@ (* the maximal number that fits *)
       Fold[Function[{was, is},
         Flatten[Function[old, Flatten[{old, #}]& /@
             Range[0, (sum - Drop[is, -1].old)/Last[is]]] /@ was, 1]],
         Array[{#}&, Floor[sum/First[summands]] + 1, 0],
         Drop[Flatten /@ FoldList[List, {}, summands], 2]]][Range[2, numberOfVars]]
```

Here is an example.

```
In[7]:= allλs[6, 3]

Out[7]= {{6, 0, 0}, {3, 0, 1}, {0, 0, 2}, {4, 1, 0}, {1, 1, 1}, {2, 2, 0}, {0, 3, 0}}
```

This is the proof.

```
In[8]:= {1, 2, 3}.#& /@ %

Out[8]= {6, 6, 6, 6, 6, 6, 6}
```

Now, it is straightforward to implement the Waring formula.

```
In[9]:= WaringFormula[k_Integer?(# >= 1&), varList_] :=
       Expand[(-1)^k Plus @@ Function[esp,
         ((-1)^(Plus @@ #) ((Plus @@ #) - 1)! k/(Times @@ (#!))*
         (Times @@ (esp^#)))& /@ (* the weights *)
                       allλs[k, Length[varList]]][
         (* the elementary symmetrical polynomials *)
         Array[ElementarySymmetricPolynomials[#, varList]&,
                        {Length[varList]}]]] == PowerSum[k, varList]
```

We test them for some examples.

```
In[10]:= Outer[WaringFormula, {2, 3, 4},
                       {{x, y}, {x, y, z}, {w, x, y, z}}, 1]
Out[10]= {{True, True, True}, {True, True, True}, {True, True, True}}
```

d) We use again the definition of the elementary symmetric polynomials from the package Algebra`SymmetricPolyno
mials`.

```
In[1]:= Needs["Algebra`SymmetricPolynomials`"]

In[2]:= Off[SymmetricPolynomial::hideg];

In[3]:= ElementarySymmetricPolynomials[k_, varList_] :=
                   SymmetricPolynomial[varList, k] /. _SymmetricPolynomial -> 0
```

This is the identity under consideration.

```
In[4]:= zero[vars_, k_, {α_, β_, γ_}] :=
       With[{n = Length[vars]},
       Sum[ElementarySymmetricPolynomials[k, Delete[vars, j]], {j, n}] -
       (α n + β k + γ) ElementarySymmetricPolynomials[k, vars] ]
```

We use various k and n to get a system of equations that allows us to uniquely determine the integers α, β, and γ.

```
In[5]:= eqs = Table[vars = Table[ξ[k], {k, n}];
                       ζερω = zero[vars, k, {α, β, γ}];
                       Last /@ Internal`DistributedTermsList[ζερω, vars][[1]],
                       {n, 2}, {k, 2}];

In[6]:= Solve[# == 0& /@ Flatten[eqs], {α, β, γ}]

Out[6]= {{α → 1, β → -1, γ → 0}}
```

Here is a quick check that the identity also holds for larger values of k and n.

```
In[7]:= Table[zero[Table[ξ[k], {k, n}], k, {1, -1, 0}] // Expand,
             {n, 10}, {k, 10}] // Flatten // Union
```

```
Out[7]= {0}
```

e) We start by defining the normalized elementary symmetric polynomials.

```
In[1]:= Needs["Algebra`SymmetricPolynomials`"]
```

```
In[2]:= NormalizedElementarySymmetricPolynomials[k_, varList_] :=
          SymmetricPolynomial[varList, k]/Binomial[Length[varList], k]
```

The function `NewtonInequality` generates the inequality for a given *k*.

```
In[3]:= NewtonInequality[k_, varList_] :=
          With[{f = NormalizedElementarySymmetricPolynomials[#, varList]&},
               f[k - 1] f[k + 1] <= f[k]^2]
```

Using the function `Resolve`, it is straightforward to verify the ten inequalities.

```
In[4]:= Table[With[{xs = Table[x[k], {k, n}]},
          Table[Resolve[ForAll[xs, Element[xs, Reals],
                                NewtonInequality[k, xs]]], {k, n - 1}]],
          {n, 2, 4}]
Out[4]= {{True}, {True, True}, {True, True, True}}
```

f) Here is an implementation in a functional style. We first determine the degree of the polynomial and using this *degree* (calculated only one time), we construct a pure function `Coefficient[poly, x, degree - #1]&`. `Fold` gets as its first argument a function implementing the definition of the s_j. The two arguments of this function are the already-calculated s_k results and the a_{k-j} (which are calculated once in the beginning) appearing in the third argument of `Fold`. Finally, we get the list of all power sums of order less than or equal to *n*.

```
In[1]:= PowerSums[poly_, x_, m_Integer] :=
          Fold[#1 ~ Join ~ {-#1.Rest[#2] - First[#2] Length[#2]}&, {-First[#]},
               Rest[Map[Reverse, MapThread[Take,
               {Table[#, {m}], Range[m]}]]]]&[Map[#, Range[m]]/#[0]&[
                 Apply[Coefficient, Function @@ {{poly, x, #1 - #2}&[
                                    Exponent[poly, x], #1]}, {1}]]]
```

Let us look at an example. We will use the following polynomial.

```
In[2]:= poly = 3 x^7 - 5 x^6 + 3 x^5 - 6 x^4 + 3 x^3 - 7 x^2 - 4 x + 9
Out[2]= 9 - 4 x - 7 x^2 + 3 x^3 - 6 x^4 + 3 x^5 - 5 x^6 + 3 x^7
```

After rationalizing, the numerical calculation gives the following result.

```
In[3]:= MapIndexed[(Plus @@ (#1^#2[[1]]))&,
          Table[Evaluate[x /. NSolve[poly == 0, x]], {8}]] // Rationalize
```

$$Out[3]= \left\{ \frac{5}{3}, \frac{7}{9}, \frac{152}{27}, \frac{643}{81}, \frac{4655}{243}, \frac{33796}{729}, \frac{112397}{2187}, \frac{517291}{6561} \right\}$$

The same result is obtained with `PowerSums`.

```
In[4]:= PowerSums[poly, x, 8]
```

$$Out[4]= \left\{ \frac{5}{3}, \frac{7}{9}, \frac{152}{27}, \frac{643}{81}, \frac{4655}{243}, \frac{33796}{729}, \frac{112397}{2187}, \frac{517291}{6561} \right\}$$

6. Generalized Lissajou Figures, Hyperspherical Harmonics, Hydrogen Orbitals

a) We first define the normalization, the weight functions, and the domains for the orthogonal polynomials implemented in *Mathematica* (this time, in comparison to Exercise 2 of this chapter in a more procedural manner).

```
In[1]:= weight[polynomial:(HermiteH  | JacobiP   | GegenbauerC |
                           LaguerreL | LegendreP | ChebyshevT | ChebyshevU),
             var_Symbol, para_:"without"] :=
          Module[{α, β, m},
             Switch[polynomial, (* the various polynomials *)
                HermiteH, Exp[-var^2],
                JacobiP, α = para[[1]]; β = para[[2]]; (1 - var)^α (1 + var)^β,
                GegenbauerC, α = para[[1]]; (1 - var^2)^α,
                LaguerreL, α = para[[1]]; Exp[-var] var^α,
                LegendreP, 1,
```

```
                    ChebyshevT, 1/Sqrt[1 - var^2],
                    ChebyshevU, Sqrt[1 - var^2]]];
In[2]:= normalization[polynomial:(HermiteH    | JacobiP   | GegenbauerC |
                              LaguerreL   | LegendreP | ChebyshevT  |
                              ChebyshevU),
                    n_Integer?(# >= 0&) | n_Symbol, para_:"without"] :=
    Module[{α, β, m, Γ = Gamma},
    Switch[polynomial,
                  (* the various polynomials *)
                  HermiteH, Sqrt[Pi] 2^n n!,
                  JacobiP, α = para[[1]]; β = para[[2]];
                            2^(α + β + 1)/(2n + α + β + 1)*
                              Γ[n + α + 1]Γ[n + β + 1]/n!/Γ[n + α + β + 1],
                  GegenbauerC, α = para[[1]]; If[α == 0, If[n == 0, 1/2,
                      2Pi/n^2], Pi 2^(1 - 2α) Γ[1 + 2α]/(α + 1)/Γ[α]^2, 2Pi/n^2],
                  LaguerreL, α = para[[1]]; Γ[n + 1 + α]/n!,
                  LegendreP, m = para[[1]]; (n + m)!/(n - m)! 2/(2n + 1),
                  ChebyshevT, If[n == 0, Pi, Pi/2],
                  ChebyshevU, Pi/2]]
In[3]:= domain[polynomial:(HermiteH    | JacobiP   | GegenbauerC |
                          LaguerreL   | LegendreP | ChebyshevT  | ChebyshevU)] :=
    Switch[polynomial, (* the various polynomials *)
                  HermiteH, {-Infinity, Infinity}, JacobiP, {-1, 1},
                  GegenbauerC, {-1, 1}, LaguerreL, {0, Infinity},
                  LegendreP, {-1, 1}, ChebyshevT, {-1, 1}, ChebyshevU, {-1, 1}]
```

Now, we look at these figures graphically. We reparametrize the polynomials that are not defined in $(-1, 1)$, so that in all cases we use the interval $(-1, 1)$ for ParametricPlot. To get a finite range of values, we include the weight factor for HermiteH, JacobiP, GegenbauerC, LaguerreL, LegendreP, and ChebyshevU, but not with ChebyshevT.

```
In[4]:= GeneralizedLissajousFigure[
            polynomial :(HermiteH    | JacobiP    | GegenbauerC |
                        LaguerreL   | LegendreP | ChebyshevT  | ChebyshevU),
                    {n1_Integer?Positive, n2_Integer?Positive}, para_:"without"] :=
    Module[{norm1, norm2, int, wt, xCoord, yCoord, picture, ε = 10^-12},
        Off[General::under]; Off[General::unfl];
        (* normalize *)
        {norm1, norm2} = N[normalization[polynomial, #, para]]& /@ {n1, n2};
        int = domain[polynomial];
        (* map interval where polys are orthogonal to {-1, 1} *)
        Which[int == {-1, 1}, rule = x -> t,
              int == {0, Infinity}, rule = x -> Tan[(t + 1)/2 Pi/2],
              int == {-Infinity, Infinity}, rule = x -> Tan[Pi/2 t]];
        If[(* avoid singularity of weight function of ChebyshevT *)
            polynomial === ChebyshevT, wt = 1,
            wt = N[weight[polynomial, x, para]]];
        (* calculate parametrized x- and y-coordinates *)
        {xCoord, yCoord} = 1/Sqrt[#2] wt polynomial[#1, Sequence @@
                              If[para === "without", {}, para], x]& @@@
                                            {{n1, norm1}, {n2, norm2}};
        (* use ParametricPlot to make a picture *)
        picture = ParametricPlot[Evaluate[N[{xCoord, yCoord} /. rule]],
                        (* avoid touching endpoints *) {t, -1 + ε, 1 - ε},
                        PlotRange -> All, Frame -> True, FrameTicks -> None,
                        Axes -> False, PlotPoints -> 60 Max[{n1, n2}],
                        DisplayFunction -> Identity];
        On[General::under]; On[General::unfl];
        picture[[1]]]
```

Now, we can give an example for each of the orthogonal polynomials.

```
In[5]:= Show[GraphicsArray[#]]& /@
        Map[Graphics[GeneralizedLissajousFigure[Sequence @@ #],
                Frame -> True, FrameTicks -> None, PlotRange -> All,
                PlotLabel -> StyleForm[ToString @ (InputForm @ #),
                            FontFamily -> "Courier", FontSize -> 5]]&,
```

```
(* a few examples, a lot of other combinations also give nice pictures *)
{{{HermiteH, {7, 5}}, {JacobiP, {4, 12}, {1/2, 3/5}},
  {GegenbauerC, {10, 13}, {1/2}}},
 {{LaguerreL, {5, 7}, {60}}, {ChebyshevT, {4, 9}},
  {ChebyshevU, {4, 11}}}}, {2}]
```

{HermiteH, {7, 5}}   {JacobiP, {4, 12}, {1/2, 3/5}}   {GegenbauerC, {10, 13}, {1/2}}

{LaguerreL, {5, 7}, {60}}   {ChebyshevT, {4, 9}}   {ChebyshevU, {4, 11}}

In the following animation, we change α in $\left\{P_{17}^{(\alpha,2-\alpha)},\ P_{16}^{(\alpha,2-\alpha)}\right\}$.

```
In[6]:= frame[α_] := Graphics[GeneralizedLissajousFigure[
                              JacobiP, {17, 16}, {α, 2 - α}]]
```

```
In[7]:= Show[GraphicsArray[#]]& /@ Partition[
          Table[Show[frame[α], PlotRange -> All, AspectRatio -> 1,
                DisplayFunction -> Identity], {α, 0, 2, 2/17}], 6]
```

```
Do[Show[frame[α], PlotRange -> All, AspectRatio -> 1],
   {α, 0, 2, 1/16}]
```

For the special properties of generalized Lissajou figures $\{T_{n-1}(t),\ T_n(t)\}$, see [202], [452].

b) The function $\text{ev}[dim, \lambda]$ calculates the eigenvalues of the operator $\hat{\Lambda}^2$ by differentiating the Gegenbauer polynomials $C_\lambda^{(\alpha)}(\mathbf{e} \cdot \mathbf{e}')$. For the vectors $\mathbf{e}$ and $\mathbf{e}'$, we use generic symbolic vectors of unit length.

```
In[1]:= ev[dim_, λ_] :=
          Module[{X = Table[x[j], {j, dim}], Ξ = Table[ξ[j], {j, dim}], ψ},
                 (* the eigenfunction *)
```

```
ψ = GegenbauerC[λ, (dim - 2)/2, X.Ξ/(Sqrt[X.X] Sqrt[Ξ.Ξ])];
(* apply angular momentum operator *)
Sum[-Nest[(x[i] D[#, x[j]] - x[j] D[#, x[i]])&, ψ, 2],
    {i, dim}, {j, i + 1, dim}]/ψ //
    (* cancel out eigenfunction *) Cancel]
```

The next inputs calculate the eigenvalues for $2 < d \le 6$ and $0 \le \lambda \le 4$. They all are of the form $\lambda(\lambda + d - 2)$.

```
In[2]:= {Table[ev[3, λ], {λ, 0, 4}], Table[λ(λ + 3 - 2), {λ, 0, 4}]}
```

```
Out[2]= {{0, 2, 6, 12, 20}, {0, 2, 6, 12, 20}}
```

```
In[3]:= {Table[ev[4, λ], {λ, 0, 4}], Table[λ(λ + 4 - 2), {λ, 0, 4}]}
```

```
Out[3]= {{0, 3, 8, 15, 24}, {0, 3, 8, 15, 24}}
```

```
In[4]:= {Table[ev[5, λ], {λ, 0, 4}], Table[λ(λ + 5 - 2), {λ, 0, 4}]}
```

```
Out[4]= {{0, 4, 10, 18, 28}, {0, 4, 10, 18, 28}}
```

```
In[5]:= {Table[ev[6, λ], {λ, 0, 4}], Table[λ(λ + 6 - 2), {λ, 0, 4}]}
```

```
Out[5]= {{0, 5, 12, 21, 32}, {0, 5, 12, 21, 32}}
```

We continue with checking the orthogonality.

```
In[6]:= (* hyperspherical coordinates *)
    ξC[d_, {r_, ϑ_}, n_] := r Product[Sin[ϑ[i]], {i, n - 1}] Cos[ϑ[n]];
    ξC[d_, {r_, ϑ_}, d_] := r Product[Sin[ϑ[i]], {i, d - 1}]
```

```
In[9]:= (* hyperspherical volume element *)
    fDet[d_, {r_, ϑ_}] := Simplify[Det[Outer[D, Table[ξC[d, {r, ϑ}, n], {n, d}],
                                       Union[Array[ϑ, d - 1], {r}]]]]
```

```
In[11]:= (* a specialized integration function; avoid any patterns with
             high complexity in matching *)
    (* separate nD integrals into 1D integrals *)
    integrate[t_, iters__] :=
    Module[{vars = First /@ {iters}, tL = If[Head[t] === Times, List @@ t, {t}],
            ints1D, δ},
           ints1D = Times @@@  (Function[v,
                               Select[tL, MemberQ[#, v, Infinity]&]] /@ vars);
           δ = Cancel[t/Times @@ ints1D];
           δ Times @@ (integrate @@@ Transpose[{ints1D, {iters}}])] /;
                                                         Length[{iters}] > 1

    (* canonicalize 1D integrals *)
    integrate[int_, iter_] := integrateC[int /. iter[[1]] -> ξ, MapAt[ξ&, iter, 1]]

    (* extract terms not depending on the integration variable *)
    integrate[t_, iter_] :=
    Module[{tL = If[Head[t] === Times, List @@ t, {t}], tI},
           tI = Times @@ Select[tL, MemberQ[#, iter[[1]], Infinity]&];
           Cancel[t/tI] integrateC[tI, iter]]

    (* cache basic 1D integrals *)
    integrateC[int_, iter_] := integrateC[int, iter] = Integrate[int, iter]
```

```
In[21]:= (* integration over hypersphere surface *)
    unitSphereIntegrate[int_, {d_, ϑ_}] :=
    Module[{ve = fDet[d, {r, ϑ}] /. r -> 1, (* integration domains *)
            iters  = Sequence @@ Table[{ϑ[i], 0, If[i == d - 1, 2, 1] Pi},
                                       {i, d - 1}]},
           (* do no use built-in Integrate; but more specialized
              and faster function integrate *)
           (* distribute integration over sums *)
           If[Head[#2] === Plus, #1 /@ #2, #1 @ #2]&[
                              integrate[# ve, iters]&, Expand[int]]]
```

```
In[23]:= orthogonalityCheck[d_, {m_, n_}] :=
    Module[{λ = (d - 2)/2, Λ, ϑ1, ϑ2, ϑ3, v1, v2, v3, integrand, ep},
           (* degenerate case d = 2 *)
```

```
          Λ = If[d === 2, Sequence @@ {}, λ];
          (* three vectors v1, v2, and v3 *)
          Do[{v1, v2, v3} = Table[ςC[d, {r, #}, k], {k, d}]& /@
                           {θ1, θ2, θ3} /. r -> 1, {j, 3}];
          (* the integrand *)
          integrand = GegenbauerC[n, Λ, v1.v2] GegenbauerC[m, Λ, v2.v3];
          (* do the integration over v2 *)
          ep = unitSphereIntegrate[integrand, {d, θ2}];
          (* right-hand side *)
          rhs = KroneckerDelta[m, n] If[λ == 0, 1, λ]/(n + λ) 2Pi^(λ + 1)/λ! *
                GegenbauerC[n, Λ, v1.v3];
          (* simplify result *) ep - rhs // Simplify]
```

In[24]:= `Table[orthogonalityCheck[d, {m, n}],`
        `{d, 2, 4}, {m, 1, 3}, {n, 1, 3}] // Flatten // Union`

Out[24]= `{0}`

Let us now define the harmonic polynomials $H_\lambda(\mathbf{r}, \mathbf{r}')$. X and Ξ are *dim*D vectors.

In[25]:= `H[λ_, dim_, X_, Ξ_] := Sqrt[X.X]^λ *`
        `GegenbauerC[λ, (dim - 2)/2, X.Ξ/(Sqrt[X.X] Sqrt[Ξ.Ξ])]/`
        `GegenbauerC[λ, (dim - 2)/2, 1]`

Verifying that the $H_\lambda(\mathbf{r}, \mathbf{r}')$ fulfill the Laplace equation and that $H_\lambda(\mathbf{e}', \mathbf{e}') = 1$ is straightforward for small λ.

In[26]:= `Table[With[{H = H[λ, dim, Table[x[k], {k, dim}], Table[ξ[k], {k, dim}]]},`
        `(* apply Laplace operator *)`
        `Sum[D[H, {x[j], 2}], {j, dim}] //`
        `(* write over common denominator *) Together],`
      `{dim, 3, 6}, {λ, 0, 5}] // Flatten // Union`

Out[26]= `{0}`

In[27]:= `Table[With[{H = H[λ, dim, Table[ξ[k], {k, dim}], Table[ξ[k], {k, dim}]]},`
        `Together[H] /. (* {ξ[1], ..., ξ[dim]} is normalized *)`
        `Sum[ξ[k]^2, {k, dim}] -> 1],`
      `{dim, 3, 9}, {λ, 0, 9}] // Flatten // Union`

Out[27]= `{1}`

Now, we turn to the multidimensional spherical harmonics. For given values of the m_j in the form of the list *ms* and variables ϑ_j and ϕ, the function Y calculates the spherical harmonics.

In[28]:= `Y[ms_, θs_, φ_] :=`
        `With[{d = Length[ms] + 1}, Exp[I ms[[-1]] φ] *`
        `Product[Sin[θs[[k + 1]]]^(ms[[k + 2]]) GegenbauerC[ms[[k + 1]] - ms[[k + 2]],`
        `(d - k - 2 + 2ms[[k + 2]])/2, Cos[θs[[k + 1]]]], {k, 0, d - 3}]]`

Here is a visualization of $Y_{61}^{30,15}(\vartheta_1, \vartheta_2, 0)$.

In[29]:= `Show[GraphicsArray[`
        `Block[{$DisplayFunction = Identity, pl3d},`
        `{(* 3D plot *)`
        `pl3d = Plot3D[Evaluate[Y[{61, 30, 15}, {θ1, θ2}, 0]],`
               `{θ1, 0, Pi}, {θ2, 0, Pi}, PlotPoints -> 240,`
               `PlotRange -> All, ColorFunction -> (Hue[0.8 #]&),`
               `Mesh -> False, Axes -> False],`
        `(* contour plot *)`
        `Show[ContourGraphics[pl3d],`
            `ContourLines -> False, FrameTicks -> False]}]]]`

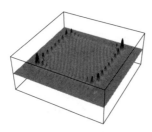 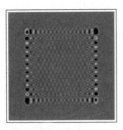

The function ΔS is the angular part of the Laplace operator (in the form of a pure function). The dimension *d* is determined by the length of the list θs.

```
In[30]:= ΔS[θs_, φ_] :=
          With[{d = Length[θs] + 2},
          (Sum[Product[Sin[θs[[j]]], {j, k}]^-2 *
             Sin[θs[[k + 1]]]^(-d + 2 + k) D[Sin[θs[[k + 1]]]^(d - 2 - k) *
                D[#, θs[[k + 1]]],
             θs[[k + 1]]], {k, 0, d - 3}] +
          Product[Sin[θs[[j]]], {j, d - 2}]^-2 D[#, {φ, 2}])&]
```

Again, it is straightforward to verify that the $Y_{m_0}^{m_1,...,m_{d-2}}$ are eigenfunctions of the angular part of the *d*D Laplace operator with eigenvalues $-m_0(m_0 + d - 2)$.

```
In[31]:= Table[(ΔS[{θ}, φ][#] + m0 (m0 + 3 - 2) #)&[Y[{m0, m1}, {θ}, φ]],
             {m0, 0, 5}, {m1, 0, m0}] // Simplify // Flatten // Union

Out[31]= {0}
```

```
In[32]:= Table[(ΔS[{θ1, θ2}, φ][#] + m0 (m0 + 4 - 2) #)&[Y[{m0, m1, m2}, {θ1, θ2}, φ]],
             {m0, 0, 5}, {m1, 0, m0}, {m2, 0, m1}] // Simplify // Flatten // Union

Out[32]= {0}
```

```
In[33]:= Table[(ΔS[{θ1, θ2, θ3}, φ][#] + m0 (m0 + 5 - 2) #)&[
                   Y[{m0, m1, m2, m3}, {θ1, θ2, θ3}, φ]],
             {m0, 0, 5}, {m1, 0, m0}, {m2, 0, m1}, {m3, 0, m2}] //
                                        Simplify // Flatten // Union

Out[33]= {0}
```

c) We start by defining $p_{12,6}(r, z)$. We use a cylindrical coordinate system with coordinates ρ and *z* (*p* is independent of the azimuthal angle) (For a discussion of the radial zeros, see [554]).

```
In[1]:= ψ[{n_, l_}, r_, θ_] := Exp[-r] r^l *
          LaguerreL[n - l - 1, 2 l + 1, 2 r] LegendreP[n, l, Cos[θ]]
```

```
In[2]:= (* compiled function for p *)
        p = Compile[{ρ, z}, Evaluate[N @ Abs[ψ[{12, 6}, Sqrt[ρ^2 + z^2],
                              ArcCos[z/Sqrt[ρ^2 + z^2]]]]^2]];
```

Here is a sketch of $p_{12,6}(r, \vartheta)$ over the ρ,*z*-plane.

```
In[4]:= R = 11; ε = 10^-8; pp = 200;
        data = #/Max[#]&[Table[p[ρ, z], {z, -R, R, R/pp},
                              {ρ, ε, R - ε, (R - ε)/pp}]];
```

```
In[6]:= ListPlot3D[data, PlotRange -> All, Mesh -> False,
               MeshRange -> {{ε, R}, {-R, R}}]
```

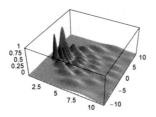

The next graphic shows $p_{12,6}(r, \vartheta) = c$ in the ρ,z-plane.

```
In[7]:= (* number of contours *) v = 60;
        dataM = Select[Sort[Flatten[data]], (0.05 <= # <= 0.45)&];
In[9]:= (* homogeneous contour spacing *)
        d = Round[Length[dataM]/v];
        cls = Append[#[[Round[d/2]]]& /@ Partition[dataM, d],
                #[[Round[Length[#]/2]]]&[Take[dataM, -Mod[Length[dataM], d]]]];
In[12]:= ListContourPlot[data, PlotRange -> All, ContourShading -> False,
                MeshRange -> {{ε, R}, {-R, R}}, Contours -> cls,
                ContourStyle -> Table[{Thickness[0.002], Hue[0.8 k/v]},
                                      {k, 0, v}]]
```

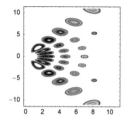

Given the $p_{12,6}(r, \vartheta) = c$ in the ρ,z-plane it is straightforward to generate the equisurfaces $p_{12,6}(r, \vartheta) = c$ in $\mathbb{R}^3$. We just rotate the 2D contour lines around the z-axis to sweep out the equisurfaces. For a given value of c, the function surfaceOfRevolutionContourPlot calculates the equisurface.

```
In[13]:= (* rotate list of points around z-axis *)
        cfR = Compile[{{xyz, _Real, 1}, {ppφ, _Integer}, φMax},
                    Table[{{Cos[φ], Sin[φ], 0.}, {-Sin[φ], Cos[φ], 0.},
                          {0., 0., 1.}}.xyz, {φ, 0, φMax, φMax/ppφ}]];
In[15]:= (* make polygons from rotated points *)
        makePolygons[m_] :=
        Table[Polygon[{m[[i, j]], m[[i + 1, j]],
                      m[[i + 1, j + 1]], m[[i, j + 1]]}],
              {i, Length[m] - 1}, {j, Length[m[[1]]] - 1}]
In[17]:= lineTo3DSurfaceOfRevolution[Line[l_], φMax_, pp_] :=
                makePolygons[cfR[Insert[#, 0, 2], pp, φMax]& /@ l]
In[18]:= (* make equisurface *)
        surfaceOfRevolutionContourPlot[c_, φMax_:3/2Pi, opts___] :=
        Module[{lcp, lines},
        (* make contour plot in ρ,z-plane *)
        lcp = ListContourPlot[data, Contours -> {c}, ContourShading -> False,
                        MeshRange -> {{ε, R}, {-R, R}},
                        DisplayFunction -> Identity];
        (* extract contour lines *)
        lines = Cases[Graphics[lcp], _Line, Infinity];
        (* make graphics of equisurfaces *)
        Graphics3D[{EdgeForm[],
                        SurfaceColor[Hue[2c], Hue[2c], 2.78],
                lineTo3DSurfaceOfRevolution[#, φMax, 32]& /@ lines},
                ViewPoint -> {3, 3, 1}, SphericalRegion -> True,
                PlotRange -> 10{{-1, 1}, {-1, 1}, {-1, 1}}, Boxed -> False]]
```

Here are some of the surfaces $p_{12,6}(r, \vartheta) = c$ for various c. To better see the inner structure, we use $0 \le \varphi \le 3/2\,\pi$.

```
In[20]:= Show[GraphicsArray[surfaceOfRevolutionContourPlot /@ #]]& /@
                Partition[Reverse[First /@ Partition[Drop[cls, -1], 10]], 3];
```

The next graphic shows six of the surfaces. We cut off the edges of the polygons to better see inside.

```
In[21]:= toDiamondPolygons[expr_] := expr /. Polygon[l_] :>
                Polygon[Plus @@@ Partition[Append[l, First[l]], 2, 1]/2]
In[22]:= Show[toDiamondPolygons[surfaceOfRevolutionContourPlot[#]& /@
            {0.6, 0.4, 0.3, 0.2, 0.1, 0.05}]]
```

Here is an animation of $p_{12,6}(r, \vartheta) = c$ for various c.

```
Do[Show @ surfaceOfRevolutionContourPlot[cls[[-k]]], {k, v}]
```

The interested could now continue with a visualization of the probability distribution in momentum space. The unnormalized momentum representation of the energy eigenfunctions contains Gegenbauer polynomials [80], [366], [446], [471].

```
In[23]:= χ[{n_, l_}, p_, θ_] := p^l/(n^2 p^2 + 1)^(l + 2) *
            GegenbauerC[n - l - 1, l + 1, (n^2 p^2 - 1)/(n^2 p^2 + 1)]*
            LegendreP[n, l, Cos[θ]]
```

For the nD version, see [271]. For the much more complicated situation of moving hydrogen atoms, see [296]. For more graphics of this kind, see [560].

7. Zeros of Hermite Polynomials, q-Hermite Polynomials, Pseudodifferential Operator, Moments of Hermite Polynomial Zeros

a) To get a rough impression of what happens with the zeros, let us look at the following graphic.

```
In[1]:= DensityPlot[(Abs[HermiteH[n, z]])^(1/8), {z, -2.5, 2.5}, {n, 1, 6},
            PlotPoints -> {100, 100}, Mesh -> False, AspectRatio -> 1]
```

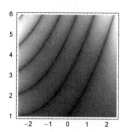

This picture suggests that the zeros approach 0 with decreasing n and two zeros meet at odd n. To look at this behavior in more detail, we construct and solve a differential equation for the zeros $z_0(n)$ in the sense of continuation methods [413], [622], [353], [609]. From $H_n(z(n), n) = 0$, a differential equation for the zeros follows by differentiation with respect to n: [138]

$$z_0'(n) = \frac{\dfrac{\partial H_n(z)}{\partial n}}{\dfrac{\partial H_n(z)}{\partial z}}\Bigg|_{z=z_0(n)}.$$

Unfortunately, we can only explicitly calculate one of the two partial derivatives.

```
In[2]:= {D[HermiteH[n, z], z], D[HermiteH[n, z], n]}

Out[2]= {2 n HermiteH[-1 + n, z], HermiteH^(1,0)[n, z]}
```

For the calculation of the derivative with respect to the index, we use the command ND from the package Numerical `Math`NLimit` (this allows more control over the numerical calculation of the derivative than does the feature of Derivative to be evaluated for approximative values).

```
In[3]:= Needs["NumericalMath`NLimit`"]
```

Now, we can implement the right-hand side of the above differential equation of the zeros.

```
In[4]:= f[z_?NumberQ, n_?NumberQ] := f[z, n] =
        -ND[HermiteH[v, z], v, n, (* found by inspection *)
            Scale -> 10^-2, Terms -> 6]/(2 n HermiteH[n - 1, z])
```

We solve the differential equation starting from higher n and move toward lower n. Here, this is done for the positive zeros of $H_6(z)$ (the negative ones follow from symmetry).

```
In[5]:= zeros = Drop[Last /@ List @@ NRoots[HermiteH[6, z] == 0, z], 3];
        (* solving from n = 6 until the roots disappear from the real axis *)
        sol3 = NDSolve[{z'[n] == f[z[n], n],
                        z[6] == Last[zeros]}, z, {n, 6, 10^-6}];
        sol2 = NDSolve[{z'[n] == f[z[n], n],
                        z[6] == zeros[[2]]}, z, {n, 6, 3}];
        sol1 = NDSolve[{z'[n] == f[z[n], n],
                        z[6] == First[zeros]}, z, {n, 6, 5}];
```

So, we can make the following picture of the location of the zeros $\tilde{z}(n)$ for noninteger n.

```
In[10]:= Show[Graphics[{Thickness[0.005],
         (* use symmetry about z == 0 to calculate one-half *)
             Line[Table[{ (z[n] /. sol3)[[1]], n}, {n, 1, 6, 5/100}]],
             Line[Table[{-(z[n] /. sol3)[[1]], n}, {n, 1, 6, 5/100}]],
             Line[Table[{ (z[n] /. sol2)[[1]], n}, {n, 3, 6, 3/60 }]],
             Line[Table[{-(z[n] /. sol2)[[1]], n}, {n, 3, 6, 3/60 }]],
             Line[Table[{ (z[n] /. sol1)[[1]], n}, {n, 5, 6, 3/20 }]],
             Line[Table[{-(z[n] /. sol1)[[1]], n}, {n, 5, 6, 3/20 }]]}],
             Frame -> True, AspectRatio -> 1]
```

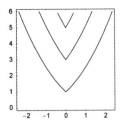

For an approximation of the zeros of the Hermite polynomials, see [525]. For some theoretical consideration of $H_n(z)$ as a function of the real parameter n, see [184].

b) We start by implementing the recursive definition of the q-Hermite polynomials.

```
In[1]:= qHermite[n_, q_, x_] := qHermite[n, q, x] =
        Collect[1/Sqrt[q] q^(-2(n - 1))2 x qHermite[n - 1, q, x] -
                2/q^2 (1 - 1/q^(2n - 2))/(1 - 1/q^2) qHermite[n - 2, q, x],
            x, Together]

        qHermite[0, q_, x_] := 1;

        qHermite[1, q_, x_] := 2/Sqrt[q] x;
```

Here are the first q-Hermite polynomials. The odd order q-Hermite polynomials have an overall $q^{1/2}$ prefactor.

```
In[5]:= qHermite[2, q, x]
```

$$Out[5]= -\frac{2}{q^2} + \frac{4 x^2}{q^3}$$

```
In[6]:= qHermite[3, q, x]
```

$$Out[6]= -\frac{4 (1 + q^2 + q^4) x}{q^{13/2}} + \frac{8 x^3}{q^{15/2}}$$

```
In[7]:= qHermite[4, q, x]
```

$$Out[7]= \frac{4 (1 + q^2 + q^4)}{q^8} - \frac{8 (1 + q^2 + 2 q^4 + q^6 + q^8) x^2}{q^{13}} + \frac{16 x^4}{q^{14}}$$

Here are two contour plots showing the behavior of $\operatorname{Re}\big(H_{12}^{(q)}(z)\big)$ for real q and q on the unit circle.

```
In[8]:= Show[GraphicsArray[
        Block[{$DisplayFunction = Identity},
         {(* for real q, 3/4 <=q <= 1 *)
          ContourPlot[Evaluate[Log[10, 1 + qHermite[12, q, x]^2]],
                      {x, -3, 3}, {q, 0.75, 1}, PlotPoints -> 100,
                      Contours -> 200, ColorFunction -> (Hue[3.8 #]&),
                      ContourLines -> False, AspectRatio -> 1/2],
          (* on the q-unit circle *)
          ContourPlot[Evaluate[Log[10, Abs[Re[qHermite[12, Exp[I φ], x]^2]]]],
                      {x, -3, 3}, {φ, 0, 2Pi}, PlotPoints -> 200,
                      Contours -> 60, ColorFunction -> (Hue[3.8 #]&),
                      ContourLines -> False, AspectRatio -> 1]}]]]
```

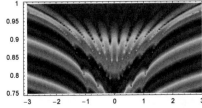 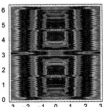

Next, we derive the differential equation for $z(q)$ being implicitly defined by $H_n^{(q)}(z) = 0$.

```
In[9]:= Do[ode[n][ζ_, {q0_, r_, φ_}] =
          With[{h = Numerator[qHermite[n, q, z]]},
             D[h /. z -> z[q], q] /. z[q] -> ζ[φ] /.
                  z'[q] :> ζ'[φ]/D[q0 + r Exp[I φ], φ] /.
                  q -> q0 + r Exp[I φ]],
        {n, 3, 6}]
```

Now let us deal with the branch points of $z(q)$. As discussed in Subsection 1.11.1 of the Numerics volume [567], a necessary condition for $z(q)$ to be a branch point is $H_n^{(q)}(z) = 0$ and $\partial H_n^{(q)}(z(q))/\partial z = 0$.

```
In[10]:= disc[p_, z_] := With[{f = Numerator[Together[p]]},
                Factor[Resultant[f, D[f, z], z]]]
```

Fortunately, for the numerical solution of the equations, the resulting univariate polynomials in q that define the branch points factor nicely.

```
In[11]:= Table[disc[qHermite[n, q, x], x], {n, 6}]
```

$$Out[11]= \left\{2, -128\,q, -16384\,q^3\,(1 - q + q^2)^3\,(1 + q + q^2)^3,\right.$$
$$67108864\,q^{10}\,(1 + q^2)^2\,(1 - q + q^2)^3\,(1 + q + q^2)^3\,(1 - q^4 + 3\,q^6 + q^{10})^2,$$
$$549755813888\,q^{22}\,(1 + q^2)^2\,(1 - q + q^2)^3\,(1 + q + q^2)^3\,(1 - q + q^2 - q^3 + q^4)^5$$
$$(1 + q + q^2 + q^3 + q^4)^5\,(1 - q^4 + 3\,q^{10} + q^{14})^2, -144115188075855872\,q^{43}$$
$$(1 + q^2)^6\,(1 - q + q^2)^5\,(1 + q + q^2)^5\,(1 - q + q^2 - q^3 + q^4)^5\,(1 + q + q^2 + q^3 + q^4)^5$$
$$(1 - q^2 - q^4 + 5\,q^6 - 6\,q^8 + 6\,q^{10} + 2\,q^{12} - 14\,q^{14} + 21\,q^{16} - 15\,q^{18} + 9\,q^{20} + 7\,q^{22} - 14\,q^{24} +$$
$$\left. 26\,q^{26} - 22\,q^{28} + 30\,q^{30} - 18\,q^{32} + 20\,q^{34} - 7\,q^{36} + 7\,q^{38} - q^{40} + q^{42})^2\right\}$$

Here are the branch points for the q-Hermite polynomials $H_2^{(q)}(z)$ to $H_{10}^{(q)}(z)$.

```
In[12]:= Show[Graphics[Reverse @
        Table[qPoly = disc[qHermite[n, q, x], x];
              (* solve for zeros of the discriminant *)
              sol = N[q /. Solve[qPoly == 0, q]];
              {Hue[(n - 2)/10], PointSize[0.01],
                    Point[{Re[#], Im[#]}]& /@ sol}, {n, 2, 10}]],
        PlotRange -> All, Frame -> True, AspectRatio -> Automatic]
```

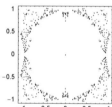

Because of the high powers of q that appear in the q-Hermite polynomials, the order of the branch points is most easily determined not symbolically, but rather numerically. Starting from an initial point q_{init}, we solve the above-calculated differential equation ode[n][ζ, {q0, r, φ}] around each branch point $q0$. After one round we return either q_{init} or to another solution of the discriminant equation. We count how many rounds we need to return to q_{init}—this is the order of the branch point. The function branchPointOrders implements this procedure.

```
In[13]:= (* make cycles from a permutation *)
        toCycles[l_] := MapIndexed[{#2[[1]], #1}&, l] //.
          {{a___, {b_, A___, c_}, d___, {c_, B___, e_}, f___} :>
             {a, {b, A, c, B, e}, f},
           {a___, {b_, A___, c_}, d___, {e_, B___, b_}, f___} :>
             {a, {e, B, b, A, c}, d, f}} //.
          {a___, {b_, c___, b_}, d___} :> {a, {b, c}, d}
In[15]:= branchPointOrders[n_] :=
        Module[{d, qPoly, q0, r, F, qBcs, nqBcs, z0s, tab},
        (* the discriminant *)
        d = disc[qHermite[n, q, z], z];
        (* the implicit definition for z[q] *)
        qPoly = Numerator[Together[qHermite[n, q, z]]];
```

```
(* ODE for z[q] around a branch point *)
ODE[q0_, r_] = D[qPoly /. z -> z[q], q] /. z[q] -> F[φ] /.
                z'[q] :> F'[φ]/D[q0 + r Exp[I φ], φ] /.
                q -> q0 + r Exp[I φ];
(* the potential branch points *)
qBcs = Union[q /. Solve[d == 0, q]];
nqBcs = N[qBcs, 30];
(* encircling each branch point on each sheet *)
{First[#], Length[#]}& /@ Split[Sort[
Table[q0 = nqBcs[[k]];
    r = Min[Abs[nqBcs[[k]] - Delete[nqBcs, k]]]/2;
    z0s = z /. {ToRules[NRoots[qPoly == 0 /. q -> q0 + r, z]]};
    (* solve differential equation around branch points *)
    tab = Table[ζ[2Pi] /.
                NDSolve[{ode[n][ζ, {q0, r, φ}] == 0, ζ[0] == z0s[[j]]},
                ζ, {φ, 2Pi, 2Pi}, PrecisionGoal -> 12][[1]], {j, n}];
(* match sheet numbers *)
Sort[Length /@ toCycles[Flatten[
                    Position[#, Min[#]]&[Abs[# - z0s]]& /@ tab]]],
    {k, Length[qBcs]}]]]]]
```

The branch points of $H_3^{(q)}(z)$ are all of square root-type.

```
In[16]:= branchPointOrders[3]

Out[16]= {{{1, 2}, 5}}
```

13 of the branch points of $H_4^{(q)}(z)$ are of square root-type, and four are of order 4.

```
In[17]:= branchPointOrders[4]

Out[17]= {{{4}, 4}, {{2, 2}, 13}}
```

21 of the branch points of $H_5^{(q)}(z)$ are of square root-type, and eight are of order 4.

```
In[18]:= branchPointOrders[5]

Out[18]= {{{1, 4}, 8}, {{1, 2, 2}, 17}, {{1, 1, 1, 2}, 4}}
```

Now let us calculate the zeros $z_0(q)$ of $H_n^{(q)}(z_0(q))$. The following graphic calculates the zeros of $H_{10}^{(q)}(z_0(q))$ for a dense set of q-values on a circle of radius 0.99. The picture shows that the zeros form a complicated pattern.

```
In[19]:= Show[Graphics[{PointSize[0.003],
        Table[Point[{Re[#], Im[#]}]& /@
        Cases[NRoots[qHermite[10, q, x] == 0 /.
                    q -> 0.99 Exp[I φ], x], _Real | _Complex, {-1}],
                    {φ, 0, 2Pi, 2Pi/5000}]}],
            PlotRange -> 3/2 {{-1, 1}, {-1, 1}}, Frame -> True,
            AspectRatio -> Automatic]
```

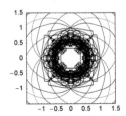

For real q, the dependence is much simpler.

```
In[20]:= Show[Graphics[Table[Point[{#, q}]& /@
        Cases[NRoots[qHermite[12, q, x] == 0 /. q -> q, x], _Real, {-1}],
        {q, 1, 0.01, -0.01}]], PlotRange -> All, Frame -> True]
```

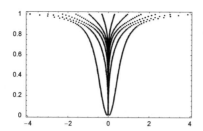

Instead of many calls to `NRoots`, we will solve the differential equation for zeros $z_0(q)$ of $H_n^{(q)}(z_0(q))$. This is a much more efficient way to obtain the zeros. The function `zerosGraphic` shows the curves $z_0(r_q \exp(i\,\varphi_q))$ for a given n and a given r_q. This function calls `NRoots` to get the zeros for $\varphi_q = 0$ and then solves the differential equation for $0 \le \varphi_q \le 2\pi$.

```
In[21]:= zerosGraphic[n_, rq_, opts___] :=
        Module[{z0s, nsols},
        (* zeros for φ == 0 *)
        z0s = Cases[NRoots[qHermite[n, q, z] == 0 /.
                            q -> rq, z, 50], _?NumberQ, Infinity];
        (* solve differential equations *)
        nsols = NDSolve[{ode[n][ζ, {0, rq, φ}] == 0, ζ[0] == #},
                    ζ, {φ, 0, 2Pi}, MaxSteps -> 10^5,
                    PrecisionGoal -> 10, MaxStepSize -> 0.01]& /@ z0s;
        (* display (z[φ] == 0)-curves *)
        ParametricPlot[Evaluate[{Re[ζ[φ]], Im[ζ[φ]]} /. (First /@ nsols)],
                    {φ, 0, 2Pi}, opts, PlotStyle -> {Thickness[0.002]},
                    PlotRange -> All, Frame -> True, Axes -> False,
                    FrameTicks -> None, PlotPoints -> 1000,
                    PlotPoints -> 1000, AspectRatio -> Automatic]]
```

The following animation shows the dependence of $z_0(r_q \exp(i\,\varphi_q))$ as a function of r_q for $H_5^{(q)}(z_0(q))$.

```
In[22]:= Show[GraphicsArray[
            zerosGraphic[5, #, DisplayFunction -> Identity]& /@ #]]& /@
                Partition[Table[r, {r, 1/2, 3/2, 1/15}], 4]
```

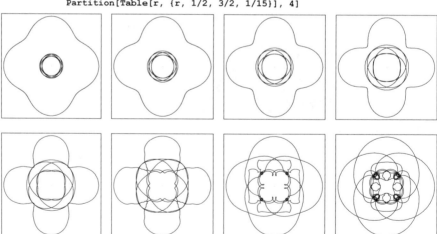

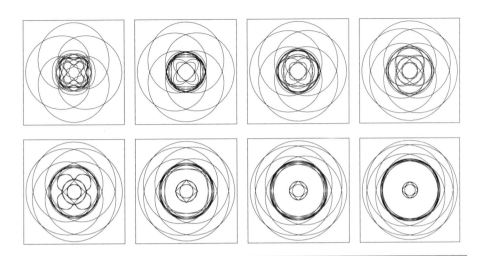

```
Do[zerosGraphic[5, r], {r, 1/2, 3/2, 1/51}]
```

c) We will proceed as we did in Section 2.10 when calculating a high-precision approximation of the quartic oscillator ground-state. The only difficulty compared to the nonrelativistic harmonic oscillator is the pseudodifferential operator $\sqrt{m^2 + \partial_{xx}}$. Differentiation transforms into multiplication by Fourier transformation [136], [439], [255]. This suggests to deal with this operator in a Fourier transformed space. Luckily, the most natural orthogonal polynomials for the harmonic oscillator problem—the Hermite polynomials—are eigenfunctions of the Fourier transform (see Exercise 44a) of Chapter 1): $\mathcal{F}_x[\phi_n(x)](\xi) = i^n \phi_n(\xi)$.

(Here the $\phi_n(\xi)$ are the normalized harmonic oscillator eigenfunctions.) This allows the straightforward (numeric) calculation of the matrix elements of $\langle \phi_n | (m^2 + \partial_{xx})^{1/2} - m | \phi_m \rangle = i^n (-i)^m \langle \phi_n | (m^2 + x^2)^{1/2} - m | \phi_m \rangle$. The condition $\psi_\varepsilon(0) = 0$ is easy to fulfill because the odd harmonic oscillator eigenfunctions are a complete system for all odd functions.

```
In[1]:= φ[n_, z_] := Exp[-z^2/2] HermiteH[n, z]/c[n]
        c[n_] := Sqrt[Sqrt[Pi] 2^n n!]
```

The matrix elements of the potential term $\langle \phi_n | x^2 | \phi_m \rangle$ can be easily evaluated exactly. The matrix elements $\langle \phi_n | \sqrt{m^2 + x^2} - m | \phi_m \rangle$ we have to calculate numerically. To get them with a sufficient precision we could either let the integrand be unevaluated and call the built-in Hermite polynomials or evaluate the integrand and use high-precision arithmetic. We choose the latter approach. So, we can define the matrix elements $\mathbb{T}[\{i, j\}] = \langle \phi_n | \sqrt{m^2 + \partial_{xx}} - m | \phi_m \rangle$ and $\mathbb{V}[\{i, j\}] = \langle \phi_n | x^2 | \phi_m \rangle$ in the following way.

```
In[3]:= opts = Sequence[PrecisionGoal -> 10, AccuracyGoal -> 10,
                        MaxRecursion -> 10, WorkingPrecision -> 32];

        T[{i_, j_}, m_] := (T[{i, j}, m] = T[{j, i}, m] =
        I^i (-I)^j NIntegrate[Evaluate[φ[i, ξ] (Sqrt[m^2 + ξ^2] - m) φ[j, ξ]],
                        {ξ, -Infinity, Infinity}, Evaluate[opts]])

        V[{i_, j_}] := (V[{i, j}] = V[{j, i}] =
        c[i]/c[j]/4 (4 (j - 1) j KroneckerDelta[i, j - 2] +
                    (4 j + 2) KroneckerDelta[i, j] + KroneckerDelta[i, j + 2]))
```

We now form the Hill determinant and calculate its lowest eigenvalue until we are sure about the first five digits.

```
In[7]:= With[{m = 1},
        Do[tab = Table[T[{i, j}, m] + V[{i, j}],
                        {i, 1, dim, 2}, {j, 1, dim, 2}];
           Print[{dim, InputForm @ Eigenvalues[tab][[-1]]}], {dim, 10, 40, 10}]]
        {10, 1.6661545551936488102`8.795715336376436}
```

{20, 1.6640165290201472651`6.943020412441538}

{30, 1.66401262155747702`5.759082319611624}

{40, 1.6640126124770890658`5.187016687217712}

This shows that the lowest eigenvalue of the asymmetric states is about 1.640126 ....

Now let us sketch the dependence of this eigenvalue on m. For such a sketch, we can reduce the precision and accuracy goals of NIntegrate to speed up the numerical integration.

```
In[8]:= opts = Sequence[PrecisionGoal -> 6, AccuracyGoal -> 6, MaxRecursion -> 8];
```

We fix the matrix dimension to be 13 and calculate the lowest eigenvalue for $10^{-2} \le m \le 10^{2}$. For $m \to 0$ we get a purely "photonic" state (with a linear dispersion relation) and, for $m \to \infty$, the matrix elements T[{i, j}] approach zero so that only the potential term survives.

```
In[9]:= dim = 25;
        data = Table[m = 10^log10m;
              tab = Table[T[{i, j}, m] + V[{i, j}],
                          {i, 1, dim, 2}, {j, 1, dim, 2}] // Chop;
              1.{m, Eigenvalues[tab][[-1]]}, {log10m, -2, 2, 4/20}];
In[11]:= ListPlot[data, PlotJoined -> True]
```

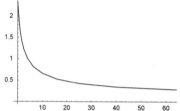

For convenient 3D system of functions that vanish at the origin, see [613], [444], [584], [250], [249], [135], [254], [346]. For a general discussion of square roots of Laplacians, see [429], [578]; for more general potentials, see [90]. For the inappropriateness of such-type Hamiltonians to generate time developments, see [540]. For fractional powers of more general operators, see [116]. For the transformation of nonrelativistic wave functions into relativistic ones, see [404].

d) For any given m and n, we can calculate $\mu_m(n)$ numerically to any desired precision. For a fixed m, calculating the moments for sufficiently many n allows determining all of the $c_j^{(m)}$. Using enough digits allows reconstructing the rational $c_j^{(m)}$ from the calculated high-precision approximation. The following function HermiteZeroMoment implements these steps; we calculate the polynomial roots by default with 50-digit precision. (By symmetry, the moments for odd m vanish.)

```
In[1]:= HermiteZeroMoment[m_Integer?(Positive[#] && EvenQ[#]&),
                          n_, prec_:50] :=
        Module[{c, ansatz, eqs, sol},
        (* ansatz for the moments *)
        ansatz[v_] = Sum[c[j] v^j, {j, 0, m/2}];
        (* m/2 + 1 equations for the m/2 + 1 unknowns c[j] *)
        eqs = Table[ansatz[n] == (* numerical moments *)
                1/n Plus @@ ((z /. {ToRules[
                    N[Roots[HermiteH[n, z] == 0, z],
                          prec]]})^m), {n, 1, m/2 + 1}];
        (* solve equations *)
        sol = Solve[(* make rational equations *)
                    Rationalize[eqs], Table[c[k], {k, 0, m/2}]];
        (* factor and return answer *)
        Factor[ansatz[n] /. sol]]
```

Here are the first six nontrivial moments.

```
In[2]:= Table[Subscript[μ, m] == HermiteZeroMoment[m, n],
              {m, 2, 12, 2}] // TraditionalForm
```

Out[2]//TraditionalForm= $\left\{ \mu_2 = \left\{ \frac{n-1}{2} \right\}, \mu_4 = \left\{ \frac{1}{4}(n-1)(2n-3) \right\}, \mu_6 = \left\{ \frac{1}{8}(n-1)(5n^2-17n+15) \right\},$

$\mu_8 = \left\{ \frac{1}{16}(n-1)(14n^3-79n^2+155n-105) \right\}, \mu_{10} = \left\{ \frac{1}{32}(n-1)(42n^4-344n^3+1106n^2-1644n+945) \right\},$

$\mu_{12} = \left\{ \frac{1}{64}(n-1)(132n^5-1454n^4+6724n^3-16226n^2+20274n-10395) \right\} \right\}$

To calculate $\mu_{100}(n)$ we have to determine 50 coefficients in a linear system of equations. To uniquely form the rational numbers corresponding to the floating-point solutions, we now calculate the zeros with 500 digits. The resulting expression for $\mu_{100}(n)$ contains quite large integers (with up to 80 digits).

```
In[3]:= μ100[n_] = HermiteZeroMoment[100, n, 500];
```

```
In[4]:= Short[μ100[n], 20]
```

Out[4]//Short= $\{ \frac{1}{1125899906842624}((-1+n)$
$(-2725392139750729502980713245400918633290796330545 80341$
$373432882344310620117 1875 +$
$1137071616538329280013286141778013117011771892830521 1392101$
$943204661907519531250 n -$
$2284081555059617595802283986060921489275718694780993 5715744$
$905708612358278125000 n^2 +$
$2950299203844805818626592546341771791228491082049863 6715887$
$475004392077542187500 n^3 -$
$2760667139231151283611005097581969335570310407130404 3159201$
$452215350277052000000 n^4 +$
$\ll 60 \gg + 78255296668214263113086400755199431 6 n^{45} -$
$9467892698771941068378701496412404 n^{46} + 8908849730153277490052195 8400096 n^{47} -$
$5814013661835764434273215796 04 n^{48} + 19782616577561606536237744 56 n^{49})) \}$

As a quick check of the result, we calculate $\mu_{100}(120)$ directly using 1000 digits and compare with the last result.

```
In[5]:= With[{m = 100, n = 120}, 1 - μ100[n]/(Plus @@ ((Last /@
          List @@ N[Roots[HermiteH[n, z] == 0, z], 1000])^m)/n)]
Out[5]= {0. × 10^-999}
```

8. Iterated Polynomial Substitution

First, the iterated substitution and the calculation of the roots of the polynomials constructed in this fashion are implemented.

```
In[1]:= iteratedRootPicture[iter_, degree_, opts___] :=
        Module[{p},
          (* the polynomials used in the replacement *)
          Do[p[j] = Sum[z^i N[Random[] Exp[2Pi I Random[]]],
                      {i, 0, j}], {j, degree}];
          (* show graphics *)
          Show[Graphics[{PointSize[0.002],
          Point[{Re[#], Im[#]}]& /@ (* root finding *)
          Cases[NRoots[# == 0, z]& /@ (* nesting the replacement *)
            NestList[Expand[# /. z^n_. :> p[n]]&,
            (* the initial polynomial *) Sum[z^i, {i, 0, degree}], iter],
            _?NumberQ, {-1}]}], opts, AspectRatio -> 1,
          PlotRange -> Automatic, Frame -> True, FrameTicks -> None]]
```

Here are a few pictures of the resulting roots for polynomials of various degrees.

```
In[2]:= Show[GraphicsArray[#, GraphicsSpacing -> 0.02]]& /@
          Partition[Table[iteratedRootPicture[100, i,
                      DisplayFunction -> Identity], {i, 4, 26, 2}], 4]
```

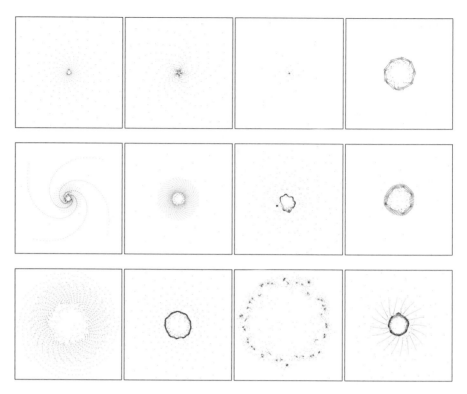

Next, we modify the definition of `iteratedRootPicture` to iterate the substitution of orthogonal polynomials.

```
In[3]:= iteratedOPRootPicture[poly_, α_, iter_, degree_, opts___] :=
        Module[{p, c = 0.},
          (* the polynomials used in the replacement *)
          Do[p[j, z_] = poly[j, z], {j, degree}];
          (* show graphics *)
          Show[Graphics[{PointSize[0.002], Point[{Re[#], Im[#]}]& /@
          Cases[(* roots *) NRoots[# == 0, z]& /@ (* nesting the replacement *)
              (nl = NestList[Expand[#/Max[Abs[CoefficientList[#, z]]]]&[
                        Expand[# /. z^n_. :>
                        p[n, (* nontrivial phase *) Exp[I (c++)^α] z]]]&,
              (* the initial polynomial *) p[degree, z], iter]),
              _?NumberQ, {-1}]}], opts, AspectRatio -> 1,
            PlotRange -> Automatic, Frame -> True, FrameTicks -> None]]
```

Here are some examples.

```
In[4]:= Show[GraphicsArray[#]]& @
        Block[{$DisplayFunction = Identity},
          iteratedOPRootPicture[##, 200, 24]& @@@ (* four examples *)
          {{ChebyshevU, 0.25}, {HermiteH, 0.5},
           {LegendreP, 0.75}, {GegenbauerC, 1.0}}]
```

9. Hermite Polynomials, Coherent States, Isospectral Potentials, Wave Packets

a) These are the normalized harmonic oscillator wavefunctions.

In[1]:= ϕ[n_, x_] := Exp[-x^2/2] HermiteH[n, x]/Sqrt[Sqrt[Pi] 2^n n!]

This is the norm of the function $\Psi(\alpha; x)$.

In[2]:= normS = Integrate[Exp[-α x^4]^2, {x, -Infinity, Infinity},
                    Assumptions -> (α > 0)]

Out[2]= $\dfrac{2^{3/4}\,\mathrm{Gamma}\left[\frac{5}{4}\right]}{\alpha^{1/4}}$

Next, we implement the overlap integral.

In[3]:= int[n_] := int[n] =
                Integrate[1/Sqrt[normS] Exp[-α x^4] ϕ[n, x],
                    {x, -Infinity, Infinity}, Assumptions -> (α > 0)]

In[4]:= int[0]

Out[4]= $\dfrac{e^{\frac{1}{32\,\alpha}}\,\mathrm{BesselK}\left[\frac{1}{4},\,\frac{1}{32\,\alpha}\right]}{2\,2^{7/8}\,\pi^{1/4}\,\alpha^{3/8}\,\sqrt{\mathrm{Gamma}\left[\frac{5}{4}\right]}}$

For $n = 0$, a plot of the overlap integral suggests that $\alpha \approx 0.2$.

In[5]:= Plot[int[0], {α, 0, 1}]

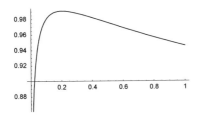

Here is the exact value for the maximal overlap integral.

In[6]:= MapAt[-#&, FindMinimum[-int[0], {α, 0.2}], 1]

Out[6]= {0.990834, {α → 0.203947}}

The following graphic shows the optional $\Psi(\alpha^*; x)$ and $\phi_0(x)$.

In[7]:= Plot[Evaluate[{1/Sqrt[normS] Exp[-α x^4] /. %[[2]], ϕ[0, x]}],
            {x, -3, 3}, PlotStyle -> {Hue[0], Hue[0.8]}]

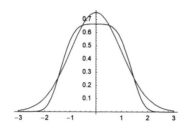

For $n = 2$, we just repeat the above calculations.

In[8]:= int[2]

Out[8]= $-\dfrac{1}{4\,2^{7/8}\,\pi^{1/4}\,\alpha^{9/8}\,\sqrt{\text{Gamma}\left[\frac{5}{4}\right]}}$

$\left(\sqrt{2}\;e^{\frac{1}{32\,\alpha}}\,\alpha^{3/4}\,\text{BesselK}\left[\frac{1}{4},\,\frac{1}{32\,\alpha}\right] - 4\,\sqrt{\alpha}\;\text{Gamma}\left[\frac{3}{4}\right]\text{Hypergeometric1F1}\left[\frac{3}{4},\,\frac{1}{2},\,\frac{1}{16\,\alpha}\right] + 2\,\text{Gamma}\left[\frac{5}{4}\right]\text{Hypergeometric1F1}\left[\frac{5}{4},\,\frac{3}{2},\,\frac{1}{16\,\alpha}\right]\right)$

The expression int[2] is much more complicated than int[0]. To avoid complicated differentiations, we use FindMin-imum with two starting values.

In[9]:= Plot[int[2], {α, 0.003, 0.01}]

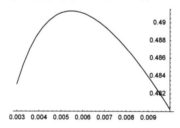

The maximal overlap integral is about 0.5.

In[10]:= MapAt[-#&, FindMinimum[-int[2], {α, 0.004, 0.006}], 1]

Out[10]= {0.4913, {α → 0.00548265}}

Here is the function of maximum overlap shown.

In[11]:= Plot[Evaluate[{1/Sqrt[normS] Exp[-α x^4] /. %[[2]], φ[2, x]}],
        {x, -5, 5}, PlotStyle -> {Hue[0], Hue[0.8]}]

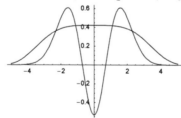

b) These are the normalized harmonic oscillator wavefunctions.

In[1]:= φ[n_, x_] := Exp[-x^2/2] HermiteH[n, x]/Sqrt[Sqrt[Pi] 2^n n!]

$\phi_1(x)$ is an antisymmetric function.

In[2]:= φ[1, x]

Out[2]= $\dfrac{\sqrt{2}\;e^{-\frac{x^2}{2}}\,x}{\pi^{1/4}}$

To maximize $\int_{-\infty}^{\infty} \Psi(x)\,\phi_k(x)\,dx$, the function $\Psi(x)$ should be also antisymmetric. This leads to the following ansatz for $\Psi(x)$.

```
In[3]:= Φ = Exp[-γ(x - ξ)^2] - Exp[-γ(x + ξ)^2]
```

$$\text{Out[3]}= \quad e^{-\gamma\,(x-\xi)^2} - e^{-\gamma\,(x+\xi)^2}$$

Next, we calculate the norm of $\Psi(x)$.

```
In[4]:= normS = Assuming[γ > 0, Sqrt[2 Integrate[Φ^2, {x, 0, Infinity}]]] //
                                                                        Simplify
```

$$\text{Out[4]}= \quad (2\,\pi)^{1/4}\,\sqrt{\frac{1 - e^{-2\,\gamma\,\xi^2}}{\sqrt{\gamma}}}$$

For the overlap integral, we obtain the following expression.

```
In[5]:= 2 Simplify[1/normS *
                Assuming[γ > 0, Integrate[Φ φ[1, x], {x, 0, Infinity}]]]
```

$$\text{Out[5]}= \quad \frac{4\,2^{3/4}\,e^{-\frac{\gamma\,\xi^2}{1+2\,\gamma}}\,\gamma\,\sqrt{\pi + 2\,\pi\,\gamma}\,\xi}{\sqrt{\pi}\,\sqrt{\dfrac{1-e^{-2\,\gamma\,\xi^2}}{\sqrt{\gamma}}}\,(1+2\,\gamma)^2}$$

```
In[6]:= int = 4 2^(3/4) γ ξ Exp[-γ ξ^2 /(1 + 2 γ)]/
            Sqrt[(1 - Exp[-2 γ ξ^2])/Sqrt[γ]]/(1 + 2 γ)^(3/2);
```

Trying to maximize the overlap integral using FindMinimum may generate messages (even though we use sufficient precision and iterations).

```
In[7]:= FindMinimum[Evaluate[int], {γ, 1}, {ξ, 1}, MaxIterations -> 100,
                AccuracyGoal -> 20, WorkingPrecision -> 30]
```

FindMinimum::nrnum : The function value $3.52355600583903760634527590405 \times 10^{-13} + 3.52355600583903760634527590405 \times 10^{-13}\,i$ is not a real number at $\{\gamma, \xi\} = \{-12.9048559283425447432655656894, -1.4697171155259250561944 9642132\}$. More…

$$\text{Out[7]}= \quad \{-0.948315263746366995426698080269,$$
$$\{\gamma \to 2.72589719038320081697971941177,\ \xi \to -1.12099508030951941076974916096\}\}$$

A plot of the function to be maximized shows that the function does not have a maximum in the domain specified. For $\xi \to 0$, the overlap integral increases monotonously.

```
In[8]:= With[{ε = 10^-5},
            ContourPlot[Evaluate[int], {γ, ε, 2}, {ξ, ε, 3/2},
                PlotPoints -> 100, Contours -> 60,
                ColorFunction -> (Hue[0.8 #]&)]]
```

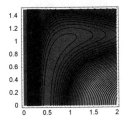

Looking at the series expansion of $\Psi(x)$ for $\xi \approx 0$ shows that for $\gamma = 1/2$, the function $\Psi(x)$ approaches $\phi_1(x)$.

```
In[9]:= ser = Series[Φ/normS, {ξ, 0, 4}] // Simplify
```

$$\text{Out[9]}= \quad 2\,e^{-x^2\gamma}\left(\frac{2}{\pi}\right)^{1/4} x\,\gamma^{3/4} + \frac{1}{3}\,e^{-x^2\gamma}\left(\frac{2}{\pi}\right)^{1/4} x\,\gamma^{7/4}\,(-3 + 4\,x^2\,\gamma)\,\xi^2 +$$
$$\frac{e^{-x^2\gamma}\,x\,\gamma^{11/4}\,(5 - 40\,x^2\,\gamma + 16\,x^4\,\gamma^2)\,\xi^4}{30\,2^{3/4}\,\pi^{1/4}} + O[\xi]^5$$

```
In[10]:= CoefficientList[ser, ξ] /. γ -> 1/2
```

$$\text{Out[10]}= \quad \left\{\frac{\sqrt{2}\,e^{-\frac{x^2}{2}}\,x}{\pi^{1/4}},\ 0,\ \frac{e^{-\frac{x^2}{2}}\,x\,(-3 + 2\,x^2)}{6\,\sqrt{2}\,\pi^{1/4}},\ 0,\ \frac{e^{-\frac{x^2}{2}}\,x\,(5 - 20\,x^2 + 4\,x^4)}{240\,\sqrt{2}\,\pi^{1/4}}\right\}$$

This means the overlap integral can be made arbitrarily close to 1.

```
In[11]:= Abs[Integrate[φ[1, x] #, {x, -Infinity, Infinity}]]& /@ %
```

$$Out[11]= \left\{1, 0, 0, 0, \frac{1}{48}\right\}$$

For the overlap integral with $\phi_3(x)$, we get a finite optimal value of about 0.27.

```
In[12]:= FullSimplify[1/normS *
            Assuming[γ > 0, Integrate[Ψ φ[3, x], {x, 0, Infinity}]]]
```

$$Out[12]= \frac{2 \, 2^{1/4} \, e^{-\frac{\gamma \xi^2}{1+2\gamma}} \, \gamma \, \xi \, (3 + 4 \, \gamma^2 \, (-3 + 2 \, \xi^2))}{\sqrt{3} \, \sqrt{\frac{1 - e^{-2\gamma\xi^2}}{\sqrt{\gamma}}} \, (1 + 2 \, \gamma)^{7/2}}$$

```
In[13]:= int = 2 2^(1/4) γ ξ (3 + 4 γ^2 (2 ξ^2 - 3)) Exp[-γ ξ^2/(1 + 2 γ)]/
            (Sqrt[3] Sqrt[(1 - Exp[-2 γ ξ^2])/Sqrt[γ]] (1 + 2 γ)^(7/2));

In[14]:= FindMinimum[Evaluate[int], {γ, 1}, {ξ, 1},
            MaxIterations -> 100, AccuracyGoal -> 10,
            WorkingPrecision -> 30, Method -> "Gradient"] // MapAt[-#&, #, 1]&
```

$$Out[14]= \{0.271009939062901523837852307891, \{γ \to 5.71914, ξ \to 0.587898\}\}$$

c) These are the time-dependent coherent states $\psi_C(\alpha; x, t)$.

```
In[1]:= ψC[α_][x_, t_] = -(-1)^(3/4) *
          Exp[(-x^2 + 2 Sqrt[2] x α Exp[- I t] -
          α^2 Exp[-2 I t] + Conjugate[α]^2 - 2 Re[α]^2)/2]/
          (Pi^(1/4) Sqrt[1 - I Cot[t]] Sqrt[Sin[t]]);
```

For $t = 0$, they reduce to simple Gaussians.

```
In[2]:= Series[ψC[α][x, t], {t, 0, 0}] // Normal // ExpandAll
```

$$Out[2]= \frac{e^{-\frac{x^2}{2} + \sqrt{2} \, x \, \alpha - \frac{\alpha^2}{2} + \frac{Conjugate[\alpha]^2}{2} - Re[\alpha]^2}}{\pi^{1/4}}$$

The coherent states $\psi_C(\alpha; x, t)$ fulfill the Schrödinger equation

$$i\hbar \frac{\partial \psi_C(\alpha; x, t)}{\partial t} = -\frac{\hbar^2}{2m} \frac{\partial^2 \psi_C(\alpha; x, t)}{\partial x^2} + \frac{m\omega^2}{2} x^2 \psi_C(\alpha; x, t).$$

(We use units $\hbar = 2$, $m = 2$, and $\omega = 1$.)

```
In[3]:= Simplify[2 I D[ψC[α][x, t], t] - (-D[ψC[α][x, t], x, x] + x^2 ψC[α][x, t])]

Out[3]= 0
```

The coherent states $\psi_C(\alpha; x, t)$ are eigenfunctions of the annihilation operator $a = 2^{-1/2} (x + \partial_x)$.

```
In[4]:= a = 1/Sqrt[2] (x # + D[#, x])&;

        a[ψC[α][x, t]]/ψC[α][x, t] // Simplify

Out[5]= e^{-it} α
```

The following two pictures visualize the coherent states $\psi_C(\alpha; x, t)$. In the x,t-plane they are periodically oscillating localized packets.

```
In[6]:= Module[{α = 4, ε = 10^-6, p3d, lcp},
          (* the 3D plot *)
          p3d = Plot3D[Re[Evaluate[ψC[5][x, t]]], {x, -12, 12}, {t, ε, 2Pi - ε},
                    PlotPoints -> 250, Mesh -> False, PlotRange -> All,
                    DisplayFunction -> Identity];
          (* the contour plot *)
          lcp = ListContourPlot[p3d[[1]], Contours -> 60,
                  ContourLines -> False, ColorFunction -> (Hue[0.8 #]&),
                  MeshRange -> {{-12, 12}, {ε, 2Pi - ε}},
                  DisplayFunction -> Identity];
          (* show both plots *)
          Show[GraphicsArray[{p3d, lcp}]]]
```

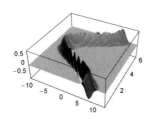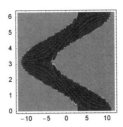

The relation between the energy

$$\int_{-\infty}^{\infty}\left(-\frac{\hbar^2}{2\,m}\frac{\partial^2\,\psi_C(\alpha;\,x,\,t)}{\partial x^2} + \frac{m\,\omega^2}{2}\,x^2\,\psi_C(\alpha;\,x,\,t)\right)\overline{\psi_C(\alpha;\,x,\,t)}\,dx$$

and α can be obtained by integration.

```
In[7]:= Integrate[Simplify[((-D[ψC[α][x, t], x, x] +  x^2 ψC[α][x, t]) *
                       (ψC[α][x, t] /.  c_Complex :> Conjugate[c]))],
             {x, -Infinity, Infinity},
             GenerateConditions -> False] // Together
```

$$\text{Out[7]= } -\frac{\mathbb{i}\,e^{-2\,\text{Im}[\alpha]^2}\,(1 + 2\,\alpha^2)\,\text{Csc}[t]}{\sqrt{\text{Csc}[t]^2}}$$

```
In[8]:= FullSimplify[I PowerExpand[%]]
```

$$\text{Out[8]= } e^{-2\,\text{Im}[\alpha]^2}\,(1 + 2\,\alpha^2)$$

Now, let us carry out the integration. We first check the stated identity numerically.

```
In[9]:= ϕ[n_, x_] = Exp[-x^2/2] HermiteH[n, x]/Sqrt[Sqrt[Pi] 2^n n!];
```

```
In[10]:= ψCInt[α_, x_, n_] :=
           NIntegrate[Evaluate[ψC[α][x, t] Exp[(n + 1/2) I t]], {t, 0, 2Pi}]
```

Here is a check for a "random" set of parameters.

```
In[11]:= With[{α = 5, x = Pi, n = 30},
              ψCInt[α, x, n]/(2 Pi Exp[-Abs[α]^2/2] α^n/Sqrt[n!] ϕ[n, x])]
Out[11]= 1. - 1.14705×10^{-14} i
```

The next graphic shows the harmonic oscillator eigenfunction $\phi_{30}(x)$ together with 241 numerically calculated values of the Fourier transform of the coherent state.

```
In[12]:= Off[NIntegrate::ploss]; Off[NIntegrate::ncvb];

        Module[{e = 30, n = 30, α1, data, pl, lp},
         α1 = Sqrt[2 e - 1.]/Sqrt[2];
         data = Table[{x, ψCInt[α1, x, n]}, {x, -15, 15, 1/8}];
         (* the calculated points *)
         lp = ListPlot[{#[[1]], Re[#[[2]]]}& /@ data, PlotJoined -> False,
                 Axes -> False, Frame -> True, DisplayFunction -> Identity,
                 PlotStyle -> {GrayLevel[0], PointSize[0.01]}];
         (* the exact harmonic oscillator eigenfunctions *)
         pl = Plot[2 Pi Exp[-α1^2/2] α1^n/Sqrt[n!] ϕ[n, x], {x, -15, 15},
                 PlotPoints -> 200, Axes -> False,
                 PlotStyle -> {Hue[0], Thickness[0.006]},
                 Frame -> True, DisplayFunction -> Identity];
         (* display both parts *)
         Show[{pl, lp}, DisplayFunction -> $DisplayFunction]]
```

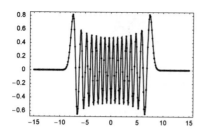

To visualize how the harmonic oscillator eigenfunctions emerge $\int_0^T \psi_C(\alpha; x, t)\, e^{(n+1/2)it}\, dt$ as a T approaches 2π, we calculate interpolating functions that approximate the integral as a function of T.

```
In[14]:= ψCIntT[α_, x_, n_] :=
         With[{ε = 10^-8},
           NDSolve[{int'[t] == ψC[α][x, t] Exp[(n + 1/2) I t],
               int[ε] == 0}, int, {t, ε, 2Pi - ε}, MaxStepSize -> 0.1]]
```

We choose α such that the expectation values of the energy of the coherent states coincides with the harmonic oscillator eigenfunction's eigenvalues $1 + 2\alpha^2 = 2n + 1$.

```
In[15]:= Module[{δx = 1/20, n = 30, pp = 200, α1, ipo, ε = 10^-8},
         α1 = Sqrt[n];
         Do[ipo[x] = ψCIntT[α1, x, n][[1, 1, 2]], {x, -15, 15, δx}];
         data = Table[ipo[x][t], {x, -15, 15, δx},
                       {t, ε, 2Pi - ε, (2Pi - 2ε)/pp}]];
```

The following pictures show 3D and contour plots of how the harmonic oscillator eigenfunctions develop as a function of T. The second row of pictures shows the imaginary parts. For $T = 2\pi$, it vanishes. For both the real and the imaginary part, we show a 3D plot and a contour plot.

```
In[16]:= Module[{(* common option settings *)
                 opts = Sequence[PlotRange -> All, DisplayFunction -> Identity,
                                 MeshRange -> {{-15, 15}, {0, 2Pi}}]},
         Function[reIm, Show[GraphicsArray[{
         (* the 3D plot *)
         ListPlot3D[reIm[Transpose[data]], opts,
                    Mesh -> False, Ticks -> False, ViewPoint -> {0, -2, 1}],
         (* the contour plot *)
         ListContourPlot[reIm[Transpose[data]], opts, FrameTicks -> False,
                    ColorFunction -> (Hue[0.8 #]&),
                    Contours -> 50, ContourLines -> False]}]]] /@ {Re, Im}]
```

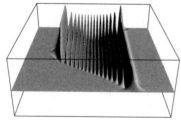

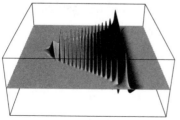

d) We will assume $n = 50$ in the following calculations. Here are the harmonic oscillator eigenfunctions and the classical probability density.

```
In[1]:= n = 50;

In[2]:= φ[n_, x_] := Exp[-x^2/2] HermiteH[n, x]/Sqrt[Sqrt[Pi] 2^n n!]
        Pcl[x_] := UnitStep[x + Sqrt[2n + 1]] *
                      UnitStep[Sqrt[2n + 1] - x] 1/Pi/Sqrt[2n + 1 - x^2];
        pcl = {{Hue[0], Thickness[0.002], Line[Table[{x, Pcl[x]} // N,
                                            {x, -12, 12, 24/500}]]}};
```

A graphic shows clearly that $P_{cl}(x)$ is the average of $P_{qu}(x)$ locally.

```
In[5]:= Plot[φ[50, x]^2, {x, -12, 12}, PlotPoints -> 500,
            Frame -> True, Axes -> False, Prolog -> pcl]
```

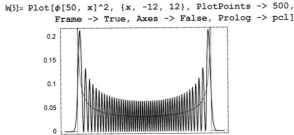

The averaged probability density $\tilde{P}_{qu}(x; \varepsilon)$ we calculate numerically from the exact indefinite integral.

```
In[6]:= H[x_] = (Integrate[#, x]& /@ Expand[φ[50, x]^2]) // Simplify;

In[7]:= PAv[x_, ε_] := With[{ξ = SetPrecision[x, 30], e = SetPrecision[ε, 30]},
                           (H[ξ + e] - H[ξ - e])/(2 e)]
```

The next graphic shows the mean difference $\delta(\varepsilon) = \int_{-\infty}^{\infty} \left(\tilde{P}_{qu}(x; \varepsilon) - P_{cl}(x)\right)^2 dx$ between $\tilde{P}_{qu}(x; \varepsilon)$ and $P_{cl}(x)$.

```
In[8]:= δ[ε_] := 2 NIntegrate[Evaluate[(
            Interpolation[Table[{x, PAv[x, ε]},
                                {x, 0, 10.1, 10.1/500}]][x] - Pcl[x])^2],
                                {x, 0, Sqrt[101] - 10^-6}]

In[9]:= Off[NIntegrate::ncvb]
        δs = Table[{10^e, δ[10^e]}, {e, -2, 1, 3/30}];
        ListPlot[δs // N, PlotJoined -> True]
```

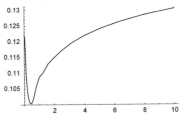

The last graphic shows that an optimal averaging ε is about 0.4. This agrees with a handwaving approximation in which we want to smooth $P_{qu}(x)$ over about one period, this means $\varepsilon \approx 2\sqrt{2/n}$. The resulting $\tilde{P}_{qu}(x; 0.4)$ agrees well with $P_{qcl}(x)$.

```
In[12]:= ipo = Interpolation[Table[{x, PAv[x, 0.4]}, {x, -12, 12, 24/1000}]];
         Plot[ipo[x], {x, -12, 12}, PlotRange -> All]
```

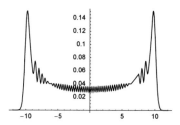

For the simultaneous limit $n \to \infty$ and $\hbar \to 0$, see [83].

e) Here, we define the functions $\Psi_n(\alpha, z)$.

```
In[1]:= φ[n_, z_] := Exp[-z^2/2] HermiteH[n, z]/Sqrt[Sqrt[Pi] 2^n n!]
```

```
In[2]:= Ψ[η_, n_, z_] = Evaluate //@ (1/Sqrt[n! LaguerreL[n, 1 - 1/η]] *
        Sum[(1/η - 1)^((n - k)/2) n!/(n - k)!/Sqrt[k!] φ[k, z], {k, 0, n}]);
```

To interpolate we use the simple function $\eta(a) = a\,n + (1-a)\,\alpha \big/ \sqrt{a}$.

```
In[3]:= IφΨ[a_, x_] := With[{n = 12, α = 2},
                          Ψ[a, Round[a n + (1 - a) α/Sqrt[a]], x]]
```

The transition between the harmonic oscillator eigenstate and the coherent state is piecewise smooth. Each time the number of summands in $\Psi_n(\alpha, z)$ increases, we get discontinuities.

```
In[4]:= Plot3D[IφΨ[a, x], {x, -6, 6}, {a, 0.03, 1 - 10^-4},
              PlotPoints -> 120, Mesh -> False,
              PlotRange -> All, ViewPoint -> {0, 3, 1}]
```

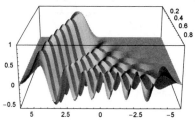

f) The following inputs implement the new eigenfunctions $\psi_n[\phi_n](z;\lambda)$ and the potential $\mathcal{V}[\phi_n](z;\lambda)$.

```
In[1]:= int[n_, z_] := int[n, z] =
        With[{I = Integrate[#, {ζ, -Infinity, z}, GenerateConditions -> False]&},
          (* thread over sums *)
          If[Head[#] === Plus, I /@ #, I[#]]&[Expand[φ[n][ζ]^2]]]
```

```
ψ[n_][λ_, z_] := ψ[n][λ, z] = Sqrt[λ (λ + 1)] φ[n][z]/(int[n, z] + λ);
```

```
V[n_][λ_, z_] := V[n][λ, z] = V[z] - 2D[Log[int[n, z] + λ], {z, 2}]
```

Here is a short verification that the $\psi_n[\phi_n](z;\lambda)$ are eigenfunctions of the potential $\mathcal{V}[\phi_n](z;\lambda)$.

```
In[6]:= -D[ψ[n][λ, z], {z, 2}] + V[n][λ, z] ψ[n][λ, z] -
         ε ψ[n][λ, z] // Simplify // Numerator
Out[6]= -√(λ (1 + λ)) ((ε - V[z]) φ[n][z] + φ[n]''[z])
```

In the limit $\lambda \to \infty$, we recover the starting eigenfunctions $\phi_n(z)$ and the starting potential $V(z)$.

```
In[7]:= Series[ψ[n][λ, z], {λ, Infinity, 3}] //
        DeleteCases[#, GenerateConditions -> False]& // Simplify
```

Out[7]= $\phi[n][z] + \dfrac{(1 - 2 \int_{-\infty}^{z} \phi[n][\zeta]^2 \, d\zeta) \, \phi[n][z]}{2\lambda} +$

$\dfrac{1}{8} \left(-1 - 4 \int_{-\infty}^{z} \phi[n][\zeta]^2 \, d\zeta + 8 \left(\int_{-\infty}^{z} \phi[n][\zeta]^2 \, d\zeta \right)^2 \right) \phi[n][z] \left(\dfrac{1}{\lambda} \right)^2 +$

$\dfrac{1}{16} \left(1 + 2 \int_{-\infty}^{z} \phi[n][\zeta]^2 \, d\zeta + 8 \left(\int_{-\infty}^{z} \phi[n][\zeta]^2 \, d\zeta \right)^2 - 16 \left(\int_{-\infty}^{z} \phi[n][\zeta]^2 \, d\zeta \right)^3 \right) \phi[n][z] \left(\dfrac{1}{\lambda} \right)^3 + O\left[\dfrac{1}{\lambda} \right]^4$

In[8]:= `Series[V[n][λ, z], {λ, Infinity, 2}] //`
`  DeleteCases[#, GenerateConditions -> False, Infinity]& // Simplify`

Out[8]= $V[z] - \dfrac{4 \, (\phi[n][z] \, \phi[n]'[z])}{\lambda} + \left(2 \, \phi[n][z]^4 + 4 \left(\int_{-\infty}^{z} \phi[n][\zeta]^2 \, d\zeta \right) \phi[n][z] \, \phi[n]'[z] \right) \left(\dfrac{1}{\lambda} \right)^2 + O\left[\dfrac{1}{\lambda} \right]^3$

Here are the explicit $\phi_n(z)$ for the harmonic oscillator.

In[9]:= `V[z_] := z^2;`
`  φ[n_][z_] = 1/Sqrt[Sqrt[Pi] 2^n n!] Exp[-z^2/2] HermiteH[n, z];`

Of course, also in this special case the $\psi_n[\phi_n](z; \lambda)$ are eigenfunctions of the potential $\mathcal{V}[\phi_n](z; \lambda)$.

In[11]:= `Table[-D[ψ[n][λ, z], {z, 2}] + V[n][λ, z] ψ[n][λ, z] -`
`    (2n + 1) ψ[n][λ, z], {n, 0, 5}] // Together`

Out[11]= `{0, 0, 0, 0, 0, 0}`

The following two graphics show $\psi_n[\phi_n](z; \lambda)$ and $\mathcal{V}[\phi_n](z; \lambda) - z^2$ for $n = 0, 1, ..., 8$.

In[12]:= `gr1 = Show[Table[`
`    Plot[Evaluate[Table[n + ψ[n][λ, z], {λ, 1/24, 2, 2/24}]],`
`      {z, -5, 5}, PlotStyle ->`
`      Table[{Thickness[0.002], Hue[λ/3]}, {λ, 1/24, 2, 2/24}],`
`      DisplayFunction -> Identity, Frame -> True, Axes -> False],`
`      {n, 0, 8}], PlotRange -> All]`

In[13]:= `gr2 = (DownValues[In][[-2]] /. (* modify last input *)`
`    HoldPattern[ψ[n][λ, z]] :> (V[n][λ, z] - V[z])/10)[[2]]`

In[14]:= `Show[GraphicsArray[{gr1, gr2}]]`

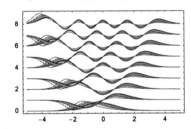 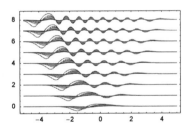

Now let us try to conjecture the values of the two mentioned integrals. Carrying out some numerical experiments for the integrand $\mathcal{V}[\phi_n](z; \lambda) - V(z)$ for various n and positive λ yields immediately the conjecture that the integral is 0.

In[15]:= `Off[NIntegrate::ploss]; Off[NIntegrate::ncvb];`
`  Table[{n, NIntegrate[Evaluate[Together[V[n][1/(1 + n), z] - V[z]]],`
`      {z, -6, 0, 6}, WorkingPrecision -> 22,`
`      AccuracyGoal -> 6]}, {n, 0, 5}]`

Out[16]= $\{\{0, 0. \times 10^{-7}\}, \{1, 0. \times 10^{-7}\}, \{2, 0. \times 10^{-7}\}, \{3, 0. \times 10^{-7}\}, \{4, 0. \times 10^{-7}\}, \{5, 0. \times 10^{-7}\}\}$

Note that the potential $\mathcal{V}[n](\lambda, z)$ is generically not an even function.

In[17]:= `{V[3][1/8, z] /. z -> 2.3, V[3][1/8, z] /. z -> -2.3}`

Out[17]= `{6.19922, 5.01627}`

Carrying out a similar experiment for the integrand $(\psi_n[\phi_n](z; \lambda) - \phi_n(z))^2$ suggests that the integral is only a function of λ and independent of n.

In[18]:= `{#, Table[NIntegrate[Evaluate[(ψ[k][#, z] - φ[k][z])^2], {z, -8, 8}],`
`    {k, 0, 4}]}& /@ (* three "random" values *) {1, E, Pi}`

Out[18]= {{1, {0.0394837, 0.0394837, 0.0394837, 0.0394837, 0.0394837}},
        {e, {0.00815439, 0.00815439, 0.00815439, 0.00815439, 0.00815439}},
        {π, {0.00634999, 0.00634999, 0.00634999, 0.00634999, 0.00634999}}}

To calculate the value of the integral we choose the simplest case, namely $n = 0$. This is the integrand.

In[19]:= Evaluate[(ψ[0][λ, z] - φ[0][z])^2] // Simplify

Out[19]= $\dfrac{e^{-z^2}\ \left(1 + 2\lambda - 2\sqrt{\lambda\ (1+\lambda)}\ + \text{Erf}[z]\right)^2}{\sqrt{\pi}\ (1 + 2\lambda + \text{Erf}[z])^2}$

Changing variables allows us to calculate the exact value of the integral.

In[20]:= Integrate[(% /. Erf[z] -> ζ) Sqrt[Pi]/2/Exp[-z^2], {ζ, -1, 1},
                Assumptions -> λ > 0]

Out[20]= $2 + 2\sqrt{\lambda\ (1+\lambda)}\ \text{Log}[\lambda] - 2\sqrt{\lambda\ (1+\lambda)}\ \text{Log}[1+\lambda]$

In[21]:= δ[λ_] = FullSimplify[%, λ > 0]

Out[21]= $2 + 2\sqrt{\lambda\ (1+\lambda)}\ \text{Log}\left[\dfrac{\lambda}{1+\lambda}\right]$

This result agrees with the above-calculated numerical values.

In[22]:= {δ[1], δ[E], δ[-Pi]} // N

Out[22]= {0.0394837, 0.00815439, 3.98782}

Here is the value of the integral as a function of λ shown.

In[23]:= Plot[Re[2 + 2 Sqrt[λ(1 + λ)]Log[λ/(λ + 1)]],
                {λ, -6, 6}, Frame -> True, Axes -> False]

For isospectral transformations for the 1D wave equation, see [234].

g) To avoid the repeated calculation of high-order Hermite polynomials, we will calculate approximations of $\phi_k(z)$, $0 \leq n \leq 200$ using 1000 points.

In[1]:= φ[n_, z_] := Exp[-z^2/2] HermiteH[n, z]/Sqrt[Sqrt[Pi] 2^n n!]

In[2]:= pp = 10^3; n0 = 10^2;
        Do[φ[n] = Table[φ[n, N[z]], {z, -20, 20, 40/pp}] //
                                Developer`ToPackedArray, {n, 0, 200}];

In[4]:= Do[Φ[100, m] = 1/Sqrt[2m + 1] Plus @@ Table[φ[n0 + k], {k, -m, m}],
        {m, 0, n0}]

Here are the resulting wave packets.

In[5]:= ListPlot3D[Table[(Φ[100, m])^2, {m, 0, n0}], Mesh -> False,
                PlotRange -> All, MeshRange -> {{-20, 20}, {0, 2n0}}]

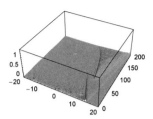

The last graphic suggests that we have optimal localization near $m \approx 10$. Calculating an approximative value for the spatial spread $\int_{-\infty}^{\infty} |z| \, \Psi_n^{(m)}(z - \bar{z}) \, dz$ (where $\bar{z} = \int_{-\infty}^{\infty} z \, \Psi_n^{(m)}(z) \, dz$), we obtain $m \approx 12$.

```
In[6]:= zExtension[l_List] :=
          Module[{sum = MapIndexed[{-20 + 40 #2[[1]]/pp, #1}&, l], xm, sumS},
                (* center *)
                xm = Plus @@ (Times @@@ sum)/Plus @@ (Last /@ sum);
                (* shifted wave packet *)
                sumS = ({First[#] - xm, Last[#]}&) /@ sum;
                (* average of Abs[x] *)
                Plus @@ (Abs[#1] #2& @@@ sumS)/Plus @@ (Last /@ sumS)]

In[7]:= ListPlot[Table[{m, zExtension[Φ[n0, m]^2]}, {m, 0, 40}]]
```

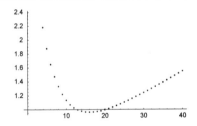

The time-dependent states $\Psi_n^{(m)}(z, t)$ show that the localization property is a function of t.

```
In[8]:= ΦtGraphics[t_, opts___] :=
          With[{Φts = Table[1/Sqrt[2m + 1] Plus @@ Table[
                          φ[n0 + k]  Exp[-I (2(n0 + k) + 1) N[t]],
                          {k, -m, m}], {m, 0, n0}]},
          ListPlot3D[Abs[Φts]^2, opts, PlotRange -> All, Mesh -> False,
                    Axes -> False, PlotRange -> {All, All, {0, 1.5}}]]

In[9]:= Show[GraphicsArray[ΦtGraphics[#, DisplayFunction -> Identity]& /@ #]]& /@
          Partition[Table[t, {t, 0, Pi/2, Pi/2/8}], 3]
```

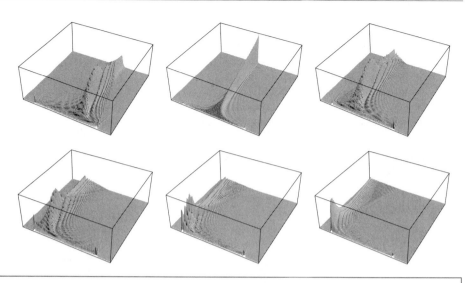

```
Do[ΨtGraphics[t], {t, 0, Pi/2, Pi/2/20}]
```

h) The implementations of the *d*-dimensional factorial, *d*-dimensional derivative, and the *d*-dimensional Hermite polynomials are straightforward.

```
In[1]:= dNumber[d_][n_] := n + (d - 1)/2 (1 - (-1)^n)
        dFactorial[d_][n_] := Product[dNumber[d][k], {k, 1, n}]
        (* == If[EvenQ[n], 2^n (n/2)! Gamma[(n + d)/2]/Gamma[d/2],
                           2^n ((n - 1)/2)! Gamma[(n + d + 1)/2]/Gamma[d/2]] *)

In[4]:= dD[d_][f_, x_] := D[f, x] + (d - 1)/2/(2x) (f - (f /. x -> -x))

In[5]:= dHermiteH[d_][n_, x_] := Factor[Together[
          n!/dFactorial[d][n] (-1)^n Exp[x^2] Nest[dD[d][#, x]&, Exp[-x^2], n]]]

In[6]:= ϕ[d_][n_, x_] := dFactorial[d][n] /(n! Sqrt[Pi^(d/2) 2^n dFactorial[d][n]])*
                         Exp[-x^2/2] dHermiteH[d][n, x]
```

Here are closed forms for the lowest eigenfunctions. The dimension *d* appears polynomially in the *d*-dimensional Hermite polynomials and nonpolynomially in the normalization.

```
In[7]:= Table[ϕ[d][n, x], {n, 0, 4}] // Simplify
```

$$
Out[7]= \left\{ \frac{e^{-\frac{x^2}{2}}}{\sqrt{\pi^{d/2}}}, \ \frac{\sqrt{2}\, e^{-\frac{x^2}{2}}\, x}{\sqrt{d\, \pi^{d/2}}}, \ -\frac{e^{-\frac{x^2}{2}}\, (1 + d - 4\, x^2)}{2\sqrt{2}\, \sqrt{d\, \pi^{d/2}}}, \right.
$$

$$
\left. -\frac{e^{-\frac{x^2}{2}}\, x\, (5 + d - 4\, x^2)}{2\sqrt{d\, (2 + d)\, \pi^{d/2}}}, \ \frac{e^{-\frac{x^2}{2}}\, (5 + d^2 - 40\, x^2 + 16\, x^4 + d\, (6 - 8\, x^2))}{8\sqrt{2}\, \sqrt{d\, (2 + d)\, \pi^{d/2}}} \right\}
$$

The next graphics show the real and imaginary part and the absolute value of the first four eigenfunctions.

```
In[8]:= Table[Show[GraphicsArray[
          Plot3D[Evaluate[#[ϕ[d][n, x]]], {x, -2, 2}, {d, -6, 6}, PlotPoints -> 80,
            ClipFill -> None, Mesh -> False, PlotRange -> {-3, 3},
            DisplayFunction -> Identity]& /@ {Re, Im, Abs}]], {n, 0, 3}]
```

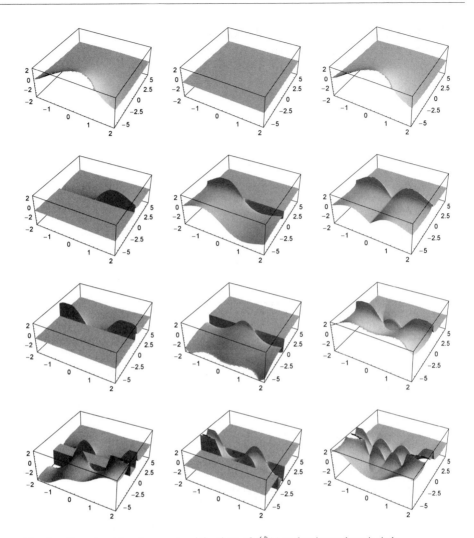

The following graphics show the real and imaginary part and the phase of $\phi_n^{(d)}(x)$ as d varies on the unit circle.

```
In[9]:= Show[GraphicsArray[
    Plot3D[Evaluate[#[φ[Exp[I φ]][8, x]]], {x, -4, 4}, {φ, 0, 2Pi},
        PlotPoints -> 80, ClipFill -> None, Mesh -> False,
        PlotRange -> All, DisplayFunction -> Identity]& /@ {Re, Im, Arg}]]
```

10. High-Order Perturbation Theory, Eigenvalue Differential Equation

a) We calculate the matrix elements for $V(z) = z^4$ in the harmonic oscillator basis.

```
In[1]:= reduceProducts[expr_] :=
          FixedPoint[Expand[# /. z_^m_. HermiteH[n_, z_] :>
            (* recursion relation for Hermite polynomials *)
            z^(m - 1)(n HermiteH[n - 1, z] + HermiteH[n + 1, z]/2)]&, expr]

In[2]:= makeh[V_, z_] :=
          Sqrt[2^m m!]/Sqrt[2^n n!] Simplify[reduceProducts[V HermiteH[n, z]] /.
          (* use orthogonality *) HermiteH[n_, z] -> KroneckerDelta[n, m]]

In[3]:= (V[n_, m_] = Sqrt[2^n n!]/Sqrt[2^m m!] makeh[z^4, z]) /.
                  KroneckerDelta[a_, b_] :> Subscript[δ, a, b]
```

$$Out[3]= \ n\,(-6 + 11\,n - 6\,n^2 + n^3)\,\delta_{m,-4+n} +$$
$$\frac{1}{16}\,(16\,n\,(1 - 3\,n + 2\,n^2)\,\delta_{m,-2+n} + 12\,(1 + 2\,n + 2\,n^2)\,\delta_{m,n} + 12\,\delta_{m,2+n} + 8\,n\,\delta_{m,2+n} + \delta_{m,4+n})$$

We write $\varepsilon[n][j]$ for $^{(n)}\varepsilon_j$ and $c[n][j][k]$ for $^{(n)}c_{j,k}$. The implementation of the above recursion relations for $^{(n)}\varepsilon_j$ and $^{(n)}c_{j,k}$ is straightforward. We choose $N_n = 2^n\,n!$ to simplify the calculations and so avoid the calculation of any factorials.

```
In[4]:= ε[n_][0] := 2n + 1

        ε[n_][j_] := ε[n][j] =
            Sum[If[V[n, k] === 0, 0, V[n, k] c[n][j - 1][k]], {k, n - 4, n + 4}]

        c[n_][j_][n_] := KroneckerDelta[j, 0]

        c[n_][0][k_] := KroneckerDelta[k, n]

        c[n_][j_][k_] := c[n][j][k] = (1/(ε[k][0] - ε[n][0])*
          (Sum[ε[n][l] c[n][j - 1][k], {l, 1, j - 1}] -
          Sum[If[V[k, l] === 0, 0, V[k, l] c[n][j - 1][l]], {l, k - 4, k + 4}]))
```

Here are the first 10 values of the $^{(0)}\varepsilon_j$.

```
In[9]:= Table[ε[0][k], {k, 0, 10}] // Timing
```

$$Out[9]= \left\{0.05\ \text{Second},\ \left\{1,\ \frac{3}{4},\ -\frac{21}{16},\ \frac{333}{64},\ -\frac{30885}{1024},\ \frac{916731}{4096},\ -\frac{65518401}{32768},\right.\right.$$
$$\left.\left.\frac{2723294673}{131072},\ -\frac{1030495099053}{4194304},\ \frac{54626982511455}{16777216},\ -\frac{6417007431590595}{134217728}\right\}\right\}$$

Analyzing the $^{(0)}c_{j,k}$ with nontrivial values shows that they fulfill the condition $k \le 4\,i$; we will use this fact later).

```
In[10]:= cValues = ({{{#[[1, 1, 1]], #[[1, 1, 0, 1]]},
                  Sign[Abs[#[[2]]]]}&) /@ Drop[SubValues[c], -4];

In[11]:= Show[Graphics[{If[#[[2]] === 0, GrayLevel[1/2], Hue[0]],
                  Rectangle[#[[1]] - 1/3, #[[1]] + 1/3]}& /@ cValues],
              AspectRatio -> 1, Frame -> True]
```

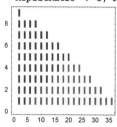

By temporarily redefining the `KroneckerDelta` function, we obtain the first few values of $^{(n)}\varepsilon_j$ for symbolic n [167], [160], [27], [237].

```
In[12]:= Unprotect[KroneckerDelta];
         KroneckerDelta[v + a_., v + b_.] := 0 /; a =!= b

In[14]:= Table[{i, ε[v][i] // Together // Expand}, {i, 0, 5}]
```

Out[14]= $\left\{\{0, 1 + 2\,v\}, \left\{1, \dfrac{3}{4} + \dfrac{3\,v}{2} + \dfrac{3\,v^2}{2}\right\}, \left\{2, -\dfrac{21}{16} - \dfrac{59\,v}{16} - \dfrac{51\,v^2}{16} - \dfrac{17\,v^3}{8}\right\},\right.$

$\left\{3, \dfrac{333}{64} + \dfrac{1041\,v}{64} + \dfrac{177\,v^2}{8} + \dfrac{375\,v^3}{32} + \dfrac{375\,v^4}{64}\right\},$

$\left\{4, -\dfrac{30885}{1024} - \dfrac{111697\,v}{1024} - \dfrac{80235\,v^2}{512} - \dfrac{71305\,v^3}{512} - \dfrac{53445\,v^4}{1024} - \dfrac{10689\,v^5}{512}\right\},$

$\left. \left\{5, \dfrac{916731}{4096} + \dfrac{3569679\,v}{4096} + \dfrac{3090693\,v^2}{2048} + \dfrac{2786805\,v^3}{2048} + \dfrac{3662295\,v^4}{4096} + \dfrac{262647\,v^5}{1024} + \dfrac{87549\,v^6}{1024}\right\}\right\}$

In[15]:= `KroneckerDelta[v + a_., v + b_.] =.`
`Protect[KroneckerDelta];`

Now let us implement a more efficient recursion for ground-state values $^{(0)}\varepsilon_j$. We optimize the following calculation for memory and avoid the storage of vanishing $^{(0)}c_{j,k}$. The function `makeDefinition` sets up the necessary definitions for calculating the $^{(0)}\varepsilon_j$ to precision *prec* (`Infinity` gives exact values). The variables ϵ and $\mathcal{V}$ are globally visible.

In[17]:= `makeDefinition[prec_] :=`
```
((* remove existing definitions *)Clear[e, V, c];
(* remember values of V *)
V[n_, m_] := V[n, m] = V[n, m];
(* the recursive definitions for e and c *)
e[0] = N[1, prec];
e[i_] := e[i] = SetPrecision[Sum[V[0, k] c[i - 1, k], {k, -4, 4, 2}], prec];
c[0, 0] = N[1, prec];
c[i_, 0] = 0;
c[0, k_] = 0;
c[i_, k_] := (c[i, k] = 0) /; k > 4i;
c[i_, k_] := c[i, k] = (* use prescribed precision at each step *)
  SetPrecision[
  (Sum[e[j] c[i - j, k], {j, i - 1}] -
  Sum[V[k, n] c[i - 1, n], {n, k - 4, k + 4, 2}])/(2k), prec])
```

In[18]:= `$RecursionLimit = Infinity;`

Now let us calculate the first 100 and 200 values of the $^{(0)}\varepsilon_j$ [60], [61], [574].

In[19]:= `makeDefinition[Infinity]`
`{Table[e[i], {i, 0, 100}]}; // Timing, MemoryInUse[]}`
Out[20]= `{{10.04 Second, Null}, 5900456}`

In[21]:= `{N[#], #}& @ e[100]`
Out[21]= `{-3.28935×10^174,`
`  -41878320957485246059475829426903698800162646287922978002634877338214172599300622290`
`  484427080138253842161250173908984941843608395640039751570160005591333761454831562`
`  856431481017572485955052988729434494787549075929269306577338553740256951507787836`
`  60618847102306340395/`
`  127314748520905380391777855525586135065716774604121015664758778084648831235208544`
`  136462336}`

In[22]:= `{Table[e[i], {i, 0, 200}]}; // Timing, MemoryInUse[]}`
Out[22]= `{{89.71 Second, Null}, 20110096}`

The calculation time is proportional to $j^{4+\varepsilon}$ and the memory use is proportional to $j^{2+\varepsilon}$. This means that within a day using less than 1 GB of memory the calculation of $^{(0)}\varepsilon_{1000}$ is feasible on a year-2005 computer. Here is an approximate value of the number of hours and the amount of memory it will take.

In[23]:= `{Round[%[[1, 1, 1]] (1000/200)^4.2/3600] hours,`
`         Round[%[[2]] (1000/200)^2.2/10.^6 MB}`
Out[23]= `{21 hours, 693.662 MB}`

We can get an estimation for the value of the divergent series using `SequenceLimit`.

In[24]:= `SequenceLimit[N[FoldList[Plus, 0, Table[e[i], {i, 0, 100}]], 400],`
`                WynnDegree -> 30]`
`         SequenceLimit::seqlim : The general form of the sequence`
`             could not be determined, and the result may be incorrect. More…`

Out[24]= 1.39235164153029

Now, we use the exact values to carry out the Padé approximation. The function `formalPowerSeriesToDiagonal` `Padé` calculates the numerical value of a (divergent) series with coefficients *as* at *x*.

```
In[25]:= formalPowerSeriesToDiagonalPadé[as_, M_, x_] :=
          Det[Table[If[j != M, as[[i + j + 2]],
                    Sum[as[[k - M + i + 1]] x^k, {k, M - i, M}]],
                  {j, 0, M}, {i, 0, M}]]/
          Det[Table[If[j != M, as[[i + j + 2]], x^(M - i)], {j, 0, M}, {i, 0, M}]]

In[26]:= Υ[x_, M_, prec_] := formalPowerSeriesToDiagonalPadé[
                    N[Table[ε[k], {k, 0, 2M + 2}], prec], M, N[x, prec]]
```

Taking $M = 100$ shows that $\varepsilon_0 = 1.39235164153029...$ [521] in agreement with the above value from `SequenceLimit`.

```
In[27]:= Table[{o, SetPrecision[Υ[1, o, 600], 15]}, {o, 20, 100, 20}] // TableForm
```

20	1.39234770518470
40	1.39235163134848
Out[27]//TableForm= 60	1.39235164142807
80	1.39235164152820
100	1.39235164153022

For multivalued Padé approximations of the anharmonic oscillator, see [514], [197]. For other methods to sum a finite number of terms of a divergent series numerically, see [618], [20], [200], [417], [473], [57], [440], and [63]; for instanton contribution-based calculations of such potentials, see [299]. For the reconstruction of the singularities of a function from the series, see [246]. For a renormalized series, see [168], [326], and [58]. For perturbations of the form λx^4 for large λ, see [400].

The following inputs use a sequence transformation [615], [616], [614] to extrapolate the value of $\sum_{j=0}^{\infty} {}^{(0)}\varepsilon_j$.

```
In[28]:= (* form partial sums E[n] *)
          MapIndexed[(E[#2[[1]] - 1] = #1)&,
            Rest @ FoldList[Plus, 0, Table[ε[k], {k, 0, 200}]]];

In[30]:= (* the sequence transformation of order k *)
          δ[k_, n_] :=
          (Sum[(-1)^j Binomial[k, j] Pochhammer[n + j + 1, k - 1]/
                  ε[n + j + 1], {j, 0, k}])^-1 *
          Sum[(-1)^j Binomial[k, j] Pochhammer[n + j + 1, k - 1] E[n + j]/
                  ε[n + j + 1], {j, 0, k}]

In[32]:= Table[{k, δ[k, 10] // N[#, 20]&}, {k, 40, 180, 20}] // TableForm
```

40	1.3923419896733785866
60	1.3923515771351775521
80	1.3923516423909829683
Out[32]//TableForm= 100	1.3923516415206794092
120	1.3923516415306481736
140	1.3923516415302869185
160 .	1.3923516415302914909
180	1.3923516415302918434

The last two results agree well with a direct diagonalization of the corresponding Hill matrix.

```
In[33]:= reduceProducts[expr_] :=
          FixedPoint[Expand[# /. z_^m_ . HermiteH[n_, z_] :>
                    (* recursion relation for Hermite polynomials *)
                    z^(m - 1) (n HermiteH[n - 1, z] + HermiteH[n + 1, z]/2)]&, expr]

In[34]:= makeHillh[V_, z_] := Sqrt[2^m m!]/Sqrt[2^n n!] Simplify[
          reduceProducts[(2n + 1) HermiteH[n, z] - z^2 HermiteH[n, z] +
                    (* the potential *) V HermiteH[n, z]] /.
          (* use orthogonality *) HermiteH[n_, z] -> KroneckerDelta[n, m]]

In[35]:= hillh[n_, m_] = makeHillh[z^2 + z^4, z];

In[36]:= (* extract lowest eigenvalue *)
          lowestEigenValue[p_, prec_] :=
            Last[Eigenvalues[N[Table[hillh[n, m], {n, 0, p, 2}, {m, 0, p, 2}], prec]]]
```

```
In[38]:= Table[{o, SetPrecision[lowestEigenValue[o, 30], 15]},
           {o, 20, 120, 20}] // TableForm
```

Out[38]//TableForm=

20	1.39235244305397
40	1.39235164153844
60	1.39235164153030
80	1.39235164153029
100	1.39235164153029
120	1.39235164153029

A quick method to obtain a rough approximation is the root approximant [630]. Taking only the first seven terms into account, we obtain the value of the lowest eigenvalue within 1%.

```
In[39]:= Module[{o = 3, A, n, rootAnsatz, series, eqs, nAs, sols},
           (* form kth-order root approximant ansatz *)
           step[{o_, k_}] := {(o + A[k] x^k)^n[k], k + 1};
           rootAnsatz = Nest[step, {1, 1}, o][[1]];
           (* series expansion for small x *)
           series = Series[rootAnsatz, {x, 0, 2 o}];
           (* solve equations for the n[k] and A[k] *)
           eqs = Rest[CoefficientList[series, x]] - Table[ε[0][k], {k, 2o}];
           nAs = Union[Cases[eqs, _A | _n, Infinity]];
           sols = NSolve[eqs, nAs];
           (* largest root approximant for x == 1 *)
           Max[Cases[rootAnsatz /. sols /. x -> 1, _Real]]]
```

Out[39]= 1.37987

Now, we will try to determine the α_k. To do this we need sufficiently many of the $^{(0)}\varepsilon_j$ to carry out a numerical fit. We calculate high-precision values for the first 500. Using 200 digits makes the calculation much faster than the exact one from above, mainly because no fractions have to be reduced.

```
In[40]:= makeDefinition[200]
           Table[ε[i], {i, 0, 500}]; // Timing
```

Out[41]= {620.24 Second, Null}

```
In[42]:= ε[500]
```

Out[42]= -5.317093881145884971799375679183193084386961628654137602204465809386378277638386073\
         312609710377743314733721304116773895678636522228373314130302067128087187740332790\
         925724506477765109791092248803750625×10^1220

After dividing out the exponentially growing prefactor and subtracting 1, we are left with a sequence that asymptotically approaches 0.

```
In[43]:= α[k_] := -(-1)^k 2^(3/2 - k) 3^(k + 1/2) (k - 1/2)!/Pi^(3/2);
```

```
In[44]:= ListPlot[Table[{k, Abs[ε[k]/α[k]] - 1}, {k, 100}],
             PlotRange -> {-1, 1}, Frame -> True, Axes -> False]
```

A numerical fit to this sequence yields the following result. We fit the sequence with the ansatz $\sum_{j=1}^{5} \alpha_j k^{-j}$.

```
In[45]:= data = Table[{k, Abs[ε[k]/α[k]] - 1}, {k, 200, 500}];
           fit1 = Fit[SetPrecision[data, 40], {1/k, 1/k^2, 1/k^3, 1/k^4, 1/k^5}, k]
```

Out[46]= $-\dfrac{360.98797738558960359265905452}{k^5} -$

$\dfrac{39.723120868839670609533283744}{k^4} - \dfrac{6.8880410248830534140908927100061}{k^3} -$

$\dfrac{1.9385587087575016207418296240273 9}{k^2} - \dfrac{1.3194444458953755759265514803927156 1}{k}$

If α_1 is a rational number (with a small denominator), then using the continued fraction form of 1.3194444458... will reveal it. The following data show that by taking five terms of the continued fraction approximation we get an excellent agreement with the numerical value, about 4 orders of magnitude better than we would get by taking only five terms. So, we naturally conjecture $\alpha_1 = 95/72$ [60], [61].

```
In[47]:= ξ = -fit1[[-1, 1]];
cf = ContinuedFraction[Abs[ξ]];
δs = {#, N[# - ξ]}& /@ Table[FromContinuedFraction @ Take[cf, k], {k, 1, 6}]
```

Out[49]= $\left\{\{1, -0.319444\}, \left\{\dfrac{4}{3}, 0.0138889\right\}, \left\{\dfrac{29}{22}, -0.00126263\right\},\right.$

$\left.\left\{\dfrac{33}{25}, 0.000555554\right\}, \left\{\dfrac{95}{72}, -1.45093 \times 10^{-9}\right\}, \left\{\dfrac{12630188}{9572353}, 6.67753 \times 10^{-15}\right\}\right\}$

Unfortunately, the coefficient α_2 does not reveal itself so easily.

```
In[50]:= data2 = Table[{k, Abs[ε[k]/α[k]] - (1 - 95/72 1/k)}, {k, 200, 500}];
fit2 = Fit[SetPrecision[data2, 40], Table[k^-α, {α, 2, 6}], k]
```

Out[51]= $-\dfrac{3423.4786814324245760502819198}{k^6} -$

$\dfrac{301.01198764179125366087565353 4}{k^5} - \dfrac{40.134765743127576073198887375 46}{k^4} -$

$\dfrac{6.886659262100286075569373774921 23}{k^3} - \dfrac{1.938560974560601285618165341715617 80}{k^2}$

```
In[52]:= ξ = -fit2[[-1, 1]];
cf = ContinuedFraction[Abs[ξ]];
δs = {#, N[# - ξ]}& /@ Table[FromContinuedFraction @ Take[cf, k], {k, 1, 14}]
```

Out[54]= $\left\{\{1, -0.938561\}, \{2, 0.061439\}, \left\{\dfrac{31}{16}, -0.00106097\right\},\right.$

$\left\{\dfrac{95}{49}, 0.000214536\right\}, \left\{\dfrac{126}{65}, -0.0000994361\right\}, \left\{\dfrac{221}{114}, 0.0000355167\right\},$

$\left\{\dfrac{347}{179}, -0.0000134885\right\}, \left\{\dfrac{568}{293}, 5.57834 \times 10^{-6}\right\}, \left\{\dfrac{915}{472}, -1.65253 \times 10^{-6}\right\},$

$\left\{\dfrac{2398}{1237}, 6.02009 \times 10^{-8}\right\}, \left\{\dfrac{24895}{12842}, -2.74936 \times 10^{-9}\right\}, \left\{\dfrac{52188}{26921}, 1.43162 \times 10^{-10}\right\},$

$\left.\left\{\dfrac{494587}{255131}, -2.43313 \times 10^{-12}\right\}, \left\{\dfrac{3019710}{1557707}, 8.3105 \times 10^{-14}\right\}\right\}$

Inspired by the presence of π in the prefactor of the asymptotic expansion, we all assume α_2 to be a product of a rational number and a fractional power of π. fitJump[x, πExp] returns the maximal ratio of two successive terms of the continued fraction-based reconstruction of $x/\pi^{\pi Exp}$ and their position.

```
In[55]:= fitJump[x_, πExp_] :=
Module[{ξ = Abs[x Pi^πExp], cf, δs, ser},
(* the continued fraction expansion *)
cf = ContinuedFraction[ξ];
(* difference to original value *)
δs = {#, N[# - ξ]}& /@ Table[FromContinuedFraction @ Take[cf, k], {k, 1, 6}];
(* ratios *)
ser = #1/#2& @@@ Partition[Abs[Last /@ δs], 2, 1];
(* optimal ratio and its position *)
{N @ Max[ser], Position[ser, Max[ser]][[1]]}]
```

Comparing the achievable approximations for various powers of π suggests $\alpha_2 = $ rational $\pi^{5/3}$.

```
In[56]:= Function[e, {e, fitJump[-fit2[[-1, 1]], e]}] /@ Table[e, {e, -2, 2, 1/3}]
```

Out[56]= $\left\{\{-2, \{77.257, \{4\}\}\}, \left\{-\dfrac{5}{3}, \{32780.5, \{3\}\}\right\}, \left\{-\dfrac{4}{3}, \{27.0306, \{4\}\}\right\},\right.$

$\left\{-1, \{3.2712, \{2\}\}\right\}, \left\{-\dfrac{2}{3}, \{25.6977, \{2\}\}\right\}, \left\{-\dfrac{1}{3}, \{108.999, \{2\}\}\right\},$

$\left\{0, \{57.9081, \{2\}\}\right\}, \left\{\dfrac{1}{3}, \{27.4058, \{2\}\}\right\}, \left\{\dfrac{2}{3}, \{121.034, \{3\}\}\right\},$

$\left.\left\{1, \{148.576, \{2\}\}\right\}, \left\{\dfrac{4}{3}, \{27.2935, \{2\}\}\right\}, \left\{\dfrac{5}{3}, \{20.678, \{1\}\}\right\}, \{2, \{28.8878, \{4\}\}\}\right\}$

Using again the continued fraction terms of the resulting real number suggests $\alpha_2 = 21/73\,\pi^{5/3}$.

```
In[57]:= ξ = -fit2[[-1, 1]] Pi^(-5/3);
        cf = ContinuedFraction[Abs[ξ]];
        δs = {#, N[# - ξ]}& /@ Table[FromContinuedFraction @
                          Take[cf, k], {k, 1, 12}]
```

$$\text{Out[59]= } \left\{\{0, -0.287671\}, \left\{\tfrac{1}{3}, 0.0456622\right\}, \left\{\tfrac{2}{7}, -0.00195689\right\}, \left\{\tfrac{21}{73}, 5.96966\times10^{-8}\right\},\right.$$

$$\left\{\tfrac{66005}{229446}, -6.45192\times10^{-12}\right\}, \left\{\tfrac{132031}{458965}, 3.04407\times10^{-12}\right\}, \left\{\tfrac{198036}{688411}, -1.20926\times10^{-13}\right\},$$

$$\left\{\tfrac{3300607}{11473541}, 5.68015\times10^{-15}\right\}, \left\{\tfrac{3498643}{12161952}, -1.48622\times10^{-15}\right\}, \left\{\tfrac{13796536}{47959397}, 2.28222\times10^{-16}\right\},$$

$$\left.\left\{\tfrac{17295179}{60121349}, -1.18593\times10^{-16}\right\}, \left\{\tfrac{31091715}{108080746}, 3.53018\times10^{-17}\right\}\right\}$$

Repeating the last steps for α_3 leads us to the conjecture $\alpha_3 = 223/149\,\pi^{4/3}$.

```
In[60]:= data3 = Table[{k, Abs[ε[k]/α[k]] -
                     (1 - 95/72 1/k - 21/73 Pi^(5/3) 1/k^2)}, {k, 100, 500}];
        fit3 = Fit[SetPrecision[data3, 40], Table[k^-α, {α, 3, 7}], k]
```

$$\text{Out[61]= } \frac{12459.9952387791209777371621368 0}{k^7} -$$

$$\frac{4487.811480753611554221301280744 2}{k^6} - \frac{283.156777533260446373859998179105}{k^5} -$$

$$\frac{40.2564825616173661346688091478894 2}{k^4} - \frac{6.88629165821254234008008227915237740}{k^3}$$

```
In[62]:= Function[e, {e, fitJump[fit3[[-1, 1]], e]}] /@ Table[e, {e, -2, 2, 1/3}]
```

$$\text{Out[62]= } \left\{\{-2, \{1035.14, \{5\}\}\}, \left\{-\tfrac{5}{3}, \{114.09, \{4\}\}\right\}, \left\{-\tfrac{4}{3}, \{3031.87, \{2\}\}\right\},\right.$$

$$\left\{-1, \{23.9212, \{1\}\}\right\}, \left\{-\tfrac{2}{3}, \{58.1592, \{3\}\}\right\}, \left\{-\tfrac{1}{3}, \{8.47413, \{2\}\}\right\},$$

$$\left\{0, \{10.0701, \{2\}\}\right\}, \left\{\tfrac{1}{3}, \{28.3707, \{4\}\}\right\}, \left\{\tfrac{2}{3}, \{15.8379, \{5\}\}\right\},$$

$$\left.\left\{1, \{4.09009, \{3\}\}\right\}, \left\{\tfrac{4}{3}, \{17.3131, \{2\}\}\right\}, \left\{\tfrac{5}{3}, \{1914.74, \{4\}\}\right\}, \{2, \{50.8543, \{2\}\}\}\right\}$$

```
In[63]:= ξ = -fit3[[-1, 1]] Pi^(-4/3);
        cf = ContinuedFraction[Abs[ξ]];
        δs = {#, N[# - ξ]}& /@ Table[FromContinuedFraction @ Take[cf, k], {k, 1, 12}]
```

$$\text{Out[65]= } \left\{\{1, -0.496645\}, \left\{\tfrac{3}{2}, 0.0033546\right\}, \left\{\tfrac{223}{149}, -1.10644\times10^{-6}\right\},\right.$$

$$\left\{\tfrac{8923}{5962}, 1.92533\times10^{-8}\right\}, \left\{\tfrac{9146}{6111}, -8.19378\times10^{-9}\right\}, \left\{\tfrac{27215}{18184}, 8.05303\times10^{-10}\right\},$$

$$\left\{\tfrac{90791}{60663}, -1.01237\times10^{-10}\right\}, \left\{\tfrac{208797}{139510}, 1.69232\times10^{-11}\right\}, \left\{\tfrac{508385}{339683}, -4.17867\times10^{-12}\right\},$$

$$\left.\left\{\tfrac{717182}{479193}, 1.96483\times10^{-12}\right\}, \left\{\tfrac{1225567}{818876}, -5.83595\times10^{-13}\right\}, \left\{\tfrac{1942749}{1298069}, 3.57177\times10^{-13}\right\}\right\}$$

So, we have the following conjectured expansion

$$^{(0)}\varepsilon_j = -(-1)^j\,\frac{2\sqrt{6}}{\pi^{3/2}}\left(\frac{3}{2}\right)^j\left(j-\frac{1}{2}\right)!\left(1-\frac{95}{72}\frac{1}{j}-\frac{21}{73}\,\pi^{5/3}\frac{1}{j^2}-\frac{223}{149}\,\pi^{4/3}\frac{1}{j^3}+\cdots\right).$$

For higher-order perturbation theory of the anharmonic oscillator based on the hypervirial theorem, see [547], [320]; for a treatment within logarithmic perturbation theory, see [41]; for a linear delta expansion based perturbation theory, see [14]; for a supersymmetry-based perturbation theory for the quartic oscillator, see [224].

In Section 2.10, we used a large initial basis of harmonic oscillator states. In this exercise, we basically used a projection into the ground state. One could now go on and mix the two approaches and use a perturbation expansion that projects into a subspace spanned by the first k states [416].

b) It is straightforward to implement the given differential equations ahoOdes[o]. For the ground state, we can restrict the system to even quantum numbers and we truncate the infinite system at $n = o$.

```
In[1]:= ahoOdes[o_, g_:g, α_:1] := Flatten @
           {Table[α D[ε[n][g], g] == V[n, n][g], {n, 0, o, 2}],
            Table[α D[V[m, n][g], g] ==
                  Sum[If[m == k, 0, V[m, k][g] V[k, n][g]/(ε[m][g] - ε[k][g])],
```

```
                    {k, 0, o, 2}] +
              Sum[If[n == k, 0, V[m, k][g] V[k, n][g]/(ε[n][g] - ε[k][g])],
                    {k, 0, o, 2}], {m, 0, o, 2}, {n, 0, o, 2}],
       (* initial condition from harmonic oscillator *)
       Table[ε[n][0] == 2n + 1, {n, 0, o, 2}],
       Table[V[m, n][0] == Vz4[n, m] , {m, 0, o, 2}, {n, 0, o, 2}]};
```

The initial conditions, meaning the quantities $\varepsilon_n(g)$ and $V_{k,l}(g)$, follow from the harmonic oscillator solutions.

```
In[2]:= reduceProducts[expr_] := (* from Section 2.10 *)
        FixedPoint[Expand[# /. z_^m_. HermiteH[n_, z_] :>
           (* recursion relation for Hermite polynomials *)
              z^(m - 1)(n HermiteH[n - 1, z] + HermiteH[n + 1, z]/2)]&, expr]
In[3]:= c[n_] := Sqrt[Sqrt[Pi] 2^n n!]

        Vz4[n_, m_] = c[m]/c[n] reduceProducts[(z^4 - z^2) HermiteH[n, z]] /.
                             HermiteH[n_, z] -> KroneckerDelta[n, m];
```

Solving now the equations numerically with only 20 symmetric states included yields 10 correct digits.

```
In[5]:= (* high-precision value from above *)
        εExact = 1.060362090484182899647046;

        (* the variables of the system of differential equations *)
        εVVars[o_] := Flatten[{Table[ε[n], {n, 0, o, 2}],
                            Table[V[m, n], {m, 0, o, 2}, {n, 0, o, 2}]}];

        Table[{o, εExact - (ε[0][1] /.
           NDSolve[ahoOdes[o], εVVars[o], {g, 1, 1}, MaxSteps -> 10^4,
                PrecisionGoal -> 12])},
              (* number of included states *) {o, 4, 40, 12}]
Out[9]= {{4, {-0.010405}}, {16, {-0.0000123741}}, {28, {-2.41126×10^-8}}, {40, {-7.69664×10^-11}}}
```

We end by solving the system of differential equations in the complex *g*-plane.

```
In[10]:= (* truncate low for faster numerical solution *) o = 8;
         R = 8; L = R/Sqrt[2]; ppr = 240; ppφ = 240;

         data = Table[
            (* solve differential equations radially outwards
               in the complex g-plane *)
            ndsol = NDSolve[ahoOdes[o, r, 1/D[r Exp[I φ], r]],
                            εVVars[o], {r, 0, R}, MaxSteps -> 10^5,
                            PrecisionGoal -> 3, AccuracyGoal -> 3];
            (* values along ray in the g-plane *)
            Table[{r, φ, ε[0][r] /. ndsol[[1, 1]]}, {r, 0, R, R/ppr}],
                 {φ, -Pi, Pi, 2Pi/ppφ}];

         (* interpolate data in polar coordinates *)
         ipo = Interpolation[Flatten[data, 1]];
```

The following two graphics show $\mathrm{Re}(\varepsilon_0(g))$ and $\mathrm{Im}(\varepsilon_0(g))$.

```
In[16]:= (* function for calculating homogeneously spaced contours *)
         homogeneousContours[data_, n_] :=
         Module[{d = Sort[Flatten[data]], λ}, λ = Length[d];
            #[[Round[λ/n/2]]]& /@ Partition[d, Round[λ/n]]]
In[18]:= (* interpolate data in polar coordinates *)
         ipo = Interpolation[Flatten[data, 1]];
In[20]:=
         Show[GraphicsArray[
         Function[reIm, Block[{cp, $DisplayFunction = Identity},
         (* contour plot in Cartesian coordinates *)
         cp = ContourPlot[reIm[ipo[Sqrt[x^2 + y^2], ArcTan[x, y]]],
                          {x, -L, L}, {y, -L, L}, PlotPoints -> 240];
         (* use homogeneous contour spacing *)
         ListContourPlot[cp[[1]], Contours -> homogeneousContours[cp[[1]], 20],
```

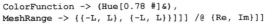

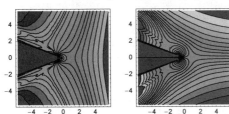

For a related approach based on a sequence of infinitesimal unitary transformations from a Hamiltonian with known spectrum to a more complicated one, see [223], [182].

11. Sextic Oscillator, Time-Dependent Calogero Potential

a) For fixed t, we will expand $\Psi(x, t)$ in eigenfunctions of the sextic oscillator $\psi_n(x, t)$. The eigenfunctions of the sextic oscillator we will expand in eigenfunctions $\phi_n(x)$ of the harmonic oscillator.

```
In[1]:= φ[n_, x_] := Exp[-x^2/2] HermiteH[n, x]/c[n]
        c[n_] := Sqrt[Sqrt[Pi] 2^n n!]
```

(The time-dependent harmonic oscillator states are given by $\Phi_n(x, t) = \exp(-i(2n+1)t)\,\phi_n(x)$. Here is a quick check.)

```
In[3]:= Table[ψ[x_, t_] = Exp[-I (2 n + 1) t] φ[n, x];
        I D[ψ[x, t], t] == -D[ψ[x, t], x, x] + x^2 ψ[x, t] // Simplify,
        {n, 0, 20}] // Union
Out[3]= {True}
```

This means we have the following formula for $\Psi(x, t) = \sum_{k=0}^{\infty} c_k \psi_n(x, t)$. The coefficients c_k follow from $\Psi(x, 0)$ as $\Psi(x, 0) = \sum_{k=0}^{\infty} c_k \psi_n(x)$ where $c_k = \int_{-\infty}^{\infty} \Psi(x, 0)\,\psi_n(x)\,dx$.

The time-independent eigenfunctions $\psi_n(x)$ of the sextic oscillator we expand in harmonic states $\phi_n(x)$ through $\psi_n(x) = \sum_{k=0}^{\infty} d_k\,\Phi_n(x)$. The d_k we obtain from diagonalizing a truncated Hill matrix.

The time-dependent eigenfunctions $\psi_n(x, t)$ have the form $\psi_n(x, t) = \sum_{k=0}^{\infty} \exp(-i\,\epsilon_n\,t)\,\psi_n(x)$ where the eigenvalues ϵ_k also follow from diagonalizing a truncated Hill matrix. Using some of the functions from Subsection 2.10, we calculate the matrix elements of x^6 in the harmonic oscillator basis.

```
In[4]:= reduceProducts[expr_] := With[{H = HermiteH},
        FixedPoint[Expand[# /. z_^m_. H[n_, z_] :>
            (* recursion relation for Hermite polynomials *)
            z^(m - 1) (n H[n - 1, z] + H[n + 1, z]/2)]&, expr]]

In[5]:= makeh[V_, z_] := c[m]/c[n] Simplify[
        reduceProducts[(2n + 1) HermiteH[n, z] - z^2 HermiteH[n, z] +
            (* the potential *) V HermiteH[n, z]] /.
        (* use orthogonality *) HermiteH[n_, z] -> KroneckerDelta[n, m]]

In[6]:= makeh[z^6, z] /. KroneckerDelta -> δ // FullSimplify
```

$$Out[6]= \frac{1}{64\sqrt{2^n\,n!}}\left(\sqrt{2^m\,m!}\right.$$

$(16\,(-1 + n)\,n\,(2\,(-3 + n)\,(-2 + n)\,(2\,(-5 + n)\,(-4 + n)\,\delta[m, -6 + n] + 3\,(-3 + 2\,n)\,\delta[m, -4 + n]) +$
$\qquad (11 + 15\,(-1 + n)\,n)\,\delta[m, -2 + n]) + 8\,(1 + 2\,n)\,(19 + 10\,n\,(1 + n))\,\delta[m, n] +$
$\qquad 4\,(41 + 15\,n\,(3 + n))\,\delta[m, 2 + n] + 6\,(5 + 2\,n)\,\delta[m, 4 + n] + \delta[m, 6 + n]))$

```
In[7]:= h[n_, m_] = Block[{δ = KroneckerDelta}, If[n - m > 6, 0, #]&[%]]
```

$$Out[7]= \text{If}\left[-m + n > 6, 0, \frac{1}{64\sqrt{2^n\,n!}}\right.$$

$\left(\sqrt{2^m\,m!}\,(16\,(-1 + n)\,n\,(2\,(-3 + n)\,(-2 + n)\,(2\,(-5 + n)\,(-4 + n)\,\text{KroneckerDelta}[m, -6 + n] +\right.$
$\qquad 3\,(-3 + 2\,n)\,\text{KroneckerDelta}[m, -4 + n]) +$
$\qquad (11 + 15\,(-1 + n)\,n)\,\text{KroneckerDelta}[m, -2 + n]) + 8\,(1 + 2\,n)\,(19 + 10\,n\,(1 + n))$
$\qquad \text{KroneckerDelta}[m, n] + 4\,(41 + 15\,n\,(3 + n))\,\text{KroneckerDelta}[m, 2 + n] +$
$\qquad 6\,(5 + 2\,n)\,\text{KroneckerDelta}[m, 4 + n] + \text{KroneckerDelta}[m, 6 + n]))\right]$

Next, we calculate the eigenvalues and eigenfunctions of a 250×250 Hill matrix and take into account the first 60 eigenstates (later we will verify that these are enough states). We denote the eigenvectors by `evecs` and the eigenvalues by `evals`.

```
In[8]:= p = 250;
        (* use machine-arithmetic *) prec = $MachinePrecision - 1;
        es = Eigensystem[N[Table[h[n, m], {n, 0, p}, {m, 0, p}], prec]];
In[11]:= d = 60;
        {evals, evecs} = Take[Reverse[#], d]& /@ es;
```

For fast plotting, we now calculate numericalized packed array versions of the discretized harmonic oscillator states. We use 300 points for the spatial discretization and $-3.5 \le x \le 3.5$ (the graphics below show that this is an appropriate range).

```
In[13]:= ppx = 360; L = 7/2;
        Do[ϕN[k] = Developer`ToPackedArray[
          Join[(-1)^k Reverse[Rest[#]], #]&[Table[ϕ[k, N[z]],
                                        {z, 0, L, L/ppx}]]],
          {k, 0, p}];
```

Using the expansion coefficients `evecs`, we can now assemble the approximate eigenstates of the sextic oscillator $\psi_n(x)$.

```
In[15]:= Needs["NumericalMath`ListIntegrate`"];

In[16]:= Do[EVS[j] = Sum[evecs[[j, k + 1]] ϕN[k], {k, 0, p}];
        (* eigenfunctions should be correctly normalized;
           the ϕN[k] and the matrix eigenvectors evecs are both normalized;
           otherwise normalize via
           EVS[j] = EVS[j]/Sqrt[ListIntegrate[Abs[EVS[j]]^2, L/ppx]] *),
        {j, d}]
```

Here are the first view eigenfunctions shown.

```
In[17]:= Show[Reverse @
        Table[ListPlot[If[j === 1, Abs[#], Identity[#]]& @ EVS[j],
                       PlotJoined -> True, Frame -> True,
                       Axes -> False, DisplayFunction -> Identity,
                       PlotStyle -> {Thickness[0.002], Hue[j/14]}],
          {j, 5}], DisplayFunction -> $DisplayFunction]
```

Discretized versions of the time-dependent wave functions $\Psi(x, t)$ are calculated using the function $H[\mathcal{K}, t]$.

```
In[18]:= H[𝒦_, t_] := Sum[cf[𝒦, j] Exp[-I evals[[j]] t] EVS[j], {j, d}];
```

The expansion coefficients $cf[\mathcal{K}, j]$ in $H[\mathcal{K}, t]$ we obtain by numerically integrating the product of the initial wave function $\Psi(x, 0)$ and the time-independent eigenfunctions $\psi_n(x)$ of the sextic oscillator.

```
In[19]:= zTab = Table[N[z], {z, -L, L, L/ppx}];

        data[𝒦_, ppt_:400, T_:10] :=
        Module[{if},
            Do[if = Interpolation[Transpose[{zTab, EVS[j]}]];
                cf[𝒦, j] = NIntegrate[(Cos[x Pi/4]^2 Exp[I 𝒦 x]) if[x], {x, -2, 2},
                                PrecisionGoal -> 5, AccuracyGoal -> 5], {j, d}];
            Table[H[𝒦, t], {t, 0, T, T/ppt}]];
```

The following autocorrelation plots [483] show how the scalar product $\int_{-\infty}^{\infty} \Psi(x, t)\, \overline{\Psi(x, 0)}\, dt$ depends on the time t. We use $\mathcal{K} = 2$, $\mathcal{K} = 3$, and $\mathcal{K} = 4$.

```
In[21]:= autoCorrelationGraphic[𝒦_, ppt_, T_] :=
    Module[{d = data[𝒦, ppt, T]},
            (* overlap integrals as a function of t *)
            ℒ = (ListIntegrate[d[[1]] #, L/ppx]& /@
                        Conjugate[Take[d, All]])/((* norm *)(3/2));
        Graphics[MapIndexed[(* color with increasing time *)
                        {Hue[0.8 #2[[1]]/ppt], Line[#1]}&,
                        Partition[{Re[#], Im[#]}& /@ ℒ, 2, 1]],
                Frame -> True, AspectRatio -> 1]]
In[22]:= Show[GraphicsArray[autoCorrelationGraphic[#, 500, 10]& /@ {2, 3, 4}]]
```

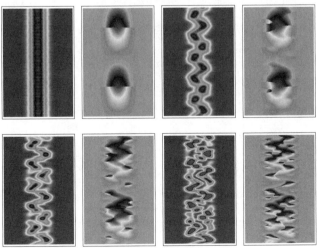

The function `absArgGraphics` finally generates density graphics of the square of the absolute value and the phase of $\Psi(x, t)$.

```
In[23]:= absArgGraphics[𝒦_, ppt_:400, T_:10] :=
    Module[{f, tab = data[𝒦, ppt, T]},
    GraphicsArray[(* show absolute value and argument *)
    ListDensityPlot[#1, PlotRange -> #2, Mesh -> False, FrameTicks -> False,
                DisplayFunction -> Identity, AspectRatio -> 3/2,
                ColorFunction -> (Hue[0.8 #]&)]& @@@
        {{Abs[tab]^2, Automatic}, (* suppress noise *)
        f = #/Max[#]&[Plus @@@ Transpose[Abs[tab]]];
        {f #& /@ Arg[tab], {-Pi, Pi}}}]]
```

Here is the animation showing $\Psi(x, t)$ as a function of $\mathcal{K}$. For increasing $\mathcal{K}$ the initial wave packet moves faster to the right, splits faster into smaller pieces, and its peak reaches larger x-values. We also get the typical quantum carpet-lines for larger $\mathcal{K}$-values [308].

```
In[24]:= frames = 9;
    Show[GraphicsArray[Flatten[absArgGraphics[#][[1]]& /@ #]]]& /@
            Partition[Table[𝒦, {𝒦, 0, 6, 6/frames}], 2]
```

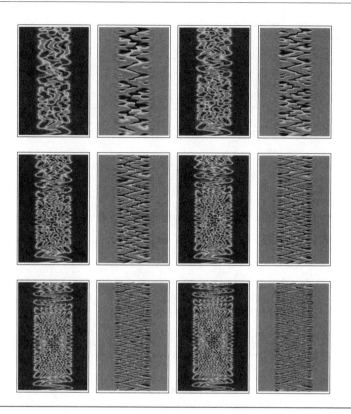

```
frames = 60;
Do[Show[absArgGraphics[𝒦]], {𝒦, 0, 6, 6/frames}]
```

It remains to show that we took enough eigenstates into account [92]. The following graphic shows that the last of the 60 eigenstates contribute on average less than 10^{-3} to the result (which is a small enough contribution for a visualization).

```
In[26]:= ListPlot3D[Table[(* display on logarithm *)
            Log[10, Abs[cf[𝒦, j]]], {𝒦, 0, 6, 6/frames}, {j, d}],
            PlotRange -> All, MeshRange -> {{1, 50}, {0, 6}},
            Mesh -> False, AxesLabel -> {"j", "𝒦", None}]
```

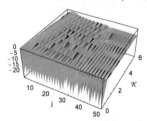

Now, we will compare the expansion results with a direct numerical evaluation of the differential equation. For a concrete example, we use $\mathcal{K} = 3$ and $0 \le t \le 1$. The calculation takes much longer than the expansion method and in addition, we get warning messages concerning the quality of the result. This shows that for the problem under consideration the expansion in eigenfunctions is the superior method [291].

```
In[27]:= (nsol =
            NDSolve[{I D[Ψ[x, t], t] == -D[Ψ[x, t], x, x] + x^6 Ψ[x, t],
```

```
              Ψ[x, 0] == Cos[x Pi/4]^2 Exp[I 3 x] If[-2 <= x <= 2, 1, 0],
               Ψ[-L, t] == 0, Ψ[+L, t] == 0},
              Ψ, {x, -L, L}, {t, 0, 1}, MaxSteps -> 10000,
            AccuracyGoal -> 3, PrecisionGoal -> 3,
            Method -> {"MethodOfLines", "SpatialDiscretization" ->
                      {"TensorProductGrid", "DifferenceOrder" -> 4,
                       "MaxPoints" -> {400}, "MinPoints" -> {400}}}];) //
                                                                  Timing

    NDSolve::eerri :
       Warning: Estimated initial error on the specified spatial grid in the direction
         of independent variable x exceeds prescribed error tolerance. More…

Out[27]=  {0.34 Second, Null}
```

The following two graphics show the two calculated $\Psi(x, t)$ agree for small times.

```
In[28]:= (* how far did the numerical solution go before reaching
          the maximum number of steps? *)
       tMax = nsol[[1, 1, 2, 1, 2, 2]];

Show[GraphicsArray[
Block[{{$DisplayFunction = Identity, pp = 200,
        opts = Sequence[Mesh -> False, ColorFunction -> (Hue[0.8 #]&)]},
   {DensityPlot[Evaluate[Abs[Ψ[x, t]]^2 /. nsol[[1]]]   (* lists at t == 0? *),
               {x, -L, L}, {t, 0, tMax}, PlotPoints -> pp,
               Evaluate[opts]],
   (* calculate data *) data[3];
   ListDensityPlot[Abs[Table[H[3, t], {t, 0, tMax, tMax/pp}]]^2,
                opts, MeshRange -> {{-L, L}, {0, tMax}}]}]]]]
```

b) These are the eigenvalues and eigenfunctions of the Calogero potential.

```
In[1]:= With[{α = 1/2 Sqrt[1 + 4 γ]},
         ε[n_] = 4n + 2 + 2α;
         ψ[n_, x_] = x^((2 α + 1)/2) Exp[-x^2/2] LaguerreL[n, α, x^2]];
```

For the expansion, we will need a couple of hundred $\psi_n(x)$. Because the $\psi_n(x)$ contain Laguerre polynomials of order n, the direct calculation of a dense set of $\psi_n(x)$ would be relatively slow. So, we use a compiled version of the recurrence relations for the associated Laguerre polynomials. The function `LaguerreLListCF[n, λ, x]` will calculate the list $\{L_0^\lambda(x), ..., L_n^\lambda(x)\}$ (which works well for the small values of λ we are using here).

```
In[2]:= LaguerreLListCF =
   Compile[{{n, _Integer}, λ, x}, Join[{1.}, Last /@
   NestList[Function[l, With[{v = l[[1]] + 1, lm2 = l[[2]], lm1 = l[[3]]},
              {v, lm1, ((2 v + λ - x - 1)/v) lm1 - ((v + λ - 1)/v) lm2}]],
              {1, LaguerreL[0, λ, x], LaguerreL[1, λ, x]}, n - 1]],
              {{_, _Real, 0}}]];
```

The following timing comparison shows a speed-up of about a factor 150 compared to the direct evaluation of the Laguerre polynomials.

```
In[3]:= {Do[ℒ1 = LaguerreLListCF[500, 2, 12.3], {150}] // Timing,
       (ℒ2 = Table[LaguerreL[k, 2, 12.3], {k, 0, 500}];) // Timing}
Out[3]=  {{0.14 Second, Null}, {0.08 Second, Null}}
```

For a graphical application, the values calculated with `LaguerreLListCF` are precise enough.

```
In[4]:= £1 - £2 // Abs // Max

Out[4]= 7.89413 × 10⁻¹²
```

We will use 400 spatial discretization points and will take into account the first 401 eigenfunctions. ψN will be a discretized eigenfunction.

```
In[5]:= o = 400; γ = 2; ppx = 400; L = 12; ppt = ppx;

In[6]:= (* the spatial part *)
        xTab = Table[N[x], {x, 0, L, L/ppx}];
        (* the Laguerre part *)
        lTab = Table[LaguerreLListCF[o, 1/2 Sqrt[1 + 4 γ], N[x^2]],
                {x, 0, L, L/ppx}] // Transpose ;
        (* the power-exponential prefactor part *)
        facTab = Table[x^((Sqrt[4γ + 1.] + 1)/2) Exp[-x^2/2], {x, 0, L, L/ppx}];
        Do[ψN[k - 1] = Developer`ToPackedArray[N[facTab lTab[[k]]]],
                {k, o + 1}];
```

We now normalize the $\psi_n(x)$. We integrate $\int_0^\infty \psi_n(x)^2 \, dx$ numerically. (For the exact symbolic normalization factor, see [26].)

```
In[13]:= Needs["NumericalMath`ListIntegrate`"]

        Do[ψif[k] = Interpolation[Transpose[{xTab, ψN[k]}]], {k, 0, o}]
        (* numerical calculation of the normalization *)
        Do[norm[k] = ListIntegrate[ψN[k]^2, L/ppx], {k, 0, o}];
        (* the normalized eigenfunctions *)
        Do[ψN[k] = ψN[k]/norm[k], {k, 0, o}];
        (* reinterpolate normalized eigenfunctions *)
        Do[ψif[k] = Interpolation[Transpose[{xTab, ψN[k]}]], {k, 0, o}]
```

The following graphic shows the first 40 normalized eigenfunctions.

```
In[21]:= Show[Graphics[{Thickness[0.002],
        Table[{Hue[0.8 k/100], Line[Transpose[{xTab, ψN[k]}]]},
            {k, 40}]}], PlotRange -> All, Frame -> True]
```

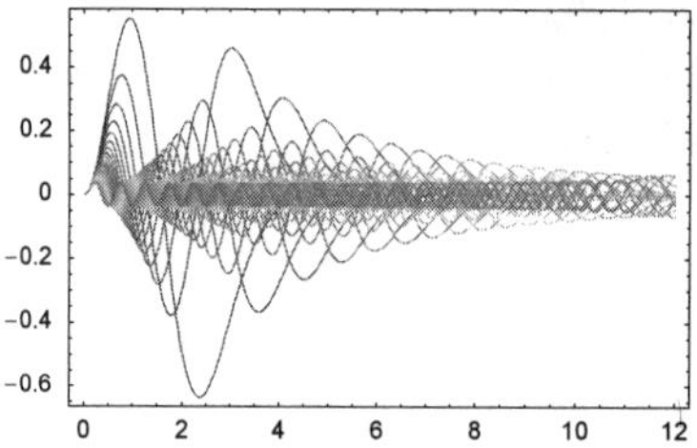

The time-dependent solution $\Psi(x, t)$ has the following form (here $\tilde{\psi}_n(x)$ are the normalized eigenfunctions):

$$\Psi(x, t) = \sum_{k=0}^{\infty} c_k \, e^{-i \, \varepsilon_k \, t} \, \tilde{\psi}_n(x).$$

The expansion coefficients c_k follow from $\Psi(x, 0)$ through $c_k = \int_{-\infty}^{\infty} \overline{\Psi(x, 0)} \, \tilde{\psi}_n(x) \, dx$.

The function ΨTab[T, ppt] calculates a discretized solution of the initial value problem under consideration for $0 \le t \le T$ using *ppt* time discretization steps.

```
In[22]:= ΨTab[T_, ppt_] := Table[Sum[coeffs[k] Exp[-1. I t (4k + 2 + Sqrt[1 + 4 γ])]]*
                        ψN[k], {k, 0, o - 1}], {t, 0, T, T/ppt}];
```

We obtain the expansion coefficients c_k again by numerical integration. The following Do loop creates the pictures for the animation. The function ΨGraphics displays the square of the absolute value and the argument.

```
In[23]:= cf = Compile[{𝒦, x}, Cos[𝒦 (x - Pi)] Sin[(x - Pi)]^(1/2)];
        Do[ψN1[k] = Table[ψif[k][x], {x, Pi, 2Pi, Pi/ppx}], {k, 0, o}];

In[25]:= (* argument that is defined at z = 0 *)
        SetAttributes[arg, Listable];
        arg[0.] = 0; arg[0. + I 0.] = 0; arg[x_] := Arg[x];
```

```
In[28]:= ΨGraphics[𝒦_] :=
    Module[{𝒦Tab},
        𝒦Tab = Table[cf[𝒦, x], {x, Pi, 2Pi, Pi/ppx}];
        (* calculate expansion coefficients *)
        Do[c[𝒦][k] = coeffs[k] = ListIntegrate[𝒦Tab ψN1[k], Pi/ppx], {k, 0, o}];
        (* display solution *)
        GraphicsArray[data = ΨTab[2, 400];
        ListDensityPlot[#[data], Mesh -> False, DisplayFunction -> Identity,
                        ColorFunction -> (Hue[0.8 #]&),
                        FrameTicks -> None]& /@ {Abs[#]^2&, arg}]]

In[29]:= frames = 5;
    Show[GraphicsArray[#]]& /@ Partition[(* show two 𝒦-values in one row *)
            Flatten[Table[ΨGraphics[𝒦][[1]], {𝒦, 3, 9, 6/frames}]], 4]
```

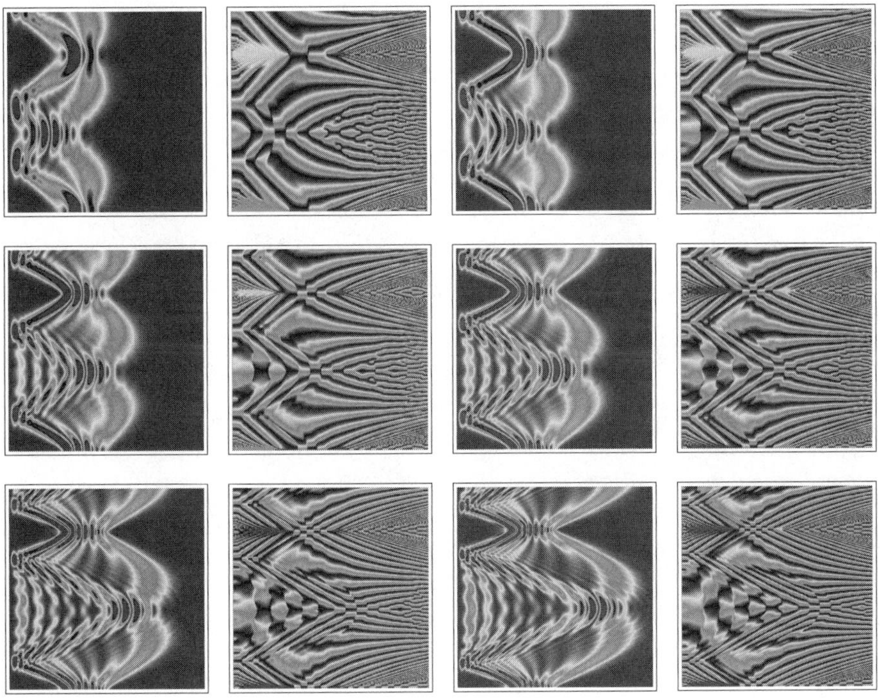

```
frames = 60;
Do[Show[ΨGraphics[𝒦]], {𝒦, 3, 9, 6/frames}]
```

The following graphic shows the logarithm of the absolute value of the expansion coefficients c_k. The picture shows that—for the goal of graphics—we took enough eigenstates into account.

```
In[31]:= SubValues[c][[1]]

Out[31]= HoldPattern[c[3][0]] :→ 0.00489985

In[32]:= ListPlot3D[Table[Log[10, Abs[c[𝒦][k]]],
                    {𝒦, 3, 9, 6/frames}, {k, 0, o}], Mesh -> False]
```

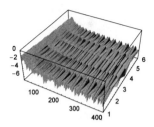

For a closed form of the time-dependent Green's function of the Calogero potential, see [607] and [608].

References

1   P. Abbott. *The Mathematica Journal* 3, n4, 28 (1993).

2   M. Abramowitz, I. A. Stegun. *Handbook of Mathematical Functions*, National Bureau of Standards, Washington, 1964.

3   L. Adawi. *J. Math. Phys.* 12, 358 (1971).

4   P. Agarwal, P. Singh. *Ganita* 54, 69 (2003).

5   B. D. Agrawal, I. K. Khanna. *Proc. Am. Math. Soc.* 22, 646 (1969).

6   S. Ahmed, M. E. Muldoon. *SIAM J. Math. Anal.* 14, 372 (1983).

7   Z. Ahmed. *Phys. Lett.* A 282, 343 (2001).

8   H. Ahmedov, I. H. Duru. *arXiv:q-alg*/9609028 (1996).

9   N. I. Akhiezer. *The Classical Moment Problem*, Oliver & Boyd, Edinburgh, 1965.

10   G. Álvarez, C. J. Howls, H. J. Silverstone. *J. Phys.* A 35, 4003 (2002).

11   G. Álvarez, C. J. Howls, H. J. Silverstone. *J. Phys.* A 35, 4017 (2002).

12   G. Alvarez, S. Graffi, H. J. Silverstone. *Phys. Rev.* A 38, 1687 (1988).

13   R. Alvarez–Nordaze, J. S. Dehesa. *Appl. Math. Comput.* 128, 167 (2002).

14   P. Amore, A. Aranda. *arXiv:quant-ph*/0310079 (2003).

15   F. Antonsen. *arXiv:quant-ph*/9809062 (1998).

16   F. Antonsen. *Phys. Rev.* A 60, 812 (1999).

17   N. Aquino. *J. Math. Chem.* 18, 349 (1995).

18   A. Arda. *arXiv:quant-ph*/0411168 (2004).

19   N. Ari in S. L. Koh, C. G. Speziale (eds.). *Recent Advances in Engineering Mathematics*, Springer-Verlag, Berlin, 1989.

20   G. A. Arteca, F. M. Fernández, E. Castro. *Large Order Perturbation Theory and Summation Methods in Quantum Mechanics*, Springer-Verlag, Berlin, 1990.

21   R. Askey. *Orthogonal Polynomials and Special Functions*, SIAM, Philadelphia, 1975.

22   R. Askey in D. Stanton (ed.). *q-Series and Partitions*, Springer-Verlag, New York, 1989.

23   R. A. Askey in R. Wong (ed.). *Asymptotic and Computational Analysis*, Marcel Dekker, New York, 1990.

24   N. M. Atakishiyev, A. Frank, K. B. Wolf. *J. Math. Phys.* 35, 3253 (1994).

25   N. M. Atakishiyev in D. Levi, O. Ragnisco (eds.). *SIDE III—Symmetries and Integrability of Differential Equations*, American Mathematical Society, Providence, 2000.

26   M. K. Atakishiyeva, N. M. Atakishiyev, L. E. Vincent. *Bol. Soc. Mat. Mex.* 6, 191 (2000).

27   G. Auberson, M. C. Peyranére. *arXiv:hep-th*/0110275 (2001).

28   J. Avery. *Hyperspherical Harmonics*, Kluwer, Dordrecht, 1989.

29   J. Avery. *J. Phys. Chem.* 97, 2406 (1993).

30   J. Avery in D. R. Herschbach, J. Avery, O. Goscinsky (eds.). *Dimensional Scaling in Chemical Physics*, Kluwer, Dordrecht, 1993.

31   J. Avery. *J. Math. Chem.* 24, 169 (1998).

32   B. Bacus, Y. Meurice, A. Soemadi. *J. Phys.* A 28, L381 (1995).

33   B. K. Bagchi. *Supersymmetry in Quantum and Classical Mechanics*, Chapman & Hall, Boca Raton, 2000.

34   B. Bagchi, S. Mallik, C. Quesne. *arXiv:quant-ph*/0106021 (2001).

35   O. V. Bagdasaryan, A. V. Dar'yan. *J. Cont. Phys.* 27, n5, 1 (1993).

36   V. G. Bagrov, D. M. Gitman. *Exact Solutions of Relativistic Wave Equations*, Kluwer, Dordrecht, 1990.

37   G. A. Baker, Jr. *Quantitative Theory of Critical Phenomena*, Academic Press, Boston, 1990.

38   G. A. Baker, P. Graves–Morris. *Padé Approximants*, Cambridge University Press, Cambridge, 1996.

39   A. B. Balentekin. *Phys. Rev.* E 64, 066105 (2001).

40   R. Balian, G. Parisi, A. Voros in G. Parisi. *Field Theory, Disorder and Simulations*, World Scientific, Singapore, 1992.

41   S. K. Bandyopadhyay, K. Bhattacharya. *Int. J. Quant. Chem.* 90, 27 (2002).

42   N. S. Banerjee, J. F. Geer. *NASA Report* CR-201751 (1997).
      http://techreports.larc.nasa.gov/icase/1997/icase-1997-56.pdf

43   D. Bar, L. P. Horwitz. *Phys. Lett.* A 296, 265 (2002).

44   F. Barra, P. Gaspard. *J. Phys.* A 32, 3357 (1999).

45   L. S. Bartell. *J. Math. Chem.* 19, 401 (1996).

46   J. L. Basdevant. *Fortschr. Phys.* 20, 283 (1972).

47   H. Bateman. *Higher Transcendental Functions* v. II, McGraw-Hill, New York, 1953.

48   K. Bay, W. Lay. *J. Math. Phys.* 38, 2127 (1997).

49   D. Baye, P.-H. Heenen. *J. Phys.* A 19, 2041 (1986).

50   D. Baye, M. Hesse, J.-M. Sparenberg, M. Vincke. *J. Phys.* B 31, 3439 (1998).

51   D. Baye, M. Vincke. *Phys. Rev.* E 59, 7195 (1999).

52   D. Baye, M. Hesse, M. Vincke. *Phys. Rev.* E 65, 026701 (2002).

53   D. Baye, J. Goldbeter, J.-M. Sparenberg. *arXiv:quant-ph*/0201021 (2002).

54   N. Bazley, D. Fox. *Phys. Rev.* 124, 483 (1961).

55   P. Beckmann. *Orthogonal Polynomials for Engineers and Physicists*, Golem, Golden, 1973.

56   S. Bell. *J. Phys.* B 3, 735 (1970).

57   B. Bellet. *arXiv:math-ph*/0208015 (2002).

58   V. V. Belokurov, Y. P. Solol'ev, E. T. Shavgulidze. *Math. Notes* 68, 22 (2000).

59   T. S. Belozerova, V. K. Henner. *Comput. Phys. Commun.* 73, 145 (1992).

60   C. M. Bender, T. T. Wu. *Phys. Rev.* 184, 1231 (1969).

61   C. M. Bender, T. T. Wu. *Phys. Rev.* D 7, 1620 (1973).

62   C. M. Bender, T. T. Wu. *Phys. Rev. Lett.* 27, 461 (1971).

63   C. M. Bender, L. R. Mead, N. Papanicolaou. *J. Math. Phys.* 28, 1016 (1987).

64   C. M. Bender, L. M. A. Bettencourt. *Phys. Rev.* D 54, 7710 (1996).

65   C. M. Bender, S. Boettcher, P. Meisinger. *J. Math. Phys.* 40, 2201 (1998).

66   C. M. Bender. *Phys. Rep.* 315, 27 (1999).

67   C. M. Bender, M. Berry, P. N. Meisinger, V. M. Savage, M. Simsek. *J. Phys.* A 34, L 31 (2001).

68   C. M. Bender, P. N. Meisinger, Q. Wang. *arXiv:quant-ph*/0211123 (2002).

69   C. M. Bender, P. N. Meisinger, Q. Wang. *arXiv:quant-ph*/0211166 (2002).

70   C. M. Bender, D. C. Brody, H. F. Jones. *arXiv:quant-ph*/0303005 (2003).

71   C. M. Bender. *arXiv:quant-ph*/0501052 (2005).

72   M. N. Berberan–Santos. *J. Math. Chem.* 37, 101 (2005).

73   C. Berg, M. E. H. Ismail. *arXiv:math.CA*/9405213 (1994).

74   C. Berg, A. Ruffing. *Commun. Math. Phys.* 323, 29 (2001).

75   F. Bergeron, G. Labelle, P. Leroux. *Combinatorial Species and Tree-like Structures*, Cambridge University Press, Cambridge, 1998.

76   C. Bernard, V. M. Savage. *arXiv:hep-lat*/0106009 (2001).

77   A. Bernardini, P. E. Ricci. *Math. Comput. Model.* 36, 1115 (2002).

78   D. R. Bès, G. G. Dussel, H. M. Sofiía. *Am. J. Phys.* 45, 191 (1977).

79   E. Besalu, R. Carbó–Dorca. *J. Chem. Edu.* 75, 502 (1998).

80   H. A. Bethe, E. E. Salpeter. *Quantum Mechanics of One- and Two-Electron Atoms*, Springer-Verlag, Heidelberg, 1957.

81   K. Bhattacharyya, J. K. Bhattacharjee. *Phys. Lett.* A 331, 288 (2004).

82   N. E. J. Bjerrum–Bohr. *arXiv:quant-ph*/0302107 (2003).

83   J. Blank, P. Exner, M. Havlíček. *Hilbert Space Operators in Quantum Physics*, American Institute of Physics, New York, 1985.

84   S. M. Blinder. *J. Math. Chem.* 14, 319 (1993).

85   S. J. Blundell. *Am. J. Phys.* 61, 1147 (1993).

86   B. D. Bojanov, O. I. Rahman. *J. Math. Anal. Appl.* 189, 781 (1995).

87   D. Bonatsos, C. Daskaloyannis, D. Ellinas, A. Faessler. *arXiv:hep-th*/9402014 (1994).

88   V. V. Borzov. *arXiv:math.CA*/0002226 (2000).

89   V. V. Borzov, E. V. Damaskinsky. *arXiv:math.QA*/0209181 (2002).

90   S. Boukraa, J.-L. Basdevant. *J. Math. Phys.* 30, 1060 (1989).

91   L. J. Boya, H. Rosu, A. J. Seguí–Santonja, F. J. Vila. *arXiv:quant-ph*/9711059 (1997).

92   J. P. Boyd. *J. Comput. Phys.* 54, 382 (1984).

93   J. P. Boyd, D. W. Moore. *Dyn. Atmosph. Oceans* 10, 51 (1986).

94   J. P. Boyd. *Chebyshev and Fourier Spectral Methods*, Springer-Verlag, Berlin, 1989.

95   J. P. Boyd in A. V. Ilin, L. R. Scott. *Proceedings of the Third International Conference on Spectral and High Order Methods*, Houston Journal of Mathematics, Houston, 1996.

96   J. P. Boyd. *J. Comput. Phys.* 204, 253 (2005).

97   M. Brack, B. P. van Zyl. *Phys. Rev. Lett.* 86, 1574 (2001).

98   M. Brack, M. V. N. Murthy. *J. Phys.* 36, 1111 (2003).

99   F. Brau. *arXiv:math-ph*/0411009 (2004).

100   D. M. Bressoud. *Proofs and Confirmations*, Cambridge University Press, Cambridge, 1999.

101   C. Brezinski. *Padé-Type Approximation and General Orthogonal Polynomials*, Birkhauser, Basel, 1980.

102   C. Brezinski, A. Draux, A. P. Magnus, P. Maroni, A. Ronveaux. *Polynomes Orthogonaux et Applications*, Springer-Verlag, Berlin, 1985.

103   C. Brezinski. *Num. Alg.* 36, 309 (2004).

104   J. W. Brown, R. V. Churchill. *Fourier Series and Boundary Value Problems*, McGraw-Hill, New York, 1993.

105   Č. Burdík, G. S. Pogosyan. *arXiv:quant-ph*/0403129 (2004).

106   R. G. Buschman. *Am. Math. Monthly* 69, 288 (1962).

107   R. G. Buschman. *Proc. Am. Math. Soc.* 13, 675 (1963).

108   W. H. Butler. *Phys. Rev.* B 14, 468 (1976).

109   V. Bykov, A. Kytmanov, M. Lazman, M. Passare (ed.). *Elimination Methods in Polynomial Computer Algebra*, Kluwer, Dordrecht, 1998.

110   F. W. Byron, R. W. Fuller. *Mathematics of Classical and Quantum Physics* v. 1, Addison Wesley, Reading, 1970.

111   E. Caliceti. *J. Phys.* A 33, 3753 (2000).

112   H. E. Camblong, L. N. Epele, H. Fanchiotti, C. A. Garciá Canal. *arXiv:hep-th*/0003267 (2000).

113   R. G. Campos, L. A. Avila. *Glasgow Math. J.* 37, 105 (1995).

114   R. Carbó, E. Besalu. *J. Math. Chem.* 13, 331 (1993).

115   J. L. Cardoso, R. Álvarzez–Nodarse. *J. Phys.* A 36, 2055 (2003).

116   C. M. Carracedo, M. S. Alix. *The Theory of Fractional Powers of Operators*, Elsevier, Amsterdam, 2001.

117   B. C. Carlson. *Special Functions of Applied Mathematics*, Academic Press, New York, 1977.

118   K. M. Case. *J. Math. Phys.* 21, 702 (1980).

119   K. M. Case. *J. Math. Phys.* 21, 709 (1980).

120   W. E. Caswell. *Ann. Phys.* 123, 153 (1979).

121   J. M. Cerverhó, A. Rodríguez. *arXiv:cond-mat*/0206486 (2002).

122   O. A. Chalykh, A. P. Veselov. *J. Nonlin. Math. Phys.* 12, S1, 179 (2005).

123   G. K.-L. Chan, M. Head-Gordon. *J. Chem. Phys.* 116, 4462 (2002).

124   S. Chaturvedi, V. Gupta. *J. Phys.* A 33, L251 (2000).

125   S. K. Chatterjea, H. M. Srivastava. *Studies Appl. Math.* 83, 319 (1990).

126   L. V. Chebotarev. *Ann. Phys.* 273, 114 (1999).

127   J.-L. Chen, L. C. Kwek, C. H. Oh. *Phys. Rev.* A 67, 012101 (2003).

128   W. Y. C. Chen, K.-W. Li, Y.-N. Yeh. *Studies Appl. Math.* 94, 327 (1995).

129   Y. Chen, N. Lawrence. *J. Phys.* A 35, 4651 (2002).

130   Y. Chen, G. Pruessner. *J. Phys.* A 38, L191 (2005).

131   Y. F. Chen, K. F. Huang. *J. Phys.* A 36, 7751 (2003).

132   T. Cheon, P. Exner, P. Šeba. *arXiv:quant-ph*/0004058 (2000).

133   M. H. Cherkani, F. Brouillard, M. Chibisov. *J. Phys.* B 34, 49 (2001).

134   T. S. Chihara. *Introduction to Orthogonal Polynomials*, Gordon and Breach, New York, 1978.

135   L.-Y. C. Chiu, M. Moharerrzadeh. *Int. J. Quant. Chem.* 73, 265 (1999).

136   T. W. Chiu. *J. Phys.* A 19, 2537 (1986).

137   J. R. Choi. *Int. J. Theor. Phys.* 42, 853 (2003).

138   A. E. Chubykalo, R. A. Flores. *arXiv:physics*/0210089 (2002).

139   N. L. Chuprikov. *Semiconductors* 30, 246 (1996).

140   J. Cigler. *Körper, Ringe, Gleichungen*, Spektrum, Heidelberg, 1995.

141   J. Cisło, J. Lopuszanski. *arXiv:quant-ph*/0012005 (2000).

142   F. Citrini, L. Malegat. *J. Phys.* B 35, 1657 (2002).

143   J. Cizek, E. R. Vrscay. *Phys. Rev.* A 30, 1550 (1984).

144   J. Čizek, E. J. Weniger, P. Bracken, V. Špirko. *Phys. Rev.* E 53, 2925 (1996).

145   E. Clarke, X. Zhao. *The Mathematica Journal* 3, n1, 56 (1993).

146   A. Corichi, J. Cortez, H. Quevedo. *Ann. Phys.* 313, 446 (2004).

147   N. Cotfas. *J. Phys.* A 35, 9355 (2002).

148   N. Cotfas. *arXiv:math-ph*/0410050 (2004).

149   C. A. Coulson, J. C. Nash. *J. Phys.* B 7, 657 (1974).

150   R. Courant, D. Hilbert. *Methods of Mathematical Physics*, v.1, v.2, Interscience, New York, 1963.

151   A. Cuyt (ed.). *Nonlinear Numerical Methods and Rational Approximations* II, Kluwer, Dordrecht, 1994.

152   M. Cvetic, L. Picman. *J. Phys.* A 14, 379 (1981).

153   J. P. Dahl, W. P. Schleich. *arXiv:quant-ph*/0110134 (2001).

154   G. Darboux. *Leçons sur la théorie des surfaces*, v. 2, Gauthier–Villars, Paris, 1915.

155   C. N. Davies, M. Aylward in F. L. Alt (ed.). *Adv. Comput.* 2, 55 (1961).

156   E. B. Davies. *Bull. Lond. Math. Soc.* 34, 513 (2002).

157   P. J. Davis. *Interpolation and Approximation*, Dover, New York, 1975.

158   R. N. Deb, A. Khare, B. D. Roy. *arXiv:quant-ph*/0211008 (2002).

159   J. Dehesa (ed.). *J. Comput. Appl. Math.* 49, n1-3 (1993).

160   E. Delabaere, H. Dillinger. *J. Math. Phys.* 38, 6126 (1997).

161   E. Delabaere, F. Pham. *Ann. Phys.* 261, 180 (1997).

162   E. Delabaere, D. T. Trinh. *J. Phys.* A 33, 8771 (2000).

163   M. Demiralp, E. Suhubi. *J. Math. Phys.* 18, 777 (1977).

164   P. Desrosiers, L. Lapointe, P. Mathieu. *arXiv:hep-th*/0305038 (2003).

165   J. J. Diamond. *Am. J. Phys.* 60, 912 (1992).

166   K. Dilcher, K. B. Stolarsky. *Trans. Am. Math. Soc.* 357, 965 (2004).

167   I. K. Dmitrieva, G. I. Plindov. *Phys. Lett.* A 79, 47 (1980).

168   I. V. Dobrovolska, R. S. Tutik. *arXiv:quant-ph*/0108142 (2001).

169   V. V. Dodonov. *J. Opt.* B 4, R1 (2002).

170   S.-H. Dong. *Int. J. Theor. Phys.* 41, 1991 (2002).

171   S.-H. Dong. *Found. Phys. Lett.* 15, 385 (2002).

172   P. Dorey, C. Dunning, R. Tateo. *J. Phys.* A 34, L 391 (2001).

173   P. Dorey, C. Dunning, R. Tateo. *arXiv:hep-th*/0201108 (2002).

174   P. Dorey, R. Tateo. *arXiv:hep-th*/9812211 (1998).

175   P. Dorey, A. Millican–Slater, R. Tateo. *arXiv:hep-th*/0410013 (2004).

176   T. A. Driscoll, B. Fornberg. *Num. Alg.* 26, 77 (2001).

177   K. Driver, P. Duren. *J. Comput. Appl. Math.* 135, 293 (2001).

178   J. E. Drummond. *J. Phys.* A 14, 1651 (1981).

179   W. Duch. *J. Phys.* A 16, 4233 (1983).

180   B. I. Dunlap. *Phys. Rev.* A 66, 032502 (2002).

181   G. V. Dunne. *arXiv:hep-th*/0207046 (2002).

182   S. Dusuel, G. S. Uhrig. *J. Phys.* A 37, 9275 (2004).

183   E. N. Economou. *Green Functions in Quantum Physics*, Springer-Verlag, Berlin, 1983.

184   A. Elbert, M. E. Muldoon. *Proc. R. Soc. Edinb.* A 129, 57 (1999).

185   M. El-Mikkawy. *Appl. Math. Comput.* 146, 759 (2003).

186   T. Endo, Y. Hirayoshi, K. Toyoshima. *J. Phys. Soc. Jpn.* 74, 547 (2005).

187   A. Erdely. *Higher Transcendental Functions* v.1–3, McGraw-Hill, New York, 1953.

188   P. Exner, M. Tater, D. Vanek. *Math. Phys. Archive* 01-101 (2001).
       http://rene.ma.utexas.edu/mp_arc/c/01/01-101.ps.gz

189   M. R. Evans, R. A. Blythe. *Physica* A 313, 110 (2002).

190   K. Ey, A. Ruffing. *Progr. Theor. Phys.* S 150, 37 (2003).

191   M. Fabre de la Ripelle. *Few Body Systems* 14, 1 (1993).

192   U. Fano, A. R. P. Rau. *Symmetries in Quantum Physics*, Academic Press, San Diego, 2001.

193   D. Felbacq, B. Guizal. *J. Math. Phys.* 39, 4604 (1998).

194   P. Falloon, J. B. Wang. *Comput. Phys. Commun.* 134, 167 (2001).

195   H. Falomir, P. A. G. Pisani, A. Wipf. *arXiv:math-ph*/0112019 (2001).

196   H. Falomir, M. A. Muschietti, P. A. G. Pisani. *J. Math. Phys.* 45, 4560 (2004).

197   T. M. Feil, H. H. H. Homeier. *Comput. Phys. Commun.* 158, 124 (2004).

198   F. M. Fernández, R. Guardiola. *J. Phys.* A 26, 7169 (1993).

199   F. M. Fernández, E. A. Castro. *Algebraic Methods in Quantum Chemistry and Physics*, CRC Press, Boca Raton, 1996.

200   F. M. Fernández. *Int. J. Quant. Chem.* 81, 268 (2001).

201   F. M. Fernández. *Introduction to Perturbation Theory in Quantum Mechanics*, CRC Press, Boca Raton, 2001.

202   G. Fischer. *Plane Algebraic Curves*, American Mathematical Society, Providence, 2001.

203   L. Fishman, M. V. de Hoop, M. J. N. van Stralen. *J. Math. Phys.* 41, 4881 (2000).

204   G. P. Flessas, G. S. Anagnostatos. *J. Phys.* A 15, L537 (1982).

205   R. Floreannini, J. LeTourneux, L. Vinet. *Ann. Phys.* 226, 331 (1993).

206   E. R. Floyd. *Int. J. Mod. Phys.* A 15, 1363 (2000).

207   P. J. Forrester, J. B. Rogers. *SIAM J. Math. Anal.* 17, 461 (1986).

208   L. Fox, I. B. Parker. *Chebyshev Polynomials in Numerical Analysis*, Oxford University Press, London, 1968.

209   B. R. Frieden. *arXiv:quant-ph*/0006012 (2000).

210   H. Fu, Y. Feng, A. I. Solomon. *J. Phys.* A 33, 2231 (2000).

211   D. Funaro. *Polynomial Approximation of Differential Equations*, Springer-Verlag, Berlin, 1992.

212   A. Gangopadhyaya, J. V. Mallow, C. Rasinariu, U. P. Sukhatme. *Chinese J. Phys.* 39, 101 (2001).

213   P. Garbaczewski, J. R. Klauder, R. Olkiewicz. *arXiv:quant-ph*/9505003 (1995).

214   P. Garbaczewski, W. Karwowski. *arXiv:math-ph*/0104010 (2001).

215   N. Gauthier. *Am. J. Phys.* 61, 857 (1993).

216   W. Gautschi. *Orthogonal Polynomials*, Oxford University Press, Oxford, 2004.

217   J. Geer, N. S. Banarjee. *J. Sc. Comput.* 12, 253 (1997).

218   A. Gelb. *J. Sc. Comput.* 20, 433 (2004).

219   J. Geronimus. *J. Lond. Math. Soc.* 6, 55 (1931).

220   F. Gesztesy, C. Macdeo, L. Streit. *J. Phys.* A 18, L503 (1985).

221   T. L. Gill, W. W. Zachary. *mp_arc* 03-109 (2001).   http://rene.ma.utexas.edu/mp_arc/c/03/03-109.pdf.gz

222   T. L. Gill, W. W. Zachary. *arXiv:quant-ph*/0405150 (2004).

223   S. D. Glazek, K. G. Wilson. *Phys. Rev.* D 49, 4214 (1994).

224   B. Gönül, N. Çelik, E. Olgar. *arXiv:quant-ph*/0412162 (2004).

225   D. Gottlieb, C.-W. Shu, A. Solomonoff, H. Vandeven. *J. Comput. Appl. Math.* 43, 81 (1992).

226   D. Gottlieb, C.-W. Shu. *SIAM Rev.* 39, 644 (1997).

227   S. Godoy, L. Andrade, E. Braun. *Physica* A 196, 416 (1993).

228   S. Godoy, S. Fujita. *J. Chem. Phys.* 97, 5148 (1992).

229   G. H. Golub, C. F. van Loan. *Matrix Computations*, Johns Hopkins University Press, Baltimore, 1989.

230   G. Gómez–Ullate, N. Kamran, R. Milson. *arXiv:quant-ph/*0308062 (2003).

231   C. Gonera, P. Kosinski, M. Majewski, P. Maslanka. *Acta Phys. Polonica* B 30, 915 (1999).

232   P. Gosselin, B. Grosdidier, H. Mohrbach. *Phys. Lett.* A 256, 125 (1999).

233   V. B. Gostev, V. S. Mineev, A. R. Frenkin. *Theor. Math. Phys.* 68, 664 (1986).

234   H. P. W. Gottlieb. *Inverse Problems* 18, 971 (2002).

235   H. W. Gould. *Fibon. Quart.* 37, 135 (1999).

236   S. Graffi, V. Grecchi. *Phys. Rev.* D 8, 3487 (1973).

237   C. G. Gray, G. Karl. *arXiv:physics/*0312071 (2003).

238   D. J. Griffith, N. F. Taussig. *Am. J. Phys.* 60, 883 (1992).

239   D. J. Griffith, C. A. Steinke. *Am. J. Phys.* 69, 137 (2001).

240   C. Grosche, F. Steiner. *Handbook of Feynman Path Integrals*, Springer-Verlag, Heidelberg 1998.

241   C. Grosche. *arXiv:nlin.SI/*0411053 (2004).

242   R. Guantes, S. C. Farantos. *J. Chem. Phys.* 113, 10429 (2000).

243   R. Guida, K. Konishi. *Ann. Phys.* 241, 152 (1995).

244   B.-y. Guo, C.-l. Xu. *Adv. Comput. Math.* 19, 35 (2003).

245   N. Gurappa, P. K. Panigrahi, T. S. Raju. *arXiv:cond-mat/*9901073 (1999).

246   A. J. Guttmann in C. Domb, J. L. Lebowitz (eds.). *Phase Transitions and Critical Phenomena* v.13, Academic Press, London, 1989.

247   W. Hackbusch. *Iterative Solution of Large Sparse Systems of Equations*, Springer-Verlag, New York, 1994.

248   Y. M. Hakobyan, M. Kibler, G. S. Pogosyan, A. N. Sissakian. *arXiv:quant-ph/*9712014 (1997).

249   R. L. Hall, W. Lucha, F. F. Schöberl. *arXiv:hep-th/*0012127 (2000).

250   R. L. Hall, W. Lucha, F. F. Schöberl. *ESI Preprint* 984 (2001).   ftp://ftp.esi.ac.at/pub/Preprints/esi984.ps

251   R. L. Hall, W. Lucha, F. F. Schöberl. *J. Phys.* A 34, 5059 (2001).

252   R. L. Hall, W. Lucha, F. F. Schöberl. *arXiv:math-ph/*0110015 (2001).

253   R. L. Hall, W. Lucha, F. F. Schöberl. *arXiv:hep-th/*0110220 (2001).

254   R. L. Hall, W. Lucha, F. F. Schöberl. *arXiv:hep-th/*0210149 (2002).

255   R. L. Hall, W. Lucha, F. Schöberl. *Int. J. Mod. Phys.* A 18, 2657 (2003).

256   R. L. Hall, Q. D. Katabeth, N. Saad. *arXiv:math-ph/*0410035 (2004).

257   K. Hallberg. *arXiv:cond-mat/*0303557 (2003).

258   I. G. Halliday, P. Suranyi. *Phys. Rev.* D 21, 1529 (1980).

259   F. R. Halpern, T. W. Yonkman. *J. Math. Phys.* 15, 1718 (1974).

260   B. Hamprecht, A. Pelster. *arXiv:cond-mat/*0105509 (2001).

261   B. Hamprecht, H. Kleinert. *arXiv:hep-th/*0302116 (2003).

262   C. R. Handy. *J. Math. Phys.* 29, 32 (1988).

263   C. R. Handy. *arXiv:math-ph/*0104035 (2001).

264   C. R. Handy. *arXiv:math-ph/*0104036 (2001).

265   C. R. Handy, D. Khan, X.-Q. Wang, C. J. Tymczak. *arXiv:math-ph*/0104037 (2001).

266   C. R. Handy, X. Q. Wang. *arXiv:math-ph*/0105019 (2001).

267   H. Hasegawa. *Open Sys. Inform. Dyn.* 4, 359 (1997).

268   T. Hatsuda, T. Kunihiro, T. Tanaka. *arXiv:hep-th*/9612097 (1996).

269   T. Hatsuda, T. Kunihiro, T. Tanaka. *Phys. Rev. Lett.* 78, 3229 (1997).

270   M. Hazewinkel. *CWI Quart.* 12, 93 (1999).

271   X.-F. He. *Phys. Rev.* B 43, 2063 (1991).

272   E. J. Heller. *J. Chem. Phys.* 62, 1544 (1974).

273   E. Heller in M.-J. Giannoni, A. Voros, J. Zinn-Justin (eds.). *Chaos and Quantum Physics*, North-Holland, Amsterdam, 1991.

274   E. J. Heller in J. A. Yeazell, T. Uzer (eds.). *The Physics and Chemistry of Wave Packets*, Wiley, New York, 2000.

275   G. Helmberg, P. Wagner. *J. Approx. Th.* 89, 308 (1997).

276   E. Hendriksen, H. van Rossum in M. Alfaro, J. S. Dehesa, F. J. Marcellan, J. L. Rubio de Francia, J. Vinuesa (eds.). *Orthogonal Polynomials and Their Applications*, Springer-Verlag, Berlin, 1988.

277   J. M. Herbert, W. C. Ermler. *Comput. Chem.* 22, 169 (1998).

278   R.-M. Hervé, M. Hervé. *J. Math. Phys.* 32, 956 (1991).

279   R. Hinterding, J. Wess. *Eur. J. Phys.* C 6, 183 (1999).

280   F. T. Hioe, E. W. Montroll. *J. Math. Phys.* 16, 1945 (1975).

281   F. T. Hioe, D. MacMillon, E. W. Montroll. *J. Math. Phys.* 17, 1320 (1976).

282   F. T. Hioe, D. MacMillan, E. W. Montroll. *Phys. Rep.* 43, 305 (1978).

283   K. C. Ho, Y. T. Liu, C. F. Lo, K. L. Liu, W. M. Kwok, M. L. Shiu. *Phys. Rev.* A 53, 1280 (1996).

284   E. W. Hobson. *The Theory of Spherical and Ellipsoidal Harmonics*, Chelsea, New York, 1955.

285   I. A. Howard, N. H. March, L. M. Nieto. *J. Phys.* A 35, 4985 (2002).

286   L. Infeld, T. E. Hull. *Rev. Mod. Phys.* 23, 21 (1951).

287   H. Ishikawa. *J. Phys.* A 35, 4453 (2002).

288   M. E. H. Ismail. *Num. Funct. Anal. Optim.* 21, 191 (2000).

289   M. E. H. Ismail. *Pac. J. Math.* 193, 355 (2000).

290   A. A. Istmest'ev, G. S. Pogosyan, A. N. Sissakian, P. Winternitz. *J. Math. Phys.* 40, 1549 (1999).

291   I. B. Ivanov. *arXiv:quant-ph*/0203019 (2002).

292   M. A. Jafarizadeh, A. Rostami. *arXiv:math-ph*/0003013 (2000).

293   M. A. Jafarizadeh, H. Goudarzi. *arXiv:math-ph*/0404054 (2004).

294   S. R. Jain, B. Grémaud, A. Khare. *Phys. Rev.* E 66, 016216 (2002).

295   G. S. Japaridze. *J. Phys.* A 35, 1709 (2002).

296   M. Järvinen. *arXiv:hep-ph*/0411208 (2004).

297   R. Jáuregui, R. Paredes, G. T. Sánchez. *arXiv:cond-mat*/0405045 (2004).

298   A. J. Jerri. *The Gibbs Phenomenon in Fourier Analysis, Splines and Wavelet Approximations*, Kluwer, Dordrecht, 1998.

299   U. Jentschura, J. Zinn–Justin. *arXiv:math-ph*/0103010 (2001).

300   M. N. Jones. *Spherical Harmonics and Tensors for Classical Field Theory*, Wiley, New York, 1985.

301   A. Joseph. *Rev. Mod. Phys.* 39, 829 (1967).

302   J.-H. Jung, B. D. Shizgal. *Int. J. Comput. Appl. Math.* 172, 131 (2004).

303　J.-P. Kahane, P.-G. Lemarié–Rieusset. *Fourier Series and Wavelets*, Gordon and Breach, Luxembourg, 1995.

304　E. G. Kalnins, W. Miller, Jr. in T. M. Rassias, H. M. Srivastava, A. Yanushauskas (eds.). *Topics in Polynomials of One and Several Variables and Their Applications*, World Scientific, Singapore, 1993.

305　T. M. Kalotas, A. R. Lee. *Eur. J. Phys.* 16, 119 (1995).

306　D. Kaminski in C. Dunkl, M. Ismail, R. Wong (eds.). *Special Functions*, World Scientific, Singapore, 2000.

307　L. W. Kantorowitsch, G. P. Akilov. *Funktionalanalysis in normiereten Räumen*, Akademie-Verlag, Berlin, 1964.

308　A. E. Kaplan, I. Marzoli, W. E. Lamb, Jr., W. P. Schleich. *Phys. Rev.* A 61, 032101 (2000).

309　J. Katriel, G. Adam. *J. Phys.* B 3, 13 (1970).

310　M. Kauers in J. Gutierrez (ed.). *ISSAC 04*, ACM Press, New York, 2004.

311　H. Kautzleben. *Kugelfunktionen* (Reihe Geomagnetismus und Aeronomie, v. 1/ Ergänzungsband), Teubner, Leipzig, 1965.

312　S. V. Kerov. *St. Petersburg Math. J.* 5, 925 (1994).

313　S. V. Kerov. *St. Petersburg Math. J.* 12, 1049 (2001).

314　S. V. Kerov. *Asymptotic Representation Theory of the Symmetric Group and Its Applications in Analysis*, American Mathematical Society, Providence, 2003.

315　A. T. Kerris, R. J. Lombard. *Phys. Lett.* A 319, 263 (2003).

316　P. B. Khan, Y. Zarmi. *J. Math. Phys.* 40, 4658 (1999).

317　A. Khare, U. Sukhatme. *J. Phys.* A 22, 2847 (1989).

318　A. Khare, U. P. Sukhatme. *Phys. Rev.* A 40, 6185 (1989).

319　A. Khare. *arXiv:quant-ph*/0409003 (2004).

320　J. P. Killingbeck, A. Grosjean, G. Jolicard. *J. Phys.* A 34, 8309 (2001).

321　A. H. E. Kinanim M. Daoud. *Phys. Lett.* A 283, 291 (2001).

322　V. V. Kisil. *Ann. Combinat.* 6, 45 (2002).

323　R. Klippert, H. C. Rosu. *Int. J. Theor. Phys.* 41, 331 (2002).

324　J. R. Klauder, B. S. Skagerstam. *Coherent States*, World Scientific, Singapore, 1985.

325　H. Kleindienst, A. Lüchow. *Int. J. Quant. Chem.* 48, 239 (1993).

326　H. Kleinert, V. Schulte–Frohlinde. *Critical Properties of ϕ^4-Theories*, World Scientific, Singapore, 2001.

327　R. Klippert, H. C. Rosu. *arXiv:quant-ph*/0005048 (2000).

328　R. Kochendörffer. *Introduction to Algebra*, Wolters-Noordhoff, Groningen, 1972.

329　R. Koekoek, R. F. Swarttouw. *arXiv:math.CA*/9602214 (1996).

330　W. Koepf in R. P. Gilbert, J. Kajiwara, Y. S. Xu (eds.). *Recent Developments in Complex Analysis and Computer Algebra*, Kluwer, Dordrecht, 1999.

331　T. Koike. *Ann. Henri Poincaré* 1, 193 (2000).

332　A. Kormányos, J. Cserti, G. Vattay. *arXiv:cond-mat*/0005407 (2000).

333　A. V. Kotikov. *arXiv:hep-ph*/0102177 (2001).

334　L. P. Kouwenhoven, F. W. J. Hekking, B. J. van Wees, C. J. P. M. Harmans. *Phys. Rev. Lett.* 65, 361 (1990).

335　V. Kowalenko, A. A. Rawlinson. *J. Phys.* A 31, L663 (1998).

336　A. M. Krall in M. Alfaro, J. S. Dehesa, F. J. Marcellan, J. L. Rubio de Francia, J. Vinuesa (eds.). *Orthogonal Polynomials and Their Applications*, Springer-Verlag, Berlin, 1988.

337　A. B. J. Kuilaars, A. Martínez–Finkelshtein, R. Orive. *arXiv:math.CA*/0301037 (2003).

338　T. Kunihiro. *arXiv:hep-th*/9710087 (1997).

339   L. C. Kwek, Y. Liu, H. C. Oh, X.-B. Wang. *arXiv:quant-ph*/0007031 (2000).

340   K. H. Kwon, J. K. Lee, B. H. Yoo. *Result. Math.* 24, 119 (1993).

341   K. H. Kwon, L. L. Littlejohn. *J. Korean Math. Soc.* 34, 973 (1997).

342   C. Lämmerzahl. *J. Math. Phys.* 34, 3918 (1993).

343   J. R. Laguna. *arXiv:cond-mat*/0207340 (2002).

344   W. Lang. *J. Comput. Appl. Math.* 89, 237 (1998).

345   O. E. Lanford III in K. Osterwalder, R. Stora (eds.). *Les Houches*, Session XLIII, Elsevier, Amsterdam, 1984.

346   F. Lanzara, V. Maz'ya, G. Schmidt. *ESI Preprint* 984 (2003).   ftp://ftp.esi.ac.at:/pub/Preprints/esi1412.ps

347   A. Lascoux, P. Pragacz. *Ann. Combinat.* 6, 169 (2002).

348   L. Lathouwers. *J. Math. Phys.* 16, 1393 (1975).

349   C. Leubner, M. Alber, N. Schupfer. *Am. J. Phys.* 56, 1123 (1988).

350   G. Lévai, B. Kónya, Z. Papp. *J. Math. Phys.* 39, 5811 (1998).

351   H. Li, D. Kusnezov. *arXiv:cond-mat*/9907202 (1999).

352   C.-T. Li, A. Klein. *Chinese J. Phys.* 39, 555 (2001).

353   T.-Y. Li. *Math. Intell.* 9, n3, 33 (1987).

354   J. F. Liebmann. *Int. J. Quant. Chem.* 101, 283 (2005).

355   J. C. Light, I. P. Hamilton, J. V. Lill. *J. Phys. Chem.* 82, 1400 (1985).

356   J. C. Light, T. Carrington, Jr. *Adv. Chem. Phys.* 114, 263 (2000).

357   R. G. Littlejohn, M. Cargo. *J. Chem. Phys.* 117, 27 (2002).

358   R. G. Littlejohn, M. Cargo. *J. Chem. Phys.* 117, 37 (2002).

359   Q.-H. Liu. *J. Phys.* A 32, L57 (1999).

360   Q. H. Liu, X. Wang, W. H. Qi, L. P. Fu, B. Hu. *J. Math. Phys.* 43, 170 (2002).

361   Q. H. Liu, W. H. Qi, T. G. Liu, Z. H. Zhu. *Int. J. Theor. Phys.* 42, 783 (2003).

362   S. S. Lo, D. A. Morales. *Int. J. Quant. Chem.* 88, 263 (2002).

363   S. S. Lo, D. A. Morales. *J. Math. Chem.* 35, 21 (2004).

364   B. F. Logani. *SIAM J. Math. Anal.* 21, 1031 (1990).

365   M. A. Lohe, A. Thilagam. *J. Phys.* A 37, 6181 (2004).

366   J. R. Lombardi. *Phys. Rev.* A 22, 797 (1980).

367   J. López-Bonilla, G. Ovando. *Irish Math. Soc. Bull.* 44, 61 (2000).

368   J. López–Bonilla, J. Morales, G. Ovando. *Bull. Allahabad Math. Soc.* 17, 45 (2002).

369   R. López–Mobilia, P. L. Nash. *J. Phys.* A 38, 227 (2005).

370   A. Lorek, A. Ruffing, J. Wess. *arXiv:hep-th*/9605161 (1996).

371   A. Losev. *J. Phys. Cond. Mat.* 15, 1007 (2003).

372   M. Luban, D. L. Pursey. *Phys. Rev.* D 33, 431 (1986).

373   D. S. Lubinsky. *Acta Appl. Math.* 61, 207 (2000).

374   W. Lucha, F. F. Schöberl. *arXiv:hep-ph*/0408184 (2004).

375   D. Lynden–Bell, R. M. Lynden–Bell. *arXiv:cond-mat*/9904019 (1999).

376   I.G. Macdonald. *Symmetric Functions and Hall Polynomials*, Oxford University Press, New York, 1995.

377   M. H. Macfarlane. *Ann. Phys.* 271, 159 (1999).

378   T. M. MacRobert. *Spherical Harmonics*, Dutton, New York, 1927.

379   W. Magnus, F. Oberhettinger, R. P. Soni. *Formulas and Theorems for the Special Functions of Mathematical Physics*, Springer-Verlag, Berlin, 1966.

380   R. Maj, S. Mrówczynski. *arXiv:physics*/0401063 (2004).

381   F. Malek, R. Vaillancourt. *Comput. Math. Appl.* 30, 37 (1995).

382   D. E. Manolopoulos, R. E. Wyatt. *Chem. Phys. Lett.* 152, 23 (1988).

383   T. Mansour. *arXiv:math.CO*/0108043 (2001).

384   F. Marcellan, A. Branquinho, J. Petronilho. *Acta Appl. Math.* 34, 283 (1994).

385   A. N. H. March, L. M. Nieto. *Phys. Rev.* A 63, 044502 (2001).

386   M. A. Martín–Delgado, G. Sierra, R. M. Noack. *J. Phys.* A 32, 6079 (1999).

387   A. Martines, A. Sarso, R. Yan'es. *Russ. Acad. Sci. Sb. Math.* 83, 483 (1995).

388   V. P. Maslov, S. A. Molchanov, A. Y. Gordan. *Russ. J. Math. Phys.* 1, 71 (1993).

389   J. C. Mason, D. C. Handscomb. *Chebyshev Polynomials*, CRC Press, Boca Raton, 2003.

390   A. Matos–Abiague. *J. Phys.* A 34, 11059 (2001).

391   V. B. Matveev. *Inverse Problems* 17, 633 (2001).

392   H. A. Mavromatis. *J. Phys.* A 32, 2601 (1999).

393   E. B. McBride. *Obtaining Generating Functions*, Springer-Verlag, Berlin, 1971.

394   M. Martens, T. Nowicki. *Invent. Math.* 144, 225 (2001).

395   R. L. Matcha, K. C. Daiker. *J. Math. Phys.* 15, 114 (1974).

396   D. McPeake, J. F. McCann. *Comput. Phys. Commun.* 161, 119 (2004).

397   S. M. McRae, E. R. Vrscay. *J. Math. Phys.* 33, 3004 (1992).

398   H. Meißner, E. O. Steinborn. *Phys. Rev.* A 56, 1189 (1997).

399   I. Mendas. *J. Phys.* A 26, L93 (1993).

400   Y. Meurice. *Phys. Rev. Lett.* 88, 141601 (2002).

401   B. Mielnik. *J. Math. Phys.* 25, 3387 (1984).

402   B. Mielnik. *J. Phys.* A 37, 10007 (2004).

403   G. V. Milovanović. *Publ. de l'Inst. Math.* 64, 53 (1998).

404   R. M. Mir–Kasimov. *Int. J. Theor. Phys.* 32, 607 (2002).

405   D. S. Mitronović, D. D. Tošić, R. R. Janić. *Specijalne Funkcije*, Naučna Knjiga, Beograd, 1978.

406   H. Miyazaki, I. Tsutsui. *arXiv:quant-ph*/0202037 (2002).

407   N. Mizutani, H. Yamada. *arXiv:physics*/97102013 (1997).

408   J. J. Monzón, L. L. Sánches–Soto. *Opt. Commun.* 162, 1 (2001).

409   J. Morales, J. Lopez-Bonilla, A. Palma. *J. Math. Phys.* 28, 1032 (1987).

410   J. Morales, A. Flores-Riveros. *J. Math. Phys.* 30, 393 (1989).

411   J. Morales, J. J. Peña, P. Portillo, G. Ovando, V. Gaftoi. *Int. J. Quant. Chem.* 65, 205 (1997).

412   J. Morales, J. J. Peña, J. L. López–Bonilla. *J. Math. Phys.* 42, 966 (2001).

413   A. Morgan. *Solving Polynomial Systems Using Continuation for Engineering and Scientific Problems*, Prentice-Hall, Englewood Cliffs, 1987.

414   M. Morifuji, K. Kato. *Phys. Rev.* B 68, 035108 (2003).

415   H. E. Moses, R. T. Prosser. *SIAM Rev.* 35, 610 (1993).

416   S. Moukouri. *Phys. Lett.* A 325, 177 (2004).

417   A. I. Mudrov, K. B. Varnashev. *arXiv:cond-mat*/9805081 (1998).

418   E. J. Mueller. *arXiv:cond-mat*/0405425 (2004).

419   C. Müller. *Analysis of Spherical Symmetries in Euclidean Spaces*, Springer-Verlag, New York, 1998.

420   R. Z. Muratov. *Tech. Phys.* 47, 380 (2002).

421   O. Mustafa, M. Odeh. *arXiv:quant-ph*/0001038 (2000).

422   O. Mustafa, M. Odeh. *Eur. J. Phys.* B 15, 143 (2000).

423   J. Naas, H. L. Schmid. *Mathematisches Wörterbuch*, Akademie-Verlag, Berlin, 1961.

424   B. Nagel. *J. Math. Phys.* 35, 1549 (1994).

425   S. M. Nagiyev, E. I. Jafarov, R. M. Imanov. *arXiv:math-ph*/0302042 (2003).

426   H. Nakamura. *J. Chem. Phys.* 97, 256 (1992).

427   K. Nakamura, M. Lakshmanan. *Phys. Rev. Lett.* 57, 1661 (1986).

428   K. Namsrai. *Int. J. Theor. Phys.* 37, 1531 (1998).

429   K. Namsrai, H. V. von Geramb. *Int. J. Theor. Phys.* 40, 1929 (2001).

430   I. P. Natanson. *Constructive Function Theory*, v 1, Fredrick Ungar, New York, 1964.

431   I. P. Natanson. *Constructive Function Theory*, v 2, Fredrick Ungar, New York, 1964.

432   P. Nevai. *SIAM J. Math. Anal.* 15, 1177 (1986).

433   P. Nevai (ed.). *Orthogonal Polynomials: Theory and Practice*, Kluwer, Amsterdam, 1990.

434   R. G. Newton. *Am. J. Phys.* 62, 1042 (1994).

435   C. P. Niculescu. *J. Inequal. Pure Appl. Math.* 1, 2, a17 (2000).

436   A. F. Nikiforov, V. B. Uvarov. *Special Functions of Mathematical Physics*, Birkhäuser, Basel, 1988.

437   A. F. Nikiforov, S. K. Suslov, V. B. Uvarov. *Classical Orthogonal Polynomials*, Springer-Verlag, Berlin, 1991.

438   T. Nowicki. *Invent. Math.* 144, 233 (2001).

439   H. N. Núñez–Yépez. *arXiv:physics*/0001030 (2000).

440   M. A. Núñez. *Phys. Rev.* E 68, 016703 (2003).

441   S. Odake, R. Sasaki. *J. Phys.* A 35, 8283 (2002).

442   A. M. Odlyzko in R. Graham, M. Grötschel, L. Lovász. *Handbook of Combinatorics*, v.2, Elsevier, Amsterdam 1995.

443   K. Okamoto in M. Kashiwara, T. Kawai (eds.). *Algebraic Analysis*, Academic Press, Boston, 1988.

444   M. G. Olsson, S. Veseli. *Phys. Rev.* D 52, 5141 (1995).

445   B. Osilenker. *Fourier Series in Orthogonal Polynomials*, World Scientific, Singapore, 1999.

446   E. Öztekin. *Int. J. Quant. Chem.* 100, 236 (2004).

447   A. Palma, L. Sandoval, J. Morales. *Int. J. Quant. Chem.* S 21, 729 (1987).

448   C. Palmer, P. N. Stavrinou. *J. Phys.* A 37, 6987 (2004).

449   J. Pappademos, U. Sukhatme, A. Pagnamenta. *Phys. Rev.* A 48, 3525 (1993).

450   M. P. Pato. *Physica* A 312, 153 (2002).

451   G. Paz. *arXiv:quant-ph*/0009046 (2000).

452   D. Pecker. *Compositio Math.* 87, 1 (1993).

453   A. Pelster, F. Weissbach. *arXiv:quant-ph*/0105095 (2001).

454   P. Pereyra. *arXiv:cond-mat*/0009064 (2000).

455   P. Pereyra. *Physica* E 17, 209 (2003).

456   F. Pérez–Bernal, I. Martel, J. M. Aria, J. Gómes–Camachao. *Phys. Rev.* A 63, 052111 (2001).

457   R. T. Perkins, L. V. Knight. *Acta Cryst.* A 40, 617 (1984).

458   O. Perron. *Algebra*, de Gruyter, Berlin, 1951.

459   P. Piecuch. *Rep. Math. Phys.* 24, 187 (1986).

460   J. Plemelj. *Problems in the Sense of Riemann and Klein*, Interscience, New York, 1964.

461   W. Polak. *J. Appl. Phys.* 74, 777 (1993).

462   A. D. Polyanin. *Handbook of Linear Partial Differential Equations for Engineers and Scientists*, Chapman & Hall, Boca Raton, 2002.

463   D. Port. *J. Combinat. Th.* A 98, 219 (2002).

464   G. Pöschl, E. Teller. *Z. Phys.* 83, 143 (1933).

465   R. E. Powell, S. M. Shah. *Summability Theory and Applications*, van Nostrand, New York, 1972.

466   R. Prus, A. Sym. *Phys. Lett.* A 336, 459 (2005).

467   R. R. Puri. *Mathematical Models of Quantum Optics*, Springer-Verlag, Berlin, 2001.

468   S. K. Rangarajan, S. P. Purushothaman. *J. Comput. Appl. Math.* 177, 461 (2005).

469   T. M. Rassias, H. M. Srivastava, A. Yanushauskas. *Topics in Polynomials of One and Several Variables and Their Applications*, World Scientific, Singapore, 1993.

470   B. Rath. *J. Phys. Soc. Jpn.* 66, 3693 (1997).

471   E. Red, C. A. Weatherford. *Int. J. Quant. Chem.* 100, 208 (2004).

472   L. B. Rédei. *Acta Sci. Math.* 37, 115 (1975).

473   C. E. Reid. *Int. J. Quant. Chem.* 1, 521 (1967).

474   G. Reinisch. *Phys. Rev.* A 56, 3409 (1997).

475   T. N. Rescigno, C. W. McCurdy. *Phys. Rev.* A 62, 032706 (2001).

476   M. A. Reyes, D. Jimenez, H. C. Rosu. *math-ph*/0301022 (2003).

477   N. Ripamonti. *J. Phys.* A 29, 5137 (1996).

478   T. J. Rivlin. *The Chebyshev Polynomials*, Wiley, New York, 1974.

479   M. M. Rizk. *Ukrainian Math. J.* 53, 155 (2001).

480   R. W. Robinett. *Am. J. Phys.* 63, 823 (1995).

481   R. W. Robinett. *Am. J. Phys.* 65, 190 (1997).

482   R. W. Robinett. *Eur. J. Phys.* 23, 165 (2002).

483   R. W. Robinett. *arXiv:quant-ph*/0401031 (2004).

484   M. A. Rodríguez, P. Winternitz. *arXiv:math-ph*/0110018 (2001).

485   C. Rorres. *SIAM J. Appl. Math.* 27, 303 (1974).

486   J. O. Rosas–Ortiz. *arXiv:quant-ph*/9812003 (1998).

487   B. Rosenstein, M. Usher. *Phys. Rev.* D 36, 2381 (1987).

488   H. Rosu. *arXiv:quant-ph*/9904007 (1999).

489   H. C. Rosu. *arXiv:quant-ph*/0104003 (2001).

490   H. C. Rosu, F. A. de la Cruz. *arXiv:quant-ph*/0107043 (2001).

491   C. C. Roy, A. Khan. *phys. stat. sol.* (b) 176, 101 (1993).

492   M. G. Rozman, P. Reineker, R. Tehver. *Phys. Rev.* A 49, 3310 (1994).

493   M. G. Rozman, P. Reineker, R. Tehver. *Phys. Lett.* A 187, 127 (1994).

494   B. Rui, R. Wong. *J. Approx. Th.* 98, 146 (1999).

495   T. W. Ruijgrok. *Acta Phys. Polonica* B 31, 1655 (2000).

496   N. Saad, R. L. Hall, Q. D. Katabeh. *arXiv:math-ph*/0410039 (2004).

497   R. A. Sack. *J. Math. Phys.* 5, 245 (1964).

498   W. Salejda, P. Szyszuk. *Physica* A 252, 547 (1998).

499   L. Šamaj, J. K. Percus, P. Kalinay. *arXiv:math-ph*/0210004 (2002).

500   B. F. Samsonov. *arXiv:quant-ph*/9805051 (1998).

501   A. M. Sánchez, J. D. Bejarno. *J. Phys.* A 19, 887 (1986).

502   S. S. Sannikov–Proskuryakov. *Russ. Phys. J.* 46, 432 (2003).

503   G. Sansone. *Orthogonal Functions*, Interscience, New York, 1959.

504   M. Sassoli de Bianchi, M. D. Ventra. *J. Math. Phys.* 36, 1753 (1995).

505   M. Sassoli de Bianchi, M. Di Ventra. *Superlattices and Microstructures* 20, 149 (1996).

506   W. P. Schleich. *Quantum Optics in Phase Space*, Wiley–VCH, Berlin, 2001.

507   J. Schliemann, F. G. Mertens. *arXiv:cond-mat*/9811371 (1998).

508   G. Schmeisser. *Lin. Alg. Appl.* 193, 11 (1993).

509   H. J. Schmidt, J. Schnack. *arXiv:cond-mat*/0209397 (2002).

510   B. I. Schneider, N. Nygaard. *Phys. Rev.* E 70, 056706 (2004).

511   U. Schollwöck. *arXiv:cond-mat*/0409292 (2004).

512   W. Schuster. *Math. Semesterber.* 48, 1 (2001).

513   W. F. Scott. *Math. Scientist* 19, 87 (1994).

514   A. V. Sergeev, D. Z. Goodson. *J. Phys.* A 31, 4301 (1998).

515   B. D. Shizgal, H. Chen. *J. Chem. Phys.* 104, 4137 (1996).

516   P. E. Shanley. *Ann. Phys.* 186, 292 (1988).

517   B. S. Shastry, A. Dhar. *arXiv:cond-mat*/0101464 (2001).

518   K. C. Shin. *arXiv:math-ph*/0201013 (2002).

519   K. C. Shin. *arXiv:hep-ph*/0207251 (2002).

520   H. J. Silverstone, J. G. Harris. *Phys. Rev.* A 32, 1965 (1985).

521   B. Simon. *Ann. Phys.* 58, 76 (1970).

522   B. Simon in D. Bessis (eds.). *Cargèse Lectures in Physics* v5, Gordon and Breach, New York, 1972.

523   L. Skála, J. Čížek, V. Kapsa, E. J. Weniger. *Phys. Rev.* A 56, 4471 (1997).

524   L. Skála, J. Čížek, E. J. Weniger, J. Zamastil. *Phys. Rev.* A 59, 102 (1999).

525   E. R. Smith. *Am. Math. Monthly* 43, 354 (1936).

526   W. L. Smith. *J. Phys.* B 2, 1 (1969).

527   I. S. Sogami, K. Koizumi. *arXiv:math-ph*/0109017 (2001).

528   D.-Y. Song, J. H. Park. *arXiv:quant-ph*/0109108 (2001).

529   W. Specht in M. Deuring, G. Köthe (eds.). *Enzyklopädie der Mathematischen Wissenschaften* I 1, n3, T.2, Teubner, Stuttgart, 1958.

530   D. E. Spencer, P. Moon. *Field Theory Handbook*, Springer-Verlag, Berlin, 1988.

531   D. W. L. Sprung, H. Wu, J. Martonell. *Am. J. Phys.* 61, 1118 (1993).

532   M. Srednicki. *arXiv:hep-th*/0409035 (2004).

533   H. Stahl, V. Totik. *General Orthogonal Polynomials*, Cambridge University Press, Cambridge, 1992.

534   W.-H. Steeb, A. Fatykhova. *Can. J. Phys.* 72, 147 (1994).

535   F. M. Stein. *SIAM Rev.* 1, 167 (1959).

536   E. O. Steinborn, K. Ruedenberg. *Adv. Quant. Chem.* 7, 1 (1973).

537   A. I. Stepanets. *Uniform Approximations by Trigonometric Polynomials*, VSP, Utrecht, 2001.

538   F. H. Stillinger. *J. Math. Phys.* 18, 1224 (1977).

539   V. Strassen. *Num. Math.* 20, 238 (1973).

540   F. Strocchi. *arXiv:hep-th*/0401143 (2004).

541   J. Sucher. *J. Math. Phys.* 4, 17 (1963).

542   N. K. Suetin. *Klassicheskie Ortogonalnie Mnogochlenni*, Nauka, Moscow, 1979.

543   S. P. Suetin. *Russ. Math. Surv.* 57, 43 (2002).

544   C. V. Sukumar. *J. Phys.* A 37, 10287 (2004).

545   I. M. Suslov. *JETP* 93, 1 (2001).

546   J. Suzuki. *J. Phys.* A 32, L183 (1999).

547   R. J. Swenson, S. H. Danforth. *J. Chem. Phys.* 57, 1734 (1972).

548   H. K. Sy, T. C. Chua. *Phys. Rev.* B 48, 7930 (1993).

549   V. Szalay. *J. Chem. Phys.* 99, 1978 (1993).

550   V. Szalay, G. Czakó, Á. Nagy, T. Furtenbacher, A. G. Császár. *J. Chem. Phys.* 119, 10512 (2003).

551   J. Szczesny. *Fortschr. Phys.* 49, 723 (2001).

552   G. Szegö. *Orthogonal Polynomials*, Am. Math. Soc., Providence, 1939.

553   R. Szmytkowski, B. Zywicka–Mozejko. *Phys. Rev.* A 62, 022104 (2000).

554   V. F. Tarasov. *Int. J. Mod. Phys.* B 16, 3939 (2002).

555   H. Taseli, A. Zafer. *Int. J. Quant. Chem.* 63, 936 (1997).

556   H. Taseli, M. B. Ersecen. *J. Math. Chem.* 34, 177 (2003).

557   M. Tashkova, S. Donev. *Helv. Phys. Acta* 67, 691 (1994).

558   M. G. Tashkova, S. G. Donev. *phys. stat. sol.* (b) 200, 75 (1997).

559   N. M. Temme. *Special Functions*, Wiley, New York, 1996.

560   B. Thaller. *Advanced Visual Quantum Mechanics*, Springer, New York, 2005.

561   W. J. Thompson. *Am. J. Phys.* 60, 378 (1992).

562   J.-P. Tignol. *Galois' Theory of Algebraic Equations*, Longman, New York, 1988.

563   A. N. Tikhonov, A. A. Samarskii. *Equations of Mathematical Physics*, Pergamon Press, Oxford, 1963.

564   D. H. Towne, C. R. Hadlock. *Am. J. Phys.* 45, 255 (1977).

565   F. G. Tricomi. *Vorlesungen über Orthogonalreihen*, Springer-Verlag, Berlin, 1955.

566   M. Trott. *The Mathematica GuideBook for Programming*, Springer-Verlag, New York, 2004.

567   M. Trott. *The Mathematica GuideBook for Numerics*, Springer-Verlag, New York, 2005.

568   A. Truman, H. Z. Zhao. *Proc. Am. Math. Soc.* 128, 1003 (2000).

569   M. Trzetrzelewski. *arXiv:hep-th*/0407059 (2004).

570   G. Tsaur, J. Wang. *Phys. Rev.* A 65, 012104 (2001).

571   G. S. Tschumper, M. R. Hoffmann. *J. Math. Chem.* 31, 105 (2002).

572   I. Tsutsui, T. Fülöp, T. Cheon. *arXiv:quant-ph*/0209110 (2001).

573   I. Tsutsui, T. Fülöp. *arXiv:quant-ph*/0312028 (2003).

574   A. V. Turbiner, A. G. Ushveridze. *J. Math. Phys.* 29, 2053 (1988).

575   A. V. Turbiner. *arXiv:hep-th*/0108160 (2001).

576   C. J. Tymczak, G. S. Japaridze, C. R. Hardy, X.-Q. Wang. *Phys. Rev. Lett.* 80, 3673 (1998).

577   C. J. Tymczak, G. S. Japaridze, C. R. Hardy, X.-Q. Wang. *Phys. Rev.* A 58, 2708 (1998).

578   T. Umeda. *arXiv:math.SP*/0310090 (2003).

579   A. Unterberger. *Ann. Inst. Fourier* 29, 201 (1979).

580   W. Van Assche. *arXiv:math.CA* /9307220 (1993).

581   W. Van Assche, R. J. Yañez, R. González-Férez, J. S. Dehesa. *J. Math. Phys.* 41, 6600 (2000).

582   B. L. van der Waerden. *Algebra*, vI, Springer-Verlag, Berlin, 1971.

583   S. J. L. van Eijndhoven, J. L. H. Meyers. *J. Math. Anal. Appl.* 146, 89 (1990).

584   M. van Iersel, C. F. M. van der Burgh, B. L. G. Bakker. *arXiv:hep-ph*/0010243 (2000).

585   J. van Mill, A. Ran. *Indag. Math.* 7, 199 (1996).

586   D. A. Varshalovich, A. N. Moskalev, V. K. Khersonskii. *Quantum Theory of Angular Momentum*, World Scientific, Singapore, 1988.

587   D. J. Vezzetti, M. M. Cahay. *J. Phys.* D 19, L53 (1986).

588   P. Vignolo, A. Minguzzi, M. P. Tosi. *Phys. Rev. Lett.* 85, 2850 (2001).

589   J. M. Vigoureux, R. Giust. *Opt. Commun.* 186, 231 (2000).

590   A. Vijay, R. E. Wyatt, G. D. Billing. *J. Chem. Phys.* 111, 10794 (1999).

591   A. Vijay, H. Metiu. *J. Chem. Phys.* 116, 60 (2002).

592   N. J. Vilenkin, A. U. Klimyk. *Representations of Lie Groups and Special Functions*, v.2, Kluwer, Dordrecht, 1991.

593   M. Vincke, L. Malegat, D. Baye. *J. Phys.* B 26, 811 (1993).

594   F. Vinette, J. Čizek. *J. Math. Phys.* 32, 3392 (1991).

595   V. S. Vladimirov. *Equations of Mathematical Physics*, Mir, Moscow, 1984.

596   A. L. W. von Bachhaus. *Le Mathematiche* L, 47 (1995).

597   A. Voros. *J. Phys.* A 27, 4653 (1994).

598   A. Voros. *J. Phys.* A 32, 5993 (1999).

599   A. Voros. *arXiv:math-ph*/0005029 (2000).

600   A. Voros. *J. Phys.* A 33, 7423 (2000).

601   A. Voros. *arXiv:math-ph*/0201052 (2002).

602   J. S. Walker, J. Gathright. *Am. J. Phys.* 62, 408 (1994).

603   G. G. Walter. *Wavelets and Orthogonal Systems with Applications*, CRC, Boca Raton, 1994.

604   L.-Y. Wang, X.-Y. Gu, Z.-Q. Ma, S.-H. Dong. *Found. Phys. Lett.* 15, 569 (2002).

605   W. Wang. *Phys. Lett.* A 303, 125 (2002).

606   Z. X. Wang, D. R. Guo. *Special Functions*, World Scientific, Singapore, 1989.

607   S. Watanabe. *Commun. Partial Diff. Equ.* 26, 571 (2001).

608   S. Watanabe in S. Saitoh, N. Hayashi, M. Yamamoto (eds.). *Analytic Extension Formulas and Their Applications*, Kluwer, Dordrecht, 2001.

609   L. T. Watson. *Arab. J. Sci. Eng.* 16, 297 (1991).

610   C. A. Weatherford, E. Red, A. Wynn III. *Int. J. Quant. Chem.* 90, 1289 (2002).

611   J. A. C. Weidman. *Num. Math.* 61, 409 (1992).

612   S. Weigert. *arXiv:quant-ph*/02305177 (2003).

613   E. J. Weniger. *J. Math. Phys.* 26, 276 (1985).

614   E. J. Weniger. *J. Comput. Appl. Math.* 32, 291 (1990).

615   E. J. Weniger, J. Čízk, F. Vinette. *Phys. Lett.* A 156, 169 (1991).

616   E. J. Weniger, J. Čízk, F. Vinette. *J. Math. Phys.* 34, 571 (1993).

617   E. J. Weniger. *Phys. Rev. Lett.* 77, 2859 (1996).

618   E. J. Weniger. *arXiv:math.CA*/0107080 (2001).

619   H. E. White. *Phys. Rev.* 37, 1416 (1931).

620   A. J. C. Wilson. *Acta Cryst.* A 42, 81 (1986).

621   B. R. Wong. *J. Phys.* A 31, 1101 (1998).

622   S. C. Woon. *arXiv:physics*/9705021 (1997).

623   H. Wu, D. W. L. Sprung, J. Martorell. *J. Appl. Phys.* 26, 798, (1993).

624   A. Wünsche. *J. Phys.* A 32, 3179 (1999).

625   A. Wünsche. *J. Opt.* B 4, 359 (2002).

626   A. Wurm, M. Berg. *arXiv:physics*/0212061 (2002).

627   H. Yamamoto, Y. Kaniwe, K. Taniguchi. *phys. stat. sol.* (b) 154, 195 (1989).

628   R. J. Yáñez, W, Van Assche, R. González–Férez, J. S. Dehesa. *J. Math. Phys.* 40, 5675 (1999).

629   V. I. Yukalov, E. P. Yukalova. *arXiv:hep-ph*/9911357 (1999).

630   V. I. Yukalov, S. Gluzman. *Int. J. Mod. Phys.* B 18, 3027 (2004).

631   A. Zarzo, J. S. Dehesa. *J. Comput. Appl. Math.* 50, 613 (1994).

632   A. I. Zayed. *Handbook of Function and Generalized Function Transformations*, CRC Press, Boca Raton, 1996.

633   Q.-H. Zhang, S. Chen, Y. Qu. *IJMMS* n1 33 (2005).

634   M. Znojil. *arXiv:quant-ph*/0102034 (2001).

635   M. Znojil. *arXiv:quant-ph*/0103054 (2001).

636   M. Znojil. *J. Nonl. Math. Phys.* 9, S2, 122 (2002).

637   A. Zvonkin in D. Krob, A. A. Mikhalev, A. V. Mikhalev (eds.). *Formal Power Series and Algebraic Combinatorics*, Springer-Verlag, Berlin, 2000.

The Classical Special Functions

3.0 Remarks

This chapter presents some of the special functions that are among the most important for technical and scientific applications. We do not attempt to discuss all special functions that are built-in *Mathematica*, not all of their important properties and applications, nor do we look at all of them graphically. Moreover, in contrast to the previous chapter, we do not discuss the corresponding differential equations, series expansions, or differentiation and integration formulas. Instead, our aim is to introduce the nomenclature, to illustrate the power of *Mathematica* as a tool for dealing with special functions, and to show how they can be used to provide effective solutions to a variety of problems. So some famous functions (like the Riemann Zeta function) will not be discussed in the following sections because it would be outside of the scope of the *GuideBooks*, but we will nevertheless refer to some of the not-discussed functions in the exercises of this chapter. For a quite complete treatment, including series, product, and integral representations, as well as visualizations of all special functions of *Mathematica*, see the comprehensive site http://functions.wolfram.com (and the forthcoming sites http://dlmf.nist.gov [811], [892] and http://algo.inria.fr/esf [885]).

In analogy with the inverse trigonometric functions, most special functions have branch cuts in the complex plane. Because these can be complicated to describe (see [846]), we do not go into detail here. Most of the special functions have interesting representations as contour integrals whose integrands are compositions of elementary functions (see [333] and [1297]). Because the associated contours typically run over several branches of the corresponding Riemann surfaces, we cannot work with these representations directly using `Integrate` and/or `NIntegrate` (but see Exercises 6 and 16). This problem can be overcome by first constructing analytic continuations. (When trying to do this, be careful, because some of these formulas from the cited books use branch cuts other than those defined in *Mathematica*.) Furthermore, we do not discuss the very interesting and important Stokes phenomenon in the asymptotic expansion of special functions [143], [200], [143], [886], [896], and [350], *q*-versions of special functions, etc.

We advise the reader not to be intimidated by special functions and their appearance in *Mathematica* output from integration. The classical special functions are extremely useful tools for obtaining closed-form as well as series solutions to a variety of problems [146], and *Mathematica* has the capabilities to use the special functions in an efficient way. (*Mathematica* can calculate numerical values for special functions, make series solutions, differentiate and integrate them, and simplify expressions involving special functions by using special identities known for these functions.) We recommend that readers, if possible, use them in solutions of their own problems. We also note that *Mathematica* cannot handle *all* cases of integration and simplification of special functions. Therefore, *Mathematica* is not an exclusive resource for scientific calculations, but it is certainly a very useful tool. But because of its rich programming language and its algebraic capabilities, working with special functions in *Mathematica* can be very successful.

In this chapter, we will mainly use *Mathematica*'s capabilities to solve problems in closed form (as compared to in truncated series) using special functions. In the exercises, we will also use *Mathematica* to investigate special functions themselves. But *Mathematica* is also a perfect environment to discover new identities and properties of special functions. It is the main spirit of the *Mathematica GuideBooks* to apply *Mathematica* to problems from engineering and physics. In this introduction, we just give a toy example of "experimental mathematics". Let us conjecture that the generalized harmonic numbers $H_k^{(r)}$ [1220] (in *Mathematica*, `HarmonicicNumber[r, n]`) defined by $H_n^{(r)} = \sum_{j=1}^{n} 1/j^r$ fulfill identities of the form

$$\sum_{a}^{m} \alpha_{r_a} H_n^{(r_a)} = \sum_{b,c=1}^{m} \beta_{r_b r_c} \sum_{j=1}^{n-1} H_{n-j}^{(r_b)} H_j^{(r_c)}.$$

Here, α_{r_a}, $\beta_{r_b r_c}$ are constants and m is a small integer. We want to find some such identities. Let $\mathbb{H}_r(z)$ be the generating function of the generalized harmonic numbers $\mathbb{H}_r(z) = \sum_{k=0}^{\infty} H_k^{(r)} z^k$. Then, the above identities can be rewritten as

$$\sum_{a}^{m} \alpha_{r_a} \mathbb{H}_{r_a}(z) = \sum_{b,c=1}^{m} \beta_{r_b r_c} \mathbb{H}_{r_b}(z) \, \mathbb{H}_{r_c}(z).$$

Using a truncated power series for the $\mathbb{H}_r(z)$ in the last equation and comparing powers of z yields a system of linear equations for the α_{r_a}, $\beta_{r_b r_c}$. Using enough terms in the power series gives an overdetermined system of equations. If such an overdetermined system has a solution, it is probably because a relation of the form of the last equation exists. `Hs[d, z, n]` represents the series of $\mathbb{H}_r(z)$ to order n.

```
In[1]:= Hs[d_, z_, n_] := Hs[d, z, n] =
          Sum[HarmonicNumber[d, k] z^k, {k, n}] + O[z]^n
```

We will use $m = 3$ in the following calculations. This gives 12 unknowns α_{r_a}, $\beta_{r_b r_c}$. Using 15 terms for the power series gives three more equations than unknowns.

```
In[2]:= eq[l_, n_] := Sum[C[l[[i]]] Hs[l[[i]], z, n], {i, Length[l]}] +
          Sum[C[l[[i]], l[[j]]] Hs[l[[i]], z, n] Hs[l[[j]], z, n],
          {i, Length[l]}, {j, i, Length[l]}];
```

Restricting ourselves to $r_a, r_b, r_c \le 10$, we get the following 20 potential identities from the 220 tried possibilities.

```
In[3]:= Off[Solve::svars];
        CData = Module[{cl, sol, C}, Table[
        (* coefficients of powers of z *)
        cl = CoefficientList[Expand[eq[{i, j, k}, 15]], z];
        (* solve overdetermined system *)
        sol = Solve[# == 0& /@ cl, Cases[cl, _C, Infinity]];
        If[(* keep a nontrivial solution *)
           Union[Last /@ sol[[1]]] =!= {0},
           {{i, j, k}, sol}, {}], {i, 10}, {j, i}, {k, j}]];

In[5]:= eqRes[l_, n_] := (* general ansatz *)
          Sum[C[l[[i]]] Subscript[H, l[[i]]], {i, Length[l]}] +
          Sum[C[l[[i]], l[[j]]] Subscript[H, l[[i]]] Subscript[H, l[[j]]],
          {i, Length[l]}, {j, i, Length[l]}];

In[6]:= gfIdentities = Union[Factor[Numerator[Together[#]]& /@
          (* write result in polynomial form *)
          Flatten[eqRes[#[[1]], Length[#[[1]]]]] /. #[[2]]& /@
             DeleteCases[Flatten[CData, 2], {}, Infinity]] /. _C -> 1],
             (* eliminate doubles *)
          SameTest -> (MatchQ[Cancel[#1/#2], _Integer | _Rational]&)] /.
             _Integer p_Plus -> p
```

Out[6]= {3 H₁ + H₁² - 2 H₂ - H₁ H₂, 5 H₁ + 2 H₁² + 4 H₂ - H₁ H₂ + H₂² - 6 H₃ - H₁ H₃ - H₂ H₃,

26 H₁ + 6 H₁² - 9 H₂ + H₁ H₂ - H₂² - 6 H₄ - 7 H₁ H₄ + H₂ H₄,

43 H₁ + 12 H₁² - 53 H₃ - 15 H₁ H₃ - 3 H₃² + 26 H₄ + 3 H₁ H₄ + 3 H₃ H₄,

61 H₂ + 8 H₂² - 159 H₃ - 17 H₂ H₃ - 9 H₃² + 96 H₄ + 9 H₂ H₄ + 9 H₃ H₄,

-H₁ - 5 H₄ - 4 H₁ H₄ + 5 H₅ + 4 H₁ H₅,

103 H₁ + 17 H₁² + 31 H₃ + 52 H₁ H₃ - 3 H₃² - 70 H₅ - 69 H₁ H₅ + 3 H₃ H₅,

-32 H₂ - 13 H₂² + 76 H₃ + 36 H₂ H₃ - 5 H₃² - 40 H₅ - 23 H₂ H₅ + 5 H₃ H₅,

4 H₃ - 9 H₄ + H₃ H₄ - H₄² + 5 H₅ - H₃ H₅ + H₄ H₅,

71 H₂ + 15 H₂² - 358 H₄ - 86 H₂ H₄ - 7 H₄² + 280 H₅ + 71 H₂ H₅ + 7 H₄ H₅,

-H₁ - 6 H₅ - 5 H₁ H₅ + 6 H₆ + 5 H₁ H₆, 5 H₄ - 11 H₅ + H₄ H₅ - H₅² + 6 H₆ - H₄ H₆ + H₅ H₆,

-H₁ - 7 H₆ - 6 H₁ H₆ + 7 H₇ + 6 H₁ H₇, 6 H₅ - 13 H₆ + H₅ H₆ - H₆² + 7 H₇ - H₅ H₇ + H₆ H₇,

-H₁ - 8 H₇ - 7 H₁ H₇ + 8 H₈ + 7 H₁ H₈, 7 H₆ - 15 H₇ + H₆ H₇ - H₇² + 8 H₈ - H₆ H₈ + H₇ H₈,

-H₁ - 9 H₈ - 8 H₁ H₈ + 9 H₉ + 8 H₁ H₉, 8 H₇ - 17 H₈ + H₇ H₈ - H₈² + 9 H₉ - H₇ H₉ + H₈ H₉,

-H₁ - 10 H₉ - 9 H₁ H₉ + 10 H₁₀ + 9 H₁ H₁₀, 9 H₈ - 19 H₉ + H₈ H₉ - H₉² + 10 H₁₀ - H₈ H₁₀ + H₉ H₁₀}

Translating the products back to convolution-type sums, we have the following results (we display only some of the identities).

```
In[7]:= hIdentities = gfIdentities /.
       (* products ⟹ convolution sums *)
       {Subscript[H, a_]^2 -> Sum[h[a, n - k] h[a, k], {k, n - 1}],
        Subscript[H, a_] Subscript[H, b_] ->
                    Sum[h[a, n - k] h[b, k], {k, n - 1}],
        Subscript[H, a_] -> h[a, n]};
```

```
In[8]:= (* format h[r, n] as harmonic numbers *)
       MakeBoxes[h[r_, n_], TraditionalForm] :=
       SubsuperscriptBox["H", ToString[n], RowBox[{"(", ToString[r], ")"}]];
```

```
       (Part[# == 0& /@ Sort[hIdentities,
                       ByteCount[#1] < ByteCount[#2]&],
           (* take only four equations *) {1, 3, 8, 12}]) //.
           (* unite sums *)
       HoldPattern[α_. Sum[a_, iter_] + β_. Sum[b_, iter_]] :>
            Sum[Evaluate[Factor[α a + β b]], iter] //.
       HoldPattern[Sum[m_Integer r_, i_]] :> m Sum[r, i] //
                          TableForm // TraditionalForm
```

Out[11]//TraditionalForm=

$$3 H_n^{(1)} - 2 H_n^{(2)} + \sum_{k=1}^{n-1} H_{-k+n}^{(1)} \left(H_k^{(1)} - H_k^{(2)} \right) = 0$$

$$-H_n^{(1)} - 9 H_n^{(8)} + 9 H_n^{(9)} - 8 \sum_{k=1}^{n-1} H_{-k+n}^{(1)} \left(H_k^{(8)} - H_k^{(9)} \right) = 0$$

$$9 H_n^{(8)} - 19 H_n^{(9)} + 10 H_n^{(10)} + \sum_{k=1}^{n-1} \left(H_{-k+n}^{(8)} - H_{-k+n}^{(9)} \right) \left(H_k^{(9)} - H_k^{(10)} \right) = 0$$

$$5 H_n^{(4)} - 11 H_n^{(5)} + 6 H_n^{(6)} + \sum_{k=1}^{n-1} \left(H_{-k+n}^{(4)} - H_{-k+n}^{(5)} \right) \left(H_k^{(5)} - H_k^{(6)} \right) = 0$$

Until now, the identities are only conjectures. *Mathematica* can evaluate all of the last sums symbolically.

```
In[12]:= hIdentities /. h -> HarmonicNumber // Expand // Simplify
Out[12]= {0, 0, 0, 0, 0, 0, 0, 0, 0, 0, 0, 0, 0, 0, 0, 0, 0, 0, 0, 0}
```

We see that all of the conjectured identities are true identities. Now that one has a set of formulas for explicit r, one could go on and search for patterns in some of the formulas, such as

$$-\mathbb{H}_1(z) - (n + 1) \mathbb{H}_n(z) - n \mathbb{H}_1(z) \mathbb{H}_n(z) + (n + 1) \mathbb{H}_{n+1}(z) + n \mathbb{H}_1(z) \mathbb{H}_{n+1}(z) = 0$$

and formulate and prove them for general r.

The last examples dealt with the possibility dealing with special functions symbolically. Equally (and sometimes even more) important is the fact that all the special functions of *Mathematica* can be calculated numerically for arbitrary complex parameters with arbitrary precision. This is a wonderful resource for visualizations for exactly solvable problems. As a first example let us consider a simple, but not so well-known potential problem from quantum mechanics: Solving the eigenvalue problem of the time-independent Schrödinger equation with a Liouville potential e^{2x} [346], [347] [348], [1079]. (For shorter formulas we use the normalized wave vector k instead of the energy ε, where $k = \varepsilon^{1/2}$.

$$-\psi_k''(x) + e^{2x}\,\psi_k(x) = k^2\,\psi_k(x)$$

The shape of the potential suggests a continuous spectrum for $\varepsilon > 0$. The normalized eigenfunctions can easily be calculated to be Bessel functions with imaginary index (the normalization is $\int_{-\infty}^{\infty}\psi_k(x)\,\overline{\psi_{k'}(x)} = \delta(k - k')/(2\,k))$:

$$\psi_k(x) = \frac{\sqrt{\sinh(\pi k)}}{\pi}\,K_{ik}(e^x).$$

The corresponding momentum-space eigenfunctions are obtained after carrying out a Fourier transform of $\psi_k(x)$.

$$\tilde{\psi}_k(p) = \int_{-\infty}^{\infty} e^{-ixp}\,\psi_k(x)\,dx = \frac{\sqrt{\sinh(\pi k)}}{4\pi}\,2^{-ip}\,\Gamma\!\left(-\frac{i}{2}\,(p+k)\right)\Gamma\!\left(-\frac{i}{2}\,(p-k)\right).$$

The next quantity of interest would be the Wigner function (see, for instance, [599], [1350], [1214], [389], [1113], [389], [390], [926], [615], [616], [918], [731]). Using regularized generalized hypergeometric functions $_0\tilde{F}_3$, it can be expressed in the following way:

$$w_k(x, p) = \int_{-\infty}^{\infty} e^{-iyp}\,\overline{\psi}_k\!\left(x - \frac{y}{2}\right)\psi_k\!\left(x + \frac{y}{2}\right)dy = \frac{i}{4}\,\pi\,\sqrt{\sinh(\pi k)}\,\times$$

$$(c_-\,4^{-i\,(k-p)}\,e^{2\,i\,(k-p)\,x}\ _0\tilde{F}_3(1 + ik,\ 1 - ip,\ 1 + i(k - p),\ e^{4x}/16) +$$
$$c_+\,4^{i\,(k+p)}\,e^{-2\,i\,(k+p)\,x}\ _0\tilde{F}_3(1 - ik,\ 1 - ip,\ 1 - i(k + p),\ e^{4x}/16) -$$
$$c_+\,4^{-i\,(k+p)}\,e^{2\,i\,(k+p)\,x}\ _0\tilde{F}_3(1 + ik,\ 1 + ip,\ 1 + i(k + p),\ e^{4x}/16) -$$
$$c_-\,4^{i\,(k-p)}\,e^{-2\,i\,(k-p)\,x}\ _0\tilde{F}_3(1 - ik,\ 1 + ip,\ 1 - i(k - p),\ e^{4x}/16))$$
$$c_\pm = \operatorname{csch}(k\,\pi)\operatorname{csch}(p\,\pi)\operatorname{csch}((k \pm p)\,\pi).$$

This means the solutions of this problem contains Bessel functions, Gamma functions, and generalized hypergeometric functions of complex variables and parameters.

Now let us visualize the position-space and momentum-space eigenfunctions as a function of k. The position-space eigenfunctions are concentrated in the region $x < 0$ where the potential is exponentially vanishing. To achieve this localization we need superpositions of left- and right-waves. The momentum-space eigenfunctions show a sharp peak at $p = \pm k$.

```
In[13]:= ψ[k_][x_] := 1/Pi Sqrt[Sinh[Pi k]] BesselK[I k, Exp[x]]

In[14]:= Ψ[k_][p_] := 1/Pi Sqrt[Sinh[Pi k]] 2^(-I p - 2)*
                      Gamma[-I(p + k)/2] Gamma[-I(p - k)/2]
```

```
In[15]:= Show[GraphicsArray[
    Block[{$DisplayFunction = Identity},
    {(* position wavefunctions *)
    Plot3D[Re[ψ[k][x]], {x, -20, 3}, {k, 0.01, 3},
        PlotPoints -> 200, Mesh -> False],
    (* momentum wavefunctions *)
    Plot3D[Abs[Ψ[k][p]], {p, -3, 3}, {k, 0.01, 2},
        PlotPoints -> 200, Mesh -> False,
        PlotRange -> {0, 3}, ClipFill -> None]}]]]
```

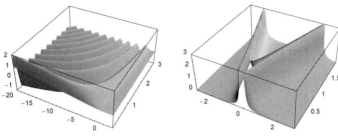

Now let us define the Wigner function.

```
In[16]:= w[k_][x_, p_] :=
    With[{cP = Csch[k Pi] Csch[(k + p) Pi] Csch[p Pi],
          cM = Csch[k Pi] Csch[(k - p) Pi] Csch[p Pi]},
    I Pi/4 Sinh[Pi k] (
    4^(-I(k - p)) Exp[2I(k - p)x] cM HypergeometricPFQRegularized[{},
                    {1 + I k, 1 - I p, 1 + I(k - p)}, Exp[4x]/16] -
    4^(+I(k - p)) Exp[-2I(k - p)x] cM HypergeometricPFQRegularized[{},
                    {1 - I k, 1 + I p, 1 - I(k - p)}, Exp[4x]/16] +
    4^(+I(k + p)) Exp[-2I(k + p)x] cP HypergeometricPFQRegularized[{},
                    {1 - I k, 1 - I p, 1 - I(k + p)}, Exp[4x]/16] -
    4^(-I(k + p)) Exp[2I(k + p)x] cP HypergeometricPFQRegularized[{},
                    {1 + I k, 1 + I p, 1 + I(k + p)}, Exp[4x]/16])]
```

The Wigner function shows the expected oscillations in the classically allowed region $p^2 + e^{2x} < k^2$. Outside the classically allowed region the Wigner function is exponentially vanishing. The following graphic shows $w_k(x, p)$ for $k = 3/2$.

```
In[17]:= Plot3D[Evaluate[Re[w[1.5][x, p]]], {x, -12, 2}, {p, -2.5, 2.5},
        PlotPoints -> 180, PlotRange -> All, Mesh -> False]
```

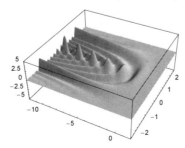

We could now go on and visualize the Titchmarsh m function or the Husimi function for the Liouville potential, but we will end here.

While many special functions are built into *Mathematica*, not all are. Hundreds of specialized special functions are used in various fields of mathematics and physics. However, *Mathematica*'s high-precision arithmetic, programming capabilities, and symbolic mathematics power make it straightforward to implement additional special functions, to work with them, and to investigate them.

As an example let us consider a recent addition to the large pool of special functions: the three Ramanujan elliptic functions $a(q)$, $b(q)$, and $c(q)$ [11], [184], [187], [329], [603], [802], [283], [803], [1170], [311], [1133]. (In a more general setting these function could be considered as the zero values of Ramanujan's theta functions; obtained by adding powers $z^n w^m$ into the sums below.) We will not "fully" incorporate these function, but rather use its numerical evaluation and its series expansion as an example.

They can be defined by the following series expansions.

$$a(q) = \sum_{m=-\infty}^{\infty} \sum_{n=-\infty}^{\infty} q^{n^2+m^2+mn}$$

$$b(q) = \sum_{m=-\infty}^{\infty} \sum_{n=-\infty}^{\infty} e^{2\pi i/3\,(m-n)}\, q^{n^2+m^2+mn}$$

$$c(q) = \sum_{m=-\infty}^{\infty} \sum_{n=-\infty}^{\infty} q^{(n+1/3)^2+(m+1/3)^2+(m+1/3)(n+1/3)}$$

From these definitions follows a straightforward implementation of their series expansions around $q = 0$. (Although single instead of double series exists for the Ramanujan elliptic theta functions, because of the quadratic exponents the double series is quite fast for $|q| < 1$.) (We associate the new definitions not with `Series`, but rather with the new functions, for which we use a typical *Mathematica* notation.)

```
In[18]:= RamanujanEllipticA /:
         Series[RamanujanEllipticA[q_^k_.], {q_, 0, o_}] :=
         Series[Sum[(q^k)^(m^2 + m n + n^2),
                      {m, -f[o, k], f[o, k]}, {n, -f[o, k], f[o, k]}], {q, 0, o}]

         RamanujanEllipticB /:
         Series[RamanujanEllipticB[q_^k_.], {q_, 0, o_}] := ExpToTrig //@
         Series[Sum[Exp[2Pi I/3 (m - n)] (q^k)^(m^2 + m n + n^2),
                      {m, -f[o, k], f[o, k]}, {n, -f[o, k], f[o, k]}], {q, 0, o}]

         RamanujanEllipticC /:
         Series[RamanujanEllipticC[q_^k_.], {q_, 0, o_}] :=
         Series[Sum[(q^k)^((m + 1/3)^2 + (m + 1/3) (n + 1/3) + (n + 1/3)^2),
                      {m, -f[o, k], f[o, k]}, {n, -f[o, k], f[o, k]}], {q, 0, o}]

         f[o_, k_] :=  Ceiling[2/Sqrt[3] Sqrt[o]] + 2
```

The three functions fulfill a beautiful identity—they parametrize a cubic (quartics are parametrized by Jacobi theta functions; for parametrized quadratics, see [279]):

$$a(q)^3 = b(q)^3 + c(q)^3.$$

Here we check this identity up to $O(q)^{1000}$.

```
In[25]:= cubicIdentity = RamanujanEllipticA[q]^3 -  RamanujanEllipticB[q]^3 -
                RamanujanEllipticC[q]^3;
```

```
In[26]:= With[{o = 1000},
            cubicIdentity /. abc:(RamanujanEllipticA |
                                  RamanujanEllipticB |
                                  RamanujanEllipticC)[q] :>
                                  Series[abc, {q, 0, o}]]
```

Out[26]= $O[q]^{1001}$

Like most special functions, the three functions $a(q)$, $b(q)$, and $c(q)$ are related to other special functions and various fulfill differential equations [802]. Here is quick check for the identity $a(q) = {}_2F_1(1/3, 2/3; 1; c(q)^3/a(q)^3)$ where ${}_2F_1(\alpha, \beta; \gamma; z)$ is the Gauss hypergeometric function (see below).

```
In[27]:= With[{o = 100},
         Series[RamanujanEllipticA[q], {q, 0, o}] -
         Hypergeometric2F1[1/3, 2/3, 1,
                Series[RamanujanEllipticC[q], {q, 0, o}]^3/
                Series[RamanujanEllipticA[q], {q, 0, o}]^3]]
```

Out[27]= $O[q]^{101}$

And here is a quick check for three differential equations containing $a(q)$ and $c(q)$.

```
In[28]:= With[{o = 100},
         Series[{2 a[q]^3 c[q]^3 - 2 c[q]^6 -  9 q a[q] a'[q] +
                 18 q^2 a'[q]^2 - 9 q^2 a[q] a''[q],
                 a[q]^4 c[q]^2 - a[q] c[q]^5 + 9 q c[q] c'[q] -
                 18 q^2 c'[q]^2 + 9 q^2 c[q] c''[q],
                 a[q]^3 c[q] - c[q]^4 + 3 q c[q] a'[q] - 3 q a[q] c'[q]} /.
            Rule @@@ Transpose[{{a, b, c},
            Function[q, Evaluate[Normal[Series[#[q], {q, 0, o}]]]]& /@
            {RamanujanEllipticA, RamanujanEllipticB, RamanujanEllipticC}}],
            {q, 0, o}]]
```

Out[28]= $\{O[q]^{301/3}, O[q]^{301/3}, O[q]^{301/3}\}$

And, as with most special functions, for special parameters it is possible to find closed-form values for the function values. The quotient $\xi_n = a(\exp(-2\pi(n/3)^{1/2}))/c(\exp(-2\pi(n/3)^{1/2}))$ is an algebraic number for integer n. The function algebraicIdentity uses the package function NumberTheory`Recognize to find this algebraic number. (For the small n we are dealing with here, we calculate the quotient numerically to 100 digits and look for algebraic numbers of degree less than 12.)

```
In[29]:= Needs["NumberTheory`Recognize`"]
```

```
In[30]:= algebraicIdentity[n_Integer?Positive, maxDegree_:12]  :=
         Module[{prec = 100, o, ξ, q, poly, ξAlg},
           (* number of terms of the series needed to obtain
              precision prec for a and c *)
           o = Ceiling[-prec/Log[10, Exp[-2 Pi Sqrt[n/3]]]] + 10;
           (* calculate ξ = a/c *)
           ξ = Divide @@ ((Normal[Series[#[q], {q, 0, o}]] /.
                       q -> N[Exp[-2 Pi Sqrt[n/3]], 100])& /@
                       {RamanujanEllipticC, RamanujanEllipticA});
           (* find polynomial with root ξ *)
           poly = Recognize[ξ, maxDegree, t];
           (* find the root *)
           ξAlg = Select[Solve[poly == 0, t], (t /. #) - ξ == 0&];
           (* form identity *)
```

```
a[Exp[-2 Pi Sqrt[n/3]]]/c[Exp[-2 Pi Sqrt[n/3]]] ==
Simplify[ξAlg[[1, 1, 2]]]]
```

Here are the algebraic values for $\xi_1, \ldots, \xi_6$.

In[31]:= `Table[algebraicIdentity[n], {n, 6}] // TraditionalForm`

Out[31]//TraditionalForm=

$$\left\{ \frac{a\left(e^{-\frac{2\pi}{\sqrt{3}}}\right)}{c\left(e^{-\frac{2\pi}{\sqrt{3}}}\right)} = \frac{1}{\sqrt[3]{2}}, \ \frac{a\left(e^{-2\sqrt{2/3}\,\pi}\right)}{c\left(e^{-2\sqrt{2/3}\,\pi}\right)} = \frac{\sqrt[3]{2-\sqrt{2}}}{2^{2/3}}, \ \frac{a\left(e^{-2\pi}\right)}{c\left(e^{-2\pi}\right)} = \frac{1}{2}\left(-1+\sqrt{3}\right), \right.$$

$$\left. \frac{a\left(e^{-\frac{4\pi}{\sqrt{3}}}\right)}{c\left(e^{-\frac{4\pi}{\sqrt{3}}}\right)} = \frac{\sqrt[3]{\frac{1}{2}\left(9-5\sqrt{3}\right)}}{3^{2/3}}, \ \frac{a\left(e^{-2\sqrt{5/3}\,\pi}\right)}{c\left(e^{-2\sqrt{5/3}\,\pi}\right)} = \frac{\sqrt[3]{\frac{1}{2}\left(25-11\sqrt{5}\right)}}{5^{2/3}}, \ \frac{a\left(e^{-2\sqrt{2}\,\pi}\right)}{c\left(e^{-2\sqrt{2}\,\pi}\right)} = \frac{1}{10}\left(4-\sqrt{6}\right) \right\}$$

The numerical calculation for nonbuilt-in (meaning user-defined) special functions can often be achieved through finite (sometimes infinite) sums of built-in special functions. From the mentioned references we can implement the following expressions involving the built-in function `EllipticTheta`.

In[32]:=
```
RamanujanEllipticA[q_Real?InexactNumberQ] :=
With[{θ = EllipticTheta},
    θ[3, 0, q] θ[3, 0, q^3] + θ[2, 0, q] θ[2, 0, q^3]]

RamanujanEllipticB[q_Real?InexactNumberQ] :=
With[{θ = EllipticTheta},
    (-θ[2, 0, q^3] (θ[2, 0, q] - 3 θ[2, 0, q^9]) -
     θ[3, 0, q^3] (θ[3, 0, q] - 3 θ[3, 0, q^9]))/2]

RamanujanEllipticC[q_Real?InexactNumberQ] :=
With[{θ = EllipticTheta},
    (θ[2, 0, q] (θ[2, 0, q^(1/3)] - θ[2, 0, q^3]) +
     θ[3, 0, q] (θ[3, 0, q^(1/3)] - θ[3, 0, q^3]))/2]
```

A quick check for a random argument confirms again the cubic identity.

In[37]:= `cubicIdentity /. q -> N[Sqrt[2]/E^Pi, 1000]`

Out[37]= $0. \times 10^{-1000}$

Slight modifications to the above definitions for the numerical evaluation allow the continuation into the complex q-plane. Here this is done for $a(q)$.

In[38]:=
```
RamanujanEllipticC[q_Complex?InexactNumberQ] :=
With[{θ = EllipticTheta},
    (* use sector-dependent different representations *)
    Which[0 <= Arg[q] < Pi/3 || -Pi/3 <= Arg[q] < 0,
        θ[3, 0, q] θ[3, 0, q^3] + θ[2, 0, q] θ[2, 0, q^3],
        Pi/3 <= Arg[q] < Pi || -Pi <= Arg[q] < -Pi/3,
        θ[3, 0, q] θ[3, 0, q^3] +
        I Sign[Arg[q]] θ[2, 0, q] θ[2, 0, q^3]]]
```

As is easily visible from the q-series, the three functions $a(q)$, $b(q)$, and $c(q)$ have the unit circle as their boundary of analyticity. The following graphic shows large oscillations near $|q| = 0.95$. They are caused by the dense set of poles situated along the unit circle. The left graphic shows the real part colored according to the (scaled) imaginary part and the right graphic shows the imaginary part colored according to the (scaled) real part.

```
In[39]:= Show[GraphicsArray[
    Block[{$DisplayFunction = Identity},
    Function[{ri, ir},(* colored 3D plot of real or imaginary part *)
    ParametricPlot3D[{rq Cos[φq], rq Sin[φq], Sequence @@
        ({ArcTan[ri[#]], (* color as a function of Im[c] *)
        {EdgeForm[], SurfaceColor[#, #, 3]}&[
                           Hue[(ArcTan[ir[#]] + Pi/2)/Pi]]}&[
    RamanujanEllipticC[rq Exp[I φq]]])},
    {rq, 10^-6, 0.95}, {φq, 0, 2Pi},
    PlotPoints -> {100, 200}, Compiled -> False, Boxed -> False,
    Axes -> False, PlotRange -> All, BoxRatios -> {1, 1, 2/3}]] @@@
                (* value and color value *) {{Re, Im}, {Im, Re}}]]]
```

Using *Mathematica*'s symbolic functions allows to derive identities for the functions $a(q)$, $b(q)$, and $c(q)$. As an example let us look for modular equations of the form

$$\sum_{a_{k_1},a_{k_e},b_{k_1},b_{k_e},c_{k_1},c_{k_e}} \alpha_{a_{k_1},a_{k_e},b_{k_1},b_{k_e},c_{k_1},c_{k_e}}\, a(q)^{a_{k_1}}\, a(q^e)^{a_{k_e}}\, b(q)^{b_{k_1}}\, b(q^e)^{b_{k_e}}\, c(q)^{c_{k_1}}\, c(q^e)^{c_{k_e}} = 0$$

with e a fixed positive integer, $a_{k_1} + a_{k_e} + b_{k_1} + b_{k_e} + c_{k_1} + c_{k_e} \le d$ and the $\alpha_{a_{k_1},a_{k_e},b_{k_1},b_{k_e},c_{k_1},c_{k_e}}$ integer.

We will search for identities of this form in the following way. First, we generate all monomials $a(q)^{a_{k_1}}\, a(q^5)^{a_{k_5}}\, b(q)^{b_{k_1}}\, b(q^5)^{b_{k_5}}\, c(q)^{c_{k_1}}\, c(q^5)^{c_{k_5}}$. Then, we form series expansions of these monomials. Taking sufficiently many series terms results in an overdetermined system of linear equations. Using NullSpace we find all nontrivial solutions of this overdetermined system. We start by constructing all monomials. (Here we choose $e = 5$ and $d = 4$.)

```
In[40]:= basis = Module[{d = 4, e = 5, vars},
        vars = {A[q], B[q], C[q], A[q^e], B[q^e], C[q^e]};
        Flatten[Table[Evaluate[Times @@ (vars^Table[ε[j], {j, 6}])],
                Evaluate[Sequence @@
                    Table[{ε[j], 0, d - Sum[ε[i], {i, j - 1}]}, {j, 6}]]]]];
```

There are 210 monomials in the basis. The set of rules serRules generates the series expansions for the monomials. To avoid fractional powers, we use the third powers of $a(q)$, $b(q)$, and $c(q)$.

```
In[41]:= (* number of terms in the q-series *)
    o = Round[1.5 Length[basis]];

    serRules =
    With[{(* avoid fractional series with zero coefficients *)
            S = Series[Normal[Series[#1, {q, 0, o}]^#2] /. q -> q^#3, {q, 0, o}]&},
        {A[q] -> S[RamanujanEllipticA[q], 3, 1],
        B[q] -> S[RamanujanEllipticB[q], 3, 1],
```

```
        C[q] -> S[RamanujanEllipticC[q], 3, 1],
        A[q^5] -> S[RamanujanEllipticA[q], 3, 5],
        B[q^5] -> S[RamanujanEllipticB[q], 3, 5],
        C[q^5] -> S[RamanujanEllipticC[q], 3, 5]}];
```

In[44]:= `basisSeries = Table[basis[[k]] /. serRules, {k, Length[basis]}];`

For each series from `basisSeries`, we extract the coefficients with respect to q and calculate the null space of the resulting matrix. (For speed reasons, we use a modular null space computation here.)

In[45]:=
```
coeffMat = If[Length[#] < o,
        Join[#, Table[0, {o - Length[#]}]], Take[#, o]]& /@
            (CoefficientList[Normal[#], q]& /@ basisSeries);

    ns = NullSpace[Transpose[coeffMat], Modulus -> Prime[10^6]];
```

153 null spaces were found. Here is their distribution as a function of their dimensions.

In[48]:= `{First[#], Length[#]}& /@ Split[Sort[Length[DeleteCases[#, 0]]& /@ ns]]`

Out[48]= `{{3, 53}, {4, 17}, {5, 16}, {6, 1},`
         `{7, 5}, {9, 1}, {19, 7}, {22, 1}, {23, 30}, {25, 22}}`

In the final step, we calculate the nonmodular null spaces, factor the resulting identities, and select the relevant factors.

In[49]:=
```
fullNullSpace[candidateNullSpace_] :=
    basis[[Flatten[Position[candidateNullSpace, 1]]]].#& /@
        (* nonmodular null space for candidates *)
        NullSpace[Transpose[coeffMat[[Flatten[Position[candidateNullSpace, 1]]]]]]
```

In[50]:=
```
allIdentities =
    Select[Union[Flatten[Cases[Factor[fullNullSpace[#]],
        _Plus, Infinity]& /@ (ns /. i_Integer?(# =!= 0&) -> 1)
            (* high-order check *)]], Normal[# /. serRules] == 0&];
```

66 nontrivial relations were found. We display the first 12 (the remaining one are quite large expressions). As expected, the first one is the cubic identity from above.

In[51]:= `Length[allIdentities]`

Out[51]= `66`

In[52]:= `TraditionalForm[Take[allIdentities, 12] /.` (* express in a, b, and c *)
                           `{A -> (a[#]^3&), B -> (b[#]^3&), C -> (c[#]^3&)}]`

Out[52]//TraditionalForm=

$$\{a(q)^3 - b(q)^3 - c(q)^3, \; a(q^5)^3 - b(q^5)^3 - c(q^5)^3, \; a(q)^3 \, a(q^5)^3 - b(q)^3 \, b(q^5)^3 - b(q^5)^3 \, c(q)^3 - b(q)^3 \, c(q^5)^3 - c(q)^3 \, c(q^5)^3,$$

$$a(q^5)^3 \, a(q)^6 - b(q^5)^3 \, c(q)^6 - b(q)^6 \, b(q^5)^3 - 2 \, b(q)^3 \, b(q^5)^3 \, c(q)^3 - b(q)^6 \, c(q^5)^3 - c(q)^6 \, c(q^5)^3 - 2 \, b(q)^3 \, c(q)^3 \, c(q^5)^3,$$

$$a(q)^3 \, a(q^5)^6 - b(q)^3 \, b(q^5)^6 - b(q)^3 \, c(q^5)^6 - c(q)^3 \, c(q^5)^6 - b(q^5)^6 \, c(q)^3 - 2 \, b(q)^3 \, b(q^5)^3 \, c(q^5)^3 - 2 \, b(q^5)^3 \, c(q)^3 \, c(q^5)^3,$$

$$23625 \, b(q^5)^9 - 23436 \, b(q)^3 \, b(q^5)^6 - 1664 \, c(q)^3 \, b(q^5)^6 + 467875 \, c(q^5)^3 \, b(q^5)^6 - 189 \, b(q)^6 \, b(q^5)^3 +$$
$$147 \, c(q)^6 \, b(q^5)^3 - 124125 \, c(q^5)^6 \, b(q^5)^3 - 6482 \, b(q)^3 \, c(q)^3 \, b(q^5)^3 - 1820 \, b(q)^3 \, c(q^5)^3 \, b(q^5)^3 +$$
$$69180 \, c(q)^3 \, c(q^5)^3 \, b(q^5)^3 - 224 \, c(q)^9 + 7625 \, c(q^5)^9 - 800 \, b(q)^3 \, c(q)^6 + 336 \, b(q)^3 \, c(q^5)^6 +$$
$$9564 \, c(q)^3 \, c(q^5)^6 + 208 \, a(q)^6 \, c(q)^3 + 19 \, b(q)^6 \, c(q^5)^3 + 1923 \, c(q)^6 \, c(q^5)^3 - 1938 \, b(q)^3 \, c(q)^3 \, c(q^5)^3,$$

$$23625 \, b(q^5)^9 - 23436 \, b(q)^3 \, b(q^5)^6 - 1664 \, c(q)^3 \, b(q^5)^6 + 467875 \, c(q^5)^3 \, b(q^5)^6 - 189 \, b(q)^6 \, b(q^5)^3 +$$
$$147 \, c(q)^6 \, b(q^5)^3 - 124125 \, c(q^5)^6 \, b(q^5)^3 - 6482 \, b(q)^3 \, c(q)^3 \, b(q^5)^3 - 1820 \, b(q)^3 \, c(q^5)^3 \, b(q^5)^3 +$$
$$69180 \, c(q)^3 \, c(q^5)^3 \, b(q^5)^3 - 16 \, c(q)^9 + 7625 \, c(q^5)^9 - 592 \, b(q)^3 \, c(q)^6 + 336 \, b(q)^3 \, c(q^5)^6 +$$
$$9564 \, c(q)^3 \, c(q^5)^6 + 208 \, a(q)^3 \, b(q)^3 \, c(q)^3 + 19 \, b(q)^6 \, c(q^5)^3 + 1923 \, c(q)^6 \, c(q^5)^3 - 1938 \, b(q)^3 \, c(q)^3 \, c(q^5)^3,$$

$$23625 \, b(q^5)^9 - 23436 \, b(q)^3 \, b(q^5)^6 - 1664 \, c(q)^3 \, b(q^5)^6 + 467875 \, c(q^5)^3 \, b(q^5)^6 - 189 \, b(q)^6 \, b(q^5)^3 +$$

$147\, c(q)^6\, b(q^5)^3 - 124125\, c(q^5)^6\, b(q^5)^3 - 6482\, b(q)^3\, c(q)^3\, b(q^5)^3 - 1820\, b(q)^3\, c(q^5)^3\, b(q^5)^3 +$
$69180\, c(q)^3\, c(q^5)^3\, b(q^5)^3 - 16\, c(q)^9 + 7625\, c(q^5)^9 - 384\, b(q)^3\, c(q)^6 + 336\, b(q)^3\, c(q^5)^6 +$
$9564\, c(q)^3\, c(q^5)^6 + 208\, b(q)^6\, c(q)^3 + 19\, b(q)^6\, c(q^5)^3 + 1923\, c(q)^6\, c(q^5)^3 - 1938\, b(q)^3\, c(q)^3\, c(q^5)^3,$
$26\, a(q)^9 - 49625\, b(q^5)^9 + 16\, c(q)^9 + 18375\, c(q^5)^9 + 45978\, b(q)^3\, b(q^5)^6 + 592\, b(q)^3\, c(q)^6 -$
$355\, b(q^5)^3\, c(q)^6 + 1822\, b(q)^3\, c(q)^6 + 1086125\, b(q^5)^3\, c(q^5)^6 + 12978\, c(q)^3\, c(q^5)^6 + 3621\, b(q)^6\, b(q^5)^3 +$
$3822\, b(q^5)^6\, c(q)^3 + 13866\, b(q)^3\, b(q^5)^3\, c(q)^3 + 189\, b(q)^6\, c(q^5)^3 - 1429875\, b(q^5)^6\, c(q^5)^3 -$
$5355\, c(q)^6\, c(q^5)^3 - 107640\, b(q)^3\, b(q^5)^3\, c(q^5)^3 - 5446\, b(q)^3\, c(q)^3\, c(q^5)^3 - 178640\, b(q)^3\, c(q^5)^3\, c(q^5)^3,$
$-222125\, b(q^5)^9 + 207348\, b(q)^3\, b(q^5)^6 + 16952\, c(q)^3\, b(q^5)^6 - 6187375\, c(q^5)^3\, b(q^5)^6 + 14673\, b(q)^6\, b(q^5)^3 -$
$1567\, c(q)^6\, b(q^5)^3 + 4468625\, c(q^5)^6\, b(q^5)^3 + 61946\, b(q)^3\, c(q)^3\, b(q^5)^3 - 428740\, b(q)^3\, c(q^5)^3\, b(q^5)^3 -$
$783740\, c(q)^3\, c(q^5)^3\, b(q^5)^3 + 184\, c(q)^9 + 65875\, c(q^5)^9 + 104\, a(q)^3\, b(q)^6 + 3064\, b(q)^3\, c(q)^6 +$
$6952\, b(q)^3\, c(q^5)^6 + 42348\, c(q)^3\, c(q^5)^6 + 737\, b(q)^6\, c(q^5)^3 - 23343\, c(q)^6\, c(q^5)^3 - 19846\, b(q)^3\, c(q)^3\, c(q^5)^3,$
$208\, b(q)^9 + 29535\, b(q^5)^3\, b(q)^6 + 1455\, c(q^5)^3\, b(q)^6 + 438132\, b(q^5)^6\, b(q)^3 + 6512\, c(q)^6\, b(q)^3 +$
$13568\, c(q^5)^6\, b(q)^3 + 130374\, b(q^5)^3\, c(q)^3\, b(q)^3 - 855660\, b(q^5)^3\, c(q^5)^3\, b(q)^3 - 37754\, c(q)^3\, c(q^5)^3\, b(q)^3 -$
$467875\, b(q^5)^9 + 384\, c(q)^9 + 124125\, c(q^5)^9 - 3281\, b(q^5)^3\, c(q)^6 + 9061375\, b(q^5)^3\, c(q^5)^6 + 75132\, c(q)^3\, c(q^5)^6 +$
$35568\, b(q^5)^6\, c(q)^3 - 12842625\, b(q^5)^6\, c(q^5)^3 - 48609\, c(q)^6\, c(q^5)^3 - 1636660\, b(q^5)^3\, c(q)^3\, c(q^5)^3,$
$-420625\, b(q^5)^9 + 391260\, b(q)^3\, b(q^5)^6 + 32240\, c(q)^3\, b(q^5)^6 - 11906875\, c(q^5)^3\, b(q^5)^6 + 29157\, b(q)^6\, b(q^5)^3 -$
$2987\, c(q)^6\, b(q^5)^3 + 8813125\, c(q^5)^6\, b(q^5)^3 + 117410\, b(q)^3\, c(q)^3\, b(q^5)^3 - 859300\, b(q)^3\, c(q^5)^3\, b(q^5)^3 -$
$1498300\, c(q)^3\, c(q^5)^3\, b(q^5)^3 + 352\, c(q)^9 + 139375\, c(q^5)^9 + 5536\, b(q)^3\, c(q)^6 + 14240\, b(q)^3\, c(q^5)^6 +$
$94260\, c(q)^3\, c(q^5)^6 + 208\, a(q)^6\, b(q)^3 + 1493\, b(q)^6\, c(q^5)^3 - 44763\, c(q)^6\, c(q^5)^3 - 41630\, b(q)^3\, c(q)^3\, c(q^5)^3\}$

We end here our short excursion about Ramanujan elliptic functions. The interested reader can continue this journey and implement differentiation, integration, simplification, and transformation rules for these functions (not many such rules are known) to be used by functions such as `D`, `Integrate`, `Sum`, `Product`, `FullSimplify`, and `FunctionExpand`.

3.1 Introduction

Most special functions arise as the solutions of some important ordinary differential equations (which frequently occur in practical problems), and also arise in connection with certain definite and indefinite integrals. Most of them were discovered in the 18th and 19th centuries, and are often named after the people who invented/discovered them. For many details on special functions, see, e.g., [830], [1109], [42], [44], [768], [3], [1217], [1164], [1222], [956], [438], [1297], [771], [1311], [784], [860], [1104], [1175], [1273], and on-line at http://functions.wolfram.com and http://dlmf.nist.gov/.

For calculations involving special functions in *Mathematica*, the following two functions are very valuable. The first is `FullSimplify`. In distinction to `Simplify`, the function `FullSimplify` knows many simplification rules for special functions. (But be aware that this additional knowledge built-in into `FullSimplify` also means that it will typically need a longer time to carry out its job).

`FullSimplify`[*expression*, *opions*]

    tries to simplify the expression *expression* under the options *options*.

Here is a first simple example.

In[1]:= **FullSimplify[Gamma[z] z]**

Out[1]= Gamma[1 + z]

FullSimplify carries all of the options of Simplify.

In[2]:= **Complement[First /@ Options[Simplify], First /@ Options[FullSimplify]]**

Out[2]= {}

Here is a first example that uses FullSimplify. FullSimplify successfully shows that $x(z) = \varphi^{2-\varphi} z^{\varphi-1}$ is a solution of the functional differential equation $x'(z) = 1/x(x(z))$ [1144], while Simplify does not.

In[3]:= **x[z_] := GoldenRatio^(2 - GoldenRatio) z^(GoldenRatio - 1)**

In[4]:= **{x'[z] - 1/x[x[z]] // Simplify, x'[z] - 1/x[x[z]] // FullSimplify}**

Out[4]= $\{(-1 + \text{GoldenRatio}) \text{GoldenRatio}^{2-\text{GoldenRatio}} z^{-2+\text{GoldenRatio}} -$

$\text{GoldenRatio}^{(-2+\text{GoldenRatio})\text{GoldenRatio}} z^{-(-1+\text{GoldenRatio})^2}, 0\}$

The second useful function when dealing with special functions is FunctionExpand.

FunctionExpand[*expression*]

   tries to rewrite the expression *expression* in simpler function and tries to canonicalize its arguments.

Here, FunctionExpand is used to simplify the following Bessel function.

In[5]:= **{BesselJ[5/2, Pi], FunctionExpand[BesselJ[5/2, Pi]]}**

Out[5]= $\left\{\text{BesselJ}\left[\frac{5}{2}, \pi\right], \frac{3\sqrt{2}}{\pi^2}\right\}$

FunctionExpand currently does not have any options.

In[6]:= **Options[FunctionExpand]**

Out[6]= {Assumptions :→ $Assumptions}

As a motivation concerning special functions and an example of the use of FullSimplify and FunctionEx : pand, let us investigate some integrals. All special functions have integral and/or series representations involving elementary functions. This fact means, on the other hand, that special functions arise as integrals or sums over (compositions of) elementary functions. Here is a list of 33 elementary functions.

In[7]:= **elementaryFunctions =**
   {Plus, Times, 1/#&, Sqrt, Power[#, C]&, Power[C, #]&,
   Exp, Log[C, #]&, Log[#, C]&,
   Sin, Cos, Sec, Csc, Tan, Cot,
   Sinh, Cosh, Sech, Csch, Tanh, Coth,
   ArcSin, ArcCos, ArcSec, ArcCsc, ArcTan, ArcCot,
   ArcSinh, Cosh, ArcSech, ArcCsch, ArcTanh, ArcCoth};

We will now try to carry out all 1089 integrals of the form $\int^z f(g(x))\,dx$, where f and g are elementary functions from the list elementaryFunctions.

In[8]:= **integrals =**
   Module[{integrand, λ = Length[elementaryFunctions]},
      Flatten[Table[integrand =
         elementaryFunctions[[i]][elementaryFunctions[[j]][z]];
         (* the pairs: {integrand, integral} *)
         {integrand, Integrate[integrand, z]},
               {i, λ}, {j, λ}], 1]];

The list allDoneIntegrals contains all functions and, when found, their antiderivatives.

In[9]:= `allDoneIntegrals = Select[integrals, FreeQ[#, _Integrate, Infinity]&];`

About 60% of all integrals were done.

In[10]:= `{Length[allDoneIntegrals],`
`    N[Length[allDoneIntegrals]]/Length[elementaryFunctions]^2 100 "%"}`

Out[10]= `{640, 58.7695 %}`

Although the integrand has a very simple form, the integrals are sometimes complicated. Here are 10 randomly chosen elements of `allDoneIntegrals`.

In[11]:= `SeedRandom[222];`

```
Apply[(Integrate[HoldForm[#1], z] == #2)&,
    allDoneIntegrals[[Nest[Function[{a, b},
                {Append[a, b[[#]]], Delete[b, #]}&[
                Random[Integer, {1, Length[b]}]]]] @@ #&,
        {{}, Range[Length[allDoneIntegrals]]}, 10][[1]]]], {1}] //
                                        TraditionalForm
```

Out[15]//TraditionalForm=

$$\left\{ \int \sec^{-1}(z)\, dz = z\sec^{-1}(z) - \log\left(z\left(\sqrt{\frac{z^2-1}{z^2}} + 1\right)\right),\right.$$

$$\int \sinh^{-1}(\cosh(z))\, dz =$$

$$\frac{1}{8}\left(8 z \sinh^{-1}(\cosh(z)) + 4 i \pi \tanh^{-1}\left(\frac{\sqrt{2}\,\cosh(z)}{\sqrt{\cosh(2z)+3}}\right) + 2 z \log\left(e^{-z}\left(-\sinh(z) + e^z - \sqrt{\cosh^2(z)+1}\right)\right) - \right.$$

$$2 \tanh^{-1}\left(\frac{\sinh(z)}{\sqrt{\cosh^2(z)+1}}\right)\log\left(e^{-z}\left(-\sinh(z) + e^z - \sqrt{\cosh^2(z)+1}\right)\right) +$$

$$i \pi \log\left(e^{-z}\left(-\sinh(z) + e^z - \sqrt{\cosh^2(z)+1}\right)\right) - 2 z \log\left(e^{-z}\left(-\sinh(z) + e^z + \sqrt{\cosh^2(z)+1}\right)\right) -$$

$$2 \tanh^{-1}\left(\frac{\sinh(z)}{\sqrt{\cosh^2(z)+1}}\right)\log\left(e^{-z}\left(-\sinh(z) + e^z + \sqrt{\cosh^2(z)+1}\right)\right) -$$

$$i \pi \log\left(e^{-z}\left(-\sinh(z) + e^z + \sqrt{\cosh^2(z)+1}\right)\right) + 2 z \log\left(e^z \sinh(z) - e^z \sqrt{\cosh^2(z)+1} + 1\right) -$$

$$2 \tanh^{-1}\left(\frac{\sinh(z)}{\sqrt{\cosh^2(z)+1}}\right)\log\left(e^z \sinh(z) - e^z \sqrt{\cosh^2(z)+1} + 1\right) +$$

$$i \pi \log\left(e^z \sinh(z) - e^z \sqrt{\cosh^2(z)+1} + 1\right) - 2 z \log\left(e^z \sinh(z) + e^z \sqrt{\cosh^2(z)+1} + 1\right) -$$

$$2 \tanh^{-1}\left(\frac{\sinh(z)}{\sqrt{\cosh^2(z)+1}}\right)\log\left(e^z \sinh(z) + e^z \sqrt{\cosh^2(z)+1} + 1\right) -$$

$$i \pi \log\left(e^z \sinh(z) + e^z \sqrt{\cosh^2(z)+1} + 1\right) +$$

$$2 \operatorname{Li}_2\left(e^z\left(\sqrt{\cosh^2(z)+1} - \sinh(z)\right)\right) + 2 \operatorname{Li}_2\left(e^{-z}\left(\sinh(z) - \sqrt{\cosh^2(z)+1}\right)\right) -$$

$$2\,\mathrm{Li}_2\!\left(e^{-z}\left(\sinh(z)+\sqrt{\cosh^2(z)+1}\,\right)\right)-2\,\mathrm{Li}_2\!\left(-e^{z}\left(\sinh(z)+\sqrt{\cosh^2(z)+1}\,\right)\right),$$

$$\int \csc^{-1}(\cosh(z))\,dz = z\,\csc^{-1}(\cosh(z)) +$$

$$\frac{i\sqrt{\frac{(-1+e^{2z})^2}{(1+e^{2z})^2}}\ (1+e^{2z})\left(z\left(\log(1-i\,e^z)-\log(1+i\,e^z)\right)-\mathrm{Li}_2(-i\,e^z)+\mathrm{Li}_2(i\,e^z)\right)}{-1+e^{2z}},$$

$$\int \operatorname{sech}^{-1}(\tan(z))\,dz = \frac{1}{4}\Bigg(2\,i\,\log\!\left(1-i\,e^{-\operatorname{sech}^{-1}(\tan(z))}\right)\operatorname{sech}^{-1}(\tan(z)) +$$

$$2\,i\,\log\!\left(1+i\,e^{-\operatorname{sech}^{-1}(\tan(z))}\right)\operatorname{sech}^{-1}(\tan(z))-2\,i\,\log\!\left(1-i\left(-1+\sqrt{2}\,\right)e^{-\operatorname{sech}^{-1}(\tan(z))}\right)\operatorname{sech}^{-1}(\tan(z))-$$

$$2\,i\,\log\!\left(1+i\left(1+\sqrt{2}\,\right)e^{-\operatorname{sech}^{-1}(\tan(z))}\right)\operatorname{sech}^{-1}(\tan(z))-2\,\pi\,\operatorname{sech}^{-1}(\tan(z))+$$

$$8\sin^{-1}\!\left(\sqrt{\frac{1}{2}+\frac{i}{2}}\,\right)\tanh^{-1}\!\left(\frac{(1+i)\tanh(\tfrac{1}{2}\,\operatorname{sech}^{-1}(\tan(z)))}{\sqrt{2}}\right)+\pi\log\!\left(1-i\,e^{-\operatorname{sech}^{-1}(\tan(z))}\right)-$$

$$\pi\log\!\left(1+i\,e^{-\operatorname{sech}^{-1}(\tan(z))}\right)-4\sin^{-1}\!\left(\sqrt{\frac{1}{2}+\frac{i}{2}}\,\right)\log\!\left(1-i\left(-1+\sqrt{2}\,\right)e^{-\operatorname{sech}^{-1}(\tan(z))}\right)+$$

$$4\sin^{-1}\!\left(\sqrt{\frac{1}{2}+\frac{i}{2}}\,\right)\log\!\left(1+i\left(1+\sqrt{2}\,\right)e^{-\operatorname{sech}^{-1}(\tan(z))}\right)+4\,\pi\log\!\left(1+e^{\operatorname{sech}^{-1}(\tan(z))}\right)-$$

$$4\,\pi\log\!\left(\cosh\!\left(\frac{1}{2}\,\operatorname{sech}^{-1}(\tan(z))\right)\right)+\pi\log\!\left(-\sin\!\left(\frac{1}{4}\,(\pi-2\,i\,\operatorname{sech}^{-1}(\tan(z)))\right)\right)-$$

$$\pi\log\!\left(\sin\!\left(\frac{1}{4}\,(2\,i\,\operatorname{sech}^{-1}(\tan(z))+\pi)\right)\right)-2\,i\,\mathrm{Li}_2\!\left(-i\,e^{-\operatorname{sech}^{-1}(\tan(z))}\right)-2\,i\,\mathrm{Li}_2\!\left(i\,e^{-\operatorname{sech}^{-1}(\tan(z))}\right)+$$

$$2\,i\,\mathrm{Li}_2\!\left(i\left(-1+\sqrt{2}\,\right)e^{-\operatorname{sech}^{-1}(\tan(z))}\right)+2\,i\,\mathrm{Li}_2\!\left(-i\left(1+\sqrt{2}\,\right)e^{-\operatorname{sech}^{-1}(\tan(z))}\right)\Bigg)+$$

$$\frac{1}{4}\Bigg(-2\,i\,\log\!\left(1-i\,e^{-\operatorname{sech}^{-1}(\tan(z))}\right)\operatorname{sech}^{-1}(\tan(z))-2\,i\,\log\!\left(1+i\,e^{-\operatorname{sech}^{-1}(\tan(z))}\right)\operatorname{sech}^{-1}(\tan(z))+$$

$$2\,i\,\log\!\left(1+i\left(-1+\sqrt{2}\,\right)e^{-\operatorname{sech}^{-1}(\tan(z))}\right)\operatorname{sech}^{-1}(\tan(z))+$$

$$2\,i\,\log\!\left(e^{-\operatorname{sech}^{-1}(\tan(z))}\left(-i\left(1+\sqrt{2}\,\right)+e^{\operatorname{sech}^{-1}(\tan(z))}\right)\right)\operatorname{sech}^{-1}(\tan(z))+2\,\pi\,\operatorname{sech}^{-1}(\tan(z))-$$

$$8\,i\sin^{-1}\!\left(\sqrt{\frac{1}{2}-\frac{i}{2}}\,\right)\tan^{-1}\!\left(\frac{(1+i)\tanh(\tfrac{1}{2}\,\operatorname{sech}^{-1}(\tan(z)))}{\sqrt{2}}\right)-\pi\log\!\left(1-i\,e^{-\operatorname{sech}^{-1}(\tan(z))}\right)+$$

$$\pi\log\!\left(1+i\,e^{-\operatorname{sech}^{-1}(\tan(z))}\right)-4\sin^{-1}\!\left(\sqrt{\frac{1}{2}-\frac{i}{2}}\,\right)\log\!\left(1+i\left(-1+\sqrt{2}\,\right)e^{-\operatorname{sech}^{-1}(\tan(z))}\right)-$$

$$4\,\pi\log\!\left(1+e^{\operatorname{sech}^{-1}(\tan(z))}\right)+4\sin^{-1}\!\left(\sqrt{\frac{1}{2}-\frac{i}{2}}\,\right)\log\!\left(e^{-\operatorname{sech}^{-1}(\tan(z))}\left(-i\left(1+\sqrt{2}\,\right)+e^{\operatorname{sech}^{-1}(\tan(z))}\right)\right)+$$

$$4\,\pi\log\!\left(\cosh\!\left(\frac{1}{2}\,\operatorname{sech}^{-1}(\tan(z))\right)\right)-\pi\log\!\left(-\sin\!\left(\frac{1}{4}\,(\pi-2\,i\,\operatorname{sech}^{-1}(\tan(z)))\right)\right)+$$

$$\pi\log\!\left(\sin\!\left(\frac{1}{4}\,(2\,i\,\operatorname{sech}^{-1}(\tan(z))+\pi)\right)\right)+2\,i\,\mathrm{Li}_2\!\left(-i\,e^{-\operatorname{sech}^{-1}(\tan(z))}\right)+2\,i\,\mathrm{Li}_2\!\left(i\,e^{-\operatorname{sech}^{-1}(\tan(z))}\right)-$$

$$2\,i\,\mathrm{Li}_2\!\left(-i\left(-1+\sqrt{2}\right)e^{-\mathrm{sech}^{-1}(\tan(z))}\right)-2\,i\,\mathrm{Li}_2\!\left(i\left(1+\sqrt{2}\right)e^{-\mathrm{sech}^{-1}(\tan(z))}\right),$$

$$\int e^{\coth(z)}\,dz=\frac{\mathrm{Ei}(\coth(z)+1)}{2\,e}-\frac{1}{2}\,e\,\mathrm{Ei}(\coth(z)-1),\quad \int z\,dz=\frac{z^2}{2},$$

$$\int C^{\cot(z)}\,dz=\frac{1}{2}\,i\,C^{-i}\left(C^{2\,i}\,\mathrm{Ei}((\cot(z)-i)\log(C))-\mathrm{Ei}((\cot(z)+i)\log(C))\right),$$

$$\int\frac{\log(\cosh(z))}{\log(C)}\,dz=\frac{z\log(\cosh(z))-i\left(-\frac{iz^2}{2}-i\log(1+e^{-2z})z+\frac{1}{2}\,i\,\mathrm{Li}_2(-e^{-2z})\right)}{\log(C)},$$

$$\int\frac{z}{\sqrt{1-z^2}}\,dz=-\sqrt{1-z^2},\quad \int\frac{\log(\sin(z))}{\log(C)}\,dz=\frac{-z\log(1-e^{2\,i\,z})+z\log(\sin(z))+\frac{1}{2}\,i\,(z^2+\mathrm{Li}_2(e^{2\,i\,z}))}{\log(C)}\Bigg\}$$

Now, let us find out which special functions were needed to express the integrals of the most simple compositions of elementary functions.

```
In[13]:= usedSpecialFunctions =
        DeleteCases[Union[Level[Last /@ allDoneIntegrals, {-1}, Heads -> True]],
            z | C | _?NumericQ | List | Power | Log |
            _?(MemberQ[elementaryFunctions, #]&)]
Out[13]= {CoshIntegral, CosIntegral, EllipticE, EllipticF,
        Erf, Erfi, ExpIntegralEi, FresnelC, FresnelS, Gamma,
        Hypergeometric2F1, LogIntegral, PolyLog, SinhIntegral, SinIntegral}
```

Here, we count the integrals containing special functions.

```
In[14]:= Count[allDoneIntegrals,
            _?(MemberQ[#, Alternatives @@ usedSpecialFunctions,
                {0, Infinity}, Heads -> True]&)]
Out[14]= 324
```

We now apply FunctionsExpand to the results of integration.

```
In[15]:= functionExpandedDoneIntegrals =
            {#[[2]], FunctionExpand[#[[2]]]}& /@ allDoneIntegrals;
```

For further use, we keep only the results that really were expanded.

```
In[16]:= interestingFunctionExpandedDoneIntegrals =
            Select[functionExpandedDoneIntegrals, (UnsameQ @@ #)&];
```

```
In[17]:= Length[interestingFunctionExpandedDoneIntegrals]
Out[17]= 331
```

In most cases, the application of FunctionExpand results in an increase of the size of the expressions, but sometimes the expressions might shrink. The following picture shows the size ratio (measured with ByteCount) of the expanded expression versus the unexpanded expression.

```
In[18]:= ListPlot[Sort[Apply[#2/#1&, Map[ByteCount,
                interestingFunctionExpandedDoneIntegrals, {2}], {1}]],
            PlotRange -> All, Frame -> True, Axes -> False]
```

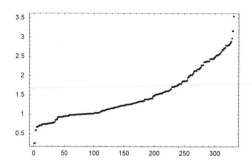

Here are the first few results of the application of `FunctionExpand`. We see that $\sqrt{product}$ was rewritten in the form $\sqrt{\prod_k factors_k}$ and $\log(product)$ as $\sum_k \log(factors_k)$. Here, the functions `Power` and `Log` remain unchanged, but their arguments get "expanded" (linearized as much as possible).

In[19]:= **Take[interestingFunctionExpandedDoneIntegrals, 5] // TableForm**

Out[19]//TableForm=

$$\sqrt{1 - z^2} + z\ \mathtt{ArcSin[z]} \qquad\qquad \sqrt{1 - z}\ \sqrt{1 + z} + z\ \mathtt{ArcSin[z]}$$

$$-\sqrt{1 - z^2} + z\ \mathtt{ArcCos[z]} \qquad\qquad -\sqrt{1 - z}\ \sqrt{1 + z} + z\ \mathtt{ArcCos[z]}$$

$$z\ \mathtt{ArcSec[z]} - \mathtt{Log}\left[z\left(1 + \sqrt{\tfrac{-1+z^2}{z^2}}\right)\right] \qquad\qquad z\ \mathtt{ArcSec[z]} - \mathtt{Log}\left[z\left(1 + \tfrac{\sqrt{(-1+z)\ (1+z)}}{\sqrt{-i\ z}\ \sqrt{i\ z}}\right)\right]$$

$$z\ \mathtt{ArcCsc[z]} + \mathtt{Log}\left[z\left(1 + \sqrt{\tfrac{-1+z^2}{z^2}}\right)\right] \qquad\qquad z\ \mathtt{ArcCsc[z]} + \mathtt{Log}\left[z\left(1 + \tfrac{\sqrt{(-1+z)\ (1+z)}}{\sqrt{-i\ z}\ \sqrt{i\ z}}\right)\right]$$

$$z\ \mathtt{ArcTan[z]} - \tfrac{1}{2}\ \mathtt{Log[1 + z^2]} \qquad\qquad z\ \mathtt{ArcTan[z]} + \tfrac{1}{2}\ (-\mathtt{Log[i\ (-i + z)]} - \mathtt{Log[-i\ (i + z)]})$$

In some cases, `FunctionExpand` also expanded special functions into simpler functions.

In[20]:= **changedInterestingFunctionExpandedDoneIntegrals =**
   Select[Select[interestingFunctionExpandedDoneIntegrals,
         DeleteCases[Union[Level[First[#], {-1}, Heads -> True]],
            z | C | _?NumericQ | List | Power | Log |
            _?(MemberQ[elementaryFunctions, #]&)] =!= {}&],
         (* any change? *)
         DeleteCases[Union[Level[First[#], {-1}, Heads -> True]],
            z | C | _?NumericQ | List | Power | Log |
            _?(MemberQ[elementaryFunctions, #]&)] =!=
         DeleteCases[Union[Level[Last[#], {-1}, Heads -> True]],
            z | C | _?NumericQ | List | Power | Log |
            _?(MemberQ[elementaryFunctions, #]&)]&];

In[21]:= **Length[changedInterestingFunctionExpandedDoneIntegrals]**

Out[21]= 51

Here are some of these cases before and after expansion.

In[22]:= **Take[changedInterestingFunctionExpandedDoneIntegrals, 3]**

Out[22]= $\left\{\left\{-\mathtt{Cos[z]\ Hypergeometric2F1}\left[\tfrac{1}{2}, \tfrac{1-C}{2}, \tfrac{3}{2}, \mathtt{Cos[z]}^2\right]\mathtt{Sin[z]}^{1+C}\ (\mathtt{Sin[z]}^2)^{\frac{1}{2}\ (-1-C)},\right.\right.$

$$\left.-\frac{\mathtt{Beta[Cos[z]}^2,\ \tfrac{1}{2},\ 1+\tfrac{1}{2}\ (-1+C)]\ \mathtt{Cos[z]\ Sin[z]}^{1+C}\ (\mathtt{Sin[z]}^2)^{\frac{1}{2}\ (-1-C)}}{2\ \sqrt{\mathtt{Cos[z]}^2}}\right\},$$

$$\left\{-\frac{\mathtt{Cos[z]}^{1+C}\ \mathtt{Hypergeometric2F1}\left[\tfrac{1+C}{2}, \tfrac{1}{2}, \tfrac{3+C}{2}, \mathtt{Cos[z]}^2\right]\mathtt{Sin[z]}}{(1+C)\ \sqrt{\mathtt{Sin[z]}^2}},\right.$$

$$-\frac{\text{Beta}[\text{Cos}[z]^2, \frac{1+C}{2}, \frac{1}{2}]\,\text{Cos}[z]^{1+C}\,(\text{Cos}[z]^2)^{\frac{1}{2}(-1-C)}\,\text{Sin}[z]}{2\,\sqrt{\text{Sin}[z]^2}}\Bigg\},$$

$$\Bigg\{\frac{\text{Csc}[z]\,\text{Hypergeometric2F1}[\frac{1-C}{2}, \frac{1}{2}, \frac{3-C}{2}, \text{Cos}[z]^2]\,\text{Sec}[z]^{-1+C}\,\sqrt{\text{Sin}[z]^2}}{-1+C},$$

$$\frac{(1-C)\,\text{Beta}[\text{Cos}[z]^2, \frac{1-C}{2}, \frac{1}{2}]\,(\text{Cos}[z]^2)^{\frac{1}{2}(-1+C)}\,\text{Csc}[z]\,\text{Sec}[z]^{-1+C}\,\sqrt{\text{Sin}[z]^2}}{2\,(-1+C)}\Bigg\}\Bigg\}$$

Using `FullSimplify`, we can check the transformation carried out by `FunctionExpand`.

```
In[23]:= Table[FullSimplify[Subtract @@
              interestingFunctionExpandedDoneIntegrals[[i]]], {i, 12}]

Out[23]= {0, 0, 0, 0, 0, 0, 0, 0, 0, 0, 0, 0}
```

Now, we come back to our list of integrals of compositions of elementary functions. We will check the correctness of the integrals. We differentiate the resulting integral, subtract the original integrand, and see if the result is equal to zero. From the first 200 integrals, more than one-half simplify to 0 by purely rational operations (meaning applying `Together`).

```
In[24]:= togetheredDifferences = Table[Together[
              D[allDoneIntegrals[[i, 2]], z] - allDoneIntegrals[[i, 1]]], {i, 200}];

In[25]:= Count[togetheredDifferences, 0, {1}]

Out[25]= 127
```

Applying `Simplify` simplifies more functions, but not all.

```
In[26]:= simplifiedDifferences = Simplify[togetheredDifferences];

In[27]:= Count[simplifiedDifferences, 0, {1}]

Out[27]= 179
```

Using the more powerful `FullSimplify` finally reduces all differences to 0.

```
In[28]:= fullSimplifiedDifferences = FullSimplify[simplifiedDifferences];

In[29]:= Count[fullSimplifiedDifferences, 0, {1}]

Out[29]= 200
```

`FullSimplify` is a powerful function and can be very beneficial for calculations involving special functions. Not only for single integrals, but also for multiple integrals, it helps often. The following direct try to calculate the volume of the supersphere [98], [1365], [290] $|x|^n + |y|^n + |z|^n < 1$ fails.

```
In[30]:= 8 Integrate[(1 - x^n - y^n)^(1/n), {x, 0, 1}, {y, 0, (1 - x^n)^(1/n)},
              GenerateConditions -> False]
```

$$\text{Out[30]}= 8\,\text{Integrate}\Big[(1-x^n)^{\frac{1}{n}}\Big(1-x^n-\big((1-x^n)^{\frac{1}{n}}\big)^n\Big)^{\frac{1}{n}}$$

$$\Big(1+\frac{\big((1-x^n)^{\frac{1}{n}}\big)^n}{-1+x^n}\Big)^{-1/n}\,\text{Hypergeometric2F1}\Big[\frac{1}{n}, -\frac{1}{n}, 1+\frac{1}{n}, -\frac{\big((1-x^n)^{\frac{1}{n}}\big)^n}{-1+x^n}\Big],$$

$$\{x, 0, 1\}, \text{GenerateConditions} \rightarrow \text{False}\Big]$$

Using `FullSimplify` after the first integration and at the end, we easily obtain the desired result.

```
In[31]:= With[{conds = Element[n, Integers] && n > 0 && 0 < x < 1},
           FullSimplify[#, conds]& @ FullSimplify[%, cond,
                 TransformationFunctions -> (* extract finite part *)
                 {(# //. (α_ a_^e_ b_^f_) :> α (a/b)^e /; e/f == -1 &&
```

```
                    ((Together //@ PowerExpand //@ {a, b}) === {0, 0}))&,
                Together}]]
```

Out[31]= $\dfrac{8 \; \mathrm{Gamma} \left[1 + \frac{1}{n}\right]^2 \; \mathrm{Gamma} \left[\frac{1}{n}\right]}{3 \; \mathrm{Gamma} \left[\frac{3}{n}\right]}$

For $n = 2$, we get from $8\,\Gamma(1 + 1/n)^2\,\Gamma(1/n)/(3\,\Gamma(3/n))$ the well-known volume of a sphere.

In[32]:= `% /. n -> 2`

Out[32]= $\dfrac{4\,\pi}{3}$

Similar remarks apply to the function `FunctionExpand`. The following sum [66] containing the function `Factorial2` stays unevaluated.

In[33]:= `Sum[Evaluate[(-1)^k (𝒦r)^(2k)/((2k)!! (d + 2k - 2)!!)],`
`        {k, 0, Infinity}]`

Out[33]= $\displaystyle\sum_{k=0}^{\infty} \dfrac{(-1)^k \, \mathcal{K}r^{2\,k}}{(2\,k)!! \; (-2 + 2\,k + d)!!}$

Rewriting the summand in Gamma functions allows to carry out the sum.

In[34]:= `MapAt[Evaluate, MapAt[Simplify[FunctionExpand[#],`
`                        Element[k, Integers]]&, %, 1], 1]`

Out[34]= $2^{-\frac{1}{4}+\frac{1}{4}\mathrm{Cos}[\pi d]} \; \pi^{\frac{1}{2}} \; \mathrm{Sin}\left[\frac{\pi d}{2}\right]^2 \; \mathcal{K}r^{1-\frac{d}{2}} \; \mathrm{BesselJ}\left[\frac{1}{2}\,(-2+d),\,\mathcal{K}r\right]$

The same happens in the following sum [1102], [1182].

In[35]:= `Sum[Evaluate[(-1)^k/(n - k)! (2k + 1)!!/(k! (k + 1)!)], {k, 0, n}]`

Out[35]= $\displaystyle\sum_{k=0}^{n} \dfrac{(-1)^k \, (1 + 2\,k)!!}{k! \; (1 + k)! \; (-k + n)!}$

In[36]:= `MapAt[Evaluate, MapAt[Simplify[FunctionExpand[#],`
`                        Element[k, Integers]]&, %, 1], 1]`

Out[36]= $\dfrac{\mathrm{Hypergeometric2F1}\left[\frac{3}{2},\,-n,\,2,\,2\right]}{\mathrm{Gamma}[1 + n]}$

`FullSimplify` tries to write the result in functions with fewer arguments. In the following the two-argument expression `PolyGamma[2,1]` is converted to the one-argument function `Zeta[3]`.

In[37]:= `Series[1/(2 ε^3) Log[Gamma[1 + ε]^3 Gamma[1 - ε]/Gamma[1 + 2ε]],`
`        {ε, 0, 0}] // (Together //@ #)&`

Out[37]= $-\dfrac{1}{2}\,\mathrm{PolyGamma}[2, 1] + O[\varepsilon]^1$

In[38]:= `FullSimplify[%]`

Out[38]= $\mathrm{Zeta}[3] + O[\varepsilon]^1$

The function `FullSimplify` has one more option than `Simplify`.

In[39]:= `Complement[Sort[First /@ Options[FullSimplify]],`
`            Sort[First /@ Options[Simplify]]]`

Out[39]= `{ExcludedForms}`

> ExcludedForms
>
> is an option of FullSimplify that determines which expressions should not be simplified
>
> **Default:**
>
> {}
>
> **Admissible:**
>
> *listOfExpressionsOrPatterns*

In the next input, FullSimplify does not attempt to simplify expressions with the head f.

In[40]:= **FullSimplify[f[(x^2 - 1)/(x - 1)] + f[(x^2 - 1)/(x - 1)],**
     ExcludedForms -> {_f}]

Out[40]= $f[1 + x] + f\left[\dfrac{-1 + x^2}{-1 + x}\right]$

In addition to the option ExcludedForms, there is one other difference with respect to the options of Sim‌plify and FullSimplify: the default setting of the value TimeConstraint. Simplify will use each single rule not longer than 5 minutes, whereas FullSimplify has no time constraints. (This means that the result of Simplify[*expr*] may depend on the computer used, but the result of FullSimplify[*expr*] does not.)

In[41]:= **Options[#, TimeConstraint]& /@ {Simplify, FullSimplify}**

Out[41]= $\{\{\text{TimeConstraint} \rightarrow 300\}, \{\text{TimeConstraint} \rightarrow \infty\}\}$

We compare the power of Simplify versus FullSimplify on a concrete example. The sinc function $\mathrm{sinc}(x) = \sin(\pi x)/(\pi x)$ fulfills the following reproducing identity for $n, m \in \mathbb{Z}$, $n \neq m$ [663], [870], [592], [1190], [423], [1189]:

$$h\,\mathrm{sinc}(m - n) = \frac{1}{h}\int_{-\infty}^{\infty}\mathrm{sinc}\!\left(\frac{x - mh}{h}\right)\mathrm{sinc}\!\left(\frac{x - nh}{h}\right)dx.$$

The direct integral gives a rather large result.

In[42]:= **Sinc[x_] := Sin[Pi x]/(Pi x)**

In[43]:= **rhsInt = Integrate[Sinc[(x - m h)/h] Sinc[(x - n h)/h]/h,**
      {x, -Infinity, Infinity}];

  rhsInt // TraditionalForm // Short[#, 6]&

Out[45]//Short=

$\dfrac{1}{\pi^2}\Big(h\,\mathrm{If}\Big[\mathrm{Im}(h) = 0 \wedge (\mathrm{Re}(h^2\,m^2) \le 0 \vee \mathrm{Im}(h^2\,m^2) \neq 0) \wedge$

  $(h^2\,\mathrm{Re}(n^2) \le 0 \vee h\,\mathrm{Re}(n^2) = 0 \vee h\,\mathrm{Im}(n^2) \neq 0),$

  $\dfrac{\pi\,(\ll 13\gg + \ll 1\gg)}{2\,(m - n)\,|h|},\ \mathrm{Integrate}\Big[\dfrac{\sin(\frac{\pi\,(x - h\,m)}{h})\,\sin(\frac{\pi\,(x - h\,n)}{h})}{(x - h\,m)\,(x - h\,n)},$

  $\{x, -\infty, \infty\}, \mathrm{Assumptions} \rightarrow (\mathrm{Im}(h^2\,m^2) = 0 \wedge \mathrm{Re}(h^2\,m^2) > 0) \vee$

  $(h\,\mathrm{Im}(n^2) = 0 \wedge h^2\,\mathrm{Re}(n^2) > 0 \wedge h\,\mathrm{Re}(n^2) \neq 0) \vee \mathrm{Im}(h) \neq 0\Big]\Big]\Big)$

We rewrite the integral and state the identity in the following form.

In[46]:= **διφφ = h Sinc[(m - n)] == 1/(2 (m - n) Pi^2) Cos[(m - n) Pi]***
   ((CosIntegral[-2 m Pi] - CosIntegral[2 m Pi] - CosIntegral[-2 n Pi] +
   CosIntegral[2 n Pi] - Log[-h m] + Log[h m] + Log[-h*n] - Log[h n]) +

```
      2 Pi Sin[(m - n) Pi]);
```

```
      LeafCount[διφφ]
```
Out[48]= 101

`Simplify` cannot reduce this expression further, `FullSimplify` can reduce the expression to about a 80% of the original size. (`Integrate` returned a result containing the special functions Ci and Si and `Simplify` makes no attempts to simplify special functions.)

In[49]:= **LeafCount[#[διφφ]]& /@ {Simplify, FullSimplify}**

Out[49]= {101, 82}

`FullSimplify` with the assumption about n and m can prove the identity to result in `True`.

In[50]:= **#[διφφ, Element[{n, m}, Integers] && n != m && h > 0]& /@**
                                           {Simplify, FullSimplify}

Out[50]= {Cos[(m - n) π] (CosIntegral[-2 m π] - CosIntegral[2 m π] - CosIntegral[-2 n π] +
          CosIntegral[2 n π] - Log[-m] + Log[m] + Log[-n] - Log[n]) == 0, True}

Let us make one remark about the `TransformationFunctions` option of `Simplify`/`FullSimplify`: We will try to show that certain Laguerre polynomials are solutions of the $\mathcal{PT}$-symmetric oscillator [71], [1368]:

$$-\frac{\partial^2 \psi_{n,c,\alpha,q}(x)}{\partial x^2} + V_{c,\alpha}(x)\,\psi_{n,c,\alpha,q}(x) = (4\,n + 2 - 2\,\alpha)\,\psi_{n,c,\alpha,q}(x)$$

$$V_{c,\alpha}(x) = (x - i\,c)^2 + \frac{\alpha^2 - \frac{1}{4}}{(x - i\,c)^2}$$

$$\psi_{n,c,\alpha,q}(x) = \sqrt{\frac{n!}{\Gamma(n - q\,\alpha + 1)\cos(\pi\,(1/2 - \alpha))}}\;e^{-(x - i\,c)^2/2}\,(x - i\,c)^{1/2 - q\,\alpha}\,L_n^{-q\,\alpha}((x - i\,c)^2)$$

Here $c > 0$ and $n \in \mathbb{N}$, $q = \pm 1$. A direct attempt to verify that the $\psi_{n,c,\alpha,q}(x)$ are solutions fails (we ignore the normalization factor and specialize to $q = 1$).

In[51]:= **V[c_, α_, x_] := (x - I c)^2 + (α^2 - 1/4)/(x - I c)^2**

```
      ψ[n_, q_, c_, α_, x_] := Exp[-(x - I c)^2/2] (x - I c)^(-q α + 1/2) *
      LaguerreL[n, -q α, (x - I c)^2]
```

```
      ε[n_, q_, α_] := (4n + 2 - 2 q α)
```

In[54]:= **δ = With[{q = 1},**
                  With[{Ψ = ψ[n, q, c, α, x]},
                       -D[Ψ, x, x] - (ε[n, q, α] Ψ - V[c, α, x] Ψ) // FullSimplify]]

Out[54]= $4\,e^{\frac{1}{2}(c + i\,x)^2}\,(-i\,c + x)^{\frac{1}{2} - \alpha}$
          $((c + i\,x)^2\,\text{LaguerreL}[-2 + n, 2 - \alpha, (-i\,c + x)^2] + (1 + (c + i\,x)^2 - \alpha)$
          $\text{LaguerreL}[-1 + n, 1 - \alpha, (-i\,c + x)^2] - n\,\text{LaguerreL}[n, -\alpha, (-i\,c + x)^2])$

Next we add a transformation function representing the missing relation for the simplification.

In[55]:= **TF[LaguerreL[n_, λ_, z_]] := ((1 - z + λ) LaguerreL[-1 + n, 1 + λ, z] -**
                                 z LaguerreL[-2 + n, 2 + λ, z])/n

Unfortunately δ still does not simplify to zero.

In[56]:= **FullSimplify[δ, TransformationFunctions -> {TF}]**

Out[56]= $4 \, e^{\frac{1}{2} \, (c+i \, x)^2} \, (-i \, c + x)^{\frac{1}{2}-\alpha}$
$((c + i \, x)^2 \, \text{LaguerreL}[-2 + n, \, 2 - \alpha, \, (-i \, c + x)^2] + (1 + (c + i \, x)^2 - \alpha)$
$\text{LaguerreL}[-1 + n, \, 1 - \alpha, \, (-i \, c + x)^2] - n \, \text{LaguerreL}[n, \, -\alpha, \, (-i \, c + x)^2])$

The reason for this is that the transformation functions are only applied "locally". This means that all three Laguerre polynomials occurring in δ are transformed according to TF, but not more. After applying TF to each of them, the resulting expression becomes larger (measured with the default option setting of the Complexity: Function option), so these transformations are discarded. To achieve the desired simplification, we must also expand the whole expression containing the transformed Laguerre polynomials. The transformation function TF1 does this.

```
In[57]:= TF1[expr_] :=
        Module[{(* all Laguerre polynomials occurring in expr *)
                Ls = Cases[expr, _LaguerreL, {0, Infinity},
                Heads -> True], newExprs},
                (* recursive call to FullSimplify;
                    use different TransformationFunctions option setting *)
                FullSimplify[expr, TransformationFunctions ->
                    (Function[r, Together[expr /. r]& /@ ((# -> TF[#])& /@ Ls))]]
```

Now δ simplifies to 0.

```
In[58]:= FullSimplify[δ, TransformationFunctions -> {TF1}]
```

Out[58]= 0

FullSimplify uses a large repertory of rules. Here is a complicated zero (the Fibonacci identity comes from [880]).

```
In[59]:= Sin[Pi/24] - FunctionExpand[Sin[Pi/24]] +
        Beta[1, 2, 3, 4] - FunctionExpand[Beta[1, 2, 3, 4]] +
        Root[#^3 + 4# - 7&, 2] - ToRadicals[Root[#^3 + 4# - 7&, 2] +
        Fibonacci[n + 2]^3 + Fibonacci[n + 1]^3 -
        Fibonacci[n]^3 - Fibonacci[3n + 3]] // (id = #)&
```

Out[59]= $\frac{49}{60} + \frac{1}{4} \sqrt{3 \, (2 - \sqrt{2})} - \frac{\sqrt{2 + \sqrt{2}}}{4} - 2 \, (1 - i \, \sqrt{3}) \left(\frac{2}{3 \, (63 + \sqrt{4737})} \right)^{1/3} +$

$\dfrac{(1 + i \, \sqrt{3}) \, (\frac{1}{2} \, (63 + \sqrt{4737}))^{1/3}}{2 \cdot 3^{2/3}} + \text{Beta}[1, 2, 3, 4] + \text{Fibonacci}[n]^3 - \text{Fibonacci}[1 + n]^3 -$

$\text{Fibonacci}[2 + n]^3 + \text{Fibonacci}[3 + 3 \, n] + \text{Root}[-7 + 4 \, \#1 + \#1^3 \, \&, \, 2] + \text{Sin}\left[\dfrac{\pi}{24}\right]$

CSC will carry out the complicated simplification of id. As a side effect, we count the number of transformations applied (count1), the number of transformations resulting in an expression not larger than the original one (count2), and the number of transformations resulting in a smaller expression (count3). Here, we measure the size of an expression with LeafCount.

```
In[60]:= CSC := (count1 = 0; count2 = 0; count3 = 0;
        lf = LeafCount[id];
        (* simplify and keep track of intermediate results *)
        FullSimplify[id, Element[n, Integers] && n > 0,
            ComplexityFunction -> ((count1++;
            Which[# === lf, count2++,
                    # < lf, lf = #; count2++; count3++]; #)&[LeafCount[#]&])])
```

FullSimplify is able to show that id equals 0. More than 5000 transformations were attempted, only 8 of them resulted in smaller expressions.

In[61]:= **CSC // Timing**

Out[61]= {70.13 Second, 0}

In[62]:= **{count1, count2, count3}**

Out[62]= {14326, 65, 8}

Because of cached results, simplifying again, CSC is much faster (this is similar to the caching of high-precision numerical values of expressions).

In[63]:= **CSC // Timing**

Out[63]= {13.54 Second, 0}

For FullSimplify to try all of its knowledge, the expression must have a certain minimal complexity. The following Hermite function returns unchanged.

In[64]:= **FullSimplify[HermiteH[v, z], ComplexityFunction ->**
                         (Count[#, _HermiteH, Infinity]&)]

Out[64]= HermiteH[v, z]

Wrapping an additional F around in another result.

In[65]:= **FullSimplify[F[HermiteH[v, z]], ComplexityFunction ->**
                         (Count[#, _HermiteH, Infinity]&)]

Out[65]= $F\left[\left(2^v \sqrt{\pi} \left(-2 z \text{ Gamma}\left[\frac{1-v}{2}\right] \text{ Hypergeometric1F1}\left[\frac{1-v}{2}, \frac{3}{2}, z^2\right] + \right.\right.\right.$
$\left.\left.\left. \text{Gamma}\left[-\frac{v}{2}\right] \text{ Hypergeometric1F1}\left[-\frac{v}{2}, \frac{1}{2}, z^2\right]\right)\right) \middle/ \left(\text{Gamma}\left[\frac{1-v}{2}\right] \text{ Gamma}\left[-\frac{v}{2}\right]\right)\right]$

By counting already-encountered functions, we can transform HermiteH[v, z] into a variety of equivalent forms. Here is one more example.

In[66]:= **FullSimplify[%, ComplexityFunction ->**
                     (Count[#, _HermiteH | _Hypergeometric1F1, Infinity]&)]

Out[66]= $F\left[\frac{1}{2 \text{ Gamma}[-v]}\left(-2 z \text{ Gamma}\left[\frac{1-v}{2}\right] \text{ HypergeometricPFQ}\left[\left\{\frac{1}{2} - \frac{v}{2}\right\}, \left\{\frac{3}{2}\right\}, z^2\right] + \right.\right.$
$\left.\left. \text{Gamma}\left[-\frac{v}{2}\right] \text{ HypergeometricPFQ}\left[\left\{-\frac{v}{2}\right\}, \left\{\frac{1}{2}\right\}, z^2\right]\right)\right]$

FullSimplify and FunctionExpand use transformations that are correct for generic values of the occurring variables. This means, the simplified expression can have different values at lower-dimensional subsets of some variable values. Here is an example: the function NCF[n, z] is a polynomial for integer n. For generic complex values of n, NCF[n, z] does not have a finite value.

In[67]:= **NCF[n_, z_] = GegenbauerC[2n, -n - 1, z]**

Out[67]= GegenbauerC[2 n, -1 - n, z]

In[68]:= **{NCF[n, z] // FullSimplify, NCF[n, z] // FunctionExpand}**

            FullSimplify::infd :
                 Expression GegenbauerC[2 n, -1 - n, z] simplified to ComplexInfinity. More…

Out[68]= {ComplexInfinity, ComplexInfinity}

As an orthogonal polynomial, the associated Gegenbauer polynomials are genuine polynomials for (positive) integers n.

In[69]:= **Table[NCF[n, z], {n, -3, 3}]**

Out[69]= $\{-2 (-1 + 6 z^2), 1 - 4 z^2, 0, 1, 2 + 4 z^2, 3 + 12 z^2, 4 + 24 z^2\}$

A numerical calculation confirms the problem for noninteger n.

In[70]:= **{NCF[1, Pi], NCF[1 + 10^-20 I, Pi]} // N[#, 50]& // N**

Out[70]= {41.4784, ComplexInfinity}

3.2 Gamma, Beta, and Polygamma Functions

The Gamma function is probably the most important special function of mathematical physics. (For a detailed treatment, see [809], [951], [1056], [61], and [172]). It is the analytical continuation in the complex plane of the standard factorial function (which was originally defined only for positive integers). Many special functions are built of series that contain terms with Gamma functions. In contrast to most other functions, it does not satisfy a second-order ordinary differential equation with polynomial coefficients [92], [987], [1083], but instead satisfies the simple functional equation $z\,\Gamma(z) = \Gamma(z + 1)$. It can be represented, for instance, in either of the following two ways:

$$\Gamma(z) \;=\; \int_0^\infty t^{z-1}\,e^{-t}\,dt \quad \mathrm{Re}(z) > 0$$

$$\frac{1}{\Gamma(z)} \;=\; z\,e^{\gamma z}\prod_{n=1}^\infty \Big(1 + \frac{z}{n}\Big)e^{-z/n}.$$

In *Mathematica*, the Gamma function is defined in the following way.

Gamma [a, z_0, z_1]

    represents the (generalized) incomplete Gamma function $\Gamma(a, z_0, z_1) = \int_{z_0}^{z_1} t^{a-1}\,e^{-t}\,dt$.

Gamma [a, z] = Gamma [a, z, ∞]

Gamma [a] = Gamma [a, 0]

Mathematica recognizes the defining integral.

In[1]:= **Integrate[t^(a - 1) Exp[-t], {t, z0, z1}] // Together**

Out[1]= $-(z0 - z1)$ If$\Big[\mathrm{Re}\Big[\dfrac{z0}{z0 - z1}\Big] \geq 1 \;||\; \mathrm{Re}\Big[\dfrac{z0}{-z0 + z1}\Big] \geq 0 \;||\; \mathrm{Im}\Big[\dfrac{z0}{-z0 + z1}\Big] \neq 0,$

$\dfrac{-\text{Gamma}[a, z0] + \text{Gamma}[a, z1]}{z0 - z1}$, Integrate$\Big[e^{(-1+t)\,z0-t\,z1}\,(z0 - t\,z0 + t\,z1)^{-1+a},$

$\{t, 0, 1\}$, Assumptions $\rightarrow \mathrm{Im}\Big[\dfrac{z0}{-z0 + z1}\Big] == 0$ && $\mathrm{Re}\Big[\dfrac{z0}{z0 - z1}\Big] < 1$ && $\mathrm{Re}\Big[\dfrac{z0}{-z0 + z1}\Big] < 0\Big]\Big]$

For certain special choices of the arguments, we get exact results. This includes the integers n in which we have $n! = \Gamma(n + 1)$. (On the other hand, this relation is used to define Factorial for noninteger arguments).

In[2]:= **Table[{i, Gamma[i + 1]}, {i, 12}]**

Out[2]= {{1, 1}, {2, 2}, {3, 6}, {4, 24}, {5, 120}, {6, 720}, {7, 5040},
    {8, 40320}, {9, 362880}, {10, 3628800}, {11, 39916800}, {12, 479001600}}

Certain fractions as arguments also give simplified results.

In[3]:= **Join[Table[{i/2, Gamma[i/2]}, {i, 1, 9, 2}],**
    Table[{i/3, Gamma[i/3]}, {i, 1, 2}],
    Table[{i/4, Gamma[i/4]}, {i, 1, 3, 2}]]

Out[3]= $\left\{\left\{\frac{1}{2}, \sqrt{\pi}\right\}, \left\{\frac{3}{2}, \frac{\sqrt{\pi}}{2}\right\}, \left\{\frac{5}{2}, \frac{3\sqrt{\pi}}{4}\right\}, \left\{\frac{7}{2}, \frac{15\sqrt{\pi}}{8}\right\}, \left\{\frac{9}{2}, \frac{105\sqrt{\pi}}{16}\right\},\right.$

$\left.\left\{\frac{1}{3}, \text{Gamma}\left[\frac{1}{3}\right]\right\}, \left\{\frac{2}{3}, \text{Gamma}\left[\frac{2}{3}\right]\right\}, \left\{\frac{1}{4}, \text{Gamma}\left[\frac{1}{4}\right]\right\}, \left\{\frac{3}{4}, \text{Gamma}\left[\frac{3}{4}\right]\right\}\right\}$

The Gamma function has a pole of order one in the complex plane at each of the negative integers and at zero.

In[4]:= **Plot3D[{Abs[Gamma[x + I y]], Hue[Arg[Gamma[x + I y]]/Pi]},**
        {x, -2, 3.2}, {y, -1, 1.2}, PlotPoints -> 125, Mesh -> False]

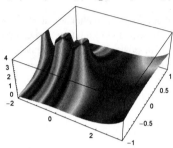

Because of the pole at $z = 0$, $\Gamma(z)$ does not have a power series expansion around 0. But $1/\Gamma(z)$ does have a power series expansion around 0.

In[5]:= **Series[1/Gamma[z], {z, 0, 5}] // Simplify**

Out[5]= $z + \text{EulerGamma } z^2 + \frac{1}{12} \left(6 \text{ EulerGamma}^2 - \pi^2\right) z^3 +$

$\frac{1}{12} \left(2 \text{ EulerGamma}^3 - \text{EulerGamma } \pi^2 - 2 \text{ PolyGamma}[2, 1]\right) z^4 +$

$\frac{\left(60 \text{ EulerGamma}^4 - 60 \text{ EulerGamma}^2 \pi^2 + \pi^4 - 240 \text{ EulerGamma PolyGamma}[2, 1]\right) z^5}{1440} + O[z]^6$

(We will discuss the function PolyGamma from the last result in a moment.) But $\Gamma(z)$ has a Laurent expansion around 0.

In[6]:= **Series[Gamma[z], {z, 0, 3}]**

Out[6]= $\frac{1}{z} - \text{EulerGamma} + \frac{1}{2} \left(\text{EulerGamma}^2 + \frac{\pi^2}{6}\right) z +$

$\frac{1}{6} \left(-\text{EulerGamma}^3 - \frac{\text{EulerGamma } \pi^2}{2} + \text{PolyGamma}[2, 1]\right) z^2 +$

$\frac{1}{24} \left(\text{EulerGamma}^4 + \text{EulerGamma}^2 \pi^2 + \frac{3\pi^4}{20} - 4 \text{ EulerGamma PolyGamma}[2, 1]\right) z^3 + O[z]^4$

Mathematica can also calculate a series expansion around ∞.

In[7]:= **Series[Gamma[z], {z, Infinity, 3}]**

Out[7]= $e^{-z} z^z \left(\sqrt{2\pi} \sqrt{\frac{1}{z}} + \frac{1}{6} \sqrt{\frac{\pi}{2}} \left(\frac{1}{z}\right)^{3/2} + \frac{1}{144} \sqrt{\frac{\pi}{2}} \left(\frac{1}{z}\right)^{5/2} + O\left[\frac{1}{z}\right]^{7/2}\right)$

To get an impression of the behavior of the Gamma function Γ at infinity, we can display a contour plot of $\Gamma(1/z)$. This accumulates all of the poles around the origin.

In[8]:= **With[{pp = 499, n = 50},**
    Module[{data, dataReIm, cls, Γ = Gamma},
    (* the function values *)
    data = Table[Γ[1./(x + I y)], {y, -1/2, 1/2, 1/pp}, {x, -1/2, 1/2, 1/pp}];
    (* show real and imaginary part *)
    Show[GraphicsArray[

```
((* make machine numbers *)
 dataReIm = Map[If[Abs[#] > $MaxMachineNumber,
               $MaxMachineNumber/2 Sign[#], #]&, #[data], {2}];
 (* values of contour lines for homogeneous spacing *)
 cls = #[[Round[pp^2/n/2]]]& /@
       Partition[Sort[Flatten[dataReIm]], Round[pp^2/n]];
 (* the contour plot *)
 ListContourPlot[dataReIm, Contours -> cls, ContourLines -> False,
                 DisplayFunction -> Identity, PlotLabel -> #,
                 ColorFunction -> (Hue[Random[]]&),
                 MeshRange -> {{-1/2, 1/2}, {-1/2, 1/2}}])& /@
                 (* show real and imaginary part *) {Re, Im}]]]]
```

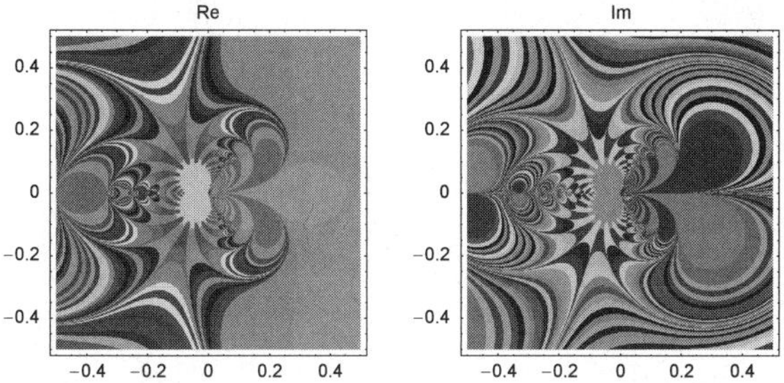

Complicated expressions involving several Gamma functions can often be simplified using `FullSimplify`. We now give a simple example.

In[9]:= `Gamma[z] Gamma[1 - z] // FullSimplify`

Out[9]= $\pi \, \mathrm{Csc}[\pi z]$

The next input calculates and simplifies the product $\prod_{k=1}^{\infty} (1 - (z/k)^{11})$ for $z > 0$.

In[10]:= `Product[1 - (z/k)^11, {k, Infinity}] // FullSimplify[#, z > 0]&`

Out[10]= $1 / (z^{10} \, \mathrm{Gamma}[1 - z] \, \mathrm{Gamma}[(-1)^{1/11} z] \, \mathrm{Gamma}[-(-1)^{2/11} z]$
$\mathrm{Gamma}[(-1)^{3/11} z] \, \mathrm{Gamma}[-(-1)^{4/11} z] \, \mathrm{Gamma}[(-1)^{5/11} z] \, \mathrm{Gamma}[-(-1)^{6/11} z]$
$\mathrm{Gamma}[(-1)^{7/11} z] \, \mathrm{Gamma}[-(-1)^{8/11} z] \, \mathrm{Gamma}[(-1)^{9/11} z] \, \mathrm{Gamma}[-(-1)^{10/11} z])$

If we want to simplify only expressions that contain Gamma functions, we can use the more specialized function `Developer`GammaSimplify`.

In[11]:= `Gamma[z] Gamma[1 - z] // Developer`GammaSimplify`

Out[11]= $\pi \, \mathrm{Csc}[\pi z]$

The incomplete Gamma function $\Gamma(a, z)$ has a branch cut along the negative real axis with respect to z. Its analytical continuation is given by the formula $\Gamma(a, \exp(2\,k\,\pi\,i)\,z) = \exp(2\,k\,\pi\,i\,a)\,\Gamma(a, z) + (1 - \exp(2\,k\,\pi\,i\,a))\,\Gamma(a)$ [1013]. The following pictures show a part of the Riemann surface of the incomplete Gamma function $\Gamma((3 + i)/10, z)$.

In[12]:= `With[{α = 0.3 + 0.1 I, ε = 10^-12},`
`  Show[GraphicsArray[`
`  Graphics3D[{EdgeForm[Thickness[0.002]],`
`    SurfaceColor[Hue[0.09], Hue[0.18], 2.3],`
`    Table[(* split whole graphics into pieces *)`
`    Last /@ Partition[Cases[`
`    ParametricPlot3D[(* the sheets *)`
`            {r Cos[φ], r Sin[φ], #[Exp[2 k Pi I α] Gamma[α, r Exp[I φ]] +`

```
                    (1 - Exp[2 k Pi I α]) Gamma[α]]},
                  {r, 0, 2}, {φ, -Pi + ε, Pi - ε},
          PlotPoints -> {30, 40}, DisplayFunction -> Identity],
        _Polygon, Infinity], 2], {k, -2, 2}]},
            BoxRatios -> {1, 1, 2.5}, PlotRange -> All]& /@
           (* show real and imaginary part *) {Re, Im}]]]
```

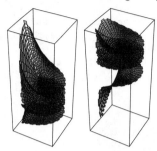

Using the incomplete Gamma function, we can find the complete solution to the following differential equation.

In[13]:= `DSolve[x y''[x] - (x + ν) y'[x] + ν y[x] == 0, y[x], x]`

Out[13]= $\{\{y[x] \rightarrow e^x C[1] - e^x C[2] \text{Gamma}[1 + ν, x]\}\}$

The incomplete Gamma function is also needed to express the following infinite sum of complete Gamma functions [833], [832].

In[14]:= `K1[z_] = Sum[Gamma[z - k], {k, 0, Infinity}]`

Out[14]= $\dfrac{(-1)^z \pi \text{Csc}[\pi z]}{e} - \dfrac{(-1)^z \pi \text{Csc}[\pi z] \text{Gamma}[-z, -1]}{e \text{Gamma}[-z]}$

Rewriting the last solution slightly and using positive integers for $ν$ reveals the interesting fact that the solutions are $\exp(x)$ and its Taylor series of order $ν - 1$ [949].

In[15]:= `dSol[n_, x_] := Subscript[α, 1] Exp[x] + Subscript[α, 2]*`
         `                Expand[FunctionExpand[Exp[x] Gamma[n, x]/(n - 1)!]];`

In[16]:= `Table[dSol[n, x], {n, 8}] // TableForm // TraditionalForm`

Out[16]//TraditionalForm=

$e^x \alpha_1 + \alpha_2$

$e^x \alpha_1 + (x + 1)\alpha_2$

$e^x \alpha_1 + \left(\frac{x^2}{2} + x + 1\right)\alpha_2$

$e^x \alpha_1 + \left(\frac{x^3}{6} + \frac{x^2}{2} + x + 1\right)\alpha_2$

$e^x \alpha_1 + \left(\frac{x^4}{24} + \frac{x^3}{6} + \frac{x^2}{2} + x + 1\right)\alpha_2$

$e^x \alpha_1 + \left(\frac{x^5}{120} + \frac{x^4}{24} + \frac{x^3}{6} + \frac{x^2}{2} + x + 1\right)\alpha_2$

$e^x \alpha_1 + \left(\frac{x^6}{720} + \frac{x^5}{120} + \frac{x^4}{24} + \frac{x^3}{6} + \frac{x^2}{2} + x + 1\right)\alpha_2$

$e^x \alpha_1 + \left(\frac{x^7}{5040} + \frac{x^6}{720} + \frac{x^5}{120} + \frac{x^4}{24} + \frac{x^3}{6} + \frac{x^2}{2} + x + 1\right)\alpha_2$

In dealing with the Gamma function, it is very useful to introduce the so-called Pochhammer symbol $(a)_n$. The Pochhammer symbol is particularly important in dealing with hypergeometric, confluent hypergeometric, and generalized hypergeometric functions (see below).

Pochhammer[a, n]

> represents the Pochhammer symbol $(a)_n$, defined as $(a)_n = a(a+1)(a+2) \cdots (a+n-1)$.

The shorter and more general definition (because it is applicable for arbitrary complex n) is $(a)_n = \Gamma(a+n)/\Gamma(a)$. When the second argument is an integer, we get an explicit value.

In[17]:= **Apply[Pochhammer, {{1, 6}, {5, 8}, {7/5 I, 8}, {5/6, 9/5}}, {1}]**

Out[17]= $\left\{720, 19958400, -\dfrac{788766524}{390625} - \dfrac{1463438704\, i}{78125}, \dfrac{Gamma\left[\frac{79}{30}\right]}{Gamma\left[\frac{5}{6}\right]}\right\}$

When the second argument is a positive integer, we have the following interesting Verde–Star identity [1266]:

$$\frac{(x)_{m+1} - (y)_{m+1}}{x - y} = \sum_{k=0}^{m} (x)_k \, (y + k + 1)_{m-k}.$$

We now examine a few special cases.

In[18]:= **VerdeStarIdentity[m_, x_, y_] := Expand[**
  (Pochhammer[x, m + 1] - Pochhammer[y, m + 1]) - (x - y)*
  Sum[Pochhammer[x, k] Pochhammer[y + k + 1, m - k], {k, 0, m}]];

  Table[VerdeStarIdentity[i, x, y], {i, 0, 8}]
Out[19]= {0, 0, 0, 0, 0, 0, 0, 0, 0}

For integer a, n, with a, $a + n < 1$ the ratio $\Gamma(a+n)/\Gamma(a)$ is not immediately defined because of the singularities of the Gamma function. (This situation is similar to the behavior of Binomial [471]). In these cases, the Pochhammer symbol is given as $(a)_n = (-1)^n (-a)!/(-(a+n))!$.

In[20]:= **Table[Pochhammer[a, n] == (-1)^n (-a)!/(-(a + n))!,**
  {a, 0, -5, -1}, {n, -5, a}] // Flatten // Union
Out[20]= {True}

Here is a picture of $(a)_n$ in the a,n-plane.

In[21]:= **With[{ε = 10^-5},**
  Plot3D[Pochhammer[a, n], {a, -5 - ε/2, 5 + ε/2}, {n, -5 - ε, 5 + ε},
    PlotPoints -> 200, PlotRange -> {-2, 2},
    AxesLabel -> {"a", "n", None}, ClipFill -> None,
    Mesh -> False, ColorFunction -> Hue]]

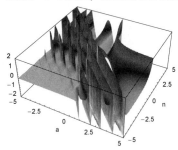

Gamma functions often arise in certain combinations. One particular arrangement leads to the introduction of Beta functions.

Beta[z_0, z_1, a, b]

   represents the (generalized) incomplete Beta function $B(z_0, z_1, a, b) = \int_{z_0}^{z_1} t^{a-1}(1-t)^{b-1}\,dt$.

Beta[z_1, a, b] = Beta[0, z_1, a, b]

Beta[a, b] = Beta[1, a, b]

Although *Mathematica* is able to compute this definite integral, it does not print out the result in the form of a Beta function.

In[22]:= **Integrate[t^(a - 1) (1 - t)^(b - 1), {t, z0, z1},**
                GenerateConditions -> False]

Out[22]= $-$Beta[z0, a, 1 + b] $-$ Beta[z0, 1 + a, b] + Beta[z1, a, 1 + b] + Beta[z1, 1 + a, b]

Using FullSimplify we can convert the last expression into a difference of three-argument Beta functions.

In[23]:= **FullSimplify[%]**

Out[23]= $-$Beta[z0, a, 1 + b] $-$ Beta[z0, 1 + a, b] + Beta[z1, a, 1 + b] + Beta[z1, 1 + a, b]

Note the order of the parameters and the limits of integration in comparison with the Gamma function. For the complete beta function, we have

$$B(a, b) = \frac{\Gamma(a)\,\Gamma(b)}{\Gamma(a+b)}.$$

The incomplete Beta function allows us to map the unit circle onto a regular *n*-gon. The map is given by $z \to \int_0^z (1 - \xi^n)^{-2/n}\,d\xi$ [405], [937], [344], [966], [176]. Here, we visualize this map for $n = 3, 4, 5, 6$. (For averaging over *n*-gons, see [91].)

In[24]:= **f[n_, z_] = z/n Beta[z^n, 1/n, 1 - 2/n]/((z^n)^(1/n));**

In[25]:= **With[{ε = 10^-8}, unitCirclePoints = Table[r Exp[I φ],**
                {r, ε, 1 - ε, (1 - 2ε)/30}, {φ, 0, 2Pi, 2Pi/60}] // N];

In[26]:= **Show[GraphicsArray[**
            Table[Graphics[(* radial and azimuthal lines *)
                {Line /@ #, Line /@ Transpose[#]}&[
                Map[{Re[#], Im[#]}&[f[n, #]]&, unitCirclePoints, {-1}]],
                AspectRatio -> Automatic], {n, 3, 6}]]]

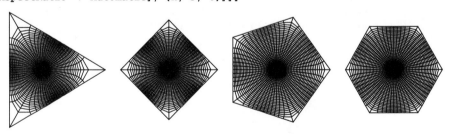

We conclude this section with a short discussion of the Polygamma functions.

PolyGamma[*n*, *z*]

    represents the Polygamma function $\Psi^{(n)}(z)$. It is defined via the Digamma function $\Psi(z) = \frac{\Gamma'(z)}{\Gamma(z)}$.

PolyGamma[*n*, *z*] $= \frac{d^n \Psi(z)}{dz^n}$, $n \in \mathbb{N}$

Polygamma functions frequently arise in closed sums of rational expressions. Here is an example.

In[27]:= **Sum[1/(n + a)^3, {n, 0, m}]**

Out[27]= $\frac{1}{2}$ (-PolyGamma[2, a] + PolyGamma[2, 1 + a + m])

Here is a slightly more complicated, infinite sum.

In[28]:= **Sum[1/(n^4 + b n + 1), {n, 0, m}]**

Out[28]= $-$RootSum$\left[1 + b\, \#1 + \#1^4\ \&,\ \dfrac{\text{PolyGamma}[0, -\#1]}{b + 4\, \#1^3}\ \&\right]\ +$

    RootSum$\left[2 + b + 4\,m + b\,m + 6\,m^2 + 4\,m^3 + m^4 + 4\,\#1 + b\,\#1 + 12\,m\,\#1\ +\right.$

        $12\,m^2\,\#1 + 4\,m^3\,\#1 + 6\,\#1^2 + 12\,m\,\#1^2 + 6\,m^2\,\#1^2 + 4\,\#1^3 + 4\,m\,\#1^3 + \#1^4\ \&,$

        $\left.\dfrac{\text{PolyGamma}[0, -\#1]}{4 + b + 12\,m + 12\,m^2 + 4\,m^3 + 12\,\#1 + 24\,m\,\#1 + 12\,m^2\,\#1 + 12\,\#1^2 + 12\,m\,\#1^2 + 4\,\#1^3}\ \&\right]$

It is possible to compute exact values for Polygamma functions at integers and half integers.

In[29]:= **{PolyGamma[0, 1], PolyGamma[3, 1/2]}**

Out[29]= {-EulerGamma, π^4}

Interestingly, $\Psi(z)$ can be expressed in elementary functions for all rational z.

In[30]:= **FunctionExpand[PolyGamma[0, 2/15]] // Simplify**

Out[30]= $\dfrac{1}{4}\left(-4\,\text{EulerGamma} - \dfrac{2\left(1 + \sqrt{5} + \sqrt{30 - 6\sqrt{5}}\,\right)\pi}{\sqrt{3} + \sqrt{15} - \sqrt{10 - 2\sqrt{5}}} - 6\,\text{Log}[3]\ -\right.$

    $4\,\text{Log}[5] + \text{Log}[64] - \text{Log}[5 - \sqrt{5}\,] - \sqrt{5}\,\text{Log}[5 - \sqrt{5}\,] - \text{Log}[5 + \sqrt{5}\,]\ +$

    $\sqrt{5}\,\text{Log}[5 + \sqrt{5}\,] + \left(1 + \sqrt{5} - \sqrt{30 - 6\sqrt{5}}\,\right)\text{Log}\left[\dfrac{1}{8}\left(\sqrt{3} + \sqrt{15} - \sqrt{10 - 2\sqrt{5}}\,\right)\right]\ +$

    $\left(1 + \sqrt{5} + \sqrt{30 - 6\sqrt{5}}\,\right)\text{Log}\left[\dfrac{1}{8}\left(\sqrt{3} + \sqrt{15} + \sqrt{10 - 2\sqrt{5}}\,\right)\right]\ -$

    $\left(-1 + \sqrt{5} - \sqrt{6\,(5 + \sqrt{5}\,)}\,\right)\text{Log}\left[\dfrac{1}{8}\left(\sqrt{3} - \sqrt{15} + \sqrt{2\,(5 + \sqrt{5}\,)}\,\right)\right]\ -$

    $\left.\left(-1 + \sqrt{5} + \sqrt{6\,(5 + \sqrt{5}\,)}\,\right)\text{Log}\left[\dfrac{1}{8}\left(\sqrt{3}\,(-1 + \sqrt{5}\,) + \sqrt{2\,(5 + \sqrt{5}\,)}\,\right)\right]\right)$

We now present a graph of the first few Polygamma functions on the real axis. The right graphics shows the lines of Re($\Psi^{(n)}(z)$) = 0 over the complex z-plane.

In[31]:= **Show[GraphicsArray[**
    Block[{$DisplayFunction = Identity},
      {(* plots along the real axis *)
      Plot[Evaluate[Table[PolyGamma[i, z], {i, 0, 10}]], {z, 1, 3.5},
        PlotStyle -> Table[Hue[i 0.8/10], {i, 0, 10}]],
      (* contour plots over the complex plane *)
      Show[Table[
      ContourPlot[Im @ PolyGamma[i, x + I y], {x, -4, 2}, {y, -3, 3},
        PlotPoints -> 60, Contours -> {0}, ContourShading -> False,

```
ContourStyle -> {Hue[i 0.8/10], Thickness[0.001]}],
   {i, 0, 10}]]}]]]
```

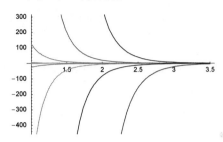

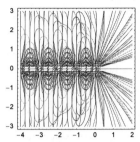

Here, we should note that `FullSimplify` also knows many identities of polygamma functions.

In[32]:= **PolyGamma[2, z + 1] + (-1)^(2 + 1) PolyGamma[2, z]**

Out[32]= $-$PolyGamma[2, z] + PolyGamma[2, 1 + z]

In[33]:= **% // FullSimplify**

Out[33]= $\dfrac{2}{z^3}$

Here is a more complicated example. This time we use the more specialized simplifier `Developer`PolyGam:
maSimplify` directly.

In[34]:= **D[Gamma[x + 1]/Gamma[x + 1/2], {x, 4}] /. x -> 0**

Out[34]= $-\dfrac{\pi^{7/2}}{4} - 3\,\pi^{3/2}\,\text{PolyGamma}\Big[0, \tfrac{1}{2}\Big]^2 + \dfrac{\text{PolyGamma}[0, \tfrac{1}{2}]^4}{\sqrt{\pi}} +$

$6\left(\text{EulerGamma}^2 + \dfrac{\pi^2}{6}\right)\left(-\dfrac{\pi^{3/2}}{2} + \dfrac{\text{PolyGamma}[0, \tfrac{1}{2}]^2}{\sqrt{\pi}}\right) +$

$\dfrac{4\,\text{PolyGamma}[0, \tfrac{1}{2}]\,\text{PolyGamma}[2, \tfrac{1}{2}]}{\sqrt{\pi}} -$

$4\,\text{EulerGamma}\left(\dfrac{3}{2}\,\pi^{3/2}\,\text{PolyGamma}\Big[0, \tfrac{1}{2}\Big] - \dfrac{\text{PolyGamma}[0, \tfrac{1}{2}]^3}{\sqrt{\pi}} - \dfrac{\text{PolyGamma}[2, \tfrac{1}{2}]}{\sqrt{\pi}}\right) -$

$\dfrac{4\,\text{PolyGamma}[0, \tfrac{1}{2}]\left(-\text{EulerGamma}^3 - \frac{\text{EulerGamma}\,\pi^2}{2} + \text{PolyGamma}[2, 1]\right)}{\sqrt{\pi}} +$

$\dfrac{\text{EulerGamma}^4 + \text{EulerGamma}^2\,\pi^2 + \frac{3\pi^4}{20} - 4\,\text{EulerGamma}\,\text{PolyGamma}[2, 1]}{\sqrt{\pi}}$

In[35]:= **Developer`PolyGammaSimplify[%] // Together**

Out[35]= $\dfrac{-3\,\pi^4 - 10\,\pi^2\,\text{Log}[4]^2 + 5\,\text{Log}[4]^4 + 240\,\text{Log}[4]\,\text{Zeta}[3]}{5\sqrt{\pi}}$

3.3 Error Functions and Fresnel Integrals

The error function is defined as the indefinite integral of the Gaussian function $\exp(-x^2)$ (up to a constant factor):

$$\mathrm{erf}(z) = \frac{2}{\sqrt{\pi}} \int_0^z \exp(-t^2)\, dt.$$

In *Mathematica*, the error function is written as follows.

`Erf[`z_0`, `z_1`]`

    represents the (generalized) error function $\mathrm{erf}(z_0, z_1) = \frac{2}{\sqrt{\pi}} \int_{z_0}^{z_1} \exp(-t^2)\, dt.$

`Erf[`z`] = Erf[0, `z`]`

Here is a graph of the error function.

```
In[1]:= Show[GraphicsArray[
    Block[{$DisplayFunction = Identity},
        {(* plot of the one-argument Erf *) Plot[Erf[z], {z, -2, 2}],
        (* 3D plot of the two-argument Erf *)
        Plot3D[Erf[z0, z1], {z0, -2, 2}, {z1, -2, 2},
            PlotPoints -> 60, Mesh -> False]}]]]
```

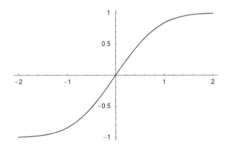 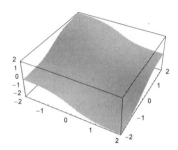

Here is a simple view of the real and imaginary parts, as well of the absolute value and the phase of erf(z) over the complex z-plane. (For a more detailed structure becomes visible for more evenly spaced contour lines.)

```
In[3]:= data = With[{pp = 61},
                Table[Erf[N[x + I y]], {y, -3, 3, 6/pp}, {x, -3, 3, 6/pp}]];
```

```
In[4]:= Show[GraphicsArray[Function[f,
    ListContourPlot[f[data], Contours -> (#[[60]]& /@
                Partition[Sort[Flatten[f[data]]], 120]),
                FrameTicks -> False, DisplayFunction -> Identity,
                ColorFunction -> (Hue[3.4 #]&),
                ContourStyle -> {Thickness[0.002]}]] /@ {Re, Im, Abs, Arg}]]
```

We get the error function by integrating $\exp(-t^2)$.

In[5]:= **Integrate[Exp[-t^2], {t, 0, z}]**

Out[5]= $\frac{1}{2}$ $\sqrt{\pi}$ Erf[z]

Mathematica is also able to find the integral of the error function [574].

In[6]:= **Integrate[Erf[t], {t, 0, z}]**

Out[6]= $\frac{-1 + e^{-z^2}}{\sqrt{\pi}}$ + z Erf[z]

It can also find higher integrals.

In[7]:= **Integrate[%, {t, 0, z}]**

Out[7]= $-\frac{z}{\sqrt{\pi}}$ + $\frac{e^{-z^2} z}{\sqrt{\pi}}$ + z² Erf[z]

In[8]:= **Nest[Integrate[#, {z, 0, z}]&, Exp[-z^2], 6]**

Out[8]= $\frac{1}{960}$ e$^{-z^2}$ $\left(8 + 18 z^2 + 4 z^4 + e^{z^2} \left(-4 \left(2 + 5 z^2 \left(2 + z^2\right)\right) + \sqrt{\pi} z \left(15 + 4 z^2 \left(5 + z^2\right)\right) \text{Erf}[z]\right)\right)$

Using the Rodriguez formula from Chapter 2 for negative integer n, we can write these iterated integrals in a compact form $H_n(z) = (-1)^n \exp(z^2) d^n \exp(-z^2)/dz^n$.

The following function is closely related to the error function.

Erfc[z]

represents the complementary error function. It satisfies Erfc[z] = 1 - Erf[z].

One interesting application of the function erfc is the following: The function

$$\psi_k(x, t) = e^{i(k x - k^2 t/2)} \text{erfc}\!\left(e^{-\frac{i\pi}{4}} (x - k t)/\sqrt{2 t}\right)$$

is a solution of the time-dependent Schrödinger equation $i \partial \psi(x, t)/\partial t = -1/2 \partial^2 \psi(x, t)/\partial x^2$ with initial conditions $\psi_k(x, 0) = \theta(-x) e^{i k x}$ [912], [963], [1260], [493], [838], [915], [913], [916], [1076], [492], [378], [110], [642], [514], [246]. Here, we check that the function fulfills the Schrödinger equation.

In[9]:= **M[x_, t_, k_] = Exp[I(k x - k^2 t/2)]*
                     Erfc[Exp[-I Pi/4] (x - k t)/Sqrt[2 t]];**

In[10]:= (* quick check *)
        **freeSchrödingerOperator[x_, t_] := (I D[#, t] + 1/2 D[#, x, x])&
        freeSchrödingerOperator[x, t] @ M[x, t, k] // Simplify**

Out[12]= 0

The next picture shows the time dependent absolute value of the wavefunction. The flow from the left half-space into the right one is clearly visible. For $t \to 0$, we see very high-frequency structures near $x = 0$. They arise from the Fourier spectrum of the initial conditions.

In[13]:= **Show[GraphicsArray[
        Block[{ε = 10^-3, $DisplayFunction = Identity},
          {Plot3D[Abs[M[x, t, 1]], {x, -6, 6}, {t, ε, 8}, PlotPoints -> 60],
          ContourPlot[Abs[M[x, t, 1]], {x, -1, 0}, {t, ε, 0.1},
                PlotPoints -> 120, ColorFunction -> (Hue[3 #]&)]}]]]**

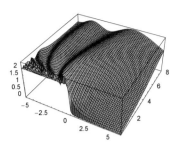 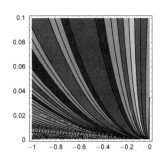

The function erfc(z) can be solved to use another interesting quantum mechanical problem. Consider the 1D free Schrödinger equation in the half space $x > 0$. In addition we assume the presence of a constant weakly absorbing potential $1 - i\, V_i$ [379], [915].

```
In[14]:= SchrödingerEquationWithAbsorption[Ψ_, {x_, t_}, Vi_] :=
             I D[Ψ, t] - (-1/2 D[Ψ, {x, 2}] + (1 - I Vi) Ψ)
```

The boundary value problem $\psi(0, t) = \exp(i\,\omega_0\, t)$, $\psi(x, 0) = 0$ (meaning a driving force on the left domain boundary and initially vanishing wave function everywhere), can then be solved in closed form through the following expression.

```
In[15]:= ψ[{x_, t_}, Vi_, ω0_] = With[{w = Function[z, Exp[-z^2] Erfc[-I z]]},
             Exp[-I (1 - I Vi) t] Exp[I x^2/(2 t)]/2 *
             (w[-(1 + I)/Sqrt[2] Sqrt[t] (+Sqrt[ω0 - 1 + I Vi] - x/(Sqrt[2] t))] +
              w[-(1 + I)/Sqrt[2] Sqrt[t] (-Sqrt[ω0 - 1 + I Vi] - x/(Sqrt[2] t))])];
```

```
In[16]:= (* correctness check of the result *)
         {SchrödingerEquationWithAbsorption[ψ[{x, t}, Vi, ω0], {x, t}, V1],
          ψ[{0, t}, V1, ω0]} // FullSimplify
```

$$
\text{Out[17]= } \left\{ \frac{1}{2}\, i\, e^{-i\,\left(t\,\omega 0 + \sqrt{2}\, x\, \sqrt{-1 + i\, Vi + \omega 0}\right)}\, (V1 - Vi) \right.
$$

$$
\left(e^{2\,i\,\sqrt{2}\, x\, \sqrt{-1 + i\, Vi + \omega 0}}\ \text{Erfc}\left[\frac{\left(\frac{1}{2} - \frac{i}{2}\right)\left(x - \sqrt{2}\, t\, \sqrt{-1 + i\, Vi + \omega 0}\right)}{\sqrt{t}} \right] + \right.
$$

$$
\left. \text{Erfc}\left[\frac{\left(\frac{1}{2} - \frac{i}{2}\right)\left(x + \sqrt{2}\, t\, \sqrt{-1 + i\, Vi + \omega 0}\right)}{\sqrt{t}} \right] \right),\ \left. e^{-i\, t\, \omega 0} \right\}
$$

The next graphic shows the time-development of $|\psi(x, t)|^2$; we use a logarithmic scale.

```
In[18]:= Plot3D[Evaluate[Log[Abs[ψ[{x, t}, 10^-4, 99/100]]^2]],
             {x, 0, 100}, {t, 10^-3, 10000}, PlotPoints -> 180, Mesh -> False]
```

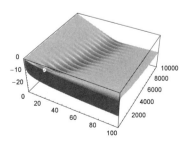

For a purely imaginary argument, the error function has a purely imaginary value. This suggests introducing the following function.

```
Erfi[z]
```

represents the imaginary complementary error function. It satisfies
`Erfi[z] =-I Erf[Iz]`.

Many initial value problems of the time-dependent Schrödinger equation of the harmonic oscillator

$$i \frac{\partial \psi(x, t)}{\partial t} = -\frac{1}{2} \frac{\partial^2 \psi(x, t)}{\partial x^2} + \frac{\omega^2}{2} x^2 \psi(x, t)$$

can be solved using the Erfi function. Using the well-known propagator

$$G(x, y; \tau) = \sqrt{\frac{\omega}{\sin(\omega\tau)}} \ \exp\left(\frac{1}{2} i\omega\left((x^2 + y^2)\cot(\omega\tau) - \frac{2xy}{\sin(\omega\tau)}\right)\right)$$

[693], [97], [931], [547], [1330], we get the following solution for the initial condition $\psi(x, 0) = \theta(x - \pi)\,\theta(\pi - x)\cos(x)$ after carrying out the corresponding integral.

In[19]:= `GF[ω_][x_, y_, τ_] :=`
`    Sqrt[ω/Sin[ω τ]] Exp[I ω/2((x^2 + y^2) Cot[ω τ] - 2 x y/Sin[ω τ])]`

In[20]:= `(* quick check *)`
`hoSchrödingerOperator[ω_][x_, t_] :=`
`                    (I D[#, t] + 1/2 D[#, x, x] - ω^2/2 x^2 #)&;`

`(* apply Schrödinger operator *)`
`hoSchrödingerOperator[ω][x, t] @ GF[ω][x, y, τ] // Simplify`
Out[24]= `0`

In[25]:= `ψ[x_, τ_] = Integrate[GF[1][x, y, τ] Cos[y], {y, -Pi, Pi},`
`                       GenerateConditions -> False] // Simplify`

Out[25]= $\left(\frac{1}{8} + \frac{i}{8}\right) e^{\frac{1}{2} i (-2-3 x^2+(2+x^2) \cos[2 \tau]) \csc[2 \tau]} \sqrt{\pi} \sqrt{\csc[\tau]}$

$\left(\sqrt{2} \ e^{i \csc[2 \tau] (x-\sin[\tau])^2} \ \mathrm{Erf}\left[\frac{(-1)^{3/4} (x - \pi \cos[\tau] + \sin[\tau])}{\sqrt{\sin[2 \tau]}}\right] - \right.$

$\sqrt{2} \ e^{i \csc[2 \tau] (x+\sin[\tau])^2} \ \mathrm{Erf}\left[\frac{(-1)^{3/4} (-x + \pi \cos[\tau] + \sin[\tau])}{\sqrt{\sin[2 \tau]}}\right] +$

$(1 - i) (-1)^{3/4} \left(e^{i \csc[2 \tau] (x+\sin[\tau])^2} \ \mathrm{Erfi}\left[\frac{(-1)^{1/4} (-x - \pi \cos[\tau] + \sin[\tau])}{\sqrt{\sin[2 \tau]}}\right] - \right.$

$\left.\left. e^{i \csc[2 \tau] (x-\sin[\tau])^2} \ \mathrm{Erfi}\left[\frac{(-1)^{1/4} (x + \pi \cos[\tau] + \sin[\tau])}{\sqrt{\sin[2 \tau]}}\right]\right)\right) \sec[\tau] \sqrt{\sin[2 \tau]}$

We use the periodicity of $\psi[x, \tau]$ with respect to x and τ to visualize the last function.

In[26]:= `Module[{ε = 10^-6, pp = 200, cp, gr},`
`        (* use symmetry with respect to x and τ *)`
`        cp = ContourPlot[Abs[ψ[N[x], N[τ]]]^2,`
`                        {x, -6, 0}, {τ, ε, Pi/2 - ε},`
`                        PlotPoints -> pp, ContourLines -> False,`
`                        DisplayFunction -> Identity, Contours -> 120,`
`                        ColorFunction -> (Hue[0.7 #]&)];`
`        (* make polygons *)`
`        gr = Graphics[cp];`
`        (* generate three other parts *)`

```
Show[{#, # /. Polygon[l_] :>
        Polygon[{#[[1]], Pi - 2 ε - #[[2]]}]& /@ l]}&[
        {gr, gr /. Polygon[l_] :> Polygon[{-1, 1}#& /@ l]}],
        FrameTicks -> None, DisplayFunction -> $DisplayFunction]]
```

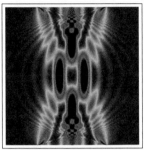

(Using the expansion of $G(x, y; \tau)$ into harmonic oscillator eigenfunction the time-evolved wave function can also be considered a fractional Fourier transform [931], [990].)

If we replace the exponential function in the integrand in the definition of the error function by the sine or cosine function, we get the Fresnel integrals.

`FresnelC[z]`
> is the Fresnel integral $C(z)$. $C(z) = \int_0^z \cos(\frac{\pi}{2} t^2)\, dt$.

`FresnelS[z]`
> is the Fresnel integral $S(z)$. $S(z) = \int_0^z \sin(\frac{\pi}{2} t^2)\, dt$.

Fresnel integrals can be used to view the boundary of a "sharp" shadow under a microscope.

Physical Application: Fresnel Diffraction

Suppose we are given a semi-infinite opaque surface that is being illuminated by a light source of intensity I_0 located at distance q from the surface boundary. Suppose a screen is set up at a distance p behind the opaque surface.

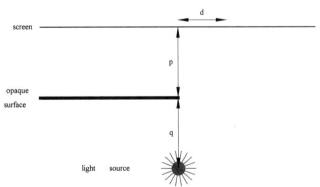

Then, the light intensity I at a point on the screen that is at a distance d from the geometric boundary of the shadow is given by

$$I = \frac{I_0}{2}\left(\left(C(w) + \frac{1}{2}\right)^2 + \left(S(w) + \frac{1}{2}\right)^2\right) \quad \text{with } w \sim \frac{\pi d k q}{p(q + p)}$$

where k is the wave vector of light (the nonzero value of the intensity I in the shadow zone arises from the scattering of the light at the edge of the surface). (See, e.g., [754], [992], [964], [569], [1153], [587], [769], and [686].)

We now look at the light–shadow boundary in more detail.

```
In[27]:= ListPlot[Table[N[{d, ((FresnelC[d] + 1/2)^2 + (FresnelS[d] + 1/2)^2)/2}],
              {d, -2., 4., 0.05}],
           PlotJoined -> True, Prolog -> {GrayLevel[1/2],
           Thickness[0.01], Line[{{-2, 0}, {0, 0}, {0, 1}, {4, 1}}]},
           PlotStyle -> {GrayLevel[0], Thickness[0.002]}]
```

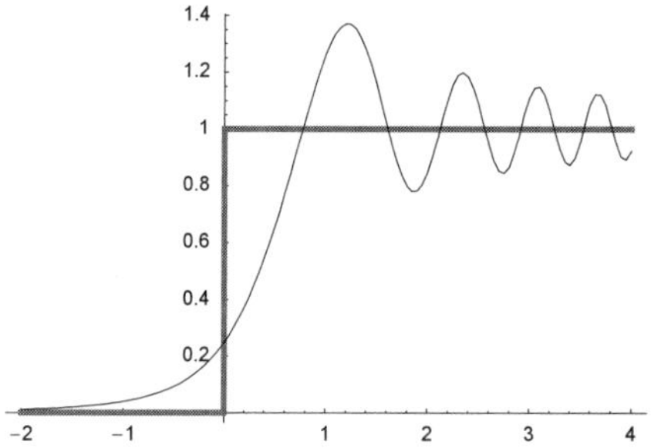

It is clear from this plot that because of the bending, the light–shadow boundary is not sharp. In the part of the screen in the shadow, the intensity decreases monotonically, whereas in the part of the screen that is directly illuminated, the intensity oscillates with increased intensity in the neighborhood of the boundary.

```
In[28]:= {d, ((FresnelC[d] + 1/2)^2 + (FresnelS[d] + 1/2)^2)/2} //.
              d -> 1.22 (* approximate first maximum *)

Out[28]= {1.22, 1.37042}
```

The Fresnel integrals $C(z)$ and $S(z)$ are also sufficient to construct the exact solution of the diffraction problem of an incoming plane wave on a half-plane [1171], [209], [813], [194], [1127], [147], [665], [285], [316], [638], [965]; (see [1081] and [314] for the time-dependent problem). Let the screen be positioned at the negative x-axis and the plane wave comes from the negative y-direction. Then the exact solution for the scattering problem is given by the following expression

```
In[29]:= (ψ[α_][x_, y_] =
        Module[{ρ = 2/Sqrt[Pi] (x^2 + y^2)^(1/4), φ = ArcTan[x, -y]/2,
              C = FresnelC, S = FresnelS},
              Exp[ I y] (1 + I + 2 C[ρ Sin[φ]] + 2 I S[ρ Sin[φ]]) +
              α Exp[-I y] (1 + I - 2 C[ρ Cos[φ]] - 2 I S[ρ Cos[φ]])]) //
                                                TraditionalForm
```

Out[29]//TraditionalForm=

$$e^{-iy}\,\alpha\left(-2C\left(\frac{2\sqrt[4]{x^2 + y^2}\,\cos(\frac{1}{2}\tan^{-1}(x, -y))}{\sqrt{\pi}}\right) - 2iS\left(\frac{2\sqrt[4]{x^2 + y^2}\,\cos(\frac{1}{2}\tan^{-1}(x, -y))}{\sqrt{\pi}}\right) + (1 + i)\right) +$$

$$e^{iy}\left(2C\left(\frac{2\sqrt[4]{x^2 + y^2}\,\sin(\frac{1}{2}\tan^{-1}(x, -y))}{\sqrt{\pi}}\right) + 2iS\left(\frac{2\sqrt[4]{x^2 + y^2}\,\sin(\frac{1}{2}\tan^{-1}(x, -y))}{\sqrt{\pi}}\right) + (1 + i)\right)$$

The parameter α takes the values $\alpha = 1$ for Dirichlet boundary conditions on the screen, $\alpha = -1$ for Neumann boundary conditions on the screen, and $\alpha = 0$ for an absorbing screen.

Let us visualize the above $\psi[\alpha][x, y]$. To avoid the relatively time-consuming calculation of the Fresnel integrals at many points in a graphic we generate interpolating functions of $C(z)$ and $S(z)$.

```
In[30]:= {C, S} = FunctionInterpolation[#[x], {x, -5, 5},
              MaxRecursion -> 8]& /@ {FresnelC, FresnelS};
```

```
In[31]:= ψIpo[α_][x_, y_] = ψ[α][x, y] /. {FresnelC -> C, FresnelS -> S};
```

data $[\alpha]$ is an array of values of $\psi[\alpha][x, y]$.

```
In[32]:= pp = 241; L = 12;
         dataS[α_] = Table[ψIpo[α][N[x], N[y]], {y, -L, L, 2L/pp}, {x, -L, L, 2L/pp}];
```

The function `diffractionGraphics` generates a contour plot of the absolute value of the sum of the incoming and diffracted wave. In addition we display the lines of zero real and imaginary part.

```
In[34]:= diffractionGraphics[α_, opts___] :=
         Module[{cpAbs, cpRe, cpIm, screen,
                 mr = MeshRange -> {{-L, L}, {-L, L}}},
         Block[{$DisplayFunction = Identity},
         (* magnitude of ψ[α][x, y] *)
         cpAbs = ListContourPlot[Abs[dataS[α]], PlotRange -> All,
                 ColorFunction -> (Hue[0.8 #]&), mr,
                 Contours -> 60, ContourLines -> False];
         (* zeros of real and imaginary part of ψ[α][x, y] *)
         {cpRe, cpIm} = ListContourPlot[#1[dataS[α]],
                 ContourShading -> False, Contours -> {0}, mr,
                 ContourStyle -> {GrayLevel[#2]}]& @@@ {{Re, 1}, {Im, 0.5}};
         (* the screen *)
         screen = Graphics[{{Thickness[0.02], GrayLevel[0],
                         Line[{{-L, 0}, {0, 0}}]}}];
         (* display all four graphic together *)
         Show[{cpAbs, cpRe, cpIm, screen}, opts, FrameTicks -> None]]
```

Here are the graphics for the three cases $\alpha = \pm 1$, $\alpha = 0$.

```
In[35]:= Show[GraphicsArray[diffractionGraphics[#,
              DisplayFunction -> Identity]& /@ {-1, 0, 1}]]
```

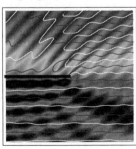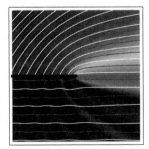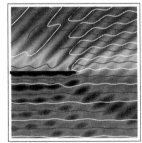

For the case of the diffraction on a slit, see [1122].

3.4 Exponential Integral and Related Functions

The antiderivatives of the $\sin(t)/t$ and $\cos(t)/t$ functions are defined by

$$\text{Si}(z) = \int_0^z \frac{\sin(t)}{t}\, dt$$

and

$$\text{Ci}(z) = -\int_z^\infty \frac{\cos(t)}{t}\, dt.$$

`SinIntegral[z]`
  represents the function $\text{Si}(z)$.
`CosIntegral[z]`
  represents the function $\text{Ci}(z)$.

Mathematica is able to find both integrals.

In[1]:= `{Integrate[Sin[t]/t, t], Integrate[Cos[t]/t, t]}`
Out[1]= `{SinIntegral[t], CosIntegral[t]}`

The following plot shows both functions along the real axis.

In[2]:= `Show[GraphicsArray[`
          `Plot[#, {z, 0.1, 3Pi}, DisplayFunction -> Identity,`
            `PlotLabel -> StyleForm[#, FontFamily -> "Courier"],`
            `PlotRange -> {{0, 3Pi}, {-2, 2}}, Frame -> True,`
            `Axes -> False]& /@ {SinIntegral[z], CosIntegral[z]}]]`

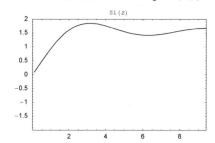 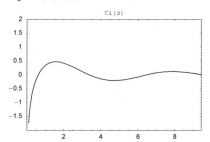

Mathematica is able to develop both functions in a series, despite the logarithmic singularity of $\text{Ci}(z)$ at $z = 0$.

In[3]:= `{Series[SinIntegral[z], {z, 0, 3}], Series[CosIntegral[z], {z, 0, 3}]}`
Out[3]= $\left\{ z - \dfrac{z^3}{18} + O[z]^4,\ (\text{EulerGamma} + \text{Log}[z]) - \dfrac{z^2}{4} + O[z]^4 \right\}$

Whenever possible, we get exact values for the sin integral and the cos integral.

In[4]:= `{SinIntegral[0], SinIntegral[Infinity]}`
Out[4]= $\left\{ 0,\ \dfrac{\pi}{2} \right\}$

Taking a linear combination of Si(z) and Ci(z), and using analytic continuation, we get the exponential integrals. As the name suggests, they arise frequently in integrals that involve exponential functions. Their explicit definitions are:

$$E_n(z) = \int_1^\infty \frac{e^{-zt}}{t^n}\, dt,$$

$$\mathrm{Ei}(z) = \mathcal{P} \int_{-z}^\infty \frac{e^{-t}}{t}\, dt$$

(here $\mathcal{P}$ denotes the principal part). In *Mathematica*, here are the exponential integral functions.

```
ExpIntegralE[n, z]
    represents the exponential integral En(z).
ExpIntegralEi[z]
    represents the exponential integral Ei(z).
```

We now plot these functions for real arguments.

In[5]:= **Needs["Graphics`Legend`"]**

In[6]:= **Plot[Evaluate[Table[ExpIntegralE[n, z], {n, 8}]], {z, 0.001, 2},**
          PlotStyle -> Table[{AbsoluteThickness[0.005],
                              Hue[(i - 1)/9]}, {i, 8}],
          PlotLegend -> (StyleForm[#, FontFamily -> "Courier"]& /@
                          Table[ExpIntegralE[n, z], {n, 8}]),
          LegendPosition -> {0, 0}, LegendSize -> {1.05, 0.5}]

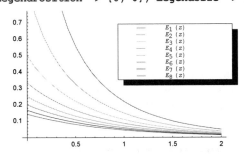

As with all special functions, the exponential integrals can also be computed for arbitrary complex arguments.

In[7]:= **ExpIntegralE[I, I] // N**

Out[7]= $-0.516028 - 0.192991\,\mathrm{i}$

Some of their symbolic properties are also available.

In[8]:= **D[ExpIntegralE[n, z], z]**

Out[8]= $-\mathrm{ExpIntegralE}[-1 + n, z]$

In[9]:= **ExpIntegralE[0, z]**

Out[9]= $\dfrac{e^{-z}}{z}$

In[10]:= **Series[ExpIntegralEi[z], {z, 0, 3}]**

Out[10]= $\frac{1}{2}\left(-2\,i\,\pi\,\text{Floor}\left[\frac{\pi + \text{Arg}[z]}{2\,\pi}\right] + \left(2\,\text{Log}[z] + O[z]^4\right)\right) + \left(\text{EulerGamma} + z + \frac{z^2}{4} + \frac{z^3}{18} + O[z]^4\right)$

The last special function we discuss in this section is li(z), the integral of the logarithmic function: li(z) = $\mathcal{P}\int_0^z 1/\ln(t)\,dt$. Modulo a differential algebraic constant, the exponential integral Ei(z) is a special case of the incomplete Gamma function.

In[11]:= **FunctionExpand[-Gamma[0, -z] - I Pi]**

Out[11]= $-i\,\pi + \text{ExpIntegralEi}[z] + \text{Log}[-z] + \frac{1}{2}\left(\text{Log}\left[\frac{1}{z}\right] - \text{Log}[z]\right)$

In[12]:= **PowerExpand[%]**

Out[12]= ExpIntegralEi[z]

Ei(z) has an unusual branch cut. Its value at the branch cut $(-\infty, 0)$ is the average of the value above and below the branch cut. The following graphic shows the exceptional behavior along the line $(-\infty, 0)$.

In[13]:= **Module[{pp = 30},**
    Off[N::meprec];
    dataExpIntegralEi =
    Table[Abs[Im[ExpIntegralEi[zr + I zi]]] // N[#, 22]&,
        {zi, -1, 1, 2/pp}, {zr, -2, 0, 2/pp}];
    On[N::meprec];
    ListPlot3D[dataExpIntegralEi, MeshRange -> {{-2, 0}, {-1, 1}}]]

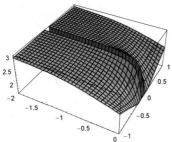

We have already encountered this function in Chapter 1 of the Numerics volume [1237] of the *GuideBooks* in connection with $\pi(x)$, the number of prime numbers less than x. The Riemann hypothesis is equivalent to the assertion that the number $\pi(x)$ of prime numbers less than x is given by: li(x) $-\pi(x) = O(\sqrt{x}\,\ln(x))$ (where $O(\ldots)$ is the Landau symbol). For more on the use of li(x) to compute the number n of prime numbers less than x, see [1105], [649], and [1292].

LogIntegral[z]
    represents the integral of the logarithm li(z).

In[14]:= **Integrate[1/Log[z], z]**

Out[14]= LogIntegral[z]

The integral of the logarithm li(z) can be expressed in terms of the exponential integral Ei(z) as li(z) = Ei(ln(z)).

In[15]:= **Integrate[1/Log[z], {z, x0, x1}, GenerateConditions -> False]**

Out[15]= -LogIntegral[x0] + LogIntegral[x1]

3.5 Bessel and Airy Functions

The four Bessel functions $J_\nu(z)$, $I_\nu(z)$, $K_\nu(z)$, and $Y_\nu(z)$ are the best known and most frequently used special functions. That is why we will devote a slightly longer section to them and present a couple of applications. Following *Mathematica*'s naming convention, they are written as follows.

BesselJ[*n*, *z*]

    represents the Bessel function $J_n(z)$.

BesselI[*n*, *z*]

    represents the Bessel function $I_n(z)$.

BesselK[*n*, *z*]

    represents the Bessel function $K_n(z)$.

BesselY[*n*, *z*]

    represents the Bessel function $Y_n(z)$ (sometimes called $N_n(z)$).

Here are plots for these functions along the real axis for various ν. We discuss some of these plots further in a moment.

```
In[1]:= Show[GraphicsArray[#]]& /@ Partition[
    Plot[Evaluate[Table[#[i, z], {i, 0, 5}]], {z, 0, 5},
        DisplayFunction -> Identity, PlotRange -> {-2, 2},
        PlotLabel -> StyleForm[ToString[#] <> "[n, z]",
            FontFamily -> "Courier"]]& /@
        {BesselJ, BesselY, BesselI, BesselK}, 2]
```

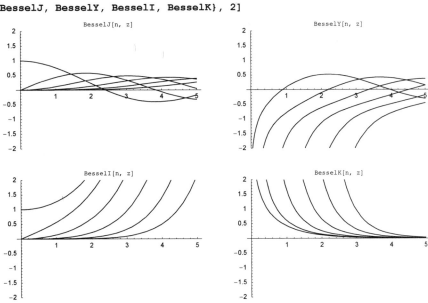

The next array of graphics shows contour plots of the real and imaginary parts of the four Bessel functions over the *x,ν*-plane.

```
ln[2]:= reImContourPlot[f_, pp_, c_, opts___] :=
    Module[{ξ = 3, data, reData, imData, recls, imcls, reLcp, imLcp, opt},
    opt = Sequence[ContourShading -> False, DisplayFunction -> Identity,
                MeshRange -> {{-ξ, ξ}, {-ξ, ξ}}];
    (* function values *)
    data = Table[N[f[ν, x]], {ν, -ξ, ξ, 2ξ/pp}, {x, -ξ, ξ, 2ξ/pp}];
    (* real and imaginary parts *)
    {reData, imData} = {Re[data], Im[data]};
    (* values for contours *)
    {recls, imcls} = (#[[Floor[pp^2/c/2]]]& /@
        Partition[Sort[Flatten[#]], Floor[pp^2/c]])& /@ {reData, imData};
    (* make contour plots *)
    reLcp = ListContourPlot[reData, opt, Contours -> recls,
                    ContourStyle -> Table[{Thickness[0.002],
                        RGBColor[α, 0, 1 - α]}, {α, 0, 1, 1/c}]];
    imLcp = ListContourPlot[imData, opt, Contours -> imcls,
                    ContourStyle -> Table[{Thickness[0.002],
                        RGBColor[0, 1 - α, α]}, {α, 0, 1, 1/c}]];
    (* display both contour plots *)
    Show[{reLcp, imLcp}, opts]]
```

```
ln[3]:= Show[GraphicsArray[#]]& /@ Map[reImContourPlot[#, 201, 30]&,
        {{BesselJ, BesselY}, {BesselI, BesselK}}, {2}]
```

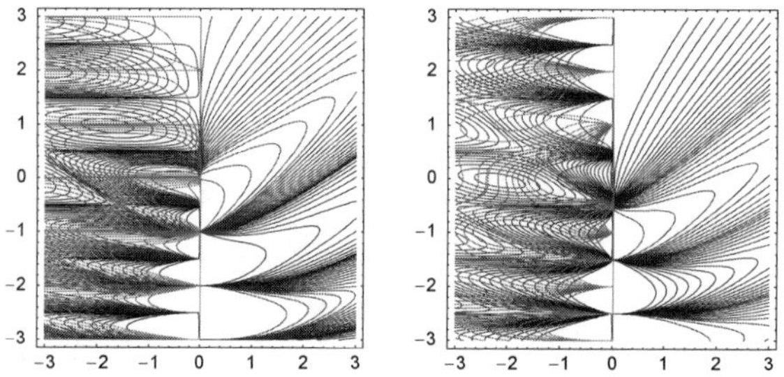

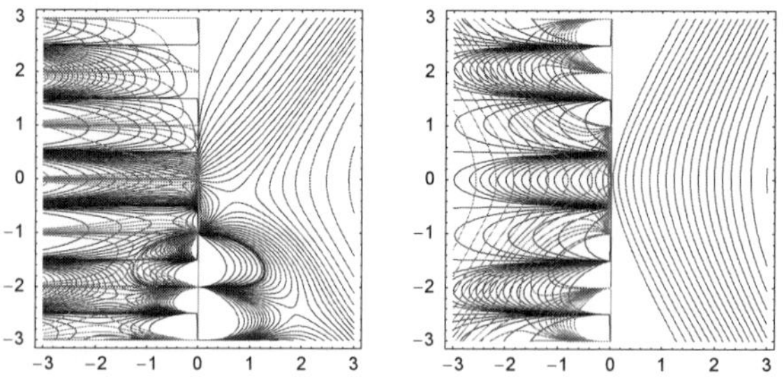

Most Bessel functions have branch cuts along the negative real axis.

```
ln[4]:= With[{ε = 10^-10},
    ParametricPlot3D[{r Cos[φ], r Sin[φ], Im[BesselJ[r Exp[I φ], r Exp[I φ]]]],
                {EdgeForm[]}}, {r, 0, 3},
```

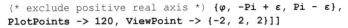

```
(* exclude positive real axis *) {φ, -Pi + ε, Pi - ε},
PlotPoints -> 120, ViewPoint -> {-2, 2, 2}]]
```

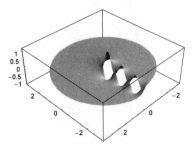

Here is little application of Bessel functions, especially of the Bessel function $I_\nu(z)$: the probability w_t for a random walker [617], [1308] who starts at $t = 0$ at the origin of a d dimensional cubic lattice to be at the point $\{n_1, n_2, \ldots, n_d\}$ at time t ($t \in \mathbb{N}$) is [126], [854]:

$$w_t = \frac{\partial^t}{\partial \xi^t} \prod_{k=1}^{d} I_n\left(\frac{\xi}{d}\right)\Big|_{\xi=0}$$

Here is this formula implemented.

```
In[5]:= w[t_, n_List] := Module[{ξ}, D[Times @@ BesselI[n, ξ/Length[n]],
                    {ξ, t}] /. ξ -> 0]
```

Here are the probabilities that a random walker in three dimensions returns to the origin after t steps.

```
In[6]:= Table[w[t, {0, 0, 0}], {t, 0, 16}]
```

$$\text{Out[6]}= \left\{1, 0, \frac{1}{6}, 0, \frac{5}{72}, 0, \frac{155}{3888}, 0, \frac{2485}{93312}, 0, \right.$$
$$\left. \frac{3619}{186624}, 0, \frac{902671}{60466176}, 0, \frac{1445015}{120932352}, 0, \frac{19046885}{1934917632}\right\}$$

The following graphic shows the probability of a random walker in two dimensions. To avoid to carrying out the differentiation of the product of Bessel functions for each lattice point, we precompute the probabilities for each integer t and symbolic n_1, n_2.

```
In[7]:= With[{(* lattice size *) m = 10},
       (* precompute derivatives *)
       Do[w[2][t][{n1_, n2_}] = Expand[w[t, {n1, n2}]], {t, 0, m + 4}];
       (* make graphic *)
       Show[GraphicsArray[#]] & /@
       Partition[Table[ListDensityPlot[
               Table[If[Sqrt[i^2 + j^2] > t, (* easy *) 0, w[2][t][{i, j}]],
                   {i, -m, m}, {j, -m, m}],
               ColorFunction -> (Hue[0.7 #] &), DisplayFunction -> Identity,
               FrameTicks -> None, MeshRange -> {{-m, m}, {-m, m}},
               MeshStyle -> {GrayLevel - 0.5, Thickness[0.002]}],
                   {t, 0, m + 4}], 5]]
```

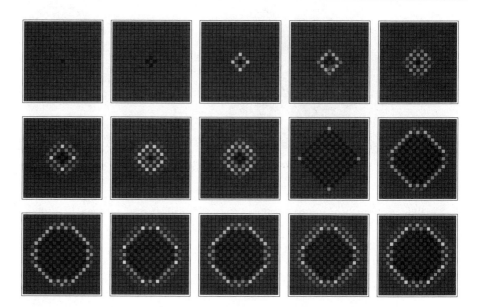

Using explicit expressions for the derivatives $I_\nu^{(n)}(0)$, we can quickly calculate the probabilities for larger values of t.

```
In[8]:= Module[{t = 64, data, opts = Sequence[FrameTicks -> None, Mesh -> False],
            BesselIDerivative, w},
    (* derivatives Derivative[0, n][BesselI][ν, 0] *)
    BesselIDerivative[ν_, n_] := BesselIDerivative[ν, n] =
     If[Element[(n - ν)/2, Integers],
        n!/(2^n (Gamma[(n - ν + 2)/2] Gamma[(n + ν + 2)/2])), 0];
    (* probabilities; use expansion for D[f[z] g[z], {z, n}] *)
    w[t_, {n1_, n2_}] := Sum[Binomial[t, k] BesselIDerivative[n1, k]*
                                        BesselIDerivative[n2, t - k],
                            {k, 0, t}]/2^t;
    (* probabilities on the lattice *)
    data = Table[w[t, {i, j}], {i, -t - 2, t + 2}, {j, -t - 2, t + 2}];
    (* visualize probabilities *)
    Show[GraphicsArray[
    Block[{$DisplayFunction = Identity},
     {(* linear scaling *)
      ListDensityPlot[data, ColorFunction -> (Hue[0.78 #]&), opts],
      (* nonlinear, multivalued scaling *)
      ListDensityPlot[data, ColorFunctionScaling -> False, opts,
                    ColorFunction -> (If[# == 0, RGBColor[1, 1, 1],
                                        Hue[Log[#]]]&)]}]]]]
```

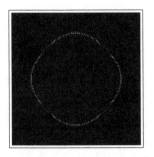

Note that for the special two-dimensional (2D) [1317], [715], [716] case under consideration, one can get a closed form for the probability as

$$w_t(n_x, n_y) = \delta_{0,((t-n_x-n_y)/2) \bmod 1} \, 4^{-t} \binom{t}{(t+n_x-n_y)/2}^2 \binom{(t+n_x-n_y)/2}{n_x} \Big/ \binom{(t+n_x+n_y)/2}{n_x}.$$

(For the use Bessel functions for quantum random walks, see [124]; and for the use of the Airy functions for quantum random walks, [695].)

Special functions can be used in the same way that elementary functions can be used—inside mathematics and inside *Mathematica*, for numeric and symbolic calculations and for graphics. Here, we use the Bessel function $J_0(z)$ to produce a fractal under the iteration $z \to J_0(z/c)$.

```
In[9]:= With[{pp = 201, a = 1},
        data = Table[c = N[cr + I ci];
        (* iterate *)
        FixedPoint[If[Abs[#] > 10.^3, Indeterminate, BesselJ[0, #/c]]&,
                    1. + 1. I, 10], {cr, -a, a, 2a/pp}, {ci, -a, a, 2a/pp}];
        (* show convergent c *)
        ListDensityPlot[Map[If[NumberQ[#], 1, 0]&, data, {2}],
                        Mesh -> False, ColorFunction -> GrayLevel,
                        FrameTicks -> None, AspectRatio -> 0.6]]
```

Mathematica can differentiate and integrate Bessel functions. Next, we integrate the famous Weber–Schafheitlin integral [899].

```
In[10]:= Clear[a, b, μ, ν, ω, x];
        int1[a_, b_, μ_, ν_, ω_] =
            Integrate[BesselJ[μ, a x] BesselJ[ν, b x] x^ω, {x, 0, Infinity},
                      Assumptions -> a > 0 && b > 0 && Re[ω] < 1 && b/a < 1]
```

Out[11]= If$\left[\text{Re}\left[\mu + \nu + \omega\right] > -1, \left(2^{\omega} a^{-1-\nu-\omega} b^{\nu} \text{Gamma}\left[\frac{1}{2} (1 + \mu + \nu + \omega)\right]\right.\right.$

$\qquad$ Hypergeometric2F1Regularized$\left[\frac{1}{2} (1 - \mu + \nu + \omega), \frac{1}{2} (1 + \mu + \nu + \omega), 1 + \nu, \frac{b^2}{a^2}\right]\right) /$

$\qquad$ Gamma$\left[\frac{1}{2} (1 + \mu - \nu - \omega)\right]$, Integrate[x$^{\omega}$ BesselJ[μ, a x] BesselJ[ν, b x],

$\qquad$ {x, 0, ∞}, Assumptions $\rightarrow$ 1 + Re$\left[\mu + \nu + \omega\right] \le 0\right]$

For $b/a > 1$, we get a result that is *not* the analytic continuation of the last result [1304].

In[12]:= int2[a_, b_, μ_, ν_, ω_] =
$\qquad$ Integrate[BesselJ[μ, a x] BesselJ[ν, b x] x^ω, {x, 0, Infinity},
$\qquad\qquad$ Assumptions -> a > 0 && b > 0 && Re[ω] < 1 && b/a > 1];

In[13]:= {int1[a, b, μ, ν, ω], int2[a, b, μ, ν, ω]} /.
$\qquad$ {a -> 2, b -> 1, μ -> 2, ν -> 3, ω -> -2}

Out[13]= $\left\{\dfrac{1}{96}, -\dfrac{5}{6}\right\}$

For the special case $\nu = \mu + 1$, $a = 1$, and $\omega = 0$ we get closed forms that are quite different [550].

In[14]:= {int1[1, b, μ, μ + 1, 0][[2]], int2[1, b, μ, μ + 1, 0][[2]]} // Simplify

Out[14]= {0, b$^{-1-\mu}$}

Bessel functions $J_{\nu}(z)$ with real index $\nu > -1$ have only real zeros. For real index $\nu < -1$, we have additional $2\lfloor -\nu \rfloor$ complex roots [920], [220]. The following sequence of graphics shows how at negative integers a new complex root forms. We display the contour lines of zero real and imaginary parts. The formation of a crossing from an "avoided crossing" situation is clearly visible.

In[15]:= zeroLinePicture[ν_, pp_, {X_, Y_}, opts___] :=
$\qquad$ Module[{ε = 10^-1, data},
$\qquad$ (* array of function values *)
$\qquad$ data = Table[BesselJ[ν, x + I y],
$\qquad\qquad\qquad$ {y, ε, Y, (Y - ε)/pp}, {x, -X, X, 2X/pp}];
$\qquad$ (* contour plots of real and imaginary parts *)
$\qquad$ Show[Apply[ListContourPlot[#1[data],
$\qquad\qquad\qquad$ MeshRange -> {{-X, X}, {0, Y}},
$\qquad\qquad\qquad$ ContourShading -> False, AspectRatio -> 0.8,
$\qquad\qquad\qquad$ PlotLabel -> "ν = " <> ToString[ν],
$\qquad\qquad\qquad$ DisplayFunction -> Identity,
$\qquad\qquad\qquad$ ContourStyle -> {Hue[#2]}, Contours -> {0}]&,
$\qquad\qquad$ {{Re, 0.}, {Im, 0.78}}, {1}], opts,
$\qquad\qquad$ DisplayFunction -> $DisplayFunction]]

In[16]:= With[{δ = 10^-4},
$\qquad\qquad$ Show[GraphicsArray[##]]& /@ Partition[
$\qquad\qquad$ Table[zeroLinePicture[N[ν], 81, {5, 4}, DisplayFunction -> Identity],
$\qquad\qquad\qquad$ {ν, -6 + δ, -6 - δ, -2δ/11}], 4]]

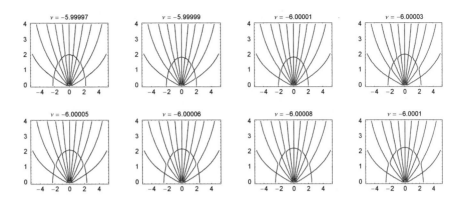

The following graphic uses the truncated Taylor series of $z^{-\nu} J_\nu(z)$ to visualize the zeros of $J_\nu(z)$ as a function of ν. At each negative integer ν two zeros coming from infinity join along the real axis. Between negative integers the already existing zeros wander into the complex plane and return at the next integer.

```
In[17]:= Module[{vs, λ, v, z},
         (* fractional part of the v-values;
            use exponentially spaced points near 0 and 1 *)
         vs = Join[Table[-10^α, {α, -20, -2, 18/60}],
                 Table[-α, {α, 1/100, 99/100, 98/100/60}],
                 Table[-1 + 10^α, {α, -2, -20, -18/60}]];
         (* number of v-values *) λ = Length[vs];
         Show[Graphics3D[Table[Table[{Hue[0.8 j/λ],
                 (* zeros as points in v,Re[z],Im[z]-space *)
                 Point[{v0 + vs[[j]], Re[#], Im[#]}]& /@
                 (* approximate BesselJ zeros *)
                 Cases[NRoots[(* Taylor polynomial around z == 0 for fixed v *)
                   (Sum[(-1)^k (z/2)^(2k)/(k! Gamma[k + v + 1]), {k, 0, 16}] /.
                   v -> v0 + vs[[j]]) == 0, z, 20], _?NumberQ, {-1}]},
                   {j, λ}], {v0, -1, -6, -1}] // N,
             PlotRange -> {All, 6.8 {-1, 1}, 6.8 {-1, 1}}, Axes -> True,
             BoxRatios -> {3, 1, 1}, AxesLabel -> {"v", "Re[z]", "Im[z]"}]]]
```

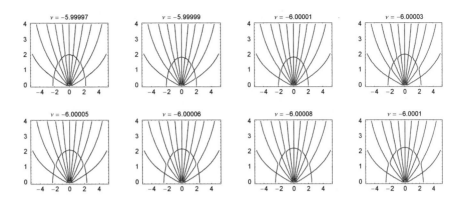

To better see how the zeros form as the common points of vanishing real and imaginary parts, we show the surfaces $\mathrm{Re}(z^{-\nu} J_\nu(z)) = 0$ and $\mathrm{Im}(z^{-\nu} J_\nu(z)) = 0$. The in the first, third, and fourth quadrant of the complex z-plane as a function of ν.

```
In[18]:= Needs["Graphics`ContourPlot3D`"]

In[19]:= makeHole[Polygon[l_]] :=
         With[{f = Partition[Append[#, First[#]], 2, 1]&},
```

```
        Function[mp, MapThread[Polygon[Join[#1, Reverse[#2]]]&,
        {f[l], f[mp + 0.78 (# - mp)& /@ l]}]][Plus @@ l/Length[l]]];
```

```
In[20]:= Module[{ε = 10^-10, pps = {36, 24, 24}, cRe, cIm},
        (* contour plot of real and imaginary parts == 0 for x, y > 0 *)
        {cRe,  cIm} = ContourPlot3D[#[(x + I y)^-v BesselJ[v, x + I y]],
                        {v, -5, 5}, {x, ε, 5}, {y, ε, 5},
                        MaxRecursion -> 0, PlotPoints -> pps,
                        DisplayFunction -> Identity]& /@ {Re, Im};
        (* extract polygons, mirror in two other quadrants *)
        {rePolys, imPolys} = N @ Flatten[{#, Map[({1, 1, -1} #)&, #, {-2}],
            Map[({1, -1, -1} #)&, #, {-2}]}&[
            Cases[#, _Polygon, Infinity]]]& /@ {cRe,  cIm};
        (* display surfaces of vanishing real and imaginary parts *)
        Show[Graphics3D[{EdgeForm[], (* cut holes in surfaces *)
        (* real part in red; imaginary part in light green *)
        {SurfaceColor[Hue[0.02], Hue[0.12], 2.4], makeHole /@ rePolys},
         SurfaceColor[Hue[0.24], Hue[0.12], 2.4], makeHole /@ imPolys}},
            BoxRatios -> {4, 1, 1}, Axes -> True,
            AxesLabel -> {"v", "Re[z]", "Im[z]"}]]
```

Certain special Bessel functions (whose indices are multiples of one-half, so-called spherical Bessel functions) reduce to trigonometric functions.

In[21]:= **BesselJ[1/2, r]**

Out[21]= $\dfrac{\sqrt{\frac{2}{\pi}}\ \text{Sin}[r]}{\sqrt{r}}$

In[22]:= **BesselI[5/2, r]**

Out[22]= $\dfrac{-\frac{6\,\text{Cosh}[r]}{r} + \left(2 + \frac{6}{r^2}\right)\,\text{Sinh}[r]}{\sqrt{2\,\pi}\ \sqrt{r}}$

Bessel functions can be easily differentiated with respect to their second argument.

In[23]:= **D[BesselY[n, z], z]**

Out[23]= $\dfrac{1}{2}\ (\text{BesselY}[-1 + n,\ z] - \text{BesselY}[1 + n,\ z])$

But they cannot be differentiated with respect to their first index, i.e., with respect to their first argument in closed form for generic complex *n*.

In[24]:= **D[BesselY[n, z], n]**

Out[24]= $\text{BesselY}^{(1,0)}[n,\ z]$

Bessel functions arise, for example, in the solution of the Helmholtz equation in cylindrical coordinates [420], [1019].

Physical Application: Oscillation of a Circular Drum

We now show how a Helmholtz equation describes the motion of the membrane surface of a circular drum. The displacement $u(\mathbf{r}, t)$, at each point $\mathbf{r}$ as a function of t, satisfies the wave equation

$$\Delta\, u(\mathbf{r},\, t) - \frac{1}{c_{ph}^2}\, \frac{\partial^2 u(\mathbf{r},\, t)}{\partial t^2} = 0.$$

Here, Δ is the Laplace operator and c_{ph}^2 (= tension per unit surface density) is the phase velocity. Assuming the solution is periodic in time, that is, $u(\mathbf{r},\, t) = \cos(\omega\, t + \alpha)\, u(\mathbf{r})$ we are led to the Helmholtz equation

$$\Delta\, u(\mathbf{r}) + \frac{\omega^2}{c_{ph}^2}\, u(\mathbf{r}) = 0.$$

In polar coordinates, the Laplace operator becomes [468]

$$\Delta\, u(\mathbf{r}) = \frac{1}{r}\, \frac{\partial}{\partial r}\left(r\, \frac{\partial u(\mathbf{r})}{\partial r}\right) + \frac{1}{r^2}\, \frac{\partial^2 u(\mathbf{r})}{\partial \varphi^2}.$$

Now, assuming (after another separation of variables [856], [709], [60], [81], [895], [1159], [1110], [198], [1362], [429], [1029], [86], [894]) that

$$u(\mathbf{r}) = \cos(m\, \varphi)\, R(r) \quad \text{or} \quad u(\mathbf{r}) = \sin(m\, \varphi)\, R(r)$$

by the 2π-periodicity of the displacement in the φ-direction, we see that $m \in \mathbb{N}$. The function $R(r)$ still has to be determined. This leads to the following ordinary differential equation for the displacement in the radial direction: $(\omega^2 / c_{ph}^2 = \lambda)$

$$\frac{1}{r}\, \frac{\partial}{\partial r}\left(\frac{r\, \partial R(r)}{\partial r}\right) - \frac{m^2}{r^2}\, R(r) + \lambda\, R(r) = 0$$

or

$$r\, (r\, R'(r))' + (r^2\, \lambda - m^2)\, R(r) = 0.$$

The boundary condition for this differential equation describing the drum are a finite value for $R(0)$ and $R(r_0) = 0$ where r_0 is the radius of the drum and the membrane of the drum is assumed to be fixed along its edge. For a very detailed discussion of the vibrations of a circular and an ellipsoidal drum, see [868], [292], [291], [1068]; for time-dependent drum sizes, see [1349].

We now solve the radial differential equation.

```
In[25]:= DSolve[r D[r D[R[r], r], r] + (r^2 λ - m^2) R[r] == 0, R[r], r]

Out[25]= {{R[r] → BesselJ[m, r √λ] C[1] + BesselY[m, r √λ] C[2]}}
```

First, we take account of the behavior of $J_m(x)$ and $Y_m(x)$ at $x \approx 0$.

```
In[26]:= {Series[BesselJ[m, x], {x, 0, 1}], Series[BesselY[m, x], {x, 0, 1}]}
```

Out[26]= $\left\{ x^m \left(\dfrac{2^{-m}}{\text{Gamma}[1 + m]} + O[x]^2 \right), \right.$

$\left. x^m \left(-\dfrac{2^{-m} \text{Cos}[m \pi] \text{Gamma}[-m]}{\pi} + O[x]^2 \right) + x^{-m} \left(-\dfrac{2^m \text{Gamma}[m]}{\pi} + O[x]^2 \right) \right\}$

We find that the solution of the above differential equation is $R(r) \sim J_{|m|}(\lambda^{1/2} r)$.

The boundary condition implies $\lambda_{k\,j} = \mu_{k\,j}^2 / r_0^2$, where μ_{kj} is the jth zero of the Bessel function $J_k(z)$. The zeros of the Bessel function can be found most quickly by using the package `NumericalMath`BesselZero``. We now calculate the first roots (first by absolute value) using `FindRoot`. We choose the starting value by looking at the above plot for $J_\nu(x)$.

```
In[27]:= {μ[0, 1], μ[0, 2], μ[1, 1], μ[1, 2], μ[3, 3]} =
          FindRoot[BesselJ[#1, z] == 0, {z, #2}][[1, 2]] & @@@
                         {{0, 2}, {0, 5}, {1, 4}, {1, 7}, {3, 13}}
Out[27]= {2.40483, 5.52008, 3.83171, 7.01559, 13.0152}
```

Next, we examine the oscillation of the drum graphically ($r_0 = 1$). (Here, the dark black line is the edge of the drum.)

```
In[28]:= Show[GraphicsArray[Map[Show[{
          (* transforming to Cartesian coordinates *)
          ParametricPlot3D[{r Cos[φ], r Sin[φ], #}, {φ, 0, 2Pi}, {r, 0, 1},
                           PlotPoints -> 18, DisplayFunction -> Identity],
          (* the borderline *)
          ParametricPlot3D[{Cos[φ], Sin[φ], 0, Thickness[0.02]}, {φ, 0, 2Pi},
                           DisplayFunction -> Identity]}, Boxed -> False,
                           Axes -> False, DisplayFunction -> Identity] &,
            (* the to be displayed eigenfunctions *)
            {BesselJ[0, μ[0, 1] r], BesselJ[0, μ[0, 2] r],
             Cos[φ] BesselJ[1, μ[1, 1] r], Cos[2φ] BesselJ[1, μ[1, 2] r]}, {1}]]]
```

To visualize the oscillation modes, it is useful to generate a plot of the lines of equidistortion. For a round drum, `ContourPlot` is not immediately applicable. However, we can use `ContourPlot` to create an equipotential-line representation in the r,φ-coordinates, and then use `Graphics[ContourGraphics[...]]` to turn this into a graphics object that consists of lines and polygons, and whose coordinates are transformed to an x,y-Cartesian coordinate system. Then, it is straightforward to create an equipotential-line plot of the oscillation of our drum.

```
In[29]:= Show[Graphics[{
          Map[If[(* leave GrayLevel untouched *) Head[#] === List,
              (* transforming to Cartesian coordinates *)
              {#[[1]] Cos[#[[2]]], #[[1]] Sin[#[[2]]]}, #] &,
              (* converting ContourGraphics into Graphics
                 for further manipulations on the 2D primitives *)
              Graphics[ (* making the ContourPlot in polar coordinates *)
                ContourPlot[Sin[3 φ] BesselJ[3, μ[3, 3] r],
                            {r, 0, 1}, {φ, 0, 2Pi},
              PlotPoints -> 45, DisplayFunction -> Identity]][[1]], {-2}],
          (* the boundary *)
          {Thickness[0.002], GrayLevel[1], Line[{{0, 0}, {1, 0}}]},
```

```
{Thickness[0.01], Circle[{0, 0}, 1]}}],
PlotRange -> All, AspectRatio -> Automatic]
```

We sidestep a moment and using the package NumericalMath`BesselZeros` have a look at the possible $\mu_{k\,j}$ for larger k and j. The left graphic shows the $\mu_{k\,j}$ as a function of $|m|$. The right graphic shows a logarithmic plot of a histogram of the sorted differences of the 30699 lowest neighboring $\mu_{k\,j}$. The straight line indicates the complete integrability of the problem [139], [652].

In[30]:= **Needs["NumericalMath`BesselZeros`"]**

In[31]:= **Show[GraphicsArray[**
```
        {(* the eigenvalues as a function of m *)
        Graphics[Table[{Point[{-m, #}], Point[{m, #}]}& /@
                        BesselJZeros[m, 35], {m, 0, 20}],
            Frame -> True, PlotRange -> {{-20, 20}, {0, 100}}],
        (* eigenvalue spacing distribution *)
        ListPlot[{First[#], Log[10, Length[#]]}& /@
        Split[Sort[Round[1000 (-Subtract @@@ Partition[Sort[
                Select[Flatten[Table[BesselJZeros[m, 200], {m, 0, 200}]],
                                            (# < 627)&]], 2, 1)]]],
            PlotJoined -> True, PlotRange -> {{0, 130}, All},
            DisplayFunction -> Identity]}]]
```

For time-dependent superpositions of such eigenfunctions, see [1069], [396]. For the visualization of the eigenmodes of an ellipse–shaped membrane, see below. For drums with radius-dependent phase velocity (this means mass density), see [834]; for drums with a circular hole, see [985]. Next, we will deal with the much more general case of an arbitrarily shaped membrane.

Physical Application: Oscillation of a General-Shaped Drum

Above, we derived that the functions $\exp(i\,m\,\varphi)\,J_m(k\,r)$ are solutions of the 2D Helmholtz equation $\Delta u(\mathbf{r}) + k^2\,u(\mathbf{r}) = 0$ (here, $\mathbf{r} = \{r,\ \varphi\}$ in polar coordinates). Because $\exp(i\,m\,\varphi)\,J_m(k\,r)$ for a complete system in $\mathbb{R}^2$ [1280], [515], [1042], every solution of the Helmholtz equation can be represented as:

$$u(r, \varphi) = \sum_{m=-\infty}^{\infty} c_m \exp(i\, m\, \varphi)\, J_m(k\, r).$$

Let the boundary of a general (compact, connected) drum be parametrized by $r(s)$ and $\varphi(s)$, where s is the curve length. If we keep the membrane fixed at the rim, we have $u(r(s), \varphi(s)) = 0$. Using the periodicity of $u(r(s), \varphi(s))$, $u(r(s), \varphi(s)) = u(r(s + \ell), \varphi(s + \ell))$ (ℓ is the length of the circumference of the drum), we decompose u along the boundary in its Fourier coefficients

$$c_n(k) = \int_0^\ell \exp(-2\,\pi\, i\, n\, s/\ell)\, u(r(s), \varphi(s))\, ds$$

[1119], [31], [1155], [1178], [320]. To ensure the Dirichlet boundary condition of the membrane, this expression should vanish. Substituting the Fourier–Bessel expansion of $u(r, \varphi)$ into the expression for the $c_n(k)$, we obtain that the determinant of the matrix with entries

$$c_{nm}(k) = \int_0^\ell \exp(-2\,\pi\, i\, n\, s/\ell) \exp(i\, m\, \varphi(s))\, J_m(k\, r(s))\, ds$$

should vanish. Truncating the m-basis yields a finite determinant that vanishes for certain k. For such values of k we can calculate the null space of the corresponding linear homogeneous system; this means the c_m and, in this way, obtain a series representation of the eigenfunctions of the vibrating membrane.

Let us carry out the just-sketched program for one particular example: A very popular shape of a drum for physicists is a stadium-shaped drum [878], [734], [161], [1208], [366], [28], [1194], [30], [304], [581], [68], [268], [645], [942]. Let the diameter of the circular parts of the stadium be $2R$ and the length of the straight middle part be $2\,a$. (For $R = a$, this stadium is called the Bunimovich stadium [235].) Here is a parametrization of the boundary of the stadium (it can be obtained from elementary geometry). The parametrization starts in the lower right "corner". Its first piece is the right arc, its second is the top straight piece, its third is the left arc, and its fourth is the lower straight part.

```
In[32]:= (* parametrizations for φ for the four segments *)
        φ[1][s_, {R_, a_}] = ArcTan[a + R Cos[s - Pi/2], R Sin[s - Pi/2]];
        φ[2][s_, {R_, a_}] = ArcTan[R Pi + a - s, R];
        φ[3][s_, {R_, a_}] = ArcTan[-a + R Cos[-((4 a + Pi R - 2 s)/(2 R))],
                            R Sin[-((4 a + Pi R - 2 s)/(2 R))]];
        φ[4][s_, {R_, a_}] = ArcTan[s - 2 Pi R - 3 a, -R];
```

```
In[37]:= (* parametrizations for r for the four segments *)
        r[1][s_, {R_, a_}] =
                Sqrt[(a + R Cos[s - Pi/2])^2 + (R Sin[s - Pi/2])^2];
        r[2][s_, {R_, a_}] = Sqrt[(R Pi + a - s)^ 2 + R^2];
        r[3][s_, {R_, a_}] =
                Sqrt[(-a + R Cos[-((4 a + Pi R - 2 s)/(2 R))])^2 +
                    (R Sin[-((4 a + Pi R - 2 s)/(2 R))])^2];
        r[4][s_, {R_, a_}] = Sqrt[(s - 2 Pi R - 3 a)^2 + (-R)^2];
```

```
In[42]:= (* curve lengths the 4 segments *)
        sLimit[1][{R_, a_}] = Pi R;
        sLimit[2][{R_, a_}] = Pi R + 2 a;
        sLimit[3][{R_, a_}] = 2 Pi R + 2 a;
        sLimit[4][{R_, a_}] = 2 Pi R + 4 a;
```

```
        (* lengths the circumference *)
        ℓ[{R_, a_}] = 2 Pi R + 4 a;
In[49]:= (* full parametrization for φ *)
        φ[s_, {R_, a_}] =  Which @@ Flatten[
        Table[{s <= sLimit[j][{R, a}], φ[j][s, {R, a}]}, {j, 4}]];

        (* full parametrization for r *)
        r[s_, {R_, a_}] =  Which @@ Flatten[
        Table[{s <= sLimit[j][{R, a}], r[j][s, {R, a}]}, {j, 4}]];
```

Here is a sketch of the Bunimovich stadium using the just-defined parametrization.

```
In[53]:= With[{R = 1, a = 1},
        ParametricPlot[
            Evaluate[r[s, {R, a}] {Cos[φ[s, {R, a}]], Sin[φ[s, {R, a}]]}],
                    {s, 0, ℓ[{R, a}]},
            PlotStyle -> {Hue[0], Thickness[0.01]}, Frame -> True,
            PlotRange -> All, Axes -> False, AspectRatio -> Automatic]]
```

In the following discussion we will calculate some possible shapes of the vibrating Bunimovich stadium. To do this, we have to calculate the integrals $c_{nm}(k)$. Instead of using the built-in function NIntegrate, we will implement a fast numerical integration based on a discretization of the boundary. Because of the periodicity of $\varphi(s)$ and $r(s)$, we use the trapezoidal integration rule. The precision we can achieve with this crude integration for the eigenvalues and eigenfunctions is sufficient for the purpose here—some graphics of the eigenfunctions.

```
In[54]:= sValues[n_, {R_, a_}] := Table[s, {s, 0, ℓ[{R, a}], ℓ[{R, a}]/n}]

        φValues[n_, {R_, a_}] := φ[#, {R, a}]& /@ sValues[n, {R, a}]

        rValues[n_, {R_, a_}] := r[#, {R, a}]& /@ sValues[n, {R, a}]
In[57]:= (* a fast, specialized numerical integration *)
        TrapezoidalIntegrateC =
        Compile[{{data, _Complex, 1}, {h, _Real}},
        Module[{sum = 0. + 0. I},
                Do[sum = sum + data[[i]], {i, Length[data]}];
                h (sum - (First[data] + Last[data])/2.)]];
```

To avoid the (time-consuming) calculation of $J_m(k\,r(s))$ for each integer m, we use the following three-term recursion relation of Bessel functions.

```
In[59]:= BesselJ[v, z] == 2 (v - 1)/z BesselJ[v - 1, z] - BesselJ[v - 2, z] //
                                                        FullSimplify
Out[59]= True
```

It is most effective to calculate the whole matrix $c_{nm}(k)$ $-o \le n$, $m \le o$ at once to avoid the recalculation of the various quantities appearing in these expressions. The function `fourierMatrix` calculates the matrix of integrals $c_{nm}(k)$. Here, $-o \le n$, $m \le o$, and for the numerical integration, *pp* points are taken into account. To speed up the calculation of the matrix, we will pack all lists.

```
In[60]:= fourierMatrix[k_, o_, {R_, a_}, pp_] :=
    Module[{hδ = N[𝓁[{R, a}]/pp], rData, φData, sData, l1, l2, l3},
      (* packed versions of the vectors needed *)
      {rData, φData, sData} =
        Developer`ToPackedArray[N[#[pp, {1, 1}]]]& /@
                              {rValues, φValues, sValues};
      (* the two exponential functions *)
      Do[l1[m] = Exp[N[I] m φData], {m, -o, o}];
      Do[l2[n] = Exp[N[-2Pi I n  sData/𝓁[{R, a}]]], {n, -o, o}];
      (* use stable recursion toward m==0 *)
      l3[-o] = Developer`ToPackedArray[BesselJ[-o, k rData]];
      l3[-o + 1] = Developer`ToPackedArray[BesselJ[-o + 1, k rData]];
      (* use recursion formulas for BesselJ *)
      Do[l3[m] = 2. (m - 1.)/(k rData) l3[m - 1] - l3[m - 2],
        {m, -o + 2, 0}];
      l3[o] = Developer`ToPackedArray[BesselJ[o, k rData]];
      l3[o - 1] = Developer`ToPackedArray[BesselJ[o - 1, k rData]];
      (* use recursion formulas *)
      Do[l3[m] = 2. (m + 1.)/(k rData) l3[m + 1] - l3[m + 2],
        {m, o - 2, 1, -1}];
      (* the table of integrals *)
      Table[TrapezoidalIntegrateC[l1[m] l2[n] l3[m], hδ],
          {m, -o, o}, {n, -o, o}]];
```

Here, we calculate a 61×61 matrix of the above integrals. The calculation is about three orders of magnitude faster than the direct use of `NIntegrate` (which for each input would compile and sample the integrand). We display the matrix element $c_{30\,30}$. The two results agree to at least six digits—surely enough for visualization purposes.

```
In[61]:= fourierMatrix[16, 30, {1, 1}, 201][[-1, -1]] // Timing
Out[61]= {1.45 Second, -0.0330124 + 0.0188618 i}
```

```
In[62]:= With[{k = 16, l = 30, n = 30, R = 1, a = 1},
    NIntegrate[Evaluate[
            Exp[-2 Pi I n s/𝓁[{R, a}]] Exp[I l φ[s, {R, a}]] *
            BesselJ[l, k r[s, {R, a}]]],
            {s, 0, Pi, Pi + 2, 2 Pi + 2, 𝓁[{R, a}]}]] // Timing
Out[62]= {0.67 Second, -0.0330124 + 0.0188618 i}
```

Here is a visualization of a typical matrix generated by `fourierMatrix`. The checkerboard pattern arises because the eigenfunctions can be grouped according to their symmetry. (We do not separate the various symmetry states here.)

```
In[63]:= Module[{k = 25, o = 50, pp = 101, R = 1, a = 1},
    ListDensityPlot[Abs[fourierMatrix[k, o, {R, a}, pp]],
                ColorFunction -> (Hue[0.8 #]&), Mesh -> False,
                MeshRange -> {{-o, o}, {-o, o}}]]
```

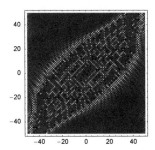

Now, let us calculate some k such that $|c_{nm}(k)|$ vanishes.

```
In[64]:= det[k_?NumericQ, o_, {R_, a_}, pp_] :=
           Det[fourierMatrix[k, o, {R, a}, pp]]
```

To find eigenvalues k^*, we have to search for zeros of the corresponding determinant. Because of the smooth shape of the stadium, we expect the main contribution of an eigenvalue k^* to arise from $J_m(\xi)$, where $2\, r(s)\, k^* \le |m|$. We take $o = 20$ and try to find some eigenvalues near $k \approx 10$. We calculate a table of values for the determinant for $8 \le k \le 12$.

```
In[65]:= Timing[mat = Table[{k, det[k, 20, {1, 1}, 301]}, {k, 8, 12, 1/30}];]

Out[65]= {65.07 Second, Null}
```

A plot of the value of the determinant shows many zeros in this range.

```
In[66]:= Show[Apply[Function[{reim, col},
              (* real part in red; imaginary part in blue *)
              ListPlot[{#[[1]], reim[#[[2]]]}& /@ mat,
                        PlotJoined -> True, PlotStyle -> col,
                        DisplayFunction -> Identity]],
              {{Re, {Hue[0]}}, {Im, {Hue[0.8], Dashing[{0.01, 0.01}]}}}, {1}],
           DisplayFunction -> $DisplayFunction,
           PlotRange -> {-1, 1}, Frame -> True, Axes -> False]
```

We will select the three zeros from the interval (10.5, 11) for visualizing the eigenfunctions. To find precise values for the corresponding k's we use `FindRoot`. The start values for the root-searching process can be extracted from `mat` by looking for adjacent values that change sign.

```
In[67]:= startPairs = Map[First,
             Select[Partition[{#[[1]], Re[#[[2]]]}& /@ mat, 2, 1],
                10.5 < #[[1, 1]] < 11.0 &&  (* sign change *)
                #[[1, 1]] > 4 && #[[1, 2]] #[[2, 2]] < 0 &], {2}] // N

Out[67]= {{10.7, 10.7333}, {10.7667, 10.8}, {10.8333, 10.8667}}
```

Here are the refined values of k^*.

```
In[68]:= kRoots = (k /. FindRoot[Re[det[k, 20, {1, 1}, 301]],
            Evaluate[{k, Sequence @@ #}],
                MaxIterations -> 30, AccuracyGoal -> 10])& /@ startPairs
Out[68]= {10.7322, 10.792, 10.847}
```

To obtain the corresponding $c_n(k^*)$, we calculate the null space of the matrix $c_{mn}(k^*)$.

```
In[69]:= fourierMatrices =
            Transpose[fourierMatrix[#, 20, {1, 1}, 301]]& /@ kRoots;
```

```
In[70]:= nontrivialHomogeneousSolutions =
            NullSpace[#, Tolerance -> 0.1][[-1]]& /@ fourierMatrices;
```

Now, we can visualize the vibrating stadium-shaped drum. The function Ψ calculates the elongation of the membrane. `eigenfunctionPicture` calculates and displays the eigenfunction belonging to the eigenvalue k and the null space v. `boundary` is a thick black line that represents the rim of the membrane.

```
In[71]:= Ψ[{r_, φ_}, k_, v_] := v.Table[BesselJ[l, k r] Exp[I l φ],
                    {l, -(Length[v] - 1)/2, (Length[v] - 1)/2}]
```

```
In[72]:= eigenfunctionPicture[k_, v_, {ppφ_, ppr_}, col_, opts___] :=
        Module[{points, polys, boundary},
        (* the stadium boundary *)
        boundary = {Thickness[0.01], GrayLevel[0],
                    Line[Table[1.006 r[s, {1, 1}]*
                            {Cos[φ[s, {1, 1}]], Sin[φ[s, {1, 1}]], 0},
                                {s, 0, 2Pi + 4, (2Pi + 4)/501}]]};
        (* the points *)
        points = Table[ρ = N[α r[s, {1, 1}]]; φ = N[φ[s, {1, 1}]];
                {ρ Cos[φ], ρ Sin[φ], Re[Ψ[{α ρ, φ}, k, v]]},
                {s, 0, ℓ[{1, 1}], ℓ[{1, 1}]/ppφ}, {α, 0, 1, 1/ppr}];
        (* make polygons *)
        polys = Table[Polygon[{#[[i, j]], #[[i, j + 1]], #[[i + 1, j + 1]],
                            #[[i + 1, j]]}&[points]], {i, ppφ}, {j, ppr}];
        (* show graphics *)
        Show[Graphics3D[{boundary, EdgeForm[], col, polys}], opts,
            PlotRange -> All, Boxed -> False, BoxRatios -> {4, 2, 1}]]
```

Here are pictures of the three selected eigenoscillations. Because we have to calculate more than 20000 values of Bessel functions for each picture, the next calculation will take a few minutes. Because the corresponding classical system (a stadium shaped 2D billiard) is chaotic, the eigenfunctions exhibit a much more complicated structure than do the ones of the circular case [69], [823], [580], [1229], [786], [581], [732]. Analyzing the number of nodal domains allows to quantify this [166]. For an experimental realizations of such eigenoscillations of a metallic plate see [1186].

```
In[73]:= Show[GraphicsArray[
        Table[eigenfunctionPicture[kRoots[[k]],
                nontrivialHomogeneousSolutions[[k]], {201, 101},
                    SurfaceColor[Hue[k/4], Hue[k/4 + 0.4], 2.8],
                        DisplayFunction -> Identity], {k, 3}]]]
```

Using the above general ansatz for the solution of the Helmholtz equation in $\mathbb{R}^2$, we could continue to investigate some more examples, numerically and symbolically. We leave it to the reader to, say, calculate the eigenvalues of a regular $n$gon in the limit of large n [155].

Now let us the Airy functions. The Airy functions Ai(z) and Bi(z) are linear combinations of Bessel functions:

$$\text{Ai}(z) = \frac{1}{\pi} \sqrt{\frac{z}{3}} \, K_{1/3}\!\left(\frac{2}{3} z^{3/2}\right)$$

$$\text{Bi}(z) = \sqrt{\frac{z}{3}} \left(I_{-1/3}\!\left(\frac{2}{3} z^{3/2}\right) + I_{1/3}\!\left(\frac{2}{3} z^{3/2}\right)\right).$$

Using `FullSimplify` with an appropriate option setting for the `ComplexityFunction` gives the following representations.

```
In[74]:= FullSimplify[Airies[AiryAi[z], AiryBi[z]], ComplexityFunction ->
                (Count[#, _AiryAi | _AiryBi, Infinity]&)]
```

$$\text{Out[74]= Airies}\Big[\frac{(z^{3/2})^{2/3} \, \text{BesselI}\big[-\frac{1}{3}, \frac{2\,z^{3/2}}{3}\big] - z \, \text{BesselI}\big[\frac{1}{3}, \frac{2\,z^{3/2}}{3}\big]}{3\,(z^{3/2})^{1/3}},$$

$$\frac{(z^{3/2})^{2/3} \, \text{BesselI}\big[-\frac{1}{3}, \frac{2\,z^{3/2}}{3}\big] + z \, \text{BesselI}\big[\frac{1}{3}, \frac{2\,z^{3/2}}{3}\big]}{\sqrt{3}\,(z^{3/2})^{1/3}}\Big]$$

And here is the hypergeometric representation of the two Airy functions.

```
In[75]:= FullSimplify[F[AiryAi[z], AiryBi[z]], ComplexityFunction ->
                (Count[#, _AiryAi | _AiryBi | _BesselI, Infinity]&)]
```

$$\text{Out[75]= F}\Big[-\frac{1}{3\cdot3^{2/3}}\Big(-3\,\text{Hypergeometric0F1Regularized}\big[\frac{2}{3}, \frac{z^3}{9}\big] +$$

$$3^{1/3}\,z\,\text{Hypergeometric0F1Regularized}\big[\frac{4}{3}, \frac{z^3}{9}\big]\Big),$$

$$\frac{1}{3^{5/6}}\Big(3^{2/3}\,\text{Hypergeometric0F1Regularized}\big[\frac{2}{3}, \frac{z^3}{9}\big] +$$

$$z\,\text{Hypergeometric0F1Regularized}\big[\frac{4}{3}, \frac{z^3}{9}\big]\Big)\Big]$$

`AiryAi[z]`
   represents the Airy function Ai(z).

`AiryBi[z]`
   represents the Airy function Bi(z).

In view of their associated differential equation, the derivatives Ai$'(z)$ and Bi$'(z)$ are also of interest.

```
AiryAiPrime[z]
```
   represents the derivative Ai$'(z)$ of the Airy function Ai(z).
```
AiryBiPrime[z]
```
   represents the derivative Bi$'(z)$ of the Airy function Bi(z).

These functions have their own names because of their independent importance, and, because $z^{3/2}$ is involved in their analytic continuation for arbitrary complex arguments, they cannot immediately be expanded in terms of the Bessel functions using the above formulas. They satisfy the differential equation $w''(z) = z\,w(z)$ [614].

```
In[76]:= DSolve[w''[z] == z w[z], w[z], z]
```
```
Out[76]= {{w[z] → AiryAi[z] C[1] + AiryBi[z] C[2]}}
```

The solution of the following differential equation is a rational function in Ai(x), Ai$'(x)$, Bi(x), and Bi$'(x)$.

```
In[77]:= DSolve[w''[x] + 2 w[x] w'[x] - 1 == 0, w, x]
```

```
                InverseFunction::ifun : Inverse functions are
                    being used. Values may be lost for multivalued inverses. More…

                InverseFunction::ifun : Inverse functions are
                    being used. Values may be lost for multivalued inverses. More…

                InverseFunction::ifun : Inverse functions are
                    being used. Values may be lost for multivalued inverses. More…

                General::stop : Further output of
                    InverseFunction::ifun will be suppressed during this calculation. More…

                Solve::tdep : The equations appear to involve the variables
                    to be solved for in an essentially non-algebraic way. More…

                Solve::tdep : The equations appear to involve the variables
                    to be solved for in an essentially non-algebraic way. More…
```

$$\text{Out[77]}= \left\{\left\{w \to \text{Function}\left[\{x\}, -\frac{\text{AiryBiPrime}[x - C[1]] + \text{AiryAiPrime}[x - C[1]]\,C[2]}{-\text{AiryBi}[x - C[1]] - \text{AiryAi}[x - C[1]]\,C[2]}\right]\right\}\right\}$$

Ai(z) exhibits a simple behavior along the real axis. Bi(z) also has complex zeros along the directions $\arg(z) \approx \pm\pi/3$. The right graphic shows these zeros.

```
In[78]:= Show[GraphicsArray[
        Block[{$DisplayFunction = Identity},
          {Plot[AiryAi[z], {z, -16, 5}],
            ContourPlot[ArcTan @ Abs[AiryBi[ x + I y]],
                        {x, -8, 8}, {y, -8, 8}, PlotPoints -> 160,
                        PlotRange -> All, ColorFunction -> (Hue[0.8#]&)]}]]]
```

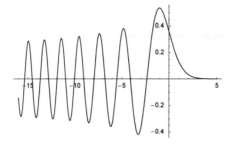
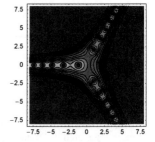

Ai(z) can also be written in integral form:

$$3^{-1/3}\,\pi\,\mathrm{Ai}(3^{-1/3}\,z) = \int_0^\infty \cos(t^3 + z\,t)\,dt.$$

This integral can be computed by *Mathematica* and expressed in terms of Bessel functions of order $1/3$ and $-1/3$. (Note that one has to use use constructions like $(z^6)^{(1/4)}$ instead of $z^{2/3}$ to avoid branch cut problems.)

Mathematica can calculate many definite integrals that contain Airy functions. Here are two examples [1255], [1054].

```
In[80]:= FourierTransform[AiryAi[x^2], x, y] // FullSimplify[#, y > 0]&
```

$$\text{Out[80]}= \frac{1}{288\sqrt{\pi}}\left(48\,2^{1/6}\,\pi\,\left(3\,\mathrm{AiryAi}\!\left[-\frac{y}{2^{2/3}}\right]\mathrm{AiryAi}\!\left[\frac{y}{2^{2/3}}\right]+\mathrm{AiryBi}\!\left[-\frac{y}{2^{2/3}}\right]\mathrm{AiryBi}\!\left[\frac{y}{2^{2/3}}\right]\right)+ \right.$$
$$\left. \sqrt{6}\,y^4\,\mathrm{HypergeometricPFQ}\!\left[\{\},\left\{\frac{4}{3},\frac{3}{2},\frac{5}{3}\right\},-\frac{y^6}{5184}\right]\right)$$

```
In[81]:= Integrate[AiryAi[x]^3, {x, 0, Infinity},
              GenerateConditions -> False] // FullSimplify
```

$$\text{Out[81]}= \frac{1}{18\,\pi^{3/2}\,\mathrm{Gamma}\!\left[\frac{1}{3}\right]\mathrm{Gamma}\!\left[\frac{7}{3}\right]}$$
$$\left(3^{1/6}\sqrt{\pi}\left(-4\,\mathrm{Gamma}\!\left[\frac{1}{3}\right]\mathrm{Hypergeometric2F1}\!\left[1,\frac{3}{2},\frac{7}{3},4\right]+\mathrm{Gamma}\!\left[\frac{7}{3}\right]\right.\right.$$
$$\left.\left(-3+6\,\mathrm{Hypergeometric2F1}\!\left[1,\frac{11}{6},\frac{8}{3},4\right]\right)\right)+$$
$$\left.\mathrm{Gamma}\!\left[\frac{1}{3}\right]\mathrm{Gamma}\!\left[\frac{7}{6}\right]\mathrm{Gamma}\!\left[\frac{7}{3}\right]\mathrm{Root}\!\left[2916+\#1^6\,\&,5\right]\right)$$

The Airy function $\mathrm{Ai}(z)$ allows for a very interesting solution of the one-dimensional (1D) Schrödinger equation: a nonspreading wave "packet" [140], [1250], [1172], [933].

```
In[82]:= ψ[x_, t_, β_] := Exp[I β^3 t/2 (x - β^3 t^2/6)] AiryAi[β (x - β^3 t^2/4)]
```

```
In[83]:= (* quick check that ψ fulfills Schrödinger equation *)
          freeParticleSchrödingerOperator[x_, t_] := (I D[#, t] + 1/2 D[#, x, x])&
          freeParticleSchrödingerOperator[x, t] @ ψ[x, t, β] // Simplify
Out[85]= 0
```

Here, the real part, the imaginary part, and the absolute value of such a wave packet are shown.

```
In[86]:= Module[{pp = 220, data},
          data = Table[ψ[x, t, 1], {t, -5, 5, 10/pp}, {x, -8, 12, 20/pp}];

          Show[GraphicsArray[ListPlot3D[#[data], Mesh -> False,
                  DisplayFunction -> Identity]& /@ {Re, Im, Abs}]]]
```

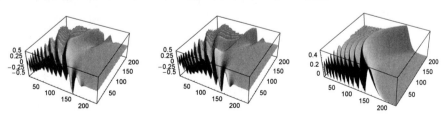

Mathematical Remark: Uniform Approximation of Linear Turning Point Problems

In applications, one often encounters the following differential equation:

$$y''(x) = \lambda^2 f(x) y(x)$$

where λ^2 is a large parameter and $f(x)$ is a real-valued function. A typical example for this type of equation is the Schrödinger equation. In many cases of interest for a given $f(x)$, an exact solution cannot be found and one has to use approximative methods. A very popular case is the so-called WKB approximation (see [893], [607], [455], [1157], [1151], [417], [289], [1163], [989], [979], [1071], [125], [52], [911], [1283], [579], [339], [1268], [138], [980], [1043], [1066], [292], [1149], [318], [1152], [482], [1192], [510], [481], [1284], [376], [1099], and [855]; for the use of Airy functions for uniform approximation of Wigner functions, see [462]). It consists of the following: We change the independent and dependent variable via

$$z(x) = \int^x \sqrt{f(x)} \, dx \quad w(z) = f^{\frac{1}{4}}(x) y(x).$$

Then, the differential equation in $w(z)$ is

$$w''(z) = \left(\lambda^2 - f^{-\frac{3}{4}}(x(z)) \frac{d^2}{dx^2} f^{-\frac{1}{4}}(x(z)) \right) w(z).$$

Neglecting the second term, we get the following two linearly independent solutions:

$$y_1(x) = f^{-\frac{1}{4}}(x) \exp\left(+\lambda \int^x \sqrt{f(x)} \, dx \right)$$

$$y_2(x) = f^{-\frac{1}{4}}(x) \exp\left(-\lambda \int^x \sqrt{f(x)} \, dx \right).$$

We see that neglecting the above term is not possible if λ^2 is not large or if x is near a zero of $f(x)$. In the case of a simple zero of $f(x)$ (the so-called linear turning point problem), a similar change of variables $y(x) \longrightarrow w(z)$

$$\frac{2}{3} z^{\frac{3}{2}}(x) = \int^x \sqrt{f(x)} \, dx, \quad w(z) = \left(\frac{f(x)}{z(x)} \right)^{\frac{1}{4}} y(x)$$

yields

$$w''(z) = \left(\lambda^2 z + \left(\frac{f(x(z))}{z} \right)^{-3/4} \frac{d^2}{dx^2} \left(\frac{f(x(z))}{z} \right)^{-1/4} \right) w(z).$$

Neglecting again the second term on the right-hand side containing a second derivative, we obtain the following two solutions this time:

$$y_1(x) = \left(\frac{z(x)}{f(x)} \right)^{\frac{1}{4}} \mathrm{Ai}\left(\lambda^{\frac{2}{3}} \int^x \sqrt{f(x)} \, dx \right)$$

$$y_2(x) = \left(\frac{z(x)}{f(x)} \right)^{\frac{1}{4}} \mathrm{Bi}\left(\lambda^{\frac{2}{3}} \int^x \sqrt{f(x)} \, dx \right).$$

To make the solutions also sensible at the turning points, we modify the transformation from x to z in the following way: If x_0 is a zero of $f(x)$, and (without loss of generality) assuming $f(x > x_0) > 0$, we choose the following:

$$\frac{2}{3} z^{\frac{3}{2}}(x) = \int_{x_0}^{x} \sqrt{f(x)}\, dx \quad x \ge x_0$$

$$\frac{2}{3} (-z)^{\frac{3}{2}}(x) = \int_{x_0}^{x} \sqrt{-f(x)}\, dx \quad x \le x_0.$$

For large λ^2, this is also a nondivergent solution around x_0. Using the asymptotic expansions for the Airy functions, the WKB solutions are the limit of appropriate linear combinations of these solutions.

Before looking at an explicit example, let us use *Mathematica* to change the independent and dependent variables in the differential equation $y''(x) = \lambda^2 f(x) y(x)$. The first of the above transformations gives the following differential equation.

In[87]:= `Remove[z, x, f, w, λ];`

`z[x_] := Integrate[Sqrt[f[x]], x]`

```
Collect[Expand[#/Coefficient[#, w''[z]]]&[
    Numerator[Together[  (* use z = z[x] *)
    (D[#, {x, 2}] - λ^2 f[x] #)&[f[x]^(-1/4) w[z[x]]]] //.
                    Integrate[Sqrt[f[x]], x] -> z]], w[z]]
```

Out[89]= $w[z] \left(-\lambda^2 + \dfrac{5\, f'[x]^2}{16\, f[x]^3} - \dfrac{f''[x]}{4\, f[x]^2}\right) + w''[z]$

The coefficient of the term $w(z)$ can be rewritten in the way given above by using the following identity.

In[90]:= `Expand[f[x]^(-3/4) D[f[x]^(-1/4), {x, 2}]]`

Out[90]= $\dfrac{5\, f'[x]^2}{16\, f[x]^3} - \dfrac{f''[x]}{4\, f[x]^2}$

In a similar way, the second of the above changes of variables yields the following result.

In[91]:= `z[x_] := (3/2 Integrate[Sqrt[f[x]], x])^(2/3)`

```
Collect[Expand[#/Coefficient[#, w''[z]]&[(* write nicely *)
    Numerator[Together[PowerExpand[ (D[#, {x, 2}] - λ^2 f[x] #)&[
        f[x]^(-1/4) z[x]^(1/4) w[z[x]]] //.
    Integrate[Sqrt[f[x]], x] -> 2/3 z^(3/2)]]]]], w[z]]
```

Out[92]= $w[z] \left(-\dfrac{5}{16\, z^2} - z\, \lambda^2 + \dfrac{5\, z\, f'[x]^2}{16\, f[x]^3} - \dfrac{z\, f''[x]}{4\, f[x]^2}\right) + w''[z]$

This can be transformed into the form above used.

In[93]:= `PowerExpand[Expand[ (f[x]/z[x])^(-3/4) *`
`                D[(f[x]/z[x])^(-1/4), {x, 2}]] //.`
`                Integrate[Sqrt[f[x]], x] -> 2/3 z^(3/2)]`

Out[93]= $-\dfrac{5}{16\, z^2} + \dfrac{5\, z\, f'[x]^2}{16\, f[x]^3} - \dfrac{z\, f''[x]}{4\, f[x]^2}$

As an example, let us investigate the well-known harmonic oscillator as an eigenvalue problem. (We set $\lambda = 1$, which can always be done, after a suitable change of variables.) The straightforward change of variables gives $y''(x) = (x^2 - \varepsilon)\, y(x)$. The eigenvalues are $\varepsilon_n = 2n + 1$ and the normalized solutions that go to zero as x goes to infinity are

$$y_n(x) = \frac{1}{\sqrt{\sqrt{\pi}\, 2^n\, n!}}\, e^{-\frac{x^2}{2}}\, H_n(x).$$

where $n = 0, 1, 2, \ldots$ and $H_n(x)$ are the Hermite polynomials, as discussed in Section 2.2. We implement this.

```
In[94]:= Remove[z, x, f, y, n, ε, w, λ, ξ];
```

```
        exactSolutionHO[n_, x_] :=
                1/Sqrt[Sqrt[Pi] 2^n n!] Exp[-x^2/2] HermiteH[n, x]
```

Let us now investigate the uniform approximation near the right turning point $x_0 = \sqrt{\varepsilon}$. We carry out the relevant integrations needed in $z = z(x)$. (We do not use definite integrals here to make sure we get a real-values branch of the integral.)

```
In[96]:= f[ε_, x_] := x^2 - ε;
```

```
In[97]:= {#, Limit[#, y -> Sqrt[ε]]}&[Integrate[Sqrt[+f[ε, y]], y]]
```

$$Out[97]= \left\{ \frac{1}{2}\, y\, \sqrt{y^2 - \varepsilon} - \frac{1}{2}\, \varepsilon\, \mathrm{Log}\left[y + \sqrt{y^2 - \varepsilon}\right],\ -\frac{1}{4}\, \varepsilon\, \mathrm{Log}[\varepsilon] \right\}$$

```
In[98]:= {#, Limit[#, y -> Sqrt[ε]]}&[Integrate[Sqrt[-f[ε, y]], y]] //
                                        FullSimplify[#, ε > 0]&
```

$$Out[98]= \left\{ \frac{1}{2}\, \left(y\, \sqrt{-y^2 + \varepsilon} + \varepsilon\, \mathrm{ArcTan}\left[\frac{y}{\sqrt{-y^2 + \varepsilon}}\right] \right),\ -\frac{\pi\, \varepsilon}{4} \right\}$$

```
In[99]:= gl = (2 ε Log[x + Sqrt[x^2 - ε]] - 2 x Sqrt[x^2 - ε] - ε Log[ε])/4;
```

```
In[100]:= gr = (Pi ε + 2 x Sqrt[ε - x^2] + 2 ε ArcTan[x/Sqrt[ε - x^2]])/4;
```

Then, we can also implement the uniform approximations.

```
In[101]:= ξ[ε_, x_] := (3/2)^(2/3) ((ε Log[Sqrt[ε]])/2 + (x Sqrt[x^2 - ε] -
                        ε Log[x + Sqrt[x^2 - ε]])/2)^(2/3) /; x >= Sqrt[ε]
```

```
        ξ[ε_, x_] := (-(3/2)^(2/3) (ε Pi/4 +
                        (ε ArcTan[(x Sqrt[ε - x^2])/(x^2 - ε)] -
                        (x Sqrt[ε - x^2]))/2)^(2/3)) /; x < Sqrt[ε]
```

To fulfill the boundary condition $y(x \to \infty) = 0$, we have to choose the solution containing the Airy function $Ai(x)$.

```
In[103]:= uniformApproximationHO[n_, x_] :=
                (ξ[2n + 1, x]/f[2n + 1, x])^(1/4) AiryAi[ξ[2n + 1, x]]
```

The expression $z(x)/f(x)$, which appeared in the neglected terms of the above differential equation, remains finite as x approaches x_0.

```
In[104]:= Simplify[Normal[Series[((3/2 gr)^(2/3)/f[ε, x])^(1/4),
                        {x, Sqrt[ε], 0}]]] // PowerExpand
```

Out[104]= $\frac{1}{2\,2^{5/12}\,(x\,\varepsilon^{3/2}-\varepsilon^2)^{1/4}}\left((x^2-4\,x\,\sqrt{\varepsilon}+7\,\varepsilon)^{1/4}\right.$

$$\left(i\,\sqrt{2}\,x\,\sqrt{x-\sqrt{\varepsilon}}\,\varepsilon^{1/4}+\left(\left(6-6\,(-1)^{\text{Floor}\left[\frac{\pi+\text{Arg}\left[x-\sqrt{\varepsilon}\right]-\text{Arg}\left[-\frac{x^2}{x+\sqrt{\varepsilon}}\right]}{2\,\pi}\right]}\,i^{2\,\text{Floor}\left[\frac{\text{Arg}\left[\frac{-2\,x^2+x\,\sqrt{\varepsilon}+\varepsilon}{x+\sqrt{\varepsilon}}\right]}{2\,\pi}\right]}\right)\pi-\right.\right.$$

$$\left.\left.\left.\frac{i\,\sqrt{2}\,\sqrt{x-\sqrt{\varepsilon}}}{\varepsilon^{1/4}}\right)\varepsilon\right)^{1/6}\right)$$

To compare the two solutions exactSolutionHO and uniformApproximationHO quantitatively, we match them at x_0.

```
In[105]:= scaling[n_] := scaling[n] =
    1/(2^(1/6) (2n + 1)^(1/12)) AiryAi[0]/exactSolutionHO[n, Sqrt[2n + 1]]
```

For comparison, let us implement the last WKB solutions. We get them by the expansion of the Airy function for large positive and negative arguments.

```
In[106]:= (* turn of message for essential singularities in Series *)
    Off[Series::esss];

    asympRight[x_] = Series[AiryAi[x], {x, Infinity, 0}] // Normal
```

Out[108]= $\dfrac{e^{-\frac{2\,x^{3/2}}{3}}\,\left(\frac{1}{x}\right)^{1/4}}{2\,\sqrt{\pi}}$

```
In[109]:= asympLeft[x_] = Series[AiryAi[x], {x, -Infinity, 0}] // Normal
```

Out[109]= $\dfrac{\text{Sin}\left[\frac{\pi}{4}+\frac{2}{3}\,(-x)^{3/2}\right]}{\sqrt{\pi}\,(-x)^{1/4}}$

```
In[110]:= WKBsolHO[n_, x_] := (ξ[2n + 1, x]/f[2n + 1, x])^(1/4) *
    If[x > Sqrt[2n + 1], asympRight[ξ[2n + 1, x]], asympLeft[ξ[2n + 1, x]]]
```

(As these formulas arise from one solution, they also solve the so-called "connection" associated with the WKB solutions, which, because of their singularity at x_0, cannot be joined smoothly together there.)

Now, let us graphically show the various solutions. (We use a form that is mostly used for visualizations of time-independent states in quantum mechanics textbooks.)

```
In[111]:= With[{nMax = 5},
    Show[Graphics[{
    (* the potential x^2 *)
    {Thickness[0.01], Plot[x^2,
                {x, -1.1 Sqrt[2nMax + 1], 1.1 Sqrt[2nMax + 1]},
                DisplayFunction -> Identity][[1, 1, 1]]},
    {Thickness[0.003], GrayLevel[0.2], Dashing[{0.002, 0.002}],
    (* the scaled WKB solutions
        (not shown to the left of the left turning point) *)
    Table[Plot[1/scaling[n] WKBsolHO[n, x] + (2n + 1),
                {x, #1 Sqrt[2n + 1], #2 Sqrt[2n + 1]},
                PlotPoints -> 60, DisplayFunction -> Identity][[1, 1, 1]],
            {n, 0, nMax}]& @@@ (* middle and right region *)
                    {{-0.99, 0.99}, {1.01, 1.60}}}},
    (* the eigenvalues *)
    {Thickness[0.002], GrayLevel[0.8],
    Table[Line[{# 1.8 Sqrt[2n + 1], 2n + 1}& /@ {-1, 1}], {n, 0, nMax}]},
```

```
(* the exact solutions *)
 {Thickness[0.002],
  Table[Plot[exactSolutionHO[n, x] + (2n + 1),
            {x, -1.8 Sqrt[2n + 1], 1.8 Sqrt[2n + 1]},
            PlotPoints -> 60, DisplayFunction -> Identity][[1, 1, 1]],
        {n, 0, nMax}]},
 (* the scaled uniform approximation solutions, not shown to
    the left of the left turning point *)
 {Thickness[0.002], GrayLevel[0.5], Dashing[{0.01, 0.01}],
  Table[Plot[1/scaling[n] uniformApproximationHO[n, x] + (2n + 1),
            {x, -0.99 Sqrt[2n + 1], 1.8 Sqrt[2n + 1]},
            PlotPoints -> 60, DisplayFunction -> Identity][[1, 1, 1]],
        {n, 0, nMax}]}}],
    Frame -> True, AspectRatio -> 0.7, FrameLabel -> None]]
```

The WKB solutions diverge at the right turning point [1045], [1191] and are not useful approximations there, whereas the uniform approximation coincides to the accuracy of the plot with the exact solution, including at the turning point itself, as long as we stay away from the second turning point $x_0 = \sqrt{\varepsilon}$. To get a solution that is also valid there, one can use a more general transformation $z = z(x)$; see the cited references for details. For the uniform approximation of the x^n-potential, see [36].

As a second example with only one turning point, let us treat the smoothed step potential [467], [15], [16], [479]:

$$y''(x) = \left(\frac{1}{1 - e^{-x}} - \varepsilon\right) y(x).$$

We implement this by proceeding exactly in the same way as above.

In[112]:= `Clear[f, gr, gl, ξ, ε, x, scaling]`

```
f[ε_, x_] := 1/(1 + Exp[-x]) - ε;
(* f[ε, x] == 0 for x -> Log[ε/(1 - ε)] *)
```

In[115]:= `{gr[ε_, x_], gl[ε_, x_]} =`
```
  With[{e = Exp[x], s = Sqrt[1 - ε]},
   {(* right side *)
    x s - Log[-1 + 1/ε] s - Log[ε/(1 - ε)] s + Log[2 - 2 ε +
    1/e - 2 ε 1/e + 2 s Sqrt[(-ε + e - ε e)/(1 + e)] + 2/e Sqrt[1 - ε]*
    Sqrt[(-ε + e - ε e)/(1 + e)]] s - ArcTan[((-1 + 2 ε + 2 ε 1/e)*
    Sqrt[(-ε + e - ε e)/(1 + e)])/(2 (-1 + ε + ε 1/e) Sqrt[ε])] Sqrt[ε] +
    (Pi (-1 + ε) ε Sqrt[1/(1 - ε)])/(2 Sqrt[(1 - ε) ε]),
    (* left side *)
    - (ArcTan[((1 - 2 ε + 2 e - 2 ε e) s Sqrt[(ε - e + ε e)/
        (1 + e)])/(2 (-1 + ε) (ε - e + ε e))] s) + Log[-1 + 2 ε +
```

```
        2ε/e + 2 Sqrt[ε] Sqrt[(ε - e + ε e)/(1 + e)] + 2/e Sqrt[ε] *
        Sqrt[(ε - e + ε e)/(1 + e)]] Sqrt[ε] + (Pi (-1 + ε) ε *
        Sqrt[1/ε])/(2 Sqrt[(1 - ε) ε])}];

In[116]:= (* ξ in the classically allowed and forbidden region *)
        ξ[ε_, x_] := ( (3/2)^(2/3) gr[ε, x]^(2/3)) /; x >= Log[ε/(1 - ε)];
        ξ[ε_, x_] := (-(3/2)^(2/3) gl[ε, x]^(2/3)) /; x <  Log[ε/(1 - ε)];

In[119]:= uniformApproximationStep[ε_, x_] :=
                    (ξ[ε, x]/f[ε, x])^(1/4) AiryAi[ξ[ε, x]];

        (* the exact solution can be expressed
            through hypergeometric functions; see below *)
        exactSolutionStep[ε_, x_] =
        ε^(-1/4) (* normalize to DiracDelta[ε - ε'] *) *
        (1/(1 + Exp[x]))^Sqrt[1 - ε] (1 - 1/(1 + Exp[x]))^(I Sqrt[ε])*
        Hypergeometric2F1[I Sqrt[ε] + Sqrt[1 - ε], I Sqrt[ε] + Sqrt[1 - ε] + 1,
                        2 Sqrt[1 - ε] + 1, 1/(1 + Exp[x])];

        (* for better comparison scale the solution *)
        scaling[ε_] := scaling[ε] =
        uniformApproximationStep[ε (1 + 10^-8)(* for the picture *),
            Log[ε/(1 - ε)]]/ exactSolutionStep[ε, Log[ε/(1 - ε)]];
```

Again, the uniform approximation agrees well with the exact solutions [228]. Here is a sketch of the potential and five solutions at different energies.

```
In[124]:= Module[{pic2, max},
        Table[(* the exact wave functions *)
        pic1[ε] = (* the exact solutions *)
        Plot[exactSolutionStep[ε, x], {x, -30, 10},
            DisplayFunction -> Identity][[1]], {ε, 1/6, 5/6, 1/6}];
        (* the potential *)
        pot = Plot[1/(1 + Exp[-x]), {x, -30, 10},
                DisplayFunction -> Identity][[1]];
        max = Max[Flatten[Abs[Table[Last /@ pic1[ε][[1, 1, 1]],
                            {ε, 1/6, 5/6, 1/6}]]]];
        (* rescale the exact solutions to fit
            in a picture together with the potential *)
        Do[pic2[ε] = Map[{0, ε} + # {1, 0.9 1/12 /max}&,
                        pic1[ε], {-2}], {ε, 1/6, 5/6, 1/6}];
        Do[pic3[ε] = (* the approximate solutions *)
            Map[{0, ε} + # {1, 0.9 1/12 /max}&,
            Plot[1/scaling[ε] uniformApproximationStep[ε, x], {x, -30, 10},
                DisplayFunction -> Identity][[1]], {-2}], {ε, 1/6, 5/6, 1/6}];
        (* show the wave functions and the potential *)
        Show[Graphics[ (* make potential a polygon *)
        {GrayLevel[0.5], Polygon[Join[pot[[1, 1, 1]],
        {{10, 0}, {10, -0.05}, {-30, -0.05}}]],
         {GrayLevel[0.5], Thickness[0.002],
          Table[Line[{{-30, ε}, {10, ε}}], {ε, 1/6, 5/6, 1/6}]},
         {GrayLevel[0], Thickness[0.002], Table[pic2[ε], {ε, 1/6, 5/6, 1/6}]},
         {Hue[0], Thickness[0.002], Table[pic3[ε], {ε, 1/6, 5/6, 1/6}]}}],
        PlotRange -> All, Frame -> True, FrameTicks -> None, AspectRatio -> 1/2]]
```

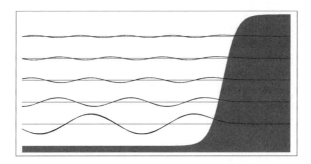

To better see the difference between the exact solution and the approximate one, let us look at the square of their difference. (The bigger the difference, the lower ε.)

```
In[125]:= (* turn off messages generated at x == Log[ε/(1 - ε)] *)
          Off[Power::infy]; Off[Infinity::indet]; Off[Plot::plnr]

          Plot[Evaluate[Table[Abs[1/scaling[ε] uniformApproximationStep[ε, x] -
                              exactSolutionStep[ε, x]]^2,
                         {ε, 1/6, 5/6, 1/6}]], {x, -8, 5},
              PlotRange -> All, Frame -> True, Axes -> False,
              FrameLabel -> {"x", None}]

          On[Power::infy]; On[Infinity::indet]; On[Plot::plnr]
```

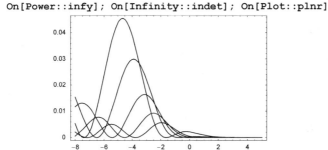

Comparing the absolute size of the error with the maximum value of the exact solution, we see that the approximation is quite good.

```
In[130]:= Table[Max[Abs[Last /@ pic1[ε][[1, 1, 1]]]], {ε, 1/6, 5/6, 1/6}]
Out[130]= {5.00095, 2.18531, 1.26949, 0.854864, 0.655851}
```

For more on Bessel and Airy functions, see the comprehensive book [1304] (and [711]). For the special role of Bessel functions as Volterra transcendentals, see [1220].

3.6 Legendre Functions

The Legendre functions $P_\nu^\mu(z)$, $Q_\nu^\mu(z)$ of the first and second kind are two linearly independent solutions of the differential equation

$$(1 - z^2)\, w''(z) - 2\, z\, w'(z) + \left(\nu(\nu + 1) - \frac{\mu^2}{1 - z^2} \right) w(z) = 0.$$

(We have already encountered this differential equation in Section 2.6.) We now try to solve this differential equation using *Mathematica*.

```
In[1]:= DSolve[(1 - z^2) w''[z] - 2z w'[z] +
            (ν(ν + 1) - μ^2/(1 - z^2)) w[z] == 0, w[z], z]
Out[1]= {{w[z] → C[1] LegendreP[ν, μ, z] + C[2] LegendreQ[ν, μ, z]}}
```

Using *Mathematica*, we get the Legendre functions $P_\nu^\mu(z)$ and $Q_\nu^\mu(z)$ as follows.

```
LegendreP[ν, μ, z]
```
   represents the Legendre function $P_\nu^\mu(z)$.

```
LegendreQ[ν, μ, z]
```
   represents the Legendre function $Q_\nu^\mu(z)$.

For ν, $\mu \in \mathbb{N}$, we get the corresponding associated Legendre polynomials, see Section 2.6. If the second variable is missing in LegendreP[ν, μ, z], the second variable is assumed to be 0, and we have the Legendre function $P_\nu(z)$. Similarly, if the second variable is missing in LegendreQ[ν, μ, z], it is assumed to be 0, and we get the Legendre function $Q_\nu(z)$.

The natural domain of the Legendre function is $-1 \le z \le 1$. To see what the Legendre functions look like on this interval, we plot $P_i^{1/2}(z)$ and $Q_i^{1/2}(z)$. (To eliminate the singularities at the endpoints, we choose the domain $-0.99 \le z \le 0.99$.)

```
In[2]:= Show[GraphicsArray[
         Plot[Evaluate[Table[#[i, 1/2, z], {i, 0, 3, 1/3}]], {z, -0.99, 0.99},
              PlotStyle -> Table[Dashing[{(i - 1)/200, (10 - i)/200}], {i, 10}],
              PlotLabel -> StyleForm[ToString[#] <> "[i, 1/2, z]",
                                 FontFamily -> "Courier", FontSize -> 7],
              DisplayFunction -> Identity]& /@ {LegendreP, LegendreQ}]]
```

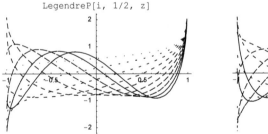
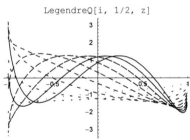

In the complex z-plane, it is easy to see the discontinuity of this function. (The discontinuity comes from a branch cut along $(-\infty, 1)$). (We use here the four argument version. See [1326] for details.)

```
In[3]:= Plot3D[Im[LegendreP[2.9, 0.4, 3, x + I y]],
            {x, -2.2, 2.2}, {y, -0.2, 0.2}, PlotPoints -> 30,
            AxesLabel -> {"Re(z)", "Im(z)", "P(z)"}]
```

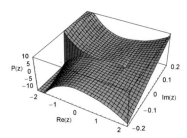

For nonnegative integers ν, μ, we get the corresponding Legendre polynomials $P_\nu^\mu(z)$. For a variety of other ν, μ, the corresponding Legendre function also reduces to a simple expression.

In[4]:= **LegendreP[1, 1/3, z]**

Out[4]= $\dfrac{(1+z)^{1/6}\,(-1+3\,z)}{2\,(1-z)^{1/6}\,\text{Gamma}[\frac{2}{3}]}$

In[5]:= **LegendreP[1/2, 7/2, z]**

Out[5]= $-\dfrac{3\,(1+z)^{7/4}\,(1+4\,z^2)}{8\,\sqrt{\pi}\,(1+\frac{1}{2}\,(-1+z))^{7/2}\,(1-z)^{7/4}}$

Here again, series expansions can be obtained.

In[6]:= **(#[n, m, z] ~ Series[#[n, m, z], {z, 0, 1}])& /@**
{LegendreP, LegendreQ} // Simplify // TraditionalForm

Out[6]//TraditionalForm=

$$\Big\{P_n^m(z) \sim \frac{2^m\,\sqrt{\pi}}{\Gamma(\tfrac{1}{2}\,(-m-n+1))\,\Gamma(\tfrac{1}{2}\,(-m+n+2))} +$$

$$\left(\frac{2^m\,\sqrt{\pi}\,m}{\Gamma(\tfrac{1}{2}\,(-m-n+1))\,\Gamma(\tfrac{1}{2}\,(-m+n+2))} + 2^{-n}\,n\,(n+1)\,{}_2\tilde{F}_1(1-n,\,-m-n;\,2-m;\,-1)\right)z + O(z^2),\ Q_n^m(z) \sim$$

$$2^{-m-1}\,\pi^{3/2}\left(\frac{4^m\,\cot(m\,\pi)}{\Gamma(\tfrac{1}{2}\,(-m-n+1))\,\Gamma(\tfrac{1}{2}\,(-m+n+2))} - \frac{\csc(m\,\pi)\,\Gamma(m+n+1)}{\Gamma(\tfrac{1}{2}\,(m-n+1))\,\Gamma(-m+n+1)\,\Gamma(\tfrac{1}{2}\,(m+n+2))}\right) + \frac{1}{2}$$

$$\pi\,\csc(m\,\pi)\Bigg(\cos(m\,\pi)\left(\frac{2^m\,\sqrt{\pi}\,m}{\Gamma(\tfrac{1}{2}\,(-m-n+1))\,\Gamma(\tfrac{1}{2}\,(-m+n+2))} + 2^{-n}\,n\,(n+1)\,{}_2\tilde{F}_1(1-n,\,-m-n;\,2-m;\,-1)\right) +$$

$$\frac{\Gamma(m+n+1)\left(\frac{2^{-m}\,m\,\sqrt{\pi}}{\Gamma(\tfrac{1}{2}\,(m-n+1))\,\Gamma(\tfrac{1}{2}\,(m+n+2))} - 2^{-n}\,n\,(n+1)\,{}_2\tilde{F}_1(1-n,\,m-n;\,m+2;\,-1)\right)}{\Gamma(-m+n+1)}\Bigg)z + O(z^2)\Big\}$$

The coefficients of the expansion at a generic point can again be expressed as Legendre functions (this follows from the recursion relation obeyed by the Legendre functions).

In[7]:= **(#[n, m, z] ~ Series[#[n, m, z], {z, Subscript[ξ, 0], 2}])& /@**
{LegendreP, LegendreQ} // Simplify // TraditionalForm

Out[7]//TraditionalForm=

$$\Big\{ P_n^m(z) \sim P_n^m(\xi_0) + \frac{(n\, P_n^m(\xi_0)\, \xi_0 - (m+n)\, P_{n-1}^m(\xi_0))\,(z - \xi_0)}{\xi_0^2 - 1} +$$

$$\frac{((n-1)\, n\, P_n^m(\xi_0)\, \xi_0^2 - (m+n)\,(2n-3)\, P_{n-1}^m(\xi_0)\, \xi_0 + (m^2 + (2n-1)\, m + (n-1)\, n)\, P_{n-2}^m(\xi_0) - n\, P_n^m(\xi_0))\,(z-\xi_0)^2}{2\,(\xi_0^2 - 1)^2}$$

$$+ O((z - \xi_0)^3),\ Q_n^m(z) \sim Q_n^m(\xi_0) + \frac{(n\, Q_n^m(\xi_0)\, \xi_0 - (m+n)\, Q_{n-1}^m(\xi_0))\,(z-\xi_0)}{\xi_0^2 - 1} + \frac{1}{2\,(\xi_0^2 - 1)^2}$$

$$(((n-1)\, n\, Q_n^m(\xi_0)\, \xi_0^2 - (m+n)\,(2n-3)\, Q_{n-1}^m(\xi_0)\, \xi_0 + (m^2 + (2n-1)\, m + (n-1)\, n)\, Q_{n-2}^m(\xi_0) - n\, Q_n^m(\xi_0))$$

$$(z - \xi_0)^2) + O((z - \xi_0)^3) \Big\}$$

The Legendre functions with index $-1/2 + i\tau$ [1275] appear frequently in various applications [778], [768]. Here is a graph showing the behavior of $P_{-1/2+i\tau}(x)$ over the τ,x-plane.

```
In[8]:= Plot3D[LegendreP[-1/2 + I τ, x], {τ, 0, 15}, {x, 1, 15},
          ViewPoint -> {1, 3, 1}, PlotPoints -> 60]
```

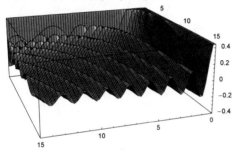

These functions occur often when applying the separation of variable method to the Laplace equation in various coordinate systems. A typical example is the following. Imagine a charge q at the z-axis at $z = R$ [768], [1309], [783], [622]. A symmetric metallic cone at zero potential is formed by $\vartheta = \vartheta_0$. Using a spherical coordinate system, the potential $\varphi(r, \vartheta)$ inside the cone is given by the following integral:

$$\varphi(r, \vartheta) \sim$$
$$\frac{1}{\sqrt{r^2 + R^2 - 2\,r\,R\cos(\vartheta)}} - \frac{1}{\sqrt{r\,R}} \int_0^\infty \frac{P_{-1/2+i\tau}(-\cos(\vartheta_0))}{P_{-1/2+i\tau}(+\cos(\vartheta_0))}\, P_{-1/2+i\tau}(\cos(\vartheta))\, \frac{\cos(\tau \log(\frac{r}{R}))}{\cosh(\pi\,\tau)}\, d\tau.$$

Let us calculate and visualize an example. We will use $\vartheta_0 = \pi/4$ and $R = 1$. Because of the quickly increasing term $\cosh(\pi\,\tau)$ in the denominator of the integrand, we carry out the numerical integration only up to $\tau = 8$. This avoids the time-consuming numerical evaluation of the Legendre functions at large imaginary indices. Because our goal here is to visualize the equipotential surfaces, we use relatively low option value settings for PrecisionGoal and AccuracyGoal.

```
In[9]:= φ[r_, ϑ_, {ϑ0_, R_}] :=
    1/Sqrt[r^2 + R^2 - 2 r R Cos[ϑ]] -
    (* carry out integral numerically *)
    1/Sqrt[R r] NIntegrate[Re[
    LegendreP[-1/2 + I τ, +Cos[ϑ]]/LegendreP[-1/2 + I τ, Cos[ϑ0]]*
    LegendreP[-1/2 + I τ, -Cos[ϑ0]] Cos[τ Log[r/R]]/Cosh[Pi τ]],
                {τ, 0, 8}, Compiled -> False,
                PrecisionGoal -> 3, AccuracyGoal -> 3]
```

φData is a list of potential values.

```
In[10]:= ppr = 16; ppϑ = 10; ε = 0.01;
        (φData = Table[φ[r, ϑ, {Pi/4, 1}], {r, ε, 2, (2 - ε)/ppr},
                        {ϑ, 0, Pi/4 - ε, (Pi/4 - ε)/ppϑ}];) // Timing
Out[11]= {27.21 Second, Null}
```

Because of the spherical symmetry, the problem is effectively a 2D one and we can use `ListContourPlot` to calculate equipotential lines.

```
In[12]:= (* make spacing of equicontour lines homogeneous *)
        ppc = 15;
        cls = #[[Round[ppr ppϑ/ppc/2]]]& /@
              Partition[Sort[Flatten[φData]], Round[ppr ppϑ/ppc]];

In[15]:= lcp = ListContourPlot[φData,
                MeshRange -> {{0, Pi/4 - ε}, {ε, 2}},
                Contours -> Drop[cls, 4], ContourShading -> False,
                DisplayFunction -> Identity];

In[16]:= (* new contour values *)
        contours = Select[Cases[Graphics[lcp], _Line, Infinity],
                          (* remove points *)
                          (Max[Flatten[Outer[#.#&[#1 - #2]&,
                                  #[[1]], #[[1]], 1]]] > ε)&];
```

By mapping from the Cartesian coordinates to spherical coordinates and by rotating the equipotential lines around the *z*-axis, we obtain the equipotential surfaces.

```
In[18]:= Needs["Graphics`Graphics3D`"]

In[19]:= to3DContours[Line[l_], ppφ_] :=
        Module[{line3D, data3D, lsp, ℛ},
          (* map lines in the x-z-plane *)
          line3D = Apply[{#2 Sin[#1], 0, #2 Cos[#1]}&, l, {1}];
          (* rotate lines around the z-axis *)
          data3D = Table[ℛ = N[{{+Cos[φ], Sin[φ], 0},
                                 {-Sin[φ], Cos[φ], 0}, {0, 0, 1}}];
                        ℛ.#& /@ line3D, {φ, 0, -Pi, -Pi/ppφ}];
          (* display equi-potential surfaces *)
          lsp = ListSurfacePlot3D[data3D, PlotRange -> All,
                                  DisplayFunction -> Identity];
          (* cut holes in polygons *)
          makeHole[#, 0.8]& /@ Cases[lsp, _Polygon, Infinity]]
```

Now, we cut holes in the equipotential surfaces to better see their nesting, and add the metallic cylinder.

```
In[20]:= makeHole[Polygon[l_], f_] :=
        Module[{mp = Plus @@ l/4, q}, q = (mp + f (# - mp))& /@ l;
              MapThread[Polygon[Join[#1, Reverse[#2]]]&,
                Partition[Append[#, First[#]], 2, 1]& /@ {l, q}]]

In[21]:= (* the metallic cone *)
        metallicCone =
        ParametricPlot3D[{x Cos[φ], -x Sin[φ], x,
                        {EdgeForm[{Thickness[0.002], GrayLevel[1]}],
                         SurfaceColor[GrayLevel[0.2]]}},
                        {x, 0, 1.65}, {φ, -Pi, 0}, DisplayFunction -> Identity];
```

Here are the resulting equipotential surfaces. The nearly spherical equipotential surfaces near the charge at *z* = 1 are nicely visible.

```
In[23]:= Show[{(* the metallic cone *) metallicCone,
         (* the equipotential surfaces *)
         Graphics3D[{EdgeForm[], MapIndexed[{SurfaceColor[Hue[#2[[1]]/22],
                   Hue[#2[[1]]/22], 3], to3DContours[#1, 12]}&, contours]}]},
         BoxRatios -> Automatic, PlotRange -> All, Boxed -> False,
         Axes -> False, Background -> GrayLevel[0.6],
         ViewPoint -> {1, -3, 1}, DisplayFunction -> $DisplayFunction]
```

For more on Legendre functions, see [604], [1067], [835], and [778]. For a detailed discussion concerning the relation of Legendre functions to hypergeometric functions, see [975].

3.7 Hypergeometric Functions

Hypergeometric functions are of extraordinary importance among the special functions. There are two reasons for this: Nearly two-thirds of all other special functions can be expressed in terms of hypergeometric functions, and hypergeometric functions can be generalized by systematically introducing additional arguments. Moreover, a lot of ordinary differential equations can be solved in hypergeometric functions [1086], [280], and using an appropriate integral transform, various special cases of partial differential equations can be reduced to hypergeometric differential equations [1050]. In addition, a wide class of partial differential equations can be solved in terms of generalized hypergeometric functions [752]. At present, mainly the hypergeometric functions of one variable have been implemented in *Mathematica*.

For details on hypergeometric functions, see [845], [8], [75], [1162], [227], [1038], [230], [1125], and [499], and the literature cited therein.

Many identities related to hypergeometric functions can today be proved by modern algorithms. We do not discuss them here; the interested reader should consult [710], [1012], [720], and [704] and the references therein.

The Gauss hypergeometric function $_2F_1(a, b; c; z)$ is defined by

$$_2F_1(a, b; c; z) = \sum_{k=0}^{\infty} \frac{(a)_k (b)_k}{(c)_k} \frac{z^k}{k!}$$

where $(d)_k$ is the Pochhammer symbol. As an immediate generalization of this function, we get the generalized hypergeometric function $_pF_q(a_1, a_2, \ldots, a_p; b_1, b_2, \ldots, b_q; z)$

$$_pF_q(a_1, a_2, \ldots, a_p; b_1, b_2, \ldots, b_q; z) = \sum_{k=0}^{\infty} \frac{(a_1)_k (a_2)_k \cdots (a_p)_k}{(b_1)_k (b_2)_k \cdots (b_q)_k} \frac{z^k}{k!}.$$

We first examine the *Mathematica* formulas for these functions.

```
Hypergeometric2F1[a, b, c, z]
```
   represents the Gauss hypergeometric function $_2F_1(a, b; c; z)$.
```
HypergeometricPFQ[{a₁, a₂, ... , aₚ}, {b₁, b₂, ... , bₚ}, z]
```
   represents the generalized hypergeometric function $_pF_q(a_1, a_2, \ldots, a_p; b_1, b_2, \ldots, b_q; z)$.

Note that P and Q are not specified in `HypergeometricPFQ`, but instead p and q are defined by the actual list lengths. Many special choices of the parameters lead to simpler functions, or can be reduced to simpler hypergeometric functions. (A short overview of such simplifications can be found in [613], and detailed tables can be found in the references listed above, especially, [1038].) We now give a few examples.

In[1]:= (* autoevaluation into elementary functions *)
   {**Hypergeometric2F1[1, 1, 1, z]**, **Hypergeometric2F1[1/2, 1/2, 3/2, z]**,
    Hypergeometric2F1[1, 1, 2, z]}

Out[2]= $\left\{ \dfrac{1}{1-z}, \dfrac{ArcSin[\sqrt{z}]}{\sqrt{z}}, -\dfrac{Log[1-z]}{z} \right\}$

In[3]:= (* autoevaluation into rational functions *)
   {**Hypergeometric2F1[6, 1/2, 1, z]**, **HypergeometricPFQ[{1, 3, 3}, {2, 2}, z]**}

Out[4]= $\left\{ \dfrac{256 - 640\,z + 960\,z^2 - 800\,z^3 + 350\,z^4 - 63\,z^5}{256\,(1-z)^{11/2}}, \dfrac{-4 + 3\,z - z^2}{4\,(-1+z)^3} \right\}$

In[5]:= (* autoevaluation into simpler hypergeometric functions *)
   HypergeometricPFQ[{2, 2, 2}, {1/2, 1}, z]

Out[6]= $\text{HypergeometricPFQ}\left[\{2, 2\}, \left\{\dfrac{1}{2}\right\}, z\right] + 8\,z\,\text{HypergeometricPFQ}\left[\{3, 3\}, \left\{\dfrac{3}{2}\right\}, z\right]$

Using the function `FunctionExpand` many hypergeometric functions can be expanded in simpler functions (like Bessel functions, error functions, or elementary functions).

In[7]:= **FunctionExpand[%]** // (* for a more compact result *) **Simplify**

Out[7]= $\dfrac{1}{8\,(-1+z)^5}$

$\left((-1+z) \left(29 + 30 \left(\dfrac{1}{z}\right)^{1/4} z^{1/4} - 51 \left(\dfrac{1}{z}\right)^{3/4} z^{3/4} + 87\,z + 30 \left(\dfrac{1}{z}\right)^{1/4} z^{5/4} - 48 \left(\dfrac{1}{z}\right)^{3/4} z^{7/4} + \right.\right.$

$\left. 4\,z^2 - 52 \left(\dfrac{1}{z}\right)^{1/4} z^{9/4} + 76 \left(\dfrac{1}{z}\right)^{3/4} z^{11/4} - 8\,z^3 - 8 \left(\dfrac{1}{z}\right)^{1/4} z^{13/4} + 16 \left(\dfrac{1}{z}\right)^{3/4} z^{15/4} \right) -$

$\left. 3\,(22\,\sqrt{1-z}\,z^{3/2} + 4\,\sqrt{1-z}\,z^{5/2} + 9\,\sqrt{-(-1+z)\,z}\,)\,ArcSin[\sqrt{z}\,] \right)$

Using `FunctionExpand` it is possible to rewrite a large amount of hypergeometric functions. Here is the last output rewritten in elementary functions. (Such transformations do not happen automatically because often the resulting expressions are quite large and are numerically not so well behaved as shorter expressions involving hypergeometric functions).

In[8]:= **FunctionExpand[%]** // **Simplify**

Out[8]= $\frac{1}{8\,(-1+z)^5}$

$$\left((-1+z)\left(29+30\left(\frac{1}{z}\right)^{1/4}z^{1/4}-51\left(\frac{1}{z}\right)^{3/4}z^{3/4}+87\,z+30\left(\frac{1}{z}\right)^{1/4}z^{5/4}-48\left(\frac{1}{z}\right)^{3/4}z^{7/4}+\right.\right.$$

$$\left.4\,z^2-52\left(\frac{1}{z}\right)^{1/4}z^{9/4}+76\left(\frac{1}{z}\right)^{3/4}z^{11/4}-8\,z^3-8\left(\frac{1}{z}\right)^{1/4}z^{13/4}+16\left(\frac{1}{z}\right)^{3/4}z^{15/4}\right)-$$

$$\left.3\,(22\,\sqrt{1-z}\,z^{3/2}+4\,\sqrt{1-z}\,z^{5/2}+9\,\sqrt{-(-1+z)\,z}\,)\,\text{ArcSin}[\sqrt{z}\,]\right)$$

For negative integer b or/and c, the hypergeometric functions degenerate into polynomials. Here is an example.

In[9]:= **Hypergeometric2F1[a, -3, c, z]**

Out[9]= $\frac{1}{c\,(1+c)\,(2+c)}\,(2\,c+3\,c^2+c^3-6\,a\,z-9\,a\,c\,z-$

$3\,a\,c^2\,z+6\,a\,z^2+6\,a^2\,z^2+3\,a\,c\,z^2+3\,a^2\,c\,z^2-2\,a\,z^3-3\,a^2\,z^3-a^3\,z^3)$

And here are two matrices $\Phi_{i,j}$ and $P_{i,j}$ whose elements are degenerate hypergeometric functions and that obey the identity $\Phi = P.P^T$ [295].

In[10]:= **Φ[i_, j_][x_, y_] :=**
  (x y)^(i + j - 2) Hypergeometric2F1[1 - i, 1 - j, 1, (y/x)^2]

  P[i_, j_][x_, y_] :=
  If[i >= j, x^(i - j) y^(i + j - 2) Binomial[i - 1, j - 1], 0]

In[12]:= **Table[Array[Φ[##][x, y]&, {d, d}] -**
        (#.Transpose[#])&[Array[P[##][x, y]&, {d, d}]] //
                      Together // Flatten // Union, {d, 12}]

Out[12]= {{0}, {0}, {0}, {0}, {0}, {0}, {0}, {0}, {0}, {0}, {0}, {0}}

There are a huge amount of cases of the hypergeometric functions when they reduce to more elementary functions. Many of them happen automatically, many more can be obtained using FunctionExpand. For special algebraic parameters hypergeometric functions sometimes reduce to algebraic numbers. Here is an example of such a case [150], [151], [1370], [54] (for a general discussion of algebraic values of transcendental functions, see [1293]).

In[14]:= **Hypergeometric2F1[1/12, 5/12, 1/2, 1323/1331] // FunctionExpand**

Out[14]= $\text{Hypergeometric2F1}\left[\frac{1}{12},\frac{5}{12},\frac{1}{2},\frac{1323}{1331}\right]$

We numerically evaluate the last expression. The fact that we get $\text{\$MaxExtraPrecisionMessages}$ indicates the fact that we have a true 0 here.

In[15]:= **% - 3/4 11^(1/4\`\`100)**

Out[15]= $0.\times10^{-101}$

More complicated functions can be expressed as hypergeometric functions, too. For instance, the following gives the partial sums of the Taylor expansion of cos.

In[16]:= **Sum[(-1)^i/(2i)! x^(2i), {i, 0, n}]**

Out[16]= $\text{Cos}\left[\sqrt{x^2}\right]+\dfrac{x^2\,(-x^2)^n\,\text{HypergeometricPFQ}\left[\{1\},\{\frac{3}{2}+n,2+n\},-\frac{x^2}{4}\right]}{\text{Gamma}[3+2\,n]}$

Here is a quick check.

In[17]:= **Expand[PowerExpand[% /. n -> 6]] == Normal[Series[Cos[x], {x, 0, 12}]]**

Out[17]= True

Using the last result and the differential equation of the hypergeometric function $_1F_2$, we can derive the differential equation

$$z \chi_n^{(3)}(z) - 2 n \chi_n''(z) + z \chi_n'(z) - 2 n \chi_n(z) = 0$$

for $\chi_n(z) = \sum_{k=0}^{n} (-1)^k z^{2k} / (2 k)!$.

```
In[18]:= Module[{x, z}, Expand @
           Table[z x'''[z] - 2 n x''[z] + z x'[z] -2 n x[z] /.
               x -> Function @@ {z, Sum[(-1)^k z^(2k)/(2k)!, {k, 0, n}]},
               {n, 0, 24, 2}]]
Out[18]= {0, 0, 0, 0, 0, 0, 0, 0, 0, 0, 0, 0, 0}
```

And here is the general solution of this third-order differential equations.

```
In[19]:= DSolve[z x'''[z] - 2 n x''[z] + z x'[z] -2 n x[z] == 0 , x[z] , z]
Out[19]= {{x[z] → e^{-i z} C[1] + e^{i z} C[2] + e^{-i z} C[3]
             ((-i z)^{-1-2 n} z^{1+2 n} Gamma[1 + 2 n, -i z] + i e^{2 i z} (i z)^{-2 n} z^{2 n} Gamma[1 + 2 n, i z])}}
```

Many related quantities, for instance, the closed form of the Padé approximation of degree (m, n) of $\cos(x)$, can be expressed in hypergeometric functions; see [526], [406], [564], and [80]. The Padé approximation of degree (m, n) of $\exp(x)$ is given as $_1F_1(-m; -m-n; x)/_1F_1(-n; -m-n; -x)$ [1313].

```
In[20]:= PadéExp[{n_, m_}, z_] = FullSimplify[
           Sum[(m + n - j)! m!/((m + n)! j! (m - j)!) z^j, {j, 0, Infinity}]/
           Sum[(m + n - j)! n!/((m + n)! j! (n - j)!) (-z)^j, {j, 0, Infinity}]]
Out[20]= Hypergeometric1F1[-m, -m - n, z]
         ─────────────────────────────────
         Hypergeometric1F1[-n, -m - n, -z]
```

```
In[21]:= Table[PadéExp[{k, k}, z], {k, 3}]
Out[21]= { 1 + z/2      1 + z/2 + z²/12      1 + z/2 + z²/10 + z³/120  }
           ───────  ,  ───────────────  ,  ────────────────────────
           1 - z/2      1 - z/2 + z²/12      1 - z/2 + z²/10 - z³/120
```

```
In[22]:= Table[Series[PadéExp[{j, j}, z] - Exp[z], {z, 0, 2j + 1}], {j, 4}]
Out[22]= { z³/12 + O[z]⁴,  -z⁵/720 + O[z]⁶,  z⁷/100800 + O[z]⁸,  -z⁹/25401600 + O[z]¹⁰ }
```

The Padé approximation of the discontinuous function $y = \text{sign}(x)$, $-1 \le x \le 1$ can also expressed through hypergeometric functions, this time the generalized hypergeometric function $_3F_2$ is needed [939].

```
In[23]:= sign[{n_, m_}, x_] := 4/Sqrt[Pi] n!/m! Gamma[m + 3/2]/Gamma[n + 1/2]*
              Gamma[n + m + 2]/Gamma[n + m + 3/2] x*
           HypergeometricPFQ[{-m, -n + 1/2, n + m + 2}, {3/2, 3/2}, x^2]/
           HypergeometricPFQ[{-n, -m - 1/2, n + m + 3/2}, {1/2, 1}, x^2]
```

For explicit integers n and m, the $\text{sign}[\{n, m\}, x]$ reduces to a rational functions in x.

```
In[24]:= Table[Subscript[sign, n, m][x] == sign[{n, m}, x],
             {n, 0, 2}, {m, 0, 2}] // Flatten // Simplify // TraditionalForm
```

Out[24]//TraditionalForm=

$$\left\{4\,x = \pi\,\mathrm{sign}_{0,0}(x),\ 16\,x^3 + 3\,\pi\,\mathrm{sign}_{0,1}(x) = 24\,x,\right.$$

$$4\,x\,(48\,x^4 - 80\,x^2 + 45) = 15\,\pi\,\mathrm{sign}_{0,2}(x),\ \mathrm{sign}_{1,0}(x) = \frac{64\,x}{15\,\pi\,x^2 + 6\,\pi},\ \mathrm{sign}_{1,1}(x) = \frac{64\,x\,(8\,x^2 + 9)}{15\,\pi\,(21\,x^2 + 2)},$$

$$\mathrm{sign}_{1,2}(x) = -\frac{128\,x\,(24\,x^4 - 100\,x^2 - 45)}{105\,\pi\,(45\,x^2 + 2)},\ \mathrm{sign}_{2,0}(x) = \frac{256\,x}{15\,\pi\left(-\frac{21\,x^4}{8} + 7\,x^2 + 1\right)},$$

$$\left.\mathrm{sign}_{2,1}(x) = \frac{8192\,x\,(10\,x^2 + 3)}{105\,\pi\,(99\,x^4 + 216\,x^2 + 8)},\ \mathrm{sign}_{2,2}(x) = \frac{4096\,x\,(56\,x^4 + 200\,x^2 + 25)}{315\,\pi\,(715\,x^4 + 440\,x^2 + 8)}\right\}$$

The following graphic shows that with increasing n and m the approximations get better and better. The right graphic shows the 6-6-approximation over the complex plane.

```
In[25]:= With[{v = 6, μ = 6},
    Show[GraphicsArray[
      Block[{$DisplayFunction = Identity},
        {(* plots of the various orders *)
        Show[Table[Plot[Evaluate[sign[{n, m}, x]], {x, -1, 1},
                    PlotStyle -> {RGBColor[n/v, 0, m/μ]},
                    DisplayFunction -> Identity],
                {n, 0, v}, {m, 0, μ}],
            Frame -> True, Axes -> False, PlotRange -> All],
        (* 3D plot of order 6-6 over the complex plane *)
        Plot3D[Re[sign[{μ, v}, x + I y]], {x, -1/2, 1/2}, {y, -1/2, 1/2},
            PlotPoints -> 161, Mesh -> False,
            ViewPoint -> {-3, -2, 3/2}]}]]]]
```

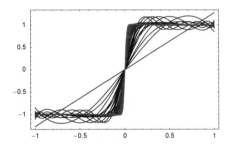 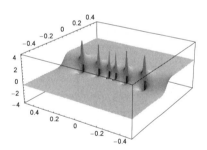

Also, the iterated integral [654] of $\exp(-z^2)$ discussed above can be expressed as a generalized hypergeometric function. We express $\exp(-z^2)$ by its series representation, integrate each term m times, and obtain the following series, which *Mathematica* can sum in closed form. (For integral representations of the iterated error function, see [501].)

```
In[27]:= iteratedErfIntegral[m_, z_] =
    Sum[(z^2)^n z^m (-1)^n Gamma[2n + 1]/Gamma[2n + m + 1]/n!,
        {n, 0, Infinity}]
```

$$\text{Out[27]}= \frac{z^m\ \mathrm{HypergeometricPFQ}\left[\left\{\frac{1}{2},\,1\right\},\,\left\{\frac{1}{2} + \frac{m}{2},\,1 + \frac{m}{2}\right\},\,-z^2\right]}{\mathrm{Gamma}[1 + m]}$$

Here is an example.

```
In[28]:= iteratedErfIntegral[6, z] // Simplify
```

$$\text{Out[28]}= \frac{1}{960}\,e^{-z^2}\left(8 + 18\,z^2 + 4\,z^4 - 4\,e^{z^2}\,(2 + 10\,z^2 + 5\,z^4) + e^{z^2}\,\sqrt{\pi}\,\sqrt{z^2}\,(15 + 20\,z^2 + 4\,z^4)\,\mathrm{Erf}\left[\sqrt{z^2}\,\right]\right)$$

Here is a quick check of the result.

In[29]:= `Table[D[iteratedErfIntegral[k, z], {z, k}] // Simplify, {k, 6}]`

Out[29]= $\left\{ e^{-z^2}, e^{-z^2}, e^{-z^2}, e^{-z^2}, e^{-z^2}, e^{-z^2} \right\}$

Many complicated indefinite and definite integrals can be expressed in terms of generalized hypergeometric functions. Here are some examples.

In[30]:= `Integrate[x^a b^x Log[x^c], x]`

Out[30]= $\dfrac{1}{(1+a)^2} (x^{1+a} (-x \, Log[b])^{-1-a}$

$(c \, x \, HypergeometricPFQ[\{1+a, 1+a\}, \{2+a, 2+a\}, x \, Log[b]] \, Log[b] \, (-x \, Log[b])^a +$
$(1+a)^2 \, (c \, Gamma[1+a] \, Log[x] - Gamma[1+a, -x \, Log[b]] \, Log[x^c])))$

In[31]:= `Integrate[x^a Sqrt[x^c - x^b], x]`

Out[31]= $-\Big(2 \, x^{1+a} \Big((2+2a+2b-c)(x^b - x^c) - (b-c) \, x^b \sqrt{1-x^{b-c}}$

$Hypergeometric2F1\Big[\dfrac{2+2a+2b-c}{2b-2c}, \dfrac{1}{2}, \dfrac{2+2a+4b-3c}{2b-2c}, x^{b-c} \Big] \Big) \Big) \Big/$

$\Big((2+2a+2b-c)(2+2a+c) \sqrt{-x^b + x^c} \Big)$

In[32]:= `Integrate[Exp[-x] Sin[x^2 + 1], {x, 0, Infinity}]`

Out[32]= $\dfrac{1}{12} \Big(-6 \, Cos[1] \, HypergeometricPFQ\Big[\{1\}, \Big\{\dfrac{3}{4}, \dfrac{5}{4}\Big\}, -\dfrac{1}{64}\Big] +$

$HypergeometricPFQ\Big[\{1\}, \Big\{\dfrac{5}{4}, \dfrac{7}{4}\Big\}, -\dfrac{1}{64}\Big] \, Sin[1] + 3\sqrt{2\pi} \Big(Cos\Big[\dfrac{5}{4}\Big] + Sin\Big[\dfrac{5}{4}\Big] \Big) \Big)$

A generalization of the Fresnel integrals, $\int_0^z \exp(i\,\xi^\alpha)\,d\xi$ can be expressed as the hypergeometric function $_1F_2$.

In[33]:= `generalizedErf[α_, z_] = Integrate[Exp[I τ^α], {τ, 0, z},`
`                     Assumptions -> (z > 0 && Re[α] > 0)]`

Out[33]= $\dfrac{1}{1+\alpha} \Big((z^{-2\alpha})^{-\frac{1+\alpha}{2\alpha}} \Big(i \, HypergeometricPFQ\Big[\Big\{\dfrac{1}{2}+\dfrac{1}{2\alpha}\Big\}, \Big\{\dfrac{3}{2}, \dfrac{3}{2}+\dfrac{1}{2\alpha}\Big\}, -\dfrac{z^{2\alpha}}{4}\Big] +$

$\sqrt{z^{-2\alpha}} \, (1+\alpha) \, HypergeometricPFQ\Big[\Big\{\dfrac{1}{2\alpha}\Big\}, \Big\{\dfrac{1}{2}, 1+\dfrac{1}{2\alpha}\Big\}, -\dfrac{z^{2\alpha}}{4}\Big] \Big) \Big)$

For $\alpha = 2$, we recover the Fresnel integrals.

In[34]:= `Sqrt[2/Pi] generalizedErf[2, Sqrt[Pi/2] z] //`
`                     FunctionExpand // Simplify[#, z > 0]&`

Out[34]= `FresnelC[z] + i FresnelS[z]`

The following graphic shows the behavior of `generalizedErf` as a function of α and z.

In[35]:= `polys = Cases[`
`    ParametricPlot3D[Evaluate[{Re[#], Im[#], α}&[generalizedErf[α, t]]],`
`                {α, 0.1, 3}, {t, 10^-6, 3}, PlotRange -> All,`
`                PlotPoints -> {50, 300}, Compiled -> False,`
`                DisplayFunction -> Identity], _Polygon, Infinity];`

In[36]:= `Show[Graphics3D[{EdgeForm[], SurfaceColor[Hue[0.12], Hue[0.8], 2.8],`
`                polys,   (* other half; use symmetry *)`
`                        Map[{-1, -1, 1}#&, polys, {-2}]} /.`
`                    (* make diamonds *)`
`                    Polygon[l_] :> Polygon[Apply[Plus,`
`                            Partition[Append[l, First[l]]/2, 2, 1], {1}]]],`
`                    BoxRatios -> {1, 1, 1.2}, Boxed -> False,`
`                    ViewPoint -> {3, 0, 2}]`

Using the generalized hypergeometric function, we can express a function $f(x)$ with the property $d^n f(x)/d x^n = f(x)$ and $d^k f(x)/d x^k \neq f(x)$ for $k < n$:

$$f(x) = {}_0F_n\!\left(; \frac{1}{n}, \frac{2}{n}, \ldots, \frac{n-1}{n}; \frac{x^n}{n^n}\right)$$

[1239]. Here are the first six of these functions. For $n = 1$ we recover the well known solution $\exp(x)$.

```
In[37]:= Table[Sum[x^(k n)/(k n)!, {k, 0, Infinity}] // Simplify,
            {n, 1, 6}] // PowerExpand
```

Out[37]= $\left\{e^x, \; \text{Cosh}[x], \; \frac{1}{3}\left(e^x + 2\, e^{-x/2}\, \text{Cos}\left[\frac{\sqrt{3}\,x}{2}\right]\right),\right.$

$\frac{1}{2}\,(\text{Cos}[x] + \text{Cosh}[x]),\; \text{HypergeometricPFQ}\left[\{\}, \left\{\frac{1}{5}, \frac{2}{5}, \frac{3}{5}, \frac{4}{5}\right\}, \frac{x^5}{3125}\right],$

$\left. \text{HypergeometricPFQ}\left[\{\}, \left\{\frac{1}{6}, \frac{1}{3}, \frac{1}{2}, \frac{2}{3}, \frac{5}{6}\right\}, \frac{x^6}{46656}\right]\right\}$

(Using similar hypergeometric functions and their inverses it is possible to find nice generalizations of the sine and cosine functions [792].)

Functions like D, Limit, and Series can operate on many special functions, especially on hypergeometric and hypergeometric-type functions. Here is a small example: The image force F of a point charge e outside a dielectric sphere with radius R and permittivity ϵ is [238]

$$F = \frac{2e\epsilon}{3+\epsilon}\left(\frac{R}{\rho}\right)^3 \frac{1}{\rho^2}\, {}_2F_1\!\left(3, \frac{\epsilon+3}{\epsilon+2}; 1 + \frac{\epsilon+3}{\epsilon+2}; \left(\frac{R}{\rho}\right)^2\right).$$

(ρ is the distance of the point charge from the sphere center, the outside of the sphere has permittivity 1 and $\rho > R$.)

```
In[38]:= F[ε_, e_, R_, ρ_] := -2 e ε/(3 + ε) (R/ρ)^3/ρ^2 *
            Hypergeometric2F1[3, (ε + 3)/(ε + 2), 1 + (ε + 3)/(ε + 2), (R/ρ)^2]
```

We will quickly derive the limits $\epsilon \to \infty$ (isolated conducting sphere), $\epsilon \to 0$ (small permittivity), $\rho \to \infty$ (large distance) and $\rho \to R$ (near distance). For a nice-looking result we apply various functions like Factor, Simplify, and FullSimplify.

```
In[39]:= {Limit[F[ε, e, R, ρ], ε -> ∞] // Factor[Numerator[#]]/Denominator[#]&,
            Series[F[ε, e, R, ρ], {ε, 0, 1}] // Normal // FunctionExpand // Simplify}
```

Out[39]= $\left\{\dfrac{R^3\, e\, (R^2 - 2\,\rho^2)}{\rho^3\, (R^2 - \rho^2)^2},\; \dfrac{e\,\epsilon\,\left(-R\rho\,(R^2 + \rho^2) + (R^2 - \rho^2)^2\, \text{ArcTanh}\left[\frac{R}{\rho}\right]\right)}{4\,(-R^2\,\rho + \rho^3)^2}\right\}$

```
In[40]:= {Series[F[ε, e, R, ρ], {ρ, Infinity, 6}] // Normal,
            (* use Series twice to get leading term only *)
```

```
Series[Normal @ Simplify[Series[F[ε, e, R, ρ], {ρ, R, -2}],
          ρ > R && R > 0], {ρ, R, -2}] // Normal // FullSimplify}
```

Out[40]= $\left\{ -\dfrac{2\,R^3\,e\,\epsilon}{(3+\epsilon)\,\rho^5}, \; -\dfrac{e\,\epsilon}{4\,(2+\epsilon)\,(R-\rho)^2} \right\}$

We recover the well known conducting sphere limit, the $(\rho - R)^{-2}$ image charge force from a plane and a ρ^{-5} large distance force.

Mathematica is not only a good tool for using special functions in problem solutions, but it is also suited for deriving old and new identities about special functions. A well-known set of identities fulfilled by the hypergeometric function $_2F_1(a, b; c; z)$ are the so-called contiguous relations [1270]. These are relations of the form

$$(\alpha_0 + \alpha_1 z)\,_2F_1(\ldots; z) + (\beta_0 + \beta_1 z)\,_2F_1(\ldots; z) + (\gamma_0 + \gamma_1 z)\,_2F_1(\ldots; z) = 0.$$

The three hypergeometric functions in the last formula have parameters that differ by ± 1. Let us derive all of these contiguous relations. Given a triple *triple*, the function `makeContiguousRelation` returns the contiguous relation if exists; else it returns 0. To determine the values of the coefficients $\alpha_0, \alpha_1, \beta_0, \beta_1, \gamma_0, \gamma_1$ we use the series expansion of the hypergeometric functions around $z = 0$. Then we compare coefficients of z to get a linear system of equations.

```
In[41]:= makeContiguousRelation[triple_] :=
    Module[{eqs, vars, ser, preSol, rhsVars, rhsRule, sol},
        (* the ansatz for the contiguous relation *)
        eqs = Sum[(C[k, 0] + C[k, 1] z) triple[[k]], {k, 3}];
        (* the variables C[…] *)
        vars = Cases[eqs, _C, ∞];
        (* make series expansion *)
        ser = CoefficientList[Series[eqs, {z, 0, 5}], z];
        (* solve for the C[…]; (NullSpace would be slightly faster) *)
        (* suppress messages and use optimal solution method *)
        Off[Solve::svars]; SetOptions[RowReduce, Method -> OneStepRowReduction];
        preSol = Solve[((# == 0)& /@ ser), vars];
        (* the undetermined C[…] *)
        rhsVars = Union[Flatten[Cases[#, _C, Infinity]& /@
                        (Last /@ preSol[[1]])]];
        (* use value 1 for the undetermined C[…] *)
        rhsRule = (# -> 1)& /@ rhsVars;
        (* values for the C[…] *)
        sol = MapAt[Together, #, 2]& /@ (preSol[[1]] /. rhsRule);
        (* return the relation in a nice form *)
        Collect[Numerator[Together[eqs /. sol /. rhsRule]],
                triple, Simplify]]
```

Here are the 35 possible triples of hypergeometric functions with arguments differing by ± 1.

```
In[42]:= allTriples =
    With[{hypergeos =
    {(* all possible hypergeometric functions with parameters ±1 *)
    Hypergeometric2F1[a, b, c, z]       , Hypergeometric2F1[a - 1, b, c, z],
    Hypergeometric2F1[a + 1, b, c, z], Hypergeometric2F1[a, b - 1, c, z],
    Hypergeometric2F1[a, b + 1, c, z], Hypergeometric2F1[a, b, c - 1, z],
    Hypergeometric2F1[a, b, c + 1, z]}},
    (* the triples *)
    Flatten[Table[{hypergeos[[i]], hypergeos[[j]], hypergeos[[k]]},
                {i, 7}, {j, i + 1, 7}, {k, j + 1, 7}], 2]];
```

We can find 25 nontrivial contiguous relations.

```
In[43]:= Off[Solve::"svars"];
        allContiguousRelations =
                  DeleteCases[makeContiguousRelation /@ allTriples, 0];
```

```
In[45]:= Length[allContiguousRelations]
```

Out[45]= 26

Here is a quick numerical check for the relations found at some "random" complex values for the parameters a, b, and c and the argument z.

```
In[46]:= Off[N::"meprec"];
        N[allContiguousRelations /. {a -> 1/5 + 9/5 I, b -> -3/2 - 4 I,
                                c -> 7/3 - 4/9 I, z -> 1/5 + 7/2 I},
           (* use high-precision arithmetic *)
           $MachinePrecision + 1] // Abs // Max
```

Out[47]= $0. \times 10^{-61}$

Here is, finally, a nicely written form of the 25 contiguous relations.

```
In[48]:= (# == 0& /@ allContiguousRelations) // TraditionalForm
```

Out[48]//TraditionalForm=

$\{(a-c)\,_2F_1(a-1, b; c; z) + (c + a\,(z-2) - b\,z)\,_2F_1(a, b; c; z) + (a - a\,z)\,_2F_1(a+1, b; c; z) = 0,$
$(c-a)\,_2F_1(a-1, b; c; z) + (b-c)\,_2F_1(a, b-1; c; z) - (a-b)(z-1)\,_2F_1(a, b; c; z) = 0,$
$(a-c)\,_2F_1(a-1, b; c; z) + (-a-b+c)\,_2F_1(a, b; c; z) + (b - b\,z)\,_2F_1(a, b+1; c; z) = 0,$
$(c-a)\,_2F_1(a-1, b; c; z) + (c-1)(z-1)\,_2F_1(a, b; c-1; z) + (a + (b-c+1)z - 1)\,_2F_1(a, b; c; z) = 0,$
$-c\,_2F_1(a-1, b; c; z) + (c - c\,z)\,_2F_1(a, b; c; z) + (c-b)\,z\,_2F_1(a, b; c+1; z) = 0,$
$(b-c)\,_2F_1(a, b-1; c; z) + (-a-b+c)\,_2F_1(a, b; c; z) + (a - a\,z)\,_2F_1(a+1, b; c; z) = 0,$
$-(a-b)(z+1)\,_2F_1(a, b; c; z) - b\,(z+1)\,_2F_1(a, b+1; c; z) + a\,(z+1)\,_2F_1(a+1, b; c; z) = 0,$
$-(c-1)(z+1)\,_2F_1(a, b; c-1; z) - (a-c+1)(z+1)\,_2F_1(a, b; c; z) + a\,(z+1)\,_2F_1(a+1, b; c; z) = 0,$
$c\,(-a-b\,z+c\,z)\,_2F_1(a, b; c; z) + (a-c)(c-b)\,z\,_2F_1(a, b; c+1; z) - a\,c\,(z-1)\,_2F_1(a+1, b; c; z) = 0,$
$(b-c)\,_2F_1(a, b-1; c; z) + (c + b\,(z-2) - a\,z)\,_2F_1(a, b; c; z) + (b - b\,z)\,_2F_1(a, b+1; c; z) = 0,$
$(c-b)\,_2F_1(a, b-1; c; z) + (c-1)(z-1)\,_2F_1(a, b; c-1; z) + (b + (a-c+1)z - 1)\,_2F_1(a, b; c; z) = 0,$
$-c\,_2F_1(a, b-1; c; z) + (c - c\,z)\,_2F_1(a, b; c; z) + (c-a)\,z\,_2F_1(a, b; c+1; z) = 0,$
$-(c-1)(z+1)\,_2F_1(a, b; c-1; z) - (b-c+1)(z+1)\,_2F_1(a, b; c; z) + b\,(z+1)\,_2F_1(a, b+1; c; z) = 0,$
$c\,(-b-a\,z+c\,z)\,_2F_1(a, b; c; z) + (a-c)(c-b)\,z\,_2F_1(a, b; c+1; z) - b\,c\,(z-1)\,_2F_1(a, b+1; c; z) = 0,$
$(c-1)\,c\,(z-1)\,_2F_1(a, b; c-1; z) +$
$\quad c\,(-2\,z\,c + c + (a+b+1)z - 1)\,_2F_1(a, b; c; z) + (a-c)(b-c)\,z\,_2F_1(a, b; c+1; z) = 0,$
$-(a-b)(a-c)\,_2F_1(a-1, b; c; z) + b\,(c + a\,(z-2) - b\,z)\,_2F_1(a, b+1; c; z) + a\,(a+b-c)\,_2F_1(a+1, b; c; z) = 0,$
$(-a^2 + 2\,c\,a - a - c^2 + c)\,_2F_1(a-1, b; c; z) +$
$\quad (c-1)(c + a\,(z-2) - b\,z)\,_2F_1(a, b; c-1; z) + a\,(a + (b-c+1)z - 1)\,_2F_1(a+1, b; c; z) = 0,$
$-c\,_2F_1(a-1, b; c; z) + c\,_2F_1(a, b-1; c; z) + (a-b)\,z\,_2F_1(a, b; c+1; z) = 0,$
$(a-c)(-b+c-1)\,_2F_1(a-1, b; c; z) -$
$\quad (a+b-c)(c-1)\,_2F_1(a, b; c-1; z) + b\,(a + (b-c+1)z - 1)\,_2F_1(a, b+1; c; z) = 0,$
$-(a-b)(b-c)\,_2F_1(a, b-1; c; z) - b\,(a+b-c)\,_2F_1(a, b+1; c; z) + a\,(-c - b\,(z-2) + a\,z)\,_2F_1(a+1, b; c; z) = 0,$
$(a-c+1)(c-b)\,_2F_1(a, b-1; c; z) -$
$\quad (a+b-c)(c-1)\,_2F_1(a, b; c-1; z) + a\,(b + (a-c+1)z - 1)\,_2F_1(a+1, b; c; z) = 0,$
$(a-b)(c-1)(z+1)\,_2F_1(a, b; c-1; z) - b\,(a-c+1)(z+1)\,_2F_1(a, b+1; c; z) +$
$\quad a\,(b-c+1)(z+1)\,_2F_1(a+1, b; c; z) = 0,\ (a-b)(a-c)(b-c)\,z\,_2F_1(a, b; c+1; z) -$
$\quad b\,c\,(a + (b-c)\,z)\,_2F_1(a, b+1; c; z) + a\,c\,(b + (a-c)\,z)\,_2F_1(a+1, b; c; z) = 0,$
$-(c-1)\,c\,(a + (b-c)\,z)\,_2F_1(a, b; c-1; z) + (b-c)(a^2 - 2\,c\,a + a + (c-1)\,c)\,z\,_2F_1(a, b; c+1; z) +$
$\quad a\,c\,(-2\,z\,c + c + (a+b+1)z - 1)\,_2F_1(a+1, b; c; z) = 0,\ (-b^2 + 2\,c\,b - b - c^2 + c)\,_2F_1(a, b-1; c; z) +$

$$(c-1)(c+b(z-2)-az)\,_2F_1(a,b;c-1;z)+b(b+(a-c+1)z-1)\,_2F_1(a,b+1;c;z)=0,$$
$$-(c-1)c(b+(a-c)z)\,_2F_1(a,b;c-1;z)+(a-c)(b^2-2cb+b+(c-1)c)z\,_2F_1(a,b;c+1;z)+$$
$$bc(-2zc+c+(a+b+1)z-1)\,_2F_1(a,b+1;c;z)=0\}$$

For a negative integer $c\ (=-m)$, the hypergeometric function $\,_2F_1(a,b;c;z)$ is infinite, although the limit value

$$\lim_{c\to-m}\frac{\,_2F_1(a,b;c;z)}{\Gamma(c)}$$

exists. We can compute this limit as follows.

> ```
> Hypergeometric2F1Regularized[a, b, c, z]
> ```
>   represents the function $\,_2F_1\,(a,b;c;z)/\Gamma\,(c)$.

$\,_2F_1(a,b;c;z)$ satisfies the following simple differential equation:

$$z(1-z)\,w''(z)+(c-(a+b+1)z)\,w'(z)-ab\,w(z)=0.$$

Let us look at this.

```
In[49]:= DSolve[z(1 - z) w''[z] + (c - (a + b + 1)z) w'[z] - a b w[z] == 0,
            w[z], z]
```
```
Out[49]= {{w[z] → C[1] Hypergeometric2F1[a, b, c, z] +
             (-1)^(1-c) z^(1-c) C[2] Hypergeometric2F1[1 + a - c, 1 + b - c, 2 - c, z]}}
```

The corresponding differential equation for the function $\,_pF_q(a_1,a_2,\dots,a_p;b_1,b_2,\dots,b_q;z)$ is [7]

$$z\frac{d}{dz}\left(z\frac{d}{dz}+b_1-1\right)\left(z\frac{d}{dz}+b_2-1\right)\cdots\left(z\frac{d}{dz}+b_q-1\right)w(z)-$$
$$z\left(z\frac{d}{dz}+a_1\right)\left(z\frac{d}{dz}+a_2\right)\cdots\left(z\frac{d}{dz}+a_p\right)w(z)=0.$$

Here, the construction of the differential equation of the generalized hypergeometric function is implemented.

```
In[50]:= makePFQODE[aList_List, bList_List, w_, z_] :=
          (* factor coefficients if possible *)
          Factor /@ Collect[ (* write ODE as ODE in w(z) *)
          Select[Factor[(* nest the differential operators *)
          z D[Fold[z D[#1, z] + (#2 - 1)#1&, w[z], Reverse[bList]], z] -
          z    Fold[z D[#1, z] + (#2)#1&, w[z], Reverse[aList]]],
                MemberQ[#, w, {0, Infinity}, Heads -> True]&],
                Table[Derivative[i][w][z],
                      {i, Max[Length[aList], Length[bList]]}]] == 0
```

When aList has two elements and bList has one element, we are back at the hypergeometric function $\,_2F_1$.

```
In[51]:= makePFQODE[{a, b}, {c}, w, z]
```
```
Out[51]= a b w[z] - (c - z - a z - b z) w'[z] + (-1 + z) z w''[z] == 0
```

```
In[52]:= DSolve[%, w[z], z]
```
```
Out[52]= {{w[z] → C[1] Hypergeometric2F1[a, b, c, z] +
             (-1)^(1-c) z^(1-c) C[2] Hypergeometric2F1[1 + a - c, 1 + b - c, 2 - c, z]}}
```

DSolve can also solve some of the differential equations of the generalized hypergeometric functions. (In the following case, we get a differential equation of order three; so the general solution is built from three linearly independent solutions.)

In[53]:= **makePFQODE[{a1, a2}, {c1, c2}, w, z]**

Out[53]= a1 a2 w[z] - (c1 c2 - z - a1 z - a2 z) w'[z] + z (-1 - c1 - c2 + z) w''[z] - z² w⁽³⁾[z] == 0

In[54]:= **DSolve[%, w[z], z]**

Out[54]= {{w[z] → C[1] HypergeometricPFQ[{a1, a2}, {c1, c2}, z] + (-1)^{1-c1} z^{1-c1} C[2]
    HypergeometricPFQ[{1 + a1 - c1, 1 + a2 - c1}, {2 - c1, 1 - c1 + c2}, z] + (-1)^{1-c2}
    z^{1-c2} C[3] HypergeometricPFQ[{1 + a1 - c2, 1 + a2 - c2}, {2 - c2, 1 + c1 - c2}, z]}}

Looking at the series representation of the hypergeometric function $_pF_q(a_1, a_2, ..., a_p; b_1, b_2, ..., b_q; z)$, one recognizes that the function has singularities at negative integer b_i. The differential equation is of course also well defined for negative integer b_i. So what will the solutions be in this case? Here is an example.

In[55]:= **makePFQODE[{a1, a2}, {-4, c2}, w, z]**

Out[55]= a1 a2 w[z] + (4 c2 + z + a1 z + a2 z) w'[z] + z (3 - c2 + z) w''[z] - z² w⁽³⁾[z] == 0

DSolve again returns three linear independent solutions, two hypergeometric functions and one Meijer G function.

In[56]:= **DSolve[%, w[z], z]**

Out[56]= {{w[z] → z⁵ C[2] HypergeometricPFQ[{5 + a1, 5 + a2}, {6, 5 + c2}, z] + (-1)^{1-c2} z^{1-c2}
    C[1] HypergeometricPFQ[{1 + a1 - c2, 1 + a2 - c2}, {-3 - c2, 2 - c2}, z] +
    C[3] MeijerG[{{1 - a1, 1 - a2}, {}}, {{0, 5}, {1 - c2}}, z]}}

The Meijer G function $G_{m,n}^{p,q}\left(z \left| \begin{array}{c} a_1, ..., a_p \\ b_1, ..., b_q \end{array} \right.\right)$ is a generalization of the hypergeometric functions $_pF_q(a_1, a_2, ..., a_p; b_1, b_2, ..., b_q; z)$. It allows us to express the logarithmic solutions of the hypergeometric differential equation.

MeijerG[{{a_1, ... , a_n}, {a_{n+1}, ... , a_p}},
　　{{b_1, ... , b_m}, {b_{m+1}, ... , b_q}}, z]
　gives the Meijer G function $G_{m,n}^{p,q}\left(z \left| \begin{array}{c} a_1, ..., a_p \\ b_1, ..., b_q \end{array} \right.\right)$.

Because of the character of this *GuideBooks*, we will not give a detailed definition of the Meijer G function here; see [845] for details.

Similarly to the generalized hypergeometric function, many indefinite and definite integrals can be expressed through Meijer G functions. Here is one example.

In[57]:= **Integrate[Exp[-x] Log[x^3 + 1], {x, 0, Infinity}]**

Out[57]= $\dfrac{\sqrt{3}\ \text{MeijerG}[\{\{0\}, \{\}\}, \{\{0, 0, \frac{1}{3}, \frac{2}{3}\}, \{\}\}, \frac{1}{27}]}{2\pi}$

Instead of generalizations of the hypergeometric function $_2F_1(a, b; c; z)$ we will now discuss a special case of it. Because so many choices for the variables are available, we do not present plots of the hypergeometric functions here. In addition to the Gaussian hypergeometric function $_2F_1(a, b; c; z)$, the so-called confluent hypergeometric functions $M(a, b, z)$ and $U(a, b, z)$ are frequently useful. They are defined by

$$_1F_1(a; b; z) = M(a, b, z) = \sum_{k=0}^{\infty} \frac{(a)_k}{(b)_k} \frac{z^k}{k!}$$

$$U(a, b, z) = \frac{\pi}{\sin(\pi b)} \left(\frac{_1F_1(a; b; z)}{\Gamma(1 + a - b) \Gamma(b)} - z^{1-b} \frac{_1F_1(1 + a - b; 2 - b; z)}{\Gamma(a) \Gamma(2 - b)} \right).$$

In *Mathematica*, these two functions are implemented under the names `Hypergeometric1F1` and `Hyper`. `geometricU`.

`Hypergeometric1F1[a, b, z]`

> represents the confluent hypergeometric function $_1F_1(a; b; z)$.

`HypergeometricU[a, b, z]`

> represents the confluent hypergeometric function $U(a, b, z)$.

Both of these functions are solutions of the confluent hypergeometric differential equation

$$z w''(z) + (b - z) w'(z) - a w(z) = 0.$$

Mathematica also solves this differential equation.

```
In[58]:= DSolve[z w''[z] + (b - z) w'[z] - a w[z] == 0, w[z], z]
Out[58]= {{w[z] → C[1] HypergeometricU[a, b, z] + C[2] LaguerreL[-a, -1 + b, z]}}
```

For special choices of the variables, confluent hypergeometric functions can also be expressed in terms of simpler functions.

```
In[59]:= {Hypergeometric1F1[a, a, z], Hypergeometric1F1[1, 2, 2z],
         Hypergeometric1F1[1/2, 3/2, -z^2]}
```
$$\text{Out[59]= } \left\{ e^z, \frac{-1 + e^{2z}}{2z}, \frac{\sqrt{\pi} \text{ Erf}[z]}{2z} \right\}$$

Again, a variety of linear combinations of hypergeometric functions of type $_1F_1(a; b; z)$ and $U(a, b, z)$ can also be reduced to simpler functions. For example, in some handbooks, one finds the formula

$$\text{Si}(z) = \frac{\pi}{2} - \frac{1}{2} i(e^{-iz} U(1, 1; iz) - e^{iz} U(1, 1; -iz)).$$

```
In[60]:= myIntegralSin[y_] =
         Pi/2 - 1/2I (Exp[-I y] HypergeometricU[1, 1, +I y] -
                      Exp[+I y] HypergeometricU[1, 1, -I y]) // FullSimplify
```
$$\text{Out[60]= } \frac{1}{2} (\pi + i \text{ Gamma}[0, -i y] - i \text{ Gamma}[0, i y])$$

```
In[61]:= % - SinIntegral[y] // FullSimplify[#, y > 0]&
Out[61]= 0
```

In Section 2.2, we showed the integrated density of states for a harmonic oscillator. Hermite polynomials (functions) can be expressed through hypergeometric $_1F_1$ functions.

```
In[62]:= HermiteH[v, z] == FullSimplify[2^v Sqrt[Pi]*
         (Hypergeometric1F1[-v/2, 1/2, z^2]/Gamma[(1 - v)/2] -
         2z Hypergeometric1F1[(1 - v)/2, 3/2, z^2]/Gamma[-v/2])]
```

Out[62]= $\text{HermiteH}[\nu, z] == \dfrac{1}{2\,\text{Gamma}[-\nu]}\left(-2\,z\,\text{Gamma}\left[\dfrac{1}{2}-\dfrac{\nu}{2}\right]\,\text{Hypergeometric1F1}\left[\dfrac{1-\nu}{2}, \dfrac{3}{2}, z^2\right]\,+\right.$

$\left.\text{Gamma}\left[-\dfrac{\nu}{2}\right]\,\text{Hypergeometric1F1}\left[-\dfrac{\nu}{2}, \dfrac{1}{2}, z^2\right]\right)$

(Not only for the harmonic oscillator, but in general are confluent hypergeometric functions very important for quantum-mechanical problems [936].) By analytically continuing α in $-\psi''(x) + \alpha/4\,x^2\,\psi(x) = \varepsilon\,\psi(x)$ from $\alpha = 1$ to $\alpha = -1$, we are led to the inverted harmonic oscillator [547], [1137], [959], [1138], [153], [1341], [557], [1139], [1141], [127], [968], [1140], [310], [772], [560], [993], [338] (for general complex α, see [356], [357]). Obviously, the inverted harmonic oscillator has a continuous spectrum. Using hypergeometric $_1F_1$ functions, it is straightforward to calculate its local density of states. We start by implementing purely real-valued wavefunctions $\psi(x)$ [3], [509].

In[63]:= `y[1][ε_, x_] := Exp[-x^2/4] Hypergeometric1F1[ε/2 + 1/4, 1/2, x^2/2]`

```
y[2][ε_, x_] := -(-1)^(3/4) x Exp[-x^2/4]*
                Hypergeometric1F1[ε/2 + 3/4, 3/2, x^2/2]
```

```
ParabolicCylinderW[+1, ε_, x_] := Cosh[-Pi ε]^(1/4)/(2 Sqrt[Pi]) *
          (Abs[Gamma[1/4 - I ε/2]] y[1][I ε, x Exp[I Pi/4]] -
     Sqrt[2] Abs[Gamma[3/4 - I ε/2]] y[2][I ε, x Exp[I Pi/4]])
```

```
ParabolicCylinderW[-1, ε_, x_] := Cosh[-Pi ε]^(1/4)/(2 Sqrt[Pi]) *
          (Abs[Gamma[1/4 - I ε/2]] y[1][I ε, x Exp[I Pi/4]] +
     Sqrt[2] Abs[Gamma[3/4 - I ε/2]] y[2][I ε, x Exp[I Pi/4]])
```

Here is a quick numerical check that these functions really satisfy the differential equation for the inverted harmonic oscillator.

In[67]:= `invertedHOSchrödingerOperator[x_, t_] := (-D[#, {x, 2}] - 1/4 x^2 # -  ε #)&`

In[68]:= `(invertedHOSchrödingerOperator[x, t] @`
`            ParabolicCylinderW[+1, ε, x]) /. {ε -> -2, x -> 4} // N[#, 22]&`
Out[68]= $0. \times 10^{-68} + 0. \times 10^{-68}\,i$

We use a δ-function normalization because of the continuous spectrum.

In[69]:= `ParabolicCylinderWNormalized[1, ε_, x_] :=`
`            ParabolicCylinderW[1, ε, x]/Sqrt[2Pi Sqrt[1 + Exp[-2Pi ε]]]`

```
ParabolicCylinderWNormalized[-1, ε_, x_] :=
            ParabolicCylinderW[-1, ε, x]/Sqrt[2Pi Sqrt[1 + Exp[-2Pi ε]]]
```

We implement symmetric and antisymmetric solutions [101] because of the reflection symmetry of the potential. Using them, it is straightforward to express the 1D local density of states.

In[72]:= `ψSymmetric[ε_, x_] := 1/Sqrt[2] (ParabolicCylinderWNormalized[+1, ε, x] +`
`                                ParabolicCylinderWNormalized[-1, ε, x])`

```
ψAsymmetric[ε_, x_] := 1/Sqrt[2] (ParabolicCylinderWNormalized[+1, ε, x] -
                                ParabolicCylinderWNormalized[-1, ε, x])
```

```
localDensityOfStates[ε_, x_] := ψSymmetric[ε, x]^2 + ψAsymmetric[ε, x]^2
```

Here is a picture of the local density of states. In the classical allowed region, one clearly sees the 1D semi-classical Thomas–Fermi contribution $(\varepsilon + x^2/4)^{-1/2}$. In the classically forbidden region, the density is very small. Near $-\varepsilon \approx x^2/4$, the density is maximal.

```
In[75]:= Plot3D[localDensityOfStates[ε, x], {x, -8., 0.}, {ε, -8., 8.},
            PlotRange -> All, PlotPoints -> {60, 120},
            Compiled -> False, ViewPoint -> {3, 1, 2}, Mesh -> False,
            AxesLabel -> {"x", "ε", None}]
```

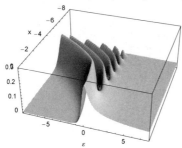

As mentioned in the beginning of this subsection, hypergeometric functions can be generalized to more than one variable. *Mathematica* has currently one of these generalizations, the Appell function $F_1(a; b_1, b_2; c; z_1, z_2)$.

AppellF1$[a, b_1, b_2, c, z]$

    represents the generalized hypergeometric function $F_1(a; b_1, b_2; c; z_1, z_2)$.

As with the generalized hypergeometric function and the Meijer G function, the Appell function F_1 allows us to express many otherwise unexpressable integrals. Here is again one example (the roots appearing in the result are the roots of the polynomial $a x^3 + b x^2 + c x + d = 0$).

```
In[76]:= Integrate[(a x^3 + b x^2 + c x + d)^e, x]
```

$$Out[76]= \frac{1}{1+e} \left((d + x (c + x (b + a x)))^e \text{ AppellF1}\left[1 + e, -e, -e, 2 + e, \right.\right.$$

$$\frac{-x + \text{Root}[d + c\, \#1 + b\, \#1^2 + a\, \#1^3\, \&,\, 1]}{\text{Root}[d + c\, \#1 + b\, \#1^2 + a\, \#1^3\, \&,\, 1] - \text{Root}[d + c\, \#1 + b\, \#1^2 + a\, \#1^3\, \&,\, 2]},$$

$$\left.\frac{-x + \text{Root}[d + c\, \#1 + b\, \#1^2 + a\, \#1^3\, \&,\, 1]}{\text{Root}[d + c\, \#1 + b\, \#1^2 + a\, \#1^3\, \&,\, 1] - \text{Root}[d + c\, \#1 + b\, \#1^2 + a\, \#1^3\, \&,\, 3]}\right]$$

$$(x - \text{Root}[d + c\, \#1 + b\, \#1^2 + a\, \#1^3\, \&,\, 1])$$

$$\left(\frac{x - \text{Root}[d + c\, \#1 + b\, \#1^2 + a\, \#1^3\, \&,\, 2]}{\text{Root}[d + c\, \#1 + b\, \#1^2 + a\, \#1^3\, \&,\, 1] - \text{Root}[d + c\, \#1 + b\, \#1^2 + a\, \#1^3\, \&,\, 2]} \right)^{-e}$$

$$\left.\left(\frac{x - \text{Root}[d + c\, \#1 + b\, \#1^2 + a\, \#1^3\, \&,\, 3]}{\text{Root}[d + c\, \#1 + b\, \#1^2 + a\, \#1^3\, \&,\, 1] - \text{Root}[d + c\, \#1 + b\, \#1^2 + a\, \#1^3\, \&,\, 3]} \right)^{-e}\right)$$

For many interesting identities related to the Appell function F_1, see [1337].

3.8 Elliptic Integrals

Elliptic integrals arise in the integration of expressions of the form

$$R\left(t, \sqrt{a_3 t^3 + a_2 t^2 + a_1 t + a_0}\right) \quad \text{and} \quad R\left(t, \sqrt{a_4 t^4 + a_3 t^3 + a_2 t^2 + a_1 t + a_0}\right)$$

where $R(x, y)$ is a rational function. When the limits of integration agree with zeros and/or poles of the integrand, the results can often be expressed in compact form. Here is an example.

```
In[1]:= Integrate[Sqrt[(1 - x)/((x - 2)(x^2 - 2x + 3))], {x, 1, 2}] //
                                                           FullSimplify
```

$$\text{Out[1]=} \quad \int_1^2 \sqrt{\frac{1-x}{-6+7x-4x^2+x^3}} \, dx$$

We distinguish the following three elliptic integrals (and present both the so-called incomplete and so-called complete elliptic integrals):

- elliptic integral of the first kind

$$F(\varphi \mid m) = \int_0^\varphi \frac{1}{\sqrt{1 - m\sin^2(\vartheta)}} \, d\vartheta \quad K(m) = \int_0^{\pi/2} \frac{1}{\sqrt{1 - m\sin^2(\vartheta)}} \, d\vartheta$$

- elliptic integral of the second kind

$$E(\varphi \mid m) = \int_0^\varphi \sqrt{1 - m\sin^2(\vartheta)} \, d\vartheta \quad E(m) = \int_0^{\pi/2} \sqrt{1 - m\sin^2(\vartheta)} \, d\vartheta$$

- elliptic integral of the third kind

$$\Pi(n; \varphi \mid m) = \int_0^\varphi \frac{1}{(1 - n\sin^2(\vartheta)) \sqrt{1 - m\sin^2(\vartheta)}} \, d\vartheta$$

$$\Pi(n \mid m) = \int_0^{\pi/2} \frac{1}{(1 - n\sin^2(\vartheta)) \sqrt{1 - m\sin^2(\vartheta)}} \, d\vartheta.$$

Here are the associated *Mathematica* commands.

```
EllipticF[φ, m]
     represents the incomplete elliptic integral F(φ | m).
EllipticK[m]
     represents the complete elliptic integral K(m).
EllipticE[φ, m]
     represents the incomplete elliptic integral E(φ | m).
EllipticE[m]
     represents the complete elliptic integral E(m).
EllipticPi[n, φ, m]
     represents the incomplete elliptic integral Π(n, φ | m).
EllipticPi[n, m]
     represents the complete elliptic integral Π(n | m).
```

We now examine two elliptic integrals for $0 \le m \le 1$ and $0 \le \varphi \le \pi/2$.

```
In[2]:= Show[GraphicsArray[{
    (* picture of EllipticF[φ, m] *)
    Plot3D[EllipticF[φ, m], {φ, 0, Pi/2 - 10^-10}, {m, 0, 1},
        PlotLabel -> StyleForm["EllipticF[φ, m]",
                FontFamily -> "Courier", FontSize -> 7],
        DisplayFunction -> Identity],
    (* picture of EllipticE[φ, m] *)
    Plot3D[EllipticE[φ, m], {φ, 0, Pi/2 - 10^-10}, {m, 0, 1},
        PlotLabel -> StyleForm["EllipticE[φ, m]",
                FontFamily -> "Courier", FontSize -> 7],
        DisplayFunction -> Identity]}]]
```

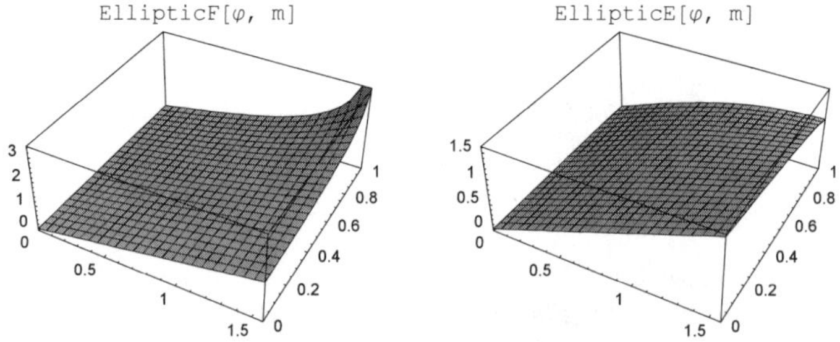

For a fixed m, the following pictures show the behavior of $\mathrm{Re}(E(\varphi\,|\,m))$ and $\mathrm{Re}(F(\varphi\,|\,m))$ over the complex φ-plane. We assemble the following pictures from five individual plots to avoid steep vertical walls from the branch cuts.

```
In[3]:= With[{ε = 10^-10},
    Show[GraphicsArray[Graphics3D[{EdgeForm[],
     SurfaceColor[Hue[0.1], Hue[0.2], 2.3],
     Cases[Block[{$DisplayFunction = Identity},
     (* make five pieces of the surface *)
     Table[Graphics3D[Plot3D[Re[#[x + I y, 0.2]],
             {x, -Pi/2 + ε + i Pi, Pi/2 - ε + i Pi},
             {y, -3/2 Pi, 3/2Pi},
             PlotPoints -> {25, 40}]], {i, -2, 2}]],
        _Polygon, Infinity]}, PlotRange -> All,
        ViewPoint -> {-1.3, -4, 1.8}, Axes -> False, Boxed -> False,
        BoxRatios -> {1, 1, 1}]& /@ {EllipticE, EllipticF}]]]
```

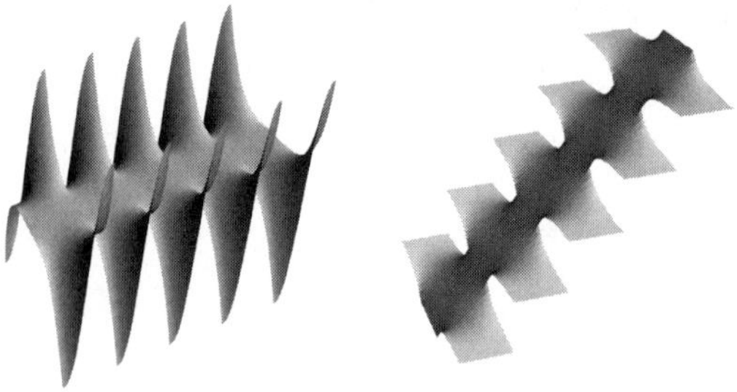

Complete elliptic integrals are special cases of hypergeometric functions.

```
In[4]:= {Hypergeometric2F1[1/2, 1/2, 1, k], Hypergeometric2F1[-1/2, 1/2, 1, k]}
```

Out[4]= $\left\{ \dfrac{2\ \text{EllipticK[k]}}{\pi},\ \dfrac{2\ \text{EllipticE[k]}}{\pi} \right\}$

Mathematica can differentiate elliptic integrals.

In[5]:= `D[EllipticK[k], k]`

Out[5]= $\dfrac{\text{EllipticE[k]} - (1 - k)\ \text{EllipticK[k]}}{2\ (1 - k)\ k}$

Also, the incomplete elliptic integrals $E(\varphi \mid m)$ and $F(\varphi \mid m)$ can be expressed through hypergeometric functions. But to do this, we need the bivariate hypergeometric Appell F_1 function.

In[6]:= `Integrate[1/Sqrt[(1 - ξ^2) (1 - m ξ^2)], {ξ, 0, x},`
         `GenerateConditions -> False] // PowerExpand`

Out[6]= $\dfrac{\sqrt{1 - x^2}\ \sqrt{1 - m\,x^2}\ \text{EllipticF[ArcSin[x], m]}}{\sqrt{-1 + x^2}\ \sqrt{-1 + m\,x^2}}$

In[7]:= `Integrate[1/Sqrt[(1 - ξ^2)] Sqrt[(1 - m ξ^2)], {ξ, 0, x},`
         `GenerateConditions -> False] // PowerExpand`

Out[7]= `EllipticE[ArcSin[x], m]`

In[8]:= `(* ±1/2 → α *)`
        `Integrate[((1 - ξ^2) (1 - m ξ^2))^α, {ξ, 0, x},`
         `GenerateConditions -> False] // PowerExpand`

Out[9]= $x\ (1 - x^2)^{-\alpha}\ (-1 + x^2)^{\alpha}\ (1 - m\,x^2)^{-\alpha}\ (-1 + m\,x^2)^{\alpha}\ \text{AppellF1}\!\left[\dfrac{1}{2},\ -\alpha,\ -\alpha,\ \dfrac{3}{2},\ x^2,\ m\,x^2\right]$

One little known application of elliptic integrals is a formula for expressing the surface area of an ellipsoid. Let $a > b > c$ be the principal axes, and then the surface area A is given by the following (see [910], [746], [766], [822], [773], [415], [1064], [196], and [1065] for the *n*D ellipsoid):

$$A = 2\pi \left(c^2 + b\sqrt{a^2 - c^2}\ F(\vartheta \mid m) + \dfrac{b\,c^2}{\sqrt{a^2 - c^2}}\ E(\vartheta \mid m) \right)$$

where

$$m = \dfrac{a^2(b^2 - c^2)}{b^2(a^2 - c^2)},\quad \vartheta = \arcsin\sqrt{1 - \dfrac{c^2}{a^2}}\ .$$

(For the surface area of an ellipsoid of revolution, see [491].)

Furthermore, the length of geodesics of a rotational ellipsoid (the earth) can be expressed in elliptic integrals of the third kind (see [635], [1314], [1272], [828], [401], [725], [474], [1233], [719], [1230], [1034], [1205], [402], [412], [1008], [452], [403], and [1206]).

Elliptic integrals (and elliptic functions) fulfill many algebraic and differential-algebraic relations. The derivatives of the functions three complete elliptic integrals $K(m)$, $E(m)$, and $\Pi(n, m)$ can be rationally expressed through $K(m)$, $E(m)$, and $\Pi(n, m)$. In the following inputs we use this fact to derive linear third-order differential equations for $\Pi(n, m)$ with respect to n and m.

In[10]:= `makeΠOde[ξ: (n | m)] :=`
         `Module[{eqs, gb, η = Complement[{n, m}, {ξ}][[1]]},`
         `(* four equations in w, w', w'', and w''' *)`
         `eqs = Table[Derivative[k][w][ξ] - D[EllipticPi[n, m], {ξ, k}],`
                    `{k, 0, 3}] // Together // Numerator;`
         `(* collect with respect to w, w', w'', and w''' *)`
         `Collect[Cases[Factor[GroebnerBasis[eqs, {},`
                `(* eliminate elliptic functions K, E, and Π *)`

```
                    {EllipticE[m], EllipticK[m], EllipticPi[n, m]}]],
                   _Plus?(MemberQ[#, w, Infinity, Heads -> True]&),
                    Infinity][[1]], Table[D[w[ξ], {ξ, k}], {k, 0, 3}]]]
```

In[11]:= `{makeΠOde[n], makeΠOde[m]}`

Out[11]= $\{-2\,w[n] + (4 + 4\,m - 16\,n)\,w'[n] + (-3\,m + 8\,n + 8\,m\,n - 13\,n^2)\,w''[n] +$
$(-2\,m\,n + 2\,n^2 + 2\,m\,n^2 - 2\,n^3)\,w^{(3)}[n],\ 3\,w[m] + (-12 + 42\,m - 6\,n)\,w'[m] +$
$(-28\,m + 44\,m^2 + 8\,n - 24\,m\,n)\,w''[m] + (-8\,m^2 + 8\,m^3 + 8\,m\,n - 8\,m^2\,n)\,w^{(3)}[m]\}$

Substitution shows the correctness of the last two differential equations.

In[12]:= `{%[[1]] /. w -> Function[n, EllipticPi[n, m]],`
        `%[[2]] /. w -> Function[m, EllipticPi[n, m]]} // Together`

Out[12]= `{0, 0}`

We will end this section with a not so well known application of elliptic integrals. Consider the differential equation (Zeilon equation):

$$\frac{\partial^3 G(x, y, z)}{\partial x^3} + \frac{\partial^3 G(x, y, z)}{\partial y^3} + \frac{\partial^3 G(x, y, z)}{\partial z^3} = \delta(x)\,\delta(y)\,\delta(z).$$

Fourier transforming this equation yields

$$\tilde{G}(k_x, k_y, k_z) = -\frac{i}{(2\pi)^{3/2}}\,\frac{1}{(k_x^3 + k_y^3 + k_z^3)}.$$

Carrying out the (nontrivial) inverse Fourier transform yields the Green's function [411], [708] of the Zeilon operator. It will contain the incomplete elliptic integral $F(u\,|\,m)$ where u is a complicated algebraic function of x, y, and z. Here is a *Mathematica* version of $G(x, y, z)$ [1290], [1291].

In[13]:= `poly[{x_, y_, z_}, ξ_] := (x^6 + y^6 + z^6 -`
              `2 (x^3 y^3 + x^3 z^3 + y^3 z^3)) ξ^3 -`
              `9 4^(1/3) (x y z)^2 ξ^2 -`
              `3 16^(1/3) x y z (x^3 + y^3 + z^3) ξ -`
              `4 (x^3 y^3 + x^3 z^3 + y^3 z^3)`

In[14]:= `polyD[{x_, y_, z_}, ξ_] = D[poly[{x, y, z}, ξ], ξ];`

In[15]:= `G[{x_, y_, z_}] :=`
         `With[{X = Abs[x], Y = Abs[y], Z = Abs[z]},`
         `(* which case? *)`
         `If[Intersection[Sign[{x, y, z}], {-1, 1}] =!= {-1, 1} &&`
             `X^(3/2) <= Y^(3/2) + Z^(3/2) &&`
             `Y^(3/2) <= X^(3/2) + Z^(3/2) &&`
             `Z^(3/2) <= X^(3/2) + Y^(3/2),`
            `-Sign[x] Beta[1/3, 1/3]/(8 Sqrt[3] Pi),`
         `(* calculate roots *)`
         `roots = Cases[NRoots[poly[{x, y, z}, ξ] == 0, ξ], _Real, {-1}];`
         `If[Length[roots] === 1, ξ = First[roots],`
            `ξ = roots[[Position[#, Max[#]]&[Abs[`
                            `polyD[{x, y, z}, #]& /@ roots]][[1, 1]]]]];`
         `2^(1/3)/(8 3^(3/4) Pi) Sign[3 2^(1/3) x y z - x^3 - y^3 - z^3]*`
         `EllipticF[ArcCos[(Sqrt[3] - 1 - ξ)/(Sqrt[3] + 1 + ξ)],`
                   `(2 + Sqrt[3])/4]]]`

Because the Zeilon operator is neither elliptic nor hyperbolic, it has an interesting structure. It has regions where it is constant (as have Green's functions of hyperbolic operators [1300], [411]) and regions where it is a smooth analytic function (as have Green's functions of elliptic operators) [887], [453], [1048], [99]. The following graphic shows $G(x, y, 0.8)$.

```
In[16]:= Needs["Graphics`Graphics3D`"]
```

```
In[17]:= ppr = 91; ppφ = 120;
        points = Table[{r Cos[φ], r Sin[φ], G[{r Cos[φ], r Sin[φ], 4/5}]} // N,
                      {r, 0, 3, 3/ppr}, {φ, Pi/4, 5Pi/4, Pi/ppφ}];
```

```
In[19]:= (* extract polygons *)
        polys = Cases[ListSurfacePlot3D[points,
                      DisplayFunction -> Identity], _Polygon, Infinity];
        (* show final graphic *)
        Show[Graphics3D[{EdgeForm[], polys, Apply[{#2, #1, #3}&, polys, {-2}]}],
            PlotRange -> {-0.13, 0.13}, BoxRatios -> {1, 1, 0.7},
            ViewPoint -> {3, 3, 2}]
```

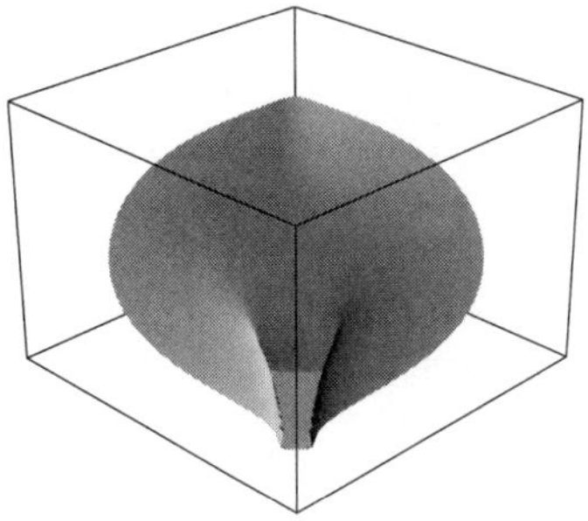

A very detailed discussion of elliptic integrals can be found in [242], [945], whereas their relation to hypergeometric functions is reviewed in [40].

After having introduced the incomplete and complete elliptic integrals, let us use the derive some of the famous modular equations for the ratio $K(1 - x)/K(x)$ [181], [130], [131], [276], [1084], [1274]. Specifically we want to derive polynomials $p_n(x, y)$, such that

$$\frac{K(1 - x)}{K(x)} = n \frac{K(1 - y)}{K(y)}$$

holds. Here n is a positive integer. To derive modular equations we will make a series expansion and compare coefficients. Because the complete elliptic integral $K(x)$ has a logarithmic singularity at $x = 1$, we exponentiate the ratio $K(1 - x)/K(x)$. To eliminate a further prefactor of $1/\pi$, it is most convenient to introduce the function $q(z)$, the elliptic nome $q(z) = \exp(-\pi K(1 - z)/K(z))$.

Rewriting the above equation in the elliptic nome, equating both side to α and solving for x and y yields the following two equations: $x = q^{-1}(\alpha)$ and $y = q^{-1}(\alpha^{1/n})$. Making a general ansatz

$$p_n(x, y) = \sum_{k,l=0}^{d(n)} c_{kl} x^k y^l$$

we will substitute the series $x(\alpha)$ and $y(\alpha)$ into $p_n(x, y)$, extract coefficients of α and use the function `Null` `Space` to extract the nontrivial c_{kl}. To avoid dealing with too large integers in intermediate calculations, we will use `NullSpace` with the `Modulus` option.

Instead of using the series of $K(x)$ for $x = 0$, 1 and deriving the series for $q^{-1}(z)$, it is more convenient to use the known product representation of $q^{-1}(z)$.

$$q^{-1}(z) = 16\,z \prod_{k=1}^{\infty} \left(\frac{z^{2k} + 1}{z^{2k-1} + 1} \right)^8$$

The function *invqSeriesPoly*[*o*][*z*] calculates, reuses, and caches the series $q^{-1}(z)$ to order *o*.

```
In[23]:= invqSeriesPoly[o_][z_] :=
        Module[{(* currently known orders *) os = #[[1, 1, 0, 1]]& /@
                 Drop[SubValues[invqSeriesPoly], -1], pos},
          (* smallest already known order > o *)
          pos = Position[os, _?(# > o&), {1}, 1];
          If[pos =!= {}, (* use known series *)
                     Normal[invqSeriesPoly[SubValues[invqSeriesPoly][[
                         pos[[1, 1]], 1, 1, 0, 1]]][z] + O[z]^o],
          invqSeriesPoly[o][z_] = (* calculate series to order o *)
          Normal[16 ζ Product[((1 + ζ^(2k))/(1 + ζ^(2k - 1)))^8,
                     {k, 1, Ceiling[o/2 + 1]}] + O[ζ]^o];
          invqSeriesPoly[o][z]]]
```

The coefficients of the series of $q^{-1}(z)$ are quickly growing.

```
In[24]:= ListPlot[Log[10, 1 + Abs[CoefficientList[
                     invqSeriesPoly[400][z], z]]] // N]
```

In the following we will derive all modular equations for $n < 10$. Using a modulus with 100 digits will turn out to be sufficient.

```
In[25]:= Needs["NumberTheory`NumberTheoryFunctions`"]
        p = NextPrime[10^30];
```

For fixed *n* and *d*, we have $(d + 1)^2$ unknown coefficients in the ansatz for $p_n(x, y)$ and a series to order *o* in α will have *no* terms. To uniquely determine the c_{kl}, we will use $o = \lceil 5/4\,(d + 1)^2/n \rceil$. This gives plenty more equations than unknowns. The following function calculateModularEquation[*n*, *d*, {*x*, *y*}, *p*] tries to find a modular equation for a fixed *d*. Its implementation is straightforward.

```
In[27]:= calculateModularEquation[n_, d_, {x_, y_}, p_:p] :=
        Module[{o = Ceiling[5/4 (d + 1)^2/n], i = invqSeriesPoly,
                 α, A, isX, isY, ser, cs, m, ns},
          (* series for lhs and rhs *)
          {isX, isY} = {i[o][α], i[n o][A] /. A -> α^(1/n)} + O[α]^o;
          (* ansatz for modular equation; linear in the c[i, j] *)
          ser = Sum[c[i, j] isX^i isY^j, {i, 0, d}, {j, 0, d}];
          (* the unknowns c[i, j] *)
          cs = Union[Cases[ser, _c, Infinity]];
          (* the coefficient matrix of the c's *)
```

```
m = Function[eq, Coefficient[eq, #]& /@ cs] /@ ser[[3]];
(* calculate modular null space *)
ns = NullSpace[m, Modulus -> p];
If[ns === {}, {}, (* lift coefficients and form equation *)
    Mod[Last[ns], p, -Floor[p/2]].cs /. c[i_, j_] :> x^i y^j]]
```

The function `calculateModularEquation[n, {x, y}]` finally starts with $d = n$ and increases d until a modular equation is found.

In[28]:= `calculateModularEquation[n_, {x_, y_}] :=`
```
        Module[{d = n, modEq},
                (* increase d (starting from n) until we find a modular equation *)
                While[(modEq = calculateModularEquation[n, d, {x, y}]) === {},
                    d = d + 1];
                (* group terms *)
                modEq //. i_Integer b_ + i_Integer c_ :> i (b + c)]
```

Here are the first nine modular equations. The smaller ones take only seconds to generate, the larger ones minutes.

In[29]:= m**E**[2] = `calculateModularEquation[ 2, {x, y}]`

Out[29]= $-16 x + 16 x y + y^2 - 2 x y^2 + x^2 y^2$

In[30]:= m**E**[3] = `calculateModularEquation[ 3, {x, y}]`

Out[30]= $x^4 - 762 x^2 y^2 + y^4 - 132 (x^3 y + x y^3) + 384 (x^2 y + x y^2 + x^3 y^2 + x^2 y^3) - 256 (x y + x^3 y^3)$

In[31]:= m**E**[4] = `calculateModularEquation[ 4, {x, y}]`

Out[31]= $-4096 (x + x^3) - 16384 x^2 y + 5632 x^2 y^2 + 2560 x^2 y^3 + y^4 + 6 x^2 y^4 + x^4 y^4 + $
$8192 (x^2 + x y + x^3 y) - 4864 (x y^2 + x^3 y^2) + 768 (x y^3 + x^3 y^3) - 4 (x y^4 + x^3 y^4)$

In[32]:= m**E**[5] = `calculateModularEquation[ 5, {x, y}]`

Out[32]= $x^6 + 691180 x^3 y^3 + y^6 + 133135 (x^4 y^2 + x^2 y^4) - 207360 (x^3 y^2 + x^2 y^3 + x^4 y^3 + x^3 y^4) - $
$133120 (x^2 y^2 + x^4 y^4) - 3590 (x^5 y + x y^5) + 43520 (x^4 y + x^5 y^2 + x y^4 + x^2 y^5) - $
$138240 (x^3 y + x y^3 + x^5 y^3 + x^3 y^5) + 163840 (x^2 y + x y^2 + x^5 y^4 + x^4 y^5) - 65536 (x y + x^5 y^5)$

In[33]:= m**E**[6] = `calculateModularEquation[ 6, {x, y}]`

Out[33]= $65536 x^4 - 262144 x^4 y + 425984 x^4 y^2 - 360448 x^4 y^3 + $
$59155456 x^4 y^4 - 118016000 x^4 y^5 + 59607296 x^4 y^6 - 615680 x^4 y^7 + $
$y^8 + 70 x^4 y^8 + x^8 y^8 - 540672 (x^3 y^2 + x^5 y^2) + 1572864 (x^2 y^2 + x^6 y^2) - $
$1048576 (x y^2 + x^7 y^2) + 1622016 (x^3 y^3 + x^5 y^3) - 4718592 (x^2 y^3 + x^6 y^3) + $
$3145728 (x y^3 + x^7 y^3) - 39868416 (x^3 y^4 + x^5 y^4) + 13764096 (x^2 y^4 + x^6 y^4) - $
$3440640 (x y^4 + x^7 y^4) + 77033472 (x^3 y^5 + x^5 y^5) - 19663872 (x^2 y^5 + x^6 y^5) + $
$1638400 (x y^5 + x^7 y^5) - 38331328 (x^3 y^6 + x^5 y^6) + 8836992 (x^2 y^6 + x^6 y^6) - $
$309312 (x y^6 + x^7 y^6) + 84928 (x^3 y^7 + x^5 y^7) + 208512 (x^2 y^7 + x^6 y^7) + $
$14400 (x y^7 + x^7 y^7) - 56 (x^3 y^8 + x^5 y^8) + 28 (x^2 y^8 + x^6 y^8) - 8 (x y^8 + x^7 y^8)$

In[34]:= m**E**[7] = `calculateModularEquation[ 7, {x, y}]`

Out[34]= $x^8 - 7639890874 x^4 y^4 + y^8 - 1905600312 (x^5 y^3 + x^3 y^5) + $
$4686427648 (x^4 y^3 + x^3 y^4 + x^5 y^4 + x^4 y^5) - 3908889600 (x^3 y^3 + x^5 y^5) + $
$133672476 (x^6 y^2 + x^2 y^6) + 63926016 (x^5 y^2 + x^6 y^3 + x^2 y^5 + x^3 y^6) - $
$916944896 (x^4 y^2 + x^2 y^4 + x^6 y^4 + x^4 y^6) + 1158348800 (x^3 y^2 + x^2 y^3 + x^6 y^5 + x^5 y^6) - $
$499580928 (x^2 y^2 + x^6 y^6) - 51464 (x^7 y + x y^7) + $
$1858304 (x^6 y + x^7 y^2 + x y^6 + x^2 y^7) - 15307264 (x^5 y + x^7 y^3 + x y^5 + x^3 y^7) + $
$50462720 (x^4 y + x y^4 + x^7 y^4 + x^4 y^7) - 78905344 (x^3 y + x y^3 + x^7 y^5 + x^5 y^7) + $
$58720256 (x^2 y + x y^2 + x^7 y^6 + x^6 y^7) - 16777216 (x y + x^7 y^7)$

In[35]:= m**E**[8] = `calculateModularEquation[ 8, {x, y}]`

Out[35]= $5368709120 \, x^4 - 4026531840 \, (x^3 + x^5) + 1610612736 \, (x^2 + x^6) -$
$268435456 \, (x + x^7) - 21474836480 \, x^4 \, y + 94355062784 \, x^4 \, y^2 - 207903260672 \, x^4 \, y^3 +$
$229027610624 \, x^4 \, y^4 - 136603762688 \, x^4 \, y^5 + 37108924416 \, x^4 \, y^6 + 121552896 \, x^4 \, y^7 +$
$y^8 + 70 \, x^4 \, y^8 + x^8 \, y^8 + 16106127360 \, (x^3 \, y + x^5 \, y) - 6442450944 \, (x^2 \, y + x^6 \, y) +$
$1073741824 \, (x \, y + x^7 \, y) - 42849009664 \, (x^3 \, y^2 + x^5 \, y^2) - 2617245696 \, (x^2 \, y^2 + x^6 \, y^2) -$
$1711276032 \, (x \, y^2 + x^7 \, y^2) + 72175583232 \, (x^3 \, y^3 + x^5 \, y^3) + 30400315392 \, (x^2 \, y^3 + x^6 \, y^3) +$
$1375731712 \, (x \, y^3 + x^7 \, y^3) - 80637394944 \, (x^3 \, y^4 + x^5 \, y^4) - 33296351232 \, (x^2 \, y^4 + x^6 \, y^4) -$
$580059136 \, (x \, y^4 + x^7 \, y^4) + 59772633088 \, (x^3 \, y^5 + x^5 \, y^5) + 8409317376 \, (x^2 \, y^5 + x^6 \, y^5) +$
$119930880 \, (x \, y^5 + x^7 \, y^5) - 20606054400 \, (x^3 \, y^6 + x^5 \, y^6) + 1927176192 \, (x^2 \, y^6 + x^6 \, y^6) -$
$9801728 \, (x \, y^6 + x^7 \, y^6) + 64647168 \, (x^3 \, y^7 + x^5 \, y^7) + 8626176 \, (x^2 \, y^7 + x^6 \, y^7) +$
$167936 \, (x \, y^7 + x^7 \, y^7) - 56 \, (x^3 \, y^8 + x^5 \, y^8) + 28 \, (x^2 \, y^8 + x^6 \, y^8) - 8 \, (x \, y^8 + x^7 \, y^8)$

In[36]:= `mE[ 9] = calculateModularEquation[ 9, {x, y}]`

Out[36]= $x^{12} - 215456569281636 \, x^6 \, y^6 + y^{12} - 149214551526168 \, (x^7 \, y^5 + x^5 \, y^7) +$
$151466682034176 \, (x^6 \, y^5 + x^5 \, y^6 + x^7 \, y^6 + x^6 \, y^7) - 70275008901120 \, (x^5 \, y^5 + x^7 \, y^7) -$
$47817388695057 \, (x^8 \, y^4 + x^4 \, y^8) + 72221649580032 \, (x^7 \, y^4 + x^8 \, y^5 + x^4 \, y^7 + x^5 \, y^8) -$
$48938555719680 \, (x^6 \, y^4 + x^4 \, y^6 + x^8 \, y^6 + x^6 \, y^8) +$
$8124308717568 \, (x^5 \, y^4 + x^4 \, y^5 + x^8 \, y^7 + x^7 \, y^8) + 5382109396992 \, (x^4 \, y^4 + x^8 \, y^8) -$
$4616302195932 \, (x^9 \, y^3 + x^3 \, y^9) + 13461347954688 \, (x^8 \, y^3 + x^9 \, y^4 + x^3 \, y^8 + x^4 \, y^9) -$
$14620704497664 \, (x^7 \, y^3 + x^9 \, y^5 + x^3 \, y^7 + x^5 \, y^9) +$
$4239280766976 \, (x^6 \, y^3 + x^3 \, y^6 + x^9 \, y^6 + x^6 \, y^9) + 2972205318144 \, (x^5 \, y^3 + x^3 \, y^5 + x^9 \, y^7 + x^7 \, y^9) -$
$1797645139968 \, (x^4 \, y^3 + x^3 \, y^4 + x^9 \, y^8 + x^8 \, y^9) + 28588376064 \, (x^3 \, y^3 + x^9 \, y^9) +$
$20502218818 \, (x^{10} \, y^2 + x^2 \, y^{10}) + 293245873152 \, (x^9 \, y^2 + x^{10} \, y^3 + x^2 \, y^9 + x^3 \, y^{10}) -$
$756142522368 \, (x^8 \, y^2 + x^{10} \, y^4 + x^2 \, y^8 + x^4 \, y^{10}) + 119541399552 \, (x^7 \, y^2 + x^{10} \, y^5 + x^2 \, y^7 + x^5 \, y^{10}) +$
$926350049280 \, (x^6 \, y^2 + x^2 \, y^6 + x^{10} \, y^6 + x^6 \, y^{10}) - 739774562304 \, (x^5 \, y^2 + x^2 \, y^5 + x^{10} \, y^7 + x^7 \, y^{10}) +$
$95428804608 \, (x^4 \, y^2 + x^2 \, y^4 + x^{10} \, y^8 + x^8 \, y^{10}) + 45097156608 \, (x^3 \, y^2 + x^2 \, y^3 + x^{10} \, y^9 + x^9 \, y^{10}) -$
$508940 \, (x^{11} \, y + x \, y^{11}) + 46550016 \, (x^{10} \, y + x^{11} \, y^2 + x \, y^{10} + x^2 \, y^{11}) -$
$818644992 \, (x^9 \, y + x^{11} \, y^3 + x \, y^9 + x^3 \, y^{11}) + 5560270848 \, (x^8 \, y + x^{11} \, y^4 + x \, y^8 + x^4 \, y^{11}) -$
$18712756224 \, (x^7 \, y + x^{11} \, y^5 + x \, y^7 + x^5 \, y^{11}) + 34527510528 \, (x^6 \, y + x \, y^6 + x^{11} \, y^6 + x^6 \, y^{11}) -$
$35634806784 \, (x^5 \, y + x \, y^5 + x^{11} \, y^7 + x^7 \, y^{11}) + 19327352832 \, (x^4 \, y + x \, y^4 + x^{11} \, y^8 + x^8 \, y^{11}) -$
$4294967296 \, (x^3 \, y + x^2 \, y^2 + x \, y^3 + x^{11} \, y^9 + x^{10} \, y^{10} + x^9 \, y^{11})$

In[37]:= `mE[10] = calculateModularEquation[10, {x, y}]`

Out[37]= $16777216 \, x^6 - 100663296 \, x^6 \, y + 264241152 \, x^6 \, y^2 - 398458880 \, x^6 \, y^3 + 3418075615068160 \, x^6 \, y^4 -$
$13672301176815616 \, x^6 \, y^5 + 22231869734322176 \, x^6 \, y^6 - 18842555650342912 \, x^6 \, y^7 +$
$8822154915289600 \, x^6 \, y^8 - 2191068209689600 \, x^6 \, y^9 + 233847292688512 \, x^6 \, y^{10} -$
$22302416512 \, x^6 \, y^{11} + y^{12} + 924 \, x^6 \, y^{12} + x^{12} \, y^{12} - 3764387840 \, (x^5 \, y^2 + x^7 \, y^2) +$
$45634027520 \, (x^4 \, y^2 + x^8 \, y^2) - 144955146240 \, (x^3 \, y^2 + x^9 \, y^2) + 171798691840 \, (x^2 \, y^2 + x^{10} \, y^2) -$
$68719476736 \, (x \, y^2 + x^{11} \, y^2) + 18821939200 \, (x^5 \, y^3 + x^7 \, y^3) - 228170137600 \, (x^4 \, y^3 + x^8 \, y^3) +$
$724775731200 \, (x^3 \, y^3 + x^9 \, y^3) - 858993459200 \, (x^2 \, y^3 + x^{10} \, y^3) +$
$343597383680 \, (x \, y^3 + x^{11} \, y^3) - 2592582416465920 \, (x^5 \, y^4 + x^7 \, y^4) +$
$1094046618419200 \, (x^4 \, y^4 + x^8 \, y^4) - 227018046177280 \, (x^3 \, y^4 + x^9 \, y^4) +$
$17235569541120 \, (x^2 \, y^4 + x^{10} \, y^4) - 719407022080 \, (x \, y^4 + x^{11} \, y^4) +$
$10370216734228480 \, (x^5 \, y^5 + x^7 \, y^5) - 4374817452851200 \, (x^4 \, y^5 + x^8 \, y^5) +$
$903723530321920 \, (x^3 \, y^5 + x^9 \, y^5) - 63788317409280 \, (x^2 \, y^5 + x^{10} \, y^5) +$
$816043786240 \, (x \, y^5 + x^{11} \, y^5) - 16876203371315200 \, (x^5 \, y^6 + x^7 \, y^6) +$
$7115470060748800 \, (x^4 \, y^6 + x^8 \, y^6) - 1442260797440000 \, (x^3 \, y^6 + x^9 \, y^6) +$
$87600712908800 \, (x^2 \, y^6 + x^{10} \, y^6) - 541463674880 \, (x \, y^6 + x^{11} \, y^6) +$
$14332930596290560 \, (x^5 \, y^7 + x^7 \, y^7) - 6035507411845120 \, (x^4 \, y^7 + x^8 \, y^7) +$
$1166794094264320 \, (x^3 \, y^7 + x^9 \, y^7) - 43150800322560 \, (x^2 \, y^7 + x^{10} \, y^7) +$
$211346784256 \, (x \, y^7 + x^{11} \, y^7) - 6723478886778880 \, (x^5 \, y^8 + x^7 \, y^8) +$
$2813719433651200 \, (x^4 \, y^8 + x^8 \, y^8) - 497891173611520 \, (x^3 \, y^8 + x^9 \, y^8) -$

$$3380639166720 \, (x^2 \, y^8 + x^{10} \, y^8) - 46191738880 \, (x \, y^8 + x^{11} \, y^8) +$$
$$1657288659128320 \, (x^5 \, y^9 + x^7 \, y^9) - 671757202278400 \, (x^4 \, y^9 + x^8 \, y^9) +$$
$$104020090695680 \, (x^3 \, y^9 + x^9 \, y^9) + 5977562145280 \, (x^2 \, y^9 + x^{10} \, y^9) +$$
$$4995153920 \, (x \, y^9 + x^{11} \, y^9) - 168184172107456 \, (x^5 \, y^{10} + x^7 \, y^{10}) +$$
$$59018375123200 \, (x^4 \, y^{10} + x^8 \, y^{10}) - 7950567090400 \, (x^3 \, y^{10} + x^9 \, y^{10}) +$$
$$192920376000 \, (x^2 \, y^{10} + x^{10} \, y^{10}) - 202645600 \, (x \, y^{10} + x^{11} \, y^{10}) -$$
$$2200531264 \, (x^5 \, y^{11} + x^7 \, y^{11}) + 10115142400 \, (x^4 \, y^{11} + x^8 \, y^{11}) + 3048452320 \, (x^3 \, y^{11} + x^9 \, y^{11}) +$$
$$186694720 \, (x^2 \, y^{11} + x^{10} \, y^{11}) + 1450080 \, (x \, y^{11} + x^{11} \, y^{11}) - 792 \, (x^5 \, y^{12} + x^7 \, y^{12}) +$$
$$495 \, (x^4 \, y^{12} + x^8 \, y^{12}) - 220 \, (x^3 \, y^{12} + x^9 \, y^{12}) + 66 \, (x^2 \, y^{12} + x^{10} \, y^{12}) - 12 \, (x \, y^{12} + x^{11} \, y^{12})$$

While we could continue to derive modular equations for larger n, because of the size of the results we will end here. Let us quickly check the correctness of the found modular equations. For a given x and a "randomly chosen" x, we solve the transcendental equation $K(1 - x)/K(x) = n \, K(1 - y)/K(y)$ numerically to high precision with respect to y and calculate $p_n(x, y)$.

```
In[38]:= (* equation to be solved *)
    KRatios[n_, {ξ_?NumericQ, y_?NumberQ}, prec_:1000] :=
    With[{η = SetPrecision[y, prec + 20], K = EllipticK},
        K[1 - ξ]/K[ξ] - n K[1 - η]/K[η]];

    (* solve the equation numerically *)
    Clear[ηN];
    ηN[n_, ξ_, prec_:2000, opts___] := ηN[n, ξ, prec, opts] =
        (η /. FindRoot[KRatios[n, {ξ, η}],
                    (* tight initial conditions for small n *)
                    {η, 10^-(3n), 1 - 10^-(3n)},
                    opts, WorkingPrecision -> prec + 20,
                    MaxIterations -> 100, AccuracyGoal -> prec] //
                        SetPrecision[#, prec/2]&)
```

A 1000-digit check for the "random" value $x = 1/e$ confirms the correctness of the found modular equations.

```
In[44]:= Table[Block[{n = v, x = 1/E}, mE[v] /. y -> ηN[v, x]], {v, 3, 10}]
Out[44]= {0. × 10^-997, 0. × 10^-996, 0. × 10^-995,
        0. × 10^-992, 0. × 10^-991, 0. × 10^-989, 0. × 10^-987, 0. × 10^-985}
```

For generalizations, see [827].

3.9 Elliptic Functions

Elliptic functions are the inverse functions of elliptic integrals. There are three families of elliptic functions: Jacobi's elliptic functions, Weierstrass's elliptic functions, and elliptic theta functions (known as Jacobi's theta functions and Neville's theta functions). *Mathematica* supports all of these three groups of elliptic functions. Here, we discuss only Jacobi's elliptic functions, the most widely used group. (The exercises contain examples to other elliptic functions.) Some properties and applications of Weierstrass's elliptic functions and elliptic theta will be dealt with in the exercises. For introductions to the subject of elliptic function, see [1047], [475], [193], [897], [1033], [419], [1213], [435], and [1281]. We have the following definition. Let

$$u(\varphi \mid m) = \int_0^\varphi \frac{1}{\sqrt{1 - m \sin^2(\vartheta)}} \, d\vartheta$$

where *m* is a parameter. Then, the Jacobi amplitude am(*u* | *m*) is defined by:

$$\varphi = \varphi(u, m) = \text{am}(u \mid m).$$

The Jacobi functions sn(*u* | *m*), cn(*u* | *m*), and dn(*u* | *m*) are defined as follows:

$$\text{sn}(u \mid m) = \sin(\text{am}(u \mid m))$$
$$\text{cn}(u \mid m) = \cos(\text{am}(u \mid m))$$
$$\text{dn}(u \mid m) = \sqrt{1 - m\sin^2(\text{am}(u \mid m))}.$$

By analogy with the trigonometric functions, we now define additional Jacobi functions via

$$\text{dc}(u \mid m) = \frac{\text{dn}(u \mid m)}{\text{cn}(u \mid m)} \quad \text{cd}(u \mid m) = \frac{\text{cn}(u \mid m)}{\text{dn}(u \mid m)} \quad \text{cs}(u \mid m) = \frac{\text{cn}(u \mid m)}{\text{sn}(u \mid m)}$$

$$\text{sc}(u \mid m) = \frac{\text{sn}(u \mid m)}{\text{cn}(u \mid m)} \quad \text{sd}(u \mid m) = \frac{\text{sn}(u \mid m)}{\text{dn}(u \mid m)} \quad \text{ds}(u \mid m) = \frac{\text{dn}(u \mid m)}{\text{sn}(u \mid m)}$$

$$\text{nc}(u \mid m) = \frac{1}{\text{cn}(u \mid m)} \quad \text{nd}(u \mid m) = \frac{1}{\text{dn}(u \mid m)} \quad \text{ns}(u \mid m) = \frac{1}{\text{sn}(u \mid m)}$$

Jacobi elliptic functions are generalizations of the trigonometric and hyperbolic functions and, in fact, reduce to them for *m* = 0 and *m* = 1, respectively. As functions of the complex variable *u*, elliptic functions are doubly periodic. In *Mathematica*, the elliptic functions are given as follows.

JacobiAmplitude[*u*, *m*]

   represents the Jacobi amplitude am(*u* | *m*).

With the elliptic function JacobiAmplitude, we can make a picture of a very interesting minimal surface: one that has six pairwise parallel edges of a cube as its boundaries. The equation of this minimal surface is *h*(*x*) *h*(*y*) = *h*(*z*), where *h*(*x*) is the inverse function of *x*(*h*). *x*(*h*) is defined via [1124], [53], [48]

$$x(h) = \int_0^h \frac{1}{\sqrt{1 + t^2 + t^4}} \, dt.$$

After writing the above integral in terms of EllipticF and inverting the relation, we can implement the following graphic.

```
In[1]:= Module[{f, g, h, ε = 10^-12, pp = 60},
    f[t_] := Which[t < 1, ArcTan[Sqrt[3] (1 + t)/(1 - t)],
                   t == 1, Pi/2,
                   t > 1, Pi + ArcTan[Sqrt[3] (1 + t)/(1 - t)]];
    (* the above integral *)
    g[t_] := 2/3N[EllipticF[f[t], 8/9] - EllipticF[Pi/3, 8/9]];
    (* the function h after inversion *)
    h[z_] := N[(Tan[#] - Sqrt[3])/(Tan[#] + Sqrt[3]) &[
             If[# < N[Pi/2], #, # - Pi] &[Chop[N[JacobiAmplitude[
                  3/2z + EllipticF[Pi/3, 8/9], 8/9]], ε]]]];
    (* display graphics *)
    Show[Graphics3D[{EdgeForm[{Thickness[0.002], GrayLevel[0.9]}],
        Graphics3D[Plot3D[g[h[x] h[y]],
          (* starting and ending an ε away from the edges of the cube *)
```

```
{x, ε, # - ε}, {y, ε, # - ε},
   PlotPoints -> pp, DisplayFunction -> Identity]][[1]],
{Thickness[0.01], GrayLevel[0],
 (* the edges of the cube that are touched *)
 Line[{{0, 0, 0}, {#, 0, 0}}], Line[{{0, 0, 0}, {0, #, 0}}],
 Line[{{#, 0, #}, {#, #, #}}], Line[{{0, #, #}, {#, #, #}}],
 Line[{{#, 0, 0}, {#, 0, #}}], Line[{{0, #, 0}, {0, #, #}}]},
{Thickness[0.01], GrayLevel[0], Dashing[{0.02, 0.02}],
 (* the other edges of the cube dashed *)
 Line[{{0, 0, 0}, {0, 0, #}}], Line[{{#, #, 0}, {#, #, #}}],
 Line[{{#, 0, 0}, {#, #, 0}}], Line[{{0, #, 0}, {#, #, 0}}],
 Line[{{0, 0, #}, {#, 0, #}}], Line[{{0, 0, #}, {0, #, #}}]}}],
Axes -> False, Boxed -> False, ViewPoint -> {1, -2, 0.9},
PlotRange -> All, BoxRatios -> {1, 1, 1}]&[
 (* the boundary coordinates *)
 2/3 (EllipticF[2Pi/3, 8/9] - EllipticF[Pi/3, 8/9]) // N]]
```

Here are five translated and rotated copies of the last surface. They fit together smoothly.

```
In[2]:= With[{L = 2/3 (EllipticF[2Pi/3, 8/9] - EllipticF[Pi/3, 8/9]) // N,
          pl = Cases[%, _Polygon | _Line, Infinity]},
     Show[Graphics3D[{EdgeForm[], Hue[0], Thickness[0.006], pl,
          Apply[({#3, #1, #2} + {1, 0, 1} L)&, pl, {-2}],
          Apply[({#1, #3, #2} + {0, 1, 1} L)&, pl, {-2}],
          Apply[({#3, #1, #2} - {1, 0, 1} L)&, pl, {-2}],
          Apply[({#1, #3, #2} - {0, 1, 1} L)&, pl, {-2}]}],
        ViewPoint -> {-3, -3, 3}, Boxed -> False]]
```

(For more complicated triple periodic minimal surfaces, see [517], [107], [808], [699].)

Jacobi*pq*[*u*, *m*] with *p*, *q* =S,C,D,N, *p* ≠ *q*

  represents the Jacobi elliptic function *pq*(*u* | *m*).

We first examine the limit cases *m* = 0, 1.

In[3]:= **allJacobis = {JacobiNS, JacobiNC, JacobiND, JacobiSN, JacobiSC, JacobiSD,
                JacobiCN, JacobiCS, JacobiCD, JacobiDN, JacobiDS, JacobiDC};**

In[4]:= **TableForm[{allJacobis, #[u, 0]& /@ allJacobis,
                    #[u, 1]& /@ allJacobis} // Transpose,
            TableHeadings -> {None, {" ", "m = 0", "m = 1"}}]**

Out[4]//TableForm=

	m = 0	m = 1
JacobiNS	Csc[u]	Coth[u]
JacobiNC	Sec[u]	Cosh[u]
JacobiND	1	Cosh[u]
JacobiSN	Sin[u]	Tanh[u]
JacobiSC	Tan[u]	Sinh[u]
JacobiSD	Sin[u]	Sinh[u]
JacobiCN	Cos[u]	Sech[u]
JacobiCS	Cot[u]	Csch[u]
JacobiCD	Cos[u]	1
JacobiDN	1	Sech[u]
JacobiDS	Csc[u]	Csch[u]
JacobiDC	Sec[u]	1

The elliptic functions satisfy equations analogous to $\sin^2 x + \cos^2 x = 1$. However, *Mathematica* does not recognize this.

In[5]:= **FullSimplify[JacobiSN[u, m]^2 + JacobiCN[u, m]^2]**

Out[5]= JacobiCN[u, m]^2 + JacobiSN[u, m]^2

The correctness of the above identity can be checked numerically for an arbitrary argument.

In[6]:= **Chop[Table[JacobiSN[#1, #2]^2 + JacobiCN[#1, #2]^2&[
                    Random[Complex], Random[Complex]], {12}]]**

Out[6]= {1., 1., 1., 1., 1., 1., 1., 1., 1., 1., 1., 1.}

Addition formulas for elliptic functions are not implemented in *Mathematica* Version 5.1. Thus, for example, we have

$$\text{sn}(u + v \mid m) = \frac{\text{sn}(u \mid m)\,\text{cn}(v \mid m)\,\text{dn}(v \mid m) + \text{sn}(v \mid m)\,\text{cn}(u \mid m)\,\text{dn}(u \mid m)}{1 - m\,\text{sn}^2(u \mid m)\,\text{sn}^2(v \mid m)}.$$

But applying Expand does not give an expansion of the following expression.

In[7]:= **JacobiSN[u + v, m] // Expand**

Out[7]= JacobiSN[u + v, m]

A function of the form EllipticExpand (doing for Jacobi*pq* what TrigExpand does for trigonometric functions) does not yet exist.

In[8]:= **JacobiSN[u + v, m] // EllipticExpand**

Out[8]= EllipticExpand[JacobiSN[u + v, m]]

But because the two fundamental identities for Jacobi elliptic functions $cn(u \mid m)^2 + sn(u \mid m)^2 = 1$ and $dn(u \mid m)^2 + m\, sn(u \mid m)^2 = 1$ are purely polynomial, when working with Jacobi elliptic functions containing formulas, one can frequently successfully use GroebnerBasis and related functions for computations and simplifications. Here is an example: The Wronskian of the three functions $\{sn(u \mid m), cn(u \mid m), dn(u \mid m)\}$ [335]. This is its immediate form.

```
In[9]:= Det[{#, D[#, u], D[#, u, u]}]&[
          {JacobiSN[u, m], JacobiCN[u, m], JacobiDN[u, m]}]
```

Out[9]= m JacobiCN[u, m]^4 JacobiDN[u, m]^2 - JacobiCN[u, m]^2 JacobiDN[u, m]^4 +
          m^2 JacobiCN[u, m]^4 JacobiSN[u, m]^2 - JacobiDN[u, m]^4 JacobiSN[u, m]^2 +
          m^2 JacobiCN[u, m]^2 JacobiSN[u, m]^4 - m JacobiDN[u, m]^2 JacobiSN[u, m]^4

Using the algebraic identities for $sn(u \mid m)$, $sn(u \mid m)$, and $sn(u \mid m)$, we obtain the simple result $m - 1$ for the Wronskian.

```
In[10]:= GroebnerBasis[{(* name Wronskian *) SNCNDNWronskian - %,
                        (* defining relations *)
                        JacobiSN[u, m]^2 + JacobiCN[u, m]^2 - 1,
                        1 - m JacobiSN[u, m]^2 - JacobiDN[u, m]^2},
                        {JacobiSN[u, m], JacobiCN[u, m], JacobiDN[u, m]}]
```

Out[10]= {1 - m + SNCNDNWronskian, -1 + m - m JacobiCN[u, m]^2 + JacobiDN[u, m]^2,
          -1 + JacobiCN[u, m]^2 + JacobiSN[u, m]^2}

The elliptic functions can be explicitly differentiated with respect to their first variable.

```
In[11]:= D[JacobiSN[u, m], u]
```

Out[11]= JacobiCN[u, m] JacobiDN[u, m]

```
In[12]:= D[JacobiAmplitude[u, m], u]
```

Out[12]= JacobiDN[u, m]

Derivatives with respect to the second variable are not explicitly calculated.

```
In[13]:= D[JacobiSN[u, m], m]
```

Out[13]= $\dfrac{1}{2\,(1-m)\,m}$ (JacobiCN[u, m] JacobiDN[u, m]
          $((1 - m)\, u$ - EllipticE[JacobiAmplitude[u, m], m] + m JacobiCD[u, m] JacobiSN[u, m]))

```
In[14]:= D[JacobiAmplitude[u, m], m]
```

Out[14]= $\dfrac{1}{2\,(-1+m)\,m}$ $(((-1 + m)\, u$ + EllipticE[JacobiAmplitude[u, m], m]) JacobiDN[u, m] -
          m JacobiCN[u, m] JacobiSN[u, m])

We now examine the Jacobi amplitude and one of the elliptic functions over the complex u-plane. The Jacobi amplitude has many branch points in the complex plane. The double periodicity of $dn(u, 1/5)$ is evident.

```
In[15]:= With[{pp = 120, m = 4/5, k = 3},
          Show[GraphicsArray[
          Plot3D[Re[#[ux + I uy, 1/5]],
                     (* the periodicity of the elliptic
                        functions is given by the elliptic integrals *)
                     {ux, -k EllipticK[m], k EllipticK[m]},
                     {uy, -k EllipticK[1 - m], k EllipticK[1 - m]},
                     MeshStyle -> {Thickness[0.002]}, PlotPoints -> pp,
                     Mesh -> False, PlotLabel ->
                     StyleForm["Re[" <> ToString[#] <> "[ux + I uy, 1/5]]",
                               FontFamily -> "Courier", FontWeight -> "Bold"],
                     DisplayFunction -> Identity]& /@ {JacobiAmplitude, JacobiDN}]]]
```

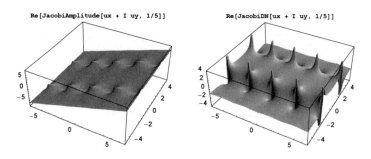

Re[JacobiAmplitude[ux + I uy, 1/5]]   Re[JacobiDN[ux + I uy, 1/5]]

Many computational rules for elliptic functions that are similar to those for trigonometric and hyperbolic functions. For example, we can find an explicit value for arguments of the form $u = a/2\,K(m) + i\,b/2\,K(1-m)$, $a, b \in \mathbb{Z}$. Here is an example.

In[16]:= **Off[Power::infy];**
Table[JacobiCD[a EllipticK[m] + I b EllipticK[1 - m], m],
    {a, -1, 1, 1/2}, {b, -1, 1, 1/2}]

Out[17]= $\left\{\left\{\text{ComplexInfinity}, -\dfrac{i}{m^{1/4}}, 0, \dfrac{i}{m^{1/4}}, \text{ComplexInfinity}\right\},\right.$

$$\left\{\frac{1}{\sqrt{1-\sqrt{1-m}}}, \frac{1-i}{\left(-i\sqrt{1-\sqrt{1-m}} + \sqrt{1+\sqrt{1-m}}\right)m^{1/4}}, \frac{1}{\sqrt{1+\sqrt{1-m}}},\right.$$

$$\frac{\sqrt{2}\left(i\sqrt{1-\sqrt{m}} + \sqrt{1+\sqrt{m}}\right)\left(\frac{1-m}{m}\right)^{1/4}}{\left(-i\sqrt{1-\sqrt{1-m}} + \sqrt{1+\sqrt{1-m}}\right)\left(i\sqrt{1-\sqrt{1-m}} + \sqrt{1+\sqrt{1-m}}\right)(1-m)^{1/4}},$$

$$\left.\frac{1}{\sqrt{1-\sqrt{1-m}}}\right\}, \left\{\frac{1}{\sqrt{m}}, \frac{1}{m^{1/4}}, 1, \frac{1}{m^{1/4}}, \frac{1}{\sqrt{m}}\right\}, \left\{\frac{1}{\sqrt{1-\sqrt{1-m}}},\right.$$

$$\frac{\sqrt{2}\left(i\sqrt{1-\sqrt{m}} + \sqrt{1+\sqrt{m}}\right)\left(\frac{1-m}{m}\right)^{1/4}}{\left(-i\sqrt{1-\sqrt{1-m}} + \sqrt{1+\sqrt{1-m}}\right)\left(i\sqrt{1-\sqrt{1-m}} + \sqrt{1+\sqrt{1-m}}\right)(1-m)^{1/4}},$$

$$\left.\frac{1}{\sqrt{1+\sqrt{1-m}}}, \frac{1-i}{\left(-i\sqrt{1-\sqrt{1-m}} + \sqrt{1+\sqrt{1-m}}\right)m^{1/4}}, \frac{1}{\sqrt{1-\sqrt{1-m}}}\right\},$$

$$\left.\left\{\text{ComplexInfinity}, \frac{i}{m^{1/4}}, 0, -\frac{i}{m^{1/4}}, \text{ComplexInfinity}\right\}\right\}$$

Elliptic functions play a major role in:
- The analytical solution of gyrating motions of rigid and deformable bodies (see [1059], [566], [826], [750], [777], [848], [533], [766], [924], [1020], [440], [234], [740], [1323], [821], [1353], and [272])
- Celestial mechanics [224], [286], [563], [1332], [225], [483], [824], [349], [1288], [222], [223], [1318], [118]
- Modelling chaotic systems [1248], [1249], [201], [202], [767], [632], [705], [685], [974], [932], [530]
- Ordinary [862], [77], [714], [516], [208], [1269] and supersymmetric quantum mechanics [414], [416], [682], [661], [370], [662], [51], [456]
- Electrodynamics [70]
- Modeling double well systems [1053]
- Semiclassical quantum mechanics [267], [1354]
- Nonlinear quantum mechanics [262], [309], [576], [875], [263]
- Quantum field theory [925], [388], [205]
- Modelling traffic jams [575], [929]

- Modelling solids [787], [788], [424], [1135], [1361], [122], [1232], [683]
- Statistical mechanics [1116]
- Closed form solutions of nonlinear partial differential equations [307], [506], [863], [408], [1035], [370], [371], [372], [218], [203], [371], [308], [1005], [858], [1089], [902], [660], [930]
- Writing the Painlevé equations in Hamiltonian form [836], [837], [1209], [1210]
- Schwarz–Christoffel maps [966]
- Quasiperiodic functions [353]
- And many more fields [121], [240], [12]

Here is an application of elliptic functions to numerical analysis. We construct an analytic functions that approximates the characteristic function of the interval $[-1, 1]$, namely $\chi_{[-1,1]}(z) = \theta(z + 1)\,\theta(1 - z)$ as $m \to 1$ [1188].

```
In[18]:= χApprox[z_, m_] :=
            (1 + m JacobiSN[2 EllipticK[m]/(I Pi) Log[(z - 1)/(z + 1)] -
                        EllipticK[m] + I EllipticK[1 - m], m])/(1 + m)
```

```
In[19]:= With[{pp = 24},
            Plot[Evaluate[Table[χApprox[z, m], {m, 1/pp, 1 - 1/pp, 1/pp}]], {z, -2, 2},
                PlotStyle -> Table[{Thickness[0.002], Hue[0.78 m]},
                                {m, 1/pp, 1 - 1/pp, 1/pp}],
                Frame -> True, Axes -> False, PlotRange -> All]]
```

One of the simplest applications of elliptic functions in this connection is the mathematical pendulum.

Physical Application: Mathematical Pendulum

Newton's law for the angle of deflection of a mathematical pendulum as a function of time can be derived from the torque operating on the body [46]: $l\,\varphi''(t) = -g\sin(\varphi(t))$. Here, l is the length of the pendulum, and g is the acceleration of the earth's gravity.

Conservation of energy (which mathematically means that the independent variable t does not appear explicitly in the differential equation) makes it easy to find an analytic solution of this nonlinear differential equation. (Without loss of generality, let $\varphi'(0) = 0$, $\varphi(0) = \varphi_0$.)

$$\varphi(t) = 2\arcsin\left(\sin\left(\frac{\varphi_0}{2}\right)\operatorname{sn}\left(\sqrt{\frac{g}{l}}\,t\,\middle|\,\sin\left(\frac{\varphi_0}{2}\right)^2\right)\right)$$

This is the solution for the oscillating case; the case of overdeflection of the pendulum can be solved in a similar way. We do not go into this case here (see [1204] and [109]). Here, φ_0 is the maximum angle of deflection. For a detailed discussion of the mathematical pendulum, see [539], [1095], [234], [1358], and [998].

Mathematica is not able to find an explicit solution for $\varphi(t)$ with initial conditions.

In[20]:= `DSolve[{1 φ''[t] == -g Sin[φ[t]], φ[0] == 0}, φ[t], t]`

> Solve::ifun : Inverse functions are being used by Solve, so some solutions
> may not be found; use Reduce for complete solution information. More…

> DSolve::bvfail :
> For some branches of the general solution, unable to solve the conditions. More…

> DSolve::bvfail :
> For some branches of the general solution, unable to solve the conditions. More…

Out[20]= `{}`

Without initial conditions, a complete solution is found.

In[21]:= `DSolve[{1 φ''[t] == -g Sin[φ[t]]}, φ[t], t]`

> Solve::ifun : Inverse functions are being used by Solve, so some solutions
> may not be found; use Reduce for complete solution information. More…

Out[21]= $\left\{\left\{\varphi[t] \to 2 \text{ JacobiAmplitude}\left[\frac{\sqrt{2 g t^2 + 1 t^2 C[1] + 4 g t C[2] + 2 1 t C[1] C[2] + 2 g C[2]^2 + 1 C[1] C[2]^2}}{2\sqrt{1}}, \frac{4 g}{2 g + 1 C[1]}\right]\right\}, \left\{\varphi[t] \to 2 \text{ JacobiAmplitude}\left[\frac{\sqrt{2 g t^2 + 1 t^2 C[1] + 4 g t C[2] + 2 1 t C[1] C[2] + 2 g C[2]^2 + 1 C[1] C[2]^2}}{2\sqrt{1}}, \frac{4 g}{2 g + 1 C[1]}\right]\right\}\right\}$

Here, the solution can be implemented directly.

In[22]:= `elongation[t_, φ0_, l_, g_] =`
`    2 ArcSin[Sin[φ0/2] JacobiSN[Sqrt[g/l] t, Sin[φ0/2]^2]]`

Out[22]= $2 \text{ ArcSin}\left[\text{JacobiSN}\left[\sqrt{\frac{g}{1}} t, \text{Sin}\left[\frac{\varphi 0}{2}\right]^2\right] \text{Sin}\left[\frac{\varphi 0}{2}\right]\right]$

In the limiting case of a small deflection, we have the known result.

In[23]:= `elongationLinearized[t_, φ0_, l_, g_] =`
`    Series[elongation[t, φ0, l, g], {φ0, 0, 1}] // Normal`

Out[23]= $\varphi 0 \text{ Sin}\left[\sqrt{\frac{g}{1}} t\right]$

We now compare the exact solution for three distinct maximum angles of deflection with the "small deviation" approximation.

In[24]:= `Show[GraphicsArray[`
`    Plot[{elongation[t, #, 1, 10],`
`        elongationLinearized[t, #, 1, 10]}, {t, 0, 2},`
`        PlotStyle -> {GrayLevel[0], GrayLevel[1/2]},`
`        PlotLabel -> StyleForm[ToString[InputForm[#]],`

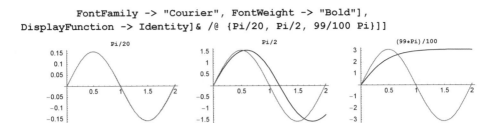

```
              FontFamily -> "Courier", FontWeight -> "Bold"],
       DisplayFunction -> Identity]& /@ {Pi/20, Pi/2, 99/100 Pi}]]
```

The narrow admissible region of the harmonic approximation can be clearly seen.

Jacobi's elliptic function also plays an important role for the solution of electrostatic problems [766], [500], [446], [1058], [70], [1312], [797]. Here, we visualize the field distribution and the current flow through a rectangular metallic plate with electrodes attached to opposite corners.

```
In[25]:= With[{m = 0.3, n = 16},
      Module[{(* the size of the plate *)
      xm = EllipticK[m], ym = EllipticK[1 - m], tab},
      (* the values of the complex potential *)
      tab = Table[Log[JacobiCN[x + I y, m]],
                     {x, 0, xm, 1/n}, {y, 0, ym, 1/n}];
      (* the equipotential and current flow lines *)
      Show[Graphics /@ {
      (* real part *)
      ListContourPlot[Re[tab], ContourShading -> False,
                      ContourStyle -> {Thickness[0.002]},
                      DisplayFunction -> Identity, Contours -> 16],
      (* imaginary part *)
      ListContourPlot[Im[tab], ContourShading -> False,
                      ContourStyle -> {Dashing[{0.01, 0.01}],
                      Thickness[0.002], GrayLevel[1/2]},
                      DisplayFunction -> Identity, Contours -> 18]},
          DisplayFunction -> $DisplayFunction, PlotRange -> All,
          FrameTicks -> None, FrameStyle -> {Thickness[0.05]},
          AspectRatio -> ym/xm]]]
```

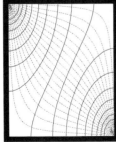

The arithmetic–geometric mean plays an important role in connection with the numerical calculation of elliptic integrals and elliptic functions (see [259], [260], [334], [536], and [233]).

The arithmetic–geometric mean of two numbers a and b is the limit value $a_\infty = b_\infty$ of the following iteration (for a detailed discussion of which root to take in case of complex a and b, see [334]):

$$
\begin{array}{ccc}
 & a_i & b_i \\[4pt]
i = 1 & a_1 = \dfrac{a+b}{2} & b_1 = \sqrt{a\,b} \\[10pt]
i = 2 & a_2 = \dfrac{a_1 + b_1}{2} & b_2 = \sqrt{a_1\,b_1} \\[10pt]
i = 3 & a_3 = \dfrac{a_2 + b_2}{2} & b_3 = \sqrt{a_2\,b_2} \\[6pt]
 & \vdots & \vdots
\end{array}
$$

ArithmeticGeometricMean[a, b]

   gives the arithmetic–geometric mean of a and b.

We have the plot.

In[26]:= **Plot3D[Re[ArithmeticGeometricMean[a, b]],**
                  {a, -3, 3}, {b, -3, 3}]

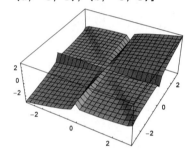

The iterations leading to the arithmetic–geometric mean converge quite fast. Starting with $a = 3$, $b = 1$, we just need 18 steps to get 100000 correct digits of the arithmetic–geometric mean. The number of correct digits doubles roughly per step. (This fast convergence makes it possible to calculate many elliptic functions fast [183].)

In[27]:= **fpl = With[{a = 3, b = 1}, Drop[**
           FixedPointList[{(#[[1]] + #[[2]])/2, Sqrt[#[[1]] #[[2]]]}&,
                               N[{a, b}, 100000]], -1]];

In[28]:= **N[First /@ ((fpl[[-1]] - #) & /@ fpl), 4]**

Out[28]= $\{-1.136,\ -0.1364,\ -0.002409,\ -7.777 \times 10^{-7},\ -8.115 \times 10^{-14},$
          $-8.833 \times 10^{-28},\ -1.047 \times 10^{-55},\ -1.470 \times 10^{-111},\ -2.897 \times 10^{-223},\ -1.126 \times 10^{-446},$
          $-1.700 \times 10^{-893},\ -3.879 \times 10^{-1787},\ -2.018 \times 10^{-3574},\ -5.465 \times 10^{-7149},$
          $-4.007 \times 10^{-14298},\ -2.154 \times 10^{-28596},\ -6.223 \times 10^{-57193},\ 0. \times 10^{-100000}\}$

Interestingly, the values after the nth iteration can be given in a closed form using elliptic theta functions [1114], [183].

In[29]:= **AGMPair[{a_, b_}, n_] :=**
           With[{τ = EllipticNomeQ[1 - (b/a)^2], ϑ = EllipticTheta},
               ArithmeticGeometricMean[a, b]*
               {ϑ[3, 0, τ^(2^(n + 1))]^2, ϑ[4, 0, τ^(2^(n + 1))]^2}] /; a > b

Here is an example.

```
In[30]:= Off[N::meprec];
    With[{a = 3, b = 1},
        Rest[NestList[Function[{a, b},
                                {(a + b)/2, Sqrt[a b]}] @@ #&, {b, a}, 4]] -
                Table[AGMPair[{a, b}, n], {n, 0, 3}]] // N[#, 22]&
Out[31]= {{0.×10⁻⁷¹, 0.×10⁻⁷¹}, {0.×10⁻⁷¹, 0.×10⁻⁷¹},
    {0.×10⁻⁷¹, 0.×10⁻⁷¹}, {0.×10⁻⁷¹, 0.×10⁻⁷¹}}
```

One of the relations mentioned above between the elliptic integral $K(m)$ and the arithmetic–geometric mean is $K(m) = \pi\left(2 \, \text{agm}\left(1, \sqrt{1-m}\,\right)\right)$.

The sequence of the a_i can be used to represent $\exp(\pi)$ in the following way [1098]:

$$e^\pi = 32 \prod_{j=0}^{\infty} \left(\frac{a_{j+1}}{a_j}\right)^{2^{j}+1}$$

where $a = a_0 = 1$ and $b = b_0 = 2^{-1/2}$. Here is a high-precision numerical check [1360] of this formula.

```
In[32]:= N[32 Times @@ MapIndexed[#1^(2^(-#2[[1]] + 2))&,
    Flatten[Map[Divide @@ Reverse[#]&, Partition[First /@
    NestList[{(#[[1]] + #[[2]])/2, Sqrt[#[[1]] #[[2]]]}&,
            {1, 1/Sqrt[N[2, 1000]]}, 30], 2, 1], {1}]]] - E^Pi,
            1000] // N
Out[32]= 0.
```

For details and further applications, see [966], [1233], [1034], [779], [422] (which employs a special notation), [539], [1228], [766], [29], and [967].

3.10 Product Log Function

In this section, we will briefly discuss the `ProductLog` function (sometimes also called the Lambert function).

```
In[1]:= ?ProductLog
            System`ProductLog

            Attributes[ProductLog] = {Listable, Protected}
```

```
ProductLog[n, z]
    gives the value of the nth solution of z e^z = w for z.
```

The defining equation autosimplifies.

```
In[2]:= z - w Exp[w] /. w -> ProductLog[z]
Out[2]= 0
```

Here is a numerical test for the equation.

```
In[3]:= ReleaseHold[Hold[z - w E^w] /.
            w -> ProductLog[-23, z] /. z -> N[1/2, 33] - 10^-12]
Out[3]= 0.×10⁻³⁴ + 0.×10⁻³⁴ i
```

The `ProductLog` function allows us to solve for many transcendental equations. Here is the prime example of the type of equations that can be solved using `ProductLog`.

In[4]:= `Solve[y == x^x, x]`

> InverseFunction::ifun : Inverse functions are
>   being used. Values may be lost for multivalued inverses. More…

> Solve::ifun : Inverse functions are being used by Solve, so some solutions
>   may not be found; use Reduce for complete solution information. More…

Out[4]= $\left\{\left\{x \to \dfrac{\text{Log}[y]}{\text{ProductLog}[\text{Log}[y]]}\right\}\right\}$

From the $z^{z^{z^{\cdot^{\cdot^{\cdot}}}}}$ iteration, rewritten in a Dyson equation like-manner.

In[5]:= `Solve[w[z] == z^w[z], w[z]]`

> InverseFunction::ifun : Inverse functions are
>   being used. Values may be lost for multivalued inverses. More…

> Solve::ifun : Inverse functions are being used by Solve, so some solutions
>   may not be found; use Reduce for complete solution information. More…

Out[5]= $\left\{\left\{w[z] \to -\dfrac{\text{ProductLog}[-\text{Log}[z]]}{\text{Log}[z]}\right\}\right\}$

This gives just $y\,\text{Exp}[y]$.

In[6]:= `Solve[ProductLog[x] == y, x, VerifySolutions -> False]`

Out[6]= $\{\{x \to e^y\,y\}\}$

Mathematica knows many integrals containing the `ProductLog` function.

In[7]:= `Integrate[x^n ProductLog[x], x]`

Out[7]= $\dfrac{1}{(1+n)^3}\,(e^{-n\,\text{ProductLog}[x]}\,x^n$

$\qquad (-(1+n)\,\text{Gamma}[2+n,\,-(1+n)\,\text{ProductLog}[x]] + \text{Gamma}[3+n,\,-(1+n)\,\text{ProductLog}[x]])$
$\qquad (-(1+n)\,\text{ProductLog}[x])^{-n})$

It also knows the series around 0.

In[8]:= `Off[Series::lss];`
        `Series[ProductLog[x], {x, 0, 6}]`

Out[9]= $x - x^2 + \dfrac{3\,x^3}{2} - \dfrac{8\,x^4}{3} + \dfrac{125\,x^5}{24} - \dfrac{54\,x^6}{5} + O[x]^7$

We can derive this series by inverting the series for $w\,\text{Exp}[w]$.

In[10]:= `InverseSeries[Series[Exp[w] w, {w, 0, 6}], x]`

Out[10]= $x - x^2 + \dfrac{3\,x^3}{2} - \dfrac{8\,x^4}{3} + \dfrac{125\,x^5}{24} - \dfrac{54\,x^6}{5} + O[x]^7$

On the principal sheet, `ProductLog` has a branch point at $-1/e$.

In[11]:= `Series[ProductLog[z], {z, -1/E, 5}]`

Out[11]= $-1 + \sqrt{2\,e}\,\sqrt{z + \dfrac{1}{e}} - \dfrac{2}{3}\,e\,\left(z + \dfrac{1}{e}\right) + \dfrac{11\,e^{3/2}\,\left(z + \frac{1}{e}\right)^{3/2}}{18\,\sqrt{2}} -$

$\qquad \dfrac{43}{135}\,e^2\,\left(z + \dfrac{1}{e}\right)^2 + \dfrac{769\,e^{5/2}\,\left(z + \frac{1}{e}\right)^{5/2}}{2160\,\sqrt{2}} - \dfrac{1768\,e^3\,\left(z + \frac{1}{e}\right)^3}{8505} + O\left[z + \dfrac{1}{e}\right]^6$

At infinity, the behavior of `ProductLog [z]` is complicated.

In[12]:= (* suppress messages *)
    Off[Series::lss]; Off[General::ivar];
    Off[Series::esss]; Off[SeriesData::csa];
    Series[ProductLog[z], {z, Infinity, 0}] // Normal

Out[15]= $\dfrac{1}{2\,\mathrm{Log}\left[\frac{1}{z}\right]^{2}}\left(-2\,\mathrm{Log}\left[\frac{1}{z}\right]^{3} - 2\,\mathrm{Log}\left[-\mathrm{Log}\left[\frac{1}{z}\right]\right]\right.-$

$\left.2\,\mathrm{Log}\left[\frac{1}{z}\right]\mathrm{Log}\left[-\mathrm{Log}\left[\frac{1}{z}\right]\right] - 2\,\mathrm{Log}\left[\frac{1}{z}\right]^{2}\mathrm{Log}\left[-\mathrm{Log}\left[\frac{1}{z}\right]\right] + \mathrm{Log}\left[-\mathrm{Log}\left[\frac{1}{z}\right]\right]^{2}\right)$

The first derivatives of the ProductLog function.

In[16]:= **Table[D[ProductLog[z], {z, i}], {i, 3}] // Simplify**

Out[16]= $\left\{\dfrac{\mathrm{ProductLog}[z]}{z + z\,\mathrm{ProductLog}[z]},\ -\dfrac{\mathrm{ProductLog}[z]^{2}\,(2 + \mathrm{ProductLog}[z])}{z^{2}\,(1 + \mathrm{ProductLog}[z])^{3}},\right.$

$\left.\dfrac{\mathrm{ProductLog}[z]^{3}\,(9 + 8\,\mathrm{ProductLog}[z] + 2\,\mathrm{ProductLog}[z]^{2})}{z^{3}\,(1 + \mathrm{ProductLog}[z])^{5}}\right\}$

Here is a differential equation, the solution of which contains the ProductLog function.

In[17]:= **DSolve[w'[z] == Exp[-w[z]]/(w[z] + 1), w, z]**

    InverseFunction::ifun : Inverse functions are
        being used. Values may be lost for multivalued inverses. More…

    Solve::ifun : Inverse functions are being used by Solve, so some solutions
        may not be found; use Reduce for complete solution information. More…

Out[17]= {{w → Function[{z}, ProductLog[z + C[1]]]}}

The ProductLog function appears also in the simplest possible delay-differential equation $x'(t) = a\,x(t - \tau)$ [299], [1057].

In[18]:= **x[t_] := Exp[ProductLog[α τ]/τ t] ξ**

In[19]:= **x'[t] == α x[t - τ] // FullSimplify**

Out[19]= True

In the left graphic, we see the branch cut along the real axis, starting at $-1/e$ of the principal sheet [634]. The branch point at the other sheets (the right graphic shows sheet number five) are also along the real axis, starting at 0.

In[20]:= **Show[GraphicsArray[**
        Plot3D[Im[#[zr + I zi]], {zr, -1, 1}, {zi, -1, 1},
            PlotPoints -> 40, PlotRange -> All,
            DisplayFunction -> Identity]& /@
                {ProductLog, ProductLog[5, #]&}]]

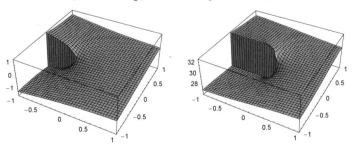

Next, we generate some pictures of the Riemann surface of `ProductLog`. We show the real and imaginary parts of `ProductLog[z]`.

```
In[21]:= χ[reIm_, i_, {r_, φ_}] :=
        {-1/2/E + 1/2/E Cos[φ] Cosh[r], 1/2/E Sin[φ] Sinh[r],
         reIm @ ProductLog[i, -1/2/E + 1/2/E Cos[φ] Cosh[r] +
                                 I 1/2/E Sin[φ] Sinh[r]]}
```

Making holes in the polygons shows better the structure of the whole surface.

```
In[22]:= makeHole[Polygon[l_], α_] :=
     Module[{λ = Function[p, (p + α (# - p))& /@ l][Plus @@ l/4]},
            {MapThread[Polygon[Join[#1, Reverse[#2]]]&,
             Partition[Append[#, First[#]]&[#], 2, 1]& /@ {l, λ}],
            Line[Append[#, First[#]]]&[λ]}]
```

```
In[23]:= With[{ε = 10^-10},
     Show[GraphicsArray[
     Graphics3D[{EdgeForm[], Thickness[0.002],
                 SurfaceColor[Hue[#2], Hue[#2], 2.4],
     makeHole[#, 0.78]& /@ Cases[Table[{
     ParametricPlot3D[(* upper half *) Evaluate @ χ[#1, i, {r, φ}],
             {φ, ε, Pi - ε}, {r, ε, 3}, DisplayFunction -> Identity],
     ParametricPlot3D[(* lower half *) Evaluate @ χ[#1, i + 1, {r, -φ}],
             {φ, ε, Pi - ε}, {r, ε, 3}, DisplayFunction -> Identity]},
             {i, -2, 2}], _Polygon, Infinity]},
         BoxRatios -> {2, 1, 2}, Boxed -> False, PlotRange -> #3]& @@@
         (* color and plotrange for real and imaginary part *)
                     {{Re, 0.0, {-8, 2}}, {Im, 0.22, {-8, 8}}}]]]
```

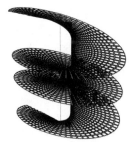

Now let us consider the iterated application of the function $W_n(z)$=`ProductLog[n, z]`. Here is the Julia set of the unit circle under the map $z \to W_n(z)$. At each step, we use three of the infinitely many solutions.

```
In[24]:= Show[Graphics[{PointSize[0.003],
     Table[{Hue[α], Point[{Im[#], Re[#]}]& /@ Flatten[NestList[
     Flatten[(Table[ProductLog[k, #], {k, -1, 1}])& /@ #]&,
             {1. Exp[I 2 Pi α]}, 5]]}, {α, 0, 1, 1/120}]}],
         Frame -> True, PlotRange -> All,
         AspectRatio -> Automatic, FrameTicks -> None]
```

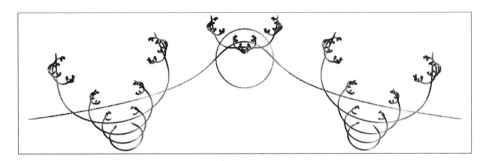

The ProductLog function arises in many applications, especially in problems related to combinatorics. Here is an example: The Stirling numbers of the second kind $S_n^{(m)}$ have a unique maximum with respect to m for a fixed n. The function mMaxStirlingS2 finds the m^*, where for a given n this maximum occurs.

```
In[25]:= mMaxStirlingS2[n_] :=
    Module[{m = 1, sOld = StirlingS2[n, 1], sNew},
        While[(sNew = StirlingS2[n, m = m + 1]) > sOld, sOld = sNew];
        m - 1]
```

We have the interesting relation $\lim_{n \to \infty} m^*/n = W(n)^{-1}$ [881], [1338], [256]. Using $n < 10^3$, we clearly see how m^*/n approaches $W(n)^{-1}$.

```
In[26]:= mMaxData = Table[{n, mMaxStirlingS2[n]/n}, {n, 1000}];
```

```
In[27]:= Plot[1/ProductLog[n], {n, 1, 1000},
        Epilog -> {PointSize[0.002], Hue[0], Point /@ mMaxData},
        PlotRange -> {0, 0.5}, Frame -> True]
```

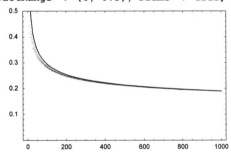

The ProductLog function also arises in the approximate solution of the equation $\sum_{k=1}^{n} 1/k = m$ with respect to m. The zeroth order solution for large m is $n \approx \exp(m - \gamma)$ and the first order solution is $n \approx 1/(2\,W(-\exp(\gamma - m)/2))$.

For many applications of the ProductLog function, see [1256], [495], [244], [644], [486], [100], [1016], [330], [1203], [551], [182], [991], [336], [1299], and [108]. For the properties of a function related to the ProductLog function, see [524].

For solutions of related equations of the form $w(x)\,f(w(x)) = x$, see [849].

3.11 Mathieu Functions

Mathieu functions are the solutions to the differential equation $w''(z) + (a - 2q\cos(2z))\,w(z) = 0$ [868], [826], [59], [1197], [747], [879].

`MathieuC[a, q, z]`

    represents the even Mathieu solution $\mathrm{Ce}_a(q, z)$ to the Mathieu differential equation.

`MathieuS[a, q, z]`

    represents the odd Mathieu solution $\mathrm{Se}_a(q, z)$ to the Mathieu differential equation.

Because Mathieu functions are solutions to second-order differential equations, the derivatives are available too.

`MathieuCPrime[a, q, z]`

    represents the derivative with respect to z of the Mathieu function $\mathrm{Ce}_a(q, z)$.

`MathieuSPrime[a, q, z]`

    represents the derivative with respect to z of the Mathieu function $\mathrm{Se}_a(q, z)$.

The parameter a is called the characteristic. The next picture shows a family of solutions for a fixed q.

```
In[1]:= Plot[Evaluate[Table[Re[MathieuC[a, 1, z]],
           {a, -2, 3, 1/4}]], {z, -10, 10},
           PlotStyle -> Table[{Hue[i/30], Thickness[0.002]}, {i, 30}],
           PlotRange -> {-20, 20}, Frame -> True, Axes -> False]
```

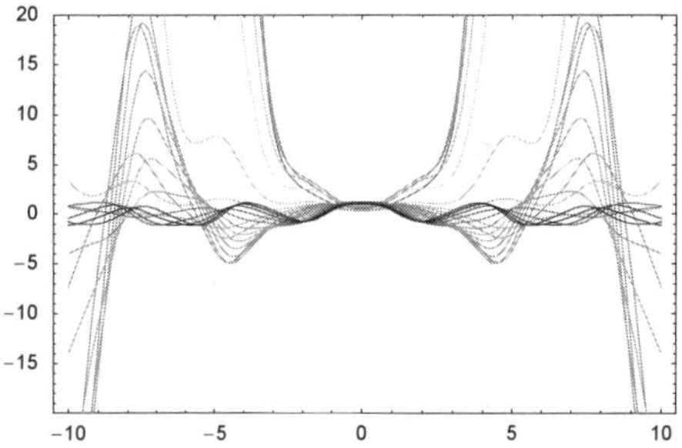

A numerical solution of the corresponding differential equation gives the same solution within the error bounds of `NDSolve`.

```
In[2]:= With[{a = 3, q = 1},
           Plot[Evaluate[Abs[MathieuC[a, q, z] -
              (w[z] /. NDSolve[{w''[z] + (a - 2q Cos[2z]) w[z] == 0,
                         w[0] == MathieuC[a, q, 0], w'[0] == 0},
                      w, {z, 0, 10}])]], {z, 0, 10}]]
```

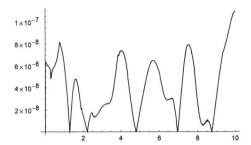

Here, the equivalent family for the odd Mathieu functions is shown.

```
In[3]:= Plot[Evaluate[Table[Re[MathieuS[a, 1, z]],
          {a, -2, 3, 1/4}]], {z, -10, 10},
          PlotStyle -> Table[{Hue[i/30], Thickness[0.002]}, {i, 30}],
          PlotRange -> {-20, 20}, Frame -> True, Axes -> False]
```

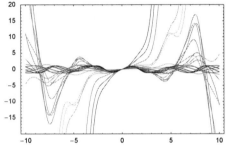

Phase space-like trajectories $\{Ce_a(q, z), \partial Ce_a(q, z)/\partial z\}$ can show interesting behavior [698].

```
In[4]:= phaseSpaceMathieuGraphics[a_, q_, T_, opts___] :=
    Module[{l, pp = ParametricPlot[
            {MathieuC[a, q, t], MathieuCPrime[a, q, t]},
                {t, 0, T}, PlotPoints -> 2000,
                DisplayFunction -> Identity]},
      {points, l} = {#, Length[#]}&[Cases[pp, _Line, Infinity][[1, 1]]];
      (* color curve *)
      Show[Graphics[{Thickness[0.002],
      MapIndexed[{Hue[0.8 #2[[1]]/l], Line[#]}&,
                Partition[points, 2, 1]]}],
                opts, AspectRatio -> 1, Frame -> True]]

In[5]:= Show[GraphicsArray[{
    phaseSpaceMathieuGraphics[8.49494, 5.60263, 300,
                            DisplayFunction -> Identity],
    phaseSpaceMathieuGraphics[2.76108, 2.14745, 300,
                            DisplayFunction -> Identity]}]]
```

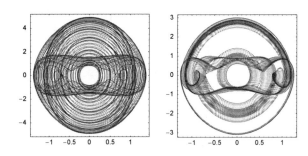

Some of the solutions grow, whereas some stay bounded with increasing z. When the solutions stay bounded, they are of the form $e^{irz} u(z)$, where r is a real rational or integer called the characteristic exponent and $u(z)$ is a periodic function of period 2π. For a given q, the value of a such that the Mathieu functions are bounded can be obtained with the following two functions.

MathieuCharacteristicA[r, q]

    gives the value of a such that MathieuC[a, q, z] is of the form $e^{irz} u(z)$ with periodic $u(z)$.

MathieuCharacteristicB[r, q]

    gives the value of a such that MathieuS[a, q, z] is of the form $e^{irz} u(z)$ with periodic $u(z)$.

Here, for $q = 1$, the value of the characteristic as a function of the characteristic exponent (assumed real here) is shown.

In[6]:= **Plot[Re[MathieuCharacteristicA[r, 1]], {r, -3, 3}]**

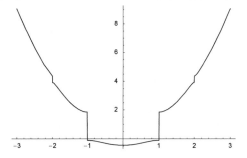

We see jumps in the last picture that indicate the regions of a where the even Mathieu function is not bounded. For $a \leqslant -q$, we get purely imaginary characteristic exponents.

In[7]:= **Plot[Re[MathieuCharacteristicA[I r, 1]], {r, -3, 3}]**

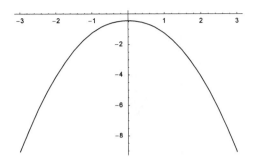

The regions in the *a,q*-plane that allow for bounded solutions are of particular practical importance because in many applications they correspond to stable oscillations (these are the so-called resonance tongues [216]). Here are these regions.

```
In[8]:= Show[Graphics[{#, Map[{-1, 1}#&, #, {-2}]}&[
        (* form polygon from points *)
        Polygon[Join[#[[1]], Reverse[#[[2]]]]]& /@
              Partition[First /@ Flatten[Plot[Evaluate @
        (* table of points *)
        Transpose[{Table[MathieuCharacteristicA[i, q], {i, 0, 8}],
                 Table[MathieuCharacteristicB[i, q], {i, 1, 9}]}],
            {q, 0, 60}, DisplayFunction -> Identity,
            PlotPoints -> 120][[1]]], 2]]],
        AspectRatio -> 1, Frame -> True, PlotRange -> {All, {-50, 100}}]
```

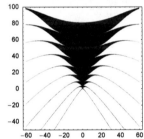

The following graphics of Ce$_a$(2, *z*) and Se$_a$(2, *z*) clearly shows the *a*-intervals where Ce$_a$(2, *z*) and Se$_a$(2, *z*) stay bounded.

```
In[9]:= ppz = 200; ppa = 150;
    data = Table[MathieuC[a, 2, N[z]], {a, -3, 8, 11/ppa}, {z, -30, 30, 60/ppz}];

    Show[GraphicsArray[
    ListPlot3D[ArcTan @ #[data], Mesh -> False, MeshRange -> {{-30, 30}, {-3, 8}},
              DisplayFunction -> Identity]& /@ {Re, Im, Abs}]]
```

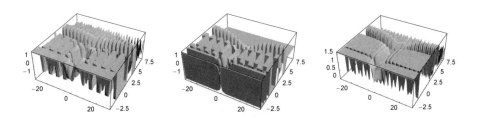

The next two graphics show $Ce_a(2, z)$ and $Se_a(2, z)$ inside one of the a-intervals of the last graphic.

```
In[12]:= Module[{ab, ε = 10^-6},
    Show[GraphicsArray[
    ParametricPlot3D[{z, ab = #1[r, 2], #2[ab, 2, z]},
              {EdgeForm[]}}, {z, -30, 30}, {r, 1 + ε, 2 - ε},
          PlotPoints -> {200, 100}, BoxRatios -> {2, 1, 1/2},
          PlotRange -> All, Compiled -> False,
          DisplayFunction -> Identity]& @@@
        (* characteristics and functions *)
      {{MathieuCharacteristicA, MathieuC},
      {MathieuCharacteristicB, MathieuS}}]]]
```

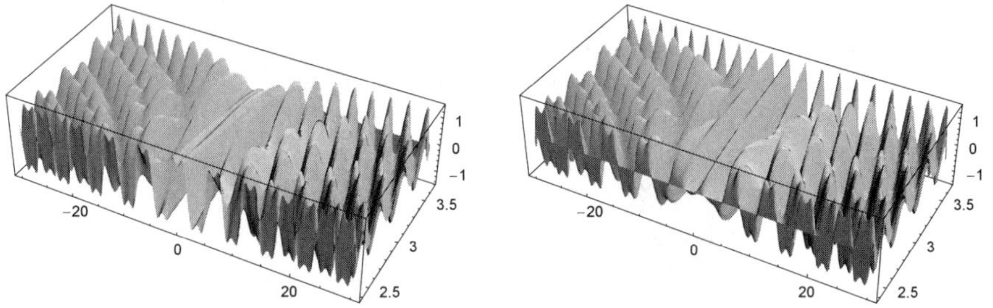

For small q, the term $2\, q \cos(2\, z)\, w(z)$ can be considered as a small perturbation to the differential equation. This means that, for small q, one expects $a \propto r^2$. This is indeed the case.

```
In[13]:= Show[GraphicsArray[
    Block[{$DisplayFunction = Identity},
        {Plot[MathieuCharacteristicA[r, 1], {r, 0, 3}],
        Plot[MathieuCharacteristicA[r, 1] - r^2, {r, 0, 3}]}]]]
```

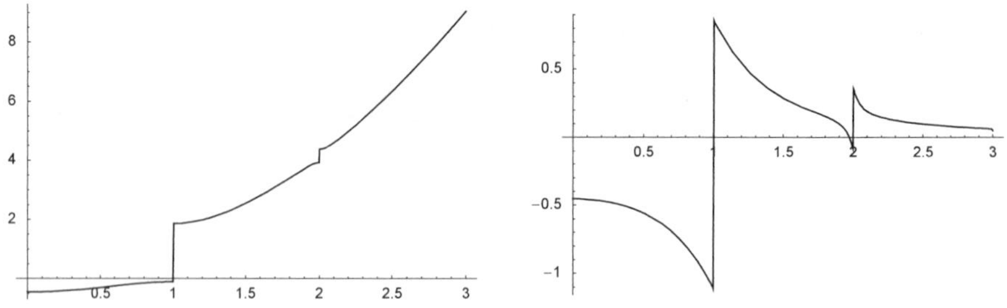

(For larger q (and small z), a periodic version of a harmonic oscillator is obtained [152].)

The following graphics show the even and odd Mathieu functions with period 2π, 4π, 6π and 8π.

```
In[14]:= Show[GraphicsArray[
    Block[{$DisplayFunction = Identity},
    Plot[Evaluate[Table[Re[#1[#2[r, 1], 1, z]], {r, 1, 4}]], {z, 0, 3 2Pi},
        PlotStyle -> Table[{Hue[i/7], Thickness[0.002]}, {i, 6}],
        PlotRange -> All, Frame -> True, Axes -> False]& @@@
        (* functions and characteristics *)
        {{MathieuC, MathieuCharacteristicA},
         {MathieuS, MathieuCharacteristicB}}]]]
```

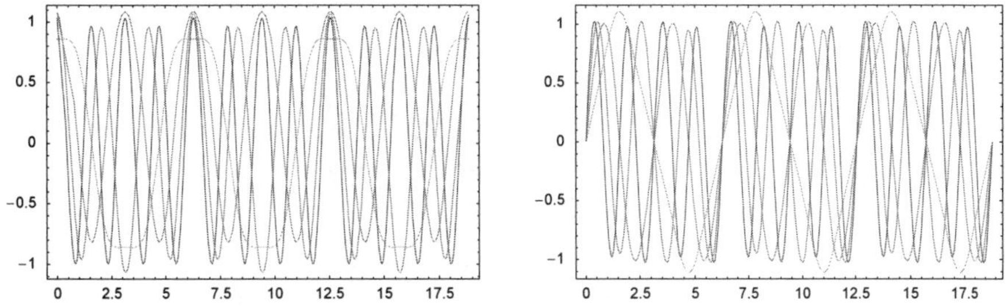

The characteristic a as a function of q can be regarded as a multivalued function with the integer r in $a_r(q)$ labelling the sheets [164], [621]. Some of the branch points of $a(q)$ appear on the imaginary axis. A plot of $a_r(q)$ shows the branch points nicely. The functions $a_r(q)$, $b_r(q)$ have a complicated branch cut structure as a function of q. The right graphic shows $a_{10}(q)$.

```
In[15]:= Show[GraphicsArray[
    Block[{$DisplayFunction = Identity},
    {Show[Plot[Evaluate[Table[Re[MathieuCharacteristicA[i, I q]],
                        {i, 0, 6}]], {q, 0, 17},
        PlotStyle -> Table[{Hue[i/9], Thickness[0.002]}, {i, 7}]] /.
        (* delete steep vertical lines *)
    Line[l_] :> Line /@ Select[Partition[l, 2, 1],
                (#[[2, 2]] - #[[1, 2]])/(#[[2, 1]] - #[[1, 1]]) < 100&]],
    ContourPlot[Abs @ MathieuCharacteristicA[10, qx + I qy],
                {qx, 0, 100}, {qy, 0, 120}, PlotPoints -> 200,
                PlotRange -> All, ColorFunction -> (Hue[0.8 #]&),
                Contours -> 50]}]]]
```

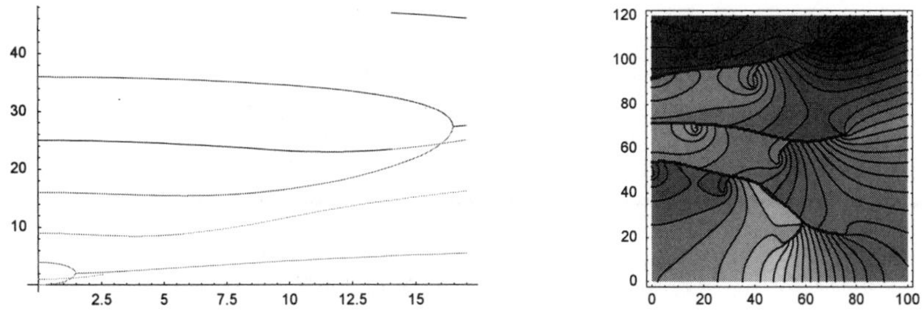

The branch cut structure of $a_r(q)$, $b_r(q)$ is not standardized. To obtain values from the appropriate sheet the functions MathieuCharacteristicA, MathieuCharacteristicB accept a third argument, which is the starting value for the numerical procedure that calculates $a_r(q)$, $b_r(q)$. The following graphics show branches of the multivalued $a_{40}(q)$, once calculated using MathieuCharacteristicA with two arguments and once as a continuous function.

```
In[16]:= Show[GraphicsArray[
    Block[{$DisplayFunction = Identity},
        {ListPlot[Table[MathieuCharacteristicA[40, q], {q, 0, 820}] // N],
         ListPlot[FoldList[MathieuCharacteristicA[40, #2, N[#1]]&,
            (* use step size 1 *)
            MathieuCharacteristicA[40, 0] // N,
            Table[q, {q, 1, 820}]]] // N}]]]
```

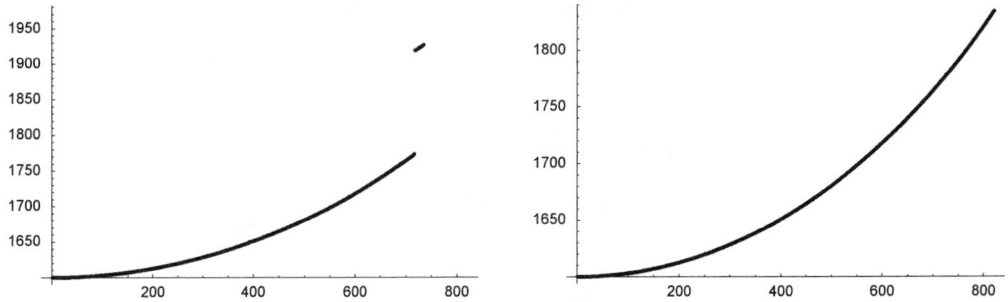

The above contour plot shows that $a_{10}(q)$ has a branch point near $q \approx 75 + 66\,i$. Encircling this branch point twice by using the three-argument version of MathieuCharacteristicA shows that this branch points is of square root-type [164].

```
In[17]:= ListPlot[N[{#1, Re[#2]}]& @@@
    FoldList[Function[{φa, φ}, {φ, MathieuCharacteristicA[10,
                            N[75 + 66 I + 8 Exp[I φ]], φa[[2]]]}],
                {-Pi/2, N[MathieuCharacteristicA[10, 75 + 58 I]]},
                Table[φ, {φ, -Pi/2, 7Pi/2, 2Pi/30}]],
        Frame -> True, Axes -> False, PlotJoined -> True]
```

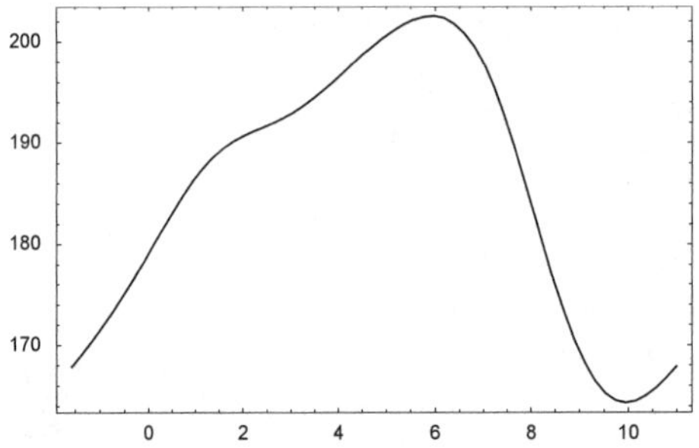

The characteristic exponent as a function of a and q can be obtained from the function MathieuCharacteristicExponent:

MathieuCharacteristicExponent[*a*, *q*]

   gives the characteristic exponent *r* for MathieuC[*a*, *q*, *z*] and MathieuS[*a*, *q*, *z*].

In the stability regions (bands) the characteristic exponent is purely real. Outside it has a nonvanishing imaginary and a constant real part. The following graphic shows the real and imaginary parts of *r* for $q = 2$.

```
In[18]:= Plot[{Re @ MathieuCharacteristicExponent[a, 2],
        Im @ MathieuCharacteristicExponent[a, 2]}, {a, -3, 12},
        PlotStyle -> {Hue[0], Hue[0.78]}, Frame -> True, Axes -> False]
```

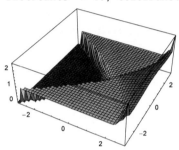

Here, the characteristic exponent as a function of a and q is shown.

```
In[19]:= Plot3D[Re[MathieuCharacteristicExponent[a, q]], {a, -3, 3}, {q, -3, 3},
            PlotPoints -> 40, ColorFunction -> (Hue[#/2]&)]
```

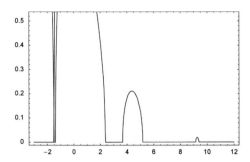

We see jumps in the last picture that indicate the regions of r, where the even Mathieu function is not bounded. The characteristic exponent as a function of a and q can be obtained from the function. The characteristic exponent can also be complex. In this case, the solutions are exponentially increasing or decreasing. Here is such a situation.

```
In[20]:= Plot[Re[MathieuS[MathieuCharacteristicB[0.2 I, 1], 1, z]],
            {z, -3 2Pi, 3 2Pi}]
```

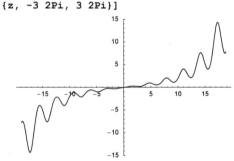

As an example for the use of Mathieu functions, let us discuss the oscillations of an ellipsoidal drum. An ellipsoidal drum is the simplest generalization of the circular drum discussed in the above section about Bessel and Airy functions.

Physical Application: Oscillation of an Ellipsoidal Drum

Let us consider a membrane inside an ellipse-shaped region with the membrane held fixed at the boundary [291], [1068]. Then the amplitude $u(\mathbf{r}, t)$ of the membrane is governed by the wave equation

$$\Delta u(\mathbf{r}) + \frac{1}{c_{ph}^2} \frac{\partial^2 u(\mathbf{r}, t)}{\partial t^2} = 0.$$

Here, Δ is the Laplace operator and c_{ph}^2 (= tension per unit surface density) is the phase velocity. Assuming the solution is periodic in time, that is, $u(\mathbf{r}, t) = \cos(\omega t + \alpha) u(\mathbf{r})$ we are led to the Helmholtz equation

$$\Delta u(\mathbf{r}) + \lambda^2 u(\mathbf{r}) = 0, \quad \lambda = \frac{\omega}{c_{ph}}.$$

The Laplace operator separates in an elliptical coordinate system. An elliptical coordinate system is related to a Cartesian coordinate system by the following equations:

$$x = c \cosh(r) \cos(\varphi)$$
$$y = c \sinh(r) \sin(\varphi) \qquad\qquad 0 \leq \varphi \leq 2\pi, \, 0 < r < \infty.$$

For a given ellipse, the relation between the half-axis a and b, the parameter c, and the maximal value r_0 of r is given by $a = c \cosh r_0$ and $b = c \sinh r_0$.

Making a separation ansatz in the form $u(\mathbf{r}) = \Phi(\varphi) R(r)$, we obtain the following two ordinary differential equations [908], [868]:

$$(a - 2q \cosh(2r)) R(r) + R''(r) = 0$$
$$(a + 2q \cos(2\varphi)) \Phi(\varphi) + \Phi''(\varphi) = 0$$

where $q = \frac{\lambda^2}{4c}$ and $2a$ is the separation constant.

The boundary conditions for $\Phi(\varphi)$ are periodicity: $\Phi(\varphi + k\, 2\pi) = \Phi(\varphi)$, $k \in \mathbb{Z}$. Because of the Dirichlet boundary condition of the whole membrane, the boundary condition for $R(r)$ is $R(r_0) = 0$.

For a more detailed discussion of an ellipsoidal drum, see [868], [292], [291], [1068], [37], [1235], [559], [282], and [840].

The following example follows [1236] closely. See this article for more details. In accordance with the symbols used in the literature, let us define the two functions $ce_n(q, z)$ and $se_n(q, z)$, which are the even and odd solutions of the Mathieu equation with periodicity $2\pi n$.

```
In[21]:= (* even functions *)
      ce[n_Integer?NonNegative][q_, z_] =
         MathieuC[MathieuCharacteristicA[n, q], q, z];
      (* odd functions *)
      se[n_Integer?Positive][q_, z_] =
         MathieuS[MathieuCharacteristicB[n, q], q, z];
```

Now, the "radial" (r dependent) equation derived above must now be used to satisfy the Dirichlet boundary condition on the outer boundary. The solutions of the radial equation are again Mathieu functions, but this time with a purely imaginary argument. Using continuity of the elongation along the line connecting the two foci, this leads to the following form of the eigenfunctions:

$$\psi_{nj}^{cs} \propto \begin{cases} ce_n(q_{nj}^c, \varphi)\, ce_n(q_{nj}^c, ir) & n = 0, \ldots, \quad j = 1, \ldots \\ se_n(q_{nj}^s, \varphi)\, se_n(q_{nj}^s, ir) & n = 1, \ldots, \quad j = 1, \ldots \end{cases} \qquad cs = c \text{ or } s.$$

The corresponding eigenvalues are given by $\lambda^{cs}_{nj} = 2\sqrt{q^{cs}_{nj}/c}$, cs $= c$ or s. To fulfill the boundary condition $R(r_0) = 0$ we must have $ce_n(q^c_{nj}, i\,r_0) = 0$ or $se_n(q^s_{nj}, i\,r_0) = 0$. For a given n and r_0, this implicitly defines a countable infinite number of values q^c_{nj} and q^s_{nj}. For the following let us fix the value of r_0 to be $1/2$.

In[25]:= **r0Example = 1/2;**

Next, we have to find the values q^{cs}_{nj}. Here, we will have to resort to numerical techniques. Plots show that for small values of n, the q^{cs}_{nj} are well separated.

In[26]:= **se[1][1, I r0Example] // N**

Out[26]= 0. + 0.309619 i

In[27]:= **Show[GraphicsArray[#]]& /@ Partition[**
  Block[{$DisplayFunction = Identity},
    {Plot[Evaluate[I se[1][q, I r0Example]], {q, 0, 100},
      PlotRange -> {-0.0001, 0.0001}],
    Plot[Evaluate[ce[3][q, I r0Example]], {q, 0, 100},
      PlotRange -> {-0.0001, 0.0001}],
    Plot[Evaluate[I se[5][q, I r0Example]], {q, 0, 100},
      PlotRange -> {-0.0001, 0.0001}],
    Plot[Evaluate[I se[12][q, I r0Example]], {q, 0, 100},
      PlotRange -> {-1, 1}]}], 2]

Next, we have to find the values q^{cs}_{nj}. Here, we will have to resort to numerical techniques. Plots show that for small values of n, the q^{cs}_{nj} are well separated. (We take into account that the functions are purely imaginary for an imaginary argument.)

In[28]:= **findQValues[f:ce | se, r0_, n_, {q0_, q1_}, pp_] :=**
  Module[{qData, zeroEnclosingIntervals},
  (* a dense set of points *)
    qData = Table[N[{q, If[f === ce, 1, I] f[n][q, I r0]}],
      {q, q0, q1, (q1 - q0)/pp}];
  (* intervals where a zero occurs *)

```
         zeroEnclosingIntervals =
         Map[First, Select[Partition[qData, 2, 1],
                           Re[#[[1, 2]] #[[2, 2]]] < 0&], {2}];
         (* finding accurate values for the zeros *)
         q /. Apply[FindRoot[Evaluate[f[n][q, I r0] == 0],
                    {q, ##}]&, zeroEnclosingIntervals, {1}]]
```

For the first few n, we calculate the corresponding first few q^{cs}_{nj}.

In[29]:= `Table[MapIndexed[(qc[i][#2[[1]]] = #1)&,`
                      `findQValues[ce, r0Example, i, {0, 100}, 100]],`
               `{i, 0, 6}];`

In[30]:= `Short[SubValues[qc], 6]`

Out[30]//Short=

```
        {HoldPattern[qc[0][1]] :→ 3.17779, HoldPattern[qc[0][2]] :→ 22.6941,
         HoldPattern[qc[0][3]] :→ 60.4245, HoldPattern[qc[1][1]] :→ 5.3837,
         HoldPattern[qc[1][2]] :→ 27.5324, HoldPattern[qc[1][3]] :→ 68.0215,
         HoldPattern[qc[2][1]] :→ 8.39973, ≪5≫, HoldPattern[qc[4][1]] :→ 17.0229,
         HoldPattern[qc[4][2]] :→ 46.2685, HoldPattern[qc[4][3]] :→ 94.8499,
         HoldPattern[qc[5][1]] :→ 22.6667, HoldPattern[qc[5][2]] :→ 54.0154,
         HoldPattern[qc[6][1]] :→ 29.2084, HoldPattern[qc[6][2]] :→ 62.552}
```

In[31]:= `Table[MapIndexed[(qs[i][#2[[1]]] = #1)&,`
                      `findQValues[se, r0Example, i, {0, 100}, 100]],`
               `{i, 1, 6}];`

In[32]:= `Short[SubValues[qs], 6]`

Out[32]//Short=

```
        {HoldPattern[qs[1][1]] :→ 10.6489 + 0. i,
         HoldPattern[qs[1][2]] :→ 39.2868 + 0. i, HoldPattern[qs[1][3]] :→ 86.1063 + 0. i,
         HoldPattern[qs[2][1]] :→ 14.1239 + 0. i, HoldPattern[qs[2][2]] :→ 45.5018 + 0. i,
         HoldPattern[qs[2][3]] :→ 95.0878 + 0. i, HoldPattern[ ≪1≫ ] :→ ≪1≫, ≪1≫,
         HoldPattern[qs[4][1]] :→ 23.2388 + 0. i, HoldPattern[qs[4][2]] :→ 59.9612 + 0. i,
         HoldPattern[qs[5][1]] :→ 28.9455 + 0. i, HoldPattern[qs[5][2]] :→ 68.2552 + 0. i,
         HoldPattern[qs[6][1]] :→ 35.4485 + 0. i, HoldPattern[qs[6][2]] :→ 77.2877 + 0. i}
```

Using these q^{cs}_{nj}, we can define the eigenfunctions in the following way. (We suppress the dependence on r_0.)

In[33]:= `ψc[n_, j_][r_, φ_] := ce[n][qc[n][j], φ] ce[n][qc[n][j], I r]`

In[34]:= `ψs[n_, j_][r_, φ_] := se[n][qs[n][j], φ] I se[n][qs[n][j], I r]`

Now, let us visualize the eigenfunctions. We start with the 3D pictures.

In[35]:= `EigenFunctionPlot3D[ψ_, r0_, {r_, φ_}, col_, pp_:{50, 100}, opts___] :=`
         `Module[{polys = Cases[`
                 `ParametricPlot3D[{Cosh[r] Cos[φ], Sinh[r] Sin[φ], ψ},`
                               `{r, 0, r0}, {φ, 0, 2Pi}, PlotPoints -> pp,`
                               `DisplayFunction -> Identity], _Polygon, Infinity],`
                 `maxZ}, (* scale the elongation *)`
                 `maxZ = Max[Abs[Transpose[Level[polys, {-2}]][[-1]]]];`
         `Show[Graphics3D[{boundary3D[r0],`
               `{col, EdgeForm[], Apply[{#1, #2, #3/maxZ 0.5}&, polys, {-2}]}}],`
               `opts, PlotRange -> All, Boxed -> False]]`

In[36]:= `boundary3D[r0_] := boundary3D[r0] =`
         `{Thickness[0.01], Line[Table[{Cosh[r0] Cos[φ], Sinh[r0] Sin[φ], 0},`
                               `{φ, 0., 2.Pi, 2Pi/200}]]}`

Here are some of the above-calculated eigenfunctions, starting with the ground state [511].

```
In[37]:= randomColor := SurfaceColor[Hue[Random[]], Hue[Random[]], 3 Random[]];
```

```
In[38]:= (* group into pairs *)
    efp3D[type_, {i1_, j1_}, {i2_, j2_}] :=
    Show[GraphicsArray[
    Block[{$DisplayFunction = Identity},
    Apply[EigenFunctionPlot3D[type[##]][r, φ], r0Example, {r, φ},
                            randomColor]&,
        {{i1, j1}, {i2, j2}}, {1}]]]]
```

```
In[40]:= efp3D[ψc, {0, 1}, {0, 2}]
```

```
In[41]:= efp3D[ψc, {0, 3}, {1, 1}]
```

```
In[42]:= efp3D[ψc, {1, 2}, {1, 3}]
```

Here are two of the higher states shown.

```
In[43]:= EigenFunctionPlot3D[ψc[4, 3][r, φ], r0Example, {r, φ}, randomColor,
                            {60, 120}]
```

Next we display some of the ψs-functions.

In[44]:= **efp3D[ψs, {1, 1}, {1, 2}]**

In[45]:= **efp3D[ψs, {1, 3}, {2, 1}]**

In[46]:= **efp3D[ψs, {2, 2}, {2, 3}]**

Here is again one of the higher states shown.

In[47]:= **EigenFunctionPlot3D[ψs[6, 1][r, φ], r0Example, {r, φ}, randomColor,
                            {60, 120}]**

Here, two much higher state are shown. We use considerably more points to resolve the more complicated structure of the eigenfunctions.

In[48]:= **{qs[12][1], qs[12][2]} = findQValues[se, r0Example, 12, {0, 180}, 200]**

Out[48]= {91.7035 + 0. i, 147.933 + 0. i}

In[49]:= **Show[GraphicsArray[
        Block[{$DisplayFunction = Identity},**

```
Table[EigenFunctionPlot3D[ψs[12, j][r, φ],
                r0Example, {r, φ}, randomColor, {160, 240}],
                {j, 2}]]]]
```

Now, we will construct contour plots of the membrane. Using ContourPlot, we will make a contour plot in the rectangular r,φ-domain and then map the rectangle to the ellipse. The function addPoints refines the outlines of the polygons generated in the contour plot. This is necessary to get a smooth mapping.

```
In[50]:= addPoints[points_, δε_] :=
        Module[{n, l}, Join @@ (
           Function[segment, (* segment too short? *)
                   If[(l = Sqrt[#. #]&[Subtract @@ segment]) < δε, segment,
                   n = Floor[l/δε] + 1; (* form segments *)
                   Table[# + i/n (#2 - #1), {i, 0, n - 1}]& @@ segment]] /@
                       Partition[Append[points, First[points]], 2, 1])]
```

```
In[51]:= toEllipse[{r_, φ_}] = {Cosh[r] Cos[φ], Sinh[r] Sin[φ]};
```

```
In[52]:= boundary2D[r0_] := boundary2D[r0] =
        {Thickness[0.01], Line[Table[{Cosh[r0] Cos[φ], Sinh[r0] Sin[φ]},
                               {φ, 0., 2.Pi, 2Pi/200}]]}
```

The function efPlotContour generates a contour plot of ψ within the ellipse.

```
In[53]:= EigenFunctionContourPlot[ψ_, r0_, {r_, φ_}, col_,
                        pp_:{60, 100}, cts_:40, opts___] :=
        Module[{cp = ContourPlot[ψ, {r, 0, r0}, {φ, 0, 2Pi},
                           PlotPoints -> pp, ColorFunction -> col,
                           ContourLines -> False,
                           DisplayFunction -> Identity],
            δε = Cosh[r0] 2Pi/100, contourValues},
           (* make homogeneously spaced contours *)
           cV = contourValues = #[[Round[(Times @@ pp)/cts/2]]]& /@
                       Partition[Sort[Flatten[Re[cp[[1]]]]],
                                   Round[(Times @@ pp)/cts]];
           Show[Graphics[cp /. (Contours -> _) -> (Contours -> 20)] /.
               {p_Polygon :> (toEllipse /@ addPoints[#, δε]&) /@ p},
               Frame -> False, AspectRatio -> Automatic,
               DisplayFunction -> $DisplayFunction,
               Epilog -> boundary2D[r0]]]
```

Here are the functions from above shown as contour plots.

```
In[54]:= (* group into pairs *)
        efcP[type_, {i1_, j1_}, {i2_, j2_}] :=
        Show[GraphicsArray[
        Block[{$DisplayFunction = Identity},
        Apply[EigenFunctionContourPlot[type[##][r, φ], r0Example,
```

```
                                                  {r, φ}, Hue[0.8 #]&]&,
                              {{i1, j1}, {i2, j2}}, {1}]]]]
```

In[56]:= **efcP[ψc, {0, 1}, {0, 2}]**

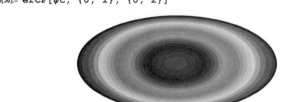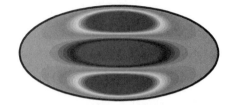

In[57]:= **efcP[ψc, {0, 3}, {1, 1}]**

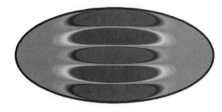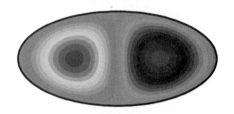

In[58]:= **efcP[ψc, {1, 2}, {1, 3}]**

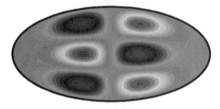

And here is again one of the higher states shown.

In[59]:= **EigenFunctionContourPlot[ψc[4, 3][r, φ],**
                              r0Example, {r, φ}, Hue[0.8 #]&]

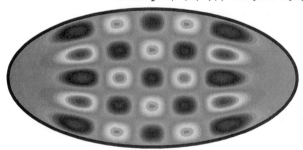

And here again some of the ψs-type functions.

In[60]:= **efcP[ψs, {1, 1}, {1, 2}]**

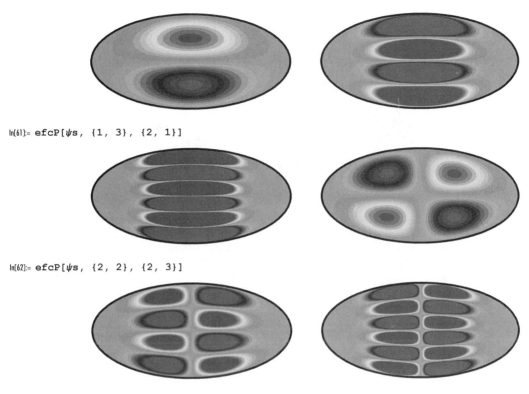

In[61]:= **efcP[ψs, {1, 3}, {2, 1}]**

In[62]:= **efcP[ψs, {2, 2}, {2, 3}]**

In[63]:= **EigenFunctionContourPlot[ψs[6, 1][r, φ],**
                      r0Example, {r, φ}, Hue[0.8 #]&]

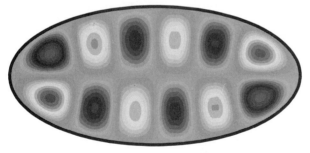

In distinction to the circular membrane, the ellipse-shaped membrane has one free parameter; in our notation, this is r_0. Varying this parameter allows for degenerate eigenvalues. For $r_0 \approx 0.495$, the two states $\psi^c_{1,3}$ and $\psi^s_{5,2}$ are degenerate.

In[62]:= **r0Degenerate = 0.49526836;**
     {{qc[1][2]}, {qs[5][1]}} =
             {findQValues[ce, r0Degenerate, 1, {50, 80}, 10],
              findQValues[se, r0Degenerate, 5, {50, 80}, 10]}
Out[63]= **{{69.3274}, {69.3274 + 0. i}}**

Here the contour plot of the two states is shown.

```
In[66]:= Show[GraphicsArray[
    Block[{$DisplayFunction = Identity},
    {EigenFunctionContourPlot[ψc[1, 2][r, φ],
                              r0Degenerate, {r, φ}, Hue[0.8 #]&],
    EigenFunctionContourPlot[ψs[5, 2][r, φ],
                              r0Degenerate, {r, φ}, Hue[0.8 #]&]}]]]
```

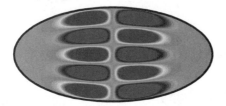 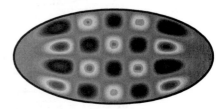

For degenerate eigenstates also linear combinations are eigenstates; this means $(1 - \alpha)\psi_{13}^c + \alpha\psi_{52}^s$ is an eigenstate, too. Here some of the resulting states for different α are shown.

```
In[67]:= DegenerateEigenFunctionContourPlot[
                    {ψ1_, ψ2_}, r0_, {r_, φ_}, frames_,
                    col_, pp_:{50, 100}, cts_:30, opts___] :=
    Module[{data1 = Table[Re[ψ1], {φ, 0, 2Pi, 2Pi/pp[[2]]},
                                   {r, 0, r0, r0/pp[[1]]}],
            data2 = Table[Re[ψ2], {φ, 0, 2Pi, 2Pi/pp[[2]]},
                                   {r, 0, r0, r0/pp[[1]]}],
            δε = Cosh[r0] 2Pi/100, contourValues, max1, max2, data},
            {max1, max2} = Max[Abs[#]]& /@ {data1, data2};
            {data1, data2} = {data1/max1, data2/max2};
            (* make homogeneously spaced contours *)
        Table[(* add elongation values *)
              data = (1 - i/frames) data1 + i/frames data2;
              (* contour values *)
              contourValues = #[[Round[(Times @@ pp)/cts/2]]]& /@
                              Partition[Sort[Flatten[data]],
                                        Round[(Times @@ pp)/cts]];
              (* show contour plot *)
              Show[Graphics[ListContourPlot[data, Contours -> contourValues,
                 ContourLines -> False, MeshRange -> {{0, r0}, {0, 2Pi}},
                 ColorFunction -> col, DisplayFunction -> Identity]] /.
                 (* map polygons to ellipse *)
                 {p_Polygon :> (toEllipse /@ addPoints[#, δε]&) /@ p},
                 opts, Frame -> False, AspectRatio -> Automatic,
                 DisplayFunction -> $DisplayFunction,
                 Epilog -> boundary2D[r0]], {i, 0, frames}]];

In[68]:= Show[GraphicsArray[#]]& /@ Partition[
    DegenerateEigenFunctionContourPlot[
                    {ψc[1, 2][r, φ], ψs[5, 2][r, φ]}, r0Degenerate,
                    {r, φ}, 20, Hue[0.8 #]&, {50, 100}, 15,
                    DisplayFunction -> Identity], 3]
```

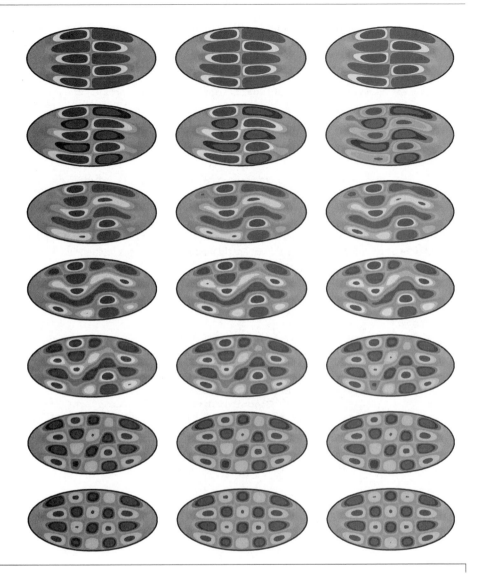

```
DegenerateEigenFunctionContourPlot[{ψc[1, 2][r, φ], ψs[5, 2][r, φ]},
        r0Degenerate, {r, φ}, 30, Hue[0.8 #]&]
```

For the corresponding quantum-mechanical eigenvalue problem, including a magnetic field, see [611] and [612]. For other applications of Mathieu functions, see [1088], [549], [434], and [787].

We will end this section with a solid-state physics view [1161], [65], [681], [513] on the Mathieu equation and functions. Because of the importance of crystalline materials, the study of periodic potentials V in the Schrödinger equation $-\Delta\psi + V\psi = \varepsilon\psi$ is of great relevance. After 1D piecewise constant potentials (Kronig–Penney potential and its limiting cases) the potential $V(z) \sim \cos(z)$ is the next simple choice (the next complicated, but still exactly solvable one is the choice of a double-periodic elliptic function like the Weierstrass function $V(z) \sim \wp(z; \{g_2, g_3\})$ [1211]). But this is just the Mathieu equation. Transforming

$w''(z) + (a - 2q \cos(2z)) w(z) = 0$ into a Schrödinger form $-\psi''(z) + V_0 \cos(z) \psi(z) = \varepsilon \psi(z)$) yields the following dictionary: characteristic $a \Leftrightarrow$ energy ε and parameter $q \Leftrightarrow$ potential strength V_0. Here is the solution of the Mathieu equation in Schrödinger form. (In the following, we will often choose $V_0 = -1$, giving a potential well at the origin.)

```
In[69]:= sol = DSolve[-ψ''[z] + V0 Cos[z] ψ[z] == ε ψ[z], ψ[z], z]
```

$\text{Out[69]}= \left\{ \left\{ \psi[z] \to C[1] \text{ MathieuC} \left[4\,\varepsilon,\, 2\,V0,\, \frac{z}{2} \right] + C[2] \text{ MathieuS} \left[4\,\varepsilon,\, 2\,V0,\, \frac{z}{2} \right] \right\} \right\}$

```
In[70]:= -ψ''[z] + V0 Cos[z] ψ[z] == ε ψ[z] /.
              ψ -> Function[z, Evaluate[ψ[z] /. sol[[1]]]] // Simplify
```

Out[70]= True

```
In[71]:= sol = DSolve[-ψ''[z] + 0 V0 Cos[z] ψ[z] == ε ψ[z], ψ[z], z]
```

$\text{Out[71]}= \left\{ \left\{ \psi[z] \to C[1] \text{ Cos}[z\sqrt{\varepsilon}] + C[2] \text{ Sin}[z\sqrt{\varepsilon}] \right\} \right\}$

A complete set of quantum numbers for a periodic potential $V_0 \cos(z)$ is the band number $n = 0, 1, \ldots$ and the quasi-wave vector $\mathcal{K}$ $(-1 \le \mathcal{K} \le 1)$ [702], [757]. The band structure of the cos-potential can be shown in a reduced or in an extended zone scheme. Here are the first few dispersions relations $\varepsilon_n(\mathcal{K})$ shown for $V_0 = -1$.

```
In[72]:= With[{V0 = -1, ε = 10^-6},
         Show[GraphicsArray[
         Function[f, Show[{#, (* mirror *) # /.
                      Line[l_] :> Line[{-1, 1}#& /@ l]}]&[
         (* the curves *)
         Table[ParametricPlot[f, {κ, j + ε, j + 1 - ε},
                      DisplayFunction -> Identity,
                      Compiled -> False], {j, 0, 3}]],
                      Frame -> True, Axes -> False], HoldAll] /@
              Unevaluated[{{κ, MathieuCharacteristicA[κ, 2 V0]/4},
                 (* fold in *) {κ - Floor[κ], MathieuCharacteristicA[
                      If[EvenQ[Floor[κ]], κ, 2 Floor[κ] + 1 - κ], 2 V0]/4}}]]]]
```

The width of the lowest energy bands depends exponentially on the potential strength V_0.

```
In[73]:= Δe[n_, V0_] := With[{ε = 10^-10},
              MathieuCharacteristicA[n + 1 - ε, 2V0] -
              MathieuCharacteristicA[n + ε, 2V0]]
```

```
In[74]:= Plot[Evaluate[Table[Log[10, Δe[j, V0]], {j, 0, 6}]],
              {V0, -30, 0}, Frame -> True, Axes -> False]
```

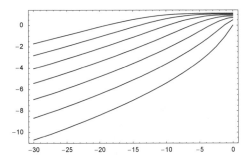

The well known Bloch theorem states that inside an allowed energy band (meaning the corresponding wave functions stay bounded as $|z| \to \infty$) the wave functions can be written in the form $\psi_{n\mathcal{K}}(z) = e^{i\mathcal{K}z} u_{n\mathcal{K}}(z)$ where $u_{n\mathcal{K}}(z)$ is periodic with the periodicity of the underlying potential and $\mathcal{K}$ is again the quasi-wave vector [702], [757], [653]. This means that the Bloch wave vector $\mathcal{K}$ is the characteristic exponent. In the free particle limit ($V_0 = 0$), the Mathieu functions reduce to trigonometric functions.

```
In[75]:= {MathieuC[MathieuCharacteristicA[κ, 0], 0, z],
         MathieuS[MathieuCharacteristicB[κ, 0], 0, z]}
```

$$\text{Out[75]=}\ \left\{ \cos\left[z\,\sqrt{\kappa^2}\right],\ \sin\left[z\,\sqrt{\kappa^2}\right] \right\}$$

Compensating for the square root in the last expression, we have the following definition for the Bloch functions $\psi_{n\mathcal{K}}(z)$ and their periodic parts $u_{n\mathcal{K}}(z)$. These functions are normalized [245] to $\int_{-\pi}^{\pi} |\psi_{n\mathcal{K}}(z)|\,dz = 2\pi$ with respect to z and orthogonal

$$\int_{-\infty}^{\infty} \psi_{n\mathcal{K}}(z)\,\overline{\psi_{n'\,\mathcal{K}'}(z)}\,dz = 2\pi\,\delta_{n,n'}\,\delta(\mathcal{K} - \mathcal{K}').$$

```
In[76]:= BlochK[κ_, V0_, z_] :=
             MathieuC[MathieuCharacteristicA[κ, 2 V0], 2 V0, z/2] +
         Sign[κ] I MathieuS[MathieuCharacteristicB[κ, 2 V0], 2 V0, z/2]

         periodicu[κ_, V0_, z_] := BlochK[κ, V0, z]/Exp[I κ z/2]
```

The following graphics show the absolute value and the argument of the periodic Bloch functions and their periodic parts. The 2π-periodicity and the shape-change of behavior at the band edges are clearly visible.

```
In[79]:= Show[GraphicsArray[
         ContourPlot[# @ periodicu[κ, -1, z], {z, -5/2Pi, 5/2Pi}, {κ, -4, 4},
                 PlotPoints -> 100, DisplayFunction -> Identity,
                 PlotRange -> All, ColorFunction -> (Hue[0.8 #]&)]& /@
                                         {Abs, Arg[#]^2&}]]
```

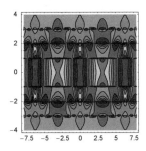 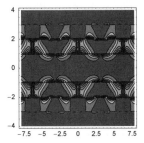

Within a band, the quasiperiodicity of the Bloch functions yields a complicated behavior when viewed over many periods. The following two graphics show the real part of the Bloch functions of the lowest two energy bands over 24 periods.

```
In[80]:= With[{ε = 10^-3},
    Show[GraphicsArray[
    Plot3D[Evaluate[Re[BlochK[κ, -3, z]]],
        {z, -24 Pi, 24 Pi}, {κ, # + ε, # + 1 - ε},
        PlotPoints -> 180, Mesh -> False, PlotRange -> All,
        DisplayFunction -> Identity, Axes -> False]& /@ {0, 1}]]]
```

This complicated structure can be interpreted in the following way: For large V_0, the periodic potential becomes a series of independent potential wells (at least for the low lying energy states). Instead of having infinitely many wells, we look now at the situation of a finite number ($d = 24$ in the following) of wells. We use periodic boundary conditions, meaning, we identify the first with the $d + 1$ well. The overlap integrals between the localized wave functions of the wells depend exponentially on the distances between the wells. This leads to the following simple model for *dim* extended states out of *dim* states localized in the wells (a circulant matrix [26], [360]).

```
In[81]:= dim = 24;
    Module[{(* small overlap constant *) α = 10^-2,
        H = Table[α^Min[Abs[i - j], dim - Abs[i - j]],
            {i, dim}, {j, dim}];
        {evals, evecs} = Eigensystem[N[H]]];
```

The eigenvalues show already the typical dispersion relation of periodic potentials in 1D. (Because of the symmetry of H, the eigenvalues appear in pairs.)

```
In[83]:= ListPlot[evals, PlotJoined -> True, Frame -> True]
```

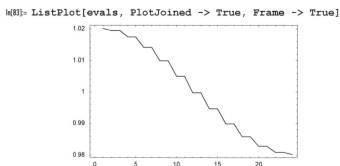

We now model the kth state inside each well with the kth state of a harmonic oscillator $\phi[k, x]$. Using the eigenvectors from above we form the eigenfunctions extending over all d wells. We use the ground-state and the first excited state, as the well functions. Similar to the above graphics of the Bloch functions, we easily recognize the totally symmetric and totally antisymmetric states as the first and last state for finite d, or as the sates at the band edges respectively.

```
In[84]:= ϕ[k_, x_] := HermiteH[k, x] Exp[-x^2]

    Function[e, Show[GraphicsArray[
      Plot[Evaluate[(* form extended states *)
          Sum[ϕ[e, x - 5 k] evecs[[#, k]], {k, dim}]],
          {x, 0, (dim + 1) 5}, PlotRange -> All, Axes -> False,
          DisplayFunction -> Identity, Frame -> True,
          FrameTicks -> False, PlotPoints -> 300,
          PlotStyle -> {Thickness[0.002]}]& /@
          (* five selected states *) {1, 6, 12, 18, 24}]]] /@ {0, 1}
```

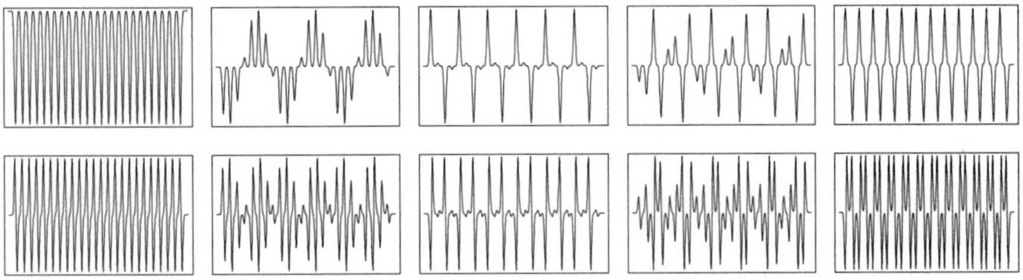

Bloch functions $\psi_{n\mathcal{K}}(z)$ are extended states. Often one wants localized states. A natural choice for real-valued states in a periodic potential localized near $z \approx Z$ are the so called Wannier functions $W_n(z - Z) = \int_{-\pi}^{\pi} \psi_{n\mathcal{K}}(z) e^{-i\mathcal{K}z} dz$ [1298]. (The choice of a possible $\mathcal{K}$-dependent phase of $\psi_{n\mathcal{K}}(z)$ is determined by $\psi_{n0}(0)$ and $\psi_{n1}(0)$ [1004], [859], [1166], [1167], [226]). (Wannier functions can be thought of as special eigendifferentials for the continuous spectrum of the allowed energy bands [882], [737], [738], [739]; for Wannier functions constructed from Gaussians, see [648].)

```
In[86]:= WannierW[n_, V0_, Z_, z_, nIntOpts___] :=
    With[{ε = 10^-8},
      NIntegrate[Evaluate[Exp[-I κ Z] BlochK[+κ, V0, z + Z] +
                          Exp[+I κ Z] BlochK[-κ, V0, z + Z]],
          {κ, n + ε, n + 1 - ε},
          (* numerical integration options *) nIntOpts]]
```

The 0th Wannier functions is a localized bump that quickly goes to zero.

```
In[87]:= wData = Table[{z, WannierW[0, -1, 0, z,
                          PrecisionGoal -> 4, MaxRecursion -> 10]},
                {z, 0, 9Pi, 9Pi/120}];

In[88]:= WannierListGraphic[wData_] :=
    With[{wDataAll = Join[Reverse[{-1, 1}#& /@ #], #]&[wData]},
        Show[GraphicsArray[ListPlot[#, PlotRange -> All, Frame -> True,
                        Axes -> False, PlotJoined -> True,
                        DisplayFunction -> Identity]& /@
          {{#1, #2}& @@@ wDataAll, {#1, Log @ Abs[#2]}& @@@ wDataAll}]]]

In[89]:= WannierListGraphic[wData // N]
```

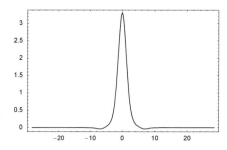
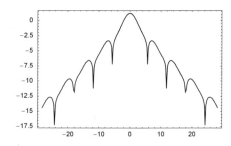

Qualitatively we have $|W_n(z - Z)| \propto \exp(-\lambda |z - Z|)$ as $|z - Z| \to \infty$ where the decay length λ is equal to the distance of the nearest branch point of $\varepsilon_{n\mathcal{K}}$ from the real $\mathcal{K}$-axis [702]. For the above data wData, we can approximate λ numerically as $\lambda \approx -0.5$.

```
In[90]:= Fit[{{#1, Log @ Abs[#2]}& @@@ wData, {1, x}, x]
Out[90]= 0.0614477 - 0.530262 x
```

The branch points of $\varepsilon_{n\mathcal{K}}$ are located at $j \pm i\zeta$, $j \in \mathbb{Z}$. The branch point of relevance for the lowest Wannier function is located at $\mathcal{K} \approx 1 + 0.48035\,i$, which is in good agreement with the above estimate.

```
In[91]:= Show[GraphicsArray[
        Plot[Evaluate[{Re[#], Im[#]}&[MathieuCharacteristicA[# + I r, -1]/4]],
            {r, -1, 1}, PlotStyle -> {GrayLevel[0], Hue[0]},
            PlotRange -> All, DisplayFunction -> Identity]& /@ {0, 1}]]
```

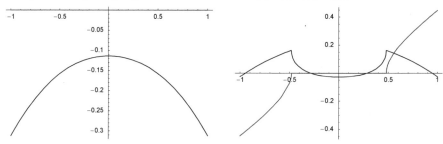

(A more detailed analysis shows that the decay is $|z|^{-3/4} \exp(-\lambda |z|)$ [577], [317].)

Higher Wannier functions are still exponentially localized, but typically show a more complicated structure. Here is $W_1(z)$ shown.

```
In[92]:= (* slightly more efficient version of WannierW *)
        WannierW[n_, V0_, Z_, z_, nIntOpts___] := With[{ε = 10^-8},
        NIntegrate[2 Re[Exp[-I κ Z] BlochK[+κ, V0, z + Z]],
            {κ, n + ε, n + 1 - ε}, nIntOpts]]

In[94]:= wData = Table[{z, WannierW[1, -1, 0, z,
                            PrecisionGoal -> 4, MaxRecursion -> 10]},
            {z, 0, 9Pi, 9Pi/120}];

        WannierListGraphic[wData // N]
```

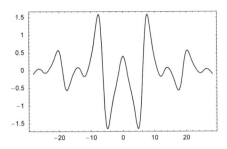

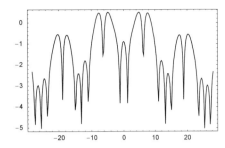

In the limit $v_0 \to 0$, we have $W_n(z - Z) \to 4\sin((z - Z)/2)/(z - Z)$. For $V_0 = -0.1$ the Wannier function $W_1(z)$ resembles $4\sin(z/2)/z$ quite well.

```
In[97]:= wData = Table[{z, WannierW[0, -0.1, 0, z]}, {z, 0, 6Pi, 6Pi/60}];
         WannierListGraphic[wData];
```

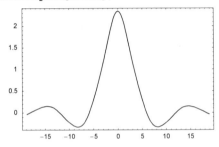

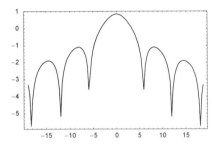

For an interpretation of $-\psi''(z) + V_0\cos(z)\,\psi(z) = \varepsilon\,\psi(z)$ as a quantum mechanical pendulum, see [25], [1032], [397], [703], [82], [18], and [399]; for a semiclassical treatment, see [89], [950]; for the double pendulum, see [1010].

3.12 Additional Special Functions

Not all special functions mentioned in the above references are implemented in Version 4. However, most of the functions that are not implemented can be expressed relatively easily in terms of other functions that are available in *Mathematica*. These include Weber functions $E_v(z)$, Anger functions $J_v(z)$, Lommel functions $s_{\mu,v}(z)$, $S_{\mu,v}(z)$, and Kelvin functions $ber_v(z)$, $bei_v(z)$, $ker_v(z)$, $kei_v(z)$.

However, a few functions are definitely not available (in *Mathematica* Version 4), including the hypergeometric functions of several variables, that is, in two variables the Kampé de Fériet function (see, for instance, [50], [958]) and Lauricella functions in more variables, Epstein Zeta functions for arbitrarily many variables, and so on.

Some of the special functions implemented in *Mathematica* and not discussed here are the following:

■ Elliptic theta functions

$$\vartheta_1(u \mid q),\ \vartheta_2(u \mid q),\ \vartheta_3(u \mid q),\ \vartheta_4(u \mid q)$$

These are solutions of partial differential equations similar to the heat equation (see [421], [1047], and [120]).

■ Weierstrass functions $\wp(u; g_2, g_3)$ and their derivatives $\wp'(u; g_2, g_3)$. These are also inverse functions for elliptic integrals and permit the solution of differential equations of the form:

$$x''(t) + a\,x(t) + b\,x(t)^2 + c = 0$$

which is Newton's law of motion for a force $\sim a\,x + b\,x^2 + c$. As an elliptic function a lot of other interesting problems from classical mechanics can also be expressed using Weierstrass functions; see the listing in Section 3.9 and [1310]. The Weierstrass function is also very useful for the construction of closed-form solutions of 1D chaotic maps [221].

■ Polylogarithm $\mathrm{Li}_\nu(z)$.

■ Zeta functions $\zeta(s, a)$. These frequently arise in summation problems and are of particular interest for solving number-theoretic problems.

■ Lerch functions $\Phi(z, s, a)$. These, too, are of great interest in number theory. In addition, they allow closed-form expressions of Fermi integrals [315]

$$\int_0^\infty \frac{x^s}{e^{x-\mu}+1}\, dx = e^\mu\,\Gamma(s+1)\,\Phi(-e^\mu, s+1, 1).$$

Although not discussed in the main part of this chapter, we will meet some of the mentioned functions in the exercises.

To conclude this section, we state the following warning: In rare cases (in general for large complex values of the variables and the parameters), the values of the special functions will be incorrectly computed when using machine accuracy. For such arguments, one should make use of *Mathematica*'s bignum arithmetic capability in the form N [*specialFunction(complexNumber)*, *numberGreaterThan$MachinePrecision*] to compare results.

3.13 Solution of Quintic Polynomials

In this section, we treat a subject that is relatively unknown, but nevertheless very interesting: The roots of a generic fifth-degree polynomial can be expressed in closed form by using hypergeometric functions. After the proofs of Abel, Ruffini, and Galois showing that it is impossible to solve a generic quintic in radicals (for a review of these early investigations, see [1018], [668], [67], [1223], and [496]), Brioschi [214], Hermite [586], and Kronecker [728] gave solutions of the quintic in closed form using elliptic modular functions, which can be expressed through elliptic theta functions. (For a modernized rephrasing of these approaches, see [1364], [667], [1017], [941], [359], [676], [677], [45], and [191], and the material from the poster "*Solving the Quintic with Mathematica*" *MathSource* 0207-122. For a solution in Poisson-type series, see [62].) Later, Felix Klein in his book *Lectures on the Icosahedron* [689] and in his article *Further Investigations on the Icosahedron* [688] expressed the roots using hypergeometric functions [590], [591], [748], [749], [1009].

Because we have discussed hypergeometric functions but not elliptic theta functions, we will outline Klein's approach. We do not follow Klein directly, because his treatment has been simplified (see [688], [692], [518], [806], [393], [1165], and [620]). The complete theory and motivation behind Klein's solution is outside the scope of this book on *Mathematica*; for details, see [322], [441], [1320], [1277], [1279], [1278], [1112], [690], [692], [544], [691], and [207]). We will discuss here only the construction of an explicit solution of a quintic equation by itself. A couple of other methods also explicitly express the roots of a quintic (see, for instance, [791], [117], [1143], [675], [1246], [407], and [1002]).

A basic sketch of Klein's approach for solving polynomials of degree five is the following: Klein, observing that the symmetry group of an icosahedron is just the alternating group of five letters, expressed the five roots of a quintic through a single root of a polynomial of degree sixty. Setting this polynomial to zero gives a special equation, the icosahedral equation. By looking at the possible monodromy groups of the hypergeometric differential equation or by analyzing the stereographic projection of an icosahedron into a half-plane, this root of the degree sixty polynomial can be expressed as the ratio of two hypergeometric functions. So the roots of a general degree five polynomial can be expressed in closed form.

Let us start with a general quintic of the form

$$x^5 + A\,x^4 + B\,x^3 + C\,x^2 + D\,x + E = 0$$

where $A, B, C, D, E \in \mathbb{C}$.

With a Tschirnhaus transformation (see [1240], [571], and [271], or illustrated in more modern terms in [519], [537], [723], [1305], [1], [2], [722], and [1193]), we can transform this general quintic into a so-called principal quintic of the form

$$z^5 + 5\,a\,z^2 + 5\,b\,z + c = 0.$$

(The coefficients 5 has historical origin and a, b, and c are functions of A, B, C, D, and E.)

The idea of the Tschirnhaus transformation is the following. We are looking for a new equation in which the coefficients of the degree 3 and degree 4 terms are 0. To achieve this, we make the following transformation with the roots x_i of the original equation $z_i(x_i) = x_i^2 + \alpha\,x_i + \beta$. Now, by requiring that the roots z_i of the transformed equation obey $\sum_{i=0}^{4} z_i = 0$ and $\sum_{i=0}^{4} z_i^2 = 0$, we force the coefficients of z^4 and z^3 to vanish (this can be seen immediately by Vieta's relations). The transformed roots z_i are, of course, still unknown, but because of Newton's relations (see Exercise 5 in Chapter 2), we can sum the roots of the transformed quintic to get its coefficients. Because we can express the sum of the old roots through the coefficients of the old equation via Newton's relations, we can express the coefficient of the new principal quintic in closed form through the coefficients of the old quintic. The two parameters α and β are determined so that the above two sums vanish. A quintic with missing degree 4 and 3 terms is called a principal quintic.

This elimination of the quartic and cubic term is implemented next. The first argument of ReduceToPrinci֙palQuintic is the general quintic, the second is the variable in which *quintic* is supposed to be a quintic, and the third argument is the variable of the transformed equation. The result of ReduceToPrincipalQuintic is a list with the Tschirnhaus transformation as its first argument (in the form of a pure function) and the transformed equation as its second. (We only check if the equation is a quintic. We do not check to see if it is already in principal form or in an even more reduced form. The restriction of the polynomial variables to symbols can, of course, be relaxed, but it is sufficient for our purposes here.)

```
In[1]:= ReduceToPrincipalQuintic[quintic_Equal,
                        oldVar_Symbol, newVar_Symbol] :=
    Module[{leftHandSide, oldCoefficient, pS, α, β, x},
        leftHandSide = Cancel[#/Coefficient[#, oldVar, 5]
                    ]&[Subtract @@ quintic];
        (* set nonexistent coefficients to 0 *)
        Do[oldCoefficient[i] = If[Evaluate[i > 5], 0,
            Coefficient[leftHandSide, oldVar, 5 - i]], {i, 10}];
        (* the recursive definition of powersums *)
        pS[k_] := pS[k] = -Expand[(Sum[pS[k - j]*
```

```
          oldCoefficient[j], {j, 1, k - 1}] + k oldCoefficient[k])];
(* α and β from the transformation x^2 + α x + β *)
α = (-pS[1] pS[2] + 5 pS[3] + Sqrt[5] Sqrt[pS[2]^3 -
    2 pS[1] pS[2] pS[3] + 5 pS[3]^2 + pS[1]^2 pS[4] -
    5 pS[2] pS[4]])/(pS[1]^2 - 5 pS[2]);
β = (-5 pS[2]^2 + 5 pS[1] pS[3] + Sqrt[5] pS[1] Sqrt[pS[2]^3 -
    2 pS[1] pS[2] pS[3] + 5 pS[3]^2 + pS[1]^2 pS[4] -
    5 pS[2 ]pS[4]])/(5(-pS[1]^2 + 5 pS[2]));
(* do the transformation; calculation of the new coefficients *)
{Function @@ {{1, α, β}.{#^2, #, 1}},
  Collect[newVar^5 + {-newVar^2/3, -newVar/4, -1/5}.(Table[Collect[
      (x^2 + α x + β)^i + 4β^i, x], {i, 3, 5}] /.
                x^n_ :> pS[n]), newVar] == 0}] /;
  (* test if quintic is really a quintic *)
  With[{diff = Subtract @@ quintic}, PolynomialQ[diff] &&
      Length[CoefficientList[diff, oldVar]] === 6]
```

Let us look at an example. We start with a quintic whose coefficients are nonzero.

In[2]:= **oldQuintic = x^5 + 5I x^4 - 5 x^3 + (5 + I) x^2 + 3 x - 4 == 0;**

Here is the Tschirnhaus transformed equation.

In[3]:= **{trafo, newQuintic} = ReduceToPrincipalQuintic[oldQuintic, x, z] //**
                                                        Simplify

Out[3]= $\left\{ \frac{1}{250} \left((-50 - 375\,i) + 25\,i\,\sqrt{181 + 40\,i} \right) + \frac{1}{50} \left((-75 + 160\,i) + 5\,\sqrt{181 + 40\,i} \right) \#1 + \#1^2 \, \&, \right.$
$(3214585 - 2403013\,i) - (310249 - 204125\,i)\,\sqrt{181 + 40\,i} +$
$(100 + 75\,i) \left((-6995 - 37308\,i) + (771 + 2630\,i)\,\sqrt{181 + 40\,i} \right) z +$
$25\,i\left((-25985 + 4662\,i) + (1351 + 110\,i)\,\sqrt{181 + 40\,i} \right) z^2 + 6250\,z^5 == 0 \Big\}$

After applying Tschirnhaus's transformation, the resulting equation typically has quite complicated coefficients, and this is true even in the case in which the original equation has integer coefficients. Let us make a numerical check of the transformation by looking at the transformed roots of the old equation.

In[4]:= **Last /@ List @@ NRoots[oldQuintic, x]**
Out[4]= {-1.10162 - 1.94615 i, -0.770325 + 0.158724 i,
    0.606354 + 0.674669 i, 0.624501 - 3.69754 i, 0.641091 - 0.189709 i}

In[5]:= **Sort[trafo /@ % // N]**
Out[5]= {-2.78278 + 2.60269 i, -1.34256 - 2.13223 i,
    -0.198047 - 2.99322 i, 0.56839 + 1.78424 i, 3.755 + 0.738514 i}

We look at the roots of the new equation.

In[6]:= **Last /@ List @@ NRoots[newQuintic, z]**
Out[6]= {-2.78278 + 2.60269 i, -1.34256 - 2.13223 i,
    -0.198047 - 2.99322 i, 0.56839 + 1.78424 i, 3.755 + 0.738514 i}

Conversely, if we have somehow obtained the solutions of the transformed equation, the determination of the corresponding roots of the new equation is unique. However, at first glance, we might think otherwise, because it is not clear which of the roots to use when inverting the $z_i(x_i)$. But this root $z_i(x_i)$ has to satisfy both the old equation and the transformed equation; so the root is uniquely defined. Let us have a look at this with a simple example.

In[7]:= **{trafo, eq} = Simplify[ReduceToPrincipalQuintic[**
        Expand[(x - 1)(x - 2)(x - 3)(x - 4)(x - 5)] == 0, x, z]]

Out[7]= $\{\frac{1}{250}(1750 + 150 \text{ i } \sqrt{35}) + \frac{1}{50}(-300 - 10 \text{ i } \sqrt{35}) \text{ #1 + #1}^2 \text{ \&},$

$1152 + 1056 \text{ z} + 440 \text{ z}^2 + 25 \text{ z}^5 == 0\}$

Next, we solve the transformed quintic. In this case, in which the original quintic factors completely over the integers (or rationals), the Tschirnhaus transformed equation looks simple in most cases and is in this case also explicitly solvable in radicals.

In[8]:= `Solve[eq, z] // Simplify`

Out[8]= $\{\{z \to -2\}, \{z \to -1 - \text{ i } \sqrt{\frac{7}{5}}\}, \{z \to 2 - 2 \text{ i } \sqrt{\frac{7}{5}}\}, \{z \to -1 + \text{ i } \sqrt{\frac{7}{5}}\}, \{z \to 2 + 2 \text{ i } \sqrt{\frac{7}{5}}\}\}$

Of course, we get the same roots by applying the transformation directly to the roots of the original quintic.

In[9]:= `(trafo /@ {1, 2, 3, 4, 5}) // Simplify`

Out[9]= $\{2 + 2 \text{ i } \sqrt{\frac{7}{5}}, -1 + \text{ i } \sqrt{\frac{7}{5}}, -2, -1 - \text{ i } \sqrt{\frac{7}{5}}, 2 - 2 \text{ i } \sqrt{\frac{7}{5}}\}$

We can explicitly find the correct root with `PolynomialGCD`.

In[10]:= `5 (Together[PolynomialGCD[` (* original and transformed equations *)
       `Expand[(x - 1)(x - 2)(x - 3)(x - 4)(x - 5)],`
       `Expand[trafo[x] - #]]]& /@ %)`

Out[10]= $\{-1 + x, -2 + x, -3 + x, -4 + x, -5 + x\}$

Remark: One can also do the above Tschirnhaus transformation calculation directly, without using Newton's relations. (But to eliminate the coefficient of x^2, which is always possible but not needed here, the above method is much faster and shorter.) Here is a direct version (it does not matter which solution we pick):

```
Collect[Subtract @@ Eliminate[{z == x^2 + α x + β,
    x^5 + A1 x^4 + B1 x^3 + C1 x^2 + D1 x + E1 == 0}, x], z];

Solve[{Coefficient[%, z, 4] == 0, Coefficient[%, z, 3] == 0}, {α, β}] //
                                                           Simplify
```

$\{\{β \to \frac{1}{20 c_1^2 - 50 c_2}$

$(5 c_2 c_1^2 + (15 c_3 - \sqrt{5} \sqrt{8 c_3 c_1^3 + (16 c_4 - 3 c_2^2) c_1^2 - 38 c_2 c_3 c_1 + 12 c_2^3 + 45 c_3^2 - 40 c_2 c_4})$

$c_1 - 20 c_2^2), α \to \frac{1}{4 c_1^2 - 10 c_2}(4 c_1^3 - 13 c_2 c_1 + 15 c_3 - \sqrt{5}$

$\sqrt{8 c_3 c_1^3 + (16 c_4 - 3 c_2^2) c_1^2 - 38 c_2 c_3 c_1 + 12 c_2^3 + 45 c_3^2 - 40 c_2 c_4})\}, \{β \to \frac{1}{20 c_1^2 - 50 c_2}$

$(5 c_2 c_1^2 + (15 c_3 + \sqrt{5} \sqrt{8 c_3 c_1^3 + (16 c_4 - 3 c_2^2) c_1^2 - 38 c_2 c_3 c_1 + 12 c_2^3 + 45 c_3^2 - 40 c_2 c_4})$

$c_1 - 20 c_2^2), α \to \frac{1}{4 c_1^2 - 10 c_2}(4 c_1^3 - 13 c_2 c_1 + 15 c_3 +$

$\sqrt{5} \sqrt{8 c_3 c_1^3 + (16 c_4 - 3 c_2^2) c_1^2 - 38 c_2 c_3 c_1 + 12 c_2^3 + 45 c_3^2 - 40 c_2 c_4})\}\}$

Having transformed the generic quintic into one of the form $z^5 + 5 a z^2 + 5 b z + c = 0$, let us look for a solution of this simpler equation. First, we introduce five functions that we will use frequently later. The quantities f, H, T, t_i, and W_i ($i = 0, 1, 2, 3, 4$) are functions depending on two parameters u and v.

In[11]:= `(*` ϵ `== Exp[2Pi I/5]; but to save space and`
        `to avoid evaluation we use just` ϵ `*)`
        `f[u_, v_] = u v (u^10 + 11 u^5 v^5 - v^10);`

```
H[u_, v_] = -u^20 - v^20 + 228(u^15 v^5 - u^5 v^15) -
            494 u^10 v^10;

T[u_, v_] = u^30 + v^30 + 522 (u^25 v^5 - u^5 v^25) -
            10005 (u^20 v^10 + u^10 v^20);

W[k_, u_, v_] = -e^(4k) u^8 + e^(3k) u^7 v - 7 e^(2k) u^6 v^2 -
                7 e^k u^5 v^3 + 7 e^(4k) u^3 v^5 -
                7 e^(3k) u^2 v^6 - e^(2k) u v^7 - e^k v^8;

t[k_, u_, v_] = e^(3k) u^6 + 2 e^(2k) u^5 v - 5 e^k u^4 v^2 -
                5e^(4k) u^2 v^4 - 2 e^(3k) u v^5 + e^(2k) v^6;
```

Here, ϵ is equal to `Exp[2Pi I/5]`, but to keep things short and readable, we just write ϵ. We will later on use the algebraic properties of ϵ, but we will not assign the explicit value in the moment. Similarly, we will sometimes write H for `H[u, v]`, and so on, to avoid evaluation.

Let us briefly discuss the relationship between these functions and the icosahedron [688], [829], [478], [867], [400], [337], [677]. The roots of `f[z, 1]` are just the locations of the stereographic projections of the corners of an icosahedron of unit radius (`myIco`) and center {0, 0, 0}. This can be visualized as follows: The squares in the x,y-plane are around the zeros of `f[z, 1]` (determined numerically).

In[17]:= **Needs["Graphics`Polyhedra`"]**

```
(* rescale the icosahedron *)
myIco = Function[x, Map[#/x&, Polyhedron[Icosahedron][[1]], {-1}]][
                    Polyhedron[Icosahedron][[1, 1, 1, 1, 3]]];


StereographicProjection[{x_, y_, z_}] := {x/(1 + z), y/(1 + z), 0}

Show[Graphics3D[{
(* the skeleton of the icosahedron; make holes in the faces *)
Function[poly,
   Module[{p = poly[[1]], mp, i, e}, mp = (Plus @@ p)/Length[p];
     i = (mp + 0.9(# - mp))& /@ p; e = (mp + (# - mp))& /@ p;
     i = Reverse /@ Partition[Append[i, i[[1]]], 2, 1];
     e = Partition[Append[e, e[[1]]], 2, 1];
     {Polygon /@ MapThread[Join, {i, e}]}]] /@ myIco,
(* a mesh in the x,y-plane *)
{Thickness[0.002], GrayLevel[0.7],
{Line /@ #, Line /@ Transpose[#]}&[
             Table[{i, j, 0}, {i, -2, 2, 1/5}, {j, -2, 2, 1/5}]]},
(* the stereographic projections of the vertices *)
{Thickness[0.008], GrayLevel[0],
 If[(* avoid point {0, 0, -1} *) Abs[#[[3]] + 1] > 10^-4,
     Line[{#, StereographicProjection[#]}], {}]& /@
                     Union[Flatten[First /@ myIco, 1]]},
(* the zeros of f[z] *)
  {Thickness[0.002], GrayLevel[0],
  Line[{# + {0.05,  0.05, 0}, # + {-0.05, 0.05, 0}, # + {-0.05, -0.05, 0},
        # + {0.05, -0.05, 0}, # + { 0.05, 0.05, 0}}]& /@
({Re[#], Im[#], 0}& /@ Last /@ (List @@ NRoots[f[z, 1] == 0, z]))}}],
Axes -> True, Boxed -> False, PlotRange -> All]
```

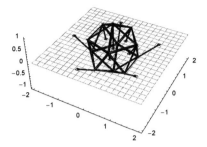

Up to a factor, H is the Hessian of f (the definition of the Hessian is immediately visible in the following input). T is the functional determinant of f and H.

```
In[22]:= {Det[{{D[f[u, v], u, u], D[f[u, v], u, v]},
              {D[f[u, v], v, u], D[f[u, v], v, v]}}]/H[u, v],
         Det[{{D[f[u, v], u], D[f[u, v], v]},
              {D[H[u, v], u], D[H[u, v], v]}}]/T[u, v]} // Cancel
Out[22]= {121, 20}
```

In analogy to the interpretation of the zeros of f, one recognizes the zeros of H are the stereographic projections of the midpoints of the faces (which is why H has degree 20) and the zeros of T are the stereographic projections of the midpoints of the 30 edges of an icosahedron with unit distance of the relevant points from the origin [647].

Now, let us interpret t_i and W_i. The roots of t are the stereographic projections of the vertices of five octahedra that can be inscribed into the icosahedron so its vertices are located at the midpoints of the edges of the icosahedron. This is demonstrated numerically using the analogous property of T.

```
In[23]:= (Last /@ (List @@ NRoots[T[z, 1] == 0, z])) -
         (Table[Last /@ (List @@ NRoots[(t[k, z, 1] /.
               ε -> Exp[2Pi I/5]) == 0, z]), {k, 0, 4}] //
                Flatten // Union) // Chop // Union
Out[23]= {0}
```

The W_i are the stereographic projections of the midpoints of the faces of the five octahedra. (We will not prove this because it is straightforward but lengthy.) After making these definitions, let us continue solving the quintics. We start by making the following ansatz for the five roots of a quintic in principal form

$$z_i = \frac{\lambda \, f}{H} \, W_i + \frac{\mu \, f^3}{H \, T} \, t_i \, W_i \quad (i = 0, \dots 4)$$

where W_i, t_i, f, H, and T as defined above are functions of $u(a, b, c)$, $v(a, b, c)$, and λ and μ are parameters to be determined that also depend on a, b, and c. The variables a, b, and c come from $z^5 + 5\,a\,z^2 + 5\,b\,z + c = 0$, and the exact dependence of u, v, λ, and μ on a, b, and c still has to be calculated. To determine λ and μ, we use Newton's relations for the power sums of the roots z_i. As a first step, we sum the jth ($j = 1, \dots, 5$) powers of the roots z_i and reexpress the resulting quantities through f, T, and H one step at a time until we have no remainder that contains u. In the following procedure, plugInAnsatz carries out this calculation. We use a couple of obvious identities for $\exp(2\,i\pi/5) = \epsilon$ here.

```
In[24]:= (* properties of Exp[2Pi I/5] *)
         ε /: ε^n_ ? (NumberQ[#] && # > 5&) = ε^Mod[n, 5];
         ε /: 1 + ε + ε^2 + ε^3 + ε^4 = 0;
         ε /: ε^5 = 1;
```

```
(* plugInAnsatz expects products as arguments *)
plugInAnsatz[expr_] :=
Module[{poly, rest, i, uInsideQ},
If[FreeQ[expr, W[k], {0, Infinity}] && FreeQ[expr, t[k], {0, Infinity}],
(* nothing to do, just five times the original expression *)5 expr,
(* the powersum *)
poly = Sum[Evaluate[expr /. {t[k] :> t[k, u, v], W[k] :> W[k, u, v]}],
          {k, 0, 4}];
If[Max[Exponent[expr, #]& /@ {t[k], W[k]}] <= 2,
           (* not much to do *) (Factor @ poly),
(* prepare the polynomial *)
poly = Factor /@ Collect[poly, {u, v}];
(* i[f], i[H], and i[T] are the powers of f, H, and T in poly *)
rest = poly; i[f] = 0; i[H] = 0; i[T] = 0;
uInsideQ = MemberQ[#, u, {0, Infinity}]&;
While[(* try to divide the rest until it is completely reexpressed
          in f, H, and T *)
      uInsideQ[rest, u, {0, Infinity}],
      Which[uInsideQ[temp = PolynomialGCD[rest, f[u, v]]],
            i[f] = i[f] + 1; rest = Cancel[rest/temp],
            uInsideQ[temp = PolynomialGCD[rest, H[u, v]]],
            i[H] = i[H] + 1; rest = Cancel[rest/temp],
            uInsideQ[temp = PolynomialGCD[rest, T[u, v]]],
            i[T] = i[T] + 1; rest = Cancel[rest/temp]]];
(* write the polynomial poly in f, H, and T *)
Cancel[poly/(f[u, v]^i[f] H[u, v]^i[H] T[u, v]^i[T])]*
     f^i[f] H^i[H] T^i[T]]]]
```

Be aware that the following is a fairly large calculation; and in Klein's original work [688], a number of pages are devoted to this problem. It is an elementary, but really large, calculation.

Here are the power sums of the roots $\sum_0^4 z_i^j$ for $j = 1, \ldots, 5$ reexpressed through f, H, T, λ, and μ.

In[30]:= `Table[plugInAnsatz /@` (* most terms fortunately cancel *)
            `Expand[(λ f/H W[k] + μ f^3/H/T t[k] W[k])^i], {i, 5}]`

Out[30]= $\left\{0, \ 0, \ -\dfrac{120 \ f^5 \ \lambda^3}{H^3} - \dfrac{15 \ f^5 \ \lambda^2 \ \mu}{H^3} - \dfrac{1080 \ f^{10} \ \lambda \ \mu^2}{H^3 \ T^2} - \dfrac{15 \ f^{10} \ \mu^3}{H^3 \ T^2}, \right.$

$-\dfrac{20 \ f^5 \ \lambda^4}{H^3} - \dfrac{360 \ f^{10} \ \lambda^2 \ \mu^2}{H^3 \ T^2} - \dfrac{20 \ f^{10} \ \lambda \ \mu^3}{H^3 \ T^2} - \dfrac{540 \ f^{15} \ \mu^4}{H^3 \ T^4},$

$\left. -\dfrac{5 \ f^5 \ \lambda^5}{H^3} + \dfrac{50 \ f^{10} \ \lambda^3 \ \mu^2}{H^3 \ T^2} - \dfrac{225 \ f^{15} \ \lambda \ \mu^4}{H^3 \ T^4} - \dfrac{5 \ f^{15} \ \mu^5}{H^3 \ T^4} \right\}$

Now, we manually refine the result by introducing the two quantities Z and V through

$$f^5(u(Z), v(Z)) - Z T^2(u(Z), v(Z)) = 0$$
$$1728 Z - Z V = 0.$$

In[31]:= `((Together[%] /. {f^5 -> T^2 Z, f^10 -> T^4 Z^2, f^15 -> T^6 Z^3}) //`
            `Simplify) /. T^2 Z/H^3 -> 1/V`

Out[31]= $\left\{0, \ 0, \ -\dfrac{15 \ (8 \ \lambda^3 + \lambda^2 \ \mu + 72 \ Z \ \lambda \ \mu^2 + Z \ \mu^3)}{V}, \right.$

$\left. -\dfrac{20 \ (-\lambda^4 + 18 \ Z \ \lambda^2 \ \mu^2 + Z \ \lambda \ \mu^3 + 27 \ Z^2 \ \mu^4)}{V}, \ -\dfrac{5 \ (\lambda^5 - 10 \ Z \ \lambda^3 \ \mu^2 + 45 \ Z^2 \ \lambda \ \mu^4 + Z^2 \ \mu^5)}{V} \right\}$

Using Newton's relations, we compare the power sums with the coefficients of the principal quintic in the form

$$z^5 + 5\,a\,z^2 + 5\,b\,z + c = 0.$$

```
In[32]:= (Clear[powerSum, coefficient];
        (* the coefficients of the above polynomial in z *)
        Evaluate[coefficient /@ {1, 2, 3, 4, 5}] = {0, 0, 5 a, 5 b, c};
        (* the general definition of power sums *)
        powerSum[k_] := powerSum[k] =
            -Expand[(Sum[powerSum[k - j] coefficient[j], {j, 1, k - 1}] +
                    k coefficient[k])];

        Array[powerSum, 5])
Out[32]= {0, 0, -15 a, -20 b, -5 c}
```

Finally, this gives the following set of equations that connect the coefficients a, b, c of the reduced quintic with the parameters λ, μ of the ansatz.

```
In[33]:= Append[Function[eq, Function[fac, #/fac[[1]]& /@
                eq][Cases[eq[[1]], _?NumberQ]]] /@
            Map[V #&, DeleteCases[MapThread[Equal, {%, %%}], True], {2}],
        1 + Z V == 1728 Z] // TableForm
```

```
Out[33]//TableForm=
```
$$a\,V == 8\,\lambda^3 + \lambda^2\,\mu + 72\,Z\,\lambda\,\mu^2 + Z\,\mu^3$$
$$b\,V == -\lambda^4 + 18\,Z\,\lambda^2\,\mu^2 + Z\,\lambda\,\mu^3 + 27\,Z^2\,\mu^4$$
$$c\,V == \lambda^5 - 10\,Z\,\lambda^3\,\mu^2 + 45\,Z^2\,\lambda\,\mu^4 + Z^2\,\mu^5$$
$$1 + V\,Z == 1728\,Z$$

Now, let us pause for a moment and inspect the result. Solving the above four equations allows us to express Z, V, λ, and μ in terms of the given a, b, and c in the principal form quintic. The only difficult part of this calculation is to determine u and v (given Z) to satisfy the equation

$$f^5(u(Z), v(Z)) - Z\,T^2(u(Z), v(Z)) = 0.$$

This equation is shown in an inhomogeneous form (sometimes this equation is written in the literature with $Z \to 1728\,Z$).

```
In[34]:= IcosahedralEquation[u_, v_, Z_] = f[u, v]^5 - Z T[u, v]^2;

        IcosahedralEquation[z_, Z_] = IcosahedralEquation[z, 1, Z]
Out[35]= z^5 (-1 + 11 z^5 + z^10)^5 - (1 + z^30 - 10005 (z^10 + z^20) + 522 (-z^5 + z^25))^2 z
```

We see that we have "reduced" the solution of a degree five polynomial to the solution of a polynomial of degree 60. Fortunately, it is possible to solve this equation completely for all 60 roots in closed form. Here the title of this chapter comes into play. One pair $u_1(Z)$, $v_1(Z)$ of solutions of the above icosahedral equation is given by the following. (For details on how to calculate them, see [688], [606], [687], [1028], [438], [1123], [861], [213], [1112], [1165], [85], and [447] for a sextic.)

```
In[36]:= SolutionIcosahedralEquation[1, X_] =
                1/X^(01/60) Hypergeometric2F1[-1/60, 29/60, 4/5, 1728 X];

        SolutionIcosahedralEquation[2, X_] =
                X^(11/60) Hypergeometric2F1[11/60, 41/60, 6/5, 1728 X];
```

For a classical proof that these are the solutions of the icosahedral equation, see the references mentioned. Here is a numerical check for five randomly chosen examples.

```
In[39]:= Array[Chop @ IcosahedralEquation[
             SolutionIcosahedralEquation[1, #],
             SolutionIcosahedralEquation[2, #], #]&[
                Random[Complex, {-10 - 10I, 10 + 10 I}]]&, {5}]
Out[39]= {0, 0, 0, 0, 0}
```

Inspecting the two hypergeometric functions, one recognizes that they are solutions of a hypergeometric differential equation of the form 15.5.7 and 15.5.8 in [3] with $a = -1/60$, $b = 11/60$, $c = 1/2$, and $1728\,Z \to 1/z$. Transforming to the variable Z, we get the following differential equation:

$$w''(Z) + \frac{15552\,Z - 5}{(10368\,Z^2 - 6\,Z)}\,w'(Z) + \frac{11}{3600\,Z^2\,(1728\,Z - 1)}\,w(Z) = 0.$$

At this point, let us check that the solutions $z(Z)$ of the icosahedral equation are given as the ratio of the two hypergeometric functions given above. From the solution of Exercise 12.4, we know that if $y_1(z)$ and $y_2(z)$ are two linear independent solution of the differential equation

$$y''(z) + p_1(z)\,y'(z) + p_2(z)\,y(z) = 0,$$

then the ratio $w(z) = y_1(z)/y_2(z)$ fulfills the following differential equation (see Chapter 1):

$$w'''(z)\,w'(z) - \frac{3}{2}\,w''(z)^2 + \left(\frac{p_1(z)^2}{2} - 2\,p_2(z) + p_1'(z)\right)w'(z)^2 = 0.$$

Using the linear differential equation, we obtain the following nonlinear differential to be fulfilled by Solu-tionIcosahedralEquation[1, Z]/SolutionIcosahedralEquation[2, Z].

```
In[40]:= Clear[w, z, ode1];
        ode1 = w'''[z] w'[z] - 3/2 w''[z]^2 +
             (p1[z]^2/2 - 2 p2[z] + p1'[z]) w'[z]^2 /.
             {p1 -> Function[Z, (15552 z - 5)/(10368 Z^2 - 6 Z)],
              p2 -> Function[Z, 11/(3600 Z^2 (1728 Z - 1))]}
```

$$Out[41]= \left(-\frac{11}{1800\,z^2\,(-1 + 1728\,z)} + \frac{(-5 + 15552\,z)^2}{2\,(-6\,z + 10368\,z^2)^2} - \right.$$
$$\left.\frac{(-5 + 15552\,z)\,(-6 + 20736\,z)}{(-6\,z + 10368\,z^2)^2} + \frac{15552}{-6\,z + 10368\,z^2}\right)w'[z]^2 - \frac{3}{2}\,w''[z]^2 + w'[z]\,w^{(3)}[z]$$

```
In[42]:= ode1 = Numerator[Together[ode1]]
```

$$Out[42]= -24\,w'[z]^2 + 35472\,z\,w'[z]^2 - 55987200\,z^2\,w'[z]^2 -$$
$$75\,z^2\,w''[z]^2 + 259200\,z^3\,w''[z]^2 - 223948800\,z^4\,w''[z]^2 +$$
$$50\,z^2\,w'[z]\,w^{(3)}[z] - 172800\,z^3\,w'[z]\,w^{(3)}[z] + 149299200\,z^4\,w'[z]\,w^{(3)}[z]$$

On the other hand, the icosahedral equation defines $w(Z)$ and we can differentiate the equation three times with respect to Z.

```
In[43]:= IcosahedralEquation[w[z], z]
```

$$Out[43]= w[z]^5\,(-1 + 11\,w[z]^5 + w[z]^{10})^5 -$$
$$z\,(1 + w[z]^{30} - 10005\,(w[z]^{10} + w[z]^{20}) + 522\,(-w[z]^5 + w[z]^{25}))^2$$

```
In[44]:= Table[ruleD[i] = Solve[D[IcosahedralEquation[w[z], z], {z, i}] == 0,
                Derivative[i][w][z]][[1]], {i, 3}];
```

Substituting now these derivatives into the Schwarz differential equation, we obtain a large polynomial in $w(z)$ and z. (The following is a trivial, but again, large calculation.

```
In[45]:= odePoly = Numerator[Together[ode1 /. ruleD[3] /. ruleD[2] /. ruleD[1]]];
```

```
In[46]:= {Length[odePoly], Exponent[odePoly, {w[z], z}]}
```

```
Out[46]= {2, {350, 6}}
```

But one of the factors of this polynomial is the icosahedral equation (which we assume to be fulfilled by $w(z)$).

```
In[47]:= PolynomialGCD[IcosahedralEquation[w[z], z], odePoly]
```

$$Out[47]= z + w[z]^5 - 1044\, z\, w[z]^5 - 55\, w[z]^{10} + 252474\, z\, w[z]^{10} + 1205\, w[z]^{15} +$$
$$10445220\, z\, w[z]^{15} - 13090\, w[z]^{20} + 100080015\, z\, w[z]^{20} + 69585\, w[z]^{25} +$$
$$10446264\, z\, w[z]^{25} - 134761\, w[z]^{30} + 199655084\, z\, w[z]^{30} - 69585\, w[z]^{35} -$$
$$10446264\, z\, w[z]^{35} - 13090\, w[z]^{40} + 100080015\, z\, w[z]^{40} - 1205\, w[z]^{45} -$$
$$10445220\, z\, w[z]^{45} - 55\, w[z]^{50} + 252474\, z\, w[z]^{50} - w[z]^{55} + 1044\, z\, w[z]^{55} + z\, w[z]^{60}$$

Although we actually need only one solution of the icosahedral equation IcosahedralEquation[z, 1, Z] == 0 for the solution of the quintic, we show how to obtain the other 59 to illustrate the special property of this equation. The routine AllRoots generates all of the roots (including the one given as an argument) by applying the symmetry group of the icosahedron to the one root we have [535].

```
In[48]:= AllRoots[root1_,
                  (* numerically or symbolically *)num_] :=
       Function[ε, (* from the symmetry of the icosahedron *)
       Flatten[{Table[ε^i root1, {i, 5}], Table[-ε^i/root1, {i, 5}],
               Table[ ε^j ((ε + ε^4) root1 + ε^i)/
                     (root1 - ε^i (ε + ε^4)), {i, 5}, {j, 5}],
               Table[-ε^j (root1 - ε^i (ε + ε^4))/
                     ((ε + ε^4) root1 + ε^i), {i, 5}, {j, 5}]}]][
                                      num[Exp[2Pi I/5]]];
```

Here, this process of the generation of the other roots is visualized: Only the gray line is actually calculated; all other lines are generated through the use of symmetry.

```
In[49]:= Function[values,
       Show[Function[data, Graphics[{
       {Thickness[0.005], GrayLevel[0], projected =
         (* all other pieces *) (Line /@ (Map[{Re[#], Im[#]}&,
       Thread[AllRoots[#, N]& /@ data], {-1}]))},
         (* one piece *)
       {Thickness[0.015], GrayLevel[0.6], Line[{Re[#], Im[#]}& /@
         data]}}]][Function[x, SolutionIcosahedralEquation[1, x]/
           SolutionIcosahedralEquation[2, x]] /@ values],
       PlotRange -> {{-5, 5}, {-5, 5}}, AspectRatio -> Automatic,
       Frame -> True]][(* the points to be mapped *)
           N @ Join[Table[10^-i/1728, {i, 52, 2, -4}],
                   Table[10^-i/1728, {i, 3, 0, -1/10}],
                   Table[10^-i/1728, {i, 0, -1, -4/100}],
                   Table[10^-i/1728, {i, -1, -51, -5}]]]]
```

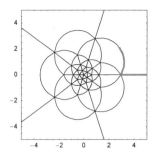

We can also look at the way the roots depend on Z varying on the 3D icosahedron by going back to the sphere from the complex plane.

```
In[50]:= InverseStereographicProjection[{x_, y_}] :=
                    {2 x, 2 y, 1 - x^2 - y^2}/(1 + x^2 + y^2);
         Show[Graphics3D[{{EdgeForm[], myIco}, Map[InverseStereographicProjection,
                                  projected, {-2}]}], Boxed -> False]
```

As is obvious from the `IcosahedralEquation` and the above discussion of the zeros of f, T, and H, the zeros of the `IcosahedralEquation` for $Z = 0$ are the stereographic projections of the vertices of the icosahedra. For $Z = \infty$, the zeros of the `IcosahedralEquation` are the stereographic projections of the midpoints of the edges. We observe that $1728\, f^5 - T^2 = H^3$.

```
In[52]:= 1728 f[u, v]^5 - T[u, v]^2 == H[u, v]^3 // ExpandAll
Out[52]= True
```

So for $Z = 1/1728$, the zeros of the icosahedra equation are the centers of the 20 faces of an icosahedron.

Now, let us check numerically that we really have all 60 roots after applying `AllRoots` to one root.

```
In[53]:= firstRoot = N[#, 40]& @ (SolutionIcosahedralEquation[1, 34/10]/
                                 SolutionIcosahedralEquation[2, 34/10]);

         allRoots = AllRoots[firstRoot, N[#, 40]&];

         Union[Chop[N[IcosahedralEquation[#, 1, 34/10]& /@ allRoots, 40]]]
Out[55]= {0, 0.×10^-3 + 0.×10^-3 i, 0.×10^-3 + 0.×10^-4 i, 0.×10^-3 + 0.×10^-4 i,
         0.×10^-3 + 0.×10^-4 i, 0.×10^-4 + 0.×10^-4 i, 0.×10^-4 + 0.×10^-5 i,
         0.×10^-4 + 0.×10^-5 i, 0.×10^-5 + 0.×10^-5 i, 0.×10^-5 + 0.×10^-6 i, 0.×10^-5 + 0.×10^-6 i}
```

Now we return to the solution of the quintic. The above ansatz simplifies for the special solution of the icosahedron equation that we chose. Using the differential equation for `SolutionIcosahedralEquation[1, Z]` and `SolutionIcosahedralEquation[2, X]` from above, the Wronskian of the two solutions given can be easily calculated to be $Z^{-5/6}(1-1728\,Z)^{-2/3}/5$. Now, we differentiate the icosahedron equation, written in the form $T^2/f^5 = Z$.

```
In[56]:= D[T[y1[Z], y2[Z]]^2/f[y1[Z], y2[Z]]^5, Z];
```

Observe that the above expression is equal to the product of the given Wronskian multiplied by $5\,T/(H\,f)\,H^3/f^5$.

```
In[57]:= Cancel[% /(5 (y2[Z] y1'[Z] - y1[Z] y2'[Z])*
                T[y1[Z], y2[Z]]/(H[y1[Z], y2[Z]] f[y1[Z], y2[Z]]) *
                H[y1[Z], y2[Z]]^3/f[y1[Z], y2[Z]]^5)]
Out[57]= 1
```

Then, we use the above-mentioned relation $T^2 + H^3 - 1728\,f^5 = 0$.

```
In[58]:= T[z[Z], 1]^2 + H[z[Z], 1]^3 - 1728 f[z[Z], 1]^5 // Together
Out[58]= 0
```

We also use $Z = T/(f\,H)$; we have from the following input that the quantity f is equal to one for the above `SolutionIcosahedralEquations`.

```
In[59]:= Factor /@ Eliminate[{Z H f == T, f^5 == X T^2,
                    T^2 + H^3 == 1728 f^5}, {H, T}]
Out[59]= f^9 X (-1 + 1728 X)^2 z^6 == f^8
```

It now remains to solve the above nonlinear system for $V = V(a, b, c)$, $Z = Z(a, b, c)$, $\lambda = \lambda(a, b, c)$, and $\mu = \mu(a, b, c)$. First, we eliminate the variable V to get a system of three equations in three unknowns.

```
In[60]:= eqs = Numerator[Together[#]]& /@ (Apply[Subtract,
         {a V == (8 λ^3 + λ^2 μ + 72 λ μ^2 Z + μ^3 Z),
          b V == (-λ^4 + 18 λ^2 μ^2 Z + λ μ^3 Z + 27 μ^4 Z^2),
          c V == (λ^5 -10 λ^3 μ^2 Z + 45 λ μ^4 Z^2 + μ^5 Z^2)}, {1}] /.
                        Solve[1 + V Z == 1728 Z, V][[1]])
Out[60]= {-a + 1728 a Z - 8 Z λ^3 - Z λ^2 μ - 72 Z^2 λ μ^2 - Z^2 μ^3,
         -b + 1728 b Z + Z λ^4 - 18 Z^2 λ^2 μ^2 - Z^2 λ μ^3 - 27 Z^3 μ^4,
         -c + 1728 c Z - Z λ^5 + 10 Z^2 λ^3 μ^2 - 45 Z^3 λ μ^4 - Z^3 μ^5}
```

To speed up the following calculations, we define a modified resultant function. `myResultant` does not take into account any algebraic dependencies between the coefficients of the two polynomials given as its arguments. (This is for the system of polynomials under consideration of advantage.)

```
In[61]:= myResultant[poly1_, poly2_, x_] :=
         Module[{c, a, t, r},
         (* replace all coefficients in poly1 and poly2 by local symbols
            without taking into account any dependencies *)
         Do[c[i] = Expand[CoefficientList[If[i == 1, poly1, poly2], x]];
            t[i] = MapIndexed[If[# =!= 0, a[i][#2[[1]] - 1], 0]&, c[i]];
            polya[i] = t[i].(x^Range[0, Length[c[i]] - 1]);
            r[i] = Apply[Rule, Transpose[{t[i], c[i]}], {1}], {i, 2}];
         (* form resultant with symbolic coefficient polynomial *)
         Resultant[polya[1], polya[2], x] //.
            (* restore original coefficients *) Flatten[{r[1], r[2]}]]
```

First, let us eliminate Z to get two equations for λ and μ.

```
In[62]:= res1 = Factor @ myResultant[eqs[[1]], eqs[[2]], Z]
Out[62]= -μ^4 (24 λ + μ)^3 (-46656 a^2 b - 5832 a^3 λ + 5184 b^2 λ^2 - 216 a b λ^3 - 108 a^2 λ^4 + 8 b λ^6 + a λ^7 +
            729 a^3 μ + 144 b^2 λ μ - 99 a b λ^2 μ + 9 a^2 λ^3 μ + b λ^5 μ + b^2 μ^2 - 2 a b λ μ^2 + a^2 λ^2 μ^2)
```

```
In[63]:= res2 = Factor @ myResultant[eqs[[1]], eqs[[3]], Z]
```
$$Out[63]= -\mu^4 \; (24\,\lambda + \mu)^3 \; (-77760\,a^2\,c\,\lambda + 5184\,c^2\,\lambda^2 + 9720\,a^3\,\lambda^3 +$$
$$792\,a\,c\,\lambda^4 - 180\,a^2\,\lambda^6 - 8\,c\,\lambda^7 + a\,\lambda^9 - 1728\,a^2\,c\,\mu + 144\,c^2\,\lambda\,\mu - 999\,a^3\,\lambda^2\,\mu +$$
$$179\,a\,c\,\lambda^3\,\mu + 8\,a^2\,\lambda^5\,\mu - c\,\lambda^6\,\mu + c^2\,\mu^2 + 18\,a^3\,\lambda\,\mu^2 + 3\,a\,c\,\lambda^2\,\mu^2 + a^3\,\mu^3)$$

Eliminating μ from `res1` and `res2`, the resulting expression contains a quadratic polynomial in λ as one of its factors (we do not need all other, larger factors for our purposes here).

```
In[64]:= res3 = Factor @ myResultant[Last[res1], Last[res2], μ];
         First[res3]
```
$$Out[65]= -64\,a^2\,b^2 + 27\,a^3\,c + b\,c^2 + 11\,a^3\,b\,\lambda + 2\,b^2\,c\,\lambda - a\,c^2\,\lambda - a^4\,\lambda^2 + b^3\,\lambda^2 - a\,b\,c\,\lambda^2$$

Using the solution for λ following from the last equation, μ, Z, and V can be calculated from the above `res1` and `eqs` and $1 + V\,Z == 1728\,Z$. The case $a = 0$ is easily treated as a limiting value of the general solution.

So we can finally implement the following solution `RootsPrincipalQuintic` of a principal quintic developed by Klein. The first lines of the program calculate λ, μ, and Z in terms of the given a, b, and c, and the second part is just the above ansatz simplified by the use of $f = 1$. The behavior of `RootsPrincipalQuin`-`tic` is similar to the behavior of (N)`Roots`. We allow an optional third argument determining the precision used in the calculations and give the result in the form of a logical `Or`. (It is always wise to make the behavior of user-implemented functions similar to the behavior of built-in functions, although we do not implement an independent `NRootsPrincipalQuintic`).

```
In[66]:= RootsPrincipalQuintic[quintic_Equal, var_Symbol, prec_:Automatic] :=
         Module[{a, b, c, n, λ, μ, mmZ, Z, H, T, u, v, sol, w},
            {a, b, c} = {1/5, 1/5, 1}Function[diff, Coefficient[diff, var, #]& /@
                               {2, 1, 0}][Subtract @@ quintic];
            (* the precision to be used *)
            n = If[prec === Infinity, Identity,
                   If[prec === Automatic, N, N[#, prec]&]];
            ε = Exp[I 2 Pi/5] // n;
            If[a =!= 0, (* the above results *)
            λ = (-#2/#1/2 + Sqrt[#2^2/#1^2/4 - #3/#1])&[
                  (a^4 + a b c - b^3), -(11a^3 b - a c^2 + 2b^2c),
                  (64 a^2 b^2 - 27a^3c - b c^2)] // n;
            mmZ = (λ b + c)/a;
            V = (λ^2 - 3mmZ)^3/(λ c - mmZ b);
            Z = 1/(1728 - V);
            μ = (V a - 8 λ^3 - 72 λ mmZ)/(λ^2 + mmZ),
            (* a == 0 as a limiting case *)
            λ = -c/b;
            w = Sqrt[(256 b^5 + c^4)/b^6];
            Z = 1/(1728 + (5c^2 - 3b^3 w)^3/(4b^5(c^2 + b^3 w))) // n;
            μ = (8c(-7c^2 + 9b^3 w))/(b c^2 + b^4 w) // n];
            (* the main part, solving the degree 60 equation
               as the ratio of two hypergeometrics *)
            {u, v} = SolutionIcosahedralEquation[#, Z]& /@ {1, 2};
            (* using the above formula for the five roots *)
            H = H[u, v]; T = T[u, v];
            sol = Table[λ W[k, u, v]/H + μ t[k, u, v] W[k, u, v]/T/H,
                      {k, 0, 4}] // n;  Clear[ε];
            (Or @@ (var == #& /@ Sort[sol])) /. Complex[r_, _?(# == 0.&)] :> r] /;
            (* test if quintic is a quintic *)
               With[{diff = Subtract @@ quintic},
```

```
            PolynomialQ[diff] && Length[CoefficientList[diff, var]] === 6 &&
            Take[CoefficientList[diff, var], {4, 5}] === {0, 0}]
```

Let us test the routine numerically with some examples.

```
In[67]:= NRoots[z^5 + 4 z^2 + 5 z + 1 == 0, z]
```

Out[67]= z == -0.997672 - 0.449727 i || z == -0.997672 + 0.449727 i ||
        z == -0.249677 || z == 1.12251 - 1.44371 i || z == 1.12251 + 1.44371 i

```
In[68]:= RootsPrincipalQuintic[z^5 + 4 z^2 + 5 z + 1 == 0, z]
```

Out[68]= z == -0.997672 - 0.449727 i || z == -0.997672 + 0.449727 i ||
        z == -0.249677 || z == 1.12251 - 1.44371 i || z == 1.12251 + 1.44371 i

Of course, the formula also works for complex coefficients and for more digits in a reasonable time.

```
In[69]:= NRoots[z^5 + 5 I z^2 + (4 - 7I) z + 4I == 0, z, 40] // Timing
```

Out[69]= {0.01 Second, z == -1.848967262991700045961082641841224843445-
        1.008274474039560948344553626584831563494i ||
        z == -0.598462623824917714703668533942208687486+
        1.720369084039852876855390890344294520648i ||
        z == 0.341997156002664477301819553253047826147-
        0.377195370621742553366617966255745590934i ||
        z == 0.876059709267730703917539738331730404177+
        0.832049216864767867443820064534991189300i ||
        z == 1.229373021546222579445391884198655300607-
        1.166948456243317242588039362038708555521i}

```
In[70]:= RootsPrincipalQuintic[z^5 + 5 I z^2 + (4 - 7I) z + 4I == 0, z, 40] //
                                                             Timing
```

Out[70]= {0.02 Second, z == -1.848967262991700045961082641841225-
        1.008274474039560948344553626584832i ||
        z == -0.598462623824917714703668533942209+
        1.720369084039852876855390890344295i || z ==
        0.341997156002664477301819553253048- 0.377195370621742553366617966255746i ||
        z == 0.876059709267730703917539738331730+
        0.832049216864767867443820064534991i || z ==
        1.229373021546222579445391884198655- 1.166948456243317242588039362038709i}

The argument *prec* of RootsPrincipalQuintic determines the precision to be used in the calculation, so the result is of lower precision.

```
In[71]:= Precision[Last[%]]
```

Out[71]= 33.2783

And because the case of a zero coefficient of z^2 is treated separately, we also get the following solutions right.

```
In[72]:= NRoots[z^5 + 5 z + 1 == 0, z, 20]
```

Out[72]= z == -1.0044974557968355185 - 1.0609465064060406436i ||
        z == -1.0044974557968355185 + 1.0609465064060406436i ||
        z == -0.1999361021712199956 ||
        z == 1.1044655068824455162 - 1.05982966915252011667i ||
        z == 1.1044655068824455162 + 1.05982966915252011667i

```
In[73]:= RootsPrincipalQuintic[z^5 + 5 z + 1 == 0, z, 20]
```

Out[73]= z == -1.004497455796836 + 1.060946506406041 i ||
    z == -1.004497455796836 - 1.060946506406041 i ||
    z == -0.199936102171220 || z == 1.104465506882446 - 1.059829669152520 i ||
    z == 1.104465506882446 + 1.059829669152520 i

We could also have a look at nonnumerical solutions, but because such solutions are very long, we do not do this here.

In[74]:= **RootsPrincipalQuintic[z^5 + 2 z^2 + (1 - I) z + 2I == 0,**
                                     z, Infinity] // ByteCount

Out[74]= 1920680

Now, we have to build the solution of a general quintic of the form $x^5 + A x^4 + B x^3 + C x^2 + D x + E = 0$. First, we apply a Tschirnhaus transformation to reduce the quintic to a principal one; then we solve this principal one with `RootsPrincipalQuintic`; and after that, we inverse the Tschirnhaus transformation to go back to z_i from the x_i. Because these multiple substitutions result in big symbolic solutions, we do not carry this out explicitly.

For the solution of more general polynomials in hypergeometric functions, see [1198].

Exercises

1.¹¹ Asymptotic Series, Carlitz Expansion, Contour Lines of Gamma Function, Bessel Zeros, Asymptotic Expansion of Gamma Function Ratio, Integrals of Function Compositions, $W_{1+i}(1+i)$

a) Special functions are often developed in asymptotic series. Taylor series are convergent, but asymptotic series are not. Investigate the convergence behavior of the following asymptotic representation of the Bessel function of order 0:

$$J_0(z) \sim \frac{1}{\pi\sqrt{2\pi z}} \left(\sum_{k=0}^{\infty} e^{-i(z-\pi/4)} \frac{\Gamma(k+\frac{1}{2})^2}{(-2iz)^k \, k!} + \sum_{k=0}^{\infty} e^{i(z-\pi/4)} \frac{\Gamma(k+\frac{1}{2})^2}{(2iz)^k \, k!} \right)$$

and the Airy function Ai(z)

$$\text{Ai}(z) \sim \frac{1}{2\sqrt{\pi}} \frac{1}{z^{1/4}} \exp\left(-\frac{2}{3} z^{2/3}\right) \sum_{k=0}^{\infty} (-1)^k \frac{\Gamma(3k+\frac{1}{2})}{54^k \, \Gamma(k) \, \Gamma(k+1)} \left(\frac{2}{3} z^{\frac{3}{2}}\right)^{-k} \qquad |\arg(z)| < \pi$$

for some specific values of z in the complex z-plane. To calculate numerical values of asymptotic series, one best (not taking into account hyperrefinements) sums only the terms as long as they are decreasing [197].

b) The asymptotic expansion of the Airy function Ai(z) at $z = \infty$ can be written in the form [417]

$$\text{Ai}(z) \sim \frac{1}{2\sqrt{\pi}} \frac{1}{z^{1/4}} \exp\left(-\frac{2}{3} z^{2/3}\right) \left(1 + \sum_{k=0}^{\infty} \frac{2}{3} \frac{a_{k+1}}{k+1} z^{3/2} \right).$$

Here the $a_k(\zeta)$ are defined through the recursion

$$a_1(\zeta) = -\frac{5}{72} \frac{1}{\zeta}$$

$$a_{k+1}(\zeta) = \left(\frac{k-1}{k}\right) a_k(\zeta) + \frac{1}{2k} \frac{\partial a_k(\zeta)}{\partial \zeta} + \frac{5}{72k} \int_{-\infty}^{\zeta} \frac{a_{k+1}(z)}{z^2} \, dz.$$

Calculate the 8 leading terms (as a Laurent series in $2/3 \, z^{3/2}$) of the first 1000 $a_k(\zeta)$. How well do these terms approximate the exact values?

c) Verify by a series expansion in z the following relation [904] to some low order in z:

$$\left(\frac{az}{2}\right)^\mu \left(\frac{bz}{2}\right)^\nu = \sum_{k=0}^{\infty} (\sigma_k - \sigma_{k-2}) J_{\mu+k}(az) J_{\nu+k}(bz)$$

$$\sigma_{-1} = \sigma_{-2} = 0$$

$$\sigma_k = \frac{\Gamma(\nu+k+1)\,\Gamma(\mu+1)}{k!} \left(\frac{a}{b}\right)^k {}_2F_1\left(-k, \mu+1; -\nu-k; \frac{b^2}{a^2}\right).$$

d) Make a picture of the lines where $\text{Re}(\Gamma(z)) = 0$ and $\text{Im}(\Gamma(z)) = 0$ in the complex z-plane.

e) Calculate the first few (in magnitude) real zeros ν_n of $J_\nu(2)$. What is remarkable here?

f) Calculate the first four coefficients $c_i(\alpha, \beta)$ of the following series expansion:

$$\frac{\Gamma(z + \alpha)}{\Gamma(z + \beta)} \sim z^{\alpha - \beta} \sum_{i=0}^{\infty} \frac{c_i(\alpha, \beta)}{z^i}, \text{ as } z \to \infty.$$

g) Find the real x^* such that the thousands partial sum $f_{1000}(x)$ of $\sin(x)$, where $f_n(x) = \sum_{i=0}^{n} (-1)^i x^{2i+1}/(2i+1)!$ deviates from $\sin(x)$ by 10^{-1000}.

h) At which dimension d are the volume $V = \pi^{d/2}/\Gamma(d/2 + 1)$ and the surface area $A = d\,\pi^{d/2}/\Gamma(d/2 + 1)$ of a d-dimensional unit sphere maximal [41], [1242], [34]?

The probability density $p_d(\rho)$ for the Euclidean distance ρ of two points chosen at random in a d-dimensional sphere of radius 1 is [1243]

$$p_d(\rho) = \frac{2d\,\rho^{d-1}}{B(\frac{d}{2} + \frac{1}{2}, \frac{1}{2})} \left({}_2F_1\!\left(\frac{1}{2}, \frac{1}{2} - \frac{d}{2}; \frac{3}{2}; 1\right) - \frac{1}{2}\,\rho\,{}_2F_1\!\left(\frac{1}{2}, \frac{1}{2} - \frac{d}{2}; \frac{3}{2}; \frac{\rho^2}{4}\right) \right)$$

What is the limit $\lim_{d\to\infty} p_d(\rho)$? What is the average distance between two points in the limit $d \to \infty$?

i) A radial symmetric potential function $V_d(|\mathbf{x} - \mathbf{y}|)$ in d dimensions can be decomposed à la Fefferman–de la Llave through [565]

$$V_d(|\mathbf{x} - \mathbf{y}|) = \int_0^{\infty} \int_{\mathbb{R}^d} \chi_{r/2}(\mathbf{z} - \mathbf{y})\, g_d(r)\, \chi_{r/2}(\mathbf{x} - \mathbf{z})\, d\mathbf{z}\, dr.$$

Here $\chi_r(\mathbf{x}) = \theta(|\mathbf{x}| - r)$ is the characteristic function of a ball around $\mathbf{x}$ of radius r. The function $g(r)$ can be expressed through

$$g_d(r) = \frac{2(-1)^{d+1}}{\Gamma(\frac{d-1}{2})\,(\pi r^2)^{\frac{d-1}{2}}} \int_r^{\infty} V^{(d+1)}(\rho)\,\rho\,(\rho^2 - r^2)^{\frac{d-3}{2}}\, d\rho$$

for smooth $V(r)$ and the convolution integral over the two balls through

$$\int_{\mathbb{R}^d} \chi_r(\mathbf{z} - \mathbf{y})\, \chi_r(\mathbf{x} - \mathbf{z})\, d\mathbf{z} = \left(\frac{\pi}{4}\right)^{\frac{d-1}{2}} \frac{\theta(2r - |\mathbf{x} - \mathbf{y}|)}{\Gamma(\frac{d+1}{2})} \int_{|\mathbf{x}|}^{2r} (4r^2 - \xi^2)^{\frac{d-1}{2}}\, d\xi.$$

Calculate $g_d(r)$ for some potentials $V_d(|\mathbf{x} - \mathbf{y}|)$ and derive integration-free formulas for $g_d(r)$ for $d = 3, 5, 7$ for potentials that vanish sufficiently fast at infinity.

j) Calculate the value of the Bessel functions $J_{10^{100}}(10^{100})$ to 1000 digits. $J_\nu(z)$ for $z \approx \nu$ and $\nu \to \infty$ can be efficiently computed by Meissel's formula [1304]:

$$J_\nu(z) \sim \frac{1}{3\pi} \sum_{m=0}^{\infty} \left(\frac{z}{6}\right)^{-\frac{m+1}{3}} B_m(z-\nu) \sin\left(\frac{m+1}{3}\pi\right) \Gamma\left(\frac{m+1}{3}\right)$$

$$B_m(\zeta) = 6^{-\frac{m+1}{3}} b_m(m)$$

$$b_n(m) = [w^n]\left\{ e^{\zeta w}\left(\frac{\sinh(w) - w}{w^3}\right)^{-(m+1)/3}\right\}.$$

Here, $[z^n]\{f(z)\}$ denotes the n-th coefficient in the Taylor expansion of $f(z)$ around $z = 0$.

k) The first zero $z_0^{(\nu)}$ of the Bessel function $J_\nu(z)$ can be bounded by [555], [680], [657]

$$\frac{1}{\left(\sigma_n^{(\nu)}\right)^n} < \left(z_0^{(\nu)}\right)^2 < \frac{\sigma_n^{(\nu)}}{\sigma_{n+1}^{(\nu)}}.$$

Here, $\sigma_n^{(\nu)}$ is the Rayleigh sum $\sigma_n^{(\nu)} = \sum_{k=1}^{\infty} \left(z_k^{(\nu)}\right)^{-2n}$. The Rayleigh sum $\sigma_n^{(\nu)}$ obeys the following recurrence relation:

$$\sigma_n^{(\nu)} = \frac{1}{\nu + n} \sum_{k=1}^{n-1} \sigma_k^{(\nu)} \sigma_{n-k}^{(\nu)}$$

$$\sigma_1^{(\nu)} = \frac{1}{4(\nu+1)}.$$

Use these formulas to calculate $z_0^{(1)}$ to 100 digits. What is the maximal n needed? How fast can one calculate $z_0^{(1)}$ to 100 digits using Rayleigh sums?

l) For all positive integers n, the number x defined implicitly by $K(1-x)/K(x) = \sqrt{n}$ is algebraic [41], [183], [1370]. Use numerical techniques to find an exact algebraic x for $n = 10$.

m) Try to calculate all 13824 indefinite integrals of the functions $f_1(f_2(f_3(x)))$, where f_i is a trigonometric or hyperbolic or inverse trigonometric or inverse hyperbolic function. How many of the doable triple integrals will contain special functions? Which special functions appear and how often?

n) The `ProductLog` function $W_k(z)$ has the following integral representation ($z \notin (-1/e, 0)$) [1146], [1147], [664]:

$$W_k(z) = 1 + (\log(z) + 2\pi i k - 1) \exp\left(\frac{i}{2\pi} \int_0^\infty \log\left(\frac{(2k-1)\pi i + t - \log(t) + \log(z)}{(2k+1)\pi i + t - \log(t) + \log(z)}\right) \frac{1}{t+1} dt\right).$$

Use this formula to calculate $W_{1+i}(1 + i)$ to 20 digits.

o) The Gumbel probability distribution $p(x)$ has the form ($a > 0$) [553], [284], [946]:

$$p(x) = c\left(\exp(b(x-\xi) - e^{b(x-\xi)})\right)^a.$$

Find the values of the parameters $b(a)$, $c(a)$, and $\xi(a)$ such that $p(x)$ is normalized to 1, has a mean of 0, and a second moment of 1.

p) Consider the command `Binomial[x, y]` as a function of two complex variables x, and y. What are the "correct" values of `Binomial[negativeInteger, anotherNegativeInteger]`?

q) Find a closed form description of the curves that separate the white and the black areas in the following graphic.

```
ContourPlot[Im[(x + I y)^(x + I y)], {x, -10, 10}, {y, -10, 10},
            PlotPoints -> 250, Contours -> {0}];
```

r) Let g_n be the geometric mean of all irreducible fractions from the unit interval with maximal denominator n [888]:

$$g_n = \sqrt[n]{\frac{1}{n} \times \frac{2}{n} \times \cdots \times \frac{n}{n}}.$$

Calculate the limit of g_n as n tends to infinity and the first correction term.

Calculate the following infinite product (known as the Wallis product [1022]) and the first correction term.

$$\frac{2}{1} \frac{2}{3} \frac{4}{3} \frac{4}{5} \frac{6}{5} \frac{6}{7} \frac{8}{7} \frac{8}{9} \frac{10}{9} \frac{10}{11} \cdots.$$

Calculate the following limit and the first correction term for large n:

$$\lim_{n \to \infty} \frac{2^{4n+2}}{(n+1)^3} \frac{(n+1)!^4}{(2n+1)!^2}.$$

Calculate the following limit and the first correction term for large n: $\lim_{n \to \infty} \sum_{k=1}^{n^2} n/(k^2 + n^2)$ [527].

s) Consider the following generalization $\Gamma_k(z)$ of the classical Gamma function $\Gamma(z)$: $\Gamma_k(z) = k^{z/k-1} \Gamma(z/k)$ [392]. Derive polynomial partial differential equations that are fulfilled by $\Gamma_k(z)$.

t) The generalized Bell numbers $B_k(n)$ are defined by [64], [323]

$$\underbrace{\exp(\exp(\cdots(\exp(z))))}_{k \text{ exp's}} = \sum_{k=0}^{\infty} \frac{B_k(n)}{n!} z^k.$$

Show by explicit calculation for $1 \le m, n \le 6$ that the following identity holds:

$$B_m(n+1) = \sum_{k_1, k_2, \ldots, k_m = 0}^{\infty} \delta_{\sum_{j=1}^{m} k_j, n} \frac{n!}{\prod_{j=1}^{m} k_j!} \prod_{j=1}^{m} B_j(k_j).$$

u) Predict if $1 - \texttt{Erfc[66.66]}$ will be a machine number.

v) For factorially divergent sums, Borel summation means expressing the factorial function through its integral representation and then exchanging summation and integration:

$$\sum_{k=1}^{\infty} f(k)(a+bk)! = \sum_{k=1}^{\infty} f(k) \int_0^{\infty} e^{-t} t^{a+bk} dt \underset{\mathcal{B}}{=} \int_0^{\infty} e^{-t} \left(\sum_{k=1}^{\infty} f(k) t^{a+bk} \right) dt$$

Sums of the form $s_j = \sum_{k=1}^{\infty} (-3)^k k^{-j} (k - 1/2)!$, $k = 0, 1, \ldots$ occur in the perturbation expansion of the quartic anharmonic oscillator [713], [528]. Calculate the first few of these sums. The summands of these sums decrease with increasing j, do the sums decrease too? Conjecture a closed form for $\lim_{j \to \infty} s_j$.

w) Calculate the normalized ground-state $\psi_N(z)$ of the Hamiltonian [952], [953], [954]

$$\hat{H} = -\frac{\partial^2}{\partial z^2} + V(z)$$

$$V(z) = \frac{\frac{19z^2}{4} - \left(\sqrt{5} - \frac{1}{2}\right)}{(z^2 + 1)^2}.$$

Calculate z^*, such that $V(z^*) = \psi_N(z^*)$.

x) Consider the two integrals $\int_{-X}^{X} H_1(x) H_{3/2}(x) \exp(-x^2/2)\, dx$ and $\int_0^X L_1(x) L_{3/2}(x)\, e^{-x}\, dx$ involving Laguerre functions $L_\nu(x)$ and Hermite functions $H_\nu(x)$. Find the leading terms of these integrals for large real X. Laguerre and Hermite functions are orthogonal for nonnegative integer indices. Does orthogonality still hold for these fractional indices?

y) Evaluate the following integral [1303], [512], [907]:

$$\frac{1}{\pi^3} \int_0^\pi \int_0^\pi \int_0^\pi \frac{1}{1 - \cos(x)\cos(y)\cos(z)}\, dx\, dy\, dz.$$

z) The function $L(z) = \mathrm{Li}_2(z) + 1/2\ln(z)\ln(1-z)$ fulfills the two identities $L(z) + L(1-z) = \pi^2/6$ and $L(z) = L(z/(z-1)) + L(z^2)/2$. In addition, for special arguments identities like [810]

$$6L\left(\frac{1}{3}\right) - L\left(\frac{1}{9}\right) = \frac{\pi^2}{3}$$

$$2L\left(2\cos\left(\frac{3}{7}\pi\right)\right) + L\left(\left(2\cos\left(\frac{3}{7}\pi\right)\right)^2\right) = \frac{4}{21}\pi^2$$

$$3L\left(\frac{1}{2\cos(\frac{\pi}{9})}\right) + 3L\left(\left(\frac{1}{2\cos(\frac{\pi}{9})}\right)^2\right) - L\left(\left(\frac{1}{2\cos(\frac{\pi}{9})}\right)^3\right) = \frac{7}{18}\pi^2$$

hold.

Write a program that searches for (and finds) such identities.

2.¹² Elliptic Integrals

The incomplete elliptic integral of the third kind is defined by the following integral representation:

$$\Pi(n; \phi \mid m) = \int_0^\phi \frac{1}{(1 - n\sin(\varphi))\sqrt{1 - m\sin(\varphi)}}\, d\varphi.$$

a) Derive an inhomogeneous, linear, third-order differential equation for $\frac{\partial\Pi(n;\phi\mid m)}{\partial m}$ [242]. (The coefficients will be polynomials in n and m.)

b) Derive an inhomogeneous, linear, third-order differential equation for $\frac{\partial\Pi(n;\phi\mid m)}{\partial n}$. (The coefficients will again be polynomials in n and m.)

c) Starting from the integral representation, derive a nonlinear, third-order differential equation for $\frac{\partial\Pi(n;\phi\mid m)}{\partial\phi}$.

d) Calculate a "nice" result for the following integral:

$$\int_{(3-2\sqrt{3})/12}^{1/12} \sqrt{\frac{3-12\,x}{(12\,x-1)\,(48\,x^2-24\,x-1)}}\ dx.$$

e) Show that in the addition theorem for elliptic integrals of the first kind

$$\int_0^x \frac{1}{\sqrt{(1-\tau^2)\,(1-k\,\tau^2)}}\ d\tau + \int_0^y \frac{1}{\sqrt{(1-\tau^2)\,(1-k\,\tau^2)}}\ d\tau = \int_0^{z(x,y)} \frac{1}{\sqrt{(1-\tau^2)\,(1-k\,\tau^2)}}\ d\tau$$

the z on the right-hand side as a function of x and y fulfills [96]

$$(k\,y^2\,z^2-1)^2\,x^4 - 2\,(((k\,(y^2+z^2-2)-2)\,z^2+1)\,y^2+z^2)\,x^2 + (y^2-z^2)^2 = 0.$$

f) Determine the magnetic field of a circular current (also beyond the axis), and examine it graphically. Then, investigate the magnetic field of a Helmholtz coil [247] in the neighborhood of a symmetry point. In which direction is the field more inhomogeneous: in the radial or in the perpendicular direction?

3.^{12} Weierstrass Function

a) Visualize the following two Weierstrass functions:

$$\wp\!\left(z;\ \frac{8}{3},\ \frac{8}{3}\right) \quad \text{and} \quad \wp'\!\left(z;\ \frac{8}{3},\ \frac{8}{3}\right)$$

in the complex z-plane. What is an appropriate z-domain? In *Mathematica*, the Weierstrass $\wp(z; g_2, g_3)$ is `WeierstrassP[z, {g2, g3}]` and its derivative $\wp'(z; g_2, g_3)$ (with respect to z) is `Weierstrass` `PPrime[z, {g2, g3}]`.

b) Make a picture of the function $\wp'(z; g_2, g_3)$ over the Riemann z-sphere. Use appropriate values for g_2 and g_3.

c) The function $\wp(z; g_2, g_3)$ has the following series expansion around $z = 0$:

$$\wp(z; g_2, g_3) = \frac{1}{z^2} + \frac{g_2}{20}\,z^2 + \frac{g_3}{28}\,z^4 + c_6\,z^6 + c_8\,z^8 + \cdots.$$

By using the differential equation of the Weierstrass function:

$$\wp'^2(z; g_2, g_3) = 4\,\wp^3(z; g_2, g_3) + g_2\,\wp'^2(z; g_2, g_3) + g_3$$

(where the derivative is with respect to z), find the coefficients c_6 to c_{20}.

d) The Weierstrass function $\wp(u; g_2, g_3)$ fulfills the differential equation (differentiation with respect to u):

$$\wp'(u; g_2, g_3)^2 = 4\,\wp(u; g_2, g_3)^3 - g_2\,\wp(u; g_2, g_3) - g_3.$$

In [3], formula 18.4.1, the following addition theorem is given:

$$\wp(u+v; g_2, g_3) = \frac{1}{4}\left(\frac{\wp'(u; g_2, g_3) - \wp'(v; g_2, g_3)}{\wp(u; g_2, g_3) - \wp(v; g_2, g_3)}\right)^2 - \wp(u; g_2, g_3) - \wp(v; g_2, g_3).$$

Derive a polynomial form of the addition theorem $p(\wp(u+v; g_2, g_3),\ \wp(u; g_2, g_3),\ \wp(v; g_2, g_3))$ (this means $p(x, y, z)$ being a polynomial in x, y, and z, symmetric in y and z) by eliminating the derivative terms. Derive the corresponding double argument formula that expresses $\wp(2\,v; g_2, g_3)$ polynomially in $\wp(v; g_2, g_3)$.

e) In [3] formula 18.4.2, the following addition theorem for the derivative of the Weierstrass function $\wp'(u; g_2, g_3)$ is given:

$$\wp'(u + v; g_2, g_3) = -(\wp(u + v; g_2, g_3)\,(\wp'(u; g_2, g_3) - \wp'(v; g_2, g_3)) +$$
$$\wp(u; g_2, g_3)\,\wp'(v; g_2, g_3) - \wp'(u; g_2, g_3)\,\wp(v; g_2, g_3))/(\wp(u; g_2, g_3) - \wp(v; g_2, g_3)).$$

Derive a polynomial form of the addition theorem $q(\wp'(u + v; g_2, g_3), \wp'(u; g_2, g_3), \wp'(v; g_2, g_3))$ (this means $q(x, y, z)$ is a polynomial in x, y, and z, symmetric in y and z) by eliminating the nondifferentiated terms. Derive the corresponding double argument formula that expresses $\wp'(2\,v; g_2, g_3)$ polynomially in $\wp'(v; g_2, g_3)$.

f) The general solution of the functional equation (Sutherland–Calogero model [249], [1202], [248], [204], [1262], [759], [241], [206])

$$\varphi(x)\,\varphi(y) + \varphi(x)\,\varphi(z) + \varphi(y)\,\varphi(z) = f(x) + f(y) + f(z)$$

for $z = x + y$ is given by

$$\varphi(x) = \alpha\,\zeta(x; g_2, g_3) + \beta\,x$$
$$f(x) = -\frac{1}{2}\left(\alpha^2\,\zeta(x; g_2, g_3)^2 + \alpha^2\,\frac{\partial\zeta(x; g_2, g_3)}{\partial x} + 2\,\beta\,x\,\zeta(x; g_2, g_3)\,\alpha + \beta^2\,x^2\right)$$

$$\varphi(x)\,\varphi(y) + \varphi(x)\,\varphi(z) + \varphi(y)\,\varphi(z) = f(x) + f(y) + f(z).$$

Here, $\zeta(x; g_2, g_3)$ is the Weierstrass Zeta function (in *Mathematica*, `WeierstrassZeta[x, {g_2, g_3}]`):

$$\zeta(x; g_2, g_3) = \frac{1}{x} + \int_0^x\left(\frac{1}{x^2} - \wp(x; g_2, g_3)\right)dx \quad \text{or} \quad \frac{\partial\zeta(x; g_2, g_3)}{\partial x} = -\wp(x; g_2, g_3).$$

Use the above addition formula for $\wp(u + v; g_2, g_3)$ to show that $\varphi(x)$ and $f(x)$ fulfill the above functional equation.

g) Use the above addition formula for $\wp(u + v; g_2, g_3)$ to show that the following identity holds for $z = x + y$ [395]:

$$\zeta(x; g_2, g_3) + \zeta(y; g_2, g_3) + \zeta(z; g_2, g_3) = \sqrt{\wp(x; g_2, g_3) + \wp(y; g_2, g_3) + \wp(z; g_2, g_3)}\,.$$

h) The n-argument multiplication formula for Weierstrass function can be expressed in the following form ($n \in \mathbb{N}$) [476], [1001], [477]:

$$\wp(n\,z; g_2, g_3) = \wp(z; g_2, g_3) - \frac{\psi_{n-1}\,\psi_{n+1}}{\psi_n^2}$$

$$\psi_1 = 1$$
$$\psi_2 = -\wp'(z; g_2, g_3)$$
$$\psi_3 = 3\,\wp(z; g_2, g_3)^4 - \frac{3}{2}\,g_2\,\wp(z; g_2, g_3)^2 - 3\,g_3\,\wp(z; g_2, g_3) - \frac{g_2^2}{16}$$
$$\psi_4 = \wp'(z; g_2, g_3)\left(-2\,\wp(z; g_2, g_3)^6 + \frac{5\,g_2}{2}\,\wp(z; g_2, g_3)^4 + 10\,g_3\,\wp(z; g_2, g_3)^3 + \right.$$
$$\left.\frac{5\,g_2^2}{8}\,\wp(z; g_2, g_3)^2 + \frac{g_2\,g_3}{2}\,\wp(z; g_2, g_3) - \frac{g_2^3}{32} + g_3^2\right)$$

$$\psi_n = \begin{cases} -\psi_{\frac{n}{2}}\left(\psi_{\frac{n}{2}+2}\,\psi_{\frac{n}{2}-1}^2 - \psi_{\frac{n}{2}-2}\,\psi_{\frac{n}{2}+1}^2\right)\Big/\wp'(z;\,g_2,\,g_3) & \text{if } n \text{ is even} \\ \psi_{\frac{n-1}{2}+2}\,\psi_{\frac{n-1}{2}}^3 - \psi_{\frac{n-1}{2}-1}\,\psi_{\frac{n-1}{2}}^3 & \text{if } n \text{ is odd.} \end{cases}$$

Use this formula to derive $\wp(5\,v;\,g_2,\,g_3) = R(\wp(v;\,g_2,\,g_3))$, where R is a rational function.

i) Show that the function $\psi_\lambda(x)$ [729], [1087], [727], [210], [22], [506], [977]

$$\psi_\lambda(x) = \frac{\sigma(\lambda - x;\,g_2,\,g_3)}{\sigma(x;\,g_2,\,g_3)\,\sigma(\lambda;\,g_2,\,g_3)}\, e^{x\,\zeta(\lambda;g_2,g_3)}$$

is a solution of the Lamé equation [1297], [1197]:

$$-\psi_\lambda''(x) + 2\,\wp(x;\,g_2,\,g_3)\,\psi_\lambda(x) = -\wp(\lambda;\,g_2,\,g_3)\,\psi_\lambda(x).$$

Here $\sigma(x;\,g_2,\,g_3)$ is the Weierstrass sigma function (in *Mathematica* WeierstrassSigma[z, {g_2, g_3}]):

$$\sigma(x;\,g_2,\,g_3) = \exp\left(\int_0^x \left(\zeta(x;\,g_2,\,g_3) - \frac{1}{x}\right) dx\right) \quad \text{or} \quad \frac{\partial\sigma(x;\,g_2,\,g_3)}{\partial x} = \sigma(x;\,g_2,\,g_3)\,\zeta(x;\,g_2,\,g_3).$$

j) Visualize the Riemann surface of the inverse Weierstrass function $\wp^{(-1)}(z;\,1+i,\,1-2\,i)$.

k) The Weierstrass sigma function $\sigma(z;\,g_2,\,g_3)$ fulfills a partial differential equation of the form

$$\frac{\partial^2\,\sigma(z;\,g_2,\,g_3)}{\partial z^2} =$$

$$G_2(z,\,g_2,\,g_3)\,\frac{\partial\sigma(z;\,g_2,\,g_3)}{\partial g_2} + G_3(z,\,g_2,\,g_3)\,\frac{\partial\sigma(z;\,g_2,\,g_3)}{\partial g_3} + G_C(z,\,g_2,\,g_3)\,\sigma(z;\,g_2,\,g_3).$$

The functions $G_2(z,\,g_2,\,g_3)$, $G_3(z,\,g_2,\,g_3)$, and $G_C(z,\,g_2,\,g_3)$ are total degree two polynomials of g_2, g_3, and z. Use the series representation:

$$\sigma(z;\,g_2,\,g_3) = \sum_{m=0}^{\infty}\sum_{n=0}^{\infty} a_{m,n}\,\frac{\left(\frac{g_2}{2}\right)^m (2\,g_3)^n}{(4\,m + 6\,n + 1)!}\,z^{4m+6n+1}$$

$$a_{0,0} = 1$$

$$a_{m,n} = \frac{16}{3}\,(n + 1)\,a_{m-2,n+1} - \frac{1}{3}\,(2\,m + 3\,n - 1)\,(4\,m + 6\,n - 1)\,a_{m-1,n} + 3\,(m + 1)\,a_{m+1,n-1} \;\text{ if } n,\, m \geq 0$$

$$a_{m,n} = 0 \;\text{ else}$$

to find the polynomials G_2, G_3, and G_C.

l) The function

$$\Sigma(\tau) = \sigma\!\left(\frac{2}{3}\,\tau;\, g_2(\omega_1(1,\,\tau),\, \omega_2(1,\,\tau)),\, g_3(\omega_1(1,\,\tau),\, \omega_2(1,\,\tau))\right)^5 \Big/$$

$$\left(\sigma\!\left(\frac{1}{3}\,\tau;\, g_2(\omega_1(1,\,\tau),\, \omega_2(1,\,\tau)),\, g_3(\omega_1(1,\,\tau),\, \omega_2(1,\,\tau))\right)^4\right.$$

$$\left.\sigma\!\left(\frac{4}{3}\,\tau;\, g_2(\omega_1(1,\,\tau),\, \omega_2(1,\,\tau)),\, g_3(\omega_1(1,\,\tau),\, \omega_2(1,\,\tau))\right)\right)$$

takes on algebraic values for certain $\tau = i\sqrt{n}$, $n \in \mathbb{N}$ [671]. Conjecture at least 10 such values for τ and express the corresponding values $\Sigma(\tau)$ in radicals. ($\{\omega_1(g_2,\, g_3),\, \omega_2(g_2,\, g_3)\}$ are the half periods corresponding to the invariants $\{g_2,\, g_3\}$. (In *Mathematica*, the half periods can be calculated from the invariants by the function `WeierstrassHalfPeriods[{g2,g3}]`.)

m) The equations of motions for a 2D periodic set of n point vertices $z_\alpha(t)$ of strength Γ_α ($\alpha = 1, \ldots, n$) are [199], [1195], [983], [1196]

$$\frac{dz_\alpha(t)}{dt} = \frac{i}{2\pi} \sum_{\substack{\beta=1 \\ \alpha \neq \beta}}^{n} \Gamma_\beta\, \overline{\phi(z_\alpha(t) - z_\beta(t))}$$

$$\phi(z) = \zeta(z;\, g_2,\, g_3) + \left(\frac{\pi}{4\,|\mathrm{Im}(\omega_1\,\overline{\omega_2})|} - \frac{\zeta(\omega_1;\, g_2,\, g_3)}{\omega_1}\right) z - \frac{\pi}{4\,|\mathrm{Im}(\omega_1\,\overline{\omega_2})|}\,\bar{z}.$$

(For 1D periodic point vertices, see .[909])

The basic lattice vectors are $2\,\omega_1$ and $2\,\omega_2$, $\zeta(z;\, g_2,\, g_3)$ is the Weierstrass Zeta function and g_2, and g_3 are the invariants corresponding to the half periods ω_1 and ω_2. Calculate and visualize orbits of a few vortices.

4.^{L2} Jacobi's Elliptic Functions

a) The Jacobi elliptic functions $\mathrm{sn}(u\mid m)$, $\mathrm{sd}(u\mid m)$, $\mathrm{sc}(u\mid m)$, $\mathrm{nd}(u\mid m)$, $\mathrm{nc}(u\mid m)$, $\mathrm{dn}(u\mid m)$, $\mathrm{dc}(u\mid m)$, $\mathrm{cn}(u\mid m)$, $\mathrm{cd}(u\mid m)$, $\mathrm{cd}(u\mid m)$, $\mathrm{ds}(u\mid m)$, and $\mathrm{ns}(u\mid m)$ fulfill nonlinear ordinary differential equations of the following type (where pq in the following formulas is any of the just-mentioned 12 Jacobi functions):

$$\frac{\partial}{\partial z}\,\mathrm{pq}(u\mid m) = \pm\sqrt{a_{\mathrm{pq}}(m) + b_{\mathrm{pq}}(m)\,\mathrm{pq}^2(u\mid m) + c_{\mathrm{pq}}(m)\,\mathrm{pq}^4(u\mid m)}$$

$$\frac{\partial^2}{\partial z^2}\,\mathrm{pq}(u\mid m) = d_{\mathrm{pq}}(m)\,\mathrm{pq}(u\mid m) + e_{\mathrm{pq}}(m)\,\mathrm{pq}^3(u\mid m).$$

By using the built-in capabilities of *Mathematica* to give series approximations of the Jacobi elliptic function, determine the coefficients $a_{\mathrm{pq}}(m)$, $b_{\mathrm{pq}}(m)$, $c_{\mathrm{pq}}(m)$, $d_{\mathrm{pq}}(m)$, and $e_{\mathrm{pq}}(m)$ for all of the above elliptic Jacobi functions.

b) The 12 Jacobi functions $\mathrm{cd}(u\mid m)$, $\mathrm{cn}(u\mid m)$, $\mathrm{cs}(u\mid m)$, $\mathrm{dc}(u\mid m)$, $\mathrm{dn}(u\mid m)$, $\mathrm{ds}(u\mid m)$, $\mathrm{nc}(u\mid m)$, $\mathrm{nd}(u\mid m)$, $\mathrm{ns}(u\mid m)$, $\mathrm{sc}(u\mid m)$, $\mathrm{sd}(u\mid m)$, and $\mathrm{sn}(u\mid m)$ satisfy many identities similar to the ones known for trigonometric functions. The equivalent formulas to $\sin^2(x) + \cos^2(x) = 1$ are

$$\mathrm{cn}(u\mid m)^2 + \mathrm{sn}(u\mid m)^2 = 1,$$
$$\mathrm{dn}(u\mid m)^2 + m\,\mathrm{sn}(u\mid m)^2 = 1.$$

Derive all possible formulas for pairs of squares of Jacobi functions $pq(u \mid m)$ and $rs(u \mid m)$ of the form

$f(m)\,pq(u \mid m)^2 + g(m)\,rs(u \mid m)^2 + h(m)\,pq(u \mid m)^2\,rs(u \mid m)^2 + l(m) = 0$, where f, g, h, and l are constant or linear polynomials in m.

c) The Jacobi functions $pq(u \mid m)$ have polynomial addition formulas $p(pq(u + v \mid m), pq(u \mid m), pq(v \mid m), m) = 0$ where p is a polynomial.

The typically shown formulas are [3]:

$$sn(u + v \mid m) = \frac{cn(v \mid m)\,dn(v \mid m)\,sn(u \mid m) + cn(u \mid m)\,dn(u \mid m)\,sn(v \mid m)}{1 - m\,sn(u \mid m)^2\,sn(v \mid m)^2}$$

$$cn(u + v \mid m) = \frac{cn(u \mid m)\,cn(v \mid m) - sn(u \mid m)\,dn(u \mid m)\,sn(v \mid m)\,dn(v \mid m)}{1 - m\,sn(u)^2\,sn(v)^2}$$

$$dn(u + v) \mid m = \frac{dn(u \mid m)\,dn(v \mid m) - m\,sn(u \mid m)\,cn(u \mid m)\,sn(v \mid m)\,cn(v \mid m)}{1 - m\,sn(u \mid m)^2\,sn(v \mid m)^2}.$$

Using the defining formulas for Jacobi functions, the above square identities, and these three addition formulas, derive the addition formulas expressed only in $pq(u \mid m)$. This means as $p(pq(u + v \mid m), pq(u \mid m), pq(v \mid m), m) = 0$ for all twelve Jacobi functions.

d) Use the following implicit definition of the Jacobi function $sn(z \mid m)$ to derive the first five terms of the series expansion of $sn(z \mid m) = \sum_{n=0}^{\infty} c_n(z)\,m^n$ around $m = 0$ [944]:

$$z = \int_0^{sn(z|m)} \frac{1}{\sqrt{1 - t^2}\,\sqrt{1 - m\,t^2}}\,dt.$$

An alternative way to calculate this series expansion is given by the following set of formulas [1329]:

$$sn(z \mid m) = \sum_{n=0}^{\infty} \frac{m^n}{(2\,n)!}\,f_{2n}(z)$$

$$f_0(z) = \sin(z)$$

$$g_{2n}(z) = \sum_{k=0}^{n} \binom{2\,n}{2\,k} f_{2k}(z)\,f_{2n-2k}(z)$$

$$g''_{2n}(z) = -4\,g_{2n}(z) - 8\,n\,(2\,n - 1)\,g_{2n-2}(z) + 12\,(2\,n - 1)\,n \sum_{k=0}^{n-1} \binom{2\,n - 2}{2\,k} g_{2k}(z)\,g_{2n-2k-2}(z)$$

$$g_{2n}(0) = g'_{2n}(0) = 0$$
$$g_0(z) = \sin(z)^2.$$

Use these formulas to calculate the $f_2(z)$, $f_4(z)$, $f_6(z)$, and $f_8(z)$ and compare with the above result.

e) The Jacobi elliptic function sn obeys the following functional equation.

$$sn(z \mid m) = -\frac{2\,(m\,sn(\frac{z}{4} \mid m)^4 - 2\,sn(\frac{z}{4} \mid m)^2 + 1)\,(m\,(sn(\frac{z}{4} \mid m)^2 - 2)\,sn(\frac{z}{4} \mid m)^2 + 1)\,sn(\frac{z}{2} \mid m)}{(m\,sn(\frac{z}{4} \mid m)^4 - 1)^2\,(m\,sn(\frac{z}{2} \mid m)^4 - 1)}.$$

Use this equation to calculate $sn(1 + i \mid 1/3 + 2i)$ to 100 digits.

f) Polynomial (and rational) functions (with z-independent coefficients) of the Jacobi functions $cn(z \mid m)$, $sn(z \mid m)$, and $dn(z \mid m)$ are frequently solutions of a Sturm–Liouville-type eigenvalue equation of the form

$$-\Psi''(m \mid z) + V(m \mid \Psi(m \mid z)) \Psi(m \mid z) = \varepsilon(m) \Psi(m \mid z)$$

where the potential V is a function of z only through $\Psi(m \mid z)$. Find some $\Psi(m \mid z)$ where the corresponding $V(m \mid \Psi(m \mid z))$ is a rational function of $\Psi(m \mid z)$ and m with no explicit z-dependence.

g) Find a solution of the form

$$u(x, t) = \sum_{j=0}^{o} \alpha_j(k, c, m; \beta_1, \beta_2, \beta_3) \, sn(k\,(x - c\,t) \mid m)^j$$

of the nonlinear wave equation (Bussinesq equation) [800], [1295]

$$\frac{\partial^2 u(x, t)}{\partial t^2} = \beta_1 \frac{\partial^2 u(x, t)}{\partial x^2} + \beta_2 \frac{\partial^4 u(x, t)}{\partial x^4} + \beta_3 \frac{\partial^2 u(x, t)^2}{\partial x^2}.$$

Find some other nonlinear evolution equations of the form $\psi_t = p(\psi, \psi_x, \psi_{xx}, \psi_{xxx}, \dots)$ or $\psi_{tt} = q(\psi, \psi_x, \psi_{xx}, \psi_{xxx}, \dots)$ or $\psi_{ttt} = \dots$ where $p, q, \dots$ are polynomials and the equations have solutions that can be expressed as a sum of powers of $sn(k\,(x - c\,t) \mid m)$ [1073], [326], [1000], [1333], [87], [293], [449], [1334], [1345], [609], [1134], [1335], [1336], [798], [799], [1296], [789].

h) Derive a second-order differential equation of the function $w(z) = sn(\alpha \log(a + b\,z) \mid m)$; view a and b as integration constants [251].

i) Solve the equation for a pendulum $\varphi''(t) = -\sin(\varphi(t))$, $\varphi(0) = 0$, $0 \le \varphi_0 \le \pi$, $\varphi'(0) = 2\sin(\varphi_0/2)$ [46] numerically along straight rays in the complex t-plane (φ_0 is the maximal elongation). Explain the resulting "complex oscillations" using visualizations of the Riemann surface of the solution $\varphi(t) = 2\arcsin(\sin(\varphi_0/2)\,sn(t \mid \sin^2(\varphi_0/2)))$.

5.^{12} Rocket with Discrete Propulsion, Neat Product, Harmonic Oscillator Spectrum

a) The discrete rocket problem is as follows: Imagine a rocket that burns its fuel in discrete masses of size $(m_i - m_f)/n$ (where m_i is the initial mass, m_f the final mass, and n is the number of burns) rather than continuously. Give an analytic formula for the final velocity as a function of n. What happens in the limiting case $n \to \infty$?

b) Calculate a nice result for the following product [302]. Find some generalizations that can be calculated in closed form.

$$\prod_{k=1}^{\infty} \exp\!\left(-\frac{1}{k}\right)\!\left(1 + \frac{1}{k} + \frac{1}{k^2}\right).$$

c) Using the functions `DSolve` and `Series`, calculate the eigenvalues $\varepsilon_n = 2n + 1$, $n \in \mathbb{N}$ of the harmonic oscillator [1324], [1306], $-\psi_n''(z) + z^2\,\psi_n(z) = \varepsilon_n\,\psi_n(z)$, $z \in (-\infty, \infty)$.

6.^{12} Contour Integral, Asymptotics of Bessel Function, Isophotes, Circular Andreev Billiard

a) The Bessel function $J_\nu(z)$ has the following integral representation [1304]:

$$J_\nu(z) = \frac{\Gamma(\frac{1}{2} - \nu)\,(\frac{z}{2})^2}{2\pi i\,\sqrt{\pi}} \int_C e^{izt}\,(t^2 - 1)^{\nu - \frac{1}{2}}\,dt.$$

A possible parametrization of the contour C in the complex t-plane is given by $t(\varphi) = a\cos(\varphi) + i\,b\sin(2\,\varphi)$, $a > 1$, $b > 0$, $0 \le \varphi \le 2\pi$. The contour C is located on the Riemann surface of the (generically multivalued) integrand. Implement this representation, and use it to calculate the numerical value of $J_e(i\,\pi)$.

b) Bessel functions of large real positive order ν and arguments $z < \nu$ can be expanded in an asymptotic series in the following way [1304]: We start from the differential equation for $J_\nu(\nu z)$ with respect to z

$$z^2\,\frac{d^2 J_\nu(\nu z)}{dz^2} + z\,\frac{d J_\nu(\nu z)}{dz} + \nu^2(z^2 - 1)\,J_\nu(\nu z) = 0$$

and introduce new (in the moment unknown) functions $u_i(z)$ via

$$J_\nu(z) = \frac{\nu^\nu}{\Gamma(\nu + 1)}\,\exp(u(z))$$

$$u(z) = \nu \int_c^z u_1(\zeta)\,d\zeta + \sum_{i=-\infty}^{0} \nu^i \int_0^z u_i(\zeta)\,d\zeta.$$

The constant c is chosen in such a way that for small z, this expansion agrees with the ordinary series expansion of $J_\nu(z)$ around $z = 0$. The functions $u_i(z)$ have to be calculated by equating the coefficients of the powers of ν to zero. Calculate symbolically the first terms of this series expansion, and use the resulting series to calculate the numerical value of $J_{1000}(100)$.

c) The intensity of a converging spherical wave, which was diffracted at a circular aperture, is in the meridonal plane given by the square of the absolute value of the following integral $\int_0^1 \rho\,J_0(v\,\rho)\,e^{-iu\rho^2/2}\,d\rho$ [793], [179], [1187], [857]. Visualize the intensity inside the domain $|u| < 10\,\pi$, $|v| < 5\,\pi$.

d) The eigenfunctions superconductor–normal metal system is described by the Bogoliubov–de Gennes equation (a generalization of the Schrödinger equation) [374], [1158], [5], [116], [1121], [1131]

$$\begin{pmatrix} -\hbar^2/2\,m(\mathbf{r})^{-1}\,\Delta + V(\mathbf{r}) - \mu(\mathbf{r}) & \delta(\mathbf{r}) \\ \overline{\delta(\mathbf{r})} & -\hbar^2/2\,m(\mathbf{r})^{-1}\,\Delta + V(\mathbf{r}) - \mu(\mathbf{r}) \end{pmatrix} \begin{pmatrix} \psi_n^{(e)}(\mathbf{r}) \\ \psi_n^{(h)}(\mathbf{r}) \end{pmatrix} = \varepsilon_n \begin{pmatrix} \psi_n^{(e)}(\mathbf{r}) \\ \psi_n^{(h)}(\mathbf{r}) \end{pmatrix}.$$

Here the quantities $m(\mathbf{r})$, $V(\mathbf{r})$, $\mu(\mathbf{r})$, and $\delta(\mathbf{r})$ are the position-dependent mass, potential, chemical potential, and pair potential (which vanishes in the nonsuperconducting state). $u_n(\mathbf{r})$ and $v_n(\mathbf{r})$ are the electron and hole components of the wave function and ε_n are the eigenvalues. For the case of a concentric superconductor–normal conductor arrangement with equal chemical potentials, no (interface) potential, and no mass mismatch, a superconductor of diameter $\rho^{(S)}$ and normal conductor of diameter $\rho^{(N)}$ the problem simplifies considerably. (Here we assume the superconductor in the center surrounded by the normal conductor, meaning $\rho^{(S)} < \rho^{(N)}$.)

After separation of variables and matching wave functions the eigenvalue problem reduces to [340] finding the roots of the following determinantal equation

$$
\begin{vmatrix}
\psi_m^{(N,e)}(\varepsilon, \rho^{(S)}) & 0 & \delta^{(e)}(\varepsilon)\,\psi_m^{(S,e)}(\varepsilon, \rho^{(S)}) & \delta^{(h)}(\varepsilon)\,\psi_m^{(S,h)}(\varepsilon, \rho^{(S)}) \\
0 & \psi_m^{(N,h)}(\varepsilon, \rho^{(S)}) & \psi_m^{(S,e)}(\varepsilon, \rho^{(S)}) & \psi_m^{(S,h)}(\varepsilon, \rho^{(S)}) \\
\psi_m^{(N,e)\prime}(\varepsilon, \rho^{(S)}) & 0 & \delta^{(e)}(\varepsilon)\,\psi_m^{(S,e)\prime}(\varepsilon, \rho^{(S)}) & \delta^{(h)}(\varepsilon)\,\psi_m^{(S,h)\prime}(\varepsilon, \rho^{(S)}) \\
0 & \psi_m^{(N,h)\prime}(\varepsilon, \rho^{(S)}) & \psi_m^{(S,e)\prime}(\varepsilon, \rho^{(S)}) & \psi_m^{(S,h)\prime}(\varepsilon, \rho^{(S)})
\end{vmatrix} = 0
$$

where (choosing appropriate units) the electron wave function components are

$$
\psi_m^{(N,e)}(\varepsilon, \rho^{(S)}) = J_m(k(\varepsilon)\,\rho^{(S)}) - J_m(k(\varepsilon)\,\rho^{(N)})\,Y_m(k(\varepsilon)\,\rho^{(S)}) / Y_m(k(\varepsilon)\,\rho^{(N)})
$$
$$
\psi_m^{(S,e)}(\varepsilon, \rho^{(S)}) = J_m(q(\varepsilon)\,\rho^{(S)})
$$

and the hole wave functions are $\psi^{(N,h)} = \psi_m^{(N,e)}(-\varepsilon, \rho^{(S)})$ and $\psi^{(S,h)} = \overline{\psi_m^{(S,e)}(-\varepsilon, \rho^{(S)})}$. Here a prime denotes differentiation with respect to the second argument $\rho^{(S)}$. The azimuthal quantum number m is integer. The wave vectors k and q are $k(\varepsilon) = (1 + \varepsilon)^{1/2}$ and $q(\varepsilon) = (1 + (\varepsilon - \Delta_0)^{1/2})^{1/2}$. Finally the electron and hole pairing potentials are $\delta^{(e)}(\varepsilon) = \Delta_0(\varepsilon - (\varepsilon^2 - \Delta_0^2)^{-1/2})$ and $\delta^{(h)}(\varepsilon) = \overline{\delta^{(e)}(\varepsilon)}$.

The eigenvalues of the resulting determinant have interesting dependence on m [340], [804]. Visualize the eigenvalues for $0 < \varepsilon_n < \Delta_0$ and $0 \le m \le 2.5\,\rho^{(S)}$ for the parameter values $\Delta_0 = 0.15$, $\rho^{(S)} = 200$, and $\rho^{(N)} = 400$.

7.^{L1} Euler's Integral for Beta Function, Beta Probability Distribution, Euler's Constant

a) The Beta function $B(p, q)$ can be represented by the following integral (see [583] and [264]):

$$
B(p, q) = -\frac{e^{i(q-p)}}{4\sin(p\pi)\sin(q\pi)} \int_C t^{p-1}(1-t)^{q-1}\,dt.
$$

Here, C is a contour that encloses the point $\{0, 0\}$ and $\{1, 0\}$ (see [765], [724], [598], [736], and [1325]), in a way that is topologically equivalent to a path parametrized by:

$$
\{x(s), y(s)\} = \{(\cos(s)^2\cosh(\sin(\pi/4 + s))^2 - \sin(s)^2\sinh(\sin(\pi/4 + s))^2),
$$
$$
\cos(s)\cos(\sin(\pi/4 + s))\sin(s)\sin(\sin(\pi/4 + s))\}.
$$

The integration has to be carried out on the Riemann surface of the integrand. (So the integrand is a continuous function along the path of integration.) Implement the numerical evaluation of the Beta function via the above integral representation.

b) Visualize a discrete realization of the probability density $w(x) \propto (1 - x^2)^\gamma$ using $x_j(\gamma)$, $0 \le x_j(\gamma) \le 1$, $j = 0, ..., n$ [1321].

c) Euler's constant γ has the well known representation $\gamma = \lim_{n\to\infty}(\sum_{k=1}^n 1/k - \ln(n))$ [1344], [387], [386]. Find a generalization of the form

$$
\gamma = \lim_{n\to\infty}\left(\left(\sum_{k=1}^n 1/k\right) - \ln\left(n + \left(\sum_{j=0}^o \frac{\alpha_j}{n^j}\right)\right)\right)
$$

that has optimal convergence properties for $o = 100$.

8.^{L2} Time-Dependence in cos-Potential, Singular Potential Eigenvalues

a) Use expansion in eigenfunctions to visualize the time-development of the following initial value problem for a triple well potential

$$i\,\frac{\partial \Psi(x,\,t)}{\partial t} = -\,\frac{\partial^2\,\Psi(x,\,t)}{\partial x^2} - V_0\cos(x)\,\Psi(x,\,t)$$

$$\Psi(-3\,\pi,\,t) = \Psi(3\,\pi,\,t)$$

$$\Psi(x,\,0) = \cos(\mathcal{K}\,x)\,\cos^2\!\left(\frac{\pi}{2}\,\frac{x}{x_M}\right)\theta(x_M^2 - x^2)$$

for $-3\,\pi \le x \le 3\,\pi$, $0 \le t \le 10$. Use $V_0 = 12$, $\mathcal{K} = 6$, $x_M = \pi/6$. How does this expansion compare with a direct numerical solution of the initial value problem using `NDSolve`?

b) Calculate bound states for the potential $V(x) = -e^x$ in the Schrödinger equation $-\psi''(x) + V(x)\,\psi(x) = \varepsilon\,\psi(x)$ [484], [696], [640], [39]. Calculate the corresponding WKB eigenvalues.

9.¹² Dependencies, Numerical Function Evaluation, Usage Messages, Derivative Definition

a) By looking at examples of mathematical functions, find out which functions are used in internal computations of those mathematical functions.

b) Find examples for functions and exact arguments such that the result of N[*specialFunction*[*arguments*]] is incorrect. (N[*specialFunction*[*arguments*], *precisionLargerThanMachinePrecision*] is, nearly always, correct).

c) Numeric functions (meaning functions that carry the `NumericFunction` attribute) have usage messages like `"BesselJ[n, z] gives the Bessel function..."`. Check which other "kinds" of usage messages exist for numeric functions.

d) Assume you have a (special) function Y(z), such that Y'(0) = c, but Y'(z) cannot be expressed in closed form for general z. How could one implement that `Y'[0]` evaluates, but avoid the constant presence of a definition for `Derivative`?

10.¹³ Perturbation Theory

a) Consider the following eigenvalue problem (where λ is the eigenvalue) [1090], [815], [457], [58]:

$$-y''(x) + \alpha\,x\,y(x) = \lambda\,y(x), \quad y(-l/2) = y(l/2) = 0.$$

For small α ($\alpha \ll \pi/l^2$), starting with the exact solution of the problem, determine the dependence of $\lambda_i = \lambda_i(\alpha)$ on α in the form:

$$\lambda_i(\alpha) \approx \lambda_i(\alpha = 0) + c_1\,\alpha + c_2\,\alpha^2.$$

b) Consider the following eigenvalue problem in $(-\infty, \infty)$ for $v_0 < 0$ (where λ is the continuous eigenvalue) [58]:

$$-y''(x) + \theta\!\left(\left(\frac{l}{2}\right)^2 - x^2\right)V_0\,y(x) + \alpha\,x\,y(x) = \lambda\,y(x).$$

(In physics terms, we are considering a quantum well in an electric field in the limit of small electric fields.) Now, the spectrum is for finite α a continuous one. For small α ($\alpha \to 0$), again starting with the exact solution of the problem, determine the dependence of the $\lambda_i = \lambda_i(\alpha)$ on α in the form

$$\lambda_i(\alpha) \approx \lambda_i(\alpha = 0) + c_1\,\alpha + c_2\,\alpha^2$$

where λ_i, $\lambda_i < v_0$ (which means it is a Gamov or quasi-bound state [377], [962], [173]) is now defined as a solution of the "eigenvalue equation" (pole condition of the corresponding Green's function [6], [1052], [171], and [741]) which one obtains by taking into account only outgoing waves to the far left and only asymptotically vanishing solutions to the far right. Check the limit $\alpha \to 0$ for the eigenvalue equation and the limit (calculated in part a) $V_0 \to \infty$ for c_2.

11.^{L2} Fermi–Dirac Integrals, Sum of All Reciprocal 9-Free Numbers, Zagier's Function

a) In [1265], the following approximation of the Fermi–Dirac integral

$$F_\alpha(z) = \frac{1}{\Gamma(1+\alpha)} \int_0^\infty \frac{t^\alpha}{1+z^{-1}e^t} \, dt$$

as

$$F_\alpha^{(n,k)}(z) \approx \frac{2\pi \csc(a\pi)}{\Gamma(a+1)} \sum_{i=0}^n (\pi^2(2i+1)^2 + \log^2(z))^{a/2} \cos\!\left(a\left(\pi - \tan^{-1}\!\left(\frac{(2i+1)\pi}{\log(z)}\right)\right)\right) +$$

$$\frac{\pi}{\Gamma(a+1)} \cos\!\left(a\left(\pi - \tan^{-1}\!\left(\frac{(2(n+1)+1)\pi}{\log(z)}\right)\right)\right) \csc(a\pi)\,(\pi^2(2(n+1)+1)^2 + \log^2(z))^{a/2} -$$

$$2\csc(a\pi)\,(\pi^2(2(n+1)+1)^2 + \log^2(z))^{\frac{a+1}{2}} \times \left(\sum_{i=0}^k \frac{\zeta(2i)}{\Gamma(a+2(1-i))}\right.$$

$$\left. (\pi^2(2(n+1)+1)^2 + \log^2(z))^{-i} \sin\!\left(a\pi - (a-2i+1)\tan^{-1}\!\left(\frac{(2(n+1)+1)\pi}{\log(z)}\right)\right)\right)$$

was given. Here $1 < z < \infty$, $-1 < \alpha < \infty$, $\alpha \neq 0, 1, 2, \ldots$, $k = 0, 1, 2, \ldots$, $2k > a+1$, $n = 0, 1, \ldots$ and $\zeta(x)$ is the Riemann zeta function.

For moderate values of k and n ($n < 8$), examine the usability of this expression (accuracy and availability of all needed functions) for other programs.

b) The asymptotic expansion for $\eta \to \infty$ of the integral (a slightly rewritten form of the integral from the last subexercise)

$$\mathcal{F}_d(\eta) = \frac{1}{\Gamma(d/2)} \int_0^\infty \frac{\varepsilon^{d/2-1}}{1+e^{\varepsilon-\eta}} \, d\varepsilon$$

is for $d \in \mathbb{N}$ given by [394], [497], [1096]

$$\mathcal{F}_d(\eta) \underset{\eta \to \infty}{=} \mathcal{F}_{d,o}^{(\text{alg})}(\eta) + \mathcal{F}_{d,o}^{(\text{exp})}(\eta)$$

$$\mathcal{F}_{d,o}^{(\text{alg})}(\eta) =$$

$$\left(\sum_{k=0}^{\lfloor d/4 \rfloor} \frac{2(1-2^{1-2k})\,\zeta(2k)}{\Gamma(d/2-2k+1)} \eta^{(d-4k)/2}\right) + \frac{\sin(d\pi/2)}{\pi} \sum_{k=\lfloor d/4 \rfloor+1}^{o} \frac{2(1-2^{1-2k})\,\zeta(2k)}{\Gamma(2k-d/2)} \eta^{(d-4k)/2}$$

$$\mathcal{F}_{d,o}^{(\text{exp})}(\eta) = \cos(\pi(d/2-1)) \sum_{k=1}^{o} \frac{(-1)^{k+1}}{k^{d/2}} e^{-k\eta}.$$

A graph of $\mathcal{F}_{d,o}^{(\text{alg})}(\eta)/\mathcal{F}_{d,o}^{(\text{exp})}(\eta)$ in the d,η-plane shows for $o \to \infty$ "special" points. Calculate the location of these points for $d \to \infty$ to 20 digits.

c) The sum [656], [625], [807]

$$
s = \sum_{\substack{k=1 \\ k \text{ is } 9\text{-free}}}^{\infty} \frac{1}{k}
$$

of all reciprocal numbers that are free of the digit 9 is a convergent sum. It is given by [463], [76]

$$
s = \beta_0 \ln(10) + \sum_{k=2}^{\infty} 10^{-k}\,\beta(k-1)\,\zeta(k)
$$

where the β_k are implicitly defined by:

$$
\sum_{k=1}^{n} \binom{n}{k}(10^{-k+n+1} - 10^k + 1)\,\beta(n-k) = 10\,(11^n - 10^n), \quad n = 1,\, 2,\, \dots.
$$

Calculate a 50-digit approximation of s.

d) Consider the "special" function $\varphi(z)$ defined as [1352], [43]:

$$
\varphi(z) = \sum_{n=0}^{\infty} \prod_{k=1}^{n} (1 - z^k).
$$

For $z = \exp(2\pi i\, p/q)$, $p, q \in \mathbb{Z}$, $q \neq 0$, the sum converges; for all other complex values of z, the sum diverges. Up to which denominators q can one use machine arithmetic to calculate $\varphi(z)$? As a function of q, the relation $\langle |\varphi(z)| \rangle \propto q^{\alpha}$ holds for large q (averaging $\langle \cdot \rangle$ is with respect to p). Find an approximative value of α. Visualize for some $\frac{p}{q}$ how the sum converges as one adds terms.

12.^{12} Heat Equation, Green's Function for a Rectangle, Theta Function Addition Formulas, Theta Function Series Expansion, Bose Gas

a) Find the solution of the following boundary-value problem [113]

$$
\frac{\partial}{\partial t} T(x, t) = a^2 \frac{\partial^2}{\partial x^2} T(x, t)
$$
$$
T(0, t) = T(l, t) = 0
$$
$$
T(x, 0) = \delta(x - y)
$$

using elliptic Theta functions:

$$
\vartheta_3(u, q) = 1 + 2 \sum_{n=1}^{\infty} q^{n^2} \cos(2\,n\,u)
$$

where $\vartheta_3(u, q)$ =EllipticTheta[3, u, q]. Examine graphically the dependence on temperature.

b) The Green's function [411], [1216] for the Laplace equation with homogeneous Dirichlet boundary conditions on a rectangle $(0, a) \times (0, b)$ usually is written in the form:

$$G_{ab}(x, y, x', y') = \frac{4}{ab\pi^2} \sum_{n=1}^{\infty} \sum_{m=1}^{\infty} \frac{\sin(\frac{m\pi x}{a}) \sin(\frac{n\pi y}{b}) \sin(\frac{m\pi x'}{a}) \sin(\frac{n\pi y'}{b})}{(\frac{m}{a})^2 + (\frac{n}{b})^2}.$$

(Sometimes, it is also written in the form of a single sum as a Fourier expansion in $\sin(n\pi x/a)\sin(n\pi x'/a)$ [102], [461], [630].)

Using elliptic Theta functions, it is possible to find a closed form for this sum (see, e.g., [966], [1219], [631], and [333]):

$$G_{ab}(x, y, x', y') = \frac{1}{2\pi} \mathrm{Re}\left(\ln\left(\frac{\vartheta_1\left(\frac{\pi(z+z')}{2a}, q\right)\vartheta_1\left(\frac{\pi(z-z')}{2a}, q\right)}{\vartheta_1\left(\frac{\pi(z+z')}{2a}, q\right)\vartheta_1\left(\frac{\pi(z-z')}{2a}, q\right)}\right)\right)$$

with $q = e^{-\pi b/a}$, $z = x + iy$, $z' = x' + iy'$. Here, $\vartheta_1(z, q)$ is the Theta function EllipticTheta[1, z, q].

As a Green's function of a 2D problem, $G_{ab}(x, y, x', y')$ must have the following three properties:

$$\left(\frac{\partial^2}{\partial x^2} + \frac{\partial^2}{\partial y^2}\right)G_{ab}(x, y, x', y') = -\frac{1}{\pi^2}\delta(x - x')\delta(y - y')$$

$$G_{ab}(x, y, x', y') = -\frac{1}{2\pi}\ln\sqrt{(x - x')^2 + (y - y')^2} + smoothFunction(x, y, x', y')$$

$$G_{ab}(x, y, x', y') = 0 \quad \text{for } \{x, y\} \text{ on the boundary of the rectangle.}$$

Can one establish symbolically these three properties directly using *Mathematica*? Examine graphically whether the boundary values are 0, whether a logarithmic singularity is present at $z = z'$, and whether the minimax principle holds (i.e., $G_{ab}(x, y, x', y')$ has no local minima or maxima in the interior).

Given the Green's function, the solution of the boundary value problem

$$\left(\frac{\partial^2}{\partial x^2} + \frac{\partial^2}{\partial y^2}\right)\psi(x, y) = 0$$

$$\psi(x, y)\,|_{\{x,y\}\in\partial R_{a,b}} = u(x, y)$$

(where $\partial R_{a,b}$ is the boundary of the rectangle) can be expressed as [1280], [1359]

$$\psi(x, y) = -\int_{\partial R_{a,b}} \frac{\partial G_{ab}(x, y, x', y')}{\partial \vec{n}}\, u(x', y')\, ds'.$$

Here $\partial./\partial\vec{n}$ is the outward directed normal derivative with respect to the primed variables and ds' is the arc length measure and the integration extends along the boundary of the rectangle $\partial R_{a,b}$.

Use this representation to numerically solve the boundary value problem

$$\left(\frac{\partial^2}{\partial x^2} + \frac{\partial^2}{\partial y^2}\right)\psi(x, y) = 0$$

$$\psi(0, y) = \psi(2, y) = 1$$
$$\psi(x, 0) = \psi(x, 4) = 2$$

in the rectangle $R = (0, 2) \times (0, 1)$ and to visualize the solution.

c) The theta functions $\vartheta_k(z, q)$ (in *Mathematica* $\texttt{EllipticTheta[k, z, q]}$), $k \in \{1, 2, 3, 4\}$, $z \in \mathbb{C}$, $q \in \mathbb{C}$, $|q| < 1$) fulfill many identities. One class of identities are addition formulas of the form [762], [830], [726], [169], [1049], [1184]

$$\vartheta_{i_1}(w, q)\, \vartheta_{i_2}(x, q)\, \vartheta_{i_3}(y, q)\, \vartheta_{i_4}(z, q) \pm \vartheta_{j_1}(w, q)\, \vartheta_{j_2}(x, q)\, \vartheta_{j_3}(y, q)\, \vartheta_{j_4}(z, q) =$$
$$\pm \vartheta_{k_1}(\omega, q)\, \vartheta_{k_2}(\xi, q)\, \vartheta_{k_3}(\psi, q)\, \vartheta_{k_4}(\zeta, q) \pm \vartheta_{l_1}(\omega, q)\, \vartheta_{l_2}(\xi, q)\, \vartheta_{l_3}(\psi, q)\, \vartheta_{l_4}(\zeta, q)$$

where

$$\omega = (w + x + y + z)/2$$
$$\xi = (w + x - y - z)/2$$
$$\psi = (w - x + y - z)/2$$
$$\zeta = (w - x - y + z)/2$$

and i_1, i_2, i_3, i_4, j_1, j_2, j_3, j_4, k_1, k_2, k_3, k_4, l_1, l_2, l_3, $l_4 \in \{1, 2, 3, 4\}$. Use numerical techniques to find all identities of the above form. Calculate a "generating" set of identities.

d) Consider the theta function $\vartheta_3(z, q)$ (in *Mathematica*, it is $\texttt{EllipticTheta[3, z, q]}$)

$$\vartheta_3(z, q) = 1 + 2 \sum_{k=1}^{\infty} q^{k^2} \cos(k\, z).$$

Use the partial differential equation for $\vartheta_3(z, q)$

$$-\frac{1}{4} \frac{\partial^2 \vartheta_3(z, q)}{\partial z^2} = q \frac{\partial \vartheta_3(z, q)}{\partial q}$$

to derive the first eight terms in the series expansion of $\vartheta_3(z, q)$. Use the ordinary differential equation for $\vartheta_3(q) = \vartheta_3(0, q)$ [1371], [940], [1339]

$$(\vartheta_3(q)^2\, (d^3\, \vartheta_3(q)) - 15\, \vartheta_3(q)\, (d\vartheta_3(q))\, (d^2\, \vartheta_3(q)) + 30\, (d\vartheta_3(q))^3)^2 +$$
$$32\, \big(\vartheta_3(q)\, (d^2\, \vartheta_3(q)) - 3\, (d\vartheta_3(q))^2\big)^3 - \vartheta_3(q)^{10}\, \big(\vartheta_3(q)\, (d^2\, \vartheta_3(q)) - 3\, (d^2\, \vartheta_3(q))^2\big)^2 = 0$$

(here, $d = q \frac{\partial}{\partial q}$ was used for brevity) to express these coefficients $c_0(q)$, ..., $c_{10}(q)$ as functions of $\vartheta_3(q)$, $\vartheta_3'(q)$, and $\vartheta_3''(q)$.

e) The canonical partition function for N identical noninteracting bosons can be expressed as $Z_N(\beta) = N^{-1} \sum_{n=1}^{N} Z_1(n\, \beta)\, Z_{N-n}(\beta)$ [1117], [180], [921], [84], [1118], [756], [425]. Here $Z_0 = 1$, $\beta = 1/(k_B\, T)$ and $Z_1(\beta)$ is the one-particle partition function. For a particle in a 3D box of dimensions $L \times L \times L$ with infinite walls, the one-particle partition function is [1116], [558], [343], [903], [381], [469], [1199], [269]

$$Z_1^{(box)}(\beta) = \left(\sum_{v=1}^{\infty} e^{-x\, v^2} \right)^3 = \left(\frac{1}{2}\, (\vartheta_3(0,\ e^{-x}) - 1) \right)^3 = \left(\sqrt{\frac{2}{2\pi}}\, K(q^{-1}(e^{-x})) - \frac{1}{2} \right)^3$$

where $x = \beta\, \varepsilon_v$, $\varepsilon_v = \hbar^2/(2\, m)\, (\pi\, n/L)^2$ (or $x = 3/8\, (\lambda(T)/L)^2$, $\lambda(T) = h\, (3\, m\, k_B\, T)^{-1/2}$ with $\lambda(T)$ the de Broglie thermal wavelength), $K(z)$ being the complete elliptic integral of the first kind, and $q^{-1}(z)$ the inverse nome (in *Mathematica* $\texttt{InverseEllipticNomeQ}$). The ground state has energy $3\, \varepsilon_v$.

The corresponding result for a 3D harmonic oscillator with identical frequencies is [219], [659], [816], [1116], [351], [425], [678] $Z_1^{(HO)}(\beta) = \left(\sum_{v=1}^{\infty} e^{-\tilde{x}(v+1/2)} \right)^3 = \left(e^{\tilde{x}/2} / (e^{\tilde{x}} - 1) \right)^3$ where $\tilde{x} = \beta \tilde{\varepsilon}_v$, $\tilde{\varepsilon}_v = \hbar \omega$. The ground state has energy $3/2\, \tilde{\varepsilon}_v$.

The probability that n out of N particles occupy the state ε at temperature T is [1307], [229], [1116], [1231], [1015]

$$p_\varepsilon^{(N)}(n; T) = \frac{1}{Z_N(\beta)} \left(e^{-n\beta\varepsilon} Z_{N-n}(\beta) - e^{-(n+1)\beta\varepsilon} Z_{N-n-1}(\beta) \right).$$

Visualize the average ground state occupation for $N = 100$ for various temperatures for the 3D box and the harmonic oscillator.

The mean energy of N particles is $U_N(T) = k_B T^2 (\partial \ln(Z_N(\beta))/\partial T)_{N,L}$ and the specific heat $c_{v,N}(T) = 1/N\, \partial U_N(T)/\partial T$. Visualize $c_{v,N}(T)$ for various temperatures and particle numbers N for the 3D box and the harmonic oscillator.

13.^{13} Scattering on Cylinder, Coulomb Scattering, Spiral Waves, Optical Black Hole, Corrugated Wall Scattering, Random Solutions of the Helmholtz Equation

a) The scattered wave $\mathcal{E}(r, \varphi; t)$ of an incoming plane electromagnetic wave $E \exp(i\omega t + ikx)$, scattered on a conducting cylinder of radius R and potential zero, is given by [768], [1245], [795], [638], [795]:

$$\mathcal{E}(r, \varphi; t) = -E \left(\frac{J_0(kR)}{H_0^{(2)}(kR)} H_0^{(2)}(kr) + 2 \sum_{n=1}^{\infty} (-i)^n \frac{J_n(kR)}{H_n^{(2)}(kR)} H_n^{(2)}(kr) \cos(n\varphi) \right).$$

Here, $H_0^{(2)}(x)$ is the Hankel function of the second kind, $H_0^{(2)}(x) = J_n(x) - iY_n(x)$. Visualize the scattering process.

b) Water waves (Poincaré waves) in constant depth water are governed by the 2D Helmholtz equation with the boundary conditions $\mathbf{n}.\mathrm{grad}\, u(\mathbf{r}) + i\beta\, \mathbf{t}.\mathrm{grad}\, u(\mathbf{r})$ on scatterers ($\mathbf{t}$ is the normalized tangent vector and $\mathbf{n}$ is the normalized normal vector). The solution of the scattering problem of a disk is [852]

$$u(r, \varphi; t) = u_0 \left(e^{ikx} - \sum_{n=-\infty}^{\infty} i^n \frac{n R J_n'(kR) - n\beta J_n(kR)}{n R H_0^{(1)}{}'(kR) - n\beta H_0^{(1)}(kR)} H_n^{(1)}(kr)\, e^{in\varphi} \right).$$

Here, $H_0^{(1)}(x)$ is the Hankel function of the first kind, $H_0^{(1)}(x) = J_n(x) + iY_n(x)$ and primes denote differentiation with respect to the argument. Visualize this scattering process for various β (the parameter taking into account the rotation of the earth).

c) The scattering of a vertically polarized electromagnetic wave $\mathcal{E}_z(r, \varphi)\, e^{i\omega t}$ impinging from the left on an infinite dielectric cylinder of radius R and refractive index n can be described by the following equations [1357], [1126], [584]:

$$\mathcal{E}_z(r, \varphi) = \begin{cases} \mathcal{E}_0 \sum_{m=-\infty}^{\infty} a_m J_m(kr)\, e^{im\varphi} & \text{for } r \le R \\ \mathcal{E}_0\, e^{ikx} + \mathcal{E}_0 \sum_{m=-\infty}^{\infty} b_m H_m^{(1)}(kr)\, e^{im\varphi} & \text{for } r \ge R \end{cases}$$

$$a_m = i^m \; \frac{J_m'(k\,R)\,H_0^{(1)}(k\,R) - J_m(k\,R)\,H_0^{(1)\prime}(k\,R)}{n\,J_m'(n\,k\,R)\,H_0^{(1)}(k\,R) - J_m(n\,k\,R)\,H_0^{(1)\prime}(k\,R)}$$

$$b_m = i^m \; \frac{J_m'(k\,R)\,J_m(n\,k\,R) - J_m(k\,R)\,J_m'(n\,k\,R)}{n\,J_m'(n\,k\,R)\,H_0^{(1)}(k\,R) - J_m(n\,k\,R)\,H_0^{(1)\prime}(k\,R)}.$$

Here $\mathcal{E}_0$ is the amplitude of the incoming wave, k is the wave vector ($k = 2\pi/\lambda$), n is the index of refraction, and $H_m^{(1)}(z)$ is again the Hankel function of the first kind.

Make an animation showing $|\mathcal{E}_z(r, \varphi)|$ for a cylinder with radius $R \approx 10^1\,\lambda$ with the index of refraction varying from frame to frame.

d) The quantum-mechanical wavefunction for the scattering process of an asymptotically plane wave $\psi(\mathbf{r}) \sim e^{ikz}$ as $|\mathbf{r}| \to \infty$ (unavoidable logarithmic terms suppressed) on a Coulomb potential $V(r) \sim \kappa/r$ is given by the following series [1247], [755], [948], [324], [17], [1215], [701], [790], [104]:

$$\psi(r, \vartheta) = \sum_{l=0}^{\infty} (2\,l + 1)\,i^l\,e^{i\,\eta_l(\kappa;\,k)}\,R_l(\kappa;\,k\,r)\,P_l(\cos(\vartheta))$$

$$\eta_l(\kappa;\,k) = \operatorname{Arg}(\Gamma(l + i\,\kappa/k + 1))$$

$$R_l(\kappa;\,k\,r) = \frac{e^{-\frac{\pi\kappa}{2k}}\,(2\,k\,r)^l}{(2\,l+1)!}\,e^{i\,k\,r}\,|\Gamma(l + i\,\kappa/k + 1)|\;{}_1F_1(l + i\,\kappa/k + 1;\,2\,l + 2;\,-2\,i\,k\,r).$$

(For the corresponding formula in d dimensions, see [1331].) Here $\eta_l(\kappa;\,k)$ is the phase shift. It contains the Gamma function $\Gamma(z)$ and the Legendre polynomials $P_l(z)$. Negative κ correspond to attracting potentials and positive κ correspond to repulsive potentials. Generate an animation that shows how $\operatorname{Re}(\psi(\mathbf{r}))$ and $|\psi(\mathbf{r})|^2$ behave as a function of the potential strength κ.

e) Make a circular contour plot of the real or the imaginary part of the function $u_n(r, \varphi)$:

$$u_n(r, \varphi) = e^{i\,n\,\varphi}\,J_n(\sqrt{\mu}\,r).$$

Here, $0 \le \varphi \le 2\pi$, $0 \le r \le R$ ($R \approx 25$ is a suitable value), $n = 1, 2, 3, \dots$, and μ (in general a complex quantity) is determined by the following equation:

$$\sqrt{\mu}\,J_{n-1}(\sqrt{\mu}\,R) - n\left(\frac{1}{R} + i\right)J_n(\sqrt{\mu}\,R) = 0.$$

The resulting pictures visualize spiral waves of scalar reaction-diffusion equations. (For details, see [380].)

f) The behavior of a light wave (impinging from the right) near a vortex can be approximately described by [780], [781]

$$F(v;\,r,\,\varphi;\,t) = \sum_{m=-\infty}^{\infty} \phi_m(v;\,r,\,\varphi)\,e^{i\,\omega\,t}$$

$$\phi_m(v;\,r,\,\varphi) = (-i)^{\sqrt{m^2 + 2\,v\,m}}\,J_{\sqrt{m^2 + 2\,v\,m}}(k\,r)\,e^{i\,m\,\varphi}.$$

Here k is the wave vector and ω the frequency of the light wave. v is the strength of the vortex. Make an animation that shows how the light wave bends around the vortex with increasing vortex strength v.

g) The following equations describe the stationary 2D scattering problem of a plane wave (with wave vector k and angle α to the y-axis) on the periodic curve $y = b \sin(\beta)$ with Dirichlet boundary conditions:

$$-\frac{\partial^2}{\partial x^2} \psi_{k,\alpha}(x, y) - \frac{\partial^2}{\partial y^2} \psi_{k,\alpha}(x, y) = k^2 \psi_{k,\alpha}(x, y)$$

$$\psi_{k,\alpha}(x, b \sin(\beta x)) = 0$$

Expanding $\psi_{k,\alpha}(x, y)$ in Bragg-reflected waves gives [1263], [1348], [1200], [1302], [889], [1301], [955], [794], [638], [633], [1257], [72], [163], [20]

$$\psi_{k,\alpha}(x, y) = e^{i(k_x x + k_y y)} + \sum_{j=-\infty}^{\infty} c_j(k, \alpha) e^{i(K_x^{(j)} x + K_y^{(j)} y)}.$$

Here the abbreviations $k_x = k \sin(\alpha)$, $k_y = k \cos(\alpha)$, $K_x^{(j)} = k_x + j\beta$, and $K_y^{(j)} = \left((k_x^2 + k_y^2) - K_x^{(j)2}\right)^{1/2}$ were used.

The $c_j(k, \alpha)$ are solution of the following infinite system of linear equations:

$$J_n(b k_y) + \sum_{j=-\infty}^{\infty} c_j(k, \alpha) J_{n-j}(b K_y^{(j)}) = 0.$$

Make an animation that shows the $|\psi_{k,\alpha}(x, y)|$, $\mathrm{Re}(\psi_{k,\alpha}(x, y)) = 0$, and $\mathrm{Im}(\psi_{k,\alpha}(x, y)) = 0$ as a function of k and α.

h) Make an animation of contour plots of $\mathrm{Re}(\psi_\tau(r, \varphi)) = 0$ and $\mathrm{Im}(\psi_\tau(r, \varphi)) = 0$ where

$$\psi_\tau(r, \varphi; k) = \sum_{n=-\infty}^{\infty} c_n(\tau) e^{in\varphi} J_n(k r)$$

is a random superposition of orthogonal solutions of the 2D Helmholtz equation $\Delta\psi = k^2 \psi$ [170], [139], [162], [1093], [905].

Here r and φ are polar coordinates, k is a real constant and the $c_n(\tau)$ are smooth random functions of τ.

Make 3D contour plots of the corresponding 3D functions

$$\psi(r, \vartheta, \varphi; k) = \sum_{l=0}^{\infty} \sum_{m=-l}^{l} c_{l,m} \frac{J_{l+1/2}(k r)}{\sqrt{k r}} Y_{lm}(\vartheta, \varphi)$$

where the $Y_{lm}(\vartheta, \varphi)$ are spherical harmonics and the $c_{l,m}$ are again random coefficients.

14.^{L1} Wronskian of Legendre Functions, Separation of Variables in Toroidal Coordinate System

a) Prove that the Wronskian for the Legendre functions is given by:

$$W(P_\nu^\mu(z), Q_\nu^m(z)) = P_\nu^\mu(z) \frac{\partial Q_\nu^\mu(z)}{\partial z} - Q_\nu^\mu(z) \frac{\partial P_\nu^\mu(z)}{\partial z}$$

$$= \frac{\Gamma(\nu + \mu + 1)}{\Gamma(\nu - \mu + 1)} \frac{1}{1 - z^2}.$$

b) The relationship between toroidal coordinates [908] and Cartesian ones is given by:

$$\{x, y, z\} = \left\{ \frac{c \, \sinh(\eta) \cos(\vartheta)}{\cosh(\eta) - \cos(\varphi)}, \; \frac{c \, \sinh(\eta) \sin(\vartheta)}{\cosh(\eta) - \cos(\varphi)}, \; \frac{c \, \sin(\varphi)}{\cosh(\eta) - \cos(\varphi)} \right\}.$$

Visualize the coordinate surfaces of the toroidal coordinate system.

Show that the Laplace equation can be solved by the method of separation of variables in toroidal coordinates and that the solution is a linear combination of functions of the form

$$\sqrt{2 \, (\cosh(\eta) - \cos(\varphi))} \; (c_1 \cos(\varphi) + c_2 \sin(\varphi)) \, (c_3 \cos(\vartheta) + c_4 \sin(\vartheta)) \, (c_5 \, P_\nu^\mu(\cosh(\eta)) + c_6 \, Q_\nu^\mu(\cosh(\eta)))$$

where the c_i are constants to be determined by the specific boundary conditions. Use the package `Calculus`` `VectorAnalysis``. For some application of the use of toroidal coordinates for the solution of electromagnetic boundary value and other problems, see [618], [119], [106], [943], [9], [10], [1264], [265], and [776].

15.¹² Riemann–Siegel Formula, Zeros of Hurwitz Zeta Function, Zeta Zeta Function, Harmonic Polylogarithms, $1 \times 2 \times 3 \times \cdots = \sqrt{2\pi}$

a) The Riemann zeta function $\zeta(z) = \sum_{k=1}^{\infty} k^{-z}$, $\mathrm{Re}(z) > 1$ is of most interest along the critical line $z = 1/2 + i\, t$ [764], [426], [629], [1226], [995], [144], [145], [305], [83]. There, it can be conveniently written as $\zeta(1/2 + i\,t) = Z(t)\, e^{i\,\vartheta(t)}$. Here, $\vartheta(t)$ is the Riemann–Siegel Theta function (in *Mathematica* `RiemannSiegel`` `Theta[t]`) and $Z(t)$ is the Riemann–Siegel function (in *Mathematica* `RiemannSiegelZ[t]`). $Z(t)$ is purely real for real t. The Riemann–Siegel Theta function can be expressed through the logarithm of the Gamma function (in *Mathematica*, `LogGamma[z]`):

$$\vartheta(t) = \mathrm{Im}\left(\log\Gamma\left(i\, \frac{t}{2} + \frac{1}{4} \right) - \frac{t}{2} \log(t) \right).$$

For large positive t, the Riemann–Siegel function $Z(t)$ has the following expansion (this is the celebrated asymptotic Riemann–Siegel formula):

$$Z(t) \underset{t \to \infty}{\sim} 2 \sum_{k=1}^{v(t)} \frac{1}{\sqrt{k}} \cos(\vartheta(t) - t \log(k)) + R(t)$$

where $(t) = \lfloor (t/(2\pi))^{1/2} \rfloor$.

The term $R(t)$ can be expanded in powers of $t^{1/4}$ [426], [1160]:

$$R(t) = (-1)^{v(t)-1} \left(\frac{t}{2\pi} \right)^{-\frac{1}{4}} \sum_{k=0}^{\infty} c_k \left(\sqrt{\frac{t}{2\pi}} - v(t) \right) \left(\frac{t}{2\pi} \right)^{-\frac{k}{2}}.$$

The coefficients $c_k(p)$ can be expressed in the following way:

$$c_k(p) = [\omega^k]\left(\exp\left(i\left(\log\left(\frac{t}{2\pi} \right) - \frac{t}{2} - \frac{\pi}{8} - \vartheta(t) \right) \right) \times [y^0]\left(\left(\sum_{j=0}^{\infty} A_j(y)\, \omega^j \right) \left(\sum_{j=0}^{\infty} \psi^{(j)}(p)\, \frac{y^j}{j!} \right) \right) \right).$$

In the last formula, $[x^n]\, (f(x))$ denotes the coefficient of x^n of the Taylor series expansion of $f(x)$ around $x = 0$. The sequence $A_j(y)$ obeys the following recursion relation:

$$A_j(y) = -\frac{1}{2} y A_{j-1}(y) - \frac{1}{32\pi^2} \frac{\partial^2}{\partial y^2} \frac{A_{j-1}(y)}{y}$$

$$A_0(y) = e^{2\pi i y^2}.$$

Finally, $\psi(p)$ is the following function:

$$\psi(p) = \frac{\cos(2\pi(p^2 - p - 1/16))}{\cos(2\pi p)}.$$

Calculate the first 10 coefficients c_k. Taking in $R(t)$ the first 10 terms into account, how many correct digits does one get for $t = 10^6$? (For more efficient ways to calculate the c_k's, see [145].)

b) The generalized zeta function (Hurwitz's Zeta function [753], [1179]) $\zeta(z, \alpha)$ is defined in *Mathematica* by the analytic continuation of the following formula (when α not a negative integer or 0):

$$\zeta(z, \alpha) = \sum_{n=0}^{\infty} \frac{1}{((n+\alpha)^2)^{z/2}}.$$

The *Mathematica* form is $\texttt{Zeta[z, }\alpha\texttt{]}$. For $\alpha = 1$, the function $\zeta(z, 1)$ is Riemann's $\zeta(z)$ function with all its nontrivial zeros (conjectured) on the line $z = 1/2 + it$, $t > 0$ (see [426] and [629]). Visualize the dependence of the zeros of $\zeta(z, \alpha)$ for $1/2 \le \alpha \le 1$ [480].

c) For positive integers s, the values of the Zeta function $Z(s)$ (of the nontrivial zeros $z_k = 1/2 + it_k$) of the Riemann Zeta function $\zeta(z)$ [774]

$$Z(s) = \sum_{k=1}^{\infty} \frac{1}{\left(\frac{1}{4} + t_k\right)^s}$$

(here conjugate roots have been grouped pairwise so that $t_k > 0$ is assumed) can be obtained in closed form by equating coefficients of s^n in the following identity [1285], [1286], [319], [1287]:

$$\sum_{k=1}^{\infty} \frac{Z(k)}{k} (s(1-s))^k =$$

$$-\left(\left(\log(2\sqrt{\pi}) + \frac{\gamma}{2} - 1\right)s + \sum_{k=2}^{\infty} \frac{((1-2^{-k})\zeta(k) - 1)}{k} s^k - \sum_{l=1}^{\infty} \frac{1}{l}\left(\sum_{k=1}^{\infty} \frac{\gamma_{k-1}}{(k-1)!} s^k\right)^l\right).$$

Here γ_k are the Stieltjes constants (in *Mathematica* $\texttt{StieltjesGamma[k]}$). Calculate the exact values for $Z(1), \ldots, Z(5)$ and compare the values with the ones obtained by summing over the first 1000 nontrivial zeros explicitly.

d) The harmonic polylogarithm functions $H(a_1, a_2, \ldots, a_n; z)$ are defined recursively through [1055], [504], [901], [505], [167], [361], [1282]

$$H(a_1, a_2, \ldots, a_n; z) = \int_0^z f(a_1; \zeta) H(a_2, \ldots, a_n; \zeta) \, d\zeta$$

$$f(-1, z) = 1/(1+z) \qquad H(-1; z) = +\ln(1+z)$$
$$f(\pm 0, z) = 1/z \qquad H(\pm 0; z) = +\ln(z)$$
$$f(-1, z) = 1/(1-z) \qquad H(+1; z) = -\ln(1-z).$$

For which $a_k = -1, 0, 1$ can *Mathematica* find exact finite values of $H(a_1, a_2; 1)$, $H(a_1, a_2, a_3; 1)$, and $H(a_1, a_2, a_3, a_4; 1)$?

e) Sums (especially infinite sums) whose summands contain special functions can often be calculated by using an integral representation for the special function, then interchanging summation and integration (assuming the conditions to do this are fulfilled) [1181], [841]. Use the integral representation

$$\psi^{(n)}(z) = \frac{\partial^n}{\partial t^n}\left(\gamma + \int_0^1 \frac{1 - t^{z-1}}{1-t}\, dt\right)$$

for the *n*th derivative of the digamma function $\psi(z)$ to calculate some sums of the form [970], [971], [972]

$$\sum_{k=1}^{\infty} R(k)\, P(\psi(z), \psi^{(1)}(z), \dots).$$

Here $R(k)$ is a rational function in k and $P(x, y, \dots)$ is a polynomial. Can one calculate some sums that the built-in `Sum` cannot find?

f) Motivate the finite result for the following divergent product: $1 \times 2 \times 3 \times \cdots = \sqrt{2\pi}$.

16.^{12} Riemann Surface of Gauss Hypergeometric Function, $K(z)/K(1-z)$, $\mathrm{erf}^{(-1)}$

a) Carry out the analytical continuation by using Kummer relations [1028], [3], [23], [674], [1037], [38], that are relevant to the analytical continuation ($1 - c, b - a, c - a - b$ not integers in all of the following formulas):

$$\begin{aligned}
{}_2F_1(a; b, c; z) =\ & (1-z)^{c-a-b}\, {}_2F_1(c-a; c-b, c; z) \\
=\ & (1-z)^{-a}\, {}_2F_1\!\left(a; c-b, c; \frac{z}{z-1}\right) \\
=\ & (1-z)^{-b}\, {}_2F_1\!\left(b; c-a, c; \frac{z}{z-1}\right) \\
=\ & \frac{\Gamma(c)\,\Gamma(c-a-b)}{\Gamma(c-a)\,\Gamma(c-b)}\, {}_2F_1(a; b, a+b-c+1; 1-z) \\
& + \frac{\Gamma(c)\,\Gamma(a+b-c)}{\Gamma(a)\,\Gamma(b)}\, (1-z)^{c-a-b}\, {}_2F_1(c-a; c-b, c-a-b+1; 1-z) \\
& \text{if } |\arg(1-z)| < \pi \\
=\ & \frac{\Gamma(c)\,\Gamma(b-a)}{\Gamma(b)\,\Gamma(c-a)}\, (-z)^{-a}\, {}_2F_1\!\left(a; 1-c+a, 1-b+a; \frac{1}{z}\right) \\
& + \frac{\Gamma(c)\,\Gamma(a-b)}{\Gamma(a)\,\Gamma(c-b)}\, (-z)^{-b}\, {}_2F_1\!\left(b; 1-c+b, b-a+1; \frac{1}{z}\right) \\
& \text{if } |\arg(-z)| < \pi \\
=\ & \frac{\Gamma(c)\,\Gamma(b-a)}{\Gamma(b)\,\Gamma(c-a)}\, (1-z)^{-a}\, {}_2F_1\!\left(a; c-b, a-b+1; \frac{1}{1-z}\right) \\
& + \frac{\Gamma(c)\,\Gamma(a-b)}{\Gamma(a)\,\Gamma(c-b)}\, (1-z)^{-b}\, {}_2F_1\!\left(b; c-a, b-a+1; \frac{1}{1-z}\right)
\end{aligned}$$

if $|\arg(1-z)| < \pi$

$$
= \frac{\Gamma(c)\,\Gamma(c-a-b)}{\Gamma(c-a)\,\Gamma(c-b)}\, z^{-a}\,{}_2F_1\!\left(a;\, a-c+1,\, a+b-c+1;\, 1-\frac{1}{z}\right)
$$

$$
+ \frac{\Gamma(c)\,\Gamma(a+b-c)}{\Gamma(a)\,\Gamma(b)}\,(1-z)^{c-a-b}\,{}_2F_1\!\left(c-a;\, 1-a,\, c-a-b+1;\, 1-\frac{1}{z}\right)
$$

if $|\arg(z)| < \pi$.

b) Construct pictures of the sheets of the Riemann surface of ${}_2F_1(1/3;\, 1/2,\, 1/6;\, z)$ that are directly connected with the principal sheet. Carry out the analytical continuation by solving the differential equation for $w(z) = {}_2F_1(a;\, b,\, c;\, z)$:

$$
z(1-z)\,w''(z) + (c-(a+b+1)z)\,w'(z) - a\,b\,w(z) = 0.
$$

c) Construct pictures of the sheets of the Riemann surface of $w(z) = K(z)/K(1-z)$ [1346] that are neighboring the principal sheet. Carry out the analytical continuation by solving the Schwarz differential equation for $K(z)/K(1-z)$. Find symbolic expressions for the neighboring sheets.

d) Construct a picture of some sheets of the Riemann surface of $w(z) = \mathrm{erf}^{(-1)}(z)$ in the neighborhood of the point $z = 1$.

17.¹² Kummer's 24 Solutions of the Gauss Hypergeometric Differential Equation, Appell Differential Equation

a) The Liouville transformation

$$
y(z) = \exp\!\left(-\frac{1}{2}\int^{z} p(\xi)\, d\xi\right) u(z)
$$

transforms the differential equation

$$
y''(z) + p(z)\,y'(z) + q(z)\,y(z) = 0
$$

into the normal form $u''(z) + g(z)\,u(z) = 0$. (See also Exercise 11 of Chapter 1)

Use the Liouville transformation to transform the hypergeometric differential equation

$$
z(1-z)\,y''(z) + (c-(a+b+1)z)\,y'(z) - a\,b\,y(z) = 0
$$

into normal form. Then, apply a change of the independent variable from z to x of the form [619]

$$
z(x) = \frac{\alpha + \beta\,x}{\gamma + \delta\,x}
$$

and transform the resulting differential equation again to normal form $w''(x) + h(x)\,w(x) = 0$.

One particular solution of the hypergeometric differential equation is $y(z) = {}_2F_1(a,\, b;\, c;\, z)$.

Determine all $\alpha(a, b, c)$, $\beta(a, b, c)$, $\gamma(a, b, c)$, and $\delta(a, b, c)$ that leave the above $g(z)$ form invariant (meaning that $h(x) = z'(x)^2\, g(z(x)) \doteq g(x)$ holds) so that the solution of $w''(x) + h(x)\,w(x) = 0$ can again be expressed as a hypergeometric function ${}_2F_1(a', b'; c'; z)$, where $a' = a'(a, b, c)$, $b' = b'(a, b, c)$ and $c' = c'(a, b, c)$. Finally, transform the resulting solution back to the original function $y(z)$ to get a new form of the solution of the hypergeometric differential equation (see [1028], [3], [258], [23], [598], [674], [1037], and [38]).

b) The bivariate hypergeometric function $w(z_1, z_2) = F_1(a; b_1, b_2; c; z_1, z_2)$ fulfills the following coupled system of partial differential equations.

$$(1 - z_1) z_1 \frac{\partial^2 w(z_1, z_2)}{\partial z_1^2} + (1 - z_1) z_2 \frac{\partial^2 w(z_1, z_2)}{\partial z_1 \partial z_2} +$$

$$(c - (a + b_1 + 1) z_1) \frac{\partial w(z_1, z_2)}{\partial z_1} - b_1 z_2 \frac{\partial w(z_1, z_2)}{\partial z_2} - a b_1 w(z_1, z_2) = 0$$

$$(1 - z_2) z_2 \frac{\partial^2 w(z_1, z_2)}{\partial z_2^2} + (1 - z_2) z_1 \frac{\partial^2 w(z_1, z_2)}{\partial z_1 \partial z_2} +$$

$$(c - (a + b_2 + 1) z_2) \frac{\partial w(z_1, z_2)}{\partial z_2} - b_2 z_1 \frac{\partial w(z_1, z_2)}{\partial z_1} - a b_2 w(z_1, z_2) = 0$$

For fixed z_2, derive an ordinary differential equation of $F_1(a; b_1, b_2; c; z_1, z_2)$ with respect to z_1 [236], [364].

18.^{12} Roots of Differentiated Polynomials

a) Visualize the following theorem: Given a polynomial $f(z)$ of arbitrary degree over the complex (or real) numbers, all roots of the polynomial $f'(z)$ lie in the convex hull of the roots of $f(z)$. (This is the so-called Gauss–Lucas theorem; see [806], [844], [843], [1176], [188], [94], [1044], [583], [1142], [842], and [898]; for a sharpening of this theorem, see [345].) Look in the standard packages for calculating the convex hull.

b) Take a random polynomial over the real or complex numbers of degree greater than 10 and show graphically its roots and all of the roots of the polynomial differentiated m times. The picture suggests that the roots lie on some curves. Try to connect the roots in the "right" order by building a (the "obvious") continuous version (as a function of α) of $d^\alpha x^n / dx^\alpha$ [761], [871], [872], [873], [1024], [874]. Visualize this generalization of differentiation by itself.

19.^{12} Coinciding Bessel Zeros, π-Formulas

a) For μ and ν nonintegers, it is possible that $J_\nu(x)$ and $J_\mu(x)$ have two (or more) zeros in common. Find numerically real values ν, μ, x, z, such that $J_\nu(x) = J_\mu(x) = 0$ *and* simultaneously $J_\nu(z) = J_\mu(z) = 0$ [128], [1013].

b) Ramanujan-like series for $1/\pi$

$$\frac{1}{\pi} = \sum_{k=0}^{\infty} \frac{a + bk}{k!^3} \left(\frac{1}{2}\right)_k \left(\frac{1}{3}\right)_k \left(\frac{2}{3}\right)_k c^k$$

can be generated based on algebraic solutions α_n (α_n is expressable in radicals) of the following transcendental equation for integer $n > 1$:

$$\frac{{}_2F_1(\frac{1}{3}, \frac{2}{3}; 1; 1 - \alpha_n)}{{}_2F_1(\frac{1}{3}, \frac{2}{3}; 1; \alpha_n)} = \sqrt{n} .$$

The parameters a, b, and c algebraic numbers dependent on n and are defined through a_n by [275], [135], [277], [278]

$$a_n = \frac{8}{9}\sqrt{\frac{n}{3}}\ \alpha_n(\alpha_n - 1)\ \frac{{}_2F_1(\frac{4}{3}, \frac{5}{3}; 2; \alpha_n)}{{}_2F_1(\frac{1}{3}, \frac{2}{3}; 1; \alpha_n)} + \frac{1}{\pi\,{}_2F_1(\frac{1}{3}, \frac{2}{3}; 1; \alpha_n)^2}$$

$$b_n = \frac{2}{3}\sqrt{3\,n}\ \sqrt{1 - 4\,\alpha_n\,(1 - \alpha_n)}$$

$$c_n = 4\,\alpha_n\,(1 - \alpha_n).$$

(The appearance of π is the formula for a does not make this a circular definition, the resulting series contains only integers, rationals, and algebraic numbers.) Use numerical techniques to calculate explicit forms of such π-series for $2 \le n \le 20$.

20.^{L1} Force-Free Magnetic Fields, Bessel Beams, Gauge Transformation

a) Calculate x in

$$B_z = x\,e^{-lz}\ J_0(r\,x)$$
$$B_\varphi = a\,e^{-lz}\ J_1(r\,x)$$
$$B_r = l\,e^{-lz}\ J_1(r\,x).$$

(B_z, B_φ, B_r are the components of the magnetic field in a cylindrical coordinate system) such that $\vec{B}$ is a force-free magnetic field. This means that div $\vec{B} = 0$ and curl $\vec{B} = a\,\vec{B}$ hold simultaneously [851], [129], [1351], [1319].

b) Consider an electromagnetic wave in vacuum with the magnetic field components (in a cylindrical coordinate system) [866], [869], [1128], [917], [820], [1074], [1107], [383], [1097], [192]

$$B_z = 0$$
$$B_\varphi = J_1(k\sin(\alpha)\,r)\exp(i(k\cos(\alpha)\,z - \omega\,t))$$
$$B_r = 0.$$

Calculate the corresponding electric field $\{E_z, E_\varphi, E_r\}$ of this wave.

c) A vector potential $\mathbf{A}_0(r, \varphi)$ of a homogeneous magnetic field of strength H in e_z-direction can be chosen in cylindrical coordinates as $A_\varphi(r, \varphi) = H\,r/2$, $A_r(r, \varphi) = 0$, $A_z(r, \varphi) = 0$. (Here $A_\varphi(r, \varphi)$ is the azimuthal part of the vector potential in cylindrical coordinates and r is the radius. For domains with polygonal symmetry, it is often useful to have vanishing normal components of the vector potential along the polygon edges [298]. After a gauge transformation $\mathbf{A}_0(r, \varphi) \to \mathbf{A}(r, \varphi) = \mathbf{A}_0(r, \varphi) + \mathbf{G}(r, \varphi)$ this can be achieved. For a square centered at the origin, with edges parallel to the coordinate axes and edge length a, the gauge transformation is [297]

$$G_r(r, \varphi) = H\,\frac{r^3}{a^6}\left(a^2(a^2 - r^2) + (a^4 - 4\,r^2\,a^2 + 2\,r^4)\exp\left(1 - \frac{2\,r^2}{a^2}\right)\mathrm{Ei}\left(\frac{2\,r^2}{a^2} - 1\right)\right)\sin(4\,\varphi)$$

$$G_\varphi(r, \varphi) = H\left(\frac{r^3}{a^4}\left(a^2 + (a^2 - 2\,r^2)\exp\left(1 - \frac{2\,r^2}{a^2}\right)\mathrm{Ei}\left(\frac{2\,r^2}{a^2} - 1\right)\right)\cos(4\,\varphi)\right)$$

$$G_z(r, \varphi) = 0.$$

Verify by explicit calculation that the vector potential $\mathbf{A}(r, \varphi)$ fulfills the stated properties. Visualize the flowlines of this vector potential and its equicontour lines.

21.^{12} Riemann Surface of the Bootstrap Equation

Visualize the Riemann surface of the function $w(z)$ that is implicitly defined by the equation $z = 2w - e^w + 1$.

22.^{11} Differential Equation of Powers of Airy Functions, Map Airy Distribution, Zeros of Airy Function

a) Find the (linear) differential equation that is obeyed by $\text{Ai}(z)^n$ for $1 \le n \le 10$, n integer.

b) Starting from the series expansion of $\text{Ai}(z)$ for large negative z, derive the first terms of the expansion of the zeros $z^{(i)}$ of $\text{Ai}(z)$ $(\text{Ai}(z^{(i)}) = 0, i = 0, 1, 2, \dots)$ for large i in terms of descending powers of $3\pi/8\,(4\,i - 1)$.

c) In [90], the "map-Airy" distribution $p(x)$ was introduced.

$$p(x) = 2\,e^{-\frac{2x^3}{3}}\,(x\,\text{Ai}(x^2) - \text{Ai}'(x^2))$$

It is a probability distribution. Calculate its asymptotics as $x \to \pm\infty$, the corresponding cumulative distribution function, and its first moment. Calculate $\int_0^\infty x^2\,p(x)\,dx$.

23.^{12} Differential Equation for Dedekind η Function, Darboux–Halphen System

a) The Dedekind Eta function $\eta(z)$ [775], [49], [1145], [148], [627] has the Fourier product representation

$$\eta(z) = e^{\frac{i\pi z}{12}} \prod_{k=1}^{\infty} (1 - e^{2ik\pi z}),$$

where $\text{Im}(z) > 0$. The function $\eta(z)$ obeys a fourth-order, nonlinear differential equation [149] $p(\eta(z), \eta'(z), \eta''(z), \eta'''(z), \eta''''(z)) = 0$, where p is a multivariate polynomial of total degree 4. Find the polynomial p.

b) Show that the functions

$$w_1(\tau) = \frac{1}{2}\,\frac{\partial \ln\left(\frac{\lambda'(\tau)}{\lambda(\tau)}\right)}{\partial \tau}, \quad w_2(\tau) = \frac{1}{2}\,\frac{\partial \ln\left(\frac{\lambda'(\tau)}{\lambda(\tau)-1}\right)}{\partial \tau}, \quad w_3(\tau) = \frac{1}{2}\,\frac{\partial \ln\left(\frac{\lambda'(\tau)}{\lambda(\tau)\,(\lambda(\tau)-1)}\right)}{\partial \tau}$$

where $\lambda(\tau) = q^{(-1)}(e^{i\pi\tau})$ are solutions of the Darboux–Halphen system [572]:

$$w_1'(z) = w_1(z)\,(w_2(z) + w_3(z)) + w_2(z)\,w_3(z)$$
$$w_2'(z) = w_2(z)\,(w_1(z) + w_3(z)) + w_1(z)\,w_3(z)$$
$$w_3'(z) = w_3(z)\,(w_1(z) + w_2(z)) + w_1(z)\,w_2(z).$$

Here $q^{(-1)}$ is the inverse of the elliptic nome q (in *Mathematica* `InverseEllipticNomeQ`). How could one calculate $q^{(-1)\prime}$?

24.^{11} Ramanujan Identities for φ and λ Function

a) For many positive integers p, q, r, the expression

$$\left(\frac{\vartheta_3(0, \exp(-p\,\pi))}{\vartheta_3(0, \exp(-q\,\pi))}\right)^r = \left(\frac{\displaystyle\sum_{k=-\infty}^{\infty} \exp(-k^2\,p\,\pi)}{\displaystyle\sum_{k=-\infty}^{\infty} \exp(-k^2\,q\,\pi)}\right)^r$$

is an algebraic number [132], [274], [133]. (Here, $\vartheta_3(z, q)$ is the function `EllipticTheta[3, z, q]`.) Use *Mathematica*'s high-precision numerics to find some such integers p, q, r and the corresponding algebraic numbers.

b) For positive integer n the expression

$$\lambda_n = \frac{1}{3\sqrt{3}} \left(\frac{\eta\left(\left(1 + i\sqrt{\frac{n}{3}}\right)/2\right)}{\eta\left(\left(1 + i\sqrt{3n}\right)/2\right)} \right)^6$$

is an algebraic number [1046], [136]. (Here, $\eta(\tau)$ is the function `DedekindEta[τ]`.) Use *Mathematica*'s high-precision numerics to find the algebraic values for $1 \le n \le 10$. Find all $n \le 100$ such that λ_n is the root of a polynomial of degree four or less.

25.¹³ Identities for Gamma Function Values, Identities for Dedekind η Function

a) Gamma functions fulfill many identities of the form

$$\frac{\prod\limits_{k=1}^{n} \Gamma(r_k)^{a_k}}{\prod\limits_{k=1}^{m} \Gamma(s_k)^{c_k}} = \pi^p \alpha.$$

Here, the r_k and s_k are rational numbers, a_k, c_k, and p are small integers, and α is an algebraic number. Examples of such identities are [258], [185]:

$$\frac{\Gamma(\frac{5}{12})^2}{\Gamma(\frac{1}{4})^2 \Gamma(\frac{2}{3})^2} = \frac{1}{2\pi} \sqrt{-9 + 6\sqrt{3}}$$

$$\frac{\Gamma(\frac{2}{5})^2 \Gamma(\frac{8}{15})^2}{\Gamma(\frac{4}{15})^2 \Gamma(\frac{2}{3})^2} = \frac{3^{7/10}}{\sqrt[6]{5}} \sqrt{\frac{2}{5 + \sqrt{5}}} \sqrt{3} (\sqrt{5} - 1) + \frac{\sqrt{2(5 + \sqrt{5})}}{\sqrt{10 - 2\sqrt{5}} + \sqrt{3} + \sqrt{15}}.$$

Using the functional equations obeyed by the Gamma functions (n being a positive integer)

$$\Gamma\left(\frac{k}{n}\right) \Gamma\left(1 - \frac{k}{n}\right) = \pi \csc\left(\pi \frac{k}{n}\right)$$

$$\Gamma(n z) = n^{nz - 1/2} (2\pi)^{(1-n)/2} \prod_{k=0}^{n} \Gamma\left(z + \frac{k}{n}\right)$$

write a program that generates such identities.

b) The Dedekind Eta function (defined for $\text{Im}(\tau) > 0$)

$$\eta(\tau) = e^{\frac{i\pi\tau}{12}} \prod_{k=1}^{\infty} (1 - e^{2i\pi\tau}) = e^{\frac{i\pi\tau}{12}} \sum_{k=-\infty}^{\infty} (-1)^k e^{k(3k-1)i\pi\tau}$$

fulfills many functional equations of the form [700], [154], [131], [134], [853], [520], [1267], [1342], [157], [938], [158]

$P(\eta(c_1 \tau), \ldots, \eta(c_n \tau)) = 0.$

Here P is a multivariate polynomial over the integers and the c_i are positive integers. Examples of such functional identities are:

$$\eta(6\,\tau)^4 \, \eta(\tau)^9 - 9\,\eta(2\,\tau)^4 \, \eta(3\,\tau)^8 \, \eta(\tau) + 8\,\eta(2\,\tau)^9 \, \eta(3\,\tau)^3 \, \eta(6\,\tau) = 0$$

$$\eta(8\,\tau)^4 \, \eta(2\,\tau)^{14} + \eta(\tau)^8 \, \eta(4\,\tau)^4 \, \eta(8\,\tau)^4 \, \eta(2\,\tau)^2 - 2\,\eta(\tau)^4 \, \eta(4\,\tau)^{14} = 0$$

$$\eta(10\,\tau)^2 \, \eta(\tau)^5 - 5\,\eta(2\,\tau)^2 \, \eta(5\,\tau)^4 \, \eta(\tau) + 4\,\eta(2\,\tau)^5 \, \eta(5\,\tau)\,\eta(10\,\tau) = 0$$

$$\eta(4\,\tau)^2 \, \eta(12\,\tau)^2 \, \eta(\tau)^8 - 4\,\eta(3\,\tau)^2 \, \eta(4\,\tau)^8 \, \eta(\tau)^2 + 3\,\eta(2\,\tau)^8 \, \eta(6\,\tau)^4 = 0$$

$$\eta(\tau)^4 \, \eta(2\,\tau)\,\eta(3\,\tau)^2 \, \eta(12\,\tau)^3 \, \eta(6\,\tau)^2 + \eta(2\,\tau)^3 \, \eta(3\,\tau)^6 \, \eta(12\,\tau)^3 - 2\,\eta(\tau)^2 \, \eta(4\,\tau)\,\eta(6\,\tau)^9 = 0$$

Conjecture at least 25 more relations of this kind.

Solutions

1. Asymptotic Series, Carlitz Expansion, Contour Lines of Gamma Function, Bessel Zeros, Asymptotic Expansion of Gamma Function Ratio, Integrals of Function Compositions, $W_{1+i}(1 + i)$

a) Here is the recursive definition of the asymptotic formula for the Bessel function $J_0(z)$.

```
In[1]:= myBessel0[z_, 0] := myBessel0[z, 0] = 1/(Pi Sqrt[2Pi z]) Gamma[1/2]^2 *
                                (Exp[-I(z - Pi/4)] + Exp[+I(z - Pi/4)]) // N;

    myBessel0[z_, ord_] := (* remember terms *) myBessel0[z, ord] =
    myBessel0[z, ord - 1] + (* new term *)
      1/(Pi Sqrt[2Pi z]) Gamma[ord + 1/2]^2 *
      (Exp[-I(z - Pi/4)]/(-2I z)^ord/ord! +
      Exp[+I(z - Pi/4)]/(+2I z)^ord/ord!) // N
```

Here is a list of the successive approximations for $z = 2 + 3i$ and a graph showing the convergence behavior in the complex plane. In addition, we show a magnified version of the part of the graph containing the most interesting point.

```
In[3]:= Function[ta,
    Show[GraphicsArray[{
    (* rough sketch *)
    Graphics[Line[{Re[#], Im[#]}& /@ ta],
      Axes -> True, DisplayFunction -> Identity],
    (* the microscopic view *)
    Graphics[{{PointSize[0.02], (* the exact value *)
          Point[{Re[#], Im[#]}&[BesselJ[0, 2 + 3I] // N]]},
          {MapIndexed[{Hue[#2[[1]]/21], Line[#1]}&,
            Partition[{Re[#], Im[#]}& /@ ta, 2, 1]]}},
      Axes -> True, (* appropriate plot range *)
      PlotRange -> {{-0.455, -0.48}, {-4.31, -4.32}}]}]]][
                Table[myBessel0[2 + 3I, o], {o, 0, 20}]]
```

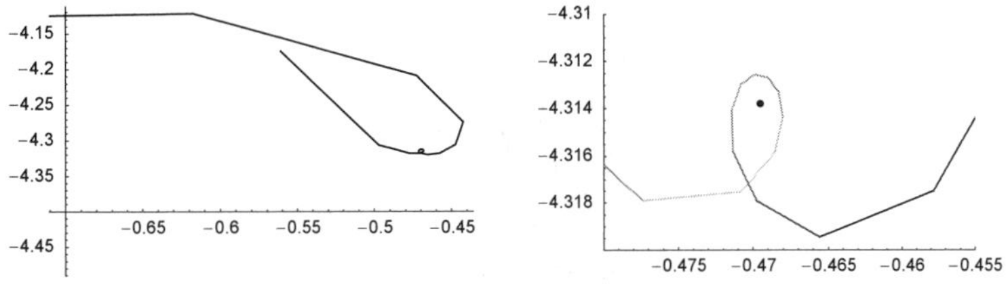

For comparison, here is the exact value.

```
In[4]:= BesselJ[0, 2 + 3I] // N
Out[4]= -0.469517 - 4.31379 i
```

Let us make the calculation for the Airy function more automatic. These are the coefficients in the sum.

```
In[5]:= c[k_] := c[k] = Gamma[3k + 1/2]/(2^k 3^(3k) Gamma[k + 1/2] Gamma[k + 1])
```

`lim` calculates how many terms of the series must be taken into account by looking at `c[k]` as a continuous function of `k`.

```
In[6]:= lim[z_] := lim[z] = Floor @
    FindMinimum[Abs[Gamma[3k + 1/2] (2/3 z^(3/2))^(-k)/
                (2^k 3^(3k) Gamma[k + 1/2] Gamma[k + 1])],
          (* appropriate for the following *) {k, 12, 100}][[2, 1, 2]]
```

The following graphic gives an idea about the size of `lim` along the positive real axis.

```
In[7]:= ListPlot[Table[lim[n], {n, 100}]]
```

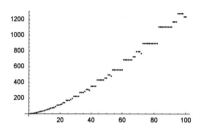

AiryAiSumList gives the list of the partial sums.

```
In[8]:= AiryAiSumList[z_] :=
        N[1/(2 Sqrt[Pi]) z^(-1/4) Exp[-2/3 z^(3/2)] FoldList[Plus, 0,
          Table[(-1)^k c[N[k]] (2/3 z^(3/2))^-k, {k, 0, lim[Abs[z]] + 5}]]]
```

Here is the behavior of 25 values in the complex plane.

```
In[9]:= Show[Graphics[{{Thickness[0.002], Hue[0],
        (* the partial sums of the asymptotic series *)
        Table[Line[{Re[#], Im[#]}& /@ Take[AiryAiSumList[x + I y], -10]],
            {x, 2.222, 2.223, 0.0002}, {y, 2.222, 2.223, 0.0002}]},
        {PointSize[0.005],
        (* the exact values *)
        Table[Point[{Re[#], Im[#]}]&[AiryAi[x + I y]],
            {x, 2.222, 2.223, 0.0002}, {y, 2.222, 2.223, 0.0002}]}}],
            Frame -> True]
```

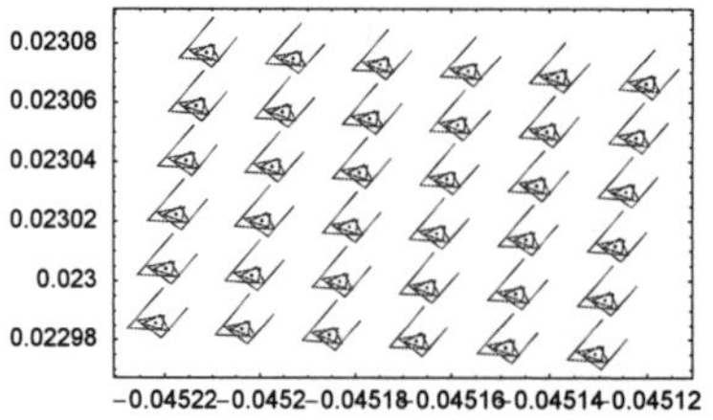

Near the negative real axis, the given asymptotic series breaks down [143]. The following picture demonstrates this clearly. We show the relative error as a function of the argument.

```
In[10]:= AiryAiSum[z_] :=
        1/(2 Sqrt[Pi]) z^(-1/4) Exp[-2/3 z^(3/2)] *
        (* cumulative sums *)
        Fold[Plus, 0, Table[(-1)^k c[k] (2/3 z^(3/2))^-k, {k, 0, lim[Abs[z]]}]]
In[11]:= (* relative error *)
        δ[z_] := AiryAiSum[z]/AiryAi[z] - 1
In[13]:= (* |z| == 3; δ as a function of Arg[z] *)
        Plot[Abs[δ[4 Exp[I φ]]], {φ, -Pi, Pi},
            Frame -> True, Axes -> False, PlotRange -> All]
```

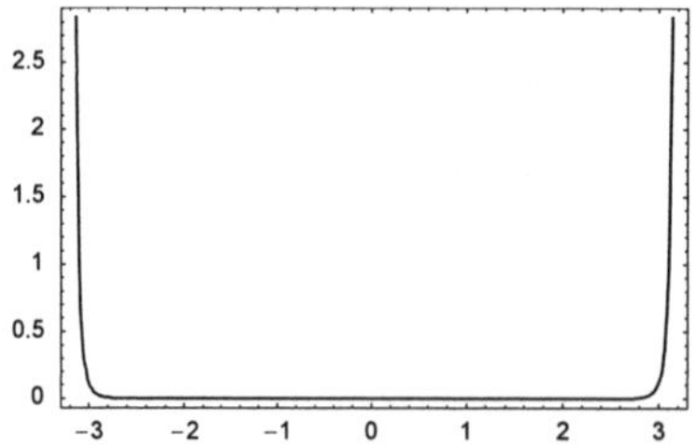

Here is the number of terms until the minimal size element is reached in dependence of z in the complex plane, calculated and then shown. (For an optimal truncation, one should sum until the smallest element occurs, but not taking the smallest element itself into account.)

```
In[15]:= numberOfTerms[z_] :=
         Module[{k = 0, old, new},
                 old = N[(-1)^k c[N[k]] z^(-3/2k)];
                 (* as long as terms are decreasing *)
                 While[k = k + 1.;
                       Abs[new = N[(-1)^k c[N[k]] (2/3 z^(3/2))^-k]] < Abs[old],
                       old = new]; k]
```

```
In[16]:= pp = 61;
         data = Table[numberOfTerms[N[x + I y]],
                      {x, -5, 5, 10/pp}, {y, -5, 5, 10/pp}];
```

```
In[18]:= {Min[#], Max[#]}&[data]
```

```
Out[18]= {1., 27.}
```

```
In[19]:= ListContourPlot[data, MeshRange -> {{-5, 5}, {-5, 5}},
                         Contours -> Range[Max[data]], ContourShading -> False]
```

For more on the properties of asymptotic expansions, see [439], [394], [979], [217], [88], and [935]; for some numerical investigations on asymptotic expansions, see [190].

As discussed in Section 1.7, Sum will sum many mildly divergent sums.

```
In[20]:= SymbolicSum`SymbolicSum[Gamma[3k + 1/2]/(2^k 3^(3k) Gamma[k + 1/2]*
                                 Gamma[k + 1]) (-1)^k (2/3 z^(3/2))^-k,
                                 {k, 0, Infinity}, GenerateConditions -> False]
```

$$Out[20]= \frac{2 e^{\frac{2 z^{3/2}}{3}} \sqrt{z^{3/2}} \text{ BesselK}\left[\frac{1}{3}, \frac{2 z^{3/2}}{3}\right]}{\sqrt{3 \pi}}$$

The last result is actually the function Ai(z) for $z > 0$ [482], [1150], [1151]. Here is a random numerical check.

```
In[21]:= Table[1/(2 Sqrt[Pi]) z^(-1/4) Exp[-2/3 z^(3/2)] % - AiryAi[z] /.
               z -> Random[Real, {0, 1}], {10}] // Chop
```

```
Out[21]= {0, 0, 0, 0, 0, 0, 0, 0, 0, 0}
```

Regularizing the sum with $1/k!$ yields a convergent sum.

```
In[22]:= Sum[Gamma[3k + 1/2]/(2^k 3^(3k) Gamma[k + 1/2] Gamma[k + 1])
             (-1)^k ζ^k/(* make convergent *)k!, {k, 0, Infinity}]
```

$$Out[22]= \text{HypergeometricPFQ}\left[\left\{\frac{1}{6}, \frac{5}{6}\right\}, \{1\}, -\frac{ζ}{2}\right]$$

Laplace transforming the result gives again the Airy function for positive z.

```
In[23]:= LaplaceTransform[% /. ζ -> ζ t, t, 1]
```

$$Out[23]= \frac{e^{\frac{1}{ζ}} \sqrt{\frac{2}{\pi}} \text{ BesselK}\left[\frac{1}{3}, \frac{1}{ζ}\right]}{\sqrt{ζ}}$$

```
In[24]:= 1/(2 Sqrt[Pi]) z^(-1/4) Exp[-2/3 z^(3/2)] (% /. ζ -> 1/(2/3 z^(3/2)))
```

$$Out[24]= \frac{\text{BesselK}\left[\frac{1}{3}, \frac{2 z^{3/2}}{3}\right]}{\sqrt{3} \pi \sqrt{\frac{1}{z^{3/2}}} z^{1/4}}$$

b) The implementation of the recursion for the $a_k(\zeta)$ is straightforward. Induction shows that the $a_k(\zeta)$ have the form $a_k(\zeta) = \sum_{j=2}^{k} c_{k,j}\, \zeta^{-k}$ with rational c_k. This form allows to replace the definite integral by a faster indefinite one. To avoid getting too many terms we represent the $a_k(\zeta)$ as a SeriesData-object.

```
In[1]:= a[1] = Series[-5/72/ζ, {ζ, Infinity, 10}]

       a[k_] := a[k] = (k - 2)/(k - 1) a[k - 1] + D[a[k - 1], ζ]/(2 (k - 1)) +
                       5/72/(k - 1) Integrate[a[k - 1]/ζ^2, ζ]
```

$$\text{Out[1]=} \quad -\frac{5}{72\,\zeta} + O\!\left[\frac{1}{\zeta}\right]^{11}$$

Here are the first few $a_k(\zeta)$.

```
In[3]:= {a[2], a[3], a[4], a[5]}
```

$$\text{Out[3]=} \quad \left\{ \frac{385\left(\frac{1}{\zeta}\right)^2}{10368} + O\!\left[\frac{1}{\zeta}\right]^{12},\ \frac{385\left(\frac{1}{\zeta}\right)^2}{20736} - \frac{85085\left(\frac{1}{\zeta}\right)^3}{4478976} + O\!\left[\frac{1}{\zeta}\right]^{12}, \right.$$

$$\frac{385\left(\frac{1}{\zeta}\right)^2}{31104} - \frac{85085\left(\frac{1}{\zeta}\right)^3}{4478976} + \frac{37182145\left(\frac{1}{\zeta}\right)^4}{3869835264} + O\!\left[\frac{1}{\zeta}\right]^{12},$$

$$\left. \frac{385\left(\frac{1}{\zeta}\right)^2}{41472} - \frac{935935\left(\frac{1}{\zeta}\right)^3}{53747712} + \frac{37182145\left(\frac{1}{\zeta}\right)^4}{2579890176} - \frac{5391411025\left(\frac{1}{\zeta}\right)^5}{1114512556032} + O\!\left[\frac{1}{\zeta}\right]^{12} \right\}$$

Calculating now 10 leading terms of the first 1000 $a_k(\zeta)$ is a matter of seconds.

```
In[4]:= o = 1000;
       Do[a[k], {k, o}]; // Timing
Out[5]= {2.36 Second, Null}
```

We extract the coefficients $c_{k,j}$ and visualize them.

```
In[6]:= coeffs = Table[Table[{k, Coefficient[Normal[a[k]], ζ, -o]},
                        {k, 1, o}], {o, 2, 10}];
```

```
In[7]:= Show[Table[
       ListPlot[(* display logarithm of absolute value *)
                Log[DeleteCases[Abs[coeffs[[k]]], {0, _} | {_, 0}]] // N,
                PlotStyle -> {PointSize[0.005], Hue[k/13]},
                DisplayFunction -> Identity],
                {k, 9}], DisplayFunction -> $DisplayFunction,
       PlotRange -> All, Frame -> True, Axes -> False]
```

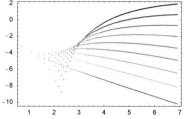

Now we will compare the $\sum_{k=0}^{o} a_{k+1}(\zeta)/(k+1)$ with the exact values. These are the exact values of the asymptotic expansion as a Laurent series in $2/3\,z^{3/2}$.

```
In[8]:= Off[Series::esss];
       seriesAi = Series[Normal[Series[AiryAi[z], {z, Infinity, 20}]]/
                  (Exp[-2/3 z^(3/2)] (z)^(-1/4)/(2 Sqrt[Pi])) /.
                  z -> (3/2 ζ)^(2/3)], {ζ, Infinity, 10}]
```

$$\text{Out[9]=} \quad 1 - \frac{5}{72\,\zeta} + \frac{385\left(\frac{1}{\zeta}\right)^2}{10368} - \frac{85085\left(\frac{1}{\zeta}\right)^3}{2239488} + \frac{37182145\left(\frac{1}{\zeta}\right)^4}{644972544} - \frac{5391411025\left(\frac{1}{\zeta}\right)^5}{46438023168} +$$

$$\frac{5849680962125\left(\frac{1}{\zeta}\right)^6}{20061226008576} - \frac{1267709431363375\left(\frac{1}{\zeta}\right)^7}{1444408272617472} + \frac{2562040760785380875\left(\frac{1}{\zeta}\right)^8}{831979165027663872} -$$

$$\frac{6653619855759634132375\left(\frac{1}{\zeta}\right)^9}{53912249893792618 9056} + \frac{4318199286388002551911375\left(\frac{1}{\zeta}\right)^{10}}{7763363984706137122 4064} + O\!\left[\frac{1}{\zeta}\right]^{11}$$

The next graphic shows the relative error and its logarithm as a function of the number of coefficients $a_k(\zeta)$.

```
In[10]:= exactCoeff[d_] := SeriesCoefficient[seriesAi, d]
```

```
In[11]:= Show[GraphicsArray[Function[idLog,
         Show[Table[ListPlot[(* function of the relative error *)
                     N @ idLog[1 - Rest[FoldList[Plus, 0,
                           (#2/#1)& @@@ coeffs[[k]]]]]/exactCoeff[k + 1]],
               PlotStyle -> {PointSize[0.003], Hue[(k - 1)/10]},
               DisplayFunction -> Identity], {k, 9}]] /@
               (* identity and logarithm *) {Identity, Log[10, #]&}]]
```

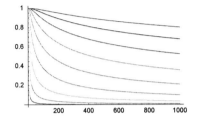
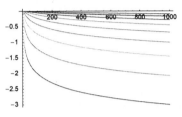

c) Because the following expressions are sometimes quite big, we suppress the output. Here is the definition of σ_k.

```
In[1]:= s[-1, n_, m_] = 0; s[-2, n_, m_] = 0;

       s[k_, n_, m_] := Gamma[n + k + 1] Gamma[m + 1]/k! (a/b)^k *
                         Hypergeometric2F1[-k, m + 1, -n - k, (b/a)^2]
```

This is the implementation of the terms of the series.

```
In[3]:= term[k_] = (* use power series at generic points *)
       PowerExpand @ Assuming[a > 0 && b > 0 && z > 0,
       Normal[Series[(s[k, n, m] - s[k - 2, n, m]) *
                         BesselJ[m + k, a z] BesselJ[n + k, b z],
                         {z, 0, 12}]] ];
```

We calculate explicitly the first summands.

```
In[4]:= sum = Expand[Sum[term[k], {k, 0, 6}]];
```

These are the ones involving the lowest powers in z.

```
In[5]:= powers = Union[Cases[sum, z^_, {0, Infinity}]]
```

$$\text{Out[5]= } \{z^{m+n},\ z^{2+m+n},\ z^{4+m+n},\ z^{6+m+n},\ z^{8+m+n},\ z^{10+m+n},$$
$$z^{12+m+n},\ z^{14+m+n},\ z^{16+m+n},\ z^{18+m+n},\ z^{20+m+n},\ z^{22+m+n},\ z^{24+m+n}\}$$

```
In[6]:= terms = Apply[Plus, Cases[sum, _ #]/#& /@ Take[powers, 5], {1}];
```

Using FullSimplify, we see that all higher powers vanish identically and that the lowest power has the correct prefactor.

```
In[7]:= FullSimplify /@ Take[terms, 2]
```

$$\text{Out[7]= } \{2^{-m-n}\, a^m\, b^n,\ 0\}$$

For the next terms, FullSimplify will need a long time; so it is faster to simplify the expression manually. Looking at the structure of one of the elements of terms, we see many Gamma functions with different arguments.

```
In[8]:= Short[terms[[3]] /. Gamma -> Γ, 4]
```

$$\text{Out[8]//Short= } \frac{2^{-5-m-n}\, a^{4+m}\, b^n}{(1+m)\,(2+m)} + \frac{2^{-4-m-n}\, a^{2+m}\, b^{2+n}}{(1+m)\,(1+n)} + \ll 24 \gg + \frac{2^{-4-m-n}\, a^{2+m}\, b^{2+n}\, n\, \Gamma[1+m]}{(1+n)\,(2+n)\,\Gamma[3+m]} +$$
$$\frac{2^{-4-m-n}\, a^{2+m}\, b^{2+n}\, m\, n\, \Gamma[1+m]}{(1+n)\,(2+n)\,\Gamma[3+m]} + \frac{2^{-5-m-n}\, a^{4+m}\, b^n\, n^2\, \Gamma[1+m]}{(1+n)\,(2+n)\,\Gamma[3+m]} - \frac{2^{-4-m-n}\, a^{2+m}\, b^{2+n}\, \Gamma[1+m]\, \Gamma[1+n]}{\Gamma[3+m]\, \Gamma[3+n]}$$

If we canonicalize the arguments of the Gamma functions by hand, it is easy to establish the identity we are looking for.

```
In[9]:= Together[# //. (* rewrite shifted Gamma functions *)
               {Gamma[n + i_Integer] -> (n + i - 1) Gamma[n + i - 1],
               Gamma[m + i_Integer] -> (m + i - 1) Gamma[m + i - 1]}]& /@ terms
```

Out[9]= $\{2^{-m-n}\, a^m\, b^n,\ 0,\ 0,\ 0,\ 0\}$

d) We use ContourPlot to look for the lines where the real or the imaginary part of $\Gamma(z)$ vanish and then combine the two graphics into one. To save some time, we use the property $\Gamma(z) = \overline{\Gamma(\bar{z})}$.

```
In[1]:= Module[{ε = 10^-3, gr11, gr2, grRe, grIm},
         (* the two ContourGraphics *)
         {gr1, gr2} = ContourPlot[(* Re or Im *) #[Gamma[x + I y]],
                            {x, -7 + ε, 7}, {y, ε, 7},
                            PlotPoints -> 100, Contours -> {0},
                            ContourShading -> False,
                            DisplayFunction -> Identity]& /@ {Im, Re};
         (* keeping only the lines *)
         {grRe, grIm} = Flatten[DeleteCases[Graphics[#][[1]],
                            AbsoluteThickness[_] |
                            Thickness[_] | GrayLevel[_],
                            {0, Infinity}]]& /@ {gr1, gr2};
         (* lines of vanishing real and imaginary parts *)
         Show[Graphics[{(* real part *)
           {Thickness[0.002], grRe,
            Map[#{1, -1}&, grRe, {-2}], Line[{{-7 + ε, 0}, {7, 0}}]},
                    (* imaginary part *)
           {Dashing[{0.015, 0.015}], Thickness[0.002], grIm,
            Map[#{1, -1}&, grIm, {-2}]}},
               Frame -> True, AspectRatio -> Automatic,
               PlotLabel -> StyleForm["Re[Gamma[z]] == 0, Im[Gamma[z]] == 0",
                            FontFamily -> "Courier", FontSize -> 7]]]
```

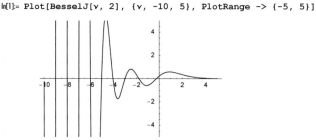

For a detailed discussion of these lines, see [1252], [174].

e) For a rough estimation of the zeros, let us have a look at the plot of $J_v(2)$.

```
In[1]:= Plot[BesselJ[v, 2], {v, -10, 5}, PlotRange -> {-5, 5}]
```

The plot shows that $v_i = -i$ is a reasonable starting value. Because *Mathematica* cannot differentiate Bessel functions with respect to the order in closed form, we give FindRoot two starting values to calculate the zeros.

```
In[2]:= Table[N[#, 15]& @
          FindRoot[BesselJ[v, 2], {v, v0 + 2/10, v0 - 2/10},
                 AccuracyGoal -> 20, WorkingPrecision -> 50], {v0, -2, -10, -1}]
Out[2]= {{v → -1.78932135266695}, {v → -2.96105888069356}, {v → -3.99604799733464},
         {v → -4.99977431981483}, {v → -5.99999184132705}, {v → -6.99999979492956},
         {v → -7.99999999619187}, {v → -8.99999999994551}, {v → -9.99999999999938}}
```

Interestingly, the zeros are *very* nearly to negative integers; for a detailed discussion of this fact, see [332] and [465].

```
In[3]:= MapIndexed[((v /. #1) + (#2[[1]] + 1))&, %]
```

Out[3]= {0.21067864733305, 0.03894111930644, 0.00395200266536, 0.00022568018517,

$8.15867295 \times 10^{-6}$, 2.0507044×10^{-7}, 3.80813×10^{-9}, 5.449×10^{-11}, 6.2×10^{-13}}

f) Unfortunately, directly entering the fraction under consideration into Series does not give a result in the form we want.

```
In[1]:= (* turn off messages;
         we know we expand at an essential singularity *)
      Off[Series::esss];
      Series[Gamma[z + α]/Gamma[z + β], {z, Infinity, 3}]
```

$$\text{Out[3]}= \left(z\left(1+\frac{\alpha}{z}\right)\right)^{-\frac{1}{2}+z+\alpha}\left(z\left(1+\frac{\beta}{z}\right)\right)^{\frac{1}{2}-z-\beta}\left(e^{-\alpha+\beta}+e^{-\alpha+\beta}\left(\frac{1}{2}\left(\frac{1}{144}-\frac{\alpha}{6}\right)+\frac{1}{2}\left(-\frac{1}{144}+\frac{\beta}{6}\right)\right)\left(\frac{1}{z}\right)^2+\right.$$
$$e^{-\alpha+\beta}\left(\frac{1}{24}\left(-\frac{1}{144}+\frac{\alpha}{6}\right)+\frac{1}{3}\left(\frac{1}{24}\left(\frac{1}{144}-\frac{\alpha}{6}\right)-\frac{\alpha}{72}+3\left(-\frac{1}{360}+\frac{\alpha^2}{12}\right)\right)\right)+$$
$$\frac{1}{12}\left(-\frac{1}{144}+\frac{1}{2}\left(\frac{1}{144}-\frac{\beta}{6}\right)\right)+\frac{1}{12}\left(\frac{1}{144}+\frac{1}{2}\left(-\frac{1}{144}+\frac{\beta}{6}\right)\right)+\frac{1}{24}\left(\frac{1}{144}-\frac{\beta}{6}\right)+$$
$$\left.\frac{1}{3}\left(\frac{1}{24}\left(-\frac{1}{144}+\frac{\beta}{6}\right)+\frac{\beta}{72}-3\left(-\frac{1}{360}+\frac{\beta^2}{12}\right)\right)\right)\left(\frac{1}{z}\right)^3+O\left[\frac{1}{z}\right]^4\right)$$

The reason for the failure is that the denominator and the numerator have essential singularities at $z = \infty$, and *Mathematica* currently does not cancel the parts outside of SeriesData. We can cancel these parts using PowwerExpand.

```
In[4]:= ser = PowerExpand[%] // Simplify
```

$$\text{Out[4]}= z^{\alpha-\beta}\left(1+\frac{-\alpha+\alpha^2+\beta-\beta^2}{2z}+\right.$$
$$\frac{1}{24}\left(-10\alpha^3+3\alpha^4+\alpha^2\left(9+6\beta-6\beta^2\right)+\alpha\left(-2-6\beta+6\beta^2\right)+\beta\left(2-3\beta-2\beta^2+3\beta^3\right)\right)\left(\frac{1}{z}\right)^2+$$
$$\frac{1}{48}\left(-7\alpha^5+\alpha^6-\alpha\left(-1+\beta\right)^2\beta\left(4+3\beta\right)+\alpha^4\left(17+3\beta-3\beta^2\right)-\left(-1+\beta\right)^2\beta^2\left(2+3\beta+\beta^2\right)+\right.$$
$$\left.\left.\alpha^3\left(-17-10\beta+10\beta^2\right)+\alpha^2\left(6+11\beta-12\beta^2-2\beta^3+3\beta^4\right)\right)\left(\frac{1}{z}\right)^3+O\left[\frac{1}{z}\right]^4\right)$$

Canceling the factor $\exp(\alpha - \beta)$ yields for the resulting series the structure $z^{(\alpha-\beta)}\sum_{k=0}^{n}c_k\,z^{-k}$. Here are the first four c_k.

```
In[5]:= Table[Coefficient[1/z^(α - β) ser, z, -k] z^k, {k, 0, 3}]
```

$$\text{Out[5]}= \left\{1, \frac{-\alpha+\alpha^2+\beta-\beta^2}{2z}, \frac{-10\alpha^3+3\alpha^4+\alpha^2\left(9+6\beta-6\beta^2\right)+\alpha\left(-2-6\beta+6\beta^2\right)+\beta\left(2-3\beta-2\beta^2+3\beta^3\right)}{24z^2}, \right.$$
$$\frac{1}{48z^3}\left(-7\alpha^5+\alpha^6-\alpha\left(-1+\beta\right)^2\beta\left(4+3\beta\right)+\alpha^4\left(17+3\beta-3\beta^2\right)-\right.$$
$$\left.\left.\left(-1+\beta\right)^2\beta^2\left(2+3\beta+\beta^2\right)+\alpha^3\left(-17-10\beta+10\beta^2\right)+\alpha^2\left(6+11\beta-12\beta^2-2\beta^3+3\beta^4\right)\right)\right\}$$

(For an analytical expression for the coefficients, see [1234], [960], [996], and [814].) These are the differences between the original expression and the four partial approximations.

```
In[6]:= diffList = Gamma[z + α]/Gamma[z + β] - z^(α - β) *
                    Rest[FoldList[Plus, 0, %]]
```

$$\text{Out[6]}= \left\{-z^{\alpha-\beta}+\frac{\text{Gamma}[z+\alpha]}{\text{Gamma}[z+\beta]}, -z^{\alpha-\beta}\left(1+\frac{-\alpha+\alpha^2+\beta-\beta^2}{2z}\right)+\frac{\text{Gamma}[z+\alpha]}{\text{Gamma}[z+\beta]}, -z^{\alpha-\beta}\left(1+\frac{-\alpha+\alpha^2+\beta-\beta^2}{2z}+\right.\right.$$
$$\left.\frac{-10\alpha^3+3\alpha^4+\alpha^2\left(9+6\beta-6\beta^2\right)+\alpha\left(-2-6\beta+6\beta^2\right)+\beta\left(2-3\beta-2\beta^2+3\beta^3\right)}{24z^2}\right)+$$
$$\frac{\text{Gamma}[z+\alpha]}{\text{Gamma}[z+\beta]}, -z^{\alpha-\beta}\left(1+\frac{-\alpha+\alpha^2+\beta-\beta^2}{2z}+\right.$$
$$\frac{-10\alpha^3+3\alpha^4+\alpha^2\left(9+6\beta-6\beta^2\right)+\alpha\left(-2-6\beta+6\beta^2\right)+\beta\left(2-3\beta-2\beta^2+3\beta^3\right)}{24z^2}+\frac{1}{48z^3}$$
$$\left(-7\alpha^5+\alpha^6-\alpha\left(-1+\beta\right)^2\beta\left(4+3\beta\right)+\alpha^4\left(17+3\beta-3\beta^2\right)-\left(-1+\beta\right)^2\beta^2\left(2+3\beta+\beta^2\right)+\right.$$
$$\left.\left.\left.\alpha^3\left(-17-10\beta+10\beta^2\right)+\alpha^2\left(6+11\beta-12\beta^2-2\beta^3+3\beta^4\right)\right)\right)+\frac{\text{Gamma}[z+\alpha]}{\text{Gamma}[z+\beta]}\right\}$$

This shows the logarithm of the absolute value of the difference of the approximations to the exact ratio for the case $\alpha = e$, $\beta = \pi$.

```
In[7]:= Block[{α = E, β = Pi},
         Plot[Evaluate[Log[Abs[diffList]]], {z, 10, 100},
             Frame -> True, AxesLabel -> {None, "z"}]]
```

For the asymptotics of the ratio of products of Gamma functions, see [231], [232], [996], and [1062].

g) To avoid calculations with huge rationals or floating point numbers with many digits, we start by calculating a closed-form expression for the partial sums.

```
In[1]:= g[x_, n_] = Sum[x^(2i + 1)/(2i + 1)! (-1)^i, {i, 0, n}]
```

$$Out[1]= \; x \left(\frac{x^2 \, (-x^2)^n \; \text{HypergeometricPFQ}\left[\{1\}, \{2+n, \frac{5}{2}+n\}, -\frac{x^2}{4}\right]}{\text{Gamma}[4+2n]} + \frac{\text{Sin}\left[\sqrt{x^2}\right]}{\sqrt{x^2}} \right)$$

Now, let us look at the difference between $\sin(x)$ and the 1000th partial sum of its Taylor series for some x from 1 to 300.

```
In[2]:= $MaxExtraPrecision = 10000;
```

```
In[3]:= {#, Log[10, Abs[N[g[#, 1000] - Sin[#], 22]]]}& /@
                  (* various n *) {1, 10, 200, 200, 300}
```

```
Out[3]= {{1, -5745.4250427682328090955450},
          {10, -3742.4250534686811620945926}, {200, -1136.4662635182518561466728},
          {200, -1136.4662635182518561466728}, {300, -783.7607903272220158145400}}
```

This means the value x^* is in the range $200 < x^* < 300$. We use `FindRoot` to find its precise value.

```
In[4]:= δ[x_ ?NumberQ] := Function[z, Log[10,
                 Abs[N[g[z, 1000] - Sin[z], 22]]]][SetPrecision[x, Infinity]]
```

```
In[5]:= FindRoot[δ[x] == -1000, {x, 200, 300}]
```

```
Out[5]= {x → 233.971}
```

For asymptotics about the zeros of g[x, n] as $n \to \infty$, see [670], [1080].

h) These are the expressions for the area and the volume of a *dim*-dimensional unit sphere.

```
In[1]:= area[ dim_] := dim Pi^(dim/2)/Gamma[1 + dim/2];
         volume[dim_] :=     Pi^(dim/2)/Gamma[1 + dim/2];
```

The closed form of the volume agrees with the following obvious recursive calculation of the volume.

```
In[3]:= Table[volume[dim], {dim, 1, 12}] -
           (* integrate next dimension over circular cross section *)
           NestList[Integrate[# /. r -> Sqrt[r^2 - z^2], {z, -r, r},
                          Assumptions -> r > 0]&, 2r, 11] /. r -> 1 // Union
```

```
Out[3]= {0}
```

A quick plot shows that both the volume and the area are maximal around dim $\approx$ 6.

```
In[4]:= Show[GraphicsArray[
           Plot[#[dim], {dim, 0, 15}, Frame -> True, Axes -> False,
               PlotLabel -> ToString[#], PlotRange -> All,
               DisplayFunction -> Identity]& /@
               (* display are and volume *) {area, volume}]]
```

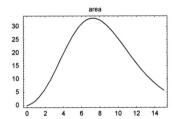

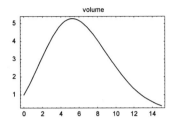

Here is a precise numerical value at which dimensions the maximum area occurs.

```
In[5]:= FindRoot[Evaluate[D[area[dim], dim] == 0], {dim, 6},
                WorkingPrecision -> 40, AccuracyGoal -> Infinity]

Out[5]= {dim → 7.25694640486057678013283838869076 9236619}
```

If we restrict the problem to positive integer dimensions, the 7-sphere has maximal area.

```
In[6]:= area[6] < area[7] > area[8]

Out[6]= True
```

And here is a precise numerical value of the dimension at which the maximum volume occurs.

```
In[7]:= FindRoot[Evaluate[D[volume[dim], dim] == 0], {dim, 6},
                WorkingPrecision -> 40, AccuracyGoal -> Infinity]

Out[7]= {dim → 5.25694640486057678013283838869076 9236619}
```

If we restrict the problem to positive integer dimensions, the 5-sphere has maximal volume.

```
In[8]:= volume[4] < volume[5] > volume[6]

Out[8]= True
```

This critical dimension depends on the radius of the hyperspheres. For non-unit spheres the critical dimension increases quickly with the radius.

```
In[9]:= (* condition for extrema; ρ is the hypersphere radius *)
        radiusAV[dim_, ρ_] = D[ρ^dim {area[dim], 1/ρ volume[dim]}, dim];

        Show[GraphicsArray[
         ContourPlot[#, {ρ, 1/10, 2}, {dim, 1/2, 20},
                     Contours -> {0}, ContourShading -> False,
                     DisplayFunction -> Identity,
                     PlotPoints -> 40]& /@ radiusAV[dim, ρ]]]
```

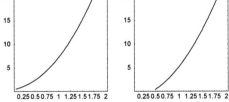

If we allow negative dimensions, we can find more maxima.

```
In[12]:= Show[GraphicsArray[
          Plot[#[dim], {dim, -20, 0}, Frame -> True, Axes -> True,
               PlotLabel -> ToString[#], PlotRange -> All,
               DisplayFunction -> Identity]& /@ {area, volume}]]
```

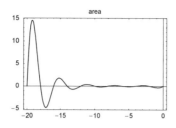

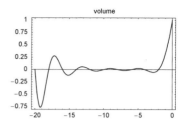

In the complex dimension plane, the area and the volume function look similar.

```
In[13]:= Show[GraphicsArray[
    ContourPlot[Abs[Evaluate[#[redim + I imdim]]],
            {redim, -10, 10}, {imdim, -3, 3}, Contours -> 50,
            PlotPoints -> 120, ContourLines -> False,
            ColorFunction -> (Hue[0.8 #]&), PlotLabel -> ToString[#],
            DisplayFunction -> Identity]& /@
            (* display are and volume *) {area, volume}]]
```

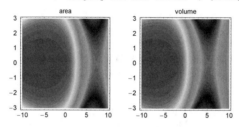

The sums of the volumes and areas of all even- and odd-dimensional unit spheres have the following values [56].

```
In[14]:= Sum[volume[dim], {dim, #, Infinity, 2}]& /@ {0, 1}
```

$$Out[14]= \{e^\pi, \ e^\pi \operatorname{Erf}[\sqrt{\pi}\,]\}$$

```
In[15]:= Sum[area[dim], {dim, #, Infinity, 2}]& /@ {0, 1}
```

$$Out[15]= \{2\,e^\pi\,\pi, \ 2\left(1 + e^\pi\,\pi\operatorname{Erf}[\sqrt{\pi}\,]\right)\}$$

A related interesting value that contains e and π follows by considering the limit of the ratio of the volumes of spheres with a dimension-dependent radius [1129].

```
In[16]:= (* sphere with dimension-dependent radius Sqrt[dim]/2 *)
        dimVolume[dim_] := (Sqrt[dim]/2)^dim volume[dim]

        Limit[dimVolume[dim + 1]/dimVolume[dim], dim -> Infinity]
```

$$Out[18]= \sqrt{\frac{e\,\pi}{2}}$$

For volumes of other compact manifolds, see [195].

Here is the probability density for the distance ρ of two points chosen at random in a d-dimensional unit sphere.

```
In[19]:= p[d_, ρ_] := 2 d/Beta[d/2 + 1/2, 1/2]*
            (Hypergeometric2F1[1/2, 1/2 - d/2, 3/2, 1] -
            ρ/2 Hypergeometric2F1[1/2, 1/2 - d/2, 3/2, ρ^2/4])ρ^(d - 1)
```

Integrating $p_d(\rho)$ yields the probability distribution.

```
In[20]:= (P[d_, ρ_] = Integrate[p[d, r], {r, 0, ρ}, Assumptions -> 0 < ρ < 2 && d > 0] //
                                    FullSimplify) // TraditionalForm
```

Out[20]//TraditionalForm=

$$\rho^d + \frac{2^d\,\mathrm{B}_{\frac{\rho^2}{4}}\left(\frac{d+1}{2}, \frac{d+1}{2}\right) - \rho^d\,\mathrm{B}_{\frac{\rho^2}{4}}\left(\frac{1}{2}, \frac{d+1}{2}\right)}{\mathrm{B}\left(\frac{d+1}{2}, \frac{1}{2}\right)}$$

The following graphic shows the distributions for $1 \le d \le 30$. The maximum of $p_d(\rho)$ seems to shift right towards a limit of $\rho \approx 1.4$ and to become narrower at the same time.

```
In[21]:= Show[GraphicsArray[
          Plot[Evaluate[Table[#[d, ρ], {d, 30}]], {ρ, 0, 2},
              PlotRange -> All, PlotStyle -> Table[Hue[0.8 d/30], {d, 30}],
              DisplayFunction -> Identity]& /@ {p, P}]]
```

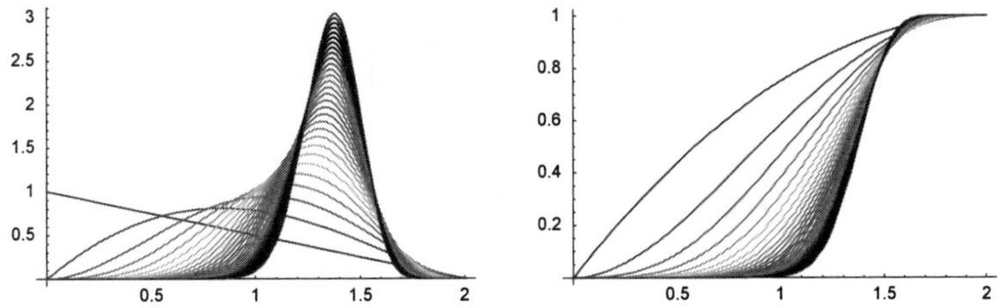

A natural conjecture (see below) is that the maximum occurs at $\rho = 2^{1/2}$. The value $p_{1000}(2^{1/2}) \approx 17.81$ suggests (keeping the normalization of $p_d(\rho)$ in mind) that $\lim_{d \to \infty} p_d(\rho) = \delta(\rho - \sqrt{2})$.

```
In[22]:= p[1000, N[Sqrt[2], 200]] // N

Out[22]= 17.8102
```

The average distance is given by the following expression.

```
In[23]:= ρAv[d_] = Integrate[ρ p[d, ρ], {ρ, 0, 2},
                            Assumptions -> d > 0] // FullSimplify
```

$$Out[23]= \frac{2^d \, d^2 \, \text{Gamma}[1 + \frac{d}{2}] \, \text{Gamma}[\frac{d}{2}]}{(1 + d) \, \sqrt{\pi} \, \text{Gamma}[\frac{3}{2} + d]}$$

In the limit $d \to \infty$ the average distance becomes $2^{1/2}$.

```
In[24]:= Limit[ρAv[d] /. Gamma[ξ_] :> Sqrt[2Pi] Sqrt[1/ξ] Exp[-ξ] ξ^ξ,
             d -> Infinity]

Out[24]= √2
```

The maximum of $p_d(\rho)$ is also at $\rho_{max} = 2^{1/2}$ in the limit $d \to \infty$. For $d = 100$, the difference is only 1% and for $d = 1000$ the difference is only 10^{-16}.

```
In[25]:= pD[d_, ρ_] = D[p[d, ρ], ρ] // FullSimplify;

        pDN[d_, ρ_?NumericQ] := pD[SetPrecision[d, 50], SetPrecision[ρ, 50]]

        zeroD[d_, ρ0_] := FindRoot[pDN[d, ρ] == 0, {ρ, ρ0, ρ0 (1 - 10^-3)},
                                    (* use high-precision *) WorkingPrecision -> 30]

In[28]:= Sqrt[2] - ρ /. zeroD[100, Sqrt[2]]

Out[28]= 0.0105636870205904930720463550507
```

The result $\overline{\rho_\infty} = \sqrt{2}$ is easily understood: It is the distance between two unit vectors in l_2. In the limit $d \to \infty$, we get more and more dimensions and probabilistically two random vectors will be orthogonal. In addition, because of the radial Jacobian ρ^{d-1} for large d, most of the volume of a d-dimensional unit sphere is concentrated in a thin shell near the surface. This means most randomly chosen vectors will have norm 1, and this, in turn, means $\lim_{d \to \infty} \overline{\rho} = \sqrt{2}$.

For the average distance of to the nth neighbor of uniformly distributed points, see [273], [156]. For the average size of the smallest component of a vector on a dD sphere, see [1063]. For further properties of higher dimensional spheres, see [1070]. For the grazing goat problem in dD, see [473].

i) Without loss of generality, we set $\mathbf{y} = \mathbf{0}$ and denote $|\mathbf{x}|$ by x. The convolution integral can be written in a compact way using a beta function.

```
In[1]:= FullSimplify[#, r > 0]& @
         (UnitStep[2r - x] 1/Gamma[(n + 1)/2] (Pi/4)^((n - 1)/2) *
```

```
Integrate[(4r^2 - y^2)^((n - 1)/2), {y, x, 2r},
        Assumptions -> n > 0 && x > 0 && 2r > x])
```

Out[1]= $\pi^{n/2} \, r^n \left(\dfrac{1}{\text{Gamma}\left[1 + \frac{n}{2}\right]} - \dfrac{x \, \text{Beta}\left[\frac{x^2}{4\,r^2}, \frac{1}{2}, \frac{1+n}{2}\right]}{\sqrt{\pi} \, \sqrt{x^2} \, \text{Gamma}\left[\frac{1+n}{2}\right]} \right) \text{UnitStep}[2\,r - x]$

In[2]:= (* use hypergeometric form versus Beta function for autoevaluation *)
```
χConvolution[n_, r_, x_] = Pi^((n - 1)/2) r^(n - 1)*
    (-x Gamma[1 + n/2] Hypergeometric2F1[1/2, 1/2 - n/2, 3/2,
            x^2/(4 r^2)]  + Sqrt[Pi] r Gamma[(1 + n)/2] Sign[r]*
        UnitStep[2 r - x])/(Gamma[1 + n/2] Gamma[(1 + n)/2])
```

Out[3]= $\left(\pi^{\frac{1}{2}(-1+n)} \, r^{-1+n} \left(-x \, \text{Gamma}\left[1 + \frac{n}{2}\right] \text{Hypergeometric2F1}\left[\frac{1}{2}, \frac{1}{2} - \frac{n}{2}, \frac{3}{2}, \frac{x^2}{4\,r^2}\right] + \right. \right.$
$\left. \left. \sqrt{\pi} \, r \, \text{Gamma}\left[\frac{1+n}{2}\right] \text{Sign}[r] \, \text{UnitStep}[2\,r - x] \right) \right) \Big/ \left(\text{Gamma}\left[1 + \frac{n}{2}\right] \text{Gamma}\left[\frac{1+n}{2}\right] \right)$

This is the general formula for calculating $g_d(r)$.

```
In[4]:= g[n_, V_, r_, intAssumptions___] :=
    2(-1)^(n + 1)/(Gamma[(n - 1)/2] (Pi r^2)^((n - 1)/2))*
    Integrate[(D[V, {r, n + 1}] /. r -> ρ) ρ (ρ^2 - r^2)^((n - 3)/2),
            {ρ, r, Infinity}, Assumptions -> And[r > 0, intAssumptions]]
```

For a Coulomb potential $V \sim |x|^{-1}$, we have $g_d(r) \sim r^{-d-2}$.

```
In[5]:= Table[g[d, 1/r, r], {d, 2, 5}] // Simplify[#, r > 0]&
```

Out[5]= $\left\{ \dfrac{3}{r^4}, \dfrac{16}{\pi\,r^5}, \dfrac{30}{\pi\,r^6}, \dfrac{192}{\pi^2\,r^7} \right\}$

Here are the functions $g_3(r)$ for a Yukawa, another exponential and an oscillating potential. (For the case $\sin(r)/r$ we add a convergence-achieving factor.)

```
In[6]:= (* for Sin[r] 1/r potential, use non-Riemann integral
        Integrate[Sin[ρ], {ρ, r, Infinity},
                GenerateConditions -> False] ==> Cos[r]
        or
        Series[g[3, Exp[-ε r] Sin[r] 1/r, r] /. If[_, a_, _] :> a,
            {ε, 0, 0}]  // Normal *)
    {g[3, Exp[-r] 1/r, r], g[3, Exp[-r^2] 1/r, r],
    Normal[Series[g[3, Exp[-ε r] Sin[r] 1/r, r, ε > 0], {ε, 0, 0}]]} //
                                FullSimplify[#, r > 0]&
```

Out[7]= $\left\{ \dfrac{2\,e^{-r}\,(2 + r)\,(4 + r\,(2 + r))}{\pi\,r^5}, \dfrac{8\,e^{-r^2}\,(2 + 2\,r^2 + r^4 + 2\,r^6)}{\pi\,r^5}, \right.$
$\left. \dfrac{2\,(r\,(-8 + r^2)\,\text{Cos}[r] - 4\,(-2 + r^2)\,\text{Sin}[r])}{\pi\,r^5} \right\}$

Using the original decomposition, we recover the first two potentials immediately. The third integral

```
In[8]:= Integrate[χConvolution[3, r/2, x] #, {r, x, Infinity},
            Assumptions -> x > 0]& /@ Take[%, 3] // FullSimplify
```

Out[8]= $\left\{ \dfrac{e^{-x}}{x}, \dfrac{e^{-x^2}}{x}, \dfrac{1}{360\,x} \right.$
$\left(-20\,(30\,x + (19 - 18\,\text{EulerGamma})\,x^3 + 18\,x\,\text{Cos}[x] - 66\,\text{Sin}[x]) + x^2 \left(-360\,x\,\text{CosIntegral}[x] + \right. \right.$
$x^3 \left(-35\,\text{HypergeometricPFQ}\left[\{1, 1\}, \{2, \frac{5}{2}, 3\}, -\frac{x^2}{4}\right] + 4\,\text{HypergeometricPFQ}\left[\{1, 1\}, \right. \right.$
$\left. \{2, 3, \frac{7}{2}\}, -\frac{x^2}{4}\right] + 10\,\text{HypergeometricPFQ}\left[\{1, 1\}, \{\frac{5}{2}, 3, 3\}, -\frac{x^2}{4}\right] -$
$\left. \left. \left. 2\,\text{HypergeometricPFQ}\left[\{1, 1\}, \{3, 3, \frac{7}{2}\}, -\frac{x^2}{4}\right] \right) + 360\,x\,\text{Log}[x] + 360\,\text{Sin}[x] \right) \right) \right\}$

The last complicated result can be simplified to $\sin(x)/x$ which recovers the above result.

```
In[9]:= Series[%[[3]] - Sin[x]/x, {x, 0, 20}]
```

Out[9]= $O[x]^{21}$

To derive integration-free formulas for $g_d(r)$, we have to carry out partial integrations. The rule set `partialIntegrate`
`Rules` implements these partial integrations. We assume that the potential and its derivatives vanish at infinity, so we only
obtain the contributions from $\rho = r$.

```
In[10]:= partialIntegrateRules =
    {Integrate[f_ Derivative[n_][V][ρ], {ρ, r, Infinity}, ___] :>
        (* do partial integration *)
        -Limit[f, ρ -> r] Derivative[n - 1][V][r] -
        Integrate[Together[D[f, ρ]] Derivative[n - 1][V][ρ], {ρ, r, Infinity}],
        (* integrate complete differential; use value at ∞ *)
        Integrate[Derivative[n_][V][ρ], {ρ, r, Infinity}, ___] :>
            -Derivative[n - 1][V][r],
        (* fast vanishing potential *) Derivative[_][V][Infinity] :> 0};
```

Here are the resulting integration-free formulas for $g_d(r)$.

```
In[11]:= Table[Factor[g[d, V[r], r] //. partialIntegrateRules], {d, 3, 11, 2}]
```

$$Out[11]= \left\{ -\frac{2\,(-V''[r] + r\,V^{(3)}[r])}{\pi\,r^2}, \quad \frac{4\,(3\,V''[r] - 3\,r\,V^{(3)}[r] + r^2\,V^{(4)}[r])}{\pi^2\,r^4}, \right.$$

$$-\frac{8\,(-15\,V''[r] + 15\,r\,V^{(3)}[r] - 6\,r^2\,V^{(4)}[r] + r^3\,V^{(5)}[r])}{\pi^3\,r^6},$$

$$\frac{16\,(105\,V''[r] - 105\,r\,V^{(3)}[r] + 45\,r^2\,V^{(4)}[r] - 10\,r^3\,V^{(5)}[r] + r^4\,V^{(6)}[r])}{\pi^4\,r^8},$$

$$\left. -\frac{32\,(-945\,V''[r] + 945\,r\,V^{(3)}[r] - 420\,r^2\,V^{(4)}[r] + 105\,r^3\,V^{(5)}[r] - 15\,r^4\,V^{(6)}[r] + r^5\,V^{(7)}[r])}{\pi^5\,r^{10}} \right\}$$

j) A direct calculation of $J_\nu(\nu)$ becomes time-consuming for larger ν.

```
In[1]:= Table[Timing[N[BesselJ[10^k, 10^k]]], {k, 4}]
```

$$Out[1]= \{\{1.66696 \times 10^{-18}\ \text{Second}, 0.207486\}, \{1.66696 \times 10^{-18}\ \text{Second}, 0.0963667\},$$
$$\{0.02\ \text{Second}, 0.0447307\}, \{3.8\ \text{Second}, 0.0207622\}\}$$

It is straightforward to implement the calculation of the $B_m(\zeta)$ from the Meissel expansion. We will pregenerate 30 terms of
the series expansion.

```
In[2]:= ser = Series[E^(z ε w)/((Sinh[w] - w)/w^3)^((m + 1)/3), {w, 0, 35}];
```

```
In[3]:= coeffs = CoefficientList[ser, w] 1/6^((1/3) (m + 1));
```

```
In[4]:= Bterms = MapIndexed[(#1 /. m -> #2[[1]] - 1)&, coeffs] // Expand;
```

Here are the first five Bs.

```
In[5]:= Take[Bterms, 5]
```

$$Out[5]= \left\{1, z\,\varepsilon, -\frac{1}{20} + \frac{z^2\,\varepsilon^2}{2}, -\frac{z\,\varepsilon}{15} + \frac{z^3\,\varepsilon^3}{6}, \frac{1}{280} - \frac{z^2\,\varepsilon^2}{24} + \frac{z^4\,\varepsilon^4}{24}\right\}$$

Making an assignment to the symbol B, we can implement the terms in the asymptotic expansion of $J_\nu(z)$ for $z \approx \nu$.

```
In[6]:= MapIndexed[(B[#2[[1]] - 1, z_] = #1)&, Bterms /. ε -> 1];
```

```
In[7]:= besselJTerms[v_, z_, o_] :=
    Table[1/(3 Pi) B[m, z - v] Sin[(m + 1) Pi/3] *
        Gamma[(m + 1)/3]/(z/6)^((m + 1)/3), {m, 0, o}];
```

The terms of the series are rapidly decreasing. Here, we check the case $\nu = 1000$ against the built-in function $J_\nu(z)$.

```
In[8]:= (J1000 = besselJTerms[N[10^3, 50], N[10^3, 50], 30]) // N
```

$$Out[8]= \{0.0447307, 0., 0., 0., -5.86929 \times 10^{-8}, 0., -1.98803 \times 10^{-10}, 0., 0., 0., 4.868 \times 10^{-15},$$
$$0., 3.10313 \times 10^{-17}, 0., 0., 0., -1.80072 \times 10^{-21}, 0., -1.58442 \times 10^{-23}, 0., 0., 0.,$$
$$1.54769 \times 10^{-27}, 0., 1.69169 \times 10^{-29}, 0., 0., 0., -2.40156 \times 10^{-33}, 0., -3.09153 \times 10^{-35}\}$$

```
In[9]:= Fold[Plus, 0, J1000] - BesselJ[1000, 1000]
```

$$Out[9]= -5.9618269147 \times 10^{-39}$$

The precision reached is in the order of the size of the last term of the series. Now, it is straightforward to calculate
$J_{10^{100}}(10^{100})$. The 30-th term is in the order 10^{-1037}, and the first term is about 10^{-34}. This means that we have enough terms
to get 1000 correct digits. Carrying out now all calculations with enough precision yields the desired result.

```
In[10]:= (JExp100 = besselJTerms[N[10^100, 1100], N[10^100, 1100], 35]) // N
```

$\text{Out[10]}= \{2.07622 \times 10^{-34}, 0., 0., 0., -1.2645 \times 10^{-169}, 0.,$
$\quad -9.22763 \times 10^{-237}, 0., 0., 0., 1.048778354945894 \times 10^{-370}, 0.,$
$\quad 1.440345284614099 \times 10^{-437}, 0., 0., 0., -3.879539557040791 \times 10^{-571}, 0.,$
$\quad -7.354207942796650 \times 10^{-638}, 0., 0., 0., 3.334398611424452 \times 10^{-771}, 0.,$
$\quad 7.852130949281306 \times 10^{-838}, 0., 0., 0., -5.173995989896592 \times 10^{-971}, 0.,$
$\quad -1.434959279500102 \times 10^{-1037}, 0., 0., 0., 1.265859087127370 \times 10^{-1170}, 0.\}$

```
In[11]:= {(res = Fold[Plus, 0, JExp100] // N[#, 1000]&) // N, Precision[res]}
```

$\text{Out[11]}= \{2.07622 \times 10^{-34}, 1000.\}$

For another expansion for large parameter and argument values, see [997].

k) Here is a straightforward implementation of the Rayleigh sums.

```
In[1]:= σ[n_, v_] := σ[n, v] = Together[1/(n + v) Sum[σ[k, v]σ[n - k, v], {k, n - 1}]]
        σ[1, v_] = 1/(4(v + 1));
```

A symbolic calculation of the Rayleigh sums is relatively time- and memory-consuming. For large n, the Rayleigh sums become large rational functions. The following code calculates $z_0^{(1)}$ to 20 digits using a symbolic v.

```
In[3]:= (n = 1;
        While[(σ[n, v]/σ[n + 1, v] - σ[n, v]^(-1/n) /. v -> 1) > 10^-20,
              n = n + 1];
        n) // Timing
```

$\text{Out[3]}= \{4.35 \text{ Second}, 41\}$

For a fixed v, the Rayleigh sums become rational numbers. For $n \geq 191$, we get $z_0^{(1)}$ to 100 correct digits.

```
In[4]:= $MaxExtraPrecision = 200;
        With[{v = 1},
        n = 1; (* until we a have precise result *)
        While[(Sqrt[σ[n, v]/σ[n + 1, v]] - σ[n, v]^(-1/n/2) /. v -> 1) > 10^-100,
              n = n + 1];
        {n, N[Sqrt[σ[n, v]/σ[n + 1, v]], 100]}] // Timing
```

$\text{Out[5]}= \{5.9 \text{ Second}, \{191,$
$\quad 3.8317059702075123156144358863081607665645452742878019287622989899188393095190114 7\backslash$
$\quad 0214112874757423127\}\}$

Using high-precision floating point numbers instead of rational numbers speeds up the calculation.

```
In[6]:= $MaxExtraPrecision = 200;
        With[{(* start with approximative number *) v = N[1, 200]},
        n = 1;
        (* iterate as long as needed *)
        While[(Sqrt[σ[n, v]/σ[n + 1, v]] - σ[n, v]^(-1/n/2) /. v -> 1) > 10^-100,
              n = n + 1];
        (* return result *)
        {n, N[Sqrt[σ[n, v]/σ[n + 1, v]], 100]}] // Timing
```

$\text{Out[7]}= \{0.2 \text{ Second}, \{191,$
$\quad 3.8317059702075123156144358863081607665645452742878019287622989899188393095190114 7\backslash$
$\quad 0214112874757423127\}\}$

A further speedup can be achieved by using ListConvolve instead of explicitly summing the terms in the Rayleigh sums.

```
In[8]:= σ1[n_, v_] := σ1[n, v] = 1/(n + v)*
                                 ListConvolve[#, #]&[Table[σ1[k, v], {k, n - 1}]][[1]]
        σ1[1, v_] = 1/(4(v + 1));

In[10]:= $MaxExtraPrecision = 200;
        With[{v = N[1, 200]},
        n = 1; (* iterate as long as needed *)
        While[(Sqrt[σ1[n, v]/σ1[n + 1, v]] -
               σ1[n, v]^(-1/n/2) /. v -> 1) > 10^-100,
              n = n + 1];
        (* return result *)
        {n, N[Sqrt[σ1[n, v]/σ1[n + 1, v]], 100]}] // Timing
```

Out[11]= {0.16 Second, {191,
    3.8317059702075123156144358863081607665645452742878019287622989899188393095190114T
    0214112874757423127}}

A numerical root finding and a plot of $J_1(z)$ show that we really have the first root to 100 digits.

In[12]:= (z /. FindRoot[BesselJ[1, z] == 0, {z, 3},
                    WorkingPrecision -> 200]) - %[[2, 2]]

Out[12]= 0. × 10⁻¹⁰⁰

For coupled equations for the zeros of Bessel functions, see [1091], [919].

l) A graph of $K(1 - x)/K(x)$ shows that the x will be near to 0.

In[1]:= Plot[EllipticK[1 - x]/EllipticK[x], {x, 0, 1},
            Frame -> True, Axes -> False]

To find an exact expression for x, we start by numerically computing a high-precision value for x.

In[2]:= g[x_?NumberQ, prec_] :=
    With[{ξ = SetPrecision[x, prec]}, EllipticK[1 - ξ]/EllipticK[ξ]]

In[3]:= ε = 10^-10;
    x = ξ /. FindRoot[g[ξ, 200] == Sqrt[10],
                    {ξ, ε, 1 - ε}, AccuracyGoal -> 100,
                    WorkingPrecision -> 200, MaxIterations -> 200] //
                    Re // SetPrecision[#, 100]&

Out[4]= 0.0007752017293532922488376453553255501692319869434272712607280923833754612812020806T
    740099251537389029166

Using the package NumberTheory`Recognize`, we can get the polynomial of which x is a root.

In[5]:= Needs["NumberTheory`Recognize`"]

In[6]:= poly = Recognize[x, 20, t] // Factor

Out[6]= (1 + t) (1 - 1292 t + 2598 t² - 1292 t³ + t⁴)

This is the algebraic form of x.

In[7]:= xAlgebraic = t /. Select[{ToRules[Roots[poly == 0, t]]},
                    N[(t /. #), 22] - N[x] == 0&][[1]] // FullSimplify

Out[7]= $323 - 228\sqrt{2} + 144\sqrt{5} - 102\sqrt{10}$

To make sure the result is correct, we check the identity to 1000 digits.

In[8]:= g[N[xAlgebraic, 1000], 1000] - Sqrt[10]

Out[8]= 0. × 10⁻¹⁰⁰⁰

m) These are the 24 functions under consideration.

In[1]:= f = {(* trigonometric functions *)
        Sin, Cos, Tan, Cot, Sec, Csc,
        (* hyperbolic functions *)
        Sinh, Cosh, Tanh, Coth, Sech, Csch,
        (* inverse trigonometric functions *)
        ArcSin, ArcCos, ArcTan, ArcCot, ArcSec, ArcCsc,
        (* inverse hyperbolic functions *)
        ArcSinh, ArcCosh, ArcTanh, ArcCoth, ArcSech, ArcCsch};

```
lf = Length[f];
```

We define the integrands and the integrals by referring to their position in the list f.

```
In[4]:= integrand[{i_, j_, k_}, x_] := f[[i]][f[[j]][f[[k]][x]]];

        tripleIntegral[{i_, j_, k_}, x_] :=
                    Integrate[integrand[{i, j, k}, x], x];
```

When *Mathematica* was able to carry out the integration, the function `containedSpecialFunction` returns a list of the special functions present in the result.

```
In[6]:= containedSpecialFunction[int_] :=
        DeleteCases[Complement[Level[int, {-1}, Heads -> True],
                        Level[int, {-1}, Heads -> False]],
                (* take out elementary functions *)
                List | Plus | Times | Power | Log | Sequence @@ f]
```

Now, we carry out all 13824 calls to `Integrate`. This calculation takes about $1\frac{1}{2}$ hours on a 2 GHz computer. If the integration succeeds, we store the result in the list bag.

```
In[7]:= (* results *)
        bag = Table[Null, {i, lf}, {j, lf}, {k, lf}];

        Off[Power::infy]; Off[Sum::div]; Off[Infinity::indet];
        Do[(* to see some progress while waiting
            CellPrint[Cell[TextData[{"∘ Now integrating integral number ",
                    ToString[(i - 1)^2 lf + (j - 1) lf + k], ": ",
                    Cell[BoxData[FormBox[MakeBoxes[#, TraditionalForm]&[
                    integrand[{i, j, k}, x]], TraditionalForm]]]}], "Text",
                                        CellTags -> "currentIntegral"]];
            NotebookLocate["currentIntegral"];
            NotebookDelete[EvaluationNotebook[]]; *)
            (* try to do the integration *)
            int = tripleIntegral[{i, j, k}, x];
            (* was the integral done? *)
            If[FreeQ[int, _Integrate], bag[[i, j, k]] = int],
            {i, lf}, {j, lf}, {k, lf}]; // Timing
Out[10]= {37873.7 Second, Null}
```

Mathematica was able to find integrals for about 22% of all integrands.

```
In[11]:= Count[bag, _?(# =!= Null&), {3}]
Out[11]= 3075
```

About 460—this means about 3%—contained special functions.

```
In[12]:= analyzedBag = Map[If[# === Null, {},
                        containedSpecialFunction[#]]&, bag, {3}];
In[13]:= Length[Flatten[analyzedBag]]
Out[13]= 460
```

These are the special functions needed in carrying out the integrations.

```
In[14]:= Union[Flatten[analyzedBag]]
Out[14]= {CoshIntegral, CosIntegral, EllipticE, EllipticF,
         EllipticPi, Hypergeometric2F1, SinhIntegral, SinIntegral}
```

The Gauss hypergeometric function $_2F_1(a, b; c; z)$ is by far the most frequently appearing function in the results.

```
In[15]:= {#, Count[analyzedBag, #, Infinity, Heads -> True]}& /@
                        Union[Flatten[analyzedBag]]
Out[15]= {{CoshIntegral, 24}, {CosIntegral, 24}, {EllipticE, 70}, {EllipticF, 115},
         {EllipticPi, 11}, {Hypergeometric2F1, 168}, {SinhIntegral, 24}, {SinIntegral, 24}}
```

Because the list f contains functions and inverse functions in a regular position, plotting the done integrals in a f_1, f_2, f_3-space shows some pattern.

In[16]:= Show[Graphics3D[MapIndexed[If[#1 =!= Null,
            Cuboid[#2 - {1, 1, 1}/3, #2 + {1, 1, 1}/3], {}]&, bag, {3}]]]

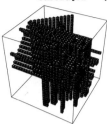

n) The function `productLog` implements the numerical calculation according to the given integral. To avoid numericalization problems of the integrand, we numericalize the integrand explicitly.

```
In[1]:= productLog[k_, z_, wp_:$MachinePrecision, opts___] :=
        1 + (Log[z] - 1 + 2 k Pi I) Exp[1/(2 Pi I) *
        NIntegrate[(* numericalize integrand *) Evaluate[N[
        Log[((2k - 1) I Pi + t + Log[z] - Log[t])/
            ((2k + 1) I Pi + t + Log[z] - Log[t])]/(t + 1), wp]],
            {t, 0, Infinity}, WorkingPrecision -> wp, opts]]
```

The values calculated with `productLog` agree with the ones of the built-in `ProductLog` for integer k.

```
In[2]:= Table[ProductLog[k, 1 + I] - productLog[k, 1 + I],
            {k, -3, 3}] // Abs // Max
Out[2]= 0. × 10⁻⁷
```

`productLog` is a sensible analytic continuation of `ProductLog` for complex k. (The continuation is not unique because we do not have a finite accumulation point of $W_k(z)$ with respect to k.) The next graphic show the real and imaginary part of along $W_{1+i}(z)$ along the lines $z = t$ and $z = 2 + t$.

```
In[3]:= data1 = Table[{k , productLog[k, 1 + I]}, {k, -10, 10, 1/3}] // N;
        data2 = Table[{ki, productLog[2 + I ki, 1 + I]}, {ki, -10, 10, 1/3}] // N;
```

```
In[5]:= Function[data, Show[GraphicsArray[
        Function[reim, ListPlot[{#[[1]], reim[#[[2]]]}& /@ data, Frame -> True,
                            Axes -> False, DisplayFunction -> Identity,
                            PlotStyle -> PointSize[0.01]]] /@ {Re, Im}]]] /@
                                                    {data1, data2}
```

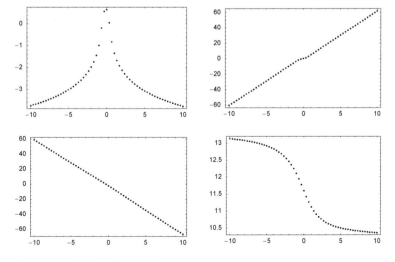

So we can use `productLog` to calculate $W_{1+i}(1 + i)$.

```
In[6]:= productLog[1 + I, 1 + I, 30, PrecisionGoal -> 20]
```
```
Out[6]= -8.1651615019266634444 + 4.4234781811167593075 i
```

o) This is the (unnormalized) Gumbel distribution.

```
In[1]:= p[x_] = c Exp[b (x - ξ) - Exp[b (x - ξ)]]^a;
```

It is straightforward to calculate the zeroth, first, and second moments. (To make the integration easier, we carry out a shift of the integration variable manually.)

```
In[2]:= m0 = Integrate[p[x + ξ], {x, -Infinity, Infinity},
              Assumptions -> b < 0 && a > 0]
```
$$Out[2]= -\frac{a^{-a} \, c \, \text{Gamma}[a]}{b}$$

```
In[3]:= m1 = Integrate[(x + ξ) p[x + ξ], {x, -Infinity, Infinity},
              Assumptions -> b < 0 && a > 0]
```
$$Out[3]= -\frac{a^{-a} \, c \, \text{Gamma}[a] \, (b \, ξ - \text{Log}[a] + \text{PolyGamma}[0, a])}{b^2}$$

```
In[4]:= m2 = Integrate[(x + ξ)^2 p[x + ξ], {x, -Infinity, Infinity},
              Assumptions -> b < 0 && a > 0]
```
$$Out[4]= \frac{a^{-a} \, c \, \text{Gamma}[a] \, (-(b \, ξ - \text{Log}[a] + \text{PolyGamma}[0, a])^2 - \text{PolyGamma}[1, a])}{b^3}$$

```
In[5]:= eqs = ({m0, m1, m2} /. If[cond_, res_, _] :> res) // FullSimplify
```
$$Out[5]= \left\{-\frac{a^{-a} \, c \, \text{Gamma}[a]}{b}, \, -\frac{a^{-a} \, c \, \text{Gamma}[a] \, (b \, ξ - \text{Log}[a] + \text{PolyGamma}[0, a])}{b^2},\right.$$
$$\left.\frac{a^{-a} \, c \, \text{Gamma}[a] \, (-(b \, ξ - \text{Log}[a] + \text{PolyGamma}[0, a])^2 - \text{PolyGamma}[1, a])}{b^3}\right\}$$

The conditions on the first moments to be 0, 1, and 0 result in three equations for the three variables c, b, and $ξ$.

```
In[6]:= (sol = Solve[Thread[eqs == {1, 0, 1}], {b, ξ, c}]) // TraditionalForm
```
$$Out[6]//TraditionalForm= \left\{\left\{ξ \to \frac{\log(a) - ψ^{(0)}(a)}{\sqrt{ψ^{(1)}(a)}}, c \to -\frac{a^a \sqrt{ψ^{(1)}(a)}}{\Gamma(a)}, b \to \sqrt{ψ^{(1)}(a)}\right\}, \left\{ξ \to \frac{ψ^{(0)}(a) - \log(a)}{\sqrt{ψ^{(1)}(a)}}, c \to \frac{a^a \sqrt{ψ^{(1)}(a)}}{\Gamma(a)}, b \to -\sqrt{ψ^{(1)}(a)}\right\}\right\}$$

The resulting distribution contains Gamma and Polygamma functions of a.

```
In[7]:= (p[a_, x_] = p[x] /. sol[[1]] // FullSimplify) // TraditionalForm
```

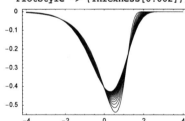

$$Out[7]//TraditionalForm= -\frac{a^a \left(e^{\sqrt{ψ^{(1)}(a)} \, x - \frac{e^{\sqrt{ψ^{(1)}(a)} \, x + ψ^{(0)}(a)}}{a} + ψ^{(0)}(a)}\right)^a \sqrt{ψ^{(1)}(a)}}{\Gamma(a)}$$

Here is a plot of a family parametrized by a.

```
In[8]:= Plot[Evaluate[Table[p[a, x], {a, 1/2, 5/2, 5/2/20}]], {x, -4, 4},
            PlotRange -> All, Frame -> True, Axes -> False,
            PlotStyle -> {Thickness[0.002]}]
```

For the application of the Gumbel distribution to football game results, see [545].

p) Expressing the Binomial functions through Gamma functions shows singular denominators for negative integer x and y.

```
In[1]:= Binomial[x, y] // FunctionExpand
```

Out[1]= $\dfrac{\text{Gamma}\,[1 + x]}{\text{Gamma}\,[1 + x - y]\ \text{Gamma}\,[1 + y]}$

For $x \geq y$ as a function of two complex variables, we do not have a removable singularity at negative integers. Depending on how the point $\{x,\ y\}$ is approached, different values result.

```
In[2]:= Table[rc := Random[Complex, 10^-19 {-1 - I, 1 + I}, 30];
         Binomial[-5 + rc, -8 + rc] // N, {5}]
Out[2]= {-152.334 + 67.5158 i, 39.0447 + 6.03727 i,
         71.9376 - 18.0395 i, -19.781 + 15.6696 i, -19.4786 + 5.14629 i}
```

This means for the situation of the last input there is no "correct" value.

For $x < y$ as a function of two complex variables we do have a removable singularity at negative integers. By continuity we infer the value is 0.

```
In[3]:= Table[rc := Random[Complex, 10^-19 {-1 - I, 1 + I}, 30];
         Binomial[-5 + rc, -3 + rc] // N, {5}]
Out[3]= {4.42749×10^-21 - 7.39199×10^-21 i,
         -4.22214×10^-21 - 1.87218×10^-20 i, 3.2522×10^-21 + 7.61639×10^-22 i,
         6.87007×10^-21 - 9.96873×10^-21 i, -5.94349×10^-21 - 2.22034×10^-20 i}
```

Approaching the negative integers $x \geq y$ in a symmetric way we get a finite limit [805].

```
In[4]:= Table[rc := Random[Complex, 10^-19 {-1 - I, 1 + I}, 30];
         Binomial[-5 + #, -8 + #]&[rc] // N // Chop, {5}]
Out[4]= {-35., -35., -35., -35., -35.}
```

Here we use series to confirm this observation.

```
In[5]:= Series[Binomial[-5 + εx, -3 + εy], {εx, 0, 1}, {εy, 0, 1}] //
                                            Normal // Simplify
Out[5]= - 1/144 (12 + 13 εx) εy
```

```
In[6]:= Series[Binomial[-5 + εx, -8 + εy], {εx, 0, 1}, {εy, 0, 1}] //
                                            Normal // Simplify
Out[6]= (- 35/4 - 35/εx - 7175 εx/144) εy
```

One can define binomial through an order dependent limit [529]. This forces $\binom{m}{n}$ to vanish for negative integer n.

```
In[7]:= binomial[m_, n_] :=
         Normal[Series[Series[μ!/(ν! (μ - ν)!), {ν, n, 0}], {μ, m, 0}]]
In[8]:= {binomial[-5, -3], binomial[-5, -8]}
Out[8]= {0, 0}
```

q) This is the graphic under consideration.

```
In[1]:= cp = ContourPlot[Im[(x + I y)^(x + I y)], {x, -10, 10}, {y, -10, 10},
                        PlotPoints -> 250, Contours -> {0}]
```

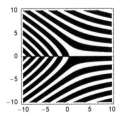

Rewriting z^z as $\exp(z \ln(z))$ shows that the curves arise from $z \ln(z) = k\, i\, \pi$.

```
In[2]:= ContourPlot[Im[(x + I y) Log[(x + I y)]], {x, -8, 8}, {y, -8, 8},
                    ContourShading -> False, PlotPoints -> 200,
                    Contours -> Table[k Pi, {k, -15, 15}]]
```

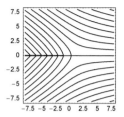

Solving $z \ln(z) = x$ yields $z_k^{(j)} = (i k \pi + \xi)/W_j(i k \pi + \xi)$ as a parametric description for the curves. Here ξ is the real parameter along a curve and the integers j and k count the curves.

```
In[3]:= Off[InverseFunction::ifun]; Off[Solve::ifun];
        Solve[z Log[z] == x, z]
```

$$Out[4]= \left\{\left\{z \to \frac{x}{\text{ProductLog}[x]}\right\}\right\}$$

Due to the branch cut structure of $\ln(z)$ and $W_j(z)$ only the branches $j = 0, \pm 1$ are needed. The following graphic overlays the parametrized curves on the original contour plot.

```
In[5]:= cLines[n_, {kMin_, kMax_}, col_] :=
        ParametricPlot[Evaluate[Table[{Re[#], Im[#]}& @
        (* the parametrizations *)
        (Function[x, x/ProductLog[n, x]][N[k I Pi + x]]),
        {k, kMin, kMax}]], {x, -50, 50},
        (* use enough plotpoints *) PlotPoints -> 500,
        Frame -> True, Axes -> False, AspectRatio -> Automatic,
        Compiled -> False, PlotStyle -> {{Thickness[0.01], col}},
        DisplayFunction -> Identity, PlotRange -> {{-10, 10}, {-10, 10}}];

In[6]:= (* lines from three branches *)
        cl1 = cLines[ 0, {-10, 10}, RGBColor[1, 0, 0]];
        cl2 = cLines[ 1, {-10, -1}, RGBColor[0, 1, 0]];
        cl3 = cLines[-1, {  0, 10}, RGBColor[0, 0, 1]];

In[10]:= Show[{cp, cl1, Graphics[{Hue[0.12], Line[{{0, 0}, {-10, 0}}]}]},
             (* make curves fit *)
             cl2 /. Line[l_] :> Line /@ DeleteCases[Partition[l, 2, 1],
                    {{_, _?Negative}, {_, _?Negative}}],
             cl3 /. Line[l_] :> Line /@ DeleteCases[Partition[l, 2, 1],
                    {{_, _?Positive}, {_, _?Positive}}],
             (* straight lines *)
             Graphics[{GrayLevel[0.5], Thickness[0.01],
                    Line[{{-10, -10}, {10, -10}, {10, 10},
                          {-10, 10}, {-10, -10}}]}]},
             PlotRange -> {{-10, 10}, {-10, 10}}]
```

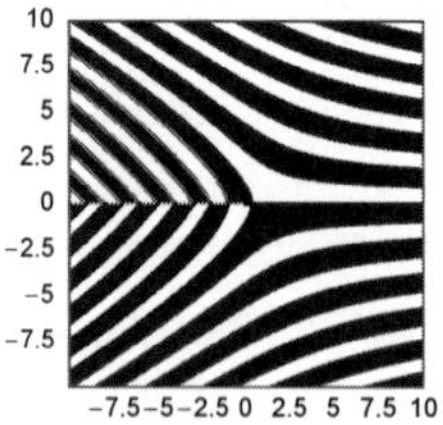

r) A quick numerical experiment shows that the limit probably exists.

```
In[1]:= cf = Compile[{{n, _Integer}}, 1/n Product[k^(1/n), {k, n}]];

In[2]:= ListPlot[Table[cf[k], {k, 1000}], PlotRange -> All]
```

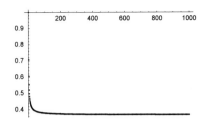

In[3]:= **cf[10^6] // Timing**

Out[3]= {0.7 Second, 0.367882}

Pulling out the *n*'s from the denominators gives the defining product for the factorial. Taking into account its leading term of the factorial function at ∞ yields immediately the answer $1/e$.

In[4]:= (* turn of messages from series at essential singularities *)
     Off[Series::esss]

In[6]:= **Series[Π = 1/n Product[k, {k, n}]^(1/n), {n, Infinity, 2}] // Simplify**

Out[6]= $\frac{1}{e} + \frac{-\text{Log}[\frac{1}{n}] + \text{Log}[2\pi]}{2\,e\,n} + \frac{\left(2 + 3\,\text{Log}[\frac{1}{2\,n\,\pi}]^2\right)\left(\frac{1}{n}\right)^2}{24\,e} + O\left[\frac{1}{n}\right]^3$

The agreement between the numerical value and the approximation is excellent for $n = 10^4$.

In[7]:= **Block[{n = 10000.},**
     {Π, 1/E + (-Log[n^(-1)] + Log[2*Pi])/(2*E n)}]

Out[7]= {0.368083, 0.368083}

A direct calculation of the infinite product gives the value $\pi/2$.

In[8]:= **Product[(2k)(2k)/((2k - 1)(2k + 1)), {k, 1, Infinity}]**

Out[8]= $\frac{\pi}{2}$

But *Mathematica* can also find the finite product.

In[9]:= **Π = Product[(2k)(2k)/((2k - 1)(2k + 1)), {k, 1, n}]**

Out[9]= $\frac{\pi\,\text{Gamma}[1 + n]^2}{2\,\text{Gamma}[\frac{1}{2} + n]\,\text{Gamma}[\frac{3}{2} + n]}$

And after substituting the leading term of $\Gamma(z)$ at $z = \infty$, the limit can be found.

In[10]:= **ΓSeries[z_] = Normal[Series[Gamma[z], {z, Infinity, 2}]];**
      ΠS = Π /. Gamma[z_] :> ΓSeries[z]

Out[11]= $\dfrac{(\frac{1}{2} + n)^{-\frac{1}{2}-n}(1 + n)^{2+2n}(\frac{3}{2} + n)^{-\frac{3}{2}-n}\pi\left(\frac{1}{6}\left(\frac{1}{1+n}\right)^{3/2}\sqrt{\frac{\pi}{2}} + \sqrt{\frac{1}{1+n}}\sqrt{2\pi}\right)^2}{2\left(\frac{1}{6}\left(\frac{1}{\frac{1}{2}+n}\right)^{3/2}\sqrt{\frac{\pi}{2}} + \sqrt{\frac{1}{\frac{1}{2}+n}}\sqrt{2\pi}\right)\left(\frac{1}{6}\left(\frac{1}{\frac{3}{2}+n}\right)^{3/2}\sqrt{\frac{\pi}{2}} + \sqrt{\frac{1}{\frac{3}{2}+n}}\sqrt{2\pi}\right)}$

In[12]:= **Series[FullSimplify[Normal[Series[%, {n, Infinity, 2}]]],**
      {n, Infinity, 1}]

Out[12]= $\frac{\pi}{2} - \frac{\pi}{8\,n} + O\left[\frac{1}{n}\right]^2$

The agreement between the exact value and the approximation is again excellent.

In[13]:= **N[{Π, Pi/2 (1 - 1/(4 n))} /. n -> N[1000, 20], 8]**

Out[13]= {1.5704039, 1.5704036}

For a nice generalization of the last product, see [986].

Now we deal with the third expression.

In[14]:= **pi[n_] := 2^(4n + 2) (n + 1)!^4/((2n + 1)!^2 (n + 1)^3)**

The limit is π.

In[15]:= `Limit[pi[n], n -> Infinity]`

Out[15]= π

To calculate the limit we first use the Stirling expansion for $n!$.

In[16]:= `Simplify[pi[n] /. (n_)! -> Normal[Series[n!, {n, Infinity, 3}]]]`

Out[16]= $\dfrac{2^{-7+4n} (1+n)^{-5+4n} (1+2n)^{1-4n} (313 + 600 n + 288 n^2)^4 \pi}{81 e^2 (313 + 1200 n + 1152 n^2)^2}$

To avoid constructs of the form $f(n)^{g(n)}$ we take the logarithm of the last expression and calculate its limit. Exponentiating the result gives the limit π and the first terms of the series in $1/n$.

In[17]:= `Exp @ Simplify[Series[Log[%], {n, Infinity, 3}]]`

Out[17]= $\pi + \dfrac{\pi}{4n} - \dfrac{7}{32} \pi \left(\dfrac{1}{n}\right)^2 + \dfrac{2623 \pi \left(\frac{1}{n}\right)^3}{13824} + O\left[\dfrac{1}{n}\right]^4$

The first two correction terms reproduce well the difference from above.

In[18]:= `Pi/(4 n) - 7/32 Pi /n^2 /. n -> SetPrecision[10^6, 7]`

Out[18]= 7.853975×10^{-7}

We continue with the sum. Mathematica can carry out the sum, and the limit.

In[19]:= `sum = Sum[n/(n^2 + k^2), {k, n^2}]`

Out[19]= $n \left(\dfrac{-1 + n \pi \operatorname{Coth}[n\pi]}{2n^2} - \dfrac{i (\operatorname{PolyGamma}[0, 1 - i n + n^2] - \operatorname{PolyGamma}[0, 1 + i n + n^2])}{2n}\right)$

In[20]:= `Limit[%, n -> Infinity]`

Out[20]= $\dfrac{\pi}{2}$

For large n, we have the following expansion.

In[21]:= `Series[sum, {n, Infinity, 2}]`

Out[21]= $\dfrac{1}{2} \left(\pi \operatorname{Coth}[n\pi] + \left(-\dfrac{3}{n} + O\left[\dfrac{1}{n}\right]^3\right)\right)$

Taking into account that $\lim_{x\to\infty} \coth(x) = 1$ and the value is approached exponentially fast, we have for the limit and its first correction term: $\pi/2 - 3/2/n$.

This results agrees favorable with a numerical value for large n.

In[22]:= `Block[{n = N[1000, 10]},`
`        {sum, Pi/2 - 3/2/n}]`

Out[22]= $\{1.569296328 + 0. \times 10^{-12} i, 1.569296326795\}$

s) This defined the generalized Gamma function.

In[1]:= `Γ[k_, z_] := k^(z/k - 1) Gamma[z/k]`

Here is a recursion relation for the generalized Gamma function.

In[2]:= `Γ[k, z + k] - z Γ[k, z] // FullSimplify`

Out[2]= 0

Mathematica recognizes the infinite product representation of the generalized Gamma function.

In[3]:= `FullSimplify[1/Γ[k, z] -`
`        z k^-(z/k) Exp[z/k EulerGamma] Product[(1 + z/(k n)) Exp[-z/(k n)], {n, Infinity}]]`

Out[3]= 0

Using only first derivatives, we find exactly one first-order PDE. We denote the function by w.

In[4]:= `Union[Cases[GroebnerBasis[{w - Γ[k, z], Subscript[w, z] - D[Γ[k, z], z],`
`                           Subscript[w, k] - D[Γ[k, z], k]}, {},`
`           {Gamma[z/k], PolyGamma[0, z/k], PolyGamma[1, z/k]},`
`           MonomialOrder -> EliminationOrder] // Factor,`

```
              _Plus?(MemberQ[#, w, Infinity]&), Infinity]] // Simplify
```

Out[4]= {w (k - z) + k² w_k + k z w_z}

In[6]:= (k - z) Γ[k, z] + k^2 D[Γ[k, z], k] + k z D[Γ[k, z] , z] // Simplify

Out[6]= 0

To derive more PDEs fulfilled by the generalized Gamma function, we start by forming all partial derivatives up to order two.

In[7]:= (eqs =
        {w - Γ[k, z], Subscript[w, z] - D[Γ[k, z], z], Subscript[w, k] - D[Γ[k, z], k],
         Subscript[w, z, z] - D[Γ[k, z], z, z], Subscript[w, k, k] - D[Γ[k, z], k, k],
         Subscript[w, k, z] - D[Γ[k, z], k, z]}) // TraditionalForm

Out[7]//TraditionalForm= $\left\{ w - k^{\frac{z}{k}-1}\Gamma\!\left(\frac{z}{k}\right),\ -\Gamma\!\left(\frac{z}{k}\right)\log(k)\,k^{\frac{z}{k}-2} - \Gamma\!\left(\frac{z}{k}\right)\psi^{(0)}\!\left(\frac{z}{k}\right)k^{\frac{z}{k}-2} + w_z,\ z\,\Gamma\!\left(\frac{z}{k}\right)\psi^{(0)}\!\left(\frac{z}{k}\right)k^{\frac{z}{k}-3} - \Gamma\!\left(\frac{z}{k}\right)\left(\frac{\frac{z}{k}-1}{k} - \frac{z\log(k)}{k^2}\right)k^{\frac{z}{k}-1} + w_k, \right.$

$-\Gamma\!\left(\frac{z}{k}\right)\log^2(k)\,k^{\frac{z}{k}-3} - \Gamma\!\left(\frac{z}{k}\right)\psi^{(0)}\!\left(\frac{z}{k}\right)^2 k^{\frac{z}{k}-3} - 2\,\Gamma\!\left(\frac{z}{k}\right)\log(k)\,\psi^{(0)}\!\left(\frac{z}{k}\right)k^{\frac{z}{k}-3} - \Gamma\!\left(\frac{z}{k}\right)\psi^{(1)}\!\left(\frac{z}{k}\right)k^{\frac{z}{k}-3} + w_{z,z},$

$-z^2\,\Gamma\!\left(\frac{z}{k}\right)\psi^{(0)}\!\left(\frac{z}{k}\right)^2 k^{\frac{z}{k}-5} - z^2\,\Gamma\!\left(\frac{z}{k}\right)\psi^{(1)}\!\left(\frac{z}{k}\right)k^{\frac{z}{k}-5} + z\,\Gamma\!\left(\frac{z}{k}\right)\left(\frac{\frac{z}{k}-3}{k} - \frac{z\log(k)}{k^2}\right)\psi^{(0)}\!\left(\frac{z}{k}\right)k^{\frac{z}{k}-3} +$

$z\,\Gamma\!\left(\frac{z}{k}\right)\left(\frac{\frac{z}{k}-1}{k} - \frac{z\log(k)}{k^2}\right)\psi^{(0)}\!\left(\frac{z}{k}\right)k^{\frac{z}{k}-3} - \Gamma\!\left(\frac{z}{k}\right)\left(\frac{\frac{z}{k}-1}{k} - \frac{z\log(k)}{k^2}\right)^2 k^{\frac{z}{k}-1} - \Gamma\!\left(\frac{z}{k}\right)\left(\frac{2\log(k)z}{k^3} - \frac{2z}{k^3} - \frac{\frac{z}{k}-1}{k^2}\right)k^{\frac{z}{k}-1} +$

$w_{k,k},\ z\,\Gamma\!\left(\frac{z}{k}\right)\psi^{(0)}\!\left(\frac{z}{k}\right)^2 k^{\frac{z}{k}-4} + z\,\Gamma\!\left(\frac{z}{k}\right)\log(k)\,\psi^{(0)}\!\left(\frac{z}{k}\right)k^{\frac{z}{k}-4} + z\,\Gamma\!\left(\frac{z}{k}\right)\psi^{(1)}\!\left(\frac{z}{k}\right)k^{\frac{z}{k}-4} + \Gamma\!\left(\frac{z}{k}\right)\psi^{(0)}\!\left(\frac{z}{k}\right)k^{\frac{z}{k}-3} -$

$\left. \Gamma\!\left(\frac{z}{k}\right)\log(k)\left(\frac{\frac{z}{k}-1}{k} - \frac{z\log(k)}{k^2}\right)k^{\frac{z}{k}-2} - \Gamma\!\left(\frac{z}{k}\right)\left(\frac{\frac{z}{k}-1}{k} - \frac{z\log(k)}{k^2}\right)\psi^{(0)}\!\left(\frac{z}{k}\right)k^{\frac{z}{k}-2} - \Gamma\!\left(\frac{z}{k}\right)\left(\frac{1}{k^2} - \frac{\log(k)}{k^2}\right)k^{\frac{z}{k}-1} + w_{k,z} \right\}$

To obtain polynomial PDEs, we have to eliminate the Gamma and polygamma functions.

In[8]:= gb = GroebnerBasis[(* form numerators *)
                          Numerator[Factor[Together[Rest[eqs] /.
                          (* eliminate gamma function *)
                          Solve[eqs[[1]] == 0, Gamma[z/k]][[1]]]]], {},
                          {PolyGamma[0, z/k], PolyGamma[1, z/k]},
                          MonomialOrder -> EliminationOrder] // Factor;

In[9]:= candidatePDEs = Sort[Union[Cases[gb, _Plus, Infinity]], (Length[#1] < Length[#2])&];
        Length[candidatePDEs]

Out[10]= 23

Here are some of the so-obtained PDEs.

In[11]:= (selectedPDEs = Collect[#, {w, Subscript[w, z], Subscript[w, k], Subscript[w, z, z],
         Subscript[w, k, k], Subscript[w, k, z]}, Factor]& /@ Take[candidatePDEs, 12]) //
                                                                        TraditionalForm

Out[11]//TraditionalForm=

$\{ w_k\,k^2 + z\,w_z\,k + w\,(k-z),\ w_{k,z}\,k^3 + w_k\,k^2 + 2\,w_z\,k^2 + z\,w_{z,z}\,k^2 - w\,z,$

$w_{k,k}\,k^3 + z\,w_{k,z}\,k^2 + (2\,k-z)\,w_k\,k + w\,z,\ -2\,w_k\,k^2 + z\,w_{k,z}\,k^2 + z^2\,w_{z,z}\,k + w\,(z-2\,k) - z^2\,w_z,$

$w_{k,k}\,k^2 + z\,w_{k,z}\,k + w + (3\,k-z)\,w_k + z\,w_z,\ w_k\,w_{z,z} - 2\,w_z^2\,k^2 + w_z\,(-w_{k,z}\,k^3 - w_k\,k^2) + w\,(z\,w_z + k\,(k-z)\,w_{z,z}),$

$w_{k,k}\,k^2 + 2\,w_z\,k + (k+z)\,w_{k,z}\,k + z\,w_{z,z}\,k + (3\,k-z)\,w_k,\ k^2\,w_k^2 - z^2\,w_z^2 + w\,(w_{k,k}\,k^2 + 2\,z\,w_{k,z}\,k + (3\,k-2\,z)\,w_k + z\,w_z + z^2\,w_{z,z}),$

$-k\,w_k^2 - z\,w_z\,w_k + w\,(2\,w_k + 2\,w_z + k\,w_{k,k} + (k+z)\,w_{k,z} + z\,w_{z,z}),$

$w\,(z\,w_{k,k} + 2\,k\,w_{k,z} + z\,w_{z,z}) + w_k\,(-w_{k,k}\,k^2 + (4\,k-z)\,w_{k,z}\,k + (3\,k-z)\,z\,w_{z,z}) + w_z\,(-2\,w_{k,k}\,k^2 + z^2\,w_{k,z} + z^2\,w_{z,z}),$

$(7\,k-2\,z)\,w_k^2 + (2\,k^2\,w_{k,k} - 2\,k\,(k-z)\,w_{k,z})\,w_k + 2\,z\,w_z^2 + w_z\,(2\,w_{k,k}\,k^2 + (6\,k+z)\,w_k) + w\,(-k\,w_{k,k} + (z-3\,k)\,w_{k,z} - z\,w_{z,z}),$

$(7\,k-2\,z)\,w_k^2 + (2\,w_{k,k}\,k^2 + 2\,w_{z,z}\,k^2 + 2\,z\,w_{k,z}\,k)\,w_k - 2\,(2\,k-z)\,w_z^2 +$

$w_z\,(3\,z\,w_k - 2\,k^2\,w_{k,z}) + w\,(-k\,w_{k,k} + (-k-z)\,w_{k,z} + (2\,k-3\,z)\,w_{z,z})\}$

We end with a quick check of their correctness.

In[12]:= FullSimplify[selectedPDEs /. Subscript[w, kz__] :> D[Γ[k, z], kz] /. w -> Γ[k, z]]

Out[12]= {0, 0, 0, 0, 0, 0, 0, 0, 0, 0, 0, 0}

t) We start by calculating the first few $B_k(n)$.

```
In[1]:= exp[k_, x_] := exp[k, x] = Nest[Exp, x, k];

     Do[MapIndexed[(B[k, #2[[1]] - 1] = Together[#1 (#2[[1]] - 1)!])&,
              CoefficientList[Series[exp[k, x], {x, 0, 7}], x]], {k, 6}]
```

sum implements the right-hand side of the identity under consideration.

```
In[3]:= sum[m_, n_] := Sum[Evaluate[KroneckerDelta[Sum[k[j], {j, m}], n]*
                    n! Product[B[j, k[j]]/k[j]!, {j, m}]],
                Evaluate[Sequence @@ Table[{k[j], 0, n}, {j, m}]]]
```

A straightforward calculation confirms the identity for $1 \le m, n \le 6$.

```
In[4]:= Table[sum[o, n] - B[o, n + 1] // Expand, {o, 6}, {n, 6}] //
                                        Flatten // Union

Out[4]= {0}
```

u) No, it will not be a machine number.

```
In[1]:= 1 - Erfc[66.66]
```

```
Out[1]= 0.999999999999999999999999999999999999999999999999999999999999999999999999999999999999999999999999999999999`.
        999999999999999999999999999999999999999999999999999999999999999999999999999999999999999999999999999999999`.
        999999999999999999999999999999999999999999999999999999999999999999999999999999999999999999999999999999999`.
        999999999999999999999999999999999999999999999999999999999999999999999999999999999999999999999999999999999`.
        999999999999999999999999999999999999999999999999999999999999999999999999999999999999999999999999999999999`.
        999999999999999999999999999999999999999999999999999999999999999999999999999999999999999999999999999999999`.
        999999999999999999999999999999999999999999999999999999999999999999999999999999999999999999999999999999999`.
        999999999999999999999999999999999999999999999999999999999999999999999999999999999999999999999999999999999`.
        999999999999999999999999999999999999999999999999999999999999999999999999999999999999999999999999999999999`.
        999999999999999999999999999999999999999999999999999999999999999999999999999999999999999999999999999999999`.
        999999999999999999999999999999999999999999999999999999999999999999999999999999999999999999999999999999999`.
        999999999999999999999999999999999999999999999999999999999999999999999999999999999999999999999999999999999`.
        999999999999999999999999999999999999999999999999999999999999999999999999999999999999999999999999999999999`.
        999999999999999999999999999999999999999999999999999999999999999999999999999999999999999999999999999999999`.
        999999999999999999999999999999999999999999999999999999999999999999999999999999999999999999999999999999999`.
        999999999999999999999999999999999999999999999999999999999999999999999999999999999999999999999999999999999`.
        999999999999999999999999999999999999999999999999999999999999999999999999999999999999999999999999999999999`.
        999999999999999999999999999999999999999999999999999999999999999999999999999999999999999999999999999999999`.
        999999999999999999999999999999999999999999999999999999999999999999999999999999999999999999999999999999999`.
        999999999999999999999999999999999999999999999999999999999999999999999999999999999999999999999999999999999`.
        999999999999999999999999999999999999999999999999999999999999999999999999999999999999999999999999999999999`.
        999999999999999999999999999999999999999999999999999999999999999999999999999999999999999999999999999999999`.
        999999999999999999999999999999986943289047535825
```

```
In[2]:= MachineNumberQ[%]

Out[2]= False
```

In the first step, `Erfc[66.66]` is calculated. The result is a very small high-precision number. Its exponent is smaller than the smallest exponent for machine numbers. So this number must be represented as a high-precision number.

```
In[3]:= Erfc[66.66]

Out[3]= 1.305671095246418 × 10^{-1932}
```

The last result is then subtracted from 1, leaving a number that is far very close to 1. Because happens because subtracting a high-precision number from a exact number yields again a high-precision number.

The same situation happens, of course, for any expression of the form *exactNormalSizeNumber - verySmallApproximateNumber*. Huge numbers of digits can be generated in this manner.

```
In[4]:= ByteCount[1 - Exp[-25000.^2]]

Out[4]= 112710616
```

v) Here we calculate the Borel sums for the first seven j.

```
In[1]:= res = Table[Integrate[Exp[-t] Sum[(-3)^k/k^j t^(k - 1/2), {k, Infinity}],
                  {t, 0, Infinity}], {j, 0, 6}];
```

The results are relatively complicated expressions containing various special functions.

```
In[2]:= TraditionalForm[Take[res, 3] // FullSimplify]
```

$$Out[2]//TraditionalForm= \left\{-\sqrt{\pi} + \frac{\sqrt[3]{e}\,\pi\,\mathrm{erfc}\left(\frac{1}{\sqrt{3}}\right)}{\sqrt{3}},\ \frac{1}{3}\,\sqrt{\pi}\left(3\gamma - 3\pi\,\mathrm{erfi}\left(\frac{1}{\sqrt{3}}\right) + 2\,{}_2F_2\left(1, 1;\ \frac{3}{2}, 2;\ \frac{1}{3}\right) + \log\left(\frac{64}{27}\right)\right),\right.$$

$$\left.-\frac{1}{12}\,\sqrt{\pi}\left(5\pi^2 + 8\left(-2\sqrt{3\pi}\,{}_2F_2\left(\frac{1}{2}, \frac{1}{2}, \frac{3}{2}, \frac{3}{2};\ \frac{1}{3}\right) + {}_3F_3\left(1, 1, 1;\ \frac{3}{2}, 2, 2;\ \frac{1}{3}\right)\right) + 6\left(\gamma + \log\left(\frac{4}{3}\right)\right)^2\right)\right\}$$

While the summands decrease for increasing j, the corresponding Borel sums seem to increase.

```
In[3]:= res // N[#, 22]& // N
```

```
Out[3]= {-0.723923, -1.21591, -1.66226, -2.01377, -2.26323, -2.42673, -2.52747}
```

It is straightforward to conjecture the inner sum to be $t^{-1/2}\,\mathrm{Li}_j(-3\,t)$.

```
In[4]:= Table[{j, Sum[(-3)^k/k^4 t^(k - 1/2), {k, Infinity}]}, {j, 4}]
```

$$Out[4]= \left\{\left\{1,\ \frac{\mathrm{PolyLog}[4, -3\,t]}{\sqrt{t}}\right\},\ \left\{2,\ \frac{\mathrm{PolyLog}[4, -3\,t]}{\sqrt{t}}\right\},\right.$$

$$\left.\left\{3,\ \frac{\mathrm{PolyLog}[4, -3\,t]}{\sqrt{t}}\right\},\ \left\{4,\ \frac{\mathrm{PolyLog}[4, -3\,t]}{\sqrt{t}}\right\}\right\}$$

Using this result we can numerically calculate values of the integral for larger j. For $j \to \infty$ the sequence seems to approach a limit.

```
In[5]:= Table[{j, NIntegrate[Exp[-t] PolyLog[j, -3 t]/Sqrt[t],
                  {t, 0, Infinity}]}, {j, 20}]
```

```
Out[5]= {{1, -1.21591}, {2, -1.66226}, {3, -2.01377}, {4, -2.26323}, {5, -2.42673},
        {6, -2.52747}, {7, -2.58658}, {8, -2.61997}, {9, -2.63827}, {10, -2.64807},
        {11, -2.65322}, {12, -2.65589}, {13, -2.65727}, {14, -2.65797}, {15, -2.65832},
        {16, -2.6585}, {17, -2.65859}, {18, -2.65864}, {19, -2.65866}, {20, -2.65867}}
```

Calculating, say s_{200} to high precision leads to the conjecture $\lim_{j \to \infty} s_j = -3/2\,\pi$.

```
In[6]:= NIntegrate[Exp[-t] PolyLog[100, -3 t]/Sqrt[t], {t, 0, 200},
                  WorkingPrecision -> 40, PrecisionGoal -> 25]
```

```
Out[6]= -2.658680776358274040947251
```

```
In[7]:= 3/2 Sqrt[Pi] // N[#, 25]&
```

```
Out[7]= 2.658680776358274040947251
```

w) Asymptotically the potential $V(z)$ approaches 0. This means the ground-state has an energy is greater or equal to zero. So let us try the smallest possible value 0. Here is the solution of the differential equation.

```
In[1]:= V[z_] := (-(Sqrt[5] - 1/2) + 19/4 z^2)/(1 + z^2)^2
```

```
        Φ = ψ[z] /. DSolve[-ψ''[z] + V[z] ψ[z] == 0, ψ[z], z][[1]]
```

$$Out[2]= (1 + z^2)^{\frac{1}{4} - \frac{\sqrt{5}}{2}} C[1] + z\,(1 + z^2)^{\frac{1}{4} - \frac{\sqrt{5}}{2}} C[2]\ \mathrm{Hypergeometric2F1}\left[\frac{1}{2}, \frac{1}{2} - \sqrt{5}, \frac{3}{2}, -z^2\right]$$

A quick visualization shows that the hypergeometric term is asymptotically increasing, so the algebraic term is the relevant one.

```
In[3]:= Show[GraphicsArray[
          Plot[Abs[Cases[Φ, _ C[#]]/C[#]], {z, -6, 6},
              DisplayFunction -> Identity]& /@ {1, 2}]]
```

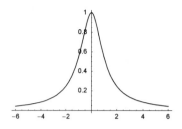 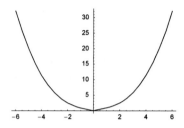

We normalize the algebraic part and obtain the ground-state $\psi_N[z]$.

In[4]:= `Cases[Ψ, _ C[1]][[1]]/C[1]`

Out[4]= $(1 + z^2)^{\frac{1}{4} - \frac{\sqrt{5}}{2}}$

In[5]:= `FullSimplify[Factor //@ Abs[%]^2, z > 0]`

Out[5]= $(1 + z^2)^{\frac{1}{2} - \sqrt{5}}$

In[6]:= `Integrate[%, {z, -Infinity, Infinity}]`

Out[6]= $\dfrac{\sqrt{\pi}\ \text{Gamma}[-1 + \sqrt{5}\ ]}{\text{Gamma}[-\frac{1}{2} + \sqrt{5}\ ]}$

In[7]:= `ψN[z_] = FullSimplify[Sqrt[%%]/Sqrt[%], z > 0]`

Out[7]= $\dfrac{(1 + z^2)^{\frac{1}{4}(1-2\sqrt{5})}\ \sqrt{\dfrac{\text{Gamma}\left[-\frac{1}{2}+\sqrt{5}\ \right]}{\text{Gamma}\left[-1+\sqrt{5}\ \right]}}}{\pi^{1/4}}$

Here is a visualization of the potential (in red) and the ground-state.

In[8]:= `Plot[Evaluate[{V[z], ψN[z]}], {z, -6, 6},`
       `    PlotStyle -> {{Hue[0], Thickness[0.02]},`
       `                    {GrayLevel[0], Thickness[0.002]}}]`

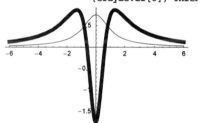

Asymptotically we have $V(z) < \psi_N(z)$.

In[9]:= `Plot[V[z]/ψN[z] - 1, {z, 0, 2000}, PlotRange -> {-1, 1}]`

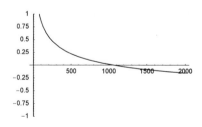

We have $z^* \approx 1068$.

In[10]:= `FindRoot[V[z]/ψN[z] - 1, {z, 1200}, WorkingPrecision -> 50]`

Out[10]= {z → 1068.7231271204161008606129460053160613748624691698}

x) We start with the integral containing the Laguerre functions. The built-in `Integrate` cannot evaluate this integral in closed form.

In[1]:= {Integrate[LaguerreL[1, x] LaguerreL[3/2, x] Exp[-x], x],
       Integrate[LaguerreL[1, x] LaguerreL[3/2, x] Exp[-x], {x, 0, X}]}

Out[1]= $\left\{ \int e^{-x} (1-x) \text{ LaguerreL}\left[\frac{3}{2}, x\right] dx, \int_0^X e^{-x} (1-x) \text{ LaguerreL}\left[\frac{3}{2}, x\right] dx \right\}$

Using the representation $L_v(z) = {}_1F_1(-v; 1; z)$, we can get a closed form of the integral.

In[2]:= intL[X_] = Integrate[Hypergeometric1F1[-2/3, 1, x]*
                  LaguerreL[1, x] Exp[-x], {x, 0, X},
                  GenerateConditions -> False]

Out[2]= $X \text{ Hypergeometric1F1}\left[\frac{5}{3}, 2, -X\right] +$

$\frac{1}{18} X^2 \left(-9 \text{ Hypergeometric1F1}\left[\frac{5}{3}, 3, -X\right] + 5 X \text{ Hypergeometric1F1}\left[\frac{8}{3}, 4, -X\right]\right)$

Here is a quick numerical check of the last result.

In[3]:= With[{X = 10, prec = 20}, {intL[X] // N[#, 22]&,
       NIntegrate[LaguerreL[1, x] LaguerreL[2/3, x]Exp[-x],
                  {x, 0, X}, WorkingPrecision -> prec + 11]}]

Out[3]= {1.427748109051749593142, 1.427748109051749593142}

To obtain the asymptotics for large X, we implement a rule `ser1F1Rule` that replaces the hypergeometric function ${}_1F_1(a; b; z)$ by its asymptotics.

In[4]:= (* turn off messages *) Off[Series::esss]
       ser1F1Rule = (HypergeometricPFQ[{a_}, {b_}, z_] |
           Hypergeometric1F1[a_, b_, z_]) ->
           Series[Hypergeometric1F1[a, b, z], {z, Infinity, 1}] // Normal

Out[5]= HypergeometricPFQ[{a_}, {b_}, z_] | Hypergeometric1F1[a_, b_, z_] → Gamma[b]

$\left(e^z \left(\frac{1}{z}\right)^{-a+b} \left(\frac{1}{\text{Gamma}[a]} + \frac{-a+a^2+b-ab}{z \text{ Gamma}[a]}\right) + \left(\frac{1}{z}\right)^a \left(\frac{(-1)^{-a}}{\text{Gamma}[-a+b]} + \frac{(-1)^{-a}(-a-a^2+ab)}{z \text{ Gamma}[-a+b]}\right)\right)$

In[6]:= asympL[x_] = intL[x] /. ser1F1Rule // FullSimplify

Out[6]= $\frac{1}{80\pi}$

$\left(3\sqrt{3} e^{-x} \left(-\frac{1}{x}\right)^{4/3} \left(60 (-2+x) x^2 \text{ Gamma}\left[\frac{4}{3}\right] + 50 x (-1+2x) \text{ Gamma}\left[\frac{7}{3}\right] + 24 (-1)^{1/3} e^x \left(-\frac{1}{x}\right)^{1/3}\right.\right.$

$\left.\left.x^2 \text{ Gamma}\left[\frac{8}{3}\right] + \frac{8 (-1)^{1/3} e^x \text{ Gamma}\left[\frac{11}{3}\right]}{(-\frac{1}{x})^{2/3}} + 5 (-1)^{1/3} e^x \left(-\frac{1}{x}\right)^{1/3} \text{ Gamma}\left[\frac{11}{3}\right]\right)\right)$

So we obtain the following result for large X.

In[7]:= Series[Expand[asympL[x]] /. Exp[-x] :> 0, {x, Infinity, 2}]

Out[7]= $\frac{9\sqrt{3} \text{ Gamma}\left[\frac{8}{3}\right]}{10\pi (\frac{1}{x})^{1/3}} - \frac{3 (\sqrt{3} \text{ Gamma}\left[\frac{11}{3}\right]) (\frac{1}{x})^{2/3}}{10\pi} + \frac{3\sqrt{3} \text{ Gamma}\left[\frac{11}{3}\right] (\frac{1}{x})^{5/3}}{16\pi} + O\left[\frac{1}{x}\right]^{7/3}$

A quick numerical check confirms the last result.

In[8]:= Block[{X = 10.^6}, {Normal[%] /. x -> X, intL[X]}]

Out[8]= {74.6564, 74.6564}

Interpreting the divergent integral in a regularized sense (meaning dropping fractional powers of X) shows that orthogonality still holds; no term of the form X^0 appears in the expansion.

Now let us deal with the Hermite-function-involving integral. This time `Integrate` gives a result.

In[9]:= intH[X_] = Integrate[HermiteH[1, x] HermiteH[3/2, x] Exp[-x^2], {x, -X, X}]

Out[9]= $-\frac{X^3 \text{ Gamma}\left[-\frac{1}{4}\right] \text{ Hypergeometric1F1}\left[\frac{7}{4}, \frac{5}{2}, -X^2\right]}{\sqrt{\pi}}$

For real X, the last result can be considerably simplified, only Bessel functions are needed.

```
In[10]:= intH1 = Simplify[FunctionExpand[intH[X]], Element[X, Reals]]
```

$$\text{Out[10]=} \quad \frac{2\,e^{-\frac{x^2}{2}}\,\sqrt{\pi}\,X^3\,\left(\text{BesselI}\left[\frac{1}{4},\,\frac{x^2}{2}\right] - \text{BesselI}\left[\frac{5}{4},\,\frac{x^2}{2}\right]\right)}{\sqrt{\text{Abs}[X]}}$$

Using the expansions of the Bessel functions $I_\nu(x)$ for large positive X, we obtain the asymptotics $3\,X^{-1/2}$. This vanishes for large X and orthogonality holds again.

```
In[11]:= intH2 = Simplify[intH1 /. BesselI[n_, x_] :>
                    (Normal[Series[BesselI[v, ξ], {ξ, Infinity, 1}]] /.
                        {v -> n, ξ -> x}), X > 0]
```

$$\text{Out[11]=} \quad \frac{e^{-x^2}\,\left(12\,e^{x^2} + (-1)^{3/4}\,(9 + 16\,X^2)\right)}{4\,\sqrt{X}}$$

```
In[12]:= intH3 = Expand[intH2] /. Exp[-X^2] -> 0
```

$$\text{Out[12]=} \quad \frac{3}{\sqrt{X}}$$

y) A direct evaluation of this triple integral fails (otherwise it would not be listed as an exercise).

```
In[1]:= TimeConstrained[
          1/Pi^3 Integrate[1/(1 - Cos[x] Cos[y] Cos[z]),
                          {x, 0, Pi}, {y, 0, Pi}, {z, 0, Pi}], 100]
```

```
Out[1]= $Aborted
```

The structure of the integrand $1/(1-\zeta)$ suggests expanding the integrand in a series $1/(1-\zeta) = \sum_{k=0}^{\infty} \zeta^k$. The resulting terms $(\cos(x)\cos(y)\cos(z))^k$ are easily integrated (using factorization in three independent integrals and symmetry of $\cos(\xi)$ around $\xi = \pi/2$).

```
In[2]:= int = Integrate[Cos[x]^(2k), {x, 0, Pi/2},
                      Assumptions -> k > 0]^3 // FullSimplify
```

$$\text{Out[2]=} \quad \frac{\pi^{3/2}\,\text{Gamma}\left[\frac{1}{2} + k\right]^3}{8\,\text{Gamma}[1 + k]^3}$$

Summing these terms gives a nice short answer.

```
In[3]:= sum = 2^3/Pi^3 Sum[Evaluate[int], {k, 0, Infinity}]
```

$$\text{Out[3]=} \quad \frac{\pi}{\text{Gamma}\left[\frac{3}{4}\right]^4}$$

A quick numerical check using Monte-Carlo integration confirms the result

```
In[4]:= {1/Pi^3 NIntegrate[1/(1 - Cos[x] Cos[y] Cos[z]),
                          {x, 0, Pi}, {y, 0, Pi}, {z, 0, Pi},
                          Method -> MonteCarlo[(* seed *) 8830911524],
                          MaxPoints -> 10^6],
          N[%]}
```

```
Out[4]= {1.39309, 1.3932}
```

z) We first remark that the arguments of the shown identities are relatively low-order algebraic numbers.

```
In[1]:= RootReduce[Together[TrigToExp[2 Cos[3/7 Pi]]]]
```

```
Out[1]= Root[1 - 2 #1 - #1^2 + #1^3 &, 2]
```

```
In[2]:= RootReduce[Together[TrigToExp[1/(2 Cos[Pi/9])]]]
```

```
Out[2]= Root[-1 + 3 #1^2 + #1^3 &, 3]
```

This suggest searching for identities of the form $\sum_{j=1}^{n} c_j\,L(\gamma^j) = d\,\pi^2$ where the c_j and d are rational numbers, and γ is an algebraic number (or even algebraic integers) arising from some "nice" trigonometric expression.

$$d\,\sum_{j=1}^{n} c_j\,L(\gamma^j) = d\,\pi^2.$$

Here is the modified dilogarithm.

```
In[3]:= L[z_] := PolyLog[2, z] + 1/2 Log[z] Log[1 - z]
```

We can numerically calculate the $L(\gamma^j)$. To determine rational numbers c_j and d, we will make use of the function `Lattice Reduce`. The function `rationalSumFactors` returns a list cs of rational numbers, such that the identity $cs.terms = rhs$ holds with the precision of the inputs *term* and *rhs*.

```
In[4]:= rationalSumFactors[terms_, rhs_] :=
         Module[{λ = Length[terms] + 1, prec = Round[Precision[{terms, rhs}]],
                  v = Prepend[terms, -rhs]},
                 (* all integers *) vExact = Round[10^prec v];
                 basis = Transpose[Prepend[IdentityMatrix[λ], vExact]];
                 (* find new basis *)
                 lrBasis = LatticeReduce[basis];
                 v = Rest[First[lrBasis]]; Rest[v]/First[v]]
```

The following two inputs show how for a given γ, the above identities are recovered.

```
In[5]:= rationalSumFactors[{L[1/3], L[1/9]} // N[#, 20]&, Pi^2]
```

```
Out[5]= {18, -3}
```

```
In[6]:= With[{γ = 2 Cos[3/7 Pi]},
             rationalSumFactors[{L[γ], L[γ^2], L[γ^3]} // N[#, 40]&, Pi^2] ]
```

$$Out[6]= \left\{ \frac{21}{2}, \frac{21}{4}, 0 \right\}$$

Given an algebraic number γ, the function `LIdentity` tries to find the rational factors making an identity of the form described above.

```
In[7]:= Clear[LIdentity];

        (* avoid redoing work *)
        LIdentity[γ_] := LIdentity[γ] =
        Module[{id, γA, d, deg, cs, den, f = 4, res},
               $MaxExtraPrecision = 500;
               (* avoid messages *)
               Internal`DeactivateMessages[
               (* explicit algebraic number *)
               γA = TimeConstrained[RootReduce[Together[TrigToExp[γ]]], 12];
               d = Which[γA === $Aborted, Infinity,
                          Head[γA] === Root, Exponent[γA[[1]][C], C],
                          True, 1];
               If[d > 10, res = $Failed,
               (* increase number of terms to be used in the sum *)
               While[deg = f d;
                     (cs = rationalSumFactors[Table[L[γ^e], {e, deg}] //
                                                N[#, 40]& // N[#, 20]&, Pi^2];
                 Not[And @@ (NumericQ /@ cs)]) && f > 1, f--];
               (* check returned coefficients *)
               res = If[(* sensible result from rationalSumFactors? *)
                        cs =!= $Aborted && FreeQ[cs, Indeterminate | ComplexInfinity] &&
                        Max[Abs[Denominator[cs]]] < 200 &&
                        (* identity holds to higher precision? *)
                        N[cs.Table[L[γ^e], {e, deg}] - N[Pi^2, 100]] == 0.,
                        den = Max[Denominator[cs]];
                        Expand[den (cs.Table[£[g^e], {e, deg}])] == den Pi^2 /; g == γ,
                        $Failed]]]; res]
```

In the next input, the last of the given three identities is found. For easier readability of the identities we express them as a condition.

```
In[10]:= LIdentity[1/(2 Cos[Pi/9])]
```

$$Out[10]= 54 \, £[g] + 54 \, £[g^2] - 18 \, £[g^3] == 7 \, \pi^2 \; /; \; g == \frac{1}{2} \, Sec\left[\frac{\pi}{9}\right]$$

Now, we can implement our search. The function `findLIdentityTry` tries to find an identity for a given seed to generate an algebraic number.

```
In[11]:= Clear[findLIdentityTry];

    findLIdentityTry[seed_, {f_:6, n_:16, exp_:6}] :=
    Module[{γ, res},
    SeedRandom[seed]; $MaxExtraPrecision = 500;
        (* do not try too long for a given number *)
        res = TimeConstrained[Internal`DeactivateMessages[
        (* random algebraic number *)
        γ = (2 Random[Integer] - 1)*
            (Random[Integer, {1, f}]/Random[Integer, {1, f}]*
            (* just three functions as candidates *)
            {Cos, Sin, Tan, Exp[I #]&}[[Random[Integer, {1, 3}]]]][
                Random[Integer, {1, n}]/Random[Integer, {1, n}] 2 Pi])^
                                            (Random[Integer, {-1, 1} exp])];
        (* or use a root directly:
        γ = Function[d, Root[Table[Random[Integer, {-3, 3}], {d}].
            (#^Range[d]), Random[Integer, {1, d}]]][Random[Integer, {3, 12}]] *)
        (* try to find identity for γ *)
        LIdentity[γ], 30];
        If[res === $Aborted, res = $Failed]; res]
```

We create 10000 "random" values of γ and check if an identity of the form conjectured above exists. We print the seed and the identity found. We found seven identities (some being simple variations of each other).

```
In[13]:= identityBag = {};

    Do[id = findLIdentityTry[k, {}];
        If[id =!= $Failed && FreeQ[identityBag, Verbatim[id]],
            Print[{k, id}]; AppendTo[identityBag, id]], {k, 10^4}]
```

$$\left\{56,\ 12\,\mathcal{L}[g] = \pi^2\ /;\ g = \frac{1}{2}\right\}$$

$$\left\{821,\ 10\,\mathcal{L}[g] - 5\,\mathcal{L}[g^2] = \pi^2\ /;\ g = \frac{2}{-1+\sqrt{5}}\right\}$$

$$\left\{1330,\ 144\,\mathcal{L}[g] - 96\,\mathcal{L}[g^2] + 12\,\mathcal{L}[g^4] = 13\,\pi^2\ /;\ g = \text{Tan}\left[\frac{3\,\pi}{8}\right]\right\}$$

$$\left\{2853,\ 18\,\mathcal{L}[g^2] - 3\,\mathcal{L}[g^4] = \pi^2\ /;\ g = -\frac{1}{\sqrt{3}}\right\}$$

$$\left\{3101,\ 24\,\mathcal{L}[g] - 6\,\mathcal{L}[g^2] = \pi^2\ /;\ g = -\frac{8}{(-1-\sqrt{5})^3}\right\}$$

$$\left\{4914,\ 18\,\mathcal{L}[g] - 3\,\mathcal{L}[g^2] = \pi^2\ /;\ g = \frac{1}{3}\right\}$$

$$\left\{7463,\ 18\,\mathcal{L}[g^2] - 3\,\mathcal{L}[g^4] = \pi^2\ /;\ g = \frac{1}{\sqrt{3}}\right\}$$

We now pack the function findLIdentity in an additional While loop that calls findLIdentityTry until an identity is found.

```
In[15]:= findLIdentity[{f_:6, n_:16, exp_:6}] :=
    Module[{id}, While[id = findLIdentityTry[Random[Integer, {1, 10^20}], {f, n, exp}];
                    id === $Failed]; id]]
```

We could now continue to search for more identities. Here are some more of the identities found using the function findLIdentity.

```
In[16]:= γValues =
    {(* rationals *)
    1/3, 1/2, -I, I,
    (* radicals *)
    -1/Sqrt[2], 1/Sqrt[2], 1/2^(1/3), 1/Sqrt[3],
    -1 + Sqrt[2], 1 + Sqrt[2], -2 + Sqrt[5],
    (-1 - Sqrt[5])/2, (1 - Sqrt[5])/2, (3 - Sqrt[5])/2,
    (-1 + Sqrt[5])/2, (1 + Sqrt[5])/2,
    (* roots of unity *)
    (-1)^(2/15), (-1)^(4/15), -(-1)^(7/15), (-1)^(8/15), -(-1)^(11/15),
    -(-1)^(13/15),
    (* trigonometric expressions *)
    2 Cos[11 Pi/7], -Cot[3 Pi/8], Cot[5 Pi/8]^2, Sec[Pi/9]/2,
```

```
        -Sec[22 Pi/7]/2, Root[-1 - # + #^3 &, 1]^2,
        (* explicit roots *)
        Root[-1 + # + #^2 + #^3 &, 1], Root[-1 + 3 # + #^2 + #^3 &, 1],
        Root[-1 + 2 # + #^2 + 2 #^3 &, 1], Root[-1 + 2 #^2 + 2 #^3 &, 1],
        Root[-1 + # + #^4 &, 2], Root[-1 - #^2 + #^4 &, 1],
        Root[-1 + #^3 + #^4 &, 1], Root[-1 + # + #^2 + #^3 + #^4 &, 2],
        Root[-1 + 2 #^3 + #^4 & , 2], Root[-1 + 2 #^4 & , 2],
        Root[-1 - # - #^4 + #^5 &, 2], Root[-1 + # + #^2 + #^4 + #^5 &, 1]};
```

In[17]:= `idList = LIdentity /@ γValues`

Out[17]= $\{18\,£[g] - 3\,£[g^2] == \pi^2 \,/;\, g == \frac{1}{3},\ 12\,£[g] == \pi^2 \,/;\, g == \frac{1}{2},\ 12\,£[g] + 12\,£[g^3] == \pi^2 \,/;\, g == -i,$

$12\,£[g] + 12\,£[g^3] == \pi^2 \,/;\, g == i,\ 12\,£[g^2] == \pi^2 \,/;\, g == -\frac{1}{\sqrt{2}},\ 12\,£[g^2] == \pi^2 \,/;\, g == \frac{1}{\sqrt{2}},$

$12\,£[g^3] == \pi^2 \,/;\, g == \frac{1}{2^{1/3}},\ 18\,£[g^2] - 3\,£[g^4] == \pi^2 \,/;\, g == \frac{1}{\sqrt{3}},$

$16\,£[g] - 4\,£[g^2] == \pi^2 \,/;\, g == -1 + \sqrt{2},\ 144\,£[g] - 96\,£[g^2] + 12\,£[g^4] == 13\,\pi^2 \,/;\, g == 1 + \sqrt{2},$

$24\,£[g] - 6\,£[g^2] == \pi^2 \,/;\, g == -2 + \sqrt{5},\ 6\,£[g] + 6\,£[g^2] == \pi^2 \,/;\, g == \frac{1}{2}\left(-1 - \sqrt{5}\right),$

$15\,£[g^2] == \pi^2 \,/;\, g == \frac{1}{2}\left(1 - \sqrt{5}\right),\ 15\,£[g] == \pi^2 \,/;\, g == \frac{1}{2}\left(3 - \sqrt{5}\right),$

$10\,£[g] == \pi^2 \,/;\, g == \frac{1}{2}\left(-1 + \sqrt{5}\right),\ 10\,£[g] - 5\,£[g^2] == \pi^2 \,/;\, g == \frac{1}{2}\left(1 + \sqrt{5}\right),$

$-15\,£[g^7] - 15\,£[g^8] == 2\,\pi^2 \,/;\, g == (-1)^{2/15},\ 15\,£[g^7] + 15\,£[g^8] == 4\,\pi^2 \,/;\, g == (-1)^{4/15},$

$5\,£[g^7] + 5\,£[g^8] == \pi^2 \,/;\, g == -(-1)^{7/15},\ 5\,£[g^7] + 5\,£[g^8] == \pi^2 \,/;\, g == (-1)^{8/15},$

$15\,£[g^7] + 15\,£[g^8] == 4\,\pi^2 \,/;\, g == -(-1)^{11/15},\ -15\,£[g^7] - 15\,£[g^8] == 2\,\pi^2 \,/;\, g == -(-1)^{13/15},$

$42\,£[g] + 21\,£[g^2] == 4\,\pi^2 \,/;\, g == 2\,\mathrm{Cos}\left[\frac{11\,\pi}{7}\right],\ 30\,£[g^2] - 6\,£[g^4] == \pi^2 \,/;\, g == -\mathrm{Cot}\left[\frac{3\,\pi}{8}\right],$

$30\,£[g^2] - 6\,£[g^2] == \pi^2 \,/;\, g == \mathrm{Cot}\left[\frac{5\,\pi}{8}\right]^2,\ 54\,£[g] + 54\,£[g^2] - 18\,£[g^3] == 7\,\pi^2 \,/;\, g == \frac{1}{2}\,\mathrm{Sec}\left[\frac{\pi}{9}\right],$

$42\,£[g] + 21\,£[g^2] == 5\,\pi^2 \,/;\, g == -\frac{1}{2}\,\mathrm{Sec}\left[\frac{22\,\pi}{7}\right],$

$10\,£[g] - 3\,£[g^2] + 10\,£[g^3] - £[g^4] - 4\,£[g^5] - 5\,£[g^6] + 2\,£[g^{10}] == 2\,\pi^2 \,/;$
$\quad g == \mathrm{Root}[-1 - \#1 + \#1^3 \,\&,\, 1]^2,$

$12\,£[g] + 6\,£[g^2] - 12\,£[g^3] == \pi^2 \,/;\, g == \mathrm{Root}[-1 + \#1 + \#1^2 + \#1^3 \,\&,\, 1],$

$12\,£[g] + 18\,£[g^2] - 6\,£[g^4] == \pi^2 \,/;\, g == \mathrm{Root}[-1 + 3\,\#1 + \#1^2 + \#1^3 \,\&,\, 1],$

$12\,£[g] + 6\,£[g^2] + 4\,£[g^3] - 4\,£[g^4] == \pi^2 \,/;\, g == \mathrm{Root}[-1 + 2\,\#1 + \#1^2 + 2\,\#1^3 \,\&,\, 1],$

$8\,£[g] + 6\,£[g^2] + 4\,£[g^3] - 6\,£[g^4] - 4\,£[g^5] - 4\,£[g^6] + 2\,£[g^{12}] == \pi^2 \,/;$
$\quad g == \mathrm{Root}[-1 + 2\,\#1^2 + 2\,\#1^3 \,\&,\, 1],\ 6\,£[g] + 6\,£[g^4] == \pi^2 \,/;\, g == \mathrm{Root}[-1 + \#1 + \#1^4 \,\&,\, 2],$

$10\,£[g^2] - 5\,£[g^4] == \pi^2 \,/;\, g == \mathrm{Root}[-1 - \#1^2 + \#1^4 \,\&,\, 1],$

$6\,£[g^3] + 6\,£[g^4] == \pi^2 \,/;\, g == \mathrm{Root}[-1 + \#1^3 + \#1^4 \,\&,\, 1],$

$12\,£[g] + 6\,£[g^3] - 12\,£[g^4] == \pi^2 \,/;\, g == \mathrm{Root}[-1 + \#1 + \#1^2 + \#1^3 + \#1^4 \,\&,\, 2],$

$2\,£[g^2] + 8\,£[g^3] + 7\,£[g^4] - 2\,£[g^8] == \pi^2 \,/;\, g == \mathrm{Root}[-1 + 2\,\#1^3 + \#1^4 \,\&,\, 2],$

$12\,£[g^4] == \pi^2 \,/;\, g == \mathrm{Root}[-1 + 2\,\#1^4 \,\&,\, 2],$

$9\,£[g^2] + 6\,£[g^3] - 3\,£[g^4] + 6\,£[g^5] - 3\,£[g^6] == \pi^2 \,/;\, g == \mathrm{Root}[-1 - \#1 - \#1^4 + \#1^5 \,\&,\, 2],$

$6\,£[g] + 9\,£[g^2] - 2\,£[g^3] + 6\,£[g^5] - 7\,£[g^6] + 2\,£[g^9] == \pi^2 \,/;$
$\quad g == \mathrm{Root}[-1 + \#1 + \#1^2 + \#1^4 + \#1^5 \,\&,\, 1]\}$

Here is a quick check of these identities.

In[18]:= `checkIdentity[Verbatim[Condition][a__ == b_, g == γ_]] :=`
   `Block[{$MaxExtraPrecision = 100},`
     `N[a - b /. g -> γ /. £ -> L, 22]]`

In[19]:= `Off[N::meprec];`
   `checkIdentity /@ idList`

Out[20]= $\{0. \times 10^{-121}, 0, 0. \times 10^{-121} + 0. \times 10^{-121}$ i, $0. \times 10^{-121} + 0. \times 10^{-121}$ i, $0, 0, 0,$
$0. \times 10^{-121}, 0. \times 10^{-120}, 0. \times 10^{-119} + 0. \times 10^{-119}$ i, $0. \times 10^{-120}, 0. \times 10^{-120} + 0. \times 10^{-120}$ i,
$0. \times 10^{-120}, 0. \times 10^{-120}, 0. \times 10^{-120}, 0. \times 10^{-120} + 0. \times 10^{-120}$ i, $0. \times 10^{-119} + 0. \times 10^{-119}$ i,
$0. \times 10^{-119} + 0. \times 10^{-120}$ i, $0. \times 10^{-111} + 0. \times 10^{-111}$ i, $0. \times 10^{-111} + 0. \times 10^{-111}$ i,
$0. \times 10^{-119} + 0. \times 10^{-120}$ i, $0. \times 10^{-119} + 0. \times 10^{-119}$ i, $0. \times 10^{-119}, 0. \times 10^{-120},$
$0. \times 10^{-120}, 0. \times 10^{-120}, 0. \times 10^{-119}, 0. \times 10^{-119} + 0. \times 10^{-119}$ i, $0. \times 10^{-120}, 0. \times 10^{-120},$
$0. \times 10^{-121}, 0. \times 10^{-120}, 0. \times 10^{-121}, 0. \times 10^{-120} + 0. \times 10^{-120}$ i, $0. \times 10^{-120} + 0. \times 10^{-120}$ i,
$0. \times 10^{-120}, 0. \times 10^{-120}, 0. \times 10^{-120}, 0. \times 10^{-120} + 0. \times 10^{-120}$ i, $0. \times 10^{-120}\}$

2. Elliptic Integrals

a) We start by making an ansatz for the differential equation under consideration.

```
In[1]:= w = EllipticPi[n, φ, m];
        ansatz = α3 D[w, {m, 3}] + α2 D[w, {m, 2}] + α1 D[w, {m, 1}] + α0 w + αC;
```

For determining the coefficients $\alpha3$, $\alpha2$, $\alpha1$, $\alpha0$, and αC, we clear denominators in `ansatz`.

```
In[3]:= eqs1 = Numerator[Together[ansatz]];
```

The resulting polynomial is quite large.

```
In[4]:= Length[eqs1]
```

Out[4]= 375

To get an impression of the form of the resulting equation, we set all *n* and *m* in the prefactors equal to 1.

```
In[5]:= TraditionalForm @ Union[(List @@ eqs1) /.
        {α3 -> 1, α2 -> 1, α1 -> 1, α0 -> 1, αC -> 1} /.
        a_Integer n^_. r_ -> r /. m^_. r_ -> r /. a_Integer r_ -> r]
```

Out[5]//TraditionalForm= $\Big\{ \sin(\phi),\ \sin^3(\phi),\ \sin^5(\phi),\ E(\phi\,|\,m)\sqrt{\sin^2(\phi)+1},\ F(\phi\,|\,m)\sqrt{\sin^2(\phi)+1},\ \sqrt{1-m\sin^2(\phi)},\ E(\phi\,|\,m)\sqrt{1-m\sin^2(\phi)},$

$F(\phi\,|\,m)\sqrt{1-m\sin^2(\phi)},\ \Pi(n;\phi\,|\,m)\sqrt{1-m\sin^2(\phi)},\ \sin^2(\phi)\sqrt{1-m\sin^2(\phi)},\ E(\phi\,|\,m)\sin^2(\phi)\sqrt{1-m\sin^2(\phi)},$

$F(\phi\,|\,m)\sin^2(\phi)\sqrt{1-m\sin^2(\phi)},\ \Pi(n;\phi\,|\,m)\sin^2(\phi)\sqrt{1-m\sin^2(\phi)},\ \sin^4(\phi)\sqrt{1-m\sin^2(\phi)},$

$E(\phi\,|\,m)\sin^4(\phi)\sqrt{1-m\sin^2(\phi)},\ F(\phi\,|\,m)\sin^4(\phi)\sqrt{1-m\sin^2(\phi)},\ \Pi(n;\phi\,|\,m)\sin^4(\phi)\sqrt{1-m\sin^2(\phi)},$

$\sqrt{m\sin^2(\phi)+1},\ E(\phi\,|\,m)\sqrt{m\sin^2(\phi)+1},\ F(\phi\,|\,m)\sqrt{m\sin^2(\phi)+1},\ \Pi(n;\phi\,|\,m)\sqrt{m\sin^2(\phi)+1},$

$\sin^2(\phi)\sqrt{m\sin^2(\phi)+1},\ E(\phi\,|\,m)\sin^2(\phi)\sqrt{m\sin^2(\phi)+1},\ F(\phi\,|\,m)\sin^2(\phi)\sqrt{m\sin^2(\phi)+1},$

$\Pi(n;\phi\,|\,m)\sin^2(\phi)\sqrt{m\sin^2(\phi)+1},\ \sin^4(\phi)\sqrt{m\sin^2(\phi)+1},\ E(\phi\,|\,m)\sin^4(\phi)\sqrt{m\sin^2(\phi)+1},$

$F(\phi\,|\,m)\sin^4(\phi)\sqrt{m\sin^2(\phi)+1},\ \Pi(n;\phi\,|\,m)\sin^4(\phi)\sqrt{m\sin^2(\phi)+1},\ \sin(2\phi),\ \sin^2(\phi)\sin(2\phi),\ \sin^4(\phi)\sin(2\phi)\Big\}$

Assuming that for generic *n*, *m*, and ϕ, the three terms `EllipticE[φ, m] Sqrt[1-mSin[φ]^2]`, `EllipticF[φ, m] Sqrt[1-mSin[φ]^2]`, and `EllipticPi[n, φ, m] Sqrt[1-mSin[φ]^2]` are linearly independent, we replace them with new variables.

```
In[6]:= eqs2 = eqs1 /. {EllipticE[φ, m] Sqrt[1 - m Sin[φ]^2] -> we,
        EllipticF[φ, m] Sqrt[1 - m Sin[φ]^2] -> wf,
        EllipticPi[n, φ, m] Sqrt[1 - m Sin[φ]^2] -> wp};
```

Based on the linear independence, we extract the coefficients of `we`, `wf`, and `wp` from `eqs2`. `ce`, `cf`, and `cp` are these coefficients; `c0` are the remaining terms.

```
In[7]:= {ce, cf, cp} = Coefficient[eqs2, (* variables *) {we, wf, wp}];
        c0 = Expand[eqs2 - (ce we + cf wf + cp wp)];
```

Solving the resulting system of four equations for the five unknowns $\alpha3$, $\alpha2$, $\alpha1$, $\alpha0$, and αC gives the following result.

```
In[9]:= sol = Factor //@ Solve[{ce == 0, cf == 0, cp == 0, c0 == 0},
        {α3, α2, α1, α0, αC}]
```

Solve::svars : Equations may not give solutions for all "solve" variables. More…

Out[9]= $\left\{\left\{\alpha 0 \to \dfrac{3\,\alpha 2}{4\,(-7\,m+11\,m^2+2\,n-6\,m\,n)},\right.\right.$

$\alpha C \to -\dfrac{3\,\alpha 2\,\mathrm{Sin}[2\,\phi]}{8\,(-7\,m+11\,m^2+2\,n-6\,m\,n)\,\sqrt{1-m\,\mathrm{Sin}[\phi]^2}\,(-1+m\,\mathrm{Sin}[\phi]^2)^2},$

$\left.\left.\alpha 1 \to \dfrac{3\,(-2+7\,m-n)\,\alpha 2}{2\,(-7\,m+11\,m^2+2\,n-6\,m\,n)},\ \alpha 3 \to \dfrac{2\,(-1+m)\,m\,(m-n)\,\alpha 2}{-7\,m+11\,m^2+2\,n-6\,m\,n}\right\}\right\}$

The last result suggests the choice $\alpha 2 \to 4\,(11\,m^2-7\,m+2\,n-6\,m\,n)$ to make the formulas as simple as possible.

In[10]:= `res = With[{α2T = 2 (-7 m + 11 m^2 + 2 n - 6 m n)},`
`            Append[sol[[1]] /. α2 -> α2T, α2 -> α2T]]`

Out[10]= $\left\{\alpha 0 \to \dfrac{3}{2},\ \alpha C \to -\dfrac{3\,\mathrm{Sin}[2\,\phi]}{4\,\sqrt{1-m\,\mathrm{Sin}[\phi]^2}\,(-1+m\,\mathrm{Sin}[\phi]^2)^2},\right.$

$\left.\alpha 1 \to 3\,(-2+7\,m-n),\ \alpha 3 \to 4\,(-1+m)\,m\,(m-n),\ \alpha 2 \to 2\,(-7\,m+11\,m^2+2\,n-6\,m\,n)\right\}$

So we finally obtain the following differential equation.

$$4\,(-1+m)\,m\,(m-n)\,\frac{\partial^3\,\Pi(n;\phi\,|\,m)}{\partial m^3} + 2\,(-7\,m+11\,m^2+2\,n-6\,m\,n)\,\frac{\partial^2\,\Pi(n;\phi\,|\,m)}{\partial m^2} +$$

$$3\,(-2+7\,m-n)\,\frac{\partial\,\Pi(n;\phi\,|\,m)}{\partial m} + \frac{3}{2}\,\Pi(n;\phi\,|\,m) = \frac{3\,\sin(2\,\phi)}{4\,\sqrt{(1-m\,\sin^2(\phi))^5}}$$

A check shows that the differential equation is correct.

In[11]:= `(ansatz /. res) // Simplify`

Out[11]= 0

b) We more or less repeat the calculation from Part a) of this exercise, but this time we differentiate with respect to n. So again we start by making an ansatz for the differential equation under consideration.

In[1]:= `w = EllipticPi[n, φ, m];`
`      ansatz = α3 D[w, {n, 3}] + α2 D[w, {n, 2}] + α1 D[w, {n, 1}] + α0 w + αC;`

For determining the coefficients $\alpha 3$, $\alpha 2$, $\alpha 1$, $\alpha 0$, and αC, we clear denominators in ansatz.

In[3]:= `eqs1 = Numerator[Together[ansatz]];`

The resulting polynomial has even more terms than the equivalent polynomial in Part a).

In[4]:= `Length[eqs1]`

Out[4]= 588

To get an impression of the form of the resulting equation, we set all n and m in the prefactors to 1.

In[5]:= `Union[(List @@ eqs1) /.`
`        {α3 -> 1, α2 -> 1, α1 -> 1, α0 -> 1, αC -> 1} /.`
`        a_Integer n^_. r_ -> r /. m^_. r_ -> r /. a_Integer r_ -> r]`

Out[5]= $\{$ m, m², m³, n⁶, n⁷, n⁸, n⁹, EllipticE[ϕ, m], EllipticF[ϕ, m], EllipticPi[n, ϕ, m], Sin[ϕ]²,

EllipticE[ϕ, m] Sin[ϕ]², EllipticF[ϕ, m] Sin[ϕ]², EllipticPi[n, ϕ, m] Sin[ϕ]², Sin[ϕ]⁴,

EllipticE[ϕ, m] Sin[ϕ]⁴, EllipticF[ϕ, m] Sin[ϕ]⁴, EllipticPi[n, ϕ, m] Sin[ϕ]⁴, Sin[ϕ]⁶,

EllipticE[ϕ, m] Sin[ϕ]⁶, EllipticF[ϕ, m] Sin[ϕ]⁶, EllipticPi[n, ϕ, m] Sin[ϕ]⁶,

Sin[ϕ] $\sqrt{1+\mathrm{Sin}[\phi]^2}$, Sin[$\phi$]³ $\sqrt{1+\mathrm{Sin}[\phi]^2}$, Sin[$\phi$]⁵ $\sqrt{1+\mathrm{Sin}[\phi]^2}$,

Sin[ϕ] $\sqrt{1+m\,\mathrm{Sin}[\phi]^2}$, Sin[$\phi$]³ $\sqrt{1+m\,\mathrm{Sin}[\phi]^2}$, Sin[$\phi$]⁵ $\sqrt{1+m\,\mathrm{Sin}[\phi]^2}$ $\}$

Assuming that for generic n, m, and ϕ, the three terms EllipticE[ϕ, m], EllipticF[ϕ, m], and EllipticPi[n, ϕ, m] are linear independent, we replace them by new variables.

In[6]:= `eqs2 = eqs1 /. {EllipticE[φ, m] -> we, EllipticF[φ, m] -> wf,`
`                EllipticPi[n, φ, m] -> wp};`

Based on the linear independence, we extract the coefficients of we, wf, and wp from eqs2.

In[7]:= `{ce, cf, cp} = Coefficient[eqs2, #]& /@ {we, wf, wp};`
`      c0 = Expand[eqs2 - (ce we + cf wf + cp wp)];`

Solving the resulting system of four equations for the five unknowns $\alpha 3$, $\alpha 2$, $\alpha 1$, $\alpha 0$, and αC gives the following result.

In[9]:= sol = Factor //@ Solve[{ce == 0, cf == 0, cp == 0, c0 == 0},
                               {α3, α2, α1, α0, αC}]

Solve::svars : Equations may not give solutions for all "solve" variables. More…

Out[9]= $\left\{\left\{\alpha 0 \to -\dfrac{\alpha 1}{2\ (1+m-4\ n)},\ \alpha C \to -\dfrac{\alpha 1\ \sqrt{1-m\ \text{Sin}[\phi]^2}\ \text{Sin}[2\ \phi]}{4\ (1+m-4\ n)\ (-1+n\ \text{Sin}[\phi]^2)^3},\right.\right.$

$\left.\left.\alpha 3 \to \dfrac{(m-n)\ (-1+n)\ n\ \alpha 1}{2\ (1+m-4\ n)},\ \alpha 2 \to \dfrac{(-3\ m+8\ n+8\ m\ n-13\ n^2)\ \alpha 1}{4\ (1+m-4\ n)}\right\}\right\}$

The last result suggests the chosen $\alpha 1 \to 4\ (m-4\ n+1)$.

In[10]:= res = With[{α1T = 4 (m - 4 n + 1)},
                    Append[(sol[[1]] /. α1 -> α1T), α1 -> α1T]]

Out[10]= $\left\{\alpha 0 \to -2,\ \alpha C \to -\dfrac{\sqrt{1-m\ \text{Sin}[\phi]^2}\ \text{Sin}[2\ \phi]}{(-1+n\ \text{Sin}[\phi]^2)^3},\right.$

$\left.\alpha 3 \to 2\ (m-n)\ (-1+n)\ n,\ \alpha 2 \to -3\ m+8\ n+8\ m\ n-13\ n^2,\ \alpha 1 \to 4\ (1+m-4\ n)\right\}$

So we finally obtain the following differential equation.

$$2\ (m-n)\ (n-1)\ n\ \frac{\partial^3\ \Pi(n;\ \phi\mid m)}{\partial n^3} + (-3\ m + 8\ n + 8\ m\ n - 13\ n^2)\ \frac{\partial^2\ \Pi(n;\ \phi\mid m)}{\partial n^2} +$$

$$4\ (1 + m - 4\ n)\ \frac{\partial\Pi(n;\ \phi\mid m)}{\partial n} - 2\ \Pi(n;\ \phi\mid m) = \frac{\sqrt{1 - m\ \sin^2(\phi)}\ \sin(2\ \phi)}{(n\ \sin^2(\phi) - 1)^3}$$

A quick check shows that the differential equation is correct.

In[11]:= (ansatz /. res) // Simplify

Out[11]= 0

c) Looking at the derivative of $\Pi(n;\ \phi\mid m)$ with respect to ϕ, one obvious possibility to derive a nonlinear differential equation is the following: Differentiate $\Pi(n;\ \phi\mid m)$ twice with respect to ϕ, and eliminate the trigonometric terms in ϕ. Here are the first two derivatives.

In[1]:= {d1 = D[EllipticPi[n, φ, m], φ], d2 = D[d1, φ]}

Out[1]= $\left\{\dfrac{1}{\sqrt{1-m\ \text{Sin}[\phi]^2}\ (1-n\ \text{Sin}[\phi]^2)},\right.$

$\left.\dfrac{2\ n\ \text{Cos}[\phi]\ \text{Sin}[\phi]}{\sqrt{1-m\ \text{Sin}[\phi]^2}\ (1-n\ \text{Sin}[\phi]^2)^2} + \dfrac{m\ \text{Cos}[\phi]\ \text{Sin}[\phi]}{(1-m\ \text{Sin}[\phi]^2)^{3/2}\ (1-n\ \text{Sin}[\phi]^2)}\right\}$

Let D1 stand for $\partial\Pi(n;\ \phi\mid m)/\partial\phi$ and D2 for $\partial^2\ \Pi(n;\ \phi\mid m)/\partial\phi^2$. Then we have the following set of equations.

In[2]:= eqs = Factor[Numerator[Together[
                {D1^2 - d1^2, D2^2 - d2^2, Cos[φ]^2 + Sin[φ]^2 - 1}]]]

Out[2]= $\{1 - \text{D1}^2 + \text{D1}^2\ m\ \text{Sin}[\phi]^2 + 2\ \text{D1}^2\ n\ \text{Sin}[\phi]^2 - 2\ \text{D1}^2\ m\ n\ \text{Sin}[\phi]^4 - \text{D1}^2\ n^2\ \text{Sin}[\phi]^4 + \text{D1}^2\ m\ n^2\ \text{Sin}[\phi]^6,$

$-\text{D2}^2 + 3\ \text{D2}^2\ m\ \text{Sin}[\phi]^2 + 4\ \text{D2}^2\ n\ \text{Sin}[\phi]^2 + m^2\ \text{Cos}[\phi]^2\ \text{Sin}[\phi]^2 +$

$4\ m\ n\ \text{Cos}[\phi]^2\ \text{Sin}[\phi]^2 + 4\ n^2\ \text{Cos}[\phi]^2\ \text{Sin}[\phi]^2 - 3\ \text{D2}^2\ m^2\ \text{Sin}[\phi]^4 - 12\ \text{D2}^2\ m\ n\ \text{Sin}[\phi]^4 -$

$6\ \text{D2}^2\ n^2\ \text{Sin}[\phi]^4 - 6\ m^2\ n\ \text{Cos}[\phi]^2\ \text{Sin}[\phi]^4 - 12\ m\ n^2\ \text{Cos}[\phi]^2\ \text{Sin}[\phi]^4 +$

$\text{D2}^2\ m^3\ \text{Sin}[\phi]^6 + 12\ \text{D2}^2\ m^2\ n\ \text{Sin}[\phi]^6 + 18\ \text{D2}^2\ m\ n^2\ \text{Sin}[\phi]^6 + 4\ \text{D2}^2\ n^3\ \text{Sin}[\phi]^6 +$

$9\ m^2\ n^2\ \text{Cos}[\phi]^2\ \text{Sin}[\phi]^6 - 4\ \text{D2}^2\ m^3\ n\ \text{Sin}[\phi]^8 - 18\ \text{D2}^2\ m^2\ n^2\ \text{Sin}[\phi]^8 - 12\ \text{D2}^2\ m\ n^3\ \text{Sin}[\phi]^8 -$

$\text{D2}^2\ n^4\ \text{Sin}[\phi]^8 + 6\ \text{D2}^2\ m^3\ n^2\ \text{Sin}[\phi]^{10} + 12\ \text{D2}^2\ m^2\ n^3\ \text{Sin}[\phi]^{10} + 3\ \text{D2}^2\ m\ n^4\ \text{Sin}[\phi]^{10} -$

$4\ \text{D2}^2\ m^3\ n^3\ \text{Sin}[\phi]^{12} - 3\ \text{D2}^2\ m^2\ n^4\ \text{Sin}[\phi]^{12} + \text{D2}^2\ m^3\ n^4\ \text{Sin}[\phi]^{14},\ -1 + \text{Cos}[\phi]^2 + \text{Sin}[\phi]^2\}$

Next, we have to eliminate $\text{Sin}[\phi]$ and $\text{Cos}[\phi]$. We start by eliminating $\text{Sin}[\phi]$ from the first two equations.

In[3]:= res1 = Factor[Resultant[eqs[[1]], -1 + Cos[φ]^2 + Sin[φ]^2, Sin[φ]]]

Out[3]= $(-1 + \text{D1}^2 - \text{D1}^2\ m - 2\ \text{D1}^2\ n + 2\ \text{D1}^2\ m\ n + \text{D1}^2\ n^2 - \text{D1}^2\ m\ n^2 + \text{D1}^2\ m\ \text{Cos}[\phi]^2 +$

$2\ \text{D1}^2\ n\ \text{Cos}[\phi]^2 - 4\ \text{D1}^2\ m\ n\ \text{Cos}[\phi]^2 - 2\ \text{D1}^2\ n^2\ \text{Cos}[\phi]^2 + 3\ \text{D1}^2\ m\ n^2\ \text{Cos}[\phi]^2 +$

$2\ \text{D1}^2\ m\ n\ \text{Cos}[\phi]^4 + \text{D1}^2\ n^2\ \text{Cos}[\phi]^4 - 3\ \text{D1}^2\ m\ n^2\ \text{Cos}[\phi]^4 + \text{D1}^2\ m\ n^2\ \text{Cos}[\phi]^6)^2$

In[4]:= res2 = Factor[Resultant[eqs[[2]], -1 + Cos[φ]^2 + Sin[φ]^2, Sin[φ]]]

Out[4]= $(D2^2 - 3\,D2^2\,m + 3\,D2^2\,m^2 - D2^2\,m^3 - 4\,D2^2\,n + 12\,D2^2\,m\,n - 12\,D2^2\,m^2\,n + 4\,D2^2\,m^3\,n + 6\,D2^2\,n^2 - 18\,D2^2\,m\,n^2 +$
$18\,D2^2\,m^2\,n^2 - 6\,D2^2\,m^3\,n^2 - 4\,D2^2\,n^3 + 12\,D2^2\,m\,n^3 - 12\,D2^2\,m^2\,n^3 + 4\,D2^2\,m^3\,n^3 + D2^2\,n^4 -$
$3\,D2^2\,m\,n^4 + 3\,D2^2\,m^2\,n^4 - D2^2\,m^3\,n^4 + 3\,D2^2\,m\,Cos[\phi]^2 - m^2\,Cos[\phi]^2 - 6\,D2^2\,m^2\,Cos[\phi]^2 +$
$3\,D2^2\,m^3\,Cos[\phi]^2 + 4\,D2^2\,n\,Cos[\phi]^2 - 4\,m\,n\,Cos[\phi]^2 - 24\,D2^2\,m\,n\,Cos[\phi]^2 + 6\,m^2\,n\,Cos[\phi]^2 +$
$36\,D2^2\,m^2\,n\,Cos[\phi]^2 - 16\,D2^2\,m^3\,n\,Cos[\phi]^2 - 4\,n^2\,Cos[\phi]^2 - 12\,D2^2\,n^2\,Cos[\phi]^2 +$
$12\,m\,n^2\,Cos[\phi]^2 + 54\,D2^2\,m\,n^2\,Cos[\phi]^2 - 9\,m^2\,n^2\,Cos[\phi]^2 - 72\,D2^2\,m^2\,n^2\,Cos[\phi]^2 +$
$30\,D2^2\,m^3\,n^2\,Cos[\phi]^2 + 12\,D2^2\,n^3\,Cos[\phi]^2 - 48\,D2^2\,m\,n^3\,Cos[\phi]^2 + 60\,D2^2\,m^2\,n^3\,Cos[\phi]^2 -$
$24\,D2^2\,m^3\,n^3\,Cos[\phi]^2 - 4\,D2^2\,n^4\,Cos[\phi]^2 + 15\,D2^2\,m\,n^4\,Cos[\phi]^2 - 18\,D2^2\,m^2\,n^4\,Cos[\phi]^2 +$
$7\,D2^2\,m^3\,n^4\,Cos[\phi]^2 + m^2\,Cos[\phi]^4 + 3\,D2^2\,m^2\,Cos[\phi]^4 - 3\,D2^2\,m^3\,Cos[\phi]^4 + 4\,m\,n\,Cos[\phi]^4 +$
$12\,D2^2\,m\,n\,Cos[\phi]^4 - 12\,m^2\,n\,Cos[\phi]^4 - 36\,D2^2\,m^2\,n\,Cos[\phi]^4 + 24\,D2^2\,m^3\,n\,Cos[\phi]^4 +$
$4\,n^2\,Cos[\phi]^4 + 6\,D2^2\,n^2\,Cos[\phi]^4 - 24\,m\,n^2\,Cos[\phi]^4 - 54\,D2^2\,m\,n^2\,Cos[\phi]^4 + 27\,m^2\,n^2\,Cos[\phi]^4 +$
$108\,D2^2\,m^2\,n^2\,Cos[\phi]^4 - 60\,D2^2\,m^3\,n^2\,Cos[\phi]^4 - 12\,D2^2\,n^3\,Cos[\phi]^4 + 72\,D2^2\,m\,n^3\,Cos[\phi]^4 -$
$120\,D2^2\,m^2\,n^3\,Cos[\phi]^4 + 60\,D2^2\,m^3\,n^3\,Cos[\phi]^4 + 6\,D2^2\,n^4\,Cos[\phi]^4 - 30\,D2^2\,m\,n^4\,Cos[\phi]^4 +$
$45\,D2^2\,m^2\,n^4\,Cos[\phi]^4 - 21\,D2^2\,m^3\,n^4\,Cos[\phi]^4 + D2^2\,m^3\,Cos[\phi]^6 + 6\,m^2\,n\,Cos[\phi]^6 +$
$12\,D2^2\,m^2\,n\,Cos[\phi]^6 - 16\,D2^2\,m^3\,n\,Cos[\phi]^6 + 12\,m\,n^2\,Cos[\phi]^6 + 18\,D2^2\,m\,n^2\,Cos[\phi]^6 -$
$27\,m^2\,n^2\,Cos[\phi]^6 - 72\,D2^2\,m^2\,n^2\,Cos[\phi]^6 + 60\,D2^2\,m^3\,n^2\,Cos[\phi]^6 + 4\,D2^2\,n^3\,Cos[\phi]^6 -$
$48\,D2^2\,m\,n^3\,Cos[\phi]^6 + 120\,D2^2\,m^2\,n^3\,Cos[\phi]^6 - 80\,D2^2\,m^3\,n^3\,Cos[\phi]^6 - 4\,D2^2\,n^4\,Cos[\phi]^6 +$
$30\,D2^2\,m\,n^4\,Cos[\phi]^6 - 60\,D2^2\,m^2\,n^4\,Cos[\phi]^6 + 35\,D2^2\,m^3\,n^4\,Cos[\phi]^6 + 4\,D2^2\,m^3\,n\,Cos[\phi]^8 +$
$9\,m^2\,n^2\,Cos[\phi]^8 + 18\,D2^2\,m^2\,n^2\,Cos[\phi]^8 - 30\,D2^2\,m^3\,n^2\,Cos[\phi]^8 + 12\,D2^2\,m\,n^3\,Cos[\phi]^8 -$
$60\,D2^2\,m^2\,n^3\,Cos[\phi]^8 + 60\,D2^2\,m^3\,n^3\,Cos[\phi]^8 + D2^2\,n^4\,Cos[\phi]^8 - 15\,D2^2\,m\,n^4\,Cos[\phi]^8 +$
$45\,D2^2\,m^2\,n^4\,Cos[\phi]^8 - 35\,D2^2\,m^3\,n^4\,Cos[\phi]^8 + 6\,D2^2\,m^3\,n^3\,Cos[\phi]^{10} + 12\,D2^2\,m^2\,n^3\,Cos[\phi]^{10} -$
$24\,D2^2\,m^3\,n^3\,Cos[\phi]^{10} + 3\,D2^2\,m\,n^4\,Cos[\phi]^{10} - 18\,D2^2\,m^2\,n^4\,Cos[\phi]^{10} + 21\,D2^2\,m^3\,n^4\,Cos[\phi]^{10} +$
$4\,D2^2\,m^3\,n^3\,Cos[\phi]^{12} + 3\,D2^2\,m^2\,n^4\,Cos[\phi]^{12} - 7\,D2^2\,m^3\,n^4\,Cos[\phi]^{12} + D2^2\,m^3\,n^4\,Cos[\phi]^{14})^2$

Next, we eliminate Cos[ϕ]. We observe that Cos[ϕ] appears only in even powers. So we introduce a new $c = $ Cos[ϕ]$\wedge 2$ variable and eliminate this variable c.

In[5]:= res1a = res1[[1]] /. Cos[ϕ]^e_ -> c^(e/2)

Out[5]= $-1 + D1^2 - D1^2\,m + c\,D1^2\,m - 2\,D1^2\,n + 2\,c\,D1^2\,n + 2\,D1^2\,m\,n - 4\,c\,D1^2\,m\,n + 2\,c^2\,D1^2\,m\,n +$
$D1^2\,n^2 - 2\,c\,D1^2\,n^2 + c^2\,D1^2\,n^2 - D1^2\,m\,n^2 + 3\,c\,D1^2\,m\,n^2 - 3\,c^2\,D1^2\,m\,n^2 + c^3\,D1^2\,m\,n^2$

In[6]:= res2a = res2[[1]] /. Cos[ϕ]^e_ -> c^(e/2)

Out[6]= $D2^2 - 3\,D2^2\,m + 3\,c\,D2^2\,m - c\,m^2 + c^2\,m^2 + 3\,D2^2\,m^2 - 6\,c\,D2^2\,m^2 + 3\,c^2\,D2^2\,m^2 - D2^2\,m^3 + 3\,c\,D2^2\,m^3 -$
$3\,c^2\,D2^2\,m^3 + c^3\,D2^2\,m^3 - 4\,D2^2\,n + 4\,c\,D2^2\,n - 4\,c\,m\,n + 4\,c^2\,m\,n + 12\,D2^2\,m\,n - 24\,c\,D2^2\,m\,n +$
$12\,c^2\,D2^2\,m\,n + 6\,c\,m^2\,n - 12\,c^2\,m^2\,n + 6\,c^3\,m^2\,n - 12\,D2^2\,m^2\,n + 36\,c\,D2^2\,m^2\,n - 36\,c^2\,D2^2\,m^2\,n +$
$12\,c^3\,D2^2\,m^2\,n + 4\,D2^2\,m^3\,n - 16\,c\,D2^2\,m^3\,n + 24\,c^2\,D2^2\,m^3\,n - 16\,c^3\,D2^2\,m^3\,n + 4\,c^4\,D2^2\,m^3\,n -$
$4\,c\,n^2 + 4\,c^2\,n^2 + 6\,D2^2\,n^2 - 12\,c\,D2^2\,n^2 + 6\,c^2\,D2^2\,n^2 + 12\,c\,m\,n^2 - 24\,c^2\,m\,n^2 + 12\,c^3\,m\,n^2 -$
$18\,D2^2\,m\,n^2 + 54\,c\,D2^2\,m\,n^2 - 54\,c^2\,D2^2\,m\,n^2 + 18\,c^3\,D2^2\,m\,n^2 - 9\,c\,m^2\,n^2 + 27\,c^2\,m^2\,n^2 -$
$27\,c^3\,m^2\,n^2 + 9\,c^4\,m^2\,n^2 + 18\,D2^2\,m^2\,n^2 - 72\,c\,D2^2\,m^2\,n^2 + 108\,c^2\,D2^2\,m^2\,n^2 - 72\,c^3\,D2^2\,m^2\,n^2 +$
$18\,c^4\,D2^2\,m^2\,n^2 - 6\,D2^2\,m^3\,n^2 + 30\,c\,D2^2\,m^3\,n^2 - 60\,c^2\,D2^2\,m^3\,n^2 + 60\,c^3\,D2^2\,m^3\,n^2 - 30\,c^4\,D2^2\,m^3\,n^2 +$
$6\,c^5\,D2^2\,m^3\,n^2 - 4\,D2^2\,n^3 + 12\,c\,D2^2\,n^3 - 12\,c^2\,D2^2\,n^3 + 4\,c^3\,D2^2\,n^3 + 12\,D2^2\,m\,n^3 - 48\,c\,D2^2\,m\,n^3 +$
$72\,c^2\,D2^2\,m\,n^3 - 48\,c^3\,D2^2\,m\,n^3 + 12\,c^4\,D2^2\,m\,n^3 - 12\,D2^2\,m^2\,n^3 + 60\,c\,D2^2\,m^2\,n^3 - 120\,c^2\,D2^2\,m^2\,n^3 +$
$120\,c^3\,D2^2\,m^2\,n^3 - 60\,c^4\,D2^2\,m^2\,n^3 + 12\,c^5\,D2^2\,m^2\,n^3 + 4\,D2^2\,m^3\,n^3 - 24\,c\,D2^2\,m^3\,n^3 +$
$60\,c^2\,D2^2\,m^3\,n^3 - 80\,c^3\,D2^2\,m^3\,n^3 + 60\,c^4\,D2^2\,m^3\,n^3 - 24\,c^5\,D2^2\,m^3\,n^3 + 4\,c^6\,D2^2\,m^3\,n^3 + D2^2\,n^4 -$
$4\,c\,D2^2\,n^4 + 6\,c^2\,D2^2\,n^4 - 4\,c^3\,D2^2\,n^4 + c^4\,D2^2\,n^4 - 3\,D2^2\,m\,n^4 + 15\,c\,D2^2\,m\,n^4 - 30\,c^2\,D2^2\,m\,n^4 +$
$30\,c^3\,D2^2\,m\,n^4 - 15\,c^4\,D2^2\,m\,n^4 + 3\,c^5\,D2^2\,m\,n^4 + 3\,D2^2\,m^2\,n^4 - 18\,c\,D2^2\,m^2\,n^4 + 45\,c^2\,D2^2\,m^2\,n^4 -$
$60\,c^3\,D2^2\,m^2\,n^4 + 45\,c^4\,D2^2\,m^2\,n^4 - 18\,c^5\,D2^2\,m^2\,n^4 + 3\,c^6\,D2^2\,m^2\,n^4 - D2^2\,m^3\,n^4 + 7\,c\,D2^2\,m^3\,n^4 -$
$21\,c^2\,D2^2\,m^3\,n^4 + 35\,c^3\,D2^2\,m^3\,n^4 - 35\,c^4\,D2^2\,m^3\,n^4 + 21\,c^5\,D2^2\,m^3\,n^4 - 7\,c^6\,D2^2\,m^3\,n^4 + c^7\,D2^2\,m^3\,n^4$

So the final differential equation is given by the sum (head Plus) term of the following resultant.

In[7]:= res3 = Factor[Resultant[res1a, res2a, c]] // Simplify

Out[7]= $-m^5\,n^{10}\,(16\,D1^{14}\,(-1 + m)\,(m - n)^6\,(-1 + n)^2 -$
$\qquad 27\,D1^2\,D2^4\,m^4\,n^2 - D2^6\,m^4\,n^2 - 8\,D1^{12}\,(m - n)^3\,(2\,m^4\,(-1 + n)^2 + 2\,n^3\,(2 - 2\,n + n^2) -$
$\qquad\qquad 2\,m\,n^2\,(6 - 5\,n + n^2 + n^3) + 3\,m^2\,n\,(13 - 20\,n + 7\,n^2 + 2\,n^3) - m^3\,(4 + 29\,n - 64\,n^2 + 33\,n^3)) -$
$\qquad D1^{14}\,D2^2\,m^2\,n\,(243\,m^2\,n + D2^2\,(m - n)^2\,(-10\,n + m\,(-8 + 9\,n))) +$
$\qquad D1^6\,n\,(-729\,m^4\,n + D2^4\,(-1 + m)\,(m - n)^4\,n -$
$\qquad\qquad 9\,D2^2\,m^2\,(-3\,m\,(-12 + n)\,n^2 - 10\,n^3 + 3\,m^2\,n\,(4 - 16\,n + 3\,n^2) + 2\,m^3\,(8 - 15\,n + 9\,n^2))) -$
$\qquad D1^8\,(2\,D2^2\,(m - n)^2\,(2\,m\,(-10 + n)\,n^3 + 4\,n^4 + m^3\,n\,(8 + 15\,n - 33\,n^2) -$
$\qquad\qquad 3\,m^2\,n^2\,(9 - 19\,n + 2\,n^2) + 4\,m^4\,(2 - 5\,n + 3\,n^2)) +$

$$27 \, m^2 \, n \, (24 \, m \, n^2 - 8 \, n^3 - 3 \, m^2 \, n \, (26 - 18 \, n + 9 \, n^2) + m^3 \, (8 + 27 \, n - 54 \, n^2 + 27 \, n^3))) +$$
$$D1^{10} \, (8 \, (-2 + D2^2 \, (-1 + n)) \, n^6 - 8 \, m^6 \, (2 + (27 + D2^2) \, n - (54 + D2^2) \, n^2 + 27 \, n^3) +$$
$$8 \, m^3 \, n^3 \, (202 - 135 \, n + 27 \, n^2 + 27 \, n^3 - 10 \, D2^2 \, (-1 + n^2)) +$$
$$8 \, m^2 \, n^4 \, (-84 + 54 \, n - 27 \, n^2 + 5 \, D2^2 \, (-2 + n + n^2)) - 8 \, m \, n^5 \, (-12 + D2^2 \, (-5 + 4 \, n + n^2)) -$$
$$m^4 \, n^2 \, (40 \, D2^2 \, (1 + n - 2 \, n^2) + 3 \, (755 - 702 \, n + 27 \, n^2 + 216 \, n^3)) +$$
$$m^5 \, n \, (D2^2 \, (8 + 32 \, n - 40 \, n^2) + 3 \, (176 + 315 \, n - 846 \, n^2 + 459 \, n^3))))$$

So we finally obtain the following differential equation.

$$16 \, (m-1) \, (m-n)^6 \, (n-1)^2 \left(\frac{\partial \Pi(n; \phi \mid m)}{\partial \phi} \right)^{14} -$$

$$8 \, (m-n)^3 \, (2 \, (n-1)^2 \, m^4 - (n \, (n \, (33 \, n - 64) + 29) + 4) \, m^3 + 3 \, n \, (n \, (n \, (2 \, n + 7) - 20) + 13) \, m^2 -$$
$$2 \, n^2 \, (n \, (n^2 + n - 5) + 6) \, m + 2 \, n^3 \, ((n-2) \, n + 2)) \left(\frac{\partial \Pi(n; \phi \mid m)}{\partial \phi} \right)^{12} +$$

$$\left(-8 \left(2 - \left(\left(\frac{\partial^2 \Pi(n; \phi \mid m)}{\partial \phi^2} \right)^2 - 27 \, n + 27 \right) (n-1) \, n \right) m^6 +$$

$$n \left(3 \, (9 \, n \, (n \, (51 \, n - 94) + 35) + 176) - 8 \left(\frac{\partial^2 \Pi(n; \phi \mid m)}{\partial \phi^2} \right)^2 (n-1) \, (5 \, n + 1) \right) m^5 -$$

$$n^2 \left(40 \, (-2 \, n^2 + n + 1) \left(\frac{\partial^2 \Pi(n; \phi \mid m)}{\partial \phi^2} \right)^2 + 3 \, (27 \, n \, (8 \, n^2 + n - 26) + 755) \right) m^4 +$$

$$8 \, n^3 \left(-10 \, (n^2 - 1) \left(\frac{\partial^2 \Pi(n; \phi \mid m)}{\partial \phi^2 \, \phi} \right)^2 + 27 \, n \, (n^2 + n - 5) + 202 \right) m^3 +$$

$$8 \, n^4 \left(5 \, (n^2 + n - 2) \left(\frac{\partial^2 \Pi(n; \phi \mid m)}{\partial \phi^2} \right)^2 - 27 \, (n-2) \, n - 84 \right) m^2 -$$

$$8 \, n^5 \left(\left(\frac{\partial^2 \Pi(n; \phi \mid m)}{\partial \phi^2} \right)^2 (n-1) \, (n+5) - 12 \right) m + 8 \left(\left(\frac{\partial^2 \Pi(n; \phi \mid m)}{\partial \phi^2} \right)^2 (n-1) - 2 \right) n^6 \right)$$

$$\left(\frac{\partial \Pi(n; \phi \mid m)}{\partial \phi} \right)^{10} -$$

$$\left(27 \, n \, ((27 \, n \, (n-1)^2 + 8) \, m^3 - 3 \, n \, (9 \, (n-2) \, n + 26) \, m^2 + 24 \, n^2 \, m - 8 \, n^3) \, m^2 + 2 \, (m-n)^2 \right.$$

$$(4 \, (n-1) \, (3 \, n - 2) \, m^4 + n \, (3 \, n \, (5 - 11 \, n) + 8) \, m^3 - 3 \, (n-9) \, n^2 \, (2 \, n - 1) \, m^2 + 2 \, (n-10) \, n^3 \, m + 4 \, n^4)$$

$$\left. \left(\frac{\partial^2 \Pi(n; \phi \mid m)}{\partial \phi^2} \right)^2 \right) \left(\frac{\partial \Pi(n; \phi \mid m)}{\partial \phi} \right)^8 +$$

$$n \left(-729 \, n \, m^4 + 9 \, (-2 \, (3 \, n \, (3 \, n - 5) + 8) \, m^3 - 3 \, n \, (n \, (3 \, n - 16) + 4) \, m^2 + 3 \, (n-12) \, n^2 \, m + 10 \, n^3) \right.$$

$$\left. \left(\frac{\partial^2 \Pi(n; \phi \mid m)}{\partial \phi^2} \right)^2 m^2 + (m-1) \, (m-n)^4 \, n \left(\frac{\partial^2 \Pi(n; \phi \mid m)}{\partial \phi^2} \right)^4 \right) \left(\frac{\partial \Pi(n; \phi \mid m)}{\partial \phi} \right)^6 -$$

$$\left(\frac{\partial^2 \Pi(n; \phi \mid m)}{\partial \phi^2} \right)^2 m^2 \, n \left(243 \, n \, m^2 + (m-n)^2 \, (m \, (9 \, n - 8) - 10 \, n) \left(\frac{\partial^2 \Pi(n; \phi \mid m)}{\partial \phi^2} \right)^2 \right) \left(\frac{\partial \Pi(n; \phi \mid m)}{\partial \phi} \right)^4 -$$

$$27 \left(\frac{\partial^2 \Pi(n; \phi \mid m)}{\partial \phi^2} \right)^4 m^4 \, n^2 \left(\frac{\partial \Pi(n; \phi \mid m)}{\partial \phi} \right)^2 - \left(\frac{\partial^2 \Pi(n; \phi \mid m)}{\partial \phi^2} \right)^6 m^4 \, n^2 = 0$$

Here is a quick numerical check that the derived differential equation is correct. (A symbolic check would be much more expensive because of the high powers in res3.)

```
In[8]:= res3[[-1]] /. {D1 -> D[EllipticPi[n, φ, m], φ],
                       D2 -> D[EllipticPi[n, φ, m], φ, φ]} /.
                      {n -> 1/2, m -> 1/3, φ -> N[3/7, 32]}
```

Out[8]= $0. \times 10^{-31}$

d) This is the integrand.

```
In[1]:= f[x_] = Sqrt[3 - 12 x]/Sqrt[(-1 + 12 x)(-1 - 24 x + 48 x^2)];
```

Unfortunately, just using `Integrate` to get the definite integral fails. (If this had worked, this would not be an exercise of the *GuideBooks*.)

```
In[2]:= lowerLimit = (3 - 2 Sqrt[3])/12;
       upperLimit = 1/12;
       ε = 10^-40;

       Integrate[f[x], {x, lowerLimit, upperLimit}]
```

Out[5]= $\displaystyle\int_{\frac{1}{12}(3-2\sqrt{3})}^{\frac{1}{12}} \frac{\sqrt{3 - 12\,x}}{\sqrt{(-1 + 12\,x)\,(-1 - 24\,x + 48\,x^2)}}\; dx$

The integrand has three singularities.

```
In[6]:= singularities = x /. NSolve[(-1 + 12 x)(-1 - 24 x + 48 x^2) == 0, x]
```

Out[6]= $\{-0.0386751, 0.0833333, 0.538675\}$

The integration region extends between the left-most two singularities.

```
In[7]:= With[{ε = 10^-6}, Show[
          Plot[{Re[f[x]], Im[f[x]]}, {x, #1 + ε, #2 - ε},
               DisplayFunction -> Identity,
               PlotStyle -> {{Thickness[0.002]},
                             {Thickness[0.002], Dashing[{0.02, 0.01}]}}]& @@@
               (* domains for continuous curve parts *)
               Partition[Flatten[{-3/4, singularities, 3/4}], 2, 1],
          DisplayFunction -> $DisplayFunction,
          Frame -> True, Axes -> False, PlotRange -> {-10, 10},
          (* x-interval of interest *)
          Prolog -> {{Hue[0], Thickness[0.02],
                      Line[{{(3 - 2 Sqrt[3])/12, 0}, {1/12, 0}}]}}]]
```

Because each singularity is of the form $(x - x_0)^{1/2}$, the integrand is integrable at the singularities and `NIntegrate` can calculate a high-precision approximation of the integral.

```
In[8]:= NIntegrate[f[x], {x, lowerLimit, upperLimit},
                    WorkingPrecision -> 50, MaxRecursion -> 9, PrecisionGoal -> 30]
```

    NIntegrate::ncvb :
        NIntegrate failed to converge to prescribed accuracy after 10 recursive
           bisections in x near x = 0.08333333333333333`. More…

Out[8]= $0.29930766349188126811700097 + 0. \times 10^{-26}\,i$

The above `Power::infy` and `Infinity::indet` messages from `Integrate` often indicate problems with determining the definite integral from the indefinite one. (For example, expressions of the form $0/0$ result from substituting limits.) Indeed, *Mathematica* has no problems calculating the corresponding indefinite integral.

In[9]:= `int = Integrate[f[x], x] // FullSimplify`

Out[9]= $\left(\sqrt{-1+4x}\ (-1+12x)^{3/2}\ \sqrt{1+\dfrac{2(-1+\sqrt{3})}{-1+12x}}\right.$

$\sqrt{-1+\dfrac{2(1+\sqrt{3})}{-1+12x}}\ \left(\text{EllipticF}\left[\text{ArcSin}\left[\dfrac{\sqrt{-(-3+\sqrt{3})\ (-1+4x)}}{\sqrt{-1+12x}}\right],\ 2+\sqrt{3}\right]-\right.$

$\left.\left.\text{EllipticPi}\left[\dfrac{1}{2}(3+\sqrt{3}),\ \text{ArcSin}\left[\dfrac{\sqrt{-(-3+\sqrt{3})\ (-1+4x)}}{\sqrt{-1+12x}}\right],\ 2+\sqrt{3}\right]\right)\right)\Big/$

$\left(6\ 3^{1/4}\ \sqrt{-1+\sqrt{3}}\ \sqrt{1-4x}\ \sqrt{1+12x\ (1+4x\ (-7+12x))}\right)$

The indefinite integral is a smooth function inside the region of integration.

In[10]:= `Plot[Re @ int, {x, lowerLimit, upperLimit},`
         `Frame -> True, Axes -> False]`

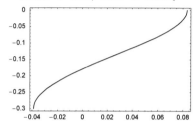

But when substituting limits, we get indeterminate expressions.

In[11]:= `(* avoid further message generation *)`
         `Internal`DeactivateMessages[`
            `{int /. x -> upperLimit, int /. x -> lowerLimit} // N[#, 22]&]`

Out[12]= `{Indeterminate, Indeterminate}`

Substituting *x*-values near to the endpoints of integration results in well-defined expressions. The so-obtained numerical value agrees with the one from the numerical integration.

In[13]:= `N[(int /. x -> upperLimit - ε) -`
         `(int /. x -> lowerLimit + ε), 200] // N`
Out[13]= `0.299308 + 0. i`

`Series` is be able to obtain expansions at the endpoints of integration.

In[14]:= `Assuming[Element[x, Reals] && x < upperLimit,`
         `Series[int, {x, upperLimit, 1}]] //`
         `FullSimplify[#, Element[x, Reals] && x < upperLimit]&`

Out[14]= $\left(\text{EllipticF}\left[\pi-\dfrac{1}{16}i\ (-1+\sqrt{3})\ (-1+12x)-\dfrac{1}{2}i\ \text{Log}\left[-\dfrac{2}{9}(-3+\sqrt{3})\right]+\dfrac{1}{2}i\ \text{Log}\left[-\dfrac{1}{12}+x\right],\right.\right.$

$2+\sqrt{3}\Big]-\text{EllipticPi}\left[\dfrac{1}{2}(3+\sqrt{3}),\ \pi-\dfrac{1}{16}i\ (-1+\sqrt{3})\ (-1+12x)-\right.$

$\left.\left.\dfrac{1}{2}i\ \text{Log}\left[-\dfrac{2}{9}(-3+\sqrt{3})\right]+\dfrac{1}{2}i\ \text{Log}\left[-\dfrac{1}{12}+x\right],\ 2+\sqrt{3}\right]\right)\ O\left[x-\dfrac{1}{12}\right]^0$

The main difficulty in obtaining the last expansion is the $\sin^{-1}((3^{1/2}+1+4/(12x-1))^{1/2}/(2^{1/2}\ 3^{1/4}))$ term inside the elliptic integrals. For $x \to 1/12$, this expression diverges logarithmically.

In[15]:= `elliptics = Cases[int, _EllipticF | _EllipticPi, Infinity]`

Out[15]= $\left\{\text{EllipticF}\left[\text{ArcSin}\left[\dfrac{\sqrt{-(-3+\sqrt{3})\ (-1+4x)}}{\sqrt{-1+12x}}\right],\ 2+\sqrt{3}\right],\right.$

$\left.\text{EllipticPi}\left[\dfrac{1}{2}(3+\sqrt{3}),\ \text{ArcSin}\left[\dfrac{\sqrt{-(-3+\sqrt{3})\ (-1+4x)}}{\sqrt{-1+12x}}\right],\ 2+\sqrt{3}\right]\right\}$

In[16]:= `Assuming[Element[x, Reals] && x < upperLimit,`
      `Series[elliptics[[1, 1]], {x, upperLimit, 1}]] //`
                `FullSimplify[#, Element[x, Reals] && x < upperLimit]&`

Out[16]= $\left(\pi - \frac{1}{2}\ i\ \text{Log}\left[-\frac{2}{9}\left(-3 + \sqrt{3}\right)\right] + \frac{1}{2}\ i\ \text{Log}\left[-\frac{1}{12} + x\right]\right) + -\frac{3}{4}\ i\ \left(-1 + \sqrt{3}\right)\ \left(x - \frac{1}{12}\right) + O\left[x - \frac{1}{12}\right]^2$

Resorting to the integral representation of $F(\varphi \mid m)$ allows us to calculate the value of $F(i\infty \mid m)$. *Mathematica* does not evaluate this expression by default.

In[17]:= `EllipticF[I ∞, m]`

Out[17]= `EllipticF[i ∞, m]`

In[18]:= `Integrate[1/(Sqrt[1 - t^2] Sqrt[1 - m t^2]), {t, 0, I Infinity},`
                `GenerateConditions -> False] // PowerExpand`

Out[18]= `i EllipticK[1 - m]`

We generate a rule to replace the incomplete elliptic integral of the first kind.

In[19]:= `FRule = EllipticF[φ_, m_] :> EllipticK[m] - EllipticK[1/m]/Sqrt[m];`

In[20]:= `(* quick numerical check for the rule *)`
      `elliptics[[1]] - FRule[[2]] /. {x -> upperLimit - ε,`
                    `m -> elliptics[[1, 2]]} // N[#, 100]& // N`

Out[21]= $1.95035 \times 10^{-20} + 0.\ i$

The prefactor of $F(i\infty \mid m)$ also results in expressions of the form 0/0 after substitution of $x \to 1/2$. But because it is an algebraic function, it can be easily calculated using `Limit`.

In[22]:= `cf = Coefficient[int, elliptics[[1]]];`

In[23]:= `{cfl, cfu} = FullSimplify[RootReduce @`
                      `{Limit[cf, x -> upperLimit, Direction -> +1],`
                        `Limit[cf, x -> lowerLimit, Direction -> -1]}]`

Out[23]= $\left\{-\frac{1}{6}\sqrt{\frac{1}{2}\left(3 + \sqrt{3}\right)}, -\frac{1}{6}\sqrt{\frac{1}{2}\left(3 + \sqrt{3}\right)}\right\}$

In[24]:= `fPart = (cfu (elliptics[[1]] /. FRule) -`
                `cfl (elliptics[[1]] /. x -> lowerLimit)) // FullSimplify`

Out[24]= $\frac{1}{6}\sqrt{\frac{1}{2}\left(3 - \sqrt{3}\right)}\ \text{EllipticK}[2 - \sqrt{3}\,]$

In[25]:= `(* quick numerical check *)`
      `{(cf elliptics[[1]] /. (x -> upperLimit - 10^-40)) -`
      `(cf elliptics[[1]] /. (x -> lowerLimit + 10^-40)) // N[#, 100]& // N,`
      `N[fPart]}`

Out[26]= $\{0.225014 + 0.\ i,\ 0.225014\}$

We proceed in a similar way for $\Pi(n; \varphi \mid m)$ and use its transformation formula for imaginary φ [201].

In[27]:= `ΠRule = EllipticPi[n_, φ_, m_] :> (* ±1 depending on ±I φ *)`
              `I (EllipticK[1 - m] - n EllipticPi[1 - n, 1 - m])/(1 - n);`

In[28]:= `plusMinus1 = Round[elliptics[[2]]/ΠRule[[2]] /. {x -> 1/12 - 10^-40,`
                `m -> elliptics[[2, 3]], n -> elliptics[[2, 1]]}];`

      `(* quick numerical check for the rule *)`
      `{elliptics[[2]] , plusMinus1 ΠRule[[2]]} /.`
                  `{x -> upperLimit - ε, m -> elliptics[[2, 3]],`
                  `n -> elliptics[[2, 1]]} // N[#, 30]& // N`

Out[31]= $\{0. - 0.522639\ i,\ 0. - 0.522639\ i\}$

In[32]:= `cp = Coefficient[int, elliptics[[2]]];`

In[33]:= `{cpl, cpu} = FullSimplify[RootReduce @`
                      `{Limit[cp, x -> upperLimit, Direction -> 1],`
                        `Limit[cp, x -> lowerLimit, Direction -> -1]}]`

Out[33]= $\left\{\frac{1}{6}\sqrt{\frac{1}{2}\left(3+\sqrt{3}\right)}, \frac{1}{6}\sqrt{\frac{1}{2}\left(3+\sqrt{3}\right)}\right\}$

In[34]:= `pPart = (cpu (plusMinus1 elliptics[[2]] /. ΠRule) -`
`        cpl (elliptics[[2]] /. x -> lowerLimit)) // FullSimplify`

Out[34]= $-\dfrac{1}{6\,(1+\sqrt{3}\,)}$

$\left(i\sqrt{\frac{1}{2}\left(3+\sqrt{3}\right)}\left(-2\,\text{EllipticK}[-1-\sqrt{3}\,]+\left(3+\sqrt{3}\right)\text{EllipticPi}\!\left[\frac{1}{2}\left(-1-\sqrt{3}\right),-1-\sqrt{3}\,\right]-\right.\right.$

$\left.\left.i\left(1+\sqrt{3}\right)\text{EllipticPi}\!\left[\frac{1}{2}\left(3+\sqrt{3}\right),2+\sqrt{3}\,\right]\right)\right)$

In[35]:= `(* quick numerical check *)`
`    {(cp elliptics[[2]] /. (x -> upperLimit - ε)) -`
`     (cp elliptics[[2]] /. (x -> lowerLimit + ε)) // N[#, 100]& // N,`
`    N[pPart]}`

Out[36]= $\{0.0742937 + 0.\,i,\ 0.0742937 + 7.9176\times10^{-16}\,i\}$

Now, we can put all obtained expressions together.

In[37]:= `res = fPart + pPart // FullSimplify`

Out[37]= $\dfrac{1}{6\sqrt{2}}\left(\sqrt{3-\sqrt{3}}\ \text{EllipticK}[2-\sqrt{3}\,]-\dfrac{1}{1+\sqrt{3}}\right.$

$\left(i\sqrt{3+\sqrt{3}}\left(-2\,\text{EllipticK}[-1-\sqrt{3}\,]+\left(3+\sqrt{3}\right)\text{EllipticPi}\!\left[\frac{1}{2}\left(-1-\sqrt{3}\right),-1-\sqrt{3}\,\right]-\right.\right.$

$\left.\left.\left.i\left(1+\sqrt{3}\right)\text{EllipticPi}\!\left[\frac{1}{2}\left(3+\sqrt{3}\right),2+\sqrt{3}\,\right]\right)\right)\right)$

In[38]:= `N[res]`

Out[38]= $0.299308 + 7.9176\times10^{-16}\,i$

Using a transformation for $K(z)$ allows us to eliminate the explicit i from the last expression.

In[39]:= `KRule[z_] = EllipticK[z] -> -I EllipticK[1 - z] + EllipticK[1/z]/Sqrt[z];`

In[40]:= `badzs = Select[First /@ Union[Cases[res, _EllipticK, Infinity]],`
`        # > 1&];`
`    res /. If[badzs =!= {}, KRule[badzs[[1]]], {}] // FullSimplify`

Out[41]= $\dfrac{1}{6\sqrt{2}}\left(\sqrt{3-\sqrt{3}}\ \text{EllipticK}[2-\sqrt{3}\,]-\dfrac{1}{1+\sqrt{3}}\right.$

$\left(i\sqrt{3+\sqrt{3}}\left(-2\,\text{EllipticK}[-1-\sqrt{3}\,]+\left(3+\sqrt{3}\right)\text{EllipticPi}\!\left[\frac{1}{2}\left(-1-\sqrt{3}\right),-1-\sqrt{3}\,\right]-\right.\right.$

$\left.\left.\left.i\left(1+\sqrt{3}\right)\text{EllipticPi}\!\left[\frac{1}{2}\left(3+\sqrt{3}\right),2+\sqrt{3}\,\right]\right)\right)\right)$

So we have the following result.

$$\frac{\sqrt{1+\sqrt{3}}}{12}\left(\sqrt{3-\sqrt{3}}\ K\!\left(\frac{1-\sqrt{3}}{2}\right)-2\sqrt[4]{3}\left(K(2-\sqrt{3})+(\sqrt{3}-3)\Pi(\sqrt{3}-2\,|\,2-\sqrt{3})\right)\right)$$

It agrees with the high-precision numerical integration value calculated above.

In[42]:= `N[%, 30]`

Out[42]= $0.299307663491881268117009677712 + 0.\times10^{-31}\,i$

e) We start by deriving partial differential equations for $z(x, y)$ by differentiating the starting equation.

In[1]:= `(* D[Integrate[f[τ], {τ, 0, T}], T] == f[T] *)`
`    integrand[τ_] := 1/Sqrt[(1 - τ^2) (1 - k τ^2)]`

In[3]:= `eqs1 = FullSimplify[{integrand[x] - integrand[z[x, y]] D[z[x, y], x],`
`                     integrand[y] - integrand[z[x, y]] D[z[x, y], y]}]`

Out[3]= $\left\{ \dfrac{1}{\sqrt{(-1+x^2)\,(-1+k\,x^2)}} - \dfrac{z^{(1,0)}\,[x,\,y]}{\sqrt{(-1+z[x,\,y]^2)\,(-1+k\,z[x,\,y]^2)}} \right.,$

$\left. \dfrac{1}{\sqrt{(-1+y^2)\,(-1+k\,y^2)}} - \dfrac{z^{(0,1)}\,[x,\,y]}{\sqrt{(-1+z[x,\,y]^2)\,(-1+k\,z[x,\,y]^2)}} \right\}$

For further algebraic manipulations we eliminate the radicals.

In[4]:= **Sx = Numerator[Together[eqs1[[1, 1]]^2 - eqs1[[1, 2]]^2]]**

Out[4]= $1 - z[x,\,y]^2 - k\,z[x,\,y]^2 + k\,z[x,\,y]^4 - z^{(1,0)}\,[x,\,y]^2 +$
$x^2\,z^{(1,0)}\,[x,\,y]^2 + k\,x^2\,z^{(1,0)}\,[x,\,y]^2 - k\,x^4\,z^{(1,0)}\,[x,\,y]^2$

In[5]:= **Sy = Numerator[Together[eqs1[[2, 1]]^2 - eqs1[[2, 2]]^2]]**

Out[5]= $1 - z[x,\,y]^2 - k\,z[x,\,y]^2 + k\,z[x,\,y]^4 - z^{(0,1)}\,[x,\,y]^2 +$
$y^2\,z^{(0,1)}\,[x,\,y]^2 + k\,y^2\,z^{(0,1)}\,[x,\,y]^2 - k\,y^4\,z^{(0,1)}\,[x,\,y]^2$

Now it is straightforward to show that $z(x,\,y)$ obeys the derived differential equations.

In[6]:= **id = ((y^2 - z^2)^2 + x^4 (k y^2 z^2 - 1)^2 -**
2 x^2 (z^2 + y^2 (1 + z^2 (k (y^2 + z^2 - 2) - 2))));

In[7]:= **(PolynomialGCD[id, Numerator[Together[**
#1 /. Solve[(D[id /. z -> z[x, y], #2]) == 0,
D[z[x, y], #2]][[1]]]] /. z[x, y] -> z]& @@@
{{Sx, x}, {Sy, y}} // Union

Out[7]= $\{x^4 - 2\,x^2\,y^2 + y^4 - 2\,x^2\,z^2 - 2\,y^2\,z^2 + 4\,x^2\,y^2\,z^2 +$
$4\,k\,x^2\,y^2\,z^2 - 2\,k\,x^4\,y^2\,z^2 - 2\,k\,x^2\,y^4\,z^2 + z^4 - 2\,k\,x^2\,y^2\,z^4 + k^2\,x^4\,y^4\,z^4\}$

Checking the cases $x = 0$ and $y = 0$ shows finally that z as a function of x and y fulfills the given identity.

f) By the Biot–Savart rule for a line current (see [114], [531], [718], [978], [1207], and [257]),

$$A(r) = \frac{\mu\,\mu_0}{4\,\pi} \int_{R^3} \frac{j(r')}{|r - r'|}\,dr'$$

we easily find the φ component $A_\varphi(r) = A\varphi$ of the vector potential $A(r)$ of a circular current in the x,y-plane with current current and radius r0 in cylindrical coordinate. i is the magnitude of the current.

In[1]:= **i μ μ0 r0/(4 Pi) Integrate[Cos[φ]/Sqrt[z^2 + r^2 + r0^2 - 2r r0 Cos[φ]],**
{φ, 0, 2Pi}, GenerateConditions -> False]

Out[1]= $\dfrac{i\,\mu\,\mu 0\,\left(-2\,((r-r0)^2 + z^2)\,\text{EllipticE}\!\left[-\frac{4\,r\,r0}{(r-r0)^2 + z^2}\right] + 2\,(r^2 + r0^2 + z^2)\,\text{EllipticK}\!\left[-\frac{4\,r\,r0}{(r-r0)^2 + z^2}\right]\right)}{4\,\pi\,r\,\sqrt{(r-r0)^2 + z^2}}$

The last can be rewritten in the following form.

In[2]:= **Aφ = i μ μ0 r0 /(4 Pi) Pi r r0 ***
Hypergeometric2F1[3/4, 5/4, 2, 4 r^2 r0^2/(r^2 + r0^2 + z^2)^2]/
(r^2 + r0^2 + z^2)^(3/2)

Out[2]= $\dfrac{r\,r0^2\,i\,\mu\,\mu 0\,\text{Hypergeometric2F1}\!\left[\frac{3}{4},\,\frac{5}{4},\,2,\,\frac{4\,r^2\,r0^2}{(r^2 + r0^2 + z^2)^2}\right]}{4\,(r^2 + r0^2 + z^2)^{3/2}}$

Note the compactness of this formula, expressed in terms of hypergeometric functions, compared with the longer formulas using elliptic integrals found in textbooks and the original literature [538]; for some similar hypergeometric expressions for inductance formulas, see [650], [651].

Using $B(r) = \text{rot}\,A(r)$ in cylindrical coordinates, this is the z-component of the magnetic field $B(r)$.

In[3]:= **Bz = 1/r D[r Aφ, r]**

Out[3]= $\dfrac{1}{r}\left(-\dfrac{3\,r^3\,r0^2\,i\,\mu\,\mu0\;\text{Hypergeometric2F1}\left[\frac{3}{4},\,\frac{5}{4},\,2,\,\frac{4\,r^2\,r0^2}{(r^2+r0^2+z^2)^2}\right]}{4\,(r^2+r0^2+z^2)^{5/2}} + \right.$

$\dfrac{r\,r0^2\,i\,\mu\,\mu0\;\text{Hypergeometric2F1}\left[\frac{3}{4},\,\frac{5}{4},\,2,\,\frac{4\,r^2\,r0^2}{(r^2+r0^2+z^2)^2}\right]}{2\,(r^2+r0^2+z^2)^{3/2}} +$

$\left. \dfrac{15\,r^2\,r0^2\left(-\frac{16\,r^3\,r0^2}{(r^2+r0^2+z^2)^3}+\frac{8\,r\,r0^2}{(r^2+r0^2+z^2)^2}\right)i\,\mu\,\mu0\;\text{Hypergeometric2F1}\left[\frac{7}{4},\,\frac{9}{4},\,3,\,\frac{4\,r^2\,r0^2}{(r^2+r0^2+z^2)^2}\right]}{128\,(r^2+r0^2+z^2)^{3/2}} \right)$

Thus, this is the radial component.

In[4]:= **Br = -D[Aφ, z]**

Out[4]= $\dfrac{3\,r\,r0^2\,z\,i\,\mu\,\mu0\;\text{Hypergeometric2F1}\left[\frac{3}{4},\,\frac{5}{4},\,2,\,\frac{4\,r^2\,r0^2}{(r^2+r0^2+z^2)^2}\right]}{4\,(r^2+r0^2+z^2)^{5/2}} +$

$\dfrac{15\,r^3\,r0^4\,z\,i\,\mu\,\mu0\;\text{Hypergeometric2F1}\left[\frac{7}{4},\,\frac{9}{4},\,3,\,\frac{4\,r^2\,r0^2}{(r^2+r0^2+z^2)^2}\right]}{8\,(r^2+r0^2+z^2)^{9/2}}$

To get series expansions of the magnetic field, we can also proceed in another way. Instead of first carrying out the integration and then making a series expansion around $r = 0$, we can first make a series expansion of the integrand around $r = 0$ and then integrate term by term. The next inputs demonstrate this process.

In[5]:= **AφIntegrand = i μ μ0 r0/(4 Pi) Cos[φ]/Sqrt[z^2 + r^2 + r0^2 - 2r r0 Cos[φ]];**

In[6]:= **BzIntegrand = Simplify /@ Series[1/r D[r AφIntegrand, r], {r, 0, 5}]**

Out[6]= $\dfrac{r0\,i\,\mu\,\mu0\,\text{Cos}[\varphi]}{4\,\pi\,\sqrt{r0^2+z^2}\,r} + \dfrac{r0^2\,i\,\mu\,\mu0\,\text{Cos}[\varphi]^2}{2\,\pi\,(r0^2+z^2)^{3/2}} + \dfrac{3\,r0\,i\,\mu\,\mu0\,\text{Cos}[\varphi]\,(r0^2-2\,z^2+3\,r0^2\,\text{Cos}[2\,\varphi])\,r}{16\,\pi\,(r0^2+z^2)^{5/2}} +$

$\dfrac{r0^2\,i\,\mu\,\mu0\,\text{Cos}[\varphi]^2\,(-3\,(r0^2+z^2)+5\,r0^2\,\text{Cos}[\varphi]^2)\,r^2}{2\,\pi\,(r0^2+z^2)^{7/2}} +$

$\dfrac{5\,r0\,i\,\mu\,\mu0\,\text{Cos}[\varphi]\,(9\,r0^4-72\,r0^2\,z^2+24\,z^4+20\,r0^2\,(r0^2-6\,z^2)\,\text{Cos}[2\,\varphi]+35\,r0^4\,\text{Cos}[4\,\varphi])\,r^3}{256\,\pi\,(r0^2+z^2)^{9/2}} +$

$\dfrac{3\,r0^2\,i\,\mu\,\mu0\,\text{Cos}[\varphi]^2\,(29\,r0^4-40\,r0^2\,z^2+120\,z^4-28\,r0^2\,(r0^2+10\,z^2)\,\text{Cos}[2\,\varphi]+63\,r0^4\,\text{Cos}[4\,\varphi])\,r^4}{128\,\pi\,(r0^2+z^2)^{11/2}} +$

$+\dfrac{1}{2048\,\pi\,(r0^2+z^2)^{13/2}}$

$(7\,r0\,i\,\mu\,\mu0\,\text{Cos}[\varphi]\,(50\,r0^6-900\,r0^4\,z^2+1200\,r0^2\,z^4-160\,z^6+105\,r0^2\,(r0^4-16\,r0^2\,z^2+16\,z^4)$

$\text{Cos}[2\,\varphi]+126\,r0^4\,(r0^2-10\,z^2)\,\text{Cos}[4\,\varphi]+231\,r0^6\,\text{Cos}[6\,\varphi])\,r^5)+O[r]^6$

In[7]:= **Integrate[#, {φ, 0, 2Pi}]& /@ Normal[BzIntegrand]**

Out[7]= $\dfrac{3\,r^2\,r0^2\,(r0^2-4\,z^2)\,i\,\mu\,\mu0}{8\,(r0^2+z^2)^{7/2}} + \dfrac{r0^2\,i\,\mu\,\mu0}{2\,(r0^2+z^2)^{3/2}} + \dfrac{45\,r^4\,r0^2\,(r0^4-12\,r0^2\,z^2+8\,z^4)\,i\,\mu\,\mu0}{128\,(r0^2+z^2)^{11/2}}$

In[8]:= **BrIntegrand = Simplify /@ Series[-D[AφIntegrand, z], {r, 0, 5}]**

Out[8]= $\dfrac{r0\,z\,i\,\mu\,\mu0\,\text{Cos}[\varphi]}{4\,\pi\,(r0^2+z^2)^{3/2}} + \dfrac{3\,r0^2\,z\,i\,\mu\,\mu0\,\text{Cos}[\varphi]^2\,r}{4\,\pi\,(r0^2+z^2)^{5/2}} + \dfrac{3\,r0\,z\,i\,\mu\,\mu0\,\text{Cos}[\varphi]\,(3\,r0^2-2\,z^2+5\,r0^2\,\text{Cos}[2\,\varphi])\,r^2}{16\,\pi\,(r0^2+z^2)^{7/2}} +$

$\dfrac{5\,r0^2\,z\,i\,\mu\,\mu0\,\text{Cos}[\varphi]^2\,(r0^2-6\,z^2+7\,r0^2\,\text{Cos}[2\,\varphi])\,r^3}{16\,\pi\,(r0^2+z^2)^{9/2}} +$

$\dfrac{15\,r0\,z\,i\,\mu\,\mu0\,\text{Cos}[\varphi]\,\left((r0^2+z^2)^2-14\,r0^2\,(r0^2+z^2)\,\text{Cos}[\varphi]^2+21\,r0^4\,\text{Cos}[\varphi]^4\right)r^4}{32\,\pi\,(r0^2+z^2)^{11/2}} +$

$\dfrac{21\,r0^2\,z\,i\,\mu\,\mu0\,\text{Cos}[\varphi]^2\,\left(5\,(r0^2+z^2)^2-30\,r0^2\,(r0^2+z^2)\,\text{Cos}[\varphi]^2+33\,r0^4\,\text{Cos}[\varphi]^4\right)r^5}{32\,\pi\,(r0^2+z^2)^{13/2}} + O[r]^6$

In[9]:= **Integrate[#, {φ, 0, 2Pi}]& /@ Normal[BrIntegrand]**

Out[9]= $\dfrac{15\,r^3\,r0^2\,z\,(3\,r0^2-4\,z^2)\,i\,\mu\,\mu0}{32\,(r0^2+z^2)^{9/2}} + \dfrac{3\,r\,r0^2\,z\,i\,\mu\,\mu0}{4\,(r0^2+z^2)^{5/2}} + \dfrac{105\,r^5\,r0^2\,z\,(5\,r0^4-20\,r0^2\,z^2+8\,z^4)\,i\,\mu\,\mu0}{256\,(r0^2+z^2)^{13/2}}$

We now look at both components in a neighborhood of the axis of symmetry.

In[10]:= **Series[Bz, {r, 0, 3}] // Simplify**

Out[10]= $\dfrac{r0^2\,i\,\mu\,\mu0}{2\,(r0^2 + z^2)^{3/2}} + \dfrac{3\,r0^2\,(r0^2 - 4\,z^2)\,i\,\mu\,\mu0\,r^2}{8\,(r0^2 + z^2)^{7/2}} + O[r]^4$

In[11]:= **Series[Br, {r, 0, 3}] // Simplify**

Out[11]= $\dfrac{3\,r0^2\,z\,i\,\mu\,\mu0\,r}{4\,(r0^2 + z^2)^{5/2}} + \dfrac{15\,r0^2\,z\,(3\,r0^2 - 4\,z^2)\,i\,\mu\,\mu0\,r^3}{32\,(r0^2 + z^2)^{9/2}} + O[r]^4$

We can very easily visualize the vector field **B(r)** in the r,z-plane ($r > 0$). To achieve this, we choose concrete values for the parameters.

In[12]:= **Needs["Graphics`PlotField`"]**

In[13]:= **valueRules = {μ -> 1, $\mu0$ -> 1, r0 -> 1, i -> 1};**
PlotVectorField[Evaluate[N[{Br, Bz} /. valueRules]],
{r, 0.1, 2.3, 0.14}, {z, -1.5, 1.5, 0.14}]

We clearly see the circular flow of the magnetic field around the wire. We could go on and use ContourPlot3D to make a picture of surfaces of constant magnitude of the magnetic field of the current under investigation.

For a Helmholtz coil, we have to superimpose the magnetic fields of two parallel circular rings, which are a distance r0 apart.

In[15]:= **BHr = (Br /. z -> z + r0/2) + (Br /. z -> z - r0/2);**

In[16]:= **BHz = (Bz /. z -> z + r0/2) + (Bz /. z -> z - r0/2);**

We now find the series expansion at the point {0, 0} of symmetry.

In[17]:= **{hz, hr} = (Series[Normal[Series[BHz, {r, 0, 5}]], {z, 0, 5}] //**
Normal // Simplify // PowerExpand)& /@ {BHz, BHr}

Out[17]= $\Big\{ \dfrac{1}{78125\,\sqrt{5}\,r0^9}\,(8\,(125\,r0^4\,(125\,r0^4 - 144\,z^4) - $
$\qquad 18\,r^4\,(375\,r0^4 - 6160\,r0^2\,z^2 + 8736\,z^4) + 240\,r^2\,(225\,r0^4\,z^2 - 616\,r0^2\,z^4))\,i\,\mu\,\mu0),$
$\qquad \dfrac{1}{78125\,\sqrt{5}\,r0^9}\,(8\,(125\,r0^4\,(125\,r0^4 - 144\,z^4) - 18\,r^4\,(375\,r0^4 - 6160\,r0^2\,z^2 + 8736\,z^4) + $
$\qquad 240\,r^2\,(225\,r0^4\,z^2 - 616\,r0^2\,z^4))\,i\,\mu\,\mu0)\Big\}$

After some minor manipulations, we obtain the following expression for the magnetic field in the vicinity of the symmetry point.

In[18]:= **$i\,\mu\,\mu0$/r0 * (Normal[Series[Normal[Series[Sqrt[hz^2 + hr^2],**
{z, 0, 5}]], {r, 0, 5}]] /.
(* remove prefactors *) Power[i^2 μ^2 $\mu0$^2/r0^2, 1/2] -> 1)

Out[18]= $\dfrac{1}{r0}\Bigg(\Bigg(\dfrac{8\,\sqrt{\tfrac{2}{5}}}{5} - \dfrac{1152\,\sqrt{\tfrac{2}{5}}\,z^4}{625\,r0^4} + r^4\Bigg(-\dfrac{432\,\sqrt{\tfrac{2}{5}}}{625\,r0^4} + \dfrac{177408\,\sqrt{\tfrac{2}{5}}\,z^2}{15625\,r0^6} - \dfrac{1257984\,\sqrt{\tfrac{2}{5}}\,z^4}{78125\,r0^8} \Bigg) + $

$\qquad r^2\Bigg(\dfrac{3456\,\sqrt{\tfrac{2}{5}}\,z^2}{625\,r0^4} - \dfrac{236544\,\sqrt{\tfrac{2}{5}}\,z^4}{15625\,r0^6} \Bigg)\Bigg)\,i\,\mu\,\mu0 \Bigg)$

We see that there are many terms of the order (radial or z-distance to the midpoint)/(ring radius)^4, not only the one of the form $(z/r_0)^4$, which are mentioned typically (e.g., [1156]).

We now graphically examine the homogeneity of the resulting magnetic field. To answer the question of which direction has the largest homogeneity, we consider $|\mathbf{B}(r, z)|$ as a function of r and z (right graphic).

```
In[19]:= Show[GraphicsArray[
            Block[{$DisplayFunction = Identity, ε = 10^-7},
                {(* vector field plot *)
                 PlotVectorField[Evaluate[N[{BHr, BHz} /. valueRules]],
                                 {r, ε, 2/3 + ε, (2/3 + 2 ε)/10},
                                 {z, -2/3, 2/3, (4/3)/10}],
                 (* 3D plot of field magnitude *)
                 Plot3D[Evaluate[Sqrt[Br^2 + Bz^2] /. valueRules],
                        {r, ε, 0.2}, {z, ε, 0.2},
                        PlotRange -> All, AxesLabel -> {"r", "z", None},
                        Compiled -> False]}]]]
```

The last plot shows the somewhat lesser homogeneity in the *z*-direction. For more on circular currents (including fields near the flow, formulas for the fields in terms of Legendre functions, etc.), see [454]; for saddle-shaped Helmholtz coils, see [570]. For the movement of particles in such magnetic fields, see [498].

3. Weierstrass Function

a) In view of the periodicity and then rotational symmetry of this function, it suffices to look at just one of these parts. Here is a detailed look at the part with the pole.

```
In[1]:= With[{pp = 90},
          Show[GraphicsArray[
          Plot3D[Re[#1[x + I y, 2.66, 2.66]], {x, -1, 1}, {y, -1, 1},
                 PlotPoints -> pp, SphericalRegion -> True,
                 Mesh -> False, ClipFill -> None,
                 (* indicate parameters *)
                 PlotLabel ->  StyleForm[#2 <> "[z, 8/3, 8/3I]",
                                         FontFamily -> "Courier", FontSize -> 6],
                 PlotRange -> {-6, 10}, DisplayFunction -> Identity]& @@@
           {{WeierstrassP, "℘"}, {WeierstrassPPrime, "℘'"}}]]]
```

Note that the half-periods for the values of the invariants $g_2 = 2.66$ and $g_3 = 2.66$ are given by $\omega = 1.18983...$ and $\omega' = -0.59491... + i\,1.23788...$. Using these values, we can implement a visualization of the real part or the imaginary part or the phase (determined via the first argument of manyPeriodParallelograms) Weierstrass function on a bigger region, consisting of $(m + 1)^2$ period parallelograms. (Here, we use a contour plot instead of Plot3D.)

```
In[2]:= manyPeriodParallelograms[reImAbsArg_, pp_, plotRange_, m_] :=
          Module[{ω, ωs, tab, gr, l, ls, h, gr1, gr2, cs = 20},
            (* the periods *)
            {ω, ωs} = WeierstrassHalfPeriods[{2.66, 2.66}];
            (* ContourPlot of reImAbsArg[WeierstrassP] == 0
               inside the period parallelogram;
               to do this transform the period parallelogram
```

```
                     into a square and then transform back *)
               gr = Graphics[ListContourPlot[
               Table[reImAbsArg @ WeierstrassP[-ω - ωs + s/pp 2ω +
                       t/pp 2ωs, 2.66, 2.66], {s, 0, pp}, {t, 0, pp}],
                   PlotRange -> plotRange, ContourShading -> False,
                   Contours -> cs, DisplayFunction -> Identity]][[1]];
               (* the back transformation *)
               l = Abs[2ω]; ls = Abs[2ωs]; h = ls Sin[Arg[ωs]];
               gr1 = Map[If[Head[#] === List,
                           {(#[[1]] - 1)/pp l, (#[[2]] - 1)/pp h}, #]&, gr, {-2}];
               gr2 = Map[If[Head[#] === List,
                           {#[[1]] + (#[[2]] - 0)/h Re[2ωs], #[[2]]}, #]&, gr1, {-2}];
               (* make more period parallelograms *)
               Show[Graphics[{Table[Map[If[Head[#] === List,
                           {#[[1]] + i Re[2ω] + j Re[2ωs],
                           #[[2]] + i Im[2ω] + j Im[2ωs]}, #]&, gr2, {-2}],
                       {i, -m, m}, {j, -m, m}]}], AspectRatio -> Automatic]];
```

This shows the contour lines of the square of the argument.

In[42]:= **manyPeriodParallelograms[Arg[#]^2&, 43, All, 1]**

b) Here is a picture of a suitable $\wp'(z; g_2, g_3)$. (Suitable means here that the period parallelogram has an appropriate extension for the picture to be constructed.)

In[1]:= **{g2, g3} = {-207.228 + 906.1 I, 658.759 + 188.243 I};**
       {ω1, ω2} = WeierstrassHalfPeriods[{g2, g3}];

In[3]:= **(* to avoid the edges of the period parallelogram *) ε = 10^-13;**
       pP[z_] = WeierstrassPPrime[z, g2, g3];

       pPP3D = ParametricPlot3D[(* use {Re[z], Im[z], Re[f[z]]} *)
                               {Re[#], Im[#], Re[pP[#]]}&[x ω1 + y ω2],
                               {x, -1 + ε, 1 - ε}, {y, -1 + ε, 1 - ε},
                       BoxRatios -> {1, 1, 1}, PlotPoints -> 40,
                       Compiled -> False, PlotRange -> Automatic]

       unitPolys1 = Cases[pPP3D, _Polygon, Infinity];

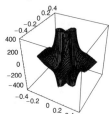

Although the picture looks good, the polygons forming the graphics are not ideal. Because of the Automatic setting of the PlotRange option of ParametricPlot3D, the picture looks decent. But some of the polygons extend far outside the displayed volume.

In[7]:= **{Min[#], Max[#]}&[Last /@ Level[unitPolys1, {-2}]]**

Out[7]= {-993122., 993122.}

To avoid the far outlying points, we cut all polygons at the heights ±500. The function splitPolygon splits a polygon *p* at the height *h*. It returns a list with two elements. The first element is a list of the polygons that lie below *h*, and the second element is a list with the polygons that lie above *h*.

```
In[8]:= (* split a polygon at height h *)
       splitPolygon[p:Polygon[l_], h_] :=
       Module[{zs = Last /@ l, mp = Plus @@ l/Length[l], newPolys},
         (* return {polygonsBelow, polygonsAbove} *)
         Which[Max[zs] <= h, {p, {}}, Min[zs] >= h, {{}, p},
             Min[zs] < h < Max[zs],
             If[Length[l] > 3,
                 (* split polygon into triangles *)
                 newPolys = (Polygon[Append[#, mp]]& /@
                   Partition[Append[l, First[l]], 2, 1]);
                 Transpose[splitPolygon[#, h]& /@ newPolys],
                 splitTriangle[p, h]]]];
In[10]:= (* split a triangle at height h *)
        splitTriangle[p:Polygon[l_], h_] :=
        Module[{zs = Last /@ l, c, p1, p2, p3, p1z, p2z, p3z},
          c = Count[zs, _?(# > h&)];
          Which[c === 3, {{}, p}, c === 0, {p, {}},
              (* one point is above h *)
              c === 1, pos = Position[zs, Max[zs]][[1, 1]];
              (* p1 is highest *)
              {p1, p2, p3} = RotateLeft[l, pos - 1];
              {p1z, p2z, p3z} = Last /@ {p1, p2, p3};
              p12 = p2 + (h - p2z)/(p1z - p2z) (p1 - p2);
              p13 = p3 + (h - p3z)/(p1z - p3z) (p1 - p3);
              {Polygon[{p2, p12, p13, p3}], Polygon[{p12, p1, p13}]},
              (* two points are above h *)
              c === 2, pos = Position[zs, Min[zs]][[1, 1]];
              {p1, p2, p3} = RotateLeft[l, pos - 1];
              {p1z, p2z, p3z} = Last /@ {p1, p2, p3};
              p12 = p1 + (h - p1z)/(p2z - p1z) (p2 - p1);
              p13 = p1 + (h - p1z)/(p3z - p1z) (p3 - p1);
              {Polygon[{p12, p1, p13}], Polygon[{p2, p12, p13, p3}]}]]
In[12]:= h = 500;
        (* cut top *)
        unitPolys2 = Transpose[splitPolygon[#, +h]& /@ unitPolys1];
        (* cut bottom *)
        unitPolys3 = Transpose[splitPolygon[#, -h]& /@
                                  Flatten[unitPolys2[[1]]]];
        finalPolys = Flatten[unitPolys3[[2]]];
```

The next graphic shows all resulting polygons. We use the All option setting for PlotRange.

```
In[18]:= Show[Graphics3D[finalPolys], BoxRatios -> {1, 1, 1},
            Axes -> True, PlotRange -> All]
```

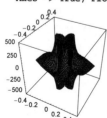

The $\wp'$ function is doubly periodic. This means we can cover the complex z-plane with copies of finalPolys in a square array-like fashion. Here we show an array of size 3×3 array.

```
In[19]:= Show[Graphics3D[{EdgeForm[], SurfaceColor[Hue[0.22], Hue[0.8], 2.93],
             Table[Apply[{#1 + Re[ix 2 ω1 + iy 2 ω2],
                          #2 + Im[ix 2 ω1 + iy 2 ω2], #3}&,
                     finalPolys, {-2}], {ix, -1, 1}, {iy, -1, 1}]}],
             PlotRange -> {-500, 500}, BoxRatios -> {2, 2, 1}]
```

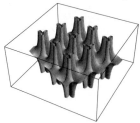

Now we have to map the so-constructed carpet of $\wp'$-parallelograms to the Riemann sphere. The function toSphere maps the point z of the complex plane onto the Riemann sphere.

```
In[20]:= toSphere[z_] := Function[{φ, θ}, {0, 0, 1/2} +
             1/2 {Cos[φ] Sin[θ], Sin[φ] Sin[θ], Cos[θ]}][
             Arg[z], Pi - 2 ArcTan[Abs[z]]]
```

And the function toDeformedSphere maps the point $\{x, y, z\}$ of the complex $(x + i\, y)$-plane onto the "squeezed" Riemann sphere. The "squeezing" is proportional to the height z of the point.

```
In[21]:= toDeformedSphere[{x_, y_, z_}] :=
             (# + 0.3 z/h #)&[toSphere[x + I y] - {0, 0, 1/2} ]
```

Applying now toDeformedSphere to a 5×5 carpet of $\wp'$-parallelograms gives the following picture. The more we approach the north pole, the denser the triple poles of $\wp'$ become.

```
In[22]:= Show[Graphics3D[{EdgeForm[],
             Table[{SurfaceColor[Hue[Random[]], Hue[Random[]], 2.6],
                Map[toDeformedSphere,
                (* carpet of 5×5 ℘'s in the period parallelogram *)
                Apply[{#1 + Re[ix 2 ω1 + iy 2 ω2],
                       #2 + Im[ix 2 ω1 + iy 2 ω2], #3}&,
                       finalPolys, {-2}], {-2}]}, {ix, -2, 2}, {iy, -2, 2}]}],
             ViewPoint -> {1, 2, -4}, PlotRange -> All, Boxed -> False]
```

Here is the last graphic with the corners of all polygons cut off.

```
In[23]:= Show[% /. Polygon[l_] :>
                Polygon[Plus @@@ Partition[Append[l, First[l]], 2, 1]/2],
             ViewPoint -> {-2, 0, -3/2}]
```

Using the elliptic ψ function $\psi_n(z; g_2, g_3) = \sigma(n\,z; g_2, g_3)\,\sigma(z; g_2, g_3)^{-n^2}$ [1311] we could generate similar pictures with higher-order poles. The following graphic shows $\psi_4(z; 1, i)$ over the period parallelogram.

```
In[24]:= ψ[n_][z_, {g2_, g3_}] :=
         WeierstrassSigma[n z, {g2, g3}]/WeierstrassSigma[z, {g2, g3}]^(n^2)

In[25]:= With[{g2 = 1, g3 = I, n = 4},
         Module[{ω1, ω2}, {ω1, ω2} = WeierstrassHalfPeriods[N[{g2, g3}]];
         ParametricPlot3D[(* display one period *)
         Evaluate[{Re[s ω1 + t ω2], Im[s ω1 + t ω2],
                 Re @ ψ[n][s ω1 + t ω2, {1, I}],
                 {EdgeForm[], SurfaceColor[Hue[s t], Hue[s t], 2.2]}}],
             {s, -1, 1}, {t, -1, 1}, PlotPoints -> 161,
             Boxed -> False, BoxRatios -> {1, 1, 0.6},
             Axes -> False, PlotRange -> {All, All, {-50, 50}}]]]
```

The interested reader can go on and display the last function over the Riemann sphere.

c) The implementation is straightforward. We plug the series into the differential equations and compare the coefficients of equal powers of z.

```
In[1]:= WeierstrassPSeries[ord_Integer?(# >= 6&)] :=
        Module[{c},
              c[2] = g2/20; c[4] = g3/28;
              ser = 1/z^2 + Sum[c[2i] z^(2i), {i, 1, ord/2}] +
                 (* make a series *) O[z]^(2Floor[ord/2]);
              Off[Solve::svars]; (* solve for the coefficients by
                                    plugging ser in the differential equation *)
              sol = Factor[Together[Solve[D[ser, z]^2 - 4 ser^3 + g2 ser + g3 == 0,
                     Table[c[2i], {i, 3, ord/2}]]]];
              On[Solve::svars]; (ser /. sol)[[1]]]
```

Here is the series up to order 20.

```
In[2]:= WeierstrassPSeries[20]
```

$$
\text{Out[2]}= \frac{1}{z^2} + \frac{g2\,z^2}{20} + \frac{g3\,z^4}{28} + \frac{g2^2\,z^6}{1200} + \frac{3\,g2\,g3\,z^8}{6160} +
$$
$$
\frac{(49\,g2^3 + 750\,g3^2)\,z^{10}}{7644000} + \frac{g2^2\,g3\,z^{12}}{184800} + \frac{g2\,(539\,g2^3 + 18000\,g3^2)\,z^{14}}{11435424000} +
$$
$$
\frac{g3\,(1421\,g2^3 + 5500\,g3^2)\,z^{16}}{29821792000} + \frac{g2^2\,(5929\,g2^3 + 363750\,g3^2)\,z^{18}}{18868449600000} + O[z]^{20}
$$

The above approach works, but it is much more efficient to first calculate (more or less by hand) a recurrence relation for the coefficients [306]. Here, this can be done easily to get the following coefficients.

In[3]:= `c[2] = g2/20; c[4] = g3/28;`
`c[n_] := c[n] = Factor[3/((n + 3)(n/2 - 2)) Sum[c[k] c[n - k - 2],`
`                                          {k, 2, n - 4, 2}]]`

The coefficients agree with the above-calculated ones.

In[5]:= `Table[c[i], {i, 2, 20, 2}]`

Out[5]= $\left\{ \dfrac{g2}{20}, \dfrac{g3}{28}, \dfrac{g2^2}{1200}, \dfrac{3\,g2\,g3}{6160}, \dfrac{49\,g2^3 + 750\,g3^2}{7644000}, \dfrac{g2^2\,g3}{184800}, \dfrac{g2\,(539\,g2^3 + 18000\,g3^2)}{11435424000}, \right.$

$\dfrac{g3\,(1421\,g2^3 + 5500\,g3^2)}{29821792000}, \dfrac{g2^2\,(5929\,g2^3 + 363750\,g3^2)}{18868449600000}, \left. \dfrac{g2\,g3\,(19061\,g2^3 + 184500\,g3^2)}{49972802880000} \right\}$

d) For brevity and easier readability, we suppress the g_2 and g_3 argument of $\wp(z; g_2, g_3)$ in the following. We will use an algebraic approach [211] in the following.

We take the given addition formula and supplement it with the two differential equations for $\wp(u; g_2, g_3)$ and $\wp(v; g_2, g_3)$. We generate three polynomial equations by clearing the denominators.

In[1]:= `eqs = Numerator[Together[#]]& /@`
`        {p[u + v] + p[u] + p[v] - ((p'[u] - p'[v])/(p[u] - p[v]))^2/4,`
`         p'[u]^2 - 4 p[u]^3 + g2 p[u] + g3, p'[v]^2 - 4 p[v]^3 + g2 p[v] + g3}`

Out[1]= $\{ 4\,\wp[u]^3 - 4\,\wp[u]^2\,\wp[v] - 4\,\wp[u]\,\wp[v]^2 + 4\,\wp[v]^3 + 4\,\wp[u]^2\,\wp[u + v] -$
$8\,\wp[u]\,\wp[v]\,\wp[u + v] + 4\,\wp[v]^2\,\wp[u + v] - \wp'[u]^2 + 2\,\wp'[u]\,\wp'[v] - \wp'[v]^2,$
$g3 + g2\,\wp[u] - 4\,\wp[u]^3 + \wp'[u]^2, \; g3 + g2\,\wp[v] - 4\,\wp[v]^3 + \wp'[v]^2 \}$

Using `GroebnerBasis` with `EliminationOrder`, it is now straightforward to derive the polynomial we are looking for.

In[2]:= `gb = Factor[GroebnerBasis[eqs, {p[u], p[v]}, {p'[u], p'[v]},`
`                MonomialOrder -> EliminationOrder]]`

Out[2]= $\{ (\wp[u] - \wp[v])^2\,(g2^2 + 16\,g3\,\wp[u] + 16\,g3\,\wp[v] + 8\,g2\,\wp[u]\,\wp[v] + 16\,\wp[u]^2\,\wp[v]^2 +$
$16\,g3\,\wp[u + v] + 8\,g2\,\wp[u]\,\wp[u + v] + 8\,g2\,\wp[v]\,\wp[u + v] - 32\,\wp[u]^2\,\wp[v]\,\wp[u + v] -$
$32\,\wp[u]\,\wp[v]^2\,\wp[u + v] + 16\,\wp[u]^2\,\wp[u + v]^2 - 32\,\wp[u]\,\wp[v]\,\wp[u + v]^2 + 16\,\wp[v]^2\,\wp[u + v]^2) \}$

Obviously, the last factor is the one that is of interest. A quick numerical check of the result shows its correctness.

In[3]:= `(gb[[1, -1]] /. p[u_] -> WeierstrassP[u, {g2, g3}]) /.`
`          {u -> N[2, 30], v -> 1, g2 -> 1/2, g3 -> I}`

Out[3]= $0. \times 10^{-21} + 0. \times 10^{-21}\,i$

The corresponding double argument formula is linear in $\wp(2v; g_2, g_3)$.

In[4]:= `Factor[gb[[1, -1]] /. u -> v]`

Out[4]= $g2^2 + 32\,g3\,\wp[v] + 8\,g2\,\wp[v]^2 + 16\,\wp[v]^4 + 16\,g3\,\wp[2\,v] + 16\,g2\,\wp[v]\,\wp[2\,v] - 64\,\wp[v]^3\,\wp[2\,v]$

In[5]:= `Solve[% == 0, p[2v]]`

Out[5]= $\left\{ \left\{ \wp[2\,v] \rightarrow \dfrac{-g2^2 - 32\,g3\,\wp[v] - 8\,g2\,\wp[v]^2 - 16\,\wp[v]^4}{16\,(g3 + g2\,\wp[v] - 4\,\wp[v]^3)} \right\} \right\}$

e) For the addition formula for $\wp'(u + v; g_2, g_3)$, we proceed in a similar way as in part c). We supplement the given formula with the three differential equations for $\wp(u; g_2, g_3)$, $\wp(v; g_2, g_3)$, and $\wp(u + v; g_2, g_3)$ and generate four polynomial equations by clearing the denominators.

In[1]:= `eqs = Numerator[Together[#]]& /@`
`        {p'[u + v] + (p[u + v] (p'[u] - p'[v]) + p[u] p'[v] - p'[u] p[v])/`
`          (p[u] - p[v]),`
`         p'[u]^2 - 4 p[u]^3 + g2 p[u] + g3, p'[v]^2 - 4 p[v]^3 + g2 p[v] + g3,`
`         p'[u + v]^2 - 4 p[u + v]^3 + g2 p[u + v] + g3}`

Out[1]= $\{ -\wp[v]\,\wp'[u] + \wp[u + v]\,\wp'[u] + \wp[u]\,\wp'[v] - \wp[u + v]\,\wp'[v] + \wp[u]\,\wp'[u + v] - \wp[v]\,\wp'[u + v],$
$g3 + g2\,\wp[u] - 4\,\wp[u]^3 + \wp'[u]^2, \; g3 + g2\,\wp[v] - 4\,\wp[v]^3 + \wp'[v]^2,$
$g3 + g2\,\wp[u + v] - 4\,\wp[u + v]^3 + \wp'[u + v]^2 \}$

Using again `GroebnerBasis` with `EliminationOrder`, it is now straightforward to derive the polynomial we are looking for. This time, the result is quite large.

In[2]:= `gb = GroebnerBasis[eqs, {p'[u], p'[v], p'[u + v]}, {p[u], p[v], p[u + v]},`
`                MonomialOrder -> EliminationOrder] // Factor;`

```
In[3]:= {Head[gb[[1]]], Length[gb[[1]]]}
Out[3]= {Times, 4}
```

```
In[4]:= Length /@ (List @@ gb[[1]])
Out[4]= {2, 2, 2, 670}
```

The last factor is the addition formula we are looking for. The first ones are the following.

```
In[5]:= Take[gb[[1]], 3]
```

$$Out[5]= (\wp'[u] - \wp'[v]) \; (\wp'[u] + \wp'[u+v]) \; (\wp'[v] + \wp'[u+v])$$

Here are the first 25 summands of the resulting polynomial.

```
In[6]:= Take[gb[[1, -1]], 25]
```

$$
\begin{aligned}
Out[6]= \; & 4 \, g2^9 \, \wp'[u]^6 - 324 \, g2^6 \, g3^2 \, \wp'[u]^6 + 8748 \, g2^3 \, g3^4 \, \wp'[u]^6 - 78732 \, g3^6 \, \wp'[u]^6 - \\
& 12 \, g2^9 \, \wp'[u]^5 \, \wp'[v] + 972 \, g2^6 \, g3^2 \, \wp'[u]^5 \, \wp'[v] - 26244 \, g2^3 \, g3^4 \, \wp'[u]^5 \, \wp'[v] + \\
& 236196 \, g3^6 \, \wp'[u]^5 \, \wp'[v] - 3 \, g2^9 \, \wp'[u]^4 \, \wp'[v]^2 - 486 \, g2^6 \, g3^2 \, \wp'[u]^4 \, \wp'[v]^2 + \\
& 32805 \, g2^3 \, g3^4 \, \wp'[u]^4 \, \wp'[v]^2 - 472392 \, g3^6 \, \wp'[u]^4 \, \wp'[v]^2 - 216 \, g2^6 \, g3 \, \wp'[u]^6 \, \wp'[v]^2 + \\
& 11664 \, g2^3 \, g3^3 \, \wp'[u]^6 \, \wp'[v]^2 - 157464 \, g3^5 \, \wp'[u]^6 \, \wp'[v]^2 + 26 \, g2^9 \, \wp'[u]^3 \, \wp'[v]^3 - \\
& 648 \, g2^6 \, g3^2 \, \wp'[u]^3 \, \wp'[v]^3 - 21870 \, g2^3 \, g3^4 \, \wp'[u]^3 \, \wp'[v]^3 + 551124 \, g3^6 \, \wp'[u]^3 \, \wp'[v]^3 + \\
& 432 \, g2^6 \, g3 \, \wp'[u]^5 \, \wp'[v]^3 - 23328 \, g2^3 \, g3^3 \, \wp'[u]^5 \, \wp'[v]^3 + 314928 \, g3^5 \, \wp'[u]^5 \, \wp'[v]^3 + \\
& 12 \, g2^6 \, \wp'[u]^7 \, \wp'[v]^3 - 648 \, g2^3 \, g3^2 \, \wp'[u]^7 \, \wp'[v]^3 + 8748 \, g3^4 \, \wp'[u]^7 \, \wp'[v]^3
\end{aligned}
$$

The formula is relatively complicated; it is a polynomial of total degree 21.

```
In[7]:= Max[Plus @@@ (First /@
         Internal`DistributedTermsList[gb, {p'[u], p'[v], p'[u + v]}][[1, 1]])]
Out[7]= 21
```

The largest factor is the one that is of interest. Here is a quick numerical check of the result.

```
In[8]:= SeedRandom[1111];
        (gb[[1, -1]] /. p'[u_] -> WeierstrassPPrime[u, {g2, g3}]) /.
          {u -> N[2, 50], v -> 1, g2 -> 1/2, g3 -> I}
Out[9]= 0. × 10^-28 + 0. × 10^-28 i
```

The double argument version of the formula is considerably simpler (again, the last factor is the one under consideration).

```
In[10]:= Factor[gb[[1, -1]] /. v -> u]
```

$$
\begin{aligned}
Out[10]= \; & 4 \, (\wp'[u] + \wp'[2\,u])^6 \\
& (g2^9 - 81 \, g2^6 \, g3^2 + 2187 \, g2^3 \, g3^4 - 19683 \, g3^6 - 54 \, g2^6 \, g3 \, \wp'[u]^2 + 2916 \, g2^3 \, g3^3 \, \wp'[u]^2 - \\
& 39366 \, g3^5 \, \wp'[u]^2 - 33 \, g2^6 \, \wp'[u]^4 + 1782 \, g2^3 \, g3^2 \, \wp'[u]^4 - 24057 \, g3^4 \, \wp'[u]^4 + \\
& 108 \, g2^3 \, g3 \, \wp'[u]^6 - 2916 \, g3^3 \, \wp'[u]^6 - 33 \, g2^3 \, \wp'[u]^8 + 891 \, g3^2 \, \wp'[u]^8 - 54 \, g3 \, \wp'[u]^{10} + \\
& \wp'[u]^{12} - 24 \, g2^6 \, \wp'[u]^3 \, \wp'[2\,u] + 1296 \, g2^3 \, g3^2 \, \wp'[u]^3 \, \wp'[2\,u] - 17496 \, g3^4 \, \wp'[u]^3 \, \wp'[2\,u] + \\
& 864 \, g2^3 \, g3 \, \wp'[u]^5 \, \wp'[2\,u] - 23328 \, g3^3 \, \wp'[u]^5 \, \wp'[2\,u] + 240 \, g2^3 \, \wp'[u]^7 \, \wp'[2\,u] - \\
& 6480 \, g3^2 \, \wp'[u]^7 \, \wp'[2\,u] + 864 \, g3 \, \wp'[u]^9 \, \wp'[2\,u] - 24 \, \wp'[u]^{11} \, \wp'[2\,u] - 240 \, g2^3 \, \wp'[u]^6 \, \wp'[2\,u]^2 - \\
& 5184 \, g3^2 \, \wp'[u]^6 \, \wp'[2\,u]^2 - 3456 \, g3 \, \wp'[u]^8 \, \wp'[2\,u]^2 + 192 \, \wp'[u]^{10} \, \wp'[2\,u]^2 - 512 \, \wp'[u]^9 \, \wp'[2\,u]^3)
\end{aligned}
$$

```
In[11]:= (%[[-1]] /. p'[u_] -> WeierstrassPPrime[u, {g2, g3}]) /.
           {u -> N[1, 40], g2 -> 1/2, g3 -> 1 + I}
Out[11]= 0. × 10^-27 + 0. × 10^-27 i
```

Note that for deriving the above addition formula for $\wp'(u + v; g_2, g_3)$, we do not actually need the given addition formula for $\wp'(u + v; g_2, g_3)$. The addition formula for $\wp(u + v; g_2, g_3)$ together with the differential equations for $\wp(v; g_2, g_3)$, $\wp(u; g_2, g_3)$, and $\wp(u + v; g_2, g_3)$ are sufficient.

f) Now, let us deal with the mentioned functional equation of the Sutherland–Calogero model. These are the two functions $\varphi(x)$ and $f(x)$ as well as the functional equation.

```
In[1]:= (* define functions φ and f *)
        φ[x_] = α WeierstrassZeta[x, {g2, g3}] + β x;
        f[x_] = -1/2 (α^2 D[WeierstrassZeta[x, {g2, g3}], x] +
                 α^2 WeierstrassZeta[x, {g2, g3}]^2 +
                 2 α β x WeierstrassZeta[x, {g2, g3}] +  + β^2 x^2);
```

In[4]:= eqs = φ[x] φ[y] + φ[y] φ[z] + φ[x] φ[z] - f[x] - f[y] - f[z] /.
          z -> -(x + y);

Here is a quick numerical check for the functional equation.

In[5]:= eqs /. ((# -> Random[Complex, {0, 1 + I}, 50])& /@ {g2, g3, α, β, x, y})

Out[5]= $0. \times 10^{-44} + 0. \times 10^{-44}$ i

To show that the two functions $\varphi(x)$ and $f(x)$ are solutions of the functional equation, we will differentiate the functional equation with respect to x and show that the resulting equation is an identity. So to show the correctness of the functional equation, we also have to show that the functional equation is correct for some point x, say, $x = 0$.

In[6]:= Series[eqs /. {WeierstrassP[z_, {g2_, g3_}] :> z^-2 + g2 z^2/20,
                      WeierstrassZeta[z_, {g2_, g3_}] :> z^-1 -g2 z^3/60},
              {x, 0, 1}]

Out[6]= $O[x]^2$

In the first step, we eliminate all Weierstrass Zeta functions to obtain an equation in Weierstrass $\wp$ functions and its derivative.

In[7]:= pRules = {WeierstrassP[ξ_, _] -> p[ξ], WeierstrassPPrime[ξ_, _] -> p'[ξ]};

In[8]:= (gb1 = Factor[GroebnerBasis[{eqs, D[eqs, x]},
              {}, {WeierstrassZeta[x, {g2, g3}], WeierstrassZeta[y, {g2, g3}],
              WeierstrassZeta[-x - y, {g2, g3}]},
              MonomialOrder -> EliminationOrder]]) /. pRules

Out[8]= $\{\alpha^2 \ (4 \ p[x]^3 - 4 \ p[x]^2 \ p[-x - y] - 4 \ p[x] \ p[-x - y]^2 + 4 \ p[-x - y]^3 + 4 \ p[x]^2 \ p[y] - 8 \ p[x] \ p[-x - y] \ p[y] + 4 \ p[-x - y]^2 \ p[y] - p'[x]^2 + 2 \ p'[x] \ p'[-x - y] - p'[-x - y]^2)\}$

To match the above addition formula, in a second step, we eliminate two derivatives $\wp'(y; g_2, g_3)$ and $\wp'(-x - y; g_2, g_3)$.

In[9]:= gb2 = Factor[GroebnerBasis[{gb1, D[gb1, x], D[gb1, y]},
              {}, {WeierstrassPPrime[y, {g2, g3}],
              WeierstrassPPrime[-x - y, {g2, g3}]},
              MonomialOrder -> EliminationOrder]] /. pRules

Out[9]= $\{\alpha^2 \ (p[x] - p[-x - y])^2 \ (p[x] + p[-x - y] + p[y])$
         $(g2^2 - 16 \ g2 \ p[x]^2 + 64 \ p[x]^4 - 8 \ g2 \ p[x] \ p[-x - y] + 64 \ p[x]^3 \ p[-x - y] + 16 \ p[x]^2 \ p[-x - y]^2 - 8 \ g2 \ p[x] \ p[y] + 64 \ p[x]^3 \ p[y] + 8 \ g2 \ p[-x - y] \ p[y] - 32 \ p[x]^2 \ p[-x - y] \ p[y] - 32 \ p[x] \ p[-x - y]^2 \ p[y] + 16 \ p[x]^2 \ p[y]^2 - 32 \ p[x] \ p[-x - y] \ p[y]^2 + 16 \ p[-x - y]^2 \ p[y]^2 - 16 \ p[x] \ p'[x]^2 - 16 \ p[-x - y] \ p'[x]^2 - 16 \ p[y] \ p'[x]^2)\}$

Using now the symmetry of Weierstrass $\wp$ function $\wp(-x; g_2, g_3) = \wp(x; g_2, g_3)$ and the above addition formula (from exercise a), we can easily show that the last factor of gb2 is equivalent to the defining differential equations for $\wp(x; g_2, g_3)$ and $\wp(y; g_2, g_3)$.

In[10]:= eqs = (* make polynomials *) Numerator[Together[#]]& /@
              {p[x + y] + p[x] + p[y] - ((p'[x] - p'[y])/(p[x] - p[y]))^2/4,
              p'[x]^2 - 4 p[x]^3 + g2 p[x] + g3, p'[y]^2 - 4 p[y]^3 + g2 p[y] + g3}

Out[10]= $\{4 \ p[x]^3 - 4 \ p[x]^2 \ p[y] - 4 \ p[x] \ p[y]^2 + 4 \ p[y]^3 + 4 \ p[x]^2 \ p[x + y] - 8 \ p[x] \ p[y] \ p[x + y] + 4 \ p[y]^2 \ p[x + y] - p'[x]^2 + 2 \ p'[x] \ p'[y] - p'[y]^2,$
          $g3 + g2 \ p[x] - 4 \ p[x]^3 + p'[x]^2, g3 + g2 \ p[y] - 4 \ p[y]^3 + p'[y]^2\}$

In[11]:= GroebnerBasis[Flatten[{gb2[[-1, -1]] /. p[-x - y] -> p[x + y],
              eqs /. pRules}],
              {}, {p[x + y]}, MonomialOrder -> EliminationOrder]

Out[11]= $\{-g3 - g2 \ p[y] + 4 \ p[y]^3 - p'[y]^2, -g3 - g2 \ p[x] + 4 \ p[x]^3 - p'[x]^2\}$

g) Using the symmetry properties for $\wp(u; g_2, g_3)$ and $\zeta(u; g_2, g_3)$, the identity

$$\zeta(x; g_2, g_3) + \zeta(y; g_2, g_3) + \zeta(z; g_2, g_3) = (\wp(x; g_2, g_3) + \wp(y; g_2, g_3) + \wp(z; g_2, g_3))^{1/2}$$

reads as follows.

In[1]:= id = With[{ζ = WeierstrassZeta[#, {g2, g3}]&, p = WeierstrassP[#, {g2, g3}]&},
              ζ[u] + ζ[v] - ζ[u + v] - Sqrt[p[u] + p[v] + p[u + v]]];

Again, after checking the identity for an appropriate set of initial conditions, the algebraic part to be shown is the following set of equations.

```
In[2]:= eqs1 = {D[id, u], D[id, v]} // Together // Numerator;
```

We eliminate the square roots and obtain a polynomial set of equations.

```
In[3]:= eliminateSquareRoot[f_] :=
         Module[{squareRoot = Plus @@ Cases[f, __ Power[_, 1/2], {1}]},
                Expand[(f - squareRoot)^2 - squareRoot^2]]
In[4]:= eqs2 = (eliminateSquareRoot /@ eqs1) /.
               {WeierstrassP[ξ_, _] -> ℘[ξ], WeierstrassPPrime[ξ_, _] -> ℘'[ξ]}
```

$$Out[4]= \{-4\,\wp[u]^3 - 4\,\wp[u]^2\,\wp[v] + 4\,\wp[u]^2\,\wp[u+v] + 8\,\wp[u]\,\wp[v]\,\wp[u+v] +$$
$$4\,\wp[u]\,\wp[u+v]^2 - 4\,\wp[v]\,\wp[u+v]^2 - 4\,\wp[u+v]^3 + \wp'[u]^2 + 2\,\wp'[u]\,\wp'[u+v] + \wp'[u+v]^2,$$
$$-4\,\wp[u]\,\wp[v]^2 - 4\,\wp[v]^3 + 8\,\wp[u]\,\wp[v]\,\wp[u+v] + 4\,\wp[v]^2\,\wp[u+v] - 4\,\wp[u]\,\wp[u+v]^2 +$$
$$4\,\wp[v]\,\wp[u+v]^2 - 4\,\wp[u+v]^3 + \wp'[v]^2 + 2\,\wp'[v]\,\wp'[u+v] + \wp'[u+v]^2\}$$

Eliminating the functions that involve $u + v$, we can reduce the resulting polynomials to zero using the defining differential equations.

```
In[5]:= gb2 = GroebnerBasis[Flatten[{eqs2,
           (* make polynomials *) Numerator[Together[
           ℘[u + v] + ℘[u] + ℘[v] - ((℘'[u] - ℘'[v])/(℘[u] - ℘[v]))^2/4]],
           ℘'[u + v]^2 - 4 ℘[u + v]^3 + g2 ℘[u + v] + g3}],
                {}, {℘'[u + v], ℘[u + v]},
                MonomialOrder -> EliminationOrder] // Factor
```

$$Out[5]= \{(\wp'[u] - \wp'[v])$$
$$(4\,g3\,\wp[u]^3 - 4\,g2\,\wp[u]^4 + 16\,\wp[u]^6 - 12\,g3\,\wp[u]^2\,\wp[v] + 8\,g2\,\wp[u]^3\,\wp[v] + 12\,g3\,\wp[u]\,\wp[v]^2 -$$
$$48\,\wp[u]^4\,\wp[v]^2 - 4\,g3\,\wp[v]^3 - 8\,g2\,\wp[u]\,\wp[v]^3 + 4\,g2\,\wp[v]^4 + 48\,\wp[u]^2\,\wp[v]^4 - 16\,\wp[v]^6 +$$
$$g2\,\wp[u]\,\wp'[u]^2 - 8\,\wp[u]^3\,\wp'[u]^2 - g2\,\wp[v]\,\wp'[u]^2 + 12\,\wp[u]\,\wp[v]^2\,\wp'[u]^2 - 4\,\wp[v]^3\,\wp'[u]^2 +$$
$$\wp'[u]^4 - 2\,g2\,\wp[u]\,\wp'[u]\,\wp'[v] + 8\,\wp[u]^3\,\wp'[u]\,\wp'[v] + 2\,g2\,\wp[v]\,\wp'[u]\,\wp'[v] -$$
$$8\,\wp[v]^3\,\wp'[u]\,\wp'[v] - 2\,\wp'[u]^3\,\wp'[v] + g2\,\wp[u]\,\wp'[v]^2 + 4\,\wp[u]^3\,\wp'[v]^2 -$$
$$g2\,\wp[v]\,\wp'[v]^2 - 12\,\wp[u]^2\,\wp[v]\,\wp'[v]^2 + 8\,\wp[v]^3\,\wp'[v]^2 + 2\,\wp'[u]\,\wp'[v]^3 - \wp'[v]^4),$$
$$(g3 - g2\,\wp[u] + 4\,\wp[u]^3 - g2\,\wp[v] + 12\,\wp[u]^2\,\wp[v] + 12\,\wp[u]\,\wp[v]^2 +$$
$$4\,\wp[v]^3 + 2\,\wp'[u]^2 - 2\,\wp'[u]\,\wp'[v] + \wp'[v]^2)$$
$$(4\,g3\,\wp[u]^3 - 4\,g2\,\wp[u]^4 + 16\,\wp[u]^6 - 12\,g3\,\wp[u]^2\,\wp[v] + 8\,g2\,\wp[u]^3\,\wp[v] + 12\,g3\,\wp[u]\,\wp[v]^2 -$$
$$48\,\wp[u]^4\,\wp[v]^2 - 4\,g3\,\wp[v]^3 - 8\,g2\,\wp[u]\,\wp[v]^3 + 4\,g2\,\wp[v]^4 + 48\,\wp[u]^2\,\wp[v]^4 - 16\,\wp[v]^6 +$$
$$g2\,\wp[u]\,\wp'[u]^2 - 8\,\wp[u]^3\,\wp'[u]^2 - g2\,\wp[v]\,\wp'[u]^2 + 12\,\wp[u]\,\wp[v]^2\,\wp'[u]^2 - 4\,\wp[v]^3\,\wp'[u]^2 +$$
$$\wp'[u]^4 - 2\,g2\,\wp[u]\,\wp'[u]\,\wp'[v] + 8\,\wp[u]^3\,\wp'[u]\,\wp'[v] + 2\,g2\,\wp[v]\,\wp'[u]\,\wp'[v] -$$
$$8\,\wp[v]^3\,\wp'[u]\,\wp'[v] - 2\,\wp'[u]^3\,\wp'[v] + g2\,\wp[u]\,\wp'[v]^2 + 4\,\wp[u]^3\,\wp'[v]^2 -$$
$$g2\,\wp[v]\,\wp'[v]^2 - 12\,\wp[u]^2\,\wp[v]\,\wp'[v]^2 + 8\,\wp[v]^3\,\wp'[v]^2 + 2\,\wp'[u]\,\wp'[v]^3 - \wp'[v]^4)\}$$

```
In[6]:= PolynomialReduce[#, {℘'[u]^2 - 4 ℘[u]^3 + g2 ℘[u] + g3,
                        ℘'[v]^2 - 4 ℘[v]^3 + g2 ℘[v] + g3},
                       {℘'[u], ℘'[v], ℘'[u], ℘[u]}][[2]]& /@ gb2
```

```
Out[6]= {0, 0}
```

h) We want to express $\wp(5\,z; g_2, g_3)$ rationally through $\wp(z; g_2, g_3)$. We start by implementing the recursively defined ψ_n.

```
In[1]:= With[{(* abbreviations *) ℘ = WeierstrassP[z, {g2, g3}],
              ℘s = WeierstrassPPrime[z, {g2, g3}]},
        (* initial terms *)
        ψ[1] = 1;
        ψ[2] = -℘s;
        ψ[3] = 3 ℘^4 - 3/2 g2 ℘^2 - 3 g3 ℘ - 1/16 g2^2;
        ψ[4] = ℘s (-2 ℘^6 + 5/2 g2 ℘^4 + 10 g3 ℘^3 + 5/8 g2^2 ℘^2 +
                  1/2 g2 g3 ℘ + g3^2 - 1/32 g2^3);
        (* recursion for even and odd order *)
        ψ[n_?EvenQ] := ψ[n] = Expand[
          -ψ[n/2]/℘s (ψ[n/2 + 2] ψ[n/2 - 1]^2 - ψ[n/2 - 2] ψ[n/2 + 1]^2)];
        ψ[n_?OddQ] := ψ[n] = Expand[ψ[(n - 1)/2 + 2] ψ[(n - 1)/2]^3 -
                                   ψ[(n - 1)/2 - 1] ψ[(n - 1)/2 + 1]^3]]
```

Here is the resulting polynomial form of the five-argument formula.

In[2]:= num = With[{n = 5}, Numerator[Together[WeierstrassP[n z, {g2, g3}] -
        (WeierstrassP[z, {g2, g3}] - ψ[n + 1] ψ[n - 1]/ψ[n]^2)]]];

num also contains derivative terms $\wp'(z; g_2, g_3)$.

In[3]:= Exponent[num, WeierstrassPPrime[z, {g2, g3}]]

Out[3]= 8

We eliminate the derivative terms using the defining differential equation of $\wp(z; g_2, g_3)$.

In[4]:= odeEqs = WeierstrassPPrime[z, {g2, g3}]^2 -
        4 WeierstrassP[z, {g2, g3}]^3 + g2 WeierstrassP[z, {g2, g3}] + g3;

In[5]:= res = Factor[Resultant[num, odeEqs, WeierstrassPPrime[z, {g2, g3}]]];

In[6]:= {Head[res], Head[First[res]]}

Out[6]= {Power, Plus}

Here is a quick numerical check of the result.

In[7]:= res[[1]] /. {g2 -> 1/2, g3 -> I, z -> 2.`100}

Out[7]= 0. × 10^{-81} + 0. × 10^{-81} i

Finally, we solve for $\wp(5 z; g_2, g_3)$ and simplify the result.

In[8]:= Exponent[res[[1]], WeierstrassP[{1, 5} z, {g2, g3}]]

Out[8]= {25, 1}

In[9]:= Solve[res[[1]] == 0, WeierstrassP[5 z, {g2, g3}]] /.
            WeierstrassP[1 z, {g2, g3}] -> ℘ // Simplify

Out[9]= {{WeierstrassP[5 z, {g2, g3}] →
        $(25\, g2^{12}\, \wp + 8\, g2^{11}\, (3\, g3 - 310\, \wp^3) + g2^{10}\, (-5440\, g3\, \wp^2 + 44704\, \wp^5) - 640\, g2^9$
        $(11\, g3^2\, \wp - 47\, g3\, \wp^4 - 1602\, \wp^7) + 4096\, g2^7\, \wp^2\, (45\, g3^3 + 4308\, g3^2\, \wp^3 - 3805\, g3\, \wp^6 - 11750\, \wp^9) -$
        $1280\, g2^8\, (3\, g3^3 - 53\, g3^2\, \wp^3 - 6520\, g3\, \wp^6 + 5485\, \wp^9) +$
        $8192\, g2^6\, \wp\, (35\, g3^4 + 2830\, g3^3\, \wp^3 + 4910\, g3^2\, \wp^6 - 63084\, g3\, \wp^9 + 55030\, \wp^{12}) +$
        $32768\, g2^5\, (5\, g3^5 + 600\, g3^4\, \wp^3 - 1930\, g3^3\, \wp^6 - 44390\, g3^2\, \wp^9 + 69870\, g3\, \wp^{12} - 42812\, \wp^{15}) -$
        $2621440\, g2^3\, \wp^4\, (235\, g3^5 + 1150\, g3^4\, \wp^3 - 3566\, g3^3\, \wp^6 + 5552\, g3^2\, \wp^9 - 4075\, g3\, \wp^{12} + 1190\, \wp^{15}) +$
        $327680\, g2^4\, \wp^2\, (30\, g3^5 - 1159\, g3^4\, \wp^3 - 4490\, g3^3\, \wp^6 + 11646\, g3^2\, \wp^9 - 12032\, g3\, \wp^{12} + 9083\, \wp^{15}) -$
        $2097152\, g2^2\, (g3^7 + 255\, g3^6\, \wp^3 + 2400\, g3^5\, \wp^6 - 15050\, g3^4\, \wp^9 + 34170\, g3^3\, \wp^{12} - 36654\, g3^2\, \wp^{15} +$
        $3610\, g3\, \wp^{18} - 1125\, \wp^{21}) + 8388608\, g2\, \wp^2\, (-35\, g3^7 - 469\, g3^6\, \wp^3 + 3900\, g3^5\, \wp^6 - 15830\, g3^4\, \wp^9 +$
        $25610\, g3^3\, \wp^{12} - 9435\, g3^2\, \wp^{15} + 583\, g3\, \wp^{18} + 50\, \wp^{21}) + 16777216\, \wp\, (-5\, g3^8 - 65\, g3^7\, \wp^3 +$
        $685\, g3^6\, \wp^6 - 3410\, g3^5\, \wp^9 + 11425\, g3^4\, \wp^{12} - 5735\, g3^3\, \wp^{15} + 3145\, g3^2\, \wp^{18} + 520\, g3\, \wp^{21} + \wp^{24})) /$
        $(g2^6 + 200\, g2^5\, \wp^2 + 400\, g2^4\, \wp\, (g3 - 5\, \wp^3) + 128\, g2^3\, (g3^2 - 10\, g3\, \wp^3 + 150\, \wp^6) -$
        $768\, g2^2\, (5\, g3^2\, \wp^2 - 58\, g3\, \wp^5 + 35\, \wp^8) - 2048\, g2\, \wp\, (5\, g3^3 - 60\, g3^2\, \wp^3 - 30\, g3\, \wp^6 + 31\, \wp^9) -$
        $4096\, (g3^4 - 25\, g3^3\, \wp^3 + 15\, g3^2\, \wp^6 + 95\, g3\, \wp^9 - 5\, \wp^{12}))^2\}\}$

i) This is the function $\psi_\lambda(x)$.

In[1]:= ψ[λ_][x_] = WeierstrassSigma[λ - x, {g2, g3}]/
        (WeierstrassSigma[λ, {g2, g3}] WeierstrassSigma[x, {g2, g3}])*
        Exp[WeierstrassZeta[λ, {g2, g3}] x];

We have to show that the following quantity δ is equal to 1.

In[2]:= δ = (D[ψ[λ][x], x, x] - 2 WeierstrassP[x, {g2, g3}] ψ[λ][x])/
        (WeierstrassP[λ, {g2, g3}] ψ[λ][x]) // Simplify

Out[2]= $-\dfrac{1}{\text{WeierstrassP}[\lambda, \{g2, g3\}]}$
        (WeierstrassP[x, {g2, g3}] + WeierstrassP[-x + λ, {g2, g3}] - (WeierstrassZeta[x,
            {g2, g3}] - WeierstrassZeta[λ, {g2, g3}] + WeierstrassZeta[-x + λ, {g2, g3}])^2)

Or we have to show that the difference of the numerator and the denominator is zero.

In[3]:= zero = Numerator[δ] - Denominator[δ]

```
Out[3]= -WeierstrassP[x, {g2, g3}] - WeierstrassP[λ, {g2, g3}] -
          WeierstrassP[-x + λ, {g2, g3}] + (WeierstrassZeta[x, {g2, g3}] -
          WeierstrassZeta[λ, {g2, g3}] + WeierstrassZeta[-x + λ, {g2, g3}])²
```

To show that `zero` is identically zero, we differentiate `zero` with respect to x and $λ$. In addition to coinciding initial conditions, we have to show that `{zero, D[zero, x], D[zero, λ]}` is identically zero. Using the symmetry of the Weierstrass $\wp$ function, we arrive at the following set of equations.

```
In[4]:= gb1 = (GroebnerBasis[{zero, D[zero, x], D[zero, λ]},
              {}, {WeierstrassZeta[x, {g2, g3}], WeierstrassZeta[λ, {g2, g3}]},
              MonomialOrder -> EliminationOrder] // Factor) //.
        {λ -> u, x -> -v} //.
        {WeierstrassP[uv_, {g2, g3}] :> ℘[uv],
         WeierstrassPPrime[uv_, {g2, g3}] :> ℘'[uv]} /. ℘'[-v] -> -℘'[v]
Out[4]= {℘[v] ℘'[u] - ℘[u + v] ℘'[u] - ℘[u] ℘'[v] + ℘[u + v] ℘'[v] - ℘[u] ℘'[u + v] + ℘[v] ℘'[u + v],
         4 ℘[u]³ + 4 ℘[u]² ℘[v] - 4 ℘[u]² ℘[u + v] - 8 ℘[u] ℘[v] ℘[u + v] - 4 ℘[u] ℘[u + v]² +
         4 ℘[v] ℘[u + v]² + 4 ℘[u + v]³ - ℘'[u]² - 2 ℘'[u] ℘'[u + v] - ℘'[u + v]²,
         -4 ℘[u]³ + 4 ℘[u] ℘[v]² + 4 ℘[u] ℘[v] ℘[u + v] - 4 ℘[v]² ℘[u + v] + 4 ℘[u] ℘[u + v]² -
         4 ℘[v] ℘[u + v]² + ℘'[u]² - ℘'[u] ℘'[v] + ℘'[u] ℘'[u + v] - ℘'[v] ℘'[u + v],
         4 ℘[u]³ + 4 ℘[v]³ - 12 ℘[u] ℘[v] ℘[u + v] + 4 ℘[u + v]³ - ℘'[u]² +
         ℘'[u] ℘'[v] - ℘'[v]² - ℘'[u] ℘'[u + v] - ℘'[v] ℘'[u + v] - ℘'[u + v]²}
```

Using now the addition theorem from part a) and the differential equations for $\wp$, we arrive at the conclusion that `zero` vanishes identically.

```
In[5]:= gb2 = Numerator[Together[gb1 /. ℘'[u + v] ->
              -(℘[u + v] (℘'[u] - ℘'[v]) + ℘[u] ℘'[v] - ℘'[u] ℘[v])/(℘[u] - ℘[v])]];

In[6]:= gbR = GroebnerBasis[(* make polynomials *) Numerator[Together[#]]& /@
              {℘[u + v] + ℘[u] + ℘[v] - ((℘'[u] - ℘'[v])/(℘[u] - ℘[v]))^2/4,
               ℘'[u]^2 - 4 ℘[u]^3 + g2 ℘[u] + g3, ℘'[v]^2 - 4 ℘[v]^3 + g2 ℘[v] + g3},
              {℘[u + v], ℘'[u], ℘[u], ℘'[v], ℘[v]}];

In[7]:= Last[PolynomialReduce[#, gbR,
              {℘[u + v], ℘'[u], ℘[u], ℘'[v], ℘[v]}]]& /@ gb2
Out[7]= {0, 0, 0, 0}
```

For a similar solution of the Jimbo–Miwa equation, see [801].

j) A generic inverse Weierstrass function has three branch points. Here are the three branch points for the invariants $g_2 = 1 + i, g_3 = 1 - 2i$.

```
In[1]:= {g2, g3} = {1 + I, 1 - 2 I};
        branchPoints = z /. Solve[4 z^3 - g2 z - g3 == 0, z] // N
Out[2]= {0.82709 - 0.165204 i, -0.783007 - 0.533101 i, -0.0440829 + 0.698306 i}
```

The following contour plot shows the branch cuts as clusters of contour lines. The white points represent are branch points.

```
In[3]:= ContourPlot[Im[InverseWeierstrassP[zr + I zi, {g2, g3}]],
              {zr, -1, 1}, {zi, -1, 1}, PlotPoints -> 160,
              ColorFunction -> (Hue[0.8 #]&), Contours -> 30,
              Epilog -> {GrayLevel[1], PointSize[0.03],
                         Point[{Re[#], Im[#]}]& /@ branchPoints}]
```

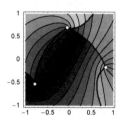

The position of the branch cuts can be described by the solutions of $4z^3 - g_2 z - g_3 = negativeRealNumber$. We could use this information to divide the z-plane into branch cut free regions. Inside these regions, we then would solve the differential equation $w''(z) + w'(z)(12 z^2 - g_2)/(2(4 z^3 - g_2 z - g_3)) = 0$ for the inverse Weierstrass function. By solving the differential equation encircling the branch points multiple times, we could model the Riemann surface of $\wp^{(-1)}(z; 1 + i, 1 - 2 i)$. Because the only goal here is a picture of the Riemann surface, we choose a more convenient way. The Riemann surface of the inverse function $z(w)$ of a function $w(z)$ is obviously given by the parametrization $\{\text{Re}(z), \text{Im}(z), f(w)\} = \{\text{Re}(z(w)), \text{Im}(z(w)), f(w)\}$, where f is a function that allows for a faithful representation of the Riemann surface—in most cases, Re or Im. Here, the parametrization variables of the surface are Re(w) and Im(w). So we can generate the following surface:

```
In[4]:= {g2, g3} = {1 + I, 1 - 2I};
       f[z_] = WeierstrassP[z, {g2, g3}];
       pp = 51;
       points = Table[{Re[f[wr + I wi]], Im[f[wr + I wi]], wr + I wi},
                     {wr, -2.5, 2.5, 5/pp}, {wi, -2.5, 2.5, 5/pp}];

In[8]:= (* form polygons from array of points *)
       prePolys = Table[Polygon[{points[[i, j]], points[[i, j + 1]],
                        points[[i + 1, j + 1]], points[[i + 1, j]]}],
                   {i, pp}, {j, pp}] // Flatten;

In[10]:= polysRe = Map[Re, prePolys, {-1}];
        polysIm = Apply[{#1, #2, Im[#3]}&, prePolys, {-2}];
```

Because some of the polygons of the last graphic "cross infinity", we eliminate these polygons. In addition, we cut away the vertices of the other polygons to allow for a better view inside the multisheeted surface.

```
In[12]:= tooBigPolygonQ[Polygon[l_], δMax_] :=
        Max[#.#& /@ Apply[Subtract, (* the edges *)
                Partition[Append[l, First[l]], 2, 1], {1}]] > δMax^2

In[13]:= makeDiamondPolygon[Polygon[l_]] :=
        Polygon[Apply[Plus, Partition[Append[l, First[l]]/2, 2, 1], {1}]]
```

So we arrive at the following pictures for the Riemann surface of the Weierstrass function $\wp^{(-1)}(z; 1 + i, 1 - 2 i)$. The left picture uses the real part and the right displays the imaginary part.

```
In[14]:= Show[GraphicsArray[
        Graphics3D[{EdgeForm[{Thickness[0.002], Hue[0.79]}],
                   SurfaceColor[Hue[0.12], Hue[0.22], 1.4],
                   makeDiamondPolygon /@ DeleteCases[#, _?(tooBigPolygonQ[#, 1]&)]}],
             BoxRatios -> {1, 1, 2}, PlotRange -> All,
             ViewPoint -> {2, -2, 0}]& /@ {polysRe, polysIm}]]
```

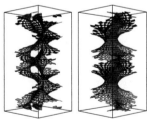

k) The equations for the $a_{m,n}$ do not immediately allow for a recursion. So we generate the first equations for the $a_{m,n}$ and solve for the lowest order $a_{m,n}$. The function `getSeries` returns the terms of the Taylor series of $\sigma(z; g_2, g_3)$.

```
In[1]:= a[0, 0] = 1;
       a[m_, n_] := 0 /; Negative[m] || Negative[n];
       eq[m_, n_] := (a[m, n] == 3 (m + 1) a[m + 1, n - 1] +
                     16/3 (n + 1) a[m - 2, n + 1] -
                     1/3 (2m + 3n -1) (4m + 6n - 1) a[m - 1, n])

In[4]:= getSeries[o_] :=
       Module[{eqs, sol},
          (* the a-equations *)
          eqs = Rest @ Flatten @ Table[eq[m, n], {m, 0, o}, {n, 0, o}];
          Off[Solve::svars];
          (* solve the a-equations *)
          sol = Solve[eqs, Cases[eqs, _a, Infinity] // Union];
          On[Solve::svars];
```

```
(* make series *)
Sum[a[m, n] (g2/2)^m (2g3)^n z^(4m + 6n + 1)/(4m + 6n + 1)!,
    {m, 0, o}, {n, 0, o}] /.
  Select[sol[[1]], FreeQ[#[[2]], _a, Infinity]&] /. _a -> 0]
```

Here is the series up to order 19.

In[5]:= `ser = getSeries[7] + O[z]^20 // Simplify`

Out[5]= $z - \dfrac{g2\,z^5}{240} - \dfrac{g3\,z^7}{840} - \dfrac{g2^2\,z^9}{161280} - \dfrac{(g2\,g3)\,z^{11}}{2217600} + \dfrac{(23\,g2^3 - 576\,g3^2)\,z^{13}}{16605388800} +$

$\dfrac{19\,g2^2\,g3\,z^{15}}{96864768000} + \dfrac{(107\,g2^4 + 52992\,g2\,g3^2)\,z^{17}}{1896999616512000} + \dfrac{(311\,g2^3\,g3 + 4416\,g3^3)\,z^{19}}{4505374089216000} + O[z]^{20}$

Using `ser` in an ansatz for the coefficients of the partial differential equations gives a linear set of equations for the coefficients G2 [*j*], G3 [*j*], and GC [*j*].

In[6]:= `diff = D[ser, z, z] - (Sum[G2[j] z^j, {j, 0, 2}] D[ser, g2] +`
                           `Sum[G3[j] z^j, {j, 0, 2}] D[ser, g3] +`
                           `Sum[GC[j] z^j, {j, 0, 2}] ser);`

Solving the system results in the following partial differential equation for $\sigma(z; g_2, g_3)$:

$$\frac{\partial^2\,\sigma(z; g_2, g_3)}{\partial z^2} = 12\,g_3\,\frac{\partial\sigma(z; g_2, g_3)}{\partial g_2} + \frac{2}{3}\,g_2^2\,\frac{\partial\sigma(z; g_2, g_3)}{\partial g_3} - \frac{z^2}{12}\,\sigma(z; g_2, g_3).$$

In[7]:= `Solve[# == 0& /@ CoefficientList[diff, z],`
        `Flatten[Table[{G2[j], G3[j], GC[j]}, {j, 0, 2}]]]`

Out[7]= $\left\{\left\{\text{G3}[2] \to 0,\ \text{G3}[0] \to \dfrac{2\,g2^2}{3},\ \text{G2}[2] \to 0,\ \text{G3}[1] \to 0,\right.\right.$

$\left.\left.\text{G2}[1] \to 0,\ \text{G2}[0] \to 12\,g3,\ \text{GC}[1] \to 0,\ \text{GC}[2] \to -\dfrac{g2}{12},\ \text{GC}[0] \to 0\right\}\right\}$

l) This implements the function $\Sigma(\tau)$.

In[1]:= `Σ[τ_?InexactNumberQ] :=`
        `With[{σ = WeierstrassSigma[#, WeierstrassInvariants[{1, τ}]]&},`
        `σ[2/3 τ]^5/(σ[τ/3]^4 σ[4/3 τ])]`

To find algebraic values for $\Sigma(i\sqrt{n}\,)$, we start by calculating a 30-digit approximation. Then we find the nearest polynomial of order four or less. (We restrict ourselves to order four to always obtain radicals when solving the polynomial.) We solve the polynomial and determine the distance of the roots to a 200-digit approximation of $\Sigma(i\sqrt{n}\,)$. If this distance is small (say less than 10^{-150}), then $\Sigma(i\sqrt{n}\,)$ and the root agree to about 150 digits, and we naturally conjecture that they actually are identical.

In[2]:= `Needs["NumberTheory`Recognize`"]`

In[3]:= `Do[(* low and high precision value for Σ[I Sqrt[n]] *)`
        `  x = SetPrecision[ξ = N[Σ[I Sqrt[n] + I 0.1`200^^-199], 200], 30];`
        `  (* integer coefficient polynomial such that x is a root *)`
        `  poly = Recognize[x, 4, t];`
        `  (* minimal distance between root and Σ[I Sqrt[n]] *)`
        `  δ = Min[Abs[(Last /@ (List @@ NRoots[poly == 0, t, 200])) - ξ]];`
        `  (* print radical value if root distance is small *)`
        `  If[δ < 10^-150, Print[Σ[n] == t /. Select[Solve[poly == 0, t],`
        `                        Abs[t - ξ /. #] < 10^-100&][[1]]]],`
        `  {n, 120}]`

$\Sigma[1] = 3 + \sqrt{3} + \sqrt{3\left(3 + 2\sqrt{3}\,\right)}$

$\Sigma[2] = 5 + 2\sqrt{6}$

$\Sigma[3] = 3\left(2 + \sqrt{3}\,\right)$

$\Sigma[4] = 3 + 2\sqrt{3} + 2\sqrt{2}\ 3^{3/4}$

$\Sigma[6] = 3\left(3 + 2\sqrt{2}\,\right)$

$\Sigma[8] = 9 + 4\sqrt{3} + 4\sqrt{2 + \sqrt{3}}$

$\Sigma[9] = 3\left(3 + \sqrt{3} + \sqrt{3\left(3 + 2\sqrt{3}\,\right)}\,\right)$

$$\Sigma[12] = 3\left(7 + 4\sqrt{3}\right)$$

$$\Sigma[15] = 18 + \frac{9\sqrt{15}}{2} + \frac{3}{2}\sqrt{5\left(31 + 8\sqrt{15}\right)}$$

$$\Sigma[18] = 9\left(5 + 2\sqrt{6}\right)$$

$$\Sigma[24] = 3\left(15 + 6\sqrt{6} + 2\sqrt{2\left(49 + 20\sqrt{6}\right)}\right)$$

$$\Sigma[30] = 3\left(27 + 12\sqrt{5} + 2\sqrt{2\left(161 + 72\sqrt{5}\right)}\right)$$

$$\Sigma[36] = 3\left(45 + 26\sqrt{3} + 2\sqrt{6\left(168 + 97\sqrt{3}\right)}\right)$$

$$\Sigma[42] = 3\left(75 + 20\sqrt{14} + 2\sqrt{6\left(449 + 120\sqrt{14}\right)}\right)$$

$$\Sigma[48] = 3\left(119 + 84\sqrt{2} + 4\sqrt{3\left(577 + 408\sqrt{2}\right)}\right)$$

$$\Sigma[60] = 3\left(279 + 72\sqrt{15} + 4\sqrt{5\left(1921 + 496\sqrt{15}\right)}\right)$$

$$\Sigma[72] = 9\left(201 + 116\sqrt{3} + 4\sqrt{5042 + 2911\sqrt{3}}\right)$$

$$\Sigma[78] = 3\left(867 + 170\sqrt{26} + 12\sqrt{2\left(5201 + 1020\sqrt{26}\right)}\right)$$

$$\Sigma[102] = 3\left(3267 + 2310\sqrt{2} + 8\sqrt{17\left(19601 + 13860\sqrt{2}\right)}\right)$$

For functions having algebraic values for all algebraic arguments, see[1103].

m) It is straightforward to implement the above formulas describing the motion of n lattice-periodicized vortices. ω_1 and ω_2 are the half periods of the lattice, $\Gamma List$ the list of the vortex strengths, $z0List$ the list of their initial conditions and T the time over which to solve the equations of motion. The optional argument *numericalSolutionCheck* implements a check of the numerical solution based on the conservation law $\sum_{\alpha=1}^{n} \Gamma_\alpha z_\alpha(t) = constant$.

```
In[1]:= vortexArrayMotionGraphics[{ω1_, ω2_}, ΓList_, z0List_, T_,
                           numericalSolutionCheck_:False] :=
    Module[{n = Length[ΓList], g2, g3, Δ, φ, equationsOfMotion,
            ndsol, ΣΓ0, ℓp, pp, pp = 600},
      (* definitions for φ *)
      {g2, g3} = WeierstrassInvariants[{ω1, ω2}] // N;
      Δ = 4 Abs[Im[ω1 Conjugate[ω2]]];
      φ[z_] = WeierstrassZeta[z, {g2, g3}] - Pi/Δ Conjugate[z] +
              (Pi/Δ - WeierstrassZeta[ω1, {g2, g3}]/ω1) z;
      (* equations of motion *)
      equationsOfMotion =
      Table[z[α]'[t] == I/(2Pi) Sum[If[α === β, 0,
          Conjugate[ΓList[[β]] φ[z[α][t] - z[β][t]]]], {β, n}], {α, n}];
      (* solve equations of motions numerically *)
      ndsol = NDSolve[Join[equationsOfMotion,
                      Thread[Table[z[α][0], {α, n}] == z0List]],
                  Table[z[α], {α, n}], {t, 0, T},
                  PrecisionGoal -> 10, MaxSteps -> 10^5];
      (* make graphics *)
      Block[{$DisplayFunction = Identity},
      If[numericalSolutionCheck, (* check conservation law *)
      ΣΓ0 = Sum[ΓList[[α]] z[α][0], {α, n}] /. ndsol[[1]];
      (* check solution quality *)
      ℓp = ListPlot[Log[10, Abs[Table[Sum[ΓList[[α]] z[α][t], {α, n}]/ΣΓ0 -
                          1 /. ndsol[[1]], {t, 0, T, T/pp}]]]],
                      PlotRange -> All];
      pp = ParametricPlot[Evaluate[Table[{Re[z[α][t]], Im[z[α][t]]}, {α, n}] /.
                  ndsol[[1]]], {t, 0, ndsol[[1, 1, 2, 1, 1, 2]]},
                  PlotPoints -> pp, PlotStyle ->
                  Table[{Thickness[0.002], Hue[0.78 (α - 1)/n]},
                    {α, 0, n - 1}], PlotRange -> All,
                  Axes -> False, Frame -> True, FrameTicks -> False]];
      (* display graphics *)
      If[numericalSolutionCheck, Show[GraphicsArray[{ℓp, pp}]], Show[pp]]]
```

To generate random instances of lattices, vortex strengths, and positions, we implement a function `randomVortexArray`·
`MotionGraphics` which accepts a seed *seed* for the random number generator as its first argument.

```
In[2]:= randomVortexArrayMotionGraphics[seed_, T_,
                                        numericalSolutionCheck_:False] :=
        Module[{n, ωs, Γs, z0s},
            (* seed random number generator *) SeedRandom[seed];
            (* number of vortices *) n = Random[Integer, {2, 5}];
            (* lattice half periods *) ωs = Table[Random[], {2}]{1, I};
            (* vortex strength; sum is zero *)
            Γs = Append[#, -Plus @@ #]&[Table[Random[Real, {-1, 1}], {n - 1}]];
            (* initial positions *)
            z0s = Table[Random[Complex, {-1 - I, 1 + I}], {n}];
            (* make graphics of vortex motion *)
            vgr = vortexArrayMotionGraphics[ωs, Γs, z0s, T,
                                            numericalSolutionCheck]]
```

Calling `randomVortexArrayMotionGraphics` with random seeds results in a variety of vortex movements similar to the nonperiodicized case. Here are two situations with a largely translational motion structure. (We color each vortex trajectory differently.)

```
In[3]:= Show[GraphicsArray[Block[{$DisplayFunction = Identity},
        randomVortexArrayMotionGraphics @@@
                {{657261560047, 1024}, {724575402766, 128}}]]]
```

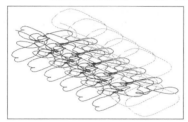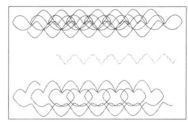

Here are two situations with a largely rotational motion structure.

```
In[4]:= Show[GraphicsArray[Block[{$DisplayFunction = Identity},
        randomVortexArrayMotionGraphics @@@
                {{491575599708, 16}, {191769440444, 32}}]]]
```

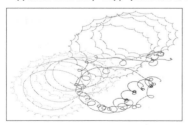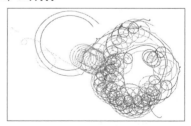

We end with two examples that show complicated trajectories. The left graphic shows $\ln(|(\sum_{\alpha=1}^{n} \Gamma_a z_\alpha(t)) - (\sum_{\alpha=1}^{n} \Gamma_a z_\alpha(0))|)$ and confirms the correctness of the numerical solutions of the trajectories.

```
In[5]:= randomVortexArrayMotionGraphics[349646403399, 256, True]
```

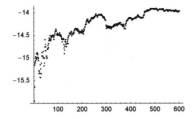
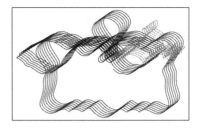

In[6]:= **randomVortexArrayMotionGraphics[808492574369, 064, True]**

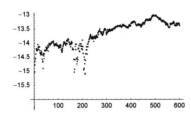

4. Jacobi's Elliptic Functions

a) Let us start with the first-order differential equations.

The following implementation is straightforward. We expand the elliptic function under consideration as a series around $z = 0$ and differentiate this series. We compare the result with the series expansion of

$$\sqrt{a_f(m) + b_f(m)\, f^2(u \mid m) + c_f(m)\, f^4(u \mid m)}\,.$$

Comparing the coefficients of equal powers of z gives a linear system of equations for $a_f(m)$, $b_f(m)$, and $c_f(m)$. The case that the series of $\partial f(u \mid m)/\partial u$ has no constant term is treated especially to avoid dividing by zero. The undetermined sign on the right-hand side of the differential equation is determined for real $0 < m < 1$ and $0 < z < K(m)/2$ by trying both possibilities and numerically choosing the right one. (The HoldForm at the end is used to avoid actually differentiating the function, and the result of makeODEForJacobis is visually a differential equation.)

```
In[1]:= makeODEForJacobis[jacobi_] :=
    Module[{func, ser, ser1, ser2, ser3, ser4, ser5, s0, a, b, c},
    func = jacobi[z, m];
    (* make the series expansion *)
    ser = Series[func, {z, 0, 5}];
    If[ (* first term of D[ser, z] equal 0 ? *)
       Coefficient[Normal[D[ser, z]], z, 0] =!= 0,
       (* plug ser in differential equation; solve for a, b, and c *)
       (* the If tests if D[ser, z] - Sqrt[a + b ser^2 + c ser^4] or
                           D[ser, z] + Sqrt[a + b ser^2 + c ser^4]
          is appropriate *)
    short1 = Hold[ser2 = Normal[ser1];
                  ser3 = Join[{0}, Exponent[#, z]& /@
                         Take[Union[Cases[ser2, z^_., {2}]], 2]];
                  ser4 = Coefficient[ser2, z, #]& /@ ser3;
                  ser5 = Simplify[PowerExpand @ Simplify @
                              Solve[# == 0& /@ ser4, {a, b, c}]];
                  ser5 = Select[ser5, (((ser4 /. #) /.
                              {m -> 1/2}) === {0, 0, 0})&]];
       If[ser1 = D[ser, z] - Sqrt[a + b ser^2 + c ser^4];
          ReleaseHold[short1]; ser5 === {}, ReleaseHold[short1]],
       s0 = Coefficient[Normal[ser], z, 0];
       (* plug ser in differential equation; solve for a, b, and c *)
    short2 = Hold[ser2 = Normal[ser1];
                  ser3 = Join[{0}, Exponent[#, z]& /@
```

```
                        Take[Union[Cases[ser2, z^_., {2}]], 2]];
            ser4 = Coefficient[ser2, z, #]& /@ ser3;
            ser5 = Solve[# == 0& /@ ser4, {b, c}];
            ser5 = Select[ser5, (((ser4 /. #) /.
                        {m -> 1/2}) === {0, 0, 0})&]];
    If[ser1 = D[ser, z] - Sqrt[-(b + c) s0 + b ser^2 + c ser^4];
        ReleaseHold[short2]; ser5 === {},
        ser1 = D[ser, z] + Sqrt[-(b + c) s0 + b ser^2 + c ser^4];
        ReleaseHold[short2]];
    ser5 = Simplify[{Join[ser5[[1]], {a -> ((-(b + c) s0) /. ser5)[[1]]}]}]]];
(* fix undetermined sign *)
If[Chop[(D[func, z] - Sqrt[a + b func^2 + c func^4] /. ser5)[[1]] /.
                        {m -> 0.3, z -> 0.5}] == 0,
(* write nicely, avoid differentiation *)
    (HoldForm[D[func1, z]] /. func1 -> func)  ==
            (+Sqrt[a + b func^2 + c func^4] /. ser5)[[1]],
    (HoldForm[D[func1, z]] /. func1 -> func)  ==
            (-Sqrt[a + b func^2 + c func^4] /. ser5)[[1]]]]
```

Now, for the functions that *Mathematica* can differentiate, the differential equations are calculated.

```
In[2]:= Off[Solve::ifun];
        (makeODEForJacobis /@
        {JacobiCD, JacobiCN, JacobiDC, JacobiDN, JacobiNC,
        JacobiND, JacobiSC, JacobiSD, JacobiSN, JacobiDC}) // TableForm
```

Out[3]//TableForm=

$$\partial_z \text{JacobiCD}[z, m] == -\sqrt{1 + (-1 - m)\, \text{JacobiCD}[z, m]^2 + m\, \text{JacobiCD}[z, m]^4}$$

$$\partial_z \text{JacobiCN}[z, m] == -\sqrt{1 - m + (-1 + 2m)\, \text{JacobiCN}[z, m]^2 - m\, \text{JacobiCN}[z, m]^4}$$

$$\partial_z \text{JacobiDC}[z, m] == \sqrt{m + (-1 - m)\, \text{JacobiDC}[z, m]^2 + \text{JacobiDC}[z, m]^4}$$

$$\partial_z \text{JacobiDN}[z, m] == -\sqrt{-1 + m + (2 - m)\, \text{JacobiDN}[z, m]^2 - \text{JacobiDN}[z, m]^4}$$

$$\partial_z \text{JacobiNC}[z, m] == \sqrt{-m + (-1 + 2m)\, \text{JacobiNC}[z, m]^2 + (1 - m)\, \text{JacobiNC}[z, m]^4}$$

$$\partial_z \text{JacobiND}[z, m] == \sqrt{-1 + (2 - m)\, \text{JacobiND}[z, m]^2 + (-1 + m)\, \text{JacobiND}[z, m]^4}$$

$$\partial_z \text{JacobiSC}[z, m] == \sqrt{1 + (2 - m)\, \text{JacobiSC}[z, m]^2 + (1 - m)\, \text{JacobiSC}[z, m]^4}$$

$$\partial_z \text{JacobiSD}[z, m] == \sqrt{1 + (-1 + 2m)\, \text{JacobiSD}[z, m]^2 + (-1 + m)\, m\, \text{JacobiSD}[z, m]^4}$$

$$\partial_z \text{JacobiSN}[z, m] == \sqrt{1 + (-1 - m)\, \text{JacobiSN}[z, m]^2 + m\, \text{JacobiSN}[z, m]^4}$$

$$\partial_z \text{JacobiDC}[z, m] == \sqrt{m + (-1 - m)\, \text{JacobiDC}[z, m]^2 + \text{JacobiDC}[z, m]^4}$$

Here is a quick numerical check of the last result.

```
In[4]:= ReleaseHold[%] /. {m -> N[2/5, 22], z -> 3/10}
Out[4]= {True, True, True, True, True, True, True, True, True, True}
```

The two Jacobi functions $ds(u \mid m)$ and $ns(u \mid m)$ do not have Taylor series expansions around $z = 0$, but Laurent expansions.

```
In[5]:= Series[JacobiDS[z, m], {z, 0, 3}]
```

$$\text{Out[5]}= \frac{1}{z} + \frac{1}{6}(1 - 2m)\, z + O[z]^2$$

```
In[6]:= Series[JacobiNS[z, m], {z, 0, 3}]
```

$$\text{Out[6]}= \frac{1}{z} + \frac{1}{6}(1 + m)\, z + O[z]^2$$

The definitions of $ds(u \mid m)$ and $ns(u \mid m)$ are $ds(u \mid m) = 1 / sd(u \mid m)$ and $ns(u \mid m) = 1 / sn(u \mid m)$. Therefore, we observe the identities and how the coefficients $a_f(m)$, $b_f(m)$, $c_f(m)$ transform under the substitution $f(z \mid m) \to 1 / f(z \mid m)$.

```
In[7]:= Collect[(D[1/f[x], x] /. f'[x] -> Sqrt[A + B f[x]^2 + C f[x]^4])^2 -
                        (Sqrt[a + b (1/f[x])^2 + c (1/f[x])^4])^2, f[x]]
```

$$\text{Out[7]}= -a + C + \frac{A - c}{f[x]^4} + \frac{-b + B}{f[x]^2}$$

Then, we can immediately obtain the missing two equations.

```
D[JacobiDS[u, m], u] ==
-Sqrt[m (m - 1) + (2m - 1) JacobiDS[u, m]^2 + JacobiDS[u, m]^4]
```

```
D[JacobiNS[u, m], u] ==
    -Sqrt[m - (m + 1) JacobiNS[u, m]^2 + JacobiNS[u, m]^4]
```

The coefficients $d_f(m)$ and $e_f(m)$ for the second-order differential equations can be easily obtained from the previously calculated coefficients of the first-order differential equations. By differentiating the first-order differential equations, we have the following result.

```
In[8]:= Cancel[D[Sqrt[a + b f[x]^2 + c f[x]^4], x] /.
                f'[x] -> Sqrt[a + b f[x]^2 + c f[x]^4]]
Out[8]=  b f[x] + 2 c f[x]^3
```

Now, we have $d_f(m) = b_f(m)$ and $e_f(m) = 2\,c_f(m)$.

b) There are 12 different Jacobi functions. We use the following obvious abbreviations and suppress the argument m.

```
In[1]:= allJacobis[u_] = {cd[u], cn[u], cs[u], dc[u], dn[u], ds[u],
                          nc[u], nd[u], ns[u], sc[u], sd[u], sn[u]}
Out[1]=  {cd[u], cn[u], cs[u], dc[u], dn[u], ds[u], nc[u], nd[u], ns[u], sc[u], sd[u], sn[u]}
```

These are the two basic identities for squares of Jacobi functions.

```
In[2]:= basicSquareSums[u_] = {cn[u]^2 + sn[u]^2 - 1, 1 - m sn[u]^2 - dn[u]^2}
Out[2]=  {-1 + cn[u]^2 + sn[u]^2, 1 - dn[u]^2 - m sn[u]^2}
```

The three basic Jacobi functions are $sn(u \mid m)$, $cn(u \mid m)$, and $dn(u \mid m)$. The other nine Jacobi functions can be easily defined in terms of these three functions.

```
In[3]:= jacobiDefinitions[u_] = (* make polynomials *) Numerator[Together[
            (* form differences *) Apply[Subtract, #, {1}]&[
            (* definitions of the Jacobi elliptic functions *)
            {cd[u] == cn[u]/dn[u], dc[u] == dn[u]/cn[u], ns[u] == 1/sn[u],
             sd[u] == sn[u]/dn[u], nc[u] == 1/cn[u], ds[u] == dn[u]/sn[u],
             nd[u] == 1/dn[u], sc[u] == sn[u]/cn[u], cs[u] == cn[u]/sn[u]}]]]
Out[3]=  {-cn[u] + cd[u] dn[u], cn[u] dc[u] - dn[u], -1 + ns[u] sn[u],
          dn[u] sd[u] - sn[u], -1 + cn[u] nc[u], -dn[u] + ds[u] sn[u],
          -1 + dn[u] nd[u], cn[u] sc[u] - sn[u], -cn[u] + cs[u] sn[u]}
```

The 66 possible pairs of different Jacobi functions are the following.

```
In[4]:= allJacobiPairs = Flatten[Table[{allJacobis[u][[i]], allJacobis[u][[j]]},
                          {i, 12}, {j, i + 1, 12}], 1]
Out[4]=  {{cd[u], cn[u]}, {cd[u], cs[u]}, {cd[u], dc[u]}, {cd[u], dn[u]},
          {cd[u], ds[u]}, {cd[u], nc[u]}, {cd[u], nd[u]}, {cd[u], ns[u]},
          {cd[u], sc[u]}, {cd[u], sd[u]}, {cd[u], sn[u]}, {cn[u], cs[u]},
          {cn[u], dc[u]}, {cn[u], dn[u]}, {cn[u], ds[u]}, {cn[u], nc[u]},
          {cn[u], nd[u]}, {cn[u], ns[u]}, {cn[u], sc[u]}, {cn[u], sd[u]}, {cn[u], sn[u]},
          {cs[u], dc[u]}, {cs[u], dn[u]}, {cs[u], ds[u]}, {cs[u], nc[u]}, {cs[u], nd[u]},
          {cs[u], ns[u]}, {cs[u], sc[u]}, {cs[u], sd[u]}, {cs[u], sn[u]}, {dc[u], dn[u]},
          {dc[u], ds[u]}, {dc[u], nc[u]}, {dc[u], nd[u]}, {dc[u], ns[u]}, {dc[u], sc[u]},
          {dc[u], sd[u]}, {dc[u], sn[u]}, {dn[u], ds[u]}, {dn[u], nc[u]}, {dn[u], nd[u]},
          {dn[u], ns[u]}, {dn[u], sc[u]}, {dn[u], sd[u]}, {dn[u], sn[u]}, {ds[u], nc[u]},
          {ds[u], nd[u]}, {ds[u], ns[u]}, {ds[u], sc[u]}, {ds[u], sd[u]}, {ds[u], sn[u]},
          {nc[u], nd[u]}, {nc[u], ns[u]}, {nc[u], sc[u]}, {nc[u], sd[u]}, {nc[u], sn[u]},
          {nd[u], ns[u]}, {nd[u], sc[u]}, {nd[u], sd[u]}, {nd[u], sn[u]}, {ns[u], sc[u]},
          {ns[u], sd[u]}, {ns[u], sn[u]}, {sc[u], sd[u]}, {sc[u], sn[u]}, {sd[u], sn[u]}}
```

We derive a possible relation of the squares of Jacobi functions by using the two basic square identities and the definitions of the Jacobi functions. To eliminate all not-needed functions, we use `GroebnerBasis`, with three arguments.

```
In[5]:= squareSumFormula[{jac1_, jac2_}] :=
            With[{gb = GroebnerBasis[Join[basicSquareSums[u], jacobiDefinitions[u]],
                          {jac1, jac2}, DeleteCases[allJacobis[u], jac1 | jac2]]},
                (* pq[u] qp[u] - 1 *)
                If[Exponent[gb[[1]], {jac1, jac2}] === {1, 1},
                    gb /. {jac1 -> jac1^2, jac2 -> jac2^2}, gb]]
```

Here are all square identities.

In[6]:= `TableForm[# == 0& /@ (squareSums = squareSumFormula /@ allJacobiPairs)]`

$\{cd[u]^2 - m\,cd[u]^2 - cn[u]^2 + m\,cd[u]^2\,cn[u]^2\} == 0$

$\{cd[u]^2 - m\,cd[u]^2 - cs[u]^2 + cd[u]^2\,cs[u]^2\} == 0$

$\{-1 + cd[u]^2\,dc[u]^2\} == 0$

$\{1 - m - dn[u]^2 + m\,cd[u]^2\,dn[u]^2\} == 0$

$\{1 - m - ds[u]^2 + cd[u]^2\,ds[u]^2\} == 0$

$\{1 - m\,cd[u]^2 - cd[u]^2\,nc[u]^2 + m\,cd[u]^2\,nc[u]^2\} == 0$

$\{-1 + m\,cd[u]^2 + nd[u]^2 - m\,nd[u]^2\} == 0$

$\{1 - m\,cd[u]^2 - ns[u]^2 + cd[u]^2\,ns[u]^2\} == 0$

$\{1 - cd[u]^2 - cd[u]^2\,sc[u]^2 + m\,cd[u]^2\,sc[u]^2\} == 0$

$\{-1 + cd[u]^2 + sd[u]^2 - m\,sd[u]^2\} == 0$

$\{1 - cd[u]^2 - sn[u]^2 + m\,cd[u]^2\,sn[u]^2\} == 0$

$\{cn[u]^2 - cs[u]^2 + cn[u]^2\,cs[u]^2\} == 0$

$\{-1 + m - m\,cn[u]^2 + cn[u]^2\,dc[u]^2\} == 0$

$\{1 - m + m\,cn[u]^2 - dn[u]^2\} == 0$

$\{1 - m + m\,cn[u]^2 - ds[u]^2 + cn[u]^2\,ds[u]^2\} == 0$

$\{-1 + cn[u]^2\,nc[u]^2\} == 0$

$\{-1 + nd[u]^2 - m\,nd[u]^2 + m\,cn[u]^2\,nd[u]^2\} == 0$

$\{1 - ns[u]^2 + cn[u]^2\,ns[u]^2\} == 0$

$\{-1 + cn[u]^2 + cn[u]^2\,sc[u]^2\} == 0$

$\{-1 + cn[u]^2 + sd[u]^2 - m\,sd[u]^2 + m\,cn[u]^2\,sd[u]^2\} == 0$

$\{-1 + cn[u]^2 + sn[u]^2\} == 0$

$\{-1 + m - cs[u]^2 + cs[u]^2\,dc[u]^2\} == 0$

$\{-1 + m - cs[u]^2 + dn[u]^2 + cs[u]^2\,dn[u]^2\} == 0$

$\{1 - m + cs[u]^2 - ds[u]^2\} == 0$

$\{-1 - cs[u]^2 + cs[u]^2\,nc[u]^2\} == 0$

$\{-1 - cs[u]^2 + nd[u]^2 - m\,nd[u]^2 + cs[u]^2\,nd[u]^2\} == 0$

$\{1 + cs[u]^2 - ns[u]^2\} == 0$

$\{-1 + cs[u]^2\,sc[u]^2\} == 0$

$\{-1 + sd[u]^2 - m\,sd[u]^2 + cs[u]^2\,sd[u]^2\} == 0$

$\{-1 + sn[u]^2 + cs[u]^2\,sn[u]^2\} == 0$

$\{-dc[u]^2 + m\,dc[u]^2 - m\,dn[u]^2 + dc[u]^2\,dn[u]^2\} == 0$

$\{-dc[u]^2 + m\,dc[u]^2 - ds[u]^2 + dc[u]^2\,ds[u]^2\} == 0$

Out[6]//TableForm= $\quad\{-m + dc[u]^2 - nc[u]^2 + m\,nc[u]^2\} == 0$

$\{-m + dc[u]^2 - dc[u]^2\,nd[u]^2 + m\,dc[u]^2\,nd[u]^2\} == 0$

$\{m - dc[u]^2 - ns[u]^2 + dc[u]^2\,ns[u]^2\} == 0$

$\{-1 + dc[u]^2 - sc[u]^2 + m\,sc[u]^2\} == 0$

$\{-1 + dc[u]^2 - dc[u]^2\,sd[u]^2 + m\,dc[u]^2\,sd[u]^2\} == 0$

$\{1 - dc[u]^2 - m\,sn[u]^2 + dc[u]^2\,sn[u]^2\} == 0$

$\{m\,dn[u]^2 - ds[u]^2 + dn[u]^2\,ds[u]^2\} == 0$

$\{-m - nc[u]^2 + m\,nc[u]^2 + dn[u]^2\,nc[u]^2\} == 0$

$\{-1 + dn[u]^2\,nd[u]^2\} == 0$

$\{m - ns[u]^2 + dn[u]^2\,ns[u]^2\} == 0$

$\{-1 + dn[u]^2 - sc[u]^2 + m\,sc[u]^2 + dn[u]^2\,sc[u]^2\} == 0$

$\{-1 + dn[u]^2 + m\,dn[u]^2\,sd[u]^2\} == 0$

$\{-1 + dn[u]^2 + m\,sn[u]^2\} == 0$

$\{-m - ds[u]^2 - nc[u]^2 + m\,nc[u]^2 + ds[u]^2\,nc[u]^2\} == 0$

$\{-m - ds[u]^2 + ds[u]^2\,nd[u]^2\} == 0$

$\{m + ds[u]^2 - ns[u]^2\} == 0$

$\{-1 - sc[u]^2 + m\,sc[u]^2 + ds[u]^2\,sc[u]^2\} == 0$

$\{-1 + ds[u]^2\,sd[u]^2\} == 0$

$\{-1 + m\,sn[u]^2 + ds[u]^2\,sn[u]^2\} == 0$

$\{nc[u]^2 - m\,nd[u]^2 - nc[u]^2\,nd[u]^2 + m\,nc[u]^2\,nd[u]^2\} == 0$

$\{-nc[u]^2 - ns[u]^2 + nc[u]^2\,ns[u]^2\} == 0$

$\{-1 + nc[u]^2 - sc[u]^2\} == 0$

$$\{-1 + nc[u]^2 - m\,sd[u]^2 - nc[u]^2\,sd[u]^2 + m\,nc[u]^2\,sd[u]^2\} == 0$$
$$\{1 - nc[u]^2 + nc[u]^2\,sn[u]^2\} == 0$$
$$\{-m\,nd[u]^2 - ns[u]^2 + nd[u]^2\,ns[u]^2\} == 0$$
$$\{1 - nd[u]^2 + sc[u]^2 - nd[u]^2\,sc[u]^2 + m\,nd[u]^2\,sc[u]^2\} == 0$$
$$\{-1 + nd[u]^2 - m\,sd[u]^2\} == 0$$
$$\{1 - nd[u]^2 + m\,nd[u]^2\,sn[u]^2\} == 0$$
$$\{-1 - sc[u]^2 + ns[u]^2\,sc[u]^2\} == 0$$
$$\{-1 - m\,sd[u]^2 + ns[u]^2\,sd[u]^2\} == 0$$
$$\{-1 + ns[u]^2\,sn[u]^2\} == 0$$
$$\{sc[u]^2 - sd[u]^2 - sc[u]^2\,sd[u]^2 + m\,sc[u]^2\,sd[u]^2\} == 0$$
$$\{-sc[u]^2 + sn[u]^2 + sc[u]^2\,sn[u]^2\} == 0$$
$$\{-sd[u]^2 + sn[u]^2 + m\,sd[u]^2\,sn[u]^2\} == 0$$

For a quick numerical check of these identities, we substitute some random numerical values.

```
In[7]:= replaceJacs = {cd[u_] -> JacobiCD[u, m], cn[u_] -> JacobiCN[u, m],
                       cs[u_] -> JacobiCS[u, m], dc[u_] -> JacobiDC[u, m],
                       dn[u_] -> JacobiDN[u, m], ds[u_] -> JacobiDS[u, m],
                       nc[u_] -> JacobiNC[u, m], nd[u_] -> JacobiND[u, m],
                       ns[u_] -> JacobiNS[u, m], sc[u_] -> JacobiSC[u, m],
                       sd[u_] -> JacobiSD[u, m], sn[u_] -> JacobiSN[u, m]}
        (* or shorter, but less readable
        (#[u_] -> (ToExpression["Jacobi" <> ToUpperCase[ToString[#]]])
         [u, m])& /@ {cd, cn, cs, dc, dn, ds, nc, nd, ns, sc, sd, sn} *)
Out[7]= {cd[u_] -> JacobiCD[u, m], cn[u_] -> JacobiCN[u, m], cs[u_] -> JacobiCS[u, m],
         dc[u_] -> JacobiDC[u, m], dn[u_] -> JacobiDN[u, m], ds[u_] -> JacobiDS[u, m],
         nc[u_] -> JacobiNC[u, m], nd[u_] -> JacobiND[u, m], ns[u_] -> JacobiNS[u, m],
         sc[u_] -> JacobiSC[u, m], sd[u_] -> JacobiSD[u, m], sn[u_] -> JacobiSN[u, m]}

In[9]:= squareSums /. replaceJacs /.
        Table[{u -> Random[], v -> Random[], m -> Random[]}, {3}] //
                                       Chop // Flatten // Union
Out[9]= {0}
```

c) We will use the following strategy here to derive all 12 addition formulas. We start with the known addition formulas and write them in polynomial form (clearing the denominator). We add to this equation the basic square formulas and the definitions of the Jacobi functions. Then, we use `GroebnerBasis` with `EliminationOrder` to derive a form that contains only the one function under consideration. The resulting polynomial might factor and we take out the factor that is the addition formula. After deriving an addition formula for $pq(u \mid m)$, we derive the addition formula for $qp(u \mid m)$ by the already derived formula and the definition of $qp(u \mid m)$ in terms of $pq(u \mid m)$ as $pq(u \mid m)\,qp(u \mid m) - 1 = 0$.

```
In[1]:= basicSquareSums[u_] = {cn[u]^2 + sn[u]^2 - 1, 1 - m sn[u]^2 - dn[u]^2}
Out[1]= {-1 + cn[u]^2 + sn[u]^2, 1 - dn[u]^2 - m sn[u]^2}
```

Here is the addition formula for $sn(u \mid m)$ written in polynomial form.

```
In[2]:= additionFormulaSN = Numerator[Together[Subtract @@ #]]&[
        sn[u + v] == (sn[u] cn[v] dn[v] + sn[v] cn[u] dn[u])/
                     (1 - m sn[u]^2 sn[v]^2)]
Out[2]= cn[v] dn[v] sn[u] + cn[u] dn[u] sn[v] - sn[u + v] + m sn[u]^2 sn[v]^2 sn[u + v]
```

We add the two basic square identities in u and v and eliminate the functions $cn(u \mid m)$, $cn(v \mid m)$, $dn(u \mid m)$, and $dn(v \mid m)$.

```
In[3]:= (* addition formula for function sn *)
        gbSN = Factor @ GroebnerBasis[
        Flatten[{additionFormulaSN, basicSquareSums[u], basicSquareSums[v]}],
        {sn[u + v], sn[u], sn[v]}, {cn[v], dn[v], cn[u], dn[u]},
        MonomialOrder -> EliminationOrder]
Out[4]= {(-1 + m sn[u]^2 sn[v]^2) (sn[u]^4 - 2 sn[u]^2 sn[v]^2 + sn[v]^4 -
         2 sn[u]^2 sn[u + v]^2 - 2 sn[v]^2 sn[u + v]^2 + 4 sn[u]^2 sn[v]^2 sn[u + v]^2 +
         4 m sn[u]^2 sn[v]^2 sn[u + v]^2 - 2 m sn[u]^4 sn[v]^2 sn[u + v]^2 - 2 m sn[u]^2 sn[v]^4 sn[u + v]^2 +
         sn[u + v]^4 - 2 m sn[u]^2 sn[v]^2 sn[u + v]^4 + m^2 sn[u]^4 sn[v]^4 sn[u + v]^4)}
```

The second factor is the polynomial we were looking for. Here is a quick numerical check of this formula.

```
In[5]:= gbSN[[1, -1]] /. sn[u_] -> JacobiSN[u, m] /.
         Table[SetPrecision[{u -> Random[], v -> Random[],
                   m -> Random[]}, 30], {6}] // Chop
Out[5]= {0, 0, 0, 0, 0, 0}
```

Having the addition formula for $sn(u \mid m)$ makes it straightforward to derive the one for $ns(u \mid m)$.

```
In[6]:= (* addition formula for function ns *)
         gbNS = Factor @ GroebnerBasis[
           Join[gbSN, {sn[u] ns[u] - 1, sn[v] ns[v] - 1, sn[u + v] ns[u + v] - 1}],
           {ns[u + v], ns[u], ns[v]}, {sn[u + v], sn[u], sn[v]},
           MonomialOrder -> EliminationOrder]
Out[7]= {-(m - ns[u]^2 ns[v]^2)
         (m^2 - 2 m ns[u]^2 ns[v]^2 + ns[u]^4 ns[v]^4 - 2 m ns[u]^2 ns[u + v]^2 - 2 m ns[v]^2 ns[u + v]^2 +
            4 ns[u]^2 ns[v]^2 ns[u + v]^2 + 4 m ns[u]^2 ns[v]^2 ns[u + v]^2 - 2 ns[u]^4 ns[v]^2 ns[u + v]^2 -
            2 ns[u]^2 ns[v]^4 ns[u + v]^2 + ns[u]^4 ns[u + v]^4 - 2 ns[u]^2 ns[v]^2 ns[u + v]^4 + ns[v]^4 ns[u + v]^4)}
```

Again, the second factor is the formula we were looking for. Here is a quick numerical check.

```
In[8]:= gbNS[[1, -1]] /. ns[u_] -> JacobiNS[u, m] /.
         Table[SetPrecision[{u -> Random[], v -> Random[],
                   m -> Random[]}, 30], {6}] // Chop
Out[8]= {0, 0, 0, 0, 0, 0}
```

In a similar way, we can derive the addition formulas for $cn(u \mid m)$ and $nc(u \mid m)$.

```
In[9]:= additionFormulaCN = Numerator[Together[Subtract @@ #]]&[
         cn[u + v] == (cn[u] cn[v] - sn[u] dn[u] sn[v] dn[v])/
                        (1 - m sn[u]^2 sn[v]^2)]
Out[9]= cn[u] cn[v] - cn[u + v] - dn[u] dn[v] sn[u] sn[v] + m cn[u + v] sn[u]^2 sn[v]^2
```

```
In[10]:= (* addition formula for function cn *)
         gbCN = Factor @ GroebnerBasis[
         Flatten[{additionFormulaCN, basicSquareSums[u], basicSquareSums[v]}],
         {cn[u + v], cn[u], cn[v]}, {sn[v], dn[v], sn[u], dn[u]},
         MonomialOrder -> EliminationOrder]
Out[11]= {(-1 + m - m cn[u]^2 - m cn[v]^2 + m cn[u]^2 cn[v]^2)
           (1 - m - cn[u]^2 + m cn[u]^2 - cn[v]^2 + m cn[v]^2 - m cn[u]^2 cn[v]^2 +
            2 cn[u] cn[v] cn[u + v] - cn[u + v]^2 + m cn[u + v]^2 - m cn[u]^2 cn[u + v]^2 -
            m cn[v]^2 cn[u + v]^2 + m cn[u]^2 cn[v]^2 cn[u + v]^2)}
```

```
In[12]:= (* quick numerical check *)
         gbCN[[1, -1]] /. cn[u_] -> JacobiCN[u, m] /.
         Table[SetPrecision[{u -> Random[], v -> Random[],
                   m -> Random[]}, 30], {6}] // Chop
Out[13]= {0, 0, 0, 0, 0, 0}
```

```
In[14]:= (* addition formula for function nc *)
         gbNC = Factor @ GroebnerBasis[
           Join[gbCN, {cn[u] nc[u] - 1, cn[v] nc[v] - 1, cn[u + v] nc[u + v] - 1}],
           {nc[u + v], nc[u], nc[v]}, {cn[u + v], cn[u], cn[v]},
           MonomialOrder -> EliminationOrder]
Out[15]= {(m - m nc[u]^2 - m nc[v]^2 - nc[u]^2 nc[v]^2 + m nc[u]^2 nc[v]^2)
           (-m + m nc[u]^2 + m nc[v]^2 + nc[u]^2 nc[v]^2 - m nc[u]^2 nc[v]^2 - 2 nc[u] nc[v] nc[u + v] +
            m nc[u + v]^2 + nc[u]^2 nc[u + v]^2 - m nc[u]^2 nc[u + v]^2 + nc[v]^2 nc[u + v]^2 -
            m nc[v]^2 nc[u + v]^2 - nc[u]^2 nc[v]^2 nc[u + v]^2 + m nc[u]^2 nc[v]^2 nc[u + v]^2)}
```

```
In[16]:= (* quick numerical check *)
         gbNC[[1, -1]] /. nc[u_] -> JacobiNC[u, m] /.
         Table[SetPrecision[{u -> Random[], v -> Random[],
                   m -> Random[]}, 30], {6}] // Chop
Out[17]= {0, 0, 0, 0, 0, 0}
```

Now, we derive the addition formulas for $dn(u \mid m)$ and $nd(u \mid m)$.

```
In[18]:= additionFormulaDN = Numerator[Together[Subtract @@ #1]]&[
            dn[u + v] == (dn[u] dn[v] - m sn[u] cn[u] sn[v] cn[v])/
                         (1 - m sn[u]^2 sn[v]^2)]
```

$$\text{Out[18]= } dn[u]\, dn[v] - dn[u+v] - m\, cn[u]\, cn[v]\, sn[u]\, sn[v] + m\, dn[u+v]\, sn[u]^2\, sn[v]^2$$

```
In[19]:= (* addition formula for function dn *)
        gbDN = Factor @ GroebnerBasis[
          Flatten[{additionFormulaDN, basicSquareSums[u], basicSquareSums[v]}],
          {dn[u + v], dn[u], dn[v]}, {sn[v], cn[v], sn[u], cn[u]},
          MonomialOrder -> EliminationOrder]
```

$$\text{Out[20]= } \{(-1 + m + dn[u]^2 + dn[v]^2 - dn[u]^2\, dn[v]^2)$$
$$(1 - m - dn[u]^2 + m\, dn[u]^2 - dn[v]^2 + m\, dn[v]^2 + dn[u]^2\, dn[v]^2 - 2\, m\, dn[u]\, dn[v]\, dn[u+v] -$$
$$dn[u+v]^2 + m\, dn[u+v]^2 + dn[u]^2\, dn[u+v]^2 + dn[v]^2\, dn[u+v]^2 - dn[u]^2\, dn[v]^2\, dn[u+v]^2)\}$$

```
In[21]:= (* quick numerical check *)
        gbDN[[1, -1]] /. dn[u_] -> JacobiDN[u, m] /.
          Table[SetPrecision[{u -> Random[], v -> Random[],
                              m -> Random[]}, 30], {6}] // Chop
```

$$\text{Out[22]= } \{0, 0, 0, 0, 0, 0\}$$

```
In[23]:= (* addition formula for function nd *)
        gbND = Factor @ GroebnerBasis[
          Join[gbDN, {dn[u] nd[u] - 1, dn[v] nd[v] - 1, dn[u + v] nd[u + v] - 1}],
          {nd[u + v], nd[u], nd[v]}, {dn[u + v], dn[u], dn[v]},
          MonomialOrder -> EliminationOrder]
```

$$\text{Out[24]= } \{(-1 + nd[u]^2 + nd[v]^2 - nd[u]^2\, nd[v]^2 + m\, nd[u]^2\, nd[v]^2)$$
$$(1 - nd[u]^2 - nd[v]^2 + nd[u]^2\, nd[v]^2 - m\, nd[u]^2\, nd[v]^2 + 2\, m\, nd[u]\, nd[v]\, nd[u+v] -$$
$$nd[u+v]^2 + nd[u]^2\, nd[u+v]^2 - m\, nd[u]^2\, nd[u+v]^2 + nd[v]^2\, nd[u+v]^2 -$$
$$m\, nd[v]^2\, nd[u+v]^2 - nd[u]^2\, nd[v]^2\, nd[u+v]^2 + m\, nd[u]^2\, nd[v]^2\, nd[u+v]^2)\}$$

```
In[25]:= (* quick numerical check *)
        gbND[[1, -1]] /. nd[u_] -> JacobiND[u, m] /.
          Table[SetPrecision[{u -> Random[], v -> Random[],
                              m -> Random[]}, 30], {6}] // Chop
```

$$\text{Out[26]= } \{0, 0, 0, 0, 0, 0\}$$

Next, we will derive the addition formulas for cd(u | m) and dc(u | m). The addition formula for cd(u | m) in terms of sn(u | m), sn(v | m), cn(u | m), cn(v | m), dn(u | m), and dn(v | m) are obtained directly from the addition formulas of cn(u | m) and dn(u | m).

```
In[27]:= additionFormulaCD = Numerator[Together[Subtract @@ #1]]&[
            cd[u + v] == (cn[u] cn[v] - sn[u] dn[u] sn[v] dn[v])/
                         (dn[u] dn[v] - m sn[u] cn[u] sn[v] cn[v])]
```

$$\text{Out[27]= } cn[u]\, cn[v] - cd[u+v]\, dn[u]\, dn[v] +$$
$$m\, cd[u+v]\, cn[u]\, cn[v]\, sn[u]\, sn[v] - dn[u]\, dn[v]\, sn[u]\, sn[v]$$

```
In[28]:= (* addition formula for function cd *)
        gbCD = Factor @ GroebnerBasis[
          Flatten[{additionFormulaCD, basicSquareSums[u], basicSquareSums[v],
                   -cn[u] + cd[u] dn[u], -cn[v] + cd[v] dn[v]}],
          {cd[u + v], cd[u], cd[v]}, {sn[u], cn[u], dn[u], sn[v], cn[v], dn[v]},
          MonomialOrder -> EliminationOrder]
```

$$\text{Out[29]= } \{(-1 + m)^2\, (-1 + m\, cd[u]^2\, cd[v]^2)$$
$$(-1 + cd[u]^2 + cd[v]^2 - m\, cd[u]^2\, cd[v]^2 - 2\, cd[u]\, cd[v]\, cd[u+v] + 2\, m\, cd[u]\, cd[v]\, cd[u+v] +$$
$$cd[u+v]^2 - m\, cd[u]^2\, cd[u+v]^2 - m\, cd[v]^2\, cd[u+v]^2 + m\, cd[u]^2\, cd[v]^2\, cd[u+v]^2)\}$$

```
In[30]:= (* quick numerical check *)
        gbCD[[1, -1]] /. cd[u_] -> JacobiCD[u, m] /.
          Table[SetPrecision[{u -> Random[], v -> Random[],
                              m -> Random[]}, 30], {6}] // Chop
```

$$\text{Out[31]= } \{0, 0, 0, 0, 0, 0\}$$

```
In[32]:= (* addition formula for function dc *)
        gbDC = Factor @ GroebnerBasis[
          Join[gbCD, {cd[u] dc[u] - 1, cd[v] dc[v] - 1, cd[u + v] dc[u + v] - 1}],
```

```
                 {dc[u + v], dc[u], dc[v]}, {cd[u + v], cd[u], cd[v]},
                 MonomialOrder -> EliminationOrder]
Out[33]=  {-(-1+m)² (m - dc[u]² dc[v]²)
          (-m + m dc[u]² + m dc[v]² - dc[u]² dc[v]² + 2 dc[u] dc[v] dc[u + v] - 2 m dc[u] dc[v] dc[u + v] +
             m dc[u + v]² - dc[u]² dc[u + v]² - dc[v]² dc[u + v]² + dc[u]² dc[v]² dc[u + v]²)}
```

```
In[34]:=  (* quick numerical check *)
          gbDC[[1, -1]] /. dc[u_] -> JacobiDC[u, m] /.
          Table[SetPrecision[{u -> Random[], v -> Random[],
                              m -> Random[]}, 30], {6}] // Chop
Out[35]=  {0, 0, 0, 0, 0, 0}
```

Next, we will derive the addition formulas for sc($u \mid m$) and cs($u \mid m$). The addition formula for sc($u \mid m$) in terms of sn($u \mid m$), sn($v \mid m$), cn($u \mid m$), cn($v \mid m$), dn($u \mid m$), and dn($v \mid m$) are obtained directly from the addition formulas of sn($u \mid m$) and cn($u \mid m$).

```
In[36]:=  additionFormulaSC = Numerator[Together[Subtract @@ #]]&[
             sc[u + v] == (sn[u] cn[v] dn[v] + sn[v] cn[u] dn[u])/
                          (cn[u] cn[v] - sn[u] dn[u] sn[v] dn[v])]
Out[36]=  cn[u] cn[v] sc[u + v] - cn[v] dn[v] sn[u] - cn[u] dn[u] sn[v] - dn[u] dn[v] sc[u + v] sn[u] sn[v]
```

```
In[37]:=  (* addition formula for function sc *)
          gbSC = Factor @ GroebnerBasis[
             Flatten[{additionFormulaSC, basicSquareSums[u], basicSquareSums[v],
                cn[u] sc[u] - sn[u], cn[v] sc[v] - sn[v]}],
             {sc[u + v], sc[u], sc[v]}, {sn[u], cn[u], dn[u], sn[v], cn[v], dn[v]},
             MonomialOrder -> EliminationOrder]
Out[38]=  {(-1 - sc[u]² - sc[v]² - sc[u]² sc[v]² + m sc[u]² sc[v]²)
          (sc[u]⁴ - 2 sc[u]² sc[v]² + sc[v]⁴ - 2 sc[u]² sc[u + v]² - 2 sc[v]² sc[u + v]² -
             8 sc[u]² sc[v]² sc[u + v]² + 4 m sc[u]² sc[v]² sc[u + v]² - 2 sc[u]⁴ sc[v]² sc[u + v]² +
             2 m sc[u]⁴ sc[v]² sc[u + v]² - 2 sc[u]² sc[v]⁴ sc[u + v]² + 2 m sc[u]² sc[v]⁴ sc[u + v]² +
             sc[u + v]⁴ - 2 sc[u]² sc[v]² sc[u + v]⁴ + 2 m sc[u]² sc[v]² sc[u + v]⁴ +
             sc[u]⁴ sc[v]⁴ sc[u + v]⁴ - 2 m sc[u]⁴ sc[v]⁴ sc[u + v]⁴ + m² sc[u]⁴ sc[v]⁴ sc[u + v]⁴)}
```

```
In[39]:=  (* quick numerical check *)
          gbSC[[1, -1]] /. sc[u_] -> JacobiSC[u, m] /.
          Table[SetPrecision[{u -> Random[], v -> Random[],
                              m -> Random[]}, 30], {6}] // Chop
Out[40]=  {0, 0, 0, 0, 0, 0}
```

```
In[41]:=  (* addition formula for function cs *)
          gbCS = Factor @ GroebnerBasis[
             Join[gbSC, {sc[u] cs[u] - 1, sc[v] cs[v] - 1, sc[u + v] cs[u + v] - 1}],
             {cs[u + v], cs[u], cs[v]}, {sc[u + v], sc[u], sc[v]},
             MonomialOrder -> EliminationOrder]
Out[42]=  {-(-1 + m - cs[u]² - cs[v]² - cs[u]² cs[v]²)
          (1 - 2 m + m² - 2 cs[u]² cs[v]² + 2 m cs[u]² cs[v]² + cs[u]⁴ cs[v]⁴ - 2 cs[u]² cs[u + v]² +
             2 m cs[u]² cs[u + v]² - 2 cs[v]² cs[u + v]² + 2 m cs[v]² cs[u + v]² - 8 cs[u]² cs[v]² cs[u + v]² +
             4 m cs[u]² cs[v]² cs[u + v]² - 2 cs[u]⁴ cs[v]² cs[u + v]² - 2 cs[u]² cs[v]⁴ cs[u + v]² +
             cs[u]⁴ cs[u + v]⁴ - 2 cs[u]² cs[v]² cs[u + v]⁴ + cs[v]⁴ cs[u + v]⁴)}
```

```
In[43]:=  (* quick numerical check *)
          gbCS[[1, -1]] /. cs[u_] -> JacobiCS[u, m] /.
          Table[SetPrecision[{u -> Random[], v -> Random[],
                              m -> Random[]}, 30], {6}] // Chop
Out[44]=  {0, 0, 0, 0, 0, 0}
```

Finally, we will derive the addition formulas for ds($u \mid m$) and sd($u \mid m$). The addition formula for ds($u \mid m$) in terms of sn($u \mid m$), sn($v \mid m$), cn($u \mid m$), cn($v \mid m$), dn($u \mid m$), and dn($v \mid m$) are obtained directly from the addition formulas of sn($u \mid m$) and dn($u \mid m$).

```
In[45]:=  additionFormulaDS = Numerator[Together[Subtract @@ #]]&[
             ds[u + v] == (dn[u] dn[v] - m sn[u] cn[u] sn[v] cn[v])/
                          (sn[u] cn[v] dn[v] + sn[v] cn[u] dn[u])]
```

Out[45]= -dn[u] dn[v] + cn[v] dn[v] ds[u + v] sn[u] +
       cn[u] dn[u] ds[u + v] sn[v] + m cn[u] cn[v] sn[u] sn[v]

In[46]:= (* addition formula for function ds *)
       gbDS = Factor @ GroebnerBasis[
         Flatten[{additionFormulaDS, basicSquareSums[u], basicSquareSums[v],
                   -dn[u] + ds[u] sn[u], -dn[v] + ds[v] sn[v]}],
         {ds[u + v], ds[u], ds[v]}, {sn[u], cn[u], dn[u], sn[v], cn[v], dn[v]},
         MonomialOrder -> EliminationOrder]

Out[47]= {(-m + m² + m ds[u]² + m ds[v]² + ds[u]² ds[v]²)
       (m² - 2 m³ + m⁴ + 2 m ds[u]² ds[v]² - 2 m² ds[u]² ds[v]² + ds[u]⁴ ds[v]⁴ +
       2 m ds[u]² ds[u + v]² - 2 m² ds[u]² ds[u + v]² + 2 m ds[v]² ds[u + v]² -
       2 m² ds[v]² ds[u + v]² + 4 ds[u]² ds[v]² ds[u + v]² - 8 m ds[u]² ds[v]² ds[u + v]² -
       2 ds[u]⁴ ds[v]² ds[u + v]² - 2 ds[u]² ds[v]⁴ ds[u + v]² +
       ds[u]⁴ ds[v]² ds[u + v]⁴ - 2 ds[u]² ds[v]² ds[u + v]⁴ + ds[v]⁴ ds[u + v]⁴)}

In[48]:= (* quick numerical check *)
       gbDS[[1, -1]] /. ds[u_] -> JacobiDS[u, m] /.
       Table[SetPrecision[{u -> Random[], v -> Random[],
                            m -> Random[]}, 30], {6}] // Chop

Out[49]= {0, 0, 0, 0, 0, 0}

In[50]:= (* addition formula for function sd *)
       gbSD = Factor @ GroebnerBasis[
         Join[gbDS, {sd[u] ds[u] - 1, sd[v] ds[v] - 1, sd[u + v] ds[u + v] - 1}],
         {sd[u + v], sd[u], sd[v]}, {ds[u + v], ds[u], ds[v]},
         MonomialOrder -> EliminationOrder]

Out[51]= {(1 + m sd[u]² + m sd[v]² - m sd[u]² sd[v]² + m² sd[u]² sd[v]²)
       (sd[u]⁴ - 2 sd[u]² sd[v]² + sd[v]⁴ - 2 sd[u]² sd[u + v]² - 2 sd[v]² sd[u + v]² +
       4 sd[u]² sd[v]² sd[u + v]² - 8 m sd[u]² sd[v]² sd[u + v]² + 2 m sd[u]⁴ sd[v]² sd[u + v]² -
       2 m² sd[u]⁴ sd[v]² sd[u + v]² + 2 m sd[u]² sd[v]⁴ sd[u + v]² - 2 m² sd[u]² sd[v]⁴ sd[u + v]² +
       sd[u + v]⁴ + 2 m sd[u]² sd[v]² sd[u + v]⁴ - 2 m² sd[u]² sd[v]² sd[u + v]⁴ +
       m² sd[u]⁴ sd[v]⁴ sd[u + v]⁴ - 2 m³ sd[u]⁴ sd[v]⁴ sd[u + v]⁴ + m⁴ sd[u]⁴ sd[v]⁴ sd[u + v]⁴)}

In[52]:= (* quick numerical check *)
       gbSD[[1, -1]] /. sd[u_] -> JacobiSD[u, m] /.
       Table[SetPrecision[{u -> Random[], v -> Random[],
                            m -> Random[]}, 30], {6}] // Chop

Out[53]= {0, 0, 0, 0, 0, 0}

For a very symmetric form of the addition theorems, see [327].

d) We have the Taylor expansion $\mathrm{sn}(z \mid m) = \sum_{k=0}^{\infty} c_k(z) m^k$, where $c_k(z) = \mathrm{sn}^{(0,k)}(z, 0)$. In the following we will calculate closed-form expressions for $\mathrm{sn}^{(0,k)}(z, 0)$. The zeroth term $c_0(z)$ autoevaluates.

In[1]:= JacobiSN[z, 0]

Out[1]= Sin[z]

The derivatives do not automatically simplify.

In[2]:=
       D[JacobiSN[z, m], m] /. m -> 0

       Power::infy : Infinite expression $\frac{1}{0}$ encountered. More…

       ∞::indet : Indeterminate expression 0 Cos[z] ComplexInfinity encountered. More…

Out[2]= Indeterminate

In[3]:= Limit[D[JacobiSN[z, m], m], m -> 0]

Out[3]= Limit[$\frac{1}{2 (1 - m) m}$
       (JacobiCN[z, m] JacobiDN[z, m] ((1 - m) z - EllipticE[JacobiAmplitude[z, m], m] +
       m JacobiCD[z, m] JacobiSN[z, m])), m → 0]

This is the defining integral relation for sn($z \mid m$). To prevent *Mathematica* from carrying out the integration, we insert a dummy function $f(t)$. In a moment we will set $f(t)$ to 1. To avoid any autoevaluation, we use `jacobiSN` instead of `JacobiSN`.

```
In[4]:= def = z - Integrate[f[t]/(Sqrt[1 - t^2] Sqrt[1 - m t^2]),
                   {t, 0, jacobiSN[z, m]},
                   GenerateConditions -> False]
```

$$\text{Out[4]=}\quad z - \text{Integrate}\left[\frac{f[t]}{\sqrt{1-t^2}\ \sqrt{1-m\,t^2}},\ \{t,\,0,\,\text{jacobiSN}[z,\,m]\},\ \text{GenerateConditions} \rightarrow \text{False}\right]$$

We differentiate the relation `def` and substitute the value $m = 0$. Setting $f(t)$ to 1 yields an equation for sn$^{(0,k)}(z, 0)$.

```
In[5]:= eq[1] = D[def, m] /. m -> 0 /. f -> (1&)
```

$$\text{Out[5]=}\quad \frac{1}{4}\left(-\text{ArcSin}[\text{jacobiSN}[z,\,0]] + \text{jacobiSN}[z,\,0]\sqrt{1-\text{jacobiSN}[z,\,0]^2}\right) - \frac{\text{jacobiSN}^{(0,1)}[z,\,0]}{\sqrt{1-\text{jacobiSN}[z,\,0]^2}}$$

Because Jacobi functions do not have branch cuts, we can simplify the last expression.

```
In[6]:= (* use value of JacobiSN[z, 0] and simplify inverse functions *)
        jacobiSNSimplify[expr_] := expr //. jacobiSN[z, 0] -> Sin[z] //.
                   {ArcSin[Sin[z]] -> z, (1 - Sin[z]^2)^n_ -> Cos[z]^(2n),
                    (Cos[z]^n_)^m_ :> Cos[z]^(n m)}

In[8]:= sol[1] = Solve[eq[1] == 0, Derivative[0, 1][jacobiSN][z, 0]][[1]] //
                                        jacobiSNSimplify
```

$$\text{Out[8]=}\quad \left\{\text{jacobiSN}^{(0,1)}[z,\,0] \rightarrow \frac{1}{4}\,\text{Cos}[z]\,(-z + \text{Cos}[z]\,\text{Sin}[z])\right\}$$

Here is a quick numerical check of the last result for a random numerical value of z.

```
In[9]:= With[{ε = 10^-20, z0 = 1/(Sqrt[2] - 1) + 1/Pi I},
           {(* numerical approximation of derivative definition *)
               (JacobiSN[z0, ε/2] - JacobiSN[z0, -ε/2])/ε,
            (* above result *)
            Derivative[0, 1][jacobiSN][z, 0] /. sol[1] /. z -> z0}] //
                (* numericalize to high precision *) N[#, 30]& // N
```

$$\text{Out[9]=}\quad \{0.576733 + 0.216977\,\mathbf{i},\ 0.576733 + 0.216977\,\mathbf{i}\}$$

For the next derivatives, we automate the above two steps.

```
In[10]:= sol[n_] := sol[n] =
           Module[{d},
              d[n] = D[def, {m, n}] /. m -> 0 /. f -> (1&) /.
                          Flatten[Table[sol[k], {k, n - 1}]] // jacobiSNSimplify;
              Solve[d[n] == 0, Derivative[0, n][jacobiSN][z, 0]][[1]] //
                          jacobiSNSimplify // Simplify]
```

Here are the Taylor coefficients $c_2(z)$ to $c_5(z)$.

```
In[11]:= sol[2]
```

$$\text{Out[11]=}\quad \left\{\text{jacobiSN}^{(0,2)}[z,\,0] \rightarrow \right.$$
$$\left. \frac{1}{32}\,\text{Cos}[z]\,(\text{Cos}[z]\,(6\,\text{Sin}[z] + \text{Sin}[3\,z]) + z\,(-9 + 12\,\text{Sin}[z]^2 - 2\,z\,\text{Tan}[z]))\right\}$$

```
In[12]:= sol[3]
```

$$\text{Out[12]=}\quad \left\{\text{jacobiSN}^{(0,3)}[z,\,0] \rightarrow \frac{1}{2048}\right.$$
$$(32\,z\,(-21 + z^2)\,\text{Cos}[z] + 3\,(-156\,z\,\text{Cos}[3\,z] - 20\,z\,\text{Cos}[5\,z] + 67\,\text{Sin}[z] -$$
$$\left. 88\,z^2\,\text{Sin}[z] + 82\,\text{Sin}[3\,z] - 72\,z^2\,\text{Sin}[3\,z] + 16\,\text{Sin}[5\,z] + \text{Sin}[7\,z]))\right\}$$

```
In[13]:= sol[4]
```

$$\text{Out[13]=}\quad \left\{\text{jacobiSN}^{(0,4)}[z,\,0] \rightarrow \right.$$
$$\frac{1}{8192}\,(4\,z\,(-1845 + 128\,z^2)\,\text{Cos}[z] + 72\,z\,(-83 + 12\,z^2)\,\text{Cos}[3\,z] - 1260\,z\,\text{Cos}[5\,z] -$$
$$84\,z\,\text{Cos}[7\,z] + 2214\,\text{Sin}[z] - 3120\,z^2\,\text{Sin}[z] + 32\,z^4\,\text{Sin}[z] + 2835\,\text{Sin}[3\,z] -$$
$$\left. 3888\,z^2\,\text{Sin}[3\,z] + 690\,\text{Sin}[5\,z] - 600\,z^2\,\text{Sin}[5\,z] + 72\,\text{Sin}[7\,z] + 3\,\text{Sin}[9\,z])\right\}$$

```
In[14]:= sol[5]
```

Out[14]= $\{$jacobiSN$^{(0,5)}$ [z, 0] $\rightarrow$

$\frac{1}{131072}$ (-8 z (55245 - 4580 z^2 + 16 z^4) Cos[z] + 5 (288 z (-272 + 69 z^2) Cos[3 z] +

40 z (-531 + 100 z^2) Cos[5 z] + 3 (-812 z Cos[7 z] - 36 z Cos[9 z] +

2 (12104 - 19208 z^2 + 544 z^4 + 8 (1925 - 3174 z^2 + 108 z^4) Cos[2 z] + (3739 - 5592 z^2)

Cos[4 z] + 475 Cos[6 z] - 392 z^2 Cos[6 z] + 33 Cos[8 z] + Cos[10 z]) Sin[z])))$\}$

A numerical check of the derived results shows good agreement.

```
In[15]:= $MaxExtraPrecision = 200;
        With[{ε = 10^-20, z0 = 1/(Sqrt[2] - 1)},
        Table[N[ε^-k Sum[(-1)^(k - j) Binomial[k, j] JacobiSN[z0, j ε],
                    {j, 0, k}], 30] -
              (Derivative[0, k][jacobiSN][z, 0] /. sol[k] /. z -> z0),
              {k, 5}]]
Out[16]= {2.49068102×10^-22, -8.0323727111×10^-21,
        -3.8160520020×10^-20, -8.772201337×10^-20, 8.835607901×10^-20}
```

Now let us deal with the second method to calculate the series terms under consideration. We start by calculating the first few $g_{2n}(z)$. Recursively creating and solving the differential equations are straightforward.

```
In[17]:= ode[o_?EvenQ] := With[{n = o/2},
        {g[2n]''[z] == -4g[2n][z] - 8n(2n - 1) g[2n - 2][z] +
         12n(2n - 1) Sum[Binomial[2n - 2, 2k] g[2k][z] g[2n - 2 - 2k][z],
                    {k, 0, n - 1}],
         g[2n][0] == 0, g[2n]'[0] == 0}]
In[18]:= g[0][z_] = Sin[z]^2;
In[19]:= Do[dsol = DSolve[ode[k], g[k][z], z] // Simplify;
         g[k][z_] = dsol[[1, 1, 2]], {k, 2, 10, 2}]
```

Now that we have the $g_{2n}(z)$, we can calculate the $f_{2n}(z)$. To do this we solve the equation that connects $g_{2n}(z)$ with the $f_{2k}(z)$ for $f_{2n}(z)$.

```
In[20]:= f[0][z_] = Sin[z];
        Off[RuleDelayed::rhs];
        f[o_?EvenQ][z_] := f[o][z_] = With[{n = o/2},
            (g[2n][z] - Sum[Binomial[2n, 2k] f[2k][z] f[2n - 2k][z],
                    {k, 1, n - 1}])/(2 f[0][z]) // Simplify]
```

This is the resulting series.

```
In[23]:= jacobiSN[z_, m_, o_] := Sum[m^n/(2n)! f[2n][z], {n, 0, o}]
```

The following plot shows how the approximation quality of the Taylor series increases with the number of terms.

```
In[24]:= Off[Plot::plnr]; Off[Pattern::patvar];

        With[{(* a "random" point *) z = 1.3 - 0.6 I},
        Plot[Evaluate @ Table[Log[10,
                    Abs[JacobiSN[z, m] - jacobiSN[z, m, o]]], {o, 0, 4}],
              {m, 0, 0.5}, PlotRange -> All, Frame -> True, Axes -> False,
              PlotStyle -> Table[{Hue[o/5]}, {o, 0, 4}]]]
```

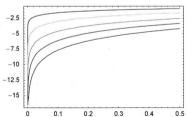

The coefficients calculated with this method agree with the above-calculated ones.

```
In[26]:= jac = jacobiSN[z, m, 5];
        Table[sol[k][[1, 2]] - k! Coefficient[jac, m, k], {k, 5}] // Simplify
```

```
Out[27]= {0, 0, 0, 0, 0}
```

e) The right-hand side functional equation expresses sn$(z \mid m)$ in sn$(z/2 \mid m)$, and sn$(z/4 \mid m)$. By repeatedly applying the functional equation, the argument of sn can be made as small as possible. We apply the functional equation until $|z| < \varepsilon$ and then use just the first term in the series expansion of sn$(z \mid m)$. Here this process is implemented.

```
In[1]:= jacobiSN[z_, m_, p_] :=
        Block[{sn, SN, ε = 10^-(p + 2), $RecursionLimit = Infinity},
         sn[ζ_, μ_] := sn[ζ, μ] =
          If[Abs[ζ] > ε, (* functional equation *)
             -((2(1 - 2 sn[ζ/4, μ]^2 + μ sn[ζ/4, μ]^4)*
             (1 + μ sn[ζ/4, μ]^2 (sn[ζ/4, μ]^2 - 2)) sn[ζ/2, μ])/
             ((μ sn[ζ/4, μ]^4 - 1)^2 (μ sn[ζ/2, μ]^4 - 1))),
             (* first term of the Taylor series *) ζ];

        SetAttributes[SN, NumericFunction];
        SN[ζ_?InexactNumberQ, μ_] := sn[ζ, μ];
        N[SN[z, m], p]]
```

After using the functional equation 337 times, we obtain a 100-digit approximation of sn$(1 + i \mid 1/3 + 2\,i)$.

```
In[2]:= jacobiSN[1 + I, 1/3 + 2 I, 100]

Out[2]= 1.3766372303219856480900018009559056232493736287899793631313559467442082832351608858⌣
        500776562656400296 -
        0.3119308883926265216698027327710572343231020791609682333016256964019877222118451115⌣
        9906915189745176540 i

In[3]:= %  - JacobiSN[1 + I, 1/3 + 2 I]

Out[3]= 0. × 10^-101 + 0. × 10^-101 i
```

f) Using the fundamental identities of the three Jacobi functions cn$(z \mid m)^2 + $ sn$(z \mid m)^2 - 1$, and dn$(z \mid m)^2 + m\,$sn$(z \mid m)^2 - 1$ it is straightforward to implement the function getPotentialPolynomial that, for a given solution function ψ, returns a polynomial connecting the potential V and function ψ.

```
In[1]:= getPotentialPolynomial[ψ_, {z_, m_}, {V_, Φ_}] :=
        GroebnerBasis[{(* basic Jacobi pq identities *)
                       JacobiCN[z, m]^2 + JacobiSN[z, m]^2 - 1,
                       JacobiDN[z, m]^2 + m JacobiSN[z, m]^2 - 1,
                       (* polynomials from ψ and Φ *)
                       Numerator[Together[Φ - ψ]],
                       Numerator[Together[-D[ψ, z, z] + V ψ]]},
                      {V, m, Φ}, {JacobiCN[z, m], JacobiSN[z, m], JacobiDN[z, m]},
                      MonomialOrder -> EliminationOrder] // Factor
```

Here are two examples.

```
In[2]:= getPotentialPolynomial[1/JacobiDN[z, m]^2, {z, m}, {V, Φ}]

Out[2]= {-2 + 8 Φ - 4 m Φ - V Φ - 6 Φ^2 + 6 m Φ^2}

In[3]:= getPotentialPolynomial[JacobiSN[z, m]/(JacobiDN[z, m] + α), {z, m}, {V, Φ}]

Out[3]= {(-1 + α)^2 (1 + α)^2 Φ
         (-1 + 4 m - 4 m^2 - 2 V + 4 m V - V^2 + α^2 + 2 m α^2 + m^2 α^2 + 2 V α^2 + 2 m V α^2 + V^2 α^2 - 4 m Φ^2 + 12 m^2 Φ^2 -
         8 m^3 Φ^2 - 4 m V Φ^2 + 4 m^2 V Φ^2 + 8 m α^2 Φ^2 - 8 m^2 α^2 Φ^2 + 5 m^3 α^2 Φ^2 + 8 m V α^2 Φ^2 - 4 m^2 V α^2 Φ^2 -
         4 m α^4 Φ^2 - 4 m^2 α^4 Φ^2 - 4 m V α^4 Φ^2 - 4 m^2 Φ^4 + 8 m^3 Φ^4 - 4 m^4 Φ^4 +
         12 m^2 α^2 Φ^4 - 16 m^3 α^2 Φ^4 + 4 m^4 α^2 Φ^4 - 12 m^2 α^4 Φ^4 + 8 m^3 α^4 Φ^4 + 4 m^2 α^6 Φ^4)}
```

For a rational solution, we must check if one of the factors of the polynomial returned by getPotentialPolynomial is linear in V. The function getRationalPotential does this and returns V rationally expressed through ψ.

```
In[4]:= getRationalPotential[potential_, {V_, Φ_}] :=
        Module[{factors, goodFactors},
         If[potential === {}, {}, (* analyze factors *)
            factors = If[Head[#] === Times, List @@ #, {#}]&[
                      Factor[potential[[1]]]]];
         (* select factors that are linear in V *)
         goodFactors = Select[factors, Exponent[#, V] === 1&];
```

```
    If[goodFactors =!= {},
        Solve[goodFactors[[1]] == 0, V][[1]], {}] // Simplify]
```

Here are again two examples.

```
In[5]:= getRationalPotential[
          getPotentialPolynomial[JacobiDN[z, m]^2, {z, m}, {V, Ψ}], {V, Ψ}]
```

$$Out[5]= \left\{V \to 8 + m\left(-4 + \frac{2}{\Phi}\right) - \frac{2}{\Phi} - 6\Phi\right\}$$

```
In[6]:= getRationalPotential[
          getPotentialPolynomial[1/JacobiDN[z, m]^2, {z, m}, {V, Ψ}], {V, Ψ}]
```

$$Out[6]= \left\{V \to 8 - \frac{2}{\Phi} - 6\Phi + m(-4 + 6\Phi)\right\}$$

We now implement a random search. As the ψ, we use sums of the form

$$\Psi(m \mid z) = \sum_{j=0}^{n} c_j \, \text{cn}(z \mid m)^{\alpha_j^{(cn)}} \, \text{sn}(z \mid m)^{\alpha_j^{(sn)}} \, \text{dn}(z \mid m)^{\alpha_j^{(dn)}}$$

where the $\alpha_j^{(pq)}$ are randomly chosen integers and the c_j are z- and m-independent constants.

```
In[7]:= r[n_] := Random[Integer, {-n, n}]

        randomJacobiPQPolynomial[maxExp_, n_] :=
        (* a random polynomial in Jacobi functions *)
        Sum[c[j] JacobiCN[z, m]^r[maxExp] JacobiSN[z, m]^r[maxExp]*
            JacobiDN[z, m]^r[maxExp], {j, n}]

In[9]:= searchPotentials[{maxExp_, n_}, o_, maxTime_:5] :=
        Module[{ψ, counter = 0},
            While[counter < o,
                ψ = randomJacobiPQPolynomial[maxExp, n];
                res = (* do not try too hard *) TimeConstrained[
                    getRationalPotential[getPotentialPolynomial[
                        ψ, {z, m}, {V, Ψ}], {V, Ψ}], maxTime];
                If[res != {} && res =!= $Aborted,
                    Print[{Φ -> ψ, res[[1]]}]; counter++]]]
```

Here are possible results of the random search.

```
In[10]:= SeedRandom[123456789];
         searchPotentials[{3, 1}, 5]
```

$$\left\{\Phi \to \frac{c[1]\, JacobiDN[z, m]^2}{JacobiSN[z, m]^2}, \ V \to -4 + \frac{6\Psi}{c[1]} + \frac{2m^2\, c[1]}{\Phi} + m\left(8 - \frac{2\,c[1]}{\Phi}\right)\right\}$$

$$\left\{\Phi \to \frac{c[1]\, JacobiCN[z, m]}{JacobiSN[z, m]}, \ V \to 2 - m + \frac{2\Phi^2}{c[1]^2}\right\}$$

$$\left\{\Phi \to \frac{c[1]\, JacobiCN[z, m]\, JacobiSN[z, m]}{JacobiDN[z, m]}, \ V \to 2\left(-2 + m + \frac{m^2\Phi^2}{c[1]^2}\right)\right\}$$

$$\left\{\Phi \to \frac{c[1]\, JacobiSN[z, m]}{JacobiCN[z, m]\, JacobiDN[z, m]}, \ V \to \frac{2((-1+m)^2\Phi^2 + (1+m)\,c[1]^2)}{c[1]^2}\right\}$$

$$\left\{\Phi \to \frac{c[1]\, JacobiCN[z, m]}{JacobiDN[z, m]\, JacobiSN[z, m]}, \ V \to 2 - 4m + \frac{2\Phi^2}{c[1]^2}\right\}$$

```
In[12]:= searchPotentials[{2, 2}, 1]
```

$$\left\{\Phi \to \frac{c[2]}{JacobiCN[z, m]^2} + \frac{c[1]\, JacobiSN[z, m]^2}{JacobiCN[z, m]^2}, \ V \to \right.$$
$$\left. \frac{2(-3(-1+m)\Phi^2 + c[1]^2 + 2(-1+m)\,c[1]\,c[2] - m\,c[2]^2 - 2\Phi((-2+m)\,c[1] + c[2] - 2m\,c[2]))}{\Phi(c[1] + c[2])}\right\}$$

Many more functions Ψ and potentials V can be found this way. For an alternative way to derive such potentials, see [733], [365], [79], [363], [532], [448], [684], and [488]; for trigonometric cases, [850]. For self-consistent potentials, see [1130], [1075]. For multidimensional generalizations, see [436]. For quasi-exactly solvable models with elliptic potentials, see [489]. For finding solutions to second-order differential equations with elliptic coefficients, see [237].

g) This is the nonlinear differential operator under consideration.

```
In[1]:= ℒ[x_, t_] = (D[#, {t, 2}] - β[1] D[#, x, x] -
                  β[2] D[#, {x, 4}] - β[3] D[#^2, {x, 2}])&;
```

We try to find a solution for $o = 2$ by applying $\mathcal{L}$ to the ansatz solution.

```
In[2]:= o = 2;
        ansatz = Sum[α[j] JacobiSN[k (x - c t), m]^j, {j, 0, o}]
```

$$Out[3]= \alpha[0] + JacobiSN[k (-c t + x), m] \alpha[1] + JacobiSN[k (-c t + x), m]^2 \alpha[2]$$

```
In[4]:= δ1 = ℒ[x, t] @ ansatz;
```

```
In[5]:= δ2 = δ1 /. (* use some abbreviations for shorter outputs *)
              {_JacobiSN -> sn, _JacobiCN -> cn, _JacobiDN -> dn} // Simplify
```

$$Out[5]= -c^2 k^2 (dn^2 + cn^2 m) sn\,\alpha[1] + 2 c^2 k^2 (-dn^2 sn^2 + cn^2 (dn^2 - m sn^2)) \alpha[2] +$$
$$k^2 (cn^2 m sn (\alpha[1] + 2 sn\,\alpha[2]) + dn^2 (sn\,\alpha[1] - 2 cn^2 \alpha[2] + 2 sn^2 \alpha[2])) \beta[1] -$$
$$k^4 (dn^4 (sn\,\alpha[1] - 8 cn^2 \alpha[2] + 8 sn^2 \alpha[2]) +$$
$$cn^2 m^2 sn (-4 sn^2 (\alpha[1] + 2 sn\,\alpha[2]) + cn^2 (\alpha[1] + 8 sn\,\alpha[2])) -$$
$$2 dn^2 m (4 cn^4 \alpha[2] + 2 sn^3 (\alpha[1] + 2 sn\,\alpha[2]) - cn^2 sn (7 \alpha[1] + 36 sn\,\alpha[2]))) \beta[2] -$$
$$2 ((cn\,dn\,k\,\alpha[1] + 2 cn\,dn\,k\,sn\,\alpha[2])^2 - k^2 (\alpha[0] + sn (\alpha[1] + sn\,\alpha[2]))$$
$$(cn^2 m sn (\alpha[1] + 2 sn\,\alpha[2]) + dn^2 (sn\,\alpha[1] - 2 cn^2 \alpha[2] + 2 sn^2 \alpha[2]))) \beta[3]$$

To equate coefficients, the power terms must be linear independent. Assuming linear independence of the $sn(k (x - c t) \mid m)^j$, we rewrite all $cn(k (x - c t) \mid m)$ and $dn(k (x - c t) \mid m)$ as $sn(k (x - c t) \mid m)$ using the defining algebraic relations between the Jacobi elliptic functions.

```
In[6]:= gb = GroebnerBasis[{δ2, cn^2 + sn^2 - 1, 1 - m sn^2 - dn^2},
                           {}, {cn, dn}];

        Short[gb, 9]
```

$$Out[7]//Short= \{-c^2 k^2 sn\,\alpha[1] - c^2 k^2 m sn\,\alpha[1] + 2 c^2 k^2 m sn^3 \alpha[1] + 2 c^2 k^2 \alpha[2] - 4 c^2 k^2 sn^2 \alpha[2] -$$
$$4 c^2 k^2 m sn^2 \alpha[2] + 6 c^2 k^2 m sn^4 \alpha[2] + k^2 sn\,\alpha[1] \beta[1] + k^2 m sn\,\alpha[1] \beta[1] -$$
$$2 k^2 m sn^3 \alpha[1] \beta[1] - 2 k^2 \alpha[2] \beta[1] + 4 k^2 sn^2 \alpha[2] \beta[1] + 4 k^2 m sn^2 \alpha[2] \beta[1] -$$
$$6 k^2 sn^3 \alpha[2] \beta[1] - k^4 sn\,\alpha[1] \beta[2] - 14 k^4 m sn\,\alpha[1] \beta[2] - k^4 m^2 sn\,\alpha[1] \beta[2] +$$
$$20 k^4 m sn^3 \alpha[1] \beta[2] + 20 k^4 m^2 sn^3 \alpha[1] \beta[2] - 24 k^4 m^2 sn^5 \alpha[1] \beta[2] + 8 k^4 \alpha[2] \beta[2] +$$
$$\ll 8 \gg + 120 k^4 m^2 sn^4 \alpha[2] \beta[2] - 120 k^4 m^2 sn^6 \alpha[2] \beta[2] + 2 k^2 sn\,\alpha[0] \alpha[1] \beta[3] +$$
$$2 k^2 m sn\,\alpha[0] \alpha[1] \beta[3] - 4 k^2 m sn^3 \alpha[0] \alpha[1] \beta[3] - 2 k^2 \alpha[1]^2 \beta[3] + 4 k^2 sn^2 \alpha[1]^2 \beta[3] +$$
$$4 k^2 m sn^2 \alpha[1]^2 \beta[3] - 6 k^2 m sn^4 \alpha[1]^2 \beta[3] - 4 k^2 \alpha[0] \alpha[2] \beta[3] + 8 k^2 sn^2 \alpha[0] \alpha[2] \beta[3] +$$
$$8 k^2 m sn^2 \alpha[0] \alpha[2] \beta[3] - 12 k^2 m sn^4 \alpha[0] \alpha[2] \beta[3] - 12 k^2 sn\,\alpha[1] \alpha[2] \beta[3] +$$
$$18 k^2 sn^3 \alpha[1] \alpha[2] \beta[3] + 18 k^2 m sn^3 \alpha[1] \alpha[2] \beta[3] - 24 k^2 m sn^5 \alpha[1] \alpha[2] \beta[3] -$$
$$12 k^2 sn^2 \alpha[2]^2 \beta[3] + 16 k^2 sn^4 \alpha[2]^2 \beta[3] + 16 k^2 m sn^4 \alpha[2]^2 \beta[3] - 20 k^2 m sn^6 \alpha[2]^2 \beta[3]\}$$

Extracting the coefficients of the powers of $sn(k (x - c t) \mid m)$ gives a linear system for the α_j that can be easily solved.

```
In[8]:= cl = CoefficientList[gb[[1]], sn];
```

```
        Off[Solve::svars];
        solr = Solve[# == 0& /@ cl, Table[α[j], {j, 0, o}]] //
                     Union // Simplify
```

$$Out[10]= \left\{\{\alpha[1] \to 0, \alpha[2] \to 0\}, \left\{\alpha[0] \to \frac{c^2 - \beta[1] + 4 k^2 (1 + m) \beta[2]}{2 \beta[3]}, \alpha[1] \to 0, \alpha[2] \to -\frac{6 k^2 m \beta[2]}{\beta[3]}\right\}\right\}$$

Here is the resulting solution.

```
In[11]:= U[x_, t_] = ansatz /. solr[[2]]
```

$$Out[11]= -\frac{6 k^2 m\, JacobiSN[k (-c t + x), m]^2 \beta[2]}{\beta[3]} + \frac{c^2 - \beta[1] + 4 k^2 (1 + m) \beta[2]}{2 \beta[3]}$$

For $m = 1$, the solution can be written in elementary functions.

```
In[12]:= U[x, t] /. m -> 1
```

$$Out[12]= -\frac{6 k^2 Tanh[k (-c t + x)]^2 \beta[2]}{\beta[3]} + \frac{c^2 - \beta[1] + 8 k^2 \beta[2]}{2 \beta[3]}$$

To find other evolution equations having solutions that can be expressed as a sum of powers of $sn(k (x - c t) \mid m)$, we will conduct a random search. We slightly modify the above steps and bundle them in a function `nonlinearEvolution`. `EquationTry`. We use now `Reduce` instead of `Solve` because a random equation will in most cases, if at all, have solutions of the wanted form only under additional restrictions on the parameters.

```
In[13]:= ψAnsatz[d_] := Sum[α[j] JacobiSN[k (x - c t), m]^j, {j, 0, d}]

In[14]:= nonlinearEvolutionEquationTry[pde_, d_] :=
        Module[{s1, s2, gb, cl, sn, cn, dn},
          (* the ansatz *)
          ansatz = ψAnsatz[d];
          (* substitute ansatz into the partial differential equation *)
          s1 = pde /. ψ -> Function[Evaluate[ansatz /. {x -> #1, t -> #2}]];
          (* rename variables *)
          s2 = s1 /. {_JacobiSN -> sn, _JacobiCN -> cn, _JacobiDN -> dn};
          (* form Gröbner basis *)
          gb = GroebnerBasis[{s2, cn^2 + sn^2 - 1, 1 - m sn^2 - dn^2},
                  {}, {cn, dn}, MonomialOrder -> EliminationOrder];
          cl = CoefficientList[gb[[1]], sn];
          (* use Reduce to get conditions on the βs *)
          Reduce[Join[# == 0& /@ cl, {m != 0, c != 0, k != 0}],
                  Table[α[j], {j, 0, d}]]]
```

Once we have found a candidate nonlinear evolution equation we use the following routine extractNontrivialNon
linearPart to verify that the differential equation is nonlinear under the restrictions generated by Reduce and that the
solution is nontrivial (meaning it contains x and t).

```
In[15]:= extractNontrivialNonlinearPart[res_] :=
        Module[{ψTerms, Φ, listΦ, ψRules, ncls, conds, resRed},
          (* conditions for a nonlinear pde *)
          ψTerms = Union[Cases[deq, Subscript[β, _] _, {1}] /.
                          Subscript[β, _] -> 1];
          listΦ = List @@ Collect[deq, ψTerms];
          ψRules = (# //. Derivative[_, _][ψ][x, t] :> Φ //. ψ[x, t] :> Φ)&;
          ncls = ψRules[Select[listΦ, (Exponent[ψRules[#], Φ] > 1)&]] //. Φ -> 1;
          conds = And[(* conditions for a nontrivial solution *)
                      Not[α[1] == 0 && α[2] == 0],
                      Not[And @@ ((# == 0)& /@ ncls)]];
          resRed = Select[res, (Reduce[And[#, conds]] =!= False)&];
          (* remove nondegeneracy conditions *)
          DeleteCases[resRed, _ != 0, Infinity]]
```

We start with an obvious generalization of the above equation and the restrictions bring us back to the form above.

```
In[16]:= (* two abbreviations *)
        Φ = ψ[x, t]; B[k_] := Subscript[β, k];

In[18]:= res = nonlinearEvolutionEquationTry[deq =
          D[Φ, t, t] - B[1] D[Φ, {x, 2}] - B[2] D[Φ, {x, 4}] +
          B[3] D[Φ, x]^2 + B[4] Φ D[Φ, {x, 2}], 2];
        extractNontrivialNonlinearPart[res]
```

$$\text{Out[19]=}\quad \beta_3 == \beta_4 \,\&\&\, \alpha[0] == \frac{-c^2 + \beta_1 - 4\,k^2\,\beta_2 - 4\,k^2\,m\,\beta_2}{\beta_4} \,\&\&\, \alpha[1] == 0 \,\&\&\, \alpha[2] == \frac{12\,k^2\,m\,\beta_2}{\beta_4}$$

Here is a nonlinear equation from [1000] investigated.

```
In[20]:= res = nonlinearEvolutionEquationTry[deq =
          D[Φ, t] + B[1] Φ^2 D[Φ, x] + B[2] D[Φ, {x, 5}], 2];
        extractNontrivialNonlinearPart[res] // Simplify
```

> Reduce::useq : The answer found by Reduce contains unsolved
> equation(s) $\left\{0 == \frac{1}{20} \left(-\sqrt{10}\,\sqrt{\beta_1}\,\alpha[0] - \sqrt{10}\,\sqrt{\ll 2\gg}\,\text{Power}[\ll 2\gg]\right),\right.$
>
> $0 == \frac{1}{20} \left(-\sqrt{10}\,\sqrt{\beta_1}\,\alpha[0] - \sqrt{10}\,\sqrt{\ll 2\gg}\,\text{Power}[\ll 2\gg]\right),$
>
> $\ll 4\gg,\ 0 == \frac{1}{20} \left(\sqrt{10}\,\sqrt{\beta_1}\,\alpha[0] - \sqrt{10}\,\sqrt{\ll 2\gg}\,\text{Power}[\ll 2\gg]\right),$
>
> $\left. 0 == \frac{1}{20} \left(\sqrt{10}\,\sqrt{\beta_1}\,\alpha[0] - \sqrt{10}\,\sqrt{\ll 2\gg}\,\text{Power}[\ll 2\gg]\right)\right\}.$ A likely reason for this is
> that the solution set depends on branch cuts of *Mathematica* functions. More…

Out[21]=
$$\Bigg(m == -1 \;\&\&\; c == -72\,k^4\,\beta_2 \;\&\&\; \alpha[0] == 0 \;\&\&$$

$$\alpha[1] == 0 \;\&\&\; \bigg(\alpha[2] == -\frac{6\,i\,\sqrt{10}\;k^2\,\sqrt{\beta_2}}{\sqrt{\beta_1}} \;||\; \alpha[2] == \frac{6\,i\,\sqrt{10}\;k^2\,\sqrt{\beta_2}}{\sqrt{\beta_1}} \bigg) \Bigg) \;||$$

$$\Bigg(m == 2 \;\&\&\; c == -72\,k^4\,\beta_2 \;\&\&\; \bigg(\alpha[0] == -\frac{6\,i\,\sqrt{10}\;k^2\,\sqrt{\beta_2}}{\sqrt{\beta_1}} \;||\; \alpha[0] == \frac{6\,i\,\sqrt{10}\;k^2\,\sqrt{\beta_2}}{\sqrt{\beta_1}} \bigg) \;\&\&$$

$$\alpha[1] == 0 \;\&\&\; \alpha[2] == -2\,\alpha[0] \bigg) \;||\; \bigg(c == -24\,k^4\,(1-m+m^2)\,\beta_2 \;\&\&$$

$$\bigg(\alpha[0] == -\frac{2\,\sqrt{10}\;\sqrt{-k^4\,(1+m)^2\,\beta_2}}{\sqrt{\beta_1}} \;||\; \alpha[0] == \frac{2\,\sqrt{10}\;\sqrt{-k^4\,(1+m)^2\,\beta_2}}{\sqrt{\beta_1}} \bigg) \;\&\&$$

$$\alpha[1] == 0 \;\&\&\; \alpha[2] == \frac{72\,k^4\,(-2+m)\,m\,\beta_2\,\alpha[0]}{c+72\,k^4\,\beta_2} \Bigg)$$

We implement a function nonlinearEvolutionEquationTry that generates a random nonlinear evolution equation and attempts to find a solution in the form of a sum of powers of sn($k\,(x-c\,t)\,|\,m$). The argument of the function nonlinearEvolutionEquationTry are the parameters for the order of derivatives, the number of terms, and so on.

```
In[22]:= nonlinearEvolutionEquationRandomTry[dt_, dx_, deg_, p_, o_, d_] :=
    Module[{Φ = ψ[x, t], rd, r, pde, solαs},
    (* generate a random pde *)
    rd := D[ψ[x, t], {x, Random[Integer, {0, dx}]}];
    r := Random[Integer, {0, deg}];
    pde = D[Φ, {t, dt}] + Sum[Subscript[β, j] Product[rd^r, {p}], {j, o}];
    (* solve the pde *)
    solαsPre = nonlinearEvolutionEquationTry[pde, d];
    (* make replacement rules from result *)
    solαs = {ToRules[Union[solαsPre]]};
    (* return pde and solutions *)
    If[solαs === {} || (Table[α[j], {j, 0, d}] /. solαs) === {{0, 0}},
        $Failed, {pde, DeleteCases[ansatz /. solαs, α[0] | 0] /.
            α[k_] :> Subscript[α, k]}]]
```

We implement one more function for the random search. The function randomTry calls the function nonlinearEvolutionEquationTry, constraining its memory and time usage. When nonlinearEvolutionEquationRandomTry succeeds, a list containing the equation and the solution is returned.

```
In[23]:= randomTry[] := MemoryConstrained[TimeConstrained[
    nonlinearEvolutionEquationRandomTry[
        (* orders of t-derivatives *) Random[Integer, {1, 3}],
        (* orders of x-derivatives *) Random[Integer, {3, 6}],
        (* powers of x-derivatives *) Random[Integer, {1, 5}],
        (* number of factors in each summand *) Random[Integer, {2, 4}],
        (* number of pde terms *) Random[Integer, {2, 7}],
        (* number of terms in ansatz *) Random[Integer, {2, 2}]], 240], 10^7]
```

Running the following input will give some potential candidates for nonlinear evolution equations.

```
Module[{rs, rt},
Do[While[rs = Random[Integer, {1, 10^10}]; SeedRandom[rs];
        rt = randomTry[rs];
        rt === $Failed || rt === $Aborted || FreeQ[rt, x], Null];
    Print["Seed for potential nonlinear pde: ", rs]; {100}]];
```

Here are the results of some of the seeds obtained with the last input.

```
In[24]:= SeedRandom[7011957036]; randomTry[] // TraditionalForm
```

Out[24]//TraditionalForm=

$$\left\{ \psi^{(0,3)}(x,\,t) + \beta_1\,\psi^{(2,0)}(x,\,t)\,\psi^{(3,0)}(x,\,t) + \beta_2\,\psi^{(1,0)}(x,\,t)\,\psi^{(4,0)}(x,\,t),\, \left\{ \frac{(2\,m^2\,c^3 - 5\,m\,c^3 - 4\,c^3)\,\text{JacobiSN}(k\,(x - c\,t),\,-1)^2}{8\,k^2\,\beta_2} + \alpha_0, \right. \right.$$

$$\left. \left. \frac{(2\,m^2\,c^3 - 5\,m\,c^3 - 4\,c^3)\,\text{JacobiSN}\!\left(k\,(x - c\,t),\,\frac{1}{2}\right)^2}{8\,k^2\,\beta_2} + \alpha_0,\; \frac{(2\,m^2\,c^3 - 5\,m\,c^3 - 4\,c^3)\,\text{JacobiSN}(k\,(x - c\,t),\,2)^2}{8\,k^2\,\beta_2} + \alpha_0 \right\} \right\}$$

In[25]:= **SeedRandom[1379006063]; randomTry[] // TraditionalForm**

Out[25]//TraditionalForm=

$$\left\{ \beta_3\,\psi(x,\,t)^3 + \beta_2\,\psi^{(4,0)}(x,\,t)^2\,\psi(x,\,t) + \beta_1\,\psi^{(4,0)}(x,\,t)^2 + \psi^{(0,2)}(x,\,t), \right.$$

$$\left. \left\{ -\frac{\sqrt{2}\,c\,k\,\text{JacobiSN}(k\,(x - c\,t),\,-1)}{\sqrt{\beta_3}},\; \frac{\sqrt{2}\,c\,k\,\text{JacobiSN}(k\,(x - c\,t),\,-1)}{\sqrt{\beta_3}} \right\} \right\}$$

These partial differential equations should now be analyzed more carefully using the functions nonlinearEvolution EquationTry and extractNontrivialNonlinearPart.

Here are some more nonlinear evaluation equations that were found using randomTry[]. We start with an equation that is third order in t.

The function $\psi(x,\,t) = 5\,(2\,c - (-1)^{1/3}\,c)\,/(48\,k^4\,\beta_1)\,\text{sn}(k\,(x - c\,t)\,|\,(-1)^{1/3})^2$ is a solution of the nonlinear partial differential equation $\psi_t - \beta_1\,\psi_{xx}\,\psi_{xxx} + 3/5\,\beta_1\,\psi_x\,\psi_{xxxx} = 0$.

In[26]:= **res = nonlinearEvolutionEquationTry[deq =**
    D[Φ, t] - B[1] D[Φ, {x, 2}] D[Φ, {x, 3}] +
    B[2] D[Φ, x] D[Φ, {x, 4}], 2];
    extractNontrivialNonlinearPart[res]

Out[27]= $\beta_1 == \dfrac{5\,\beta_2}{3}$ && (m == $(-1)^{1/3}$ || m == $-(-1)^{2/3}$) && $\alpha[1] == 0$ && $\left(\alpha[2] == 0 \;||\; \alpha[2] == \dfrac{2\,c - c\,m}{16\,k^4\,\beta_2} \right)$

The function $\psi(x,\,t) = 5\,c\,/(12\,k^2\,(m^2 - m + 1)\,\beta_1)\,(3\,m\,\text{sn}(k\,(x - c\,t)\,|\,m)^2 - m - 1)$ is a solution of the nonlinear partial differential equation $\psi_t + \beta_1\,\psi\,\psi_{xxx} - c/(24\,k^4\,(m^2 - m + 1))\,\psi_{xxxxx} = 0$.

In[28]:= **res = nonlinearEvolutionEquationTry[deq =**
    D[Φ, t] + B[1] Φ D[Φ, {x, 3}] + B[2] D[Φ, {x, 5}], 2];
    extractNontrivialNonlinearPart[res]

Out[29]= $\left(m == -1\ \&\&\ c == -72\,k^4\,\beta_2\ \&\&\ \alpha[0] == 0\ \&\&\ \alpha[1] == 0\ \&\&\ \alpha[2] == \dfrac{30\,k^2\,\beta_2}{\beta_1} \right)$ ||

$\left(m == 2\ \&\&\ c == -72\,k^4\,\beta_2\ \&\&\ \alpha[0] == \dfrac{30\,k^2\,\beta_2}{\beta_1}\ \&\&\ \alpha[1] == 0\ \&\&\ \alpha[2] == -2\,\alpha[0] \right)$ ||

$\left(c == -24\,(k^4\,\beta_2 - k^4\,m\,\beta_2 + k^4\,m^2\,\beta_2)\ \&\&\ \alpha[0] == \dfrac{10\,(k^2\,\beta_2 + k^2\,m\,\beta_2)}{\beta_1}\ \&\& \right.$

$\left. \alpha[1] == 0\ \&\&\ \alpha[2] == \dfrac{72\,(-2\,k^4\,m\,\beta_2\,\alpha[0] + k^4\,m^2\,\beta_2\,\alpha[0])}{c + 72\,k^4\,\beta_2} \right)$

And the function $\psi(x,\,t) = \big(\alpha_2\,\beta_4\,(3\,\text{sn}(k\,(x - c\,t)\,|\,(-1)^{1/3})^2 + (-1)^{1/3} - 2) - 3\,\beta_2\big)\big/(3\,\beta_4)$ is a solution of the nonlinear partial differential equation $\psi_t + \psi_x\,(c - 2\,\beta_4\,\psi_{xx}) + (\beta_2 + \beta_4\,\psi)\,\psi_{xxx} = 0$.

In[30]:= **res = nonlinearEvolutionEquationTry[deq =**
    D[Φ, t] + D[Φ, x] (c - 2 B[4] D[Φ, {x, 2}]) +
    D[Φ, {x, 3}] (B[2] + B[4]Φ), 2];
    extractNontrivialNonlinearPart[res]

Out[31]= $(\beta_4 == 0\ \&\&\ \beta_2 == 0)$ || $(\beta_4 == 0\ \&\&\ \beta_2 == 0\ \&\&\ \alpha[1] == 0)$ ||

$(\beta_4 == 0\ \&\&\ \beta_2 == 0\ \&\&\ \alpha[2] == 0)$ || $\left((m == (-1)^{1/3}\ ||\ m == -(-1)^{2/3})\ \&\& \right.$

$\left. \alpha[1] == 0\ \&\&\ \left(\alpha[2] == 0\ ||\ \alpha[2] == \dfrac{-\beta_2 - m\,\beta_2 - \beta_4\,\alpha[0] - m\,\beta_4\,\alpha[0]}{\beta_4} \right) \right)$

For a method to modify differential equations in such a way that they become periodic solutions, see [250]. For differential equations with doubly-periodic coefficients, see [212].

h) This is the function under consideration.

In[1]:= **w = JacobiSN[α Log[a + b z], m];**

We form the first two derivatives with respect to z and supplement the resulting equations with the two fundamental identities of the Jacobi elliptic functions.

```
In[2]:= (eqs = Join[Numerator[Together[{w, w', w''} - {w, D[w, z], D[w, z, z]}]],
            {JacobiCN[α Log[a + b z], m]^2 + JacobiSN[α Log[a + b z], m]^2 - 1,
             1 - m JacobiSN[α Log[a + b z], m]^2 - JacobiDN[α Log[a + b z], m]^2}]) //
            TraditionalForm
```

Out[2]//TraditionalForm= $\{w - \text{JacobiSN}(\alpha \log(a+b z), m), -b \alpha \operatorname{cn}(\alpha \log(a+b z) \mid m) \operatorname{dn}(\alpha \log(a+b z) \mid m) + a w' + b z w',$

$w'' a^2 + 2 b z w'' a + b^2 \alpha \operatorname{cn}(\alpha \log(a+b z) \mid m) \operatorname{dn}(\alpha \log(a+b z) \mid m) +$

$b^2 m \alpha^2 \operatorname{cn}(\alpha \log(a+b z) \mid m)^2 \operatorname{sn}(\alpha \log(a+b z) \mid m) + b^2 \alpha^2 \operatorname{dn}(\alpha \log(a+b z) \mid m)^2 \operatorname{sn}(\alpha \log(a+b z) \mid m) + b^2 z^2 w'',$

$\operatorname{cn}(\alpha \log(a+b z) \mid m)^2 + \operatorname{sn}(\alpha \log(a+b z) \mid m)^2 - 1, -\operatorname{dn}(\alpha \log(a+b z) \mid m)^2 - m \operatorname{sn}(\alpha \log(a+b z) \mid m)^2 + 1\}$

Now we have five equations and we have to eliminate the five terms $\operatorname{sn}(\lambda \log(a + b z) \mid m)$, $\operatorname{cn}(\lambda \log(a + b z) \mid m)$, $\operatorname{dn}(\lambda \log(a + b z) \mid m)$, a, and b. Because this is generically not possible, we start by eliminating the first four of these terms.

```
In[3]:= GroebnerBasis[eqs, {}, {JacobiSN[α Log[a + b z], m], JacobiCN[α Log[a + b z], m],
                JacobiDN[α Log[a + b z], m], a},
            MonomialOrder -> EliminationOrder] // Factor // FullSimplify
```

Out[3]= $\{b^2 \alpha ((-1 + w^2 (1 + m - m w^2 + (-1 + m (-1 + 2 w^2))^2 \alpha^2)) (w')^4 -$

$2 w (-1 + w^2) (1 + m - m (3 + m) w^2 + 2 m^2 w^4) \alpha^2 (w')^2 w'' + (-1 + w^2)^2 (-1 + m w^2)^2 \alpha^2 (w'')^2)\}$

Fortunately we were lucky and the last factor of the last result does not contain b. So, we have the following differential equation for $w(z) = \operatorname{sn}(\alpha \log(a + b z) \mid m)$:

$$\alpha^2 w''(z)^2 (w(z)^2 - 1)^2 (m w(z)^2 - 1)^2 + w'(z)^4 (w(z)^2 ((m \alpha^2 (2 w(z)^2 - 1) - 1)^2 - m w(z)^2 + m + 1) - 1) -$$
$$2 \alpha^2 w(z) w'(z)^2 w''(z) (w(z)^2 - 1) (2 m^2 w^4 - m (m + 3) w(z)^2 + m + 1) = 0.$$

i) We start with the numerical solution of the pendulum equation along straight rays in the complex t-plane. The function `pendulumNDSolve[α, φ0, T]` solves the differential equation along the ray $t = \tau \exp(i \alpha)$ for $0 \le \tau \le T$ and the function `complexElongationN` plots the real part of $\varphi(t)$ along this ray. We color the resulting curve according to the maximal elongation $\varphi0$.

```
In[1]:= (* numerical solution of the pendulum equation
         along the ray t == τ Exp[I α] *)
        pendulumNDSolve[α_, φ0_, T_, opts___] :=
          NDSolve[{Exp[-2I α] φ''[τ] == -Sin[φ[τ]],
                φ[0] == 0, φ'[0] == Exp[I α] 2 Sin[φ0/2]},
                φ, {τ, 0, T}, opts, MaxSteps -> 10^5,
                PrecisionGoal -> 12, MaxStepSize -> Pi/20];

In[3]:= (* plot oscillations along the ray t == τ Exp[I α] *)
        complexElongationN[α_, φ0_, T_, opts___] :=
        Module[{ndsol},
        (* solve pendulum equation along the ray t == τ Exp[I α] *)
        ndsol = pendulumNDSolve[α, φ0, T];
        (* plot real part of φ[t] *)
        Plot[Evaluate[Re[φ[τ]] /. ndsol[[1]]], {τ, 0, T}, opts,
            PlotPoints -> 20 T, PlotStyle -> Hue[0.8 Abs[φ0]/Pi],
            AspectRatio -> 1/3]]
```

Here are the resulting oscillations for $\alpha = 0$, $\alpha = \pi/6$, $\alpha = \pi/5$, $\alpha = \pi/4$, and $\alpha = e$. Complex time t runs downwards and each plot shows the oscillations for 25 initial φ_0 from the interval $[-5\pi/6, 5\pi/6]$. For nonvanishing α, we get much more complicated oscillations that for the classical case of purely real time.

```
In[5]:= Show[GraphicsArray[(* let time run downwards *)
        Function[α, Show[# /. Line[l_] :> Line[Reverse[{-1, 1}#]]& /@ l],
                    Frame -> True, FrameTicks -> False,
                    Axes -> False, AspectRatio -> 3]& @
        Table[complexElongationN[α, φ0, 60, DisplayFunction -> Identity],
            {φ0, -5/6 Pi, 5/6 Pi, 10/6 Pi/24}]] /@
            (* list of t-directions *) {0, Pi/6, Pi/5, Pi/4, E}]]
```

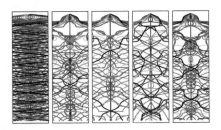

In the next two plots we fix φ_0 (we use a small and a large value) and show the resulting oscillations for 26 different directions (this time the coloring is according to the direction). Again, for nonvanishing α, we get quite complicated oscillations.

```
In[6]:= Show[GraphicsArray[Function[φ0,
         Show[Table[complexElongationN[α, φ0, 60,
             DisplayFunction -> Identity] /. _Hue :> Hue[0.8 Abs[α]],
           {α, -Pi, Pi, 2Pi/26}], Axes -> False, Frame -> True,
           PlotRange -> All, FrameTicks -> False]] /@
         (* small and large maximal elongation *) {0.01 Pi, 0.99 Pi}]]
```

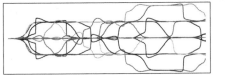 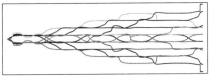

We remark that the observed behavior of the solution could be, in principle, reproduced using the following pair of purely real differential equations.

```
In[7]:= Thread[{x''[t], y''[t]} ==
         {ComplexExpand[Re[#]], ComplexExpand[Im[#]]}&[
           Expand[-ExpToTrig[Exp[2 I α]] TrigExpand[Sin[x[t] + I y[t]]]]]]
Out[7]= {x''[t] == -Cos[2 α] Cosh[y[t]] Sin[x[t]] + Cos[x[t]] Sin[2 α] Sinh[y[t]],
         y''[t] == -Cosh[y[t]] Sin[2 α] Sin[x[t]] - Cos[2 α] Cos[x[t]] Sinh[y[t]]}
```

To understand the complex oscillations for complex times, we consider the exact solution $\varphi P[t, \varphi 0]$ of the pendulum equation.

```
In[8]:= φP[t_, φ0_] := 2 ArcSin[Sin[φ0/2] JacobiSN[t, Sin[φ0/2]^2]]
```

The inner Jacobi elliptic function $\operatorname{sn}(t \mid \sin^2(\varphi_0/2))$ is a meromorphic function of t. But the outer inverse sin function adds a complicated branch cut structure to $\varphi(t)$. The function complexElongations calculates and visualizes the solution along the ray $t = \tau \exp(i\alpha)$ for $0 \le \tau \le T$ on the main sheet of arcsin and $2kM + 1$ neighboring sheets. We again color the resulting curve according to the maximal elongation $\varphi 0$.

```
In[9]:= complexElongations[α_, φ0_, T_, kM_:1, opts___] :=
         Module[{pl, lines},
         Show[Graphics[{Thickness[0.002], Hue[0.8 Abs[φ0]/Pi],
         (* elongation on the main sheet *)
         pl = Plot[Evaluate[Re[φP[Exp[I α] τ, φ0]]],
               {τ, 0, T}, DisplayFunction -> Identity,
               PlotPoints -> 20 T][[1, 1, 1, 1]];
         (* elongation on the neighboring sheet;
            use analytic continuations of ArcSin *)
         lines = {Table[{Line[({0, 4k Pi} + #)& /@ pl],
                     Line[({0, (4k + 2) Pi} + {1, -1}#)& /@ pl]},
                   {k, -kM, kM}]}], opts]]
```

The following three graphics show the continued exact solution and the numerical solution for the parameters $\alpha = 2$, $\varphi_0 = \pi/3$, $\alpha = E$, $\varphi_0 = \pi/2$, and a complex $\alpha = \exp(i\, 2\pi/5)$, $\varphi_0 = \pi/12$. The numerical solution winds in a complicated manner through the various sheets of the exact solutions.

```
In[10]:= complexElongationPlot[paramList_, opts___] :=
         Show[GraphicsArray[
```

```
Show[{(* multiple sheets from the symbolic form *)
       complexElongations[#1, #2, #3, #4,
       DisplayFunction -> Identity] /. _Thickness :> Thickness[0.02],
       (* one path from the numeric solution *)
       complexElongationN[#1, #2, #3,
       opts, DisplayFunction -> Identity] /. _Hue :> GrayLevel[0]}]& @@@
                                                  paramList]]
```

In[11]:= complexElongationPlot[{{2, Pi/3, 80, 2}, {E, Pi/2, 80, 2},
                                {Exp[I 2Pi/5], Pi/12, 80, 2}}]

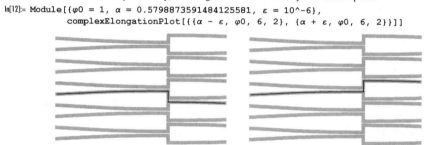

The next two graphics show that, for a fixed φ_0, small variations in α can result in greatly different paths. We see that at this special α, the two sheets meet nearly vertically, meaning we are in the vicinity of a branch point.

In[12]:= Module[{$\varphi 0$ = 1, α = 0.5798873591484125581, ε = 10^-6},
            complexElongationPlot[{{α - ε, $\varphi 0$, 6, 2}, {α + ε, $\varphi 0$, 6, 2}}]]

The sheets of the exact solutions can form quite complicated patterns along a ray in the complex plane. Here are three examples for $\alpha = \pi/3$, $\alpha = \pi/4$, and $\alpha = 99/100\,\pi/2$. Each picture shows the sheets for 13 different maximal elongations φ_0.

In[13]:= Show[GraphicsArray[Function[α, Show[
            Table[complexElongations[α, $\varphi 0$, 60, 1, DisplayFunction -> Identity],
               {$\varphi 0$, -5/6 Pi, 5/6 Pi, 10/6 Pi/12}]]] /@ {Pi/3, Pi/4, 99/100 Pi/2}]]

To understand why the numerical solution of the pendulum equations moves from sheet to sheet in such a complicated manner, we look at the branch points of $\varphi(t)$. The arcsin function has branch points at ± 1 and at infinity. For $0 \le \varphi_0 \le \pi/2$, the inner $\sin(\varphi_0/2)\,\mathrm{sn}(t \mid \sin^2(\varphi_0/2))$ never becomes ± 1. But the Jacobi elliptic function sn has poles and these poles give rise to branch points of $\varphi(t)$.

Next, we solve the pendulum equation radially outwards for a dense set of α and visualize the resulting oscillations in the complex t-plane using a contour plot. Sudden color variations in the contour plot indicate a discontinuous behavior of the oscillations, meaning branch cuts in the resulting numerical $\varphi_N(t)$. We indicate the branch points (located at multiples of the half periods) through small black points.

In[14]:= halfPeriods[$\varphi 0$_] := 2 EllipticK[{Sin[$\varphi 0$/2]^2, 1 - Sin[$\varphi 0$/2]^2}]

In[15]:= Module[{pp = 361, T = 24, $\varphi 0$ = 1, c = Ceiling[#]&,
            di = DisplayFunction -> Identity},

```
data = Table[
(* solve differential equation radially for many α's *)
    ndsol = pendulumNDSolve[α, φ0, T];
    Table[Evaluate[Re[φ[τ]] /. ndsol[[1]]], {τ, 0, T, T/pp}],
        {α, 0, 2Pi, 2Pi/pp}];
(* make contour plot in polar coordinates *)
lcp = ListContourPlot[data, PlotRange -> All, Contours -> 60,
        ColorFunction -> (Hue[1.6 (# - 1/2)]&),
        MeshRange -> {{0, T}, {0, 2Pi}}, di];
(* add intermediate points to a polygon with long edges *)
addPoints[p_, δε_] := Module[{n, l}, Join @@ (
    Function[s, If[(l = Sqrt[#. #]&[Subtract @@ s]) < δε, s,
        n = Floor[l/δε] + 1;
        Table[# + i/n (#2 - #1), {i, 0, n - 1}]& @@ s]] /@
                Partition[Append[p, First[p]], 2, 1])];
(* add points to polygon edges and  map to Cartesian t-coordinates *)
lcpR = Show[DeleteCases[Graphics[lcp], _Line, Infinity] /. Polygon[l_] :>
        Polygon[{#1 Cos[#2], #1 Sin[#2]}& @@@ addPoints[l, 0.1]], di];
(* half-periods and branch point positions *)
{δx, δy} = halfPeriods[N @ φ0];
branchPoints = Select[#, #[[1]].#[[1]] <= T^2&]& @
    Flatten[Table[Point[{i δx, j δy + δy/2}],
            {i, -c[T/δx], c[T/δx]}, {j, -c[T/δy], c[T/δy]}]];
(* show contour plot and branch points *)
Show[{lcpR, Graphics[{PointSize[0.01], GrayLevel[0], branchPoints}]},
    AspectRatio -> Automatic, Frame -> False, FrameTicks -> False,
    DisplayFunction -> $DisplayFunction]]
```

The last graphic allows now easily to explain the complicated oscillations of the pendulum in the complex plane. The poles of the Jacobi elliptic function sn are located in a rectangular array [950]. These poles are branch points of the elongation $\varphi(t)$ and when coming from a straight line from the center the solution continues on the different sheet. (After a second branch point is passed in the same direction, the two sheet might coincide again.) And the farther one is from the origin, the more branch points one has traversed to the right or left and solutions for any two values of α (also very nearby ones) can deviate arbitrarily much.

We continue with a visualization of the Riemann surface of $\varphi(t)$. Because only branch points are "real", and branch cuts can be drawn arbitrarily (or come to existence due to following a solution along different path), we now use the symbolic solution $\varphi(t)$ in Cartesian coordinates. The following plot of $\varphi(t)$ has straight line branch cuts parallel to the real axis. (We choose the number of plotpoints to emphasize the cuts).

```
In[16]:= makeHole[Polygon[l_], f_] := (* cut a hole in a polygon *)
    Module[{mp = Plus @@ l/Length[l], L}, L = (mp + f(# - mp))& /@ l;
        {MapThread[Polygon[Join[#1, Reverse[#2]]]&,
                Partition[Append[#, First[#]], 2, 1]& /@ {l, L}]}]

In[17]:= Module[{φ0 = Pi/3., kM = 1, δx, δy, polys, polysH,
            coloredPolys, coloredPolysR, coloredSheets},
    {δx, δy} = 2 halfPeriods[φ0];
    (* 3D plot of the main sheet *)
    polys = Cases[Graphics3D @
        Plot3D[Re @ φP[τx + I τy, φ0], {τx, -δx, +δx}, {τy, -δy, +δy},
                PlotPoints -> {48, 49}, PlotRange -> All,
                DisplayFunction -> Identity], _Polygon, Infinity];
    (* make holes in polygons *)
```

```
polysH = Flatten[makeHole[#, 0.8]& /@ polys];
(* color polygons according to real part of function value *)
coloredPolys = {SurfaceColor[#, #, 2.3]&[
                  Hue[0.8 Abs[Last[Plus @@ #[[1]]/4]]/Pi]], #}& /@ polysH;
(* mirror  and shift polygons *)
coloredPolysR = coloredPolys /.
              Polygon[l_] :> Polygon[({1, 1, -1}# + {0, 0, 2Pi})& /@ l];
(* polygons of the first sheets *)
coloredSheets =
Table[Map[If[Head[#] === List, (# + {0, 0, 4k Pi}), #]&,
          {coloredPolys, coloredPolysR}, {-2}], {k, -kM, kM}];
(* show all polygons *)
Show[Graphics3D[{EdgeForm[], coloredSheets }],
    BoxRatios -> {1, 1, 1}, ViewPoint -> {4, -2, 1}, Boxed -> False]]
```

We could now continue and, say, solve the pendulum equation along circles. The following three functions `pendulumND` `SolveAzimuthal`, `complexElongationsAzimuthal`, and `complexElongationPlotAzimuthal` follow closely their ray counterparts.

```
In[18]:= (* numerical solution of the pendulum equation
            along the circle with radius ρ *)
        pendulumNDSolveAzimuthal[ρ_, φ0_, αMax_, opts___] :=
         NDSolve[{-Exp[-2I α]/ρ^2 (φ''[α] - I φ'[α]) == -Sin[φ[α]],
              φ[0] == φP[ρ, φ0], φ'[0] == I ρ D[φP[t, φ0], t] /. {t -> ρ}},
             φ, {α, 0, αMax}, opts, MaxSteps -> 10^5,
             PrecisionGoal -> 12, MaxStepSize -> 2Pi/60];

In[20]:= (* plot oscillations along the circle of radius ρ around the origin *)
        complexElongationNAzimuthal[ρ_, φ0_, αMax_, opts___] :=
        Module[{ndsol},
        (* solve pendulum equation along the circle *)
        ndsol = pendulumNDSolveAzimuthal[ρ, φ0, αMax];
        (* plot real part of φ[t] *)
        Plot[Evaluate[Re[φ[α]] /. ndsol[[1]]], {α, 0, αMax}, opts,
            PlotPoints -> Round[60 αMax/(2Pi)], PlotStyle -> Hue[0.8 Abs[φ0]/Pi],
            AspectRatio -> 1/3]]

In[22]:= complexElongationsAzimuthal[ρ_, φ0_, αMax_, kM_:1, opts___] :=
        Module[{pl, lines},
        Show[Graphics[{Thickness[0.002], Hue[0.8 Abs[φ0]/Pi],
         (* elongation on the main sheet *)
         pl = Plot[Evaluate[Re[φP[ρ Exp[I α], φ0]]], {α, 0, αMax},
                  DisplayFunction -> Identity,
                  PlotPoints -> Round[60 αMax/(2Pi)]][[1, 1, 1, 1]];
         (* elongation on the neighboring sheet *)
         lines = {Table[{Line[({0, 4k Pi} + #)& /@ pl],
                         Line[({0, (4k + 2) Pi} + {1, -1}#)& /@ pl]},
                       {k, -kM, kM}]}}], opts]];

In[23]:= complexElongationPlotAzimuthal[paramList_, opts___] :=
        Show[GraphicsArray[
        Show[{(* multiple sheets from the symbolic form *)
             complexElongationsAzimuthal[#1, #2, #3, #4,
              DisplayFunction -> Identity] /. _Thickness :> Thickness[0.02],
```

```
(* one path from the numeric solution *)
complexElongationNAzimuthal[#1, #2, #3,
  opts, DisplayFunction -> Identity] /. _Hue :> GrayLevel[0]}]& @@@
                                                      paramList]]
```

Here are again three examples.

```
In[24]:= complexElongationPlotAzimuthal[
           {{21, 1, 2Pi, 1}, {78/5, 2/3Pi, 2Pi, 1}, {37, 1/2, 2Pi, 1}}]
```

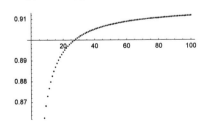

We could now go on and investigate the number of visited sheets as a function of the distance for a given ray direction and maximal elongation, or see what happens along spiral-shaped paths, but we end here. For applications of unusual pendulum solutions, see [79], [239]. For the analytic structure of the pendulum equation with a periodic forcing, see [994]. For genuinely complex mechanical models, see [932].

5. Rocket with Discrete Propulsion, Neat Product, Harmonic Oscillator Spectrum

a) We easily see from the conservation of momentum law that the final velocity v_f is [385], [1021]:

$$v_f = u \, \frac{(m_i - m_f)}{n} \sum_{k=1}^{n} \frac{1}{m_f + k\,(m_i - m_f)/n}.$$

To find an analytical expression for this sum as a function of n, we use Sum.

```
In[1]:= vf = Sum[((mf - mi) u)/(k (mf - mi) - mf n), {k, n}]
```

$$\text{Out[1]= } -u \left(\text{PolyGamma}\left[0, \, 1 - \frac{mf\,n}{mf - mi}\right] - \text{PolyGamma}\left[0, \, 1 + n - \frac{mf\,n}{mf - mi}\right]\right)$$

Substituting $m_i/m_f \to n$, and normalizing the nozzle velocity, we get a unitless result.

```
In[2]:= vf = Simplify[vf/u /. mf -> mi/v]
```

$$\text{Out[2]= } -\text{PolyGamma}\left[0, \, \frac{-1 + n + v}{-1 + v}\right] + \text{PolyGamma}\left[0, \, \frac{-1 + v + n\,v}{-1 + v}\right]$$

We now graphically examine the behavior of the final velocity v_f as a function of the number of impulses. We use the value $v = 5/2$.

```
In[3]:= Block[{v = 5/2}, ListPlot[Table[vf, {n, 1, 100}]]]
```

In order to find the limiting value for an infinite number of impulses, we look at the behavior of PolyGamma[0, x] at infinity.

```
In[4]:= ser[x_] = Normal[Series[PolyGamma[0, x], {x, Infinity, 2}]]
```

$$\text{Out[4]= } -\frac{1}{12\,x^2} - \frac{1}{2\,x} - \text{Log}\left[\frac{1}{x}\right]$$

```
In[5]:= ser[(-1 + v + n v)/(-1 + v)] - ser[(-1 + n + v)/(-1 + v)]
```

Out[5]= $\dfrac{(-1+v)^2}{12\ (-1+n+v)^2} + \dfrac{-1+v}{2\ (-1+n+v)} - \dfrac{(-1+v)^2}{12\ (-1+v+n\ v)^2} -$

$\dfrac{-1+v}{2\ (-1+v+n\ v)} + \text{Log}\Big[\dfrac{-1+v}{-1+n+v}\Big] - \text{Log}\Big[\dfrac{-1+v}{-1+v+n\ v}\Big]$

In[6]:= **Series[%, {n, Infinity, 0}] // Simplify[#, v > 0]&**

Out[6]= $\text{Log}[v] + O\Big[\dfrac{1}{n}\Big]^1$

This means the limiting velocity is $\log(v/(v-1)) + \log(v-1) = \log(v)$ [847]. Here is an asymptotic value of the above mass relationship.

In[7]:= **Log[5/2] // N**

Out[7]= 0.916291

Here it is with units: $v_f = u \ln(m_i/m_f)$. For further details on this problem, see [847] and [189]. For a similar problem, see [1030].

(V5.1 on byblis2) Out[10]=

TMGBs`SaveNotebookAndQuitKernel[]

b) A direct calculation of the infinite product gives a closed-form result.

In[1]:= **Product[Exp[-1/k] (1 + 1/k + 1/k^2), {k, Infinity}]**

Out[1]= $e^{-\text{EulerGamma}-\text{LogGamma}\big[\frac{3}{2}-\frac{i\sqrt{3}}{2}\big]-\text{LogGamma}\big[\frac{3}{2}+\frac{i\sqrt{3}}{2}\big]}$

Here is a simplified form of the result.

In[2]:= **FullSimplify[%]**

Out[2]= $\dfrac{e^{-\text{EulerGamma}}\ \text{Cosh}\big[\frac{\sqrt{3}\ \pi}{2}\big]}{\pi}$

And here is a generalization of the product that contains the two parameters a and b.

In[3]:= **Product[Exp[-1/k] (1 + a/k + b/k^2), {k, Infinity}]**

Out[3]= $\dfrac{e^{-\text{EulerGamma}}}{\text{Gamma}\big[\frac{1}{2}\ \big(2+a-\sqrt{a^2-4b}\ \big)\big]\ \text{Gamma}\big[\frac{1}{2}\ \big(2+a+\sqrt{a^2-4b}\ \big)\big]}$

Further generalizations can be calculated too, but the results become more messy.

In[4]:= **Product[Exp[-1/k] (1 + a/k + b/k^2 + c/k^3), {k, Infinity}] //.**
    (* introduce abbreviations *)
    -a^2 b^2 + 4 b^3 + 4 a^3 c - 18 a b c + 27c^2 -> D //.
    -2 a^3 + 9 a b - 27 c + 3 Sqrt[3] Sqrt[D] -> E //.
    -4 a^3 + 18 a b -54 c + 6 Sqrt[3] Sqrt[D] -> 2 E

Out[4]= $e^{-\text{EulerGamma}} \Big/ \Big(\text{Gamma}\Big[\dfrac{1}{6}\ \Big(6+2\ a - \dfrac{2\ 2^{1/3}\ a^2}{E^{1/3}} + \dfrac{6\ 2^{1/3}\ b}{E^{1/3}} - 2^{2/3}\ E^{1/3}\Big)\Big]\ \text{Gamma}\Big[$

$\dfrac{2\ 2^{1/3}\ (1+i\ \sqrt{3})\ a^2 - 6\ 2^{1/3}\ b - 6\ i\ 2^{1/3}\ \sqrt{3}\ b + 12\ E^{1/3} + 4\ a\ E^{1/3} + 2^{2/3}\ E^{2/3} - i\ 2^{2/3}\ \sqrt{3}\ E^{2/3}}{12\ E^{1/3}}\Big]$

$\text{Gamma}\Big[$

$\dfrac{2\ 2^{1/3}\ (1-i\ \sqrt{3})\ a^2 - 6\ 2^{1/3}\ b + 6\ i\ 2^{1/3}\ \sqrt{3}\ b + 12\ E^{1/3} + 4\ a\ E^{1/3} + 2^{2/3}\ E^{2/3} + i\ 2^{2/3}\ \sqrt{3}\ E^{2/3}}{12\ E^{1/3}}\Big]\Big)$

For similar products containing π, e, and γ, see [301], [186].

c) This is the general solution of the differential equation under consideration. (We divide out an unimportant numeric prefactor.)

In[1]:= **odeSol = FullSimplify[First[ψ[z] /.**
    DSolve[{-ψ''[z] + z^2 ψ[z] == ε ψ[z]}, ψ[z], z]]]/(1/2 - I/2) /.
    (* write all parts as hypergeometric functions *)
    HermiteH[n_, z_] :> 2^n HypergeometricU[-n/2, 1/2, z^2]

Out[1]= $(1+i)\ e^{-\frac{z^2}{2}}$

$\Big(\text{C[2] Hypergeometric1F1}\Big[\dfrac{1-\varepsilon}{4},\ \dfrac{1}{2},\ z^2\Big] + 2^{\frac{1}{2}\ (-1+\varepsilon)}\ \text{C[1] HypergeometricU}\Big[\dfrac{1-\varepsilon}{4},\ \dfrac{1}{2},\ z^2\Big]\Big)$

For the eigenfunctions to be square integrable in $(-\infty, \infty)$, they have to vanish faster than $z^{-1/2}$ as $z \to \infty$. So we have a look at the series of odeSol at $z = \infty$.

```
In[2]:= odeSol /. {Hypergeometric1F1[a_, b_, x_] ->
              Normal[Series[Hypergeometric1F1[a, b, x], {x, Infinity, 0}]],
          HypergeometricU[a_, b_, x_] ->
              Normal[Series[HypergeometricU[a, b, x], {x, Infinity, 0}]]}
```

Series::esss : Essential singularity encountered in $e^{\frac{1}{x} + O\left[\frac{1}{x}\right]^2}$. More...

Series::esss : Essential singularity encountered in $e^{\frac{1}{x} + O[x]^2}$. More...

$$
\text{Out[2]=} \quad (1 + i)\, e^{-\frac{z^2}{2}} \left[2^{\frac{1}{2}(-1+\varepsilon)} \left(\frac{1}{z^2}\right)^{\frac{1-\varepsilon}{4}} C[1] + \right.
$$

$$
\left((-1)^{\frac{1}{4}(-1+\varepsilon)} \sqrt{\pi} \left(\frac{1}{z^2}\right)^{\frac{1}{4}(-1+\varepsilon)} C[2] \left((-1)^{\frac{1-\varepsilon}{4}} e^{z^2} \sqrt{\frac{1}{z^2}}\, \text{Gamma}\left[\frac{1}{2} + \frac{1}{4}(-1+\varepsilon)\right] + \right. \right.
$$

$$
\left. \left. \left(\frac{1}{z^2}\right)^{\frac{1-\varepsilon}{2}} \text{Gamma}\left[\frac{1-\varepsilon}{4}\right] \right) \right) \middle/ \left(\text{Gamma}\left[\frac{1}{2} + \frac{1}{4}(-1+\varepsilon)\right] \text{Gamma}\left[\frac{1-\varepsilon}{4}\right] \right) \right]
$$

The exponential growing term vanishes when $\text{Gamma}[3/4 - \varepsilon/4]$ becomes infinity; this means $\varepsilon_n = 4n + 3, n \in \mathbb{N}$. The second possibility to cancel the exponential growing terms is $C[2] = 0$.

```
In[3]:= odeSol /. {C[2] -> 0, C[1] -> 1}
```

$$
\text{Out[3]=} \quad (1 + i)\, 2^{\frac{1}{2}(-1+\varepsilon)}\, e^{-\frac{z^2}{2}}\, \text{HypergeometricU}\left[\frac{1-\varepsilon}{4}, \frac{1}{2}, z^2\right]
$$

Let us now consider the origin $z = 0$.

```
In[4]:= % /. {Exp[-z^2/2] -> Normal[Series[Exp[-z^2/2], {z, 0, 3}]],
          HypergeometricU[a_, b_, z_] ->
              Normal[Series[HypergeometricU[a, b, z], {z, 0, 3}]]} /.
          (* for positive real z *)
          {(z^2)^(1/2) -> z, (z^2)^(-1/2) -> 1/z} //
              (CoefficientList[#, z]&) // Simplify
```

$$
\text{Out[4]=} \quad \left\{ \frac{(1+i)\, 2^{\frac{1}{2}(-1+\varepsilon)} \sqrt{\pi}}{\text{Gamma}\left[\frac{3}{4} - \frac{\varepsilon}{4}\right]}, \; -\frac{(1+i)\, 2^{\frac{1+\varepsilon}{2}} \sqrt{\pi}}{\text{Gamma}\left[\frac{1-\varepsilon}{4}\right]}, \; -\frac{(1+i)\, 2^{\frac{1}{2}(-3+\varepsilon)} \sqrt{\pi}\, \varepsilon}{\text{Gamma}\left[\frac{3}{4} - \frac{\varepsilon}{4}\right]}, \right.
$$

$$
\frac{\left(\frac{1}{3} + \frac{i}{3}\right) 2^{\frac{1}{2}(-1+\varepsilon)} \sqrt{\pi}\, \varepsilon}{\text{Gamma}\left[\frac{1-\varepsilon}{4}\right]}, \; \frac{\left(\frac{1}{3} + \frac{i}{3}\right) 2^{\frac{1}{2}(-7+\varepsilon)} \sqrt{\pi}\, (-1 + \varepsilon^2)}{\text{Gamma}\left[\frac{3}{4} - \frac{\varepsilon}{4}\right]},
$$

$$
-\frac{\left(\frac{1}{15} + \frac{i}{15}\right) 2^{\frac{1}{2}(-5+\varepsilon)} \sqrt{\pi}\, (-9 + \varepsilon^2)}{\text{Gamma}\left[\frac{1-\varepsilon}{4}\right]}, \; -\frac{\left(\frac{1}{45} + \frac{i}{45}\right) 2^{\frac{1}{2}(-9+\varepsilon)} \sqrt{\pi}\, (30 - 31\varepsilon + \varepsilon^3)}{\text{Gamma}\left[\frac{3}{4} - \frac{\varepsilon}{4}\right]},
$$

$$
\frac{\left(\frac{1}{315} + \frac{i}{315}\right) 2^{\frac{1}{2}(-7+\varepsilon)} \sqrt{\pi}\, (210 - 79\varepsilon + \varepsilon^3)}{\text{Gamma}\left[\frac{1-\varepsilon}{4}\right]}, \; \frac{\left(\frac{1}{45} + \frac{i}{45}\right) 2^{\frac{1}{2}(-11+\varepsilon)} \sqrt{\pi}\, (-9 + \varepsilon)\, (-5 + \varepsilon)\, (-1 + \varepsilon)}{\text{Gamma}\left[\frac{3}{4} - \frac{\varepsilon}{4}\right]},
$$

$$
-\frac{\left(\frac{1}{315} + \frac{i}{315}\right) 2^{\frac{1}{2}(-9+\varepsilon)} \sqrt{\pi}\, (-11 + \varepsilon)\, (-7 + \varepsilon)\, (-3 + \varepsilon)}{\text{Gamma}\left[\frac{1-\varepsilon}{4}\right]} \right\}
$$

When $\text{Gamma}[1/4 - \varepsilon/4]$ (meaning $\varepsilon_n = 4n + 1, n \in \mathbb{N}$) becomes infinity, the odd terms vanish and $\psi(z)$ becomes an even function with $\psi'(0) = 0$, making $\psi(z)$ continuous at $z = 0$. At $\varepsilon_n = 4n + 3, n \in \mathbb{N}$, the even terms vanish and $\psi(0) = 0$, making $\psi(z)$ a smooth odd function Putting these two results together, we arrive at $\varepsilon_n = 2n + 1, n \in \mathbb{N}$, the well-known oscillator spectrum.

Sometimes the solution of the differential equation $z^2 \psi(z) - \psi''(z) = \varepsilon \psi(z)$ is written in the form [1221]

$$
c_1 \exp(-z^2/2)\, H_{(\varepsilon-1)/2}(z) + c_2 \exp(-z^2/2)\; {}_1F_1(-(\varepsilon - 1)/4; 1/2; z^2).
$$

Taking into account the asymptotics of ${}_1F_1$ and $H_{(\varepsilon-1)/2}$ yields the same result: $\varepsilon_n = 2n + 1, n \in \mathbb{N}$.

For a derivation of the spectrum through the Titchmarsh–Weyl m function, see [1324]. The interested reader could now continue and derive the energy spectrum for the harmonic oscillator in a box [760], [13], [877], [288], [961] (meaning $\psi_n(-L) = \psi_n(+L) = 0$ for large L) by using higher order terms in the series expansion of the hypergeometric functions. For a similar approach to the hydrogen spectrum, see [300], [253]. For a difference equation-based approach to eigenvalue problems, see [215].

For the most general canonical commutation relation-based quantization, see [973]. For nonclassical, canonoid, Newton-equivalent quantizations of the harmonic oscillator and the corresponding eigenvalues, see [373], [1218], and [437], [254]. For the (supersymmetric) half-space oscillator, see [354], [355], [487], [1369], [427]. For harmonic oscillators with a position-dependent mass, see [27]. For nonlinear harmonic oscillators, see [252].

6. Contour Integral, Asymptotics of Bessel Function, Isophotes, Circular Andreev Billiard

a) Let us have a look at the branch cuts of the integrand and how the integration contour crosses them.

```
In[1]:= With[{ε = 10^-6},
        (* picture of the integrand *)
        Show[{Apply[Plot3D[Im[((x + I y)^2 - 1)^(1/Pi)], ##,
                            DisplayFunction -> Identity, PlotPoints -> 20]&,
           (* use many parts to avoid branch cuts *)
               {{{x, -2, -ε}, {y, -2, -ε}}, {{x, ε, 2}, {y, -2, -ε}},
                {{x, -2, -ε}, {y, ε, 2}}, {{x, ε, 2}, {y, ε, 2}}}, {1}],
           (* the integration contour *)
            Graphics3D[{Thickness[0.01], Line[Table[{3/2 Cos[p], Sin[2p],
                        Im[((3/2Cos[p] + I Sin[2p])^2 - 1)^(1/Pi)] +
                           (* lift slightly above the surface *) 0.05},
                    {p, 0, N[2Pi], N[2Pi/300]}]]}]},
              DisplayFunction -> $DisplayFunction, AxesLabel -> {"x", "y", None}]]
```

We see that the complex t-plane is cut along the line (interval) $(-1, 1)$ and along the imaginary axis. (That this is a branch cut follows easily from the branch cut of the Power function along the negative real axis and solving the equation $(t^2 - 1) = \xi$ for $-\infty < \xi < 0$.) These locations of the branch cuts mean that, fortunately, our contour never crosses any branch cut; it always stays on the "right" sheet. The integrand is continuous everywhere along the path and does not have to explicitly continue the integrand analytically. (If the contour of integration would not exactly go from the upper half-plane to the lower half-plane and vice versa in the point $\{0, 0\}$, we would have much more to do.) So we can implement immediately.

```
In[2]:= t[φ_] = a Cos[φ] + b I Sin[2φ]; dt = D[t[φ], φ];
        a = 12/10; b = 1;

        myBesselJ[n_, z_] := Gamma[1/2 - n] (z/2)^n/(2Pi I Sqrt[Pi]) *
             NIntegrate[Evaluate[Exp[I z t[φ]] (t[φ]^2 - 1)^(n - 1/2) D[t[φ], φ]],
                        {φ, 0, 2Pi}]
```

Now, we calculate and compare the result with the built-in Bessel function.

```
In[5]:= {myBesselJ[E, I Pi] // N, BesselJ[E, I Pi] // N}

Out[5]= {-0.639505 - 1.34955 i, -0.639505 - 1.34955 i}
```

b) First, we calculate the differential equation for $u(z)$.

```
In[1]:= BesselJAnsatz = n^n/Gamma[n + 1] Exp[Integrate[u[zs], z]];

        ((z^2 D[BesselJAnsatz, {z, 2}] + z D[BesselJAnsatz, z] -
           n^2 (1 - z^2) BesselJAnsatz) // Factor)*
                 Gamma[n + 1]/n^n/Exp[Integrate[u[zs], z]]
```

Out[2]= $-n^2 + n^2 z^2 + z\, u[zs] + z^2\, u[zs]^2$

Now, we determine $u_0(z)$ and c. To do this, we take the first terms of the series expansion of $u(z)$.

In[3]:= ```ser = Sum[n^i u[i][z], {i, 1, -1, -1}] +```
    ```(* make a Series; because of Infinity we cannot just use O[Infinity] *)```
    ```SeriesData[n, Infinity, {1}, 3, 3, 1]```

Out[3]= $\dfrac{u[1][z]}{\frac{1}{n}} + u[0][z] + \dfrac{u[-1][z]}{n} + O\!\left[\dfrac{1}{n}\right]^3$

We equate the coefficient of v^2 to zero.

In[4]:= ```Coefficient[Normal[z^2 (ser^2 + D[ser, z]) + z ser + n^2 (z^2 - 1)], n, 2]```

Out[4]= $-1 + z^2 + z^2\, u[1][z]^2$

Then, we solve the resulting equation for $u_1(z)$.

In[5]:= ```DSolve[% == 0, u[1], z]```

Out[5]= $\left\{\left\{u[1] \to \text{Function}\!\left[\{z\},\ -\dfrac{\sqrt{1-z^2}}{z}\right]\right\},\ \left\{u[1] \to \text{Function}\!\left[\{z\},\ \dfrac{\sqrt{1-z^2}}{z}\right]\right\}\right\}$

To decide which of the two solutions we need and to determine the constant c, we compare the resulting series solutions with the exact series of $J_v(z)$.

In[6]:= ```Normal[Series[BesselJ[n, n z], {z, 0, 1}]]```

Out[6]= $\dfrac{2^{-n}\, e^{2\, i\, n\, \pi\, \text{Floor}\left[\frac{\pi - \text{Arg}[n] - \text{Arg}[z]}{2\pi}\right]}\, n^n\, z^n}{\text{Gamma}[1+n]}$

In[7]:= ```n^n/Gamma[n + 1] Exp[Expand[n Normal[```
    ```Series[Integrate[-Sqrt[1 - z^2]/z, z] + C,```
    ```{z, 0, 1}]]]] //. Exp[a_ Log[b_] + c_] -> b^a Exp[c]```

Out[7]= $\dfrac{2^n\, e^{-n+C\,n}\, n^n\, z^{-n}}{\text{Gamma}[1+n]}$

This shows that $u_0(z) = (1 - z^2)/z$ is the appropriate solution and the constant c has to be chosen in such a way that the constant C becomes -1. (There is no need to explicitly calculate c, which is given implicitly by $c\exp\!\left((1 - c^2)^{1/2} - 1\right)$ $= 1 + (1 - c^2)^{1/2}$.) Having calculated the first term of the expansion of $u(z)$, the further terms require no manipulation by hand, and we can implement the calculation of $u_0(z)$ to, say, $u_{-9}(z)$ at once. Here is the initial series up to order 9.

In[8]:= ```ser = Sum[n^i u[i][z], {i, 1, -8, -1}] +```
    ```SeriesData[n, Infinity, {1}, 9, 9, 1];```

    ```odeSer = z^2 (ser^2 + D[ser, z]) + z ser + n^2 (z^2 - 1);```

Given $u_1(z)$, we recursively determine the coefficients of v^i, equate them to zero, and solve the resulting differential equation for $u_{i-1}(z)$.

In[10]:= ```Clear[u];```
    ```u[1] = Function[z, Sqrt[1 - z^2]/z];```

    ```Do[u[i] = Function[z, ##]& @@ {Together[DSolve[Coefficient[```
        ```Normal[odeSer], n, i + 1] == 0, u[i], z][[1, 1, 2]][z]]},```
    ```{i, 0, -5, -1}]```

In[13]:= ```??u```

    ```Global`u```
    ```

$$u[-5] = \text{Function}\left[z, \ -\frac{z \left(512 + 27264 \, z^2 + 156416 \, z^4 + 184176 \, z^6 + 44684 \, z^8 + 1073 \, z^{10}\right)}{1024 \sqrt{1-z^2} \ (-1+z^2)^8}\right]$$

$$u[-4] = \text{Function}\left[z, \ -\frac{z \left(16 + 368 \, z^2 + 924 \, z^4 + 374 \, z^6 + 13 \, z^8\right)}{32 \, (-1+z^2)^7}\right]$$

$$u[-3] = \text{Function}\left[z, \ \frac{z \left(64 + 560 \, z^2 + 456 \, z^4 + 25 \, z^6\right)}{128 \sqrt{1-z^2} \ (-1+z^2)^5}\right]$$

$$u[-2] = \text{Function}\left[z, \ \frac{z \left(4 + 10 \, z^2 + z^4\right)}{8 \, (-1+z^2)^4}\right]$$

$$u[-1] = \text{Function}\left[z, \ -\frac{z \left(4 + z^2\right)}{8 \sqrt{1-z^2} \ (-1+z^2)^2}\right]$$

$$u[0] = \text{Function}\left[z, \ -\frac{z}{2 \, (-1+z^2)}\right]$$

$$u[1] = \text{Function}\left[z, \ \frac{\sqrt{1-z^2}}{z}\right]$$

So we finally have the following expansion.

In[14]:= `myAsymptoticBesselJ[n_, z_] =`
        `Module[{x}, (n^n /Gamma[n + 1] Exp[n (Integrate[u[1][x], x] - 1) +`
                `Sum[Together[Integrate[n^i u[i][z], {z, 0, x}]],`
                `{i, 0, -5, -1}]]) /. {x -> z/n}];`

`Short[myAsymptoticBesselJ[n, z], 12]`

Out[16]//Short=
$$\frac{1}{\text{Gamma}[1+n]} \left(n^n \exp\left[-\frac{z^2 \left(4 + \frac{z^2}{n^2}\right)}{16 \, n^4 \left(-1 + \frac{z^2}{n^2}\right)^3} + \frac{z^2 \left(32 + \frac{288 \, z^2}{n^2} + \frac{232 \, z^4}{n^4} + \frac{13 \, z^6}{n^6}\right)}{128 \, n^6 \left(-1 + \frac{z^2}{n^2}\right)^6} + \right.\right.$$
$$\left.\left. \ll 2 \gg + \frac{1}{4} \, i \left(\pi + i \, \text{Log}\left[-1 + \frac{z^2}{n^2}\right]\right) + n \left(-1 + \sqrt{1 - \frac{z^2}{n^2}} + \text{Log}\left[\frac{z}{n}\right] - \text{Log}\left[1 + \sqrt{1 - \frac{z^2}{n^2}}\right]\right) \right] \right)$$

Here is the value calculated with this expansion. It agrees well with the value from the built-in command.

In[18]:= `{N[myAsymptoticBesselJ[1000, 100], 22], N[BesselJ[1000, 100.]]}`

Out[18]= `{1.902621040949654807218×10^-870 + 0.×10^-892 i, 1.90262104095027×10^-870}`

c) This is a straightforward definition for the intensity. Because we are interested in a picture only, the relatively low precision goal of 4 is enough.

In[1]:= `int[u_, v_] := Abs[NIntegrate[BesselJ[0, v ρ] Exp[-I u ρ^2/2] ρ, {ρ, 0, 1},`
        `Compiled -> False, PrecisionGoal -> 4]]^2`

We use the symmetry of the intensity $I(u, v) = I(-u, v) = I(u, -v)$. Here is a contour plot of the intensity in the first quadrant.

In[2]:= `ε = 10.^-6; pp = 50;`
        `cp = ContourPlot[Abs[int[u, v]], {u, ε, 10Pi}, {v, ε, 5Pi},`
                `PlotPoints -> pp, ColorFunction -> Hue,`
                `AspectRatio -> Automatic, Compiled -> False]`

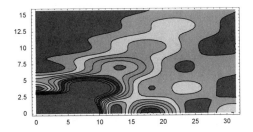

To get a smoother picture, we space the contour lines more evenly.

```
In[4]:= (lcp = ListContourPlot[cp[[1]], Contours ->
           (#[[pp/2]]& /@ Partition[Sort[Flatten[cp[[1]]]], pp]),
         ColorFunction -> (Hue[0.7 #]&), ContourLines -> False,
         MeshRange -> {{ε, 10Pi}, {ε, 5Pi}}, DisplayFunction -> Identity];)
```

Generating the graphics in the other three quadrants by reflection, we get the following picture for the intensity distribution.

```
In[5]:= With[{gr = Graphics[lcp]},
         Show[Function[δ, gr /. Polygon[l_] :> Polygon[δ #& /@ l]] /@
                {{1, 1}, {-1, 1}, {1, -1}, {-1, -1}},
           DisplayFunction -> $DisplayFunction,
           FrameTicks -> None, AspectRatio -> Automatic]]
```

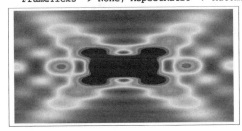

d) It is straightforward to implement the calculation of the determinant described. For the range $0 < \varepsilon_n < \Delta_0$, we can carry out the complex conjugation explicitly by the changing signs of the square roots.

```
In[1]:= k   = Sqrt[1 + ε];
        {q, qc}   = Sqrt[1 + # Sqrt[ε^2 - Δ0^2]]& /@ {+1, -1};

        {δe, δh} = Δ0/(ε + # Sqrt[ε^2 - Δ0^2])& /@ {-1, 1};
In[4]:= ψNe = BesselJ[m, k r] -
              BesselJ[m, k ρN]/BesselY[m, k ρN] BesselY[m, k r];
        ψNh = ψNe /. ε -> -ε;
        {ψSe, ψSh} = {BesselJ[m, q r], BesselJ[m, qc r]};
In[7]:= det1 = Det[{{ψNe, 0, δe ψSe, δh ψSh},
                    {0, ψNh, ψSe, ψSh},
                    {D[ψNe, r], 0, δe D[ψSe, r], δh D[ψSh, r]},
                    {0, D[ψNh, r], D[ψSe, r], D[ψSh, r]}} /. r -> ρS];
```

The calculation of the energy spectrum is now, in principle, straightforward. In practice, we have to do some more programming to yield the energy eigenvalues in a reasonable amount of time. The presence of 22 different Bessel functions in det1, the size of det1, and the ε-dependent denominators with zeros in the interval $0 < \varepsilon_n < \Delta_0$ need to be addressed. The 22 Bessel functions and the size imply a relatively slow calculation of the determinant; the poles with respect to ε make numerical root finding more difficult. We start with removing the denominators from the expanded determinant.

```
In[8]:= det2 = Factor[Numerator[Together[det1]]]/Δ0/2;
```

det2 is a relatively large expression.

```
In[9]:= det2 // LeafCount
Out[9]= 22053
```

We introduce unique variables for all Bessel functions to avoid their repeated time-consuming calculation in det2 for a given ε. (We introduce a few more abbreviations than needed because later we will get the Bessel functions for all first arguments anyway.)

```
In[10]:= (* auxiliary function;
          m stands for m - 1, 0 for m + 0, p for m + 1 *)
       aux[x_] := Which[x === m - 1, "m", x === m, "0", x === m + 1, "p"]
In[12]:= (* all BesselJ[_, _] that occur *)
       allBesselJs = Union[Cases[det2, _BesselJ, Infinity]];

       (* introduce abbreviations *)
       allBesselJsAbbr = allBesselJs /.
       ((BesselJ[m_, #1] :> ToExpression[ToString[#2] <> aux[m]])& @@@
              {{ρN Sqrt[1 - ε], J1}, {ρS Sqrt[1 - ε], J2},
               {ρN Sqrt[1 + ε], J3}, {ρS Sqrt[1 + ε], J4},
               {ρS Sqrt[1 - Sqrt[-Δ0^2 + ε^2]], J5},
               {ρS Sqrt[1 + Sqrt[-Δ0^2 + ε^2]], J6}});
In[16]:= (* all BesselY[_, _] that occur *)
       allBesselYs = Union[Cases[det2, _BesselY, Infinity]];

       (* introduce abbreviations *)
       allBesselYsAbbr = allBesselYs /.
       ((BesselY[m_, #1] :> ToExpression[ToString[#2] <> aux[m]])& @@@
              {{ρN Sqrt[1 - ε], Y1}, {ρS Sqrt[1 - ε], Y2},
               {ρN Sqrt[1 + ε], Y3}, {ρS Sqrt[1 + ε], Y4}});
```

The resulting expression is still large and we simplify it using the function OptimizeExpression.

```
In[20]:= besselRules = Join[Rule @@@ Transpose[{allBesselJs, allBesselJsAbbr}],
                           Rule @@@ Transpose[{allBesselYs, allBesselYsAbbr}]];
In[21]:= (* substitute abbreviations *)
       det3 = det2 //. besselRules;
In[23]:= optimize = Experimental`OptimizeExpression[#,
                    OptimizationLevel -> 1, ExcludedForms -> {}]&;
In[24]:= det4 = optimize[det3];
       (* length of det3 and size of the optimized expression *)
       {Length[Expand[det3]], LeafCount[det4]}
Out[26]= {160, 1805}
```

To further speed up the computations, we compile the resulting expression. This gives the function cfDet.

```
In[27]:= cfDet = ReleaseHold[Hold[Compile[vars, body]] /.
           vars -> Join[{{ε, _Complex}, {Δ0, _Complex}}, {#, _Complex}& /@
           (* the Bessel functions *)
           {J1m, J10, J1p, J2m, J20, J2p, J3m, J30, J3p, J4m, J40, J4p,
            J5m, J50, J5p, J6m, J60, J6p, Y1m, Y10, Y1p, Y2m, Y20, Y2p,
            Y3m, Y30, Y3p, Y4m, Y40, Y4p}] /. (* the determinant *)
               body -> det4 /. OptimizedExpression[e_] :> e];
```

The compilation was successful. For being able to calculate high-precision values of the determinant, we extract the last argument from the compiled function and call it ucfDet.

```
In[28]:= {Union[Head /@ Flatten[cfDet[[4]]]], (ucfDet = cfDet[[5]]) // Head}
Out[28]= {{Integer, Real}, Function}
```

A direct definition of the determinant so far is the following.

```
In[29]:= SetDelayed @@ Flatten[Hold[det5[m_, ε_, Δ0_, {ρS_, ρN_}], rhs] /.
           rhs -> (Hold[ucfDet @@ args] /. args -> (ucfDet[[1]] //.
                                             (Reverse /@ besselRules)))]
```

Calculating a single value of the determinant takes seconds, even at this point .

```
In[30]:= det5[300, 0.1, 0.15, {200, 400}] // Timing
Out[30]= {2.07 Second, -1.86762 × 10^{-24} - 2.42842 × 10^{-7} i}
```

We have to find the zeros for hundreds of values of *m* and this would yield calculation times in the order of days. We can gain a dramatic speed-up by dealing with all *m* at once for a given numerical value of ε. By using the recursion relations $J_v(z) = 2(v + 1)/z\, J_{v+1}(z) - J_{v+2}(z)$ and $Y_v(z) = 2(v - 1)/z\, Y_{v-1}(z) - Y_{v-2}(z)$, we can quickly calculate hundreds of Bessel function values with only two calls to the built-in Bessel functions. This is done with the two functions BesselJList and BesselJList. Here we assume *v1* and *v2* and $v2 - v1$ to be positive integers

```
In[31]:= BesselJList[{v1_, v2_}, z_] :=
    (* special case for z = 0 *)
    If[z == 0, Table[If[v == 0, 1, 0], {v, v1, v2}],
    (* downward recursion for stability *)
    Module[{Jv2 = BesselJ[v2, z], Jv2m1 = BesselJ[v2 - 1, z], v},
      Reverse[Join[{Jv2, Jv2m1}, #[[2]]& /@ Rest[
        NestList[(Function[{Jh, Jl, v}, {Jl, 2(v + 1)/z Jl - Jh, v - 1}] @@ #)&,
          {Jv2, Jv2m1, v2 - 2}, Round[v2 - v1 - 1]]]]]]]
```

```
In[32]:= BesselYList[{v1_, v2_}, z_] :=
    (* upward recursion for stability *)
    Module[{Yv1 = BesselY[v1, z], Yv2p1 = BesselY[v1 + 1, z]},
    Join[{Yv1, Yv2p1}, #[[2]]& /@ #[[2]]& /@ Rest[
      NestList[(Function[{Yl, Yh, v}, {Yh, 2(v - 1)/z Yh - Yl, v + 1}] @@ #)&,
        {Yv1, Yv2p1, v1 + 2}, Round[v2 - v1 - 1]]]]]
```

The following check shows that this resulted in a hundreds-of-time faster calculation of the Bessel function values. At the same time, the precision of the result is sufficient for our visualization purpose.

```
In[33]:= Module[{v1 = -1, v2 = 500, z = N[500]},
            {Timing[JL1 = BesselJList[{v1, v2}, z];],
             Timing[JL2 = Table[BesselJ[k, z], {k, v1, v2}];],
             Max[Abs[JL1 - JL2]]}]
Out[33]= {{0.02 Second, Null}, {1.19 Second, Null}, 6.52256 × 10^{-14}}
```

```
In[34]:= Module[{v1 = -1, v2 = 500, z = N[500]},
            {Timing[YL1 = BesselYList[{v1, v2}, z];],
             Timing[YL2 = Table[BesselY[k, z], {k, v1, v2}];],
             Max[Abs[YL1 - YL2]]}]
Out[34]= {{0.02 Second, Null}, {1.17 Second, Null}, 7.58282 × 10^{-14}}
```

Using now the lists of Bessel functions and the optimized expressions from above, we can implement the function detε· List. Depending on the input, detεList either uses the compiled or the pure function to calculate the determinant.

```
In[35]:= detεList[ε_, Δ0_, ρS_, ρN_, {mMin_, mMax_}] :=
    Module[{J1, J2, J3, J4, J5, J6, Y1, Y2, Y3, Y4},
      (* list of Bessel J functions for fixed ε *)
      {J1, J2, J3, J4, J5, J6} = BesselJList[{mMin - 1, mMax + 1}, #]& /@
        {ρN Sqrt[1 - ε], ρS Sqrt[1 - ε], ρN Sqrt[1 + ε], ρS Sqrt[1 + ε],
         ρS Sqrt[1 - Sqrt[-Δ0^2 + ε^2]], ρS Sqrt[1 + Sqrt[-Δ0^2 + ε^2]]};
      (* list of Bessel Y functions for fixed ε *)
      {Y1, Y2, Y3, Y4} = BesselYList[{mMin - 1, mMax + 1}, #]& /@
        {ρN Sqrt[1 - ε], ρS Sqrt[1 - ε], ρN Sqrt[1 + ε], ρS Sqrt[1 + ε]};
      (* all arguments for cfDet *)
      JY = Join[{ε, Δ0}, #]& /@ Flatten /@ Transpose[
        Partition[#, 3, 1]& /@ {J1, J2, J3, J4, J5, J6, Y1, Y2, Y3, Y4}];
      (* use the compiled cfDet only for machine numbers;
         else use ucfDet *)
      If[MatrixQ[JY, MachineNumberQ], cfDet @@@ JY, ucfDet @@@ JY]]
```

A computation of the values of the determinant for relatively large *m* shows numerical quality of detεList.

```
In[36]:= Module[{ε = 1/10, Δ0 = 15/100, ρS = 200, ρN = 400,
             mMin = 400, mMax = 402, res1, res2},
            res1 = detεList[0.1, 0.15, 200, 400, {mMin, mMax}];
            (* use high-precision numbers for comparison *)
            res2 = Table[det5[m, N[ε, 22], N[Δ0, 22], {ρS, ρN}],
                    {m, mMin, mMax}];
            Max[Abs[1 - res1/res2]]]
Out[36]= 4.50973 × 10^{-12}
```

Now, we calculate the value of the determinant at a dense ε-grid for all m. It takes about half an hour for more than 10^5 values of m and ε.

```
In[37]:= (εdata = With[{Δ0 = 0.15, ρS = 200, ρN = 400,
              mMin = 0, mMax = 450, ppε = 300},
              Table[{ε, #}& /@ detεList[N[ε], Δ0, ρS, ρN, {mMin, mMax}],
                  {ε, 0, Δ0, Δ0/ppε}]];) // Timing
Out[37]= {620.98 Second, Null}
```

The imaginary part is of relevance only—a nonzero real part results exclusively from rounding errors.

```
In[38]:= Max[Abs[Re[Map[Last, εdata, {2}]]]]
Out[38]= 1.9476 × 10^-11
```

To find the zeros for a given m, we transpose the array εdata, and search for zeros with respect to ε for fixed m. We interpolate the data linearly (we could use a higher-order interpolation, but for the visualization of the eigenvalues this is sufficient). We see a dramatic change in the functional behavior of the eigenvalues near $m \approx \rho^{(S)}$.

```
In[39]:= (* interpolate zero *)
        interpolatedZero[{{ε1_, v1_}, {ε2_, v2_}}] := (ε1 v2 - ε2 v1)/(v2 - v1)
In[41]:= imData = Transpose[εdata] /. Complex[r_, i_] :> i;

        theZeros = Table[Point[{μ - 1, #}]& /@ Select[interpolatedZero /@
                      (* interpolate between sign changing ε-values *)
                      Select[Partition[imData[[μ]], 2, 1],
                          #[[1, 2]] #[[2, 2]] < 0. &], 0 < # < 0.15&],
                      {μ, Length[imData]}];

        Show[Graphics[{PointSize[0.002], theZeros}],
            Frame -> True, AspectRatio -> 1, PlotRange -> All]
```

The left graphic shows that the number of zeros as a function of m becomes a complicated function around $m \approx \rho^{(S)}$. The right graphic shows the average distance between two zeros in the interval $0 < \varepsilon < \Delta0$.

```
In[44]:= Show[GraphicsArray[
        Block[{$DisplayFunction = Identity},
          {(* plot of the number of eigenvalues in (0, Δ0) *)
            ListPlot[MapIndexed[{#2[[1]] - 1, #1}&, Length /@ theZeros],
                PlotRange -> All, Axes -> False, Frame -> True],
            (* plot of average eigenvalue spacing *)
            ListPlot[MapIndexed[{#2[[1]] - 1, #1}&,
            (Plus @@ #/If[# === {}, 1, Length[#]])&[-Subtract @@@
              Partition[Sort[#[[1, 2]]& /@ #], 2, 1]]& /@ theZeros],
                PlotRange -> All, Axes -> False, Frame -> True]}]]]
```

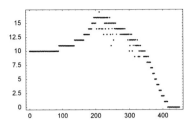 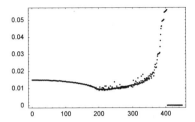

If we relax the restriction of m being an integer, we can observe the zeros as a function of m. We again use the function detεList, but this time repeatedly with noninteger limits for *mMin* and *mMax*.

```
In[45]:= detailedGraphics[Δ0_, {εMin_, εMax_}, ρS_, ρN_,
                      {mMin_, mMax_}, {ppε_, ppδm_}, opts___] :=
       Module[{ℓ},
              (* calculate values of the determinant *)
              ℓ = Transpose /@ Im[Table[
                 Table[detεList[N[ε], N[Δ0], ρS, ρN, {mMin + δ, mMax + δ}],
                    {ε, εMin, εMax, (εMax - εMin)/ppε}],
                 {δ, 0, (ppδm - 1)/ppδm, 1/ppδm}]];
              (* show zeros using ContourPlot *)
              ListContourPlot[(* rearrange data for increasing m *)
              Transpose @ Flatten[Table[ℓ[[i, j]],
                            {j, Length[ℓ[[1]]]}, {i, Length[ℓ]}], 1],
                opts, MeshRange -> {{mMin, mMax - 1/ppδm}, {εMin, εMax}},
                Contours -> {0}, ContourShading -> False]]
```

Here are two detailed graphs of the zeros for noninteger m for $0 < m < 10$ and $300 < m < 310$, where $0 < \varepsilon < \Delta0/3$. We see that for larger $m \in \mathbb{R}$, $\varepsilon(m)$ becomes an oscillating function. With an nonintegral period, the $\varepsilon(m)$ for integer m appear "random".

```
In[46]:= Show[GraphicsArray[
       Block[{$DisplayFunction = Identity},
          detailedGraphics[0.15, {0., 0.05}, 200, 400, #, {24, 24}]& /@
          {{0, 10}, {300, 310}}]]]
```

The interested reader could now continue to calculate and visualize the eigenfunctions (or its components). To avoid the time-consuming calculation of the Bessel functions involved, one could, for instance, work in polar coordinates and solve the differential equation of the Bessel functions. For the circular Andreev billiard with applied magnetic field, see [341]; for an angular sector of a circular Andreev billiard, see [342].

7. Euler's Integral for Beta Function, Beta Probability Distribution, Euler's Constant

a) Let us first have a look at the integration path.

```
In[1]:= ParametricPlot[{Cos[s]^2 Cosh[Sin[Pi/4 + s]]^2 -
                      Sin[s]^2 Sinh[Sin[Pi/4 + s]]^2,
                      Cos[s] Cos[Sin[Pi/4 + s]] Sin[s] Sin[Sin[Pi/4 + s]]},
                    {s, 0, 2Pi}, Frame -> True, Axes -> False]
```

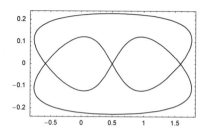

The points 0 and 1 are branch points, and the corresponding branch cuts are $(-\infty, 0)$ and $(1, \infty)$, respectively. Because the direct evaluation of the integrand exhibits discontinuities when crossing the branch cuts, we split the integration path into four pieces. Here, this is visualized.

```
In[2]:= Show[GraphicsArray[#]]& /@
        Partition[(* the wavy lines of the branch cuts *)
        Graphics[{Line[Table[{1 + 0.02 i, (-1)^i 0.02}, {i, 0, 42}]],
                Line[Table[{  - 0.02 i, (-1)^i 0.02}, {i, 0, 42}]],
        (* showing the direction of the curve *)
              Apply[Function[{p1, p2},
                Module[{d = p2 - p1, n, s},
                      n = d/Sqrt[d.d]; s = Reverse[n]{-1, 1};
                      Polygon[{p1 + s d, p1 - s d, p2}]]],
          Partition[#[[1, 1, 1, 1]], 2, 1], {1}]},
        PlotRange -> {{-0.85, 1.85}, {-0.3, 0.3}},
        AspectRatio -> 1/2, Frame -> True, Axes -> True]& /@
        (ParametricPlot[ (* the pieces of the integration path *)
        {Cos[s]^2 Cosh[Sin[Pi/4 + s]]^2 - Sin[s]^2 Sinh[Sin[Pi/4 + s]]^2,
        Cos[s] Cos[Sin[Pi/4 + s]] Sin[s] Sin[Sin[Pi/4 + s]]},
        Evaluate[Flatten[{s, #}]],
        DisplayFunction -> Identity, Axes -> False]& /@
                (* the four segments *)
                Partition[{0, Pi/2, Pi, 3Pi/2, 2Pi}, 2, 1]), 2]
```

Now, we have to analytically continue the integrand across the branch cuts. When crossing the negative real axis from above, the analytic continuation of t^p is $\exp(2\,i\,p\,\pi)\,t^p$. This can be seen by rewriting t^p in the following way:

$$t^p = (e^{\ln(t)})^p = e^{\ln(t)\,p}$$

and by taking into account $(\varepsilon \to 0)$

$\ln(negativeRealNumber + i\,\varepsilon) - \ln(negativeRealNumber - i\,\varepsilon) = 2\,i\,\pi.$

Treating the other crossings of the branch cuts in a similar way, we can implement the path over the whole parameter range.
As a sketch, here is the path on the Riemann surface of $z^{4/3}(1-z)^{3/2}$.

```
In[3]:= Show[GraphicsArray[Function[reIm,
          Module[{p = 4/3, q = 3/2, ε = 10^-5, r1 = 3/2, r2 = 1/2,
                  intPath, sheets},
          (* the integration path *)
          intPath = ParametricPlot3D[Function[t,
          {Re[t], Im[t], 0.01 + reIm @ Which[
            (* the analytic continuations across the branch cuts *)
               0 <= N[s] <= N[Pi/2], t^(p - 1) (1 - t)^(q - 1),
               N[Pi/2] <= N[s] <= N[Pi],
                  Exp[2I Pi (p)] t^(p - 1) (1 - t)^(q - 1),
               N[Pi/2] <= N[s] <= N[3/2 Pi],
                  Exp[2I Pi (p - q)] t^(p - 1) (1 - t)^(q - 1),
               N[3Pi/2] <= N[s] <= N[2Pi],
                  Exp[2I Pi (-q)] t^(p - 1) (1 - t)^(q - 1)]}][N[
            Cos[s]^2 Cosh[Sin[Pi/4 + s]]^2 - Sin[s]^2 Sinh[Sin[Pi/4 + s]]^2 +
          I Cos[s] Cos[Sin[Pi/4 + s]] Sin[s] Sin[Sin[Pi/4 + s]]]],
            {s, ε, N[2Pi] - ε}, Compiled -> False, PlotPoints -> 360,
            DisplayFunction -> Identity];
          (* all sheets of the Riemann surface of t^(p - 1) (1 - t)^(q - 1) *)
          sheets = Table[ParametricPlot3D[
          {r1 r Cos[φ] + 1/2, r2 r Sin[φ],
           reIm[(1/2 +  r (r1 Cos[φ] + I r2 Sin[φ]))^(p - 1) *
              Exp[I 2Pi/3 i] (1 - (1/2 +  r (r1 Cos[φ] +
                         I r2 Sin[φ])))^(q - 1) Exp[I 2Pi/2 j]]},
          {r, ε, 1}, #, PlotPoints -> {13, 16}, DisplayFunction -> Identity],
          {i, 0, 2}, {j, 0, 1}]& /@ (* stay away from branch cuts *)
                {{φ, ε, Pi - ε}, {φ, Pi + ε, 2Pi - ε}};
          (* the picture *)
          Graphics3D[{{Thickness[0.012], Hue[0], intPath[[1]]},
            {EdgeForm[], SurfaceColor[Hue[0.22], Hue[0.12], 2.6],
              (* making holes in the polygons to see
                 the integration path better *)
            Function[{fac, points}, Function[mp, Function[newPoints,
            {{EdgeForm[], MapThread[Polygon[Join[#1, Reverse[#2]]]&,
                            Partition[Append[#, First[#]]&[#], 2, 1]& /@
                            {points, newPoints}]}}][
              mp + fac(# - mp)& /@ points]][(Plus @@ points)/
                Length[points]]][0.8, First @ #]& /@
                    Flatten[First /@ Flatten[sheets]]]},
            PlotRange -> All, BoxRatios -> {1, 1/2, 3/2},
            Boxed -> False, Axes -> False, ViewPoint -> {0, -2.4, 2.2}]]] /@
            (* use real part and imaginary part *) {Re, Im}]]
```

So we can now implement the following function `myBeta` (for especially tailored integration formulas for the integrands
occurring here, see [503]).

```
In[4]:= t =  Cos[s]^2 Cosh[Sin[Pi/4 + s]]^2 - Sin[s]^2 Sinh[Sin[Pi/4 + s]]^2 +
          I Cos[s] Cos[Sin[Pi/4 + s]] Sin[s] Sin[Sin[Pi/4 + s]];
```

```
dtds = D[t, s]; (* change of the integration variable *)
```

```
myBeta[p_, q_] := -Exp[I Pi (-p + q)]/(4 Sin[p Pi] Sin[q Pi]) *
(NIntegrate[t^(p - 1) (1 - t)^(q - 1) dtds, {s, 0, Pi/2}] +
    Exp[2I Pi p]          NIntegrate[t^(p - 1) (1 - t)^(q - 1) dtds,
                                        {s, Pi/2, Pi}] +
    Exp[2I Pi (p - q)]  NIntegrate[t^(p - 1) (1 - t)^(q - 1) dtds,
                                        {s, Pi, 3Pi/2}] +
    Exp[2I Pi (  - q)]  NIntegrate[t^(p - 1) (1 - t)^(q - 1) dtds,
                                        {s, 3Pi/2, 2Pi}]) // N
```

This is a small test for some random values.

```
In[7]:= SeedRandom[1234];
        Table[{p = Random[Complex, {-2 - 2 I, 2 + 2 I}],
               q = Random[Complex, {-2 - 2 I, 2 + 2 I}],
               N[{myBeta[p, q], Beta[p, q]}]}, {5}]
```

Out[8]= {{0.724648 + 1.41687 i, -0.271648 + 1.58976 i,
        {-1.93965 - 0.226004 i, -1.93965 - 0.226004 i}}, {1.56018 - 0.0578006 i,
        0.220496 - 1.6652 i, {-0.177619 + 0.341665 i, -0.177619 + 0.341665 i}},
        {-1.95644 - 0.762574 i, 0.507089 - 1.07702 i,
        {-4.74595 - 0.723191 i, -4.74595 - 0.723191 i}},
        {-0.277333 + 1.05752 i, 1.23668 - 0.0937346 i,
        {-0.362214 - 0.635235 i, -0.362214 - 0.635235 i}}, {0.572929 + 0.170776 i,
        0.262934 - 1.16784 i, {0.918492 + 0.609069 i, 0.918492 + 0.609069 i}}}}

```
In[9]:= Apply[Subtract, Last /@ %, {1}] // Abs
```

Out[9]= {1.26598 × 10^{-11}, 5.7435 × 10^{-12}, 2.16295 × 10^{-13}, 6.20241 × 10^{-13}, 4.29979 × 10^{-12}}

b) We first calculate the normalization factor of the probability distribution $(1 - x^2)^\gamma$.

```
In[1]:= norm = Integrate[1/(1 - ξ^2)^γ, {ξ, 0, 1}, Assumptions -> Re[γ] < 1]
```

Out[1]= $\dfrac{\sqrt{\pi}\ \text{Gamma}[1 - \gamma]}{2\ \text{Gamma}[\frac{3}{2} - \gamma]}$

```
In[2]:= w[γ_][x_] = 1/norm Integrate[1/(1 - ξ^2)^γ, {ξ, 0, x},
                                       GenerateConditions -> False]
```

Out[2]= $\dfrac{2\ x\ \text{Gamma}[\frac{3}{2} - \gamma]\ \text{Hypergeometric2F1}[\frac{1}{2},\ \gamma,\ \frac{3}{2},\ x^2]}{\sqrt{\pi}\ \text{Gamma}[1 - \gamma]}$

Carrying out the integration with respect to x gives the cumulative probability distribution.

```
In[3]:= W[γ_][x_] = FullSimplify[w[γ][x], 0 <= x <= 1,
           ComplexityFunction ->
           (If[MemberQ[#, _Hypergeometric2F1 |
                           _HypergeometricPFQ, Infinity], 1, 0]&)]
```

Out[3]= $\dfrac{\text{Beta}[x^2,\ \frac{1}{2},\ 1 - \gamma]\ \text{Gamma}[\frac{1}{2}\ (3 - 2\ \gamma)]}{\sqrt{\pi}\ \text{Gamma}[1 - \gamma]}$

We have to invert this relation into the form $x(\gamma) = x(W(\gamma))$. *Mathematica* does not have an inverse Beta function built in, but it has a near relative of it.

```
In[4]:= Names["*Inverse*Beta*"]
```

Out[4]= {InverseBetaRegularized}

To make use of the function `InverseBetaRegularized`, we transform the `Beta` function into a `BetaRegularized` function.

```
In[5]:= W[γ_][x_] = W[γ][x] /. Beta[x_, a_, b_] :> BetaRegularized[x, a, b] Beta[a, b]
```

Out[5]= $\dfrac{\text{Beta}[\frac{1}{2},\ 1 - \gamma]\ \text{BetaRegularized}[x^2,\ \frac{1}{2},\ 1 - \gamma]\ \text{Gamma}[\frac{1}{2}\ (3 - 2\ \gamma)]}{\sqrt{\pi}\ \text{Gamma}[1 - \gamma]}$

Now, `Solve` can invert the relation and we obtain the following result, which is a nice and short answer.

```
In[6]:= sol = Solve[W[γ][x] == ξ, x, InverseFunctions -> True] // FullSimplify
```

Out[6]= $\left\{\left\{x \to -\sqrt{\text{InverseBetaRegularized}\left[\xi, \frac{1}{2}, 1 - \gamma\right]}\right\},\right.$

$\left.\left\{x \to \sqrt{\text{InverseBetaRegularized}\left[\xi, \frac{1}{2}, 1 - \gamma\right]}\right\}\right\}$

Defining a continuous version of the $x_j(\gamma)$, we can generate a graphic showing the x_j as a function of γ. For $\gamma = 0$, the $x_j(\gamma)$ are equidistant. For $\gamma \to 1$, they cluster around 1. For $\gamma \to \infty$, they cluster around 0.

In[7]:= `x[γ_][w_] = Sqrt[InverseBetaRegularized[w, 1/2, 1 - γ]];`

In[8]:= `Module[{ppγ = 50, pps = 30, εs = 10^-2, εγ = 10^-3},`
`    Show[Graphics[`
`        Table[N @ Line[Table[{{x[γ][w], γ}, {γ, -5, 1 - εγ, (6 - εγ)/ppγ}]],`
`            {w, εs, 1, (1 - εs)/pps}]], Frame -> True, AspectRatio -> 2]]`

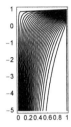

c) The sum $\sum_{k=1}^{n} 1/k$ can be expressed in closed form through $\psi(n) + 1/n + \gamma$. Expanding this expression around $n = \infty$ allows to calculate the α_j by matching terms of the same order in n^{-j}.

In[1]:= `o = 3;`
`    γSer = Series[PolyGamma[0, n], {n, Infinity, o + 1}] + 1/n + EulerGamma -`
`        Log[n + Sum[α[j] n^-j, {j, 0, o + 1}]] // ExpandAll`

Out[2]= $\text{EulerGamma} + \dfrac{\frac{1}{2} - \alpha[0]}{n} + \left(-\dfrac{1}{12} + \dfrac{\alpha[0]^2}{2} - \alpha[1]\right)\left(\dfrac{1}{n}\right)^2 + \left(-\dfrac{1}{3}\alpha[0]^3 + \alpha[0]\,\alpha[1] - \alpha[2]\right)\left(\dfrac{1}{n}\right)^3 +$

$\left(\dfrac{1}{120} + \dfrac{\alpha[0]^4}{4} - \alpha[0]^2\,\alpha[1] + \dfrac{\alpha[1]^2}{2} + \alpha[0]\,\alpha[2] - \alpha[3]\right)\left(\dfrac{1}{n}\right)^4 + O\left[\dfrac{1}{n}\right]^5$

αEqs are the nonlinear equations in the α_j.

In[3]:= `αEqs = Rest[γSer[[3]]];`

We make use of the fact that the jth equation of αEqs is linear in α_{j-1} and solve recursively for the highest α_j.

In[4]:= `αSols = {};`
`    Do[sol = Solve[(αEqs[[k]] /. αSols) == 0, α[k - 1]];`
`        AppendTo[αSols, sol[[1, 1]]], {k, o}];`
`    αSols`

Out[6]= $\left\{\alpha[0] \to \dfrac{1}{2}, \alpha[1] \to \dfrac{1}{24}, \alpha[2] \to -\dfrac{1}{48}\right\}$

To calculate the α_j up to order 100 quickly and without excessive memory requirements, we make a slight change to the last code. Instead of generating the equations for all the α_j at once, we recursively use already calculated α_j to obtain only univariate equations.

In[7]:= `o = 101; αSols = {};`
`    Do[AppendTo[αSols, Solve[`
`                (Series[PolyGamma[0, n], {n, Infinity, k + 1}] + 1/n + EulerGamma -`
`                 (Log[n + Sum[α[j] n^-j, {j, 0, k + 1}]] /. αSols))[[3, k + 1]] == 0,`
`                α[k - 1]][[1, 1]]], {k, o}];`

The following graphic shows the logarithmic difference γ – approximation for $n = 1$, $n = 10$ and $n = 100$. For finite n, an optimal order of the series exists. From $n = 1$, we obtain the approximation $\gamma \approx 1 - \ln(885877/580608)$.

In[9]:= `$MaxExtraPrecision = Infinity;`

`    Show[GraphicsArray[`
`    ListPlot[Table[{j, Log[10, Abs[#]]& @`

```
N[(Sum[1/k, {k, n}] - Log[n + Sum[α[k] n^-k, {k, 0, j}]]]) -
   EulerGamma /. αSols /. n -> #, 22]}, {j, 100}],
   DisplayFunction -> Identity]& /@ {1, 10, 100}]]
```

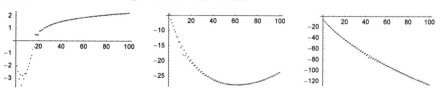

8. Time-Dependence in cos-Potential, Singular Potential Eigenvalues

a) The expansion in eigenfunctions, is given by $\Psi(x, t) = \sum_{j=0}^{\infty} c_j e^{-i\varepsilon_j t} \psi_j(x)$ where the $\psi_j(x)$ fulfill

$$-\psi_j''(x) - V_0 \cos(x) \psi_j(x) = \varepsilon_j \psi_j(x)$$

and are normalized according to $\int_{-3\pi}^{3\pi} |\psi_j(x)|^2 dx = 1$ and the expansion coefficients are defined through $c_j = \int_{-3\pi}^{3\pi} \psi_j(x) \Psi(x, 0) dx$.

To carry out an expansion in eigenfunctions we first need the eigenfunctions and eigenvalues. The time-independent differential equation is a Mathieu equation.

In[1]:= `DSolve[-1/2 y''[x] - V0 Cos[x] y[x] == ε y[x], y[x], x]`

Out[1]= $\left\{\left\{y[x] \to C[1] \text{ MathieuC}\left[8 \varepsilon, -4 V0, \frac{x}{2}\right] + C[2] \text{ MathieuS}\left[8 \varepsilon, -4 V0, \frac{x}{2}\right]\right\}\right\}$

For the (yet to be determined) eigenvalue ε, we define the symmetric eigenfunctions $\psi[x, \varepsilon]$ normalized to $\psi[x, \varepsilon]=1$. (Because of the symmetry of the initial conditions, we only need the symmetric eigenfunctions [24].)

In[2]:= `V0 = 12;`
      `ψ[x_, ε_] := MathieuC[8ε, -4 V0, x/2]/MathieuC[8ε, -4 V0, 0]`

Taking the definition of the function `MathieuCharacteristicA` into account, the eigenvalues can be straightforwardly obtained as $a_{k/3}(-4 V_0)$, $2 k \in \mathbb{N}$. To resolve the low-lying states correctly, we use high-precision arithmetic here. We will take the first 160 states for the expansion; for the visualization we are interested here this will be enough.

In[4]:= `o = 160;`
      `εData = Table[MathieuCharacteristicA[k/3, -4 V0]/8 // N[#, 40]&,`
      `   {k, 0, 2o, 2}];`

The higher eigenvalues ε_j quickly approach the asymptotic value $\tilde{\varepsilon}_j = (2 j - 2)^2 / 72$ because of the decreasing influence of the periodic potential. The first graphic shows $\log_{10}(\varepsilon_{j+1} - \varepsilon_j)$ and the second shows $\log_{10}(|1 - \varepsilon_j / \tilde{\varepsilon}_j|)$. Observe the extremely small energy differences between some of the low-states. They are caused by tunneling effects of the corresponding strongly localized eigenfunctions within each of the three equivalent wells. Each well contributes one energy level, and symmetry considerations select pairs and singles from the resulting triplets.

In[6]:= `Off[Power::infy]; Off[Graphics::gptn];`
      `Show[GraphicsArray[`
      `Block[{$DisplayFunction = Identity},`
      `{(* consecutive eigenvalue difference *)`
      `ListPlot[Log[10, -Subtract @@@ Partition[εData, 2, 1]],`
      `        PlotRange -> All],`
      `(* approach to asymptotic value *)`
      `ListPlot[Log[10, Abs[1 - MapIndexed[#1/((2 #2[[1]] - 2)^2/9/8)&,`
      `                              εData]]],`
      `        PlotRange -> Automatic]}]]]`

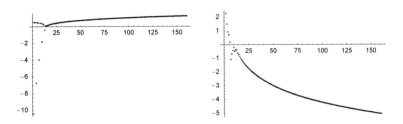

We quickly check that the symmetric eigenfunctions fulfill the periodic boundary conditions (meaning $\psi'_j(3\pi) = 0$) at the calculated ε_j.

```
In[8]:= TimeConstrained[Derivative[1, 0][ψ][3Pi, #], 5]& /@
                    εData // Abs // Chop // Union

Out[8]= {0}
```

Now we have to calculate the eigenfunctions. We calculate them on a dense grid for positive x and then form interpolating functions. (Because of the symmetry of the initial conditions, we will generate the values for negative x by reflection.)

```
In[9]:= pp = 600;
        ψData = Table[Table[{x, ψ[x, N[εData[[k]]]]},
                        {x, 0, 3Pi, 3Pi/pp}], {k, o}];

In[11]:= Do[ψIpo[k] = Interpolation[ψData[[k]], InterpolationOrder -> 16], {k, o}]
```

Because we need normalized eigenfunctions for the expansion, we calculate the norm of the ψIpo[k]. Because of the periodicity of the eigenfunctions, we use the option setting Trapezoidal for NIntegrate.

```
In[12]:= cTab = Table[
         1/Sqrt[NIntegrate[Evaluate[ψIpo[k][Abs[x]]^2], {x, -3Pi, 3Pi},
                    MaxRecursion -> 10, Method -> Trapezoidal]],
              {k, o}];
```

The following graphic shows the first 18 normalized eigenfunctions "at" their energies together with the triple well potential. The then nearly degenerate eigenfunctions deep inside the wells are clearly visible. (The third, sixth, ... are not degenerate because its degenerate partner would be antisymmetric.)

```
In[13]:= Plot[Evaluate[Table[εData[[k]] + 4 cTab[[k]] ψIpo[k][Abs[x]],
                    {k, 24}] // N], {x, -3Pi, 3Pi},
            PlotStyle -> Table[{Thickness[0.002],
                            Hue[0.8 (k - 1)/24]}, {k, 24}],
            PlotRange -> All, Frame -> True, Axes -> False,
            AspectRatio -> 1/4, (* the triple well potential *)
            Prolog -> {Thickness[0.002], GrayLevel[0],
            Line[Table[{x, -V0 Cos[x]}, {x, -3. Pi, 3. Pi, 6Pi/300}]]}]
```

Now we calculate the expansion coefficients c_j (including the normalization coefficients).

```
In[14]:= Module[{𝒦 = 6, xM = Pi/6},
        αTab = Table[2 cTab[[k]] NIntegrate[Evaluate[
                    ψIpo[k][Abs[x]] Cos[𝒦 x] Cos[x/xM Pi/2]^2],
                {x, 0, xM}, MaxRecursion -> 8,
                AccuracyGoal -> 5], {k, o}]];
```

The following plot of the logarithm of absolute value of the expansion coefficients confirms that we took enough states into account for the purpose of a visualization.

```
In[15]:= ListPlot[Log[10, Abs[αTab]], PlotRange -> All]
```

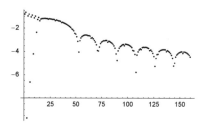

A quick check of Parseval's theorem shows that we did not miss any important eigenstate. (The 1% difference is caused by taking into account the first 160 states only; for the visualization purpose here this is sufficient.)

```
In[16]:= {(* norm of initial wave function *)
        Module[{𝒦 = 6, xM = Pi/6},
               2 Integrate[(Cos[𝒦 x] Cos[x/xM Pi/2]^2)^2, {x, 0, xM}]] // N,
        (* sum of modes *) Sum[αTab[[k]]^2, {k, o}]}
Out[16]= {0.229074, 0.226713 + 1.35381×10^-14 i}
```

Now we are ready to calculate the time-dependent $\Psi(x, t)$. For each t, we have to form the sum $\sum_{j=0}^{o} c_j\, e^{-i\,\varepsilon_j\, t}\,\psi_j(x)$. To do this in a time-efficient manner, we pack the lists of eigenfunction values.

```
In[17]:= ψDataP = Developer`ToPackedArray[Last /@
                  (* reduce number of spatial points *)
                  Append[First /@ Partition[#, 3], #[[-1]]]]& /@ ψData;
In[18]:= Φ[t_] := Sum[cTab[[k]] αTab[[k]] Exp[-I t εData[[k]]] ψDataP[[k]],
               {k, o}]
```

The following two plots show $|\Psi(x, t)|$ and $\arg(\Psi(x, t))$. Because the mean energy of the initial $\Psi(x, 0)$ is much below the maximum of the potential, the majority of the wave packet stay inside the middle well. Some high-frequency components leak into the two neighboring wells. Because the energies ε_j are not (small) rational multiples of each other, the resulting pattern in the middle well is aperiodic.

```
In[19]:= T = 10; ppt = 600;
        ΦData = Table[Φ[t], {t, 0, T, T/ppt}];
In[21]:= (* add values for negative x *)
        ΦDataAll = Transpose[Join[Reverse[Rest[#]], #]& /@ ΦData];
In[23]:= ListDensityPlot[#[ΦDataAll], Mesh -> False, FrameTicks -> False,
                      PlotRange -> #2, ColorFunction -> (Hue[0.8 #]&),
                      AspectRatio -> 1/3]& @@@
            {{Abs, {0, 2/3}}, {Arg[#]^2&, {0, Pi^2}}}
```

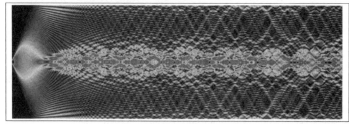

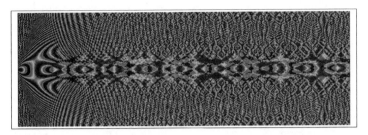

It remains to calculate of $\Psi(x, t)$ using NDSolve. We will use the "Pseudospectral" method because of the periodic boundary condition. The result from NDSolve agrees visually with the result from above.

```
In[24]:= Module[{𝒦 = 6, xM = Pi/6, X = 3Pi, pp = 120, T = 10, V0 = 12},
          (* solve differential equation *)
          NDSolve[{I D[Φ[x, t], {t, 1}] ==
                    -1/2 D[Φ[x, t], {x, 2}] - V0 Cos[x] Φ[x, t],
                  Φ[x, 0] == Cos[𝒦 x] If[Abs[x] < xM, Cos[x/xM Pi/2]^2, 0],
                  Φ[-X, t] == Φ[+X, t]}, Φ[x, t], {x, -X, X}, {t, 0, T},
                  (*  for plotting purposes use small precision goal *)
                  AccuracyGoal -> 3, PrecisionGoal -> 3,
                  Method -> {"MethodOfLines", "SpatialDiscretization" ->
                              {"TensorProductGrid", DifferenceOrder -> "Pseudospectral",
                               "MaxPoints" -> pp, "MinPoints" -> pp}}] //
                                                    (nsol = #)&;
          (* display square absolute value *)
          DensityPlot[Evaluate[Abs[Φ[x, t]] /. nsol[[1]]],
                    {t, 0, nsol[[1, 1, 2, 0, 1, 2, 2]]}, {x, -X, X},
                    PlotPoints -> {600, 400}, Mesh -> False,
                    PlotRange -> {0, 2/3}, AspectRatio -> 1/3,
                    ColorFunction -> (Hue[0.78 #]&)]]
```

For the time-evolution of similar wave packets in the cos-potential, see [525].

b) The potential $V(x) = -e^x$ is singular at $x = \infty$; it approaches ∞ faster than any power of x.

In the first moment, one might not expect the Schrödinger equation with the potential $V(x) = -e^x$ to have any bound states. For $\varepsilon > 0$, a particle could basically move freely in the left half space and is accelerated in the right half space. So, one expects a continuous spectrum. For $\varepsilon < 0$, a particle cannot enter the classically forbidden left half space and is accelerated in the right half space. Again one would expect a continuous spectrum (like for the linear potential $V(x) = -F x$.) The possibility for a particle to move freely in an unbounded domain would under most circumstances still imply a continuous spectrum and no bound states. But in a physical picture, bound states are caused by suitable interferences of travelling waves. And for the $V(x) = -e^x$ potential, a classical particle will reach infinity in finite time (and will become reflected there and return). So, in this fast decaying potential interference of left- and right-moving waves and as a result bound states are possible [1051].

For clarity, we calculate the $x(t)$ dependence for a classical particle starting with zero velocity and zero potential energy.

```
In[1]:= V[x_] := -Exp[x]

In[2]:= x[t] /. DSolve[{x''[t] == -D[V[x[t]], x[t]],
                        x[0] == Log[-ε], x'[0] == 0}, x[t], t]
```

```
Solve::ifun : Inverse functions are being used by Solve, so some solutions
    may not be found; use Reduce for complete solution information. More…

Solve::incnst :
Inconsistent or redundant transcendental equation. After reduction, the
```
$$\text{bad equation is } 4\,\text{ArcTanh}\Big[\text{Tanh}\Big[\frac{1}{2}\,\sqrt{C[\ll 1\gg]}\ \text{Power}[\ll 2\gg]\Big]\Big]^{2} == 0. \text{ More…}$$

```
Solve::verif :
Potential solution {C[2] → 0, C[1] → 2 ε} (possibly discarded by verifier)
    should be checked by hand. May require use of limits. More…

Solve::ifun : Inverse functions are being used by Solve, so some solutions
    may not be found; use Reduce for complete solution information. More…

Solve::ifun : Inverse functions are being used by Solve, so some solutions
    may not be found; use Reduce for complete solution information. More…

General::stop :
Further output of Solve::ifun will be suppressed during this calculation. More…
```

Out[2]= $\Big\{\text{Log}\Big[\varepsilon\Big(-1+\text{Tanh}\Big[\frac{\sqrt{t^2\,\varepsilon}}{\sqrt{2}}\Big]^{2}\Big)\Big]\Big\}$

In[3]:= `X[t_, ε_] = Log[ε (Tanh[(t Sqrt[ε])/Sqrt[2]]^2 - 1)]`

Out[3]= $\text{Log}\Big[\varepsilon\Big(-1+\text{Tanh}\Big[\frac{t\,\sqrt{\varepsilon}}{\sqrt{2}}\Big]^{2}\Big)\Big]$

In[4]:= `(* equation of motion and initial conditions are fulfilled *)`
```
    {D[X[t, ε], t, t] - Exp[X[t, ε]] // Simplify,
    X[0, ε], Derivative[1, 0][X][0, ε]}
```
Out[5]= `{0, Log[-ε], 0}`

The classical solution has a singularity at $t = \pi/(-2\,\varepsilon)^{1/2}$ (at this time the particle reaches infinity). This is shown in the left picture. The right picture shows solutions for complex initial conditions.

In[6]:= `Show[GraphicsArray[`
```
    Block[{$DisplayFunction = Identity, nsol},
    {(* x-t solution *)
    Plot[Evaluate[X[t, -3]], {t, 0, 2}],
    (* solutions in the complex x-plane *)
    Off[NDSolve::ndsz];
    Show[Table[(* solve differential equation *)
                nsol = NDSolve[{x''[t] == -D[V[x[t]], x[t]],
                        (* start at the x-unit circle *)
                        x[0] == Exp[I φ], x'[0] == 0},
                        x[t], {t, 0, 20}];
                (* parametric plot of the solution *)
        ParametricPlot[Evaluate[{Re[x[t]], Im[x[t]]} /. nsol],
                {t, 0, nsol[[1, 1, 2, 0, 1, 1, 2]]},
                DisplayFunction -> Identity,
                PlotStyle -> {{Thickness[0.002], Hue[φ/(2Pi)]}}],
        {φ, 0, 2Pi, 2Pi/96}], Frame -> True, Axes -> False,
    PlotRange -> {5 {-1, 1}, 10 {-1, 1}}]}]]]
```

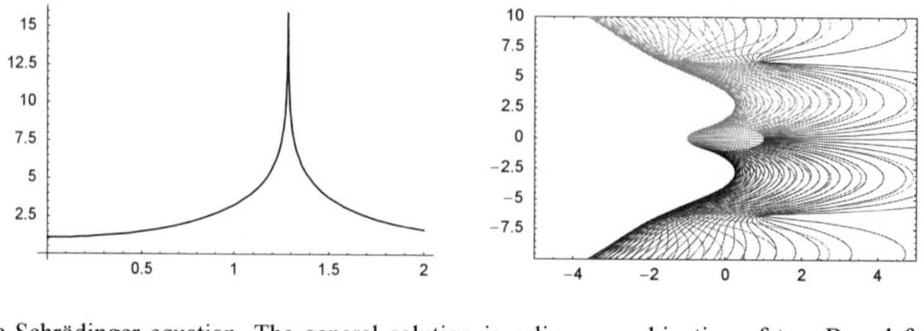

Now we solve the Schrödinger equation. The general solution is a linear combination of two Bessel functions with real indices for negative ε.

```
In[7]:= ψ[x_, ε_] = DSolve[-ψ''[x] - Exp[x] ψ[x] == ε ψ[x], ψ[x], x][[1, 1, 2]]
```

```
Out[7]= BesselJ[-2 i √ε, 2 √e^x] C[1] Gamma[1 - 2 i √ε] +
        BesselJ[2 i √ε, 2 √e^x] C[2] Gamma[1 + 2 i √ε]
```

Taking the asymptotics of the Bessel functions at negative infinity into account, we find that the solution $\psi_\varepsilon(x) = J_{-2i(-\varepsilon)^{1/2}}(\exp(x/2))$ is a bounded solution for all $\varepsilon < 0$. Here is a family of such solutions.

```
In[8]:= Plot[Evaluate[Table[BesselJ[-2 I Sqrt[ε], 2 Sqrt[Exp[x]]],
            {ε, -1, -10, -1/2}]], {x, -6, 6},
        Frame -> True, Axes -> False, PlotStyle ->
        Table[{Thickness[0.002], Hue[(ε + 10)/12]}, {ε, -1, -10, -1/2}]]
```

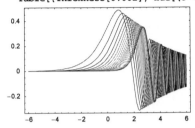

For brevity, we introduce $v(\varepsilon) = -2i(-\varepsilon)^{1/2}$. Then the normalization integral $\int_{-\infty}^{\infty} |\psi_\varepsilon(x)|\, dx$ is finite, namely $1/v(\varepsilon)$. This means we have found true bound states.

```
In[9]:= Integrate[v BesselJ[v, y]^2 2/y, {y, 0, Infinity},
            Assumptions -> v > 0]
```

```
Out[9]= 1
```

And the overlap integral becomes $4\sin(\pi/2\,(\mu - v))/(\pi\,(\mu - v)\,(\mu + v))$.

```
In[10]:= Integrate[BesselJ[v, y] BesselJ[μ, y] 2/y, {y, 0, Infinity},
            Assumptions -> Re[v + μ] > 0]
```

$$Out[10]= \frac{4\,\mathrm{Sin}[\frac{1}{2}\,\pi\,(\mu - v)]}{\pi\,\mu^2 - \pi\,v^2}$$

Because different bound states must have orthogonal wave functions, we obtain the additional constraint $v(\varepsilon_k) - v(\varepsilon_l) = 2j$, $j \in \mathbb{N}$. This now gives a family of countable infinite bound states parametrized, for instance, by the highest state ε_0. As for other highly singular potentials [464], [142], [999], [742], [112], ε_0 cannot be fixed uniquely, but corresponds to different self-adjoint extensions of the original symmetric differential operator [177], [1051], [444], [165], [1241].

Now we will calculate the corresponding WKB eigenvalues. This means we have to solve the equation $\int_{-\ln(-\varepsilon_n)}^{\infty} (\varepsilon_n - V(x))^{1/2}\, dx = (n + 1/2)\,\pi$ with respect to ε_n. The integral is obviously strongly divergent. So we regularize it (using the `GenerateConditions -> False` option setting).

```
In[11]:= Integrate[Sqrt[ε - V[x]], {x, Log[-ε], Infinity},
            Assumptions -> Im[ε] == 0 && ε < 0,
            GenerateConditions -> False]
```

$$Out[11]= 2\sqrt{\varepsilon}\ \mathrm{ArcSinh}[\sqrt{\varepsilon}] - i\sqrt{-\varepsilon}\left(\mathrm{Log}\left[1 - i\sqrt{-\frac{1+\varepsilon}{\varepsilon}}\right] - \mathrm{Log}\left[1 + i\sqrt{-\frac{1+\varepsilon}{\varepsilon}}\right]\right)$$

```
In[12]:= int = 2 Sqrt[ε] (ArcSinh[Sqrt[ε]] - ArcTanh[1/Sqrt[ε/(1 + ε)]])
```

$$Out[12]= 2\sqrt{\varepsilon}\left(\mathrm{ArcSinh}[\sqrt{\varepsilon}] - \mathrm{ArcTanh}\left[\frac{1}{\sqrt{\frac{\varepsilon}{1+\varepsilon}}}\right]\right)$$

A slightly nicer written form of this quantization condition is obtained by calculating the indefinite integral and explicitly removing the infinite part.

```
In[13]:= intIndef = Integrate[Sqrt[ε - V[x]], x];
         ser = Series[intIndef /. Exp[x] -> y, {y, ∞, 1}] // FullSimplify
```

$$Out[14]= \frac{2}{\sqrt{\frac{1}{y}}} + \pi\sqrt{-\frac{1}{\varepsilon}}\,\varepsilon - \varepsilon\sqrt{\frac{1}{y}} + O\left[\frac{1}{y}\right]^{3/2}$$

This yields the WKB eigenvalues $\varepsilon_n = (2n-1)^2/4$.

```
In[15]:= εWKB[n_] := -(1 + 2n)^2/4
```

Here is a quick check that they fulfill the quantization condition obtained from the regularized definite integral.

```
In[16]:= Table[FullSimplify[TrigToExp[
                int + (k + 1/2) Pi /. ε -> εWKB[k]]], {k, 0, 12}]
Out[16]= {0, 0, 0, 0, 0, 0, 0, 0, 0, 0, 0, 0, 0}
```

These WKB eigenvalues also fulfill the orthogonality conditions.

```
In[17]:= Table[4/Pi Sin[1/2 Pi(μ - ν)]/((μ - ν)(μ + ν)) /.
                {μ -> -2 I Sqrt[εWKB[m]], ν -> -2 I Sqrt[εWKB[n]]},
                {m, 0, 6}, {n, m + 1, 6}] // Flatten // Union
Out[17]= {0}
```

Interestingly, the WKB eigenvalues are the ones where the second solution of the differential equation $J_{2i(-\varepsilon)^{1/2}}(\exp(x/2))$ degenerates into the first solution $J_{-2i(-\varepsilon)^{1/2}}(\exp(x/2))$ and the second linear independent solution is logarithmic.

For strongly singular potentials in 3D and peratization, see [472], [33]. For a 2D polynomial potential without local minima that supports bound states, see [1254].

9. Dependencies, Numerical Function Evaluation, Usage Messages, Derivative Definition

a) Using On[ ], we can see "everything that is going on in a calculation", but the "output" may be very extensive. (We could direct the "output" to a file via

```
$MessagePrePrint = Identity;
$Messages = {OpenWrite["fileName", FormatType -> InputForm]}
```

which we could then read into *Mathematica* and (string) analyze.) On the other hand, Trace often does not provide enough information. Thus, we take the following approach: We associate an extra rule with all mathematical functions, so that when they are called, the corresponding function is entered in the list callList. The condition for the applicability of the extra rule (according to the standard evaluation strategy, the user-defined rule is used before the built-in rule) is never satisfied, and thus, the computation of all expressions proceeds normally. The disadvantage of this approach is that the additional rule has to be checked at each time and so increases the running time of most inputs. The association of the result proceeds as follows:

```
Unprotect[ mathFunction];
HoldPattern[ mathFunction[___]] := nothing /; (callList = {callList, mathFunction}; False);
Protect[ mathFunction];
```

Because we do not want to write this procedure for every mathematical function individually, we implement it.

```
In[1]:= (Unprotect[#];
        (* add the new rule *)
        ToExpression @ ("HoldPattern[" <> ToString[#] <>
            "[___]] := nothing /; (callList = {callList, " <> ToString[#] <> "}; False)");
        Protect[#])& /@
          (functionList =
          (* all functions with the NumericFunction attribute *)
          ToExpression /@ Select[Names["*"], MemberQ[Attributes[#], NumericFunction]&]);
```

Note that some functions may call other routines internally without passing through our rules; we will not see those calls in our list. We define a function analyze that gives the number of times the new added rules were used. The function must have the attribute HoldAll to avoid the evaluation of its argument before the function really comes to work. The function calls we collect in the globally visible list callList.

```
In[2]:= SetAttributes[analyze, HoldAll]

        analyze[x_] := (callList = {}; x; {#[[1]], Length[#]}& /@
                        Split[Sort[Flatten[callList]]])
```

We now give a few examples. In computations involving special functions, many other functions are called.

```
In[4]:= analyze[PolyLog[4, N[5, 100]]]
```

In[4]:= `{{Divide, 1}, {Gamma, 12}, {Log, 1}, {PolyLog, 2}, {Zeta, 2}}`

In[5]:= `analyze[JacobiAmplitude[4, 0.5]]`

Out[5]= `{{Abs, 2}, {ArcSin, 1}, {Ceiling, 2}, {EllipticF, 2}, {EllipticK, 2}, {Exp, 1},`
`{Floor, 6}, {Im, 6}, {Mod, 5}, {Power, 16}, {Re, 9}, {Round, 3}, {Times, 24}}`

In[6]:= `analyze[HypergeometricU[1.89I, 0.6, 1.03]]`

Out[6]= `{{Hypergeometric1F1Regularized, 2}, {HypergeometricU, 1}, {Power, 2}}`

In[7]:= `analyze[SphericalHarmonicY[5, 2, t, p]]`

Out[7]= `{{Cos, 1}, {Csc, 1}, {Divide, 4}, {Exp, 1}, {Gamma, 3}, {Hypergeometric2F1, 1},`
`{Im, 3}, {Log, 1}, {Max, 8}, {Plus, 112}, {Pochhammer, 1}, {Power, 93},`
`{Sec, 5}, {Sin, 1}, {SphericalHarmonicY, 1}, {Sqrt, 1}, {Times, 816}}`

In[8]:= `analyze[HypergeometricPFQ[{1, 2}, {3, 5, 7}, 2]]`

Out[8]= `{{Abs, 11}, {BesselI, 4}, {BesselJ, 2}, {BesselK, 2}, {BesselY, 2}, {Gamma, 2},`
`{Im, 4}, {Plus, 168}, {Power, 22}, {Sign, 1}, {Sqrt, 2}, {Times, 168}}`

In[9]:= `analyze[Erf[13.7 + 7.9 I]]`

Out[9]= `{{Erf, 1}, {Gamma, 2}}`

In[10]:= `analyze[HermiteH[23, ξ]]`

Out[10]= `{{HermiteH, 1}, {Plus, 8188}, {Power, 12}, {Times, 48}}`

In[11]:= `analyze[HermiteH[23.9 I, 2.34 I + 6]]`

Out[11]= `{{HermiteH, 1}, {Hypergeometric1F1Regularized, 4}, {HypergeometricU, 1}, {Power, 4}}`

In[12]:= `analyze[BesselJ[1/2, z]]`

Out[12]= `{{BesselJ, 1}, {Cos, 3}, {Im, 1}, {Log, 1}, {Plus, 4}, {Power, 16}, {Sin, 3}, {Times, 60}}`

In[13]:= `analyze[Hypergeometric2F1[a, b, c, z]]`

Out[13]= `{{Abs, 2}, {Hypergeometric2F1, 1}, {Im, 6},`
`{Log, 3}, {Plus, 152}, {Power, 6}, {Times, 124}}`

`FunctionExpand` and `FullSimplify` carry out many calls to special and elementary functions.

In[14]:= `analyze[FunctionExpand[PolyGamma[2/5]]]`

Out[14]= `{{Abs, 5}, {Arg, 4}, {Ceiling, 1}, {Cos, 3}, {Exp, 2}, {Floor, 4}, {Im, 14},`
`{Log, 113}, {Max, 2}, {Mod, 1}, {Plus, 7404}, {Power, 383}, {Sin, 3}, {Times, 2240}}`

In[15]:= `analyze[FullSimplify[HypergeometricPFQ[{1, 2, 4, 1/3}, {2, 3, 4, 1/7}, 1/2]]]`

Out[15]= `{{Abs, 2}, {ArcCos, 4}, {ArcSec, 1}, {ArcSin, 1}, {ArcSinh, 1}, {ArcTan, 2}, {ArcTanh, 3},`
`{Divide, 15}, {Gamma, 24}, {Hypergeometric1F1, 1}, {Hypergeometric1F1Regularized, 12},`
`{Plus, 608}, {Power, 100}, {Round, 12}, {Times, 1336}}`

Because we have introduced additional rules for a rather large number of functions, we should not compute more than necessary, as the processing speed clearly is slower than normal. To cancel the above definition, we use the following code (or start a new *Mathematica* session.)

In[16]:= `??Sin`

```
System`Sin

Attributes[Sin] = {Listable, NumericFunction, Protected}
HoldPattern[Sin[___]] := nothing /; (callList = {callList, Sin}; False)
```

In[17]:= `((Unprotect[#];`
`ToExpression["Unset[HoldPattern[" <> ToString[#] <> "[___]]]"];`
`Protect[#]; Null)& /@ functionList);`

In[18]:= `??Sin`

```
System`Sin

Attributes[Sin] = {Listable, NumericFunction, Protected}
```

We compare the above-obtained information with the one we would obtain with `Trace`. We use again the $_2F_1(a, b; c; z)$ example.

```
In[19]:= (tr = Trace[Hypergeometric2F1[a, b, c, z], TraceInternal -> True]) // ByteCount

Out[19]= 105952
```

And here is the corresponding result of `On[]`.

```
Block[{nb = NotebookPut[Notebook[{
  Cell["\<Block[{$MessagePrePrint = Identity},
  On[]; Hypergeometric2F1[a, b, c, z]; Off[]]\>", "Input"]}]]},
  (* evaluate the input *)
  SelectionMove[nb, Next, All];
  FrontEndTokenExecute[nb, "SelectAll"];
  FrontEndTokenExecute[nb, "Evaluate"]]
```

The next input carries out a quick count of the functions carrying the `NumericFunction` attribute in the nested list `tr`.

```
In[20]:= Off[Attributes::notfound]; Off[Attributes::ssle];
        {First[#], Length[#]}& /@ Split[Sort[Cases[tr, _?(MemberQ[Attributes[#],
                            NumericFunction]&), {-1}, Heads -> True]]]

Out[21]= {{Abs, 3}, {ArcTan, 2}, {ArcTanh, 2}, {BesselI, 6}, {BesselJ, 6},
         {BesselK, 6}, {BesselY, 6}, {Binomial, 2}, {Factorial, 6}, {Gamma, 16},
         {Hypergeometric2F1, 5}, {Im, 9}, {LerchPhi, 1}, {Log, 21}, {Plus, 223},
         {Pochhammer, 4}, {Power, 78}, {Re, 1}, {Sqrt, 5}, {Times, 302}}
```

When functions have symbolic variables, we can manage with somewhat fewer rules. In this case, we can associate the call to the counter with the symbolic variable of the input rather than the called function. One problem with this is that the rule for the symbolic variable cannot immediately be nested arbitrarily deeply in a rule, and thus we restrict ourselves here to calls on the function of the form

```
x /: f_?(# =!= collection&) [___, x, ___]
```

and

```
x /: f_?(# =!= collection&) [___, ___ x, ___].
```

We now implement this.

```
In[22]:= SetAttributes[informAboutWhatIsUsed, HoldAll]

        informAboutWhatIsUsed[expr_] :=
        Module[{buildingBlocks, nameList, erg, a, b, ls, calledFunctions, checkedContexts},
              (* turn off some messages *)
              Off[Context::ssle]; Off[Context::notfound]; Off[AppendTo::rvalue];
              (* the atoms of expr which are of interest here *)
              buildingBlocks = Complement[ToString /@ Cases[
               Union[Level[Hold[expr], {-1}, Heads -> True]], _Symbol],
              (* the system commands *)
              nameList = Names["System`*"]];
              (* add the "watching rules" to the buildingBlocks; as argument *)
              ToExpression[StringJoin[#, "/: f___?(# =!=
               " <> ToString[b] <> "&)[___, ", #, ", ___] := Null /;
              (AppendTo[" <> ToString[a] <> ", f]; False)"]]& /@ buildingBlocks;
              (* add the "watching rules" to the buildingBlocks; as head *)
              ToExpression[StringJoin[#, "/: f___?(# =!=
               " <> ToString[b] <> "&)[___, ___" #, ", ___] := Null /;
              (AppendTo[" <> ToString[a] <> ", f]; False)"]]& /@ buildingBlocks;
              a = b[]; (* our local containers; do not use List - it is from System` *)
              (* now evaluate the expression *)
              erg = Evaluate[expr];
              (* analyze the content of a *)
```

```
          ls = ToString /@ List @@ Union[a];
          calledFunctions = Intersection[nameList, ls];
          checkedContexts = Complement[Union[Context /@ ls], {"System`"}];
          (* delete nonsensical information *)
          checkedContexts = DeleteCases[checkedContexts, Context[_]];
          (* remove rules *)
          ToExpression[StringJoin[#, "/: f___?(# =!= " <> ToString[b] <> "&)[
              ___, ", #, ", ___] =."]]& /@ buildingBlocks;
          ToExpression[StringJoin[#, "/: f___?(# =!=
              " <> ToString[b] <> "&)[___, ___", #, ", ___] =."]]& /@ buildingBlocks;
          On[AppendTo::rvalue];
          (* print out the result *)
          Print["The following functions were used:\n", calledFunctions];
          Print[" "];
          Print["The following contexts were run through:\n", checkedContexts];
          erg]
```

The heart of the following is an analog of the above string construction:

```
StringJoin[#, "/: f_?(# =!= " <> ToString[b] <> "&)[___, ", #, ", ___] := Null /;
    (AppendTo[" <> ToString[a] <> ", f]; False)"]& /@ …
```

which comes from evaluating `informAboutWhatIsUsed` for all nonbuilt-in names `buildingBlocks`. The called functions will be collected in a$n. (They will not be collected in a list in order not to interfere with the call of the function `List`.) After the variables of `informAboutWhatIsUsed` have been computed, the request for `buildingBlocks` will again be deleted to get back to the previous state of *Mathematica*. The used functions are printed out as a side effect inside the calculation. In addition to the called functions, we also print out the contexts involved. We now look at a few examples.

```
In[24]:= informAboutWhatIsUsed[BesselJ[1/2, z]]
```

```
The following functions were used:
{BesselJ, Cos, Log, Plus, Power, Sin, Times}
```

```
The following contexts were run through:
{}
```

$$Out[24]= \frac{\sqrt{\frac{2}{\pi}}\, Sin[z]}{\sqrt{z}}$$

The rules temporarily associated with the variable z are no longer visible.

```
In[25]:= ??z
```

```
Global`z
```

```
In[26]:= informAboutWhatIsUsed[LegendreQ[-1/2, z]]
```

```
The following functions were used:
{LegendreQ}
```

```
The following contexts were run through:
{}
```

$$Out[26]= LegendreQ\left[-\frac{1}{2}, z\right]$$

```
In[27]:= informAboutWhatIsUsed[LegendreP[3, 5, z]]
```

```
The following functions were used:
{LegendreP}
```

```
The following contexts were run through:
{}
```

$$Out[27]= 0$$

```
In[28]:= informAboutWhatIsUsed[Gamma[z] Gamma[1 - z] // FullSimplify]
```

```
The following functions were used:
{Expand, Gamma, Head, List, Plus, Power, SameQ, Simplify, Times, UnsameQ}
```

```
                The following contexts were run through:
                {Assumptions`, Developer`, Integrate`, System`GammaDump`, System`Private`}
Out[28]=  π Csc[π z]
```

Naturally, other *Mathematica* functions can be investigated for their mutual dependence. The functions Integrate, Solve, Series, Limit, Sum, ... are good candidates for an interesting result. Here are some examples.

```
In[29]:= informAboutWhatIsUsed[Integrate[Exp[-x^2], {x, -Infinity, Infinity}]]
                The following functions were used:
                {Alternatives, Apart, Attributes, Collect, D, Expand,
                  FactorList, FreeQ, Greater, Head, If, Less, List, Log, Normal, NumericQ,
                  Plus, PolynomialQ, Power, Rule, SeriesData, Times, Together, Variables}

                The following contexts were run through:
                {Integrate`, Integrate`ExponentialsDump`,
                  Integrate`FindIntegrandDump`, Integrate`ImproperDump`,
                  Integrate`NLtheoremDump`, Internal`, Limit`, System`Private`}
Out[29]=  √π

In[30]:= informAboutWhatIsUsed[Limit[Csc[x]/Log[x] x^x, x -> Infinity]]
                The following functions were used:
                {Cos, Csc, Divide, Exp, Expand, FactorList,
                  FactorSquareFree, Greater, Integrate, Less, LessEqual, List, Log, Plus,
                  Power, Rule, SeriesData, Sin, Times, Together, Unequal, Variables}

                The following contexts were run through:
                {Integrate`, Limit`, System`Private`}
Out[30]=  Limit[ (x^x Csc[x]) / Log[x] , x → ∞]

In[31]:= informAboutWhatIsUsed[Solve[{x^2 + y^2 == 9, x^2 - y^2 == 5}, {x, y}];]
                The following functions were used:
                {Alternatives, Equal, List, Power, ReplaceAll, Roots, Rule}

                The following contexts were run through:
                {Integrate`, Solve`}
```

We could use similar techniques to collect, not all functions encountered in a calculation, but say all real numbers. We attach an upvalue to the symbol Real that can collect every real number seen in the evaluation process.

```
In[32]:= SetAttributes[myContainer, Flat]

          Unprotect[Real]
          Real /: (f_ /; f =!= myContainer)[___, x_Real, ___] :=
                          Null /; (bag = myContainer[bag, x]; False)
Out[33]=  {Real}
```

Here is a numerical calculation for a hypergeometric function carried out.

```
In[35]:= bag = myContainer[];

          N[HypergeometricPFQ[{1, 1}, {2, 3, 4}, Pi^4]]
Out[36]=  566.348
```

bag now contains all of the real numbers calculated.

```
In[37]:= {Length[Union @ bag], Length[bag]}
Out[37]=  {188, 541}
```

The high-precision version of the last input carries out much more work.

```
In[38]:= bag = myContainer[];

      N[HypergeometricPFQ[{1, 1}, {2, 3, 4}, Pi^4], 22]
Out[39]= 566.3482972729391555980
```

bag now contains all of the real numbers calculated.

```
In[40]:= {Length[Union @ bag], Length[bag]}
Out[40]= {657, 1224}
```

Here an autonumericalization example.

```
In[41]:= bag = myContainer[];

      Sin[Pi/24] < (-Sqrt[3 (2 - Sqrt[2])]/4 + Sqrt[2 + Sqrt[2]]/4
```

N::meprec : Internal precision limit \$MaxExtraPrecision = 50.`

reached while evaluating $\frac{1}{4}\sqrt{3\left(2-\sqrt{2}\right)}-\frac{\sqrt{2+\sqrt{2}}}{4}+\mathrm{Sin}\left[\frac{\pi}{24}\right]$. More...

```
Out[42]= Sin[ π/24 ] < - 1/4 √(3 (2 - √2)) + √(2 + √2)/4
```

```
In[43]:= {Length[Union @ bag], Length[bag]}
Out[43]= {4, 4}
```

b) To find such examples, we have to conduct automated random searches. Here is a list of some of the functions discussed in this chapter, together with the number of arguments for each function.

```
In[1]:= functions = (* functions and number of arguments *)
      {{Gamma, 1, 2, 3}, {Pochhammer, 2}, {Beta, 2, 3, 4},
       {PolyGamma, 2}, {Erf, 1, 2}, {Erfc, 1, 2}, {Erfi, 1, 2},
       {FresnelC, 1}, {FresnelS, 1}, {SinIntegral, 1}, {CosIntegral, 1},
       {ExpIntegralE, 2}, {ExpIntegralEi, 1}, {LogIntegral, 1},
       {BesselJ, 2}, {BesselI, 2}, {BesselK, 2}, {BesselY, 2},
       {AiryAi, 1}, {AiryBi, 1}, {AiryAiPrime, 1}, {AiryBiPrime, 1},
       {LegendreP, 2, 3}, {LegendreQ, 2, 3},
       {Hypergeometric1F1, 3}, {Hypergeometric2F1, 4},
       {Hypergeometric2F1Regularized, 4}, {HypergeometricU, 3},
       {EllipticF, 2, 3}, {EllipticK, 1}, {EllipticE, 1, 2}, {EllipticPi, 2, 3},
       {JacobiAmplitude, 2}, {JacobiSN, 2}, {JacobiCN, 2}, {JacobiDN, 2},
       {ArithmeticGeometricMean, 2}, {ProductLog, 1},
       {MathieuC, 3}, {MathieuS, 3}};
```

Now, we must generate random arguments. We do not use exact rational arguments to avoid that N[*rationalNumber*] and N[*rationalNumber*, *precisionLargerThanMachinePrecision*] are actually different numbers because of different nonleading digits. Instead, we generate a list of random digits together with their scale and sign. (Naturally, we would expect machine precision to fail for "large" arguments of special functions. This is the reason we randomly select a scale of the numbers.)

```
In[2]:= preRandomNumber[digits_, scale_] :=
      {(* sign *) {"+", "-"}[[ Random[Integer, {1, 2}]]],
       (* first digit *)         Random[Integer, {1, 9}],
       (* next digits *) Table[Random[Integer, {1, 9}], {digits}],
       (* scale in 10^scale *) Random[Integer, {1, digits}]}
```

Later, we will use complex numbers as arguments; so we need two preRandomNumbers. preRandomNumberPair generates such a pair.

```
In[3]:= preRandomNumberPair[digits_, scale_] :=
              Table[preRandomNumber[digits, scale], {2}]
```

The function makeMachineNumber converts our prerandom numbers into machine numbers.

```
In[4]:= makeMachineNumber[{l1_List, l2_List}] :=
          makeMachineNumber[l1] + I makeMachineNumber[l2]
In[5]:= makeMachineNumber[{sign_String, firstDigit_, otherDigits_, scale_}] :=
          StringJoin[Flatten[{sign, ToString[firstDigit], ".",
```

```
                               ToString /@ otherDigits, "*^",
                               ToString[scale]}]] // ToExpression
```

Similarly, the function `makeHighPrecisionNumber` converts prerandom numbers into high-precision numbers of precision *prec*.

```
In[6]:= makeHighPrecisionNumber[{l1_List, l2_List}, prec_] :=
           makeHighPrecisionNumber[l1, prec] + I makeHighPrecisionNumber[l2, prec]
```

```
In[7]:= makeHighPrecisionNumber[{sign_String, firstDigit_,
                                otherDigits_, scale_}, prec_] :=
        StringJoin[Flatten[{sign, ToString[firstDigit], ".",
                            ToString /@ otherDigits, "`",
                            ToString[prec],"*^", ToString[scale]}]] // ToExpression
```

Here is an example for the three functions `preRandomNumberPair`, `makeMachineNumber`, and `makeHighPrecisionNumber`.

```
In[8]:= preRandomNumberPair[10, 6]
```

```
Out[8]= {{+, 8, {2, 6, 2, 8, 5, 3, 4, 5, 1, 1}, 1}, {+, 3, {4, 3, 9, 9, 2, 7, 1, 4, 6, 4}, 7}}
```

```
In[9]:= {makeMachineNumber[%], makeHighPrecisionNumber[%, $MachinePrecision + 1]}
```

```
Out[9]= {82.6285 + 3.43993 × 10^7 i, 82.628534511000000 + 3.4399271464000000 × 10^7 i}
```

`randomFunction` randomly selects one of the functions from the set of special functions `functions` with a possible number of arguments.

```
In[10]:= randomFunction[ζ_] := #[[1]][Sequence @@ Table[ζ[i], {i, #[[2]]}]]& @
                              functions[[Random[Integer, {1, Length[functions]}]]]
```

Here are three instances of `randomFunction`.

```
In[11]:= Table[randomFunction[ξ], {3}]
```

```
Out[11]= {EllipticPi[ξ[1], ξ[2]], ExpIntegralE[ξ[1], ξ[2]], AiryAi[ξ[1]]}
```

The function `test` takes as arguments a random function *func*, the arguments *numbers* of the function *func* in the form generated by `preRandomNumberPair`, and a precision *prec*. `test` tests if the machine precision value of *func* at *numbers* agrees with the high-precision value. When they agree `test` returns `True`. If they do not agree, `test` returns the function and the numbers that caused the disagreement, and in case the evaluation did not result in an explicit number, `test` returns `$Failed`. (This means an overflow or underflow or an indeterminate result was calculated.)

```
In[12]:= test[{func_, ζ_}, numbers_, prec_] :=
         Module[{l = Length[numbers], ζ, machineResult, bignumResult},
         (* convert to machine numbers *)
         Do[ζ[i] = makeMachineNumber[numbers[[i]]], {i, l}];
         machineResult = func;
         (* convert to high-precision numbers *)
         Do[ζ[i] = makeHighPrecisionNumber[numbers[[i]], prec], {i, l}];
         bignumResult = N[func, prec];
         If[(* are the results numbers? *)
            NumberQ[machineResult] && NumberQ[bignumResult] &&
            machineResult =!= Underflow[] && bignumResult =!= Underflow[] &&
            machineResult =!= Overflow[]  && bignumResult =!= Overflow[],
            If[(* do machine and high-precision result agree? *)
               machineResult == bignumResult,
               True, (* return function and arguments *)
                    {{func[[0]], numbers},
                     {machineResult, N[bignumResult]}}], $Failed]]
```

Here is a test that shows agreement between the machine and the high-precision version.

```
In[13]:= SeedRandom[1000];
         args = {randomFunction[ξ], Table[preRandomNumberPair[10, 6], {2}]}
```

```
Out[14]= {Pochhammer[ξ[1], ξ[2]],
          {{{+, 6, {3, 3, 7, 9, 7, 1, 4, 9, 3, 2}, 1}, {+, 1, {1, 5, 4, 7, 8, 7, 8, 5, 5, 6}, 6}},
           {{-, 3, {2, 6, 6, 3, 3, 9, 8, 5, 5, 5}, 3}, {+, 7, {6, 9, 9, 4, 2, 8, 4, 1, 5, 7}, 7}}}}
```

```
In[15]:= test[{args[[1]], ξ}, args[[2]], 40]
```

```
Out[15]= {{Complex,
         {{{+, 6, {3, 3, 7, 9, 7, 1, 4, 9, 3, 2}, 1}, {+, 1, {1, 5, 4, 7, 8, 7, 8, 5, 5, 6}, 6}},
          {{-, 3, {2, 6, 6, 3, 3, 9, 8, 5, 5, 5}, 3}, {+, 7, {6, 9, 9, 4, 2, 8, 4, 1, 5, 7}, 7}}}},
         {5.99078915736834×10⁻⁵²⁵⁵⁰²⁵⁷ - 8.9966966559407×10⁻⁵²⁵⁵⁰²⁵⁸ i,
          5.990788874276513×10⁻⁵²⁵⁵⁰²⁵⁷ - 8.996710487903841×10⁻⁵²⁵⁵⁰²⁵⁸ i}}
```

To avoid the generation of a large number of messages, we turn off the following messages.

```
In[16]:= Off[General::ovfl]; Off[General::unfl]; Off[General::dbyz];
         Off[N::meprec]; Off[Infinity::indet];
```

Now, we search for an example in which the machine results deviates slightly from the corresponding high-precision result. Because the high-precision evaluation of special functions can be very time-consuming, we use TimeConstrained here. If no difference between the machine arithmetic result and the high-precision result exists, the function randomTest will return $Failed (meaning that it failed to find an example where the machine number result was incorrect).

```
In[18]:= randomTest[r_, δ_] :=
         Module[{func, numbers, res, ratio},
           SeedRandom[r];
           func = randomFunction[ξ];
           numbers = Table[preRandomNumberPair[10, 6], {Length[func]}];
           (* wait one second only *)
           res = TimeConstrained[test[{func, ξ}, numbers,
                             $MachinePrecision + 10], 1];
           (* analyze res *)
           If[res =!= $Failed && res =!= True && res =!= $Aborted,
             ratio = res[[2, 1]]/res[[2, 2]];
             If[ratio =!= Underflow[] && ratio =!= Overflow[] &&
               Abs[Abs[ratio] - 1] > δ, {r, func, res},
               $Failed], $Failed]]
```

For the vast majority of functions and arguments, the machine precision version will be correct and the function random⁚ Test will fail to find a function and an argument that result in a wrong numerical value.

```
In[19]:= Table[randomTest[k, 1/10], {k, 100}] // Union
```

> Divide::infy : Infinite expression $\dfrac{1}{\text{Underflow[]}}$ encountered.

> Divide::infy : Infinite expression $\dfrac{1}{\text{Underflow[]}}$ encountered.

> Divide::infy : Infinite expression $\dfrac{1}{\text{Underflow[]}}$ encountered.

> General::stop :
>   Further output of Divide::infy will be suppressed during this calculation. More…

```
Out[19]= {$Failed}
```

To find examples of functions and arguments in which the machine values are wrong, we can conduct random searches. By running such a test for a long enough time, we can find some examples. Instead of collecting the function and the arguments, we collect the (much shorter) random number seed.

```
While[randomTest[r = Random[Integer, {1, 10^20}], 1/10] === $Failed, Null]; r
```

Here are three such examples.

```
In[20]:= randomTest[#, 1/10]& /@ {816995408349, 777945441789, 441683551009}
```

> Power::infy : Infinite expression $\dfrac{1}{0. + 0. i}$ encountered. More…

```
Out[20]= {{816995408349, JacobiAmplitude[ξ[1], ξ[2]], {{Complex,
           {{{+, 8, {3, 9, 8, 8, 2, 8, 1, 4, 8, 4}, 10}, {-, 8, {3, 8, 7, 7, 9, 7, 8, 8, 8, 7}, 1}},
            {{+, 4, {8, 9, 2, 3, 7, 9, 1, 4, 4, 3}, 9}, {-, 1, {3, 8, 3, 7, 9, 6, 8, 1, 6, 7}, 10}}}},
           {0.135613 + 0.036035 i, -0.163632 + 0.54043 i}}},
         {777945441789, HypergeometricU[ξ[1], ξ[2], ξ[3]], {{Complex,
           {{{+, 8, {8, 9, 6, 2, 7, 2, 1, 3, 9, 9}, 3}, {+, 4, {5, 5, 4, 4, 7, 5, 5, 7, 6, 4}, 2}},
```

```
        {{-, 5, {9, 1, 3, 6, 7, 4, 4, 1, 6, 3}, 1}, {+, 7, {2, 5, 6, 1, 7, 7, 5, 1, 5, 8}, 6}},
         {{-, 2, {2, 1, 4, 2, 6, 2, 9, 5, 2, 9}, 1}, {-, 1, {2, 8, 2, 4, 8, 2, 7, 5, 4, 5}, 9}}}},
```
$\{-2.045487191331937 \times 10^{-81334} + 4.602689111925826 \times 10^{-81334}\, \mathrm{i},\ 0.+0.\, \mathrm{i}\}\}\}$,
```
   {441683551009, JacobiSN[ξ[1], ξ[2]], {{Complex,
      {{{-, 4, {2, 1, 1, 2, 3, 7, 3, 6, 6, 8}, 9}, {-, 5, {3, 5, 2, 7, 4, 8, 3, 6, 7, 7}, 1}},
        {{+, 8, {5, 6, 1, 9, 1, 2, 5, 5, 4, 2}, 7}, {-, 7, {2, 4, 1, 8, 4, 5, 5, 8, 4, 8}, 9}}}},
       {-0.0865041 + 0.18907 i, -0.0780553 + 0.0884771 i}}}}
```

c) Here are two different kinds of usage messages. One is of type "*function* [*z, args*] is the ... " and one is of type "*function* [*z, args*] gives the ... ".

In[1]:= **??BesselJ**

System`BesselJ

Attributes[BesselJ] = {Listable, NumericFunction, Protected}

In[2]:= **??Exp**

System`Exp

Attributes[Exp] = {Listable, NumericFunction, Protected, ReadProtected}

To investigate if there are further kinds of message phrasings, we first must get all usage messages. We call all functions with arguments (to make sure autoloaded functions are loaded) and then explicitly read in all messages to do this.

```
In[3]:= Get[ToFileName[{$TopDirectory, "SystemFiles", "Kernel",
                  "TextResources", $Language}, "Usage.m"]];

In[4]:= systemCommands = Names["System`*"];
        (* autoload functions *)
        (C @@ ToHeldExpression[#]) & /@ systemCommands;
        (* clear the ReadProtected attribute *)
        If[MemberQ[Attributes[#], ReadProtected],
           ClearAttributes[#, ReadProtected]] & /@
             Apply[Unevaluated, ToHeldExpression /@
                    DeleteCases[systemCommands, "I"], {1}];
```

156 numeric functions exist.

```
In[9]:= functionAndMessages =
        Map[{#, MessageName[Evaluate[ToExpression[#]], "usage"]} &,
            (* the numeric functions *)
            (Select[DeleteCases[systemCommands, "I"],
                    MemberQ[Attributes[#], NumericFunction] &]), {1}];

In[10]:= Length[functionAndMessages]

Out[10]= 156
```

We extract the first words from the usage messages and join similar ones.

```
In[11]:= getFirstWords[{f_, s_String}] :=
         Module[{firstPos, secondPos},
           If[MatchQ[StringPosition[s, ToString[f]], {{1, _}, ___}],
             (* argument ending position *)
             firstPos = StringPosition[s, "]"][[1, 1]];
             (* next spaces *)
             secondPos = Select[First /@ StringPosition[s, " "], # > firstPos &][[3]];
             (* the phrase *)
             {StringTake[s, {firstPos + 2, secondPos - 1}], f}]]

In[12]:= {#[[1, 1]], Length[#]} & /@ Split[Sort[DeleteCases[
            getFirstWords /@ functionAndMessages, Null]], #1[[1]] == #2[[1]] &]

Out[12]= {{{Abs, Abs::usage}, 1}, {{AiryAi, AiryAi::usage}, 1},
          {{AiryAiPrime, AiryAiPrime::usage}, 1}, {{AiryBi, AiryBi::usage}, 1},
          {{AiryBiPrime, AiryBiPrime::usage}, 1}, {{AppellF1, AppellF1::usage}, 1},
          {{ArcCos, ArcCos::usage}, 1}, {{ArcCosh, ArcCosh::usage}, 1},
          {{ArcCot, ArcCot::usage}, 1}, {{ArcCoth, ArcCoth::usage}, 1},
          {{ArcCsc, ArcCsc::usage}, 1}, {{ArcCsch, ArcCsch::usage}, 1},
          {{ArcSec, ArcSec::usage}, 1}, {{ArcSech, ArcSech::usage}, 1},
```

```
{{ArcSin, ArcSin::usage}, 1}, {{ArcSinh, ArcSinh::usage}, 1},
{{ArcTan, ArcTan::usage}, 1}, {{ArcTanh, ArcTanh::usage}, 1}, {{Arg, Arg::usage}, 1},
{{ArithmeticGeometricMean, ArithmeticGeometricMean::usage}, 1},
{{BesselI, BesselI::usage}, 1}, {{BesselJ, BesselJ::usage}, 1},
{{BesselK, BesselK::usage}, 1}, {{BesselY, BesselY::usage}, 1},
{{Beta, Beta::usage}, 1}, {{BetaRegularized, BetaRegularized::usage}, 1},
{{Binomial, Binomial::usage}, 1}, {{Ceiling, Ceiling::usage}, 1},
{{ChebyshevT, ChebyshevT::usage}, 1}, {{ChebyshevU, ChebyshevU::usage}, 1},
{{Clip, Clip::usage}, 1}, {{Conjugate, Conjugate::usage}, 1}, {{Cos, Cos::usage}, 1},
{{Cosh, Cosh::usage}, 1}, {{CoshIntegral, CoshIntegral::usage}, 1},
{{CosIntegral, CosIntegral::usage}, 1}, {{Cot, Cot::usage}, 1},
{{Coth, Coth::usage}, 1}, {{Csc, Csc::usage}, 1}, {{Csch, Csch::usage}, 1},
{{DedekindEta, DedekindEta::usage}, 1}, {{Divide, Divide::usage}, 1},
{{EllipticE, EllipticE::usage}, 1}, {{EllipticF, EllipticF::usage}, 1},
{{EllipticK, EllipticK::usage}, 1}, {{EllipticNomeQ, EllipticNomeQ::usage}, 1},
{{EllipticPi, EllipticPi::usage}, 1}, {{Erf, Erf::usage}, 1},
{{Erfc, Erfc::usage}, 1}, {{Erfi, Erfi::usage}, 1},
{{Exp, Exp::usage}, 1}, {{ExpIntegralE, ExpIntegralE::usage}, 1},
{{ExpIntegralEi, ExpIntegralEi::usage}, 1}, {{Factorial, Factorial::usage}, 1},
{{Factorial2, Factorial2::usage}, 1}, {{Fibonacci, Fibonacci::usage}, 1},
{{Floor, Floor::usage}, 1}, {{FractionalPart, FractionalPart::usage}, 1},
{{FresnelC, FresnelC::usage}, 1}, {{FresnelS, FresnelS::usage}, 1},
{{Gamma, Gamma::usage}, 1}, {{GammaRegularized, GammaRegularized::usage}, 1},
{{GegenbauerC, GegenbauerC::usage}, 1}, {{HarmonicNumber, HarmonicNumber::usage}, 1},
{{HermiteH, HermiteH::usage}, 1}, {{Hypergeometric0F1, Hypergeometric0F1::usage}, 1},
{{Hypergeometric0F1Regularized, Hypergeometric0F1Regularized::usage}, 1},
{{Hypergeometric1F1, Hypergeometric1F1::usage}, 1},
{{Hypergeometric1F1Regularized, Hypergeometric1F1Regularized::usage}, 1},
{{Hypergeometric2F1, Hypergeometric2F1::usage}, 1},
{{Hypergeometric2F1Regularized, Hypergeometric2F1Regularized::usage}, 1},
{{HypergeometricU, HypergeometricU::usage}, 1},
{{Im, Im::usage}, 1}, {{IntegerPart, IntegerPart::usage}, 1},
{{InverseBetaRegularized, InverseBetaRegularized::usage}, 1},
{{InverseEllipticNomeQ, InverseEllipticNomeQ::usage}, 1},
{{InverseErf, InverseErf::usage}, 1}, {{InverseErfc, InverseErfc::usage}, 1},
{{InverseGammaRegularized, InverseGammaRegularized::usage}, 1},
{{InverseJacobiCD, InverseJacobiCD::usage}, 1},
{{InverseJacobiCN, InverseJacobiCN::usage}, 1},
{{InverseJacobiCS, InverseJacobiCS::usage}, 1},
{{InverseJacobiDC, InverseJacobiDC::usage}, 1},
{{InverseJacobiDN, InverseJacobiDN::usage}, 1},
{{InverseJacobiDS, InverseJacobiDS::usage}, 1},
{{InverseJacobiNC, InverseJacobiNC::usage}, 1},
{{InverseJacobiND, InverseJacobiND::usage}, 1},
{{InverseJacobiNS, InverseJacobiNS::usage}, 1},
{{InverseJacobiSC, InverseJacobiSC::usage}, 1},
{{InverseJacobiSD, InverseJacobiSD::usage}, 1},
{{InverseJacobiSN, InverseJacobiSN::usage}, 1},
{{JacobiAmplitude, JacobiAmplitude::usage}, 1}, {{JacobiCD, JacobiCD::usage}, 1},
{{JacobiCN, JacobiCN::usage}, 1}, {{JacobiCS, JacobiCS::usage}, 1},
{{JacobiDC, JacobiDC::usage}, 1}, {{JacobiDN, JacobiDN::usage}, 1},
{{JacobiDS, JacobiDS::usage}, 1}, {{JacobiNC, JacobiNC::usage}, 1},
{{JacobiND, JacobiND::usage}, 1}, {{JacobiNS, JacobiNS::usage}, 1},
{{JacobiP, JacobiP::usage}, 1}, {{JacobiSC, JacobiSC::usage}, 1},
{{JacobiSD, JacobiSD::usage}, 1}, {{JacobiSN, JacobiSN::usage}, 1},
{{JacobiZeta, JacobiZeta::usage}, 1}, {{KleinInvariantJ, KleinInvariantJ::usage}, 1},
{{LaguerreL, LaguerreL::usage}, 1}, {{LerchPhi, LerchPhi::usage}, 1},
{{Log, Log::usage}, 1}, {{LogGamma, LogGamma::usage}, 1},
{{LogIntegral, LogIntegral::usage}, 1}, {{MathieuC, MathieuC::usage}, 1},
{{MathieuCharacteristicA, MathieuCharacteristicA::usage}, 1},
```

```
{{MathieuCharacteristicB, MathieuCharacteristicB::usage}, 1},
{{MathieuCharacteristicExponent, MathieuCharacteristicExponent::usage}, 1},
{{MathieuCPrime, MathieuCPrime::usage}, 1}, {{MathieuS, MathieuS::usage}, 1},
{{MathieuSPrime, MathieuSPrime::usage}, 1}, {{Max, Max::usage}, 1},
{{Min, Min::usage}, 1}, {{Minus, Minus::usage}, 1},
{{Mod, Mod::usage}, 1}, {{ModularLambda, ModularLambda::usage}, 1},
{{Multinomial, Multinomial::usage}, 1}, {{NevilleThetaC, NevilleThetaC::usage}, 1},
{{NevilleThetaD, NevilleThetaD::usage}, 1},
{{NevilleThetaN, NevilleThetaN::usage}, 1},
{{NevilleThetaS, NevilleThetaS::usage}, 1}, {{Plus, Plus::usage}, 1},
{{Pochhammer, Pochhammer::usage}, 1}, {{PolyLog, PolyLog::usage}, 1},
{{Power, Power::usage}, 1}, {{Quotient, Quotient::usage}, 1}, {{Re, Re::usage}, 1},
{{Rescale, Rescale::usage}, 1}, {{RiemannSiegelTheta, RiemannSiegelTheta::usage}, 1},
{{RiemannSiegelZ, RiemannSiegelZ::usage}, 1}, {{Round, Round::usage}, 1},
{{Sec, Sec::usage}, 1}, {{Sech, Sech::usage}, 1},
{{Sign, Sign::usage}, 1}, {{Sin, Sin::usage}, 1}, {{Sinh, Sinh::usage}, 1},
{{SinhIntegral, SinhIntegral::usage}, 1}, {{SinIntegral, SinIntegral::usage}, 1},
{{SphericalHarmonicY, SphericalHarmonicY::usage}, 1}, {{Sqrt, Sqrt::usage}, 1},
{{StruveH, StruveH::usage}, 1}, {{StruveL, StruveL::usage}, 1},
{{Subtract, Subtract::usage}, 1}, {{Tan, Tan::usage}, 1}, {{Tanh, Tanh::usage}, 1},
{{Times, Times::usage}, 1}, {{UnitStep, UnitStep::usage}, 1}, {{Zeta, Zeta::usage}, 1}}
```

We see that the usage messages come in four major groups: "*function*[*z*, *args*] is the … ", "*function*[*z*, *args*] gives the … ", "*function*[*z*, *args*] yields the … ", and "*function*[*z*, *args*] represents the … ".

d) Unfortunately we cannot associate the identity $Y'(0) = c$ with Y.

In[1]:= Y /: Y'[0] = c;

> TagSet::tagpos : Tag Y in Y'[0] is too deep for an assigned rule to be found. More…

Because we want to avoid adding a permanent definition to Derivative, the only remaining possibility is to add a rule to Y. This rule itself should then generate a rule such that $Y'[0]$ evaluates to c. Here are two SetDelayed definitions for YD and YO that set up the generation of two such rules.

In[2]:= YD := (Derivative[1][Y][0] =.; YO; c)

```
YO := (Y /: Derivative[1][Y] :=
         ((* remove current definition *)
          Y /: Derivative[1][Y] =.;
          (* make new definition for Derivative *)
          Derivative[1][Y][0] := YD;
          (* return original head *)
          Derivative[1][Y]))
```

At the moment $Y'[0]$ does only evaluate trivially.

In[4]:= Y'[0]

Out[4]= Y'[0]

Evaluating YO generates a definition for Y'.

In[5]:= YO

In[6]:= ??Y

> Global`Y

> Y' ^:= (Y /: Y' =.; Y'[0] := YD; Y')

In the next input, the upvalue definition for Y' generates a definition for $Y'[0]$ (associated with Derivative) and removes the original definition for Y'. The execution of the definition for $Y'[0]$ restores the original definition for Y' and removes the temporarily present definition for $Y'[0]$.

In[7]:= Y'[0]

Out[7]= c

Here are the currently active definitions.

```
In[8]:= (* no additional definitions for Derivative *)
       #[Derivative]& /@ {OwnValues, DownValues, UpValues, SubValues}

Out[9]= {{}, {}, {},
       {HoldPattern[InverseLaplaceTransform^(0, 0, System`LaplaceTransformDump`m_Integer?Positive) [
           System`LaplaceTransformDump`f_, System`LaplaceTransformDump`s_,
           System`LaplaceTransformDump`t_]] :>
         Module[{System`LaplaceTransformDump`inverse, System`LaplaceTransformDump`t1},
           (∂{System`LaplaceTransformDump`t1, System`LaplaceTransformDump`m}
             System`LaplaceTransformDump`inverse /.
           System`LaplaceTransformDump`t1 -> System`LaplaceTransformDump`t) /;
         FreeQ[System`LaplaceTransformDump`inverse = InverseLaplaceTransform[
           System`LaplaceTransformDump`f, System`LaplaceTransformDump`s,
           System`LaplaceTransformDump`t1], InverseLaplaceTransform]] /;
       FreeQ[System`LaplaceTransformDump`t, System`LaplaceTransformDump`s]}}

In[10]:= (* an upvalue for Υ *)
       #[Υ]& /@ {OwnValues, DownValues, UpValues, SubValues}

Out[11]= {{}, {}, {HoldPattern[Υ'] :> (Υ /: Υ' =.; Υ'[0] := ΥD; Υ')}, {}}
```

Evaluating Υ' instead Υ'[0] does, of course, generate a definition for Derivative that is not automatically removed.

10. Perturbation Theory

a) Physicists may think of second-order time-independent perturbation theory [669], [568]

$$c_{i2} = \sum_{\substack{j=1 \\ j \neq i}}^{\infty} \frac{\left| \int_{-l/2}^{l/2} y_i(x, \alpha = 0) \, x \, y_j(x, \alpha = 0) \right|^2}{\lambda_i(\alpha = 0) - \lambda_j(\alpha = 0)}$$

but we have an exercise for this chapter on special functions; so we do not use the above formula. Such an approach is difficult to generalize to eigenvalue problems with continuous spectra (the resulting integrals over the continuum states require some care (see [1251], [331], and [666]), as treated in Part b) of this exercise [1225], [1052], or to higher order corrections in α. (A general overview of perturbation theory for eigenvalue problems can be found in [1154].) We begin with the exact solution of the differential equation for arbitrary α.

```
In[1]:= DSolve[{-y''[x] + α x y[x] == λ y[x]}, y[x], x, GeneratedParameters -> a]
```

$$\text{Out[1]= } \left\{\left\{y[x] \to a[1] \, \text{AiryAi}\left[\frac{x\alpha - \lambda}{\alpha^{2/3}}\right] + a[2] \, \text{AiryBi}\left[\frac{x\alpha - \lambda}{\alpha^{2/3}}\right]\right\}\right\}$$

The coefficients a[1] and a[2] are determined from the boundary condition at $x = \pm l/2$.

$$a_2 \, \text{Ai}\left(\frac{-\lambda + \frac{\alpha l}{2}}{\alpha^{2/3}}\right) + a_1 \, \text{Bi}\left(\frac{-\lambda + \frac{\alpha l}{2}}{\alpha^{2/3}}\right) = 0$$

$$a_2 \, \text{Ai}\left(\frac{-\lambda - \frac{\alpha l}{2}}{\alpha^{2/3}}\right) + a_1 \, \text{Bi}\left(\frac{-\lambda - \frac{\alpha l}{2}}{\alpha^{2/3}}\right) = 0.$$

This linear system of equations in a_1, a_2 has a nontrivial solution only if the associated coefficient determinant vanishes.

```
In[2]:= s1[x_] = AiryAi[(-λ + α x)/α^(2/3)];
       s2[x_] = AiryBi[(-λ + α x)/α^(2/3)];
In[4]:= deter = Det[{{s1[-1/2], s2[-1/2]}, {s1[1/2], s2[1/2]}}]
```

$$\text{Out[4]= } -\text{AiryAi}\left[\frac{\frac{l\alpha}{2} - \lambda}{\alpha^{2/3}}\right] \text{AiryBi}\left[\frac{-\frac{l\alpha}{2} - \lambda}{\alpha^{2/3}}\right] + \text{AiryAi}\left[\frac{-\frac{l\alpha}{2} - \lambda}{\alpha^{2/3}}\right] \text{AiryBi}\left[\frac{\frac{l\alpha}{2} - \lambda}{\alpha^{2/3}}\right]$$

At this point, it remains "only" to develop a series expansion for the solution $\lambda_i(\alpha)$ of this equation. We still do not need an explicit solution, because

$$\lambda_i(\alpha) \approx \lambda_i(\alpha = 0) + c_{1i}\,\alpha + c_{2i}\,\alpha^2 = \approx \lambda_i(\alpha = 0) + \left.\frac{\partial \lambda_i(\alpha)}{\partial \alpha}\right|_{\alpha=0} \alpha + \frac{1}{2!}\left.\frac{\partial^2 \lambda_i(\alpha)}{\partial \alpha^2}\right|_{\alpha=0} \alpha^2$$

and so the first and the second derivative of $\lambda_i(\alpha)$ with respect to α can be obtained immediately by implicit differentation of the exact eigenvalue equation.

In[5]:= `Solve[D[lhsEigenvalueEquation[λ[α], α] == 0, {α, 2}], λ''[α]]`

Out[5]= $\{\{\lambda''[\alpha] \to$

$(-\text{lhsEigenvalueEquation}^{(0,2)}[\lambda[\alpha], \alpha] - 2\,\lambda'[\alpha]\,\text{lhsEigenvalueEquation}^{(1,1)}[\lambda[\alpha], \alpha] -$

$\lambda'[\alpha]^2\,\text{lhsEigenvalueEquation}^{(2,0)}[\lambda[\alpha], \alpha])\,/\,\text{lhsEigenvalueEquation}^{(1,0)}[\lambda[\alpha], \alpha]\}\}$

All of the derivatives involved can be found directly by differentiation of the remaining `lhsEigenvalueEqua`‹ `tion[λ[α], α]` (= deter) or, alternatively, from the following result.

In[6]:= `Solve[D[lhsEigenvalueEquation[λ[α], α] == 0, α], λ'[α]]`

Out[6]= $\left\{\left\{\lambda'[\alpha] \to -\dfrac{\text{lhsEigenvalueEquation}^{(0,1)}[\lambda[\alpha], \alpha]}{\text{lhsEigenvalueEquation}^{(1,0)}[\lambda[\alpha], \alpha]}\right\}\right\}$

In[7]:= `%% /. %`

Out[7]= $\Big\{\Big\{\lambda''[\alpha] \to \Big(-\text{lhsEigenvalueEquation}^{(0,2)}[\lambda[\alpha], \alpha] +$

$\dfrac{2\,\text{lhsEigenvalueEquation}^{(0,1)}[\lambda[\alpha], \alpha]\,\text{lhsEigenvalueEquation}^{(1,1)}[\lambda[\alpha], \alpha]}{\text{lhsEigenvalueEquation}^{(1,0)}[\lambda[\alpha], \alpha]} -$

$\dfrac{\text{lhsEigenvalueEquation}^{(0,1)}[\lambda[\alpha], \alpha]^2\,\text{lhsEigenvalueEquation}^{(2,0)}[\lambda[\alpha], \alpha]}{\text{lhsEigenvalueEquation}^{(1,0)}[\lambda[\alpha], \alpha]^2}\Big)\Big/$

$\text{lhsEigenvalueEquation}^{(1,0)}[\lambda[\alpha], \alpha]\Big\}\Big\}$

This takes care of preparations that do not involve _Mathematica_, and in principle one might expect _Mathematica_ to finish the job. However, this is unfortunately not the case.

In[8]:= `% /.` (* for better readability *)
`Derivative[i_, j_][lhsEigenvalueEquation][λ[α], α] :>`
`(HoldForm[D[deter, {λ, ii}, {α, jj}]] /. {ii -> i, jj -> j})`

Out[8]= $\Big\{\Big\{\lambda''[\alpha] \to$

$\dfrac{-\partial_{\{\lambda,0\},\{\alpha,2\}}\,\text{deter} + \dfrac{2\,\partial_{\{\lambda,0\},\{\alpha,1\}}\text{deter}\,\partial_{\{\lambda,1\},\{\alpha,1\}}\text{deter}}{\partial_{\{\lambda,1\},\{\alpha,0\}}\text{deter}} - \dfrac{(\partial_{\{\lambda,0\},\{\alpha,1\}}\text{deter})^2\,\partial_{\{\lambda,2\},\{\alpha,0\}}\text{deter}}{(\partial_{\{\lambda,1\},\{\alpha,0\}}\text{deter})^2}}{\partial_{\{\lambda,1\},\{\alpha,0\}}\,\text{deter}}\Big\}\Big\}$

First, the resulting expression is very large, and thus further manipulation of it (if it makes sense at all) would take a very long time.

In[9]:= `(toBeSeried = %[[1, 1, 1, 2]] // ReleaseHold);`
`LeafCount[toBeSeried]`

Out[10]= `2808`

Moreover, the built-in limit cannot work with this horrendous expression. `Limit[toBeSeried, α->0]` would never return a result. Using the package `Calculus`Limit`` does not do any good either.

This means that we have to do some further work on the function manually. We go back to deter, and we look more carefully at its behavior for $\alpha \to 0$. Because the $\lambda_i(\alpha = 0)$ are finite and positive, all variables in the Airy function go to $-\infty$ as $\alpha \to 0$. We now make use of the first two terms of the asymptotic formula for the two Airy functions.

In[11]:= `ord1 = 2;`

(* turn off messages *) `Off[Series::"esss"];`
`asymptoticRules =`
`{AiryAi[x_] -> Normal[Series[AiryAi[x], {x, -Infinity, ord1}]],`
`AiryBi[x_] -> Normal[Series[AiryBi[x], {x, -Infinity, ord1}]]}`

Out[13]= $\Big\{\text{AiryAi}[x\_] \to \dfrac{-\frac{5}{48}\,i\,(\frac{1}{x})^{3/2}\,\text{Cos}[\frac{\pi}{4} + \frac{2}{3}\,(-x)^{3/2}] + \text{Sin}[\frac{\pi}{4} + \frac{2}{3}\,(-x)^{3/2}]}{\sqrt{\pi}\,(-x)^{1/4}},$

$\text{AiryBi}[x\_] \to \dfrac{\text{Cos}[\frac{\pi}{4} + \frac{2}{3}\,(-x)^{3/2}] + \frac{5}{48}\,i\,(\frac{1}{x})^{3/2}\,\text{Sin}[\frac{\pi}{4} + \frac{2}{3}\,(-x)^{3/2}]}{\sqrt{\pi}\,(-x)^{1/4}}\Big\}$

```
In[14]:= asymptoticRules =   (* the explicit form of the expansion *)
    Block[{arg = Pi/4 + 2/3 (-x)^(3/2), α = (-x)^(1/4), β = (-1/x)^(3/2)},
        {AiryAi[x_] -> (-5/48 β Cos[arg] +          Sin[arg])/(Sqrt[Pi] α),
         AiryBi[x_] -> (          Cos[arg] + 5/48 β Sin[arg])/(Sqrt[Pi] α)}];
```

In the interest of readability, we suppress some output in the following, and we name the intermediate results. (For this kind of calculation, protecting the variables is usually not necessary; we do it interactively just once.) Substituting `asymptotic` Rules in `deter` gives the following expression.

```
In[15]:= deterAsymp1 = deter /. asymptoticRules;
```

Because this expression is zero, we should eliminate the denominator.

```
In[16]:= (deterAsymp2 = Numerator[deterAsymp1 // Together]) // InputForm
```

```
Out[16]//InputForm= -120*l*α^(5/3)*Sqrt[-(α^(2/3)/(l*α - 2*λ))]*Cos[Pi/4 + (2*(-((-(l*α)/2 -
    λ)/α^(2/3)))^(3/2))/3]*
    Cos[Pi/4 + (2*(-(((l*α)/2 - λ)/α^(2/3)))^(3/2))/3] -
    240*α^(2/3)*Sqrt[-(α^(2/3)/(l*α - 2*λ))]*λ*
    Cos[Pi/4 + (2*(-((-(l*α)/2 - λ)/α^(2/3)))^(3/2))/3]*Cos[Pi/4 + (2*(-(((l*α)/2 -
    λ)/α^(2/3)))^(3/2))/3] -
    120*l*α^(5/3)*Sqrt[α^(2/3)/(l*α + 2*λ)]*Cos[Pi/4 + (2*(-((-(l*α)/2 -
    λ)/α^(2/3)))^(3/2))/3]*
    Cos[Pi/4 + (2*(-(((l*α)/2 - λ)/α^(2/3)))^(3/2))/3] +
    240*α^(2/3)*λ*Sqrt[α^(2/3)/(l*α + 2*λ)]*
    Cos[Pi/4 + (2*(-((-(l*α)/2 - λ)/α^(2/3)))^(3/2))/3]*Cos[Pi/4 + (2*(-(((l*α)/2 -
    λ)/α^(2/3)))^(3/2))/3] +
    288*Sqrt[2]*l^2*α^2*Cos[Pi/4 + (2*(-(((l*α)/2 - λ)/α^(2/3)))^(3/2))/3]*
    Sin[Pi/4 + (2*(-((-(l*α)/2 - λ)/α^(2/3)))^(3/2))/3] -
    1152*Sqrt[2]*λ^2*Cos[Pi/4 + (2*(-((-(l*α)/2 - λ)/α^(2/3)))^(3/2))/3]*
    Sin[Pi/4 + (2*(-(((l*α)/2 - λ)/α^(2/3)))^(3/2))/3] -
    25*Sqrt[2]*α^(4/3)*Sqrt[-(α^(2/3)/(l*α - 2*λ))]*
    Sqrt[α^(2/3)/(l*α + 2*λ)]*Cos[Pi/4 + (2*(-(((l*α)/2 - λ)/α^(2/3)))^(3/2))/3]*
    Sin[Pi/4 + (2*(-((-(l*α)/2 - λ)/α^(2/3)))^(3/2))/3] -
    288*Sqrt[2]*l^2*α^2*Cos[Pi/4 + (2*(-((-(l*α)/2 - λ)/α^(2/3)))^(3/2))/3]*
    Sin[Pi/4 + (2*(-(((l*α)/2 - λ)/α^(2/3)))^(3/2))/3] +
    1152*Sqrt[2]*λ^2*Cos[Pi/4 + (2*(-((-(l*α)/2 - λ)/α^(2/3)))^(3/2))/3]*
    Sin[Pi/4 + (2*(-(((l*α)/2 - λ)/α^(2/3)))^(3/2))/3] +
    25*Sqrt[2]*α^(4/3)*Sqrt[-(α^(2/3)/(l*α - 2*λ))]*
    Sqrt[α^(2/3)/(l*α + 2*λ)]*Cos[Pi/4 + (2*(-((-(l*α)/2 - λ)/α^(2/3)))^(3/2))/3]*
    Sin[Pi/4 + (2*(-(((l*α)/2 - λ)/α^(2/3)))^(3/2))/3] -
    120*l*α^(5/3)*Sqrt[-(α^(2/3)/(l*α - 2*λ))]*
    Sin[Pi/4 + (2*(-((-(l*α)/2 - λ)/α^(2/3)))^(3/2))/3]*Sin[Pi/4 + (2*(-(((l*α)/2 -
    λ)/α^(2/3)))^(3/2))/3] -
    240*α^(2/3)*Sqrt[-(α^(2/3)/(l*α - 2*λ))]*λ*Sin[Pi/4 + (2*(-((-(l*α)/2 -
    λ)/α^(2/3)))^(3/2))/3]*
    Sin[Pi/4 + (2*(-(((l*α)/2 - λ)/α^(2/3)))^(3/2))/3] -
    120*l*α^(5/3)*Sqrt[α^(2/3)/(l*α + 2*λ)]*
    Sin[Pi/4 + (2*(-((-(l*α)/2 - λ)/α^(2/3)))^(3/2))/3]*Sin[Pi/4 + (2*(-(((l*α)/2 -
    λ)/α^(2/3)))^(3/2))/3] +
    240*α^(2/3)*λ*Sqrt[α^(2/3)/(l*α + 2*λ)]*Sin[Pi/4 + (2*(-((-(l*α)/2 -
    λ)/α^(2/3)))^(3/2))/3]*
    Sin[Pi/4 + (2*(-(((l*α)/2 - λ)/α^(2/3)))^(3/2))/3]
```

The products of the angle functions oscillate infinitely in the limiting case $\alpha \to 0$. Hence, we rewrite the angle functions with a single variable to extract a nonoscillating part.

```
In[17]:= (deterAsymp3 = TrigReduce /@ deterAsymp2);
```

We again expand the arguments of the resulting angle functions in a series.

```
In[18]:= ord2 = 2;
    (deterAsymp4 = deterAsymp3 /.
        {Cos[x_] :> Cos[Normal[Series[x, {α, 0, ord2}]]]},
```

```
Sin[x_] :> Sin[Normal[Series[x, {α, 0, ord2}]]]});
Take[deterAsymp4, 3]
```

Out[20]= $144 \left(\sqrt{2} \; 1^2 \; \alpha^2 \; \text{Cos} \left[\frac{1^2 \; \alpha}{8 \sqrt{\lambda}} + \frac{4 \; \lambda^{3/2}}{3 \; \alpha} \right] - \sqrt{2} \; 1^2 \; \alpha^2 \; \text{Sin} \left[\frac{1^3 \; \alpha^2}{96 \; \lambda^{3/2}} - 1 \sqrt{\lambda} \right] \right) -$

$\quad 144 \left(\sqrt{2} \; 1^2 \; \alpha^2 \; \text{Cos} \left[\frac{1^2 \; \alpha}{8 \sqrt{\lambda}} + \frac{4 \; \lambda^{3/2}}{3 \; \alpha} \right] + \sqrt{2} \; 1^2 \; \alpha^2 \; \text{Sin} \left[\frac{1^3 \; \alpha^2}{96 \; \lambda^{3/2}} - 1 \sqrt{\lambda} \right] \right) -$

$\quad 576 \left(\sqrt{2} \; \lambda^2 \; \text{Cos} \left[\frac{1^2 \; \alpha}{8 \sqrt{\lambda}} + \frac{4 \; \lambda^{3/2}}{3 \; \alpha} \right] - \sqrt{2} \; \lambda^2 \; \text{Sin} \left[\frac{1^3 \; \alpha^2}{96 \; \lambda^{3/2}} - 1 \sqrt{\lambda} \right] \right)$

Next, we simplify the roots (without loss of generality, we assume $\alpha > 0$).

```
In[21]:= rootRules =
        {Sqrt[α^(2/3)/(2λ + α 1)] -> α^(1/3)(2λ + α 1)^(-1/2),
         Sqrt[(-2 α^(2/3))/(-2λ + α 1)] -> Sqrt[2] α^(1/3)(2λ - α 1)^(-1/2)};
```

```
In[22]:= (deterAsymp5 = deterAsymp4 //. rootRules // Expand)
```

Out[22]= $-120 \; 1 \; \alpha^{5/3} \sqrt{-\frac{\alpha^{2/3}}{1 \; \alpha - 2 \; \lambda}} \; \text{Cos} \left[\frac{1^3 \; \alpha^2}{96 \; \lambda^{3/2}} - 1 \sqrt{\lambda} \right] - 240 \; \alpha^{2/3} \sqrt{-\frac{\alpha^{2/3}}{1 \; \alpha - 2 \; \lambda}} \; \lambda \; \text{Cos} \left[\frac{1^3 \; \alpha^2}{96 \; \lambda^{3/2}} - 1 \sqrt{\lambda} \right] -$

$\quad \frac{120 \; 1 \; \alpha^2 \; \text{Cos} \left[\frac{1^3 \; \alpha^2}{96 \; \lambda^{3/2}} - 1 \sqrt{\lambda} \right]}{\sqrt{1 \; \alpha + 2 \; \lambda}} + \frac{240 \; \alpha \; \lambda \; \text{Cos} \left[\frac{1^3 \; \alpha^2}{96 \; \lambda^{3/2}} - 1 \sqrt{\lambda} \right]}{\sqrt{1 \; \alpha + 2 \; \lambda}} - 288 \sqrt{2} \; 1^2 \; \alpha^2 \; \text{Sin} \left[\frac{1^3 \; \alpha^2}{96 \; \lambda^{3/2}} - 1 \sqrt{\lambda} \right] +$

$\quad 1152 \sqrt{2} \; \lambda^2 \; \text{Sin} \left[\frac{1^3 \; \alpha^2}{96 \; \lambda^{3/2}} - 1 \sqrt{\lambda} \right] + \frac{25 \sqrt{2} \; \alpha^2 \; \text{Sin} \left[\frac{1^3 \; \alpha^2}{96 \; \lambda^{3/2}} - 1 \sqrt{\lambda} \right]}{\sqrt{-1 \; \alpha + 2 \; \lambda} \; \sqrt{1 \; \alpha + 2 \; \lambda}}$

`additionalRootRules` calculates series expansions for $(\lambda \pm \alpha \, l)^{1/2}$.

```
In[23]:= ord3 = 2;
        additionalRootRules =
        {(2 λ + α 1)^(-1/2) -> Normal[Series[(2 λ + α 1)^(-1/2), {α, 0, ord3}]],
         (2 λ - α 1)^(-1/2) -> Normal[Series[(2 λ - α 1)^(-1/2), {α, 0, ord3}]]}
```

Out[24]= $\left\{ \frac{1}{\sqrt{1 \; \alpha + 2 \; \lambda}} \to \frac{3 \; 1^2 \; \alpha^2}{32 \sqrt{2} \; \lambda^{5/2}} - \frac{1 \; \alpha}{4 \sqrt{2} \; \lambda^{3/2}} + \frac{1}{\sqrt{2} \; \sqrt{\lambda}}, \right.$

$\quad \left. \frac{1}{\sqrt{-1 \; \alpha + 2 \; \lambda}} \to \frac{3 \; 1^2 \; \alpha^2}{32 \sqrt{2} \; \lambda^{5/2}} + \frac{1 \; \alpha}{4 \sqrt{2} \; \lambda^{3/2}} + \frac{1}{\sqrt{2} \; \sqrt{\lambda}} \right\}$

```
In[25]:= deterAsymp6 = Expand[deterAsymp5 //. additionalRootRules]
```

Out[25]= $-120 \; 1 \; \alpha^{5/3} \sqrt{-\frac{\alpha^{2/3}}{1 \; \alpha - 2 \; \lambda}} \; \text{Cos} \left[\frac{1^3 \; \alpha^2}{96 \; \lambda^{3/2}} - 1 \sqrt{\lambda} \right] - \frac{45 \; 1^3 \; \alpha^4 \; \text{Cos} \left[\frac{1^3 \; \alpha^2}{96 \; \lambda^{3/2}} - 1 \sqrt{\lambda} \right]}{4 \sqrt{2} \; \lambda^{5/2}} +$

$\quad \frac{105 \; 1^2 \; \alpha^3 \; \text{Cos} \left[\frac{1^3 \; \alpha^2}{96 \; \lambda^{3/2}} - 1 \sqrt{\lambda} \right]}{2 \sqrt{2} \; \lambda^{3/2}} - \frac{90 \sqrt{2} \; 1 \; \alpha^2 \; \text{Cos} \left[\frac{1^3 \; \alpha^2}{96 \; \lambda^{3/2}} - 1 \sqrt{\lambda} \right]}{\sqrt{\lambda}} +$

$\quad 120 \sqrt{2} \; \alpha \sqrt{\lambda} \; \text{Cos} \left[\frac{1^3 \; \alpha^2}{96 \; \lambda^{3/2}} - 1 \sqrt{\lambda} \right] - 240 \; \alpha^{2/3} \sqrt{-\frac{\alpha^{2/3}}{1 \; \alpha - 2 \; \lambda}} \; \lambda \; \text{Cos} \left[\frac{1^3 \; \alpha^2}{96 \; \lambda^{3/2}} - 1 \sqrt{\lambda} \right] -$

$\quad 288 \sqrt{2} \; 1^2 \; \alpha^2 \; \text{Sin} \left[\frac{1^3 \; \alpha^2}{96 \; \lambda^{3/2}} - 1 \sqrt{\lambda} \right] + \frac{225 \; 1^4 \; \alpha^6 \; \text{Sin} \left[\frac{1^3 \; \alpha^2}{96 \; \lambda^{3/2}} - 1 \sqrt{\lambda} \right]}{1024 \sqrt{2} \; \lambda^5} +$

$\quad \frac{25 \; 1^2 \; \alpha^4 \; \text{Sin} \left[\frac{1^3 \; \alpha^2}{96 \; \lambda^{3/2}} - 1 \sqrt{\lambda} \right]}{8 \sqrt{2} \; \lambda^3} + \frac{25 \; \alpha^2 \; \text{Sin} \left[\frac{1^3 \; \alpha^2}{96 \; \lambda^{3/2}} - 1 \sqrt{\lambda} \right]}{\sqrt{2} \; \lambda} + 1152 \sqrt{2} \; \lambda^2 \; \text{Sin} \left[\frac{1^3 \; \alpha^2}{96 \; \lambda^{3/2}} - 1 \sqrt{\lambda} \right]$

This is our simplified result for `deter`. First, we now consider the limiting case $\alpha \to 0$.

```
In[26]:= Limit[deterAsymp6, α -> 0, Direction -> -1]
```

Out[26]= $-1152 \sqrt{2} \; \lambda^2 \; \text{Sin}[1 \sqrt{\lambda}]$

Thus, $\sqrt{\lambda} \, l = j \pi$, $j = 1, 2, \ldots$, which can also be seen more or less directly from the differential equation for $\alpha = 0$. Now, we turn to the computation of c_{i1}, c_{i2}. Because the first derivative of `deterAsymp6` with respect to α vanishes, we have $c_{i1} = 0$ (this was to be expected from the symmetry of the problem), and the computation of c_{i2} simplifies.

```
In[27]:= Simplify[Normal[Series[D[deterAsymp6, α], {α, 0, 0}]]] // PowerExpand
```

Out[27]= 0

```
In[28]:= PowerExpand[Normal[
         Series[-1/2 (D[deterAsymp6, {α, 2}]/D[deterAsymp6, λ]), {α, 0, 0}]]]
```

$$Out[28]= -\frac{-\frac{360\sqrt{2}\,1\,Cos[1\sqrt{\lambda}]}{\sqrt{\lambda}}+24\sqrt{2}\,1^3\sqrt{\lambda}\,Cos[1\sqrt{\lambda}]+576\sqrt{2}\,1^2\,Sin[1\sqrt{\lambda}]-\frac{25\sqrt{2}\,sin[1\sqrt{\lambda}]}{\lambda}}{2\,(-576\sqrt{2}\,1\,\lambda^{3/2}\,Cos[1\sqrt{\lambda}]-2304\sqrt{2}\,\lambda\,Sin[1\sqrt{\lambda}])}$$

We now take account of the eigenvalue equation for $\alpha = 0$ [47], [1174], [398], [485], [1244].

```
In[29]:= % //. {Sin[1 Sqrt[λ]] -> 0} // Simplify
```

$$Out[29]= \frac{-15+1^2\,\lambda}{48\,\lambda^2}$$

We finally obtain the following expression for the eigenvalue of $\alpha = 0$.

```
In[30]:= % /. λ -> (i Pi/1)^2
```

$$Out[30]= \frac{1^4\,(-15+i^2\,\pi^2)}{48\,i^4\,\pi^4}$$

We now examine the quality of the approximation for the three smallest eigenvalues in comparison with the exact results (computed numerically). Let $l = 1$.

```
In[31]:= Show[GraphicsArray[Table[
         (* the approximations *)
         Plot[i^2 Pi^2 + (-15 + i^2 Pi^2)/(48 i^4 Pi^4) α^2, {α, 0, 101},
             Epilog -> {PointSize[0.03], (* the exact values *)
             Point /@ Table[{α,
             λ /. FindRoot[(deter /. 1 -> 1) == 0, {λ, i^2 N[Pi^2]}]},
                            {α, 0.01, 100.01, 10.0}]},
             PlotRange -> All, DisplayFunction -> Identity,
             AxesLabel -> (StyleForm[#, "Input", FontSize -> 5]& /@ {α, λ}),
             PlotLabel -> StyleForm["i = " <> ToString[i], FontSize -> 6]],
             {i, 1, 3}], GraphicsSpacing -> 0]]
```

Let us compare the result $l^4\,(i^2\,\pi^2-15)/(48\,i^4\,\pi^4)$ with the perturbation theory result obtained from the sum given above. This is a simplified form of the matrix element.

```
In[32]:= Clear[a, 1, m, n];

         int1 = Simplify[2 Integrate[
                   Expand[Sin[n Pi x/1] (a x - a 1/2) Sin[m Pi x/1]],
                   {x, 0, 1}], Element[{n, m}, Integers]]
```

$$Out[33]= \frac{4\,(-1+(-1)^{m+n})\,a\,1^2\,m\,n}{(m^2-n^2)^2\,\pi^2}$$

ev is the eigenvalue of the unperturbed problem.

```
In[34]:= ev[n_] := (n^2 Pi^2)/1^2
```

term[n, m] are the summands in the infinite sum.

```
In[35]:= term[n_, m_] = int1^2/(ev[n] - ev[m]) /. {a -> 1, 1 -> 1}
```

$$Out[35]= \frac{16\,(-1+(-1)^{m+n})^2\,m^2\,n^2}{(m^2-n^2)^4\,\pi^4\,(-m^2\,\pi^2+n^2\,\pi^2)}$$

To evaluate the infinite sum symbolically, we split the sum part in the terms less than and greater than n.

```
In[36]:= sum[n_] := Together[Sum[Evaluate[term[n, m]], {m, n - 1}] +
                       (Sum[#, {m, n + 1, Infinity}]& /@ Apart[term[n, m]])]
```

The classical perturbation theory result agrees with the one from above.

In[37]:= `f[i_] := Cancel[((-15 + i^2 Pi^2))/(48 i^4 Pi^4)];`

In[38]:= `Table[{sum[i], f[i]}, {i, 3}]`

Out[38]= $\left\{\left\{\frac{-15+\pi^2}{48\,\pi^4},\ \frac{-15+\pi^2}{48\,\pi^4}\right\},\ \left\{\frac{-15+4\,\pi^2}{768\,\pi^4},\ \frac{-15+4\,\pi^2}{768\,\pi^4}\right\},\ \left\{\frac{-5+3\,\pi^2}{1296\,\pi^4},\ \frac{-5+3\,\pi^2}{1296\,\pi^4}\right\}\right\}$

By increasing ord1, ord2, and ord3, and the corresponding expressions for $\partial^{2n}\lambda_i(\alpha)/\partial\alpha^{2n}|_{\alpha=0}\,\alpha^{2n}$ and applying repeated implicit differentiation of the exact eigenvalue equation, it is straightforward to find a correction term of higher order in α. (The odd order terms all vanish identically for the example treated here because of symmetry.)

b) Here, we follow the same spirit as in Part a). This means we calculate an implicit equation for the eigenvalue and then get the coefficient c_2 by differentiation with respect to α and λ. This ignores exponentially small lifetime effects [197], [35], [377], [981].

These are the three solutions of the governing differential equation in the three regions $z < -l/2$, $-l/2 < z < l/2$, and $z > l/2$.

In[1]:= `(* -1/2 < z < 1/2 *)`
`DSolve[-y''[z] + α z y[z] == λ y[z], y[z], z,`
`                GeneratedParameters -> B][[1, 1, 2]]`

Out[2]= `AiryAi`$\left[\frac{z\,\alpha-\lambda}{\alpha^{2/3}}\right]$`B[1]` + `AiryBi`$\left[\frac{z\,\alpha-\lambda}{\alpha^{2/3}}\right]$`B[2]`

We rewrite the solution in a form slightly easier to use later.

In[3]:= `solMiddle = B[1] AiryBi[α^(1/3) (z - λ/α)] +`
`               B[2] AiryAi[α^(1/3) (z - λ/α)];`

In[4]:= `(* z > 1/2 *)`
`DSolve[-y''[z] + α z y[z] + v0 y[z] == λ y[z], y[z], z,`
`               GeneratedParameters -> C][[1, 1, 2]] /.`
`               (* use at +Infinity vanishing solution only *) _AiryBi -> 0`

Out[5]= `AiryAi`$\left[\frac{v0+z\,\alpha-\lambda}{\alpha^{2/3}}\right]$`C[1]`

In[6]:= `solRight = C[2] AiryAi[α^(1/3) (z + (v0 - λ)/α)];`

In[7]:= `(* z < -1/2 *)`
`solLeft = DSolve[-y''[z] + α z y[z] + v0 y[z] == λ y[z], y[z], z,`
`                GeneratedParameters -> A][[1, 1, 2]]`

Out[8]= `A[1] AiryAi`$\left[\frac{v0+z\,\alpha-\lambda}{\alpha^{2/3}}\right]$ + `A[2] AiryBi`$\left[\frac{v0+z\,\alpha-\lambda}{\alpha^{2/3}}\right]$

Taking into account the asymptotics of the Airy functions, we achieve a purely outgoing wave (Sommerfeld's radiation condition) in the far left by using the following linear combination of AiryAi and AiryBi (see [458], [296], [914], [19], [35], [270], [459], [1041], [1111], [610]).

In[9]:= `solLeft = A[1] ( AiryBi[α^(1/3) ((-λ + v0)/α + z)] -`
`               I AiryAi[α^(1/3) ((-λ + v0)/α + z)])`

Out[9]= `A[1]`$\left(-\mathrm{i}\ \mathrm{AiryAi}\left[\alpha^{1/3}\left(z+\frac{v0-\lambda}{\alpha}\right)\right]+\mathrm{AiryBi}\left[\alpha^{1/3}\left(z+\frac{v0-\lambda}{\alpha}\right)\right]\right)$

Integrating the differential equation under consideration over an infinitesimal neighborhood around $z = -l/2$ or $z = l/2$, we obtain the condition that the function $y(z)$ as well its first derivative $y'(z)$ must be continuous at $z = -l/2$ and $z = l/2$. So we have the following equations connecting the three solutions solLeft, solMiddle, and solRight at $z = -l/2$ and $z = l/2$.

In[10]:= `eq1 = (solLeft    /. z -> -1/2) - (solMiddle /. z -> -1/2)`

Out[10]= `A[1]`$\left(-\mathrm{i}\ \mathrm{AiryAi}\left[\alpha^{1/3}\left(-\frac{1}{2}+\frac{v0-\lambda}{\alpha}\right)\right]+\mathrm{AiryBi}\left[\alpha^{1/3}\left(-\frac{1}{2}+\frac{v0-\lambda}{\alpha}\right)\right]\right)$ -
$\mathrm{AiryBi}\left[\alpha^{1/3}\left(-\frac{1}{2}-\frac{\lambda}{\alpha}\right)\right]$`B[1]` - $\mathrm{AiryAi}\left[\alpha^{1/3}\left(-\frac{1}{2}-\frac{\lambda}{\alpha}\right)\right]$`B[2]`

In[11]:= `eq2 = (solMiddle /. z -> +1/2) - (solRight  /. z -> +1/2)`

Out[11]= $\mathrm{AiryBi}\left[\alpha^{1/3}\left(\frac{1}{2}-\frac{\lambda}{\alpha}\right)\right]$`B[1]` + $\mathrm{AiryAi}\left[\alpha^{1/3}\left(\frac{1}{2}-\frac{\lambda}{\alpha}\right)\right]$`B[2]` - $\mathrm{AiryAi}\left[\alpha^{1/3}\left(\frac{1}{2}+\frac{v0-\lambda}{\alpha}\right)\right]$`C[2]`

In[12]:= `eq3 = (D[solLeft,  z] /. z -> -1/2) - (D[solMiddle, z] /. z -> -1/2)`

Out[12]= A[1] $\left(-i\,\alpha^{1/3}\,\text{AiryAiPrime}\left[\alpha^{1/3}\left(-\frac{1}{2}+\frac{v0-\lambda}{\alpha}\right)\right]+\alpha^{1/3}\,\text{AiryBiPrime}\left[\alpha^{1/3}\left(-\frac{1}{2}+\frac{v0-\lambda}{\alpha}\right)\right]\right)-$

$\quad\alpha^{1/3}\,\text{AiryBiPrime}\left[\alpha^{1/3}\left(-\frac{1}{2}-\frac{\lambda}{\alpha}\right)\right]\,B[1]-\alpha^{1/3}\,\text{AiryAiPrime}\left[\alpha^{1/3}\left(-\frac{1}{2}-\frac{\lambda}{\alpha}\right)\right]\,B[2]$

In[13]:= eq4 = (D[solRight, z] /. z -> +1/2) - (D[solMiddle, z] /. z -> +1/2)

Out[13]= $-\alpha^{1/3}\,\text{AiryBiPrime}\left[\alpha^{1/3}\left(\frac{1}{2}-\frac{\lambda}{\alpha}\right)\right]\,B[1]-$

$\quad\alpha^{1/3}\,\text{AiryAiPrime}\left[\alpha^{1/3}\left(\frac{1}{2}-\frac{\lambda}{\alpha}\right)\right]\,B[2]+\alpha^{1/3}\,\text{AiryAiPrime}\left[\alpha^{1/3}\left(\frac{1}{2}+\frac{v0-\lambda}{\alpha}\right)\right]\,C[2]$

These are four equations in the four unknown coefficients of the various Airy functions. These four equations have a nontrivial solution if and only if the corresponding coefficient determinant vanishes.

In[14]:= (* we extract the coefficient in such a complicated manner because the
          numbering 1 or 2 of the constants might depend on the Mathematica version *)
       abc = Cases[{solLeft, solMiddle, solRight}, _A | _B | _C,
                {0, Infinity}] // Union

Out[15]= {A[1], B[1], B[2], C[2]}

In[16]:= mat = Outer[Coefficient, Expand /@ {eq1, eq2, eq3, eq4}, abc]

Out[16]= $\left\{\left\{-i\,\text{AiryAi}\left[\alpha^{1/3}\left(-\frac{1}{2}+\frac{v0-\lambda}{\alpha}\right)\right]+\text{AiryBi}\left[\alpha^{1/3}\left(-\frac{1}{2}+\frac{v0-\lambda}{\alpha}\right)\right]\right.\right.,$

$\quad\left.-\text{AiryBi}\left[\alpha^{1/3}\left(-\frac{1}{2}-\frac{\lambda}{\alpha}\right)\right],\,-\text{AiryAi}\left[\alpha^{1/3}\left(-\frac{1}{2}-\frac{\lambda}{\alpha}\right)\right],\,0\right\},$

$\quad\left\{0,\,\text{AiryBi}\left[\alpha^{1/3}\left(\frac{1}{2}-\frac{\lambda}{\alpha}\right)\right],\,\text{AiryAi}\left[\alpha^{1/3}\left(\frac{1}{2}-\frac{\lambda}{\alpha}\right)\right],\,-\text{AiryAi}\left[\alpha^{1/3}\left(\frac{1}{2}+\frac{v0-\lambda}{\alpha}\right)\right]\right\},$

$\quad\left\{-i\,\alpha^{1/3}\,\text{AiryAiPrime}\left[\alpha^{1/3}\left(-\frac{1}{2}+\frac{v0-\lambda}{\alpha}\right)\right]+\alpha^{1/3}\,\text{AiryBiPrime}\left[\alpha^{1/3}\left(-\frac{1}{2}+\frac{v0-\lambda}{\alpha}\right)\right],\right.$

$\quad\left.-\alpha^{1/3}\,\text{AiryBiPrime}\left[\alpha^{1/3}\left(-\frac{1}{2}-\frac{\lambda}{\alpha}\right)\right],\,-\alpha^{1/3}\,\text{AiryAiPrime}\left[\alpha^{1/3}\left(-\frac{1}{2}-\frac{\lambda}{\alpha}\right)\right],\,0\right\},$

$\quad\left\{0,\,-\alpha^{1/3}\,\text{AiryBiPrime}\left[\alpha^{1/3}\left(\frac{1}{2}-\frac{\lambda}{\alpha}\right)\right],\,-\alpha^{1/3}\,\text{AiryAiPrime}\left[\alpha^{1/3}\left(\frac{1}{2}-\frac{\lambda}{\alpha}\right)\right],\right.$

$\quad\left.\left.\alpha^{1/3}\,\text{AiryAiPrime}\left[\alpha^{1/3}\left(\frac{1}{2}+\frac{v0-\lambda}{\alpha}\right)\right]\right\}\right\}\right\}$

In[17]:= deter = Factor[Det[mat] // PowerExpand]/ α^(2/3);

This coefficient determinant is the implicit equation for the eigenvalues $\lambda(\alpha)$. Because the problem under consideration has a continuous spectrum for λ in $(-\infty, \infty)$, the solutions of deter for a given α are, in general, complex valued. In the limit $\alpha \to 0$, some of them (at least one) become real. These real solutions are the solutions we are looking for, an expansion dependent of α.

In[18]:= Short[deter, 12]

Out[18]//Short= $i\,\text{AiryAi}\left[\alpha^{1/3}\left(\frac{1}{2}-\frac{\lambda}{\alpha}\right)\right]\,\text{AiryAiPrime}\left[\alpha^{1/3}\left(-\frac{1}{2}+\frac{v0-\lambda}{\alpha}\right)\right]$

$\quad\text{AiryAiPrime}\left[\alpha^{1/3}\left(\frac{1}{2}+\frac{v0-\lambda}{\alpha}\right)\right]\,\text{AiryBi}\left[\alpha^{1/3}\left(-\frac{1}{2}-\frac{\lambda}{\alpha}\right)\right]-$

$\quad i\,\text{AiryAi}\left[\alpha^{1/3}\left(\frac{1}{2}+\frac{v0-\lambda}{\alpha}\right)\right]\,\text{AiryAiPrime}\left[\alpha^{1/3}\left(-\frac{1}{2}+\frac{v0-\lambda}{\alpha}\right)\right]\,\text{AiryAiPrime}\left[\alpha^{1/3}\left(\frac{1}{2}-\frac{\lambda}{\alpha}\right)\right]$

$\quad\text{AiryBi}\left[\alpha^{1/3}\left(-\frac{1}{2}-\frac{\lambda}{\alpha}\right)\right]-i\,\text{AiryAi}\left[\alpha^{1/3}\left(-\frac{1}{2}-\frac{\lambda}{\alpha}\right)\right]\,\text{AiryAiPrime}\left[\alpha^{1/3}\left(-\frac{1}{2}+\frac{v0-\lambda}{\alpha}\right)\right]$

$\quad\text{AiryAiPrime}\left[\alpha^{1/3}\left(\frac{1}{2}+\frac{v0-\lambda}{\alpha}\right)\right]\,\text{AiryBi}\left[\alpha^{1/3}\left(\frac{1}{2}-\frac{\lambda}{\alpha}\right)\right]+\ll13\gg+$

$\quad i\,\text{AiryAi}\left[\alpha^{1/3}\left(\frac{1}{2}+\frac{v0-\lambda}{\alpha}\right)\right]\,\text{AiryAi}\left[\alpha^{1/3}\left(-\frac{1}{2}-\frac{\lambda}{\alpha}\right)\right]\,\text{AiryAiPrime}\left[\alpha^{1/3}\left(-\frac{1}{2}+\frac{v0-\lambda}{\alpha}\right)\right]$

$\quad\text{AiryBiPrime}\left[\alpha^{1/3}\left(\frac{1}{2}-\frac{\lambda}{\alpha}\right)\right]-i\,\text{AiryAi}\left[\alpha^{1/3}\left(-\frac{1}{2}+\frac{v0-\lambda}{\alpha}\right)\right]$

$\quad\text{AiryAi}\left[\alpha^{1/3}\left(\frac{1}{2}+\frac{v0-\lambda}{\alpha}\right)\right]\,\text{AiryAiPrime}\left[\alpha^{1/3}\left(-\frac{1}{2}-\frac{\lambda}{\alpha}\right)\right]\,\text{AiryBiPrime}\left[\alpha^{1/3}\left(\frac{1}{2}-\frac{\lambda}{\alpha}\right)\right]+$

$\quad\text{AiryAi}\left[\alpha^{1/3}\left(\frac{1}{2}+\frac{v0-\lambda}{\alpha}\right)\right]\,\text{AiryAiPrime}\left[\alpha^{1/3}\left(-\frac{1}{2}-\frac{\lambda}{\alpha}\right)\right]\,\text{AiryBi}\left[\alpha^{1/3}\left(-\frac{1}{2}+\frac{v0-\lambda}{\alpha}\right)\right]$

$\quad\text{AiryBiPrime}\left[\alpha^{1/3}\left(\frac{1}{2}-\frac{\lambda}{\alpha}\right)\right]-\text{AiryAi}\left[\alpha^{1/3}\left(\frac{1}{2}+\frac{v0-\lambda}{\alpha}\right)\right]\,\text{AiryAi}\left[\alpha^{1/3}\left(-\frac{1}{2}-\frac{\lambda}{\alpha}\right)\right]$

$\quad\text{AiryBiPrime}\left[\alpha^{1/3}\left(-\frac{1}{2}+\frac{v0-\lambda}{\alpha}\right)\right]\,\text{AiryBiPrime}\left[\alpha^{1/3}\left(\frac{1}{2}-\frac{\lambda}{\alpha}\right)\right]$

To manipulate the terms of deter individually and to avoid possible reorderings caused by the attribute Orderless, we change the head of deter from Plus to List.

```
In[19]:= deter = List @@ deter;
```

Similar to part a) of this solution, we now need the derivative of this implicit equation for $\lambda = \lambda(\alpha)$ for vanishing α. Again, a direct calculation via implicit differentiation of `deter` with respect to α fails. So we have to take care of the singularities by hand that come up in taking this limit to eliminate all oscillatory terms and to get smooth dependencies on α. First, we construct a replacement table of all Airy functions that appear in `deter` by their asymptotic terms. We write all terms of the asymptotic expansion in a form appropriate for later calculation by collecting them with respect to trigonometric or exponential functions.

```
In[20]:= (* turn off messages *) Off[Series::"esss"];

    asympAiryRules =
    Function[airy,
    airy -> Module[{asy, ll, z, coll, pm},
            (* is the expansion point +Infinity or -Infinity? *)
            pm = If[MemberQ[airy, v0, {-1}], +1, -1];
            asy = Expand[Normal[Series[Head[airy][z],
                {z, pm Infinity, If[pm == 1, 3, 2]}]]];
            (* collect with respect to Cos, Sin, and Exp functions *)
            ll = Union[Cases[asy, _Cos | _Sin | Power[E, _], {0, Infinity}]];
            coll = Plus @@ (Coefficient[asy, #] #& /@ ll);
            coll /. z -> airy[[1]]]] /@
            (* all Airy functions occurring in deter *)
                Union[Cases[deter, AiryAi[_] | AiryAiPrime[_] |
                                    AiryBi[_] | AiryBiPrime[_], {0, Infinity}]];

In[22]:= asympAiryRules = (* explicit form of the asymptotic expansion *)
    Module[Evaluate[Table[ToExpression["c" <> ToString[k]], {k, 70}]],
        c1 = α^(1/3); c2 = -1/2; c3 = 1/α; c4 = -λ; c5 = v0 + c4; c6 = c3 c5;
        c7 = c2 + c6; c8 = c1 c7; c9 = Sqrt[c8]; c10 = c8 c9; c11 = -2 c10/3;
        c12 = E^c11; c13 = Sqrt[Pi]; c14 = 1/c13; c15 = 1/c1; c16 = 1/c7;
        c17 = c15 c16; c18 = c17^(1/4); c19 = 1/2; c20 = c19 + c6; c21 = c1 c20;
        c22 = Sqrt[c21]; c23 = c21 c22; c24 = -2 c23/3; c25 = E^c24; c26 = 1/c20;
        c27 = c15 c26; c28 = c27^(1/4); c29 = -λ c3; c30 = c2 + c29; c31 = c1 c30;
        c32 = c1^2; c33 = 1/c32; c34 = 1/c30; c35 = -c15 c34; c36 = Sqrt[c35];
        c37 = c30^2; c38 = 1/c37; c39 = -c1 c30; c40 = c39^(1/4); c41 = c40^2;
        c42 = c40 c41; c43 = Pi/4; c44 = Sqrt[c39]; c45 = c39 c44; c46 = 2 c45/3;
        c47 = c43 + c46; c48 = Cos[c47]; c49 = Sin[c47]; c50 = c19 + c29;
        c51 = c1 c50; c52 = 1/c50; c53 = -c15 c52; c54 = Sqrt[c53]; c55 = c50^2;
        c56 = 1/c55; c57 = -c1 c50; c58 = c57^(1/4); c59 = c58^2; c60 = c58 c59;
        c61 = Sqrt[c57]; c62 = c57 c61; c63 = 2 c62/3; c64 = c43 + c63;
        c65 = Cos[c64]; c66 = Sin[c64]; c67 = 1/c18; c68 = c17 c18;
        c69 = 2 c10/3; c70 = E^c69; (* the 14 rules *)
        {AiryAi[c8] -> c12 c14 c18/2, AiryAi[c21] -> c14 c25 c28/2,
         AiryAi[c31] -> -5/48 c14 c33 c36 c38 c42 c48 - c14 c15 c34 c42 c49,
         AiryAi[c51] -> -5/48 c14 c33 c54 c56 c60 c65 - c14 c15 c52 c60 c66,
         AiryAiPrime[c8] -> c12 (-c14 c67/2 - 7 c14 c68/96),
         AiryAiPrime[c21] -> c25 (-c14/2/c28 - 7 c14 c27 c28/96),
         AiryAiPrime[c31] -> -c14 c40 c48 - 7/48 c14 c15 c34 c36 c40 c49,
         AiryAiPrime[c51] -> -c14 c58 c65 - 7/48 c14 c15 c52 c54 c58 c66,
         AiryBi[c8] -> c14 c18 c70,
         AiryBi[c31] -> -c14 c15 c34 c42 c48 + 5/48 c14 c33 c36 c38 c42 c49,
         AiryBi[c51] -> -c14 c15 c52 c60 c65 + 5/48 c14 c33 c54 c56 c60 c66,
         AiryBiPrime[c8] -> (c14 c67 - 7 c14 c68/48) c70,
         AiryBiPrime[c31] -> -7/48 c14 c15 c34 c36 c40 c48 + c14 c40 c49,
         AiryBiPrime[c51] -> -7/48 c14 c15 c52 c54 c58 c65 + c14 c58 c66}];
```

Applying `asympAiryRules`, we have the following form of the determinant.

```
In[23]:= deter1 = deter //. asympAiryRules;
```

We have two kinds of exponential functions inside of `deter1`.

```
In[24]:= Cases[deter1, Power[E, _], {0, Infinity}] // Union
```

Out[24]= $\left\{ e^{-\frac{2}{3} \alpha^{1/3} \left(-\frac{1}{2} + \frac{v0-\lambda}{\alpha}\right)} \sqrt{\alpha^{1/3} \left(-\frac{1}{2} + \frac{v0-\lambda}{\alpha}\right)} - \frac{2}{3} \alpha^{1/3} \left(\frac{1}{2} + \frac{v0-\lambda}{\alpha}\right) \sqrt{\alpha^{1/3} \left(\frac{1}{2} + \frac{v0-\lambda}{\alpha}\right)} \right.$,

$\left. e^{\frac{2}{3} \alpha^{1/3} \left(-\frac{1}{2} + \frac{v0-\lambda}{\alpha}\right)} \sqrt{\alpha^{1/3} \left(-\frac{1}{2} + \frac{v0-\lambda}{\alpha}\right)} - \frac{2}{3} \alpha^{1/3} \left(\frac{1}{2} + \frac{v0-\lambda}{\alpha}\right) \sqrt{\alpha^{1/3} \left(\frac{1}{2} + \frac{v0-\lambda}{\alpha}\right)} \right\}$

In the limit $\alpha \to 0$, the second of them vanishes exponentially fast. (Physically speaking, this means that finite lifetime effects are nonperturbative and cannot be caught in a perturbation expansion done here.) We now approximate the arguments of the exponential functions by their series.

In[25]:= `% /. {Exp[x_] :> Exp[Expand[Normal[Series[x, {α, 0, 2}]]]]}`

Out[25]= $\left\{ e^{-\frac{1^2 \alpha}{8 \sqrt{v0-\lambda}} - \frac{4 \, v0 \sqrt{v0-\lambda}}{3 \alpha} + \frac{4 \sqrt{v0-\lambda} \, \lambda}{3 \alpha}}, \ e^{\frac{1^3 \alpha^2}{96 \, (v0-\lambda)^{3/2}} - \frac{1}{\alpha} \sqrt{v0-\lambda}} \right\}$

Because of the exponential decay of these terms, we can replace all such terms by 0 because after the differentiation with respect to α at the end to get c_2, these terms would vanish anyway.

In[26]:= `deter2 = deter1 //. {(* the term of the form Exp[-1/α +...] vanishes *)`
        `Select[%%, Exponent[Normal[Series[#[[2]],`
                           `{α, 0, 0}]], α] === -1&][[1]] -> 0};`

This reduces the number of nontrivial terms in the determinant to 8.

In[27]:= `deter2 = DeleteCases[deter2, 0];`

In[28]:= `Length[deter2]`

Out[28]= 8

Starting now, let us massage each term in the remaining sum individually by introducing the variables $aux[i]$, $i = 1, \ldots, 8$.

In[29]:= `Do[aux1[i] = deter2[[i]], {i, Length[deter2]}]`

All terms have one factor in common, the second exponential term. So we can drop this one, too, because we have to equate the whole `deter2` to 0.

In[30]:= `Cases[Table[aux1[i], {i, Length[deter2]}], Exp[_], {0, Infinity}] // Union`

Out[30]= $\left\{ e^{\frac{2}{3} \alpha^{1/3} \left(-\frac{1}{2} + \frac{v0-\lambda}{\alpha}\right)} \sqrt{\alpha^{1/3} \left(-\frac{1}{2} + \frac{v0-\lambda}{\alpha}\right)} - \frac{2}{3} \alpha^{1/3} \left(\frac{1}{2} + \frac{v0-\lambda}{\alpha}\right) \sqrt{\alpha^{1/3} \left(\frac{1}{2} + \frac{v0-\lambda}{\alpha}\right)} \right\}$

In[31]:= `Intersection @@ Table[List @@ aux1[i], {i, Length[deter2]}]`

Out[31]= $\left\{ e^{\frac{2}{3} \alpha^{1/3} \left(-\frac{1}{2} + \frac{v0-\lambda}{\alpha}\right)} \sqrt{\alpha^{1/3} \left(-\frac{1}{2} + \frac{v0-\lambda}{\alpha}\right)} - \frac{2}{3} \alpha^{1/3} \left(\frac{1}{2} + \frac{v0-\lambda}{\alpha}\right) \sqrt{\alpha^{1/3} \left(\frac{1}{2} + \frac{v0-\lambda}{\alpha}\right)} \right\}$

In[32]:= `Do[aux2[i] = aux1[i]/%[[1]], {i, Length[deter2]}]`

The resulting terms contain the following trigonometric functions.

In[33]:= `allTrigs = Cases[Table[aux2[i], {i, Length[deter2]}], _Cos | _Sin,`
                          `{0, Infinity}] // Union`

Out[33]= $\left\{ \mathrm{Cos}\left[\frac{\pi}{4} - \frac{2}{3} \alpha^{1/3} \left(-\frac{1}{2} - \frac{\lambda}{\alpha}\right) \sqrt{-\alpha^{1/3} \left(-\frac{1}{2} - \frac{\lambda}{\alpha}\right)}\right], \ \mathrm{Cos}\left[\frac{\pi}{4} - \frac{2}{3} \alpha^{1/3} \left(\frac{1}{2} - \frac{\lambda}{\alpha}\right) \sqrt{-\alpha^{1/3} \left(\frac{1}{2} - \frac{\lambda}{\alpha}\right)}\right] \right.$,

$\left. \mathrm{Sin}\left[\frac{\pi}{4} - \frac{2}{3} \alpha^{1/3} \left(-\frac{1}{2} - \frac{\lambda}{\alpha}\right) \sqrt{-\alpha^{1/3} \left(-\frac{1}{2} - \frac{\lambda}{\alpha}\right)}\right], \ \mathrm{Sin}\left[\frac{\pi}{4} - \frac{2}{3} \alpha^{1/3} \left(\frac{1}{2} - \frac{\lambda}{\alpha}\right) \sqrt{-\alpha^{1/3} \left(\frac{1}{2} - \frac{\lambda}{\alpha}\right)}\right] \right\}$

In the limit of vanishing α, these trigonometric functions oscillate more and more because of the α in the denominators of their arguments. To get a sensible limit, we have to combine appropriate combinations of these trigonometric functions to cancel the singular denominators.

For shortness of the following code, let us define a function `trigRule` that expands the arguments of trigonometric functions into a series.

In[34]:= `ord3 = 2;`

        `trigRule = {Sin[x_] :> Sin[Normal[Series[x, {α, 0, ord3}]]],`
                   `Cos[x_] :> Cos[Normal[Series[x, {α, 0, ord3}]]]};`

For a later application of `PowerExpand` to simplify all expressions and to write them in a canonical form, we define some rules that make sure that all arguments of power functions are positive. (We are only interested in the regime $\lambda < v_0$, so all powers with bases $\lambda - v_0$ should be transformed.)

```
In[36]:= rule1 = {
    a_ (λ - v0)^el_ -> a (-1)^el (v0 - λ)^el,
    a_ (-(1/(λ - v0)))^el_ -> a (v0 - λ)^(-el),
    a_ (λ - v0)^el_ (v0 - λ)^e2_ -> a (-1)^el (v0 - λ)^(el + e2),
    a_ ((-λ + v0)^(-1))^el_ (-λ + v0)^e2_ -> a (-λ + v0)^(e2 - el),
    a_ (-1/(λ - v0))^el_ (-λ + v0)^e2_ -> a (-λ + v0)^(-el + e2),
    a_ (-1/(λ - v0))^el_ (λ - v0)^e2_ -> a (-λ + v0)^(-el + e2) (-1)^e2};
```

The two functions f1 and f2 will do all of the expansion for small α in the following main calculation.

```
In[37]:= (* for dealing with the trigonometric functions *)
    f1[x_] := TrigReduce[x] //. trigRule // PowerExpand

In[39]:= ord4 = 3;
    (* for dealing with the remaining (prefactor) terms *)
    f2[x_] := Together[(Normal[Series[Expand[x], {α, 0, ord4}]] //
                        Expand) //. rule1];
```

Now comes the main calculation. Inside this main calculation, we do the following: For every term from `deter2`, we extract the trigonometric functions, combine all products of two trigonometric functions into one, and expand the resulting argument for small α. The same expansion for α is done for the corresponding prefactors of the resulting trigonometric functions.

```
In[42]:= Module[{sc1, sc2, withoutSC, sinCos1, sinCos2, coeffs1, coeffs2,
             preFac11, preFac12, preFac21, preFac22, tr11, tr12, tr21, tr22},
        Do[(* extract trigonometric functions *)
           {sc1, sc2} = Cases[aux2[i], _?(MemberQ[#, Cos[_] | Sin[_],
                              {0, Infinity}]&), {1}];
           (* the terms not containing trigonometric functions *)
           withoutSC = aux2[i]/(sc1 sc2);
           (* the prefactors of the trigonometric functions *)
           sinCos1 = Cases[sc1, #[_], {0, Infinity}][[1]]& /@ {Cos, Sin};
           sinCos2 = Cases[sc2, #[_], {0, Infinity}][[1]]& /@ {Cos, Sin};
           coeffs1 = Coefficient[sc1, #]& /@ sinCos1;
           coeffs2 = Coefficient[sc2, #]& /@ sinCos2;
           (* combine trigonometric functions into one and expand in series *)
           tr11 = f1[sinCos1[[1]] sinCos2[[1]]];
           tr12 = f1[sinCos1[[1]] sinCos2[[2]]];
           tr21 = f1[sinCos1[[2]] sinCos2[[1]]];
           tr22 = f1[sinCos1[[2]] sinCos2[[2]]];
           (* here, most time is spent in evaluating these four series *)
           preFac11 = f2[coeffs1[[1]] coeffs2[[1]] withoutSC];
           preFac12 = f2[coeffs1[[1]] coeffs2[[2]] withoutSC];
           preFac21 = f2[coeffs1[[2]] coeffs2[[1]] withoutSC];
           preFac22 = f2[coeffs1[[2]] coeffs2[[2]] withoutSC];
           (* add all contributions together *)
           aux3[i] = (preFac11 tr11 + preFac12 tr12 +
                     preFac21 tr21 + preFac22 tr22), {i, 1, Length[deter2]}]];
```

After this calculation, the following trigonometric expressions are present in the result.

```
In[43]:= trigos1 =
    Union[Flatten[Table[Cases[aux3[i], Sin[_] | Cos[_], {0, Infinity}],
                    {i, Length[deter2]}]]]
```

$$\text{Out[43]= } \left\{ \cos\left[\frac{1^3 \alpha^2}{96 \lambda^{3/2}} - 1 \sqrt{\lambda}\right], \cos\left[\frac{1^2 \alpha}{8 \sqrt{\lambda}} + \frac{4 \lambda^{3/2}}{3 \alpha}\right], \sin\left[\frac{1^3 \alpha^2}{96 \lambda^{3/2}} - 1 \sqrt{\lambda}\right], \sin\left[\frac{1^2 \alpha}{8 \sqrt{\lambda}} + \frac{4 \lambda^{3/2}}{3 \alpha}\right] \right\}$$

Here, we still have trigonometric functions with highly oscillating terms in the limit $\alpha \to 0$. After collecting with respect to these trigonometric functions, we see that these (bounded) terms do not matter because of the vanishing of their prefactor.

```
In[44]:= eve1 = Together /@ Collect[
                  Expand[Sum[aux3[i], {i, 1, Length[deter2]}]], trigos1];
```

We are left with trigonometric functions that do not have α in the denominator of their arguments. All remaining terms behave smoothly in the limit $\alpha \to 0$.

In[45]:= **Union[Cases[eve1, Sin[_] | Cos[_], {0, Infinity}]]**

Out[45]= $\left\{ \text{Cos}\left[\frac{13\,\alpha^2}{96\,\lambda^{3/2}} - 1\,\sqrt{\lambda}\right], \text{Sin}\left[\frac{13\,\alpha^2}{96\,\lambda^{3/2}} - 1\,\sqrt{\lambda}\right]\right\}$

So this is how the result looks in the moment.

In[46]:= **Short[eve1, 12]**

Out[46]//Short= $\dfrac{71\,v0\,\alpha^2\,(\frac{1}{v0-\lambda})^{7/2}\,\text{Cos}\left[\frac{13\,\alpha^2}{96\,\lambda^{3/2}} - 1\,\sqrt{\lambda}\right]}{48\,\pi^2} - \dfrac{v0^3\,\text{Cos}\left[\frac{13\,\alpha^2}{96\,\lambda^{3/2}} - 1\,\sqrt{\lambda}\right]}{\pi^2\,(v0-\lambda)^3} + \dfrac{35\,v0\,\alpha^2\,\text{Cos}\left[\frac{13\,\alpha^2}{96\,\lambda^{3/2}} - 1\,\sqrt{\lambda}\right]}{2304\,\pi^2\,(v0-\lambda)\,\lambda^3} -$

$\dfrac{51\,v0^3\,\alpha^2\,(\frac{1}{v0-\lambda})^{5/2}\,\sqrt{\frac{1}{\lambda}}\,\text{Cos}\left[\frac{13\,\alpha^2}{96\,\lambda^{3/2}} - 1\,\sqrt{\lambda}\right]}{64\,\pi^2\,\lambda^{5/2}} - \dfrac{13\,v0^3\,\alpha^2\,\text{Cos}\left[\frac{13\,\alpha^2}{96\,\lambda^{3/2}} - 1\,\sqrt{\lambda}\right]}{32\,\pi^2\,(v0-\lambda)^3\,\lambda^2} -$

$\dfrac{35\,\alpha^2\,\text{Cos}\left[\frac{13\,\alpha^2}{96\,\lambda^{3/2}} - 1\,\sqrt{\lambda}\right]}{2304\,\pi^2\,(v0-\lambda)\,\lambda^2} + \dfrac{13\,v0^2\,\alpha^2\,(\frac{1}{v0-\lambda})^{5/2}\,\sqrt{\frac{1}{\lambda}}\,\text{Cos}\left[\frac{13\,\alpha^2}{96\,\lambda^{3/2}} - 1\,\sqrt{\lambda}\right]}{8\,\pi^2\,\lambda^{3/2}} -$

$\dfrac{71\,v0^2\,\alpha^2\,(\frac{1}{v0-\lambda})^{7/2}\,\text{Cos}\left[\frac{13\,\alpha^2}{96\,\lambda^{3/2}} - 1\,\sqrt{\lambda}\right]}{192\,\pi^2\,\lambda} + \dfrac{13\,v0^2\,\alpha^2\,\text{Cos}\left[\frac{13\,\alpha^2}{96\,\lambda^{3/2}} - 1\,\sqrt{\lambda}\right]}{32\,\pi^2\,(v0-\lambda)^3\,\lambda} -$

$\dfrac{13\,v0\,\alpha^2\,(\frac{1}{v0-\lambda})^{5/2}\,\sqrt{\frac{1}{\lambda}}\,\text{Cos}\left[\frac{13\,\alpha^2}{96\,\lambda^{3/2}} - 1\,\sqrt{\lambda}\right]}{64\,\pi^2\,\sqrt{\lambda}} - \dfrac{13\,\alpha^2\,(\frac{1}{v0-\lambda})^{5/2}\,\sqrt{\frac{1}{\lambda}}\,\sqrt{\lambda}\,\text{Cos}\left[\frac{13\,\alpha^2}{96\,\lambda^{3/2}} - 1\,\sqrt{\lambda}\right]}{32\,\pi^2} -$

$\dfrac{71\,\alpha^2\,(\frac{1}{v0\,\ll1\gg\,\ll1\gg})^{7/2}\,\lambda\,\text{Cos}\left[\frac{13\,\alpha^2}{96\,\lambda^{3/2}} - 1\,\sqrt{\lambda}\right]}{64\,\pi^2} + \dfrac{3\,\ll3\gg}{\pi^2\,(\ll1\gg)^3} - \dfrac{3\,\ll3\gg}{\ll1\gg\,\ll1\gg} + \ll1\gg + \dfrac{\ll1\gg}{\ll1\gg} -$

$\dfrac{\ll1\gg}{192\,\pi^2} - \dfrac{1\,v0^2\,\alpha^2\,(\ll1\gg)^{5/2}\,\text{Sin}\left[\frac{13\,\alpha^2}{96\,\lambda\,\ll1\gg} - 1\,\sqrt{\lambda}\right]}{32\,\pi^2\,(v0-\lambda)^2} + \dfrac{25\,v0\,\alpha^2\,\sqrt{\frac{1}{v0-\lambda}}\,\text{Sin}\left[\frac{13\,\alpha^2}{96\,\lambda^{3/2}} - 1\,\sqrt{\lambda}\right]}{4608\,\pi^2\,\lambda^{7/2}} -$

$\dfrac{37\,\alpha^2\,\sqrt{\frac{1}{v0-\lambda}}\,\text{Sin}\left[\frac{13\,\alpha^2}{96\,\lambda^{3/2}} - 1\,\sqrt{\lambda}\right]}{2304\,\pi^2\,\lambda^{5/2}} + \dfrac{13\,v0^3\,\alpha^2\,(\frac{1}{v0-\lambda})^{5/2}\,\text{Sin}\left[\frac{13\,\alpha^2}{96\,\lambda^{3/2}} - 1\,\sqrt{\lambda}\right]}{32\,\pi^2\,\lambda^{5/2}} -$

$\dfrac{13\,v0^2\,\alpha^2\,(\frac{1}{v0-\lambda})^{5/2}\,\text{Sin}\left[\frac{13\,\alpha^2}{96\,\lambda^{3/2}} - 1\,\sqrt{\lambda}\right]}{16\,\pi^2\,\lambda^{3/2}} + \dfrac{v0^3\,(\frac{1}{v0-\lambda})^{5/2}\,\text{Sin}\left[\frac{13\,\alpha^2}{96\,\lambda^{3/2}} - 1\,\sqrt{\lambda}\right]}{2\,\pi^2\,\sqrt{\lambda}} -$

$\dfrac{49\,\alpha^2\,(\frac{1}{v0-\lambda})^{5/2}\,\text{Sin}\left[\frac{13\,\alpha^2}{96\,\lambda^{3/2}} - 1\,\sqrt{\lambda}\right]}{4608\,\pi^2\,\sqrt{\lambda}} - \dfrac{71\,\alpha^2\,\text{Sin}\left[\frac{13\,\alpha^2}{96\,\lambda^{3/2}} - 1\,\sqrt{\lambda}\right]}{64\,\pi^2\,(v0-\lambda)^2\,\sqrt{\lambda}} -$

$\dfrac{2\,v0^2\,(\frac{1}{v0-\lambda})^{5/2}\,\sqrt{\lambda}\,\text{Sin}\left[\frac{13\,\alpha^2}{96\,\lambda^{3/2}} - 1\,\sqrt{\lambda}\right]}{\pi^2} + \dfrac{5\,v0\,(\frac{1}{v0-\lambda})^{5/2}\,\lambda^{3/2}\,\text{Sin}\left[\frac{13\,\alpha^2}{96\,\lambda^{3/2}} - 1\,\sqrt{\lambda}\right]}{2\,\pi^2} -$

$\dfrac{(\frac{1}{v0-\lambda})^{5/2}\,\lambda^{5/2}\,\text{Sin}\left[\frac{13\,\alpha^2}{96\,\lambda^{3/2}} - 1\,\sqrt{\lambda}\right]}{\pi^2} + \dfrac{13\,\alpha^2\,\sqrt{\frac{1}{\lambda}}\,\text{Sin}\left[\frac{13\,\alpha^2}{96\,\lambda^{3/2}} - 1\,\sqrt{\lambda}\right]}{32\,\pi^2\,(-v0+\lambda)^2}$

We again apply PowerExpand to canonicalize the form of the eigenvalue equation.

In[47]:= **eve2 = PowerExpand[eve1];**

Taking the numerator, we finally get the eigenvalue equation up to second order in α. (Because we are only interested in corrections of the "eigenvalues" to second order in α, we only keep terms up to second order in α here.)

In[48]:= **Short[eve3 = Numerator[Factor[eve2]], 4]**

Out[48]//Short= $-360\,1\,v0^6\,\alpha^2\,\sqrt{\lambda}\,\text{Cos}\left[\frac{13\,\alpha^2}{96\,\lambda^{3/2}} - 1\,\sqrt{\lambda}\right] + 70\,v0^2\,\alpha^2\,(v0-\lambda)^{7/2}\,\sqrt{\lambda}\,\text{Cos}\left[\frac{13\,\alpha^2}{96\,\lambda^{3/2}} - 1\,\sqrt{\lambda}\right] + \ll46\gg +$

$57600\,v0^2\,\lambda^7\,\text{Sin}\left[\frac{13\,\alpha^2}{96\,\lambda^{3/2}} - 1\,\sqrt{\lambda}\right] - 25344\,v0\,\lambda^8\,\text{Sin}\left[\frac{13\,\alpha^2}{96\,\lambda^{3/2}} - 1\,\sqrt{\lambda}\right] + 4608\,\lambda^9\,\text{Sin}\left[\frac{13\,\alpha^2}{96\,\lambda^{3/2}} - 1\,\sqrt{\lambda}\right]$

Let us check the resulting eigenvalue equation in the limit $\alpha \to 0$.

In[49]:= **eveAlphaZero1 = Simplify[eve3 /. α -> 0]**

Out[49]= $-2304\,(v0-\lambda)^2\,\lambda^3\,\left(2\,(v0-\lambda)^{7/2}\,\sqrt{\lambda}\,\text{Cos}[1\,\sqrt{\lambda}] + (v0-2\,\lambda)\,(v0-\lambda)^3\,\text{Sin}[1\,\sqrt{\lambda}]\right)$

In the tradition typically used in physics, we use abbreviations for the wave vectors k and κ in favor of λ and v_0 for the classically allowed and forbidden regions via the following substitutions: $k^2 \longrightarrow \lambda$ and $\kappa^2 \longrightarrow v_0 - \lambda$. So we can derive the following replacement rules.

In[50]:= **waveVectorRule =**
  {λ^e_. -> k^(2 e), (v0 - λ)^e_. -> κ^(2 e), v0^e_. -> k^(2 e) + κ^(2 e)};

Now, the above eveAlphaZero1 can be rewritten as follows.

```
In[51]:= eveAlphaZero2 = eveAlphaZero1 /. waveVectorRule // Expand // Factor
```

$$Out[51]= 2304\, k^6\, \kappa^{10}\, (-2\, k\, \kappa\, \text{Cos}[k\, l] + k^2\, \text{Sin}[k\, l] - \kappa^2\, \text{Sin}[k\, l])$$

The last factor of this equation $-2\, k\, \kappa \cos(k\, l) + (k^2 - \kappa^2) \sin(k\, l) = 0$ is the well-known eigenvalue equation for the bound states of a quantum well (see [1003], [57], [490], [467], [900], [1148], [409], [410], [1367], [988], [352], [546], [115], [785], [812], and [362]): $-y''(x) + \theta((l/2)^2 - x^2)\, v_0\, y(x) = \lambda\, y(x)$.

Now, let us analyze the coefficient c_2 in the Taylor expansion of the solution $\lambda(\alpha)$ of eveAlphaZero1. As in part a), the first derivative of the eigenvalue equation with respect to α vanishes identically because of the symmetry of the potential under consideration.

```
In[52]:= D[eve2, α] /. α -> 0
```

$$Out[52]= 0$$

To simplify the following expressions, we make use of the eigenvalue equation derived above.

```
In[53]:= eveRule = {a_ Cos[k l] k^e1_. κ^e2_. ->
                    -a k^(e1 - 1) κ^(e2 - 1)(κ^2 - k^2) Sin[k l]/2};
```

So we have the numerator and denominator of the ratio needed for c_2.

```
In[54]:= num1 = ((D[eve3, {α, 2}] /. α -> 0) //. waveVectorRule) // Factor;
```

```
In[55]:= num2 = num1 /. eveRule // Together // Factor
```

$$Out[55]= -2\, \kappa^2\, (-1320\, k^{11}\, l\, \text{Cos}[k\, l] + 264\, k^{13}\, l^3\, \text{Cos}[k\, l] + 130\, k^{10}\, \text{Sin}[k\, l] +$$
$$1188\, k^{10}\, l\, \kappa\, \text{Sin}[k\, l] - 300\, k^{12}\, l^3\, \kappa\, \text{Sin}[k\, l] + 4\, k^8\, \kappa^2\, \text{Sin}[k\, l] + 288\, k^{10}\, l^2\, \kappa^2\, \text{Sin}[k\, l] -$$
$$2772\, k^8\, l\, \kappa^3\, \text{Sin}[k\, l] + 660\, k^{10}\, l^3\, \kappa^3\, \text{Sin}[k\, l] - 367\, k^6\, \kappa^4\, \text{Sin}[k\, l] -$$
$$936\, k^8\, l^2\, \kappa^4\, \text{Sin}[k\, l] + 2604\, k^6\, l\, \kappa^5\, \text{Sin}[k\, l] - 600\, k^8\, l^3\, \kappa^5\, \text{Sin}[k\, l] + 522\, k^4\, \kappa^6\, \text{Sin}[k\, l] +$$
$$1224\, k^6\, l^2\, \kappa^6\, \text{Sin}[k\, l] - 2352\, k^4\, l\, \kappa^7\, \text{Sin}[k\, l] + 228\, k^6\, l^3\, \kappa^7\, \text{Sin}[k\, l] -$$
$$304\, k^2\, \kappa^8\, \text{Sin}[k\, l] - 648\, k^4\, l^2\, \kappa^8\, \text{Sin}[k\, l] + 864\, k^2\, l\, \kappa^9\, \text{Sin}[k\, l] - 48\, k^4\, l^3\, \kappa^9\, \text{Sin}[k\, l] +$$
$$60\, \kappa^{10}\, \text{Sin}[k\, l] + 72\, k^2\, l^2\, \kappa^{10}\, \text{Sin}[k\, l] - 180\, l\, \kappa^{11}\, \text{Sin}[k\, l] + 12\, k^2\, l^3\, \kappa^{11}\, \text{Sin}[k\, l])$$

```
In[56]:= denom1 = ((D[eve3, λ] /. α -> 0) //. waveVectorRule) // Factor;
```

```
In[57]:= denom2 = denom1 /. eveRule // Together // Factor
```

$$Out[57]= 576\, k^4\, \kappa^2$$
$$(22\, k^{11}\, l\, \text{Cos}[k\, l] + 352\, k^{10}\, \text{Sin}[k\, l] - 25\, k^{10}\, l\, \kappa\, \text{Sin}[k\, l] - 700\, k^8\, \kappa^2\, \text{Sin}[k\, l] + 55\, k^8\, l\, \kappa^3\, \text{Sin}[k\, l] +$$
$$692\, k^6\, \kappa^4\, \text{Sin}[k\, l] - 50\, k^6\, l\, \kappa^5\, \text{Sin}[k\, l] - 322\, k^4\, \kappa^6\, \text{Sin}[k\, l] + 19\, k^4\, l\, \kappa^7\, \text{Sin}[k\, l] +$$
$$48\, k^2\, \kappa^8\, \text{Sin}[k\, l] - 4\, k^2\, l\, \kappa^9\, \text{Sin}[k\, l] + 2\, \kappa^{10}\, \text{Sin}[k\, l] + l\, \kappa^{11}\, \text{Sin}[k\, l])$$

Multiplying by $-1/2$, we finally arrive at the following value for the coefficient c_2.

```
In[58]:= c2 = -1/2 Collect[num2  , {Sin[k l], Cos[k l]}, Simplify]/
                  Collect[denom2, {Sin[k l], Cos[k l]}, Simplify]
```

$$Out[58]= -(-528\, k^{11}\, l\, (-5 + k^2\, l^2)\, \kappa^2\, \text{Cos}[k\, l] -$$
$$2\, \kappa^2\, (-300\, k^{12}\, l^3\, \kappa + 60\, \kappa^{10}\, (1 - 3\, l\, \kappa) + 4\, k^2\, \kappa^8\, (-76 + 216\, l\, \kappa + 18\, l^2\, \kappa^2 + 3\, l^3\, \kappa^3) -$$
$$6\, k^4\, \kappa^6\, (-87 + 392\, l\, \kappa + 108\, l^2\, \kappa^2 + 8\, l^3\, \kappa^3) - 4\, k^8\, \kappa^2\, (-1 + 693\, l\, \kappa + 234\, l^2\, \kappa^2 + 150\, l^3\, \kappa^3) +$$
$$k^6\, \kappa^4\, (-367 + 2604\, l\, \kappa + 1224\, l^2\, \kappa^2 + 228\, l^3\, \kappa^3) +$$
$$2\, k^{10}\, (65 + 594\, l\, \kappa + 144\, l^2\, \kappa^2 + 330\, l^3\, \kappa^3))\, \text{Sin}[k\, l])\, /$$
$$(2\, (12672\, k^{15}\, l\, \kappa^2\, \text{Cos}[k\, l] + 576\, k^4\, \kappa^2\, (2\, k^6\, \kappa^4\, (346 - 25\, l\, \kappa) + k^{10}\, (352 - 25\, l\, \kappa) - 4\, k^2\, \kappa^8$$
$$(-12 + l\, \kappa) + \kappa^{10}\, (2 + l\, \kappa) + 5\, k^8\, \kappa^2\, (-140 + 111\, l\, \kappa) + k^4\, \kappa^6\, (-322 + 191\, l\, \kappa))\, \text{Sin}[k\, l]))$$

In the limit of infinitely large v_0, we are back to part a). This limit can be modeled as the limit $\kappa \to 0$ [177]. Making use of this and the corresponding limit of the k, we come back to the result of part a) of this exercise.

```
In[59]:= Function[maxκExp, Coefficient[Numerator[ c2], κ, maxκExp]/
                  Coefficient[Denominator[c2], κ, maxκExp] /.
                  {k -> i Pi/l} // Simplify] @@
                  Union[Exponent[#, κ]& /@ {Numerator[c2], Denominator[c2]}]
```

$$Out[59]= \frac{l^4\, (-15 + i^2\, \pi^2)}{48\, i^4\, \pi^4}$$

For the perturbation theory of a constant field between two δ-potentials, see [32]. For the physical relevance of such calculations, see [1077], [105], [460], [573], [636], [507], [296], [19], [1041], [628], [494], [1106], [1294], [1289], and [906] and the references therein.

11. Fermi Integrals, Sum of All Reciprocal 9-Free Numbers, Zagier's Function

a) Here is the implementation `fermiIntApprox` of the approximation (including the bounds on the variables), along with the exact result `fermiIntExact`.

```
In[1]:= fermiIntApprox[a_?(((TrueQ[-1 < # < Infinity] && !IntegerQ[#]) ||
                    Head[a] == Symbol)&),
                z_?(TrueQ[1 < # < Infinity] || Head[z] == Symbol &),
                n_Integer?(# >= -1&), k_Integer] :=
        (2Pi/(Sin[Pi a] Gamma[1 + a]) *
        Sum[((2i + 1)^2 Pi^2 + Log[z]^2)^(a/2) *
         Cos[a(Pi - ArcTan[(2i + 1)Pi/Log[z]])], {i, 0, n}] +
        Pi/(Sin[Pi a] Gamma[1 + a]) (Log[z]^2 + (2(n + 1) + 1)^2 Pi^2)^(a/2)*
         Cos[a(Pi - ArcTan[(2(n + 1) + 1)Pi/Log[z]])] -
        2/Sin[Pi a] (Log[z]^2 + (2(n + 1) + 1)^2 Pi^2)^((a + 1)/2)*
         Sum[Zeta[2i]/Gamma[a + 2(1 - i)]*
          (Log[z]^2 + (2(n + 1) + 1)^2 Pi^2)^-i*
          Sin[a Pi - (a + 1 - 2i) ArcTan[(2(n + 1) + 1)Pi/Log[z]]],
          {i, 0, k}]) /; If[NumberQ[a], N[2k - a] > 1, True]

In[2]:= fermiIntExact[a_, z_] = z LerchPhi[-z, a + 1, 1];
```

The most useful case for applications is $\alpha = 1/2$. (The replacement rule `Cos[x_] :> Cos[ExpandAll[x]]` serves to simplify terms of the form `Cos[(Pi + ⋯)/2]`.)

```
In[3]:= (fermiIntApprox[1/2, z, 5, 5] /.
            {Cos[x_] :> Cos[ExpandAll[x]]}) // Short[#, 12]&
```

$$
\text{Out[3]//Short= } -2\,(169\,\pi^2 + \text{Log}[z]^2)^{3/4} \left(-\frac{2\,\text{Cos}\left[\frac{3}{2}\,\text{ArcTan}\left[\frac{13\,\pi}{\text{Log}[z]}\right]\right]}{3\,\sqrt{\pi}} + \frac{65\,\pi^{19/2}\,\text{Cos}\left[\frac{17}{2}\,\text{ArcTan}\left[\frac{13\,\pi}{\text{Log}[z]}\right]\right]}{768\,(169\,\pi^2 + \text{Log}[z]^2)^5} + \right.
$$

$$
\frac{11\,\pi^{15/2}\,\text{Cos}\left[\frac{13}{2}\,\text{ArcTan}\left[\frac{13\,\pi}{\text{Log}[z]}\right]\right]}{640\,(169\,\pi^2 + \text{Log}[z]^2)^4} + \frac{\pi^{11/2}\,\text{Cos}\left[\frac{9}{2}\,\text{ArcTan}\left[\frac{13\,\pi}{\text{Log}[z]}\right]\right]}{144\,(169\,\pi^2 + \text{Log}[z]^2)^3} +
$$

$$
\left. \frac{\pi^{7/2}\,\text{Cos}\left[\frac{5}{2}\,\text{ArcTan}\left[\frac{13\,\pi}{\text{Log}[z]}\right]\right]}{120\,(169\,\pi^2 + \text{Log}[z]^2)^2} + \frac{\pi^{3/2}\,\text{Cos}\left[\frac{1}{2}\,\text{ArcTan}\left[\frac{13\,\pi}{\text{Log}[z]}\right]\right]}{6\,(169\,\pi^2 + \text{Log}[z]^2)} \right) + 4\,\sqrt{\pi}
$$

$$
\left((\pi^2 + \text{Log}[z]^2)^{1/4}\,\text{Sin}\left[\frac{1}{2}\,\text{ArcTan}\left[\frac{\pi}{\text{Log}[z]}\right]\right] + (9\,\pi^2 + \text{Log}[z]^2)^{1/4}\,\text{Sin}\left[\frac{1}{2}\,\text{ArcTan}\left[\frac{3\,\pi}{\text{Log}[z]}\right]\right] + \right.
$$

$$
(25\,\pi^2 + \text{Log}[z]^2)^{1/4}\,\text{Sin}\left[\frac{1}{2}\,\text{ArcTan}\left[\frac{5\,\pi}{\text{Log}[z]}\right]\right] +
$$

$$
(49\,\pi^2 + \text{Log}[z]^2)^{1/4}\,\text{Sin}\left[\frac{1}{2}\,\text{ArcTan}\left[\frac{7\,\pi}{\text{Log}[z]}\right]\right] + (81\,\pi^2 + \text{Log}[z]^2)^{1/4}
$$

$$
\text{Sin}\left[\frac{1}{2}\,\text{ArcTan}\left[\frac{9\,\pi}{\text{Log}[z]}\right]\right] + (121\,\pi^2 + \text{Log}[z]^2)^{1/4}\,\text{Sin}\left[\frac{1}{2}\,\text{ArcTan}\left[\frac{11\,\pi}{\text{Log}[z]}\right]\right] \right) +
$$

$$
2\,\sqrt{\pi}\,(169\,\pi^2 + \text{Log}[z]^2)^{1/4}\,\text{Sin}\left[\frac{1}{2}\,\text{ArcTan}\left[\frac{13\,\pi}{\text{Log}[z]}\right]\right]
$$

```
In[4]:= FreeQ[%, _Zeta | _Gamma, {0, Infinity}]

Out[4]= True
```

All γ and Zeta functions have been computed, and now nothing stands in the way of their further use outside of *Mathematica*. Finally, we examine the quality of the approximation over a wide parameter range.

```
In[5]:= TableForm[Apply[{#1, #3 - #2, (#3 - #2)/#3}&,
                Table[{E^z, fermiIntExact[1/2, Exp[z]],
                fermiIntApprox[1/2, Exp[z], 5, 4]} // N // Chop,
            {z, 0.1, 5.1, 0.4}], {1}],
            TableHeadings -> {{}, {"z", "abs. error ", "rel. error "}}]
```

z	abs. error	rel. error
1.10517	1.24484×10^{-10}	1.50423×10^{-10}
1.64872	1.33853×10^{-10}	1.19797×10^{-10}
2.4596	1.42203×10^{-10}	9.63981×10^{-11}
3.6693	1.49375×10^{-10}	7.84605×10^{-11}
5.47395	1.55418×10^{-10}	6.46637×10^{-11}
8.16617	1.60237×10^{-10}	5.39114×10^{-11}
12.1825	1.63763×10^{-10}	4.54018×10^{-11}

Out[5]//TableForm=

18.1741	1.66066×10^{-10}	3.85841×10^{-11}
27.1126	1.67097×10^{-10}	3.30262×10^{-11}
40.4473	1.66894×10^{-10}	2.84324×10^{-11}
60.3403	1.65342×10^{-10}	2.45618×10^{-11}
90.0171	1.62626×10^{-10}	2.12805×10^{-11}
134.29	1.58868×10^{-10}	1.84768×10^{-11}

The approximation is excellent.

b) We start by implementing the two expansions $\mathcal{F}_{d,o}^{(alg)}(\eta)$, $\mathcal{F}_{d,o}^{(exp)}(\eta)$.

```
In[1]:= FAsympAlg[d_, η_, o_] :=
          Sum[(2 (1 - 2^(1 - 2k)) η^(1/2(d - 4k)) Zeta[2k])/
            Gamma[1 + d/2 - 2k], {k, 0, Floor[d/4]}] -
          Sin[(d Pi)/2]/Pi Sum[(2(1 - 2^(1 - 2k))η^(1/2(d - 4k)) Zeta[2k])/
                          Gamma[-d/2 + 2k], {k, 1 + Floor[d/4], o}]


        FAsympExp[d_, η_, o_] :=
          Cos[Pi (d/2 - 1)] Sum[(-1)^(k + 1)/(k^(d/2)) Exp[-k η], {k, o}]


        FAsymp[d_, η_, o_] := FAsympAlg[d, η, o] + FAsympExp[d, η, o]
```

Because for even d the part $\mathcal{F}_{d,o}^{(alg)}(\eta)$ vanishes, the expansion is most accurate in these cases.

```
In[6]:= (* closed form of the exact result *)
        F[d_, η_] = 1/Gamma[d/2] Integrate[ε^(d/2 - 1)/(1 + Exp[ε - η]),
                                  {ε, 0, Infinity}, Assumptions -> d > 0]
```

$$Out[7]= -PolyLog\left[\frac{d}{2}, -e^{\eta}\right]$$

```
In[8]:= h[d_, η_, o_] := (FAsymp[d, η, o] - F[d, η])/F[d, η]

In[9]:= Plot[Log[10, Abs[h[d, 6, 8]]], {d, 0, 10}, PlotRange -> All]
```

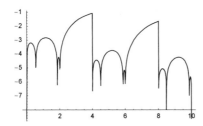

Here are two typical expansions. One is purely algebraic and one is purely exponential in nature.

```
In[10]:= FAsymp[1, η, 5]
```

$$Out[10]= -\frac{\frac{146 \pi^{19/2}}{460550215125 \eta^{19/2}} + \frac{127 \pi^{15/2}}{638512875 \eta^{15/2}} + \frac{62 \pi^{11/2}}{893025 \eta^{11/2}} + \frac{7 \pi^{7/2}}{675 \eta^{7/2}} + \frac{\pi^{3/2}}{3 \eta^{3/2}}}{\pi} + \frac{2\sqrt{\eta}}{\sqrt{\pi}}$$

```
In[11]:= FAsymp[2, η, 5]
```

$$Out[11]= \frac{e^{-5\eta}}{5} - \frac{e^{-4\eta}}{4} + \frac{e^{-3\eta}}{3} - \frac{e^{-2\eta}}{2} + e^{-\eta} + \eta$$

A contour plot of $\mathcal{F}_{d,o}^{(alg)}(\eta) \big/ \mathcal{F}_{d,o}^{(exp)}(\eta)$ over the η,d-plane shows that the "special" points occur for $d = 1 + 4k$, $k \in \mathbb{N}$.

```
In[12]:= g[{d_, η_}, o_] := FAsympExp[d, η, o]/FAsympAlg[d, η, o]

In[13]:= With[{ε = 10^-6},
          ContourPlot[Evaluate[g[{d, η}, 5]], {η, ε, 3/2}, {d, ε, 9},
                  PlotPoints -> 200, PlotRange -> {-1, 1},
                  Contours -> 40, ColorFunction -> (Hue[0.8#]&),
                  ContourStyle -> {Thickness[0.002]}]]
```

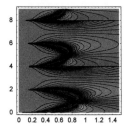

At the special points $\mathcal{F}_{d,o}^{(alg)}(\eta)$ vanishes. It is straightforward to calculate high-precision approximations of these points for a given k.

```
In[14]:= ηSpecial[d_] := FindRoot[Evaluate[𝓕AsympAlg[d, η, 2 d + 100] == 0],
                                  {η, 0.9}, WorkingPrecision -> 100] // N[#, 30]&
```

Here are the η-values for the first few points.

```
In[15]:= Table[{c, ηSpecial[1 + 4c]}, {c, 0, 4}]

Out[15]= {{0, {η → 0.869292163309593920883438143793}},
          {1, {η → 0.731131644870916378502124488633}},
          {2, {η → 0.696619639988909704490663500553}},
          {3, {η → 0.686821448211657580235757424992}},
          {4, {η → 0.684104146213182408556561001752}}}
```

The last numbers suggest that for $k \to \infty$ these numbers approach a limit. At $k = 34$ we obtain a 20-digit approximation of this limit.

```
In[16]:= ηSO = η /. ηSpecial[1]; k = 1;
         While[ηSN = η /. ηSpecial[1 + 4 k];
              (* until we have a 20 digit approximation for η *)
              Abs[ηSN - ηSO] > 10^-20, k++; ηSO = ηSN];
         {k, ηSN}
Out[18]= {34, 0.68312946455902666316407848634З}
```

c) The sum under consideration converges very slowly. The following plot shows the first 50000 partial sums.

```
In[1]:= ListPlot[FoldList[Plus, 0, Table[If[MemberQ[IntegerDigits[k], 9], 0, 1./k],
                          {k, 5 10^4}]], PlotRange -> All]
```

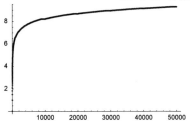

To use the closed-form expansion $\beta_0 \ln(10) + \sum_{k=2}^{\infty} 10^{-k} \beta(k-1) \zeta(k)$ for the sum under consideration, we must first determine the β_k's. β_0 can be calculated immediately to $\beta_0 = 10$.

```
In[2]:= eq[n_] := Sum[Binomial[n, k] (10^(n - k + 1) -
                     10^k +1) β[n - k], {k, n}] == 10 (11^n - 10^n)

In[3]:= eq[1]

Out[3]= β[0] == 10
```

We calculate the higher β_k by recursively solving the equation and backsubstitution of the already-calculated β_k. The next input calculates the first 60 β_k's.

```
In[4]:= βList = {β[0] -> 10};
        Do[AppendTo[βList, Solve[eq[k + 1] /. βList, β[k]][[1, 1]]], {k, 1, 60}]
```

Using these β_k's we calculate high-precision approximations for s.

```
In[6]:= sums = FoldList[Plus, 10 Log[10],
                        -Table[N[10^-k β[k - 1] Zeta[k] /.  βList, 200], {k, 2, 60}]];
```

The relative change of the sums from the last list shows that the last sum has about 59 correct digits.

```
In[7]:= ListPlot[Reverse[Log[10, N[Abs[Last[sums] - Rest[Reverse[sums]]]]]]]
```

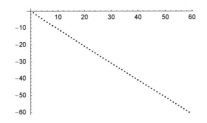

Here is the sum calculated to the required 50 digits.

```
In[8]:= N[sums[[-1]], 50]
Out[8]= 22.920676619264150348163657094375931914944762436998
```

For the similar sum $\sum_{k=1}^{\infty} n^{-1+\alpha/n}$, $\alpha \in \mathbb{R}$, see [1031].

d) Here is a direct implementation of $\varphi(z)$. We truncate the sum at n terms.

```
In[1]:= φ[z_, nMax_] := Sum[Product[1 - z^k, {k, 1, n}], {n, 0, nMax}]
```

It works, but it is relatively slow even for floating point numbers.

```
In[2]:= φ[Exp[2Pi I 11/99.], 100] // Timing
Out[2]= {0.02 Second, -0.0783152 - 27.3733 i}
```

Avoiding the repeated calculations of the products and compiling the function speeds up the calculations by about a factor 500.

```
In[3]:= φc = Compile[{{n, _Integer}, {d, _Integer}},
         Module[{q = Exp[2. Pi I n/d], sum = 1.0 + 0.0 I, prod = 1.0 + 0.0 I},
             Do[sum = sum + (prod = prod (1 - q^k)), {k, 1, d}]; sum]];
In[4]:= Table[φc[11, 99], {1000}] // Union // Timing
Out[4]= {0.07 Second, {-0.0783152 - 27.3733 i}}
```

Of course, we must make sure that using machine arithmetic gives correct results. Comparing the machine arithmetic results with a high-precision calculation shows that for denominators less than or equal 200 the resulting error is less than 0.02%. For denominators of order 300, the error is already 100%.

```
In[5]:= (* high-precision version of φ *)
        φHP = Function[{n, d, p},
        Module[{q = N[Exp[2 Pi I n/d], p], sum = 1, prod = 1},
            Do[sum = Expand[sum + (prod = prod (1 - q^k))], {k, 1, d}];
            sum]];
In[7]:= (* all irreducible fractions with denominator j *)
        fract[j_] := Select[Table[i/j, {i, j}], Denominator[#] === j&]
In[9]:= (* relative error *)
        δ[j_] := Abs[(#1 - #2)/#1]&[φc[Numerator[#], Denominator[#]],
                          φHP[Numerator[#], Denominator[#], 100]]& /@
                                      fract[j] // Abs // Max
In[11]:= {δ[100], δ[200], δ[300]}
Out[11]= {2.20223×10^-11, 0.00023656, 1.00049}
```

For further use, we calculate the values of $\varphi(\exp(2 \pi i p/q))$ for the first 19024 irreducible fractions.

```
In[12]:= n = 200;
         irreducibleFractions = Table[fract[j], {j, n}];
```

```
data = ({#, φc[Numerator[#], Denominator[#]]}& /@ #)& /@
                                    irreducibleFractions;
```

Omitting the first denominators shows that $\langle |\varphi(z)| \rangle \propto q^{3/2}$ to a very high precision.

In[15]:= `fitData = {Log[Denominator[#[[1, 1]]]], Log[(Plus @@ #/Length[#])&[`
        `Abs[#[[2]]]& /@ #]]}& /@ Drop[data, 50];`

In[16]:= `ListPlot[fitData // N]`

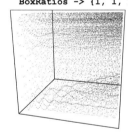

In[17]:= `Fit[fitData, {1, k}, k]`

Out[17]= $0.0515768 + 1.50012 \, k$

Rescaling $\varphi(\exp(2\pi i \, p/q))$ to $\tilde{\varphi}(\exp(2\pi i \, p/q)) = q^{3/2}\varphi(\exp(2\pi i \, p/q))$ allows all function values to be of order 10^0. Here, we display the absolute value of the $\tilde{\varphi}(\exp(2\pi i \, p/q))$ over the p,q-plane.

In[18]:= `Show[Graphics3D[{PointSize[0.002],`
        `Map[Point[{Denominator[#[[1]]], Numerator[#[[1]]],`
            `Abs[#[[2]]]/Denominator[#[[1]]]^(3/2) - 1}]&, data, {2}]}],`
        `BoxRatios -> {1, 1, 1}, ViewPoint -> {0.5, -2, 0.5}]`

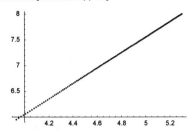

The next graphic shows $\tilde{\varphi}(\exp(2\pi i \, p/q))$ as a function of p/q.

In[19]:= `makePoint[{f_, z_}] := Point[{f, Abs[z] Denominator[f]^(-3/2)}]`

In[20]:= `Show[Graphics[{PointSize[0.002],`
        `MapIndexed[{Hue[0.8 #2[[1]]/n], makePoint[#1]}&, data, {2}]}],`
        `Frame -> True, PlotRange -> {0.95, 1.15}, AspectRatio -> 1/3]`

The Fourier transform of the pairs $\{p/q, \tilde{\varphi}(\exp(2\pi i \, p/q))\}$ shows an interesting structure.

In[21]:= `ListPlot[Abs[Fourier[{#[[1]], #[[2]]/Denominator[#[[1]]]^(3/2)}& /@`
        `Sort[Flatten[data, 1], #1[[1]] < #2[[1]]&]]],`
        `PlotStyle -> {PointSize[0.002]}]`

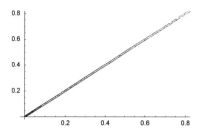

Now, let us visualize how the individual terms form the sum. The function φcL calculates a list of all partial sums.

```
In[22]:= φcL = Compile[{{n, _Integer}, {d, _Integer}},
            Module[{q = Exp[2. Pi I n/d], sum = 1.0 + 0.0 I, prod = 1.0 + 0.0 I, tab},
               tab = Table[prod = prod (1 - q^k), {k, 1, d}];
               Rest[FoldList[Plus, 0. + 0. I, tab]]]];
```

Because of the $\langle|\varphi(z)|\rangle \propto q^{3/2}$ dependence of the function values, we use a polar coordinate system with a logarithmic scale for the radial coordinate. The next picture shows how the partial sums approach the value of $\varphi\!\left(\exp\!\left(2\,\pi\,i\,\frac{p}{q}\right)\right)$.

```
In[23]:= data = Map[φcL[Numerator[#], Denominator[#]]&,
             Take[irreducibleFractions, 50], {2}];

In[24]:= (* avoid unevaluated Arg[0] *)
         Unprotect[Arg]; Arg[0.0 + 0.0 I] = 0;

         Show[Graphics[{Thickness[0.002], Reverse[MapIndexed[Function[{l, pos},
               {Hue[0.8 pos[[1]]/30],  (* log r scale *)
                  Line[({Re[#], Im[#]}&[Log[1 + Abs[#]] Exp[I Arg[#]]])& /@ l]}],
                  data, {2}]]}], Frame -> True,
            PlotRange -> 10 {{-1, 1}, {-1, 1}}, AspectRatio -> Automatic]
```

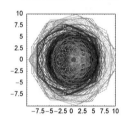

12. Heat Equation, Green's Function for a Rectangle, Theta Function Addition Formulas, Theta Function Series Expansion, Bose Gas

a) Unfortunately, attempting to get a closed-form sum for the original series fails.

```
In[1]:= Sum[Exp[-a^2 n^2] Sin[n x] Sin[n y], {n, 1, Infinity}]
```

$$\text{Out[1]}= \sum_{n=1}^{\infty} e^{-a^2 n^2} \operatorname{Sin}[n\,x] \operatorname{Sin}[n\,y]$$

In fact, even the defining series $\vartheta_3(q, u) = 1 + 2 \sum_{n=1}^{\infty} q^{n^2} \cos(2\,nu)$ is not recognized by *Mathematica*.

```
In[2]:= Sum[q^(n^2) Cos[2 n u], {n, Infinity}]
```

$$\text{Out[2]}= \sum_{n=1}^{\infty} q^{n^2} \operatorname{Cos}[2\,n\,u]$$

After some manipulation by hand, we have

$$T(x, t) = \frac{1}{2\,l}\left(\vartheta_3\!\left(\exp\!\left(-\frac{\pi^2\,a^2}{l^2}\,t\right), \frac{\pi(x - y)}{2\,l}\right) - \vartheta_3\!\left(\exp\!\left(-\frac{\pi^2\,a^2}{l^2}\,t\right), \frac{\pi(x + y)}{2\,l}\right)\right).$$

This is the usually presented series expansion in the references (see, e.g., [1168], [1224], [543], [1100], [646], [102], [78], and [758]). Here is the *Mathematica* implementation.

```
In[3]:= temperature[a_, l_, x_, y_, t_] =
        1/(21) (EllipticTheta[3, Pi/(21) (x - y), Exp[-Pi^2 a^2/l^2 t]] -
              EllipticTheta[3, Pi/(21) (x + y), Exp[-Pi^2 a^2/l^2 t]]);
```

To illustrate the usefulness of the closed formula, we now present a numerical attempt to compute the temperature (without changing the default option settings in NSum) from the series.

```
In[4]:= temperatureSeries[a_, l_, x_, y_, t_, opts___Rule] :=
        2/1  NSum[Exp[-Pi^2 a^2/l^2 t n^2] Sin[n Pi x/l] Sin[n Pi y/l],
              {n, Infinity}, opts] // N
```

For typical arguments, temperatureSeries will run quite long (for the efficient numerical evaluation of such-type series, see [796]).

```
In[5]:= temperatureSeries[1, N[Pi], 0.1, 1, 0.1] // Timing
```

```
Out[5]= {1.45 Second, 0.0744301}
```

The formula using elliptic Theta functions is much faster.

```
In[6]:= Do[temperature[1, N[Pi], x, 1, 0.1], {x, 0, 1, 1/1000}] // Timing
```

```
Out[6]= {0.11 Second, Null}
```

Here is a plot of the temperature for $t = 0.1$. (Let $l = \pi$ and $a = 1$.) The right graphic shows a 3D plot of $T(x, t)$.

```
In[7]:= Show[GraphicsArray[
        Block[{$DisplayFunction = Identity},
        (* plot of T for t = 0.1 and 3D plot of T as a function of t *)
        {Plot[temperature[1, Pi, 0.5, y, 0.1], {y, 0, Pi},
            PlotRange -> All, Compiled -> False, AxesLabel -> {"x", "T"},
            Prolog -> {Thickness[0.01], Line[{{0.5, 0}, {0.5, 0.6}}]},
            PlotStyle -> {Thickness[0.002]}];
        Plot3D[temperature[1, Pi, 0.1, x, t], {x, 0, Pi/3}, {t, 0.05, 0.2},
            PlotRange -> All, PlotPoints -> 25, AxesLabel -> {"x", "t", "T"}]}]]]
```

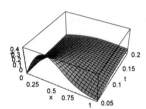

Not only to linear, but also solutions to many nonlinear partial differential equation can be expressed in Theta functions [408]. For a solution of the initial value problem of the 1D heat equation using Meijer G functions, see [1183].

b) Here is the direct implementation of Green's function. The four variables x, y, xs, and ys are all real.

```
In[1]:= GreenRectangle[{x_, y_}, {xs_, ys_}, a_, b_] :=
        Module[{q = Exp[-Pi b/a], t, C = Conjugate, z = x + I y, zs = xs + I ys},
              t[arg_] = EllipticTheta[1, arg/(2a) Pi, q];
              1/(2Pi) Re[Log[(t[z + C[zs]] t[z - C[zs]])/(t[z + zs] t[z - zs])]]] /.
              Conjugate[z_] :> (z /. i_Complex :> Conjugate[i])
```

For convenience, we now define the Green's function for symbolic arguments.

```
In[2]:= gf = GreenRectangle[{x, y}, {xs, ys}, a, b]
```

$$\text{Out[2]= } \frac{\text{Re}\left[\text{Log}\left[\dfrac{\text{EllipticTheta}\left[1, \frac{\pi (x+xs+i\,y-i\,ys)}{2a}, e^{-\frac{b\pi}{a}}\right]\, \text{EllipticTheta}\left[1, \frac{\pi (x-xs+i\,y+i\,ys)}{2a}, e^{-\frac{b\pi}{a}}\right]}{\text{EllipticTheta}\left[1, \frac{\pi (x-xs+i\,y-i\,ys)}{2a}, e^{-\frac{b\pi}{a}}\right]\, \text{EllipticTheta}\left[1, \frac{\pi (x+xs+i\,y+i\,ys)}{2a}, e^{-\frac{b\pi}{a}}\right]}\right]\right]}{2\pi}$$

The Green's function satisfies the Laplace equation. (We do not get the isolated point where the right-hand side does not equal 0 because differentiation is generic. This means single-point singularities are ignored by *Mathematica* by default.)

```
In[3]:= D[gf, {x, 2}] + D[gf, {y, 2}] // Simplify
```

Out[3]= 0

Unfortunately, the asymptotic behavior for $x \approx x'$ and $y \approx y'$ cannot be computed directly. (Because there is currently no mechanism to tell Series that the variables involved are all real, some problems with differentiating Re are expected.)

In[4]:= Off[EllipticThetaPrime::argr];
ser = Series[gf, {x, xs, 1}, {y, ys, 0}] // Short[#, 4]&

$$\text{Out[5]//Short=} \quad \frac{\mathrm{Re}\left[\mathrm{Log}\left[\dfrac{\text{EllipticTheta}\left[1, \frac{\pi\,(x+xs+i\,y-i\,ys)}{2\,a}, e^{-\frac{b\pi}{a}}\right]\text{EllipticTheta}\left[1, \frac{\pi\,(\ll 1\gg)}{2\,a}, e^{-\frac{\ll 1\gg}{a}}\right]}{\text{EllipticTheta}\left[1, \frac{\pi\,(x-xs+i\,y-i\,ys)}{2\,a}, e^{-\frac{b\pi}{a}}\right]\text{EllipticTheta}\left[1, \frac{\pi\,(\ll 1\gg)}{2\,a}, e^{-\frac{\ll 1\gg}{a}}\right]}\right]\right]}{2\,\pi}$$

In[6]:= Off[EllipticThetaPrime::argr];
Series[gf /. Re[ζ_] :> ζ, {x, xs, 0}, {y, ys, 0}] // TraditionalForm

$$\text{Out[7]//TraditionalForm=} \quad \left(\frac{\log\left|-\dfrac{2\,i\,a\vartheta_1\left(\frac{\pi\,xs}{a}, e^{-\frac{b\pi}{a}}\right)\vartheta_1\left(\frac{i\pi\,ys}{a}, e^{-\frac{b\pi}{a}}\right)}{\pi\vartheta_1\left(\frac{\pi\,xs}{a}+\frac{i\pi\,ys}{a}, e^{-\frac{b\pi}{a}}\right)\vartheta_1'\left(0, e^{-\frac{b\pi}{a}}\right)}\right| - \log(y - ys)}{2\,\pi} + O((y - ys)^1)\right) + O((x - xs)^1)$$

In[8]:= Off[EllipticThetaPrime::argr];
Series[gf /. Re[ζ_] :> ζ, {y, ys, 0}, {x, xs, 0}] // TraditionalForm

$$\text{Out[9]//TraditionalForm=} \quad \left(\frac{\log\left|\dfrac{2\,a\vartheta_1\left(\frac{\pi\,xs}{a}, e^{-\frac{b\pi}{a}}\right)\vartheta_1\left(\frac{i\pi\,ys}{a}, e^{-\frac{b\pi}{a}}\right)}{\pi\vartheta_1\left(\frac{\pi\,xs}{a}+\frac{i\pi\,ys}{a}, e^{-\frac{b\pi}{a}}\right)\vartheta_1'\left(0, e^{-\frac{b\pi}{a}}\right)}\right| - \log(x - xs)}{2\,\pi} + O((x - xs)^1)\right) + O((y - ys)^1)$$

Similarly, the homogeneous Dirichlet conditions on the boundary cannot be verified symbolically.

In[10]:= ((GreenRectangle[#, {xs, ys}, a, b] // FullSimplify) === 0)& /@
{{0, y}, {2a, y}, {x, 0}, {x, 2b}}
Out[10]= {False, False, False, False}

We now present a plot of the Green's function. One sees that on the boundary, the Green's function vanishes and that the minimax principle holds. A smooth function remains after subtracting the singular part from the Green's function. Here, this is shown in the vicinity of the singularity.

In[11]:= Show[GraphicsArray[
Block[{$DisplayFunction = Identity},
{(* overall 3D view *)
Plot3D[Evaluate[GreenRectangle[{x, y}, {2, 2}, 4, 3]],
{x, 0, 4}, {y, 0, 3}, PlotPoints -> 120,
Mesh -> False, PlotRange -> {0, 1/3}, ClipFill -> None],
(* overall contour view *)
ContourPlot[Evaluate[GreenRectangle[{x, y}, {2, 2}, 4, 3]],
{x, 0, 4}, {y, 0, 3}, PlotPoints -> 40, Contours -> 30,
ContourShading -> False, AspectRatio -> Automatic],
(* zoom-in view *)
Plot3D[Evaluate[GreenRectangle[{x, y}, {2, 2}, 4, 3] -
(-1/(2Pi) Log[Sqrt[(x - 2)^2 + (y - 2)^2]])],
{x, 1.99, 2.01}, {y, 1.99, 2.01}, PlotPoints -> 20]}]]];

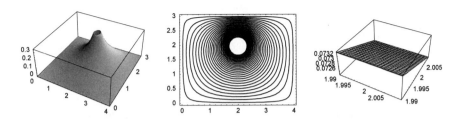

So all properties of a Green's function seem to be satisfied.

(Because of the zero boundary conditions, GreenRectangle[φ_1, φ_2, φ_1', φ_2', π, π] can be nicely visualized on a torus. We leave this to the reader.)

We will now use the Green's function to solve the boundary value problem. We start by calculating the normal derivative of the Green's function.

In[12]:= `gf = GreenRectangle[{x, y}, {xs, ys}, a, b];`
`(dgf[{x_, y_}, {xs_, ys_}, a_, b_] = (* do nor differentiate Re *)`
`             1/a Re[D[a gf /. Re[ζ_] :> ζ, xs] // Simplify]) // TraditionalForm`

Out[13]//TraditionalForm=

$$
\mathrm{Re}\left(\frac{\vartheta_1'\left(\frac{\pi(x-xs+i(y-ys))}{2a}, e^{-\frac{b\pi}{a}}\right)}{\vartheta_1\left(\frac{\pi(x-xs+i(y-ys))}{2a}, e^{-\frac{b\pi}{a}}\right)} + \frac{\vartheta_1'\left(\frac{\pi(x+xs+i(y-ys))}{2a}, e^{-\frac{b\pi}{a}}\right)}{\vartheta_1\left(\frac{\pi(x+xs+i(y-ys))}{2a}, e^{-\frac{b\pi}{a}}\right)} - \frac{\vartheta_1'\left(\frac{\pi(x-xs+i(y+ys))}{2a}, e^{-\frac{b\pi}{a}}\right)}{\vartheta_1\left(\frac{\pi(x-xs+i(y+ys))}{2a}, e^{-\frac{b\pi}{a}}\right)} - \frac{\vartheta_1'\left(\frac{\pi(x+xs+i(y+ys))}{2a}, e^{-\frac{b\pi}{a}}\right)}{\vartheta_1\left(\frac{\pi(x+xs+i(y+ys))}{2a}, e^{-\frac{b\pi}{a}}\right)}}{4a}\right)
$$

To minimize the numerical integrations needed, we subtract 1 from the boundary values. This gives two sides of the rectangle with homogeneous boundary conditions where the integral vanishes. We reduce the default values for the precision and accuracy goals of NIntegrate because we are only interested in a visualization here.

In[14]:= `SetOptions[NIntegrate, PrecisionGoal -> 4,`
`                          AccuracyGoal -> 4, MaxRecursion -> 8];`

`(* integral along the rectangle boundary *)`
`ψ[{x_, y_}, a_, b_] := (* integrand is peaked near ys ≈ y near edges *)`
`      NIntegrate[dgf[{x, y}, {0, ys}, a, b], {ys, 0, y, b}] +`
`      NIntegrate[dgf[{x, y}, {a, ys}, a, b], {ys, b, y, 0}]`

`(* no integration needed here *)`
`ψ[{x_, y_}, a_, b_] := 0. /; y == 0 || y == a`

We use the symmetry of the problem under consideration to calculate the values of $\psi(x, y)$ in only one quarter of the rectangle. Now we add the value 1 to the solution to fulfill the original boundary conditions.

In[20]:= `ψDataQuarter = Flatten[#, 1]& @`
`      With[{a = 2, b = 1, ε = 10^-3, app = 50, bpp = 50},`
`          Table[{x, y, ψ[{x, y}, a, b] + 1},`
`                  {x, ε, a/2, (a/2 - ε)/app}, {y, 0, b/2, b/2/bpp}]];`

Mirroring the data on the two symmetry lines of the rectangle yields the list of data ψDataQuarter.

In[21]:= `ψDataAll = Union @ With[{a = 2, b = 1},`
`      Join[ψDataQuarter, {a - #1, #2, #3}& @@@ ψDataQuarter,`
`          {#1, b - #2, #3}& @@@ ψDataQuarter,`
`          {a - #1, b - #2, #3}& @@@ ψDataQuarter]];`

Now we interpolate the data and calculate the Laplace operator applied to the resulting function. For a numerically reliable differentiation we use a higher interpolation order in Interpolation.

In[22]:= `ipo = Interpolation[ψDataAll, InterpolationOrder -> 12];`
`Δ[x_, y_] = D[ipo[x, y], x, x] + D[ipo[x, y], y, y];`

The following graphics show the function $\psi(x, y)$ and $(\partial^2/\partial x^2 + \partial^2/\partial y^2)\psi(x, y)$. In average, the right graphic shows that the residuals are mostly less than 10^{-5}. Taking into account that we performed the numerical integration to only 4 digits no better result was to be expected. Near the boundaries where the integrand becomes sharply peaked and the derivatives become one-sided and near the corners where we have discontinuous boundary conditions, the error of the solution is larger.

```
In[24]:= With[{a = 2, b = 1, ε = 10^-3},
         Show[GraphicsArray[
          Block[{$DisplayFunction = Identity},
           {(* 3D plot of the solution of Laplace equation *)
            Plot3D[ipo[x, y], {x, ε, a - ε}, {y, 0, b},
                  PlotPoints -> 60, Mesh -> False],
            (* contour plot of the solution of Laplace equation *)
            ContourPlot[ipo[x, y], {x, ε, a - ε}, {y, 0, b},
                       PlotPoints -> 160, ColorFunction -> (Hue[0.78 #]&),
                       Contours -> Append[Table[c, {c, 1, 2, 1/20}], 1.1097698],
                       PlotRange -> All, AspectRatio -> Automatic],
            (* error of the solution *)
            Plot3D[Log[10, Abs[Δ[x, y]]], {x, ε, a - ε}, {y, 0, b},
                  PlotPoints -> 60, PlotRange -> All, Mesh -> False]}]]]];
```

We end by remarking that for constant boundary values on the edges no numerical integration is actually necessary. The integral $\psi(x, y) = -\int \partial G_{ab}(x, y, x', y')/\partial x' \times constant \, dy'$ can be carried out in closed form by partial integration.

```
In[25]:= (idgf[{x_, y_}, {xs_, ys_}, a_, b_] = Re[Simplify[
          Integrate[D[gf /. Re -> Identity /.
                  (* use symbolic indefinite integration *) EllipticTheta -> θ,
                  xs], ys] /. θ -> EllipticTheta]]) // TraditionalForm
```

Out[25]//TraditionalForm=

$$-\frac{1}{2\pi}\mathrm{Im}\left(\log\left(\vartheta_1\left(\frac{\pi(x - xs + i(y - ys))}{2a}, e^{-\frac{b\pi}{a}}\right)\right) + \log\left(\vartheta_1\left(\frac{\pi(x + xs + i(y - ys))}{2a}, e^{-\frac{b\pi}{a}}\right)\right) + \log\left(\vartheta_1\left(\frac{\pi(x - xs + i(y + ys))}{2a}, e^{-\frac{b\pi}{a}}\right)\right) + \log\left(\vartheta_1\left(\frac{\pi(x + xs + i(y + ys))}{2a}, e^{-\frac{b\pi}{a}}\right)\right)\right)$$

```
In[26]:= Ψ[{x_, y_}, a_, b_] :=
          idgf[{x, y}, {0, b}, a, b] - idgf[{x, y}, {0, 0}, a, b] +
          1 - (idgf[{x, y}, {a, b}, a, b] - idgf[{x, y}, {a, 0}, a, b])
```

Here is a check for "random" values of x and y showing that $\psi[\{x, y\}, a, b]$ agrees with $\Psi[\{x, y\}, a, b]$.

```
In[27]:= With[{a = 1, b = 2, x = 1/Pi, y = 1/E},
          {ψ[{x, y}, a, b], Ψ[{x, y}, a, b] // N}]
Out[27]= {0.655828 + 0. i, 0.655828}
```

For the Green's function of a thin strip with homogeneous Dirichlet boundary conditions (free water surface over a finite height basin), see [1011].

c) To find all identities, we cannot carry out an exhaustive search over all possible combinations of $i_1, i_2, i_3, i_4, j_1, j_2, j_3, j_4$, $k_1, k_2, k_3, k_4, l_1, l_2, l_3, l_4$ and the three $\pm$ (this would mean to check $4^{16} \, 2^3 \approx 3.5 \, 10^{10}$ cases) in a reasonable time. (Although on a modern year 2002 computer this is in principle certainly doable.) So we will proceed in the following way.

We fix "random" values for the variables x, y, z, w, and q and calculate all possible 131072 values of the left-hand side and all possible 262144 of the right-hand side of the identities. Then we unite both lists and group them into sublists of equal numerical values. After deleting multiple left-hand and right-hand sides, we are left with candidates of the identities. Each left-hand and right-hand side is stored in the form C[*numericalValue*, {*indices1*, *signs*, *indices2*}]. Using random values for the variables x, y, z, w, and q does not guarantee that it will be a correct identity; we will have to check again afterwards. On the other hand, we will not miss any identity in this way. If it is an identity, then it certainly holds for selected random values.

Here this approach is implemented.

```
In[1]:= {x, y, z, w} = Table[1/Prime[k], {k, 4}];
        {ω, ξ, ψ, ζ} = {w + x + y + z, w + x - y - z, w - x + y - z, w - x - y + z}/2;

In[3]:= (* numerical values of the theta functions *)
        Do[(ϑ[i, #] = N[EllipticTheta[i, #, 1/E], $MachinePrecision + 5])& /@
                                                {w, x, y, z, ω, ξ, ψ, ζ}, {i, 4}]

In[5]:= (* product of the left-hand side *)
        ϑΠx[i_, j_, k_, l_] := ϑΠx[i, j, k, l] = ϑ[i, w] ϑ[j, x] ϑ[k, y] ϑ[l, z]

        (* product of the right-hand side *)
        ϑΠξ[i_, j_, k_, l_] := ϑΠξ[i, j, k, l] = ϑ[i, ω] ϑ[j, ξ] ϑ[k, ψ] ϑ[l, ζ]

In[9]:= (* all possible left-hand side *)
        tab1 = Flatten[
        Table[C[ϑΠx[i1, i2, i3, i4] + (-1)^β ϑΠx[j1, j2, j3, j4],
                {{i1, i2, i3, i4}, {β}, {j1, j2, j3, j4}}],
              {i1, 4}, {i2, 4}, {i3, 4}, {i4, 4},
              {j1, 4}, {j2, 4}, {j3, 4}, {j4, 4}, {β, 0, 1}]];

In[11]:= (* all possible right-hand side *)
        tab2 = Flatten[
        Table[C[(-1)^α ϑΠξ[k1, k2, k3, k4] + (-1)^β ϑΠξ[l1, l2, l3, l4],
                {{k1, k2, k3, k4}, {α, β}, {l1, l2, l3, l4}}],
              {k1, 4}, {k2, 4}, {k3, 4}, {k4, 4},
              {l1, 4}, {l2, 4}, {l3, 4}, {l4, 4}, {α, 0, 1}, {β, 0, 1}]];
```

We now create a list called `tab3`. It contains all the values of the differences of the products.

```
In[13]:= tab3 = MapAt[N, #, 1]& /@ Flatten[{tab1, tab2}];
        Length[tab3]
Out[14]= 393216
```

`tab3` is a list with 393216 elements. We split `tab3` into sublist with equal numerical values and delete the sublists that are not candidates for identities.

```
In[15]:= splits = Split[Sort[tab3], (First[#1] == First[#2])&];

        splits1 = Select[splits, (* both sides occur? *) ((Length[#] > 1) &&
                        (Length[Union[Length /@ (#[[2, 2]]& /@ #)]] === 2))&];
```

Next we remove trivial cases that arise from commutativity.

```
In[17]:= splits2 = DeleteCases[DeleteCases[splits1,
                        C[_, {a_, {1}, a_}] | C[_, {a_, {1, 0}, a_}] |
                        C[_, {a_, {0, 1}, a_}], {2}], {}];

        uniteRules = Join[Function[{m1, m2},
            {s___, c:C[_, {a_, m1, b_}], t___,
                        C[_, {b_, m2, a_}], u___} :> {s, c, t, u}] @@@
                {{{0}, {0}}, {{1, 0}, {0, 1}}, {{0, 1}, {1, 0}}},
                                Function[m,
            {s___, c:C[_, {a_, m, b_}], t___,
                        C[_, {b_, m, a_}], u___} :> {s, c, t, u}] /@ {{0, 0}, {1, 1}}];
In[20]:= splits3 = (#  //. uniteRules)& /@ splits2;
```

Now we have 288 candidates, each representing one identity.

```
In[21]:= {Length[splits3], Union[Length /@ splits3]}

Out[21]= {288, {2}}
```

`splits5` contains a sorted version of the result.

```
In[22]:= splits4 = Sort[#, Length[#1[[2, 2]]] < Length[#2[[2, 2]]]]& /@ splits3;

In[23]:= splits5 = Sort[Sort[splits4, OrderedQ[{#1[[1, 2, 2]], #2[[1, 2, 2]]}]&],
                                OrderedQ[{#1[[1, 2, 1]], #2[[1, 2, 1]]}]&];
```

`makeEquation` generates a form of the corresponding identity that has head `Equal`.

```
In[24]:= makeEquation[{C[_, {{a___}, {α_}, {b___}}],
                C[_, {{c___}, {β_, γ_}, {d___}}]}] :=
    (Subscript[Π, a]) + (-1)^α Subscript[Π, b] ==
    (-1)^β Subscript[OverTilde[Π], c] + (-1)^γ Subscript[OverTilde[Π], d]
In[25]:= allIdentities = makeEquation /@ splits5;
```

To display the 288 identities as concisely as possible, we make use of the common notation [762], [830]:

$$\vartheta_{i_1}(w, q)\, \vartheta_{i_2}(x, q)\, \vartheta_{i_3}(y, q)\, \vartheta_{i_4}(z, q) \equiv [i_1, i_2, i_3, i_4]$$
$$\vartheta_{i_1}(\omega, q)\, \vartheta_{i_2}(\xi, q)\, \vartheta_{i_3}(\psi, q)\, \vartheta_{i_4}(\zeta, q) \equiv \{i_1, i_2, i_3, i_4\}.$$

```
In[26]:= writeIdentityInShortForm[id_] := id /.
    {Subscript[Π, a___] :> "[" <> StringJoin[ToString /@ {a}] <> "]",
    Subscript[OverTilde[Π], a___] :> "{" <> StringJoin[ToString /@ {a}] <> "}"}
```

Here are all 288 identities found.

```
In[27]:= (* add +s for alignment *)
    (StringReplace[(#1 <> "=" <> #2)& @@ (
      If[StringTake[#, {1, 1}] === "-", #, "+" <> #]&[
        ToString[#]]& /@ (List @@ #)), " " -> ""])& /@
          (writeIdentityInShortForm /@ allIdentities)
Out[28]= {+[1111]+[4444]==-{2222}+{3333}, +[1111]+[2222]==+{3333}-{4444},
    +[1111]+[3333]==+{1111}+{3333}, +[1111]-[3333]==-{2222}-{4444},
    +[1111]-[2222]==+{1111}-{2222}, +[1111]-[4444]==+{1111}-{4444},
    +[1122]+[2211]==-{3344}+{4433}, +[1122]+[3344]==-{2211}+{4433},
    +[1122]+[4433]==+{1122}+{4433}, +[1122]-[4433]==-{2211}-{3344},
    +[1122]-[3344]==+{1122}-{3344}, +[1122]-[2211]==+{1122}-{2211},
    +[1133]+[3311]==-{2244}+{4422}, +[1133]+[2244]==-{3311}+{4422},
    +[1133]+[4422]==+{1133}+{4422}, +[1133]-[4422]==-{2244}-{3311},
    +[1133]-[2244]==+{1133}-{2244}, +[1133]-[3311]==+{1133}-{3311},
    +[1144]+[4411]==-{2233}+{3322}, +[1144]+[2233]==+{3322}-{4411},
    +[1144]+[3322]==+{1144}+{3322}, +[1144]-[3322]==-{2233}-{4411},
    +[1144]-[2233]==+{1144}-{2233}, +[1144]-[4411]==+{1144}-{4411},
    +[1212]+[2121]==-{3434}+{4343}, +[1212]+[4343]==+{1212}+{4343},
    +[1212]+[3434]==-{2121}+{4343}, +[1212]-[3434]==+{1212}-{3434},
    +[1212]-[4343]==-{2121}-{3434}, +[1212]-[2121]==+{1212}-{2121},
    +[1221]+[4334]==+{1221}+{4334}, +[1221]+[2112]==-{3443}+{4334},
    +[1221]+[3443]==-{2112}+{4334}, +[1221]-[3443]==+{1221}-{3443},
    +[1221]-[2112]==+{1221}-{2112}, +[1221]-[4334]==-{2112}-{3443},
    +[1234]+[4321]==+{1234}+{4321}, +[1234]+[3412]==+{1234}+{3412},
    +[1234]+[2143]==+{1234}+{2143}, +[1234]-[2143]==+{3412}+{4321},
    +[1234]-[3412]==+{2143}+{4321}, +[1234]-[4321]==+{2143}+{3412},
    +[1243]+[4312]==+{1243}+{4312}, +[1243]+[3421]==+{1243}+{3421},
    +[1243]+[2134]==+{1243}+{2134}, +[1243]-[2134]==+{3421}+{4312},
    +[1243]-[3421]==+{2134}+{4312}, +[1243]-[4312]==+{2134}+{3421},
    +[1313]+[3131]==-{2424}+{4242}, +[1313]+[4242]==+{1313}+{4242},
    +[1313]+[2424]==-{3131}+{4242}, +[1313]-[2424]==+{1313}-{2424},
    +[1313]-[4242]==-{2424}-{3131}, +[1313]-[3131]==+{1313}-{3131},
    +[1324]+[4231]==+{1324}+{4231}, +[1324]+[2413]==+{1324}+{2413},
    +[1324]+[3142]==+{1324}+{3142}, +[1324]-[3142]==+{2413}+{4231},
    +[1324]-[2413]==+{3142}+{4231}, +[1324]-[4231]==+{2413}+{3142},
    +[1331]+[4224]==+{1331}+{4224}, +[1331]+[3113]==-{2442}+{4224},
    +[1331]+[2442]==-{3113}+{4224}, +[1331]-[2442]==+{1331}-{2442},
    +[1331]-[3113]==+{1331}-{3113}, +[1331]-[4224]==-{2442}-{3113},
    +[1342]+[4213]==+{1342}+{4213}, +[1342]+[2431]==+{1342}+{2431},
    +[1342]+[3124]==+{1342}+{3124}, +[1342]-[3124]==+{2431}+{4213},
    +[1342]-[2431]==+{3124}+{4213}, +[1342]-[4213]==+{2431}+{3124},
    +[1414]+[4141]==-{2323}+{3232}, +[1414]+[3232]==+{1414}+{3232},
    +[1414]+[2323]==+{3232}-{4141}, +[1414]-[2323]==+{1414}-{2323},
```

$+[1414]-[3232]=-\{2323\}-\{4141\},\ +[1414]-[4141]=+\{1414\}-\{4141\},$
$+[1423]+[3241]=+\{1423\}+\{3241\},\ +[1423]+[2314]=+\{1423\}+\{2314\},$
$+[1423]+[4132]=+\{1423\}+\{4132\},\ +[1423]-[4132]=+\{2314\}+\{3241\},$
$+[1423]-[2314]=+\{3241\}+\{4132\},\ +[1423]-[3241]=+\{2314\}+\{4132\},$
$+[1432]+[2341]=+\{1432\}+\{2341\},\ +[1432]+[3214]=+\{1432\}+\{3214\},$
$+[1432]+[4123]=+\{1432\}+\{4123\},\ +[1432]-[4123]=+\{2341\}+\{3214\},$
$+[1432]-[3214]=+\{2341\}+\{4123\},\ +[1432]-[2341]=+\{3214\}+\{4123\},$
$+[1441]+[4114]=-\{2332\}+\{3223\},\ +[1441]+[3223]=+\{1441\}+\{3223\},$
$+[1441]+[2332]=+\{3223\}-\{4114\},\ +[1441]-[2332]=+\{1441\}-\{2332\},$
$+[1441]-[3223]=-\{2332\}-\{4114\},\ +[1441]-[4114]=+\{1441\}-\{4114\},$
$+[2112]+[4334]=+\{2112\}+\{4334\},\ +[2112]+[3443]=-\{1221\}+\{4334\},$
$+[2112]-[3443]=+\{2112\}+\{3443\},\ +[2112]-[4334]=-\{1221\}-\{3443\},$
$-[1221]+[2112]=-\{1221\}+\{2112\},\ +[2121]+[4343]=+\{2121\}+\{4343\},$
$+[2121]+[3434]=-\{1212\}+\{4343\},\ +[2121]-[3434]=+\{2121\}-\{3434\},$
$+[2121]-[4343]=-\{1212\}-\{3434\},\ -[1212]+[2121]=-\{1212\}+\{2121\},$
$+[2134]+[4312]=+\{1243\}-\{3421\},\ +[2134]+[3421]=+\{1243\}-\{4312\},$
$+[2134]-[3421]=+\{2134\}-\{3421\},\ +[2134]-[4312]=+\{2134\}-\{4312\},$
$-[1243]+[2134]=-\{3421\}-\{4312\},\ +[2143]+[4321]=+\{1234\}-\{3412\},$
$+[2143]+[3412]=+\{1234\}-\{4321\},\ +[2143]-[3412]=+\{2143\}-\{3412\},$
$+[2143]-[4321]=+\{2143\}-\{4321\},\ -[1234]+[2143]=-\{3412\}-\{4321\},$
$+[2211]+[3344]=-\{1122\}+\{4433\},\ +[2211]+[4433]=+\{2211\}+\{4433\},$
$+[2211]-[4433]=-\{1122\}-\{3344\},\ +[2211]-[3344]=+\{2211\}-\{3344\},$
$-[1122]+[2211]=-\{1122\}+\{2211\},\ +[2222]+[4444]=-\{1111\}+\{3333\},$
$+[2222]+[3333]=+\{2222\}+\{3333\},\ +[2222]-[3333]=-\{1111\}-\{4444\},$
$+[2222]-[4444]=+\{2222\}-\{4444\},\ -[1111]+[2222]=-\{1111\}+\{2222\},$
$+[2233]+[4411]=-\{1144\}+\{3322\},\ +[2233]+[3322]=+\{2233\}+\{3322\},$
$+[2233]-[3322]=-\{1144\}-\{4411\},\ +[2233]-[4411]=+\{2233\}-\{4411\},$
$-[1144]+[2233]=-\{1144\}+\{2233\},\ +[2244]+[3311]=-\{1133\}+\{4422\},$
$+[2244]+[4422]=+\{2244\}+\{4422\},\ +[2244]-[4422]=-\{1133\}-\{3311\},$
$-[1133]+[2244]=-\{1133\}+\{2244\},\ +[2244]-[3311]=+\{2244\}-\{3311\},$
$+[2314]+[3241]=+\{1423\}-\{4132\},\ +[2314]+[4132]=+\{1423\}-\{3241\},$
$+[2314]-[4132]=+\{2314\}-\{4132\},\ -[1423]+[2314]=-\{3241\}-\{4132\},$
$+[2314]-[3241]=+\{2314\}-\{3241\},\ +[2323]+[4141]=-\{1414\}+\{3232\},$
$+[2323]+[3232]=+\{2323\}+\{3232\},\ +[2323]-[3232]=-\{1414\}-\{4141\},$
$+[2323]-[4141]=+\{2323\}-\{4141\},\ -[1414]+[2323]=-\{1414\}+\{2323\},$
$+[2332]+[4114]=-\{1441\}+\{3223\},\ +[2332]+[3223]=+\{2332\}+\{3223\},$
$+[2332]-[3223]=-\{1441\}-\{4114\},\ +[2332]-[4114]=+\{2332\}-\{4114\},$
$-[1441]+[2332]=-\{1441\}+\{2332\},\ +[2341]+[3214]=+\{1432\}-\{4123\},$
$+[2341]+[4123]=+\{1432\}-\{3214\},\ +[2341]-[4123]=+\{2341\}-\{4123\},$
$+[2341]-[3214]=+\{2341\}-\{3214\},\ -[1432]+[2341]=-\{3214\}-\{4123\},$
$+[2413]+[4231]=+\{1324\}-\{3142\},\ +[2413]+[3142]=+\{1324\}-\{4231\},$
$+[2413]-[3142]=+\{2413\}-\{3142\},\ +[2413]-[4231]=+\{2413\}-\{4231\},$
$-[1324]+[2413]=-\{3142\}-\{4231\},\ +[2424]+[3131]=-\{1313\}+\{4242\},$
$+[2424]+[4242]=+\{2424\}+\{4242\},\ +[2424]-[4242]=-\{1313\}-\{3131\},$
$+[2424]-[3131]=+\{2424\}-\{3131\},\ -[1313]+[2424]=-\{1313\}+\{2424\},$
$+[2431]+[4213]=+\{1342\}-\{3124\},\ +[2431]+[3124]=+\{1342\}-\{4213\},$
$+[2431]-[3124]=+\{2431\}-\{3124\},\ +[2431]-[4213]=+\{2431\}-\{4213\},$
$-[1342]+[2431]=-\{3124\}-\{4213\},\ +[2442]+[4224]=+\{2442\}+\{4224\},$
$+[2442]+[3113]=-\{1331\}+\{4224\},\ +[2442]-[3113]=+\{2442\}-\{3113\},$
$+[2442]-[4224]=-\{1331\}-\{3113\},\ -[1331]+[2442]=-\{1331\}+\{2442\},$
$+[3113]+[4224]=+\{3113\}+\{4224\},\ -[2442]+[3113]=-\{2442\}+\{3113\},$
$+[3113]-[4224]=-\{1331\}-\{2442\},\ -[1331]+[3113]=-\{1331\}+\{3113\},$
$+[3124]+[4213]=+\{1342\}-\{2431\},\ -[2431]+[3124]=-\{2431\}+\{3124\},$
$+[3124]-[4213]=+\{3124\}-\{4213\},\ -[1342]+[3124]=-\{2431\}-\{4213\},$
$+[3131]+[4242]=+\{3131\}+\{4242\},\ -[2424]+[3131]=-\{2424\}+\{3131\},$
$+[3131]-[4242]=-\{1313\}-\{2424\},\ -[1313]+[3131]=-\{1313\}+\{3131\},$
$+[3142]+[4231]=+\{1324\}-\{2413\},\ -[2413]+[3142]=-\{2413\}+\{3142\},$
$+[3142]-[4231]=+\{3142\}-\{4231\},\ -[1324]+[3142]=-\{2413\}-\{4231\},$
$+[3214]+[4123]=+\{1432\}-\{2341\},\ +[3214]-[4123]=+\{3214\}-\{4123\},$
$-[1432]+[3214]=-\{2341\}-\{4123\},\ -[2341]+[3214]=-\{2341\}+\{3214\},$

```
          +[3223]+[4114]=+{3223}+{4114},  -[2332]+[3223]=+{1441}+{4114},
          +[3223]-[4114]=+{1441}+{2332},  -[1441]+[3223]=+{2332}+{4114},
          +[3232]+[4141]=+{3232}+{4141},  -[2323]+[3232]=+{1414}+{4141},
          +[3232]-[4141]=+{1414}+{2323},  -[1414]+[3232]=+{2323}+{4141},
          +[3241]+[4132]=+{1423}-{2314},  +[3241]-[4132]=+{3241}-{4132},
          -[2314]+[3241]=-{2314}+{3241},  -[1423]+[3241]=-{2314}-{4132},
          +[3311]+[4422]=+{3311}+{4422},  +[3311]-[4422]=-{1133}-{2244},
          -[2244]+[3311]=-{2244}+{3311},  -[1133]+[3311]=-{1133}+{3311},
          +[3322]+[4411]=+{3322}+{4411},  -[2233]+[3322]=+{1144}+{4411},
          +[3322]-[4411]=+{1144}+{2233},  -[1144]+[3322]=+{2233}+{4411},
          +[3333]+[4444]=+{3333}+{4444},  -[2222]+[3333]=+{1111}+{4444},
          +[3333]-[4444]=+{1111}+{2222},  -[1111]+[3333]=+{2222}+{4444},
          +[3344]+[4433]=+{3344}+{4433},  +[3344]-[4433]=-{1122}-{2211},
          -[1122]+[3344]=-{1122}+{3344},  -[2211]+[3344]=-{2211}+{3344},
          +[3412]+[4321]=+{1234}-{2143},  -[2143]+[3412]=-{2143}+{3412},
          +[3412]-[4321]=+{3412}-{4321},  -[1234]+[3412]=-{2143}-{4321},
          +[3421]+[4312]=+{1243}-{2134},  -[2134]+[3421]=-{2134}+{3421},
          +[3421]-[4312]=+{3421}-{4312},  -[1243]+[3421]=-{2134}-{4312},
          +[3434]+[4343]=+{3434}+{4343},  +[3434]-[4343]=-{1212}-{2121},
          -[2121]+[3434]=-{2121}+{3434},  -[1212]+[3434]=-{1212}+{3434},
          +[3443]+[4334]=+{3443}+{4334},  -[2112]+[3443]=-{2112}+{3443},
          +[3443]-[4334]=-{1221}-{2112},  -[1221]+[3443]=-{1221}+{3443},
          -[2332]+[4114]=-{2332}+{4114},  -[3223]+[4114]=-{1441}-{2332},
          -[1441]+[4114]=-{1441}+{4114},  -[3214]+[4123]=-{3214}+{4123},
          -[1432]+[4123]=-{2341}-{3214},  -[2341]+[4123]=-{2341}+{4123},
          -[2314]+[4132]=-{2314}+{4132},  -[1423]+[4132]=-{2314}-{3241},
          -[3241]+[4132]=-{3241}+{4132},  -[2323]+[4141]=-{2323}+{4141},
          -[3232]+[4141]=-{1414}-{2323},  -[1414]+[4141]=-{1414}+{4141},
          -[3124]+[4213]=-{3124}+{4213},  -[2431]+[4213]=-{2431}+{4213},
          -[1342]+[4213]=-{2431}-{3124},  -[2442]+[4224]=+{1331}+{3113},
          -[3113]+[4224]=+{1331}+{2442},  -[1331]+[4224]=+{2442}+{3113},
          -[3142]+[4231]=-{3142}+{4231},  -[2413]+[4231]=-{2413}+{4231},
          -[1324]+[4231]=-{2413}-{3142},  -[2424]+[4242]=+{1313}+{3131},
          -[3131]+[4242]=+{1313}+{2424},  -[1313]+[4242]=+{2424}+{3131},
          -[2134]+[4312]=-{2134}+{4312},  -[3421]+[4312]=-{3421}+{4312},
          -[1243]+[4312]=-{2134}-{3421},  -[2143]+[4321]=-{2143}+{4321},
          -[3412]+[4321]=-{3412}+{4321},  -[1234]+[4321]=-{2143}-{3412},
          -[3443]+[4334]=+{1221}+{2112},  -[2112]+[4334]=+{1221}+{3443},
          -[1221]+[4334]=+{2112}+{3443},  -[3434]+[4343]=+{1212}+{2121},
          -[2121]+[4343]=+{1212}+{3434},  -[1212]+[4343]=+{2121}+{3434},
          -[3322]+[4411]=-{1144}-{2233},  -[2233]+[4411]=-{2233}+{4411},
          -[1144]+[4411]=-{1144}+{4411},  -[2244]+[4422]=+{1133}+{3311},
          -[1133]+[4422]=+{2244}+{3311},  -[3311]+[4422]=+{1133}+{2244},
          -[3344]+[4433]=+{1122}+{2211},  -[1122]+[4433]=+{2211}+{3344},
          -[2211]+[4433]=+{1122}+{3344},  -[3333]+[4444]=-{1111}-{2222},
          -[2222]+[4444]=-{2222}+{4444},  -[1111]+[4444]=-{1111}+{4444}}
```

Now we check them by making random substitutions of high-precision numbers.

```
In[29]:= checkIdentity[id_] :=
           Block[{x, y, z, w}, (id) /. (* restore original arguments *)
             {Subscript[Π, a___] :> Times @@ EllipticTheta[{a}, {w, x, y, z}, q],
              Subscript[OverTilde[Π], a___] :>
               Times @@ EllipticTheta[{a}, {w + x + y + z, w + x - y - z,
                                            w - x + y - z, w - x - y + z}/2, q]} /.
             (* random substitutions *)
             Apply[Rule, Transpose[{{w, x, y, z, q},
                         Table[Random[Real, {1/10, 9/10}], 30], {5}]}], {1}]]
```

Next we check them again using 10 sets of high-precision numbers.

```
In[30]:= SeedRandom[111];
         Table[checkIdentity[allIdentities] // Union, {5}]
```

Out[31]= {{True}, {True}, {True}, {True}, {True}}

We observe that 288 identities in 128 different monomials are surely not minimal. To calculate a minimal set we generate a Gröbner basis.

In[32]:= allProducts = Cases[allIdentities, Subscript[Π, ___] |
                  Subscript[OverTilde[Π], ___], Infinity] // Union;

The Gröbner basis contains only 64 equations.

In[33]:= Length[gb = GroebnerBasis[Subtract @@@ allIdentities, allProducts]]

Out[33]= 64

In minimal form, the identities contain five terms instead of four. (We suppress the = 0 part). Here are the first 12 of these identities shown.

In[34]:= writeIdentityInShortForm /@ Take[gb, 12] //
                    TableForm[#, TableAlignments -> Right]&

Out[34]//TableForm=
$$
\begin{aligned}
&\{1111\} + \{2222\} - \{3333\} - \{4444\} + 2\,[4444] \\
-&\{1122\} - \{2211\} - \{3344\} - \{4433\} + 2\,[4433] \\
&\{1133\} + \{2244\} + \{3311\} + \{4422\} - 2\,[4422] \\
&\{1144\} + \{2233\} - \{3322\} - \{4411\} + 2\,[4411] \\
&\{1212\} + \{2121\} + \{3434\} + \{4343\} - 2\,[4343] \\
-&\{1221\} - \{2112\} - \{3443\} - \{4334\} + 2\,[4334] \\
-&\{1234\} + \{2143\} + \{3412\} - \{4321\} + 2\,[4321] \\
-&\{1243\} + \{2134\} + \{3421\} - \{4312\} + 2\,[4312] \\
&\{1313\} + \{2424\} + \{3131\} + \{4242\} - 2\,[4242] \\
-&\{1324\} + \{2413\} + \{3142\} - \{4231\} + 2\,[4231] \\
-&\{1331\} - \{2442\} - \{3113\} - \{4224\} + 2\,[4224] \\
-&\{1342\} + \{2431\} + \{3124\} - \{4213\} + 2\,[4213]
\end{aligned}
$$

Taking the symmetry of the original identity between the second, third, and fourth index into account, we can reduce the last set to 32.

In[35]:= (* generate all equivalent identities *)
        makeEquivalents[id_] :=
        Function[r, id /. Rule @@ {Subscript[p:(Π | OverTilde[Π]), α_, β_, γ_, δ_],
                          Subscript[p, Sequence @@ r]}] /@
          (* exchange and rename of variables symmetries *)
          {{α, β, γ, δ}, {α, β, δ, γ}, {α, γ, β, δ},
           {α, γ, δ, β}, {α, δ, β, γ}, {α, δ, γ, β}}

In[37]:= (* keep one copy of each identity *)
        posis = First /@ Union[Sort /@ Table[Flatten[
                  Position[gb, #]& /@ makeEquivalents @ gb[[k]]], {k, Length[gb]}]];

In[39]:= (* align corresponding terms *)
        orderedResult[p_Plus] := Function[lastSummand,
            Apply[Plus, HoldForm @@ {writeIdentityInShortForm[
                Append[List @@ (p - lastSummand), lastSummand]]},
                        {1}]][Cases[p, (-2 | 2)_][[1]]]

In[41]:= Sort[orderedResult /@ (gb[[#]]& /@ posis)] //
                TableForm[#, TableAlignments -> Right]&

$$
\begin{aligned}
-&\{1111\} + \{2222\} - \{3333\} + \{4444\} + 2\,[1111] \\
&\{1212\} - \{2121\} - \{3434\} + \{4343\} - 2\,[1212] \\
-&\{1221\} + \{2112\} + \{3443\} - \{4334\} + 2\,[1221] \\
&\{1313\} - \{2424\} - \{3131\} + \{4242\} - 2\,[1313] \\
-&\{1331\} + \{2442\} + \{3113\} - \{4224\} + 2\,[1331] \\
&\{1423\} + \{2314\} + \{3241\} + \{4132\} - 2\,[1423] \\
-&\{1432\} - \{2341\} - \{3214\} - \{4123\} + 2\,[1432] \\
-&\{1441\} + \{2332\} - \{3223\} + \{4114\} + 2\,[1441] \\
-&\{1212\} + \{2121\} - \{3434\} + \{4343\} - 2\,[2121] \\
&\{1122\} - \{2211\} + \{3344\} - \{4433\} + 2\,[2211] \\
&\{1111\} - \{2222\} - \{3333\} + \{4444\} + 2\,[2222] \\
&\{1423\} + \{2314\} - \{3241\} - \{4132\} - 2\,[2314] \\
&\{1441\} - \{2332\} - \{3223\} + \{4114\} + 2\,[2332] \\
-&\{1313\} + \{2424\} - \{3131\} + \{4242\} - 2\,[2424]
\end{aligned}
$$

Out[41]//TableForm=

$$-\{1342\} - \{2431\} + \{3124\} + \{4213\} + 2 \ [2431]$$
$$\{1331\} - \{2442\} + \{3113\} - \{4224\} + 2 \ [2442]$$
$$\{1331\} + \{2442\} - \{3113\} - \{4224\} + 2 \ [3113]$$
$$\{1423\} - \{2314\} + \{3241\} - \{4132\} - 2 \ [3241]$$
$$-\{1133\} - \{2244\} + \{3311\} + \{4422\} - 2 \ [3311]$$
$$-\{1144\} - \{2233\} - \{3322\} - \{4411\} + 2 \ [3322]$$
$$-\{1111\} - \{2222\} - \{3333\} - \{4444\} + 2 \ [3333]$$
$$-\{1243\} + \{2134\} - \{3421\} + \{4312\} + 2 \ [3421]$$
$$-\{1212\} - \{2121\} + \{3434\} + \{4343\} - 2 \ [3434]$$
$$\{1221\} + \{2112\} - \{3443\} + \{4334\} + 2 \ [3443]$$
$$\{1423\} - \{2314\} - \{3241\} + \{4132\} - 2 \ [4132]$$
$$-\{1331\} - \{2442\} - \{3113\} - \{4224\} + 2 \ [4224]$$
$$-\{1234\} + \{2143\} + \{3412\} - \{4321\} + 2 \ [4321]$$
$$\{1212\} + \{2121\} + \{3434\} + \{4343\} - 2 \ [4343]$$
$$\{1144\} + \{2233\} - \{3322\} - \{4411\} + 2 \ [4411]$$
$$\{1133\} + \{2244\} + \{3311\} + \{4422\} - 2 \ [4422]$$
$$-\{1122\} - \{2211\} - \{3344\} - \{4433\} + 2 \ [4433]$$
$$\{1111\} + \{2222\} - \{3333\} - \{4444\} + 2 \ [4444]$$

d) The first term c_0 is obviously $\vartheta_3(q)$. Substituting the series in the partial differential equation and equating powers of z yields linear equations for the c_k. The next input calculates the first few c_k.

```
In[1]:= Clear[c];
        c[0][q] = θ[q];

        Do[sum = Sum[c[2j][q] z^(2 j), {j, 0, o}];
           coeff = Coefficient[-1/4 D[sum, {z, 2}] - q D[sum, q], z, o - 2];
           c[o][q] = c[o][q] /. Solve[coeff == 0, c[o][q]][[1]] // Expand,
           {o, 2, 14, 2}]
```

In[4]:= ??c

Global`c

$$c[0][q] = \vartheta[q]$$
$$c[2][q] = -2 \, q \, \vartheta'[q]$$
$$c[4][q] = \tfrac{2}{3} \, q \, \vartheta'[q] + \tfrac{2}{3} \, q^2 \, \vartheta''[q]$$
$$c[6][q] = -\tfrac{4}{45} \, q \, \vartheta'[q] - \tfrac{4}{15} \, q^2 \, \vartheta''[q] - \tfrac{4}{45} \, q^3 \, \vartheta^{(3)}[q]$$
$$c[8][q] = \tfrac{2}{315} \, q \, \vartheta'[q] + \tfrac{2}{45} \, q^2 \, \vartheta''[q] + \tfrac{4}{105} \, q^3 \, \vartheta^{(3)}[q] + \tfrac{2}{315} \, q^4 \, \vartheta^{(4)}[q]$$
$$c[10][q] = -\tfrac{4 \, q \, \vartheta'[q]}{14175} - \tfrac{4}{945} \, q^2 \, \vartheta''[q] - \tfrac{4}{567} \, q^3 \, \vartheta^{(3)}[q] - \tfrac{8 \, q^4 \, \vartheta^{(4)}[q]}{2835} - \tfrac{4 \, q^5 \, \vartheta^{(5)}[q]}{14175}$$
$$c[12][q] = \tfrac{4 \, q \, \vartheta'[q]}{467775} + \tfrac{124 \, q^2 \, \vartheta''[q]}{467775} + \tfrac{8 \, q^3 \, \vartheta^{(3)}[q]}{10395} + \tfrac{52 \, q^4 \, \vartheta^{(4)}[q]}{93555} + \tfrac{4 \, q^5 \, \vartheta^{(5)}[q]}{31185} + \tfrac{4 \, q^6 \, \vartheta^{(6)}[q]}{467775}$$
$$c[14][q] = -\tfrac{8 \, q \, \vartheta'[q]}{42567525} - \tfrac{8 \, q^2 \, \vartheta''[q]}{675675} - \tfrac{344 \, q^3 \, \vartheta^{(3)}[q]}{6081075} - \tfrac{16 \, q^4 \, \vartheta^{(4)}[q]}{243243} - \tfrac{32 \, q^5 \, \vartheta^{(5)}[q]}{1216215} - \tfrac{8 \, q^6 \, \vartheta^{(6)}[q]}{2027025} - \tfrac{8 \, q^7 \, \vartheta^{(7)}[q]}{42567525}$$

Using the ordinary differential equation for $\vartheta_3(q)$ we will now eliminate $\vartheta_3^{(3)}(q)$ from c_6 and $\vartheta_3^{(3)}(q)$ and $\vartheta_3^{(4)}(q)$ from c_8.

```
In[5]:= θode = 1/q^2 Factor @ With[{θ = θ3[q], d = q D[#, q]&},
        (θ^2 d[d[d[θ]]] - 15 θ d[θ] d[d[θ]] + 30 d[θ]^3)^2 +
        32 (θ d[d[θ]] - 3 d[θ]^2)^3 - θ^10 (θ d[d[θ]] - 3 d[θ]^2)^2]
```

Out[5]= $\vartheta3[q]^4 \, \vartheta3'[q]^2 - \vartheta3[q]^{12} \, \vartheta3'[q]^2 + 2 \, q \, \vartheta3[q]^3 \, \vartheta3'[q]^3 + 6 \, q \, \vartheta3[q]^{11} \, \vartheta3'[q]^3 - $
$3 \, q^2 \, \vartheta3[q]^2 \, \vartheta3'[q]^4 - 9 \, q^2 \, \vartheta3[q]^{10} \, \vartheta3'[q]^4 - 36 \, q^3 \, \vartheta3[q] \, \vartheta3'[q]^5 + 36 \, q^4 \, \vartheta3'[q]^6 + $
$6 \, q \, \vartheta3[q]^4 \, \vartheta3'[q] \, \vartheta3''[q] - 2 \, q \, \vartheta3[q]^{12} \, \vartheta3'[q] \, \vartheta3''[q] - 24 \, q^2 \, \vartheta3[q]^3 \, \vartheta3'[q]^2 \, \vartheta3''[q] + $
$6 \, q^2 \, \vartheta3[q]^{11} \, \vartheta3'[q]^2 \, \vartheta3''[q] + 54 \, q^3 \, \vartheta3[q]^2 \, \vartheta3'[q]^3 \, \vartheta3''[q] - 36 \, q^4 \, \vartheta3[q] \, \vartheta3'[q]^4 \, \vartheta3''[q] + $
$9 \, q^2 \, \vartheta3[q]^4 \, \vartheta3''[q]^2 - q^2 \, \vartheta3[q]^{12} \, \vartheta3''[q]^2 + 6 \, q^3 \, \vartheta3[q]^3 \, \vartheta3'[q] \, \vartheta3''[q]^2 - $
$63 \, q^4 \, \vartheta3[q]^2 \, \vartheta3'[q]^2 \, \vartheta3''[q]^2 + 32 \, q^4 \, \vartheta3[q]^3 \, \vartheta3''[q]^3 + 2 \, q^2 \, \vartheta3[q]^4 \, \vartheta3'[q] \, \vartheta3^{(3)}[q] - $
$30 \, q^3 \, \vartheta3[q]^3 \, \vartheta3'[q]^2 \, \vartheta3^{(3)}[q] + 60 \, q^4 \, \vartheta3[q]^2 \, \vartheta3'[q]^3 \, \vartheta3^{(3)}[q] + $
$6 \, q^3 \, \vartheta3[q]^4 \, \vartheta3'[q] \, \vartheta3^{(3)}[q] - 30 \, q^4 \, \vartheta3[q]^3 \, \vartheta3'[q] \, \vartheta3''[q] \, \vartheta3^{(3)}[q] + q^4 \, \vartheta3[q]^4 \, \vartheta3^{(3)}[q]^2$

Here is a quick check of the differential equation.

```
In[6]:= θ3Ser[q_] = 1 + 2 Sum[q^(n^2), {n, 1, 10}] + O[q]^100;
        θode /. θ3 -> θ3Ser
```

Out[7]= $O[q]^{99}$

$\vartheta_3''(q)$ appears quadratically in the differential equation $\mathcal{O}$Ode. So it is straightforward to express $\vartheta_3^{(3)}(q)$ in $\vartheta_3(q)$, $\vartheta_3'(q)$, and $\vartheta_3''(q)$. We use the series $\mathcal{O}$3Ser to select one of the two possible square roots.

```
In[8]:= sol3 = Select[Solve[𝒪Ode == 0, 𝒪3'''[q]],
                (((𝒪3'''[q] /. #) == D[𝒪3[q], {q, 3}]) /.
                   𝒪3 -> 𝒪3Ser)&][[1]] // Simplify
```

$$\text{Out[8]}= \left\{ 𝒪3^{(3)}[q] \to \right.$$
$$\frac{1}{q^4\,𝒪3[q]^4}\left(-q^2\,𝒪3[q]^4\,𝒪3'[q] + 15\,q^3\,𝒪3[q]^3\,𝒪3'[q]^2 - 30\,q^4\,𝒪3[q]^2\,𝒪3'[q]^3 - 3\,q^3\,𝒪3[q]^4\,𝒪3''[q] + \right.$$
$$15\,q^4\,𝒪3[q]^3\,𝒪3'[q]\,𝒪3''[q] + \sqrt{\left(q^4\,𝒪3[q]^4\,(-3\,q\,𝒪3'[q]^2 + 𝒪3[q]\,(𝒪3'[q] + q\,𝒪3''[q]))\right)^2}$$
$$\left.\left.\left(𝒪3[q]^{10} + 96\,q^2\,𝒪3'[q]^2 - 32\,q\,𝒪3[q]\,(𝒪3'[q] + q\,𝒪3''[q])\right)\right)\right)\right\}$$

To express $\vartheta_3^{(4)}(q)$ from through $\vartheta_3(q)$, $\vartheta_3'(q)$, and $\vartheta_3''(q)$, we differentiate the differential equation, eliminate $\vartheta_3^{(3)}(q)$, and solve for $\vartheta_3^{(4)}(q)$.

```
In[9]:= res = Resultant[𝒪Ode, D[𝒪Ode, q], 𝒪3'''[q]] // Factor // Last;
```

```
In[10]:= sol4 = Select[Solve[res == 0, 𝒪3''''[q]],
                 (((𝒪3''''[q] /. #) == D[𝒪3[q], {q, 4}]) /.
                    𝒪3 -> 𝒪3Ser)&][[1]] // Simplify
```

$$\text{Out[10]}= \left\{ 𝒪3^{(4)}[q] \to \right.$$
$$\frac{1}{q^6\,𝒪3[q]^6}\left(5\,q^3\,𝒪3[q]^6\,𝒪3'[q] + q^3\,𝒪3[q]^{14}\,𝒪3'[q] - 123\,q^4\,𝒪3[q]^5\,𝒪3'[q]^2 - 3\,q^4\,𝒪3[q]^{13}\,𝒪3'[q]^2 + \right.$$
$$588\,q^5\,𝒪3[q]^4\,𝒪3'[q]^3 - 822\,q^6\,𝒪3[q]^3\,𝒪3'[q]^4 + 11\,q^4\,𝒪3[q]^6\,𝒪3''[q] + q^4\,𝒪3[q]^{14}\,𝒪3''[q] -$$
$$156\,q^5\,𝒪3[q]^5\,𝒪3'[q]\,𝒪3''[q] + 408\,q^6\,𝒪3[q]^4\,𝒪3'[q]^2\,𝒪3''[q] - 33\,q^6\,𝒪3[q]^5\,𝒪3''[q]^2 -$$
$$2\,\sqrt{\left(q^6\,𝒪3[q]^6\,(3\,𝒪3[q] - 14\,q\,𝒪3'[q])^2\,(-3\,q\,𝒪3'[q]^2 + 𝒪3[q]\,(𝒪3'[q] + q\,𝒪3''[q]))^2\right)}$$
$$\left.\left.\left(𝒪3[q]^{10} + 96\,q^2\,𝒪3'[q]^2 - 32\,q\,𝒪3[q]\,(𝒪3'[q] + q\,𝒪3''[q])\right)\right)\right)\right\}$$

Expressing $\vartheta_3^{(5)}(q)$ through $\vartheta_3(q)$, $\vartheta_3'(q)$, and $\vartheta_3''(q)$ is slightly more expensive. We substitute the just-calculated expressions for $\vartheta_3^{(3)}(q)$ and $\vartheta_3^{(4)}(q)$ into the twice-differentiated differential equation and solve for $\vartheta_3^{(5)}(q)$. We start by eliminating the two square roots introduced after substituting $\vartheta_3^{(3)}(q)$ and $\vartheta_3^{(4)}(q)$.

```
In[11]:= num = Expand[Numerator[Together[D[𝒪Ode, q, q] /. sol3 /. sol4]]];
```

```
In[12]:= roots = Cases[num, Power[_, _Rational], Infinity] // Union
```

$$\text{Out[12]}= \left\{ \sqrt{\left(q^4\,𝒪3[q]^4\,(-3\,q\,𝒪3'[q]^2 + 𝒪3[q]\,(𝒪3'[q] + q\,𝒪3''[q]))\right)^2} \right.$$
$$\left(𝒪3[q]^{10} + 96\,q^2\,𝒪3'[q]^2 - 32\,q\,𝒪3[q]\,(𝒪3'[q] + q\,𝒪3''[q]))\right),$$
$$\sqrt{\left(q^6\,𝒪3[q]^6\,(3\,𝒪3[q] - 14\,q\,𝒪3'[q])^2\,(-3\,q\,𝒪3'[q]^2 + 𝒪3[q]\,(𝒪3'[q] + q\,𝒪3''[q]))^2\right)}$$
$$\left.\left(𝒪3[q]^{10} + 96\,q^2\,𝒪3'[q]^2 - 32\,q\,𝒪3[q]\,(𝒪3'[q] + q\,𝒪3''[q]))\right) \right\}$$

```
In[13]:= num1 = num /. {roots[[1]] -> ℛ1, roots[[2]] -> ℛ2};
         num2 = Expand[#^2 - (num1 - #)^2]&[Plus @@ Cases[num1, _ ℛ1, {1}]];
         num3 = Expand[#^2 - (num2 - #)^2]&[Plus @@ Cases[num2, _ ℛ2, {1}]];
         num4 = Expand[num3 /. ℛ1 -> roots[[1]] /. ℛ2 -> roots[[2]]];
```

The resulting polynomial is quite large and of high degree.

```
In[17]:= {Length[num4],
         Max[Plus @@@ First /@ Internal`DistributedTermsList[num4,
                    {q, 𝒪3[q], 𝒪3'[q], 𝒪3''[q], 𝒪3'''''[q]}][[1]]]}
```

$$\text{Out[17]}= \{2616, 144\}$$

We factor the polynomial and select the relevant factor for expressing $\vartheta_3^{(5)}(q)$.

```
In[18]:= (* easy part of the factorization *)
         num5 = FactorSquareFree[num4];
         {Head[num5], num5 // Length}
```

$$\text{Out[20]}= \{\text{Times}, 6\}$$

```
In[21]:= num6 = Select[List @@ num5, (MemberQ[#, 𝒪3, {0, Infinity}, Heads -> True] &&
                    Normal[# /. 𝒪3 -> 𝒪3Ser] === 0)&][[1]];
```

```
In[22]:= (* hard part of the factorization *)
         Length[num7 = Factor[num6]]
```

$$\text{Out[23]}= 2$$

```
In[24]:= num8 = Select[List @@ num7, (MemberQ[#, θ3, {0, Infinity}, Heads -> True] &&
                Normal[# /. θ3 -> θ3Ser] === 0)&][[1]];
```

So we finally arrive at the following expression for $\vartheta_3^{(5)}(q)$.

```
In[25]:= sol5 = Select[Solve[num8 == 0, θ3'''''[q]],
            (((θ3'''''[q] /. #) == D[θ3[q], {q, 5}]) /.
                θ3 -> θ3Ser)&][[1]] // Simplify
```

$$
\text{Out[25]= } \Big\{ \vartheta3^{(5)}[q] \to \frac{1}{q^8\,\vartheta3[q]^8}
$$

$$
\big(-26\,q^4\,\vartheta3[q]^8\,\vartheta3'[q] - 10\,q^4\,\vartheta3[q]^{16}\,\vartheta3'[q] + 855\,q^5\,\vartheta3[q]^7\,\vartheta3'[q]^2 + 75\,q^5\,\vartheta3[q]^{15}\,\vartheta3'[q]^2 -
$$

$$
6615\,q^6\,\vartheta3[q]^6\,\vartheta3'[q]^3 - 135\,q^6\,\vartheta3[q]^{14}\,\vartheta3'[q]^3 + 20280\,q^7\,\vartheta3[q]^5\,\vartheta3'[q]^4 -
$$

$$
21870\,q^8\,\vartheta3[q]^4\,\vartheta3'[q]^5 - 50\,q^5\,\vartheta3[q]^8\,\vartheta3''[q] - 10\,q^5\,\vartheta3[q]^{16}\,\vartheta3''[q] +
$$

$$
1185\,q^6\,\vartheta3[q]^7\,\vartheta3'[q]\,\vartheta3''[q] + 45\,q^6\,\vartheta3[q]^{15}\,\vartheta3'[q]\,\vartheta3''[q] - 7050\,q^7\,\vartheta3[q]^6\,\vartheta3'[q]^2\,\vartheta3''[q] +
$$

$$
12060\,q^8\,\vartheta3[q]^5\,\vartheta3'[q]^3\,\vartheta3''[q] + 330\,q^7\,\vartheta3[q]^7\,\vartheta3''[q]^2 - 1485\,q^8\,\vartheta3[q]^6\,\vartheta3'[q]\,\vartheta3''[q]^2 +
$$

$$
\sqrt{ \big(q^8\,\vartheta3[q]^8\,(\vartheta3[q]^{10} + 96\,q^2\,\vartheta3'[q]^2 - 32\,q\,\vartheta3[q]\,(\vartheta3'[q] + q\,\vartheta3''[q])) \big) }
$$

$$
(3\,q\,\vartheta3[q]^{10}\,\vartheta3'[q]^2 + 2232\,q^3\,\vartheta3'[q]^4 - 35\,\vartheta3[q]^3\,(\vartheta3'[q] + q\,\vartheta3''[q]) -
$$

$$
\vartheta3[q]^{11}\,(\vartheta3'[q] + q\,\vartheta3''[q]) - 6\,q^2\,\vartheta3[q]\,\vartheta3'[q]^2\,(283\,\vartheta3'[q] + 143\,q\,\vartheta3''[q]) +
$$

$$
q\,\vartheta3[q]^2\,(423\,\vartheta3'[q]^2 + 356\,q\,\vartheta3'[q]\,\vartheta3''[q] + 38\,q^2\,\vartheta3''[q]^2))^2) \big) \Big\}
$$

We succeeded in expressing $\vartheta_3^{(3)}(q)$, $\vartheta_3^{(4)}(q)$, and $\vartheta_3^{(5)}(q)$ through $\vartheta_3(q)$, $\vartheta_3'(q)$, and $\vartheta_3''(q)$.

For further relations between the derivatives of the theta functions, see [103].

e) In the following, we use the convenient units $\tilde{\varepsilon}_v = \varepsilon_v = k_B = 1$. While this will yield different quantitative results of the 3D box and the harmonic oscillator, it allows for an easy qualitative comparison. Here are the two one-particle partition functions. (For the box, we use the form involving the third Jacobi theta function. While it cannot be differentiated with respect to x in closed form, it is numerically faster to evaluate than the form containing the elliptic nome.)

```
In[1]:= {Z13DBox[x_] = Sum[Exp[-x v^2], {v, 1, Infinity}]^3,
        Z13DHO[ x_] = Sum[Exp[-x (v + 1/2)], {v, 0, Infinity}]^3}
```

$$
\text{Out[1]= } \Big\{ \frac{1}{8}\,(-1 + \text{EllipticTheta}[3, 0, e^{-x}])^3, \; \frac{e^{3\,x/2}}{(-1 + e^x)^3} \Big\}
$$

While it is possible to use the above recursion relation to obtain closed-form expressions for Z_N, for larger N this is not practical because Z_N will have $p(N)$ terms where $p(n)$ is the partitions function. This means for $N = 100$, Z_N has about 200 million terms and is so not practical for any numerical computation. Here are the first few Z_N.

```
In[2]:= Z[0, β_] = 1; Z[1, β_] = Subscript[Z, 1][β];
        Z[n_, β_] := Z[n, β] = Factor[Sum[Z[1, k β] Z[n - k, β], {k, n}]/n]
        Table[Z[j, β], {j, 0, 5}]
```

$$
\text{Out[4]= } \Big\{ 1,\; Z_1[\beta],\; \frac{1}{2}\,(Z_1[\beta]^2 + Z_1[2\beta]),\; \frac{1}{6}\,(Z_1[\beta]^3 + 3\,Z_1[\beta]\,Z_1[2\beta] + 2\,Z_1[3\beta]),
$$

$$
\frac{1}{24}\,(Z_1[\beta]^4 + 6\,Z_1[\beta]^2\,Z_1[2\beta] + 3\,Z_1[2\beta]^2 + 8\,Z_1[\beta]\,Z_1[3\beta] + 6\,Z_1[4\beta]),
$$

$$
\frac{1}{120}\,(Z_1[\beta]^5 + 10\,Z_1[\beta]^3\,Z_1[2\beta] + 15\,Z_1[\beta]\,Z_1[2\beta]^2 +
$$

$$
20\,Z_1[\beta]^2\,Z_1[3\beta] + 20\,Z_1[2\beta]\,Z_1[3\beta] + 30\,Z_1[\beta]\,Z_1[4\beta] + 24\,Z_1[5\beta]) \Big\}
$$

For any explicit numerical value of β, we can calculate numerical values $Z_N^{(\text{box})}(\beta)$ and $Z_N^{(\text{HO})}(\beta)$ much faster after the values of $Z_1(k\,\beta)$ are known. The functions zkβListCF and zkβListHP take the $Z_1(k\,\beta)$ as arguments and calculate machine and high-precision values of $Z_N(\beta)$.

```
In[5]:= (* compiled version *)
        zkβListCF = Compile[{{Z1βList, _Real, 1}},
            Module[{ZkβList = Table[1., {k, Length[Z1βList] + 1}]},
                (* carry out recursion numerically *)
                ZkβList[[2]] = Z1βList[[1]];
                Do[ZkβList[[n + 1]] = Sum[Z1βList[[k]] ZkβList[[n - k + 1]],
                                {k, n}]/n, {n, 2, Length[Z1βList]}];
                ZkβList]];

        (* high-precision version (extract code from compiled version) *)
        OwnValues[zkβListHP] = OwnValues[zkβListCF] /. 1. -> 1 /.
                HoldPattern[zkβListCF] :> zkβListHP /.
```

```
HoldPattern[CompiledFunction[_, _, _, _,
                  body_, ___]] :> body;
```

The following graphics show zkβListCF at work. The two graphics use the (unphysical) values $Z_1(k \beta) = tan(k)$ and $Z_1(k \beta) = random[0, 1]$.

```
In[10]:= Show[GraphicsArray[SeedRandom[111];
         Block[{$DisplayFunction = Identity,
             opts = Sequence[PlotRange -> All, Frame -> True, Axes -> False]},
            {(* linear scale *)
             ListPlot[zkβListCF[Table[Tan[k], {k, 2000}]] // N, opts],
             (* logarithmic scale *)
             ListPlot[Log[MapIndexed[{#2[[1]], #1}&,
                         zkβListCF[Table[Random[], {k, 2000}]]]] // N, opts]}]]]
```

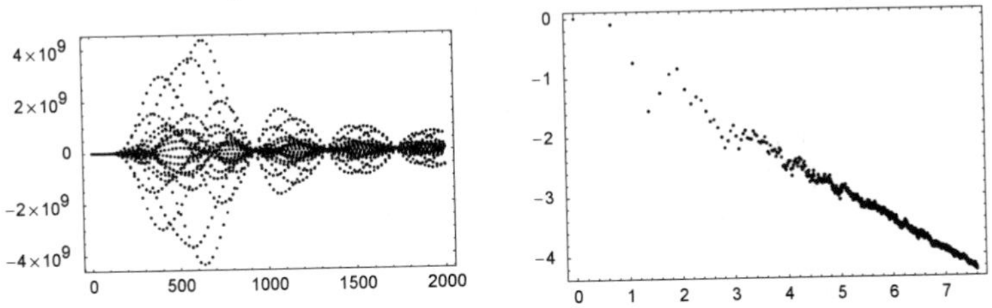

The next two graphics show the N-particle partition functions of the box and the harmonic oscillator for relatively small T.

```
In[11]:= Show[GraphicsArray[
         Block[{$DisplayFunction = Identity},
             ListPlot[zkβListCF[Table[N[#, 60], {k, 1, 100}]], PlotRange -> All]& /@
             (* use different β *) {Z13DBox[k 1/20], Z13DHO[k 1/4]}]]]
```

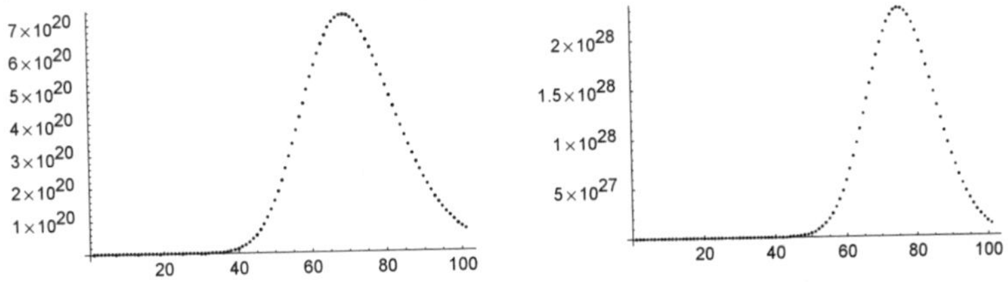

Given the partition N-particle partition functions, we can now calculate the occupation probabilities $p_\varepsilon^{(N)}(n; T)$.

```
In[12]:= (* to allow precise calculation of the theta function *)
         $MaxExtraPrecision = 1000;
         pnList[ε_, N_, T_, Z_] :=
         Module[{β = 1/T, zkList = Table[N[Z[k 1/T], 60], {k, 1, N}], Z},
             (* use compiled or high-precision version *)
             zkβList = If[Min[Abs[zkList]] < $MinMachineNumber,
                         zkβListHP, zkβListCF][zkList];
             Z[-1] = 0; Z[k_] := zkβList[[k + 1]];
             Table[(Exp[-n β ε] Z[N - n] - Exp[-(n + 1) β ε] Z[N - n - 1])/
                 Z[N], {n, 0, N}]]
```

The occupation probabilities $p_\varepsilon^{(N)}(n; T)$ allow to calculate the average occupancy of the level ε as an expectation value. The following graphics show the probability that all particles will be in the ground state as a function of temperature for $N = 10$ [589].

```
In[15]:= With[{N = 10},
         Show[GraphicsArray[
             ListPlot[#, PlotRange -> All, Frame -> True, Axes -> False,
                     DisplayFunction -> Identity]& /@
             (* 3D harmonic oscillator and 3D box *)
             {Table[{T, pnList[3  , N, T, Z13DBox][[-1]]}, {T, 1/20, 5, 1/20}],
              Table[{T, pnList[3/2, N, T, Z13DHO ][[-1]]}, {T, 1/20, 3/2, 1/20}]}]]]
```

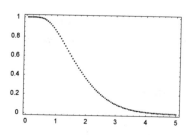
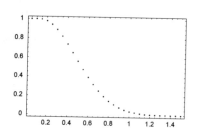

The following graphics show $p_\varepsilon^{(100)}(n; T)$ for various temperatures for the 3D box and the harmonic oscillator. Observe the different scales for T. We use the values $\varepsilon = 3/2$ for the ground state of the 3D harmonic oscillator and $\varepsilon = 3$ for the ground state of the 3D box.

```
In[16]:= Show[GraphicsArray[
            Block[{pp = 60, N = 100, $DisplayFunction = Identity},
                ListPlot3D[Chop[#[[1]]], Mesh -> False, PlotRange -> {0, 0.2},
                    MeshRange -> {{1, 100}, #[[2]]},
                    ViewPoint -> {-1, -2, 1}, ClipFill -> None]& /@
            (* 3D harmonic oscillator and 3D box *)
            {{Table[pnList[3  , N, T, Z13DBox], {T, 1, 30, (30 - 1)/pp}], {1, 30}},
             {Table[pnList[3/2, N, T, Z13DHO ], {T, 1/2, 5, (5 - 1/2)/pp}], {1/2, 1}}}]]]
```

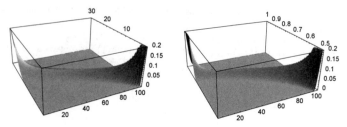

The next two graphics show the average fraction of particles in the ground state as a function of temperature. The Bose–Einstein condensation—meaning a finite fraction of all particles being in the lowest available energy state—is clearly visible. For small T, the relative fraction of particles in the ground state is of the form $1 - (T/T_c)^\alpha$ for larger N [375], [548], [74], [351], [1366], [730], [1201], [418].

```
In[17]:= averageGroundstateOccupation[N_, T_, Z_] := Plus @@
            MapIndexed[(#2[[1]] - 1) #1&, pnList[If[Z === Z13DBox, 3, 3/2], N, T, Z]]/N
In[18]:= With[{N = 100, f = averageGroundstateOccupation},
            Show[GraphicsArray[
            ListPlot[#, PlotRange -> All, Frame -> True, Axes -> False,
                    DisplayFunction -> Identity, PlotJoined -> True]& /@
            (* 3D harmonic oscillator and 3D box *)
            {Table[{T, f[100, T, Z13DBox]}, {T, 1/20, 30, 1/3}],
             Table[{T, f[100, T, Z13DHO ]}, {T, 1/20, 6, 1/10}]}]]]
```

For the calculation of the specific heat, we have to differentiate the N-particle partition function twice. In our units, we have $c_{V,N} = (Z_N''(1/T) Z_N(1/T) - Z_N'^2(1/T))/Z_N^2(1/T)$. We calculate the derivatives $Z_N'(1/T)$ and $Z_N''(1/T)$ numerically. (For very small N, we could use the exact formulas and differentiate symbolically.)

```
In[19]:= cv[N_, T_, Z1_] :=
    Module[(* use high-precision arithmetic *)
        {ε = 10^-20, Thp = SetPrecision[T, 120], ZN},
        ZN[β_] := zkβListHP[Table[Z1[k β], {k, 1, N}]][[-1]];
        (* numerically calculate derivatives *)
        ZN0 = ZN[1/Thp]; ZNεP = ZN[1/Thp + ε]; ZNεM = ZN[1/Thp - ε];
        ZNε2P = ZN[1/Thp + 2ε]; ZNε2M = ZN[1/Thp - 2ε];
        ZND1 = (ZNεP - ZNεM)/(2ε);
        ZND2 = (ZNε2P + ZNε2M - 2ZN0)/(4ε^2);
        (* specific heat *) (ZND2 ZN0 - ZND1^2)/(T^2 ZN0^2)/N];
```

The next two graphics show the specific heat for increasing particle numbers from $N = 1$ to $N = 100$. For the specific heat, we see the typical bump structure indicating a phase transition in the $N \to \infty$ limit [287]. Due to the stronger confinement of the box, the transition temperature increases faster with the number of particles (for fixed volume) compared with the harmonic oscillator [375], [1307], [548].

```
In[20]:= cvData[{minT_, maxT_, δT_}, N_Integer, Z_] :=
        Table[{T, cv[N, T, Z]}, {T, minT, maxT, δT}]

    Show[GraphicsArray[Function[data, Show[Function[
    ListPlot[data, PlotJoined -> True, PlotRange -> All,
        AxesOrigin -> {0, 0}, DisplayFunction -> Identity,
        PlotStyle -> {#2}]] @@@ (* N and color values *)
    {{1, Hue[0]}, {2, Hue[0.1]}, {5, Hue[0.2]} , {10, Hue[0.3]},
    {50, Hue[0.4]}, {100, Hue[0.8]}}]] /@
    (* 3D harmonic oscillator and 3D box *)
    {cvData[{1/10, 30, 1/6}, #1, Z13DBox],
    cvData[{11/10, 7, 1/10}, #1, Z13DHO]}]]
```

The interested reader could now continue and calculate the mean square fluctuations of the number of particles in the ground state via $\delta^2 n_0 = z^2 Y''(z) + z Y'(z)|_{z=1}$ where $Y(z) = Z_N(T) \sum_{n=0}^{N} z^n p_{\varepsilon_0}^{(N)}(n; T)$ [1307], [1026], [623], but we end here. For similar calculations in the microcanonical, grand canonical, and fourth ensemble, see [541], [1307], [608], [921], [588], [589], [934], [84], [1366]; for the limit $L \to \infty$, see [1258], [1259]; for the 2D case, see [1203].

13. Scattering on Cylinder, Coulomb Scattering, Spiral Waves, Optical Black Hole, Corrugated Wall Scattering, Random Solutions of the Helmholtz Equation

a) The implementation of the formula for the scattered field strength is straightforward.

```
In[1]:= (* define Hankel functions through Bessel functions *)
    HankelH[2][n_, x_] := BesselJ[n, x] - I BesselY[n, x]

In[3]:= ε[r_, φ_, {k_, R_}, m_] :=
    If[r < R, (* inside cylinder *) 0,
        (* incoming wave *) Exp[-I k r Cos[φ]] - (* scattering wave *)
        BesselJ[0, k R]/HankelH[2][0, k R] HankelH[2][0, k r] -
        2 Sum[(-I)^n BesselJ[n, k R]/HankelH[2][n, k R] *
            HankelH[2][n, k r] Cos[n φ], {n, m}]]
```

We will fix the radius of the scattering cylinder as 1 and use a wave vector $k = 5$. Numerical experiments with some random values for r and φ show that taking into account about 12 terms of the sum yields about 10 correct digits, plenty for a visualization.

```
In[4]:= εScatteredTerms[r_, φ_, {k_, R_}, m_] :=
        Table[(-I)^n BesselJ[n, k R]/HankelH[2][n, k R] *
            HankelH[2][n, k r] Cos[n φ], {n, m}]
In[5]:=
Show[Graphics[{Thickness[0.002],
        Table[Line[MapIndexed[{#2[[1]], #1}&,
            Log[10, Abs[εScatteredTerms[Random[Real, {1, 5}],
                2Pi Random[], {5, 1}, 20]]]]], {50}]}], Frame -> True]
```

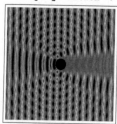

We will show a contour plot of the scattering process. To use the built-in function `ContourPlot`, we calculate the field values on a rectangular grid. We use the symmetry with respect to the x-axis.

```
In[6]:= cp = ContourPlot[Re[ε[Sqrt[x^2 + y^2], ArcTan[x, y], {5, 1}, 8]],
            {x, -10, 10}, {y, 10^-6, 10},
            PlotPoints -> 60 {2, 1}, DisplayFunction -> Identity,
            ColorFunction -> (Hue[0.8 #]&), Compiled -> False,
            Contours -> 30, FrameTicks -> None, ContourLines -> False,
            (* the circular scatterer *)
            Epilog -> {{GrayLevel[0], Disk[{0, 0}, 1]}}];
In[7]:= Show[{cp, Graphics[cp] /. Polygon[l_] :> Polygon[{1, -1} #& /@ l]},
            DisplayFunction -> $DisplayFunction]
```

For a finite potential inside the cylinder, see [1355]. For the more interesting scattering process on vortices and Aharonov–Bohm scattering, see [141], [1276], [582], [1094], [567], [368], and [508]; for some interesting 3D scattering wave functions, see [443]. For arrays of circular scatterers, see [294]. For rotating cylinders, see [600] and [601].

b) c is the r-independent factor in the doubly infinite sum. We evaluate it for symbolic n to carry out the differentiations involved. Because we want to visualize the scattering for various β we also make a definition for the Hankel functions that remembers already-calculated values.

```
In[1]:= HankelH[1][n_, x_] := HankelH[1][n, x] = BesselJ[n, x] + I BesselY[n, x];
In[2]:= (c[kR_, n_, β_] =
        (kR Derivative[0, 1][BesselJ][n, kR] - n β BesselJ[n, kR])/
        (kR Derivative[0, 1][HankelH[1]][n, kR] - n β HankelH[1][n, kR]) // Simplify)
Out[2]= (-kR BesselJ[-1 + n, kR] + 2 n β BesselJ[n, kR] + kR BesselJ[1 + n, kR]) /
        (-kR BesselJ[-1 + n, kR] + 2 n β BesselJ[n, kR] + kR BesselJ[1 + n, kR] -
            i kR BesselY[-1 + n, kR] + 2 i n β BesselY[n, kR] + i kR BesselY[1 + n, kR])
```

For numeric values of kR, n, and β, we use the function $c1$ to calculate and store the scattering coefficient.

```
In[3]:= c1[kR_, n_, β_] := c1[kR, n, β] = c[kR, n, β];

    uS[r_, φ_, {β_, k_, R_}, o_] := Module[{sum = 0.},
```

```
              Do[sum = sum + I^n c1[k R, n, β] HankelH[1][n, k r] Exp[I n φ],
                 {n, -o, o}]; sum]
 In[5]:= u[r_, φ_, {β_, k_, R_}, o_] :=
           If[r <= R, 0, Exp[I k r Cos[φ]] - uS[r, φ, {β, k, R}, o]]
```

The function `scatterGraphics` finally makes a pair of contour plots, one for the real part and one for the absolute value. For a graphic visualization it is sufficient to neglect terms with $|n| > 12$ for the parameters used below. (We use $R = 1$ and $k \approx 10^0$.)

```
 In[6]:= scatterGraphics[{β_, k_}, pp_, opts___] :=
           With[{R = 1.},
           Show[GraphicsArray[Function[reAbs,
           (* mirror to lower half *)
           Show[{#, # /. Polygon[l_] :> Polygon[# {1, -1}& /@ l],
                 Graphics[{GrayLevel[0], Disk[{0, 0}, R]}]}]&[
           (* the contour plot in the upper half plane *)
           Graphics[ContourPlot[reAbs[u[Sqrt[x^2 + y^2], ArcTan[x, y],
                                 {β, k, R}, 12]], {x, -10, 10}, {y, 0, 10},
                 opts, PlotPoints -> pp{2, 1}, PlotRange -> All,
                 FrameTicks -> None, ContourLines -> False, Contours -> 30,
                 DisplayFunction -> Identity, Compiled -> False,
                 ColorFunction -> (Hue[0.8 #]&)]]]]] /@ {Re, Abs}]]]
```

Here are three visualizations of the scattering process. We use the parameters $R = 1$, $k = 4$, and $\beta = 0$, $\beta = 1/2$, and $\beta = 1$ and display the real part and the absolute value.

```
 In[7]:= scatterGraphics[{#, 4}, 90]& /@ {0, 1/2, 1}
```

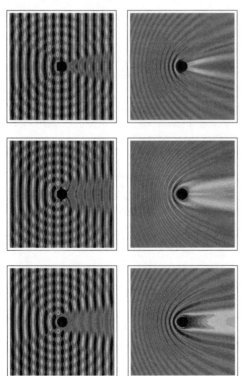

For the scattering of a wave packet on a cylinder, see [1356].

c) For the calculation of the expansion coefficients, as well as for the expansion itself, we will need Bessel function values of the form $\{J_{-o}(z), J_{-o+1}(z), \ldots, J_{o-1}(z), J_o(z)\}$ and $\{Y_{-o}(z), Y_{-o+1}(z), \ldots, Y_{o-1}(z), Y_o(z)\}$. Generating a list of Bessel functions by direct calls to `BesselJ` and `BesselY` surely works, but is relatively time-consuming. Using the recursion relations with respect to the index of the Bessel functions, we need only two calls to the built-in functions to generate $\{J_0(z), ..., J_{o-1}(z), J_o(z)\}$ and $\{Y_0(z), ..., Y_{o-1}(z), Y_o(z)\}$. The negative indices values can be obtained using $J_{-n}(z) = (-1)^n J_n(z)$ and $Y_{-n}(z) = (-1)^n Y_n(z)$. The following two functions BesselJList and BesselYList implement the recursion relations. To guarantee numerical stability, we use backward recursion for the $J_n(z)$ and forward recursion for the $Y_n(z)$ [502].

```
In[1]:= BesselJList[o_Integer, z_] :=
          If[z == 0., Table[If[m == 0, 1., 0.], {m, -o, o}],
              Module[{JO = BesselJ[o, z], JN = BesselJ[o - 1, z], m = o - 2,
                  JList = Table[1., {o + 1}]},
                  (* backward iteration for stability *)
                  JList[[o + 1]] = JO; JList[[o]] = JN;
                  Do[{JO, JN} = {JN, 2 (m + 1)/z JN - JO};
                      m--; JList[[m + 2]] = JN, {o - 1}];
                  Join[Reverse[Table[(-1)^m, {m, o}] Rest[JList]], JList]]]
```

```
In[2]:= BesselYList[o_Integer, z_] :=
          Module[{YO = BesselY[0, z], YN = BesselY[1, z], m = 2,
              YList = Table[1., {o + 1}]},
              (* forward iteration for stability *)
              YList[[1]] = YO; YList[[2]] = YN;
              Do[{YO, YN} = {YN, 2/z (m - 1) YN - YO};
                  m++; YList[[m]] = YN, {o - 1}];
              Join[Reverse[Table[(-1)^m, {m, o}] Rest[YList]], YList]]
```

The following timings show the dramatic speed-ups resulting from the use of BesselJList and especially BesselYList.

```
In[3]:= {(tJ1 = Table[BesselJ[m, 2.], {m, -500, 500}];) // Timing,
          (tJ2 = BesselJList[500, 2.];) // Timing, Max[Abs[(tJ1 - tJ2)/tJ1]]}
Out[3]= {{0.07 Second, Null}, {0.01 Second, Null}, 3.88 × 10^{-13}}
```

```
In[4]:= {(tY1 = Table[BesselY[m, 2.], {m, -400, 400}];) // Timing,
          (tY2 = BesselYList[400, 2.];) // Timing, Max[Abs[(tY1 - tY2)/tY1]]}
Out[4]= {{4.85 Second, Null}, {0. Second, Null}, 1. × 10^{-15}}
```

Compiling the two functions BesselJList and BesselYList would yield a further speed-up, but we will not use the compiled versions in the following because for some cylinder parameters R and n, some elements of the resulting list might not be machine numbers, but high-precision numbers due to large exponents.

```
In[5]:= {BesselJListC, BesselYListC} =
          (Function[body, (* avoid evaluation *)
          Compile[{{o, _Integer}, {z, _Real}}, body], {HoldAll}] @@
          (DownValues[#][[1]] /. (* (re)use existing definition *)
          (Verbatim[HoldPattern][_] :> f_) :> Hold[f]))& /@
                                            {BesselJList, BesselYList}
```

$$Out[5]= \left\{ \text{CompiledFunction}\Big[\{o, z\}, \right.$$
$$\text{If}\Big[z == 0., \text{Table}[\text{If}[m == 0, 1., 0.], \{m, -o, o\}], \text{Module}\Big[\{JO = BesselJ[o, z],$$
$$JN = BesselJ[o - 1, z], m = o - 2, JList = Table[1., \{o + 1\}]\}, JList[[o + 1]] = JO;$$
$$JList[[o]] = JN; Do\Big[\{JO, JN\} = \Big\{JN, \frac{2 (m + 1) JN}{z} - JO\Big\}; m--; JList[[m + 2]] = JN, \{o - 1\}\Big];$$
$$\text{Join}[\text{Reverse}[\text{Table}[(-1)^m, \{m, o\}] \text{ Rest}[JList]], JList]\Big], -\text{CompiledCode-}\Big],$$
$$\text{CompiledFunction}\Big[\{o, z\}, \text{Module}\Big[\{YO = BesselY[0, z], YN = BesselY[1, z],$$
$$m = 2, YList = Table[1., \{o + 1\}]\}, YList[[1]] = YO; YList[[2]] = YN;$$
$$Do\Big[\{YO, YN\} = \Big\{YN, \frac{2 (m - 1) YN}{z} - YO\Big\}; m++; YList[[m]] = YN, \{o - 1\}\Big];$$
$$\text{Join}[\text{Reverse}[\text{Table}[(-1)^m, \{m, o\}] \text{ Rest}[YList]], YList]\Big], -\text{CompiledCode-}\Big]\right\}$$

```
In[6]:= {Do[BesselYList[ 250, 2.], {100}] // Timing,
          Do[BesselYListC[250, 2.], {100}] // Timing}
          CompiledFunction::cfn : Numerical error encountered
              at instruction 30; proceeding with uncompiled evaluation. More…

          CompiledFunction::cfn : Numerical error encountered
              at instruction 30; proceeding with uncompiled evaluation. More…
```

```
CompiledFunction::cfn : Numerical error encountered
    at instruction 30; proceeding with uncompiled evaluation. More…

General::stop : Further output of
    CompiledFunction::cfn will be suppressed during this calculation. More…
```

Out[6]= {{0.34 Second, Null}, {0.44 Second, Null}}

Now we implement the calculation of the coefficients a_m and b_m. First we express them in Bessel functions J_k and Y_k. In the following, we will measure length in units of λ, meaning $k = 2\pi$.

```
In[7]:= J[m_, z_] = BesselJ[m, z];
        H[m_, z_] = BesselJ[m, z] + I BesselY[m, z];
        d = Derivative;  k = 2Pi;


        αβ[m_, {R_, n_}] =
          I^m {d[0, 1][J][m, k R] H[m, k R] - J[m, k R] d[0, 1][H][m, k R],
              d[0, 1][J][m, k R] J[m, n k R] - n  J[m, k R] d[0, 1][J][m, n k R]}/
              (n H[m, k R] d[0, 1][J][m, n k R] - J[m, n k R] d[0, 1][H][m, k R]);
```

The function abLists calculates the coefficients a_m and b_m for values of the radius R and refractive index n by taking partial waves between $-o$ and o into account.

```
In[11]:= abLists[o_, {R_, n_}] :=
         Module[{αβList = Table[αβ[m, {R, n}], {m, -o, o}], NBesselRules, m},
            (* rules for replacing exact values by numerical values *)
            NBesselRules = Dispatch[Rule @@@ (Join @@
            (Transpose[{Table[#1, {m, 0, o}], Drop[#2, o]}]& @@@
            {{BesselJ[m,   k R], BesselJList[o, N[  k R]]},
             {BesselJ[m, n k R], BesselJList[o, N[n k R]]},
             {BesselY[m,   k R], BesselYList[o, N[  k R]]}}))];
            Transpose[αβList /. NBesselRules]]
```

The natural coordinates for the problem at hand are polar coordinates. In a polar coordinate system, the problem factors in an r-dependent part and an φ-dependent part and in the calculation of field amplitude values the r- and φ-dependent parts can be reused. On the other hand, Cartesian coordinates are most suited for the visualization. We bridge the two coordinate systems by constructing an interpolating function in polar coordinates and convert the Cartesian coordinates needed in the plots to the then-everywhere-defined function in polar coordinates. The following function torφBounds calculates the r,φ-bounds for a rectangular plotting area $\{\{\xi_1, \xi_2\}, \{\eta_1, \eta_2\}\}$ taking into account symmetry.

```
In[12]:= torφBounds[{{x1_, x2_}, {y1_, y2_}}] :=
         Module[{ξ1, ξ2, η1, η2, points, rMin, rMax, φMin, φMax},
            {ξ1, ξ2} = Sort[N[{x1, x2}]]; {η1, η2} = Sort[N[{y1, y2}]];
            points = {{ξ1, η1}, {ξ1, η2}, {ξ2, η1}, {ξ2, η2}};
            {rMin, rMax} = {Min[#], Max[#]}&[Sqrt[#.#& /@ points]];
            {φMin, φMax} = {Min[#], Max[#]}&[ArcTan @@@ points];
            Which[(* contains the origin *)
                  ξ1 < 0 < ξ2 && η1 < 0 < η2, {{0, rMax}, {0, Pi}},
                  (* above or below the real line *)
                  η1 >= 0 || y2 <= 0, {{rMin, rMax}, {φMin, φMax}},
                  (* does intersect positive real line *)
                  ξ1 >= 0, {{ξ1, rMax}, {0, φMax}},
                  (* does intersect negative real line *)
                  η1 < 0 < η2, {{-ξ2, rMax}, {φMax, Pi}}]]
```

As mentioned, in polar coordinates, the radial and azimuthal values of the Bessel and exponential functions can be reused many times. So, for each set of parameters R and n, we will cache the values of φL, JaL, and HbLvia a SetDelayed[. Set[...]] construction.

```
In[13]:= makeCachedJYDefinitions[] :=
         (Clear[φL, JaL, HbL];
          With[{tpa = Developer`ToPackedArray},
          φL[φ_] := φL[φ] = tpa @ Table[Exp[I m N[φ]], {m, -o, o}];
          JaL[r_] := JaL[r] = tpa @ (aL BesselJList[o, N[r]]);
          HbL[r_] := HbL[r] = tpa @ (bL (BesselJList[o, N[r]] +
                                      I BesselYList[o, N[r]]))])
```

In the last definitions of the list of Bessel function values `Jal` and `HaL`, we incorporated the expansion coefficients a_m and b_m. This allows us now to define the field strength as a simple scalar product between the list of radial function values and the list of azimuthal function values.

```
In[14]:= ε[{r_, φ_}, {R_, n_}] := If[r <= R, JaL[N[n k r]].φL[N[φ]],
                                     Exp[I k r Cos[φ]] + HbL[N[1 k r]].φL[N[φ]]]
```

Now we are ready to implement the main function `makeIpoAndDensityGraphic`. It will generate a density plot of $f(\mathcal{E}_z(r, \varphi))$ over the Cartesian domain $\{\{x_1, x_2\}, \{y_1, y_2\}\}$ using *ppInt* points for the interpolation and *ppPlot* points for the plotting.

```
In[15]:= makeIpoAndDensityGraphic[f_, {R_, n_}, {{x1_, x2_}, {y1_, y2_}},
                         {ppInt_, ppPlot_}, α_:10, plotOpts___] :=
        ((* truncation order *)
        o = Ceiling[α R];
        (* partial wave coefficients *)
        {aL, bL} = abLists[o, {R, n}];
        (* definitions for caching lists of J-, Y-, and H-values *)
        makeCachedJYDefinitions[];
        (* Cartesian domain => polar domain *)
        {{rMin, rMax}, {φMin, φMax}} = torφBounds[{{x1, x2}, {y1, y2}}];
        (* ε in the polar domain *)
        rφData = Table[Table[{r, φ}, ε[{r, φ}, {R, n}]},
                              {φ, φMin, φMax, (φMax - φMin)/ppInt}],
                       {r, rMin, rMax, (rMax - rMin)/ppInt}];
        (* interpolate ε in the polar domain *)
        ipo = Interpolation[Flatten[rφData, 1]];
        (* make density plot(s) *)
        DensityPlot[#[ipo[Sqrt[x^2 + y^2], Abs[ArcTan[x, y]]]] ,
                    {x, x1, x2}, {y, y1, y2}, plotOpts, PlotRange -> All,
                    Mesh -> False, PlotPoints -> ppPlot, ColorFunction -> Hue,
                    Epilog -> {GrayLevel[0.5], Thickness[0.002],
                               Line[R Table[{Cos[φ], Sin[φ]},
                                            {φ, 0, 2Pi, 2Pi/360.}]]},
                    AspectRatio -> Automatic, PlotLabel -> N[n],
                    FrameTicks -> None]& /@ Flatten[{{f}}])
```

For a cylinder of radius $R = 12\lambda$ and for the refractive index values $n = 2.5$, $n = 1.708$, and $n = 1.25$ we now show $|\mathcal{E}_z(r, \varphi)|$.

```
In[16]:= Show[GraphicsArray[
        Block[{$DisplayFunction = Identity}, Flatten[
        makeIpoAndDensityGraphic[Abs, {12, #},
            24 {{-1, 1}, {-1, 1}}, {400, 400}]& /@
                              {2.5, 1.708, 1.25}]]]]
```

2.5   1.708   1.25

 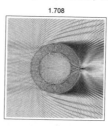 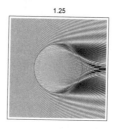

The last graphics show a variety of interesting physics: wave catastrophes as the equivalents classical caustics [1185], the lasing mode due to internal resonance for the $n = 1.708$ case [658], [1060], [751], [1363], [1343], [93], [1315], [552], whispering gallery modes [717], and more [597], [1136], [1173], [1036], [281], [540], [178].

Here is the corresponding animation.

```
Do[makeIpoAndDensityGraphic[Abs, {12, n}, 24 {{-1, 1}, {-1, 1}},
                           {400, 400}],
   {n, 3, 11/10, -19/10/50}]
```

Next we show the field strength for the last graphic across the boundary of the cylinder. The field strength is a continuous function with a discontinuous derivative at $r = R$.

In[17]:= Show[GraphicsArray[Plot[{Re[ipo[r, #]], Im[ipo[r, #]]}, {r, 11, 13},
                      DisplayFunction -> Identity]& /@ {0, 1}]];

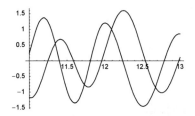 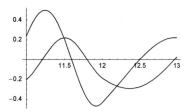

To easily check that we took enough partial waves into account, we define a function checkConvergence[] that displays the logarithm of the absolute value of $a_m J_m(k r)$ and $b_m H_m^{(1)}$. For the above graphic the highest partial waves contribute less than 10^{-20} and are surely small enough to be neglected for a visualization.

In[18]:= checkConvergence[] :=
        Show[Block[{$DisplayFunction = Identity,
                   opts = Sequence[PlotRange -> All, PlotJoined -> True]},
              {(* listplots of logarithm of absolute values of coefficients *)
               ListPlot[Log[Abs[JaL[N[1.25 k rMax]]]], opts,
                       PlotStyle -> {RGBColor[1, 0, 0]}],
               ListPlot[Log[Abs[HbL[N[k rMax]]]], opts,
                       PlotStyle -> {RGBColor[0, 0, 1]}]}]]

In[19]:= checkConvergence[]

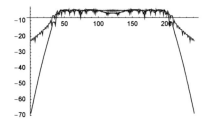

The next graphics compare the absolute value, the real part, and the argument of the field strength. We use the parameters $R = 12 \lambda$ and $n = 1.25$.

In[20]:= Show[MapAt[Transpose, GraphicsArray[
        Block[{$DisplayFunction = Identity},
        makeIpoAndDensityGraphic[{Abs, Re, Arg}, {12, 1.25}, {{6, 30}, {-6, 6}},
                            {400, 400}, 6, AspectRatio -> 3,
                            ColorFunction -> (Hue[2.8 #]&)]]],
                       {{1, 1, 1}, {1, 2, 1}}]]
```
```

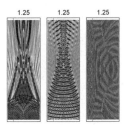

The structure of the field strength inside the cylinder is a very sensitive function of the refractive index. The following three graphics show the absolute value of the field strength for $R = \lambda$ and nearly identical values of n.

```
In[21]:= Show[GraphicsArray[
    Block[{$DisplayFunction = Identity}, Flatten[
    makeIpoAndDensityGraphic[Abs, {1, #}, 2 {{-1, 1}, {-1, 1}},
        {300, 300}, 20]& /@ {2.085, 2.088, 2.091}]]]]
```

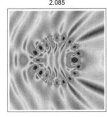
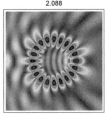
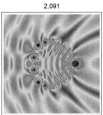

Let us compare the just-calculated intensity with the intensity of a ray approximation. The function `rayPath` gives the path of a light ray through the dielectric cylinder. The path follows easily from elementary geometry and Snell's law. We use a cylinder of radius 1.

```
In[22]:= rayPath[n_, y0_] :=
    Module[{x0 = -3, x1, y1, x2, y2, x3, y3, α, β},
        (* hitting the cylinder *)
        x1 = -Sqrt[1 - y0^2]; y1 = y0; α = ArcTan[x1, y1];
        m12 = Tan[α - Pi + ArcSin[Sin[α]/n]];
        (* leaving the cylinder *)
        {x2, y2} = {m12^2 x1 - m12 y1 + #, y1 + m12 (-x1 + #)}&[
            Sqrt[1 - m12^2 (x1^2 - 1) + 2 m12 x1 y1 - y1^2]]/(1 + m12^2) // N;
        β = ArcTan[x2, y2]; m23 = Tan[β - ArcSin[Sin[β] n]];
        x3 = -x0; y3 = y2 + m23 (x3 - x2);
        (* the internal path *)
        {{x0, y0}, {x1, y1}, {x2, y2}, {x3, y3}}]
```

Here are the paths for 50 rays and cylinders with $n = 1.1$, $n = 1.5$, and $n = 3$.

```
In[23]:= Off[Graphics::gptn];
    With[{o = 25},
    Show[GraphicsArray[
    Graphics[{Table[{Hue[0.8 y0^2], Thickness[0.002],
                    (* the ray *) Line[rayPath[#, y0 // N]]},
                {y0, -(o - 1)/o, (o - 1)/o, 1/o}],
            {GrayLevel[0], Thickness[0.01], Circle[{0, 0}, 1]}},
            AspectRatio -> Automatic, PlotLabel -> "n = " <> ToString[#],
            PlotRange -> 1.5{{-1, 1}, {-1, 1}}]& /@ {1.1, 1.5, 3}]]]
```

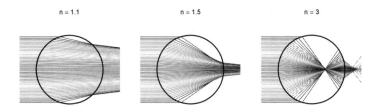

n = 1.1   n = 1.5   n = 3

For an incoming homogeneous density of rays, the intensity I at a point $\{x, y\}$ is given as $I(x, y) = \int_{-R}^{R} \delta(\text{path}(x, y, y_0)) \, dy_0$. Inside the cylinder we have the following expression for $\text{path}(x, y, y_0)$.

```
In[25]:= Clear[n]; rayPath[n, y0];
         (* the path inside the cylinder *)
         Y[{x_, y_}, y0_, n_] =
         Module[{x1, y1, x2, α, R = 1},
               x1 = -Sqrt[R^2 - y0^2]; y1 = y0; α = ArcTan[x1, y1];
               y - y1 - m12 (x - x1)]
```

$$Out[27]= \ y - y0 - \left(x + \sqrt{1 - y0^2}\right) \operatorname{Tan}\left[\operatorname{ArcSin}\left[\frac{\operatorname{Sin}\left[\operatorname{ArcTan}\left[-\sqrt{1 - y0^2}, y0\right]\right]}{n}\right] + \operatorname{ArcTan}\left[-\sqrt{1 - y0^2}, y0\right]\right]$$

Using the relation $\int_{-R}^{R} \delta(f(y_0)) \, dy_0 = \theta(R - \tilde{y}_0) \, \theta(\tilde{y}_0 - R) \, |f'(\tilde{y}_0)|^{-1}$ where $\tilde{y}_0$ is a simple zero of $f(y_0)$ we can implement the calculation of the ray density inside the cylinder. We solve the equation $Y[\{x, y\}, y_0, n] == 0$ numerically.

```
In[28]:= (* the derivative with respect to y0 *)
         dY[{x_, y_}, y0_, n_] = D[Y[{x, y}, y0, n], y0];
In[30]:= (* numerical root finding *)
         y0Root[{x_, y_}, n_] := If[Head[#] === List, #, {}]& @
               FindRoot[Evaluate[Y[{x, y}, y0, n] == 0], {y0, -1, 1},
                         MaxIterations -> 50]
In[32]:= (* the density of the rays *)
         δRay[{x_, y_}, n_] := With[{y0 = y0 /. y0Root[{x, y}, n]},
                                    If[NumericQ[y0] && Im[y0] == 0.,
                                       1/Abs[dY[{x, y}, y0, n]], 0.]]
```

The following graphics show the ray density inside the cylinder for $n = 1.1$, $n = 1.5$, and $n = 3$. The last two cases show pronounced singularity at the caustics where rays intersect.

```
In[34]:= (* turn off some FindRoot messages arising
            from caustic singularities *)
         Off[FindRoot::frsec]; Off[FindRoot::frmp]; Off[FindRoot::cvmit];

         Show[GraphicsArray[
         Show[{#, # /. Polygon[l_] :> Polygon[{1, -1, 1}#& /@ l]}]&[
         ParametricPlot3D[{r Cos[φ], r Sin[φ],
                           Log[δRay[{r Cos[φ], r Sin[φ]}, #] + 1], EdgeForm[]},
                          {r, 0, 0.999}, {φ, 0, Pi}, PlotRange -> All,
                          PlotPoints -> 121, BoxRatios -> {1, 1, 0.6},
                          DisplayFunction -> Identity]& /@ {1.1, 1.5, 3.0}]]]
```

The ray density in the region to the right of the cylinder could be analyzed in a similar manner. For the reconstruction of the rays from the solution of the Helmholtz equation, see [123].

We could no continue with left-handed materials (negative *n*) [1132], [1006], [1007], [865], [1340] or negative dielectric permittivity (complex *n*) [712] values. Here are some such examples.

```
In[37]:= Show[GraphicsArray[#]]& /@
          Block[{$DisplayFunction = Identity},
          Partition[
            makeIpoAndDensityGraphic[Abs, {#1, #2}, 2 #1 {{-1, 1}, {-1, 1}},
                                     {160, 160}, 25][[1]]& @@@
            (* radius and refractive index *)
            {{1, -2.275}, {1, -2.85}, {1, -1.01},
             {0.003763447, -0.99996851 - 0.0003650372 I},
             {0.1062547,    5.53477900 - 0.4851389 I},
             {1.8000000,   -0.99080166 + 0.13532207 I},
             {5.3644691,    0.37637662 - 0.01305735 I},
             {0.6007915,    0.59520844 - 0.08625483 I},
             {3.2902513,    0.40114371 - 0.05842709 I}}, 3]]
```

For dielectric cylinders with an inner hole, see [585], [561]; for enclosed cylinders, see [369], [1253]; for cylinders with a hexagonal cross section, see [1316]; for cylinders with a spiral-shaped cross section, see [770]; for arrays of cylinders, see [470]; for the time-dependent problem, see [876].

d) Because of the cylindrical symmetry of the problem around the *z*-axis, it suffices to calculate and display $\psi(\mathbf{r})$ in the *x,z*-plane. In this plane we will use the projection of the corresponding 3D spherical coordinate system. This has the advantage that the summands forming $\psi(r, \vartheta)$ factor and so dramatically reduce the number of $R_l(\kappa; k\, r)$ and $P_l(\cos(\vartheta))$ function values that have to be calculated (compared to a square grid in the *x,z*-plane, say).

The function makeDefinitions implements the definitions of $R_l(\kappa; k\, r)$ and $P_l(\cos(\vartheta))$. To avoid building up large amounts of cached values of the R[κ, *l*, *k*, *r*] and Θ[*l*, ϑ] we add the Clear[R, Θ] statement so that for each value of κ, we calculate and store a new set of R and Θ values.

```
In[1]:= makeDefinitions :=
          (Clear[R, θ];
```

```
(* the radial part *)
R[κ_, l_, k_, r_] := R[κ, l, k, r] =
        (2 l + 1) I^l Exp[I Arg[Gamma[l + 1 + I κ/k]]]*
        (Exp[-1/2 Pi κ/k] Abs[Gamma[l + 1 + I κ/k]]/(2l + 1)!*
        (2 k r)^l Exp[I k r] Hypergeometric1F1[l + 1 + I κ/k, 2l + 2, -2 I k r]);
(* the azimuthal part *)
Θ[l_, ϑ_] := Θ[l, ϑ] = LegendreP[l, Cos[ϑ]])
```

Now we implement the calculation of the sum $\psi(r, \vartheta)$. For each set of κ, l, k, r values, we will sum the first 20 terms explicitly and additional terms until the fourth digit of the sum does not change anymore.

```
In[2]:= ψ[k_, κ_, r_, ϑ_] :=
    Module[{Σ, l, ε = 10^-4, L = 20},
            Σ = Sum[R[κ, l, k, r] Θ[l, ϑ], {l, 0, L}];
            l = L + 1;
            (* add terms until desired precision is reached *)
            While[t = R[κ, l, k, r] Θ[l, ϑ]; Abs[t/Σ] > ε,
                    Σ = Σ + t; l++];
            (* return sum *) Σ]
```

The function makeGraphics generates a 3D graphic of $f(\psi(r, \vartheta))$.

```
In[3]:= Needs["Graphics`Graphics3D`"]
```

```
In[4]:= makeGraphics[f_, data_, opts___] :=
    Module[{d = Apply[{#1, #2, f[#3]}&, data, {2}], polys},
        (* the polygons *)
        polys = Cases[ListSurfacePlot3D[d, DisplayFunction -> Identity],
                    _Polygon, Infinity];
        (* the final graphics *)
        Graphics3D[{EdgeForm[], polys, Map[{1, -1, 1}#&, polys, {-2}]}, opts,
                    PlotRange -> All, BoxRatios -> {1, 1, 1/4}, Axes -> True]]
```

The function picturePair finally generates a GraphicsArray of Re($\psi(r)$) and $|\psi(r)|^2$ for the wave number k and the potential strength κ.

```
In[5]:= picturePair[k_, κ_] :=
    With[{ε = 10.^-6, rmax = 16, ppr = 80, ppφ = 120},
    Module[{data},
    (* clear old values of R and Θ *) makeDefinitions;
    (* calculate data; exclude real line *)
    data = Table[{r Cos[ϑ], r Sin[ϑ], ψ[k, κ, N[r], N[ϑ]]},
                    {r, ε, rmax, (rmax - ε)/ppr}, {ϑ, ε, Pi - ε, (Pi - 2ε)/ppφ}];
    (* display pair of graphics; real part and absolute value *)
    Show[GraphicsArray[{makeGraphics[Re, data], makeGraphics[Abs[#]^2&, data,
                    PlotRange -> {0, 1.1 Max[Map[Abs[#[[3]]]&, data, {2}]^2]}]}]]]]
```

The graphics show nicely for attractive potentials $|\psi(\{0, 0, 0\})|^2 \sim |\kappa|/k$ and for repulsive potentials $|\psi(\{0, 0, 0\})|^2 \sim \kappa/k \exp(-2\kappa/k)$. (For the scattering on the $1/r^2$ potential, see [111], [325].)

```
In[6]:= Do[picturePair[1, κ], {κ, -8, 8, 2}]
```

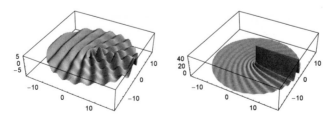

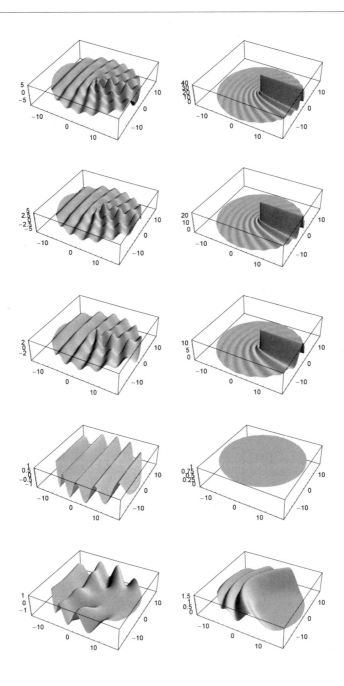

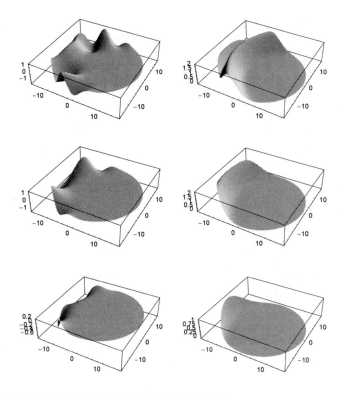

```
Do[picturePair[1, κ], {κ, -8, 8, 1/2}]
```

e) This is the equation to be solved for μ.

```
In[1]:= eq[n_, R_, μ_] = Sqrt[μ] BesselJ[n - 1, Sqrt[μ] R] -
                n (1/R + I) BesselJ[n, Sqrt[μ] R]
```

$$\text{Out[1]=} \quad \sqrt{\mu} \; \text{BesselJ}[-1+n, R\sqrt{\mu}] - n\left(i + \frac{1}{R}\right)\text{BesselJ}[n, R\sqrt{\mu}]$$

By making pictures of the lines of the real part and the imaginary part equal to zero, for instance, by using running the following inputs we see that there are many of solutions to this equation.

```
Module[{pp = 200, data},
 Show[GraphicsArray[Table[
 (* calculate function value only once *)
 data = Table[eq[n, 25, μx + I μy],
              {μy, -1/2., 1/2., 1/pp}, {μx, -1/2., 1/2., 1/pp}];
 Show[ListContourPlot[#1[data],
                ContourShading -> False, Contours -> {0},
                MeshRange -> {{-1, 1}, {-1, 1}}/2,
                ContourStyle -> {{Thickness[0.002], #2}},
                DisplayFunction -> Identity]& @@@
              (* real and imaginary part *)
              {{Re, Hue[0]}, {Im, Hue[0.78]}}]], {n, 1, 3}]]]]
```

Because we are not interested in an exhaustive study of all solutions here, we just numerically find a few of them using `FindRoot`. (We could also search for all intersections of all lines in the above picture and use these intersections as starting points for `FindRoot`.) To get different values for μ, we start with randomly chosen starting values. Sometimes the search will not converge. To avoid printing out the corresponding warnings while still being able to catch the generation of a message with `Check`, we use the following construction. We temporarily redirect the message `FindRoot::cvnwt`, so that it is not printed out.

```
In[2]:= Unprotect[Message];
```

```
        HoldPattern[Message[FindRoot::cvnwt, iter_]] :=
        Block[{$Messages = sentNowhere},
                Message[FindRoot::cvnwt, iter]] /; $Messages =!= sentNowhere;
```

Here, the search for $n = 1, 2, 3$ is carried out.

```
In[4]:= Do[For[μBag[n] = {}, (* three values for μ for every n *)
                Length[μBag[n]] < 3,
                Null, (* find next value of μ *)
                If[NumberQ[#], newμValue = #; (* new value of μ found? *)
                    If[Min[Abs[# - newμValue]& /@ μBag[n]] > 10^-5 &&
                        Abs[newμValue] > 10^-3,
                    AppendTo[μBag[n], newμValue]]]&[(* numerical root finding *)
                Check[μ /. FindRoot[Evaluate[eq[n, 25, μ]],
                        {μ, Random[Complex, {-1 - I, 1 + I}]}], "noConvergence"]]],
        {n, 1, 3}]
```

These are the μ-values collected.

```
In[5]:= ??μBag
```

```
        Global`μBag
```

```
        μBag[1] = {0.023414 - 0.00189104 i, 0.810664 - 0.104472 i, 0.282693 - 0.025253 i}
        μBag[2] = {0.0421652 - 0.00169257 i, 0.515604 - 0.0215858 i, 0.215849 - 0.00879449 i}
        μBag[3] = {0.270933 - 0.0072965 i, 0.152389 - 0.00408565 i, 0.420963 - 0.0114023 i}
```

Now, we must make the picture. Using the technique for making the circular contour plot for the drum in Section 3.5, we first make a `ListContourPlot` in r,φ-coordinates and then transform into polar coordinates. Because some of the polygons of the contour graphics will have only a few edge points with large differences in their φ-coordinates, we add additional points along these edges to get a smooth boundary after the transformation into polar coordinates. The routine `addEdgePoints` adds these points, so that the distance between two points becomes less than δ.

```
In[6]:= addEdgePoints[δ_] :=
        Module[{φDist, δ1, fl},
        Polygon[li_] :> Polygon[Join @@ ((* test if new points are needed *)
        If[(φDist = Abs[#[[1, 2]] - #[[2, 2]]]) < δ, #,
            δ1 = #[[2]] - #[[1]]; (* add new points *)
            Table[#[[1]] + 1/fl δ1, {1, 0, fl = Floor[φDist/δ]}]]& /@
                Partition[Append[li, First[li]], 2, 1])]]
```

Now, we can finally implement the function `circlePicture`. To avoid the multiple (time-consuming) calculation of the Bessel functions, we use the construction `Outer[Im[#1 #2]&, Table[...], Table[...]]`.

```
In[7]:= circlePicture[n_, R_, μ_, pp_, opts___] :=
        Module[{data, lcp},
        (* data for the ListContourPlot *)
        data = Outer[Im[#1 #2]&, Table[Exp[I n φ], {φ, 0, N[2Pi], N[2Pi/pp]}],
                Table[BesselJ[n, Sqrt[μ] r], {r, 0., N[R], N[R]/pp}]];
        (* ListContourPlot in r,φ-coordinates *)
        lcp = ListContourPlot[data, PlotRange -> All,
            MeshRange -> {{0, R}, {0, 2Pi}}, DisplayFunction -> Identity];
        Show[Graphics[{Thickness[0.002],
        (* transform into polar coordinates *)
        Map[If[Head[#] === List, #[[1]] {Cos[#[[2]]], Sin[#[[2]]]}, #]&,
            Graphics[lcp][[1]] /. addEdgePoints[0.1], {-2}]}],
            AspectRatio -> Automatic, PlotRange -> All]]
```

Using `circlePicture`, we can now look at the circular contour plots of the μ-values collected in the μBag.

```
In[8]:= Show[GraphicsArray[
            circlePicture[1, 25, #, 56, DisplayFunction -> Identity]& /@ μBag[1]]]
```

```
In[9]:= Show[GraphicsArray[
            circlePicture[2, 25, #, 66, DisplayFunction -> Identity]& /@ μBag[2]]]
```

```
In[10]:= Show[GraphicsArray[
            circlePicture[3, 25, #, 86, DisplayFunction -> Identity]& /@ μBag[3]]]
```

Using the package function `ListSurfacePlot3D`, we can generate the corresponding 3D plots.

```
In[11]:= Needs["Graphics`Graphics3D`"]

In[12]:= circlePicture3D[n_, R_, μ_, pp_, opts___] :=
         Module[{data, lcp, polys},
         (* data for the ListSurfacePlot3D *)
         data = Outer[{#2[[1]] Sin[#1[[1]]], #2[[1]] Cos[#1[[1]]],
                       Im[#1[[2]] #2[[2]]]}&,
               Table[{φ, Exp[I n φ]}, {φ, 0, N[2Pi], N[2Pi/pp]}],
               Table[{r, BesselJ[n, Sqrt[μ] r]}, {r, 0., N[R], N[R]/pp}], 1];
         (* make polygons *)
         polys = Cases[ListSurfacePlot3D[data, DisplayFunction -> Identity],
                       _Polygon, Infinity];
         (* the picture *)
         Show[Graphics3D[{EdgeForm[],
               SurfaceColor[Hue[Random[]], Hue[Random[]], 3 Random[]], polys}],
               Boxed -> False, BoxRatios -> {1, 1, 1/4}, opts]]

In[13]:= Show[GraphicsArray[
            circlePicture3D[1, 25, #, 60, DisplayFunction -> Identity]& /@ μBag[1]]]
```

In[14]:= Show[GraphicsArray[
          circlePicture3D[2, 25, #, 60, DisplayFunction -> Identity]& /@ μBag[2]]]

In[15]:= Show[GraphicsArray[
          circlePicture3D[3, 25, #, 60, DisplayFunction -> Identity]& /@ μBag[3]]]

f) Using some ideas from the last subexercises, we will employ a polar coordinate system. makeDefinitions generates the definitions to compute $F(v; r, \varphi; t)$.

```
In[1]:= makeDefinitions[v_, k_, mMax_] :=
        (Clear[rF, φF];
        (* pack lists for fast scalar multiplication *)
        rF[r_] := rF[r] = Developer`ToPackedArray @
        Table[0. I + (-I)^Sqrt[m^2 + 2m v]*
              BesselJ[Sqrt[m^2 + 2m v], N[k r]], {m, -mMax, mMax}];
        φF[φ_] := φF[φ] = Developer`ToPackedArray @
                          Table[0. I + Exp[I m N[φ]], {m, -mMax, mMax}];
        F[r_, φ_] := rF[r].φF[φ])
```

Because of the polar coordinate system, we will implement a function circularListContourPlot.

```
In[2]:= circularListContourPlot[data_, R_, opts___] :=
        Module[{δε = 0.1},
        (* make contour plot in r,φ-coordinates *)
        prod = Times @@ Dimensions[data];
        cls = #[[Round[prod/50/2]]]& /@
                 Partition[Sort[Flatten[data]], Round[prod/50]];
        (* the color function *)
        colFc = Compile[{x}, #[x]]&[Function[Evaluate[
         Which @@ Flatten[Append[Function[x, {# < x[[1]], x[[2]]}] /@
                MapIndexed[{#1, (#2[[1]] - 1)/50 0.8}&, cls], {True, 0.815}]]]]];
        colF[x_] := Hue[colFc[x]];
        (* the contour plot *)
        lcp = ListContourPlot[data, Contours -> cls, ContourLines -> False,
                MeshRange -> {{0, 2Pi}, {0, R}}, DisplayFunction -> Identity,
                ContourStyle -> {Thickness[0.002]}, PlotRange -> All,
```

```
                ColorFunctionScaling -> False, ColorFunction -> colF];
         (* make Graphics-object from contour plot *)
         gr = Graphics[lcp];
         (* transform to polar coordinates *)
         Show[{gr /. (lp:(Polygon | Line))[l_] :> lp[{#2 Cos[#1], #2 Sin[#1]}& @@@
                      (* add points *) addPoints[lp][l, δε]],
               Graphics[{Thickness[0.01], GrayLevel[0], Circle[{0, 0}, R]}]}, opts,
              DisplayFunction -> $DisplayFunction, Frame -> False]];
```

```
In[3]:= (* add points to get smooth curves *)
        addPoints[lp_][points_, δε_] :=
        Module[{n, l}, Join @@ (Function[pair,
                If[(* additional points needed? *)
                   (l = Sqrt[#. #]&[Subtract @@ pair]) < δε, pair,
                   n = Floor[l/δε] + 1;
                   Table[# + i/n (#2 - #1), {i, 0, n - 1}]& @@ pair]] /@
                Partition[If[lp === Polygon,
                             Append[#, First[#]], #]&[points], 2, 1])]
```

The function makePicture generates a graphic of the light wave amplitude as a function of the vortex strength n and the wave vector k.

```
In[5]:= makePicture[{ν_, k_, R_, mMax_}, {ppr_, ppφ_}, opts___] :=
        Module[{data, ε = 10^-5},
               makeDefinitions[ν, k, mMax];
               data = Table[F[r, φ], {r, ε, R, (R - ε)/ppr}, {φ, 0, 2Pi, 2Pi/ppφ}];
               circularListContourPlot[Im[data], R, opts]]
```

```
In[6]:= Show[GraphicsArray[makePicture[{#, 3, 7, 50}, {120, 240},
                                        DisplayFunction -> Identity]& /@ #]]& /@
                     Partition[Table[ν, {ν, 1/100, 2, (2 - 1/100)/14}], 3]
```

```
Do[makePicture[{v, 3, 7, 50}, {120, 240}], {v, 1/100, 2, (2 - 1/100)/36}]
```

g) To determine the $c_j(k, \alpha)$ we have to truncate the system of equations. To construct the resulting finite-dimensional linear system requires the calculation of many values of Bessel functions of different order, but equal arguments. To speed up these Bessel function calculations we implement a recursive definition of Bessel functions [1322].

```
In[1]:= makeBesselJDefinitions[o_] :=
        ((* remove existing definitions *)Clear[besselJ];
        (* initial conditions *)
        besselJ[0, z_] := besselJ[0, z] = BesselJ[0, z];
        besselJ[o, z_] := besselJ[o, z] = BesselJ[o, z];
        besselJ[o - 1, z_] := besselJ[o - 1, z] = BesselJ[o - 1, z];
        besselJ[-o, z_] := besselJ[-o, z] = BesselJ[-o, z];
        besselJ[-o + 1, z_] := besselJ[-o + 1, z] = BesselJ[-o + 1, z];
        (* recursion relations *)
        besselJ[n_?Positive, z_] := besselJ[n, z] =
                2/z (n + 1)besselJ[n + 1, z] - besselJ[n + 2, z];
        besselJ[n_?Negative, z_] := besselJ[n, z] =
                2/z (n - 1)besselJ[n - 1, z] - besselJ[n - 2, z];)
```

The following quick timing comparison shows that we easily speed the calculation up by a factor of ten and more.

```
In[2]:= makeBesselJDefinitions[100]
        (tab1 = Table[besselJ[o, -40.], {o, -100, 100}]); // Timing
Out[3]= {0.01 Second, Null}
```

```
In[4]:= (tab2 = Table[BesselJ[o, -40.], {o, -100, 100}]); // Timing
Out[4]= {0.08 Second, Null}
```

Within the precision needed for the following calculation, the two results `tab1` and `tab2` agree.

```
In[5]:= Max[Abs[tab1 - tab2]]
Out[5]= 1.79023 × 10^{-14}
```

Now for the direct calculation of the $c_j(k, \alpha)$: To obtain a numerically stable solution, we generate more equations than we face unknowns $c_j(k, \alpha)$. This means we use `PseudoInverse` instead of `LinearSolve` for the linear algebra problem. We start with 41 of the $c_j(k, \alpha)$ assumed to be nonvanishing and increase their number until we are sure that we took enough of the $c_j(k, \alpha)$ into account.

```
In[6]:= makeψ[k_, α_, {b_, β_}] :=
    Module[{kx = N[k Sin[α]], ky = -N[k Cos[α]], Kx, Ky, m = 20, μ},
        Kx[j_] = kx + j β; Ky[j_] = Sqrt[k^2 - Kx[j]^2];
        (* increase m until we have enough c's *)
        While[μ = m + 10;
                (* make definitions for Bessel functions *)
                makeBesselJDefinitions[2μ + 2];
                (* set up linear algebra problem *)
                c = Table[-besselJ[j, b ky], {j, -μ, μ}];
                A = Table[besselJ[n - j, b Ky[j]], {n, -μ, μ}, {j, -m, m}];
                (* calculate the c's *)
                cs = PseudoInverse[A].c // Chop;
                (* was m large enough? *)
                Count[cs, _?(# =!= 0&)] < 2m, m = m + 10];
        (* put together ψ *)
        expVector = Table[Exp[I (Kx[j] x + Ky[j] y)], {j, -m, m}];
        ψIncoming = Exp[I (kx x + ky y)];
        ψScattering = cs.expVector;
        (* set ψ to 0 inside the wall *)
        UnitStep[y - b Sin[β x]] (ψIncoming + ψScattering)]
```

The following graphic shows $|\psi_{k,\alpha}(x, y)|$ along the wall $y = b \sin(\beta x)$. $|\psi_{k,\alpha}(x, y)|$ is quite small ($< 10^{-6}$) and the solution quality is sufficient for a visualization.

```
In[7]:= With[{k = 2, α = 0.5, b = 1/4, β = 2},
    Plot[Evaluate[Abs[makeψ[k, α, {b, β}] /.
                _UnitStep -> 1 /. y -> b Sin[β x]]], {x, 0, 2Pi}]]
```

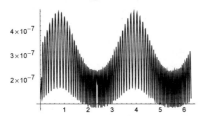

For the visualization of $|\psi_{k,\alpha}(x, y)|$, $\text{Re}(\psi_{k,\alpha}(x, y)) = 0$, and $\text{Im}(\psi_{k,\alpha}(x, y)) = 0$ as a function of k and α, we finally implement a function `scatteringGraphic`. It displays the three quantities over the x-region $(0, p\,2\pi)$.

```
In[8]:= scatteringGraphic[ψ_, {x_, y_}, {b_, β_}, {Y_, p_}, {ppx_, ppy_}] :=
        ((* compile ψ for speed *)
        ψc = Compile[{x, y}, ψ];
        cdps = Table[(* data of ψ *)
        data = Table[ψc[x, y], {y, -b, Y, (Y + b)/ppy},
                                {x, k Pi/β, (k + 1) Pi/β, Pi/β/ppx}];
        opts = Sequence[MeshRange -> {{k Pi/β, (k + 1) Pi/β}, {-b, Y}},
                        DisplayFunction -> Identity];
        {(* density plot of Abs[ψ]^2 *)
        ListDensityPlot[Abs[data]^2, opts, Mesh -> False],
        (* contour plot of Re[ψ] == 0 and Im[ψ] == 0 *)
        ListContourPlot[#1[data], opts, Contours -> {0},
                        DisplayFunction -> Identity, ContourLines -> True,
                        ContourShading -> False, ContourStyle -> {#2}]& @@@
                    {{Re, RGBColor[1, 0, 0]}, {Im, RGBColor[0, 0, 1]}}}],
                    {k, 0, 2p - 1}];
        (* the corrugated wall in green *)
        wall = Graphics[{RGBColor[0, 1, 0],
                Polygon[Join[Table[{x, b Sin[β x]}, {x, 0, p 2Pi/β, p 2Pi/β/120}],
                        {{p 2Pi/β, -b - Y/10}, {0, -b - Y/10}}]] // N}];
        (* display all prepared elements *)
        Show[{cdps, wall}, PlotRange -> All, AspectRatio -> 0.4,
            FrameTicks -> None, DisplayFunction -> Identity])
```

Now we have all ingredients together to make the animation. First we fix α and vary k. We use $b = 0.25$ and a period of the wall of π. Looking at the picture one notices that—in analogy with the Bloch theorem for a periodic potential—the wave function itself is not periodic, only its absolute value is.

```
In[9]:= Show[GraphicsArray[#]]& /@ Partition[
          (ψ[x_, y_] = makeψ[#, 0.3, {0.25, 2}];
                    scatteringGraphic[ψ[x, y], {x, y}, {0.25, 2},
                        {7, 3}, {50, 60}])& /@ Table[k, {k, 2, 8, 6/7}], 2]
```

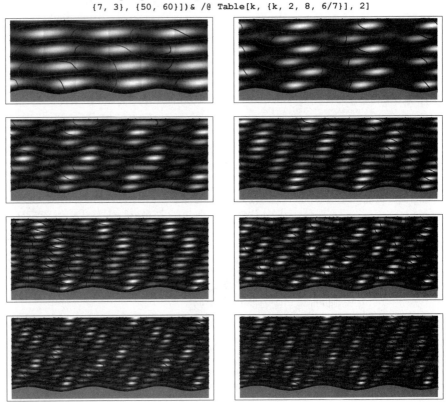

And here is the corresponding animation.

```
Do[ψ[x_, y_] = makeψ[k, 0.3, {0.25, 2}];
    Show[scatteringGraphic[ψ[x, y], {x, y}, {0.25, 2}, {7, 3}, {50, 60}],
        DisplayFunction -> $DisplayFunction], {k, 1.1, 8, 0.1}]
```

Now fix k and vary α. Again we use $b = 0.25$ and a period of the wall of π.

```
In[10]:= Show[GraphicsArray[#]]& /@ Partition[
          (ψ[x_, y_] = makeψ[6, #, {0.25, 2}];
            scatteringGraphic[ψ[x, y], {x, y}, {0.25, 2},
            {7, 3}, {50, 60}])& /@
            Table[φ, {φ, 1/8 Pi/2, 7/8 Pi/2, 6/8 Pi/2/7}], 2]
```

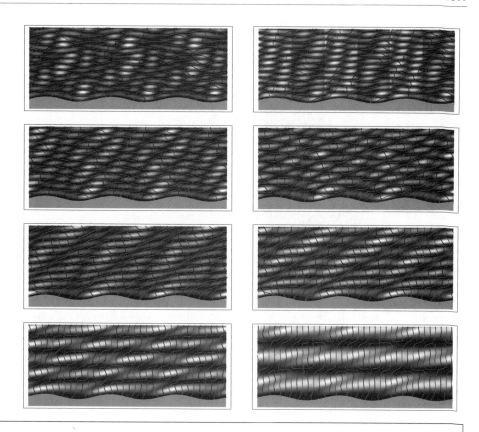

```
Do[ψ[x_, y_] = makeψ[6, φ, {0.25, 2}];
   Show[scatteringGraphic[ψ[x, y], {x, y}, {0.25, 2}, {7, 3}, {50, 60}],
                          DisplayFunction -> $DisplayFunction],
   {φ, 1/8 Pi/2, 7/8 Pi/2, 6/8 Pi/2/60}]
```

Here we dealt with a half space bounded by a corrugated wall; for channels with corrugated walls, see [817], [818], [21], and [819]; for more complicated corrugation patterns, see [982]. For the corrugated wall-reflection on negative phase-velocity materials, see [384]. For an interesting periodic potential with no classical analogue, see [1027].

h) We start by implementing the smooth random functions $c_n(\tau)$. We use random trigonometric polynomials for the absolute value and the argument of the $c_n(\tau)$. (For visualization, the exact specification of the probability distribution of the c_n does not matter.)

```
In[1]:= rF[n_, τ_] :=  Sum[Random[Real, {-1, 1}] *
                          Cos[j 2Pi τ + 2Pi Random[]], {j, n}]

In[2]:= rAbs[n_, τ_] :=
        Module[{min, max, f = rF[n, τ]}, (* normalize to Abs[f] <= 1 *)
               {min, max} = {Min[#], Max[#]}&[Table[f, {τ, 0, 1, 1/200}]];
               (f - min)/(max - min)]

In[3]:= rArg[n_, τ_] :=  rF[n, τ]
```

We will truncate the double infinite n-sum at $o = 300$. We will use $k = 1$ in the following. Any other choice of k is possible, and just sets the scale for the resulting structures. For $k = 1$, most of the $J_o(k\,r)$ are sufficiently small for the purpose of a graphics.

```
In[4]:= o = 300; k = 1;
        SeedRandom[111];
        Do[c[k, τ_] = rAbs[3, τ] Exp[2Pi I rArg[2, τ]], {k, -o, o}]
```

Here are some of the functions shown as parametric curves.

```
In[7]:= Show[Table[
        ParametricPlot[Evaluate[{Re[#], Im[#]}& @ c[k, τ]], {τ, 0, 1},
                Frame -> True, Axes -> False,
                DisplayFunction -> Identity,
                AspectRatio -> Automatic, PlotStyle -> {Hue[k/12]}],
            {k, 0, 10}], DisplayFunction -> $DisplayFunction]
```

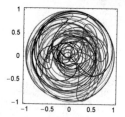

We cannot make a contour plot in polar coordinates directly. So, we first calculate $\psi_\tau(r, \varphi)$ on a r,φ-grid in polar coordinates and then interpolate these data to obtain $\psi_\tau(x, y)$. This avoids the time-consuming evaluation of Bessel functions through $\psi_\tau\left((x^2 + y^2)^{1/2}, \arctan(y/x)\right)$. The compiled function ψC calculates $\psi_\tau(r, \varphi)$ for a given value of φ and lists of values $\{J_n(k\,r)\}$ and $\{c_n(\tau)\}$ for $n = -o, \dots, 0, \dots, o$.

```
In[8]:= ψC = Compile[{φ, {js, _Complex, 1}, {cs, _Complex, 1}},
            Module[{λ = Length[js] - 1},
                Sum[cs[[j + λ + 1]] js[[Abs[j] + 1]] Exp[I j φ],
                {j, -λ, λ}]]];
```

For each r, we calculate the values $\{J_n(k\,r)\}$ only once and store them.

```
In[9]:= jList[r_] := jList[r] = If[Abs[#] < $MinMachineNumber, 0., #]& /@
                Table[BesselJ[m, N[k r]], {m, 0, o}];
```

Now, we have all ingredients together so we can implement a function cPlot that generates a contour plot of the real and imaginary parts of $\psi_\tau(r, \varphi)$.

```
In[10]:= (* define arctan at origin *)
         arcTan[x_, y_] := If[x == y == 0, 0., ArcTan[x, y]];


In[12]:= cPlot[τ_, {ppr_, ppφ_, ppxy_}, L_, opts___] :=
         Module[{R = Sqrt[2] L, cList, data, ipo, dataC, cpRe, cpIm,
                commonOpts = Sequence[MeshRange -> {{-L, L}, {-L, L}},
                                Contours -> {0}, DisplayFunction -> Identity]},
             (* c-values *)
             cList = Table[c[k, τ], {k, -o, o}];
             (* r,φ-data *)
             data  = Table[Table[{φ, r, ψC[φ, jList[r], cList]},
                            {φ, -Pi, Pi, 2Pi/ppφ}], {r, 0, R, R/ppr}];
             (* interpolate r,φ-data *)
             ipo = Interpolation[Flatten[data, 1]];
             (* x,y-data *)
             dataC = Table[ipo[arcTan[x, y], Sqrt[x^2 + y^2]],
                        {y, -L, L, 2.L/ppxy}, {x, -L, L, 2.L/ppxy}];
             (* colored contour plot of the real part *)
             cpRe = ListContourPlot[Re[dataC], ContourLines -> False,
                        MeshRange -> {{-L, L}, {-L, L}}, ColorFunction ->
                        (If[# < 0, RGBColor[1, 0, 0], RGBColor[0, 0, 1]]&),
                        ColorFunctionScaling -> False, commonOpts];
             (* contour lines of vanishing imaginary part *)
             cpIm = ListContourPlot[Im[dataC], ContourShading -> False,
                        ContourStyle -> {{GrayLevel[1]}}, commonOpts];
```

```
(* unite real and imaginary part contour plots *)
Show[{cpRe, cpIm}, opts, FrameTicks -> False]]
```

Here are three of the resulting contour plots for various τ. We see typical "random" structures of the length scale $k = 1$.

In[13]:= `Show[GraphicsArray[cPlot[#, {150, 300, 301}, 40]& /@ {0, 1/3, 2/3}]]`

And here is the corresponding animation.

```
frames = 60;
Do[cPlot[τ, {150, 300, 301}, 50, DisplayFunction -> $DisplayFunction],
   {τ, 0, 1 - 1/frames, 1/frames}]
```

For the model use of such pictures in physics, see [170] and [139].

Now let us deal with the 3D case. We now define random values $c[l, m]$.

In[14]:= `c[l_, m_] := c[l, m] = Random[] Exp[2Pi I Random[]]`

To avoid the repeated numerical calculation of the Bessel functions and the Legendre functions forming the nontrivial part of the spherical harmonics, we cache their values.

In[15]:= `P[l_, r_] := P[l, r] = If[r == 0, 0.,`
`                            If[Abs[#] < $MinMachineNumber, 0., #]&[`
`                                N[BesselJ[l + 1/2, r]/r]]];`

```
θ[l_, m_, θ_] := θ[l, m, θ] =  Sqrt[(2l + 1) (l - m)!/
                 (4Pi (l + m)!)] LegendreP[l, m, 2, Cos[θ]] // N;

(* no caching needed for this simple function *)
Φ[m_, φ_] = Exp[I m φ] // N;
```

Here is the definition of the $\psi(r, \vartheta, \varphi)$.

In[21]:= `ψ[r_, θ_, φ_] :=`
`         Sum[c[l, m] P[l, r] θ[l, m, θ] Φ[m, φ], {l, 0, o}, {m, -l, l}]`

We now proceed similarly as above: we calculate $\psi(r, \vartheta, \varphi)$ on a tensor product grid in spherical coordinates and interpolate the resulting function. Because of the double sum occurring in the definition of $\psi(r, \vartheta, \varphi)$, we will use fewer points than in the 2D case to avoid a very time-consuming calculation. (Repeating the calculation with more points yields qualitatively similar pictures.)

In[22]:= `R = 20; o = 20; ppr = 36; ppθ = 36; ppφ = 60;`
`        SeedRandom[123];`

```
rθφData = Table[{r, θ, φ, ψ[r, θ, φ]}, {φ, -Pi, Pi, 2Pi/ppφ},
                {θ, 0, Pi, Pi/ppθ}, {r, 0, R, R/ppr}];
```

In[25]:= `ipo = Interpolation[Flatten[rθφData, 2]]`

Out[25]= `InterpolatingFunction[{{0., 20.}, {0., 3.14159}, {-3.14159, 3.14159}}, <>]`

$\psi C[x, y, z]$ calculates the random superposition in Cartesian coordinates.

In[26]:= `Clear[ψC];`
`        ψC[x_?NumberQ, y_, z_] := If[x == y == z == 0, 0,`

```
         ipo[Sqrt[x^2 + y^2 + z^2],
             ArcCos[z/Sqrt[x^2 + y^2 + z^2]], arcTan[x, y]]]
```

We again display the surfaces of vanishing real and imaginary parts. This time we use the function `ContourPlot3D` from the package `Graphics`ContourPlot3D``.

```
In[28]:= Needs["Graphics`ContourPlot3D`"]
```

```
In[29]:= (* cut hole in polygons *)
         outlinePolygon[Polygon[l_], α_] := Function[mp,
           Polygon[Join[#1, Reverse[#2]]]& @@@ Transpose[
             Partition[Append[#, First[#]], 2, 1]& /@
             {1, mp + α (# - mp)& /@ 1}]][(Plus @@ l)/Length[l]]
```

```
In[31]:= (* cube of maximal size *) L = R/Sqrt[3.];
```

```
         Show[GraphicsArray[Graphics3D[{EdgeForm[], #2, outlinePolygon[#, 0.6]& /@
           (* extract polygons from contour plots *)
           Cases[ContourPlot3D[#1[ψC[x, y, z]], {x, -L, L}, {y, -L, L}, {z, -L, L},
                      PlotPoints -> {24, 3}, MaxRecursion -> 1,
                      Compiled -> False, DisplayFunction -> Identity],
                      _Polygon, Infinity]}]& @@@
           (* function and coloring *)
           {{Re, SurfaceColor[Hue[0.02], Hue[0.12], 2.6]},
            {Im, SurfaceColor[Hue[0.22], Hue[0.12], 2.6]}}]]
```

14. Wronskian of Legendre Functions, Separation of Variables in Toroidal Coordinate System

a) For a second-order differential equation of the form $w''(z) + f_1(z) w'(z) + f_0(z) w(z) = 0$, Liouville's theorem states the following relation for the Wronskian $W(z)$:

$$W(z) = W(z_0) \exp\left(-\int_{z_0}^{z_1} f_1(z') \, dz'\right).$$

Using the differential equations of the Legendre functions, the second factor is easily calculated. (We use $z_0 = 0$.)

```
In[1]:= zDependentPart = Exp[Integrate[2ζ/(1 - ζ^2), {ζ, 0, z},
                            GenerateConditions -> False]] // Simplify
```

$$Out[1]= \frac{1}{1 - z^2}$$

So the calculation of $W(0)$ remains to be carried out. We get this expression by the following manipulations.

```
In[2]:= Normal[Series[ LegendreP[n, m, z],    {z, 0, 0}]]*
        Normal[Series[D[LegendreQ[n, m, z], z], {z, 0, 0}]] -
        Normal[Series[ LegendreQ[n, m, z],    {z, 0, 0}]]*
        Normal[Series[D[LegendreP[n, m, z], z], {z, 0, 0}]]
```

$$Out[2]= \frac{2^{-1+m} (m + n) \pi^{3/2} \text{Csc}[m\pi] \left(\frac{2^m \sqrt{\pi} \, \text{Cos}[m\pi]}{\text{Gamma}[1-\frac{m}{2}-\frac{n}{2}] \, \text{Gamma}[\frac{1}{2}-\frac{m}{2}+\frac{n}{2}]} - \frac{2^{-m} \sqrt{\pi} \, \text{Gamma}[m+n]}{\text{Gamma}[1+\frac{m}{2}-\frac{n}{2}] \, \text{Gamma}[\frac{1}{2}+\frac{m}{2}+\frac{n}{2}] \, \text{Gamma}[-m+n]} \right)}{\text{Gamma}[\frac{1}{2}-\frac{m}{2}-\frac{n}{2}] \, \text{Gamma}[1-\frac{m}{2}+\frac{n}{2}]} -$$

$$\frac{2^{-1+m} (m + n) \pi^{3/2} \text{Csc}[m\pi] \left(\frac{2^m \sqrt{\pi} \, \text{Cos}[m\pi]}{\text{Gamma}[\frac{1}{2}-\frac{m}{2}-\frac{n}{2}] \, \text{Gamma}[1-\frac{m}{2}+\frac{n}{2}]} - \frac{2^{-m} \sqrt{\pi} \, \text{Gamma}[1+m+n]}{\text{Gamma}[\frac{1}{2}+\frac{m}{2}-\frac{n}{2}] \, \text{Gamma}[1+\frac{m}{2}+\frac{n}{2}] \, \text{Gamma}[1-m+n]} \right)}{\text{Gamma}[1-\frac{m}{2}-\frac{n}{2}] \, \text{Gamma}[\frac{1}{2}-\frac{m}{2}+\frac{n}{2}]}$$

Now, we must simplify this part to $\Gamma(n + m + 1)/\Gamma(n - m + 1)$. Here, this is done.

In[3]:= **FullSimplify[%] // TrigToExp // RootReduce // ToRadicals**

Out[3]= $\dfrac{\text{Gamma}[1 + m + n]}{\text{Gamma}[1 - m + n]}$

b) First we load in the package `Calculus`VectorAnalysis``.

In[1]:= **Needs["Calculus`VectorAnalysis`"]**

The command `CoordinatesToCartesian` gives the required transformation from Cartesian to toroidal coordinates.

In[2]:= **??CoordinatesToCartesian**

> CoordinatesToCartesian[pt] gives the Cartesian coordinates of the point pt given in
>   the default coordinate system. CoordinatesToCartesian[pt, coordsys] gives the
>   Cartesian coordinates of the point given in the coordinate system coordsys. More...
>
> Attributes[CoordinatesToCartesian] = {Protected, ReadProtected}

In[3]:= **CoordinatesToCartesian[{η, φ, ϑ}, Toroidal]**

Out[3]= $\left\{ \dfrac{\text{Cos}[\vartheta]\,\text{Sinh}[\eta]}{-\text{Cos}[\varphi] + \text{Cosh}[\eta]},\ \dfrac{\text{Sin}[\vartheta]\,\text{Sinh}[\eta]}{-\text{Cos}[\varphi] + \text{Cosh}[\eta]},\ \dfrac{\text{Sin}[\varphi]}{-\text{Cos}[\varphi] + \text{Cosh}[\eta]} \right\}$

Because in the following, we will only work in toroidal coordinates, we change globally to this coordinate system.

In[4]:= **SetCoordinates[Toroidal[η, ϑ, φ, 1]]**

Out[4]= Toroidal[η, ϑ, φ, 1]

The package also provides information about the ranges of the allowed η, ϑ, φ-values and the range of the parameter c.

In[5]:= **{CoordinateRanges[Toroidal], ParameterRanges[]}**

Out[5]= $\{\{0 \le \eta < \infty,\ -\pi < \vartheta \le \pi,\ -\pi < \varphi \le \pi\},\ 0 < \#1 < \infty\}$

Toroidal coordinates are orthogonal coordinates, meaning that at every point in the space, the tangent planes to *coordinate = constant* are pairwise orthogonal. Here is a visualization of the surfaces $\eta = constant$, $\varphi = constant$, and $\vartheta = constant$.

In[6]:= **g[η_, ϑ_, φ_] = CoordinatesToCartesian[{η, φ, ϑ}, Toroidal];**

In[7]:= **(* set of coordinate hypersurfaces *)**
tab = Apply[ParametricPlot3D[##, PlotPoints -> 25,
**   DisplayFunction -> Identity]&,**
**   (* just a couple of coordinate surfaces *)**
**   {{g[2, ϑ, φ],   {ϑ, -Pi/2, Pi}, {φ, -Pi, Pi}},**
**   {g[3/2, ϑ, φ], {ϑ, -Pi/2, Pi}, {φ, -Pi, Pi}},**
**   {g[3, ϑ, φ],   {ϑ, -Pi/2, Pi}, {φ, -Pi, Pi}},**
**   {g[η, ϑ, Pi/2], {η, 0, 4},   {ϑ, -Pi/2, Pi}},**
**   {g[η, ϑ, Pi],   {η, 0, 4},   {ϑ, -Pi/2, Pi}},**
**   {g[η, ϑ, Pi/4], {η, 0, 4},   {ϑ, -Pi/2, Pi}},**
**   {g[η, ϑ, -Pi/4], {η, 0, 4},   {ϑ, -Pi/2, Pi}},**
**   {g[η, ϑ, -Pi/2], {η, 0, 4},   {ϑ, -Pi/2, Pi}},**
**   {g[η, -Pi/2, φ],   {η, 0.8, 3}, {φ, -Pi, Pi}},**
**   {g[η, 0, φ]   ,   {η, 0.8, 3}, {φ, -Pi, Pi}},**
**   {g[η, Pi/2, φ],   {η, 0.8, 3}, {φ, -Pi, Pi}}}, {1}];**

In[9]:= **Show[Graphics3D[{EdgeForm[{GrayLevel[1], Thickness[0.002]}],**
**   (* color each set of hypersurfaces differently *)**
**   {SurfaceColor[RGBColor[1, 0, 0], RGBColor[1, 0, 0], 2.5],**
**   Cases[Take[tab, {1, 3}], _Polygon, Infinity]},**
**   {SurfaceColor[RGBColor[0, 1, 0], RGBColor[0, 1, 0], 2.5],**
**   Cases[Take[tab, {4, 8}], _Polygon, Infinity]},**
**   {SurfaceColor[RGBColor[0, 0, 1], RGBColor[0, 0, 1], 2.5],**
**   Cases[Take[tab, {9, 11}], _Polygon, Infinity]}}],**
**   PlotRange -> All, ViewPoint -> {-3, -2, 2}, Boxed -> False]**

Now, let us tackle the more complicated system of the separation of variables method for the Laplacian in toroidal coordinates. The package also provides the Laplace operator for us.

In[10]:= **??Laplacian**

    Laplacian[f] gives the Laplacian of the scalar- or vector-valued
      function f in the default coordinate system. Laplacian[f, coordsys]
      gives the Laplacian of f in the coordinate system coordsys. **More...**

    Attributes[Laplacian] = {Protected, ReadProtected}

In[11]:= **SetCoordinates[Toroidal[η, ϑ, φ, c]]**

Out[11]= Toroidal[η, ϑ, φ, c]

We apply the Laplacian to $\sqrt{2(\cosh(\eta) - \cos(\varphi))}\; f(\eta, \vartheta, \varphi)$. (Note the prefactor; it is necessary to achieve the separation.)

In[12]:= **(Δ = Laplacian[Sqrt[2(Cosh[η] - Cos[ϑ])] f[η, ϑ, φ]])**

Out[12]= $\frac{1}{c^3}\left((-\cos[\vartheta] + \cosh[\eta])^3 \operatorname{Csch}[\eta] \left(\frac{\sqrt{2}\, c\, \operatorname{Csch}[\eta]\, f^{(0,0,2)}[\eta, \vartheta, \varphi]}{\sqrt{-\cos[\vartheta] + \cosh[\eta]}} - \right. \right.$

$\left. \frac{c \sin[\vartheta] \sinh[\eta] \left(\frac{f[\eta,\vartheta,\varphi] \sin[\vartheta]}{\sqrt{2}\sqrt{-\cos[\vartheta]+\cosh[\eta]}} + \sqrt{2}\sqrt{-\cos[\vartheta]+\cosh[\eta]}\, f^{(0,1,0)}[\eta, \vartheta, \varphi]\right)}{(-\cos[\vartheta]+\cosh[\eta])^2} + \right.$

$\frac{1}{-\cos[\vartheta]+\cosh[\eta]}\left(c \sinh[\eta] \left(\frac{\cos[\vartheta] f[\eta, \vartheta, \varphi]}{\sqrt{2}\sqrt{-\cos[\vartheta]+\cosh[\eta]}} - \frac{f[\eta, \vartheta, \varphi]\sin[\vartheta]^2}{2\sqrt{2}(-\cos[\vartheta]+\cosh[\eta])^{3/2}} + \right.\right.$

$\left.\left. \frac{2\sin[\vartheta] f^{(0,1,0)}[\eta, \vartheta, \varphi]}{\sqrt{2}\sqrt{-\cos[\vartheta]+\cosh[\eta]}} + \sqrt{2}\sqrt{-\cos[\vartheta]+\cosh[\eta]}\, f^{(0,2,0)}[\eta, \vartheta, \varphi] \right)\right) +$

$\frac{c \cosh[\eta] \left(\frac{f[\eta,\vartheta,\varphi]\sinh[\eta]}{\sqrt{2}\sqrt{-\cos[\vartheta]+\cosh[\eta]}} + \sqrt{2}\sqrt{-\cos[\vartheta]+\cosh[\eta]}\, f^{(1,0,0)}[\eta, \vartheta, \varphi]\right)}{-\cos[\vartheta]+\cosh[\eta]} -$

$\frac{c \sinh[\eta]^2 \left(\frac{f[\eta,\vartheta,\varphi]\sinh[\eta]}{\sqrt{2}\sqrt{-\cos[\vartheta]+\cosh[\eta]}} + \sqrt{2}\sqrt{-\cos[\vartheta]+\cosh[\eta]}\, f^{(1,0,0)}[\eta, \vartheta, \varphi]\right)}{(-\cos[\vartheta]+\cosh[\eta])^2} +$

$\frac{1}{-\cos[\vartheta]+\cosh[\eta]}\left(c \sinh[\eta] \left(\frac{\cosh[\eta] f[\eta, \vartheta, \varphi]}{\sqrt{2}\sqrt{-\cos[\vartheta]+\cosh[\eta]}} - \frac{f[\eta, \vartheta, \varphi]\sinh[\eta]^2}{2\sqrt{2}(-\cos[\vartheta]+\cosh[\eta])^{3/2}} + \right.\right.$

$\left.\left.\left.\left. \frac{2\sinh[\eta] f^{(1,0,0)}[\eta, \vartheta, \varphi]}{\sqrt{2}\sqrt{-\cos[\vartheta]+\cosh[\eta]}} + \sqrt{2}\sqrt{-\cos[\vartheta]+\cosh[\eta]}\, f^{(2,0,0)}[\eta, \vartheta, \varphi] \right)\right)\right)\right)\right)$

Now we simplify this expression. Because we are interested in the Laplace equation (the right-hand side vanishes), we can concentrate on the numerator.

In[13]:= **(Δ1 = Numerator[Together[Δ]]) // Short[#, 12]&**

Out[13]//Short= $-\operatorname{Csch}[\eta]\, \left(-2\sqrt{2}\cos[\vartheta]^3 f[\eta, \vartheta, \varphi]\sinh[\eta] + \right.$

$6\sqrt{2}\cos[\vartheta]^2\cosh[\eta]^2 f[\eta, \vartheta, \varphi]\sinh[\eta] - 4\sqrt{2}\cosh[\eta]^3 f[\eta, \vartheta, \varphi]\sinh[\eta] -$

$3\sqrt{2}\cos[\vartheta] f[\eta, \vartheta, \varphi]\sin[\vartheta]^2 \sinh[\eta] + 3\sqrt{2}\cosh[\eta] f[\eta, \vartheta, \varphi]\sin[\vartheta]^2 \sinh[\eta] -$

$3\sqrt{2}\cos[\vartheta] f[\eta, \vartheta, \varphi]\sinh[\eta]^3 + 3\sqrt{2}\cosh[\eta] f[\eta, \vartheta, \varphi]\sinh[\eta]^3 -$

$12\sqrt{2}\cos[\vartheta]^2\coth[\eta] f^{(0,0,2)}[\eta, \vartheta, \varphi] + 12\sqrt{2}\cos[\vartheta]\cosh[\eta]\coth[\eta] f^{(0,0,2)}[\eta, \vartheta, \varphi] -$

$4\sqrt{2}\cosh[\eta]^2\coth[\eta] f^{(0,0,2)}[\eta, \vartheta, \varphi] + 4\sqrt{2}\cos[\vartheta]^3\operatorname{Csch}[\eta] f^{(0,0,2)}[\eta, \vartheta, \varphi] +$

$4\sqrt{2}\cos[\vartheta]^3\sinh[\eta] f^{(0,2,0)}[\eta, \vartheta, \varphi] - 12\sqrt{2}\cos[\vartheta]^2\cosh[\eta]\sinh[\eta] f^{(0,2,0)}[\eta, \vartheta, \varphi] +$

$12\sqrt{2}\cos[\vartheta]\cosh[\eta]^2\sinh[\eta] f^{(0,2,0)}[\eta, \vartheta, \varphi] - 4\sqrt{2}\cosh[\eta]^3\sinh[\eta] f^{(0,2,0)}[\eta, \vartheta, \varphi] +$

$4\sqrt{2}\cos[\vartheta]^3\cosh[\eta] f^{(1,0,0)}[\eta, \vartheta, \varphi] - 12\sqrt{2}\cos[\vartheta]^2\cosh[\eta]^2 f^{(1,0,0)}[\eta, \vartheta, \varphi] +$

$$12 \sqrt{2} \, \text{Cos}[\vartheta] \, \text{Cosh}[\eta]^3 \, f^{(1,0,0)}[\eta, \vartheta, \varphi] - 4 \sqrt{2} \, \text{Cosh}[\eta]^4 \, f^{(1,0,0)}[\eta, \vartheta, \varphi] +$$
$$4 \sqrt{2} \, \text{Cos}[\vartheta]^3 \, \text{Sinh}[\eta] \, f^{(2,0,0)}[\eta, \vartheta, \varphi] - 12 \sqrt{2} \, \text{Cos}[\vartheta]^2 \, \text{Cosh}[\eta] \, \text{Sinh}[\eta] \, f^{(2,0,0)}[\eta, \vartheta, \varphi] +$$
$$12 \sqrt{2} \, \text{Cos}[\vartheta] \, \text{Cosh}[\eta]^2 \, \text{Sinh}[\eta] \, f^{(2,0,0)}[\eta, \vartheta, \varphi] - 4 \sqrt{2} \, \text{Cosh}[\eta]^3 \, \text{Sinh}[\eta] \, f^{(2,0,0)}[\eta, \vartheta, \varphi])$$

In[14]:= `Δ2 = Δ1/Csch[η];`

We look for all different partial derivatives of $f(\eta, \vartheta, \varphi)$ (and $f(\eta, \vartheta, \varphi)$ itself).

In[15]:= `allfs = Join[Union[Cases[Δ1, Derivative[__][f][__], {0, Infinity}]],`
        `{f[η, ϑ, φ]}]`

Out[15]= $\{ f^{(0,0,2)}[\eta, \vartheta, \varphi], f^{(0,2,0)}[\eta, \vartheta, \varphi], f^{(1,0,0)}[\eta, \vartheta, \varphi], f^{(2,0,0)}[\eta, \vartheta, \varphi], f[\eta, \vartheta, \varphi] \}$

Now, we extract all prefactors of these terms.

In[16]:= `Δ3 = Simplify[Factor[Plus @@ Cases[Δ2, # _]]]& /@ allfs`

Out[16]= $\{ -4 \sqrt{2} \, (\text{Cos}[\vartheta] - \text{Cosh}[\eta])^3 \, \text{Csch}[\eta] \, f^{(0,0,2)}[\eta, \vartheta, \varphi],$
$\quad -4 \sqrt{2} \, (\text{Cos}[\vartheta] - \text{Cosh}[\eta])^3 \, \text{Sinh}[\eta] \, f^{(0,2,0)}[\eta, \vartheta, \varphi],$
$\quad -4 \sqrt{2} \, (\text{Cos}[\vartheta] - \text{Cosh}[\eta])^3 \, \text{Cosh}[\eta] \, f^{(1,0,0)}[\eta, \vartheta, \varphi],$
$\quad -4 \sqrt{2} \, (\text{Cos}[\vartheta] - \text{Cosh}[\eta])^3 \, \text{Sinh}[\eta] \, f^{(2,0,0)}[\eta, \vartheta, \varphi],$
$\quad -\sqrt{2} \, (\text{Cos}[\vartheta] - \text{Cosh}[\eta])^3 \, f[\eta, \vartheta, \varphi] \, \text{Sinh}[\eta] \}$

We simplify the last result $\Delta 3$.

In[17]:= `Δ4 = Sqrt[2]/(8 Sinh[η]) Cancel[#/(Cosh[η] - Cos[ϑ])^3]& /@ (Plus @@ Δ3)`

Out[17]= $\frac{1}{4} f[\eta, \vartheta, \varphi] + \text{Csch}[\eta]^2 \, f^{(0,0,2)}[\eta, \vartheta, \varphi] +$
$\quad f^{(0,2,0)}[\eta, \vartheta, \varphi] + \text{Coth}[\eta] \, f^{(1,0,0)}[\eta, \vartheta, \varphi] + f^{(2,0,0)}[\eta, \vartheta, \varphi]$

It is now obvious that this equation allows us to separate variables. So we make the following ansatz:
$f[\eta, \vartheta, \varphi] = \eta\text{Func}[\eta] \, \vartheta\text{Func}[\vartheta] \, \varphi\text{Func}[\varphi]$.

In[18]:= `Δ5 = #/(ηFunc[η] ϑFunc[ϑ] φFunc[φ])& /@`
        `(Δ4 /. {Derivative[d1_, d2_, d3_][f][η, ϑ, φ] :>`
        `       D[ηFunc[η] ϑFunc[ϑ] φFunc[φ], {η, d1}, {ϑ, d2}, {φ, d3}],`
        `       f[η, ϑ, φ] -> ηFunc[η] ϑFunc[ϑ] φFunc[φ]})`

Out[18]= $\frac{1}{4} + \frac{\text{Coth}[\eta] \, \eta\text{Func}'[\eta]}{\eta\text{Func}[\eta]} + \frac{\eta\text{Func}''[\eta]}{\eta\text{Func}[\eta]} + \frac{\vartheta\text{Func}''[\vartheta]}{\vartheta\text{Func}[\vartheta]} + \frac{\text{Csch}[\eta]^2 \, \varphi\text{Func}''[\varphi]}{\varphi\text{Func}[\varphi]}$

Introducing separation constants $-m^2$ and $-n^2$ via
$-\varphi\text{Func}''[\varphi] == -m\text{^}2 \, \varphi\text{Func}''[\varphi]$
$-\vartheta\text{Func}''[\vartheta] == -n\text{^}2 \, \vartheta\text{Func}''[\vartheta]$
gives cos, sin as solutions of the differential equations in ϑ and φ, and the remaining differential equation in η transforms into the following.

In[19]:= `Δ6 = Δ5 /. {φFunc''[φ]/φFunc[φ] -> - m^2, ϑFunc''[ϑ]/ϑFunc[ϑ] -> - n^2}`

Out[19]= $\frac{1}{4} - n^2 - m^2 \, \text{Csch}[\eta]^2 + \frac{\text{Coth}[\eta] \, \eta\text{Func}'[\eta]}{\eta\text{Func}[\eta]} + \frac{\eta\text{Func}''[\eta]}{\eta\text{Func}[\eta]}$

In[20]:= `Δ7 = ηFunc[η] Δ6 // Expand`

Out[20]= $\frac{\eta\text{Func}[\eta]}{4} - n^2 \, \eta\text{Func}[\eta] - m^2 \, \text{Csch}[\eta]^2 \, \eta\text{Func}[\eta] + \text{Coth}[\eta] \, \eta\text{Func}'[\eta] + \eta\text{Func}''[\eta]$

We change the independent variable via $s = \cosh(\eta)$ in this differential equation.

In[21]:= `Δ8 = Δ7 /. {ηFunc[η]         ->                sFunc[s],`
        `           ηFunc'[η]  -> D[Cosh[η], η]        sFunc'[s],`
        `           ηFunc''[η] -> D[Cosh[η], {η, 2}] sFunc'[s] +`
        `                         D[Cosh[η], η]^2    sFunc''[s]}`

Out[21]= $\frac{s\text{Func}[s]}{4} - n^2 \, s\text{Func}[s] - m^2 \, \text{Csch}[\eta]^2 \, s\text{Func}[s] + 2 \, \text{Cosh}[\eta] \, s\text{Func}'[s] + \text{Sinh}[\eta]^2 \, s\text{Func}''[s]$

We rewrite all terms depending on η as terms in s of this differential equation.

In[22]:= `Δ9 = Δ8 /. {Cosh[η] -> s, Sinh[η]^2 -> s^2 - 1, Csch[η]^2 -> 1/(s^2 - 1)}`

Out[22]= $\frac{s\text{Func}[s]}{4} - n^2 \, s\text{Func}[s] - \frac{m^2 \, s\text{Func}[s]}{-1 + s^2} + 2 \, s \, s\text{Func}'[s] + (-1 + s^2) \, s\text{Func}''[s]$

Reordering the terms shows that it is a Legendre differential equation.

```
In[23]:= Δ10 = Collect[-Δ9, {sFunc[s], sFunc'[s], sFunc''[s]}]
```

$$Out[23]= \left(-\frac{1}{4} + n^2 + \frac{m^2}{-1+s^2}\right) sFunc[s] - 2 s\, sFunc'[s] + (1-s^2)\, sFunc''[s]$$

The solutions of the last differential equation are Legendre functions of lower index $n - 1/2$ and upper index $|m|$.

15. Riemann–Siegel Formula, Zeros of Hurwitz Zeta Function, Zeta Zeta Function, Harmonic Polylogarithms, $1 \times 2 \times 3 \times \cdots = \sqrt{2\pi}$

a) *Mathematica* cannot calculate the series expansion of `RiemannSiegelZ` directly.

```
In[1]:= Series[RiemannSiegelZ[t], {t, Infinity, 3}]
```

```
Out[1]= RiemannSiegelZ[t]
```

So we will use the given formulas to calculate the coefficients c_k. We will first calculate the series expansion of $\exp(i(\log(t/(2\pi)) - t/2 - \pi/8 - \vartheta(t)))$ and then the series expansion of $(\sum_{j=0}^{\infty} A_j(y)\, \omega^j)\,(\sum_{j=0}^{\infty} \psi^{(j)}(p)\, y^j / j!)$.

Mathematica also cannot calculate the series expansion of `RiemannSiegelTheta`.

```
In[2]:= Series[RiemannSiegelTheta[t], {t, Infinity, 3}]
```

```
Out[2]= RiemannSiegelTheta[t]
```

But *Mathematica* can compute the series expansion of `LogGamma[z]` around $z = \infty$.

```
In[3]:= Series[LogGamma[z], {z, Infinity, 5}]
```

$$Out[3]= \frac{-1 - Log[\frac{1}{z}]}{\frac{1}{z}} + \left(\frac{1}{2} Log[2\pi] + \frac{1}{2} Log[\frac{1}{z}]\right) + \frac{1}{12 z} - \frac{1}{360}\left(\frac{1}{z}\right)^3 + \frac{\left(\frac{1}{z}\right)^5}{1260} + O\left[\frac{1}{z}\right]^6$$

Using this series, it is straightforward to calculate the series expansion of the Riemann–Siegel Theta function.

```
In[4]:= RiemannSiegelThetaSeries[t_, o_] :=
         Module[{aux1, aux2, aux3, aux4},
                (* series for LogGamma *)
                aux1 = Normal[Simplify[PowerExpand[
                               Series[LogGamma[z], {z, Infinity, o + 4}]]]];
                (* change variables *)
                aux2 = aux1 /. z -> I t/2 + 1/4;
                (* simplify logarithmic terms *)
                aux3 = Expand[PowerExpand[Normal[Series[aux2,
                               {t, Infinity, o}]]]] - I t/2 Log[Pi];
                (* take imaginary part *)
                aux4 = (Plus @@ (Cases[aux3, _Complex _]/I)) + O[t]^2
```

Here are the first 20 terms of the series expansion.

```
In[5]:= rsts = RiemannSiegelThetaSeries[t, 20]
```

$$Out[5]= \frac{91546277357}{131491430400\, t^{19}} + \frac{5749691557}{64012419072\, t^{17}} + \frac{118518239}{8021606400\, t^{15}} +$$
$$\frac{8191}{2555904\, t^{13}} + \frac{1414477}{1476034560\, t^{11}} + \frac{511}{1216512\, t^9} + \frac{127}{430080\, t^7} + \frac{31}{80640\, t^5} +$$
$$\frac{7}{5760\, t^3} + \frac{1}{48 t} - \frac{\pi}{8} + \left(-\frac{1}{2} - \frac{Log[2]}{2} - \frac{Log[\pi]}{2} + \frac{Log[t]}{2}\right) t + O[t]^2$$

The following table shows the size of the terms for $t = 100$.

```
In[6]:= Block[{t = N[100, 100],
              ser = FoldList[Plus, 0, Drop[Insert[#, Last[#], 2]&[
                              Reverse[List @@ Normal[rsts]]], -1]]},
              ser - RiemannSiegelTheta[t]] // N
```

```
Out[6]= {-87.9722, 0.392491, -0.000208335, -1.21532×10⁻⁹, -3.84454×10⁻¹⁴,
         -2.95336×10⁻¹⁸, -4.20149×10⁻²², -9.58616×10⁻²⁶, -3.20622×10⁻²⁹,
         -1.47839×10⁻³², -8.98912×10⁻³⁶, -6.96886×10⁻³⁹, -6.70928×10⁻⁴²}
```

Given the series expansion of $\vartheta(t)$, it is straightforward to calculate the series expansion of $\exp(i(\log(t/(2\pi)) - t/2 - \pi/8 - \vartheta(t)))$.

```
In[7]:= RiemannSiegelThetaExpSeries[ω_, o_] :=
          Module[{aux1, aux2},
                 aux1 = Normal[RiemannSiegelThetaSeries[t, o]];
                 aux2 = Expand[PowerExpand[t/2 Log[t/(2Pi)]] -
                                 t/2 - Pi/8 - aux1]] /. t -> 2Pi/ω^2;
                 Series[Exp[I aux2], {ω, 0, o}]]
```

Here are the first few terms of this series.

```
In[8]:= RiemannSiegelThetaExpSeries[ω, 8]
```

$$Out[8]= 1 + \frac{i\,\omega^2}{96\,\pi} - \frac{\omega^4}{18432\,\pi^2} + \frac{4027\,i\,\omega^6}{26542080\,\pi^3} - \frac{16123\,\omega^8}{10192158720\,\pi^4} + O[\omega]^9$$

Now, let us deal with the series expansion of the second factor, namely, $(\sum_{j=0}^{\infty} A_j(y)\,\omega^j)\,(\sum_{j=0}^{\infty} \psi^{(j)}(p)\,y^j/j!)$. We start with the calculation of the $A_j(y)$. We implement the recursion relation and the series expansion independently. Because the $\psi^{(j)}(p)$ involving series contains only nonnegative powers of y, we need all nonpositive powers of y in the series for the $A_j(y)$.

Be aware of the Expand in the definitions of the $\mathcal{A}[j]$. Because we differentiate, nested expressions arise, and expanding the results is essential to get expressions of manageable size.

```
In[9]:= 𝓐[0] = Exp[I 2Pi y^2];
        𝓐[j_] := 𝓐[j] = Expand[-1/2 y 𝓐[j - 1] - 1/32 /Pi^2 D[1/y 𝓐[j - 1], {y, 2}]];
        A[j_] := 𝓐[j] + O[y]
```

Taking now enough terms with respect to ω in all series expansions, we arrive at the following function RiemannSiegel⌐ ExtraTermsList that calculates the coefficients c_k.

```
In[12]:= RiemannSiegelExtraTermsList[ω_, n_] :=
           Module[{aux1, aux2},
                  aux1 = RiemannSiegelThetaExpSeries[ω, n + 1];
                  aux2 = Series[Coefficient[Normal[Sum[A[j] ω^j, {j, 0, 3n + 4}] *
                                  Sum[Derivative[k][ψ][p] y^k/k!, {k, 0, 3n + 4}]], y, 0],
                                  {ω, 0, n + 1}];
                  Take[Expand[CoefficientList[aux1 aux2, ω]], n]]
```

Here is a list of the first ten coefficients. We do not substitute the actual form of $\psi(p)$ because the high-order derivatives would give quite large results.

```
In[13]:= res = RiemannSiegelExtraTermsList[ω, 10]
```

$$Out[13]= \Big\{\psi[p],\ -\frac{\psi^{(3)}[p]}{96\,\pi^2},\ \frac{\psi''[p]}{64\,\pi^2} + \frac{\psi^{(6)}[p]}{18432\,\pi^4},\ \frac{\psi'[p]}{64\,\pi^2} - \frac{\psi^{(5)}[p]}{3840\,\pi^4} - \frac{\psi^{(9)}[p]}{5308416\,\pi^6},$$

$$\frac{\psi[p]}{128\,\pi^2} + \frac{19\,\psi^{(4)}[p]}{24576\,\pi^4} + \frac{11\,\psi^{(8)}[p]}{5898240\,\pi^6} + \frac{\psi^{(12)}[p]}{2038431744\,\pi^8},$$

$$-\frac{5\,\psi^{(3)}[p]}{3072\,\pi^4} - \frac{901\,\psi^{(7)}[p]}{82575360\,\pi^6} - \frac{7\,\psi^{(11)}[p]}{849346560\,\pi^8} - \frac{\psi^{(15)}[p]}{978447237120\,\pi^{10}},$$

$$\frac{5\,\psi'[p]}{2048\,\pi^4} + \frac{367\,\psi^{(6)}[p]}{7864320\,\pi^6} + \frac{18889\,\psi^{(10)}[p]}{237817036800\,\pi^8} + \frac{17\,\psi^{(14)}[p]}{652298158080\,\pi^{10}} + \frac{\psi^{(18)}[p]}{563585608581120\,\pi^{12}},$$

$$-\frac{5\,\psi'[p]}{2048\,\pi^4} - \frac{407\,\psi^{(5)}[p]}{2621440\,\pi^6} - \frac{6649\,\psi^{(9)}[p]}{11890851840\,\pi^8} - \frac{2131\,\psi^{(13)}[p]}{5707608883200\,\pi^{10}} - \frac{\psi^{(17)}[p]}{15655155793920\,\pi^{12}} -$$

$$\frac{\psi^{(21)}[p]}{378729528966512640\,\pi^{14}},\ \frac{41\,\psi[p]}{32768\,\pi^4} + \frac{427\,\psi^{(4)}[p]}{1048576\,\pi^6} + \frac{26405\,\psi^{(8)}[p]}{8455716864\,\pi^8} + \frac{88651\,\psi^{(12)}[p]}{22830435532800\,\pi^{10}} +$$

$$\frac{11153\,\psi^{(16)}[p]}{8766887244595200\,\pi^{12}} + \frac{23\,\psi^{(20)}[p]}{180347394745958400\,\pi^{14}} + \frac{\psi^{(24)}[p]}{290864278246281707520\,\pi^{16}},$$

$$-\frac{2603\,\psi^{(3)}[p]}{3145728\,\pi^6} - \frac{3781\,\psi^{(7)}[p]}{264241152\,\pi^8} - \frac{21543701\,\psi^{(11)}[p]}{669692775628800\,\pi^{10}} - \frac{66727\,\psi^{(15)}[p]}{3652869685248000\,\pi^{12}} -$$

$$\frac{8503\,\psi^{(19)}[p]}{2524863526443417600\,\pi^{14}} - \frac{13\,\psi^{(23)}[p]}{60596724634642022400\,\pi^{16}} - \frac{\psi^{(27)}[p]}{25130673640478739529280\,\pi^{18}}\Big\}$$

Showing all prefactors of the last result in factored form yields a slightly nicer looking result.

```
In[14]:= showIntegersFactored[Rational[p_, q_]] :=
                        showIntegersFactored[p]/showIntegersFactored[q];

         showIntegersFactored[1] = 1;

         showIntegersFactored[p_PrimeQ] = p;
```

```
showIntegersFactored[i_Integer?Negative] = -showIntegersFactored[-i];

showIntegersFactored[i_Integer?Positive] :=
    HoldForm @@ {Times @@ Apply[HoldForm[Power[##]]&, FactorInteger[i], {1}]}
```

In[21]:= `Take[res, 6] /. r_Rational :> showIntegersFactored[r] // TableForm`

Out[21]//TableForm=

$\psi[p]$

$-\frac{\psi^{(3)}[p]}{\pi^2\,(2^5\,3^1)}$

$\frac{\psi''[p]}{\pi^2\,2^6} + \frac{\psi^{(6)}[p]}{\pi^4\,(2^{11}\,3^2)}$

$-\frac{\psi'[p]}{\pi^2\,2^6} - \frac{\psi^{(5)}[p]}{\pi^4\,(2^8\,3^1\,5^1)} - \frac{\psi^{(9)}[p]}{\pi^6\,(2^{16}\,3^4)}$

$\frac{\psi[p]}{\pi^2\,2^7} + \frac{19^1\,\psi^{(4)}[p]}{\pi^4\,(2^{13}\,3^1)} + \frac{11^1\,\psi^{(8)}[p]}{\pi^6\,(2^{17}\,3^2\,5^1)} + \frac{\psi^{(12)}[p]}{\pi^8\,(2^{23}\,3^5)}$

$-\frac{5^1\,\psi^{(3)}[p]}{\pi^4\,(2^{10}\,3^1)} - \frac{(17^1\,53^1)\,\psi^{(7)}[p]}{\pi^6\,(2^{18}\,3^2\,5^1\,7^1)} - \frac{7^1\,\psi^{(11)}[p]}{\pi^8\,(2^{21}\,3^4\,5^1)} - \frac{\psi^{(15)}[p]}{\pi^{10}\,(2^{28}\,3^6\,5^1)}$

For a quick numerical check of the derived coefficients c_k, we substitute the actual value of $\psi(p)$.

In[22]:= `evalR = res /. ψ -> Function[p, Cos[2Pi(p^2 - p - 1/16)]/Cos[2 Pi p]];`

This results in a large expression.

In[23]:= `ByteCount[evalR]`

Out[23]= `5113400`

We use the package `NumericalMath`OptimizeExpression`` to shrink the expression to a smaller size.

In[24]:= `optimize = Experimental`OptimizeExpression[#,`
`        OptimizationLevel -> 1, ExcludedForms -> {}]&;`

In[25]:= `evalRO = optimize[evalR];`

The next table shows the size of the various correction terms for $t = 10^6$.

In[26]:= `Block[{t = N[10^6, 50], v, p},`
`        v = Floor[Sqrt[t/(2 Pi)]]; p = Sqrt[t/(2 Pi)] - v;`
`        mainTerm = 2 Sum[Cos[RiemannSiegelTheta[t] - t Log[k]]/Sqrt[k], {k, v}];`
`        resTerms = (-1)^(v - 1) (t/(2 Pi))^(-1/4)*`
`                    MapIndexed[(t/(2Pi))^(-(#2[[1]] - 1)/2) #1&, First[evalRO]];`
`        (* differences of approximate and exact values *)`
`        (mainTerm + FoldList[Plus, 0, resTerms]) - RiemannSiegelZ[t]] // N`

Out[26]= $\{0.0397462, 1.17633\times10^{-6}, 4.58937\times10^{-10}, 1.32087\times10^{-13}, 5.21417\times10^{-17}, 3.69943\times10^{-19},$
$5.17449\times10^{-23}, 3.13999\times10^{-25}, 4.43323\times10^{-29}, 4.8514\times10^{-31}, 5.47272\times10^{-35}\}$

The Riemann–Siegel function $Z(t)$ is of order 10^0 ($Z(10^6) = -2.806133\ldots$). This means we get around 35 correct digits.

b) To get the dependence of the zeros $z_0(\alpha)$, we deduce from $\zeta(z_0(\alpha), \alpha) = 0$ by implicit differentiation the following differential equation for the zeros [312]:

$$\frac{\partial z_0(\alpha)}{\partial \alpha} = -\frac{\frac{\partial \zeta(z,\alpha)}{\partial \alpha}}{\frac{\partial \zeta(z,\alpha)}{\partial z}}\Bigg|_{z=z_0(\alpha)}$$

The derivative with respect to α can be evaluated.

In[1]:= `D[Zeta[z, α], α]`

Out[1]= `-z Zeta[1 + z, α]`

However, it is not possible to express the derivative with respect to z in a closed form.

In[2]:= `D[Zeta[z, α], z]`

Out[2]= `Zeta`$^{(1,0)}$`[z, α]`

So we implement this differentiation numerically. (We do not use the built-in feature of `Derivative` to evaluate for numeric quantities because we do not have control concerning the precision of the result).

```
In[3]:= Needs["NumericalMath`NLimit`"];

     SetOptions[ND, Scale -> 10^-5, Terms -> 5];

     rhs[z_?NumberQ, α_?NumberQ] := (z Zeta[z + 1, α]/ND[Zeta[ζ, α], ζ, z])
```

As initial conditions for the differential equations, we use the zeros at $\alpha = 1$. A plot of $|\zeta(y + \frac{1}{2}\,i)|$ along the critical line shows many minima—this means potential zeros.

```
In[6]:= pl = Plot[Abs[Zeta[1/2 + I y]], {y, 10., 200},
              PlotPoints -> 500, AspectRatio -> 1/5]
```

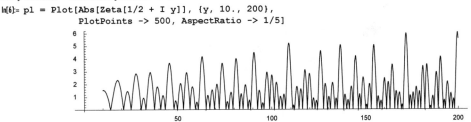

More than 3000 points were sampled to get this curve.

```
In[7]:= Length[pl[[1, 1, 1, 1]]]

Out[7]= 3323
```

Here is a rough search for the first 50 zeros.

```
In[8]:= pre = Select[Partition[pl[[1, 1, 1, 1]], 3, 1],
                    (* search for minima *)
                    #[[1, 2]] > #[[2, 2]] && #[[3, 2]] > #[[2, 2]]&];
In[9]:= zeroIntervals = Take[Map[First, Drop[#, {2}]& /@ pre, {2}], 50];

In[10]:= Show[pl, (* zero-intervals *)
              Graphics[{Hue[0], PointSize[0.01], Point[{#, 0}]& /@
                  (Apply[Plus, Map[First, pre, {2}], {1}]/3)}]]
```

Here is a refinement for their actual values.

```
In[11]:= zeros = FindRoot[Zeta[z, 0] == 0, {z, 1/2 + I #[[1]], 1/2 + I #[[2]]},
                    AccuracyGoal -> 12, MaxIterations -> 50][[1, 2]]& /@
                                                                zeroIntervals
Out[11]= {0.5 + 14.1347 i, 0.5 + 21.022 i, 0.5 + 25.0109 i, 0.5 + 30.4249 i, 0.5 + 32.9351 i,
         0.5 + 37.5862 i, 0.5 + 40.9187 i, 0.5 + 43.3271 i, 0.5 + 48.0052 i, 0.5 + 49.7738 i,
         0.5 + 52.9703 i, 0.5 + 56.4462 i, 0.5 + 59.347 i, 0.5 + 60.8318 i, 0.5 + 65.1125 i,
         0.5 + 67.0798 i, 0.5 + 69.5464 i, 0.5 + 72.0672 i, 0.5 + 75.7047 i, 0.5 + 77.1448 i,
         0.5 + 79.3374 i, 0.5 + 82.9104 i, 0.5 + 84.7355 i, 0.5 + 87.4253 i, 0.5 + 88.8091 i,
         0.5 + 92.4919 i, 0.5 + 94.6513 i, 0.5 + 95.8706 i, 0.5 + 98.8312 i, 0.5 + 101.318 i,
         0.5 + 103.726 i, 0.5 + 105.447 i, 0.5 + 107.169 i, 0.5 + 111.03 i, 0.5 + 111.875 i,
         0.5 + 114.32 i, 0.5 + 116.227 i, 0.5 + 118.791 i, 0.5 + 121.37 i, 0.5 + 122.947 i,
         0.5 + 124.257 i, 0.5 + 127.517 i, 0.5 + 129.579 i, 0.5 + 131.088 i, 0.5 + 133.498 i,
         0.5 + 134.757 i, 0.5 + 138.116 i, 0.5 + 139.736 i, 0.5 + 141.124 i, 0.5 + 143.112 i}
```

Now, we solve the above differential equation for each of the first 50 zeros.

```
In[12]:= Do[sol[i] = NDSolve[{z'[a] == rhs[z[a], a], z[1] == zeros[[i]]},
                    z, {a, 1, 0.5}, PrecisionGoal -> 6,
                    AccuracyGoal -> 6, MaxStepSize -> 0.01,
```

```
                              Compiled -> False, MaxSteps -> 3000][[1, 1, 2]],
          {i, Length[zeros]}]
```

Here is a quick check of the quality of the zeros. For a visualization the zeros are precise enough. We display the logarithm of the absolute difference.

```
In[13]:= Show[Graphics[Table[{Hue[i/70],
                          Line[Table[{a, Log[10, Abs[Zeta[sol[i][a], a]]]},
                                     {a, 0.5, 1, 0.01}]]}, {i, Length[zeros]}]],
          Frame -> True]
```

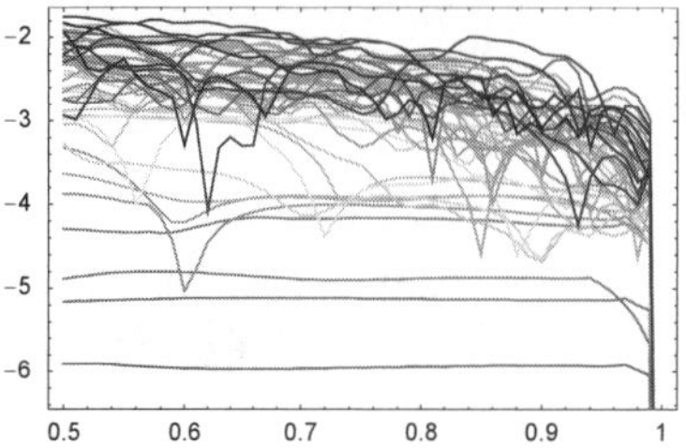

To better differentiate between the various curves, we generate 50 colors in random order.

```
In[14]:= randomColors = (* randomly reorder a list of 50 colors *)
          Hue /@ (Nest[Function[{a}, {Flatten[{a[[1]], #}],
          DeleteCases[a[[2]], #[[1]]]}&[Take[a[[2]], {Random[Integer,
          {1, Length[a[[2]]]}]}]]]]], {{}, Range[Length[zeros]]},
          Length[zeros]][[1]]/Length[zeros]);
```

Now, we can make a picture of how the zeros move in the complex z_0-plane in parametric dependence of α. In the following graphic the imaginary axis runs to the right.

```
In[15]:= Show[Graphics[
          {{Thickness[0.002], GrayLevel[1/2], Line[{{1/2, 0}, {1/2, 150}}]},
           Line[{{0, 0}, {0, 150}}]},
          {Thickness[0.002], Table[{randomColors[[i]],
          (* path of a zero as a function of a *)
          ParametricPlot[Evaluate[{Re[sol[i][a]], Im[sol[i][a]]}],
                         {a, 0.5, 1}, DisplayFunction -> Identity][[1]]},
                         {i, Length[zeros]}]}}] /. (* {x, y} <-> {y, x} *)
                                       Line[l_] :> Line[Reverse /@ l],
          Frame -> True, PlotRange -> All, Axes -> False, AspectRatio -> 1/4]
```

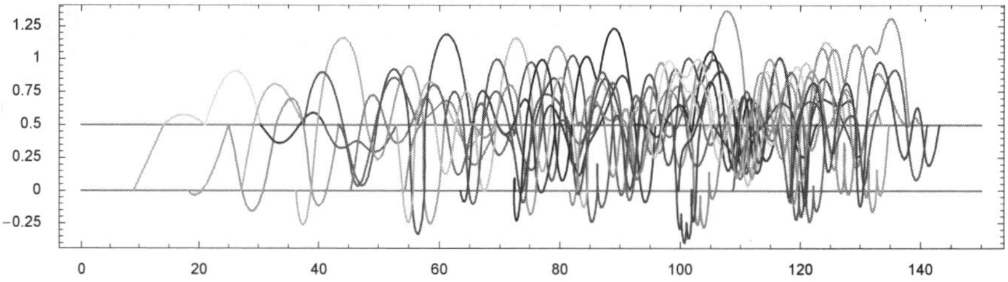

The zeros $z_0(1)$ are mapped onto themselves or into multiples of $2\,i\pi/\ln(2)$. This can be seen from the functional equation that connects the values of $\zeta(z, 1)$ and $\zeta(z, 1/2)$: $(2^z - 1)\,\zeta(z, 1) = \zeta(z, 1/2)$.

The following graphic shows the contours $\operatorname{Re}(\zeta(z, a)) = 0$ and $\operatorname{Re}(\zeta(z, a)) = 0$ as a varies from 0.8 to 1.0. The black dots are the zeros.

```
In[16]:= reImZeroGraphic[a_, opts___] :=
          With[{ppx = 90, ppy = 200},
           data = Table[Evaluate[Zeta[x + I y, a]],
                   {x, -1, 2, 3/ppx}, {y, 1, 50, 49/ppy}];
           Show[(* contours Re[Zeta[z, a]] == 0 and Im[Zeta[z, a]] *)
                ListContourPlot[#1[data], Contours -> {0}, ContourStyle -> {#2},
                ContourShading -> False, DisplayFunction -> Identity,
                MeshRange -> {{1, 50}, {-1, 2}}]& @@@
```

```
        {{Re, {Thickness[0.005], Hue[0.0]}}, {Im, {Thickness[0.005], Hue[0.8]}}},
        DisplayFunction -> $DisplayFunction,
        (* critical line *)
        Prolog -> {GrayLevel[0.6], Thickness[0.005], Line[{{1, 1/2}, {50, 1/2}}]},
        (* zeros as points *)
        Epilog -> Table[{GrayLevel[0], PointSize[0.012],
                Point[{Im[#], Re[#]}&[sol[k][a]]]}, {k, 12}],
        AspectRatio -> 0.3, FrameTicks -> None, Frame -> True,
        PlotRange -> {{1, 50}, {-1, 2}}]]]
In[17]:= Do[reImZeroGraphic[a], {a, 0.8, 1.00, 0.2/4}]
```

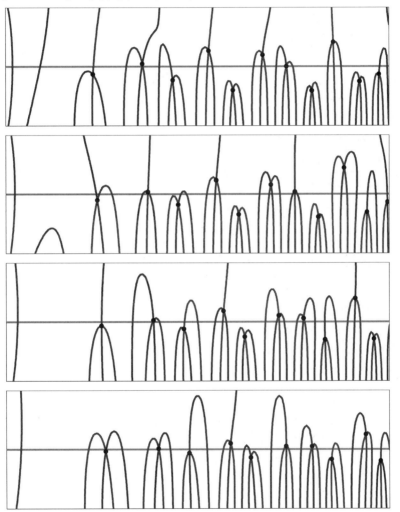

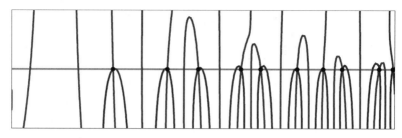

Here is the corresponding animation that uses more frames.

```
Do[reImZeroGraphic[a], {a, 0.8, 1.00, 0.2/100}]
```

Finally, let us have a 3D look at how the zeros evolve in dependence of α.

```
In[18]:= Show[Graphics3D[
           {(* critical lines *)
            {Thickness[0.005], GrayLevel[1/2],
             Line[{{1/2, 0, 1/2}, {1/2, 150, 1/2}}],
             Line[{{1/2, 0, 1}, {1/2, 150, 1}}]},
            {Thickness[0.002], GrayLevel[3/4], Line[{{0, 0, 1/2}, {0, 150, 1/2}}]},
            Table[{Thickness[0.002], randomColors[[i]],
                   (* let the parameter a be the z-coordinate *)
                   ParametricPlot3D[Evaluate[{Re[sol[i][a]], Im[sol[i][a]], a}],
                                    {a, 0.5, 1}, PlotPoints -> 200,
                                    DisplayFunction -> Identity][[1]]},
                  {i, Length[zeros]}]}],
           DisplayFunction -> $DisplayFunction, PlotRange -> All, Axes -> True,
           BoxRatios -> {1, 5, 3}, AxesLabel -> {"x", "y", "a"},
           ViewPoint -> {-1.3, -1.1, 0.9}]
```

With a straightforward change of variables from α to ρ via $\alpha \to \rho\,e^{i\varphi}$ we could solve the above differential equation $\partial\zeta(z_0(\alpha),\alpha)/\partial\alpha = 0$ for fixed φ and so obtain the zero-surfaces for complex α.

For the zeros of the incomplete Zeta function in the complex plane, see [706], [707].

c) Truncating the infinite series and extracting coefficients yields a linear system for the $Z(k)$ that can easily be solved.

```
In[1]:= o = 6;
        S = Sum[Z[k]/k (s (1 - s))^k, {k, 1, o}] +
            ((EulerGamma/2 + Log[2 Sqrt[Pi]] - 1) s +
            Sum[((1 - 2^-k) Zeta[k] - 1)/k s^k, {k, 2, o}] -
            Sum[1/1 Sum[StieltjesGamma[k - 1]/(k - 1)! s^k, {k, 1, o}]^1,
                {1, 1, o}]) + O[s]^6;
In[3]:= (Zs = (Solve[# == 0& /@ CoefficientList[S, s],
                     Table[Z[k], {k, o - 1}]] // FullSimplify)) // TraditionalForm
```

Out[3]//TraditionalForm= $\left\{\left\{Z(1) \to \frac{1}{2}(2+\gamma-\log(4\pi)), Z(2) \to 2\gamma_1 - \log(4\pi) - \frac{\pi^2}{8} + \gamma^2 + \gamma + 3,\right.\right.$

$$Z(3) \to 6\gamma_1 + \gamma(3\gamma_1 + \gamma^2 + 3\gamma + 3) + \frac{3\gamma_2}{2} - \frac{7\zeta(3)}{8} - 3\log(4\pi) - \frac{3\pi^2}{8} + 10,$$

$$Z(4) \to 2\gamma_1(\gamma_1 + 10) + 2\gamma^2(2\gamma_1 + 5) + 6\gamma_2 + 2\gamma(6\gamma_1 + \gamma_2 + 5) + \frac{2\gamma_3}{3} - \frac{7\zeta(3)}{2} - 10\log(4\pi) - \frac{1}{96}\pi^2(120+\pi^2) +$$

$$\gamma^4 + 4\gamma^3 + 35, Z(5) \to 5\gamma^3(\gamma_1 + 3) + \frac{5}{2}\gamma^2(8\gamma_1 + \gamma_2 + 14) + \frac{5}{6}\gamma(6(\gamma_1(\gamma_1 + 9) + 2\gamma_2 + 7) + \gamma_3) +$$

$$\frac{1}{96}(240\gamma_1(4\gamma_1 + \gamma_2 + 28) + 4(540\gamma_2 + 80\gamma_3 + 5\gamma_4 - 315\zeta(3) - 840\log(4\pi) + 3024) - 93\zeta(5) - 5\pi^2(84 + \pi^2)) +$$

$$\left.\left.\gamma^5 + 5\gamma^4\right\}\right\}$$

Numericalizing the last expression yields the following values.

In[4]:= `NZs = Zs /. (z_Z -> r_) :> (z -> N[N[r, 50], 20])`

```
NIntegrate::slwcon :
Numerical integration converging too slowly; suspect one of the
    following: singularity, value of the integration being 0, oscillatory
    integrand, or insufficient WorkingPrecision. If your integrand is
    oscillatory try using the option Method->Oscillatory in NIntegrate. More...

NIntegrate::slwcon :
Numerical integration converging too slowly; suspect one of the
    following: singularity, value of the integration being 0, oscillatory
    integrand, or insufficient WorkingPrecision. If your integrand is
    oscillatory try using the option Method->Oscillatory in NIntegrate. More...

NIntegrate::slwcon :
Numerical integration converging too slowly; suspect one of the
    following: singularity, value of the integration being 0, oscillatory
    integrand, or insufficient WorkingPrecision. If your integrand is
    oscillatory try using the option Method->Oscillatory in NIntegrate. More...

General::stop : Further output of
    NIntegrate::slwcon will be suppressed during this calculation. More...
```

Out[4]= $\{\{Z[1] \to 0.023095708966121033814,$
$Z[2] \to 0.00003710063643746487153, Z[3] \to 1.4367786028869177485\times 10^{-7},$
$Z[4] \to 6.5982791454240115269\times 10^{-10}, Z[5] \to 3.1938918608673242323\times 10^{-12}\}\}$

For comparison we calculate numerical approximations of the first 1000 zeros of $\zeta(z)$. We do this by generating a fictitious plot, extracting intervals containing the roots and then using a numerical root-finding procedure. Using $t < 2000$ yields more than 1500 roots.

In[5]:= `pl = Plot[Abs[Zeta[1/2 + I y]], {y, 10., 2000},`
`            PlotPoints -> 30000, AspectRatio -> 1/5,`
`            DisplayFunction -> Identity];`

In[6]:= `Length[pre = Select[Partition[pl[[1, 1, 1, 1]], 3, 1],`
`            (* search for minima *)`
`            #[[1, 2]] > #[[2, 2]] && #[[3, 2]] > #[[2, 2]]&]]`

Out[6]= 1517

In[7]:= `zeros = FindRoot[Zeta[z, 0] == 0, {z, 1/2 + I #[[1]], 1/2 + I #[[2]]},`
`            AccuracyGoal -> 12, MaxIterations -> 50][[1, 2]]& /@`
`            Take[Map[First, Drop[#, {2}]& /@ pre, {2}], 500];`

This yields the following values for the $Z(k)$.

In[8]:= `Table[Fold[Plus, 0, ((1/4 + Im[#]^2)& /@ zeros)^-s], {s, 5}]`

Out[8]= $\{0.0219469, 0.0000371001, 1.43678\times 10^{-7}, 6.59828\times 10^{-10}, 3.19389\times 10^{-12}\}$

The relative error is a quickly decreasing function of k.

In[9]:= `(# - %)/#&[Last /@ NZs[[1]]]`

Out[9]= $\{0.0497404, 0.0000138704, 3.17488\times 10^{-9}, 7.40426\times 10^{-13}, 3.92023\times 10^{-15}\}$

In a similar way one can calculate the sums $\sum_{k=1}^{\infty} (1/2 \pm i\,t_k)_k^{-n}$ of the nontrivial zeros $z_k = 1/2 + i\,t_k$, using [160]

$$\sum_{k=-\infty}^{\infty} 1 - \left(1 - \frac{1}{\rho_k}\right)^n = -\sum_{j=1}^{n} \eta_{j-1} \binom{n}{j} - \sum_{j=2}^{n} (-1)^{j-1}\binom{n}{j}(1 - 2^{-j})\zeta(j) + 1 - \frac{1}{2}(\log(4\pi) + \gamma)\,n$$

where the η_k defined through $\log(s\,\zeta(s+1)) = -\sum_{k=0}^{\infty} \eta_k\, s^{k+1}/(k+1)$.

```
In[10]:= η[n_] := η[n] = -(n + 1) Coefficient[
                        Series[Log[s Zeta[1 + s]], {s, 0, n + 2}], s, n + 1]
In[11]:= lhs[n_] := 1 - (1 - 1/ρ)^n


        rhs[n_] := -Sum[η[j - 1] Binomial[n, j],
                       {j, 1, n}] + 1 - (Log[4 Pi] + EulerGamma) n/2 -
                   Sum[(-1)^(j - 1) Binomial[n, j] (1 - 2^(-j)) Zeta[j],
                       {j, 2, n}]
```

Here are explicit expressions for small integer n.

```
In[13]:= With[{o = 6}, (* Solve for Sum[zeros^-n] *)
            Solve[Table[Expand[lhs[j]] == rhs[j], {j, o}] /. ρ^k_ :> z[-k],
                Table[z[j], {j, 6}]] // Simplify] // TraditionalForm
```

Out[13]//TraditionalForm=

$$\left\{\left\{ z(1) \to \frac{1}{2}(2 + \gamma - \log(4\pi)),\ z(2) \to 2\gamma_1 - \frac{\pi^2}{8} + \gamma^2 + 1, \right.\right.$$

$$z(3) \to 3\gamma\,\gamma_1 + \frac{3\gamma_2}{2} - \frac{7\zeta(3)}{8} + \gamma^3 + 1,\ z(4) \to 2\gamma_1^2 + 4\gamma^2\,\gamma_1 + 2\gamma\,\gamma_2 + \frac{2\gamma_3}{3} - \frac{\pi^4}{96} + \gamma^4 + 1,$$

$$z(5) \to \frac{5\gamma_2\,\gamma_1}{2} + 5\gamma^3\,\gamma_1 + \frac{5\gamma^2\,\gamma_2}{2} + \frac{5}{6}\gamma(6\gamma_1^2 + \gamma_3) + \frac{5\gamma_4}{24} - \frac{31\zeta(5)}{32} + \gamma^5 + 1,$$

$$\left.\left. z(6) \to 2\gamma_1^3 + \gamma_3\,\gamma_1 + 6\gamma^4\,\gamma_1 + \frac{3\gamma_2^2}{4} + 3\gamma^3\,\gamma_2 + \gamma^2(9\gamma_1^2 + \gamma_3) + \frac{1}{4}\gamma(24\gamma_1\,\gamma_2 + \gamma_4) + \frac{\gamma_5}{20} - \frac{\pi^6}{960} + \gamma^6 + 1 \right\}\right\}$$

Again, the so-calculated values agree well with the explicit sums over the sums.

```
In[14]:= N[%]
Out[14]= {{z[1.] → 0.0230957, z[2.] → -0.0461543, z[3.] → -0.000111158,
          z[4.] → 0.0000736272, z[5.] → 7.15093×10^-7, z[6.] → -2.81436×10^-7}}

In[15]:= Table[Plus @@ (((1/#)^n + (1/Conjugate[#])^n) &) /@ zeros,
            {n, 1, 6}]
Out[15]= {0.0219469 + 0. i, -0.0438567 + 0. i, -0.000111157 + 0. i,
          0.0000736262 + 0. i, 7.15093×10^-7 + 0. i, -2.81436×10^-7 + 0. i}
```

d) On a first glance the question seems straightforward to answer. Just implement the recursive definition and then evaluate the values under consideration.

```
In[1]:= {f[-1, x_], f[ 0, x_], f[+1, x_]} = {1/(x + 1), 1/x, 1/(1 - x)};

        {H[{-1}, x_], H[{0}, x_], H[{1}, x_]} = {Log[1 + x], Log[x], -Log[1 - x]};
```

Unfortunately there are two problems with such an approach. Because we are ultimately interested in $H(a_1, a_2, \ldots, a_n; 1)$ instead of $H(a_1, a_2, \ldots, a_n; z)$ we should use definite integrations which, for special parameter values, will have a higher chance to succeed than indefinite integration (at least in the last integration step). But to avoid the intermediate generation of If statements in multivariate definite integration, the option GenerateConditions is set by default to False. This setting has the unpleasant side effect (in the context of this exercise) that divergent integrals will be Hadamard regularized. The following comparison between nested and unnested and numerical integration shows the problem.

```
In[3]:= Off[NIntegrate::slwcon]; Off[Integrate::idiv]; Off[NIntegrate::ncvb];


        {Integrate[f[-1, y] f[0, x] H[{0}, x], {y, 0, 1}, {x, 0, y}],
         Integrate[f[-1, y] Integrate[f[0, x] H[{0}, x], {x, 0, y},
                  GenerateConditions -> False], {y, 0, 1}],
         Integrate[f[-1, y] Integrate[f[0, x] H[{0}, x], {x, 0, y},
                  GenerateConditions -> True], {y, 0, 1}],
         Integrate[f[-1, y] Integrate[f[0, x] H[{0}, x], {x, 0, y}], {y, 0, 1}],
```

```
(* numerical integration *)
NIntegrate[f[-1, y] f[0, x] H[{0}, x], {y, 0, 1}, {x, 0, y}]}
```

$$\text{Out[4]= }\left\{\frac{3\ \text{Zeta}[3]}{4},\ \frac{3\ \text{Zeta}[3]}{4},\ \int_0^1 \frac{\text{Integrate}[\frac{\text{Log}[x]}{x},\ \{x,\ 0,\ y\},\ \text{GenerateConditions} \to \text{True}]}{1+y}\,dy,\right.$$

$$\left.\int_0^1 \frac{\int_0^y \frac{\text{Log}[x]}{x}\,dx}{1+y}\,dy,\ -4.98937 \times 10^{10}\right\}$$

To ensure that we try all possibilities of nesting and both possible settings for the GenerateConditions option, we implement the following two functions H and $\mathcal{H}$.

```
In[5]:= H[{a_, b_}, 1, opts___] :=
           Integrate[Evaluate[f[a, x] H[{b}, x]], {x, 0, 1}, opts]


       H[{a_, b_, c_}, 1, opts___] :=
           Integrate[Evaluate[f[a, y] f[b, x] H[{c}, x]], {y, 0, 1}, {x, 0, y}, opts]


       H[{a_, b_, c_, d_}, 1, opts___] :=
           Integrate[Evaluate[f[a, z] f[b, y] f[c, x] H[{d}, x]],
                               {z, 0, 1}, {y, 0, z}, {x, 0, y}, opts]


       H[{a_, b_}, 1, opts___] := H[{a, b}, 1, opts]


       H[{a_, b_, c_}, 1, opts___] :=
           Integrate[f[a, y] Integrate[f[b, x] H[{c}, x], {x, 0, y}, opts], {y, 0, 1}]


       H[{a_, b_, c_, d_}, 1, opts___] :=
           Integrate[f[a, z] Integrate[f[b, y]*
                       Integrate[f[c, x] H[{d}, x], {x, 0, y}, opts],
                               {y, 0, z}, opts], {z, 0, 1}]
```

To detect divergent integrals and potential integration errors, we implement the function ⊩H for a numerical integration.

```
In[14]:= SetOptions[NIntegrate,
                    WorkingPrecision -> 30, PrecisionGoal -> 3, AccuracyGoal -> 3];


        H[{a_, b_}, 1, opts___] := With[{ε = 10^-20},
          NIntegrate[Evaluate[f[a, x] H[{b}, x]], {x, ε, 1 - ε}]]


        H[{a_, b_, c_}, 1, opts___] := With[{ε = 10^-20},
          NIntegrate[Evaluate[f[a, y] f[b, x] H[{c}, x]],
                     {y, ε, 1 - ε}, {x, ε, y}]]


        H[{a_, b_, c_, d_}, 1, opts___] := With[{ε = 10^-22},
          NIntegrate[Evaluate[f[a, z] f[b, y] f[c, x] H[{d}, x]],
                     {z, ε, 1 - ε}, {y, ε, z}, {x, ε, y}]]
```

For a given n, the function allTuples generates all possible $\{a_1, a_2, ..., a_n\}$.

```
In[19]:= allTuples[n_] := Flatten[Outer[List, Sequence @@ Table[{-1, 0, 1}, {n}]], n - 1]
```

The function findIntegral tries to calculate $H(a_1, a_2, ..., a_n; z)$. It uses the functions H and $\mathcal{H}$ and compares the results with the numerical integration result. When agreement is found, the corresponding exact result is returned.

```
In[20]:=
       f[i_, h_, tf_] := TimeConstrained[formatResult @ h[i, 1,
                               GenerateConditions -> tf], 120]


       formatResult[r_] := If[MemberQ[r, _Integrate, {0, Infinity}], "I", r]


       findIntegral[as_, maxTime_: 120] :=
       Module[{resH1, resH2, resH1, resH2, resH, sel},
               (* integrate symbolically in various ways *)
               {resH1, resH2, resH1, resH2} =
               {f[as, H, False], f[as, H, True], f[as, H, False], f[as, H, True]};
               (* integrate numerically; do not wait forever *)
               resH = TimeConstrained[formatResult @ H[as, 1], maxTime];
               (* select symbolic result that agrees with numeric result *)
```

```
sel = Select[{resH1, resH1, resH1, resH2}, Abs[N[#/resH - 1]] < 1/10&];
(* return simplified result *)
If[sel =!= {}, Subscript[H, Sequence @@ as] == Simplify[sel[[1]]],
   Sequence @@ {}]]
```

Calculating all $9 + 27 + 81$ possible harmonic polylogarithms yields $6 + 16 + 11$ results for the nondivergent values at $z = 1$.

$$H_{-1,-1} = \frac{\log^2(2)}{2}, \ H_{-1,0} = -\frac{\pi^2}{12}, \ H_{-1,1} = \frac{1}{12}(\pi^2 - 6\log^2(2)), \ H_{0,-1} = \frac{\pi^2}{12}, \ H_{0,1} = \frac{\pi^2}{6}, \ H_{1,0} = -\frac{\pi^2}{6}$$

$$H_{-1,-1,-1} = \frac{\log^3(2)}{6}, \ H_{-1,-1,0} = \frac{1}{24}(3\zeta(3) - \pi^2\log(4)), \ H_{-1,-1,1} = \frac{1}{24}(3\zeta(3) - 4\log^3(2)),$$

$$H_{-1,0,-1} = \frac{1}{12}(\pi^2\log(2) - 3\zeta(3)), \ H_{-1,0,1} = \frac{1}{6}\pi^2\log(2) - \frac{5\zeta(3)}{8}, \ H_{-1,1,-1} = \frac{1}{12}(-2\log^3(2) + \pi^2\log(2) - 3\zeta(3)),$$

$$H_{-1,1,0} = \frac{1}{12}\pi^2\log(2) - \zeta(3), \ H_{-1,1,1} = \frac{1}{24}(4\log^3(2) - \pi^2\log(4) + 21\zeta(3)),$$

$$H_{0,-1,-1} = \frac{\zeta(3)}{8}, \ H_{0,-1,0} = -\frac{1}{2}(3\zeta(3)), \ H_{0,-1,1} = \frac{1}{8}(13\zeta(3) - \pi^2\log(4)), \ H_{0,0,-1} = \frac{3\zeta(3)}{4},$$

$$H_{0,0,1} = \zeta(3), \ H_{0,1,-1} = \frac{1}{4}\pi^2\log(2) - \zeta(3), \ H_{0,1,0} = -2\zeta(3), \ H_{0,1,1} == \zeta(3)$$

$$H_{-1,-1,-1,-1} = \frac{\log^4(2)}{24}, \ H_{-1,-1,1,-1} = \frac{1}{120}\left(-15\pi^2\log^2(2) - 4\pi^4 + 10\left(\log^4(2) + 33\zeta(3)\log(2) + 36\text{Li}_4\left(\frac{1}{2}\right)\right)\right),$$

$$H_{-1,0,0,-1} = \frac{3}{4}\log(2)\zeta(3) - \frac{\pi^4}{288},$$

$$H_{-1,1,-1,-1} = \frac{1}{24}\left(-4\log^4(2) + \pi^2\log(16)\log(2) - 69\zeta(3)\log(2) + \frac{4\pi^4}{5} - 72\text{Li}_4\left(\frac{1}{2}\right)\right),$$

$$H_{-1,1,1,1} = \text{Li}_4\left(\frac{1}{2}\right), \ H_{0,-1,-1,-1} = \frac{1}{360}\left(15\pi^2\log^2(2) + 4\pi^4 - 15\left(\log^4(2) + 21\zeta(3)\log(2) + 24\text{Li}_4\left(\frac{1}{2}\right)\right)\right),$$

$$H_{0,0,-1,0} = -\frac{1}{240}(7\pi^4), \ H_{0,0,0,-1} = \frac{7\pi^4}{720}, \ H_{0,0,0,1} = \frac{\pi^4}{90}, \ H_{0,1,1,0} = -\frac{\pi^4}{72}, \ H_{0,1,1,1} = \frac{\pi^4}{90}, \ H_{1,1,-1,0} = \frac{\pi^4}{90}$$

```
In[24]:= HPLValues[1] = {Subscript[H, -1] == Log[1 + x], Subscript[H, 0] == Log[x],
          Subscript[H, +1] == -Log[1 - x]};
```

```
In[25]:= HPLValues[2] = findIntegral /@ allTuples[2]
```

Out[25]= $\left\{H_{-1,-1} == \frac{\text{Log}[2]^2}{2}, \ H_{-1,0} == -\frac{\pi^2}{12}, \right.$

$\left. H_{-1,1} == \frac{1}{12}(\pi^2 - 6\,\text{Log}[2]^2), \ H_{0,-1} == \frac{\pi^2}{12}, \ H_{0,1} == \frac{\pi^2}{6}, \ H_{1,0} == -\frac{\pi^2}{6}\right\}$

```
In[26]:= HPLValues[3] = findIntegral /@ allTuples[3]
```

Out[26]= $\left\{H_{-1,-1,-1} == \frac{\text{Log}[2]^3}{6}, \ H_{-1,-1,0} == \frac{1}{24}(-\pi^2\,\text{Log}[4] + 3\,\text{Zeta}[3]), \ H_{-1,-1,1} == -\frac{1}{6}\,\text{Log}[2]^3 + \frac{\text{Zeta}[3]}{8}, \right.$

$H_{-1,0,-1} == \frac{1}{12}(\pi^2\,\text{Log}[2] - 3\,\text{Zeta}[3]), \ H_{-1,0,1} == \frac{1}{6}\pi^2\,\text{Log}[2] - \frac{5\,\text{Zeta}[3]}{8},$

$H_{-1,1,-1} == \frac{1}{12}(\pi^2\,\text{Log}[2] - 2\,\text{Log}[2]^3 - 3\,\text{Zeta}[3]), \ H_{-1,1,0} == \frac{1}{12}\pi^2\,\text{Log}[2] - \text{Zeta}[3],$

$H_{-1,1,1} == \frac{1}{24}(4\,\text{Log}[2]^3 - \pi^2\,\text{Log}[4] + 21\,\text{Zeta}[3]), \ H_{0,-1,-1} == \frac{\text{Zeta}[3]}{8},$

$H_{0,-1,0} == -\frac{3\,\text{Zeta}[3]}{2}, \ H_{0,-1,1} == \frac{1}{8}(-\pi^2\,\text{Log}[4] + 13\,\text{Zeta}[3]), \ H_{0,0,-1} == \frac{3\,\text{Zeta}[3]}{4},$

$\left. H_{0,0,1} == \text{Zeta}[3], \ H_{0,1,-1} == \frac{1}{4}\pi^2\,\text{Log}[2] - \text{Zeta}[3], \ H_{0,1,0} == -2\,\text{Zeta}[3], \ H_{0,1,1} == \text{Zeta}[3]\right\}$

The various harmonic polylogarithms are not independent from each other. For instance, they fulfill the following identity [168].

```
In[27]:= HPLIds[as_, HIds_] :=
          With[{n = Length[as]},
          Subscript[H, Sequence @@ as] ==
```

```
                 Sum[(-1)^(k + 1) Subscript[H, Sequence @@ Reverse[Take[as, k]]]*
                     If[k == n, 1, Subscript[H, Sequence @@ Take[as, {k + 1, n}]]],
                     {k, n}]] /. Rule @@@ HIds
```

Here this identity is shown explicitly for order 3.

In[28]:= `HPLIds[{a1, a2, a3}, {}]`

Out[28]= $H_{a1,a2,a3} == -H_{a3} H_{a2,a1} + H_{a1} H_{a2,a3} + H_{a3,a2,a1}$

The just-calculated values fulfill this identity.

In[29]:= `HPLIds[#, Join @@ Table[HPLValues[k], {k, 3}]]& /@`
`        {{-1, -1, -1}, {-1, 0, -1}, {-1, 1, -1}, {0, -1, 0}, {0, 0, 0},`
`         {0, 1, 0}, {1, -1, 1}, {1, 0, 1}, {1, 1, 1}}`

Out[29]= {True, True, True, True, True, True, True, True, True}

We end with the more time-consuming calculation of the harmonic polylogarithms of order 4.

In[30]:= `HPLValues[4] = findIntegral /@ allTuples[4]`

Out[30]= $\Big\{ H_{-1,-1,-1,-1} == \dfrac{\text{Log}[2]^4}{24},$

$H_{-1,-1,-1,0} == \dfrac{\pi^4}{90} + \dfrac{1}{12} \pi^2 \text{Log}[2]^2 - \dfrac{1}{6} i \pi \text{Log}[2]^3 - \text{PolyLog}[4, 2] + \text{Log}[2] \text{Zeta}[3],$

$H_{-1,-1,0,-1} == -\dfrac{\pi^4}{30} - \dfrac{1}{12} \pi^2 \text{Log}[2]^2 - \text{Log}[2] \text{PolyLog}[3, 2] + 3 \text{PolyLog}[4, 2] - \text{Log}[4] \text{Zeta}[3],$

$H_{-1,0,-1,-1} == \dfrac{\pi^4}{30} + \dfrac{3}{8} \pi^2 \text{Log}[2]^2 - \dfrac{1}{2} i \pi \text{Log}[2]^3 - 3 \text{PolyLog}[4, 2] + \dfrac{11}{4} \text{Log}[2] \text{Zeta}[3],$

$H_{-1,0,0,-1} == -\dfrac{\pi^4}{288} + \dfrac{3}{4} \text{Log}[2] \text{Zeta}[3], \quad H_{-1,1,1,1} == \text{PolyLog}\Big[4, \dfrac{1}{2}\Big],$

$H_{0,-1,-1,-1} == \dfrac{\pi^4}{90} + \dfrac{1}{24} \pi^2 \text{Log}[2]^2 - \dfrac{\text{Log}[2]^4}{24} - \text{PolyLog}\Big[4, \dfrac{1}{2}\Big] - \dfrac{7}{8} \text{Log}[2] \text{Zeta}[3],$

$H_{0,-1,-1,0} == -\dfrac{\pi^4}{288}, \quad H_{0,0,-1,0} == -\dfrac{7 \pi^4}{240}, \quad H_{0,0,0,-1} == \dfrac{7 \pi^4}{720}, \quad H_{0,0,0,1} == \dfrac{\pi^4}{90},$

$H_{0,0,1,0} == -\dfrac{\pi^4}{30}, \quad H_{0,1,0,1} == \dfrac{\pi^4}{120}, \quad H_{0,1,1,0} == -\dfrac{\pi^4}{72}, \quad H_{0,1,1,1} == \dfrac{\pi^4}{90} \Big\}$

e) We start by implementing a function *Sum* that will deal with summands that are polynomial in $\psi^{(k)}(z)$.

In[1]:= `Sum[f_, {k_, k0_, k1_}] :=`
`        With[{sum = polyGammaSum[f, {k, k0, k1}]},`
`            sum /; sum =!= $Failed] /;`
`                 (* are summands polynomial in the PolyGammas? *)`
`                 PolynomialQ[f, Union[Cases[f, _PolyGamma, Infinity]]]`

`    Sum[f_, {k_, k1_}] := Sum[f, {k, 1, k1}]`

The function `polyGammaSum` substitutes dummy variables $I[o, z]$ for the $\psi^{(o)}(z)$ after expanding the summand.

In[4]:= `polyGammaSum[f_, {k_, k0_, k1_}] :=`
`        Module[{f1, f2, res},`
`          (* expand and introduce dummy variables *)`
`          f1 = Expand[Expand[f] /. PolyGamma[o_, z_] :>`
`                      (I[o, z] - KroneckerDelta[o, 0] EulerGamma)];`
`          (* make list of summands *)`
`          f2 = If[Head[f1] === Plus, List @@ f1, {f1}];`
`          (* sum and integrate *)`
`          res = Plus @@ (useIntegralRepresentation[#, {k, k0, k1}]& /@ f2);`
`          If[FreeQ[res, _Integrate, {0, Infinity}, Heads -> True], res, $Failed]]`

`    polyGammaSum[f_, {k_, k1_}] := polyGammSum[f, {k, 1, k1}]`

The function `useIntegralRepresentation` uses the integral representation for the derivatives of the Digamma function and carries out the summation and then the integration (if possible). We will not check the conditions for interchanging summation and integration, but will later use a numerical check of the results.

In[7]:= `int[o_, z_, t_] := Derivative[o][(1 - t^(# - 1))/(1 - t) &][z];`

`    useIntegralRepresentation[expr_, {k_, k0_, k1_}] :=`

```
Module[{preFactor, I1, I2, I3, I4, I5},
  (* polygamma independent prefactor *)
  preFactor = expr /. _I -> 1;
  I1 = expr/preFactor;
  (* use integral representation for each polygamma function *)
  I2 = If[Head[I1] === Times, List @@ I1, {I1}];
  I3 = Flatten[I2 /. i_I^e_ :> Table[i, {e}]];
  I4 = MapIndexed[(#1 /. I[o_, z_] :> Product[int[o, z, t[#2[[1]]]]])&, I3];
  (* do the summation *)
  I5 = Sum[Evaluate[preFactor (Times @@ I4)], {k, k0, k1}];
  If[FreeQ[I5, _Sum, {0, ∞}, Heads -> True],
    (* do the integration(s) *)
    Fold[Integrate[##, GenerateConditions -> False (* True *)]&,
      Normal[I5], Table[{t[j], 0, 1}, {j, Length[I4]}]],
    $Failed]]
```

We start by looking at the series expansions of $\psi^{(n)}(z)$ at infinity to find out how $R(k)$ must behave to achieve convergence.

In[9]:= `Table[Series[PolyGamma[k, n], {n, Infinity, k + 1}] // Normal, {k, 0, 3}]`

Out[9]= $\left\{ -\dfrac{1}{2n} - \text{Log}\left[\dfrac{1}{n}\right], \dfrac{1}{2n^2} + \dfrac{1}{n}, -\dfrac{1}{n^3} - \dfrac{1}{n^2}, \dfrac{3}{n^4} + \dfrac{2}{n^3} \right\}$

We start with the two examples $n^{-2}(\gamma + \psi(n))$ and $n^{-1}(\psi^{(1)}(n))$. *Sum* and Sum give the same results.

In[10]:= `summand = 1/n^2 (PolyGamma[n] + EulerGamma);`
      `{Sum[#, {n, Infinity}], Sum[#, {n, Infinity}]}&[summand] // Expand`
Out[11]= `{Zeta[3], Zeta[3]}`

In[12]:= `summand = 1/n PolyGamma[1, n];`
      `{Sum[#, {n, Infinity}], Sum[#, {n, Infinity}]}&[summand] // Expand`

      ∞::indet : Indeterminate expression 0 (-∞) encountered. More…

      ∞::indet : Indeterminate expression 0 (-∞) encountered. More…

      ∞::indet : Indeterminate expression 0 (-∞) encountered. More…

      General::stop :
        Further output of ∞::indet will be suppressed during this calculation. More…

Out[13]= $\left\{ 2\,\text{Zeta}[3],\ \displaystyle\sum_{n=1}^{\infty} \dfrac{\text{PolyGamma}[1, n]}{n} \right\}$

The two examples $1/(n(n+1))(\gamma + \psi(n))^2$ and $(n+1)/(n(n+3)(n+5))(\gamma + \psi(n))(\gamma + \psi(n+2))$ are not summed by the built-in Sum function.

In[14]:= `summand = 1/(n (n + 1)) (PolyGamma[n] + EulerGamma)^2;`
      `{Sum[#, {n, Infinity}], Sum[#, {n, Infinity}]}&[summand] // Expand`
Out[15]= $\left\{ 1 + \dfrac{\pi^2}{6},\ \displaystyle\sum_{n=1}^{\infty} \dfrac{(\text{EulerGamma} + \text{PolyGamma}[0, n])^2}{n\,(1+n)} \right\}$

In[16]:= `{N[%[[1]]], (* numerical check *) NSum[Evaluate[summand], {n, Infinity}]}`
Out[16]= `{2.64493, 2.64493}`

In[17]:= `summand = (n + 1)/(n (n + 3) (n + 5)) (PolyGamma[n] + EulerGamma)*`
                                          `(PolyGamma[n + 2] + EulerGamma);`
      `{Sum[#, {n, Infinity}], Sum[#, {n, Infinity}]}&[summand] // Expand`
      ∞::indet : Indeterminate expression -∞ + ∞ encountered. More…

      ∞::indet : Indeterminate expression 0 (-∞) encountered. More…

      ∞::indet : Indeterminate expression 0 (-∞) encountered. More…

      General::stop :
        Further output of ∞::indet will be suppressed during this calculation. More…

Out[18]= {Indeterminate,

$$\sum_{n=1}^{\infty} \frac{(1+n)\ (\text{EulerGamma} + \text{PolyGamma}[0, n])\ (\text{EulerGamma} + \text{PolyGamma}[0, 2+n])}{n\ (3+n)\ (5+n)}\ \}$$

In[19]:= **{N[%[[1]]],** (* numerical check *) **NSum[Evaluate[summand], {n, Infinity}]}**

Out[19]= {Indeterminate, 1.90383}

So we obtain the following results.

$$\sum_{n=1}^{\infty} \frac{1}{n(n+1)}\ (\gamma + \psi(n))^2 = 1 + \frac{\pi^2}{6}$$

$$\sum_{n=1}^{\infty} \frac{(n+1)}{n\ (n+3)\ (n+5)}\ (\psi(n) + \gamma)\ (\psi(n+2) + \gamma) = \frac{\zeta(3)}{15} + \frac{34\ \pi^2}{675} + \frac{11939}{9000}$$

f) A direct use of Product gives—not unexpectedly—the result ∞.

In[1]:= **Product[k, {k, Infinity}]**

Product::div : Product does not converge. More...

Product::div : Product does not converge. More...

Out[1]= $\prod_{k=1}^{\infty} k$

We can make the product convergent by replacing k with $k^{\exp(-\varepsilon k)}$ and letting $\varepsilon \to +0$ at the end. For finite ε, *Mathematica* gives a closed form for the regularized product.

In[2]:= **Product[k^Exp[-ε k], {k, Infinity}]**

Out[2]= $e^{-\text{PolyLog}^{(1,0)}[0, e^{-\varepsilon}] + e^{-\varepsilon}\ \text{LerchPhi}^{(0,1,0)}[e^{-\varepsilon}, 0, 1]}$

In the limit $\varepsilon \to +0$, this can easily be simplified to $(2\pi)^{1/2}$.

In[3]:= **% /. ε -> 0**

Out[3]= $e^{-\text{PolyLog}^{(1,0)}[0, 1] + \text{LerchPhi}^{(0,1,0)}[1, 0, 1]}$

In[4]:= **% /. Derivative[1, 0][PolyLog][0, 1] :> (D[PolyLog[v, 1], v] /. v -> 0)**

Out[4]= $e^{-\text{PolyLog}^{(1,0)}[0, 1] + \text{LerchPhi}^{(0,1,0)}[1, 0, 1]}$

This is the same result as obtained by the Zeta-regularization of products [1039], [639], [243], [673], [63], [428], [602], [442], [382], [266], [624], [626]:

$$\prod_{k=1}^{\infty} a_k = \exp\left(\left.\left(\frac{\partial}{\partial s} \sum_{k=1}^{\infty} a_k^{-s}\right)\right|_{s=0}\right)$$

In[5]:= **Exp[-D[Sum[k^-s, {k, Infinity}], s] /. s -> 0]**

Out[5]= $\sqrt{2\pi}$

The slightly more general result $\prod_{k=0}^{\infty} (k+x) = (2\pi)^{1/2}/\Gamma(x)$ goes back to M. Lerch, 1894 [782]. The product of all prime numbers is $4\pi^2$ [922], [923]. For similar products, see [743], [1347], [744], and [745].

16. Riemann Surface of Gauss Hypergeometric Function, $K(z)/K(1-z)$, $erf^{(-1)}$

a) The defining sum of the hypergeometric function $_2F_1(a; b, c; z)$ converges generically only inside the unit circle. The Kummer relations allow the analytic continuation to the whole complex plane. In this process, $_2F_1(a; b, c; z)$ gets a branch cut along $(1, \infty)$. On the other hand, by looking at the differential equation, we recognize that $z = 0$, $z = 1$, and $z = \infty$ are regular singular points. Because $_2F_1(1/3; 1/2, 1/6; z)$ is regular at the origin, the principal sheet of $_2F_1(1/3; 1/2, 1/6; z)$ has branch points at 1 and at ∞. These are the two points where we must be especially careful.

First, we load the relevant graphics package and define the parameters of the problem.

```
In[1]:= Needs["Graphics`Graphics3D`"]
```

```
In[2]:= ppφ = 32; ppr = 12;          (* plot points *)
        a = 1/3; b = 1/2; c = 1/6;   (* parameter of 2F1 *)
        ε = 10^-8;                    (* to have a small distance to the branch cut *)
```

The relation

$$_2F_1(a; b, c; z) = \frac{\Gamma(c)\,\Gamma(c-a-b)}{\Gamma(c-a)\,\Gamma(c-b)}\, _2F_1(a; b, a+b-c+1; 1-z) + \frac{\Gamma(c)\,\Gamma(a+b-c)}{\Gamma(a)\,\Gamma(b)}\, (1-z)^{c-a-b}\, _2F_1(c-a; c-b, c-a-b+1; 1-z)$$

gives, starting with z inside the unit circle, the analytic continuation to the circle with center $z = 0$ and radius 1. For z-values from this disk $|1-z| < 1$, we encounter no new branch cuts because of the arguments of the two hypergeometric functions, which are $1-z$. The factor $(-z)^{c-a-b}$ shows that $_2F_1(1/3; 1/2, 1/6; z)$ behaves near $z = 1$ like $(1-z)^{-\frac{2}{3}}$. This shows that $z = 1$ is a branch point of order three. If we now replace $(-z)^{c-a-b}$ by $\exp(j\,2\pi i/3)(-z)^{c-a-b}$, $j = 0, 1, 2$ in the above formula for the analytic continuation, we get all sheets that are glued together at $z = 1$. To avoid computing three times (once for each sheet) the time-consuming calculation of the values of the hypergeometric functions on the principal sheet, we compute them only once and multiply later with $\exp(j\,2\pi i/3)$. Here are the $z = x+iy$-values in the region $2/10 \le |z-1| \le 8/10$.

```
In[5]:= xyTab3 = Table[{1 + r Cos[φ], r Sin[φ]} // N,
                    {φ, ε, 2Pi - ε, (2Pi - 2ε)/ppφ}, {r, 1/5, 1, 4/5/ppr}];
```

These are the corresponding values needed in the above formula on the principal sheet.

```
In[6]:= hyperTab3 = Apply[Function[{x, y}, N @
                    {Gamma[c] Gamma[c - a - b]/(Gamma[c - a] Gamma[c - b])*
                    Hypergeometric2F1[a, b, a + b - c + 1, 1 - (x + I y)],
                    (1 - (x + I y))^(c - a - b) Gamma[c] Gamma[a + b - c]/
                    (Gamma[a] Gamma[b])*
                    Hypergeometric2F1[c - a, c - b, c - a - b + 1, 1 - (x + I y)]}],
                        xyTab3, {2}];
```

Using `MapThread`, we combine `xyTab3` and the real part of `hyperTab3` in points and display the resulting surfaces with `ListSurfacePlot3D`. We show the real and the imaginary parts.

```
In[7]:= (* frequently used options for inner (optsi) and
            outer (optso) functions in the following *)
        optsi = Sequence[Axes -> True, PlotRange -> All,
                        DisplayFunction -> Identity];
        optso = Sequence[ViewPoint -> {3, -1, 1.5}, BoxRatios -> {1, 1, 1.8},
                        AxesLabel -> {"x", "y", None}];
In[10]:= Show[GraphicsArray[
        Block[{$DisplayFunction = Identity},
        Function[reIm, Show[Table[
        ListSurfacePlot3D[MapThread[Append[#1, reIm[#2]]&,
            {xyTab3, Map[N[{1, Exp[2Pi I j/3]}.#]&, hyperTab3, {-2}]}, 2],
                        optsi], {j, 0, 2}], optso]] /@ {Re, Im}]]]
```

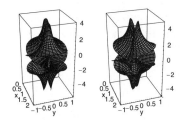

Concerning the branch point $z = \infty$, we use

$$_2F_1(a; b, c; z) = \frac{\Gamma(c)\,\Gamma(b-a)}{\Gamma(b)\,\Gamma(c-a)}\,(-z)^{-a}\,{}_2F_1\!\left(a; 1 - c + a, 1 - b + a; \frac{1}{z}\right) + \frac{\Gamma(c)\,\Gamma(a-b)}{\Gamma(a)\,\Gamma(c-b)}\,(-z)^{-b}\,{}_2F_1\!\left(b; 1 - c + b, b - a + 1; \frac{1}{z}\right)$$

for the analytical continuation. Now, we make the replacements $(-z)^{-a} \rightarrow \exp(j\,2\,\pi\,i/3)\,(-z)^{-a}$, $j = 0, 1, 2$ and $(-z)^{-b} \rightarrow \exp(i\,2\,\pi\,k/2)\,(-z)^{-a}$, $k = 0, 1$ to get all possible sheets that are glued together at ∞. As the z-region, we chose an elliptical ring.

```
In[11]:= xyTab6 = Table[{1/2 + (1/2 + r) Cos[φ], r Sin[φ]} // N,
                {φ, ε, 2Pi - ε, (2Pi - 2ε)/ppφ}, {r, 1/5, 4/5, 3/5/ppr}];
```

These are the relevant values from the principal sheet.

```
In[12]:= hyperTab6 = Apply[Function[{x, y}, N @
            {Gamma[c] Gamma[b - a]/(Gamma[b] Gamma[c - a])*
            (-(x + I y))^(-a) Hypergeometric2F1[a, 1 - c + a, 1 - b + a, 1/(x + I y)],
            Gamma[c] Gamma[a - b]/(Gamma[a] Gamma[c - b])(-(x + I y))^(-b) *
            Hypergeometric2F1[b, 1 - c + b, 1 - a + b, 1/(x + I y)]}],
                xyTab6, {2}];
```

The picture is constructed in complete analogy to the above one.

```
In[13]:= Show[GraphicsArray[
            Block[{$DisplayFunction = Identity},
            Function[reIm, Show[Table[
            ListSurfacePlot3D[MapThread[Append[#1, reIm[#2]]&,
                {xyTab6, Map[N[{Exp[2Pi I j/3], Exp[2Pi I k/2]}.#]&,
                        hyperTab6, {-2}]}], 2], optsi], {j, 0, 2}, {k, 0, 1}],
                        optso]] /@ {Re, Im}]]]
```

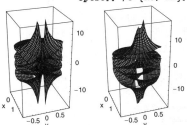

b) Now, let us reconstruct the above pictures by the solution of the hypergeometric differential equation. We first repeat the definitions from Part a).

```
In[1]:= Needs["Graphics`Graphics3D`"]

In[2]:= ppφ = 32; ppr = 12;              (* plot points *)
        a = 1/3; b = 1/2; c = 1/6;       (* parameter of 2F1 *)
        ε = 10^-8;                       (* to have a small distance to the branch cut *)

In[5]:= xyTab3 = Table[{1 + r Cos[φ], r Sin[φ]} // N,
                    {φ, ε, 2Pi - ε, (2Pi - 2ε)/ppφ}, {r, 1/5, 1, 4/5/ppr}];
```

```
In[6]:= xyTab6 = Table[{1/2 + (1/2 + r) Cos[φ], r Sin[φ]} // N,
                 {φ, ε, 2Pi - ε, (2Pi - 2ε)/ppφ}, {r, 1/5, 4/5, 3/5/ppr}];
```

First, we change the independent variable from z to $z = mp + r_x \cos(t) + i\, r_y \sin(t)$. (This form allows the description of the above-used, ellipse-shaped region with the midpoint mp.) Then, the differential equation in the variable t has the following form.

```
In[7]:= z1 = mp + rx Cos[t] + I ry Sin[t];
        k1 = D[z1, t]  (* for changing derivatives *);

        (* the transformed differential equation *)
        (eq1 = Collect[Numerator[Together[
            z1 (1 - z1)(1/k1 D[1/k1 w'[t], t]) +
            (c - (a + b + 1) z1)1/k1 w'[t] - a b w[t]]],
            {w''[t], w'[t], w[t]}] == 0) // Short[#, 12]&
```

Out[10]//Short=
$(-ry^3 \, Cos[t]^3 - 3\, i\, rx\, ry^2\, Cos[t]^2\, Sin[t] + 3\, rx^2\, ry\, Cos[t]\, Sin[t]^2 + i\, rx^3\, Sin[t]^3)\, w[t] +$
$\quad(6\, i\, mp\, rx\, Cos[t] - 6\, i\, mp^2\, rx\, Cos[t] + 6\, i\, rx^2\, Cos[t]^2 - 12\, i\, mp\, rx^2\, Cos[t]^2 -$
$\quad\quad i\, ry^2\, Cos[t]^2 + 11\, i\, mp\, ry^2\, Cos[t]^2 - 6\, i\, rx^3\, Cos[t]^3 + 11\, i\, rx\, ry^2\, Cos[t]^3 -$
$\quad\quad 6\, mp\, ry\, Sin[t] + 6\, mp^2\, ry\, Sin[t] - 10\, rx\, ry\, Cos[t]\, Sin[t] + 2\, mp\, rx\, ry\, Sin[t] -$
$\quad\quad 4\, rx^2\, ry\, Cos[t]^2\, Sin[t] - 11\, ry^3\, Cos[t]^2\, Sin[t] + i\, rx^2\, Sin[t]^2 -$
$\quad\quad 11\, i\, mp\, rx^2\, Sin[t]^2 - 6\, i\, ry^2\, Sin[t]^2 + 12\, i\, mp\, ry^2\, Sin[t]^2 - 11\, i\, rx^3\, Cos[t]\, Sin[t]^2 -$
$\quad\quad 4\, i\, rx\, ry^2\, Cos[t]\, Sin[t]^2 + 11\, rx^2\, ry\, Sin[t]^3 - 6\, ry^3\, Sin[t]^3)\, w'[t] +$
$\quad(-6\, mp\, ry\, Cos[t] + 6\, mp^2\, ry\, Cos[t] - 6\, rx\, ry\, Cos[t]^2 + 12\, mp\, rx\, ry\, Cos[t]^2 +$
$\quad\quad 6\, rx^2\, ry\, Cos[t]^3 - 6\, i\, mp\, rx\, Sin[t] + 6\, i\, mp^2\, rx\, Sin[t] - 6\, i\, rx^2\, Cos[t]\, Sin[t] +$
$\quad\quad 12\, i\, mp\, rx^2\, Cos[t]\, Sin[t] - 6\, i\, ry^2\, Cos[t]\, Sin[t] + 12\, i\, mp\, ry^2\, Cos[t]\, Sin[t] +$
$\quad\quad 6\, i\, rx^3\, Cos[t]^2\, Sin[t] + 12\, i\, rx\, ry^2\, Cos[t]^2\, Sin[t] + 6\, rx\, ry\, Sin[t]^2 - 12\, mp\, rx\, ry\, Sin[t]^2 -$
$\quad\quad 12\, rx^2\, ry\, Cos[t]\, Sin[t]^2 - 6\, ry^3\, Cos[t]\, Sin[t]^2 - 6\, i\, rx\, ry^2\, Sin[t]^3)\, w''[t] = 0$

First, we solve this differential equation by encircling $z = 1$ three times (because of the branch point of order 3). Because the real axis is part of the lower lip of the branch cut while we integrate the differential equation clockwise, we choose a value just above the branch cut for the initial conditions of the differential equation. For various values of r, we solve the differential equation.

```
In[11]:= mp = 1;
        Table[sol1[r] = NDSolve[N[{eq1,
                    w[0] == Hypergeometric2F1[a, b, c, z1] /. t -> ε,
                    w'[0] == D[Hypergeometric2F1[a, b, c, z1],
                        t] /. t -> ε]} /. {rx -> r, ry -> r},
                    {w}, {t, 0, 6 2Pi}, MaxSteps -> 10000,
                    PrecisionGoal -> 10, AccuracyGoal -> 8],
            {r, 1/5, 4/5, 3/5/ppr}];
```

We extract the function values.

```
In[13]:= Table[hyperTab3DE[j] = Table[sol1[r][[1, 1, 2]][N[φ]],
                    {φ, j 2Pi + ε, j 2Pi + 2Pi - ε, (2Pi - 2ε)/ppφ},
                    {r, 1/5, 4/5, 3/5/ppr}],
            {j, 0, 2}];
```

Finally, we display the result.

```
In[14]:= (* frequently used options for inner (optsi) and
            outer (optso) functions in the following *)
        optsi = Sequence[Axes -> True, PlotRange -> All,
                    DisplayFunction -> Identity];
        optso = Sequence[ViewPoint -> {3, -1, 1.5}, BoxRatios -> {1, 1, 1.8},
                    AxesLabel -> {"x", "y", None}];

In[17]:= Show[GraphicsArray[
        Block[{$DisplayFunction = Identity},
        Function[reIm, Show[Table[ListSurfacePlot3D[MapThread[Append[#1, Re[#2]]&,
            {xyTab3, hyperTab3DE[j]}, 2], optsi], {j, 0, 2}], optso]] /@ {Re, Im}]]]
```

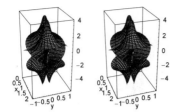

The picture is the same as the one above (as expected). For the branch point at infinity, we solve the differential equation along ellipses with center $1/2 + 0\,i$.

```
In[18]:= mp = 1/2;
         Table[sol1[r] = NDSolve[N[{eq1,
                 w[0] == Hypergeometric2F1[a, b, c, z1] /. t -> ε,
                 w'[0] == D[Hypergeometric2F1[a, b, c, z1],
                     t] /. t -> ε}] //. {rx -> 1/2 + r, ry -> r},
                 {w}, {t, 0, 6 2Pi}, MaxSteps -> 10000,
                 PrecisionGoal -> 10, AccuracyGoal -> 8],
             {r, 1/5, 4/5, 3/5/ppr}];
```

We extract the function values from the resulting `InterpolatingFunction`-objects.

```
In[20]:= Table[hyperTab6DE[j] =
         Table[sol1[r][[1, 1, 2]][N[φ]],
             {φ, j 2Pi + ε, j 2Pi + 2Pi - ε, (2Pi - 2ε)/ppφ},
             {r, 1/5, 4/5, 3/5/ppr}], {j, 0, 5}];
```

Finally, we display the result.

```
In[21]:= Show[GraphicsArray[
         Block[{$DisplayFunction = Identity},
         Function[reIm, Show[Table[ListSurfacePlot3D[MapThread[Append[#1, Re[#2]]&,
             {xyTab6, hyperTab6DE[i]}, 2], optsi], {i, 0, 5}], optso]] /@ {Re, Im}]]]
```

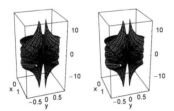

Again, the picture is the same as the one constructed using the Kummer relations.

The so-constructed sheets of the Riemann surface of $_2F_1(1/3; 1/2, 1/6; z)$ also have a branch point at $z = 0$. By using some more complicated paths for solving the differential equation, we could go on and also make a picture of these sheets. We end here. The reader, if interested in this subject, should be able to carry out these further manipulations.

As a third possible way to get the various sheets of the hypergeometric function, we could use the integral representation of the hypergeometric function and deform the integration path in dependence of z appropriately. We do not carry out this approach here; for details, see [687]. For further discussions of the structure of the Riemann surface of the hypergeometric function $_2F_1(a, b; c; z)$, see [137], [735], [556], and [1227].

c) Here are graphics of the real and imaginary parts of $K(z)/K(1 - z)$.

```
In[1]:= f[z_] = EllipticK[z]/EllipticK[1 - z];

In[2]:= Show[GraphicsArray[
         Plot3D[#[f[x + I y]], {x, -2, 2}, {y, -2, 2},
             DisplayFunction -> Identity, Mesh -> False,
             PlotPoints -> 120, PlotRange -> All]& /@ {Re, Im}]]
```

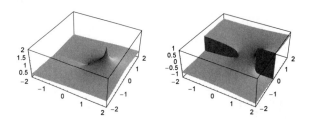

The right branch cut for $z > 1$ shows the typical shape of a logarithmic branch point. Taking into account the series expansions of $K(z)$ near $z = 0$ and $z = 1$, namely $K(z) \underset{z \to 0}{\approx} \pi/2\,(1 + z/4 + 9/64\,z^2 + \cdots)$ and $K(z) \underset{z \to 1}{\approx} -1/2 \ln(1 - z) + \ln(4)$ we recognize that the points $z = 0$ and $z = 1$ are both logarithmic branch points. The numerator of $K(z)/K(1 - z)$ gives rise to the branch point at $z = 1$ and the denominator to the branch point at $z = 0$. This means the Riemann surface of $K(z)/K(1 - z)$ has double infinite many sheets. Both functions, $K(z)$, and $K(1 - z)$ are (linearly independent) solutions of the following hypergeometric differential equation.

```
In[3]:= a = 1/2; b = 1/2; c = 1;
        ode = z (1 - z) w''[z] + (c - (a + b + 1)z) w'[z] - a b w[z]
```
$$\text{Out[4]}= \ -\frac{w[z]}{4} + (1 - 2 z) w'[z] + (1 - z) z w''[z]$$

```
In[5]:= {ode /. w -> Function[z, EllipticK[z]],
         ode /. w -> Function[z, EllipticK[1 - z]]} // Simplify
```
$$\text{Out[5]}= \ \{0, 0\}$$

The corresponding Schwarz differential equation for the ratio $K(z)/K(1 - z)$ is then given by the following expression (see Exercise 4 of Chapter 1).

```
In[6]:= f[z_] := (1 - 2z)/(z (1 - z))
        g[z_] := -1/(4 z (1 - z))
```

```
In[8]:= odeRatio = Collect[w'''[z] w'[z] - 3/2 w''[z]^2 +
          (f[z]^2/2 - 2 g[z] + f'[z]) w'[z]^2,
          {w'''[z] w'[z], w''[z]^2, w'[z]^2}, Factor[Together[#]]&]
```
$$\text{Out[8]}= \ -\frac{(1 - z + z^2)\, w'[z]^2}{2\,(-1 + z)^2\, z^2} - \frac{3}{2}\, w''[z]^2 + w'[z]\, w^{(3)}[z]$$

Here is a quick check that $K(z)/K(1 - z)$ really fulfills this differential equation.

```
In[9]:= odeRatio /. w -> f // Simplify
```
$$\text{Out[9]}= \ 0$$

To construct the Riemann surface by solving the Schwarz differential equation, we have to encircle the branch points. We will do this recursively and repeatedly. First we will encircle the $z = 0$ branch points. We will use the so-obtained functions values and derivative values as initial conditions for encircling the $z = 1$ branch point. In a similar way, we will encircle the $z = 1$ branch points and then use the so-obtained functions values and derivative values will be used as initial conditions for encircling the $z = 0$ branch point. To encircle the branch points, we change independent and dependent variables in the differential equation: $z = mp + r_x \cos(\varphi) + i\, r_y \sin(\varphi)$, $w(z) = \omega(\varphi)$. (This is the parametrization of an elliptical path around mp.)

```
In[10]:= odeω[{ω_, φ_}, {mp_, rx_, ry_}] =
         Module[{ξ = mp + rx Cos[φ] + I ry Sin[φ], k}, k = D[ξ, φ];
         Simplify[Numerator[Together[
           (1/k D[1/k D[1/k ω'[φ], φ], φ]) (1/k ω'[φ]) -
           3/2 (1/k D[1/k ω'[φ], φ])^2 -
           (1 - ξ + ξ^2)/(2 ξ^2 (ξ - 1)^2) (1/k ω'[φ])^2]]]];
```

The functions dChange["*oldVar -> newVar*"] transforms the derivatives needed for the initial conditions.

```
In[11]:= (* transform w'[z] and w''[z] to ω'[φ] and ω''[φ] *)
         dChange["z -> φ"][{f0_, f1_, f2_}, φ0_, {mp_, rx_, ry_}] :=
         Module[{φ, ξ, k, κ},
            ξ = mp + rx Cos[φ] + I ry Sin[φ];
```

```
            k = D[ξ, φ]; κ = D[1/k, φ];
            {f0, k f1, k^2 (f2 - κ f1)} /. φ -> φ0]
In[13]:= (* transform ω'[φ] and ω''[φ] to w'[z] and w''[z] *)
        dChange["φ -> z"][{f0_, f1_, f2_}, φ0_, {mp_, rx_, ry_}] :=
        Module[{φ, ξ, k, κ},
              ξ = mp + rx Cos[φ] + I ry Sin[φ];
              k = D[ξ, φ]; κ = D[1/k, φ];
              {f0, 1/k f1, 1/k^2 f2 + κ/k f1} /. φ -> φ0]
```

The function solveODEAroundBranchPoint encircles the branch point *mp o* times by solving the Schwarz differential equation along this path.

```
In[15]:= solveODEAroundBranchPoint[{mp_, rx_, ry_},
                                {f0_, f1_, f2_}, φ0_, o_] :=
        Module[{ξ, eqs}, (* ξ = mp + rx Cos[φ] + I ry Sin[φ]; *)
          eqs = Flatten[{odeω[{ω0, φ}, {mp, rx, ry}] == 0 /.
                        {ω0' -> ω1, ω0'' -> ω2, ω0''' -> ω2'},
          ω0'[φ] == ω1[φ], ω1'[φ] == ω2[φ],
          Thread[{ω0[φ0], ω1[φ0], ω2[φ0]} ==
                    dChange["z -> φ"][{f0, f1, f2}, φ0, {mp, rx, ry}]]}];
          NDSolve[eqs, {ω0, ω1, ω2}, {φ, -o 2Pi, o 2Pi},
                    PrecisionGoal -> 12, AccuracyGoal -> 12, MaxSteps -> 10^4]];
```

Here are the function values and the first two derivatives of the four sheets directly neighboring to the principal sheet.

```
In[16]:= o = 1;

In[17]:= nsol11 = solveODEAroundBranchPoint[{1, 1/2, 1/2},
                        {f[1/2], f'[1/2], f''[1/2]}, Pi, o + 1];

In[18]:= tab11 = dChange["φ -> z"][#, Pi, {1, 1/2, 1/2}]& /@
              Table[{ω0[Pi + j 2Pi], ω1[Pi + j 2Pi], ω2[Pi + j 2Pi]} /.
                    nsol11[[1]], {j, -o, o}] // Chop
Out[18]= {{1. + 2. i, 0.913893, 0.835201}, {1., 0.913893, 0.835201}, {1. - 2. i, 0.913893, 0.835201}}

In[19]:= nsol01 = solveODEAroundBranchPoint[{0, 1/2, 1/2},
                        {f[1/2], f'[1/2], f''[1/2]}, 0, o + 1];

In[20]:= tab01 = dChange["φ -> z"][#, 0, {0, 1/2, 1/2}]& /@
              Table[{ω0[0 + j 2Pi], ω1[0 + j 2Pi], ω2[0 + j 2Pi]} /.
                    nsol01[[1]], {j, -o, o}] // Chop
Out[20]= {{0.2 - 0.4 i, -0.109667 - 0.146223 i, -0.0467712 + 0.160359 i},
         {1., 0.913893, 0.835201}, {0.2 + 0.4 i, -0.109667 + 0.146223 i, -0.0467712 - 0.160359 i}}
```

Using the last function values and derivatives as initial values for encircling the other branch point we end up with a set of 13 initial conditions.

```
In[21]:= (startData = Union[Join[
        Flatten[(dChange["φ -> z"][#, 0, {0, 1/2, 1/2}]& /@
              Table[{ω0[0 + j 2Pi], ω1[0 + j 2Pi], ω2[0 + j 2Pi]} /.
              solveODEAroundBranchPoint[{0, 1/2, 1/2}, #, 0, o + 1][[1]],
                                        {j, -o, o}])& /@ tab11, 1],
        Flatten[(dChange["φ -> z"][#, Pi, {1, 1/2, 1/2}]& /@
              Table[{ω0[Pi + j 2Pi], ω1[Pi + j 2Pi], ω2[Pi + j 2Pi]} /.
              solveODEAroundBranchPoint[{1, 1/2, 1/2}, #, Pi, o + 1][[1]],
                                        {j, -o, o}])& /@ tab01, 1],
              SameTest -> ((Max[Abs[#1 - #2]] < 10^-3)&)]) // (First /@ #&)
Out[21]= {0.0344828 + 0.413793 i, 0.0344828 - 0.413793 i,
         0.0769231 - 0.615385 i, 0.0769231 + 0.615385 i, 0.2 - 0.4 i, 0.2 + 0.4 i,
         0.2 + 1.6 i, 0.2 + 2.4 i, 0.2 - 1.6 i, 0.2 - 2.4 i, 1. + 0. i, 1. + 2. i, 1. - 2. i}
```

To calculate the values of the function on the various sheets, we will solve the Schwarz differential equation along the ray $(1/2, 1/2 + it)$ using the initial conditions startData. Then we will use these solutions as initial conditions for encircling both branch points together by ellipse-shaped integration paths. The values of the solution of the Schwarz differential equation along these ellipse-shaped integration paths constitute the sheets of the Riemann surface. To solve the Schwarz differential equation along the ray $1/2 + it$, we carry out the following change of variables in $z = 1/2 + I y$, $w(z) = Y(y)$.

```
In[22]:= odeY[{Y_, y_}] =
    Module[{ξ = 1/2 + I y, k}, k = D[ξ, y];
        Simplify[Numerator[Together[
            (1/k D[1/k D[1/k Y'[y], y], y]) (1/k Y'[y]) -
            3/2 (1/k D[1/k Y'[y], y])^2 -
            (1 - ξ + ξ^2)/(2 ξ^2 (ξ - 1)^2) (1/k Y'[y])^2]]]]
```

$$Out[22]= \; -4 \, (-3 + 4 \, y^2) \, Y'[y]^2 - 3 \, (1 + 4 \, y^2)^2 \, Y''[y]^2 + 2 \, (1 + 4 \, y^2)^2 \, Y'[y] \, Y^{(3)}[y]$$

Here are the two corresponding instances of dChange.

```
In[23]:= (* transform w'[z] and w''[z] to ω'[φ] and ω''[φ] *)
    dChange["z -> y"][{f0_, f1_, f2_}, y0_] :=
    Module[{y, ξ, k, κ}, ξ = 1/2 + I y;
        k = D[ξ, y]; κ = D[1/k, y];
        {f0, k f1, k^2 (f2 - κ f1)} /. y -> y0]
In[25]:= dChange["y -> z"][{f0_, f1_, f2_}, y0_] :=
    Module[{φ, ξ, k, κ, y}, ξ = 1/2 + I y;
        k = D[ξ, y]; κ = D[1/k, y];
        {f0, 1/k f1, 1/k^2 f2 + κ/k f1} /. y -> y0]
```

The function solveODENorthwards solves the Schwarz differential equation the ray $(1/2, \; 1/2 + I\,t)$.

```
In[26]:= solveODENorthwards[{f0_, f1_, f2_}, y0_, yMax_] :=
    Module[{eqs},
    eqs = Flatten[{odeY[{Y0, y}] == 0 /.
                    {Y0' -> Y1, Y0'' -> Y2, Y0''' -> Y2'},
        Y0'[y] == Y1[y], Y1'[y] == Y2[y],
        Thread[{Y0[y0], Y1[y0], Y2[y0]} ==
                dChange["z -> y"][{f0, f1, f2}, y0]]}];
    NDSolve[eqs, {Y0, Y1, Y2}, {y, y0, yMax},
                PrecisionGoal -> 12, AccuracyGoal -> 12, MaxSteps -> 10^4]];
```

Now we have all functions to actually solve the Schwarz differential equation on a grid and construct the corresponding sheets of the Riemann surface.

```
In[27]:= ppy = 6; ppφ = 24; ε = 10^-3;
    insideIterator  = {y, ε, 1/2 - ε, (1/2 - 2 ε)/ppy};
    outsideIterator = {y, 1/2 + ε, 1, (1/2 - ε)/ppy};

    Needs["Graphics`Graphics3D`"]

    Do[(* solve differential equation along (1/2, 1/2 + I) *)
        yIFs = YIF[k] = solveODENorthwards[startData[[k]], 0, 1];
        (* construct function values and derivative values *)
        {startDataInside, startDataOutside} =
        Table[{y, dChange["y -> z"][{Y0[y], Y1[y], Y2[y]} /.
                        yIFs[[1]], y]}, #]& /@ {insideIterator, outsideIterator};
        (* encircle both branch points *)
        ((R[#1] = solveODEAroundBranchPoint[{1/2, #1, 1/2 + #1},
          #2, Pi/2, 1/2])& @@@ #)& /@ {startDataInside, startDataOutside};
        (* points of the sheets *)
        {pointsInside, pointsOutside} =
        Table[r = R[y][[1, 1, 2]];
            Table[{1/2., 0., 0.} + {y Cos[φ], y Sin[φ], r[φ]},
                    {φ, -Pi, Pi, 2Pi/ppφ}], #]& /@
                                {insideIterator, outsideIterator};
        (* construct polygons *)
        sheetPoly[k] = Function[reIm, Cases[
        ListSurfacePlot3D[Apply[{#1, #2, reIm[#3]}&, #, {-2}],
                        DisplayFunction -> Identity], _Polygon, Infinity]& /@
                            {pointsInside, pointsOutside}] /@ {Re, Im},
        {k, Length[startData]}]
```

Coloring the sheets differently and cutting the edges of the polygons (to have a better view of the other sheets) yields the following graphic of the sheets that are nearest to the principal sheet of $K(z)/K(1-z)$.

```
In[32]:= Do[sheetColor[k] = SurfaceColor[Hue[Random[]], Hue[Random[]], 2.3],
         {k, Length[startData]}]

In[33]:= makeDiamondPolygon[Polygon[l_]] :=
         Polygon[Plus @@@ Partition[Append[l, First[l]], 2, 1]/2]

In[34]:= Show[Graphics3D[
         Table[{sheetColor[k], sheetPoly[k][[2]]},
              {k, Length[startData]}] /. p_Polygon :> makeDiamondPolygon[p],
              BoxRatios -> {1, 1, 3/4}, ViewPoint -> {0, -3, 1},
              PlotRange -> {-1/2, 1}]]
```

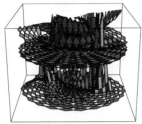

Now let us construct symbolic expressions for the other sheets. From the integral representation $K(z) = \int_0^{\pi/2} (1 - z\sin^2(\vartheta))^{1/2}\, d\vartheta$, the following jump behavior along the branch cut $(1, \infty)$ of $K(z)$ follows: $\mathrm{Im}(K(z \pm 0^+ i)) = \pm K(1 - z)$. This formula allows the implementation of the function KKPExtensionStep. KKPExtensionStep[K, z][f] will return a list of the neighboring sheets of f, obtained by continuing the function K.

```
In[35]:= KKPExtensionStep[K_, z_][f_] :=
         Module[{c = 1, Kp = DeleteCases[{EllipticK[z],
                                          EllipticK[1 - z]}, K][[1]], f, k},
              (* replace all Ks by extended ones *)
              f = f /. K :> K + I k[c++] Kp;
              (* reduce fraction *)
              ExpandNumerator /@ Together[ (* the neighboring sheets *)
                Table[Evaluate[f], Evaluate[Sequence @@ Table[{k[j], -1, 1}, {j, c - 1}]]]]]
```

Using the abbreviations $K = K(z)$ and $K' = K(1 - z)$ we show KKPExtensionStep at work.

```
In[36]:= abbreviate[expr_] :=
         expr /. {EllipticK[z] -> K, EllipticK[1 - z] -> K'}

In[37]:= (ext1 = KKPExtensionStep[EllipticK[z], z][f[z]]) // abbreviate
```
$$Out[37]= \left\{ \frac{K - i\,K'}{K'},\ \frac{K}{K'},\ \frac{K + i\,K'}{K'} \right\}$$

```
In[38]:= (ext2 = KKPExtensionStep[EllipticK[1 - z], z][f[z]]) // abbreviate
```
$$Out[38]= \left\{ \frac{K}{-i\,K + K'},\ \frac{K}{K'},\ \frac{K}{i\,K + K'} \right\}$$

The sheets of the lists ext1 and ext2 are solutions of the Schwarz differential equation for $K(z)/K(1 - z)$.

```
In[39]:= (odeRatio /.  w -> Function[z, #])& /@ Flatten[{ext1, ext2}] // Simplify

Out[39]= {0, 0, 0, 0, 0, 0}
```

We unite the application of KKPExtensionStep with respect to $K(z)$ and $K(1 - z)$ in the function KKPExtensions.

```
In[40]:= KKPExtensions[n_, z_] := KKPExtensions[n, z] =
         DeleteCases[Union[Flatten[KKPExtensionStep[EllipticK[1 - z], z] /@
                Union[Flatten[KKPExtensionStep[EllipticK[z], z] /@
                                          KKPExtensions[n - 1, z]]]]],
                _?NumberQ]

         KKPExtensions[0, z_] := {EllipticK[z]/EllipticK[1 - z]}

In[42]:= KKPExtensions[1, z] // abbreviate
```

$$\text{Out[42]=} \left\{ -\frac{i\,K'}{-i\,K+K'}, \; \frac{i\,K'}{-i\,K+K'}, \; -\frac{i\,K'}{i\,K+K'}, \; \frac{i\,K'}{i\,K+K'}, \; \frac{K}{K'}, \; \frac{K}{-i\,K+K'}, \; \frac{K}{i\,K+K'}, \; \frac{K-i\,K'}{K'}, \; \frac{K-i\,K'}{-i\,K+K'}, \right.$$
$$\left. \frac{K+i\,K'}{K'}, \; \frac{K+i\,K'}{i\,K+K'}, \; \frac{2\,K-i\,K'}{K'}, \; \frac{2\,K-i\,K'}{-i\,K+K'}, \; \frac{2\,K-i\,K'}{i\,K+K'}, \; \frac{2\,K+i\,K'}{K'}, \; \frac{2\,K+i\,K'}{-i\,K+K'}, \; \frac{2\,K+i\,K'}{i\,K+K'} \right\}$$

Let us now use the symbolic descriptions of the other sheets for generating a graphic. All sheets generated by KKPExten∴ sions are of the form $(a\,K(z) + b\,K(1 - z))/(c\,K(z) + c\,K(1 - z))$, where a, b, c, d are Gaussian integers. This means the branch cuts of all these sheets are $(-\infty, 0)$ and $(1, \infty)$. The function makeSheetPolygons excludes the branch cuts to avoid steep vertical walls.

```
In[43]:= makeSheetPolygons[{KKs_, z_}, f_, pp_] :=
        Module[{ε = 10^-12, x, y},
        Cases[Graphics3D[#], _Polygon, Infinity]& /@
        (* above and below the real axis *)
        (Plot3D[Evaluate[f[KKs] /. z -> x + I y],
                {x, -2, 2}, {y, ##}, PlotPoints -> {pp, Round[pp/2]},
                DisplayFunction -> Identity]& @@@ {{ -2, -ε}, {ε, 2}})]
```

```
In[44]:= Show[Graphics3D[makeSheetPolygons[{f[z], z}, Im, 30]],
        Axes -> True, PlotRange -> All]
```

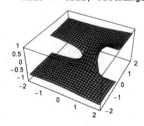

The function graphicsPair shows the sheets *KKs* from two different viewpoints, from the positive and negative *x*-directions.

```
In[45]:= graphicsPair[{KKs_, z_}, pp_] :=
        With[{gr = makeSheetPolygons[{#, z}, Im, 30]& /@ KKs},
        Show[GraphicsArray[
          Graphics3D[{EdgeForm[Thickness[0.002]],
                      SurfaceColor[Hue[0.12], Hue[0], 2.7],
                      gr}, ViewPoint -> #]& /@ {{+3, -1, 2}, {-3, -1, 2}}] /.
          p_Polygon :> makeDiamondPolygon[p]]]
```

The following two pairs of graphics show that the sheets of ext1 join smoothly along the positive *x*-axis and the sheets of ext2 join smoothly along the negative *x*-axis.

```
In[46]:= graphicsPair[{ext1, z}, 30]
```

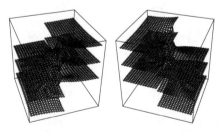

```
In[47]:= graphicsPair[{ext2, z}, 30]
```

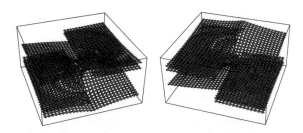

Carrying out the continuation repeatedly yields more and more sheets. KKPExtensions[3, z] contains 2232 sheets, including the ones that contain the above numerically calculated startData. Here is a quick check of this statement. (The maximal difference of 5 10^{-11} agrees with the PrecisionGoal option setting used for NDSolve above.)

```
In[48]:= Function[1, Min[Abs[1 - #]]& /@ (First /@ startData)][
           KKPExtensions[3, z] /. z -> 1/2.] // Max

Out[48]= 7.96381×10^-9
```

We end with a graphic of the sheets of KKPExtensions[1, z].

```
In[49]:= graphicsPair[{KKPExtensions[1, z], z}, 20]
```

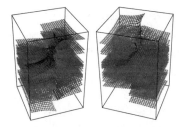

d) Starting with the differential equation $w''(z) + 2 z w'(z) = 0$ for $w(z) = \mathrm{erf}(z)$, it is straightforward to derive the differential equation $w''(z) - 2 w(z) w'(z)^2 = 0$ for $w(z) = \mathrm{erf}^{(-1)}(z)$.

```
In[1]:= w''[z] - 2 w[z] w'[z]^2 /. w -> InverseErf

Out[1]= 0
```

Now we adapt to a coordinate system that is appropriate for displaying the Riemann surface near $z = 1$. Changing now variables according to $z = 1 + r \exp(i\varphi)$ and $w(z) = \Phi(\varphi)$, we obtain the differential equation $\Phi''(\varphi) - 2 \Phi(\varphi) \Phi'(\varphi)^2 - i \Phi'(\varphi) = 0$ for $\Phi(\varphi)$.

```
In[2]:= z = 1 + r Exp[I φ];
        {w0 = Φ[φ], w1 = D[Φ[φ], φ] 1/D[z, φ], w2 = D[w1, φ] 1/D[z, φ]};

        -Numerator[Together[w2 - 2 w0 w1^2]]

Out[4]= - i Φ'[φ] - 2 Φ[φ] Φ'[φ]^2 + Φ''[φ]
```

Now we solve this differential equation for various values of r over a φ-range corresponding to multiple surroundings of the point $z = 1$.

```
In[5]:= ε = 10^-3;
        Do[nsol[r] = NDSolve[{Φ''[φ] - 2 Φ[φ] Φ'[φ]^2 - I Φ'[φ] == 0,
                        (* initial conditions on the real axis *)
                        Φ[Pi] == 0, Φ'[Pi] == -I InverseErf'[1 - r]},
                    Φ, {φ, -6Pi, 8Pi}][[1, 1, 2]],
                {r, ε, 7/4 - ε, (7/4 - 2ε)/40}]
In[7]:= data = Table[{1 + r Cos[φ], r Sin[φ], nsol[r][φ]},
                    {φ, -6Pi, 8Pi, 2Pi/30},
                    {r, ε, 7/4 - ε, (7/4 - 2ε)/40}];
```

The visualization of the so-obtained values of $\mathrm{erf}^{(-1)}(z)$ at the sheets next to the principal sheet is straightforward. We display $\mathrm{Re}(\mathrm{erf}^{(-1)}(z))$ and $\mathrm{Im}(\mathrm{erf}^{(-1)}(z))$.

```
In[8]:= makePolys[data_] :=
        Table[Polygon[{data[[i, j]], data[[i, j + 1]],
                       data[[i + 1, j + 1]], data[[i + 1, j]]}],
              {i, Length[data] - 1}, {j, Length[data[[1]]] - 1}];
In[9]:= With[{λ = Length[data]},
        Show[GraphicsArray[
        Function[reIm, (* display real and imaginary part *)
        Graphics3D[{EdgeForm[], (* color according to φ *)
            MapIndexed[{SurfaceColor[Hue[#2[[1]]/λ], Hue[#2[[1]]/λ], 2.4],
            #1}&, makePolys[Map[MapAt[reIm, #, 3]&, data, {2}]]] // N},
            BoxRatios -> {1, 1, 1.75}, ViewPoint -> {-3, 0, 1}]] /@ {Re, Im}]]]
```

17. Kummer's 24 Solutions of the Gauss Hypergeometric Differential Equation, Appell Differential Equation

a) Before treating the hypergeometric differential equation in detail, let us first perform the change of the independent variable $z \to x$ in a generic differential equation. Denote the function $u(z)$ expressed as a function of x by $U(x) = u(z(x))$.

We can then transform derivatives of u with respect to z to derivatives of U with respect to x in the following manner $u[z]$ $\to U[x]$.

```
In[1]:= Derivative[1 ][u][z] := 1/z'[x] Derivative[1][U][x];
        Derivative[n_][u][z] := Together[1/z'[x] D[Derivative[n - 1][u][z], x]];
```

So the left-hand side of our second-order differential equations reads as a differential equation in x.

```
In[3]:= u''[z] + g[x] U[x]
```

$$Out[3]= g[x] U[x] + \frac{z'[x] U''[x] - U'[x] z''[x]}{z'[x]^3}$$

It is slightly rewritten. We write everything above a common denominator and keep only the numerator.

```
In[4]:= Numerator[Together[%]]/z'[x] // Expand
```

$$Out[4]= g[x] U[x] z'[x]^2 + U''[x] - \frac{U'[x] z''[x]}{z'[x]}$$

It now shows the form of the Liouville transformation to apply.

```
In[5]:= U[x_] = Exp[-1/2 Integrate[-z''[x]/z'[x], x]] w[x]
```

$$Out[5]= w[x] \sqrt{z'[x]}$$

After some cosmetic rearrangements, the differential equation for $w(x)$ is given here.

```
In[6]:= Collect[Numerator[Together[%%]]/(4z'[x]^2), {w[x], w''[x]}]
```

$$Out[6]= w''[x] + \frac{w[x] (4 g[x] z'[x]^4 - 3 z''[x]^2 + 2 z'[x] z^{(3)}[x])}{4 z'[x]^2}$$

For the transformation $z(x) = (\alpha + \beta\, x)/(\gamma + \delta\, x)$, the last two difficult-looking terms (also called the Schwarz derivative) fortunately vanish identically.

```
In[7]:= With[{z = (α + β x)/(γ + δ x)},
        Together[-3/4 D[z, {x, 2}]^2/D[z, x]^2 + 1/2 D[z, {x, 3}]/D[z, x]]]
Out[7]= 0
```

The above equation thus shows the relation $h(x) = z'(x)^2\, g(z(x))$.

In[8]:= (* do not need any special rule for derivatives of u anymore *)
        Derivative[1][u][z] =.
        Derivative[n_][u][z] =.

Now, let us go to the hypergeometric differential equation. The Liouville transformation is now given by the following.

In[11]:= Simplify[ExpandAll[Exp[-1/2 Integrate[
                    (c - (1 + a + b) z)/(z (1 - z)), z]]]] u[z]

Out[11]= $(-1 + z)^{\frac{1}{2}(-1-a-b+c)} z^{-c/2} u[z]$

In[12]:= y[z_] = ((1 - z)^((-1 - a - b + c)/2) u[z])/z^(c/2)

Out[12]= $(1 - z)^{\frac{1}{2}(-1-a-b+c)} z^{-c/2} u[z]$

So this is the normal form of the hypergeometric differential equation.

In[13]:= (z (1 - z) D[y[z], {z, 2}] + (c - (1 + a + b) z) D[y[z], z] -
                                        a b y[z]) // Together

Out[13]= $-\dfrac{1}{4(-1+z)}$

$\left((1-z)^{-\frac{1}{2}-\frac{a}{2}-\frac{b}{2}+\frac{c}{2}} z^{-1-\frac{c}{2}} (2\,c\,u[z] - c^2\,u[z] - 4\,a\,b\,z\,u[z] - 2\,c\,z\,u[z] + 2\,a\,c\,z\,u[z] + 2\,b\,c\,z\,u[z] + \right.$
$\left. z^2\,u[z] - a^2\,z^2\,u[z] + 2\,a\,b\,z^2\,u[z] - b^2\,z^2\,u[z] + 4\,z^2\,u''[z] - 8\,z^3\,u''[z] + 4\,z^4\,u''[z])\right)$

Here is some rewriting done with the last expression to bring it to polynomial form.

In[14]:= Cancel[(Factor /@ Numerator[%])/(1 - z)^(c/2)]

Out[14]= $-2\,c\,u[z] + c^2\,u[z] + 4\,a\,b\,z\,u[z] + 2\,c\,z\,u[z] - 2\,a\,c\,z\,u[z] - 2\,b\,c\,z\,u[z] -$
$z^2\,u[z] + a^2\,z^2\,u[z] - 2\,a\,b\,z^2\,u[z] + b^2\,z^2\,u[z] - 4\,z^2\,u''[z] + 8\,z^3\,u''[z] - 4\,z^4\,u''[z]$

So now it can be cast in the following form.

In[15]:= #/(-4 (z - 1)^2 z^2) & /@ (Factor /@ Collect[%, {u[z], u'[z], u''[z]}])

Out[15]= $-\dfrac{(-2\,c + c^2 + 4\,a\,b\,z + 2\,c\,z - 2\,a\,c\,z - 2\,b\,c\,z - z^2 + a^2\,z^2 - 2\,a\,b\,z^2 + b^2\,z^2)\,u[z]}{4(-1+z)^2 z^2} + u''[z]$

The function $g(z)$ is then given by the following.

In[16]:= g[z_, a_, b_, c_] = Coefficient[%, u[z]]

Out[16]= $-\dfrac{-2\,c + c^2 + 4\,a\,b\,z + 2\,c\,z - 2\,a\,c\,z - 2\,b\,c\,z - z^2 + a^2\,z^2 - 2\,a\,b\,z^2 + b^2\,z^2}{4(-1+z)^2 z^2}$

We use the explicit form of $z(x)$.

In[17]:= z = (α x + β)/(γ x + δ);

One could think that after making the definitions, the equality $g(x) = h(x)$ can now be easily solved via these definitions with a call to Reduce.

In[18]:= g1 = g[x, a, b, c];
        h1 = D[z, x]^2 g[z, as, bs, cs];
        (* Reduce[# == 0& /@ CoefficientList[
                            Numerator[Together[g1 - h1]], x],
                    {α, β, γ, δ, as, bs, cs}] *)

Unfortunately, this input never returns. So we must help *Mathematica* here. We could start with the last equation and see what could happen.

In[21]:= CoefficientList[Numerator[Together[g1 - h1]], x][[-1]] // Factor

Out[21]= $-(-1 + a - b)(1 + a - b)\,α^2\,(α - γ)^2\,γ^2$

This means $α = 0$ or $γ = 0$ or ... Instead of giving nine equations to Reduce, or analyzing backward starting with the last equation of all possibilities we will go another route and split the equality g1 == h1 in two groups by comparing numerator and denominator independently [605]. Because the numerators and denominators are only determined up to a constant, we multiply with an auxiliary variable auxFac.

We look at the numerators and the denominators.

`In[22]:=` `Together[g1] == Together[h1]`

`Out[22]=`
$$\frac{2c - c^2 - 4abx - 2cx + 2acx + 2bcx + x^2 - a^2x^2 + 2abx^2 - b^2x^2}{4(-1+x)^2 x^2} ==$$

$-((-\beta\gamma + \alpha\delta)^2 (-x^2\alpha^2 + as^2 x^2\alpha^2 - 2 as\, bs\, x^2\alpha^2 + bs^2 x^2\alpha^2 - 2x\alpha\beta + 2 as^2 x\alpha\beta - 4 as\, bs\, x\alpha\beta +$
$2 bs^2 x\alpha\beta - \beta^2 + as^2\beta^2 - 2 as\, bs\,\beta^2 + bs^2\beta^2 + 4 as\, bs\, x^2\alpha\gamma + 2 cs\, x^2\alpha\gamma - 2 as\, cs\, x^2\alpha\gamma -$
$2 bs\, cs\, x^2\alpha\gamma + 4 as\, bs\, x\beta\gamma + 2 cs\, x\beta\gamma - 2 as\, cs\, x\beta\gamma - 2 bs\, cs\, x\beta\gamma - 2 cs\, x^2\gamma^2 +$
$cs^2 x^2\gamma^2 + 4 as\, bs\, x\alpha\delta + 2 cs\, x\alpha\delta - 2 as\, cs\, x\alpha\delta - 2 bs\, cs\, x\alpha\delta + 4 as\, bs\,\beta\delta +$
$2 cs\,\beta\delta - 2 as\, cs\,\beta\delta - 2 bs\, cs\,\beta\delta - 4 cs\, x\gamma\delta + 2 cs^2 x\gamma\delta - 2 cs\,\delta^2 + cs^2\delta^2)) /$
$(4(x\alpha + \beta)^2 (x\alpha + \beta - x\gamma - \delta)^2 (x\gamma + \delta)^2)$

These suggest that it would be easiest to begin the computation with the terms that do not have all of the variables, for example, the denominators.

`In[23]:=` `h1 = Together[h1];`

`eqDen1 = Denominator[g1] - auxFac Denominator[h1]`

`Out[24]=` $4(-1+x)^2 x^2 - 4\,auxFac\,(x\alpha + \beta)^2 (x\alpha + \beta - x\gamma - \delta)^2 (x\gamma + \delta)^2$

Now, we have only seven equations to be solved.

`In[25]:=` `(eqDen2 = CoefficientList[eqDen1, x]) // Short[#, 12]&`

`Out[25]//Short=` $\{-4\,auxFac\,\beta^4\delta^2 + 8\,auxFac\,\beta^3\delta^3 - 4\,auxFac\,\beta^2\delta^4, -8\,auxFac\,\beta^4\gamma\delta - 16\,auxFac\,\alpha\beta^3\delta^2 +$
$24\,auxFac\,\beta^3\gamma\delta^2 + 24\,auxFac\,\alpha\beta^2\delta^3 - 16\,auxFac\,\beta^2\gamma\delta^3 - 8\,auxFac\,\alpha\beta\delta^4,$
$4 - 4\,auxFac\,\beta^4\gamma^2 - 32\,auxFac\,\alpha\beta^3\gamma\delta + 24\,auxFac\,\beta^3\gamma^2\delta - 24\,auxFac\,\alpha^2\beta^2\delta^2 +$
$72\,auxFac\,\alpha\beta^2\gamma\delta^2 - 24\,auxFac\,\beta^2\gamma^2\delta^2 + 24\,auxFac\,\alpha^2\beta\delta^3 - 32\,auxFac\,\alpha\beta\gamma\delta^3 - 4\,auxFac\,\alpha^2\delta^4,$
$-8 - 16\,auxFac\,\alpha\beta^3\gamma^2 + 8\,auxFac\,\beta^3\gamma^3 - 48\,auxFac\,\alpha^2\beta^2\gamma\delta + 72\,auxFac\,\alpha\beta^2\gamma^2\delta - 16\,auxFac\,\beta^2\gamma^3\delta -$
$16\,auxFac\,\alpha^3\beta\delta^2 + 72\,auxFac\,\alpha^2\beta\gamma\delta^2 - 48\,auxFac\,\alpha\beta\gamma^2\delta^2 + 8\,auxFac\,\alpha^3\delta^3 - 16\,auxFac\,\alpha^2\gamma\delta^3,$
$4 - 24\,auxFac\,\alpha^2\beta^2\gamma^2 + 24\,auxFac\,\alpha\beta^2\gamma^3 - 4\,auxFac\,\beta^2\gamma^4 - 32\,auxFac\,\alpha^3\beta\gamma\delta +$
$72\,auxFac\,\alpha^2\beta\gamma^2\delta - 32\,auxFac\,\alpha\beta\gamma^3\delta - 4\,auxFac\,\alpha^4\delta^2 + 24\,auxFac\,\alpha^3\gamma\delta^2 - 24\,auxFac\,\alpha^2\gamma^2\delta^2,$
$-16\,auxFac\,\alpha^3\beta\gamma^2 + 24\,auxFac\,\alpha^2\beta\gamma^3 - 8\,auxFac\,\alpha\beta\gamma^4 - 8\,auxFac\,\alpha^4\gamma\delta +$
$24\,auxFac\,\alpha^3\gamma^2\delta - 16\,auxFac\,\alpha^2\gamma^3\delta, -4\,auxFac\,\alpha^4\gamma^2 + 8\,auxFac\,\alpha^3\gamma^3 - 4\,auxFac\,\alpha^2\gamma^4\}$

Reduce returns the answer immediately. (Here, the value of *auxFac* also becomes determined.)

`In[26]:=` `solDen = Reduce[# == 0& /@ eqDen2, {α, β, γ, δ, auxFac}] // Union`

`Out[26]=` $\left(\alpha == 0\ \&\&\ \gamma == -\beta\ \&\&\ \delta == \beta\ \&\&\ \beta \neq 0\ \&\&\ auxFac == \frac{1}{\beta^6}\right) ||$

$\left(\alpha == 0\ \&\&\ \gamma == \beta\ \&\&\ \delta == 0\ \&\&\ \beta \neq 0\ \&\&\ auxFac == \frac{1}{\beta^6}\right) ||$

$\left(\beta == 0\ \&\&\ \gamma == 0\ \&\&\ \delta == \alpha\ \&\&\ \alpha \neq 0\ \&\&\ auxFac == \frac{1}{\alpha^6}\right) ||$

$\left(\beta == 0\ \&\&\ \gamma == \alpha\ \&\&\ \delta == -\alpha\ \&\&\ \alpha \neq 0\ \&\&\ auxFac == \frac{1}{\alpha^6}\right) ||$

$\left(\beta == -\alpha\ \&\&\ \gamma == 0\ \&\&\ \delta == -\alpha\ \&\&\ \alpha \neq 0\ \&\&\ auxFac == \frac{1}{\alpha^6}\right) ||$

$\left(\beta == -\alpha\ \&\&\ \gamma == \alpha\ \&\&\ \delta == 0\ \&\&\ \alpha \neq 0\ \&\&\ auxFac == \frac{1}{\alpha^6}\right)$

Inserting the solutions calculated into the expression for z, we get the explicit form of the transformations.

`In[27]:=` `variableTrafos = Cancel /@ ((α x + β)/(γ x + δ) /. {ToRules[solDen]})`

`Out[27]=` $\left\{\dfrac{1}{1-x},\ \dfrac{1}{x},\ x,\ \dfrac{x}{-1+x},\ 1-x,\ \dfrac{-1+x}{x}\right\}$

Now, we prepare the three equations resulting from the numerators.

`In[28]:=` `eqNum1 = Numerator[g1] - auxFac Numerator[h1]`

`Out[28]=` $2c - c^2 - 4abx - 2cx + 2acx + 2bcx + x^2 - a^2x^2 + 2abx^2 - b^2x^2 +$
$auxFac\,(-\beta\gamma + \alpha\delta)^2 (-x^2\alpha^2 + as^2 x^2\alpha^2 - 2 as\, bs\, x^2\alpha^2 + bs^2 x^2\alpha^2 - 2x\alpha\beta + 2 as^2 x\alpha\beta -$
$4 as\, bs\, x\alpha\beta + 2 bs^2 x\alpha\beta - \beta^2 + as^2\beta^2 - 2 as\, bs\,\beta^2 + bs^2\beta^2 + 4 as\, bs\, x^2\alpha\gamma + 2 cs\, x^2\alpha\gamma -$
$2 as\, cs\, x^2\alpha\gamma - 2 bs\, cs\, x^2\alpha\gamma + 4 as\, bs\, x\beta\gamma + 2 cs\, x\beta\gamma - 2 as\, cs\, x\beta\gamma - 2 bs\, cs\, x\beta\gamma -$
$2 cs\, x^2\gamma^2 + cs^2 x^2\gamma^2 + 4 as\, bs\, x\alpha\delta + 2 cs\, x\alpha\delta - 2 as\, cs\, x\alpha\delta - 2 bs\, cs\, x\alpha\delta +$
$4 as\, bs\,\beta\delta + 2 cs\,\beta\delta - 2 as\, cs\,\beta\delta - 2 bs\, cs\,\beta\delta - 4 cs\, x\gamma\delta + 2 cs^2 x\gamma\delta - 2 cs\,\delta^2 + cs^2\delta^2)$

In[29]:= `(eqNum2 = CoefficientList[Numerator[Together[eqNum1]], x]) // Short[#, 12]&`

Out[29]//Short= $\{2\,c - c^2 - \text{auxFac}\,\beta^4\,\gamma^2 + \text{as}^2\,\text{auxFac}\,\beta^4\,\gamma^2 - 2\,\text{as}\,\text{auxFac}\,\text{bs}\,\beta^4\,\gamma^2 + \text{auxFac}\,\text{bs}^2\,\beta^4\,\gamma^2 +$
$2\,\text{auxFac}\,\alpha\,\beta^3\,\gamma\,\delta - 2\,\text{as}^2\,\text{auxFac}\,\alpha\,\beta^3\,\gamma\,\delta + 4\,\text{as}\,\text{auxFac}\,\text{bs}\,\alpha\,\beta^3\,\gamma\,\delta - 2\,\text{auxFac}\,\text{bs}^2\,\alpha\,\beta^3\,\gamma\,\delta +$
$4\,\text{as}\,\text{auxFac}\,\text{bs}\,\beta^3\,\gamma^2\,\delta + 2\,\text{auxFac}\,\text{bs}\,\beta^3\,\gamma^2\,\delta - 2\,\text{as}\,\text{auxFac}\,\text{cs}\,\beta^3\,\gamma^2\,\delta - 2\,\text{auxFac}\,\text{bs}\,\text{cs}\,\beta^3\,\gamma^2\,\delta -$
$\text{auxFac}\,\alpha^2\,\beta^2\,\delta^2 + \text{as}^2\,\text{auxFac}\,\alpha^2\,\beta^2\,\delta^2 - 2\,\text{as}\,\text{auxFac}\,\text{bs}\,\alpha^2\,\beta^2\,\delta^2 + \text{auxFac}\,\text{bs}^2\,\alpha^2\,\beta^2\,\delta^2 -$
$8\,\text{as}\,\text{auxFac}\,\text{bs}\,\alpha\,\beta^2\,\gamma\,\delta^2 - 4\,\text{auxFac}\,\text{cs}\,\alpha\,\beta^2\,\gamma\,\delta^2 + 4\,\text{as}\,\text{auxFac}\,\text{cs}\,\alpha\,\beta^2\,\gamma\,\delta^2 +$
$4\,\text{auxFac}\,\text{bs}\,\text{cs}\,\alpha\,\beta^2\,\gamma\,\delta^2 - 2\,\text{auxFac}\,\text{cs}\,\beta^2\,\gamma^2\,\delta^2 + \text{auxFac}\,\text{cs}^2\,\beta^2\,\gamma^2\,\delta^2 +$
$4\,\text{as}\,\text{auxFac}\,\text{bs}\,\alpha^2\,\beta\,\delta^3 + 2\,\text{auxFac}\,\text{cs}\,\alpha^2\,\beta\,\delta^3 - 2\,\text{auxFac}\,\text{cs}\,\alpha^2\,\beta\,\delta^3 - 2\,\text{auxFac}\,\text{bs}\,\text{cs}\,\alpha^2\,\beta\,\delta^3 +$
$4\,\text{auxFac}\,\text{cs}\,\alpha\,\beta\,\gamma\,\delta^3 - 2\,\text{auxFac}\,\text{cs}^2\,\alpha\,\beta\,\gamma\,\delta^3 - 2\,\text{auxFac}\,\text{cs}\,\alpha^2\,\delta^4 + \text{auxFac}\,\text{cs}^2\,\alpha^2\,\delta^4,$
$-4\,a\,b - 2\,c + \ll 52 \gg + 2\,\text{auxFac}\,\text{cs}^2\,\alpha^2\,\gamma\,\delta^3, \; 1 - a^2 + 2\,a\,b - b^2 - \text{auxFac}\,\alpha^2\,\beta^2\,\gamma^2 +$
$\text{as}^2\,\text{auxFac}\,\alpha^2\,\beta^2\,\gamma^2 - 2\,\text{as}\,\text{auxFac}\,\text{bs}\,\alpha^2\,\beta^2\,\gamma^2 + \text{auxFac}\,\text{bs}^2\,\alpha^2\,\beta^2\,\gamma^2 + 4\,\text{as}\,\text{auxFac}\,\text{bs}\,\alpha\,\beta^2\,\gamma^3 +$
$2\,\text{auxFac}\,\text{cs}\,\alpha\,\beta^2\,\gamma^3 - 2\,\text{as}\,\text{auxFac}\,\text{cs}\,\alpha\,\beta^2\,\gamma^3 - 2\,\text{auxFac}\,\text{bs}\,\text{cs}\,\alpha\,\beta^2\,\gamma^3 - 2\,\text{auxFac}\,\text{cs}\,\beta^2\,\gamma^4 +$
$\text{auxFac}\,\text{cs}^2\,\beta^2\,\gamma^4 + \ll 10 \gg + 4\,\text{as}\,\text{auxFac}\,\text{cs}\,\alpha^2\,\beta\,\gamma\,\delta + 4\,\text{auxFac}\,\text{bs}\,\text{cs}\,\alpha^2\,\beta\,\gamma\,\delta +$
$4\,\text{auxFac}\,\text{cs}\,\alpha\,\beta\,\gamma^3\,\delta - 2\,\text{auxFac}\,\text{cs}^2\,\alpha\,\beta\,\gamma^3\,\delta - \text{auxFac}\,\alpha^4\,\delta^2 + \text{as}^2\,\text{auxFac}\,\alpha^4\,\delta^2 -$
$2\,\text{as}\,\text{auxFac}\,\text{bs}\,\alpha^4\,\delta^2 + \text{auxFac}\,\text{bs}^2\,\alpha^4\,\delta^2 + 4\,\text{as}\,\text{auxFac}\,\text{bs}\,\alpha^3\,\gamma\,\delta^2 + 2\,\text{auxFac}\,\text{cs}\,\alpha^3\,\gamma\,\delta^2 -$
$2\,\text{as}\,\text{auxFac}\,\text{cs}\,\alpha^3\,\gamma\,\delta^2 - 2\,\text{auxFac}\,\text{bs}\,\text{cs}\,\alpha^3\,\gamma\,\delta^2 - 2\,\text{auxFac}\,\text{cs}\,\alpha^2\,\gamma^2\,\delta^2 + \text{auxFac}\,\text{cs}^2\,\alpha^2\,\gamma^2\,\delta^2\}$

For every one of the six solutions of solDen, we join them with the equations eqNum2 and calculate $a' = a'(a, b, c)$, $b' = b'(a, b, c)$, and $c' = c'(a, b, c)$. Because the hypergeometric differential equation (and so also $g(x)$) is symmetric in a and b, a' and b', respectively, we sort out solutions that result from this symmetry in the following calculation. In this manner, we arrive at the following 24 solutions of $g(x) = z'(x)^2\,g(z(x))$.

In[30]:= `Clear[z];`

```
allSolutions = Flatten[Table[
(* join numerator equations with denominator equations *)
eq1 = Join[# == 0& /@ eqNum2, List @@ solDen[[i]]];
(* solve for as, bs and cs *)
preSol = (Or @@ (And @@@ Apply[Equal, Solve[eq1, {as, bs, cs},
                  {auxFac, α, β, γ, δ}], {2}])) // Union;
(* sort out solutions that result from
    a ↔ b or as ↔ bs symmetry *)
bag = {Sort @ First[preSol]}; rest = Sort /@ Rest[List @@ preSol];
Do[If[MemberQ[bag, (Sort @ (First[rest] /. {as -> bs, bs -> as})) |
              (Sort @ (First[rest] /. {a  -> b, b  -> a}))],
      Null, AppendTo[bag, Sort @ First[rest]]]; rest = Rest[rest],
   {Length[preSol] - 1}];
   (* prepare solutions in form of Rules *)
Map[Prepend[#, z -> variableTrafos[[i]]]&, {ToRules[Or @@ bag]}],
                    {i, Length[solDen]}], 1];
```

In[32]:= `TableForm[%]`

Out[32]//TableForm=

$z \to \frac{1}{1-x}$	$\text{as} \to 1 - a$	$\text{bs} \to 1 + b - c$	$\text{cs} \to 1 - a + b$
$z \to \frac{1}{1-x}$	$\text{as} \to a$	$\text{bs} \to -b + c$	$\text{cs} \to 1 + a - b$
$z \to \frac{1}{1-x}$	$\text{as} \to 1 + a - c$	$\text{bs} \to 1 - b$	$\text{cs} \to 1 + a - b$
$z \to \frac{1}{1-x}$	$\text{as} \to -a + c$	$\text{bs} \to b$	$\text{cs} \to 1 - a + b$
$z \to \frac{1}{x}$	$\text{as} \to 1 - a$	$\text{bs} \to -a + c$	$\text{cs} \to 1 - a + b$
$z \to \frac{1}{x}$	$\text{as} \to a$	$\text{bs} \to 1 + a - c$	$\text{cs} \to 1 + a - b$
$z \to \frac{1}{x}$	$\text{as} \to 1 + b - c$	$\text{bs} \to b$	$\text{cs} \to 1 - a + b$
$z \to \frac{1}{x}$	$\text{as} \to -b + c$	$\text{bs} \to 1 - b$	$\text{cs} \to 1 + a - b$
$z \to x$	$\text{as} \to 1 - a$	$\text{bs} \to 1 - b$	$\text{cs} \to 2 - c$
$z \to x$	$\text{as} \to a$	$\text{bs} \to b$	$\text{cs} \to c$
$z \to x$	$\text{as} \to 1 + a - c$	$\text{bs} \to 1 + b - c$	$\text{cs} \to 2 - c$
$z \to x$	$\text{as} \to -a + c$	$\text{bs} \to -b + c$	$\text{cs} \to c$
$z \to \frac{x}{-1+x}$	$\text{as} \to 1 - a$	$\text{bs} \to 1 + b - c$	$\text{cs} \to 2 - c$
$z \to \frac{x}{-1+x}$	$\text{as} \to a$	$\text{bs} \to -b + c$	$\text{cs} \to c$
$z \to \frac{x}{-1+x}$	$\text{as} \to 1 + a - c$	$\text{bs} \to 1 - b$	$\text{cs} \to 2 - c$

$z \to \frac{x}{-1+x}$	as $\to -a + c$	bs $\to b$	cs $\to c$
$z \to 1 - x$	as $\to 1 - a$	bs $\to 1 - b$	cs $\to 1 - a - b + c$
$z \to 1 - x$	as $\to a$	bs $\to b$	cs $\to 1 + a + b - c$
$z \to 1 - x$	as $\to 1 + a - c$	bs $\to 1 + b - c$	cs $\to 1 + a + b - c$
$z \to 1 - x$	as $\to -a + c$	bs $\to -b + c$	cs $\to 1 - a - b + c$
$z \to \frac{-1+x}{x}$	as $\to 1 - a$	bs $\to -a + c$	cs $\to 1 - a - b + c$
$z \to \frac{-1+x}{x}$	as $\to a$	bs $\to 1 + a - c$	cs $\to 1 + a + b - c$
$z \to \frac{-1+x}{x}$	as $\to 1 + b - c$	bs $\to b$	cs $\to 1 + a + b - c$
$z \to \frac{-1+x}{x}$	as $\to -b + c$	bs $\to 1 - b$	cs $\to 1 - a - b + c$

Now, reversing the transformations of the normal form x to z and constructing the normal form of the original equation in the solutions, we arrive at the following 24 solutions of the hypergeometric differential equation (Kummer's 24 solutions [605]).

```
In[33]:= ((PowerExpand[Together //@
            (* transform back to solution original differential equation for 2F1 *)
            (Unevaluated[Hypergeometric2F1[as, bs, cs, x]*
                    x^+(cs/2)(1 - x)^-((cs - as - bs - 1)/2)*
                    z^-(c/2)(1 - z)^+((c - a - b - 1)/2)/Sqrt[D[x, z]]] /. #)] /.
            (* change (z - 1)^exp to the form (1 - z)^exp
               to be uniform and to have standard form;
               ignore z-independent phase factor *)
            {(z - 1)^exp_ -> (1 - z)^exp} /.
            (* expand exponents to make them collapse down *)
            {(zz_)^(exp_) :> zz^Expand[exp]}) //.
            (* get rid of prefactors not containing z *)
            {numFac_?(FreeQ[#, z]&) zFacs___ hyp_Hypergeometric2F1 ->
                zFacs hyp})& /@ (allSolutions /. {z -> x, x -> z}) //
                            TableForm[#, TableSpacing -> {3}]&
```

Out[33]//TableForm=

$(1 - z)^{-1-b+c} z^{1-c}$ Hypergeometric2F1$[1 - a, 1 + b - c, 1 - a + b, \frac{1}{1-z}]$

$(1 - z)^{-a}$ Hypergeometric2F1$[a, -b + c, 1 + a - b, \frac{1}{1-z}]$

$(1 - z)^{-1-a+c} z^{1-c}$ Hypergeometric2F1$[1 + a - c, 1 - b, 1 + a - b, \frac{1}{1-z}]$

$(1 - z)^{-b}$ Hypergeometric2F1$[-a + c, b, 1 - a + b, \frac{1}{1-z}]$

$(1 - z)^{-a-b+c} z^{a-c}$ Hypergeometric2F1$[1 - a, -a + c, 1 - a + b, \frac{1}{z}]$

z^{-a} Hypergeometric2F1$[a, 1 + a - c, 1 + a - b, \frac{1}{z}]$

z^{-b} Hypergeometric2F1$[1 + b - c, b, 1 - a + b, \frac{1}{z}]$

$(1 - z)^{-a-b+c} z^{b-c}$ Hypergeometric2F1$[-b + c, 1 - b, 1 + a - b, \frac{1}{z}]$

$(1 - z)^{-a-b+c} z^{1-c}$ Hypergeometric2F1$[1 - a, 1 - b, 2 - c, z]$

Hypergeometric2F1$[a, b, c, z]$

z^{1-c} Hypergeometric2F1$[1 + a - c, 1 + b - c, 2 - c, z]$

$(1 - z)^{-a-b+c}$ Hypergeometric2F1$[-a + c, -b + c, c, z]$

$(1 - z)^{-1-b+c} z^{1-c}$ Hypergeometric2F1$[1 - a, 1 + b - c, 2 - c, \frac{z}{1-z}]$

$(1 - z)^{-a}$ Hypergeometric2F1$[a, -b + c, c, \frac{z}{1-z}]$

$(1 - z)^{-1-a+c} z^{1-c}$ Hypergeometric2F1$[1 + a - c, 1 - b, 2 - c, \frac{z}{1-z}]$

$(1 - z)^{-b}$ Hypergeometric2F1$[-a + c, b, c, \frac{z}{1-z}]$

$(1 - z)^{-a-b+c} z^{1-c}$ Hypergeometric2F1[1 - a, 1 - b, 1 - a - b + c, 1 - z]

Hypergeometric2F1[a, b, 1 + a + b - c, 1 - z]

z^{1-c} Hypergeometric2F1[1 + a - c, 1 + b - c, 1 + a + b - c, 1 - z]

$(1 - z)^{-a-b+c}$ Hypergeometric2F1[-a + c, -b + c, 1 - a - b + c, 1 - z]

$(1 - z)^{-a-b+c} z^{a-c}$ Hypergeometric2F1[1 - a, -a + c, 1 - a - b + c, $\frac{-1+z}{z}$]

z^{-a} Hypergeometric2F1[a, 1 + a - c, 1 + a + b - c, $\frac{-1+z}{z}$]

z^{-b} Hypergeometric2F1[1 + b - c, b, 1 + a + b - c, $\frac{-1+z}{z}$]

$(1 - z)^{-a-b+c} z^{b-c}$ Hypergeometric2F1[-b + c, 1 - b, 1 - a - b + c, $\frac{-1+z}{z}$]

For the "automatic" derivation of more complicated hypergeometric identities based on the differential equation of $_2F_1(a, b; c; z)$, see [720]. For q-analogues of these equations, see [328].

b) These are the two differential equations obeyed by $F_1(a; b_1, b_2; c; z_1, z_2)$. (We use the symbol F1 instead of the built-in symbol AppellF1 to avoid the evaluation of the derivatives.)

```
In[1]:= eqs0 = With[{w = F1[z1, z2]},
         {z1 (1 - z1) D[w, z1, z1] + z2 (1 - z1) D[w, z1, z2] +
          (c - (1 + a + b1) z1) D[w, z1] - b1 z2 D[w, z2] - a b1 w,
          z2 (1 - z2) D[w, z2, z2] + z1 (1 - z2) D[w, z1, z2] +
          (c - (1 + a + b2) z2) D[w, z2] - b2 z1 D[w, z1] - a b2 w}]
```

```
Out[1]= {-a b1 F1[z1, z2] - b1 z2 F1^(0,1)[z1, z2] + (c - (1 + a + b1) z1) F1^(1,0)[z1, z2] +
         (1 - z1) z2 F1^(1,1)[z1, z2] + (1 - z1) z1 F1^(2,0)[z1, z2],
         -a b2 F1[z1, z2] + (c - (1 + a + b2) z2) F1^(0,1)[z1, z2] + (1 - z2) z2 F1^(0,2)[z1, z2] -
         b2 z1 F1^(1,0)[z1, z2] + z1 (1 - z2) F1^(1,1)[z1, z2]}
```

To derive an ordinary differential equation with respect to z_1, we must eliminate all terms containing derivatives with respect to z_2. So we need more equations than just the two equations eqs0 to eliminate the three derivatives with respect to z_2. Differentiating the two differential equations eqs0 repeatedly with respect to z_1 and z_2 gives us more equations (and more derivatives). The pair of partial differential equations is of second order, so we expect the ordinary differential equation to be at least of order three. eqs1 is a list of all second and third order equations following from eqs0.

```
In[2]:= eqs1 = {eqs0[[1]], D[eqs0[[1]], z1],
               eqs0[[2]], D[eqs0[[2]], z1], D[eqs0[[1]], z2],
               D[eqs0[[2]], z1, z1], D[eqs0[[1]], z1, z2]};
```

These are the seven derivatives with respect to z_2 that we must eliminate.

```
In[3]:= Union[Cases[eqs1, Derivative[_, _?(# =!= 0&)][F1][_, _], Infinity]]
```

```
Out[3]= {F1^(0,1)[z1, z2], F1^(0,2)[z1, z2], F1^(1,1)[z1, z2],
         F1^(1,2)[z1, z2], F1^(2,1)[z1, z2], F1^(2,2)[z1, z2], F1^(3,1)[z1, z2]}
```

Because in intermediate stages of the elimination process some equations might factor nontrivially, we do not use a built-in function for the elimination, but the following function eliminate. eliminate[*eqsOld*, *eqNumber*, *var*] eliminates the variable *var* from the set of equations *eqsOld* by using the *eqNumber*th equation from *eqsOld*.

```
In[4]:= eliminate[eqsOld_, eqNumber_, var_] :=
        Factor[Numerator[Together[Delete[eqsOld, eqNumber] /.
          Solve[eqsOld[[eqNumber]] == 0, var][[1]]]]] /.
        (* remove not relevant factors *)
        _Plus?(FreeQ[#, F1, {0, Infinity}, Heads -> True]&) -> 1
```

Applying now eliminate repeatedly and starting with eqs1 yields after six steps the ordinary differential equation we were looking for. (In the first elimination step, two derivatives are eliminated at once.)

```
In[5]:= eqsT = eqs1;
        While[(derivativesToEliminate =
               Union[Cases[eqsT, Derivative[_, _?(# =!= 0&)][F1][_, _],
                 Infinity]]) =!= {},
          (* next derivative to eliminate *)
```

```
             nextDerivative = derivativesToEliminate[[-1]];
             (* new equations *)
             eqsT = eliminate[eqsT, Union[First /@
                        Position[eqsT, nextDerivative]][[-1]],
                    nextDerivative]]
```

Collecting and simplifying terms yields the following linear third order differential equation for $F_1(a; b_1, b_2; c; z_1, z_2)$.

```
In[7]:= (ode = Collect[eqsT, Table[D[w[z1, z2], {z1, k}], {k, 0, 3}],
                    FullSimplify]) /.
        {Derivative[d1_, 0][w][z1, z2] :> Derivative[d1][W][z1],
         w[z1, z2] -> W[z1]} /. (* for a nicer-looking result *)
                    {b1 -> Subscript[b, 1], b2 -> Subscript[b, 2],
                     z1 -> Subscript[z, 1], z2 -> Subscript[z, 2],
                     F1 -> Subscript[F, 1]} //  TraditionalForm
```

Out[7]//TraditionalForm= $\{a\,b_1\,(b_1+1)\,F_1(z_1, z_2) + (b_1+1)\,(-c + (2a+b_1+2)\,z_1 + (-a+b_2-1)\,z_2)\,F_1^{(1,0)}(z_1, z_2) +$
$((a + 2\,b_1 + 4)\,z_1^2 - (c + b_1 + (a + b_1 - b_2 + 3)\,z_2 + 2)\,z_1 + (c - b_2 + 1)\,z_2)\,F_1^{(2,0)}(z_1, z_2) + (z_1 - 1)\,z_1\,(z_1 - z_2)\,F_1^{(3,0)}(z_1, z_2)\}$

Here is a quick numerical check for the correctness of the derived differential equation.

```
In[8]:= SeedRandom[111];
        ode /. {Derivative[d1_, d2_][F1][z1, z2] :>
           D[AppellF1[a, b1, b2, c, z1, z2], {z1, d1}, {z2, d2}],
              F1[z1, z2] :> AppellF1[a, b1, b2, c, z1, z2]} /.
           (* numerical values for parameters *)
           Rule @@@ Transpose[{{a, b1, b2, c, z1, z2},
                    Table[Random[Real, {-1/2, 1/2}, 50], {6}]}]]
```

Out[9]= $\{0. \times 10^{-40}\}$

18. Roots of Differentiated Polynomials

a) The implementation is straightforward: We choose a random polynomial, calculate its roots, differentiate it, calculate its roots, ... For the calculation of the convex hull, we use the command `ConvexHull` from the package `DiscreteMath\`ComputationalGeometry\``. It returns an ordered list with points that span the convex hull.

```
In[1]:= Needs["DiscreteMath`ComputationalGeometry`"]
```

```
In[2]:= GaussLucasTheoremVisualization[ord_, maxCoeff_, opts___] :=
        Show[Graphics[Line[Append[#, First[#]]]& /@
        (Function[points, points[[#]]]& /@
        (* making the convex hull *) ConvexHull[points]] /@
        (Map[{Re[#], Im[#]}&, #]& /@ ((#[[1, 2]]& /@ (List @@
        (* the zeros *)
        NSolve[# == 0, C]))& /@ (* all derivatives of the polynomial *)
        Function[poly, Array[D[poly,
        {C, #}]&, ord - 1, 0]][(* the randomly chosen polynomial *)
        Array[C^#&, ord].Array[Random[Head[maxCoeff],
                                  {-maxCoeff, maxCoeff}]&, ord]]))),
         PlotRange -> All, AspectRatio -> Automatic,
         Axes -> False, Frame -> True, opts]
```

Here are two examples with different coefficient ranges.

```
In[3]:= Show[GraphicsArray[GaussLucasTheoremVisualization[12, #,
                    DisplayFunction -> Identity]& /@ {3, 5 + 5I}]]
```

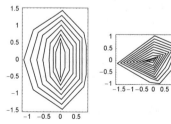

Note that polynomials, although in some sense the simplest of all functions, show a rich behavior in many respects. A couple of other, interesting "root-pictures" related to the above images could be produced by using, for instance, a visualization of Sturm's theorem on the number of real roots of a polynomial in a given interval via [1212].

```
SturmPicture[poly_, var_] :=
Module[{localPoly, x, degree, polyTab, zerosTab, mi, ma, intervalTab,
        polyFuncTab, intervalMidPointsTab, negPositionsTab, negRectangles},
  (* a local version of the polynomial *)
  localPoly = Expand[poly //. var -> x];
  (* degree of the polynomial *)
  degree = Exponent[localPoly, x];
  (* the derivatives up to degree -1 of the polynomial *)
  polyTab = Table[D[localPoly, {x, i}], {i, 0, degree - 1}];
  (* the zeros of all derivatives *)
  zerosTab = Union[Cases[NRoots[# == 0, x], _Real, {-1}]]& /@ polyTab;
  (* x-interval in which all sign changes are happening *)
  {mi, ma} = {-1, 1} Abs[Subtract @@ #]/8 +
                      #&[{Min[#], Max[#]}&[Flatten[zerosTab]]];
  (* intervals in which the polynomials have equal signs *)
  intervalTab = If[# === {}, {}, Partition[Join[{mi}, #, {ma}], 2, 1]]& /@
                                                           zerosTab;
  (* all polynomials as Functions *)
  polyFuncTab = Apply[Function[##, Listable]&, {x, #}& /@ polyTab, {1}];
  (* the midpoints of all intervals *)
  intervalMidPointsTab = Apply[Plus[##]/2&, intervalTab, {2}];
  (* position of the intervals in which the polynomials are negative *)
  negPositionsTab = Flatten[Position[#, _?(# < 0&)]]& /@
  (#[[1]][#[[2]]]& /@ Transpose[{polyFuncTab, intervalMidPointsTab}]);
  (* the intervals in which the polynomials are negative *)
  negIntervalsTab = Apply[Part, Transpose[{intervalTab, negPositionsTab}], {1}];
  (* black rectangles in which the polynomials are negative *)
  negRectangles = MapIndexed[Rectangle[{#[[1]], #2[[1]] - 1}, {#[[2]], #2[[1]]}]&,
                             negIntervalsTab, {2}];
  (* show all intervals in which the polynomials are negative *)
  Graphics[negRectangles,
       PlotRange -> {{mi, ma}, {0, degree + 1}}, Frame -> True]];

(* an example *)
SturmPicture[Product[y - i, {i, -20, 20}], y]
```

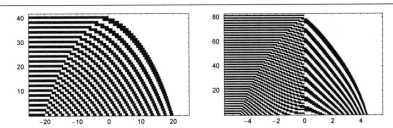

b) Let us start with the roots of a random polynomial, the roots of the polynomial differentiated, the roots of the polynomial differentiated twice, and so on. In the case of real coefficients, the picture has the obvious mirror symmetry between the lower and the upper half-planes, which is generically lost in the case of complex coefficients.

```
In[1]:= SeedRandom[999];

        Show @ GraphicsArray @ (Function[coeffs,
        Graphics[MapIndexed[{PointSize[#2[[1]]/Length[coeffs] 0.02], #1}&,
            (* points at the roots *)
            Map[Point[{Re[#[[1, 2]]], Im[#[[1, 2]]]}]}&, (NSolve[# == 0, x]& /@
```

```
(NestList[(* differentiate the polynomials *) N[D[#, x]]&,
          (* form the polynomials *)
          coeffs.Array[x^#&, Length[coeffs], 0],
                            Length[coeffs] - 2])), {2}]],
  PlotRange -> All, AspectRatio -> Automatic, Axes -> False,
    Frame -> True]] /@ (* two random lists of coefficients *)
    {Array[Random[Integer, {-8, 8}]&, 25],
     Array[Random[Complex, {-12 - 12I, 12 + 12I}]&, 25]})
```

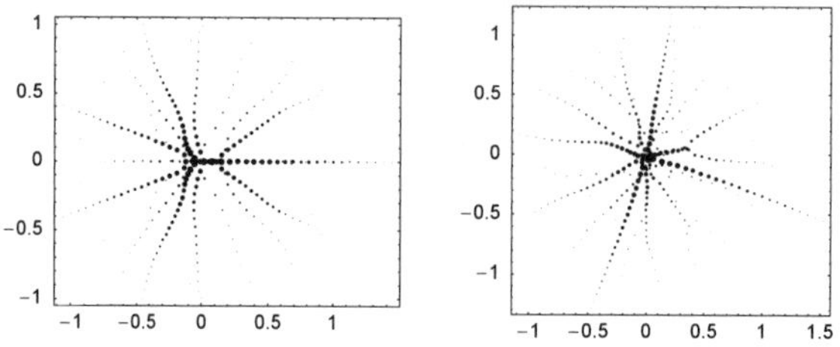

The roots seem to lie on curves. To connect them in the "right" order, we would have to know the expression $d^\alpha polynomial(x)/dx^\alpha$. Taking into account only powers of the independent variable and using linearity, we should look for a continuous form of $d^\alpha x^n/dx^\alpha$. Looking at integer values of α, one obvious possibility is as follows: $d^\alpha x^n/dx^\alpha = n!/(n-\alpha)!\,x^{n-\alpha}$. (Here Pochhammer functions come into play, which is why this exercise belongs in this chapter.)

Interpreting $\alpha!$ not only for integer α (which *Mathematica* does automatically, because every function (if possible) is defined for arbitrary complex arguments), we have the desired generalization. (Such generalizations of the order of differentiation go back to Liouville; for a detailed discussion, see [1101], [1180], [255], [957], [884], [1061], [641], [404], [643], [891], and [765]; for applications in physics, see [595], [431], [321], [367], [831], [1072], [1169], [1025], [1115], [1108], [697], [1328], [578], [433], [655], [95], [694], [721], [1082], [1085], [175], [883], [594], [976], [55], [1014], [1092], [73], [261], [4], [763], [430], [593], [466], [542], [432], [523], [1040], and [313], for the relation between special functions and fractional differentiation, see [679], [1327], and [1078]. For fractional differences, see [890], [391], [927], [534], [596], and [928]. For the differentiation of discrete functions, see [358].)

The following picture shows how this formula interpolates between integer values of α. The black lines represent x^3 differentiated and integrated an integral number of times.

```
In[3]:= With[{(* n in x^n *) n = 3}, Show[
        (* the x^3 n - times differentiated *)
        {Plot3D[Re[n!/(n - a)! x^(n - a)], {x, -3, 3}, {a, 7, -2},
                PlotPoints -> {60, 55}, ClipFill -> None,
                MeshStyle -> Thickness[0.002], PlotRange -> {-20, 30},
                DisplayFunction -> Identity],
         Graphics3D[{Thickness[0.01], ParametricPlot3D[
         (* the prefactor*x^m from integer times differentiating *)
         Evaluate[# + {0, 0, 0.3}& /@
                 {{x, 0, x^3}, {x, 1, 3x^2}, {x, 2, 6x}, {x, 3, 6},
                  {x, 4, 0}, {x, -1, x^4/4}, {x, -2, x^5/20}, {x, 5, 0},
                  {x, 6, 0}, {x, 7, 0}}], {x, -3, 3},
         DisplayFunction -> Identity][[1]]]}],
        DisplayFunction -> $DisplayFunction, AxesLabel -> {"x", "a", None}]]
```

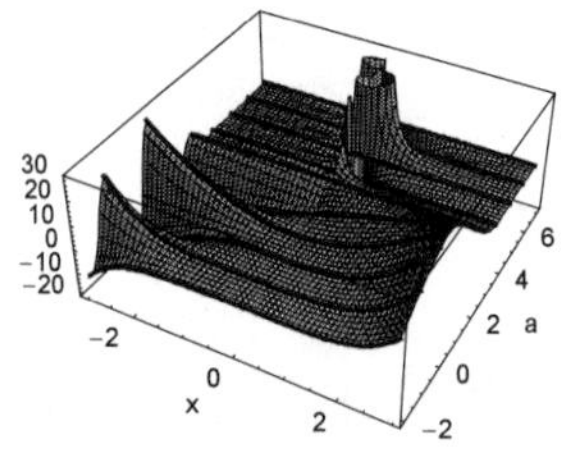

Using the above relation for every term of a polynomial, we can define a polynomial differentiated a noninteger number of times. (Because we are interested only in the zeros of such expressions, we use Cancel[...] to pull out the term y^α, which can never be zero.)

```
In[4]:= DifferentiatedPolynomial[poly_, var_, y_, α_] :=
        Cancel[# y^α]& /@ Plus @@ MapIndexed[#1 y^(#2[[1]] - 1 - α)
                               (#2[[1]] - 1)!/(#2[[1]] - 1 - α)!&,
                                        CoefficientList[poly, var]]
```

Here is an example.

```
In[5]:= DifferentiatedPolynomial[a1 x^5 + a2 x^3 - a3 x^2 - a4, x, x, a]
```

$$Out[5]= -\frac{2\,a3\,x^2}{(2-a)!} + \frac{6\,a2\,x^3}{(3-a)!} + \frac{120\,a1\,x^5}{(5-a)!} - \frac{a4}{(-a)!}$$

This is iterated here.

```
In[6]:= NestList[DifferentiatedPolynomial[#, x, x, a]&,
                 a1 x^5 + a2 x^3 - a3 x^2 - a4, 4] // TableForm
```

$$Out[6]//TableForm=$$
$$-a4 - a3\,x^2 + a2\,x^3 + a1\,x^5$$
$$-\frac{2\,a3\,x^2}{(2-a)!} + \frac{6\,a2\,x^3}{(3-a)!} + \frac{120\,a1\,x^5}{(5-a)!} - \frac{a4}{(-a)!}$$
$$-\frac{4\,a3\,x^2}{((2-a)!)^2} + \frac{36\,a2\,x^3}{((3-a)!)^2} + \frac{14400\,a1\,x^5}{((5-a)!)^2} - \frac{a4}{((-a)!)^2}$$
$$-\frac{8\,a3\,x^2}{((2-a)!)^3} + \frac{216\,a2\,x^3}{((3-a)!)^3} + \frac{1728000\,a1\,x^5}{((5-a)!)^3} - \frac{a4}{((-a)!)^3}$$
$$-\frac{16\,a3\,x^2}{((2-a)!)^4} + \frac{1296\,a2\,x^3}{((3-a)!)^4} + \frac{207360000\,a1\,x^5}{((5-a)!)^4} - \frac{a4}{((-a)!)^4}$$

To avoid spurious lines in the picture (without changing the topology of the actual result in the sense of which root of the polynomial $p(x)$ is continuously "connected" with which root of the polynomial $p'(x)$), we drop all terms in which $\alpha > n+1$, n being a positive integer, by defining a ReducedGeneralizedPolynomial. These terms vanish at negative integer values of $n - \alpha$, but are not equal to 0 between these values. Omitting these terms simplifies the ongoing numerical tasks.

```
In[7]:= ReducedDifferentiatedPolynomial[poly_, var_, y_, α_, n_] :=
        Drop[CoefficientList[#, y], n].Array[y^#&,  (* avoid y -> 0 solutions *)
                     Exponent[#, y] + 1 - n, 0]&[
                       DifferentiatedPolynomial[poly, var, y, α]]
```

```
In[8]:= Array[ReducedDifferentiatedPolynomial[
              a1 x^5 + a2 x^3 - a3 x^2 - a4, x, x, a, #]&, 5, 0] // TableForm
```

$$Out[8]//TableForm=$$
$$-\frac{2\,a3\,x^2}{(2-a)!} + \frac{6\,a2\,x^3}{(3-a)!} + \frac{120\,a1\,x^5}{(5-a)!} - \frac{a4}{(-a)!}$$
$$-\frac{2\,a3\,x}{(2-a)!} + \frac{6\,a2\,x^2}{(3-a)!} + \frac{120\,a1\,x^4}{(5-a)!}$$
$$-\frac{2\,a3}{(2-a)!} + \frac{6\,a2\,x}{(3-a)!} + \frac{120\,a1\,x^3}{(5-a)!}$$
$$\frac{6\,a2}{(3-a)!} + \frac{120\,a1\,x^2}{(5-a)!}$$
$$\frac{120\,a1\,x}{(5-a)!}$$

Because these so-calculated expressions are again polynomials in the unknown, we use NSolve for generating a dense set of zeros between integer values of the number of differentiations. (We could, of course, also use a less-dense set and apply Interpolation.) Here, we implement this. (The routine ZerosOfDifferentiatedPolynomialGraphics shows the roots of the polynomial differentiated an integer number of times in big points and the roots of the polynomial differentiated a nonintegral number of times in small points.)

```
In[9]:= ZerosOfDifferentiatedPolynomialGraphics[poly_, var_, steps_] :=
        Module[{a},
               Show[Graphics[{{PointSize[0.008],
                 Table[Point[{Re[#], Im[#]}]& /@ Cases[
                 (* integer times differentiated — the bigger dots *)
                 NRoots[D[poly, {var, i}] == 0, var], _?NumberQ, {-1}],
                                {i, 0, Exponent[poly, var] - 1}]},
                 {PointSize[0.003], Table[Function[polya, Table[Point[{Re[#],
                 Im[#]}]]& /@ (* solve the polynomial *)
                 Cases[NRoots[N[polya] == 0, var], _?NumberQ, {-1}],
                      {a, i, i + 1. - 10^-6, (* between two integers *)
                 (1. - 10^-6)/steps}]][
```

```
                        (* the polynomial between two integer times differentiated *)
                        ReducedDifferentiatedPolynomial[poly, var, var, a, i]],
                                              {i, 0, Exponent[poly, var] - 2}]}}],
                        PlotRange -> All, Frame -> True, Axes -> False,
                        AspectRatio -> Automatic]]
```

Here is an example.

```
In[10]:= ZerosOfDifferentiatedPolynomialGraphics[
             8 z^12 - 6 z^11 + z^10 - 4 z^9 - z^8 - 3 z^7 + 8 z^6 +
             6 z^5 + 4 z^4 - z^3 + z^2 + 5 z - 7, z, 120]
```

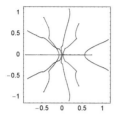

Now, we recognize which roots "belong together". Another possibility would have been to calculate the paths of the roots via the solution of a nonlinear first-order differential equation. Such an ordinary differential equation could be easily derived from the corresponding polynomial ReducedDifferentiatedPolynomial[...]. By starting at the known values of the roots of polynomials differentiated an integral number of times, we could solve the differential equations numerically. In this case, care has to be taken when dividing by 0 and singular points of the differential equation. For just seeing which root path belongs to which, the above pointwise approach is appropriate. For a more detailed picture of the paths of the roots, the differential equation approach is more suitable. In addition, the differential equation approach does easily allow for a complex-valued order of differentation. (For the averaging properties of multiple differentiation, see [451].)

Note that this picture looks different from what one would get by a linear interpolation (the use of another interpolation does not change the picture dramatically) of the coefficients between *polynomial*(z) and *polynomial*'(z).

```
In[11]:= Module[{p, po, a},
                (* the original polynomial *)
                p[0] = 8 x^12 - 6 x^11 + x^10 - 4 x^9 - x^8 - 3 x^7 +
                       8 x^6 + 6 x^5 + 4 x^4 - x^3 + x^2 + 5 x - 7;
                    (* the polynomial p[0] differentiated i times *)
                Do[p[i] = D[p[0], {x, i}], {i, 12}];
                    (* linear interpolations between i times and
                       (i + 1)-times differentiated polynomial *)
                Do[po[i, a_] = ((1 - a) CoefficientList[p[i], x] +
                            a Append[CoefficientList[p[i + 1], x], 0]).
                            Table[x^j, {j, 0, 12 - i}], {i, 0, 11}];
            Show[Graphics[(* the roots of an integer times differentiated *)
                {{PointSize[0.008], Map[Point[{Re[#], Im[#]}]&,
                  Table[Cases[NRoots[p[i] == 0, x],
                        _?NumberQ, {-1}], {i, 0, 11}], {-1}]},
                    (* the roots of the interpolants *)
                {PointSize[0.003], Map[Point[{Re[#], Im[#]}]&,
                  Table[Cases[NRoots[po[i, a] == 0, x], _?NumberQ, {-1}],
                      {i, 0, 11}, {a, 0, 1, 1/250}], {-1}]}}],
                        Frame -> True, AspectRatio -> Automatic]]
```

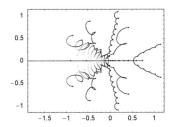

19. Coinciding Bessel Zeros, π-Formulas

a) We search for v, μ, x, z by starting from random values until we find such values that the two identities hold. We might get some messages FindMinimum::regex if no minimum can be found within the specified search region.

```
In[1]:= r := Random[Real, {0, 100}];
```

```
In[2]:= SeedRandom[1111];
       While[Check[(* search for minimum; catch messages *)
                   Not[FindMinimum[Re[(BesselJ[v, x] - BesselJ[μ, x])^2 +
                                      (BesselJ[v, z] - BesselJ[μ, z])^2],
                                   Evaluate[Sequence @@
                                   (pre = {#, r, r, 0, 100}& /@
                                    {v, μ, x, z})]]][[1]] < 10^-10], True],
           Null];
       FindMinimum::regex :
         Reached the point {100.398, 68.8359, 83.3, 91.0552} which is outside
            the region {{0., 0., 0., 0.}, {100., 100., 100., 100.}}. More…
```

Here are the starting values.

```
In[4]:= pre
```

```
Out[4]= {{v, 2.87865, 64.7165, 0, 100}, {μ, 70.0279, 4.94753, 0, 100},
        {x, 64.7728, 15.3376, 0, 100}, {z, 68.5022, 80.2957, 0, 100}}
```

Now, we use the starting values to search with higher precision for the minima.

```
In[5]:= FindMinimum[(BesselJ[v, x] - BesselJ[μ, x])^2 +
                    (BesselJ[v, z] - BesselJ[μ, z])^2,
                    Evaluate[Sequence @@ SetPrecision[pre, 40]],
                    WorkingPrecision -> 30, AccuracyGoal -> 20]
```

```
Out[5]= {0, {v → 2.18810937453033197719862923552, μ → 78.4210614132455312738251981096,
        x → 65.4516359547319078845469661486, z → 68.5720145475780063099891824436}}
```

b) To obtain algebraic numbers, we will make heavy use of the function Recognize from the package NumberTheory` ` Recognize`.

```
In[1]:= Needs["NumberTheory`Recognize`"]
```

Let us start by implementing some of the formulas needed. $\mathcal{KR}$ is the ratio of hypergeometric functions in the transcendental equation. a and b are the definition of the constants a, b, and c through α_n.

```
In[2]:= 𝒦R[x_] := Hypergeometric2F1[1/3, 2/3, 1, 1 - x]/
                  Hypergeometric2F1[1/3, 2/3, 1, x]
```

A quick plot shows that the solutions for the transcendental equation for α_n lie in the interval $(0, 1)$.

```
In[3]:= Plot[𝒦R[SetPrecision[x, 30]], {x, 0, 1}, Frame -> True, Axes -> False]
```

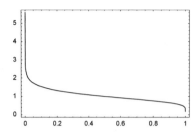

For further use we define three functions that express the variables a, b, and c through α_n.

```
In[4]:= a[n_, αn_] = 1/(Pi Hypergeometric2F1[1/3, 2/3, 1, αn]^2) +
                (8 Sqrt[n] (-1 + αn) αn Hypergeometric2F1[4/3, 5/3, 2, αn])/
                (9 Sqrt[3] Hypergeometric2F1[1/3, 2/3, 1, αn]);

In[5]:= b[n_, αn_] := 2 Sqrt[3 n]/3 Sqrt[1 - 4 αn (1 - αn)]

In[6]:= c[n_, αn_] := 4 αn (1 - αn)
```

A numerical root finding-based technique for determining α_n, a, and b is straightforward. We first calculate high-precision solutions (using enough digits in all intermediate calculations) of α_n and then find a "nice" algebraic number that is identical to the numerical approximation of α_n to more than 100 digits. To make sure that we found the correct algebraic number we verify the original identity to twice as many digits. Then we proceed similarly to find an algebraic number for a. Algebraic values of the two quantities b and c follow directly from the algebraic form of α_n.

```
In[7]:= k[x_?InexactNumberQ, prec_] := With[{ξ = SetPrecision[x, prec]}, 𝒦R[ξ]]

In[8]:= findSmallAlgebraicIntegerSecant[n_, prec_] :=
    Block[{$MaxExtraPrecision = 1000, ε = 10^-(3n), o = 2, polyα, α1, α2},
    α1 = Re[α /. FindRoot[k[α, 2prec] == Sqrt[n], {α, ε, 1 - ε},
                    AccuracyGoal -> 2prec, WorkingPrecision -> 2prec,
                    MaxIterations -> 2prec]] // SetPrecision[#, prec]&;
    (* find smallest degree polynomial *)
    While[polyα = Recognize[α1, o, t];
        α2 = Select[t /. Union[{ToRules[Roots[polyα == 0, t]]}],
                Abs[# - α1] == 0.&];
        If[α2 =!= {},
            (* high-precision check of original equation *)
            N[𝒦R[α2[[1]]], 2prec] - Sqrt[n] =!= 0., True],
        (* increase degree *) o = o + 2]; α2]
```

Here is an example.

```
In[9]:= findSmallAlgebraicIntegerSecant[10, 100] // Timing
```

$$\text{Out[9]= } \left\{0.93 \text{ Second, } \left\{\frac{223 - 70 \sqrt{10}}{54 \left(54 + \sqrt{10} \left(247 + 14 \sqrt{10}\right)\right)}\right\}\right\}$$

From the above graphic we see that the values for α_n for larger n converge to 0. This means instead of using a secant-based root finding method, we could use the one-starting value version of FindRoot and a good starting value. We obtain a good starting value by using the series expansion of the ratio of hypergeometric functions.

```
In[10]:= Off[InverseFunction::ifun]; Off[Solve::ifun];
    Select[Solve[Normal[FullSimplify[Series[𝒦R[x], {x, 0, 1}]]] == Sqrt[n], x],
        Im[x /. # /. n -> 2.] == 0&]
```

$$\text{Out[11]= } \left\{\left\{x \to \frac{9}{5} \text{ ProductLog}\left[15 \left(e^{-6 \sqrt{3} \sqrt{n} \pi}\right)^{1/9}\right]\right\}\right\}$$

```
In[12]:= findSmallAlgebraicInteger[n_, prec_, δprec_:20] :=
    Block[{$MaxExtraPrecision = 1000, ε = 10^-20, o = 2, polyα, α1, sel, α2},
    α1 = α /. FindRoot[𝒦R[α] == Sqrt[n], (* use good starting value *)
                {α, 9/5 ProductLog[15 Exp[-2 Pi Sqrt[n]/Sqrt[3]]]},
                WorkingPrecision -> 2 prec, AccuracyGoal -> prec + δprec] //
                            SetPrecision[#, prec]&;
    (* find smallest degree polynomial *)
    While[polyα = Recognize[α1, o, t];
        sel = Select[t /. Union[{ToRules[Roots[polyα == 0, t]]}],
```

```
                        Abs[# - α1] == 0.&];
          If[sel =!= {}, α2 = sel[[1]]];
          (* high-precision check of original equation *)
          N[𝒦R[α2], 2prec] - Sqrt[n] =!= 0.,
          (* increase degree *) o = o + 2]; α2]
```

Here is again the $n = 10$ example from above.

```
In[13]:= findSmallAlgebraicInteger[10, 100] // Timing
```

$$\text{Out[13]}= \left\{0.63 \text{ Second}, \frac{223 - 70\sqrt{10}}{54\left(54 + \sqrt{10}\left(247 + 14\sqrt{10}\right)\right)}\right\}$$

Now now continue to construct the terms of the infinite sums.

```
In[14]:= makeπTerm[n_, k_, prec_:150] :=
         Module[{α, α1, α2, polyα, e1, e2, polye, a, b, h, o},
         (* α as an algebraic number *)
         α2 = Simplify[findSmallAlgebraicInteger[n, prec]];
         (* numerical value for a *)
         a1 = N[a[n, α2], prec];
         (* a as an algebraic number *)
         o = 2;
         While[polyα = Recognize[a1, o, t];
               a2Pre = Select[t /. Union[{ToRules[Roots[polyα == 0, t]]}],
                      Abs[# - a1] == 0.&];
               If[a2Pre === {}, True, a2 = a2Pre[[1]];
                   (* high-precision check of original equation *)
                   N[a[n, α2], 2prec] - a2 =!= 0.],
                   (* increase degree *) o = o + 2];
         (* a, b, c as an algebraic numbers *)
         {a, b, c} = Simplify[#]& /@ {a2, b[n, α2], c[n, α2]};
         (* return summand *)
         (a + b k) Pochhammer[1/2, k] Pochhammer[1/3, k]*
            Pochhammer[2/3, k]/k!^3 c^k // Simplify]
```

The calculation of the Ramanujan-type π-formulas can now be done quickly.

```
In[15]:= $MaxExtraPrecision = 1000;
         makeπTerm[17, k]
```

$$\text{Out[16]}= \frac{1}{(k!)^3}$$

$$\left(\left(-\left(\left(74172\,2^{2/3} + 6661\left(9657941 + 2343501\sqrt{17}\right)^{1/3} - 21\,2^{1/3}\left(9657941 + 2343501\sqrt{17}\right)^{2/3}\right)\left(74172\right.\right.\right.\right.$$

$$2^{2/3} - 3165\left(9657941 + 2343501\sqrt{17}\right)^{1/3} - 21\,2^{1/3}\left(9657941 + 2343501\sqrt{17}\right)^{2/3} -$$

$$34\sqrt{\left(17\left(-74172\,2^{2/3}\left(9657941 + 2343501\sqrt{17}\right)^{1/3} -\right.\right.}$$

$$\left.\left.1748\left(9657941 + 2343501\sqrt{17}\right)^{2/3} + 21\,2^{1/3}\left(9657941 + 2343501\sqrt{17}\right)\right)\right)\right)\right)\Bigg/$$

$$\left(289\left(289\left(9657941 + 2343501\sqrt{17}\right)^{1/3} + \sqrt{\left(17\left(-74172\,2^{2/3}\left(9657941 + 2343501\sqrt{17}\right)^{1/3} -\right.\right.}\right.\right.$$

$$\left.\left.\left.1748\left(9657941 + 2343501\sqrt{17}\right)^{2/3} + 21\,2^{1/3}\left(9657941 + 2343501\sqrt{17}\right)\right)\right)^2\right)\right)^k$$

$$\left(\frac{2}{51}\sqrt{\left(\frac{2}{3}\left(-1244 - \frac{855788}{\left(2934393569 + 737141607\sqrt{17}\right)^{1/3}} + \left(2934393569 + 737141607\sqrt{17}\right)^{1/3}\right)\right)} +$$

$$\left(2\left(-74172\,2^{2/3} - 1748\left(9657941 + 2343501\sqrt{17}\right)^{1/3} + 21\,2^{1/3}\left(9657941 + 2343501\sqrt{17}\right)^{2/3} +\right.$$

$$17\sqrt{\left(17\left(-74172\,2^{2/3}\left(9657941 + 2343501\sqrt{17}\right)^{1/3} -\right.\right.}$$

$$\left.\left.\left.1748\left(9657941 + 2343501\sqrt{17}\right)^{2/3} + 21\,2^{1/3}\left(9657941 + 2343501\sqrt{17}\right)\right)\right)\right) k\right) \Bigg/$$

$$\left(\sqrt{51}\left(289\left(9657941 + 2343501\sqrt{17}\right)^{1/3} + \sqrt{\left(17\left(-74172\,2^{2/3}\left(9657941 + 2343501\sqrt{17}\right)^{1/3} -\right.\right.}\right.\right.$$

$$\left.\left.\left.\left.1748\left(9657941 + 2343501\sqrt{17}\right)^{2/3} + 21\,2^{1/3}\left(9657941 + 2343501\sqrt{17}\right)\right)\right)\right)\right)\right)$$

$$\text{Pochhammer}\left[\frac{1}{3}, k\right]\text{Pochhammer}\left[\frac{1}{2}, k\right]\text{Pochhammer}\left[\frac{2}{3}, k\right]\right)$$

```
In[17]:= $MaxExtraPrecision = 1000;

        Do[πTerm[j] = makeπTerm[j, k], {j, 2, 20}] // Timing
Out[18]= {36.86 Second, Null}
```

Here are the summands for a few selected values of j shown.

```
In[19]:= (* smallest few j *)
        Table[FullSimplify[πTerm[j]], {j, 2, 5}] // TraditionalForm
Out[20]//TraditionalForm=
```

$$\left\{ \frac{2^{-k}(6k+1)\left(\frac{1}{3}\right)_k\left(\frac{1}{2}\right)_k\left(\frac{2}{3}\right)_k}{3\sqrt{3}\,(k!)^3}, \; -\frac{\left(-18+\frac{21\sqrt{3}}{2}\right)^k\left((-7+3\sqrt{3})k+\sqrt{3}-2\right)\left(\frac{1}{3}\right)_k\left(\frac{1}{2}\right)_k\left(\frac{2}{3}\right)_k}{(k!)^3}, \right.$$

$$\left. \frac{\left(\frac{2}{27}\right)^k\left(\frac{20k}{9}+\frac{8}{27}\right)\left(\frac{1}{3}\right)_k\left(\frac{1}{2}\right)_k\left(\frac{2}{3}\right)_k}{(k!)^3}, \; \frac{2^{2k+1}5^{-3k-1}(33k+4)\left(\frac{1}{3}\right)_k\left(\frac{1}{2}\right)_k\left(\frac{2}{3}\right)_k}{3\sqrt{3}\,(k!)^3} \right\}$$

```
In[21]:= (* largest here calculated j *)
        πTerm[20] // FullSimplify // TraditionalForm
Out[22]//TraditionalForm=
```

$$\frac{1}{\sqrt{\pi}\,(k!)^3\,\Gamma(k)}$$

$$\left(2^{3k+2}\,3^{-3k-1}\left(205692+118755\sqrt{3}+91988\sqrt{5}+53109\sqrt{15}\right)^k\left(205694+118755\sqrt{3}+91988\sqrt{5}+53109\sqrt{15}\right)^{-2k-1}\right.$$

$$\left(15\left(159327+91988\sqrt{3}+71253\sqrt{5}+41138\sqrt{15}\right)k+4\left(36828+21263\sqrt{3}+16470\sqrt{5}+9509\sqrt{15}\right)\right)$$

$$\left.\Gamma(3k)\,\Gamma\!\left(k+\frac{1}{2}\right)\right)$$

We end by calculating the average number of digits are gained for each summand. We calculate this number by explicitly summing the first 100 terms and comparing the result with $1/\pi$.

```
In[23]:= digitsPerSummand[term_] :=
        With[{termN = N[term, 1000], l = 100},
             (* logarithm of absolute value of error *)
             -Log[10, Abs[N[1/Pi - Sum[termN, {k, 0, 100}]]]]]/100]
In[24]:= ListPlot[Table[{j, digitsPerSummand[πTerm[j]]}, {j, 2, 20}]]
```

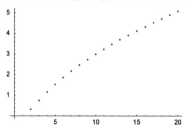

20. Force-Free Magnetic Fields, Bessel Beams, Gauge Transformation

a) We load the package `Calculus`VectorAnalysis`` to carry out the `Div` and `Curl` operation easily.

```
In[1]:= Needs["Calculus`VectorAnalysis`"]

        SetCoordinates[Cylindrical[r, φ, z]]
Out[2]= Cylindrical[r, φ, z]

In[3]:= Div[{Br, Bφ, Bz}, Cylindrical] // FullSimplify
Out[3]= Br
        ──
        r

In[4]:= Curl[{Br, Bφ, Bz}, Cylindrical] / {Br, Bφ, Bz} // FullSimplify
Out[4]= {0, 0, Bφ }
             ──
             Bz r
```

Substituting the given expressions, we get the following expressions for the components of $\mathbf{B}$.

In[5]:= {Bz, Bφ, Br} = {x, a, 1} Exp[-1 z] BesselJ[{0, 1, 1}, x r];

The divergence vanishes identically.

In[6]:= Div[{Br, Bφ, Bz}, Cylindrical] // FullSimplify

Out[6]= 0

For the curl we get the following result.

In[7]:= Curl[{Br, Bφ, Bz}, Cylindrical]/{Br, Bφ, Bz} // FullSimplify

Out[7]= $\left\{a, \dfrac{-1^2 + x^2}{a}, a\right\}$

This means the Div-equation is already fulfilled and the Curl-equation gives the following value for *x*.

In[8]:= Solve[%[[2]] == a, x]

Out[8]= $\left\{\left\{x \to -\sqrt{a}\ \sqrt{\dfrac{a^2 + 1^2}{a}}\right\}, \left\{x \to \sqrt{a}\ \sqrt{\dfrac{a^2 + 1^2}{a}}\right\}\right\}$

In[9]:= % /. Power[a_, 1/2] Power[b_, 1/2] :> Power[Expand[a b], 1/2]

Out[9]= $\left\{\left\{x \to -\sqrt{a^2 + 1^2}\right\}, \left\{x \to \sqrt{a^2 + 1^2}\right\}\right\}$

b) We start by defining our default coordinate system to be a cylindrical one.

In[1]:= Needs["Calculus`VectorAnalysis`"];
        SetCoordinates[Cylindrical[r, φ, z]];

These are the magnetic field components.

In[3]:= Br = 0;
        Bφ = BesselJ[1, k Sin[α] r] Exp[I (k Cos[α] z - ω t)];
        Bz = 0;

They fulfill the equation div **B**(**r**) = 0.

In[6]:= Div[{Br, Bp, Bz}, Cylindrical] // FullSimplify

Out[6]= 0

Using the Maxwell equation curl **B**(**r**, *t*) = ∂**E**(**r**, *t*)/∂*t* we can determine the electric field components up to an integration constant.

In[7]:= Integrate[Curl[{Br, Bφ, Bz}, Cylindrical], t] ==
        {Er[r, φ, z, t], Eφ[r, φ, z, t], Ez[r, φ, z, t]} // FullSimplify

Out[7]= $\Big\{\dfrac{e^{-i t \omega + i k z \cos[\alpha]}\ k\ \mathrm{BesselJ}[1, k r \sin[\alpha]]\ \cos[\alpha]}{\omega}$,

        $0, \dfrac{i\ e^{-i t \omega + i k z \cos[\alpha]}\ k\ \mathrm{BesselJ}[0, k r \sin[\alpha]]\ \sin[\alpha]}{\omega}\Big\} ==$

        {Er[r, φ, z, t], Eφ[r, φ, z, t], Ez[r, φ, z, t]}

In[8]:= Er = k/ω Cos[α] BesselJ[1, k Sin[α] r] Exp[I (Cos[α] k z - ω t)];
        Eφ = 0;
        Ez = I k Sin[α]/ω BesselJ[0, k Sin[α] r] Exp[I (Cos[α] k z - ω t)];

The electric field fulfills the equation div **E**(**r**, *t*) = 0.

In[11]:= Div[{Er, Ep, Ez}, Cylindrical] // FullSimplify

Out[11]= 0

The last remaining Maxwell equation curl **E**(**r**, *t*) = -∂**B**(**r**, *t*)/∂*t* is fulfilled if ω = ±*k* holds.

In[12]:= Curl[{Er, Eφ, Ez}, Cylindrical] + D[{Br, Bφ, Bz}, t] // FullSimplify

Out[12]= $\Big\{0, \dfrac{i\ e^{-i t \omega + i k z \cos[\alpha]}\ (k - \omega)\ (k + \omega)\ \mathrm{BesselJ}[1, k r \sin[\alpha]]}{\omega}, 0\Big\}$

c) We start by implementing the vector potential after the gauge transformation. We consider the vector potential in a plane and ignore the trivial *z*-component.

In[1]:= (* radial component of A in cylindrical coordinates *)
        Ar[r_, φ_] := H (r^3/a^6 (a^2 (a^2 - r^2) +

```
        (a^4 - 4 a^2 r^2  + 2 r^4) Exp[1 - 2r^2/a^2]*
        ExpIntegralEi[2 r^2/a^2 - 1]) Sin[4 φ])

   (* azimuthal component of A in cylindrical coordinates *)
   Aφ[r_, φ_] := H (r/2 + r^3/a^4 (a^2 + (a^2 - 2 r^2)*
      Exp[1 - 2r^2/a^2] ExpIntegralEi[2 r^2/a^2 - 1]) Cos[4 φ]);
```

In[6]:=
```
   (* A in cylindrical coordinates *)
   A[{r_, φ_}] := Ar[r, φ] { Cos[φ], Sin[φ]} +
                  Aφ[r, φ] {-Sin[φ],  Cos[φ]}
```

A direct calculation shows that we have curl $\mathbf{A}(r, \varphi) = H\,\mathbf{e}_z$.

In[8]:=
```
   Needs["Calculus`VectorAnalysis`"]
   Curl[{Ar[r, φ], Aφ[r, φ], 0}, Cylindrical[r, φ, ζ]] // Simplify
```
Out[9]= {0, 0, H}

In cylindrical coordinates, one edge of the square can be parametrized as $r = a/(2\cos(\varphi))$. Again a direct calculation shows that the normal component of $\mathbf{A}(r, \varphi)$ vanishes along this edge of the square. Because of the obvious fourfold rotational symmetry of $\mathbf{G}(r, \varphi)$, it follows that this property holds on all four edges of the square.

In[10]:=
```
   A[{a/2/Cos[φ], φ}].{1, 0} // FullSimplify[#, -Pi/4 < φ < Pi/4]&
```
Out[10]= 0

Next, we visualize the vector potential $\mathbf{A}(r, \varphi)$ and the flow lines of this potential. The left graphic shows the direction of the vector potential at discrete points. In the center the radial symmetry of $\mathbf{A}_0(r, \varphi)$ is clearly visible and, near the edges of the square, the vector potential becomes parallel to the edges. The right graphic shows the flow lines. They change from nearly circular in the center to square near the edges.

In[11]:=
```
   (* a short line segment indicating the direction
       of the vector potential *)
   direction[{r_, φ_}, α_] :=
   Line[{{r Cos[φ], r Sin[φ]}, {r Cos[φ], r Sin[φ]} + α A[{r, φ}]}]
```
In[13]:=
```
   (* differential equation for the flow lines *)
   odes = Block[{(* without loss of generality *) a = 2, H = 1},
           Thread[{x'[s], y'[s]} ==
               A[{Sqrt[x[s]^2 + y[s]^2], ArcTan[x[s], y[s]]}]]];
```
In[15]:=
```
   Block[{a = 2, H = 1, pp = 60, n = 36, directions, border},
      (* the square *)
   border = {Hue[0], Thickness[0.01],
               Line[{{-1, -1}, {1, -1}, {1, 1}, {-1, 1}, {-1, -1}}]}];
      (* options for both graphics *)
   commonOpts = Sequence[Axes -> False, AspectRatio -> Automatic,
                         PlotRange -> 1.001 {{-1, 1}, {-1, 1}}];
      (* direction of the vector potential at (pp + 1)^2 points *)
   directions = Table[direction[{2. ri/pp, φ +
                            If[EvenQ[ri], 0, 2Pi/pp/2]}, 0.08],
                 {ri, 0, pp}, {φ, 0, 2Pi, 2Pi/pp}];
      (* show both graphics *)
   Show[GraphicsArray[{
      (* directions *)
      Graphics[{directions, border}, commonOpts],
      (* the flow lines *)
      Show[{Table[
      ParametricPlot[Evaluate[{x[s], y[s]} /.
         (* solve differential equations numerically  *)
         NDSolve[Join[odes, {x[0] == x0, y[0] == 0}],
                    {x, y}, {s, -3Pi, 3Pi}]], {s, -3Pi, 3Pi},
                    DisplayFunction -> Identity, PlotPoints -> 120],
         {x0, 1/n, 1 - 1/n, 1/n}], Graphics[border]}, commonOpts]}]]]
```

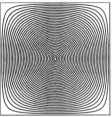

We end with visualizing $|\mathbf{A}(r, \varphi)|$. The graphics show the $|\mathbf{A}(r, \varphi)|$ becomes maximal at the centers of the edges.

```
In[16]:= Show[GraphicsArray[
          Block[{a = 2, H = 1, ε = 10^-8, absA, $DisplayFunction = Identity},
            absA = Sqrt[#.#]&[A[{Sqrt[x^2 + y^2], ArcTan[x, y]}]];
            {Plot3D[Evaluate[absA], {x, -1 + ε, 1 - ε}, {y, -1 + ε, 1 - ε},
                PlotPoints -> 60, Mesh -> False],
             ContourPlot[Evaluate[absA], {x, -1 + ε, 1 - ε}, {y, -1 + ε, 1 - ε},
                PlotPoints -> 60, ColorFunction -> (Hue[0.8 #]&)]}]]]
```

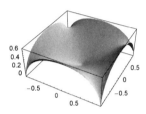
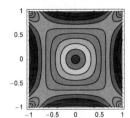

21. Riemann Surface of the Bootstrap Equation

Mathematica can solve the implicit equation $z = 2w - \exp(w) + 1$ with respect to w by using the ProductLog function.

```
In[1]:= Solve[z == 2 w - Exp[w] + 1, w]
```

> InverseFunction::ifun : Inverse functions are
>   being used. Values may be lost for multivalued inverses. More…
>
> Solve::ifun : Inverse functions are being used by Solve, so some solutions
>   may not be found; use Reduce for complete solution information. More…

$$Out[1]= \left\{\left\{w \to \frac{1}{2}\left(-1 + z - 2\,\text{ProductLog}\left[-\frac{1}{2}\,e^{-\frac{1}{2}+\frac{z}{2}}\right]\right)\right\}\right\}$$

Because Solve does not produce all possible solutions for multivalued inverse functions, we only get one sheet in this way. The other sheets are easy to obtain by using the analytic continuation for the ProductLog function.

```
In[2]:= w[k_, z_] = (z - 1)/2 - ProductLog[k, -Exp[(z - 1)/2]/2];
```

Here is a quick numerical check that these solutions satisfy the original equation.

```
In[3]:= Table[Chop[z -( 2 w[k, z] - Exp[w[k, z]] + 1) /.
              {k -> Random[Integer, {-10, 10}],
               z -> Random[Complex, {-5 - 5I, 5 + 5I}]}], {100}] // Union
Out[3]= {0}
```

Now, we will calculate the branch points. The principal sheet of the ProductLog function has a branch point at $z = -1/e$. Using the argument of $w[k, z]$, we get as the branch point(s) for $w(z)$ the following z-value.

```
In[4]:= Solve[-Exp[(z - 1)/2]/2 == -1/E, z]
```

> Solve::ifun : Inverse functions are being used by Solve, so some solutions
>   may not be found; use Reduce for complete solution information. More…

$$Out[4]= \left\{\left\{z \to 1 + 2\,\text{Log}\left[\frac{2}{e}\right]\right\}\right\}$$

Again, Solve does not give us all solutions. We obtain the other ones by adding the other sheets of the Log functions: $z_m = 1 + 2 (\log(2/e) + 2 i \pi m)$.

```
In[5]:= branchPointPrincipalSheet[m_] := 1 + 2 (Log[2/E] + 2 Pi I m)
```

The other sheets of the ProductLog function do not contribute to branch points in the finite part of the complex plane.

```
In[6]:= Solve[-Exp[(z - 1)/2]/2 == 0, z]
Out[6]= {}
```

The branch cut of the principal sheet of the ProductLog function extends along the negative real axis.

```
In[7]:= Solve[-(Exp[(z - 1)/2]/2) == -positiveReal, z]
```

```
        Solve::ifun : Inverse functions are being used by Solve, so some solutions
            may not be found; use Reduce for complete solution information. More...
```

```
Out[7]= {{z → 1 + 2 Log[2 positiveReal]}}
```

This means the branch cuts of w[k, z] are straight lines starting at $1 + 2(\log(2/e) + 2 i \pi m) \approx 0.386 + 4 m i \pi$ and going to the right, being parallel to the real axis. The following picture clearly shows these branch cuts.

```
In[8]:= Plot3D[Evaluate[Im[w[0, zr + I zi]]], {zr, -12, 12}, {zi, -20, 20},
            PlotPoints -> 60]
```

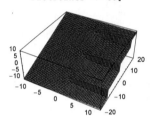

Although the other sheets of these ProductLog functions do not have branch points in the finite part of the complex z-plane, nevertheless, branch cuts are present. When *positiveReal* in $1 + 2 \ln(2 \, positiveReal)$ varies over the positive reals, the resulting branch cut is the whole real axis. In a similar way, the branch cuts at $real + 4 \, m \, i \, \pi$ are generated. Here, the branch cuts for the second sheet are shown.

```
In[9]:= Plot3D[Evaluate[Im[w[2, zr + I zi]]], {zr, -12, 12}, {zi, -20, 20},
            PlotPoints -> 60]
```

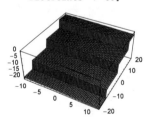

Now, we can generate a picture of the Riemann surface by dividing the z-plane in parallel stripes and using some neighboring sheets of the principal sheet of the ProductLog function.

```
In[10]:= ε = 10^-8;
        Show[Graphics3D[{EdgeForm[],
            SurfaceColor[Hue[Random[]], Hue[Random[]], 3 Random[]],
            Cases[Table[Graphics3D[(* make the graphic *)
            Plot3D[Evaluate[Re[w[k, zr + I zi]]], {zr, -20, 20},
                {zi, j 4 Pi + ε, (j + 1) 4 Pi - ε}, PlotPoints -> {40, 20},
                DisplayFunction -> Identity]],
                (* some sheets and some stripes *)
                {k, -2, 2}, {j, -2, 2}], _Polygon, Infinity]}],
            BoxRatios -> {1, 1, 0.6}, ViewPoint -> {3, -0.9, 0.7}]
```

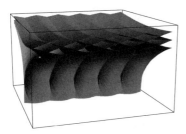

The imaginary part gives a good view of the branch points, all of which connect two sheets.

```
In[12]:= (DownValues[In][[$Line - 1]] /.
   (* change to imaginary part; change view point *)
   {Re -> Im, {3, -0.9, 0.7} -> {-3, 1, 0.3}})[[2]]
```

For a more detailed discussion of the Riemann surface of this function, see [562], [1261], and [825].

22. Differential Equation of Powers of Airy Functions, Map Airy Distribution, Zeros of Airy Function *

a) Motivated by the differential equation for the Airy function itself, we make for the differential equation for the powers of Airy functions an ansatz of the form $\sum_{i=0}^{n+1} c_i(z)\, w^{(i)}(z) = 0$ and determine the coefficients $c_i(z)$ by observing that the functions $Ai^i(z)$, $Ai'^i(z)$ are linearly independent [864]. This is implemented in a compact form below. *pow* is the power, w is the independent, and z is the dependent variable of the resulting differential equation.

```
In[1]:= ODEOfAiryPowers[α_, w_, z_] :=
   (* the ansatz *) (Array[D[w[z], {z, #}]&, α + 2, 0].Array[C, α + 2, 0] //.
   Map[Factor, Function[ansatz,
   (* set coefficients, and solve for the C[i] *)
   Flatten[{Solve[# == 0& /@ Apply[Plus, Cases[ansatz, _ #]/#& /@
   Array[AiryAi[z]^(α - #) AiryAiPrime[z]^#&, α + 1, 0], {1}],
   {C[pow + 1] -> 1}}]]][Expand[(* the ansatz *)
   Array[C, α + 2, 0].Array[D[AiryAi[z]^α, {z, #}]&, α + 2, 0]]], {2}]) == 0
```

Here are the differential equations for the first 10 powers of Ai(z).

```
In[2]:= Off[Solve::svars];
   Array[ODEOfAiryPowers[#, w, z]&, 10]
```

$$Out[3]= \{-z\, C[2]\, w[z] + C[2]\, w''[z] = 0, \ -2\, C[3]\, w[z] - 4\, z\, C[3]\, w'[z] + C[3]\, w^{(3)}[z] = 0,$$
$$9\, z^2\, C[4]\, w[z] - 10\, C[4]\, w'[z] - 10\, z\, C[4]\, w''[z] + C[4]\, w^{(4)}[z] = 0,$$
$$64\, z\, C[5]\, w[z] + 64\, z^2\, C[5]\, w'[z] - 30\, C[5]\, w''[z] - 20\, z\, C[5]\, w^{(3)}[z] + C[5]\, w^{(5)}[z] = 0,$$
$$-5\, (-26 + 45\, z^3)\, C[6]\, w[z] + 518\, z\, C[6]\, w'[z] +$$
$$259\, z^2\, C[6]\, w''[z] - 70\, C[6]\, w^{(3)}[z] - 35\, z\, C[6]\, w^{(4)}[z] + C[6]\, w^{(6)}[z] = 0,$$
$$-3456\, z^2\, C[7]\, w[z] - 4\, (-295 + 576\, z^3)\, C[7]\, w'[z] + 2352\, z\, C[7]\, w''[z] +$$
$$784\, z^2\, C[7]\, w^{(3)}[z] - 140\, C[7]\, w^{(4)}[z] - 56\, z\, C[7]\, w^{(5)}[z] + C[7]\, w^{(7)}[z] = 0,$$
$$7\, z\, (-2776 + 1575\, z^3)\, C[8]\, w[z] - 38748\, z^2\, C[8]\, w'[z] - 4\, (-1485 + 3229\, z^3)\, C[8]\, w''[z] +$$
$$7896\, z\, C[8]\, w^{(3)}[z] + 1974\, z^2\, C[8]\, w^{(4)}[z] - 252\, C[8]\, w^{(5)}[z] - 84\, z\, C[8]\, w^{(6)}[z] + C[8]\, w^{(8)}[z] =$$
$$0, \ 256\, (-155 + 1152\, z^3)\, C[9]\, w[z] + 256\, z\, (-925 + 576\, z^3)\, C[9]\, w'[z] -$$
$$236160\, z^2\, C[9]\, w''[z] - 20\, (-1095 + 2624\, z^3)\, C[9]\, w^{(3)}[z] + 21840\, z\, C[9]\, w^{(4)}[z] +$$
$$4368\, z^2\, C[9]\, w^{(5)}[z] - 420\, C[9]\, w^{(6)}[z] - 120\, z\, C[9]\, w^{(7)}[z] + C[9]\, w^{(9)}[z] = 0,$$
$$-243\, z^2\, (-13084 + 3675\, z^3)\, C[10]\, w[z] + 44\, (-11870 + 96111\, z^3)\, C[10]\, w'[z] +$$
$$11\, z\, (-141740 + 96111\, z^3)\, C[10]\, w''[z] - 1036860\, z^2\, C[10]\, w^{(3)}[z] -$$
$$110\, (-600 + 1571\, z^3)\, C[10]\, w^{(4)}[z] + 52668\, z\, C[10]\, w^{(5)}[z] + 8778\, z^2\, C[10]\, w^{(6)}[z] -$$

$$660\, C[10]\, w^{(7)}[z] - 165\, z\, C[10]\, w^{(8)}[z] + C[10]\, w^{(10)}[z] == 0,$$
$$-1280\, z\, (-12731 + 28800\, z^3)\, C[11]\, w[z] - 256\, z^2\, (-190111 + 57600\, z^3)\, C[11]\, w'[z] +$$
$$264\, (-13995 + 122624\, z^3)\, C[11]\, w''[z] + 176\, z\, (-41795 + 30656\, z^3)\, C[11]\, w^{(3)}[z] -$$
$$3669600\, z^2\, C[11]\, w^{(4)}[z] - 220\, (-783 + 2224\, z^3)\, C[11]\, w^{(5)}[z] + 114576\, z\, C[11]\, w^{(6)}[z] +$$
$$16368\, z^2\, C[11]\, w^{(7)}[z] - 990\, C[11]\, w^{(8)}[z] - 220\, z\, C[11]\, w^{(9)}[z] + C[11]\, w^{(11)}[z] == 0\}$$

A test confirms that these are the correct differential equations.

```
In[4]:= MapIndexed[Function[{ode, pow}, ExpandAll[
            ode /. {w[z] -> AiryAi[z]^pow[[1]],
            Derivative[n_][w][z] -> D[AiryAi[z]^pow[[1]], {z, n}]}]], %]
Out[4]= {True, True, True, True, True, True, True, True, True, True}
```

b) We start with the series expansion of Ai(z) for large negative arguments.

```
In[1]:= ser = Series[AiryAi[z], {z, -Infinity, 12}]
```

Series::esss :

Essential singularity encountered in $\text{Cos}\left[\dfrac{2\,\mathbf{i}}{3\,(\frac{1}{z})^{3/2}} + \dfrac{\pi}{4} + O\left[\dfrac{1}{z}\right]^{27/2}\right]$. More...

Series::esss : Essential singularity encountered in $\text{Cos}\left[\dfrac{2\,\mathbf{i}}{3\,z^{3/2}} + \dfrac{\pi}{4} + O[z]^{27/2}\right]$. More...

Series::esss :

Essential singularity encountered in $\text{Sin}\left[\dfrac{2\,\mathbf{i}}{3\,(\frac{1}{z})^{3/2}} + \dfrac{\pi}{4} + O\left[\dfrac{1}{z}\right]^{27/2}\right]$. More...

General::stop :

Further output of Series::esss will be suppressed during this calculation. More...

$$Out[1]= \frac{1}{\sqrt{\pi}\,(-z)^{1/4}}\left(\text{Cos}\left[\frac{\pi}{4} + \frac{2}{3}\,(-z)^{3/2}\right]\left(-\frac{5}{48}\,\mathbf{i}\left(\frac{1}{z}\right)^{3/2} - \frac{85085\,\mathbf{i}\,(\frac{1}{z})^{9/2}}{663552} - \right.\right.$$
$$\left.\frac{5391411025\,\mathbf{i}\,(\frac{1}{z})^{15/2}}{6115295232} - \frac{1267709431363375\,\mathbf{i}\,(\frac{1}{z})^{21/2}}{84537841287168} + O\left[\frac{1}{z}\right]^{25/2}\right) +$$
$$\left(1 + \frac{385\,(\frac{1}{z})^3}{4608} + \frac{37182145\,(\frac{1}{z})^6}{127401984} + \frac{5849680962125\,(\frac{1}{z})^9}{1761205026816} + \right.$$
$$\left.\left.\frac{2562040760785380875\,(\frac{1}{z})^{12}}{32462531054272512} + O\left[\frac{1}{z}\right]^{13}\right)\text{Sin}\left[\frac{\pi}{4} + \frac{2}{3}\,(-z)^{3/2}\right]\right)$$

This expression should be zero. So we can drop the $\pi^{1/2}\,(-z)^{-1/4}$ term. Further on, we switch from $-z$ to y to avoid intermediate imaginary expressions.

```
In[2]:= approx = ((ser // Normal) Sqrt[Pi] (-z)^(1/4)) /. {z -> -y};

In[3]:= Shallow[approx, 8]
```

$$Out[3]//Shallow= \left(-\frac{5}{48}\,\mathbf{i}\,(-\text{Power}[\ll 2\gg])^{3/2} - \frac{85085\,\mathbf{i}\,(-\text{Power}[\ll 2\gg])^{9/2}}{663552} - \frac{5391411025\,\mathbf{i}\,(-\text{Power}[\ll 2\gg])^{15/2}}{6115295232} - \right.$$
$$\left.\frac{1267709431363375\,\mathbf{i}\,(-\text{Power}[\ll 2\gg])^{21/2}}{84537841287168}\right)\text{Cos}\left[\frac{\pi}{4} + \frac{2\,y^{3/2}}{3}\right] +$$
$$\left(1 + \frac{2562040760785380875}{32462531054272512\,y^{12}} - \frac{5849680962125}{1761205026816\,y^9} + \frac{37182145}{127401984\,y^6} - \frac{385}{4608\,y^3}\right)\text{Sin}\left[\frac{\pi}{4} + \frac{2\,y^{3/2}}{3}\right]$$

The resulting expression has the form $a_s \sin(\pi/4 + 2/3\, y^{3/2}) - a_c \cos(\pi/4 + 2/3\, y^{3/2})$. We extract the coefficients of the two trigonometric functions.

```
In[4]:= {sinCoeff, cosCoeff} = { Coefficient[#, Sin[Pi/4 + 2/3 y^(3/2)]],
            -Coefficient[#, Cos[Pi/4 + 2/3 y^(3/2)]]}&[approx]
```

$$Out[4]= \left\{1 + \frac{2562040760785380875}{32462531054272512\,y^{12}} - \frac{5849680962125}{1761205026816\,y^9} + \frac{37182145}{127401984\,y^6} - \frac{385}{4608\,y^3}, \right.$$
$$\left.\frac{5}{48}\,\mathbf{i}\left(-\frac{1}{y}\right)^{3/2} + \frac{85085\,\mathbf{i}\left(-\frac{1}{y}\right)^{9/2}}{663552} + \frac{5391411025\,\mathbf{i}\left(-\frac{1}{y}\right)^{15/2}}{6115295232} + \frac{1267709431363375\,\mathbf{i}\left(-\frac{1}{y}\right)^{21/2}}{84537841287168}\right\}$$

The condition $a_s \sin(\pi/4 + 2/3\, y^{3/2}) - a_c \cos(\pi/4 + 2/3\, y^{3/2}) = 0$ means $\tan(\pi/4 + 2/3\, y^{3/2}) = a_c/a_s$ or $\pi/4 + 2/3\, y^{3/2} = \arctan(a_c/a_s) + k\,\pi$, $k = -1, -2, -3, \ldots$ (by taking into account that $\mathrm{Ai}(z)$ has only zeros for negative real z). The term a_c/a_s is small for large positive y, and we approximate it by its first series terms.

```
In[5]:= shouldBeZero = Pi/4 + 2/3 y^(3/2) + k Pi -
            Normal[Series[ArcTan[Cancel[cosCoeff/sinCoeff]], {y, Infinity, 18}]]
```

$$Out[5]= \frac{\pi}{4} + k\,\pi - \frac{5}{48}\left(\frac{1}{y}\right)^{3/2} + \frac{1105\left(\frac{1}{y}\right)^{9/2}}{9216} - \frac{82825\left(\frac{1}{y}\right)^{15/2}}{98304} + \frac{1282031525\left(\frac{1}{y}\right)^{21/2}}{88080384} + $$

$$\frac{138827178967109332375\left(\frac{1}{y}\right)^{27/2}}{14023813415445725184} - \frac{120757654567430948374337 5\left(\frac{1}{y}\right)^{33/2}}{7898211715579032423628 8} + \frac{2\,y^{3/2}}{3}$$

Now, we have a relation of the form $f(y, k) = 0$, which we would like to solve for $y = y(c)$. We carry out this inversion by first changing variables from y to x via $y = (3/2\,x)^{2/3}$ and then inverting the resulting series for $d_k = -3\,(k\,\pi + \pi/4)/2$ ($= 3/8\,\pi\,(4\,k - 1)$, $k = 1, 2, \ldots$). We suppress the k-dependence of d in the following.

```
In[6]:= shouldBeZero = shouldBeZero //.
            {y -> (3/2 x)^(2/3), Pi/4 + k Pi -> -2/3 d} // PowerExpand
```

$$Out[6]= -\frac{2\,d}{3} - \frac{120757654567430948374337 5}{6831760306541400667717632\,x^{11}} + $$
$$\frac{138827178967109332375}{53912249893792618905 6\,x^{9}} + \frac{1282031525}{1504935936\,x^{7}} - \frac{82825}{746496\,x^{5}} + \frac{1105}{31104\,x^{3}} - \frac{5}{72\,x} + x$$

```
In[7]:= inv = InverseSeries[Series[shouldBeZero + 2/3 d,
                                    {x, Infinity, 6}]] /. {x -> - 2/3 d}
```

$$Out[7]= \frac{1}{-\frac{1}{\frac{2\,d}{3}}} + \frac{5}{72} \frac{1}{\frac{2\,d}{3}(-2\,d)} - \frac{1255\left(-\frac{1}{\frac{2\,d}{3}}\right)^{3}}{31104} + \frac{272075\left(-\frac{1}{\frac{2\,d}{3}}\right)^{5}}{2239488} + O\left[-\frac{1}{\frac{2\,d}{3}}\right]^{7}$$

Now, changing back from x to the original variable z, and writing the resulting expression in descending powers of d, we finally have the following result.

```
In[8]:= z = -((Factor[Normal[(3/2 inv)^(2/3)]] // PowerExpand) //. {(-1)^(1/3) -> -1})
```

$$Out[8]= -\frac{77125 - 11520\,d^{2} + 8640\,d^{4} + 82944\,d^{6}}{82944\,d^{16/3}}$$

```
In[9]:= (* to avoid reordering of the sum in Plus *)
        (HoldForm @@ {Sort[List @@ Expand[-z/d^(2/3)],
            Exponent[#1, d] > Exponent[#2, d]&]}) /. List[l__] :> -d^(2/3) Plus[l]
```

$$Out[10]= -d^{2/3}\left(1 + \frac{5}{48\,d^{2}} - \frac{5}{36\,d^{4}} + \frac{77125}{82944\,d^{6}}\right)$$

Now we can put all of the above steps together to calculate some more terms of the zeros.

```
In[11]:= res =
         Module[{z, y, x, ser, approx, sinCoeff, cosCoeff, shouldBeZero,
                  inv, k, ord = 26},
         (* the series *)
         ser = Series[AiryAi[z], {z, -Infinity, ord}];
         approx = ((ser // Normal) Sqrt[Pi] (-z)^(1/4)) /. {z -> -y};
         (* coefficients of sin and cos *)
         {sinCoeff, cosCoeff} = { Coefficient[#, Sin[Pi/4 + 2/3 y^(3/2)]],
                          -Coefficient[#, Cos[Pi/4 + 2/3 y^(3/2)]]}&[approx];
         (* the expression that has to vanish *)
         shouldBeZero = Pi/4 + 2/3 y^(3/2) + k Pi -
            Normal[Series[ArcTan[Cancel[cosCoeff/sinCoeff]], {y, Infinity, ord}]];
         shouldBeZero = PowerExpand[shouldBeZero //.
                 {y -> (3/2 x)^(2/3), Pi/4 + k Pi -> 2/3 d}];
         (* invert series *)
         inv = InverseSeries[Series[shouldBeZero - 2/3 d,
                                    {x, Infinity, ord}]] /. {x -> 2/3 d};
         z = Factor[Normal[(3/2 inv)^(2/3)]] // PowerExpand;
         (* some cosmetic refinements for a better look *)
         (HoldForm @@ {Sort[
         DeleteCases[List @@ Expand[z/d^(2/3)], _ d^(_?(# < -10&))],
                          Exponent[#1, d] > Exponent[#2, d]&]}) /.
                                    List[l__] :> -d^(2/3) Plus[l]]
```

This result agrees with [3], 10.4.105 and [1023]. Let us have a brief look at the quality of the derived series. Here are the exact zeros of Ai(z), starting at the approximations just calculated.

```
exactZeros = Table[FindRoot[AiryAi[ζ] == 0,
        Evaluate[{ζ, ReleaseHold[res]} /. (* rename k --> i *)
                         {d -> 3Pi/8 (4i - 1)}}],
            WorkingPrecision -> 30, AccuracyGoal -> 20][[1, 2]], {i, 8}]
```

We collect in `tab` the partial sums for the various orders of `res`.

```
tab = -d^(2/3) Rest[
    FoldList[Plus, 0, Table[d^i Coefficient[res[[1, 3]], d, i],
                     {i, 0, -10, -2}]]];
```

Here are the differences between the approximations and the exact zeros for the first eight zeros.

```
TableForm[N[Table[N[(tab /. {d -> 3Pi/8 (4i - 1)}) -
                    exactZeros[[i]], 20], {i, 8}], 1]]
```

For a derivation of such expansions using the Lagrange–Bürmann theorem, see [445]. For the zeta function of the Airy zeros, see [182].

c) This is the map-Airy distribution.

```
In[1]:= p[x_] := 2 Exp[-2/3 x^3] (x AiryAi[x^2] - AiryAiPrime[x^2])
```

A direct call to `Series` does not succeed in giving the asymptotic.

```
In[2]:= Series[p[x], {x, Infinity, 1}]
```

$$\text{Out[2]= } e^{-\frac{2x^3}{3}}\left(-\frac{2}{\frac{1}{x}}+O\left[\frac{1}{x}\right]^2\right)\left(e^{-\frac{2}{3}(x^2)^{3/2}}\left(-\frac{\sqrt{\frac{1}{x}}}{2\sqrt{\pi}}+O\left[\frac{1}{x}\right]^{3/2}\right)+e^{-\frac{2}{3}(x^2)^{3/2}}\left(-\frac{\sqrt{\frac{1}{x}}}{2\sqrt{\pi}}+O\left[\frac{1}{x}\right]^{5/2}\right)\right)$$

Substituting the series expansions for the Airy functions gives the expansions

$$p(x)\underset{x\to\infty}{\longrightarrow}\frac{2}{\sqrt{\pi}}\sqrt{x}\,e^{-\frac{4}{3}x^3}$$

$$p(x)\underset{x\to-\infty}{\longrightarrow}\frac{7}{48\sqrt{\pi}}\frac{1}{\sqrt{x^5}}.$$

```
In[3]:= p[x] /. ({AiryAi[x_] -> Normal[Series[AiryAi[ξ], {ξ, Infinity, 1}]],
          AiryAiPrime[x_] -> Normal[Series[AiryAiPrime[ξ],
                               {ξ, Infinity, 1}]]} /. ξ -> x)
```

$$\text{Out[3]= } 2\,e^{-\frac{2x^3}{3}}\left(-e^{-\frac{2}{3}(x^2)^{3/2}}\left(-\frac{1}{2\sqrt{\pi}\,(\frac{1}{x^2})^{1/4}}-\frac{7\,(\frac{1}{x^2})^{5/4}}{96\sqrt{\pi}}\right)+e^{-\frac{2}{3}(x^2)^{3/2}}\left(\frac{(\frac{1}{x^2})^{1/4}}{2\sqrt{\pi}}-\frac{5\,(\frac{1}{x^2})^{7/4}}{96\sqrt{\pi}}\right)x\right)$$

```
In[4]:= Simplify[%, x > 0]
```

$$\text{Out[4]= } \frac{e^{-\frac{4x^3}{3}}(1+48x^3)}{24\sqrt{\pi}\,x^{5/2}}$$

```
In[5]:= p[-x] /. ({AiryAi[x_] -> Normal[Series[AiryAi[ξ], {ξ, Infinity, 2}]],
          AiryAiPrime[x_] -> Normal[Series[AiryAiPrime[ξ],
                               {ξ, Infinity, 2}]]} /. ξ -> x)
```

$$\text{Out[5]= } 2\,e^{\frac{2x^3}{3}}\left(-e^{-\frac{2}{3}(x^2)^{3/2}}\left(-\frac{1}{2\sqrt{\pi}\,(\frac{1}{x^2})^{1/4}}-\frac{7\,(\frac{1}{x^2})^{5/4}}{96\sqrt{\pi}}\right)-e^{-\frac{2}{3}(x^2)^{3/2}}\left(\frac{(\frac{1}{x^2})^{1/4}}{2\sqrt{\pi}}-\frac{5\,(\frac{1}{x^2})^{7/4}}{96\sqrt{\pi}}\right)x\right)$$

```
In[6]:= Simplify[%, x > 0]
```

$$\text{Out[6]= } \frac{1}{4\sqrt{\pi}\,x^{5/2}}$$

Now, let us calculate the cumulative distribution function. When dealing with complicated expressions in special functions, it is often of advantage to have the argument of the special functions as simple as possible. So we introduce a second distribution $\mathbb{p}(x) = p(x)/(2\,x^{1/2})$. (The denominator arises from the change of variables.)

```
In[7]:= ℘[y_] = p[Sqrt[y]]/(2 Sqrt[y]);
```

Mathematica can calculate the indefinite integral needed for the cumulative distribution function $P(x) = \int_{-\infty}^{x} p(x)\, dx$.

In[8]:= (int = Integrate[p[y], y]) // TraditionalForm

Out[8]//TraditionalForm=
$$-\frac{2\,_2F_2\left(\frac{7}{6},\frac{5}{3};\frac{7}{3},\frac{8}{3};-\frac{4y^{3/2}}{3}\right)y^{5/2}}{15\cdot3^{2/3}\,\Gamma\left(\frac{5}{3}\right)} - \frac{_2F_2\left(\frac{5}{6},\frac{4}{3};\frac{5}{3},\frac{7}{3};-\frac{4y^{3/2}}{3}\right)y^2}{6\sqrt[3]{3}\,\Gamma\left(\frac{4}{3}\right)} +$$

$$\frac{_2F_2\left(\frac{1}{6},\frac{2}{3};\frac{1}{3},\frac{5}{3};-\frac{4y^{3/2}}{3}\right)y}{3^{2/3}\,\Gamma\left(\frac{2}{3}\right)} + \frac{2\,_2F_2\left(-\frac{1}{6},\frac{1}{3};-\frac{1}{3},\frac{4}{3};-\frac{4y^{3/2}}{3}\right)\sqrt{y}}{\sqrt[3]{3}\,\Gamma\left(\frac{1}{3}\right)}$$

In[9]:= cdf[x_] = PowerExpand[int /. y -> x^2];

Here is the proof of the correctness of cdf as the indefinite integral.

In[10]:= FullSimplify[D[cdf[x], x] - p[x]]

Out[10]= 0

A plot shows that we must fix the integration constant to have a nonzero value.

In[11]:= ListPlot[Table[{x, cdf[x]}, {x, -2.8, 2.8, 5.6/30}],
         PlotJoined -> True, Frame -> True]

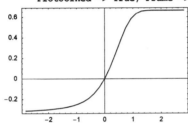

Using the asymptotics of the hypergeometric functions $_2F_2$, we find that the integration constant needed is $1/3$.

In[12]:= cdfAsymp = cdf[x] /. HypergeometricPFQ[{a1_, a2_}, {b1_, b2_}, z_] ->
         With[{Γ = Times @@ Gamma[{##}]&},
             E^z z^(a1 + a2 - b1 - b2) Γ[b1, b2]/Γ[a1, a2] +
             (-z)^-a1 Γ[a2 - a1, b1, b2]/Γ[a2, b1 - a1, b2 - a1] +
             (-z)^-a2 Γ[a1 - a2, b1, b2]/Γ[a1, b1 - a2, b2 - a2]];

In[13]:= FullSimplify[cdfAsymp, x > 0]

Out[13]= $\dfrac{2}{3} - \dfrac{e^{-\frac{4x^3}{3}}}{2\sqrt{\pi}\,x^{3/2}}$

So we have for $P(x)$ the following result.

$$p(x) = \frac{1}{3} + \frac{2x}{\sqrt[3]{3}\,\Gamma\left(\frac{1}{3}\right)}\,_2F_2\left(-\frac{1}{6},\frac{1}{3};-\frac{1}{3},\frac{4}{3};-\frac{4x^3}{3}\right) + \frac{x^2}{3^{2/3}\,\Gamma\left(\frac{2}{3}\right)}\,_2F_2\left(\frac{1}{6},\frac{2}{3};\frac{1}{3},\frac{5}{3};-\frac{4x^3}{3}\right) -$$

$$\frac{x^4}{6\sqrt[3]{3}\,\Gamma\left(\frac{4}{3}\right)}\,_2F_2\left(\frac{5}{6},\frac{4}{3};\frac{5}{3},\frac{7}{3};-\frac{4x^3}{3}\right) - \frac{2x^5}{15\cdot3^{2/3}\,\Gamma\left(\frac{5}{3}\right)}\,_2F_2\left(\frac{7}{6},\frac{5}{3};\frac{7}{3},\frac{8}{3};-\frac{4x^3}{3}\right)$$

The first moment $\int_{-\infty}^{\infty} x\, p(x)\, dx$ is 0. This follows immediately from the corresponding indefinite integral and the asymptotics of the Airy functions.

In[14]:= intM1y = Integrate[p[y] Sqrt[y], y] // FullSimplify

Out[14]= $\dfrac{3^{5/6}\,\text{Gamma}\left[\frac{4}{3}\right]}{2\,\pi}$ -

$\dfrac{1}{27}\,e^{-\frac{2\,y^{3/2}}{3}}\left(27\,y\,\text{AiryAiPrime}[y] - 27\,3^{1/3}\,\text{HypergeometricOF1Regularized}\left[-\frac{1}{3},\,\frac{y^3}{9}\right] +\right.$

$\left.3^{2/3}\,y^4\,\text{HypergeometricOF1Regularized}\left[\frac{7}{3},\,\frac{y^3}{9}\right]\right)$

In[15]:= `intM1 = PowerExpand[intM1y /. y -> x^2]`

Out[15]= $\dfrac{3^{5/6}\,\text{Gamma}\left[\frac{4}{3}\right]}{2\,\pi}$ -

$\dfrac{1}{27}\,e^{-\frac{2\,x^3}{3}}\left(27\,x^2\,\text{AiryAiPrime}[x^2] - 27\,3^{1/3}\,\text{HypergeometricOF1Regularized}\left[-\frac{1}{3},\,\frac{x^6}{9}\right] +\right.$

$\left.3^{2/3}\,x^8\,\text{HypergeometricOF1Regularized}\left[\frac{7}{3},\,\frac{x^6}{9}\right]\right)$

In[16]:= `D[intM1, x] - x p[x] // FullSimplify`

Out[16]= $-\dfrac{1}{27\,3^{1/3}}$

$\left(2\,e^{-\frac{2\,x^3}{3}}\,x^7\left(-9\,\text{HypergeometricOF1Regularized}\left[\frac{4}{3},\,\frac{x^6}{9}\right] + 12\,\text{HypergeometricOF1Regularized}\left[\right.\right.\right.$

$\left.\left.\left.\frac{7}{3},\,\frac{x^6}{9}\right] + x^6\,\text{HypergeometricOF1Regularized}\left[\frac{10}{3},\,\frac{x^6}{9}\right]\right)\right)$

In[17]:= `Simplify[intM1 /. ({AiryAi[x_] ->`
`Normal[Series[AiryAi[ξ], {ξ, Infinity, 1}]]} /. ξ -> x), x > 0]`

Out[17]= $\dfrac{3^{5/6}\,\text{Gamma}\left[\frac{4}{3}\right]}{2\,\pi}$ -

$\dfrac{1}{27}\,e^{-\frac{2\,x^3}{3}}\left(27\,x^2\,\text{AiryAiPrime}[x^2] - 27\,3^{1/3}\,\text{HypergeometricOF1Regularized}\left[-\frac{1}{3},\,\frac{x^6}{9}\right] +\right.$

$\left.3^{2/3}\,x^8\,\text{HypergeometricOF1Regularized}\left[\frac{7}{3},\,\frac{x^6}{9}\right]\right)$

In[18]:= `Simplify[intM1 /. ({AiryAi[x_] ->`
`Normal[Series[AiryAi[ξ], {ξ, -Infinity, 1}]]} /. ξ -> x), x < 0]`

Out[18]= $\dfrac{3^{5/6}\,\text{Gamma}\left[\frac{4}{3}\right]}{2\,\pi}$ -

$\dfrac{1}{27}\,e^{-\frac{2\,x^3}{3}}\left(27\,x^2\,\text{AiryAiPrime}[x^2] - 27\,3^{1/3}\,\text{HypergeometricOF1Regularized}\left[-\frac{1}{3},\,\frac{x^6}{9}\right] +\right.$

$\left.3^{2/3}\,x^8\,\text{HypergeometricOF1Regularized}\left[\frac{7}{3},\,\frac{x^6}{9}\right]\right)$

Because of the asymptotics of $p(x)$ at $-\infty$, the second moment $\int_{-\infty}^{\infty}x^2\,p(x)\,dx$ does not exist.

In[19]:= `intM2 = PowerExpand[(Integrate[p[y] y, y] // FullSimplify) /. y -> x^2]`

Out[19]= $\dfrac{3\,3^{1/6}\,\text{Gamma}\left[\frac{5}{3}\right]}{4\,\pi}$ +

$\dfrac{1}{54}\,e^{-\frac{2\,x^3}{3}}\left(27\,x^4\,\text{AiryAi}[x^2] + 27\,3^{2/3}\,\text{HypergeometricOF1Regularized}\left[-\frac{2}{3},\,\frac{x^6}{9}\right] -\right.$

$\left.3^{1/3}\,x^{10}\,\text{HypergeometricOF1Regularized}\left[\frac{8}{3},\,\frac{x^6}{9}\right]\right)$

In[20]:= `D[intM2, x] - x^2 p[x] // FullSimplify`

Out[20]= 0

The integral $\int_0^{\infty}x^2\,p(x)\,dx$ evaluates to $1/(3^{1/3}\,\Gamma(1/3))$.

In[21]:= `(intM2[[2]]) - (intM2 /. x -> 0)`

Out[21]= $-\dfrac{3^{2/3}}{2\ \text{Gamma}\left[-\frac{2}{3}\right]}-\dfrac{3\ 3^{1/6}\ \text{Gamma}\left[\frac{5}{3}\right]}{4\ \pi}+$

$\dfrac{1}{54}\ e^{-\frac{2\ x^3}{3}}\left(27\ x^4\ \text{AiryAi}\left[x^2\right]+27\ 3^{2/3}\ \text{HypergeometricOF1Regularized}\left[-\frac{2}{3},\ \frac{x^6}{9}\right]-\right.$

$\left.3^{1/3}\ x^{10}\ \text{HypergeometricOF1Regularized}\left[\frac{8}{3},\ \frac{x^6}{9}\right]\right)$

23. Differential Equation for Dedekind η Function, Darboux–Halphen System

a) To find the differential equation for $\eta(z)$, we will make polynomial ansatz for p in the form

$$p(\eta(z),\eta'(z),\eta''(z),\eta'''(z),\eta''''(z))=\sum_{i_0,i_1,i_2,i_3,i_4=1}^{4}c_{i_0\,i_1\,i_2\,i_3\,i_4}\ \eta(z)^{i_0}\ \eta'(z)^{i_1}\ \eta''(z)^{i_2}\ \eta'''(z)^{i_3}\ \eta''''(z)^{i_4}$$

and then we will substitute the first Fourier series terms of $\eta(z)$ into this ansatz. As a result, we obtain a linear system for the $c_{i_0\,i_1\,i_2\,i_3\,i_4}$ that can be solved.

To avoid unnecessary multiplications, we will consider $\eta(z)$ as a series in $X=\exp(\frac{\pi i z}{12})$ and use SeriesData as the data structures to manipulate. The function ηSeries[n] produces a power series of order n in X.

In[1]:= `ηSeries[n_] := X Product[1 - X^(24 k), {k, 1, n/24 + 1}] + O[X]^n`

The series is a very sparse one. Only 11 terms out of the first 1000 powers possible are nonvanishing.

In[2]:= `ηSeries[1000]`

Out[2]= $X-X^{25}-X^{49}+X^{121}+X^{169}-X^{289}-X^{361}+X^{529}+X^{625}-X^{841}-X^{961}+O[X]^{1000}$

To calculate the differentiated functions $\eta'(z)$, $\eta''(z)$, $\eta'''(z)$, and $\eta''''(z)$, we backsubstitute $X\longrightarrow\exp(\frac{\pi i z}{12})$.

In[3]:= `expηSeries[n_] := Normal[ηSeries[n]] /. X -> Exp[Pi I z/12];`

Here are the first few even terms of the Fourier series of $\eta(z)$.

In[4]:= `expηSeries[500]`

Out[4]= $e^{\frac{i\pi z}{12}}-e^{\frac{25\,i\pi z}{12}}-e^{\frac{49\,i\pi z}{12}}+e^{\frac{121\,i\pi z}{12}}+e^{\frac{169\,i\pi z}{12}}-e^{\frac{289\,i\pi z}{12}}-e^{\frac{361\,i\pi z}{12}}$

The function $d\eta$Series[d][n] implements the d-times differentiated series for $\eta(z)$ of order n and express all terms through X.

In[5]:= `dηSeries[d_][n_] := dηSeries[d][n] = (D[expηSeries[n], {z, d}] /.`
`                          Exp[exp_] -> X^(12 exp/(I Pi z))) + O[X]^(n - d)`

In[6]:= `dηSeries[2][300]`

Out[6]= $-\dfrac{\pi^2\,X}{144}+\dfrac{625\,\pi^2\,X^{25}}{144}+\dfrac{2401\,\pi^2\,X^{49}}{144}-\dfrac{14641\,\pi^2\,X^{121}}{144}-\dfrac{28561\,\pi^2\,X^{169}}{144}+\dfrac{83521\,\pi^2\,X^{289}}{144}+O[X]^{298}$

Let us substitute our differentiated expressions into the ansatz for the differential equation. As we will see later, taking the first six nonvanishing terms of the series into account will generate enough terms to solve uniquely for all coefficients $c_{i_0\,i_1\,i_2\,i_3\,i_4}$.

In[7]:= `o = 300;`
`      ansatz =`
`      Sum[If[i0 + i1 + i2 + i3 > 4, 0, (* the terms *)`
`              c[i0, i1, i2, i3, i4] dηSeries[0][o]^i0 *`
`              dηSeries[1][o]^i1 dηSeries[2][o]^i2  *`
`              dηSeries[3][o]^i3 dηSeries[4][o]^i4],`
`              {i0, 0, 4}, {i1, 0, 4}, {i2, 0, 4}, {i3, 0, 4}, {i4, 0, 4}];`

Equating the coefficients of the powers of X to zero gives us 96 nontrivial equations.

In[9]:= `(eqs1 = DeleteCases[CoefficientList[Normal[ansatz], X], 0, {1}]) // Length`

Out[9]= 96

On the other hand, we have 350 unknowns $c_{i_0\,i_1\,i_2\,i_3\,i_4}$.

In[10]:= `(vars = Cases[eqs1, _c, Infinity] // Union) // Length`

In[10]:= 350

Because in the process of differentiation π appeared in various powers, the resulting equations eqs1 also contain π in various powers. Here, this is exemplified in the 11th equation.

In[11]:= **eqs1[[11]]**

Out[11]= $-\dfrac{390625\,\pi^8\,c[0,\,0,\,0,\,0,\,2]}{214990848} + \dfrac{203125\,i\,\pi^7\,c[0,\,0,\,0,\,1,\,1]}{17915904} + \dfrac{15625\,\pi^6\,c[0,\,0,\,0,\,2,\,0]}{1492992} +$

$\dfrac{195625\,\pi^6\,c[0,\,0,\,1,\,0,\,1]}{1492992} - \dfrac{8125\,i\,\pi^5\,c[0,\,0,\,1,\,1,\,0]}{124416} - \dfrac{625\,\pi^4\,c[0,\,0,\,2,\,0,\,0]}{10368} -$

$\dfrac{195325\,i\,\pi^5\,c[0,\,1,\,0,\,0,\,1]}{124416} - \dfrac{7825\,\pi^4\,c[0,\,1,\,0,\,1,\,0]}{10368} + \dfrac{325}{864}\,i\,\pi^3\,c[0,\,1,\,1,\,0,\,0] +$

$\dfrac{25}{72}\,\pi^2\,c[0,\,2,\,0,\,0,\,0] - \dfrac{195313\,\pi^4\,c[1,\,0,\,0,\,0,\,1]}{10368} + \dfrac{7813}{864}\,i\,\pi^3\,c[1,\,0,\,0,\,1,\,0] +$

$\dfrac{313}{72}\,\pi^2\,c[1,\,0,\,1,\,0,\,0] - \dfrac{13}{6}\,i\,\pi\,c[1,\,1,\,0,\,0,\,0] - 2\,c[2,\,0,\,0,\,0,\,0]$

Assuming that the powers of π are linear independent, we equate the coefficients of all powers of π for all equations. This gives us many more nontrivial equations, namely, 1246.

In[12]:= **(eqs2 = DeleteCases[Flatten[CoefficientList[#, pi]& /@**
           (eqs1 /. Pi -> pi)], 0]) // Length

Out[12]= 1246

Solving these 1246 equations gives the following result (for space reasons, we display only the nonzero solutions).

In[13]:= **sol = Solve[# == 0& /@ eqs2, vars];**
         DeleteCases[sol, _c -> 0, {2}]

          Solve::svars : Equations may not give solutions for all "solve" variables. More...

Out[14]= {{c[0, 4, 0, 0, 0] → -18 c[3, 0, 0, 0, 1], c[0, 4, 0, 0, 1] → -18 c[3, 0, 0, 0, 2],
         c[0, 4, 0, 0, 2] → -18 c[3, 0, 0, 0, 3], c[0, 4, 0, 0, 3] → -18 c[3, 0, 0, 0, 4],
         c[1, 2, 1, 0, 0] → 12 c[3, 0, 0, 0, 1], c[1, 2, 1, 0, 1] → 12 c[3, 0, 0, 0, 2],
         c[1, 2, 1, 0, 2] → 12 c[3, 0, 0, 0, 3], c[1, 2, 1, 0, 3] → 12 c[3, 0, 0, 0, 4],
         c[2, 0, 2, 0, 0] → 33 c[3, 0, 0, 0, 1], c[2, 0, 2, 0, 1] → 33 c[3, 0, 0, 0, 2],
         c[2, 0, 2, 0, 2] → 33 c[3, 0, 0, 0, 3], c[2, 0, 2, 0, 3] → 33 c[3, 0, 0, 0, 4],
         c[2, 1, 0, 1, 0] → -28 c[3, 0, 0, 0, 1], c[2, 1, 0, 1, 1] → -28 c[3, 0, 0, 0, 2],
         c[2, 1, 0, 1, 2] → -28 c[3, 0, 0, 0, 3], c[2, 1, 0, 1, 3] → -28 c[3, 0, 0, 0, 4]}}

Only a few coefficients are not equal to zero. After backsubstitution, we get the following result.

In[15]:= **ode1 = Sum[If[i0 + i1 + i2 + i3 > 4, 0, c[i0, i1, i2, i3, i4] ***
             η[z]^i0 η'[z]^i1 η''[z]^i2 η'''[z]^i3 η''''[z]^i4],
             {i0, 0, 4}, {i1, 0, 4}, {i2, 0, 4},
             {i3, 0, 4}, {i4, 0, 4}] /. sol[[1]] // Factor

Out[15]= $-(18\,\eta'[z]^4 - 12\,\eta[z]\,\eta'[z]^2\,\eta''[z] - 33\,\eta[z]^2\,\eta''[z]^2 + 28\,\eta[z]^2\,\eta'[z]\,\eta^{(3)}[z] - \eta[z]^3\,\eta^{(4)}[z])$

$(c[3,\,0,\,0,\,0,\,1] + c[3,\,0,\,0,\,0,\,2]\,\eta^{(4)}[z] +$

$\quad c[3,\,0,\,0,\,0,\,3]\,\eta^{(4)}[z]^2 + c[3,\,0,\,0,\,0,\,4]\,\eta^{(4)}[z]^3)$

The first factor looks like the result we are looking for:

$$18\,\eta'(z)^4 - 12\,\eta(z)\,\eta''(z)\,\eta'(z)^2 + 28\,\eta(z)^2\,\eta^{(3)}(z)\,\eta'(z) - 33\,\eta(z)^2\,\eta''(z)^2 - \eta(z)^3\,\eta^{(4)}(z) = 0.$$

In[16]:= **ode2 = Cases[ode1, _Plus?(FreeQ[#, _c, Infinity]&)][[1]]**

Out[16]= $18\,\eta'[z]^4 - 12\,\eta[z]\,\eta'[z]^2\,\eta''[z] - 33\,\eta[z]^2\,\eta''[z]^2 + 28\,\eta[z]^2\,\eta'[z]\,\eta^{(3)}[z] - \eta[z]^3\,\eta^{(4)}[z]$

To verify the correctness of the result, we carry out a numerical check. d[f, z0, ε] implements a quick numerical differentiation of the function f at point z0, and η[i][x, ε] are approximations for the numerical derivatives. (We do not use N[D[...]] to have a better control over the precision and points used.)

In[17]:= **d[f_, z0_, ε_] = (f[z0 + ε] - f[z0 - ε])/(2ε);**

In[18]:=   **ηD[0][z_, ε_] = DedekindEta[z];**
          Do[ηD[i][z_, ε_] = Together[d[ηD[i - 1][#, ε]&, z, ε]], {i, 1, 5}]

In[20]:= **??ηD**

Global`ηD

ηD[0][z_, ε_] = DedekindEta[z]

ηD[1][z_, ε_] = $\frac{-\text{DedekindEta}[z-\varepsilon]+\text{DedekindEta}[z+\varepsilon]}{2\,\varepsilon}$

ηD[2][z_, ε_] = $\frac{-2\,\text{DedekindEta}[z]+\text{DedekindEta}[z-2\,\varepsilon]+\text{DedekindEta}[z+2\,\varepsilon]}{4\,\varepsilon^2}$

ηD[3][z_, ε_] = $\frac{-\text{DedekindEta}[z-3\,\varepsilon]+3\,\text{DedekindEta}[z-\varepsilon]-3\,\text{DedekindEta}[z+\varepsilon]+\text{DedekindEta}[z+3\,\varepsilon]}{8\,\varepsilon^3}$

ηD[4][z_, ε_] = $\frac{6\,\text{DedekindEta}[z]+\text{DedekindEta}[z-4\,\varepsilon]-4\,\text{DedekindEta}[z-2\,\varepsilon]-4\,\text{DedekindEta}[z+2\,\varepsilon]+\text{DedekindEta}[z+4\,\varepsilon]}{16\,\varepsilon^4}$

ηD[5][z_, ε_] =
$\frac{-\text{DedekindEta}[z-5\,\varepsilon]+5\,\text{DedekindEta}[z-3\,\varepsilon]-10\,\text{DedekindEta}[z-\varepsilon]+10\,\text{DedekindEta}[z+\varepsilon]-5\,\text{DedekindEta}[z+3\,\varepsilon]+\text{DedekindEta}[z+5\,\varepsilon]}{32\,\varepsilon^5}$

Two numerical checks of the differential equations at random points in the upper half-plane strongly confirm that the derived differential equation is correct.

In[21]:= quickCheck[z0_, ε0_, prec_] :=
    {Max[Abs[#]], Plus @@ #}&[(* substitute numerical values *)
      (List @@ ode2) /. {η[z] -> N[ηD[0][z0, ε0], prec],
        Derivative[k_][η][z] -> N[ηD[k][z0, ε0], prec]}] // N[#, 22]&

In[22]:= quickCheck[2/3 + 3/4 I, 10^-100, 500]

Out[22]= {9.333661071319466658021, 0. × 10^{-100} + 0. × 10^{-100} i}

In[23]:= quickCheck[7/11 + 13/9 I, 10^-100, 500]

Out[23]= {0.07422974187168967869175, 0. × 10^{-101} + 0. × 10^{-101} i}

Also, symbolically, the differential equation still holds when more series terms are taken into account.

In[24]:= With[{o = 1000},
    ode2 /. {η[z] -> dηSeries[0][o], Derivative[k_][η][z] -> dηSeries[k][o]}]

Out[24]= O[X]^{999}

Note that introducing a new function $u(z) = \eta'(z)/\eta(z)$ allows us to reduce the order of the differential equation by one and gives the following nice differential equation $u^{(3)}(z) - 24\,u(z)\,u''(z) + 36\,u'(z)^2 = 0$.

In[25]:= Factor[ode2 //. Derivative[k_][η][z] :> D[u[z] η[z], {z, k - 1}]]

Out[25]= η[z]^4 (-36 u'[z]^2 + 24 u[z] u''[z] - u$^{(3)}$[z])

b) We start by defining the three solutions w_1, w_2, and w_3.

In[1]:= λ[τ_] = InverseEllipticNomeQ[Exp[I Pi τ]];
    w1[τ_] = 1/2 D[Log[λ'[τ]/λ[τ]], τ];
    w2[τ_] = 1/2 D[Log[λ'[τ]/(λ[τ] - 1)], τ];
    w3[τ_] = 1/2 D[Log[λ'[τ]/(λ[τ] (λ[τ] - 1))], τ];

Although InverseEllipticNomeQ is a complicated modular function, *Mathematica* can find the derivative of EllipticNomeQ.

In[5]:= InverseEllipticNomeQ'[q]

Out[5]= (2 EllipticK[InverseEllipticNomeQ[q]]^2
    (-1 + InverseEllipticNomeQ[q]) InverseEllipticNomeQ[q]) /
    (π q (-EllipticE[1 - InverseEllipticNomeQ[q]] EllipticK[InverseEllipticNomeQ[q]] +
      EllipticK[1 - InverseEllipticNomeQ[q]]
      (-EllipticE[InverseEllipticNomeQ[q]] + EllipticK[InverseEllipticNomeQ[q]]))))

InverseEllipticNomeQ is the inverse of the function EllipticNomeQ. So to calculate $q^{(-1)'}$ we use the following identity.

In[6]:= InverseFunction[f]'

Out[6]= $\frac{1}{f'[f^{(-1)}[\#1]]}$ &

The function EllipticNomeQ can be expressed through complete elliptic integrals.

In[7]:= eq = EllipticNomeQ[m] // FunctionExpand

Out[7]= $e^{-\frac{\pi\,\text{EllipticK}[1-m]}{\text{EllipticK}[m]}}$

The differentiation of the complete elliptic integrals is straightforward. Replacing $\exp(-\pi K(1-m)/K(m))$ by $q(m)$, substituting $q^{(-1)}(q)$ for m, and using $q(q^{(-1)}(z)) = z$, we arrive at the following derivative.

```
In[8]:= 1/D[eq, m] /.
        Exp[Pi EllipticK[1 - m]/EllipticK[m]] -> 1/EllipticNomeQ[m] /.
          m -> InverseEllipticNomeQ[q] /.
            EllipticNomeQ[InverseEllipticNomeQ[q]] -> q // Simplify
Out[8]= -(2 EllipticK[InverseEllipticNomeQ[q]]^2
         (-1 + InverseEllipticNomeQ[q]) InverseEllipticNomeQ[q]) /
        (π q (EllipticE[InverseEllipticNomeQ[q]] EllipticK[1 - InverseEllipticNomeQ[q]] +
           (EllipticE[1 - InverseEllipticNomeQ[q]] - EllipticK[1 - InverseEllipticNomeQ[q]])
           EllipticK[InverseEllipticNomeQ[q]]))
```

This result agrees with the built-in derivative.

```
In[9]:= % - InverseEllipticNomeQ'[q] // Together
Out[9]= 0
```

The actual proof that w_1, w_2, and w_3 are solutions is straightforward. We evaluate the derivatives of the $w_i(\tau)$ and writing the result over a common denominator yields the desired 0.

```
In[10]:= {D[w1[τ], τ] - (w1[τ](w2[τ] + w3[τ]) - w2[τ] w3[τ]),
         D[w2[τ], τ] - (w2[τ](w1[τ] + w3[τ]) - w1[τ] w3[τ]),
         D[w2[τ], τ] - (w2[τ](w1[τ] + w3[τ]) - w1[τ] w3[τ])} // Together
Out[10]= {0, 0, 0}
```

For generalizations, see [839].

24. Ramanujan Identities for φ and λ Function

a) This is the expression under consideration.

```
In[1]:= f[{p_, q_, r_}] := (EllipticTheta[3, 0, Exp[-p Pi]]/
                           EllipticTheta[3, 0, Exp[-q Pi]])^r
```

To find the polynomial a number is a root of, we use the function `Recognize` from the package `NumberTheory`Recog᾽ nize᾽`.

```
In[2]:= Needs["NumberTheory`Recognize`"]
```

We will carry out the search for the triples $\{p, q, r\}$ in the following way. We choose the smallest possible values for $\{p, q, r\}$ and calculate a 200-digit approximation for `f[{p, q, r}]`. We use `Recognize` to find a polynomial of maximal degree 25 for this number. If the polynomial found has a lower degree, we use this polynomial and check if one root is within 2000 digits identical to a 2500-digits approximation of `f[{p, q, r}]`. The function `verify` carries out this "verification".

```
In[3]:= verify[{p_, q_, r_}, poly_] :=
        Module[{s1, t1},
               s1 = N[f[{p, q, r}], 2500];
               (* calculate polynomial root numerically *)
               t1 = t /. FindRoot[Evaluate[poly], {t, s1},
                                  AccuracyGoal -> 3000, WorkingPrecision -> 5000];
               Abs[s1 - t1] < 10^-2000]
```

Now, we carry out the actual search for the triples $\{p, q, r\}$. We will search for 12 triples.

```
In[4]:= resultBag = {};
        maxDegree = 16;
        (* suppress messages *)
        Off[FindRoot::cvnwt];

        SeedRandom[1234567890];

        Module[{p = 2, q = 1, r = 0, s, poly},
        While[Length[resultBag] < 12,
              (* increase p, q, r values *)
              If[r < 5, r = r + 1,
                 If[q == p - 1, p = p + 1; q = 1; r = 1, q = q + 1; r = 1]];
```

```
    If[(* a new triple? *)
       MemberQ[resultBag, {{p, q, _}, _}], Null,
       s = N[f[{p, q, r}], 200];
       (* find polynomial *) poly = Recognize[s, maxDegree, t];
       If[(* is it a good polynomial? *)
          Exponent[poly, t] < maxDegree,
          If[verify[{p, q, r}, poly],
             AppendTo[resultBag, {{p, q, r}, poly}]]]]]]]
```

In the final step, we form a Root-object, verify that it agrees with the ratio of the Theta functions to 3000 digits (which is not a proof, but a very strong argument in favor), and try to rewrite the Root-object in radicals.

```
In[10]:= makeAlgebraicNumber[{{p_, q_, r_}, poly_}] :=
         Module[{s1, polyRoots, diffs, alg},
                (* high-precision value for f *)
                s1 = N[f[{p, q, r}], 3100];
                (* the root of the polynomial *)
                polyRoots = t /. Solve[poly == 0, t];
                diffs = Abs[polyRoots - s1];
                pos = Position[diffs, Min[diffs]][[1, 1]];
                (* Root → radicals *)
                alg = Together[ToRadicals[polyRoots[[pos]]]];
                {{p, q, r}, alg, N[alg, 3100] - s1}]
```

Here are the {p, q, r} triples and the corresponding algebraic numbers.

```
In[11]:= makeAlgebraicNumber /@ resultBag;
```

Having found some of the numbers we are interested in, we format the result in a nice way.

```
In[12]:= niceForm[{{p_, q_, r_}, alg_, δ_?(Abs[#] < 10^-3000&)}] :=
         Apply[HoldForm[Equal[##]]&,
               {EllipticTheta[3, 0, Exp[-p Pi]]/
                EllipticTheta[3, 0, Exp[-q Pi]], alg^(1/r)}]
In[13]:= TraditionalForm[niceForm /@ %%]
```

$$\left\{ \frac{\vartheta_3(0, e^{-2\pi})}{\vartheta_3(0, e^{-\pi})} = \frac{\sqrt{2 + \sqrt{2}}}{2}, \; \frac{\vartheta_3(0, e^{-3\pi})}{\vartheta_3(0, e^{-\pi})} = \frac{\sqrt[4]{3 + 2\sqrt{3}}}{\sqrt{3}}, \; \frac{\vartheta_3(0, e^{-3\pi})}{\vartheta_3(0, e^{-2\pi})} = \frac{2^{3/4} \sqrt[4]{9 + 6\sqrt{3} - 2\sqrt{6(7 + 4\sqrt{3})}}}{\sqrt{3}}, \right.$$

$$\frac{\vartheta_3(0, e^{-4\pi})}{\vartheta_3(0, e^{-\pi})} = \frac{1}{4}(2 + 2^{3/4}), \; \frac{\vartheta_3(0, e^{-4\pi})}{\vartheta_3(0, e^{-2\pi})} = \sqrt{\frac{17 + 12\sqrt{2} + 2\sqrt{140 + 99\sqrt{2}}}{2(17 + 12\sqrt{2})}}, \; \frac{\vartheta_3(0, e^{-5\pi})}{\vartheta_3(0, e^{-\pi})} = \sqrt{\frac{1}{5}(2 + \sqrt{5})},$$

$$\frac{\vartheta_3(0, e^{-5\pi})}{\vartheta_3(0, e^{-2\pi})} = \sqrt{\frac{2}{5}\left(4 + 2\sqrt{5} - \sqrt{2(9 + 4\sqrt{5})}\right)}, \; \frac{\vartheta_3(0, e^{-5\pi})}{\vartheta_3(0, e^{-3\pi})} = \frac{\sqrt[4]{3\left(-27 + 18\sqrt{3} + 4\sqrt{15(7 - 4\sqrt{3})}\right)}}{\sqrt{5}},$$

$$\frac{\vartheta_3(0, e^{-5\pi})}{\vartheta_3(0, e^{-4\pi})} = \sqrt{\text{Root}[390625\,\#1^8 - 50000000\,\#1^7 - 310000000\,\#1^6 + 396800000\,\#1^5 -}$$
$$1573120000\,\#1^4 + 1392640000\,\#1^3 + 121241600\,\#1^2 + 20971520\,\#1 + 1048576\,\&, \, 3],$$

$$\frac{\vartheta_3(0, e^{-6\pi})}{\vartheta_3(0, e^{-\pi})} = \sqrt{\text{Root}[2985984\,\#1^8 - 1492992\,\#1^6 - 331776\,\#1^5 - 38016\,\#1^4 - 82944\,\#1^3 - 31392\,\#1^2 - 3648\,\#1 + 1\,\&, \, 4]},$$

$$\frac{\vartheta_3(0, e^{-6\pi})}{\vartheta_3(0, e^{-2\pi})} = \sqrt{\text{Root}[729\,\#1^8 - 972\,\#1^6 + 14256\,\#1^5 + 270\,\#1^4 - 9504\,\#1^3 - 3996\,\#1^2 - 528\,\#1 + 1\,\&, \, 4]},$$

$$\frac{\vartheta_3(0, e^{-6\pi})}{\vartheta_3(0, e^{-3\pi})} = \frac{1}{2}\sqrt{\left(\left(194 + 112\sqrt{3} + 4\sqrt{2(2340 + 1351\sqrt{3})} + \right.\right.}$$

$$\left.\left.\sqrt{2\left(74977 + 43288\sqrt{3} + 774\sqrt{2(2340 + 1351\sqrt{3})} + 448\sqrt{6(2340 + 1351\sqrt{3})}\right)}\right)\right/$$

$$\left.\left(97 + 56\sqrt{3} + 2\sqrt{2(2340 + 1351\sqrt{3})}\right)\right)\right\}$$

b) This is the definition of λ_n.

```
In[1]:= λ[n_] := (DedekindEta[(1 + I Sqrt[n/3])/2]/
            DedekindEta[(1 + I Sqrt[3*n])/2])^6/(3 Sqrt[3])
```

We will again use the package `NumberTheory`Recognize`` to calculate an algebraic number of degree d that is near to λ_n.

```
In[2]:= Needs["NumberTheory`Recognize`"]

In[3]:= nearestRoot[λ_, d_] :=
        Module[{poly = Recognize[λ, d, C], rs},
                (* all roots of best polynomial *)
                rs = C /. {ToRules[Roots[poly == 0, C]]};
                (* select nearest root *)
                rs[[Position[#, Min[#]]&[Abs[rs - λ]][[1, 1]]]]]
```

By iteratively raising the degree, the function λAlg finds an algebraic number that agrees with λ_n to at least 5000 digits.

```
In[4]:= λAlg[n_, wp_] :=
        Module[{k = 1, λN = Re[N[λ[n], wp]], r, δ},
        While[(* best possible root *)
                r = nearestRoot[λN, 8 k];
                (* high-precision check *)
                δ = N[r - λ[n], 5000];
                δ != 0, (* raise precision *)k++]; r]
```

Using now sufficient precision for the numericalization of $\lambda[n]$, the following algebraic values can be calculated in seconds.

```
In[5]:= Off[N::meprec];
        λAlg[1, 100]
Out[6]= 1

In[7]:= λAlg[2, 100]
```

Out[7]= $\sqrt{-264 + 153\sqrt{3} + \sqrt{2\left(69961 - 40392\sqrt{3}\right)}}$

In[8]:= $\lambda Alg[3, 100]$

Out[8]= $\sqrt{3\left(-3 + 2\sqrt{3}\right)}$

In[9]:= $\lambda Alg[4, 100]$

Out[9]= $\sqrt{-18550 + 15147\sqrt{\frac{3}{2}} + 45\sqrt{\frac{1}{2}\left(679753 - 277508\sqrt{6}\right)}}$

In[10]:= $\lambda Alg[5, 100]$

Out[10]= $\frac{1}{2}\left(1 + \sqrt{5}\right)$

In[11]:= $\lambda Alg[6, 100]$

Out[11]= $3\sqrt{3\left(-17918 + 12670\sqrt{2} + 5\sqrt{3\left(8561499 - 6053894\sqrt{2}\right)}\right)}$

In[12]:= $\lambda Alg[7, 100]$

Out[12]= $\sqrt{-135 + 52\sqrt{7} + 6\sqrt{6\left(172 - 65\sqrt{7}\right)}}$

In[13]:= $\lambda Alg[8, 200]$

Out[13]= $\sqrt{Root[1 + 30248592 \#1 - 188475440 \#1^2 + 42675408 \#1^3 - }$
$953184234 \#1^4 + 42675408 \#1^5 - 188475440 \#1^6 + 30248592 \#1^7 + \#1^8 \&, 4]$

In[14]:= $\lambda Alg[9, 200]$

Out[14]= 3

In[15]:= $\lambda Alg[10, 200]$

Out[15]= $\sqrt{Root[1 + 340904160 \#1 - 3585031924 \#1^2 - 7530593760 \#1^3 + }$
$22278443046 \#1^4 - 7530593760 \#1^5 - 3585031924 \#1^6 + 340904160 \#1^7 + \#1^8 \&, 8]$

Now let us look for the values of n, such that λ_n is an algebraic number of degree four or less.

```
In[16]:= simpleλ[n_, (* maximal degree *) d_, wp_] :=
        Module[{λN = Re[N[λ[n], wp]], poly, rs, r, δ},
              poly = Recognize[λN, d, C];
              (* small coefficients? *)
              If[Max[Abs[CoefficientList[poly, C]]] < 10^10,
                 rs = C /. {ToRules[Roots[poly == 0, C]]};
                 r = rs[[Position[#, Min[#]]&[Abs[rs - λN]][[1, 1]]]];
                 (* high-precision check *)
                 δ = N[r - λ[n], 2000]; If[δ == 0, r, {}], {}]]

In[17]:= DeleteCases[Table[Subscript[λ, n] ==
                    simpleλ[n, 4, 100], {n, 100}], _ == {}] //
                    TableForm // TraditionalForm
```

Out[17]//TraditionalForm=

$\lambda_1 = 1$

$\lambda_3 = \sqrt{3\left(-3 + 2\sqrt{3}\right)}$

$\lambda_5 = \frac{1}{2}\left(1 + \sqrt{5}\right)$

$\lambda_9 = 3$

$\lambda_{13} = \frac{7}{4} + \frac{\sqrt{13}}{4} + \frac{1}{2}\sqrt{\frac{23}{2} + \frac{7\sqrt{13}}{2}}$

$\lambda_{17} = 4 + \sqrt{17}$

$\lambda_{21} = \frac{9}{4} + \frac{3\sqrt{21}}{4} + \frac{1}{2}\sqrt{\frac{171}{2} + \frac{39\sqrt{21}}{2}}$

$$\lambda_{25} = 9 + 4\sqrt{5}$$

$$\lambda_{33} = 3\left(6 + \sqrt{33}\right)$$

$$\lambda_{41} = 32 + 5\sqrt{41}$$

$$\lambda_{49} = 55 + 12\sqrt{21}$$

$$\lambda_{57} = 45 + 6\sqrt{57} + \sqrt{6\left(687 + 91\sqrt{57}\right)}$$

$$\lambda_{65} = 72 + 9\sqrt{65} + 4\sqrt{5\left(129 + 16\sqrt{65}\right)}$$

$$\lambda_{73} = 112 + 13\sqrt{73} + 4\sqrt{1555 + 182\sqrt{73}}$$

$$\lambda_{81} = 3\left(75 + 52\sqrt[3]{3} + 36\,3^{2/3}\right)$$

$$\lambda_{89} = 500 + 53\sqrt{89}$$

$$\lambda_{97} = 364 + 37\sqrt{97} + 2\sqrt{2\left(33161 + 3367\sqrt{97}\right)}$$

25. Identities for Gamma Function Values, Identities for Dedekind η Function

a) We start by generating all identities for a given denominator n. To generate a closed set of equations, we generate equations also for the divisors of n. We use Γ for the Gamma functions to avoid numerical equality tests (this would slow down all calculations considerably). The $\csc(\pi k/n)$ can be transformed into algebraic numbers later by using FunctionExpand.

```
In[1]:= allIΓdentities[n_] :=
        Module[{t1, t2, divs = Rest[Divisors[n]]},
        (* the equations following from the functional identities *)
        t1 = Table[Γ[k/n] Γ[1 - k/n] - Pi Csc[Pi k/n],
                   {k, 1, n - 1}];
        t2 = Flatten[Function[v, Function[z,
                   Γ[v z] - v^(v z - 1/2) (2Pi)^((1 - v)/2)*
                   Product[Γ[z + k/v], {k, 0, v - 1}]] /@ (1/divs)] /@ divs];
        (* evaluate integer and half-integer values and canonicalize arguments *)
        Flatten[{t1, t2}] /. Γ -> Gamma /. Gamma -> Γ /. cscRule /. gammaRule]
```

We reduce the number of cosecants and the Gamma functions with arguments greater than 1 with the following two rules.

```
In[2]:= cscRule = Csc[z_] :> Csc[Pi - z] /; z > Pi/2;
        gammaRule = Γ[z_?(# > 1&)] :> (z - 1) Γ[z - 1];
```

For $n = 5$, we obtain the following set of three equations.

```
In[4]:= Γ5Eqs = allIΓdentities[5]
```

$$\text{Out[4]= } \left\{ -2\sqrt{\frac{2}{5 - \sqrt{5}}}\ \pi + \Gamma\!\left[\tfrac{1}{5}\right]\Gamma\!\left[\tfrac{4}{5}\right],\ -2\sqrt{\frac{2}{5 + \sqrt{5}}}\ \pi + \Gamma\!\left[\tfrac{2}{5}\right]\Gamma\!\left[\tfrac{3}{5}\right], \right.$$

$$\left. -2\sqrt{\frac{2}{5 + \sqrt{5}}}\ \pi + \Gamma\!\left[\tfrac{2}{5}\right]\Gamma\!\left[\tfrac{3}{5}\right],\ -2\sqrt{\frac{2}{5 - \sqrt{5}}}\ \pi + \Gamma\!\left[\tfrac{1}{5}\right]\Gamma\!\left[\tfrac{4}{5}\right],\ 1 - \frac{\sqrt{5}\ \Gamma\!\left[\tfrac{1}{5}\right]\Gamma\!\left[\tfrac{2}{5}\right]\Gamma\!\left[\tfrac{3}{5}\right]\Gamma\!\left[\tfrac{4}{5}\right]}{4\,\pi^2} \right\}$$

Now, we divide all Gamma values occurring in the identities in two classes: the ones we want to keep and the ones we want to eliminate. Given a set of equations *eqs*, the function ΓsToEliminate returns a list of the Gamma values to eliminate.

```
In[5]:= ΓsToEliminate[eqs_, ΓsToKeep_] :=
                Complement[Union[Cases[eqs, _Γ, Infinity]], ΓsToKeep]
```

The function step will eliminate one (randomly chosen) Gamma value from *eqs*. It will try to keep the Gamma values from ΓsToKeep and return a reduced set of equations. For a fast execution, we deal only with Gamma values that appear linearly in the equations. (For the nonlinear case, we could use the more time-consuming GroebnerBasis command.)

```
In[6]:= step[eqs_, ΓsToKeep_] :=
        Module[{remainingGammas, exponents, posis, nextΓ, eqNumber, eqsNew},
        remainingGammas = ΓsToEliminate[eqs, ΓsToKeep];
        (* look for Gamma values that appear linearly *)
        exponents = Exponent[eqs, #]& /@ remainingGammas;
        posis = Position[exponents, 1, {2}];
        If[posis =!= {},
           (* select a Gamma value to eliminate *)
           r = Random[Integer, {1, Length[posis]}];
```

```
          {nextΓ, eqNumber} = posis[[r]];
          eqsNew = Union[DeleteCases[Numerator[Together[
                    eqs /. Solve[eqs[[eqNumber]]] == 0,
                        remainingGammas[[nextΓ]]][[1]]]], 0]];
          (* remove pure trig identities in csc[…] *)
          eqsNew = DeleteCases[eqsNew, _?(FreeQ[#, Γ]&)];
          eqsNew = Factor[eqsNew];
          eqsNew = If[MatchQ[#, _ _Γ], 0, #]& /@ eqsNew;
          eqsNew = Replace[#, _?NumericQ p_Plus -> p]& /@ eqsNew;
          eqsNew = Union[DeleteCases[eqsNew, 0]], eqs]]
```

Keeping $\Gamma[1/5]$, $\Gamma[4/5]$ in Γ5Eqs, we obtain the following (trivial) identities.

In[7]:= step[Γ5Eqs, {Γ[1/5], Γ[4/5]}]

Out[7]= $\left\{-2\sqrt{\dfrac{2}{5-\sqrt{5}}}\ \pi + \Gamma\left[\dfrac{1}{5}\right]\Gamma\left[\dfrac{4}{5}\right],\ 2\pi - \sqrt{\dfrac{10}{5+\sqrt{5}}}\ \Gamma\left[\dfrac{1}{5}\right]\Gamma\left[\dfrac{4}{5}\right]\right\}$

After the repeated application of step, we might be ending up with identities of the form we are looking for. The function makeGammaRatioIdentity determines if this is the case and formats the result nicely.

```
In[8]:= makeGammaRatioIdentity[expr_] :=
          Module[{aux1 = Expand[expr], aux2, aux3, aux4, aux5, α, β, A},
          If[Head[aux1] === Plus,
              aux2 = List @@ aux1;
              aux3 = DeleteCases[Union[#/(# /. _Γ -> 1)& /@ aux2], 1];
              If[Length[aux3] === 2 &&
                  (* different Gamma values? *)
                  ((Numerator[#] =!= 1 && Denominator[#] =!= 1)&[aux3[[1]]/aux3[[2]]]),
                  {α, β} = aux3;
                  aux4 = aux1 /. {aux3[[1]] -> A};
                  aux5 = Cancel[Solve[aux4 == 0, A][[1, 1, 2]]/β];
                  α/β == aux5]]]
```

To generate identities randomly, we implement a function randomPermutation that mixes the lists of Gamma values, so that we keep a random set of them.

```
In[9]:= randomPermutation[l_List] :=
          Module[{lTemp = l, j, λ = Length[l]},
              Do[j = Random[Integer, {i, λ}];
                  {lTemp[[i]], lTemp[[j]]} = {lTemp[[j]], lTemp[[i]]}, {i, λ}];
              lTemp]
```

Now, we have everything together to implement the function randomIdentity that tries to generate a random identity of Gamma functions of the form wanted.

```
In[10]:= randomIdentity[n_] :=
          Module[{eqs = allΓdentities[n], allΓs, Γs},
              allΓs = Union[Cases[eqs, _Γ, Infinity]];
              (* the Gamma values to keep; make a random choice *)
              Γs = Γ /@ Take[randomPermutation[allΓs],
                          {Random[Integer, {2, Min[6, Length[allΓs]]}]}];
              (* apply step until all unwanted Gamma values are eliminated *)
              FixedPoint[step[#, Γs]&, eqs, Length[allΓs] + 5]]
```

For a given seed *seed* of the random number generator, the function γId will return the identities for Gamma values found.

```
In[11]:= γId[seed_] :=
          (SeedRandom[seed];
          Union[DeleteCases[makeGammaRatioIdentity /@
                  randomIdentity[Random[Integer, {5, 48}]], Null],
              SameTest -> (First[#1] === First[#2]&)])
```

With a proper seed, we can obtain many identities of the form we were looking for.

In[12]:= γId[90938315823026141192]

Out[12]= $\left\{\dfrac{\Gamma\left[\frac{5}{12}\right]^2}{\Gamma\left[\frac{1}{4}\right]^2\Gamma\left[\frac{2}{3}\right]^2} == \dfrac{3^{3/4}\left(-\sqrt{2}+\sqrt{6}\right)}{4\pi}\right\}$

In[13]:= γId[52339558529598473094]

Out[13]= $\left\{ \dfrac{\Gamma[\frac{1}{5}]^2 \, \Gamma[\frac{4}{15}]^2}{\Gamma[\frac{2}{15}]^2 \, \Gamma[\frac{1}{3}]^2} == \dfrac{3^{1/10} \, 5^{1/6} \, \text{Csc}[\frac{\pi}{15}] \, \text{Csc}[\frac{4\pi}{15}]^2 \, \text{Sin}[\frac{2\pi}{15}]}{2\sqrt{2} \, (5 + \sqrt{5})} \right\}$

As remarked above, using `FunctionExpand`, we can convert the trigonometric expressions in nested radicals.

In[14]:= FunctionExpand[%] // FunctionExpand // Simplify

Out[14]= $\left\{ \dfrac{\Gamma[\frac{1}{5}]^2 \, \Gamma[\frac{4}{15}]^2}{\Gamma[\frac{2}{15}]^2 \, \Gamma[\frac{1}{3}]^2} == \dfrac{\sqrt{2} \; 3^{1/10} \, 5^{1/6} \, (\sqrt{3} + \sqrt{15} - \sqrt{10 - 2\sqrt{5}})}{2\sqrt{10} + 3\sqrt{3}\,(5+\sqrt{5}) - \sqrt{15}\,(5+\sqrt{5})} \right\}$

The right-hand side of the following identity autoevaluates to radicals.

In[15]:= γId[72121807168444733700]

Out[15]= $\left\{ \dfrac{\Gamma[\frac{1}{12}]^2}{\Gamma[\frac{1}{4}]^2 \, \Gamma[\frac{1}{3}]^2} == -\dfrac{3\sqrt{2} \; 3^{1/4}}{(-3+\sqrt{3})\,\pi} \right\}$

Here are some more examples of such identities.

In[16]:= γId[8382603719166208017]

Out[16]= $\left\{ \dfrac{\Gamma[\frac{3}{16}]^4 \, \Gamma[\frac{17}{48}]^2 \, \Gamma[\frac{7}{16}]^2 \, \Gamma[\frac{5}{16}]^2 \, \Gamma[\frac{11}{16}]^4 \, \Gamma[\frac{41}{48}]^2 \, \Gamma[\frac{15}{16}]^2}{\Gamma[\frac{23}{48}]^2 \, \Gamma[\frac{47}{48}]^2} == \right.$

$\left. \dfrac{1}{512} \left(\dfrac{3}{2} \right)^{3/4} \pi^7 \, \text{Csc}[\frac{\pi}{16}]^2 \, \text{Csc}[\frac{7\pi}{48}]^2 \, \text{Csc}[\frac{3\pi}{16}]^4 \, \text{Csc}[\frac{5\pi}{16}]^4 \, \text{Csc}[\frac{17\pi}{48}]^2 \, \text{Csc}[\frac{3\pi}{8}]^2 \, \text{Csc}[\frac{7\pi}{16}]^2 \right\}$

In[17]:= γId[31354816994988410858]

Out[17]= $\left\{ \dfrac{\Gamma[\frac{2}{3}]^2}{\Gamma[\frac{1}{4}]^2 \, \Gamma[\frac{11}{24}]^2 \, \Gamma[\frac{23}{24}]^2} == \dfrac{2 \; 2^{5/6} \, \text{Sin}[\frac{\pi}{24}] \, \text{Sin}[\frac{11\pi}{24}]}{3^{1/4} \, \pi^2} \right\}$

In[18]:= γId[50618374844506072572]

Out[18]= $\left\{ \dfrac{\Gamma[\frac{1}{3}]^2}{\Gamma[\frac{1}{12}]^2 \, \Gamma[\frac{3}{4}]^2} == \dfrac{3\sqrt{2} \; 3^{1/4} - 2\sqrt{2} \; 3^{3/4}}{6\,(-3+\sqrt{3})\,\pi} \right\}$

In[19]:= γId[88395267401272261046]

Out[19]= $\left\{ \dfrac{\Gamma[\frac{13}{30}]^2}{\Gamma[\frac{1}{6}]^2 \, \Gamma[\frac{3}{10}]^2 \, \Gamma[\frac{29}{30}]^2} == -\dfrac{3^{3/5} \, 5^{5/6} \, \text{Csc}[\frac{13\pi}{30}] \, \text{Sin}[\frac{\pi}{30}]}{2\,(-5+\sqrt{5})\,\pi^2} \right\}$

In[20]:= γId[38487837109080339099]

Out[20]= $\left\{ \dfrac{\Gamma[\frac{2}{9}]^2 \, \Gamma[\frac{8}{9}]^2}{\Gamma[\frac{1}{4}]^2 \, \Gamma[\frac{4}{9}]^2 \, \Gamma[\frac{11}{12}]^2} == \dfrac{2\sqrt{2}\,(-1+\sqrt{3})\,\text{Sin}[\frac{4\pi}{9}]^2}{3^{7/12}\,\pi} \right\}$

In[21]:= γId[80171190622502344844]

Out[21]= $\left\{ \dfrac{\Gamma[\frac{1}{5}]^2}{\Gamma[\frac{1}{20}]^2 \, \Gamma[\frac{13}{20}]^2} == \dfrac{\sqrt{5-\sqrt{5}} \, \text{Csc}[\frac{9\pi}{20}] \, \text{Sin}[\frac{7\pi}{20}]}{4 \; 2^{3/10} \, 5^{3/4} \, \pi} \right\}$

In[22]:= γId[24115537869945002262]

Out[22]= $\left\{ \dfrac{\Gamma[\frac{1}{4}]^2 \, \Gamma[\frac{5}{12}]^2}{\Gamma[\frac{1}{9}]^2 \, \Gamma[\frac{5}{36}]^2 \, \Gamma[\frac{13}{36}]^2 \, \Gamma[\frac{17}{36}]^4 \, \Gamma[\frac{29}{36}]^4} == \dfrac{3^{1/12} \, \text{Csc}[\frac{2\pi}{9}]^2 \, \text{Sin}[\frac{7\pi}{36}]^2 \, \text{Sin}[\frac{17\pi}{36}]^2}{4 \; 2^{1/6}\,(3+\sqrt{3})\,\pi^5} , \right.$

$\left. \dfrac{\Gamma[\frac{1}{4}]^2}{\Gamma[\frac{1}{9}]^2 \, \Gamma[\frac{13}{36}]^2 \, \Gamma[\frac{17}{36}]^2 \, \Gamma[\frac{29}{36}]^2} == \dfrac{\text{Csc}[\frac{5\pi}{36}] \, \text{Csc}[\frac{2\pi}{9}]^2 \, \text{Sin}[\frac{7\pi}{36}] \, \text{Sin}[\frac{17\pi}{36}]}{8 \; 2^{2/3} \, 3^{1/4} \, \pi^3} \right\}$

In[23]:= γId[7221486915064620684]

Out[23]= $\left\{ \dfrac{\Gamma[\frac{3}{10}]^2 \, \Gamma[\frac{2}{5}]^4 \, \Gamma[\frac{19}{40}]^2}{\Gamma[\frac{1}{40}]^2 \, \Gamma[\frac{11}{40}]^2 \, \Gamma[\frac{31}{40}]^2} == \dfrac{16 \; 2^{3/5}\,(-3+\sqrt{5})\,\pi \, \text{Sin}[\frac{\pi}{40}]^2 \, \text{Sin}[\frac{9\pi}{40}]^2 \, \text{Sin}[\frac{11\pi}{40}]^2}{-5+\sqrt{5}} \right\}$

Using the following program, we could now automatically search for Gamma identities.

```
Module[{ΓBag, rand, n, res, res1, finalRes},
ΓBag = {};
Do[(* a random seed *)
   rand = Random[Integer, {1, 10^20}];
   SeedRandom[rand];
   n = Random[Integer, {5, 48}];
   (* do not waste too much time on one example *)
   res = TimeConstrained[randomIdentity[n], 100];
   If[res =!= {} && res =!= $Aborted,
     res1 = Cases[makeGammaRatioIdentity /@ res, _Equal];
     (* print not already encountered identities *)
     If[res1 =!= {},
        finalRes = Select[res1, (FreeQ[ΓBag, #[[1]]] &&
                                 FreeQ[ΓBag, 1/#[[1]]])&];
        If[finalRes =!= {},
           ΓBag = Flatten[{ΓBag, First /@ finalRes}];
           (* potentially write right-hand sides as algebraic numbers and powers of π:
              finalRes = Union[TimeConstrained[finalRes //
                            FunctionExpand // RootReduce // ToRadicals, 300]]; *)
           Print[{rand, n, finalRes}]]]],
   (* many trials *) {10^5}]]
```

For similar calculations, see [1271].

b) A quick look at the examples shows two features:

a) all polynomials P are homogeneous

b) the coefficients c_i are the divisors of the largest of the c_i.

So we will look for identities of the form $P(\eta(c_1 \tau), \ldots, \eta(c_n \tau)) = 0$ with these three properties. To find such polynomials P we will implement the following strategy:

Given a positive integer c_{max} we form all monomials (so called η-products [522], [159], [413], [521], [984]) $\eta(c_1 \tau)^{e_1} \cdots \eta(c_n \tau)^{e_n}$ with $0 \le e_k \le s$, $e_1 + \cdots + e_n = s$ (s being a fixed positive integer) and the c_k are divisors of c_{max}. The fractional part of the Puiseux expansions of the monomials allow to naturally group the list of exponents $\{e_k\}$. Inside these groups we look for linear relations between the monomials

$$\sum_{\substack{e_1,\ldots,e_n=0 \\ e_1+\cdots+e_n=s}}^{s} \alpha_{e_1,\ldots,e_n} \, \eta(c_1 \tau)^{e_1} \cdots \eta(c_n \tau)^{e_n} = 0.$$

Here we expect the $\alpha_{e_1,\ldots,e_n}$ to be integers. The main function for finding such linear relations is NullSpace.

The product and series representations of $\eta(\tau)$ strongly suggest the use of the variable $q = e^{2i\pi\tau}$ instead of τ. The following two functions convert from $\eta(c\,\tau)$ to the corresponding factors $1 - q^c$ from the product representation and back. For brevity, here and in the following discussions, we will just use the factor $1 - q^c$ instead of the whole product. We also will always use η with the argument τ and the products with the variable q.

```
In[1]:= ηToΠ[expr_] := expr /. η[c_. τ] :> q^(c/24) (1 - q^(c k))
```

```
ΠToη[expr_] := PowerExpand[expr /. (1 - q^(c_. k)) :>
                     Exp[-c 2 Pi I τ/24] η[c τ] /. q :> Exp[2 Pi I τ]]
```

Here is an example.

```
In[3]:= ηToΠ[η[3τ]^3 η[5τ]^4]
```

```
Out[3]= q^(29/24) (1 - q^(3k))^3 (1 - q^(5k))^4
```

```
In[4]:= ΠToη[%]
```

```
Out[4]= η[3 τ]^3 η[5 τ]^4
```

The next step is the generation of all exponent lists $\{e_k\}_{k=1,...,n}$ for a given positive integer s. This means we need all partitions of s into n nonnegative integers. This is easily implemented—here is one possibility.

```
In[5]:= makeAllPartitions[s_, n_] := Flatten[
          Table[Evaluate[Append[Table[e[k], {k, n - 1}],
              s - Sum[e[j], {j, n - 1}]]],
              Evaluate[Sequence @@ Table[{e[k], 0, s - Sum[e[j],
              {j, k - 1}]}, {k, n - 1}]]], n - 2]

        makeAllPartitions[s_, 1] = {{s}};
```

Here is an example. We use $n = 3$ and $s = 6$.

```
In[8]:= makeAllPartitions[3, 6]
```

```
Out[8]= {{0, 0, 0, 0, 0, 3}, {0, 0, 0, 0, 1, 2}, {0, 0, 0, 0, 2, 1}, {0, 0, 0, 0, 3, 0},
         {0, 0, 0, 1, 0, 2}, {0, 0, 0, 1, 1, 1}, {0, 0, 0, 1, 2, 0}, {0, 0, 0, 2, 0, 1},
         {0, 0, 0, 2, 1, 0}, {0, 0, 0, 3, 0, 0}, {0, 0, 1, 0, 0, 2}, {0, 0, 1, 0, 1, 1},
         {0, 0, 1, 0, 2, 0}, {0, 0, 1, 1, 0, 1}, {0, 0, 1, 1, 1, 0}, {0, 0, 1, 2, 0, 0},
         {0, 0, 2, 0, 0, 1}, {0, 0, 2, 0, 1, 0}, {0, 0, 2, 1, 0, 0}, {0, 0, 3, 0, 0, 0},
         {0, 1, 0, 0, 0, 2}, {0, 1, 0, 0, 1, 1}, {0, 1, 0, 0, 2, 0}, {0, 1, 0, 1, 0, 1},
         {0, 1, 0, 1, 1, 0}, {0, 1, 0, 2, 0, 0}, {0, 1, 1, 0, 0, 1}, {0, 1, 1, 0, 1, 0},
         {0, 1, 1, 1, 0, 0}, {0, 1, 2, 0, 0, 0}, {0, 2, 0, 0, 0, 1}, {0, 2, 0, 0, 1, 0},
         {0, 2, 0, 1, 0, 0}, {0, 2, 1, 0, 0, 0}, {0, 3, 0, 0, 0, 0}, {1, 0, 0, 0, 0, 2},
         {1, 0, 0, 0, 1, 1}, {1, 0, 0, 0, 2, 0}, {1, 0, 0, 1, 0, 1}, {1, 0, 0, 1, 1, 0},
         {1, 0, 0, 2, 0, 0}, {1, 0, 1, 0, 0, 1}, {1, 0, 1, 0, 1, 0}, {1, 0, 1, 1, 0, 0},
         {1, 0, 2, 0, 0, 0}, {1, 1, 0, 0, 0, 1}, {1, 1, 0, 0, 1, 0}, {1, 1, 0, 1, 0, 0},
         {1, 1, 1, 0, 0, 0}, {1, 2, 0, 0, 0, 0}, {2, 0, 0, 0, 0, 1}, {2, 0, 0, 0, 1, 0},
         {2, 0, 0, 1, 0, 0}, {2, 0, 1, 0, 0, 0}, {2, 1, 0, 0, 0, 0}, {3, 0, 0, 0, 0, 0}}
```

The following graphic shows that the number of exponent lists increases quickly with n and s.

```
In[9]:= ListPlot3D[Table[Length @ makeAllPartitions[n, s], {s, 8}, {n, 8}],
          AxesLabel -> {"n", "s", None}, PlotRange -> All]
```

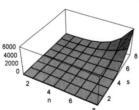

The product representation of the Dedekind Eta functions shows that for small q we have $\eta(c\,\tau) \propto q^{\frac{c}{24}}$. From this follows $\eta(c_1\,\tau)^{e_1} \cdots \eta(c_n\,\tau)^{e_n} \propto q^{(c_1\,e_1 + \cdots + c_n\,e_n)/24}$ for small q. This means that, for a set of coefficient–exponent lists $\{\{c_{i_1}, e_{i_1}\}, ..., \{c_{i_n}, s_{i_n}\}\}$ be able to form a linear relation, the fractional part of all $(c_{i_1}\,e_{i_1} + \cdots + c_{i_n}\,e_{i_n})/24$. The function splitProductGroupLists splits a list of coefficient–exponent lists pl into groups of equal fractional part of the $(c_{i_1}\,e_{i_1} + \cdots + c_{i_n}\,e_{i_n})/24$.

```
In[10]:= leadingqFactor[l_] :=
              (Times @@ (q^(-#/24)& /@ Apply[Times, 1, {1}])) /. q^e_ :> e
```

```
In[11]:= splitProductGroupLists[pl_] := Map[(Last /@ #)&,
           Split[Sort[{Mod[leadingqFactor[#], 1], #}& /@ pl], #1[[1]] === #2[[1]]&]]
```

Here is an example for $s = 8$, $n = 4$, and $\{c_1, c_2, c_3, c_4\} = \{1, 3, 7, 21\}$. We obtain 12 groups.

```
In[12]:= Length /@ splitProductGroupLists[
           Transpose[{{1, 3, 7, 21}, #}]& /@ makeAllPartitions[8, 4]]
```

```
Out[12]= {14, 14, 14, 13, 15, 13, 14, 14, 14, 13, 14, 13}
```

While theoretically it is not necessary to split the list of coefficient–exponent lists into subgroups, it is an essential step from a computational point of view. The resulting monomial lists are much smaller. This allows to treat examples with larger n and s.

To convert coefficient–exponent lists to products we use the function coefficientExponentListToProducts.

In[13]:= coefficientExponentListToProducts[l_] := Times @@ ηToΠ[η[#1 τ]^#2& @@@ l];

Here is again a simple example.

In[14]:= coefficientExponentListToProducts[{{1, 2}, {3, 3}, {5, 7}}]

Out[14]= $q^{23/12} (1 - q^k)^2 (1 - q^{3k})^3 (1 - q^{5k})^7$

Now let us implement the conversion of a product of Eta functions into their series around $q = 0$. The function qSeries[ηproduct, o] generates the series to order o. As mentioned above, for brevity we identify the product $\prod_{k=1}^{\infty} (1 - q^{ck})^e$ with $(1 - q^{ck})^e$. For efficiency, we spend some care in the implementation of qSeries.

In[15]:= (* recursive, pattern-based calculation of the series *)
(* products *)
qSeries𝐼[t_Times, o_] := Series[qSeries[#, o]& /@ t, {q, 0, o}]
(* single powers of q *)
qSeries𝐼[q^e_., o_] := q^e
(* powers *)
qSeries𝐼[Power[p_Plus, e_], o_] := qSeries𝐼[p, o]^e
(* the q products *)
qSeries𝐼[1 - q^(c_ k), o_] :=
    Series[Normal[qSeries𝐼[1 - q^k, o]] /. q -> q^c, {q, 0, o}];
(* the basis product; use sum representation *)
qSeries𝐼[1 - q^k, o_] := qSeries𝐼[1 - q^k, o] =
    Series[Sum[(-1)^k q^(k (3k - 1)/2), {k, -#, #}]&[
            Ceiling[(Sqrt[24o + 1] + 1)/6] + 2], {q, 0, o}];

(* guarantee right order in q *)
qSeries[expr_, o_] := qSeries𝐼[expr, o] + O[q]^(o + 1)

Here is again an example, the series of $q^4 \prod_{k=1}^{\infty} (1 - q^{2k})^3 (1 - q^{7k})^5$.

In[29]:= qSeries[q^4 (1 - q^(2k))^3 (1 - q^(7k))^5, 100] // Timing

Out[29]= {0.02 Second, $q^4 - 3 q^6 + 5 q^{10} - 5 q^{11} + 15 q^{13} - 7 q^{16} - 25 q^{17} + 5 q^{18} - 15 q^{20} + 35 q^{23} + 34 q^{24} + 10 q^{25} - 30 q^{27} - 35 q^{30} + 5 q^{31} - 15 q^{32} + 34 q^{34} - 70 q^{37} - 30 q^{38} - 6 q^{39} + 73 q^{41} + 105 q^{44} + 60 q^{45} + 8 q^{46} - 40 q^{48} + 42 q^{51} - 160 q^{52} - 40 q^{53} - 185 q^{55} + 35 q^{58} + 71 q^{59} + 65 q^{60} + 120 q^{62} - 175 q^{65} + 30 q^{66} + 185 q^{67} + 126 q^{69} - 105 q^{72} + 125 q^{73} - 261 q^{74} + 45 q^{76} + 140 q^{79} + 180 q^{80} - 273 q^{81} - 225 q^{83} - 63 q^{86} - 405 q^{87} + 155 q^{88} - 65 q^{90} + 315 q^{93} + 37 q^{94} + 440 q^{95} + 315 q^{97} + 35 q^{100} + O[q]^{101}$}

Because later we have to calculate many series of Eta products, we did optimize qSeries (for instance, by using the series representation of the Dedekind Eta function and by caching the results of already-calculated expansions of powers). A direct calculation of the series expansion of the product works, of course, but is much slower.

In[30]:= (Series[q^4 Product[(1 - q^(2k))^3 (1 - q^(7k))^5,
                    {k, 100}], {q, 0, 100}] == %[[2]]) // Timing

Out[30]= {0.08 Second, True}

For the null space computation we do not need the series, but rather the list of coefficients of the q-series. Given a q-product $(1 - q^{ck})^e$, the function coefficientList calculates a vector of the coefficients of length o. To be able to deal with larger examples we store the list of coefficients as a packed array.

In[31]:= coefficientList[prod_, o_] := coefficientList[prod, o] =
    (* to save memory *) Developer`ToPackedArray @
            (* add trailing zeros if needed *)
            Join[#, Table[0, {o + 1 - Length[#]}]]& @
                    CoefficientList[Normal[qSeries[prod, o]], q]

Here is the coefficient list corresponding to the last example.

In[32]:= coefficientList[q^4 (1 - q^(2k))^3 (1 - q^(7k))^5, 14]

Out[32]= {0, 0, 0, 0, 1, 0, -3, 0, 0, 0, 5, -5, 0, 15, 0}

Taking the coefficient list vectors of all Eta products of one group yields the matrix whose null space is to be determined. The function modularSearch makes a primarily computation of this null space. Because a full null space computation is quite expensive for larger matrices (for an $n \times n$ matrix it is $O(n^3)$), we start with a modular null space computation.

Although it has the same complexity, its absolute runtime is much smaller. This modular null space eliminates many Eta products. We will determine the null space corresponding to the remaining monomials in a moment more carefully.

```
In[33]:= modularSearch[listOfProducts_, o_, prime_:Prime[10^4]] :=
            NullSpace[Transpose[coefficientList[#, o]& /@
                        listOfProducts], Modulus -> prime]
```

If the length of the list of products is small enough, we can also afford to calculate the exact null space.

```
In[34]:= exactSearch[listOfProducts_, o_] := NullSpace[
            Transpose[coefficientList[#, o]& /@ listOfProducts]]
```

Frequently it will happen that the monomials that form a null space returned by `modularSearch` have common factors. Because these factors will unnecessary slow down the computations, we eliminate them with the function `eliminateCommonFactors`.

```
In[35]:= eliminateCommonFactors[listOfProducts_] :=
            Module[{pl = listOfProducts /. 1 - q^(c_. k) :> C[c]},
                    pl/(PolynomialGCD @@ pl) /. C[c_] :> 1 - q^(c k)]
```

Here is an example of `eliminateCommonFactors` at work.

```
In[36]:= eliminateCommonFactors[
            {q^2 (1 - q^(2k))^3 (1 - q^(3k))^3,
             q^5 (1 - q^(2k))^4 (1 - q^(3k))^1 (1 - q^(5k))^3}]
Out[36]= {(1 - q^(3 k))^2, q^3 (1 - q^(2 k)) (1 - q^(5 k))^3}
```

We do one more optimization: The monomials forming a nontrivial null space might be monomials not only in q, but in a power of q. Using the function `reducePowers`, we rewrite such monomials as monomials in q.

```
In[37]:= reducePowers[listOfProducts_] :=
            If[FreeQ[listOfProducts /. (1 - q^_) :> 1, q, Infinity],
            With[{gcd = GCD @@ (Last /@ Cases[listOfProducts, q^_, Infinity]/k)},
                 listOfProducts /. q^e_ :> q^(e/gcd)], listOfProducts]
```

In the following example effectively the substitution $q^3 \to q$ is carried out.

```
In[38]:= reducePowers[{(1 - q^(3k))^5, (1 - q^(6k))^5, 1 - q^(9k)}]
Out[38]= {(1 - q^k)^5, (1 - q^(2 k))^5, 1 - q^(3 k)}
```

Now we have nearly everything together to find linear relations between a set products of Eta functions.

```
In[39]:= makeProductIdentities[listOfProducts_] :=
            Module[{ms, ms1, pl, pl1, es},
                (* start with a modular null space computation *)
                ms = modularSearch[listOfProducts,
                                    (* use more terms than unknowns *)
                                    Round[1.5 Length[listOfProducts]]];
                If[ms === {}, {},
                    (* nonempty null space was found; keep relevant monomials *)
                    ms1 = Map[If[# =!= 0, 1, 0]&, ms, {-1}];
                    (* for each vector that spans the null space *)
                    Table[(* keep only relevant monomials *)
                          pl = DeleteCases[ms1[[j]] listOfProducts, 0];
                          If[Length[pl] > 1,
                              (* simplify resulting products *)
                              pl1 = reducePowers[eliminateCommonFactors[pl]];
                              (* exact null space computation;
                                 use again more terms than unknowns *)
                              es = exactSearch[pl1, Round[2 Length[pl1]]];
                              If[es === {}, {},
                                  (* make η products *)
                                  ηProducts = eliminateCommontFactors[ΠToη[pl1]];
                                  (#.ηProducts)& /@ es], {}],
                      {j, Length[ms]}]]]
```

In the last step of the calculations done by `makeProductIdentities`, products are converted to Eta functions. For a nicer-looking result, the function `eliminateCommontFactors` is applied to extract common powers of τ.

In[40]:= `eliminateCommonτFactors[listOfProducts_] :=`
        `listOfProducts/First[Union[listOfProducts /. _η :> 1]]`

We test the function `findModularEquations` using the monomials of an example given in the exercise.

In[41]:= `eliminateCommonFactors[ηToΠ[{η[10τ]^2 η[τ]^5, η[2τ]^2 η[5τ]^4 η[τ],`
                    `η[2τ]^5 η[5τ]η[10τ]}]]`

Out[41]= $\{(1-q^k)^5 (1-q^{10 k})^2, (1-q^k)(1-q^{2 k})^2 (1-q^{5 k})^4, (1-q^{2 k})^5 (1-q^{5 k})(1-q^{10 k})\}$

In[42]:= `makeProductIdentities[%]`

Out[42]= $\{\{-5 \eta[\tau] \eta[2 \tau]^2 \eta[5 \tau]^4 + 4 \eta[2 \tau]^5 \eta[5 \tau] \eta[10 \tau] + \eta[\tau]^5 \eta[10 \tau]^2\}\}$

Before searching for further functional identities we implement one more function. `unionηIdentities[ηIdentities]` removes doubles from the list *ηIdentities*.

In[43]:= `unionηIdentities[l_] :=`
    `Union[Cases[Union[Factor[Flatten[{l, -1}]]], _Plus, Infinity],`
        `SameTest -> (Expand[#1 - #2] === 0 || Expand[#1 + #2] === 0&)]`

In[44]:= `findModularEquations[exponentDegree_, cMax_] :=`
    `Module[{cList, numberOfηa, aps, ces, cesGroups, res},`
      `(* list of the ci's *)`
      `cList = Divisors[cMax];`
      `numberOfηa = Length[cList];`
      `(* list of all monomials *)`
      `aps = makeAllPartitions[exponentDegree, numberOfηa];`
      `ces = Transpose[{cList, #}]& /@ aps;`
      `(* split monomials in groups *)`
      `cesGroups = splitProductGroupLists[ces];`
      `(* deals with each group individually *)`
      `(* eliminate doubles *) unionηIdentities @`
      `Table[(* simplify monomials in groups *)`
          `actProductGroups = eliminateCommonFactors[`
          `coefficientExponentListToProducts /@ cesGroups[[β]]];`
          `(* try to find identities inside group *)`
          `If[Length[actProductGroups] > 1,`
            `aux = makeProductIdentities[actProductGroups];`
            `If[Flatten[aux] =!= {}, aux, {}]],`
        `{β, Length[cesGroups]}]]`

We test the function `findModularEquations` using the values $s = 13, c_{max} = 6$.

In[45]:= `ME[13, 6] = findModularEquations[13, 6]`

Out[45]= $\{9 \eta[\tau] \eta[2 \tau]^4 \eta[3 \tau]^8 - 8 \eta[2 \tau]^9 \eta[3 \tau]^3 \eta[6 \tau] - \eta[\tau]^9 \eta[6 \tau]^4,$
$\eta[\tau]^6 \eta[3 \tau]^6 - \eta[\tau]^5 \eta[2 \tau]^5 \eta[6 \tau] + 9 \eta[\tau] \eta[2 \tau] \eta[3 \tau]^5 \eta[6 \tau]^5 - 8 \eta[2 \tau]^6 \eta[6 \tau]^6,$
$\eta[2 \tau]^9 \eta[3 \tau]^4 - \eta[\tau]^9 \eta[3 \tau] \eta[6 \tau]^3 - 9 \eta[\tau]^4 \eta[2 \tau] \eta[6 \tau]^8,$
$-\eta[\tau] \eta[2 \tau]^3 \eta[3 \tau]^9 + \eta[2 \tau]^8 \eta[3 \tau]^4 \eta[6 \tau] - \eta[\tau]^4 \eta[6 \tau]^9,$
$\eta[2 \tau]^4 \eta[3 \tau]^9 - \eta[\tau]^8 \eta[3 \tau] \eta[6 \tau]^4 - 8 \eta[\tau]^3 \eta[2 \tau] \eta[6 \tau]^9\}$

Let us check the found identities by using high-precision values for the Dedekind Eta functions at "random" arguments.

In[46]:= `ηN[τ_?NumericQ] := ηN[τ] = DedekindEta[τ];`
    `NCheck[ηIdentities_, prec_:1000, z0_: 1/Pi + E/2 I] :=`
        `ηIdentities /. η -> ηN /. τ -> N[z0, prec]`

Here is a 1000-digit check of the five found identities.

In[48]:= `NCheck[ME[13, 6]] // Abs // Max`

Out[48]= $0. \times 10^{-1002}$

The first of the found identities was mentioned in the exercise part. Because some of the identities might factor we implement a function `factorIdentities`.

In[49]:= `factorIdentities[ids_, o_:100] :=`
    `Select[Union[Cases[Factor[ids], _Plus, Infinity]],`
        `(* check the factors *) NCheck[#] == 0.&]`

The identities $ME[13, 6]$ did not factor.

In[50]:= `factorIdentities[M$\mathcal{E}$[13, 6]] === M$\mathcal{E}$[13, 6]`

Out[50]= True

Now it is straightforward to search 25 additional modular equations: we just try increasing values for s and c_{max}.

In[51]:= `$\mathcal{F}\eta I$[exponentDegree_, cMax_] := factorIdentities[`
       `findModularEquations[exponentDegree, cMax]]`

To select only identities that are algebraically independent we implement a function `reduceIdentities`. Given a set of identities *ids* it will first calculate a Gröbner basis of *ids* and then tries to find a subset of *ids* that generates the same ideal.

In[52]:= `reduceIdentities[ids_] :=`
       `Module[{mo = MonomialOrder -> DegreeReverseLexicographic,`
             `vars = Union[Cases[ids, _$\eta$, Infinity]],`
             `idsNew = ids, idsTemp, gbTemp},`
           `(* try if an identity follows from the other ones *)`
           `Do[idsTemp = DeleteCases[idsNew, ids[[k]]];`
             `gbTemp = GroebnerBasis[idsTemp, vars, mo];`
             `If[Last[PolynomialReduce[ids[[k]], gbTemp, vars, mo]] === 0,`
               `idsNew = idsTemp], {k, Length[ids]}];`
           `idsNew]`

All equations from $M\mathcal{E}$[13, 6] are algebraically independent.

In[53]:= `reduceIdentities[M$\mathcal{E}$[13, 6]] == M$\mathcal{E}$[13, 6]`

Out[53]= True

Here are some of infinitely possible identities we were looking for (we are not exhaustive here).

In[54]:= `$\mathcal{F}\eta I$[24, 4] // TraditionalForm`

Out[54]//TraditionalForm=

$$\{\eta(2\,\tau)^{24} - 16\,\eta(\tau)^8\,\eta(4\,\tau)^{16} - \eta(\tau)^{16}\,\eta(4\,\tau)^8\}$$

In[55]:= `$\mathcal{F}\eta I$[18, 6] // reduceIdentities // TraditionalForm`

Out[55]//TraditionalForm=

$$\{-\eta(6\,\tau)^4\,\eta(\tau)^9 + 9\,\eta(2\,\tau)^4\,\eta(3\,\tau)^8\,\eta(\tau) - 8\,\eta(2\,\tau)^9\,\eta(3\,\tau)^3\,\eta(6\,\tau),$$
$$\eta(6\,\tau)^3\,\eta(\tau)^{14} - \eta(2\,\tau)^9\,\eta(3\,\tau)^3\,\eta(\tau)^5 + 81\,\eta(2\,\tau)^5\,\eta(3\,\tau)^7\,\eta(6\,\tau)^4\,\eta(\tau) - 72\,\eta(2\,\tau)^{10}\,\eta(3\,\tau)^2\,\eta(6\,\tau)^5,$$
$$\eta(\tau)^6\,\eta(3\,\tau)^6 + 9\,\eta(\tau)\,\eta(2\,\tau)\,\eta(6\,\tau)^5\,\eta(3\,\tau)^5 - \eta(\tau)^5\,\eta(2\,\tau)^5\,\eta(6\,\tau)\,\eta(3\,\tau) - 8\,\eta(2\,\tau)^6\,\eta(6\,\tau)^6,$$
$$8\,\eta(3\,\tau)^3\,\eta(2\,\tau)^{14} + \eta(\tau)^9\,\eta(6\,\tau)^3\,\eta(2\,\tau)^5 - 81\,\eta(\tau)^5\,\eta(3\,\tau)^4\,\eta(6\,\tau)^7\,\eta(2\,\tau) - 9\,\eta(\tau)^{10}\,\eta(3\,\tau)^5\,\eta(6\,\tau)^2,$$
$$-\eta(3\,\tau)\,\eta(6\,\tau)^3\,\eta(\tau)^9 - 9\,\eta(2\,\tau)\,\eta(6\,\tau)^8\,\eta(\tau)^4 + \eta(2\,\tau)^9\,\eta(3\,\tau)^4,$$
$$-\eta(\tau)\,\eta(2\,\tau)^3\,\eta(3\,\tau)^9 + \eta(2\,\tau)^8\,\eta(6\,\tau)\,\eta(3\,\tau)^4 - \eta(\tau)^4\,\eta(6\,\tau)^9, \eta(2\,\tau)^4\,\eta(3\,\tau)^9 - \eta(\tau)^8\,\eta(6\,\tau)^4\,\eta(3\,\tau) - 8\,\eta(\tau)^3\,\eta(2\,\tau)\,\eta(6\,\tau)^9,$$
$$\eta(2\,\tau)^3\,\eta(3\,\tau)^{14} + \eta(\tau)^3\,\eta(6\,\tau)^9\,\eta(3\,\tau)^5 - \eta(\tau)^7\,\eta(2\,\tau)^4\,\eta(6\,\tau)^5\,\eta(3\,\tau) - 8\,\eta(\tau)^2\,\eta(2\,\tau)^5\,\eta(6\,\tau)^{10},$$
$$-8\,\eta(\tau)^3\,\eta(6\,\tau)^{14} + \eta(2\,\tau)^3\,\eta(3\,\tau)^9\,\eta(6\,\tau)^5 - \eta(\tau)^4\,\eta(2\,\tau)^7\,\eta(3\,\tau)^5\,\eta(6\,\tau) + \eta(\tau)^5\,\eta(2\,\tau)^2\,\eta(3\,\tau)^{10}\}$$

In[56]:= `$\mathcal{F}\eta I$[22, 8] // reduceIdentities // TraditionalForm`

Out[56]//TraditionalForm=

$$\{-\eta(4\,\tau)^2\,\eta(8\,\tau)^4\,\eta(\tau)^{12} - 4\,\eta(2\,\tau)^2\,\eta(8\,\tau)^8\,\eta(\tau)^8 - 8\,\eta(4\,\tau)^{10}\,\eta(8\,\tau)^4\,\eta(\tau)^4 + \eta(2\,\tau)^{10}\,\eta(4\,\tau)^8,$$
$$-\eta(2\,\tau)^{12} - 32\,\eta(8\,\tau)^8\,\eta(2\,\tau)^4 + 8\,\eta(4\,\tau)^{12} + \eta(\tau)^8\,\eta(4\,\tau)^4, \eta(4\,\tau)^{12} - \eta(\tau)^4\,\eta(2\,\tau)^2\,\eta(8\,\tau)^4\,\eta(4\,\tau)^2 - 4\,\eta(2\,\tau)^4\,\eta(8\,\tau)^8,$$
$$-32\,\eta(\tau)^4\,\eta(2\,\tau)^2\,\eta(8\,\tau)^{12} - 4\,\eta(\tau)^8\,\eta(4\,\tau)^2\,\eta(8\,\tau)^8 - \eta(\tau)^4\,\eta(2\,\tau)^{10}\,\eta(8\,\tau)^4 + \eta(2\,\tau)^8\,\eta(4\,\tau)^{10}\}$$

In[57]:= `$\mathcal{F}\eta I$[20, 9] // reduceIdentities // TraditionalForm`

Out[57]//TraditionalForm=

$$\{\eta(3\,\tau)^{12} - 27\,\eta(\tau)^3\,\eta(9\,\tau)^9 - 9\,\eta(\tau)^6\,\eta(9\,\tau)^6 - \eta(\tau)^9\,\eta(9\,\tau)^3\}$$

In[58]:= `$\mathcal{F}\eta I$[12, 10] // reduceIdentities // TraditionalForm`

Out[58]//TraditionalForm=

$$\{-\eta(10\,\tau)^2\,\eta(\tau)^5 + 5\,\eta(2\,\tau)^2\,\eta(5\,\tau)^4\,\eta(\tau) - 4\,\eta(2\,\tau)^5\,\eta(5\,\tau)\,\eta(10\,\tau),$$
$$\eta(10\,\tau)\,\eta(\tau)^8 - \eta(2\,\tau)^5\,\eta(5\,\tau)\,\eta(\tau)^3 + 25\,\eta(2\,\tau)^3\,\eta(5\,\tau)^3\,\eta(10\,\tau)^2\,\eta(\tau) - 20\,\eta(2\,\tau)^6\,\eta(10\,\tau)^3,$$
$$-\eta(5\,\tau)\,\eta(10\,\tau)\,\eta(\tau)^5 - 5\,\eta(2\,\tau)\,\eta(10\,\tau)^4\,\eta(\tau)^2 + \eta(2\,\tau)^5\,\eta(5\,\tau)^2,$$
$$\eta(\tau)^4\,\eta(5\,\tau)^4 + 5\,\eta(\tau)\,\eta(2\,\tau)\,\eta(10\,\tau)^3\,\eta(5\,\tau)^3 - \eta(\tau)^3\,\eta(2\,\tau)^3\,\eta(10\,\tau)\,\eta(5\,\tau) - 4\,\eta(2\,\tau)^4\,\eta(10\,\tau)^4,$$
$$-\eta(\tau)\,\eta(2\,\tau)\,\eta(5\,\tau)^5 + \eta(2\,\tau)^4\,\eta(10\,\tau)\,\eta(5\,\tau)^2 - \eta(\tau)^2\,\eta(10\,\tau)^5, \eta(2\,\tau)^2\,\eta(5\,\tau)^5 - \eta(\tau)^4\,\eta(10\,\tau)^2\,\eta(5\,\tau) - 4\,\eta(\tau)\,\eta(2\,\tau)\,\eta(10\,\tau)^5,$$
$$\eta(2\,\tau)\,\eta(5\,\tau)^8 + \eta(\tau)\,\eta(10\,\tau)^5\,\eta(5\,\tau)^3 - \eta(\tau)^3\,\eta(2\,\tau)^2\,\eta(10\,\tau)^3\,\eta(5\,\tau) - 4\,\eta(2\,\tau)^3\,\eta(10\,\tau)^6,$$

$-4\,\eta(\tau)\,\eta(5\,\tau)\,\eta(2\,\tau)^8 - \eta(\tau)^6\,\eta(10\,\tau)\,\eta(2\,\tau)^3 + 25\,\eta(5\,\tau)^6\,\eta(10\,\tau)\,\eta(2\,\tau)^3 - 100\,\eta(\tau)\,\eta(5\,\tau)\,\eta(10\,\tau)^6\,\eta(2\,\tau)^2 + 5\,\eta(\tau)^7\,\eta(5\,\tau)^3,$

$-4\,\eta(5\,\tau)\,\eta(2\,\tau)^9 - \eta(\tau)^5\,\eta(10\,\tau)\,\eta(2\,\tau)^4 - 100\,\eta(5\,\tau)\,\eta(10\,\tau)^6\,\eta(2\,\tau)^3 + 25\,\eta(5\,\tau)^9\,\eta(2\,\tau) + 5\,\eta(\tau)^6\,\eta(5\,\tau)^3\,\eta(2\,\tau) +$

$25\,\eta(\tau)\,\eta(5\,\tau)^4\,\eta(10\,\tau)^5, -4\,\eta(\tau)\,\eta(10\,\tau)^8 + \eta(2\,\tau)\,\eta(5\,\tau)^5\,\eta(10\,\tau)^3 - \eta(\tau)^2\,\eta(2\,\tau)^3\,\eta(5\,\tau)^3\,\eta(10\,\tau) + \eta(\tau)^3\,\eta(5\,\tau)^6\}$

In[59]:= $\mathcal{F}\eta\mathcal{I}$[07, 12] // reduceIdentities // TraditionalForm

Out[59]//TraditionalForm=

$\{-\eta(6\,\tau)^2\,\eta(12\,\tau)\,\eta(\tau)^4 + 4\,\eta(3\,\tau)\,\eta(4\,\tau)^3\,\eta(6\,\tau)^2\,\eta(\tau) - 3\,\eta(2\,\tau)^2\,\eta(3\,\tau)^4\,\eta(12\,\tau),$

$-\eta(4\,\tau)\,\eta(6\,\tau)^2\,\eta(\tau)^4 - 4\,\eta(2\,\tau)^2\,\eta(3\,\tau)\,\eta(12\,\tau)^3\,\eta(\tau) + \eta(2\,\tau)^2\,\eta(3\,\tau)^4\,\eta(4\,\tau),$

$-\eta(\tau)\,\eta(6\,\tau)^2\,\eta(4\,\tau)^4 + \eta(2\,\tau)^2\,\eta(3\,\tau)^3\,\eta(12\,\tau)\,\eta(4\,\tau) - \eta(\tau)\,\eta(2\,\tau)^2\,\eta(12\,\tau)^4,$

$\eta(3\,\tau)\,\eta(6\,\tau)^2\,\eta(4\,\tau)^4 - \eta(\tau)^3\,\eta(6\,\tau)^2\,\eta(12\,\tau)\,\eta(4\,\tau) - 3\,\eta(2\,\tau)^2\,\eta(3\,\tau)\,\eta(12\,\tau)^4\}$

In[60]:= $\mathcal{F}\eta\mathcal{I}$[08, 12] // reduceIdentities // TraditionalForm

Out[60]//TraditionalForm=

$\{-\eta(6\,\tau)^2\,\eta(12\,\tau)\,\eta(\tau)^4 + 4\,\eta(3\,\tau)\,\eta(4\,\tau)^3\,\eta(6\,\tau)^2\,\eta(\tau) - 3\,\eta(2\,\tau)^2\,\eta(3\,\tau)^4\,\eta(12\,\tau),$

$\eta(3\,\tau)\,\eta(2\,\tau)^7 - 9\,\eta(3\,\tau)\,\eta(6\,\tau)^6\,\eta(2\,\tau) - 4\,\eta(\tau)^3\,\eta(4\,\tau)^2\,\eta(6\,\tau)\,\eta(12\,\tau)^2 - \eta(\tau)^5\,\eta(3\,\tau)^2\,\eta(6\,\tau) + 8\,\eta(3\,\tau)\,\eta(4\,\tau)^5\,\eta(6\,\tau)\,\eta(12\,\tau),$

$\eta(12\,\tau)\,\eta(2\,\tau)^7 + 9\,\eta(6\,\tau)^6\,\eta(12\,\tau)\,\eta(2\,\tau) - 8\,\eta(4\,\tau)^5\,\eta(6\,\tau)\,\eta(12\,\tau)^2 - 2\,\eta(\tau)^2\,\eta(3\,\tau)^2\,\eta(4\,\tau)^3\,\eta(6\,\tau) + \eta(\tau)^5\,\eta(3\,\tau)\,\eta(6\,\tau)\,\eta(12\,\tau),$

$-\eta(4\,\tau)\,\eta(6\,\tau)^2\,\eta(\tau)^4 - 4\,\eta(2\,\tau)^2\,\eta(3\,\tau)\,\eta(12\,\tau)^3\,\eta(\tau) + \eta(2\,\tau)^2\,\eta(3\,\tau)^4\,\eta(4\,\tau),$

$-\eta(3\,\tau)^2\,\eta(12\,\tau)\,\eta(\tau)^5 - 2\,\eta(4\,\tau)^2\,\eta(12\,\tau)^3\,\eta(\tau)^3 + \eta(3\,\tau)^3\,\eta(4\,\tau)^3\,\eta(\tau)^2 + 8\,\eta(3\,\tau)\,\eta(4\,\tau)^5\,\eta(12\,\tau)^2 - 9\,\eta(2\,\tau)\,\eta(3\,\tau)\,\eta(6\,\tau)^5\,\eta(12\,\tau),$

$-\eta(3\,\tau)^4\,\eta(4\,\tau)^4 + 4\,\eta(\tau)\,\eta(3\,\tau)\,\eta(12\,\tau)^3\,\eta(4\,\tau)^3 + \eta(\tau)^3\,\eta(3\,\tau)^3\,\eta(12\,\tau)\,\eta(4\,\tau) - \eta(\tau)^4\,\eta(12\,\tau)^4,$

$-\eta(\tau)\,\eta(6\,\tau)^2\,\eta(4\,\tau)^4 + \eta(2\,\tau)^2\,\eta(3\,\tau)^3\,\eta(12\,\tau)\,\eta(4\,\tau) - \eta(\tau)\,\eta(2\,\tau)^2\,\eta(12\,\tau)^4,$

$\eta(3\,\tau)\,\eta(6\,\tau)^2\,\eta(4\,\tau)^4 - \eta(\tau)^3\,\eta(6\,\tau)^2\,\eta(12\,\tau)\,\eta(4\,\tau) - 3\,\eta(2\,\tau)^2\,\eta(3\,\tau)\,\eta(12\,\tau)^4,$

$-3\,\eta(\tau)\,\eta(6\,\tau)^7 - \eta(\tau)\,\eta(2\,\tau)^6\,\eta(6\,\tau) + \eta(\tau)^2\,\eta(2\,\tau)\,\eta(3\,\tau)^5 - 8\,\eta(\tau)\,\eta(2\,\tau)\,\eta(4\,\tau)\,\eta(12\,\tau)^5 + 4\,\eta(2\,\tau)\,\eta(3\,\tau)\,\eta(4\,\tau)^2\,\eta(12\,\tau)^2,$

$-\eta(6\,\tau)\,\eta(2\,\tau)^7 - 12\,\eta(4\,\tau)\,\eta(12\,\tau)^5\,\eta(2\,\tau)^2 + 9\,\eta(6\,\tau)^7\,\eta(2\,\tau) + \eta(\tau)^5\,\eta(3\,\tau)\,\eta(6\,\tau)^2 - 4\,\eta(4\,\tau)^5\,\eta(6\,\tau)^2\,\eta(12\,\tau),$

$-\eta(6\,\tau)\,\eta(2\,\tau)^7 + \eta(\tau)\,\eta(3\,\tau)^5\,\eta(2\,\tau)^2 - 4\,\eta(4\,\tau)\,\eta(12\,\tau)^5\,\eta(2\,\tau)^2 - 3\,\eta(6\,\tau)^7\,\eta(2\,\tau) + 4\,\eta(4\,\tau)^5\,\eta(6\,\tau)^2\,\eta(12\,\tau),$

$-\eta(\tau)\,\eta(4\,\tau)\,\eta(6\,\tau)\,\eta(2\,\tau)^5 - 8\,\eta(\tau)\,\eta(4\,\tau)^2\,\eta(12\,\tau)^5 - \eta(\tau)^3\,\eta(3\,\tau)^2\,\eta(12\,\tau)^3 + 2\,\eta(3\,\tau)^3\,\eta(4\,\tau)^3\,\eta(12\,\tau)^2 + \eta(\tau)^2\,\eta(3\,\tau)^5\,\eta(4\,\tau),$

$3\,\eta(4\,\tau)\,\eta(6\,\tau)^7 - \eta(2\,\tau)^6\,\eta(4\,\tau)\,\eta(6\,\tau) - 8\,\eta(2\,\tau)\,\eta(4\,\tau)^2\,\eta(12\,\tau)^5 - 2\,\eta(\tau)^2\,\eta(2\,\tau)\,\eta(3\,\tau)^2\,\eta(12\,\tau)^3 + \eta(\tau)\,\eta(2\,\tau)\,\eta(3\,\tau)^5\,\eta(4\,\tau)\}$

In[61]:= $\mathcal{F}\eta\mathcal{I}$[12, 14] // reduceIdentities // TraditionalForm

Out[61]//TraditionalForm=

$\{\eta(14\,\tau)^4\,\eta(\tau)^8 - 56\,\eta(2\,\tau)^3\,\eta(7\,\tau)\,\eta(14\,\tau)^7\,\eta(\tau) - 8\,\eta(2\,\tau)^7\,\eta(7\,\tau)\,\eta(14\,\tau)^3\,\eta(\tau) + 7\,\eta(2\,\tau)^4\,\eta(7\,\tau)^8,$

$-\eta(7\,\tau)^4\,\eta(2\,\tau)^8 + 7\,\eta(\tau)^3\,\eta(7\,\tau)^7\,\eta(14\,\tau)\,\eta(2\,\tau) + \eta(\tau)^7\,\eta(7\,\tau)\,\eta(14\,\tau)^3\,\eta(2\,\tau) - 7\,\eta(\tau)^4\,\eta(14\,\tau)^8\}$

In[62]:= $\mathcal{F}\eta\mathcal{I}$[12, 15] // reduceIdentities // TraditionalForm

Out[62]//TraditionalForm=

$\{\eta(3\,\tau)^6\,\eta(5\,\tau)^6 - \eta(\tau)^5\,\eta(3\,\tau)\,\eta(15\,\tau)\,\eta(5\,\tau)^5 - 5\,\eta(\tau)^3\,\eta(3\,\tau)^3\,\eta(15\,\tau)^3\,\eta(5\,\tau)^3 - 9\,\eta(\tau)\,\eta(3\,\tau)^5\,\eta(15\,\tau)^5\,\eta(5\,\tau) - \eta(\tau)^6\,\eta(15\,\tau)^6,$

$18\,\eta(5\,\tau)\,\eta(15\,\tau)^2\,\eta(3\,\tau)^8 + \eta(\tau)^2\,\eta(5\,\tau)^3\,\eta(3\,\tau)^6 + 2\,\eta(\tau)^5\,\eta(15\,\tau)^3\,\eta(3\,\tau)^3 - 25\,\eta(\tau)\,\eta(5\,\tau)^8\,\eta(15\,\tau)\,\eta(3\,\tau) -$

$\eta(\tau)^7\,\eta(5\,\tau)^2\,\eta(15\,\tau)\,\eta(3\,\tau) - 25\,\eta(\tau)^2\,\eta(5\,\tau)^3\,\eta(15\,\tau)^6, -9\,\eta(\tau)\,\eta(5\,\tau)\,\eta(15\,\tau)\,\eta(3\,\tau)^8 - \eta(\tau)^6\,\eta(15\,\tau)^2\,\eta(3\,\tau)^3 +$

$5\,\eta(5\,\tau)^6\,\eta(15\,\tau)^2\,\eta(3\,\tau)^3 - 45\,\eta(\tau)\,\eta(5\,\tau)\,\eta(15\,\tau)^7\,\eta(3\,\tau)^2 + 10\,\eta(\tau)^2\,\eta(5\,\tau)^8\,\eta(3\,\tau) + 10\,\eta(\tau)^3\,\eta(5\,\tau)^3\,\eta(15\,\tau)^5,$

$-\eta(3\,\tau)\,\eta(5\,\tau)\,\eta(15\,\tau)\,\eta(\tau)^8 - 5\,\eta(5\,\tau)^2\,\eta(15\,\tau)^6\,\eta(\tau)^3 + \eta(3\,\tau)^6\,\eta(5\,\tau)^2\,\eta(\tau)^3 - 5\,\eta(3\,\tau)\,\eta(5\,\tau)^7\,\eta(15\,\tau)\,\eta(\tau)^2 -$

$90\,\eta(3\,\tau)^2\,\eta(15\,\tau)^8\,\eta(\tau) + 10\,\eta(3\,\tau)^3\,\eta(5\,\tau)^5\,\eta(15\,\tau)^3, -2\,\eta(5\,\tau)^2\,\eta(15\,\tau)\,\eta(\tau)^8 - \eta(3\,\tau)^2\,\eta(15\,\tau)^3\,\eta(\tau)^6 +$

$2\,\eta(3\,\tau)^5\,\eta(5\,\tau)^3\,\eta(\tau)^3 - 225\,\eta(3\,\tau)\,\eta(5\,\tau)\,\eta(15\,\tau)^8\,\eta(\tau) - 9\,\eta(3\,\tau)^7\,\eta(5\,\tau)\,\eta(15\,\tau)^2\,\eta(\tau) + 25\,\eta(3\,\tau)^2\,\eta(5\,\tau)^6\,\eta(15\,\tau)^3\}$

In[63]:= $\mathcal{F}\eta\mathcal{I}$[11, 16] // reduceIdentities // TraditionalForm

Out[63]//TraditionalForm=

$\{-\eta(16\,\tau)^2\,\eta(2\,\tau)^6 + 2\,\eta(\tau)^2\,\eta(8\,\tau)^5\,\eta(2\,\tau) - \eta(\tau)^4\,\eta(4\,\tau)^2\,\eta(16\,\tau)^2,$

$\eta(8\,\tau)\,\eta(2\,\tau)^6 - 4\,\eta(\tau)^2\,\eta(4\,\tau)^2\,\eta(16\,\tau)\,\eta(2\,\tau) - \eta(\tau)^4\,\eta(4\,\tau)^2\,\eta(8\,\tau),$

$-\eta(\tau)^2\,\eta(8\,\tau)^6 + \eta(2\,\tau)^5\,\eta(16\,\tau)^2\,\eta(8\,\tau) - 2\,\eta(\tau)^2\,\eta(4\,\tau)^2\,\eta(16\,\tau)^4,$

$\eta(2\,\tau)\,\eta(8\,\tau)^6 - \eta(\tau)^2\,\eta(4\,\tau)^2\,\eta(16\,\tau)^2\,\eta(8\,\tau) - 2\,\eta(2\,\tau)\,\eta(4\,\tau)^2\,\eta(16\,\tau)^4,$

$\eta(4\,\tau)^{10} - \eta(\tau)^4\,\eta(2\,\tau)^2\,\eta(8\,\tau)^4 - 8\,\eta(2\,\tau)^4\,\eta(8\,\tau)^2\,\eta(16\,\tau)^4 - 4\,\eta(\tau)^2\,\eta(2\,\tau)^3\,\eta(8\,\tau)^3\,\eta(16\,\tau)^2\}$

In[64]:= $\mathcal{F}\eta\mathcal{I}$[08, 21] // reduceIdentities // TraditionalForm

Out[64]//TraditionalForm=

$\{-\eta(3\,\tau)^4\,\eta(7\,\tau)^4 + 7\,\eta(\tau)\,\eta(3\,\tau)\,\eta(21\,\tau)^3\,\eta(7\,\tau)^3 + 3\,\eta(\tau)^2\,\eta(3\,\tau)^2\,\eta(21\,\tau)^2\,\eta(7\,\tau)^2 + \eta(\tau)^3\,\eta(3\,\tau)^3\,\eta(21\,\tau)\,\eta(7\,\tau) - \eta(\tau)^4\,\eta(21\,\tau)^4,$

$-\eta(21\,\tau)\,\eta(\tau)^7 - 14\,\eta(7\,\tau)^4\,\eta(21\,\tau)\,\eta(\tau)^3 - 91\,\eta(21\,\tau)^7\,\eta(\tau) - 28\,\eta(3\,\tau)^4\,\eta(21\,\tau)^3\,\eta(\tau) +$

$7\,\eta(3\,\tau)\,\eta(7\,\tau)^7 - 14\,\eta(3\,\tau)^3\,\eta(7\,\tau)\,\eta(21\,\tau)^4 + \eta(3\,\tau)^7\,\eta(7\,\tau), \eta(7\,\tau)\,\eta(3\,\tau)^7 - \eta(\tau)\,\eta(21\,\tau)^3\,\eta(3\,\tau)^4 +$

$13\,\eta(7\,\tau)\,\eta(21\,\tau)^4\,\eta(3\,\tau)^3 - 3\,\eta(7\,\tau)^7\,\eta(3\,\tau) - \eta(\tau)^4\,\eta(7\,\tau)^3\,\eta(3\,\tau) - 10\,\eta(\tau)\,\eta(21\,\tau)^7 - \eta(\tau)^3\,\eta(7\,\tau)^4\,\eta(21\,\tau)\}$

In[65]:= $\mathcal{F}\eta\mathcal{I}$[05, 30] // TraditionalForm

Out[65]//TraditionalForm=

$$\{-\eta(3\,\tau)\,\eta(5\,\tau)\,\eta(6\,\tau)\,\eta(10\,\tau)+2\,\eta(2\,\tau)\,\eta(6\,\tau)\,\eta(30\,\tau)\,\eta(10\,\tau)+\eta(\tau)\,\eta(3\,\tau)\,\eta(5\,\tau)\,\eta(15\,\tau)-\eta(\tau)\,\eta(2\,\tau)\,\eta(15\,\tau)\,\eta(30\,\tau)\}$$

In[66]:= $\mathcal{F}\eta\mathcal{I}$[07, 32] // reduceIdentities // TraditionalForm

Out[66]//TraditionalForm=

$$\begin{aligned}
\{&\eta(8\,\tau)\,\eta(2\,\tau)^6-4\,\eta(\tau)^2\,\eta(4\,\tau)^2\,\eta(16\,\tau)^2\,\eta(2\,\tau)-\eta(\tau)^4\,\eta(4\,\tau)^2\,\eta(8\,\tau),\\
&\eta(2\,\tau)\,\eta(8\,\tau)^6-\eta(\tau)^2\,\eta(4\,\tau)^2\,\eta(16\,\tau)^2\,\eta(8\,\tau)-2\,\eta(2\,\tau)\,\eta(4\,\tau)\,\eta(16\,\tau)^4,\\
&-4\,\eta(32\,\tau)^2\,\eta(4\,\tau)^5-\eta(2\,\tau)^2\,\eta(16\,\tau)\,\eta(4\,\tau)^4+8\,\eta(2\,\tau)^2\,\eta(16\,\tau)^5+4\,\eta(\tau)^2\,\eta(2\,\tau)\,\eta(8\,\tau)\,\eta(16\,\tau)^3+\eta(\tau)^4\,\eta(8\,\tau)^2\,\eta(16\,\tau),\\
&\eta(16\,\tau)\,\eta(4\,\tau)^6-4\,\eta(2\,\tau)^2\,\eta(8\,\tau)^2\,\eta(32\,\tau)^2\,\eta(4\,\tau)-\eta(2\,\tau)^4\,\eta(8\,\tau)^2\,\eta(16\,\tau),\\
&-\eta(16\,\tau)^2\,\eta(2\,\tau)^5-4\,\eta(4\,\tau)\,\eta(16\,\tau)\,\eta(32\,\tau)^2\,\eta(2\,\tau)^3-8\,\eta(4\,\tau)^2\,\eta(32\,\tau)^4\,\eta(2\,\tau)+2\,\eta(8\,\tau)^4\,\eta(16\,\tau)^2\,\eta(2\,\tau)+\eta(\tau)^2\,\eta(8\,\tau)^5,\\
&\eta(4\,\tau)\,\eta(16\,\tau)^6-\eta(2\,\tau)^2\,\eta(8\,\tau)^2\,\eta(32\,\tau)^2\,\eta(16\,\tau)-2\,\eta(4\,\tau)\,\eta(8\,\tau)^2\,\eta(32\,\tau)^4\}
\end{aligned}$$

In[67]:= $\mathcal{F}\eta\mathcal{I}$[14, 34] // TraditionalForm

Out[67]//TraditionalForm=

$$\begin{aligned}
\{&\eta(2\,\tau)\,\eta(17\,\tau)^3\,\eta(34\,\tau)^2\,\eta(\tau)^8+\eta(17\,\tau)^4\,\eta(34\,\tau)^3\,\eta(\tau)^7-\\
&\eta(2\,\tau)^3\,\eta(17\,\tau)^7\,\eta(\tau)^4-4\,\eta(2\,\tau)^4\,\eta(34\,\tau)^7\,\eta(\tau)^3-17\,\eta(2\,\tau)^2\,\eta(17\,\tau)^8\,\eta(34\,\tau)\,\eta(\tau)^3-\\
&68\,\eta(2\,\tau)^3\,\eta(17\,\tau)\,\eta(34\,\tau)^8\,\eta(\tau)^2+4\,\eta(2\,\tau)^8\,\eta(17\,\tau)^2\,\eta(34\,\tau)^3\,\eta(\tau)+4\,\eta(2\,\tau)^7\,\eta(17\,\tau)^3\,\eta(34\,\tau)^4\}
\end{aligned}$$

In[68]:= $\mathcal{F}\eta\mathcal{I}$[08, 39] // reduceIdentities // TraditionalForm

Out[68]//TraditionalForm=

$$\begin{aligned}
\{&-\eta(3\,\tau)^4\,\eta(13\,\tau)^4+13\,\eta(\tau)\,\eta(3\,\tau)\,\eta(39\,\tau)^3\,\eta(13\,\tau)^3+3\,\eta(\tau)\,\eta(3\,\tau)^3\,\eta(39\,\tau)\,\eta(13\,\tau)^3+\\
&3\,\eta(\tau)^2\,\eta(3\,\tau)^2\,\eta(39\,\tau)^2\,\eta(13\,\tau)^2+3\,\eta(\tau)^3\,\eta(3\,\tau)\,\eta(39\,\tau)^3\,\eta(13\,\tau)+\eta(\tau)^3\,\eta(3\,\tau)^3\,\eta(39\,\tau)\,\eta(13\,\tau)-\eta(\tau)^4\,\eta(39\,\tau)^4\}
\end{aligned}$$

In[69]:= $\mathcal{F}\eta\mathcal{I}$[08, 49] // reduceIdentities // TraditionalForm

Out[69]//TraditionalForm=

$$\begin{aligned}
\{&\eta(7\,\tau)^8-49\,\eta(\tau)\,\eta(49\,\tau)^3\,\eta(7\,\tau)^4-35\,\eta(\tau)^2\,\eta(49\,\tau)^2\,\eta(7\,\tau)^4-7\,\eta(\tau)^3\,\eta(49\,\tau)\,\eta(7\,\tau)^4-343\,\eta(\tau)\,\eta(49\,\tau)^7-\\
&343\,\eta(\tau)^2\,\eta(49\,\tau)^6-147\,\eta(\tau)^3\,\eta(49\,\tau)^5-49\,\eta(\tau)^4\,\eta(49\,\tau)^4-21\,\eta(\tau)^5\,\eta(49\,\tau)^3-7\,\eta(\tau)^6\,\eta(49\,\tau)^2-\eta(\tau)^7\,\eta(49\,\tau)\}
\end{aligned}$$

The reader can now go on and derive (with sufficient computational power) as many identities as wanted. All these "experimentally" found identities can be proved using modular forms [450], [1120], [969], [554], [303], [14], [947].

The interested reader could now go on and search for more complicated identities where the arguments of the Dedekind Eta functions are of the form $k\tau+p/q$. Simple example of such identities are $\eta(i/\tau)^2=\tau\,\eta(i\,\tau)^2$ [637], [672] and $\eta(\tau)\,\eta(4\,\tau)\,\eta(\tau+1/2)-\exp(i\,\pi/24)\,\eta(2\,\tau)^3=0$, $\exp(i\,\pi/3)\,\eta(\tau-1/2)^8-\eta(\tau)^8-16\,\eta(4\,\tau)^8=0$ [1267]. Another possible extension is the use of derivatives of the Dedekind Eta function [1339]. For the application of this method to find new modular equations of the Rogers–Ramanujan continued fraction, see [1238].

References

1 S. S. Abhyankar, T. T. Moh. *J. reine angew. Math.* 260, 47 (1973).

2 S. S. Abhyankar, T. T. Moh. *J. reine angew. Math.* 261, 29 (1973).

3 M. Abramowitz, I. A. Stegun. *Handbook of Mathematical Functions*, National Bureau of Standards, Washington, 1964.

4 B. N. Achar, J. W. Hanneken, T. Enck, T. Clarke. *Physica* A 297, 361 (2001).

5 I. Adagideli, P. M. Goldbart. *arXiv:cond-mat*/0108102 (2001).

6 J. A. Adam. *Phys. Rep.* 142, 263 (1986).

7 V. Adamchik, O. I. Marichev, V. K. Tuan. *Dokl. Akad. Nauk BSSR* 30, 876 (1986).

8 V. Adamchik, A.D. Lizarev. *Diff. Equ.* 23, 858 (1987).

9 G. N. Afanasiev. *J. Comput. Phys.* 69, 196 (1987).

10 G. N. Afanasiev. *J. Phys.* A 26, 731 (1993).

11 R. P. Agarwal. *Resonance of Ramanujan's Mathematics*, New Age International, New Delhi, 1999.

12 M. Agop, V. Griga, C. G. Buzea, N. Rezlescu, C. Buzea, A. Zacharias, I. Petreus. *Austral. J. Phys.* 51, 21 (1998).

13 V. C. Aguilera–Navarro, E. L. Koo, A. H. Zimerman. *J. Phys.* A 13, 3585 (1980).

14 S. Ahlgren. *J. Number Th.* 89, 222 (2001).

15 Z. Ahmed. *J. Phys.* A 32, 2767 (1999).

16 Z. Ahmed. *J. Phys.* A 33, 3161 (2000).

17 Z. Ahmed. *arXiv:quant-ph*/0310019 (2003).

18 H. Ahmedov, I. H. Duru. *arXiv:quant-ph*/0108041 (2001).

19 D. Ahn, S. L. Chang. *Phys. Rev.* B 34, 9034 (1986).

20 G. B. Akguc, L. E. Reichl. *Phys. Rev.* A 67, 046202 (2003).

21 G. B. Akguc, L. E. Reichl. *Int. J. Quant. Chem.* 98, 173 (2004).

22 A. Akhmetshin, Y. Volvovsky. *arXiv:hep-th*/0007163 (2000).

23 M. A. Al-Bassam in G. M. Rassias (ed.). *The Mathematical Heritage of C. F. Gauss*, World Scientific, Singapore, 1991.

24 H. A. Alhendi, E. I. Lashin. *arXiv:quant-ph*/0406075 (2004).

25 R. Aldrovandi, P. L. Ferreira. *Am. J. Phys.* 48, 660 (1980).

26 R. Aldrovandi. *Special Matrices of Mathematical Physics*, World Scientific, Singapore, 2001.

27 A. D. Alhaidari. *Int. J. Theor. Phys.* 42, 2999 (2003).

28 D. Alonso, P. Gaspard. *J. Phys.* A 27, 1599 (1994).

29 G. Almkvist, B. Berndt. *Am. Math. Monthly* 98, 585 (1988).

30 H. Alt, H.-D. Gräf, H. L. Harney, R. Hoffebert, H. Lengeler, A. Richter, P. Schardt, H. A. Weidenmüller. *Phys. Rev. Lett.* 74, 62 (1994).

31 H. Alt, C. Dembowski, H.-D. Gräf, R. Hoffebert, H. Rehfeld, A. Richter, C. Schmit. *arXiv:chao-dyn*/9906032 (1999).

32 G. Álvarez, B. Sundaram. *J. Phys.* A 37, 9735 (2004).

33 H. H. Aly, Riazuddin, A. H. Zimmermann. *Phys. Rev.* 136, B 1174 (1964).

34 H. Alzer. *J. Math. Anal. Appl.* 252, 353 (2000).

35   G. Alvarez. *Phys. Rev.* A 37, 4079 (1988).

36   G. Álvarez, C. Casares. *J. Phys.* A 33, 2499 (2000).

37   S. Ancey, A. Folacci, P. Gabrielli. *J. Phys.* A 34, 1341, (2001).

38   A. Anderson. *J. Math. Phys.* 35, 6018 (1994).

39   A. Anderson. *Ann. Phys.* 232, 292 (1994).

40   G. D. Anderson, M. K. Vamanamurthy, M. Vuorinen in H. M. Srivastava, S. Owa (eds.). *Current Topics in Analytic Function Theory*, World Scientific, Singapore, 1992.

41   G. D. Anderson, M. K. Vamanamurthy, M. K. Vuorinen. *Conformal Invariants, Inequalities, and Quasiconformal Maps*, Wiley & Sons, New York, 1997.

42   G. E. Andrews, R. Askey, R. Roy. *Special Functions*, Cambridge University Press, Cambridge, 1999.

43   G. E. Andrews, J. Jiménez–Urroz, K. Ono. *Duke Math. J.* 108, 395 (2001).

44   L. C. Andrews. *Special Functions for Engineers and Applied Mathematicians*, MacMillan, New York, 1985.

45   I. A. Antipova. *Sib. Math. J.* 44, 757 (2003).

46   S. S. Antman. *SIAM Rev.* 40, 927 (1998).

47   J.-P. Antoine, J.-P. Gazeau, P. Monceau, J. R. Klauder, K. A. Penson. *arXiv:math-ph*/0012044 (2000).

48   H. Aoki, M. Koshino, D. Takeda, H. Morise. *arXiv:cond-mat*/0306729 (2003).

49   T. Apostol. *Modular Functions and Dirichlet Series in Number Theory*, Springer-Verlag, New York, 1976.

50   P. Appell, F. Kampé de Fériet. *Fonctions Hypergéométriques et Hypersphériques*, Gauthier-Villars, Paris, 1926.

51   V. Aquilanti, A. Caligiana, S. Cavalli. *Int. J. Quant. Chem.* 92, 99 (2003).

52   T. Aoki, T. Kawai, Y. Takei. *Sugaku Exp.* 8, 217 (1995).

53   H. Aoki, M. Koshino, D. Takeda, H. Morise. *arXiv:cond-mat*/0109512 (2001).

54   N. Archinard. *J. Number Th.* 101, 244 (2003).

55   P. Arena, R. Caponetto, L. Fortuna, D. Porto. *Nonlinear Noninteger Order Circuits and Systems*, World Scientific, Singapore, 2000.

56   J. Arndt, C. Haenel. *π Unleashed*, Springer-Verlag, Berlin, 2001.

57   D. L. Aronstein, C. R. Stoud, Jr. *J. Math. Phys.* 41, 8349 (2000).

58   G. P. Arrighini, N. Durante, C. Guidotti. *J. Math. Chem.* 25, 93 (1997).

59   F. M. Arscott. *Periodic Differential Equations*, MacMillan, New York, 1964.

60   F. M. Arscott in W. N. Everitt, B. D. Sleeman (eds.). *Ordinary and Partial Differential Equations*, Springer-Verlag, Berlin, 1981.

61   E. Artin. *The Gamma Function*, Holt, Rinehart and Winston, New York, 1964.

62   M. Artzouni. *arXiv:math.GM*/0410548 (2004).

63   A. Asada in K. Tas, D. Krupka, O. Krupková, D. Beleanu (eds.). *Global Analysis and Applied Mathematics*, American Institure of Physics, College Park, 2004.

64   N. Asai, I. Kubo, H.-H. Kuo. *arXiv:math.CO*/0104137 (2001).

65   N. W. Ashcroft, N. D. Mermin. *Solid State Physics*, Saunders, Philadelphia, 1976.

66   J. Avery. *J. Math. Chem.* 24, 169 (1998).

67   R. G. Ayoub. *Arch. Hist. Exact Sci.* 23, 253 (1980).

68   A. Bäcker, R. Schubert. *J. Phys.* A 32, 4795 (1999).

69   A. Bäcker. *arXiv:nlin.CD*/0204061 (2002).

70   S. M. Badalyan, F. M. Peeters. *Physica* B 316/317 216 (2001).

71   B. Bagchi, C. Quesne, M. Znojil. *arXiv:quant-ph*/0108096 (2001).

72   M. Bagieu, D. Maystre. *J. Opt.* A 1, 537 (1999).

73   R. L. Bagley, P. J. Torvik. *J. Appl. Mech.* 51, 294 (1984).

74   V. Bagnato, D. E. Pritchard, D. Kleppner. *Phys. Rev.* A 35, 4354 (1987).

75   W. N. Bailey. *Generalized Hypergeometric Series*, Cambridge University Press, Cambridge, 1935.

76   R. Baillie. *Am. Math. Monthly* 86, 372 (1979).

77   J. Bajer, A. Miranowicz. *arXiv:quant-ph*/0109075 (2001).

78   S. D. Bajpai, S. Mishra. *Jnanabha* 21, 47 (1991).

79   I. Bakas, C. Sourdis. *arXiv:hep-th*/0205007 (2002).

80   B. A. Baker, Jr., P. Graves-Morris. *Padé Approximants*, Cambridge University Press, Cambridge, 1996.

81   B. J. Baker. *J. Lond. Math. Soc.* 42, 385 (1967).

82   G. L. Baker, J. A. Blackburn, H. J. T. Smith. *Am. J. Phys.* 70, 525 (2002).

83   E. P. Balanzario. *Bol. Soc. Mat. Mex.* 10, 1 (2004).

84   N. L. Balazs, T. Bergman. *Phys. Rev.* A 58, 2359 (1998).

85   F. Baldarassi, B. Dwork. *Am. J. Math.* 101, 42 (1979).

86   G. R. Baldock, T. Bridgeman. *Mathematical Theory of Wave Motion*, Ellis Horwood, New York, 1981.

87   D. Baldwin, W. Hereman, J. Sayers. *arXiv:nlin.SI*/0311028 (2003).

88   R. Balian, G. Parisi, A. Voros. *Phys. Rev. Lett.* 41, 1141 (1978).

89   B. Balzer, S. Dilthey, G. Stock, M. Thoss. *J. Chem. Phys.* 119, 5795 (2003).

90   C. Banderier, P. Flajolet, G. Schaeffer, M. Soria. *Preprint* (2000). http://pauillac.inria.fr/algo/flajolet/Publications/coal.ps.gz

91   D. Bang, B. Elmabsout. *J. Phys.* A 36, 11453 (2003).

92   S. B. Bank, R. P. Kaufman. *Funk. Ekvacioj* 19, 53 (1976).

93   G. Bao, X. Fang, W. Tan, T. Van. *J. Math. Chem.* 27, 251 (2000).

94   E. J. Barbeau. *Polynomials*, Springer-Verlag, New York, 1989.

95   E. Barkai, R. Silbay. *arXiv:cond-mat*/0002020 (2000).

96   H. H. Barnum. *Ann. Math.* 11, 103 (1910).

97   F. A. Barone, H. Boschi–Filho, C. Farina. *arXiv:quant-ph*/0205085 (2002).

98   A. H. Barr in D. Kirk (ed.). *Graphics Gems* III, Academic Press, Boston, 1992.

99   J. Barros–Neto, F. Cardoso *arXiv:math.CA*/0101065 (2001).

100   D. A. Barry, J.-Y. Parlange, L. Li, H. Prommer, C. J. Cunningham, F. Stagnitti. *Math. Comput. Simul.* 53, 95 (2000).

101   G. Barton. *Ann. Phys.* 166, 322 (1986).

102   G. Barton. *Elements of Green's Functions and Propagators: Potentials, Diffusion, Waves*, Clarendon Press, Oxford, 1989.

103   A. Baruch. *Ueber die Differentialrelationen zwischen den Thetafunctionen eines Arguments.*, Ph. D. thesis, Halle, 1910.

104   V. G. Baryshevskii, I. D. Feranchuk, P. B. Kats. *arXiv:quant-ph*/0403050 (2004).

105   G. Bastard, E. E. Mendez, L. L. Chang, L. Esaki. *Phys. Rev.* B 28, 3241 (1983).

106   J. W. Bates. *J. Math. Phys.* 38, 3679 (1997).

107   V. R. Batista. *Pacific J. Math.* 212, 347 (2003).

108   M. Bauer, O. Golinelli. *arXiv:cond-mat*/0003049 (2000).

109   G. Baumann. *Mathematica in the Theoretischen Physik*, Springer-Verlag, Berlin, 1993.

110   A. D. Baute, I. L. Egusquiza, J. G. Muga. *arXiv:quant-ph*/0007066 (2000).

111   M. Bawin, S. A. Coon. *arXiv:quant-ph*/0012039 (2000).

112   M. Bawin, S. A. Coon. *quant-ph*/0302199(2003).

113   J. V. Beck, K. D. Cole, A. Haji-Sheikh, B. Litkouhi. *Heat Conduction Using Green's Function*, Hemisphere Publishing, London, 1992.

114   R. Becker, F. Sauter. *Theorie der Elektrizität* v.1, Teubner, Stuttgart, 1962.

115   J. L. Beeby. *Proc. Camb. Phil. Soc.* 59, 607 (1963).

116   C. W. J. Beenakker. *arXiv:cond-mat*/0406018 (2004).

117   G. Belardinelli. *Memorial des Sciences Mathematiques* CXLV (1960).

118   V. V. Beletsky. *Essays on the Motion of Celestial Bodies*, Birkhäuser, Basel, 2001.

119   V. Belevitch, J. Boersma. *Philips J. Res.* 38, 79 (1983).

120   R. Bellmann. *A Brief Introduction to Theta Functions*, Holt, Rinehart and Winston, New York, 1961.

121   E. D. Belokolos, A. I. Bobenko, V. Z. Enol'skii, A. R. Its, V. B. Mateev. *Algebro-Geometric Approach to Nonlinear Integrable Equations*, Springer-Verlag, Berlin, 1994.

122   E. D. Belokolos, V. Z. Enolskii, M. Salerno. *arXiv:cond-mat*/0401440 (2004).

123   J.-D. Benamou, F. Collino, O. Runborg. *J. Comput. Phys.* 199, 717 (2004).

124   D. Ben–Avraham, . M. Bollt, C. Tamon. *arXiv:cond-mat*/0409514 (2004).

125   C. M. Bender, S. A. Orszag. *Advanced Mathematical Methods for Scientists and Engineers*, McGraw-Hill, New York, 1978.

126   C. M. Bender, S. Boettcher, L. R. Mead. *J. Math. Phys.*, 35, 368 (1994).

127   V. A. Benderskii, E. V. Vetoshkin, E. I. Kats. *JETP* 95, 645 (2002).

128   T. C. Benton, H. D. Knoble. *Math. Comput.* 32, 533 (1978).

129   M. A. Berger. *Phys. Rev. Lett.* 70, 705 (1993).

130   B. Berndt, A. J. Biagioi, J. M. Purtilo. *Indian J. Math.* 29, 215 (1987).

131   B. C. Berndt. *Ramanujan's Notebooks* v.3, v.4, Springer-Verlag, New York, 1994.

132   B. C. Berndt, H. H. Chan. *Mathematika* 42, 278 (1995).

133   B. C. Berndt, H. H. Chan, L.-C. Zhang in M. E. H. Ismail, D. R. Masson, M. Rahman (eds.). *Special Functions, q-Series and related Topics*, American Mathematical Society, Providence, 1997.

134   B. C. Berndt in R. P. Bambah, V. C. Dumir, R. J. Hans–Gill. *Number Theory*, Birkhäuser, Basel, 2000.

135   B. C. Berndt, H. H. Chan. *Ill. J. Math.* 45, 75 (2001).

136   B. C. Berndt, H. H. Chan, S.-Y. Kang, L.-C. Zhang. *Pac. J. Math.* 202, 267 (2002).

137   K. Berry, M. Tretkoff in W. Abikoff, J. S. Birman, K. Kuiken (eds.). *The Mathematical Legacy of Wilhelm Magnus Groups, Geometry and Special Functions*, American Mathematical Society, Providence, 1994.

138   M. V. Berry, K. E. Mount. *Rep. Prog. Phys.* 35, 315 (1972).

139   M. V. Berry. *J. Phys.* A 10, 2083 (1977).

140   M. V. Berry. *Am. J. Phys.* 47, 264 (1979).

141   M. V. Berry, R. G. Chambers, M. D. Large, C. Upstill, J. C. Walmsley. *Eur. J. Phys.* 1, 154 (1980).

142   M. Berry, Z. V. Lewis. *Proc. R. Soc. London* A 370, 459 (1980).

143   M. V. Berry in H. Segur, S. Tanveer, H. Levine (eds.). *Asymptotic beyond All Orders*, Plenum Press, New York, 1991.

144   M. V. Berry, J. P. Keating. *Proc. R. Soc. Lond.* A 437, 151 (1992).

145   M. V. Berry. *Proc. R. Soc. Lond.* A 450, 439 (1995).

146   M. V. Berry. *Phys. Today* 42, n4, 11 (2001).

147   M. V. Berry. *Phil. Trans. R. Soc. Lond.* 360, 1023 (2002).

148   M. J. Bertin, A. Decomps-Guilloux, M. Grandet-Hugot, M. Pathiaux-Delefosse, J. P. Schreiber. *Pisot and Salem Numbers*, Birkhäuser, Basel, 1992.

149   D. Bertrand, W. Zudilin. *arXiv:math.NT/0006176* (2000).

150   F. Beukers in F. Q. Gouvêa, N. Yui (eds.). *Advances in Number Theory*, Clarendon, Oxford, 1991.

151   F. Beukers. *J. reine angew. Math.* 434, 45 (1993).

152   V. B. Bezerra, M. A. Rego–Monteiro. *arXiv:hep-th/0409134* (2004).

153   R. K. Bhaduri, A. Khare, J. Law. *Phys. Rev.* E 52, 486 (1995).

154   S. Bhargava, D. D. Somashekara. *J. Math. Anal. Appl.* 176, 554 (1993).

155   J. K. Bhattacharjee, K. Banerjee. *J. Phys.* A 20, L759 (1987).

156   P. Bhattacharyya, B. K. Chakrabarti. *arXiv:math.PR/0212230* (2002).

157   S. Bhargava, C. Adiga, M. S. M. Naika. *Adv. Stud. Contemp. Math.* 5, 37 (2002).

158   S. Bhargava, C. Adiga, M. S. M. Naika. *Indian J. Math.* 45, 23 (2003).

159   A. J. F. Biagioli. *Glasgow Math. J.* 35, 307 (1993).

160   P. Biane, J. Pitman, M. Yor. *Bull. Amer. Math. Soc.* 38, 435 (2001).

161   W. E. Bies, L. Kaplan, M. R. Haggerty, E. J. Heller. *arXiv:nlin.CD/0004024* (2000).

162   W. E. Bies, E. J. Heller. *J. Phys.* A 35, 5673 (2002).

163   J. F. Bird. *Optics Comm.* 136, 349 (1997).

164   G. Blanch, D. S. Clemm. *Math. Comput.* 23, 97 (1969).

165   J. Blank, P. Exner, M. Havlíček. *Hilbert Space Operators in Quantum Physics*, American Institute of Physics, New York, 1985.

166   G. Blum, S. Gnutzman, U. Smilansky. *arXiv:nlin.CD/0109029* (2001).

167   J. Blümlein. *arXiv:hep-ph/0311046* (2003).

168   J. Blümlein. *Comput. Phys. Comm.* 159, 19 (2004).

169   K. Bobek. *Einleitung in die Theorie der elliptischen Funktionen*, Teubner, Leipzig, 1884.

170   E. Bogomolny, C. Schmit. *arXiv:nlin.CD/0110019* (2001).

171   A. Bohm, M. Gadella, B. B. Mainland. *Am. J. Phys.* 57, 1103 (1989).

172   P. E. Böhmer. *Differenzengleichungen und bestimmte Integrale*, Koehler, Leipzig, 1939.

173   A. Bohm, M. Gadella. *Dirac Kets, Gamow Vectors and Gel'fand Triplets*, Springer-Verlag, Berlin, 1989.

174   J. Bohmann, C.-E. Froberg. *Math. Comput.* 58, 315 (1992).

175   M. Bologna, C. Tsallis, P. Grigolini. *arXiv:cond-mat/0003482* (2000).

176   D. Bonciani, F. Vlacci. *Complex Variables* 49, 271 (2004).

177   G. Bonneau, J. Faraut, G. Valent. *Am. J. Phys.* 69, 288 (2001).

178   A. V. Boriskin, A. I. Nosich. *IEEE Trans. Antennas Prop.* 50, 1245 (2002).

179   M. Born, E. Wolf. *Principles of Optics*, Pergamon Press, Oxford, 1970.

180   P. Borrmann, G. Franke. *J. Chem. Phys.* 98, 2484 (1993).

181   J. Borwein. *Proc. Am. Math. Soc.* 95, 365 (1985).

182   J. Borwein, D. Bailey, R. Girgensohn. *Experimentation in Mathematics*, A K. Peters, Natick, 2004.

183   J. M. Borwein, P. B. Borwein. *Pi and the AGM*, Wiley, New York, 1987.

184   J. M. Borwein, P. B. Borwein. *Trans. Am. Math. Soc.* 323, 691 (1991).

185   J. M. Borwein, I. J. Zucker. *IMA J. Numer. Anal.* 12, 519 (1992).

186   P. Borwein, W. Dykshoorn. *J. Math. Anal. Appl.* 179, 203 (1993).

187   J. M. Borwein, P. B. Borwein, F. G. Garvan. *Trans. Am. Math. Soc.* 343, 35 (1994).

188   P. Borwein, T. Erdély. *Polynomials and Polynomial Inequalities*, Springer-Verlag, New York, 1995.

189   S. K. Bose. *Am. J. Phys.* 51, 463 (1983).

190   D. L. Bosley. *SIAM Rev.* 38, 128 (1996).

191   U. Bottazzini in I. Grattan-Guiness (ed.). *Companion Encyclopedia of the History and Philosophy of the Mathematical Sciences* 1, Routledge, London, 1994.

192   Z. Bouchal. *arXiv:physics*/0309109 (2003).

193   F. Bowman. *Introduction to Elliptic Functions with Application*, Dover, New York, 1961.

194   J. J. Bowman, T. B. A. Senior in J. J. Bowman, T. B. A. Senior, P. L. E. Uslenghi (eds.) *Electromagnetic and Acoustic Scattering by Simple Shapes*, Hemisphere, New York, 1987.

195   L. J. Boya, E. C. G. Sudarshan, T. Tilma. *arXiv:math-ph*/0210033 (2002).

196   J. H. Boyd. *Ann. Math.* 7, 1 (1892).

197   J. P. Boyd. *Acta Appl. Math.* 56, 1 (1999).

198   C. P. Boyer, E. G. Kalnins, W. Miller, Jr. *Nagoya Math. J.* 60, 35 (1976).

199   P. Boyland, M. Stremler, H. Aref. *Physica* D 175, 69 (2003).

200   B. L. J. Braaksma, G. K. Immink, M. van der Put. *The Stokes Phenomena and Hilbert's 16th Problem*, World Scientific, Singapore, 1996.

201   M. Brack, M. Mehta, K. Tanaka. *arXiv:nlin.CD*/0105048 (2001).

202   M. Brack, S. N. Fedotkin, A. G. Magner, M. Mehat. *arXiv:nlin.CD*/0207043 (2002).

203   P. Bracken, A. M. Grundland. *J. Math. Phys.* 42, 1250, (2001).

204   H. W. Braden. *arXiv:nlin.SI*/0005046 (2000).

205   H. W. Braden, A. Gorsky, A. Odesskii, V. Rubtsov. *arXiv:hep-th*/0111066 (2001).

206   H. W. Braden, J. G. B. Byatt–Smith. *Phys. Lett.* A 295, 208 (2002).

207   R. Brauer. *Math. Annalen* 110, 473 (1934).

208   P. A. Braun, V. I. Sarichev. *Phys. Rev.* A 49, 1704 (1994).

209   W. Braunbek. *Optik* 9, 174 (1952).

210   Y. V. Brezhnev. *arXiv:nlin.SI*/0007028 (2000).

211   Y. V. Brezhnev. *Rep. Math. Phys.* 48, 39 (2001).

212   Y. V. Brezhnev. *J. Math. Phys.* 45, 696 (2004).

213   F. Brioschi. *Math. Annalen* 13, 109 (1878).

214   M. F. Brioschi. *Math. Annalen* 11, 111 (1877).

215   J. T. Broad. *Phys. Rev.* A 26, 3078 (1982).

216   H. Broer, C. Simó. *J. Diff. Eq.* 166, 290 (2000).

217   T. J. I'A. Bromwich. *An Introduction to the Theory of Infinite Series*, MacMillan, London, 1965.

218   J. C. Bronski, L. D. Carr, B. Deconinck, J. Nathan Kutz. *Phys. Rev. Lett.* 86, 1402 (2001).

219   F. Brosens, J. T. Devreese, L. F. Lemmens. *Solid State Comm.* 100, 123 (1996).

220   B. M. Brown, S. P. Eastham. *Proc. R. Soc. Lond.* 459, 2431 (2003).

221   R. Brown, L. O. Chua. *Int. J. Bifurc. Chaos* 6, 219 (1996).

222   V. A. Brumberg. *Celest. Mech. Dynam. Astron.* 59, 1 (1994).

223   V. A. Brumberg, T. Fukushima. *Celest. Mech. Dynam. Astron.* 60, 69 (1994).

224   V. A. Brumberg. *Analytical Techniques of Celestial Mechanics*, Springer-Verlag, Heidelberg, 1995.

225   V. A. Brumberg, E. V. Brumberg. *Celestial Dynamics at High Eccentricities*, Gordon and Breach, Amsterdam, 1999.

226   A. Bruno–Alfonso, G.- Hai. *J. Phys.* CM 15, 6701 (2003).

227   H. Buchholz. *The Confluent Hypergeometric Functions: With Special Emphasis on Its Applications*, Springer-Verlag, Berlin, 1969.

228   R. A. Buckingham in D. R. Bates (ed.). *Quantum Theory*, v.1, Academic Press, New York, 1961.

229   E. Buffet, J. V. Pulè. *J. Math. Phys.* 24, 1608 (1983).

230   W. K. Bühler. *Math. Intell.* 7, n2, 35 (1985).

231   W. Bühring. *Int. J. Math. Math. Sci.* 24, 505 (2000).

232   W. Bühring. *arXiv:math.CA*/0206200 (2002).

233   S. Bullett. *Topology* 30, 171 (1991).

234   S. Bullett, J. Stark. *SIAM Rev.* 35, 631 (1993).

235   L. A. Bunimovich. *Funct. Anal. Appl.* 8, 254 (1974).

236   J. L. Burchnall. *Quart. J. Math.* 13, 90 (1942).

237   R. Burger, G. Labahn, M. van Hoeij. *Technical Report University Waterloo* CS-2002-25 (2002). ftp://cs-archive.uwaterloo.ca/cs-archive/CS-2002-25/

238   L. M. Burko. *arXiv:physics*/0201024 (2002).

239   L. M. Burko. *Eur. J. Phys.* 24, 125 (2003).

240   J. G. B. Byatt–Smith, H. W. Braden. *arXiv:math-ph*/0110012 (2001).

241   J. G. B. Byatt–Smith, H. W. Braden. *arXiv:nlin.SI*/0201014 (2002).

242   P. F. Byrd, M. D. Friedman. *Handbook of Elliptic Integrals for Engineers and Scientists*, Springer-Verlag, Berlin, 1971.

243   A. A. Bytsenko, G. Cognola, E. Elizalde, V. Moretti, S. Zerbini. *Analytic Aspects of Quantum Fields*, World Scientific, New Jersey, 2003.

244   J.-M. Caillol. *arXiv:cond-mat*/0306562 (2003).

245   J.-L. Calais. *Int. J. Quant. Chem.* 63, 223 (1997).

246   G. G. Calderón, J. Villavicencio, F. Delgado, J. G. Muga. *arXiv:quant-ph*/0206020 (2002).

247   R. C. Calhoun. *Am. J. Phys.* 64, 1399 (1996).

248   F. Calogero. *Lett. Nuov. Cim.* 13, 507 (1975).

249   F. Calogero. *Classical Many-Body Problems Amenable to Exact Treatments*, Springer-Verlag, Berlin, 2001.

250   F. Calogero. *J. Phys.* A 35, 985 (2002).

251   F. Calogero, J.-P. Francoise. *J. Nonlin. Math. Phys.* 9, 99 (2002).

252   F. Calogero, S. Graffi. *Phys. Lett.* A 313, 356 (2003).

253   F. Calogero. *Phys. Lett.* A 319, 240 (2003).

254   F. Calogero. *J. Nonl. Math. Phys.* 11, 1 (2004).

255   L. M. B. C. Campos. *IMA J Appl. Math.* 33, 109 (1984).

256   E. R. Canfield, C. Pomerance. *Integers* 2, A2 (2002).   http://www.integers-ejcnt.org/vol2.html

257   J. Cantarella, D. DeTurck, H. Gluck. *Am. Math. Monthly* 109, 409 (2002).

258 C. Carathéodory. *Theory of Functions of a Complex Variable*, Birkhäuser, Basel, 1950.

259 B. C. Carlson. *SIAM Rev.* 12, 332 (1970).

260 B. C. Carlson. *Am. Math. Monthly* 78, 496 (1971).

261 A. Carpinteri, F. Mainardi (eds.). *Fractals and Fractional Calculus in Continuum Mechanics*, Springer-Verlag, Wien, 1997.

262 L. D. Carr, C. W. Clark, W. P. Reinhardt. *arXiv:cond-mat*/9911177 (1999).

263 L. D. Carr, K. W. Mahmud, W. P. Reinhardt. *Phys. Rev.* A 64, 033603 (2001).

264 G. F. Carrier, M. Krook, C. E. Pearson. *Functions of a Complex Variable*, Hod Books, New York, 1983.

265 N. J. Carron. *Am. J. Phys.* 63, 717 (1995).

266 P. Cartier. *Preprint* IHES/M/00/48 (2000). http://www.ihes.fr/PREPRINTS/M00/Resu/resu-M00-48.html

267 C. A. A. de Carvalho, R. M. Cavalcanti, E. S. Fraga, S. E. Jorás. *Ann. Phys.* 273, 146 (1999).

268 G. Casati, T. Prosen. *Physica* D 131, 293 (1999).

269 H. B. G. Casimir in E. G. D. Cohen (ed.). *Fundamental Problems in Statistical Mechanics* v. 2, North-Holland, Amsterdam, 1968.

270 R. M. Cavalcanti, P. Giacconi, R. Soldati. *arXiv:quant-ph*/0307232 (2003).

271 A. Cayley. *Philos. Trans.* CLII, 561 (1862).

272 E. Cerda, L. Mahadevan, J. M. Pasini. *Proc. Natl. Acad. Sci.* 101, 1807 (2004).

273 N. J. Cerf, J. Boutet de Monvel, O. Bohigas, O. C. Martin, A. G. Percus. *J. Phys. France* I, 117 (1997).

274 H. H. Chan, V. Tan in B. C. Berndt, F. Gesztesy (eds.). *Continued Fractions: From Analytic Number Theory to Constructive Approximations*, American Mathematical Society, Providence, 1999.

275 H. H. Chan, W.-C. Liaw. *Pacific J. Math.* 192, 219 (2000).

276 H. H. Chan, W.-C. Liaw. *Can. J. Math.* 52, 31 (2000).

277 H. H. Chan, W.-C. Liaw, V. Tan. *J. Lond. Math. Soc.* 64, 93 (2001).

278 H. H. Chan in M. A. Bennett, B. C. Berndt, N. Boston, H. G. Diamond, A. J. Hildebrand, W. Philipp. *Number Theory for the Millennium*, A K Peters, Natick, 2002.

279 H. H. Chan, K. S. Chua, P. Solé. *J. Number Th.* 99, 361 (2003).

280 L. Chan, E. S. Cheb–Terrab. *arXiv:math-ph*/0402063 (2004).

281 R. K. Chang, A. J. Campillo (eds.). *Optical Processes in Microcavities*, World Scientific, Singapore, 1996.

282 L. Chaos-Cador, E. Ley-Koo. *Rev. Mex. Fis.* 48, 67 (2002).

283 R. Chapman. *arXiv:math.NT*/0009187 (2000).

284 S. C. Chapman, G. Rowlands. *arXiv:cond-mat*/0007275 (2000).

285 S. J. Chapman, J. M. H. Lawry, J. R. Ockendon, V. H. Saward. *Wave Motion* 33, 41 (2001).

286 J. Chapront, J.-L. Simon. *Celest. Mech. Dynam. Astron.* 63, 171 (1996).

287 K. C. Chase, A. Z. Mekjian, L. Zamick. *arXiv:cond-mat*/9708070 (1997).

288 R. N. Chaudhuri, B. Mukherjee. *J. Phys.* A 16, 3193 (1983).

289 L. V. Chebotarev. *Ann. Phys.* 255, 305 (1997).

290 C. Chen, F. Chen, Y. Feng. *Graphical Models* 63, 1 (2001).

291 G. Chen, P. J. Morris, J. Zhou. *SIAM Rev.* 36, 453 (1994).

292 G. Chen, J. Zhou. *Vibration and Damping in Distributed Systems*, v.II CRC Press, Boca Raton, 1993.

293 H. Chen, H. Zhang. *Chaos, Solitons, Fractals* 15, 585 (2003).

294 Y.-Y. Chen, Z. Ye. *Phys. Rev.* E 64, 036616 (2002).

295   G.-S. Cheon, M. El-Mikkawy. *Appl. Math. Comput.* 158, 159 (2004).

296   Y. Chiba, S. Ohnishi. *Phys. Rev.* B 41, 6065 (1990).

297   L. F. Chibotaru, A. Ceulemans, V. Teniers, V. Bruyndoncx, V. V. Moshchalkov. *Eur. J. Phys.* B 27, 341 (2002).

298   L. F. Chibotaru, A. Ceulemans, M. Lorenzini, V. V. Moschalkov. *Europhys. Lett.* 63, 159 (2003).

299   C. Chicone, S. M. Kopeikin, B. Mashhoon, D. G. Retzloff. *arXiv:gr-qc*/0101122 (2001).

300   A. G. Chirkov, A. Y. Berdnikov. *Tech. Phys.* 46, 368 (2001).

301   J. Choi, T. Y. Seo. *Indian J. Pure Appl. Math.* 30, 649 (1999).

302   J. Choi, J. Lee, H. M. Srivastava. *Kodai Math. J.* 26, 44 (2003).

303   Y.-S. Choi. *Adv. Math.* 156, 180 (2000).

304   K. M. Christoffel, P. Brumer. *Phys. Rev.* A 33, 1309 (1986).

305   B. K. Choudhoury. *Proc. R. Soc. Lond.* A 450, 477 (1995).

306   R. Chouikha. *CRM Proc. Lecture Notes* 22, 53 (1999).

307   K. W. Chow. *J. Phys. Soc. Japan* 69, 1313 (2000).

308   K. W. Chow. *Wave Motion* 35, 71 (2001).

309   P. L. Christiansen, Y. B. Gaididei, S. F. Mingaleev. *arXiv:cond-mat*/0003146 (2000).

310   D. Chruscinski. *arXiv:math-ph*/0307047 (2003).

311   K. S. Chua, P. Solé. *Eur. J. Combinat.* 25, 179 (2004).

312   A. E. Chubykalo, R. A. Flores. *arXiv:physics*/0210089 (2002).

313   K. V. Chukbar. *JETP* 81, 1025 (1995).

314   A. Ciarkowski, B. Atamaniuk. *arXiv:math-ph*/0412003 (2004).

315   S. Ciccariello. *J. Math. Phys.* 45, 3353 (2004).

316   P. C. Clemmow. *The Plane Wave Spectrum Representation of Electromagnetic Fields*, Pergamon Press, Oxford, 1966.

317   S. D. Clow, B. R. Johnson. *Phys. Rev.* B 68, 235107 (2003).

318   D. Cocolicchio, M. Viggiano. *arXiv:quant-ph*/9710004 (1997).

319   M. W. Coffey. *J. Comput. Appl. Math.* 166, 525 (2004).

320   D. Cohen, N. Lepore, E. J. Heller. *arXiv:nlin.CD*/0108014 (2001).

321   C. F. M. Coimbra. *Ann. Phys.* 12, 692 (2003).

322   F. N. Cole. *Am. J. Math.* 9, 45 (1887).

323   C. B. Collins. *J. Comput. Appl. Math.* 131, 195 (2001).

324   A. K. Common, T. Stacey. *J. Phys.* A 11, 259 (1978).

325   S. A. Coon, B. R. Holstein. *arXiv:quant-ph*/0202091 (2002).

326   F. Cooper, A. Khare, U. Sukhatme. *arXiv:nlin.SI*/0203018 (2002).

327   S. Cooper. *Aequ. Math.* 56, 69 (1998).

328   S. Cooper. *J. Math. Anal. Appl.* 272, 43 (2002).

329   S. Cooper. *J. Comput. Appl. Math.* 160, 77 (2003).

330   R. M. Corless, D. J. Jeffrey in J. Calmet, B. Benhamou, O. Caprotti, L. Henocque, V. Sorge (eds.). *Artificial Intelligence, Automated Reasoning, and Symbolic Computation*, Springer-Verlag, Berlin, 2002.

331   M. Coronado, N. Dominguez, J. Flores, C. de la Portilla. *Am. J. Phys.* 50, 27 (1982).

332   J. Coulomb. *Bull. Sci. Math.* 60, 297 (1936).

333   R. Courant, D. Hilbert. *Methods of Mathematical Physics*, v.1, v.2, Interscience, New York, 1962.

334   D. A. Cox. *L'Enseignement Math.* 30, 275 (1984).

335   T. Craig. *Am. J. Math.* 5, 355 (1882).

336   S. R. Cranmer. *Am. J. Phys.* 72, 1397 (2004).

337   S. Crass. *arXiv:math.DS*/9903054 (1999).

338   D. S. F. Crothers. *J. Phys.* B 9, L513 (1976).

339   B. J. B. Crowley. *J. Phys.* A 13, 1227 (1980).

340   J. Cserti, A. Bodor, J. Koltai, G. Vattay. *arXiv:cond-mat*/0105472 (2001).

341   J. Cserti, P. Polinák, G. Palla, U. Zülicke, C. J. Lambert. *cond-mat*/0303246 (2003).

342   J. Cserti, B. Béri, P. Pollner, Z. Kaufmann. *arXiv:cond-mat*/0405404 (2004).

343   J. A. Cuesta, R. P. Sear. *Phys. Rev.* E 65, 021406 (2002).

344   L. M. Cureton, J. R. Kuttler. *J. Sound Vib.* 220, 83 (1999).

345   B. Ćurgus, V. Mascioni. *Proc. Am. Math. Soc.* 132, 2973 (2004).

346   T. Curtright, D. Fairlie, C. Zachos. *arXiv:hep-th*/9711183 (1997).

347   T. Curtright, T. Uematsu, C. Zachos. *arXiv:hep-th*/0011137 (2000).

348   T. Curtight. *arXiv:quant-ph*/0011101 (2000).

349   L. Cveticanin. *Physica* A 317, 83 (2003).

350   A. B. O. Daalhuis. *Proc. R. Soc. Edinb.* 123A, 731 (1993).

351   F. Dalfavo, S. Giorgini, L. P. Pitaevskii, S. Stringari. *Rev. Mod. Phys.* 71, 463 (1999).

352   M. G. E. daLuz, E. J. Heller, B. K. Cheng. *J. Phys.* A 31, 2975 (1998).

353   I. Dana, V. E. Chernov. *J. Phys.* A 35, 10101 (2002).

354   A. Das, S. A. Pernice. *Nucl. Phys.* B 561, 357 (1999).

355   A. Das, S. Pernice. *arXiv:hep-th*/0207112 (2002).

356   E. B. Davies. *Proc. R. Soc. Lond.* A 455, 585 (1999).

357   E. B. Davies, A. B. J. Kuijlaars. *J. Lond. Math. Soc.* 70, 420 (2004).

358   M. Davio, J.-P. Deschamps, A. Thayse. *Discrete and Switching Functions*, McGraw-Hill, New York, 1978.

359   H. T. Davis. *Introduction to Nonlinear Differential and Integral Equations*, United States Atomic Energy Commission, Washington, 1960.

360   P. J. Davis. *Circulant Matrices*, Wiley, New York, 1979.

361   A. I. Davydychev, M. Y. Kalmykov. *arXiv:hep-th*/0303162 (2003).

362   M. A. M. deAguir. *Phys. Rev.* A 48, 2567 (1993).

363   N. Debergh, J. Ndimubandi, B. Van den Bossche. *Ann. Phys.* 298, 361 (2002).

364   A. Debiard, B. Gaveau. *Bull. Sci. Math.* 126, 773 (2002).

365   N. Debergh, B. Van den Bossche. *Ann. Phys.* 308, 605 (2003).

366   S. De Biévre. *J. Phys.* A 25, 3399 (1992).

367   L. Debnath. *Int. J. Math. Math. Sci.* 54, 3413 (2003).

368   Y. Décanini, A. Folacci. *Phys. Rev.* A 67, 042704 (2003).

369   Y. Décanini, A. Folacci. *arXiv:nlin.CD*/0305034 (2003).

370   B. Deconinck, B. A. Frigyik, J. N. Katz. *arXiv:cond-mat*/0101478 (2001).

371   B. Deconinck, B. A. Frigyik, J. N. Kutz. *Phys. Lett.* A 283, 177 (2001).

372   B. Deconinck, B. A. Frigyik, J. N. Kutz. *arXiv:cond-mat*/0106231 (2001).

373   A. Degasperis, S. N. M. Ruijsenaars. *Ann. Phys.* 293, 92 (2001).

374   P. G. de Gennes. *Superconductivity of Metals and Alloys*, Addison-Wesley, Reading, 1989.

375   S. R. de Groot, G. J. Hooyman, C. A. ten Seldam. *Proc. R. Soc. Lond.* A 203, 266 (1950).

376   E. Delabaere, H. Dillinger, F. Pham. *J. Math. Phys.* 38, 6126 (1997).

377   R. de la Madrid, M. Gadella. *arXiv:quant-ph*/0201109 (2002).

378   F. Delgado, H. Cruz, J. G. Muga. *arXiv:quant-ph*/0207159 (2002).

379   F. Delgado, J. G. Muga, A. Ruschhaupt. *Phys. Rev.* A 69, 022106 (2004).

380   M. Dellnitz, M. Golubitski, A. Hohmann, I. Stewart. *Int. J. Bifurc. Chaos* 5, 1487 (1995).

381   M. Demetrian. *arXiv:physics*/0303110 (2003).

382   C. Deninger. *Invent. Math.* 107, 135 (1992).

383   E. C. de Oliveira, W. A. Rodrigues, Jr. *Ann. Physik* 7, 654 (1998).

384   R. A. Depine, A. Lakhtakia. *arXiv:physics*/0408050 (2004).

385   C. A. de Sousa, V. H. Rodrigues. *arXiv:physics*/0211075 (2002).

386   D. W. DeTemple, S. H. Wang. *J. Math. Anal. Appl.* 160, 149 (1991).

387   D. W. DeTemple. *Am. Math. Monthly* 100, 468 (1993).

388   E. D'Hoker, D. H. Phong. *arXiv:hep-th*/9912271 (1999).

389   N. C. Dias, J. N. Prata. *Ann. Phys.* 311, 120 (2004).

390   N. C. Dias, J. N. Prata. *Ann. Phys.* 313, 110 (2004).

391   J. B. Diaz, T. J. Osler. *Math. Comput.* 28, 185 (1974).

392   R. Díaz, E. Pariguan. *arXiv:math.CA*/0405596 (2004).

393   L. E. Dickson. *Modern Algebraic Theories*, Sanborn, Chicago, 1926.

394   R. B. Dingle. *Asymptotic Expansions: Their Derivation and Interpretation*, Academic Press, London, 1973.

395   A. C. Dixon. *The Elementary Properties of the Elliptic Functions*, Macmillan, New York, 1894.

396   M. A. Doncheski, S. Heppelmann,R. W. Robinett, D. C. Tussey. *Am. J. Phys.* 71, 541 (2003).

397   M. A. Doncheski, R. W. Robinett. *arXiv:quant-ph*/0307079 (2003).

398   S.-H. Dong, Z.-Q. Ma. *Am. J. Phys.* 70, 520 (2002).

399   J. Dorignac, S. Flach. *arXiv:cond-mat*/0112168 (2001).

400   P. Doyle and C. McMullen. *Acta Math.* 163, 151 (1989).

401   V. Dragovich. *J. Phys.* A 29, L 317 (1996).

402   V. Dragović. *arXiv:math-ph*/0008009 (2000).

403   V. Dragović. *J. Phys.* A 35, 2213 (2002).

404   D. W. Dreisigmeyer, P. M. Young. *J. Phys.* A 36, 8297 (2003).

405   K. Driver, P. Duren. *J. Math. Anal. Appl.* 239, 72 (1999).

406   K. A. Driver, N. M. Temme. *Quaest. Math.* 22, 7 (1999).

407   R. J. Drociuk. *arXiv:math.GM*/0005026 (2000).

408   B. A. Dubrovin. *Russian Math. Surv.* 36, 11 (1981).

409   M. Dudek, S. Giller, P. Milczarski. *arXiv:quant-ph*/9712041 (1997).

410   M. Dudek, S. Giller, P. Milczarski. *J. Math. Phys.* 40, 1163 (1999).

411   D. G. Duffy. *Green's Functions with Applications*, Chapman & Hall, Boca Raton, 2001.

412   H. R. Dullin, P. H. Richter, A. P. Veselov, H. Waalkens. *Physica* D 155, 159 (2001).

413   D. Dummit, H. Kiselevsky, J. McKay. *Contemp. Math.* 45, 89 (1985).

414   G. Dunne, J. Feinberg. *arXiv:hep-th*/9706012 (1997).

415   C. F. Dunkl, D. E. Ramirez. *ACM Trans. Math. Softw.* 20, 413 (1994).

416   G. Dunne, J. Feinberg in H. Aratyn, T. D. Imbo, W.-Y. Keung, U. Sukhatme (eds.). *Supersymmetry and Integrable Models*, Springer, Berlin, 1998.

417   T. M. Dunster. *Studies Appl. Math.* 103, 293 (2001).

418   L. Durand. *Am. J. Phys.* 72, 1082 (2004).

419   H. Durége. *Theorie der elliptischen Funktionen*, Teubner, Leipzig, 1908.

420   J. Dutka. *Arch. Hist. Exact Sci.* 49, 105 (1995).

421   P. Du Val. *Elliptic Functions and Elliptic Curves*, Cambridge University Press, Cambridge, 1973.

422   A. Eagle. *The Elliptic Functions as They Should Be*, Galloway and Porter, Cambridge, 1958.

423   R. Easther, G. Guralnik, S. Hahn. *arXiv:hep-ph*/9903225 (1999).

424   W. Ebeling, P. S. Landa, V. G. Ushakov. *Phys. Rev.* E 63, 046601 (2001).

425   B. Eckhardt in D. A. Hejhal, F. Friedman, M. C. Gutzwiller, A. M. Odlyzko (eds.). *Emerging Applications of Number Theory* , Springer-Ver;lag, New York, 1999.

426   H. M. Edwards. *Riemann's Zeta Function*, Academic Press, Boston, 1974.

427   H. I. Elim. *arXiv:quant-ph*/9901009 (1999).

428   E. Elizalde. *arXiv:hep-th*/9906229 (1999).

429   W. C. Elmore, M. E. Heald. *Physics of Waves*, McGraw-Hill, New York, 1969.

430   S. A. Elwakil, M. A. Zahran. *Chaos, Solitons & Fractals* 9, 1545 (1999).

431   N. Engheta. *J. Electrom. Waves Appl.* 9, 1179 (1995).

432   N. Enghata. *IEEE Antennas Prop. Mag.* 39, n4, 35 (1997).

433   N. Enghata in D. H. Werner, R. Mittra (eds.). *Frontiers in Electromagnetics*, IEEE Press, New York, 2000.

434   R. Englman, A. Yahalom, M. Baer. *Int. J. Quant. Chem.* 90, 266 (2002).

435   A. Enneper. *Elliptische Funktionen, Theorie und Geschichte*, Louis Nebert, Halle, 1890.

436   V. Z. Enolskii in F. K. Abdullaev, V. V. Konotop (eds.). *Nonlinear Waves: Classical and Quantum Aspects*, Kluwer, Dordrecht, 2004.

437   E. Ercolessi, G. Morandi, G. Marmo. *Int. J. Mod. Phys.* A 17, 3779 (2002).

438   A. Erdély. *Higher Transcendental Functions*, v.I, v.II, v.III, McGraw-Hill, New York, 1953.

439   A. Erdély. *Asymptotic Expansions*, Dover, New York, 1956.

440   T. Ertl, H. Ruder, R. Allrutz, K. Gruber, M. Günther, F. Hospach, M. Ruder, J. Subke, K. Widmayer. *Visual Comput.* 9, 453 (1993).

441   J.-H. Eschenburg, L. Hefendehl–Hebeker. *Math. Semesterber.* 47, 193 (2000).

442   R. Estrada, J. M. Garcia-Bondía, J. C. Várilly. *arXiv:funct-an*/9702001 (1997).

443   P. Exner, P. Šeba. *Phys. Lett.* A 245, 35 (1998).

444   P. Exner, P. Štóviček, P. Vytras. *J. Math. Phys.* 43, 2151 (2002).

445   B. R. Fabijonas, F. W. J. Olver. *SIAM Rev.* 41, 762 (1999).

446   V. I. Fabrikant. *Int. J. Eng. Sci.* 29, 1425 (1991).

447   W. Fakler. *mathPAD* 9, n1, 34 (1999).

448   E. Fan, J. Zhang. *Phys. Lett.* A 305, 383 (2002).

449   E. Fan. *Chaos, Solitons, Fractals* 16, 819 (2003).

450   H. M. Farkas, I. Kra. *Theta Constants, Riemann Surfaces and the Modular Group*, American Mathematical Society, Providence, 2001.

451   D. W. Farmer, R. C. Rhoades. *arXiv:math.NT*/0310252 (2003).

452   Y. N. Fedorov. *Functional Analysis Appl.* 35, 199 (2001).

453   J. Fehrman. *Arkiv. Math.* 13, 209 (1975).

454   S. Fenster. *Am. J. Phys.* 43, 683 (1975).

455   C. R. Fernández–Pousa, F. Mateos, M. T. Flores–Arias, C. Bao, M. V. Pérez. *J. Mod. Opt.* 51, 367 (2004).

456   D. J. Fernández, B. Mielnik, O. Rosas–Ortiz, B. F. Samsonov. *quant-ph*/0302204 (2003).

457   F. M. Fernández, E. A. Castro. *Physica* A 11, 334 (1982).

458   F. M. Fernández, E. A. Castro. *Am. J. Phys.* 53, 757 (1985).

459   F. M. Fernández. *Introduction to Perturbation Theory in Quantum Mechanics*, CRC Press, Boca Raton, 2001.

460   R. Ferreira, G. Bastard in E. Burstein, C. Weisbuch (eds.). *Confined Electrons and Photons*, Plenum Press, New York, 1995.

461   J. G. Fikioris, J. L. Tsalamengas. *J. Franklin Inst.* 324, 1 (1987).

462   S. Filippas, G. N. Makrakis. *Multiscale Model. Simul.* 1, 674 (2003).

463   H.-J. Fischer. *Elem. Math.* 48, 100 (1993).

464   W. Fischer, H. Leschke, P. Müller. *J. Phys.* A 30, 5579 (1997).

465   P. Flajolet, R. Schott. *Eur. J. Combinat.* 11, 421 (1990).

466   E. Flores, T. J. Osler. *Am. J. Phys.* 67, 718 (1999).

467   S. Flügge, H. Marshall. *Practical Quantum Mechanics*, Springer-Verlag, Berlin, 1974.

468   G. B. Folland. *Fourier Analysis and Its Applications*, Wadsworth & Brooks, Pacific Grove, 1992.

469   D. I. Ford. *Am. J. Phys.* 39, 215 (1971).

470   S. Foteinopoulou, E. N. Economou, C. M. Soukoulis. *arXiv:cond-mat*/0210346 (2002).

471   D. Fowler. *Am. Math. Monthly* 103, 1 (1996).

472   W. M. Frank, D. J. Land. *Rev. Mod. Phys.* 43, 36 (1971).

473   M. Fraser. *Coll. Math. J.* 15, 126 (1984).

474   R. Fricke. *Kurzgefasste Vorlesungen über verschiedene Gebiete der höheren Mathematik mit Berücksichtigung der Anwendungen*, Teubner, Leipzig, 1900.

475   R. Fricke. *Analytisch-funktionentheoretische Vorlesungen*, Teubner, Leipzig, 1900.

476   R. Fricke. in *Encyklopädie der mathematischen Wissenschaften* v.3/1/2, Teubner, Leipzig 1904-1916.

477   R. Fricke. *Die elliptischen Funktionen and ihre Anwendungen* v. 2, Teubner, Leipzig, 1922.

478   R. Fricke. *Lehrbuch der Algebra*, v.II, Vieweg, Braunschweig, 1926.

479   H. Friedrich, J. Trost. *Phys. Rev. Lett.* 76, 4869 (1996).

480   H. Frisk, S. de Gossen. *arXiv:math-ph*/0102007 (2001).

481   N. Fröman, P. O. Fröman. *Phase Integral Method*, Springer-Verlag, Heidelberg, 1996.

482   D. N. Fröman, P. O. Fröman. *J. Math. Phys.* 39, 4417 (1998).

483   T. Fujiwara, H. Fukuda, H. Ozaki. *arXiv:math-ph*/0210044 (2002).

484   T. Fülöp. *arXiv:hep-th*/9502145 (1995).

485   T. Fülöp, T. Cheon, I. Tsutsui. *Phys. Rev.* A 66, 052102 (2002).

486   I. N. Galidakis. *Complex Variables* 49, 759 (2004).

487 A. Gangopadhyaya, J. V. Mallow. *arXiv:hep-th*/0206133 (2002).

488 A. Ganguli. *Mod. Phys. Lett.* A 15, 1923 (2000).

489 A. Ganguly. *J. Math. Phys.* 43, 5310 (2002).

490 P. Garbaczewski, W. Karwowski. *arXiv:math-ph*/0310023 (2003).

491 E. J. Garboczi, K. A. Snyder, J. F. Douglas, M. F. Thorpe. *Phys. Rev.* E 52, 819 (1995).

492 G. García-Calderón, A. Rubio, J. Villavicencio. *arXiv:math.DG*/0006183 (2000).

493 G. García–Calderón, J. Villavicencio. *Phys. Rev.* E 66, 032104 (2002).

494 F. Garcia-Moliner, V. R. Velasco. *Theory of Single and Multiple Interfaces*, World Scientific, Singapore, 1992.

495 E. Gardi, G. Grunberg, M. Karliner. *arXiv:hep-ph*/9806462 (1998).

496 L. Gärding, C. Skau. *Arch. Hist. Exact Sci.* 48, 181 (1994).

497 T. M. Garoni, N. E. Frankel, M. L. Glasser. *J. Math. Phys.* 42, 1860 (2001).

498 F. G. Gascón, D. Peralta–Salas. *Phys. Lett.* A 333, 72 (2004).

499 G. Gaspar, M. Rahman. *Hypergeometric Series*, Cambridge University Press, Cambridge, 1990.

500 A. K. Gautesen, F. J. Sabina. *Quart. J. Mech. Appl. Math.* 43, 363 (1990).

501 W. Gautschi. *Math. Comput.* 15, 227 (1961).

502 W. Gautschi. *SIAM Rev.* 9, 24 (1967).

503 W. Gautschi. *J. Comp. Appl. Math.* 139, 173 (2002).

504 T. Gehrman, E. Remiddi. *arXiv:hep-ph*/0107173 (2001).

505 T. Gehrman, E. Remiddi. *arXiv:hep-ph*/0111255 (2001).

506 F. Gesztesy, R. Weikard. *Bull. Am. Math. Soc.* 35, 271 (1998).

507 A. K. Ghatak, K. Thyagarajan, M. R. Shenoy. *IEEE Quant. Electron.* 24, 1524 (1988).

508 P. Giacconi. *Phys. Rev.* D 53, 952 (1996).

509 A. Gil, J. Segura, N. M. Temme. *arXiv:math.NA*/0401131 (2004).

510 S. Giller. *arXiv:quant-ph*/9903097 (1999).

511 G. M. L. Gladwell, N. B. Willms. *J. Sound Vib.* 188, 419 (1995).

512 M. L. Glasser, I. J. Zucker. *Proc. Natl. Acad. Sc.* 74, 1800 (1977).

513 M. Glück, A. R. Kolovsky, H. J. Korsch. *arXiv:quant-ph*/0111132 (2001).

514 S. Godoy. *Phys. Rev.* A 65, 042111 (2002).

515 A. Góngora-T, J. V. José, S. Schaffner, P. H. E. Tiesinga. *arXiv:nlin.CD*/0008037 (2000).

516 J. A. González, M. A. del Olmo. *J. Phys.* A 31, 8841 (1998).

517 C. Goodman–Strauss, J. M. Sullivan. *arXiv:math.MG*/0205145 (2002).

518 P. Gordan. *Math. Annalen* 13, 375 (1878).

519 P. Gordan. *Math. Annalen* 28, 152 (1887).

520 B. Gordon, D. Sinor in K. Alladi (ed.). *Number Theory, Madras, 1987*, Springer-Verlag, Berlin, 1989.

521 B. Gordon, K. Hughes. *Contemp. Math.* 143, 415 (1993).

522 B. Gordon, S. Robins. *Glasgow Math. J.* 37, 1 (1995).

523 R. Gorenflo, F. Mainardi, D. Moretti, G. Pagnini, P. Paradisi. *Phys. Lett.* A 305, 106 (2002).

524 R. W. Gosper, Jr. *SIGSAM Bull.* 32, 8 (1998).

525 L. Gosse, P. A. Markovich. *J. Comput. Phys.* 197, 387 (2004).

526   X. Gou-liang, L. Jia-kai. *Chin. J. Numer. Math. Appl.* 17, 13 (1995).

527   H. W. Gould, T. A. Chapman. *Am. Math. Monthly* 69, 651 (1962).

528   S. Graffi, V. Greechi. *J. Math. Phys.* 19, 1002 (1978).

529   R. L. Graham, D. E. Knuth, O. Patashnik. *Concrete Mathematics*, Addison-Wesley, Reading, 1994.

530   B. Grammaticos, A. Ramani, C. M. Viallet. *arXiv:math-ph*/0409081 (2004).

531   P. Graneau, N. Graneau. *Newtonian Electrodynamics*, World Scientific, Singapore, 1996.

532   S. Gravel, P. Winternitz. *J. Math. Phys.* 43, 5902 (2002).

533   A. Gray. *A Treatise on Gyrostatics and Rotational Motion*, Dover, New York, 1959.

534   H. L. Gray, N. F. Zhang. *Math. Comput.* 50, 513 (1988).

535   J. J. Gray. *Linear Differentaial Equations and Group Theory*, Birkhäuser, Boston, 2000.

536   D. Grayson. *Arch. Math.* 52, 507 (1989).

537   M. L. Green. *Compos. Math.* 37, 233 (1978).

538   A. G. Greenhill. *Trans. Am. Math. Soc.* 8, 447 (1907).

539   A. G. Greenhill. *The Applications of the Elliptic Functions*, Dover, New York, 1959.

540   A. D. Greenwood, J.-M. Jin. *IEEE Antennas Prop.* 41, 9 (1999).

541   M. Grether, M. Fortes, M. de Llano, J. L. de Río, F. J. Sevilla, M. A. Solís, A. V. Valladares. *arXiv:cond-mat*/0205468 (2002).

542   P. Grigolini, A. Rocco, B. J. West. *arXiv:cond-mat*/9809075 (1998).

543   G. Grillo. *Expo. Math.* 14, 181 (1996).

544   J. J. Gray. *Linear Differential Equations and Group Theory from Riemann to Poincaré*, Birkhäuser, Basel, 1999.

545   J. Greenhough, P. C. Birch, S. C. Chapman, G. Rowlands. *arXiv:cond-mat*/0110605 (2001).

546   C. Grosche. *Phys. Rev. Lett.* 71, 1 (1993).

547   C. Grosche, F. Steiner. *Handbook of Feynman Path Integrals*, Springer-Verlag, Heidelberg, 1998.

548   S. Grossmann, M. Holthaus. *Optics Express* 1, 262 (1997).

549   Z.-Y. Gu, S.-W. Qian. *J. Phys.* A 21, 2573 (1988).

550   E. Gubler. *Math. Ann.* 48, 37 (1896).

551   F. Guil, M. Mañas, L. M. Alonso. *arXiv:nlin.SI*/0209051 (2002).

552   B. Guizal, D. Felbacq. *Phys. Rev.* E 66, 026602 (2002).

553   E. J. Gumbel. *Statistics of Extremes*, Columbia University Press, New York, 1958.

554   R. C. Gunning. *Lectures on Modular Forms*, Princeton University Press, Princeton, 1962.

555   D. P. Gupta, M. E. Muldoon. *arXiv:math.CA*/9910128 (1999).

556   V. Gurarii, V. Katsnelson. *Preprint NTZ* 3/2000(2000).   http://www.uni-leipzig.de/~ntz/abs/abs0300.htm

557   A. H. Guth, S.-Y. Pi. *Phys.Rev.* D 32, 1899 (1985).

558   G. Gutiérrez, J. M. Yáñez. *Am. J. Phys.* 65, 739 (1997).

559   J. C. Gutiérrez–Vega, R. M. Rodríguez–Dagnino. *Am. J. Phys.* 71, 222330 (2003).

560   S. Habib. *Phys Rev.* D 42, 2566 (1990).

561   G. Hackenbroich, J. U. Nöckel. *arXiv:chao-dyn*/9806020 (2000).

562   R. Hagedorn, J. Rafelski. *Commun. Math. Phys.* 83, 563 (1982).

563   Y. Hagihara. *Celestial Mechanics*, v.1, MIT Press, Cambridge, 1972.

564   N. H. S. Haidar. *J. Comput. Anal. Appl.* 4, 389 (2002).

565   C. Hainzl, R. Seiringer. *Lett. Math. Phys.* 61, 75 (2002).

566   G. Hamel. *Theoretische Mechanik*, Springer-Verlag, Heidelberg, 1949.

567   J. Hamilton. *Aharonov-Bohm and other Cyclic Phenomena*, Springer-Verlag, Berlin, 1997.

568   K. Hannabuss. *An Introduction to Quantum Theory*, Clarendon Press, Oxford, 1997.

569   J. H. Hannay. *Proc. R. Soc. Lond.* A 450, 51 (1995).

570   H. Hanssum. *J. Phys.* A 16, 3385 (1983).

571   R. Harley. *Quart. J. Pure Appl. Math.* VI, 38 (1864).

572   J. Harnad, J. McKay in D. Levi, O. Ragnisco (eds.). *SIDE III—Symmetries and Integrability of Differential Equations*, American Mathematical Society, Providence, 2000.

573   P. Harrison. *Quantum Wells, Wires, and Dots*, Wiley, Chichester, 2000.

574   D. R. Hartree. *Manchester Lit. Phil. Soc.* 80, 85 (1936).

575   K. Hasebe, A. Nakayama, Y. Sugiyama. *arXiv:patt-sol/*9812003 (1998).

576   I. A. Hassanien, R. A. Zait, A.-B. Abdel–Salam. *Physica Script.* 67, 457 (2003).

577   L. He, D. Vanderbilt. *Phys. Rev. Lett.* 86, 5341 (2001).

578   X.-F. He. *Phys. Rev.* B 42, 11751 (1990).

579   J. Heading. *An Introduction to Phase-Integral Methods*, Methuen, London, 1962.

580   E. J. Heller. *Phys. Rev. Lett.* 53, 1515 (1984).

581   E. Heller in M.-J. Giannoni, A. Voros, J. Zinn-Justin (eds.). *Chaos and Quantum Physics*, North-Holland, Amsterdam, 1991.

582   W. C. Henneberger. *Adv. Imaging Electron Phys.* 112, 56 (2000).

583   P. Henrici. *Applied and Computational Complex Analysis*, v.1, Wiley, New York, 1974.

584   M. Hentschel, H. Schomerus, R. Schubert. *arXiv:physics/*0208006 (2002).

585   M. Hentschel, K. Richter. *arXiv:physics/*0210002 (2002).

586   C. Hermite. *Compt. Rend.* 46, 508 (1858).

587   J. S. Hersch, M. R. Haggerty, E. J. Heller. *arXiv:nlin.CD/*0003023 (2000).

588   C. Herzog, M. Olshanii. *arXiv:cond-mat/*9609002 (1996).

589   C. Herzog, M. Olshanii. *Phys. Rev.* A 55, 3254 (1997).

590   W. Heymann. *Zeitschrift Math. Phys.* 39, 162, 193, 257, 321 (1894).

591   W. Heymann. *Zeitschrift Math. Phys.* 42, 81, 113 (1897).

592   J. R. Higgins. *Bull. Am. Math. Soc.* 12, 45 (1985).

593   R. Hilfer in B. Dubrulle, F. Graner, D. Sornette (eds.). *Scale Invariance and Beyond*, Springer-Verlag, Berlin, 1997.

594   R. Hilfer. *arXiv:cond-mat/*0006427 (2000).

595   R. Hilfer. *Applications of Fractional Calculus in Physics*, World Scientific, Singapore, 2000.

596   S. Hilger. *Nonlin. Anal Theor. Methods Appl.* 30, 2683 (1997).

597   S. C. Hill, R. E. Benner in P. W. Barber, R. K. Chuang (eds.). *Optical Effects Associated With Small Particles*, World Scientific, Singapore, 1988.

598   E. Hille. *Ordinary Differential Equations in the Complex Domain*, Wiley, New York 1976.

599   M. Hillery, R. O'Connell, M. Scully, E. Wigner. *Phys. Rep.* 106, 121 (1984).

600   P. Hillion. *Rep. Math. Phys.* 41, 223 (1998).

601   P. Hillion. *Rep. Math. Phys.* 41, 235 (1998).

602   M. Hirano, N. Kurokawa, M. Wakayama. *J. Ramanujan Soc.* 18, 195 (2003).

603 M. Hirschhorn, F. Garvan, J. Borwein. *Can. J. Math.* 45, 673 (1993).

604 E. W. Hobson. *The Theory of Spherical and Ellipsoidal Harmonics*, Cambridge University Press, Cambridge, 1931.

605 H. Hochstadt. *The Functions of Mathematical Physics*, Wiley, New York, 1971.

606 J. Hodgkinson. *Proc. Lond. Math. Soc.* 24, 71 (1925).

607 M. H. Holmes. *Introduction to Perturbation Methods*, Springer-Verlag, New York, 1995.

608 M. Holthaus, E. Kalinowski. *Ann. Phys.* 270, 198 (1998).

609 Y. C. Hon, E. Fan. *Appl. Math. Comput.* 146, 813 (2003).

610 P. Hoodbhoy. *arXiv:quant-ph*/0411031 (2004).

611 K. Hornberger, U. Smilansky. *arXiv:chao-dyn*/9912022 (1999).

612 K. Hornberger, U. Smilansky. *Phys. Rep.* 367, 249 (2002).

613 Y. Hsu. *Int. J. Mod. Phys.* C 4, 805 (1993).

614 J. H. Hubbard, J. M. McDill, A. Noonburg, B. H. West. *Coll. Math. J.* 25, 419 (1994).

615 M. Hug, C. Menke, W. P. Schleich. *Phys. Rev.* A 57, 3188 (1998).

616 M. Hug, C. Menke, W. P. Schleich. *Phys. Rev.* A 57, 3206 (1998).

617 B. D. Hughes. *Random Walks and Random Environments*, v.1, Clarendon Press, Oxford, 1995.

618 A. Hulme. *Proc. Camb. Phil. Soc.* 92, 183 (1982).

619 M. Humi. *arXiv:math-ph*/0202020 (2002).

620 B. Hunt. *The Geometry of Some Special Arithmetic Quotients*, Springer-Verlag, Heidelberg, 1996.

621 C. Hunter, B. Guerrieri. *Studies Appl. Math.* 64, 113 (1981).

622 M. Idemen. *Wave Motion* 38, 251 (2003).

623 Z. Idziaszek, K. Rzazewski. *arXiv:cond-mat*/0304361 (2003).

624 G. Illies. *Commun. Math. Phys.* 220, 69 (2001).

625 F. Irwin. *Am. Math. Monthly* 23, 149 (1916).

626 H. Itoyama, T. Oota. *arXiv:hep-th*/0206123 (2002).

627 C. Itzykson, J.-M. Drouffe. *Statistical Field Theory*, v.2, Cambridge University Press, Cambridge, 1989.

628 E. L. Ivchenko, G. Pikus. *Superlattices and Other Heterostructures*, Springer-Verlag, Berlin, 1995.

629 A. Ivic. *The Riemann Zeta-Function*, Wiley, New York, 1985.

630 J. D. Jackson. *Am. J. Phys.* 67, 107 (1999).

631 A. W. Jacobson. *Proc. Am. Math. Soc.* 1, 682 (1950).

632 M. A. Jafarizadeh, S. Behnia. *arXiv:nlin.CD*/0208024 (2002).

633 V. Jamnejad–Dailami, R. Mittra, T. Itoh. *IEEE Trans. Antennas Prop.* 20, 392 (1972).

634 D. J. Jeffrey, D. E. G. Hare, R. M. Corless. *Math. Scientist* 21, 1 (1996).

635 F. Joachimsthal. *Anwendung der Differential- und Integralrechnung auf die allgemeine Theorie der Flächen un der Linien doppelter Krümmung*, Teubner, Leipzig, 1890.

636 W. Jasolski. *Phys. Rep.* 271, 1 (1996).

637 H.-G. Jeon. *Far East J. Math. Sci.* 5, 81 (2002).

638 D. S. Jones. *Acoustic and Electromagnetic Waves*, Clarendon, Oxford, 1986.

639 J. Jorgenson, S. Lang. *Basic Analysis of Regularized Series and Product*, Springer-Verlag, New York 1994.

640 G. Jorjadze, G. Weigt. *arXiv:hep-th*/0207041 (2002).

641 G. Jumarie. *Choas, Solitons, Fractals* 22, 907 (2004).

642   G. Kaelbermann. *J. Phys.* A 34, 6465 (2001).

643   R. N. Kalia (ed.). *Recent Advances in Fractional Calculus*, Global Publishing Company, Sauk Rapids, 1993.

644   D. Kalman. *Coll. Math. J.* 32, 2 (2001).

645   S. O. Kamphorst, S. P. de Carvalho. *arXiv:math.DS*/0009210 (2000).

646   L. W. Kantorowitsch, W. I. Krylow. *Näherungsmethoden der höheren Analysis*, Deutscher Verlag der Wissenschaften, Berlin, 1956.

647   Y.-M. Kao, P.-G. Luan. *arXiv:cond-mat*/0210338 (2002).

648   H. Karabulut, E. L. Sibert III. *J. Math. Phys.* 38, 4815 (1997).

649   A. A. Karatsuba, S. M. Voronin. *The Riemann Zeta-Function*, De Gruyter, Berlin, 1992.

650   P. W. Karlsson. *J. Comput. Appl. Math.* 37, 171 (1991).

651   P. W. Karlsson. *J. Comput. Appl. Math.* 118, 215 (2000).

652   D. L. Kaufman. *Am. J. Phys.* 67, 133 (1999).

653   P. E. Kaus, W. K. R. Watson. *Phys. Rev.* 120, 44 (1960).

654   J. Kaye. *J. of Math. Phys.* 34, 119 (1955).

655   S. Kempfle, I. Schäfer, H. Beyer. *Nonl. Dyn.* 29, 99 (2002).

656   A. J. Kempner. *Am. Math. Monthly* 21, 48 (1914).

657   M. K. Kerimov. *J. Vich. Mat. Mat. Fis.* 39, 1962 (1999).

658   M. Kerker. *The Scattering of Light.*, Academic Press, New York, 1969.

659   W. Ketterle, N. J. van Druten. *Phys. Rev.* A 54, 656 (1996).

660   P. G. Kevrekidis, V. V. Konotop, A. R. Bishop, S. Takeno. *J. Phys.* A 35, L641 (2002).

661   D. Khare. *arXiv:quant-ph*/0105030 (2001).

662   A. Khare, U. Sukhatme. *arXiv:quant-ph*/0105044 (2001).

663   K. Khare, N. George. *J. Phys.* A 36, 10011 (2003).

664   A. Kheyfits. *Fract. Calc. Appl. Anal.* 7, 177 (2004).

665   A. I. Khizhnyak, S. P. Anokhov, R. A. Lymarenko, M. S. Soskin, M. V. Vasnetsov. *J. Opt. Soc. Am.* A 17, 2199 (2000).

666   D. Kiang *Am. J. Phys.* 45, 308 (1977).

667   L. Kiepert. *J. Math.* 87, 114 (1878).

668   M. Kiernan. *Arch. Hist. Exact Sciences* 8, 40 (1971).

669   J. Killingbeck. *Rep. Progress Phys.* 40, 963 (1977).

670   S. S. Kim. *Am. Math. Monthly* 106, 968 (1999).

671   D. Kim, J. K. Koo. *Bull. Korean Math. Soc.* 37, 675 (2000).

672   D. Kim, J. K. Koo. *J. Korean Math. Soc.* 40, 977 (2003).

673   K. Kimoto, M. Wakayama. *Int. Math. Res. Notes* 17, 855 (2003).

674   T. Kimura, K. Shima. *Memoirs Faculty of Science, Kyushu University* A 46, 137 (1992).

675   R. B. King. *Beyond the Quartic Equation*, Birkhäuser, Boston 1996.

676   R. B. King, E. R. Cranfield. *J. Math. Phys.* 32, 823 (1991).

677   R. B. King, E. R. Cranfield. *Comput. Math. Applic.* 24, 13 (1992).

678   K. Kirsten, D. J. Toms. *Phys. Rev.* A 54, 4188 (1996).

679   V. Kiryakova. *J. Phys.* A 30, 5085 (1997).

680 N. Kishore. *Proc. Am. Math. Soc.* 14, 527 (1963).

681 C. Kittel. *Introduction to Solid State Physics*, Wiley, New York, 1986.

682 A. Khare, U. Sukhatme. *arXiv:quant-ph*/9906044 (1999).

683 A. Khare, U. Sukhatme. *arXiv:quant-ph*/0402206 (2004).

684 A. Khare, U. Sukhatme. *Phys. Lett.* A 324, 406 (2004).

685 K. Kimura, H. Yahagi, R. Hirota, A. Ramani, B. Grammaticos, Y. Ohta. *J. Phys.* A 35, 9205 (2002).

686 M. Kleber. *Phys. Rep.* 236, 331 (1994).

687 F. Klein. *Vorlesungen über die hypergeometrische Funktion*, Springer-Verlag, Berlin, 1933.

688 F. Klein. *Math. Annalen* 121, 503 (1877).

689 F. Klein *Lectures on the Icosahedron and the Solution of Equations of the Fifth Degree*, Dover, New York, 1956, republished with commentaries by P. Slodowy, Birkhäuser, Basel, 1993.

690 F. Klein. *Lectures on Mathematics*, Am. Math. Soc., New York, 1911.

691 F. Klein. *Gesammelte mathematische Abhandlungen* II, Springer-Verlag, Berlin, 1922.

692 F. Klein. *Vorlesungen über die Entwicklung der Mathematik im 19. Jahrhundert*, Springer-Verlag, Berlin, 1926.

693 H. Kleinert. *Path Integrals in Quantum Mechanics, Statistics and Polymer Physics*, World Scientific, Singapore, 1995.

694 M. Klimek. *Czech. J. Phys.* 52, 1247 (2002).

695 P. L. Knight, E. Roldán, J. E. Sipe. *J. Mod. Opt.* 51, 1761 (2004).

696 H. Kobayashi, I. Tsutsui. *Nucl. Phys.* B 472, 409 (1996).

697 L. Y. Kobelev. *arXiv:physics*/0001035 (2000).

698 R. Kobes, S. Pelěs. *arXiv:nlin.CD*/0005005 (2000).

699 E. Koch. *Z. Krist.* 215, 386 (2004).

700 G. Köhler. *Math. Scand.* 66, 147 (1990).

701 W. Kohn. *Rev. Mod. Phys.* 26, 292 (1954).

702 W. Kohn. *Phys. Rev.* 115, 809 (1959).

703 H. D. Koenig. *Phys. Rev.* 44, 657 (1933).

704 W. Koepf. *arXiv:math.CA*/9603215 (1996).

705 T. Kohda, H. Fujisaki. *Physica* D 148, 242 (2001).

706 K. S. Kolbig. *Math. Comput.* 24, 679 (1970).

707 K. S. Kolbig. *Math. Comput.* 26, 551 (1972).

708 H. Konig. *Proc. Am. Math. Soc.* 120, 1315 (1994).

709 T. H. Koornwinder in R. Martini (ed.). *Geometrical Approaches to Differential Equations*, Springer-Verlag, Berlin, 1980.

710 T. H. Koornwinder. *J. Comput. Appl. Math.* 99, 449 (1998).

711 B. G. Korenev. *Bessel Functions and Their Applications*, Taylor & Francis, London, 2002.

712 J. P. Kottmann, O. J. F. Martin, D. R. Smith, S. Schultz. *J. Micros.* 202, 60 (2001).

713 V. Kowalenko, A. R. Rawlinson. *J. Phys.* A 31, L663 (1998).

714 K. Kowalski, J. Rembielinski, L. C. Papaloucas. *J. Phys.* A 29, 8841 (1996).

715 G. Kozma, E. Schreiber. *arXiv:math.PR*/0212156 (2002).

716 G. Kozma, E. Schreiber. *Electr. J. Prob.* 9, 1 (2004).

717 J. C. Knight, N. Dubreuil, V. Sandoghdar, J. Hare, V. Lefèvre–Seguin, J. M. Raimond, S. Haroche. *Optics Lett.* 21, 698 (1996).

718   H. E. Knoepfel. *Magnetic Fields*, Wiley, New York, 2000.

719   H. Knörrer. *Invent. Math.* 59, 119 (1980).

720   W. Koepf. *Hypergeometric Summation*, Vieweg, Braunschweig, 1998.

721   K. M. Kolwankar, A. D. Gangal. *arXiv:chao-dyn* 9711010 (1997).

722   M. Kracht, E. Kreyszig. *Historia Mathematica* 17, 16 (1990).

723   H. Kraft. *arXiv:math.AC*/0403323 (2004).

724   A. Kratzer, W. Franz. *Transzendente Funktionen*, Geest & Portig, Leipzig, 1960.

725   N. N. Kravchenko. *Vestnik Mosk. Univ* s.1, n4, 69 (1996).

726   A. Krazer. *Lehrbuch der Thetafunktionen*, Teubner, Leipzig, 1903.

727   I. M. Krichever. *Funct. Anal. Appl.* 15, 282 (1981).

728   L. Kronecker. *Monatsberichte der Königlich Preuss. Akad. der Wiss.* 609, (1861); reprinted in K. Hensel (ed.). *Kroneckers Werke*, Teubner, Leipzig, 1929.

729   I. Krichever, S. P. Novikov. *arXiv:math-ph*/0003004 (2000).

730   D. A. Krueger. *Phys. Rev.* 172, 211 (2002).

731   J. G. Krüger, A. Poffyn. *Physica* A 85, 84 (1976).

732   C. A. Kruelle, A. Kittel, J. Peinke, R. Richter. *Z. Naturf.* 52a, 581 (1997).

733   G. Krylov, M. Robnik. *J. Phys.* A 34, 5403 (2001).

734   A. Kudrolli, M. C. Abraham, J. P. Gollub. *arXiv:nlin.CD*/0002045 (2000).

735   K. Kuiken, J. T. Masterson in W. Abikoff, J. S. Birman, K. Kuiken (eds.). *The Mathematical Legacy of Wilhelm Magnus Groups*, Geometry and Special Functions, American Mathematical Society, Providence, 1994.

736   A. B. J. Kuilaars, A. Martínez–Finkelshtein, R. Orive. *arXiv:math.CA*/0301037 (2003).

737   V. I. Kukulin, O. A. Rubtsova. *Theor. Math. Phys.* 130, 54 (2002).

738   V. I. Kukulin, O. A. Rubtsova. *Theor. Math. Phys.* 134, 404 (2003).

739   V. I. Kukulin, O. A. Rubtsova. *Theor. Math. Phys.* 139, 693 (2004).

740   I. M. Kulić, H. Schiessel. *arXiv:physics*/0206084 (2002).

741   K. Kunze, P. Ziesche. *Wiss. Zeitschrift TU Dresden* 37, 201 (1988).

742   P. Kurasov. *Am. J. Phys.* A 29, 1767 (1996).

743   N. Kurokawa, M. Wakayama. *Kyushu University Preprint Series* 2002-7 (2002).

744   N. Kurokawa, M. Wakayama. *Kyushu University Preprint Series* 2002-8 (2002).

745   N. Kurokawa, M. Wakayama. *Kyushu University Preprint Series* 2003-19 (2003).

746   L. S. Kwok. *J. Theor. Biol.* 139, 573 (1989).

747   N. W. Lachlan. *Theory and Application of Mathieu Functions*, Clarendon Press, Oxford 1951.

748   L. K. Lachtin. *Math. Sbornik* 16, 597 (1891).

749   L. K. Lachtin. *Math. Sbornik* 17, 1 (1893).

750   M. Lakshmanan, S. Rajasekar. *Nonlinear Dynamics*, Springer-Verlag, Berlin, 2003.

751   C. C. Lam, P. T. Leung, K. Young. *J. Am. Opt. Soc.* B 9, 1585 (1992).

752   E. Lanckau. *Complex Integral Operators in Mathematical Physics*, Johann Ambrosius Barth, Berlin, 1993.

753   E. Landau. *Bibliotheca Mathematica* 3, 69 (1907).

754   L. D. Landau, E. M. Lifschitz. *Course of Theoretical Physics, Electrodynamics*, v. II, Pergamon Press, Oxford, 1982.

755   L. D. Landau, E. M. Lifshitz. *Quantum Mechanics*, Pergamon Press, Oxford, 1977.

756   P. T. Landsberg. *Thermodynamics—With Quantum Statistical Illustrations*, Interscience, New York, 1961.

757   P. T. Landsberg in P. T. Landsberg (ed.). *Solid State Theory*, Wiley, London, 1969.

758   S. Lang. *El. Math.* 51, 17 (1996).

759   E. Langmann. *arXiv:math-ph*/0102005 (2001).

760   U. Larsen. *J. Phys.* A 16, 2137 (1983).

761   R. E. Larson. *Coll. Math. J.* 5, 68 (1974).

762   W. Láska. *Sammlung von Formeln der reinen und angewandten Mathematik*, Vieweg, Braunschweig, 1894.

763   N. Laskin. *arXiv:quant-ph*/0206098 (2002).

764   A. Laurincikas. *Limit Theorems for the Riemann Zeta-Function*, Kluwer, Dordrecht, 1996.

765   J. L. Lavoie, T. J. Osler, R. Tremblay. *SIAM Rev.* 18, 240 (1976).

766   D. F. Lawden. *Elliptic Functions and Applications*, Springer-Verlag, New York, 1989.

767   P. G. L. Leach, G. P. Flessas. *J. Phys.* A 34, 6013 (2001).

768   N. N. Lebedev. *Special Functions and Their Applications*, Prentice-Hall, Englewood Cliffs, 1965.

769   S. W. Lee, J. Boersma. *J. Math. Phys.* 16, 1746 (1975).

770   S.-Y. Lee, S. Rim, J.-W. Ryu, T.-Y. Kwon, M. Choi, C.-M. Kim. *Phys. Rev. Lett.* 93, 164102 (2004).

771   T.-C. Lee. *Mathematical Methods in Physical Sciences and Engineering*, Vantage Press, New York, 1995.

772   R. Lefebvre, A. Palma. *Int. J. Quant. Chem.* 65, 487 (1997).

773   D. H. Lehmer. *Can. J. Math.* 2, 267 (1950).

774   D. H. Lehmer. *Math. Comput.* 50, 265 (1988).

775   J. Lehner. *Discontinuous Groups and Automorphic Functions*, American Mathematical Society, Providence, 1964.

776   R. Lehoucq, J.-P. Uzan, J. Weeks. *Kodai Math. J.* 26, 119 (2003).

777   E. Leimanis. *The General Problem of the Motion of Coupled Rigid Bodies about a Fixed Point*, Springer-Verlag, Berlin, 1965.

778   J. Lense. *Sphere Functions*, Geest & Portig, Leipzig, 1950.

779   H. Lenz. *Math. Z.* 67, 153 (1957).

780   U. Leonhardt, M. Wilkens. *Europhys. Lett.* 42, 365 (1998).

781   U. Leonhardt, P. Piwnicki. *arXiv:physics*/0009093 (2000).

782   M. Lerch. *Rozpravy České Akad.* 3, n 28, 1 (1894).

783   L. C. Lew Yan Voon, M. Willatzen. *Europhys. Lett.* 62, 299 (2003).

784   L. Lewin. *Polylogarithms and Associated Functions*, North Holland, Amsterdam, 1981.

785   X. Leyronas, M. Combescot. *arXiv:cond-mat*/0201553 (2002).

786   B. Li, B. Hu. *arXiv:cond-mat*/9712082 (1997).

787   H. Li, D. Kusnezov. *arXiv:cond-mat.*/9907202 (1999).

788   Y.-Q. Li, B. Chen. *arXiv:cond-mat*/9907171 (1999).

789   P. Li, Z. Pan. *Phys. Lett.* A 332, 39 (2004).

790   Q. Lin. *arXiv:quant-ph*/0010078 (2000).

791   F. Lindemann. *Nachrichten der Königlichen Gesellschaft der Wissenschaften und der Georg-August-Universität zu Göttingen* 1892, 292 (1892).

792   P. Lindqvist. *Ric. Mat.* 44, 269 (1995).

793   H. H. Linfoot, E. Wolf. *Proc. Phys. Soc.* B 69, 823 (1956).

794   C. M. Linton. *J. Eng. Math.* 33, 377 (1998).

795   C. M. Linton, P. McIver. *Hanbook of Mathematical Techniques for Wave/Structure Interactions*, Chapman & Hall, Boca Raton, 2001.

796   C. M. Linton. *Proc. R. Soc. Lond.* A 455, 1767 (2004).

797   H. J. Lippmann, F. Kuhrt. *Z. Naturf.* 13 a, 462 (1958).

798   J. Liu, L. Yang, K. Yang. *Chaos, Solitons, Fractals* 20, 1157 (2004).

799   J. Liu, L. Yang, K. Yang. *Phys. Lett.* A 325, 268 (2004).

800   S. Liu, Z. Fu, S. Liu, Q. Zhao. *Phys. Lett.* A 289, 69 (2001).

801   X.-Q. Liu, S. Jiang. *Appl. Math. Comput.* 158, 177 (2004).

802   Z.-G. Liu. *J. Number Th.* 85, 231 (2000).

803   Z.-G. Liu in F. Garvan, M. Ismail (eds.). *Symbolic Computation, Number Theory, Special Functions, Physics and Combinatorics*, Kluwer, Dordrecht, 2001.

804   A. Lodder, Y. V. Nazarov. *Phys. Rev.* B 58, 5783 (1998).

805   D. Loeb. *Adv. Math.* 91, 64 (1992).

806   A. Loewy in P. Epstein (ed.). *Repertorium der höheren Analysis*, Teubner, Leipzig, 1920.

807   C. T. Long. *Coll. Math. J.* 12, 320 (1981).

808   E. A. Lord, A. L. Mackay. *Curr. Sc.* 85, 346 (2003).

809   F. Lösch, F. Schoblik. *Die Fakultät*, Teubner, Leipzig, 1951.

810   J. H. Loxton. *Acta Arithm.* 43, 155 (1984).

811   D. Lozier in C. Dunkl, M. Ismail, R. Wong (eds.). *Special Functions*, World Scientific, Singapore, 2000.

812   R. Luck, J. W. Stevens. *SIAM Rev.* 44, 227 (2002).

813   E. Lüneburg in E. Meister (ed.). *Modern Mathematical Methods in Diffraction Theory and its Applications in Engineering*, Peter Lang, Franfurt, 1997.

814   Y. L. Luke. *The Special Functions and Their Approximations I*, Academic Press, New York, 1969.

815   T. Lukes, G. A. Ringwood, B. Suprato. *Physica* A 84, 421 (1976).

816   S. Lumb, S. K. Muthu. *Int. J. Mod. Phys.* B 17, 5855 (2004).

817   G. A. Luna–Acosta, K. Na, L. E. Reichl. *Phys. Rev.* E 53, 3271 (1996).

818   G. A. Luna–Acosta, J. A. Méndez–Bermúdez, F. M. Izrailev. *Phys. Lett.* A 274, 192 (2000).

819   G. A. Luna–Acosta, J. A. Méndez–Bermúdez, F. M. Izrailev. *Phys. Rev.* E 64, 036206 (2001).

820   J. T. Lunardi. *arXiv:physics/0105097* (2001).

821   P. Lynch, C. Houghton. *arXiv:nlin.SI/0303038* (2003).

822   L. R. M. Maas. *J. Comput. Appl. Math.* 51, 237 (1994).

823   S. W. MacDonald, A. N. Kaufman. *Phys. Rev. Lett.* 42, 1189 (1979).

824   G. Z. Machabeli, A. D. Rogava. *Phys. Rev.* A 50, 98 (1994).

825   W. D. MacMillan. *Ann. Math.* 19, 26 (1917).

826   W. D. MacMillan. *Dynamics of Rigid Bodies*, Dover, New York, 1936.

827   H. S. Madhusudhan, M. S. M. Naika, K. R. Vasuki. *Hardy–Ramanujan J.* 24, 3 (2001).

828   A. G. Magner, K. Arita, S. N. Fedotkin, K. Matsuyanagi. *arXiv:nlin.SI/0208005* (2002).

829   N. Magot, A. Zvonkin. *Discr. Math.* 217, 249 (2000).

830   W. Magnus, F. Oberhettinger, R. P. Soni. *Formulas and Theorems for the Special Functions of Mathematical Physics*, Springer-Verlag, Berlin, 1966.

831   F. Mainardi. *Chaos, Solitons & Fractals* 7, 1461 (1996).

832   B. J. Malešević. *Univ. Beograd Publ. Elektrotehn. Fak.* 14, 26 (2003).

833   B. J. Malešević. *arXiv:math.NT*/0406235 (2004).

834   S. S. Malu, A. Siddharthan. *arXiv:math-ph*/0001030 (2000).

835   B. N. Mandal, N. Mandal. *Integral Expansions Related to Mehler-Fock Type Transforms*, Longman, Essex, 1997.

836   Y. Manin. *arXiv:alg-geom*/9605010 (1996).

837   Y. I. Manin. *AMS Transl.* 186, 131 (1998).

838   V. Man'ko, M. Moshinsky, A. Sharma. *arXiv:quant-ph*/9902075 (1999).

839   T. Mano. *J. Math. Kyoto Univ.* 42, 41 (2002).

840   H. C. Manoharan, C. P. Lutz, D. M. Eigler. *Nature* 403, 512 (2000).

841   T. Mansour. *Adv. Appl. Math.* 28, 196 (2002).

842   M. Marden. *Am. Math. Monthly* 92, 643 (1985).

843   M. Marden. *Geometry of Polynomials*, American Mathematical Society, Providence, 1966.

844   M. Marden. *The Geometry of the Zeros of a Polynomial in a Complex Variable*, Math. Surveys 3, American Mathematical Society, New York, 1949.

845   O. I. Marichev. *Handbook of Integral Transforms of Higher Transcendental Functions*, Wiley, New York, 1983.

846   O. Marichev. *Function Cuts and Continuity*, MathSource 0205-007 (1992).

847   R. Markert, B. Zastrau. *ZAMM* 67, T430 (1987).

848   A. I. Markushevich. *Remarkable Sine Functions*, Elsevier, New York, 1966.

849   V. E. Markushin, R. Rosenfelder, A. W. Schreiber. *arXiv:math-ph*/0104019 (2001).

850   M. G. Marmorino. *J. Math. Chem.* 32, 303 (2003).

851   G. E. Marsh. *Force-Free Magnetic Fields*, World Scientific, Singapore, 1996.

852   P. A. Martin. *Math. Meth. Appl. Sc.* 24, 913 (2001).

853   Y. Martin. *Trans. Am. Math. Soc.* 348, 4825 (1996).

854   N. Martzel, C. Aslangul. *arXiv:cond-mat*/0009275 (2000).

855   V. P. Maslov. *The Complex WKB Method for Nonlinear Equations 1: Linear Theory*, Birkhäuser, Basel, 1994.

856   M. H. Martin. *J. Rat. Mech. Anal.* 2, 315 (1953).

857   M. Martínez–Corral, L. Muñoz–Excrivá, A. Pons, M. T. Caballero. *Eur. J. Phys.* 22, 361 (2001).

858   K. Maruno, A. Ankiewicz, N. Akhmediev. *arXiv:nlin.PS*/0209045 (2002).

859   N. Marzari, D. Vanderbilt. *Phys. Rev.* B 56, 12847 (1997).

860   A. M. Mathai. *A Handbook of Generalized Special Functions for Statistical and Physical Sciences*, Clarendon Press, Oxford, 1993.

861   M. Matsuda. *Lectures on Algebraic Solutions of Hypergeometric Differential Equations*, Kinokuniya, Kyoto, 1985.

862   S. Matsutani. *Rev. Math. Phys.* 9, 943 (1997).

863   S. Matsutani. *arXiv:math.DG*/0008153 (2000).

864   P. A. Maurone, A. J. Phares. *J. Math. Phys.* 20, 2191 (1979).

865   M. W. McCall, A. Lakhtakia, W. S. Weiglhofer. *Eur. J. Phys.* 23, 353 (2002).

866   K. T. McDonald. *arXiv:physics*/0006046 (2000).

867   H. McKean, V. Moll. *Elliptic Curves*, Cambridge University Press, Cambridge, 1997.

868   N. M. McLachlan. *Theory and Applications of Mathieu Functions*, Clarendon Press, Oxford, 1947.

869   J. H. McLeod. *J. Opt. Soc. Am.* 44, 592 (1954).

870   J. McNamee, F. Stenger, E. L. Whitney. *Math. Comput.* 25, 141 (1971).

871   A. McNaughton. *J. Fract. Calculus* 8, 119 (1995).

872   A. McNaughton. *J. Fract. Calculus* 10, 91 (1996).

873   A. McNaughton. *J. Fract. Calculus* 16, 55 (1999).

874   A. McNaughton. *J. Fract. Calculus* 19, 35 (2001).

875   K. W. Mahmud, J. N. Kutz, W. P. Reinhardt. *Phys. Rev.* A 66, 063607 (2002).

876   L. Méès, G. Gousbet, G. Gréhan. *J. Optics* A 4, S150 (2002).

877   W. N. Mei, Y. C. Lee. *J. Phys.* A 16, 1623 (1983).

878   J. D. Meiss. *Chaos* 2, 267 (1999).

879   J. Meixner, F. W. Schäfke, G. Wolf. *Mathieu Funktionen and Spheroidal Functions and Their Mathematical Foundations,* Springer, Berlin, 1980.

880   R. S. Melham. *Fib. Quart.* 37, 305 (1999).

881   V. V. Menon. *J. Comb. Th.* 15, 11 (1973).

882   A. Messiah. *Quantum Mechanics*, North Holland, Amsterdam, 1961.

883   R. Metzler, J. Klafter. *Europhys. Lett.* 51. 492 (2000).

884   R. Metzler. *Phys. Rep.* 339, 1 (2000).

885   L. Meunier, B. Salvy in J. R. Sendra (ed.). *ISSAC 2003*, ACM Press, New York, 2003.

886   R. E. Meyer. *SIAM Rev.* 31, 435 (1989).

887   S. G. Mikhlin. *Mathematical Physics, an Advanced Course*, North Holland, Amsterdam, 1970.

888   M. Mikolás. *Acta Sci. Math.* 14, 5 (1951).

889   R. F. Millar. *Radio Sc.* 8, 785 (1973).

890   K. S. Miller, B. Ross in H. M. Srivastava, S. Owa (eds.). *Univalent Functions, Fractional Calculus, and their Applications*, Horwood, Chichester, 1989.

891   K. S. Miller. *An Introduction to the Fractional Calculus and Fractional Differential Equations*, Wiley, New York, 1993.

892   B. R. Miller, A. Youssef. *Ann. Math. Artif. Intell.* 38, 121 (2003).

893   S. C. Miller, R. H. Good. *Phys. Rev.* 91, 174 (1953).

894   W. Miller, Jr. *Symmetry and Separation of Variables*, Addison-Wesley, Reading, 1977.

895   W. Miller, Jr. *IMA Preprint* 1687 (2000).   http://www.ima.umn.edu/preprints/feb2000/1687.ps

896   G. Millington. *Radio Science* 4, 95 (1969).

897   L. M. Milne-Thompson. *Die elliptischen Funktionen von Jacobi*, Springer-Verlag, Berlin, 1931.

898   G. V. Milovanovic, D. S. Mitrinovic, T. M. Rassias. *Topics in Polynomials: Extremal Problems, Inequalities, Zeros*, World Scientific, Singapore, 1994.

899   R. N. Miroshin. *Math. Notes* 5, 682 (2001).

900   M. Miyamoto. *arXiv:quant-ph*/0105033 (2001).

901   S. Moch, P. Uwer, S. Weinzierl. *arXiv:hep-ph*/0110083 (2001).

902   Y. Mochimaru, M.-W. Bae. *Int. J. Diff. Eq. Appl.* 5, 275 (2002).

903   M. I. Molina. *arXiv:physics*/9704006 (1997).

904   M. Möller. *Results in Math.* 24, 147 (1993).

905   A. G. Monastra, U. Smilansky, S. Gnutzmann. *arXiv:nlin.CD*/0212006 (2002).

906   B. S. Monozon, A. N. Shalaginov. *Solid State Commun.* 89, 167 (1994).

907   E. W. Montroll. *J. Soc. Indust. Appl. Math.* 4, 241 (1956).

908   P. Moon, D. E. Spencer. *Field Theory Handbook*, Springer-Verlag, Berlin, 1988.

909   J. Montaldi, A. Soulière, T. Tokieda. *arXiv:math.DS*/0210028 (2002).

910   P. A. P. Moran in G. Kallianpur, P. R. Krishnaiah, J. K. Gosh (eds.). *Statistics and Probablility: Essays in Honor of C. R. Rao*, North Holland, Amsterdam, 1982.

911   H. Moriguchi. *J. Phys. Soc. Jap.* 14, 1771 (1959).

912   M. Moshinsky. *Phys. Rev.* 88, 625 (1952).

913   M. Moshinsky, D. Schuch. *Comput. Math. Appl.* 41, 641 (2001).

914   C. Moyer. *J. Phys.* C 6, 1461 (1973).

915   J. G. Muga, M. Büttiker. *arXiv:quant-ph*/0001039 (2000).

916   J. G. Muga. *arXiv:quant-ph*/0105081 (2001).

917   D. Mugnai. *Phys. Lett.* A 278, 6 (2000).

918   S. Mukamel. *Principles of Nonlinear Optical Spectroscopy*, Oxford University Press, Oxford, 1995.

919   M. E. Muldoon. *Lett. Nuov. Cim.* 23, 447 (1978).

920   M. E. Muldoon. *J. Comput. Appl. Math.* 65, 299 (1995).

921   W. J. Mullin, J. P. Fernández. *arXiv:cond-mat*/0211115 (2002).

922   E. Muñoz–Garcia, R. Pérez–Marco. *IHES Preprint* M03-24 (2003).
      http://www.ihes.fr/PREPRINTS/M03/Resu/resu-M03-34.html

923   E. Muñoz–Garcia, R. Pérez–Marco. *IHES Preprint* M03-52 (2003).
      http://www.ihes.fr/PREPRINTS/M03/Resu/resu-M03-52.html

924   H. Murrell. *The Mathematica Journal* 2, n1, 61 (1992).

925   G. Mussardo, S. Penati. *arXiv:hep-th*/9907039 (1999).

926   H. Nachbagauer. *arXiv:hep-th*/9703105 (1997).

927   A. Nagai. *J. Phys. Soc. Jpn.* 72, 2181 (2003).

928   A. Nagai. *J. Nonlin. Math. Phys.* 10, 133 (2003).

929   K. Nakanishi. *arXiv:patt-sol*/9909005 (1999).

930   K. Nakayama, M. Wadati. *J. Phys. Soc. Jpn.* 62, 473 (1993).

931   V. Namias. *J. Inst. Maths. Appl.* 25, 241 (1980).

932   A. Nanayakkara. *J. Phys.* A 37, 4321 (2004).

933   A. B. Nassar, J. M. F. Bassalo, P. d. T. S. Alencar. *Am. J. Phys.* 63, 849 (1995).

934   P. Navez, D. Bitouk, M. Gajda, Z. Idziaszek, K. Rzazewski. *Phys. Rev. Lett.* 79, 1789 (1997).

935   A. H. Nayfeh. *Introduction to Perturbation Techniques*, Wiley, New York, 1981.

936   J. Negro, L. M. Nieto, O. Rosas–Ortiz. *J. Math. Phys.* 41, 7964 (2000).

937   Z. Nehari. *Conformal Mapping*, McGraw-Hill, New York, 1952.

938   E. Neher. *Jahresber. DMV* 87, 164 (1985).

939   G. Németh, G. Páris. *J. Math. Phys.* 26, 1175 (1985).

940   Y. V. Nesterenko, P. Philippon. *Introduction to Algebraic Independence Theory*, Springer-Verlag, Berlin, 2001.

941   E. Netto. *Vorlesungen über Algebra*, Teubner, Leipzig, 1900.

942   B. Neuberger, J. W. Neuberger, D. W. Noid. *arXiv:math.SC*/0105217 (2001).

943   W. Neutsch. *Coordinates*, de Gruyter, Berlin, 1996.

944 E. H. Neville. *J. London Math. Soc.* 15, 113 (1940).

945 E. H. Neville. *Can. J. Math.* 11, 175 (1959).

946 V. B. Nevzorov. *Records: Mathematical Theory*, American Mathematical Society, Providence, 2000.

947 M. Newman. *Trans. Am. Math. Soc.* 73, 313 (1952).

948 R. G. Newton. *Scattering Theory of Waves and Particles*, Springer-Verlag, New York, 1982.

949 T. A. Newton. *Am. Math. Monthly* 81, 592 (1974).

950 D. A. Nicole, P. J. Walters. *J. Phys.* A 21, 2351 (1988).

951 N. Nielson. *Handbuch der Theorie der Gammafunktion*, Chelsea, New York, 1965.

952 M. M. Nieto. *arXiv:hep-th*/0005281 (2000).

953 M. M. Nieto. *arXiv:hep-th*/0106223 (2001).

954 M. M. Nieto. *arXiv:quant-ph*/0112142 (2001).

955 M. Nieto–Vesperinas. *Scattering and Diffraction in Physical Optics*, Wiley, New York, 1991.

956 A. F. Nikiforov, V. B. Uvarov. *Special Functions of Mathematical Physics*, Birkhäuser, Basel, 1988.

957 K. Nishimoto. *Fractional Calculus*, Descartes Press, Koriyama, 1991.

958 A. W. Niukkanen. *J. Phys.* A 16, 1813 (1983).

959 S. Nonnenmacher, A. Voros. *J. Phys.* A 30, 295 (1997).

960 N. E. Nörlund. *Rend. Circ. Mat. Palermo* 10, 27 (1961).

961 M. A. Núñez. *Int. J. Quant. Chem.* 62, 449(1997).

962 H. M. Nussenzweig. *Causality and Dispersion Relations*, Academic Press, London, 1972.

963 H. M. Nussenzweig in A. Frank, B. Wolf (eds.). *Symmetries in Physics*, Springer-Verlag, Berlin, 1992.

964 J. F. Nye, J. H. Hannay, W. Liang. *Proc. Royal Soc. Lond.* A 449, 515 (1995).

965 J. F. Nye. *Proc. R. Soc. Lond.* A 458, 401 (2002).

966 F. Oberhettinger, W. Magnus. *Anwendungen der elliptischen Funktionen in Physik and Technik*, Springer-Verlag, Berlin, 1949.

967 K. Odagiri. *Eureka* 53, 36 (1994).

968 D. H. J. O'Dell. *arXiv:quant-ph*/0011105 (2000).

969 A. Ogg. *Modular Forms and Dirichlet Series*, Benjamin, New York, 1969.

970 O. M. Ogreid, P. Osland. *J. Comput. Appl. Math.* 98, 245 (1998).

971 O. M. Ogreid, P. Osland. *J. Comput. Appl. Math.* 136, 389 (2001).

972 O. M. Ogreid, P. Osland. *J. Comput. Appl. Math.* 140, 659 (2002).

973 Y. Ohnuki, S. Kamefuchi. *J. Math. Phys.* 19, 67 (1978).

974 Y. Ohta, A. Ramani, B. Grammaticos. *J. Phys.* A 35, L635 (2002).

975 R. Olbricht. *Verhandlungen Kaiserl. Leopold.-Carol. Deutschen Akademie*, 52, 7 (1888).

976 A. I. Olemskoi. *arXiv:cond-mat*/9906367 (1999).

977 M. A. Olshanetsky, A. M. Perelomov. *Phys. Rep.* 94, 313 (1983).

978 M. H. Oliveira, J. A. Miranda. *arXiv:physics*/0011015 (2000).

979 F. W. J. Olver. *Asymptotics and Special Functions*, Academic Press, New York, 1974.

980 F. W. J. Olver. *Proc. Roy. Soc.* A 289, 501 (1978).

981 R. E. O'Malley, Jr. *Introduction to Singualr Perturbations*, Academic Press, New York, 1974.

982 J. A. Méndez–Bermúdez, G. A. Luna–Acosta, F. M. Izrailev. *arXiv:cond-mat*/0309595 (2003).

983 K. A. O'Neil. *J. Math. Phys.* 30, 1373 (1989).

984 K. Ono. *The Web of Modularity: Arithmetic of the Coefficients of Modular Forms and q-Series*, American Mathematical Society, Providence, 2004.

985 J. S. E. Ortiz, R. E. de Carvalho. *Brazilian J. Phys.* 31, 538, (2001).

986 T. J. Osler. *Am. Math.Monthly* 106, 774 (1999).

987 A. Ostrowski. *Math. Ann.* 94, 248 (1925).

988 Y. C. Ou, Z. Cao, Q. Shen. *J. Chem. Phys.* 121, 8175 (2004).

989 K. E. Oughstun, G. C. Sherman. *Electromagnetic Pulse Propagation in Causal Dielectrics*, Springer-Verlag, Berlin, 1994.

990 H. M. Ozaktas, Z. Zalevsky, M. Alper Kutay. *The Fractional Fourier Transform*, Wiley, Chichester, 2001.

991 E. Packel, D. Yuen. *Coll. Math. J.* 35, 337 (2004).

992 H. J. Pain. *The Physics of Vibrations and Waves*, John Wiley & Sons, Chichester, 1993.

993 A. Palma, L. Sandoval, M. Martin, R. Lefebre. *Int. J. Quant. Chem.* 99, 484 (2004).

994 S. Parthasarathy, M. Lakhsmanan. *J. Phys.* A 23, L1223 (1990).

995 R. B. Paris. *Proc. Royal Soc. Lond.* A 446, 565 (1994).

996 R. B. Paris, D. Kamisnki. *Asymptotics and the Mellin–Barnes Integrals*, Cambridge University Press, Cambridge, 2001.

997 R. B. Paris. *Proc. R. Soc. Lond.* A 460, 2737 (2004).

998 D. Park. *Classical Dynamics and Its Quantum Analogues*, Springer-Verlag, Berlin 1990.

999 D. K. Park, S.-K. Yoo. *arXiv:hep-th*/9712134 (1997).

1000 E. J. Parkes, B. R. Duffy, P. C. Abbott. *Phys. Lett.* A 295, 280 (2002).

1001 E. Pascal. *Repertorium der höheren Mathematik*, v.1/1, Teubner, Leipzig 1900.

1002 S. J. Patterson. *Hist. Math.* 17, 132 (1990).

1003 P. Paul, D. Nkemzi. *J. Math. Phys.* 41, 4551 (2000).

1004 F. B. Pedersen, G. T. Einevoll, P. C. Hemmer. *Phys. Rev.* B 44, 5470 (1991).

1005 L. A. Peletier, W. C. Troy. *Spatial Patterns*, Birkhäuser, Boston, 2001.

1006 J. B. Pendry. *Phys. Rev. Lett.* 85, 3966 (2000).

1007 J. B. Pendry, S. A. Ramakrishnan. *J. Phys.* CM 15, 6345 (2003).

1008 A. M. Perelomov. *Reg. Chaotic Dynamics* 5, 89 (2000).

1009 A. M. Perelomov. *arXiv:math-ph*/0303016 (2003).

1010 L. Perotti. *arXiv:nlin.CD*/0406018 (2004).

1011 M. A. Peter, M. H. Meylan. *Wave Motion* 40, 1 (2004).

1012 M. Petkovšek, H. S. Wilf, D. Zeilberger. *A=B*, A. K. Peters, Wellesley, 1996.

1013 E. N. Petropoulou, P. D. Siafarikas, I. D. Stabolas. *J. Comput. Appl. Math.* 153, 387 (2003).

1014 T. Pfitzenreiter. *ZAMM* 84, 284 (2004).

1015 F. Philippe, J. Arnaud, L. Chusseau. *arXiv:math-ph*/0211029 (2002).

1016 G. Pickett, Y. Millev. *J. Phys.* A 35, 4485 (2002).

1017 J. Pierpont. *Monatshefte Math. Phys.* VI, 15 (1895).

1018 P. Pierpont. *Bull. Am. Math. Soc.* 2, 200 (1896).

1019 V. P. Pikulin, S. I. Pohozaev. *Equations in Mathematical Physics*, Birkhäuser, Basel, 2001.

1020 E. Pina. *Rev. Mex. Fis.* 39, 10 (1993).

1021 M. J. Pinheiro. *arXiv:physics/*0304113 (2003).

1022 N. Pippenger. *Am. Math. Monthly* 87, 391 (1980).

1023 G. Pittaluga, L. Sacripante. *SIAM J. Math. Anal.* 22, 260 (1991).

1024 I. Podlubny. *Fractional Differential Equations*, Academic Press, San Diego, 1999.

1025 I. Podlubny. *Fract. Calc. Appl. Anal.* 5, 1 (2002).

1026 H. D. Politzer. *Phys. Rev.* A 54, 5048 (1996).

1027 I. V. Ponomarev. *Austral. J. Phys.* 52, 859 (1998).

1028 E. G. C. Poole. *Introduction to the Theory of Linear Differential Equations*, Clarendon Press, Oxford, 1936.

1029 R. Portugal, L. Golebiowski, D. Frenkel. *Am. J. Phys.* 67, 534 (1999).

1030 T. Pöschel, N. V. Brilliantov. *arXiv:cond-mat/*9906138 (1999).

1031 E. Pourreza, K. Azizi. *Int. J. Diff. Eq. Appl.* 4, 207 (2002).

1032 T. Pradhan, A. V. Khare. *Am. J. Phys.* 41, 59 (1973).

1033 G. Prasad. *An Introduction to the Theory of Elliptic Functions and Higher Transcendentals*, Calcutta, 1928.

1034 V. Prasolov, Y. Solovyev. *Elliptic Functions and Elliptic Integrals*, American Mathematical Society, Providence, 1997.

1035 G. M. Pritula, V. E. Vekslerchik. *arXiv:nlin.SI/*0009043 (2000).

1036 J. R. Probert–Jones. *J. Am. Opt. Soc.* A 1, 822 (1984).

1037 R. T. Prosser. *Am. Math. Monthly* 101, 535 (1994).

1038 A. P. Prudnikov, Y. A. Brychkov, O. I. Marichev. *Integrals and Series*, v. 1-3, Gordon and Breach, New York, 1990.

1039 J. R. Quine, S. H. Heydari, R. Y. Song. *Trans. Am. Math. Soc.* 338, 213 (1993).

1040 E. M. Rabei, T. S. Alhalholy, A. Rousan. *Int. J. Mod. Phys.* A 19, 3083 (2004).

1041 A. Rabinovitch, R. Thieberger, M. Friedman. *J. Phys.* B 18, 393 (1985).

1042 F. Raciti, E. Venturino in A. M. Anile, V. Capasso, A. Greco (eds.). *Progress in Industrial Mathematics at ECMI 2000*, Springer-Verlag, Berlin, 2002.

1043 P. M. Radmore. *J. Phys.* A 13, 173 (1979).

1044 Q. I. Rahman, G. Schmeisser. *Analytic Theory of Polynomials*, Oxford University Press, Oxford, 2002.

1045 M. J. Rakovic, E. A. Solov`ev. *Phys. Rev.* A 40, 6692 (1989).

1046 K. G. Ramanathan. *J. Indian Math. Soc.* 52, 71 (1987).

1047 H. E. Rauch, A. Lebowitz. *Elliptic Functions, Theta Functions and Riemann Surfaces*, Williams & Wilkins, Baltimore, 1973.

1048 J. Rauch. *Partial Differential Equations*, Springer-Verlag, New York, 1997.

1049 O. Rausenberger. *Lehrbuch der Theorie der periodischen Funktionen einer Variablen*, Teubner, Leipzig, 1884.

1050 K. Rawer. *Annalen Physik* 35, 385 (1939).

1051 M. Reed, B. Simon. *Methods of Modern Mathematical Physics* v. II, Academic Press, New York, 1975.

1052 M. Reed, B. Simon. *Methods of Modern Mathematical Physics* v. IV, Academic Press, New York, 1978.

1053 L. E. Reichl, W. M. Zheng. *Phys. Rev.* A 29, 2186 (1984).

1054 W. H. Reid. *Z. angew. Math. Phys.* 48, 646 (1997).

1055 E. Remiddi, J. A. M. Vermaseren. *arXiv:hep-ph/*9905237 (1999).

1056 R. Remmert. *Classical Topics in Function Theory*, Springer, New York, 1998.

1057 H.-S. Ren. *Appl. Math. E-Notes* 1, 40 (2001).

1058 R. W. Rendell, S. M. Girvin. *Phys. Rev.* B 23, 6610 (1981).

1059   F. J. Richelot. *J. Math.* 45, 233 (1852).

1060   R. D. Richtmyer. *J. Appl. Phys.* 10, 391 (1939).

1061   F. Riewe. *Phys. Rev.* E 55, 3581 (1997).

1062   T. D. Riney. *Trans. Am. Math. Soc.* 88, 214 (1958).

1063   I. Rivin. *arXiv:math.PR*/0305252 (2003).

1064   I. Rivin. *arXiv:math.MG*/0306387 (2003).

1065   I. Rivin. *arXiv:math.MG*/0403375 (2004).

1066   F. Robicheaux, U. Fano, M. Cavagnero, D. A. Harmin. *Phys. Rev.* A 35, 3619 (1987).

1067   L. Robin. *Fonctions Sphériques de Legendre et Fonctions Sphéroidales*, v.1, v.2, v.3, Gauthier-Villars, Paris, 1959.

1068   R. W. Robinett. *Am. J. Phys.* 64, 440 (1996).

1069   R. W. Robinett, S. Heppelmann. *Phys. Rev.* A 65, 062103 (2002).

1070   M. Robnik. *J. Phys.* A 13, L349 (1980).

1071   M. Robnik, L. Salasnich. *J. Phys.* A 30, 1711 (1997).

1072   A. Rocco, B. J. West. *Physica* A 265, 535 (1999).

1073   M. Rodrigo, M. Mimura. *Jpn. J. Industr. Appl. Math.* 18, 657 (2001).

1074   W. A. Rodrigues Jr., D. S. Thober, A. L. Xavier Jr. *Phys. Lett.* A 284, 217 (2001).

1075   F. J. Romeiras, G. Rowlands. *Phys. Rev.* A 33, 3499 (1986).

1076   R. Romo, J. Villavicencio. *arXiv:quant-ph*/0006052 (2000).

1077   E. Rosencher, A. Fiore, B. Vinter, V. Berger, P. Bois, J. Nagle. *Science* 271, 168 (1996).

1078   B. Ross. *Methods of Summation*, Descartes Press, Kaguike, 1987.

1079   H. Rosu, J. Socorro. *arXiv:gr-qc*/9610018 (1996).

1080   F. Rothe. *Nieuw Archief Wiskunde* 5/1 397 (2000).

1081   K. Rottbrand. *ZAMM* 5, 321 (1998).

1082   A. A. Rousan, E. Malkawi, E. M. Rabei, H. Wydan. *Fract. Calc. Appl. Anal.* 5, 155 (2002).

1083   L. A. Rubel. *Am. Math. Monthly* 96, 777 (1989).

1084   R. Russel. *Proc. Lond. Math. Soc.* 19, 91 (1888).

1085   R. S. Rutman. *Theor. Math. Fiz.* 105, 393 (1995).

1086   A. Rubinowicz. *Sommerfeldsche Polynommethode*, Springer-Verlag, Berlin, 1972.

1087   S. N. M. Ruijsenaars. *Commun. Math. Phys.* 110, 191 (1987).

1088   L. Ruby. *Am. J. Phys.* 64, 39 (1996).

1089   G. A. Rudykh, É. I. Semenov. *Math. Notes* 70, 714 (2001).

1090   K. C. Rustagi, J. Ducuing. *Optic Comm.* 10, 258 (1974).

1091   P. C. Sabatier. *Lett. Nuov. Cim.* 21, 41 (1978).

1092   A. I. Saichev, G. M. Zaslavsky. *Chaos* 7, 753 (1997).

1093   A. I. Saichev, K.-F. Bergren, A. F. Sadreev. *arXiv:nlin.CD*/0012019 (2000).

1094   S. Sakoda, M. Omote. *Advances in Imaging and Electron Physics* 110, 101 (1999).

1095   K. L. Sala. *SIAM J. Math. Anal.* 20, 1514 (1989).

1096   L. Salasnich. *arXiv:math-ph*/0008030 (2000).

1097   J. Salo, M. M. Salomaa. *J. Opt.* A 3, 366 (2001).

1098   E. Salamin. *Math. of Comput.* 135, 565 (1976).

1099   V. V. Samarin, S. M. Samarina. *J. Vis. Mat. Fiz.* 41, 1099 (2001).

1100   A. A. Samarski. *Equations of Mathematical Physics*, Pergamon Press, Oxford, 1963.

1101   S. G. Samko, A. A. Kilbas, O. I. Marichev. *Fractional Integrals and Derivatives*, Gordon & Breach, 1993.

1102   A. A. Samoletov. *J. Comput. Appl. Math.* 131, 503 (2001).

1103   D. Sato. *Proc. Am. Math. Soc.* 14, 996 (1963).

1104   J. Saurer. *Bases of Special Functions and Their Domain of Convergence*, Akademie-Verlag, Berlin, 1993.

1105   J. W. Sander. *Math. Semesterber.* 39, 185 (1992).

1106   H. Sari, Y. Ergün, I. Sökmen, M. Tomak. *Superlattices and Microstructures* 20, 163 (1996).

1107   T. Sauter, F. Paschke. *Phys. Lett.* A 285, 1 (2001).

1108   I. Schäfer, H. J. Seifert. *ZAMM* 82, 423 (2002).

1109   F. W. Schäfke. *Einführung in die Theorie der speziellen Funktionen der mathematischen Physik*, Springer-Verlag, Berlin, 1963.

1110   H. Scheffé. *Technometrics* 12, 388 (1970).

1111   T. B. Scheffler. *Proc. R. Soc. Lond.* A 336, 475 (1974).

1112   W. Scheibner. *Beiträge zur Theorie der linearen Transformationen*, Teubner, Leipzig, 1907.

1113   W. P. Schleich. *Quantum Optics in Phase Space*, Wiley–VCH, Berlin, 2001.

1114   L. Schlesinger. *Handbuch der Theorie der linearen Differentialgleichungen* v. II/2, Tuebner, Leipzig, 1898..

1115   A. Schmidt, L. Gaul. *ZAMM* 83, 26 (2003).

1116   H.-J. Schmidt, J. Schnack. *Physica* A 260, 479 (1998).

1117   H.-J. Schmidt, J. Schnack. *arXiv:cond-mat*/9803151 (1998).

1118   H.-J. Schmidt, J. Schnack. *arXiv:cond-mat*/0104293 (2001).

1119   C. Schmit in M.-J. Giannoni, A. Voros, J. Zinn-Justin (eds.). *Chaos and Quantum Physics*, North-Holland, Amsterdam, 1991.

1120   B. Schoeneberg. *Elliptic Modular Functions*, Springer-Verlag, Berlin, 1974.

1121   H. Schomerus, C. W. J. Beenakker. *Phys. Rev. Lett.* 82, 2951 (1999).

1122   H. F. Schouten, T. D. Visser, G. Gbur, D. Lenstra, H. Blok. *J. Opt.* A 6, S277 (2004).

1123   H. A. Schwarz. *J. reine angew. Math.* 75, 292 (1873).

1124   H. A. Schwarz. *Gesammelte Mathematische Abhandlungen* v.1, Springer-Verlag, Berlin, 1890.

1125   J. B. Seaburn. *Hypergeometric Functions and Their Applications*, Springer-Verlag, New York, 1991.

1126   T. B. A. Senior, P. L. E. Uslenghi in J. J. Bowman, T. B. A. Senior, P. L. E. Uslenghi (eds.) *Electromagnetic and Acoustic Scattering by Simple Shapes*, Hemisphere, New York, 1987.

1127   S. Servadio. *Phys. Rev.* A 24, 793 (1981).

1128   A. Shaarawi, I. M. Besieris. *J. Phys.* A 33, 7227 (2000).

1129   Shabot–Marcos, A. Sandoval–Villalbazo. *arXiv:math.MG*/0411106 (2004).

1130   B. S. Shastry. *Phys. Rev. Lett.* 50, 633 (1983).

1131   A. Shelankov, M. Ozana. *arXiv:cond-mat*/9907230 (1999).

1132   J. Q. Shen. *arXiv:cond-mat*/0402213 (2004).

1133   L.-C. Shen. *Proc. Am. Math. Soc.* 120, 1131 (1994).

1134   S. Shen, Z. Pan. *Phys. Lett.* A 308, 143 (2003).

1135   L. A. Shenyavskiĭ. *Prikl. Mat. Mekh.* 43, 1079 (1979).

1136 N. V. Shepelevich, V. V. Lopatin, V. P. Maltsev, V. N. Lopatin. *J. Opt.* A 1, 448 (1999).

1137 T. Shimbori, T. Kobayashi. *arXiv:math-ph*/9910009 (1999).

1138 T. Shimbori. *arXiv:quant-ph*/9912073 (1999).

1139 T. Shimbori. *Phys. Lett.* A 273, 37 (2000).

1140 T. Shimbori, T. Kobayashi. *arXiv:quant-ph*/0006019 (2000).

1141 T. Shimbori, T. Kobayashi. *arXiv:hep-th*/0010039 (2000).

1142 M. Shub, D. Tischler, R. F. Williams. *SIAM J. Math. Anal.* 19, 246 (1988).

1143 J. Shurman. *Geometry of the Quintic*, Wiley, New York 1997.

1144 J. Si, W. Zhang, G.-H. Kim. *Appl. Math. Comput.* 150, 647 (2004).

1145 C. L. Siegel. *Advanced Analytic Number Theory*, Tata Bombay, 1980.

1146 C. E. Siewert, E. E. Burniston. *J. Math. Anal. Appl.* 43, 626 (1973).

1147 C. E. Siewert, E. E. Burniston. *J. Math. Anal. Appl.* 46, 329 (1973).

1148 C. E. Siewert. *J. Math. Phys.* 19, 434 (1978).

1149 H. J. Silverstone. *Phys. Rev. Lett.* 55, 2523 (1985).

1150 H. J. Silverstone, S. Nakai, J. G. Harris. *Phys. Rev.* A 32, 1341 (1985).

1151 H. J. Silverstone, J. G. Harris, J. Čížek, J. Paldus. *Phys. Rev.* A 32, 1965 (1985).

1152 H. J. Silverstone. *Int. J. Quant. Chem.* 29, 261 (1986).

1153 M. P. Silvermann, W. Strange. *Am. J. Phys.* 64, 773 (1996).

1154 B. Simon. *Am. Math. Soc.* 24, 303 (1991).

1155 I. B. Simonenko. *Dokl. Akad. Nauk* 367, 603 (1999).

1156 K. Simony. *Theoretische Elektrotechnik*, Deutscher Verlag der Wissenschaften, Berlin, 1977.

1157 A. Sinha, R. Roychoudhury. *Int. J. Quant. Chem.* 73, 497 (1999).

1158 O. Šipr, B. L. Györffy. *J. Phys.* CM 8, 169 (1996).

1159 E. K. Sklyanin. *arXiv:solv-int*/9504001 (1995).

1160 S. L. Skorokhodov. *Programming Comput. Software* 29, 75 (2003).

1161 J. C. Slater. *Phys. Rev.* 87, 807 (1952).

1162 L. J. Slater. *Generalized Hypergeometric Functions*, Cambridge University Press, Cambridge, 1966.

1163 S. Y. Slavyanov. *Asymptotic Solutions to the One-Dimensional Schrödinger Equation*, American Mathematical Society Providence, 1996.

1164 S. Y. Slavyanov, W. Lay, A. Seeger. *Special Functions*, Oxford University Press, Oxford, 2000.

1165 P. Slodowy in H. Knörrer, C.-G. Schmidt, J. Schwermer, P. Slodowy. *Mathematische Miniaturen* 3, Birkhäuser, Basel, 1986.

1166 V. P. Smirnov, D. E. Usvyat. *Phys. Rev.* B 59, 9695 (1999).

1167 V. P. Smirnov, D. E. Usvyat. *Phys. Rev.* B 64, 245108 (2001).

1168 W. I. Smirnow. *Lehrgang der höheren Mathematik* v.2, Verlag der Wissenschaften, Berlin, 1955.

1169 M. Sokolov, J. Klafter, A. Blumen. *Phys. Today* 55, n11, 48 (2002).

1170 P. Solé, P. Loyer. *Eur. J. Combinat.* 19, 222 (1998).

1171 A. Sommerfeld. *Vorlesungen über Theoretische Physik: Optik* v. 4, Harri Deutsch, Frankfurt, 1994.

1172 D.-Y. Song. *arXiv:quant-ph*/0211107 (2002).

1173 C. M. Sorensen. *Optic Comm.* 173, 145 (2000).

1174 Y. Sosov, C. E. Theodosiou. *J. Math. Phys.* 43, 2831 (2002).

1175 J. Spanier, K. B. Oldham. *An Atlas of Functions*, Hemisphere, New York, 1987.

1176 W. Specht in M. Deuring, G. Köthe (eds.). *Enzyklopädie der Mathematischen Wissenschaften* I 1, n3, T.2, Teubner, Stuttgart, 1958.

1177 J. Spieß. *Math. Comput.* 55, 839 (1990).

1178 M. E. Spina, M. Saraceno. *arXiv:nlin.CD*/0011022 (2000).

1179 R. Spira. *Math. Comput.* 30, 863 (1976).

1180 H. M. Srivastava, R. K. Saxena. *Appl. Math. Comput.* 118, 1 (2001).

1181 H. M. Srivastava, J. Choi. *Series Associated with the Zeta and Related Functions*, Kluwer, Dordrecht, 2001.

1182 H. M. Srivastava. *J. Comput. Appl. Math.* 142, 441 (2002).

1183 S. S. Srivastava, B. M. L. Srivastava. *Jñānābha* 28, 129 (1998).

1184 J. Stalker. *Complex Analysis*, Birkhäuser, Boston, 1998.

1185 J. J. Stamnes. *Waves in Focal Regions*, Institute of Physics, Bristol, 1986.

1186 A. J. Starobin, S. W. Teitsworth. *arXiv:nlin.CD*/0010028 (2000).

1187 A. H. K. Stelzer, F.-M. Haar. *Adv. Imaging Elect. Phys.* 106, 293 (1999).

1188 F. Stenger. *SIAM J. Num. Anal.* 12, 239 (1975).

1189 F. Stenger. *SIAM Rev.* 23, 165 (1981).

1190 F. Stenger. *Numerical Methods Based on Sinc and Analytic Functions*, Springer-Verlag, New York, 1993.

1191 S. A. Stepin, A. A. Arshanov. *Dokl. Akad. Nauk.* 378, 18 (2001).

1192 B. Y. Sternin, V. E. Shatalov. *Borel-Laplace Transform and Asymptotic Theory*, CRC Press, Boca Raton, 1996.

1193 J. Stillwell. *Math. Intell.* 17, n2, 58 (1995).

1194 H. J. Stockman, J. Stein. *Phys. Rev. Lett.* 64, 2215 (1990).

1195 M. A. Stremler, H. Aref. *J. Fluid Mech.* 392, 101 (1999).

1196 M. A. Stremler. *J. Math. Phys.* 45, 3584 (2004).

1197 M. J. O. Strutt. *Lamésche-, Mathieusche und verwandte Funktionen in Physik und Technik*, Springer, Berlin, 1932.

1198 B. Sturmfels. *Disc. Math.* 210, 171 (2000).

1199 C. Stutz. *Am. J. Phys.* 36, 826 (1968).

1200 S. V. Sukhinin. *J. Appl. Math. Mech.* 63, 863 (1999).

1201 A. Sütő. *arXiv:cond-mat*/0311658 (2003).

1202 B. Sutherland. *Phys. Rev.* A 4, 2019 (1971).

1203 A. Swarup, B. Cowan. *J. Low Temp. Phys.* 134, 881 (2004).

1204 J. L. Synge in S. Flügge (ed.). *Handbuch the Physik III.1, Prinzipien der klassischen Mechanik and Feldtheorie*, Springer-Verlag, Berlin, 1960.

1205 S. Tabachnikov. *Moscow Math. J.* 2, 183 (2002).

1206 M. B. Tabanov. *Russian Acad. Sci. Dokl. Math.* 48, 438 (1994).

1207 Y. Takahashi. *Found. Phys.* 23, 739 (1993).

1208 T. Takami. *Progr. Theor. Phys.* Suppl. 116, 303 (1994).

1209 K. Takasaki. *arXiv:math.QA*/0004118 (2000).

1210 K. Takasaki. *J. Math. Phys.* 42, 1443, (2001).

1211 K. Takemura. *arXiv:math.CA*/0112179 (2001).

1212 W. L. Tanner. *Messenger of Math.* 17, 95 (1888).

1213 J. Tannery, J. Molk. *Éléments de la théorie des fonctions elliptiques* VIII, Chelsea, New York, 1972.

1214 V. I. Tatarskii. *Sov. Phys. Usp.* 26, 311 (1983).

1215 J. R. Taylor. *Nuov. Cim.* B 23, 313 (1974).

1216 R. Tazzioli. *Historia Math.* 28, 232 (2001).

1217 N. M. Temme. *Special Functions*, Wiley, New York, 1996.

1218 P. Tempesta, E. Alfinito, R. A. Leo, G. Solani. *arXiv:quant-ph*/0203121 (2002).

1219 R. Terras. *Math. Proc. Camb. Phil. Soc.* 89, 331 (1981).

1220 H. P. Thielman. *Ann. Math.* 31, 193 (1930).

1221 K. Thirulogasanthar, N. Saad, A. B. von Keviczky. *arXiv:math-ph*/0404036 (2004).

1222 W. J. Thompson. *Atlas for Computing Mathematical Functions*, Wiley, New York, 1997.

1223 J.-P. Tignol. *Galois' Theory of Algebraic Equations*, Longman, Harlow, 1980.

1224 T. N. Tikhonov, A. A. Samarski. *Equations of Mathematical Physics*, Pergamon Press, New York, 1963.

1225 E. C. Titchmarsh. *Eigenfunction Expansions Associated with Second-Order Differential Equations*, Oxford University Press, Oxford, 1946.

1226 E. C. Titchmarsh. *The Theory of the Riemann Zeta Function*, Clarendon Press, Oxford, 1986.

1227 P. G. Todorov. *Mathematica Balkanica* 8, 375 (1994).

1228 F. Tölke. *Praktische Funktionenlehre* v. 5, Springer-Verlag, Berlin, 1966.

1229 M. Tomiya, N. Yoshinaga. *J. Stat. Phys.* 83, 215 (1996).

1230 J. A. Toth. *Ann. Phys.* 130, 1 (1995).

1231 M. N. Tran, R. K. Bhaduri. *Phys. Rev.* E 68, 026105 (2003).

1232 I. Travěnec. *Phys. Rev.* B 69, 033104 (2004).

1233 F. Tricomi. *Elliptische Funktionen*, Geest & Portig, Leipzig, 1948.

1234 F. G. Tricomi, A. Erdélyi. *Pacific J. Math.* 1, 133 (1951).

1235 B. A. Troesch, H. R. Troesch. *Math. Comput.* 27, 755 (1973).

1236 M. Trott. *The Mathematica Journal* 6, n4, 55 (1996).

1237 M. Trott. *The Mathematica GuideBook for Numerics*, Springer-Verlag, New York, 2005.

1238 M. Trott. *The Mathematica Journal* 9, 314 (2004).

1239 T. T. Truong, P. Audit. *J. Phys.* A 23, 2795 (1990).

1240 W. V. Tschirnhaus. *Acta Eruditorum* 2, 204 (1683).

1241 I. Tsutsui, T. Fülöp. *arXiv:quant-ph*/0209110 (2002).

1242 S.-J. Tu, E. Fischbach. *arXiv:math-ph*/0004021 (2000).

1243 S.-J. Tu and E. Fischbach. *J. Phys.* A 35, 6557 (2002).

1244 L. Turko. *J. Math. Phys.* 45, 3659 (2004).

1245 K. Umashankar, A. Taflove. *Computational Electromagnetics*, Artech House, Boston, 1993.

1246 H. Umemura in D. Mumford. *Tata Lectures on Theta* II, Birkhäuser, Basel, 1984.

1247 H. Umezawa, G. Vitello. *Quantum Mechanics*, Bibliopolis, Napoli 1986.

1248 K. Umeno. *arXiv:chao-dyn*/9704007 (1997).

1249 K. Umeno. *Phys. Rev.* E 55, 5280 (1997).

1250 K. Unnikrishnan, A. R. P. Rau. *Am. J. Phys.* 64, 1034 (1996).

1251   V. Urumov, G. J. Ivanovski. *Am. J. Phys.* 51, 950 (1983).

1252   A. A. Utzinger. *Die reellen Züge der Riemann'schen Zetafunktion*, Leemann & Co, Zürich, 1934.

1253   N. K. Uzunoglu, J. G. Fikioris. *J. Phys.* A 12, 825 (1979).

1254   T. Vachaspati. *Phys. Rev.* A 66, 014104 (2002).

1255   O. Vallée, M. Soares, C. de Izarra. *Z. angew. Math. Phys.* 48, 156 (1997).

1256   S. R. Valluri, D. J. Jeffrey, R. M. Corless. *Can. J. Phys.* 78, 823 (2000).

1257   P. M. van den Berg, J. T. Fokkema. *J. Opt. Soc. Am.* 69, 27 (1979).

1258   M. van den Berg, J. T. Lewis. *Physica* A 110, 550 (1982).

1259   M. van den Berg, J. T. Lewis, J. V. Pulè. *Helv. Phys. Acta* 59, 1271 (1986).

1260   W. van Dijk, Y. Nogami. *Phys. Rev. Lett.* 83, 2867 (1999).

1261   A. van Hameren, R. Kleiss. *arXiv:physics/9804022* (1998).

1262   J. F. van Diejen, L. Vinet. *Calogero–Moser–Sutherland Models*, Springer-Verlag, New York, 2000.

1263   J. Vaníček, E. J. Heller. *arXiv:nlin.CD/0101055* (2001).

1264   B. P. van Milligen, A. Lopez Fraguas. *Comput. Phys. Commun.* 81, 74 (1994).

1265   F. J. F. Velicia. *Phys. Rev.* A 34, 4387 (1986).

1266   L. Verde-Star. *Stud. Appl. Math.* 85, 215 (1991).

1267   H. A. Verril. *CRM Proc.* 30, 253 (2001).

1268   M. D. Verweij. *J. Acoust. Soc. Am.* 92, 2223 (1992).

1269   A. P. Veselov, A. B. Shabat. *Funct. Anal. Appl.* 27, 1 (1993).

1270   R. Vidunas. *arXiv:math.CA/0109222* (2001).

1271   R. Vidunas. *arXiv:math.CA/0403510* (2004).

1272   H. Viesel. *Archiv Math.* 22, 106 (1971).

1273   N. J. Vilenkin, A. V. Klimyk. *Representation of Lie Groups and Special Functions* v. 1-3, Kluwer, Amsterdam, 1993.

1274   M. B. Villarino. *arXiv:math.NT/0308028* (2003).

1275   N. Virchenko, I. Fedotova. *Generalized Associated Legendre Functions and Their Applications*, World Scientific, Singapore, 2001.

1276   F. Vivanco, F. Melo, C. Coste, F. Lund. *Phys. Rev. Lett.* 83, 1966 (1999).

1277   G. Vivanti. *Elementi della teoria delle funzioni poliedriche e modulari*, Hoepli, Milano, 1906.

1278   G. Vivanti. *Archiv Math.* 8, 53 (1908).

1279   G. Vivanti. *Les functions polédriques et modulaires*, Gauthier-Villars, 1910.

1280   V. S. Vladimirov. *Equations of Mathematical Physics*, Mir, Moscow, 1984.

1281   S. G. Vlădut. *Kronecker's Jugendtraum and Modular Functions*, Gordon Breach, Luxembourg, 1995.

1282   J. Vollinga, S. Weinzierl. *arXiv:hep-ph/0410259* (2004).

1283   A. Voros. *Algebra i Analys* 8, 103 (1996).

1284   A. Voros. *arXiv:math-ph/9902016* (1999).

1285   A. Voros. *arXiv:math.CV/0104051* (2001).

1286   A. Voros. *Ann. Inst. Fourier* 53, 665 (2003).

1287   A. Voros. *arXiv:math.NT/0404213* (2004).

1288   H. Waalkens, J. Wiersig, H. D. Dullin. *arXiv:chao-dyn/9812012* (1998).

1289   M. Wagner, H. Mizuta. *Phys. Rev.* B 48, 14393 (1993).

1290 P. Wagner. *Math. Scand.* 86, 273 (2000).

1291 P. Wagner in H. Florain, N. Ortner, F. J. Schnitzer, W. Tutschke (eds.). *Functional-Analytic and Complex Methods, their Interactions, and Applications to Partial Differential Equations*, World Scientific, Singapore, 2001.

1292 S. Wagon. *Mathematica in Action*, W. H. Freeman, New York, 1990.

1293 M. Waldschmidt. *J. Comput. Anal. Appl.* 160, 323 (2003).

1294 C.-S. Wang, D.-S. Chu. *Physica* B 129, 227 (1993).

1295 J. P. Wang. *J. Nonlin. Math. Phys.* 9, 212 (2002).

1296 Q. Wang, Y. Chen, Z. Hongqing. *Chaos, Solitons, Fractals.* 23, 477 (2004).

1297 Z. X. Wang, D. R. Guo. *Special Functions*, World Scientific, Singapore, 1989.

1298 G. H. Wannier. *Phys. Rev.* 52, 191 (1937).

1299 R. D. H. Warburton, J. Wang. *Am. J. Phys.* 72, 1404 (2004).

1300 H. A. Warchall. *Electr. J. Diff. Eq.* Conf. 04, 245 (2000). http://ejde.math.unt.edu/conf-proc/04/w3/abstr.html

1301 T. Watanabe, Y, Choyal, K. Minami, V. L. Granatstein. *Phys. Rev.* E 69, 056606 (2004).

1302 P. C. Watermann. *J. Acoust. Soc. Am.* 57, 791 (1975).

1303 G. N. Watson. *Quart. J. Math.* 10, 266 (1939).

1304 G. N. Watson. *A Treatise on the Theory of Bessel Functions*, Cambridge University Press, Cambridge, 1958.

1305 H. Weber. *Lehrbuch der Algebra*, Vieweg, Braunschweig, 1898.

1306 T. A. Weber, C. L. Hammer. *J. Math. Phys.* 21, 24 (1980).

1307 C. Weiss, M. Wilkens. *Optics Express* 1, 272 (1997).

1308 G. H. Weiss. *Aspects and Applications of the Random Walk*, North Holland, Amsterdam, Oxford, 1994.

1309 G. Wendt in S. Flügge (ed.). *Handbuch der Physik* v.16, Springer-Verlag, Berlin, 1958.

1310 E. T. Whittaker. *A Treatise on the Analytical Dynamics of Particles and Rigid Bodies*, Dover, New York, 1944.

1311 E. T. Whittaker, G. N. Watson. *A Course on Modern Analysis*, Cambridge University Press, Cambridge, 1962.

1312 R. F. Wick. *J. Appl. Phys.* 25, 741 (1954).

1313 F. Wielonky in B. C. Berndt, Y.-S. Choi, S.-Y. Kinag in B. C. Berndt, F. Gesztesy (eds.). *Continued Fractions: From Analytic Number Theory to Constructive Approximations*, American Mathematical Society, Providence, 1999.

1314 J. Wiersig, P. H. Richter. *Z. Naturf.* 51a, 219 (1996).

1315 J. Wiersig. *arXiv:physics/0206018* (2002).

1316 J. Wiersig. *arXiv:physics/0210052* (2002).

1317 F. van Wijland, S. Caser, H. J. Hilhorst. *arXiv:cond-mat/9609018* (1996).

1318 C. A. Williams, T. van Flandern, E. A. Wright. *Celest. Mech.* 40, 367 (1987).

1319 B. L. Willis. *J. Math. Phys.* 43, 2610 (2002).

1320 A. Wiman in *Encyclopädie der mathematischen Wissenschaften* 1/1, Teubner, Leipzig, 1904.

1321 G. Wimmer, G. Altmann. *Thesaurus of Univariate Discrete Probability Distributions*, Stamm, Esssen, 1999.

1322 J. Wimp. *Computation With Recurrence Relations*, Pitman, Boston, 1984.

1323 J. Wittenburg. *Dynamics of Systems of Rigid Bodies*, Teubner, Stuttgart, 1977.

1324 M. R. M. Witwit, N. A. Gordon, J. P. Killingbeck. *J. Comput. Appl. Math.* 106, 131 (1999).

1325 J. Wohlfart. *Invent. Math.* 92, 187 (1988).

1326 S. Wolfram. *The Mathematica Book*, Cambridge University Press and Wolfram Media, 1996.

1327 S. C. Woon. *arXiv:Math.CA/9812147* (1998).

1328   S. C. Woon. *arXiv:hep-th*/9812234 (1998).

1329   S. Wrigge. *Math. Comput.* 37, 495 (1981).

1330   M. Xiao. *Phys. Rev.* E 60, 6226 (1999).

1331   D. Yafaev. *J. Phys.* A 30, 6981 (1997).

1332   R. M. Yamaleev. *Ann. Phys.* 277, 1 (1999).

1333   Z. Yan. *Comput. Phys. Comm.* 148, 30 (2002).

1334   Z. Yan. *J. Phys.* A 36, 1961 (2003).

1335   Z. Yan. *Comput. Phys. Comm.* 153, 1 (2003).

1336   Z.-.Y. Yan. *Int. J. Mod. Phys.* C 14, 225 (2003).

1337   L. Yang. *arXiv:math.NT*/0309415 (2003).

1338   Y. Yang. *Electr. J. Combinatorics.* 8, #19 (2000).

1339   Y. Yang. *Math. Z.* 246, 1 (2004).

1340   Z. Ye. *arXiv:cond-mat*/0408477 (2004).

1341   K. H. Yeon, S. Zhang, Y. D. Kim, C. I. Um, T. F. George. *Phys. Rev.* A 61, 042103 (2000).

1342   J. Yi. *Ramanujan J.* 5, 377 (2001).

1343   S. Yi. *Optics Comm.* 159, 7 (1999).

1344   L. Yingying. *Am. Math. Monthly* 109, 845 (2002).

1345   E. Yomba. *Preprint MPI Mathematik* MPI-2003-94 (2003).
       http://www.mpim-bonn.mpg.de/cgi-bin/preprint/preprint_search.pl/MPI-2003-94.ps?ps=MPI-2003-94

1346   M. Yoshida. *Hypergeometric Functions, My Love*, Vieweg, Braunschweig, 1997.

1347   M. Y. Yoshimoto in C. Jia, K. Matsumoto (eds.). *Analytic Number Theory*, Kluwer, Dordrecht, 2002.

1348   K. W. Yu, J. T. K. Wan. *arXiv:cond-mat*/0102059 (2001).

1349   C. Yüce. *Phys. Lett.* A 327, 107 (2004).

1350   C. Zachos. *arXiv:hep-th*/0110114 (2001).

1351   H. A. Zaghloul, O. Barajas in A. Lakhtakia (ed.). *Essays on the Formal Aspects of Electromagnetic Theory*, World
       Scientific, Singapore, 1993.

1352   D. Zagier. *MPIM Preprint 78/1999*/0002022 (1999).
       http://www.mpim-bonn.mpg.de/cgi-bin/preprint/preprint_search.pl/MPI-1999-78-b.ps?ps=MPI-1999-78-b

1353   J. Zagrodzinski. *Phys. Rev.* E 51, 2566 (1995).

1354   J. Zak. *J. Phys.* A 36, L553 (2003).

1355   W. Zakowicz. *J. Phys.* A 36, 4445 (2003).

1356   W. Zakowicz. *J. Phys.* B 37, L153 (2004).

1357   W. Zakowski. *Phys. Rev.* E 64, 066610 (2001).

1358   S. J. Zaroodny. *Am. Math. Monthly* 68, 593 (1961).

1359   E. Zauderer. *Partial Differential Equations of Applied Mathematics*, Wiley, New York, 1989.

1360   D. Zeilberger. *Math. Intell.* 16, n4, 11 (1994).

1361   P. Zeiner, R. Dirl, B. L. Davies. *Phys. Rev.* B 58, 7681 (1998).

1362   R. Zhdanov, A. Zhalij. *arXiv:math-ph*/9911018 (1999).

1363   X. H. Zheng, D. Carroll. *J. Opt.* A 1, 168 (1999).

1364   A. Zhivkov. *Preprint Humboldt University* 2002/P-11 (2002).
       http://www.mathematik.hu-berlin.de/publ/pre/2002/P-11-.ps

1365 L. Zhou, C. Kambhamettu. *Graphical Models* 63, 1 (2001).

1366 R. M. Ziff, G. E. Uhlenbeck, M. Kac. *Phys. Rep.* 32, 169 (1977).

1367 M. Znojil. *arXiv:quant-ph*/9812027 (1998).

1368 M. Znojil. *arXiv:hep-th*/0209062 (2002).

1369 M. Znojil. *arXiv:hep-th*/0209262 (2002).

1370 I. J. Zucker, G. S. Joyce. *Math. Proc. Camb. Phil. Soc.* 131, 309 (2001).

1371 V. V. Zudilin. *Mat. Sbornik* 191, 1827 (2000).

Index

The alphabetization is character-by-character, including spaces. Numbers and symbols come first, with the exception of $. All fonts are treated equally.

The index entries refer to the sections or subsections and are hyperlinked. The index entry for a subject from within the exercises and solutions are hyperlinked mostly to the exercises and not to the corresponding solutions.

"*subject* in action" refers to examples or solutions of exercises making very heavy use of *subject*, or could be considered archetypical use of *subject*.

Index entries are grouped at most one level deep. Index entries containing compound names, such as Riemann–Siegel, are mentioned on their own and not as a subentry under the first name.

Built-in functions are referenced to the section in which they are first discussed. Built-in functions and functions defined in the standard packages appear in the font Courier bold (example: **Plot**); functions defined in *The Mathematica GuideBooks* appear in the font Courier plain (example: DistributionOfBends).

Numbers and Symbols

120-cell 367
27 lines on the Clebsch surface 357
28 bitangents 357
9-free numbers, sum of ~ 1140
e **E** 349
π **Pi** 1150, 1081, 1128
Γ **Gamma** 1001
B **Beta** 1006
ψ **PolyGamma** 1007
Π **EllipticPi** 1063
ϑ **EllipticTheta** 1140, 1141
ζ **WeierstrassZeta** 1131
σ **WeierstrassSigma** 1132
ψ **PolyGamma** 1007
ζ **Zeta** 1147
η **DedekindEta** 1152
ϑ **EllipticTheta** 1140, 1141
θ **UnitStep** 266

δ **DiracDelta** 268

Mathematica **input forms and output forms**

∂ **D** 129
'   **Derivative** 139
$\in$ **Element** 8
$\exists$ **Exists** 62
!   **Factorial** 359
$\forall$ **ForAll** 62
d   **Integrate** 156

A

Abel
   differential equations 243
   type differential equations 255
Abel–Plana formula 479
Abel–Ruffini theorem 96

Aborts
   because of time constraints 1263
   intentionally induced ~ 4, 581
Absolute value
   differentiating the ~ 131
   of expressions 91
   of integrands 299
   of polynomial roots 96
Accelerated charges 358
Accumulation, of singularities 1002
Addition
   exact ~ of polynomial roots 99
   of series 207
   of Taylor series 207
Addition theorems
   for elliptic functions 1130, 1134
   for elliptic integrals 1130
   for Hermite polynomials 885
   for Jacobi functions 1134
   for Laguerre polynomials 838